Books are to be returned on or before
the last date below.

19/11/05

LIBREX —

# HANDBOOK OF
# OPTICAL COMPONENTS
# AND ENGINEERING

# Handbook of Optical Components and Engineering

*Editor-in-Chief*

## Kai Chang

Department of Electrical Engineering
Texas A&M University
College Station, Texas

*Members of Editorial Board*

# HANDBOOK OF OPTICAL COMPONENTS AND ENGINEERING

*Edited by*
**KAI CHANG**

**⊗WILEY-INTERSCIENCE**

A JOHN WILEY & SONS, INC., PUBLICATION

*Library of Congress Cataloging-in-Publication Data:*

Handbook of optical components and engineering / Kai Chang, Editor.
    p. cm.
  ISBN 0-471-39055-0 (Cloth)
  1. Optical instruments–Handbooks, manuals, etc. I. Chang, Kai S.
  TA 1520.H365 2003
  621.36–dc22
                                       2003016770

Printed in the United States of America.

10 9 8 7 6 5 4 3 2 1

# HANDBOOK OF OPTICAL COMPONENTS AND ENGINEERING

1. Optical Wave Propagation
2. Infrared Techniques
3. Optical Lenses
4. Optical Resonators
5. Spatial Filters and Fourier Optics
6. Semiconductor Lasers
7. Solid-State Lasers
8. Liquid Lasers
9. Gas Lasers
10. Optical Fiber Transmission Technology
11. Optical Channel Waveguides and Waveguide Couplers
12. Planar Optical Waveguides and Applications
13. Optical Attenuators, Isolators, Circulators, and Polarizers
14. Optical Filters for Telecommunication Applications
15. Wavelength Division Multiplexers and Demultiplexers
16. Wide-Bandwidth Optical Intensity Modulators
17. Optical Modulation: Acousto-Optical Devices
18. Optical Modulation: Magneto-Optical Devices
19. Optical Detectors
20. Acousto-Optic Modulators and Switches
21. Optical Amplifiers

# HANDBOOK OF RF/MICROWAVE COMPONENTS AND ENGINEERING

1. Transmission Lines
2. Transmission-Line Discontinuities
3. Filters, Hybrids and Couplers, Power Combiners, and Matching Networks
4. Cavities and Resonators
5. Ferrite Control Components
6. Surface Acoustic Wave Devices
7. Quasi-Optical Techniques
8. Components for Surveillance and Electronic Warfare Receivers
9. Antennas I: Fundamentals and Numerical Methods
10. Antennas II: Reflector, Lens, Horn, and Other Microwave Antennas of Conventional Configuration

11. Antennas III: Array, Millimeter Wave, and Integrated Antennas

12. Antennas IV: Microstrip Antennas

13. Antennas V: Active Integrated Antennas

14. Mixers and Detectors

15. Multipliers and Parametric Devices

16. Semiconductor Control Devices: PIN Diodes

17. Semiconductor Control Devices: Phase Shifters and Switches

18. Transferred Electron Devices

19. IMPATT and Related Transit-Time Devices

20. Microwave Silicon Bipolar Transistors and Monolithic Integrated Circuits

21. FETs: Power Applications

22. FETs: Low-Noise Applications

23. High-Electron-Mobility Transistors: Principles and Applications

24. Heterojunction Bipolar Transistors and Applications

25. Oscillators and Frequency Synthesizers

26. RF Components

27. Microwave Superconductors

28. Microwave MEMS and Micromachining

# CONTENTS

**Preface**                                                                          **xv**

**Contributors**                                                                     **xvii**

**1. Optical Wave Propagation**                                                      **1**

*Andrew K. Chan*

1.1 Introduction and Organization of the Chapter   1
1.2 Electromagnetic Wave Propagation   2
1.3 Geometrical Optics   15
1.4 Propagation of Gaussian Beam   29
1.5 Diffraction   40
    References   51

**2. Infrared Techniques**                                                           **53**

*Asu R. Jha and Peter N. Kupferman*

2.1 Introduction   53
2.2 Definition of the Infrared   54
2.3 History of Infrared Detectors   54
2.4 Device Physics   58
2.5 Mercury Cadmium Telluride   70
2.6 Indium Antimonide   86
2.7 Schottky-Barrier Detectors   91
2.8 Sprite Detectors   96
2.9 Pyroelectric Detectors   103
2.10 Extrinsic Silicon   107
2.11 Systems Analysis   109
     References   113

## 3. Optical Lenses                                                                115

*John R. Rogers*

3.1  Introduction   115
3.2  First-Order Optics   118
3.3  Aberrations   134
3.4  Real Ray-Trace Techniques   151
3.5  Diffraction Effects   155
3.6  Modular Optical Design   160
     References   162
     Bibliography   162

## 4. Optical Resonators                                                            163

*Kenichi Iga*

4.1  Introduction   163
4.2  Waveguide-Type Fabry–Perot Resonators   164
4.3  Gaussian Beams and Propagation Matrices   171
4.4  Fabry–Perot Open Resonators with Concave Mirrors   176
4.5  Distributed Feedback/Reflector Resonators   188
4.6  Resonator for Vertical Cavity Surface Emitting Laser   195
4.7  Summary   196
     References   196

## 5. Spatial Filters and Fourier Optics                                            199

*Francis T. S. Yu*

5.1   Introduction   199
5.2   Fresnel–Kirchhoff Theory and System Transforms   199
5.3   Analysis of Coherent and Incoherent Optical Systems   205
5.4   Coherent Optical Processing   210
5.5   Complex Spatial Filter Synthesis   212
5.6   Partially Coherent Optical Processing   215
5.7   White Light Optical Processing   218
5.8   Spatial Encoding and Image Sampling   220
5.9   Real-Time Optical Processing   222
5.10  Adaptive Processing   228
5.11  Large-Capacity Information Processing   229
5.12  Optical Neural Networks   232
5.13  Fundamental Approaches   234
5.14  Concluding Remark   236
      References   236

**6. Semiconductor Lasers**                                                    **239**

*Jane J. Yang and Bing W. Liang*

6.1   Introduction   239
6.2   Applications   240
6.3   Basic Operation   241
6.4   Concepts of High-Power Operation   271
6.5   Surface-Emitting Semiconductor Lasers   290
6.6   Semiconductor Tunable Laser Diodes   292
6.7   High-Speed Semiconductor Lasers   308
6.8   Vertical-Cavity Surface-Emitting Lasers (VCSELs)   323
6.9   Other Quantum Devices: Quantum Cascade, Quantum Wire, and Quantum DOT Lasers   347
6.10  Conclusion   350
       References   352


**7. Solid-State Lasers**                                                      **369**

*John M. McMahon*

7.1   Introduction   369
7.2   Excitation of Solid-State Lasers   375
7.3   Laser Oscillators   386
7.4   Laser Amplifiers   399
7.5   Optical Modulators   412
7.6   System Strategies for Oscillator–Amplifier Laser Systems   419
7.7   Average Power Effects   424
7.8   Miscellaneous Topics   428
       References   437


**8. Liquid Lasers**                                                           **443**

*B. Wilhelmi, E. Döpel, and F. Weidner*

8.1   Introduction   443
8.2   Parameters of the Active Media: Comparison with Other Lasers   446
8.3   Excitation of the Active Medium, Frequency Selection, and Tuning   460
8.4   Operational Principles, Design, and Parameters of Liquid Lasers   471
8.5   Selected Applications   496
       Bibliography   501
       References   502

## 9. Gas Lasers                                                511

*T. Lehecka, N. C. Luhmann, Jr., W. A. Peebles, J. Goldhar, and S. P. Obenschain*

9.1  Introduction   511
9.2  Far-Infrared Lasers   541
9.3  Excimer Lasers   624
     References   647

## 10. Optical Fiber Transmission Technology                    659

*Chinlon Lin*

10.1 Introduction   659
10.2 Optical Fibers for Transmission   660
10.3 Optical Fiber Transmission Technology   679
10.4 Transmission Limitations   686
10.5 Future Directions in Optical Transmission Technology   691
10.6 Future Applications: Fiber to the Home and Intelligent
     Buildings   692
     References   694

## 11. Optical Channel Waveguides and Waveguide Couplers        697

*Talal K. Findakly*

11.1 Optical Planar Waveguides   697
11.2 Optical Channel Waveguides   702
11.3 Optical Directional Couplers   707
     References   716

## 12. Planar Optical Waveguides and Applications               719

*Shi-Kay Yao*

12.1 Planar Optical Waveguides   720
12.2 Passive Waveguide Optical Components   753
12.3 Waveguide Measurement Techniques and Applications   809
     References   830

## 13. Optical Attenuators, Isolators, Circulators, and Polarizers   837

*Feng Qing Zhou and J. J. Pan*

13.1 Introduction   837
13.2 Optical Polarizers   838
13.3 Polarization Rotators and Phase Retarders   846

13.4 Optical Attenuators   850

13.5 Optical Isolators   863

13.6 Optical Circulators   874
     References   884

**14. Optical Filters for Telecommunication Applications**                **887**

*C. K. Madsen*

14.1 Introduction   887

14.2 Optical Filter Theory   890

14.3 Finite Impulse Response Optical Filters   898

14.4 Infinite Impulse Response Optical Filters   913

14.5 Thin Film Filters   917

14.6 Bragg Gratings   921

14.7 Conclusion   925
     References   925

**15. Wavelength Division Multiplexers and Demultiplexers**                **935**

*José Capmany and Salvador Sales*

15.1 Introduction   935

15.2 General Issues on WDM MUX/DEMUX and OADM
     Devices   936

15.3 Thin-Film MUX/DEMUX Devices   954

15.4 Free-Space MUX/DEMUX Devices   961

15.5 Optical Fiber-Based MUX/DEMUX Devices   970

15.6 Integrated Optics-Based MUX/DEMUX
     Devices   979

15.7 Epilogue   998
     References   998

**16. Wide-Bandwidth Optical Intensity Modulators**                **1009**

*G. L. Li and P. K. L. Yu*

16.1 Introduction   1009

16.2 Physical Effects and Modulator Materials   1010

16.3 Modulation Efficiency and Modulator Waveguide
     Design   1027

16.4 Lumped-Element Modulators   1041

16.5 Traveling-Wave Modulators   1050

16.6 A Comparison of the Available Modulators   1069
     References   1072

**17. Optical Modulation: Acousto-Optical Devices**                    **1079**

*Chen S. Tsai*

17.1  Introduction    1079
17.2  Bulk-Wave Acousto-Optical Bragg Diffraction and
      Applications    1080
17.3  Guided-Wave Acousto-Optical Bragg Diffraction in Planar
      Waveguides    1084
17.4  Key Performance Parameters of Planar Guided-Wave
      Acousto-Optical Bragg Modulator and Deflector    1099
17.5  Construction of Wide-Band Guided-Wave Acousto-Optical
      Bragg Modulators and Deflectors    1102
17.6  Applications of Guided-Wave Acousto-Optical Bragg Cells
      and Modules    1111
      References    1121


**18. Optical Modulation: Magneto-Optical Devices**                    **1131**

*Alan E. Craig*

18.1  Introduction    1131
18.2  Units    1132
18.3  Material Physics of Magnetics    1135
18.4  Spin Waves    1149
18.5  Material Physics of Optics    1160
18.6  Magneto-Optics    1164
18.7  Real Materials    1169
18.8  Modern Magneto-Optical Devices    1187
      References    1207
      Suggestions for Further Reading    1211


**19. Optical Detectors**                                              **1215**

*P. K. L. Yu and H. D. Law*

19.1  Introduction    1215
19.2  Terminology in Optical Detection and Material
      Considerations    1216
19.3  Photodiodes    1220
19.4  Photoconductive Devices    1239
19.5  Nonlinearity in Photodetectors    1243
      References    1249

**20. Acousto-Optic Modulators and Switches**    **1255**

*Shi-Kay Yao*

20.1 Introduction   1255
20.2 Basic Properties of Acousto-Optic Modulators   1257
20.3 Design Procedures of Acousto-Optic Modulators   1277
20.4 Fabrication of Acousto-Optic Modulators   1287
20.5 Summary   1296
     References   1298

**21. Optical Amplifiers**    **1301**

*Chin B. Su*

21.1 Erbium-Doped Fiber Amplifier   1301
21.2 Raman Fiber Amplifier   1312
21.3 Semiconductor Optical Amplifier   1316
     References   1320

**Index**    **1321**

# PREFACE

The four-volume *Handbooks of Microwave and Optical Components* were published in 1989 and 1990. Since then, we have witnessed a rapid development in the microwave and optical areas. Radio frequency and microwave wireless personal communications has become one of the hottest growth areas. Optical devices have been used in a wide variety of applications from compact disk players to communication systems. With these expanding applications, it is time to update the handbooks with new material in response to the new developments.

In the second edition, Volume 1 and Volume 2 of the first edition have been combined into one book entitled *Handbook of RF/Microwave Components and Engineering*. Volume 3 and Volume 4 have been combined into one book entitled *Handbook of Optical Components and Engineering*. New chapters have been added, and many old chapters have been revised. This handbook is intended to serve as a compendium of principles and design data for practicing microwave and optical engineers. Although it is expected to be most useful to engineers actively engaged in designing microwave and optical systems, it should also be of considerable value to engineers in other disciplines who have a desire to understand the capabilities and limitations of microwave and optical systems.

To achieve these goals, this handbook covers almost all important components. Theoretical discussions and mathematical formulations are given only where essential. Whenever possible, design results are presented in graphic and tabular form; references are given for further study. The book provides, in practical fashion, a wealth of essential principles, methods, design information, and references to help solve problems in high-frequency spectra.

Each chapter is written as a self-contained unit with its own list of references. Some overlap is inevitable among chapters, but it has been kept to a minimum. It is hoped that this comprehensive handbook will offer the type of detailed information necessary for use in today's complex and rapidly changing high-frequency engineering.

The authors who have contributed chapters to this handbook have done an excellent job of condensing mountains of material into readable accounts of their respective areas.

The emphasis throughout has been to provide an overview and practical information of each subject.

I would like to thank all members of the editorial board for their advice and suggestions, Dr. Felix Schwering for organizing the antenna chapters, Ms. Cassie Craig of Wiley for managing this project, and Mr. George Telecki, our Wiley editor, for his constant encouragement. I wish especially to thank my wife, Suh-jan, for her patience and support.

**Kai Chang**
College Station, Texas

# CONTRIBUTORS

**José Capmany**
Optical Communications Group
IMCO2 Research Institute
Universidad Politécnica de Valencia
Spain

**Andrew K. Chan**
Department of Electrical Engineering
Texas A&M University
College Station, Texas

**Alan E. Craig**
U.S. Naval Research Laboratory
Washington, D.C.

**E. Döpel**
University of Applied Sciences
Jena, Germany

**Talal K. Findakly**
Research Division
Hoechst Celanese Corporation
Summit, New Jersey

**J. Goldhar**
University of Maryland
College Park, Maryland

**Kenichi Iga**
Japan Society for the Promotion of
Science
Tokyo, Japan

**Asu R. Jha**
Jha Technical Consulting Services
Cerritos, California

**Peter N. Kupferman**
Jet Propulsion Laboratory
California Institute of Technology
Pasadena, California

**H. D. Law**
Agura Hill, California

**T. Lehecka**
University of California
Los Angeles, California

**G. L. Li**
Semiconductor Photonics Department
Agere Systems, Inc.
Breinigsville, Pennsylvania

**Bing W. Liang**
Lytek Inc.
Phoenix, Arizona

**Chinlon Lin**
Bell Communications Research
(Bellcore)
Red Bank, New Jersey

**N. C. Luhmann, Jr.**
University of California
Los Angeles, California

**C. K. Madsen**
Bell Laboratories
Lucent Technologies
Murray Hill, New Jersey

**John M. McMahon**
Optical Sciences Division
Naval Research Laboratory
Washington, D.C.

**S. P. Obenschain**
Laser Plasma Branch
Naval Research Laboratory
Washington, D.C.

**J. J. Pan**
Lightwaves 2020, Inc.
Milpitas, California

**W. A. Peebles**
University of California
Los Angeles, California

**John R. Rogers**
Institute of Optics
University of Rochester
Rochester, New York

**Salvador Sales**
Optical Communications Group
IMCO2 Research Institute
Universidad Politécnica de Valencia
Spain

**Chin B. Su**
Department of Electrical Engineering
Texas A&M University
College Station, Texas

**Chen S. Tsai**
Department of Electrical and Computer
Engineering
and Institute for Surface and Interface
Science
University of California
Irvine, California

**F. Weidner**
MLC GmbH
Jena, Germany

**B. Wilhelmi**
JENOPTIK AG
Jena, Germany

**Jane J. Yang**
Advanced Photonic Sciences
Los Angeles, California

**Shi-Kay Yao**
Optech Laboratory
Rowland Heights, California

**P. K. L. Yu**
Department of Electrical and Computer
Engineering
University of California, San Diego
La Jolla, California

**Francis T. S. Yu**
Department of Electrical Engineering
The Pennsylvania State University
University Park, Pennsylvania

**Feng Qing Zhou**
Lightwaves 2020, Inc.
Milpitas, California

# 1

# OPTICAL WAVE PROPAGATION

ANDREW K. CHAN
*Department of Electrical Engineering*
*Texas A&M University*
*College Station, Texas*

## 1.1 INTRODUCTION AND ORGANIZATION OF THE CHAPTER

The subject of optical wave propagation is of fundamental importance to optical technology. It concerns the movement of optical energy from one space point to another. As light moves through an optical system, the general concerns at the output are the amplitude and phase of the optical field. As new materials are developed and new systems are designed, wave propagation problems are continuously being generated to challenge the mind of the researcher. It is impossible to cover every aspect of the subject even at the minimal level. In this introductory chapter, only the most basic topics in optical wave propagation are presented.

Section 1.2 is an introduction to the theory of electromagnetic waves. It serves as a quick review for those who are already familiar with the wave aspect of light. At the same time, it provides a foundation for further study in wave optics. Only topics pertinent to optical technology are included in this chapter. Section 1.3 covers the fundamentals of geometrical optics. It aims to give the other aspect of light propagation, which considers the traveling path of light in a medium called a ray. The ray equation and the eikonal equation, which completely specifies the ray path and the surface of the geometric wavefront, are derived. Applications of geometrical optics to propagation in an inhomogeneous medium are discussed. Ray tracing and the *ABCD* matrix transformation are emphasized. Section 1.4 treats the subject of the Gaussian beam in some detail. The paraxial wave equation is developed. Only the lowest-order solution is discussed, with special emphasis placed on the physical significance of the beam parameters. The Gaussian modes of lens waveguides and graded-index medium are presented without derivation, and the transformation of a Gaussian beam using complex representation is developed. The problem of coupling from the output of a laser

*Handbook of Optical Components and Engineering,*   Edited by Kai Chang
ISBN 0-471-39055-0   © 2003 John Wiley & Sons, Inc.

to a graded-index waveguide is treated by mode matching. Some properties of the Hermite — Gaussian modes are discussed briefly at the end of the section. In Section 1.5 the diffraction integral is obtained through the use of a free-space Green's function. The paraxial approximation is made to obtain the Fresnel and Fraunhofer diffraction formulas. Since diffraction theory plays an important role in Fourier optics, the linear system representation of diffraction and the convolution integrals are identified. Amplitude and phase diffraction grating are treated briefly.

## 1.2   ELECTROMAGNETIC WAVE PROPAGATION

Optical wave propagation is a dynamic phenomenon in classical electromagnetism. It concerns changes in amplitude and phase of a wave, and the transfer of energy from one space point to another. Many optical phenomena can be explained based on electromagnetic wave theory. Reflection and transmission at a dielectric interface, total reflection and total transmission, propagation in a layered medium, and waves in anisotropic crystals are a few of the fundamental subjects discussed in this introductory section.

### 1.2.1   Maxwell Equations

The basic equations in electromagnetic wave theory are those developed by James C. Maxwell. These equations, written in the time domain, are

$$\nabla \times \overline{\mathscr{E}} = -\partial \overline{\mathscr{B}}/\partial t \tag{1.1a}$$

$$\nabla \times \overline{\mathscr{H}} = \overline{\mathscr{J}} + \partial \overline{\mathscr{D}}/\partial t \tag{1.1b}$$

$$\nabla \cdot \overline{\mathscr{D}} = \rho \tag{1.1c}$$

$$\nabla \cdot \overline{\mathscr{B}} = 0 \tag{1.1d}$$

where $\overline{\mathscr{E}}, \overline{\mathscr{D}}, \overline{\mathscr{H}}$, and $\overline{\mathscr{B}}$ are time-domain vectors corresponding to electric field intensity, displacement vector, magnetic field intensity, and magnetic flux density respectively. $\overline{\mathscr{J}}$ is the conduction current density. These fields are further related to one another through the electric and magnetic characteristic parameters of the material medium by

$$\overline{\mathscr{B}} = \overline{\overline{\mu}} \, \overline{\mathscr{H}} \tag{1.2}$$

$$\overline{\mathscr{D}} = \overline{\overline{\epsilon}} \, \overline{\mathscr{E}} \tag{1.3}$$

$$\overline{\mathscr{J}} = \sigma \overline{\mathscr{E}} \tag{1.4}$$

where $\overline{\overline{\epsilon}}, \overline{\overline{\mu}}$, and $\sigma$ are, respectively, the electric permittivity tensor, the magnetic permeability tensor, and the conductivity of the medium. These equations are known as the constitutive relations. In a homogeneous and isotropic medium, these parameters are scalar constants. For inhomogeneous materials, the parameters are scalar functions of the coordinates. In anisotropic crystals, the permittivity is a tensor. More than one wave can propagate in anisotropic crystals, in which the $\overline{\mathscr{D}}$ and $\overline{\mathscr{E}}$ fields are not parallel to each other. Wave propagation in crystals is discussed in a later section. In what follows, a homogeneous and isotropic medium is assumed. In addition, instead of $\epsilon$, the optical refractive index $n^2 = \epsilon/\epsilon_0$ will be used.

In time-harmonic case where the time variation of the fields is assumed to be $e^{j\omega t}$, the field vectors can be written in the form

$$\overline{\mathscr{E}} = \text{Re } \{\overline{\mathbf{E}}e^{j\omega t}\} \tag{1.5}$$

where Re denotes the real part; the Maxwell equations in the frequency domain are given by

$$\nabla \times \overline{\mathbf{E}} = -j\omega\overline{\mathbf{B}} \tag{1.6a}$$

$$\nabla \times \overline{\mathbf{H}} = \overline{\mathbf{J}} + j\omega\overline{\mathbf{D}} \tag{1.6b}$$

$$\nabla \cdot \overline{\mathbf{D}} = \rho \tag{1.6c}$$

$$\nabla \cdot \overline{\mathbf{B}} = 0 \tag{1.6d}$$

where $\overline{\mathbf{E}}$, $\overline{\mathbf{H}}$, $\overline{\mathbf{D}}$, $\overline{\mathbf{B}}$, and $\overline{\mathbf{J}}$ are complex vector functions of frequency $\omega$.

### 1.2.2   Boundary Conditions

A solution to the Maxwell equations in (1.6) describes the fields in a region where the characteristic parameters of the medium are uniform or continuous. If the parameters change abruptly across a boundary surface, a set of conditions is necessary to specify the discontinuities of the fields. These conditions are expressed as

$$\hat{n} \cdot (\overline{\mathbf{B}}_1 - \overline{\mathbf{B}}_2) = 0 \tag{1.7a}$$

$$\hat{n} \cdot (\overline{\mathbf{D}}_1 - \overline{\mathbf{D}}_2) = \rho_s \tag{1.7b}$$

$$\hat{n} \times (\overline{\mathbf{E}}_1 - \overline{\mathbf{E}}_2) = 0 \tag{1.7c}$$

$$\hat{n} \times (\overline{\mathbf{H}}_1 - \overline{\mathbf{H}}_2) = \overline{\mathbf{J}}_s \tag{1.7d}$$

where $\overline{\mathbf{J}}_s$ and $\rho_s$ are the surface current density and surface charge density, respectively, with $\hat{n}$ being the unit normal to the boundary surface. In optics, where most of the boundary surfaces are interfaces of dielectrics, $\rho_s$ and $\overline{\mathbf{J}}_s$ in general are zero. The boundary conditions can be expressed in terms of the tangential and normal components of the fields as

$$B_{1n} = B_{2n} \tag{1.8a}$$

$$D_{1n} = D_{2n} \tag{1.8b}$$

$$E_{1t} = E_{2t} \tag{1.8c}$$

$$H_{1t} = H_{2t} \tag{1.8d}$$

where $n$ and $t$ are subscripts for the normal and tangential components.

### 1.2.3   The Poynting Theorem

The energy contained in an electromagnetic field is given in terms of the Poynting vector by the equation

$$\overline{\mathscr{P}} = \overline{\mathscr{E}} \times \overline{\mathscr{H}} \tag{1.9}$$

The unit of $\overline{\mathscr{P}}$ is watts/meter$^2$ and $\overline{\mathscr{P}}$ represents the power density of the field. In the time-domain, the Poynting theorem is expressed mathematically as

$$\frac{\partial}{\partial t} \int_V \frac{\overline{\mathscr{E}} \cdot \overline{\mathscr{D}}}{2} \, dv + \frac{\partial}{\partial t} \int_V \frac{\overline{\mathscr{H}} \cdot \overline{\mathscr{B}}}{2} \, dv + \int_V \overline{\mathscr{J}} \cdot \overline{\mathscr{E}} \, dv = - \int_S \overline{\mathscr{P}} \cdot d\overline{\mathbf{s}} \tag{1.10}$$

where the quantities $\int_V \frac{\overline{\mathscr{E}} \cdot \overline{\mathscr{D}}}{2} \, dv$ and $\int_V \frac{\overline{\mathscr{H}} \cdot \overline{\mathscr{B}}}{2} \, dv$ are the electric and magnetic stored energy in the volume $V$. The third term in the equation is the total power lost in the volume, and the term $- \int_S \overline{\mathscr{P}} \cdot d\mathbf{s}$ is the total power supplied to the volume $V$. Hence the Poynting theorem states that the total power transfer into a volume bounded by a closed surface must be equal to the power loss plus the power stored in the volume. In the time-harmonic case, the complex Poynting theorem in differential form is stated as

$$\nabla \cdot (\overline{\mathbf{E}} \times \overline{\mathbf{H}}^*) = -j\omega(\overline{\mathbf{B}} \cdot \overline{\mathbf{H}}^* - \overline{\mathbf{E}} \cdot \overline{\mathbf{D}}^*) - \overline{\mathbf{E}} \cdot \overline{\mathbf{J}}^* \tag{1.11}$$

### 1.2.4   The Wave Equation and Its Solution

For an isotropic and homogeneous medium, the time-domain equation for the electric field in a source free region is the wave equation given by

$$\nabla^2 \overline{\mathscr{E}} = \frac{1}{c^2} \frac{\partial^2 \overline{\mathscr{E}}}{\partial t^2} \tag{1.12}$$

where $c$ is the velocity of light in the medium. The $\overline{\mathscr{H}}$ field satisfies the same equation. For the time-harmonic case, (1.12) is reduced to the Helmholtz equation

$$(\nabla^2 + k^2)\overline{\mathbf{E}} = 0 \tag{1.13}$$

where

$$k = \omega/c \tag{1.14}$$

is the wave number and $\omega$ the angular frequency.

The simplest solution to (1.13) is the uniform plane wave solution given by

$$\overline{\mathbf{E}}(\overline{\mathbf{r}}) = \overline{\mathbf{E}}_0 e^{-j\overline{k} \cdot \overline{r}} \tag{1.15}$$

and the associated magnetic field $\overline{\mathbf{H}}$ is

$$\overline{\mathbf{H}}(\overline{\mathbf{r}}) = \frac{k}{\omega\mu} \hat{k} \times \overline{\mathbf{E}} \tag{1.16}$$

where $\bar{r}$ is the position vector and $\overline{\mathbf{E}}_0$ is a constant vector. The wave number vector $\bar{\mathbf{k}}$ is given by

$$\bar{\mathbf{k}} = k_x \hat{x} + k_y \hat{y} + k_z \hat{z} \tag{1.17}$$

with $k^2 = k_x^2 + k_y^2 + k_z^2$. Since

$$\bar{\mathbf{k}} \cdot \bar{\mathbf{r}} = \text{constant} \tag{1.18}$$

is an equation of a plane, the wave is called a plane wave. This constant phase plane coincides with the plane of constant amplitude in a uniform plane wave. The scalar wave theory is usually adequate to treat most problems in optical technology. The vector solution in (1.15) may be reduced to

$$E(r) = E_0 e^{-j\bar{k}\cdot\bar{r}} \tag{1.19}$$

In a spherical coordinate system, the scalar wave equation is simplified to

$$\frac{1}{r}\frac{\partial^2}{\partial r^2}(rE) + k^2(E) = 0 \tag{1.20}$$

by using radial symmetry, and the solution corresponding to an outward-going wave is

$$E = E_0 \frac{e^{-jkr}}{r} \tag{1.21}$$

where $r$ is the radial coordinate of the spherical system. The constant-amplitude and constant-phase surfaces are the same spherical surface of radius $r$.

In a cylindrical coordinate system, the scalar wave equation is

$$\frac{1}{\rho}\frac{\partial}{\partial \rho}\left(\rho \frac{\partial E}{\partial \rho}\right) + \frac{1}{\rho^2}\frac{\partial^2 E}{\partial \phi^2} + \frac{\partial^2 E}{\partial z^2} + k^2 E = 0 \tag{1.22a}$$

Assuming a cylindrical wave propagating along the $z$-axis such that the $z$-dependent of $E$ is $e^{\pm jk_z z}$, (1.22a) is reduced to

$$\frac{1}{\rho}\frac{\partial}{\partial \rho}\left(\rho \frac{\partial E}{\partial \rho}\right) + \frac{1}{\rho^2}\frac{\partial^2 E}{\partial \phi^2} + k_\rho^2 E = 0 \tag{1.22b}$$

where $k_\rho^2 = k^2 - k_z^2$. Equation (1.22b) is simpler and more suitable for studies in fiber optics. Its solution is given by

$$E(\rho, \phi) = \sum_m E_m B_m(k_\rho \rho) e^{jm\phi} \tag{1.23}$$

where $E_m$ is a constant and $B_m$ is a Bessel function of order $m$ appropriately chosen for a given boundary value problem.

### 1.2.5 Polarization

At a fixed point in space, the electric field vector $\overline{\mathscr{E}}$ of a time-harmonic electromagnetic wave varies with time. The polarization of a wave is described by the locus of the tip of the $\overline{\mathscr{E}}$ vector as time progresses. The three types of polarization are linear, circular, and elliptical polarization. There are also the left-handed and right-handed polarization for the circular and elliptical cases.

Suppose that a uniform plane wave is propagating in the $z$ direction. The electric field of the wave can be written as

$$\overline{\mathbf{E}} = (E_x \hat{x} + E_y \hat{y}) e^{-jkz} \tag{1.24}$$

where $E_x$ and $E_y$ are complex amplitudes of the $x$ and $y$ components of $\overline{\mathbf{E}}$. The complex ratio

$$E_y / E_x = A e^{j\psi} \tag{1.25}$$

completely specifies the polarization of the wave. The types of polarization and their corresponding values for $A$ and $\psi$ are listed below.

|  | Linear | Circular | | Elliptical | |
| --- | --- | --- | --- | --- | --- |
|  |  | Right Handed | Left Handed | Right Handed | Left Handed |
| $A$ |  | 1 | 1 |  |  |
| $\psi$ | $0, \pi$ | $-\pi/2$ | $\pi/2$ | $-\pi < x < 0$ | $0 < x < \pi$ |

A polarization chart is shown in Fig. 1.1. It is important to note that the sense of rotation is determined by assuming that the wave is approaching the observer.

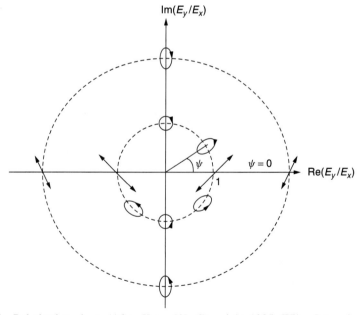

**Figure 1.1**   Polarization chart. (After Kong [3]. Copyright 1985, Wiley-Interscience. Reproduced by permission.)

If the observer points his right thumb in the direction of the propagating wave and the electric field vector rotates in the direction of his fingers, the wave is right-handed polarized; left-handed polarized otherwise. The equation for $\mathscr{E}_x$ and $\mathscr{E}_y$ for determining the polarization of the wave is

$$\left(\frac{\mathscr{E}_x}{\mathscr{E}_{x0}}\right)^2 - 2\left(\frac{\mathscr{E}_x}{\mathscr{E}_{x0}}\right)\left(\frac{\mathscr{E}_y}{\mathscr{E}_{y0}}\right)\cos\psi + \left(\frac{\mathscr{E}_y}{\mathscr{E}_{y0}}\right)^2 = \sin^2\psi \tag{1.26}$$

where $\mathscr{E}_x = \mathscr{E}_{x0}\cos(\omega t + \phi_x)$, $\mathscr{E}_y = \mathscr{E}_{y0}\cos(\omega t + \phi_y)$, and $\psi = \phi_y - \phi_x$. This is an equation of an ellipse for an arbitrary value of $\psi$. If $\psi$ is 0 or $\pi$, (1.26) represents the equation of a straight line. When $\psi$ is $\pm\pi/2$ and $\mathscr{E}_{x0} = \mathscr{E}_{y0}$, (1.26) is an equation of a circle.

### 1.2.6 Reflection and Transmission at a Boundary

When a traveling plane wave encounters a boundary separating two media with different material parameters, energy is being partially reflected and transmitted at the boundary. The coefficients of reflection and transmission can be expressed as functions of the refractive indexes and the angle of incidence. For a linearly polarized wave whose electric field is perpendicular to the plane of incidence (perpendicular polarized wave), the amplitude reflection coefficient and transmission coefficients are given by

$$R_\perp = \frac{n_1\cos\theta_i - n_2\cos\theta_t}{n_1\cos\theta_i + n_2\cos\theta_t} \tag{1.27}$$

and

$$T_\perp = 1 + R_\perp = \frac{2n_1\cos\theta_i}{n_1\cos\theta_i + n_2\cos\theta_t} \tag{1.28}$$

If the wave is linearly polarized so that the electric field is parallel to the plane of incidence, $R_\parallel$ and $T_\parallel$ are given by

$$R_\parallel = \frac{n_2\cos\theta_i - n_1\cos\theta_t}{n_2\cos\theta_i + n_1\cos\theta_t} \tag{1.29}$$

$$T_\parallel = \frac{2n_1\cos\theta_i}{n_2\cos\theta_i + n_1\cos\theta_t} \tag{1.30}$$

The geometry of these two cases is shown in Fig. 1.2.

Sometimes it is more convenient to study the reflection and transmission of light intensity rather than the wave amplitude. The reflectivity and transmissivity are expressed in terms of the incident angle and the transmitted angles as [1]

$$r_\perp = |R_\perp|^2 = \frac{\sin^2(\theta_i - \theta_t)}{\sin^2(\theta_i + \theta_t)} \tag{1.31}$$

$$t_\perp = \frac{n_2\cos\theta_t}{n_1\cos\theta_i}|T_\perp|^2 = \frac{\sin 2\theta_i \sin 2\theta_t}{\sin^2(\theta_i + \theta_t)} \tag{1.32}$$

$$r_\parallel = |R_\parallel|^2 = \frac{\tan^2(\theta_i - \theta_t)}{\tan^2(\theta_i + \theta_t)} \tag{1.33}$$

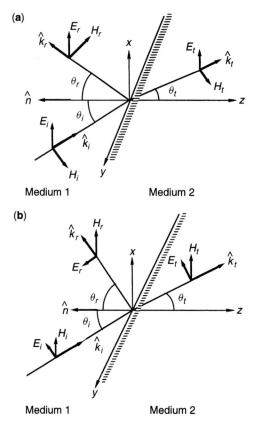

**Figure 1.2**   Geometry for the reflection and transmission of waves. **(a)** Perpendicular polarization, **(b)** Parallel polarization. (After Paris and Hurd. *Basic Electromagnetic Theory*, Copyright 1969, McGraw Hill Book Company. Reproduced by permission.)

$$t_\parallel = \frac{n_2 \cos\theta_t}{n_1 \cos\theta_i}|T_\parallel|^2 = \frac{\sin 2\theta_i \sin 2\theta_t}{\sin^2(\theta_i + \theta_t)\cos^2(\theta_i - \theta_t)} \tag{1.34}$$

where Snell's law, $n_1 \sin\theta_i = n_2 \sin\theta_t$, has been used to eliminate the refractive indexes in the expressions. The law of conservation of energy requires that the sum of the reflected energy and the transmitted energy equals the incident energy. Hence

$$r_\perp + t_\perp = 1 \tag{1.35}$$

$$r_\parallel + t_\parallel = 1 \tag{1.36}$$

### 1.2.7   Critical Angle and Brewster Angle

If the refractive index of the incident medium ($n_1$) is larger than that of the transmitting medium ($n_2$), there exists a critical angle $\theta_c$ such that

$$\sin\theta_c = n_2/n_1 \tag{1.37}$$

and the transmitted angle $\theta_t$ is $\pi/2$. If the angle of incidence is greater than $\theta_c$, no energy is transferred into the transmitting medium in the direction normal to the boundary. The field amplitudes decay exponentially into the transmitting medium and the reflection coefficients $R_\perp$ and $R_\parallel$ are complex quantities with unit magnitude. The reflectivity $r$ is unity and the transmissivity $t$ is zero for both polarizations. In this case, the transmitted wave propagates along the boundary surface, with its amplitude diminishing along the direction normal to the surface. This is an example of an inhomogeneous plane wave.

If a parallel polarized wave is incident on the boundary at an angle $\theta_b$ such that $R_\parallel = 0$, this angle is called the Brewster angle (polarization angle). Total transmission of energy occurs at this angle. The value of the Brewster angle is determined by

$$\tan \theta_b = n_2/n_1 \tag{1.38}$$

and the transmitted angle

$$\theta_t = \frac{\pi}{2} - \theta_b \tag{1.39}$$

In this case it is evident from (1.33) and (1.34) that $r_\parallel = 0$ and $t_\parallel = 1$. The incident energy is totally transferred into the transmitting medium.

### 1.2.8  Reflection and Transmission through Thin Film

A thin film may be defined as a thin layer of uniform dielectric sandwiched between two semi-infinite dielectric media of different refraction indexes. Consider a dielectric film of thickness $d$ with refractive index $n_2$ sandwiched between semi-infinite media with refractive indexes $n_1$ and $n_3$ as shown in Fig. 1.3. For the perpendicular polarization, the reflection coefficient $R$ and transmission coefficient $T$ are given by [1]

$$R = \frac{R_{12} + R_{23}e^{j2u}}{1 + R_{12}R_{23}e^{j2u}} \tag{1.40}$$

$$T = \frac{T_{12}T_{23}e^{ju}}{1 + R_{12}R_{23}e^{j2u}} \tag{1.41}$$

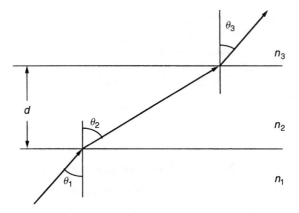

**Figure 1.3**  Geometry for a thin film.

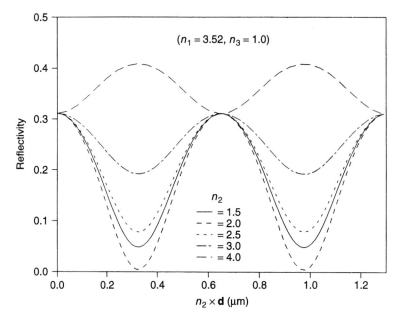

**Figure 1.4** Reflectivity of a thin film.

where $R_{12}$, $R_{23}$, $T_{12}$, and $T_{23}$ are the reflection and transmission coefficients at the first and second interfaces, respectively. The quantity $u$ is given by

$$u = n_2 k_0 d \cos \theta_2 \tag{1.42}$$

and $\theta_2$ can be calculated from Snell's law. The reflectivity and transmissivity of the thin film are expressed as

$$r = |R|^2 \tag{1.43}$$

$$t = \frac{n_3 \cos \theta_3}{n_1 \cos \theta_1} |T|^2 \tag{1.44}$$

Several graphs on the reflectivity are shown in Fig. 1.4.

One should note that the reflection coefficient in (1.40) is a periodic function of $u$. By proper choice of the refractive index of $n_2$ and the thickness $d$ of the film, it is possible to reduce the reflection in medium 1 substantially. In particular, for the case of normal incidence, if the thickness $d$ is odd multiples of quarter wavelength, the reflectivity is zero when

$$n_2 = \sqrt{n_1 n_3} \tag{1.45}$$

### 1.2.9 Reflection and Transmission from Layered Dielectric

Assuming that a plane wave in region 0 is incident on the boundary at $z = 0$ as shown in Fig. 1.5, the reflection and transmission coefficients can be obtained through matching the conditions at each boundary. The simplest approach is to formulate the

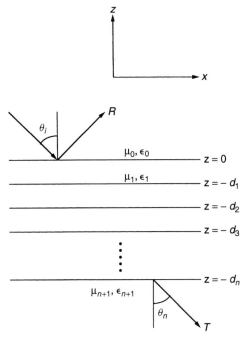

**Figure 1.5**  Multilayer medium. (After Kong [3]. Copyright 1986, Wiley-Interscience. Reproduced by permission.)

problem in terms of a propagation matrix. For a perpendicular polarized wave, the electric field in the $q$th layer is given by [3]

$$E_{qy} = (A_q e^{-jk_{qz}z} + B_q e^{jk_{qz}z}) e^{-jk_x x} \tag{1.46}$$

The associated magnetic field in the same layer can be obtained using Maxwell equations. $k_{qz}$ is related to the wave number of the $q$th layer by

$$k_{qz}^2 + k_x^2 = n_q^2 k_0^2 \tag{1.47}$$

where $n_q$ is the refractive index of the $q$th layer and $k_x$ is the wave number in the $x$ direction for all layers through Snell's law.

The wave amplitudes $A_q$ and $B_q$ are related to the amplitudes in the next layer by the boundary conditions at $z = -d_q$. In terms of matrices, the relationship is given by

$$\begin{bmatrix} A_{q+1} e^{jk_{(q+1)z}d_{q+1}} \\ B_{q+1} e^{-jk_{(q+1)z}d_{q+1}} \end{bmatrix} = [P_{q+1,q}] \begin{bmatrix} A_q e^{jk_{qz}d_q} \\ B_q e^{-jk_{qz}d_q} \end{bmatrix} \tag{1.48}$$

where the propagation matrix $P$ is

$$[P_{q+1,q}] = \frac{1}{2}(1 + C_{q+1,q}) \begin{bmatrix} e^{jk_{(q+1)z}(d_{q+1}-d_q)} & R_{q+1,q} e^{jk_{(q+1)z}(d_{q+1}-d_q)} \\ R_{q+1,q} e^{-jk_{(q+1)z}(dq+1-dq)} & e^{-jk_{(q+1)z}(dq+1-dq)} \end{bmatrix} \tag{1.49}$$

with

$$C_{q+1,q} = \frac{\mu_{q+1}k_{qz}}{\mu_q k_{(q+1)z}} \tag{1.50}$$

and

$$R_{q+1,q} = \frac{1 - C_{q+1,q}}{1 + C_{q+1,q}} \tag{1.51}$$

is the reflection coefficient at the boundary $z = -d_q$. The first of the two subscripts denotes the region with the incident wave.

From (1.49), the amplitudes in a given layer can be related to the amplitudes in any other layer through simple matrix multiplications. For $p > q$ we obtain

$$\begin{bmatrix} A_p e^{jk_{pz}d_p} \\ B_p e^{-jk_{pz}d_p} \end{bmatrix} = [P_{p,p-1}][P_{p-1,p-2}]\dots[P_{q+1,q}] \begin{bmatrix} A_q e^{jk_{qz}d_q} \\ B_q e^{-jk_{qz}d_q} \end{bmatrix} \tag{1.52}$$

Assuming a unity amplitude on the incident wave, the reflection coefficient $R = A_0$ and the transmission coefficient $T = B_{n+1}$ can be expressed in matrix form:

$$\begin{bmatrix} 0 \\ T \end{bmatrix} = [P_{n+1,0}] \begin{bmatrix} R \\ 1 \end{bmatrix} \tag{1.53}$$

with

$$[P_{n+1,0}] = [P_{n+1,n}][P_{n,n-1}]\dots[P_{2,1}][P_{1,0}] \tag{1.54}$$

and

$$[P_{n+1,n}] = \frac{1}{2}(1 + C_{n+1,n}) \begin{bmatrix} e^{-jk_{(n+1)z}d_n} & R_{n+1,n}e^{-jk(n+1)_z d_n} \\ R_{n+1,n}e^{jk_{(n+1)z}d_n} & e^{jk_{(n+1)z}d_n} \end{bmatrix} \tag{1.55}$$

The values of $R$ and $T$ can be obtained by solving (1.53) using (1.54) and (1.55).

### 1.2.10 Waves in Anisotropic Crystals

In this section the propagation of wave in an anisotropic medium is explained briefly. An electrically anisotropic medium is characterized by its permittivity tensor $\bar{\bar{\epsilon}}$:

$$\bar{\bar{\epsilon}} = \begin{bmatrix} \epsilon_{xx} & \epsilon_{xy} & \epsilon_{xz} \\ \epsilon_{yx} & \epsilon_{yy} & \epsilon_{yz} \\ \epsilon_{zx} & \epsilon_{zy} & \epsilon_{zz} \end{bmatrix} \tag{1.56}$$

and the constitutive relation for the electric field becomes

$$\bar{D} = \bar{\bar{\epsilon}}\,\bar{E} \tag{1.57}$$

All elements of $\bar{\bar{\epsilon}}$ are nonzero for a general anisotropic medium. However, the permittivity tensor in (1.56) must be a symmetric tensor as can be deduced from the argument of conservation of energy. In addition, the energy stored in an electric field must be positive. This condition requires that [1]

$$\epsilon_{xx} E_x^2 + \epsilon_{yy} E_y^2 + \epsilon_{zz} E_z^2 + 2\epsilon_{xy} E_x E_y + 2\epsilon_{xz} E_x E_z + 2\epsilon_{yz} E_y E_z > 0 \qquad (1.58)$$

If the right-hand side is set to be a positive constant, (1.58) becomes an equation of an ellipsoid. By the properties of a quadric, (1.58) can always be transformed so that it is expressed in terms of a set of three principal axes. Hence there exist a coordinate system in an anisotropic medium such that the energy stored in an electric field is given by

$$\epsilon_x E_x^2 + \epsilon_y E_y^2 + \epsilon_z E_z^2 > 0 \qquad (1.59)$$

and $\overline{\overline{\epsilon}}$ in (1.56) becomes

$$\overline{\overline{\epsilon}} = \begin{bmatrix} \epsilon_x & 0 & 0 \\ 0 & \epsilon_y & 0 \\ 0 & 0 & \epsilon_z \end{bmatrix} \qquad (1.60)$$

Anisotropic crystals can be classified into three groups, depending on the relationships of the elements in (1.60). These groups are:

(i) Isotropic medium if $\epsilon_x = \epsilon_y = \epsilon_z$

(ii) Uniaxial medium if any two of the three nonzero elements in (1.60) have the same value

(iii) Biaxial medium if all three of the nonzero elements in (1.60) are distinct

An anisotropic crystal can be demonstrated graphically by the indicatrix shown in Fig. 1.6. The three major axes of the indicatrix represent the principal dielectric constants of the medium. If an arbitrary line $OP$ is drawn from the origin, the cross-sectional plane perpendicular to the line $OP$ is an ellipse. There are two waves propagating in directions normal to $OP$ and their phase velocities are different.

For isotropic medium where $\epsilon_x = \epsilon_y = \epsilon_z$, the indicatrix degenerates into a sphere. Since any cross-sectional plane through the center of a sphere is a circle, waves propagating through an isotropic medium have only one phase velocity and the $\overline{\mathbf{D}}$ field is parallel to the $\overline{\mathbf{E}}$ field. For an uniaxial medium, two of the three principal axes in the indicatrix have the same length and the ellipsoid degenerates into a spheroid. The axis that is different in length from the other two is called the optic axis. Any plane

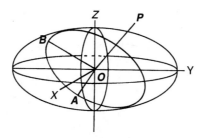

**Figure 1.6**   Indicatrix. (After Nye. *Physical Properties of Crystals*, 1957, Oxford University Press. Reproduced by permission.)

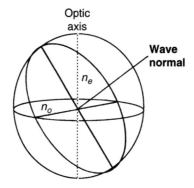

**Figure 1.7**   Indicatrix for uniaxial medium. (After Nye. Reproduced by permission.)

perpendicular to the optic axis is a circle and the polarization of a wave propagating along the optic axis is preserved. The indicatrix for this case is shown in Fig. 1.7. If a plane wave is incident on the boundary of a uniaxial crystal, two refracted waves occur inside the crystal: the ordinary wave and the extraordinary wave. This phenomenon is called double refraction or birefringence. If the thickness of a uniaxial crystal is chosen such that

$$d = \lambda_B/4 \tag{1.61}$$

and $\lambda_B$ is the birefringent wavelength given by

$$\lambda_B = \frac{2\pi}{|k_B|} = \frac{2\pi}{|k_o - k_e|} \tag{1.62}$$

then the polarization of the wave at the output plane of a uniaxial crystal is circular for a linearly polarized incident wave. Such a device is called a quarter-wave plate. $k_B$ in (1.62) is the birefringent wave number which is the difference of the wave numbers of the two waves in the crystal.

When the diagonal elements of the matrix in (1.60) are all different, the indicatrix is a triaxial ellipsoid and the crystal is biaxial. There are two directions from the origin where the cross-sectional plane is a circle. These are the two optic axes of a biaxial crystal as shown in Fig. 1.8. The three groups of crystal classified according to the crystal system are listed as follow:

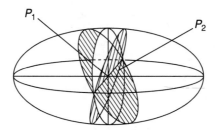

**Figure 1.8**   Indicatrix for biaxial medium. (After Nye. Reproduced by permission.)

| Crystal Classification | Crystal Systems |
|---|---|
| Isotropic | Cubic |
| Uniaxial | Hexogonal, trigonal, tetragonal |
| Biaxial | Monoclinic, orthorhombic, triclinic |

## 1.3  GEOMETRICAL OPTICS

To solve problems in optics using the physical optic approach as discussed in Section 1.2, the Maxwell equations should be solved and the solutions are uniquely determined by applying the appropriate initial and boundary conditions. However, the existence of closed-form solutions is limited to only a few problems which process geometrical symmetry and homogeneity of the medium. Approximation methods or numerical methods are needed to solve the more general problems. For those problems where the wavelength can be considered very small compared to the dimensions of the object involved (or equivalently, the operating frequency is very high), approximate solutions are much easier to obtain using geometrical optics.

Geometrical optics is a branch in optics in which the wavelength of the light wave is assumed to be zero. In this approximation, no information is furnished by the solutions on certain wave phenomena of light such as diffraction, polarization, and interference. On the other hand, the propagation law can be described by such geometrical language as the ray (a curve in space) and the eikonal (a surface in space). This simplification allows the light ray to be traced from point to point in space (ray tracing). It is found to be most advantageous for problems consisting of an inhomogeneous medium, multiple-lens systems, and optical instrument designs.

In geometrical optics, solutions to the wave equation are expressed in terms of the phase function (the eikonal) and the path of the propagating wave (the ray). After a few preliminary concepts have been introduced, both the ray equation and the eikonal equation are derived. The ray equations in other coordinate systems are listed as reference. Applications of these equations in an inhomogeneous medium, graded-index fiber, and certain types of graded-index lens are shown. The method of ray tracing and the *ABCD* matrix method are also discussed.

### 1.3.1  Some Preliminary Concepts

*Fermat's Principle.* Fermat's principle states that a propagating light ray always chooses the path that minimizes the optical path length. The differential optical path length is defined as

$$dl = n\,ds \tag{1.63}$$

where $ds$ is the differential length and $n$ is the refractive index. Mathematically, Fermat's principle assumes the form

$$\int_{P_1}^{P_2} n(x, y, z)\,ds = \text{minimum} \tag{1.64}$$

where $P_1$ and $P_2$ are two points in space.

This principle plays a fundamental role in geometrical optics. Snell's law, the ray equation, and the eikonal equation can be derived from this law via variational calculus [2]. The geometrical movements of the light ray are ultimately governed by this principle.

***Snell's Law.***   Snell's law is the most frequently used physical law in geometrical optics. It governs the directional change of a light ray when it encounters a change in the refractive index of the medium. It states that the direction of the refracted ray is related to the direction of the incident ray by the equation

$$n_1 \sin \theta_1 = n_2 \sin \theta_2 \tag{1.65}$$

where $n_1$ and $n_2$ are the refractive indexes of the first and second media, respectively. $\theta_1$ is the angle between the incident ray and the boundary surface normal, and $\theta_2$ is the angle between the refracted ray and the normal (see Fig. 1.9). If the boundary surface is that of a mirror, Snell's law of reflection is

$$\theta_1' = \pi - \theta_1 \tag{1.66}$$

where $\theta_1'$ is the angle between the reflected ray and the normal. If the light ray travels from a medium with high refractive index to a medium with lower refractive index, there exists a critical angle $\theta_c$ such that there will be no transmitted ray if $\theta_1 > \theta_c$.

***Unit Vectors and the Radius of Curvature.***   Two mathematical definitions are useful to the understanding of geometrical optics. If $C$ is a curve in space as shown in Fig. 1.10, and $P$ is a point on the curve with corresponding position vector $\overline{\mathbf{R}} = x\hat{x} + y\hat{y} + z\hat{z}$, the unit tangent vector at $P$ is

$$\hat{s} = \frac{d\overline{\mathbf{R}}}{ds} = \frac{dx}{ds}\hat{x} + \frac{dy}{ds}\hat{y} + \frac{dz}{ds}\hat{z} \tag{1.67}$$

and the unit normal vector $\hat{N}$ is

$$\hat{N} = -\rho \frac{d\hat{s}}{ds} \tag{1.68}$$

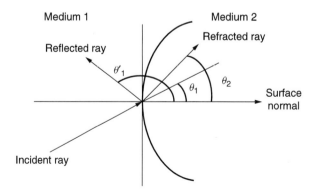

**Figure 1.9**   Snell's law.

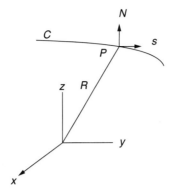

**Figure 1.10**   Unit vectors along a curve in space.

where $\rho$ is the radius of curvature defined as

$$\frac{1}{\rho} = \left|\frac{d\hat{s}}{ds}\right| = \left|\frac{d^2\overline{\mathbf{R}}}{ds^2}\right| \tag{1.69}$$

and $\hat{N}$ is on the convex side of the curve. $ds$ is the differential arc length along the curve $C$.

### 1.3.2   Ray Equation and Eikonal Equation

The ray equation and the eikonal equation can be directly derived from the Fermat's principle using variational calculus, or they may be obtained from the Maxwell equations. Another simple and perhaps the most informative way is to obtain these equations from a one-dimensional wave equation [5].

Let the scalar wave equation in a source-free region be written as

$$[\nabla^2 + k_0^2 n^2(x, y, z)]\psi = 0 \tag{1.70}$$

where $\psi$ represents any component of the $\overline{\mathbf{E}}$ or $\overline{\mathbf{H}}$ field. $k_0$ is the free-space wave number and $n(x, y, z)$ is the refractive index. The solution of (1.70) can be written in the form

$$\psi = A(x, y, z)e^{-jk_0 L(x,y,z)} \tag{1.71}$$

where $A(x, y, z)$ is the amplitude function and $L(x, y, z)$ is the phase function. Upon substitution of (1.71) into (1.70) and retaining only the terms containing the zeroth order of $k_0^{-1}$, we obtain the eikonal equation

$$|\nabla L|^2 = n^2(x, y, z) \tag{1.72}$$

The surface

$$L(x, y, z) = \text{constant} \tag{1.73}$$

is called the geometrical wavefront. If the unit normal to the geometrical wavefront is defined as

$$\hat{s} = \nabla L / |\nabla L| = \nabla L / n \qquad (1.74)$$

then $\hat{s}$ represents the direction of the light ray.

The ray equation, whose solution represents the path of the light ray, can be obtained from (1.72) by rewriting it as

$$dL/ds = n(x, y, z) \qquad (1.75)$$

where $dL/ds$ is the derivative in the direction of $\nabla L$. Taking the gradient of (1.75) and using (1.67) and (1.74), the ray equation is

$$\frac{d}{ds}\left(n\frac{d\overline{\mathbf{R}}}{ds}\right) = \nabla n \qquad (1.76)$$

where $\overline{\mathbf{R}}$ is the position vector and $ds$ is the incremental ray path length normal to the eikonal.

From the derivations above, one may observe that the solution to the vector equation (1.76) gives the position of the ray as a function of the coordinates, while the solutions to (1.72) and (1.74) give the direction of the ray and an equation of a surface which describe the geometrical wavefront. In view of (1.74), the ray path can be considered as the orthogonal trajectory to the geometrical wavefront. In addition, it has been shown that the direction of $\hat{s}$ coincides with the direction of the Poynting vector in an isotropic medium [3]. Hence the ray trajectory is the path through which optical energy is transported in the medium.

### 1.3.3    The Ray Equation in Three Coordinate Systems

In optical system designs, one may need to determine the ray path through optical components with cylindrical or spherical geometry. The problem can be made simpler if the ray equations are expressed in the appropriate coordinate system. These equations are listed below for the three coordinate systems most commonly used in optical technology [4].

(a)  Cartesian coordinate system $(x, y, z)$

$$\frac{d}{ds}\left(n\frac{dx}{ds}\right) = \frac{\partial n}{\partial x} \qquad (1.77a)$$

$$\frac{d}{ds}\left(n\frac{dy}{ds}\right) = \frac{\partial n}{\partial y} \qquad (1.77b)$$

$$\frac{d}{ds}\left(n\frac{dz}{ds}\right) = \frac{\partial n}{\partial z} \qquad (1.77c)$$

(b) Cylindrical coordinate system $(\rho, \phi, z)$

$$\frac{d}{ds}\left(n\frac{d\rho}{ds}\right) - n\rho\left(\frac{d\phi}{ds}\right)^2 = \frac{\partial n}{\partial \rho} \qquad (1.78a)$$

$$n\frac{d\rho}{ds}\frac{d\phi}{ds} + \frac{d}{ds}\left(n\rho\frac{d\phi}{ds}\right) = \frac{1}{\rho}\frac{d}{ds}\left(n\rho^2\frac{d\phi}{ds}\right) = \frac{1}{\rho}\frac{\partial n}{\partial \phi} \qquad (1.78b)$$

$$\frac{d}{ds}\left(n\frac{dz}{ds}\right) = \frac{\partial n}{\partial z} \qquad (1.78c)$$

(c) Spherical coordinate system $(r, \theta, \phi)$

$$\frac{d}{ds}\left(n\frac{dr}{ds}\right) - nr\sin^2\theta\left(\frac{d\phi}{ds}\right)^2 - nr\left(\frac{d\theta}{ds}\right)^2 = \frac{\partial n}{\partial r} \qquad (1.79a)$$

$$\frac{d}{ds}\left(nr\frac{d\theta}{ds}\right) - nr\sin\theta\cos\theta\left(\frac{d\phi}{ds}\right)^2 + n\frac{dr}{ds}\frac{d\theta}{ds} = \frac{1}{r}\frac{\partial n}{\partial \theta} \qquad (1.79b)$$

$$\frac{d}{ds}\left(nr\sin\theta\frac{d\phi}{ds}\right) + nr\cos\theta\frac{d\theta}{ds}\frac{d\phi}{ds} + n\sin\theta\frac{dr}{ds}\frac{d\theta}{ds} = \frac{1}{r\sin\theta}\frac{\partial n}{\partial \phi} \qquad (1.79c)$$

### 1.3.4  Basic Properties of Light Rays

As a direct solution to the ray equation, three basic properties of light rays are given as follow:

(i) Rays in a homogeneous medium are straight lines. From (1.76), the ray equation in a homogeneous medium is reduced to

$$d^2\overline{\mathbf{R}}/ds^2 = 0 \qquad (1.80)$$

and the solution is

$$\overline{\mathbf{R}} = s\overline{\mathbf{A}} + \overline{\mathbf{B}} \qquad (1.81)$$

where $\overline{\mathbf{A}}$ and $\overline{\mathbf{B}}$ are constant vectors. Equation (1.81) is the vector form of a straight-line equation.

(ii) Rays in a spherical symmetric medium are curves on a plane through the point of symmetry and the launch point. The refractive index of a spherical symmetric medium is written as

$$n(r, \theta, \phi) = n(r) \qquad (1.82)$$

Let a vector $\overline{\mathbf{U}}$ be defined as

$$\overline{\mathbf{U}} = \overline{\mathbf{R}} \times (n\hat{s}) \qquad (1.83)$$

where $\overline{\mathbf{R}}$ is the position vector and $\hat{s}$ is the direction of the incident ray. The quantity

$$\frac{d\overline{\mathbf{U}}}{ds} = \frac{d\overline{\mathbf{R}}}{ds} \times \left(n\frac{d\overline{\mathbf{R}}}{ds}\right) + \overline{\mathbf{R}} \times \frac{d}{ds}\left(n\frac{d\overline{\mathbf{R}}}{ds}\right) \qquad (1.84)$$

is identically zero, which indicates that the vector $\overline{U}$ remains constant with respect to $s$. That is, the ray path remains on the plane containing $\overline{R}$ and $\hat{s}$.

(iii) Boundary condition for the ray trajectory. The boundary condition for the ray can be obtained by considering the ray vector as given in (1.74)

$$n\frac{d\overline{R}}{ds} = n\hat{s} = \nabla L \qquad (1.85)$$

It follows that the integral

$$\int_A (\nabla \times n\hat{s}) \cdot \hat{a}\, dA \qquad (1.86)$$

vanishes for arbitrary area $A$ with unit normal $\hat{a}$. If the area is chosen to be a rectangular shape across the boundary between two different media as shown in Fig. 1.11, applying Stokes's theorem to (1.86) yields

$$(n_1\hat{s}_1 - n_2\hat{s}_2) \cdot \hat{l} = 0 \qquad (1.87)$$

where $\hat{l}$ is the unit vector tangent to the boundary. Equation (1.87) is the boundary condition for the light ray.

### 1.3.5  Ray Tracing in an Inhomogeneous Media

The solutions to the eikonal equation (1.72) and the ray equation (1.76) describe the wavefront and the ray trajectory, respectively. Closed-form solutions are not readily available if the problems involve complicated geometry or refractive index inhomogeneity. Several practical problems with index inhomogeneity in one spatial variable are used here to illustrate the solution approach. Other cases of interest are listed.

*Axial Inhomogeneous Medium.* Assuming a dielectric half-space whose index of refraction varies only with $x$ as shown in Fig. 1.12, the solution to the eikonal

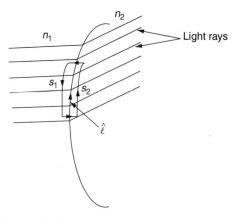

**Figure 1.11**  Boundary condition for the ray.

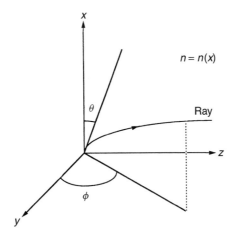

**Figure 1.12**   Ray in a planar inhomogeneous medium.

equation (1.72) using the method of separation of variables is

$$L(x, y, z) = \pm \int_K^x \sqrt{n^2(x) - (b^2 + c^2)} \, dx \pm by \pm cz \qquad (1.88)$$

where $b$ and $c$ are constants to be determined by the launch conditions of the ray, and $K$ is a constant. The launch conditions include the position at which the ray enters the medium and the direction of the ray at the entry point. The signs in (1.88) should be chosen to indicate the direction of the wave propagation.

The solution to the ray equation (1.76) for this problem is

$$y = \int \frac{b \, dx}{\sqrt{n^2(x) - (b^2 + c^2)}} \qquad (1.89a)$$

$$z = \int \frac{c \, dx}{\sqrt{n^2(x) - (b^2 + c^2)}} \qquad (1.89b)$$

As an example, let the launch coordinate be the origin and the launch angle be $(\theta_0, \phi_0)$; then the plane containing the ray is

$$z = y \tan \phi_0 \qquad (1.90)$$

and the equation for $z$ is

$$z = \int_0^x \frac{n_0 \sin \theta_0 \sin \phi_0 \, dx}{\sqrt{n^2(x) - n_0^2 \sin^2 \theta_0}} \qquad (1.91)$$

with $n_0$ being the index at $x = 0$.

It is important to note that there exists a turning point in the ray path if $n(x)$ is monotonic decreasing with increasing $x$. The turning point $x_0$ satisfies the equation

$$n(x_0) = n_0 \sin \theta_0 \qquad (1.92)$$

Equation (1.91) is a general solution for arbitrary index variation $n(x)$. Three different index functions will be considered here and their ray solutions given without derivation.

(i) If the index variation with respect to $x$ is [5]

$$n^2(x) = n^2_{max}(1 - a^2x^2)$$  (1.93)

with $\alpha$ a constant, the ray trajectory is given by

$$z = \frac{\sin\theta_0}{a}\sin^{-1}\left(\frac{a}{\cos\theta_0}x\right)$$  (1.94)

where the launch point is assumed to be the origin and $\theta_0$ is the launch angle. A graph of the ray trajectory is shown in Fig. 1.13.

(ii) If the index varies with respect to $x$ as [4]

$$n(x) = c/|x|$$  (1.95)

where $c$ is a constant, the equation of the trajectory is

$$(zn^2_0\sin^2\theta_0 + b)^2 = c^2 - n_0x^2\sin^2\theta_0$$  (1.96)

with $b$ is a constant of integration, $n_0$ is the index at the launch point $(x_0, 0)$, and $\theta_0$ is the launch angle.

Equation (1.96) is that of an ellipse, which means that the ray trajectory is a closed path. For some particular values of $c$, the equation can be reduced to that of a circle. The ellipses from different launch conditions always have one of the major axes coincide with the $z$-axis.

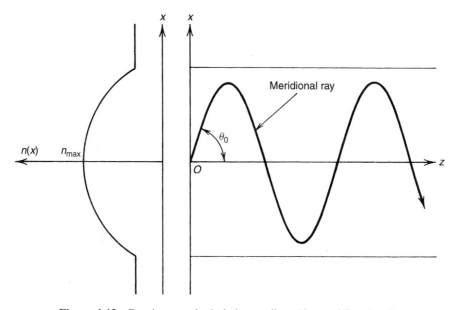

**Figure 1.13**   Ray in a quadratic index medium (the meridional ray).

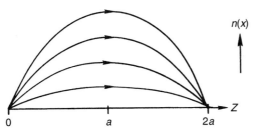

**Figure 1.14** Ray in a medium with hypobolic secant index profile. (After Cornbleet [4]. Copyright 1984, Wiley. Reproduced by permission.)

(iii) An inhomogeneous medium with hyperbolic secant index profile has the property to refocus the light rays. This index profile is given by [4]

$$n(x) = n_0 \sec h(\pi x/2a) \tag{1.97}$$

The equation for the trajectory is

$$\sin(\pi z/2a) = \tan \theta_0 \sin h(\pi x/2a) \tag{1.98}$$

A graph of the ray trajectory is shown in Fig. 1.14.

***Cylindrically Inhomogeneous Medium.*** The refractive index of a cylindrically inhomogeneous medium varies with respect to the transverse coordinates orthogonal to the axial direction. For a general cylindrical medium, Cornbleet [4] uses a conformal map to transform the ray equations from cylindrical coordinates to rectangular coordinates. Solutions to these equations in rectangular coordinates can be obtained for those cases where the rays are confined to surfaces with one of the coordinates being held constant. As a special case, the ray solutions in a circular cylindrical medium with a quadratic index variation is given here. For a circular cylindrical structure having cylindrically symmetric index variation $[n = n(\rho)]$, the general solution for the optical ray is

$$\phi = \int \frac{c \, d\rho}{\rho^2 \sqrt{n^2(\rho) - (c^2/\rho^2 + b^2)}} \tag{1.99}$$

$$z = \int \frac{b \, d\rho}{\sqrt{n^2(\rho) - (c^2/\rho^2 + b^2)}} \tag{1.100}$$

where $b$ and $c$ are constants to be determined from launch conditions. When $c = 0$, the ray path stays on a plane containing the $z$-axis and the ray is called a meridional ray. The ray expression is the same as (1.91) for the inhomogeneous half-space. In particular, if the index profile is quadratic in $\rho$, such as that of a graded-index fiber,

$$n^2(\rho) = n_{\max}^2 (1 - a^2 \rho^2) \tag{1.101}$$

the ray trajectory is given by

$$\rho = \frac{\cos \theta_0}{a} \sin \left( \frac{az}{\sin \theta_0} \right) \tag{1.102}$$

where $\theta_0$ is the launch angle with respect to the radial axis.

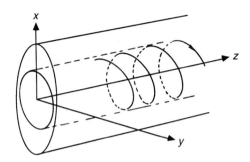

**Figure 1.15**   Skew ray in a graded-index fiber.

If $c \neq 0$, the ray path is an elliptical helix. Such a ray is a skew ray and the equation for the ray trajectory is given by [5]

$$\frac{\rho^2 \cos^2(\phi - \phi_0 - \pi/4)}{\left(\dfrac{c}{n_{\max} a \sqrt{p - q}}\right)^2} + \frac{\rho^2 \sin^2(\phi - \phi_0 - \pi/4)}{\left(\dfrac{c}{n_{\max} a \sqrt{p + q}}\right)^2} \tag{1.103}$$

where $p$ and $q$ are constants given by

$$p = \frac{n_{\max}^2 - b^2}{2 n_{\max}^2 a^2} \tag{1.104a}$$

$$q^2 = p^2 - \left(\frac{c}{n_{\max} a}\right)^2 \tag{1.104b}$$

An illustration of the skew ray is shown in Fig. 1.15. Other applications using inhomogeneous cylindrical medium for lens are given in Refs. 4 and 6.

***Spherically Symmetric Inhomogeneous Medium.***   If the inhomogeneity of a medium varies with respect to the radial coordinate of a spherical coordinate system alone, the medium is spherically symmetric. A ray entering this medium will always be on a plane containing the incident ray and the point of symmetry.

Assuming that a medium is spherically symmetric and the equatorial plane contains the ray trajectory, Eqs. (1.79a)–(1.79c) are reduced to

$$\frac{d}{ds}\left(n \frac{dr}{ds}\right) - nr \left(\frac{d\phi}{ds}\right)^2 = \frac{\partial n}{\partial r} \tag{1.105a}$$

$$\frac{d}{ds}\left(nr \frac{d\phi}{ds}\right) + n \frac{dr}{ds}\frac{d\phi}{ds} = 0 \tag{1.105b}$$

The equation of the ray is given by

$$\phi = \int_K^r \frac{A\, dr}{r[r^2 n^2(r) - A^2]^{1/2}} \tag{1.106}$$

where $A$ is $nr \sin \Psi$, with $\Psi$ the angle between $\overline{R}$ and $\overline{s}$ at a point on the ray and $K$ an integration constant. The solution to this equation for a given index profile $n(r)$ requires the Abels integral given in Ref. [4]. The ray trajectories of several different spherical symmetric lenses are listed below.

(i) Maxwell's fish-eye

Index profile:   $n(r) = \dfrac{2}{1 + r^2}$     (1.107a)

Ray equation:   $r = (1 + \sin^2 \phi \cot^2 \alpha)^{1/2} + \sin \phi \cot \alpha$     (1.107b)

(ii) Luneberg collimating lens

Index profile:   $n(r) = (2 - r^2)^{1/2}$     (1.108a)

Ray equation:   $r^2 = \sin^2 \alpha / [1 - \cos \alpha \cos(2\phi + \alpha)]$     (1.108b)

(iii) Eaton lens

Index profile:   $n(r) = \left( \dfrac{2}{r} - 1 \right)^{1/2}$     (1.109a)

Ray equation:   $r = \sin^2 \alpha / (1 - \cos \alpha \cos \phi)$     (1.109b)

The angle $\alpha$ in the equations above is the launch angle with respect to the normal of the spherical surface at the launch point. The ray pictures are shown in Fig. 1.16.

### 1.3.6  *ABCD* Transformation

In many optical system designs, it is important to know how the ray path will be altered as the ray encounters a boundary. It is well understood that the ray path is fully specified by the position and the slope of the ray path with respect to an arbitrary chosen axis. A simple $2 \times 2$ matrix, called the *ABCD* matrix, can be formulated for each optical system to relate the direction and position of the output ray to those of the input ray. Moreover, the equivalent matrix of two optical systems in cascade is equal to the product of the matrices of the individual systems. Using this approach, the ray path can be traced from the source plane to the image plane by simple matrix multiplications. The matrices of several simple systems are shown in this section. For more complicated systems and the derivation of the *ABCD* matrices, the reader is referred to Refs. 7–9.

Let the input ray at plane 1 be located at $r_1$, and the ray slope is $r_1'$, as shown in Fig. 1.17. In the same manner, the output ray is specified by $r_2$ and $r_2'$ at the output

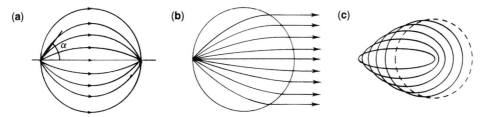

**Figure 1.16**  Rays in spherical symmetric index media: **(a)** Maxwell's fish-eye; **(b)** Luneberg lens; **(c)** Eaton lens. (After Cornbleet [4]. Reproduced by permission.)

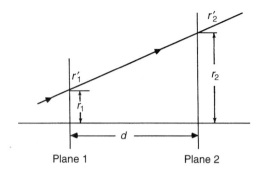

**Figure 1.17**   Ray geometry for the *ABCD* transformation.

plane 2. Assuming that the refractive index of the medium is unity and the distance separating the two planes is $d$, using simple geometry, we obtain

$$\begin{bmatrix} r_2 \\ r_2' \end{bmatrix} = \begin{bmatrix} 1 & d \\ 0 & 1 \end{bmatrix} \begin{bmatrix} r_1 \\ r_1' \end{bmatrix} \tag{1.110}$$

The square matrix in (1.110) is the *ABCD* matrix or the transformation matrix $T$ for the space between planes 1 and 2. If the space between the two planes is filled with material with refractive index $n$, the transformation matrix for a dielectric slab of thickness $d$ is

$$T = \begin{bmatrix} 1 & d/n \\ 0 & 1 \end{bmatrix} \tag{1.111}$$

If the space between planes 1 and 2 is only half-filled with dielectrics whose refractive indices are $n_1$ and $n_2$, the composite system transformation matrix is

$$T = \begin{bmatrix} 1 & d/2n_1 \\ 0 & 1 \end{bmatrix} \begin{bmatrix} 1 & d/2n_2 \\ 0 & 1 \end{bmatrix} = \begin{bmatrix} 1 & \dfrac{d}{2}\left(\dfrac{1}{n_1} + \dfrac{1}{n_2}\right) \\ 0 & 1 \end{bmatrix} \tag{1.112}$$

For a convex lens of focal length $f$, the input plane 1 may be placed on one side of the lens and plane 2 on the other side. The *ABCD* matrix of a simple convex lens is

$$T = \begin{bmatrix} 1 & 0 \\ -1/f & 1 \end{bmatrix} \tag{1.113}$$

For image formation, one must have $r_2 = Mr_1$ with $M$ the magnification, and the position $r_2$ is independent $r_1'$. If plane 1 is placed at a distance $d_1$ in front of a lens and plane 2 is placed at $d_2$ behind the lens, the transformation matrix becomes

$$T = \begin{bmatrix} 1 - d_2/f & d_1 + d_2(1 - d_1/f) \\ -1/f & 1 - d_1/f \end{bmatrix} \tag{1.114}$$

The magnification $M$ is given by

$$M = |d_2/d_1| \tag{1.115}$$

and $r_2$ is independent of $r_1'$ if

$$\frac{1}{d_1} + \frac{1}{d_2} = \frac{1}{f} \tag{1.116}$$

which is the well-known Gaussian lens formula.

The elements of the *ABCD* matrix can be used for the transformation of Gaussian beam parameters in lens waveguides and for stability study of optical resonators. These discussions will be taken up in the next section. A list of the transformation matrices is included in Table 1.1.

### 1.3.7  *ABCD* Matrix in Quadratic Index Medium [9]

When the refractive index of an optical system varies continuously with respect to the spatial coordinates, a more general definition of the slope of the ray path is needed to account for the index variations. Let the ray slope $r(z)$ be defined as

$$r'(z) = n(z)\frac{dr(z)}{dz} \tag{1.117}$$

so that the *ABCD* matrix is given by

$$\begin{bmatrix} r_2 \\ r_2' \end{bmatrix} = \begin{bmatrix} A & B \\ C & D \end{bmatrix} \begin{bmatrix} r_1 \\ r_1' \end{bmatrix} \tag{1.118}$$

The general expression of a medium with quadratic variation can be written in the form

$$n(r, z) = n_0(z) - \tfrac{1}{2}n_q(z)r^2(z) \tag{1.119}$$

where $n_0(z)$ refers to the refractive index on the $z$-axis and $n_q(z)$ is the curvature of the index profile at the axis. The ray equation similar to (1.76) is written as

$$\frac{d}{dz}\left[n_0(z)\frac{dr(z)}{dz}\right] = -n_q(z)r(z) \tag{1.120}$$

where the paraxial approximation has been made. If the ray slope is defined as in (1.117) to be

$$r'(z) = n_0(z)\frac{dr(z)}{dz} \tag{1.121}$$

then (1.120) becomes

$$\frac{d}{dz}r'(z) = -n_q(z)r(z) \tag{1.122}$$

If $n_0(z)$ and $n_q(z)$ are both constants, as in the case of a graded-index fiber, the two equations can be combined to give

$$\frac{d^2}{dz^2}r(z) = -\frac{n_q}{n_0}r(z) \tag{1.123}$$

**TABLE 1.1   Ray Matrices for Some Common Optical Elements and Media. (After Yariv [8], Copyright 1975, Wiley. Reproduced by permission)**

| | | |
|---|---|---|
| (1) | Straight section length $d$ $n = 1$ | 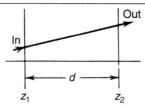 |

$$\begin{bmatrix} 1 & d \\ 0 & 1 \end{bmatrix}$$

| | |
|---|---|
| (2) Thin lens: focal length $f$ ($f > 0$, converging; $f < 0$, diverging) | 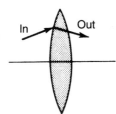 |

$$\begin{bmatrix} 1 & 0 \\ \dfrac{-1}{f} & 1 \end{bmatrix}$$

| | |
|---|---|
| (3) Dielectric interface: refractive indices $n_1, n_2$ | 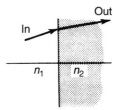 |

$$\begin{bmatrix} 1 & 0 \\ 0 & \dfrac{n_1}{n_2} \end{bmatrix}$$

| | |
|---|---|
| (4) Curved dielectric interface: radius $R$ | 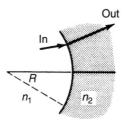 |

$$\begin{bmatrix} 1 & 0 \\ \dfrac{n_2 - n_1}{n_2} \dfrac{1}{R} & \dfrac{n_1}{n_2} \end{bmatrix}$$

| | |
|---|---|
| (5) Curved mirror: radius of curvature $R$ | 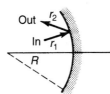 |

$$\begin{bmatrix} 1 & 0 \\ \dfrac{-2}{R} & 1 \end{bmatrix}$$

| | |
|---|---|
| (6) A medium with a quadratic index profile $n$ | 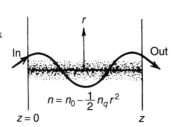 |

$$\begin{bmatrix} \cos\left(\sqrt{\dfrac{n_q}{n_0}}\,z\right) & \sqrt{\dfrac{n_0}{n_q}}\sin\left(\sqrt{\dfrac{n_q}{n_0}}\,z\right) \\ -\sqrt{\dfrac{n_q}{n_0}}\sin\left(\sqrt{\dfrac{n_q}{n_0}}\,z\right) & \cos\left(\sqrt{\dfrac{n_q}{n_0}}\,z\right) \end{bmatrix}$$

$$n = n_0 - \frac{1}{2}n_q r^2$$

The solution to (1.123) is

$$r(z) = r_0 \cos\left(\sqrt{\frac{n_q}{n_0}}z\right) + \frac{1}{\sqrt{n_0 n_q}} \sin\left(\sqrt{\frac{n_q}{n_0}}z\right) r_0' \qquad (1.124)$$

If the definition of (1.121) is used in (1.124), it becomes

$$r(z) = r_0 \cos\left(\sqrt{\frac{n_q}{n_0}}z\right) + \sqrt{\frac{n_0}{n_4}} \sin\left(\sqrt{\frac{n_q}{n_0}}z\right) \frac{dr_0}{dz} \qquad (1.125)$$

Similarly, the equation for the ray slope is

$$\frac{dr(z)}{dz} = -\sqrt{\frac{n_q}{n_0}} \sin\left(\sqrt{\frac{n_q}{n_0}}z\right) r_0 + \cos\left(\sqrt{\frac{n_q}{n_0}}z\right) \frac{dr_0}{dz} \qquad (1.126)$$

Hence the *ABCD* matrix is represented by

$$T = \begin{bmatrix} \cos\left(\sqrt{\frac{n_q}{n_0}}z\right) & \sqrt{\frac{n_0}{n_q}} \sin\left(\sqrt{\frac{n_q}{n_0}}z\right) \\ -\sqrt{\frac{n_q}{n_0}} \sin\left(\sqrt{\frac{n_q}{n_0}}z\right) & \cos\left(\sqrt{\frac{n_q}{n_0}}z\right) \end{bmatrix} \qquad (1.127)$$

Equation (1.125) points out that the ray position will vary sinusoidally with $z$ for $n_q$ being positive ($n_0$ is assumed to be positive); that is, the refractive index decreases from the $z$-axis. The medium is called a stable quadratic duct. However, if $n_q$ is negative, the ray will diverge away from the axis, and the medium is an unstable quadratic index medium.

## 1.4 PROPAGATION OF GAUSSIAN BEAM

In previous discussions on geometrical optics, the optical ray is considered as a path along which optical energy is transported. The ray cannot be considered as a field because it lacks amplitude, phase, and other field properties. An optical beam wave such as one produced by a laser can be treated more accurately by the wave nature of light. The field of an optical beam can be approximately represented by the Hermite–Gaussian functions.

The Hermite–Gaussian functions developed in this section are exact solutions to the paraxial wave equation. They form a complete set of orthogonal functions and are often used as basis functions in normal-mode expansion. Some characteristics of this set of functions are listed at the end of this section.

The Hermite–Gaussian modes occur as solutions to the wave equation in a quadratic index medium. They are also the normal modes of a lens waveguide. Because the mathematical functions used in the expressions of the Gaussian beam and the Gaussian modes are very similar, they are treated together in this section.

After the paraxial wave equation is introduced, the lowest-order beam and its properties will be discussed. Propagation of the Gaussian beam in various optical systems will be summarized. Finally, some properties of the Hermite–Gaussian function are presented briefly.

### 1.4.1  Paraxial Wave Equation and Its Solution [11]

If one considers the propagation of light nearly parallel to an arbitrary chosen axis (say the $+z$-axis), the solution to the scalar wave equation may be expressed in the form

$$\psi(x, y, z) = u(x, y, z)e^{-jkz} \tag{1.128}$$

Since $k = 2\pi/\lambda$ is in general a large number for small optical wavelengths, the approximation

$$k\frac{\partial u}{\partial z} \gg \frac{\partial^2 u}{\partial z^2} \tag{1.129}$$

is valid. Upon substitution of (1.128) into the scalar wave equation and using (1.129), we obtain the paraxial wave equation (reduced wave equation)

$$\nabla_t^2 u - 2jk\frac{\partial u}{\partial z} = 0 \tag{1.130}$$

where $\nabla_t$ is the del operator for the transverse coordinates only. The solution to this equation can be written as

$$\psi(x, y, z) = A\frac{w_0}{w(z)} H_n\left(\sqrt{2}\frac{x}{w(z)}\right) H_m\left(\sqrt{2}\frac{y}{w(z)}\right)$$

$$\cdot \exp\left\{-j\left[kz - (m+n+1)\tan^{-1}\frac{\lambda z}{\pi w_0^2} + \frac{\pi(x^2+y^2)}{\lambda R(z)}\right]\right\} e^{-(x^2+y^2)/w^2(z)} \tag{1.131}$$

which is the expression of the $(n, m)$th order Gaussian beam. Various parameters contained in this complicated expression are explained below.

$A$ is a normalization constant.

$w(z)$ is the spot size of the beam, which varies with distance $z$. It is defined to be the distance between the two half-power points of the fundamental mode. In terms of $w_0$ it is given by

$$w^2(z) = w_0^2\left[1 + \left(\frac{z}{z_0}\right)^2\right] \tag{1.132}$$

with $w_0$ the minimum spot size of the beam, which occurs at $z = 0$. $z_0 = \pi w_0^2/\lambda$ is a point on the $z$-axis at which the spot size is equal to $\sqrt{2}w_0$.

$R(z)$ is the wavefront curvature of the beam, which is defined as

$$R(z) = z\left[1 + \left(\frac{z_0}{z}\right)^2\right] \tag{1.133}$$

$H_m$ and $H_n$ are the $m$th- and $n$th-order Hermite polynomials which satisfy the Hermite equation

$$\frac{d^2}{dx^2}H_m(x) - 2x\frac{d}{dx}H_m(x) + 2mH_m(x) = 0 \qquad (1.134)$$

Several low-order Hermite polynomials are as follows:

$$H_0(x) = 1 \qquad (1.135a)$$

$$H_1(x) = 2x \qquad (1.135b)$$

$$H_2(x) = 4x^2 - 2 \qquad (1.135c)$$

$$H_3(x) = 8x^3 - 12x \qquad (1.135d)$$

Higher-order polynomials can be generated by the recursion formula

$$H_{n+1}(x) - 2xH_n(x) + 2nH_{n-1}(x) = 0 \qquad (1.136)$$

The properties of the Gaussian beam can be understood from the lowest-order beam, $TEM_{0,0}$ as in Fig. 1.18, where the beam profile is given in terms of circular coordinate $r^2 = x^2 + y^2$. From (1.131), the amplitude of the field is seen to vary with $r$ and $z$ as $w_0/w(z)\exp[-r^2/w^2(z)]$. The amplitude decays in the transverse direction away from the $z$-axis, and the rate of decay decreases as $w(z)$ increases with increasing $z$. This corresponds to the spreading of the beam as the beam propagates in $+z$ direction. The spreading angle $\theta$ is approximately equal to $2\lambda/\pi w_0$. The factor $w_0/w(z)$ in (1.131) serves as a weighting factor so that the total power carried by the beam remains constant for all $z$. In other words, the divergence of the beam requires the reduction of the peak amplitude for the conservation of total energy in the beam.

The longitudinal phase factor

$$\exp\{-j[kz - \tan^{-1}(z/z_0)]\} \qquad (1.137)$$

determines the phase velocity of the beam. The phase velocity

$$v_p = \frac{c}{1 - (1/kz)\tan^{-1}(z/z_0)} \qquad (1.138)$$

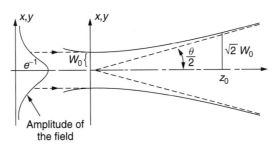

**Figure 1.18** $TEM_{0,0}$ order of Gaussian beam. (After Verdeyen [11]. Copyright 1981, Prentice Hall. Reproduced by permission.)

is slightly greater than the velocity of light. The radial phase factor

$$\exp\left\{-j\frac{\pi(x^2+y^2)}{\lambda R(z)}\right\}$$ (1.139)

indicates that the surface of constant phase is a spherical surface with its radius of curvature increasing with $z$. This observation is justified because the phase of spherical wave initiated at the origin of the coordinate system can be approximately written as

$$e^{-jkz}e^{-jk(x^2+y^2)/2R}$$ (1.140)

for $z \gg (x^2+y^2)$. The transverse variation of the phase in (1.139) and (1.140) are the same except that $R(z)$ in (1.139) is not a constant. Hence, from the phase expression, the Gaussian beam appears to have been initiated at the origin when it is observed at a large distance. As the observation point moves closer to the origin, the center of the spherical surface for the phase moves in the $-z$ direction. When the observation point is at $z = 0$, the wavefront curvature $R(z) = \infty$, the surface of constant phase is a plane, and the spot size is at its minimum.

The amplitude profile of this lowest-order Gaussian beam has only one main lobe. It is circular symmetric and has the maximum amplitude at the center of the beam. The constant phase surface of the beam is a plane at the point where the spot size is at its minimum. These are desirable properties for coupling of the beam into an optical fiber. Figure 1.19 is an illustration of the cross section of this fundamental mode and other higher-order modes.

### 1.4.2    Power Transmission through a Circular Aperture [9, 11]

The percentage of Gaussian beam power transmitted through a finite aperture is useful information for optical system design. Considering only the lowest-order beam, the

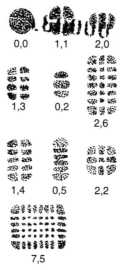

**Figure 1.19**    Cross-sectional view of higher-order Gaussian beam. (After Siegman [9]. Copyright 1986, University Science Book. Reproduced by permission.)

intensity profile is given by

$$I = \frac{2}{\pi w^2} e^{-2(x^2+y^2)/w^2} \tag{1.141}$$

where the total power of the beam has been normalized to unity. The power transmitted through a circular aperture of radius $a$ is

$$P_t = \iint I \, ds = 1 - e^{-2a^2/w^2} \tag{1.142}$$

If $a$ is the same as the spot size $w$, only 86% of the total power is transmitted. For $a = 1.5w$, approximately 99% of the power in the beam is transmitted.

Although the percentage of transmitted power is large for $a = 1.5w$, the sharp-edged aperture will cause diffraction loss and deterioration of the beam profile. The radius $a$ must be increased to $4.6w$ to reduce the ripple on the intensity profile to less than 1%.

### 1.4.3 Complex Representation of Gaussian Beam

The parameters in the Gaussian beam expression (1.131) are real and measurable quantities. They are useful for the interpretation of the behavior of the beam. These parameters are changed as the beam propagates from one optical system to another. It is much simpler to represent these parameters by a single complex radius of curvature $q(z)$ such that

$$\frac{1}{q(z)} = \frac{1}{R(z)} - j \frac{2}{kw^2(z)} \tag{1.143}$$

It has been shown [9] that a complete set of normalized Hermite-Gaussian beam solutions can be written in terms of $q(z)$ as

$$\psi(x, y, z) = \left(\frac{2}{\pi}\right)^{1/2} \left(\frac{1}{w_0}\right) \left(\frac{1}{2^{n+m}n!m!}\right)^{1/2} \left(\frac{q_0}{q(z)}\right) \left(\frac{q_0 q^*(z)}{q_0^* q(z)}\right)^{(m+n)/2}$$

$$\times H_n\left(\frac{\sqrt{2}x}{w(z)}\right) H_m\left(\frac{\sqrt{2}y}{w(z)}\right) \exp\left[-j\frac{k(x^2+y^2)}{2q(z)}\right] \tag{1.144}$$

This two-dimensional Gaussian beam expression is in fact a product solution of two one-dimensional ones with $q_0$ being the complex radius of curvature at a given reference plane. This set of Gaussian beam solutions is normalized so that the total power is unity. It is widely used to represent the physical solutions of stable lasers.

### 1.4.4 Transformation of Gaussian Beam [8]

When a Gaussian beam is propagating through an optical system, the parameters of the beam are changed according to the ABCD law given below. The parameters A, B, C, D are the elements of the transformation matrix in geometrical optics. The complex Gaussian parameter $q$, given by

$$\frac{1}{q(z)} = \frac{1}{R(z)} - j \frac{\lambda}{\pi w^2(z)} \tag{1.145}$$

is transformed according to the relation

$$q_2 = \frac{Aq_1 + B}{Cq_1 + D} \tag{1.146}$$

where $q_1$ is the complex Gaussian parameter at the input plane of an optical system and $q_2$, at the output plane. A simple example will demonstrate the simplicity of (1.146).

Let the optical system be a thin lens of focal length $f$; the $ABCD$ elements are $A = D = 1, C = -1/f$, and $B = 0$. Assuming an input beam with complex parameter $q_1$ is located at the front of the lens (plane 1), as shown in Fig. 1.20, the output parameter $q_2$ at the back plane (plane 2) of the lens is given by

$$\frac{1}{q_2} = -\frac{1}{f} + \frac{1}{q_1} \tag{1.147}$$

If the input beam at plane 1 has a planar wavefront ($R_1 = \infty$) and spot size $w_1$ so that

$$\frac{1}{q_1} = \frac{1}{\infty} - j\frac{\lambda}{\pi w_1^2} \tag{1.148}$$

the output parameter is

$$\frac{1}{q_2} = -\frac{1}{f} - j\frac{\lambda}{\pi w_1^2} \tag{1.149}$$

where the spot size remains the same across the lens but the beam is converged toward the focal point of the lens.

If this beam is allowed to propagate a distance $d$ from plane 2, the complex parameter at the plane $z = d$ (plane 3) becomes

$$q_3 = q_2 + d \tag{1.150}$$

according to (1.146). If $d$ is chosen so that the spot size is at its minimum (i.e., $R_3 = \infty$), one finds that

$$d = \frac{f}{1 + (f/z_c)^2} \tag{1.151}$$

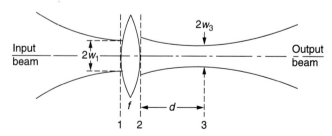

**Figure 1.20**   Transformation of Gaussian beam by a thin lens. (After Yariv [8]. Copyright 1975, Wiley. Reproduced by permission.)

where $z_c$ is the confocal length given by

$$z_c = \pi w_1^2/\lambda \qquad (1.152)$$

The ratio of the input and output spot sizes is given by [8]

$$\frac{w_3}{w_1} = \frac{f/z_c}{\sqrt{1 + (f/z_c)^2}} \qquad (1.153)$$

Since the ratio is positive and less than unity for positive $f$, the spot size at plane 3 is reduced and the beam is said to have been focused.

### 1.4.5  Gaussian Mode in Lens Waveguide

The focusing effect of a lens on a Gaussian beam has been demonstrated in the preceding section. The beam begins to diverge again after it passes the point where the spot size is minimum. To guide the beam and minimize the beam spreading, it is necessary to place lenses periodically along the path of propagation to reconverge the beam. Such a sequence of lenses forms a lens waveguide. A special case of a lens waveguide is a confocal guide where identical lenses of focal length $f$ are placed on a straight line at a distance $2f$ apart, as shown in Fig. 1.21. The modes of this lens waveguide are represented by the Hermite–Gaussian functions. The normalized expression of the ($m$, $n$)th mode is [2]

$$\psi_{mn} = \frac{1}{(2^{m+n}m!n!\lambda f)^{1/2}} H_m\left(\sqrt{\frac{\pi}{\lambda f}}x\right) H_n\left(\sqrt{\frac{\pi}{\lambda f}}y\right) e^{-(\pi/2\lambda f)(x^2+y^2)} \qquad (1.154)$$

For less restricted forms of lens waveguides, the eigenmodes are given by (1.131). Propagation of a Gaussian beam through irregular placement of lenses should be treated by the transformation law discussed above.

### 1.4.6  Gaussian Modes in Quadratic Index Medium

If a lens waveguide is idealized so that it consists of a large number of thin lenses and the spacing between them approaches zero, one obtains a continuous focusing medium whose refractive index varies according to

$$n^2(x, y) = n_0^2 - n_0 n_q (x^2 + y^2) \qquad (1.155)$$

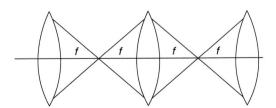

**Figure 1.21**  Confocal lens waveguide.

The medium is a quadratic index medium or square-law medium. Waves propagating in this medium satisfy the scalar wave equation

$$\nabla^2 \psi + k_0^2 n^2 \psi = 0 \tag{1.156}$$

The normalized modal solution is given by [2]

$$\psi_{lm}(x, y, z) = \left(\frac{2}{\pi} \frac{1}{2^{m+1} m! l!}\right)^{1/2} \frac{1}{w} H_l\left(\sqrt{2}\frac{y}{w}\right) H_m\left(\sqrt{2}\frac{x}{w}\right) e^{-(x^2+y^2)/w^2} e^{-j\beta_{lm}z}$$

$$\tag{1.157}$$

with

$$w^2 = \frac{2}{k_0 \sqrt{n_0 n_q}} \tag{1.158}$$

and

$$\beta_{lm} = \{n_0^2 k_0^2 - k_0 \sqrt{n_0 n_q}[(2l + 1) + (2m + 1)]\}^{1/2} \tag{1.159}$$

Notice that the phase factor $\beta_{lm}$ is independent of the transverse coordinates. This means that the waves in this medium are inhomogenous plane waves. In addition, the spot size $w$ depends only on the index coefficients $n_0$ and $n_q$. Hence the eigenmodes of a quadratic index medium, which are the exact solutions to (1.156), are Hermite–Gaussian modes with a plane wavefront and a constant spot size.

### 1.4.7   Gaussian Mode Matching [2]

If a Gaussian beam is introduced into a quadratic medium whose eigenmodes do not match with the amplitude profile and the wavefront curvature of the incoming beam, the self-focusing property of the medium and the spreading of the beam cause the spot size of the beam to be modulated as the beam propagates through the medium. This effect can be explained by the interference of many eigenmodes excited by the input beam traveling through the medium with different velocities.

It is possible to transform the Gaussian beam from the output of a laser by a lens to match the eigenmodes of a quadratic index medium. Let the input laser beam be characterized by the minimum spot size $w_1$, and its distance $d_1$ from the lens of focal length $f$. The minimum spot size $w_2$ of the output beam is located at $d_2$ as shown in Fig. 1.22.

The problem of mode matching is to determine the distances $d_1$ and $d_2$ so that the output beam profile matches the eigenmodes of the quadratic index medium for a given $w_1$. The transformation matrix is given by a product of three matrices as

$$\begin{bmatrix} A & B \\ C & D \end{bmatrix} = \begin{bmatrix} 1 & d_2 \\ 0 & 1 \end{bmatrix} \begin{bmatrix} 1 & 0 \\ -1/f & 1 \end{bmatrix} \begin{bmatrix} 1 & d_1 \\ 0 & 1 \end{bmatrix} \tag{1.160}$$

Since the elements in the transformation matrix are real and the complex Gaussian parameters of the input and output beam are purely imaginary, two equations for the distances $d_1$ and $d_2$ can be obtained. The solutions are

$$d_1 = f \pm \frac{w_1}{w_2} \sqrt{f^2 - f_1^2} \tag{1.161a}$$

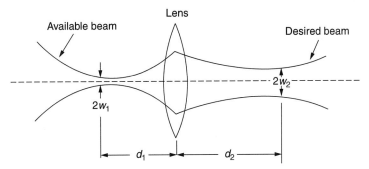

**Figure 1.22**   Gaussian mode matching. (After Marcuse [2]. Copyright 1982, Van Nostrand Reinhold. Reproduced by permission.)

$$d_2 = f \pm \frac{w_2}{w_1}\sqrt{f^2 - f_1^2} \qquad (1.161b)$$

with

$$f_1 = \frac{\pi}{\lambda}w_1 w_2 \qquad (1.162)$$

The signs on the right side of (1.161a) and (1.161b) must be simultaneously positive or negative and $f - f_1$ must be nonnegative.

### 1.4.8   Beam Propagation in a Lens Waveguide

When the spacing of the lenses in a lens waveguide are irregular, the arrangement can be viewed as a cascade of subsystems. Each subsystem consists of a lens and a separation distance before the next lens, as shown in Fig. 1.23. The transformation matrix for $M$ sections of subsystem is

$$\begin{bmatrix} A & B \\ C & D \end{bmatrix} = \prod_{k=1}^{M} \begin{bmatrix} A_k & B_k \\ C_k & D_k \end{bmatrix} \qquad (1.163)$$

The matrix for the $k$th subsystem is given by

$$\begin{bmatrix} A_k & B_k \\ C_k & D_k \end{bmatrix} = \begin{bmatrix} 1 & 0 \\ -1/f_k & 1 \end{bmatrix} \begin{bmatrix} 1 & d_k \\ 0 & 1 \end{bmatrix} \qquad (1.164)$$

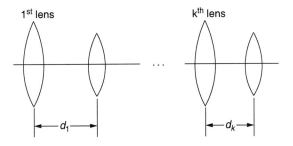

**Figure 1.23**   General lens waveguide.

with $f_k$ the focal length of the $k$th lens and $d_k$ the distance between the $k$th and $(k + 1)$st lenses. The complex Gaussian parameter $q_M$ of the output beam is given by

$$q_M = \frac{Aq_0 + B}{Cq_0 + D} \tag{1.165}$$

where $A$, $B$, $C$, $D$ are given in (1.163).

Propagation of a Gaussian beam in a biperiodic lens sequence is an interesting example since it represents the beam propagation in a laser cavity. The lenses are arranged in a biperiodic fashion as shown in Fig. 1.24. The spacing between the lenses is $d$. The elements of the transformation matrix for each period is given by [8]

$$
\begin{aligned}
A &= 1 - d/f_2 \\
B &= d\left(2 - \frac{d}{f_2}\right) \\
C &= -\left[\frac{1}{f_1} + \frac{1}{f_2}\left(1 - \frac{d}{f_2}\right)\right] \\
D &= -\left[\frac{d}{f_1} - \left(1 - \frac{d}{f_1}\right)\left(1 - \frac{d}{f_2}\right)\right]
\end{aligned}
\tag{1.166}
$$

In order that the beam is confined within the cavity to set up a cavity mode, the beam must reproduce itself after each period. The requirements on the focal lengths of the lenses and the spacing are given by

$$0 < \left(1 - \frac{d}{2f_1}\right)\left(1 - \frac{d}{2f_2}\right) < 1 \tag{1.167}$$

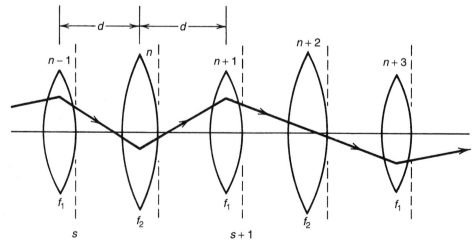

**Figure 1.24** Biperiodic lens waveguide. (After Yariv [8]. Copyright 1975, Wiley. Reproduced by permission.)

### 1.4.9  Properties of Hermite–Gaussian Function

Only a partial list of the mathematical properties of the Hermite–Gaussian function is given here for convenient reference. The reader is referred to Ref. 10 for further details. Several low-order Hermite–Gaussian functions are shown in Fig. 1.25.

Differential equation

$$\frac{d^2}{dx^2} H_m(x)e^{-x^2/2} + (\lambda_m - x^2)H_m(x)e^{-x^2/2} = 0 \tag{1.168}$$

Orthogonality

$$\int_{-\infty}^{\infty} H_m(x)H_n(x)e^{-x^2}\,dx = 0 \quad \text{for } m \neq n \tag{1.169}$$

Recursion formula for Hermite polynomial

$$H_{m+1}(x) - 2xH_m(x) + 2mH_{m-1}(x) = 0 \tag{1.170}$$

Fourier transform

$$\mathscr{F}[H_m(x)e^{-x^2/2}] = \frac{1}{\sqrt{2\pi}}(j)^m H_m(k)e^{-k^2/2} \tag{1.171}$$

Normalization

$$\int_{-\infty}^{\infty} H_m^2(x)e^{-x^2}\,dx = m!2^m\sqrt{\pi} \tag{1.172}$$

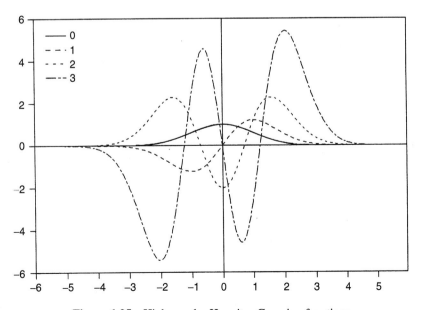

**Figure 1.25**  Higher-order Hermite–Gaussian functions.

## 1.5   DIFFRACTION

Diffraction is an important subject associated with the wave nature of light. In contrast with geometrical optics, which predicts sharp shadows behind an opaque obstacle, the diffraction theory confirms no such sharpness in the shadows. Based on diffraction theory, the far-field pattern of an antenna can be predicted and the divergence angle of a light beam can be calculated using the Fraunhofer diffraction formula. In addition, using linear system theory and the Fresnel diffraction formula, two-dimensional optical information can be processed using thin lenses and spatial frequency filters.

In this section, diffraction is treated based on the scalar wave theory. The diffraction field represented by the Kirchhoff–Fresnel integral is developed using the free-space Green's function. Huygen's principle and the Fresnel and Fraunhofer diffraction formulas are obtained through various approximations. Several closed-form diffraction patterns are given. Linear system representation of diffraction is briefly described. Diffraction by amplitude and phase gratings are presented.

### 1.5.1   Scalar Diffraction Theory

The formal derivation of the Kirchhoff diffraction integral is quite involved and will not be reproduced here. Instead, only an outline of the derivation is given so that the physical significance of the integral can be meaningfully discussed.

Let the diffracted field $\psi$ be a solution to the source free scalar wave equation

$$\nabla^2\psi + k^2\psi = 0 \tag{1.173}$$

within a given volume $V$ bounded by a closed surface $s$. $k$ is the wave number given by $\omega\sqrt{\mu\epsilon}$. A Green's function within the same region satisfies the equation

$$\nabla^2 g + k^2 g = \delta(\bar{\mathbf{r}} - \overline{\mathbf{r}'}) = \delta(x - x')\delta(y - y')\delta(z - z') \tag{1.174}$$

with the unit point source located at $\overline{\mathbf{r}'}$. After several algebraic steps and making use of the divergence theorem, the diffracted fields $\psi$ at the point $\mathbf{r}'$ can be expressed as

$$\psi(\overline{\mathbf{r}'}) = \int_s (\psi\nabla g - g\nabla\psi)\cdot\hat{n}\,ds \tag{1.175}$$

where $s$ is the closed boundary surface of the volume $V$. Equation (1.175) says that the diffraction field $\psi$ at any point $(\overline{\mathbf{r}'})$ can be obtained based on the knowledge of the field at the boundary with the help of a properly chosen Green's function. Figure 1.26 graphically demonstrate the integral representation of $\psi$ in (1.175).

The free-space Green's function is chosen to be

$$g = -\frac{1}{4\pi}\frac{e^{-jkr}}{r} \tag{1.176}$$

so that (1.175) becomes

$$\psi(\overline{\mathbf{r}'}) = \frac{1}{4\pi}\int_s \left(\nabla\psi\frac{e^{-jkr}}{r} - \psi\nabla\frac{e^{-jkr}}{r}\right)\cdot\hat{n}\,ds \tag{1.177}$$

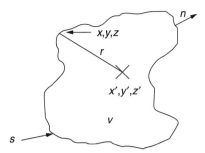

**Figure 1.26** Geometry of the integral representation. (After Marcuse [2]. Copyright 1982, Van Nostrand and Reinhold. Reproduced by permission.)

Equation (1.177) is the Kirchhoff integral, which states that the field at any interior point within the volume can be obtained once the field and its derivative are known on the boundary. In fact, the integral can be simplified if the Green's function is chosen so that it satisfies either the Dirichlet condition ($g = 0$) or the Neumann condition ($\partial g / \partial n = 0$) on the boundary. The advantage is that only one integral is nonzero in (1.177). However, these Green's functions are difficult to find for complicated boundaries, and the advantage may not outweigh the effort to secure such a Green's function.

### 1.5.2 Fresnel–Kirchhoff Integral

The diffraction integral in (1.177) is a general result by which the diffraction field can be obtained. For diffraction by a two-dimensional planar aperture, (1.177) should be simplified so that the integral may be evaluated in closed form or by numerical methods.

Assuming an aperture $A(x, y)$ is in an infinite opaque screen, the diffraction field in the $+z$ half-space is sought. Let the aperture $A(x, y)$ be illuminated by a spherical wave originated at a point $P_2$ from behind the screen and the point of observation $P$ be in the volume $V$. The surface integral in (1.177) can be separated into two integrals over the surface $S_s$ and $S_\infty$, where the surface $S_s$ is the infinite screen plus the aperture and $S_\infty$ is a portion of the spherical surface with infinite radius as shown in Fig. 1.27. The integral over $S_\infty$ is equal to zero provided that the optical field $\psi$ satisfies the Sommerfeld radiation condition

$$\lim_{R \to \infty} R(\nabla \psi \cdot \hat{n} + jk\psi) = 0 \qquad (1.178)$$

The diffracted field $\psi$ expressed in rectangular coordinates is

$$\psi(x', y', z') = \frac{1}{4\pi} \int_A \left( \nabla \psi \frac{e^{-jkr}}{r} - \psi \nabla \frac{e^{-jkr}}{r} \right) dx\, dy \qquad (1.179)$$

where the integral is over the aperture since the screen is assumed to be opaque. Given that

$$r^2 = (x - x')^2 + (y - y')^2 + z'^2 \qquad (1.180)$$

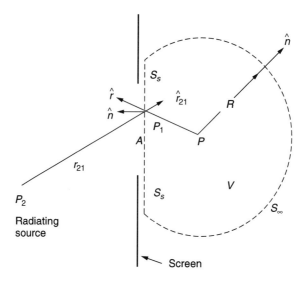

**Figure 1.27**   Separate domains for integral in (1.177).

and using the spherical wave assumption, the field $\psi(x, y)$ at point $P_1$ on the aperture is given by

$$\psi = K \frac{e^{-jkr_{21}}}{r_{21}} \tag{1.181}$$

with $K$ the amplitude of the source. The integral representation of the diffraction field at $P$ is

$$\psi(x', y', z') = \frac{jK}{2\lambda} \int_A \frac{e^{-jk(r_{21}+r)}}{r_{21}r} (\hat{r} \cdot \hat{n} - \hat{r}_{21} \cdot \hat{n}) \, dx \, dy \tag{1.182}$$

This is the Fresnel–Kirchhoff diffraction formula. The quantity $(\hat{r} \cdot \hat{n} - \hat{r}_{21} \cdot \hat{n})$ is known as the obliquity factor, which relates the diffracted angle to the incident angle.

### 1.5.3   Fraunhofer and Fresnel Diffraction [2, 5]

The integral of (1.182), although simple in form, can be evaluated in closed form for only a few cases where the geometry of the problem is simple. In some cases the closed-form expressions may be too complicated to shed light on the physical meaning of the results. It is desirable to further simplify the integral by approximation.

***Huygen's Principle.***   Before approximations are made, it is instructive to set a unit point source on the $-z$-axis and the field point on $+z$-axis so that the obliquity factor is equal to 2. The diffraction formula becomes

$$\psi(x', y', z') = \frac{j}{\lambda} \int_A \psi \frac{e^{-jkr}}{r} \, dx \, dy \tag{1.183}$$

This expression can be viewed as a convolution integral where the source function $\psi$ is convolved with the Green's function to produce the field at $(x', y', z')$. Since the Green's function represents the response due to a point source, (1.183) represents the total field due to a summation of an infinite number of point sources located at the aperture. This interpretation was first proposed by C. Huygens and is known as Huygens' principle. The geometry for plane wave diffraction is shown in Fig. 1.28.

***Fraunhofer Diffraction.***  Assuming that the field point $(x', y', z')$ is sufficiently far away from the aperture that $r$ can be approximated by a binomial series as

$$r \cong z' + \frac{1}{2}\frac{x'^2 + y'^2}{z'} - \frac{xx' + yy'}{z'} + \frac{1}{2}\frac{x^2 + y^2}{z'} \tag{1.184}$$

the diffraction formula becomes

$$\psi(x', y', z') = \frac{j}{\lambda z'}e^{-jkz'}\exp\left[-j\frac{k}{2}\left(\frac{x'^2 + y'^2}{z'}\right)\right]\int\!\!\int_A \psi(x, y)\exp\left[\frac{jk}{z'}(xx' + yy')\right.$$
$$\left. -j\frac{k}{2}\left(\frac{x^2 + y^2}{z'}\right)\right]dx\,dy \tag{1.185}$$

If $z'$ is large compared with the size of the aperture so that the last exponential term in the integral is near unity, the diffraction formula is given by

$$\psi(x', y', z') = \frac{j}{\lambda z'}e^{-jkz'}\exp\left[-j\frac{k}{2}\left(\frac{x'^2 + y'^2}{z'}\right)\right]\int\!\!\int_A \psi(x, y)$$
$$\times \exp\left[\frac{jk}{z'}(xx' + yy')\right]dx\,dy \tag{1.186}$$

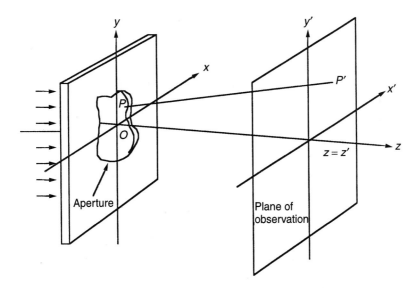

**Figure 1.28**  Diffraction of a plane wave by an aperture.

The limits of integration may be extended to infinity since the field on the screen is assumed to be zero. The diffraction field is proportional to the two-dimensional Fourier transform of the aperture distribution. This is the Fraunhofer or far-zone diffraction. The criterion for this approximation to be valid is that

$$k \frac{(x^2 + y^2)_{\text{max}}}{2z'} < \frac{\pi}{2} \tag{1.187}$$

If $A$ is a square aperture with edge dimension $d$, the Fraunhofer region is for

$$z' > d^2/\lambda \tag{1.188}$$

For the case where the aperture field varies only with respect to the $x$-axis, such as a long slit with uniform width, the diffraction formula in (1.183) is further simplified to

$$
\psi(x', y', z') = \frac{j}{\lambda} \int_{-\infty}^{\infty} \psi(x) \left[ \int_{-\infty}^{\infty} \frac{e^{-jkr}}{r} \, dy \right] dx
$$
$$
= \frac{\pi}{\lambda} \int_{-\infty}^{\infty} \psi(x) H_0^{(2)}(k\rho) \, dx \tag{1.189}
$$

where $\rho = \sqrt{z'^2 + (x - x')^2}$. Using the large argument approximation for Hankle function, (1.189) can be rewritten as

$$
\psi(x', y', z') = \frac{e^{-jkz'} e^{j\pi/4}}{\sqrt{\lambda z'}} \int_{-\infty}^{\infty} \psi(x) \exp \left[ \frac{-jk(x' - x)^2}{2z'} \right] dx \tag{1.190}
$$

If the field point is sufficiently far from the aperture, the one-dimensional Fraunhofer diffraction formula is

$$
\psi(x', y', z') = \frac{1}{\sqrt{z'\lambda}} \exp \left[ -j \left( kz' - \frac{\pi}{4} + \frac{kx'^2}{2z'} \right) \right] \int_{-\infty}^{\infty} \psi(x) e^{jkx'x/z'} \, dx \tag{1.191}
$$

Several Fraunhofer diffraction formulas of practical interest are shown in Section 1.5.6.

***Fresnel Diffraction.*** If the field point is sufficiently close to the aperture, the last term of the binomial expansion for $r$ in (1.184) must be included in the evaluation of the integral (1.185). The integral may be recognized as a convolution integral of the form

$$\psi(x', y', z') = \psi(x, y) * K(x, y, z') \tag{1.192}$$

where

$$K(x, y, z') = \frac{j}{\lambda z'} e^{-jkz'} e^{-j(k/2z')(x^2 + y^2)} \tag{1.193}$$

is the convolution kernel. Equation (1.192) can also be written in terms of a two-dimensional Fourier transform as

$$
\psi(x', y', z') = \frac{j}{\lambda z'} \exp \left[ -j \frac{k(x'^2 + y'^2)}{2z'} \right] e^{-jkz'} \mathscr{F} \left[ \psi(x, y) \exp \left[ -j \frac{k(x^2 + y^2)}{2z'} \right] \right]
$$
$$\tag{1.194}$$

Closed-form representations of the integral in either (1.192) or (1.194) are difficult to obtain. Numerical approaches must be used to calculate the near-field diffraction of an aperture. Equation (1.194) is a much better choice for numerical calculations because computer codes for fast Fourier transform (FFT) are readily available.

One classic example in Fresnel diffraction is the diffraction of a single slit illuminated by a uniform plane wave. The sharp edges of the slit excite a series of spherical waves which create complex optical interferences in the near-field region. Assuming that the width of the slit is $2a$, the Fresnel–Kirchhoff integral in one dimension is

$$\psi(x', y', z') \sim \sqrt{\frac{j}{\lambda z'}} \int_{-a}^{a} \psi(x) \exp\left[-j\frac{\pi}{z'\lambda}(x' - x)^2\right] dx \tag{1.195}$$

If the variables are normalized so that $u = x/a$ and $N = a^2/z'\lambda$, (1.195) takes on the form

$$\psi(u') \sim \sqrt{jN} \int_{-1}^{1} e^{j\pi N(u'-u)^2} du \tag{1.196}$$

$N$ is the Fresnel number, which measures the number of Fresnel zones visible at a distance $z'$. A larger value of $N$ indicates more ripples in the diffraction pattern. Equation (1.196) is a normalized expression of the diffraction field, with the Fresnel number $N$ as a parameter.

The single-slit diffraction is best expressed in terms of the complex Fresnel integral function $F(y)$, defined by

$$F(y) = \int_{0}^{y} e^{j\pi(u^2/2)} du \tag{1.197}$$

The real and imaginary parts of (1.197) are the Fresnel cosine and sine integrals, which are defined as

$$\begin{aligned} F(y) &= C(y) + jS(y) \\ &= \int_{0}^{y} \cos\frac{\pi u^2}{2} du + j\int_{0}^{y} \sin\frac{\pi u^2}{2} du \end{aligned} \tag{1.198}$$

If the complex values of the Fresnel integral function are plotted on a complex plane, they form a contour with $y$ as a parameter known as the Cornu spiral, shown in Fig. 1.29. The asymptotic value of this curve tends to $\pm(1 + j)/2$ as $y$ approaches $\pm\infty$.

The Fresnel diffraction of the single slit can be written in terms of the Fresnel integral functions as

$$\psi(u') \sim \sqrt{\frac{j}{2}}[F\sqrt{2N}(1 - u') - F(-\sqrt{2N'}(1 + u'))]e^{-jkz'} \tag{1.199}$$

### 1.5.4  Diffraction Gratings [2]

Previous discussions in Sections 1.5.2 and 1.5.3 have set the mathematical background for the diffraction theory. However, the diffractions of single apertures have little practical value. A direct application of diffraction theory to practical devices is diffraction gratings. The grating is a repetitive array of diffraction elements that has the effect

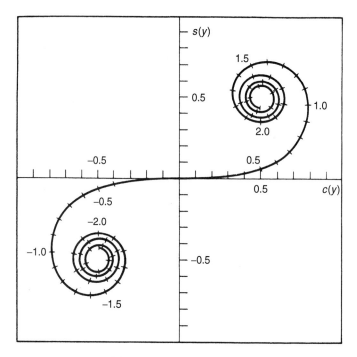

**Figure 1.29**   Cornu spiral.

on an emergent wave of producing periodic alterations in the phase, amplitude, or both. These devices are useful for the spectral analysis of light and incoherent optical processing.

***Transmission Amplitude Grating.***   A simple example of an amplitude grating is a series of parallel slits. A plane wavefront, in passing through the grating, encounters alternative opaque and transparent regions so that it undergoes a modulation in amplitude. Mathematically, the multiple slits can be expressed as

$$t(x) = \sum_{n=-N/2}^{N/2} \text{rect}\left(\frac{x}{a}\right) * \delta\left(\frac{x}{T} - n\right) \tag{1.200}$$

with rect $(\cdot)$ being a rectangular pulse function defined by

$$\text{rect}(x) = \begin{cases} 1 & |x| \leqslant \frac{1}{2} \\ 0 & \text{otherwise} \end{cases} \tag{1.201}$$

and $*$ denotes convolution. Equation (1.201) represents a series of $(N+1)$ periodic pulses of unit amplitude and width $a$; and the pulses are separated by a distance $T$ with the total grating size $D = (N+1)T$. The diffraction pattern of the transmission function (1.200) is

$$\psi \sim \frac{\sin(kax'/2z')}{kax'/2z'} \frac{\sin(kDx'/2z')}{\sin(kTx'/2z')} \tag{1.202}$$

The first of the two contributing terms in (1.202) is the diffraction of an individual slit, while the second term is the array factor due to the grating. When the sine function in the denominator vanishes at $x' = m\lambda z'/T$, the numerator also vanishes and the ratio is at a relative maximum. These are the grating lobes of the pattern. The 0th-order lobe is located on the $z$-axis while the $\pm m$th grating lobes are located at $x' = m\lambda z'/T$, respectively. For $T \gg \lambda$, the spacings between the lobes are very small and the diffraction pattern consists of alternate light and dark bands along the $x'$-axis. The pattern may be interpreted as a series of near plane waves propagating in the direction of

$$\sin\theta = m\lambda/T \tag{1.203}$$

Since the direction of propagation is dependent on the wavelength, an amplitude grating may be used for spectral analysis of light.

Another amplitude grating which is greatly used in incoherent optical processing and white light holography has the transmission function given by

$$t(x) = \tfrac{1}{2}[1 + h\cos(Hx)] \tag{1.204}$$

with $H = 2\pi/T$, where $T$ is the period of the cosine function. Unlike the periodic slit grating, which produces an infinite number of grating lobes, the diffraction pattern of the sinusoidal amplitude grating consists of the zeroth and the $\pm$ first-order grating lobes only. The pattern may be expressed in terms of $\theta \simeq x'/z'$ as

$$\psi \sim \left\{ \frac{\sin(k\theta T)(N/2)}{k\theta} + h\frac{\sin(k\theta T + 2\pi)(N/2)}{k\theta + H} + h\frac{\sin(k\theta T - 2\pi)(N/2)}{k\theta - H} \right\} \tag{1.205}$$

The second and third terms represent light diffracted away from the axis and separating from the on-axis zeroth-order diffraction. These off-axis diffractions are useful in optical processing for separating the signal from the noise.

***Phase Grating.*** A phase grating can be made out of a glass slab whose thickness is modulated in a certain fashion. Because of the higher refractive index of the glass, the phase of a plane wave passing through this glass slab is modulated by the variable thickness. A common transmission function for a phase grating is

$$t(x) = e^{j[1 + h\cos(Hx)]} \tag{1.206}$$

The calculation process for the diffraction pattern of (1.206) requires several approximations to evaluate the integral in closed form. It is sufficient to point out that the diffraction pattern of the phase grating has an infinite number of grating lobes, and the amplitude of the $m$th-order grating lobe is proportional to the $m$th-order Bessel function $J_m(h)$.

### 1.5.5 Linear System Representation of Diffraction

It is well known that the diffraction theory plays an important role in optical information processing because the diffraction fields can be represented in terms of a two-dimensional Fourier transform of the aperture field distribution. It is natural to relate optical diffractions to linear system theory since the Fourier transform is central to spectral analysis of linear systems.

Recall from equation (1.185) that the Fresnel diffraction of an aperture field $\psi(x, y)$ is given by

$$\psi(x', y', z') = \frac{j}{\lambda z'} e^{-jkz'} \int_A \int \psi(x, y) e^{-j(k/2z')[(x'-x)^2+(y'-y)^2]}\, dx\, dy \qquad (1.207)$$

This integral is recognized as a two-dimensional convolution of $\psi(x, y)$ with the function $K(x, y, z')$ as given by (1.192) and (1.193). This relationship can be seen as a linear system having $K(x, y, z')$ as the spatial system response function and the aperture function $\psi(x, y)$ as the system input. The output transform of the system $\mathscr{F}[\psi(x', y', z')]$ can be obtained from the product of the transforms of the input $\mathscr{F}[\psi(x, y)]$ and the transfer function $\mathscr{F}[K(x, y, z')]$ where

$$\mathscr{F}[K(x, y, z')] = e^{-jkz'} e^{j\pi\lambda z'(f_x^2+f_y^2)}. \qquad (1.208)$$

with $f_x = x'/\lambda z'$ and $f_y = y'/\lambda z'$ as the spatial frequencies mentioned before.

For aperture distributions whose transforms cannot be calculated in closed form, the linear system approach is an easier way to calculate the diffractions using FFT. In addition, this linear system concept is useful in holographic analysis, computer-generated holography, and spatial frequency filter designs. With the help of the phase transformation property of lenses, the input signal $\psi(x, y)$ can be processed at the spatial frequency plane by specially designed filters.

### 1.5.6    Examples of Fraunhofer Diffraction

In this section the far-field diffraction patterns of several simple apertures are given without the derivation. Only the amplitude profiles are given here. Proportional constants and the phase profiles are omitted for simplicity.

*Rectangular Aperture.* The diffraction pattern of a rectangular aperture of dimension $a$, $b$ is given by

$$\psi(x', y', z') \sim \frac{\sin(\pi a x'/\lambda z')}{\pi a x'/\lambda z'} \frac{\sin(\pi b y'/\lambda z')}{\pi b y'/\lambda z'} \qquad (1.209)$$

where the edges $a$, $b$ are oriented parallel to $x$, $y$ axes, respectively. The one-dimensional plot of the pattern is shown in Fig. 1.30.

*Circular Aperture.* A circular aperture of radius $a$ illuminated by a plane wave gives the diffraction pattern

$$\psi(\rho', z') \sim \frac{J_1(2\pi a \rho'/\lambda z')}{2\pi a \rho'/\lambda z'} \qquad (1.210)$$

with $\rho'$ the radial coordinate in the diffraction plane, and $J_1$ is the Bessel function of first order. For large value of $\rho'$, $J_1$ behaves approximately as a sine function and (1.210) shows behavior similar to (1.209). The pattern is shown in Fig. 1.31.

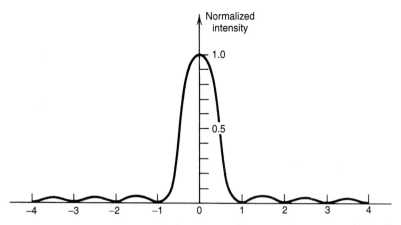

**Figure 1.30** Cross section of the Fraunhofer pattern of a rectangular aperture. (After Goodman. *Introduction to Fourier Optics*, Copyright 1968, McGraw Hill Book Company. Reproduced by permission.)

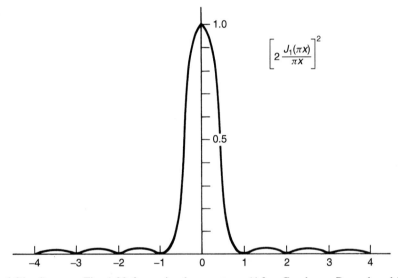

**Figure 1.31** Same as Fig. 1.30 for a circular aperture. (After Goodman. Reproduced by permission.)

*Gaussian Aperture.* A Gaussian aperture illuminated by a plane wave will produce a circular symmetric field distribution given by

$$\psi(\rho) = e^{-\rho^2/w^2} \tag{1.211}$$

and the diffraction pattern can be calculated to give

$$\psi(\rho', z') \sim e^{-(\pi w/\lambda z')^2 \rho'^2} \tag{1.212}$$

Hence the diffraction of a Gaussian distribution is still Gaussian.

**Figure 1.32**  Fresnel zone plate.

Notice that the Gaussian function in (1.211) is positive over the entire aperture plane. If the distribution is of finite extent so that

$$\psi(\rho) = \begin{cases} e^{-\rho^2/w^2}, & |\rho| \leqslant a \\ 0, & |\rho| > a \end{cases} \tag{1.213}$$

the diffraction pattern will contain many ripples due to scattering by the discontinuity at the edge. The reader is referred to Ref. 9 for more details.

***Fresnel Zone Plate.***  A Fresnel zone plate consists of a series of alternating opaque and transparent concentric annular rings of different radii and thicknesses (see Fig. 1.32). The transmission function is given by [5]

$$\psi(\rho) = \sum_{n=-\infty}^{\infty} \frac{\sin(n\pi/2)}{n\pi} e^{jn(\pi/\lambda f)\rho^2} \tag{1.214}$$

where $f$ is called the focal length of the Fresnel zone plate. The diffraction of (1.214) is an infinite series of index $n$ with the typical term given by

$$\psi_n \sim \frac{\sin(n\pi/2)}{n\pi} \int_0^\infty \rho \exp\left[j\frac{k\rho^2}{2}\left(\frac{1}{z'} - \frac{n}{f}\right)\right] J_0\left(\frac{2\pi\rho'\rho}{\lambda z'}\right) d\rho \tag{1.215}$$

The stationary phase points of (1.215) are given by the equation

$$\frac{1}{z'} - \frac{n}{f} = 0 \tag{1.216}$$

For $n = 1$, the integral is maximum at $z' = f$ and the light passing through the Fresnel zone plate is said to be focused. Other relative maximum points are

$$z' = \frac{f}{3}, \frac{f}{5}, \dots, \frac{f}{2m+1} \tag{1.217}$$

## REFERENCES

1. M. Born and E. Wolf, *Principles of Optics*, 6th ed., Pergamon Press, Elmsford, NY, 1980.
2. D. Marcuse, *Light Transmission Optics*, 2nd ed., Van Nostrand Reinhold, New York, 1982.
3. J. A. Kong, *Electromagnetic Wave Theory*, Wiley, New York, 1986.
4. S. Cornbleet, *Microwave and Optical Ray Geometry*, Wiley, New York, 1984.
5. K. Iizuka, *Engineering Optics*, Springer-Verlag, Berlin, 1985.
6. E. W. Marchand, *Graded Index Optics*, Academic Press, New York, 1978.
7. H. A. Haus, *Waves and Fields in Optic Electronics*, Prentice-Hall, Englewood Cliffs, NJ, 1984.
8. A. Yariv, *Quantum Electronics*, 2nd ed, Wiley, New York, 1975.
9. A. E. Siegman, *Laser*, University Science Books, Secaucus, NJ, 1986.
10. M. Abramwitz and I. A. Stegun, *Handbook of Mathematical Functions*. Dover, New York, 1976.
11. J. T. Verdeyen, *Laser Electronics*, Prentice Hall, Englewood Cliffs, NJ, 1981.

# 2

# INFRARED TECHNIQUES

Asu R. Jha
*Jha Technical Consulting Services*
*Cerritos, California*

Peter N. Kupferman
*Jet Propulsion Laboratory*
*California Institute of Technology*
*Pasadena, California*

## 2.1 INTRODUCTION

The field of infrared technology has evolved rapidly since the publication of the most significant and comprehensive handbook covering this subject was published: *The Infrared Handbook* [1]. No attempt has been made to redo this work. Rather, a more limited approach is taken where those aspects of the technology that have become central to those working in the field are highlighted. Perhaps the single most significant development since the publication of *The Infrared Handbook* has been the maturation of line and area array detector technology. These devices have allowed workers in the infrared to use the same type of instruments and techniques that those working in visible wavelengths have traditionally employed. Until this time, most infrared instruments have been limited to those designed around a small linear array of detectors. Now there exist staring two-dimensional cameras and Fourier transform spectrometers using arrays of several hundred detectors on a side and conventional spectrometers incorporating linear arrays of more than 500 detectors.

The approach taken here is to emphasize those characteristics common to most infrared detectors and illustrate their use for calculating and predicting sensor performance. No detailed discussion of detector physics will be undertaken because this aspect of technology has been covered extensively in appropriate literatures. Further, actual signal processing techniques will not be covered, since they belong more properly to the general and separate field of image and data processing.

*Handbook of Optical Components and Engineering,*  Edited by Kai Chang
ISBN 0-471-39055-0  © 2003 John Wiley & Sons, Inc.

There are infrared applications in almost all the physical sciences. These include, for example, geologic exploration for minerals from space, oceanographic and atmospheric studies, thermography, surveillance, remote sensing, missile tracking, biological threat detection, and astronomical research. This chapter is intended for those working in such fields who have specific application and detection requirements already defined and need to select those infrared detectors most appropriate to their needs.

## 2.2 DEFINITION OF THE INFRARED

The infrared wavelength range will be defined from 1.0 to 1000 $\mu$m for this chapter. Traditionally, the infrared was considered to begin beyond 0.7 $\mu$m, where the human eye loses sensitivity and at a time when "red"-sensitive photographic emulsions actually peaked in the blue-green. However, with the maturing of charge-coupled device (CCD) technology and subsequent development of visible detectors that are very sensitive beyond 0.7 $\mu$m, the term "infrared" has come to mean that portion of the electromagnetic spectrum longward of 1.0 $\mu$m. Although 1000 $\mu$m is considered to be the long-wave cutoff for the infrared, most common applications involve detectors sensitive below 100 $\mu$m. The atmospheric transmission of the earth [1] from 0.55 to 24 $\mu$m is given in Fig. 2.1. Most applications require a clear atmospheric window or placement above the atmosphere if the wavelength region of interest lies within absorption bands. However, there are some applications where imaging within these bands are required to provide uniform backgrounds. The three most commonly considered regions, namely, 1–2.5 $\mu$m, 3–5 $\mu$m, 8–14 $\mu$m, and 14–100 $\mu$m are labeled as short-wavelength Infrared (SWIR), mid-wavelength Infrared (MWIR), long-wavelength Infrared (LWIR), and very long wavelength Infrared (VLWIR), respectively.

## 2.3 HISTORY OF INFRARED DETECTORS

Modern infrared technology started with the development of polycrystalline films placed on substrates through vapor deposition [2]. During the 1950s discrete solid-state detectors were manufactured from lead salts (PbTe, PbSe, and PbS). The 1960s saw the development of bulk crystal growth and the appearance of single-crystal detectors from indium antimonide (InSb), mercury cadmium telluride (HgCdTe), and lead tin telluride (PbSnTe). During the 1960s and 1970s extrinsic technology was developed, principally in silicon and germanium. Initially, all detectors functioned as discrete units and planar arrays of thousands of PbS detectors, for example, would be arranged in arrays for imaging applications. Each detector would require its own preamplifier and amplifier, with resulting problems of high power requirements, uniformity, and reliability. The invention of the CCD in the early 1970s led the way to integrated arrays of detectors mated to or formed within common processing units or multiplexers. The two most common types of arrays that evolved were (1) monolithic, where the detector and signal processor used the same crystal; and (2) hybrid, where different materials, each optimized for its particular detection and signal processing functions, were mated using techniques such as indium bump bonding. Array technology has enabled imaging in the infrared region with technique similar to earlier-developed visible imaging technologies. It has also challenged sensor system design because of the required wider fields of view, cryogenic temperatures needed to operate the arrays, and image processing and calibration techniques used to remove fixed pattern noise due to inherent variations in detector sensitivity. A summary of the original most common materials is given in Table 2.1 [2]. Because of the many varied materials and operating parameters, the concept of detectivity Parameter $D^*$ was developed [3]. It is basically a means to characterize the sensitivity normalized to unit collecting area and electrical bandwidth and is most relevant to single detectors, where the incoming radiation is optically or electrically chopped. It is defined in Eq. (2.1) and discussed more fully in Section 2.4.3. Representative values versus wavelength are given in Figs. 2.2 and 2.3 [2] for the

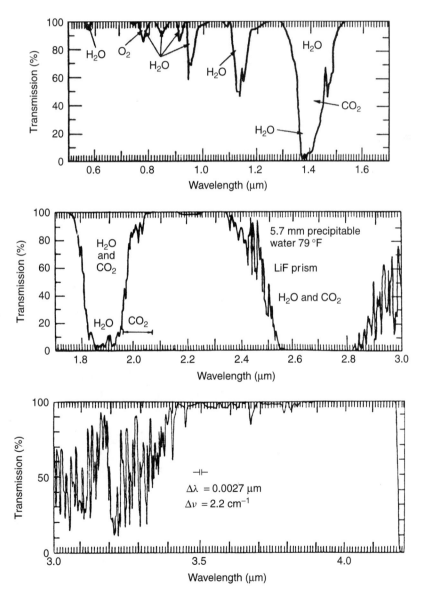

**Figure 2.1**   Atmospheric transmission at sea level over a 0.3-km path. (From Ref. 1.)

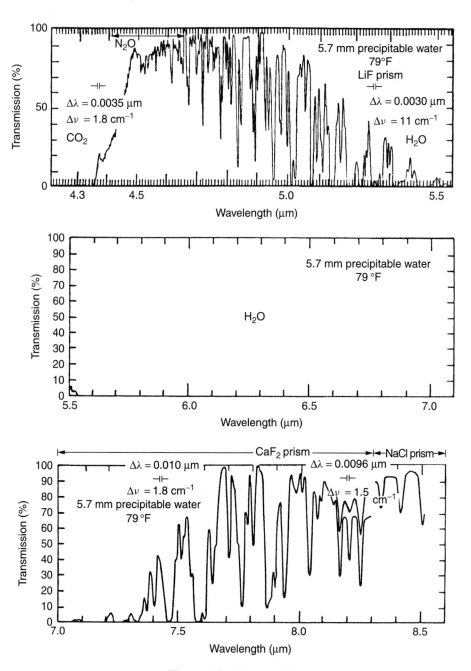

**Figure 2.1**   (*Continued*)

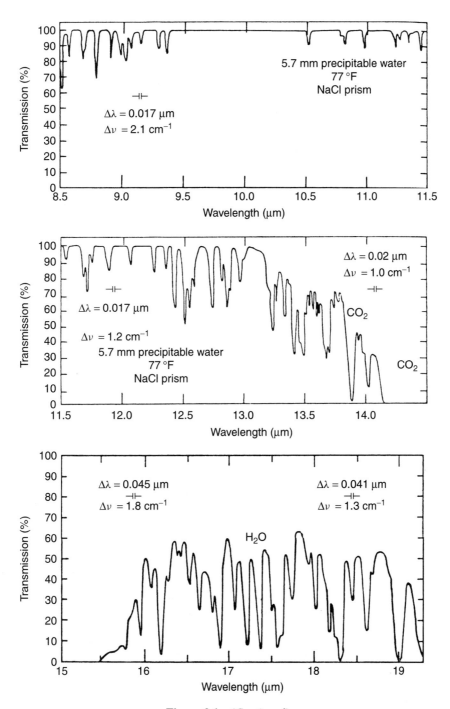

**Figure 2.1**   *(Continued)*

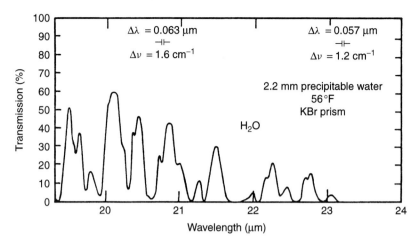

**Figure 2.1**   (*Continued*)

detectors mentioned previously. Figure 2.4 [2] illustrates the dependence of $D^*$ on the detector field-of-view (FOV) that limits unwanted background radiation.

## 2.4   DEVICE PHYSICS

The physics of semiconductors and their application to photon detection has been covered extensively in the literature [4]. A brief review will be given of those topics specific to the performance of these devices.

### 2.4.1   Intrinsic and Extrinsic

Photon detectors may be considered intrinsic if the absorbed photon excites a valence-band electron to the conduction band or extrinsic if the absorption occurs at impurity centers where either excitation from the valence band to a level within the bandgap takes place or excitation from a bandgap level to the conduction band occurs. Depending on how they are doped, either may be $n$- or $p$-type.

### 2.4.2   Photoconductive and Photovoltaic Detection

Both intrinsic and extrinsic materials may be used for photon detection. When they are operated as a photoconductive device (PC), the increase in conductivity due to an increase in electrons in the conduction band or increase in holes in the valence band is proportional to the incident photon flux. The increase in conductivity is characterized by a change in the voltage of a load resistor placed in series with the detector, which is biased with an external voltage.

By selective doping, junctions or diodes may be formed and used for detection. These devices are referred to as photovoltaic (PV) detectors because they rely on the internal junction field set up by the generation and subsequent separation of electron–hole pairs proportional to the incident photon flux. A common method to measure this internally generated voltage is in the so-called short-circuit mode, where an external

**TABLE 2.1  Common IR Detector Materials and Properties**

| Material | Maximum Temperature for Background Limited Operation | Long-Wavelength Cutoff (50%) ($\mu$m) | Peak Wavelength ($\mu$m) | Absorption Coefficient (cm$^{-1}$) | Quantum Efficiency | Resistance ($\Omega$) | $D^*$ peak (cm $\sqrt{\mathrm{Hz}}/W$) | Approximate Response Time (s) |
|---|---|---|---|---|---|---|---|---|
| InAs | | 3.6 | 3.3 | $\sim 3\times10^{3}$ | | | $3\times10^{11}$ | $5\times10^{-7}$ |
| InSb | 110 | 5.6 | 5.3 | $\sim 3\times10^{3}$ | 0.5–0.8 | $10^{3}$–$10^{4}$ | $6\times10^{10}$ $-1\times10^{11}$ | $5\times10^{-6}$ |
| Ge:Au | 60 | 9 | 6 | $\sim 2$ | 0.2–0.3 | $4\times10^{5}$ | $3\times10^{9}$ – $10^{10}$ | $3\times10^{-8}$ |
| Ge:Au(Sb) | 60 | 9 | 6 | | | $10^{4}$ | $6\times10^{9}$ $7\times10^{9}$ $-4\times10^{10}$ | $1.6\times10^{-9}$ |
| Ge:Hg | 35 | 14 | 11 | $\sim 3$ | 0.2–0.6 | $1$–$4\times10^{4}$ | $4\times10^{10}$ | } $3\times10^{-8}$ – $10^{-9}$ |
| | | 14 | 10.5 | $\sim 4$ | 0.62 | $1.2\times10^{5}$ | | |
| Ge:Hg(Sb) | 35 | 14 | 11 | | | $5\times10^{9}$ | $1.8\times10^{10}$ | $3\times10^{-10}$ – $2\times10^{-9}$ $3\times10^{-10}$ – $3\times10^{-9}$ |
| Ge:Cu | 17 | 27 | 23 | $\sim 4$ | 0.2–0.6 | $2\times10^{4}$ | $2$–$4\times10^{10}$ | $3\times10^{-4}$ – $10^{-8}$ $4\times10^{-9}$ – $1.3\times10^{-7}$ |
| Ge:Cu(Sb) | 17 | 27 | 23 | | | $2\times10^{5}$ | $2\times10^{10}$ | $<2.2\times10^{-9}$ |
| Hg:Cd, Te $x = 0.2$ | | 14 | 12 | $\sim 10^{3}$ | 0.05–0.3 | 60–400 | $10^{10}$ | $<10^{-8}$ |
| | | | | | | 20–200 | $6\times10^{10}$ | $<4\times10^{-4}$ |
| Pb:Sn, Te $x = 0.17$–$0.2$ | | 11 | 10 | $\sim 10^{4}$ | | 42 | $3\times10^{8}$ | $1.5\times10^{-8}$ |
| | | 15 | 14 | | | 52 | $1.7\times10^{10}$ | $1.2\times10^{-4}$ |

*Source:* Ref. 2.

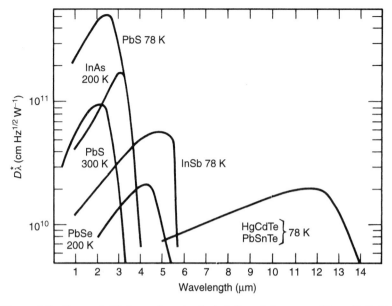

**Figure 2.2** Spectral $D^*$ for various intrinsic infrared detectors. (From Ref. 2.)

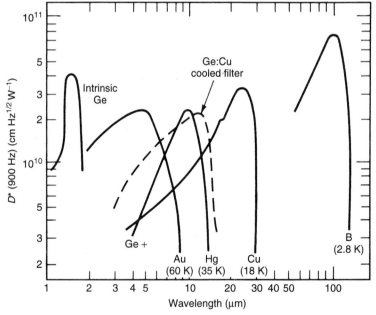

**Figure 2.3** Spectral $D^*$ for intrinsic Ge and various doped Ge extrinsic detectors. (From Ref. 2.)

current proportional to his voltage is sensed. This detection process is often augmented by an externally applied bias usually reversed from the junction field. An analytic expression relating the diode current to the external bias and the incident photon flux may be derived by considering the process to be a boundary value problem. The dark

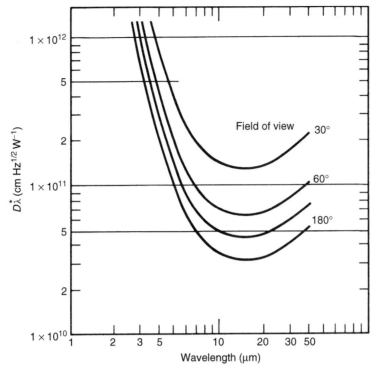

**Figure 2.4** BLIP $D^*(\lambda)$ as a function of long-wavelength cutoff for various fields of view. (From Ref. 2.)

current is the internally generated signal that is referred to as the diffusion current and adds to the noise and limits the detector dynamic range. It can be reduced by lowering the operating temperature. One fundamental difference between these two detection techniques is that PC detectors have gain, whereas PV detectors do not. Research studies performed indicate [5] that for certain emerging applications PV-HgCdTe detectors are preferred over PC-HgCdTe detectors because of superior overall performance (see Table 2.2). A typical configuration of a PC and PV detector and a simplified equivalent circuit are given in Figs. 2.5 and 2.6, respectively [4].

### 2.4.3  Figures of Merit

One of the most common figures of merit is the normalized detectivity, $D^*$, the reciprocal of the noise equivalent power normalized to unit detector size and electrical bandwidth. $D^*$ is defined in Eq. (2.1), where $A$ is the detector active area in square centimeters, $df$ is the electrical bandwidth in square-root hertz [equal to $1/(2 \cdot$ dwell time)], and NEP is the noise equivalent power in watts.

$$D^* = \sqrt{A(df)}/\text{NEP} \quad \text{cm}\sqrt{\text{Hz}}/\text{W} \tag{2.1}$$

At low temperatures one of the fundamental limits to performance is often the photoelectron shot noise due to the background photon flux. This background-limited infrared photodetector (BLIP) figure-of-merit (FOM) is given by Eq. (2.2) for PC and

**TABLE 2.2   Performance Comparison between PC- and PV-HgCdTe Detectors**

| Detector Type<br><br>Parameters | PC | PV |
|---|---|---|
| optical chopping requirement | YES | NO |
| DC bias requirement | YES | NO |
| $\left(\dfrac{1}{f}\right)$ noise component | High | Very Low |
| Impedance | Moderate | Very High |
| Dynamic Resistance $(R_0 A)$ | Moderate | High |
| Inherent gain | YES | NO |
| Linear Response Over | Limited Photon Flux Level | Very High Photon Flux Level |

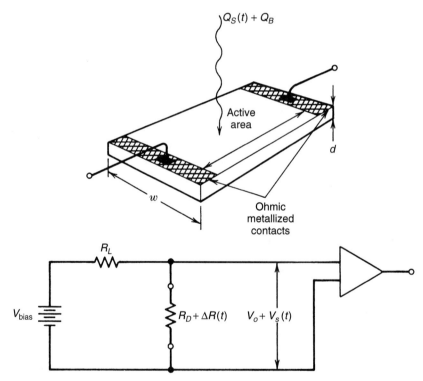

**Figure 2.5**   Configuration and geometry of a HgCdTe photoconductor (PC) detector and the bias circuit usually employed. (From Ref. 4.)

Eq. (2.3) for PV detectors. The PV-BLIP performance is better by a factor of root 2 over PC detectors. The recombination noise in PC detectors is often given as an explanation, although this is an area of current interest and research and may not be correct. Values of $D^*$ should always be accompanied by the background temperature

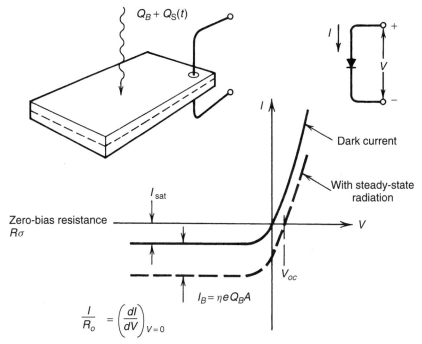

**Figure 2.6**   Geometry of a simple *p-n* junction Photovoltaic (PV) detector and its *I-V* characteristics. (From Ref. 4)

and field-of-view (FOV) used during the measurement. For a single detector with a cold shield subtending a cone given by *F/N*, this solid angle, $\Omega$, may be calculated by Eq. (2.4) in units of steradians.

$$D^*(\text{PC-BLIP}) = (\lambda/2hc)\sqrt{n/Q_b}$$

$$= 2.515 \times 10^{18} \lambda \sqrt{n/Q_b} \quad \text{cm}\sqrt{\text{Hz}}/\text{W} \tag{2.2}$$

$$D^*(\text{PV-BLIP}) = (\lambda/\sqrt{2}hc)\sqrt{n/Q_b}$$

$$3.557 \times 10^{18} \lambda \sqrt{n/Q_b} \quad \text{cm}\sqrt{\text{Hz}}/\text{W} \tag{2.3a}$$

$$\left[\frac{D^*(\text{PV-BLIP})}{D^*(\text{PC-BLIP})}\right] = \sqrt{2} \tag{2.3b}$$

$$\Omega = \pi/(1 + 4F/N^2) \qquad \text{steradian} \tag{2.4}$$

In these equations $Q_b$ is the photon irradiance in 1/s-cm², $n$ the quantum efficiency, $\lambda$ the wavelength in micrometers, $h$ Planck's constant ($6.626 \times 10^{-34}$ J-s), and $c$ the speed of light ($2.998 \times 10^{14}$ μm/s). For example, a single PV-BLIP detector uniformly sensitive from 8 to 12 μm with $n = 0.50$ within a warm background equivalent to a 300 K blackbody with unit emissivity filling an *F/2* cone will be limited by an irradiance of $3.70 \times 10^{16}$ 1/s-cm² giving a $D^*$ at 10 μm of $1.308 \times 10^{11}$ cm$\sqrt{\text{Hz}}$/W. In this example, the background is integrated from 8 to 12 μm while the detection is taking place over a narrow band centered at 10 μm.

The noise in a PC is due most often to a combination of generation–recombination (G-R) noise, Johnson noise, and $1/f$ noise as given in

$$V_n^2 = V_{gr}^2 + V_j^2 + V_l^2/f \tag{2.5}$$

The $1/f$ noise, which is not well understood, can be greatly reduced by proper crystal growth and by signal processing techniques such as correlated double sampling that eliminate it from the signal chain. Ignoring $1/f$ noise, which is proportional to the generation–recombination current, the detectivity may be written in terms of the other two sources characterized by the detector dynamic resistance-area product, $R_0A$, in ohm-cm$^2$, @zerobias and the incident photon irradiance $Q_b$ in 1/s-cm$^2$ in Eq. (2.6). The detector temperature, $T$, is in K and $q$ is the charge per electron ($1.602 \times 10^{-12}$C). Detectivity of a photovoltaic detector can be written as

$$D^*(\text{PV}) = (\lambda/hc)nq/\sqrt{4kT/R_0A + 2q^2nQ_b}$$

$$(\lambda/1.241)n/\sqrt{5.52 \times 10^{-23}T/R_0A + 5.13 \times 10^{-38}nQ_b} \tag{2.6}$$

The product $R_0A$ determines the thermal limit to detectivity and is often tabulated versus reciprocal detector temperature, where a high value of $R_0A$ implies good performance [4].

$R_0$ is defined as the dynamic resistance and an analytical expression at zero bias derived by differentiating the photodiode detector current equation is given by [6].

$$I_{\text{det}} = I_{\text{sat}}[\exp(qV/kT) - 1] + I_{gr} + V/R_{\text{shunt}} - qnQ_bA \quad \text{amperes} \tag{2.7}$$

where    $V$ = detector bias voltage
$I$ = zero-bias saturation current (dark current),  $A$
$I_{gr}$ = generation–recombination current
$R_{\text{shunt}}$ = junction shunt resistance
$A$ = photo-sensivitive detector area,  cm$^2$

The first term determines $R_0$ when $V$ is set to zero (diffusion term), and to a first approximation the quantum efficiency, $n$, is considered independent of the bias. The zero-bias resistance $R_0$ is then defined by Eq. (2.8) in ohms.

$$1/R_0 = (dI_{\text{det}}/dV)_{v=0} \quad \text{mho} \tag{2.8}$$

It is also assumed that $R_0$ is independent of the detector temperature, which is only approximately correct since there is a dependence on $I_{gr}$ and subsequently on $V/R_{\text{shunt}}$ with lower temperatures. To first order $R_0$ is given by Eq. (2.9) in ohms.

$$R_0 = 8.62 \times 10^5 T/I_{\text{sat}} \quad \text{ohms} \tag{2.9}$$

$R_0$ is graphically equivalent to the reciprocal of the slope of dark current versus voltage at zero bias, as shown in Fig. 2.6 [4].

### 2.4.3.1  *HgCdTe Photovoltaic Detectors for Various Operating Wavelengths.* Uncooled HgCdTe detectors are most attractive for high-temperature applications due to lower fundamental dark current at any given wavelength and operating temperature.

Such detectors with moderate cooling (225 K) and with cutoff wavelengths from near IR (NIR, not exceeding 1 micron) to short-wavelength IR (SWIR, 1–2 micron) to mid-wavelength IR (MWIR, 1.5–4.0 micron) are widely used for low light level (LLL) imaging and thermal imaging applications. Research studies performed on HgCdTe detectors [5] indicate that it is possible to achieve a background-limited IR pho-todetector (BLIP) at room temperature for both the NIR-LLL and thermal imaging applications. The studies further indicate that a photovoltaic detector with double-layer planar heterostructure (DLPH) device geometry shown in Fig. 2.10 eliminates potential sources of surface currents and tunneling currents. It is important to men-tion that uncooled detectors with small junction areas and operating in NIR, SMIR, and MWIR regions suffer from high-diffusion currents, dark currents, tunneling cur-rents, and surface generation-recombination (G-R) currents. Some applications require a single-chip–multispectral detector using a simple n-p-n structure integrated with a HgCdTe double-layer p-n heterojunction. Such a detector offers simultaneous and inde-pendent detection and integration of two photon currents. The MWIR-cutoff wavelength is determined by the p-HgCdTe film. The potential barrier in p-n heterojunction can be formed during the growth of p-HgCdTe film on N-HgCdTe layer as p-type dopants diffuse into N-HgCdTe layer. Note the SWIR cutoff wavelength is determined by the band gap of the p-HgCdTe layer material.

*2.4.3.2   Design and Performance Parameters of MWIR Photovoltaic Detec-tors.* MWIR-HgCdTe photovoltaic (PV) detectors when fabricated in the three-layer P-n-N configurations offer optimum performance in the 3–5-micron spectral region. Typical one-dimensional internal quantum efficiencies vary from 85% to 99% when operated at optimum cryogenic temperatures. The optical collection lengths are typi-cally 25 microns. The I-V characteristics of these detectors indicate that the diffusion current is the most dominant junction current mechanism at operating temperatures greater than 100 Kith detector figure-of-merit (FOM) or the dynamic resistance-area product ($R_0 A_d$) varies from $10^6$ ohm-cm$^2$ at a cutoff wavelength ($\lambda_c$) of 4 microns to about $10^4$ ohm-cm$^2$ at $\lambda_c$ equals to 5 microns, when operated at a cryogenic tempera-ture of 140 K. The low-frequency ($1/f$) noise current is as low as $5 \times 10^{-15}$ A/$\sqrt{\text{Hz}}$ at a wavelength of 5 microns and at 120 K.

*2.4.3.3   Critical Performance Parameters of LWIR-PV Detectors.* Low-frequency noise current, surface leakage current, dark current, and FOM ($R_0 A_d$) are the most critical parameters of the LWIR-PV detector operating in the 8–14-micron region. The presence of various noise mechanisms in the detector junction can significantly degrade the performance of the LWIR-Focal Planar Array (FPA). The low-frequency noise current in the LWIR-Hectic, double-layer planar heterojunction (DLPH) grown on a lattice-matched substrate is of critical importance. The FOM of LWIR-HgCdTe detectors operating in the 8- to-14-micron region varies from $10^2$ to $10^6$ ohm-cm$^2$, when the detectors are operating between 20 and 120 K. Note LWIR-HgCdTe detectors with FOMs greater than 1000 omh-cm$^2$ at 40 K have theoretically diffusion-limited performance down to 78 K. The $R_0 A$ product is used as an indicator of both the dark current and the coupling efficiency to external circuit. It is important to mention that this product of a photodiode with wide-gap band layer is at least 3 to 5 times higher over a detector diode without a wideband gap layer, regardless of diode's cutoff wavelength.

*2.4.3.3.1 Low-Frequency Noise and Dark Current levels in LWIR-PV Detectors.* Studies performed on LWIR detectors [5] indicate that the low-frequency noise in the intermediate performance LWIR detectors with $R_0 A$ products ranging from $10^3$ to $10^4$ ohm-cm$^2$ and operating at a cryogenic temperature of 40 K is higher than the noise in the high-performance detectors with $R_0 A$ products ranging from $10^5$ to $10^7$ ohm-cm$^2$ at the same operating temperature. This means higher values of this product are necessary for low-frequency noise levels. Note excessive low-frequency noise and dark current levels are observed with an FOM close to 10 ohm-cm$^2$ at reverse bias level of 100 mV. Note the low-frequency noise in LWIR photodiodes is limited by the trap-assisted tunneling currents. Post-implantation annealing technique is considered to be the most effective way to achieve simultaneously high $R_0 A$ product and low-frequency noise current. According to studies performed [5], the low-frequency noise current is directly proportional to dark current, but inversely proportional to square root of the electrical bandwidth. The same studies on LWIR detectors with cut of 14 microns at 60 K reveal that the excessive low-frequency noise at 1 Hz and $-100$ mV bias varies from $10^{-10}$ A/$\sqrt{\text{Hz}}$ at a dark current of 1 microampere to $10^{-8}$ A/$\sqrt{\text{Hz}}$ at a dark current of 100 microamperes. The studies further reveal that under constant operating temperature and bias level, the dark current and low-frequency noise current are dependent on detector implant area. Note the excessive low-frequency noise component is proportional to total dark current level, which includes G-R current, surface leakage current, and tunneling current.

*2.4.3.3.2 Techniques to Reduce Dark Current Levels in LWIR-PV Detectors.* The semiconductor compound $Hg_{1-x}Cd_x Te$ is the most promising detector material for the detection of IR signals and IR images. The $Hg_{0.8}Cd_{0.2}$ Te detector composition is best suited for the LWIR detectors operating in $8-14$ micron spectral region because of high detectivity and responsivity. Structural details of a HgCdTe photodiode passivated with a wideband gap epitaxial layer grown on the surface and using liquid phase epitaxy (LPE) fabrication technology are shown in Fig. 2.11. The dark current ($I_d$) is limited by the diffusion current at high operating temperatures, G-R current at low temperatures, and surface leakage current. Surface leakage current is generated by through surface states and fixed charges. Note lower reverse bias currents reduce surface leakage current levels in the wideband gap layer. Ion-implantation technique is widely used during the fabrication process, which introduces lattice damage and increases the dark current levels that can be reduced by an annealing process at 50°C over a duration ranging from 40 to 60 minutes. LWIR-HgCdTe photodiodes with good p-n junctions exhibit low dark current that is directly related to low-frequency noise current and high dynamic resistance-area product or FOM.

The surface G-R current is the major contributor to the dark current. The magnitude of this current is dependent on intrinsic carrier concentration, surface recombination velocity, surface area of the depletion region, surface state density, and surface passivation quality. The surface leakage current due to surface fixed charges is proportional to channel width and surface area, but inversely proportional to carrier life time. It is important to mention that surface passivation is essential to minimize both the dark current and surface leakage current. It is equally important to mention that surface leakage current is exponentially proportional to the band gap energy ($E_g$). Note the dark current level is independent of operating or cutoff wavelength. However, the cutoff wavelength of a LWIR-HgCdTe detector is strictly dependent on the compositional profile of the first layer as illustrated in Fig. 2.12. It is evident from Fig. 2.12 that the composition of the first layer determines the highest cutoff wavelength.

The dark current level is strictly dependent on the operating parameters, namely, bias and temperature. Under optimum reverse bias conditions, the dark current can be reduced close to 1 μA at an operating temperature of 78 K as shown in Fig. 2.13. However, higher quantum efficiencies are possible at lower cryogenic temperatures in the LWIR and very LWIR regions, as illustrated in Fig. 2.14.

### *2.4.3.4   Applications of very-LWIR Detectors For Space and Military Systems.*  Very-LWIR (VLWIR) detectors are best suited for certain military and space applications. VLWIR-focal planar array detectors are widely used in ballistic missile defense (BMD) applications, such as surveillance of long-range tactical and strategic missiles, target detection, target tracking during launch, cruise and terminal phases, and discrimination between the decoy and the hostile missile. These functions can be provided from the IR sensors located on interceptors, satellites, airborne platforms, or fixed platforms. It is important to mention [5] that most advanced BMD targets require IRFPAs operating in MWIR (3–5 micron), LWIR (8–14 micron), or VLWIR (14 micron and up) spectral regions. Research studies performed on LWIR and VLWIR detectors [5] indicate that IRFPA detectors for BMD applications require large format, high sensitivity or detectivity, low $(1/f)$ noise, good uniformity, minimum dark current, multicolor tunability, long life, and high reliability, while operating under severe environments. The studies further indicate that due to the cooler environments in space applications, VLWIR detectors will offer optimum performance because of their superior characteristics. Note multicolor IR sensing is extremely important to eliminate the adverse effects of earthshine in exoatmospheric discrimination. VLWIR-FPA detectors can play a key role in national missile defense (NMD), theater missile defense (TMD), and strategic missile defense (SMD) systems.

The missile defense systems can be used either in an endoatmosphere or exoatmosphere environment. In general, endoatmosphere interceptors and airborne surveillance systems are used for tactical missile applications to observe warm targets with high background irradiance from heated windows, scattered sunlight, and earth's surface reflections.

However, such applications require accurate measurements and subtraction of IR background irradiance to detect the target. In contrast, exoatmospheric intercepts and space-based IR surveillance sensors used for strategic applications normally engage cool targets with low background irradiance level. Both the LWIR and VLWIR detectors are best suited for strategic applications involving tracking and detection of intercontinental ballistic missiles (ICBMs), where scene is a space background and the targets are at relatively very low temperatures. In case of tactical missile applications, the most important wave bands are determined by the atmospheric transmission characteristics.

Multicolor discrimination capability is the principal requirement for IR detectors to detect and track theater and tactical missiles. Brief investigation on InSb, PtSi, SiAs, HgCdTe, and QWIP detectors indicates that HgCdTe-based QWIP detectors offer high sensitivity, improved wavelength flexibility in MWIR, LWIR, and VLWIR regions (Fig. 2.12), and multicolor tunable capability for discrimination. The MWIR-FPA detectors are barely adequate for current threats facing the IR sensors or missiles. However, to cope with increased threats posed by rouge nations in future,

LWIR-FPA detectors and VLWIR-FPA detectors with multicolor capability will be required.

Although many BMD functions can be accomplished with only one color, multicolor capability of VLWIR detectors offers much better performance in operating environments, when the target and/or background undergo changes during a missile engagement phase. Two or three spectral bands (or colors) will significantly improve the performance of the IR tracking sensor. Note in case of target detection and tracking of long-range ballistic missile during the booster burnout phase, reliable tracking and detection will result, if two colors are used, one before and one after burnout. Discrimination of real target from the decoys and debris is greatly improved by using simultaneously multicolor HgCdTe-FPA detectors. Note estimation of target thermal characteristics requires two to four colors. The $CO_2$-blocking band (14 to 16 micron) offers the best combination of spectral width and earth-sunshine blockage. Optimum tracking accuracy occurs, only when the color in the blocking band is between the other two colors. The $CO_2$-blocking band is frequently used due to its wide spectrum and good blocking characteristics.

*2.4.3.4.1 Operating Requirements for LWIR and VLWIR HgCdTe Detectors.* The HgCdTe-LWIR detector exhibits high sensitivity when operated at a cryogenic temperature of 40 K. Note a HgCdTe material suffers from tunneling dark currents caused by material impurities, particularly, when a HgCdTe detector is operated at this low temperature in the LWIR or VLWIR band. However, a QWIP detector when operated at 40 K offers high uniformity, but suffers from high dark current levels at longer wavelengths. This detector will not able to see very dim targets at long distances due to its poor quantum efficiency and low gain product. Note weight, size, and power of the cryocooler operating at 40 K or lower [7] are the basic constraints of a cryogenically cooled IR sensor used by an interceptor or satellite. For space-based surveillance and tracking systems, cryocoolers with life times exceeding 10 years are required to meet low operating costs and high reliability. Even though the Si:As-FPA detectors operating at 10 K are best suited for space-based BMD applications; nevertheless, they suffer from very short life time, poor reliability, and excessive power consumption.

*2.4.3.4.2 Performance Capabilities of Large HgCdTe VLWIR Detectors.* Large HgCdTe-VLWIR detectors operating in the 14–17-micron spectral region are best suited for remote-sensing sounding applications. Such sounding detectors provide temperature, pressure, and moisture profiles of the atmosphere for accurate weather prediction. As stated earlier, HgCdTe-VLWIR photoconductive (PC) detectors suffer from high nonlinearity and nonuniformity compared to photovoltaic detectors. Large VLWIR photovoltaic (PV) detectors with cutoff wavelength of 17 micron and diameter of 1000 micron seem to offer excellent response uniformity at 78 K. In addition, large PV detectors permit the use of passive radiators in spacecraft to cool the detector elements. Excellent response uniformity as a function of spot size implies that the low-frequency spatial response variations are absent. The 1000-micron diameter HgCdTe-VLWIR detectors [5] with cutoff wavelength of 17 micron at 78 K have demonstrated dark current levels as low as 160 µA at a reverse bias of 100 mV. These detectors exhibit nonlinearity less than 0.15% under high flux levels exceeding

$3 \times 10^{17}$ photons/cm$^2$.sec. This nonlinearity value of the PV detector is an order of magnitude better than that of a PC detector. Dark current and quantum efficiency are the most important performance parameters of a large VLWIR detector rather than the dynamic resistance or the $R_0A$ product. When a large VLWIR detector operates at a reverse bias of 50 mV, this product is close to 30 ohm-cm$^2$, the dark current is less than 100 μA and the quantum efficiency is close to 58% at 78 K [8]. Note the 1000-micron diameter PV detector is diffusion-limited at bias levels less than −50mV and at a cryogenic temperature of 78 K. The same detector is dominated by tunneling currents at operating temperatures less than 78 K. However, PV detectors with diameters close to 8 microns are diffusion-limited at temperatures exceeding 63 K and at reverse bias levels close to 200 mV. In small VLWIR-PV detectors, the tunneling commences even at 40 K and at a reverse bias of 80 mV. However, this particular detector is tunneling limited, when operated below 30 K.

### 2.4.4   Focal Plane Architecture

The goal of current technology is to mosaic a large number of detectors into arrays with a minimum of electrical leads. One of the most successful approaches has been the development of the hybrid focal plane array illustrated in Fig. 2.7 [6]. This shows a typical backsided illuminated array in which the incident radiation passes through a transparent substrate to the active layer (junction) and the resulting photocurrent passes through the individual contacts (indium bumps) to the readout circuitry (multiplexer). In many applications the multiplexer is a silicon CCD whose architecture and performance have reached a high state of development. A typical parallel-to-serial readout scheme is illustrated in Fig. 2.8 [6]. A current limitation to the general availability of such arrays has been their low yield, typically a few percent. One major reason has been the mechanical integrity and reliability of the bump bonding. Failure of these interconnections will result in "dead" or inactive pixels.

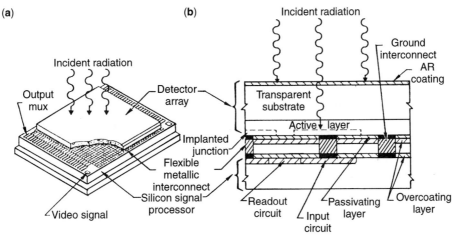

**Figure 2.7**   Architecture of a planar hybrid focal plane array: **(a)** perspective drawing; **(b)** expanded cross section. (From Ref. 6.)

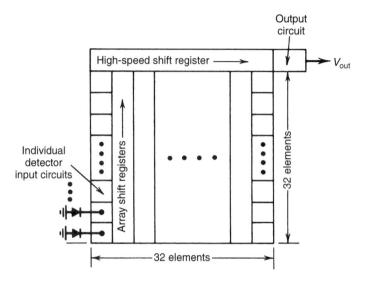

**Figure 2.8**   Parallel-to-serial readout architecture of a CCD multiplexer. (From Ref. 6.)

## 2.5   MERCURY CADMIUM TELLURIDE

Mercury cadmium telluride (HgCdTe) is a pseudo-binary alloy semiconductor [4] that is used in PC and PV detectors. These detectors are available singly, in line arrays, and arranged in two-dimensional area arrays. The alloy is often abbreviated $Hg_{1-x}Cd_xTe$, where the $x$ represents the fraction of cadmium telluride and $1 - x$ the remaining fraction of mercury telluride. The long-wave cutoff of this material is tunable by varying the value of $x$ and temperature. This variable wavelength may range from 2 to 30 $\mu$m, making HgCdTe suitable for a wide variety of infrared applications and competitive with indium antimonide in the SWIR. An important advantage of HgCdTe over extrinsic silicon, for example, is the modest cooling requirements (80 to 195 K). MWIR applications often require cooling to only 195 K, LWIR and VLWIR applications require endogenic cooling @ 80 K and 35 K, respectively, optimum performance, is the principal requirement.

### 2.5.1   HgCdTe Performance Characteristics

A plot of cutoff wavelength versus temperature is given in Fig. 2.9 [4] where by inspection it is apparent that for most applications below 12 $\mu$m the cutoff wavelength is a weak function of temperature. However, as discussed below, performance degrades rapidly with increasing temperature.

The detectivity given by $D^*$ is a function of several parameters, including electrical bandwidth, background photon flux, and $R_0A$. The degradation with frequency is shown in Fig. 2.15 [4], where the root 2 improvement of PV over PC detectors is apparent. The variation with $R_0A$ is given in Fig. 2.16 [4], which may be verified using Eq. (2.6) with the background irradiance set to zero. Degradation with increasing cutoff wavelength is illustrated by plotting $R_0A$ versus cutoff wavelength and reciprocal detector temperature in Figs. 2.17 and 2.18, respectively [4]. For a given detector

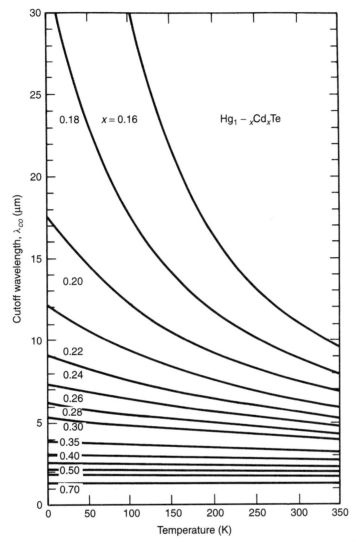

**Figure 2.9**  Cutoff wavelength for $Hg_{1-x}Cd_xTe$ as a function of temperature and alloy composition. (From Ref. 4.)

temperature, the $R_0A$ product (proportional to dark current) decreases with increasing cutoff. Figure 2.18 shows the improvement in performance for a fixed cutoff (2.15 μm) with decreasing temperature. The $D^*$ is a function of relative response whose typical variations with wavelength are given in Fig. 2.19 [10] for material tailored to the LWIR region. This curve suggests that HgCdTe cannot be made uniformly responsive over all wavelengths but only optimized over a limited region by the appropriate selection of $x$ and application of antireflection coatings over the detector.

At low temperatures (<66 K) the performance will be limited by the background. This is illustrated by the flattening of the $R_0A$ product (see Section 2.4.3) versus reciprocal temperature as shown in Fig. 2.20 [10]. The case for the detector within a hemisphere of background indicates a decrease of $R_0A$ of several orders of magnitude

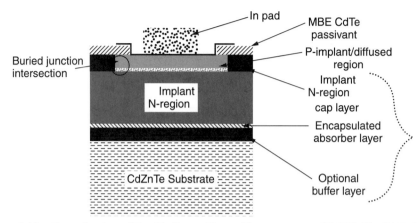

**Figure 2.10**   Cross-Section of Double-Layer Planar Heterostructure (DLPH) Hg:Cd:Te Photovoltaic Detector.

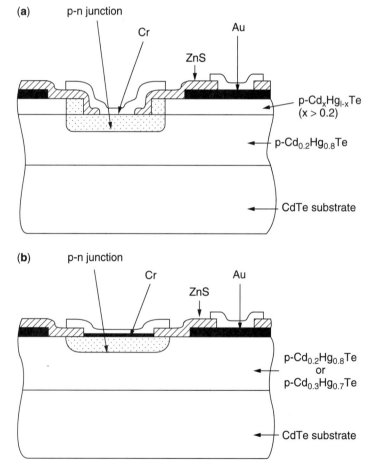

**Figure 2.11**   Structures of a Hg:Cd:Te Photodiode passivated with **(a)** wideband gap epitaxial layer and **(b)** without the wideband gap layer.

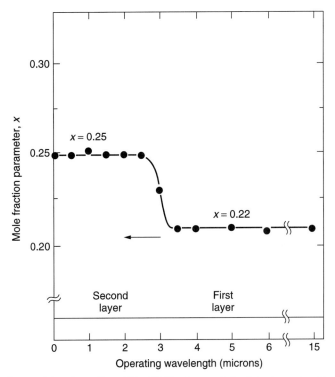

**Figure 2.12**   Compositional profile of Hg:Cd:Te photovoltaic detector using the ternary alloy $Hg_{1-x}Cd_xTe$.

over the dark-current-limited case measured with no background. The dark current as a function of the reverse bias for diodes and representative variations with voltage are given in Fig. 2.21 [10] and closely follows the diffusion term of Eq. (2.7).

### 2.5.2   HgCdTe Focal Plane Arrays

The hybrid array is one of the most common array architectures for this material. This approach to detector area arrays consists of mating the photon-sensitive material to a silicon multiplexer using individual indium bumps to electrically connect each pixel. The majority of work has been on backsided illuminated arrays because of the lack of obscuration and high pixel fill factor. A representative curve of quantum efficiency versus wavelength is given in Fig. 2.22 [6], measured $R_0A$ products in Fig. 2.23 [6], and the equivalent relative NEDT (noise equivalent delta temperature) in Figs. 2.24 and 2.25 [6] for a gray body at 300 K. An example of the use of curves such as those shown in Fig. 2.23 follows: Suppose that the detector is cooled to 65 K or $1/T \sim 15 \text{ K}^{-1}$. An application where the background is cut off beyond 10 μm gives $R_0A > 10\Omega$ -cm$^2$. However, using Eq. (2.6) for a 300-K hemispherical background gives $R_0A > 0.143$ Ω-cm$^2$ to provide background-limited performance. Therefore, under these background conditions, cooling below 65 K would be unnecessary.

The minimum near 8 μm in Fig. 2.24 can be derived from the functional relationship between NEDT and the derivative of the Planck function, $B(\lambda, T)$, which is given

**(a)**

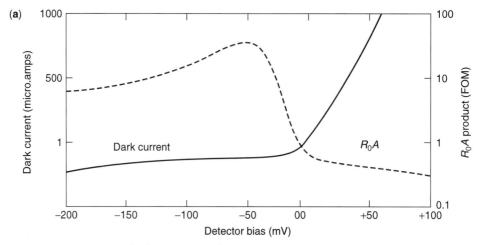

Dark current and FOM as a function of bias voltage
for a 1000-micron diameter detector

**(b)**

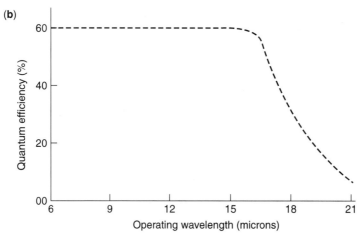

Quantum efficiency as a function of wavelength for 1000-micron
detector with cut off wavelength of 17 microns

**Figure 2.13**   **(a)** Dark current and FOM and **(b)** quantum efficiency at various wavelength for a 1000-micron diameter detector at 78 K.

by Eq. (2.10), where SNR is the signal-to-noise ratio. For 300 K backgrounds, this expression has a minimum near 8 µm for a fixed SNR.

$$\text{NEDT} = B(\lambda, T)/dB(\lambda, T)/dT/\text{SNR} \qquad (2.10)$$

Expected variations in $D^*$ for a 32 × 32 array is given by the histograms in Fig. 2.26 [6]. The order of magnitude shift toward higher $D^*$ in the right-hand histogram is due to the reduction in background because of the slower F/16 focal ratio. The spread in the histograms in Fig. 2.26 is an indicator of the array nonuniformity. HgCdTe arrays usually require some type of nonuniformity correction to remove this dominant source of fixed pattern noise. Typical variations in detector element response range from 10

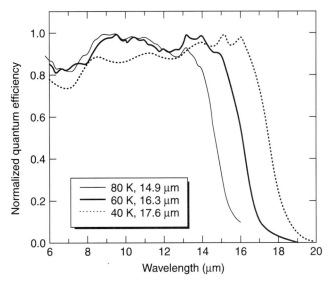

Note: As the operating wavelength is increased, the quantum efficiency decreases, regardless of cryogenic temperature and Hg:Cd:Te detector size.

**Figure 2.14** Normalized quantum efficiency of a 80-micron diameter detector as a function of wavelength at three distinct cryogenic temperatures.

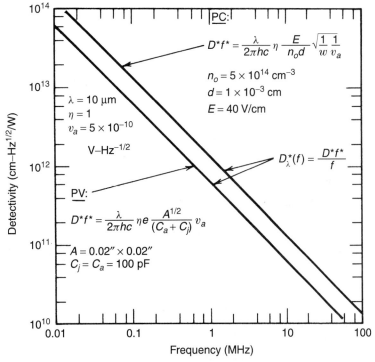

**Figure 2.15** $D^*f^*$ products for PC and PV HgCdTe detectors for representative values of detector parameters. (From Ref. 4.)

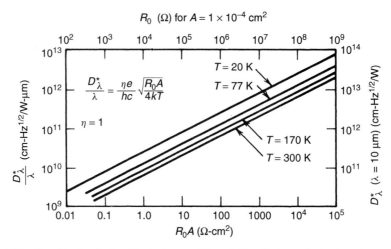

**Figure 2.16**   Graphical relationship between $D^*$ and the $R_0 A$ product at various operating temperatures. (From Ref. 4.)

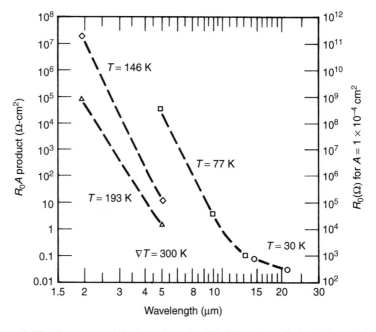

**Figure 2.17**   Summary of $R_0 A$ products for HgCdTe photodiodes. (From Ref. 4.)

to 20%. The corresponding variation of $D^*$ with reciprocal temperature appears in Fig. 2.22 [5], while histograms of dc signal and noise are given in Figs. 2.28 and 2.29 [6], respectively.

A range of $R_0 A$ curves for different cutoff wavelengths is given in Fig. 2.30 [11] illustrating typical degradation with increasing wavelength. Corresponding histograms are shown in Fig. 2.31 [11], which shows that shorter cutoff wavelength (2.3 μm)

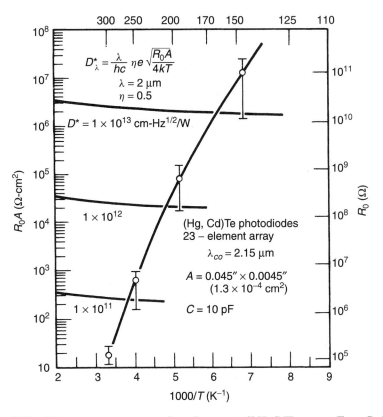

**Figure 2.18**   $R_0A$ versus temperature for a 2-μm cutoff HgCdTe array. (From Ref. 4.)

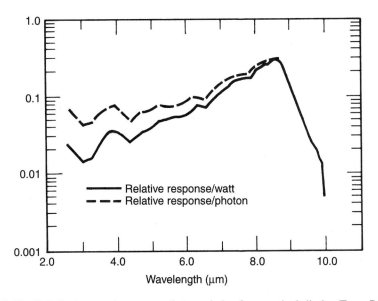

**Figure 2.19**   Relative spectral response characteristics for a typical diode. (From Ref. 10.)

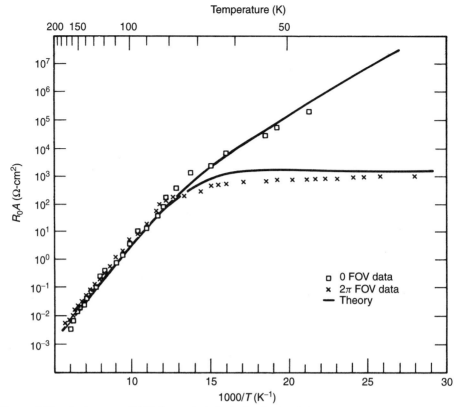

**Figure 2.20**    $R_0A$ versus $100/T$ for typical diode for fields of view 0 and $2\pi$ steradians. (From Ref. 10.)

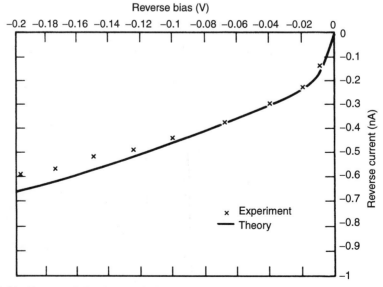

**Figure 2.21**    Reverse $I–V$ characteristic at 67 K for diode with field of view 0 steradians. (From Ref. 10.)

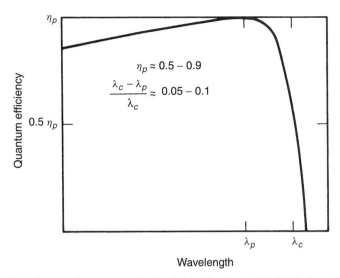

**Figure 2.22**  Model spectral response of a backside-illuminated HgCdTe detector. The maximum quantum efficiency, $\eta_p$, occurs at wavelength $\lambda_p$. The cutoff wavelength, $\lambda_c$, is defined as the wavelength at which the quantum efficiency has fallen to 50% of the peak value. (From Ref. 6.)

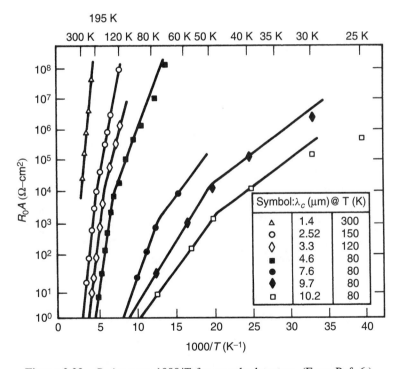

**Figure 2.23**  $R_0A$ versus $1000/T$ for sample detectors. (From Ref. 6.)

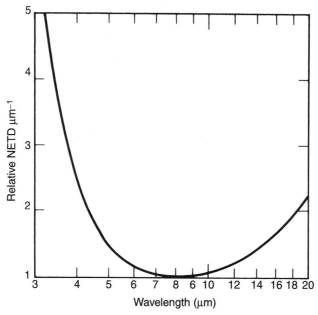

**Figure 2.24** Relative NETD per micrometer versus wavelength for a gray-body temperature of 300 K. (From Ref. 6.)

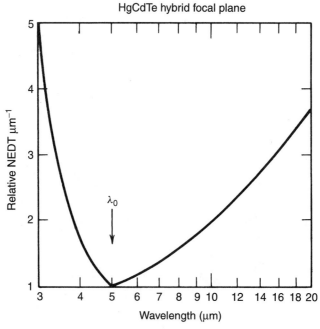

**Figure 2.25** Relative NETD per micrometer versus wavelength for $\lambda = 5$ μm for a gray-body temperature of 300 K. (From Ref. 6.)

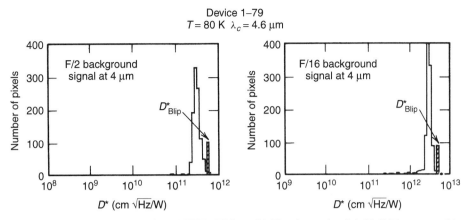

**Figure 2.26**  $D^*$ histogram of an MWIR FPA at 80 K using epitaxial HgCdTe on sapphire. (From Ref. 6.)

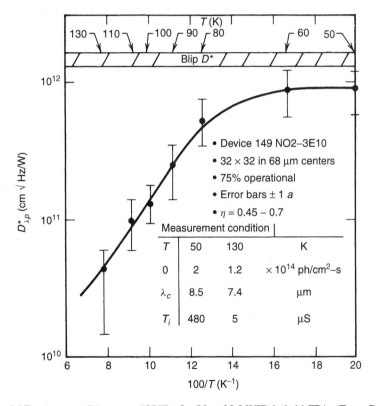

**Figure 2.27**  Average $D^*$ versus $100/T$ of a $32 \times 32$ LWIR hybrid FPA. (From Ref. 6.)

devices may be operated at higher temperatures (200 K) and still exhibit high values of $R_0A$ (>1000 $\Omega$-cm$^2$). Table 2.3 [11] summarizes the performance of these arrays where there is $R_0A$ variation with increasing operating temperatures for a range of device cutoff wavelengths. $D^*$ curves for three representative detectors as a function

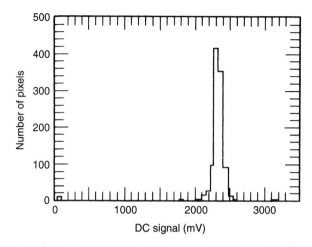

**Figure 2.28** Signal histogram of 32 × 32 LWIR hybrid FPA. (From Ref. 6.)

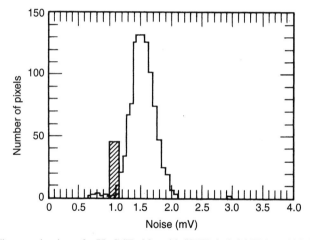

**Figure 2.29** Characterization of a HgCdTe 32 × 32 SWIR hybrid FPA at 120 K in background of $10^{13}$ photon/cm$^2$/s. (From Ref. 6.)

of temperature are given in Fig. 2.32 [12]. These curves again illustrate that these detectors may be nearly background limited performance (BLIP) and can be operated at higher temperatures for shorter cutoff wavelengths. The temperature dependence for a 4.3-μm cutoff array is given in Fig. 2.33 [11], which suggests that cooling below 60 K no longer improves performance.

### 2.5.3 Requirements For HgCdTe-Infrared Focal Planar Arrays

One single cryogenically cooled infrared focal planar array (IRFPA) is capable of meeting most of the needs of high-performance IR detectors for military, space, and meteorological applications [7]. It is important to mention that HgCdTe semiconductor is a unique ternary compound capable of detecting wavelengths over 1–14-micron region

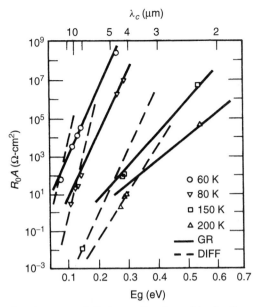

**Figure 2.30**  Calculated and measured $R_0A$ as a function of $\lambda_c$ for four selected temperatures. Optimized diodes with only G–R and diffusion dark currents were assumed in the model. (From Ref. 11.)

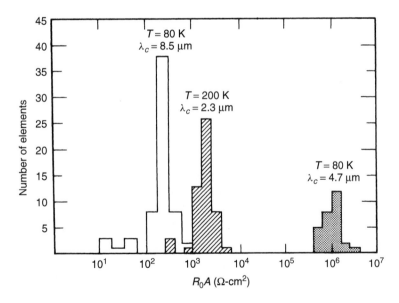

**Figure 2.31**  $R_0A$ histograms of SWIR, MWIR, and LWIR detector hybrids. (From Ref. 11.)

at 80 K. In other words, this ternary detector compound offers reasonably good performance in all bands, including SWIR, MWIR, and LWIR. Note the HgCdTe-FPA can operate at higher operating temperatures in certain applications, thereby significantly reducing the cooling constraints in terms of size, weight, cost, and power consumption

**TABLE 2.3**

| $\lambda_c$ at 80 K($\mu$m) | Structure | $R_0 A (\Omega\text{-cm}^2)$ | | |
|---|---|---|---|---|
| | | $T = 60$ K | $T = 80$ K | $T = 200$ K |
| 2.3 | Single-layer heterojunction | — | — | $8 \times 10^4$ |
| 4.7 | Planar implanted | — | $1 \times 10^7$ | 8 |
| 4.8 | Double-layer heterojunction | $3 \times 10^7$ | $3 \times 10^6$ | 1 |
| 8.5 | Single-layer heterojunction | $4 \times 10^4$ | $6 \times 10^2$ | — |
| 10.3 | Single-layer heterojunction | $3 \times 10^3$ | $2 \times 10^2$ | — |
| 11.5 | Double-layer heterojunction | $2 \times 10^2$ | 10 | — |

*Source*: Ref. 8.

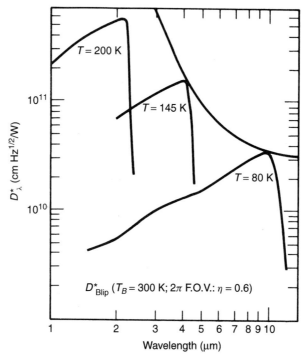

**Figure 2.32** Spectral detectivity for three representative detectors. The SWIR and the MWIR are implanted while the LWIR device is a double-layer heterojunction. (From Ref. 11.)

by using small Stirling-cycle cryocoolers or thermoelectric (TE) coolers [7]. An IRFPA using a well-controlled ion-implantation process, LPE growth technique, and optimum compositional profile can offer high sensitivity, long-term stability, and accurate control of pixel size needed for high-quality images.

**2.5.3.1  *Applications of IRFPAs in Various Bands.*** IRFPA detectors have potential applications in SWIR, MWIR, LWIR, and very-LWIR (VLWIR) bands. The SWIR band (1 to 3 micron) has several applications, including space applications for military target observation, agricultural monitoring, meteorological studies, and astronomical

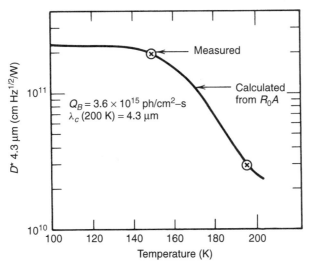

**Figure 2.33**  Calculated and measured temperature dependence of $D^*$ for an implanted MWIR hybrid under moderate-background radiation. (From Ref. 11.)

research. Such applications require low dark current, high quantum efficiency, and low readout circuit noise to achieve high detector sensitivity under low optical flux levels.

The MWIR-FPAs are of highest interest due to atmospheric transmission characteristics. This band is split into two sub-bands, namely, 3.0 to 4.2 micron and 3.7 to 4.8 micron sub-band, each having the lowest dark current and optimum detector sensitivity over the sub-band region. Studies performed on IRFPAs [13] indicate that the lower portion of the SWIR spectrum band is normally used in applications where high performance is of critical importance under cryogenic environments (typically over 80–140 K temperature range) and low photonic flux levels. The high sub-band (3.7 to 4.8 micron) is best suited for IRFPAs operating over 180–220 K temperature range. This particular sub-band when operated below 80 K offers improved performance most attractive for certain military systems such as Joule–Thomson-cooled missile seekers or Stirling-cycle-cooled forward looking IR (FLIR) imaging sensors for tactical applications. Note a MWIR-FPA detector operating above 180 K generally use thermoelectric cooler (TE), which yields minimum cost, high reliability, and noise/vibration-free operation. Such IRFPA detectors are widely used for space surveillance and commercial thermography. Note detectors operating at higher temperatures have shorter cooling times and require less input power. It is important to mention that the cryogenic power requirement for a rotary microcooler needed to cool down the MWIR-FPA at 120 K is only 55% of the one needed to cool down at 77 K. Latest microcooler research studies [7] indicate that the pulse tube refrigerator (PTR) offers significantly higher life time over the microcoolers currently being used.

*2.5.3.2  Capabilities of LWIR-FPA Detectors.*  LWIR-FPA detectors operating over 8- to 14-micron spectral region have potential applications in military systems, such as FLIRs, IR Search and Track sensors (IRSTs), missile seekers, and anti-tank weapons. These detectors are also best suited for certain scientific and space applications involving surveillance and spectrometry. LWIR-FPAs are optimized for three distinct categories:

1.  Linear or Staring arrays operating in the 12- to 14-micron spectral region.

2.  Scanning arrays operating in the 8- to 10.6-micron spectral range.

3.  Staring arrays operating in the 8- to 9.5-micron spectral region.

It is important to mention that LWIR-FPA detectors used in a space system require very low operating temperatures (below 60 K) to ensure low dark current, high cutoff wavelength, and optimum signal-to-noise (S/N) ratio. High-performance FLIRs deployed in military applications use linear arrays with wide spectral bandwidth, high quantum efficiency, and enhanced S/N ratio. The performance of LWIR-FPA staring arrays is limited by three critical parameters, namely, the maximum amount of charges storable in the read-out circuit, the integration time compatible with required frame time, and the dark current of the detector material. Regardless of the operating wavelength of the LWIR-FPA, the dark current and the detectivity are the most critical performance parameters and are strictly dependent on the operating temperatures. The dark current levels for a LWIR-FPA are much higher than those for SWIR-FPA or MWIR-FPA detectors. For the SWIR-FPA, the dark current varies from 10 nA at 180 K to 100 nA at 220 K and at 3.16 micron, whereas for a MWIR-FPA, the dark current varies from 50 nA at 80 K to 500 nA at 120 K and at 5.5 micron. However, for a LWIR-FPA detector [Ref 8], the dark current varies from 1 $\mu$A to 200 $\mu$A at 16-micron cutoff wavelength, depending on the operating temperature, reverse bias level, and the detector junction diameter. The peak detectivity for LWIR-FPA and MWIR-FPAs is better than $10^{13}$cm/$\sqrt{\text{Hz}}$. W at their recommended operating temperatures. Note the peak detectivity of a LWIR-FPA detector with 16-micron cutoff wavelength varies from $7 \times 10^{12}$ at 40 K to $4 \times 10^{11}$ at 60 K to $3 \times 10^{10}$cm/$\sqrt{\text{Hz}}$. W at 80 K operating temperature. It is evident from these statements that both the low dark current and high peak detectivity in LWIR-FPA detectors with cutoff wavelength of 16 micron and up are possible at lower cryogenic temperatures ranging from 40 to 80 K.

## 2.6    INDIUM ANTIMONIDE

Indium antimonide (InSb) is an intrinsic narrow bandgap crystal sensitive from 1.0 to 5.5 $\mu$m. Although it has been fabricated into both PC and PV detectors, photodiodes have seen the most use. InSb diodes are characterized by a high degree of linearity, stability, and uniformity. They are usually operated at or below 77 K to reduce dark current and are available in long line arrays (up to 512 elements) and moderate-sized area arrays (128 × 128). Several different readout architectures are employed, the most common being (1) charge injection devices (CIDs), (2) charge-coupled devices (CCDs), and (3) coupling to switched FET multiplexers.

### 2.6.1    InSb Performance Characteristics

Representative $I$–$V$ (current–voltage) curves are given in Fig. 2.34 [12] for different operating temperatures. Of particular interest for low-light-level (LLL) applications requiring long integration time (>0.1 s) are the low values of the dark current for these operating temperatures. These curves are used to derive the $R_0 A$ curve for near-zero bias shown in Fig. 2.35 [12], which is a common mode of operation where low noise is required such as in astronomical applications. The maximum $R_0 A$ versus

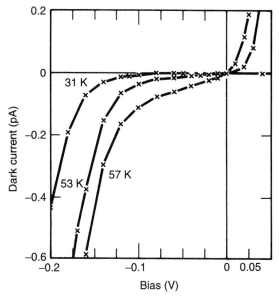

**Figure 2.34** $I-V$ characteristics of 0.5-mm-diameter InSb diode shown on several scales for various temperatures. (From Ref. 12.)

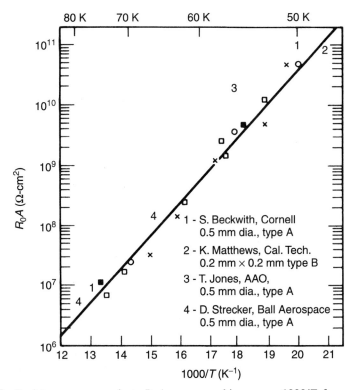

**Figure 2.35** Resistance–area product, $R_0A$, near zero bias versus $1000/T$ for several diodes of different size and type. (From Ref. 12.)

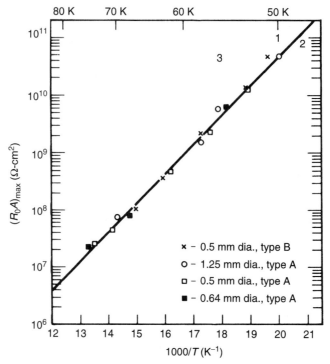

**Figure 2.36**  Maximum resistance–area product plotted versus $1000/T$. Numbers plotted correspond to sources listed in Fig. 2.35. (From Ref. 12.)

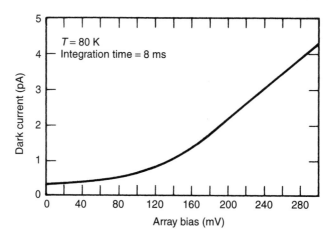

**Figure 2.37**  Dark current versus array bias. (From Ref. 14.)

reciprocal temperature is plotted in Fig. 2.36, where high values ($>10^6$ $\Omega$-cm$^2$) occur even at liquid nitrogen temperatures (80 K). Representative dark current values versus reverse bias, derived from $R_0 A$ versus reciprocal temperature, are given in Fig. 2.37 [14] for a typical operational 8-ms integration time. The quantum efficiency of the device, shown in Fig. 2.38 [14], depends partly on the antireflection coating in use.

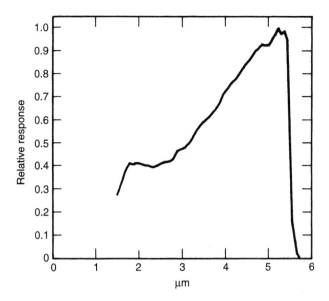

**Figure 2.38**   InSb relative spectral response. (From Ref. 14.)

**TABLE 2.4   Selected Diode Responsivity**

| Element | Responsivity (A/W) |
|---------|--------------------|
| 1 | 2.81 |
| 2 | 2.81 |
| 3 | 2.78 |
| 4 | 2.89 |
| 5 | 2.89 |
| 6 | 2.89 |
| 7 | 2.89 |
| 8 | 2.85 |
| 9 | 2.89 |
| 10 | 2.78 |

*Source*: Ref. 14.

This curve illustrates a device optimized for the MWIR near 5 μm, which is close to the intrinsic long-wave cutoff of InSb (ca. 5.5 μm). The absolute maximum quantum efficiency here is approximately 70%, which can be used to convert Fig. 2.38 to absolute values through multiplication. Typical responsivities are given in Table 2.4 [14] for 10 randomly selected diodes at 4.3 μm indicating their high degree of uniformity.

### 2.6.2   InSb Line Arrays

The characteristics of a typical 128-element line array are given in Table 2.5 [14]. Some items of note are the large full well ($2 \times 10^7 e$), high fill factor (90%), extreme linearity (2%), moderate readout noise ($2000e$), and high maximum quantum efficiency (80%).

### 2.6.3   InSb Area Arrays

One common technique for manufacturing area arrays is to use CID technology whereby metal insulator semiconductor (MIS) capacitors are formed within the InSb wafer [15].

**TABLE 2.5   128-Element FPA Mechanical/Electrical Characteristics**

Detector material: InSb
Pixel size: $200 \times 200$ μm
Number of pixels: 128
Dead space: 30 μm between pixels
Array package size: $1\frac{3}{4} \times 1 \times \frac{1}{8}$ in.
Power dissipation: 2 mW
Number of package leads: 15
Diode leakage at 80 K: 4 Pa at 0.24 V reverse bias
Dark current density at 80 K: $7 \times 10^{-9}$ A/cm$^2$ at 0.24 V reverse bias
Quantum efficiency: $\geqslant 0.8$ (1 to 5 μm)
Responsivity uniformity: 3%
Dark nonuniformity: 3% at 0.24 V reverse bias
Maximum integration time: 80 K = 0.4 s, 62 K = 28 s
Dark current temperature coefficient: drops in half every 3.5 K
Maximum charge storage: $2 \times 10^7$ electrons at 0.24 V reverse bias
Ac signal-to-noise ratio: 4000: 1 at 80 K (simple pixel reset)
Readout noise: $\sim$1200 electrons (simple pixel reset)
Spectral noise density: white, 100 Hz to 100 kHz
Maximum readout speed: 80 K–1 MHz, 62 K–200 kHz
JFET preamp freeze-out: $\approx$50 K

*Source*: Ref 10.

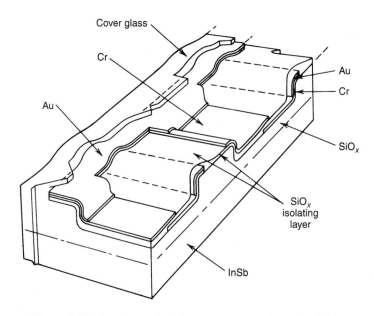

**Figure 2.39**   Structure of dual-gate area array. (From Ref. 15.)

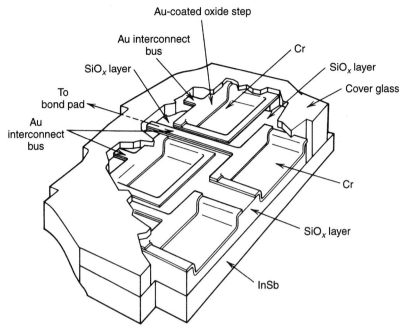

**Figure 2.40**   Structure of InSb sensing sites of line array with staggered elements. (From Ref. 15.)

An advantage of this approach is the capability to address individual or groups of pixels, called *x-y* addressing. Examples of gate structures are shown in Figs. 2.39 and 2.40 [15], and a typical circuit configuration is shown in Fig. 2.41 [15].

## 2.7   SCHOTTKY-BARRIER DETECTORS

Schottky-barrier detectors make use of the Schottky-barrier junction formed at the interface of a semiconductor such as silicon and a metal such as platinum or paladium. These two metals are the most commonly applied and result in PtSi and PdSi Schottky-barrier diodes, respectively. The long-wave cutoff for PtSi is about 5 μm, while the practical upper limit for PdSi is near 3 μm. The longer cutoff arrays are usually cooled to 77 K or below, while the shorter cutoff arrays can operate as warm as 200 K. This construction, involving the deposition of a thin film of metal, results in detector arrays with a very high level of uniformity, typically less than 1%. For many applications, the high uniformity often compensates for the relatively low quantum efficiency that is usually about 10% at 3 μm and drops below 1% at 4 μm. Because the most common of these arrays are silicon based, they are easily mated to silicon CCDs. These monolithic IRCCDs are presently available in sizes up to $160 \times 244$ pixels from Sarnoff Laboratories.

### 2.7.1   Schottky-Barrier Characteristics

A simplified diagram of the band structure showing the photoemissive process characteristic of the PtSi or PdSi Schottky-barrier diodes is shown in Fig. 2.42 [16]. Photons

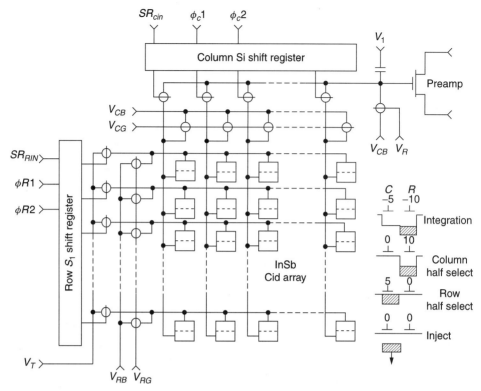

**Figure 2.41** Schematic diagram of two-dimensional focal plane configurations, including an InSb CID array and two shift registers, along with idealized potential well diagrams for different select conditions. (From Ref. 15.)

longer than 1 μm in wavelength that pass unattenuated through the silicon cause the emission of a so-called "hot hole" into this substrate. The corresponding electron moves to the CCD input gate, where it is sensed. Because the emission probability is a strong function of the threshold potential, as opposed to most photon detectors, where it is a strong function of thickness, the quantum efficiency is very similar from pixel to pixel, thus assuring high uniformity.

A typical cross section of a PtSi diode is shown in Fig. 2.43 [16]. The aluminum serves as a reflector to increase the relatively low quantum efficiency whose typical variation with wavelength is given in Fig. 2.44 [16]. Note that above 2 μm the quantum efficiency is always below 10%. These low values are the result of the very thin (20 to 100 Å) silicide layer whose shallow depth is required to ensure passage of the excited holes across the barrier and to limit optical "crosstalk." Typical dark current densities versus reverse bias voltage and reciprocal temperature are shown in Figs. 2.45 and 2.46 [16], respectively, for PtSi. The indicated low values at 77 K allow integration times from 10 to 100 ms, making these devices ideally matched to video rates of 30 to 60 frames per second. They are also very linear, as indicated by the output voltage versus input energy flux given in Fig. 2.47 [17]. This curve also illustrates the large dynamic range (about 2000) exhibited by these monolithic devices.

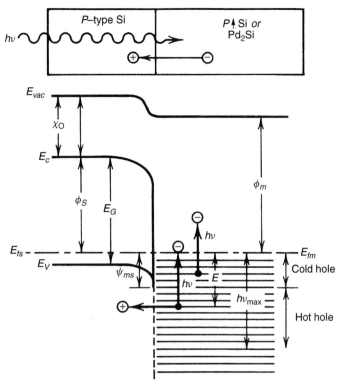

**Figure 2.42** Simplified energy band diagram illustrating the photoemission in a PtSi or PdSi Schottky-barrier detectors. Escape (emission) probability of hot holes for thick-film devices. (From Ref. 16.)

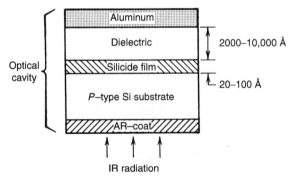

**Figure 2.43** Cross-sectional view of thin-PtSi Schottky-barrier detector with optical cavity. (From Ref. 16.)

### 2.7.2 Schottky-Barrier Array Architecture and Figures of Merit

A typical structure of a single pixel from a $64 \times 128$ array is shown in Fig. 2.48 [17]. This illustrates the usual low fill factor (here 22%) due to the buried channel CCD placed alongside each active element. This array makes use of interline transfer readout

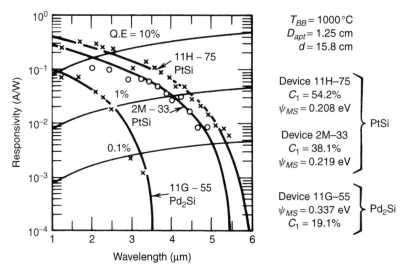

$T_{BB} = 1000\,°C$
$D_{apt} = 1.25$ cm
$d = 15.8$ cm

Device 11H–75
$C_1 = 54.2\%$
$\psi_{MS} = 0.208$ eV

Device 2M–33
$C_1 = 38.1\%$
$\psi_{MS} = 0.219$ eV

} PtSi

Device 11G–55
$\psi_{MS} = 0.337$ eV
$C_1 = 19.1\%$

} $Pd_2Si$

**Figure 2.44** Measured responsivity of thin PtSi and PdSi high-performance Schottky-barrier detectors. (From Ref. 16.)

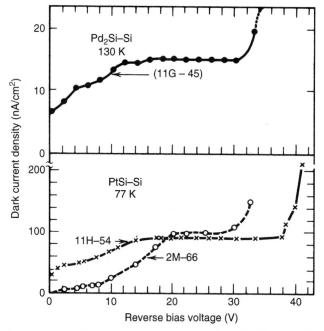

**Figure 2.45** Effect of reverse bias on the dark current density of PdSi and PtSi Schottky-barrier detectors. (From Ref. 16.)

scheme sketched in Fig. 2.49 [17]. It is part of a group of commercially available arrays listed in Table 2.6 [16]. Some relevant performance parameters for the $32 \times 63$ array is given in Table 2.7 [16], where the large dynamic range (2000:1) is due to the relatively low integrated dark current and readout noise coupled with a usable full well

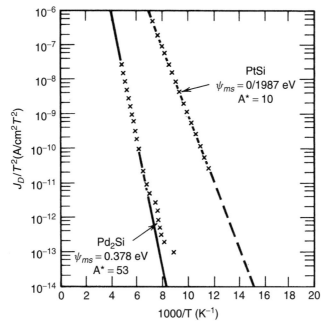

**Figure 2.46** Dark current characteristics of PtSi and PdSi Schottky-barrier detectors. (From Ref. 16.)

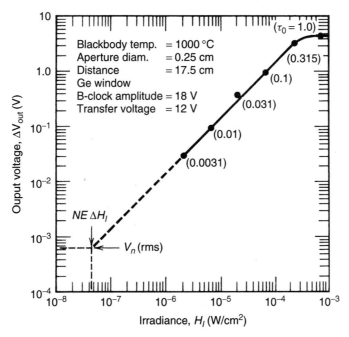

**Figure 2.47** Optical transfer characteristics of an $32 \times 63$-element IR-CCD image sensor. (From Ref. 17.)

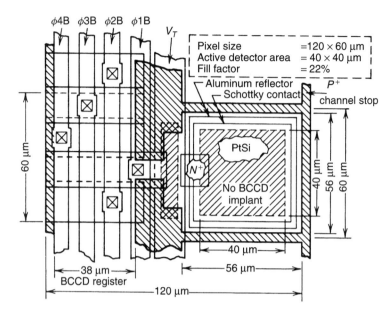

**Figure 2.48**    Layout of one pixel of an 64 × 128-element IR-CCD image sensor. (From Ref. 17.)

about $1 \times 10^6$ electrons. Most of these arrays require only a dc-type offset (subtraction of non-photondependent signals such as dark current responsible for so-called fixed pattern noise) to provide a uniformity correction. These noise sources are illustrated in Fig. 2.50 [17], which shows the dominance of 700 electrons of fixed pattern noise up to background levels of $1 \times 10^6$ electrons. Because this is a SWIR and MWIR detector, shot noise due to typical 300 K backgrounds do not limit performance. However, this curve indicates that for 300 K backgrounds, the detector is very close to BLIP because these other noise sources are relatively low.

## 2.8   SPRITE DETECTORS

The SPRITE detector ("signal processing in the element"), developed by C. T. Elliott [18], makes use of a technique almost equivalent to continuous time delay and integration. A long thin strip or filament of PC *n*-type HgCdTe is biased such that the drift velocity of a charge packet equals the scan velocity of an image. Thus, if the carriers do not recombine too quickly, the effective dwell time is greater than that of a single detector because the charge packet is continuously integrating while traversing the filament to a narrow readout region. However, the required long lifetimes lead to diffusion spreading of charge packets, which can degrade the imaging properties of the detector. There has been experimentation with a so-called "meander" technique to limit the effects of this diffusion, but these more complex SPRITEs will not be discussed here.

This technique does not require a line arrangement of several discrete detectors with individual signal processing electronics to provide the time delay and integration function. The minority carrier diffusion length must be kept low to ensure good

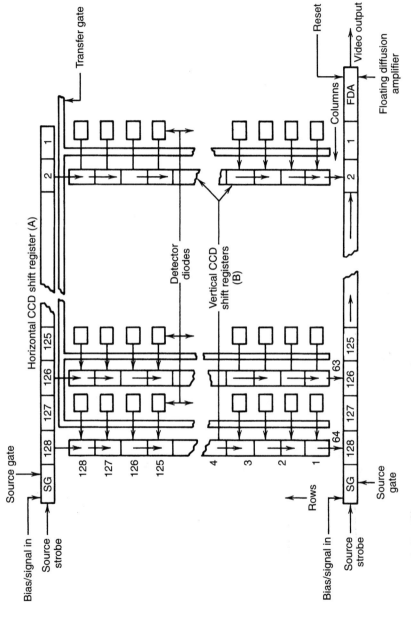

**Figure 2.49** Block diagram of an 64 × 128-element IR-CCD FPA. (From Ref. 17.)

97

**TABLE 2.6  Reported Schottky-Barrier IR-CCD Focal Plane Arrays**

| Type of FPA | Pixel Size $(\mu m)^2$ | Fill Factor (%) | | Type of SBDs | Year | Company |
|---|---|---|---|---|---|---|
| 256 × 1 | 40 (H) × 320 (V) line sensor | 50 | Thick-PtSi | $\psi_{MS} - 0.27$ eV | 1978 | RCA/RADC |
| | | | Thin-PtSi | | 1980 | RCA/RADC |
| 25 × 50 interline transfer (IT) | 160 (H) × 80 (V) | 17 | Thick-PtSi | $\psi_{MS} = 0.27$ eV | 1978 | RCA/RADC |
| | | | Thin-PtSi | | 1980/1981 | RCA/RADC |
| 32 × 63 SPSIT | 160 (H) × 80 (V) | 25 | Thin-PtSi | $\psi_{MS} = 0.208$ to 0.22 eV | 1981/1982 | RCA |
| | | | Thin-Pd$_2$Si | $\psi_{MS} = 0.337$ eV | 1982 | RCA |
| 64 × 128 IT | 120 (H) × 60 (V) | 22 | Thin-PtSi | $\psi_{MS} = 0.208$ to 0.22 eV | 1981/1982 | RCA |
| 32 × 64 IT | 133 (H) × 80 (V) | 19 | Thin-PtSi | $\psi_{MS} = 0.277$ eV | 1981 | Mitsubishi |
| 64 × 64 meander-channel IT | 130 (H) × 70 (V) | 23 | Thin-PtSi | $\psi_{MS} = 0.23$ eV | 1983 | Fujitsu |
| 256 × 256 IT | 37 (H) × 31 (V) | 25 | Thin-PtSi | $\psi_{MS} = 0.26$ eV | 1983 | Mitsubishi |

*Source*: Ref. 16.

**TABLE 2.7   Measured Performance of 32 × 63 PtSi IR-CCD FPA**[a]

| | |
|---|---|
| CCD noise | = 180 to 250 rms electrons/pixel |
| Dark current | = $10^5$ electrons/pixel |
| Dynamic range | = 70 dB |
| Response | = $4 \times 10^4$ electrons/K |
| NE $\Delta T$ | = 0.033 K |
| MRT at $f \approx \frac{1}{5} f_N$ | < 0.1 K (FPN limited) |
| (with electronic uniformity compensation) | |

[a] $160 \times 80$ μm pixels, 25% fill factor, 60 frames/s, $f/2.0$ optics.
*Source*: Ref. 17.

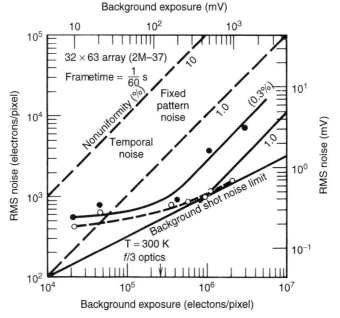

**Figure 2.50**   Noise-data analysis for an $32 \times 63$-element IR-CCD image sensor. (From Ref. 17.)

spatial MTF. Otherwise, the charge packets spread during scanning, causing a degradation in image quality. Long recombination times are also required and low carrier density materials may exhibit lifetimes as long as 5 μs for 8- to 12-μm detectors and 20 μs for 3- to 5-μm detectors cooled to 195 K by thermoelectric coolers. HgCdTe has a relatively long diffusion length of 25 μm, allowing effective use of long transit times. These devices are very compact, with well-cold shielded arrays of up to eight parallel filaments available for use in FLIR (forward-looking infrared) systems.

The detectivity is given by [18]

$$D_\lambda^* \simeq \sqrt{2n}\, D_\lambda^*(\text{BLIP})\sqrt{s\tau} \quad \text{cm}\sqrt{\text{Hz}}/\text{W} \tag{2.11}$$

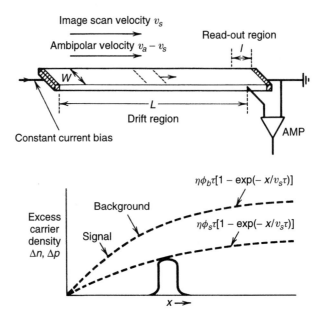

**Figure 2.51**   Schematic illustration of the operation of a SPRITE detector. (From Ref. 18.)

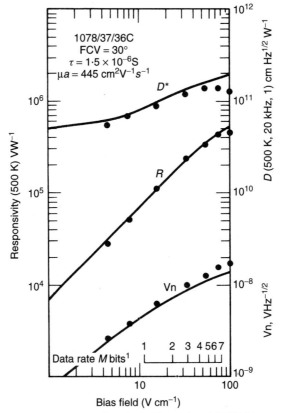

**Figure 2.52**   Performance parameters of an 8- to 13-μm band SPRITE operated at 77 K in 30° field of view. (From Ref. 18.)

where $D_\lambda^*(\text{BLIP})$ is equivalent to the background-limited detectivity of a conventional single detector, $n$ the quantum efficiency, $\tau$ the integration time, and $s$ the pixel rate. Typical values of $\tau$ are 2 μs with $s$ values of $5 \times 10^5$ s$^{-1}$.

The number of equivalent discrete detectors is given by [18]

$$N_{eq}(\text{BLIP}) \sim 2s\tau \qquad (2.12)$$

An example of a typical device 50 μm wide and 65 μm long for the 8- to 14-μm region operated at 77 K is shown in Fig. 2.51 [18]. This figure illustrates the movement of the charge packet corresponding to a single imaging pixel moving toward the readout region. It also suggests that the major component of the signal is the background—a common situation for the LWIR in particular.

### 2.8.1   SPRITE Performance

The variation of $D^*$, responsivity, and noise voltage against bias voltage is given in Figs. 2.52 and 2.53 [18] for 8- to 14-μm and 3- to 5-μm devices, respectively. In addition to high detectivities, responsivities of these devices are higher than conventional HgCdTe detectors, which are limited by sweep-out effects to $2 \times 10$ V/W. The short-wave SPRITE tends to saturate in both $D^*$ and $R$ because of short integration times imposed by its finite length.

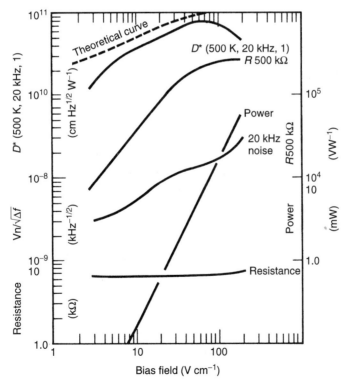

**Figure 2.53**  Performance parameters for a SPRITE with a cutoff wavelength of 4.5 μm at 183 K in 55° field of view. (From Ref. 18.)

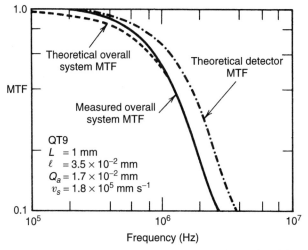

**Figure 2.54** Measured and theoretical MTF data for an experimental serial scan imager. (From Ref. 18.)

The optical imaging performance is an important parameter that needs to be characterized. A typical curve of MTF (modulation transfer function) versus frequency is given in Fig. 2.54 [18] for a two-row 8- to 14-μm device. The theoretical curve is based on Eq. (2.13) [18] and compares well with experiment.

$$\mathrm{MTF} = [1/(1 + K_s^2 Q_a^2)] \sin(K_s l/2)/(K_s l/2) \qquad (2.13)$$

where $K_s$ is the spatial frequency, $l$ the readout length, and $Q_a$ the minority-carrier diffusion length.

One major limitation to increasing performance is Joule heating of the material at high bias voltages. Increased heating also results from the limited cooling capacity of

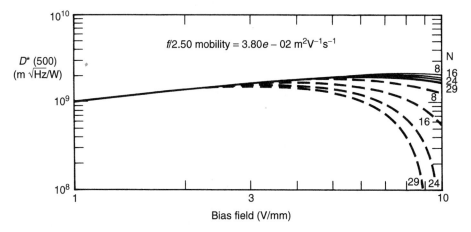

**Figure 2.55** $D^*(500)$ versus bias field for $N$ detectors. (From Ref. 20.)

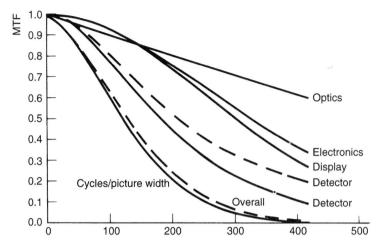

**Figure 2.56** Solid line represents MTF for baseline system. Dashed line represents MTF for an improved detector. (From Ref. 20.)

compact dewars when more than eight filaments are arranged in parallel. The decrease in performance with number of detectors versus bias voltage is given in Fig. 2.55 [19].

The different contributions to total MTF is illustrated in Fig. 2.56 [19], where detector MTF dominates performance.

## 2.9  PYROELECTRIC DETECTORS

This type of device is a thermal detector whose principle of operation is fundamentally different from the photon detectors described in this chapter. Rather than responding to individual photons, pyroelectrics respond to a change in temperature resulting from changes in input photon flux. The material is basically a capacitor creating changes in voltage due to changes in polarization. Because the principle of operation relies on changes in temperature, the input photon signal must be optically chopped [20]. A typical material is triglycine sulfate (TGS), which responds from 1.0 to 30.0 μm. TGS belongs to a subgroup of piezoelectric crystals that possess a spontaneous polarization [21]. Local charge from the neighboring atmosphere usually neutralizes this polarization. However, these surface charges cannot stay in phase with rapidly changing polarization.

The first imaging devices were pyroelectric vidicons, which take advantage of the ability to operate uncooled. One inherent disadvantage of these vidicons is their low sensitivity and often poor image quality due to high thermal diffusion during each frame time [22].

Self-scanning arrays are replacing vidicons because the chopped signal creates low heat levels, allowing compatible mating of the pyroelectric with solid-state multiplexers via flip-chip bonding without the need for exotic insulating materials. These arrays have many industrial applications, because cooling is not required and the spectral response is very uniform [23]. Line arrays are often used for the focal plane of spectrometers, while area arrays up to 16 × 16 elements have been used in direct imaging systems and provide an NEDT of about 0.1 K per picture element at a 10-Hz bandwidth.

Pyroelectrics are not well suited for use in scanning sensors that employ only one detector because they have poor high-frequency response [24] and give NEDTs of several Kelvin.

### 2.9.1 Pyroelectric Performance

Pyroelectric materials usually perform better than most other thermal detectors. A comparison between the detectivity of a pyroelectric (PE), Golay cell (G), thermopile (T), and immersed thermistor (IT) is given in Fig. 2.57 [21].

The typical spectral response of a pyroelectric vidicon is shown in Fig. 2.58 [25], which has been optimized for the 8- to 14-$\mu$m window. A representative sectional view of a lead zirconate (PZ) linear array is shown in Fig. 2.59 [26], with corresponding detectivity, responsivity, and noise shown in Fig. 2.60 [26]. The individual array elements are 200 $\mu$m square and 30 $\mu$m thick. The roll-off at high frequencies is due to the limited penetration of signal in the pyroelectric resulting from shallow diffusion depths, whereas the low-frequency roll-off is limited by the total thermal resistance.

A sketch of a typical flip-chip bonding to a silicon CCD is shown in Fig. 2.61 [26] for a 16 × 16 array. This CCD operates via direct injection as illustrated in Fig. 2.62 [26]. The NEDT for three different combinations of thickness and pitch is plotted versus

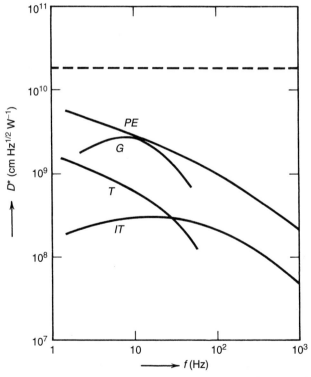

**Figure 2.57** The re-normalized detectivity as a function of frequency for various types of thermal detectors: the pyroelectric element (PE), the Golay cell (G), the thermopile (T), and an immersed thermistor (IT). The dashed line represents the limiting case of the ideal thermal detector. (From Ref. 21.)

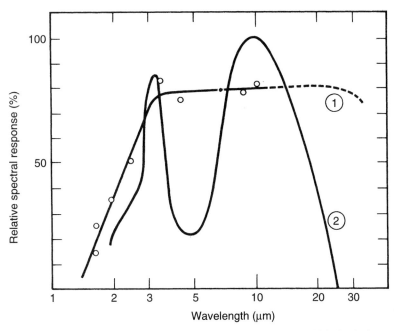

**Figure 2.58**   Typical spectra response of pyroelectric vidicons: (1) a KRS-5 window and (2) a Ge window coated for optimum transmission in the range 8 to 14 μm. The dotted circles represent absolute spectral response values. (From Ref. 24.)

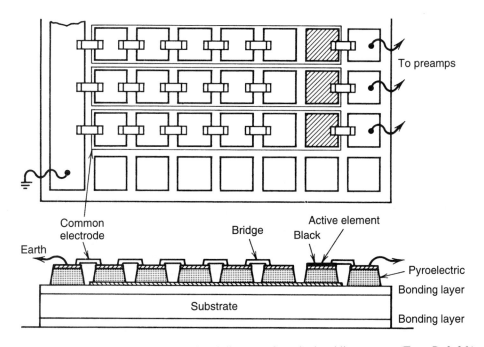

**Figure 2.59**   Schematic plan and sectional diagram of a reticulated linear array. (From Ref. 26.)

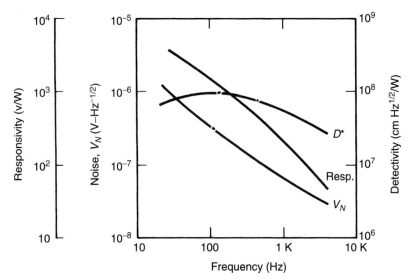

**Figure 2.60** Performance of a lead zirconate (PZ) pyroelectric array, 200-μm square elements, 30 μm thick. (From Ref. 26.)

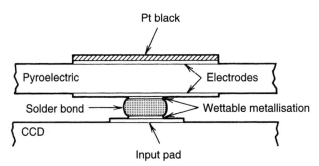

**Figure 2.61** Flip-chip solder bond interconnection of pyroelectric detector element to CCD input pad. schematic. (From Ref. 26.)

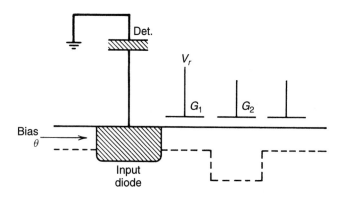

**Figure 2.62** Direct injection interface for a pyroelectric-CCD hybrid. (From Ref. 26.)

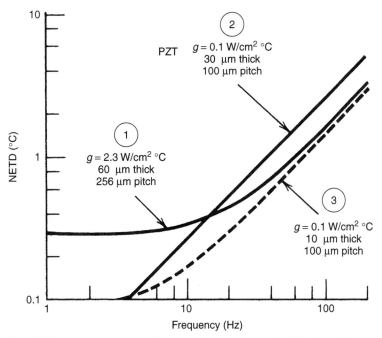

**Figure 2.63** NETD versus chopping frequency for pyroelectric-CCD hybrid arrays. (From Ref. 26.)

chopping frequency in Fig. 2.63 [26]. For this device, the dominant noise results from the MOSFET and not the more common $1/f$ source. This noise is given by Eq. (2.14), where $C_e$ is pixel capacitance in nanofarads, $a$ the injection efficiency, and $T$ the array temperature in Kelvin. The corresponding signal is given by Eq. (2.15), where $p$ is the pyroelectric coefficient, $A_e$ the element area, and $I_0$ the modulated incident radiation chopped at angular frequency $\omega_c$.

$$Q(g_m \text{ noise}) = (4kTCea/3)^{1/3} \quad \text{coulombs} \tag{2.14}$$

$$qs = apA_eI_0/\sqrt{g^2 + \omega_c^2 C^2 d^2} \quad \text{coulombs} \tag{2.15}$$

The NEDT plotted is given by the ratio of Eqs. (2.14) and (2.15).

## 2.10 EXTRINSIC SILICON

Much effort has been devoted to making silicon sensitive to the infrared in order to take advantage of the mature integrated-circuit technology associated with this material. In fact, silicon CCDs sensitive to visible light are commercially available in arrays larger then $500 \times 500$ pixels. The intrinsic long-wave cutoff of silicon is about 1 μm. However, by introducing deep-level impurities, it may be made extrinsically sensitive out to the LWIR [26]. One major remaining disadvantage of doped silicon for infrared area arrays is the low operating temperatures required. These usually range from below 20 K in the MWIR and below 40 K in the LWIR. In contrast to HgCdTe, impurities are not available to sensitive silicon specifically to these wavelength regions. The thermal ionization of deep-level impurities is mainly responsible for the low operating temperatures required. It is actually the residual boron impurities, present in all silicon, that gives rise to this substantial dark current even at temperatures as cold as 50 K. The two most common dopants in use are indium (In) and galium (Ga) for devices sensitive to the MWIR and LWIR, respectively.

### 2.10.1   Readout Structures

Two current readout techniques are in use. One is the so-called accumulation mode, making use of majority carriers, while the other more common approach makes use of minority carriers, with both considered to be monolithic devices. The accumulation mode is illustrated in Fig. 2.64 [27], which shows the equivalent surface channel structure and the linear variation with distance of the silicon energy bands caused by the insulating property of the silicon operated at low temperatures. Holes drift to the silicon oxide interface and are subsequently read out of the device.

By making use of epitaxial growth techniques, CCD shift registers of opposite conductivity are fabricated on the same chip. A sketch of this structure along with the resultant energy bands are shown in Fig. 2.65 [27]. In this architecture, the CCD is

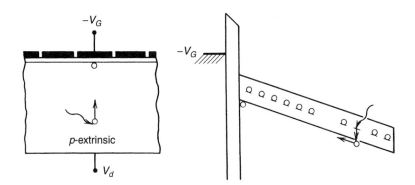

**Figure 2.64**   Accumulation Mode Extrinsic CCD. (From Ref. 27.)

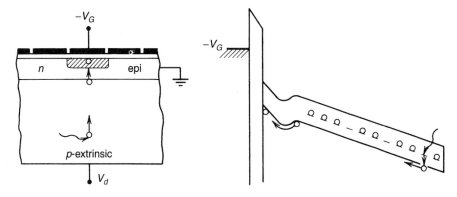

**Figure 2.65**   Extrinsic Silicon Detector With Minority Carrier CCD Readout. (From Ref. 27.)

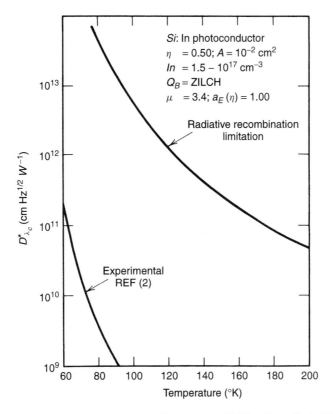

**Figure 2.66**  Detectivity Versus Temperature (Si:In). (From Ref. 28.)

a conventional minority carrier device and the photo-generated majority carriers are sensed by the conventional techniques of direct injection or indirect using diffusion at the interface.

### 2.10.2  Extrinsic Silicon Performance Characteristics

Detectivity versus temperature is plotted for both indium-doped silicon (Si:In) and Gallium-doped silicon (Si:Ga) in Figs. 2.66 and 2.67, respectively [28]. Both curves indicate that these devices do not approach their theoretical radiation limit of performance under zero background conditions. It is apparent that Si:In operated at liquid nitrogen temperatures will give very low (about $1 \times 10^9$ cm$\sqrt{\text{Hz}}$/W)$D^*$ and that cooling below 60 K is necessary. The curve for Si:Ga suggests the need for even lower temperatures for useful performance, near 20 K. Reference 28 gives some very useful comparisons with competitive devices of HgCdTe, PbSnTe, and InAsSb. All three of these intrinsic materials exhibit a $D^*$ at least two orders of magnitude higher at 80 K than Si:In or Si:Ga.

### 2.11  SYSTEMS ANALYSIS

Systems analysis is a general term used to describe an analytical approach for calculating the performance of an instrument in some quantitative terms. These terms may be

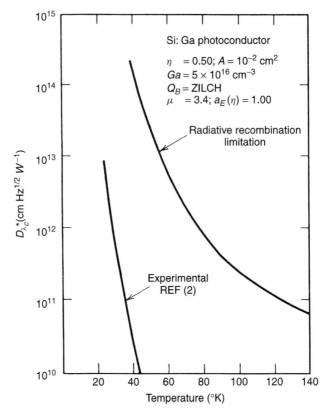

**Figure 2.67**   Detectivity Versus Temperature (Si:Ga). (From Ref. 28.)

a single signal-to-noise ratio in one spectral channel or a complete table of minimum resolvable temperature differences versus object frequency for many spectral channels. The calculations usually require knowledge of basic instrument parameters, such as optical collecting area, throughput, detector sensitivity, and noise. In addition, a target model is required consisting of its radiant properties, size, range, and absorptive and emissive effects of any intervening atmosphere. The means by which the resultant signal is encoded often requires consideration of quantitization noise and errors due to sampling. This level of analysis usually does not require a detailed optical or mechanical design of the instrument. However, it is often the first crucial step used to determine what the basic instrument parameters must be to meet specified performance and design requirements. A parametric analysis is one common technique whereby one parameter is taken as a fundamental variable while all others are fixed and family of curves plotted with some measure of performance against this variable.

### 2.11.1   Parametric Equations

A basic performance parameter is the sensor signal-to-noise ratio (SNR or *S/N*), which is the ratio of the detected signal to the rms noise in the same units. The SNR may be related to the response to a single target pixel or related to its neighbors, taking into account the total modulation transfer function of the instrument and atmosphere.

The approach to the noise calculation depends on the type of detector and the physics used for its characterization. For example, if the concept of $D^*$ is used, a logical approach would be to calculate the noise equivalent power (NEP) based on the definition of $D^*$ applicable to those operating conditions and given detector size and electric bandwidth.

Difficulties or ambiguities may arise when the specified $D^*$ was not measured under the same conditions of background or operating temperature representative of the system of interest. For these cases, explicit formulation of the contributors to the noise, such as dark current, background, read noise, and so on, in parametric form, will be more useful. However, such formulations are often not possible when knowledge of the operational physics of detectors of interest are not completely known.

Because of the multitude of parameters that affect the performance of detectors and the myriad combinations of optical components and signal processing techniques, it is not possible to give a small group of general equations applicable to all systems. Below are listed several representative parametric equations and approaches which may serve as examples for a particular system of interest.

Range equations are often required and an example is given by [29]

$$\text{SNR} = (J/R^2)(\pi D^2 R_L/4V_n)\sqrt{\omega n/k\dot\theta} \, T_0 K_b K_e \overline{R} \qquad (2.16)$$

where
$J$ = target radiance intensity in W/sr
$R$ = range
$T_0$ = optical transmission
$K_b, K_e$ = blur and electrical collection efficiencies, respectively
$\overline{R}$ = detector responsivity weighted over the band of interest
$R_L$ = load resistance (PC detector)
$D$ = fore-optics diameter
$V_n$ = noise per square-root hertz
$\dot\theta$ = scan rate
$n$ = number of detectors
$\omega$ = instantaneous field of view in the scan direction.

For infrared detectors, diffraction effects often are fundamental limiters to performance. For a given fore-optics diameter, the minimum angular resolution is fixed. This angular resolving capability is given by [29]

$$N \sim \omega/\pi(1.22\lambda/D_0)^2 \quad \text{rad} \qquad (2.17)$$
$$\sim (4/1.22\pi)^2 (A_0\omega/\lambda^2)$$

where $\omega$ is the solid angle instantaneous field of view in steradians, $A_0$ the entrance pupil area (collecting aperture), and $D_0$ the entrance pupil diameter.

Another commonly used parameter is the noise equivalent temperature (NET), given by [31].

$$\text{NET} = 4R\,dR F/N\sqrt{dF}/(\pi K_o K_e D D^*) \quad \text{kelvin} \qquad (2.18)$$

where
$R$ = range
$dR$ = detector dimension in object space
$K_o, K_e$ = optical and electrical efficiencies, respectively

$D^*$ = detectivity

$D$ = fore-optics diameter.

Physically, NET is the change in target temperature that gives a signal-to-noise ratio of one. The expression for the electrical bandwidth, derivable from the sampling or Nyquist theorem, is given in

$$df = 1/2T \quad \text{hertz} \tag{2.19}$$

where $T$ is the integration or dwell time per detector [31].

A similar expression for NET considering the target to be at temperature $T$ is given by [32]

$$\text{NET} = \pi \sqrt{dF} \sqrt{A_d} / [D^*_{300}(4\sigma T^3)] \, d\theta \, d\phi \, A_{\text{opt}} T_{\text{opt}} T_{\text{atm}} \quad \text{kelvin} \tag{2.20}$$

where
$\sigma$ = Stefan-Boltzmann constant.

$A_{\text{opt}}$ = optics collecting area

$T_{\text{opt}}, T_{\text{atm}}$ = optical and atmospheric transmissions, respectively

$d\theta$ and $d\phi$ = angular fields of view for a square detector with area $A_d$

Some representative curves of $D^*$ versus wavelength useful for calculating NET for a set of detectors is given in Fig. 2.68 [32].

Some equations that may be used in a sensor optimization program are given in Table 2.8 [33]. This table is included to convey some of the parameters that should be included in a program designed to arrive at key detector characteristics based on general system requirements.

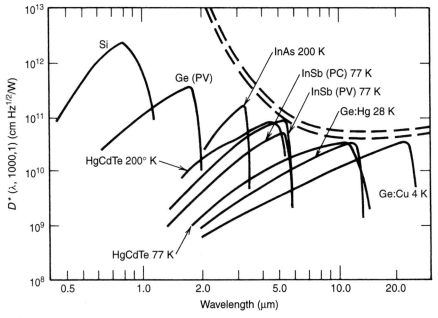

**Figure 2.68**  Spectral detectivities of photo sensors. (From Ref. 30, courtesy of Texas Instruments.)

**TABLE 2.8   Sensor Trade-off Equations[a,b]**

| | |
|---|---|
| (1) Optics diameter | $D = \dfrac{2.44\ K_1 \lambda R}{L}$ |
| (2) Number of detectors | $n = K_2 \dfrac{SR \times S}{V_T L}$ |
| (3) Computing speed | $C = K_3 \dfrac{n V_T}{L}$ |
| (4) Detector dynamic range (e.g. NEP/power) | $DR = K_4 \dfrac{J/N}{L^2 \times REQSNR}$ |
| (5) Smear reduction factor[c] | $F_{SR} = $ smear-induced drift[d]/acceptable drift acceptable |
| | $drift = K_5 V_T \left( \dfrac{J/N_\sigma}{REQSNR \times L^2} \right)^{1/3} \left( \dfrac{YR - 1975}{5} \right)$ |
| (6) Required jitter angle[c] | $\theta_J = \dfrac{K_6 (J/N_\sigma)}{R \times REQSNR \times L} \left( \dfrac{YR - 1975}{5} \right)$ |

[a]Notes and definitions

| | | | |
|---|---|---|---|
| $\lambda$ | = wavelength | REQSNR | = required SNR |
| $R$ | = range | $J$ | = target intensity (W/sr) |
| $L$ | = footprint | $N$ | = background radiance (W/sr-m$^2$) |
| SR | = search rate | $N_\sigma$ | = standard deviation of background radiance |
| $V_T$ | = target velocity | YR | = year |
| $S$ | = target detector crossings per stare | | |

[b]Constants $K_1$, $K_2$, $K_3$, $K_4$, $K_5$, and $K_6$ are dimensionless and of order unity.
[c]Equations 5 and 6 have time dependence due to expected improvements in algorithms.
[d]The equation for this is given as Eq. 24 of Valdes, "Modern Utilization of Infrared Technology V," *Proc. SPIE, 1979,* Vol. 197, p. 45.
*Source*: Ref. 33.

# REFERENCES

1. W. L. Wolfe and G. J. Zissis, *The Infrared Handbook, 1975*, Office of Naval Research, Washington, DC, 1978, pp. 5-88–5-91–.

2. H. Levinstein and J. Mudar, "Infrared Detectors in Remote Sensing," *Proc. IEEE*, **63**(1), pp. 6–14, January 1975.

3. R. C. Jones, "Phenomenological Description of the Response and Detecting Ability of Radiation Detectors," *Proc. IRE*, **47**, pp. 1495–1502, September 1959.

4. M. B. Reine and R. M. Broudy, "A Review of HgCdTe Infrared Detector Technology," *Proc. Soc. Photo-Opt. Instrum. Eng.*, **124**, pp. 80–89, 1977.

5. A. R. Jha, "Technical Report on Applications involving HgCdTe-LWIR and Very-LWIR Detectors," Jha Technical Consulting Services, Cerritos, CA, 2001, pp. 9–14.

6. J. P. Rode, "HgCdTe Hybrid Focal Plane," *Infrared Phys.*, **24**(5), pp. 443–453, 1984.

7. A. R. Jha, *"SUPERCONDUCTOR TECHNOLOGY: Applications to Microwave, Electro-optics, Electrical Machines and Propulsion systems.* New York: Wiley, 1998, pp. 81–83.

8. A. I. D'Souza et al., "Large Very-LWIR HgCdTe Photovoltaic Detectors," *J. Electron. Mater.*, **29**(6), pp. 630–635, January 2000.

9. J. Johnson, Aerospace Corp. (private communication), 1987.

10. J. P. Rosbeck, R. E. Starr, S. L. Price, and K. J. Riley, "Background and Temperature Dependent Current-Voltage Characteristics of HgCdTe Photodiodes," *J. Appl. Phys.*, **53**(9), pp. 6430–6440, September 1982.

11. M. Lanir and K. J. Riley, "Performance of PV HgCdTe Arrays for 1–14 um Applications," *IEEE Trans. Electron Devices*. **ED-29**(2), February 1982.

12. J. T. Wimmers and D. S. Smith, "Characteristics of InSb Photovoltaic Detectors at 77 K and SEC B Below," *Proc. SPIE–Int. Soc. Opt. Eng.*, **364**, pp. 123–131, 1983.

13. A. R. Jha, *"INFRARED TECHNOLOGY: Applications to Electro-optics, Photonic Devices and Infrared Sensors*. New York: Wiley, 2000, pp. 155–159.

14. G. Bailey, "Integrating 128 Element InSb Array: Recent Results," *Proc. Soc. Photo-Opt. Instrum. Eng.*, **345**, pp. 185–191, 1982.

15. M. D. Giobbons and S. C. Wang, "Status of CID InSb Detector Technology," *Proc. SPIE–Int. Soc. Opt. Eng.*, **443**, pp. 151–166, 1984.

16. W. F. Kosonocky and H. Elabd, *Schottky-Barrier Infrared Charge-Coupled Device Focal Plane Arrays*, *Proc. SPIE–IR Detectors*, **443**, pp. 167–188, 1983.

17. W. F. Kosonocky, H. Elabd, H. G. Erhardt, F. V. Shallcross, G. M. Meray, T. S. Villani, J. V. Groppe, R. Miller, V. L. Frantz, M. J. Cantella, J. Klein, and N. Roberts, "Design and Performance of 64 × 128 Element PtSi Schottky-Barrier Infrared Charge-Coupled Devices (IRCCD) Focal Plane Array," *Proc. SPIE–Int. Soc. Opt. Eng.*, **344**, pp. 66–77, 1982.

18. C. T. Elliott, "The SPRITE Detector," *International Conference on Advanced Infrared Detectors and systems*, The Institution of Electrical Engineers, pp. 1–12, 1981.

19. D. B. Webb and S. P. Brain, "TI System Trade Offs with SPRITE's," *International Conference on Advanced Infrared Detectors and Systems*, The Institution of Electrical Engineers, pp. 13–17, 1981.

20. T. S. Moss, "Infrared Detectors," *Infrared Phys.*, **6**, pp. 29–36, 1976.

21. R. L. Kroes and D. Reiss, "The Pyroelectric Properties of TGS for Applications in Infrared Detection," *NASA Tech. Memo.*, **NASA-TM-82394**, January 1981.

22. C. B. Roundy, "Pyroelectric Self-Scanning Infrared Detector Arrays," *Appl. Opt.*, **18**(7), pp. 943–945, April 1, 1979.

23. C. B. Roundy, "Operation of Pyroelectric Self-Scanning Infrared(IR) Detector Arrays," *Proc. Soc. Photo-Opt. Instrum. Eng.*, **244**, pp. 132–138, 1980.

24. J. E. Loveluck and S. G. Porter, "A Charge-Coupled (CCD) Pyroelectric Array," *Proc. SPIE–Int. Soc. Photo-Opt. Eng.*, **396**, pp. 86–90, 1983.

25. Y. Talmi, "Pyroelectric Vidicon: A New Multichannel Spectrometric Infrared (1.0–30 μm) Detector", *Applied Optics*, **17**(16), pp. 2489–2501, 1978.

26. R. Walton, F. Ainger, D. Porter, and J. Gooding, "Technologies and Performance for Linear and Two Dimensional Pyroelectric Arrays," *Proc. SPIE–Int. Soc. Opt. Eng.*, **510**, pp. 139–148, 1984.

27. D. R. Lamb and N. A. Foss, "The Application of Charge-Coupled Devices to Infra-red Image Sensing Systems," *Radio Electron. Eng.*, **50**(5), pp. 226–236, 1980.

28. N. Sclar, "Temperature Limitations of IR Extrinsic and Intrinsic Photodetectors," *IEEE Trans. Electron Devices*, **ED-27**(1), pp. 109–118, January 1980.

29. W. H. Flaugh, "Derivation of a Sensor Specification from System Requirements," *Proc. Soc. Photo-Opt. Instrum. Eng.*, **256**, pp. 2–7, 1980.

30. J. A. Jamieson, "Limitations on the Performance of Passive Infrared Sensor," *Proc. Soc. Photo-Opt. Instrum. Eng.*, **62**, pp. 269–300, 1975.

31. M. L. Fee, A. C. Liang, and R. G. Nishinaga, "Design Methodology for Mosaic Infrared Sensors," *Proc. Soc. Photo-Opt. Instrum. Eng.*, **62**, pp. 7–12, 1975.

32. G. R. Pruett, "System Limitations of Infrared Detectors," *Proc. Soc. Photo-Opt. Instrum. Eng.*, **124**, pp. 77–79, 1977.

33. T. J. Janssens and S. F. Valdes, "A Sensor Optimization Program," *Proc. Soc. Photo-Opt. Instrum. Eng.*, **253**, pp. 24–30, 1980.

# 3

# OPTICAL LENSES

JOHN R. ROGERS
*Institute of Optics*
*University of Rochester*
*Rochester, New York*

## 3.1  INTRODUCTION

Virtually all optical systems built today are designed to be symmetric about a longitudinal axis, and consist of regions of homogeneous isotropic refractive (or reflective) media. The properties of such systems may be described through the use of geometrical optics (ray optics), with diffraction effects accounted for through Fourier theory. Even systems with slight perturbations of their axial symmetry may be understood qualitatively through the equations developed for the axially symmetric case. Readers interested in optical systems with inhomogeneous media ("gradient index" optics) are referred to Marchand [1].

Properties of optical systems may be divided into *first-order* properties (i.e., size and location of an idealized image), and the *aberrations*, which describe the differences between the actual image and the ideal image. An "ideal" image is taken as one in which (1) the image is sharply focused across its entire diameter, (2) a planar object is mapped to a planar image (both normal to the axis), and (3) the image is a linearly scaled version of the object. There are, of course, instances in which deviations from this "ideal" are actually desired; however, the aberrations may conveniently be defined as deviations from this "ideal" if it is recognized that nonzero values of the aberrations will be desirable in certain instances.

### 3.1.1  Fundamental Concepts

Geometric optics refers to those properties of optical systems that may be derived from the ray theory of light. There are several fundamental concepts that act as a basis for this theory. The first of these is that light travels along a well-defined path, known

*Handbook of Optical Components and Engineering*,   Edited by Kai Chang
ISBN 0-471-39055-0   © 2003 John Wiley & Sons, Inc.

as a *ray*. The speed at which light travels is governed by the *refractive index*, which is the ratio of the speed of light in vacuum to its speed in the medium. A mirror may be correctly represented by reversing the sign of the index after the reflection. A *wavefront* is a surface of constant transit time from the source, corresponding to a surface of constant phase in wave theory. Wavefronts are assumed (in homogeneous, isotropic media) to be everywhere perpendicular to the ray paths. Properties of optical systems may be described with equal validity through descriptions of the rays or the wavefronts of the system. A *point source* is an infinitesimal point in space which emits uniformly distributed rays of light and spherical wavefronts. The *optical path length* (OPL) from point *A* to point *B* is the distance in vacuum that could be traversed by light in the time it takes light to travel from *A* to *B*:

$$\text{OPL}_{AB} = \int_B^A n \, ds$$

*Optical path difference* (OPD) refers to the difference between the optical path lengths along a given ray and a standard or "reference" ray. *Fermat*'s principle states that a ray connecting any two points will take a path such that the optical path length is stationary (zero derivative) with respect to all neighboring paths. *Snell*'s law may be derived from Fermat's principle, and states that refraction at an interface will occur such that $n' \sin i' = n \sin i$, and that the incident and emergent rays are coplanar. The wavefront description and the ray description of an optical system's performance are equivalent; however, there are instances in which one method or the other is preferred. Fermat's principle is useful in deriving the exact closed-form solution to a given imaging problem, while ray-trace algorithms are based on Snell's law. *Total internal reflection* (TIR) is the phenomenon that rays incident at high angle upon a surface at which there is a drop in index are reflected entirely.

### 3.1.2   Summary of Definitions

In addition to the basic concepts given above, there are a number of quantities that are useful in the description of optical systems. An object and its image are said to be *conjugate* to one another. *Object space* and *image space* are the "before" and "after" descriptions of the rays and wavefronts entering and leaving the optical system. It is important to realize that these are distinguished temporally rather than spatially, since the rays entering or leaving an optical system may be extended to include all space. For instance, if the object presented to an optical system is actually the image formed by some previous optical system (located to the left), that object may very well lie to the right of the optical system in question. This object presented to the second system is said to be *virtual*, since the rays must be extended through object space (along the paths they would take if there were no second optical system), in order to reach the object. A *virtual image* is one in which the rays in image space must be extended back along their paths in image space in order to reach the image.

The *optical axis* is the axis along which the system has axial symmetry, and is a ray path if there are no obscurations in the system. A ray is in the *paraxial region* if its ray angles and angles of incidence satisfy the first-order small-angle approximation: $\sin \alpha \approx \tan \alpha \approx \alpha$ (in radians); $\cos \alpha \approx 1$.

Often, and quite purposefully, this *paraxial approximation* is made for rays that do not actually satisfy the approximation. *Paraxial rays* are rays that have been traced using the paraxial laws of optics, whether or not they actually lie in the paraxial region. Rays that have been traced using the exact form of Snell's law will be known as *real rays*.

The *aperture stop*, or *stop*, is that aperture or boundary which limits the transverse extent of a bundle of rays propagating from a point source on the optical axis.

The *marginal (or axial) ray* is that ray beginning at the axial object point and passing through the extreme edge of the stop. The *chief (or principal) ray* is that ray which begins at the edge of the object and passes through the center of the aperture stop.

The *entrance pupil* is the image of the aperture stop in object space, and the *exit pupil* is its image in image space. A system is *telecentric* in object (or image) space if the chief ray is parallel to the axis in that space, which means that the pupil in that space is located at infinity.

The *field stop* is the aperture that limits transverse extent of the image or object.

The *rear focal point* (denoted $F'$) is the location where a paraxial ray traced from an infinitely remote object point through the system at any height would again cross the axis. The plane of apparent refraction for this paraxial ray is the *rear principal plane* of the system, and the distance from the rear principal plane to the rear focal point is the *rear focal length*, $f'$. The *front focal point* is the point at which an object must be placed if its paraxial rays are to be refracted parallel to the axis by the system. The apparent plane of refraction in this case is the *front principal plane*, and the distance from the front focal point to the front principal plane is the front focal length, $f$. The *nodal points* of a system are defined by the property that any ray in object space passing through the *front nodal point* exits from the *rear nodal point* at the same angle. The focal points, principal points, and nodal points are known collectively as the *cardinal points* of the optical system.

A spherical surface has a well-defined radius of curvature. The *curvature* of a surface is defined to be the reciprocal of the radius of curvature. When applied to an aspheric surface, the terms *radius* and *curvature* refer to the paraxial values of those quantities. The *vertex* of a surface is the point at which the surface crosses the optical axis, and the plane through this point perpendicular to the axis is termed the *tangent plane*. The *sagitta*, or "*sag*" of a surface refers to the function $z(r)$, where $r$ is the transverse position on the surface and $z$ is the axial position of that point measured from the tangent plane.

An aspheric surface is specified either by its *conic constant*, which is the negative of the square of the eccentricity of the (conic) surface, or by the *aspheric coefficients*, which are the coefficients of the polynomial that describes the departure of the surface from a sphere.

The *clear aperture* of an optical surface or element is the diameter of the unobstructed portion through which light may pass. The term *semiaperture* will be used to denote half the clear aperture; the word *radius* could be confused with radius of curvature. *Vignetting* is the condition in which some of the rays from off-axis object points are blocked by other apertures than the stop. The *reduced distance* between two points is defined in this chapter as the physical distance divided by the refractive index of the medium.

### 3.1.3   Notational Conventions

Figure 3.1 shows the coordinate system and the sign conventions used in this chapter. The coordinate system is right-handed, with the $z$-axis forming the axis of rotational

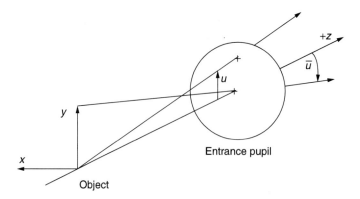

**Figure 3.1**    Coordinate system and ray angles.

symmetry of the optical system. Light is assumed to propagate (in the absence of mirrors) in the direction of increasing $z$. The intersections of rays with planes and curved surfaces are denoted by their $x$, $y$, and $z$ coordinates. The angles made by the rays with the optical axis are *ray angles* and will be taken as positive if they represent a counterclockwise rotation from the axis.

At a curved surface, the *angle of incidence* is the angle between the ray and the surface normal and is taken as positive if the angle represents a counterclockwise rotation from the normal. Axial distances (except for focal lengths) are taken as positive toward the right. The distance following an odd number of mirrors is negative. The front focal lengths points from the front focal point toward the lens (i.e., the directed distance $FP$) while the rear focal length is the distance $P'F'$. This convention gives both the front focal length and rear focal length the same sign as the system power, $\phi$. The *radius* and curvature of a surface are taken as positive if the center of curvature lies to the right of the surface.

Unless otherwise noted, subscripts in this chapter are used to denote individual surfaces or subsystems, as appropriate. Unsubscripted quantities pertain to the entire system under consideration. An overbar is used to denote quantities relating to the chief ray; unbarred ray angles and heights refer to the marginal ray. Paraxial values of quantities such as ray heights, angles, and angles of incidence are denoted by lowercase letters $(u, \bar{u}, i)$. In contrast, real ray quantities are denoted by uppercase letters $(U, \bar{U}, I)$. Primed quantities refer to the space following the surface or system in question. For instance, $\bar{u}'_1$ denotes the paraxial chief ray angle following surface 1, and is equal to $\bar{u}_2$.

## 3.2 FIRST-ORDER OPTICS

The first-order properties of a system (magnification, focal length, and image location) may be found by tracing two specific paraxial rays through the system. The paraxial equations provide a good approximation to the actual ray-trace results only when the ray angles are small; however, it is easily verified that the paraxial equations continue to predict the ideal image regardless of the ray angles involved. Thus it is recognized that the image will converge to the paraxially predicted image as the aberrations are removed from the system. For the purpose of finding the location and size of the ideal image (independent of the aberrations of the system), one may imagine scaling the ray angles down so that the small-angle condition is satisfied and an ideal image is obtained, tracing the rays to find the image properties, then invoking linearity to rescale the image back up to proper size. Since the paraxial equations are themselves linear in $y$ and $u$, the same result is achieved if the paraxial equations are formally applied to the large-angle rays directly; what results is a rough approximation to the actual ray data but an exact description of the ideal image. Such formal rays, which have been traced with the paraxial equations (regardless of whether or not they fall in the paraxial region), will be termed *paraxial rays*, and their associated quantities denoted by lowercase variables.

Although one may occasionally speak of nonparaxial values for first-order properties ("the focal length of the system at the 0.7 zone in the pupil," for instance), these properties are defined by their paraxially determined values, with the difference between any particular ray and its first-order prediction being considered as an aberration. Because only the paraxial limit is being considered, the first-order properties depend on the spherical terms of the surfaces only, and the aspherics play no role.

### 3.2.1 The Paraxial Ray-trace Equations

The laws of paraxial optics are found by assuming that both the object height and the ray heights in the optical system are small, so that the small-angle approximation may be used. This leads to the paraxial form of Snell's law:

$$n'i' = ni \tag{3.1}$$

To apply this to refraction at a curved surface, we note from Fig. 3.2 that the incidence angle is given by

$$i = u - \alpha = u + yc \tag{3.2}$$

It is simplest to consider the change in the quantity $nu$ at the refracting surface, and we find that

$$n_i'u_i' = n_iu_i - y_i\phi_i \tag{3.3}$$

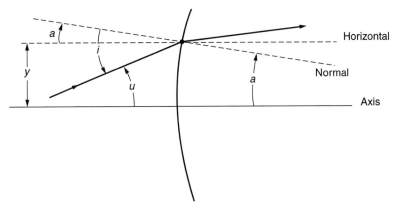

**Figure 3.2**  Incidence angles.

where the *optical power* is defined as

$$\phi_i = c_i(n'_i - n_i) \tag{3.4}$$

Optical power, having dimensions of inverse length, is often specified in inverse millimeters, or in *diopters*, which are units of inverse meters.

Although the surfaces is curved, it appears flat in the paraxial region, since the sag of the surface is proportional to $1 - \cos\theta$, which is zero in the small-angle approximation. Noting this, the paraxial transfer equation is simply

$$y_{i+1} = y_i + \tau'_i(n'_i u'_i) \tag{3.5}$$

where $\tau'_i$ is the reduced distance between the surfaces, defined as

$$\tau'_i = t'_i/n'_i \tag{3.6}$$

Since the paraxial ray-trace equations are to be applied formally to rays that do not actually satisfy the small-angle approximation, it is important to interpret correctly the meaning of the paraxial ray heights and angles. Careful examination of Eqs. (3.5) and (3.6) indicates that the paraxial ray angle $u$ is the tangent of the actual ray angle $U$ (if there are no aberrations) and that the height $y$ is measured in the tangent plane of the surface. Formal paraxial rays therefore behave in the somewhat unusual manner depicted in Fig. 3.3. If the system is perfectly corrected for aberrations, the real and paraxial rays will meet at the image plane.

### 3.2.2 First-Order Properties of Optical Systems

The rear focal point $F'$ of a system is found by tracing a ray from an object at infinity (i.e., one that enters the system parallel to the axis) at any height, and observing where it crosses the axis in image space. The rear principal plane is found by extending the object space ray and the image space ray and finding their intersection, as shown in Fig. 3.4. Because of the linearity of the paraxial equations, this point of apparent

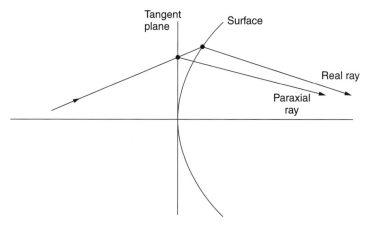

**Figure 3.3** Refraction of real and paraxial rays.

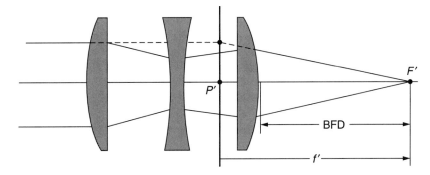

**Figure 3.4** Rear focal length and principal plane.

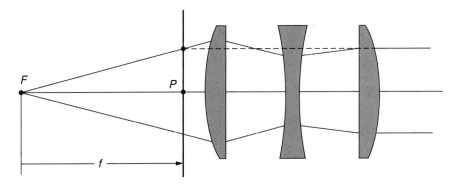

**Figure 3.5** Front focal length and principal plane.

refraction occurs at the same axial position, regardless of the height at which the ray enters the system. The rear principal point $P'$ is the intersection of the rear principal plane with the axis. The rear focal length is given by the distance from $P'$ to $F'$. This differs from the back focal distance $BFD$, which is measured from the rear surface. Similarly, the front focal point $F$, the front principal point $P$, and the front focal length $F$ may be found by tracing a ray backward through the system from an image at infinity to its conjugate object at $F$, as shown in Fig. 3.5. The front and rear focal lengths are equal only when the refractive index in object space matches that in image space; however, they are related and it is possible to define a single optical power for the system:

$$\phi = n'/f' = n/f \tag{3.7}$$

The principal planes are conjugate to one another with a magnification of $+1$. This means that any ray that strikes the front principal plane must leave the rear principal plane at the same height. It is possible, using only the rays through the front and rear focal points, to determine graphically the location and size of the image of any object, as shown in Fig. 3.6. If one defines the object and image distances $l$ and $l'$ as in the figure, the conjugate distance relationship is given by

$$\phi = n'/l' - n/l \tag{3.8}$$

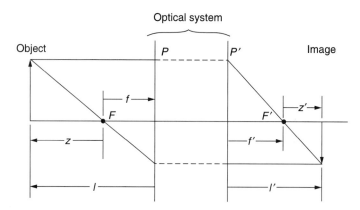

**Figure 3.6**   Graphical raytracing.

This form of the conjugate distance relationship is correct regardless of the signs of $\phi$, $l$, and $l'$. Here $n$ and $n'$ refer to the refractive indices in object and image space, respectively. It is important to note that the focal lengths, object distances, and image distances are measured relative to the principal planes, not the surfaces of the system.

Once the system power and the principal planes have been found, Eqs. (3.3) and (3.5) may be applied to trace rays through the entire system at once; one simply uses the power of the system instead of the surface power, and the ray heights at the principal planes rather than at the surfaces.

Another important pair of points for an optical system are the front and rear nodal points. Any ray in object space passing through the *front nodal point* exits from the *rear nodal point* at the same angle, as shown in Fig. 3.7. This property is important in the analysis and design of scanning systems. The nodal points are located with respect to the principal points by

$$\overline{PN} = \overline{P'N'} = f' - f \tag{3.9}$$

An important special case is when the optical system is immersed in a single medium, so that the refractive index in object space matches that in image space. Under this condition, the front and rear focal lengths are equal and the nodal points coincide with the principal points.

Although it is always possible to treat a system as a collection of surfaces, and to find the cardinal points from the ray-trace data, it is often useful to regard an

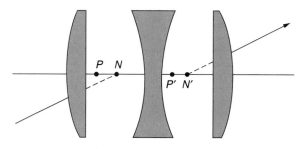

**Figure 3.7**   Nodal points of an optical system.

optical system as a collection of two constituent subsystems. The subsystems might be individual surfaces or collections of surfaces; in the latter case the subsystems could themselves be decomposed into constituents. Denoting the properties of the two subsystems with the subscripts 1 and 2, and the properties of the entire system with unsubscripted symbols, the power of the system is given by

$$\phi = \phi_1 + \phi_2 - \tau_1' \phi_1 \phi_2 \tag{3.10}$$

where $\tau_1'$ is the reduced distance $P_1' P_2 / n_1'$. The principal points of the system may be found algebraically by the relations

$$P' P_2' = n' \tau_1' (\phi_1 / \phi) \tag{3.11}$$

and

$$P_1 P = n \tau_1' (\phi_2 / \phi) \tag{3.12}$$

Considering Fig. 3.3, one recognizes that the principal planes of a single surface coincide and are located at the tangent plane of the surface. The nodal points of a single surface are located at the center of curvature of the surface, since a ray through the center strikes the surface normally and is undeviated.

The magnification of an optical system (the ratio of image height to object height) may be expressed in several ways:

$$m = \frac{l'}{n'} \frac{n}{l} = \frac{-f'}{z'} = \frac{z}{f} \tag{3.13}$$

according to Fig. 3.6. Note that all distances are referenced to the principal planes rather than to the surfaces themselves.

An important limiting case is that of a "thin lens", which is a hypothetical lens with nonzero power yet zero thickness. In such a case, Eqs. (3.10)–(3.12) (with $\tau = 0$) indicate that the principal planes and nodal points for the lens coincide at the plane of the lens, and that the power of the lens is simply the sum of the powers of the two surfaces. In the usual case that the lens is immersed in air,

$$\phi_{\text{thin lens}} = n_{\text{glass}} (c_1 - c_2) \tag{3.14}$$

Although inherently approximate, the thin lens approach is frequently used in the early stages of design because it allows two refractions and a transfer to be replaced with a single refraction.

### 3.2.3  Stops, Pupils, and the Paraxial Invariant

The *aperture stop* (or simply, the *"stop"*) is that aperture which limits the transverse extent of ray bundle originating from an axial object point, as shown in Fig. 3.8. An

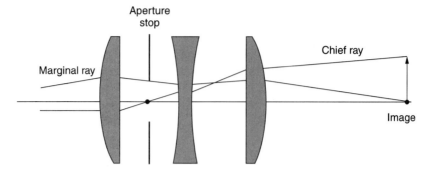

**Figure 3.8** Aperture stop and chief and marginal rays.

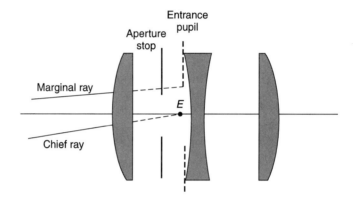

**Figure 3.9** Entrance pupil.

image is formed wherever the marginal ray crosses the axis, and the convergence angle of the marginal ray determines the limit of resolution of the system.

The chief ray connects the tips of the object and image with the center of the stop. The entrance pupil is located where the chief ray (or its extension) crosses the optical axis in object space, as Fig. 3.9 shows. To an observer looking into the system from object space, this represents the window in space through which the rays must be aimed, in order to pass through the aperture stop. Similarly, the *exit pupil* is the image of the aperture stop in image space, and is located where the chief ray crosses the axis in image space. To an observer in image space, this is the window in space from which all rays appear to emerge. When two systems are combined, the exit pupil of the first system should coincide with the entrance pupil of the second.

If the location and diameter of the entrance pupil are known, the paraxial marginal ray may simply be aimed at its edge; since this is conjugate to the stop and the exit pupil, the marginal ray will graze their boundaries as well. Similarly, the chief ray may be aimed at the center of the entrance pupil, so that it also pierces the centers of the stop and exit pupil. Unfortunately, it is usually the location and diameter of the stop which are given, rather than of the entrance pupil. In principle, the location and size of the entrance pupil may be found by tracing rays backward from the stop into object space, and observing where they cross the axis. This is more efficiently accomplished by using the fact that any ray may be written as a linear combination of the chief and marginal rays as

$$y^* = Cy + \overline{C}\overline{y} \tag{3.15}$$

$$u^* = Cu + \overline{C}\overline{u} \tag{3.16}$$

where $C$ and $\overline{C}$ are constants determined by the specific choice of the third ray. By tracing provisional marginal and chief rays and using the linear scaling properties [Eqs. (3.15) and (3.16)] it is possible to obtain the correct ray data. Beginning at the center of the object, a provisional marginal ray is traced at an arbitrary starting ray angle. (Because the paraxial refractions occur at the tangent planes, it is immaterial if this provisional ray passes outside the apertures of the elements.) Observing the height of the provisional ray at the stop, the ray heights and angles are scaled to yield the correct marginal ray data. A provisional chief ray is traced from the tip of the object through the center of an arbitrarily located "pupil." (The vertex of the first element is usually a convenient starting point.) Identifying this provisional chief ray with the starred quantities in Eqs. (3.15) and (3.16), and considering the fact that the provisional ray and the chief ray have the same height at the object plane, it is clear that $\overline{C}$ is equal to unity. Applying Eqs. (3.15) and (3.16) now to the stop, one notes that $\overline{y}$ is zero, and one may solve for the remaining two unknowns, $C$ and $\overline{u}$. Once $C$ and $\overline{C}$ are known, the correct chief ray data may be found at any plane in the system by application of Eqs. (3.15) and (3.16).

If the clear apertures of all the elements are sufficiently large, the aperture stop is the only aperture that limits the extents of the ray bundles, even for object points removed from the optical axis. *Vignetting* occurs when this condition is not met, and elements far from the stop block parts of the ray bundles for off-axis object points, causing a decrease in image intensity near the edge of the field. To the extent that the system behaves according to the paraxial approximations, freedom from vignetting is achieved if the semiaperture exceeds $|y| + |\overline{y}|$ for all the elements in the system. Because of aberrations, these estimates of the required apertures should be checked by tracing real rays.

A system is telecentric in object space or image space if the chief ray in that space is parallel to the axis, which means that the pupil in that space is located at infinity. If a system is telecentric (in image space, for example), and there is no vignetting, the ray bundles for the off-axis image points are traveling parallel to the axis, and an error in focal position of the detector does not cause a change in the apparent size of the detected image. To be sure, the detected image becomes blurry, but the centers of the blurs do not expand out from the center as they would if the pupil were near. This is useful in measurement systems in which the ability to focus is impeded, yet a highly accurate transverse measurement is desired.

The *field stop* is defined as that aperture which limits the extent of the image. In most systems, the boundary of the film or detector serves as the field stop. In some visual systems, an aperture is inserted at an internal image specifically for the purpose of providing a sharp boundary to the image. If this is not done, the field is limited by vignetting.

It is easily verified that the quantity

$$H = n\overline{u}y - nu\overline{y} \tag{3.17}$$

is invariant under both refraction and transfer. It is therefore a constant throughout the optical system and is known as the *paraxial invariant* or *Lagrange invariant*. It may be shown that the paraxial invariant is proportional to the maximum number of resolvable spots across a diameter of the image.

The *numerical aperture in image space* of a system is defined in terms of the angle of the emergent marginal ray:

$$\text{N.A.} = n' \sin U' \tag{3.18}$$

The *F-number* will be defined as the ratio of the focal length to the diameter of the entrance pupil of the system:

$$F^{\#} = f/2y_E \tag{3.19}$$

Note that (according to this definition) the *F*-number is a function of the system geometry only, whereas the numerical aperture depends on a ray angle, and therefore depends on the object position as well as the system. For systems in which the object is not at infinity, the "effective *F*-number" may be defined as

$$F^{\#}_{\text{eff}} = 1/2u' \tag{3.20}$$

[Another commonly encountered definition of the *F*-number of a system is 1/(2 N.A.), which not only varies with object position but scales with the sine of the ray angle and differs substantially from Eq. (3.20) whenever the convergence angle is large.]

Interestingly, from the marginal ray alone it is possible to find not only the image position but also the magnification. From the Lagrange invariant, it is easy to show that the magnification of image-forming systems is given by

$$m = nu/n'u' \tag{3.21}$$

The expressions given above hold for all *focal* or image-forming systems, but are meaningless for certain systems that do not form images. Such systems are discussed in the next section.

### 3.2.4   Elementary Optical Systems

One of the most common uses of optical system is to provide a visually "magnified" image of an object. Ordinary magnifying glasses, loupes, and eyepieces all serve this purposes, imaging a near object to infinity as depicted in Fig. 3.10, so that the eye can focus on it comfortably. In this case the usual definition of magnification is meaningless, because the image is infinite in extent, yet the object is of finite size. In such cases the "magnifying power" or eyepiece magnification may be defined as the ratio of the angle subtended by the magnified object to that subtended by the object if it were viewed at the usual reading distance of 250 mm:

$$m_{\text{eyepiece}} = 250 \text{ mm}/f \tag{3.22}$$

Note in the figure that an imaginary "exit pupil" has been shown external to the system, so that the pupil of the eye (which actually is the top of the system) may be placed there. This allows the viewer to see not only the light from the axial object

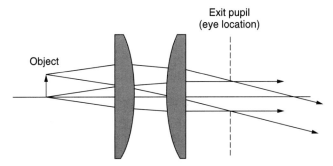

**Figure 3.10**  Simple eyepiece.

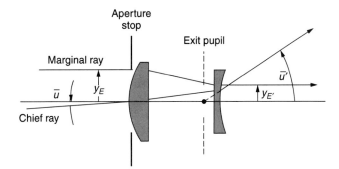

**Figure 3.11**  Galilean telescope.

point, but the images of the field points as well. In general, when coupling two optical systems together, the exit pupil of the first system should fall on the entrance pupil of the second, to avoid vignetting. The smaller of the two coincident pupils becomes the stop of the whole system.

The Galilean telescope of Fig. 3.11 is an example of an *afocal* system, which maps an object at infinity to an image at infinity. For two-element afocal systems, the separation between the elements is equal to the sum of the focal lengths, and it may also be shown that the optical power of any afocal system is zero.

Again, the magnification cannot be defined as in Eqs. (3.13) or (3.21), since no image is being formed; nevertheless, it is possible to define the "magnifying power" in a meaningful way. For afocal systems, the magnification perceived by the observer is the ratio of the chief ray angles with and without the optical system:

$$m_{\text{afocal}} = n'\bar{u}'/n\bar{u} \tag{3.23}$$

This equation is deceptively similar to Eq. (3.21), but should not be confused with it.

The Lagrange invariant yields another expression for the afocal magnification:

$$m_{\text{afocal}} = y_E/y_{E'} \tag{3.24}$$

where $y_E$ and $y_{E'}$ represent the marginal ray heights at the entrance and exit pupils, respectively.

In Fig. 3.11, the aperture stop has been shown at the front element, so that the element does not have to be larger than the axial ray bundle. It can be seen that Galilean systems have a significant problem in that the exit pupil is internal to the system and therefore the chief rays that exit from this point at any appreciable angle will miss the pupil of the eye. If one prefers to define the eye pupil to be the stop of the system, the entrance pupil of the system lies to the right of the telescope, and rays aimed at the pupil at any appreciable field angle will miss the front element.

Figure 3.12 shows a Keplerian telescope, which is identical to the Galilean system, except that the power of the rear element is reversed in sign and the spacings of the elements readjusted for afocality. The magnifying power [given by Eq. (3.23) or (3.24)] is now negative, indicating an inverted image. Because the sum of the focal lengths is now larger, the system is much longer.

Note that in the Keplerian system, the exit pupil is conveniently formed outside the system, where it is accessible to the viewer's eye. Alternatively, the telescope may be thought of as a lens that forms an image inside the device, plus a magnifying glass to observe the internal image.

A field stop is shown at the plane of the internal image, to sharply define the boundary of the image. If a measurement reticle or cross-hair were desired in the system, it must also be placed at the internal image, so that it is simultaneously in focus with the image. If no field stop were included, the field of view would be limited by vignetting at the eyepiece. The field of view may be increased by the inclusion of a *field lens* at or near the internal image plane, as shown in Fig. 3.13. Note that

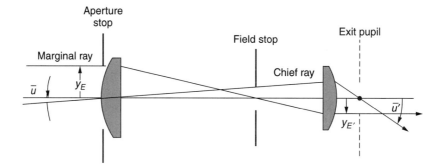

**Figure 3.12** Keplerian telescope.

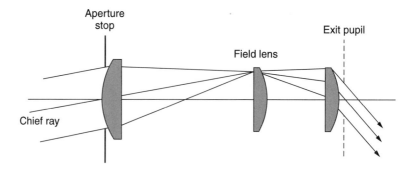

**Figure 3.13** Effect of a field lens.

the gain in field of view has been made at the expense of the *eye relief*, which is the separation between the exit pupil and the last surface. If the field lens is "thin" and is placed exactly at the internal image, the marginal ray will be unaffected, and the magnification of the system will not change. In practice, field lenses are usually placed some distance from the image, so that defects on the surface (such as scratches or dust) are not sharply focused on the image.

A microscope consists of a short focal length "objective" lens which forms a real image inside the microscope tube, plus an eyepiece to facilitate observation of the image, as shown in Fig. 3.14. Since microscope objectives are usually specified in terms of their magnifications rather than their focal lengths, it is useful to know the relation

$$m_{\text{microscope objective}} = \text{optical tube length}/f \qquad (3.25)$$

The *optical tube length* is the distance between the rear focal point of the objective and the image presented to the eyepiece. This is approximately equal to the mechanical tube length, which is the distance from the flange on the objective to that on the eyepiece. Most objectives are designed for use with a tube length of 160 or 170 mm, although some may be found for a 200-mm tube. The microscope system may also be broken down into parts: an "objective" lens that forms a real (internal) image of a real object, and a magnifier to observe that image. The magnifying power of the microscope system is the product of the objective magnification and the eyepiece magnification given by Eq. (3.22).

Systems that contain their own illumination systems (such as a 35 mm slide projector) usually operate in two stages, as shown in Fig. 3.15. The light source is imaged

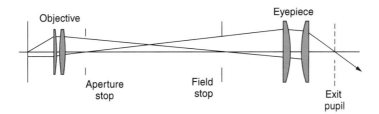

**Figure 3.14**  Compound microscope.

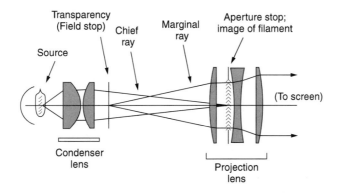

**Figure 3.15**  Projection system, with condenser.

by a *condenser lens* into the entrance pupil of the *projection lens*. When the source is a lamp filament, a concave mirror is often placed concentric with it, so that the rays emitted out the back of the lamp are redirected toward the condenser; however, this may cause overheating of the filament. If the filament is placed slightly to one side of the center of curvature of the reflector, the filament image will be shifted slightly from the filament itself, and both contribute to the illumination of the object. The concave mirror is usually a "cold mirror," which reflects the visible light but not the infrared, to protect the object from unnecessary heating. To collect as much light as possible, the condenser lens must have as high a numerical aperture as possible in the space of the source, and for this reason spherical aberration can become a problem. Fortunately, the condenser system serves only to illuminate the object, and its aberrations do not detract from the quality of the final image.

### 3.2.5 Prisms

Many complex optical manipulations may be accomplished using prisms. Dispersion, low-scatter reflection, constant-angle deviation, beam displacement, retroreflection, beam splitting, image rotation, and image inversion are all possible uses of prisms. In working with prism systems, the concepts of the *reduced thickness*, *tunnel diagram*, and *parity* are important.

A tunnel diagram is an unfolded view of the prism, and may be found by "flipping" the prism over each reflecting surface as the rays strike them. Figure 3.16 demonstrates this for the case of a 90° prism. Nondispersing prisms will have tunnel diagrams with parallel entrance and exit faces; furthermore, in prisms designed to work in convergent light, they should be perpendicular to the optical axis.

The first-order (monochromatic) properties of a prism may be found by imagining the prism to be replaced by an equivalent "block" of air whose thickness is equal to the reduced thickness of the prism. This enables one to complete the first-order design for the system without a prism, knowing that part of an air space is to be replaced with glass. To evaluate the aberrations of the system as well as the first-order properties, it is only necessary to insert a block of glass as thick as the tunnel diagram for the prism. To insert a prism of thickness $t$ and index $n$, we remove a distance $t/n$ of air and inserts a thickness $t$ of the appropriate glass. It is only necessary to model the tilts of the surfaces explicitly if the input and exit faces of the tunnel diagram are not perpendicular to the beam.

The parity of an image refers to the "handedness" of the image, that is, whether or not the image may be returned to its original state by a rotation. The parity of an image

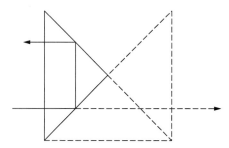

**Figure 3.16** A 90° prism, with its tunnel diagram.

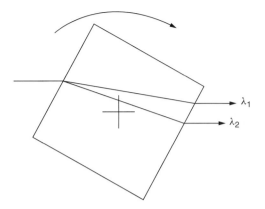

**Figure 3.17** Scanning beam displacement prism.

is *odd* (or "left-handed") for optical systems having an odd number of reflections, and *even* ("right-handed") for systems with even numbers of reflections.

Simple spectrometer and monochromator systems use dispersing prisms with non-parallel entrance and exit faces to deviate the various wavelengths by differing amounts. If the entrance and exit faces are parallel, but oblique to the beam, no angular deviation will occur; rather, the beam will be displaced laterally. Figure 3.17 shows the use of a rotating prism to produce a straight-line scan. Because of the dispersion of the glass, the displacement will be a function of wavelength, but the emergent rays will be parallel to the incident rays regardless of the wavelength.

A total internal reflection inside a prism may be used in place of a simple mirror, when efficiency and low scatter are required. Total internal reflection will occur whenever the beam is incident at angles higher than the *critical angle*. The critical angle may conveniently be found by calculating the angles of incidence for which the emergence angle reaches 90°:

$$\theta_c = \arcsin(n'/n)$$

For crown glass of index 1.52, the critical angle is approximately 41°.

If two surfaces of the prism are used for reflection, the total beam deviation angle is fixed by the angle between the two surfaces, and is independent of the angle of incidence. The *penta prism*, shown in Fig. 3.18, produces a 90° deviation of the beam,

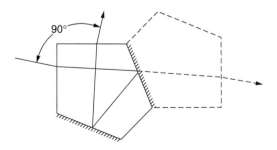

**Figure 3.18** Penta prism and tunnel diagram.

regardless of the orientation of the prism to the beam. Note that the two reflecting surfaces must be silvered, since the incidence angle is less than the critical angle.

Retroreflection in a single plane may be achieved through the use of a *right-angle prism*, used as shown in Fig. 3.16. (This is a special case of the constant angle deviation application noted above.) True retroreflection occurs with a three-dimensional corner, known as a *cube corner* or *corner cube*.

Beam splitters may be of a simple plate type, but the cube type is usually preferred, since both the reflected and transmitted beams pass through the same thickness of glass and therefore suffer the same aberrations and phase delay. Alignment is also simplified since no transverse displacement of the beam occurs. Amplitude beam splitting may be done using a broad band, partially transparent coating on the diagonal surface. Special dielectric coatings may be used instead to separate the two beams by polarization. Dichroic coatings may be used to separate the beams by color instead, although a 30° reflection is in this case preferable to 45°, due to the difficulty of separating chromatic and polarization effects at a high incidence angle. Figure 3.19 illustrates a three-color dichroic beam splitter.

Any prism system with an odd number of reflections (such as the *reversion prism* of Fig. 3.20) may be used as a beam rotation device. As the prism is rotated about the axis shown, the image rotates about that axis twice as fast as the prism. The second reflection inside the reversion prism occurs at less than the critical angle, which means that the surface must be silvered.

The *Dove prism* shown in Fig. 3.21 is an alternative to the reversion prism. If the beam is converging or diverging, the oblique faces of the dove prism introduce aberrations that cannot be compensated for by a rotationally symmetric lens system.

The addition of a "roof" onto a reversion prism turns it into an *Abbe prism*, which inverts the image in both the $x$ and $y$ directions, as indicated in Fig. 3.22. Because

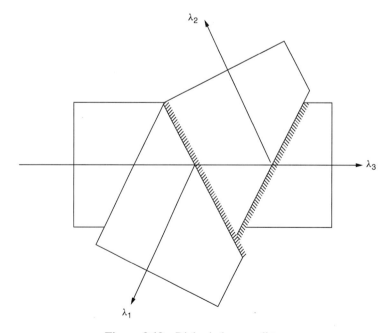

**Figure 3.19**  Dichroic beam splitter.

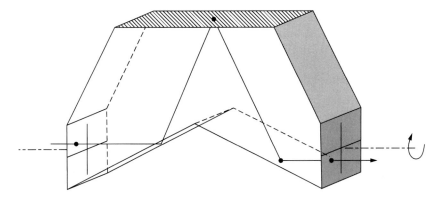

**Figure 3.20**  Reversion prism.

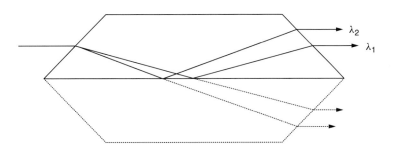

**Figure 3.21**  Dove prism, with its tunnel diagram.

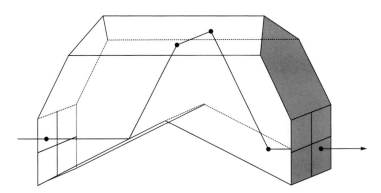

**Figure 3.22**  Abbe prism.

of the even number of reflections, the image no longer rotates with the prism. This prism is frequently used in Keplerian telescope systems to present the viewer an erect image.

Another common system for this is a pair of 90° prisms forming a *Porro system*, as shown in Fig. 3.23. The Porro system displaces the beam but does not deviate it, and is responsible for the "knuckle" shape in many binocular systems.

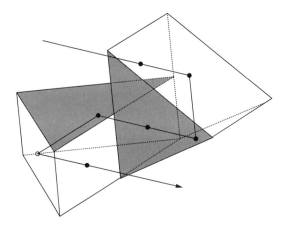

**Figure 3.23** Porro image-inversion system.

## 3.3 ABERRATIONS

The aberrations of optical systems may be regarded in two ways: the deviation of the wavefront from the ideal case, or the deviations of the rays from their paraxially predicted intersections with the detector surface. These two representations are equivalent, yet it is important to understand both. The wavefront aberration function is proportional to the OPD of the system, which is the quantity measured by an interferometric test of the system. The transverse ray deviations are usually the quantities inspected by a designer in the course of designing a system.

Regardless of whether the wavefront aberration function or the transverse ray deviation is chosen to represent the aberration, the aberration varies with position in the pupil as well as with the position of the source point in the plane of the object. For a rotationally symmetric optical system, it is sufficient to examine the imagery along any radial line in the object (or image) plane. The aberrations of object points along other lines are simply rotated versions of the aberrations along this line. Traditionally, this line is denoted as the $y$-axis and defines the principal meridian of the system. It is convenient to use polar coordinates and to normalize both the object coordinate and the radial pupil coordinate to their maximum values, so that

$$h = \frac{\text{position in object plane } (y \text{ direction})}{\text{object radius}}$$

and

$$\rho = \frac{\text{radial position in the pupil}}{\text{pupil radius}}$$

It would be reasonable to expand the wavefront aberration function as a series in the variables $h$, $\rho_x$, and $\rho_y$; however, the rotational symmetry of the system dictates that only combinations and powers of the terms $h^2$, $\rho^2$, and $h\rho_y$ are possible. Writing $h\rho_y$ as $h\rho \cos\phi$, according to Fig. 3.24, the wavefront aberration function is then

$$W(h, \rho, \cos\phi) = \sum_p \sum_q \sum_m W_{klm}(h^k \rho^l \cos^m \phi) \qquad (3.26)$$

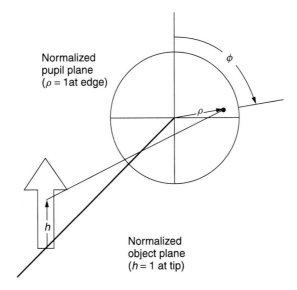

**Figure 3.24** Object and pupil coordinates.

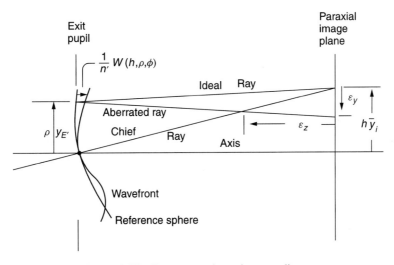

**Figure 3.25** Transverse aberration coordinates.

where $p$, $q$, and $m$ are integers, and $k = 2p + m$, $l = 2q + m$. Note that the subscripts of $W$ have been chosen to match the powers of $h$, $\rho$, and $\cos \phi$.

Figure 3.25 shows the geometry relating the wavefronts to the transverse aberrations. The wavefront aberration function $W$ is equal to the negative of the OPD between the aberrated ray and the idealized chief ray; this means that the distance along the aberrated ray from the "reference sphere" of the figure to the wavefront is $W/n'$.

Since the rays propagate normally to the wavefront, it is not surprising that the transverse ray aberrations are proportional to the derivative of the wavefront aberration function:

$$\epsilon_y = \frac{1}{n'u'}\frac{\partial W}{\partial \rho_y}$$

$$\epsilon_x = \frac{1}{n'u'}\frac{\partial W}{\partial \rho_x} \tag{3.27}$$

The "longitudinal aberration" of the rays is given by

$$\epsilon_z = \frac{-\epsilon_y}{u'\rho_y} \tag{3.28}$$

There are several common ways of representing the aberrations graphically. Perhaps the most obvious is to trace a large number of rays through the system and plot their intersections with the film plane. This is known as a "spot diagram." For rotationally symmetric systems, it is unnecessary to trace this many rays, so only "fans" of rays from the object point in question through vertical and horizontal strips in the pupil are traced, as in Fig. 3.26. These are known as the *x-fan* or *meridional* fan, and the *y-fan*

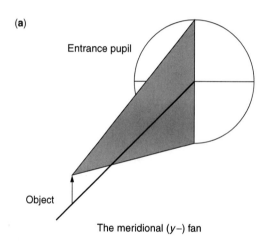

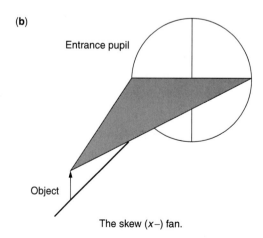

**Figure 3.26** Ray fans for a symmetrical optical system.

or *skew* fan, respectively. For the meridional fan, one plots the $\epsilon_y$ values against $\rho_y$; for the skew fan, the $\epsilon_x$ values are plotted against $\rho_x$. From the figure it is easily seen that the wavefront cross section determined by tracing the skew fan must be symmetric and that the $\epsilon_x$ data must be antisymmetric. For this reason, only the right-hand side of the skew fan need be plotted. Because the aberrations vary not only with pupil coordinate but also with field coordinate, a separate plot must be generated for each field position under study.

A third way of depicting the aberrations is to draw the contours of the wave aberration function. This corresponds to the interferogram produced by a Twyman–Green interferogram, in which the fringes are contour lines with an OPD spacing of one wavelength. Alternatively, a perspective drawing may be made of the surface of the wavefront aberration function.

The conventional orders of the terms of the wavefront aberration function are the sums of the exponents of $h$ and $\rho$ for each term. Equation (3.26) shows three second-order terms in the wavefront, six fourth-order terms, ten sixth-order terms, and so on. Because of the derivatives involved in Eq. (3.27), these same aberrations may be regarded as transverse aberrations of orders 1, 3, 5, and so on. The aberrations are usually referred to by the orders of their transverse errors; however, it should be understood that "fourth-order" and "third-order" refer to the wavefront and transverse representations of the same aberrations.

The lowest-order terms of the wavefront aberration function are $h^2$, $\rho^2$, and $h\rho_y$. The first of these represents a uniform temporal advance or delay of the wavefront for off-axis image points and has no effect on the image quality. This term is called *piston error* and is most often neglected. The $W_{020}\rho^2$ term represents a *defocus* of the wavefront, which is uniform over the field. The transverse error associated with defocus is a linear function of the pupil coordinate, and the spot diagram is a scaled version of the pupil itself. The term $W_{111}h\rho_y$ is a tilt of the wavefront, which is proportional to field position. The transverse error associated with a wavefront tilt is a transverse shift of the image, and being also a linear function of $h$, the $W_{111}h\rho_y$ term may therefore be regarded as a magnification error of the system. Defocus and magnification error are known as first-order properties of the system. Since these properties may be fully controlled by proper placement of the object and detector, they are not normally considered to be aberrations, but rather parameters that may be manipulated at will.

### 3.3.1 Chromatic Aberrations

Related to the first-order properties are the *chromatic aberrations*, which are chromatic variations of the first-order properties, due to the fact that the refractive index of glass varies with wavelength.

*Longitudinal chromatic aberration*, or "axial color," is a variation in image position with wavelength. Although it may loosely be considered as a chromatic variation in focal length, this is somewhat imprecise, as it also depends on the variation in principal plane position for the various colors.

*Transverse chromatic aberration*, or "lateral color," is a variation with wavelength of the chief ray height as it is projected onto the film. This may loosely be regarded as a chromatic difference of magnification, but the equivalence is strictly true only if either the system is telecentric in image space or there is no longitudinal color.

The traditional means of describing the chromatic aberrations is to examine the system at three wavelengths; for visible light systems these are usually the hydrogen

$F$ line ($\lambda = 0.4861$ $\mu$m), the helium $d$ line ($\lambda = 0.5876$ $\mu$m), and the hydrogen $C$ line ($\lambda = 0.6563$ $\mu$m). It is common to speak of "primary" and "secondary" values of both longitudinal and lateral color, thus giving a total of four chromatic aberrations. The primary terms represent the difference between the focal positions and image heights at two specified wavelengths (usually the $F$ and $C$ wavelengths). The secondary terms represent the differences between the parameters at the ends of the spectrum and the center of the spectrum, for instance, between $F$-light and $d$-light. Two parameters useful for the purpose of optical design are the *Abbe nu number*,

$$v_d = \frac{n_d - 1}{n_F - n_C} \tag{3.29}$$

and the *partial dispersion ratio*,

$$P_d = \frac{n_d - n_F}{n_F - n_C} \tag{3.30}$$

The significance of the parameters $v_d$ and $P_d$ may be seen by examining the axial color of a thin lens:

$$\phi_F - \phi_C = \phi_d / v_d \tag{3.31}$$

$$\phi_d - \phi_F = P_d \phi_d / v_d \tag{3.32}$$

The primary chromatic variation in focus may be represented by a wavefront term of the form $\delta_{F-C}(W_{020})\rho^2$. From this, Eqs. (3.27) and (3.28) show that the longitudinal measure of the axial chromatic aberration (traditionally denoted "Lch") is given by

$$\text{Lch} = \epsilon_z = \frac{-2\delta_{F-C}(W_{020})}{n'u'^2} \tag{3.33}$$

If the primary chromatic variation in image size is represented by a wavefront error of the form $\delta_{F-C}(W_{111})h\rho_y$, the transverse measure (traditionally, "Tch") of the lateral color at the edge of the field is given by

$$\text{Tch} = \epsilon_y = \frac{\delta_{F-C}(W_{111})}{n'u'} \tag{3.34}$$

The secondary color contributions may be represented in a similar manner, by considering the difference between the $d$ and $F$ wavelengths.

### 3.3.2   The "Monochromatic" Aberrations

The next six terms of the wavefront aberration function are fourth order in the wavefront or third order in the transverse aberrations. These terms are

$$W(h, \rho\cos\phi) = W_{040}\rho^4 \qquad \text{(spherical aberration)}$$
$$+ W_{131}h\rho^3\cos\phi \qquad \text{(coma)}$$
$$+ W_{222}h^2\rho^2\cos^2\phi \qquad \text{(astigmatism)}$$

$$+ W_{220}h^2\rho^2 \qquad \text{(field curvature)}$$
$$+ W_{311}h^3\rho\cos\phi \qquad \text{(distortion)}$$
$$+ W_{400}h^4 \qquad \text{(piston error)} \qquad (3.35)$$

From the wavefront aberration polynomial, the expressions for the transverse ray aberrations may be derived using Eqs. (3.27):

$$\epsilon_y = \sigma_1\rho^3\cos\phi + \sigma_2 h\rho^2(2 + \cos 2\phi) + (3\sigma_3 + \sigma_4)h^2\rho\cos\phi + \sigma_5 h^3$$
$$\epsilon_x = \sigma_1\rho^3\sin\phi + \sigma_2 h\rho^2\sin 2\phi + (\sigma_3 + \sigma_4)h^2\rho\sin\phi \qquad (3.36)$$

where

$$\sigma_1 = 4W_{040}/n'u'$$
$$\sigma_2 = W_{131}/n'u'$$
$$\sigma_3 = W_{222}/n'u'$$
$$\sigma_4 = (2W_{220} - W_{222})/n'u'$$
$$\sigma_5 = W_{311}/n'u' \qquad (3.37)$$

***Spherical Aberration.*** Spherical aberration is a variation in image position with radial zone in the pupil. Undercorrected spherical aberration (typically found in a simple positive lens with a remote object) causes the rays at the edge of the pupil to focus at a shorter distance than the paraxial rays. Figure 3.27 shows the wavefront aberration function, $y$-fan, and spot diagram for a spherically aberrated system, with the object point on axis. (The system is an F/6, 400-mm-focal-length planoconvex lens, with the curved side facing the object, which is located "at infinity.") The reference sphere has been centered on the paraxial image point, and the $y$-fan demonstrates the cubic form characteristic of spherical aberration. It is readily apparent that the image would be substantially improved by "stopping the system down" so that only the central part

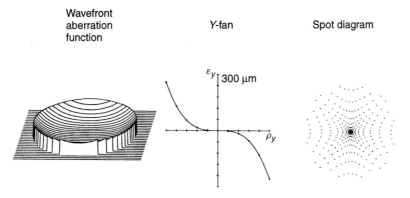

Wavefront aberration function       Y-fan       Spot diagram

**Figure 3.27** Spherical aberration, measured at paraxial focus.

of the pupil is used. The spot diameter for spherical aberration (at paraxial focus) is given by

$$\text{spot diameter} = 8W_{040}/n'u' \tag{3.38}$$

and varies with the cube of the aperture size as the system is "stopped down."

Because the wavefront shape for spherical aberration resembles that for defocus, it is possible to improve the image quality by "adding defocus" to the wavefront. Physically, this is accomplished by shifting the detector slightly from the paraxial image plane. If only third-order spherical aberration is present, the blur diameter will be a minimum three-fourths of the distance from the paraxial focus to the focus of the marginal rays. This situation is represented in Fig. 3.28. The center of the reference sphere has been shifted to the position at which the spot diameter is a minimum, which adds a quadratic term to the wavefront and a linear term to the ray fan. Because the image quality depends not simply with the diameter of the spot, but also on the distribution of rays within it, a better choice of focal position is found approximately halfway from paraxial to marginal focus. Figure 3.29 shows the wavefront aberration function, y-fan, and spot diagram at the focal position which minimizes the *rms wavefront variance*,

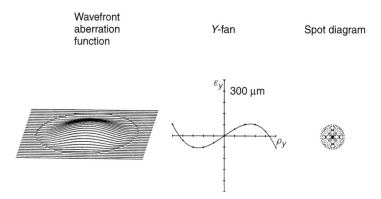

**Figure 3.28**   Spherical aberration, measured at focus of minimum blur circle.

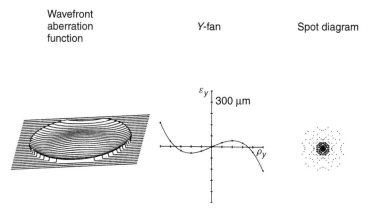

**Figure 3.29**   Spherical aberration, measured at focus of minimum rms wavefront variance.

defined as

$$W_{rms} = \left\{ \frac{1}{\text{pupil area}} \int_{\text{Pupil}} [W(h, \rho, \phi) - \overline{W}(h, \rho, \phi)]^2 \, dA \right\}^{1/2} \qquad (3.39)$$

where $\overline{W}(h, \rho, \phi)$ is the mean value of $W(h, \rho, \cos\phi)$ in the pupil. Although the outer diameter of the spot is larger than in Fig. 3.28, most of the rays are contained in the central core of the spot.

*Coma.* Coma may be regarded as a variation in magnification as a function of radial position in the pupil. This aberration varies linearly with object position, which is a highly disruptive property, since it means that there is no region within which the image is stationary in the mathematical sense. Figure 3.30 shows spot diagrams at nine points in a square field for an F/6 lens which has been aspherized to eliminate spherical aberration. The focal length is 400 mm and the field of view is 1° from side to side. (The spots are not plotted at the same scale as the field.) Although the imagery is perfect at the center of the field, no attempt has been made to correct coma, which is the dominant aberration. It can be seen that the blur function is highly asymmetric, making measurement of the exact position of an image point difficult. Because the blur is directed radially away from the center of the image field, and because it varies linearly with position within the image, the visual effect is that the image has become larger, as well as blurry. The blur function has an apex angle of 60° and a length (in the radial, or sagittal, direction) given by

$$\text{sagittal coma} = \frac{3W_{131}h}{n'u'} \qquad (3.40)$$

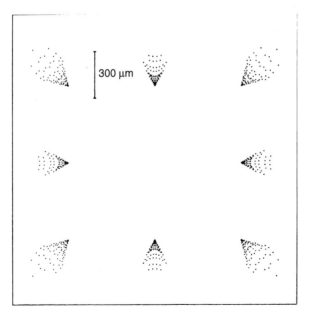

**Figure 3.30** Spot diagrams for a comatic system (Super Oslo plot). Super Oslo is a registered trademark of Sinclair Optics, Inc., Fairport, NY.

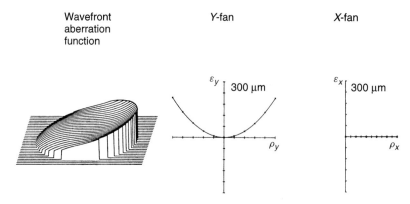

**Figure 3.31**   Wavefront and ray fans for coma.

and varies with the square of the aperture, as the system is stopped down. The width of the blur (the "tangential coma") is one-third of the sagittal coma. Figure 3.31 shows the wavefront and ray fans for this system. Note that the $x$-fan is flat, since the central cross section of the wavefront has no slope errors.

*Field Curvature and Astigmatism.* Field curvature refers to the condition in which the image is formed on a curved surface rather than the paraxial image plane. The third-order term of field curvature $(h^2\rho^2)$ represents a defocus which varies as the square of the field, so that image surface is parabolic, at least close to the optic axis. Farther from the axis, higher-order field curvature terms may dominate. In the absence of astigmatism, the surface on which the image falls is termed the "Petzval surface," and this type of field curvature is known as "Petzval field curvature."

Astigmatism refers to the condition in which the wavefront from an off-axis object point has different curvatures in the horizontal and vertical sections. This causes the rays to focus into a horizontally oriented line at one distance, and then form a vertical line image a short distance downstream, as shown in Fig. 3.32. For a rotationally symmetric system, the third-order astigmatism term depends on the square of the field angle and vanishes at the center of the image field. (This differs from the use of the word *astigmatism* in the ophthalmic profession, in which a cylindrical defect of the eye causes an error that appears across the entire field of view.) The length of the astigmatic focal lines is given by

$$\text{astigmatic blur length} = 2W_{222}h^2/n'u' \qquad (3.41)$$

and varies linearly with the aperture as the system is stopped down. The longitudinal separation between the astigmatic foci is given by

$$\text{longitudinal astigmatism} = 2W_{222}h^2/n'u'^2 \qquad (3.42)$$

Halfway between the sagittal and tangential foci lies the *medial* focus, the spot diagram for which is a circular blur whose diameter is half the length of the line foci. Although the spot diagram is identical to that for defocus, the wavefront is not. Figure 3.33 shows spot diagrams and wavefronts for astigmatism at the sagittal, medial, and tangential image locations.

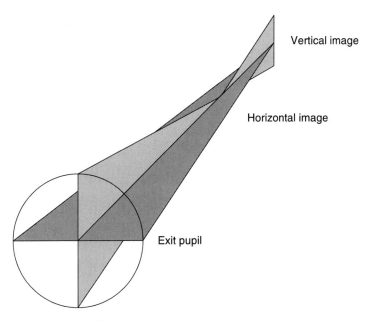

**Figure 3.32**   Line foci formed by an astigmatic system.

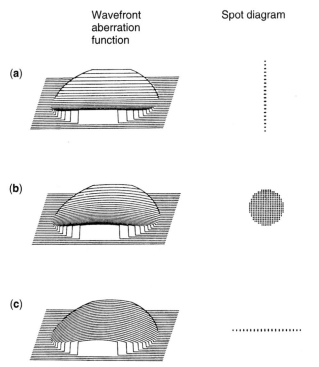

**Figure 3.33**   Wavefronts and spot diagrams for astigmatism: **(a)** at sagittal focus; **(b)** at medial focus; **(c)** at tangential focus.

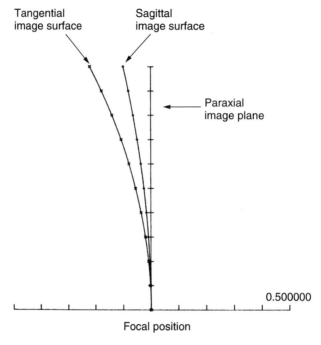

Tangential
image surface

Sagittal
image surface

← Paraxial
image plane

0.500000

Focal position

**Figure 3.34**    Field curvatures for an astigmatic system.

Note that the existence of an $h^2 \rho^2 \cos^2 \phi$ term alone implies that the tangential image "pulls away" from the sagittal image, as in Fig. 3.34. If there were no $h^2 \rho^2$ term, the sagittal image surface would remain at the paraxial image plane; however, this is not normally the case in lens systems. If there is both astigmatism and field curvature, the "field curvature" term $W_{220} h^2 \rho^2$ refers to the curvature of the sagittal image surface, and the astigmatism gives the separation between the sagittal and tangential surfaces. It will later be seen [Eqs. (3.48)] that the presence of astigmatism changes the value of the field curvature coefficient $W_{220}$, and therefore that neither the sagittal nor the tangential surface lies at the Petzval image surface. In any case, the longitudinal separation of the sagittal image field from the paraxial image plane is given by

$$\text{sagittal field sag} = -2W_{220}h^2/n'u'^2 \tag{3.43}$$

The negative sign indicates that for undercorrected field curvature (position $W_{220}$), the sagittal image surface lies to the left of the paraxial image plane. The sags of the medial and tangential surfaces are given by

$$\text{medial field sag} = \frac{-2\left(W_{220} + \frac{1}{2}W_{222}\right)h^2}{n'u'^2} \tag{3.44}$$

$$\text{tangential field sag} = \frac{-2(W_{220} + W_{222})h^2}{n'u'^2} \tag{3.45}$$

***Distortion.*** Distortion is a variation in magnification across the field of view. This causes no blurring of the image, only a shift of the image points from their paraxial

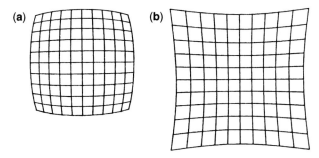

**Figure 3.35** Distorted images of a square grid: **(a)** "barrel" distortion; **(b)** "pincushion" distortion.

("ideal") image locations. Figure 3.35**a** shows the image of a rectilinear grid in the presence of 10% "barrel" distortion ($W_{311} > 0$), and Fig. 3.35**b** shows 10% "pincushion" distortion ($W_{311} < 0$). The third-order distortion term produces a shift of the image point which is proportional to the cube of the position in the object. Because there is no blurring but only a shift of all the rays for each image point, the effect on the ray fans is that the $y$-fans are displaced upward or downward by an amount proportional to the cube of $h$, and there is no change in the $x$-fans at any value of $h$. Because distortion is a fundamentally different type of aberration than the blur-inducing aberrations, it is usually depicted on a plot of percent distortion (the image shift divided by the image position) versus field height, rather than on the ray fans. In such cases the ray fans are measured from the position of the real chief ray, rather than the ideal (paraxial) chief ray.

*Higher-Order Aberrations.* The higher-order aberrations may be considered as modifications of the third-order aberrations. At low $F$-number we encounter the higher orders of spherical aberration and coma, while wide-angle systems suffer higher-order astigmatism and distortion. Higher-order spherical aberration terms simply vary with higher powers of $\rho$, causing the ray fans to deviate from the cubics of Figs. 3.27 to 3.29. Higher-order dependences of coma with $h$ can cause the coma in a system to develop a "ring node" at the field radius where the third and higher orders balance; the same is true of astigmatism. Higher-order distortion represents an image shift which is proportional to a higher power of $h$. "Oblique spherical" is a form of spherical aberration which varies with (even powers of) $\rho$. There is also "elliptical coma," which is a combination of comatic terms which causes a comatic blur having an apex angle other than $60°$. "Spherochromatism" is a term given to the variation with wavelength with spherical aberration.

While the speed of computer ray-trace algorithms has made calculation of the higher-order aberrations unnecessary for the most part, a qualitative understanding is essential to the design process. Once a general type of lens has been chosen, the higher-order aberrations are essentially fixed and only the third-order aberrations change during optimization. For this reason, some residual third-order aberrations are intentionally left in the design to offset the higher-order aberrations of similar types. Figure 3.36 shows ray fans for an achromatic doublet designed to be used at F/3. Third-order spherical aberration and defocus have been used to offset the higher-order spherical

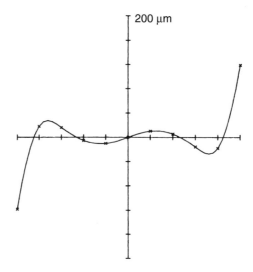

**Figure 3.36**   Ray fans for an F/3 Achromat, showing "zonal" spherical aberration.

aberration at the edge of the pupil. The aberration pattern that remains is known as *zonal spherical aberration*.

### 3.3.3   Calculation of the Third-Order Aberrations

As mentioned earlier, the third-order aberrations of an optical system may be calculated directly from the paraxial ray-trace data; it is not necessary to perform an exact ray trace. In that the wavefront aberration function represents the optical phase delay as a function of transverse position, the wavefront aberration coefficients ($W_{040}$, $W_{131}$, etc.) for the system are simply the sums of the contributions introduced by the individual surfaces. The aberrations introduced by spherical surfaces are calculated directly with the equations given below. For aspheric surfaces, the aberration contribution is decomposed into that part caused by the spherical component of the surface sag and that part introduced by the aspheric departure of the surface from the sphere.

*Surface Contributions.*  Before computing the aberration coefficients, it is necessary to compute for each surface the paraxial ray heights $y$ and $\bar{y}$, as well as the quantities $A$ and $B$, which are defined as

$$A = ni = n(u + yc) = n'(u' + yc) = n'i' \tag{3.46}$$

and

$$B = n\bar{\imath} = n(\bar{u} + \bar{y}c) = n'(\bar{u}' + \bar{y}c) = n'\bar{\imath}' \tag{3.47}$$

Using the notation $\Delta(\cdot)$ to indicate the change at the surface of the quantity in parentheses (i.e., the primed quantity minus the unprimed quantity), the spherical surface contributions to the "third-order" coefficients of the wavefront aberration function are given by

$$W_{040} = \frac{-1}{8} \sum_{\text{surfaces}} A^2 y \Delta \left( \frac{u}{n} \right)$$

$$W_{131} = \frac{-1}{2} \sum_{\text{surfaces}} A B y \Delta \left( \frac{u}{n} \right)$$

$$W_{222} = \frac{-1}{2} \sum_{\text{surfaces}} B^2 y \Delta \left( \frac{u}{n} \right) \qquad (3.48)$$

$$W_{220} = \frac{1}{2} W_{222} + \frac{H^2}{4} \sum_{\text{surfaces}} \frac{\phi}{n n'}$$

$$W_{311} = \frac{1}{2} \sum_{\text{surfaces}} \left[ \bar{y} B (2 y B - \bar{y} A) c \Delta \left( \frac{1}{n} \right) - B^3 y \Delta \left( \frac{1}{n^2} \right) \right]$$

The chromatic variations of the defocus and tilt coefficients are given by

$$\delta_{F-C}(W_{020}) = \frac{1}{2} \sum_{\text{surfaces}} A y \Delta \left( \frac{n_d - 1}{n_d \nu_d} \right)$$

$$\delta_{F-C}(W_{111}) = \sum_{\text{surfaces}} B y \Delta \left( \frac{n_d - 1}{n_d \nu_d} \right) \qquad (3.49)$$

The secondary chromatic coefficients [i.e., $\delta_{d-F}(W_{020})$ and $\delta_{d-F}(W_{111})$] may be obtained simply by replacing the quantity $\nu_d$ with the ratio $\nu_d / P$.

***Aspheric Contributions.*** To study the effects of aspheric surface shapes on the aberrations, it is first necessary to examine the forms of aspherics commonly used in optics. These may be expressed by the following sag function:

$$z(\rho) = \frac{c \rho^2}{1 + \sqrt{1 - (1 + \kappa) c^2 \rho^2}} + A_4 \rho^4 + A_6 \rho^6 + A_8 \rho^8 + \cdots \qquad (3.50)$$

where $c$ represents the curvature of the surface and $\kappa$ represents the "conic constant" of the surface. If the $\kappa$ and the $A$ coefficients are all zero, the surface is spherical. If only the $A$ coefficients are zero, the surface is a conic section revolved about the optical axis, with $\kappa$ equal to the negative of the square of the eccentricity of the conic. The departure of a conic surface from a sphere may be expanded into terms in all even powers of $\rho$, but the relationships among the various terms are fixed by the single parameter $\kappa$. The addition of each polynomial term $A_i \rho^i$ gives the designer another degree of freedom. Of these, only the fourth-order term $A_4 \rho^4$ affects the "third-order" aberrations. Conic surfaces are generally preferred to polynomial aspheres because (when tested in reflection) the spherical aberration is zero to all orders if the proper conjugates are chosen.

To compute the additional aberration contributed by the aspheric departure of the surface, we first calculate the quantity

$$a = \kappa c^3 y^4 \Delta(n) + 8A_4 y^4 \Delta(n) \tag{3.51}$$

The aspheric aberration contributions are then given by

$$\delta_{\text{asph}}(W_{040}) = \frac{1}{8}a$$

$$\delta_{\text{asph}}(W_{131}) = \frac{1}{2}\left(\frac{\bar{y}}{y}\right)a$$

$$\delta_{\text{asph}}(W_{222}) = \frac{1}{2}\left(\frac{\bar{y}}{y}\right)^2 a \tag{3.52}$$

$$\delta_{\text{asph}}(W_{220}) = \frac{1}{2}\left(\frac{\bar{y}}{y}\right)^2 a$$

$$\delta_{\text{asph}}(W_{311}) = \frac{1}{2}\left(\frac{\bar{y}}{y}\right)^3 a$$

***The Effect of Stop Shift.*** The equations above give the third-order aberrations for the system, with $y$ and $\bar{y}$ calculated with the stop in its intended position. It is useful to examine the behavior of the system as the stop position is altered. Shifting the stop essentially redefines another ray to be the chief way; however, careful study of the paraxial ray-trace equations shows that the quantity

$$\Delta S = \Delta\bar{y}/y \tag{3.53}$$

is the same for all surfaces throughout the system. The changes of the aberrations with "stop shift" are given by

$$\delta_{\text{SS}}(W_{040}) = 0$$

$$\delta_{\text{SS}}(W_{131}) = 4(\Delta S)W_{040}$$

$$\delta_{\text{SS}}(W_{222}) = 2(\Delta S)W_{131} + 4(\Delta S)^2 W_{040} \tag{3.54}$$

$$\delta_{\text{SS}}(W_{220}) = (\Delta S)W_{131} + 2(\Delta S)^2 W_{040}$$

$$\delta_{\text{SS}}(W_{311}) = 2(\Delta S)(W_{222} + W_{220}) + 3(\Delta S)^2 W_{131} + 4(\Delta S)^3 W_{040}$$

The stop-shift dependences of the chromatic aberrations are given by

$$\delta_{\text{SS}}(\delta_{F-C}(W_{020})) = 0$$

$$\delta_{\text{SS}}(\delta_{F-C}(W_{111})) = 2(\Delta S)\delta_{F-C}(W_{020}) \tag{3.55}$$

***Aberrations of Thin Lenses.*** Many optical systems consist of groups of relatively thin, widely spaced elements. In such cases, the lenses enclosed by the pairs of surfaces may be considered as the elemental units of refractive power, rather than the surfaces themselves. The aberration expressions of thin lenses are derived from the surface contribution equations, in the limit that the thicknesses of the lenses vanish. This is, of course, only approximately correct, but because it halves the number of elements in the system, it is often a worthwhile approximation. (This is particularly useful in the early stages of design; analysis is most easily carried out by ray tracing.) The thin-lens aberration expressions that follow are not in the most compact form possible; rather, they are formulated so as to give the most information about the dependences on the lens shape, object position, and stop position.

We begin the calculation by computing the index-related quantities $A_i$ through $F_i$ for each lens in the system, according to the equations of Table 3.1. From the front and rear curvatures $c$ and $c'$ and the ray angles $u$ and $u'$ for each lens, we compute the "bending factors" $X_i$ and the "conjugate factors" $Y_i$, according to Table 3.2. From these, the quantities $\alpha$ through $\eta$ are computed for each element. These give the bending and conjugate dependences for the lenses under the assumption that each element is located at a pupil of the system. Next, the parameter $T$ (which is a measure of the separation of the element from the pupil in its space) is calculated for each element, and the stop-shifted versions of $\alpha$ through $\eta$ are found from Table 3.3. Finally, the aberration coefficients may be calculated from the equations given in Table 3.4.

Much can be learned about the behavior of thin lenses through examination of Tables 3.1 to 3.4. For instance, from Table 3.2, it can be seen that the spherical aberration and coma of a thin lens with the stop in contact vary with both the bending and conjugate parameters, whereas the astigmatism depends on neither. Similarly, from

**TABLE 3.1   Thin-Lens Index Factors**

$$A_i = \frac{n_i + 2}{n_i(n_i - 1)^2} \qquad\qquad D_i = \frac{n_i^2}{(n_i - 1)^2}$$

$$B_i = \frac{4(n_i + 1)}{n_i(n_i - 1)} \qquad\qquad E_i = \frac{n_i + 1}{n_i(n_i - 1)}$$

$$C_i = \frac{3n_i + 2}{n_i} \qquad\qquad F_i = \frac{2n_i + 1}{n_i}$$

*Source*: Reference 2.

**TABLE 3.2   Bending and Conjugate Dependences for Thin Lenses**

$$\alpha_i = A_1 X_i^2 - B X_i Y_i + C Y_i^2 + D$$

$$X_i = \frac{c_i + c_i'}{c_i - c_i'} \qquad \beta_i = E_i X_i - F_i Y_i$$

$$\gamma_i = 1$$

$$Y_i = \frac{u' + u}{u' - u} \qquad \delta_i = \frac{1}{n_i}$$

$$\eta_i = 0$$

*Source*: Adapted from Ref. 3.

**TABLE 3.3   Stop-Shift Dependences for Thin Lenses**

$$\alpha_i^\dagger = \alpha_i$$

$$\beta_i^\dagger = \beta_i + T_i\alpha_i$$

$$T_i = \frac{\phi_i y_i \overline{y}_i}{2H} \qquad \gamma_i^\dagger = \gamma_i + 2T_i\beta_i + T_i^2\alpha_i$$

$$\delta_i^\dagger = \delta_i$$

$$\eta_i^\dagger = \eta_i + T_i(3\gamma_i + \delta_i) + 3T^2\beta_i + T_i^3\alpha_i$$

*Source*: Adapted from Ref. 3.

**TABLE 3.4   Thin-Lens Aberration Sums**

$$W_{040} = \frac{1}{32}\sum_{\text{lenses}} y_i^4\phi_i^3\alpha_i^\dagger \qquad W_{220} = \frac{1}{2}W_{222} + \frac{H^2}{4}\sum_{\text{lenses}}\phi_i\delta_i^\dagger$$

$$W_{131} = \frac{H}{4}\sum_{\text{lenses}} y_i^2\phi_i^2\beta_i^\dagger \qquad W_{311} = H^3\sum_{\text{lenses}} y_i^{-2}\chi_i^\dagger$$

$$W_{222} = \frac{H^2}{2}\sum_{\text{lenses}}\phi_i\gamma_i^\dagger \qquad\qquad —$$

$$\delta_{F-C}(W_{020}) = \frac{1}{2}\sum_{\text{lenses}} y_i^2\frac{\phi_i}{v_{d_i}} \quad \delta_{F-C}(W_{111}) = \sum_{\text{lenses}} y_i\overline{y}_i\frac{\phi_i}{v_{d_i}}$$

*Source*: Adapted from Ref. 3.

Table 3.3 it can be seen that the astigmatism will vary as the stop position is moved if the system has either spherical aberration or coma.

### 3.3.4   The Optical Sine Theorem

A very powerful conclusion may be reached upon closer examination of the Lagrange invariant and its relationship to the "brightness theorem" of radiative transfer. It can be shown (see, e.g., Boyd [3] from the laws of thermodynamics that the quantity $L/n^2$ is conserved throughout the optical system, where $L$ is the radiance of the system. The differential radiant flux entering the system from an infinitesimal object of area $dA$ is given by

$$d\Phi = L\,dA\,d\Omega\cos\alpha \tag{3.56}$$

where $d\Omega$ is the solid angle subtended at the object by an infinitesimal element of the entrance pupil, and $\alpha$ is the angle between the axis and the radial position in the pupil. Integrating over $\alpha$ from zero to a ray angle of $U$ gives

$$\Phi = 2\pi L\,dA\sin^2 U \tag{3.57}$$

where $U$ is the half-angle subtended by the entrance pupil at the object. Similarly, in image space one finds that

$$\Phi = 2\pi L'\,dA'\sin^2 U' \tag{3.58}$$

Conservation of energy and the brightness theorem then combine to give the expression

$$n'^2\, dA'\sin^2 U' = n^2\, dA\sin^2 U \tag{3.59}$$

This expression (or its square root) may be thought of as a nonparaxial version of the Lagrange invariant. Recognizing the ratio of image area to object area as the square of the magnification, one obtains a "real ray" expression for the magnification:

$$m_{(\text{real})} = n\sin U / n'\sin U' \tag{3.60}$$

Note that this reduces to the paraxial expression [Eq. (3.17)] in the small-angle limit. This implies that the magnification for the marginal rays will not be the same as that for rays near the axis unless the so-called Abbe sine condition is met:

$$\sin U'/u' = \sin U/u \tag{3.61}$$

Recalling that coma was the condition of having a magnification that varied as a function of pupil, it is clear that Eq. (3.61) must be satisfied if there is to be no coma in the image. Indeed, this holds not only for third-order coma, but all aberrations that are linear in the field.

From the sine condition, it can easily be shown that the principal "planes" (more precisely, the loci of equivalent refraction) are not planes as predicted by the paraxial equations, but rather spherical surfaces centered on the object and image. An interesting conclusion to be reached is that no system can simultaneously satisfy the paraxial laws of optics in terms of both the ray angles and the ray locations. Since it can usually be expected that coma has been removed (to the extent possible) from any commercially available optical system, one sees that the numerical aperture $n'\sin U'$ is a more natural measure of the "speed" of an optical system than is the $F$-number, which (according to our definition) is proportional to $\tan U'$.

## 3.4   REAL RAY-TRACE TECHNIQUES

The procedure given here for spherical surfaces is as described in References 4 and 5. The equations given for aspheric surfaces are essentially those of Reference 6, except that they are generalized to allow for different starting points.

To fully examine the aberrations seen by the real rays, it is now necessary to examine not only rays that intersect the axis ("meridional rays"), but also "skew rays" as well. It suffices to examine only one meridian of the object, provided that it is possible to examine rays that do not remain in this meridian. Although algorithms exist specifically for tracing meridional rays, modern computers are now sufficiently fast that it is quite reasonable to write only the skew ray code, and apply it to both meridional and skew rays.

As with the paraxial trace, the tasks of tracing real rays may be decomposed into the transfer of a ray from one surface to the next, and refracting at the next surface to obtain the new ray angles. In this way the ray may be propagated from the first surface to the image plane of the system. A ray will be represented by its direction cosines $(L, M, N)$ and its intersections with the surfaces $(x, y, z)$, measured from coordinate axes centered on the vertices of the surfaces. As the initial conditions for a ray trace are usually specified in terms of the ray locations at the object and entrance pupil, it is convenient to define the first surface to be a "dummy" (i.e., a surface with the same index on both sides) at the location of the entrance pupil. This allows the ray trace to begin at the first surface, which is treated like all the others. The fact that the entrance pupil might lie to the right of the first real surface is of no consequence if the thickness and indices are entered properly. Also, beginning the ray trace at the entrance pupil eliminates the problem with calculational errors that would occur in propagating a ray from an infinitely remote object to the first surface. As the ray-trace procedure is the same for all surfaces, the process of transferring from the $i$th surface to the next will be described.

### 3.4.1 Transfer between Spherical Surfaces

One begins by computing the intermediate quantity $e$:

$$e = t_i' N_i' - (L_i' x_i + M_i' y_i + N_i' z_i) \tag{3.62}$$

From this, components of the vector $\mathbf{Q}$, which is the closest approach of the skew ray to the vertex of the next surface, are computed:

$$Q_{i+1_z} = z_i + e N_i' - t_i' \tag{3.63}$$

$$Q_{i+1}^2 = x_i^2 + y_i^2 + z_i^2 - e^2 + t_i'^2 - 2t_i' z_i \tag{3.64}$$

The cosine of the real ray angle of incidence is now

$$\cos I_{i+1} = [N_i'^2 - c_{i+1}(c_{i+1} Q_{i+1}^2 - 2Q_{i+1_z})]^{1/2} \tag{3.65}$$

If the argument of the square root is negative, it means that the ray does not intersect the (spherical) surface, and the ray-trace procedure must end for this ray.

The distance along the ray between surfaces $i$ and $i+1$ is

$$A_i' = e + \frac{c_{i+1} Q_{i+1}^2 - 2Q_{i+1_z}}{N_i' + \cos I_{i+1}} \tag{3.66}$$

This is a useful quantity, since the optical path length for a ray is just the sum over all the surfaces of the quantity $n_i A_i$. The ray intersections at the next surface are given by

$$\begin{aligned} x_{i+1} &= x_i + A_i' L_i' \\ y_{i+1} &= y_i + A_i' M_i' \\ z_{i+1} &= z_i + A_i' N_i' - t_i' \end{aligned} \tag{3.67}$$

### 3.4.2 Refraction at a Spherical Surface

From Snell's law it may be seen that the cosine of the emergence angle is given by

$$\cos I_{i+1}' = \left[ 1 - \left( \frac{n_{i+1}}{n_{i+1}'} \right)^2 (1 - \cos^2 I_{i+1}) \right]^{1/2} \tag{3.68}$$

If the argument of the square root is negative, it means that the ray undergoes total internal reflection at surface $i + 1$. Defining

$$g_{i+1} = \cos I'_{i+1} - \frac{n_{i+1}}{n'_{i+1}} \cos I_{i+1} \tag{3.69}$$

we have

$$L'_{i+1} = \frac{n_{i+1}}{n'_{i+1}} L_{i+1} - g_{i+1} c_{i+1} x_{i+1} \tag{3.70}$$

$$M'_{i+1} = \frac{n_{i+1}}{n'_{i+1}} M_{i+1} - g_{i+1} c_{i+1} y_{i+1} \tag{3.71}$$

$$N'_{i+1} = \frac{n_{i+1}}{n'_{i+1}} N_{i+1} - g_{i+1} c_{i+1} z_{i+1} + g_{i+1} \tag{3.72}$$

### 3.4.3   Transfer between Aspheric Surfaces

The ray-trace equations may be written in closed form for spherical surfaces (above) and for conic aspheres. For other surfaces an iterative procedure is necessary to find the surface intersections; as a matter of programming convenience, this procedure is usually employed for conics as well. One method that converges rapidly is to begin with an estimate $\mathbf{P}_0 = (x_0, y_0, z_0)$ for the surface intersection and construct an auxiliary plane which lies tangent to the actual asphere at the point $\hat{\mathbf{P}}_0$, as shown in Fig. 3.37. The ray intersection with this auxiliary plane is found, which serves as the starting point $\mathbf{P}_1$ for the next iteration. As an initial estimate of the surface intersection, one can simply use the ray intersection with the base sphere, computed using equations given above; however, there are cases in which a ray fails to intersect the base sphere even though it intersects the asphere. For this reason, it is safer to use the ray intersection with the surface tangent plane (normal to the axis).

In what follows, it is not necessary that the surface be of any particular form, or even that it be rotationally symmetric. Although the initial estimate of the surface intersection may be taken on any convenient surface, it is assumed that the function describing the sag of the surface from a flat plane is known:

$$z = Z(x, y) \tag{3.73}$$

It is also necessary that the partial derivatives of $Z$ are known; however, these may be found numerically if they are not known explicitly. The iterative process begins with the ray direction cosines $(L, M, N)$, and an initial estimate $(x_0, y_0, z_0)$ of the surface intersection. It is important that the ray actually pass through this initial estimate.

Once the iteration has begun, all the cycles are carried out the same way. The process of computing the $(i + 1)$st estimate from the $i$th estimate will be given here. The estimate $\mathbf{P}_i$ is projected using Eq. (3.73) to find the point $\hat{\mathbf{P}}_i = (x_i, y_i, \hat{z}_i)$ on the surface, and the direction cosines of the normal vector $(\alpha_i, \beta_i, \gamma_i)$ through this point and pointing toward the center of curvature are computed using

$$(\alpha_i, \beta_i, \gamma_i) = \frac{(-\partial Z/\partial x, -\partial Z/\partial y, 1)}{[(\partial Z/\partial x)^2 + (\partial Z/\partial y)^2 + 1]^{1/2}} \tag{3.74}$$

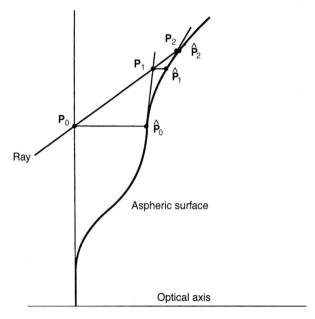

**Figure 3.37**   Finding the aspheric surface intersection points.

From these and the ray direction cosines, the intersection of the ray with the auxiliary plane gives the new estimate for the axial position of the surface intersection:

$$z_{i+1} = \frac{z_0(\alpha_i L + \beta_i M) + N[\alpha_i(x_i - x_0) + \beta_i(y_i - y_0) + \gamma_i \hat{z}_i]}{\alpha_i L + \beta_i M + \gamma_i N} \tag{3.75}$$

This gives the ray oblique ray path from $\mathbf{P}_0$ to $\mathbf{P}_{i+1}$:

$$A_{0,i+1} = \frac{z_{i+1} - z_0}{N} \tag{3.76}$$

which in turn yields new estimates for the transverse coordinates:

$$x_{i+1} = x_0 + L A_{0,i+1} \tag{3.77}$$

$$y_{i+1} = y_0 + M A_{0,i+1} \tag{3.78}$$

This completes the iteration cycle; the value of $\hat{z}_{i+1}$ can be computed and compared with $z_{i+1}$. If the two do not agree to the required accuracy (usually about six digits), another cycle can be run beginning with Eq. (3.74). After convergence, $n A_{0,i+1}$ represents the optical path length from $\mathbf{P}_0$ to the surface intersection and should be added to the path length sum if it is being computed.

### 3.4.4   Refraction at an Aspheric Surface

The dot product of the surface normal vector with the ray vector gives the cosine of the incidence angle:

$$\cos I = \alpha L + \beta M + \gamma N \tag{3.79}$$

and Eq. (3.68) may be used to obtain the cosine of the emergence angle. To convert the emergence angle into direction cosines, we note that the change in the scaled ray direction cosine vectors (scaled by the refractive indices) is parallel to the surface normal, with a proportionality constant of

$$K = n' \cos I' - n \cos I \tag{3.80}$$

Using this, the new direction cosines are given by:

$$L' = (nL + K\alpha)/n' \tag{3.81}$$

$$M' = (nM + K\beta)/n' \tag{3.82}$$

$$N' = (nN + K\gamma)/n' \tag{3.83}$$

### 3.4.5  Distortion of the Entrance Pupil

It was stated at the outset that the rays could be specified by their points of origin on the object and their locations on the entrance pupil. Since the pupil usually subtends a small angle from the object location, the radiant energy density at the pupil plane is essentially uniform. Thus, if one traces a uniform grid of rays through the entrance pupil, each ray represents the same amount of energy. Unfortunately, because of vignetting and the fact that the pupil is an *aberrated* image of the stop, some of the rays that pass through the pupil near its edge fail to pass through the stop. This is a small effect in many systems, but it becomes significant at large field angles.

There are two ways to deal with this problem. One is to define the rays by their locations on the stop rather than on the pupil, and the other is to find the exact shape of the entrance pupil, for the purpose of limiting the rays. The first method requires an iterative ray-trace method to ensure that the rays strike their desired locations on the stop, and has the effect that the rays no longer carry the same amount of energy. The second method preserves the uniformity of intensity, but requires considerable sampling of the pupil shape, which must be done for each object point under study.

## 3.5  DIFFRACTION EFFECTS

Although a detailed discussion of scalar diffraction theory goes beyond the scope of this chapter, a brief discussion of the results relevant to optical lens systems is necessary, in order to demonstrate the interrelationship of geometrical and physical optics. References 8–11 develop the diffraction theory in detail.

According to the Huygens–Fresnel principle, the scalar field amplitude $E$ at a point $\mathbf{p}'$ near the image plane may be computed using the coherent sum of "wavelets" propagating from the wavefront as it emerges from the exit pupil. According to the geometry of Fig. 3.38, this gives

$$E(\mathbf{p}') = \iint_{\text{pupil}} E(\mathbf{p}) e^{i(2\pi/\lambda)\overline{\mathbf{p}\mathbf{p}'}} \, dA \tag{3.84}$$

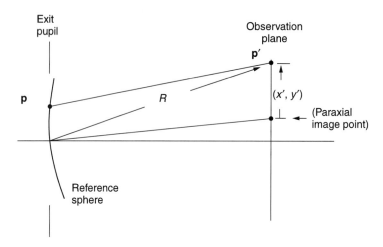

**Figure 3.38**    Coordinates for diffraction calculations.

where $E(\mathbf{p})$ is the complex field amplitude at the reference sphere centered in the pupil and $\overline{\mathbf{pp}'}$ is the oblique distance from $\mathbf{p}$ to the observation point $\mathbf{p}'$. The phase of the complex quantity $E(\mathbf{p})$ is determined by the aberrations of the system:

$$E(\mathbf{p}) = |E(\mathbf{p})|\, e^{i(2\pi/\lambda)W(h,\rho,\phi)} \tag{3.85}$$

where $W(h, \rho, \phi)$ is the wave aberration function evaluated at the appropriate field point $h$ and the normalized coordinates $(\rho, \cos\phi)$ which correspond to the pupil point $\mathbf{p}$. The function $|E(\mathbf{p})|$ represents the square root of the energy density in the exit pupil, and to first approximation is a constant within a circular boundary and zero outside it. In fact, the shape of the boundary is affected by vignetting and distortion of the exit pupil at large field angles. The distribution of energy within the pupil is affected chiefly by the energy distribution entering the system (e.g., a truncated Gaussian, for laser systems) and by variations in the transmission function of the system. The energy density is also affected by the redistribution of the rays due to aberrations of the pupil.

For points sufficiently close to the paraxial image plane of the system, the Fraunhofer approximation is valid and an integral reduces (except for a spherical phase factor) to a two-dimensional Fourier transform of the pupil function:

$$E(\mathbf{p}') = a e^{-i(2\pi/\lambda)R}\, \mathscr{F}[E(\mathbf{p})]\Big|_{\substack{\xi=x/\lambda z \\ \eta=y/\lambda z}} \tag{3.86}$$

where the frequency variables $\xi$ and $\eta$ are replaced by the spatial quantities $x/\lambda z$ and $y/\lambda z$. The quantities $x$ and $y$ are measured from the center of the reference sphere, as shown in the figure. The proportionality constant $a$ allows for conservation of energy, $R$ is the oblique distance from the center of the pupil to the observation point $\mathbf{p}$, and $z$ is the axial distance from the exit pupil to the observation plane. (The range over which the Fraunhofer approximation may be used is discussed later.) The intensity at

**p** is termed the *point spread function* PSF and is the square of the field amplitude:

$$\text{PSF}(\mathbf{p}') = a^2 \mathscr{F}[E(\mathbf{p})]^2 \Big|_{\substack{\xi = x/\lambda z \\ \eta = y/\lambda z}} \tag{3.87}$$

For a circular pupil with no aberrations, the so-called "Airy disk" is obtained:

$$\text{PSF}_{\text{Airy}}(r) = a^2 \left[ \frac{2J_1(2\pi r u'/\lambda)}{2\pi r u'/\lambda} \right]^2 \tag{3.88}$$

where $J_1(x)$ is the first order Bessel function of the first kind.

Figure 3.39 compares a cross section of an Airy disk with the spread functions for a wave of spherical aberration at paraxial focus and best (rms wavefront) focus.

The significance of the point spread function is that for linear, shift-invariant systems, the image is the convolution of (a scaled version of) the object and the PSF. The convolution of two functions $f(x, y)$ and $g(x, y)$ is given by the integral:

$$f(x, y) * g(x, y) = \iint_{-\infty}^{\infty} f(\xi, \eta) h(x - \xi, y - \eta) \, d\xi \, d\eta \tag{3.89}$$

A system is linear and shift invariant if the response function scales linearly with input intensity and is uniform across the image. Slowly varying field aberrations may be allowed if one consider the system to be "locally shift invariant." Notable exceptions are those systems that produce discretely sampled imagery, such as produced by an array detector. System that suffer image degradation due to "stray" light, or dark current or other nonlinearities of the detector violate the assumption of linearity.

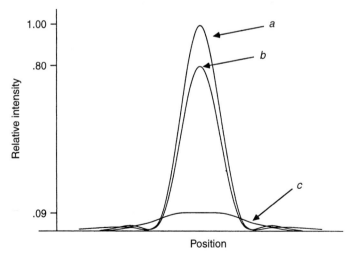

**Figure 3.39** Diffraction point spread functions: *a*, unaberrated; *b*, with one wave of spherical aberration; *c*, with one wave of spherical aberration at minimum rms wavefront focus.

One may define the *optical transfer function* (OTF) of the system to be the Fourier transform of the PSF. Since the process of Fourier transforming, squaring, and transforming again is equivalent to the process of autocorrelation, the OTF is also proportional to the autocorrelation of the pupil function. The OTF is a complex function whose arguments are the spatial frequencies in the $x$ and $y$ directions, and it is useful to decompose it into the *modulation transfer function* (MTF) and *phase transfer function* (PTF), respectively:

$$\text{OTF}(\xi, \eta) = \text{MTF}(\xi, \eta)e^{i\text{PSF}(\xi,\eta)} \qquad (3.90)$$

For linear shift-invariant systems, the MTF is the ratio of the image contrast to that of the object as a function of (sinusoidal) spatial frequency. The PTF gives the spatial phase or shift of each frequency component with respect to the corresponding frequency component in the object. Figure 3.40 gives the OTF functions for the three PSFs of Fig. 3.39. Due to the rotational symmetry of the PSFs in all three cases, the OTF functions must all be real. This restricts the PTFs to values of zero or $\pi$. Note that the abrupt jumps of the PTF in case (*c*) occurs at the points where the MTF curve passes through zero. This is a phenomenon known as *contrast reversal*.

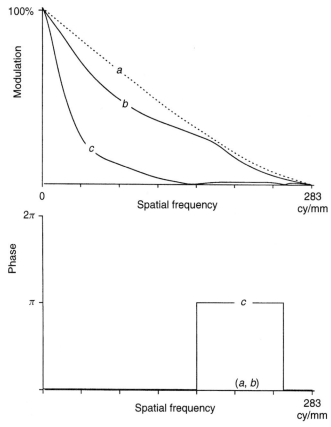

**Figure 3.40**  Optical transfer functions: *a*, unaberrated; *b*, with one wave of spherical aberration; *c*, with one wave of spherical aberration at minimum rms wavefront focus.

In all three cases, the fundamental upper limit of the MTF curve occurs at a frequency of $2u'/\lambda$.

There are some simple but useful rules of thumb for approximating diffraction effects without resorting to long calculations. A quick measure of the quality of an optical system is to compare the geometrical spot radius of a system with the Airy disk radius. This is in fact the basis for "Rayleigh's criterion," which states that (for a diffraction-limited optical system) if two points are so close to one another that their images are separated by less than one Airy disk radius, they cannot sensibly be resolved as being two distinct object points. Computing the angular separation of the object points by dividing through by the system focal length gives Rayleigh's estimate for the angular resolution of the system:

$$\alpha_{\text{Rayleigh resolution}} = \frac{1.22\lambda}{D_{\text{entrance pupil}}} \tag{3.91}$$

Another rule of thumb (the "Sparrow criterion") omits the factor of 1.22:

$$\alpha_{\text{Sparrow resolution}} = \frac{\lambda}{D_{\text{entrance pupil}}} \tag{3.92}$$

For consideration of the wavefront aberrations, the Rayleigh "quarter-wave criterion" is useful: If the peak-to-valley wavefront aberration does not exceed one quarter-wavelength, the imagery will not be sensibly aberrated. The reasoning is that if the various rays are not more than one quarter-wave out of phase at the image point, they will interfere with each other in a manner more constructive than destructive, and good imagery will result. Although "pathological" systems can be found for which this rule of thumb fails (e.g., a system with high-frequency oscillations of the phase front), it holds true in most practical cases.

Another useful measure is the *Strehl ratio*, which is the ratio of the peak value of the PSF to the peak intensity of the Airy disk. By the central ordinate theorem, this quantity is the ratio of the area under the MTF curve to that under the MTF of the unaberrated system. The term "diffraction limited" is used to mean that the system satisfies the quarter-wave criterion, or sometimes (*Maréchal's criterion*) that the Strehl ratio exceeds 80%, which is the value for one quarter-wave of defocus.

For the amount of defocus that is allowable in an image, a simple rule of thumb may be derived by assuming a wavefront error of one quarter-wave of defocus, and applying Eqs. (3.27) and (3.28) to obtain

$$\Delta z_{\text{max}} = \pm \frac{\lambda}{2n'u'^2} \tag{3.93}$$

This becomes quite simple if the object is at infinity, the image is formed in air, and the wavelength is 0.5 μm. Under these conditions, the tolerable "depth of focus" becomes simply the square of the $F$-number, in micrometers. This is precisely the criterion that determines the axial range over which the Fraunhofer approximation is valid. Simply put, the difference between a reference sphere centered on the observation point and the sphere that fits the wavefront paraxially must be a small fraction of a wavelength. Fourier methods may still be used to examine the imagery in planes other than the paraxial image (Fraunhofer) plane if the appropriate defocus term is added to the wavefront aberration function. When this is done, the Fourier transform operation calculates not the Fraunhofer integral but the Fresnel integral. The Fresnel approximation is valid provided that the quadratic approximation for defocus is valid.

## 3.6 MODULAR OPTICAL DESIGN

The large number of "off-the-shelf" lenses available today allows many optical systems to be designed and assembled far more cheaply than if each component had to be specially fabricated. Some thought should be put into the selection of the components, as the aberrations of a camera lens differ dramatically from those of an eyepiece. For this reason it is useful to group the available lenses into the functional categories of objectives and collimators, eyepieces, and relay lenses. Almost any optical system can be visualized as a combination of one or more of these elements.

### 3.6.1 Objectives and Collimators

The task of the objective lens is simply to image a remote object onto a detector at the rear focal plane, and a collimator performs the opposite task of imaging a near object to infinity. Although both operate at "infinite conjugates," collimators (and laser focusing lenses) are typically designed to give the best possible wavefront over a very narrow field, while photographic objectives are designed to yield the best possible compromise over an extended field.

For narrow field use, the simplest system is the planoconvex lens, which should be oriented with the curved side facing the long conjugate. The singlet suffers essentially all the aberrations, notably spherical aberration and color. The longitudinal chromatic aberration may be conveniently estimated as the focal length divided by the Abbe value of the glass, which is about 60, for crown glasses. Aspheric singlets are available that have been corrected for spherical aberration, and these may in principal be used to focus laser beams; however, the lens should also be "bent" to correct for coma as well, so that small misalignments of the lens do not introduce coma into the beam. For $F$-numbers above about F/5, "achromatic doublets" make excellent collimators, being corrected for spherical aberration, color, and usually coma as well. At lower $F$-numbers, higher-order spherical aberration becomes a serious problem, as Fig. 3.36 demonstrated. Off-the-shelf achromats are corrected for primary color only, and the remaining secondary longitudinal color is typically about $\frac{1}{2200}$ of the focal length. Telescope objectives of the air-spaced and three-element varieties have less zonal spherical aberration than cemented doublets, and are therefore "faster," but require more care in mounting because the two elements may become misaligned. Caution should be used when using surplus binocular objectives, as these are usually designed to operate with an image erecting prism and will suffer large amounts of axial color and spherical aberration if used without a prism. Microscope "objectives" are commonly used as laser focusing lenses, although most are designed to be used with a tube length of 160 or 170 mm rather than infinite conjugates. Special "infinity-corrected" objectives are available which may be used to focus laser beams without introducing spherical aberration. Another problem with microscope objectives is that most are designed to compensate for the spherical aberration and color introduced by a cover glass between the specimen and the objective. This may be accounted for either by introducing a 0.18-mm piece of crown glass into the space of the sample, or by selecting a "metallurgical" objective, which has been designed to work without a cover glass.

For imaging an extended object ("photographic" use), achromats are limited by astigmatism and field curvature to a useful field of about 2°. When a larger field is required, ordinary 35 mm camera objectives may be used; however, it is important to realize that the lens has been optimized for wide field coverage, and that the axial behavior is similar to that of a doublet. These objectives are optimized for distant objects and yield poor image quality when used near 1:1 magnification.

### 3.6.2 Relay Lenses

For systems operating at 1:1 conjugates (i.e., having a magnification of −1), lateral color, third-order coma, and all orders of distortion will be eliminated if the system has front-to-back symmetry. The equiconvex lens is therefore the preferred singlet shape, and for better performance, symmetric cemented-triplet lenses are available from many optics houses. Symmetric copier lenses are also available from surplus houses at a higher cost. For narrow-field applications, an excellent 1:1 relay may be constructed from two identical achromats properly oriented for collimated light between them. (While the lateral color, coma, and distortion will be canceled by the symmetry, the axial color, zonal spherical, and astigmatism will be twice as much as found in a single achromat.) Magnifications other than one may be obtained using achromats of dissimilar focal lengths. As in the 1:1 case, the lenses should be oriented and the object placed so as to give collimated light between the lenses. Wider fields may be obtained with photographic darkroom enlarger lenses.

### 3.6.3 Eyepieces

Eyepieces, loupes, and magnifying glasses are all designed to image an extended object to infinity. Although it may seem that a collimator might be used for this purpose, it must be remembered that these visual systems require an external pupil, which dramatically changes the field aberrations [see, e.g., Eqs. (3.54) or Table 3.3]. Usually, some attempt is made to control lateral color and to balance astigmatism against field curvature. Distortion is usually accepted "as is," and spherical aberration and axial color are either tolerated or compensated for by overcorrecting the objective. Coma may be corrected in the eyepiece or in the objective, depending on the intended use. (Again, care should be used when using one part of a "matched" system.) Simple two-element eyepieces and jeweller's loupes give adequate image quality over a 15° semifield angle. More expensive eyepieces are available which cover semifields up to 40°; the primary improvement in these lenses is the reduction of the field curvature, which invariably increases the number of elements and their diameters. Although the selection of available eyepieces is sufficiently large to cover essentially all eyepiece demands, one occasionally requires an "eyepiece-like" element (e.g., a lens with an external pupil covering a wide field at infinite conjugates) with a focal length that is incompatible with standard eyepiece designs. A possible alternative to a custom design is to use two closely spaced achromatic doublets oriented as if for collimated light between them. This is a good approximation to a standard eyepiece of the *Plössl* type and although spherical aberration in uncorrected, the astigmatism is relatively well balanced against field curvature over a 20° semifield.

## REFERENCES

1. E. Marchand, *Gradient Index Optics*, Academic Press, New York, 1978.
2. H. H. Hopkins, *The Wave Theory of Aberrations*, Oxford University Press (Clarendon), London, 1950. (Available from University Microfilms International, Ann Arbor, MI.)
3. R. V. Shack, *Geometrical Optics Course Notes*, University of Arizona, Tucson, 1983.
4. R. W. Boyd, *Radiometry and the Detection of Optical Radiation*, Wiley, New York, 1983.
5. D. P. Feder, "Optical Calculations with Automatic Computing Machinery," *Journal of the Optical Society of America* **41**, pp. 630–635 (1951).
6. D. P. Feder, "Calculation of an Optical Merit function and its Derivatives with Respect to the System Parameters," *Journal of the Optical Society of America*, **47**, pp. 913–925 (1957).
7. W. T. Welford, *Aberrations of the Symmetrical Optical System*, Academic Press, London, 1974.
8. J. D. Gaskill, *Linear Systems, Fourier Transforms, and Optics*, Wiley, New York, 1978.
9. J. W. Goodman, *Introduction to Fourier Optics*, McGraw-Hill, New York, 1968.
10. W. B. Wetherell, "The Calculation of Image Quality," in R. R. Shannon and J. C. Wyant, Eds., *Applied Optics and Optical Engineering*, Vol. 8, Academic Press, New York, 1980.
11. M. Born and E. Wolfe, *Principles of Optics*, 6th ed., Pergamon Press, New York, 1980.


## BIBLIOGRAPHY

J. R. Benford, "Microscope Objectives," in R. Kingslake, Ed., *Applied Optics and Optical Engineering*, Vol. 3, Academic Press, New York, 1965.

G. H. Cook, "Photographic Objectives," in R. Kingslake, Ed., *Applied Optics and Optical Engineering*, Vol. 3, Academic Press, New York, 1965.

R. E. Hopkins, "Mirror and Prism Systems," in R. Kingslake, Ed., *Applied Optics and Optical Engineering*, Vol. 3, Academic Press, New York, 1965.

S. Rosin, "Eyepieces and Magnifiers," in R. Kingslake, Ed., *Applied Optics and Optical Engineering*, Vol. 3, Academic Press, New York, 1965.

W. J. Smith, *Modern Optical Engineering*, McGraw Hill, New York, 1966.

W. Wittenstein, J. C. Fontanella, A. R. Newberry, and J. Baars, "The Definition of the OTF and the Measurement of Aliasing for Sampled Imaging Systems," *Opt. Acta*, Vol. 29, pp. 41–50, 1982.

# 4

# OPTICAL RESONATORS

Kenichi Iga

*Japan Society for the Promotion of Science*
*Tokyo, Japan*

## 4.1 INTRODUCTION

The optical resonator is a medium or space that can confine optical wave in a three-dimensional fashion to achieve laser oscillation on amplification, including active medium, in it.

Laser oscillators and resonance-type laser amplifiers both require a laser medium or an active medium to amplify light with the help of stimulated emission and a resonator that corresponds to a resonance circuit as in electronic circuits to feed the light back for the purpose of satisfying the well-known power and frequency condition in the oscillator. Since an electric circuit uses a discrete component or a one-dimensional transmission line, a spatial distribution of the electromagnetic field is not as important, while in the microwave and lightwave regions, three-dimensional resonators are used, and not only the resonant frequency, but also a spatial distribution of an electromagnetic field, must be considered. This is called a "resonant mode" or "eigenmode." The purpose of this chapter is to summarize fundamental issues on the optical resonator and to introduce a general idea of the resonant mode necessary for understanding optical resonators.

As for an optical resonator, a resonator called a Fabry–Perot interferometer, which consists of a pair of parallel reflecting mirrors, is known [1]. In the first ruby laser a pair of plane-parallel mirrors was employed [2]. A mathematical treatment of the resonant mode of a resonator with a pair of plane-parallel mirrors of finite size was initiated by Fox and Li [3]. Along with the indication that a resonator with a pair of concave mirrors had a small diffraction loss, the so-called Gaussian mode was introduced [4]. Since then various types of Fabry–Perot resonators have been devised.

A semiconductor laser is activated with the help of carrier injection through a *p-n* junction and features the utilization of a dielectric waveguide as a part of a resonator. At

*Handbook of Optical Components and Engineering,* Edited by Kai Chang
ISBN 0-471-39055-0 © 2003 John Wiley & Sons, Inc.

**TABLE 4.1   Types of Optical Resonators**

| Model | Waveguide | Structure of Reflector | Laser |
|---|---|---|---|
| Open resonator | Free space | Fabry–Perot type<br>Ring type | Gas laser<br>Solid-state laser<br>Dye laser<br>Surface emitting laser |
| Waveguide-type resonator | Hollow waveguide | Fabry–Perot type | Gas laser |
|  | Dielectric waveguide | Fabry–Perot type<br><br>Distributed feedback type<br>Distributed Bragg reflector type | Semiconductor laser |
| Traveling-wave-type resonator | Active medium | One-side mirror | High-gain gas laser |

the initial stage, a homojunction was employed for its active region and some waveguiding effects were discussed [5]. To confine light more effectively in the active medium, the double heterostructure [6] has come to be introduced, thus enabling continuous operation at room temperature. Usually, the Fabry–Perot resonator structure is used by cleaving crystals [7–9] or etching [10–13]. The distributed feedback (DFB) [14, 15] or distributed Bragg reflector (DBR) [16] is also studied, aiming at single-frequency operation and an integration-oriented structure [17].

A laser oscillator requires both an active medium to amplify light and a resonant part to feed light back. The resonator consists of a waveguide to contain an optical mode to minimize an optical loss and a reflector for reflection. Table 4.1 shows the classification of some typical optical resonators. Open-type resonators contain a laser active medium independent of the resonator formation, while the active medium of a semiconductor laser functions as a waveguide as well.

The dimensions of resonators differ according to the type of laser. Usually, resonator dimensions for a gas or solid-state laser range from about 10 cm to several meters, and that for a semiconductor laser is ranging from 1 $\mu$m to 300 $\mu$m.

## 4.2   WAVEGUIDE-TYPE FABRY–PEROT RESONATORS

### 4.2.1   Waveguides for Optical Resonators

*Modes of Two-Dimensional Dielectric Waveguide.* The semiconductor laser guides light by using a dielectric waveguide where the core with the refractive index $n_1$ is put between the cladding layers with the refractive index $n_2(n_1 > n_2)$ as shown in Fig. 4.1a, and is based on a resonator with a reflector. In a so-called mode-controlled laser, where the distributed refractive index exists in the transverse direction as seen from Fig. 4.5b, a three-dimensional waveguide is utilized. However, since fundamentals of waveguiding phenomena are introduced from a two-dimensional waveguide, we shall describe here modes of the two-dimensional waveguide.

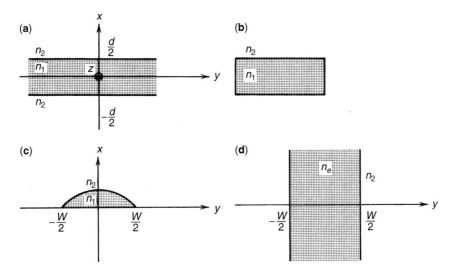

**Figure 4.1** Various dielectric waveguides.

In Fig. 4.1, the modes are divided into a TE mode and TM mode because they are uniform in the $y$ direction. Their mode functions can be obtained by solving wave equations [18], and expressed as in Table 4.2. By substituting $k_0 = 2\pi/\lambda$ for the normalized frequency $V$, we can define

$$V = k_0 \frac{d}{2} \sqrt{n_1^2 - n_2^2} \cong k_0 n_1 \frac{d}{2} \sqrt{2\Delta} \tag{4.1}$$

$$\Delta = \frac{n_1 - n_2}{n_1} \tag{4.2}$$

From the determinating equations (1) and (2) [or (2′)] and (1) and (3) [or (3′)] in Table 4.2, $\kappa$ and $\gamma$ can be obtained in the terms normalized by the thickness $d$ of the waveguide. Then the distributed intensity of light to each mode in Table 4.2 can be calculated. On the other hand, the propagation constant $\beta$ of the light traversing in the $z$ direction is written as

$$\beta^2 = n_1^2 k_0^2 - \gamma^2 \tag{4.3}$$

or, as a normalized value,

$$b \equiv \frac{\beta - n_2 k_0}{n_1 k_0 - n_2 k_0} \cong \frac{\beta^2 - n_2^2 k_0^2}{n_1^2 k_0^2 - n_2^2 k_0^2} = \frac{(\gamma d/2)^2}{V^2} \tag{4.4}$$

Figure 4.2 shows the confinement factor $\xi$ versus $V$ with other parameters $\kappa a$, $\gamma a$, and $b$.

As the value of the normalized frequency is $\beta \to n_2 (\gamma \to 0)$ in Eq. (4.4), when the TE$_1$ mode is cut off, and $\kappa d/2 \to \pi/2$ from expression (2′) in Table 4.2, the following equation is obtained:

$$V = \pi/2 \tag{4.5}$$

**TABLE 4.2  Mode Function of Two-Dimensional Dielectric Waveguide**

**TE**

$$E_y = \begin{cases} A_e \cos \kappa x & (|x| \le d/2) \\ A_e \cos\left(\dfrac{\kappa d}{2}\right) e^{-\gamma(|x|-d/2)} & (|x|) \ge d/2 \end{cases}$$

$$E_y = \begin{cases} A_0 \sin \kappa x & (|x| \le d/2) \\ \dfrac{x}{|x|} A_0 \sin\left(\dfrac{\kappa d}{2}\right) e^{-\gamma(|x|-d/2)} & (|x| \ge d/2) \end{cases}$$

$$\left(\frac{\kappa d}{2}\right)^2 + \left(\frac{\gamma d}{2}\right)^2 = V^2 \qquad (1)$$

$$\tan\left(\frac{\kappa d}{2}\right) = \frac{\gamma d/2}{\kappa d/2} \qquad (2)$$

$$\tan\left(\frac{\kappa d}{2}\right) = -\frac{\kappa d/2}{\gamma d/2} \qquad (2')$$

**TM**

$$H_y = \begin{cases} B_e \cos \kappa x & (|x| \le d/2) \\ B_e \cos\left(\dfrac{\kappa d}{2}\right) e^{-\gamma(|x|-d/2)} & (|x|) \ge d/2 \end{cases}$$

$$H_y = \begin{cases} B_0 \sin \kappa x & (|x| \le d/2) \\ \dfrac{x}{|x|} B_0 \sin\left(\dfrac{\kappa d}{2}\right) e^{-\gamma(|x|-d/2)} & (|x| \ge d/2) \end{cases}$$

$$\left(\frac{\kappa d}{2}\right)^2 + \left(\frac{\gamma d}{2}\right)^2 = V^2 \qquad (1)$$

$$\tan\left(\frac{\kappa d}{2}\right) = \left(\frac{n_1}{n_2}\right)^2 \frac{\gamma d/2}{\kappa d/2} \qquad (3)$$

$$\tan\left(\frac{\kappa d}{2}\right) = -\left(\frac{n_2}{n_1}\right)^2 \frac{\kappa d/2}{\gamma d/2} \qquad (3')$$

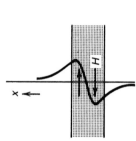

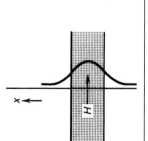

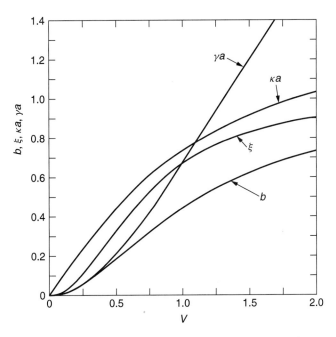

**Figure 4.2**  Confinement factor $\xi$ and other parameters $n$, $\kappa a$, and $\gamma a$ versus $V$.

and this value gives a single-mode condition of two-dimensional waveguides (a single mode at $V < \pi/2$).

***Mode Confinement Factor.***  The core region of semiconductor lasers often serves as the active layer, and consequently the ratio of the optical power confined in the core region is related to the mode gain. The mode confinement factor $\xi$ of the TE mode is calculated as

$$\xi = \frac{\displaystyle\int_0^{d/2} |E_y|^2\,dx}{\displaystyle\int_0^{\infty} |E_y|^2\,dx}$$

$$= \frac{1 + 1/\gamma d \sin \gamma d}{1 + 1/\gamma d \sin \gamma d + 2/\kappa d \cos^2(\gamma d/2)} = \frac{V + \sqrt{b}}{V + \sqrt{b}} \tag{4.6}$$

The mode confinement factor $\xi$ versus the active layer thickness $d$ is shown in Fig. 4.2.

***Modes of Three-Dimensional Waveguide.***  It is convenient for the three-dimensional waveguide, which has refractive index distribution also in the transverse direction and whose layer thickness changes in the core region as shown in Fig. 4.1**b** and **c**, to be substituted by the two-dimensional waveguide, which is uniform in the $x$ direction, by using an equivalent refractive index method. That is, the value $\beta/k_0$ (i.e., the propagation constant $\beta$ of the mode in the $x$ direction is normalized by the wave number $k_0$) becomes the equivalent refractive index $n_e$ which the mode has realized. The axis of ordinates $b$ in Fig. 4.2 also shows the increment of the equivalent refractive index $n_2$ normalized by $n_1 - n_2$, and $n_e$ has values in the range of $n_2 \leqq n_e \leqq n_1$. When the mode in the $x$ direction is designated, the change of the waveguide thickness in the $y$ direction and the index difference can be expressed in two dimensions as the refractive index distribution, as shown in Fig. 4.1**d**.

Let us first consider the case in Fig. 4.1**b**, where the mode has the index difference in the $y$ direction. We can use the two-dimensional problems described above as for two-dimensional waveguides substituted by the equivalent refractive index. Secondly in the case where the refractive index or waveguide thickness is distributed, the mode distribution becomes as follows: For the arbitrary refractive index distribution, numerical calculation is required; in the case of square distribution, and where the width of a waveguide is relatively large, approximate analytical solutions can be utilized. Setting the distribution of the equivalent refractive index as

$$n_e^2(y) = \begin{cases} n_e^2(0)[1 - (gy)^2] & y \leqslant W/2 \\ n_2^2 & y > W/2 \end{cases} \tag{4.7}$$

$$g \doteqdot \frac{\sqrt{2(n_e - n_2)/n_e}}{W/2} \tag{4.8}$$

Accordingly, the mode function $u_p(y)$ becomes

$$u_p(y) = \frac{1}{(2^p p! \sqrt{\pi} w_0)^{1/2}} H_p\left(\frac{y}{w_0}\right) e^{-(1/2)(y/w_0)^2} \tag{4.9}$$

where $H_p(\cdot)$ is the Hermite polynomial. The characteristic spot size $w_0$ is evaluated as

$$w_0 = \frac{1}{\sqrt{k_0 n_e(0)g}} \tag{4.10}$$

On the other hand, the propagation constant $\beta_p$ is

$$\beta_p = \sqrt{k_0^2 n_e^2 - g^2(2p + 1)}$$
$$\doteqdot k_0 n_e - g\left(p + \tfrac{1}{2}\right) \tag{4.11}$$

where if $\beta_1 = k_0 n_e$, the single-mode cutoff condition can be obtained.

If the following is the normalized frequency in the $y$ direction,

$$V^e = k_0 n_e \frac{W}{2} \sqrt{\frac{2(n_e - n_2)}{n_e}} \tag{4.12}$$

the cutoff $V$ value becomes $V^e = 3$. However, Eq. (4.12) denotes the approximate expression, and to be exact, $V^e = 2.4$.

## 4.2.2   Resonant Frequency

Let us first consider the resonant frequency and mode in the optical resonator in terms of a distributed-index waveguide, where analytic formulation for a propagation constant is possible. Figure 4.3 shows a Fabry–Perot waveguide-type resonator, where

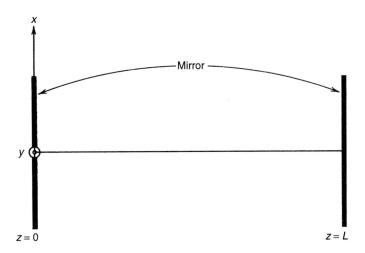

**Figure 4.3**   Model of a Fabry–Perot waveguide-type resonator.

a distributed-index waveguide is terminated with length $L$. The refractive index of the waveguide is expressed by

$$n^2(x, y) = n^2(0)[1 - g^2(x^2 + y^2)] \tag{4.13}$$

In this structure the light is reflected at both end faces and confined between them to make a resonant mode.

A guided wave in the positive $z$ direction is expressed by the following equation:

$$\psi_{pq}^+ = A_{pq}^+ u_p(x, w_{01}) u_q(y, w_{02}) \exp(-j\beta_{pq} z) \tag{4.14}$$

while a conjugate wave traversing in the negative $z$ direction is

$$\psi_{pq}^- = A_{pq}^- u_p(x, w_{01}) u_q(y, w_{02}) \exp(+j\beta_{pq} z) \tag{4.15}$$

Therefore, the standing wave in the resonator is written as

$$
\begin{aligned}
\psi(x, y, z) &= \sum_{p,q} (\psi_{pq}^+ + \psi_{pq}^-) \\
&= \sum_{p,q} [A_{pq}^+ \exp(-j\beta_{pq} z) + A_{pq}^- \exp(+j\beta_{pq} z)] u_p(x, w_{01}) u_q(y, w_{01})
\end{aligned}
$$

$$\tag{4.16}$$

Assuming that the light is completely reflected at $z = 0$ and $z = L$, and the boundary condition is $\psi(x, y, 0) = \psi(x, y, L) = 0$, we can write

$$A_{pq}^+ + A_{pq}^- = 0 \tag{4.17}$$

Therefore,

$$\sin \beta_{pq} L = 0 \tag{4.18}$$

According to Eq. (4.18),

$$\beta_{pq} L = n\pi \quad n = 1, 2, 3, \ldots \tag{4.19}$$

When we substitute the propagation constant $\beta_{pq}$ for the Hermite-Gaussian mode of the distributed index waveguide [18],

$$\beta_{pq} = k = k_0 n(0) - g(p + q + 1) \tag{4.20}$$

From Eq. (4.19) we can write, after some manipulation,

$$kL - \left(p + \tfrac{1}{2}\right) g_1 L - \left(q + \tfrac{1}{2}\right) g_2 L = n\pi \tag{4.21}$$

where $n(0)$ is the refractive index at the center axis, $g$ the focusing parameter of the waveguide, and $p$ and $q$ denote the transverse mode number in the $x$ and $y$ directions, respectively.

Since $k = n(0)(\omega/c)$ and $\omega = 2\pi f$, the frequency $f(n, p, q)$ which satisfies Eq. (4.21) can be written as

$$f(n, p, q) = \frac{c}{2n(0)L} \left[ n + \frac{g_1 L}{\pi} \left(p + \tfrac{1}{2}\right) + \frac{g_2 L}{\pi} \left(q + \tfrac{1}{2}\right) \right] \tag{4.22}$$

On the other hand, the corresponding wavelength $\lambda(n, p, q)$ is equal to $c/f(n, p, q)$ when being expressed in terms of wavelength.

$$\lambda(n, p, q)/2n(0) = L \left/ \left[ n + \frac{gL}{\pi} \left(p + \tfrac{1}{2}\right) + \frac{gL}{\pi} \left(q + \tfrac{1}{2}\right) \right] \right. \tag{4.23}$$

From Eq. (4.23), $gL < 1$ and we can understand that the integer $n$ denotes the number of half-wavelengths $\lambda/2n(0)$ in the medium of length $L$. Therefore, the number $n$ is called a longitudinal mode number. On the contrary, $p$ and $q$, which denote the number of zeros of the mode function in the directions of $x$ and $y$, respectively, are called transverse mode numbers.

Next, let us find the frequency spacing $\Delta f$ or wavelength spacing $\Delta \lambda$ in the case where the mode number is different by $\Delta n$, $\Delta p$, and $\Delta q$. From Eq. (4.22) we have

$$\begin{aligned} \Delta f &= \frac{\partial f}{\partial n} \Delta n + \frac{\partial f}{\partial p} \Delta p + \frac{\partial f}{\partial q} \Delta q \\ &= \frac{c}{2n_1 L} \Delta n + \frac{c}{2n_1} \frac{g_1}{\pi} \Delta p + \frac{c}{2n_1} \frac{g_2}{\pi} \Delta q \end{aligned} \tag{4.24}$$

The longitudinal mode spacing $\Delta f_l$ can be obtained by putting $\Delta n = 1$,

$$\Delta f_l = c/2n_1 L \tag{4.25}$$

The transverse mode spacing $\Delta f_t$ is expressed as

$$\Delta f_t = cg_i/2n_1\pi \quad i = 1, 2 \tag{4.26}$$

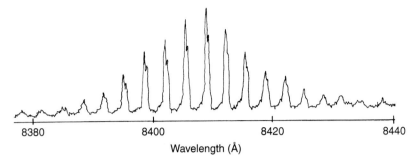

**Figure 4.4**   Spectrum of a semiconductor laser.

For example, when $L$ is 1 m and $\Delta n$ is 1, then $\Delta f_l$ is equal to 150 MHz.

The wavelength spacing can be calculated in the same manner, and the longitudinal mode spacing $\Delta \lambda_1$ is expressed as follows:

$$\Delta \lambda_l = \lambda^2 / 2 n_1 L \tag{4.27}$$

The transverse mode spacing $\Delta \lambda_t$ is

$$\Delta \lambda_t = g_i \lambda^2 / 2\pi n_1 \quad i = 1, 2 \tag{4.28}$$

Let us, for example, suppose that $n_1 = 3.5$, $L = 300$ μm, $\lambda = 0.84$ μm, and $g_i = 10^3$ m$^{-1}$; then $\Delta \lambda_l$ is 3.4 Å and $\Delta \lambda_t$ is 0.3 Å. Figure 4.4 shows a spectrum of a semiconductor laser. A series of the highest peaks is a longitudinal mode group with width approximately 4 Å. The peak seen next to its right is considered another transverse mode and is approximately 0.3 Å from the fundamental transverse mode.

## 4.3   GAUSSIAN BEAMS AND PROPAGATION MATRICES

### 4.3.1   Gaussian Beams

The field distribution $f(x', y', 0)$ of light at $z = 0$ in a coordinate system of Fig. 4.5 is transformed as

$$f(x, y, z) = \frac{j}{\lambda z} e^{-jkz} \iint dx' \, dy' \, f(x', y', 0)$$

$$\cdot \exp\left\{ -\frac{jk}{2z} [(x - x')^2 + (y - y')^2] \right\} \tag{4.29}$$

where the light propagates uniformly in the medium by the distance $z$ and $e^{j\omega t}$ is the temporal factor.

However, by assuming that the wavelength in the uniform medium is $\lambda$, the propagation constant $k$ is written as

$$k = \frac{2\pi}{\lambda} \tag{4.30}$$

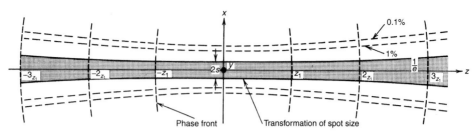

**Figure 4.5**    Transformation of Gaussian beam spot size $w$ and phase front. (From *Laser Handbook*, originally published in Japan by Ohmsha Ltd., Tokyo, 1982; reprinted with permission.)

This equation is known as the simplified Fresnel–Kirchhoff integral and expresses the diffraction of light in the numerical expression, and further is written in cylindrical coordinates as

$$f(r, \theta, z) = \frac{j}{\lambda z} e^{-jkz} \iint r' \, dr' d\theta' f(r', \theta', 0)$$

$$\cdot \exp\left\{ -\frac{jk}{2z}[r'^2 - 2rr'\cos(\theta - \theta') + r^2] \right\} \tag{4.31}$$

From Eqs. (4.29) and (4.31), the diffraction wave versus the arbitrary incident wave can be calculated. The integral to the so-called Gaussian beams having Gaussian distribution against the transverse distance $r$ is carried out analytically.

Let us assume that the field distribution with spot size $s$ of the Gaussian beams is

$$f(x', y', 0) = E_0 \exp\left( -\frac{1}{2}\frac{x'^2 + y'^2}{s^2} \right) \tag{4.32}$$

and integrate by substituting into Eq. (4.29), the following is obtained as

$$f(x, y, z) = E_0 e^{-jkz} \frac{s}{w} \exp\left[ -\frac{1}{2}P(x^2 + y^2) + j\phi \right] \tag{4.33}$$

Here the spot size $w$ and radius $R$ of the phase front are given by

$$w = s\sqrt{1 + (z/ks^2)^2} \tag{4.34}$$

$$R = z[1 + (ks^2/z)^2] \tag{4.35}$$

Parameters $P$ and $\phi$ are defined by

$$P = 1/w^2 + j(k/R) \tag{4.36}$$

$$\phi = \tan^{-1}(z/ks^2) \tag{4.37}$$

where $P$ is the waveform coefficient and $\phi$ the phase shift.

From Eq. (4.33) it can be seen that the transformed beam is still Gaussian, although there are changes in the spot size and radius phase front. It is clear that $R$ expresses the phase front if considering the phase condition

$$kz + \frac{k}{2R}r^2 = \text{constant} \tag{4.38}$$

With this equation, the functional dependence of the phase front $z = -(1/2R)r^2$ can be reduced. When $R$ is positive, the phase front is convex, as seen from $z = +\infty$.

Let us examine the parameter $z/ks^2$, which appeared in the previous equations. When this parameter is rewritten as

$$\frac{z}{ks^2} = \frac{1}{2\pi}\left(\frac{s^2}{\lambda z}\right)^{-1} \tag{4.39}$$

and the Fresnel number $N$ is defined as

$$N = s^2/\lambda z \tag{4.40}$$

where the Fresnel number $N$ is a function of the spot size $s$, distance $z$, and wavelength $\lambda$, and expresses normalized distance. Regions can be characterized according to $N$ such that

$$N \ll 1 \quad \text{Fraunhofer region}$$

$$N \geqslant 1 \quad \text{Fresnel region}$$

When the point of observation is located at a point some distance from the origin ($N \ll 1$), the spot size $w$ can be approximated from Eq. (4.34) as

$$w \simeq z/ks \tag{4.41}$$

The spreading angle $\Delta\theta$ of the beam is, therefore,

$$2\Delta\theta = 2w/z = 0.64(\lambda/2s) \tag{4.42}$$

This is analogous to the spreading angle of a diffracted plane wave from a circular aperture with $D$ in diameter, given by

$$2\Delta\theta = 1.22(\lambda/D) \tag{4.43}$$

Figure 4.5 shows the transformation of the Gaussian beam spot size and the positions of 1% and 0.1% of the electric power.

### 4.3.2  Transformation of Waveform Coefficients

Figure 4.6 represents waveform coefficients $P_0$, $P_1$, and $P_2$ at $z = 0$, $z_1$, and $z_2$, respectively. If the spot sizes and curvature radii of the wavefront are given by $s$, $w_1$, and

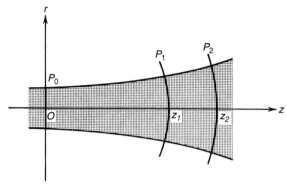

**Figure 4.6** Transformation of waveform coefficients. (From K. Iga, Y. Kokubun, and M. Oikawa, *Fundamentals of Microoptics*; copyright 1984, Ohmsha, Ltd., Tokyo, reprinted with permission.)

$w_2$, and $\infty$, $R_1$, and $R_2$, respectively, the coefficients can be expressed as [21]

$$P_0 = \frac{1}{s^2} \tag{4.44}$$

$$P_1 = \frac{1}{w_1^2} + \frac{jk}{R_1} \tag{4.45}$$

$$P_2 = \frac{1}{w_2^2} + \frac{jk}{R_2} \tag{4.46}$$

From the equations above,

$$\frac{1}{P_0} = \frac{1}{P_1} + j\frac{z_1}{k} \tag{4.47}$$

$$\frac{1}{P_0} = \frac{1}{P_2} + j\frac{z_2}{k} \tag{4.48}$$

When $P_0$ is eliminated, the relationship between $P_1$ and $P_2$ is reduced to

$$P_1 = \frac{P_2}{1 + j(1/k)(z_2 - z_1)P_2} \tag{4.49}$$

This is a special case of the linear transform

$$P_1 = \frac{AP_2 + B}{CP_2 + D} \tag{4.50}$$

It is very convenient to utilize the matrix form

$$\tilde{F} = \begin{bmatrix} A & B \\ C & D \end{bmatrix} \tag{4.51}$$

to calculate the transform for a system composed of many tandem components. It is then possible to realize a total $F$ matrix with the product of the matrices, expressed as

$$F = F_1 \cdot F_2 \cdot F_3 \cdots \tag{4.52}$$

**TABLE 4.3   Waveform Matrices for Various Optical Systems**

| Optical System | Waveform Matrix, $F$ |
|---|---|
| 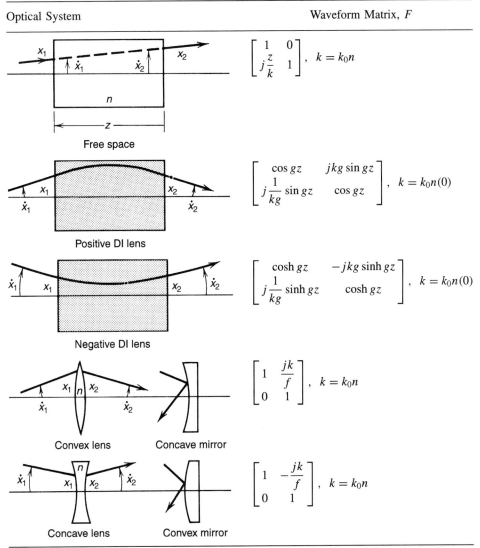 | $\begin{bmatrix} 1 & 0 \\ j\dfrac{z}{k} & 1 \end{bmatrix}, \quad k = k_0 n$ |
| | $\begin{bmatrix} \cos gz & jkg \sin gz \\ j\dfrac{1}{kg} \sin gz & \cos gz \end{bmatrix}, \quad k = k_0 n(0)$ |
| | $\begin{bmatrix} \cosh gz & -jkg \sinh gz \\ j\dfrac{1}{kg} \sinh gz & \cosh gz \end{bmatrix}, \quad k = k_0 n(0)$ |
| | $\begin{bmatrix} 1 & \dfrac{jk}{f} \\ 0 & 1 \end{bmatrix}, \quad k = k_0 n$ |
| | $\begin{bmatrix} 1 & -\dfrac{jk}{f} \\ 0 & 1 \end{bmatrix}, \quad k = k_0 n$ |

*Source*: K. Iga, Y. Kokubun, and M. Oikawa, *Fundamentals of Microoptics*; copyright 1984, Ohmsha, Ltd., Tokyo, reprinted with permission.

Table 4.3 shows a tabulation of the waveform matrices associated with some optical components. It is not difficult to obtain these matrix forms by calculating the change of a Gaussian beam when it passes through these optical components.

### 4.3.3   Ray Matrix

Figure 4.7 shows that the ray position $x_1$ and ray slope $\dot{x}_1$ at the incident position are related to $x_2$ and $\dot{x}_2$ by the same matrix representation; that is,

$$\begin{bmatrix} jk\dot{x}_1 \\ x_1 \end{bmatrix} = \begin{bmatrix} A & B \\ C & D \end{bmatrix} \begin{bmatrix} jk\dot{x}_2 \\ x_2 \end{bmatrix} \tag{4.53}$$

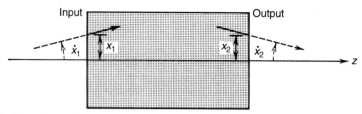

**Figure 4.7** Transformation of light beams. (From K. Iga, Y. Kokubun, and M. Oikawa, *Fundamentals of Microoptics*; copyright 1984, Ohmsha, Ltd., Tokyo, reprinted with permission.)

The propagation constant $k$ is included in the equation above to make it possible to treat a tandem connection of optical components having different refractive indices.

## 4.4    FABRY–PEROT OPEN RESONATORS WITH CONCAVE MIRRORS

### 4.4.1    Spot Size

Figure 4.8 shows some kinds of Fabry–Perot resonators using various reflecting mirrors. Among these resonators, a so-called unstable resonator which utilizes a convex mirror as shown in **(c)** is very special, that is, the diffraction loss is very high since any stable resonant mode cannot exist. This type of cavity is employed in a relatively high gain laser such as a solid-state YAG laser, where the optical field spreads in an entire volume of the resonator and a higher-gain portion near the periphery can be effectively utilized. This kind of resonator was introduced into a ruby laser and YAG laser later.

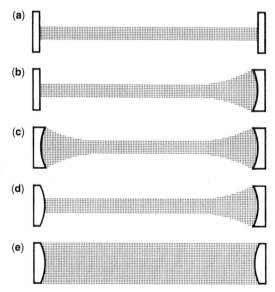

**Figure 4.8** Fabry–Perot resonator. (From *Laser Handbook*, originally published in Japan by Ohmsha Ltd., Tokyo, 1982; reprinted with permission.) The type **(e)** is an unstable resonator for enlarging the beam width [19], [20].

Among them a resonator with a plane and concave mirrors **(b)** is most widely used for gas or solid-state lasers. Taking it as an example, let us now discuss a resonant mode.

In Fig. 4.8b, a wavefront should be plane at $z = 0$. The wavefront coefficient $P_1$ is expressed as

$$P_1 = 1/s^2 \qquad (4.54)$$

Here $s$ is the spot size which is the radius of the Gaussian beam where the power is $1/e$ of the center axis. By referring to Eqs. (4.51) and (4.52), the total matrix $F$ starting from $z = 0$, reflecting by the concave mirror and then returning again to $z = 0$, can be expressed as follows:

$$
F = \begin{bmatrix} 1 & 0 \\ j\dfrac{z}{k_0} & 1 \end{bmatrix}
\begin{bmatrix} 1 & j\dfrac{k}{f} \\ 0 & 1 \end{bmatrix}
\begin{bmatrix} 1 & 0 \\ j\dfrac{z}{k_0} & 1 \end{bmatrix}
$$

$$
= \begin{bmatrix} 1 - z/f & j\dfrac{k}{f} \\ j\dfrac{z}{k}\left(2 - \dfrac{z}{f}\right) & 1 - \dfrac{z}{f} \end{bmatrix}
\equiv \begin{pmatrix} A & B \\ C & D \end{pmatrix} \qquad (4.55)
$$

According to the resonance condition, the wavefront coefficient must be the same value as that of the initial value at $z = 0$. So we must write

$$P_1 = (AP_1 + B)/(CP_1 + D)$$

Then

$$P_1 = \sqrt{B/C} \qquad (4.56)$$

Accordingly,

$$\frac{1}{s^4} = \frac{k^2/fz}{2 - z/f}$$

Therefore

$$s^4 = \frac{fz}{k^2}\left(2 - \frac{z}{f}\right) \qquad (4.57)$$

### 4.4.2  Stability of Resonators

Let us consider the stability of the optical resonator in terms of how well we can confine light by the structure of a pair of concave mirrors in Fig. 4.8c by referring to the stability diagram shown in Fig. 4.10. Let us denote the radius of curvature of the left and right mirrors as $b_1$ and $b_2$, respectively, and the cavity length as $L$, as seen from Fig. 4.9. Then we can judge the stability of the resonator according to the stability diagram in Fig. 4.10. This can easily be calculated by the matrix $F$ mentioned above. For simplification, let $b_1 = b_2 = 2f$ ($f$: focal length) and find the spot size in the center of the resonator and on the reflecting mirror. The matrix $F$ against the wave starting from the center of the resonator and returning again after being reflected by

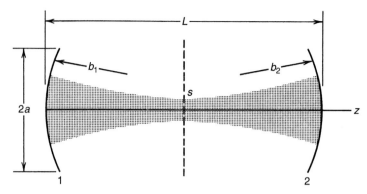

**Figure 4.9** Fabry–Perot resonator with a pair of concave mirrors. (From *Laser Handbook*, originally published in Japan by Ohmsha Ltd., Tokyo, 1982; reprinted with permission.)

the reflecting mirror 2 is expressed with the help of factors of free space and concave mirrors of Table 4.3 as

$$F = \begin{bmatrix} 1 & 0 \\ j\dfrac{L}{2k} & 1 \end{bmatrix} \begin{bmatrix} 1 & j\dfrac{k}{f} \\ 0 & 1 \end{bmatrix} \begin{bmatrix} 1 & 0 \\ j\dfrac{L}{2k} & 1 \end{bmatrix}$$

$$= \begin{bmatrix} 1 - \dfrac{L}{2f} & j\dfrac{k}{f} \\ j\dfrac{L}{2k}\left(2 - \dfrac{L}{2f}\right) & 1 - \dfrac{L}{2f} \end{bmatrix} \equiv \begin{pmatrix} A & B \\ C & D \end{pmatrix} \tag{4.58}$$

As a phase of the wave that returns after being reflected is plane and equal to the initial spot size, the wavefront coefficient $P$ can be written as

$$P = \frac{AP + B}{CP + D} \tag{4.59}$$

When $P = 1/s^2$, $A = D$, and $P = B/C$, $s^2$ can be expressed as

$$s^2 = \frac{1}{k}\sqrt{\frac{fL}{2}\left(2 - \frac{L}{2f}\right)} \tag{4.60}$$

where $L = 2f$ is called a confocal condition and $s$ obtains the maximum value $s_0 = \sqrt{f/k}$. Therefore, the following equation is obtained.

$$\frac{s}{s_0} = \left[\frac{L}{2f}\left(2 - \frac{L}{2f}\right)\right]^{1/4} \tag{4.61}$$

The spot size on the reflecting mirror is obtained by substituting $z = L/2$ for Eq. (4.61). Figure 4.11 shows $s/s_0$ and $w/s_0$ against $L/2f$.

$$\frac{w}{s_0} = \left[\frac{4(L/2f)}{(2 - L/2f)}\right]^{1/4} \tag{4.62}$$

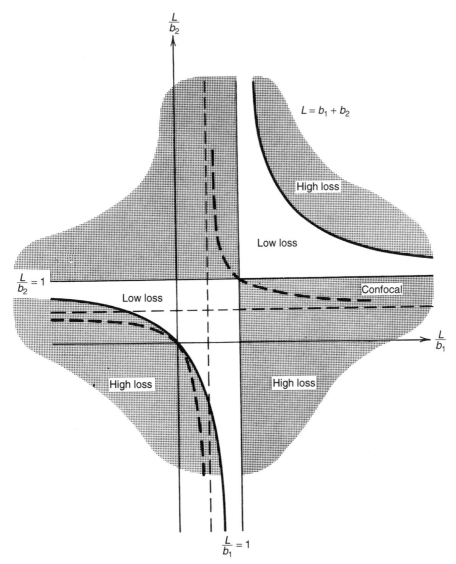

**Figure 4.10**  Stability diagram. (From G. D. Boyd and H. Kogelnik, *Bell System Technical Journal*, **41**, March 1962.)

### 4.4.3  Mode and Diffraction Loss in Fabry–Perot Resonators

When a spot size on a reflecting mirror is quite a bit smaller than a radius of the reflecting mirror, what we have discussed so far holds true and the leakage of light energy from the reflecting mirror (i.e., a diffraction loss) can be ignored. On the other hand, when the size of the reflecting mirror becomes comparable to the spot size, the diffraction loss and the distribution of the resonant mode may be different from the large mirror configuration.

To clarify such an effect, let us consider the mode in the case where the size of the reflecting mirror is finite, as shown in Fig. 4.12.

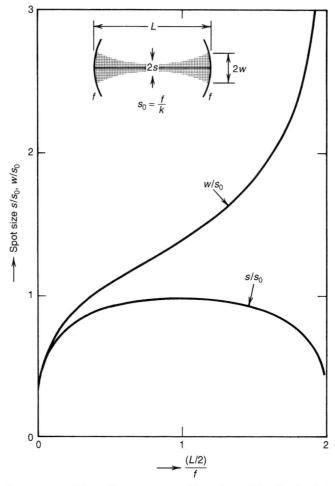

**Figure 4.11**  Spot size in a Fabry–Perot resonator. (From *Laser Handbook*, originally published in Japan by Ohmsha Ltd., Tokyo, 1982; reprinted with permission.)

***Analysis by Cartesian Coordinates.***  The field distribution on the reflecting mirror $M_1$ is expressed as follows:

$$f_1(x, y) = \phi(x)e^{+j(k/2R)x^2}\psi(y)e^{+j(k/2R)y^2} \tag{4.63}$$

Then the light that reaches the reflecting mirror $M_2$ after propagating in the resonator is expressed as

$$f_2(x, y) = \frac{j}{\lambda L}e^{-jkL} \iint f_1(x', y')e^{-(jk/2L)[(x-x')^2+(y-y')^2]}\,dx'\,dy' \tag{4.64}$$

Since $f_1$ and $f_2$ are resonant modes, we must write

$$f_2(x, y) = \gamma^2 f_1^*(x, y)e^{-jkL} \tag{4.65}$$

where $\gamma^2 = \gamma_x\gamma_y$ is a complex number.

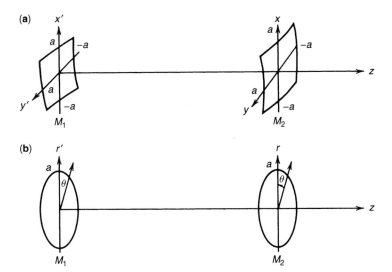

**Figure 4.12**  Fabry–Perot resonator models: **(a)** Cartesian coordinate; **(b)** cylindrical coordinate.

From the equations above,

$$\phi(x)\psi(y) = \gamma^{-2}\frac{j}{\lambda L} \iint \phi(x')\psi(y')e^{+j(k/L)(xx'+yy')} \, dx' \, dy' \tag{4.66}$$

As Eq. (4.66) can be separated into two independent equations with respect to $x$ and $y$, the following integral equation can be obtained.

$$\phi(x) = \gamma_x^{-1}\sqrt{j/\lambda L} \int \phi(x') \exp\left(j\frac{k}{L}xx'\right) dx' \tag{4.67}$$

Regarding $y$, the following equation can also be obtained:

$$\sqrt{j/\lambda L}\sqrt{L/k} = \sqrt{j/2\pi} \tag{4.68}$$

Here, with the help of normalization by $\sqrt{L/k}$, we can express the integral equation by rather simpler variables $\xi = x/\sqrt{L/k}$ and $\xi' = x'/\sqrt{L/k}$.

$$\phi(\xi) = C \int_{-\sqrt{2\pi N}}^{\sqrt{2\pi N}} \phi(\xi') \exp(j\xi\xi') \tag{4.69}$$

$$C = \gamma_x^{-1}\sqrt{j/2\pi} \tag{4.70}$$

$$N = a^2/\lambda L \tag{4.71}$$

In the case $N \gg 1$, we may consider the lower and upper bounds to be $-\infty$ and $\infty$, respectively. Then we can write

$$\phi(\xi) = C \int_{-\infty}^{\infty} \phi(\xi') \exp(j\xi\xi') \, d\xi' \tag{4.72}$$

If $\phi(\xi)$ is an even function, the following equation can be obtained:

$$\phi^{(E)}(\xi) = 2C^{(E)} \int_0^\infty \phi^{(E)}(\xi') \cos(\xi\xi') \, d\xi' \tag{4.73}$$

Let us now solve the integral equation (4.73) by using a formula on Hermite polynomials.

$$\int_0^\infty \exp(-x^2) \cos 2\beta x \, H_{2n}(\alpha x) \, dx = \frac{1}{2}\sqrt{\pi}(1-\alpha^2)^n \exp(-\beta^2) H_{2n} \frac{\alpha\beta}{\sqrt{\alpha^2-1}} \tag{4.74}$$

If we put $\alpha = \sqrt{2}$, $\sqrt{2}x = \xi'$ and $\sqrt{2}\beta = \xi$, we must write

$$\int_0^\infty \exp\left(-\tfrac{1}{2}\xi'^2\right) H_{2n}(\xi) \cos(\xi\xi') d\xi' = \sqrt{\pi/2}\, j^{2n} \exp\left(-\tfrac{1}{2}\xi^2\right) H_{2n}(\xi) \tag{4.75}$$

Accordingly, the solution is

$$\psi(\xi) = e^{-(1/2)\xi^2} H_{2n}(\xi) \tag{4.76}$$

$$\frac{1}{2c} = \frac{\sqrt{2}}{2}\sqrt{\pi}\, j^{2n} \tag{4.77}$$

$$\gamma_x^{-1} = \frac{1}{c}\sqrt{\frac{j}{2\pi}} = \sqrt{\pi}(j)^{2n}\sqrt{\frac{j}{2\pi}}$$

$$= j^{2n}\sqrt{j}$$

$$= e^{j(\pi/4+n\pi)}$$

$$= e^{j\pi(n+1/4)} \tag{4.78}$$

As a normal mode, the following equation can be obtained:

$$\phi(E)(\xi) = \frac{1}{[2^{2n}(2n)!\sqrt{\pi}]^{1/2}} e^{-(1/2)\xi^2} H_{2n}(\xi) \tag{4.79}$$

$$\gamma_x = e^{j\pi(n+1/4)} \tag{4.80}$$

where $\phi(E)(\xi)$ is normalized.

Similarly,

$$\psi(\eta) = \frac{1}{[2^{2m}(2m)!\sqrt{\pi}]^{1/2}} e^{-(1/2)\eta^2} H_{2m}(\eta) \tag{4.81}$$

$$\eta = y/\sqrt{L/k} \tag{4.82}$$

$$\gamma_y = e^{j\pi(m+1/4)} \tag{4.83}$$

$$\gamma^2 = \gamma_x\gamma_y = e^{j\pi(n+m+1/2)} = e^{j(\pi/2)(2n+2m+1)} \tag{4.84}$$

The spot size $w$ is expressed as

$$w = \sqrt{L/k} = \sqrt{2f/k} \tag{4.85}$$

In the case of an odd function, the following equation may be used as well:

$$\int_0^\infty \exp\left(-\tfrac{1}{2}\xi^2\right) H_{2n-1}(\xi) \sin \xi\xi' \, d\xi' = \sqrt{\frac{\pi}{2}}(-1)^{2n} \exp\left(-\tfrac{1}{2}\xi^2\right) H_{2n-1}(\xi) \qquad (4.86)$$

Accordingly, the eigenvalue of the integral equation is given by

$$\gamma = e^{j(\pi/2)(p+q+1)} \qquad (4.87)$$

The form of the Hermite–Gaussian mode function, including the distribution in the resonator, is shown below.

Hermite–Gaussian mode: $HG_{pq}$

$$\psi_{pq}(x, y, z) = N_{pq} \frac{s}{w} H_p\left(\frac{x}{w}\right) H_q\left(\frac{y}{w}\right)$$

$$\cdot \exp\left[-\frac{1}{2} + \left(\frac{1}{w^2} + j\frac{k}{R}\right)(x^2 + y^2)\right] e^{-jkz + j(p+q+1)\phi} \qquad (4.88)$$

$$N_{pq} = \left(\frac{1}{2^p\, p!\, 2^q\, q!\, \pi}\right)^{1/2} \qquad (4.89)$$

where $H_p(\cdot)$ is a Hermite polynomial of $p$th order and the spot size $s$ at the center of the resonator can be obtained from Eq. (4.61). With the help of calculated $s$, the spot size $w$ at an arbitrary distance, the radius of curvature of the wavefront $R$, and phase shift $\phi$ can be calculated. In this case, however, $z$ is zero at the center of the resonator. Some mode patterns are illustrated in Fig. 4.13.

Next, let us consider the case where a size of a reflecting mirror cannot be ignored compared with a spot size of a beam. In this case an integral equation taking account of the upper and lower bounds of the integral must be solved. But the eigenvalue $\gamma$ can be obtained approximately using a variational method; that is, by multiplying both sides of the equation by $\phi(x)$, $\gamma_x$ can be obtained as follows:

$$\gamma_x = \sqrt{\frac{j}{2\pi}} \frac{\displaystyle\int_{-\sqrt{2\pi N}}^{\sqrt{2\pi N}} \int_{-\sqrt{2\pi N}}^{\sqrt{2\pi N}} \phi(x)\phi(x') \exp(j\xi\xi') \, d\xi \, d\xi'}{\displaystyle\int_{-\sqrt{2\pi N}}^{\sqrt{2\pi N}} \phi^2(x) \, dx} \qquad (4.90)$$

All we have to do is to carry out the integral by substituting some suitable trial function. In general, by substitution of a mode function for $N \to \infty$ we obtain an analytic formula.

***Analysis by Cylindrical Coordinates.***   To obtain the eigenmode in the cylindrical coordinate, let us limit the size of the reflecting mirror in Fig. 4.9 to a circle with a diameter of $2a$. The electric field distribution on the mirror $M_1$ is taken as

$$\psi_1(r, \theta) = S_m(r) e^{+j(k/f)r^2} e^{-jm\theta} \qquad (4.91)$$

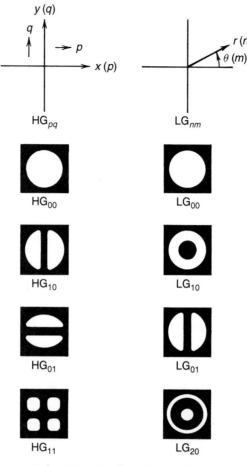

HG$_{pq}$: Hermite–Gaussian modes
LG$_{nm}$: Laguerre–Gaussian modes

**Figure 4.13**  Hermite–Gaussian modes. (From *Laser Handbook*, originally published in Japan by Ohmsha Ltd., Tokyo, 1982; reprinted with permission.)

Next the light that is reflected by the mirror $M_2$ after propagating in the resonator is expressed as

$$\psi_2(r, \theta) = \frac{j}{\lambda L} e^{-jkL} \iint r'\, dr'\, d\theta'\, \psi_1(r', \theta')$$

$$\cdot \exp\left\{ -\frac{jk}{2L}[r'^2 - 2rr'\cos(\theta' - \theta) + r^2] \right\} \qquad (4.92)$$

For $\psi_1$ and $\psi_2$ to be resonant modes, $\gamma^2$ must be the complex constant and the following can be obtained:

$$\psi_2(r, \theta) = \gamma^2 \psi_1^*(r, \theta) e^{-jkL} \qquad (4.93)$$

Then

$$S_m(r)e^{+jm\theta} = \gamma^{-2}\frac{j}{\lambda L}\iint r'\,dr'\,d\theta'\,S_m(r')e^{-jm\theta}$$

$$\cdot \exp\left[+j\frac{k}{L}rr'\cos(\theta - \theta')\right] \tag{4.94}$$

where $2f = L$ so as to simplify the equation and as a confocal condition under which a diffraction loss becomes smallest. When we integrate with respect to $\theta'$, we obtain the following integral equation with a symmetric kernel [3]:

$$S_m(r)\sqrt{r} = \gamma_m^{-2}\int_0^\alpha \sqrt{r'}\,dr'\,K_m(r, r')S_m(r') \tag{4.95}$$

$$K(r, r') = j^{n+1}\frac{k}{L}J_m\left(k\frac{rr'}{d}\right)\sqrt{rr'} \tag{4.96}$$

When $S_{nm}(r)$ is the eigenfunction and $\gamma_{nm}^2$ the eigenvalue of the integral equation, $n(= 0, 1, 2, \ldots)$ denotes the mode order in the radius direction. From the definition of the constant $\gamma$, the diffraction loss $\alpha_{nm}$ and the phase shift $\phi_{nm}$ are given as

$$\alpha_{nm} = 1 - |\gamma_{nm}^2|^2 \tag{4.97}$$

$$\phi_{nm} = \arg(\gamma_{nm}^2) \tag{4.98}$$

Fox and Li [3] found an eigenfunction and eigenmode by solving the integral equation using an iterative integration. The mode in the Fabry–Perot resonator is approximately considered a transverse electromagnetic (TEM) wave, and expressed as TEM$_{nm}$ using the mode numbers $n$ and $m$. The diffraction loss and phase shift obtained by Fox and Li are shown in Figs. 4.14 and 4.15, respectively. The abscissa represents a Fresnel number defined as

$$N = a^2/L\lambda \tag{4.99}$$

and in this case the radius of a mirror $a$ is a normalized parameter to show the size of light source in the transverse direction.

When the radius of the reflecting mirror $a$ is considerably larger than the spot size of the resonant mode, the solution of the integral equation Eq. (4.92) is analytically obtained. The solution is expressed with a Laguerre–Gaussian function in a cylindrical coordinate expression and Hermite–Gaussian function in a Cartesian coordinate [4].

The mode function in the resonator in the case where $a \to \infty$ is shown as

Laguerre–Gaussian mode: $LG_{nm}$

$$\psi_{nm}(r, \theta, z) = N_{nm}\frac{s}{w}\left(\frac{r}{w}\right)^m L_n^m\left(\frac{r^2}{w^2}\right)$$

$$\cdot \begin{pmatrix}\cos m\theta \\ \sin m\theta\end{pmatrix}\exp\left[-\frac{1}{2}\left(\frac{1}{w^2} + j\frac{k}{R}\right)r^2\right]e^{-jkz+j(2n-m+1)\phi} \tag{4.100}$$

$$N_{nm} = \left[\frac{(n-m)!}{(n!)^3\pi}\right]^{1/2} \tag{4.101}$$

where $L_n^m(\cdot)$ is the associated Laguerre polynomial.

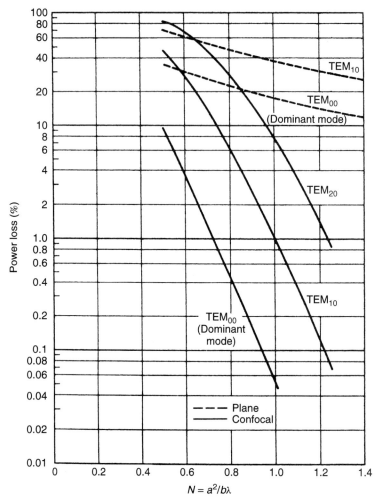

**Figure 4.14** Diffraction loss/propagation. Solid line, circular plane mirrors; dashed line, confocal spherical mirrors. (From A. G. Fox and T. Li, *Bell System Technical Journal*, **40**, 1961.)

### 4.4.4 Resonant Frequencies

Let us find a resonant frequency in the Fabry–Perot resonator. When the resonator consists of a plane mirror and a concave mirror as shown in Fig. 4.8b, the Gaussian beam traveling in the positive $z$ direction is written as

$$\psi^+(r, z) = E_0 e^{-jkz}(s/w) \exp\left(-\tfrac{1}{2}pr^2 + j\phi\right) \tag{4.102}$$

where $s$ is the spot size on the plane mirror and $w$, $P$, and $\phi$ are given from Eqs. (4.34), (4.36), and (4.37), respectively. While the conjugate wave $\psi^-$ traveling in the negative $z$ direction is given as

$$\psi^-(r, z) = E_0 e^{+jkz}(s/w) \exp\left(-\tfrac{1}{2}Pr^2 - j\phi\right) \tag{4.103}$$

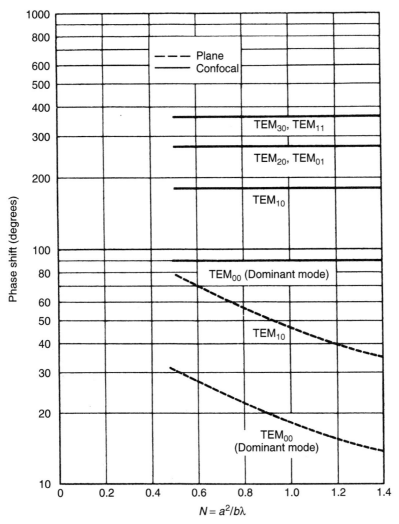

**Figure 4.15**   Phase shift in the Fabry–Perot resonator. Solid line, circular plane mirrors; dashed line, confocal spherical mirrors. (From A. G. Fox and T. Li, *Bell System Technical Journal*, **40**, 1961.)

the wave in the resonator is written similarly to Eq. (4.16) as

$$\psi(r, z) = E_0(A^+\psi^+ + A^-\psi^-) \tag{4.104}$$

When $z = 0$, from the condition $\psi = 0$, $A^-$ becomes $-A^+$. Also, when $z = L$, from $\psi = 0$.

$$\sin(kL - \phi) = n\pi \quad n = 1, 2, 3, \ldots \tag{4.105}$$

From Eq. (4.105),

$$kL - \tan^{-1}(L/ks^2) = n\pi \tag{4.106}$$

Since $s$ is determined from the spatial resonance condition, the following equation is obtained:

$$f_n = \frac{c}{2L} \left( n + \frac{1}{\pi} \tan^{-1} \sqrt{\frac{L/f}{2 - L/f}} \right) \tag{4.107}$$

where the number $n$ is a mode number of the longitudinal mode similar to the preceding section, and corresponds to $p = 0$ and $q = 0$ in the transverse mode. When we develop the argument to the Hermite–Gaussian mode of the $p$th order or $q$th order, obtain an equation similar to Eq. (4.107), and find the resonance condition, we can obtain an equation similar to Eq. (4.22).

$$f(n, p, q) = \frac{c}{2L} \left[ n + \left( p + \frac{1}{2} \right) \frac{1}{\pi} \tan^{-1} \sqrt{\frac{L/f}{2 - L/f}} \right.$$

$$\left. + \left( q + \frac{1}{2} \right) \frac{1}{\pi} \tan^{-1} \sqrt{\frac{L/f}{2 - L/f}} \right] \tag{4.108}$$

From (4.108) the longitudinal mode spacing $\Delta f$ is readily found to be

$$\Delta f = c/2L \tag{4.109}$$

When $L/2f = 1$, $\tan^{-1} \sqrt{L/(2f)/[2 - L/(2f)]}$ becomes $\pi/4$, and therefore the expression in parentheses on the right side of Eq. (4.108) becomes $n + (p + q + 1)/4$, thus causing a different longitudinal mode and degeneracy according to the combination of $p$th- and $q$th-order modes. The resonator with such a confocal structure ($L = 2f$) becomes extremely unstable and may cause a hopping between higher-order transverse modes. This sort of instability can be eliminated by off setting the cavity length $L$ to be a little bit different from $2f$.

## 4.5    DISTRIBUTED FEEDBACK/REFLECTOR RESONATORS

### 4.5.1    Resonant Frequencies

Let us discuss the distributed feedback (DFB) resonator in which a grating with the periodicity $\Lambda$ is provided on or near the surface of a waveguide as shown in Fig. 4.16.

Kaminow attempted the resonator with such a periodic structure [23], and Kogelnik and Shank [14] succeeded in the laser oscillation by means of optically pumped organic material waveguide including dyes. After that, Nakamura and co-workers [15] have realized the resonator using semiconductor materials. This resonator features a strong reflection to which the phase of a scattered wave is added in the case where the periodicity $\Lambda$ is integral multiple of $\lambda_g/2$, where the wavelength in the guide is $\lambda_g$.

It also features lasers with the single frequency because, on the contrary, the resonance wavelength depends on the periodicity $\Lambda$. From the reasons stated above, the resonance condition becomes

$$\lambda_0 \simeq 2\Lambda n_{eq}/N \quad N = 1, 2, 3, \ldots \tag{4.110}$$

where $n_{eq}$ is the equivalent refractive index and $\lambda_g = \lambda_0/n_{eq}$.

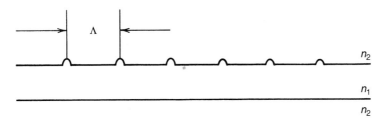

**Figure 4.16**  Distributed feedback (DFB) resonator.

By differentiating Eq. (4.110), the mode spacing $\Delta\lambda$ is written

$$\Delta\lambda = \frac{\partial\lambda_0}{\partial N}\Delta N + \frac{\partial\lambda_0}{\partial\lambda}\Delta\lambda$$

$$= -\frac{\lambda}{N}\Delta N + \frac{\lambda_N}{n_{\text{eq}}}\frac{\partial n_{\text{eq}}}{\partial\lambda}\Delta\lambda$$

Therefore,

$$|\Delta\lambda| = \frac{\lambda_0/N}{(1 - \lambda_N/n_{\text{eq}})/\partial n_{\text{eq}}/\partial\lambda} = \frac{\lambda_N}{n_{\text{eff}}}\frac{1}{N}$$

$$= \frac{\lambda_0^2}{2n_{\text{eff}}\Lambda} \tag{4.111}$$

## 4.5.2  Scattered Waves

A diffraction grating can be considered to be inhomogeneous in a waveguide, so a light wave is scattered. If the inhomogeneity, however, is periodic, the phases of scattered waves become even and are strongly scattered (in this case, diffracted) in the specific direction. Let us calculate the angle of these scattered waves. Figure 4.17 shows the diffraction of the incident light from the left bottom at the angle $\theta_i$ upon a guidewave-type diffraction grating. The light is scattered in various directions at the convex part of the diffraction grating. Now the angle is expressed as $\theta_m$ and from Fig. 4.17, the phase of the wavefront is constant between $AA'$, where the wavefront is scattered, and $BB'$, where it traverses.

$$\overline{A'B} - \overline{AB'} = \pm m\lambda_g \quad m = 0, 1, 2, \ldots \tag{4.112}$$

As $\lambda_g$ is the wavelength in the medium with the refractive index $n_2$, we can write

$$\lambda_g = \lambda_0/n_2 \tag{4.113}$$

where $\lambda_0$ is the wavelength in a vacuum. We can also write

$$\overline{A'B} = \Lambda\cos\theta_i \tag{4.114}$$

$$\overline{AB'} = \Lambda\cos\theta_m \tag{4.115}$$

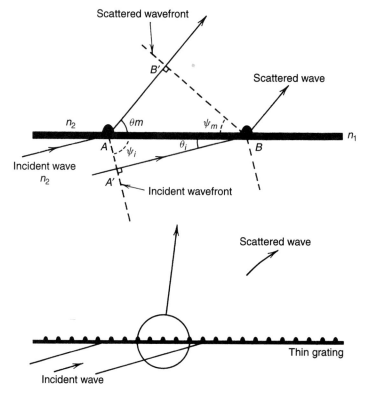

**Figure 4.17**   Scattering of waves from a waveguide grating.

From Eq. (4.110)

$$\Lambda = \lambda_0/(2n_{eq})N \tag{4.116}$$

Accordingly, Eq. (4.112) becomes

$$\cos\theta_m = \cos\theta_i \mp (2\ m/N)(n_{eq}/n_2) \tag{4.117}$$

When $\theta_i = 0$ and $n_{eq} \simeq n_2$,

$$\cos\theta_m = 1 \mp 2\ m/N \tag{4.118}$$

Table 4.4 shows the grating order $N$ and the diffraction angle to the diffraction order $m$. Pay attention to the fact that the light is scattered in both the waveguide and perpendicular directions.

### 4.5.3   Stopbands

We showed in Eq. (4.110) that the resonant frequency depends on its periodic structure. It is known, however, from a strict calculation that the Bragg wavelength does not resonate, but rather locates in the center of a stopband [24]. Let us consider the stopband in this section. Figure 4.18 shows a waveguide with a diffraction grating. For simplification, we assume that the equivalent refractive index $n$ of the waveguide

**TABLE 4.4  Light Scattering from a Grating**

| $N$ | $\cos\theta_m$ | $m$ | $\cos\theta$ | $\theta$ (deg) | Direction |
|---|---|---|---|---|---|
| 1 | $1 \mp 2m$ | 0 | 1 | 0 | •——→ |
|   |   | 1 | $-1$ | 180 | ←——• |
| 2 | $1 \mp m$ | 0 | 1 | 0 | •——→ |
|   |   | 1 | 0 | $\pm 90$ | ↕ |
|   |   | 2 | $-1$ | 180 | ←——• |
| 3 | $1 \mp 2m/3$ | 0 | 1 | 0 | •——→ |
|   |   | 1 | $\frac{1}{3}$ | $\pm 70.5$ | ⟨ |
|   |   | 2 | $-\frac{1}{3}$ | $\pm 109.5$ | ⟩ |
|   |   | 3 | $-1$ | 180 | ←——• |

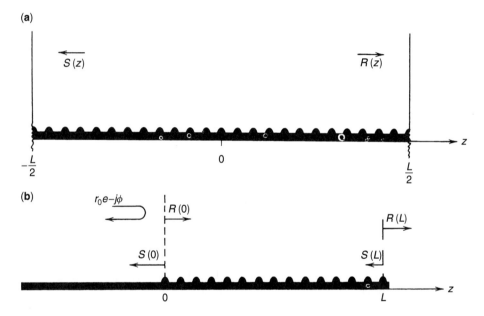

**Figure 4.18**  (a) DFB; (b) DBR.

changes periodically as the following equation. It can be easily understood that difference of a waveguide in thickness is expressed in terms of the change of the equivalent refraction index.

$$n = n_0 + \Delta n \cos(2\pi z/\Lambda) \tag{4.119}$$

where $\Lambda$ is the frequency and $\Delta n$ the change of refractive index. The coupling coefficient $\kappa$ is defined as

$$\kappa = \pi\,\Delta n/\lambda_0 = k_0\Delta n/2 \tag{4.120}$$

We also assume that the waveguide has a net gain [or loss ($<0$)]. The electric field $E(z)$ in the waveguide satisfies the following wave equation:

$$\partial^2 E/\partial z^2 + k^2 E = 0 \tag{4.121}$$

where

$$k^2 = k_0^2 n^2 + jk_0 ng \tag{4.122}$$

Now $\beta_0 = \pi/\Lambda$, and then $E(z)$ is written as the sum of a forward wave and a backward wave as follows:

$$E(z) = R(z)e^{-j\beta_0 z} + S(z)e^{j\beta_0 z} \tag{4.123}$$

By substituting Eq. (4.123) for Eq. (4.121) and multiplying by $e^{j\beta_0 z}$ or $e^{-j\beta_0 z}$, the following coupling equation can be obtained:

$$-\frac{dR}{dz} + (g/2 - j\delta)R = j\kappa S \tag{4.124a}$$

$$\frac{dS}{dz} + (g/2 - j\delta)S = j\kappa R \tag{4.124b}$$

Then

$$\delta = \beta - \beta_0 = k_0 n_0 - \pi/\Lambda \tag{4.125}$$

where $\delta$ is the offset of the frequency from the Bragg frequency, and $d^2 R/dz^2$ and $d^2 S/dz^2$ are ignored. Then the solutions of the coupling equation (4.124) are written as

$$R = r_1 e^{\gamma z} + r_2 e^{-\gamma z} \tag{4.126a}$$

$$S = s_1 e^{\gamma z} + s_2 e^{-\gamma z} \tag{4.126b}$$

By substituting Eq. (4.126) for Eq. (4.124), the following equation can be obtained:

$$\gamma^2 = \kappa^2 + (g - j\delta)^2 \tag{4.127}$$

Assuming that the reflection at the both ends of the boundary condition of DFB is ignored, we can write

$$R(-L/2) = 0 \tag{4.128a}$$

$$S(L/2) = 0 \tag{4.128b}$$

Then

$$\gamma_1 = \pm S_2, \gamma_2 = \pm S_1 \tag{4.129}$$

$$\gamma_1/\gamma_2 = S_2/S_1 = -e^{\gamma L} \tag{4.130}$$

Therefore, $R$ and $S$ are written as

$$R(z) = \sinh \gamma (z + L/2) \tag{4.131a}$$

$$S(z) = \pm \sinh \gamma (z - L/2) \tag{4.131b}$$

By substituting Eqs. (4.131a) and (4.131b) for Eqs. (4.124a) and (4.124b) and re-arranging them, the following equations are obtained:

$$\gamma + (g - j\delta) = \pm j\kappa e^{\gamma L} \qquad (4.132a)$$

$$\gamma - (g - j\delta) = \mp j\kappa e^{-\gamma L} \qquad (4.132b)$$

From the sum and the subtraction of the equations above, we can write

$$\kappa = \pm j\gamma / \sinh \gamma L \qquad (4.133)$$

$$g - j\delta = \gamma \coth \gamma L \qquad (4.134)$$

The values of $\alpha$ and $\delta$ can be obtained by finding the $\gamma$ which satisfies the condition that $\kappa$ in the eigenvalue equation is a real number and by substituting $\gamma$ for Eq. (4.134).

Figure 4.18 shows the eigenvalues $gL$ to $SL$. As seen from this figure, $\delta = 0$ (i.e., the vicinity of the Bragg frequency becomes a stopband).

### 4.5.4   Distributed Bragg Reflector-Type Resonators

The distributed Bragg reflector (DBR)-type resonator shown in Fig. 4.18**b** has a structure that a DBF region with no gain is provided on both sides or one side of the active layer. Now we may only discuss the resonance conditions of a Fabry–Perot resonator taking reflectivity of the DBR and phase shift into consideration. The DBR features the independent design of an active part and DBR part as well as connection with an output terminal at the end of the DBR. In the DFB structure the reflection at the end was ignored, as seen from the previous discussion, but when the reflection is to be considered, some influence will be given on the output side.

Assuming that the reflectivity of the right side at $z = 0$ in Fig. 4.18**b** is

$$r = r_0 e^{-j\phi} \qquad (4.135)$$

let us obtain $r_0$ and $\phi$ when we see the right from $z = 0$. Now the reflection at the right end is ignored, that is,

$$S(L) = 0$$

An wave equation of DBR becomes equal to that of DFB discussed in Section 4.5.3. However, in this case the frequency offset $\delta$ of light is previously given, and then the reflectivity $r_0$ and phase shift $\phi$ are obtained.

The complex reflectivity $r$ can be obtained by using the boundary conditions as

$$r = \frac{S(0)}{R(0)} = \frac{-j\kappa \tanh \gamma L}{\gamma + j\delta \tanh \gamma L} \qquad (4.136)$$

where the gain $g = 0$ and $\gamma$ is obtained as follows:

$$\gamma^2 = \kappa^2 - \delta^2 \qquad (4.137)$$

In the region of $\delta < \kappa$, $\gamma$ is a real number and $r_0$ and $\phi$ are written as

$$r_0 = \frac{\kappa \tanh \sqrt{\kappa^2 - \delta^2}L}{\sqrt{\kappa^2 - \delta^2} + \delta^2 \tanh \sqrt{\kappa^2 - \delta^2}L} \tag{4.138}$$

$$\phi = \frac{\pi}{2} + \tan^{-1}\left(\frac{\delta}{\sqrt{\kappa^2 - \delta^2}} \tanh \sqrt{\kappa^2 - \delta^2}L\right) \tag{4.139}$$

When $r_0$ and $\phi$ are written by using the Bragg frequency ($\delta = 0$), we have

$$r_0(0) = \tanh \kappa L \tag{4.140}$$

$$\phi(0) = \pi/2 \tag{4.141}$$

For example, when $\kappa L = 2$, $|r_0|^2 = 0.92$.

### 4.5.5 $\lambda_B/4$ Phase Shift

As is known from the preceding section, the phase shift of the right half of the DBR at the center of the Bragg frequency is $\pi/2$. If there is the similar DBR symmetrically at the left half, it becomes DFB; thus the phase shift at the left of the DBR is $\pi/2$. Then the phase shift becomes $\pi$ in total to cause a stopband. When the region with the phase shift $\psi$ is inserted as illustrated in Fig. 4.19, the sum of the phase shift of one round becomes $2\psi$. Consequently, the phase shift in total at the Bragg frequency is written as

$$\underset{\text{(DBR 1)}}{2\psi +} \quad \underset{\text{(DBR 2)}}{\pi/2} \quad + \quad \pi/2 \quad = 2\pi q \quad q = 1, 2, 3, \ldots \tag{4.142}$$

The minimum phase shift can be obtained from $q = 1$ and the following equation is written:

$$\psi = \pi/2 \tag{4.143}$$

The phase shift converted into the length is equivalent to a quarter-wavelength and is called the $\lambda_B/4$ shifter. Equation (4.142) shows that the phase shift may utilize not only a $\lambda_B/4$ length spacing but also a longer region with a slightly different propagation constant.

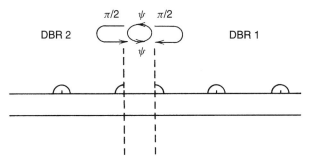

**Figure 4.19** Phase-shifted DFB.

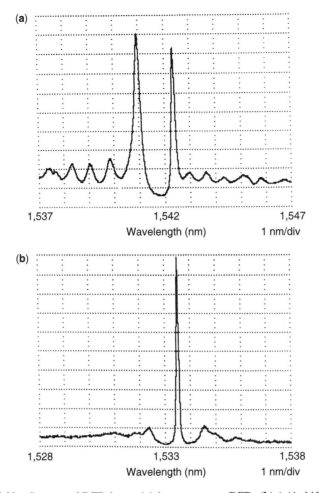

**Figure 4.20**   Spectra of DFB laser: **(a)** homogeneous DFB; **(b)** $\lambda/4$-shifted DFB.

Figure 4.20**b** shows the spectrum of the DFB laser with the phase shifter [25]. Just at the Bragg frequency $\lambda_B$, the laser oscillates in contrast to Fig. 4.20**a**, which is a spectrum of a conventional homogeneous DFB laser.

## 4.6   RESONATOR FOR VERTICAL CAVITY SURFACE EMITTING LASER

In the vertical cavity surface emitting laser (VCSEL) [26], a couple of multilayer Bragg reflectors are used in its Fabry–Perot cavity formation. In Fig. 4.21, we show the model of the VCSEL resonator [27]. The cavity length can be multiple of half wavelength $\lambda/2n$, where $\lambda$ is laser wavelength and $n$ is the refractive index of medium. The active layer is placed to match the maximum point of standing wave that can maximize the overlap of the field and gain distribution.

The field distribution is depicted in Fig. 4.21. By using this configuration, we can make the laser volume $V_a$ two or three orders of magnitude smaller than stripe lasers.

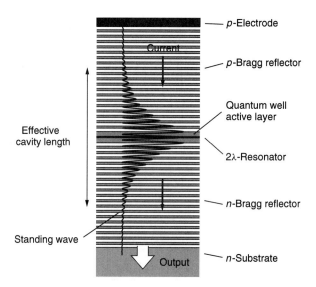

**Figure 4.21**   A model of resonator for vertical cavity surface emitting lasers.

This is advantageous to realize a high-performance semiconductor laser when we consider the scaling laws:

$$I_{th} \propto V_a \qquad (4.144)$$

$$f_r \propto \sqrt{\frac{1}{V_a}} \qquad (4.145)$$

Where $I_{th}$ is the threshold current and $f_r$ is the relaxation frequency.

## 4.7   SUMMARY

Various types of lasers have now been introduced not only into the optoelectronics area but also into the industrial and medical fields. The basic idea of an optical resonator was developed in the early 1960s, and a periodic structure was considered to be a resonator in 1970. After fundamental research and development of materials and fabrication technology, suitable optical resonators have been designed to meet the requirements of the laser of interest. The present chapter may be helpful for such purposes and to give an idea of the laser as well.

## REFERENCES

1. A. L. Schawlow and C. H. Townes, *Phys. Rev.*, **112**, p. 1958, 1940.
2. T. H. Maiman, *Nature (London)*, **187**, p. 493, 1960.
3. A. G. Fox and T. Li, *Bell Syst. Tech. J.*, **40**, p. 453, 1961.
4. G. D. Boyd and J. P. Gordon, *Bell Syst. Tech. J.*, **40**, p. 489, 1961.
5. Y. Suematsu and T. Ikegami, *J. Inst. Electron. Commun. Eng. Jpn.*, **49**, p. 1091, 1966.

6. M. B. Panish, I. Hayashi, and S. Sumski, *Appl. Phys. Lett.*, **16**, p. 326, 1970.

7. M. I. Nathan, W. P. Dumke, G. Burns, F. H. Dill, Jr., and G. Lasher, *Appl. Phys. Lett.*, **1**, p. 62, 1962.

8. T. M. Quist, R. H. Rediker, R. J. Keyes, W. E. Krag, B. Lax, A. L. McWorter, and H. J. Zeiger, *Appl. Phys. Lett.*, **1**, p. 91, 1962.

9. R. N. Hall, G. E. Fenner, J. D. Kingsley, T. J. Soltys, and R. O. Carlson, *Phys. Rev. Lett.*, **9**, p. 366, 1962.

10. A. S. Dobkin, O. N. Korekov, G. A. Lapitskaya, A. A. Pleskikon, O. N. Prozorov, L. A. Rivlin, G. A. Sukhareva, V. S. Shildyaev, and S. D. Yakubovich, *Sov. Phys.–Semicond. (Engl. Transl.)*, **4**, p. 515, 1970.

11. Y. Tarui, Y. Komiya, T. Sakamoto, H. Iida, and A. Shoji, *Jpn. J. Appl. Phys.*, **15**, Suppl., p. 293, 1976.

12. K. Iga, T. Kanbayashi, K. Wakao, and Y. Sakamoto, *Jpn. J. Appl. Phys.*, **18**, p. 2035, 1979.

13. K. Iga and B. I. Miller, *IEEE J. Quantum Electron.*, **QE-18**, p. 22, 1981.

14. H. Kogelnik and C. V. Shank, *Appl. Phys. Lett.*, **18**, p. 152, 1971.

15. M. Nakamura, A. Yariv, H. W. Yen, S. Somekh, and H. L. Garvin, *Appl. Phys. Lett.*, **22**, p. 515, 1973.

16. S. Wang, *IEEE J. Quantum Electron.*, **QE-10**, p. 413, 1970.

17. Y. Suematsu, M. Yamada, and K. Hayashi, *Proc. IEEE (Lett.)*, **63**, p. 208, 1975.

18. K. Iga, Y. Kokubun, and M. Oikawa, *Fundamentals of Microoptics*, Academic Press, New York, 1984.

19. Y. Suematsu and K. Iga, unpublished work, 1963.

20. A. E. Siegman, *Proc. IEEE*, **53**, p. 277, 1965.

21. Y. Suematsu and Y. Fukinuki, *J. Inst. Electron. Commun. Eng. Jpn.*, **48**, p. 1684, 1965.

22. G. D. Boyd and H. Kogelnik, *Bell Syst. Tech. J.*, **41**, p. 1347, 1962.

23. I. P. Kaminow, H. P. Weber, and E. A. Chandros, *Appl. Phys. Lett.*, **18**, p. 498, 1971.

24. H. Kogelnik and C. V. Shank, *J. Appl. Phys.*, **43**, p. 2327, 1972.

25. K. Utaka, S. Akiba, K. Sakai, and Y. Matsushima, *IEEE J. Quantum Electron.*, **QE-22**(7), p. 1042, July 1986.

26. K. Iga, F. Koyama, and S. Kinoshita, *IEEE J. Quant. Electron.*, **QE-24**, p. 1845, 1988.

27. K. Iga and F. Koyama, *Surface Emitting Laser*, Kyoritsu Pub. Co., Tokyo, 1999.

# 5

# SPATIAL FILTERS AND FOURIER OPTICS

FRANCIS T. S. YU
*Department of Electrical Engineering*
*The Pennsylvania State University*
*University Park, Pennsylvania*

## 5.1 INTRODUCTION

Advances in quantum electronics have brought into use the infrared and visible range of electromagnetic waves. The invention of intensive coherent light sources has permitted us to build more efficient optical systems for communication and signal processing. Most of the optical processing architectures to date have confined themselves to the cases of complete coherence or complete incoherence. However, a continuous transition between these two extremes is possible.

Recent advances in real-time spatial light modulators and electro-optic devices have brought optical signal processing to a new height. Much attention has been focused on high-speed and high-data-rate optical signal processing and computing.

In this chapter we discuss the basic principles of Fourier optics, matched spatial filters, and optical signal processing under the coherent, incoherent, and partially coherent regimes.

## 5.2 FRESNEL–KIRCHHOFF THEORY AND SYSTEM TRANSFORMS

Let us now begin the basic principle of Fresnel–Kirchhoff theory. To derive Kirchhoff's integral from a scalar wave theory is rather mathematically involved. However, it can be approached with ease using the simple *linear system* concept.

According to the Huygens principle, the complex amplitude observed at a point $p'$ of a coordinate system $\sigma(\alpha, \beta, \gamma)$, due to a monochromatic light field located in another

*Handbook of Optical Components and Engineering*,   Edited by Kai Chang
ISBN 0-471-39055-0   © 2003 John Wiley & Sons, Inc.

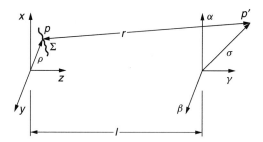

**Figure 5.1**   Coordinate system.

coordinate system $\rho(x, y, z)$, as shown in Fig. 5.1, may be calculated by assuming that each point of light source is an infinitesimal, spherical radiator. Thus the complex light amplitude $h_l(\rho; k)$ contributed by a point $p$ in the $\rho$ coordinate system can be considered to be that from an unpolarized monochromatic point source, such as

$$h_l = -\frac{i}{\lambda r} \exp[i(kr - \omega t)] \tag{5.1}$$

where $\lambda$, $k$, and $\omega$ are the wavelengths, wave number, and angular frequency, respectively, of the point source, and $r$ is the distance between the point source and the point of observation:

$$r = [(l + \gamma - z)^2 + (\alpha - x)^2 + (\beta - y)^2]^{1/2} \tag{5.2}$$

If the separation $l$ of the two coordinate systems is assumed to be large compared to the regions of interest in the $\rho$ and $\sigma$ coordinate system, $r$ may be approximated by $l$ in the denominator of Eq. (5.1) and by

$$r = l + \gamma - z + \frac{(\alpha - x)^2}{2l} + \frac{(\beta - y)^2}{2l} \tag{5.3}$$

known as *paraxial approximation* in the exponent. Equation (5.1) can be written as

$$h_l(\sigma - \rho; k) \simeq -\frac{i}{\lambda l} \exp\left\{ ik\left[ l + \gamma - z + \frac{(\alpha - x)^2}{2l} + \frac{(\beta - y)^2}{2l} \right] \right\} \tag{5.4}$$

where the time-dependent exponent has been dropped for convenience. Since Eq. (5.4) represents the free-space radiation from a monochromatic point source, it is known as free space or *spatial impulse response*. In other words, the complex amplitude produced at the $\sigma$ coordinate system by a monochromatic radiating surface located in the $\rho$ coordinate system can be written as

$$g(\sigma) = \iint_{\Sigma} f(\rho) h_l(\sigma - \rho; k) \, d\Sigma \tag{5.5}$$

where $f(\rho)$ is the complex light field of the monochromatic radiating surface, $\Sigma$ denotes the surface integral, and $d\Sigma$ is the incremental surface element. We note that

Eq. (5.5) is essential to the well-known Kirchhoff's integral derived from the scalar wave theory.

As an illustration, given a complex monochromatic radiating field $f(x, y)$, the complex light disturbances at $(\alpha, \beta)$ can be obtained by the following convolution integral:

$$g(\alpha, \beta) = f(x, y) * h_l(x, y) \tag{5.6}$$

where the asterisk denotes the convolution operation,

$$h_l(x, y) = C \exp\left(i\frac{k}{2l}\rho^2\right) \tag{5.7}$$

is the spatial impulse response between the spatial coordinate systems $(x, y)$ and $(\alpha, \beta)$, $C = -i/\lambda l \exp(ikl)$ a complex constant, and $\rho^2 = x^2 + y^2$. Consequently, Eq. (5.6) can be represented by a block box system diagram as shown in Fig. 5.2. In other words, the complex wave field distributed over the $\sigma$ coordinate system can be evaluated by a two-dimensional *convolution integral* of Eq. (5.5). The output response is equal to the input excitation convolves with the spatial impulse response.

On the other hand, if the complex light disturbances at $(\alpha, \beta)$ are provided, the monochromatic radiating field of $f(x, y)$ can be determined:

$$f(x, y) = q(\alpha, \beta) * h_l^*(\alpha, \beta) \tag{5.8}$$

where the superscript asterisk denotes the complex conjugate,

$$h_l^*(\alpha, \beta) = C^* \exp\left(-i\frac{k}{2l}\rho^2\right) \tag{5.9}$$

In other words, the input excitation can be determined by convolving the output response with the conjugate of the spatial impulse response.

It is very useful that a two-dimensional Fourier transform can be obtained with a positive lens. Fourier transform operations usually require complicated electronic spectrum analyzers or digital computers. However, this complicated transform can be performed extremely simply with a coherent optical system.

To perform Fourier transformations in optics, it is required that a positive lens is inserted in monochromatic wave field of Fig. 5.1. Let us assume that if a point source is located at the front focal length of a positive lens, the complex light field passing through the lens would be collimated into a plane wave, as shown Fig. 5.3. The output response is obviously the spatial Fourier transform of a point object. Thus a point source is equivalent to a *spatial delta function* $\delta(x, y)$. Since the separation between

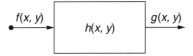

**Figure 5.2**  Linear system representation.

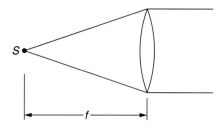

**Figure 5.3**   Fourier transform of a monochromatic point source.

the point source and the lens is $f$, the complex light distribution at the front of the lens can be shown, such as

$$
\begin{aligned}
g(\alpha, \beta) &= \delta(x, y) * h_l(x, y) \\
&= h_l(x, y) \\
&= C \exp\left[i\frac{\pi}{\lambda f}(\alpha^2 + \beta^2)\right]
\end{aligned}
\tag{5.10}
$$

The action of the lens is to change the spherical wave of Eq. (5.10) into a plane wave. It is, therefore, that the lens must induce a phase transformation, such as

$$
T(\alpha, \beta) = \exp\left[-i\frac{\pi}{\lambda f}(\alpha^2 + \beta^2)\right]
\tag{5.11}
$$

The complex light field behind the lens is

$$
g_1(\alpha, \beta) = C
\tag{5.12}
$$

which is apparently a plane wave field parallel to the $(\alpha, \beta)$ coordinate system of the lens.

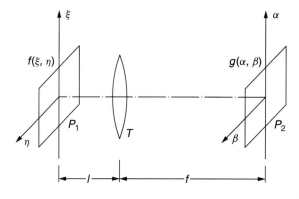

**Figure 5.4**   Fourier transform property of a positive lens.

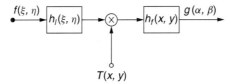

**Figure 5.5**   Linear system representation of Fig. 5.4.

We shall now demonstrate the Fourier transform property of a lens. With reference to the optical setup of Fig. 5.4, we assume that a monochromatic complex wave field at $P_1$ is $f(\xi, \eta)$. Then by applying the Fresnel–Kirchhoff theory of Eq. (5.6), the complex light distribution at $P_2$ can be written as

$$g(\alpha, \beta) = C\{[f(\xi, \eta) * h_l(\xi, \eta)]T(x, y)\} * h_f(x, y) \tag{5.13}$$

where $C$ is a proportionality constant, $h_l(\xi, \eta)$ and $h_f(x, y)$ are the corresponding spatial impulse responses, and $T(x, y)$ is the phase transform of the lens, as given in Eq. (5.11).

A linear system representation of Eq. (5.13) is shown in Fig. 5.5. Equation (5.13) can be written in integral form, such as

$$g(\alpha, \beta) = C \iint\limits_{S_1} \left[ \iint\limits_{S_2} \exp\left( i\frac{k}{2}\Delta \right) dx\, dy \right] f(\xi, \eta)\, d\xi\, d\eta \tag{5.14}$$

where $S_1$ and $S_2$ denote the surface integration over $P_1$ and the lens $T$, respectively, and

$$\Delta \triangleq \left\{ \frac{1}{l}[(x - \xi)^2 + (y - \eta)^2] + \frac{1}{f}[(\alpha - x)^2 + (\beta - y)^2 - (x^2 + y^2)] \right\} \tag{5.15}$$

It may be noted that the surface integral of the lens is assumed to be of infinite extent, since the lens is paraxially large compared to the spatial regions of interest at $P_1$ and $P_2$. Equation (5.15) can be written as

$$\Delta = \frac{1}{f}(v\xi^2 + vx^2 + \alpha^2 - 2v\xi x - 2x\alpha + v\eta^2 + vy^2 + \beta^2 - 2v\eta^2 y - 2y\beta) \tag{5.16}$$

where $v = f/l$.

By completing the square, Eq. (5.16) can be shown as

$$\Delta = \frac{1}{f} \left[ (v^{1/2}x - v^{1/2}\xi - v^{-1/2}\alpha)^2 - \alpha^2 \frac{1 - v}{v} - 2\xi\alpha \right.$$
$$\left. + (v^{1/2}y - v^{1/2}\eta - v^{-1/2}\beta)^2 - \beta^2 \frac{1 - v}{v} - 2\eta\beta \right] \tag{5.17}$$

By substituting Eq. (5.17) into Eq. (5.14), we have

$$g(\alpha, \beta) = C \exp\left[-i\frac{k}{2f}\frac{1-v}{v}(\alpha^2 + \beta^2)\right] \iint\limits_{S_1} f(\xi, \eta) \exp\left[-i\frac{k}{f}(\alpha\xi + \beta\eta)\right] d\xi\, d\eta$$

$$\cdot \iint\limits_{S_2} \exp\left\{i\frac{k}{2f}[(v^{1/2}x - v^{1/2}\xi - v^{-1/2}\alpha)^2\right.$$

$$\left. + (v^{1/2}y - v^{1/2}v - v^{-1/2}\beta)^2]\right\} dx\, dy \tag{5.18}$$

We recognize that the integral over $S_2$ is a Fresnel integral. Since $S_2$ is assumed to be infinitely extended, the Fresnel integral will converge to a complex constant, which can be incorporated with $C$. Thus Eq. (5.8) reduces to the following form:

$$g(\alpha, \beta) = C_1 \exp\left[-i\frac{k}{2f}\frac{1-v}{v}(\alpha^2 + \beta^2)\right] \cdot \iint\limits_{S_1} f(\xi, \eta) \exp\left[-i\frac{k}{f}(\alpha\xi + \beta\eta)\right] d\xi\, d\eta$$

$$\tag{5.19}$$

From this result we see that except with a quadratic phase variation, $g(\alpha, \beta)$ is essentially the Fourier transform of $f(\xi, \eta)$. In fact, the quadratic phase factor vanishes if $l = f$. Evidently, if the signal plane $P_1$ is placed at the front focal plane of the lens, the quadratic phase factor disappears, which leaves an exact Fourier transformation, such as

$$G(p, q) = C_1 \iint\limits_{S_1} f(\xi, \eta) \exp[-i(p\xi + q\eta)] d\xi\, d\eta \quad \text{for } v = 1 \tag{5.20}$$

where $p = k\alpha/f$ and $q = k\beta/f$ are the angular spatial frequency coordinates.

It must be emphasized that the exact Fourier transform takes place under the condition $l = f$. However, if $l \neq f$, a quadratic phase factor will be included. Furthermore, it can easily be shown that a quadratic phase factor also results if the signal plane $P_1$ is placed behind the lens.

In conventional Fourier transform theory, the transformation from the spatial domain to the spatial frequency domain requires a transform kernel $\exp[-i(px + qy)]$ and the transformation from the spatial frequency domain to the spatial domain requires a conjugate transform kernel $\exp[i(px + qy)]$. Since a positive lens always introduces the kernel $\exp[-i(px + qy)]$, therefore, an optical system would take only successive Fourier transformations, rather than a transformed followed by its inverse, as shown in Fig. 5.6.

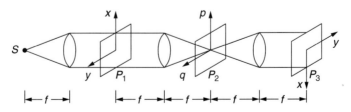

**Figure 5.6**   Successive Fourier transformations of lenses.

## 5.3  ANALYSIS OF COHERENT AND INCOHERENT OPTICAL SYSTEMS

Formulation for a general optical signal processing system under the strictly coherent and incoherent regimes will be discussed. Let a hypothetical optical system be shown in Fig. 5.7. Assume that the light emitted by the source $\Sigma$ is monochromatic, and suppose that an output light distribution of the input signal is formed at the output plane of the optical system. To demonstrate the complex light distribution of the optical system, we let $u(x, y)$ be the complex light amplitude distribution at the input signal plane due to an incremental light source $d\Sigma$. If the complex amplitude transmittance of the input plane is $f(x, y)$, the complex light field immediately behind the signal plane would be $u(x, y) f(x, y)$.

If it is assumed that the optical system in the black box is linearly spatially invariant with a spatial impulse response of $h(x, y)$, the complex light field at the output plane of the system due to $d\Sigma$ can be determined by the convolution equation

$$g(\alpha, \beta) = [u(x, y) f(x, y)] * h(x, y) \tag{5.21}$$

where the asterisk denotes the convolution operation.

We emphasized that the assumption of linearity is generally valid for small-amplitude disturbances; however, the spatial-invariance condition may be applicable only over a small region of the signal plane.

From Eq. (5.21), the intensity distribution at the output plane due to $d\Sigma$ is

$$dI(\alpha, \beta) = g(\alpha, \beta) g^*(\alpha, \beta) \, d\Sigma \tag{5.22}$$

where the superscript asterisk represents the complex conjugate. The overall output intensity distribution is therefore

$$I(\alpha, \beta) = \iint |g(\alpha, \beta)|^2 \, d\Sigma \tag{5.23}$$

which can be written in the following convolution integral:

$$I(\alpha, \beta) = \int\!\!\!\int\!\!\!\int\!\!\!\int_{-\infty}^{\infty} \Gamma(x, y; x', y') h(\alpha - x, \beta - y) h^*(\alpha - x', \beta - y')$$

$$\cdot f(x, y) f^*(x', y') \, dx \, dy \, dx' \, dy' \tag{5.24}$$

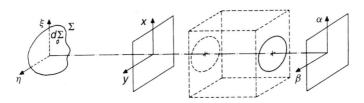

**Figure 5.7**  Hypothetical optical processing system.

where

$$\Gamma(x, y; x', y') = \iint\limits_{\Sigma} u(x, y)u^*(x', y') \, d\Sigma \qquad (5.25)$$

is the *spatial coherence function*, also known as *mutual intensity function* [1], at input plane $(x, y)$.

Let us now choose two arbitrary points $Q_1$ and $Q_2$ at the input plane. If $r_1$ and $r_2$ are the respective distances from $Q_1$ and $Q_2$ to $d\Sigma$, the complex light disturbances at $Q_1$ and $Q_2$ due to $d\Sigma$ can be written as

$$u_1(x, y) = \frac{[I(\xi, \eta)]^{-1/2}}{r_1} \exp(ikr_1) \qquad (5.26)$$

and

$$u_2(x', y') = \frac{[I(\xi, \eta)]^{1/2}}{r_1} \exp(ikr_2) \qquad (5.27)$$

where $I(\xi, \eta)$ is the intensity distribution of the light source. By substituting Eqs. (5.26) and (5.27) in Eq. (5.25), we have

$$\Gamma(x, y; x', y') = \iint\limits_{\Sigma} \frac{I(\xi, \eta)}{r_1 r_2} \exp[ik(r_1 - r_2)] \, d\Sigma \qquad (5.28)$$

In the paraxial case, $r_1 - r_2$ may be approximated by

$$r_1 - r_2 \simeq \frac{1}{r}[\xi(x - x') + \eta(y - y')] \qquad (5.29)$$

where $r$ is the separation between the source plane and the signal plane. Then Eq. (5.28) can be reduced to

$$\Gamma(x, y; x', y') = \frac{1}{r^2} \iint I(\xi, \eta) \exp\left\{i\frac{k}{r}[\xi(x - x') + \eta(y - y')]\right\} d\xi \, d\eta \qquad (5.30)$$

which is known as the *Van Cittert–Zernike theorem*. Notice that Eq. (5.30) forms an inverse Fourier transform of the source intensity distribution.

Now one of the two extreme cases is by letting the light source become infinitely large. We assume that the light source is uniform [i.e., $I(\xi, \eta) \simeq K$], Eq. (5.30) becomes

$$\Gamma(x, y; x', y') = K_1 \delta(x - x', y - y') \qquad (5.31)$$

where $K_1$ is an appropriate constant. This equation describes a completely *incoherent* optical processing system.

On the other hand, if the light source is vanishingly small [i.e., $I(\xi, \eta) \simeq K\delta(\xi, \eta)$], Eq. (5.30) becomes

$$\Gamma(x, y; x', y') = K_2 \qquad (5.32)$$

where $K_2$ is a proportionality constant. This equation describes a completely *coherent* optical processing system. In other words, a monochromatic point source describes a strictly coherent processing regime, while an extended source describes a strictly incoherent system. Furthermore, an extended monochromatic source is also known as a *spatially incoherent* source.

Referring to Eq. (5.24), for the completely incoherent case

$$\Gamma(x, y; x', y') = K_1 \delta(x - x', y - y'),$$

the intensity distribution at the output plane is

$$I(\alpha, \beta) = \iiiint\limits_{-\infty}^{\infty} \delta(x' - x, y' - y) h(\alpha - x, \beta - y)$$

$$\cdot h^*(\alpha - x', \beta - y') f(x, y) f^*(x', y') \, dx \, dy \, dx' \, dy' \quad (5.33)$$

which can be reduced to

$$I(\alpha, \beta) = \iint\limits_{-\infty}^{\infty} |h(\alpha - x, \beta - y)|^2 |f(x, y)|^2 \, dx \, dy \quad (5.34)$$

It is, therefore, apparent that for the incoherent case the output intensity distribution is the convolution of the input signal intensity with respect to the intensity impulse response. In other words, for the completely incoherent case, the optical signal processing system is linear in *intensity*, that is,

$$I(\alpha, \beta) = |h(x, y)|^2 * |f(x, y)|^2 \quad (5.35)$$

where the asterisk denotes the convolution operation.

By Fourier transformation, Eq. (5.35) can be expressed in the angular spatial frequency domain:

$$I(p, q) = [H(p, q) * H^*(p, q)][F(p, q) * F^*(p, q)] \quad (5.36)$$

where: $I(p - q)$, $H(p, q)$, and $F(p, q)$ are the Fourier transforms of $I(\alpha, \beta)$, $h(x, y)$, and $f(x, y)$, respectively; and the superscript asterisk represent the complex conjugate.

In a more convenient form, Eq. (5.35) can be written as

$$I(\alpha, \beta) = h_i(x, y) * f_i(x, y) \quad (5.37)$$

where $h_i(x, y) = |h(x, y)|^2$ and $f_i(x, y) = |f(x, y)|^2$ are the intensities of the impulse response and the input signal, respectively. Then Eq. (5.36) may be written as

$$I(p, q) = H_i(p, q) F_i(p, q) \quad (5.38)$$

where $H_i(p, q)$ and $F_i(p, q)$ are the Fourier transforms of $h_i(x, y)$ and $f_i(x, y)$, respectively, and $H_i(p, q)$ can be written as

$$H_i(p, q) = \iint\limits_{-\infty}^{\infty} h(x, y) h^*(x, y) \exp[-i(px + qy)] \, dx \, dy \quad (5.39)$$

With reference to the Fourier multiplication theorem, Eq. (5.39) can be written as

$$H_i(p, q) = \frac{1}{4\pi^2} \iint_{-\infty}^{\infty} H(p', q') H^*(p' - p, q' - q) \, dp' \, dq' \qquad (5.40)$$

which is the convolution of the complex transfer function with respect to its conjugate. On the other hand, for the completely coherent case $[\Gamma(x, y; x', y') = K_2]$, Eq. (5.24) becomes

$$I(\alpha, \beta) = g(\alpha, \beta) g^*(\alpha, \beta) = \iint_{-\infty}^{\infty} h(\alpha - x, \beta - y) f(x, y) \, dx \, dy$$

$$\cdot \iint_{-\infty}^{\infty} h^*(\alpha - x', \beta - y') f^*(x', y') \, dx' \, dy' \qquad (5.41)$$

Thus it is apparent that the optical signal processing system is linear in *complex amplitude*, that is,

$$g(\alpha, \beta) = \iint_{-\infty}^{\infty} h(\alpha - x, \beta - y) f(x, y) \, dx \, dy \qquad (5.42)$$

Again by Fourier transformation, Eq. (5.42) becomes

$$G(p, q) = H(p, q) F(p, q) \qquad (5.43)$$

where $G(p, q)$, $H(p, q)$, and $F(p, q)$ are the corresponding Fourier transforms of $g(\alpha, \beta)$, $h(x, y)$, and $f(x, y)$, respectively.

We further note that a coherence-preserving optical signal processing system that makes $\Gamma(x, y; x', y') = K$ (a constant) can easily be achieved simply by replacing the extended source $\Sigma$ with a monochromatic point source.

Let us now define the following normalized quantities:

$$\tilde{I}(p, q) = \frac{\displaystyle\iint_{-\infty}^{\infty} I(\alpha, \beta) \exp[-i(p\alpha + q\beta)] \, d\alpha \, d\beta}{\displaystyle\iint_{-\infty}^{\infty} I(\alpha, \beta) \, d\alpha \, d\beta} \qquad (5.44)$$

$$\tilde{F}_i(p, q) = \frac{\displaystyle\iint_{-\infty}^{\infty} f_i(x, y) \exp[-i(px + qy)] \, dx \, dy}{\displaystyle\iint_{-\infty}^{\infty} f_i(x, y) \, dx \, dy} \qquad (5.45)$$

and

$$\tilde{H}_i(p, q) = \frac{\displaystyle\iint_{-\infty}^{\infty} h_i(x, y) \exp[-i(px + qy)]\, dx\, dy}{\displaystyle\iint_{-\infty}^{\infty} h_i(x, y)\, dx\, dy} \tag{5.46}$$

These normalized quantities should give a set of convenient mathematical forms and also provide a concept of image contrast interpretation. Since the quality of a visual image, to a large extent, depends on the contrast (or the relative irradiance) of the image, a normalized intensity function would certainly enhance the information-bearing capacity. With the application of the Fourier convolution theorem, the following relationship can be written:

$$\tilde{I}_i(p, q) = \tilde{F}_i(p, q)\tilde{H}_i(p, q) \tag{5.47}$$

where $\tilde{H}_i(p, q)$ is commonly referred to as the *optical transfer function* (OTF) of the optical system, and the modulus of $|\tilde{H}(p, q)|$ is known as the *modulation transfer function* (MTF) of the optical system.

We further note that Eq. (5.46) can also be written as

$$\tilde{H}_i(p, q) = \frac{\displaystyle\iint_{-\infty}^{\infty} H(p', q')H^*(p' - p, q' - q)\, dp'\, dq'}{\displaystyle\iint_{-\infty}^{\infty} |H(p', q')|^2\, dp'\, dq'} \tag{5.48}$$

Upon changing the variables $p'' = p' + \frac{1}{2}p$, $q'' = q' + \frac{1}{2}q$, Eq. (5.48) results in the following symmetrical form:

$$\tilde{H}_i(p, q) = \frac{\displaystyle\iint_{-\infty}^{\infty} H(p'' + \frac{1}{2}p, q'' + \frac{1}{2}q)H^*(p'' - \frac{1}{2}p, q'' - \frac{1}{2}q)\, dp''\, dq''}{\displaystyle\iint_{-\infty}^{\infty} |H(p'', q'')|\, dp''\, dq''} \tag{5.49}$$

We note that the definition of OTF is valid for any linearly spatial invariant optical system regardless of whether the system is with or without aberrations. Furthermore, Eq. (5.49) serves as the primary link between the strictly coherent and strictly incoherent systems. There are, however, threefold limitations associated with the strictly coherent processing. First, coherent processing requires the dynamic range of a spatial light modulator to be about 100,000:1 for the input wave intensity, or a photographic density range of 4.0. Such a dynamic range is often quite difficult to achieve in practice. Second, coherent processing systems are susceptible to coherent artifact noise, which frequently degrades the image quality. Third, in coherent processing, the signal being processed is carried out by a complex wave field. However, what is actually measured at the output plane of the optical processor is the output wave intensity. The loss of the output phase distribution may seriously limit the applicability of the optical system in some applications. To overcome the drawbacks of coherent processing, it is useful to reduce either the temporal coherence or the spatial coherence, and there are several techniques available for this purpose (see Section 5.6).

## 5.4   COHERENT OPTICAL PROCESSING

We shall now consider a coherent optical signal processor as depicted in Fig. 5.8. We assume that a monochromatic point source is located at the front focal length of a collimating lens, then a monochromatic plane wave will illuminate the input signal plane $P_1$. If an object transparency $f(x, y)$ is inserted at $P_1$, the complex light field at the output plane $P_3$ can be written as

$$g(\alpha, \beta) = Kf(x, y) * h(x, y) \tag{5.50}$$

where $K$ is a proportionality constant and $h(x, y)$ is the spatial impulse response of the processor.

If we assume that the spatial frequency of the signal $f(x, y)$ lies within the spatial frequency limit of the processor, the spatial impulse response $h(x, y)$ may be approximated by the Dirac delta function, such that the output light field is

$$g(\alpha, \beta) = Kf(x, y) \tag{5.51}$$

which is proportional to the input object transparency. With reference to the Fourier transform properties of lenses (as shown in the preceding section) the complex light amplitude distribution on the plane $P_2$ is proportional to $F(p, q)$, where $F(p, q)$ is the Fourier transform of the input processing signal $f(x, y)$. The intensity distribution at $P_2$ is therefore proportional to $|F(p, q)|^2$.

Let us now assume that the coherent optical processor shown in Fig. 5.8 is terminated at $P_2$ as shown in Fig. 5.9; then the system is essentially a two-dimensional spectrum analyzer. The complex light field at $P_2$ is

$$E(p, q) = K \iint\limits_{S} f(x, y) \exp[-i(xp + yq)] \, dx \, dy = KF(p, q) \tag{5.52}$$

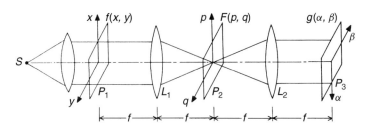

**Figure 5.8**   Coherent optical signal processor.

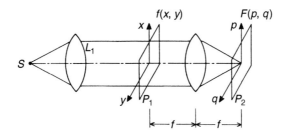

**Figure 5.9**  Coherent spectrum analyzer.

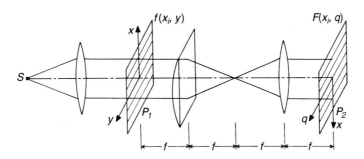

**Figure 5.10**  Multichannel spectrum analyzer.

where        $K$ = proportionality constant

$S$ = surface integration over the input plane $P_1$

$F(p, q)$ = two-dimensional Fourier transform of the input object $f(x, y)$

$p$ and $q$ = angular spatial frequency coordinates

The corresponding irradiance is therefore

$$I(p, q) = E(p, q)E^*(p, q) = K^2|F(p, q)|^2 \qquad (5.53)$$

which is proportional to the power spectrum of the input object transparency $f(x, y)$.

The optical spectrum analyzer of Fig. 5.9 can be modified to yield a multichannel one-dimensional spectrum analyzer by the addition of a cylindrical lens, as shown in Fig. 5.10. If the input multichannel signal is $f(x_i, y)$ the complex light field at the output plane of the analyzer is

$$E(x_i, q) = K \exp\left(-i\frac{k}{2f}\frac{1-v}{v}q^2\right)\int f(x_i, y)e^{-iqy}\,dy, \qquad i = 1, 2, \ldots, n \quad (5.54)$$

where $v = f/l$, and subscript $i$ denotes the corresponding channel in the $x$-axis. But since $l = 3f$ (i.e., $v = \frac{1}{3}$), Eq. (5.54) becomes

$$E(x_i, q) = K \exp\left(-i\frac{k}{f}q^2\right)F(x_i, q), \qquad n = 1, 2, \ldots, n \quad (5.55)$$

where $F(x_i, q)$ is the one-dimensional Fourier transform of $f(x_i, y)$.

The corresponding output intensity distribution is therefore

$$I(p, q) = E(x_i, q)E^*(x_i, q) = K^2 |F(x_i, q)|^2, \qquad i = 1, 2, \ldots, n \qquad (5.56)$$

Thus we see that a coherent optical spectrum analyzer can indeed perform $N$-channel one-dimensional analysis. It is possible to synthesize a spatial filter in the spatial frequency domain. Such a spatial filter consists of a transparency located in the spatial frequency domain of a coherent optical system. We stress that the desired filter can be synthesized by direct manipulation of the complex-amplitude transmittance over the frequency domain. For instance, if the processing signal $f(x, y)$ is inserted at the spatial domain $P_1$, a Fourier transform of the input signal would be distributed at $P_2$. If a spatial filter of transparency $H(p, q)$ is inserted at $P_2$, the complex light field immediately behind $P_2$ is given by

$$E(p, q) = K F(p, q) H(p, q) \qquad (5.57)$$

where $K$ is a proportionality constant.

Since the second lens $L_2$ would take an inverse Fourier transformation of the complex light field $E(p, q)$ to the output plane $P_3$, the complex-amplitude light distribution across $P_3$ would be

$$g(\alpha, \beta) = K \iint F(p, q) H(p, q) \exp[p\alpha + q\beta] \, dp \, dq \qquad (5.58)$$

where the surface integration is taken over the spatial frequency domain $P_2$.

Alternatively, by the Fourier multiplication theorem, Eq. (5.57) can be written as

$$g(\alpha, \beta) = K \iint f(x, y) h(\alpha - x, \beta - y) \, dx \, dy = K f(x, y) * h(x, y) \qquad (5.59)$$

where the integral is taken over the input spatial domain, and $h(x, y)$ is the spatial impulse response of the filter, that is,

$$h(x, y) = \mathscr{F}^{-1}[H(p, q)] \qquad (5.60)$$

We shall stress that the spatial filter $H(p, q)$ can consist of apertures or slits of any shape. Depending on the arrangement of apertures, it can represent low-pass, high-pass, or bandpass spatial filters. It is clear that any opaque portion in the filter represents a spatial frequency-band rejection. We also note that the inclusion of a phase plate with the filter would produce a phase delay. Since we have the complete freedom of synthesizing the amplitude and the phase filters independently, in principle, any complex spatial filter can be constructed. There is, however, a technique developed by Vander Lugt [2], that a complex spatial filter can actually be synthesized with a holographic technique, as will be shown in the following section.

## 5.5 COMPLEX SPATIAL FILTER SYNTHESIS

In general, a spatial filter can be described by a complex-amplitude transmittance such as

$$H(p, q) = |H(p, q)| \exp[i\phi(p, q)] \qquad (5.61)$$

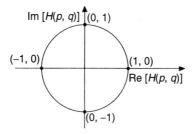

**Figure 5.11**  Complex amplitude transmittance.

In practice, the optical spatial filters are generally of the passive type. The physically realizable conditions of optical spatial filters are

$$|H(p, q)| \leqslant 1 \tag{5.62}$$

and

$$0 \leqslant \phi(p, q) \leqslant 2\pi \tag{5.63}$$

We note that such a transmittance function can be represented by a set of points within or on a unit circle in the complex plane, as shown in Fig. 5.11. The amplitude transmission of the filter changes with the optical density, and the phase delay varies with the thickness. Thus we see that a complex spatial filter can be synthesized with a combination of an amplitude and a phase delay filter.

Let us now discuss a technique of synthesizing a complex spatial filter with an interferometric method as shown in Fig. 5.12. The complex light field over the spatial frequency plane $P_2$ is

$$E(p, q) = F(p, q) + \exp(i\alpha_{0p}) \tag{5.64}$$

where $\alpha_0 = f \sin\theta$, $f$ is the focal length of the transform lens, and $F(p, q) = |F(p, q)| \exp[i\phi(p, q)]$

The corresponding intensity distribution over the recording medium would be

$$I(p, q) = 1 + |F(p, q)|^2 + 2|F(p, q)| \cos[\alpha_0 p + \phi(p, q)] \tag{5.65}$$

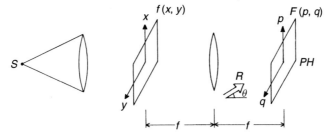

**Figure 5.12**  Synthesis of a complex spatial filter, $f(x, y)$, input signal transparency; $R$, reference beam; $PH$, recording medium.

We assumed that the recording is linear in amplitude transmission, the corresponding transmittance function of the spatial filter is

$$H(p, q) = K\{1 + |F(p, q)|^2 + 2|F(p, q)| \cos[\alpha_0 p + \phi(p, q)]\} \quad (5.66)$$

which is, in fact, a real positive function. We note that to accommodate the amplitude and phase variations of the filter, there is a price we paid at the expense of a wider bandwidth.

If this complex spatial filter is inserted in the spatial frequency plane of a coherent optical processing system as shown in Fig. 5.8, the wave field immediately behind the spatial filter is

$$E(p, q) = F(p, q)H(p, q) \quad (5.67)$$

By substitution of Eq. (5.66) into Eq. (5.67), we have

$$E(p, q) = K[F(p, q)F(p, q) + F(p, q)|F(p, q)|^2 + F(p, q)F(p, q)$$
$$\cdot \exp(ip\alpha_0) + F(p, q)F^*(p, q) \exp(-ip\alpha_0)] \quad (5.68)$$

where the superior asterisk denotes the complex conjugate. Since the complex light field at the output plane is the inverse Fourier transform of Eq. (5.68),

$$g(\alpha, \beta) = \iint E(p, q) \exp[-i(\alpha p + \beta q)] \, dp \, dq \quad (5.69)$$

which can be written as

$$g(\alpha, \beta) = K[f(\alpha, \beta) + f(\alpha, \beta) * f(\alpha, \beta) * f^*(-\alpha, -\beta) + f(\alpha, \beta) * f(\alpha + \alpha_0, \beta)$$
$$+ f(\alpha, \beta) * f^*(-\alpha + \alpha_0, -\beta)] \quad (5.70)$$

The first and second terms of Eq. (5.70) represent the zero-order diffraction, which appears at the origin of the output plane; the third and fourth terms are the convolution and cross-correlation terms, which are diffracted in the neighborhood of $\alpha = \alpha_0$ and $\alpha = -\alpha_0$, respectively. The zero-order and convolution terms are of no particular interest here; it is the cross-correlation term that is used in signal detection.

Now if the input signal is assumed to be embedded in an additive white Gaussian noise $n$, that is,

$$f'(x, y) = f(x, y) + n(x, y) \quad (5.71)$$

then the correlation term of Eq. (5.70) would be

$$R(\alpha, \beta) = K[f(x, y) + n(x, y)] * f^*(-x + \alpha_0, -y) \quad (5.72)$$

Since the cross-correlation between $n(x, y)$ and $f^*(-x + \alpha_0, -y)$ can be shown to be approximately equal to zero, Eq. (5.72) reduces to

$$R(\alpha, \beta) = f(x, y) * f^*(-x + \alpha_0, -y) \quad (5.73)$$

which is proportional to the autocorrelation of $f(x, y)$.

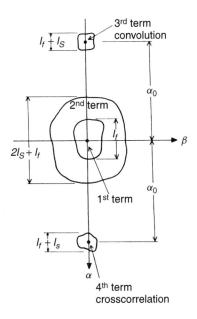

**Figure 5.13**   Sketch of the output diffraction of a complex spatial filter.

To ensure that the zero-order and the first-order diffraction terms will not overlap, a spatial carrier frequency $\alpha_0$ is required that

$$\alpha_0 > l_f + \tfrac{3}{2}l_s \qquad (5.74)$$

where $l_f$ and $l_s$ are the spatial lengths in the $x$ direction of the input object transparency and the detecting signal $f(x, y)$, respectively. To show that this is true, we consider the length of the various output terms of $g(\alpha, \beta)$ as illustrated in Fig. 5.13.

Since lengths of the first, second, third, and fourth terms of Eq. (5.70) are $l_f, 2l_s + l_f, l_f + l_s$, and $l_f + l_s$, respectively, to achieve complete separation the spatial carrier frequency $\alpha_0$ must satisfy the inequality of (5.74).

## 5.6   PARTIALLY COHERENT OPTICAL PROCESSING

Although coherent optical processors can perform a variety of complex signal operations [3], coherent processing systems are usually plagued with coherent artifact noise. These difficulties have prompted us to look at optical processing from a new standpoint, and to consider whether it is necessary for all optical processing operations to be carried out by pure coherent sources. We have found that many types of optical processing can be carried out by partially coherent sources or white light sources [4–9]. The basic advantages of partially coherent processing are: (1) it can suppress the coherent artifact noise; (2) partially coherent sources are usually inexpensive; (3) the processing environment is generally very relaxed; (4) partially coherent systems are relatively easy and economical to operate; and (5) partially coherent processors are particularly suitable for color image processing.

### 5.6.1 Spatially Partially Coherent Processing

Techniques of utilizing spatially partially coherent light to perform complex data processing have been proposed by Lohmann [4], Rhodes [5], and subsequently by Stoner [9]. These techniques share a basic concept: The optical system is characterized by use of the point spread function (PSF), which is not constrained to the class of nonnegative real functions used in conventional incoherent processing. The output intensity distribution can be adjusted by changing the PSF, that is,

$$I(x', y') = \iint O(x, y)h(x', y'; x, y)\,dx\,dy \tag{5.75}$$

where $h(x, y; x', y')$ is the PSF, and $I(x', y')$ and $O(x, y)$ are the image and object intensity distributions, respectively. If the pupil function of an optical system is $P(\alpha, \beta)$, the PSF is equal to the square of the Fourier spectrum of the pupil function. Therefore, the output intensity distribution can be adjusted by selecting an adequate pupil function. A typical processing system, as proposed by Rhodes, is shown in Fig. 5.14. This optical processing system, which is characterized by an extended pupil region, has two input pupil functions, $P_1(\alpha, \beta)$ and $P_2(\alpha, \beta)$. With the optical path lengths of the two arms of the system being equal, the overall pupil function is given by the sum of $P_1(\alpha, \beta)$ and $P_2(\alpha, \beta)$. If the path length in one arm is changed slightly, however (e.g., by moving mirror $M_2$ a small distance), a phase factor is introduced in one of the component pupil functions, with the result that

$$P(\alpha, \beta) = P_1(\alpha, \beta) + P_2(\alpha, \beta)\exp[i\phi(\alpha, \beta)] \tag{5.76}$$

It is evident that if the pupil transparencies $P_1(\alpha, \beta)$ and $P_2(\alpha, \beta)$ are recorded holographically, arbitrary PSFs can then be synthesized using the 0 to 180° phase switching operation. Thus we see that this optical system is capable of performing complex data processing with spatially partially coherent light.

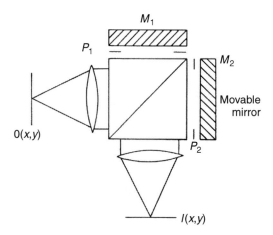

**Figure 5.14** Two-pupil incoherent processing systems.

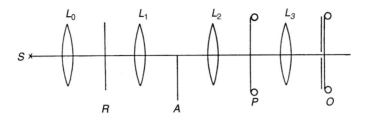

**Figure 5.15**   Achromatic optical processing system.

### 5.6.2   Achromatic Optical Processing

An achromatic optical signal processing technique with broadband source (e.g., white light) was proposed by Leith and Roth [6], as shown in Fig. 5.15. Lens $L_0$ collimates the white light point source $S$, which illuminates hologram $R$. The diffraction wavefront is spatially filtered by the aperture $A$, which removes the zero-order term and the lower sideband. The upper sideband is imaged onto the signal plane $P$; lens $L_3$ Fourier transforms the resulting wavefront; and observation in the transform plane is confined to the optical axis by a slit in the output plane $O$. The convolution of the demodulated input signal with the desired reference function is recorded by synchronously translating the signal and output films. Such a system, termed an achromatic system, thus has the flexibility of a coherent optical signal processor along with the potential for the noise immunity of an incoherent system.

### 5.6.3   Bandlimited Partially Coherent Processing

A technique using a matched filter for operation with bandlimited illumination was proposed by Morris and George [8], as shown in Fig. 5.16. The matched filter consists of a frequency-plane holographic filter, an achromatic-fringe interferometer, a color-compensating grating, and an achromatic doublet. The matched filter, a Fourier hologram, is made by recording the object spectrum and a collimated reference beam with an exposure wavelength $\lambda_0$. The reference-beam angle is at $\theta_0$ with respect to the object beam axis. In the correlation operation, the object is illuminated using a board spectral source. The various spectral components are dispersed in angle due to the grating-like structure of the matched filter. A lateral dispersed correlation signal is generated in plane $c$, which is the conventional correlation plane. These spectral

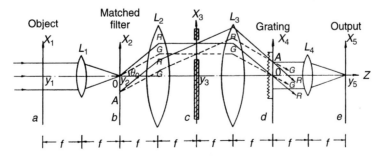

**Figure 5.16**   Bandlimited partially coherent processing system.

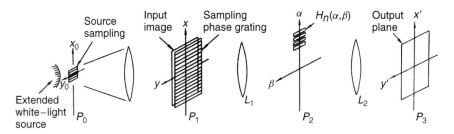

**Figure 5.17** Achromatic partially coherent optical processing system.

components are recombined by imaging the matched filter into plane $d$, where a compensation grating produces a color dispersion that is equal but opposite to that introduced at plane $b$. The color-corrected correlation signal is observed in the output plane. However, the compensated signal is still wavelength dependent, since the illumination wavelength changes the scale of the object spectrum at plane $b$. Insertion of a slit in plane $c$ provides a convenient way to bandlimit the correlation signal to $\Delta\lambda$. Bandlimiting improves the signal-to-noise ratio (SNR) of the compensated correlation output. Broadband illumination can be used for automatic scale search and object size determination. Since spatially incoherent light as well as coherent light can be used to perform matched filtering, an extended white light source is used together with a slit at plane $c$ to provide a bandlimited partially coherent processing.

### 5.6.4 Achromatic Partially Coherent Processing

An achromatic partially coherent optical processing technique with a white light source was introduced by Yu [7], Fig. 5.17 illustrates the processing system, where all the transform lenses are assumed achromatic. A high diffraction efficiency phase grating, with an angular spatial frequency $p_0$, is used at the input plane to disperse the input object spectrum into rainbow color spectrum in the Fourier plane $p_2$. Thus it permits a stripwise design of a complex spatial filter for each narrow spectral band in the Fourier plane. The achromatic output image irradiance can therefore be observed at the output plane $P_3$. The advantages of this technique are that each channel (e.g., each spectral band filter) behaves as a partially coherent channel, while the overall output noise performance is due primarily to the incoherent addition of each channel, which behaves as if under incoherent illumination. Thus, this partially coherent system has the capability for coherent noise suppression.

## 5.7 WHITE LIGHT OPTICAL PROCESSING

We now describe an achromatic partially coherent processing technique that can be carried out by a white light source [10, 11], as shown in Fig. 5.17. The partially coherent processing system is similar to a coherent processing system, except for the following: It uses an extended white light source, a source encoding mask, a signal sampling grating, multispectral band filters, and achromatic transform lenses. For example, if we place an input object transparency $s(x, y)$ in contact with an image sampling phase grating, the complex wave field, for every wavelength $\lambda$, at the Fourier plane $P_2$ would be (assuming a white light point source)

$$E(p, q; \lambda) = \iint s(x, y) \exp(ip_0 x) \exp[-i(px + qy)] \, dx \, dy = S(p - p_0, q) \quad (5.77)$$

where the integral is over the spatial domain of the input plane $P_1$, $(p, q)$ denotes the angular spatial frequency coordinate system, $p_0$ is the angular spatial frequency of the sampling phase grating, and $S(p, q)$ is the Fourier spectrum of $s(x, y)$. If we write Eq. (5.77) in the form of a spatial coordinate system $(\alpha, \beta)$, we have

$$E(\alpha, \beta; \lambda) = S\left(\alpha - \frac{\lambda f}{2\pi} p_0, \beta\right) \quad (5.78)$$

where $p = (2\pi/\lambda f)\alpha, q = (2\pi/\lambda f)\beta$, and $f$ is the focal length of the achromatic transform lens. Thus we see that the Fourier spectra would disperse into rainbow colors along the $\alpha$-axis, and each Fourier spectrum for a given wavelength $\lambda$ is centered at $\alpha = (\lambda f/2\pi)p_0$.

In complex signal filtering, we assume that a set of narrow spectral band complex spatial filters is available. In practice, all the input objects are spatial frequency limited; the spatial bandwidth of each spectral band filter $H(p_n, q_n)$ is therefore

$$H(p_n, q_n) = \begin{cases} H(p_n, q_n), & \alpha_1 < \alpha < \alpha_2 \\ 0, & \text{otherwise} \end{cases} \quad (5.79)$$

where $p_n = (2\pi/\lambda_x f)\alpha, q_n = (2\pi/\lambda_x f)\beta, \lambda_n$ is the main wavelength of the filter, $\alpha_1 = (\lambda_x f/2p)(p_0 + \Delta p)$ and $\alpha_2 = (\lambda_x f/2\pi)(p_0 - \Delta p)$ are the upper and lower spatial limits of $H(p_n, q_n)$, and $\Delta p$ is the spatial bandwidth of the input objects $s(x, y)$.

Since the limiting wavelengths of each $H(p_n, q_n)$ are

$$\lambda_l = \lambda_n \frac{p_0 + \Delta p}{p_0 - \Delta p} \quad \text{and} \quad \lambda_h = \lambda_n \frac{p_0 - \Delta p}{p_0 + \Delta p} \quad (5.80)$$

its spectral bandwidth can be approximated by

$$\Delta \lambda_n = \lambda_n \frac{4 p_0 \Delta p}{p^2 - (\Delta p)^2} \simeq \frac{4 \Delta p}{p_0} \lambda_n \quad (5.81)$$

If we place this set of spectral band filters side by side and position them properly over the smeared Fourier spectra, the intensity distribution of the output light field can be shown as

$$I(x, y) = \sum_{n=1}^{N} \Delta \lambda_n |s(x, y; \lambda_n) * h(x, y; \lambda_n)|^2 \quad (5.82)$$

where $h(x, y; \lambda_n)$ is the spatial impulse response of $H(p_n, q_n)$ and $*$ denotes the convolution operation. Thus the proposed partially coherent processor is capable of processing the signal in a complex wave field. Since the output intensity is the sum of the mutually incoherent narrow-band spectral irradiances, the disturbing coherent artifact noise can therefore be immunized. It is also apparent that the white light source emits all the visible wavelengths; the partially coherent processor is very suitable for color image processing.

## 5.8   SPATIAL ENCODING AND IMAGE SAMPLING

In this section we discuss a linear transform relationship between the spatial coherence (i.e., the mutual intensity function) and the source encoding. Since the spatial coherence depends on the image processing operation, a more relaxed coherence requirement may be used for specific image processing operations. The concept of source encoding is to alleviate the stringent coherence requirement so that an extended source can be used. In other words, source encoding is capable of generating an appropriate spatial coherence for a specific optical signal processing application, such that the available light power from the source may be efficiently utilized.

### 5.8.1   Source Encoding

We begin our discussion with Young's experiment under an extended source illumination, as shown in Fig. 5.18. First, we assume that a narrow slit is placed in the source plane $P_0$ behind an extended monochromatic source. To maintain a high degree of coherence between the slits $Q_1$ and $Q_2$ at plane $P_2$, the source size should be very narrow. If the separation between $Q_1$ and $Q_2$ is large, a narrower slit size $S_1$ is required. Thus the slit width should be

$$w \leqslant \lambda R / 2h_0 \qquad (5.83)$$

where $R$ is the distance between the planes $P_0$ and $P_1$, and $2h_0$ is the separation between $Q_1$ and $Q_2$. Let us now consider two narrow slits $S_1$ and $S_2$ located in the source plane $P_0$. We assume that the separation between $S_1$ and $S_2$ satisfies the following path-length relation:

$$r_1' - r_2' = (r_1 - r_2) + m\lambda \qquad (5.84)$$

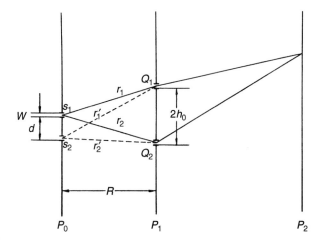

**Figure 5.18**   Young's experiment with extended source illumination.

where the $r$'s are the respective distances from $S_1$ and $S_2$ to $Q_1$ and $Q_2$, $m$ an arbitrary integer, and $\lambda$ the wavelength of the extended source. Then the interference fringes due to each of the two source slits $S_1$ and $S_2$ will be in phase and a brighter fringe pattern will be seen at plane $P_2$. To further increase the intensity of the fringes, one would simply increase the number of slits in appropriate locations in plane $P_0$, such that the separation between slits satisfies the fringe condition of Eq. (5.84). If the separation $r$ is large, that is, if $R \gg d$ and $R \gg 2h_0$, the spacing $d$ becomes

$$d = m(\lambda R/2h_0) \tag{5.85}$$

Thus, by properly encoding an extended source, it is possible to maintain a high degree of coherence between $Q_1$ and $Q_2$, and at the same time, to increase the intensity of the fringes.

To encode an extended source, we first search for the coherence function for a specific image processing operation. With reference to the optical processor shown in Fig. 5.17, the mutual intensity function at the input plane $P_1$ can be written as

$$J(x_1, x_1') = \iint \gamma(x_0) K(x_0, x_1) K(x_0, x_1) \, dx_0 \tag{5.86}$$

where the integration is over the source plane, $\gamma(x_0)$ is the intensity distribution of the encoding mask, and $K(x_0, x_1)$ is the transmittance function between the source plane $P_0$ and the input plane $P_1$, which can be written as

$$K(x_0, x_1) \approx \exp[i2\pi(x_0 x_1/\lambda f)] \tag{5.87}$$

By substituting $K(x_0, x_1)$ into Eq. (5.86), we have

$$J(x_1 - x_1') = \iint \gamma(x_0) \exp[i2\pi(x_0/\lambda_f)(x_1 - x_1')] \, dx_0 \tag{5.88}$$

From (5.88) we see that the spatial coherence and source encoding intensity from a Fourier transform pair, that is,

$$\gamma(x_0) = \mathscr{F}[J(x_1 - x_1')] \tag{5.89}$$

and

$$J(x_1 - x_1') = \mathscr{F}^{-1}[\delta(x_0)] \tag{5.90}$$

where $\mathscr{F}$ denotes the Fourier transformation operation. It is evident that the relationship of Eqs. (5.89) and (5.90) is the well-known *Van Cittert–Zernike theorem*. In other words, if a specific spatial coherence function is provided, a source encoding transmittance can be obtained through the Fourier transformation. In practice, however, the source encoding transmittance should be a positive real quantity that satisfies the physically realizable condition

$$0 \leqslant \gamma(x_0) \leqslant 1 \tag{5.91}$$

### 5.8.2   Image Sampling

There is, however, a temporal coherence requirement for partially coherent processing. Since the scale of the Fourier spectrum varies with the wavelength, a temporal coherence requirement should be imposed. If we restrict the Fourier spectra, due to wavelength spread, within a small fraction of the fringe spacing $d$ of a narrow spectral band filter $H_n(\alpha, \beta)$, we have

$$P_m f \Delta\lambda_n / 2 \ll d \tag{5.92}$$

where $1/d$ = highest spatial frequency of the filter

$P_m$ = angular spatial frequency limit of the input image transparency

$f$ = focal length of the achromatic transform lens

$\Delta\lambda_n$ = spectral bandwidth of $H_n(\alpha, \beta)$

The temporal coherence requirement of the spatial filter is, therefore

$$\Delta\lambda_n / \lambda_n \ll \pi / h_0 P_m \tag{5.93}$$

where $\lambda_n$ is the central wavelength of the $n$th narrow spectral band filter, and $2h_0 = \lambda_x f / d$ is the size of the input image transparency.

From this result it is apparent that if the spatial frequency of the input image transparency is low, a broader spectral width can be used. In other words, if a higher spatial frequency is required, a narrower spectral width is needed. Evidently, a narrower spectral spread $\Delta\lambda_n$ corresponds to a higher temporal coherence requirement, which can be obtained by increasing the image sampling frequency $p_0$. However, if a higher image sampling frequency is used, larger apertures may be required for the transform lenses in the optical system, which tend to be more expensive. Nevertheless, in practice, high-quality images have been obtained with relatively low cost lenses.

## 5.9   REAL-TIME OPTICAL PROCESSING

Two-dimensional optical processing has been used in a wide variety of applications in which the large capacity and parallel processing capability of optics are exploited. The areas of particular interest are image correlation for pattern recognition, image subtraction for machine vision, and numerical processing for optical computing. Although optical correlators that rely on general holographic filtering techniques have been introduced, the concept of programmability in optical signal processing is rather new. It is, however, primarily because of the recent development of many sophisticated spatial light modulators [12, 13] that allows us to construct such an optical system. The system can eventually be compactly packaged, and therefore it can be useful for on-board processing in space stations.

There are various available techniques to implement real-time optical processing. However, we will restrict ourselves to discuss a microcomputer-based optical signal processor. We note that this hybrid optical processor is capable of performing a variety of signal processings with high speed.

### 5.9.1   Complex Signal Detection

Let us now illustrate a technique to perform a real-time optical correlation. Figure 5.19 shows an optical setup for this purpose. A programmable magneto-optic spatial light modulator (MOSLM) is used for real-time input image write-in with a set of reference images. The set of reference images is stored in the computer memory beforehand, while the input object is sensed by a TV camera. A liquid-crystal light valve (LCLV) is used as a square-law detector for power spectrum convertion in a joint-transform correlation configuration. With a coherent readout, various image correlations can be obtained. The main feature of this proposed system is the real-time programmability, where a large set of reference images can be cross-correlated with the input image for pattern recognition and identification. The proposed system is rather economical and simple to operate. Let us consider that $K$ images are generated by a MOSLM at the input plane, as shown in Fig. 5.19. This input images can be written as

$$f(x, y) = \sum_{k=1}^{K} f_k(x - a_k, y - b_k) \tag{5.94}$$

where $(a_k, b_k)$ denotes position of the $k$th object. Since the MOSLM has an inherent grating structure, the amplitude transmittance function of the encoded MOSLM would be

$$t(x, y) = f(x, y)g(x, y) \tag{5.95}$$

where $g(x, y)$ represents a two-dimensional grating structure of the MOSLM. The corresponding joint Fourier transform at the input end of the LCLV can be written as

$$T(p, q) = F(p, q) * G(p, q) \tag{5.96}$$

where the asterisk denotes the convolution operation. The output end of the LCLV is illuminated by a beam of coherent light, as shown in Fig. 5.19. The complex amplitude

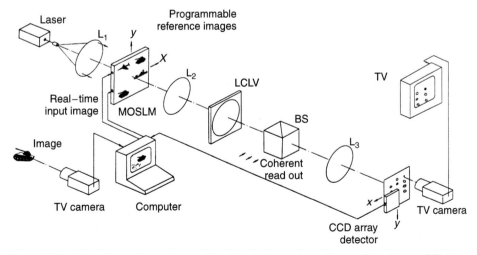

**Figure 5.19**   Real-time microcomputer-based optical correlator. L, transform lenses; BS, beam splitter.

distribution of the reflected light field is

$$A(p, q) = C_0 + C|T(p, q)|^2 \tag{5.97}$$

where $C_0$ and $C$ are the appropriate proportionality constants. The complex light field at the output plane can be written as

$$a(x, y) = C_0 \delta(x, y) + C \mathscr{F}^{-1}[|T(p, q)|^2] \tag{5.98}$$

where $\mathscr{F}^{-1}[\cdot]$ denotes the inverse Fourier transformation. By a straightforward calculation, Eq. (5.98) becomes [14]

$$a(x, y) = C_0 \delta(x, y) + C \sum_{m,n} \text{Sinc}^2 \frac{md}{l} \text{Sinc}^2 \frac{nd}{l}$$

$$\cdot \exp\left[-i2\pi \left(\frac{mx}{l} + \frac{ny}{l}\right)\right] \sum_{k}^{K} R_{kk}(x, y) + C \sum_{m,n} \text{Sinc}^2 \frac{md}{l} \text{Sinc}^2 \frac{nd}{l}$$

$$\cdot \exp\left[-i2\pi \left(\frac{mx}{l} + \frac{ny}{l}\right)\right]$$

$$\cdot \sum_{\substack{j,k \\ j \neq k}}^{K} \frac{1}{2}\{R_{jk}[x - (a_j - a_k), y - (b_j - b_k)]$$

$$+ R_{kj}[x + (a_j - a_k), y + (b_j - b_k)]\} \tag{5.99}$$

where   $l$ = period of the inherent grating structure of the MOSLM
$d$ = pixel size
$R_{kk}$ = autocorrelation of the $k$th object
$R_{jk}$ = cross-correlation of the $j$th and $k$th objects
[i.e., $R_{jk} \Delta f_j(x, y) * f_k^*(x, y)$]

We further note that the $\delta$ and $R_{kk}$ terms of Eq. (5.99) are the zero-order terms that would be appeared around the origin of the output plane, and the last two terms represent the cross-correlation terms that would be diffracted around the coordinates $x = a_j - a_k, y = b_j - b_k$, and $x = -(a_j - a_k), y = -(b_j - b_k)$, respectively, in the output plane. To ensure nonoverlapping cross-correlation distribution at the output plane, the separation between the image functions should be $|a_k - a_j| > K W_x$ or $|b_k - b_j| > K W_y$, and

$$\|a_{j1} - a_{k1}| - |a_{j2} - a_{k2}\| > 2W_x \quad \text{or} \quad \|b_{j1} - b_{k1}| - |b_{j2} - b_{k2}\| > 2W_y$$

where $W_x$ and $W_y$ denote the spatial extensions of the image function in the $x$ and $y$ directions, respectively, $K$ is the total number of the input image functions, and the subscripts 1 and 2 represent the locations of two adjacent images.

To avoid the overlapping cross-correlation terms, the images generated by the MOSLM must be set in different locations such that the geometrical symmetry is broken. For instance, the input function can be generated as

$$f(x, y) = f_1(x - a, y - b) + f_2(x, y - b) + f_3(x - a, y) + f_4(x, y + b) \tag{5.100}$$

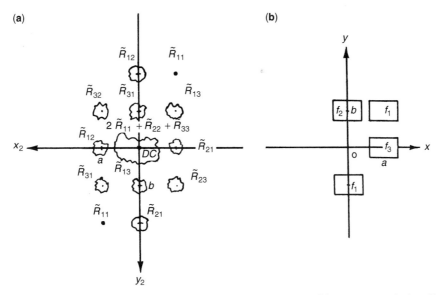

**Figure 5.20**  Correlation distribution; **(a)** input image functions; **(b)** output correlation distributions.

where we may assume that $f_4$ is a real-time object image sensed by the TV camera while $f_1$, $f_2$, and $f_3$ are references generated by the computer memory. Figure 5.20**a** shows the configuration of the input objects where $f_4$ is assumed to be identical to $f_1$. Thus, as sketched in Fig. 5.20**b**, all the first-order correlation distributions can be made mutually separated.

Since the cycle (write plus erase) time of the LCLV and MOSLM are in the order of 20 ms and <1 ms, respectively, the speed of the processing rate of the proposed system would be dependent on the cycle time of the LCLV. The resolution of the current MOSLM and LCLV is about 14 lines/mm and 30 lines/mm, respectively, measured at the 50% modulation transfer function (MTF). Thus the resolution of the overall system would depend more on the MOSLM. Even if a 50% resolution reduction is used for the overall system performance, this would correspond to a resolution of about 7 lines/mm. It is certainly a good-quality image for image correlation. Thus, within the current state of the art, a real-time microcomputer-based programmable optical correlator as applied to automatic pattern recognition and identification is feasible.

Furthermore, this proposed technique also alleviates the disadvantage and difficulty of generating a real-time programmable matched filter in the Fourier plane (as generally suggested). It is extremely difficult, if it is not impossible, to generate a complex filter for real-world object with a MOSLM. It is, however, rather simple to generate an input image with the current MOSLM.

The MOSLM is a binary transparent-type spatial light modulator. The contrast could be as high as a 400 to 1000:1 ratio under coherent illumination. It is rather suitable for input object generation. The LCLV is an image transducer spatial light modulator. This device will be used in the Fourier plane as a square-law detector, for Fourier to power spectrum conversion. However, the contrast ratio of the LCLV is about 100:1 under coherent readout. In our experience, this dynamic range of LCLV is quite adequate for the coherent readout correlation.

### 5.9.2 Optical Matrix Multiplication

To perform matrix multiplication optically, we would first decompose a higher-dimension matrices into finite number of $2 \times 2$ elementary matrices, and then utilize the *outer product* technique to evaluate the multiplication of the elementary matrices [15]. As an illustration, let us assume that an $n \times m$ matrix is given as

$$A = \begin{bmatrix} a_{1,1} & a_{1,2} & \cdots & a_{1,m} \\ a_{2,1} & a_{2,2} & \cdots & a_{2,m} \\ \vdots & \vdots & & \vdots \\ a_{n,1} & a_{n,2} & \cdots & a_{n,m} \end{bmatrix} \tag{5.101}$$

which we can decompose into the following components:

$$A_{11} = \begin{bmatrix} a_{1,1} & \cdots & a_{1,m_1} \\ \vdots & & \vdots \\ a_{n_1,1} & \cdots & a_{n_1,m_1} \end{bmatrix}$$

$$A_{12} = \begin{bmatrix} a_{1,m_1+1} & \cdots & a_{1,m} \\ \vdots & & \vdots \\ a_{n_1,m_1+1} & \cdots & a_{n_1,m} \end{bmatrix} \tag{5.102}$$

$$A_{21} = \begin{bmatrix} a_{n_1+1,1} & \cdots & a_{n_1+1,m_1} \\ \vdots & & \vdots \\ a_{n,1} & \cdots & a_{n,m_1} \end{bmatrix}$$

$$A_{22} = \begin{bmatrix} a_{n_1+1,m_1+1} & \cdots & a_{n_1+1,m} \\ \vdots & & \vdots \\ a_{n,m_1+1} & \cdots & a_{n,m} \end{bmatrix}$$

Thus matrix $A$ can be written as

$$A = \begin{bmatrix} A_{11} & A_{12} \\ A_{21} & A_{22} \end{bmatrix} \tag{5.103}$$

Similarly, an $m \times l$ matrix $B$ can also be written as

$$B = \begin{bmatrix} B_{11} & B_{12} \\ B_{21} & B_{22} \end{bmatrix} \tag{5.104}$$

The product of $A$ and $B$ would be

$$AB = \begin{bmatrix} A_{11}B_{11} + A_{12}B_{21} & A_{11}B_{12} + A_{12}B_{22} \\ A_{21}B_{11} + A_{22}B_{21} & A_{21}B_{12} + A_{22}B_{22} \end{bmatrix} \tag{5.105}$$

Thus we see that multiplication of two arbitrary matrices can be carried out by multiplication of $2 \times 2$ elementary matrices. We shall now apply this matrix multiplication technique to binary-digit matrices.

Let us assume that two vectors **a** and **b** are given in the following binary matrix forms:

$$[\mathbf{a}] = [1 \quad 0 \quad 1] \tag{5.106}$$

$$[\mathbf{b}] = [0 \quad 1 \quad 1] \tag{5.107}$$

The outer product matrix of these two vectors can be obtained as

$$C = [\mathbf{a}]^{T}[\mathbf{b}] = \begin{bmatrix} 1 \\ 0 \\ 1 \end{bmatrix} [0 \quad 1 \quad 1] = \begin{bmatrix} 0 & 1 & 1 \\ 0 & 0 & 0 \\ 0 & 1 & 1 \end{bmatrix} \tag{5.108}$$

It is apparent that the $ij$th element of the matrix $C$ represents a logic operation AND between the $i$th element of the column vector $[\mathbf{a}]^{T}$ and the $j$th element of the row vector $[\mathbf{b}]$.

Since the vectors **a** and **b** represent two 3-bit binary numbers (i.e., $\mathbf{a} = 5$, $\mathbf{b} = 3$), then by adding up the diagonal elements of matrix $C$, a five-element vector $[\mathbf{c}] = (01111)$ would be obtained. Since this vector represents a 5-bit binary number, the final result would be $0 \times 2^4 + 1 \times 2^3 + 1 \times 2^2 + 1 \times 2^1 + 1 \times 2^0 = 15 = 5 \times 3$. It is therefore apparent that the matrix multiplication can be obtained with an outer product operation technique.

Since a multiplication between two single bits is equivalent to a logic operation AND, an outer product operation can be carried out with a MOSLM. In other words, column electrodes of the MOSLM can be addressed with two binary vectors **a** and **b**, and the outer product $C$ can be directly evaluated by the MOSLM.

To exploit the parallelism for matrix multiplication, a hybrid optical architecture is shown in Fig. 5.21. We note that the grating in Fourier plane causes constructive

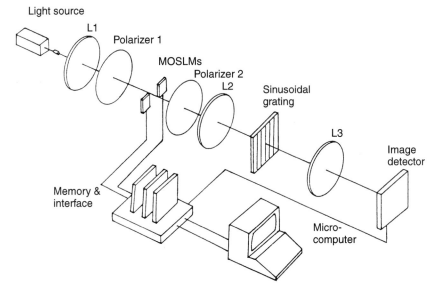

**Figure 5.21**   Hybrid optical architecture for matrix multiplication.

$$A = \begin{pmatrix} 16.75 & 22.50 \\ 22.50 & 56.50 \end{pmatrix} \qquad B = \begin{pmatrix} 50.25 & 53.25 \\ 51.50 & 10.25 \end{pmatrix}$$

$$C = A\,B = \begin{pmatrix} 2000.4375 & 1122.5625 \\ 4040.375 & 1777.25 \end{pmatrix}$$

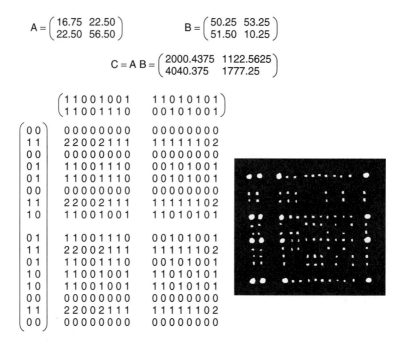

**Figure 5.22**   Matrix multiplication experiment.

interference of the outer products of *AB*, from the MOSLM, to form an elementary $2 \times 2$ matrix around the origin of the output plane. Figure 5.22 shows an experimental result obtained with matrix multiplication processing. Two $2 \times 2$ decimal matrices *A* and *B* and their product matrix *C* are shown. Each decimal matrix element is converted into an 8-bit binary number matrix, as shown in the lower part of the figure. For example, the binary, conversion of the decimal number 16.75 is (0 1 0 0 0 0 1 1). The $2 \times 2$ product matrix of *C* (i.e., four $8 \times 8$ extended binary matrix components) is also illustrated, represented by a $16 \times 16$ extended binary matrix. The output image irradiance of this matrix multiplication is depicted in the photograph on the right. Each highest-intensity spot represents a digit of 2, each intermediate-intensity spot represents a digit of 1, and each no spot represents a digit of 0. From this result we see that the proposed optical system is indeed able to perform sophisticated matrix multiplications.

## 5.10   ADAPTIVE PROCESSING

There are, however, two major architects for hybrid optical information processing depicted in Figs. 5.23 and 5.24, respectively. To illustrate the adaptivity of the joint-transform processor (JTP), we provide an interesting application to autonomous target tracking. The idea is to detect a target in the current frame with respect to the target in the previous frame. This makes the system adaptive by simply constantly updating the reference target with the dynamic input scene.

As an example, two sequential scenes of a moving object are displayed, on the input spatial light modulator (SLM) with the previous and the current frames positioned in the upper and lower half. If angular and scale tolerances are assumed smaller than

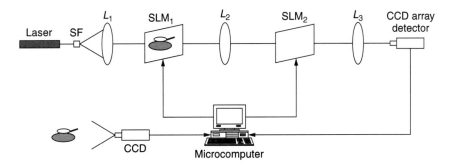

**Figure 5.23**   A microcomputer-based FDP.

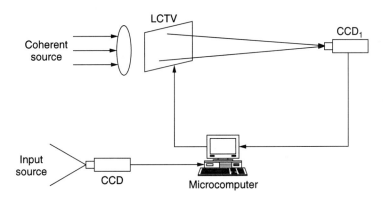

**Figure 5.24**   An ingle SLM JTP.

$\pm 5°$ and $\pm 10\%$ changes, and the motion of the target is relatively slow as compared to the processing cycle, then two high-intensity correlation peaks can be observed on the output plane.

By using a simple C language program as an example, the target position can be evaluated. The advantage of this technique is its adaptivity. For example, under a hypothetical situation, a camera is mounted on a moving space vehicle and it is assumed focusing at a fixed target on the ground terrain. As the space vehicle approaches the target, the detected scene changes continuously, the target size appears larger, and its orientation and shape change due to the motion of the vehicle. Using computer-aided design graphics, a three-dimensional tree-like model is created as a simulated target on the ground, illustrated in Fig. 5.25**a**. The JTP tracking system has little difficulty in correlating targets from different frames, even though the target in the first and the last frame look different, as shown in Fig. 5.25**b**.

## 5.11   LARGE-CAPACITY INFORMATION PROCESSING

Classic spatial matched filters are sensitive to rotational and scale variances. A score of approaches to developing composite-distortion-invariant filters have been reported. Among them, the synthetic discriminant function filter (SDF) [16] has played a central

(a)

(b)

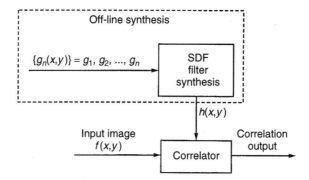

**Figure 5.25** (a) A sequence of nine simulated images as captured by a camera mounted on a moving space vehicle. Notice that only #1, #5, and #9 frames are shown. (b) Tracked positions of the ground target as seen from the vehicle's coordinate frame.

┌─────────────────────────────────────────────┐
 Off-line synthesis

$\{g_n(x,y)\} = g_1, g_2, ..., g_n$ ⟶ ┌─────────┐
                                      │   SDF   │
                                      │ filter  │
                                      │synthesis│
                                      └─────────┘
└─────────────────────────────────────────────┘
                                          │ $h(x,y)$
                                          ↓
Input image                          ┌──────────┐   Correlation
$f(x,y)$              ⟶              │Correlator│ ⟶    output
                                     └──────────┘

**Figure 5.26** A block diagram representation of the off-line SDF filters synthesis.

role. The idea of SDF can be viewed as a linear combination of classic matched filters. Figure 5.26 shows a block diagram for an SDF filtering system, where the SDF filter is off-line synthesized while the correlation is on-line processed. Techniques are proposed to improve the performance of the classic SDF filter. Nevertheless, it is not the intention here to describe all those techniques. Instead, two of the simplest approaches will be mentioned: (1) Higher noise tolerance can be achieved if the variance of the SDF filter is minimized, and (2) Sharper correlation peaks can be obtained if the average correlation energy of the SDF filter is minimized.

On the other hand, simulated annealing algorithms (SAA) have also recently been applied to pattern recognition [17]. A spatial domain bipolar filter obtained by SAA can be directly implemented on an input phase-modulating SLM in a JTP. To demonstrate the performance of a bipolar composite filter (BCF), a set of out-of-plane oriented (T72) tanks shown in Fig. 5.27**a** is used as a training set. The constructed BCF using the SAA is shown in Fig. 5.27**b**. An input scene to the JTP is shown in Fig. 5.28**a**, of which

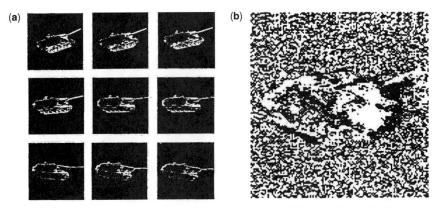

**Figure 5.27**  **(a)** Out-of-plane rotation training images of a T72 tank. **(b)** A bipolar composite filter obtained by SAA synthesis.

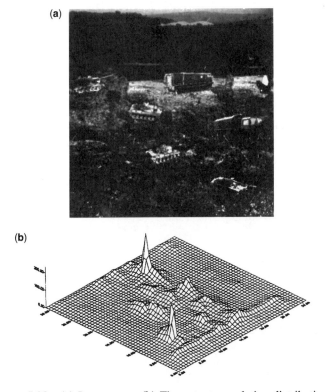

**Figure 5.28**  **(a)** Input scene. **(b)** The output correlation distributions.

the BCF of Fig. 5.27**b** is used, and the corresponding output correlation distribution as obtained is plotted in Fig. 5.28**b**. We see that target T72 tanks can indeed be extracted from the noise background.

Although SLMs can be used to display complex spatial filters, current state-of-the-art SLMs are low-resolution and low-capacity devices. On the other hand, photorefractive

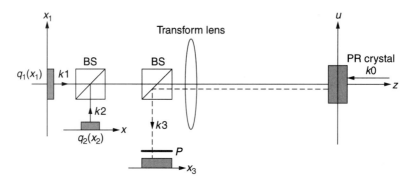

**Figure 5.29**    Wavelength-multiplexed reflection-type matched filter correlator.

(PR) materials offer real-time recording, high resolution, and massive storage capacity. However, thick PR material has limited shift invariant property due to Bragg diffraction [18]. Although thinner crystal can be used, the diffraction efficiency and storage capacity will be substantially reduced. Nevertheless high storage capacity, high diffraction efficiency, and large shift invariance can be achieved by using a reflection-type wavelength-multiplexed PR matched filter, as shown in Fig. 5.29. The matched filter is recorded by combining the Fourier spectrum of an object beam $q_1(x_1)$ with a reference plane wave from the opposite direction. Correlation can be done by inserting an input object $q_2(x_2)$ at plane $x$, by which the output correlation can be observed at plane $x_3$. To separate the reading beam $q_2(x_2)$ from the writing beam, the reading beam can be made orthogonally polarized to the writing beam by a polarized beam splitter. By using the coupled mode wave theory, the correlation peak intensity as a function of shift variable S can be shown to be

$$R(S) = \left| q_1^*(x_1) q_1(x_1) \operatorname{sinc}\left[ -\frac{\pi}{n} \frac{D}{\lambda} \frac{S(x_1 - S)}{f^2} \right] dx_1 \right|^2 \tag{5.109}$$

where $D$ is the thickness of the crystal. The shift tolerance figure-of-merit (FOM) can be shown as

$$(FOM)_{RC} = X S_{\max} = \frac{4n\lambda}{D} f^2 - S_{\max}^2 \tag{5.110}$$

where $S_{\max}$ denotes the maximum allowable shift. Plots of FOMs along with the VLC and the JTC are shown in Fig. 5.30, in which we see that the reflection-type PR correlator performs better. It has a higher shift tolerance, approximately one order higher than the VLC. The wavelength-multiplexed filter also offers higher and more uniform wavelength selectivity as compared to the angular-multiplexed technique.

## 5.12    OPTICAL NEURAL NETWORKS

Electronic computers can solve computational problems thousands of times faster and more accurately than can the human brain. However, for cognitive tasks, such as pattern recognition, understanding spoken language, and so on, the human brain is much

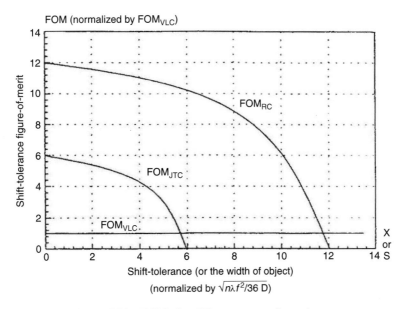

FOM (normalized by FOM$_{VLC}$)

**Figure 5.30**  FOMs for different types of correlators.

more efficient. In fact, these tasks are still beyond the reach of modern electronic computers. A neural network (*NN*) consists of a collection of processing elements, called neurons. Each neuron has many input signals, but only one fanned out signal to many pathways connected to other neurons. These pathways interconnect with other neurons to form a network. The operation of a neuron is determined by a transfer function that defines the neuron's output as a function of the input signals. Every connection entering a neuron has an adaptive coefficient called *weight* assigned to it. The weight determines the interconnection strength between neurons, and they can be changed through a learning rule that modifies the weights in response to input signals and the transfer function. The learning rule allows the response of the neuron to change with time, depending on the nature of the input signals. This means that the network adapts to the environment and organizes the information within itself, a type of learning.

Generally speaking, a one-layer neural network of $N$ neurons has $N^2$ interconnections. The transfer function of a neuron can be described by a nonlinear relationship such as a step function, making the output of a neuron either 0 or 1 (binary), or a sigmoid function, which gives rise to analog values. The operation of the $I$'th neuron in the network can be represented by a retrieval equation, as given by

$$u_i = f\left\{\sum_{j=1}^{N} T_{ij} u_j - \theta_i\right\} \qquad (5.111)$$

where $u_j$ is the activation potential of the $i$'th neuron; $T_{ij}$ is the interconnection weight matrix (IWM), between the $j$'th neuron; and $i$'th neuron, and $f$ is a nonlinear processing operator.

Light beams propagating in space will not interfere with each other, and optical systems generally have large space-bandwidth-products. These are the primary

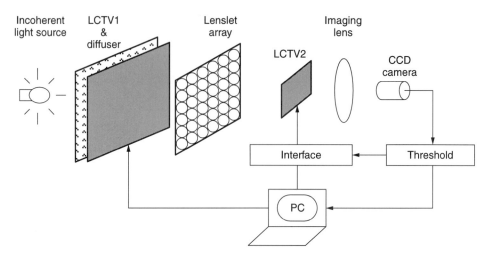

**Figure 5.31**   An LCTV-based optical neural network.

features that prompted the optical implementation of neural networks [19]. As shown in Fig. 5.31, the lenslet array is used to simulate the neuron interconnection. As the data flow is primarily controlled by the microcomputer, the hybrid optical NN is indeed an adaptive neural network.

Mention must be made that one of the most frequently used neural net models is the Hopfield model [20]. The model utilizes an associative memory retrieval process equivalent to an iteratively thresholded matrix-vector outer-product expression. Although Hopfield NN is capable of retrieving distorted and partial patterns, it is only emphasized in the intrapattern association, which ignores the association among the stored exemplars. An alternative approach can be achieved using interpattern association (IPA), by which simple logic operation can be used for the NN construction [21].

Pattern translation NN can be constructed by using the hetero-association IPA. To illustrate an example, an input–output training set is shown in Fig. 5.32**a**, and the corresponding hetero-association IWM is shown in Fig. 5.32**b**. If a partial Arabic numeral 4 is presented to the optical NN, a translated Chinese numeral can be obtained, as illustrated in Fig. 5.32(**c**). The hetero-association NN can indeed translate patterns.

## 5.13   FUNDAMENTAL APPROACHES

There are two basic approaches to optimum information processing. One is by maximizing the output signal-to-noise ratio that leads to matched filtering, and the other by minimizing the mean-square errors (MSE) that lead to a Wiener–Hopf solution. In other words, if we have a prior knowledge of the target (or signal), by maximizing the output signal-to-noise ratio and under the assumption of additive white-Gaussian noise with zero mean, then the solution for the filter synthesis is the result of a matched filter. A matched filter gives rise to autocorrelation detection, which results in highest correlation peak intensity detection. This approach has been widely used in a radar system as well as in optical pattern recognition and detection.

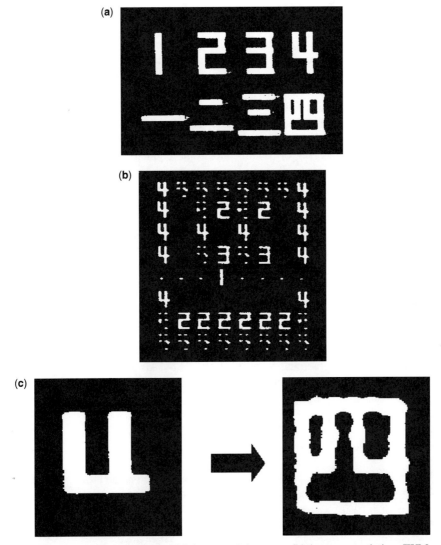

**Figure 5.32** An ANN translation, **(a)** input training set, **(b)** hetero-association IWM, and **(c)** partial input to translated output.

On the other hand, if the incoming signal (or target) is unknown, then the restoration of the incoming signal (or target) is very essential for the receiver. For optimum restoration or reconstruction of the detected target (or signal), an optimum filter can be realized by minimizing the MSE of the actual output signal (i.e., the restored target) with respect to the desired output signal, under the constrain of the physical realizable condition. The end result of the minimized MSE will generally lead us to an open-ended Wiener–Hopf solution. However, under specific conditions, a unique solution may be obtained. Nevertheless, either case for obtaining optimum information processing is not free. In most cases, it will require an excessive amount of *entropy* to accomplish it. A fundamental question is can we afford it? The answer is that sometimes we can afford it and sometimes we cannot.

## 5.14 CONCLUDING REMARK

There is, however, a common starting point between computer imaging processing and optical information processing. It is basically spatial domain processing. In other words, a temporal signal has to be first converted to a spatial signal before the optical information processing can be carried out. One of the examples must be the synthesis aperture radar (SAR) processing, by which a temporal microwave signal is converted into a spatial scanned format for optical processing. As optical information processing offers high resolution and parallel processing capacity, SAR processing has been very successful until the emerging of a better electronic counterpart in recent years, which has been totally taken over by the conventional optical processing.

Although a digital computer is essential for a sequential processor, its flexibility for implementation has overshadowed the optical technique. The major concern must be due to various physical constrains for the development of suitable electro-optic devices for optical processing. In other words, just taking advantage of optical parallelism, high resolution, and massive connectivity may not be sufficient to compete with the electronic counterpart. An idea to overcome this shortcoming is to develop better optical-digital interface devices, by which both the merits of optics and electronic counterparts can be simultaneously exploited [22].

### Acknowledgments

We acknowledge the gracious support of the U.S. Air Force Office of Scientific Research and Rome Air Development Center at Hanscom Air Force Base in the area of optical signal processing.

### REFERENCES

1. M. Born and E. Wolf, *Principles of Optics*, 2nd rev. ed., Pergamon Press, New York, 1964.
2. A. Vander Lugt, "Signal Detection by Complex Signal Filtering," *IRE Trans. Inf. Theory*, **IT-6**, p. 386, 1960.
3. L. J. Cutrona, E. N. Leith, C. J. Palermo, and L. J. Porcello, "Optical Data Processing and Filtering Systems," *IRE Trans. Inf. Theory*, **IT-16**, p. 386, 1960.
4. A. W. Lohmann, "Incoherent Optical Processing of Complex Data," *Appl. Opt.*, **16**, p. 261, 1977.
5. W. T. Rhodes, "Bipolar Pointspread Function by Phase Switching," *Appl. Opt.*, **16**, p. 265, 1977.
6. E. Leith and J. Roth, "White-Light Optical Processing and Holography," *Appl. Opt.*, **16**, p. 2565, 1977.
7. F. T. S. Yu, "A New Technique of Incoherent Complex Signal Detection," *Opt. Commun.*, **27**, p. 23, 1978.
8. G. M. Morris and N. George, "Matched Filtering Using Band-Limited Illumination," *Opt. Lett.*, **5**, p. 202, 1978.
9. W. Stoner, "Incoherent Optical Processing via Spatially Offset Pupil Masks," *Appl. Opt.*, **17**, p. 2454, 1978.
10. F. T. S. Yu, *Optical Information Processing*, Wiley-Interscience, New York, 1983.
11. F. T. S. Yu, *White-Light Optical Signal Processing*, Wiley-Interscience, New York, 1985.

12. W. P. Bleha, L. P. Lipton, E. Wiener-Avner, J. Grinberg, P. G. Reif, D. Casasent, H. B. Brown, and B. V. Markevitch, "Application of the Liquid Crystal Light Valve to Real-Time Optical Data Processing," *Opt. Eng.*, **17**, p. 371, 1978.

13. W. E. Ross, D. Psaltis, and R. H. Anderson, "Two-Dimensional Magneto-Optic Spatial Light Modulator for Signal Processing," *Opt. Eng.*, **22**, p. 485, 1983.

14. F. T. S. Yu and X. J. Lu, "A Programmable Joint Transform Correlator," *Opt. Commun.*, **52**, p. 10, 1984.

15. F. T. S. Yu and S. Jutamulia, *Optical Signal Processing, Computing and Neural Networks*, Krieger Publishing Co., Melbourne, Australia, 2000.

16. C. F. Hester and D. Casasent, "Multivariant Technique for Multiclass Pattern Recognition," *Appl. Opt.*, **19**, p. 1758, 1980.

17. S. Yin et al., "Design of a Bipolar Composite Filter Using Simulated Annealing Algorithm," *Opt. Lett.*, **20**, p. 1409, 1996.

18. F. T. S. Yu and S. Yin, "Bragg Diffraction-Limited Photorefractive Crystal-Based Correlators," *Opt. Eng.*, **34**, p. 2225, 1995.

19. D. Psaltis and N. Farhat, "Optical Information Processing Based on An Associative-Memory Model of Neural Nets with Thresholding and Feedback," *Opt. Lett.*, **10**, p. 98, 1985.

20. J. J. Hopfield, "Neural Network and Physical System with Emergent Collective Computational Abilities," *Proc. Natl. Acad. Sci.*, **79**, p. 2554, 1982.

21. T. W. Lu, X. Xu, S. Wu, and F. T. S. Yu, "A Neural Network Model Using Inter-Pattern Association (IPA)," *Appl. Opt.*, **29**, p. 284, 1990.

22. F. T. S. Yu, "Optical Information Processing Awaits Optoelectronic Devices," *Laser Focus World*, **82**, p. 71, 2002.

# 6

# SEMICONDUCTOR LASERS*

JANE J. YANG
*Advanced Photonic Sciences*
*Los Angeles, California*

BING W. LIANG
*Lytek Inc.*
*Phoenix, Arizona*

## 6.1 INTRODUCTION

The invention of the semiconductor laser dates back to 1962, when it was discovered that stimulated emission from a GaAs *p-n* junction could be obtained. The reader is referred to an early review paper that recalls the early development period [1]. Journals such as the *IEEE Special Issue on Semiconductor Lasers*, started in 1968 and published every other year, describe the early and more recent experiments. Between 1962 and 1971, many improvements in semiconductor laser technology were made, the most notable being the concept of the double-hetero-structure (DH) configuration [2, 3], stripe geometry configuration [4], and the demonstration of room temperature continuous wave (cw) operation [5]. With new and sophisticated crystal growth techniques such as molecular beam epitaxy (MBE) [6, 7] and metal organic chemical vapor deposition (MOCVD) [8–10] developed in the late 1970s, a new generation of semiconductor lasers has been designed and built that encompass an ultra-thin active layer (<200 A) and unique wave-guide structures. Such laser structures, referred to as quantum well (QW) lasers, have significantly improved output characteristics compared to those of conventional semiconductor lasers: low threshold current [11–18], high efficiency with high output power [19, 20], enhanced modulation response [21–23], reduced line width [24, 25], and the potential for incorporation of a builtin nonabsorbing mirror structure near the facets and the use of strained quantum well layers [20] for relieving

---

* This Chapter was revised from the chapter written by Dr. Luis Figueroa in the first edition.

*Handbook of Optical Components and Engineering*, Edited by Kai Chang
ISBN 0-471-39055-0 © 2003 John Wiley & Sons, Inc.

lattice mismatch. Such laser structures represent a technological discontinuity in the development of semiconductor laser technology and have had a major impact on the advancement of opto-electronics technology.

In this chapter, we provide a summary of the basic principles involved in the design and operation of various semiconductor laser structures for a variety of applications. This chapter is divided into ten sections. In Section 6.2, we describe many applications of laser diodes and future needs. In Section 6.3, we discuss basic operation, including the concept of in-plane diode lasers, semiconductor materials for diode lasers, material growth technologies, quantum well (QW) lasers, strained quantum well lasers, stripe geometry and current confinement techniques, basic characteristics of laser diodes, and single and multiple lateral and longitudinal mode lasers. In Section 6.4, we discuss concepts of high-power operation, including the effect of catastrophic damage and techniques of decreasing the facet intensity. The use of nonabsorbing mirror (NAM) technology to revolutionize the characteristics of high-power lasers is also included. Design curves relating to far field angle to laser geometry and the detailed thermal behavior that limits the output power of (GaAl)As/GaAs single-mode lasers are presented. In Section 6.5, we discuss surface emitting lasers and their potential for two-dimensional integration and high output powers. In Section 6.6, we discuss tunable laser diodes. Section 6.7 focuses on high-speed characteristics of laser diodes. In Section 6.8, detailed discussion of vertical cavity surface emitting laser diodes (VCSELs) is provided. In Section 6.9, we discuss quantum cascade laser diodes emitting in the mid-infrared region of the spectrum as well as nano-lasers, which include quantum wire and quantum dot diode lasers. In Section 6.10, we conclude the chapter with a short glimpse of what the future may bring in this important technology.

## 6.2   APPLICATIONS

The initial demand for semiconductor lasers was in fiber-optic communications at relatively low powers: 3 to 5 mW, delivered in a single, stable beam. Such lasers, cheaper, compact, and more reliable than gas lasers and easy to modulate by switching an input electric current on and off, have been installed in short- and long-range communications systems around the world and now are also extensively used in digital video and audio disk players.

Since 1980, however, advancement in key technologies such as material growth (MBE and MOCVD), micro-machining (reactive ion etching, ion beam etching, etc.), and packaging and cooling technologies has improved semiconductor laser device performance with a quantum leap. High output power (over 1 W from a single GaAs-based device) and high modulation rate can be easily achieved. In addition, very thin layer growth, which became possible using MBE and MOCVD techniques, allows some level of lattice mismatch. These so-called strained layers [26] not only improve the opto-electronic properties, but also they broaden the emission spectrum due to more flexible material compositions. Applications such as high data rate optical recording, high-speed printing, single- and multimode data-bus distribution systems, long distance transmission, free space optical communications, local area networks (LANs), Doppler optical radar, optical signal processing, high-speed optical-microwave sources, material processing, marking/cutting/drilling/welding, illumination, countermeasure, medical applications, and pumping sources for other solid state and fiber lasers are now widely pursued. New device concepts and advanced technologies have been developed to meet these demands.

Depending on the intended application, diode laser structures can be tailored to optimize the features such as coherence, power, and efficiency. In addition, when application and structure are matched, further device optimization can be addressed to meet the cost/yield, lifetime and stability (power and wavelength) requirements.

## 6.3  BASIC OPERATION

### 6.3.1  Semiconductor Laser Device Concepts — Simple In-Plane Diode Lasers

There are two practical semiconductor diode laser structures at present: one with in-plane cavities and the other with vertical cavities. The in-plane (or edge-emitting) lasers were the exclusive type of diode lasers since invention in the 1960s. The surface emitting vertical cavity lasers (VCSELs), although developed in the university laboratories in late 1970s [27], were not practical for applications until the early 1990s. The concept and operation of in-plane lasers are illustrated in this section, while the VCSELs will be described in Section 6.8.

A simple in-plane diode laser structure is shown in Fig. 6.1. This laser device can be defined by three directions: transverse, lateral, and longitudinal. In the **transverse direction**, all diode lasers are essentially a multilayered structure composed of several types of semiconductor materials. The most widely used diode lasers are red and near-infrared with wavelengths of 0.65 to 1.0 μm and long wavelengths of 1.3 to 1.67 μm. Red and near-infrared lasers are composed of layers on gallium arsenide (GaAs), aluminum gallium arsenide (AlGaAs), indium aluminum gallium arsenide (InAlGaAs), and indium gallium aluminum phosphide (InGaAlP), and they are grown on GaAs substrates. Long wavelength devices are made out of indium gallium arsenide phosphide (InGaAsP), indium gallium aluminum arsenide (InGaAlAs), and indium phosphide (InP) on InP substrates. A spectral emission range and materials used for diode lasers are summarized in Fig. 6.2.

The semiconductor materials are chemically doped with impurities to give them either an excess of electrons (n-type material) or an excess of electron vacancies, called holes, which characterize p-type material. When the diode is in operation, electrons from the n-type layer and holes from the p-type layer (carriers) are injected into the active layer where they recombine and emit light.

As the optical index of refraction of the active-layer material is larger than that of cladding layers above and below it, the light is trapped in the dielectric wave guide formed by the two cladding layers and the active layer and propagates primarily within the active layer but with some distribution into the cladding layers.

For a simple diode laser structure of Fig. 6.1, the active region (0.1 to 0.2 μm thick) confines both the injected carriers and the optical mode. This is called a double-hetero-structure-type device [2, 3] (Fig. 6.3). To improve the laser efficiency, a separate confinement hetero-structure (SCH) single or multiple quantum-well (SWQ or MQW) structure [21–23] has been developed (Fig. 6.4).

In the **longitudinal direction**, crystal facets at each end of the diode act as mirrors that confine the light within the device, which is then substantially amplified as it travels back and forth longitudinally. Lasing action occurs when enough carriers are injected into the active layer to provide the optical gain needed to overcome the

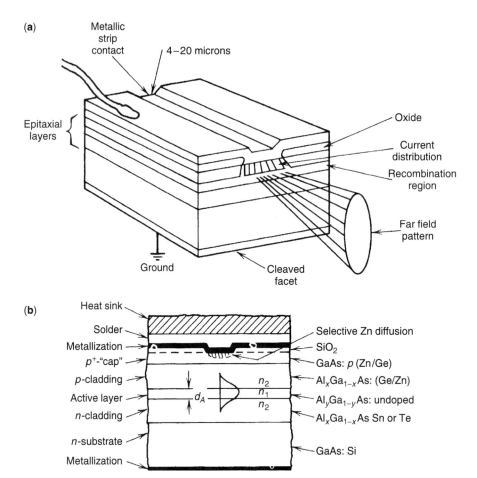

**Figure 6.1** (a) Schematic diagram of a typical stripe-geometry DH laser structure; (b) cross-sectional view showing the various epitaxial layers.

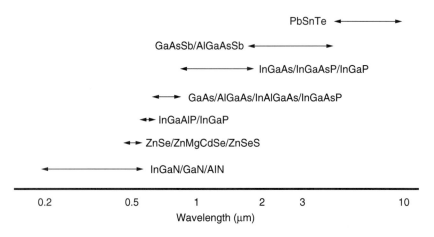

**Figure 6.2** Spectral emission range and semiconductor materials used for diode lasers.

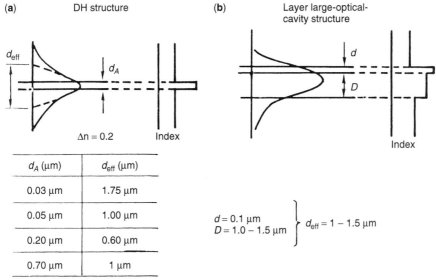

**Figure 6.3**  Schematic diagram of the two most commonly used hetero-structure configurations for high-power laser diodes.

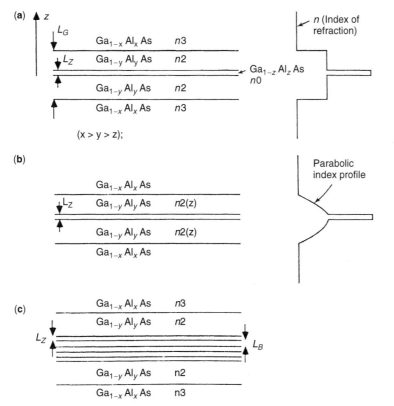

**Figure 6.4**  Schematic diagrams of various quantum well configurations: **(a)** separate carrier and optical confinement hetero-structure (SCH); **(b)** graded-index and separate carrier and optical confinement hetero-structure (GRIN-SCH); **(c)** multiple quantum-well (MQW) hetero-structure.

cavity's internal and external losses. To improve the efficiency and maximum output power of a semiconductor laser, dielectric coatings are applied to both facets. Typically, a combination of high-reflectivity ($\sim$90%) and low-reflectivity ($<$10%) dielectric coatings is applied on back and front facets. A thin $Al_2O_3$ layer is normally used for the front facet and several ($\lambda$/4) stacks of $Al_2O_3-Si$ combination for the rear facet. Other coatings, such as SiO/Si, $SiO_2$/Si, and $Si_3N_4$/Si, have also been used. It is important to use materials with similar thermal expansion coefficients to reduce internal stress. The optimum design maximizes the output power while meeting the constraints of keeping the threshold current relatively constant and maximizing the catastrophic power level.

The performance of a diode laser is determined by both its chemical composition and its physical geometry. It was recognized early in the development of injection lasers that confinement of the current and lasing emission to a narrow-stripe geometry ($<$10 $\mu$m) in the **lateral direction**, as shown in Fig. 6.1, is important in order to obtain continuous wave (cw) operation.

To achieve a cw operation at room temperature, low threshold current is needed, which in turn requires a thin active layer ($<$0.3 $\mu$m). The lasing spot size along transverse direction perpendicular to the junction plane is rather small (0.5 to 1 $\mu$m), resulting in a divergent beam of 30° to 40° full-width half-power. In the junction plane, light confinement is controlled by the current flow. The lasing spot is comparable in size to the contact stripe width (Fig. 6.1), which results in a beam with 10° to 15° of divergence angle. Thus, the laser emission pattern is elliptical.

As MBE [6, 7] and MOCVD [8–10] technologies matured and the quantum-well structures were established, threshold current density became so low that high-power broad area lasers up to 5 W with 500 $\mu$m width have become commercially available for many applications.

To better understand the diode laser performance, several key technologies associated with device fabrication will be discussed in the next subsections.

### 6.3.2 Semiconductor Materials for Diode Lasers

As shown in Fig. 6.2, several semiconductor material systems can be used to make high-quality lasers. For DH-structures, at least two compatible materials with different bandgaps are needed for their active region and cladding layers. For more complex separated confinement hetero (SCH)-structures, more layers are required. To build a high-performance laser, these materials need to have the same crystal structure and nearly the same lattice constant. Figure 6.5 shows the bandgap versus lattice constant for some established III-V compound semiconductor materials. These materials are ideal for lasers in the range of 0.7 to 1.6 $\mu$m. This range covers applications in fiber-optic communication at 1.31 $\mu$m and 1.55 $\mu$m, pumps for fiber amplifiers at 0.98 $\mu$m and 1.48 $\mu$m, pumps for Nd-doped YAG at 0.808 $\mu$m and other solid state materials, as well as compact disk (CD) players at 0.78 $\mu$m.

As seen in Fig. 6.5, the AlGaAs ternary line is almost vertical; replacing Al in GaAs does not change the lattice constant significantly. Thus, AlGaAs can be deposited on GaAs substrate without a lattice mismatch problem. To date, AlGaAs/GaAs material systems are considered to be the most mature type for high-power semiconductor lasers.

In traditional AlGaAs laser diodes, the presence of aluminum in the active region may limit device performance especially at high-power operation. Aluminum-free (Al-free) material systems were thus developed to replace AlGaAs materials [28–40].

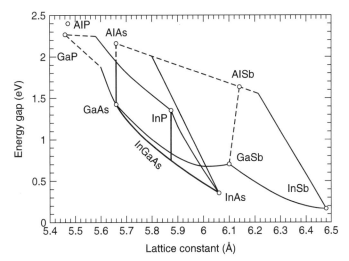

**Figure 6.5**  Energy gap vs lattice constant of ternary compounds defined by curves that connect the illustrated binaries.

Al-free materials were realized only after quantum-well and strained quantum-well technologies were established. The advantages of the Al-free system over the traditional AlGaAs systems are as follows: (1) No degradation due to oxidation of aluminum during fabrication process and laser operation [28]; (2) the InGaAsP mass transport technique can be applied for the fabrication of a buried hetero-structure [29], and very low recombination velocity at the GaAs/InGaP interface is expected [41]. Using aluminum-free systems, many high-quality, high-power devices were fabricated [30–36].

A significant problem for the Al-free compounds has been the difficulty in fabricating device structures. Varying the alloy ratio in layers of InGaAsP structures results in smaller bandgap differences than for layers containing aluminum. This allows greater leakage of current carriers and results in lower efficiency and higher device temperature. To solve these problems, Mawst et al. developed a new concept. They use Al-free compounds only in the active layer while retaining high bandgap aluminum compounds in cladding layers. With this approach, a 100-μm device operating at 805 nm with a cw output power greater than 8 W was demonstrated [37]. Short wavelength (0.7–0.78 μm) Al-free laser diodes have also been developed recently [38, 39].

For long-distance fiber optics, InGaAsP/InP materials are used. This system provides emission wavelengths in the range of 1.0–1.6 μm, at which silica fiber has minima in loss (1.55 μm) and dispersion (1.3 μm). Using InP as a substrate, a series of lattice matched quaternaries ranging from a combination of InP, InAs, GaAs, and GaP, as indicated by vertical line from InGaAs to InP in Fig. 6.5, may be fabricated. For better understanding of these materials, the reader is referred to a book edited by Wada and Hasegawa [42]. Similar approaches can be applied to other material systems.

In addition to the traditional material systems mentioned above for long wavelength laser diodes, recent development of temperature-insensitive InGaAsN quantum-well devices grown on GaAs substrates demonstrates potential as a new group of materials for emission in 1.3–1.6 μm range [43–49].

Visible lasers using material systems of InAlGaP/GaAs (red) and AlGaN/InGaN (blue/green) (not shown in Fig. 6.5) have attracted a lot of attention because of their applications in high-density data storage, display, and medical areas. For more information in these areas, the reader is referred to the papers [50–61] and books by Coldren and Corzine [26] and Nakamura and Fasol [62], respectively.

### 6.3.3  Material Growth Technologies

Semiconductor epitaxial layers can be obtained by liquid phase epitaxy (LPE) [63], molecular-beam epitaxy (MBE) [6, 7], and metal-organic chemical vapor deposition (MOCVD) [8–10]. MOCVD is often referred to as organo-metallic vapor-phase epitaxy (OMVPE). Early development of semiconductor diode lasers relied on LPE. However, MOCVD and MBE have been used, in recent years, to produce high-quality diode lasers because of their advantages in ability to produce accurate layer thickness and composition control, as well as abrupt interfaces, which are very important for quantum-well structures. Furthermore, MBE and MOCVD provide features of large area uniformity and reproducibility that are essential for mass production requirement.

LPE is essentially a process of precipitation of material from a cooling solution onto an underlying substrate. The solution and the substrate are in separate boats in a growth apparatus in which the desired temperature is accurately controlled. The solution is in saturated condition. When the desired temperature is reached, the solution is brought into contact with the surface of the substrate, and the temperature is then slowly reduced at a controlled rate and over a time interval that is appropriate for the layer designed. When the substrate is a single crystal and the lattice constant of the precipitating material is the same or nearly the same as that of the substrate, the precipitating material forms an epitaxial layer on the substrate surface.

LPE has the advantage of simplicity and high deposition rate. However, LPE suffers from drawbacks of poor reproducibility, poor layer thickness control, and uniformity, which make the technology undesirable for mass production of advanced designs such as QWs.

The MOCVD process for the formation of semiconductor compounds makes use of the following simplified reaction to produce GaAs:

$$\text{Trimethylgallium} + \text{Arsine} \xrightarrow[\text{600–800°C}]{} \text{Gallium Arsenide} + \text{Methane}$$

The metal trimethylgallium (TMGa) is liquid at the source temperature typically employed. High-purity hydrogen gas is bubbled through the liquid, and the reactant vapors are mixed with arsine and pyrolyzed (chemically combined at high temperature) at or near the substrate. Similarly, by mixing TMGa with trimethylaluminum (TMAl) and arsine, AlGaAs is obtained upon pyrolysis; the chemical composition of the resulting alloy is controlled by the ratio of the reactants. The $p$-type and $n$-type materials are obtained by incorporating a $p$-type doping such as diethylzinc (DEZn) and a $n$-type doping such as hydrogen selenide ($H_2Se$), respectively, into the reactor during growth. The doping levels are similarly controlled by the doping partial pressure. Layer thickness is controlled by growth rate and time.

The MBE is carried out in an ultra-high vacuum chamber. Elements such as Ga and As are evaporated from heated cells and epitaxially grown on a heated substrate. Each source has a shutter. With reflection high-energy electron diffraction (RHEED) and an ion-beam gauge, it is much easier to perform in situ monitoring of the growth condition. Thus, MBE has the advantage of providing very accurate control of the composition and the growth rate.

Some hybrid forms of MOCVD and MBE techniques have also been developed, i.e., gas source MBE (GSMBE), metal-organic MBE (MOMBE), and chemical beam epitaxy (CBE) [64, 65]. For more information on these technologies, readers are referred to published papers [64, 65].

### 6.3.4  Quantum-Well Lasers

A quantum-well (QW) double hetero-structure laser is formed when the active layer thickness, Lz, becomes comparable to the electron de Broglie wavelength ($\lambda = h/p$, where $h$ is Planck's constant and $p$ is the electron momentum) [66]. In such a situation, propagation of electrons is restricted in the transverse direction, and the density of states function, $g_c$, becomes a series of step function as shown in Fig. 6.6. It is readily shown that the energy of an electron confined to an infinite square well potential can be expressed as

$$E = E_n + \frac{h^2}{2m_e^*}(K_x^2 + K_y^2)$$

$$E_n = \frac{h^2}{2m_e^*}\left(\frac{n\pi}{L_z}\right)^2 \qquad \text{where } n = 1, 2, 3, \ldots$$

(6.1)

The injected electron density can be calculated using Fermi–Dirac statistics, and the electron density per unit energy is given by

$$n_c(E) = \int f_c(E)g_c(E)\,dE \tag{6.2}$$

where

$$f_c(E) = \frac{1}{1 + \exp[(E - E_{fn})/kT} \tag{6.3}$$

and $E_{fn}$ is the quasi Fermi level for injected electrons.

Figure 6.6 shows a schematic diagram of $n_c(E)$. Note how the distribution of injected electrons is shifted to higher energies for a QW laser compared to a conventional DH laser as a result of the quantization of the energy. As the gain of the laser is proportional to $n_{inj}$, the gain of the QW laser is higher than that of the conventional DH lasers [Fig. 6.6c]. This is a consequence of the narrower spectral gain line width, which results in a more sharply peaked carrier density distribution [67]. This phenomenon leads to a higher gain constant ($\beta$) and operation at a shorter wavelength for the QW laser. If the carrier injection level is high enough, the light emission could be shifted from the $n = 1$ to the $n = 2$ sub-bands. The threshold current density can be reduced compared to a conventional DH laser by reducing the active-layer thickness and keeping the confinement factor, $\Gamma$, high as we reduce the active-layer thickness. In general, separate carrier and optical confinement (SCH) structures are designed for this purpose, and the two basic types used are shown in Fig. 6.4. The local gain for a QW laser can be expressed as [21]

$$g = \frac{\eta_i \beta(L_z)J}{NL_z} - \alpha_0(L_z) \tag{6.4}$$

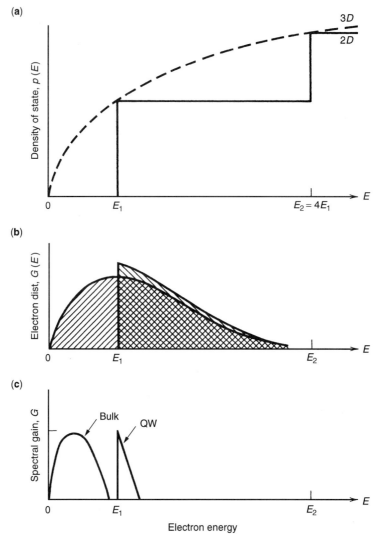

**Figure 6.6**  **(a)** Schematic diagrams of density of states for bulk material (3D) and QW het-ero-structures (2D). **(b)** The distribution of injected carriers in bulk and QW structures needed to achieve the same peak gain spectra as shown in **(c)**. Carrier densities of $n = 2 \times 10^{18}$ cm$^{-3}$ and $1.4 \times 10^{18}$ cm$^{-3}$ are required for the bulk and QW structure to reach the same peak gain value, respectively (from Ref. 20).

where $N$ = number of quantum wells; $\beta$ = gain constant, which is a function of well thickness $L_z$ ($L_Z$ in micrometers); $\alpha_0(L_Z)$ = loss constant, which is also a function of the well thickness; and $\eta_i$ = internal efficiency.

The expression (6.4) holds (i.e., remains linear for $g_{\text{max}} \lesssim 333$ cm$^{-1}$ as determined from the data of Arakawa and Yariv [21] for $N = 1$, $L_Z = 0.01$ μm, and $L_G = 0.1$ μm; we use

$$\Gamma = 0.3L_z/L_G \tag{6.5}$$

where $L_G$ is the thickness of the guide layer ($L_G \sim 0.1$ μm for the example considered).

In general, it would be helpful to use a more general expression for the confinement factor of a symmetric SCH laser. We have derived such an expression, which is accurate for $L_G \geqslant 0.5$ μm:

$$\Gamma_{\mathrm{SCH}} = \frac{L_z}{L'_G} \frac{D'^2}{D'^2 + 2} \tag{6.6}$$

where

$$L'_G = \frac{L_G}{2} + \left[ \frac{2\pi}{\lambda_0} (n_{\mathrm{eff}}^2 - n_3^2)^{1/2} \right]^{-1}$$

$$D' = \frac{2}{\lambda_0} L_G (n_2^2 - n_3^2)^{1/2} \tag{6.7}$$

and

$$n_{\mathrm{eff}} \simeq n_3 + \frac{D'^2}{4} \left( \frac{\sqrt{9 + 4D'^2}}{2 + D^2} - 1 \right)^2 (\Delta n) \quad \Delta n = n_2 - n_3 \tag{6.8}$$

Furthermore, if we assume that a parabolic index profile can be represented by an equivalent step-indexed distribution [Fig. 6.4**b**] with index difference, $\Delta n_{\mathrm{eq}}$, given by

$$\Delta n_{\mathrm{eq}} = \tfrac{2}{3} \Delta n \qquad \Delta n \text{ is the peak index difference for the parabolic index profile}$$

then we can also estimate the confinement factor for a GRIN-SCH laser by replacing $\Delta n$ in Eqs. (6.6)–(6.8) with $\Delta n_{\mathrm{eq}}$.

Arakawa and Yariv have pointed out that in order to minimize the threshold current density for a QW laser, we must optimize the number of quantum wells. This arises because $n_{\mathrm{inj}} \propto 1/NL_Z$ and we expect gain saturation at relatively low modal gains ($\Gamma g_{\mathrm{th}}$). Thus, the gain will increase sublinearly with current for peak gains greater than a maximum value. If the required modal gain is high, as for the case of a laser with high losses resulting from either high mirror or internal losses, then in order to remain in the linear region of the $g$ versus $J$ curve, we must have $N > 1$. An example of the calculation for $J_{\mathrm{th}}$ is given below. The modal gain at threshold is given by ($L_G = 0.1$ μm)

$$\Gamma g_{\mathrm{th}} = G_{\mathrm{th}} = \alpha_s + \frac{1}{L} \ln \frac{1}{R} \tag{6.9}$$

$$J_{\mathrm{th}} = \frac{G_{\mathrm{th}} L_G}{0.3\beta} + \frac{\alpha_0 L_G}{0.3\beta} \tag{6.10}$$

$$N > \frac{G_{\mathrm{th}}}{0.3 g_{\mathrm{max}}} \frac{L_G}{L_z} \quad \left( \text{or more generally, } \frac{G_{\mathrm{th}}}{\Gamma g_{\mathrm{max}}} \right) \tag{6.11}$$

If we have a laser with $L_z = 0.01$ μm, $L_G = 0.1$ μm, and $G_{\mathrm{th}} = 50$ cm$^{-1}$, $g_{\mathrm{max}} \approx 333$ cm$^{-1}$; obtained from the data of Ref. 21, we obtain $N = 5$ from Eq. (6.11); five quantum wells are required to achieve the minimum threshold current density.

The values of $\alpha_0(L_z)$ and $\beta(L_z)$ can be estimated from Fig. 1 of Arakawa and Yariv [21]. We estimate

$$\beta(L_z = 0.01 \text{ μm}) = 0.075 \text{ cm} \cdot \text{A/μm}$$

$$\alpha_0 = 19 \text{ cm}^{-1} \tag{6.12}$$

Thus, a quantum-well laser has a significantly higher gain constant (approximately twice that of a conventional laser). In addition, the value of $\alpha_0$ is significantly reduced compared to a conventional laser ($\alpha_0 = 210$ cm$^{-1}$). Using Eq. (6.12) in Eq. (6.10), we find that

$$J_{\text{th}}(L_z = 0.01 \ \mu\text{m}) = 302 \ \text{A/cm}^2 \tag{6.13}$$

which is in reasonable agreement with the value obtained in Fig. 2 of Arakawa and Yariv ($J_{\text{th}} \sim 270$ A/cm$^2$) and is reproduced in our Fig. 6.7. We should note that calculations for other values of $L_z$ require the determination of the gain versus current relationship as a function of $L_z$ and the calculation of $\beta(L_z)$ and $\alpha_0(L_z)$ from these graphs. The value of $\beta(L_z)$ can be deduced from the variation of the resonance frequency, $f_0$, as a function of $L_z$ because

$$f_0 \propto (\beta)^{1/2} \tag{6.14}$$

and thus QW lasers are expected to have a significantly higher frequency response than that of conventional lasers. Using the results of Arakawa and Yariv, we estimate

$$\beta(L_z) = 0.0123(L_z)^{-0.399} \quad (0.005 \ \mu\text{m} < L_z > 0.05 \ \mu\text{m}) \tag{6.15}$$

$L_z$ is given in micrometers and $\beta$ is in cm $\cdot$ $\mu$m/A.

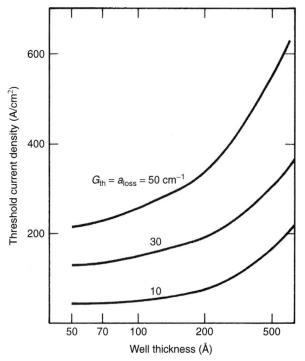

**Figure 6.7**    Threshold current as a function of the quantum-well thickness with total absorption as a parameter. The number of quantum wells is optimized so that the threshold current is a minimum (after Ref. 21).

It has also been observed [68–71] that unlike conventional double hetero-structure (DH) lasers, the threshold current ($J_{th}$) of single quantum-well (SQW) lasers increases anomalously as the laser length $L$ decreases. In DH lasers, gain increases with current linearly, resulting in $J_{th}$, which first decreases as $L$ decreases, and, at very short $L$, approaches a constant value that is dependent on mirror losses. For QW lasers, a linear gain/current relation predicts that at short $L$, $J_{th}$ of noninteracting multiple quantum-well (MQW) lasers (modal confinement factor $\Gamma_{MQW} = N_z\Gamma_{SQW}$) approaches $\ln(1/R_1R_2/2b\Gamma_{SQW})^2$ (where b is gain/current proportionality constant), which in addition to being independent of $L$, is also independent of the number of wells $N_z$. Instead, not only does $J_{th}$ increase as $L$ decrease, but the increase is also more rapid [72] and its onset occurs at longer $L$ in SQW than in MQW lasers [73]. This indicates that high electron concentration effects start to kick in for SQWs in longer cavities than for MQWs. Even though in long lasers, $J_{th}$ is lower for SQW than for their MQW counterparts, this tendency is reversed in lasers shorter than some critical length, which is dependent on QW structure. Figure 6.8 shows the calculated $j_{th}$ and $J_{th}$ ($J_{th} = j_{th}L$), respectively, as functions of laser effective length $L_{eff} = 2L/\ln(1/R_1R_2)$, which encompasses both the laser length $L$ and facet reflectivities $R_1R_2$, for $N_z = 1, 2, 4$, and $8$.

For more QW laser diode properties and their performance, readers are referred to published papers [74–86].

In closing this section on QW lasers, we would like to discuss one of their most useful characteristics for future coherent communication systems. In such systems, the laser line width, $\Delta f$, has to be kept to a value approximately 0.01 of the bit rate [87, 88]. Thus, it is important to choose a semiconductor laser with a narrow $\Delta f$. The line width $\Delta f$ can be expressed as [89]

$$\Delta f = \frac{\beta_s}{4\pi\tau_p[(J-1)/J_{th}]}(1+\alpha^2) \tag{6.16}$$

where $\beta_s$ = spontaneous emission factor, $\tau_p$ = cavity lifetime, and $\alpha$ = line-width enhancement factor, which is expressed as

$$\alpha = -\frac{\partial n_r/\partial N_{inj}}{\partial n_I/\partial N_{inj}} \cdot \left(\frac{4\pi}{\lambda_L}\right) \tag{6.17}$$

where $n_r$ and $\left(\frac{g \cdot \lambda_L}{4\pi}\right)$ are the real and imaginary portions of the indices of refraction, $N_{inj}$ is the injected carrier density, and $\lambda_L$ is the lasing wavelength.

In a recent publication, Lee et al. [88] calculated the variation of the line-width enhancement factor, $\alpha$, as a function of active layer thickness (Fig. 6.9). As the extreme case of a thin active layer is a QW structure, it is clear that QW lasers will have reduced $\alpha$ and line width. For both QW and conventional DH laser, $(dg/dN)$ increases with the increase of photon energy and has a maximum value far above the lasing photon energy. Hence, $\alpha$ and the line width of QW lasers may be reduced by increasing the threshold carrier density in the same way as in conventional DH lasers. QW lasers have a large magnitude of the density of states at the step-like conduction band edge, which leads to enhanced $dg/dN$. The value of $\alpha$ in a quantum-well laser is estimated to be ~50 to 60% (from 5 to 2) lower than a DH laser, and this implies that the line width could be reduced by a factor of 5 to 6 [87, 88].

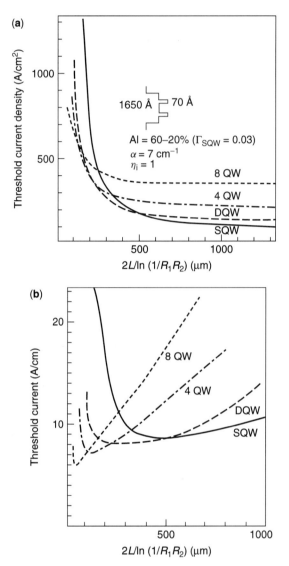

**Figure 6.8** Predicted (a) threshold current density and (b) threshold current for 70 A thick GaAs noninteracting MQWs lasers with 20% and 60% Al concentration in barrier and confining SCH layers, respectively. Al $= 60-20\%$ ($\Gamma_{SQW} = 0.03$); $\alpha = 7$ cm$^{-1}$; $\eta_i = 1$; ——— SQW; − − − DQW; −·−· 4QW; − − − − 8QW.

## 6.3.5 Strained Quantum-Well Lasers

As listed in Fig. 6.2, several sets of material systems are potentially feasible for high-performance semiconductor lasers. The most fundamental requirement for different materials to be used to make high-quality lasers is that they must have the same crystal structure and nearly the same lattice constant, so that single crystal defect-free material can be deposited on one another. It is well known that a small lattice mismatch ($\Delta a/a \sim 1\%$) can be tolerated up to a certain thickness ($\sim 20$ nm) without

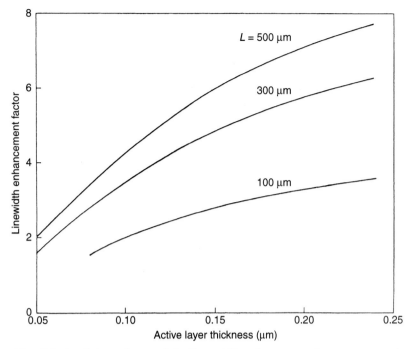

**Figure 6.9**  Calculated dependence of the line-width enhancement factor, $\alpha$, on active layer thickness at different laser length.

dislocation generated. Thus, for a thin active region, one can move slightly left or right of the lattice-matching condition. In this case, the lattice of the deposited layer distorts somewhat to fit the substrate. This causes distortion in the perpendicular direction in order to maintain the same volume of the unit cell. These distorted structures are referred to as strained layers. As we mentioned earlier, QW active regions are desirable in diode lasers for reduced threshold and improved performance. Thus, the strained QW can also be used in all diode lasers without introducing any undesired effects.

If the QW native lattice constant is larger than the surrounding lattice constant, the QW lattice will compress in the growth plane, so a compressive strain is built into the QW layer. On the other hand, if the QW lattice constant is smaller than the surrounding layers, the QW is under tensile strain (Fig. 6.10). The strain of either type increases the curvature of valence band structure, which greatly reduces the effective mass. Smaller

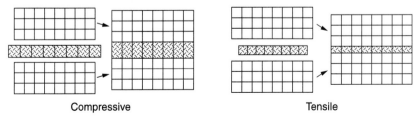

**Figure 6.10**  Schematic of sandwiching a QW with a larger or a smaller lattice constant than barrier layers to generate a compressive and a tensile strain, respectively.

valence band effective mass allows the quasi-Fermi levels to separate more symmetrically. The transparency carrier density as well as transparency current is thus reduced substantially in the strained QW structures. In addition, a better alignment of quasi-Fermi levels with the band edges in the strained materials enhances differential gain substantially. All of these effects improve laser performance a great deal compared to nonstrained quantum-well lasers. Many high-quality laser diodes based on the strained quantum-well structures have now been developed [50–61, 90–95].

### 6.3.6   Stripe Geometry and Current Confinement Techniques

As mentioned earlier, it is important to confine the current and lasing emission to a narrow stripe geometry to obtain cw operation. Confining the current to a small geometry makes it possible to control the type and number of spatial modes of the laser. In some cases, it is observed that the laser operates in a single, nearly Gaussian mode. There are various ways to fabricate lasers with a stripe geometry configuration. The most common techniques for the current confinement are dielectric isolation, proton bombardment, impurity (such as Si and Zn) diffusion, and etching/re-growth. The purpose of these processes is to create an index of refraction difference between the guiding stripe and the surrounding region (lateral refractive index difference).

The lateral refractive index difference, $\Delta n$, is composed of the following: (1) contribution from the injected carriers, $\Delta n_{fc}$ (plasma effect) [96]; (2) contribution from the nonuniform temperature profile, $\Delta n_T$; (3) contribution from the geometry, $\Delta n_B$; and (4) contribution from the dispersion of the gain profile, $\Delta n_G$ [63]. The lateral index difference can be expressed as

$$\Delta n = \underbrace{-\Gamma(1.86 \times 10^{-21})\,\Delta N}_{\Delta n_{fc}} + \underbrace{(4.9 \times 10^{-4})\,\Delta T}_{\Delta n_T} + \Delta n_B - \underbrace{\Gamma(2.2 \times 10^{20})\,\Delta N}_{\Delta n_G} \qquad (6.18)$$

where $\Delta N$ is the difference in the injected carrier density between the center and the edges of the stripe. $\Delta N$ can be calculated from the analysis of Hakki [97, 98] and is nominally 1 to $2 \times 10^{18}$ cm$^{-3}$ ($\Delta N \sim J_{th}\tau_s \eta_i / q d_A$, where $J_{th}$ is the threshold current density, $\tau_s$ is the spontaneous lifetime, $d_A$ is the active layer thickness, and $\eta_i$ is the internal efficiency); $\Delta n_B$ is the builtin index difference due to the geometry. It can be estimated by using the effective index method, which is summarized in Ref. 99, for the case of laser diodes and discussed further in this section and in Section 6.6. The confinement factor, $\Gamma$, can be estimated for a symmetric DH structure by [100]

$$\Gamma = \frac{D^2}{2 + D^2} \qquad D = \frac{2\pi}{\lambda_0}(n_1^2 - n_2^2)^{1/2} d_A \qquad (6.19)$$

where $n_1$ and $n_2$ are the index of refraction of the active and cladding layers.

To estimate the builtin index of refraction difference, $\Delta n_b$, researchers have used the concept of an effective index of refraction. Consider a symmetric two-dimensional wave guide. Figure 6.11a and **b** represent ridge-type wave guides and have been commonly used in the fabrication of mode-stabilized semiconductor lasers [101, 102]. Figure 6.11c represents a two-dimensional wave guide that utilizes the loss at the edges of the wave guide in order to stabilize the optical mode, and it is typical of the channel substrate planar (CSP) structure. To determine the variation of effective index of refraction in the junction plane, we model the two-dimensional wave guide

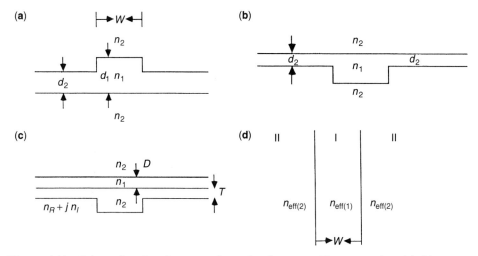

**Figure 6.11**   Schematic of various two-dimensional wave-guide geometries: **(a)** ridge wave guide; **(b)** inverted ridge wave-guide; **(c)** channel substrate planar (CSP) geometry; **(d)** effective index model.

as three distinct one-dimensional wave guides, as shown in Fig. 6.11**d** for the case of the ridge wave guide. The propagation constant, $\beta_i$, for the $i$th wave-guide region can be calculated by solving Maxwell's equation for the appropriate slab wave-guide geometry. The effective index of refraction, $n_{\text{eff}}$, for each region is obtained from ($n_{\text{eff}}$ is generally complex)

$$\beta_i = (2\pi/\lambda_0)n_{\text{eff}(i)} \tag{6.19a}$$

and we can define a lateral index difference

$$\Delta n_B = n_{\text{eff}(1)} - n_{\text{eff}(2)} \tag{6.19b}$$

It has been shown [103, 104] that the effective index method can be used to predict the wave-guiding property in the lateral direction for the lowest order mode. This method is used to analyze the optical properties of single-mode semiconductor lasers [105]. For the geometry of Fig. 6.11**a** and **b**, $\Delta n_{\text{eff}}$ can be approximated by [106]

$$\Delta n_{\text{eff}} = \frac{2D^2}{n_1(2+D)^2}\left(n_1^2 - n_2^2\right)^{1/2}\frac{\Delta d}{d}$$
$$D = \frac{2\pi}{\lambda_0}d\left(n_1^2 - n_2^2\right)^{1/2} \tag{6.19c}$$

where $\Delta d$ is the change of thickness of the ridge wave guide ($\Delta d = d_1 - d_2$).

For the case of the CSP type of wave guide, we can obtain an approximate expression for $\Delta n_{\text{eff}}$ if we assume that the absorption coefficient at the substrate goes to infinity [107]:

$$\Delta n_{\text{eff}} = (1/2n_1)(\lambda_0^2/D_{\text{eff}}^2)\Delta(1/P_s) \tag{6.19d}$$

where $D_{\text{eff}} \triangleq d + 2/P_s$ (in the region of the channel) and $P_s$ is the penetration depth of the field. For the CSP laser, we approximate $\Delta(1/P_s)$ by

$$\Delta(1/P_s) = (1/P_s) - \exp(-P_s T) \qquad (6.19e)$$

The penetration depth $(1/P_s)$ can be estimated from the confinement factor because

$$D_{\text{eff}} \cong d/\Gamma = d + 2/Ps \qquad (6.19f)$$

Thus, for the symmetric guide outside the CSP channel, we find that

$$1/P_s \simeq (d/2)(1/\Gamma - 1) \qquad (6.19g)$$

and $\Gamma$ can be estimated from Eq. (6.19).

If the geometry of the wave guide varies in the lateral direction, it is still possible to use the effective index method by subdividing the lateral wave guide into a series $(N)$ of very narrow wave guides where we assume that the geometry is constant over the dimensions of each wave guide (Fig. 6.12). By determining the propagation constant, $\beta_i$, in each of the one-dimensional wave guides, we are able to calculate the corresponding $n_{\text{eff}}$'s. The lateral optical guiding properties of the two-dimensional wave guides can then be evaluated by solving the $N$-layer slab-wave-guide problem in the lateral direction using the appropriate effective indices. In general, the solution to this problem

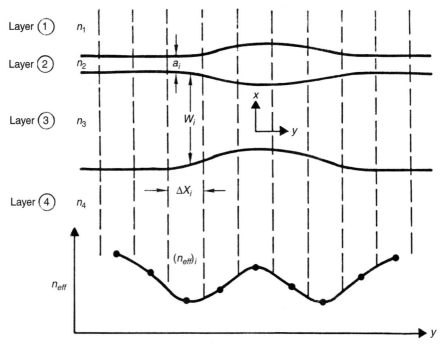

**Figure 6.12**  Schematic diagram shows the breakup of a dielectric wave guide with nonuniform lateral wave-guide thickness variation for analysis using the MODEIG wave-guide computer program.

in the case of laser diodes leads to a complex propagation constant. One appropriate numerical computational technique that can be used is a model named MODEIG developed at TRW, Inc. (G. Evans and T. Holcomb, unpublished calculations) [108] and the University of Florida [109]. This model has been used in determining the optical properties of multielement semiconductor laser arrays as discussed in Section 6.9.

In a demonstration of the effect of lateral optical guiding, researchers from Hitachi built the channel substrate planar (CSP) [110] laser, which provides a large builtin $\Delta n_B$ ($\Delta n_B > 5 \times 10^{-3}$). One useful criterion for preserving mode stability in the case of positive index guiding is [107]

$$\Delta n_B / n_1 > \Delta g / g_{\max} \tag{6.20}$$

where $g_{\max}$ is the maximum gain and $\Delta g$ is the gain dip. A further restriction on $\Delta n_B$ to provide single lateral mode operation requires that

$$W_0 < \lambda_0 / (8 n_1 \Delta n_B)^{1/2} \tag{6.21}$$

where $W_0$ is the width of the step-index wave guide (see Fig. 6.13**b** and $n_1$ is the index of refraction for the active layer; $n_1 \sim 3.5$ for GaAlAs/GaAs structures) in order to preserve single-mode operation.

To maximize the output power in the fundamental mode, a builtin mechanism for mode-dependent losses in addition to a builtin index is necessary. Two methods to

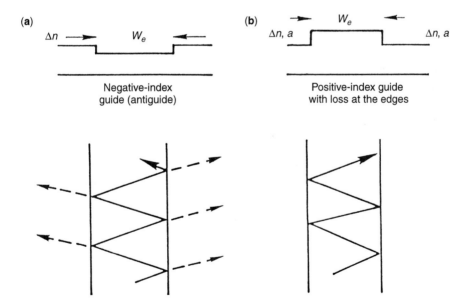

**Figure 6.13**   (a) Schematic diagram shows the guiding mechanism for a structure having a negative index profile (higher index in the wings). The profile leads to an enlarged fundamental mode and increasing losses for higher order modes. This guiding mechanism is similar to that of CDH-LOC-type structures (from Ref. 99). (b) Schematic diagram shows the guiding mechanism for a structure having a positive index profile and excess losses in the wings. The profile leads to increasing absorption losses for higher order modes. This guiding mechanism is typical of CSP-type structures (from Ref. 110).

achieve mode stability at high power levels are shown schematically in Fig. 6.13. The first method involves the formation of a lateral wave guide with high absorption losses at the edges of the stripe, such as the CSP laser (Fig. 6.13**b**), and the second involves creating a negative-index wave-guide profile ("leaky mode") as found in (Fig. 6.13**a**) in the constricted DH-structure large optical cavity (CDH-LOC) laser. Both of these structures are discussed in detail in Section 6.4.

As mentioned earlier, the simplest current confinement techniques involve either dielectric isolation or proton bombardment. However, these designs have some major drawbacks when used in combination with a mode-stabilized geometry (e.g., an etched channel). The two most important drawbacks are the possibility of misalignment, which can lead to lateral mode instability, and spectral broadening. Secondly, the amount of current spreading and lateral carrier out-diffusion can significantly increase the threshold current density. Current spreading results from the finite conductivity of the epitaxial layers outside the stripe region in semiconductor lasers with dielectric isolation. The results of the analysis performed by several workers indicates that the consequence of current spreading is the increase in the threshold current density $J_{th}$ as the stripe width is reduced. Hence, it is highly desirable to design a laser structure with a self-aligned geometry.

For a laser with a dielectric isolated stripe, there are two current-spreading layers [the $p$-GaAs contact layer and the $p$-(GaAl)As cladding layer]. The current spreading can be reduced by increasing the sheet resistance of the epitaxial layers outside the stripe region. This can usually be accomplished by either a selective Zn diffusion or proton bombardment. The use of proton bombardment is most effective because it can increase the layer resistivity to very high values ($\sim 10^7 \Omega$-cm). Proton bombardment has an additional advantage of providing a broad-area contact for heat removal. A more thorough discussion on the thermal property of stripe geometry lasers is given in Section 6.4. The second characteristic of interest in stripe geometry lasers is the lateral out-diffusion of injected carriers. This phenomenon also increases the threshold current density as the stripe width is reduced. Hakki [97, 98] solved the carrier diffusion equation in the plane of the junction and found that carrier out-diffusion is a strong function of the carrier diffusion length, $L_d$. His analysis predicted a rapid increase of $J_{th}$ when $S/L_d < 1$, where $2S$ is the stripe width. Tsang [111] combined the effects of carrier diffusion and current spreading into one formulation. A typical calculation is given in Fig. 6.14. Note that for devices with $2S < 8$ $\mu$m, the threshold current density rises rapidly. The most common method to eliminate the effects of carrier out-diffusion is to embed the active layer using a material with a larger band gap than GaAs, such as (GaAl)As. Such a stripe configuration is referred to as a buried hetero (BH)-structure. The typical threshold current is in the range 5 to 30 mA for DH lasers. By using QW structures, the threshold current is reduced to less than 1 mA.

### 6.3.7 Basic Characteristics of Diode Lasers

Light within semiconductor lasers is substantially amplified as it travels back and forth in the longitudinal direction. Lasing occurs when enough carriers are injected into the active layer, which provides the optical gain needed to overcome the cavity's internal and external losses. The operating characteristics of a typical laser diode is shown in Fig. 6.15. In the low bias current range, spontaneous emission is dominant because the carrier density in the active layer is not high enough to form a population inversion. As

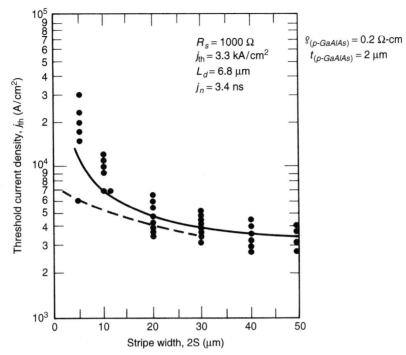

**Figure 6.14** Variation of $J_{th}$ with $2S$ for planar-stripe-geometry DH lasers at room temperature, (from Ref. 111).

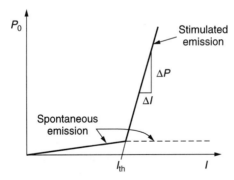

**Figure 6.15** Illustration of output power versus current for a diode laser. Below threshold, only spontaneous emission is important; above threshold, the stimulated emission power increases while the spontaneous emission is clamped at its threshold value.

the bias is increased, the population inversion occurs. Stimulated emission becomes dominant at a certain bias current. This bias current is called the threshold current. The output power above threshold current is thus linearly proportional to the injection current as shown in Fig. 6.15.

To characterize the performance of a diode laser, the output power ($P_o$) as a function of injection current ($I$) is measured. Below threshold current, only spontaneous emission is important. Above threshold current, the stimulated emission power increases while the spontaneous emission is clamped at its threshold value. Threshold current is the most important parameter for diode laser performance. It is determined by several factors due to basic material properties and device geometry, i.e., material gain, confinement factor and all losses such as absorption loss, mirror loss, scattering loss and radiation loss.

The efficiency of a diode laser after lasing is usually expressed by the external differential quantum efficiency, which is the ratio of the increase in the number of photons emitted from the laser diode to the increase in the number of carriers injected; i.e., the differential quantum efficiency is measured by the slope efficiency $[\Delta P_o / \Delta I]$ in watts/ampere above threshold and then multiplied by $[q / h\nu]$ in Coulomb/joule. For more details, the reader is referred to Coldren and Corzine [26] and Fukuda [112].

Other important characteristics of lasers are their spectral output and beam profile. When the dc bias on a Fabry–Perot laser increases, the peak wavelength, which corresponds to the highest mode in the wavelength spectrum, shifts to a shorter wavelength before lasing because of the band-filling effect [112]. This is due to the shift of peak gain or gain profile. In addition, each mode also shifts to a shorter wavelength because of the change in the refractive index. Once lasing begins, the band-filling and the plasma effect are nearly constant because the injected carrier density is nearly constant. The peak wavelength (or lasing peak wavelength) starts to red shift gradually as the injected current increases because of the heating. The lasing wavelength changes with temperature. The temperature dependence of each mode is about 0.08 nm/°C for AlGaAs/GaAs lasers, 0.1 nm/°C for 1300-nm band, and 0.12 nm/°C for 1.55-nm InGaAsP/InP lasers [112]. The average wavelength shift due to the temperature dependence of the band gap is about 0.25 nm/°C for AlGaAs/GaAs lasers, 0.4 nm/°C for 1300-nm band and 0.6 nm/°C for 1550-nm band InGaAsP/InP lasers. The peak gain shifts smoothly according to the temperature dependence of the band-gap energy, whereas the lasing wavelength change with temperature is a step-like function because it is determined by both gain profile and the FP modes. The mode hopping will not happen until the gain in the adjacent mode becomes higher than the one currently lasing.

Beam divergence is another critical parameter due to its impact on fiber coupling efficiency and the laser beam spot size and shape. It is often necessary to perform a near-field pattern and a far-field scan to measure and evaluate the spatial properties of the diode's output power. As discussed above, the output beam of edge emitting laser diodes is elliptical and highly divergent. This is because the laser's emission aperture is a narrow slit with a long and short axis, causing diffraction effects to be stronger in one direction (transverse) than the other (lateral). For more details about this aspect, the reader is referred to Ref. 112.

### 6.3.8  Single-Mode and Multimode Lasers

As discussed in Section 6.3.1, there are transverse, lateral, and longitudinal modes in diode laser operations. The transverse mode is in the perpendicular direction to the active layer, whereas the lateral mode is in a direction parallel to the active layer. The transverse and lateral modes in laser diodes are conventionally called the TM (transverse magnetic) and TE (transverse electric) modes. A laser diode usually lases in the TE-mode because the threshold for the TE-mode is lower than that for the TM-mode.

The transverse mode reflects the standing wave between the two hetero-junctions. The confinement of optical fields to the active layer is determined by the thickness of the active layer, the index difference between the active and cladding layers, and the wavelength. The active layer is usually very thin (<0.2 μm), and the index of refraction difference is less than 10%. Thus, the diode laser can only lase in the fundamental mode.

A lateral mode in the direction parallel to the active layer is much more complicated bcecause the output power of a laser diode is proportional to its emission area. Depending on application requirements, lateral emission width ranges from 1 μm to hundreds of microns. The lateral mode operation becomes very complicated when the active region is wider than 1 μm. For some applications that require only optical power but not good beam quality, such as pumping sources for solid state lasers, broad area lasers (>10 μm) or evanescent coupled arrays are widely used. These lasers operate in multilateral modes.

***Single Lateral Mode High-Power Diode Lasers.*** For many applications such as long-distance signal transmission, however, high-power, stable, single-mode, good beam quality lasers are required. Although the light in a diode laser is tightly confined in the transverse direction, this is not the case in the lateral direction because light confinement via the current flow is weak and not stable with increased drive current. As this lateral stable mode is strongly influenced by the structure of the active region, many studies have been performed on designing the dielectric wave-guide structure in the lateral direction. Detailed consideration of this topic has been described in Section 6.3.6.

In the past 20 years, researchers have concentrated on controlling the lateral modes by introducing dielectric wave-guide structures in the lateral direction. Because the light in these structures is guided by variations in the refractive index of the various materials, the corresponding devices are known as index-guided lasers. Index-guided lasers that support only the fundamental transverse mode and the fundamental lateral mode are called single-mode lasers; they emit a single, well-collimated beam of light whose intensity profile is a bell-shaped Gaussian curve.

There are several important high-power laser structures that have either become commercially available or have undergone significant development. Figure 6.16 shows schematic diagrams of the commonly used designs for high-power CW lasers.

As mentioned earlier, MBE and MOCVD, with their inherent uniform epitaxial growth and the possibility of using large-area wafers for laser diode production, have been used to fabricate lasers with high optical power in a single spatial and longitudinal mode. The excellent layer uniformity leads to a reduced spectral width and more uniform threshold characteristics. The use of MOCVD and MBE technologies for fabricating single-element QW structures has produced many single-mode high-power devices [90–93, 118]. By using three QW structure, Imafuji et al. has achieved output powers in excess of 600 mW cw in a single spatial and spectral mode with ridge wave-guide geometry [118–120]

***Multispatial Mode Diode Laser — Broad Area Lasers and Arrays.*** One of the methods used to get high power is to increase the width of the emitting region. However, as the width increases, the occurrence of multilateral modes, filamentations, and lateral-mode instability become significant. This produces a far-field pattern that is not diffraction limited and reduces the brightness. Fortunately, some applications need only high output power with minimum beam quality requirements. Broad area lasers with MOCVD and MBE grown materials offer uniform and partially coherent emission over

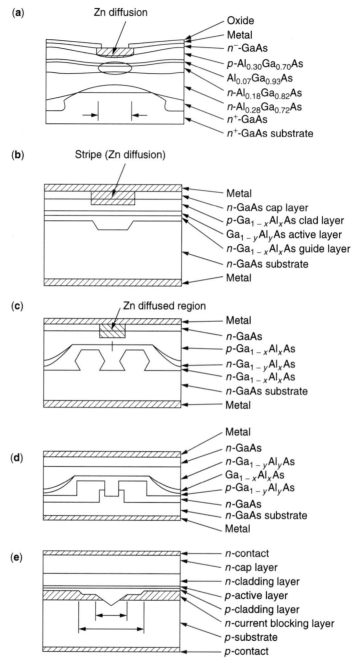

**Figure 6.16** Schematic diagram of representative high-power GaAlAs/GaAs high-power laser diodes: (**a**) constricted double heterostructure (CDH)–LOC (from Ref. 113); (**b**) channel substrate planar (CSP) (from Ref. 110); (**c**) twin-ridge structure (TRS) (from Ref. 114); (**d**) broad-area twin-ridge structure (BTRS) (from Ref. 110); (**e**) BVSIS channel V-channel inner stripe (VSIS) (from Ref. 115); (**f**) twin-channel substrate mesa guide (TCSM) laser (from Ref. 116); (**g**) inverted channel substrate planar (ICSP) laser (from Ref. 117); (**h**) RIE-etched ridge wave-guide laser (from Ref. 91); (i) TQW laser with real refractive index guided structure (from Ref. 120).

**(f)**

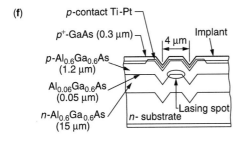

**(g)**

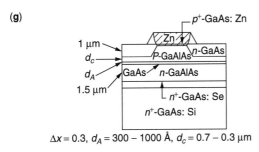

$\Delta x = 0.3$, $d_A = 300 - 1000$ Å, $d_c = 0.7 - 0.3$ μm

**(h)**

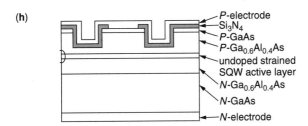

**(i)**

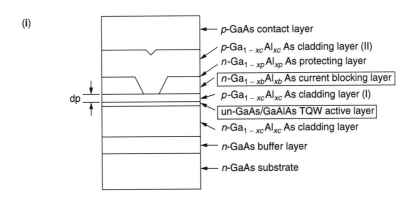

**Figure 6.16**   (*Continued.*)

the emitting aperture. It operates in multilongitudinal modes with a spectral envelope width of approximately 2-nm FWHM. These broad area devices ranging from 50 to 500 μm with cw output power up to 5 W are commercially available.

In order to achieve more stable laser operation, a broad active stripe can be divided into several nominally single-mode stripes as shown in Fig. 6.17 [19, 121–136]. These

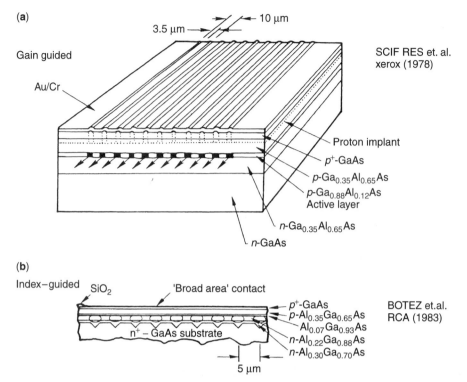

**(a)**

Gain guided

Au/Cr

3.5 μm

10 μm

SCIF RES et. al.
xerox (1978)

Proton implant

$p^+$-GaAs

$p$-Ga$_{0.35}$Al$_{0.65}$As

$p$-Ga$_{0.88}$Al$_{0.12}$As
Active layer

$n$-Ga$_{0.35}$Al$_{0.65}$As

$n$-GaAs

**(b)**

Index–guided   SiO$_2$   'Broad area' contact

$p^+$-GaAs

$p$-Al$_{0.35}$Ga$_{0.65}$As

Al$_{0.07}$Ga$_{0.93}$As

$n$-Al$_{0.22}$Ga$_{0.88}$As

$n$-Al$_{0.30}$Ga$_{0.70}$As

$n^+$ – GaAs substrate

5 μm

BOTEZ et.al.
RCA (1983)

**Figure 6.17**  Schematic diagram of two types of laser-array structures. **(a)** Gain-guided phased array using QW active layers grown by MOCVD (after Ref. 121). **(b)** Index-guided phase-locked array using CSP-LOC structures and grown by LPE (after Ref. 138).

stripes are formed by a spatial variation of injected current (gain-guided) or material composition (index-guided) [137]. Each stripe operates in the fundamental transverse mode and is evanescently coupled to the adjacent stripes. Gain-guided array structures have been used to achieve the output power level of over 5-W cw from a 500-μm emitting aperture.

The phase-locked array was first demonstrated by Scifres et al. [121] in 1978. The original coupling scheme had branched wave guides, but, it was quickly abandoned in favor of evanescent field coupling by placing the individual elements of the array in close proximity (Fig. 6.17). Later, index-guided laser arrays were also fabricated (Fig. 6.17) [137, 138].

In the past decade, work has emphasized the development of phase-locked arrays as a means of achieving mode-stabilized devices that can operate reliably at a power level of 0.5- to 1-W range [134–136, 139–160].

Initially, researchers sought to create phase-locked arrays by simply placing single-element sources in close proximity one to another. Such devices, called positive index-guided arrays, generally oscillate in several array modes above 1.5-x laser threshold, or 50 mW. None of the positive index-guided arrays reported to date has demonstrated diffraction-limited beam operation beyond 50 mW, because of either weak overall coupling or weak optical-mode confinement. A common misconception is that strong nearest-neighbor coupling implies a strong overall coupling. But Fig. 6.18

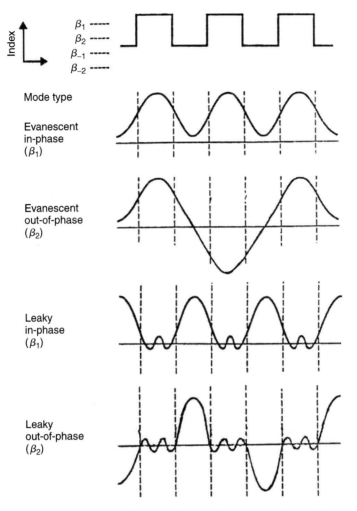

**Figure 6.18**  Arrays of both positive- and negative-index guides.

shows that nearest-neighbor coupling is, in fact, a series coupling, a scheme plagued by weak overall coherence and poor intermodal discrimination. A strong overall coupling happens only when each element equally couples to the rest — that is a parallel coupling — where intermodel discrimination is maximized and the full coherence is a system characteristic.

Botez et al. [161] at TRW have developed a whole new class of arrays: phase-locked arrays with negative-index guides or anti-guides. Such devices posses a unique feature of having both strong optical-mode confinement (within each element) and a strong overall inter-element coupling (parallel coupling).

A single anti-guided device has been discussed in Section 6.3.1 (Fig. 6.13).

For arrays, a periodic variation of the refractive index (Fig. 6.18, top) represents arrays of both positive-index guides and negative-index guides. The supported array modes are evanescent-wave type when the fields are peaked in high-index regions, and leaky-wave type when the fields are peaked in low-index regions (Fig. 6.18, bottom). Depending on preferential gain placement, one or the other type of array mode is favored to lase. For anti-guided array modes to prevail, gain is placed in the low-index regions, with the high-index regions being transparent or deliberately made lossy.

A practical way to suppress evanescent-wave array modes, which thus allows only leaky-mode oscillation, is shown in Fig. 6.19 for GaAs/AlGaAs structures. The inter-element regions contain high-index passive-guide layers placed in close proximity (0.1

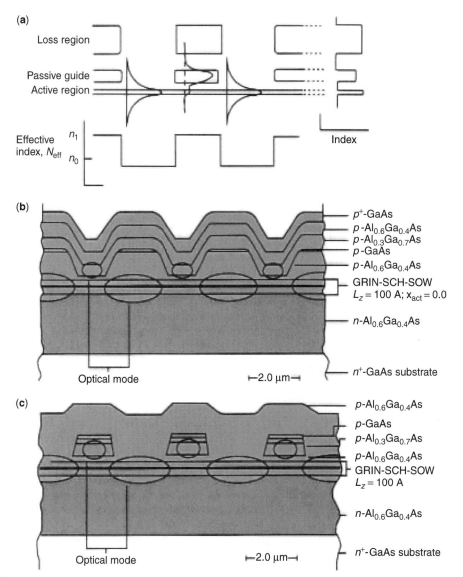

**Figure 6.19**  Means of suppressing evanescent-wave array modes: placing a high-index passive guided layer and optical absorbing material between elements (**a**) using preferential chemical etch and regrowth in the inter-element regions (**b**), or building in inter-element regions during initial growth (**c**).

to 0.2 μm) to the active layer. The fundamental transverse mode is primarily confined to the passive-guided layer, which means, between elements, the model gain is low. To further suppress oscillation of evanescent-wave modes, an optical observing material can be placed between elements as shown in Fig. 6.19**a**. Two types of anti-guided array have been fabricated by the TRW group: the complementary self-aligned stripe (CSA) array (Fig. 6.19**b**) and the self-aligned–stripe (SAS) array (Fig. 6.19**c**). In CSC-type of arrays, a preferential chemical etch and regrowth occurs in the inter-element regions. For SAS-type arrays, the inter-element regions are builtin during the initial growth, and then etch and regrowth occurs in the element regions.

A remarkable property of leaky-wave coupling is the presence of a lateral resonance condition. Because of this, lateral radiation leakage can be used to couple multiple anti-guides together. All elements are resonantly coupled in phase or out of phase when the inter-element spacing corresponds to an odd or even integral number of (lateral) half-wavelength, respectively (Fig. 6.20). At resonance, the inter-element spacing behaves as a half-wave plate. The resonant devices are called resonant-optical-waveguide (ROW) arrays. ROW arrays have uniform near-field intensity profiles, because each element is equally coupled to the rest (parallel coupling).

ROW arrays have demonstrated excellent performance orders of magnitude higher than other array types [162–165]. In cw operation, diffraction-limited beam with output power over 0.5 W is obtained. At 0.5 W, the wall-plug efficiency is as high as 25%. Overall, >2000 hours of cw operation at 0.5 W was observed with little degradation.

There are other approaches such as the use of unstable resonators to achieve a high-power stable operation. For this topic, the readers are referred to published papers [166–169]

***Single Longitudinal Mode Lasers.*** The longitudinal mode spacing depends on the cavity length, wavelength, and the gain spectral width. The number of modes decreases as the cavity length decreases. Consequently, a single longitudinal mode operation is often observed in laser diodes with a short cavity (<100-μm). However, threshold current density for short cavity lasers are high due to a large mirror loss [TRW]. Typical Fabry–Perot diode lasers usually operate at multiple-longitudinal mode because of its longer than 100-μm cavity length.

It has been observed that in index-guided lasers, the output is concentrated primarily in one single longitudinal mode at currents slightly above the threshold [170].

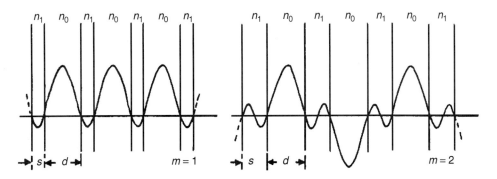

**Figure 6.20**  Lateral resonance condition via leaky-wave coupling.

Suppression ratios between the main and side modes are normally greater than $10^2$. However, as the laser power output increases beyond a critical power level (50 to 80 mW) [171, 172], the number of longitudinal modes tends to increase dramatically. The highest reported power for a single-element device that operates primarily in a single longitudinal mode is about 50 to 65 mW [171, 172]. In this section, we discuss important parameters that affect single longitudinal-mode operation at high power levels.

The laser spectrum is determined by a combination of the gain profile and the FP resonant modes. A mode-stabilized laser is considered a homogeneously broadened system. When one of the longitudinal modes reaches threshold, gain of the other modes will be clamped below threshold. This situation remains until a spatial hole burning occurs in the gain profile that produces a large nonuniformity in the junction plane. There is direct evidence that a nonuniform lateral gain profile leads to multilongitudinal mode operation [107] at high power levels.

The effect of the spatial hole burning on the spectral output of mode-stabilized lasers has been studied by Wang et al. [107]. The gain for a particular longitudinal mode can be expressed as

$$G_{m,p,q} = \frac{\int_{-\infty}^{\infty} \Gamma(x) g(x, \lambda_q) |E_{p,m}(xy)|^2 dx dy}{\int_{-\infty}^{\infty} |E_{pm}(x, y)|^2 dx dy} \tag{6.22}$$

where $p, m$ = lateral mode numbers, $q$ = longitudinal mode number set up by the resonant cavity, $g(x, \lambda_q)$ = spatial gain of the lateral mode at the wavelength, $\lambda_q$, and $G$ = mode gain.

The gain difference among different longitudinal modes comes from the dependence of the spectral gain profile $g(\lambda)$ on $\lambda$. For a uniform spatial gain (near threshold) profile, the carrier distribution is also uniform. Therefore, only one $\lambda_q$ is closest to the spectral gain peak, and this mode will be the dominant mode at low power levels. As the power level increases, spatial hole burning leads to a nonuniform gain profile. This implies a nonuniform quasi-Fermi level separation that in turn gives rise to a nonuniform spectral gain profile with a spread of the peak gain and peak wavelength. The magnitude of the spectral broadening, $\Delta\lambda$, can be written as [173]:

$$\Delta\lambda \simeq K\lambda_{\max} \frac{\Delta g}{g_{\max}} \quad K = 5 \times 10^{-3} \text{ for GaAs}$$

$$K = 1 \times 4 \times 10^{-2} \text{for GaInAsP}(\lambda = 1.3 \ \mu\text{m}) \tag{6.23}$$

(based on the experimental results of 173)

$$\Delta g = \text{spatial non-uniformity}$$

where $\lambda_{\max}$ is the peak gain wavelength.

It is clear from Eq. (6.23) that spatial variation in current density, active layer thickness, and confinement factor can lead to a nonuniform gain profile. A simple example will help to understand the magnitude of the effect. A gain nonuniformity of 10 cm$^{-1}$ for a laser with a peak gain of 100 cm$^{-1}$ would cause a $\sim$4-Å spread of the peak wavelength that for a 300-$\mu$m laser diode would lead to an excitation of two longitudinal modes. Equation (6.23) provides a useful design criterion for high-power single longitudinal mode lasers. High-power laser diodes should have a large $g_{max}$ and a reduced $\Delta g$. The large $g_{max}$ can readily be obtained with a small transverse confinement factor but large absorption loss at the edge of the guiding region (CDH or CSP type). A reduced $\Delta g$ can be achieved with either a relatively large spot size and a thin active layer or with narrow geometry and good lateral carrier and optical confinement. Finally, all high-power single-longitudinal mode lasers should strive for uniform current distribution and maximal product of active-layer thickness and the inverse of the confinement factor (i.e., $\Gamma/d_A$). It is also clear that single-longitudinal-mode operation is especially sensitive in low threshold density lasers (small $g_{max}$). In addition, from Eq. (6.23), it is evident that the achievement of single-longitudinal-mode operation at high power levels for GaInAsP/InP lasers will be much more difficult than for GaAlAs/GaAs.

Laser diodes used in optical communication systems are required to have a stable operation in a single longitudinal mode under modulation and in severe environment conditions. To meet this requirement, distributed feedback (DFB) and distributed Bragg reflector (DBR) laser diodes are commonly used.

The DFB laser structure is quite similar to a Fabry–Perot laser's, as shown in Fig. 6.1, except for the grating in the active region. Typical DFB laser diodes have a periodic sinusoidal index wave in one of the cladding layers, as shown in Fig. 6.21; this is called a grating. The grating introduces a periodic refractive index change in the active region as

$$N_r(z) = n_{eq} + \Delta n_r \cos(2\beta_o z + \Omega_g) \tag{6.24}$$

where $n_{eq}$ is an equivalent refractive index (an overall refractive index of multilayers that influences the light propagation and is a function of the layers and structures), and $\Delta n_r$ is the refractive index difference between the top and bottom layer of the grating. The term $\Omega_g$ is the phase of the grating at the center of the laser cavity and is equal

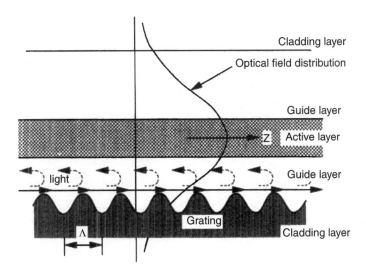

**Figure 6.21**  A cross-sectional view of a wave guide with a Bragg reflector [112].

to zero in a homogeneous grating, and $\beta_o$ is given by

$$\beta_o = 2\pi n_{eq}/\lambda_B = m\pi/\Lambda \qquad (6.25)$$

where $\Lambda$ is the grating pitch. The refractive index periodically changes by the grating pitch. This grating selectively reflects a certain wavelength (Bragg wavelength) in the gain spectrum; i.e., this wavelength corresponds to one of the multiple longitudinal modes. The lasing wavelength of a DFB laser diode can be set at an arbitrary wavelength within the gain spectrum by adjusting the grating pitch and the equivalent refractive index. The lasing wavelength shift from the gain peak is called wavelength detuning. The grating selected wavelength, i.e., the lasing wavelength in DFB laser diodes, $\lambda_{DFB}$, is given by modifying Eq. (6.25) and with $\lambda_{DFB}$ replacing $\lambda_B$.

$$\lambda_{DFB} = 2n_{eq}\Lambda/m \quad m = 1, 2, 3, \ldots \qquad (6.26)$$

The grating pitches corresponding to m = 1, 2, and 3 is called the first-order, second-order, and third-order grating, respectively. The gratings commonly used in 1.3-$\mu$m and 1.55-$\mu$m InGaAsP/InP DFB laser diodes are first, and second-order gratings, while the gratings used in 800-nm band AlGaAs/GaAs laser diodes are usually second, and third-order gratings because of the grating pitch and the fabrication difficulty.

The reflected light intensity is determined by the grating height and shape and distance between the grating and the active layer. The grating height and shape are associated with the magnitude of refractive index change and the spatial gradient of the change. The intensity of reflected light gradually increases as the gradient increases. The distance of the grating from the active layer is a measure of how strongly the optical field in the active layer is influenced by the grating. This strength can be expressed with the so-called coupling constant (in cm$^{-1}$). For a sinusoidal shape grating, the coupling constant, $\kappa$, is given by

$$\kappa = \pi \Delta n_r/\lambda_B \qquad (6.27)$$

The first-order grating's coupling constant for a 1.55-$\mu$m laser diode is about an order of magnitude larger than the second-order grating's because the propagating light wavelength coincides with the grating pitch of the first-order grating. The first-order grating also has advantages in terms of the ease of fabrication and crystal growth. It is not easy to fabricate high-order gratings while keeping crystal quality high. The total reflected light is also influenced by the length of the distributed reflecting mirror. The magnitude of the optical feedback is therefore expressed by the product of the coupling constant, $\kappa$, and the laser cavity length, L. The $\kappa$L is ordinarily designed to be about 1.

A schematic diagram of a DBR laser is shown in Fig. 6.22. The structure of a DBR laser diode is one in which one or both of the mirror facets in a Fabry–Perot laser diode has been replaced with a DBR that reflects the light emitted in the active region. The distributed reflectivity can be controlled in a manner similar to that used to control the coupling constant, $\kappa$, in DFB laser diodes. For more information on DBR laser operations, please refer to *Optical Semiconductor Devices* by Fukuda [112].

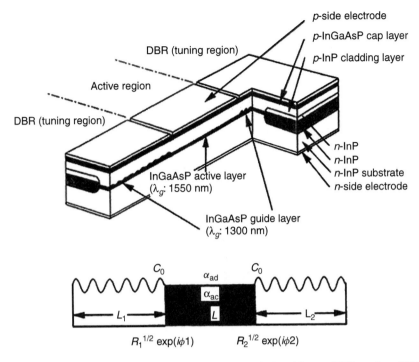

**Figure 6.22**   A schematic diagram of a DBR laser diode with two DBR regions [112].

## 6.4   CONCEPTS OF HIGH-POWER OPERATION

For many practical applications, in addition to stabilizing the lateral mode of an injection laser, it is critical to maintain the operation in a single spatial and spectral mode at high power level. Several physical mechanisms limit the output power of the injection laser. They are:

(1) Spatial hole burning effect, which leads to multispatial mode operation and is intimately related to multispectral mode operation.

(2) Temperature rise in the active layer, which will eventually cause the output power to reach a maximum.

(3) Catastrophic facet damage, which will limit the ultimate power of the laser diode.

Thus, the high-power laser designer must optimize these three physical mechanisms to achieve the maximum power. In this section, we discuss design criteria for optimizing the laser power.

### 6.4.1   Catastrophic Facet Damage (COD)

The ultimate limit in output power for GaAlAs/GaAs lasers is produced by catastrophic damage at the laser facets [175–178]. This occurs when the light intensity in the active layer increases beyond a critical level. A schematic of the physical process involved is illustrated in Fig. 6.23. It has been observed [175–178] that the

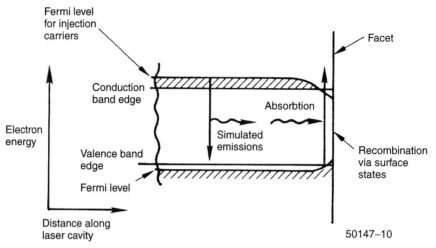

**Figure 6.23** Schematic diagram of the energy-band structure near the laser facets. Note that there is significant band bending near the laser facets, resulting from high surface recombination. Thus, light can actually be absorbed in a region several diffusion lengths away from the facet. The absorption leads to large heating at the facets, which eventually leads to catastrophic damage (after Ref. 175).

catastrophic damage is caused by a localized facet melting. The melting arises from significant absorption of the light at the facet as a result of the high-surface recombination velocity at the air — semiconductor interface — and thus, the effect is strongly dependent on the type of semiconductor materials used. For example, GaAlAs/GaAs lasers have a critical intensity of 2 to 3 $\times 10^6$ W/cm$^2$ for an uncoated facet in a cw operation [178]. On the other hand, catastrophic damage has not been observed for GaInAsP/InP lasers (i.e., optical output power is limited by heating) and the power needed to cause the catastrophic damage is believed to be an order of magnitude higher.

Over the years, researchers have discovered various methods to increase the catastrophic intensity. One of them is spreading the optical power over a larger area at the facet. Several structures have been successfully designed to widen the optical mode along the transverse direction that include the large optical cavity (LOC) (See Fig. 6.3) [178–182], the thin active layer (TAL) [183], and the QW separate confinement hetero-structure (QW-SCH) (Fig. 6.4) [21, 22]. In all of these structures, the optical wave is weakly guided by a thin active layer, causing the mode to spread into the cladding layers. In addition to the material structures, one of the most important methods involves the use of dielectric coatings on the laser facets. The use of dielectric coatings such as Al$_2$O$_3$ [184], Si$_3$N$_4$ [184], and SiO$_2$ [186] have several beneficial effects. First, a protective coating with thickness $\lambda_0/2n_{\text{die}}$ (i.e., one that does not affect the facet reflectivity) can be used to increase the intensity where catastrophic damage occurs. It is believed that the improvement (twofold to threefold improvement for Al$_2$O$_3$) might be related to a reduction in the surface recombination velocity at the interface. Another reason might be related to an increased heat flow provided by the coating with a reasonably high thermal conductivity. Of the many coatings that have been used, Al$_2$O$_3$ appears to be the most useful, due to its close thermal expansion match to GaAs, high thermal conductivity, and good protection from moisture. A dielectric coating can also be used to reduce the facet reflectivity. It was shown early [187] that a decreased facet reflectivity can be used to increase the catastrophic power damage level, $P_c$, of the laser diode.

It can be shown that [188]

$$\frac{P_c(\text{coating})}{P_c(\text{no coating})} = \frac{n(1 - R)}{(1 + R^{1/2})^2} \tag{6.28}$$

where $n$ is the index of refraction of the semiconductor (3.6 for GaAs) and $R$ is the power reflectivity of the facet. As $R \rightarrow 0$, $P_c$ can be increased by $(n)$ over an uncoated facet. Figure 6.24 provides a graph of Eq. (6.28).

More recently, other methods have been developed to increase the catastrophic intensity by reducing the absorption near the laser facets. Such structures, commonly referred to as nonabsorbing mirror (NAM) lasers, are discussed in Section 6.4.3.

From a reliability point of view, operating a laser diode at high intensities can lead to significantly reduced lifetimes. High-power diodes suffer from two degradation mechanisms: (a) high optical flux intensity and (b) high drive current.

In Fig. 6.25, we show plots of the maximum power as limited by catastrophic damage versus active-layer thickness and a corresponding plot of threshold current density versus active-layer thickness. Note a rapid increase in the threshold current for active-layer thickness of less than 0.05 µm. We should also point out that $P_{\text{max}}$ will be limited for very thin active thickness by Auger effects (not included in Fig. 6.25), which reduces the nonradiative recombination lifetime and both decreases the internal efficiency and increases the threshold current density.

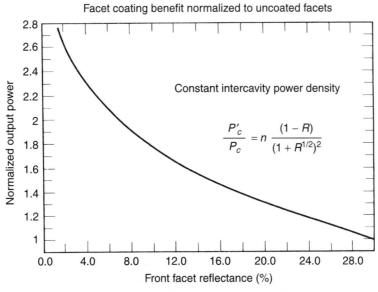

**Figure 6.24** Calculation shows the normalized maximum power (limited by catastrophic damage) versus front facet reflectivity.

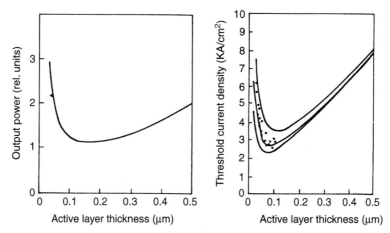

**Figure 6.25**   Calculation of the maximum power and threshold current density versus active-layer thickness. The results do not include the effects of Auger recombination and heating (from Ref. 188).

### 6.4.2   High Power Mode Stabilized Lasers with Reduced Facet Intensity

To achieve high-power operation, two major requirements have to be satisfied: high-power conversion efficiency and a large emitting area. The first requirement can be achieved by making quantum-well layers in the active region. The second requirement stems from the reliability of diode lasers, which is limited by the emitted-power flux density and thus calls for the use of broad-area or phased-locked one-dimensional arrays. The broad-area lasers and phased-locked arrays have been discussed in Section 6.3.8.

As discussed in the preceding section, one of the most effective ways to secure laser diodes with high-power operation and high reliability is to reduce the facet intensity and provide a method of stabilizing the laser's lateral mode. Over the years, researchers have developed primarily three approaches for performing this task:

(1) Increase the lasing spot size in both transverse and lateral directions while introducing a mechanism for providing lateral-mode–dependent absorption loss to discriminate against higher order modes.

(2) Eliminate or reduce the facet absorption by using structures with non absorbing mirrors (NAM).

(3) Use laser arrays, unstable resonator, and surface-emitting laser configurations to increase the mode volume.

Technique 1 is commonly used and has been discussed in Section 6.3.8. Technique 2 will be discussed in Section 6.4.3. Technique 3 is discussed in Section 6.3.8 for edge emitting lasers and in Section 6.5 for surface emitting lasers.

Given proper heat sinking (Sections 6.4.4 and 6.4.5), in order to increase the output power of a semiconductor laser without catastrophic damage, we have to increase the beam size and thus reduce the power density at the facets at a given power level. The first step is increasing the spot size in the transverse direction (perpendicular to the junction). The simplest approach to accomplish this while meeting the constraint of keeping threshold current low is to reduce the active layer thickness.

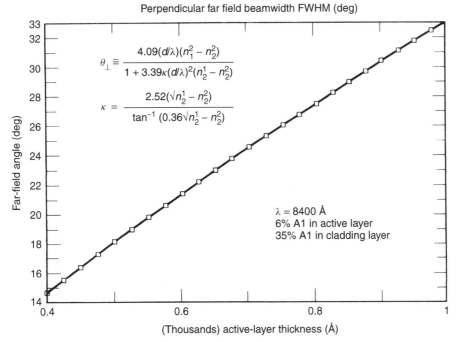

Figure 6.26   Calculation of far-field angle in transverse direction versus active-layer thickness (from Ref. 182).

An important consequence of a thin active layer is a reduced far-field beam width in the transverse direction. The beam width is reduced for a thin active layer as a result of the increasing $d_{eff}$. In Fig. 6.26, we provide a calculation of the far-field beam width versus active layer thickness using the equation derived by Botez [183]. As MOCVD and MBE growth techniques matured, QW layers are used in all laser structures, and thin active layers are common practice for all semiconductor diode lasers.

### 6.4.3   Nonabsorbing Mirror Technology

We have pointed out in Section 6.4.1 that the catastrophic facet damage level is one of the factors in limiting the power in a semiconductor laser. To eliminate this effect, we have to create a region with a higher-energy band gap, and lower surface recombination at the laser facets. This is the concept of a laser with a nonabsorbing mirror (NAM). The first NAM structure was demonstrated by Yonezu [190], who selectively diffused zinc along the length of the stripe except near the facets. This created a band-gap difference between the facet and bulk regions and achieved a threefold to fourfold increase in the CW facet damage threshold and a fourfold to fivefold increase in pulse operation. A commercial-type structure based on this technology is the CRANK TJS structure, shown schematically in Fig. 6.27**a**. In Fig. 6.27, we show a NAM-BH-LOC. A BH-LOC device [191] is first grown, and the channels are etched in the direction perpendicular to the buried mesa. LPE regrowth in the trenches provides a passive waveguide for light generated in the active laser region. Cleaving the device in the passive NAM region results in a NAM device becase the mirror regions are optically passive.

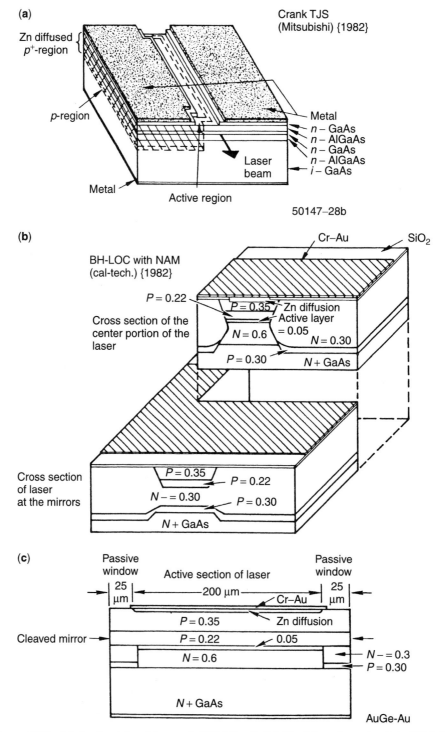

**(a)**

Zn diffused
$p^+$-region

Crank TJS
(Mitsubishi) {1982}

$p$-region

Metal
$n$ – GaAs
$n$ – AlGaAs
$n$ – GaAs
$n$ – AlGaAs
$i$ – GaAs

Laser
beam

Metal

Active region

50147–28b

**(b)**

BH-LOC with NAM
(cal-tech.) {1982}

Cr–Au          SiO$_2$

$P = 0.22$
$P = 0.35$  Zn diffusion
Active layer
$= 0.05$

Cross section of the
center portion of the
laser

$N = 0.6$          $N = 0.30$

$P = 0.30$          $N +$ GaAs

Cross section
of laser
at the mirrors

$P = 0.35$
$P = 0.22$
$N – = 0.30$          $P = 0.30$

$N +$ GaAs

**(c)**

Passive
window          Active section of laser

Passive
window

25
µm          200 µm          25
µm

Cr–Au

$P = 0.35$          Zn diffusion
$P = 0.22$          0.05

Cleaved mirror          $N = 0.6$          $N – = 0.3$
$P = 0.30$

$N +$ GaAs

AuGe-Au

**Figure 6.27**    (a) A schematic of the crank TJS (after Ref. 190); (b) a schematic of the BH-LOC
(after Ref. 191); (c) a side view of the structure.

Incorporation of the NAM structure is strongly device dependent. For example, in diffused device structures, such as the deep diffused stripe (DDS) [190] and transverse junction stripe (TJS) lasers (Fig. 6.27**a**), NAM structures have been formed by selective zinc diffusion in the cavity direction. The $n$-type region will have a wider band gap than that of the diffused region, and thus, little absorption occurs near the facets. However, most index-guided structures require an additional growth step to form the NAM region [192, 193]. More recently, a fabrication of "pseudo" NAM structures have been demonstrated [194, 195]. In these types of structures, the active layer thickness is varied along the length of the laser such that the thickness near the facets is significantly reduced. This produces two effects: (1) It increases the spot size transverse to the junction, which reduces the intensity, and (2) the thinner active layer has higher carrier injection levels, which tends to increase the quasi Fermi-level separation relative to a thick active layer device and thus reduce the absorption near the facets. The NAM structures at the present time suffer from several problems. These include the following:

(1) Due to the complexity in fabrication, they tend to have low yields.

(2) Cleaving must be carefully controlled for NAM structures in order to avoid excessive radiation losses in the NAM region. The radiation loss is a strong function of the NAM lengths, $L_1$ (see Fig. 6.28). The required NAM length is a function of the spot size, $S_0$, at the facets. Thus, a method of registering the cleavages is a requirement.

(3) The effect of the NAM structure on lateral mode control has not been documented but could lead to excess scattering and a rough far-field pattern. Recently, NAM structures have appeared in commercial products. The CRANK TJS laser (Fig. 6.27**a**) can operate reliably at $P = 15$ mW (CW), whereas the TJS laser without the NAM can only operate at $P = 3$ mW (CW). In a recent publication [196], it was reported that the Ortel Corporation has developed a buried heterostructure laser structure similar to Fig. 6.27**b** with significantly improved output power characteristics compared to the conventional BH laser. The NAMBH laser is rated at 30 mW (CW) compared to 3 to 10 mW for the conventional BH/LOC device.

Lately, the use of alloy disordering, whereby the band gap of the QW laser can be increased by diffusion of various types of impurities (e.g., zinc or silicon), can lead to a very effective technique to fabricate NAM structures [197–201] (Fig. 6.29). Such structures have enhanced a pulsed peak power by a factor of 3 or 4, and they have significantly improved GaAlAs/GaAs lasers in high-power and high-reliable operation [201].

By now, it is clear a NAM structure is required for reliable high-power GaAlAs/GaAs laser diodes, and it can possibly extend the lifetime of low-power devices. Recent experimental results [202] appear to indicate that laser structures without a NAM region show a decrease in the catastrophic power level as the device degrades. However, most of the approaches currently being implemented require elaborate processing and/or crystal growth steps. A much better approach would involve a coating deposition to reduce the surface recombination velocity and, thus, enhance the catastrophic intensity level.

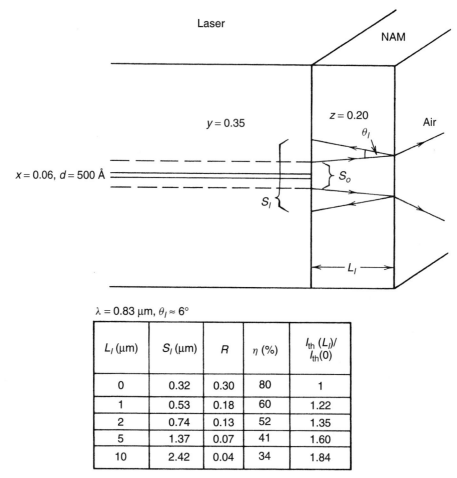

Figure 6.28 | $S_l$ (µm) | $R$ | $\eta$ (%) | $I_{th}(L_l)/I_{th}(0)$

$\lambda = 0.83$ µm, $\theta_l \approx 6°$

| $L_l$ (µm) | $S_l$ (µm) | $R$ | $\eta$ (%) | $I_{th}(L_l)/I_{th}(0)$ |
|---|---|---|---|---|
| 0 | 0.32 | 0.30 | 80 | 1 |
| 1 | 0.53 | 0.18 | 60 | 1.22 |
| 2 | 0.74 | 0.13 | 52 | 1.35 |
| 5 | 1.37 | 0.07 | 41 | 1.60 |
| 10 | 2.42 | 0.04 | 34 | 1.84 |

**Figure 6.28** Schematic diagram and calculation of diffraction effects in the NAM region. The author is grateful to Dr. C.S. Hong (Boeing HTC) for this calculation.

### 6.4.4   Thermal Considerations (GaAlAs/GaAs Semiconductor Diode Lasers)

For a laser diode to emit the maximum amount of power as limited by catastrophic damage, good heat sinking must be provided. In this section, the important aspects of thermal properties in semiconductor laser are discussed based on a simple thermal model. Several models have been presented to study the thermal properties of lasers [203]. We will follow the approach first published by Yonezu [203], but with several important modifications to provide more accurate results. These modifications are important in particular for lasers with a low $T_0$ and a rapid efficiency variation with temperature. A schematic of the laser structures used in the calculation is shown in Fig. 6.30. Two factors contribute to the temperature rise in the active layer.

(1) Conversion of a portion of the injected current in the active layer to heat

(2) Ohmic heating in the top cladding and contact layers

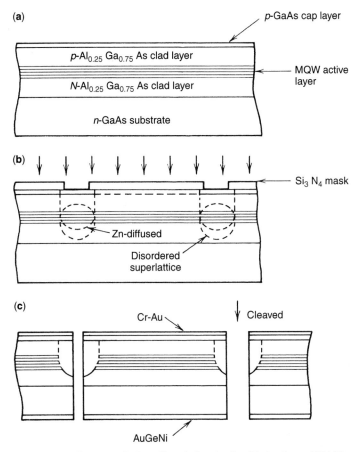

**(a)**

p-GaAs cap layer

p-Al$_{0.25}$Ga$_{0.75}$As clad layer

MQW active layer

N-Al$_{0.25}$Ga$_{0.75}$As clad layer

n-GaAs substrate

**(b)**

Si$_3$N$_4$ mask

Zn-diffused

Disordered superlattice

**(c)**

Cr-Au

Cleaved

AuGeNi

**Figure 6.29**  Schematic diagram of alloy disordering in the fabrication of NAM structures: **(a)** epitaxial growth; **(b)** selective zinc diffusion; **(c)** cleavage (from Ref. 197).

The assumptions for the model are as follows:

(1) One-dimensional heat flow.

(2) No heat flow toward the substrate; no ohmic heating in the epilayer below the substrate, and no heating in the substrate.

(3) Below threshold, all the current is going into heating.

(4) The temperature dependence of the threshold current is described by $J_{th} = J_0 \exp(\Delta T/T_0)$, where $J_0$ is the pulsed threshold current and $\Delta T$ is the junction temperature rise. $T_0$ is assumed to be 160°C for GaAlAs/GaAs.

(5) The temperature dependence for the internal efficiency is described by $\eta = \eta_0 \exp[-(\Delta T/T_1)]$, where $T_1$ is an experimentally determined parameter. In our calculations, $T_1$ is assumed to be 100°C for GaAlAs/GaAs [203].

(6) The model neglects heating at the laser facets. This assumption will tend to overestimate the maximum available power. However, as discussed in the previous section, high-power (GaAl)As/GaAs lasers with nonabsorbing mirrors will be limited only by the heating.

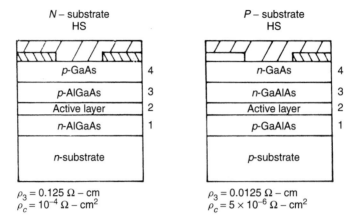

**Figure 6.30**   Schematic diagram of the various simplified DH structures used in the thermal calculations.

The output power per facet is

$$P_{\text{out}} = \tfrac{1}{2} V_j \times W \times L \times (J - J_{\text{th}}) \times \eta \tag{6.29}$$

where $V_J$ is the junction voltage, $W$ is the width of the emitting region, $L$ is the length, and $J$ is the current density at the contact. The junction temperature rise, $\Delta T$, is

$$\Delta T = R_{\text{th}} \times J_{\text{TE}} \times V_J + C \times J^2 \tag{6.30}$$

where

$$J_{\text{TE}} = J \text{ for } J < J_{\text{th}}$$

$$J_{\text{TE}} = (1 - \eta)J \text{ for } J > J_{\text{th}}$$

$$R_{\text{th}} = \frac{d_3}{K_3} + \frac{d_c}{K_4} + \frac{d_M}{K_M} + \frac{\ln(4L/W)}{\pi K_0} W \tag{6.31}$$

$$C = \rho_3 d_3 \left[ \frac{d_3}{2K_3} + \frac{d_4}{K_4} + \frac{d_M}{K_M} + \frac{\ln(4L/W)}{\pi K_0} W \right] + \rho_c \left[ \frac{d_M}{K_M} + \frac{\ln(4L/W)}{\pi K_0} \right]$$

and the various layers are described in Fig. 6.30; $d_M$ and $K_M$ are the thickness and thermal conductivity of the soldering layer. The various $\rho$'s represent the resistivities of various layers and $\rho_c$ represents the contact resistance.

With the formulation above, we can calculate the maximum output power as a function of several laser parameters. For the purpose of this section, we consider only two important cases: (1) the effect of heat sink material on maximum output power and (2) the effect of substrate type on maximum power. Figure 6.31 shows the maximum output power as limited by heating as a function of pulsed threshold current density, $J_0$, for various heat sink materials. The use of ceramic and semiconductor heat sinks has gained increased significance over the last several years [115, 204]. In many situations, it is advantageous to use the ceramic heat sink because the thermal expansion coefficient is similar to the laser diode and thus thermal stress can be minimized. Many of the

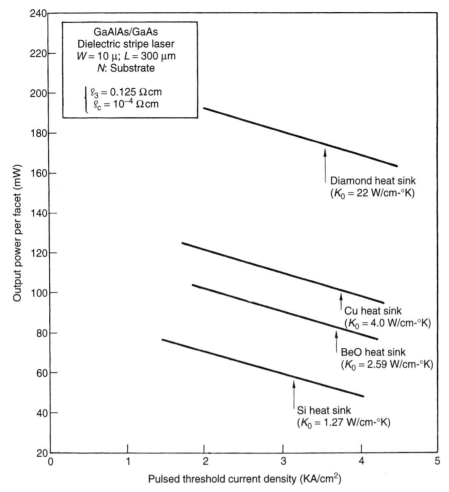

**Figure 6.31** Calculated plot of the maximum output power versus pulsed threshold current density with heat sink material as a parameter.

high-power laser results reported by the Japanese vendors are for lasers mounted on silicon heat sinks.

Figure 6.32 describes the effect of substrate type on the maximum output power. The interesting result is that it is possible to achieve significantly higher power by fabricating lasers on $p$-substrates. At high current densities, Eq. (6.30) indicates ohmic heating can play a significant role, especially if the contact and the cladding layer resistances are high. For the case of lasers fabricated on $n$-substrates, the top two layers are $p$-type and have relatively low conductivity (i.e., low mobility). Furthermore, the contact resistance in a $p$-layer is relatively high ($\sim 10^{-4}$ to $10^{-5} \Omega$-cm$^2$). By switching to a $p$-type substrate (with the use of current blocking layers), the top two layers are $n$-type and both the AlGaAs layer resistance and the top contact resistances are significantly reduced. Thus, the laser diode gets less ohmic heating in the region above the active layer. This allows the laser to produce higher power as shown in Fig. 6.32. We should point out that in our modeling, we have neglected any ohmic

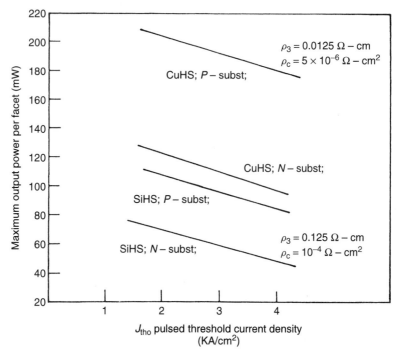

**Figure 6.32** Calculated plots of maximum output power versus pulsed threshold current for lasers having different substrate types $(n/p)$. The higher output power for $p$-substrates results from the improved electrical resistance of the contact resistance and the top cladding layer.

heat generation from epitaxial layers, below the active layer. This assumption is valid for $n$-type substrates. However, it is questionable for $p$-type substrates. More recent calculations that incorporate the effect of heat generation in these layers (Section 6.4.5) confirm that the major difference between the $n$- and $p$-type substrate lasers is the low contact resistance of the top $n$-layer in the case of $p$-type substrates.

### 6.4.5    Thermal Considerations (GaInAsP/InP Diode Lasers)

In this section, we describe a simple thermal model that can be used to predict the higher power performance of GaInAsP/InP lasers. We believe an accurate thermal model will be more useful for these laser diodes because facet absorption is expected to play a minor role, and thus bulk heating will be the ultimate limiting factor in obtaining high power.

The model used is similar to that for (GaAl)As/GaAs lasers except that we have made a greater attempt at determining the important parameters using first principles [205]. First the pulsed threshold current, $J_0$, is related to temperature by

$$J_0(T) = J_0 \exp(\Delta T / T_0) \tag{6.32}$$

where $T_0$ is an empirically determined parameter and is assumed to be 70°C for GaInAsP/InP lasers.

The quantum efficiency, $\eta_{\mathrm{tot}}$, of the laser diode can be expressed as the product of three terms, and it is given as

$$\eta_{\mathrm{tot}} = \eta_{\mathrm{ext}}\eta_{\mathrm{int}}\eta_e \tag{6.33}$$

where $\eta_{\mathrm{ext}}$ = external quantum efficiency = $[(1/L)\ln(1/R)] \div G_{\mathrm{th}}$, $\eta_{\mathrm{int}}$ = stimulated radiative efficiency, and $\eta_e$ = efficiency due to current injection.

In GaInAsP/InP devices, $\eta_e$ can be significantly lower than 1 if either the current leaks through the current blocking layers or nonradiative currents are present. We will take $\eta_e$ as a variable parameter in our model, but nominally at $\sim 0.85$. The internal efficiency can be expressed as

$$\eta_{\mathrm{int}} = \tau_{\mathrm{eff}}/\tau_{\mathrm{stim}} \tag{6.34}$$

where $\tau_{\mathrm{eff}}$ and $\tau_{\mathrm{stim}}$ are the effective and stimulated lifetimes, respectively. The stimulated lifetime is smaller than the spontaneous lifetime, but it is finite and must be continuous across the threshold. We can estimate its value from the expression relating resonance frequency and dc drive current, which is typically used to determine the resonance frequency of an analog-modulated semiconductor laser [206]. We estimate the stimulated lifetime by the following expression:

$$\tau_{\mathrm{stim}} = \begin{cases} \tau_{\mathrm{rad}}(J_0/J) & J > J_0 \\ \tau_{\mathrm{rad}} & J \leqslant J_0 \end{cases} \tag{6.35}$$

where $\tau_{\mathrm{rad}}$ is the spontaneous lifetime, $J_0$ is the pulsed threshold current density, and $J$ is the drive current. The effective lifetime is given as follows:

$$\tau_{\mathrm{eff}} = \frac{1}{1/\tau_{\mathrm{stim}} + 1/\tau_{n_{\mathrm{rad}}}} \qquad \tau_{\mathrm{rad}} = \frac{1}{Bn_{\mathrm{th}}} \qquad \tau_{n_{\mathrm{rad}}} = \frac{1}{Cn_{\mathrm{th}}^2} \tag{6.36}$$

where $B$ = radiative recombination coefficient, $C$ = Auger recombination coefficient, and $n_{\mathrm{th}}$ = threshold carrier density.

We assume that $B$ is independent of temperature. The variation of $C$ with temperature has been obtained by fitting the data from Ref. 207. The threshold carrier density was computed from the expression used by Haug [207]. We should note that our use of Eq. (6.34) for the internal efficiency is important because it relates the variation of efficiency with temperature to the important laser device parameters. A summary of all the parameters used in the calculation is given in Table 6.1. The formula used for the thermal analysis is

$$\Delta T = R_{\mathrm{th}} J_{\mathrm{te}} V_j + C' J^2 \tag{6.37}$$

where $R_{\mathrm{th}}$ = thermal resistance $\times$ area [Eq. (6.23)], $J_{te}$ = current density which heats the active region, $V_j$ = voltage across the junction, $C'$ = term due to ohmic heating of the various layers, and $J$ = applied current density.

$C'$ is expressed as

$$C' = p_1 d_1 \left( \frac{d_1}{2K_1} + d_3 K_3 + d_4 K_4 + d_m K_m + \frac{\ln(4L/W)}{\pi K_0} W \right)$$

$$+ p_3 d_3 \left[ \frac{d_3}{2K_3} K_3 + \frac{d_4}{K_4} + \frac{d_m}{K_m} + \frac{\ln(4L/W)}{\pi K_0} W \right) + p_4 d_4 \left( \frac{d_4}{2K_4} + \frac{\ln(4L/W)}{\pi K_0} W \right) \right] \tag{6.38}$$

**TABLE 6.1  Parameters Used in the Calculations**

| Parameter and Units | Values Used in InGaAsP-InP Laser |
|---|---|
| $T_0$ (°C) | 70 |
| $W$ (μm) | 2–10 |
| $L$ (μm) | 300–1800 |
| $V_j$ (V) | 0.95 |
| $d_3$ (μm) | 1 |
| $d_4$ (μm) | 2 |
| $d_m$ (μm) | 1 |
| $K_3$ (W/cm/deg) | 0.7 |
| $K_4$ (W/cm/deg) | 0.35 |
| $K_m$ (W/cm/deg) | 2.65 |
| $\rho_2$ (n-substrate) Ω-cm | 0.00924 |
| $\rho_3$ (n-substrate) Ω-cm | 0.219 |
| $\rho_4$ (n-substrate) Ω-cm | 0.4001 |
| $\rho_c$ (n-substrate) Ω-cm$^2$ | $5 \times 10^{-5}$ |
| $\rho_2$ (p-substrate) Ω-cm | 0.219 |
| $\rho_3$ (p-substrate) Ω-cm | 0.00924 |
| $\rho_4$ (p-substrate) Ω-cm | 0.01064 |
| $\rho_c$ (p-substrate) Ω-cm$^2$ | $5 \times 10^{-6}$ |
| $n_1$ | 3.5 |
| $n_2$ | 3.23 |
| $\lambda$ (μm) | 1.3 |
| $d_a$ (μm) | 0.05–0.5 |
| $\beta$ (cm · μm/kA) | 40 |
| $\alpha_0$ (cm$^{-1}$) | 35 |
| $\alpha_i$ (cm$^{-1}$) | 20 |
| $\mu_p-$ (InP) (cm$^2$/V · s) | 142 |
| $\mu_n-$ (InP) (cm$^2$/V · s) | 3378 |
| $\mu_p-$ (InGaAsP) (cm$^2$/V · s) | 78 |
| $\mu_n-$ (InGaAsp) (cm$^2$/V · s) | 2933 |
| $A$ (s$^{-1}$) | $1 \times 10^8$ |
| $B$ (cm$^3$/s) | $1 \times 10^{-10}$ |
| $C'$ (cm$^6$/s) (at 300 K) | $6.219 \times 10^{-29}$ |
| $\eta_e$ | 0.85 (except for Fig. 6.38, where $\eta_e = 1$) |

[Note the modifications from Eq. (6.31) as a result of the inclusion of ohmic heating for the epitaxial layer below the active layer.] As pointed out previously, the $C'$ term is very significant for semiconductor lasers operating at high power levels (high current density). The output power is obtained using Eq. (6.29).

We consider three cases of interest. The first case involves the variation of maximum power per facet for both $n/p$ substrate lasers as a function of active-layer thickness and heat sink material (Fig. 6.33). The geometry used is similar to that of the DC-PBH

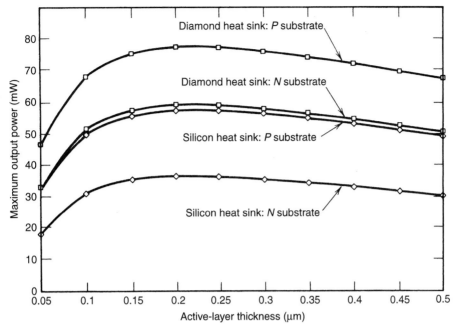

**Figure 6.33** Calculation of $P_{max}$ per facet as a function of active-layer thickness for different heat sinks. Calculations are for both $n/p$-substrates: $W = 2$ μm; $L = 300$ μm; $\eta_e = 0.85$.

laser ($W = 2$ μm; $L = 300$ μm; $T_0 = 70$ K). Several points should be noted. First, the maximum output is obtained for an optimum active layer thickness (~0.15 to 0.2 μm). Second, significantly higher output powers (~25 to 60%) are obtained for lasers fabricated on $p$-substrates compared to $n$-substrates. Third, the use of a diamond heat sink increases the output power by ~60% compared to a silicon heat sink. The output power decreases for thin active layers as a result of Auger recombination and increasing threshold current density. For thicker active layers, the output power drops as a result of the increased threshold current density. In Fig. 6.34, we show the variation of maximum power versus diode length. It is clear that increasing the length from the standard 300-μm device to 700 μm can increase the output power by ~100% (60 to 120 mW) for $n$-substrates using a diamond heat sink and a similar percentage increase for $p$-substrates. For longer lengths, the output power increases as a result of decreased thermal resistance and threshold current density. However, a counterbalance to this effect is the decreased external efficiency, which eventually leads to saturation of the output power.

Experimental results have demonstrated that an increase of output power from ~30 mW to 50 mW per facet is possible by increasing the laser cavity length from 250 to 500 μm [208]. In addition, we have used our model to predict the experimental results from Oki of $P_{max}$ as a function of $L$ [209]; the results are also shown in Fig. 6.34. Finally, in Fig. 6.35a and **b**, we plot maximum cw power as a function of maximum cw operating temperature, and $P_{max}$ versus leakage efficiency. Such relationships will be useful to determine maximum cw power from the measurements of maximum operating temperature, and to provide an estimate of the leakage current. From the graph, it is evident that lasers mounted on a diamond heat sink can operate

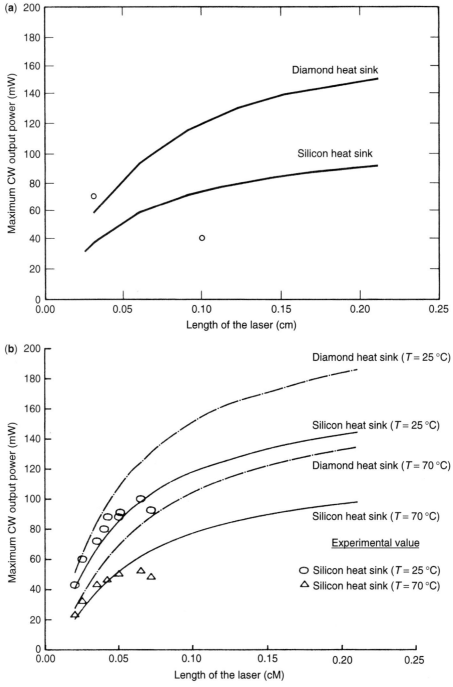

**Figure 6.34** Calculation of $P_{\text{max}}$ per facet as a function of device length for an $n$-substrate device: $W = 2\ \mu\text{m}$; $d_A - 0.2\ \mu\text{m}$; $\eta_e = 0.85$ (from Ref. 210). Calculation of $P_{\text{max}}$ as a function of device length for a $p$-substrate device. The dimensions have been chosen to match the experimental results of Oki: $W = 1.6\ \mu\text{m}$; $d_A = 0.15\ \mu\text{m}$; $\eta_e = 1.0$ (from Ref. 209).

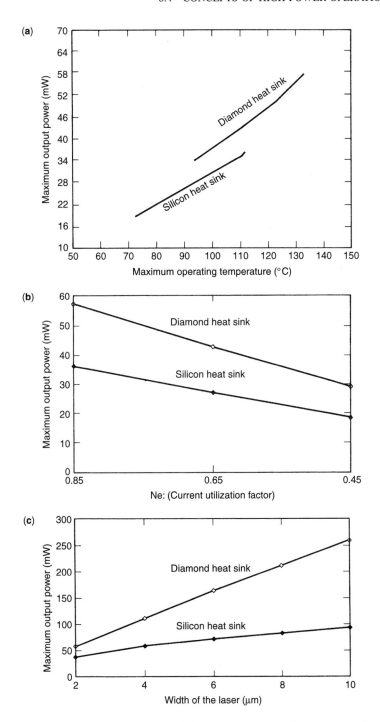

**Figure 6.35** (a) $P_{max}$ as a function of the maximum temperature of cw operation for a $n$-substrate device. (b) $P_{max}$ as a function of leakage efficiency for a $n$-substrate device. $W = 2$ μm, $d_A = 0.2$ μm, $L = 300$ μm. The graph can be used to determine the magnitude of device leakage currents. (c) Variation of maximum cw power with stripe width for an $n$-substrate (InGaAsP-InP) device.

cw to ~134°C, while devices mounted on silicon can operate cw up to 110°C. We should note also that these results agree reasonably well with published experimental data [208–210]. Lastly, it is clear that there is a correlation between high power and maximum cw temperature of operation at low power levels. This implies that lasers designed for maximum cw power can also be used in very high temperature environments when operating at lower power levels.

### 6.4.6 Reliability of Diode Lasers

Semiconductor laser diodes have been widely used in many applications such as telecommunication, military, and commercial industries. The reliability of these diodes is important for almost all applications, especially in telecommunication areas. In order to understand and implement proper screening procedures, qualification programs, and reliability predictions, it is advantageous to have some knowledge of the potential failure modes and related design issues.

Some of the common failure modes of a laser diode are as follows [211–213]:

(1) Facet damage

(2) Dark line defects

(3) Current confinement problems

(4) Normal aging

(5) Laser diode–submount interface problems

(6) Electrostatic discharge damage (ESD)

Facet damage was discussed in Section 6.4.1. Dark line defects are usually originated from laser diode crystal. They can be reduced by using a dislocation-free substrate and minimizing strains or stresses on the devices. Current confinement problems can occur from poor process control in manufacturing, poor design, and normal aging. Normal aging is characterized by an increase in threshold current. This slow deterioration of the laser diode is thought to be due to the energy released in nonradiative recombination, which aids in the creation or displacement of point defects in the crystal lattice. These localized point defects introduce charge-trapping sites into the "forbidden" energy gap. As a result, there is a further increase in nonradiative recombination and, hence, a reduction in internal quantum efficiency. The mounting and bonding of a laser diode to a submount is very important in assuring the reliability of the laser. Potential failure modes associated with the laser–submount interface include stress-induced defects, solder migration, poor solder wetting, and excessive overhang of the laser diode over the submount. Electrostatic discharge damage is a constant threat to the reliability and performance of lasers. Catastrophic damage to a laser may occur as low as 500 volts, which would not be felt by most humans (~3,000 volts). Normal handling (without ESD precautions) can therefore damage or destroy a laser [214].

Laser degradation is often measured in terms of changes in laser threshold current or in drive current required to maintain a specific optical output power. In the early stages of a laser's life, its threshold or drive current increases much more rapidly than after it has been operating for a period of time. Thus, a period of time is required, not only to stabilize the degradation, but also to stabilize the laser's electrical and optical characteristics. Key parameters are tested before and after the burn-in to monitor "infant mortality" and other reliability problems.

The purposes of the burn-in are to stabilize the device, to test if the device has sufficiently stabilized and the degradation rate is not excessive, and to weed out most failures due to poor processing and handling. It is also important to note that "harsher" burn-in conditions can frequently improve the total yield from burn-in. The burn-in process is therefore always performed at an increased temperature.

After the burn-in process, laser diode endurance tests are performed at high temperature. To obtain measurable degradation in specific laser parameters, a 5,000-hour test at 70°C and maximum-rated optical output power is needed. The lifetime of the laser diode is then extrapolated from the test data. For detailed lifetime calculation, the reader is referred to the Bellcore life-test report [215].

### 6.4.7  Master Oscillator Power Amplifiers

One approach to produce a high-power device operates in a diffraction-limited single spatial and a single spectral mode is to use a master oscillator power amplifier (MOPA). This approach takes all high-power operation concepts discussed above into consideration. The most efficient devices demonstrated with MOPA concept are the tapered amplifier MOPA [216–220]. More recently, an angled-facet, tapered amplifier (Fig. 6.36) has demonstrated high-power single mode (near-diffraction–limited beam quality) operation up to 1.9 W cw with less than 5 mW of injection power. [221, 222]

By using monolithically integrated flared-amplifier master oscillator power amplifiers (MFA-MOPA), output power greater than 2.0 W cw in a single diffraction-limited output beam has been demonstrated for InGaAs/GaAs devices emitting at ~975 nm [223]. The latest development of AlGaInAs/InP tapered high-power lasers with a distributed Bragg reflector emitting near 1550 nm also demonstrates cw power of 0.6 W with 20-dB side-mode suppression. Both devices operate in a single spatial and spectral mode throughout its operating range [224].

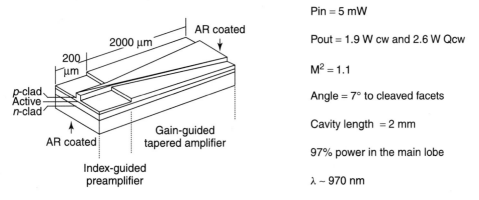

**Figure 6.36**   Schematic design of an angeled-facet tapered amplifier.

## 6.5 SURFACE-EMITTING SEMICONDUCTOR LASERS

In the last several years, there has been significant interest in the fabrication of surface-emitting lasers [225–230]. Such lasers may lead to several important advantages over existing edge-emitting devices:

(1) Potential for large-scale integration of both active and passive devices using two-dimensional arrays

(2) Potential for use in optical interconnects (e.g., between two printed circuit boards)

(3) Potential for generating of high-power, broad-area emission with a narrow far-field beam width, which could be used for free-space communications and optical pumping of solid-state lasers

There are several approaches to achieve surface emission:

(1) Two-dimensional stacked arrays — This design takes the advantages of matured edge-emitting laser bars to produce high-efficiency, high output power. Figure 6.37**a** shows the schematic drawing of the stacked array concept. A primary application of this two-dimensional stacked array is diode pumping of solid-state lasers [231–233]. Output power density of 2.5 KW/cm$^2$ with 20% duty factor is commercial available. [234].

(2) Vertical cavity surface emitting lasers (VCSELs) — The first surface-emitting laser design incorporated a vertically emitting geometry as shown in Fig. 6.37**b** [225]. The vertical cavity requires highly reflecting mirrors to increase the cavity $Q$, and thus, the output power level is somewhat limited. Detailed description of VCSELs will be provided in Section 6.8.

(3) Grating surface-emitting lasers — In this configuration (Fig. 6.37**c**), the optical cavity lies in the plane of the active layer. This is a standard diode laser configuration; hence, high gain-length products can be achieved. Surface emission is obtained by deflecting light with an appropriate designed grating [228–230]. Kojima et al., by using distributed Bragg reflector (DBR) and distributed feedback (DFB) approaches, demonstrated a cw operation of grating surface-emitting lasers at room temperature [235–236]. These DBR and DFB surface-emitting lasers also provide very narrow beam divergence, 0.1 degree for DBR and 0.13 degree for DFB, respectively.

(4) Deflecting mirror surface-emitting lasers — In this configuration, a light deflecting mirror is built-in with the cavity facet such that light can be deflected in the direction perpendicular to the wafer surface. This laser cavity has the same geometry as the edge-emitting lasers. Hence, the output power and efficiency are expected to be comparable to those of the conventional diode lasers. This structure has been implemented in both GaInAsP/InP [226] and GaAlAs/GaAs [227] semiconductor laser systems. Liau et al., by using the mass transport process (for GaInAsP/InP), fabricated very smooth parabolic deflector devices with performance equivalent to cleaved edge emitters [237, 238]. Power in excess of 1.3 W (cw) from a 164-element array of GaInAsP/InP lasers has been achieved. Unfortunately, the mass transport process is only applicable to GaInAsP materials. The same process could not be adapted to GaAs-based devices. Ion beam

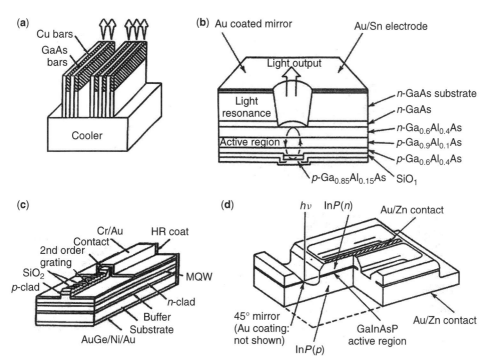

**Figure 6.37** Schematic diagram of four types of surface-emitting lasers: **(a)** stacked arrays (from Ref. 234); **(b)** vertical cavity surface-emitting design (from Ref. 225); **(c)** grating surface-emitting lasers (from Ref. 230); and (d) etched mirror design (from Refs. 226 and 227).

etching and ion beam–assisted etching (IBAE) have been used to fabricate deflectors for GaAs/AlGaAs surface emitting lasers [239–242].

There are two approaches to produce the deflector surface-emitting laser (Fig. 6.38). In the first one (Fig. 6.38**a**, called the junction up device, the laser cavity is formed by two vertical etched facets). Light is emitted from the epi-side by means of the etched 45-degree deflecting mirror. The second approach consists of one 45-degree mirror, one vertical mirror, and one etched-stop window facet. Light is emitted from the substrate via the etched hole. This junction-down device is shown in Fig. 6.38**b**.

In low-power-density applications, junction-up devices are advantageous because they are relatively easy to fabricate and have more flexibility for addressability. A $10 \times 10$ junction-up surface-emitting laser has been demonstrated at TRW [243].

As laser power increases, thermal dissipation becomes more critical. Lasers emitting from the epi-side suffer a substantial temperature rise around the active layer because of the relatively thick (100 $\mu$m), low-thermal-conducting GaAs substrate that represents a thermal barrier between the heat-creating area and the heat exchanger. The junction down approach is designed to place the heat-generating active layer within microns of the heat exchanger, thereby minimizing the thermal path.

Monolithic surface emitting through substrate arrays offer not only thermal advantages, but also cost advantages because the monolithic approach minimizes labor-intensive fabrication and assembly costs and is compatible with batch-fabrication techniques.

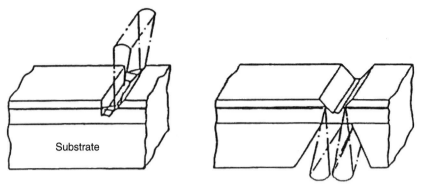

**Figure 6.38**  Two-dimensional surface emitting configurations using 45-degree deflecting micro-mirror.

By properly combining the individual device emissions, coherent and incoherent high output power light sources can be produced. For example, >10 W of coherent power can be achieved by coupling several arrays of anti-guided lasers monolithically. A similar approach can be used to produce >100-W incoherent power by simply fabricating a large number of broad area (100–200-μm wide stripe) surface-emitting lasers on the same wafer.

## 6.6  SEMICONDUCTOR TUNABLE LASER DIODES

Wavelength tunability is a very valuable feature of single-mode (SM) lasers for many applications [244]. For example, in fiber-optics communications [245], 16 or 32 channels are used in a dense-wavelength division multiplexing (DWDM) [246] system. It will be very expensive to place an instant backup for each laser with different wavelengths. With tunable lasers, however, the number of spare lasers required can be reduced substantially. Tunable lasers promise to revolutionize the optical networking industry by keeping a lid on the problem of the escalating cost of buying, stocking, and managing spares for DWDM equipment. And that is just for starters. Other applications for tunable lasers in this area stretched from enabling automated network provisioning to reconfigurable network architectures. In addition, wavelength tunability is crucial for many other applications, such as optical sensing [247], fiber measurements [248], spectroscopy [248], and reflectometry [250]. For most of these applications, a narrow spectral linewidth and continuous tunability are essential.

A semiconductor laser wavelength can be tuned electronically either by applying current or by temperature control. It can also be tuned optically and mechanically. In this section, we will outline the principles and mechanisms of the tunable SM semiconductor lasers and briefly discuss several configurations for continuous and discontinuous wavelength tuning.

### 6.6.1  Physical Mechanisms for Wavelength Tuning

For tunable semiconductor lasers, the fundamental equation [251] is

$$\frac{\Delta\lambda}{\lambda_0} = \frac{\Delta n}{\bar{n}_{g,\text{eff}}}\Gamma = \frac{\Delta n_{\text{eff}}}{\bar{n}_{g,\text{eff}}} \tag{6.39}$$

where $\Delta\lambda$ is tuning wavelength, $\lambda_0$ is the lasing wavelength before tuning, $\Delta n$ is the real part of the refractive index change, $n_{g,\text{eff}}$ is the group effective refractive index, $\Gamma$ is the optical confinement factor of the tuning region, and $\Delta n_{g,\text{eff}}$ is the real part of the effective refractive-index change. The basic mechanisms of wavelength tuning electronically are as follows:

(1) Refractive index control by carrier injection [252]: Refractive index control by electro-optical effects such as the quantumconfined Stark effect (QCSE) [253].

(2) Gain wavelength shifting and refractive index control through a junction temperature control [254].

In the case of refractive index control, the cavity mode spectrum can be tuned in a smooth manner (continuous tuning). However, if the gain wavelength shifts, it would cause mode jumps and result in a discontinuous tuning.

Due to plasma effect, the refractive index changes when free carriers are injected into a double hetero-structure [255]. This refractive index change, $\Delta n$, results from the polarization of free carriers and an absorption edge shift resulting from the free carrier injection. The former is usually the dominant factor. The real part of the refractive index change due to the carrier polarization is given by [256]

$$\Delta n = -\frac{e^2\lambda^2}{8\pi^2 c^2 n\varepsilon_0}\left(\frac{\Delta N}{m_e} + \frac{\Delta P}{m_h}\right) \tag{6.40}$$

If $\Delta N = \Delta P$ (an ambipolar injection),

$$\Delta n = -\frac{e^2\lambda^2}{8\pi^2 c^2 n\varepsilon_0}\left(\frac{1}{m_e} + \frac{1}{m_h}\right)\Delta N = \beta_{pl}\Delta N \tag{6.41}$$

where $m_e$, $m_h$, $\Delta N$, and $\Delta P$ are the effective mass and the density of the injected electrons and holes, respectively, and $n$ is the refractive index of the semiconductor.

Because according to Eq. (6.39) the wavelength tuning, $\Delta\lambda$, is proportional to the real-part change of the effective refractive index, we have

$$\Delta\lambda = \lambda_0 \frac{\Delta n}{\overline{n}_{g,\text{eff}}}\Gamma = \frac{\beta_{pl}\Gamma\lambda_0}{\overline{n}_{g,\text{eff}}}\Delta N \tag{6.42}$$

Because $\beta_{pl}$ is negative, both the refractive index and the wavelength will be reduced by the tuning with the free-carrier plasma effect. For 1550-nm InGaAsP lasers [257], if $\lambda = 1550$ nm, $n = 3.3$, $m_e = 0.05\,m_o$, $m_h = 0.5\,m_o$, $\Delta N = \Delta P = 3 \times 10^{18}\text{cm}^{-3}$, then, $\beta_{pl} = -6.75 \times 10^{-21}\text{cm}^3$, and the refractive-index change would be $-0.022$. Besides the plasma effect, carrier injection also changes the spectral shape of the band-to-band optical absorption through band filling and band-gap shrinkage [258]. This also contributes to the change of the refractive index. Therefore, $\beta_{pl}$ is as large as $-1.3 \times 10^{-20}\text{cm}^3$, and the refractive-index change can be as much as $-0.04$ for 1550 nm of wavelength with injection carrier concentration, $\Delta N = \Delta P = 3 \times 10^{18}\text{cm}^{-3}$. If we take the confinement factor, $\Gamma = 0.5$, and the effective group refractive index, $n_{g,\text{eff}} = 4$, the wavelength-tuning range $\Delta\lambda$, would be $-7.75$ nm.

For the tuning mechanism of the free-carrier plasma effect, it should be noted that for a given device structure the product of tuning range and tuning speed is fixed. They can not be optimized independently [259].

The free-carrier plasma effect not only changes the real part of the refractive index, but it changes the imaginary part also. The imaginary part change means there is an additional optical loss in this process. Again, if $\Delta N = \Delta P$ (an ambipolar injection), the optical loss due to a change of the imaginary part of the refractive index is

$$\alpha_{pl} = -2\kappa_0 \Delta n' = \frac{e^3 \lambda^2}{4\pi^2 c^3 n \varepsilon_0} \left( \frac{1}{m_e^2 \mu_e} + \frac{1}{m_h^2 \mu_h} \right) \Delta N = \kappa_{pl} \Delta N \qquad (6.43)$$

where $\mu_e$ and $\mu_h$ are the mobilities of electrons and holes, respectively. If using the example above and $\mu_e = 2000$ cm$^2$/Vs, $\mu_h = 200$ cm$^2$/Vs, then $\kappa_{pl}$ is $7.8 \times 10^{-18}$cm$^2$ and $\alpha_{pl} = 24$ cm$^{-1}$. This indicates that the carrier injection for tuning results in a large amount optical loss. In addition, the intervalence-band-absorption (IVBA) [260] by the injected holes also causes significant losses [261], which can be written as

$$\alpha_{IVBA} = \kappa_{IVBA} \Delta P = \kappa_{IVBA} \Delta N \qquad (6.44)$$

For InGaAsP at 1550-nm wavelength, $\kappa_{IVBA}$ is in the range of 2 to $4 \times 10^{-17}$cm$^2$. Assuming $\Delta N = \Delta P = 3 \times 10^{18}$cm$^{-3}$, $\alpha_{IVBA}$ is about 90 cm$^{-1}$, which is three to four times larger than $\alpha_{pl}$. IVBA is a major portion of the absorption and has to be considered for the total optical losses.

The QCSE is another approach to electronically tune the laser wavelength. However, because the QCSE is weak in the III-V semiconductors [262], by using QCSE, the refractive-index change is only in the order of $10^{-3}$ to $10^{-2}$. It really depends on how closely the band-gap wavelength of the quantum-well structure is used for tuning matches to the laser wavelength [263]. In addition to the relatively small change of the refractive index in the QCSE, the optical confinement achievable in these quantum wells (QWs) is much smaller than the case of the bulk semiconductor structure used for the free-carrier plasma effect. To calculate the wavelength tuning range through the QCSE, for 1550-nm SM lasers, we take the real part of the refractive index change, $\Delta n = -0.01$, and the effective group index, $n_{g,\text{eff}} = 4$. If the confinement factor, $\Gamma = 0.2$, we have the wavelength tuning range, $\Delta \lambda = -0.78$ nm. Therefore, the largest tuning ranges have been achieved so far with the plasma effect. On the other hand, in the QCSE, the semiconductor PN-junction is reversely biased so that no current flows and no heating occurs. The advantage of tuning with the QCSE is that no carrier lifetime limitation exists for the tuning speed.

Due to the effect of junction temperature on light-emission wavelength and refractive index of the cavity, the wavelength tuning can be achieved by temperature control. For instance, the wavelength's temperature coefficient of the 1550-nm Fabry–Perot (F-P) lasers is about 0.5-nm/K and about 0.1-nm/K for DFB and DBR lasers [264]. Thus, the wavelength tuning can be done through either an external temperature control, such as a thermal-electrical controller, or the laser driving current control. The tuning through temperature control is continuous for the DFB or DBR lasers. However, the tuning range is limited by maximum working temperature of the laser diodes. The wavelength tuning through temperature control is continuous in a small range ($\sim$0.1 nm/K) and discontinuous in a large range ($\sim$0.5 nm/K) for the 1550-nm F-P type of SM

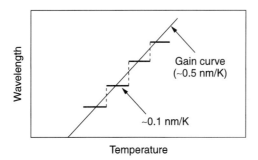

**Figure 6.39**  Schematic of lasing wavelength as a function of temperature for 1550-nm F-P type of SM semiconductor lasers. It shows a continuous and a discontinuous behavior. Within a certain temperature range (very small), the wavelength tuning through temperature is continuous. When it is beyond the range, wavelength jumps due to a mode jump.

lasers, as shown in Fig. 6.39. This type of tuning has two major disadvantages. One of them is potential wavelength instability because a very tight and precise temperature control for a long period of time is required. Any temperature perturbation would cause laser wavelength to drift. The second one is the tuning speed, which is relative slow due to the laser diode's long thermal-response time (in an order of 10 to 100 ms of range). Nevertheless, the wavelength tuning through temperature control is simple and straightforward. It is suitable for a DC-type of wavelength-tuning applications, such as measurements or sensing.

A comparison of the three types of physical tuning mechanisms discussed above is in Table 6.2. Here, we consider both the basic material aspects and device-related parameters, such as the optical confinement factor that influence the suitability of the mechanisms for different applications.

As most interests in semiconductor tunable lasers are related to the fiber-optic communication applications and the tuning speed is essential, in the following discussions, emphasis will be placed on the methods of the refractive-index control for wavelength tuning.

A continuous wavelength tuning is always preferred in most applications. This is due to its simplicity and unambiguity of the wavelength setting without interruption. In this wavelength-tuning scheme, the semiconductor laser emits in the same longitudinal mode within the entire tuning range without mode changes and jumps. As a result, the tuning range can not be larger than that of a longitudinal laser mode.

In the discontinuous tuning mode, wavelength tuning is done by mode jumps from one longitudinal mode to another. Consequently, the ultimate limitation of the tuning range in this tuning scheme is set by gain-function spectral width of the active region.

### 6.6.2  Basic Tunable Laser Structures

Ideal semiconductor tunable lasers should have two electronic controls that allow for the independent adjustment of the output power and wavelength [266], as shown in Fig. 6.40. In practice, however, a complete separation between power and wavelength controls can hardly be achieved because the tuning function by carrier injection also introduces optical losses that reduce the output power. In addition, the change of the laser current induces temperature change, which affects the emission wavelength [267].

**TABLE 6.2 A Comparison of Important Aspects of the Three Different Mechanisms for Wavelength-Tunable Laser Diodes [265]**

| Parameter | Plasma Effect | The QCSE | Temperature Tuning | Comment |
|---|---|---|---|---|
| $\Delta n$ (real part) | −0.04 | −0.01 | 0.01 | Refractive index |
| $\Gamma$ | 0.5 | 0.2 | 1 | Optical confinement factor in the tuning region |
| $\Delta\lambda$ (tuning range) | −8 nm | −1 nm | +5 nm | Wavelength |
| $f_{3dB}$ | 100 MHz | >10 GHz | <1 MHz | Laser modulation speed |
| $\alpha_H$ | −20 | −10 | Large | Linewidth enhancement factor |
| Heat generation | Large | Negligible | Very large | |
| Technology | Moderate | Demanding | Simple | |
| Tuning speed | Fast | Very fast | Slow | |
| Wavelength Stability | Good | Very good | Fair | |
| Tuning Repeatability | Good | Very good | Fair | |

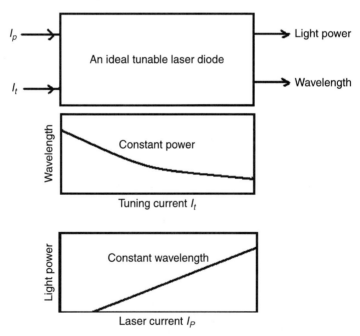

**Figure 6.40** Schematic of an ideal semiconductor tunable laser, of which the optical output power and wavelength can be controlled independently.

Nevertheless, as far a separation as possible of these two functions should be aimed at in a semiconductor-tunable laser design because it would significantly improve the handling convenience and suitability of the lasers.

As indicated in Table 6.2, the largest wavelength-tuning range is obtained from the free-carrier plasma effect. The maximum electronic-tuning range (excluding thermal heating) is limited to about 15 nm at 1550 nm wavelength [268]. Despite this limit in the tuning range, the continuously tunable lasers are still well suitable for applications such as optical communications and DWDM, in which with 1–2 nm of channel spacing, a moderate number of channels (4–8) can be covered by a single laser device.

A variety of structures has been developed to achieve a wide continuous tuning and yet maintain a narrow spectral line-width. Among those are the multisection DBR devices [269–271] as shown in Fig. 6.44 and tunable twin-guide (TTG) lasers [272], and multisection DFB lasers [273]. In both former cases, the tuning is performed by electrical–current–induced refractive-index changes in passive regions. However, in the multisection DFB lasers, a longitudinally varying bias to the active region is applied to achieve the wavelength tuning. All of these three types of lasers are integrated optoelectronic devices, each comprising an active region, a Bragg grating filter, and a tuning region for the optical gain, single mode selection, and electronic tuning function, respectively.

The combination of these three functions has been done so far either by the transverse or by the longitudinal integration as shown in Fig. 6.41. Obviously, the transversely integrated structure resembles the DFB lasers, while the longitudinally integrated structures evolve from the DBR lasers. Thus, significantly different tuning behaviors are obtained even with a similar operation principle. As to the longitudinal integration, the great number of longitudinal modes (in an order of $10^3$ to $10^4$) means that the mode discrimination is weak. Mode jumps and its related wavelength change may readily occur during a tuning process. This can be taken as a consequence of lack of synchronizing between Bragg wavelength and comb-mode spectrum in the DBR lasers. Nevertheless, integrating a phase-shift section within the composite laser cavity can synchronize them. This way the optical cavity length would match to the

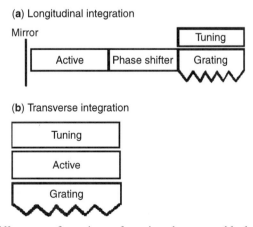

**(a)** Longitudinal integration

**(b)** Transverse integration

**Figure 6.41** Two different configurations of semiconductor-tunable lasers: longitudinal and transverse integration.

Bragg wavelength so that within a certain wavelength range (the continuous-tuning range), the same longitudinal mode is exactly at the Bragg wavelength.

Semiconductor lasers usually operate in the fundamental transverse mode. With the higher order transverse modes being cut off, transverse mode jumps are prohibited. Hence, the wavelength tuning of the transversely integrated devices inherently shows a continuous behavior. This results from the clearly defined single-mode selection of a properly designed DFB laser, which is not deteriorated by tuning the Bragg wavelength.

It was predicted and has been confirmed that the transversely integrated lasers would achieve the largest continuous-tuning ranges [274]. Up to a factor of two to three larger continuous tuning ranges for the transversely integrated lasers has been revealed. However, the longitudinally integrated lasers have proved superior in other important aspects of laser characteristics, such as spectral line-width and optical output power.

### 6.6.3   Longitudinally Integrated Structure

Depending on the tuning method, a longitudinally integrated tunable laser can provide both continuous, including quasi-continuous, and discontinuous wavelength tunings as shown in Fig. 6.42. Even though the structure change between these two is minor, the functionality has a huge difference. First, we will talk about the discontinuous tuning method.

***Discontinuous Wavelength Tuning.***   A discontinuous tuning can be achieved by a modified DBR-type of laser structure by using two different Bragg-type grating reflectors, R1 and R2, at each end of the laser cavity as shown in Fig. 6.43**a** [275]. If we design the grating reflectors in a way that they both have comb-reflection spectra but with different spacing of the reflection peaks as shown in Fig. 6.43**b**, lasing is possible only for those wavelengths that both spectra simultaneously exhibit a reflection peak. This is because a closed-feedback loop exists only at these resonance wavelengths. Similar to the DBR lasers, lasing occurs at the wavelengths of those Fabry–Perot modes that match closest with the resonance wavelengths.

The spectra shift of the reflectors can be done through carrier injection into the Bragg reflector regions, $R1$ and $R2$. If the $R1$ current is fixed at $I_{t1}$, and change the $I_{t2}$

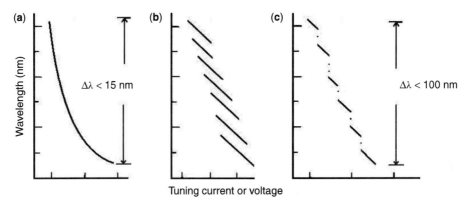

**Figure 6.42**   Three tuning modes for tunable lasers: (**a**) continuous (**b**) quasi-continuous and (**c**) discontinuous wavelength tuning.

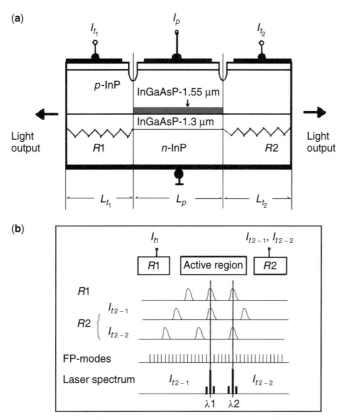

**Figure 6.43** **(a)** A schematic of three-section wavelength-tunable DBR lasers. R1 and R2 stand for two Bragg reflectors. By changing the applied currents on R1 and R2, the wavelength can be tuned discontinuously from λ1 to λ2 **(b)**.

from $I_{t2-1}$ to $I_{t2-2}$, the $R2$'s comb-modes start to shift. When one of the comb-modes of $R2$ overlaps with one of the $R1$'s, the lasing wavelength jumps from λ1 to λ2. By doing this, we can achieve a discontinuous wavelength tuning. It should be noted that this wavelength change from λ1 to λ2 is much larger than the R2's reflection spectra shift. This means the discontinuity is extremely large for this case. The comb-reflection spectra can be made through a spatial modulation of the Bragg grating. This modulation may be in the form of amplitude modulation (AM) [276] or frequency modulation (FM) [277] of the gratings. The corresponding devices are called sampled grating (SG) DBR (AM) and superstructure grating (SSG) DBR (FM) tunable lasers, respectively.

Another way to tune the laser wavelength discontinuously is using a simpler two-section DBR or DFB laser structure. Longitudinal two-section tunable DBR [278] and DFB [279] laser structures are displayed in Fig. 6.44. The simplest approach is a two-electrode DBR or DFB laser with the top contact, which is longitudinally separated into two individually biased sections. By differently biasing the laser sections, wavelength tuning can be obtained. Figure 6.45 shows laser wavelength as a function of the tuning current for both tunable two-section DBR [280] and DFB [279] lasers shown in

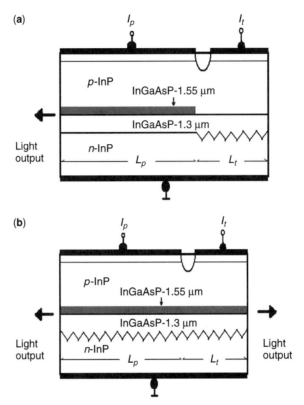

**Figure 6.44**   Schematic of longitudinally integrated wavelength-tunable two-section 1550-nm
(**a**) DBR and (**b**) DFB lasers.

Fig. 6.44**a** and **b**, respectively. The active-section length, $L_p$, is 298 μm and the Bragg-
tuning section length, $L_t$, is 250 μm for the DBR-tunable laser. The total tuning range
is about 9.4 nm. The active and Bragg-section lengths, $L_p$ and $L_t$, are 600 μm and
300 μm, respectively, for the DFB-tunable lasers. During the measurement of curve
**b**, the current $I_p$ was kept constant at 60 mA. The total tuning range is about 7.8 nm.

Comparing Fig. 6.45**a** and **b**, we can see that for the two-section tunable DBR
lasers with 70-mA tuning current, the wavelength tuning range is about 9.4 nm, which
covers nine longitudinal modes. However, for the DFB tunable laser with 50-mA
tuning current, the wavelength tuning reaches 7.8 nm but covers about 20 different
longitudinal modes. In addition, there are some unaccessible wavelengths occurring
due to a mode skip.

Therefore, in the discontinuous-tuning mode, it is of particular importance to characterize the laser carefully over a wide range of currents and temperatures in order
to know exactly under which conditions mode jumps occur and in which operation
regimes stable modes exit. Major challenges resulting from the discontinuous tuning
are the single mode operation and the wavelength access to as many wavelengths as
possible within the tuning range.

***Continuous Wavelength Tuning.***   As the two-section structure is limited to discontinuous wavelength-tuning applications, more complicated structures have been developed,

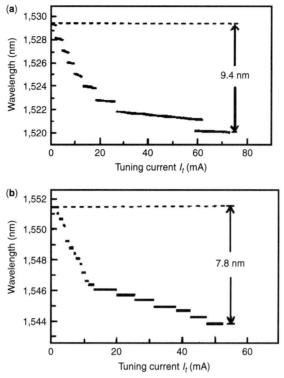

**Figure 6.45** (a) Laser wavelength as a function of tuning current for a tunable two-section DBR laser shown in Fig. 6.44a. Its active-section length, $L_p$, is 298 μm, and its Bragg-tuning section length, $L_t$, is 250 μm. The total tuning range is about 9.4 nm. (Reprinted with permission from *Appl. Phys. Lett.*, **53**, pp. 1036–1038, ©1998 American Institute of Physics.) (b) Laser wavelength as a function of the tuning current for the tunable two-section DFB lasers, as shown in Fig. 6.44b. Active and Bragg-section lengths, $L_p$ and $L_t$, are 600 μm and 300 μm, respectively, and the current, $I_p$, is kept constant at 60 mA. The total tuning range is about 7.8 nm. (Reprinted with permission from *Appl. Phys. Lett.*, **52**, pp. 1285–1287, ©1998 American Institute of Physics.)

such as three-section longitudinal DFB and DBR structures, as shown in Fig. 6.46, to obtain a continuous wavelength tuning. The key difference between the two-section and three-section ones is that there is a phase shifter in the three-section structures, which performs the comb-mode spectrum shift. A continuous tuning can be achieved by this kind of multisection-tunable DFB or DBR laser. Taking the three-section DBR structure as an example, this structure includes gain region, phase shifter, and Bragg reflector section. With three electrodes for $I_p$, $I_{ph}$, and $I_t$, this structure allows almost independent tunings of optical output power via $I_p$, of the position of the comb-mode spectrum via $I_{ph}$ and of the Bragg wavelength via $I_t$. By adjusting $I_{ph}$ and $I_t$ simultaneously, we can get a continuous wavelength tuning as shown in Fig. 6.48 [178]. The three-section DBR laser provides a convenient handling because an effective separation between the power control and tuning function is achieved.

Figure 6.47 gives a longitudinal mode contour for the three-section DBR laser operation in the $I_{ph}$–$I_t$ dimension plane. The solid lines show the mode boundaries at which wavelength/mode jump occurs. It indicates the continuous single-mode tuning

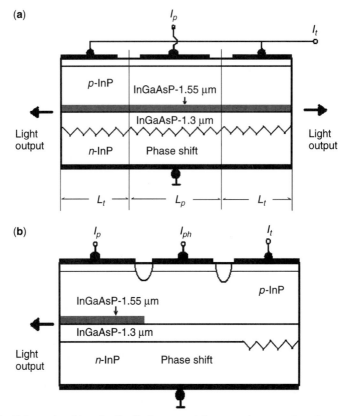

**Figure 6.46** Schematic of longitudinally integrated three-section wavelength-tunable 1550-nm DFB (**a**) and DBR lasers (**b**). The active layer is the dark area.

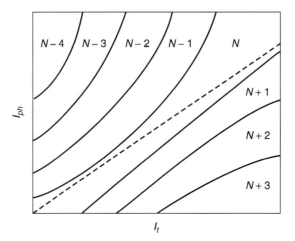

**Figure 6.47** Longitudinal mode contours for tunable laser operations in the $I_{ph} - I_t$ dimension plane. The solid lines show the mode boundaries at which wavelength mode jump occurs. It indicates the continuous single-mode tuning is achievable by following the boundary lines within a specific mode.

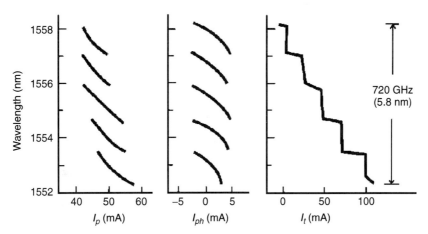

**Figure 6.48**   Effects of all three applied currents ($I_p$, $I_{ph}$, and $I_t$) on the lasing wavelength of the three-section tunable DBR lasers in Fig. 6.46(**b**).

is achievable by following the boundary lines within a specific mode. Especially for the Mode N, people can operate laser along a straight dashed line, which makes the tuning much simpler.

It should be noted that due to heating, an additional wavelength change would occur during $I_p$ change. Therefore, the lasing wavelength is a function of all three currents. A lot of work [282–285] has been performed on optimizing three-section tunable DBR laser. Up to 9 nm of discontinuous tuning and 4.4 nm of continuous tuning ranges have been achieved, which fairly well corresponds to the theoretical limit.

For three-section tunable DFB lasers, because its tuning mechanism by nonuniform current injection is complex (comprising spatial hole-burning effects, thermal heating, and for the case of a quantum-well active layer, the gain-levering effect), we will not discuss the details here.

A disadvantage in the practical application of tunable DFB lasers is that the output power and wavelength are not independently controlled. Both wavelength and output power are affected similarly by all control currents. This is in strong contrast to the idealized tunable lasers. On the other hand, these lasers yield the smallest spectral line-widths among all the tunable laser diodes. With tunable range around 1.3 nm and 1.9 nm, spectral line-widths of 98 kHz and 900 kHz, respectively, were achieved with three-section DFB lasers, as shown in Fig. 6.46**a**. Moreover, placing an electrically tunable phase-control section in the center of a two-section DFB laser proved efficient in keeping the spectral line-width constant throughout the tuning range.

*Quasi-Continuous Wavelength Tuning.*   Wavelength-tuning coverage larger than 15-nm at 1550-nm wavelength cannot be achieved continuously with one single transverse and longitudinal mode. If a large tuning range is necessary, one alternative is using discontinuous tuning. Its maximum tuning range is usually limited only by the gain bandwidth of the active region, which is on the order of 100 nm at 1550 nm. With the discontinuous tuning, however, not all the wavelengths within the covered range can be addressed individually owing to the mode jumps. Thus, it is extremely important in practice to ensure that the laser can access the set of wavelengths needed for the application.

Another alternative tuning technique is the so-called quasi-continuous tuning. In these lasers, two controls are used, one for continuous tuning in each longitudinal mode over a small wavelength range, and the other allowing discontinuous tuning over a large wavelength range via longitudinal mode changes, as shown in Fig. 6.42**b**. If the continuous tuning range exceeds the longitudinal mode spacing, overlapped continuous tuning ranges are obtained, yielding a complete coverage of the entire tuning range. However, in order to avoid instability or mode jumps, an extremely precise and sophisticated device control is required to operate each wavelength near the center of the continuous tuning ranges. This may be done through a three (or multi-) dimension figure describing the influences of various control currents and temperature on the wavelength. However, even a slight degradation or drift of the device parameters can be sufficient to make a readjustment or a new mapping of the wavelength control necessary.

### 6.6.4   Transversely Integrated Structure

The transversely integrated structure is important due to its capability in continuous wavelength tuning. The operation principle of the tunable twin-guide (TTG) DFB laser [272] can be seen in Fig. 6.49. The TTG laser essentially represents a DFB laser with an electronically tunable Bragg wavelength. Stable single-mode operation may be provided, for instance, by a quarter-wavelength shifted grating, anti-reflection (AR)/high-reflection (HR) coating on the end facets or favorable mirror to grating phases of an as-cleaved device. A uniform longitudinal stretching of the whole DFB laser does the Bragg wavelength tuning.

It is obvious that the stretching factor is equal for both laser length and grating period, which can be written as

$$\frac{L_2}{L_1} = \frac{\Lambda_2}{\Lambda_1} \tag{6.45}$$

Although the lasing wavelength is proportional to the grating period, the number of half wavelength (the mode number) in the laser cavity remains constant during stretching.

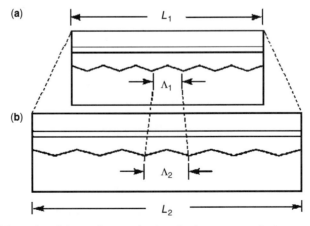

**Figure 6.49**   Schematic of the tuning mechanism in the transversely integrated tunable DFB laser. The homogeneous stretching of the optical cavity and grating results in a continuous tuning of the Bragg and laser wavelength, respectively.

So a continuous wavelength tuning with the same longitudinal mode is obtained. No mode jumps occur in this case, and the tuning is smooth.

Even though mechanically the semiconductor laser cavity cannot be stretched in significance, this operation principle can be applied because the stretching of the optical cavity length is the essential feature, and this parameter can be varied electronically via refractive index changes. This can be accomplished by placing a tuning region in one of the grading layers, which is in parallel to the active layer of the DFB lasers, as shown in Fig. 6.50**a**. By injecting electrical current through the tuning region, the effective refractive index of the relevant transverse mode is altered so the purpose of wavelength tuning is achieved. Applying Eq. (6.39) to this situation, we have the electronic wavelength tuning $\Delta\lambda_{\text{elec}}$ [286]

$$\frac{\Delta\lambda_{\text{elec}}}{\lambda_0} = \frac{\Delta n_t}{n_{g,\text{eff}}}\Gamma_t\left(1 - \frac{\alpha_{H,a}}{\alpha_{H,t}}\right) \tag{6.46}$$

Considering a transversely integrated laser at 1550-nm wavelength, if $\alpha_{H,a} = 5$, $\alpha_{H,t} = -10$, $n_{g,\text{eff}} = 4$, $\Gamma_t = 0.4$, and $\Delta n_t = -0.03$, we obtain a wavelength tuning of about $-7$ nm. This is much larger than the continuous tuning range of the longitudinally integrated structures ($\sim -3.5$ nm).

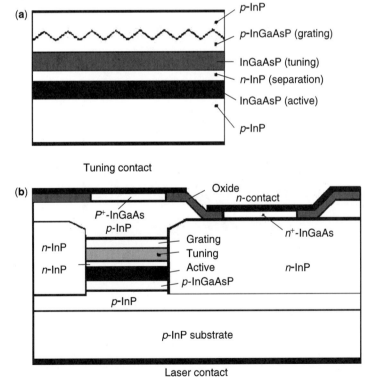

**Figure 6.50**  Schematic of (**a**) longitudinal view and (**b**) cross section of the TTG InGaAsP/InP laser structure for 1550-nm wavelength.

The TTG DFB laser was first developed in 1989 [272]. It represents the realization of a transversely integrated wavelength-tunable laser [Fig. 6.41**b**]. This structure is grown on a *p*-type InP substrate and electronically consists of two decoupled *pn*-heterojunctions in a format of a *pnp* junction, as shown in Fig. 6.50. The *n*-InP separation layer is the common electrode for both laser drive current, $I_a$, and the wavelength tuning current, $I_t$. However, optically both regions are strongly coupled as described by large confinement factors $\Gamma_t$ and $\Gamma_a$ for the tuning and active region, respectively, as shown in Fig. 6.51.

A representative experimental tuning and optical power characteristics of a 400-μm-long BH TTG DFB laser is shown in Fig. 6.52. Using tuning current of $-100$ to 100 mA, a continuous wavelength tuning of 13 nm is obtained at a constant laser drive current of 100 mA. The thermal tuning under reverse bias contributes about 4 nm, and the electronic tuning contributes about 9 nm under forward bias between 0 and 100 mA. Figure 6.52**b** shows the optical power drops almost linearly with an increase in the wavelength shift at both forward and reverse bias. An optical power above 1 mW can be maintained over the 9 nm tuning range.

Table 6.3 shows a comparison between the longitudinally and transversely integrated structure in terms of basic performance and parameters. Obviously, if the continuous wavelength tuning was essential, the TTG-type of DFB lasers would have more advantage over the longitudinal ones. However, if the continuous tuning is not required, the longitudinally integrated structure may have advantage due to its simple approach. In practice, it usually depends on the requirements.

### 6.6.5 Other Types of Tunable Lasers

As the maximum continuous tuning range is about 15 nm, as shown in Table 6.3, this is not sufficient for many applications. Thus, people have been looking for other

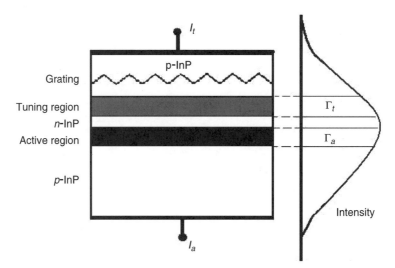

**Figure 6.51** Schematic longitudinal view of the transverse mode confinement factors for both the tuning and active region in the TTG InGaAsP/InP laser structure for 1550-nm wavelength. The electronically induced refractive index changes in the tuning region cause a change of the lasing wavelength.

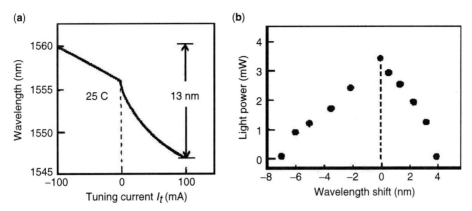

**Figure 6.52** Laser wavelength as a function of tuning current (**a**) and light output power as a function of wavelength shift (**b**) at a constant laser drive current of 100 mA for a 400 μm-long ambipolar-tunable BH TTG InGaAsP/InP DFB laser.

**TABLE 6.3  Comparison of Important Technology Aspects of Longitudinally and Transversely Integrated Tunable Lasers**

|  | Longitudinal Integration (DFB, DBR) | Transverse Integration (TTG) | Comment |
|---|---|---|---|
| Max. tuning range | ~100 nm | ~15 nm | Wavelength |
| Max. continuous-tuning range (CTR) | 7 nm | 13 nm | Wavelength |
| Typical CTR | 2–3 nm | 3–5 nm | Wavelength |
| Number of tuning electrodes | 2 | 1 |  |
| Tuning difficulty | Difficult | Easy |  |
| Ease of integration | Good | Moderate |  |
| Design flexibility | Flexible | Not very flexible |  |
| Handling | Complicate | Simple |  |
| Substrate type | N or P | P |  |
| Typical tuning current | +100 mA | +/−100 mA |  |

solutions than standard DFB and DBR lasers. Several types of structures have been studied and developed. External-cavity tunable (ECT) lasers and tunable VCSELs are some of them on the list. As mentioned above, the purpose of developing these tunable lasers is mainly to extend the wavelength-tuning range to achieve so-called widely continuous tunable lasers (>15 nm at 1550-nm wavelength). We can only brief these two structures in the following. For the details, please refer to the original papers.

External cavity tunable (ECT) is another important structure for widely tunable lasers. As shown in Fig. 6.53 [287], it comprises two external lenses and a silicon mirror with a FP laser chip. The silicon mirror is controlled by a MEMS actuator.

Tunable VCSEL is another promising tunable laser structure. By integrating an adjustable top mirror, the VCSEL can work as a tunable laser as shown in Fig. 6.54. The top mirror made of a GaAs/AlGaAs DBR is controlled by MEMS, which is similar

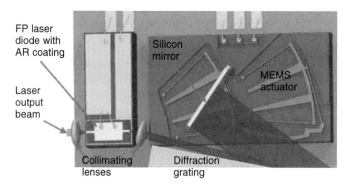

**Figure 6.53**   Schematic of ECT laser, which comprises two external lenses and a silicon mirror with a FP laser chip. The silicon mirror is controlled by a MEMS actuator.

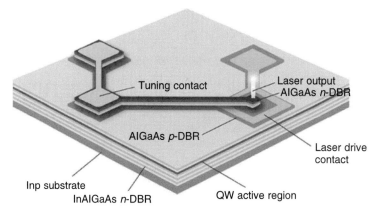

**Figure 6.54**   Schematic of tunable VCSELs, which comprises an adjustable top DBR mirror with an active region and a fixed bottom DBR mirror. The top DBR mirror is controlled by MEMS.

to the one in the ECT lasers. By adjusting the top DBR mirror position relative to the laser active region, the wavelength is tuned.

## 6.7   HIGH-SPEED SEMICONDUCTOR LASERS

The advance in compound semiconductor material (e.g., GaAs and InP system) growth technologies, such as MBE and OMVPE or MOCVD, has brought high-quality quantum-well (QW) structure into reality. With high-quality quantum wells and band-gap engineering, design and production of high-speed semiconductor lasers are realized. There is no need to mention how important the high-speed lasers are in today's communication world. Millions of faster than gigahertz, directly-modulated lasers are used in today's fiber-optic communication systems. Several companies have also commercialized semiconductor lasers faster than 10 Gbps.

The frequency response of semiconductor lasers has been studied and documented by many researchers [288–293]. The understandings of the semiconductor laser dynamics have gotten deeper and better. The effects of spontaneous emission [294], lateral diffusion [293], laser geometry [291, 292], nonlinear damping [295], and junction parasitics [296] have all been considered unlike in the early days.

In this section, the key factors will be outlined and summarized for the high-speed performance of the semiconductor QW lasers.

### 6.7.1  High-Frequency Rate Equations

The frequency response of a QW semiconductor laser can be analyzed by using three simple rate equations that describe the injected carrier and photon densities for a single longitudinal mode [297].

$$\frac{dN_B}{dt} = \frac{I}{eV_{CB}} - \frac{N_B}{\tau_r} - \frac{N_B}{\tau_{nb}} + \frac{N_W(V_W/V_{CB})}{\tau_e}$$

$$\frac{dN_W}{dt} = \frac{N_B(V_{CB}/V_W)}{\tau_r} - \frac{N_W}{\tau_n} - \frac{N_W}{\tau_{nr}} - \frac{N_W}{\tau_e} - \frac{v_g G S}{1+\varepsilon S} \qquad (6.47)$$

$$\frac{dS}{dt} = \frac{v_g G S}{1+\varepsilon S} - \frac{S}{\tau_p} - \Gamma\beta\frac{N_W}{\tau_n}$$

where  $N_{W,B}$ = carrier density in the quantum well and in the barrier layer, respectively
$\quad\quad S$ = photon density
$\quad\quad I$ = laser DC drive current
$\quad\quad e$ = electronic charge
$\quad\quad d$ = thickness of the active region
$\quad\quad G$ = gain (a function of carrier density),
$\quad\quad V_{W,CB}$ = volume of the quantum well and the confinement barrier layer
$\quad\quad v_g$ = the mode velocity
$\quad\quad \varepsilon$ = intrinsic gain compression factor
$\quad\quad \tau_p$ = photon lifetime
$\quad\quad \tau_n$ = radiative recombination lifetime in the QW
$\quad\quad \tau_{nr}$ = non-radiative recombination lifetime in the QW
$\quad\quad \tau_{nb}$ = total recombination lifetime in the confinement barrier region
$\quad\quad \tau_r$ = carrier transport time across the SCH to the QW active region
$\quad\quad \tau_e$ = carrier thermionic emission time from the QW into the barrier
$\quad\quad \Gamma$ = transverse optical confinement factor
$\quad\quad \beta$ = spontaneous emission factor

This last factor gives us the fraction of the total spontaneous emission going into the lasing mode. The equations above have neglected current spreading, lateral out-diffusion of injected carriers, and junction parasitics. We will comment on some of these effects and how they affect our results after preliminary analysis.

### 6.7.2  Small-Signal Modulation

Because the equations in Eq. (6.47) are nonlinear, exact analytical solutions are difficult to obtain. However, approximations can be made to calculate the small-signal modulation response of the semiconductor laser. To calculate the small-signal AC response, we expand $I$, $N_W$, and $N_B$, $S$ and $G$ into a DC and an AC portion [297]:

$$I = I_0 + i e^{j\omega t}$$

$$N_B = N_{B0} + n_B e^{j\omega t}$$

$$N_W = N_{W0} + n_W e^{j\omega t} \tag{6.48}$$

$$S = S_0 + s e^{j\omega t}$$

$$G = G_0 + \frac{\partial g}{\partial n} n_W e^{j\omega t}$$

Substituting Eq. (6.48) into Eq. (6.47) and neglecting second-order terms, we can get a normalized modulation $F(\omega)$ as $s(\omega)/(ie)$ and given by [298]

$$F(\omega) = \left( \frac{1}{1 + j\omega\tau_r} \right) \frac{A}{\omega_r^2 - \omega^2 + j\omega\gamma} \tag{6.49}$$

$$A = \frac{\Gamma \left( \upsilon_g \dfrac{\partial g}{\partial n} / \chi \right) S_0}{V_W (1 + \varepsilon S_0)} \tag{6.50}$$

$$\omega_r^2 = \frac{\left( \upsilon_g \dfrac{\partial g}{\partial n} / \chi \right) S_0}{\tau_p (1 + \varepsilon S_0)} \left( 1 + \frac{\varepsilon}{\upsilon_g \dfrac{\partial g}{\partial n} \tau_n} \right) \tag{6.51}$$

$$\gamma = \frac{\left( \upsilon_g \dfrac{\partial g}{\partial n} / \chi \right) S_0}{\tau_p (1 + \varepsilon S_0)} + \frac{(\varepsilon/\tau_p) S_0}{\tau_p (1 + \varepsilon S_0)} + \frac{1}{\chi \tau_n} \tag{6.52}$$

where a transport factor $\chi = 1 + (\tau_r/\tau_e)$ has been introduced into the equations. To get these equations, an assumption of $|\omega\tau_r| << 1$ has been made. This is normally the case in the frequency region where the laser speed performance is limited by the carrier transport time. There are two important terms for practical purpose. One is resonance angular frequency ($\omega_r = 2\pi f_r$), and the other is maximum $-3$dB bandwidth ($\omega_{\text{max}-3\text{dB}}$), which equals

$$\omega_{\text{max}-3\text{dB}} = \frac{\omega_r^2}{\sqrt{(\gamma(\gamma - \gamma_0) - \omega_r^2)}} \tag{6.53}$$

$$f_{\text{max}-3\text{dB}} = \frac{\omega_{\text{max}-3\text{ dB}}}{2\pi} \tag{6.54}$$

If we define, in Eq. (6.52), that $\gamma = K f_r^2 + \gamma_0$ ($K$ is the $K$ factor) [298], from Eq. (6.54) and assuming both $\gamma_0/(K f_{\text{max}-3\text{dB}}^2)$ and $(\gamma_0 K)/(4\pi^2) << 1$, we can have

$$f_{\text{max}-3\text{dB}} \approx \frac{2\sqrt{2}\pi}{K} + \frac{\gamma_0}{2\sqrt{2}\pi} \tag{6.55}$$

$$K = 4\pi^2 \left[ \tau_p + \frac{\varepsilon}{\left( v_g \dfrac{\partial g}{\partial n} / \chi \right)} \right] \tag{6.56}$$

$$\gamma_0 = \frac{1}{\chi \tau_n} \tag{6.57}$$

The $\gamma_0$ is a measure of the effective carrier recombination lifetime. As the transport factor, $\chi$, becomes a part of $\gamma_0$, it is a structure-dependent parameter. The $K$ factor is also affected by the carrier transport. When the resonance frequency is larger than $f_{\max-3dB}$, the transport delay may become the intrinsic limit of the maximum $-3$dB modulation bandwidth.

For the resonance frequency in Eq. (6.51), if we ignore the carrier transport effects and the minor terms of $\varepsilon$, it can be expressed in terms of experimentally measured parameters such as the output power and the bias current as

$$\omega_r^2 = \frac{v_g \dfrac{\partial g}{\partial n}}{\tau_p} S_0 \tag{6.58}$$

One of the key factors in Eq. (6.58) is the photon density, $S_0$, which is related to the total output power, $P_o$, and drive current, $I_o$, as

$$S_0 = \left( \frac{1}{V/\Gamma} \right) \frac{P_0}{h v v_g \alpha_m} = \left( \frac{1}{V/\Gamma} \right) \frac{\eta_i}{e v_g (\alpha_i + \alpha_m)} (I - I_{th}), \tag{6.59}$$

where $I_{th}$ is the threshold current, $\eta_i$ is the internal quantum efficiency, $\alpha_m$ is the mirror loss, $\alpha_i$ is the internal loss, $V$ is the volume of the active region, and $V/\Gamma$ is the optical mode volume. Now the resonance frequency, $f_r$, can be written as [299]

$$f_r = \frac{\omega_r}{2\pi} = \frac{1}{2\pi} \sqrt{ \left( \frac{1}{V/\Gamma} \right) \frac{v_g \dfrac{\partial g}{\partial n} \eta_i}{e} (I - I_{th})^{1/2} } \tag{6.60}$$

Even though the resonance frequency is not the limit for most of the high-speed lasers, it is a very useful tool as a starting point to optimize the laser's speed performance.

Equations (6.59) and (6.60) reveal several approaches to increase the laser modulation speed. For the photon-density limited devices, reducing the photon lifetime, $\tau_p$, will lead to the enhancement of the resonance frequency, which can be done through low-reflectivity (leaky) facet coatings. The assumption is that the drive current can be increased to obtain the optical power needed without a big negative effect. Reducing the laser's optical mode volume, such as the cavity length, can also help achieve high speed in this case. However, there are tradeoffs. One of them is the thermal dissipation problem with short cavity devices, and another one is that it is hard to produce short cavity, such as less than 200-$\mu$m lasers with high yield. For current limited or thermal dissipation-limited devices, reducing the threshold current and/or increasing the internal quantum efficiency, will enhance the laser's high-speed performance. High internal quantum efficiency is always desirable. Another important factor is the differential gain. By increasing differential gain, one can enhance the laser's modulation speed significantly. Usually, a high differential gain needs more QWs and good carrier confinement. It is clear from Eq. (6.60) that the drive current is another nub. In order to get a high-speed performance, the laser has to be driven at a current density level much higher than its threshold, minimum of three to five times. One of drawbacks of driving a laser at high current density level is the reliability concern. It is well known that the higher the drive current density is, the poorer the reliability of a laser diode demonstrates.

According to Eq. (6.60), a plot of $f_r$ versus $(I/I_{th}-1)^{1/2}$ or $P^{1/2}$ gives a straight line with slope $\dfrac{1}{2\pi}\sqrt{\dfrac{I_{th}\left(v_g\dfrac{\partial g}{\partial n}\eta_i\right)\Gamma}{eV}}$ if the parameters in the slope term do not change with drive current or the output power. As will be shown later, this is true only at low injection levels.

Figure 6.55 shows a representative plot of the small-signal normalized modulation response of a SQW $In_{0.2}Ga_{0.8}As/Al_{0.1}Ga_{0.9}As$ FP semiconductor laser with a 76-nm–wide SCH layer [298]. The relaxation oscillation resonance can be observed. Furthermore, a dip (damping) in the modulation response prior to the relaxation oscillation resonance is also observed. In Fig. 6.56, a representative plot of $f_r$ versus $P^{1/2}$ ($P^{1/2}$ is proportional to $(I/I_{th}-1)^{1/2}$) for this laser is shown. The data show a linear dependence for $f_r$ over a relatively low range of the power (<9 mW). At higher power levels, the resonance frequency showed a saturation behavior due to a decrease in the differential gain.

### 6.7.3 Key Parameters for High-Speed Semiconductor Laser Design

In order to get high-speed lasers, there are several parameters to be optimized as list in Table 6.4. They are (1) high differential gain, (2) short carrier transport time from confinement layer to the QWs, (3) long carrier-escape time from QWs to barriers or

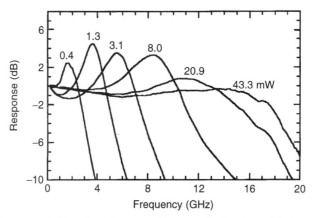

**Figure 6.55** A representative plot of a normalized small-signal modulation response for an $In_{0.2}Ga_{0.8}As/Al_{0.1}Ga_{0.9}As$ FP semiconductor laser at different output power.

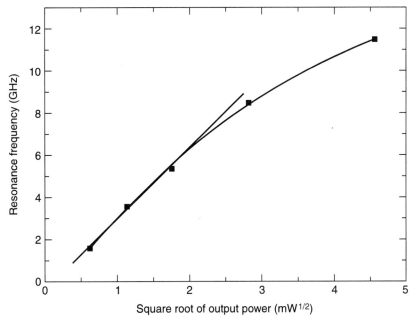

**Figure 6.56** A representative plot of $f_r$ versus $P^{1/2}$ ($P^{1/2}$ is proportional to $(I/I_{th} - 1)^{1/2}$) for the laser shown above. At low powers, the $f_r$ varies linearly with the square root of the output power. However, at high power levels, the $f_r$ shows a saturation behavior.

**TABLE 6.4   Key Parameters for High-Speed Lasers and Related Physical Parameters**

| Key Parameters for High-Speed Lasers | Related Physical Parameters |
|---|---|
| High differential gain | More QWs; strain; P-doping; wavelength detuning |
| Short carrier transport time | Confinement layer doping location |
| Short carrier capture time | Number of QWs; QW width; barrier height and width |
| Long carrier escape time from QWs | Large band offset; thick QWs and barriers; |
| Low device parasitics | Low series resistance and capacitance |
| Short photon lifetime | A leaky light-exit mirror |
| High internal quantum efficiency | Small amount of non-radiative recombination centers; good carrier confinement; small lateral leak current |
| Strained QWs | Small amount allowable lattice and/or thermal mismatch |
| Large optical confinement factor | Large index difference between active layer and cladding layer; maximum active layer and optical field overlap |
| Low junction temperature | Low series resistance; low thermal resistance and good heat dissipation |

confinement layers, (4) low device parasitics, (5) short photon lifetime, (6) high photon density, (7) high internal quantum efficiency, (8) strained QWs, (9) large optical-confinement factor, and (10) low junction temperature. In the following, we will discuss how to optimize some of these factors one by one.

***Differential Gain.*** Differential gain is the slope of the gain-carrier density curve, $\frac{\partial g}{\partial n}$. It is one of the most important parameters for high-speed lasers. Using the relaxation broadening model, the optical gain as a function of the emission energy $E_i$ can be written as [300]

$$g(E_i) = \frac{\pi \hbar e^2}{m_0^2 n_r \varepsilon_0 c E_i} \int M^2 \rho_r(E)[f_c(E) - f_v(E)]L(E)dE \qquad (6.61)$$

where $m_0$ is the free electron mass, $n_r$ is the refractive index of the gain medium, $\varepsilon_0$ is the permittivity of free space, $c$ is the speed of light, $M^2$ is the momentum matrix element that determine the relative strength of the various optical transitions, $\rho_r$ is the reduced density of states function, $f_c$ and $f_v$ are the Fermi occupation probabilities in the conduction and valence bands, respectively, and $L(E)$ is the gain broadening function. It is clear that the major influence on the gain function comes from the $M^2$ and $(\rho_r(E)[f_c(E) - f_v(E)])$ term, which is a product of the reduced density of states function and the difference in the conduction and valence band Fermi occupation probabilities.

Low-dimension materials, such as QWs, enhance the optical gain and differential gain, as shown in Fig. 6.57, due to a function profile modification of the density of

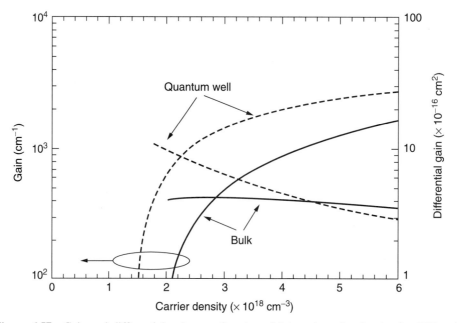

**Figure 6.57**   Gain and differential gain as a function of injected carrier density for QW and bulk structures [301].

states [302–304]. Unfortunately, there is a more rapid gain-saturation at high carrier injections due to its limited states density.

To increase the differential gain, one can (1) use multiple quantum wells, (2) use strain to remove the degeneracy of the heavy and light-hole bands [305–307], (3) use $p$-type doping in the active region [308–310], and (4) use wavelength detuning with DFB or DBR lasers.

***Multiple QWs and Stain.***  As mentioned above, for a SQW structure, the differential gain shows a rapid saturation with carrier injection. To overcome this, multiple QWs (MQWs) are used. As shown in Fig. 6.58, with more QWs, the carrier density at transparency increases and so does the differential gain [311].

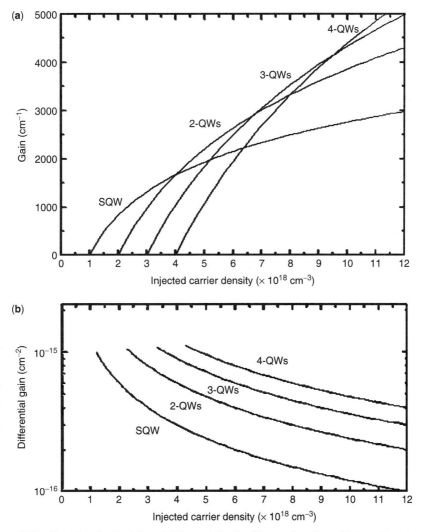

**Figure 6.58**  Calculated gain (**a**) and differential gain (**b**) as a function of injected carrier density for one, two, three, and four QWs. More QWs cause higher threshold but offer much higher differential gain.

The strain can remove the degeneracy of the valence band [312, 313]. Compressive strain moves the light-hole band edge down, and tensile strain moves the light-hole band edge up relative to the heavy-hole band. The calculation results, as shown in Fig. 6.59, indicate that the strain can drastically reduce the transparency carrier density and increase the differential gain. So, the modulation speed can be enhanced significantly [314, 315].

***P-Type Doing in the Cavity.***  *p*-type doping in the cavity can increase the gain and differential gain, which in turn improves laser's dynamic performance [317, 318]. The doping effect on the gain can be attributed to the fact that any doping in the cavity can lead to somewhat separation of the quasi-Fermi levels even without any carrier injection. However, the *n*-type doping has more a significant effect on the reduction of the transparent carrier density because the conductance band has much smaller density of states. As a result, the *p*-type doping helps the differential gain, but not the *n*-type. The *p*-type doping in the cavity also has negative effect, such as increasing the internal loss. For this, a modulation doping method can be used to take the *p*-type doping advantage but not the disadvantage by selectively putting the doping in the barrier and/or the confinement layers, but not in QWs. This avoids the increase in free carrier absorption loss. Some of the fastest lasers to date are MQW structures with *p*-type doping in the cavity.

***Wavelength Detuning.***  The spectral variation of gain and differential gain shows that the differential gain has the maximum on the shorter wavelength side of the gain peak. Thus, to take advantage of this feature, one can design a laser with lasing wavelength blue-shifted relative to the gain peak [319, 320]. However, this only works for DFB or DBR lasers because its gain peak and lasing wavelength can be different.

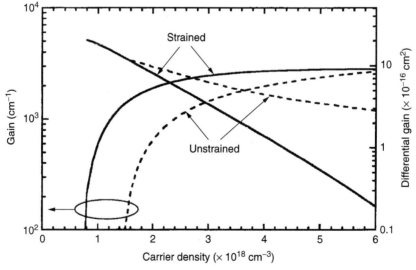

**Figure 6.59**  Calculated strain effect on the gain and differential gain for a SQW structure. It is shown that strain helps reducing the transparent carrier density, dramatically increasing the gain, and enhancing the differential gain at low level of carrier injection [316].

***Carrier Transport Time.*** There are two steps [321, 322] involved for injected carriers (electrons and holes) moving from the *p*- and *n*-type doped cladding layers into the QWs. First, the electrons and holes transport through the confinement layers by diffusion and/or drift under electric field. Second, the electrons and holes are captured by the QWs through carrier-phonon and/or carrier-carrier scattering process [323]. For the carriers in the quantum wells, there are two mechanisms then to escape. One is through the thermionic emission [324], and the other is through tunneling [325, 326]. All of these four steps are important for the laser dynamic properties, as shown in Fig. 6.60. The following is a brief discussion for these four processes.

***Carrier Transport Time in SCH Layers.*** In the SCH layers, the current continuity conditions are [327]

$$\frac{\partial n}{\partial t} = \frac{1}{e}\frac{\partial J_n}{\partial x} - U(n, p)$$

$$\frac{\partial p}{\partial t} = -\frac{1}{e}\frac{\partial J_p}{\partial x} - U(n, p) \tag{6.62}$$

Considering both the diffusion and drift, and the Einstein relation of $D/\mu = kT/e$, the electron and hole current densities can be expressed as

$$J_n = eD_n\left(\frac{enE}{kT} + \frac{\partial n}{\partial x}\right)$$

$$J_p = eD_p\left(\frac{epE}{kT} - \frac{\partial p}{\partial x}\right) \tag{6.63}$$

where *n* and *p* is electron and hole concentration, $D_{n,p}$ is electron and hole diffusion coefficient, E is electric field intensity, and U(*n*, *p*) is the net carrier recombination rate. Under high-injection ($n \approx p$) and steady-state ($\partial n/\partial t = \partial p/\partial t = 0$) conditions, according to charge neutrality ($\partial E/\partial x = 0$), Eqs. (6.62) and (6.63) can be combined as

$$\frac{d^2 n}{dx^2} - \frac{U(n)}{D_a} = 0 \tag{6.64}$$

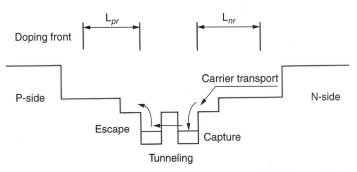

**Figure 6.60**  Four carrier transport processes: transport through diffusion and/or drift, capture, tunneling, and escape through thermionic emission. $L_{nr}$ and $L_{pr}$ are the undoped regions in N-side and P-side SCH layers, respectively.

where $D_a (= 2D_n D_p / (D_n + D_p))$ is an ambipolar diffusion coefficient. Because $U(n) = n/\tau_a$ ($\tau_a$ is the ambipolar lifetime) and the ambipolar diffusion length $L_a = \sqrt{D_a \tau_a}$, Eq. (6.64) becomes

$$\frac{\partial^2 n}{\partial x^2} - \frac{n}{L_a^2} = 0 \tag{6.65}$$

By solving this equation and assuming the $L_{nr}$ and $L_{pr} << L_a$ in Fig. 6.61, we can get the carrier transport time,

$$\tau_r = \frac{1}{2}\left(\frac{L_{nr}^2}{2D_n} + \frac{L_{pr}^2}{2D_p}\right) = \frac{1}{2}(\tau_{r,electrons} + \tau_{r,holes}) \tag{6.66}$$

This means that the total ambipolar diffusion time is equivalent to an average of the electron and hole's diffusion time. As the hole's diffusion time usually is much longer than the electron's, normally the hole's transport property is the dominant factor to the high-speed performance of a laser. One way to reduce the hole's transport time is to reduce the $L_{pr}$ so that its transport time is close to the electron's. However, the tradeoff may be a poor reliability, lower slope efficiency, or higher threshold current.

***Carrier Capture Time.*** Quantum mechanical calculations have predicted that for a SQW structure, with the QW width increase, the carrier-capture time becomes shorter [328, 329]. This is because as the well width increases, the states within the well become more bounded and the virtual high-energy states above the QW become bound to the well. This results in the final state within the QW moving in and beyond the reach of any state in the SCH separated by a phonon energy. For MQW systems, the calculations have predicted that the carrier capture time is in the range of 1 ps, which is close to the experimental results [330]. The increase in number of QWs results in the total number of final states to which the carriers get scattered, and this corresponds a decrease in the capture time. The capture time is nearly inversely proportional to the number of QWs. The capture time is a function of QW width, barrier height, and width. Due to its larger density of states in the valence band, hole capture times are smaller than are the electron's. For InGaAs/InP material system, the electron capture time is about 1 ps and hole's from 0.2 to 0.3 ps [331, 332].

The use of linear and parabolic graded index SCH layer (L-GRINSCH and P-GRINSCH) has shown a significant enhancement to the carrier capture in the QWs

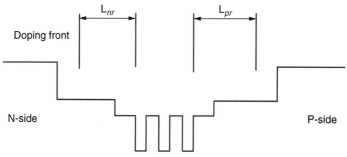

**Figure 6.61** A schematic QW active region for a laser structure. The $L_{nr}$ and $L_{pr}$ are doping recess distance in N-side and P-side SCH layers, respectively.

[320, 321]. At low temperatures, the QWs with the L-GRINSCH have exhibited almost 100% collection efficiency, whereas an ungraded SCH only showed about 50%. The collection efficiency of QWs with P-GRINSCH is in between these two limits. The capture and escape times are also influenced by the energy-level separation of the electronic states in the SCH layer relative to the QW's. The grading in the SCH layers has a profound effect on the availability and the energy-level separation of these states.

In a laser structure with L-GRINSCH at room temperature, the drifting in the SCH layer is the slowest process. However, at low temperature, the carrier capture by the QWs is more important. This is in contrast to the ungraded SCH layer, where the variation in transport time is essentially due to the change in the mobility as a function of temperature. Thus, at room temperature, the L-GRINSCH has been predicted to be only twice as fast as the ungraded SCH structures. As almost twice of SCH layer thickness is needed to get a good optical confinement, the L-GRINSCH does not help the speed performance at room temperature.

***Thermionic Emission Time.*** The thermionic emission lifetime from a quantum well can be expressed as [323]

$$\tau_e = \left(\frac{2\pi m^* L_w^2}{kT}\right)^{1/2} \exp\left(\frac{E_B}{kT}\right) \tag{6.67}$$

where $E_B$ is the effective barrier height, m* is the effective carrier mass, $k$ is the Boltzmann constant, and T is temperature. This is derived under assumptions of that the carriers in the barriers have bulklike properties and obey Boltzmann statistics. It is clear that the thermionic emission lifetime is very sensitive to temperature and the effective barrier height.

***Tunneling Time.*** In a MQW structure, if the barrier is thin enough, the electron and hole wave function is not completely isolated anymore. There is a coupling between the two wells. This interwell coupling removes the energy degeneracy that exists in an infinite barrier width. For a two QW system, the overall linear superposition of the wave functions corresponds to an electron or a hole oscillation between the wells at a frequency of $\Delta E/h$, where $h$ is Planck constant and $\Delta E$ is the energy difference between the lowest odd ($E_1$) and even ($E_2$) bound states of the two well system. In this case, the tunneling time is between the two wells given by [324, 325]

$$\tau_t = \frac{h}{2\Delta E} \tag{6.68}$$

Even though this is derived from a two-well system, this is generally valid for a MQW structure, assuming there is a weak coupling between the wells.

A MQW structure can increase the laser differential gain, which will enhance the high-speed performance significantly. However, there are some complications due to carrier transport between QWs in the MQW structure. There are two paths for the carriers in a QW that go to another. One is through thermionic emission and then captured by another QW, and the other is tunneling through the barrier in between. These two paths are competing processes, and the faster one will be dominant. Also, the transport behaviors for electrons and holes are different. We only need to consider the slower one, which does not affect the final result. If we use $\tau_b$ to represent the barrier transport time, then

$$\frac{1}{\tau_b} = \frac{1}{\tau_e} + \frac{1}{\tau_t} \qquad (6.69)$$

Usually, for thin barriers (<5 nm), the tunneling is the faster process and the hole's tunneling time is the dominant one. For intermediate barrier thickness (6 to 12 nm), the hole's thermionic emission is dominant.

*Optical Confinement Factor.*   In reality, there are three different designs of SCH layers in currently commercialized semiconductor lasers, as shown in Fig. 6.62.

For the linear grading SCH, also called the L-GRINSCH structure, the optical confinement is not as good as the steplike function SCH with the same physical thickness because the optical thickness of the L-GRINSCH layer is much thinner. However, one can increase the linear grading layer thickness to get an equivalent optical confinement. Comparing Structure **b** and **c**, it is obvious that **b** has a better optical confinement than **c**. If considering the carrier escape time, then Structure **a** is equal to or better than **c**, and **c** is better than **b** because the thermionic emission time in Structure **a** and **c** is much longer than Structure **b**, which results in a lower differential gain for Structure **b**. In turn, the reduced differential gain in Structure **b** offsets its higher optical confinement factor. Thus, in terms of the overall high-speed performance, Structure **c** has a much higher resonance-frequency than Structure **b** does if all other parameters are kept the same. So quite often, the optical confinement factor is an important factor, but not the dominant one compared to differential gain and carrier transport properties.

*Low Device Parasitics.*   In many cases, the modulation speed of semiconductor lasers is not limited by the resonance frequency, but by the device's parasitics. There are three parasitic parameters: inductance (L) [333], capacitance (C), and resistance (R). The following is a simple circuit model to simulate an edge emitting laser's AC behavior for the frequency range of 100 MHz to 6 GHz (Fig. 6.63). In this circuit model, $R_j$ and $C_j$ are the junction diffusion resistance and capacitance, while $C_p$ and $L_s$ are the capacitance and inductance from bonding pad and wire, respectively. If the bonding wire inductance is negligible [334], the equivalent circuit becomes a simple RC circuit.

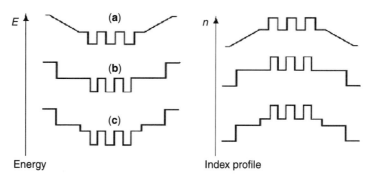

**Figure 6.62**   Three different SCH layer designs: **(a)** linear grading; **(b)** steplike SCH; **(c)** with higher barrier.

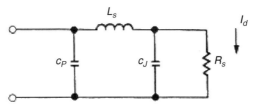

**Figure 6.63**  A simple equivalent circuit for injection lasers at a frequency ranging from 100 MHz to 6 GHz (©1982, IEEE).

To calculate a laser's normalized small-signal modulation response, we must take into account both the photon response [Eq. (6.49)] and the parasitics effect, $H(\omega)$. The amplitude square of total normalized response $|F(\omega)|^2_N$ can be expressed as

$$|F(\omega)|^2_N = \left| \frac{F(\omega)H(\omega)}{F(0)H(0)} \right|^2 \tag{6.70}$$

The photon lifetime is the inverse of the total cavity loss rate and can be written as [335]

$$\tau_p = \frac{1}{v_g(\alpha_i + \alpha_m)} = \frac{1}{v_g\left(\alpha_i + \left(\frac{1}{2L_c}\ln\left(\frac{1}{R_f R_b}\right)\right)\right)} \tag{6.71}$$

where $\alpha_i$ ($\sim 10$ cm$^{-1}$) is internal distributed losses in the cavity, $\alpha_m$ is the mirror loss due to sub-unity reflectivity for both front and back facets, $L_c$ is laser cavity length ($= 300$ µm), $R_{f,b}$ is facet reflectivity ($R_f R_b = 0.3$), and $v_g$ is the group velocity ($3 \times 10^{10}$ cm/s).

Considering the parasitic effect on the small-signal modulation response and the $K$ factor expression, Eq. (6.39) can be rewritten as [336]

$$F(\omega) = \left(\frac{1}{1 + j\omega RC}\right)\left(\frac{1}{1 + j\omega\tau_r}\right)\frac{A}{\omega_r^2 - \omega^2 + j\omega\gamma} \tag{6.72}$$

$$A = \frac{\Gamma(v_g g_0/\chi)S_0}{V_W(1 + \varepsilon S_0)} \tag{6.73}$$

$$\omega_r^2 = \frac{(v_g g_0/\chi)S_0}{\tau_p(1 + \varepsilon S_0)}\left(1 + \frac{\varepsilon}{v_g g_0\tau_n}\right) \tag{6.74}$$

$$\gamma = \frac{1}{2\pi}K\omega_r^2 + \frac{1}{\chi\tau_n} \tag{6.75}$$

$$|F(\omega)|^2_N = \frac{1}{\left(1 + \left(\frac{\omega}{RC}\right)^2\right)}\frac{1}{\left(1 + \left(\frac{\omega}{\tau_r}\right)^2\right)}\frac{\omega_r^4}{((\omega_r^2 - \omega^2) + \gamma^2\omega^2)} \tag{6.76}$$

In Fig. 6.64, there is a dip (damping) in the magnitude of the amplitude response prior to the resonance. It is evident from Fig. 6.64 that the dip in the modulation response is related to the parasitics. (It could be related to the carrier transport time too in other

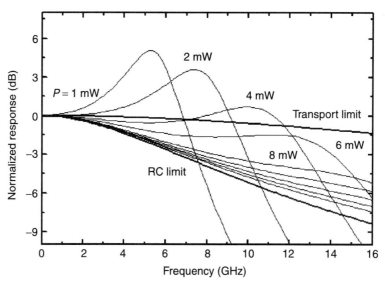

**Figure 6.64** Calculated normalized small-signal modulation response for an edge-emitting laser. The calculations include both the effect of RC parasitics and carrier transport. The parameters used are typical values for a FP laser, $\tau_p = 1.17 \times 10^{-12}$ s, $\gamma = Kf_r^2 + \gamma_0 = 0.19f_r^2 + 9.4$ ($GHz$). The parameters for the parasitics are $R_s = 10\Omega$ and $c_p + c_j = 2.4$ pf.

cases because it is not distinguishable between the RC and carrier transport limit). In some conventional semiconductor laser structures, current confinement generally is accomplished by using reverse-biased junctions (BH laser is a good example). Such junctions can lead to a relatively large capacitance [337]. People have developed laser structures with reduced capacitances ($C_j \sim 1$ pF; $R \sim 10\Omega$).

In a real laser device, the resistance includes three major parts [338]: junction diffusion resistance, $R_j$, the contact resistance, $R_c$, and the series resistance, $R_s$. The junction diffusion resistance can be expressed by $\eta kT/(eI_0)$, where $I_o$ is the DC current and $\eta$ is the ideality factor (typically for a laser diode, $\eta = 2$). When the laser operates for a high-speed application, a forward-biased current is normally high, such as 50 to 100 mA. The junction resistance, $R_j$, would be about 0.5 to 1$\Omega$, which is small compared to the other two. The major task is then to reduce the series resistance and the parasitic capacitance. For a ridge waveguide laser structure, the ridge/mesa width ($L_w$) is typically 2 µm and the height ($L_m$) is about 1.2 µm. As the doping level in the cladding layer (a typical of $7 \times 10^{17}$cm$^{-3}$) needs a good control to reduce unnecessary free-carrier absorption, the resistance from this cladding layer, $R_s = L_m/(en\mu L_w L_c)$, is about 0.5$\Omega$ if we assume the cavity length ($L_c$) is 150 µm and the mobility ($\mu$) of this cladding layer is 100 cm$^2$/(Vs). The typical contact resistivity ($r_c$) is about $0.9 \times 10^{-6}\Omega$cm$^2$. Then, the contact resistance, $R_c = r_c/(L_w L_c)$, is 3$\Omega$. The total resistance would be $R_s = R_s + R_c = 0.5 + 3 = 3.5\Omega$. 4 to 5$\Omega$ is typical total resistance for high-speed edge-emitting lasers. It is clear that the contact resistance is a major part of the total resistance. Increasing the doping level in the contact layer will reduce it. For instance, if a laser structure is grown on an $n$-type substrate, the top $p$-type contact layer should be doped as high as $5 \times 10^{19}$ to $1 \times 10^{20}$ cm$^{-3}$ to get a low contact resistance. For the $n$-type contact layer (the structure is grown on a $p$-type substrate), a doping level of higher than $2 \times 10^{18}$ cm$^{-3}$ should be enough to get a low contact resistance due to a much higher mobility of electrons. Because the junction capacitance is proportional to the junction area ($L_w L_c$) and the series resistance is inversely proportional to the junction area, a reduction of the junction area does not help much on the laser's high-speed performance due to a constant RC product for the junction part.

Another source of parasitic capacitance is from the bonding pad ($C_p$), and it plays a major role in the overall high-speed performance. The pad size and the material and dimension underneath it are key factors. Considerable effort has been employed to reduce parasitic capacitance [31, 339, 340]. Several processes have been developed to achieve a low parasitic capacitance from the bonding pad. Some of these are (1) use of isolation channels in a buried hetero-structure, (2) polyamide buried ridge waveguide geometry, (3) undercut mesa type structures, and (4) Fe-doped semi-insulating InP-based structures.

The inductance from the bonding wire is additional parasitics, even though strictly speaking, it is not a device parameter. This affects the laser high-speed performance significantly, especially for faster than 10-GHz devices. Ways of reducing inductance's effect on the high-speed performance are using shorter wire length, using thicker wires and/or using meshes or flip-chip bonding.

***Photon Density.*** From the resonance frequency expression, we know that the photon density is important for achieving a high-speed laser operation. From Eq. (6.49), we see that the high drive current and large optical mode volume are two key parameters for high photon density. The large optical mode volume requires a good optical confinement. High current operation also needs a good thermal dissipation. However, even in a good thermal environment, the high current operation is still a big concern. As discussed in Section 6.4.6, the laser reliability is severely degraded at high power levels, and thus, significant margins should be allowed in real systems applications. In this respect, quaternary 1.3-μm and 1.55-μm lasers should be the more suitable candidates for a variety of reasons. First, the smaller photon energy leads to a higher photon density for a given power level [Eq. (6.49)] and thus a higher $f_r$. Second, catastrophic damage does not appear to be a problem in this laser system. Third, the thermal conductivity of GaInAsP/InP lasers is significantly better than that of (GaAl)As devices (Sections 6.4.6 and 6.8.1), and higher power can be achieved. Finally, these laser structures are inherently more reliable than are (GaAl)As/GaAs laser devices.

## 6.8   VERTICAL-CAVITY SURFACE-EMITTING LASERS (VCSELs)

Compared to conventional edge-emitting laser diodes, VCSELs have many unique features. These features are user-friendly and can bring costs down significantly from laser chip manufacturing to packaging, integration, and fiber coupling. VCSELs have become the choice of engineering in many areas. The major advantages of VCSELs are

(1) Circularly shaped, low-numerical-aperture output beams (more directional beams resulting in high coupling efficiency to optical fiber)

(2) Narrow spectral line-width (e.g., about 0.5-nm RMS for 850-nm VCSELs)

(3) Ease of two-dimensional array fabrications

(4) Ease of chip fabrication (no facet coating) and on-wafer testing

(5) High-speed modulation up to 12.5 Gbps for now (in 2002)

(6) Temperature stability of output powers and lasing wavelength

(7) Low threshold and high efficiency in low power range

(8) Good reliability

Like everything else, VCSEL also has some disadvantages. These disadvantages are

(1) Low output power (several mW in most cases) due to high thermal resistance

(2) Very complicated epitaxial layer growth and high demand on the precise control of layer thickness, composition and doping level as well as uniformity of these.

Currently, the known and potential applications of VCSELs are

(1) Fiber-optics data communications (multimode fiber, short distance, several hundred meters) for Gigabit Ethernet and Fiber Channel

(2) Free-space optics for communications

(3) Printing and scanning

(4) Sensing, coding, and decoding

(5) Data recording and storage

(6) Medical applications, such as renal scanning

So far, commercial or close-to-commercial VCSELs cover visible and infrared wavelength range. The major wavelength bands are:

$$640 \text{ to } 690 \text{ nm } [341]$$

$$760 \text{ to } 1000 \text{ nm}[342]$$

$$1150 \text{ to } 1300 \text{ nm}[343]$$

$$\text{around}1550 \text{ nm}[344]$$

Of these wavelengths, the 850 nm is the most popular one due to its application for local-area data communications, such as Gigabit Ethernet and Storage-Area Network (SAN). Some tens of millions of this kind of VCSELs have already been produced and used in data-communication systems.

Like conventional semiconductor lasers, VCSEL has one pair of mirrors and a cavity in between. In the cavity, there are active layers, usually formed by quantum wells (QWs), barriers and confinement and cladding layers for confinement of the injected carriers. Some key aspects of VCSELs will be discussed in following sections. Due to a limited scope of this section, we will only outline and summarize the understandings and performance achieved so far. For more detailed technical discussions, one should refer to the original papers or publications listed at the end of this section.

### 6.8.1   VCSEL Basics and 850-nm VCSELS

*Top- and Bottom- DBRs.* A typical laser structure is formed by a cavity with two reflectors at the ends. For a VCSEL, because of its short gain region compared to

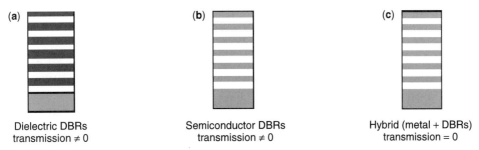

Figure 6.65   Three typed of reflectors for semiconductor VCSELs. (a) Dielectric DBRs, such as $SiO_2/TiO_2$; (b) semiconductor DBRs, such as GaAs/AlAs; and (c) hybrid reflectors: metal + dielectric DBR or semiconductor DBR.

edge-emitting lasers, a high Q (cavity quality factor) cavity is required. Thus, high-reflectivity (>99.5%) mirrors are necessary. To achieve such high reflectivity, only several types of reflectors can be used. Most reflectors with a reflectivity higher than 99% are multilayer stacks of alternating high and low refractive index materials. The optical thickness of each layer is one-quarter of the wavelength. Generally, this type of reflector is called the distributed Bragg reflector (DBR). Figure 6.65 shows three types of commonly used reflectors: dielectric DBR (such as $SiO_2/TiO_2$), semiconductor DBR (such as GaAs/AlAs), and hybrid DBR (a metal combined with a semiconductor DBR or a metal combined with a dielectric DBR). It should be noted that a metal mirror such as silver (Ag) could provide about 95% reflectivity. Among these reflectors, the hybrid DBR is considered being best choice because it can provide high reflectivity without very many layers. The major advantage of the dielectric DBRs over the semiconductor ones is the large refractive index difference between low refractive index ($n_L$) and high refractive index ($n_H$) so that few pairs are required to achieve high reflectivity. However, dielectric DBRs have many disadvantages: (1) It is not electrical conducting; (2) its thermal impedance is high; and (3) dielectric layers usually have large thermal expansion mismatch to semiconductor materials, which could cause cracking over temperature cycling. Semiconductor DBR plus metal could be the best choice if all the aspects are considered. However, it can't be used as a light-exit reflector since a metal is highly absorbing.

Multiple layer reflectors can be analyzed by a transmission matrix approach [345]. For a nonabsorbing DBR, there is a simple analytical formula to calculate its reflectivity for the Bragg wavelength, as shown below [347]

$$R = \left( \frac{\left( 1 - \left( \frac{n_1}{n_2} \right)^{2m} \right)}{\left( 1 + \left( \frac{n_1}{n_2} \right)^{2m} \right)} \right)^2 \sim \tanh^2 \left( \frac{m \Delta n}{n} \right) \qquad (6.77)$$

Where $m$ is the number of period, $n_1$ is the low index, and $n_2$ is the high index. To derive Eq. (6.77), one has to assume that the incident and exit media have the same index and that it is in between $n_1$ and $n_2$. For an absorbing DBR mirror, one can use the complex index in the equation. In practice, the DBRs in a VCSEL structure are not constructed with two abrupt layers. In order to reduce series resistance caused by a hetero-junction compositional, graded interfaces are commonly used [347]. In that case, an average index for each layer can be used to calculate its reflectivity. For a reflector design, the incident and exit materials have to be included to get correct reflectivity. In a top-emitting VCSEL structure, the incident material to a DBR is the cladding layer next to it and the exit material is air for the top DBR and for the bottom one, respectively. So it is hard to simply use this equation to do a real VCSEL design.

Most of today's commercial 850 nm VCSELs are top-emitting ones. The bottom DBR is usually $n$-type doped (such as Si), and the top is $p$-type doped (such as C or Mg). The typical top- and bottom-DBR's reflectivity is about 99.5% and 99.95%, respectively, as shown in Figs. 6.66 and 6.67. There is a strong interest in the bottom-emitting VCSELs due to (1) ease of integration with IC drivers and (2) a better over-temperature performance with flip-chip bonding. In bottom-emitting VCSELs, the number of pairs for the top- and bottom-DBR is reversed. If the lasing wavelength

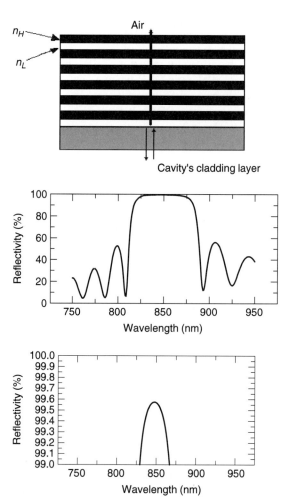

**Figure 6.66**   Schematic of a top-DBR with a cladding layer in a top-emitting VCSEL. Its typical reflectivity is 99.5%.

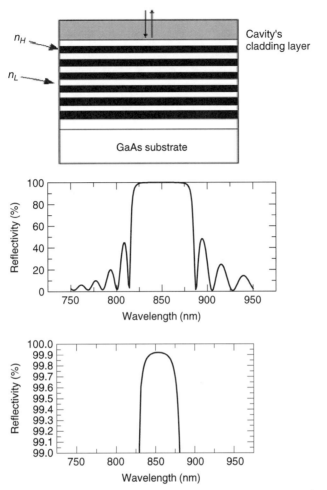

**Figure 6.67**   A bottom-DBR with the cladding layer grown on GaAs substrate for a top-emitting VCSEL. Its typical reflectivity is 99.95%.

is 980 nm, GaAs substrate does not need to be removed. However, for 850-nm bottom-emitting VCSELs, GaAs substrate must be removed owing to its strong light absorption, as shown in Fig. 6.68.

As to material composition in 850-nm DBRs, the low index ($n_L$) layer is usually AlAs or $Al_xGa_{1-x}As$ where x = 90% to 95%, and x = 12% to 18% of $Al_xGa_{1-x}As$ may be used for the high index ($n_H$) layer. With this kind of top-DBRs design, about 20 pairs are needed to obtain 99.5% reflectivity, whereas 35.5 to 40.5 pairs of layers are used to form the bottom-DBR depending on designs of reflectivity and electrical and thermal resistance [348].

In a VCSEL structure, the DBR mirrors play two important roles. One is to form an optical cavity and the other is to conduct electrical current to pump (or inject carriers into) the active region. Electrical series resistance is an important issue to address when designing the DBRs. As mentioned above, because the mirror stacks are formed by many alternating layers with different band-gap materials, there are

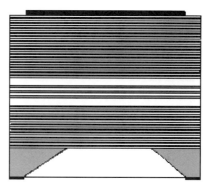

**Figure 6.68** A schematic of an 850-nm bottom-emitting VCSEL with part of GaAs substrate removed so that the light can be emitted from the bottom.

many hetero-junctions with potential barriers in each stack [349]. These barriers will contribute greatly to the total electrical resistance. In fact, it is usually the dominant factor for the series resistance issue. There are two ways to reduce series resistance resulting from these hetero-junctions: (1) reducing the potential barriers by using compositional grading [350] or some kind of compositional profiles, such as parabolic or step-like function and (2) using heavier doping at junction interfaces. A heavy doping, however, will introduce additional optical loss due to free carrier absorption in the DBR layers. There is an approximate equation [351] for the optical free-carrier loss in the 850- to 980-nm wavelength range, which is $\alpha_f$ (cm$^{-1}$) $= 5 \cdot n + 11 \cdot p$, where $n$ and $p$ is electron and hole concentration, respectively, in the unit of $1 \times 10^{18}$ cm$^{-3}$. Thus, one can see that due to a much larger free-carrier absorption coefficient of holes, the optical loss in the $p$-type DBRs is more severe. This is one of the major optical-loss factors in VCSELs, especially at long wavelength, such as 1310 nm or 1550 nm [352].

Several tradeoffs exist in VCSEL designs, such as: (1) the electrical series resistance and optical loss due to free-carrier absorption, (2) the room-temperature threshold current and high-temperature performance, (3) slope efficiency and threshold current, and (4) threshold and output power. In order to solve the series resistance (doping) and absorption conflict, one can use intracavity contact or tunnel junction to reduce the free-carrier absorption in the P-DBRs. However, there are also prices to pay for these alternatives [353].

*Quantum Wells, Confinement Layer, Cavity, Gain, and Losses.* Between the top-and bottom-DBRs, several layers exist. These layers, are quantum wells, barriers, confinement layers, and cladding layers on each side of the QWs. As a whole, it is called a cavity. For existing commercial VCSELs, the average refractive index in the cavity is larger than $n_L$. The quantum wells typically comprise of 7-nm GaAs as a well and 7 nm of ~30% AlGaAs as a barrier for 850-nm VCSELs. The number of quantum wells depends on the design, with a typical number of three.

The active layer is the engine of laser diodes. Its design and quality determine the device performance. One of the critical factors is the band-gap alignment, $\Delta E_c$ and $\Delta E_v$, as shown in Fig. 6.69. When electrons and holes are injected through current, the carriers are confined in the QWs by the potential difference between the QW and the

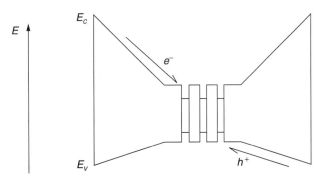

**Figure 6.69**   Schematic of band-gap offset of quantum well, barrier, confinement, and cladding layer in a VCSEL cavity.

barrier. If the confinement is small, the electrons (holes) will overflow into the confinement layer or even the cladding layer before combining with holes (electrons). These carriers outside the QWs usually recombine through nonradiative means and generate only heat, not light. More carriers tend to overflow as the temperature increases. In consequence, the output power drops and lower slope efficiency falls. Thus, a small carrier confinement causes poor over-temperature performance.

The interface quality between the QWs and confinement layers is another important factor. A lower quality interface could lead to the existence of a large amount of interface states that serve as nonradiative recombination centers. As carriers flow through the interface, some of the electrons and holes may recombine through the interface states, resulting in an increase in the threshold current and reduced laser efficiency. It requires special attention during the active region growth to obtain high-quality QWs. Several growth parameters, such as V/III ratio, growth rate, and growth temperature, can be adjusted to improve the interface quality.

Because both bottom and top mirror in VCSELs are built with DBRs, the electromagnetic field intensity actually penetrates into the DBR pairs. The effective cavity length ($L_{eff}$) is longer than the physical cavity-layer thickness. Figure 6.70 shows a normalized field-intensity profile in an 850-nm VCSEL cavity and DBR regions close to the cavity on each side. If we take $1/e$ of the intensity as a cutoff point for the effective cavity length, it extends into four DBR pairs on each side of the cavity. The total effective cavity length becomes about 1310 nm instead of about 270 nm of the physical thickness. For those few DBR layers where the field intensity is still high, special attention should be paid on deciding the doping levels. As light entering these layers is reflective backward most of the time, any absorption loss will be amplified by the number of round trips. In practice, those layers are treated differently during epitaxial growth, usually with much lower overall doping, especially at peaks of the field intensity so that the free-carrier absorption loss can be reduced significantly.

Like any other type of semiconductor lasers, the optical gain and losses are the most important factors for VCSELs. They determine the lasing behaviors such as threshold current, peak power output, and slope efficiency. Quite often, gain can be simply modeled as a two-parameter fitting equation [354]

$$g = g_0 \ln\left(\frac{J}{J_0}\right), \qquad (6.78)$$

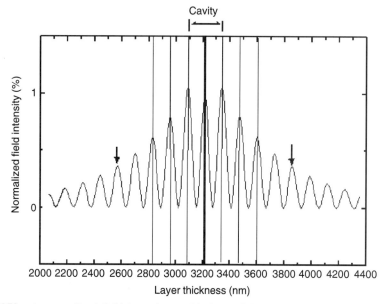

**Figure 6.70** A normalized field-intensity profile in an 850-nm VCSEL cavity and the DBR regions close to it. If taking 1/e of the intensity as a cutoff point, the effective cavity length extends to four DBR pairs on each side of the cavity.

where $g_0$ is a gain parameter factor and $J_0$ is the transparency current density. The unit of gain is $cm^{-1}$ and $A/cm^2$ for current density. $J_0$ is composed of several aspects of the current losses, such as current for spontaneous emission, carrier out-diffusion, Auger, and other types of nonradiative recombination. Table 6.5 shows typical $J_0$ and $g_0$ values for several active materials at 27°C for the wavelength of maximum gain under moderate pumping.

One can see that a different material system has different gain. Some material systems are better than others. Due to the band structure and a steplike function of the density of states, QWs usually have a much higher gain than do bulk materials.

The counterpart of the gain is losses. For VCSELs to be able to lase, the optical gain has to overcome all the losses. There are two major types of losses. One is distributed internal loss, such as those due to free-carrier absorption, and the other is mirror loss. To make an efficient laser, one has to do everything possible to increase its gain through a better structure design and material quality control and at the same time to reduce its

**TABLE 6.5  Two-Parameter Fits to Theoretical Gain versus Current Density Relationship for Three Material Systems [351]**

| Active material | $J_0$ | $g_0$ |
|---|---|---|
| Bulk GaAs | 80 | 700 |
| GaAs/Al$_{0.2}$Ga$_{0.8}$As 8-nm QW | 110 | 1300 |
| In$_{0.2}$Ga$_{0.8}$As/GaAs 8-nm QW | 50 | 1200 |

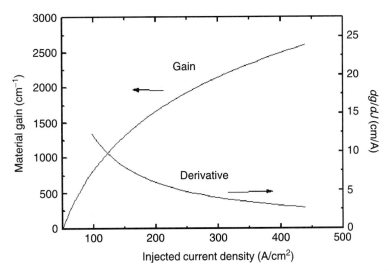

**Figure 6.71**   A calculated material gain and its derivative against the injected current density for the $In_{0.2}Ga_{0.8}As/GaAs$ 8-nm QW system in Table 6.5.

losses mentioned above. Figure 6.71 shows a calculated gain versus injected current density curve for the $In_{0.2}Ga_{0.8}As/GaAs$ 8-nm QW system in Table 6.5.

From this curve, one can see that the gain starts to increase with the injected current density dramatically in the low current range. But it starts to show a significant saturation behavior at about 150 A/cm$^2$ of current density. Also, the gain changes with temperature, and this is important because the self-heating will change the active region temperature even though the ambient temperature remains the same. Figure 6.72 shows the temperature effect on the maximum gain under a constant injected current density for three $In_{0.2}Ga_{0.8}As/GaAs$ QWs with 8-nm GaAs barriers and 10-nm GaAs confinement layers on each side. Thus, high current density will deteriorate the gain due to both the self-heating effect and the gain saturation.

***Current and Optical Confinements.***   There are three basic types of VCSEL structures based on different ways the current is confined. Even though many variations [355–360] have been explored, they are all based on one or a combination of the basic ones. Figure 6.73 shows schematics of three commonly used VCSEL structures. They are **(a)** etched mesa (or pillar) type of VCSEL, **(b)** proton-implanted VCSEL, and **(c)** oxide-confined VCSEL.

The mesa-type VCSEL is similar to a ridge wave-guide edge-emitting lasers. The etching is usually stopped just above the active layer. Thus, electrical current is confined or channeled to the mesa dimensions laterally [361, 362]. The top-emitting mesa-type of VCSELs is hard to fabricate due to the lack of space for the top electrical contact. But the mesa-type VCSELs are usually designed as bottom emission devices. By doing this, the mesa top surface can be used as a bonding pad for good electrical contact. For bottom emission, the substrate must be transparent to the emitted light. Thus, InGaAs/GaAs structures with emission wavelengths in the 0.9- to 1.0-$\mu$m range have been chosen for GaAs substrates. For the emission range of 0.78 to 0.86 $\mu$m, the GaAs substrate has to be removed in order to get the light out. Due to a small area of the

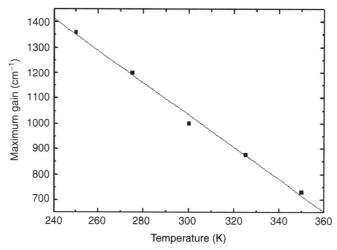

**Figure 6.72** The maximum gain as a function of temperature for the $In_{0.2}Ga_{0.8}As/GaAs$ 8-nm QW system in Table 6.5. The line is a linear fit to the data.

**Figure 6.73** Schematics of three types of commonly used VCSEL structures: (**a**) etched mesa (pillar) type of top-emitting VCSEL, (**b**) proton-implanted top-emitting VCSEL, and (**c**) oxide-confined top-emitting VCSEL.

mesa, the thermal resistance is usually high. A thick metal layer such as gold plating would help to remove the heat and improve its thermal conductance.

One of the problems with etched-mesa VCSELs is its optical loss due to roughness of the mesa wall if the mesa diameter is small [359, 360]. The smaller the mesa is etched, the lower differential efficiency gets. Therefore, this type of VCSELs is usually fabricated in research labs.

The second type of current confinement is through ion implantation. By implanting proton into GaAs/AlGaAs, the selected areas defined by a photoresist mask in the process can be made nonconductive, thus confining the current flow to the center area of the VCSEL aperture. Millions of implanted VCSEL chips have been produced and used in sensing and fiber optic applications because of the relatively mature implantation process. Optically, the implanted device is gain guided. Proton implant defines the current path to provide a desired current confinement [363]. Compared with the mesa-type VCSEL the proton-implanted VCSEL has lower thermal resistance. As the overall process is planar and implantation process is well controlled, the production yield is usually high and cost is low. Its disadvantage is low optical efficiency and kinky optical behavior because its top metal contact blocks a significant part of output light, as shown in Fig. 6.74.

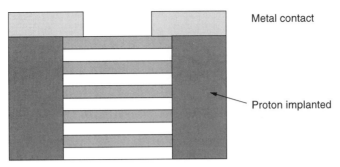

Metal contact

Proton implanted

**Figure 6.74**   A schematic of metal contact-ring on the top of a proton-implanted VCSEL chip. The metal contact ring partially blocks the light to achieve low for the sake of low series resistance.

In designing implanted VCSELs, the metal contact ring is usually made to cover part of the unimplanted areas [364]. This kind of metal contact arrangement is mainly used to reduce series resistance of the VCSEL. However, because of the partial blockage of the light, the optical efficiency of the VCSEL is low and the optical mode can be unstable. Also, the implanted proton cannot really confine the current path very well due to a scattering of ions during the implantation process. The proton profile is not abrupt and well-defined, so a current leakage path may exist. Also, because reflectivity is higher underneath the metal contact-ring, the threshold current there is lower than the center part of the aperture. This can cause proton-implanted VCSELs to start lasing under or close the ring area first. As the metal contact-ring blocks the aperture, the light-only part of light can be emitted and detected. This could cause an unstable feature in L-I curves, and it is believed to be the major cause for its kinky behavior [365].

Another issue with proton implant is the control of implant depth. Due to an energy limitation of commercially available ion-implanters, the proton implantation depth is limited to about three to four micron meters, which require about 400 KeV of energy. This places an upper limit on the top-DBR thickness. A high refractive index contrast in the DBR is needed to have a thin top-DBR and still achieve required reflectivity, such as 12% AlGaAs and AlAs for 850-nm VCSELs.

The third type of VCSEL is oxide-confined [366]. In this VCSEL structure, the current is confined to the center aperture by a very thin oxide layer, which is grown with wet oxidation of AlGaAs with very high Al content. Oxide VCSEL is the most popular type of VCSELs. Its advantages are high efficiency, high speed, low series resistance, and ease of control for small apertures (such as $<10$ $\mu$m). Its disadvantages are relatively low yield, low reproducibility, and poor reliability. The way the oxidation layer is formed is to oxidize one of the low-index layers with higher Al composition in the first two to three pairs of P-type DBRs above the cavity. For example, if the low index layer in the DBRs contains 92% of Al, the to-be-oxidized layer is 98% to 100% of Al. After epitaxial layer growth, an air post or a trench is usually formed by etching to expose the higher Al composition layer. The wafer is then inserted into a furnace with $N_2$ saturated with $H_2O$ vapor flowing. The high Al composition layer is oxidized much faster, and an aperture is formed for current confinement. This oxidation layer also provides an index guiding to the light. As this layer is nonconductive, and located very close to the active layer, the lateral current spreading is small compared to that in

**TABLE 6.6    A Comparison of Three Types of Top-Emitting VCSELs (Ranked as 1 to 5 and Referred to a Proton-Implanted VCSEL, 5 is the Best)**

| Type of VCSELs | Threshold Current | Slope Efficiency | Kinky Behavior | Power Output | Thermal Conductance | Process Cycle | Rel.* |
|---|---|---|---|---|---|---|---|
| Mesa (pillar) | 3 | 3 | 3 | 2 | 2 | 5 | 2 |
| Proton-implanted | 3 | 3 | 2 | 3 | 4 | 2 | 4 |
| Oxide-confined | 5 | 5 | 4 | 4 | 3 | 4 | 3 |

*Reliability.

the implanted VCSEL structure. So far, oxide-confined VCSEL has shown the lowest threshold current density and the highest slope efficiency.

Table 6.6 summarizes the difference of these three types of VCSELs discussed so far. A relative ranking is also provided in some key aspects.

***Thermal Properties.*** Thermal property is one of the most important properties for semiconductor lasers. In most cases, a semiconductor laser's output power is limited by the self-heating. The rollover point on the L-I curve is determined by junction temperature. A good thermal property is, especially, essential to VCSELs due to its short gain region and thick DBR layers on both top and bottom. Even though the VCSEL heat flow modeling is a three-dimensional problem, its average temperature in the active layer can be calculated and measured through a simple method. The power dissipation ($P_d$) in the VCSEL is [367]

$$P_d = P_{\text{in}} - P_o = P_{\text{in}}(1 - \eta) \qquad (6.79)$$

where $P_{\text{in}}$ is total power in, $P_o$ is optical power output, and $\eta$ is the wall-plug efficiency, and the temperature rise is

$$\Delta T = P_d \cdot Z_T \qquad (6.80)$$

where $Z_T$ is the thermal impedance. Although the VCSEL thermal modeling is a complicated three-dimensional problem, for a small-aperture VCSEL with a very thick substrate underneath, $Z_T$ can be approximated expressed as [368]

$$Z_T = 1/(4 \cdot \sigma_T \cdot a_{\text{eff}}) \qquad (6.81)$$

Here, $\sigma_T$ is the thermal conductivity of the material beneath the heat-generating junction and $a_{\text{eff}}$ is the VCSEL's effective radius. For the etched-mesa type, the $a_{\text{eff}}$ is approximately equal to the mesa diameter; for other types, such proton-implant or oxide-confined VCSELs, $a_{\text{eff}}$ is larger than the current injection radius but smaller than twice of the current injection radius due to heat spreading in surrounding epitaxial or other materials. If we take $\sigma_T = 24$ W/(mK) (an average of AlGaAs' thermal conductivity) and $a_{\text{eff}} = 10$ μm, $Z_T$ can be estimated to be 1000°C/W, which is a typical thermal impedance value for 850-nm VCSELs.

For a flip-chip bonded VCSEL, $Z_T$ can be expressed as

$$Z_T = h/(\sigma_T \cdot A) \qquad (6.82)$$

where $A$ is the heat flow area and $h$ is the distance between the heating spot and the heat sink. If we take $h = 3.5$ μm and $\sigma_T = 16$ W/(m°C) with 10 μm diameter of the heat flow area, $Z_T$ will be 400°C/W. This is a factor of 2.5 times smaller than the regularly mounted one. If we assume an input power of 40 mW ($I = 20$ mA and $V = 2$ V), for the regular mount VCSEL, the junction-temperature increase would be 40°C. But with flip-chip bonding, the junction temperature only increases 16°C. We can see a clear advantage of the flip-chip bonding on the thermal dissipation.

With increasing temperature, the overall VCSEL performance will usually degrade. The reasons are as follows:

(1) The gain decreases as temperature increases.

(2) Carrier confinement is poorer at high temperature as thermal energy increases. This causes higher threshold current, lower slope efficiency, and earlier output power rollover.

(3) Gain peak is farther away from FP wavelength due to a more sensitive relationship between gain peak wavelength and temperature (for 850-nm VCSELs, 0.3 nm/°C vs. ~0.07 nm/°C). This causes higher threshold current and lower slope efficiency.

(4) Current spreading leakage loss is larger.

(5) More free carriers at elevated temperatures and band-gap shrinking for the layers in DBRs cause increased free-carrier and band-tail absorption.

(6) Lower differential gain and earlier power rollover worsen the VCSEL's high-speed performance.

For #3 mentioned above, it is possible to detune the gain wavelength (blue shift) to FP wavelength at room temperature, so that a better match can be found for a better high-temperature performance. For all other aspects, there is no easy solution. As the junction temperature depends on both ambient temperature and the self-heating, reducing heat generation and both thermal impedance and electrical resistance is very important for a better overall performance in an extended temperature range, such as room temperature to 85°C.

***Reliability.*** During the last several years, the VCSEL reliability has been improved significantly [369, 370]. Figure 6.75 shows the improvement progress for Honeywell-made 850-nm proton-implanted VCSELs during 1995 through 2000. By now the VCSEL reliability is no longer a big issue any more, as claimed by several VCSEL manufacturers [371, 372]. However, better reliability is achieved with a lot of efforts in material and structural design, as well as in process control. So far, proton-implanted VCSEL has the most data to show good reliability due to its relatively long history in production. Greater than 2.4 million cumulative hours of operation at 70°C has been exhibited. A mean-time-to-failure (MTTF) of $3 \times 10^7$ hours for proton-implanted 850-nm VCSELs has been predicted by several vendors.

For oxide VCSELs, reliability data are limited compared to the proton-implanted one. Nevertheless, more than $1 \times 10^6$ hours of lifetime have been published as shown in Fig. 6.76. For larger aperture devices, the reliability may not be a big problem. However, the small aperture device's reliability is still a challenging issue. Figure 6.76 indicates that with smaller oxide-aperture size, such as 14 μm diameter, the device's

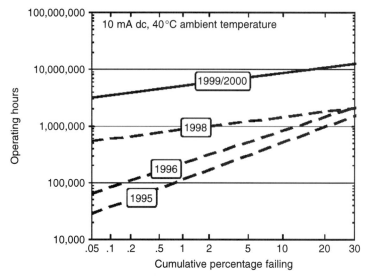

**Figure 6.75** Improvement of Honeywell's proton-implant VCSELs during a period of six years (1995 to 2000). The improvement is most significant in the low failure rate range resulting from improved uniformity of the wafer parameters.

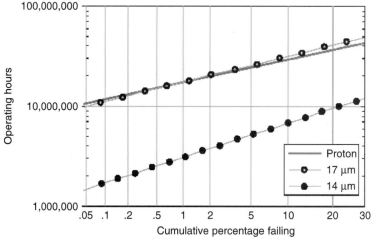

**Figure 6.76** Reliability data for Honeywell's oxide-confined VCSELs. The improvement is most significant in the large aperture VCSELs.

reliability is about a factor of five worse than that of the larger aperture (17 μm) devices, which is comparable similar to the proton-implant ones. These curves were taken under a condition of 25°C and normal operation current for each device (10 mA, 8 mA, and 6.5 mA for proton, 17-μm oxide and 14-μm oxide VCSEL, respectively). Also, as mentioned above, unlike the proton-implant process, the oxidation process is not reproducible yet. Daily quality control and monitoring are still needed to guarantee the reliability from batch to batch and from wafer to wafer.

The causes of poor reliability have been under extensive study [373, 374], and the understandings have been improving [375]. The following is a partial list of causes for poor reliability. Some of them apply to both proton-implanted and oxide-confined VCSELs, such as #1, 3, 4, and 5. But, #2 only applies to oxide VCSELs.

(1) Poor QW and active layer quality.

(2) Excessive doping in the cavity and regions close to the cavity area.

(3) Improper oxidation-layer design, including its thickness, doping level, and location.

(4) Improper oxidation process control, such as temperature, temperature ramping-down speed, and oxidation rate.

(5) Improper doping level at the interfaces of the DBRs.

(6) High electrical resistivity of DBRs.

(7) High thermal impedance of the chips and/or packaging.

(8) Small aperture sizes and high drive current density.

(9) Poor ESD handling and operation conditions.

For VCSEL arrays, the reliability would be worse than that of single element ones, as expected. Figure 6.77 shows how the number of elements affects the operating hours for Honeywell's oxide arrays.

### 6.8.2   850, 650, 780, 980, 1310, and 1550 Nm VCSELs

Table 6.7 summarizes VCSELs at different wavelengths. As discussed earlier, the key components of forming a VCSEL are QWs, barriers, confinement layer, cladding layer,

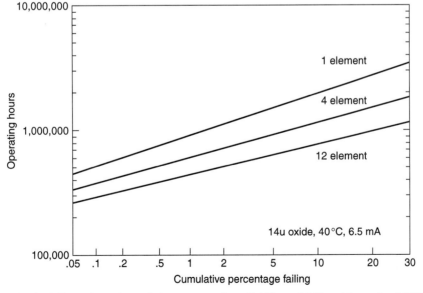

**Figure 6.77**   Effect of number of elements on the reliability of 1-D oxide-confined VCSEL arrays (from Honeywell's data).

**TABLE 6.7   A Summary of VCSEL Structures at Different Wavelengths**

| $\lambda$ (nm) | QW/Barrier/ Substrate | CMF*-Layer C**-Layer | DBRs | Dopant |
|---|---|---|---|---|
| **670** | $In_{0.54}Ga_{0.46}P/$ $In_{.5}(Ga_{.5}Al_{.5})_{.5}P/On$ GaAs | $In_{.5}(Ga_{.5}Al_{.5})_{.5}P/$ $In_{.5}(Ga_{.3}Al_{.7})_{.5}P$ | nH***- $Al_{.5}Ga_{.5}As/nL****-$ $Al_{.95}Ga_{.05}As$ | P = C, Mg, Zn N = Si, Se |
| | | | Typical numbers of pairs for HR/LR DBRs-55.5/34 | |
| **780** | $Al_{.15}Ga_{.85}As/$ $Al_{.36}Ga_{.64}As/On$ GaAs | $Al_{.36}Ga_{.64}As/$ $Al_{.6}Ga_{.4}As$ | nH-$Al_{0.3}Ga_{0.7}As/nL-$ $Al_{.95}Ga_{.05}As$ | P = C, Mg, Zn N = Si, Se |
| | | | Typical numbers of pairs for HR/LR DBRs-40.5/30 | |
| **850** | GaAs or $In_{.06}Ga_{0.94}As/$ $Al_{0.3}Ga_{0.7}As/On$ GaAs | $Al_{0.3}Ga_{0.7}As/$ $Al_{0.5}Ga_{0.5}As$ | nH-$Al_{0.15}Ga_{0.85}As/nL-$ $Al_{.95}Ga_{.05}As$ | P = C, Mg, Zn N = Si, Se |
| | | | Typical numbers of pairs for HR/LR DBRs-35.5/25 | |
| **980** | $In_{.15}Ga_{.85}As/GaAs$ or $Al_{0.1}Ga_{0.9}As/On$ GaAs | $Al_{0.3}Ga_{0.7}As/$ $Al_{0.5}Ga_{0.5}As$ | nH-GaAs/nL- $Al_{.95}Ga_{.05}As$ | P = C, Mg, Zn N = Si, Se |
| | | | Typical numbers of pairs for HR/LR DBRs-30.5/20 | |
| **1310** | $In_{.7}Ga_{.3}N_{.007}As_{.993}/$ GaAsN/On GaAs | $GaAsN/Al_{0.3}Ga_{0.7}As$ | nH-GaAs/nL-AlGaAs | P = C, Mg, Be |
| | | $GaAs/Al_{0.3}Ga_{0.7}As$ | Typical numbers of pairs for HR/LR | N = Si, Se |
| | GaAsSb/GaAs/On GaAs | $GaAs/Al_{0.3}Ga_{0.7}As$ | DBRs-35.5/25 | |
| | InAs QD/GaAs/On GaAs | | | |
| **1550** | InGaAs/AlInGaAs/ On InP | AlInGaAs/InP | nH-AlInGaAs/nL-InP or nH-InGaAlAs/nL- AlInAs or amorphors nH-GaAs/nL-AlGaAs | P = Zn, C, Mg N = Si, S |
| | | | Typical numbers of pairs for HR/LR DBRs-60.5/50 | |

*CMF—confinement, **C—cladding, ***nH—high index layer, ****nL—low index layer, HR—high reflectivity, and LR—low reflectivity.

DBRs, and substrate. Whether we can make a VCSEL at a certain wavelength depends on several factors. There must be a material system that can be grown on a suitable substrate and has enough gain at the emitting light with that wavelength. Suitable DBR stack materials must exist that can be grown with high quality on the same substrate. Preferably, the DBR stack can also be heavily doped into both *p*- and *n*-types, so electrical current can pass through the DBRs on both sides of the cavity. In addition, the materials for active layer and DBR stack should be easily processable. This is a necessary condition, but not sufficient. For a certain wavelength, the material systems are known, but they may not be well developed yet, such as GaN/AlGaN for 470-nm VCSELs. It would be some time before a working device is made, not to mention its commercialization.

In the table, we have listed several wavelengths that have been studied and developed for many years. Even though some of them are not ready for production yet, significant progress has been made during the past several years. We will talk in more detail about 850-, 780-, and 980-nm VCSELs and only briefly discuss the 650-nm and long wavelength VCSELs. For long wavelength VCSELs, the applications are ready but the technologies are not.

Overall, 850- and 980-nm VCSELs are relatively mature now. The research work on 850- and 980-nm VCSELs started in the early 1980s, and since 1980, many research groups around the world have studied these devices. The advancement on GaAs and AlGaAs thin-layer growth technology, such as MBE and MOCVD, as well as wafer (chip) fabrication processes during the past 20 years has enabled the commercialization of VCSELs. InGaAs/GaAs/AlGaAs and GaAs/AlGaAs material systems have provided a platform for researchers to understand and improve VCSEL device and its fabrication process. The demand on a high-speed and low-cost solution for short-distance fiber-optic communication was driving 850- and 980-nm VCSEL development. Prior to VCSELs, 810- to 850-nm LEDs were the key components in this field. However, due to LED's spontaneous emission, its modulation speed is limited by its carrier lifetime to several hundred Mbps, which is not enough for high-speed demand. Also, because of its broad wavelength spectral width ($\sim$20 nm), the chromatic dispersion limits its traveling distance in the multimode fibers. At that moment, VCSELs found its market in gigabit per-second (Gbps) data communication applications, such as Gigabit Ethernet, GBIC, and Fiber Channels for SAN.

The VCSEL performance is determined by both the epistructure and the chip processing design and, most importantly, the realization of the designs. Quite often, a good structural design cannot be realized with the processes and/or software and hardware tools, such as MOCVD reactors, dry-etching machines, or oxidation furnaces, due to some limitations of these tools. A good understanding of these limitations and knowing what those machines can and cannot do is necessary for the designers. Among these tools, MOCVD may be the most critical one because most of the design ideas are realized in the epilayer growth. MOCVD offers an excellent control in uniformity, accuracy, and reproducibility, which is the most important factor to achieve high-yield (so low-cost) products.

There are several important parameters to evaluate a VCSEL performance: (1) Threshold current ($I_{th}$); (2) slope efficiency (SE); (3) operational forward current and voltage ($I_f$ and $V_f$); (4) dynamic electrical resistance (dV/dI); (5) maximum light output power ($P_{max}$); and (6) modulation speed ($f_{-3db}$). Unfortunately, not all of these can be optimized at the same time. A design compromise usually has to be made according to a particular application. To design a good VCSEL, one has to fully understand the application needs, such as which is the most important parameter and which can be relaxed, because the parameters mentioned above may have conflict with each other. For instance, the low-threshold VCSEL design is totally different from the high-power one. The high-speed VCSEL is different from the high-power one also. Some of the performance parameters, however, may be commonly desired, such as low-resistance and high-slope efficiency. Other factors that need to be considered in a VCSEL design are the regulation on the eye safety, the applications for single-mode or multimode fibers, and so on. For high-power VCSEL design, because the power is the most important parameter threshold current, $I_f$ and $V_f$ may have to be scarified. To obtain high-power VCSELs, several design considerations can be made:

(1) Use lower reflectivity of the top-DBRs, such as 99.2% instead of 99.5%

(2) Use larger aperture, e.g., 20 $\mu$m vs. 12 $\mu$m

(3) Dope some areas aggressively to reduce series resistance further to increase the rollover point

(4) Use bigger offset between the gain peak and the FP mode wavelength to achieve higher power at high-temperature operation

(5) Use flip-chip bonding or better heat sinks

Figure 6.78 shows a typical L-I-V curve of a VCSEL chip designed for high-power applications, such as free-space communication. This chip has an oxide-confined square-shape aperture ($24 \times 24$ $\mu$m$^2$). Its wavelength is 845-nm, dV/dI is 38 (mV/mA), and threshold current is 5 mA. When the drive current reaches 20 mA (at 2.3 V), light output is 10 mW. The output power is over 20 mW at 45 mA of drive current. This device shows that with a proper design, VCSELs can deliver a significant amount of power. On the other hand, some applications require only low-threshold currents in which the design criteria are totally different. Figure 6.79 shows a typical L-I-V curve designed for low-threshold current applications. Its aperture size is 12 $\mu$m, and threshold current is 1.5 mA.

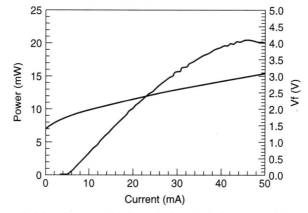

**Figure 6.78** A typical L-I-V curve of 850-nm VCSELs designed for high-power applications. This chip has an oxide-confined square-shape aperture ($24 \times 24$ $\mu$m$^2$). Its threshold current is about 5 mA.

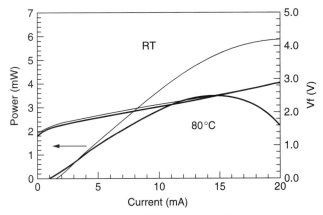

**Figure 6.79**   A typical L-I-V curve of a VCSEL designed for low-threshold current applications. Its aperture size is 12 $\mu$m, and the threshold current is about 1.5 mA.

**TABLE 6.8   Temperature Effect on Lasing Wavelength for 770-nm to 860-nm VCSELs**

| Lasing Wavelength $\lambda$ (nm) | $\Delta\lambda/\Delta T$ (nm/°C) |
|---|---|
| 777 | 0.045 |
| 798 | 0.041 |
| 808 | 0.046 |
| 811 | 0.037 |
| 823 | 0.045 |
| 835 | 0.050 |
| 853 | 0.057 |
| 860 | 0.050 |

Similar to edge-emitting FP lasers, the lasing wavelength of VCSELs changes with temperature. However, their origins are different. For edge-emitting FP lasers, the gain-peak controls lasing wavelength. For example, the wavelength temperature-coefficient of 1310-nm FP lasers is ~0.34 nm/°C. For VCSELs, however, the lasing wavelength is determined by the FP mode, which is in turn determined by optical thickness of each pair in the DBRs. The optical thickness equals physical layer thickness multiplied by its refractive index. Table 6.8 shows experimental results of temperature effect on VCSEL lasing wavelength for wavelengths from 777 to 860 nm. The results were obtained in the temperature range of 25°C to 85°C. As physical layer thickness change is determined by its thermal expansion coefficient, which is $6.8 \times 10^{-6}$/°C for GaAs and $5.2 \times 10^{-6}$/°C for AlAs, the optical thickness change due to the layer thickness change is small (on the order of 0.005 nm/°C). So the temperature effect on VCSEL wavelength is primarily due to refractive-index change versus temperature.

Visible VCSELs are always interesting and have large amount of applications and potential applications. These applications include plastic optical fiber (POF) communication, medical, sensing, DVD read-write heads, full color displays, lighting, and so on. To date, red color VCSELs (the wavelength ranges from 640 to 690 nm) are the most advanced visible VCSELs. Yet there still is a big performance gap to fill for becoming commercial uses. A significant amount of effort has been put into its studies and development.

There are two kinds of material systems that can be used to produce red color VCSELs. One is AlGaAs on GaAs, and the other is AlGaInP on GaAs. Due to the band-gap structure and energy, the AlGaAs material can only produce deep red color, about 700 nm of wavelength. And its efficiency is low. AlGaInP on GaAs is a better material system, even if it also has some intrinsic problems, such as low gain, lack of carrier confinement for the quantum wells, and high thermal impedance. AlGaInP material has been used in some commercialized optoelectronic devices, including 650-nm semiconductor DVD lasers, laser pointers, and high brightness (HB) LEDs for several years. However, even though this material is mature for those devices because of its low gain, poor carrier confinement, and thermal conductance, it has been a great challenge to use this material system to produce VCSELs with good high-temperature performance (say 70°C). Several groups have studied this material system for VCSEL development. One of the structures is shown in Table 6.7 [376]. The epitaxial structure includes 55.5 pairs of 50% AlGaAs as N-DBR and 34 pairs of P-DBR, 6-nm 56% InGaP as QWs, and 6-nm 50% (AlGa)InP barrier layers. Due to a large number of layers in the DBRs, the electrical as well as thermal resistance is high, which is a big issue for red VCSEL performance. From this structure, people have achieved 680-nm cw lasers and temperatures below 55°C have been reported. The power is relatively low even at room temperature. Further development effort is clearly needed before red VCSELs find commercial applications.

Because 1310 nm and 1550 nm are the wavelengths widely used in today's fiber-optic communications, it is expected that VCSELs at these wavelengths have huge commercial potentials. The success of 850-nm VCSELs has stimulated a great interest of LW VCSELs in the last several years. Thus, tremendous efforts have been poured into LW VCSEL development [377–379]. Owing to a larger distance-bandwidth product of the silica-based fiber at 1310 nm, the maximum data transmission rate and point-to-point distance in these links is expected to increase more than twice with multimode 1310-nm VCSELs as compared to using 850-nm VCSELs. Single-mode LW VCSELs are even more important because they are viable choices of optical emitters for the telecommunication and wavelength-division multiplexing (WDM) applications. The advantages of LW VCSELs over edge emitters, such as DFB and FP lasers, are low cost, ease of 2D array fabrication, and integrations.

Even though lots of efforts have been put into LW VCSEL development, the progress has been relatively slow due to several technology hurdles. One of the key issues is lattice mismatch between good active layer materials and good DBR mirror materials. For instance, a traditionally good active layer for 1310 nm is InGaAs, which matches InP substrate. However, there are no good DBR stack materials available. The index difference for a InGaAsP/InP [380], InGaAlAs/InP [381], and InGaAlAs/InAlAs [382] material system is not big enough to form good DBRs [383]. On the other hand, GaAs/AlAs materials can be used to build good DBRs for a 1310-nm wavelength. But no good active layer material matches GaAs substrate. So most efforts have turned to either wafer bonding [384, 385] or the dielectric mirrors approach [386]. Until recently, several new materials and technologies have appeared to be feasible to produce 1310-nm VCSELs with one epitaxial growth. These materials and technologies are (1) InGaAsN active layer with AlGaAs/GaAs DBRs on GaAs substrate [387, 388]; (2) InGaAsP active layer with air/semiconductor DBRs on InP substrate [389]; (3) InGaAs QD active layer with AlGaAs/GaAs DBRs on GaAs [390]; and (4) GaAsSb with AlGaAs/GaAs DBRs on GaAs [391, 392]. So far, the most promising material systems are GaAsSb, InGaAs QD, and InGaAsN on GaAs and InGaAsP with air/semiconductor DBRs on InP.

For traditional InGaAsP on the InP material system, there are several intrinsic issues against high performance of LW VCSELs. They are Auger recombination, carrier leakage, and carrier-related optical losses. Because for telecommunications the devices have to work normally up to 85°C, this requirement places a very high demand on the device design and material development. However, the new material systems, such as InGaAsN and GaAsSb or InAs QD, provide solutions to those issues mentioned above. Room temperature cw operation of VCSELs operating at 1.3 μm has been reported using InGaAsN and GaAsSb active layers grown on GaAs [387, 392].

### 6.8.3  High-Speed VCSELs

The high-speed properties of VCSELs are primarily determined by three factors: (1) Relaxation resonance frequency, (2) the RC (resistance and capacitance) constant, and (3) the carrier transport time ($\tau_r$), as shown in Eq. (6.76) [393]

$$\left|\frac{F(\omega)}{F(0)}\right|^2 = \frac{1}{\left(1+\left(\frac{\omega}{RC}\right)^2\right)}\frac{1}{(1+\left(\frac{\omega}{\tau_r}\right)^2}\frac{\omega_r^4}{((\omega_r^2-\omega^2)+\gamma^2\omega^2)}$$

where $F(\omega)$ is the small-signal optical frequency response, $\omega$ is angular frequency, $RC$ is the product of series resistance and parasitic capacitance, $\tau_r$ is carrier transport time, $\omega_r = 2\pi f_r$ is angular resonance frequency, $\gamma = \frac{1}{2\pi}K\omega_r^2 + \frac{1}{\chi\tau_n}$, $K$ is $K$ factor [393], and $\chi = 1 + (\tau_r/\tau_e)$ is transport factor. For biases far above threshold current, the resonance frequency is [394]

$$f_r = \frac{1}{2\pi}\left[\eta_i\frac{\Gamma v_g}{qV}\frac{\partial g}{\partial N}(I-I_{th})\right]^{1/2} = \frac{1}{2\pi}\left[\eta_i\frac{\Gamma_{xy}\varsigma v_g}{qL}\frac{\partial g}{\partial N}(J-J_{th})\right]^{1/2} \qquad (6.83)$$

where the differential gain, $\dfrac{\partial g}{\partial N}$, is obtained from the gain versus carrier density at the bias point, $\eta_i$ is the internal quantum efficiency, $v_g$ is the group velocity, $L$ is the cavity length, and $\Gamma$ is the optical confinement factor. So the key parameters determining the VCSEL speed are (1) differential gain, (2) quantum efficiency, (3) optical confinement factor, (4) cavity length, and (5) the driving condition. For VCSELs, the cavity length normally is one optical wavelength thickness. The model coupling efficiency is defined by the overlap of the QWs with the optical field in the cavity. The quantum efficiency is determined by the material system, epilayer quality, and QW design. To achieve high-speed VCSELs, people need to pay more attention on the differential gain and driving condition. Differential gain can be affected by several factors, including strain in the QWs, number of QWs, gain peak and FP wavelength offset, carrier confinement, and $p$-type doping in the active layer. Heating can also affect the high-speed modulation performance in two ways. One is heating tends to cause an early output power rollover, which limits the driving current, and the other is heating usually reduces the differential gain.

*RC* constant is another important factor for VCSEL's high-speed performance. Figure 6.80 shows a simple equivalent AC circuit of a typical VCSEL device. There, Rs is series resistance, including the contact resistance and resistance from DBRs and cladding and confinement layers. Rj and Cj are junction resistance and capacitance, respectively. Cp is parasitic capacitance from the bonding pad. Large RC constant could limit the −3dB frequency. For a commercial oxide-confined VCSEL, the typical series resistance R is about 40 Ω. If the capacitance is 0.6 pF, the RC limited −3dB frequency is about 6.6 GHz. For 10 GHz of −3dB VCSEL design, the parasitic capacitance should be controlled to be less than 0.4 pF. Figure 6.81 shows a calculated result of the small-signal frequency response for an oxide-confined VCSEL with 40 Ω of series resistance and 0.4 pF of parasitic capacitance. Its carrier transport time is 10 ps. Its threshold current is 1.5 mA with an aperture size about 9 μm. One can see that with higher drive current (higher power), the resonance frequency increases with the power, as indicated by Eq. (6.50). However, due to a large parasitic RC constant, when the power reaches 3.5 mW, the response curve reveals a rolloff behavior. The −3dB frequency shows a rollover versus square root of the output power due to the parasitic RC limitation. Figure 6.82 shows calculations of the −3dB frequency as a function of the square root of the output power for 850-nm VCSELs at different RC constants. It is obvious that if the RC is small, the −3dB frequency is proportional to the square root of the output power [so as the square root of the (I-I$_{th}$)], as shown in Eq. (6.50). However, as RC increases, larger than 16 ps for this case, the −3dB frequency has a rollover behavior due to the parasitic limitation. It is therefore very important to keep the parasitics small for high-speed VCSELs.

Another key factor for high-speed VCSEL is the carrier transport time. Like RC constant, the carrier transport time affects the −3dB frequency through a damping behavior, as shown in Eq. (6.64). It is interesting to note that the effect of RC constant and carrier transport time on the high-speed response is not distinguishable because they behave the same way. As discussed in Section 6.7, the carrier transport time

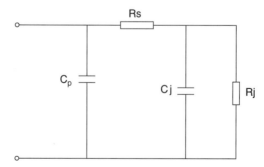

**Figure 6.80**   An equivalent AC circuit for a typical VCSEL device. Rs is series resistance, including the contact resistance and resistance from DBRs and cladding and confinement layers. Rj and Cj are junction resistance and capacitance, respectively. Cp is parasitic capacitance from the bonding pad.

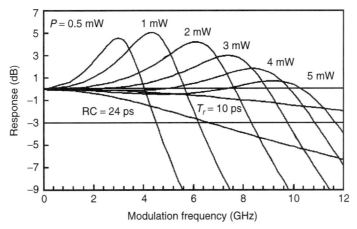

**Figure 6.81**  Calculated small-signal response as a function of modulation frequency at different output power. The RC (= 24 ps) and carrier transport time (= 10 ps) damping curves are plotted out also.

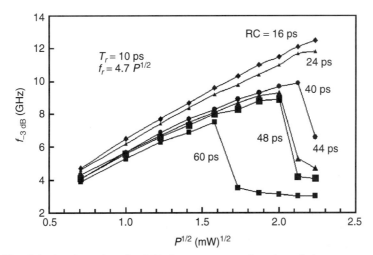

**Figure 6.82**  Calculated results of −3dB frequency as a function of the square root of the output power at different RC constants when the carrier transport time is 10 ps. Large enough RC constant starts to make the −3dB curve show a damping behavior.

consists of four major parts; carrier diffusion and/or drift under electric field from the doped cladding layer to the QWs and carrier captured by the QWs and escaping from the QWs through thermionic emission and/or tunneling process. For details, the reader is referred to Section 6.7.

By optimizing the VCSEL design and reducing the parasitic capacitance and series resistance, people have achieved up to 12.5-Gbps speed performance. Figure 6.83 shows eye diagrams of a proton-implant VCSEL, which works at 1.25 Gbps up to 10 Gbps at room temperature [395]. VCSEL operation at a data rate higher than 12.5 Gbps in general is difficult because more efforts are needed to overcome many

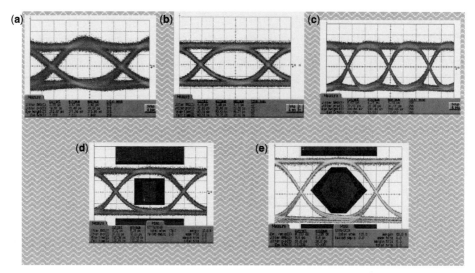

**Figure 6.83** Eye diagrams of proton-implant VCSELs operating at **(a)** 10 Gbps, **(b)** 5 Gbps, **(c)** 3.125 Gbps, **(d)** 2.5 Gbps, and **(e)** 1.25Gbps.

barriers, such as parasitic capacitance, junction RC constant, even higher differential gain, and other packaging-related issues.

### 6.8.4 VCSEL Arrays

One of the advantages of VCSELs is ease of array fabrication. VCSEL arrays have found many important applications, such as two-dimensional free-space communication [396], short reach parallel optical link [397, 398], and high-speed scanning. The demand for more bandwidth and faster speed increases significantly for these applications every day. Another array application is to generate high-power light output from the VCSEL chips.

Because VCSELs emit light in the vertical direction, they can be processed from start to finish in wafer form. Making VCSEL array involves only cutting an area of several individual chips instead of putting several chips together and aligning them to micron accuracy like the edge-emitting lasers. VCSEL arrays can be made with a simple photo-lithography process and very accurate dimensions. Figure 6.84 shows an 8 × 8 two-dimensional array made out of an 850-nm VCSEL wafer [399]. The circular dots are the light-emitting windows with metal contacts attached. The black dots at the edge are the *p*-side gold wires bonding with a square shape of bonding pads underneath. The *n*-side metal contact is on the back of the chip. For this array, all of the VCSELs can be addressed individually. The performance is very uniform, as shown by the L-I-V curves to the right side of it.

Currently, 1 × 4 and 1 × 12 one-dimensional (1D) arrays are commercialized formats for short-reach parallel optical communications. With each device delivering up to 3.35 Gbps of bandwidth, 1 × 4 and 1 × 12 can provide up to 12.5-Gbps and 40-Gbps bandwidth, respectively. This kind of massive bandwidth is achieved with relatively simple electronics and packaging, as compared to a single device providing 12.5- or 40-Gbps bandwidth with very complicated high-speed electronics, specialized high-speed packaging material, and its integration for a single device. The overall package

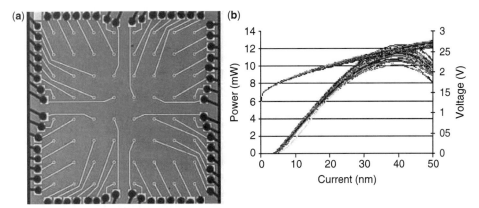

**Figure 6.84** (a) A top view of an 8 × 8 two-dimensional proton-implant 850-nm VCSEL array. The circular dots are the light-emitting windows with metal contacts attached. The black dots at the edge are the P-side gold wire bonding with a square shape of bonding pad underneath. The N-side metal contact is on the back of the chip. (b) The corresponded L-I-V curves for the 64 VCSEL devices.

size and power consumption are also smaller than a serial device operating at the same data rate. However, the room or space needed is much smaller for a single device compared to an array.

Special attention must be paid in using VCSEL arrays. One of the issues with 2-D arrays is thermal cross talk. Because the individual VCSEL cell is close to each other on the array, when one device is driven, the device next to it will also get heated up. As VCSEL characteristics is heavily dependent on the temperature, the operating condition of the devices next it might affect its performance. For this reason, it is best to use the actual overall array operation condition for testing and characterizing the individual VCSEL's electrical and optical performance. A different location has a different thermal environment. When one operates the array simultaneously, the device performance would be different. This has to be considered when a specific requirement is needed.

## 6.9   OTHER QUANTUM DEVICES: QUANTUM CASCADE, QUANTUM WIRE, AND QUANTUM DOT LASERS

### 6.9.1   Quantum Cascade Lasers

Quantum cascade lasers are semiconductor lasers based on wide band-gap materials such as InP or GaAs. [400, 401] These devices are based on an entirely new concept. The generation of photons relies on optical transitions between states created by confinement in a quantum-well thickness while keeping the same hetero-structure material. It also allows large numbers of identical active region cells to be stacked in series.

Quantum cascade lasers operate under a strong electric field, which could produce a conduction band profile resembling a staircase. Cascading electrons potentially emit a photon at each step. In each cell of the active region, for either a GaAs or InP material system, a combination of tunneling and optical phonon resonance provides population inversion between level 3 and level 2 of the active region. Super-lattices provide another efficient approach to obtaining population inversion in inter-sub-band transitions. Laser based on super-lattice active regions show high peak powers at room temperature and are suitable for very long wavelength operation ($\lambda > 12 \mu$m) [402]. For more information on these devices, the reader is referred to published papers [403–407].

### 6.9.2   Quantum Wire and Quantum Dot Lasers

As discussed in Sections 6.3.4 and 6.3.5, quantum-well semiconductor laser diodes, in which the charge carriers are quantum confined in one dimension in extremely thin active layers, exhibit improved performance compared to conventional hetero-structure laser diodes. Further improvement in laser performance is expected with the introduction of quantum confinement in more than one dimension [408]. The resulting quantum wire and quantum dot laser structures employ quasi-1D and 0D carriers, which are quantum confined to wire-like and dot-like wells of very small dimensions (in nanometer range). These devices are expected to have extremely low threshold currents in $\mu$A range [409, 410], higher modulation bandwidth and narrower spectral line-widths [411], and reduced temperature sensitivity [408] compared to their quantum-well counterpart. Laser diodes with these properties are useful in applications involving integration of a large number of densely packed laser arrays and monolithic integration of lasers with low-power electronics, such as optical interconnects, optical computing, and integrated opto-electronic circuits. In addition, recent development of quantum dots formed by colloidal synthesis makes quantum dots [412] very attractive in many biological and other optics applications such as biosensors, light-emitting diodes (LEDs) for potential white light, and tunable lasers.

Confinement of charge carriers in semiconductors within potential wells of sufficiently small dimensions can lead to significant modification of the energy band structure and the density of states distribution in these materials. [413, 414] The discrete energy levels arising from quantum confinement lead to modification of the density of states functions. The density of states distributions acquire sharper features as the carrier dimension is reduced, particularly in the case of 1D and 0D, as shown in Fig. 6.85.

Fabrication of quantum-wire and quantum-dot laser diodes pose major challenges to nanoprocessing and crystal growth technologies. The realization of these novel structures requires the preparation of uniform arrays of 2D and 3D wells with lateral dimensions in the few tens-of-nanometer range. The interfaces of these wells need to be defect-free in order to avoid nonradiative recombination effects that might reduce carrier lifetime and quantum efficiency and thus prevent lasing. Furthermore, the arrays of wells should be uniform in size and shape so that inhomogeneous broadening effects would not degrade the effects of quantum confinement. As the volume of the quantum-wire and quantum-dot wells is extremely small, a practical approach for these low-dimensional semiconductor lasers should incorporate high-density arrays of quantum wires or quantum dots, and employ low loss, tight optical confinement wave guides and cavities in order to compensate for the low modal gain.

Attempts have been made to fabricate single or multiple quantum-wire lasers by etching and regrowth [415–417], growth on vicinal substrates [418], growth on non-planar patterned substrates [419–422], and cleaved-edge-over-growth [423]. Further improvement in the fabrication techniques is required to achieve a better understanding of this low-dimensional semiconductor device.

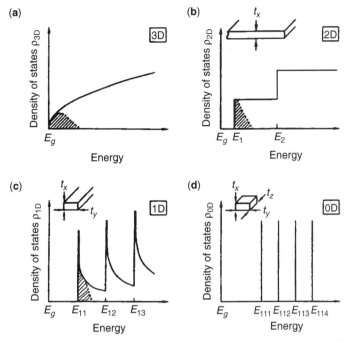

**Figure 6.85**  A schematic description of the density of states versus dimensionality. The insets illustrate rectangular potential well configurations of the corresponding quantum-confined structure: **(a)** bulk (3D), **(b)** quantum well (2D), **(c)** quantum wire (1D), and **(d)** quantum dot (0D). The crossed area indicates occupied states for sililar carrier densities.

The development of quantum-dot laser diodes has mainly focused in two areas:

(1) Self-organized (the Stransky–Krastanov process) quantum dots embedded in well-established quantum-well structures, both edge-emitting and vertical cavity lasers. Adding quantum dots to the conventional laser structures offers, potentially, a significant reduction in threshold current density, improved temperature performance, increased modulation rate, increased slope efficiency, broader gain spectrum, and less susceptibility to optical feedback [424–434]. Further research efforts are required to fully understand the characteristics of these lasers and to produce commercial viable devices.

(2) Quantum dot devices with nanometer size and length. It has been a desire to have ultra-small optical devices in the field of quantum communication and quantum computing based on semiconductor nanostructures. As an example, efficient and compact single photon emitters might be realized from single semiconductor quantum dots [435–438] if a way is found to isolate and efficiently inject current into a single quantum dot. Several approaches have been reported to realize these ultra-small devices, such as quantum-well interdiffusion [439], segmented quantum wells [440], and self-assembled quantum dots [441–443]. The results are still far from the desired single quantum-dot performance. A new approach has been reported recently to the realization of ultra-small light-emitting devices with active areas well below 1 $\mu m^2$ [444]. Using optical lithography, selective oxidation, and an active region consisting of InAs quantum dots, light-emitting

diodes (LEDs) were fabricated with light-current-voltage characteristics that scale well with nominal device area down to 600 nm diameter at room temperature. The scaling behavior provides evidence for strong carrier confinement in the quantum dots and shows the potential for the realization of high-efficiency single-photon LEDs operating at room temperature. Similar approaches may be able to realize single quantum-dot laser diodes.

Many issues must be resolved before quantum-dot–based devices can compete with existing technologies. Thus, exactly where quantum-dot devices will find their biggest commercial breakthrough remains to be seen.

## 6.10   CONCLUSION

We have discussed the present state of the art for semiconductor lasers. In this final section, we would like to take a short glimpse of what the future may bring.

With the advent of the material growth, processing, packaging, testing, and characterization technologies, semiconductor lasers with different size, shape, and performing characteristics are being realized. The lasing spectrum has been widened tremendously in the past few years. As long as there are applications, novel materials and sophisticated processing capabilities will be designed and developed. For example, lasers emitting at shorter wavelength are in demand for optical data storage because the storage density is inversely proportional to light wavelength squared. By using a blue laser at ~400 nm, the storage density is expected to increase up to a factor 4 compared to presently used red and infrared lasers. Sony has, recently, worked with Nichia to develop 405-nm (blue) diode lasers for optical video-recording applications [440]. This product features a capacity of 27 GB on a 12-cm single-layer disk, enabling users to record more than 13 hours of standard video and may be in the market as early as fall 2003. In addition to blue lasers mentioned above and the matured red lasers, green lasers are highly desirable for full color display in entertainment industries.

Lasers using indirect band-gap materials such as Si, Ge, and combinations of them [445] as well as II-VI compound semiconductor materials may someday become a reality. The use of pseudomorphic laser structures with QW and QD active layer lattice mismatch to the substrates will provide significant flexibility in tailoring the energy band structure for ultra-low-threshold currents and high-frequency operation. Furthermore, a monolithic integration of lasers with electro-optic wave guides and drivers may be used for controlling the phase of laser array structures, beam scanning, and frequency modulation. Additional components such as prisms, gratings, beam splitters, and lenses can also be integrated on the same substrate to reduce total package size.

High-power lasers are expected to deliver more power as material growth, processing, packaging, and thermal management technologies continue to improve. QD lasers, using dots-in-a well or the DWELL process, offer a significant reduction in threshold current density, improved temperature performance, reduced chirp that enables 10-GB/s direct modulation, increased slope efficiency, less susceptibility to optical feedback, and a broad gain spectrum. Taking advantage of these properties, QD semiconductor optical amplifiers may challenge erbium-doped fiber amplifiers in the near future [446].

The tunable lasers have achieved limited success during the past several years. Many issues need to be resolved for tunable lasers to really penetrate the market. Some of the key issues are wavelength tuning stability, repeatability and accessibility of the wavelength available in the tuning range, and the ease of handling. The tunable laser yield is extremely low at this moment, and the availability of a complete set of products is still problematic. The challenge for further development is to solve all of the issues mentioned here. More and continuous efforts are expected in the tunable laser's further development, including external cavity tunable (ECL) lasers and tunable VCSELs.

High-speed lasers for telecommunication and data transmission will always be needed. Currently, up to 10-GHz directly modulated lasers are available commercially. However, even higher speed, such as 40-GHz, will be needed in the near future due to the continuous increase in the bandwidth demand. Thus, the development of even higher speed lasers has been and will continue to be one of the intensive research areas for semiconductor lasers. Furthermore, direct modulated lasers and monolithic integration of opto-electronic devices that offer cost advantage over the external cavity counterpart, have also attracted a lot of attention. As mentioned in Section 6.7, there are still technical challenges to be overcome to achieve super high-speed operation. QWs and QDs with excellent carrier confinement structures are expected to play key roles in achieving the high differential gain, which is the fundamental factor in high-speed laser design.

Overall, 850-nm VCSELs have been developed and manufactured extensively for the past several years. Lots of fundamental issues have been studied and understood. This device has matured enough for large volume manufacturing. However, great challenges remain. The most significant challenge is to realize VCSELs' theoretical potential and the cost reduction for widespread commercial applications, such as printing, scaning, sensing, data storage, and medicals. Yield improvement on epitaxial layer growth and chip fabricationis is the key to reduce the cost and enable new cost-sensitive applications that in turn will increase manufacturing volume and reduce cost.

In addition to 850-nm VCSELs, long wavelength VCSELs of 1310 nm and 1550 nm will also be released for long-distance, local, and storage network applications. Shorter wavelength VCSELs (780–850 nm) may also find their positions in applications such as position translation and rotation, sensing, laser printing, atomic clocks, and scanning.

In terms of technology, single transverse mode, high speed ($\geqslant 10$ Gbps), and high power are the three major directions for VCSEL's future development. Once they are developed as mature products, VCSELs will find tremendous new applications. The future of VCSEL is indeed bright in many years ahead.

Measuring 1–100 nm across, QDs are semiconductor structures in which the electron wave function is confined in all three dimensions by the potential energy barriers that form the QD's boundaries [447]. QDs are attractive because their size, shape, and composition can all be tailored to create a variety of desired properties. These QDs can, in turn, be positioned and assembled into complexes that serve as new materials. Because of their size, they fall into the category of nanocrystals. In addition to the high-power QD lasers mentioned above for telecommunication application, ultra-small optical devices based on semiconductor nanostructures are also desired for quantum communication and quantum computing. Single QD lasers may serve the purpose of single photon emitters required in those applications.

## Acknowledgments

We would like to extend our thanks to many colleagues at TRW, Agilent, APT/Uniphase, AXT Optoelectronics, and Lytek who have either directly or indirectly contributed to this chapter through many discussions, publications, and internal reports. The authors gratefully acknowledge the contributions of D. Botez, D. Bullock, G. Evans, L. Figueroa, P. Hayashida, C. S. Hong, L. J. Mawst, S. S. Ou, G. Peterson, T. J. Roth, C. C. Shih, J. Wilcox, L. Zinkiewicz, and C. A. Zmudzinski of TRW, F. Fang of APT/Uniphase, K. L. Chen, Charlie Wang, Denny Houng, Chun Lei, Bob Weissman, Michael Tan, Hongyu Deng, Synyem Hu, Scott Corzine, Al Yuan and Steve Wang of Agilent, Frank Xiong, Zhihong Yang, Xiaobo Zhang, Wei Sun, Donghai Zhu, A. Ou, and Xiaoguang He of AXT Optoelectronics, S. Feld, Phil Dowd, Shane Johnson (also ASU) and Prof. YH Zhang (also ASU) of Lytek. Special thanks goes to Dr. Charlie Wang of Agilent and Drs. T. T. Yang and David H. Burde of Boeing for their careful review.

## REFERENCES

1. R. N. Hall, *IEEE Trans. Electron. Devices*, **ED-23**, p. 700, 1976.
2. I. Hayashi, M. B. Panish, and P. W. Foy, *IEEE J. Quantum Electron.*, **QE-5**, pp. 211–213, 1969.
3. Z. I. Alferov, V. M. Andreev, D. Z. Garbuzov, Y. V. Zhilyaev, E. D. Morozov, E. L. Portnoi, and V. G. Troffin, *Sov. Phys.–Semicond. (Engl. Transl.)*, **4**, p. 1573, 1971.
4. J. C. Dyment and L. A. D'Asaro, *Appl. Phys. Lett.*, **11**, p. 292, 1967.
5. I. Hayashi, M. B. Panish, P. W. Foy, and S. Sumski, *Appl. Phys. Lett.*, **17**, p. 109, 1970.
6. A. Y. Cho, *J. Vac. Sci. Technol.*, **16**, p. 275, 1979.
7. W. T. Tsang, *Appl. Phys. Lett.*, **34**, p. 473, 1979.
8. H. M. Manasevit and W. I. Simpson, *J. Electrochem. Soc.*, **116**, p. 1968, 1979.
9. R. D. Dupuis and P. D. Dapkus, *Appl. Phys. Lett.*, **32**, p. 406, 1978.
10. R. D. Dupuis, P. D. Daplus, R. Kolbas, and N. Holonyak, *IEEE J. Quantum Electron.*, **QE-15**, p. 756, 1979.
11. W. T. Tsang, R. A. Logan, and J. A. Ditzenberger, *Electron. Lett.*, **18**, p. 845, 1982.
12. W. T. Tsang, *IEEE J. Quantum Electron.*, **QE-20**(10), p. 1119, 1984.
13. R. L. Thornton, R. L. Burnham, T. L. Paoli, N. Holonyak, N. Jun, and D. G. Deppe, *Appl. Phys. Lett.*, **47**, p. 1239, 1985.
14. A. Kurobe, H. Furuyama, S. Maritsuka, Y. Kokobun, and M. Nakamura, *Electron. Lett.*, **22**, p. 1118, 1986.
15. H. Z. Chen, A. Ghaffari, H. Morkoc, and A. Yariv, *Electron. Lett.*, **23**, p. 1334, 1987.
16. H. K. Choi and C. A. Chen, *Appl. Phys. Lett.*, **57**, p. 321, 1990.
17. E. Kapon, S. Simhony, J. P. Harbison, L. T. Florez, and P. Worland, *Appl. Phys. Lett.*, **56**, p. 1825, 1990.
18. R. S. Geels and L. A. Coldren, *Appl. Phys. Lett.*, **57**, p. 1605, 1990.
19. D. R. Scifres, C. Lindstrom, R. D. Burnham, W. Streifer, and T. L. Paoli, *Appl. Phys., Lett.*, **19**, p. 160, 1983.
20. D. F. Welch, R. L. Thornton, R. D. Burnham, P. S. Cross, and T. L. Paoli, *Proc. Int. Semicond. Laser Conf.*, Kanazawa, Japan, October 1986, p. 16.
21. Y. Arakawa and A. Yariv, *IEEE J. Quantum Electron.*, **QE-22**(9), 1887, 1986.

22. R. Dingle, "Festkoerper probleme", *Adv. Solid State Phys.*, **15**, p. 21, 1975.

23. K. Uomi, M. Mishima, and N. Chinone, *Appl. Phys. Lett.*, **51**, p. 78, 1987.

24. T. Yuasa, T. Yamada, M. Uchida, K. Asakawa, and M. Ishii, *Proc. Int. Semicond. Laser Conf.*, Kanazawa, Japan, October 1986, p. 2.

25. P. L. Derry, T. R. Chen, Y. H. Zhuang, J. Paslaski, M. Mittlestein, K. Vahala, and A. Yariv, *Appl. Phys. Lett.*, **53**, p. 271, 1988.

26. L. A. Coldren and S. W. Corzine, *Diode Lasers and Photonic Integrated Circuits*, Wiley Series in Microwave and Optical Engineering, K. Chang, Ed., Wiley, New York, 1995.

27. H. Soda, K. Iga, C. Kitahara, and Y. Suematsu, *Japan. J. Appl. Phys.*, **18**, p. 2329, 1979.

28. D. Z. Garbuzov, I. N. Arsentyev, A. V. Ovchinnikov, and I. S. Tarasov, in *Tech. Dig., Conf. Lasers Electro-Opt.*, Opt. Amer., Washington, DC, 1988, p. 396, Paper THU44.

29. T. Ijichi, M. Phkubo, N. Matsumoto, and H. Okamoto, in *Conf. Dig., 12 IEEE Int. Semiconductor Laser Conf.*, Davos, Switzerland, 1990, p. 44, Paper D.2.

30. J. N. Warpole, S. H. Groves, Z. L. Liau, S. C. Palmateer, and D. Z. Tsang, in *Conf. Dig., 12 IEEE Int. Semiconductor Laser Conf.*, Davos, Switzerland, 1990, p. 152, Paper J.5.

31. S. L. Yellen, A. H. Shepard, C. M. Harding, J. A. Baumann, R. G. Waters, D. Z. Garbuzov, V. Pjataev, V. Kochergin, and P. S. Zory, *IEEE Photon. Techno. Lett.*, **4**, p. 1328, 1992.

32. Y. S. Sin, H. Horikawa, and T. Kamijoh, *Electron. Lett.*, **29**, p. 240, 1993.

33. G. Zhang, J. Nappi, A. Ovtchinnikov, and H. Asonen, *Electron. Lett.*, **29**, p. 429, 1993.

34. D. Z. Garbuzov, N. ju. Antonishkis, S. N. Zhigulin, N. D. Il'inskaya, A. V. Kochergin, D. A. Lifshitz, E. U. Rafailov, and M. V. Fuksman, *Appl. Phys. Lett.*, **62**, p. 1062, 1993.

35. M. Ohkubo, T. Ijichi, A. Iketani, and T. Kikuta, *IEEE J. Quantum Electron.*, **30**, p. 408, 1994.

36. M. Razeghi, *Nature*, **369**, p. 631, 1994.

37. J. Diaz, I. Eliashevich, X. Hem, H. Yi, L. Wang, E. Kolev, D. Garbuzov, and M. Razeghi, *Appl. Phys. Lett.*, **65**, p. 22, 1994.

38. S. Rusli, A. Al-Muhanna, T. Earles, and L. J. Mawst, *Electron. Lett.*, **36**, p. 630, 2000.

39. L. J. Mawst, S. Rusli, A. Al-Muhanna, and J. K. Wade, *IEEE J. Select. Topics Quantum Electron.*, **5**, p. 785, 1999.

40. L. J. Mawst, A. Bhattacharya, J. Lopez, D. Botez, D. Z. Garbuzov, L. DeMarco, J. C. Connolly, M. Jansen, F. Fang, and R. F. Nabiev, *Appl. Phys. Lett.*, **69**, p. 1532, 1996.

41. J. K. Wade, L. J. Mawst, D. Botez, M. Jansen, F. Fang, and R. F. Nabiev, *Appl. Phys. Lett.*, **70**, p. 149, 1997.

42. O. Wada and H. Hasegawa, *InP-based Materials and Devices, Physics and Technology*, Wiley Series in Microwave and Optical Engineering, K. Chang, Ed., Wiley, New York, 1999.

43. M. Kondow, T. Kitatani, S. Nakatsuka, M. C. Larson, K. Nakahara, Y. Yazawa, M. Okai, and K. Uomi, *IEEE J. Select. Topic Quantum Electron.*, **3**, p. 719, 1997.

44. C. S. Peng, T. Jouhti, P. Laukkanen, E. M. Pavelescu, J. Konrrinen, W. Li, and M. Pessa, *IEEE Photon. Technol. Lett.*, **14**, p. 275, 2002.

45. N. Tansu and L. J. Mawst, *IEEE Photon. Technol. Lett.*, **14**, pp. 444–1052, 2002.

46. W. Ha, V. Gambin, M. Wistey, S. Bank, S. Kim, and J. S. Harris, Jr., *IEEE Photon. Tech. Lett.*, **14**, p. 591, 2002.

47. E. Gouardes, T. Miyamoto, M. Kawaguchi, K. Kondo, F. Koyama, and K. Iga, *IEEE Photon. Tech. Lett.*, **14**, p. 896, 2002.

48. N. Tansu, Y. L. Chang, T. Takeuchi, D. P. Bour, S. W. Corzine, M. R. T. Tan, and L. J. Mawst, *IEEE J. Quantum Electron.*, **38**, p. 640, 2002.

49. N. Tansu, N. J. Kirsch, and L. J. Mawst, *Appl. Phys. Lett.*, **81**, p. 2523, 2002.

50. J. M. Olson, R. K. Ahrenkiel, D. J. Dunlavy, B. Keyes, and A. E. Kibber, *Appl. Phys. Lett.*, **55**, p. 1208, 1990.

51. D. F. Welch, T. Wang, and D. R. Scifres, *Electron. Lett.*, **27**, p. 693, 1991.

52. G. Hatakoshi, K. Itaya, M. Ishikawa, M. Okajima, and Y. Uematsu, *IEEE J. Quantum Electron.*, **27**, p. 1476, 1991.

53. A. Valster, C. J. Van der Poel, M. N. Finke, and M. J. B. Boermans, *Electron. Lett.*, **28**, p. 144, 1992.

54. D. P. Bour, D. W. Treat, R. L. Thornton, T. L. Paoli, R. D. Gringans, B. S. Krusor, R. S. Geels, D. W. Welch, and T. Y. Wang, *Appl. Phys. Lett.*, **60**, p. 1927, 1992.

55. S. S. Ou, J. J. Yang, R. J. Fu, and C. J. Hwang, *Appl. Phys. Lett.*, **61**, p. 892, 1992.

56. T. Tanaka, H. Yanagisawa, S. Yano, and S. Minagawa, *Electron. Lett.*, **29**, p. 606, 1993.

57. S. S. Ou, M. Jansen, J. J. Yang, R. J. Fu, and C. J. Hwang, *Electron. Lett.*, **29**, p. 233, 1993.

58. S. S. Ou, J. J. Yang, and M. Jansen, *Electron. Lett.*, **30**, p. 1303, 1994.

59. D. P. Bour, D. W. Treat, R. D. Bringans, R. S. Geels, and D. F. Welch, *Proc. InP Related Materials Conf.* (IPRM'94), Santa Barbara, CA, p. 255, 1994.

60. T. Tanaka, H. Yanagisawa, S. Kawanaka, and S. Minagawa, *IEEE Photon. Technol. Lett.*, **7**, p. 136, 1995.

61. N. M. Johnson, A. V. Nurmikko, and S. P. Denbaars, *Physics Today*, American Institute of Physics, p. 31, Oct. 2000.

62. S. Nakamura and G. Fasol, *The Blue Laser Diode, GaN Based Light Emitters and Lasers*, Springer, New York, 1997.

63. M. B. Hayashi, P. Panish, O. Foy, and S. Sumski, *Appl. Phys. Lett.*, **17**, p. 109, 1970.

64. M. B. Panish, H. Temkin, and S. Sumski, *J. Vac. Sci. Technol.*, **B3**, p. 657, 1985.

65. W. T. Tsang, in *Beam Processing Technologies*, N. G. Einspruch, S. S. Cohen, and R. N. Singh, Eds., Academic Press, New York, 1989.

66. W. T. Tsang, *The Technology and Physics of Molecular Beam Epitaxy*, E. H. C. Parder, Ed., Plenum Press, New York, 1985, p. 467.

67. C. Henry, in *Semiconductor and Semimetals*, W. T. Tsang, Ed., Vol. 22, Part B, Academic Press, New York, 1985.

68. P. S. Zory, A. R. Reisinger, R. G. Waters, L. J. Mawst, C. A. Zmudzinski, M. A. Emanuel, M. E. Givens, and J. J. Coleman, *Appl. Phys. Lett.*, **49**, p. 16, 1986.

69. P. S. Zory, A. R. Reisinger, L. J. Mawst, G. Costrini, C. A. Zmudzinski, M. A. Emanuel, M. E. Givens, and J. J. Coleman, *Electron. Lett.*, **22**, p. 476, 1986.

70. R. Reisinger, P. S. Zory, and R. G. Waters, *IEEE J. Quantum Electron.*, **23**, p. 993, 1987.

71. J. Z. Wilcox, G. L. Peterson, S. Ou, J. J. Yang, M. Jansen, and D. Schechter, *Electron. Lett.*, **24**, p. 1218, 1988.

72. S. S. Ou, J. J. Yang, J. Z. Wilcox, and M. Jansen, *Electron. Lett.*, **24**, p. 952, 1988.

73. T. Yuasa, T. Yamada, K. Asakawa, and M. Ishii, *J. Appl. Phys.*, **63**, p. 1321, 1988.

74. P. Blood, E. D. Fletcher, P. J. Hulyer, and P. M. Smowton, *Appl. Phys. Lett.*, **48**, p. 1111, 1986.

75. J. R. Shealy, *Appl. Phys. Lett.*, **50**, p. 1634, 1987.

76. K. Y. Lau, P. L. Ferry, and A. Yariv, *Appl. Phys. Lett.*, **52**, p. 88, 1988.

77. A. Kurobe, H. Furuyama, S. Naritsuka, N. Sugiyama, Y. Kokubun, and M. Nakamura, *IEEE J. Quantum Electron.*, **24**, p. 635, 1988.

78. T. Hayakawa, T. Suyama, K. Takahashi, M. Kondo, S. Yamamoto, and T. Hijikata, *Appl. Phys. Lett.*, **52**, p. 339, 1988.

79. R. G. Waters and R. K. Bertaska, *Appl. Phys. Lett.*, **52**, p. 179, 1988.

80. J. Z. Wilcox, G. L. Peterson, S. S. Ou, J. J. Yang, M. Jansen, and D. Schechter, *Appl. Phys. Lett.*, **53**, p. 2272, 1988.

81. J. Z. Wilcox, G. L. Peterson, S. S. Ou, J. J. Yang, M. Jansen, and D. Schechter, *J. Appl. Phys.*, **64**, p. 6564, 1988.

82. D. K. Wagner, R. G. Waters, P. L. Tihanyi, D. S. Hill, A. J. Roza, Jr., H. J. Vollmer, and M. M. Leopold, *IEEE J. Quantum Electron.*, **24**, p. 1258, 1988.

83. G. R. Hadley, J. P. Hohimer, and A. Owyoung, *IEEE J. Quantum Electron.*, **24**, p. 2138, 1988.

84. S. R. Chinn, P. S. Zory, and A. R. Reisinger, *IEEE J. Quantum Electron.*, **24**, p. 2191, 1988.

85. A. Behfar-Rad, J. R. Shealy, S. R. Chinn, and S. S. Wong, *IEEE J. Quantum Electron.*, **26**, p. 1476, 1990.

86. H. Jung, E. Schlosser, and R. Deufel, *Appl. Phys. Lett.*, **60**, p. 401, 1992.

87. M. Osinsky and J. Buus, *IEEE J. Quantum Electron.*, **QE-23**(1), p. 9, 1987.

88. S. Lee, L. Figueroa, and R. Ramaswamy, *IEEE J. Quant. Elect.*, **QE-25**(5), p. 862, 1989.

89. R. Fisher, D. Neuman, H. Zabel, H. Morkoc, C. Choi, and N. Otsaka, *Appl. Phys. Lett.*, **48**, p. 1223, 1986.

90. A. Larsson, S. Forouhar, J. Cody, and R. J. Lang, *IEEE Photon. Technol. Lett.*, **2**, p. 307, 1990.

91. J. S. Major, W. E. Plano, D. F. Welch, and D. Scifres, *Electron.*, **27**, p. 540, 1991.

92. S. S. Ou, J. J. Yang, M. Jansen, C. Hess, M. Sergant, C. Tu, F. Alvarez, and L. J. Lembo, *Electron. Lett.*, **28**, p. 2345, 1992.

93. S. S. Ou, J. J. Yang, C. Hess, and M. Jansen, *Electron. Lett.*, **29**, p. 542, 1993.

94. A. Wang, H. K. Choi, J. N. Walpole, G. A. Evans, W. F. Reichert, W. W. Chow, and C. T. Fuller, *Electron. Lett.*, **30**, p. 646, 1994.

95. M. Fukuda, M. Okayasy, J. Temmyo, and J. Nakano, *IEEE J. Quantum Electron.*, **30**, p. 471, 1994.

96. C. H. Henry, R. A. Logan, and K. A. Bertness, *J. Appl. Phys.*, **51**, p. 4457, 1981.

97. B. W. Hakki, *J. Appl. Phys.*, **44**, p. 5021, 1973; *J. Appl. Phys.*, **46**, p. 292, 1975.

98. D. Cook and F. R. Nash, *J. Appl. Phys.*, **46**, p. 1660, 1975.

99. D. Botez, *IEEE J. Quantum Electron.*, **QE-17**, p. 2290, 1981.

100. D. Botez, *IEEE J. Quantum Electron.*, **QE-17**, p. 230, 1978.

101. L. Figueroa and S. Wang, *Appl. Phys. Lett.*, **32**, p. 85, 1978.

102. I. Kaminow, L. Stulz, J. Ko, A. Dentai, R. Nahory, J. DeWinter, and R. Hartman, *IEEE Quantum Electron.*, **QE-19**, p. 1313, 1983.

103. H. Kogelnik, *IEEE Trans. Microwave Theory Tech.*, **MTT-23**, p. 1, 1975.

104. R. V. Ramaswamy, *Bell Syst. Tech. J.*, **53**, p. 113, 1977.

105. W. Striefer and E. Kapon, *Appl. Opt.*, **18**, p. 3724, 1979.

106. S. Wang, in *Semiconductors and Semimetals*, W. T. Tsang, Ed., Vol. 22, Part E, Academic Press, New York, 1985, p. 116.

107. S. Wang, C. Y. Chen, A. S. Liao, and L. Figueroa, *IEEE J. Quantum Electron.*, **QE-17**, p. 453, 1981.

108. R. Smith and G. Mitchell, EE Tech. Rep. 206, Dept. of Electronics, University of Washington, Seattle, December 1977.

109. L. Figueroa, T. Holcomb, K. Burghard, D. Bullock, R. Wagner, C. Morrison, and L. Zinkiewicz, *IEEE J. Quantum Electron.*, **QE-22**(11), p. 2141, 1986.

110. K. Aiki, N. Nakamura, T. Kuroda, and J. Umeda, *Appl. Phys. Lett.*, **30**, p. 649, 1977.

111. W. T. Tsang, *J. Appl. Phys.*, **49**, p. 1031, 1978.

112. M. Fukuda, *Optical Semiconductor Devices*, Wiley Series in Microwave and Optical Engineering, K. Chang, Ed., Wiley, New York, 1999.

113. H. C. Casey, Jr. and M. B. Panish, *Heterostructure Semiconductor Lasers*, Part B, Academic Press, New York, 1978, p. 231.

114. D. Ackley, *Electron. Lett.*, **20**, p. 509, 1984.

115. K. Fujiwara, T. Fujiwara, K. Hori, and M. Takusagawa, *Appl. Phys. Lett.*, **34**, p. 668, 1979.

116. D. Ackley, *Electron. Lett.*, **20**, p. 509, 1984.

117. J. J. Yang, C. S. Hong, J. Niesen, and L. Figueroa, *Electron. Lett.*, **21**, p. 751, 1985.

118. W. P. Risk, W. J. Kozlovsky, S. D. Lau, G. L. Bona, H. Jaeckel, and J. Webb, *Appl. Phys. Lett.*, **63**, p. 3134, 1993.

119. O. Imafuji, T. Takayama, H. Sugiura, M. Yuri, H. Naito, M. Kume, and K. Itoh, *13th IEEE Int. Semiconductor Laser Conf.*, PD-13, 1992.

120. O. Imafuji, T. Takayama, H. Sugiura, M. Yuri, H. Naito, M. Kume, and K. Itoh, *IEEE J. Quantum Electron.*, **29**, p. 1889, 1993.

121. D. R. Scifres, R. D. Burnham, and W. Streifer, *Appl. Phys. Lett.*, **33**, p. 1015, 1978.

122. D. E. Ackley and R. H. Engelmann, *Appl. Phys. Lett.*, **39**, p. 27, 1981.

123. D. Scifres, R. D. Burnham, W. Streifer, and M. Bernstein, *Appl. Phys. Lett.*, **41**, p. 614, 1982.

124. D. Botez and J. C. Connally, *Appl. Phys. Lett.*, **43**, p. 1097, 1983.

125. F. Kappeler, H. Westmeier, R. Gessner, M. Druminski, and K. H. Zschauer, *Proc. IEEE Int. Semicond. Laser Conf*. Rio de Janeiro, 1984, p. 90.

126. J. P. Van der Ziel, H. Temkin, and R. D. Dupuis, *Proc. IEEE Int. Semicond. Laser Conf.* Rio de Janeiro, 1984, p. 92.

127. D. F. Welch, D. Scifres, P. Cross, H. Kung, W. Streiffer, R. D. Burnham, and J. Yaeli, *Conf. Lasers Electron.*, Baltimore, 1985, ThZ3-2.

128. N. Dutta, L. A. Kozzi, S. G. Napholtz, B. P. Seger, *Proc. Conf. Lasers Electro-opt.* Baltimore, 1985, p. 44.

129. M. Taneya, M. Matsumoto, S. Matsui, Y. Yano, and T. Hijikata, *Appl. Phys. Lett.*, **47**, p. 341, 1985.

130. M. Taneya, M. Matsumoto, H. Kawanishi, S. Matsui, S. Yano, and T. Hijikata, *Proc. Conf. Lasers Electro-opt. (CLEO)*, San Francisco, 1986, Tu M5.

131. J. Ohsawa, S. Himota, T. Aoyagi, T. Kadowaki, N. Kaneno, K. Ikeda, and W. Susaki, *Electron. Lett.*, **21**, p. 779, 1985.

132. D. F. Welch, P. S. Cross, D. R. Scifres, W. Streifer, and R. D. Burnham, *Proc. Conf. Lasers Electro-opt. (CLEO)*, San Francisco, 1986, p. 66.

133. D. F. Welch, R. L. Thornton, R. D. Burnham, P. S. Cross, and T. L. Paoli, *Proc. Int. Semicond. Laser Conf.*, Kanazawa, Japan, October 1986.

134. C. Morrison, L. Zinkiewicz, A. Burghard, and L. Figueroa, *Electron. Lett.*, **21**, p. 337, 1985.

135. Y. Twu, A. Dienes, S. Wang, and J. R. Whinnery, *Appl. Phys. Lett.*, **45**, p. 709, 1984.

136. S. Mukai, C. Lindsey, J. Katz, E. Kapon, Z. Rav-Noy, S. Margalit, and A. Yariv, *Appl. Lett.*, **45**, p. 834, 1984.

137. D. E. Aceley, *Appl. Phys. Lett.*, **41**, p. 118, 1982.

138. T. Kuroda, M. Nakamura, K. Aiki, and J. Umeda, *Appl. Opt.*, **17**, p. 3264, 1978.

139. S. Lee and L. Figueroa, *Proc. Top. Meet. Semicond. Lasers*, Albuquerque, NM, February 1987.

140. S. J. Lee, L. Figueroa, and R. Ramaswamy, *IEEE J. Quant. Electron.*, **QE-25**(7), p. 1632, 1989.

141. J. K. Butler, D. E. Ackley, and D. Botez, *Appl. Phys. Lett.*, **44**, p. 293, 1984.

142. E. Kapon, J. Katz, and A. Yariv, *Opt. Lett.*, **10**, p. 125, 1984.

143. L. Figueroa, C. Morrison, H. D. Law, and F. Goodwin, *Proc. Int. Electron Devices Meet.*, 1983, p. 760.

144. L. Figueroa, C. Morrison, H. D. Law, and F. Goodwin, *J. Appl. Phys.*, **56**, p. 3357, 1984.

145. W. Streifer, A. Hardy, R. D. Burnham, and D. R. Scifres, *Electron. Lett.*, **21**, p. 118, 1985.

146. L. Figueroa, T. Holcomb, K. Burghard, D. Bullock, K. Wagner, C. Morrison, L. Zinkiewicz, and G. A. Evans, *IEEE J. Quantum Electron.*, **QE-22**(11), p. 2141, 1986.

147. Y. Twu, K. L. Chen, A. Dienes, S. Wang, and J. R. Whinnery, *Electron. Lett.*, **21**, p. 325, 1985.

148. J. Katz, E. Kappon, C. Lindsey, S. Margalit, U. Shreter, and A. Yariv, *Appl. Phys. Lett.*, **42**, p. 521, 1983.

149. D. Botez, *IEEE J. Quantum Electron.*, **QE-21**, p. 1752, 1985.

150. T. L. Paoli, W. Streifer, and R. D. Burnham, *Proc. IEEE Int. Semicond. Laser Conf.*, Rio de Janerio, 1984, p. 86.

151. K. L. Chen and S. Wang, *IEEE J. Quantum Electron.*, **QE-21**, p. 264, 1985.

152. K. L. Chen and S. Wang, *Elect. Lett.*, **21**, p. 347, 1985.

153. G. P. Agrawal, *IEEE J. Lightwave Technol.*, **LT-2**, p. 537, 1984.

154. J. Katz, E. Kapon, C. Lindsey, S. Margalit, U. Shreter, and A. Yariv, *Appl. Phys. Lett.*, **42**, p. 521, 1983.

155. D. Ackley, *Electron. Lett.*, **20**, p. 695, 1984.

156. D. Ackley, J. K. Butler, and M. Ettenberg, *IEEE J. Quantum Electron.*, **QE-22**, p. 2204, 1986.

157. E. Kapon, C. P. Lindsey, J. S. Smith, S. Margalit, and A. Yariv, *Appl. Phys. Lett.*, **45**, p. 1257, 1984.

158. D. Bullock, R. Wagner, T. Holcomb, and L. M. Frantz, *Proc. Int. Lasers Conf.*, 1985.

159. E. Marom, O. G. Ramer, and S. Ruschim, *IEEE J. Quantum Electron.*, **QE-20**, p. 1311, 1984.

160. T. R. Ranganath and S. Wang, *IEEE J. Quantum Electron.*, **QE-13**, p. 290, 1977.

161. D. Botez and J. J. Yang, *Quest, Technology Report at TRW*, Summer 1988 and Winter 1991.

162. D. Botez and G. Peterson, *Electron. Lett.*, **24**, p. 1044, 1988.

163. L. J. Mawst, D. Botez, T. J. Roth, G. Peterson, and J. J. Yang, *Electron. Lett.*, **24**, p. 958, 1988.

164. D. Botez, L. Mawst, P. Hayashida, G. Peterson, and T. J. Roth, *Appl. Phys. Lett.*, **53**, p. 464, 1988.

165. L. J. Mawst, D. Botez, and T. J. Roth, *Appl. Phys. Lett.*, **53**, p. 1236, 1988.

166. D. Botez et al., *IEEE Photonics Technol. Lett.*, **2**, p. 249, 1990.

167. J. Saltzman, T. Venketesen, R. Lang, M. Fittelstein, and A. Yariv, *Appl. Phys. Lett.*, **46**, p. 218, 1985.

168. R. Craig, L. W. Casperson, O. M. Stafsud, J. J. Yang, G. A. Evans, and R. A. Davidheiser, *Electron. Lett.*, **21**, p. 62, 1985.

169. J. Saltzman, R. Lang, M. Fittelstein, and A. Yariv, *Proc. Conf. Lasers Electro-opt. (CLEO)*, Baltimore, NW, 1985, p. 40.

170. A. E. Siegman, *Lasers*, University Science Books, Mill Valley, CA, 1986.

171. M. Nakamura, *IEEE Trans. Circuits Syst.*, **CS-26**, p. 1055, 1979.

172. K. Hamada, M. Wada, H. Shimuzu, M. Kume, A. Yoshikawa, F. Tajiri, K. Itoh, and G. Kano, *Proc. IEEE Int. Semicond. Laser Conf.*, Rio de Janeiro, 1984, p. 34.

173. F. Stern, *J. Appl. Phys.*, **47**, p. 5382, 1976.

174. J. Walpole, T. A. Lind, J. J. Hsieh, and J. P. Donnelly, **17**, p. 186, 1981.

175. C. H. Henry, P. M. Petroff, R. A. Logan, and F. R. Merritt, *J. Appl. Phys.*, **50**, p. 3721, 1979.

176. A. Moser, *Appl. Phys. Lett.*, **59**, p. 522, 1991.

177. P. W. Epperlein, P. Buchmann, and A. Jakubowicz, *Appl. Phys. Lett.*, **62**, p. 455, 1993.

178. R. W. H. Engelmann and D. Kerps, *Proc. IEEE Int. Semicondu. Laser Conf.*, Ottawa–Hull, Canada, 1982, p. 26.

179. H. Yonezu, I. Sakuma, T. Kamejima, M. Ueno, K. Iwamoto, I. Hino, and I. Hayashi, *Appl. Phys. Lett.*, **34**, p. 637, 1979.

180. L. Figueroa and S. Wang, *Appl. Phys. Lett.*, **32**, p. 85, 1978.

181. G. H. B. Thompson, *Physics of Semiconductor Laser Devices*, Wiley, New York, 1980, Chap. 5.

182. D. Botez, J. C. Connolly, M. Ettenberg, and D. B. Gilbert, *Electron. Lett.*, **19**, p. 882, 1983.

183. D. Botez, *RCA Rev.*, **39**, p. 577, 1978.

184. I. Ladany, M. Ettenberg, F. Lockwood, and H. Kressel, *Appl. Phys. Lett.*, **30**, p. 87, 1976.

185. H. Namazaki, S. Takamija, M. Ishii, and W. Susaki, *J. Appl. Phys.*, **50**, p. 3743, 1978.

186. K. Mitsuishi, N. Chinone, H. Sata, and K. Aiki, *IEEE J. Quantum Electron.*, **QE-16**, p. 728, 1980.

187. M. Ettenberg, H. S. Sommers, H. Kressel, and H. D. Lockwood, *Appl. Phys. Lett.*, **18**, p. 571, 1971.

188. H. C. Casey, Jr and M. B. Panish, *Heterostructure Semiconductor Lasers*, Part B, Academic Press, New York, 1978, p. 282.

189. *Optronic* (Japanese Publications), **2**, p. 34, 1983.

190. H. Yonezu, M. Ueno, T. Kamejima, and I. Hayashi, *IEEE J. Quantum Electron.*, **QE-15**, p. 775, 1979.

191. H. Kumabe, T. Tanuka, S. Nita, Y. Seiwa, T. Sogo, and S. Takamiya, *Jpn. J. Appl. Phys.*, **21**, p. 775, 1982.

192. H. Blauvelt, S. Margalit, and A. Yariv, *Appl. Phys. Lett.*, **40**, p. 1029, 1982.

193. D. Botez and J. C. Connally, *Proc. IEEE Int. Semicond. Laser Conf.*, Rio de Janeiro, 1984, p. 36.

194. Y. Yamamato, N. Miyauchi, S. Muci, T. Morimoto, O. Yammamoto, S. Yano, and T. Hijikata, *Appl. Phys. Lett.*, **46**, p. 319, 1985.

195. T. Shibutami, M. Kume, K. Hamada, H. Shimizu, K. Itoh, G. Kano, and I. Teramoto, *Proc. Semicond. Laser Conf.*, Kanazawa, Japan, October 1986.

196. J. Ungar, N. Bar-Chaim, and I. Ury, *Electron. Lett.*, **22**, p. 210, 1986.

197. W. D. Laidig, N. Holonyak, M. D. Camras, K. Hess, J. J. Coleman, P. Dapkus, and J. Bardeen, *Appl. Phys. Lett.*, **38**, p. 776, 1981.

198. K. Meehan, J. M. Brown, N. Holonyak, R. D. Burnham, T. L. Paoli, and W. Streifer, *Appl. Phys. Lett.*, **44**, p. 700, 1984.

199. Y. Suzuki, Y. Horikoshi, M. Kobayashi, and H. Okamoto, *Electron. Lett.*, **20**, p. 384, 1984.

200. R. L. Thornton, R. D. Burnham, T. L. Paoli, N. Holonyak, and D. G. Deppe, *Appl. Phys. Lett.*, **48**(1), p. 7, 1986.

201. R. D. Burnham, R. L. Thornton, T. L. Paoli, and N. Holonyak, *Proc. Conf. Lasers Electro-opt. (CLEO)*, San Francisco, 1986, p. 370.

202. H. Matsubara, K. Isshiki, H. Kumabe, H. Namazaki, and W. Susaki, *Proc. Conf. Lasers Electro-opt. (CLEO)*, Baltimore, 1985, p. 180.

203. H. Yonezu, T. Yauasa, T. Shinohara, T. Kamejino, and I. Sakuma, *Jpn. J. Appl. Phys.*, **15**, p. 2393, 1976.

204. K. Mitsuishi, *J. Appl. Phys.*, **55**, p. 289, 1984.

205. D. P. Wilt, T. Long, W. C. Smith, M. W. Focht, T. M. Sher, and R. L. Hartman, *Electron. Lett.*, **22**, p. 8069, 1986.

206. L. Figueroa, C. W. Slayman, and H. W. Yen, *IEEE J. Quantum Electron.*, **QE-18**, p. 1718, 1982.

207. A. Haug, *IEEE J. Quantum Electron.*, **QE-21**, p. 716, 1985.

208. E. Rezek, D. Tran, N. Adachi, and L. Yow, *Proc. Top. Meet. Semicond. Lasers*, Albuquerque, NM, February 1987.

209. M. Kawahara, S. Oshiga, A. Matoba, Y. Kawai, and Y. Tamura, *Proc. Opt. Fiber Conf.*, January 1987, Paper ME1.

210. M. Yamaguchi, H. Nishimoto, M. Kitamura, S. Namazaki, I. Moto, and K. Kobayashi, *Proc. Conf. Lasers Electro-opt. (CLEO)*, 1985, p. 180.

211. J. A. Baumann, A. H. Shepard, R. G. Waters, S. L. Yellen, C. M. Harding, and H. B. Serreze, *Proceedings SPIE 1991*, Los Angeles, CA, p. 1418.

212. M. Fukuda, M. Okayasu, J. Temmyo, and J. Nakano, *IEEE J. Quantum Electron.*, **30**, p. 471, 1994.

213. J. Haden, J. Endriz, M. Sakamoto, D. Dawson, G. Browder, K. Anderson, D. Mundinger, P. Worland, E. Wolak, and D. Scifres, *Proceedings SPIE 1995*, San Jose, CA.

214. S. P. Sim, M. J. Robertson, and R. G. Plumb, *J. Appl. Phys.*, **55**, p. 3950, 1984.

215. Introduction to Reliability of Laser Diodes and Modules, Bellcore Publication SR-TSY-001368, 1989.

216. J. N. Walpole, E. S. Kintzer, S. R. Chinn, C. A. Wang, and L. J. Missaggia, *Appl. Phys. Lett.*, **61**, p. 740, 1992.

217. E. S. Kintzer et al., *IEEE Photonics Technol. Lett.*, **5**, p. 605, 1993.

218. D. Mehuys, D. F. Welch, and L. Goldberg, *Electron. Lett.*, **28**, p. 1944, 1992.

219. S. O'Brien, D. F. Welch, R. A. Parke, D. Mehuys, K. Dzurko, R. J. Lang, R. Waarts, and D. Scifres, *IEEE J. Quantum Electron.*, **29**, p. 2052, 1993.

220. R. J. Lang, A. Hardy, R. Parke, D. Mehuys, S. O'Brien, J., Major, and D. Welch, *IEEE J. Quantum Electron.*, **29**, p. 2044, 1993.

221. P. S. Yeh, I. F. Wu, S. Jiang, and M. Dagerais, *Electron. Lett.*, **29**, p. 1981, 1993.

222. M. Dagenais, S. H. Cho, S. A. Merrit, P. J. S. Helm, S. Fox, R. Prakasam, B. Gopalan, S. Kareenahalli, V. Vusirikala, and J. J. Yang, *Air Force Diode Laser Technol. Conf.*, Walton, FL, April 1995.

223. R. Parke, D. F. Welch, A. Hardy, R. Lang, D. Mehuys, St. O'Brien, K. Dzurko, and D. Scifres, *IEEE Photonics Technol. Lett.*, **5**, p. 297, 1993.

224. S. R. Selmic, G. A. Evans, T. M. Chou, J. B. Kirk, J. N. Walpole, J. P. Donnelly, C. T. Harris, and L. J. Missaggia, *IEEE Photonics Technol. Lett.*, **14**, p. 890, 2002.

225. K. Iga, S. Ishikawa, S. Ohkoochi, and T. Nishimura, *Appl. Phys. Lett.*, **45**, p. 348, 1984.

226. J. N. Walpole and Z. L. Liau, *Proc. Conf. Laser Electro-opt.* (*CLEO*), San Francisco, 1986, Paper Tu BZ.

227. J. J. Yang, M. Jansen, and M. Sargent, *Electron. Lett.*, **22**, p. 439, 1986.

228. R. D. Burham, D. R. Scifres, and W. Streifer, *IEEE J. Quantum Electron.*, **QE-11**, p. 439, 1975.

229. P. Zory and L. D. Comerford, *IEEE J. Quantum Electron.*, **QE-11**, p. 431, 1975.

230. G. A. Evans, J. M. Hammer, N. W. Carlson, F. R. Elia, E. A. James, and J. B. Kirk, *Appl. Phys. Lett.*, **49**, p. 314, 1986.

231. F. Rosen, W. Stabile, J. C. Janton, A. McShea, J. C. Rosenberg, H. G. Petheram, J. W. Miller, Sprague, and J. M. Gilman, *IEEE Photonics Technol. Lett.*, **1**, p. 43, 1989.

232. G. L. Harnagel, M. Vakili, K. R. Anderson, D. P. Worland, J. G. Endriz, and D. R. Scifres, *Electron. Lett.*, **28**, p. 1702, 1992.

233. G. L. Harnagel, M. Vakili, D. R. Anderson, D. P. Worland, J. G. Endriz, and D. R. Scifres, *Electron. Lett.*, **29**, p. 1008, 1993.

234. J. Haden, J. Endriz, M. Sakamoto, D. Dawson, G. Brewder, K. Anderson, D. Mundinger, P. Worland, E. Wolak, and D. Scifres, *OE-Lase'95, SPIE*, San Jose, CA, 1995.

235. K. Kojima, S. Noda, K. Mitsunaga, K. Kyuma, and K. Hamanaka, *Appl. Phys. Lett.*, **50**, p. 1705, 1987.

236. K. Mitsunaga, M. Kameya, K. Kojima, S. Noda, K. Kyuma, K. Hamanaka, and T. Nakayama, *Appl. Phys. Lett.*, **50**, p. 1788, 1987.

237. Z. L. Liau and J. N. Warpole, *Appl. Phys. Lett.*, **46**, p. 115, 1985.

238. Z. L. Liau and J. N. Warpole, *Appl. Phys. Lett.*, **50**, p. 528, 1987.

239. J. J. Yang, M. Jansen, and M. Sergant, *Electron. Lett.*, **22**, p. 438, 1986.

240. J. J. Yang, M. Sergant, M. Jansen, S. S. Ou, L. Eaton, and W. W. Simmons, *Appl. Phys. Lett.*, **49**, p. 1138, 1986.

241. T. H. Windhorn and W. D. Goodhue, *Appl. Phys. Lett.*, **48**, p. 1675, 1986.

242. J. P. Donnelly, W. D. Goodhue, T. H. Windhorn, R. J. Bailey, and S. A. Lambert, *Appl. Phys. Lett.*, **51**, p. 1138, 1987.

243. J. J. Yang, L. Lee, M. Jansen, M. Sergant, S. Ou, and J. Wilcox, *Proceedings of the SPIE OE-Lase'88*, Vol. 893, Paper 893-18.

244. K. Kobayashi and I. Mito, *IEEE J. Lightwave Technol.*, **6**, pp. 1623–1633, 1988.

245. T. L. Koch and U. Koren, *IEEE J. Lightwave Technol.*, **8**, pp. 274–279, 1990.

246. K. C. Reichmann, P. D. Magill, U. Koren, B. I. Miler, M. Young, M. Newkirk, and M. D. Chien, *IEEE Photon. Technol. Lett.*, **5**, p. 1098, 1993.

247. E. M. Strzelecki, D. A. Cohen, and L. A. Coldren, *IEEE J. Lightwave Technol.*, **6**, pp. 1610–1618, 1988.

248. A. Ebber and R. Noe, *Electron. Lett.*, **26**, p. 2009, 1990.

249. I. Schneider, G. Nau, T. V. V. King, and I. Aggarwal, *IEEE Photon. Technol. Lett.*, **7**, p. 87, 1995.

250. C. W. Lee, E. T. Peng, and C. B. Su, *IEEE Photon. Technol. Lett.*, **7**, p. 664, 1995.

251. M.-C. Amann and J. Buus, *Tunable Laser Diodes*, Artech House, Inc., New York, 1998, p. 106.

252. B. R. Bennett, R. A. Soref, and J. A. Del Alamo, *IEEE J. Quantum Electron.*, **26**, p. 113, 1990.

253. N. Susa and T. Nakahara, *Appl. Phys. Lett.*, **60**, p. 2457, 1992.

254. K. Y. Lau, *Appl. Phys. Lett.*, **57**, p. 2632, 1990.

255. L. D. Westbrook, *IEE Proc., Pt. J*, **133**, p. 135, 1986.

256. R. A. Soref and J. P. Lorenzo, *IEEE J. Quantum Electron.*, **22**, p. 873, 1986.

257. M.-C. Amann and J. Buus, *Tunable Laser Diodes*, Artech House, Inc., New York, p. 87, 1998.

258. J. P. Weber, *IEEE J. Quantum Electron.*, **30**, p. 1801, 1994.

259. M.-C. Amann, in *Semiconductor Lasers II*, E. Kapon, Ed., Academic Press, New York, 1999, p. 196.

260. G. N. Childs, S. Brans, and R. A. Adams, *Semicon. Sci. Technol.*, **1**, p. 116, 1986.

261. M. Asada, A. Kameyama, and Y. Suemetsu, *IEEE J. Quantum Electron.*, **20**, p. 745, 1984.

262. S. L. Chuang, *Physics of Optoelectronic Devices*, Wiley, Chichester, U.K., 1995.

263. J. E. Zucker, I. Bar-Joseph, B. I. Miller, U. Koren, and D. S. Chemla, *Appl. Phys. Lett.*, **54**, pp. 10–12, 1988.

264. K. Chinen, K. Gen-Ei, H. Suhara, A. Tanaka, T. Matsuyama, K. Konno, and Y. Muto, *Appl. Phys. Lett.*, **51**, pp. 273–275, 1987.

265. M.-C. Amann and J. Buus, *Tunable Laser Diodes*, Artech House, Inc., New York, 1998, p. 93.

266. L. A. Coldren and S. W. Corzine, *IEEE J. Quantum Electron.*, **23**, p. 903, 1987.

267. M.-C. Amann, *Optoelectronics — Dev. Technol.*, **10**, p. 27, 1995.

268. Y. Kotaki and H. Ishikawa, *IEE Proc. Pt. J.*, **138**, p. 171, 1991.

269. X. Pen, H. Olesen, and B. Tromborg, *IEEE J. Quantum Electron.*, **24**, pp. 2423–2432, 1988.

270. N. P. Caponio, M. Goano, I. Maio, M. Meliga, G. P. Bava, G. Destefanis, and I. Montrosset, *IEEE J. Select. Areas Commun.*, **8**, pp. 1203–1213, 1990.

271. B. Stoltz, M. Dasler, and O. Sahlen, *Electron. Lett.*, **29**, pp. 700–702, 1993.

272. M.-C. Amann, S. Illek, C. Schanen, and W. Thulke, *Appl. Phys. Lett.*, **54**, pp. 2532–2533, 1989.

273. M. Okai, S. Sakano, and N. Chinone, *Proc. 15th Europ. Conf. Opt. Commun.*, Gothenburg, Sweden, 1989, p. 122.

274. M.-C. Amann and W. Thulke, *IEEE J. Select Areas Commun.*, **8**, pp. 1169–1177, 1990.

275. M.-C. Amann, in *Semiconductor Lasers II*, E. Kapon, Ed., Academic Press, New York, 1999, p. 224.

276. V. Jayaraman, A. Mathur, L. A. Coldren, and P. D. Dupkus, *IEEE Photon. Technol. Lett.*, **5**, p. 489, 1993.

277. Y. Tohmori, Y. Yoshikuni, F. Kano, H. Ishii, T. Tamamura, and Y. Kondo, *Electron Lett.*, **29**, p. 352, 1993.

278. Y. Tohmori, K. Komori, S. Arai, Y. Suematsu, and H. Oohashi, *Trans. IECE Japan*, **E68**, pp. 788–790, 1985.

279. B. Broberg and S. Nilsson, *Appl. Phys. Lett.*, **52**, pp. 1285–1287, 1988.

280. T. L. Koch, U. Koren, and B. I. Miller, *Appl. Phys. Lett.*, **53**, pp. 788–790, 1985.

281. S. Murata, I. Mito, and Kobayashi, *Electron Lett.*, **23**, pp. 403–405, 1987.

282. T. J. Reid, C. A. Park, P. J. Williams, A. K. Wood, and J. Buus, *12th IEEE Int. Semiconductor Laser Conf.*, Dovos, Switzerland, 1990, pp. 242–243.

283. T. L. Koch, U. Koren, R. P. Gnall, C. A. Burrus, and B. I. Miller, *Electron. Lett.*, **24**, pp. 1431–1433, 1988.

284. Y. Kotaki, M. Matsuda, H. Ishikawa, and H. Imai, *Electron. Lett.*, **24**, pp. 503–505, 1988.

285. P. I. Kuindersma, *Int. Conf. on Integrated Optics and Optical Fiber Communication*, Kobe, Japan, 1989, p. 19A2-1.

286. M.-C. Amann and J. Buus, *Tunable Laser Diodes*, Artech House, Inc., New York, 1998, p. 120.

287. P. Rigby, "Tunable Lasers Revisited," *Light Reading*, Jan. 10, 2003.

288. T. Ikegami and Y. Suematsu, *Proc. IEEE*, **55**, p. 122, 1967.

289. T. L. Paoli and J. E. Ripper, *Proc. IEEE*, **58**, p. 1457, 1970.

290. E. Bourkoff, D. Kerps, and R. W. H. Engelmann, *IEEE Trans. Electron. Devices*, **ED-27**, p. 2180, 1980.

291. L. Figueroa, C. W. Slayman, and H. W. Yen, *IEEE J. Quantum Electron.*, **QE-18**, p. 1718, 1982.

292. K. Lau and A. Yariv, *IEEE J. Quantum Electron.*, **QE-21**, p. 121, 1985.

293. G. H. B. Thompson, *Physics of Semiconductor Laser Devices*, Wiley, New York, 1980, Ch. 7.

294. Y. Suematsu, S. Akiba, and T. Hong, *IEEE J. Quantum Electron.*, **QE-13**, p. 596, 1977.

295. J. E. Bowers, B. R. Hemingway, A. H. Gnauck, and D. P. Wilt, *IEEE J. Quantum Electron.*, **QE-22**, p. 833, 1986.

296. J. M. Dumant, Y. Gvillausseau, and M. Monerie, *Opt. Commun.*, **33**, p. 188, 1980.

297. R. Nagarajan and J. Bowers, in *Semiconductor Laser-I*, E. Kapon, Ed., Academic Press, New York, 1999, p. 180.

298. R. Nagarajan, T. Fukushima, M. Ishikawa, J. E. Bowers, R. S. Geels, and L. A. Coldren, *IEEE Photon. Technol. Lett.*, **4**, p. 121, 1992.

299. R. Nagarajan and J. Bowers, in *Semiconductor Laser-I*, E. Kapon, Ed., Academic Press, New York, 1999, p. 248.

300. M. Asada, A. Kameyama, and Y. Suematsu, *IEEE J. Quantum Electron.*, **20**, p. 745, 1984.

301. R. Nagarajan and J. Bowers, in *Semiconductor Laser-I*, E. Kapon, Ed., Academic Press, New York, 1999, p. 208.

302. M. Asada, Y. Miyamoto, and Y. Suematsu, *IEEE J. Quantum Electron.*, **22**, p. 1915, 1986.

303. Y. Arakawa and A. Yariv, *IEEE J. Quantum Electron.*, **22**, p. 1887, 1986.

304. I. Suemune, *Phys. Rev.*, **B43**, p. 14099, 1991.

305. E. Yablonovitch and E. O. Kane, *IEEE J. Lightwave Technol.*, **4**, p. 504, 1986.

306. A. R. Adams, *Electron. Lett.*, **22**, p. 249, 1986.

307. E. D. Jones, S. K. Lyo, I. J. Fritz, J. F. Klem, J. E. Schirber, C. P. Tigges, and T. J. Drummond, *Appl. Phys. Lett.*, **54**, p. 2227, 1989.

308. C. B. Su and V. Lanzisera, *Appl. Phys. Lett.*, **45**, p. 1302, 1984.

309. K. Uomi, T. Mishima, and N. Chinone, *Jpn. J. Appl. Phys.*, **29**, p. 88, 1990.

310. I. F. Lealman, D. M. Cooper, S. D. Perrin, and M. J. Harlow, *Electron. Lett.*, **28**, p. 1032, 1992.

311. R. Nagarajan, T. Fukushima, S. W. Corzine, and J. E. Bowers, *Appl. Phys. Lett.*, **59**, p. 1835, 1991.

312. I. Suemune, *J. Quantum Electron.*, **27**, p. 1149, 1991.

313. L. F. Lester, S. D. Offsey, B. K. Ridley, W. J. Schaff, B. A. Foreman, and L. F. Eastman, *Appl. Phys. Lett.*, **59**, p. 1162, 1991.

314. S. W. Corzine and L. A. Coldren, *Appl. Phys. Lett.*, **59**, p. 588, 1991.

315. T. Fukushima, R. Nagarajan, M. Ishikawa, and J. E. Bowers, *Jpn. J. Appl. Phys.*, **32**, p. 89, 1993.

316. R. Nagarajan and J. Bowers, in *Semiconductor Laser-I*, E. Kapon, Ed., Academic Press, New York, 1999, p. 211.

317. W. H. Cheng, K. D. Buehring, A. Appelbaum, D. Renner, S. Chin, C. B. Su, A. Mar, and J. E. Bowers, *IEEE J. Quantum Electron.*, **27**, p. 1642, 1991.

318. C. E. Zah, R. Bhat, S. G. Menocal, F. Favire, N. C. Andreadakis, M. A. Koza, C. Caneau, S. A. Schwarz, Y. Lo, and T. P. Lee, *IEEE Photon. Technol. Lett.*, **2**, p. 231, 1990.

319. K. Kamite, H. Sudo, M. Yano, H. Ishikawa, and H. Imai, *IEEE J. Quantum Electron.*, **23**, p. 1054, 1987.

320. P. A. Morton, T. Tanbun-Ek, R. A. Logan, N. Chand, K. W. Wecht, A. M. Sergent, and P. F. Sciortino, Jr., *Electron. Lett.*, **30**, p. 2044, 1994.

321. S. Morin, B. Deveaud, F. Clerot, K. Fujiwara, and K. Mitsunaga, *IEEE J. Quantum Electron.*, **27**, p. 1669, 1991.

322. H. J. Polland, K. Leo, K. Rother, K. Ploog, J. Feldman, G. Peter, and E. O. Gobel, *Phys. Rev.*, **B38**, p. 7635, 1988.

323. M. Preisel, J. Mork, and H. Huang, *Phys. Rev.*, **B49**, p. 14478, 1994.

324. S. M. Sze, *Physics of Semiconductor Devices*, 2d ed., John Wiley, New York, 1981, p. 255.

325. H. Kroemer and H. Okamoto, *Jpn. J. Appl. Phys.*, **23**, p. 970, 1984.

326. G. Bastard, *Wave Mechanics Applied to Semiconductor Heterostructures*, John Wiley, New York, 1988, p. 14.

327. N. R. Howard and G. W. Johnson, *Solid State Electron.*, **8**, p. 275, 1965.

328. M. Babiker and B. K. Ridley, *Superlattice Microstructure*, **2**, p. 287, 1986.

329. J. A. Brum and G. Bastard, *Phys. Rev.*, **B33**, p. 1420, 1986.

330. P. W. M. Blom, J. E. M. Haverkort, and J. H. Wolter, *Appl. Phys. Lett.*, **58**, p. 2767, 1991.

331. H. Hirayama, J. Yoshida, Y. Miyake, and M. Asada, *Appl. Phys. Lett.*, **61**, p. 2398, 1992.

332. R. Kersting, R. Schwedler, K. Wolter, K. Leo, and H. Kurz, *Phys. Rev.*, **B46**, p. 1639, 1992.

333. J. E. Bowers, B. R. Hemenway, A. H. Gnauck, and D. P. Wilt, *IEEE J. Quantum Electron.*, **22**, p. 833, 1986.

334. S. D. Offsey, W. J. Schaff, P. J. Tasker, and L. F. Eastman, *IEEE Photon. Technol. Lett.*, **2**, p. 9, 1990.

335. R. Nagarajan and J. Bowers, in *Semiconductor Laser-I*, E. Kapon, Ed., Academic Press, New York, 1999, p. 182.

336. R. Nagarajan, M. Ishikawa, T. Fukushima, R. S. Geels, and J. E. Bowers, *IEEE J. Quantum Electron.*, **28**, p. 1990, 1992.

337. S. Y. Wang, *Proc. Int. Electron, Devices Meet. (IEDM)*, San Francisco, 1984, p. 712.

338. R. Nagarajan and J. Bowers, in *Semiconductor Laser-I*, E. Kapon, Ed., Academic Press, New York, 1999, p. 254.

339. R. Nagarajan, T. Fukushima, J. E. Bowers, R. S. Geels, and L. A. Coldren, *Appl. Phys. Lett.*, **58**, p. 2326, 1991.

340. T. R. Chen, P. C. Chen, C. Gee, and N. Bar-Chiam, *IEEE Photon. Technol. Lett.*, **5**, p. 1, 1993.

341. R. P. Schneider, Jr., M. Hagerott Crawford, K. D. Choquette, K. L. Lear, S. P. Kilcoyne, and J. J. Figiel, *Appl. Phys. Lett.*, **67**(31), p. 329, 1995.

342. Y. M. Houng, M. R. T. Tan, B. W. Liang, S. Y. Wang et al., *J. Crystal Growth*, **136**(1–4), p. 216, 1994.

343. J. J. Dudley, D. I. Basic, R. Mirin, L. Yang, B. I. Miller, R. J. Ram, T. Reynolds, E. L. Hu, and J. E. Bowerd, *Appl. Phys. Lett.*, **64**(12), pp. 1463–1465, 1994.

344. D. I. Basic, K. Streubel, R. P. Mirin, N. M. Margalit, E. L. Hu, J. E. Bowerd, D. E. Mars, L. Yang, and K. Carey, *IEEE Photon. Technol. Lett.*, **7**(11), pp. 1225–1227, 1995.

345. L. R. Coldren and S. W. Coezine, in *Diode Lasers and Photonic Integrated Circuits*, Wiley, New York, 1995, Chap. 3.

346. L. R. Coldren and S. W. Coezine, *Diode Lasers and Photonic Integrated Circuits*, Wiley, New York, 1995.

347. S. W. Corzine, R. H. Yan, and L. A. Coldren, *IEEE J. Quantum Electron.*, **27**(6), pp. 2086–2090, 1991.

348. C. Lei, L. A. Hoge, J. J. Dudley, M. R. Keever, B. Liang, J. R. Bhagat, and A. Liao, *Vertical Cavity Surface Emitting lasers-I*, **3003**, pp. 28–33, 1997.

349. R. S. Geels, S. W. Corzine, and L. A. Coldren, *IEEE J. Quantum Electron.*, **27**(6), p. 1359, 1991.

350. K. L. Lear, R. P. Schneider, K. D. Choquette, S. P. Kilcoyne, J. J. Figiel, and J. C. Zolper, *IEEE Photonics Technol. Lett.*, **6**(9), p. 1053, 1994.

351. J. W. Scott, Design, *Fabrication and Characterization of High Speed Intra-Cavity Contacted Vertical Cavity Lasers*, Ph.D. Dissertation, ECE Technical Report #95-06m University of California, Santa Barbara, 1995.

352. D. I. Dabic, *Double-Fused Long-Wavelength Vertical Cavity Lasers*, Ph.D. Dissertation, ECE Technical Report #95-06m University of California, Santa Barbara, 1995.

353. J. W. Scott, B. J. Thibeault, D. B. Young, L. A. Coldren, and F. H. Peters, *IEEE Photonics Technol. Lett.*, **6**(6), pp. 678–680, 1994.

354. F. De Martini, G. Innocenti, G. R. Jacobovitz, and P. Mataloni, *Phys. Rev. Lett.*, **59**, p. 2955, 1987.

355. J. L. Jewell, A. Sherer, S. L. McCall, Y. H. Lee, S. Walker, J. P. Harbison, and L. T. Florez, *Electron. Lett.*, **25**, pp. 1123–1124, 1989.

356. K. D. Choquette, M. Hong, R. S. Freund, S. N. G. Chu, J. P. Mannaerts, R. C. Wetzel, and R. E. Leibenguth, *IEEE Photon. Technol. Lett.*, **4**, pp. 284–287, 1993.

357. C. J. Chang-Hasnain, Y. A. Wu, G. S. Li, G. Hasnain, K. D. Choquette, C. Caneau, and L. T. Florez, *Appl. Phys. Lett.*, **63**, pp. 1307–1309, 1993.

358. D. L. Huffaker, D. G. Deppe, K. Kumar, and T. J. Rogers, *Appl. Phys. Lett.*, **65**, pp. 97–99, 1994.

359. K. D. Choquette, R. P. Schneider, Jr., K. L., Lear, and K. M. Geib, *Electron. Lett.*, **30**, pp. 2043–2044, 1994.

360. K. Tai, R. J. Fischer, K. W. Wang, S. N. G. Chu, and A. Y. Cho, *Electron. Lett.*, **25**, pp. 1644–1645, 1989.

361. Y. H. Lee, J. L. Jewell, B. Tell, K. F. Brown-Goebeler, A. Sherer, J. P. Harbison, and L. T. Florez, *Electron. Lett.*, **26**, pp. 225–227, 1990.

362. B. J. Thibeault, T. A. Strand, T. Wipiejewski, M. G. Peters, D. B. Young, S. W. Corzine, L. A. Coldren, and J. W. Scott, *J. Appl. Phys.*, **78**, pp. 5871–5875, 1995.

363. Y. H. Lee, B. Tell, K. F. Brown-Goebeler, and J. L. Jewell, *Electron. Lett.*, **26**, pp. 710–711, 1990.

364. G. Hasnain, K. Tai, L. Yang, Y. H. Wang, R. J. Fischer, J. D. Wynn, B. Weir, N. K. Dutta, and A. Y. Cho, *IEEE J. Quantum Electron.*, **27**, pp. 1377–1385, 1991.

365. K. L. Lear, R. P. Schneider, Jr., K. D. Choquette, and K. P. Kilcoyne, *IEEE Photon. Technol. Lett.*, **8**, pp. 740–742, 1996.

366. K. D. Choquette, K. L. Lear, R. P. Schneider, Jr., K. M. Geib, J. J. Figiel, and R. Hull, *IEEE Photon. Technol. Lett.*, **7**, pp. 1237–1239, 1995.

367. L. A. Codren and E. R. Hegblom, in *Vertical Cavity Surface Emitting Lasers*, C. Wilmsen, H. Temkin, and L. A. Coldren, Eds., Cambridge, New York, 1999, Chap. 2.

368. S. S. Kutateladze and V. M. Borishanski, *A Concise Encyclopedia of Heat Transfer*, Permagon, Oxford, 1966.

369. R. W. Herrick, in *Vertical-Cavity Surface-Emitting Lasers VI*, Vol. 4946, C. Lei, and S. P. Kilcoyne, Eds., SPIE, San Jose, 2000, p. 130.

370. J. K. Guenter, R. A. Hawthorn, D. N. Granville, M. K. Hibbs-Brenner, and R. A. Morgan, in *Fabrication, Testing, and Reliability of Semiconductor Lasers*, Vol. 2683, M. Fallahi and S. C. Wang, Eds., SPIE, San Jose, 1996.

371. X. Zhang, F. Xiong, W. Sun, D. Zhu, Z. Yang, J. Liu, A. Ou, and B. Liang, in *Vertical-Cavity Surface-Emitting Lasers VI*, C. Lei and S. P. Kilcoyne, Eds., SPIE, San Jose, 2000, p. 111.

372. T. D. Lowes, in *Vertical-Cavity Surface-Emitting Lasers VI*, C. Lei, and S. P. Kilcoyne, Eds., SPIE, San Jose, 2000, p. 121.

373. C. C. Wu, K. Tai, T. C. Huang, and K. F. Huang, *IEEE Photon. Technol. Lett.*, **4**(6), pp. 37–39, 1994.

374. D. Vakhshoori, J. D. Wynn, R. E. Leibenguth, and R. A. Novotny, *Electron. Lett.*, **29**, pp. 2118–2119, 1993.

375. S. Xie, R. W. Herrick, G. N. De Brabander, W. H. Widjaja, U. Koelle, A. Cheng, L. M. Giovane, F. Z. Hu, M. R. Keever, T. Osentoski, S. A. Mchugo, M. S. Mayonte, S. M. Kim, D. R. Chamberlin, S. J. Rosner, and G. Girolami, Technical Program, PhotonicsWest, San Jose, January 25–31, 2003, p. 149.

376. R. P. Schneider, Jr., M. Hagerott Crawford, K. D. Choquette, K. L. Lear, S. P. Kilcoyne, and J. J. Figiel, *Appl. Phys. Lett.*, **67**, pp. 329–331, 1995.

377. T. Tadokoro, T. H. Okamoto, Y. Kphama, T. Kawakami, and T. Kurokawa, *IEEE Photon. Technol. Lett.*, **4**(5), pp. 409–411, 1992.

378. T. Uchida, T. Miyamoto, N. Yokoushi, Y. Inaba, F. Koyama, and K. Iga, *IEEE J. Quantum Electron.*, **29**(6), pp. 1975–1980, 1993.

379. Y. Ohiso, C. Amano, Y. Itoh, K. Tateno, and Tadokoro, *Electron. Lett.*, **32**(16), pp. 1483–1484, 1996.

380. Y. Imajo, A. Kasukawa, S. Kashiwa, and H. Okamoto, *Jpn. J. Appl. Phys.*, **29**(7), pp. L1130–1132, 1990.

381. A. J. Moseley, J. Thompson, D. J. Robbins, and M. Q. Kearley, *Electron. Lett.*, **25**(25), pp. 1717–1718, 1989.

382. S. W. Choi and H. M. Park, *Jpn. J. Appl. Phys.* Part 2, **36**(6B), pp. L740–742, 1997.

383. M. J. Mondry, D. I. Basic, J. E. Bowers, and L. A. Coldren, *IEEE Photon. Technol. Lett.*, **4**(6), pp. 627–630, 1992.

384. J. J. Dudley, D. I. Basic, R. Mirin, L. Yang, B. I. Miller, R. J. Ram, T. Reynolds, E. L. Hu, and J. E. Bowerd, *Appl. Phys. Lett.*, **64**(12), pp. 1463–1465, 1994.

385. D. I. Basic, K. Streubel, R. P. Mirin, N. M. Margalit, E. L. Hu, J. E. Bowerd, D. E. Mars, L. Yang, and K. Carey, *IEEE Photon. Technol. Lett.*, **7**(11), pp. 1225–1227, 1995.

386. K. Streubel, S. Rapp, J. Andre, and J. Wallin, *IEEE Photon. Technol. Lett.*, **8**(9), pp. 1121–1123, 1996.

387. M. Kondow, K. Uomi, A. Niwa, T. Kitatani, S. Watahiki, and Y. Yazawa, *Jpn. J. Appl. Phys.*, **35**, pp. 1273–1275, 1996.

388. M. C. Larson, M. Kondow, T. Kitatani, Y. Yazawa, and M. Okai, *Electron. Lett.*, **33**, pp. 959–960, 1997.

389. Pauline Rigby, "Agilent claims VCSEL Breakthrough," *Light Reading*, Dec. 2, 2002.

390. H. Huang, D. G. Deppe, and O. B. Shchekin, Technical Program, Photonics West, San Jose, January 25–31, 2003, p. 148.

391. T. Anan, K. Nishi, S. Sugou, and K. Kasahra, *Postdeadline Abstract for the 1997 Annual Meeting of the Laser and Electro-Optics Soc.* (LEOS), 1997.

392. M. Adamcyk, S. Chaparro, P. Dowd, S. Feld, K. Hilgers, S. R. Johnson, J. Joseph, B. Liang, K. Shiralagi, S. Yu, and Y. Zhang, Technical Program Photonics West, San Jose, January 25–31, 2003, p. 149.

393. R. Nagarajan, T. Fukushima, M. Ishikawa, J. E. Bowers, R. S. Geels, and L. A. Coldren, *IEEE Photon. Technol. Lett.*, **4**, p. 121, 1992.

394. R. Nagarajan and J. Bowers, in *Semiconductor Laser-I*, E. Kapon, Ed., Academic Press, New York, 1999, p. 248.

395. X. Zhang, F. Xiong, W. Sun, D. Zhu, Z. Yang, J. Liu, A. Ou, and Bing Liang, in *Vertical Cavity Surface Emitting Lasers-VI*, C. Lei and S. P. Kilcoyne, Eds., SPIE, San Jose, 2002, p. 115.

396. K. S. Urquhart, P. Marchand, Y. Fainman, and S. H. Lee, *Appl. Opt.*, **33**, pp. 3670–3682, 1994.

397. D. B. Schwartz, C. K. Y. Chun, B. M. Foley, D. H. Hartman, M. Lebby, H. C. Lee, C. L. Shieh, S. M. Kuo, S. G. Shook, and B. Webb, *45th Electronic Components and Technology Conf.*, Las Vegas, NV, 1995, pp. 376–379.

398. K. H. Hahn, *45th Electronic Components and Technology Conf.*, Las Vegas, NV, 1995, pp. 368–375.

399. X. Zhang, F. Xiong, W. Sun, D. Zhu, Z. Yang, J. Liu, A. Ou, and Bing Liang, in *Vertical Cavity Surface Emitting lasers-VI*, C. Lei and S. P. Kilcoyne, Eds., SPIE, San Jose, p. 117, 2002.

400. J. Faist, F. Capasso, D. L. Sivco, C. Sirtori, A. L. Hutchinson, and A. Y. Cho, *Science*, **264**, p. 553, 1994.

401. C. Sirtori et al., *Pll. Phys. Lett.*, **73**, p. 3486, 1998.

402. F. Capasso, A. Tredicucci et al., *Select. Topics Quantum Electron.*, **5**, p. 792, 1999.

403. H. Pages, P. Kruck et al., *Electron. Lett.*, **35**, p. 1848, 1998.

404. J. Faist, C. Gmachi, F. Capasso, C. Sirtori, D. L. Sivco, J. N. Baillargeon, and A. Y. Cho, *Appl. Phys. Lett.*, **70**, p. 2670, 1997.

405. A. Tredicucci et al., *Nature*, **396**, p. 350, 1999.

406. C. Gmachi et al., *Science*, **286**, p. 749, 1999.

407. A. Muller, J. Faist et al., *Appl. Phys. Lett.*, **75**, p. 1509, 1999.

408. Y. Arakawa and H. Sakaki, *Appl. Phys. Lett.*, **40**, p. 939, 1982.

409. A. Yariv, *Appl. Phys. Lett.*, **53**, p. 1033, 1988.

410. Y. Miyamoto, Y. Miyake, M. Asada, and Y. Suematsu, *IEEE J. Quantum Electron.*, **QE-25**, p. 2001, 1989.

411. Y. Arakawa, K. Vahala, and A. Yariv, *Appl. Phys. Lett.*, **45**, p. 950, 1984.

412. L. E. Brus, A. L. Efros, and T. Itoh, *J. Lumin.*, **70**, 1996.

413. E. Kapon, *Proc. IEEE 80*, p. 398, 1992.

414. D. Bimberg, M. Grundmann, and N. N. Ledentsov, *Quantum Dot Heterostructures*, Wiley, Chichester, UK, 1999.

415. M. Cao, Y. Miyake, S. Tamura, H. Hirayama, S. Arai, Y. Suematsu, and Y. Miyamoto, *Trans. IEICE E*, **73**, p. 63, 1990.

416. Y. Miyamoto, M. Cao, Y. Shingai, K. Furuya, Y. Suematsu, K. G. Ravikumar, and S. Arai, *Jpn. J. Appl. Phys.*, **26**, p. L225, 1987.

417. M. Cao, P. Daste, M. Miyamoto, Y. Miyake, S. Nogiwa, S. Arai, K. Furuya, and Y. Suematsu, *Electron. Lett.*, **24**, p. 824, 1988.

418. M. Tsuchiya, L. A. Coldren, and P. , M. Petroff, *Proc. Seventh Int. Conf. Integrated Optics and Optical Fiber Communication (IOOC'89)*, Kobe, Japan, July 18–21, 1989, Paper 19C1-1, p. 104.

419. E. Kapon, D. M. Hwang, and R. Bhat, *Phys. Rev. Lett.*, **63**, p. 430, 1989.

420. E. Kapon, S. Simhony, R. Bhat, and D. M. Hwang, *Appl. Phys. Lett.*, **55**, p. 2715, 1989.

421. S. Simhony, E. Kapon, E. Colas, R. Brat, N. G. Stoffel, and D. M. Hwang, *IEEE Photon. Technol. Lett.*, **2**, p. 305, 1990.

422. S. Simhony, E. Kapon, E. Colas, D. M. Hwang, N. G. Stoffel, and P. Worland, *Appl. Phys. Lett.*, **59**, p. 2225, 1991.

423. Y. Hayamizu, M. Yoshita, S. Watanabe, H. Akiyama, L. N. Pfeiffer, and K. W. West, *Appl. Phys. Lett.*, **81**, p. 4937, 2002.

424. D. Bimberg, M. Grundmann, and N. N. Ledentsov, *Quantum Dot Hetero-structures*, Wiley, Chichester, 1998.

425. M. Grundmann, *Nano-Optoelectronics*, Springer, Berlin, 2002.

426. D. Bimberg, N. N. Ledentsov, N. Kirstaedter, O. Schmidt, M. Grundmann, V. M. Ustinov, A. YU. Egorov, A. E. Zhuikov, M. V. Maximov, P. S. Kop'ev, Zh. I. Alferov, S. S. Ruvimov, U. Gosele, and J. Heydenreich, *Jpn. J. Appl. Phys.*, **B 35**, p. 1311, 1996.

427. A. E. Zhukov, A. R. Kovsh, N. A. Maleev, S. S. Mikhrin, V. M. Ustinov, A. F. Tsatsunikov, M. V. Mzximov, B. V. Volovik, D. A. Bedarev, Yu. M. Shernyakov, P. S. Kop'ev, Zh. I. Alferov, N. N. Ledenstov, and D. Binberg, *Appl. Phys. Lett.*, **75**, p. 1926, 1999.

428. G. T. Liu, A. Stinz, T. C. Newell, A. L. Gray, P. M. Varangis, K. J. Malloy, and L. F. Lester, *IEEE J. Quantum Electron.*, **36**, p. 1272, 2000.

429. F. Klopf, J. P. Reithmaier, and A. Forchel, *Appl. Phys. Lett.*, **77**, p. 1419, 2000.

430. J. A. Lott, N. N. Ledentsov, V. M. Ustinov, N. A. Maleev, A. E. Zhukov, A. R. Kovsh, M. V. Maximov, B. V. Volovik, Z. H. I. Alferov, and D. Bimberg, *Electron. Lett.*, **36**, p. 1384, 2000.

431. R. L. Sullin, C. Ribbat, R. Grundmann, N. N. Ledentsov, and D. Bimberg, *Pll. Phys. Lett.*, **78**, p. 1207, 2001.

432. R. L. Sullin, C. Ribbat, D. Bimberg, F. Rinner, H. Konstanzer, M. T. Kelemen, and M. Mikulla, *Electron. Lett.*, **38**, p. 883, 2002.

433. A. R. Kovsh, N. A. Maleev, A. E. Zhukov, S. S. Mikhrin, A. P. Vasil'ev, Yu. M. Shernyakov, M. V. Maximov, D. A. Livshits, V. M. Ustinov, Zh. I. Alferov, N. N. Ledentsov, and D. Bimberg, *Electron. Lett.*, **38**, p. 1104, 2003.

434. R. L. Sellin, I. Kaiander, D. Ouyang, T. Kettler, U. W. Pohl, D. Bimberg, N. D. Zakharov, and P. Werner, *Appl. Phys. Lett.*, **82**, p. 841, 2003.

435. O. Benson, C. Santori, M. Pelton, and Y. Yamamoto, *Phys. Rev. Lett.*, **84**, p. 2513, 2000.

436. P. Michler, A. Imamoglu, M. D. Mason, P. J. Carson, G. F. Strouse, and S. K. Buratto, *Nature* (London), **406**, p. 968, 2000.

437. P. Michler, A. Kiraz, C. Becher, W. V. Schoenfeld, P. M. Petroff, L. Zhang, E. Hu, and A. Imamoglu, *Science*, **290**, p. 2282, 2000.

438. C. Santori, M. Pelton, G. Solomon, Y. Dale, and Y. Yamamoto, *Phys. Rev. Lett.*, **86**, p. 1502, 2001.

439. R. L. Naone, P. D. Floyd, D. B. Young, E. R. Hegblom, T. A. Strand, and L. A. Coldren, *IEEE J. Select. Top. Quantum Electron.*, **4**, p. 706, 1998.

440. T. A. Strand, B. J. Thibeault, and L. A. Coldren, *J. Appl. Phys.*, **81**, p. 3377, 1997.

441. J. M. Garcia, T. Mankad, P. O. Holtz, P. J. Wellman, and P. M. Petroff, *Appl. Phys. Lett.*, **72**, p. 3173, 1998.

442. J. K. Kim, R. L. Naone, and L. A. Coldren, *IEEE J. Select. Top. Quantum Electron.*, **6**, p. 504, 2000.

443. L. Zhang, and E. Hu, *Appl. Phys. Lett.*, **82**, p. 319, 2003.

444. A. Flore, J. X. Chen, and M. Ilegems, *Appl. Phys. Lett.*, **81**, p. 1757, 2002.

445. *Photonics Spectra*, February, 2003, pp. 36, 62.

446. T. Tumolillo, Jr, *Photonics Spectra*, January, 2003, p. 102.

447. D. Gammon, and D. G. Steel, *Physics Today*, October, 2002, p. 36.

# 7

# SOLID-STATE LASERS

JOHN M. MCMAHON
*Optical Sciences Division*
*Naval Research Laboratory*
*Washington, D.C.*

## 7.1 INTRODUCTION

The solid-state ruby laser ($Cr^{3+}$ doped into sapphire, $Al_2O_3$) was the first laser and was demonstrated by T. H. Maiman in 1960 [1]. This demonstration of the stimulated emission process, first postulated by Einstein in 1916 [2], has led to advances in many areas of science and engineering as well as to devices and technologies which have made the optical sciences one of the most rapidly growing areas of modern technology. The exact number of lasers (distinct laser ion–host combinations) at this time is in the hundreds, with new ones reported regularly. The comprehensive review of the solid-state literature by D. Ross in 1966 counted over 3100 references [3]; conservatively, one would expect that at the present time more than 15,000 publications have appeared in print.

A short handbook article such as this one cannot cover all the topics and concepts important to understanding solid-state lasers other than rather superficially. I have had to choose to omit entirely some related topics, such as nonlinear optics, dye lasers, and color center lasers. In choosing the references I have attempted to strike a reasonable balance between scholarly priority of discovery and later articles which give a more complete exposition of the topic in a mature state. I have chosen to add Section 7.8, which attempts to highlight some of the experimental pitfalls of working with these coherent sources, which may not be obvious from a reading of the more traditional optics literature.

A number of recent books and compilations exist which give much more thorough treatments of a number of topics. I recommend these both for a more comprehensive treatment of particular topics and as guides to the larger body of literature.

*Handbook of Optical Components and Engineering,*   Edited by Kai Chang
ISBN 0-471-39055-0   © 2003 John Wiley & Sons, Inc.

### 7.1.1   Basic Concepts: Einstein Relations

Processes that involve the interaction of radiation and atoms in matter include absorption and fluorescence or spontaneous emission. In absorption, a photon whose energy is equal to the difference in energy between two states or energy levels of the atom, $h\nu_{ij} = E_j - E_i$, is absorbed by the atom raising it from the lower to the upper energy level. For a group of identical atoms in thermal equilibrium at a temperature $T$, the relative population of two different energy levels $n$ and $m$ is given by

$$\frac{N_n}{N_m} = \exp\left(\frac{E_n - E_m}{kT}\right)$$

At ordinary room temperature, 300 K, most atoms will be in their lowest energy level or ground state. The Boltzmann constant $k = 1.38 \times 10^{-23}$ W · s/K. For an energy level $E_1$, whose energy above the ground state corresponds to a midvisible wavelength ($\sim$ green at $\lambda = 5000$ Å), the number of atoms in $N$ compared to the ground state will be only about 1 in $10^{42}$. As most solids have a density on the order of $10^{22}$ atoms/cm$^3$, the number of atoms in energy states substantially above the ground will ordinarily be extremely small.

In the process of fluorescence or spontaneous emission, an atom in energy state $E_j$ above the ground state decays to a lower energy level $E_i$ (not necessarily the ground state) by emitting a photon of energy $E_j - E_i$ with arbitrary phase and direction. In the process of stimulated emission, by contrast, a photon of the radiation field of energy $E_j - E_i$ causes the atom to emit a photon of the same energy and phase while simultaneously going from energy level $E_j$ to $E_i$.

Einstein was the first to derive the fact that in a classical (Planckian) radiation field at a temperature $T$ for an atom to be in equilibrium with the field, the ratio of the rates of spontaneous to stimulated emission is

$$\frac{A_{ji}}{B_{ji}} = \frac{8\pi h\nu_{ji}^3}{c^3}$$

where $c$ is the speed of light in the medium ($= c_0/n$, where $c_0 = 2.998 \times 10^8$ m/s and $n$ is the index of refraction) and $h = 6.6 \times 10^{-34}$ J · s is Planck's constant. The ratio of the rate of stimulated emission to absorption is $g_j B_{ji} = g_i B_{ij}$, where the degeneracy factors $g_m$ identify the number of states with energy $E_m$.

In a real solid at a finite temperature the energy levels of the atoms will not be infinitely sharp, but there will be a broadening due to thermal lattice vibrations and also by effects of defects in the lattice. The line shapes in crystalline lasers such as ruby ($Cr^{3+}$ doped into sapphire, $Al_2O_3$) or Nd:YAG (neodymium doped into yttrium-aluminum garnet, $Y_3Al_5O_{12}$) are good example of thermal broadening and have a Lorentzian line shape. In such cases the cross section for stimulated emission at the center of the line is

$$\sigma_{21} = \frac{A_{21}\lambda_{21}^2}{4\pi^2 n^2 \Delta\lambda}$$

where $\Delta\lambda$ is the width between half-intensity points in the line. The cross section for stimulated emission $\sigma_{21}$ is related to the absorption cross section $\sigma_{12}$ by the ratio of degeneracy factors $g_1/g_2$. Normally, rather than the spontaneous emission rate $A_{21}$ its inverse, $t_{21}$, the fluorescence lifetime, is used in the literature.

If atoms could exist only in equilibrium distributions, stimulated emission would be only an academic curiosity. However, by the use of powerful optical pumping techniques, some of the atoms in a solid can be raised by absorption to the upper level faster than they can decay back to the lower level, and a partial population inversion can be made to exist (i.e., an excess of population over that given by Boltzmann statistics). If the population of the upper level $N_2 > g_1 N_1/g_2$, then an actual population inversion exists and the rate of stimulated emission will exceed the rate of absorption. Rather than being attenuated by absorption, an incident radiation field will then be amplified by stimulated emission. (The acronym "LASER" stands for "light amplification by stimulated emission of radiation.") More rigorous and complete derivations of these relations can be found in a number of sources, such as Refs. 4 and 5.

### 7.1.2   Active Ions Exhibiting Laser Action in Solid Hosts

A large number of ions have been found to exhibit laser action when doped as an impurity into crystals and glasses. In most cases a divalent or trivalent ion will substitute for an ion of the same valence state in the host structure. In general, the energy-level structure of the particular ion will determine the general features of the laser emission, while the host matrix will determine the finer scale features of the process, such as exact positions and widths of the energy levels and the radiative and nonradiative rates between them. In some cases the laser transition can occur between an ionic level and phonon levels characteristic of vibronic modes of the crystal. Figure 7.1 illustrates the major types of level scheme encountered and the ions exhibiting such behavior.

Three level lasers are so-called because the ion is pumped to an upper state (2) and nonradiatively decays to the upper laser level (3). Laser action can occur only when the population in level 3 exceeds that in the ground state (1). In general, a rather high rate of pumping is necessary to get a population inversion ($N_3 > N_1$). In addition, since the population, $N_3 - N_1$, will change by two for every photon emitted, the energy extraction efficiency cannot exceed $E_{31}/2E_{21}$ (for a system where all degeneracy factors $g_1 = 1$). Nonetheless, the best known three-level laser, the ruby laser ($Cr^{3+}$ in $Al_2O_3$), is a reasonably efficient laser at 0.69 $\mu$m. The reasons are that the pump rate requirement is mitigated by a long 3-ms upper laser level lifetime and there are a number of strong shorter-wavelength absorption bands which all channel energy into the upper laser level.

Four level lasers are characterized by laser emission to a level well above the ground state, whose population is normally very small. In this case the population inversion will change by one for each photon emitted and the energy extraction efficiency can be as high as the ratio of the laser wavelength to the pumping frequency. The best known example of this type is the neodymium laser, in which laser action occurs most strongly from the $^4F_{3/2}$ energy level to the $^4I_{11/2}$ level at around 1.06 $\mu$m. Laser action from the same upper level is also possible, but usually weaker because of a smaller cross section, to the $^4I_{13/2}$ level ($\sim$1.3 $\mu$m) or $^4I_{15/2}$ level ($\sim$1.8 $\mu$m). Laser action to the top of the ground-state manifold, the $^4I_{9/2}$ level, generally requires reduced temperature operation to achieve a population inversion.

Figure 7.2 indicates the many energy levels involved in the laser process for the rare-earth ions [6]. Erbium probably represents one of the most complex laser ions in

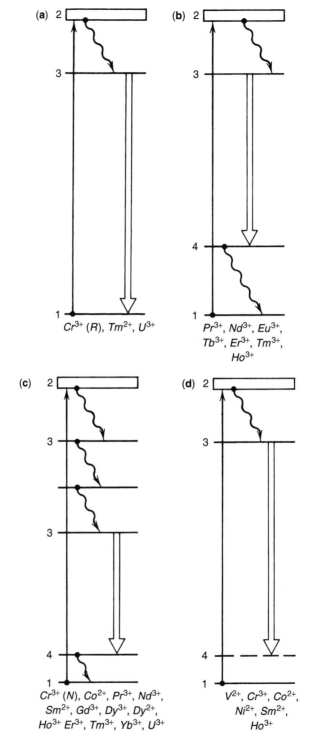

**Figure 7.1** Simplified energy-level diagrams characteristic of solid-state lasers: **(a)** three-level laser; **(b)** and **(c)** two types of four-level lasers; **(d)** a vibronic laser. (From Ref. 5.)

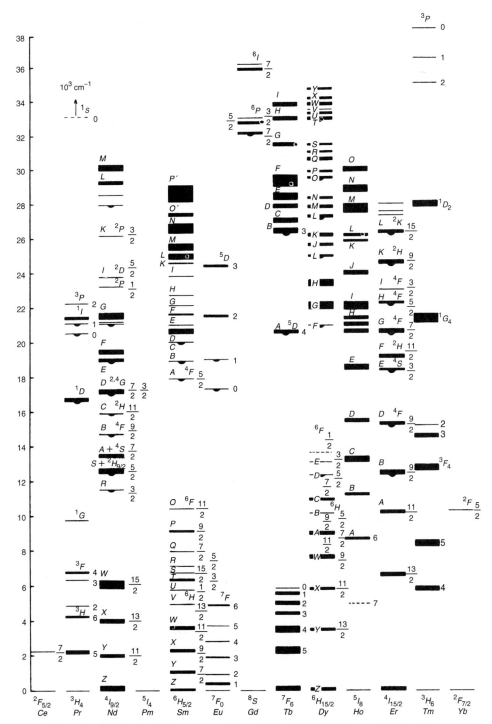

**Figure 7.2** Energy levels of the rare-earth ions. (From Ref. 6.) Those exhibiting fluorescence are denoted by a semicircle.

the number of different levels between which laser action can occur, while ytterbium is the simplest. The figure also indicates another fact with some practical importance: The various rare-earth ions have numbers of levels which are at about the same energy above the ground state. This suggests the possibility that by co-doping a host with more than one type of ion, energy can be transferred from one species to another. This has, in fact, been found to be possible in several cases. Such a transfer can potentially be used to enhance the efficiency of optical pumping. The holmium laser at 2.05 μm is generally not an efficient optically pumped laser. However, if a host such as YAG is co-doped with other rare earths, such as $Er^{3+}$ and $Tm^{3+}$, much more efficient laser action can be obtained because of energy transfer from the other ions [7]. A number of efforts have been made to find sensitizers to increase the efficiency of the neodymium laser. The most successful of these to date is co-doping with $Cr^{3+}$ in gadolinium scandium gallium garnet (GSGG) [8]. The authors report a 100% increase in efficiency for this laser compared to the Nd:YAG.

Vibronic lasers offer the potential for broadly tunable laser action, in many ways analogous to dye lasers because of the broad width over which fluorescence is observed. The broad line width, however, leads to a problem. The product of the induced emission coefficient and the fluorescence lifetime is

$$\sigma\tau = \lambda^2/(4\pi^2 n^2 \Delta\lambda)$$

from the Einstein relations. If $\Delta\lambda$ is quite broad, $\sigma\tau$ will be smaller than is typical for solid-state lasers. This can lead to problems in pumping if $\tau$ is rather small, or in extraction, if $\sigma$ is small, giving a low gain and high saturation energy density. The alexandrite laser ($Cr^{3+}$ in beryllium aluminate) [9] and the titanium-doped sapphire laser [10] are two systems in which there is much current interest. Recent developments in this area are summarized in Ref. 11.

### 7.1.3  Solid-State Laser Host Materials

A very large number of crystalline and glass hosts have been reported for laser ions. Of the crystals, only a small fraction have been developed to the point where laser rods of high optical quality are available commercially. This is largely because the development of synthetic crystals to a state of high quality is a slow and expensive process. Even if an undoped version of the crystal represents a well-developed technology, the doped crystal may require extensive modification of the growth process. Generally, the investment of time and resources necessary to develop a material from a research phase to commercial practice will not occur unless the research results indicate that the new material will have significant advantages over existing materials. The book by Kaminskii [5] documents the extensive research literature through 1981. The laser host crystals commercially available at the present time and their more noteworthy features are described next.

*$Al_2O_3$ (Sapphire).* Sapphire is a hard, strong, durable material. Doped with $Cr^{3+}$ at 0.03 to 0.05 wt% in the ruby laser [1]; more recently, the titanium sapphire laser [10] has been under active development because of its potential for wide wavelength tunability from 0.7 to 1.0 μm.

$Y_3Al_5O_{12}$ *(YAG)*.  Yttrium aluminum garnet is a hard, strong durable material which has been the most popular host material for neodymium [12] and the other rare-earth ions because of its high gain and efficiency in flashlamp-pumped operation.

$YAlO_3$ *(YALO)*.  Yttrium aluminate continues to enjoy popularity as a host for continuous neodymium lasers [13]. It also has the highest relative gain for operation on the 1.34-$\mu$m $^4F_{3/2}-^4I_{13/2}$ neodymium transition compared to the $^4F_{3/2}-^4I_{11/2}$ transition of the well-developed host crystals.

$BeAl_2O_4$.  Beryllium aluminate when doped with chromium is known as alexandrite [9] and has been developed extensively by Allied-Signal Corp. as an efficient long-pulse tunable vibronic laser. As a neodymium host in the correct orientation it has been reported to be capable of operation as an athermal material in which there is no optical distortion resulting from a thermal lens [14].

$LiYF_4$ *(YLF)*.  Lithium yttrium fluoride has found some favor as a host for neodymium [15] and the other rare earths even though it is weaker than most of the other commercially developed laser hosts. The neodymium $^4F_{3/2}-^4I_{11/2}$ transition lases at 1.053 $\mu$m in one orientation which closely matches the peak gain of the phosphate laser glasses. It also exhibits near-athermal behavior.

$Gd_3Sc_2Ga_3O_{12}$ *(GSGG)*.  The relatively large thermo-optic distortion of this material compared to YAG made it appear not particularly interesting when first reported in 1976 [16]. The more recent report of the high efficiency of this material when co-doped with chromium as a sensitizer for neodymium [8] has led to extensive investigation of this and other variants of GGG ($Gd_3Ga_5O_{12}$), both as neodymium hosts and as vibronic lasers when doped with chromium alone [17, 18].

The recent history of GSGG is a precaution that it is certainly premature to claim that a complete set of the desirable crystalline laser hosts exists. Further research or the demands of particular applications will undoubtedly result in additions to the list.

The number of distinct glass laser hosts is, if anything, more extensive than the crystalline materials. The compositions investigated and their laser and other properties were reported extensively by Lawrence Livermore National Laboratory (LLNL) as part of the laser fusion program in the series of volumes on laser glass [19]. A compilation of the properties of Soviet laser glasses is reported in Avakyants [20]. Most of the well-developed laser glasses are silicates or phosphates; the leading commercial glasses are described further in Section 7.4.4.

## 7.2  EXCITATION OF SOLID-STATE LASERS

The metastable ionic or vibronic upper laser levels could in principle be excited by a variety of mechanisms. In practice, considerations of damage to the laser host materials has resulted in almost exclusive use of optical pumping as the excitation mechanism. In the 1960s a variety of incoherent and coherent light sources were demonstrated as solid-state laser pumps. The reason why a wide variety of pump sources have resulted in operable lasers is because the values of the fluorescent lifetimes characteristic of solid-state lasers, typically $10^{-6}$ to $10^{-2}$ s, allow net optical gains to be achieved with a variety of continuous-wave (CW) or pulsed pump sources, and the moderate-sized induced emission cross sections, $10^{-19}$ to $10^{-21}$ cm$^2$, can allow efficient energy extraction either CW or pulsed. In addition, most solid-state laser ions have optical absorptions at various wavelengths leading to reasonable efficiency with sources that are broadband thermal radiators.

### 7.2.1   Flashlamp Excitation

Powerful electrically excited xenon flashlamps were developed by H.E. Edgerton in the United States [21] and I. S. Marshak in the USSR [22] for applications such as high-speed strobe lights and underwater photography. When lasers came on the scene, this technology already existed and was pressed into service immediately. There were a number of studies performed to understand the radiation characteristics of flashlamps as well as their electrical performance as circuit elements. The 1965 work of Markiewicz and Emmett [23] deserves special mention, as it pulled together much of the earlier work into a circuit description of flashlamps. In this work a voltage–current characteristic of a flashlamp was assumed as

$$V = k_0 I^{1/2}$$

which was a good empirical fit to flashlamp data. $k_0$, the impedance parameter of the lamp, is approximately $1.33l/d$ for xenon flashlamps, where $l$ is the arc length and $d$ the diameter. In their numerical simulation of single mesh capacitor–inductor circuits containing such elements, Markiewicz and Emmett found that the circuit elements were related as

$$c^3 = 2E_0(\alpha/k_0)^4 T^2$$

and that critically damped operation was obtained in the simulations for $\alpha = 0.8$. [In this equation, $C$ is the capacitance in farads; $E_0$, the energy stored initially in the capacitor, is in joules; and $T = (LC)^{1/2}$, where $L$ is the inductance in henries.] It was recognized early that operation in the region from critically damped to overdamped was important. High-efficiency lamps had polarized electrodes which would degrade rapidly with current reversal if the circuit was underdamped. Consequently, for substantially overdamped operation, the electrical pulse would have a long temporal tail. Figures 7.3 and 7.4 illustrate their results as a function of the damping constant. More complete recent calculations [24] suggest that inclusion of resistive losses in the circuit will shift the critically damped point to somewhat lower values of the damping parameter in many practical cases.

For a pulsed storage laser where the laser ion essentially integrates the pumping radiation, the peak stored energy density will clearly depend on the ratio of the pulse duration of the flashlamp and the fluorescence lifetime. The problem, however, is much more complicated than just the ratio of time constants. At constant total energy, as the flashlamp pulse is shortened, the current density, lamp spectrum, and the opacity of the lamp will all change. This will change the overlap of the radiation with the absorption spectrum of the laser ion. Other components of the pumping enclosure which may have spectrally varying reflection (transmission) in the different pump bands can also affect the results. An effort to model this rather complicated situation numerically was begun at NRL in the early 1970s [25] and carried on more recently at LLNL for the laser fusion program [26]. Figures 7.5 and 7.6 show some of these results. Figure 7.5 shows the variation in relative stored energy with population inversion and the variation in lamp life in pumping ED-2 silicate glass with a product of doping $x$ thickness of

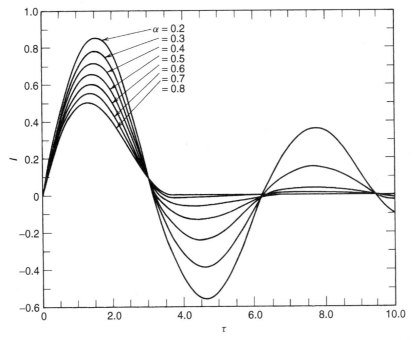

**Figure 7.3** Single-mesh-network flashlamp current as a function of damping parameter (under-damped cases). (From Ref. 23; copyright 1966 IEEE, reproduced by permission.)

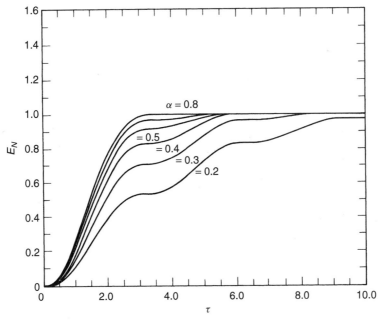

**Figure 7.4** Energy dissipated in a flashlamp as a function of damping parameter. (From Ref. 23; copyright 1966 IEEE, reproduced by permission.)

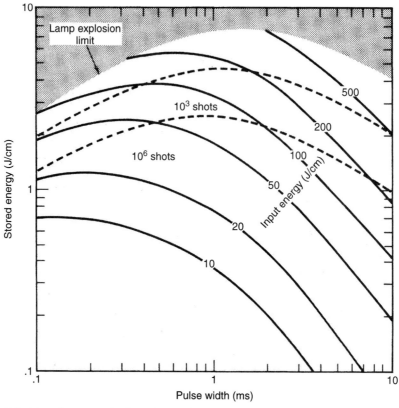

**Figure 7.5**   Pumping curve for 2 wt%-cm of ED-2 silicate laser glass. (From Refs. 25 and 79; copyright 1970 and 1971 IEEE, reproduced by permission.)

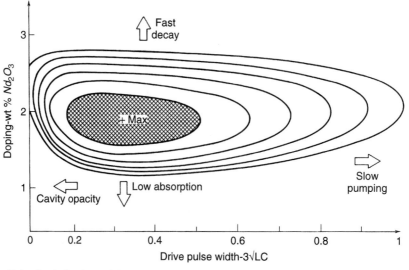

**Figure 7.6**   Optimization of ED-2 silicate glass for maximum gain versus pulse width and doping level. (From Ref. 26; copyright 1971 IEEE, reproduced by permission.)

2 wt%-cm. By accounting for the change in laser medium absorption and fluorescence with doping levels, a family of such curves can be generated and optimal performance regions determined. Figure 7.6 schematically shows such a curve for ED-2 laser glass.

An issue with xenon flashlamps is the relation between flashlamp lifetime and the relative electrical loading. In flashlamp testing in isolated circuits a body of data exists which suggests a lifetime versus loading relation of the form

$$\text{lifetime} = (Ex/E)^n$$

with the explosion energy $E_x = 2 \times 10^4 \, ld(T)^{1/2}$ (Joules). Typically, the lifetime exponent is suggested as $n = 8.5$ to $8.8$ for linear lamps [21, 22, 27]. The only data available on large helical lamps [28] seem to follow the same trend as a function of loading but with a lower explosion energy, possibly caused by mechanical flexing due to the repulsive currents between adjacent turns. Several precautions are necessary in using these data and extrapolating to service in a laser pump. Lamp radiation in a pump cavity that is emitted and then reabsorbed by the lamp plasma should be equivalent to additional electrical loading. Multiple-lamp pump geometries may also introduce additional stresses not present in single-lamp geometries. In most multiple-lamp disk amplifiers used in the fusion program a geometry is used such that the current is reversed in adjacent lamps to reduce the not-inconsiderable magnetic forces.

In pursuing flashlamp scale up issues for the laser fusion program data were developed which indicated that up to 15-mm bore diameter, quartz envelope lamps generally follow extrapolations from data on smaller lamps [29]; 20-mm-bore lamps appeared to have anomalously high catastrophic failure rates [30]. Catastrophic failure is certainly a concern with multiple-lamp pump geometries. The capacitor bank energies may typically be 5 kJ per lamp. If a lamp is broken before it is fired or breaks during a shot a low resistance, a high current arc may occur, causing substantial physical damage to other lamps and components of the laser amplifier. Although lamps are tested at the manufacturer, shipping damage, sometimes concealed, does occur. A retest before installation is certainly recommended. At NRL a further degree of protection was implemented. The metal case of laser amplifiers was isolated from the system ground by a 30-k$\Omega$ high-voltage resistor. The capacitor bank modules that fire two lamps per module use spark gap switches. If a short circuit develops because of lamp failure, the resistor limits the module current to a low-enough value that the switch opens detaching the module. This system worked in worst-case testing (installed electrodes, no lamp envelope) and has worked in practice over the past decade. A similar strategy is strongly recommended. For nonstorage laser applications, data strongly suggest that the lower the peak current density for xenon lamps, the higher the efficiency, at least for Nd:glass lasers. Figure 7.7 shows some LLNL results which suggest that the softer the spectrum from xenon lamps, the higher the pump efficiency [31]. There are two caveats for this type of laser in terms of currently undefined or poorly understood areas. As the pulse is stretched out in time, the optimum circuit may change and flashlamp damage or lifetime issues may change. For electrical pulses into flashlamps beyond a few milliseconds in duration, the larger circuit inductor becomes a high-cost item for single-mesh circuits. As moderately high voltage (2 kV), high-current solid-state diodes exist, for long pulses a multimesh switched circuit may be economically preferable in some applications.

Flashlamp failure models in existence are for rather high current density flashlamps. As this translates to relatively high optical thickness flashlamp plasmas, most of the

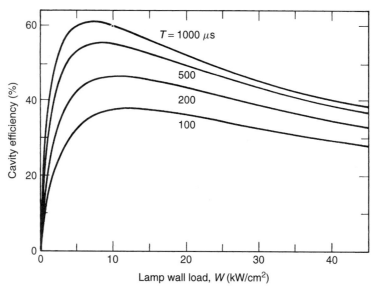

**Figure 7.7**  Cavity efficiency as a function of lamp power density for various flashlamp pulse durations. (From Ref. 31.)

dissipated heat load is into the lamp envelope; the electrodes are heated only by the fraction of the plasma within an optical depth of the end of the lamp. Long-pulse lamps for multi-millisecond lasers may have lamp plasmas which are optically rather thin; a larger fraction of the energy dissipation may be via absorption in the electrode structure. This could lead to different failure modes and statistics than in the high-current-density regime, but there are few relevant pulsed data.

Measurements have been reported on the overall radiation efficiency of xenon flash-lamps. For small lamps such as are typically used with Nd:YAG lasers, efficiencies of 50 to 60% from capacitor bank to radiation are typical [32]; for the larger-bore, longer lamp's used for large neodymium–glass lasers, somewhat higher efficiencies, on the order of 80%, have been reported [33]. Reliable triggering of xenon flashlamps was a problem in the early days of flashlamp-pumped lasers. Improved manufacturing procedures and the development of reliable triggering circuitry have eliminated these problems for most small lamp applications, and the manufacturer can provide useful information to the user in this area. The large higher-voltage lamps used in large glass lasers generally require no additional trigger source. When the bank is switched into the circuit with the lamp, the lamp initially will appear to be an open circuit. This will cause a ring-up of the voltage sufficient to trigger the lamp for bank voltages above 8 kV. One additional simplification was found for multiple-lamp arrays. The ultraviolet radiation from the first lamp module to fire will cause adjacent lamps to fire very promptly even if they have inferior triggering characteristics on their own [34].

### 7.2.2  Arclamp and Filament Lamp Operation

Arclamp and tungsten filament lamp operation has been reported for a wide variety of solid-state lasers both at room temperature and at reduced temperatures for the laser host. With these continuous pump sources, stable long-lived operation can be obtained, but efficiencies are generally not very high, typically 0.1% for CW Nd:YAG and somewhat higher for cryogenic lasers such as Er:Tm:Ho:YAG (in which erbium and thulium are sensitizers and $Ho^{3+}$ is the laser species). The issues with CW lasers are somewhat different than for pulsed lasers. It is not primarily spectral overlap of the pump with the laser species absorption bands which limits efficiency but issues related to spontaneous emission and power extraction loss for these lasers with much lower gain than pulsed lasers.

For CW laser extraction as stimulated emission to dominate over fluorescence radiation, the power density must be greater than a CW saturation flux defined as

$$P_s = h\nu/\sigma T_f$$

where $T_f$ is the fluorescence lifetime. For high-gain solid-state lasers such as neodymium YAG, $P_s = 2000$ W/CM$^2$, and indeed multistage CW YAG lasers have been built at the kilowatt levels with reasonable overall efficiencies in the 2% range [35].

To achieve this in a low-power CW oscillator is much more difficult. To have a watt out and a kilowatt or more per square centimeter circulating power in the cavity requires not only high-reflectivity resonator mirrors (typically >99.9%) but very small internal resonator losses (typically, <0.1% per pass). Components of the internal loss that must be controlled include absorption loss from the lower level,

$$\gamma_L = \sigma g N_D e^{-h\nu_g/kT}$$

where $g$ is a degeneracy factor between lower and upper laser levels, $\sigma$ is the absorption cross-section, $N_D$ the doping of lasant species, and $v_g$ the separation of the lower level from the ground state. Host absorption and scattering as well as coating losses can also strongly affect the efficiency.

### 7.2.3   Diode Laser Pumping

GaAlAs laser diodes which emit near 0.8-μm in wavelength were demonstrated to be a possible pump source for neodymium lasers some years ago [36]. An appealing feature of diode pumping was the high quantum efficiency and the small amount of thermal heating expected for this scheme. Several factors have kept this technology a laboratory curiosity and prevented broader-scale application. These include both cost and applicability issues.

To date, the cost of laser diode pump sources has been prohibitive compared to that of other pump sources. Typical single-mode laser diodes emit average powers of 10 to 20 mW of radiation. Until very recently such diodes were also very expensive and in very limited production. Adding to the cost issue is an issue of the pump duty cycle. For neodymium ions in most host materials the fluorescence lifetime is 200 to 250 μs, and pumping for longer than this would not increase the population inversion. For the low repetition rates typical of solid-state lasers, 10 to 100 Hz, the duty cycle will be 1% or much less. Additional factors that make the economics difficult include (1) the necessity to select only that fraction of laser diodes whose output wavelength matches the absorber pump bands, and (2) inefficiencies in absorbing and extracting the diode pump photons.

These factors present a superficially compelling economic case not to consider laser diodes as laser pumps. Over the long term this would probably be a mistake comparable to dismissing the transistor in the same way that 15 years ago, one could have argued that tubes were superior since it generally takes several transistors to replace one tube. The fallacy is that solid-state components have the overriding advantage that once they are produced with the right characteristics, further production is amenable to automation. The first large-scale commercial applications for laser diodes are only now becoming real. Fiberoptic communications have been the application that has led to most research on development of structures for semiconductor laser diodes and techniques such as liquid-phase epitaxy (LPE) and metal-organic chemical vapor deposition (MOCVD) have been used to produce the necessary semiconductor laser structures. Reasonably efficient diodes with a 20 to 25% wall plug efficiency are commercially available, and even higher efficiencies of 45 to 52% have been reported recently for MOCVD fabrication [37, 38]. In addition to GaAlAs laser diodes that lase at 0.8 μm, InGaAsP diodes have been developed that lase in the region 1.3 to 1.5 μm, where silica fiber-optics have maximum transparency. A much stronger factor in reducing laser diode cost than communications is the emerging market for laser video disc recorders. For this application, where sales may shortly be expected to be in the millions of units per month worldwide, many more laser diodes are needed than for communications and at as low a price as possible. Current cost goals appear to be in the range of $8 to 10 per diode, a significant decrease from earlier prices. Future-generation video disc recorders with random-access capability and read/write capability will require order-of-magnitude higher laser diode powers but may have about the same cost. A number of companies in the United States, Japan, and Europe are working on developing linear laser diode arrays to meet this need with potentially similar cost goals for arrays to those currently found for single diodes. The largest arrays currently available commercially have 40 elements and produce 500 mW CW, although a chip with 400 elements arranged in 10 adjacent arrays has recently been reported [39].

The arrays reported to date in self-excited operation do not in general have optimal spatial or spectral outputs. Typically, the lowest loss mode for an array is one in which adjacent diodes lase 180° out of phase with each other. The frequency spectrum is also typically complex. By use of injection locking, much more ideal operation can be obtained. Figure 7.8 compares the self-excited and injection-locked spectral and spatial properties of a 100-mW array [40]. Much work is ongoing to obtain well-controlled self-excited arrays.

In addition to neodymium lasers, some of the other rare-earth lasers may be even more attractive candidates for diode laser pumping. Holmium (2.05 μm) and thulium (2.02 μm) lasers have rather long upper-laser-level lifetimes in solid-state hosts, more than 10 ms in most cases. With co-doping with erbium or thulium, absorption of laser diode radiation and subsequent energy transfer could result in CW or repetitive action with a very high duty cycle for utilizing laser diode radiation. Recently, CW laser operation was demonstrated in laser diode-pumped $Ho^{3+}$ (2.06 μm) at 77 K [41], erbium (2.8 μm) at room temperature [42], and CW laser-pumped $Ho^{3+}$ laser operation at room temperature [43].

In short, technology appears to be advancing in directions which indicate that laser diode pumping will emerge over the next few years as an economically feasible pump technology as well as a very efficient and reliable technology. Although it certainly must be labeled "work in progress," it is an area that appears to have a significant future.

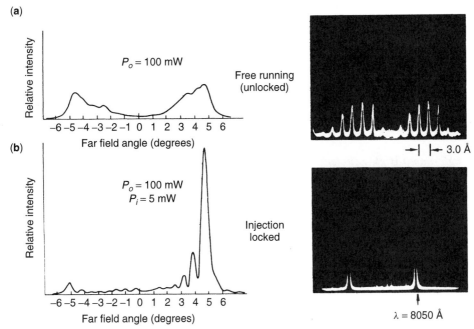

**Figure 7.8**   Injection locking of a diode laser array. (*a*) Free running far-field pattern and spectrum of a 10-stripe diode array. (*b*) Injection-locked patterns from the same array. (From L. S. Goldberg and J. F. Weller, Ref. 40.)

### 7.2.4   Pulsed or CW Laser Pumping

Laser pumping is an excellent technique for understanding the physics of new laser systems because the high spectral brightness of the pump laser allows excellent energy accounting. With dye lasers, vibronic solid-state lasers, and color center lasers, tunable radiation can be obtained between 0.4 and 4.0 $\mu$m. Short pulses can be obtained from these sources, which allow essentially instantaneous excitation of a given energy level. Subsequent radiative or nonradiative decays can then be readily studied with very accurate energy accounting. The high brightness of the sources allows small but well-defined volumes to be excited to relatively high gains. This can allow good characterization of relatively imperfect developmental laser materials.

The approach that has been used in most reported laser pumping experiments is axial pumping, where the optical axis of the pump beam and of the sample resonator are coincident or nearly so. The advantages of this approach are that relatively complete absorption of the pump radiation may easily be obtained; at the same time the transverse spatial distribution of the resultant gain will not be a function of the absorption strength but of the spatial pattern of the pump on the test crystal. If the crystal is in or near the far field of the pump (e.g., at the focus of a lens), the pump pattern will approximate a Gaussian pattern even if the pump is far from diffraction limited. The principal disadvantage of axial excitation is that it does not scale well to higher average powers, as the longitudinal heat distribution is very uneven.

### 7.2.5 Pumping Cavity: Geometrical Considerations

Over the past two and half decades a number of different optical pumping cavities have been explored for the pumping of solid-state lasers with flashlamps and CW lamps. Virtually every fabricatable geometry of lamp and pumping enclosure has been tried at one time or another. A wide variety of coolants and dopants for the coolants have been employed for various purposes. Rather than describe the history of this field in detail, we will treat three particular cases that cover most of the systems in use, which have been relatively successful.

***Single-Lamp High-Efficiency Nd:YAG Lasers.*** For most of these lasers where single-mode or high-brightness performance is not required, the pump enclosure in most current usage is as shown in cross section in Fig. 7.9, a close-coupled pump geometry [44]. The main cavity member for storage lasers is a block of samarium-doped glass with holes drilled in it (and polished) for flashlamp and laser rod and their associated cooling channels. The $Sm^{3+}$-doped glass serves to suppress transverse lasing in parasitic modes, which was found to limit the energy storage otherwise to 150 mJ for 5- to 6.6-mm-diameter Nd:YAG laser rods [45]. The coolants in most successful usage are distilled water or water–ethylene glycol mixtures, depending on the temperature range required for operation. Silicone rubber O-rings are generally used because of their superior resistance to charring when irradiated by flashlamps, in comparison to Viton or rubber O-rings. The reflector applied to the external barrel is generally a barium sulfate compound applied by proprietary processes. Typical long-pulse laser performance with neodymium YAG for such a cavity is about 3% overall efficiency at 5 to 10 times threshold. The near-field radiation pattern is generally somewhat "egg-shaped," with the broader, more intense part of the pattern on the side of the rod closest to the flashlamp.

***Single-Lamp Single-Mode Nd:YAG Lasers.*** The pump cavity most typically used for this type of laser is elliptical in cross section (Fig. 7.10) with the flashlamp along one foci and the rod along the other foci [46–48]. Typically, for good azimuthal symmetry a ratio of major axis to minor axis around 1.2:1 was found to be a good choice which also allowed ease of access to rod and flashlamp ends. An 8-cm major axis ellipse of this eccentricity can be fabricated in one piece up to 6 cm long by tilting a circular cutting tool at the appropriate angle. The end plates and elliptical barrel are then polished to a

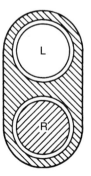

**Figure 7.9**  Close-coupled pump cavity.

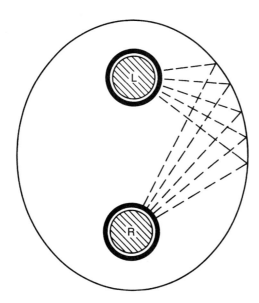

**Figure 7.10**  Cross section of an elliptical pump cavity.

smooth surface and plated, generally with gold. This procedure works reasonably well for aluminum alloys or brass cavity members. In studying the influence of features of the construction of the flashlamp and laser rod cooling jackets on the reproducibility of laser performance, several subtleties were observed [49].

With 4-mm-bore flashlamps at pump energies of 10 to 15 J, the flashlamp arc did not completely fill the lamp bore, giving a pulse-to-pulse wander of the arc center and hence its image in the laser rod. With a single-mode selecting aperture in the resonator cavity, this decentering could give significant variation in pulse-to-pulse diffraction losses and hence output energies. Use of a 3-mm-bore lamp significantly improved this situation. Operation of the lamp in a "simmer" or conducting mode reportedly achieves the same result [50].

A "thermal damper" sleeve [51] of fused silica was placed around the laser rod inside the water jacket. The objective was to reduce lateral movement of the thermal center of curvature of the rod caused by time-dependant changes in the turbulent flow pattern of the coolant around the laser rod. Without the sleeve the output energy would oscillate through a 10% variation with a characteristic period of a few seconds and little sensitivity to repetition rate. With the sleeve this oscillation was essentially eliminated.

*Large-Volume Nd:Glass Rod Amplifiers.*  Several approaches to multiple-lamp pump geometries have been developed for large-volume (100 to 1200 $cm^3$) neodymium–glass lasers. The most successful of these use laser rods in cylindrical water jackets and arrays of either linear or helical flashlamps. Experience suggests that there are several variations which should be avoided because of potentially disastrous operating consequences. These include:

- Immersed or flooded cavities. In the event of a cracked flashlamp, spectacular damage can result on the next pulse because of the relatively large mechanical forces produced by water displaced by an arc in the cavity.

- Irradiation of the laser glass with a pump spectrum beyond the ultraviolet absorption edge of the laser glass can result in very large thermal stresses at the edge of the rod, leading to permanent birefringence being established in a few tens of pulses. While the stress-induced birefringence can be annealed, the rod must be repolished, a sometimes risky process.

- Some of the phosphate laser glasses are somewhat water soluble. Use of water as a coolant can lead to measurable erosion of the barrel of the laser rod over time. At the University of Rochester it was found that this effect could effectively be prevented if cooling solutions were used which were more than 50% ethylene glycol by volume [52]. Although this solution is recommended, there are two caveats. In humid environments ethylene glycol solutions are hygroscopic and will take up water from the air. The system should be sealed. Over a period of time, ethylene glycol solutions irradiated with strong ultra-violet sources will become less transparent in the ultraviolet. If the neodymium pump band in the near ultraviolet is being used, the solution should be regularly checked on a spectrophotometer and replaced as indicated.

- Although not a catastrophe of the same magnitude as the above, parasitic oscillation of large storage mode laser amplifiers has been shown to limit the achievable gain unless active measures are taken to prevent this [53]. [With the newer phosphate laser glasses that have very low water (OH ion) content, parasitic oscillation for neodymium lasers on the $^4F_{3/2}-^4I_{13/2}$ transition at 1.35 μm, as well as on the stronger $^4F_{3/2}-^4I_{11/2}$ transition at 1.054 μm, is possible.]

Multiple-helical-lamp pump geometries were pioneered by the Companie Générale d'Electricite in France [54]. An advantage of helical lamps was that a standard capacitor bank module could be defined around available reliable 12-kV capacitors and switch gear. A family of electrically identical lamps (in bore and arc length) could be made with varying numbers of turns to pump laser amplifiers from 16 mm diameter to 90 mm diameter. A potential weakness of this approach, the necessity to disassemble the amplifier to replace a flashlamp, was minimized by developing rather long-lived flashlamps. A number of improvements were made at NRL to increase lamp life.

Probably the earliest and best documented approach to multiple-linear-lamp pump geometries occurred in the Soviet Union [55, 56]. Work undertaken during the development of the GOS-1000 laser amplifier revealed rather good azimuthal symmetry for geometries with six or more symmetrically arrayed linear flashlamps. In more recent work both in France and at LLNL [57, 58], geometries have been identified in which lamp and reflector placement and shape have been optimized to give very uniform gain profiles. In the latter cases the design trade-off is between the direct (essentially unfocused) pattern and the reflected (essentially focused) pattern for the flashlamp.

## 7.3 LASER OSCILLATORS

A laser oscillator is, in its simplest realization, a laser rod with mirrors to provide optical feedback, as in Fig. 7.11. At some point in time, $t_0$, the flashlamp is energized. At a later point in time, $t_1$, the stored population inversion is large enough that the gain exceeds the losses and laser action begins. For much of the 1960s, scientists studying solid-state lasers were studying the wide variety of experimental phenomena which

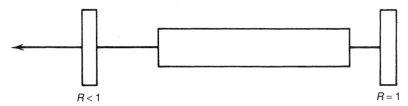

**Figure 7.11** Simple laser resonator with one totally reflecting mirror and one partially reflecting mirror.

were observed in this seemingly simple case, or others which appeared not much more complicated.

In contrast to the fairly low gain CW gas lasers then known which appeared well behaved and orderly in their output characteristics, solid-state lasers appeared to have very chaotic and unpredictable properties. Much of the confusion evident in the early literature was traceable to the poor optical quality of early solid-state laser materials and optical elements as well as high-power effects which are generally not observed with continuous lasers.

This situation is much better understood now than in the early days, due to the contributions of many researchers. What has come to be realized is that there are a hierarchy of different processes which coexist in even such a simple oscillator as that shown in the figure. By invoking them in the proper order, relatively well controlled and characterizable laser outputs can be obtained.

The first point to recognize is that our simple oscillator is really not so simple if it is a solid-state laser. In addition to the cavity reflectors, one presumably a 100% reflector and the other a partially transparent coupling mirror, $R = R_c$, there will be a variety of losses present. These could include scattering and absorption in the laser rod. This loss could also include spatially ordered distortions in the laser medium; many early ruby laser rods produced by the Verneuil process had lenslike aberrations frozen into them by the thermal gradients present in the growth technique [59, 60]. A large amount of small-angle scattering caused by many small dislocations in the crystal could also effectively mix the low-order spatial modes, making spatial mode selection in the sense understood from gas lasers hard to achieve. Adding to the richness of the phenomenology was the fact that the resonator mode spacing $k = (2L)^{-1}$, where $L$, the "optical length" of the resonator, is generally much smaller than the line width, $k_0$, for solid-state lasers, and the cavity Frensnel number, $N = A^2/\lambda L$, where $A$ is the beam radius, is frequently a large number. Unlike the situation with gas lasers where $k_0$ was small and a long narrow resonator (small $a$, large $L$) was used, with a solid-state laser many spatial and spectral modes of the resonator could be excited.

One of the more powerful discoveries of this period was that in almost all cases it was possible to separate the problems of mode selection and energy extraction from solid-state laser oscillators and find mutually compatible solutions which gave both good extraction and desirable mode properties. This is true because of the finite time over which the phenomena occurs. The desired mode need not be the only one over threshold in an absolute sense; as long as it is the fastest-growing mode, it will dominate the performance. A corollary observation is worth noting; losses can be very high when the intracavity flux is low without much affecting the efficiency, as long as they are low at high cavity fluxes.

### 7.3.1 Continuous and Long-Pulsed Oscillators

Often, the performance of some new or improved solid-state laser material is reported in the scientific literature in terms of its performance relative to some other more widely known laser material. A recent example would be the reports of the performance of Nd:Cr:GSGG versus that of Nd:YAG [8, 7]. In this particular case, a relatively complete report was published showing that because of the chromium co-doping, the neodymium $^4F_{3/2}$ level could be populated both by direct flashlamp pumping of neodymium and by nonradiative transfer from chromium ions pumped by the flashlamp. Lasing performance confirmed that about twice the overall efficiency could be obtained as in neodymium:YAG. Even with relatively complete reporting it may be difficult to compare a new material to other more established materials because of other factors. Nd:Cr:GSGG has been reported to have much greater thermo-optic lensing than Nd:YAG, for example [61]. Many reports in the literature on new materials are much more fragmentary and can cause confusion rather than enlightenment.

Most fundamentally, this is because in a laser oscillator we are trying to simultaneously optimize more parameters than there are controls. To optimize the extraction of the pump energy converted to population inversion, the intracavity photon flux should be made as large as possible to increase the stimulated emission rate compared to fluorescence decay. Increasing the photon flux corresponds to increasing the feedback by decreasing the cavity loss. However, the total loss rate $\gamma$ per transit is composed of two parts, $\gamma_L$, caused by losses in the laser, and $\gamma_R$, the coupling loss ($\gamma_R = -\frac{1}{2}\ln R$); the total useful output will be $\gamma_R/(\gamma_R + \gamma_L)$ and the better the coupling, the poorer the feedback.

Schematically, if a long pulsed oscillator is operated over a range of pump energies, the output energy will vary with excitation level as shown in Fig. 7.12. There are three regimes of operation. In region I, until the oscillator reaches threshold, it is essentially operating as a storage laser. If different $R_c$ are used, it is possible to obtain the gain as a function of pump energy. If the gain is relatively uniform, $\alpha(E = 0) = -\gamma_L$ and the passive loss can be determined (Fig. 7.13). In region III, far over threshold, the overall efficiency is approximately constant as a function of pump energy, and by varying $R_c$ the best efficiency is obtained by trading off coupling fraction against feedback. Anan'ev et al. present numerical results for optimizing typical long pulse oscillators in tabular form [62]. The intermediate region II between threshold and quasi-CW behavior is one in which the efficiency is changing rapidly. The rate of change of output in this region, often called the slope efficiency, is frequently reported as an efficiency measure. This is probably the worst measure to use, as it has the least straightforward relationship to intrinsic laser properties.

The output coupler is typically a partially reflecting mirror which reflects an optimum fraction of the incident radiation back to maintain the feedback. It can assume other forms. Anan'ev has shown that in curved mirror resonators a small mirror can be used in an unstable or telescopic resonator which will act as though it had a reflectivity of $M^{-2}$, where $M$ is the geometrical magnification [63]. This type of resonator is discussed in more detail in Section 7.3.5.

Another coupling approach is to use 100% feedback mirrors but have an intracavity frequency converter, such as a frequency-doubler crystal. Initially, as the oscillation builds up, there is almost no outcoupling or frequency conversion because the second harmonic intensity scales as the square of the intracavity intensity. If the intracavity

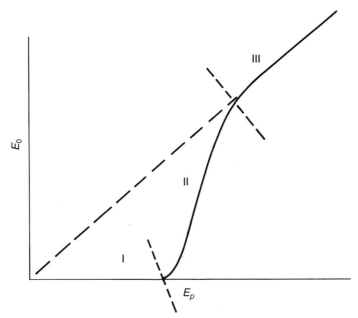

**Figure 7.12**  Output energy versus pumping energy for a long-pulse laser oscillator. In region I, up to threshold, the laser is operating in a storage mode. In region III, the laser is essentially a CW laser. In the transition, region II, the efficiency changes rapidly.

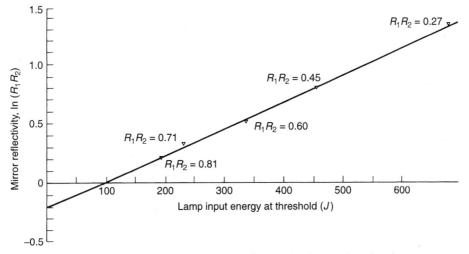

**Figure 7.13**  Threshold versus cavity loss. Both the passive loss and peak gain versus pump energy can be determined. (From Ref. 4.)

converter is picked to give a conversion which is about the same as the optimum cavity reflector, about the same efficiency can be obtained, but with an output at the second harmonic. In one of the first demonstrations of this process, 100% harmonic conversion efficiency was reported under conditions where the external conversion per pass would have been only a few percent [64].

By insertion of a Pockels cell electro-optic switch in the cavity, oscillation can be held off for a time up to about the fluorescence lifetime and the stored energy dumped as one or more sequential short pulses. This can have the result of increasing the ratio of peak to average power by a factor as large as $T_{Fl}/\Delta t$; where $\Delta t$, the duration of the $Q$-switched pulse, is typically tens of nanoseconds. As $T_{Fl}$ is typically $10^{-3}$ to $10^{-4}$ s for solid-state lasers, this enhancement can be very large, a factor of $10^5$ to $10^6$. This case is discussed in more detail in the next section.

### 7.3.2   $Q$-Switch and Short-Pulse Oscillators

In 1963, Hellwarth and McClung demonstrated the first $Q$-switched oscillator [65]. In such an oscillator, by contrast to a long pulse oscillator, a switch is used to prevent laser oscillation until a large amount of energy is stored as population inversion. If the switch is then turned on, oscillation can quickly build up and all the energy can be emitted in a brief single pulse.

Because of the high intensities that could be generated in this fashion, this type of laser oscillator had a profound effect on the development of the field of nonlinear optics. Intensities could be generated which were high enough to allow the observation of processes such as stimulated Raman scattering [66], stimulated Brilloin scattering [67], and parametric processes such as harmonic and sum and difference frequency generation. It was also possible to create and study laser-produced plasmas [68] and generate optical self-focusing [69] of intense laser beams.

The optimization of $Q$-switched lasers was studied by a number of researchers, including Wagner and Lengyel [70], Szabo and Stein [71], and Menat [72]. The models that were developed were useful for optimizing such oscillators and as a means of understanding the operation of this important kind of oscillator. In their 1963 paper, Wagner and Lengyel showed that a simple closed-form analysis could be done for several key features of $Q$-switched laser pulses. Imagine a laser resonator with single-pass gain $\alpha_0$ and total losses $\gamma_T = \gamma_0 + \gamma_C$, a transit time of the cavity $t_1$ and ask what will be the inversion remaining at the peak of the pulse and what will be the final inversion at the end. It was shown that compared to the initial inversion $n_0$, the inversion at the peak, $n_p$, was $n_p = (\gamma_T/\alpha_0)n_0$, and the inversion at the end of the pulse $n_f$ was a function of $n_p/n_0$ only.

Rewriting $p = n_p/n_0$ and $r = n_f/n_0$, several results can be obtained. The intracavity power scales as $p(1 - p + p \ln p)$ and $r = \exp[(r - 1)/p]$. Figure 7.14 shows the evolution of the $Q$-switched pulse, and the inversion and Fig. 7.15 shows $r$ as a function of $p$. Figure 7.16 shows how the intracavity pulse shape varies as a function of $p$ for several values of $p$. It can be seen that as $p$ decreases (i.e., cases with high extraction), the pulse becomes more peaked toward the front (i.e., short leading edge and long trailing edge).

In analyzing pulse shapes, a useful parameter to analyze is the pulse width. Figure 7.17 shows the ratio of the pulse energy to the peak power as a function of $p$. A striking feature of this result is that the minimum value for duration of the $Q$-switched pulse occurs for values $0.25 < p < 0.30$, or for $3.3 < \alpha_0/\gamma_T < 4$. If this range of parameters can be achieved experimentally, the pulse duration will be $9.2t_1/\alpha_0$ and will be quite stable (i.e., it will only vary as $\alpha_0^{-1}$, not as $p^n$). Away from this region, especially in the region toward $p \sim 1$, the pulse duration varies strongly as $p$ varies. The extracavity power is simply related to the intracavity power by the ratio $\gamma_C/\gamma_T$. As this ratio equals $\gamma_C/(\gamma_C^+ \gamma_L^i)$ and is a constant during the pulse, the $Q$-switched case has some interesting features:

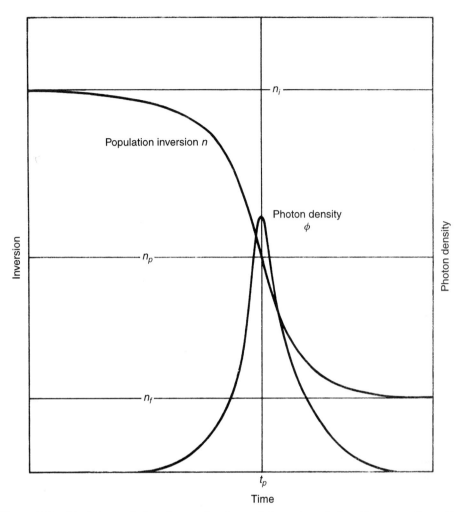

**Figure 7.14**  Evolution of the photon density and inversion during the generation of a $Q$-switched pulse. (From Ref. 70.)

a. While the best extraction, $r = 0$, occurs for very small total losses (i.e., $p = 0$), this case gives very asymmetric (fast rise time, slow fall time) and broad laser pulses.

b. The minimum-duration laser pulses occur for a value of $0.25 < p < 0.33$ even for no loss except for coupling ($\gamma_L = 0$). Near this point the pulse duration is proportional to $t_1/\alpha_0$ and can be shortened by shortening the cavity or increasing the gain.

As a historical note, with these physical results in mind, it is easy to see why neodymium: YAG became a popular oscillator material in the 1960s by comparison to ruby and Nd: glass. With YAG, high gain coefficients of $\alpha_0 = 1$ to $3$ ($G_0 = 2.7$ to $20$) and low internal losses $\gamma_L < 0.1$ could easily be achieved at stored energy densities of $0.15$ to $0.5$ J/cm$^2$. A YAG oscillator could be tuned up to give short duration and rather

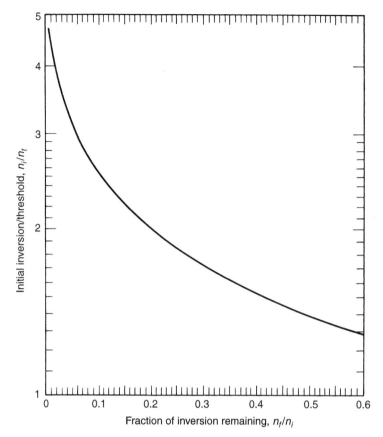

**Figure 7.15**   Final (remaining) inversion fraction $n_f/n_i$ as a function of the ratio of initial inversion to inversion at the peak of the $Q$-switched pulse. (From Ref. 70.)

stable operation at moderate energies consistent with damage-free operation. Glass and ruby oscillators, on the other hand, could operate only in the region of $p < 0.5$, with a much greater risk of damage because of the much higher saturation flux and also higher intracavity flux (i.e., less favorable ratios of $\gamma_c/\gamma_0$).

Another interesting feature of this analysis which was to have a strong impact on efforts to do mode control of laser oscillators was that although when the $Q$-switch is opened, oscillation begins to build up from the quantum noise, virtually none of the stored energy is extracted until within several pulse durations of the center of the pulse. This suggested that fairly dramatic perturbation of the early time buildup could be done with little effect on the shape of the pulse during the energy-extraction part of the $Q$-switched phase.

### 7.3.3   Mode-Locked Oscillators

It was noted earlier in this chapter that solid-state lasers frequently have a bandwidth considerably greater than the cavity axial-mode spacing such that a number of axial modes oscillated simultaneously. In contrast to the complicated mode beating observed when modes would add with random phases, a simpler case results when the modes are

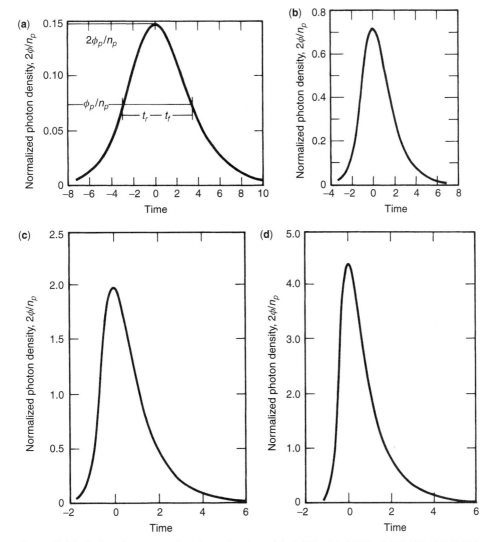

**Figure 7.16** Pulse shape as a function of $n_i/n_p$: **(a)** 1.649; **(b)** 2.713; **(c)** 4.482; **(d)** 7.389. (From Ref. 70.)

all in phase at one location in the resonator. DeMaria was the first to demonstrate that if a cavity element were used, such as a saturable absorber, which would have higher transmission for phase relations between modes such that they were all in phase at the absorber, such a phase-locked condition could be produced [73]. In this case, once every transit of the cavity all the modes would constructively interfere, generating an intense pulse whose duration was the inverse of the bandwidth. The rest of the time the modes would destructively interfere. Fleck [74] and Letokhov [75] numerically and analytically studied the noise statistics of laser oscillators to determine how such in-phase pulses could dominate the noise statistics.

For CW mode-locked gas or solid-state oscillators, mode-locking behavior was consistent with theoretical models. The work of Siegman and Kuizenga on acousto-optic

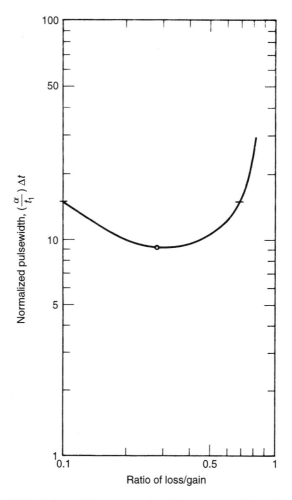

**Figure 7.17**   Pulse width versus ratio of loss to gain. (From Ref. 79.)

mode locking [76] was elegantly verified in experiments on the 1.3-μm photolytic iodine laser by Jones et al. [77], for example.

One group of lasers did not seem to follow these predictions very well, however, and that was the pulsed solid-state mode-locked lasers, especially neodymium glass mode-locked oscillators. While the early part of a mode-locked train from a glass oscillator might resemble a modulated $Q$-switched pulse envelope, the envelope became irregular at high intensity. The spectral and temporal characteristics of early pulses followed theory, but later pulses broadened enormously spectrally. Attempts to measure pulse duration using autocorrelation techniques gave values down to 100 fs time scales but generally with poor signal-to-background levels, indicating that only a fraction of the energy was in these ultrashort pulses. Using recently developed fast streak cameras with infrared sensitive photocathodes, Carman was able to explain qualitatively what was occurring [78]. Initially, an orderly train of nominal 1-ps pulses was generated; as this pulse was sequentially amplified, self-phase modulation would begin to broaden the spectral bandwidth; simultaneously, self focusing would increase the on-axis intensity, further broadening the spectrum by self-phase modulation. In a few pulses the spectrum could broaden to the full 200-cm$^{-1}$ line width of the neodymium glass and the pulse would break up into a burst of subpicosecond noise. The saturable absorber, with a relaxation time of 5 to 10 ps, could not suppress this tendency.

Mode-locked Nd:YAG oscillators, on the other hand, were generally well behaved. The gain narrowed fluorescence linewidth of 1 cm$^{-1}$ was of the right magnitude that the saturable absorber could control further broadening. In this case one could effectively design the mode-locked pulse train envelope and the width of the mode-locked pulses (by controlling the bandwidth with etalons), and well-controlled trains of 30- to 1200-ps pulses could be generated [79]. To avoid damage to optical components, envelope parameters were generally chosen to give no more than 10 to 20% of the energy in the strongest pulse.

For fusion studies this pulse was selected from the pulse train using a fast Pockels cell shutter, and the pulse was then amplified to the desired final energy. Superficially, this technique appeared to give clean subnanosecond pulses. However, in plasma studies, 20 to 30% of the experiments appeared to give anomalously high plasma temperatures as measured by the high-energy x-ray spectrum of the laser-produced plasma. Use of an extra-cavity additional saturable absorber all but eliminated such occurrences, [80]. This result was interpreted as showing that some fraction of the time, weak pre-pulses (<5% of the main pulse intensity) occurred, a not-unexpected result from the analysis of Letokhov [75].

In more recent years two different techniques come into favor for laser fusion pulse generators because of the absence of prepulses. In the first a mode-locked CW oscillator is used to generate a pulse of a desired temporal and spectral shape. This pulse is then switched into a regenerative amplifier and amplified from the 10$^{-9}$-J level to the 10$^{-3}$-J level in this device. This pulse with appropriate delay paths is then used to synthesize desired pulse shapes [81]. The second technique uses a very short cavity resonator to generate single-mode $Q$-switched oscillator pulses with a duration of 2 to 4 ns. [82]. Both Pockels cells and saturable absorber $Q$-switches have been used. The latter technique is somewhat limited by the available $Q$-switching dyes for neodymium, whose absorption only partially bleaches.

### 7.3.4  Mode Selection I: Frequency Mode Selection

As noted early in this section, the fluorescence line width of solid-state lasers is frequently much broader than the resonator axial-mode spacing. For example, for an optical resonator with an optical spacing between the mirrors of 30 cm, the axial modes are 0.0166 cm$^{-1}$ apart. With a crystalline-doped laser such as neodymium in yttrium aluminum garnet [Nd:YAG], the $^4F_{3/2}-^4I_{11/2}$ laser transition has a line width of 7 cm$^{-1}$, so potentially 400 axial modes could oscillate within this line width. It was noted early on in the history of the laser that generally much narrower and simpler spectra are observed. In some cases one or only a few axial modes were observed.

Sooy was the first to explain the narrow line width of saturable absorber $Q$-switched lasers compared to active $Q$-switched lasers [83]. The mechanism he proposed is extremely simple but also very powerful. In an actively $Q$-switched laser the intracavity flux builds up from the fluorescence level as rapidly as the exponentiation of the gain will allow to a flux that saturates the energy stored in the laser rod. There are only a small number of passes required for the flux to build up, typically 40 to 100. In a saturable absorber $Q$-switched laser, in contrast, the flashlamp pumps the laser rod over threshold gradually. Initially, oscillation begins with a very low gain per pass, just over threshold. It takes many passes for the gain to slowly increase the intracavity flux to the intensity at which the absorber begins to bleach, or saturate. Only once this has happened will the full gain be experienced, with a fast buildup to high intensity. During all of these many transits, only the parts of the spectrum near the line center will experience full gain and will build up at the expense of the wings of the spectral distribution. If the laser has a fluorescence spectrum with a Lorentzian line shape with a width $DK_0$, it is easy to show that after $n$ gain passes it will have narrowed to $DK' = DK_0 (\ln 2/n)^{1/2}$. In a saturable absorber $Q$-switched oscillator, $n$ is typically 1000 or so, compared to 40 to 100 for an active $Q$-switched laser. Only a few modes can oscillate with high gain.

Other line-narrowing mechanisms may be used to further narrow the line so that only one mode results. If the cavity has other Fabry–Perot resonances than between the end mirrors, these subsidiary resonances can be used to narrow the spectrum. If fluorescence narrowing reduces the spectral width by a factor of 40, as in the previous example, use of an intracavity etalon with about this free spectral range will give further narrowing.

It was realized that active $Q$-switches could be made to operate in a fashion analogous to saturable absorbers by opening slowly or in a two-stage fashion in which a long low transmission phase is followed by a high transmission during energy extraction [84, 85]. These techniques also allowed the optical pulse to be timed to external events much more accurately than with a saturable $Q$-switch.

A more recent approach to obtaining single-mode behavior has been to "seed" the oscillator with energy from a single-mode continuous oscillator. As long as the CW oscillator intensity is much stronger than the fluorescent emission into the laser mode volume, the pulsed laser spectrum will follow the injected laser. A recent and very compact form of such a seed generator was reported in which a GaAlAs laser diode was used to axially pump a short, 0.1-cm Nd:YAG etalon [86]. Because the etalon was so short, it could oscillate on only one of its axial modes with a line width of 10 kHz.

In all of these cases, frequency-narrowed operation can be obtained without any penalty in extraction efficiency. While this is obvious for the injected cases, it is also true for the saturable absorber case. As Fig. 7.18 illustrates, during the long buildup, the laser is oscillating at such a low intensity that effectively no energy is wasted during the precursor phase.

### 7.3.5 Mode Selection II: Spatial Mode Selection

Much of the early theory of spatial modes in lasers had little apparent connection with many of the early observations of the output radiation patterns from solid-state lasers. In the early and still relevant work of Fox and Li [87] and of Boyd and Gordon [88], radiation patterns were found which would reproduce themselves in phase and amplitude as they bounced back and forth in resonators with mirrors of various shapes and curvatures. The losses per transit for various patterns was found to depend on the Fresnel number of the resonator, $N$, where $N = a^2/\lambda l$, with $a$ the mirror radius, $l$ the mirror separation, and $\lambda$ the wavelength. Figure 7.19 compares the loss per pass for the lowest-order mode distributions. In direct analogy to the microwave region the modes were designated the $TEM_{nm}$ modes, with the Gaussian $TEM_{00}$ mode being the lowest-order pattern.

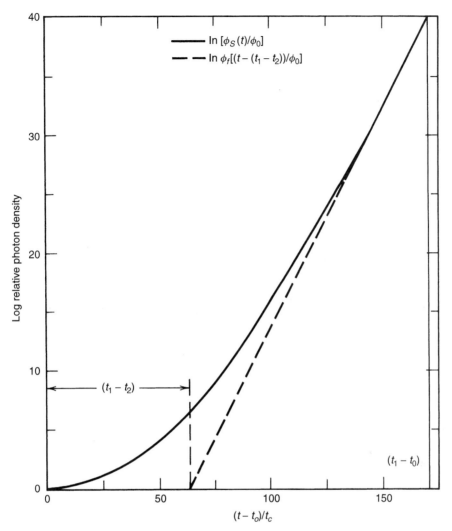

**Figure 7.18** Pulse buildup in a slowly $Q$-switched laser (solid line) and a fast $Q$-switched laser (dashed line). (From Ref. 84.)

The output from early long-pulse or $Q$-switched lasers did not bear much apparent similarity to this orderly picture, in contrast to the case with gas lasers, which often appeared to be textbook reproductions of the theory. The near-field patterns were frequently irregular and the far-field patterns frequently showed a divergence far above the diffraction limit for a $TEM_{00}$ mode. One reason was that many early solid-state laser resonators had Fresnel numbers far above unity, so that little mode selection or differentiation would be expected. Many modes could compete with very low loss. The very poor optical quality of early solid-state laser hosts such as synthetic ruby could also scramble the phase and amplitude distributions of the cavity modes. In addition, in many $Q$-switched lasers, oscillation built up so rapidly from spontaneous emission that there was not a sufficient number of cavity transits to set up cavity modes [84]. The spatial pattern in these cases was simply the gain-narrowed fluorescence pattern

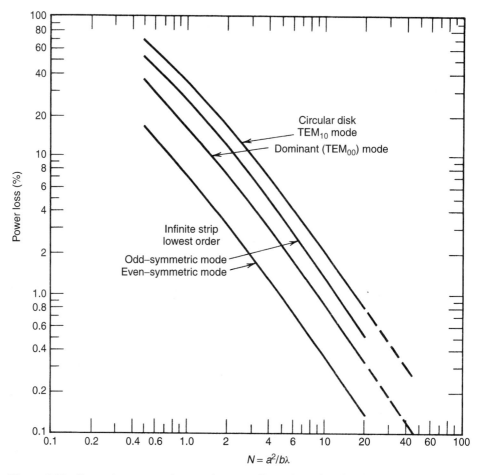

**Figure 7.19** Power loss per cavity transit versus Fresnel number for the lowest-order cavity modes. (From Ref. 87; copyright 1961 AT&T, reprinted with permission from the *Bell System Technical Journal*.)

apertured by the resonator. With slowly opening $q$-switches (to obtain frequency-mode selection), and with improved-quality laser hosts it became possible to observe $TEM_{00}$-like spatial patterns from solid-state lasers in the late 1960s. However, it proved difficult to obtain mode sizes greater than about 1 or 2 mm from most lasers. Osterink and Foster showed that this could be due to thermal lensing, causing the mode radius $a = (b\lambda/\pi)^{1/2}$ (where $b$ is the radius of mirror curvature) to have these characteristic dimensions, and also by correcting this thermal curvature they obtained large mode diameters [89]. Such a correction, of course, applies only to a steady-state condition unless an adaptive optic is used.

In the later 1960s several groups began to investigate the properties of unstable resonators, also called telescopic resonators. In such a resonator all of the rays appear to originate at a common imaginary focal point. This topic is covered most thoroughly in the publications of Anan'ev in the Soviet literature, [90, 91]. Figure 7.20 illustrates several types of such resonator. With the laser glasses in use at that time, which were

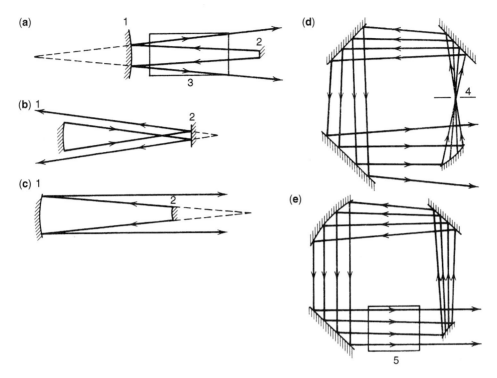

**Figure 7.20** Different mirror configurations for unstable resonators. (From Ref. 90.)

all variants of American Optical laser glass type 3669A, *dn/dT* had a negative value. This resulted in a resonator curvature that was slightly unstable. Oscillators made of this glass with high-quality optics would have a far-field pattern which was nearly diffraction limited, with a near-field pattern that resembled a Fresnel pattern from an aperture [92]. This really was an unstable resonator with a transparent coupling mirror but with a magnification only slightly greater than unity, so that mirror spillover was a small loss. In the more traditional unstable resonator treated by Anan'ev, where the two resonator mirrors are totally reflecting but of different diameters, the major loss is the geometric loss at the smaller mirror.

As was the case with frequency-mode selection, a diffraction-limited seed pulse can be injected into an oscillator to control the spatial pattern of the radiation.

## 7.4 LASER AMPLIFIERS

Fairly early on it was realized that, especially for higher-power applications, a single unit could not supply high power and all the desired spatial and frequency mode selection. The concept of the MOPA [or master oscillator; power amplifier(s)] seems to date back to at least 1964 in terms of conception [93]. In a certain sense laser amplifiers are very simple devices; in another sense they are, if not complex, at least unfamiliar in their operating modes to those who tend to refer by analogy to more familiar electrical circuits. There is no feedback or its direct analog in the operation of these laser devices. The performance at high efficiency or utilization will produce output signals which are highly distorted in any sense that a circuit engineer would understand. At the same time a laser designer might claim that all was well under control. Frantz and Nodvik first articulated an expression relating the output photon density (in time and space) to the input photon density [94]. This expression was

$$\phi(x, t) = \phi_0\{1 - [1 - \exp(-\sigma n x)] \exp[-\gamma\sigma\phi_0(t - x/c)]\}^{-1}$$

where $\gamma = 1 + g_2/g_1$ is a degeneracy factor ($\gamma \equiv 1$ for a four-level system)
    $\sigma$ = induced emission cross section
    $n$ = inversion/unit volume

This picture has withstood the test of time for laser amplifiers operated in the storage mode. That this is so for solid-state lasers is interesting and perhaps a bit curious, as the Frantz–Nodvik picture assumes homogeneous broadening of the gain line spectrally and a rather simple picture of the gain media in the sense of being represented by lumped gain and loss parameters. The reality may be much more complicated, especially for rare-earth-doped solid-state laser, where the lasant ion may exist in the host in a variety of characteristic sites, each with its own characteristic absorption and excitation spectra, induced emission spectra, and cross section as a function of wavelength and cross-relaxation time to other sites. There is a body of carefully done experiments at Lawrence Livermore National Laboratory [95–97] which suggests that the situation is indeed more complex. Fortunately, the regions where large departures from Frantz–Nodvik behavior would occur are precluded for other reasons (such as self-focusing or surface damage). There is no guarantee that this will be the case for solid-state systems other than those explored previously. A Frantz–Nodvik picture is very convenient for computation on digital computers, and many of the possible more complex physical situations can be modeled by straightforward extension of the formalism.

Historically, the extension of the Frantz–Nodvik formalism to explain the significant role that losses could play in both pulse shape distortion and energy extraction came very early. This is because the first solid-state medium in which people tried to execute the oscillator–amplifier idea at high power, ruby [$Cr^{3+}$ doped into sapphire ($Al_2O_3$)], was a rather lossy medium. Avizonos and Grotbeck reported a sequence of oscillator–amplifier experiments in which beyond a certain point adding additional amplifiers did not increase the pulse energy [98]. Their explanation for this phenomenon was rather clever. They noted that while the Frantz-Nodvik expression always predicts gain for the leading edge in time of a laser pulse, integration of the Frantz-Nodvik equation over the pulse envelope yields

$$\frac{dE(x)}{dx} = E_s g_0 \left[1 - \exp\left(\frac{-E(x)}{E_s}\right)\right] - \alpha E(x)$$

where $E_s = h\upsilon/\gamma\sigma$ is the saturation energy density.

The gain for the whole pulse will in fact go to zero at any energy density relative to the saturation flux, which is equal to the gain-to-loss ratio for the amplifier. If a pulse is introduced with an energy density above this value, it will in fact be attenuated while a weaker pulse would see amplification. Basov et al. [99] observed another manifestation of the same phenomenon. In a lossy amplifier the leading edge of the pulse will always see gain. If the temporal shape is a rising exponential, as is the case for the leading edge

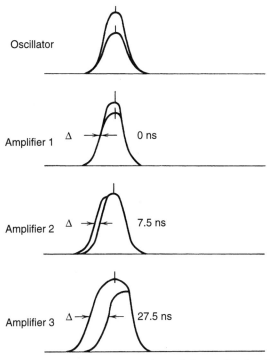

**Figure 7.21**  Advance of the leading edge of a pulse as it is amplified well above saturation in a sequence of neodymium glass amplifiers. (From Ref. 100.)

of simple $Q$-switched laser oscillator pulses, one can encounter situations in which the pulse shape may appear undistorted but the peak of the pulse may appear to exit the amplifier before the peak of the input pulse enters the amplifier. Figure 7.21 illustrates this phenomenon for a relatively high gain-to-loss ratio amplifier [100]. This "faster than light propagation" does not really violate causality, as it depends on a precursor signal to be amplified. In cases where pulse shape is important, the relevant calculation should be done, as the level where such effects may occur will be a function of the loss-to-gain ratio. In extreme cases, the relation between input and output pulses may be so strongly modified as to be nonapparent. Most of the successful Frantz–Nodvik codes use algorithms in which the input pulse is temporally split up into energy bins where the largest bin contains no more than 0.1 of the saturation energy density and, in effect, these bins are successively propagated through the amplifier using correct arithmetic on the computer to account for the influence of previous segments of the pulse. With this approach it is in fact possible to "reverse engineer" a system and build a code to define the prescription for an oscillator pulse to give a desired output pulse shape [101].

In Sections 7.4.3 and 7.4.4 we examine the processes that may limit laser amplifier performance. Some of these limits, especially those involving nonlinear phenomena such as self-focusing, self-phase modulation and optical damage, may be so dramatic as to preclude operation in certain regimes.

An important and oft-neglected rule of thumb is to design laser amplifier systems from the output back rather than the reverse procedure (i.e., from the oscillator into the amplifiers). Such an approach is much more time effective in terms of computer modeling. It is also much more cost-effective in most cases. The output stage of an oscillator–amplifier system generally costs about 50% of the total system price; it is generally where all the technological gambles are (and should) be taken. It makes no sense, economic or otherwise, to be as adventurous in designing the preliminary stages. Generally, the early stages can be designed in a very conservative fashion with little economic impact. As a practical matter, this engineering aspect is very important.

### 7.4.1   Single- and Multiple-Pass Amplification

In the case of most of the high-energy lasers that have been developed for laser fusion, single-pass amplification has almost always been the approach selected. The reason for this was largely to maximize the self-focusing limited performance. Additionally, given that the majority of costs of a system are at the output stage, the cost advantage of multiple-pass amplifier systems was viewed as minimal. This amplifier systems was viewed as minimal. This judgment was probably correct at the time it was made in terms of available materials, but should not be generalized to involve further materials or longer pulse operation. In cases where beam quality is an issue (and generally fusion lasers have had a large margin between needed and achievable beam quality), a rather strong case can be made in favor of two-pass amplification. As discussed in greater detail in Section 7.7, a two-pass system directly addresses a key issue in terms of beam quality. That is the issue of birefringence. As could be inferred from the work of Quelle [102] (1966), Anan'ev and Grishmanova [103] (1968), or the French patent of de Metz [104] (1970), a two-pass geometry offers the possibility of total correction of birefringence effects for all host media, crystalline or solid. There may be a residual uncorrected path-length distortion (i.e., a thermal lens), but for a pulsed laser this can be corrected by a suitable passive corrector. Efforts to make true "athermal" materials have been far more successful in reducing thermal lensing than in reducing birefringence effects. To exploit this opportunity for birefringence compensation, the polarization must be rotated by 90° on each successive pass of the active medium. This may be accomplished by various means, including quarter-wave plates and active modulators such as the 45° Faraday rotator. Large (i.e., 20 cm) Pockels' cells with plasma electrodes have been developed by LLNL recently and may offer the possibility of $2n$-pass amplifiers with low or eliminated birefringence [105]. An issue that is not totally resolved is the correctability of lensing effects. If the multiple-pass amplifier is used in an optical relay system in an image plane on successive passes, in principal a total correction is possible; if not, the issue is more complicated. Figures 7.22 and 7.23 contrast single- and multiple-pass amplifiers. The advantages of multipass amplifiers in energy extraction are generally large for all cases except short pulse cases, in which vacuum spatial filters must be used to suppress self-focusing. (Plasma closure of the spatial filter pinhole may preclude this application.)

### 7.4.2   Amplifier Geometries

A number of laser amplifier geometries have been investigated for optically pumped solid-state lasers. These include various types of rod, slab, and disk amplifier geometries. Which type is most appropriate for a given application depends very strongly on the application. The most notable distinction is that of single-pulse or multiple-pulse applications. In addition to the literal quasi-CW case as an example of the latter,

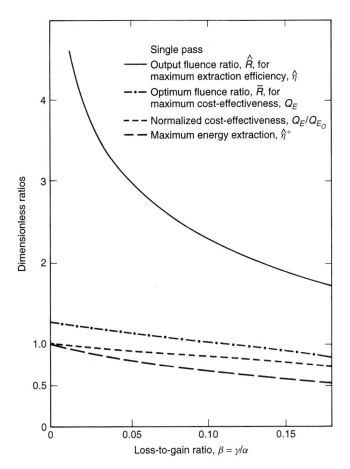

**Figure 7.22**  Single-pass amplifier optimizations versus loss-to-gain ratios. (From Ref. 162.)

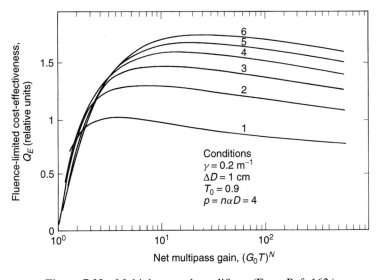

**Figure 7.23**  Multiply passed amplifiers. (From Ref. 162.)

there is also potentially a multiple-pulse "burst"-mode application. This is because many laser host materials, notably glasses and some crystals, such as yttrium lithium floride (YLF), have rather poor thermal diffusivity and structural strength but can have a very large heat capacity and very desirable thermo-optical properties. The optical distortion on a single pulse may be small enough with some of these materials (see Section 7.7.1) to the point where the limit on the number of sequential pulses in a burst may be physical destruction of the laser element rather than optical distortion.

To discuss amplifier geometries in an orderly fashion we will consider the most successful examples of the three cases: amplifiers for single-pulse operation as exemplified by laser fusion amplifiers, true average power systems, and burst-mode systems.

***Fusion Amplifiers.*** It was recognized fairly early in the laser fusion program that achieving the ultimate needs of a laser fusion reactor would require a laser technology far beyond that which existed in the early 1970s. In addition to an energy per pulse approximating $10^6$ J for a high-gain pellet, a repetition rate of 1 to 10 pulses a second, high (e.g., $>10\%$) overall efficiency, and very good coupling to the target pellet all appeared to be necessary [106]. While solid-state laser technology available at the time might not really appear to address the second, third, or fourth requirements, the high-energy storage of solid-state lasers (500 to 1000 J/L) and the great flexibility of the solid-state format made it a good and cost-effective candidate for investigation of the physics of laser fusion. Laser systems up to the LLNL Nova laser (100 KJ in a nanosecond pulse) have been built [107] and concepts developed for systems producing single pulses in the megajoule range [108].

Laser rod amplifiers were perhaps most successfully developed by the Companie Générale d'Electricité in France in the 1960s [54]. Their amplifiers used helical flashlamps to pump successively large rod amplifiers, where all of the lamps were electrically equivalent. One problem with this approach is that while reasonable pumping efficiency can be obtained, as the size is scaled up, the doping of the neodymium glass element must decrease to maintain an approximately constant doping-diameter product to get an acceptable gain profile. For the CGE amplifiers, a diameter-doping product of about 45 mm-wt% was found most useful with ED-2 silicate laser glass [34]. A necessary consequence of this trade-off is that the peak gain coefficient will decrease as the amplifier is scaled up as $r^{-1}$. As the safe power density in self-focusing-dominated situations is on the scale of the gain coefficient, this approach runs into serious problems when self-focusing begins to be a significant limit on energy extraction.

The radial pump uniformity of laser rod amplifiers has been investigated by several groups. For effectively diffuse pumping, a case approximated by helical lamps, the two competing physical effects are the near-exponential attenuation of absorbed pump light which would give the strongest deposition near the surface and the least in the center of the rod and the dielectric focusing of the laser rod (and its cooling jacket), which trends in the other direction. For a material with a rather high index of refraction, such as ruby ($N = 1.76$), the latter effect can be quite strong, and uniform energy deposition has been found to occur for a doping diameter product of 0.03 wt%-cm [109]. For neodymium-doped glass, with a much lower index of refraction, $N = 1.50$ to $1.55$, this focusing effect is much weaker and helically pumped rod amplifiers generally have nonuniform gain with peak gains at the edge of the laser rod.

With linear lamp pump geometries, an additional mechanism can be used to achieve better pump uniformity. In addition to the diffuse direct pump wave, a more strongly

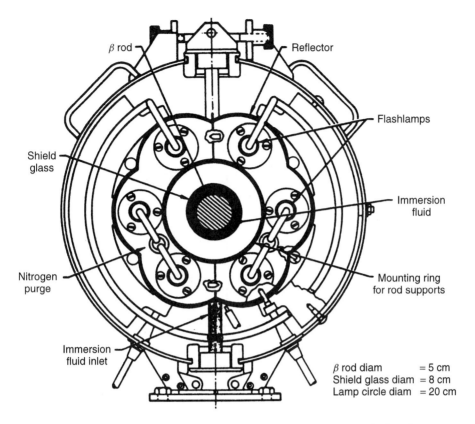

β rod

Reflector

Flashlamps

Shield glass

Immersion fluid

Nitrogen purge

Mounting ring for rod supports

Immersion fluid inlet

β rod diam        = 5 cm
Shield glass diam  = 8 cm
Lamp circle diam  = 20 cm

**Figure 7.24**   Shiva 5-cm rod amplifier in cross section. (From Ref. 110.)

focused (but generally weaker) pump wave can be focused by a reflector around the outside of the flashlamp. Figures 7.24 and 7.25 show the geometry developed by LLNL for a 5-cm-diameter amplifier using an involute reflector [110] and the radial gain uniformity obtained using this geometry [57].

With disk amplifiers the radial and azimuthal gain uniformity tends to be excellent irrespective of doping because the dominant nonuniformity is through the disk in almost the same direction as the optical propagation. Figure 7.26 shows results calculated at LLNL on gain uniformity for a disk amplifier developed at NRL with a planar reflector [111], and measurements (Fig. 7.27) at LLNL on a disk amplifier with a convoluted (focusing) reflector [112]. Little difference in radial or azimuthal uniformity is apparent in either case. Figure 7.28 is a cross-sectional view of a LLNL 20-cm amplifier and illustrates typical dimensions and proportions for these amplifiers [112].

***Medium Average Power and Burst-Mode Solid-State Lasers.***   The earliest extensively developed solid-state lasers had the active element typically in the form of a rod or cylinder. When attempts were made to operate these devices in a multiple-pulse fashion, it was found that the beam spread changed from pulse to pulse even if the pumping was very uniform. The reason was that all these materials had large thermo-optic coefficients, such that while pump uniformity might determine the thermal lens early in the pulse train, for later pulses the temperature gradient caused by edge cooling dominates.

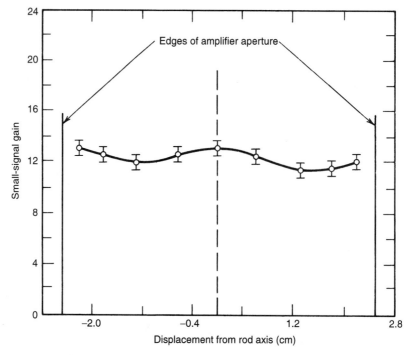

**Figure 7.25**    Radial gain variation of Shiva 5-cm-diameter rod amplifier. (From Ref. 110.)

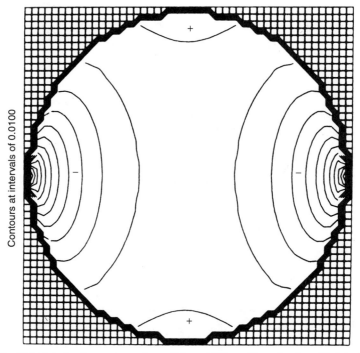

**Figure 7.26**    Calculated gain contours for a disk amplifier with a simple cylindrical flashlamp reflector. (From Ref. 111.)

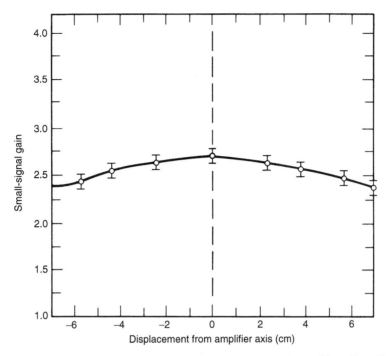

**Figure 7.27**   Measured gain on Shiva gamma (15-cm aperture) disk amplifier. (From Ref. 110.)

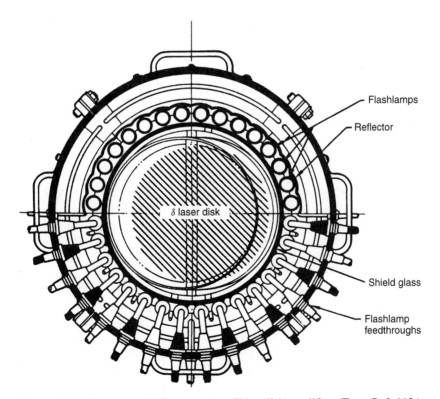

**Figure 7.28**   Geometry of 20-cm aperture Shiva disk amplifier. (From Ref. 112.)

Three concepts were proposed in the 1960s for laser amplifier geometries which might be expected to have small optical distortion even at moderate average powers; the face-cooled and pumped disk amplifier [113] and two slab amplifier geometries, the zigzag slab [114, 115] laser and the straight-through slab amplifier [116]. The optical distortion aspects of these geometries are analyzed in Section 7.7.1; in this section we deal with the energy extraction from these geometries.

The face-cooled slab laser for reasonably long (i.e., tens of nanoseconds) pulse durations is not much different from the uncooled slab, except that there may be additional loss at the extra interfaces. For short pulses and high intensities, this geometry can have a low self-focusing threshold unless the additional path length in cooling channels and windows can be minimized. Additionally, optical distortion in the coolant is a concern [117, 118].

The edge-pumped slabs will generally have much poorer gain uniformity than rods or disks. The effect on energy extraction can range from minor to significant, depending on the particular case and mode of usage. In the zigzag case, gain averaging during the bounces should minimize pump uniformity effects. With either slab case in a storage mode, the limit on absorption profile may not be extraction but transverse parasitic oscillation (see Section 7.4.3). In long-pulse relaxation-mode cases, these are potentially both very workable geometries.

### 7.4.3 Limiting Processes: Amplified Spontaneous Emission and Parasitic Oscillation

In scaling up lasers to large sizes, a concern is whether the laser radiation will preferentially be emitted only in the desired direction. Large in this context of storage lasers refers to the net gain, $\alpha L$, more than the physical size of the system. High-cross-section crystalline lasers may encounter such effects at a much smaller size than lower-gain glass lasers [45]. The two effects, amplified spontaneous emission (ASE) and parasitic oscillation, are physically distinct mechanisms. As the gain of an amplifier is increased by pumping, amplification of the photons emitted by spontaneous fluorescent decay will begin to represent a significantly larger loss than the unamplified fluorescence. The geometry of the laser medium can affect this loss since it will define the solid angles over which the fluorescence is amplified. Parasitic oscillation can occur if the medium is of a shape such that at some gain a mode will exist which is over threshold. From the definitions one would expect parasitic oscillation to have a rather distinct and sharp threshold, while ASE would be evidenced by a softer onset with a roll-off in the pump efficiency as the gain is increased further. In practice, it may be harder to make this distinction, as long path whisper or organ-pipe-mode paths without a closed-mode path may have a soft onset rather than the sharp threshold normally expected for "modes." In large-volume Nd:glass rod amplifiers surrounded by a cylindrical water jacket this was found to be the case. A solution in this particular case was to use a zinc chloride ($ZnCl_2$) solution that index matched the laser glass and water jacket with a weak addition of samarium chloride to keep modes that TIR'd from the water jacket below threshold [119]. If a larger-diameter water jacket is used, such that these mode paths do not intersect the rod, simpler to handle ethylene glycol solutions can be used to suppress rod modes [52]. An alternative approach is to use a Pyrex with a graded-index surface [120] as the water jacket, with an ethylene glycol coolant [121]. This procedure has the advantage of somewhat improved pumping efficiency because of improved flashlamp coupling.

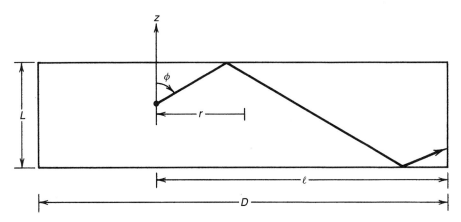

**Figure 7.29**    Parasitic mode path in a laser disk. (From Ref. 123.)

These problems were first considered for disk lasers by Swain et al. [122]. Fluorescent radiation at angles greater than $\sin^{-1}(1/n)$ to the face normal will TIR from the faces until it reaches the edge of the disk (see Fig. 7.29). If it is reflected from the edge, a parasitic mode can develop and clamp the gain. To prevent such oscillation it was necessary to put an absorbing layer at the edge which was index matched to the disk. Both liquid [122] and solid [34] edge treatments have been developed to stabilize such parasitic modes. The prescription for such a cladding depends on the properties of the laser glass but should have the general properties that the index of refraction should be equal to or slightly greater than that of the laser glass; the absorption should have sufficient optical depth to stabilize the parasitic and a low-enough absorption coefficient that such heating of the cladding (by pumping and fluorescence decay) will not cause fracture at the laser glass–cladding interface.

The issue of ASE in laser disks was considered by Trenholme [123], who used a Monte Carlo ray-tracing code, ZAP, to model this problem. Figure 7.30 illustrates the results obtained. Until a transverse gain of $\alpha D > 3$ is exceeded, ASE is not a significant factor. Above this level it can become increasingly important. As the figure indicates, the result is sensitive to details of the fluorescent line shape. In an experiment at LLNL with an index-matched liquid edge coating, it proved possible to pump a laser disk to $\alpha D = 3.6$ [124].

### 7.4.4  Effect of Laser Glass Properties on Gain and Propagation Effects

The first type of Nd:glass laser host materials to be developed extensively were the silicate glasses [125]. These host materials found favor as hosts for neodymium because of a number of factors. Among them were long fluorescent lifetimes and high quantum efficiency for neodymium. Not the least of their attraction for glass technologists was their familiarity. Most optical glasses in production were of this general class. The glasses produced by Snitzer and his colleagues at the American Optical Company when produced in the platinum-lined crucibles in common use in the United States proved to have a very low threshold for damage caused by the melting and explosion of internal inclusions or aggregates of platinum. This led by direct and indirect paths to the export and reappearance of AO composition No. 3669 to countries whose glass

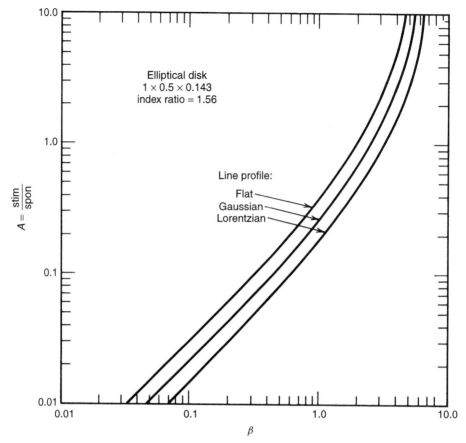

**Figure 7.30** Amplified spontaneous emission rate versus disk transverse gain coefficient. (From Ref. 123.)

industries used the older ceramic crucible technology. Melted in a ceramic crucible this glass has a damage threshold in excess of 25 J/cm$^2$ for $Q$-switched pulses. This glass, or close relatives of it, is produced as Schott LG-56, Soveril MG-915, and GLS-1 in the Soviet Union. A similar glass is produced by the Shanghai Institute of Optics and Fine Mechanic. Typically, this glass has a wavelength of peak fluorescence emission for the $^4F_{3/2}$–$^4I_{11/2}$ transition near 1.058 μm and a peak-induced emission cross-section of 1.6 to $1.8 \times 10^{-20}$ cm$^2$. It is still used widely for long-pulse commercial lasers because of its ease of production, excellent optical homogeneity, and attractive mechanical properties.

A different Li-Mg-aluminosilicate glass was developed by Owens-Illinois (OI) in conjunction with the Naval Research Laboratory (NRL) under DARPA sponsorship in the late 1960s for $Q$-switched laser applications. This glass, denoted as type ED-2 by OI, had excellent $Q$-switched damage properties with a threshold in excess of 24 J/cm$^2$ [126]. It was melted in platinum crucibles by proprietary techniques which precluded the formation of platinum aggregates [127]. The property that made this glass attractive for $Q$-switched applications was a very high gain-to-loss ratio, giving very efficient energy extraction. The peak induced emission cross section was $2.7 \times 10^{-20}$ cm$^2$, 50%

higher than the AO composition. The passive loss characteristics were even more superior. Typically, the loss was $<1 \times 10^{-3}$ cm$^{-1}$, compared to $6 - 8 \times 10^{-3}$ cm$^{-1}$ for the ceramic melted AO composition. The wavelength of peak fluorescence was 1.062 μm, quite close to the 1.064-μm wavelength of Nd:YAG.

When the focus of glass development efforts in the United States shifted to even shorter pulses to address the laser fusion problem, the technology of first choice was to use mode-locked oscillators to produce the short pulse followed by a cascade of rod and disk amplifiers to reach high peak power [34, 79]. This approach reached full maturity in the two-beam TW Argus laser [128] and the 20-beam Shiva laser [129] at Lawrence Livermore Laboratory. The output aperture of both devices was 20 cm, about the parasitic and ASE limits for this glass.

Even during the design and construction of these devices it was recognized that self-focusing posed severe limitations with the available glass (see Section 7.6.1) and a search was under way for laser glasses that would significantly ease these limitations. An extensive investigation of a variety of glass compositions was undertaken at LLL (now LLNL). Figure 7.31 is representative of what was discovered. Glasses of a beryllium fluoride–like composition have the lowest nonlinear index of refraction, followed by the fluorophosphates, phosphates, and then the silicates. A large number of glasses were melted and documented in this investigation [19]. Difficulties in producing the fluoroberyllates led the greatest interest for the 100-kJ Nova project at LLNL to center on the fluorophosphate compositions. In testing of phosphate compositions at NRL it was found that two glasses, Kigre Q-88 and OI EV-2, had markedly higher

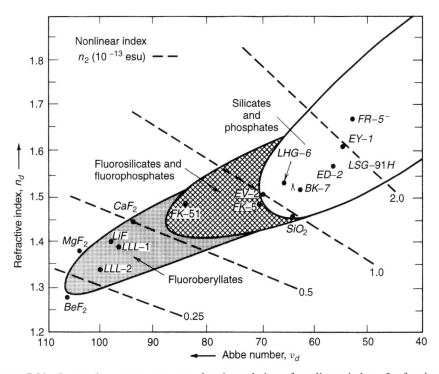

**Figure 7.31** Laser glass parameter map showing relation of nonlinear index of refraction to linear optical properties. (From Ref. 129.)

gains in a disk amplifier than would be expected solely on the basis of the larger induced emission coefficient (3.6 to $4 \times 10^{-20}$ cm$^2$) compared to ED-2 [130]. The systems implications of this result were examined and it was realized that this could give equivalent leverage on the self-focusing problem to the fluorophosphate strategy, at the penalty of requiring small apertures. An alternative strategy to approach the Nova project goals was identified [101] and reduced to practice in the two-aperture Pharos II device [121]. In practice it did not prove possible to produce the fluorophosphate glass desired for Nova with the requisite combination of optical quality and damage threshold and modified design for Nova evolved using phosphate glass and segmented apertures in the largest-diameter (46 cm) amplifiers [131]. In operation at very high fluences ($>10$ GW/cm$^2$) a problem with inclusion damage was encountered with Nova [132]. This problem was not anticipated with the phosphate glasses because of the high solubility of platinum in the molten glass, and was not noticed in earlier testing on the Pharos II and Novette lasers and was exacerbated by the very large volume pieces of glass necessary for Nova ($>3000$ cm$^3$). By careful modification of the laser glass production process it has recently proved possible to overcome this problem, and routine full-power Nova operation is expected in the near future [133]. In the course of developing the phosphate glasses it proved possible to modify the glass compositions to achieve near athermal behavior without sacrificing desirable self-focusing and gain properties, at a substantial advantage in glass fabrication and laser operations. The glasses with these desirable properties include Hoya LHG-8, Kigre Q-98, and Scott LG-750.

Laser glass development in the Soviet Union proceeded along a different path to approximately the same type of glass. The early emphasis was on developing glasses with good thermo-optic properties for long- pulse lasers, following the lead of Anan'ev [103]. Early on, the emphasis shifted to phosphate compositions, as it was found that near-athermal silicates had very small induced emission cross sections, leading to poor energy extraction [134]. By 1978, a number of phosphate glasses with near-athermal properties and reasonably high gain were apparently available in large-volume (4 cm $\times$ 24 cm $\times$ 72 cm) rectangular slabs [20].

## 7.5 OPTICAL MODULATORS

A number of functions in laser systems require optical modulators, devices whose transmission is different under different conditions. These include saturable absorbers, passive polarization switches, and electro-optic and magneto-optic polarization switches. Acousto-optical switches may also find application in some cases. Among the factors that determine the most appropriate modulator for a given application are the system aperture, damage issues, the intensities to be switched, switch losses, and high-voltage requirements. Various factors will also limit the practical as well as theoretical performance for each type.

### 7.5.1 Saturable Absorber Switches

A saturable absorber switch is a material that has the property of having a low transmission when irradiated at low intensity but a high transmission at high intensity. If a suitable material is available for a particular application, a saturable absorber switch can be a simple device that requires no external synchronization or applied high voltage. Such switches have found application both as $Q$-switches for laser oscillators and as isolators for laser amplifier systems. For ruby lasers operating at 0.694 μm, the dye crypto-cyanine in a methanol solution was found to be a very useful $Q$-switch, as the saturation intensity was relatively low, about $I_S = 50$ kW/cm$^2$, and the absorption versus intensity followed a relation of the form

$$\gamma(I) = \gamma_0/(1 + I/I_S)$$

As the peak intensity in a $Q$-switched ruby laser is typically 1 to 10 MW/cm$^2$, the loss can be quite small during the high-intensity part of the pulse when $I \gg I_S$.

For neodymium lasers a more restricted range of choices is commercially available. The Eastman Kodak 9740 and 9860 dyes have saturation fluxes in the range 50 to 70 MW/cm$^2$ and relaxation times on the order of 10 ps [135]. For simultaneously $Q$-switched and mode-locked oscillators with subnanosecond pulses, the dye can be strongly saturated and good overall efficiency can be obtained. For $Q$-switching but not mode-locking neodymium lasers, the Eastman 14015 dyes can be used because of their long relaxation time [136]. The efficiency suffers, however, because of the nonideal nature of the saturation (the absorption coefficient decreases to only about $0.5\gamma_0$, even for $I \gg I_S$).

NRL once used a two-stage saturable absorber (EK9740) optical isolator in a short-pulse amplifier train to suppress prepulses and ASE on fragile fusion targets [80]. The best performance obtained, 30 dB suppression of weak pulses and ASE for a 3-dB reduction in large-signal output from the amplifier cascade, is a typical result. Even for nanosecond pulses, at high energy densities, nontrivial pressure pulses can be generated in the absorber cell. Care in fabrication and use is advised.

### 7.5.2 Passive Polarization Switches

A quarter-wave plate plus an optical polarizer can in some cases be used as an optical isolator. In principle, if the reflection has the same polarization as the outgoing beam, very good isolation can result. This scheme can also be relatively inexpensive for beam apertures up to 20 cm using mica quarter-wave plates, such as those available from Tropel Inc. [137, 138]. In practice such approaches need to be implemented most carefully. The polarization of the target return may not be complete. If the target return has frequency-shifted components, the mica plates which are actually $m + \lambda/4$ plates, may not cleanly render linear polarization.

### 7.5.3 Electro-optic Modulators

Two types of electro-optic modulators have been used for laser $Q$-switches, Pockels cell switches and Kerr cell switches. The latter type have fallen into disfavor for several reasons: a very high modulation voltage, difficulty in handling Kerr cell liquids such as nitrobenzene, and nonlinear effects occurring in the $Q$-switch liquid. Stimulated Raman scattering was in fact discovered by analyzing the complex spectral output from a Kerr cell $Q$-switched ruby laser [66].

Development of useful Pockels cell Q-switches required the development of suitable high-quality nonlinear optical crystals with high transmission at the laser wavelength.

Typical of the Pockels cells that have been developed are the transverse field lithium niobate ($LiNbO_3$) and potassium dihydrogen phosphate (KDP) switches shown in Fig. 7.32 and the longitudinal field cylindrical ring electrode potassium dideuterium phosphate ($KD^*P$) switch shown in Fig. 7.33. The advantage of the $LiNbO_3$ switch is a very low drive voltage for half-wave operation, 300 V compared to 9000 V for the $KD^*P$ switch at 1.06 $\mu$m. The advantage of the $KD^*P$ switch is that it can be scaled up to very large apertures at the same voltage with excellent extinction: Commercial Pockels cells are available up to 90 mm aperture [140], and experimental units have been built up to 200 mm aperature using plasma electrodes [87]. $LiNbO_3$ Pockels cells are limited to 10 mm or smaller apertures by crystal availability.

As an electrical circuit component a Pockels' cell can be configured to appear as a capacitor with an associated lead inductance driven by a pulser (equivalent to a charged capacitor plus a series switch). For small cells with reasonable minimization of lead inductance, $C$ and $L$ are small enough that rise times of 1 ns are not difficult to obtain. For the larger devices used as isolators, more care must be used to make the device appear electrically to be a section of transmission line of the same impedance as the pulse driver. The pulse is then terminated into a suitable matched load. Typically, these devices are built to have a characteristic impedance of 50 $\Omega$ because of the availability of high-voltage cables, connectors, and so on, for this impedance. The limit on switching speed for this type of device is the propagation of the electrical wave across the optical aperture. With $KD^*P$, typically 200 ps/cm of aperture is a good value.

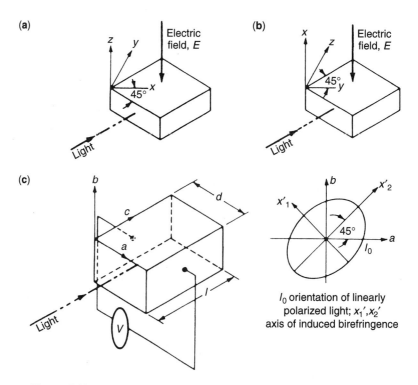

**Figure 7.32** Transverse field Pockels cell arrangements. (From Ref. 4.)

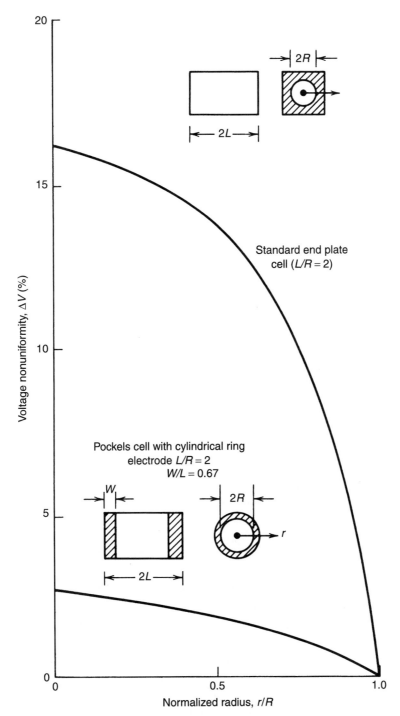

**Figure 7.33** Comparison of Pockels cell voltage nonuniformity for conventional flat plate and cylindrical ring electrode Pockels cells. (From Ref. 139.)

In most cases the optical extinction ratio of this type of device can be quite high, with several caveats. Any nonuniform strain caused by internal strains in the crystal or strains induced in mounting the crystal can cause nonuniform mechanically induced depolarization (uniform strain can be electrically biased out). The angular acceptance must be restricted to be well within the center lobe of the characteristic Maltese cross pattern. The laser beam and crystal axis should be nearly, collinear, for the same reason. Polarizers can also affect the extinction ratio achieved. For broad-line-width laser sources such as Nd:glass, the color dispersion of the Pockels coefficient may be the most fundamental limit. In most well-engineered cases, extinction ratios of 35 to 40 dB can be achieved, with an insertion loss of 20% or less (chiefly from the polarizers).

### 7.5.4   Magneto-optic Switches

Magneto-optic polarization switches using Faraday rotation in isotropic media were among the early approaches investigated for laser $Q$-switching and optical isolation [84, 141]. An advantage was that glasses with reasonable Verdet constants and high optical damage were readily available. A second advantage for $Q$-switches was that the relatively long buildup time (microseconds) of the magnetic field gave $Q$-switched laser action, with many of the mode-selection advantages of saturable absorber switching. In addition, magnetic fusion research in the previous decade had developed suitable pulsed power technology to produce the large pulsed magnetic fields necessary, 10 to 100 kG, reliably. Pockels cell technology has made this approach to $Q$-switches obsolete, especially with two-stage Pockels cells [85].

In the early 1970s the limits on Faraday rotator isolators for neodymium lasers were explored at NRL and LLNL. Using terbium-doped glasses, extinction ratios in excess of 37 dB were achieved on 90-mm-aperture devices [142] and lower-extinction-ratio devices were built at up to 200-mm apertures [143]. Figure 7.34 shows the extinction ratio as a function of rotation angle for one such system.

At the present time a comparison of Pockels cell and Faraday rotator isolator technology would tend to favor Pockels cells as the technology emerging into full maturity:

- The magnetic field required (for a constant extinction ratio) scales up as the aperture is cubed. As this roughly equates to capacitor bank size, this quickly becomes a significant cost issue for Faraday rotators. The energy required for Pockels cell pulsers is so much less that the pulser is essentially an aperature-independent cost.

- Faraday rotators are not fail-safe devices, whereas pulsed Pockels cells are fail-safe. In practice this can be a less valid point than it appears, as the laser system can do real-time current measurements and disable generation of a pulse if an anomaly in rotator operation is detected [144].

- If operation is desired at some new wavelength where Pockels cells may be hard to realize until suitable crystals are developed for the application, a workable Faraday rotator isolator approach probably can be implemented quickly.

- Cryogenic Faraday rotators may superficially appear to resolve the difficulties, but as a practical matter the operating costs of maintaining a liquid helium system with the thermal losses tied to the need for large-optical-throughput windows in the cryostat are excessive. High-temperature superconductors could, of course, change this in the future.

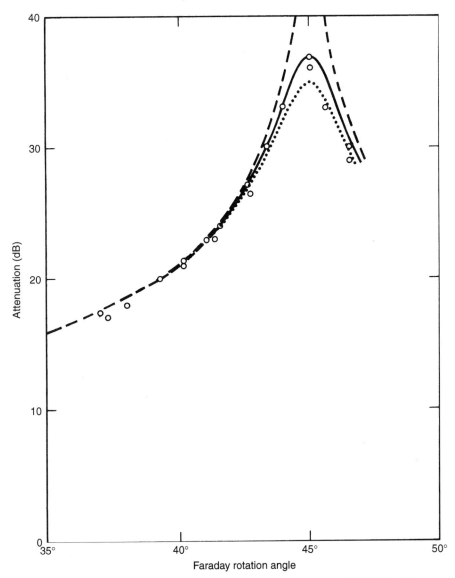

**Figure 7.34** Measured extinction ratio versus angle of rotation. The dashed line shows a calculation for an ideal case, and the solid and dotted lines show calculations for 37-dB and 35-dB extinction systems, respectively.

### 7.5.5 Acousto-optic Switches

In an acousto-optic switch a transducer is electrically driven to produce a standing sound wave in an optically transparent medium at the carrier frequency. An incident optical wave can be scattered from the sound wave at the Bragg angle,

$$\theta \sim \lambda_0 / \Delta$$

where $\lambda_0$ and $\Delta$ are the optical and acoustic wavelengths.

This approach has in general been limited to small apertures by several effects. Even for sound waves at the limit of sound propagation in solids ($Cs = 9000$ m/s) the Bragg angle is small; for a large-aperture system this would require excessive length for adequate beam walk-off of the diffracted beam. The extent to which edge effects and multiple reflections interfere with a true plane standing acoustic wave limits the device extinction ratio to 2 to 3 dB in commercial devices. Thermal dissipation in the boundary layer between the transducer and the optical element can pose difficult materials bonding and reliability problems at high duty cycles. In general, this approach has been successful only in rather low gain oscillators as a $Q$-switch, but it seems to be an area that could evolve [145].

### 7.5.6    Polarizing Optical Elements

In the prelaser era, several types of polarization-sensitive optical elements were known. These included polarizing prisms, polarized reflections from Brewster-angled surfaces and oriented film polarizers such as Polaroid. Polarizing prisms of various types function by using nonisotropic crystals, typically uniaxial materials such as calcite, oriented such that one polarization will be an ordinary ray and the other an extraordinary ray. At a tilted interface the angle for total internal reflection is $\sin \theta = 1/n$. If the ordinary and extraordinary indices are sufficiently different, a $\theta$ can be chosen such that one wave is largely transmitted and the other largely reflected. The so-called Glan–Thompson polarizer [146], which gives no lateral beam deviation, has proven to be a useful polarizer for lasers in many cases. The classic type has two identical sections cemented together with Canada balsam; for laser usage this has been replaced with an air-spaced version to avoid damage. Such prisms are available up to about a 25-mm aperture.

A single surface placed at a Brewster angle will give a totally polarized reflection ($E$-plane of incidence), but only about 10% of this component is reflected per interface. To achieve greater throughput, the classic "pile of plates polarizer" evolved. With sources with a spatial coherence length shorter than the plate thickness or spacing, this is not an unreasonable choice. Lasers, however, typically have coherence lengths of centimeters to meters, and cancellation of reflections by interference is a distinct possibility for compact devices, where reflections from successive plates can overlap spatially and the degree of cancellation depends on the plate spacing to a fraction of an optical wavelength. Two approaches have evolved to overcome this difficulty.

If plates are used which are wedged rather than with parallel surfaces, by carefully fanning the plates, interference effects can be avoided. The best reported performance for such units, 10 dB attenuation and 95% transmission for 10 Bk-7 plates and 20 dB attenuation and 90% transmission for 20 Bk-7 plates [147], compares favorably with prism polarizers without the aperture restrictions imposed by the use of calcite. A sealed device with escape windows can have a very high damage threshold and very reliable performance. Cost is the major issue with this approach. Two design issues are the avoidance of vignetting in the fanned and wedged stacks and the intolerance of these designs for other than collimated beams. For divergences greater than about 3 mrad, severe astigmatism was produced on the transmitted beams with the referenced designs.

Another approach to this problem has been to use a stack of dielectric coatings to function as an ordered and precisely spaced "pile of plates." As coating thicknesses can be held to a small fraction of an optical wavelength per layer, interference effects can be taken account of by careful computer design. The major practical fabrication issue for larger-aperture coatings is that common to other larger-aperture coated optics; unless the optic diameter is small compared to the coating gun to optic spacing, a single coating gun will systematically give thicker coatings in the center of the workpiece. For many-layer coatings this error is additive and can give appreciable wavefront distortion [148]. One solution is to use a number of "satellite" deposition guns to avoid excessively large evaporator sizes. Very high quality coatings have been produced for the laser fusion program at up to a 20-cm working aperture, essentially a $20 \times 40 \text{ cm}^2$ coating. Typical performance is 20 dB extinction with 95% transmission for the best coatings. Damage thresholds typically exceed those of the coatings on the associated Pockels cell or Faraday rotator elements.

Polaroid-type polarizers have not proven useful other than for optical testing at low power in the visible region of the spectrum. The most basic reason is that these materials absorb the "nonpass" polarization, which leads to optical distortion and damage at the power densities characteristic of solid-state lasers. It is not fundamentally clear that improved materials could not be developed which would be competitive in certain roles; it is more an issue that other approaches have evolved very successfully to meet the requirements.

## 7.6  SYSTEM STRATEGIES FOR OSCILLATOR–AMPLIFIER LASER SYSTEMS

A laser system is generally built for some specific purpose rather than as an end in itself, for example, for specific laser fusion or laser matter interaction experiments. The first principle to follow in deciding what is needed is to understand the needs of the application and necessary losses between the laser and target, such as focusing optics, beam diagnostics, and so on. It is not unusual for such losses to reduce the energy on target to 75 to 80% of the output energy for carefully designed optics and to 60 to 65% for less well optimized systems. The second principle that should be followed for sound design is to iterate the design enough to understand its robustness to degradations, which can be expected in normal use. Success in this requires a knowledge of the limiting processes and how degradations, for example in amplifier gains as the pump system ages, affect these limitations. Care in this procedure can generally allow building in some performance margin so that the system will degrade gracefully. It can also guide the designer in where to place laser diagnostics to flag necessary maintenance for the operator. It can make a large difference in whether a system is perceived as reliable, reproducible, and available or whether it appears that the operators are always fixing the laser rather than using it.

The most strongly limiting mechanisms that drive the laser design can frequently be distinguished by pulse duration. For well-designed systems with pulse lengths shorter than a few nanoseconds, self-focusing generally is the dominant mechanism. For longer pulses, damage to optical coatings is generally the limiting mechanism.

### 7.6.1  Self-Focusing Limited Operation

Initially, self-focusing estimates based on the early work of Kelley [149] and Marburger [150] predicted that self-focusing in the sense of beam collapse would not be much of a problem, especially if a geometric divergence was induced on the beam [151].

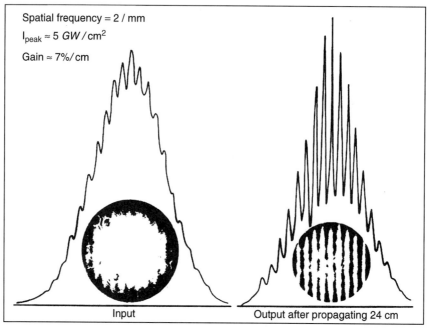

Spatial frequency = 2 / mm

$I_{peak} \approx 5\ GW/cm^2$

Gain $\approx$ 7%/cm

Input

Output after propagating 24 cm

**Figure 7.35**   Growth of a modulation on a beam due to self-focusing gain in a 24-cm-long laser rod. (From Ref. 153.)

Experimentally, this was found most decidedly not to be true. Beams collapsed not into one filament but into many small filaments. The locus of these filaments frequently coincided with Fresnel diffraction rings on the beam or diffraction patterns around damage spots on upstream optics. Bespalov and Talanov were the first to point out the source of this phenomena [152]. High-spatial-frequency ripples on the beam can collapse to a filament in a much smaller propagation length than do low-spatial-frequency ripples. At very high spatial frequencies diffraction will spread the ripples more rapidly than self-focusing can collapse them. Qualitative confirmation was soon obtained (see Fig. 7.35) and the concept of a self-focusing growth rate, or $B$-integral, at these most dangerous spatial frequencies was formulated [153]. For passive optical elements this expression is

$$B_p = \frac{8\pi^2}{c}\frac{n_2}{n_0}Il$$

where  $n_0$ = ordinary index of refraction:
  $N_2$ = nonlinear index of refraction, esu:
  $I$ = intensity W/cm$^2$
  $l$ = thickness, cm

For an amplifier in the small-signal gain regime,

$$B_a = \frac{8\pi^2}{c}\frac{n_2}{n_0}\frac{(I_o - I_{in})}{\alpha_0}$$

where $\alpha_0$ is the gain coefficient ($cm^{-1}$) and $I_o$ and $I_{in}$ are the output and input intensities.

The net self-focusing growth can be estimated by summing the contribution from individual components. An issue that was controversial at the time and which does not admit of a blanket answer is that of "how much self-focusing growth can be tolerated" before self-focusing effectively dominates the situation. The answer depends strongly on how much "noise" is present on the beam at the most dangerous spatial frequencies. The test that came to be a standard was to image the far-field pattern onto the slit of an ultrafast streak camera and watch for the appearance of a "dip" in the temporal profile as the laser intensity was increased. By this test, the lasers developed at Livermore would typically show the onset of self-focusing at a $B$-integral value of 2.5 to 4.0 [154]; the NRL group, by use of very careful filtering and beam shaping, was able to raise this level to 7.5 [155].

A significant advance in the control of small-scale self-focusing was the demonstration by LLNL [156] that image relaying and spatial filtering could largely be used to decouple the self-focusing growth in different segments and allow much higher total end-to-end self-focusing growth factors than in un-relayed systems. The Novette and Nova lasers at Livermore with short pulses ($>100$ ps) can operate stably with total $B$ integrals on the order of 12 [157]. This procedure is not totally without penalty. The spatial filters themselves will cause very low frequency spatial ripples on the beam, which will see growth and degrade the beam quality. Figure 7.36 shows an example

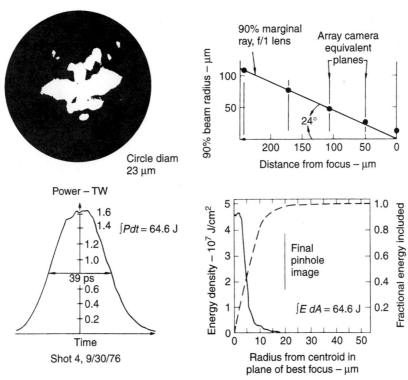

**Figure 7.36**   Beam far-field pattern and temporal pulse shape of the Argus laser at 1.6-TW peak power. (From Ref. 157.)

of such degradation. In most cases this is not an important limitation for applications such as laser fusion, as a beam spread well above diffraction limitation is still quite small and acceptable.

For nanosecond pulses spatial filtering is a more difficult option to implement, as pinhole closure by plasmas has to be confronted. This has generally been done on a case-by-case basis.

In any case, most successful laser systems operating in regimes where self-focusing is a concern have employed all or most of a variety of strategies:

1. Minimize self-focusing growth in passive components by making them have the minimum feasible thickness (i.e., minimize $n_2 l / n_0$).

2. Maximize the gain of laser amplifiers [i.e., minimize $n_2/(n_0 \alpha_0)$].

3. Have a definite strategy for controlling and measuring the beam near- and far-field pattern at a number of points.

4. Have very well controlled laser oscillators with a very small pulse-to-pulse energy dispersion.

5. When economics allows it, operate early (smaller) stages more conservatively.

The first three points are obvious; the last two may not be so obvious. When a disk amplifier is overdriven in intensity, a beam will be produced with dramatically degraded coherence, but generally no physical damage will occur. The thickness of the laser disks is usually small enough that self-focusing damage will not occur. The same is not true of laser rod amplifiers; if one is overdriven, a degraded beam and massive internal damage will occur in the form of self-focusing tracks. Oscillator statistics can dramatically affect the economics of operation of a system with rod amplifiers near the safe limits.

### 7.6.2 Damage-Limited Operation

Generally, breakdown damage will occur first at coated surfaces, then at a fluence three to four times higher at uncoated surfaces and then in the bulk of high-quality materials [158]. The presence of inclusions can radically reduce bulk levels, and control of the production process (i.e., crystal growth or glass melting, depending on the material) to prevent inclusion formation is always an issue in developing new materials. Generally, long-pulse systems are designed so that successive stages operate at the same fluence. This has the advantage that the laser designer essentially has only one damage issue for which to find a solution, and can then use this solution over and over.

What are the causes of the reduced damage levels at surfaces, coated or uncoated, compared to bulk levels? There appear to be many possible contributors; their role and relative importance is the subject of a sizable body of literature. Dirt or dust on the surface can certainly cause damage. The optical polishing itself may introduce failure points by embedding polishing compound in the surface. Bloembergen has pointed out that local field enhancement can occur in small cracks [159]. A multilayer optical coating is very complex. Milam has reported measurements on coated layers where the absorption coefficient was found to be $10^4$ times larger than expected from the bulk absorption coefficient of the same material, and ascribed the effect to very large stress in the coated layer [160]. Rather than a more extended discussion, we will present practical results for surfaces and coatings produced by major vendors.

The laser oscillator–amplifier systems produced by CGE in France in the 1960s were the first commercial solid-state lasers to produce reasonably large amounts of energy and power. The VD640 laser, for example, produced 500 J of energy in a 30-n pulse from an oscillator followed by five amplifiers, with the final amplifier having a 6.4-cm aperture. All output stages operated at 16 J/cm$^2$ average (and about 25 J/cm$^2$ peak) and used antireflection coatings. Tests at Livermore preparatory to building the Nova fusion laser with 1-ns pulses showed that most good antireflection coatings would be damaged at levels above 3.8 J/cm$^2$ (other types of coating had even higher levels) and clean bare surfaces would be damaged at 12 to 15 J/cm$^2$ [161]. For locations that can be sealed or where abrasion resistance is not important, other techniques to reduce surface reflectivity which have comparable damage properties to bare surfaces have been found. Graded-index surfaces can be produced on some materials and Sol-Gel coating techniques have also shown promise [120, 162]. For intermediate pulse widths the literature would suggest that the damage level scales as the square root of the pulse width [159].

These levels are representative of what can be achieved in the lab with great care in regard to cleanliness and control of the beam spatial profile. Many lasers must operate in less benign environments. A derating by a factor of 2 to 3 is probably prudent in those cases.

An area of special concern is reflections, especially from lens elements, as these can focus to very high powers. These should all be carefully traced out to ensure that "ghost" foci do not occur inside some other optical element. This problem can be especially severe with multiply-image-relayed systems.

### 7.6.3  Design Rules and Risk Reduction

The most important rule in developing point designs for lasers which have resulted in successful systems is to understand the sensitivity of the design to variations from the design assumptions, the impact of the most likely variations, and the costs of building in some assurance that such variations can be accommodated.

Examples of gain degradations that may be encountered where the appropriate remedy is to have some excess pumping capability include degradation of pump efficiency caused by solarization of flashlamp envelopes or oxidization of reflectors. It may be very cost-effective to be able to defer maintenance, as these degradations do not affect performance as long as greater pumping allows the original performance to be recovered.

Increased loss at surfaces in the beam due to films caused by pollution in the laboratory environment may also be acceptable if the films do not cause localized scattering and subsequent damage. If the preamplifier stages of a design are made to operate at lower fluxes than the output stages, it may be possible to operate with increased losses at much reduced maintenance levels for a very small additional initial cost.

On the other hand, surface or internal damage represents a degradation phenomena which almost always should be remedied by replacing or refinishing the damaged optic, as subsequent damage to later (and more expensive) optics in the train is highly probable.

The statistics of the laser oscillator need to be documented, carefully especially in regard to any propensities to emit pulses of greater than the expected intensity or with unusual spatial or temporal-mode properties. In some early work at NRL it was found with a system of rod amplifiers that there was a correlation between intensity level and the number of shots before internal self-focusing damage was noted, even though the nominal level was below expected single-shot levels [145]. Later examination of the statistics of the mode-located oscillator in use at that time [49] showed that the amplifier damage results were in fact quite deterministic when the oscillator statistics were taken into account. This was not a particularly bad oscillator; the standard deviation in pulse energy was 5%, somewhat better than was commercially available at the time. Later improvements reduced this dispersion by a factor of 3 and the oscillators developed at LLNL were probably even more stable. The point is to be very careful about mating a poorly characterized oscillator to an expensive set of laser amplifiers.

A design issue of some consequence is the degree of isolation to be provided against back reflection. By judicious use of Faraday rotator and Pockels cell isolators, protection can be provided against back reflections as high as 100% of the outgoing energy. The cost of doing this may, however, be dismayingly large, especially if such protection is not really needed. The reason that this is so is that a 100% protected system has to accommodate the worst-case situation on every shot for the outgoing pulse, no matter what the level of actual back reflection, and disk amplifiers can tolerate substantially high energy or intensity levels better than isolators, whether in self-focusing-limited or damage-limited regimes.

## 7.7  AVERAGE POWER EFFECTS

Many of the effects and phenomena considered in the solid-state laser literature are poised in terms appropriate for the absence of transient thermal effects. In many cases this is an appropriate approximation. The solid-state lasers developed for fusion research are currently single-pulse devices; pumping-induced thermal profiles are small in magnitude and conduction effects are generally ignorable. At the other extreme, small-aperture crystalline lasers such as Nd:YAG range finders quickly reach quasi-static conduction-limited thermal profiles, where constant values for thermal lensing and birefringence can be used by the designer.

There are a number of situations intermediate between these limits which may assume increasing practical interest. This intermediate regime is much more complex than the limiting cases of single-pulse or quasi-CW. Although aspects of it have been considered over many years, it has developed rather slowly. Although this is traceable partially to support levels for research and development in this area, it more probably reflects the complexity of the problem. The geometry of the pumped solid-state laser element, the optical path in the medium, and the thermal and thermo-optic properties of the host material can all contribute to both birefringence and wavefront distortion. To simplify the discussion and illustrate the main issues, we will confine the discussion to isotropic host materials. More complex uniaxial and biaxial hosts are known and create an even greater richness of effects.

### 7.7.1  Thermal Distortion Effects in Various Laser Media Configurations for Isotropic Laser Media

In an idealized laser, the only waste heat transferred to the host lattice might come from the quantum defect (i.e., the extent to which the emission wavelength is more than the pump wavelength). This simple picture might obtain for a diode-pumped solid-state laser. In a flashlamp-pumped solid-state laser, on the other hand, there will be a large number of pump frequencies and a different quantum defect for each. The quantum efficiency can also affect heating, especially if there are strong nonradiative decay modes for the laser ion. Other processes, such as excited-state absorption or up-conversion, can also result in losses which ultimately show up as heat in the lattice. All of these effects will be present to a greater or lesser extent in single-pulse or repetitive-pulse lasers, but as a practical matter the two cases are very distinct. The reason is that solid-state laser host materials are generally materials with a high heat capacity but with a low thermal diffusivity. Figure 7.37 illustrates the development of the thermal profile when a cylindrical laser rod is pulsed. The rod distortion relaxes slowly.

A corollary of this observation is that an optimum approach to a single pulse laser may be very nonoptimal for repetitively operated lasers. Generally, the single-pulse optimization is for overall efficiency. In a rod laser, a UV absorber may be in the coolant to prevent solarization or color-centering of the host. Most (i.e., to 70%) of the electrical input to the flashlamp will be absorbed somewhere other than in the laser rod. None of these heat inputs to the system are relevant to single-pulse operation other than as efficiency issues, but all are relevant to quasi-CW repeated-pulse operation. The engineering issue for quasi-CW lasers is primarily waste heat management; efficiency is a somewhat secondary concern.

In this light we will examine two familiar single-pulse geometries, the cooled cylindrical rod and the face-pumped disk; a quasi-CW geometry, the zigzag slab; and the single-pulse or burst-mode geometry, the thick straight-through slab.

***Cooled Cylindrical Rods.*** Generally this is the simplest of all laser rod geometries to fabricate. If pumping and cooling are radially symmetric, optical distortions can be described in a simple (cylindrical) coordinate system in terms of materials parameters and thermal field $T(r, t)$. In general, for glass or crystalline hosts one can find an arrangement of rod, ion dopant, flashlamps, and reflectors to give uniform gain. For a given material and heat load there will be an upper bound on radius for which quasi-CW operation can be maintained without thermal rupture of the material.

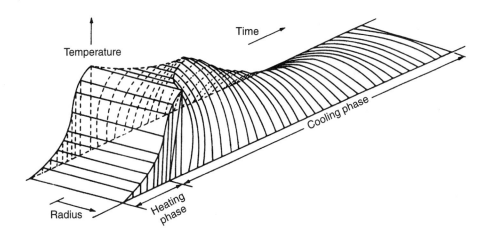

**Figure 7.37** Development and relaxation of the thermal profile in a laser rod. (From Ref. 163.)

***Face-Pumped Disk.***  As used in fusion lasers there is no cooling other than conduction to the disk holder. Nonextracted energy also results in significant fluorescence heating at the edge of the aperture. The propagation direction is almost perpendicular to the thermal gradiant due to pumping, so optical distortion is fairly insensitive to materials properties. For quasi-CW operation, face cooling would have to be arranged, a non-trivial issue, as the beam would have to pass undisturbed through the coolant. The edge heating would also add a transverse thermal gradient absent in the single-pulse case unless the edge could be very effectively cooled.

***Zigzag Slab.***  The zigzag slab was envisioned from the start to be a quasi-CW geometry. The zigzag ray paths would average out thermal gradients and the cooling issue would be about the same as for a rod of the same thickness (while the transverse dimension could be much larger, giving higher energy/pulse or power). That this has not yet emerged a clear winner is perhaps because it has the problems of the other geometries plus several uniquely its own. All of the problems of thermal management for quasi-CW rods apply here as well as the heating issues for edge claddings on disks. An issue unique to the zigzag and thick slabs is the difficulty of getting good pumping uniformity and energy extraction. Unique to the zigzag is the presence of large-area optical surfaces from which the beam bounces and through which the heat must be extracted and optical pumping must be accomplished. In addition, the distortion cancellation depends strongly on geometrical symmetry in pumping and cooling.

***Thick Slabs.***  These are very similar to thick rods (i.e., a burst mode but not really a true quasi-CW geometry). The rectangular geometry gives a greater emitting area, a different and potentially simpler thermal lensing issue, and a more difficult pumping uniformity issue. From the Soviet literature, lasers with multiple slabs $4 \times 24$ cm$^2$ in emitting area and 72 cm long have been operated [164].

### 7.7.2  Effects of Materials Properties and Configuration on Thermal Distortion and Birefringence

Quelle first analyzed thermal distortion effects for cylindrical glass laser rods pumped and cooled through the barrel which could be cast in terms of the materials properties of the host [102]. Somewhat later, Anan'ev and Grishmanova broadened this to include straight-through slab as well as rod geometries. An interesting feature of the Soviet analysis was that while the same materials parameters are important in determining the distortion, the conditions for minimizing path-length distortion are different in the two cases and may more easily be achieved in the thick slab than in the rod case. Using the more familiar Western nomenclature, these results are [103]

$$n(r)_{r,0} = n_0 + (P \pm Q)(T(r) - (T_0)) \mp [QT_{av}(r) - T_0]$$

for rods, where $r$ and $\theta$ refer to polarization direction (radial or azimuthal), and

$$n(x)_{1,2} = n_0 + (P \pm Q)[T(x) - T_0]$$

for slabs, where 1 denotes polarization in the $x$ direction and 2 denotes polarization perpendicular to $x$. In both cases,

$T(r)$ = temperature at radius ($x$ position) $r$ compared to the reservoir (coolant) temperature $T_0$

$T_{\mathrm{av}}(r)$ = average temperature from $r' = 0$ to $r$

$T_0$ = average temperature over the whole element

$$P = \frac{dn}{dT} + \frac{\alpha E}{2(1 - \nu)}(B_{\parallel} + 3B_{\perp})$$

$$Q = \frac{\alpha E}{2(1 - \nu)}(B_{\perp} - B_{\parallel})$$

$B_{\parallel}$ = stress optic coefficient for polarization parallel to the stress

$B_1$ = stress optic coefficient for polarization perpendicular to the stress

$\nu$ = Poisson's ratio

$E$ = Young's modulus

$dn/dt$ = change in refractive index with temperature

$\alpha$ = coefficient of linear thermal expansion

Several interesting situations can be determined by examination of these relations. For rods, true athermal behavior can occur only if $P = Q = 0$. For slabs, however, the requirement is that $P = \pm Q$, depending on the polarization state, a potentially much easier condition to satisfy. From the data in the paper by Avakyants [20], Soviet scientists have developed a number of laser glasses that meet this criterion.

The laser beam geometry can also influence the beam distortion and birefringence. De Metz noted that in geometries in which the element is double passed with a 90° rotation of the polarization between passes, birefringence is canceled [104]. More recently, it was recognized that in such cases the wavefront distortion will depend only on the value of $P$, not at all on $Q$ [165]. Schavelev et al. examined the temperature dependence of $P$ and $Q$ and found that while $dQ/dT < 10^{-9}/°\mathrm{C}$, $dP/dT$ was typically $10^{-8}/°\mathrm{C}$ [166]. If $P$ is sufficiently small at room temperature, the possibility exists of temperature tuning to an athermal condition. Figure 7.38 shows the results of such an experiment from the Soviet literature.

An ideal zigzag slab laser should not show lensing effects. Lensing accumulated during the first half of the path should be compensated during the second half. Achieving the necessary symmetry in alignment pumping and thermal conditions has not proven easy to do in practice, and materials with small thermo-optical coefficients can reduce the difficulty of achieving absolute symmetry.

### 7.7.3  Materials Limits for Solid-State Laser Materials

Most analyses of this problem approach it from the point of view of a quasi-static thermal problem and use the heat conduction equation to relate the thermal field as a function of position to an equivalent mechanical stress field. In principle, thermo-mechanical stresses produced by optical pumping and cooling should be related to

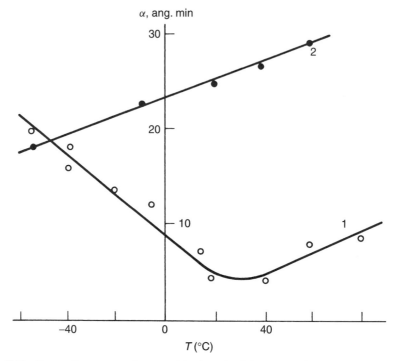

**Figure 7.38** Beam divergence of a repetitively pulsed laser as a function of base (coolant) temperature for one type GLSS-21 athermal silicate glass and two type GLS-1 silicate glass. (From Ref. 167.)

the breaking stresses encountered in purely mechanical stress tests, at least for similar surface preparation techniques. Such an approach would result in a safe stress level, which would be proportional to $E/(1 - \nu)$. For laser glasses or weak crystalline laser hosts, this may be an optimistic approach. There is a Soviet test in which a 10-mm-diameter laser rod is pulsed repetitively at 1 Hz, and the thermal gradient measured at which the rod shatters as the pump energy is slowly increased. Figure 7.39 plots these results for Soviet glasses as a function of Young's modulus from the data in Avakyants [20]. Below a value of $E$ of 4000 kg/mm$^2$, apparently no stable glasses existed, and only for $E > 8000$ kg/mm$^2$ would there be any temptation to fit a straight line to these results.

The most likely explanation for this deviation from simple models for the thermal stresses is that mounting stresses for laser rods may use up a good deal of the "safe" loading, at least for glasses and the weaker crystals. This situation can be insidious with athermal materials, as there may be little or no indication of stress up to the failure point. Use of a nonathermal "dummy" test piece may be useful to evaluate mounting stresses (and procedures).

## 7.8 MISCELLANEOUS TOPICS

In this section we cover several topics that recommend themselves as areas where experimentation with solid-state lasers may present unusual or different situations than

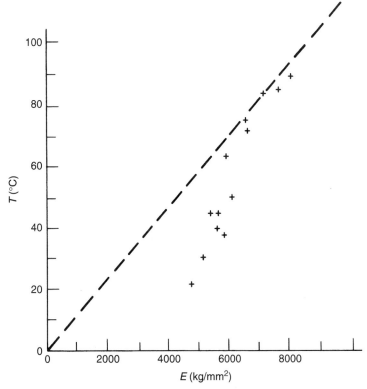

**Figure 7.39** Breakage level (average power) of 10-mm Soviet laser rods as a function of Young's modulus. The dashed line represents linear dependence. (Data from Ref. 20.)

those normally encountered in the experimental research lab and with optics as presented in typical introductory classical optics textbooks.

### 7.8.1 Laser Safety: Diffuse and Specular Reflection Hazards

Much has been written over the years about the possible hazards from laser radiation, particularly in regard to eye damage. This is a serious issue and is quite germane to any discussion of solid-state lasers, as many of them operate at wavelengths where the lens and fluid in the eye are transparent and the lens can concentrate the light in the retina. At most visible wavelengths a safe level is about 0.1 $\mu$J/cm$^2$ incident on the unprotected eye [168]. As in-beam intensities may be 1 to 10 J/cm$^2$ for commercial or research lasers, substantial attenuation is necessary before a beam can be safely viewed. A recent report of damage to a person at LLNL caused by average power damage to laser safety goggles followed by retinal damage reinforces this concern and argues for "indirect" viewing and against "direct" viewing unless the situation is totally controlled and understood [169]. Adding to the hazardous aspect of solid-state lasers is the fact that pulse durations are generally shorter than physiological response times (the "blink" reflex) by many orders of magnitude. There is no time to respond to avoid dangerous exposure other than planning ahead so that it does not occur.

With early sources which operated only at the wavelength of a particular laser transition, eye protection in the form of safety glasses or goggles were a reasonably simple and convenient means of protection. Although still useful where appropriate, there are many current situations where reliance on safety glasses alone is not adequate. In many experiments the laser beam is frequency converted to one or more other wavelengths by nonlinear optical means. The radiation of these wavelengths may also be at hazardous intensities. Tunable vibronic lasers such as the alexandrite or titanium-doped sapphire lasers can lase over a wide fraction of the spectrum. In any case, where eye protection is relied on for the primary protection, actual measured transmission of the units used should be kept on file and reviewed for appropriateness for the intended application before the fact.

Most, if not all, recent commercial solid-state laser equipment is labeled by the manufacturer as a class III or class IV laser and designed so that when used as intended, with all covers in place, only the primary beam and reflections from surfaces it encounters pose a hazard. This should not lull the user into false confidence. Reflections from external optics can be fully as hazardous as any of those protected against by the manufacturer and are the user's responsibility. Safeties or interlocks should be examined before laser activation to ensure that they are, in fact, operational. Commercial laser equipment manufactured before about 1975 did not necessarily meet current labeling or safety rules — in some cases most assuredly not. What this means is that someone who is contemplating use of a solid-state laser for some purpose has a significant responsibility to themselves and others to ensure that safe operation in fact occurs. In addition to the ocular hazard, there may also be hazards of an electrical nature as well as concerns with solvents, coolants, and fire hazards. Although each case will have its own concerns, a general checklist of procedures to result in a safe operation can be outlined. The recommended steps would be as follows:

- The suitability of the room or space that will be the intended workplace should be examined carefully. With any class III or IV laser the beam should be positively contained within the enclosure. Entrance and exit from the area should have appropriate warning signs, and procedures to gain entrance clearly spelled out. All safety-related systems, such as sprinkler or $CO_2$ systems, should be checked for proper operation. All existing (and any new) electrical systems, especially the grounds, should be explicitly checked to avoid any electrical hazards. Any work indicated as necessary by this survey should be completed and inspected before facility activation.

- The design of the experiment should be reviewed carefully to identify any potential hazards from specular reflections by some form of ray tracing. Although each setup is unique, several general rules should be followed:

  a. All specular or focused beam reflections should be traced out and steps taken to ensure no personnel hazard.

  b. When possible the beam height should be well below eye level (i.e., preferably 130 cm or less above the floor) and all specular reflections contained in this plane.
  Care should be exercised to minimize procedures that would cause a person to have his or her eyes in this beam plane.

- Plan procedures for troubleshooting and repairs. Some expected routine maintenance procedures, such as changing flashlamps or optical resonator components, will require some disassembly of the equipment. This could lead to potentially unsafe situations. A workable and safe plan should be developed in advance of the problem.

- Educate all those involved in the operation as to safety-related issues and plans.

- Follow the procedures and do not take unsafe shortcuts.


### 7.8.2  Risks in Extrapolation of Data from the Laser Literature

The twenty-fifth anniversary of the invention of the laser occurred in 1985. Although 1960 may seem a long time ago, it is not in terms of the evolution of lasers and the optical industry in general. During this period there has been a dramatic revolution in the development of precision apparatus for optical measurements. In addition to lasers per se, a whole new discipline, nonlinear optics, has grown up involving phenomena previously unknown to classical optics. In any sense the past several decades have been a very stimulating time to be involved in optical research.

At the same time, the rapid and parallel evolution of these closely related fields has created a literature with a significant number of erroneous, incomplete, and irrelevant publications, where at times it appears that both forest and trees were missed. The solid-state laser literature certainly can lay claim to its fair share of publications which the authors, present company not excepted, would like to recall in retrospect. At the same time, there are publications from very early in the life of the laser which improve with age. This situation transcends national boundaries and is probably not unique to lasers but is in fact characteristic of science and technology at work in a new area.

This can create problems for someone who is new to the area or who comes from a more traditional or better developed area of science or engineering. Probably the best guidance in terms of evaluating the published literature is to develop one's own critical judgment by doing through literature searches and intercomparisons of the results obtained by various authors. In the solid-state laser field several recent books and publications are worthy of mention. *Solid State Laser Engineering* by W. Koechner (Berlin: Springer-Verlag, 1976) and *Solid State Laser Amplifiers* by D. Brown (Berlin: Springer-Verlag, 1981) are good summaries of the physics and components. *CRC Handbook of Laser Science and Technology Handbook* Volume 4, by M. Weber (Boca Raton, FL: CRC Press, 1986), *Laser Crystals* by A. A. Kaminskii (Berlin: Springer-Verlag, 1981), and *Tunable Solid State Lasers* by A. B. Budgor and A. Pinto (Berlin: Springer-Verlag, 1985) contain compilations of most of the source data on laser crystals. Weber and Kaminskii reproduce essentially the same data base. Unfortunately, the methodology used in the Soviet experiments does not allow other than a qualitative estimate of the relative strengths of the various laser transitions. The recent series of monographs from Lawrence Livermore Laboratory on the Physics of Laser Fusion provide a valuable archive of data and methodology developed by this program.


### 7.8.3  Optical Diagnostics with Highly Coherent Sources

Many laser sources have spatial coherence lengths of centimeters to meters. This can complicate the design of optical diagnostics for sampling the beam and measuring properties such as power, energy, and beam profiles.

Beam splitters such as a glass plate with two polished optical surfaces can distort the spatial pattern even if the plate and the surfaces are optically perfect. With a coherent source, the plate becomes a wavefront shearing interferometer, and any spherical divergence on the beam will result in a interference pattern of linear fringes on the reflected beam. Although useful in other contexts as a collimation check, such fringes can distort the near-field spatial pattern. A wedged plate can be used if the reflections from the two surfaces are allowed to propagate far enough to be spatially separated. Care must also be taken to use as low an angle of incidence as possible and as collimated a beam as possible to avoid introducing some degree of coma and astigmatism on the beam.

Beam attenuators can also induce distortions on beams being diagnosed. The functions such attenuators are used for include both the reduction of the beam intensity so that the detectors work in their useful dynamic range and the reduction of spurious signals from sources such as flashlamps. For the latter function visible-blocking, infrared-transmitting filter glasses such as Hoya IR-85 or IR-90 are potentially useful, as they will block almost all the shorter-wavelength radiation from flashlamps. The decreasing response of film or vidicon detectors beyond the infrared absorption edge of these glasses generally allows for operation where residual lamp light will be below the detection threshold, while the beam to be diagnosed can use the full dynamic range. There are several potential problems in the use of broadband filters. While the attenuation in the visible at which most production measurements are carried out in the optics industry is very large, at the laser wavelength for solid-state lasers, 0.69 $\mu$m or longer, it may be small. Filter material should be checked at or near the operating wavelength to ensure that the material is optically homogeneous and the surfaces are reasonably well polished. If the internal absorption is small it may be desirable to anti-reflection coat the surfaces to avoid interference effects on the transmitted beam. With attenuating filters for the laser wavelength, one high-attenuation filter is referable to a number of separate filters of equal net attenuation to minimize interference effects causing extraneous interference patterns. For either near- or far-field detectors it is generally best to have filters placed as near the detector plane as possible. Even if the filter induces a phase distortion, the beam may have to propagate a fair distance before significant distortion of the amplitude pattern will result. Ordinary commercial filters, which consist of an absorption filter plus an interference filter, may be adequate for power or energy monitors, but in our experience, frequently demonstrate many of the foregoing problems. At the very least they need to be checked.

Silicon vidicon television cameras are frequently used as the detector elements for near-field cameras with solid-state lasers. Compact cameras with 2.54-cm vidicons are readily available, inexpensive, and have very reasonable linearity and uniformity of response. The camera output can be read out on a TV monitor or converted to a digital signal for further processing by computer. There can be problems, however, with coherent sources. The silicon element in the typical camera of this type is typically 30 $\mu$m thick and has a very strong absorption for wavelengths below 1 $\mu$m. Above 1 $\mu$m the absorption coefficient for silicon begins to decrease rapidly. It was found that at 1.06 $\mu$m, the wavelength of a Nd:YAG laser, very strong irregular modulation patterns were produced even with a demonstrably $TEM_{00}$ spatial mode laser [170]. Figures 7.40 and 7.41 show video picture and a line scan through the silicon vidicon picture. Tilting the vidicon produced a reversal of the modulation, showing that it was in interference phenomena. Further investigation revealed that the silicon wafer was chemically polished on one surface to achieve the desired final thickness. For

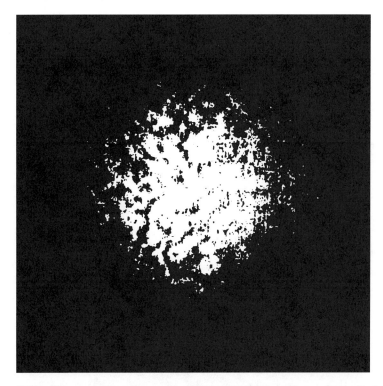

**Figure 7.40**    Silicon vidicon picture of a TEM$_{00}$ Nd:YAG oscillator. (From Ref. 170.)

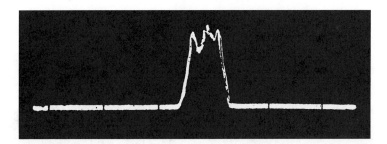

**Figure 7.41**    Line scan through the picture in Fig. 7.40. (From Ref. 170.)

wavelengths where silicon is transparent, the wafer appears to be a Fabry–Perot interferometer whose thickness varies locally. With a different type of vidicon, a lead oxide vidicon, which is not transparent at 1.064 μm, these effects were absent (Figs. 7.42 and 7.43). Very analogous effects were found with linear silicon diode arrays.

When silicate laser glasses were supplanted by higher-gain phosphate laser glasses, it was desirable to use a shorter-wavelength oscillator to match the gain of the laser glass. Neodymium-doped yttrium lithium fluoride (Nd:YLF), which lases at 1.053 μm, was found to be an almost perfect match [171]. The absorption of silicon detectors at this wavelength is enough higher than at 1.064 μm that the interference effects found with Nd:YAG were suppressed and silicon vidicons, and photodiodes are useful for the newer fusion research lasers.

**Figure 7.42**   Lead oxide vidicon picture of Nd:YAG TEM$_{00}$ oscillator. (From Ref. 170.)

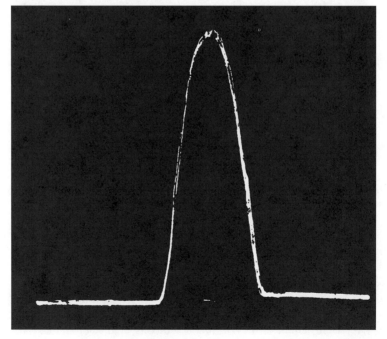

**Figure 7.43**   Line scan through the lead oxide vidicon picture. (From Ref. 170.)

Film detection of neodymium laser patterns is a somewhat limited technique. At 1.064 μm only special films such as the Kodak spectrographic 1-Z emulsions and the no-longer-manufactured Polaroid IR film demonstrated any significant response. At 1.054 μm, Polaroid positive/negative HP/N film has a limited but useful response [172]. Our measurements at this wavelength indicated a threshold for exposure of 0.03 J/cm$^2$ and a damage threshold of about 0.35 J/cm$^2$. As this film is available in 10 cm × 12 cm(4 in. × 5 in.) format, it can be used for direct in-beam sampling with a flashlamp blocking filter such a Hoya IR-90, for a "quick look," but the sensitivity is marginal for an off-line monitor.

The optics systems for either near- or far-field beam monitors needs to be approached carefully with highly coherent laser sources. For a near-field monitor the simplest and best situation is one in which the beam size is small enough that it can be put onto the vidicon or film camera (after appropriate attenuation). If this is not the case, a beam-size-reduction telescope must be used. Care is necessary in design to avoid the introduction of artifacts into the beam image by this telescope. The two major problems to avoid are designs in which the lens elements cause circular interference rings between the primary image and rays multiply reflected in the lens, and detector placement such that the images of imperfections (dust, etc.) on the primary lens or mirror are optically relayed to the detector plane.

The discussion to this point in this section has assumed a worst-case situation, which might be typical of a multi-nanosecond laser with very high temporal coherence. Any interference phenomena that occur in this case will generally achieve full fringe visibility, as the pulse duration is long compared to the light transit time through thin optical components. There are three cases of some practical significance where significant simplification may occur:

- A number of the interference effects may come from waves which are off-axis or which correspond to a wave with different collimation than that if the main beam. Subsequent spatial filtering may reduce or eliminate these beams.

- With ultrashort pulses (<30 ps) secondary reflections may lag the main pulse by a sufficient amount that full fringe visibility cannot be developed. If these after-pulses do not cause problems with the phenomena under investigation, optical systems may be useful, which would not be safe or prudent for coherent longer pulses.

- Recently, researchers at NRL have been investigating the efficacy of lasers with deliberately reduced temporal coherence for inertial confinement fusion. This "induced spatial incoherence" could also substantially reduce interference effects traceable to time-delayed reflections [173].

### 7.8.4 Prismatic Effects and Optical Aberration Control

A number of the optical layouts used in early laser experiments, especially in the 1960s, can produce undesirable effects on the laser beam quality. In some cases their use reflected the sparse availability of optical components suitable for use with high-power lasers; in many other cases the poor quality of available solid-state laser materials created even greater intrinsic beam divergence such that prismatic effects were a minor contributor to the total beam spread.

If a diverging beam is used through optical components in a laser beam, any Brewster angle rods or plates can induce coma and astigmatism on the transmitted beam. These effects become especially noticeable for geometrical divergence angles greater than about 3 mrad. Components that can contribute in this regard include laser rods, disk amplifiers, and polarizers, either dielectric film or stacked plate types.

A stacked plate polarizer can cause other problems, even with a collimated input beam. For optimum extinction for a given number of plates with coherent light, a fanned array of wedged plates is desirable to avoid any parallel surfaces and consequent interference effects. One problem with such a design is that as the beam propagates through a stack, the lateral diameter will change in the plane of incidence. If a symmetric design is used for the input and output stack, this effect will be minimized, as long as no vignetting occurs.

Minor accumulated coma and astigmatism can be corrected by having one lens in the system with a tiltable mount. Diagnosis can be accomplished several ways on a multicomponent oscillator–amplifier system. The master oscillator can be used with a far-field monitor and the lens tilted to obtain the most symmetric pattern. Testing with a Hartman plate in the near field will also reveal geometrical effects.

One other effect can occur with vacuum spatial filters or relays that will give astigmatism on the beam which looks very similar to these other problems. If the lens seat is not perfectly flat because of distortion during welding operations or burrs on the lens seat, the lens may be distorted when the assembly is evacuated, giving some net cylindrical power. Even with a distorted seat, O-rings may still hold a good vacuum. *This condition could be quite hazardous to personnel if the potentially high stresses cause the lens to rupture.* There is a good and simple test for such a condition and it is suggested that it be performed on initial assembly and any subsequent reassembly of the vacuum assembly. If a gas laser with a beam expanded to approximately fill the entrance aperture is directed through the lens and the back focus (toward the laser from the lens) is examined as the vacuum pump is started, this condition will readily be revealed. As the rays that form this image go through the lens, twice the astigmatic effect will be accumulated for this image compared to the forward focus.

One issue was raised by Soviet authors in regard to disk amplifiers and the large numbers of surfaces they introduce into a laser system. Succinctly put, it was that if each surface has a figure of $\lambda/n$ and there are $m$ surfaces, won't the best resultant figure error be on the order of $(m)^{1/2}\lambda/n$? If $n$ is on the order of 10, (i.e., $\lambda/10$ surfaces), from $m > 200$ won't the beam quality be rather low? [174]. Although this argument appears plausible, it turns out to be specious in practice. If the deviations from perfect flatness are localized to a size scale much less than the aperture and they are random, this argument would hold. On the other hand, if the deviations from flatness are large scale such that an exact description of the surface can be made down to the scale of the intrinsic roughness by use of the Seidel aberration coefficients [175], the overall wavefront error can be calculated by superposition and it in fact can be corrected by a suitable corrector. The experiment reported by LLNL in which a laser system with $164\lambda/10$ surfaces was corrected to better than $\lambda/2$ argues strongly for the latter viewpoint [176]. Our own observations on this issue appear to be consistent with LLNL; other than tilt error (which is taken out by the alignment system), and coma and astigmatism (which are minimized by design and actively corrected), the major penalty of many optical surfaces is a slowly developing spherical aberration of a magnitude that could readily be corrected [121].

## 7.8.5  Advantages of Optical versus Numerical Simulation

A number of practical situations arise in which numerical simulation of all aspects of laser systems may be difficult because the numerical solutions do not converge rapidly, or the solutions may be quite sensitive to the exact value of component focal lengths or curvature. Two examples of this sort which are frequently encountered with laser systems are the near-field amplitude pattern of a beam after it is truncated by an aperture or a series of apertures, and the exact position of secondary back foci of compound spatial filter and optical relay systems.

This type of problem can frequently be best worked out in the laboratory using a lower-power CW laser to simulate the actual optical system. This may appear a trite observation, but as a practical matter even a rather commonplace helium–neon laser may be the equivalent or better of exceedingly fast computers such as an optical processor. Another advantage of laboratory simulation is that the effects of misalignment can be accurately measured.

## REFERENCES

1. T. H. Maiman, *Nature (London)*, **187**, p. 493, 1960.
2. A. Einstein, *Verh. Dtsch. Phys. Ges.*, **18**, p. 318, 1916.
3. D. Ross, *Laser Lichtverstä*rker-used Oszillatoren, Akademische Verlagsgesellschaft, Frankfurt am Main, 1966.
4. W. Koechner, *Solid State Laser Engineering*, Springer-Verlag, Berlin, 1976.
5. A. A. Kaminskii, *Laser Crystals*, Springer-Verlag, Berlin, 1981.
6. G. H. Dicke, *Advances in Quantum Electronics*, Columbia University Press, New York, 1961, p. 170.
7. L. F. Johnson, J. E. Geusic, and L. G. Van Uitert, *Appl. Phys. Lett.*, **7**, p. 127, 1965.
8. E. V. Zharikov et al., *Sov. J. Quantum Electron. (Engli. Ed.)*, **13**, p. 1274, 1983.
9. J. C. Walling, O. G. Peterson, H. P. Jenssen, R. C. Morris, and E. W. O'Dell, *IEEE J. Quantum Electron.*, **QE-16**, p. 1302, 1980.
10. P. F. Moulton, *Conf. Lasers Electro-Opt. (CLEO)*, Tech. Dig., WA2, June 1984.
11. P. Hammerling, A. B. Budgor, and A. Pinto, Eds., *Tunable Solid State Lasers*, Springer-Verlag, Berlin, 1985.
12. J. E. Geusic, H. M. Marcos, and L. G. Van Uitert, *Appl. Phys. Lett.*, **4**, p. 182, 1964.
13. M. Birnbaun and A. W. Tucker, *IEEE J. Quantum Electron.*, **QE-9**, p. 46, 1973.
14. M. Birnbaun, private communication, 1986.
15. A. L. Harmer, A. Linz, and D. R. Gabbe, *J. Phys. Chem. Solids*, **30**, p. 1483, 1969.
16. A. A. Kaminskii, Kh. S. Bagdasarov, G. A. Bogomolova, M. M. Gritsenko, A. M. Kevorkov, and S. E. Sarkisov, *Phys. Status Solidi* **(A) 34**, p. K109, 1976.
17. B. Struve, G. Huber, V. V. Laptev, I. A. Shcherbakov, and E. V. Zharikov, *Appl. Phys. [Part] B*, **B28**, 235, 1982.
18. B. Struve et al., *Appl. Phys.*, **B30**, p. 117, 1983.
19. S. E. Stokowski, R. A. Saroyan, and M. J. Weber, *Nd-Doped Laser Glass Spectroscopic and Physical Properties*, M-95, Lawrence Livermore Lab., Livermore, CA, November 1978.
20. L. I. Avakyants, et al., *Sov. J. Quantum Electron (Engl. Transl.)* **8**, No. 4, p. 423, 1978.

21. H. E. Edgerton, *Electronic Flash Strobe*, McGraw-Hill, New York, 1920; H. E. Edgerton, H. H. Goncz, and D. W. Jameson, in H. O. Tjeck, Ed., *High Speed Photography*, Willink & Zoon, Harlem, The Netherlands, 1963.

22. I. S. Marshak, *N Impulsnoye Istochniki Sveta (Pulsed Lights Sources)*, Gosenergoizdat, Moscow and Leningrad, 1963 (NRL Transl. 1210, Ad704–944).

23. J. P. Markiewicz and J. L. Emmett, *IEEE J. Quantum Electron.*, **2**, 1966.

24. D. C. Brown and N. Nee, *IEEE Trans. Electron Devices*, **ED-24**, 1977.

25. J. B. Trenholme and J. L. Emmett, in W. G. Hyzer and W. C. Chace, Eds., *Proc. 9th Int. Cong. High Speed Photography*, SMPTE, 1970.

26. J. B. Trenholme, *Laser Program Annual Report*, UCRL-50021-75, Lawrence Livermore Lab., Livermore, CA, 1975.

27. ILC Technology, *An Introduction to Flash Tubes*, Tech Bull No. 1, An ILC Technology, Sunnyvale, CA, 1970.

28. J. D. Sturel, *Memorandum Report*, Centre des Rescherches, Companie Générale d'Electricité, Marcoussis, France, 1966.

29. L. Noble, C. Kretschmer, R. Maynard, H. Flentz, and L. Reed, *Optical Pumps for Lasers*, Rep. ECOM-0239-F, ILC Technology, Sunnyvale, CA, 1973.

30. P. B. Newell and A. P. Benson, *E. G. & G. Tech. Rep.* 8-4426, 1975.

31. J. B. Trenholme, *Laser Program Annual Report*, UCRL -50021-85, Sect. 9, Lawrence Livermore Lab., Livermore, CA, 1985.

32. J. H. Goncz and P. B. Newell, *J. Opt. Soc. Am.*, **56**, p. 87, 1966.

33. J. F. Holtzrichter and T. R. Donich, UCID-16860, Lawrence Livermore Lab., Livermore, CA, 1973.

34. J. M. McMahon, J. L. Emmett, J. F. Holtzrichter, and J. B. Trenholme, *IEEE J. Quantum Electron.*, **QE-9**, p. 992, 1972.

35. J. H. Boyden and E. G. Erickson, *Second Harmonic Generations*, Rep. AD 729681, Holobeam Inc., Paramus, NJ, 1971.

36. R. L. Keyes and T. M. Quist, *Appl. Phys. Lett.*, **4**, p. 50, 1964.

37. G. L. Harnagel et al., *Appl. Phys. Lett.*, **49**, p. 1418, 1986.

38. M. E. Givens, L. J. Mawst, C. A. Zmudzinski, M. A. Emanuel, and J. J. Coleman, *Appl. Phys. Lett.*, **50**, p. 301, 1987.

39. P. S. Cross, *Proc. Conf. Lasers Electro-Opt. (CLEO)*, Baltimore, 1987, pap. WS-1.

40. L. Goldberg and J. F. Weller, *Appl. Phys. Lett.*, **50**, p. 1713, 1987.

41. R. Allen, L. Esterowitz, L. Goldberg, J. F. Weller, and M. Storm, *Electron. Lett.*, **22**, p. 947, 1986.

42. G. L. Kintz, R. Allen, and L Esterowitz, *Proc. Conf. Lasers Electro-Opt. (CLEO)*, Baltimore, 1987, pap. FL-2.

43. T. Y. Fan, G. Huber, and R. Byer, *Proc Conf. Lasers Electro-Opt. (CLEO)*, Baltimore, 1987, pap. FL-1.

44. W. Koechner. *Solid State Laser Engineering*, Springer-Verlag, Berlin, 1976.

45. T. G. Crow and T. J. Snyder, *Techniques for Achieving High Power Q-switched Operation in* YAG:Nd, Final Tech Rep AFAL-TR-70, 69, Wright-Patterson Airforce Base, Ohio, 1970; *Laser J.*, **18**, 1970.

46. S. B. Schuldt and R. L. Aagard, *Appl. Opt.*, **6**, p. 509, 1967.

47. V. Evtuhev and J. K. Neeland, *Appl. Opt.*, **6**, p. 437, 1967.

48. J. G. Edwards, *Appl. Opt.*, **6**, p. 837, 1967.

49. J. P. Letellier, *NRL Memo Rep.*, *NRL-MR*-2684, 1973.

50. L. Noble, J. Moffat, L. Reed, and J. Richter, *Optical Pumps for Lasers*, Final Rep. TR ECOM-0035-*F*, U.S. Army Electron. Command, Ft. Monmouth, NJ, 1971.

51. R. B. Chesler, *Appl. Opt.*, **9**, p. 2190, 1970.

52. *Laboratory for Laser Energetics, Annual Report*, VI, 43, University of Rochester, College of Engineering and Applied Science, Rochester, NY, 1977.

53. J. M. McMahon, *NRL Memo Rep., NRL-MR*-7838, 1974; D. C. Brown, *High Peak Power Nd:Glass Laser Systems*, Springer-Verlag, Berlin, 1981.

54. Companie Industrielle des Lasers, *Recent Developments in Cilas Glass Lasers, Brochure*; 1973; *Neodymium Glass VD-VK*, Marcoussis, France, 1971.

55. Yu. A. Anan'ev, I. M. Buzhinskiy, M. P. Van Yokov, E. F. Dauengauer, and O. A. Shorokhov, *Opt.-Mekh. Promst.*, **9**, p. 26, 1968.

56. V. N. Alekseev, Yu. A. Anan'ev, and E. F. Dauengauer, *Sov. Phys.-Dokl.* **19** (*Engl. Transl*), p. 1, 1974.

57. G. J. Linford and S. Yarema, *Laser Program Annual Report*, UCRL-50021-75, Lawrence Livermore Lab., Livermore, CA, 1975.

58. A. Bettinger, Companie Générale d'Electricités Marcoussis, France (private communication), 1982; M. Bedu and J. Jeanjean, *Rev. Phys. Appl.*, **18**, p. 191, 1983.

59. J. M. McMahon, *Appl. Opt.*, **6**, p. 2191, 1967.

60. Yu. K. Danielieko, A. A. Manenkov, A. M. Prokhorov, and U. Ya. Khaimov-Mai'kov, *Fiz. Tverd. Tela*, **10**, No. 9, p. 2738, 1968; *IEEE J. Quantum Electron.*, **QE-5**, p. 87, 1969.

61. W. F. Krupke, M. D. Shinn, J. E. Marion, J. A. Caird, and S. E. Stokowski, *J. Opt. Soc. Am., B: Opt. Phys.*, **3**, p. 102, 1986.

62. Yu. A. Anan'ev and O. A. Shorokhov, *Sov. J. Opt. Tech. (Engl Transl)*, **44**, p. 652, 1977.

63. A. E. Siegman, *Proc IEEE*, p. 277, March 1965.

64. J. E. Geusic, H. J. Levinstein, S. Singh, R. G. Smith, and L. G. Van Uitert, *Appl. Phys. Lett.*, **12**, p. 306, 1968.

65. F. J. McClung and R. W. Hellwarth, *Proc. IRE*, **51**, p. 46, 1963.

66. E. J. Woodbury and W. K. Ng, *Proc. IRE*, **50**, p. 2347, 1962.

67. R. Y. Chiao, C. H. Townes, and B. P. Stoicheff, *Phys. Rev. Lett.*, **12**, p. 592, 1964.

68. A. Haught and L. Meyerand, *Phys. Rev. Lett.*, **11**, p. 401, 1963.

69. R. Y. Chiao, E. Garmire, and C. H. Townes, *Phys. Rev. Lett.*, **13**, p. 479, 1964.

70. W. G. Wagner and B. A. Lengyel, *J. Appl. Phys.* **34**, p. 2040, 1963.

71. A. Szabo and R. A. Stein, *J. Appl. Phys.*, **36**, p. 1562, 1965.

72. M. Menat, *J. Appl. Phys.*, **36**, p. 73, 1965; erratum: *J. Appl. Phys.*, **36**, p. 936, 1965.

73. H. W. Mocker and R. J. Collins, *Appl. Phys. Lett.*, **7**, p. 270, 1965; A. J. DeMaria, D. A. Stetser, and H. Heyman, *ibid.*, **8**, p. 174, 1966.

74. J. A. Fleck, *Phys. Rev. B: Solid State*, **1**, p. 84, 1970.

75. V. S. Letokhov, *Sov. Phys.-JETP (Engl. Transl.)* **28**, p. 562, 1969; *Sov. Phys.-JETP* **28**, p. 1026, 1969. P. G. Kryukov and V. S. Letokhov, *IEEE J. Quantum Electron.*, **QE-8**, p. 766, 1972.

76. A. E. Siegman and D. J. Kuizenga, *Appl. Phys. Lett.*, **14**, p. 181, 1969.

77. E. D. Jones, M. A. Palmer, and F. R. Franklin, *Opt. Quantum Electron.*, **8**, p. 231, 1976.

78. R. L. Carman, J. Fleck, and L. James, *IEEE J. Quantum Electron.*, **QE-8**, p. 586, 1972; J. A. Fleck and R. L. Carman, *Appl. Phys. Lett.*, **22**, p. 546, 1973.

79. J. M. McMahon and J. L. Emmett, *Proc. 11th Symp. Electron. Ion Laser Beam Technol.*, Boulder, CO, 1971.

80. B. H. Ripin et al., *Phys. Rev. Lett.*, **39**, p. 611, 1977.

81. D. J. Kuizenga, *Generation of Short Optical Pulses for Laser Fusion*, Rep. UCRL-13651, Lawrence Livermore Lab., Livermore, CA, 1975.

82. *Quantel SA*, Orsay, France, 1980.

83. W. R. Sooy, *Appl. Phys. Lett.*, **7**, p. 36, 1965.

84. J. M. McMahon, *IEEE J. Quantum Electron.*, **QE-5**, p. 489, 1969.

85. D. C. Hanna, *Opto-electronics (London)*, **3**, p. 163, 1971.

86. E. Zhou, T. Kane, G. Dixon, and R. Byer, *Opt. Lett.*, **10**, p. 62, 1985.

87. A. G. Fox and T. Li, *Bell Syst. Tech. J.*, **40**, p. 453, 1961.

88. G. D. Boyd and J. P. Gordon, *Bell Syst. Tech. J.*, **40**, p. 489, 1961.

89. L. M. Osterink and J. D. Foster, *Appl. Phys. Lett.*, **12**, p. 128, 1968.

90. Yu. A. Anan'ev, N. A. Sventsitskaya, and V. E. Sherstobitov, Sov. Pat. 274, 254, March 1968; *Zh. Eksp. Teor. Fiz.*, **55**, p. 13, 1968.

91. Yu. A. Anan'ev, *Kvantovaya Elektron. (Moscow)*, **6**, p. 3, 1971.

92. J. de Metz, A. Terneand, and P. Veyrie, *Appl. Opt.*, **5**, p. 819, 1966; J. de Metz, *Onde Electri.* **50**, p. 572, 1970.

93. V. Smiley, *Proc. IEEE*, **51**, p. 120, 1963.

94. L. M. Frantz and J. S. Nodvik, *J. Appl. Phys.*, **34**, p. 2346, 1963.

95. W. E. Martin and D. Milam, *Gain Saturation in Nd Doped Laser Materials*, LLNL Intern. Memo UCID-18868, Lawrence Livermore Lab., Lawrence, CA, 1980.

96. E. Snitzer and C. G. Young, *Lasers*, **2**, Dekker, New York, 1968.

97. J. M. Pellegrino, W. N. Yen, and M. J. Weber, *J. Appl. Phys.*, **51**, p. 6332, 1980.

98. P. V. Avizones and R. L. Grotbeck, *J. Appl. Phys.* **37**, p. 687, 1966.

99. N. G. Basov, R. V. Ambartsumyan, V. S. Zuev, P. G. Kryukov, and V. S. Letokhov, *Exp. Theor. Phys.*, **23**, p. 16, 1966.

100. T. H. DeRieux and J. M. McMahon, *NBS Spec. Publ. (U.S.)*, No. 341, 1970.

101. J. M. McMahon, *NRL Memo. Rep.*, NRL-MR-3411, 1976.

102. F. W. Quelle, Jr., *Appl. Opt.* **5**, p. 633, 1966.

103. Yu. A. Anan'ev and N. I. Grishmanova, *Zh. Prikl. Spectrosk.* **12**, p. 668, 1970.

104. J. de Metz, Thèse de Docteur Ingénieur, Université Pierre et Marie Curie, Paris, 1976.

105. J. A. Goldhar and M. A. Henesian, *Laser Program Annual Report*, UCRL-50021-83, Lawrence Livermore Lab., Livermore, CA, 1983; *Opt. Lett.*, **9**, 1984.

106. J. H. Nuckolls, L. L. Wood, A. R. Thiessen, and G. B. Zimmerman, *Nature (London)*, **239**, p. 139, 1972.

107. *Laser Program Annual Report*, UCRL-50021-84, Lawrence Livermore Lab., Livermore, CA, 1984.

108. R. Haas, *Zeus Laser Project*, Laser Program Annu. Rep., UCRL 50021-82, Lawrence Livermore Lab., Livermore, CA, 1982.

109. W. R. Sooy and J. L. Stitch, *J. Appl. Phys.*, **34**, p. 1719, 1963.

110. G. J. Linford, S. M. Yarema, and W. T. Crothers, *Laser Program Annual Report*, UCRL-50021-76, Lawrence Livermore Lab., Livermore, CA, 1976.

111. J. B. Trenholme, *Laser Program Annual Report*, UCRL 50021-73, Lawrence Livermore Lab., Livermore, CA, 1973.

112. C. B. McFann, W. D. Fountain, C. A. Hurley, and G. J. Linford, *Laser Program Annual Report*, UCRL-50021-76, Lawrence Livermore Lab., Livermore, CA, 1976.

113. J. C. Almasi, J. P. Chernoch, W. S. Martin, and K. Tomiyasu, *Face Pumped Laser*, GE Rep. to Office of Naval Research, 1966; J. P. Chernoch, W. S. Martin, and J. C. Almasi, Tech. Rep. AFAL-TR-71-3, Air Force Avionics Lab., Wright Patterson AFB, OH, 1971.

114. N. G. Bondarenko, I. V. Ereminaand, and B. I. Talanov, *JETP Lett.*, Vol. 6, p. 1, 1967; *Zh. Eksp. Teor. Fiz.*, Pis'ma Red., **6**, 1967.

115. W. B. Jones, L. M. Goldman, J. P. Chernoch, and W. S. Martin, *IEEE J. Quantum Electron.*, **QE-8**, p. 534, 1972.

116. E. M. Dianov and A. M. Prokhorov, *Sov. Phys. Dokl. (Engl. Transl.)*, **15**, p. 481, 1970.

117. W. F. Hagen, C. G. Young, J. Keefe, and D. W. Cuff, *Proc. DOD Conf.*, San Diego, CA, p. 363, 1970.

118. W. Koechner, *Solid State Laser Engineering*, Springer-Verlag, Berlin, 1976.

119. J. M. McMahon, R. P. Burns, and T. H. DeRieux, in *Laser Induced Damaged in Optical Materials Sympo.*, Boulder, CO, 1974.

120. J. J. Minot, *J. Opt. Soc. Am.*, **66**, p. 515, 1976.

121. J. M. McMahon, R. P. Burns, T. H. DeRieux, R. A. Hunzicker, and R. H. Lehmberg, *IEEE J. Quantum Electron.*, **QE-17**, p. 1629, 1981.

122. J. Z. Swain, R. E. Kidder, K. Pettipice, R. Rainer, E. D. Baird, and B. Loth, *J. Appl. Phys.*, **40**, p. 3973, 1969.

123. J. B. Trenholme, *NRL Memo. Rep.*, NRL-MR-2480, 1972.

124. G. Linford et al., *Laser Program Annual Report*, UCRL-50021-77, Lawrence Livermore Lab., Livermore, CA, 1977.

125. E. Snitzer, *Phys. Rev. Lett.*, **7**, p. 444, 1961.

126. J. M. McMahon, in *Damage in Laser Glass Symp.*, Boulder, CO, p. 49, 1969.

127. H. A. Lee, in *Damage in Laser Glass Symp.*, Boulder, Co, 1969.

128. *Laser Program Annual Report*, UCRL-50021-76, Lawrence Livermore Lab., Livermore, CA, 1976.

129. *Laser Program Annual Report*, UCRL-50021-77, Lawrence Livermore Lab., Livermore, CA, 1977.

130. J. M. McMahon et al., *Int. Quantum Electron. Conf.*, Amsterdam, 1976, paper V-5.

131. *Laser Program Annual Report*, UCRL-50021-81, Lawrence Livermore Lab., Livermore, CA, 1981.

132. *Laser Program Annual Report*, UCRL-50021-85, Lawrence Livermore Lab., Livermore, CA, 1986.

133. L. Coleman, private communication, 1987.

134. E. M. Dianov, private communication, 1978.

135. G. Girard and M. Michon, *IEEE J. Quantum Electron.*, **QE-9**, p. 979, 1973; D. von der Linde and K. F. Rogers, *ibid.*, p. 960.

136. K. H. Drexhage and G. A. Reynolds, *IEEE J. Quantum Electron.*, **QE-10**, p. 720, 1974.

137. Tropel, Inc., Rochester, NY.

138. J. F. Stowers, L. P. Bradley, and C. B. McFann, *Top. Meet. Inertial Confinement Fusion*, San Diego, CA, 1980, pap. TAF 11.

139. L. L. Steinmetz, T. W. Pouliot, and B. C. Johnson, *Appl. Opt.*, **12**, 1973.

140. Cleveland Crystals Inc., Cleveland, OH.

141. G. A. Kimber and P. J. Bateman, *A Faraday Rotation Isolator System for Use at Wavelengths between 0.4 and 0.9 $\mu m$*, Tech. Rep. 66153, Royal Aircraft Estab., UK, 1966.

142. O. C. Barr, J. M. McMahon, and J. B. Trenholme, *IEEE J. Quantum Electron.*, **QE-9**, p. 1124, 1973.

143. G. Leppelmeier and W. Simmons, *Semi-annual Reports*, UCRL-50021-73-1 and UCRL 50021-73-2, Lawrence Livermore Lab., Livermore, CA, 1973.

144. J. M. McMahon and O. C. Barr, *Proc. 17th Annu. Meet. SPIE*, San Diego, CA, 1973.

145. W. Koechner, Op. cit.

146. A. Hardy and F. Perrin, *The Principles of Optics*, McGraw-Hill, New York, 1932.

147. O. C. Barr and J. M. McMahon, *NRL Memo. Rep.*, NRL-MR-2830, 1974.

148. Optical Coating Laboratory, Inc., Tech. Note.

149. P. L. Kelley, *Phy. Rev. Lett.*, **15**, p. 1005, 1965.

150. J. H. Marburger, *NBS Spec. Publ. (U.S.)* No. 356, 1971.

151. E. S. Bliss, *IEEE J. Quantum Electron.*, **QE-8**, p. 273, 1972.

152. V. I. Bespalov and V. I. Talanov, *JEPT Lett. (Engl. Transl.)*, **3**, p. 307, 1966.

153. E. S. Bliss, D. R., Speck, J. F. Holzrichter, J. H. Erkkila, and A. J. Glass, *Appl. Phys. Lett.*, **25**, p. 448, 1974.

154. J. A. Glaze, *Opt. Eng.*, **15**, p. 136, 1976.

155. J. M. McMahon, O. C. Barr, and R. P. Burns, *Tech. Dig. Int. Electron Devices Meet.*, Washington, DC, 1973.

156. P. A. Renard and W. W. Simmons, *Appl. Opt.*, **16**, p. 779, 1977.

157. *Laser Program Annual Report*, UCRL 50021-83, Lawrence Livermore Lab., Livermore, CA, 1983.

158. W. H. Lowdermilk, D. Milam, W. L. Smith, M. J. Weber, and A. J. de Groot, *Laser Program Annual Report*, UCRL-50021-77, Lawrence Livermore Lab., Livermore, CA, 1977.

159. N. Bloembergen, *Appl. Opt.*, **12**, 1973.

160. C. K. Carniglia, J. M. Apfel, T. H. Allen, T. A. Tuttle, W. H. Lowdermilk, D. Milam, and F. Rainer, *NBS Spec. Publ. (U.S.)*, No. 568, 1980.

161. W. W. Simmons, *Laser Program Annual Report*, UCRL 50021-81, Lawrence Livermore Lab., Livermore, CA, 1981.

162. B. E. Yoldas, *Appl. Opt.*, **21**, p. 2960, 1982.

163. W. Koerchner, *Solid State Laser Engineering*, Springer-Verlag, Berlin, 1976.

164. J. N. Burdansky, E. V. Zhuzhakalo, N. E. Kovalsky, A. N. Kolomiisky, M. I. Pergament, Yu. P. Rudnitsky, and R. V. Smirnov, *Appl. Opt.*, **15**, p. 1450, 1976.

165. J. M. McMahon, S. Obenschain, H. Hellfeld, J. Meyers, C. Vollers, C. Rapp, and D. Ricks, "Current Status of Athermal Laser Glasses," *Laser Conf.*, Big Sky, MT, 1982.

166. O. S. Shchavelev, V. M. Miťkin, V. A. Babinka, N. N. Bunkina, and A. I. Stepanov, *Sov. J. Opt. Technol. (Engl. Transl.)*, **42**, p. 22, 1975.

167. K. P. Vakhmyanin, A. A. Mak, V. M. Miťkin, L. G. Popova, I. V. Raba, L. N. Soms, and A. I. Stepanov, *Sov. J. Quantum Electron. (Engl. Transl.)*, p. 106, 1976.

168. *American National Standard for the Safe Use of Lasers*, Am. Nat. Stand. Inst., New York, 1973.

169. *Laser Focus Electro-Opt. Mag.*, Aug. 1987, p. 26; also *Serious Accidents*, No. 11, DoE EH-0007, January 1986.

170. *Laser Fusion Studies at NRL July 1975–September 1976*, Rep. to Energy Research and Development Agency, 1976.

171. A. L. Harmer, A. Linz, and D. R. Gabbe, *J. Phys. Chem Solids*, **30**, p. 1483, 1969.

172. S. Kumpan, *University of Rochester*, private communication, 1980.

173. R. H. Lehmberg and S. P. Obenschain, *Opt. Commun.*, **46**, p. 27, 1983.

174. N. G. Basov et al., *FIAN (P. N. Lebedev Inst.), Prepr.*, 1974.

175. M. Born and E. Wolff, *Principles of Optics*, 2nd ed., Pergamon Press, Oxford, 1964.

176. E. S. Bliss, J. A. Glaze, K. R. Manes, J. E. Murray, and F. Rainer, *Laser Program Annual Report*, UCRL-50021-75, Lawrence Livermore Lab., Livermore, CA, 1975.

# 8

# LIQUID LASERS

B. WILHELMI
*JENOPTIK AG*
*Jena, Germany*

E. DÖPEL
*University of Applied Sciences*
*Jena, Germany*

F. WEIDNER
*MLC GmbH*
*Jena, Germany*

## 8.1  INTRODUCTION

The first liquid lasers were built only a short time after the invention of the laser in general. The use of dye molecules and metal-organic compounds as active media was proposed already in 1961 [l, 2]. One year later, lasers based on such compounds, in the frozen state, however, were set into operation [3, 4]. The real breakthrough came in 1966, when Sorokin and Lankard [5], and Schäfer, Schmidt, and Volse [6] observed stimulated emission from organic dyes. Also in 1966, Heller constructed a high-gain room-temperature laser using trivalent neodymium in the anorganic liquid selenium oxychloride [7]. In 1970 the first continuous-wave (cw) liquid laser was put into use [8]. This rhodamine 6G dye laser was pumped by a cw argon-ion laser.

It became apparent that liquid lasers have certain advantages over solid lasers and gas lasers, and that it depends on the particular aim of investigation or application whether these advantages outweigh the disadvantages. Criteria for such a comparison are discussed in Section 8.2. Advantages stem first of all from high beam quality and easy change of wavelength in conjunction with simple, flexible, and rather cheap equipment. However, compared with diode lasers and diode-pumped all-solid-state lasers, which have already achieved more than 20,000 hours operation time, the necessity of

*Handbook of Optical Components and Engineering*,  Edited by Kai Chang
ISBN 0-471-39055-0  © 2003 John Wiley & Sons, Inc.

frequently replacing active media is a disadvantage when aiming at permanent use and particularly industrial use.

From the beginning, the type of liquid laser most used has been the dye laser, in which fluorescent organic dyes in solution serve as active media. The advantage of the dye laser is its broad tunability. By using one dye under appropriate excitation, a spectral range as large as 100 nm can be covered. When several dyes are applied, such lasers can be operated from the near-ultraviolet to the near-infrared spectral region.

On one hand, highly monochromatic radiation can be obtained, even with dye lasers whose active media exhibit very broad gain profiles, by use of one or several suitable spectral filters in the cavity. This possibility, combined with tunability, enables the dye laser to be a valuable tool for high-resolution spectroscopy [R2, R9, R14, 9–13]. By frequency conversion in nonlinear optical interaction processes, the light from one or several tunable dye lasers can be transformed to other spectral regions, where no simple light sources with broad tunability are available [14, 15]. In this way, ultraviolet (UV) light with wavelengths of some tens of nanometers as well as mid-infrared light have been produced, starting with lasers in the visible. With tunable lasers, the efficiency of high-order processes can be enhanced by tuning one or several lasers to the vicinity of intermediate resonances [14–16]. If several nonlinear optical conversion processes are used "in series," the overall efficiency can be substantially increased by the use of an amplifier following some conversion steps. To achieve high gain in such a device, the input radiation must be tuned to the amplifying transition; for example, the second harmonic light from excimer laser pumped dye lasers can be amplified considerably in excimer amplifiers (see Ref. R9 and 17 and Section 8.4.3). This amplified UV radiation may then serve as pump radiation in generating high-order harmonics in gases.

On the other hand, extremely short light pulses can be obtained by locking the cavity modes in a broad spectral range by special measures (see Section 8.4.3). Such mode-locked dye lasers are applied advantageously in time-resolving spectroscopy [R9, R11, 18–25].

Moreover, dye lasers are used successfully in many other fields, where cw or pulse wave (pw) radiation is needed either at particular wavelengths or at tunable wavelengths. In many application fields, mainly where broad tunability or short light pulses are required, dye lasers had been applied from the very beginning and have paved later on the way to use all-solid-state devices by employing the same or similar procedures for tuning the wavelength or generating ultrashort light pulses (see Section 8.4).

As with other active materials, one may ask whether organic dyes can also be applied in aggregate states other than that of liquid and with an alternative type of excitation other than optical pumping.

Generally, organic dyes can be employed as laser-active media in different aggregation states: in gas phase, liquid solution, and in various types of solids, e.g., embedded in glass-like, ceramic, or polymeric host materials, as pure dye crystals or as pure organic dye polymers. The liquid dye laser was the first dye laser set into operation, and for a long time, it had been the only one with large application potential. Later, on the one hand, gas-phase dye lasers were built [26–28], but they never have gained broader application and importance. On the other hand, already since the 1960s [29], there have been strong research activities aiming at the development of solid-state dye lasers in order to combine the advantages of dye molecules (first of all, high gain, large bandwidth, and tunability) with those of typical solid lasers (robustness, easy operability, almost no service, long operation time without changing the active medium).

These advantages should mainly stem from the rigid surrounding of the light-emitting molecular groups compared to strongly fluctuating potential at such sites in liquid solutions. Organic dyes incorporated in solid host media or dye crystals have been applied with the usual optical excitation [30–32]. Nowadays, particularly such lasers are already in broad use because of simple and cheap design, production and operation as well as because of their miniaturization potential (see, e.g., [33–37]). These lasers use solid solutions of organic dyes in organic [38–46] and anorganic [47–50] materials as well as dye crystals. Ref. 51 studies the thermal and optical properties of polymer hosts for solid dye lasers, and Ref. 52 compares the laser performance of dye molecules in sol–gel glasses, polycom, and poly(methylmethacrylate) host media. Laser action has been reported from NUV [53, 54] to NIR [55]. Polymeric microcavities [56] and waveguides [57] have been designed and built for solid-state dye lasers. An additional advantage of polymeric dye lasers consists in cheap and reliable manufacturing processes, particular for making fine-structured micro devices. Large and cheap polymeric thin films can easily be obtained from solution, e.g., by spin coating [33]. Small structures can be produced by procedures applied in semiconductor chip manufacturing, first of all optical lithography and optical contact replication combined with chemical and chemophysical measures, injection molding, and hot embossing [58, 59] or can directly be written by ink jet printing [60, 61].

Dye lasers in all described aggregation states can easily be pumped by light absorption in conjunction with fluorescent emission. With this method, high gain can easily be obtained with liquid lasers (see Section 8.3.1). Until now, this has remained the mostly applied pump method also for solid-state dye lasers. Driven by the great success of electrically pumped gas lasers and semiconductor lasers, there have been efforts to use direct electric excitation also with dye lasers in all aggregation states. Experiments with special dye solutions [62–64] and dye vapors [65–67] demonstrated electroluminescence and, in the case of vapors, even gain and lasing. At present, electroluminescent polymer materials are used as incoherent light sources and displays and promise broad application. Work aimed at enhancing gain and reducing losses is in progress; see, e.g., [33, 34].

Whereas organic luminescence diodes (OLED) are already manufactured by several companies and applied in various fields, until now, there has been no real breakthrough with stimulated emission, gain, and lasing to applications until now. However, refined solid dye systems, i.e., organic polymers, show very strong electroluminescence, some at very low voltage [35–37] and are promising candidates for simple, electrically pumped, organic all-solid-state laser devices. With some selected materials, stimulated emission has been observed. Work aimed at enhancing gain and reducing losses is in progress; see, e.g., [68, 33, 34]. Lasing in solid-state electrically pumped organic systems was first achieved in dye crystals, which exhibit large carrier mobility [69].

At present, these alternatively pumped dye laser versions are still not in the state of broad application, and constructions that can really compete with other solid-state lasers and meet the requirements of specific applications are yet to come.

Other liquid lasers, based, for example, on laser-active rare-earth ions in metal-organic compounds or in anorganic solutions, are nowadays by no means superior to lasers where the same ions are used in glass or crystal. They are, therefore, rarely applied. So the dye laser is at present the only liquid laser, used in various applications, whereas other liquid lasers play only a minor role. Therefore, we concentrate mainly on dye lasers.

After some basic discussion concerning the comparison of active media and important cavity arrangements in Sections 8.2 and 8.3, respectively, we present an overview of important types and classes of dye lasers in Section 8.4. Finally, Section 8.5 is devoted to selected applications of dye lasers in science and technology.

## 8.2 PARAMETERS OF THE ACTIVE MEDIA: COMPARISON WITH OTHER LASERS

In this section, we compare some decisive properties of liquid laser media with those of gaseous and solid media.

### 8.2.1 Classes of Active Compounds and their Absorption and Emission Spectra

Liquid lasers are almost exceptionally pumped by light, and therefore both the emission spectrum and the absorption spectrum are decisive. Figures 8.1 and 8.2 present typical examples. In general, spectral lines broaden and shift when the particular atomic system (atom, molecule, or ion) is solved, where the bandwidth and the spectral shift depend on the solvent.

*Organic Dyes.* Dye lasers are operated by utilizing allowed transitions of conjugated electrons in organic molecules (see, e.g., R2, R5, R6). Nowadays the term "lasing dye" is by no means restricted to substances that appear colored; molecules are used too, which absorb and fluoresce (see Fig. 8.3) preferentially in the ultraviolet or in the infrared spectral region and which therefore may be uncolored. Table 8.1 gives some important classes and examples of lasing dyes. Obviously, many types of organic compounds, which are well known from other applications, such as coloring of textiles and food, and whiteners in washing agents and pharmaceuticals, can be used advantageously. Furthermore, some biomolecules, such as some chlorophylls, may be utilized. The lasing wavelengths of the dyes listed in Table 8.1 range from about 310 nm in the ultraviolet (2.2″-dimethyl-p-terphenyl [R6]) to 1.8 μm in the infrared (dye 26 [71]).

With organic dye molecules, almost structureless broad spectral bands appear (see Fig. 8.1) which originate from many vibronic transitions. The relative intensities of the various vibronic transitions belonging to one electronic transition are roughly given by the Franck-Condon factors.

At room temperature, the coupling of the solute to its surrounding is so strong in most cases that the single vibrational lines cannot be resolved. The spectral width of the whole absorption and emission band which contains the manifold of unresolved or almost unresolved vibronic lines is typically on the order of 10 to 100 nm. This has to be compared with the sharp lines of gas lasers, the bandwidth of which is given by the Doppler width ($\sim$1 GHz) or, at high pressure, by collision broadening, where the ratio between broadening and pressure is typically approximately 1 MHz/Pa (e.g., for the He-Ne laser transition at 0.6328 μm). In dye solutions we have line-broadening effects of a homogeneous as well as an inhomogeneous nature. Homogeneous line broadening can roughly be explained by binary collisions, which occur here at a rate of about $10^{14}$ s$^{-1}$.

Inhomogeneous broadening arises when each individual solute molecule experiences a surrounding that differs from that of other solutes for a period longer than its lifetime (or measuring time). In both liquids and solids, such site effects are in general

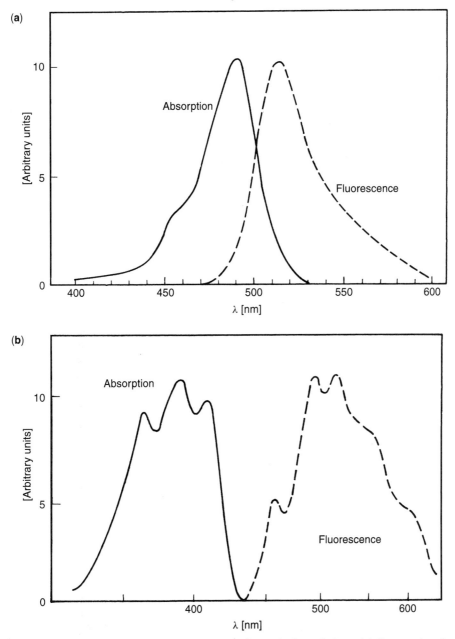

**Figure 8.1**  Absorption and fluorescence spectra of organic dye solutions; **(a)** fluorescein solved in water (from Ref. R2, p. 20); **(b)** diphenyloctateraence solved in xylene (from Ref. 70).

of anisotropic nature. In dye solutions the contributions of homogeneous and inhomogeneous line broadening can be on the same order of magnitude. Often, the value of the lifetime (or the measuring time) is decisive whether a certain process leads to homogeneous or inhomogeneous broadening. As the integral transition probability is comparatively insensitive with respect to changes in the surroundings, strong line

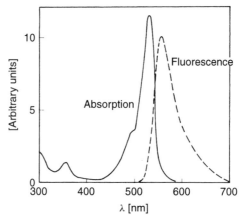

**Figure 8.2** Absorption and fluorescence spectrum of rhodamine 6G dissolved in water. (From Ref. 9, p. 338.)

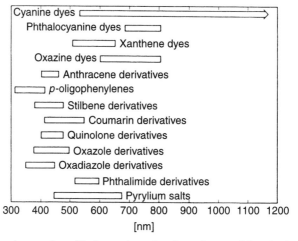

**Figure 8.3** Spectral extension of lasing action of various classes of dyes in the infrared, visible, and ultraviolet. (From Ref. R5, p. 5.)

broadening is connected with a large decrease in the maximum absorption (emission) cross section. Typical values of the maximum cross section of strong transitions in solved dyes (i.e., with oscillator strength of about 1) are on the order of $10^{-16}$ cm$^2$ (compared to $10^{-12}$ cm$^2$ in gases at low pressure).

The broad absorption bands of dye solutions are suitable for broadband optical pumping (e.g., by flashlamps; see Section 8.4.1). The broad emission bands allow tuning over large spectral ranges (see Section 8.3.2). Moreover, the frequency shift of spectral lines by changing the solvent can be used to vary the lasing range of one particular dye (see Table 8.2).

The absorption, or gain, of radiation in the liquid depends not only on the properties of the single molecules, as are the cross sections $\sigma_{ij}$, but also on the concentration of the solute and the relative occupation of various levels. If one electronic transition

**TABLE 8.1  Classes of Lasing Dyes and Some Representatives**[a]

| Class | Representatives and Range of their Laser Action |
|---|---|
| Cyanine<br>**1**<br> | 3,3'-Diethylthiadicarbocyanine iodide (DTDC) in DMSO<br>$\lambda_L = 710$–760 nm<br>3,3'-Diethyl-4,5,4',5'-dibenzothiatricarbocyanine iodide (DDTTC), (dibenzocyanine 45) in DMSO<br>$\lambda_L = 905$–970 nm<br>1,1'-Diethyl-4,4'-quinotricarbocyanine iodide (NK 124) in DMSO<br>$\lambda_L = 900$–1100 nm |
| Merocyanine<br>**2**<br> | 4-Dicyanomethylene-2-methyl-6p-dimethylaminostyryl-4H-pyran (DCM) in DMSO<br>$\lambda_L = 610$–710 nm |
| Phthalocyanine<br>**3**<br> | Magnesiumphthalocyanine in DMF<br>$\lambda_L = 682$ nm |

(continued)

**TABLE 8.1** (*Continued*)

| Class | | | | | | | Representatives and Range of their Laser Action |
|---|---|---|---|---|---|---|---|

## Xanthene

**4**

| $R_1$ | $R'_1$ | $R_2$ | $R'_2$ | $R_3$ | $R'_3$ | $R_4$ |
|---|---|---|---|---|---|---|
| Acridine red in ethanol $\lambda_L = 580{-}600$ nm | | | | | | |
| $NHMe^+$  $Cl^-$ | NHMe | H | H | H | H | H |
| Rhodamine 110 in ethanol $\lambda_L = 550{-}590$ nm | | | | | | |
| $N^+H_2$  $Cl^-$ | $NH_2$ | H | H | H | H | 2-Carboxyphenyl |
| Rhodamine B in ethanol $\lambda_L = 580{-}640$ nm | | | | | | |
| $NEt_2^+$  $Cl^-$ | $NEt_2$ | H | H | H | H | 2-Carbethoxyphenyl |
| Rhodamine 6G in ethanol $\lambda_L = 545{-}630$ nm | | | | | | |
| $NHEt^+$  $Cl^-$ | NHEt | H | H | Me | Me | 2-Carbethoxyphenyl |
| Fluorescein in ethanol $\lambda_L = 540{-}580$ nm | | | | | | |
| O | OH | H | H | H | H | 2-Carboxyphenyl |

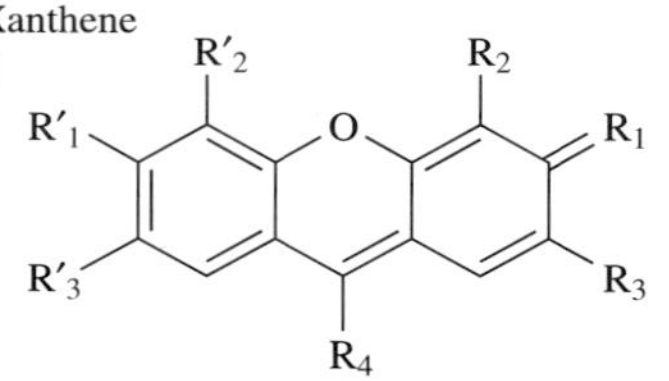

## Triarylmethane

**5**

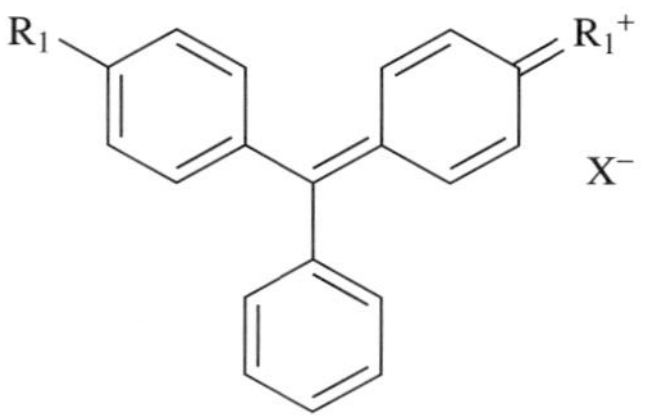

Brilliant green in glycerin
$\lambda_L = 759$ nm
$R_1 := NEt_2$

450

## Acridine

**6**

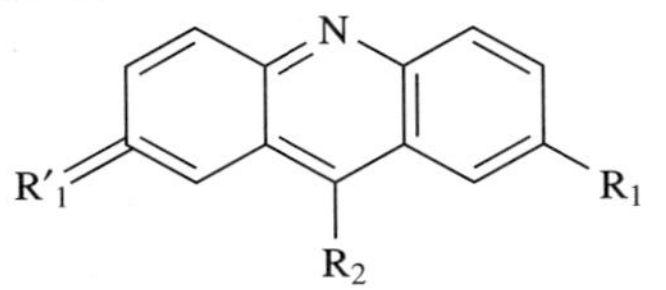

Carbazine 122 (red 7) in ethanol
$\lambda_L = 660\text{–}710$ nm

## Azine

**7**

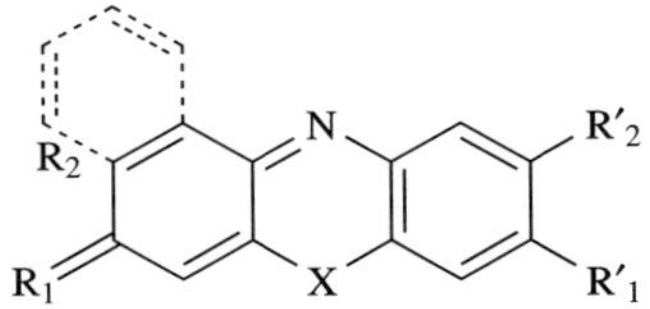

oxazine: X := O
thiazine: X := S
diazine: X := N

Oxazine

| X | $R_1$ | $R'_1$ | $R_2$ | $R'_2$ |
|---|---|---|---|---|
| | Oxazine 1 (red 8) in ethanol $\lambda_L = 690\text{–}770$ nm | | | |
| O | $N^+H_2$  $ClO_4^-$ | $NH_2$ | H | H |
| | Nile blue A (red 5) in ethanol $\lambda_L = 680\text{–}705$ nm | | | |
| O | $N^+(NH_2)(NEt_2)$  $ClO_4^-$ | $N(NH_2)(NEt_2)$ | phenyl | H |
| | Cresyl violet (oxazine 9/red 4) in ethanol $\lambda_L = 645\text{–}710$ nm | | | |
| O | $N^+(NH_2)_2$  $Cl^-$ | $N(NH_2)_2$ | phenyl | H |

## Chlorophylls

Chlorophyll $a$ in ethanol
$\lambda_L = 677$ nm, 681 nm
Bacteriochlorophylla $a$ in pyridine
$\lambda_L = 800$ nm

(continued)

**TABLE 8.1** (*Continued*)

| Class | Representatives and Range of their Laser Action |
|---|---|
| Condensed aromatic rings Example: naphthalene **8** | 9-Methylanthracene in methanol/ethanol $\lambda_L = 414$ nm |
| Fluorene **9** | |
| Oligophenylenes **10** | 2′, 2″-Dimethyl-*p*-terphenyl in cyclohexane $\lambda_L = 310$–350 nm |
| Conjugated dienes stilbene **11** | 4, 4′-Diphenylstilbene (DPS) in benzene $\lambda_L = 400$–420 nm |
| Coumarin and derivatives **12** | 7-Ethylamino-4,6-dimethylcoumarin, (coumarin 2, blue 2) in methanol $\lambda_L = 430$–485 nm |

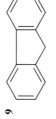

Azacoumarin and
derivatives

**13**

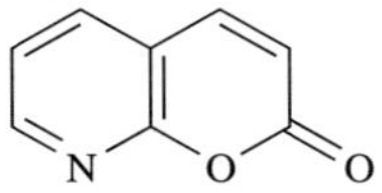

7-Morpholino-4-methyl-8-azacoumarin, (violet 3) in ethanol
$\lambda_L = 408\text{--}448$ nm

Quinolone

**14**

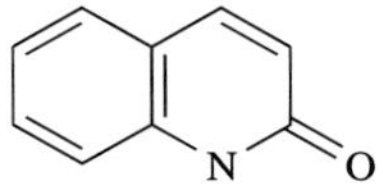

7-Dimethylamino-4-methylquinolone-2, (carbostyril 165) in EG
$\lambda_L = 414\text{--}490$ nm

Oxazole
derivatives

**15**

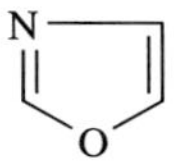

1,4-Bis(4-methyl-5-phenyloxazolyl)benzene, (dimethyl POPOP) in dioxane
$\lambda_L = 418\text{--}465$ nm

Pyrylium salts

**16**

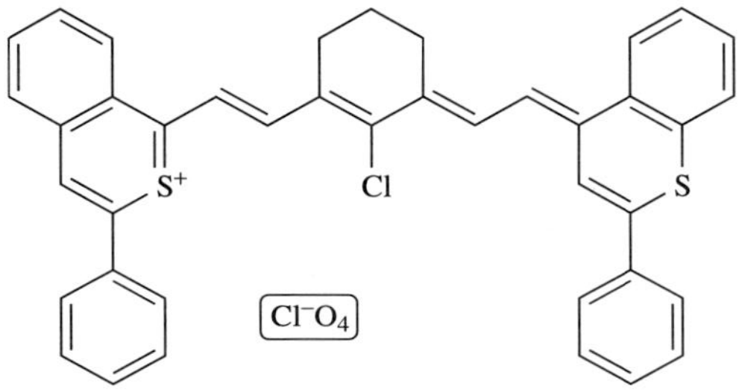

IR dye 26
$\lambda_L = 1.6\text{--}1.85$ μm

$\boxed{Cl^-O_4}$

[a]For more details, see, for example, R1-R3 and references cited there.

**453**

**TABLE 8.2   Influence of Solvent on the Absorption and Lasing Wavelengths of Rhodamine 6G (Flashlamp pumped, untuned)**

| Solvent | $\lambda_{abs}$ (nm) | $\lambda_{las}$ (nm) |
| --- | --- | --- |
| Hexafluoroisopropanol (HFIP) | 514 | 570 |
| Trifluoroethanol (TFE) | 516 | 575 |
| Ethanol (EtOH) | 530 | 590 |
| N,N-dipropylacetamide (DPA) | 537 | 595 |
| Dimethyl sulfoxide (DMSO) | 540 | 600 |

*Source*: Ref. 72.

$j \leftrightarrow i$ is in resonance with the laser radiation, the gain coefficient $g_{ij}(\nu)$ is given by

$$g_{ji}(\nu) = N(\rho_{jj} - \rho_{ii})\sigma_{ji}(\nu) \qquad (8.1a)$$

where $\rho_{jj}(\rho_{ii})$ is the probability of finding a particular molecule in state $j(i)$, and $N$ is the number density. If there are several transitions in resonance with the radiation field, the total gain coefficient $g(\nu)$ is given by

$$g(\nu) = \sum_i \sum_{j>i} g_{ji}(\nu) \qquad (8.1b)$$

The occupation probabilities $\rho_{jj}$ can be changed by radiative and nonradiative processes (compare Section 8.2.2). With unexcited samples (see, e.g., the absorption spectra in Fig. 8.1) only the electronic ground level is populated ($\rho_{00} = 1$). As an example, Fig. 8.4 compares the absorption spectrum from the ground state (unexcited sample) and the absorption spectrum from the lowest triplet level, where because of selection rules, the first one is mainly given by singlet-singlet transitions and the second one by triplet-triplet transitions. There is strong triplet-triplet absorption at long wavelengths, where the unexcited sample is highly transparent and where fluorescence radiation from the first excited singlet level appears. Thus the fluorescence radiation is strongly absorbed by triplet-triplet absorption when the $T_1$ level is occupied.

The concentration or number density of active molecules in liquid lasers can be varied to a rather large degree and, in this way, gain as high as in high-concentration crystals can be achieved. One limitation to concentration is the solubility of molecules, which can be increased by changing the solvent or by modifying parts of the solute that only slightly effect the spectral behavior but strongly influence the interaction between solute and solvent. A second limitation originates from the tendency of solute molecules to interact strongly at high concentrations and to build up dimers and aggregates. Equations 8.1a and 8.1b are only valid as long as the individual molecules act independently from one another. In most cases, the change of the spectra by solute interaction must be avoided. The tendency of the solute molecules to interact and form aggregates can be influenced by using appropriate solvents [R5, R6]. The solvents used

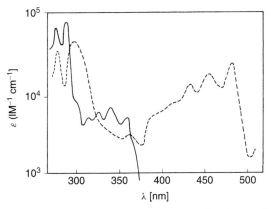

**Figure 8.4** Absorption spectrum of 1,2-benzanthracene solved in hexane from the ground level (solid line) and from the lowest triplet level (dashed line). (From Ref. 73.)

**TABLE 8.3  Some Properties of Solvents Used in Dye Lasers ($\lambda_{UV}$ is the wavelength, where 50% of the radiation transmits a sample of 1 cm length)**

| Solvent | Density (at 25°C) (g/cm$^3$) | Viscosity (at 25°C) (cP) | Low Wavelength Limit, $\lambda_{UV}{}^a$ (nm) | Refractive Index (at 25°C) |
|---|---|---|---|---|
| Methanol | 0.7866 | 0.5506 | 220 | 1.3265 |
| Ethanol | 0.785 | 1.078 | 217 | 1.3594 |
| Ethylene glycol | 1.11 | 26.09 | 235 | 1.4306 |
| DMSO | 1.0958 | 1.996 | 277 | 1.4773 |
| Dioxane | 1.028 | 1.439 | 240 | 1.42025 |
| Cyclohexane | 0.7785 | 0.898 | 222 | 1.4235 |
| Hexane | 0.6548 | 0.2985 | 207 | 1.3723 |
| Toluene | 0.867 | 0.5516 | 290 | 1.4941 |
| Dichlormethane | 1.3168 | 0.449 | 238 | 1.4212 |

*Source*: Ref. R6.

for dye lasers must show high transmission in the absorption and fluorescence band of the active molecules used (see Table 8.3).

***Rare-Earth Ions.*** With liquid lasers based on rare-earth ions the spectral properties of the active particles are far less influenced by the solute-solvent interaction or even the structure of the molecule to which the ion might be attached. This insensitivity results from the fact that the electrons taking part in the laser transitions (e.g., $4f$ electrons) are well shielded by other electrons, for example, two $5s$ electrons and six $5p$ electrons. In most cases, the effect of the surroundings on the laser transitions is smaller than the spin-orbit coupling. Thus the lines appearing are rather narrow in comparison with organic dyes; these lines are only slightly shifted when the solvent is changed. The laser transitions in such rare-earth ions are weak; their oscillator strengths are typically on the order of $10^{-6}$. Roughly speaking, the spectral properties of rare-earth ions in liquids are similar to those in solid matrices.

There are two main classes of liquid rare-earth laser media: metal-organic and anorganic solutions of rare-earth ions. Typical representatives of the first class are various chelates. They take the form

$$(RE^{3+})(L^-)_3 \quad \text{or} \quad (RE^{3+})(L^-)_4(K^+)$$

where $RE^{3+}$, $L^-$, and $K^|$ denote the rare-earth ion, the ligand of the metal-organic complex, and an additional organic cation, respectively [R7, 74–77]. In chelate solutions, the active ions are excited mainly by the inner-molecule energy transfer from the organic part of the complex, which absorbs the pump light via broad and strong singlet-singlet transitions. In most cases the excited organic ligand passes to its lowest triplet state by intersystem crossing, and then the energy goes over to the rare-earth ion by triplet energy transfer.

The main advantage of such complexes is the combination of broad absorption bands ($\Delta\tilde{\nu}_A \sim 10^3$ to $10^4$ cm$^{-1}$) and narrow emission bands ($\Delta\tilde{\nu}_E \sim 10^1$ to $10^2$ cm$^{-1}$). Disadvantages are the limited efficiency of energy transfer, the unavoidable energy loss in such relaxational and transfer processes, and the absorption losses by vibrational transitions of the solvent.

Table 8.4 comprises some representatives of the second class of rare-earth liquid laser materials: rare-earth ions directly solved in anorganic solvents. In most cases hydrogen-free solvents are used, in particular in combination with $Nd^{3+}$, to avoid the absorption losses through overtones or combination bands of the high-frequency valence vibrations of hydrogen.

## 8.2.2 Radiative and Nonradiative Processes

The probability of finding the active particle in a certain level is influenced by relaxational processes. These are decisive for the efficiency of lasers and for their temporal behavior.

**TABLE 8.4 Active Media and Solvents of Anorganic Liquid Lasers**

| Active Ion | Solvent | Admixture for Improved Solubility | Comment | Reference |
|---|---|---|---|---|
| $Nd^{3+}$ | $SeOCl_2$ | $SnCl_4$ | Very toxic, corrosive, high tenacity | R7, 7 |
| $Nd^{3+}$ | $POCl_3$ | $SnCl_4$ | Less toxic, lower tenacity | R7, 78 |
| $Nd^{3+}$ | $SOCl_2$ | $GaCl_3$ | Low toxicity, low tenacity | R7 |
| $Ce^{3+}$, $Ce^{4+}$, $En^{3+}$, $Tb^{3+}$, $Gd^{3+}$ | $H_{n+2}(P_nO_{3n+1})$ $N = 1$ to $10^4$ | | Polyphosphoric acids tend to the formation of complexes with metallic ions | R7, 79 |

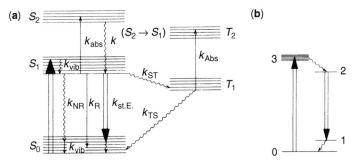

**Figure 8.5**  Level scheme of organic dye molecules with radiative and nonradiative transitions: (a) Jablonski scheme, (b) effective four-level scheme.

Liquid dye lasers can be considered as four-level lasers in most cases. Figure 8.5a shows the Jablonski scheme with the lowest electronic levels of the singlet and triplet manifold. The pump radiation excites the molecules to higher vibrational states of the $S_1$ level. Through very fast vibrational relaxation, the molecules lose most of this vibrational excess energy, that is, they attain a quasi-equilibrium distribution among the vibrational states in the $S_1$ level very rapidly (in less than 1 ps). At room temperature most molecules are near the lower edge of the energy band.

The molecules pass by radiative and nonradiative processes from these states to the ground state or other electronic levels (see Table 8.5). The lasing process we are interested in consists of spontaneous and stimulated emission by transitions from the excited level into vibronic states of the electronic ground level. For "classical" laser dyes the total rate parameter $k_R$ of spontaneous emission is larger than that of nonradiative processes and attains values of some $10^8$ s$^{-1}$. Thus the quantum yield of fluorescence from the first excited singlet state

$$\Phi_F = k_R/(k_R + k_{NR}) \qquad (8.2)$$

is high. The rate parameter $k_{NR}$ of nonradiative decay of the $S_1$ level is composed of terms describing the internal conversion $S_1 \rightarrow S_0$, intersystem crossing $S_1 \rightarrow T$, transitions to other electronic states (e.g., the $S_1$ state of conformers), and concentration-dependent quenching of $S_1$ by other solutes. The sum of the rate parameters of

**TABLE 8.5  Rate Parameters of Typical Lasing Dye Molecules in Solution**

| Process | Rate Parameter |
|---|---|
| Radiative decay $S_1 \rightarrow S_0$ | $k_R \sim 10^8 - 10^9$ s$^{-1}$ |
| Nonradiative decay $S_1 \rightarrow S_0$ | $k_{NR} \lesssim 10^8$ s$^{-1}$ |
| Nonradiative decay $S_x \rightarrow S_1$ | $k_{NR}(S_x \rightarrow S_1) > 10^{11}$ s$^{-1}$ |
| Intersystem crossing $S_1 \rightarrow T_x, T_1$ | $k_{ST} \lesssim 10^8$ s$^{-1}$ |
| Intersystem crossing $T_1 \rightarrow S_0$ | $k_{TS} \sim 10^3 - 10^6$ s$^{-1}$ |
| Vibrational relaxation | |
| Within $S_0$ | $k_{vib}(S_0) \sim 10^{11} - 10^{12}$ s$^{-1}$ |
| Within $S_1$ | $k_{vib}(S_1) \lesssim 10^{12}$ s$^{-1}$ |

nonradiative transitions from $S_1$ to all other electronic levels takes on small values, one or two orders of magnitude below $k_R$, only under particularly suitable conditions.

First, the molecule should not have any effective isomerization channels in the excited state. Internal rotation into other, nonfluorescent configurations diminishes the fluorescence quantum yield in many molecules. The rigidization of the molecular frame by additional molecular groups serves as a defence against this effect [72, 80]. Moreover, one may apply viscous solvents to hinder such internal motion [81]. Furthermore, combinations of solute and solvent should be selected, the singlet-triplet transition rate of which is rather small.

In spontaneous emission the relative probabilities of transitions between various vibronic sublevels and thus the spectral profile of fluorescence is given by the corresponding Franck-Condon factors. In stimulated emission, the transition probability depends on, in addition to the atomic parameters, the light intensity at this particular frequency. Thus the rate parameter $k_{\text{stE}}$ of the stimulated process is given by

$$k_{\text{stE}}(\nu) = \sigma(\nu)I \tag{8.3}$$

where $I$ is the photon flux at this frequency. At high radiation intensity, stimulated emission is the dominating process.

The molecules relax again very rapidly ($k_{\text{vib}} \sim 10^{12}$ s$^{-1}$) from the lower laser level, which is a vibrational state of the electronic ground level, the energy of which is determined by the radiation frequency selected. Because of the rapid relaxation from the pump levels to the upper laser level and from the lower laser level into the vibrationless ground state, the occupation of the pump levels and of the lower laser level can be neglected in most cases. This is very suitable for laser operation. Only with subpicosecond pulses may the situation change. For the generation of ultrashort light pulses, dyes for which $k_R \ll k_{NR}$ holds can also be applied advantageously, since under strong irradiation the effective lifetime of molecules in the excited state is determined primarily by the rate of stimulated emission $k_{\text{stE}}$ instead of by $k_R + k_{NR}$. This means that the stimulated radiation processes dominate for

$$\sigma(\nu)I > k_{NR} + k_R \tag{8.4}$$

This consideration is most important with near-infrared dye lasers, where $k_{NR}$ is typically on the order of $10^{11}$ s$^{-1}$ [82, 83]. Hence (for $\sigma \sim 10^{-16}$ cm$^2$), photon fluxes being larger or approximately $10^{27}$ cm$^{-2}$ s$^{-1}$ are required, which at a wavelength of 1 μm corresponds roughly to $2 \times 10^8$ W cm$^{-2}$. In typical mode-locked dye lasers, the intensities inside the active medium exceed this value by one or two orders of magnitude (compare Section 8.4.3).

From Fig. 8.5 it is obvious that several other transitions competing with the laser process may occur. Singlet-triplet relaxation decreases the occupation of the upper laser level (e.g., by transitions from $S_1$ into $T_1$ or $T_2$), and what is far more disadvantageous, such molecules may stay for a rather long time in the lowest triplet level and hence are not able to take part in further pump and laser processes during that period. Moreover, by triplet-triplet absorption, these molecules in the $T_1$ level are responsible for additional losses at the laser wavelength (compare Fig. 8.4). Therefore, dye molecules with very small triplet yield are used preferentially. Snavely and Schäfer [84] were the first to additionally use triplet quenchers, which increase the transition rate from $T_1$

to $S_0$. Since then, such means have often been applied, in particular with long pump pulses and continuous pump radiation. Note that the singlet-triplet transition rates also depend to some extent on the solvent [R2, R5, R6].

In general, we have to take into account the losses for the laser radiation originating from excited-state absorption. The gain coefficient of the active medium at the laser frequency is then given by

$$g(\nu) = -\sum_i N_i \Delta\sigma_i(\nu) \qquad (8.5)$$

where $N_i = \rho_{ii} N$ is the occupation density of level $i$ at energy level $E_i$, and $\Delta\sigma_i$ is the difference between the absorption cross section $\sigma_{ij}^A(\nu)$ with $E_j = E_i + h\nu$ and the cross section of stimulated emission $\sigma_{ik}^{stE}(\nu)$ with $E_k = E_i - h\nu$; the summation is extended over all occupied levels. This means that the effective gain coefficient depends not only on the occupation of the upper and lower laser levels, but via the excited-state absorption also on the population of other levels. First,

$$\sigma_{21}^{stE} > \sigma_{2x}^A(\nu)$$

must be required to obtain positive gain. Second, the population of all levels $i$ with $\Delta\sigma_i(\nu) > 0$ should be kept as low as possible.

Excitation energy can efficiently be transferred from donor molecules to acceptor molecules if there is sufficient overlap of the fluorescence band of the donor and an absorption band of the acceptor. Such transfer can be used to increase the tuning range, which can be roughly said to be composed of the fluorescence bands of the donor and the acceptor [85, 86]. The efficiency of the transfer depends critically on the distance between donor and acceptor. If, for example, the so-called Förster mechanism [87] is responsible for the energy transfer, the transfer rate is proportional to $R^{-6}$, where $R$ is the distance between donor and acceptor. Therefore, it is of advantage to use small distances, which can be achieved by linking donor and acceptor together by chemical bonds. This has been used in Ref. 88, for instance, to transfer excitation energy within the rhodanile molecule from the rhodamine moiety to the nile blue moiety. Thus, in comparison with the simple nile blue molecule, the fluorescence of the rhodanile molecule is fed by the two superimposed absorption bands of the nile blue group and the rhodamine one.

Furthermore, the fluorescence of solved organic molecules can be affected strongly by intra- and intermolecular donor-acceptor charge transfer. Most important for dye lasers is the formation of solute-solvent complexes in the ground state and/or in the excited state [89]. If the excited complex is nonfluorescent, the intermolecular charge transfer in the excited state represents an additional quenching mechanism. On the other hand, in some complexes — for example, in the case of dimethylaniline and anthracene — a new fluorescent band appears.

### 8.2.3  Exchange of the Active Medium

A most important advantage of gas and liquid lasers is that the active medium can easily be exchanged under operation. There are several reasons for applying this method.

First, heat has to be transported from the volume of laser operation to the surroundings. At least the energy fraction of every pump photon that corresponds to vibrational excess energy in the $S_1$ level and in the $S_0$ level is transferred to heat in the active volume. Often the energy loss is much higher, in particular, if broadband pump radiation (i.e., from flashlamps) is used. With lasers of low repetition frequency and small average power, heat conduction within the solution as well as between solution and the walls surrounding the sample can be sufficient to establish thermal equilibrium at a low temperature level. At a high repetition frequency and high average power, and in particular with continuous-wave (cw) lasers, exchange of the active solution is far more suitable.

Second, by rapidly exchanging the active medium one can get rid of molecules in the $T_1$ level or in other metastable states which do not take part in the laser process and are responsible for additional losses.

Third, active media of liquid lasers are in many cases rather complex molecular systems, which can easily be destroyed or undergo irreversible changes under irradiation, and hence the exchange of solution is required if we are always to have fresh (i.e., undamaged) molecules under operation or, at least, to decrease the concentration of nonactive molecules.

There are mainly three methods used to exchange the solution. First, self-sustained transport can be used with pw lasers of low average power by applying cuvettes, the volume of which is much larger than the active volume. The transport is maintained principally by temperature differences between the active volume and the surrounding, which results via density differences in a flow of solution. This transport can be made efficient by proper design, in which the low-density solution can rise freely from the active volume. Second, the exchange of solution in cuvettes can be forced by applying micro pumps. In cuvettes the flow velocities are in most cases below 1 m/s. Third, to obtain high exchange rates, free liquid jets are used. The solution is pumped through thin nozzles under pressure of approximately $10^6$ Pa. With appropriate nozzles the jet exhibits high optical quality, where the flow velocity is about 10 m/s. This means that with tight focusing of the pump radiation onto a spot of 10 $\mu$m diameter, the solution in the active volume is completely exchanged after a time of 1 $\mu$s, which is for many molecules shorter than the lifetime of the lowest triplet state and which guarantees that the number of destroyed or irreversibly changed molecules remains at a rather low level even under strong cw operation. Such nozzles and jet streams are applied, for example, in all commercial cw dye lasers.

The fast replacement of particles in the active volume combined with efficient cooling had also been main arguments for using liquid rare-earth lasers instead of solid-state lasers in high average-power applications. However, in recent years, improved solid materials and refined cooling systems made all-solid-state approaches superior in most applications.

## 8.3   EXCITATION OF THE ACTIVE MEDIUM, FREQUENCY SELECTION, AND TUNING

In addition to the physical and chemical properties of the liquid active laser media mentioned in the preceding section, the emission properties of liquid lasers are determined by the manner in which the active medium is excited, the resonator geometry, and possible intracavity elements (e.g., for frequency narrowing, tuning, and mode locking). All liquid lasers known up to now have been optically pumped.

### 8.3.1 Pump Sources for Dye Lasers

We want to start with a very simple and rough estimation of the pump energies necessary to reach the laser threshold and the laser's performance. Only processes in the singlet system between the electronic ground state and the first excited state will be considered (compare Fig. 8.5b). Below laser threshold, the rate equation for the population of the upper laser level $N_2$ is

$$\frac{dN_2}{dt} = \sigma_p N_0 \frac{I_p}{\hbar \omega_p} - \frac{N_2}{T_{21}} \tag{8.6}$$

where  $\sigma_p$ = absorption cross section for the pump radiation

$N_0$ = ground-state population

$I_p$ = intensity of the pump radiation

$\hbar \omega_p$ = energy of a pump photon

$T_{21}$ = energy relaxation time from the upper laser level to the ground state

Laser threshold is achieved when the gain $G = \exp\{2Lg\}$ during one roundtrip in the resonator compensates for the losses described by $2\gamma$ during the same passage:

$$\ln G_{\text{thr}} = 2g_{\text{thr}}L = 2\gamma \tag{8.7}$$

where $g_{\text{thr}} = N_{2,\text{thr}}\sigma_L$ is the gain coefficient, $\sigma_L$ is the cross section for stimulated emission, and $L$ is the length of the active medium. (When the losses originate solely from the deviation of the reflectivity of the resonator mirrors $R_1$ and $R_2$, for simplicity, we take $R_1 = R_2 = R$, we obtain $\gamma = -\ln R$.)

First, we consider Eq. (8.6) for the case of short pump pulses, for which the influence of the energy relaxation during pumping is negligible. Then the threshold inversion $N_{2,\text{thr}}$ is determined only by the pump energy $E_{p,\text{thr}}$ per unit area up until the threshold point in time. Using Eq. (8.7) and assuming almost constant $N_0$ below threshold, we find that

$$E_{p,\text{thr}} = \frac{\hbar \omega_p \gamma}{N_0 L \sigma_p \sigma_L} \tag{8.8}$$

For an estimation of $E_{p,\text{thr}}$, we use the following parameters (which can differ greatly for the various dye lasers): $\gamma = 0.3$, $\lambda_p = 400$ nm, $N_0 = 10^{-3}$ mol/l $= 5 \times 10^{-17}$ cm$^{-3}$, $L = 0.2$ cm, $\sigma_L = \sigma_p = 10^{-16}$ cm$^2$. Using this, we find $E_{p,\text{thr}} = 5 \times 10^{-5}$ J/cm$^2$.

We consider the stationary regime as a second case. The time derivative in Eq. (8.6) is then negligible, and by using Eq. (8.7), we obtain the pump intensity necessary to reach the threshold:

$$I_{p,\text{thr}} = \frac{\hbar \omega_p \gamma}{N_0 L \sigma_p \sigma_L T_{21}} \tag{8.9}$$

The losses and the length of the active medium are to be assumed smaller for the cw laser. We choose $\gamma = 0.05$ and $L = 0.2$ mm. A typical value for the relaxation time in dyes is $T_{21} = 5$ ns. The threshold pump intensity is then $I_{p,\text{thr}} = 50$ kW/cm$^2$.

The assumptions upon which this estimation is based have often to be modified in application to real lasers. This is because the processes that take place within laser dyes are very complex, and phenomena such as intersystem crossing into the triplet system, internal conversion, absorption to higher excited electronic states, loss processes within the solvent, integral and frequency-selective resonator losses, and relaxation during pumping also play a role. The estimated values for the threshold intensity, or energy, can therefore be considered only rough estimates. These estimations clarify the fundamentally differing types of excitation used for dye lasers: namely, the pulsed excitation and the continuous excitation.

The most important pulsed pump sources are flashlamps, free running pw- and Q-switched solid-state lasers, and pulsed gas discharge lasers. For continuous excitation cw-gas lasers, mostly argon or krypton ion lasers, and cw-Nd:YAG lasers as well as their higher harmonics are generally used. Their beams are focused into the active medium in order to achieve the necessary intensity. Since 1974, there have been successful attempts to employ semiconductor diode lasers in dye laser pumping, which is of particular interest in the near infrared spectral region (see, e.g., [90, 91]). The main advantages consist in the high overall efficiency and miniaturization of the whole device.

The pump rays can be transmitted into the medium perpendicular to the direction of radiation of the laser (transverse pumping) or in the direction of laser emission (longitudinal pumping). Transversal pumping, which is most often used for pulsed excitation, can be used to excite large volumes of dye. Because the absorption within the dye solution is high at the wavelength of the pump radiation (typical depths of penetration lie between 10 and 100 $\mu$m) the laser beam has a no-axial-symmetric spatial intensity profile. Figures 8.6 and 8.7 show two typical transversal pump arrangements.

Longitudinal pumping demands a very limited thickness of the active medium due to the high pump radiation absorption, so that only very small amounts of dye can be stimulated. An often used pump arrangement which allows the collinear propagation of the pump and laser beams within the active medium is depicted in Fig. 8.8a. In

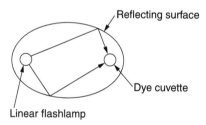

**Figure 8.6**   Transversal pump arrangement. The reflecting surface is shaped elliptically. The dye cuvette and the pump source are placed in the focus points of the ellipse. The pump source in this case is a linear flashlamp.

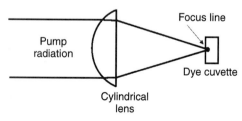

**Figure 8.7**   Transversal pump arrangement. The pump radiation is focused by a cylindrical lens to a focus line within the dye.

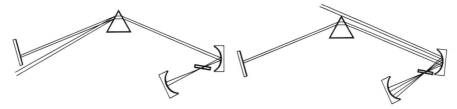

**Figure 8.8**   Longitudinal pump arrangement; (**a**) collinear pumping; the pump beam is coupled in the cavity by the dispersive prism. Pump and laser radiation are focused into the dye jet by the curved cavity mirrors; (**b**) noncollinear pumping.

an other often used arrangement (see Fig. 8.8**b**) the pump beam is not deflected by the prism and is adjusted parallel to the laser beam. The two beams are focused by the pump-mirror and intersect within the dye jet. In both cases the pump-mirrors must have broadband reflection-coatings. The dye-jet is oriented under the Brewster angle in order to prevent losses due to reflection. Arrangements of this type or with separate pump mirror (see, e.g., Fig. 8.30**b**) are typical for cw excitation.

### 8.3.2   Frequency Selection and Tuning of Dye Lasers

Dyes possess wide fluorescent bands. Without frequency selection, the spectral width of dye laser emission increases with increasing pump power. Typical values for the bandwidth are 1 to some 10 nm. In order to select a narrow frequency range within the fluorescence band of the dye, cavities with frequency-selective losses are used. To tune the laser frequency, it should be possible to drive the spectral point of minimum losses over the entire fluorescence band of the dye. The introduction of frequency-selective losses is always connected to a reduction in laser power. Because of the predominately homogenous broadening of the laser line, the reduction in power is not proportional to frequency narrowing of the laser emission. It is determined by the effective losses of the frequency-selective elements and the effective cross section for the laser emission at the chosen laser frequency.

Certain principles of frequency-selection elements are used:

1. Dispersion of the index of refraction
2. Diffraction in periodic structures
3. Multiple-beam interference
4. Change in polarization state.

In the following, the most important elements belonging to these principles are depicted.

***Dispersion of the Refraction Index.***   The most important component here is the prism. The main advantage of using the prism for the frequency narrowing and tuning of dye lasers lies in its low additional resonator losses. When using a prism whose roof surfaces face the ray path and are placed under the Brewster angle, the losses do not exceed 2 to 3% in general. Because the incoming radiation is in the vicinity of the Brewster angle, the resulting laser ray is predominantly linearly polarized. The disadvantages of prisms rest in their relatively limited spectral selectivity. Because of

**TABLE 8.6   Bandwidth of a Resonator Containing Four Prisms for Various Values of the Dispersion of the Material of the Prism (from [92])**

| Dispersion of the Prism Material ($\times 10^{-5}$/nm) | Resonator Bandwidth (nm) |
|---|---|
| 3.8 | 3.29 |
| 17.0 | 0.74 |
| 31.3 | 0.40 |
| 74.6 | 0.17 |

*Source*: Ref. 92.

the low prism losses, several prisms can be used simultaneously to improve spectral narrowing. The dispersion of the prism material is directly connected to the bandwidths that are actually achievable. The bandwidth of a resonator containing four prisms was calculated in Ref. 92. The ray hits the prism near the Brewster angle and a beam divergence of 1 mrad was assumed (see Table 8.6).

Several prisms may also be arranged in ring resonators, which have been investigated in Refs. 93–95.

Due to the relatively large bandwidth of ultrashort light pulses, prisms are well suited for frequency tuning in mode-locked lasers. The angular dispersion is also utilized in the pump arrangement shown in Fig. 8.8. For pump lasers operating at several laser lines, this arrangement allows for the selection of a single line for pumping the dye.

***Diffraction in Periodic Structures.*** The most important element in this group of tuning elements is the diffraction grating. The first spectral selective cavity for wavelength narrowing and tuning of a dye laser was constructed by Soffer and McFarland [96] using a diffraction grating. They replaced one cavity mirror with a reflective grating in a Littrow mount with 2160 lines/mm blazed at 500 nm first order with an approximately constant first-order reflectivity of $80 \pm 3\%$ for $\lambda > 552$ nm. The emission of the rhodamine 6G dye laser was narrowed from 6 nm, for which an all-dielectric reflector cavity was used, to 0.06 nm when the grating was employed.

The bandwidth of a dye laser with a nonexpanded beam which is tuned by a grating is estimated in Ref. 92. Typical values are on the order of 1 nm. Thus the grating is clearly more advantageous than the prism in the attainable bandwidth. However, one disadvantage of the grating is that it lowers the quality of the resonator because it usually causes higher losses. The efficiency of a reflection grating is below 90% in most cases. This complicates the use of gratings for cw dye lasers. Only high-quality holographic gratings are suitable for such lasers. In Ref. 97 a holographic grating with a sinusoidal groove profile was used; a spectral efficiency up to 94% was achieved. But this efficiency holds only for the polarization component perpendicular to the groove profile, causing the laser radiation to be linearly polarized. For an output of more than 400 mW, an emission bandwidth of less than 1 GHz was obtained in Ref. 97.

Pulsed lasers are operated mostly well above threshold, thereby preventing grating losses from causing major difficulties for the laser. Nevertheless, the danger of destroying the metal film of the reflection grating with the high peak powers of the laser pulses exists here. This problem may be circumvented by the use of a high-power beam-expanding telescope, as proposed by Hänsch [98] (see Fig. 8.9). Here, by using the telescope, the intensity on the grating surface is diminished and simultaneously, the effective line number of the grating itself is increased. This leads to an improvement in the spectral resolution of the grating. Disadvantages of this procedure are: (1) high demands on the chromatic correction of the telescope should the laser be tuned over a large spectral range, and (2) an increase in laser resonator length, which diminishes the efficiency, particularly for short pump pulses (e.g., from a nitrogen laser).

Less complicated and shorter resonators are possible by means of a prism beam expander [99–101] or a grazing incidence grating [102–104]. As shown in Figs. 8.10 and 8.11, both methods involve a one-dimensional beam expansion which is achieved by grazing incidence on to a prism or grating.

As a consequence of the grazing incidence, resonator losses of considerable size occur. A grazing incidence of 89° on the prism was used in Ref. 99. Here 85% of the beam was reflected, whereby no reflection-reducing coatings were used on the prism

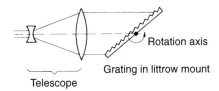

**Figure 8.9** Telescope-grating configuration. Tuning is obtained by turning the grating around the rotation axis.

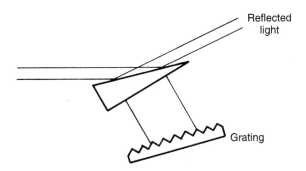

**Figure 8.10** Prism beam expander. The reflected light can be used as laser output.

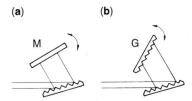

**Figure 8.11** Diffraction grating with grazing beam incidence. Tuning is obtained by rotating (a) the mirror M or (b) the second grating G in a Littrow mount.

(index of refraction 1.57). This reflection is utilized as laser output. An expansion factor of 40 was obtained using this incidence angle.

When using a grating with grazing incidence, no other optical elements are necessary, making possible very compact arrangements. Bandwidths obtained with a grating but without additional components for frequency narrowing, such as a Fabry-Perot etalon, are commonly between 1 and 10 pm. The effect of an intracavity lens and a spherical end mirror on the improvement in bandwidth and efficiency of a grazing incidence grating cavity is discussed in Ref. 105. A special form of frequency selection by means of periodic structures are lasers with distributed feedback which is discussed in Section 8.4.1.

***Multiple-Beam Interference.*** Elements using this effect for frequency narrowing are the Fabry-Perot etalon (often called simply "etalon") and the tuning wedge. The action of a Fabry-Perot etalon is based on the multiple-beam interference caused by reflection between two reflecting surfaces. Basic types of Fabry-Perot etalons are shown in Fig. 8.12.

The transmission of such an etalon is a periodic function of the angle between the etalon and the incident light, the spatial distance between the reflecting areas, and the wavelength. The wavelength distance between two neighboring transmission maxima is called the free spectral range. The spectral half-width of the transmission function of the etalon around the selected wavelength depends on the reflectivity $R$ of the reflecting surfaces. Usually, we use the finesse factor $F$ as the ratio of the free spectral range to the spectral half-width of the etalon transmission function. When the finesse of the etalon is limited by reflection only, we have $F = \pi \sqrt{R}/(1 - R)$. Tuning is accomplished by tilting the etalon or by changing the distance between the two reflecting surfaces. Hänsch [98] used a Fabry-Perot etalon between the beam expanding telescope and the grating (see Fig. 8.13).

The employment of an etalon after the beam expansion is particularly advantageous because the divergence of the laser beam reduces itself according to the expansion factor. Thus the spectral selectivity of the etalon is enhanced. Bandwidths smaller than 1 pm can be obtained using the Fabry–Perot etalon.

In dye lasers with a very narrow emission bandwidth (i.e., for high resolving spectroscopy), a first wavelength-selective step is usually followed by a second. In the latter, a fine-tuning element, such as a Fabry–Perot etalon or a similar device of small free spectral range relative to the broad fluorescence band of the active dye, is inserted into the cavity. The preselection system may consist of a prism, a filter, a grating, or a Fabry–Perot etalon with a large free spectral range, providing in most cases a bandwidth on the order of 0.1 nm. Cavities with two or three etalons for wavelength

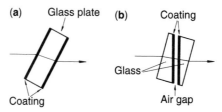

**Figure 8.12**   Basic etalon types: **(a)** glass plate with partially reflecting surfaces; **(b)** air gap etalon.

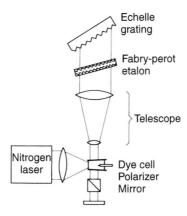

**Figure 8.13**   Hänsch-type cavity. (From Ref. 98.)

selection were used in Refs. 106–108. To avoid mode hopping during tuning the transmission peak of the etalons needs to be adjusted exactly to the frequency of the oscillating mode. Self-adapted filters using photorefractive interferometers avoid this requirement (see, e.g., [109]).

The tuning wedge shown in Fig. 8.14 is an special type of interference filter (Fabry–Perot etalon) with a variation of the transmission wavelength (of several hundred nanometers), which is achieved by a slightly variable thickness of the spacer (wedge) between the two reflective coatings. Tuning is accomplished by shifting the filter perpendicular to the beam. Without further elements for frequency narrowing in the cavity, a laser emission bandwidth below 1 nm is obtained.

***Change of Polarization State.***   When a laser beam passes a birefringent crystal [e.g., quartz or potassium dihydrogen phosphate (KDP)] in a direction, which is not the direction of its optic axis, the index of refraction is different for two directions of polarization which are oriented orthogonally to each other. The ordinary beam, whose electric field vector oscillates vertically to the plane, which is spread by the optic axis and the direction of propagation, possesses an index of refraction $n_0$. The index of

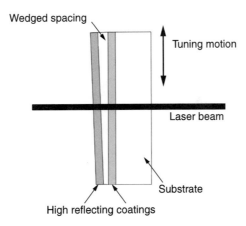

**Figure 8.14**   Tuning wedge.

refraction for the extraordinary beam is $n_e$ and depends on the direction of propagation with respect to that of the optic axis. After passage through the crystal of length $L$, there is a phase difference $\delta$ (retardation) between these two components, which depends on the wavelength: $\delta = 2\pi (n_0 - n_e)L/\lambda$. This means that the polarization state of the light behind the crystal changes with the wavelength of the incident light. If a planparallel crystal plate is cut in this way, that the optic axis lies within its surface and if this plate is placed between two parallel polarizers, the transmission of this arrangement is modulated sinusoidally with optical frequency. Maximum modulation depth is achieved if the optic axis is oriented at 45° to the direction of maximum transmission of the polarizers. These filters are called Lyot filters. In Ref. 110, this effect was used for the first time for narrowing the frequency of a dye laser.

Similar to Fabry–Perot etalons, the frequency resolution of such arrangements increases with crystal length $L$, but the free spectral range decreases. To reach high resolution in combination with a large free spectral range, several Lyot filters, with an integer ratio of their respective crystal lengths (mostly 1:2:4) can be combined. The free spectral range of the complete filter is determined by the thinnest plate, and the bandwidth is determined by the thickest plate. In Ref. 110, several KDP filters are combined with a grating and a spectral bandwidth below 1 pm is reached.

For tuning of the Lyot filters, the phase difference $\delta$ must be changed. By applying a high voltage to the electro-optical crystal, a tuning range of 0.4 nm was obtained. A simpler and more convenient tuning method was found by placing the birefringent filter at the Brewster angle as shown in Fig. 8.15. A laser with a low gain active medium is very sensitive to additional losses. Therefore a 100% modulation of the transmission is not necessary and the polarization discriminating action of the Brewster angle surfaces even can replace the polarizers. Tilting the filter (see Fig. 8.15a) or rotating (see Fig. 8.15b) the filter about its surface normal leads to a change in the angle between the optic axis and the laser light beam; this in turn changes the difference $n_0 - n_e$. In Ref. 111 a single quartz plate of 0.35 mm thickness was used as the only means for frequency narrowing. By tilting this plate between 35 and 50°, tuning between 570 and 600 nm was obtained, with a spectral bandwidth of 1 nm. Today usually three stage filters at Brewster angle with a thickness ratio of 1:2:4 are used. Tuning is accomplished by rotation about the surface normal.

Lyot filters have a relatively broad transmission function. Their spectral selectivity is therefore insufficient for dye lasers pumped by pulsed lasers, whose lasing process is limited to a modest number of cavity round trips of the light. However, low

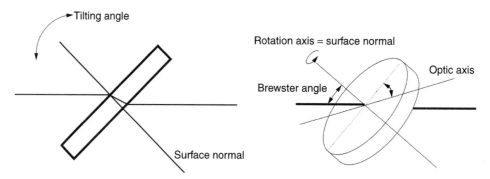

**Figure 8.15**  Lyot-filter. Tuning is obtained by **(a)** tilting or **(b)** rotating the birefringent crystal.

nonselective losses (less than 5%) and the uncomplicated possibilities for tuning make Lyot filters a well-suited and widely used component of frequency selection (or pres-election) in flashlamp pumped or cw dye lasers. Bandwidth values typically measure around 0.02 nm for cw lasers and 0.1 nm for flashlamp pumped lasers.

The Lyot filter is superior to wedge, etalon, or prism tuning in a number of important aspects: (1) Tuning keeps the alignment of the laser cavity. Therefore, the output power and the beam direction is not influenced. (2) The filter tunes continuously and avoids mode-hopping. (3) The wavelength is repeatable to less than 1 Å.

*Microresonators and Wave-Guiding Structures.* Because of the high gain in concentrated dye solutions, laser action requires only very small dimensions of the gain medium at moderate feedback. Thus, Fabry–Perot resonators with micrometer length can be employed and deliver modes of very low longitudinal order. Lasing from microspheres have also been reported, both with liquid droplets of dye solution [112–115] and with dye-doped polymer spheres [116–123]. Such small dielectric spheres act as high-quality cavities. Optical feedback is provided by light waves reflected at the boundary of the sphere. The emission spectrum shows sharp peaks, which correspond to the cavity modes, within the broad fluorescence band. The characteristics of whispering-gallery-mode dye lasers utilizing dye-doped microspheres were investigated in [124]. Microlasers can also be built by arranging scattering dielectric spheres in a highly amplifying medium. The simplest case of two microspheres positioned in a dye solution corresponds to a laser with unstable resonator [125, 126].

Wave-guiding structures supporting only certain propagating wave modes have advantageously been used for designing super-radiant dye lasers or dye lasers with distributed axial feedback (see [R2]). For example, Ippen et al. [127] constructed guided wave dye lasers by filling the dye solution into a thin, hollow glass fiber with the refractive index of the fiber being smaller than that of the solution. H. P. Weber et al. [128] obtained strong lasing in waveguides consisting of a very thin film (0.8 μm), which allows only the propagation of $TE_0$ and $TM_0$ modes. The gain was measured to be about 100 dB/cm. In other approaches, the amplifying solution was positioned outside the wave-guiding structure. The gain medium was excited by the evanescent field of a pump wave propagating in the waveguide, and the evanescent field of the laser beam experiences amplification [129].

*Distributed Feedback.* As early as 1971, it was shown by Kogelnik and Shank [130] that the feedback needed for laser generation could be produced without mirror resonators or other lumped optical components by employing Bragg reflection for coupling back laser radiation. An overview can be found in [R2]. In optics, Bragg gratings result from periodic structures of refraction index and/or gain in the active medium. This Bragg reflection leads to feedback, which makes an outer resonator superfluous. In Ref. 131, this effect was also explained theoretically, and the spatial behavior of the radiation field in a laser with distributed feedback for the stationary case near threshold was calculated. The simplest way to generate a periodic structure in a distributed feedback dye laser (DFDL) is the incidence of two pump beams (with wavelength $\lambda_p$) under the angles $\theta$ and $-\theta$ with respect to the normal axis of the active medium leading to an interference pattern with line distance $\Lambda = \lambda_p/(2 \sin \theta)$ (compare with Section 8.4.2, particularly Fig. 8.23). The spatial modulation of the population inversion determines the laser wavelength in vacuo to be $\lambda_L = 2n_L\Lambda$, where $n_L$ is the index of refraction. Tuning can be achieved by variation of the index of refraction (change of solvent or dye concentration) or of the angles of incidence of the pump radiation. The mode configuration of distributed feedback lasers can be modified by spatial variation of the feedback. For instance, A. N. Rubinov and coworkers were able to enhance the performance of such lasers by designing holographic distributed feedback lasers [132]. In [133], the polarization dynamics of distributed feedback lasers has been investigated. Distributed feedback dye lasers are generally pumped by other lasers, either cw or pw. Such lasers can simply be miniaturized and made robust and stable, broadly tunable versions included. Therefore, they have found application in spectroscopy and spectral diagnostics from ultraviolet to near-infrared [R2]. Visible and UV spectrometers based on such lasers are described in [134]. One great advantage of distributed feedback dye laser consists in the simple and efficient basic design, which allows, e.g., the simultaneous pumping of two dye lasers emitting synchronized pulses at two different wavelengths by one pump source in a compact device [135]. Particularly, distributed feedback dye lasers have the potential for simply generating short light pulses (see Section 8.4.1). Distributed feedback lasers can also be built by using solid-state dye materials (see Section 8.3.2).

***Randomly Distributed Feedback.*** In 1994, Lawandy et al. [136] observed another type of laser-like emission from optically pumped mirror-less liquid or solid solutions of the dye rhodamine 640 containing dispersed nano particles of titanium dioxide with diameter smaller then the wavelength of optical emission. Such systems show threshold behavior. Below threshold, broadband dye fluorescence is observed, where the fluorescence intensity is proportional to the laser intensity. Above threshold, one observes a sharp rise of the emitted power as a function of pump power and strong narrowing of the spectral bandwidth. The bandwidth becomes smaller than could be explained by employing the simple model of amplified spontaneous emission without taking feedback into account. The threshold for this laser-like emission decreases with increasing concentration of nano particles.

Extensive temporal and spectral studies of this emission were carried out to understand these phenomena [137–142]. In [143, 144], the observed features have been explained by taking into account the feedback of the randomly distributed dielectric particles. The expansion of the spatially varying refractive index of the solution with randomly distributed nano particles into a Fourier series provides a variety of Bragg gratings, which differ in period, direction and reflectivity. With given pump level, all modes with sufficiently high feedback (and low threshold) start to lase. In this way the phenomena observed so far could be quantitatively described, particularly the strong spectral narrowing, which could not be satisfactorily be explained by approaches which take only energetic (i.e., non-phase-sensitive) feedback into account.

Already in their early papers, Lawandy et al. [145] proposed various applications of such simple laser systems, for which they coined the term "laser paint," e.g., for marking products with secure identification and for producing patterns to be recognized and identified in rescue operations. M. Siddique et al. investigated such mirror-less dye laser action also in animal tissue [146].

The first random distributed-feedback laser that uses liquid crystals to tune its emission wavelength had been set into operation by D. Wiersma and St. Cavalieri in 2001 [147]. The feedback is provided by ground glass powder in a mixture of liquid crystal and laser dye, which supply wavelength tuning by changing temperature and gain, respectively. This laser is recommended for remote temperature measurement.

J. McNeil et al. [148] proposed lasers that start with randomly distributed feedback where the electromagnetic field of the strongest mode forces the particles to form an ordinary (nonrandom) Bragg grating. Thus, some time after start, the laser is going to operate in one stable Bragg mode.

***Photonic Crystals.*** (See, e.g., [149–152].) With the arguments of efficient wave guiding in all dimensions and sufficient distributed feedback, photonic crystals penetrated by dye solutions or doped with dye molecules, which provide the high gain needed per unit length, are candidates for promising microlasers. In [153], laser-like emission from dye-infiltrated opal photonic crystals has been reported. The photonic crystals were grown from crystallizing colloidal suspensions of $SiO_2$ spheres with diameters around 200 nm. Single crystals (size some millimeter) and poly-crystalline material (with crystallite sizes of 20–100 μm) had been obtained and used, where the dye solutions were infiltrated into the space in between the spheres.

## 8.4 OPERATIONAL PRINCIPLES, DESIGN, AND PARAMETERS OF LIQUID LASERS

In this section, we discuss certain types of liquid lasers classified with respect to excitation and the way of operation. We treat pulsed liquid lasers, cw dye lasers, and mode-locked dye lasers.

### 8.4.1 Pulse-Pumped Liquid Lasers

The two most important excitation sources of pulsed liquid lasers are flashlamps and pulsed pump lasers. The physical processes within the active medium, the laser design, and the characteristic parameters of the laser emission differ for both of these excitation sources; thus they are discussed separately.

***Flashlamp-Pumped Dye Lasers.*** Flashlamps distinguish themselves as pump sources for dye lasers in that the excitation of relatively large volumes of dye is possible by using highly energetic, incoherent light pulses. The expense involved in the above is less than that of laser pump sources. Various pump arrangements have been conceived for the different lamp geometries. The two most important lamp types are the linear (see Fig. 8.16) and coaxial flashlamps (see Fig. 8.17).

In lasers with linear flashlamps, the usual elliptical or double elliptical pump arrangements are used. Coaxial flashlamps fully include the active medium. They produce a

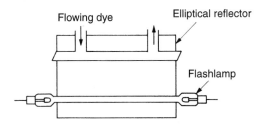

**Figure 8.16** Linear flashlamp in elliptical pump arrangement (see also Fig. 8.6).

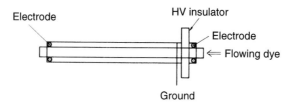

**Figure 8.17**   Coaxial flashlamp.

**TABLE 8.7   Characteristic Flashlamp Data**

|  | Linear<br>Flashlamp | Coaxial<br>Flashlamp |
|---|---|---|
| Duration of the optical pulse ($\mu$s) | 1–100 | 0.1–10 |
| Peak power (MW) | 0.1–5 | 1–50 |
| Pulse energy (kJ) | 0.1–10 | |
| Repetition rate (Hz) | 0.5–100 | |

uniform sheet discharge and have a high pump efficiency (1 to 2% from electrical input to laser output).

The electrical excitation of the flashlamps proceeds at low inductivity over a spark gap or a thyratron. A rapid increase in lamp electrode voltage is thus obtained, which makes possible operation above the static breakthrough voltage. Due to the smaller discharge inductivity, the voltage increase is faster and the available peak powers are higher in coaxial flashlamps than in linear lamps. Linear flashlamps are less expensive and easier to handle. The decisive flashlamp data are summarized in Table 8.7.

To achieve effective operations in flashlamps, some problems must be solved. Fluorescence quenching via triplet-state population in dyes exists with a time constant of about 50 ns. The efficiency obtained in lasing dye depends very greatly on the rate at which it is excited. For short pumping pulses (50 ns or less), the triplet population is not important. For most lasers pumped by flashlamps, the pump pulses last longer, so that with increasingly long pump pulses a considerable portion of the molecules collect in the triplet state as a result of intersystem crossing (compare Section 8.2.2). The number of molecules in the singlet system which are to be used for laser operation is thus reduced. As a consequence of triplet-triplet absorption bands, which extend into the fluorescence band, losses in laser operations are substantially increased [111]. Effective laser operation using flashlamp pump pulses with a duration on the order of 1 $\mu$s is possible only upon application of triplet quenchers [84, 154–156].

The high intensity of the pump beam increases the dye temperature. To avoid large thermal gradients within the dye (they can severely degrade the resonator characteristics) and to dispose of destroyed molecules, the dye is continuously exchanged by streaming the solution through the active volume (see Section 8.2.3). Because of the thermal losses of the lamp and the speed at which the dye flows, the maximal repetition rate is limited.

For low repetition rates, coaxial systems deliver high output energy per pulse. Single pulse energies 110 J [157] and 400 J [158] were achieved with an active volume of 150 cm$^3$. However, in such systems thermal effects and shock waves affect beam quality and reproducibility. For higher repetition rates the linear flashlamp is generally superior with respect to average power, lifetime, and reliability [159–161]. Pulse energies up to 40 J were reported for dye lasers with linear flashlamps [162].

Special flashlamps with gas flow were developed for lamps with high repetition rates (>200 Hz) and pump pulse energies of several joules [163]. Because of the broadband emission of flashlamps, filters for photochemically active wavelengths in the lamp spectrum which might decompose the dye are often used.

Dye lasers pumped with flashlamps are widely used and are commercially available in various configurations. The pump arrangements are compact, and the laser emission parameters can be varied within a wide range by using various lamps. Decisive parameters of commercially available flashlamp-pumped dye lasers are compiled in Table 8.8.

In addition to high-energy pulses, progress was also made in the coherence qualities of laser emission in flashlamp-pumped lasers. Flashlamp-pumped dye lasers with stable resonators and lacking additional intracavity elements emit in a relatively broad bandwidth. Bandwidths are usually on the order of 1 nm. In general, higher transversal modes also oscillate. The divergence of a multimode flashlamp-pumped dye laser is usually on the order of 2 to 5 mrad. For many applications, the TEM$_{00}$ mode is desirable, since it has a diffraction-limited divergence and the laser emission does not suffer from fluctuations due to transversal mode competition. The simplest way to overcome this problem would be the insertion of a mode aperture in the cavity. This leads to a considerable reduction in output energy, since for typical cavities the aperture size required is on the order of 1 mm, whereby utilization of the total volume of the active dye is usually no longer possible. In Ref. 164 a cavity configuration is proposed, in which the natural size of the TEM$_{00}$ mode is large and consequently fills up the cross section of the active medium more completely. As can be seen in Fig. 8.18, this arrangement combines the magnification of the TEM$_{00}$ mode volume by the telescope with Hänsch-type tuning (see Section 8.2.2). A diffraction-limited divergence of 0.1 mrad was measured for an output energy of 10 mJ (with rhodamine 6G as the laser dye) and a bandwidth below 0.1 nm. As discussed in Section 8.2.2. such an arrangement is fairly expensive. A simpler way to generate a near-diffraction-limited beam in the far field is to use an unstable resonator. (Because of the hole in the center the near-field intensity distribution is not a TEM$_{00}$ beam.) In Ref. 165 a planoconvex unstable resonator with a coaxial flashlamp is presented. With 0.1 mrad the divergence is approximately

**TABLE 8.8  Characteristic Parameters of Flashlamp-Pumped Dye Lasers**

| | |
|---|---|
| Pulse energy | 0.05–60 J |
| Average power | 0.25–100 W |
| Repetition rate | 0.5–100 Hz |
| Pulse duration | 0.2–100 μs |
| Efficiency (overall) | 0.2–2% |
| Beam diameter | 3–25 mm |
| Tuning range (with several dyes) | 350–950 mm |

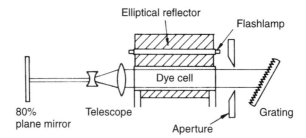

**Figure 8.18**  Flashlamp-pumped dye laser in telescopic cavity configuration. (From Ref. 164.)

1.5 times larger than diffraction limited. Pulse energies of up to several joules were obtained. By the use of a grating instead of the plane mirror or prism pair within the cavity, line widths of 1 pm were reached. Extreme reduction of the line width down to a single longitudinal mode was proven for the first time for a flashlamp-pumped dye laser in Ref. 166. In the experimental arrangement, multiprism-grating combinations were used for frequency narrowing and tuning. The spectral line width obtained was in the range of 260 MHz (0.2 pm) at a pulse energy of 50 to 60 mJ and a duration of 200 to 250 ns.

*Flashlamp-Pumped Rare-Earth Liquid Lasers.*  As discussed in Section 8.1.2. the main classes of active material in rare-earth liquid lasers are (1) solutions of metal-organic compounds (in most cases rare-earth chelates), and (2) rare-earth salts in inorganic solvents. Rare-earth liquid solutions may be effectively excited by optical means, especially by high-power flashlamps. The main pump arrangements are the linear and the coaxial lamp geometries, as described in the preceding sections (see Figs. 8.16 and 8.17).

The advantage of liquid rare-earth lasers compared with solid-state rare-earth lasers is the possibility of circulating the active liquid between the laser flashes as proposed for the first time in Ref. 1, thus solving the problem of heat removal. The laser energy emitted from each unit volume of the active medium is limited by the possible con-centration of the active ions in the solvent. The viscosity of the solvent increases greatly with the concentration of active ions. Since the flow velocity of the active medium circulation decreases with increasing viscosity, the necessary flow velocity determines the active ion concentration. Although the rare-earth solution is optically homogeneous at the beginning of each pump pulse due to the circulation of the active medium, a number of optical inhomogenities may arise during the pumping process. This effect is caused by the temperature dependence of the index of refraction, which is two orders of magnitude stronger in rare-earth solutions than in solid-state matri-ces. Therefore, typical values of the beam divergence in rare-earth liquid lasers are on the order of 100 mrad. There are some important differences in lasing behavior between the main classes of active media mentioned above. The active ions in the metal-organic compounds are $Eu^{3+}$ [167–170], $Tb^{3+}$ [75], and $Nd^{3+}$ [76, 171]. The strong pump beam absorption by the organic complexes in solutions of rare-earth chelates results in a low depth of penetration of the pump radiation (about 1 mm), so that only cuvettes of small diameters can be used. The cuvette diameters transversal to the laser beam lie between 1 and 4 mm. Therefore, the available laser energies are also limited. Due to the complex inner molecular energy transport, losses occur as a consequence of radiationless processes which diminish pump efficiency. In addition to this, the active materials are photochemically unstable. Using these materials, only low-energy relaxation oscillations could be obtained with spectral widths of several nanometers [75, 76, 167–171].

So far in the class of rare-earth liquid lasers, only the ion of neodymium ($Nd^{3+}$) dissolved in an anorganic liquid has shown satisfactory laser action. Originally, selenium oxychloride ($SeOCl_2$) was used as a solvent. However, because of its corrosivity and high viscosity, it has been replaced almost completely by the less corrosive and viscous phosphorus oxychloride ($POCl_3$). The lasing properties are almost identical for both laser solutions. To bring the Nd ion into solution, metallic Chlorides ($AlCl_3$, $SnCl_4$, $GaCl_3$, $ZrCl_4$) acting as Lewis acids have to be added. A comparison between the lasing properties of rare-earth liquid lasers and those of neodymium solid-state lasers is given in Refs. 172 and 173. With these types of liquid lasers, $Q$-switching [172, 174, 175] and mode locking [172, 176, 177] are possible. These types of lasers are also used for the amplification of ultrashort light pulses [178]. Typical laser parameters are shown in Table 8.9.

*Laser-Pulse Pumped Dye Lasers.* Important pump lasers for pulsed dye lasers and their typical parameters are shown in Table 8.10. These data should be compared with the flashlamp parameters in Table 8.7. No special means for triplet quenching are necessary for pump pulses in the nanosecond region.

The excimer laser is the most intense source of UV radiation known at present. The emission wavelength depends on the laser gas, the active compounds of which are noble-gas halides. The main laser lines are at 193 nm (ArF), 248 nm (KrF), and 308 nm (XeCl). For dye laser pumping the XeCl laser is used frequently since many dyes possess absorption bands around 300 nm. Generally, the corresponding absorption cross sections are lower than in the visible spectral region, leading to the necessity of higher dye concentrations. The UV-absorption bands often correspond to transition to higher singlet states, especially in red or infrared emitting dyes. The upper laser level is populated from these higher states by fast internal conversion. This means that in long-wavelength emitting dyes, a considerable part of the excimer laser pump energy is converted into heat. Therefore, efficient dye exchange is necessary in the active region. The effect of the UV-pump photons on the dye molecule can lead to photo-dissociation or photo-fragmentation. Laser dye stability under XeCI-laser pumping was measured

**TABLE 8.9   Parameters of Liquid Nd Lasers Using Inorganic Solvents**

| | |
|---|---|
| Concentration of the | |
| Nd Salt | 0.1–0.4  mol/l |
| Pump energy | 0.3–3  kJ |
| Repetition rate | 1–10  Hz |
| Laser output energy | 0.5–8  J |
| Laser output | |
| Spiking regime | Relaxation oscillations of 0.3–1 $\mu$s duration; bandwidth 1 nm |
| $Q$-switching | Single pulses of 10–30 ns duration; bandwidth 2–3 nm |
| Mode locking | Train of 20–100 pulses with single pulse durations smaller than 0.5 ns; bandwidth 5 nm |

**TABLE 8.10    Characteristic Parameters of Pulsed Pump Lasers**

| | Excimer (XeCl) Laser | Nitrogen Laser | Copper Vapor Laser | Nd:YAG Laser | | |
| --- | --- | --- | --- | --- | --- | --- |
| | | | | Fundamental Wave | SHG | THG |
| Wavelength (nm) | 308 | 337.1 | 510.6 578.2 | 1064 | 532 | 355 |
| Pulse duration (ns) | 10–30 | 0.2–5 | 20–50 | | 5–20 | |
| Repetition rate (Hz) | 0–1000 | 0–500 | 1000–10.000 | | 0–1000 | |
| Single pulse energy (mJ) | 10–1000 | 0.1–10 | 5–100 | Up to 1000 | Up to 400 | Up to 200 |
| Average power (W) | 10–1000 | Up to 1 | 10–1000 | Up to 1000 | Up to 400 | Up to 200 |

in Refs. 179–181. By defining the stability by the total energy of pump pulses leading to a 50% reduction of the dye laser output energy per pulse, we find typical values of the dye stability between 10 and 50 Wh per liter of dye solution. This corresponds to a number of 1 to $5 \times 10^5$ laser shots [181]. Therefore, the dye stability under XeCl-laser pumping is still fairly high (for comparable dyes the stability under a visible pump source is on the order of 50 Wh per liter). For pump wavelengths below 300 nm, dye stability decreases further, for example. multiphoton processes in the solute may lead to photochemical degradation [182].

The nitrogen laser is also suitable for pumping many dyes emitting from the near-UV up to the near-IR spectral region. Nitrogen lasers are a very economic light source due to their simple construction and low operational costs. The average power of commercial nitrogen lasers is limited to the region below 1 W. Therefore, they are used only for low-power applications.

Copper-vapor lasers (CVLs) emit on lines at 510 and 578 nm. Also, if the dye is pumped only with the shorter-wavelength line, dye emission is not achievable below 530 nm without using nonlinear optical methods of frequency conversion. CVLs are especially suited to pumping rhodamine dyes. The high repetition rate enhances the signal-to-noise ratio, thus speeding up experiments (e.g., in spectroscopy).

Nd:YAG lasers emit in the near IR. Only IR dyes can be pumped with the ground wave of these lasers. The emission wavelength of this group of dyes is already extended up to 1.8 $\mu$m [71]. In order to use Nd:YAG lasers as pump sources in the visible or UV region, it is necessary to double (532 nm), triple (355 nm) or quadruple (266 nm) their frequency. Even though these processes can be performed rather efficiently, (>80% for SHG), they increase the complexity of the pump arrangement considerably.

Since the Nd:YAG and excimer lasers are comparable with respect to the pulse repetition rate, average power and achievable single pulse energy, the decision as to what kind of pump laser should be applied depends mainly on the absorption spectrum of the dye used. A series of red emitting dyes (e.g., rhodamine 6G) may be excited very effectively by the second harmonic of the Nd:YAG laser. With a longitudinal pump arrangement energy conversion efficiencies of more than 40% are possible [181]. Using the Nd:YAG laser's fourth harmonic, dyes emitting below 320 nm can be pumped. This spectral region cannot be reached with XeCl lasers. The desired dye laser parameters, especially the spectral bandwidth, the tuning range, and the energy, determine the pump arrangement and cavity configuration of the dye laser system.

First we want to discuss some typical arrangements where the dye emission in its time behavior essentially follows the pump pulse (as a rule, the dye laser emission is somewhat shorter than the pump pulse). Such dye laser arrangements are also commercially available in several versions.

A basic requirement for the cavities of dye lasers pumped by nanosecond laser pulses is compactness. During the pump pulse duration, the dye laser radiation develops from noise and must complete a sufficient number of cavity round trips in order to find enough gain for a high peak power level. If the laser should not work in a broadband mode, strongly selective methods of frequency narrowing have to be employed, since the laser light passes the frequency-selective element only a few times. In simple systems prisms are used as dispersive elements, leading to bandwidths typically on the order of 1 nm. Stronger narrowing is achieved by gratings. The gratings are in a Littrow mount (see Fig. 8.9) or a grazing incidence configuration (see Fig. 8.11). A beam expander is often employed also for arrangements possessing grazing incidence. This enables the tilt of the grating to be reduced. This reduces the demands on the mechanics of the grating tilting and improves the reproducibility of the laser wavelength. Due to the lower selectivity of the grating in a Littrow mount, in many arrangements a Fabry–Perot etalon is inserted in the cavity between the beam expander and the grating. Fine tuning is achieved by tilting the etalon or pressure tuning. Here the tuning unit of the laser is located in a pressure chamber. By changing the inner pressure of the chamber, the light path within the tuning unit is varied.

In most configurations the active medium is pumped transversely (Fig. 8.7). The width of the excited region is on the order of several hundred micrometers; the length is typically in the centimeter region. The scheme of such cavity configurations is shown in Fig. 8.19. The cavity lengths are between 30 and 50 cm.

The energy supply of the pump lasers (except nitrogen lasers) is generally sufficiently high that the laser threshold can be greatly exceeded. The laser emits a fairly broadband radiation in this regime. The laser threshold is surpassed for higher transversal modes, too, so that the divergence of the laser beam increases. Spectral selection and mode apertures for the improvement of spectral and spatial emission characteristics diminish the efficiency of the entire system. For this reason, a master oscillator followed

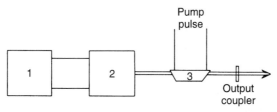

**Figure 8.19**  Scheme of a transversally pumped pulsed dye laser. The basic units are: l, frequency narrowing und tuning elements; 2, beam expansion unit; 3, active dye.

by a power amplifier (MOPA) is often used at high overall pump energies for the generation of intense narrowband and nearly diffraction-limited dye laser pulses [183, 184]. The scheme of such a MOPA configuration is presented in Fig. 8.20. The optical delay line for the amplifier pump radiation ensures that no amplified spontaneous emission (ASE) is generated in the amplifier before the laser pulse from the dye oscillator enters the amplifier. The influence of the oscillator ASE in the amplifier is reduced by a large spatial separation between oscillator and amplifier and the insertion of diaphragms in the beam pass, since the ASE divergence is much larger than the divergence of the laser beam. In Ref. 185 the ASE was suppressed below the detection limit by use of a stimulated Brillouin scattering phase conjugating reflector. In most arrangements, the dye cells are tilted to avoid parasitic oscillations; enough dye circulation must be ensured to suppress thermal degradation (heating) and photo-fragmentation of the dye. An overview of characteristic parameters of such arrangements is given in Table 8.11.

In the second half of this section, we discuss resonator configurations emitting dye laser pulses, which are short in comparison with the pump pulses. Using these methods, it is possible to generate short pulses with a variable repetition rate and energy without having to use the expensive technology of mode synchronization described in Section 8.4.3.

A method for the generation of short laser pulses using resonator transients was described as early as 1966 [186]. The principle is based on the generation of a population inversion $N$ far above the threshold inversion $N_{thr}$, in a cavity with a photon decay time $t_c$, that is short compared with the pump duration $T_p$. Thus proper values of the two parameters $N/N_{thr}$ and $T_p/t_c$ have to be chosen. Large $N/N_{thr}$ causes a steep rise in the

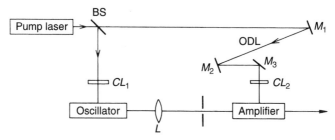

**Figure 8.20**  Scheme of a MOPA configuration. BS, beam splitter; CL, cylindrical lens; L, spherical lens; M, mirror; ODL, optical delay line.

**TABLE 8.11  Characteristic Data of Pulsed Dye Lasers**

| Pump Laser | Tuning Range (with Several Dyes) | Energy Conversion Efficiency with One Amplifier Stage (Dye Dependent) (%) |
|---|---|---|
| XeCl | 330 nm – 1 μm | 8–17 |
| SHG Nd: YAG CVL | 550 nm – 1 μm | 15–35 |

leading edge of the pulse, and $T_p/t_c$ decisively determines the possible pulse duration and should be as large as possible to obtain short pulses. If $T_p/t_c$ is too small, relaxation oscillations occur, which means that multiple pulses are emitted.

Extremely short resonators, typically between 10 and 100 µm, are used to obtain a short cavity decay time $t_c$. The reflection of the output mirror at the laser wavelength normally lies between 70 and 90%. In Refs. 187 and 188 dye laser pulses between 10 and 17 ps were generated by pumping with nitrogen laser pulses of duration between 300 and 330 ps. An energy conversion efficiency of 10% was obtained for rhodamine 6G.

Frequency tuning for this laser is usually possible only by varying the dye concentration within the dye solutions flowing perpendicular to the emission direction. Despite the extremely limited length of the resonator, numerous axial modes are still oscillating within the gain profile of the dye. The laser emits over a relatively broad band, possessing typical bandwidths of several nanometers. Spectral quality can be improved, but only at the cost of the efficiency of the laser. The resonator is further shortened to such a degree that the axial-mode distance becomes larger than the bandwidth of the dye, thereby allowing only one mode to oscillate. Single-mode operation with a bandwidth of 1.5 nm was obtained using resonator lengths of less than 5 µm. However, the energy conversion efficiency only reached 0.3% with rhodamine 6G. Tuning was accomplished by piezoelectric translation of a resonator mirror [189].

A modified method that utilizes resonator transients for the generation of short pulses is based on the idea developed in Ref. 190, where the active medium is placed into two resonators which differ greatly in their optical design (see Fig. 8.21). A resonator with a very short photon lifetime is already formed by the cuvette with two windows which possess a Fresnel reflection of 4%. A longer resonator is formed by using two highly reflective mirrors. Despite its higher threshold, the shorter resonator starts to oscillate before a significant laser intensity is reached in the long outer resonator as a consequence of the light's longer traveling time. However, the longer resonator will reach higher intracavity intensities because of its higher quality. The population inversion in the dye is then rapidly reduced below the threshold value of the shorter resonator. Laser oscillation ceases in the short cavity. When the qualities of both resonators are properly chosen, the shorter resonator emits a single light pulse which is short in comparison to the pump pulse. In Ref. 190 a dye laser pulse with a duration

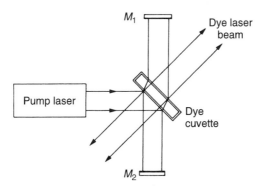

**Figure 8.21** Dye laser with quenched resonator transients. M1 and M2 are highly reflecting mirrors of the high-quality cavity. (From Refs. 190 and 191.)

of 2 ns was obtained by pumping with a 20-ns ruby laser pulse. An optimization of this method was suggested in Refs. 191 and 192. As shown in Ref. 186, extremely short resonators are necessary for the production of single pulses using resonator transients. A longer resonator leads to an emission of a train of relaxation spikes. In Ref. 192 the short resonator was chosen in such a manner that relaxation spikes were emitted for the used pump pulse. The quality of the long resonator is specifically determined so that the short resonator, of lesser quality, ceases to oscillate precisely after the first relaxation peak. This class of lasers has the name "quenched transient dye lasers" (QTDLs).

An experimental arrangement using such a laser is shown in Fig. 8.22 [192]. The XeCl pump pulse with a duration of 20 ns is split into three partial beams. The QTDL emits at a pumping level of 5- to 7-mJ pulses with an energy of 100 to 150 nJ and a duration of less than 0.5 ns. The emission wavelength of the dye $p$-terphenyl dissolved in dioxane is 340 nm. The emission pulse bandwidth of about 1 nm is narrowed to 0.2 nm by sending the pulse through a grating spectral selector. After two-stage amplification with 5- and 15-mJ pump energy, respectively, a pulse of 100 μJ is obtained. Various other possible forms of QTDL are described in Ref. 191.

Another method to produce short light pulses is that of distributed feedback in the laser (see Fig. 8.23, as well as Section 8.3 and [R2]). The nonstationary behavior of DFDL was investigated in Refs. 193–195. In general, a pw pumped DFDL emits a train of relaxation peaks. Only when the threshold is exceeded by no more than 20% will single pulses occur. Those pulses produced in DFDL can be 10 to 100 times shorter than pump pulses. With a 500-ps pulse of a nitrogen pump laser, dye laser pulses of 6 ps were generated [196]; using the third harmonic of a mode-locked Nd:YAG laser pulse with 18-ps duration, single DFDL pulses between 1.6 and 3.5 ps were generated [197]. The shorter pulses correspond to a higher pumping power. Further shortening of the DFDL pulses can be attained by traveling-wave excitation as proposed in Ref. 198. A pump pulse of about 5 ps could be shortened to 1 ps or less. With a special achromatic pump arrangement DFDL pulses of 350 fs were generated from 8-ps pump pulses [199]. In [200], tunable subpicosecond light pulses have been produced with a traveling wave dye laser. Such pulses can be used in time-resolved excite- and probe-absorption spectroscopy (see, e.g., [R11, 19, 201]).

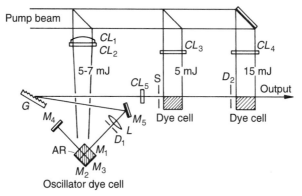

**Figure 8.22** QTDL with subsequent amplification stages. The abbreviations are: AR, antireflection-coated dye cell wall; $M_1$, uncoated dye cell wall; $M_{2-5}$, highly reflecting mirrors; L, quartz lens; G, grating: CL, cylindrical lens; D, aperture; S, slit. (From Ref. 192.)

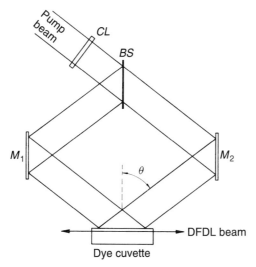

**Figure 8.23**  Scheme of DFDL. The abbreviations are: CL, cylindrical lens; BS, beam splitter; M, mirror; $\theta$, angle of incidence of the pump beams.

The outstanding advantage of the various methods described above is that single pulses of various repetition rates up to the kilohertz region possessing high energy can be produced.

These pulses can be amplified in further amplification stages to extremely high peak powers. In papers 202 and 203, all these methods have been applied to produce picosecond pulses with energies in the region of a few millijoules. Szatmàri, Racz, and Schäfer [204] were successful in generating femtosecond pulses by applying a setup where the DFDL laser was pumped with picosecond pulse from an excimer-laser QTDL. Intense femtosecond UV light pulses were then obtained by amplifying the second harmonic of these DFDL pulses in another excimer tube. Various design principles and results for pulsed dye laser amplifiers are outlined in [R11, 19, 205–207].

***Examples of a Typical Modular Pulsed Laser System for Application.*** As already mentioned, the main advantage of dye lasers is the rather simple way in which broad spectral tunability combined with miniaturization can be obtained. One of the smallest (and cheapest) configurations that has been used in many applications from NUV to NIR consists of a pw nitrogen pump laser ($\lambda = 337$ nm) and a miniaturized dye laser containing exchangeable dye cuvettes (for various gain media) and spectral tuning elements. Such configurations have been introduced to the market by several companies. As an example, we describe the highly automated version of LTB [208] named ATM (automated tunable module), which, because of its large tuning range, small pulse width, reliability, ruggedness, and simple operation and service has been applied in atomic and molecular emission, fluorescence, and absorption spectroscopic investigations in many fields of science and technology, e.g., materials sciences, chemistry, chemophysics, as well as in biology (particularly genomics and protenomics) and medical diagnostics (see Table 8.12). The dye laser modules can be directly connected to the nitrogen laser pump modules MNL 200 (100 kW pulse power) or MNL 800 (800 kW pulse power).

**TABLE 8.12  Parameters of Typical Miniaturized Automated PW Dye Lasers [208]**

The Dye Laser Modules are Preferentially Pumped by Miniaturized Nitrogen Lasers. The
    Wavelengths below 400 nm are Generated by Second Harmonic Generation in a Nonlinear
    Optical Crystal.

| Type of Dye Laser | Unit | ATM 100 | ATM 200 | ATM 300 |
|---|---|---|---|---|
| dimensions (LxWxH) | mm | | $114 \times 250 \times 168$ | |
| weight | kg | | 1.5 | |
| tuning range | nm | 400–900 ($\lambda$ fixed) | 400–900 ($\lambda$ tunable) | 400–900 ($\lambda$ tunable) |
| pump wavelength | nm | 337 | 337 | 337 |
| spectral bandwidth | nm | 5–8 | <2 | <0.1 |
| maximum conversion efficiency | % | 30 | 20 | 5 |
| resonator configuration | | mirror/mirror | mirror/grating | mirror/grating (in grazing incidence) |
| max. repetition rate | Hz | 50 | 50 | 50 |
| beam divergence | mrad | <1.5 | <1.5 | <1.5 |
| **UV1 (SHG)** | | | | |
| tuning range | nm | 225–400 | 225–400 | 225–400 |
| typical pulse width | ps | | 700 | |
| conversion efficiency with respect to dye laser | % | 8 | 8 | 10 |
| **UV2 (SHG)** | | | | |
| tuning range | nm | | 205–225 | |
| conversion efficiency with respect to dye laser | % | 8 | 8 | 10 |

Higher tuning ranges for the fundamental wave (320–1036 nm), pulse power, and repetition rates (200 Hz) can be obtained, e.g., with the LPD and SCANmate dye lasers from Lambda Physik pumped by the LPX220i excimer lasers. These modular systems can also be used in conjunction with an Nd:YAG pumped amplifiers, and then they deliver high-energy pulses above 100 mJ (at 500-mJ pump energy) [209]. The tuning accuracy for the wavelength is about 5 pm. Nonlinear optical conversion modules for generating second harmonic radiation are also offered from this source.

Another system is offered by Spectra-Physics [210], which combines a Quanta-Ray Nd:YAG pump-laser with a Sirah (Kaarst, Germany) dye laser.

## 8.4.2  Continuously Pumped Dye Lasers

To obtain a continuously working dye laser, the following problems have to be solved: pumping with a continuously working light source at power high enough to keep the laser above threshold; constant optical properties of the active medium during laser operation; suppression of unwanted molecular repopulation processes within the dye.

In typical cavity configurations the threshold pump intensity for the often used dye, rhodamine 6G, is approximately 50 kW/cm$^2$. With continuously working lamps in the usual pumping geometries, such an intensity level cannot be reached. By focusing the radiation of a 3-W cw laser on a spot of size 20 μm or less, this power can be obtained or even surpassed in the focus region. Peterson, Tuccio. and Snavely built the first cw dye laser in 1970 [8], using a focused argon laser beam in a longitudinal pump arrangement.

When the active volume is limited to a region with a diameter of approximately 10 μm, it is possible to reduce the heating up of the dye solvent and the undesired triplet occupation in the dye molecules by rapidly exchanging the dye.

In most dyes the thermal processes that lead to a spatial inhomogenity of the index of refraction proceed slower than the intersystem crossing to the triplet system. Thus, above all, the rate parameter for triplet occupation determines the necessary exchange speed of the dye. Assuming constant pump intensity and neglecting the influence of other relaxation processes, the triplet population $N_T$ increases linearly with pumping time,

$$N_T = N_2 k_{ST} t \qquad (8.10)$$

where $k_{ST}$ is the rate parameter for intersystem crossing (see Table 8.5) and $N_2$ the population density in the first excited singlet state. When the triplet occupation reaches the value

$$\sigma_T N_T = \sigma_L N_2 \qquad (8.11)$$

the absorption in the triplet system compensates for the gain in the singlet system. In Eq. (8.11), $\sigma_T$ and $\sigma_L$ are the cross section for triplet absorption and induced emission of the laser transition, respectively.

With Eqs. (8.10) and (8.11), one finds a critical time

$$t_{\text{crit}} = \frac{\sigma_L}{\sigma_T} \frac{1}{k_{ST}} \qquad (8.12)$$

For greater times, significant losses by triplet absorption have to be expected. With the ratio $\sigma_L / \sigma_T \approx 10$ the critical time $t_{\text{crit}}$ is below 1 μs. The dye laser intensity and random mean square noise as a measure of the signal-to-noise ratio of the laser emission as a function of the traveling time of a dye molecule through the pumped dye region is shown in Fig. 8.24 [211], emphasizing the necessity of a fast dye exchange. From these data one can conclude that the necessary flowing speed of the dye solution is approximately 10 m/s. In Ref. 212 the application of a stream of free-flowing dye was suggested. This jet stream technique is used today in nearly all cw dye lasers (see Section 8.2.3).

The extremely small active volume within a cw dye laser demands special cavity configurations since the mode diameter in the dye is in the micrometer range and the low gain per pass makes very low loss cavity components necessary. A linear cavity configuration especially suited for cw dye lasers was proposed in Ref. 213 (see Fig. 8.25). The center mirror in this folded three-mirror arrangement operates off-axis. This introduces astigmatism. The focus point of the rays in the paper plane is different from the focus point of the rays lying in the plane perpendicular to paper. The dye jet is arranged under the Brewster angle to the radiation beam in order to

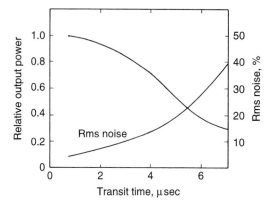

**Figure 8.24**   Relative laser output power and rms output fluctuations versus transit time of a dye molecule through the pump beam. Data obtained with a $3 \times 10^{-4}$ M solution of rhodamine 6G in water with 5% ammonyx and a pump intensity of 1 MW/cm$^2$ at 514.5 nm. (From Ref. 211.)

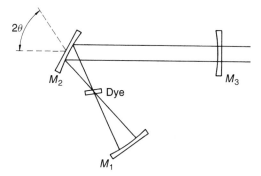

**Figure 8.25**   Astigmatically compensated three-mirror cavity (for an explanation, see the text).

minimize reflection losses. The angular position of the dye jet with respect to the beam may also lead to astigmatism. As shown in Ref. 213, the dye thickness and the folding angle $\theta$ of the cavity can be adjusted so that the astigmatism of the mirror and the dye jet cancel each other out. The parameters of the compensated cavity may be chosen in such a way that the focal spot in the dye is small and is positioned so that it is relatively insensitive to distance variations of the mirrors $M_2$ and $M_3$. The space between these two mirrors may be used for the insertion of further optical elements. A typical example proposed in Ref. 214 is shown in Fig. 8.8. The prism within the cavity serves as a frequency-narrowing element and input for the pump beam.

As discussed in Section 8.3.2, the prism is one of the optical elements inserted in the cavity to narrow the spectral bandwidth and to tune the laser output. The dispersive elements in cw dye lasers should introduce minimal additional losses in the cavity, as mentioned above. Coarse tuning over the whole tuning range of a dye with a medium bandwidth on the order of 0.1 nm and below can be obtained with Lyot filters, prisms, or wedged interference filters. For narrower bandwidths, stronger frequency selective elements have to be used. In Ref. 97 a holographic grating was used instead of one of the end mirrors. The spatially optimized grating introduced only about 8% additional loss in the cavity. With an emission linewidth below 1 GHz, tuning was possible over the whole dye bandwidth by tilting the grating. For single-longitudinal-mode oscillation, similar bandwidths can be attained by a combination of a low dispersive element [e.g., a wedged interference filter (see Fig. 8.14)] with a highly dispersive element such as a Fabry–Perot etalon. Tuning is again achieved by translating the wedged filter perpendicular to the beam and tilting the etalon. In Ref. 215 the design used was a Lyot filter and two etalons with free spectral ranges balanced on each other. If only one etalon is tuned synchronously with the cavity length, tuning ranges of about 30 GHz can be covered without mode hopping. The tuning range without mode hops could be extended only by a very expensive tuning mechanism which also employs the other etalon and the Lyot filter.

When only a few modes or even one mode oscillate in a linear cavity configuration the problem of spatial hole burning in the dye occurs. This means that a structure with a periodicity of half the laser wavelength originates in the dye. The standing cavity wave depopulates the inversion within the dye in the regions of the field strength maxima. At the field strength nodes, no induced emission takes place. This effect leads to an enhanced sensitivity to mode-hopping processes. Ring lasers with traveling waves running in only one direction overcome this drawback. The wave traveling in the undesired direction would lead to competition effects in the dye with the wave traveling in the selected direction. The suppression of this undesired wave is accomplished with an optical diode. This diode usually consists of a Faraday rotator and a birefringent crystal. For the selected wave this crystal rotates the polarization direction, and after passage the Faraday rotator returns to its original state, while for the unwanted wave the polarization rotations in the Faraday rotator and the crystal add. The ring laser is more suited to the single-mode regime (with respect to output power and stability) than the linear cavity. In the one-way regime described above, where spatial hole burning is avoided, the homogeneously broadened line favours single mode operation. Compensation of astigmatism is achieved in ring cavities in the same way as in linear cavities. However, the ring cavities require a greater deal of adjustment than linear cavities. Further, the intracavity losses in a ring cavity are usually higher than in a linear one, leading to a higher laser threshold for cw ring dye lasers. Parameters of cw dye lasers are summarized in Table 8.13.

Single-mode dye lasers with very stable output in frequency, bandwidth, and power are necessary, especially for the demands of high-resolution spectroscopy. Time variations

**TABLE 8.13  Characteristic Parameters of CW Dye Lasers**

| | |
|---|---|
| Pump power | 2–8 W |
| Conversion efficiency (dye dependent) | 5–15% |
| Laser power (without tuning element) | 0.1–1.5 W |
| Bandwidth (without tuning element) | 1 nm |

are caused primarily by pump fluctuations. The pump laser should be stabilized on a certain power level for such applications.

The spectral fluctuations of a cw dye laser are determined primarily by density fluctuations of the dye jet and by mechanical instability of the optical components. The latter effects lead to bandwidth fluctuations on the order of 10 to 100 MHz. A better stability can be obtained by active means such as locking the laser to a passive external reference cavity, which can be much better stabilized than its own laser cavity. A frequency stability better than 100 kHz was obtained in actively stabilized cw dye lasers [216, 217].

An extension of the wavelength range of cw dye lasers is possible by application of nonlinear optical methods. Because of the higher intracavity intensity, the insertion of nonlinear optical elements into the cavity seems to be advantageous. In Refs. 218–220, experimental results of the intracavity second harmonic generation are reported. Here, radiation in the regions 285 to 315 nm [218], 292 to 302 nm [219], and 240 to 250 nm [220] was generated. With an argon pump laser of 4.5 W power, 30-mW UV radiation at 300 nm was generated. The active dye was rhodamine 6G and SHG crystal ADA was used. Some problems are connected with intracavity SHG. The insertion of an intracavity crystal increases cavity losses significantly, thus strongly reducing the intracavity intensity. Since the efficiency of the SHG is frequency dependent, the position of the crystal with respect to the laser beam has to be optimized for every frequency. In Ref. 221 the SHG of cw dye laser emission in an external passive ring resonator is reported, thus decoupling the laser and the nonlinear optical process. The intensity obtained nears that of the intracavity SHG.

***Pumping by CW Diode Lasers.***   CW infrared dye lasers can also be pumped by diode lasers. Generally, semiconductor diode lasers offer substantial advantages in dye laser pumping: high overall yield (electrical input to dye laser output), miniaturization of devices, simple scaling up of output power, long lifetime of pump sources, and low costs.

As early as 1974, pw diode lasers were employed as pump sources for dye lasers, where IR dyes had been excited by 50-ns pulses from AlGaAs diode lasers at 820 nm [90]. Efficient cw pumping of infrared dye lasers by diode lasers has become possible in the 1990s after substantial improvements in such pump sources. In [91], various dyes, among them rhodamine 700, oxazine 750, oxazine 1, and DOTCI, were pumped by applying AlGaInP diode lasers, emitting at wavelengths between 670 and 690 nm. The dye lasers emitted radiation in the near-infrared spectral range between 740 and 800 nm. The optical conversion efficiency and the overall efficiency can be as high as 50% and 10%, respectively.

### 8.4.3   Mode Locking of Dye Lasers

Since 1970, considerable progress has been made in the generation of ultra-short light pulses [R11, 18, 19, 23] and their application in time-resolved spectroscopy and optoelectronic and electro-optic switching, and in the manipulation of photophysical, photochemical, and photobiological processes [R9, R11, 18–22, 24, 25]. In the first two decades since 1970, dye lasers had played a dominant role in this progress. In 1975, Ippen and Shank [222] obtained subpicosecond light pulses directly from passively mode-locked cw dye lasers. In 1981 the first light pulses with a duration of less than 0.1 ps (100 fs) were generated by improvement of the colliding ring configuration [223]. Utilizing additional dispersive elements for intra-cavity pulse compression in such ring resonators, pulses of some 10 fs were achieved [224–226]. Starting with dye laser pulses, the shortest light pulses obtained had a duration of 6 fs, and they consist of only three wave cycles of the center wavelength of 0.6 μm [227, 228]. These light pulses were produced by amplifying the ultrashort light pulses from a passively mode-locked laser and by passing the amplified pulses through a nonlinear optical element — and through an appropriate linear optical dispersive device that compresses the frequency-swept (chirped) light pulses up to the Fourier limit.

Since 1990 all-solid-state lasers with broad gain spectrum (e.g., Ti:sapphire) have taken the lead also in generating femtosecond light pulses [R11, 23, 229–230]. High gain with large spectral bandwidth as well as robustness, stability, and long operation time without service are the dominant reasons for their preference over dye lasers. In these femtosecond solid-state lasers, use has advantageously been made of linear and nonlinear optical methods developed in connection with dye lasers, particularly pulse narrowing by saturable absorption and gain depletion as well as intracavity compensation of group velocity dispersion. With solid-state lasers, some additional nonlinear optical means for intracavity pulse compression have been introduced; first of all, the great success of Kerr lens mode locking [229] has to be mentioned.

These novel measures have then been employed also with dye lasers. For instance, Kerr-lens modelocking has advantageously applied with dye lasers, and by use of this measure, very short and transform-limited light pulses have been obtained [231].

Also with ultrashort light pulses, the main advantage of dye lasers consists in the easy way to change the operation wavelength in wide spectral ranges from ultraviolet to near infrared.

To generate short light pulses, the coherent coupling of all Fourier components within a broad spectral range is necessary. With single-mode operation, the minimum achievable pulse duration, which requires very short laser cavities, is given by

$$T_{L\,min} \approx \frac{10^2}{N'c\sigma_L} \tag{8.13}$$

where $N'$ is the number density of active molecules, and $\sigma_L$ is the cross section of the laser transition [232]. As the value of $N'$ is restricted by undesired interactions between the active particles at high concentration, the minimum duration is on the order of 10 ps. This is the lower limit of pulse duration in non-mode-locked pulsed dye lasers such as a short cavity or quenched dye laser (see Section 8.4.1).

Hence, subpicosecond and femtosecond light pulses can only be produced with lasers in the multimode regime. In mode-locked lasers we have

$$T_{L\,min} = K/\Delta\nu \tag{8.14}$$

where the frequency interval $\Delta\nu$ in which the longitudinal cavity modes of the laser are locked is only restricted by the bandwidth $\Delta\nu_{21}$ of the laser transition ($\Delta\nu \leqslant \Delta\nu_{21}$). The factor $K$ is the minimum value of the pulse-duration bandwidth product, which depends on the particular pulse shape and attains 0.441 and 0.315 for Gaussian-shaped and sech$^2$-shaped pulses, respectively. In dye solutions there are very broad fluorescence bands ($\Delta\nu_{21} \leqslant 0.5 \times 10^{14}$ Hz at $\nu_{21} \sim 5 \times 10^{14}$ Hz), about one-fifth of which can be mode locked by refined techniques, which are dealt with in this section. Thus the minimum pulse duration is given by

$$T_{L\,min} \approx \frac{1.5}{\Delta \nu_{21}} \approx 30 \text{ fs} \tag{8.15}$$

(Note that with solid state lasers, the minimum achievable pulse duration is by about 5 times smaller than with dye lasers because of broader spectrum and higher gain.)

Two principal methods are used for mode locking of dye lasers: synchronous pumping and passive mode locking.

***Synchronously Pumped Dye Lasers.*** Mode locking can be achieved through suitable modulation of the gain. This can be accomplished by pumping a laser with the continuous pulse train of another laser that is already mode locked (see Figs. 8.26 and 8.27). Actively mode locked argon lasers and Nd:YAG lasers can be used advantageously with this aim. If the optical length of the laser cavity is nearly equal to that of the

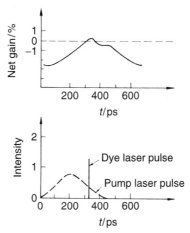

**Figure 8.26**   Temporal evolution of the net gain and the intensities of the pump laser and the dye laser in a synchronously pumped laser.

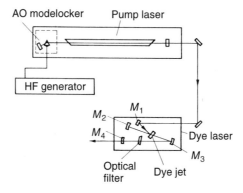

**Figure 8.27**   Scheme of a synchronously pumped dye laser.

pump laser or an integral multiple of it, then under certain conditions the gain is temporally modulated with a modulation period that is equal to the resonator-round-trip time. As in the case of active mode locking, a short pulse builds up in the time range of the maximum gain, whose pulse duration under optimum conditions can be two or even three orders of magnitude smaller than that of the pump pulses (see Fig. 8.26). By using a frequency-selective optical filter (e.g., a Fabry–Perot etalon, a Lyot filter, or a prism), which considerably narrows the bandwidth of the laser radiation in the resonator and makes it possible to change the frequency of the resulting gain maximum, the frequency of the dye laser can be tuned over a certain range (compare Section 8.3.2). In contrast to cw operation, spectral width of the filter must not be too small since otherwise the pulses will be considerably lengthened. As opposed to passively mode-locked lasers (compare next subsection), synchronous pumping has the advantage that nearly the total spectral width of the laser transition can be used for tuning.

In synchronous pumping with a continuous pump pulse train from another laser, pulse generation takes place in the following manner. (Synchronous pumping was first accomplished by using the finite pulse train of a ruby laser [233]. The method of pumping by cw lasers was introduced in Ref. 234. Subpicosecond lasers were first reported in Ref. 235.) The occupation inversion strongly increases during the action of the pump pulses, whose durations are typically on the order of 100 ps with argon lasers and Nd:YAG lasers, whereas the lifetime $T_{21}$ of the upper laser level of efficient laser dyes is some nanoseconds (compare Section 8.2.2), the duration of the pump pulses $T_p$ is very small compared to the lifetime ($T_p \gg T_{21}$). Under these conditions, the occupation inversion of the active medium depends on the pump energy supplied up to this moment. (This situation changes if one uses fast-relaxing dyes in the near-infrared region [236].) As shown in Fig. 8.26, the gain coefficient increases monotonically due to pumping until it exceeds the level of the resonator losses, that is, until the laser threshold is reached. Then in the steady-state regime, the energy of the laser pulse increases rapidly and reaches the "saturation energy" of the active medium. As a result, the occupation inversion of the dye is reduced by stimulated emission, whereby the gain drops quickly to a value below the level of losses. Thus positive gain occurs only within a small interval during the pump pulse duration, so that the generated laser radiation is concentrated in this time range. It is, of course, important that the laser pulse always travel exactly synchronously with the pump pulse through the active medium.

Starting from the noise of spontaneous emission, the laser pulse builds up always at the right position. Whether or not it experiences net gain in the following round trips depends on the matching of the resonator lengths of both lasers.

A typical arrangement of a synchronously pumped dye laser is depicted schematically in Fig. 8.27. As a pump source, an argon, krypton ion laser or a Nd:YAG laser which is actively mode locked by means of an acousto-optical modulator is normally used. A high stability of frequency and phase of the high-frequency generator, which supplies the electric signal for the modulator, is required. The relative fluctuation of the frequency should be less than $10^{-7}$. Various pump laser lines in the near ultraviolet and visible spectral range have been used for synchronous pumping. In place of noble-gas ion lasers, cw-pumped actively mode-locked solid-state lasers (e.g., Nd:YAG lasers) can be employed, and thus the fundamental wave ($\lambda = 1.06$ μm) as well as higher harmonics ($\lambda_2 = 0.53$ μm, $\lambda_3 = 0.353$ μm) of the laser radiation can be applied. Advantages of the application of solid-state lasers result from the fact that the fluctuations of the pulse parameters due to the long relaxation times of the active materials at high frequencies are smaller than in noble-gas ion lasers (for a comparison of the applications of both laser types, see Ref. 237).

In general, a continuous, stable train of pulses with durations of 30 to 300 ps, an average power of 0.1 to 10 W, and a repetition frequency on the order of $10^8$ Hz is applied for pumping the dye solution. In Fig. 8.27 the mirror $M_1$ serves to couple the pump pulse train into the dye laser resonator. Pump and laser radiation pass at a small angle through the dye. It is important that there be a very good overlapping of the waists of the pump beam and the dye laser beam in the active material. The laser radiation is deflected by the folding mirror and is partially reflected and partially coupled out at the output mirror $M_4$. In order to have a fast exchange of the active solutions, a free-flowing jet stream of good optical homogeneity is used in most cases (compare Section 8.2.3). In the setup shown in Fig. 8.27, the wavelength of the laser can be continuously changed using a tuning element. With the use of such wavelength-selective elements and of various dyes, it is possible to generate ultrashort light pulses, whose wavelengths can be tuned over a wide spectral range of about 400 to 1800 nm.

With high-quality mode locking the laser pulse duration $T_L$ is inversely proportional to the bandwidth of the frequency-selective element in the resonator and decreases monotonically with increasing pump power. Furthermore, $T_L$ decreases roughly proportional to the square root of the pump pulse duration. To achieve very short pulses, a great effort must be made to improve the stabilization of the components of the synchronously pumped laser. The length of the dye laser has to be matched to the frequency of the mode locker of the pump laser, and the influence of the surrounding air must be reduced. If the cavity length is changed slightly, the parameters of the laser pulse and its position with respect to the pulse maximum change considerably. Beyond a certain negative value of the cavity mismatch, the laser leaves the single-pulse regime and double pulses appear. Above a certain positive value of the cavity mismatch, the laser regime becomes nonstationary. It proves to be useful not to adjust the modulator frequency and the dye laser length independently, but to obtain the modulator frequency from the pulse repetition frequency of the dye laser. Furthermore, automatic control loops are applied. The pulse duration is measured where this signal is used to control the oscillator frequency [238]. Using such automatic controls, pulses that are stable over a long time and have subpicosecond duration ($\cong 0.5$ ps) can be generated. Synchronously pumped dye lasers are also constructed in ring resonator arrangements, where one direction of pulse circulation is favored by additional measures (compare next subsection).

The energy of the light pulses from synchronously pumped dye lasers is typically several nanojoules. It can be increased by more than one order of magnitude if instead of the output mirror of the laser cavity, a so-called cavity dumper is used (see Fig. 8.28). It may, for example, consist of an acousto-optical modulator that deflects the pulse out of the resonator only on the $n$th round trip, while on the other round trips the radiation passes undeflected through the modulator and remains in the resonator. In this manner energy storage occurs, and the pulse repetition frequency is reduced by the factor $n$. The Bragg cell in the cavity dumper must have small losses during the energy storage and high diffraction efficiency ($\geqslant 0.7$) after applying the driving signal. The rise time of the high-frequency power and the transit time of the ultrasonic wave through the region of the beam waist must be shorter than the resonator round-trip time. With cavity dumpers.

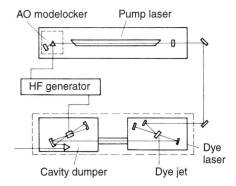

**Figure 8.28** Scheme of a synchronously pumped dye laser with cavity dumper.

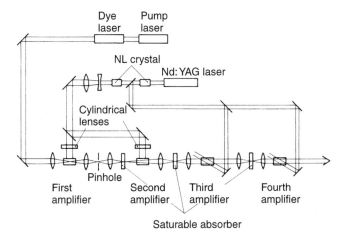

**Figure 8.29** Synchronously pumped dye laser with subsequent four-stage amplifier. (From Ref. 238.)

the pulse repetition frequency can be chosen between 0 Hz and several megahertz. The peak power of the light pulses can be as high as some $10^4$ W. In most cases the pulse duration is somewhat longer with than without the cavity dumper.

However, certain applications require even higher power than that achieved by means of cavity dumping. Figure 8.29 shows the scheme of a laser amplification system. The amplification occurs here in four consecutive dye cells that are pumped by the second harmonic of a pw pumped $Q$-switched Nd:YAG laser. The pulse repetition frequency finally achieved is given by that of the laser, which pumps the amplifier. To increase this repetition frequency, excimer lasers, nitrogen lasers, or copper-vapor lasers may be used instead of the Nd:YAG laser (compare with Table 8.10).

***Passively Mode-Locked Dye Lasers.*** In 1968, Schmidt and Schäfer observed the formation of a train of short light pulses in a rhodamine 6G laser pumped by a flash lamp after insertion of a dye cell that acted as a saturable absorber into the resonator [239]. Such flashlamp-pumped dye lasers operate in a quasi-continuous regime, whose duration is limited by that of the flashlamp pulse. A real continuous operation, which is

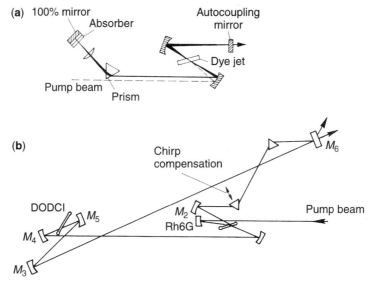

**Figure 8.30** Schemes of passively mode-locked dye lasers; **(a)** linear resonator; **(b)** ring resonator for CPM with chirp compensation. (From Ref. 242.)

most commonly used at present (cw regime) was first achieved by Ippen et al. [240] and by O'Neill [241], who continuously pumped dye lasers with argon ion lasers. By using such lasers (see Fig. 8.30), unlimited pulse trains were generated, the minimum pulse duration of which was decreased almost exponentially from hundreds of femtoseconds to tens of femtoseconds between 1975 and 1990. The shortest pulses were obtained by intracavity pulse compression in ring lasers with counterpropagating light pulses (Fig. 8.30**b**).

(By the use of pulse compression outside the laser cavity, the light pulses from such dye lasers have been further compressed after amplification [227, 228], where durations of about 5 fs have been obtained.) Fig. 8.30 shows several types of laser cavities, which contain the gain medium that is pumped by a cw laser, the saturable absorber, and possibly spectral filters and dispersive elements as well. In most cases, the gain medium and the saturable absorber consists of free-flowing jet streams (compare Sections 8.2.3). Alternatively, one may use a dye flow through a thin cell as shown for the saturable absorber in Fig. 8.30**a**.

Furthermore, the saturable absorber and the amplifying dye may be mixed in one solvent and passed through one jet stream. The employment of this mixture simplifies the adjustment and the operation of the laser, since only a few optical components are necessary. On the other hand, it is a disadvantage of the arrangement that the beam waists in the gain medium and in the saturable absorber cannot be independently varied.

Figure 8.30**b** presents a ring configuration, in which the distance between the gain jet and the absorber jet is one-fourth of the ring perimeter. In such lasers, two counterpropagating pulses build up. Their exact timing is automatically regulated by the system, since counterpropagating pulses that precisely overlap in the absorber maintain the optimum generation conditions. This is explained below.

General disadvantages of passively mode-locked dye lasers are their small tuning range, for one amplifier-absorber combination, and the necessity to find appropriate

**TABLE 8.14  Combinations of Laser Dyes and Saturable Absorbers Used in CW Passively Mode Locked Dye Lasers**

| Laser Dye | Saturable Absorber | Tuning Range (nm) | Reference |
|---|---|---|---|
| Rhodamine 6G | DODCI | 595–635 | 243 |
| | DQOCI | 580–613 | 243 |
| | DASBTI | 570–600 | 244 |
| Rhodamine B | DODCI | 610–630 | 243 |
| | DQOCI | 600–620 | 243 |
| | Cresyl Violet | 610–620 | 243 |
| | DQTCI | 616–658 | 245 |
| Sodium fluorescein | Rhodamine 6G | 546 | 243 |
| Rhodamine 700 | DOTCI | 727–740 | 246 |
| | HITCI | 762–778 | 246 |

combinations of gain medium and absorber for the desired wavelength region. Table 8.14 presents such combinations that have been used successfully.

As with active mode locking, the mechanism of passive mode locking rests on the temporal modulation of the losses and/or the gain in the resonator. In passive mode locking, however, the system itself determines the moment at which the net gain reaches a maximum. The pulse formation proceeds as follows. After the pump radiation has exceeded the laser threshold, the laser radiation is amplified from spontaneous emission in the resonator, whereby in the multimode regime dealt with here, the radiation field consists of a statistical superposition of many fluctuation peaks at the beginning. Due to high gain in the laser dye, the radiation is amplified by stimulated emission to a value at which the saturation of the absorber becomes important after a rather short time interval of only 20 to 30 cavity round trips. After this time, the absorber favors these fluctuations or groups of fluctuations that possess maximum energy, since these experience smaller losses due to the saturation of absorption. Typically, the upper-state lifetimes of the gain medium $(T_g)$ and of the absorber $(T_a)$ are much longer than the mean duration of the fluctuation peaks. Hence the saturation of both media is controlled almost only by the energy of the fluctuation peaks (and later by that of the light pulses) and does not significantly depend on the peculiar shape of the fluctuations and pulses. The saturation of the absorber and the depletion of the gain medium, respectively, begin to play a role when the pulse energy per unit area reaches around

$$E_{sa} = \hbar\omega_L/\sigma_a \tag{8.16}$$

and

$$E_{sg} = \hbar\omega_L/\sigma_g \tag{8.17}$$

where $\sigma_a$ and $\sigma_g$ are the interaction cross sections of the absorber and gain medium. In an absorber of this type, the leading edge of the fluctuation peak (and later that of the pulse) is strongly absorbed until the energy that has already passed reaches a value at which the absorption is greatly diminished due to saturation. For this reason, the trailing edge of the pulse remains almost unweakened, and thus the pulse obtains a

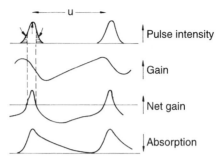

**Figure 8.31**   Formation of ultrashort light pulses in passively mode locked dye lasers by saturation of absorption and depletion of gain.

steeper leading edge in the absorber. This behavior remains in the steady-state regime as well (see Fig. 8.31). The suppression of the trailing edge occurs in the gain medium. When the pulse energy exceeds a certain value given by the saturation energy of Eq. (8.17), the amplification in the gain medium is decreased due to the depletion of the occupation inversion and falls rapidly below the value of the threshold amplification. Hence the trailing edge of the pulse experiences less gain and becomes suppressed. Due to the combined action of the absorber and the amplifier, conditions arise under which the fluctuation peak, having the highest energy, is preferred with respect to the net amplification and continuous suppression of the leading and trailing edges. Thus a single ultrashort light pulse can ultimately build up and circulate in the resonator. To saturate the absorber with less energy than the gain medium and to have higher gain at the pulse peak than at the edges, it is necessary that the absorber be saturated earlier than the gain medium. Thus if the radiation energy per area is the same in both media, we have to require $\sigma_a > \sigma_g$. By focusing the radiation tighter into the absorber than the gain medium, this requirement can be circumvented, and the process of saturation and depletion can therefore be controlled.

Note that the collision of the counterpropagating pulses within the absorber (see Fig. 8.30**b**) has approximately the same influence as an increase of the absorber cross section by a factor of 3 [19]. Furthermore, it is necessary that the occupation inversion in the amplifier not be completely built up during one round trip of the pulse, so that the losses exceed the gain at the leading edge in each round trip. This determines an upper limit for the ratio of the resonator round-trip time and the lifetime of the gain medium. On the other hand, the resonator must not be too short; otherwise, the time between subsequent pulse passage will be insufficient for the recovery of the occupation inversion by means of pumping. These requirements lead to the conditions for the parameters of the resonator and the media it contains, which define a so-called stability range within which the generation of a single ultrashort light pulse per round trip occurs.

Because of saturation of the absorber and depletion of the gain medium, the light-pulse experiences suppression of its leading and trailing edges in each round trip respectively, as explained before. Hence the pulse decreases its duration and increases its spectral width during each passage. After the pulse shortening has reached a certain value, the bandwidth-limiting action of the optical filters in the laser cavity begins to play an important part. When all additional bandwidth-limiting elements are carefully removed, the influences of the finite spectral widths of the absorber and gain media come into play. Within certain parameter ranges, a steady-state pulse shape ultimately evolves due to the balance of frequency broadening induced by the nonlinearities and frequency selection provided by linear optical filters.

In general, the frequency-selective elements in the resonator, such as gain medium, saturable absorber, laser mirrors, and additional filters, do not have the same center wavelength. For example, for all amplifier-absorber combinations presented in Table 8.14, the transition wavelength of the absorber deviates from that of the amplifier. Thus the carrier frequency $\omega_L$ of the light pulse deviates from at least one transition frequency. Off-resonance passage of light pulses through active and passive elements changes not only the temporal behavior of the modulus, but also that of the phase of the light pulse. With passively mode-locked femtosecond lasers, such effects — in particular, a decrease of the carrier frequency during the pulse, which is called a down chirp — have been observed [247]. Under such circumstances, the minimum pulse duration is obtained when the chirp that arises by the passage of light pulse through the saturable absorber, the amplifier, and other elements is compensated [224, 226]. This can be achieved by using a glass path of appropriate length or any other optical element, such as interferometers, specially designed mirrors, and grating pairs, the group-velocity dispersion of which can be controlled. In Fig. 8.30b the glass path length is varied by changing the position of one of the prisms in the cavity. By changing the path length in glass, chirp-free pulses as well as down-chirped pulses can be produced [242] (see Fig. 8.32). (For experimental and theoretical details of the optimum generation of ultrashort light pulses with passively mode-locked lasers, see Ref. 19, Chap. 6.) By carefully compensating the chirp, pulses shorter than 30 fs have been obtained directly from such lasers [225].

As with synchronously pumped dye lasers, the weak light pulses from passively mode-locked dye lasers, whose energy is in most cases below 0.1 nJ, can be amplified

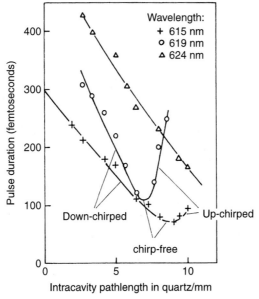

**Figure 8.32** Pulse duration versus intracavity glass path for $\lambda = 615$ nm (SQ 1 glass), $\lambda = 619$ nm (flint glass) and $\lambda = 624$ nm (SQ 1 glass). (From Ref. 242.)

[R11, 243, 248], whereby the peak power may attain terawatts [249]. The optical arrangement of the amplifier can be chosen similar to that shown in Fig. 8.29. The sequence of amplifying dye cells was pumped by use of light pulses from frequency-doubled $Q$-switched Nd:YAG lasers [250], excimer lasers [251], and copper-vapor lasers [252], the latter distinguished by their high repetition rate of about 10 kHz. By compensating the chirp resulting from various resonant and nonresonant optical components in the amplifier chain again, high-power pulses as short as some ten fs have been obtained. Such amplified pulses can then be further compressed by chirp generation in glass fibers and subsequent chirp compensation in linear optical devices with group-velocity dispersion.

***Hybrid Mode Locking.*** The shortest pulses were generated by passively mode-locked lasers. This method, however, has several disadvantages. such as the very critical adjusting of the pump and the resonator parameters for the generation of stable pulses and the restricted tuning range. To combine the advantages of passive mode locking and synchronous pumping. the so-called hybrid mode-locking regime was developed [253, 254]. In this regime, synchronous pumping of the gain medium as well as an additional passive mode locking by means of a saturable absorber were employed simultaneously. Thus, femtosecond pulses could be generated over a rather large tuning range (574 to 611 nm) [255], the duration of which was only 70 fs [237].

## 8.5   SELECTED APPLICATIONS

Nowadays dye lasers are mainly used in applications where one or several light beams with (independently) tunable wavelength are required. Although there is no fundamental progress in the improvement of dye laser systems during the last decade, there exist a great variety of such pw and cw laser applications in science, technology, and medicine. Powerful, precisely tunable sources of narrow bandwidth are available in the visible and, after nonlinear frequency conversion, also in the UV and IR spectral regions. The technology is well adopted, and there exist a great variety of experiences in their application. Here, only a few examples have been selected that demonstrate the capabilities and advantages of dye lasers.

***Application to Spectroscopy, Photochemical Reactions, and Isotope Separation.*** In these application fields, advantage is taken from the selective excitation of specific atomic or molecular transitions by one or several tunable dye lasers.

If the amount of probe molecules and as a consequence the absorption is low, methods are applied that detect the absorption not directly in the change of probe-transmission but background-free by fluorescence (e.g., laser excited atomic fluorescence spectrometry), ionization (e.g., laser induced ionization spectrometry), thermal effects (e.g., photo-thermal and photo-acoustic spectrometry), and other indirect detection principles. To reach the detection limit, high average power (for cw excitation or excitation of fast relaxing transitions) or high pulse energy (for pw excitation of slowly relaxing transitions) are required additionally to small bandwidth for achieving saturation of the transition (see, e.g., [R14] and [256]). Such principles are used as alternative detectors in chromatography [257].

To achieve high sensitivity in absorption spectrometry of probes with low concentration of target molecules (e.g., in atmospheric pollution measurement), either the absorption path is made very long [258] or the probe is placed inside the laser cavity [259].

The requirements to be met in frequency stability and line width of the lasers employed depend strongly on the aggregation state and the density of the molecules to be selectively excited and that of other molecules being in the vicinity of the active ones. In liquid and disordered solid materials at room temperature, the homogeneous linewidth of typical electronic and vibronic transitions resulting from molecular collisions (or phonon–electron interaction) is on the order of some $10^{13}$ Hz, which corresponds in the visible range to some 10 nm. That means, relative line width $\Delta\nu/\nu = |\Delta\lambda/\lambda|$ and frequency stability $\delta\nu/\nu = |\delta\lambda/\lambda|$ of the laser radiation on the order of $10^{-3}$ to $10^{-2}$ is appropriate. In gases and vapors, the inhomogeneous broadening (originating from the Doppler effect) amounts to about $10^9$ Hz, corresponding to $\Delta\nu/\nu = |\Delta\lambda/\lambda| \sim 10^{-6}$; the homogenous broadening resulting from radiation damping amounts for strong transitions to some $10^8$ Hz, and can be far smaller for weak transitions; thus, at atmospheric pressure, the (homogeneous) collision broadening, which is proportional to pressure and takes on values of about $10^{10}$ Hz at this pressure, is larger than the Doppler width, but it can be decreased far below the Doppler value, even for rather weak transitions, by decreasing the pressure adequately. Hence, for high-precision applications in gases, it makes sense to decrease the laser linewidth down to the megahertz and even kilohertz range (see, e.g., [R14]).

In many spectroscopic applications, the lasers may be operated at rather low power, particularly in transmission measurements if the relative changes of radiation power in the sample are high. However, when very small absorption losses are not measured via comparison between input and output power, but by detecting the power absorbed in the sample via photo-thermal or photo-acoustic methods, the measuring signal is directly proportional to the input power (see, e.g., [260]). By using such procedures, it is possible to measure relative absorption changes on the order of $10^{-6}$ with laser input power on the order of some 100 mW.

Also in fluorescence spectroscopy aiming at extreme detection sensitivity, the laser power is decisive, particularly in single molecule detection. As an example, we consider dye molecules in solution and use Eq. (8.6) in conjunction with the level scheme of Fig. 8.5**b**, where because of very fast vibrational relaxation within the two electronic bands, only the levels 0 and 2 are populated, and the total number of molecules in the volume under excitation and observation is given by $N_{\text{tot}} = N_0 + N_2$. The power of fluorescence light being proportional to the number of excited molecules $N_2$ is given by

$$P_F = \hbar\omega_F k_R N_2 = \hbar\omega_F \eta_F \frac{1}{T_{21}} N_2 \qquad (8.18)$$

where $k_R$ is the rate parameter of spontaneous radiation decay from level 2 and $\eta_F$ is the fluorescence quantum yield. Under stationary irradiation, we obtain from Eqs. (8.6) and (8.18) for the fluorescence quantum flux

$$\Phi_F = \frac{P_F}{\hbar\omega_F} = \eta_F N_{\text{tot}} \left[\frac{\sigma_P I_P}{\hbar\omega_P}\right] \frac{1}{1 + T_{21}\dfrac{\sigma_P I_P}{\hbar\omega_P}}. \qquad (8.19)$$

Far below and high above saturation pump intensity $I_{P\text{sat}} = (\hbar\omega_P)/(T_{21}\sigma_P)$, this equation respectively yields:

$$\Phi_F = \eta_F N_{\text{tot}} \left[\frac{\sigma_P I_P}{\hbar\omega_P}\right] \text{ and} \tag{8.20a}$$

$$\Phi_F = [1/T_{21}]\eta_F N_{\text{tot}} \tag{8.20b}$$

With $\sigma_F = 10^{-16}$ cm$^2$, $\lambda_P = 500$ nm, and $T_{21} = 10$ ns, the saturation pump intensity amounts to about 0.4 MW/cm$^2$. And, with a cross section of the pump beam (beam waist area) in the interaction volume of about 1 $\mu$m$^2$, obtained by tight focusing, the saturation power is 4 mW. That means, at a pump power of some 10 mW, the condition for applying Eq. (8.20b) is fulfilled. Even with only one molecule in the interaction volume and a quantum yield $\eta_F = 1$, the fluorescent photon flux and the fluorescent power take on the rather high values of $\Phi_L = 10^8$ s$^{-1}$ and $P_F \approx 40$ pW. Employing a photon-counting detector with quantum yield $\eta_{\text{count}} = 0.1$ and an optical system that transmits the whole fluorescence radiation emitted into the space angle $\Delta\Omega$ onto the photo detector, where $[\Delta\Omega/(4\pi)] = 0.1$, the average count rate n$_{\text{count}}$ amounts to $10^6$ s$^{-1}$. Thus, single molecules can be detected with good signal-to-noise ratio and high spatial resolution. The superiority of laser excited fluorescence markers over radioactive ones originates mainly from such high count rates even with single particles, which is decisive first of all in life sciences (see, e.g., [261]). Particularly, it can be clearly distinguished whether there is no or at least one radiative molecule in the interaction volume. And by carefully gauging the device for a specific fluorescent marker, even the number of radiative molecules $N_{\text{tot}}$ can be determined for rather small numbers. As a consequence, fluctuations of the number of molecules in the active volume can be measured down to the 10-ms range, which provides the opportunity to observe diffusion processes or the "birth" and "death" of fluorescent molecular entities. The combination of spectroscopy and microscopy provides the possibility to take pictures of molecular distribution in space. First images of molecules in solid matrices taken at low temperature by employing an intensified CCD camera were presented in [262]. Later on, even fast sequences of images (or slow motions visible in "movies") of molecular ensembles in solution and cells have been obtained (see, e.g., [263]). The laser fluorescence correlation spectroscopy, which rests on such measurements, has become a widely applied tool since the beginning of the nineties [R14, R16, 262, 264–266].

Tunable lasers, with similarly small bandwidth, but with far higher average power, are required in laser photochemistry and laser isotope separation. In both fields, the processing of 1 mol of a substance typically (chain reactions and avalanche processes excluded) requires at least the same number of photons. (For photons, one uses the unit Einstein instead of the unit mole; at a wavelength of 600 nm, 1 Einstein photons corresponds to an energy of 200 kJ.) Hence, for efficient laser processing, high average power is required. The necessary frequency stability and linewidth depend here also on the aggregation state and the width and density of absorption lines. We discuss as an example the requirements with isotope separation in gases. The laser separation rests on the isotope shift of atomic or molecular energy levels. The gas is strongly irradiated by one or several laser beams, where at least one beam is exactly tuned to an absorption line of one of the various isotopic species; after excitation to the upper level, such atoms or molecules can be ionized by additional photon absorption or the action of strong electric fields (see, e.g., [267]); the ionized entities are then extracted by electric fields.

Alternatively, the primarily excited species can be extracted by chemical reactions that start only from the excited state. The linewidth of the laser radiation required in the decisive step of selective excitation has to be far smaller than the distance between the isotope-shifted lines, otherwise one does not obtain complete separation, and smaller than the linewidth of the absorption line, otherwise the yield becomes reduced. The frequency stability should be better than about one tenth of the linewidth. When one aims only at the efficient processing of one atomic species in a specialized device, the tuning range needed for exciting various isotopes is rather small, typically smaller than ±30 GHz. Suitable measures for precise and simple tuning of high average power dye lasers within such small intervals and achieving bandwidth below 200 MHz by employing grazing incident diffraction gratings are described in [268]. Dye lasers with average power of 1 kW have been designed, set into operation, and applied for isotope separation at the Lawrence Livermore National Laboratory [269, 270]. These lasers, which consist of oscillators and amplifier chains, are pumped by copper-vapor lasers. Design considerations and simulations for a dye laser system with 10-kW average power pumped by a copper vapor laser are given in [271].

*Medical Applications.* Lasers have been used in medical diagnostics and therapy almost since their advent. The selection of the appropriate laser for a specific application is mainly determined by the dependence of the interaction between the human tissue (or organ) and laser radiation on wavelength, temporal, and spatial distribution of the light field and light power [272, 273]. We discuss this for simple examples: The penetration depth of electromagnetic radiation is determined by absorption and scattering, which strongly depend on wavelength (see Fig. 8.33). At rather low light intensity, these processes are characterized by the absorption coefficient and scattering coefficients of ordinary single photon processes within the limits of linear optics. Energy from Nd:YAG radiation ($\lambda = 1.06$ μm) or from typical NIR diode lasers (e.g., $\lambda = 0.85$ μm)

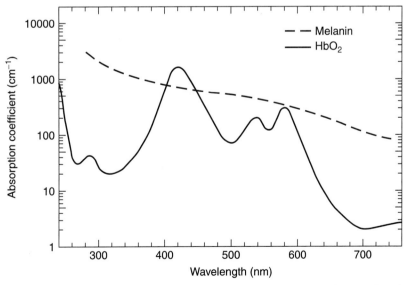

**Figure 8.33** Absorption spectrum of melanin and hemoglobin. (From [273], p. 17—data according to [274].)

is deposited in muscular tissue up to a depth of some mm, whereas the radiation from the ArF-excimer laser ($\lambda = 0.193$ $\mu$m) and of the $CO_2$ laser ($\lambda = 10.6$ $\mu$m) is absorbed within a comparatively thin layer of 10 $\mu$m and 100 $\mu$m thickness, respectively, because of strong absorption by electronic transitions in the UV and vibrational transitions in the IR. Dye laser radiation can be selectively absorbed by certain biological dye molecules, e.g., melanin, hemoglobin, porphyrin, hematoporphyrin, bile, and rhodopsin, when its wavelength matches with an absorption peak of the particular biological molecule. For very intense laser radiation, the attenuation of the input beam does not only depend on the wavelength, but because of nonlinear optical processes, also on the light intensity. The cornea of the human eye, for instance, is almost free from absorption and scattering for radiation of the visible range at low intensities; and the absorption takes on large values only in the UV, and in the IR. However, intense laser pulses at wavelengths outside the spectral ranges of linear optical absorption may experience strong multiphoton absorption and other nonlinear optical processes that lead to ionization and eventually to the formation of plasma, which reflects light arriving later on. Thus, whereas the main part of weak green light passes the cornea and reaches the retina, already the leading edge of an intense light pulse at the very same wavelength, which is focused into the front part of the cornea, built up plasma reflection and, as a consequence, the main part of the pulse does not arrive at the retina.

Lasers are applied in medical sciences for detection, characterization, and destruction of abnormal tissue. Because of the strong absorption and/or scattering of the photons in the human tissue, photophysical and photochemical treatment generally is restricted to objects near the surface where the penetration depth strongly depends on the wavelength. The selective action on a certain target without impairment of the surrounding can be realized by a proper choice of wavelength and time regime of excitation. The selective and controlled deposition of laser energy in certain parts of the human body can first be achieved by properly aligned irradiation of the respective part from outside or through endoscopes into the inside. Second, the selectivity of laser energy deposition in abnormal cells can be enhanced by matching the laser wavelength with absorption peaks of biomolecules, the concentration of which is higher in the target cells (or tissue) compared with the surrounding. Third, special drugs can be applied before laser irradiation (either orally, or by penetration through the skin or injection), which experience strong absorption at the laser wavelength and which are preferentially deposited in abnormal cells (or tissue). If the application aims at diagnostics, drugs will be chosen that exhibit strong fluorescence after laser irradiation. Then the spatial distribution of abnormal cells can be determined by fluorescence imaging.

In dermatologic surgery, the dye laser is often used for the treatment of port-wine stains, abnormal scars, and other abnormal blood vessels [275–278]. Because of the high absorbance of blood in the yellow spectral region (oxyhemoglobin with absorption peak at 578 nm), a (selective) photothermolysis of the vessels can be achieved by powerful irradiation at this wavelength. On the other hand, the chromophores of the dermis and epidermis (melanin) exhibit a low absorbance at this wavelength. Further, if the pulse duration is smaller than the thermal relaxation time (several milliseconds for typical vessels), the blood vessels are destroyed before there is sufficient time for thermal diffusion to cause heat damage of the surrounding tissue. Mainly these two facts make possible a selective interaction with the target tissue and lower the risk of scarring, which is higher after treatment with Ar-lasers (488 nm or 514 nm), for instance. A recent overview on lasers in dermatology can be found in [279]. Different lasers are compared concerning side effects in [280].

In photodynamic therapy, malignant tumor cells are destroyed by photochemical reactions, induced by absorption of laser radiation. The selectivity of the action is achieved by the application of drugs, which enrich in tumor cells due to their increased metabolism. There are also known natural molecules, which can be found in tumor cells in a much higher concentration than in normal cells [281]. These molecules (e.g., derivates of porphyrine) absorb in the red spectral region (630 nm–650 nm), a region that can be accessed easily by powerful dye lasers. However, more comfortable and cheaper diode lasers are also available in this spectral region. Therefore and because of the expected large market, great effort is made to construct powerful illumination devices on the basis of laser diodes [282].

In ophthalmology, age-related macular degeneration is treated by using dye lasers at 585 nm [283–285]. One reason for this disease is that the macula is penetrated by vessels from the choroid. This leads first to a reduced resolution and later to a loss of parts of the visual field. These vessels can be destroyed by photothermolysis, but great effort must be made in the preservation of the macula itself. As in dermatologic surgery, the strong absorption of the vessels in the yellow spectral region and the low absorption of the surrounding tissue serves for the selectivity of this process.

## BIBLIOGRAPHY

R1. R. J. Pressley, Ed., *Handbook of Lasers with Selected Data on Optical Technology, Chemical Rubber*, Cleveland, OH, 1971; Extended Translation: A. M. Prokhorov, Ed., Sovetskoye Radio, Moscow, 1978.

R2. F. P. Schäfer, Ed., *Dye Lasers*, Springer-Verlag. Berlin, 1978.

R3. R. Wallenstein, "Pulsed Dye Lasers," in M. L. Stitch, Ed., *Laser Handbook*, Vol. 3, North-Holland, Amsterdam, 1979.

R4. R. Steppel, "Organic Dye Lasers," in M. J. Weber, Ed., *Handbook of Laser Science and Technology*, Vol. l, *Lasers and Masers*, CRC Press, Boca Raton, FL, 1982.

R5. M. Maeda, *Laser Dyes*, Academic Press, New York, 1984.

R6. U. Brackmann, *Lambdachrome Laser Dyes*, Lambda Physik GmbH, Göttingen, 1986.

R7. Ju. G. Anikiyev, M. E. Zhabotinskij, and W. B. Kravchenko, *Lasers on Inorganic Liquids (in Russian)*, Moscow, Russia, 1986.

R8. F. Duarte, Ed., *High-Power Dye Lasers*, Springer, Berlin, 1991.

R9. M. Stuke, Ed., *Dye Lasers: 25 Years*, Springer, Berlin, 1992.

R10. F. J. Duarte, Ed., *Tunable Lasers Handbook*, Academic Press, San Diego, 1995.

R11. J.-C. Diels, and W. Rudolph, *Ultrashort Laser Pulse Phenomena*, Academic Press, San Diego, 1996.

R12. D. L. Andrews, *Lasers in Chemistry*, Springer, Berlin, 1997.

R13. J. Hecht, *Laser Guidebook*, McGraw-Hill, 1999.

R14. W. Demtröder, *Laser Spectroscopy*, Springer, Berlin, 2000.

R15. M. J. Weber, Ed., *Handbook of Lasers*, CRC Press, Boca Raton, 2000.

R16. R. Rigler, E. S. Elson, *Fluorescence Correlation Spectroscopy Theory and Application*, *Springer Series in Chemical Physics*, Vol. 65, Springer 2001.

## REFERENCES

1. S. G. Rautian and I. I. Sobelman, *Opt. Spektrosk.*, **10**, p. 134, 1961.
2. E. G. Brock, P. Czavinsky, E. Hormats, H. C. Nedderman, D. Stripe, and F. Unterleitner, *J. Chem. Phys.*, **35**, p. 759, 1961.
3. E. J. Schimitschek and E. G. K. Schwarz, *Nature (London)*, **196**, p. 832, 1962.
4. A. Lempicki and H. Samelson, *Phys. Lett.*, **4**, p. 133, 1963.
5. P. P. Sorokin and J. R. Lankard, *IBM J. Res. Dev.*, **10**, p. 162, 1966.
6. F. P. Schäfer, W. Schmidt, and I. Volse, *Appl. Phys. Lett.*, **9**, p. 309, 1966.
7. A. A. Heller, *Appl. Phys. Lett.*, **9**, p. 106, 1966.
8. O. G. Peterson, S. A. Tuccio, and B. B. Snavely, *Appl. Phys. Lett.*, **17**, p. 245, 1970.
9. W. Demtröder, *Laser Spectroscopy*, Springer-Verlag, Berlin, 1981, Chapters 8 and 10.
10. H. Walter, Ed., *Laser Spectroscopy of Atoms and Molecules*, Springer-Verlag, Berlin 1976.
11. D. S. Klinger, Ed., *Ultrasensitive Laser Spectroscopy*, Academic Press, New York, 1983.
12. V. S. Letokhov and V. P. Chebotayev, *Principles of Nonlinear Laser Spectroscopy (in Russian)*, Nauka, Moscow, Russia, 1975.
13. F. T. Arecchi, F. Strumia, and W. Walther, Eds., *Advances in Laser Spectroscopy*, Plenum Press, New York, 1983.
14. Y. R. Shen, *The Principles of Nonlinear Optics*, Wiley, New York, 1984.
15. M. Schubert and B. Wilhelmi, *Nonlinear Optics and Quantum Electronics*, Wiley, New York, 1986.
16. R. B. Miles and S. E. Marris, *IEEE J. Quantum Electron.*, **QE-9**, p. 470, 1973.
17. S. Szatmàri and F. P. Schäfer, *Opt. Commun.*, **48**, p. 279, 1983.
18. S. L. Shapiro, Ed., *Ultrashort Light Pulses*, Springer-Verlag, Berlin, 1977.
19. J. Herrmann and B. Wilhelmi, *Lasers for Ultrashort Light Pulses*, Elsevier, Amsterdam, 1987.
20. E. Klose and B. Wilhelmi, Eds., *Ultrafast Phenomena in Spectroscopy*, Teubner, Leipzig, 1986.
21. G. R. Fleming and A. E. Siegman, Eds., *Ultrafast Phenomena V*, Springer-Verlag, Berlin, 1986.
22. J. Manz, and L. Wöste, *Femtosecond Chemistry*, VCH, Weinheim, 1995.
23. C. Rulliere, Ed., *Femtosecond Laser Pulses: Principles and Experiments*, Springer, Berlin 1998.
24. W. Kaiser, Ed., *Ultrashort Laser Pulses: Generation and Applications*, Springer, Berlin, 1993.
25. E. Schreiber, *Femtosecond Real-Time Spectroscopy of Small Molecules and Clusters*, Springer, Berlin, 1998.
26. N. A. Borisevich, I. I. Kalosa, and V. A. Talkacev, *Zh. Prikl. Spektrosk.*, **19**, p. 1108, 1973.
27. B. Steyer, and F. P. Schäfer, *Opt. Commun.*, **10**, p. 219, 1974.
28. Yu. Yu. Stoilov, *Appl. Phys.*, **B33**, p. 63, 1984.
29. B. H. Soffer, and B. B. McFarland, *Appl. Phys. Lett.*, **10**, p. 266, 1967.
30. G. N. Dulnev, V. I. Zemskij, and B. B. Krynetzkij, *JETP Lett.*, **4**, p. 1041, 1978.
31. G. B. Altshuler, E. G. Dulneva, and I. K. Meshkovskij, *Zh. Prikl. Spektrosk.*, **36**, p. 592, 1982.
32. G. B. Altshuler, E. G. Dulneva, and K. I. Krylov, *Kvantovaya Elektron. (Moscow)*, **10**, p. 1222, 1983.

33. R. Levenson and J. Zyss, "Polymer based optoelectronics," in M. Quillec, Ed., *Materials for Optoelectronics*, Kluwer, Dortrecht, 1996, pp. 341–374.

34. G. Kranzelbinder and G. Leising, *Reports on Progress in Physics*, **63**, pp. 729–762, 2000.

35. W. Holzer, A. Penzkofer, H. Tillmann, E. Klemm, and H.-H. Hörold, *Synthetic Metals*, **124**, pp. 455–465, 2001.

36. H.-H. Hörold, H. Tillmann, C. Bader, R. Stockmann, J. Nowony, E. Klemm, W. Holzer, and A. Penzkofer, *Synthetic Metals*, **119**, pp. 199–200, 2001.

37. D. Braun and A. J. Heeger, *Appl. Phys. Lett.*, **58**, p. 1982, 1991.

38. G. S. He, J. D. Bhawalkar, C. F. Zhao, C. K. Park, and P. N. Prasad, *Appl. Phys. Lett.*, **68**, p. 3549, 1996.

39. L. K. Denisov, I. G. Kytina, V. G. Kytin, S. A. Tsogoeva, L. G. Saprykin, and B. A. Konstantinov, *Quantum Electronics*, **27**, p. 115, 1997.

40. K. C. Yee, T. Y. Tou, and S. W. Ng, *Appl. Optics*, **37**, p. 6381, 1998.

41. S. Popov, *Appl. Optics*, **37**, p. 6449, 1998.

42. Y. Oki, K. Ohno, and M. Maeda, *Jap. J. Appl. Phys.*, **34**, p. 6403, 1998.

43. A. J. Finlayson, N. Peters, P. V. Kolinsky, and M. R. W. Venner, *Appl. Phys. Lett.*, **72**, p. 2153, 1998.

44. M. D. Rahn and T. A. King, *J Modern Optics*, **45**, p. 1259, 1998.

45. G. Somasundaram and A. Ramalingam, *Optics Laser Technology*, **31**, p. 351, 1999.

46. A. Costela, I. García-Moreno, J. Barroso, and R. Sastre, *Appl. Phys.*, **B70**, p. 367, 2000.

47. C. Ye, K. S. Lam, K. P. Chik, D. Lo, and K. H. Wong, *Appl. Phys. Lett.*, **69**, p. 3800, 1996.

48. R. Gvishi, G. Ruland, and P. N. Prasad, *Optics Commun.*, **126**, p. 66, 1996.

49. D. Lo, S. K. Lam, C. Ye, and K. S. Lam, *Optics Commun.*, **156**, p. 316, 1998.

50. S. Wu and C. Zhu, *J. Materials Sci./Lett.*, **18**, p. 281, 1999.

51. W. J. Wadsworth, S. M. Giffin, I. T. McKinnie, J. C. Sharpe, A. D. Woolhouse, T. G. Haskell, and G. J. Smith, *Applied Optics*, **38**, p. 2504, 1999.

52. M. D. Rahn and T. A. King, *Applied Optics*, **34**, p. 8260, 1995.

53. X. L. Zhu and D. Lo, *Appl. Phys. Lett.*, **77**, p. 2647, 2000.

54. S. Wu and C. Zhu, *J. Materials Sci./Lett.*, **18**, p. 281, 1999.

55. V. I. Bezrodnyi and A. A. Ishchenko, *Technical Phys. Lett.*, Vol. 27, p. 740, 2001.

56. M. Sasaki, Y. Li, Y. Akatu, T. Fujii, and K. Haner, *Japanese J. Applied Physics*, **39**, p. 7145, 2000.

57. X. Peng, L. Liu, J. Wu, Y. Li, Z. Hou, L. Xu, W. Wang, F. Li, and M. Ye, *Optics Lett.*, **25**, p. 314, 2000.

58. J. A. Rogers, Z. Bao, M. Meier, A. Dodabalapur, O. J. A. Schueller, and G. M. Whitesides, *Synthetic Metals*, **115**, p. 5, 2000.

59. Y. Xia, J. J. McClelland, R. Gupta, D. Quin, X.-M. Zhao, L. L. Sohn, R. J. Celotta, and G. M. Whiteside, *Adv. Materials*, **9**, p. 147, 1997.

60. J. Barathan and Y. Yang, *Appl. Phys. Lett.*, **72**, p. 2660, 1998.

61. T. R. Hebner, C. C. Wu, D. Marcy, M. H. Lu, and J. C. Sturm, *Appl. Phys. Lett.*, **72**, p. 519, 1998.

62. R. M. Measures, *Appl. Optics*, **13**, p. 1121, 1974.

63. R. M. Measures, *Appl. Optics*, **14**, p. 909, 1975.

64. C. P. Keszthelyi, *Appl. Optics*, **14**, p. 1710, 1975.

65. B. Steyer and F. P. Schäfer, *Appl. Phys.*, **7**, p. 113, 1975.

66. G. Marowsky, R. Cordray, F. K. Tittel, W. L. Wilson, and J. W. Keto, *J. Chem. Phys.*, **67**, p. 4845, 1977.

67. N. A. Borisevich. G. B. Tolstorozhev, V. A. Tugbuyev, and D. M. Chalimanovich, *Zh. Prikl. Spektrosk.*, **23**, p. 1098, 1975; Ultrafast Relaxation Secondary Emiss., *Proc. Int. Symp. Ultrafast Phenom. Spectrosc.*, 1979, **2**, p. 8–20.

68. M. Reufer, S. Riechel, J. Crewett, J. Feldmann, T. Benstem, and W. Kowalsky, "Organic laser structures with metallic contacts" in Proceedings of the European Conference on Organic Electronics and Related Phenomena, p. 219, Potsdam, 2001.

69. J. H. Schoen, Ch. Kloc, A. Dodapalapur, and B. Batlog, *Science*, **289**, p. 599, 2000.

70. K. W. Hausser, R. Kuhn, and E. Kuhn. *Z. Phys. Chem., Abt.*, **B29**, p. 417, 1935.

71. H. J. Polland, T. Elsaesser, A. Seilmeier, and W. Kaiser, *Appl. Phys.*, **B32**, p. 53, 1983.

72. K. H. Drexhage, *Laser Focus*, Vol. 9. p. 35, 1973; "Structure and Properties of Laser Dyes," in F. P. Schäfer, Ed., *Dye Lasers*, Springer-Verlag, Berlin, p. 148, 1978.

73. H. Labhart, *Helv. Chim. Acta*, **47**, p. 2279, 1964.

74. M. E. Zhabotinskij and L. V. Ljovkin, "Rare Earth Liquid Lasers," in A. M. Prokhorov, Ed., *Handbook of Lasers with Selected Data on Optical Technology (in Russian)* Sovetskoye Radio, Moscow, Russia, 1978, pp. 346–360.

75. S. Bjorklund, G. Kellermeyer, and C. R. Hurt. *Appl. Phys. Lett.*, **10**, p. 160. 1967.

76. A. Heller, *J. Am. Chem. Soc.*, **89**, p. 167, 1967.

77. B. Wittacker, *Nature (London)*, **228**, p. 157, 1970.

78. N. E. Alekseyev, *Izv. Akad. Nauk SSSR*, **5**, p. 1038, 1965.

79. I. V. Mochalov, N. P. Bondareva, A. S. Bondarev, and S. A. Markosov, *Kvantovaya Elektron. (Moscow)*, **9**, p. 1024, 1982.

80. T. Förster, *Fluorescence of Organic Compounds (in German)*, Vandenhoeck & Ruprecht, Göttingen, 1951.

81. M. Kaschke, S. Rentsch, and B. Wilhelmi, Comments At. Mol. Spectrosc., **17**, p. 309, 1986.

82. B. Kopainsky, P. Qiu, W. Kaiser. B. Sens, and K. H. Drexhage, *Appl. Phys.*, **B29**, p. 15, 1982.

83. A. Seilmeier, W. Kaiser, B. Sens, and K. H. Drexhage, *Opt. Lett.*, **8**, p. 205, 1983.

84. B. B. Snavely and F. P. Schäfer, *Phys. Lett.*, **A28**, p. 728, 1969.

85. R. K. Jain, *Appl. Phys. Lett.*, **40**, p. 295, 1982.

86. M. Kaschke, U. Stamm, and K. Vogler, *Appl. Phys.*, **B39**, p. 183, 1986.

87. T. Förster, *Ann. Phys. (Leipzig)*, **2**, p. 55, 1948; *Z. Naturforsch.*, **A4**, p. 321, 1949.

88. F. P. Schäfer, "Principles of Dye Laser Operation," F. P. Schäfer, Ed., *Dye Lasers*, Springer-Verlag, Berlin, 1978, pp. 1–90.

89. H. Knibbe, D. Rehm, and A. Weller, *Ber. Bunsenges. Phys. Chem.*, **72**, p. 257, 1968.

90. G. Wang, *Opt. Commun.*, **10**, pp. 149–153, 1974.

91. R. Scheps, *IEEE J. Quantum Electronics*, **QE31**, p. 126, 1995.

92. G. Marowsky, *Opt. Acta*, **23**, p. 855, 1976.

93. G. Marowsky, L. Ringwelski, and F. P. Schäfer, *Z. Naturforsch.*, **A27**, p. 711, 1972.

94. G. Marowsky and F. Zaraga, *IEEE J. Quantum Electron.*, **QE-10**, p. 832, 1974.

95. G. Marowsky, *Z. Naturforsch.*, **A29**, p. 536, 1974.

96. B. H. Soffer, and B. B. McFarland, *Appl. Phys. Lett.*, **10**, p. 266, 1967.

97. J. Kuhl, G. Marowsky. and H. A. Obermayer, *Appl. Phys.*, **16**, p. 297, 1978.

98. T. W. Hänsch, *Appl. Opt.*, **11**, p. 895, 1972.

99. D. C. Hanna, P. A. Kärkkäinen, and R. Wyatt, *Opt. Quantum Electron.*, **7**, p. 115, 1975.

100. F. J. Duarte and J. A. Piper, *Am. J. Phys.*, **51**, p. 1132, 1983.

101. J. R. M. Barr, *Opt. Commun.*, **51**, p. 41, 1984.

102. I. Shoshan, N. N. Danon, and U. P. Oppenheim, *J. Appl. Phys.*, **48**, p. 4495, 1977.

103. M. G. Littmann and H. J. Metcalf, *Appl. Opt.*, **17**, p. 2224, 1978.

104. M. G., Littmann, *Opt. Lett.*, **3**, p. 138, 1978.

105. R. S. Smith and L. F. DiManre, *Appl. Opt.*, **26**, p. 855. 1987.

106. D. J. Bradley, W. G. I. Gaughey, and J. I. Vukusic, *Opt. Commun.*, **4**, p. 150, 1971.

107. I. M. Gale, *Opt. Commun.*, **7**, p. 86, 1973.

108. A. Moriarty, W. Heaks, and D. D. Davis, *Opt. Commun.*, **16**, p. 324, 1976.

109. L. Meilhac, N. Dubreuil, G. Pauliat, G. Roosen, *Optical Materials*, **18**, p. 37, 2001.

110. H. Walter and J. L. Hall, *Appl. Phys. Lett.*, **17**, p. 239, 1970.

111. W. Schmidt and F. P. Schäfer, *Z. Naturforsch.*, **A22**, 1563, 1967.

112. A. Biswas, H. Latifi, R. L. Armstrong, and R. G. Pinnick, *Opt. Lett.*, **14**, p. 214, 1989.

113. H. Latifi, A. Biswas, R. L. Armstrong, and R. G. Pinnick, *Appl. Opt.*, **29**, p. 5382, 1990.

114. A. J. Campillo, J. D. Aversale, H. B. Lin, *Phys. Rev. Lett.*, **67**, p. 437, 1991.

115. H. Taniguchi, H. Tomisawa, *Opt. Lett.*, **18**, pp. 1403–1405, 1994.

116. R. E. Benner, P. W. Barber, J. F. Owen, and R. K. Chang, *Phys. Rev. Lett.*, **44**, pp. 475–478, 1980.

117. M. K. Gonokami, K. Takeda, H. Yasuda, and K. Ema, *Jpn. J. Phys.*, **32**, p. L99, 1992.

118. H. Taniguchi, H. Yamada, T. Fujiwara, S. Tanosaki, H. Ito, H. Morozumi, and M. Baba, *Jpn. J. Appl. Phys.*, **32**, p. L58, 1993.

119. H. Taniguchi, S. Tanosaki, H. Yamada, T. Fujiwara, and M. Baba, *J. Appl. Phys.*, **73**, p. 7957, 1993.

120. H. Taniguchi and H. Tomisawa, *J. Appl. Phys.*, **78**, p. 6864, 1995.

121. S. Tanosaki, H. Taniguchi, K. Tsujita, and H. Inaba, *Appl. Phys. Lett.*, **69**, p. 719, 1996.

122. H. Taniguchi, S. Tanosaki, K. Tsujita, and H. Inaba, *IEEE J. Quantum Electronics*, **QE32**, p. 1864, 1996.

123. S. Shibata, M. Yamane, K. Kamada, K. Ohta, K. Sasaki, and H. Masuhara, *J. Sol Gel Sci. Technol.*, **8**, p. 959, 1997.

124. H. Taniguchi and S. Tanosaki, *Optical Quantum Electronics*, **26**, p. 1003, 1994.

125. B. Wilhelmi, *Exp. Techn. Phys.*, **42**, pp. 117–125, 1996.

126. B. Wilhelmi, *Microwave Optical Technol. Lett.*, **17**, pp. 111–117, 1998.

127. E. P. Ippen, C. V. Shank, and A. Dienes, *IEEE J. Quant. Electr.*, **QE7**, p. 178, 1971.

128. H. P. Weber, and R. Ulrich, *Appl. Phys. Lett.*, **19**, p. 38, 1971.

129. E. P. Ippen, and C. V. Shank, *Appl. Phys. Lett.*, **21**, p. 301, 1972.

130. H. Kogelnik and C. V. Shank, *Appl. Phys. Lett.*, **18**, p. 152, 1971.

131. H. Kogelnik and C. V. Shank, *J. Appl. Phys.*, **43**, p. 2327, 1971.

132. A. N. Rubinov, T. Sh. Efendiev, and S. A. Ryzhechkin, *J. Appl. Spectroscopy*, **67**, pp. 812–817, 2000.

133. A. N. Rubinov, Ya. A. Rubinov, V. M. Katarkevich, and T. Sh. Efendiev, *J. Appl. Spectroscopy*, **67**, pp. 973–976, 2000.

134. P. P. Yaney, D. A. V. Kliner, P. E. Schrader, and R. L. Farrow, *Rev. Scientific Instruments*, **71**, pp. 1296–1305, 2000.

135. A. Müller, *Appl. Phys.*, **B63**, p. 443, 1996.

136. N. M. Lawandy, R. M. Balachandran, A. S. L. Gomes, and E. Sauvain, *Nature (London)*, **368**, pp. 436–438, 1994.

137. A. Z. Genack and J. M. Drake, *Nature (London)*, **368**, p. 400, 1994.

138. W. Zhang, N. Cue, and K. M. Yoo, *Opt. Lett.*, **20**, pp. 1023–1025, 1995.

139. N. A. Noginov,, H. J. Caulfield, N. E. Noginov, and P. Venkatesvarlu, *Opt. Commun.*, **118**, pp. 430–437, 1995.

140. R. M. Balachandran and N. M. Lawandy, *Opt. Lett.*, **20**, pp. 1271–1273, 1995.

141. N. M. Lawandy and R. M. Balachandran, *Nature*, **373**, p. 204, 1995.

142. H. Taniguchi, M. Nishiya, S. Tanosaki, H. Inaba, *Opt. Lett.*, **21**, p. 263, 1996.

143. J. Herrmann, and B. Wilhelmi, "Randomly-distributed-feedback lasers with dispersed scattering nano particles" in P. Tomanec, Ed., *Photonics '95*, Prague 1995, Vol. 2B, pp. 476–477.

144. J. Herrmann, and B. Wilhelmi, *Appl. Phys.*, **B66**, pp. 305–312, 1998.

145. N. M. Lawandy, *Photonics Spectra*, pp. 119–124, July 1994.

146. M. Siddique, L. Yang, Q. Z. Wang, and R. R. Alfano, *Optics Commun.*, **117**, pp. 475–479, 1995.

147. D. Wiersma, and St. Cavalieri, *Nature*, **414**, pp. 708–709, 2001.

148. B. W. J. McNeil, G. R. M. Robb, R. Bonifacio, and N. Piovella, *Europhys. Lett.*, **49**, pp. 316–321, 2000.

149. E. Yablonovich, *Phys. Rev. Lett.*, **58**, pp. 2059–2062, 1987.

150. S. John, *Phys. Rev. Lett.*, **58**, pp. 2486–2489, 1987.

151. J. D. Joannopoulos, R. Meade, and J. Winn, *Photonic Crystals*, Princeton Press, Princeton, NJ, 1995.

152. J. D. Joannopoulos, P. R. Villeneuve, and S. Fan, *Nature (London)*, **386**, pp. 143–149, 1997.

153. S. V. Frolov, Z. V. Vardeny, A. A. Zakhidov, and R. H. Baughman, *Optics Commun.*, **162**, pp. 241–246, 1999.

154. F. P. Schäfer and C. Ringwelski, *Z. Naturforsch.*, **A28**, p. 792, 1973.

155. J. B. Marling, D. W. Gregg, and C. Wood, *Appl. Phys. Lett.*, **17**, p. 527, 1970.

156. R. Pappalarda, H. Samelson, and A. Lempecki, *IEEE J. Quantum Electron.*, **QE-6**, p. 716, 1970.

157. F. N. Baltakov, B. A. Barikhin. V. G. Kornilov, A. N. Rubinov, S. A. Mikhniv, and L. V. Sukhanov, *Sov. J. Tech. Phys. (Engl. Transl.)*, **17**, p. 1161, 1973; *Sov. J. Quantum Electron., (Engl. Transl.)*, **5**, p. 456, 1974.

158. F. N. Baltakov, B. A. Barikhin, and L. V. Sukhanov, *JETP Lett. (Engl. Transl.)*, **19**, p. 174, 1974.

159. C. K. Miller. J. W. Lavasek, and E. D. Jones, *Appl. Opt.*, **21**, p. 1764, 1982.

160. J. Jethwa, S. St. Anufrik, and F. Docchio, *Appl. Opt.*, **21**, p. 2778. 1982.

161. P. Mazzinghi, V. Rivano, and P. Burlamacchi, *Appl. Opt.*, **22**, p. 3335, 1983.

162. J. Fort and C. Moulin, *Appl. Opt.*, **26**, p. 1246, 1987.

163. W. W. Morey and W. H. Glenn, *Opt. Acta*, **23**, p. 873. 1976.

164. P. A. Routledge, A. J. Berry, and T. A. King, *Opt. Acta*, **33**, p. 445, 1986.

165. S. E. Neister, *Proc. SPIE—Int. Soc. Opt. Eng.*, **335**, p. 36, 1982.

166. F. J. Duarte and R. W. Courad, *Appl. Opt.*, **25**, p. 663, 1986.

167. E. J. Schimitschek, R. B. Nehrich, and J. A. Trias, *Appl. Phys. Lett.*, **9**, p. 103, 1966.

168. H. Samelson, C. Brecher, and A. Lempicki, *J. Chem. Phys.*, **64**, p. 165, 1967.

169. D. L. Ross, J. Blanc, and R. J. Pressley, *Appl. Phys. Lett.*, **8**, p. 101, 1966.

170. E. J. Schimitschek, R. B. Nehrich. and J. A. Trias, *J. Chem. Phys.*, **64**, p. 173, 1967.

171. E. M. Goryaeva, A. W. Schablya, and A. P. Serov, *Zh. Prikl. Spektrosk.*, **28**, p. 75, 1978.

172. H. Brinkschulte, E. Fill, and R. Lang, *J. Appl. Phys.*, **43**, p. 1807, 1972.

173. D. Andreou and V. I. Little. *J. Phys.*, **D6**, p. 390, 1973.

174. D. Andreou, V. I. Little, A. C. Seiden, and J. Katzenstein, *J. Phys.*, **D5**, p. 59, 1972.

175. H. Brinkschulte, J. Perschermeier, and E. Schimitschek, *J. Phys.*, **D7**, p. 1361, 1974.

176. H. Samelson and A. Lempicki, *J. Appl. Phys.*, **39**, p. 6115, 1968.

177. E. E. Fill, *J. Appl. Phys.*, **11**, p. 4749, 1970.

178. M. Green, D. Andreou, V. I. Little, and A. C. Selden, *J. Phys.*, **D9**, p. 701, 1976.

179. V. S. Antonov and K. L. Hohla, *Appl. Phys.*, **B32**, p. 9, 1983.

180. R. S. Taylor and S. Mihailov, *Appl. Phys.*, **B38**, p. 131, 1985.

181. K. L. Hohla and R. Veherenkamp, *Laser Electro-Opt.*, **14**, p. 31, 1982.

182. M. G. Raymer, J. Mostowski, and J. L. Carlstein, *Phys. Rev.*, **A 19**, p. 2304, 1979.

183. S. Lavi, M. Amit, E. Miron and L. A. Levin, *Appl. Opt.*, **24**, p. 1905, 1985.

184. S. Lavi, L. A. Levin, J. Liran, and E. Miron, *Appl. Opt.*, **18**, p. 525, 1979.

185. C. K. Ni and A. H. Kung, *Review Scientific Instruments*, **71**, p. 3309, 2000.

186. D. Roess, *J. Appl. Phys.*, **37**, p. 2004, 1966.

187. G. W. Liesegang, *Appl. Opt.*, **16**, p. 2405, 1983.

188. P. H. Chiu, S. Hsu, S. J. C. Box, and H. S. Kwok, *IEEE J. Quantum Electron.*, **QE-20**, p. 652, 1984.

189. A. J. Cox, C. D. Merrit, and G. W. Scott, *Appl. Phys. Lett.*, **40**, p. 664, 1982.

190. A. Eramian, P. Dezauzier, and O. DeWitte, *Opt. Commun.*, **7**, p. 150, 1973.

191. F. P. Schäfer, *Laser Optoelektron.*, **16**, p. 95, 1984.

192. F. P. Schäfer, L. Wenchong, and S. Szatmari, *Appl. Phys.*, **B32**, p. 123, 1983.

193. Zs. Bor, *IEEE J. Quantum Electron.*, **QE-16**, p. 517, 1980.

194. I. N. Duling and M. G. Raymer, *IEEE J. Quantum Electron.*, **QE-20**, p. 1202, 1984.

195. K. E. Süsse and F. Weidner, *Appl. Phys.*, **B37**, p. 99, 1985.

196. Zs. Bor, B. Racz. G. Szabo, and A. Müller, "The Pulse Duration of a DFDL under Single Pulse Conditions," in K. B. Eisenthal, R. M. Hochstrasser, W. Kaiser, and A. Laubereau, Eds., *Picosecond Phenomena III*, Springer-Verlag, Berlin, 1982, p. 62.

197. G. Szabo and Zs. Bor, *Appl. Phys.*, **B31**, p. 1, 1983.

198. G. Szabo, B. Racz, A Müller, R. Nikolaus, and Zs. Bor, *Appl. Phys.*, **B34**, p. 145, 1984.

199. S. Szatmàri and B. Racz, *Appl Phys.*, **B43**, p. 93, 1987; *Appl. Phys.*, **B43**, p. 173, 1987.

200. T. Schmidt-Uhlig, S. Szatmàri, G. Marowsky, and P. Simon, *Appl. Phys.*, **B68**, p. 61, 1999.

201. R. Khare, S. R. Daulatabad, K. K. Sharangpani, and R. Bhatnagar, *Optics Commun.*, **153**, p. 68, 1998.

202. S. Szatmari and F. P. Schäfer, *Appl. Phys.*, **B33**, p. 95, 1984.

203. Zs. Bor and B. Racz, *Appl. Opt.*, **24**, p. 1910, 1985.

204. S. Szatmari, B. Racz. and F. P. Schäfer. Proc. Conf. Lasers Electro-Opt. (CLEO), Baltimore, April 1987, pap. ThB 1.

205. M. Zitelli, E. Fazio, and M. Bertolotti, *IEEE J Quantum Electronics*, **QE34**, p. 609, 1998.

206. E. S. Lee and J. W. Hahn, *Optics Lett.*, **21**, p. 1836, 1996.

207. M. Wittmann, A. Penzkofer, G. Gössl, *Appl. Opt.*, **34**, p. 5287, 1995.

208. LTB Lasertechnik Berlin GmbH, http://www.ltb-berlin.de, 2001.

209. Lambda Physik AG, *http://www.lambdaphysik.com*, 2001.

210. Spectra-Physics, *http://www.spectra-physics.com*, 2001.

211. S. A. Tuccio and F. C. Strome, Jr., *Appl. Opt.*, **11**, p. 64, 1972.

212. P. K. Runge and R. Rosenberg, *IEEE J. Quantum Electron.*, **QE-8**, p. 910, 1972.

213. H. Kogelnik, E. P. Ippen, A. Dienes, and C. V. Shank, *IEEE J. Quantum Electron.*, **QE-8**, p. 373, 1972.

214. A. Dienes,. E. P. Ippen, and C. V. Shank. *IEEE J. Quantum Electron.*, **QE-8**, p. 388, 1972.

215. W. Demtröder, *Laser Spectroscopy*, Springer-Verlag, Berlin, 1981, p. 346.

216. H. Gerhardt and A. Timmermann, *Opt. Commun.*, **21**, p. 343, 1970.

217. G. Meisel, *Laser Optoelektron.*, **15**, p. 245, 1983.

218. A. I. Ferguson, M. H. Dunn, and A. Maitland, *Opt. Commun.*, **19**, p. 10, 1976.

219. A. I. Ferguson and M. H. Dunn, *Opt. Commun.*, **23**, p. 177, 1977.

220. S. Bastow and M. H. Dunn, *Opt. Commun.*, **35**, p. 259, 1980.

221. A. Renn, A. Hese, and H. Büsener, *Laser Optoelektron.*, **14**, p. 11, 1982.

222. E. P. Ippen and C. V. Shank, *Appl. Phys. Lett.*, **27**, p. 488, 1975.

223. R. L. Fork, B. I. Greene, and C. V. Shank, *Appl. Phys., Lett.*, **38**, p. 197, 1981.

224. W. Dietel, J. Fontaine, and J. C. Diels, *Opt. Lett.*, **8**, p. 4, 1983.

225. J. A. Valdmanis and R. L. Fork, *J. Opt. Soc. Am.*, **A1**, p. 1337, 1984.

226. B. Wilhelmi, W. Rudolph, E. Döpel, and W. Dietel, *Opt. Acta*, **32**, p. 1175, 1985.

227. C. V. Shank, W. Knox, R. Fork, M. Downer, and R. Stolen, *Appl. Phys. Lett.*, **46**, p. 1120, 1985.

228. C. H. Brito-Cruz, R. L. Fork, and C. V. Shank, Proc. Conf. Lasers Electro-Opt. (CLEO), Baltimore, April 1987, pap. MD 1.

229. D. E. Spence, P. N. Kean, and W. Sibbett, *Opt. Lett.*, **16**, p. 42, 1991.

230. U. Keller, "Semiconductor nonlinearities for modelocking and Q-switching," in *Semiconductors and Semimetals*, Vol. 59, Academic Press, New York, 1999, pp. 211–285.

231. Y. F. Chou, C. H. Lee, and J. Wang, *Opt. Lett.*, **19**, p. 975, 1994.

232. B. Wilhelmi, Ed., *Scientific Instrumentation at the Physics Department of Jena University*, Friedrich-Schiller-Universität, Jena, 1986, p. 7.

233. D. J. Bradley and A. J. F. Durrant, *Phys. Lett.*, **A27**, p. 73, 1968.

234. C. K. Chan and S. O. Sari, *Appl. Phys. Lett.*, **25**, p. 403, 1974.

235. J. Heritage and R. Jain, *Appl. Phys. Lett.*, **32**, p. 101, 1978.

236. U. Stamm, F. Weidner, and B. Wilhelmi, *Opt. Commun.*, **63**, p. 179, 1987.

237. G. A. Mourou and T. Sizer, *Opt Commun.*, **41**, p. 47, 1982.

238. S. R. Rotman, C. B. Roxlo, O. Bebelaar, T. K. Yee, and M. M. Salour, in R. M. Hochstrasser, W. Kaiser, and C. V. Shank, Eds., *Picosecond Phenomena II*, Springer-Verlag, Berlin, 1980, p. 50.

239. W. Schmidt and F. P. Schäfer, *Phys. Lett.*, **A26**, p. 558, 1968.

240. E. P. Ippen, C. V. Shank, and D. Dienes, *Appl. Phys. Lett.*, **21**, p. 348, 1972.

241. F. O'Neill, *Opt. Commun.*, **6**, p. 360, 1972.

242. J. C. Diels, W. Dietel, J. J. Fonlaine, W. Rudolph, and B. Wilhelmi, *J. Opt. Soc. Am. B: Opt. Phys.*, **2**, p. 680, 1985.

243. D. J. Bradley, in S. L. Shapiro, Ed., *Ultrashort Light Pulses*, Springer-Verlag, Berlin, 1977, p. 57.

244. P. M. W. French, M. D. Dawson, and J. R. Taylor, *Opt. Commun.*, **56**, p. 430, 1986.

245. P. M. W. French and J. R. Taylor, *Opt. Commun.*, **58**, p. 53, 1986.

246. K. Smith, N. Langford, W. Sibbett, and J. R. Taylor, *Opt. Acta*, **33**, p. 453, 1986.

247. W. Dietel, E. Döpel, D. Kühlke, and B. Wilhelmi, *Opt. Commun.*, **43**, p. 433, 1982.

248. B. Wilhelmi, "Propagation of femtosecond light pulses throug dye amplifiers," D. L. Andrews, Ed., *Lasers in Chemistry*, Springer, Berlin, 1997, pp. 111–127.

249. S. Szatmàri, "Terawatt-class hybrid dye/excimer lasers," M. Stuke, Ed., *Dye Lasers: 25 Years*, Springer, Berlin, 1992, pp. 129–140.

250. R. L. Fork, C. V. Shank, and R. T. Yen, *Appl. Phys. Lett.*, **41**, p. 223, 1982.

251. W. H. Knox, M. C. Downer, R. L. Fork, and C. V. Shank, *Opt. Lett.*, Vol **9**, p. 552, 1984.

252. P. B. Corkhum and R. S. Taylor, *IEEE J. Quantum Electron.*, **QE-18**, p. 1962, 1982.

253. J. P. Ryan, L. S. Goldberg, and D. J. Bradley, *Opt. Commun.*, **27**, p. 127, 1978.

254. G. W. Fehrenbach, K. J. Gruntz, and R. G. Ulbricht, *Appl. Phys. Lett.*, **33**, p. 159, 1978.

255. Y. Ishida, T. Yayima, and R. Naganuma, *Jpn. J. Appl. Phys.*, **19**, p. 717, 1980.

256. P. Stchur, K. X. Yang, X. Hou, T. Sun, and R. G. Michel, *Spectrochimica Acta*, **B56**, p. 1565, 2001.

257. C. B. Ke, K. D. Su, and K. C. Lin, *J. Chromatogr.*, **A921**, p. 247, 2001.

258. M. F. Lagutin, A. A. Zaprudnyi, V. L. Basetskii, and V. G. Pletnev, *Telecommun. Radio Eng.*, **52**, p. 100, 1998.

259. V. M. Baev, T. Latz, and P. E. Toschek, *Appl. Phys.*, **B69**, p. 171, 1999.

260. V. P. Zharov and V. S. Letokhov, *Laser Photoacoustic Spectroscopy*, Springer, Berlin, 1986.

261. O. Wolbeis, Ed., *Fluorescence Spectroscopy*, Springer, Berlin, 1992.

262. F. Güttler, T. Irmgartinger, T. Plakhovnik, A. Renn, and U. P. Wild, *Chem. Phys. Lett.*, **217**, p. 393, 1994.

263. Th. Schmidt, G. J. Schütz, W. Baumgartner, H. J. Gruber, and H. Schindler, *Proc. Nat. Acad. Sci. USA*, **93**, pp. 2926–2929, 1996.

264. D. Bates and B. Benderson, Eds., *Advances in Atomic, Molecular and Optical Physics*, Vol. 31, Academic Press, San Diego, 1993.

265. R. Rigler and J. Windengren, in O. Wolbeis, Ed., *Fluorescence Spectroscopy*, Springer, Berlin, 1992.

266. Th. Schmidt, G. J. Schütz, W. Baumgartner, H. J. Gruber, and H. Schindler, *Proc. Nat. Acad. Sci. USA*, **93**, pp. 2926–2929, 1996.

267. U. Köpf, *Laser in der Chemie (in German)*, Salle Verlag und Sauerländer Verlag, Frankfurt, 1979.

268. S. A. Kostritsa and V. A. Mishin, *Quantum Electronics*, **24**, p. 464, 1994.

269. I. L. Bass, R. E. Bonnano, P. P. Hackel, and P. R. Hammond, *Appl. Optics*, **31**, pp. 6993–7006, 1992.

270. P. P. Hackel and B. E. Warner, in J. A. Paisner, Ed., *Laser Isotope Separation, Proc. Soc. Photo-Opt. Instr. Eng.*, pp. 120–129, 1993.

271. K. Takehisa, *Appl. Optics*, **33**, p. 6360, 1994.

272. A. J. Welch and M. J. C. Van Germert, *Optical-Thermal Response of Laser-Irradiated Tissue*, Plenum Press, New York, 1995.

273. M. H. Niemz, *Laser-Tissue Interactions: Fundamentals and Applications*, Springer Berlin, Heidelberg, New York, 1996.

274. J. L. Boulnois, *Lasers Med. Sci.*, **1**, p. 47, 1986.

275. R. A. Waters, R. M. Clement, M. N. Kiernan, and G. J. Griffiths, Proceedings of medical applications of lasers (Budapest 1993), SPIE proceedings series 2086, p. 218, 1994.

276. P. Paquet, J.-F. Hermanns, G. E. Piérard, and K. Nouri, *Dermatologic Surgery*, **27**, p. 171, 2001.

277. H. Maier, *Acta Dermatovenerologica Croatica*, **9**, p. 83, 2001.

278. C. A. J. M. De Borgie, P. M. M. Bossuyt, C. M. A. M Van der Horst, and M. J. C. Van Germert, *Lasers in Surgery and Medicine*, **28**, p. 182, 2001.

279. T. S. Alster and J. R. Lupton, *Am. J. Clin. Dermatol.*, **2**, pp. 291–303, 2001.

280. M. Haersdal, *Acta Dermato-Venereologica*, Suppl. 207, p. 5, 1999.

281. A. Casas, H. Fukuda, and A. Batlle, *Proc. SPIE*, **3909**, pp. 114–123, 2000.

282. T. Zoepf, R. Jakobs, J. C. Arnold, D. Apel, A. Rosenbaum, and J. F. Riemann, *Am. J. Gastroenterol.*, **96**, p. 2093, 2001.

283. M. R. Beintema, J. A. Oosterhuis, and F. Hendrikse, *Brit. J. Ophthalmol.*, **85**, p. 708, 2001.

284. M. Schmidt-Erfurth, J. Miller, and M. Sickenberg, *Graefe's Arch. Clin. Exp. Ophthalmol.*, **236**, pp. 365–374, 1988.

285. B. Jurklies, A. Hintzmann, and N. Bornfeld, *Innovation, Carl Zeiss Oberkochen and Carl Zeiss Jena GmbH*, *Jena*, Vol. 1999/1, pp. 26–31, 1999.

# 9

# GAS LASERS

T. Lehecka, N. C. Luhmann, Jr., and W. A. Peebles
*University of California*
*Los Angeles, California*

J. Goldhar
*University of Maryland*
*College Park, Maryland*

S. P. Obenschain
*Laser Plasma Branch*
*Naval Research Laboratory*
*Washington, DC*

## 9.1 INTRODUCTION

Light amplification by stimulated emission of radiation has been a practical reality for over 25 years, while the microwave counterpart or, masers, were demonstrated six years earlier. During this time, research regarding this method of producing intense, monochromatic, coherent radiation has headed in two main directions: extending the frequency spectrum covered by laser sources, and applying laser technology to commercial and scientific techniques. To date, lasing action has been observed from the x-ray region through microwave frequencies, with applications in manufacturing, medicine, communications and basic research.

In this chapter we look at one of the most common and widely used type of laser, the gas laser. The advantages that lead to the gas lasers usefulness are discussed along with the basic theory of laser operation and the types and applications of gas lasers available today. Emphasis is placed on the longer-wavelength submillimeter lasers and ultraviolet (UV) excimer lasers, while also providing more limited information on other optical, infrared, and UV lasers. The latter choice has been made since considerable information on common lasers such as He–Ne, argon ion, and $CO_2$ can be found in the many excellent laser textbooks available as well as the references contained

*Handbook of Optical Components and Engineering*, Edited by Kai Chang
ISBN 0-471-39055-0 © 2003 John Wiley & Sons, Inc.

therein [1–4]. We begin with a brief history of laser development, followed by a discussion of basic laser theory. Section 9.1 concludes with a brief description of some of the more common gas laser systems. A more thorough review of far-infrared lasers is presented in Section 9.2, while Section 9.3 provides a detailed description of excimer lasers.

### 9.1.1   Laser History

The principles underlying maser action are grounded in quantum mechanics, involving the discrete energy levels of atoms, ions, and molecules. The possibility of amplifying devices was postulated about 1950, with the first operational maser demonstrated in 1954 [5]. This maser utilized two energy levels in the tetrahedrally structured ammonia molecule. The population inversion, a requirement for amplification, was achieved by passing a beam of ammonia molecules through a quadrupole field, which focuses the higher-energy molecules while diverging the molecules in the lower state. The focused molecules are passed through a high-$Q$ microwave cavity where oscillations at the cavity's resonant frequency of 23,870 MHz are amplified by stimulated emission in the ammonia molecules.

While this excitation, or pumping, technique is not useful for the majority of laser materials, this first maser demonstrated the possibility of obtaining amplification through the interaction of electromagnetic fields and a gain medium. In simplest terms, the two components needed for lasing action are a medium with optical gain, and a positive feedback system to support and couple the oscillations. At optical frequencies, cavities such as those employed in the ammonia maser cannot be constructed, but in the late 1950s it was proposed [6] that two optical reflectors or mirrors could be used to form an optical cavity. Using this principle, lasing action was first demonstrated in 1960 [7] with a ruby crystal providing the amplifying medium at a wavelength of 6943 Å. Later in that same year, lasing was achieved with a mixture of helium and neon gases with $\lambda = 1.15$ μm [8]. In both cases, and most subsequent lasers, pumping to inverted states was achieved by distributing the atoms to three or more energy levels rather than separating two energy levels in space as in the ammonia maser.

In 1963, strong lasing action was observed in the molecular gas carbon dioxide [9] with $\lambda = 10.6$ μm. The $CO_2$ laser is still one of the most efficient and widely used lasers. Since that time, thousands of other lasing wavelengths in various media have been discovered. These include the gas lasers, such as noble-gas ion lasers, excimer lasers, and metal vapor lasers. Throughout the remainder of this chapter, as is common in current usage, devices that achieve amplification by stimulated emission of radiation will be termed lasers whether the output wavelength is in the microwave, infrared, visible, ultraviolet, or x-ray region of the spectrum.

Before going on to look at specific types and applications of gas lasers, we briefly examine the basic theory behind laser operation.

### 9.1.2   Basic Laser Theory

As mentioned earlier, two components are required for laser operation, a cavity to confine the radiation and a medium that can provide amplification of this radiation via stimulated emission. Several laser characteristics, including directionality and spatial uniformity, are determined by the cavity properties, while attributes such as gain and output power are governed by the amplifying media.

It is the intent of this section to provide a brief review of the basic principles of laser operation. This will include a discussion of laser rate equations, a semiclassical description of the processes involved in the amplifying medium. Rate-equation models are quite adequate in predicting steady-state laser operation and can also provide much insight into transient phenomena. More complete quantum mechanical descriptions, particularly the density matrix method of laser theory, are available in several excellent books and articles [1, 10, 11] and are briefly discussed in Section 9.2. Such methods provide more accurate predictions of laser transient behavior, but are generally beyond the scope needed or intended for this review chapter.

***Rate Equations.*** It is well known that atoms, ions, and molecules can emit radiation when a transition is made from a higher-energy state to a lower one. When this process occurs due to "natural" causes (i.e., radiative decay, collisions, etc.), it is called spontaneous emission. Spectroscopists have used this principle since the early twentieth century to study properties of materials and the nature of the atom. Different particles will emit (or absorb) radiation at different characteristic wavelengths, and by studying this emission (absorption) the internal energy levels of the atom, ion, or molecule can be determined.

It was Einstein who first realized that the processes of absorption and spontaneous emission were not adequate to explain a particle's steady-state condition in an electromagnetic field [12]. For a steady-state condition to occur, the number of particles excited to higher energy levels by the absorption of electromagnetic radiation must be balanced by deexcitation by some type of emission. However, absorption is dependent on material properties and the intensity of the radiation, while the spontaneous emission rate is determined only by the internal properties of the material. To reach equilibrium, Einstein postulated that some type of emission that is dependent on the intensity of the radiation must exist. This is the stimulated emission that has made laser operation possible.

Let us consider qualitatively the stimulated and spontaneous processes. As an example, we take a collection of atoms with a simplified two-energy-level system as shown in Fig. 9.1. Under normal thermodynamic equilibrium, the populations of the two levels will be governed by the Boltzmann relation,

$$N_2/N_1 = e^{-(E_2-E_1)/kT} = e^{-h\nu/kT}$$

where $N_1 (N_2)$ is the population of level 1 (2), $E_1$ $(E_2)$ the energy of level 1 (2), $k = 1.38 \times 10^{-23}$ J/K is Boltzmann's constant, $T$ is the temperature in kelvin, and $h\nu$ is the energy associated with a photon of frequency $\nu$ $(h = 6.63 \times 10^{-34}$ J $\cdot$ s $=$ Planck's constant) equivalent to the energy difference $E_2 - E_1$. Under these conditions, the population of level 2 is smaller than that of the ground state, level 1. When a photon

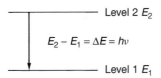

**Figure 9.1** Simplified two-energy-level system.

of frequency $\nu$ (actually, a band within the range $\nu \pm \Delta\nu$, which will be discussed later) is incident on this material, it may either be absorbed or stimulate emission (along with other processes not of interest here). The probability for these processes depends on the internal properties of the atoms, the intensity of the radiation (number of photons), and the population difference between the two energy levels. When the majority of atoms are in the ground state ($h\nu/kT \gg 1$), the probability of absorption is much greater than that for stimulated emission. As $h\nu/kT$ decreases, the probability of absorption decreases while that for stimulated emission grows, until at the limit $T \to \infty$ the two processes are equally likely and the populations in the two levels are equal. Now if it were possible in some way to create a situation where $N_2 > N_1$ or $h\nu/kT < 0$ (implying that $T < 0$), the probability of stimulated emission becomes greater than that for absorption and amplification of the incident radiation may result. This condition, $N_2 > N_1$, is known as a population inversion, and while the concept of a negative absolute temperature may seem unrealizable, the situation has been discussed in several references [4, 13–15] and is frequently used to discuss laser operation.

From this brief qualitative discussion, we have learned that the requirement for amplification in a laser is the creation of a population inversion. A semiquantitative description considering only radiative processes can be made with the help of Fig. 9.2. A photon flux of intensity $I$ (of the proper frequency) incident from the left on a slab of matter with thickness $dz$ will undergo both absorption and stimulated emission whose amplitudes depend on $\sigma$ (the cross section for reaction determined by internal properties), $I$ (the intensity), and $N_1$ or $N_2$. Also, there will be spontaneous emission at a rate determined by $A_{21}$ and $N_2$. Here, $dI$ is equal to the gain due to emission minus the loss due to absorption and can be written

$$dI = \sigma I N_2 \, dz - \sigma I N_1 \, dz + A_{21} N_2 \, dz$$

If we assume that the spontaneous emission is sufficiently small, we have

$$dI/dz = \sigma I (N_2 - N_1)$$

which is positive when $N_2 - N_1$ is positive. Thus, a population inversion is required for amplification.

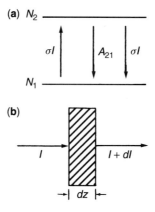

**Figure 9.2**  Amplification in a lasing medium.

Let us now see what type of system is required to obtain this inversion. The simplest is the two energy level system we have seen discussing. It is easily shown that it is not possible to obtain a population inversion for this case, under the conditions assumed, so let us now do that to gain an understanding of the rate-equation model. Utilizing Fig. 9.2a with $\sigma I = W$, we can write

$$dN_1/dt = -WN_1 + AN_2 + WN_2$$

and

$$dN_2/dt = WN_1 - AN_2 - WN_2 = -dN_1/dt$$

or after combining equations

$$d\Delta N/dt = -2W(N_2 - N_1) - 2AN_2 = W\Delta N - 2AN_2$$

This is a simple linear first-order differential equation with boundary conditions

$$N_2(0) = 0$$

$$N_1(0) = N_0$$

and

$$N_1(t) + N_2(t) = N_0 \quad \text{(total number of atoms is constant)}$$

whose solution is

$$\frac{\Delta N(t)}{N_0} = -\frac{1 + (2W/A)e^{-(2W+A)t}}{1 + 2W/A}$$

From the above, we see that $\Delta N = N_2 - N_1$ is always negative. Therefore, we cannot obtain a population inversion in this system, and any signal is attenuated as it propagates through the media.

The next system to consider is the three-level atom (note that while the term "atom" is used in these discussions, the arguments hold for ions and molecules as well). The levels and transitions for this system are shown in Fig. 9.3. Now we have involved the nonradiative rates $S_{32}$ and $S_{21}$, as well as absorption (or pumping) and stimulated emission (lasing) terms for both the 1–3 and 1–2 transitions. For the gas lasers of interest in this chapter, the major contribution to the nonradiative rates will be collisions between two particles or particles and the container walls. The $A_{mn}$'s are the radiative

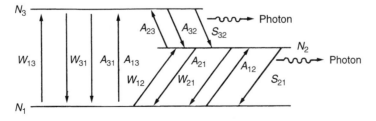

**Figure 9.3** Three-energy-level system.

transition probabilities, or the Einstein $A$ coefficients. These are constants that can be measured spectroscopically and will be discussed later when we consider individual laser systems.

If we define $\tau_{rm}$ as the radiative lifetime, $\tau_{nm}$ as the nonradiative lifetime, and $\tau_m$ the fluorescence lifetime for the $m$th level, we can write $S_m = 1/\tau_{nm}$, $A_m = 1/\tau_{rm}$ and

$$H_m = A_m + S_m = 1/\tau_m$$

Also it is known that $W_{nm} = -W_{mn}$ and the rate equations for this system are

$$dN_1/dt = -W_{13}(N_1 - N_3) - W_{12}(N_1 - N_2) - A_{13}N_1 - A_{12}N_1 + A_{31}N_3 + H_{21}N_2$$

$$dN_2/dt = -W_{12}(N_2 - N_1) + H_{32}N_3 - A_{23}N_2 - H_{21}N_2 + A_{12}N_1$$

$$dN_3/dt = -W_{13}(N_3 - N_1) - A_{31}N_3 + A_{13}N_1 - H_{32}N_3 + A_{23}N_2$$

These equations, however, are not independent, because we must satisfy the condition

$$N_1 + N_2 + N_3 = N_0 = \text{constant}$$

or

$$\frac{dN_1}{dt} + \frac{dN_2}{dt} + \frac{dN_3}{dt} = 0$$

Using two of the differential equations and this relationship allows the equations to be solved for $N_1$, $N_2$, and $N_3$ in terms of $N_0$ and the rates $A_{mn}$ and $W_{mn}$.

By making a number of assumptions that are justifiable in good laser media (namely, $S_{32} \gg A_{32}, A_{31}$; $W_{12} \ll A_{21}, S_{21}$; and $W_{13}, \gg A_{13}, A_{31}$), we can simplify these equations to

$$dN_1/dt = -W_{13}N_1 + H_{21}N_2 - W_{13}N_3$$

$$dN_2/dt = -H_{21}N_2 + S_{32}N_3$$

and

$$dN_3/dt = W_{13}N_1 - (W_{31} + S_{32})N_3$$

Letting $S_{32} \to \infty$ (fast relaxation), we find that

$$N_2(t) - N_1(t) = N_0 \left[ \frac{1 - H_{21}/W_{13}}{1 + H_{21}/W_{13}} - \frac{2\exp[-(H_{21} + W_{13})t]}{1 + H_{21}/W_{13}} \right] = \Delta N(t)$$

and

$$-1 \leqslant \frac{\Delta N(t)}{N_0} \leqslant 1$$

For this three-level system, it is possible to obtain a population inversion and hence gain. It is interesting to note that this is the scheme utilized by the first operational laser, the ruby laser, with the laser amplification occurring on the 2–1 transition. Several

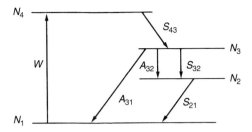

**Figure 9.4**  Simplified four-energy-level system.

of the gas lasers discussed in this chapter also involve three energy levels; however, lasing is obtained on the 3–2 transition.

The final system to be discussed is the four-level atom. A simplified diagram is shown in Fig. 9.4, where relevant assumptions regarding the dominant rates and populations have already been made.

$$\frac{dN_1}{dt} = -WN_1 + S_{21}N_2 + A_{31}N_3$$

$$\frac{dN_2}{dt} = -S_{21}N_2 + (H - A_{31})N_3$$

$$\frac{dN_3}{dt} = -HN_3 + S_{43}N_4$$

and

$$\frac{dN_4}{dt} = WN_1 + S_{43}N_4$$

with

$$H = A_{31} + A_{32} + S_{32}$$

The solution of these equations is lengthy but shows that a population inversion is obtainable with

$$-1 \leqslant \frac{\Delta N(t)}{N_0} = \frac{N_3(t) - N_2(t)}{N_0} \leqslant 1$$

To maximize the inversion on the 3–2 transition, we would like to have $S_{43} \to \infty$ and $S_{21} \to \infty$. This is called the quasi-two-level atom and its energy-level diagram is shown in Fig. 9.5. Because of the infinite rates, $N_4$ and $N_2$ will equal zero and the rate equations are

$$dN_3/dt = WN_1 - (A_{31} + A_{32})N_3$$

and

$$dN_1/dt = (A_{31} + A_{32})N_3 - WN_1$$

Solving for $\Delta N(t) = N_3(t) - N_2(t)$ gives

$$\frac{\Delta N(t)}{N_0} = \frac{W}{H + W}[1 - e^{-(H+W)t}]$$

where $H = A_{31} + A_{32}$. Thus, $0 \leqslant \Delta N(t)/N_0 \leqslant 1$.

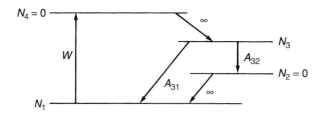

**Figure 9.5**   Quasi-two-energy-level system.

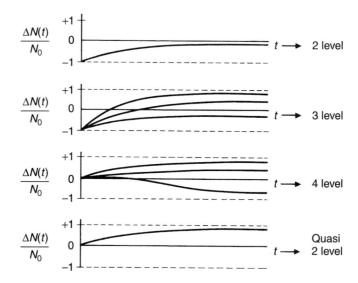

**Figure 9.6**   Energy-level population time development for various laser systems.

Qualitative plots of the variation of $\Delta N(t)/N_0$ versus time for the four atomic systems discussed are shown in Fig. 9.6. Summarizing this, no inversion can be obtained in a two-level system, while it is possible in three- or four-level systems. Four-level systems are more efficient than three-level systems (i.e., a larger inversion is obtained for equal pump powers) when the relative rates are the same, with the quasi-two-level idealization providing the most efficient operation. The rate at which the populations develop, as well as the final state, depends on the ratio of $H/W$.

While this brief section has presented only the most basic principles of laser theory, it will provide background information for the more practical sections that follow. Readers who require a more thorough understanding of the subject will find excellent treatments in a number of references [1–4]. We now restrict our attention to the subject of this chapter, lasers whose amplifying media are in a gaseous state or simply gas lasers.

### 9.1.3   Gas Laser Properties

*Advantages.* The gas laser has several properties that make it superior to liquid- or solid-state lasers for some applications. The gaseous lasing medium has an index of refraction much closer than unity than these other media; thus defects in the lasing media have less effect on the optical cavity. In most cases, the nonlinear index of gaseous medium is also very small. This small nonlinear index simplifies operation of gas lasers at very high intensities by preventing nonlinear phenomena such as filamentation. The low index of lasers allows one to employ high-velocity flows for cooling in gaseous lasers without causing optical distortion. Much higher average powers have been obtained with gaseous lasers than with solid-state or liquid lasers. There are also more avenues for pumping population inversion with gaseous lasers. In addition to the optical pumping generally used with solid-state and liquid lasers, there are gaseous lasers pumped by electrical discharges, high-energy electron beams, gas dynamics, and chemical reactions. Below we provide a review of commonly used gas lasers and follow that with a detailed examination of a few specific lasers.

### Overview of Common Gas Lasers

*He–Ne Laser.* Probably the most commonly known and widely used gas laser is the He–Ne laser. This gas laser produces the familiar small, low-power red beam of radiation used for bar code readers in supermarkets, optical disk readers for audio and video digital disks, and as an alignment tool for scientific experiments.

The He–Ne laser transitions occur in the neon atom, while the helium, which has a much simpler energy-level structure, is present to provide selective excitation. The energy-level diagrams for both He and Ne are shown in Fig. 9.7. Both the $2^3S_1$ and $2^1S_0$ levels in He are metastable and atoms can be efficiently pumped to these levels. The He $2^3S_1$ level lies within 0.04 eV of the Ne $2S_2$, and the He $2^1S_0$ level is within 0.05 eV of Ne $3S_2$. Because the neon is close to resonance conditions, selective energy transfer from the helium to neon is quite effective.

The natural lifetimes of the $2S_2$ and $3S_2$ neon levels are on the order of 10 to 20 ns due to a fast vacuum ultraviolet (VUV) transition. However, these levels are

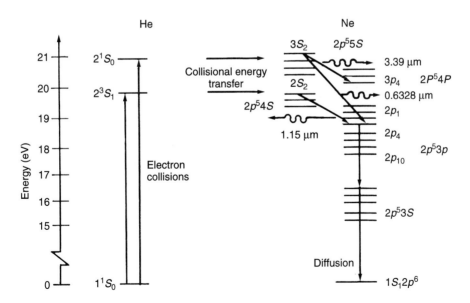

**Figure 9.7** He–Ne laser energy levels.

repopulated by radiation trapping, or reabsorption, of these VUV photons, and the effective lifetime is determined by the slower visible transitions. For the $2S_2$ level, $\tau_{\mathrm{eff}} \approx 96$ ns, while for the $3S_2$, $\tau_{\mathrm{eff}} \approx 110$ ns. The return to the ground state is a fast UV transition with $\tau_1 \approx 20$ ns. As shown in the energy-level diagram, several lasing wavelengths are available, including the visible red 6328 Å, and the near infrared 1.15-$\mu$m and 3.39-$\mu$m lines.

Because of its low efficiency ($\approx 0.1\%$) and low single-pass gain, the He–Ne laser is limited to low power operation, usually less than 10 mW. The typical arrangement for a low power ($\approx 1$ mW) He–Ne laser is shown in Fig. 9.8. The atomic pumping is provided by an electrical discharge, with the discharge current in the range 10 to 50 mA and voltages of $\approx 1$ to 5 kV. The discharge tube is $\approx 1$ to 2 mm in diameter and 10 to 30 cm in length. A ratio of 6:1 He:Ne is used with a fill pressure and tube diameter product of $\approx 3.5$ torr mm. On most inexpensive He–Ne lasers, the mirrors are glued into place on the discharge tube. Alternatively, one could use Brewster angle windows and external mirrors. A large-area cold cathode (usually of aluminum alloy) is used to limit the current density to approximately 2 mA/cm$^2$ to reduce the loss of gas due to burial of ions accelerated into the cathode. Some of the practical problems include the diffusion of He through the tube walls and accumulation of outgassed or leaked contaminants.

The He–Ne laser, as with most lasers, is finding ever-increasing application. Along with the applications already mentioned, He–Ne lasers are being used for accurate distance measuring systems [16], laser printing [17], and medical applications [18].

*Ion Lasers.* Another important category of gas laser that will be discussed only briefly here is the ion laser. These include the argon laser as well as neon, krypton, and xenon lasers. We restrict our attention to the argon ion laser. Operation of the Ne$^+$, Kr$^+$, and Xe$^+$ lasers is very similar.

The lasing medium is a gas of singly ionized Ar$^+$ ions, and laser emission is primarily in the blue-green region of the spectrum with lower-level emissions in the UV and infrared. Figure 9.9 illustrates the relative emission spectra from argon. The laser is capable of operating in the continuous-wave (CW) mode, with output powers of $\approx 20$ W available commercially. An electrical discharge in the pure argon gas is usually used as the pumping mechanism. Although the pumping is quite inefficient, a CW population inversion can be created because of the favorable ratio of upper- to lower-level lifetimes.

The energy-level diagram for the Ar$^+$ ion is shown in Fig. 9.10. Decay of the upper levels occurs only by emission of visible photons, while the lower levels relax via VUV

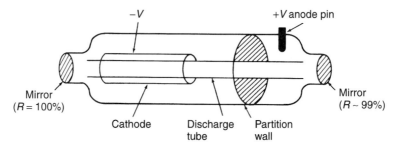

**Figure 9.8**   Simplified He–Ne laser design.

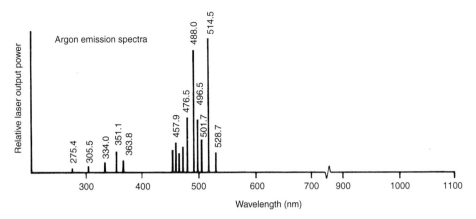

**Figure 9.9**   Relative emission spectra from argon. (Courtesy of Spectra-Physics.)

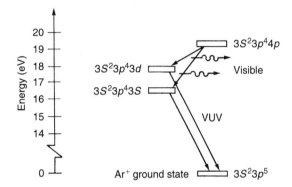

**Figure 9.10**   $Ar^+$ ion energy levels.

photon emission. Because the transition rates vary as $\tau \alpha 1/\nu^3$, the upper-level lifetimes are on the order of 10 ns while the lower-level lifetimes are $\approx 0.4$ ns. Therefore, a population inversion can build provided that the lower levels are not pumped more than $\approx 20$ times faster than the upper levels. Figure 9.11 displays the relative gain and output power for several argon lines as a function of discharge tube current and focusing magnetic field. In the visible, maximum gain typically occurs for current density $<800$ $A/cm^2$ while no saturation is observed for the UV.

The overall efficiency of the argon ion laser is poor, typically in the range 0.05 to 0.2%. This implies that as much as 80 kW of electrical power is required to produce $\approx 20$ W of laser output. The excess heat is deposited primarily on the vessel walls. Because of the large heat loads and high temperatures, refractory materials such as silica, alumina, or beryllia are usually used for the discharge tubes. Figure 9.12 contains a photograph of such a metal–ceramic plasma tube that is employed in high-power ion lasers. The interior of the tube is comprised of a number of metal webs brazed to a ceramic envelope as displayed in Fig. 9.13. The metal elements are comprised of a sputter-resistant tungsten disk that is brazed to a copper heat web (see Fig. 9.14), which in turn is brazed to the ceramic. The copper heat webs serve to transfer heat efficiently from the tungsten disks to the ceramic vacuum envelope.

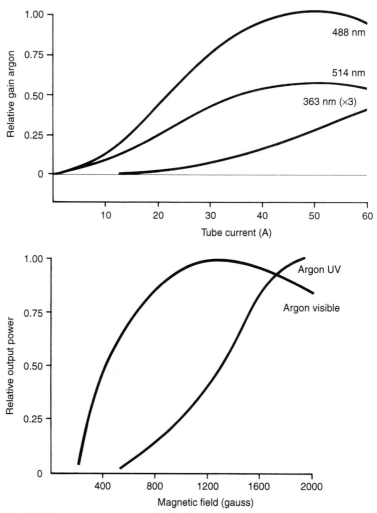

**Figure 9.11**   Relative output power and gain for argon as a function of discharge tube current and longitudinal magnetic field strength. (Courtesy of Spectra-Physics.)

**Figure 9.12**   Metal–ceramic plasma tube used in ion lasers. (Courtesy of Spectra-Physics.)

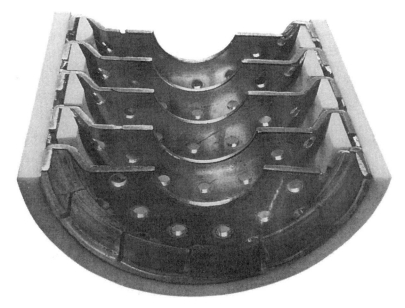

**Figure 9.13**   Photograph of interior of plasma tube. (Courtesy of Spectra-Physics.)

**Figure 9.14**   Photograph of tungsten disk welded to copper heat web. (Courtesy of Spectra-Physics.)

A typical argon laser with $\approx$10 W of output power is $\approx$1.5 m in length with a 2.5- to 4-mm-diameter discharge tube. A discharge current of $\approx$25 to 40 A is required with a voltage of $\approx$300 to 500 V. Additionally, an axial magnetic field of $\approx$ 0.1 T is applied to help confine the argon ions. The cold filling pressure is approximately 0.1 to 0.5 torr of argon gas. Water cooling at a rate of 5 to 10 L/min is also required. These discharge currents are too large for a cold cathode, so a thermionic emitting

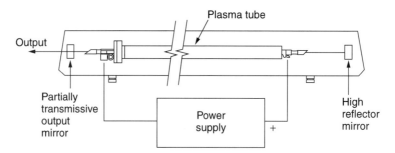

**Figure 9.15**   Basic ion laser configuration. (Courtesy of Spectra-Physics.)

cathode with a low-work-function material (e.g., BaO) coating is used. This cathode must be sealed to prevent flaking due to moisture in the air. Brewster angle windows are usually employed to allow external tuning of the laser. This tuning is accomplished with a prism system, which is tilted with respect to the longitudinal axis of the system, while the use of Brewster windows in the cavity ensures that the laser output will be plane polarized. Figure 9.15 schematically illustrates the typical ion laser configuration.

Argon ion lasers are currently being used in the fields of atomic physics and chemistry either directly for Raman spectroscopy or as a pump for a tunable dye laser [19]. Display applications, such as the now-popular laser light shows, are another area where the argon laser is useful [20]. In the medical fields, ion lasers are beginning to see widespread use for ophthalmic surgery and the treatment of glaucoma or myopil and senile subretina new vessels (the most common cause of blindness in patients over 65 years) [21].

*Cu Vapor Laser.* The copper vapor laser, first demonstrated in 1966 [22], represents a class of self-terminating metal vapor lasers. Here, the lasing transition occurs from an excited state to a low-lying metastable state. Because the lower lasing level is a true metastable state, only pulsed operation is possible, with pulse durations of $\approx$50 to 80 ns. However, the efficiency of these lasers is quite high ($\approx$1%), resulting in average power output of up to 100 W at a repetition frequency of 5 kHz [23].

An energy-level diagram for the Cu vapor laser is shown in Fig. 9.16. The upper laser levels are strongly coupled to the ground levels by fully allowed strong "resonance" transitions. The quantum efficiency, determined by the ratio of the lasing photon energy to the pump photon energy is high, in this case 328/578 and 325/510. Also, direct excitation from the $^2S_{1/2}$ level to the $^2D_{3/2}$ and $^2D_{5/2}$ levels is about a factor of 50 less than the transition to $^2P_{1/2}$ and $^2P_{3/2}$. The operating temperature of the Cu vapor laser must be maintained at about 1500°C to ensure adequate supply of vapor from the solid copper metal. Because of this, discharge tubes of alumina (a refractory material) are used.

A typical Cu vapor laser might have average power output of 100 W with peak powers in the range of 100 kW. The pulse repetition frequency can be as high as 10 kHz, with pulse duration of 30 to 50 ns and energy of 5 to 10 mJ. Figure 9.17 schematically illustrates the electrical discharge circuitry employed for a representative 6-cm-diameter 60-W laser. Neon is typically employed as a buffer gas. The overall efficiency is about 1% and the ratio of the intensity at 510 nm to that at 578 nm is about 2:1. A beam diameter of $\approx$5 cm is typical, and while peak power can be

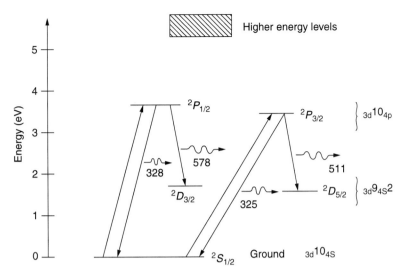

**Figure 9.16**   Cu vapor laser energy levels.

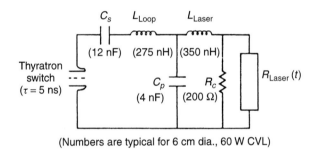

(Numbers are typical for 6 cm dia., 60 W CVL)

**Figure 9.17**   Copper vapor laser discharge circuit. (From Ref. 23.)

increased by increasing the tube diameter, and hence the volume, the pulse repetition frequency decreases with increasing tube diameter. The scaling for output with tube volume is about 50 mJ/L. Figure 9.18 displays the predicted average output power as a function of discharge tube diameter. Shown also are the measured range of current 6- to 8-cm-bore devices [23]. Table 9.1 contains the discharge and laser parameters for typical 6- to 8-cm-bore vapor lasers.

A number of applications are being found for this relatively new laser. It is used to pump a dye laser used for laser isotope enrichment of uranium. Underwater communications and mapping, semiconductor materials processing, combustion research, and medical applications are also making use of the Cu vapor laser.

*$CO_2$ Laser.* The carbon dioxide laser provides a combination of high power output and high efficiency which is superior to that of any other laser. Continuous-wave outputs in the range 10 to 20 kW and pulsed output energies of 10 J, with ~1-$\mu$s pulse widths and ~50-Hz pulse repetition frequencies, are routinely available. The overall efficiency of the $CO_2$ laser is in the 20% range. In addition, extremely high output power gas dynamic $CO_2$ lasers, with CW output powers of ~1 MW, have been developed for

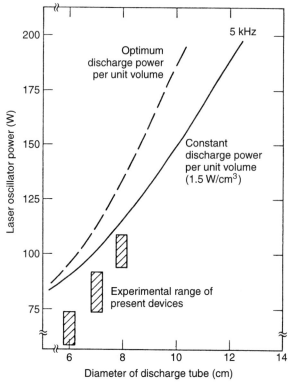

**Figure 9.18**  Computed copper vapor laser power (oscillator) as a function of discharge tube diameter. The solid line is for constant discharge power per unit volume (1.5 W/cm$^3$). The dashed line is laser power optimized with respect to discharge power per unit volume, illustrating the severity of gas-heating effects in the larger lasers. The typical range of output power in present devices is also shown. (From Ref. 23.)

**TABLE 9.1    Nominal Operating Parameters for 6- to 8-cm Large-Bore CVLs**

|  | 6 cm | 7 cm | 8 cm |
|---|---|---|---|
| Electrode spacing (cm) | 122 | 122 | 152 |
| Active length (cm) | 90 | 90 | 122 |
| Operating temperature (°C) | 1500 | 1450 | 1450 |
| Neon pressure (torr) | 32 | 25 | 25 |
| Copper partial pressure (torr) | 0.3 | 0.15 | 0.15 |
| Nominal repetition rate (kHz) | 5.0 | 5.0 | 5.0 |
| Charge voltage (kV) | 14 | 16.3 | 18.5 |
| Storage capacitance (nF) | 13 | 12 | 10.5 |
| Input power density (W/cm$^1$) | 1.85 | 1.57 |  |
| Peak discharge current (A) | 1300 | 1600 | 2100 |
| Average total power (W) | 60 | 80 | 100 |
| Code prediction | 66 | 88 | 115 |
| Light-pulse length (ns) | 62 | 70 | 85 |
| Code prediction | 57 | 62 | 75 |

*Source*: Ref. 23.

military applications. Because of the very specific applications for these gasdynamic lasers, they are not discussed further here [24].

The carbon dioxide laser provides radiation in the range 9 to 11 μm by emission from various vibrational energy levels of the linear $CO_2$ molecule. The excited vibrational states involve a stretching or bending of the atomic bonds from their equilibrium positions. A variety of processes may lead to inversion, including direct inelastic electron collisions, resonant transfer from other excited diatomic molecules, or optical pumping. Pumping to the excited state is usually produced by a dc (or pulsed) electrical discharge, or in the case of waveguide $CO_2$ lasers, by radio-frequency (RF) excitation. Tuning is provided by a wavelength-selective device, such as a diffraction grating, in the laser cavity.

Carbon dioxide is a linear symmetric molecule whose equilibrium configuration is shown in Fig. 9.19. This triatomic molecule has three vibrational degrees of freedom and the motion involved in each of the three vibrational modes is shown in Fig. 9.19. The state of the molecule is described by a set of three quantum numbers written $(v_1, v_2, v_3)$ with $v_1$, $v_2$, and $v_3$ corresponding to the symmetric stretch, bending, and asymmetric stretch, respectively. The ground state is represented by (000).

The energy levels involved in the lasing process are shown in Fig. 9.20. Lasing in the 10-μm branch occurs on the (001)–(100) transition, while the 9-μm lasing occurs between the (001)–(020) levels. The various levels in each vibrational manifold are the rotational energy levels. The numbers next to each of these correspond to quanta of rotational momenta in units of $h/2\pi$. Quantum mechanical selection rules allow only transitions that involve a change of rotational angular momentum of $\pm h/2\pi$. Transitions involving a change of $+h/2\pi$ are called P-branch transitions, while those with $-h/2\pi$ change are the R-branch transitions. An example of each is shown in Fig. 9.20. The full terminology for the transition is then determined by the number associated with the lower lasing level. For example, the 10-μm lasing line shown would be labeled the 10P6 transition. The energy levels for the $N_2$ molecule are also shown in

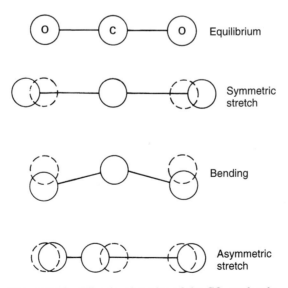

**Figure 9.19**   Vibrational modes of the $CO_2$ molecule.

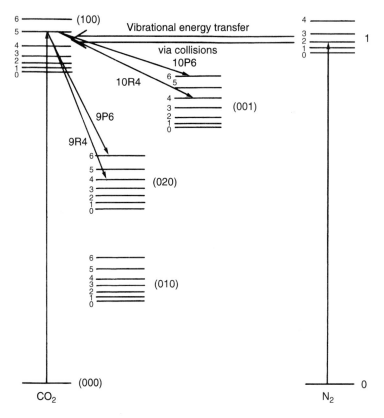

**Figure 9.20**   Low-lying vibrational–rotational energy levels for the $CO_2$ and $N_2$ molecules.

Fig. 9.20. This is a diatomic molecule with only one vibrational degree of freedom and one vibrational quantum number. Because of its simplicity, the nitrogen molecule can be efficiently excited to the upper energy level. Then, because this energy level is nearly equal to the (001) level in the $CO_2$ molecule, energy is efficiently resonantly transferred between the two molecules via collisions. This is the main excitation channel for the 001 level of the $CO_2$ laser.

Helium is added to the discharge to enhance collisional deexcitation of the $CO_2$ molecules from the lower lasing level to the ground state. The helium's high thermal velocity provides many collisions and thus the increased deexcitation rate. The helium also increases the rate of excitation of nitrogen molecules to the upper vibrational energy level [25]. Gases such as $N_2$ and He which aid the lasing process but do not emit photons at the lasing wavelength are known as buffer gases.

$CO_2$ lasers are generally divided into three groups: low power (<100 W), medium power (100 to 1000 W), and high power (>1000 W). The low-power units are usually the sealed-beam variety, with the combination of $CO_2$, $N_2$, and He gas contained in a sealed electrical discharge tube. The $CO_2/N_2/He$ ratio is approximately 1:4:15. Tube diameter is typically 5 to 10 mm, with an output of $\sim 50$ W/m. Waveguide $CO_2$ lasers are becoming increasingly popular in low-power applications. A small-diameter ($\sim 1$ to 2 mm) discharge tube confines, or guides, the radiation, and because of the improved heat diffusion to the walls, these lasers can support higher power densities. Pumping by RF excitation is usually employed in waveguide $CO_2$ lasers. The lifetime of these sealed lasers is usually limited to $\leqslant 5000$ h by contamination, escape, or breakdown ($CO_2 \rightarrow CO$) of the gas mixture.

At medium-power levels, heat removal from the lasing gas must be improved, so a slow axial flow of laser gas is produced. A typical medium-power $CO_2$ laser is shown schematically in Fig. 9.21. For a 100-W CW laser, a discharge length of approximately 1.5 m and a tube diameter of 10 mm is required. Typically, the discharge is maintained with ~10 kV and 100 to 200 mA of current. The Brewster angle windows provide a low-loss method for vacuum sealing the discharge tube. The Brewster windows also introduce losses that are polarization dependent, ensuring that the output will be linearly polarized. However, the losses also lead to heating and instability. Therefore, designs that eliminate the need for Brewster windows are becoming more commonly used as we shall soon see.

If a mirror is utilized in place of the diffraction grating, lasing will occur at the dominant wavelength, 10.58 μm. By using a grating, the operating wavelength is governed

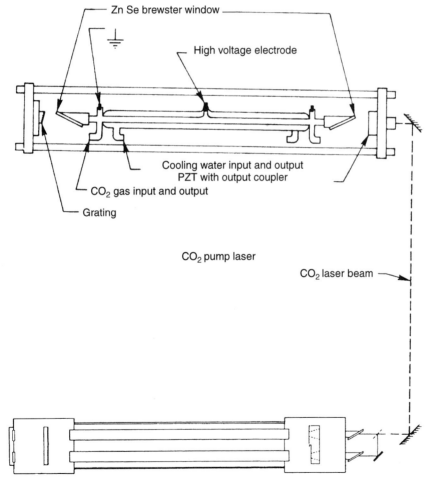

**Figure 9.21**  Typical moderate power CW $CO_2$ laser design.

by the familiar relation

$$n\lambda = d \sin\theta$$

where $n$ is the grating order ($= 1$), $\theta$ the angle between the laser beam and the grating normal, and $d$ the ruling spacing. The tuning provided by the grating is not continuous. However, discrete tuning over the region 9 to 11 $\mu$m is possible. Typical output powers, from a laser to be discussed shortly, in the 9- and 10-$\mu$m branches are shown in Figs. 9.22 and 9.23. A number of discrete wavelengths in the range 9.2 to 10.8 $\mu$m can

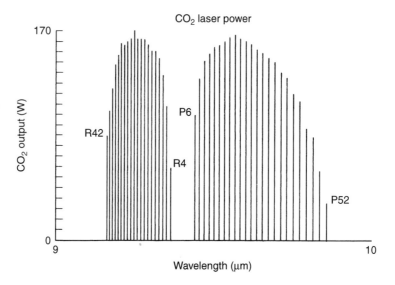

**Figure 9.22**   $CO_2$ laser output power for the 9- to 10-$\mu$m branch.

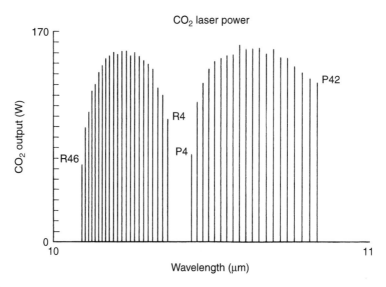

**Figure 9.23**   $CO_2$ laser output power for the 10- to 11-$\mu$m branch.

be selected. The grating also ensures linearly polarized outputs. Typical $CO_2$ diffraction gratings have 125 to 150 rulings per millimeter.

Water cooling is provided to dissipate both excess electrical energy and molecular energy from molecules striking the cavity walls. The water also cools the optics, which absorb a small amount of the radiation. Any heat buildup in the optics would cause instabilities in the cavity.

The output coupler (normally, ZnSe with $T \sim 20\%$) can be mounted in a piezoelectric transducer (PZT) which allows a small variation in the cavity length to be made ($\Delta\lambda \sim \lambda/2$ to 5 μm). This length adjustment allows fine tuning of the $CO_2$ laser frequency to match the absorption in an FIR molecular lasing medium when performing optical pumping (see Section 9.2). The tuning range is limited by the longitudinal mode spacing, which is given approximately by

$$\Delta\nu = c/2L$$

For a cavity length of 1.5 m, tuning of about 100 MHz from line center is obtainable.

Figure 9.24 schematically illustrates a significantly improved design of a grating tuned $CO_2$ laser which has found extensive application as a pump source for far-infrared

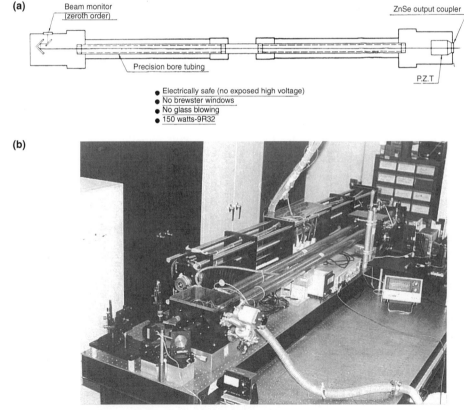

**Figure 9.24**    (a) Schematic of high power four-electrode $CO_2$ laser developed for optical pumping FIR lasers (b) Photograph of $CO_2$ laser system together with FIR laser assembly. (From Refs. 26 and 27.)

molecular lasers (see Section 9.2). This is a unique four-electrode design developed by Alain Semet at Apollo Lasers in collaboration with the UCLA group [26, 27]. The high voltage (15 kV) is isolated at the center with the entire laser structure electrically grounded. This provides electrically safe operation. This design also eliminates the need for Brewster windows in the laser cavity. Previous two- or three-electrode designs required Brewster windows to provide isolation of the discharge section from the optical mounts (see Fig. 9.24). Window imperfections induced thermal instabilities in the cavity as well as reducing the laser output power.

For the laser shown in Fig. 9.24, a grating with 135 rulings per millimeter was found to give adequate dispersion of lines while allowing coverage of all wavelengths from 9.2 to 10.8 $\mu$m. The grating also serves two other purposes. First, the zeroth-order reflection is used to monitor the laser output. Second, reflection for radiation polarized perpendicular to the rulings is higher than that for the parallel polarization. This, combined with the large gain of the $CO_2$ laser, ensures linear polarization of the laser output.

A 25-mm-diameter zinc selenide output coupler is utilized. The rear surface is antireflection coated for 10.6 $\mu$m, while the concave inner surface is reflection coated on a 20-m radius of curvature. This 20-m radius in a <3-m-long cavity, together with the grating, provides a stable resonator configuration. Reflectivities varying from 60 to 90% were tested with an 80% coupler, providing the highest output power levels while covering the full wavelength range. For a cavity length of 2.7 m, tuning of about 55 MHz from line center is obtainable, using piezoelectric translation of the output coupler.

The laser discharge tube is a 10-mm-inner-diameter-bore precision Pyrex tube which is sufficiently small to ensure single transverse-mode operation. The virtue of this construction is that no complicated and expensive glass blowing is required with the water cooling provided by an O-ring sealed tube surrounding the precision bore tube. A variety of gas mixtures with different partial pressure ratios of $CO_2/N_2/He$ were tested. While the best results were obtained with 4%:16%:80%, a mixture of 6%:18%:76% proved to provide only slightly less output power but at greatly reduced operating cost.

Output powers obtained with this design in the range 9 to 111 $\mu$m were shown in Figs. 9.22 and 9.23. Notice that the peak power of 170 W occurs in the 9R branch rather than the usual 10P branch. This is because the output coupler employed for these measurements was optimized for the shorter wavelengths, which were desired for the specific applications. Output power levels in excess 100 W are available on virtually every line, with powers greater than 130 W at almost all wavelengths of interest.

In high-power CW $CO_2$ lasers, heat removal is improved by increasing gas flow rates. Supersonic axial flow or fast transverse flow are two common techniques utilized. Operation of these high power lasers is similar to that described for medium-power lasers. For many applications, $CO_2$ lasers rated at very high peak power are required. Examples include pump sources for high-power optically pumped far-infrared lasers for fusion plasma diagnostics [28], laser-matter interaction studies [29–31], and laser-plasma accelerators [32]. The problem with the longitudinal-type discharge described earlier is that they are limited to relatively low pressures ($\approx$10 to 20 torr), so that the discharge energy is limited to $\approx$0.05 J/L [33]. This led to the development of the so-called TEA or transverse excitation, atmospheric-pressure laser [34–36]. These new lasers employed a technique to stabilize a controlled Townsend avalanche electric discharge in a high-pressure gas mix in a transverse geometry [33]. The stability was achieved either by resistive stabilization or through the use of secondary ultraviolet preionization light sources to initiate and maintain a uniform glow discharge. These are also sometimes referred to as double-discharge lasers. The laser output energy and maximum aperture are related with the size limited to a gap-pressure product of $\approx$20-cm atmospheres [37].

The solution to the foregoing energy/size constraint is to inject an externally generated electron beam into the cavity containing the large-aperture discharge region [33, 37–41]. In this approach, the electron beam provides the initial preionization pulse, which is followed by a second pulsed electric field to provide the discharge energy. Here, both of the electron beams are employed in a transverse geometry. Using this technique, operating pressures >1800 torr have been employed with energy storage densities in excess of 50 J/L. For complete details on short-pulse, high-power $CO_2$ lasers, the reader is referred to the excellent reviews [33, 40–42]. Here, we restrict our attention solely to reviewing two of the highest-peak-power facilities, which were developed for laser fusion applications, the Los Alamos National Laboratory Helios and Antares lasers [37, 43, 44].

The Helios laser system was based on a single-beam system originally developed at Los Alamos in 1971 [42]. Figure 9.25 schematically illustrates the electron beam discharge control system which was employed in all of the amplification stages [37]. Complete details of the single-beam system are contained in Table 9.2. In this system, electron guns utilizing tungsten filaments were employed with the $\approx$200-kV negative pulses provided by Marx banks which $RC$ decayed up to termination via diverter switches [37]. For the final amplification stage, an $LC$ generator (also diverter switch terminated) was utilized.

The single-beam technology outlined in Table 9.2 was modified into a dual-beam module (Gemini) in 1974 [37, 45]. Figure 9.26 schematically illustrates the configuration,

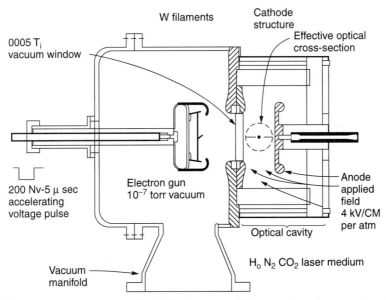

**Figure 9.25**  Electron-beam-controlled $CO_2$ laser amplifier configuration employed in Los Alamos single beam system. (From Ref. 37.)

**TABLE 9.2 Characteristic Features of the Los Alamos Single-Beam System**

| Parameter | Stages 1 and 2 | Stage 3 | Stage 4 |
|---|---|---|---|
| Electron beam | | | |
| Voltage (kV) | 120 | 155 | 250 |
| Current (A) | 100 | 500 | 1500 |
| Current density (A/cm$^2$) | 0.12 | 0.60 | 0.27 |
| Gas | | | |
| Pressure (torr) | 600 | 1800 | 1400 |
| Electric field (kV/cm-atm) | 4.3 | 3.8 | 3.5 |
| Current (kA) | 5 | 16 | 50 |
| Current density (A/cm$^2$) | 6.3 | 20 | 9 |
| Gain (P-20, cm$^{-1}$) | 0.051 | 0.049 | 0.03 |
| J/L-atm | 150 | 150 | 85 |
| Efficiency [$g_0(J)E_s$/J/L] | | | 3.2% $\left(\times \frac{1}{5}\right)$ |

*Source*: Ref. 37.

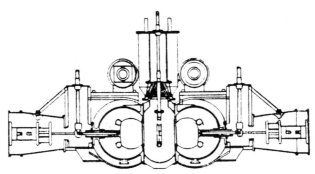

**Figure 9.26** Cross-sectional view of dual-beam module as used in the LASL Helios system. (From Ref. 37.)

which utilizes a single-electron-beam gun for the two pumping chambers. Here, triple passing is employed in the gain region and an increased aperture size is utilized ($\approx$36 cm versus 25 cm). Another change from the previous system was the use of cold cathodes [37, 46], which eliminated problems with heat loading and thermal distortion. In this scheme, thin tantalum foils ignite and generate plasma, which serves as the electron emitter. The parameters of the Gemini laser are listed in Table 9.3.

The dual-beam technology outlined in Table 9.3 was utilized in the Helios laser system with the exception that the *LC* generators were powered by Marx generators using two-mesh type C pulse-forming networks [37]. This was configured as four dual-beam modules which were combined into eight beams. Following its initial operation in April 1978, the system was measured to deliver a subnanosecond pulse of 10.7 kJ into a calorimeter [37, 43]. Figure 9.27 is a photograph of the Helios system showing the laser modules as well as the inertial confinement fusion test chamber.

The final development in LASL high-power $CO_2$ laser systems was the Antares facility [33, 37, 44]. This was originally planned in the 1970s as a 100-kJ/72-beam system but was downgraded in 1980 to a 40-kJ/24-beam system that was to deliver 10.6-$\mu$m radiation in a 1-ns pulse [33, 44]. The beam quality and optics were to be of

**TABLE 9.3 Performance Data of a Los Alamos National Laboratory Helios Dual-Beam Module**

| | |
|---|---|
| Optical design (each beam) | |
| Aperture diameter (cm) | 34 |
| Gain length (cm) | 200 |
| Operating pressure (torr) | 1800 |
| Gas mixture | $N_2/CO_2/He \frac{1}{4}:1:3$ |
| Gain | 4%/cm (P-20, 10 $\mu$m) |
| Energy output (J) | 1250 |
| Electrical design | |
| Discharge voltage (kV) | 300 |
| Discharge current (kA) | 100 |
| Pulse duration ($\mu$s) | 3 |
| Energy (J/L-atm) | 150 |
| Electron-beam voltage (kV) | 250 |
| Electron-beam current density (A/cm$^2$) | 0.3 |
| Pulse duration ($\mu$s) | 5.0 |
| Emitter | 0.013-cm-thick Ta foil |

*Source*: Ref. 37.

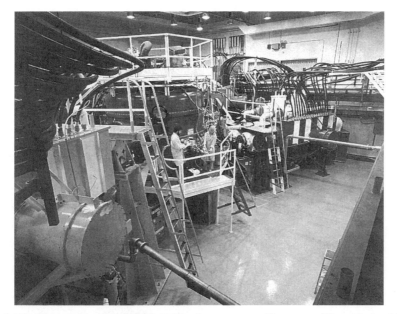

**Figure 9.27** Photograph of the LASL Helios laser system. (Courtesy of Los Alamos National Laboratory.)

sufficient quality so that the beams could be focused to a 300-$\mu$m spot with a pointing accuracy of 25 $\mu$m.

Figure 9.28 schematically illustrates the Antares front end [33]. Here, four grating-tuned, independent, single-line, injection-locked oscillators are employed. Using this configuration, the proper initial signal is generated with an output pulse selected with

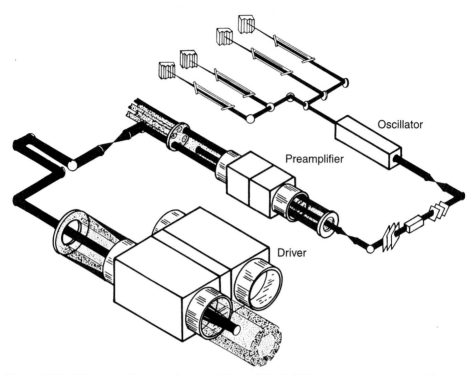

**Figure 9.28** Master oscillator and preamplifiers of the LASL Antares laser system. The multiline oscillator and electro-optical switch generate a single 10.6-μm radiation pulse which is further amplified by the preamplifier and driver shown. (From Ref. 33.)

an electro-optical Pockell's cell gate. This is then introduced into the preamplifier, which is followed by the triple-pass Cassegrain mirror module, which employs the dual-beam Helios module design. Both the preamplifier and driver-amplifiers use $SF_6$-based mixtures in gas isolation cells to suppress parasitic oscillations. At the output, a 100-J annular beam (ID = 9 cm, OD = 15 cm) of 450-cm$^2$ area is produced.

The power amplifier design departed from the designs described previously. For Antares, 12 laser beams were combined in an annulus around a single common electron beam, resulting in only two amplifier modules for the entire system. Figure 9.29 contains an artist's sketch of an amplifier module. Figure 9.30 is an actual photograph of the two amplifier modules employed in the Antares facility.

In the Antares power amplifier design, an optimum $E/P$ value of 10 V/cm-torr was chosen for the gas mixture of $CO_2/N_2 = 4:1$ to provide good efficiency and discharge stability [44]. Combined with an operating pressure of 1800 torr dictated by vessel and NaCl window safety considerations, this resulted in a discharge voltage of ≈550 kV and a gun voltage of ≈525 kV. Since the design value of the cavity gain was 2.7 m$^{-1}$, a total active gain medium length of 3 m was required for a total single-pass gain of 8.0. Electron-beam deflection considerations resulted in the amplifier cavities being subdivided into four axial sections (15 × 30.5 × cm$^3$) separated by 100 cm from adjacent sections as shown in Figs. 9.29 and 9.31. The laser gas ionization of 50 mA/cm$^2$ is provided from the central electron gun through 48 Kaptron-aluminum foil windows of 50 μm thickness. A main discharge current density of 7 A/cm$^2$ is employed.

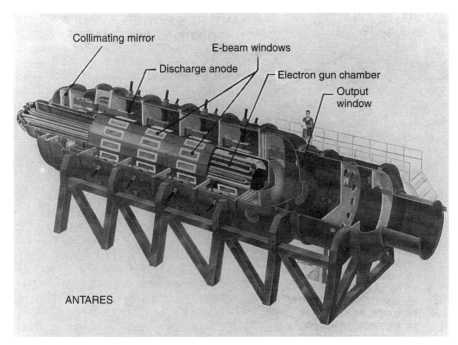

**Figure 9.29**  Artist's sketch of an Antares laser power amplifier module. (Courtesy of Los Alamos National Laboratory.)

**Figure 9.30**  Photograph of the two power amplifier modules installed in the Antares laser facility. (Courtesy of Los Alamos National Laboratory.)

Referring to Fig. 9.31, the output beam from the $CO_2$ front end is divided into 12 beamlets by the polyhedron mirror, with each beamlet entering a separate amplifier. Using a double-pass Cassegrain expansion telescope, the initial 10-cm$^2$ beamlet is expanded to the final output size of 1000 cm$^2$. The measured output of a single Antares

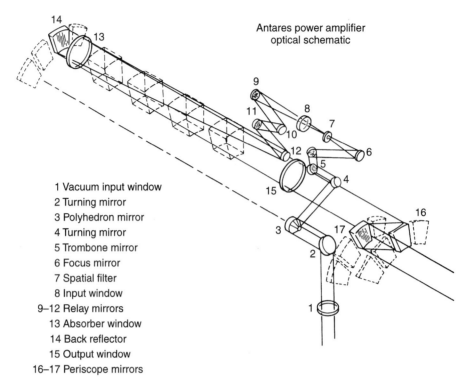

Antares power amplifier
optical schematic

1 Vacuum input window
2 Turning mirror
3 Polyhedron mirror
4 Turning mirror
5 Trombone mirror
6 Focus mirror
7 Spatial filter
8 Input window
9–12 Relay mirrors
13 Absorber window
14 Back reflector
15 Output window
16–17 Periscope mirrors

**Figure 9.31**  Optical schematic of a single beam line for the LASL Antares power amplifier. The beam from the master oscillator enters through optic 2 and finally exits through mirror system 16/17. (From Ref. 33.)

power amplifier module was 17 kJ in a 1-ns pulse of 10.5 $\mu$m radiation [33, 47]. This demonstrated the feasibility of systems for delivery pulses of $\approx$1 ns duration in the region 25 to 100 kJ.

To conclude this subsection, we note that the primary application area for high-power CW $CO_2$ lasers is in materials processing. This includes processes such as welding, cutting, heat treating, and material shaping. The laser's primary advantage in these functions is that a precise amount of energy can be delivered to a localized region. One of the major uses for pulsed TEA lasers is mask marking [48].

Low- to medium-power $CO_2$ lasers find application primarily in the medical and pure research fields. Because the 10.6-$\mu$m radiation is strongly absorbed by human tissue, the $CO_2$ laser is finding extensive use as a self-cauterizing laser scalpel. Many research and application areas, such as optical radar, plasma diagnostics, spectroscopy, and optical pumping of far-infrared lasers, among others, take advantage of the $CO_2$ laser's unique properties [16, 18].

*Excimer Lasers.*  Rare gas halide excimer lasers are perhaps the most powerful laboratory sources for ultraviolet radiation. Discharge excimer lasers are relatively inexpensive sources for high average power coherent light in the region 350 to 190 nm. These lasers are found in many research and industrial applications. Electron-beam pumped excimer lasers have attained very high energies and are a candidate for the laser fusion application. We discuss in some detail the technology and applications for these lasers.

*Far-Infrared Lasers.* The far-infrared (FIR) laser is another class of molecular laser, with operation similar to the $CO_2$ laser. However, in the FIR laser, the lasing transition occurs between two *rotational* energy levels rather than *vibrational* levels. The far-infrared region is roughly defined by 100 μm $< \lambda <$ 2000 μm.

Far-infrared lasers can be separated into two classes: electrical discharge pumped and optically pumped lasers. Optical pumping is provided by another laser, usually the $CO_2$ laser. An energy-level diagram for an optically pumped FIR laser is shown in Fig. 9.32. Operation in the discharge pumped laser is identical except that excitation is provided by electron impact rather than direct absorption of the $CO_2$ photon. Because the frequency of the $CO_2$ laser can be tuned to match an absorption line in the FIR molecule, the optical pumping process is more efficient than discharge excitation.

The lasing process is basically a three-level type, but for proper modeling one must include collisional energy transfer between the lasing levels and other rotational levels in the vibrational manifolds. This collision rate, along with the infrared absorption, determines the operating pressure of the FIR laser; at low pressures, absorption of the $CO_2$ radiation is low while at high pressures, collisions destroy the population inversion. After the lasing process, the molecules are returned to the ground state via diffusion and collisions with the container walls or vibrational-to-translational energy transfer.

For optical pumping, the laser efficiency is limited by the Manley–Rowe condition. This states that two pump photons are required to produce one FIR photon, and may

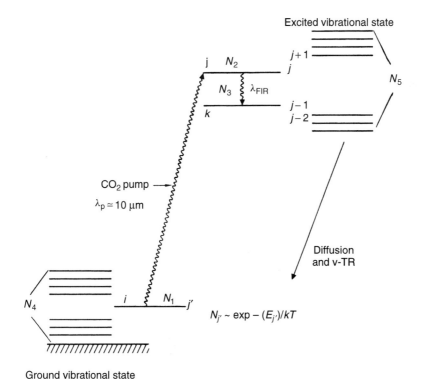

**Figure 9.32** Energy-level diagram for an optically pumped far-infrared laser.

be written as

$$P_{\text{FIR}} = \frac{1}{2} \frac{\lambda_{\text{pump}}}{\lambda_{\text{FIR}}} P_{\text{pump}}$$

(An exception to this rule are the Raman lasers discussed in Section 9.2.) For example, taking $\lambda_{\text{pump}} = 10$ μm, $\lambda_{\text{FIR}} = 500$ μm, and $P_{\text{pump}} = 100$ W predicts an FIR power of 1 W. This is a theoretical prediction and actual output powers obtained are approximately 10% of these values.

The $CO_2$ laser provides radiation at many wavelengths in the range 9 to 11 μm, and a large number of suitable FIR molecules have been found that absorb this radiation. Tabulations of FIR wavelengths and output power have been provided in the literature [49–52]. In total, these provide a choice of over 2000 wavelengths in the range 100 to 2000 μm, with several of these delivering high output power.

A simple design for a CW, optically pumped FIR laser is shown in Fig. 9.33. Here, the $CO_2$ pump radiation is coupled into the FIR cavity by focusing through a small hole in the mirror. At these long wavelengths, the optics required for a Fabry–Perot cavity are prohibitively large. For this reason, a waveguide cavity is usually employed, with either a dielectric or a metallic waveguide [53]. The output coupler reflects 100% of the $CO_2$ radiation and approximately 90% of the FIR radiation.

Application of far-infrared lasers include use as a local oscillator source for heterodyne radiometry, a technique for the determination of elements and molecules present in the upper atmosphere, other planets, or stellar bodies [54]. High atmospheric attenuations of FIR radiation, which make it unattractive for long-range communications, are advantageous for covert or secure short-range communication [55]. Small-scale laboratory modeling of microwave radar systems can be performed with the shorter-wavelength FIR radiation [56]. Because most clothing and luggage is transparent to FIR radiation with $\lambda > 1000$ μm, an FIR detection device can identify concealed weapons or possible contraband. The power levels needed for this are not harmful and such a

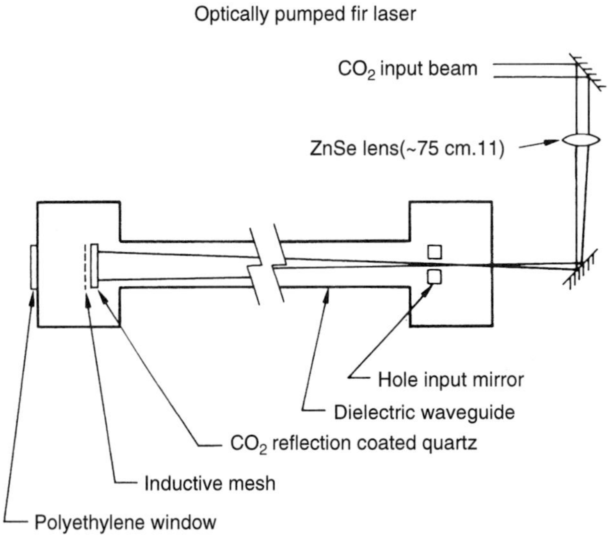

**Figure 9.33** Optically pumped far-infrared laser.

system has been designed and tested [57]. The transparency of materials to FIR radiation is useful in another application as well. Mie scattering, which occurs at voids in the plastic insulation of high-voltage cables, can be used to detect contamination or manufacturing defects before the cables are installed [58]. The far-infrared wavelength region is also ideal for a number of magnetic confinement fusion plasma diagnostics, such as collective scattering [28, 59] and interferometry [28, 60].

This section is intended only as a brief introduction to far-infrared lasers. This type of laser is discussed in greater detail in Section 9.2.

## 9.2  FAR-INFRARED LASERS

### 9.2.1  Introduction

The far-infrared portion of the spectrum extends roughly from 100 to 2000 $\mu$m. Here, one of the primary sources has been the far-infrared laser, with CW power levels up to a watt having been produced at certain frequencies. Two completely different laser types have been utilized. The first is the so-called electrical discharge type, which utilizes an electrical discharge in the appropriate gas mixture to provide the population inversion. Early work in extending lasing action to the submillimeter wave region was performed by McFarland et al. [61], who extended the output wavelength of the He–Ne laser to 28 $\mu$m followed by CW noble gas and Ne laser operation at 85 $\mu$m. Patel further extended this work to 133 $\mu$m [9].

Following the work above, the development of practical FIR sources was made possible by the investigation of polyatomic molecules such as $H_2O$ and HCN. For example, Crocker et al. described the observation of 78-$\mu$m stimulated emission in $H_2O$ [62] followed by the report of 118-$\mu$m emission by Gebbie et al. [63] as well as 72-$\mu$m $D_2O$ emission. The practical use of the $H_2O$ laser was made possible by the achievement of CW operation [64].

Perhaps the most widely employed FIR electrical discharge lasers have been the HCN and DCN. The first observations of HCN lasing were reported by Gebbie et al. [65]. In Section 9.2.1, we therefore concentrate primarily on these types. However, for completeness, Table 9.4 reproduced from the excellent FIR electrical discharge laser review by Kneubuhl and Sturzenegger, contains a tabulation of lasing media beyond 100 $\mu$m [66].

Referring back to Table 9.4, one is struck with the relative paucity of laser lines as well as the fact that they are restricted primarily to wavelengths in the region 100 to 200 $\mu$m. However, this situation was eliminated with the discovery of the optically pumped far-infrared laser by Chang and Bridges [71]. In this work, the P(20), 9.55-$\mu$m line of a $CO_2$ laser was employed to pump a vibrational transition in a molecular gas ($CH_3F$) with lasing occurring on a rotational transition. Following this work, there were a spate of publications and research efforts, culminating in a rich spectrum of lines in the region 10 to 2000 $\mu$m [49–52, 72].

A variety of both CW and pulsed pump sources have been used to provide far-infrared emission from molecules. However, the most commonly utilized source has been the $CO_2$ laser. Typical applications for the high-power pulsed FIR lasers have been in the diagnostics of high-temperature fusion plasma diagnostics [28, 60, 73]. In contrast, the lower-power CW types have found a diversity of applications, including radio astronomy, nondestructive testing, molecular spectroscopy, plasma diagnostics, and radar modeling.

**TABLE 9.4   Electrically Excited Gas Lasers Working at Wavelengths beyond 100 $\mu m^a$**

| Emitter | Vacuum Wavelength | Operation[b] | Transition | |
|---|---|---|---|---|
| He | 216.3 | CW | $4^1_p p - 4d^1 D$ | |
| Ne | 106.07 | CW | $10p\left[\dfrac{1}{2}\right]_0 - 9d\left[\dfrac{3}{2}\right]_1^0$ | |
| | 124.52 | | $9p\left[\dfrac{3}{2}\right]_1 - 8d\left[\dfrac{5}{2}\right]_2^0$ | |
| | or | CW | or | |
| | 124.76 | | $9p\left[\dfrac{3}{2}\right]_2 - 8d\left[\dfrac{5}{2}\right]_3^0$ | |
| | 126.1 | CW | ? | |
| | 132.8 | CW | ? | |
| $H_2O$ | 115.32 | p and CW | $(020)8_{35}-(020)8_{26}$ | |
| | 118.591 | p and CW | $(001)6_{42}-(020)6_{61}$ | |
| | 120.08 | p | $(001)6_{42}-(001)6_{33}$ | |
| | 220.230 | p and CW | $(100)5_{23}-(020)5_{50}$ | |
| $D_2O$ | 103.33 | p | ? | |
| | 107.731 | p and CW | $(100)11_{66}-(020)11_{75}$ | |
| | 107.91 | p | $(100)13_{68}-(100)13_{59}$ | |
| | 108.88 | p | $(100)11_{65}-(020)11_{74}$ | |
| | 110.49 | p | $(100)12_{66}-(020)12_{75}$ | |
| | 111.74 | p | $(100)13_{68}-(020)13_{77}$ | |
| | 170.08 | p | $(020)11_{47}-(020)11_{38}$ | |
| | 171.67 | p and CW | $(100)11_{0.11}-(020)11_{38}$ | |
| | 218.5 | p | ? | |
| $H_2S$ | 103.3 | p | ? | |
| | 108.8 | p | ? | |
| | 116.8 | p | ? | |
| | 126.2 | p | ? | |
| | 129.1 | p | ? | |
| | 130.8 | p | ? | |
| | 135.5 | p | ? | |
| | 140.6 | p | ? | |
| | 162.4 | p | ? | |
| | 192.9 | p | ? | |
| | 225.4 | p | ? | |
| $SO_2$ | 139.80 | p | $(020)27_{16}-(020)26_{15}$ | |
| | 140.78 | p and CW | $(100)27_{14}$ | $(020)26_{15}$ |
| | 140.88 | p and CW | $(100)26_{14}-(020)25_{15}$ | |
| | 141.98 | p | $(020)26_{16}-(020)25_{15}$ | |
| | 149.99 | p | $(100)28_{15}-(100)27_{14}$ | |
| | 151.19 | p and CW | $(100)28_{15}-(020)27_{16}$ | |
| | 151.31 | p and CW | $(100)27_{15}-(020)26_{16}$ | |
| | 192.71 | p and CW | $(100)28_{16}-(100)28_{15}$ | |
| | 206.44 | p | $(100)26_{15}-(100)26_{14}$ | |
| | 215.33 | p and CW | $(001)27_{10}-(020)27_{16}$ | |

**TABLE 9.4** (*Continued*)

| Emitter | Vacuum Wavelength | Operation[b] | Transition | |
|---|---|---|---|---|
| OCS | 123 | p | ? | |
| | 132 | p | ? | |
| HCN | 126.164 | p | $(12^20)-(05^10)$ | R(26) |
| | 128.629 | p and CW | $(12^20)-(05^10)$ | R(25) |
| | 130.838 | p | $(12^00)-(05^10)$ | R(25) |
| | 134.932 | p | $(12^00)-(05^10)$ | R(24) |
| | 284 | p and CW | $(11^10)-(11^10)$ | R(11) |
| | 309.7140 | p and CW | $(11^10)-(11^10)$ | R(10) |
| | 310.8870 | p and CW | $(11^10)-(04^00)$ | R(10) |
| | 335.1831 | p and CW | $(04^00)-(04^00)$ | R(9) |
| | 336.5578 | p and CW | $(11^10)-(04^00)$ | R(9) |
| | 372.5283 | p and CW | $(04^00)-(04^00)$ | R(8) |
| HCN or CN | 101.257 | p | ? | |
| | 112.066 | p | ? | |
| | 116.132 | p | ? | |
| or? | 201.059 | p | ? | |
| | 211.001 | p and CW | ? | |
| | 222.949 | p | ? | |
| DCN | 181.789 | p | $(22^00)-(22^00)$ | R(22) |
| | 189.9490 | p and CW | $(22^00)-(09^10)$ | R(21) |
| | 190.0080 | p and CW | $(22^00)-(22^00)$ | R(21) |
| | 194.7027 | p and CW | $(22^00)-(09^10)$ | R(20) |
| | 194.7644 | p and CW | $(09^10)-(09^10)$ | R(21) |
| | 204.3872 | p and CW | $(09^10)-(09^10)$ | R(19) |
| HCN[15] | 110.240 | p | ? | |
| or | 113.311 | p | ? | |
| | 138.768 | p | ? | |
| | 165.150 | p | ? | |
| ICN | 538.2 | p | ? | |
| or HCN | 545.4 | p | ? | |
| or CN | 676 | p | ? | |
| or? | 773.5 | p | ? | |
| $H_2O$ | 101.9 | p | ? | |
| | 119.6 | p | ? | |
| | 122.8 | p | ? | |
| | 125.9 | p | ? | |
| | 155.1 | p | ? | |
| | 157.6 | p | ? | |
| | 159.5 | p | ? | |
| | 163.8 | p | ? | |
| | 170.2 | p | ? | |
| | 184.4 | p | ? | |

*Source*: From Ref. 66.
[a]Data from Refs. 67–70.
[b]p, pulsed; CW, continuous wave.

### 9.2.2 Electrical Discharge Lasers

As mentioned above, electrically excited far-infrared lasers provide coherent output at discrete wavelengths from ~28 μm [74] to ~774 μm [75]. A comprehensive review of electrically excited submillimeter wave lasers has been given by Kneubuhl and Sturzenegger [66]. Details about specific lasers may be found in the literature [76–85]. Of the lasers above, perhaps the most widely used have been the HCN, DCN, and $H_2O$ lasers. Therefore, the following discussion will be restricted to these types.

We begin with the HCN laser, whose energy-level diagram is shown in Fig. 9.34. The transitions have been found to involve the $11^1 0$ and $04^0 0$ vibrational states, which are mixed by Coriolis perturbation [66, 87]. As noted by Kneubuhl and Sturzenegger, the lasing process in HCN (and DCN) is extremely complex and not yet precisely known [66]. A complete understanding is complicated by the fact that there are a variety of processes leading to HCN formation and molecular state inversion. In addition, the electrical discharge process is extremely complex and prevents the separate observation of the formation and inversion mechanisms. The interested reader is directed to the detailed discussion of Kneubuhl and Sturzenegger [66] together with the references contained therein. A complicated picture arises involving molecular dissociation due to ≈10-eV free electrons in the plasma together with chemical reactions in the laser plasma which both lead to HCN formation as well as excitation.

The lasing emissions in DCN have a similar explanation to those in HCN, where the transition levels are mixed by Coriolis resonance. Figure 9.35 displays the energy-level diagram corresponding to the four wavelengths suitable for CW operation near 200 μm [79]; $\lambda_1 = 189.95$ μm, $\lambda_1' = 190.01$ μm, $\lambda_2 = 194.70$ μm, and $\lambda_2' = 194.76$ μm.

Following the discovery of the $H_2O$ laser in 1964 by Crocker et al. [62], more than 100 lasing transitions have been observed in the region 2.279 to 220.23 μm. The most powerful of the CW lines is at ≈28 μm (27.972 μm), with the line at ≈118 μm also of importance. It has been found that the $H_2O$ molecule itself is responsible for the lasing action and that it is directly excited by electron impact excitation [66, 88, 89].

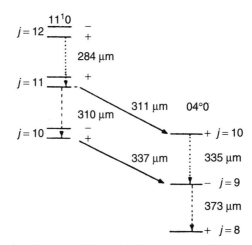

**Figure 9.34** Energy-level diagram of the submillimeter wave HCN laser, according to Hocker and Javan [86]. (From Ref. 66.)

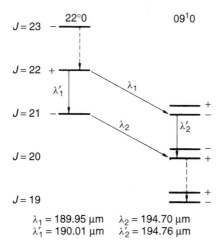

**Figure 9.35**  Energy-level diagram for the CW DCN laser lines. (From Ref. 79.)

The water molecule is a planar, bent, triatomic molecule with an equilibrium bond angle of 104.5° between molecules. The equilibrium separation distance between hydrogen and oxygen atoms is 0.956 Å. For rotation, the total angular momentum $M^2$ is given by

$$M^2 = J(J+1)h^2 \quad J = 0, 1, 2, 3, \ldots$$

and the projection on the $z$-axis by $M_z = K\hbar$, where $K = 0, \pm 1, \ldots \pm J$. The rotational energy is then given by

$$E_{\text{rot}} = (M_x^2/I_x) + (M_y^2/I_y) + (M_z^2/I_z)$$

where the $I$'s correspond to the principal moments of inertia. The $H_2O$ molecule can rotate as a rigid body with the bond lengths and angles fixed, although it is not a symmetric top molecule. Therefore, its rotational levels are bracketed between those of a prolate top and an oblate top molecule [90] as illustrated in Fig. 9.36. Here, the nonequal moments of the $H_2O$ molecule destroy the $2J + 1$ degeneracy of the comparable symmetric top molecule. Common notation results in these levels being denoted by $J_\tau$, where $\tau = -J, -J + 1, \ldots 0, \ldots J$ in increasing order of energy despite the fact that they do not correspond to a specific quantum number. The rotational selection rule is that $\Delta J = 0, \pm 1$. A number of rotational energy levels corresponding to FIR transitions are displayed in Fig. 9.37 together with their wavelengths and symmetry classes [90].

In addition to rotational modes, the $H_2O$ molecule possesses vibrational modes, which in the harmonic oscillator model are quantized as

$$E = h\nu_i \left(\nu_i + \tfrac{1}{2}\right) \quad i = 1, 2, 3$$

As illustrated in Fig. 9.38, these are the symmetric stretch mode, the bending mode, and the asymmetric stretch mode, with quantum numbers $(\nu_1, 0, 0)$, $(0, \nu_2, 0)$, and $(0, 0, \nu_3)$, respectively. The corresponding vibrational energy levels are displayed in

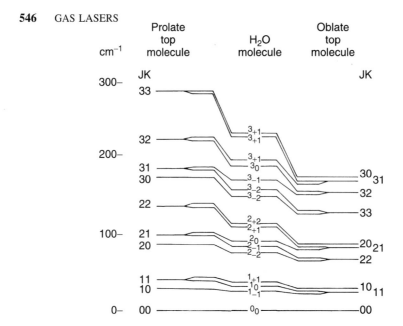

**Figure 9.36**   Rotational levels for prolate, oblate, and $H_2O$ molecule. (From Ref. 90.)

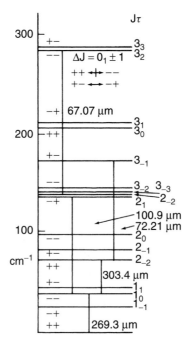

**Figure 9.37**   Rotational energy levels with symmetry classes for $H_2O$ molecule. (From Ref. 90.)

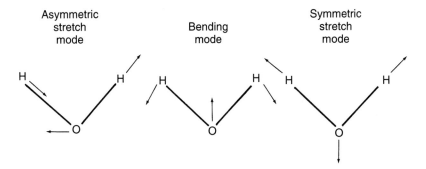

**Figure 9.38** Vibrational modes of $H_2O$ molecule. (From Ref. 90.)

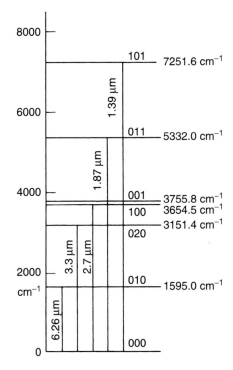

**Figure 9.39** Vibrational energy levels of an $H_2O$ molecule. (From Ref. 90.)

Fig. 9.39. In the simple harmonic oscillator approximation, the selection rules for the $i$th normal mode are given by

$$\Delta v_i = \pm 1 \quad (i = 1, 2, 3) \qquad \Delta v_j = 0 \quad (j \neq i)$$

Combining the above information, one can construct a model of the energy exchange in an $H_2O$ molecule as illustrated in Fig. 9.40. Here, the three vibrational modes of $H_2O$ are displayed together with the lower vibrational modes of $H_2$. The dashed lines correspond to lasing emission at 33, 28, and 118 μm, while the arrows represent

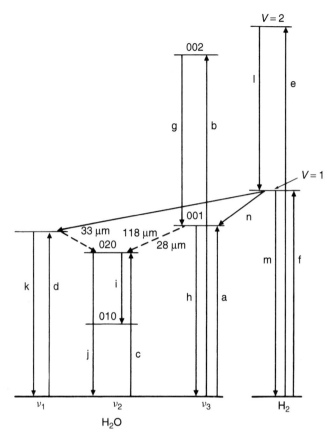

**Figure 9.40**   Model of the energy exchange in an $H_2O$ molecule. (From Ref. 90.)

**TABLE 9.5   Molecular Processes in $H_2O$**

| Process | Rate Constant (cm³/s) | Characteristic Time (atm-s) |
|---|---|---|
| (a) $e + H_2O(000) \rightarrow e = H_2O(001)$ | $k_{v3}(1)$ | |
| (b) $e + H_2O(000) \rightarrow e + H_2O(002)$ | $k_{v3}(2)$ | |
| (c) $e + H_2O(000) \rightarrow e + H_2O(020)$ | $k_{v2}(2)$ | |
| (d) $e + H_2O(000) \rightarrow e + H_2O(100)$ | $k_{v1}(1)$ | |
| (e) $e + H_2(0) \rightarrow e + H_2(2)$ | $k_{H2}(2)$ | |
| (f) $e + H_2(0) \rightarrow e + H_2(1)$ | $k_{H2}(1)$ | |
| (g) $H_2O(002) + M \rightarrow H_2O(001) + M$ | | $\tau_{v3}(2)$ |
| (h) $H_2O(001) + M \rightarrow H_2O(000) + M$ | | $\tau_{v3}(1)$ |
| (i) $H_2O(020) + M \rightarrow H_2O(010) + M$ | | $\tau_{v2}(2)$ |
| (j) $H_2O(020) + M \rightarrow H_2O(000) + M$ | | $\tau_{v2}(1)$ |
| (k) $H_2O(100) + M \rightarrow H_2O(000) + M$ | | $\tau_{v1}(1)$ |
| (l) $H_2(2) + M \rightarrow H_2(1) + M$ | | $\tau_{H2}(1)$ |
| (m) $H_2(1) + M \rightarrow H_2(0) + M$ | | $\tau_{H2}(1)$ |
| (n) $H_2(1) + H_2O(000) \rightarrow H_2O + H_2O(001)$ | $k_{vv}(1)$ | $\tau_{vv}(1)$ |
| (o) $H_2(1) + H_2O(000) \rightarrow H_2O(100)$ | $k_{vv}(2)$ | $\tau_{vv}(2)$ |

the various energy exchange processes shown in Table 9.5, where $M$ represents the collision partner $H_2O$, $H_2$, or $N_2$.

It is obvious from the discussion above that the emission spectrum of $H_2O$ is quite rich. Figure 9.41 displays the lasing assignments by Benedict et al., where the transitions are given in $cm^{-1}$ [91]. Here, the stronger CW transitions are denoted by an asterisk, while a question mark denotes a probable identity. For each level, the quantum numbers $K_a$ and $K_c$ are provided, while the columns indicate the appropriate vibrational state and the $J$ value. Mixing of two vibrational states due to a near coincidence in energy is indicated by the wavy lines. Finally, the symmetry classes $+ +$ and $+ -$ are shown as solid lines, and classes $- +$ and $- -$ are shown as dashed lines.

In the following, we discuss briefly some of the CW electrical discharge laser configurations that have been successfully employed. A particular application that has resulted in high-power CW lasers has been magnetic fusion plasma diagnostics. Researchers have reported $\approx 250$ mW at 195 $\mu$m using DCN [77, 79], $\approx 50$ mW at 118.6 $\mu$m using $H_2O$ [80], and $\approx 170$ mW at 337 $\mu$m using HCN [76, 92].

The results above have been obtained in longitudinal discharge lasers which are represented by the thermally stabilized DCN laser shown in Fig. 9.42. There are several features to note. First, the temperature of the laser waveguide tube wall is regulated by means of a jacket through which heated ($140°$C) oil flows. This has the desired effect of providing discharge stability as well as eliminating the formation of the polymer deposits, which would otherwise degrade the laser performance. The laser resonator consists of a 3.2-cm-diameter, 70-cm-long overmoded Pyrex waveguide together with two plane reflecting optics mounted nearly flush with the ends of the waveguide. The output coupler consists of a copper mesh as shown in Fig. 9.42, with the output and vacuum window constructed from 50-$\mu$m-thick Mylar. The discharge path flows between a tantalum cathode heated up to $\sim 3000$ K by the discharge and a hollow brass anode. As mentioned above, the dielectric waveguide is overmoded and can therefore support a variety of modes. The modes in a hollow dielectric circular guide can be divided into three categories: transverse electric $TE_{om}$, transverse magnetic $TM_{om}$, and hybrid $EH_{nm}$ modes [93–95]. The linearly polarized $EH_{11}$ mode is particularly attractive for FIR lasers, as it possesses extremely low loss and couples well to a free-space Gaussian mode. The measured radial intensity profile of the DCN laser described by Bruneau et al. [78] is shown in Fig. 9.43 and is seen to fit the calculated $EH_{11}$ profile quite well. The quasi-Gaussian nature of this mode is clearly seen from this figure and is obviously responsible for the excellent propagation characteristics. Figure 9.44 shows the measured spot diameter at the $e^{-1}$ intensity points of this, together with a best fit to the predictions of Gaussian beam theory:

$$d^2(z) = d_0^2[1 + (2\lambda_0 z/\pi d_0^2)^2]$$

where $d_0$ is the spot diameter at the output coupler. Beam focusing and propagation characteristics are extremely important since applications such as fusion plasma diagnostics and radio astronomy often involve extremely long unguided propagation distances. Here, it should be noted that the excellent laser mode quality was a result of the use of dielectric waveguide together with output coupling over the full aperture of the tube. Early lasers with small coupling holes located at the center of one of the resonator mirrors had considerably poorer mode qualities. More details are given in the following section, concerning optically pumped FIR laser output coupler design.

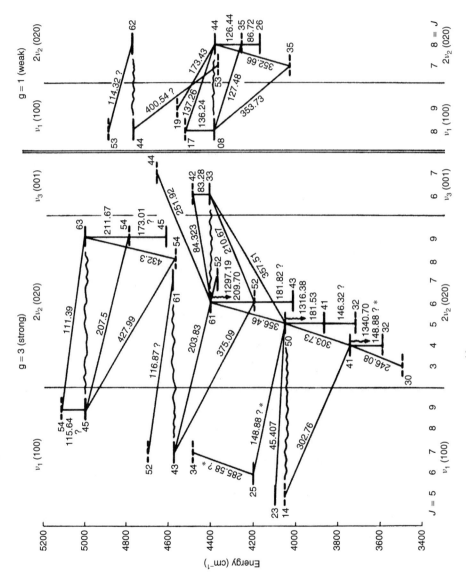

**Figure 9.41** Energy-level diagram of $H_2^{16}O$ showing most of the observed laser transition. (From Ref. 91.)

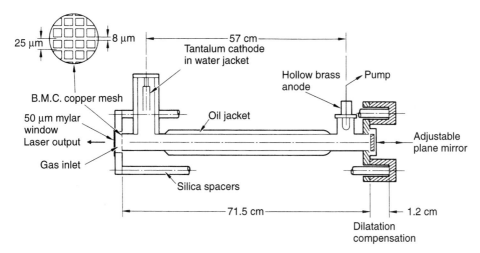

**Figure 9.42**   Schematic of the DCN electrical discharge waveguide laser. (After Ref. 78.)

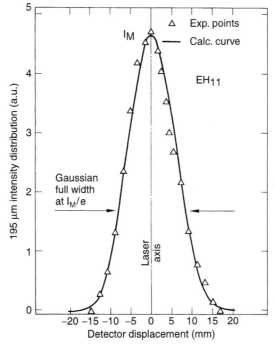

**Figure 9.43**   Radial intensity profile of the EH$_{11}$ waveguide at 23 cm from the mesh couple of the DCN laser shown in Fig. 9.41. A Gaussian curve calculated with the experimental full width of 15.7 mm at $e^{-1}$ of the intensity is shown for comparison. (After Ref. 78.)

Using a configuration similar to that shown in Fig. 9.42, Belland and Veron have performed detailed optimization of a DCN laser system [79]. To carry out this study, a variety of lasers were employed as listed in Table 9.6. As can be seen from the table, two different coupling techniques were employed. In one approach (denoted the

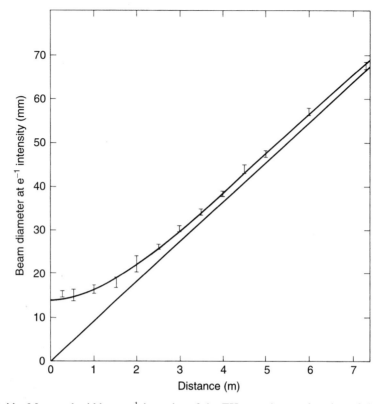

**Figure 9.44**   Measured width at $e^{-1}$ intensity of the $EH_{11}$ mode as a function of the distance from the mesh output coupler of the DCN laser shown in Fig. 9.41. The full curve is the best fit spot size to the experimental points. (After Ref. 78.)

**TABLE 9.6   Geometrical Details of CW DCN Lasers Supporting Their Study[a]**

|  | Laser | | | | |
|---|---|---|---|---|---|
|  | 1 | 2 | 3 | 4 | 5 |
| $L$ (m) | 0.57 | 1.0 | 1.0 | 1.0 | 3.0 |
| $D$ (cm) | 3.2 | 3.0 | 5.0 | 7.0 | 5.4 |
| Cavity | WG | FP | FP | FP | WG |
| Reflector | P | P | P | P | P |
|  | M | C (3 m) | C (3 m) | C (3 m) | M |
| Coupling | M (1000) | BS | BS | BS | M (500) |
| $t$ (%) | 0.9 | 1.8 | 1.8 | 1.8 | 6.0 |
| $a$ (%) | 2.2 | 6.2[b] | 2.2 | 2.0 | 2.0 |
| $S$(cm$^2$) | 2.0 | 1.8 | 1.8 | 1.8 | 5.7 |

*Source*: Ref. 79.
[a]Cavity length $\sim L + 0.4$ m; WG, waveguide; FP, Fabry–Perot; P, plane mirror; M, metal mesh (Buckbee Mears Co.); C (3 m), concave mirror with a 3-m curvature radius; BS, 45°–thin polyethylene beam splitter; $S = e^{-1}$ intensity cross-sectional area of the laser beam.
[b]This high value of diffraction loss is related to the small Fresnel number.

**TABLE 9.7 Operating Conditions, Measured Parameters, and Similarity Laws for Optimized CW DCN Laser Discharges**[a]

| | Laser | | | | | |
|---|---|---|---|---|---|---|
| | 1 | 2 | 3 | 4 | 5 | D law |
| $L$ (m)/$D$ (cm) | 0.57:3.2 | 1:3 | 1:5 | 1:7 | 3:5.4 | |
| $N_2$/$CD_4$/He | 1:3:12 | 4:9:34 | 2:5:23 | 3:7:27 | 2:5:20 | 2:5:20 |
| Flow rate (cm$^3$/min) | 53 | 47 | 89 | 92 | 80 | $15D$ |
| $P$ (torr) | 1.4 | 1.4 | 1.8 | 2.0 | 1.8 | $0.8D^{1/2}$ |
| $I$ (A) | 1.2 | 1.1 | 1.6 | 1.7 | 1.6 | $0.68D^{1/2}$ |
| $V$ (kV) | 1.7 | 2.2 | 1.9 | 1.8 | 3.6 | |
| $E$ (V/cm) | 10.0 | 10.0 | 7.7 | 6.9 | 8.2 | $18D^{-1/2}$ |
| $EI$ (W/cm) | 12.0 | 11.0 | 12.3 | 11.7 | 13.1 | 12 |

*Source*: Ref. 79.
[a]Tube wall temperature = 440 K (constant value); $E$, axial electric field = $(V - V_E)/L_D$; $L_D$, electrical discharge length $\sim L + 0.3$ m; $V_E$, electrode voltage drop $\sim 0.9$ kV.

FP cavity), a thin polyethylene beam splitter was utilized, while in the second (WG cavity), a metal mesh coupler (see next section) was utilized. Table 9.7 displays the results of their power output optimization studies for a number of discharge lengths and discharge tube diameters. The last column in Table 9.7 provides their scaling laws. An average gas mixture of $N_2$/$CD_4$/He = 2:5:20 was found optimum with only a slight dependence on the diameter $D$. The optimum flow rate is found to be proportional to $D$, while both $p$ and $I$ are proportional to $D^{1/2}$. Since the optimum axial electric field $E$ is a function of $D^{-1/2}$, they find that $EI = 12$ W/cm.

Since the homogeneous line width of DCN is comparable to HCN ($6 \pm 1$ MHz) and the total measured line width was $\approx 8.5$ MHz, Belland and Veron conclude that the line is homogeneously broadened, so that the laser output power is given by [79]

$$P = tSW_s \left( \frac{g_0 L}{a + t} - 1 \right)$$

where $g_0$ is the unsaturated gain per unit length, $W_s$ the saturation intensity, $S$ the beam cross-sectional area, and $t$ and $a$ are, respectively, the output coupling and the loss per single pass. Belland and Veron find that $g_0$ and $W_s$ can be written as [79]

$$g_0(\text{m}^{-1}) = K/D$$

and

$$W_s(\text{mW/cm}^2) = K'/D$$

where $K$ and $K'$ are as given in Table 9.8.

For optimized output coupling, the laser power is given by

$$P_0 = SW_s[(g_0 L)^{1/2} - a^{1/2}]^2$$

where $a$ is the single-pass loss coefficient. The latter is comprised of two terms, with the first, $a_0$, due to mirror losses ($\approx 2\%$ for the Belland and Veron lasers) and independent

**TABLE 9.8   Coefficients $K$ and $K'$ for Unsaturated Gain $g_0$ and Saturation Intensity $W_s$, for each Wavelength of the CW DCN Laser[a]**

| Tuning for: | Wavelength | $K$ (%/m-cm) | $K'$ (mW/cm) |
|---|---|---|---|
| $\lambda_1$ and $\lambda_2'$ | $\lambda_1$ | 44.3 | 1690 |
|  | $\lambda_2'$ | 46.4 | 1065 |
| $\lambda_2$ and $\lambda_1'$ | $\lambda_2$ | 41.6 | 2020 |
|  | $\lambda_1'$ | 38.3 | 1410 |
| $\lambda_1$ and $\lambda_2$ | $\lambda_1$ | 44.3 | 1160 |
|  | $\lambda_2$ | 41.6 | 1385 |

*Source*: Ref. 79.
[a] $\lambda_1 = 189.95$ μm, $\lambda_2 = 194.70$ μm, $\lambda_1' = 190.01$ μm, $\lambda_2' = 194.76$ μm.

of tube length and diameter. The second term, $a_1$, is associated with the dielectric waveguide losses. For the $EH_{11}$ mode in Pyrex waveguide, the loss coefficient is given by

$$a_1 = \frac{0.136(L + L')(\text{cm})}{[D\ (\text{cm})]^3}$$

where $L + L'$ is the total length of the waveguide tube. In the case of the $EH_{11}$ mode, the effective beam area $S_1$ is approximately

$$S_1 = \pi(0.5D)^2/4$$

The above then yields an optimization expression given by

$$P_o = \frac{\pi(0.5)^2}{4\ K'}\left\{(KL)^{1/2} - \left(a_0D + k_1(\hat{\eta}, \lambda)\frac{L + L'}{D^2}\right)^{1/2}\right\}^2$$

where $L$ and $L'$ are expressed in meters, $D$ is in centimeters, $P_o$ is in milliwatts, and $k_1(\hat{\eta}, \lambda) = 2\alpha D^3$, where $\hat{\eta}$ is the complex refractive index of the wall material and $\alpha$ is the electric field attenuation coefficient. Shown in Fig. 9.45 are predicted output power curves for a variety of tube diameters, discharge lengths, and $a_0$ values for a 195-μm DCN laser. Belland and Veron further point out that for fixed $L$ there is an optimum tube diameter $D_0$ (corresponding to $a_1 = a_0/2$) given by

$$D_0 = \left[2k_1(\hat{\eta}, \lambda)\frac{L + L'}{a_0}\right]^{1/3}$$

The relation above results in an optimum output power given by

$$P_o = \pi\frac{0.5^2}{4}K'[(KL)^{1/2} - (3a_0D_0/2)^{1/2}]^2$$

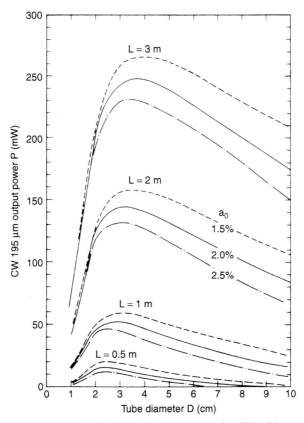

**Figure 9.45** Output power as a function of tube diameter of a CW 195-$\mu$m DCN waveguide laser with loss $a_0$ of 1.5, 2, and 2.5% per single pass for discharge lengths of 0.5, 1, 2, and 3. The Pyrex tube is always assumed to be 0.4 m longer than the discharge. The values given by the curves correspond to optimum gas conditions and cavity coupling for each combination of discharge length and tube diameter. Simultaneous tuning on the 190-$\mu$m line is achieved.

Figure 9.46 displays the predictions for the expression above. Finally, assuming a value of $a_0 = 2\%$ gives rise to the predictions displayed in Fig. 9.47 for both DCN and HCN. Note that a length of $\approx 8$ to 9 m is required to obtain 1 W of output.

The laser systems described above have involved longitudinal electrical discharges. However, excellent results have been reported from capacitively coupled RF-excited systems [96]. Figure 9.48 schematically illustrates the various electrodeless configurations that were investigated. In these arrangements, the RF electric fields are applied in the circumferential direction in case a, transversely with respect to the tube axis (case b) and along the tube axis (case c). Of the configurations above, the latter denoted $C$-coupled LERF (longitudinal electrodeless RF) was found to produce the highest output power. Figure 9.49 provides a diagram of the experimental arrangement. Laser tubes with diameters of 3.0, 4.0, and 5.5 cm were tested with parallel-plane mirrors comprising the 1-m-long cavity. As with the laser described previously, the output mode is principally the $EH_{11}$. The power is coupled out a 2.5- to 7.0-mm-diameter hole located at the center of one of the mirrors. The electrodes had a diameter of 8.0 cm and a width of 14.0 cm. A gas mixture of $CH_4/N_2 = 1:1$ was employed with a flow

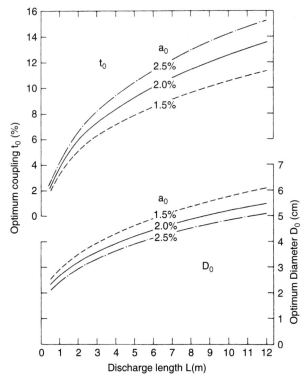

**Figure 9.46** Optimum tube diameter and corresponding optimum coupling as a function of discharge length of a CW 195-$\mu$m DCN waveguide laser for loss $a_0$ of 1.5, 2, and 2.5% per single pass.

rate of 40 cm$^3$/min STP for the largest-diameter tube. The $\pi$-type matching network typically consisted of $L_1 = 0.41$ $\mu$H and $C_1 = C_2 = 20$ to 100 pF. Using 1.5 kW drive at 27 MHz, a maximum CW output of 260 mW was obtained at 337 $\mu$m using a 1-m-long tube. An interesting feature of these studies was that the authors found that the $g_0$ value was 1.8 times that of the dc discharge while $I_s$ was essentially the same.

The discussion above concentrated solely on CW operation since it is the most commonly employed. However, it is of interest to touch basically on the pulsed case. The first studies were concerned with pulsed longitudinal discharges [97–99]. The highest reported power for such longitudinal excitation types of laser has been 1 kW with long-term reproducible repetitive operation at the 450-W level [98]. These results were obtained using a 3.6-m-long Fabry–Perot resonator with two plane 14-cm-diameter mirrors. Jassby et al. reported on an interesting approach where an auxiliary dc discharge was employed to increase the output of their pulsed HCN laser [99] (see Fig. 9.50). Using this approach in their 7.0-m-long (15-cm-diameter) laser, the laser output power was increased by a factor of 7 to $\approx$300 W. An obvious approach to high pulsed power is to employ transverse excitation. This was first reported by Lam et al. [100] using the arrangement shown in Fig. 9.51. Here, the authors reported output powers comparable to those achieved with axially pumped discharges of similar volume and current density. This approach was further investigated [101, 102] using the configuration shown in Fig. 9.52. Here, $\approx$2200 currentlimiting resistors (1 k$\Omega$,

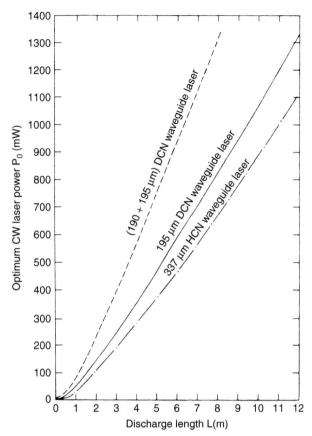

**Figure 9.47** Optimum output power as a function of discharge length of a CW 195-μm DCN waveguide laser, for a practical loss of 2%. The same assumptions as those of Fig. 9.38 are made. The total output power both obtainable on the two wavelengths 195 and 190 μm is also drawn. The optimum output power of a CW 337-μm HCN waveguide laser is reproduced for comparison.

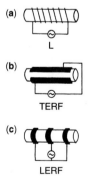

**Figure 9.48** Three systems giving electrodeless discharges: **(a)** L-coupled; **(b)** C-coupled TERF (transverse electrodeless RF); **(c)** C-coupled LERF (longitudinal electrodeless RF). (From Ref. 96.)

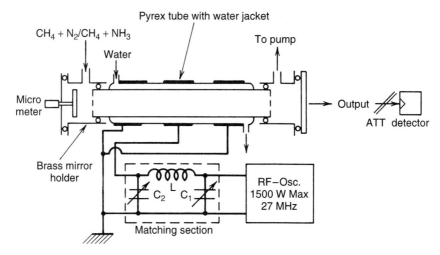

**Figure 9.49**   Outline of experimental apparatus. (From Ref. 96.)

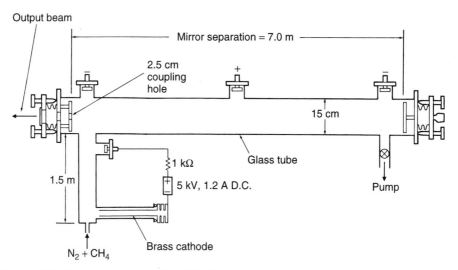

**Figure 9.50**   Schematic design of the HCN laser showing laser cavity and auxiliary discharge. (From Ref. 99.)

2 W) comprised the cathode as in the early pin-type $CO_2$ lasers. A flat aluminum surface resonator comprised the anode. The laser resonator consisted of a plexiglass enclosure as illustrated in Fig. 9.53. Both the cathode and anode assemblies were attached to the plexiglass with epoxy. Details of the mode properties of this structure have been provided by Adam and Kneubuhl [102]. Peak power outputs of at least 100 W were reported from this configuration [101]. Significant improvements were reported by Sturzenegger et al. [103], who added UV preionization as shown in Fig. 9.54. This was further improved by the use of a flowing system with a flow rate HCN/CH$_4$/He = 5:30:40 torr/L-min [104]. Typical operating pressures were 2 to 4 torr with pulse durations of 27 to 8.5 μs, respectively, with pulse energies between

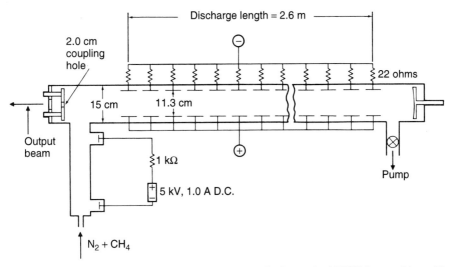

**Figure 9.51** Schematic drawing of the transverse-excitation pulsed HCN laser with auxiliary discharge. (From Ref. 100.)

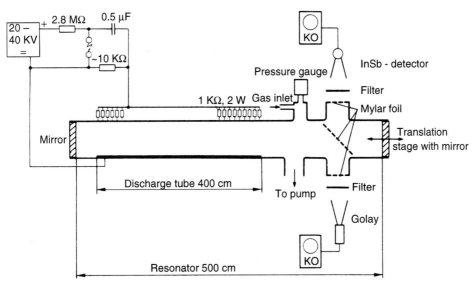

**Figure 9.52** Experimental arrangement of a transversely excited submillimeter wave HCN laser. (From Ref. 102.)

1 and 15 mJ. The results above have been summarized by Kneubuhl and Sturzenegger, who have taken account of the actual discharge volumes to provide normalized input and output energies for ease of comparison [66]. Table 9.9 summarizes their results.

### 9.2.3 Optically Pumped Fir Lasers

*Introduction and Theory.* As mentioned in the preceding, the pioneering work in this area was performed by Chang and Bridges [71]. This was followed by numerous

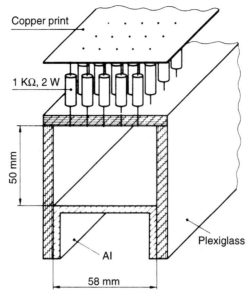

**Figure 9.53** HCN laser waveguide structure with pin-type electrodes for transverse excitation. (From Ref. 102.)

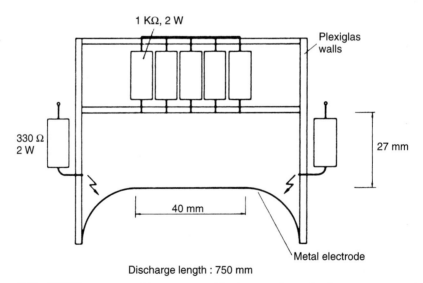

**Figure 9.54** HCN laser waveguide structure with pin-type electrodes for transverse excitation with UV preionization. (From Ref. 103.)

physics and technology advances which have made this laser extremely useful in fields ranging from spectroscopy to fusion plasma diagnostics. Excellent reviews of this field have been written by a number of authors [105–111]. In addition, a number of authors have provided tables of FIR emission and pump wavelengths for a variety of gases [49–52, 72].

**TABLE 9.9   Pulse Energy, Average Pulse Powers, and Efficiency of Electrically Excited HCN Lasers**

| Authors | Excitation | Discharge Volume (L) | Input Energy | | Output Energy | | | Power (W) | Efficiency |
|---|---|---|---|---|---|---|---|---|---|
| | | | J | J/L | mJ | μJ/L | | | |
| Turner and Poehler [97] | Longitudinal | ≈39 | 160 | ≈4 | | | | 25 | ≈$1.5 \times 10^{-4}$ |
| Sharp and Wetherell [98] | Longitudinal | ≈60 | ≈110 | ≈1.8 | ≈17 | 280 | | 1000 | ≈$2 \times 10^{-5}$ |
| Jassby et al. [99] | Longitudinal | ≈106 | ≈156 | ≈1.5 | 3 | 30 | | 300 | ≈$2.5 \times 10^{-7}$ |
| Lam et al. [100] | Transverse | ≈22 | ≈180 | ≈8 | ≈0.045 | ≈2 | | 9 | > $3 \times 10^{-7}$ |
| Adam et al. [101] | Transverse | 10 | 312 | 31 | >0.1 | >10 | | >10 | < $3 \times 10^{-6}$ |
| | | | | | <1 | <100 | | <100 | |
| Sturzenegger et al. [104] | Transverse | 1.6 | 90 | 56 | >1 | >620 | | >100 | > $10^{-5}$ |
| | | | | | <15 | <9400 | | <1500 | < $1.5 \times 10^{-4}$ |

*Source*: After Ref. 66.

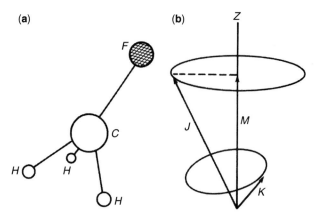

**(a)** **(b)**

**Figure 9.55** Schematic of symmetric-top molecule (CH$_3$F) and classical representation of angular momentum quantum numbers $J$, $K$, and $M$ ($J$ represents total rotational angular momentum, $K$ its projection along figure axis, and $M$ its projection along a space-fixed direction). (From Ref. 108.)

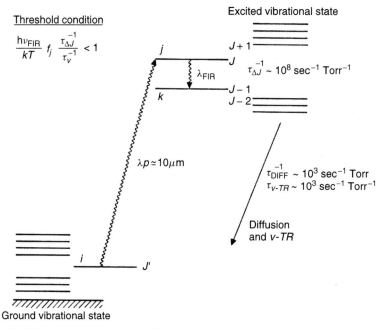

**Figure 9.56** Schematic energy-level diagram of a polar molecule, illustrating the optical pumping process. (From Ref. 107.)

The earliest optically pumped FIR laser studies involved the CH$_3$F molecule. This is a polar symmetric top molecule as illustrated in Fig. 9.55, together with a representation of the appropriate angular momenta. The optical pumping process can be understood by referring to Fig. 9.56, which schematically illustrates the vibrational–rotational energy-level diagram for a polar molecule. Here, the near coincidence of an infrared pump photon ($\lambda_p \approx 10$ $\mu$m) leads to a transition from a ground vibrational state to an excited vibrational state where the vibrational energy is given by

$$E_v^{\text{vib}} = h\nu v(v + \tfrac{1}{2})$$

where $v$ is the vibrational quantum number. Assuming a nondegenerate vibrational mode, the rotational energy (including both angular momentum and centrifugal rotation terms) can be written as [108, 112]

$$E_v^{\text{rot}} = B_v J(J + 1) + (C_v - B_v)K^2 - D_{jv}J^2(J + 1)^2$$
$$- D_{jkv} J(J + 1)K^2 - D_{kv} K^4 + \cdots$$

In the above, $J$ is the total angular momentum quantum number and $K$ its projection (see Fig. 9.55) with $0 \leqslant K \leqslant J$ and $B_v$, $C_v$, $D_{jv}$, $D_{jkv}$, and $D_{kv}$ are vibrational–rotational constants. The near pump coincidence is then seen to be a result of the multiplicity of energy levels. In the case of a pure rotational transition as appropriate for molecules possessing a permanent dipole moment, the selection rules $\Delta J = -1$ and $\Delta K = 0$ lead to a transition frequency between inverted rotational levels in the upper vibrational state given by

$$\nu = -2B_v(J + 1) - 4D_{jv}(J + 1)^3 - 2D_{jkv}(J + 1)^2 K^2$$

In addition, cascade transitions in the upper state can also occur as well as refilling transitions in the lower vibrational state. Another interesting point to note is that due to the permanent dipole moment $\mu$, the molecules exhibit a linear Stark splitting given by

$$\Delta E = \mu K M E_s [J(J + 1)]^{-1}$$

where $E_s$ is the strength of the applied electric field. The result above leads to the possibility of a small Stark tuning for frequency tuning, modulation, and optimization as well as the generation of new laser emissions [108, 113–118].

Another feature of note in FIR lasers is the output polarization, which can either be parallel or perpendicular to the pump polarization. The nature of the output polarization is determined by the quantum number $M$, which governs the dependence of the dipole matrix elements. The calculated polarizations [108, 119] are shown in Table 9.10. As pointed out by Tobin, the simplest case corresponds to the symmetric top molecule where the pump transition may be P, Q, or R($\Delta J = -1, 0, +1$), while the primary FIR emission must be of the R-type where $\Delta J = -1$.

We will now proceed to a discussion of the laser kinetics, which is essentially a summary of the detailed discussion provided by DeTemple and Danielewicz [105]. Here, the authors first begin with a simple rate equation description which corresponds to the incoherent pumping case. This approach provides a description of a number of features including the cutoff pressures and bimodal peaks [105]. However, a more sophisticated coherent pumping model is required to include multiphoton effects, which lead to effects such as gain asymmetries.

The incoherent pump model is readily introduced through consideration of the $^{12}$CH$_3$F lasing transition at 496 $\mu$m [105, 120, 121]. Figure 9.57 schematically depicts a partial energy diagram illustrating the relevant states and collision processes. As

**TABLE 9.10 Degree of Polarization of Fluorescence and Small-Signal Gain for Linearly Polarized Pump Radiation**

| Transition Sequence (Pump ↑, Emission ↓) | Degree of Polarization, $P^a$ (Quantum Mechanical) | Classical Limit of $p^b$ $J \to \infty$ | Preferred Gain |
|---|---|---|---|
| $(Q\uparrow, Q\downarrow)$ | $\dfrac{(2J+3)(2J-1)}{8J^2+8J-1}$ | $\dfrac{1}{2} \to \dfrac{I_\parallel}{I_\perp} = 3$ | $\parallel$ |
| $(P\uparrow, P\downarrow)$ | $\dfrac{J(2J-1)}{14J^2+33J+20}$ | $\dfrac{1}{7} \to \dfrac{I_\parallel}{I_\perp} = \dfrac{4}{3}$ | $\parallel$ |
| $(R\uparrow, R\downarrow)$ | $\dfrac{(2J+3)(J+1)}{14J^2-5J+1}$ | $\dfrac{1}{7}$ | $\parallel$ |
| $(Q\uparrow, P\downarrow)$ or $(P\uparrow, Q\downarrow)$ | $\dfrac{-(2J-1)}{6J+7}$ | $-\dfrac{1}{3} \to \dfrac{I_\parallel}{I_\perp} = \dfrac{1}{2}$ | $\perp$ |
| $(Q\uparrow, R\downarrow)$ or $(R\uparrow, Q\downarrow)$ | $\dfrac{-(2J+3)}{6J-1}$ | $-\dfrac{1}{3}$ | $\perp$ |
| $(P\uparrow, R\downarrow)$ or $(R\uparrow, P\downarrow)$ | $\dfrac{1}{7}$ | $\dfrac{1}{7} \to \dfrac{I_\parallel}{I_\perp} = \dfrac{4}{3}$ | $\parallel$ |

*Source*: After Ref. 108.
[a]The degree of polarization, $P = (I_\parallel - I_\perp)/(I_\parallel + I_\perp)$, where $I_\parallel, I_\perp$ are the fluorescence intensity with polarization $\parallel, \perp$ to that of the pump.
[b]The expression for $P$ can be transposed to give $I_\parallel/I_\perp = (I + P)/(I - P)$.

illustrated, the dominant absorption and emission are associated with the $K = 2$ rotational manifold. In the presence of the pump, population is promoted from the ground state ($J = 12, K = 2, v = 0$) to the upper lasing level ($J = 12, K = 2, v = 1$). Far-infrared lasing then occurs due to the $^QR(11, 12)$ rotational transition ($J = 12 \to 11, K = 2$). For sufficiently low pressure, the absorption is Doppler broadened, which results in the excitation of states with a specific axial velocity $v_z$ and magnetic quantum number $m$, giving rise to the polarization properties of the laser. The magnitude of the broadening $\Delta v_0$ is given by

$$\Delta v_D = 7.162 \times 10^{-7}[T(\text{K})/M_m]^{1/2}v_0 \quad \text{MHz}$$

where $v_0$ is the frequency and $M_m$ is the molecular weight of the molecule. This yields a value of $\Delta v_D^P = 67$ MHz for the pump transition and $\Delta v_D^{\text{FIR}} = 1.29$ MHz for the emission. The above can be compared with the pressure-broadening coefficient of $\approx 40$ MHz/torr, which leads to a pressure broadening of $\approx 2$ MHz for the typical operating pressures of 50 mtorr. Therefore, the pump transition is essentially completely Doppler broadened.

DeTemple and Danielewicz obtain the FIR gain, $\alpha_{\text{FIR}}$, by assuming that the FIR transition is homogeneously broadened and solving the coupled rate equations [105]. This yields a generalized expression given by

$$\alpha_{\text{FIR}} = \frac{1}{1+(I_{\text{FIR}}/I_{\text{FIR}}^3)L(v_{\text{FIR}}-v_{21})}\left\{\frac{\alpha_{\text{IR}}I_{\text{IR}}}{\Gamma_r}\sigma_{\text{FIR}} + \left(f_2 - \frac{g_2}{g_1}f_1\right)[n_K^{\text{exc}}\sigma_{\text{FIR}}]\right\}$$

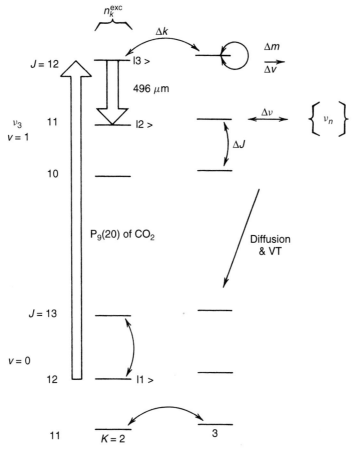

**Figure 9.57** Partial energy-level diagram of $^{12}CH_3F$ showing the states and kinetic processes that are thought to dominate for the 496-$\mu$m line.

where $I_{FIR}$ and $I_{IR}$ are the FIR and pump flux (photons/cm$^2$-s), respectively, $\Gamma_r$ the rotational relaxation rate, $\alpha_{IR}$ the saturated pump absorption coefficient, and $f_1$ the fractional rotational equilibrium population of degeneracy $g_1$ within the $K$ manifold, which has nonequilibrium population $n_k^{exc}$. In the above, DeTemple and Danielewicz have defined the detuning $L$, the stimulated emission cross section $\sigma_{FIR}$ and the FIR saturation intensity $I_{FIR}^s$ as [105]

$$L(\nu_{FIR} - \nu_{21}) = \frac{(\Delta\nu_H/2)^2}{(\nu_{FIR} - \nu_{21})^2 + (\Delta\nu_H/2)^2}$$

$$\sigma_{FIR} = \sigma_{FIR}^0 L(\nu_{FIR} - \nu_{21})$$

and

$$I_{FIR}^s = \Gamma_r/(1 + g_2/g_1)\sigma_{FIR}^0$$

where

$$\sigma_{FIR}^0 = \lambda_{21}^2 A_{21}/4\pi^2 \Delta\nu_H$$

and $\Delta\nu_H$ is the pressure-broadened line width (full width at half maximum, FWHM) [121, 122]. As pointed out by DeTemple and Danielewicz, the FIR gain expression can be understood relatively simply [121]. The first term in the square brackets represents the gain due to the pump and is proportional to the pump rate $\sigma_{IR} I_{IR}$. This gain is countered by the second term, which is negative and corresponds to absorption due to the equilibrated rotational background population. As they point out, the second term can exceed the thermal value since the excited state population $n_k^{exc}$ is also pumped. Finally, the term outside the square brackets is the familiar saturation expression corresponding to a homogeneously broadened system.

To proceed to calculate FIR output power, one requires the excited-state population, which is given by

$$n_k^{exc} = n_k^e + (\alpha_{IR} I_{IR} / \Gamma) f_k$$

In the above, $n_k^e$ is the thermal-equilibrium population density, $\Gamma$ the appropriate effective relaxation rate [vibrational–translational (VT) or diffusion], and $f_k$ the fractional population of the total excited-state manifold that is in the $K = 2$ rotational bath. The infrared pump intensity is given by

$$I_{IR} = I_{in} (\gamma_{IR} + \alpha_{IR} L_c)^{-1}$$

where $L_c$ is the cavity length, $\gamma_{IR}$ the bare cavity losses, and $\alpha_{IR}$ the absorption coefficient. In the above,

$$\alpha_{IR} = \alpha_{IR}^{(0)} (1 + I_{IR}/I_{IR}^s)^{-1/2}$$

where the saturated IR intensity

$$I_{IR}^s = \Gamma (2\sigma_{IR}^{(0)})^{-1}$$

with $\sigma_{IR}^{(0)}$ the homogeneously broadened IR cross section and $\alpha_{IR}^{(0)}$ is the unsaturated absorption coefficient. Table 9.11 lists the absorption coefficients and pump saturation intensities for a variety of common molecular gases [108, 123].

Using the results above, DeTemple and Danielewicz arrive at the familiar expression for maximum FIR output power given by [105]

$$P_{FIR} = \frac{Q_c Q_q}{1 + (g_2/g_1)} \left[ \frac{\alpha_{IR} L_c}{\gamma_{IR} + \alpha_{IR} L} \right] P_{IR}$$

where $Q_c$ is the cavity efficiency and $Q_q = \overline{h}\omega_{FIR}/\overline{h}\omega_{IR}$ is the quantum efficiency. The term in the square brackets typically ranges between 10 and 50% and corresponds to the power absorbed in the molecular gas.

For low pressures, one has $\alpha_{IR} \alpha p$, so that increased pressure results in increased output power. However, eventually one reaches an optimum pressure beyond which the power decreases. This effect is due to the existence of a cutoff pressure $p_c$ associated either with diffusion or vibrational–translational relaxation (see Ref. 105). Figure 9.58 illustrates the theoretical and experimental power dependences on pressure for the 496-μm emission line of $^{12}CH_3F$ [121]. The agreement is seen to be quite good.

A detailed example of the power of the incoherent pump model [105, 121] is shown in Fig. 9.59. Here the authors have included effects such as the fact that the pump

**TABLE 9.11** **Absorption Coefficients and Pump Saturation Intensities Reported by Weiss [123]**

| NMMW Laser Molecule | $CO_2$ Pump Line | FIR Laser Wavelength ($\mu$m) | Absorption Coefficient[a] ($cm^{-1}$/torr) | Saturation Intensity, $I_s$ (W/$cm^2$-$torr^2$) | Uncertainty ($\pm\%$) |
|---|---|---|---|---|---|
| HCOOH | 9R(16) | 446 | $5.40 \times 10^{-3}$ | 73 | 10 |
| HCOOH | 9R(18) | 393 | $4.90 \times 10^{-2}$ | 48 | 10 |
| HCOOH | 9R(20) | 433 | $1.12 \times 10^{-1}$ | 37 | 10 |
| $CH_3OH$ | 9P(36) | 118 | $1.10 \times 10^{-2}$ | 169 | 20 |
| $CH_3OH$ | 9P(16) | 570 | $3.6 \times 10^{-2}$ | $58^b$ | 20 |
| $CH_3I$ | 10P(32) | 1254 | $9.61 \times 10^{-3}$ | 2950 | 10 |
| $CH_3I$ | 10P(18) | 448 | $1.06 \times 10^{-3}$ | $6 \times 10^{-1}$ | 10 |
| $CH_3I$ | 9R(16) | 377 | $1.1 \times 10^{-4}$ | 15 | 20 |
| $CH_3I$ | 9P(34) | 508 | $3.2 \times 10^{-4}$ | 760 | 20 |
| $CH_3F$ | 10R(32) | 193 | $6 \times 10^{-3}$ | $3 \times 10^{-4}$ | 20 |
| $CH_3F$ | 9P(20) | 496 | $1.75 \times 10^{-2}$ | $2300^b$, $70^c$ | 20 |

[a]Uncertainty $< \pm 1\%$.
[b]$p < 75$ mtorr.
[c]$p > 0.5$ torr.

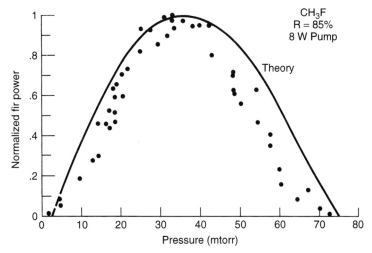

**Figure 9.58** Theoretical and experimental power versus pressure curves for $^{12}CH_3F$ at 496 $\mu$m, $R = 85\%$; 8-W pump. (From Ref. 121; copyright 1976 IEEE, reproduced by permission.)

comprises a standing wave in the usual linear case, radial variations of appropriate quantities, and molecular kinetics. First, it is noted that the standing-wave pump results in separate gain peaks associated with hole burning in the axial velocity distribution. In addition, there is an absorbing background due to the thermalized population contained in the $K = 2$ manifold. Combining the separate gain peaks and absorption curve results in the net gain curve shown in Fig. 9.59. Other effects of interest include power broadening, velocity splitting, and the transferred Lamb dip.

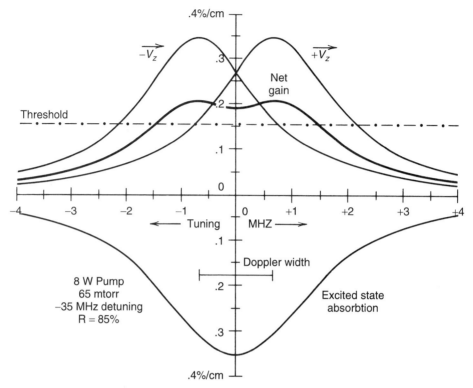

**Figure 9.59** Calculated gain versus cavity tuning for the system shown in Fig. 2.25 showing the gain peaks associated with a nonresonant standing wave. The peaks are located at a FIR frequency detuning equal to $h\omega_{FIR}$ divided by $h\omega_{IR}$ times the pump-frequency detuning (−35 MHz), $R = 85\%$, 8-W pump, pressure 65 mtorr. (From Ref. 121; copyright 1976 IEEE, reproduced by permission.)

As mentioned previously, there arise occasions when a coherent pumping model must be employed. DeTemple and Danielewicz provide the simple example of two waves interacting with a three-level system [105]. In this case, gain may occur either to the usual single-photon interaction, which involves states $|2\rangle$ and $|1\rangle$ (see Fig. 9.57) or to the Raman two-photon interaction involving states $|0\rangle$ and $|1\rangle$. The reader is referred to their work and the references contained therein for details concerning processes such as gain asymmetries and the CW Raman effect.

Due to the dearth of CW sources in the region 100 to 1500 μm, there has been considerable interest in the optimization of FIR lasers and the search for new, efficient lasing transitions. This has led to a desire to characterize those properties that lead to "winner" molecules [105, 111]. Here we will briefly review the relevant characteristics as discussed in detail by DeTemple and Danielewicz. The first is fairly obvious and is just the requirement that the IR absorption coefficient be large. The second constraint involves the smallness of the excited-state absorption. Here, one wishes that the excited rotational states should be coupled to a large number of rotational states in other vibrational levels with fast relaxation for all of them. The result of the above is a large $\alpha_{FIR}$. Another aid in diluting the population in the terminal lasing level is the use of buffer gases.

To quantify the above characterization, two parameters, $X$ and $Y$, can be defined [105]. The quantity $X$ characterizes the absorption strength and is defined as the ratio of the band strength $S$ to the rotational partition function $Z_{rot}$. The second parameter $Y$ depends on the degree of permissible high-pressure operation (and hence high conversion efficiency) and is defined to be the product of $X$ and the relative VT relaxation rate of the molecule. Table 9.12 contains the characteristics of a number of common FIR laser molecules ranging from $^{12}CH_3{}^{79}Br$, which does not possess any true CW lines, to $^{12}CH_3OH$, $^{12}CH_3F_2$, $NH_3$, and $D_2O$, which are "winner" molecules and possess large $X$ and $Y$ values.

DeTemple and Danielewicz have summarized the properties of the, better FIR molecules [105] and their results are reproduced in Table 9.13. The FIR output power levels have been exceeded in a number of cases, as we shall see later. However, the numbers are still of importance as guides to the relative strengths of the lines. For those cases where no powers are listed, the authors note that optimization studies have not been performed, but the $X$ and $Y$ parameter values promise efficient output.

***FIR Laser Design.*** We are now ready to turn to experimental details concerning FIR laser design and begin with the CW case.

*CW Optically Pumped Lasers.* As with all lasers, the heart of an FIR laser is the resonator. Figure 9.60 illustrates many of the more commonly employed resonator configurations [107]. Note that in all of the cases shown, the infrared pump is introduced into the cavity by means of a tiny hole in one of the end mirrors. As we shall see later, one problem with this approach is that feedback to the $CO_2$ laser can cause stability problems. Referring back to Table 9.9, we note that the absorption coefficients are quite small ($\sim 10^{-4}$ to $10^{-2}$ $cm^{-1}$/torr), so that the power density can be quite high

**TABLE 9.12   General Molecular Characteristics**

| Molecule | Mode | $S^a$ | $Z_{rot}{}^b$ | $X(\times 10^3)^c$ | $Y(\times 10^6)^d$ |
|---|---|---|---|---|---|
| $^{12}CH_3{}^{79}Br$ | $\nu_6$ | 29 | 7,290 | 4 | 2.5 |
| $^{12}CH_3I$ | $\nu_6$ | 36 | 9,294 | 4 | 3.8 |
| $^{12}CH_3F$ | $\nu_3$ | 440 | 2,730 | 161 | 4.2 |
| HCOOH | $\nu_6$ | $200^e$ | 8,728 | 23 | 18 |
| $^{12}C_2H_2F_2$ | $\nu_9$ | $200^e$ | 34,772 | 6 | 30 |
| HFCO | $\nu_4$ | 645 | 8,179 | 79 | 35 |
| $^{12}CF_2{}^{35}Cl_2$ | $\nu_6$ | 810 | 175,845 | 5 | 47 |
| $^{12}CH_2F_2$ | $\nu_{q(2)}$ | 1,196(269) | 12,404 | 96(22) | 159(36) |
| $^{12}CH_3OH$ | $\nu_6$ | $832^e$ | 14,280 | 58 | 1280 |
| $D_2O$ | $\nu_2$ | 400 | 225 | $2 \times 10^3$ | $9 \times 10^4$ |
| $^{14}NH_3$ | $\nu_2$ | 590 | 210 | $3 \times 10^3$ | $1.7 \times 10^5$ |

*Source*: After Ref. 105.

[a] Band intensity in units of $cm^{-3}$/atm at 300 K [124, 125].

[b] Rotational partition function calculated from spectroscopic constants exclusive of nuclear-spin statistics.

[c] $X = S/Z_{rot}$.

[d] $Y = X\gamma_{vt}$ is assumed to scale according to $10^{\alpha\nu_{min}/\sqrt{m}}$, where $\nu_{min}$ is lowest vibrational mode frequency and $m$ the molecular mass. The scaling, which is appropriate to carbonbearing molecules [126], was referenced to $^{12}CH_3F$ [127], using relaxation data for $NH_3$ [128] and $D_2O$ [129, 130].

[e] Estimated.

**TABLE 9.13  Characteristics of CW FIR Laser Lines**

| Species | Line | Mode | Transition | Offset (MHz) | $\lambda_{FIR}$ (μm)[a] | $\lambda_{FIR}$ (cm$^{-1}$)[b] | State | FIR-Output Polarization[c] | Power (mW) |
|---|---|---|---|---|---|---|---|---|---|
| D$_2$O[d] | R$_9$(17) | $\nu_2$ | $10_{-8} \to 9_{-6}$ | 32 | 113 | $88.8269_C$ | $9_{-8}$ | ⊥ | 6 |
| SO$_2$[e] | R$_9$(14) | $\nu_2$ | $37_{16,22} \to 36_{15,21}$ | 23 | 140 | $71.5834_C$ | $35_{14,22}$ | = | 3.5 |
| | R$_9$(18) | $\nu_3$ | $25_{18,8} \to 24_{17,7}$ | $-51$ | 142 | $70.3152_C$ | $23_{16,8}$ | = | 3 |
| $^{14}$NH$_3$[f] | R$_9$(30) | $\nu_2$ | 5,0,s → 6,0,a | 166 | 67[g] | $148.9718_C$ | 5,0,a | = | — |
| | R$_{10}$(2) | $\nu_2$ | 9,5,s → 9,5,a | $-180$ | 370[g] | $27.0013_C$ | 9,5,s | = | 1 |
| | R$_9$(17) | $\nu_2$ | 3,3,s → 4,3,a | 155 | 87 | $114.8181_C$ | 3,3,s | = | — |
| | P$_{10}$(13) | $\nu_2$ | 8,7,a → 8,7,s | <43 | 81 | $122.7335_C$ | 7,7,a | ⊥ | 40 |
| $^{15}$NH$_3$[h] | R$_{10}$(18) | $\nu_2$ | 4,4,a → 5,4,s | — | 153 | $65.471_C$ | 4,4,a | ⊥ | 180 |
| | R$_{10}$(42) | $\nu_2$ | 2,0,a → 3,0,s | — | 375 | $26.785_C$ | 2,0,a | = | 23 |
| HFCO[i] | R$_9$(20) | $\nu_4$ | | — | 196 | — | — | — | — |
| | R$_9$(28) | $\nu_4$ | | — | 280 | — | — | — | — |
| | R$_9$(32) | $\nu_4$ | | — | 432 | — | — | — | — |
| $^{12}$CH$_3$F[j] | P$_9$(20) | $\nu_3$ | 12,2 → 12,2 | $-44$ | 496 | $20.1574_M$ | 11,2 | ⊥ | 10 |
| | P$_9$(15) | $\nu_3$ | 6,2 → 6,2 | $-26$ | 992 | $10.0836_C$ | 5,2 | ⊥ | 1 |
| $^{13}$CH$_3$F[k] | P$_9$(32) | $\nu_3$ | 4,3 → 5,3 | 24 | 1222 | $8.1840_C$ | 4,3 | = | 10 |
| $^{12}$CD$_3$F[l] | R$_{10}$(48) | $\nu_3$ | 20,5 → 20,5 | 50 | 368 | $27.1703_C$ | 19,5 | ⊥ | 10 |
| | P$_9$(16) | $\nu_3$? | | — | 206 | $48.54_M$ | — | = | 25 |
| $^{13}$CD$_3$F[m] | R$_9$(22) | | | — | 326 | $30.684_M$ | — | ⊥ | >5 |
| | P$_{10}$(34) | | | — | 470 | $21.281_M$ | — | = | >5 |
| $^{12}$CH$_3$I[n] | P$_{10}$(18) | $\nu_6$ | 44,5 → 45,6 | 50 | 447 | $22.364_M$ | 44,6 | = | 40 |
| $^{12}$CDF$_3$[o] | R$_{10}$(10) | $\nu_5$ | 24,18 → 23,17 | — | 659 | $15.20_C$ | 22,17 | = | 3 |
| HCOOH | R$_9$(20) | $\nu_?$ | 34$_?$ → 33$_?$ | — | 433 | $23.1143_M$ | 32$_?$ | = | 35 |
| | R$_9$(22) | $\nu_?$ | 33? → 32? | — | 419 | $23.8884_M$ | 31,? | = | 30 |
| | R$_9$(18) | $\nu_?$ | $35_{-18} \to 34_{-17}$ | — | 394 | $25.4045_M$ | 33$_?$ | = | 50 |
| CH$_2$F$_2$[8] | R$_9$(18) | $\nu_?$ | $18_9 \to 18_{10}$ | — | 215 | $46.6029_M$ | 17$_?$ | ⊥ | — |
| | R$_9$(34) | $\nu_?$ | $18_9 \to 18_{10}$ | — | 288 | $34.7624_M$ | $18_8$ | = | 10 |

| Pump line | Mode | Transition | Offset | Freq. | Wavelength | Assignment | Pol. | Rel. Int. |
|---|---|---|---|---|---|---|---|---|
| R$_9$(32) | $\nu_9$ | 18$_{15}$ → 18$_{16}$ | — | 184 | 54.2576$_M$ | 17$_{14}$ | ⊥ | 150 |
| | | | — | 236 | 42.4351$_M$ | 18$_{14}$ | = | 5 |
| R$_9$(22) | $\nu_4 + \nu_9$ | 33$_{12}$ → 33$_{13}$ | — | 123 | 81.6554$_M$ | 32$_{11}$ | ⊥ | 2 |
| R$_9$(20) | $\nu_4 + \nu_9$ | 38$_?$ → 38$_?$ | — | 118 | 84.9419$_M$ | 37$_?$ | ⊥ | 70 |
| | | | — | 167 | 60.0128$_M$ | 38$_?$ | = | 3 |
| R$_9$(6) | $\nu_9$ | — | — | 237 | 42.2653$_M$ | — | = | 17 |
| P$_9$(4) | $\nu_9$ | 36$_{-25}$ → 35$_{-24}$ | — | 290 | 34.5423$_M$ | 34$_{-24}$ | = | 3 |
| | | | — | 725 | 13.7946$_M$ | 35$_{-36?}$ | ⊥ | — |
| P$_9$(6) | ? | — | — | 395 | 25.3356$_M$ | — | = | 6 |
| P$_9$(10)$'$ | $\nu_9$ | 41$_{-11}$ → 40$_{-10}$ | — | 159 | 63.0861$_M$ | 39$_{-12}$ | = | 15 |
| | | | — | 272 | 36.7189$_M$ | 40$_{-12}$ | ⊥ | 1 |
| P$'_{10}$(10) | ? | — | — | 383 | 26.1343$_M$ | — | = | 10 |
| P$_9$(22) | ? | — | — | 657 | 15.2152$_M$ | — | ⊥ | 1 |
| | | | — | 134 | 74.6315$_M$ | — | ⊥ | 1 |
| P$_9$(24) | ? | — | — | 192 | 52.1246$_M$ | — | = | 1 |
| | | | — | 109 | 91.4948$_M$ | — | = | 1 |
| | | | — | 135 | 73.9266$_M$ | — | = | 5 |
| | | | — | 256 | 39.0584$_M$ | — | ⊥ | 3 |
| $^{12}$C$_2$H$_2$F$_2^r$  P$_{10}$(22) | $\nu_9$ | 24$_{-7}$ → 23$_{-7}$ | 21 | 889 | 11.2475$_M$ | 22$_{-6}$ | = | >1 |
| | | 24$_{-5}$ → 23$_{-3}$ | -24 | 889 | 11.2503$_M$ | 22$_{-1}$ | = | >1 |
| $^{12}$CH$_3$OH$^5$  P$_9$(36) | $\nu_8$ | 16(0,1,8) → 16(0,1,8) | ~25 | 171 | 58.6248$_M$ | 16(0,2,7) | = | — |
| | | | | 119 | 84.1509$_M$ | 15(0,2,7) | ⊥ | 400 |
| R$_{10}$(38) | $\nu_8$ | 26(0,3,4) → 25(0,3,4) | 36 | 163 | 61.3371$_M$ | 24(0,1,3) | = | 18 |
| R$_9$(10) | $\nu_8$ | 26(0,2,10) → 27(0,2,10) | 38 | 97 | 103.6029$_C$ | 26(0,3,9) | = | 300 |
| P$_9$(34) | $\nu_8$ | 9(1,2,5) → 9(1,2,5) | | 71 | 141.8206$_M$ | 10(0,1,6) | ⊥ | 100 |
| P$_9$(32) | $\nu_8$ | — | | 42 | 237.1968$_M$ | 10(0,1,6) | = | 50 |
| P$_9$(16) | $\nu_8$ | 10(0,1,0) → 11(0,1,0) | 63 | 571 | 17.5264$_M$ | 10(0,1,0) | = | 38 |
| R$_{10}$(34) | | — | | 163 | — | — | | 18 |
| R$_9$(18) | | — | | 65 | — | — | | 22 |

*(Continued)*

571

**TABLE 9.13** (Continued)

| Species | Line | Mode | Transition | Offset (MHz) | $\lambda_{FIR}$ (μm)[a] | $\lambda_{FIR}$ (cm$^{-1}$)[b] | State | FIR-Output Polarization[c] | Power (mW) |
|---|---|---|---|---|---|---|---|---|---|
| $^{12}CD_3OH^t$ | $R_{10}(18)$ | $\nu_8$ | $10(1,,) \rightarrow 9(1,,)$ | — | 41 | $241.8095_M$ | $9(0,,)$ | ⊥ | >1 |
| | | | | | 44 | $228.8471_M$ | $10(0,,)$ | ∥ | >1 |
| | | | | | 858 | $11.6516_M$ | $8(1,,)$ | ∥ | 6.2 |
| | $R_{30}(36)$ | $\nu_8$ | $12(0,3,9) \rightarrow 12(0,3,9)$ | — | 419 | $23.8828_M$ | $12(0,1,8)$ | ∥ | 3.8 |
| | | | | | 254 | $39.4136_M$ | $11(0,1,8)$ | ⊥ | >1 |
| | $P_{10}(18)$ | $\nu_8$ | $28(0,3,13) \rightarrow 27(0,13,13)$ | — | 287 | $34.8059_M$ | $26(0,3,13)$ | ∥ | 4.4 |
| | $R_{10}(36)$ | $\nu_8$ | | — | 255 | $39.22_M$ | — | ⊥ | — |
| | $R_{10}(24)$ | $\nu_8$ | $28(0,2,7) \rightarrow 28(0,2,7)$ | — | 299 | $34.44_M$ | $27(0,2,7)$ | ⊥ | — |
| | | | | | 184 | $54.35_M$ | $27(0,3,6)$ | ⊥ | 60 |
| $^{12}CD_3OD^M$ | $R_{10}(18)$ | $\nu_8$ | — | — | 41 | $243.9_M$ | — | ⊥ | — |
| | $R_{10}(12)$ | $\nu_8$ | — | — | 406 | $24.63_M$ | — | ⊥ | |
| | $R_{10}(10)$ | $\nu_8$ | — | — | 312 | $32.05_M$ | — | ⊥ | |
| $1,2\text{-}C_2H_3F_2$ | $P_{10}(6)$ | $2\nu_t$ | $34_{19,15} \rightarrow 33_{16,16}$ | — | 327 | $30.62_M$ | $32_{17,17}$ | ∥ | 7.6 |
| | $R_{10}(14)$ | $\nu_4$ | $51_{34,27} \rightarrow 50_{23,?}$ | — | 243 | $41.13_M$ | $49_{22,29}$ | ∥ | 8.0 |
| | $R_{10}(16)$ | $\nu_4$? | $35_{15,20} \rightarrow 35_{14,21}$ | — | 377 | $26.55_M$ | $34_{13,22}$ | ⊥ | 1.5 |
| | $R_{10}(20)$ | $\nu_4$ | $37_{26,11} \rightarrow 36_{25,12}$ | — | 261 | $38.39_M$ | $35_{24,13}$ | ∥ | 20.0 |
| | | | | | 389 | $25.69_M$ | $36_{24,13}$ | ⊥ | 3.0 |
| | $R_{10}(30)$ | $\nu_4$ | $29_{23,6} \rightarrow 28_{22,7}$ | — | 308 | $32.47_M$ | $27_{21,8}$ | ∥ | 8.7 |
| | | | | | 442 | $22.62_M$ | $38_{21,8}$ | ∥ | 0.9 |
| | $R_{10}(38)$ | $\nu_4$ | $44_{29,16} \rightarrow 44_{27,17}$ | — | 231 | $43.18_M$ | $43_{26,18}$ | ⊥ | 7.3 |
| | | | | | 360 | $27.78_M$ | $44_{26,18}$ | ⊥ | 1.6 |
| | $P_9(34)$ | $\nu_4$ | $39_{17,22} \rightarrow 39_{18,21}$ | — | 311 | $32.14_M$ | $38_{17,22}$ | ⊥ | 5.8 |
| | | | | | 543 | $18.41_M$ | $39_{17,22}$ | ⊥ | 0.7 |
| | $P_9(28)$ | $\nu_4$ | $32_{22,10} \rightarrow 32_{23,9}$ | — | 286 | $34.93_M$ | $31_{22,10}$ | ∥ | 5.8 |
| | | | | | 423 | $23.67_M$ | $32_{22,10}$ | ∥ | 1.2 |
| | $P_9(20)$ | $\nu_4$ | $73_{17,36} \rightarrow 73_{18,55}$ | — | 228 | $43.84_M$ | $72_{17,56}$ | ⊥ | 1.8 |

| CD₂F | Pump line | | | Wavelength (µm) | Frequency (cm⁻¹) | | | | Pol. | Rel. int. |
|---|---|---|---|---|---|---|---|---|---|---|
| | P$_9$(40) | — | — | 31.5406$_M$ | 317 | — | — | — | $\parallel$ | 0.5 |
| | | — | — | 30.9426$_M$ | 323 | — | — | — | $\parallel$ | 0.4 |
| | | — | — | 31.1918$_M$ | 321 | — | — | — | $\parallel$ | 0.3 |
| | R$_{10}$(44) | — | — | — | 456 | — | — | — | $|$ | 0.4 |
| | R$_{10}$(42) | — | — | 48.1151$_M$ | 208 | — | — | — | $\parallel$ | 1.5 |
| | R$_{10}$(38) | — | — | 45.8155$_M$ | 218 | — | — | — | $\parallel$ | 0.3 |
| | | — | — | 52.6781$_M$ | 190 | — | — | — | $\parallel$ | 10.0 |
| | R$_{10}$(34) | — | — | 28.3365$_M$ | 353 | — | — | — | $\perp$ | 0.4 |
| | | — | — | 19.9769$_M$ | 501 | — | — | — | $\parallel$ | 1.3 |
| | R$_{10}$(24) | — | — | 53.2427$_M$ | 188 | — | — | — | $\perp$ | 1.0 |
| | R$_{10}$(14) | — | — | 27.2184$_M$ | 367 | — | — | — | $\parallel$ | 0.2 |

*Source*: Ref. 105.

[a] Rounded to the nearest micrometer.

[b] Rounded to the nearest $10^{-4}$ cm; $C$, based on calculation; $M$, based on measurement.

[c] Polarization relative to pump polarization.

[d] Refs. 131 and 132. Sequence-band line.

[e] Ref. 133.

[f] Refs. 134 and 135. R$_9$(17) line is sequence band. P$_{10}$(13) line is a N$_{20}$ laser.

[g] Continuous-wave Raman lines.

[h] Refs. 135 and 136. R$_{10}$(18) line pumped by $^{13}$C$^{16}$O laser.

[i] Ref. 137.

[j] Refs. 138 and 139. P$_9$(15) line sequence band.

[k] Ref. 138.

[l] Ref. 140.

[m] Ref. 141.

[n] Refs. 138, 142, and 143.

[o] Ref. 144.

[p] Refs. 142 and 145.

[q] Refs. 146–149.

[r] Ref. 150.

[s] Refs. 136, 138, and 151–153.

[t] Refs. 131 and 154.

[u] Refs. 136 and 155–157.

[v] Ref. 158.

[w] Ref. 159.

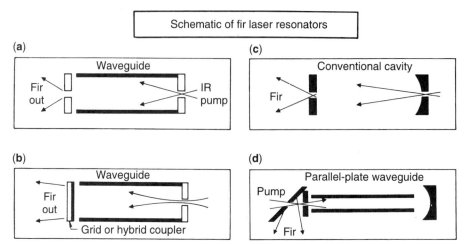

**Figure 9.60**   Four resonators commonly employed for CW operation. The hollow circular waveguide structures illustrated in **(a)** and **(b)** are widely used because of their relative compactness, favorable dimensions for wall deexcitation collisions, and allowable variety of output coupling techniques for metal waveguides transverse dimensions are typically a few wavelengths, whereas hollow dielectric guides must have diameters greater than 100 wavelengths for low propagation loss. The latter geometry provides best mode quality as discussed in the text. **(c)** illustrates conventional Fabry–Perot resonators. Transverse dimensions are 10 to 15 cm. The parallel-plate waveguide **(d)** suitable for Stark tuning is a very compact structure. Plate separation is only a few wavelengths and the small mode volume restricts available output power. (From Ref. 107.)

on the FIR output mirror. This has led to considerable work on a variety of output coupler designs. A number of commonly employed output coupling techniques are displayed in Fig. 9.61. In Fig. 9.61**a**, a Michelson arrangement is depicted where mirrors $M_1$ and $M_2$ are utilized together with a mesh oriented at 45°. Shown in Fig. 9.61**b** is an output coupler comprised of two silicon etalons forming a Fabry–Perot coupler [161]. The infrared input pump and FIR output are separated in one case using a silicon plate and making use of Brewster's angle and the polarization properties of the medium [162]. An alternative approach is to employ a grating and make use of the disparate wavelengths. To obtain good mode quality, one wishes to extract the beam over the full aperture of the laser tube. One approach that permits this is the hybrid metal mesh-dielectric mirror [163, 164] shown in Fig. 9.61**c**. Here, the fragile mirror structure is protected from the infrared pump by a dielectric coating. For example, a four-layer coating of Ge and $CaF_2$ provides a specular reflectivity of greater than 90% over a 2-μm range in the near IR. The final configuration shown in Fig. 9.61**d** has been referred to as a giant hole coupler [151]. This coupler consists basically of a silicon window coated with a dielectric to reflect (~98%) the IR pump radiation (~9 to 11 μm). To serve properly as an FIR output coupler, the window is masked to produce a coupling hole of the required diameter and then overcoated with an ~500-Å-thick layer of gold. The FIR output is taken from the hole in the gold film. For typical FIR wavelengths ($\geqslant$ 100 μm), the dielectric film produces negligible attenuation. An important feature to note is that due to the partial FIR reflectivity associated with the high index substrate, relatively large coupling holes can be employed without significantly perturbing the modes of the resonator. Therefore, better behaved output beams can be obtained as compared to the simple hole in a mirror.

Referring back to the metal mesh mirror configuration shown in Fig. 9.61c, it is of interest to examine its design in detail. As noted by Danielewicz and Coleman [163], one begins with the freestanding metal mesh calculations derived by Ulrich [165]. Figure 9.62 schematically illustrates the geometry assumed for a mesh of thickness $t$. To calculate the transmission properties, an equivalent circuit model is employed where the freestanding mesh is treated as a uniform transmission line shunted by a normalized admittance $Y$. For the lossless case, one has [165]

$$Y = -j(2/Z_0)(\lambda/g) = -2jb$$

where

$$Z_0 = 2\ln[\csc(\pi a_{\text{eff}}/g)]$$

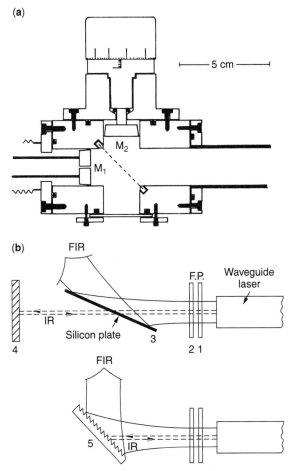

**Figure 9.61** Output couplers used in far-infrared lasers: **(a)** Michelson coupler comprised of $M_1$, $M_2$, and the mesh (from Ref. 160); **(b)** multiple silicon etalons (F.P.) (from Ref. 161); **(c)** hybrid metal mesh-dielectric mirrors (from Ref. 164); left inductive grid, right capacitive grid (after Ref. 105); **(d)** giant hole coupler (from Ref. 151.)

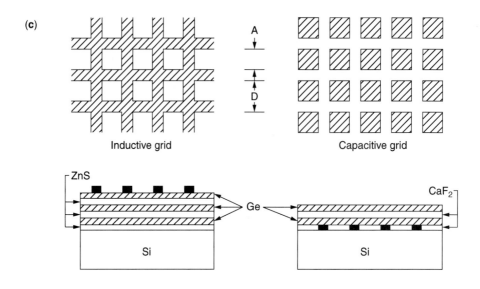

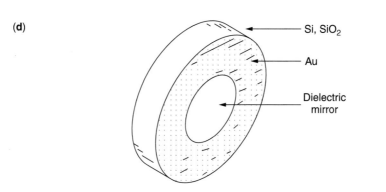

**Figure 9.61**   (*Continued*)

and

$$a_{\text{eff}} = a + (t/2\pi)[1 + \ln(8\pi a/t)]$$

The transmission and reflection coefficients are then given, respectively, by

$$T = \frac{1}{1 + b^2}$$

and

$$R = \frac{b^2}{1 + b^2}$$

Including the effect of a substrate of refractive index $n$ results in a transmission coefficient $T'$ given by

$$T' = \frac{nT}{R + (n + 1/2)^2 T}$$

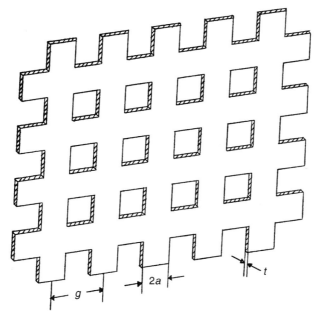

**Figure 9.62** Schematic sketch of a metal mesh and definition of the mesh parameters $g$, $2a$, and $t$. (From Ref. 163.)

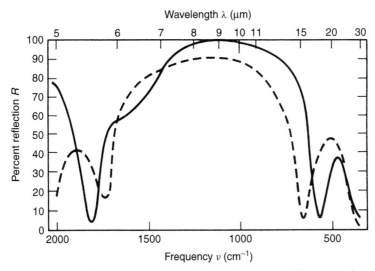

**Figure 9.63** Measured reflection spectra of a high-reflectance multilayer coating of Ge and $CaF_2$ and a Si substrate (solid line) and a metal mesh dielectrically coated mirror (dashed line). (From Ref. 163.)

An example of the infrared properties of a typical FIR mirror is shown in Fig. 9.63. This was designed to operate with a 10-$\mu$m pump and consisted of a four-layer coating of Ge and $C_aF_2$ on a Si substrate. For operation at $\sim$9 $\mu$m wavelength, the reflectivity of a bare coated substrate was $>$99.5% while the mesh mirror case dropped to $\sim$90%.

As with the case of electrical discharge FIR lasers, optically pumped FIR lasers typically employ waveguide structures. Both conventional metal waveguides have been employed as well as hollow dielectric waveguides [93, 95, 166]. Figure 9.64 displays the theoretical attenuation coefficient as a function of wavelength for a variety of commonly employed low-loss waveguides [167], where the nomenclature for the dielectric waveguide will be described shortly. Usually, dielectric waveguides are employed in preference to smooth-bore metallic waveguides since the linearly polarized $EH_{11}$ mode couples well to a low-order Gaussian mode, thereby permitting the laser optics to be located outside the waveguide [95].

It is now appropriate to begin an examination of actual FIR laser systems and experimental results. Figure 9.65 contains a photograph of an optically pumped far-infrared laser system that was designed for fusion plasma diagnostics [168]. A relatively compact system is realized by embedding both the $CO_2$ pump laser and the FIR laser cavities in a single Invar support frame with dimensions of $2 \times 0.4 \times 0.4$ m. The low-temperature coefficient of expansion of the Invar is essential for long-term stability. In addition, the laser system shown in Fig. 9.65 utilizes a temperature-controlled circulator

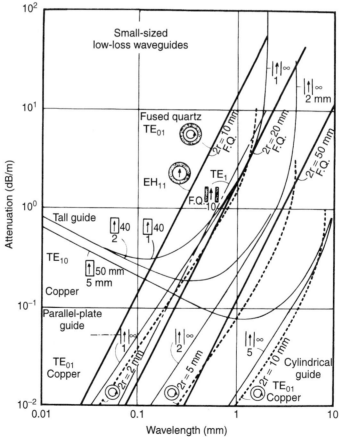

**Figure 9.64**   Attenuation coefficient as a function of wavelength for both copper and quartz waveguides. (After Ref. 167.)

**Figure 9.65** Photograph of the Aerospace Corp. far-infrared laser package. (Courtesy of E. Fletcher.)

to flow coolant throughout the structural components of the laser for improved stability. Referring again to the photograph, the grating tuned $CO_2$ pump laser is located above the dielectric waveguide FIR laser. Notice that in addition to the two Brewster angle ZnSe windows on the $CO_2$ pump laser, there is a third Brewster window located at the $CO_2$ pump input end at the FIR laser. However, even with cooling of the windows, absorption of the $CO_2$ pump can degrade the overall performance of the system. Therefore, as we shall see later, there has been considerable interest in configurations that eliminate the Brewster windows.

For many applications, laser stability is extremely important. The first consideration is amplitude stability. For the laser shown in Fig. 9.65, a long-term stability of $\pm 3\%$ is obtained without active stabilization of the $CO_2$ pump beam as shown in Fig. 9.66. As we shall see in the following, this can be significantly improved by means of active stabilization. In addition to amplitude stability, the frequency stability of the laser is vital for many applications, such as receiver local oscillator service or active probing of plasmas.

Typical frequency stability data for the laser shown in Fig. 9.65 are illustrated in Fig. 9.67, where the heterodyne beat signal between two $CH_2F_2$ lasers whose frequency difference was held at 2 MHz is displayed on a spectrum analyzer. The free-running frequency stability for a 2-min exposure is seen to be better than 20 kHz, as determined from the full width at half maximum of the signal [168].

Wattenbach et al. have devoted considerable effort toward developing a compact and stable FIR laser to serve as a local oscillator for radio astronomy studies [170]. Figure 9.68 schematically illustrates their FIR laser configuration. By using Invar stabilizing rods, the thermal expansion is limited to 2 $\mu$m/K, which greatly contributes to passively stabilizing the FIR laser. As can be seen in the diagram, a hybrid metal

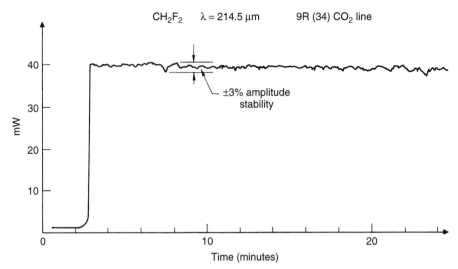

**Figure 9.66**   Amplitude stability of a $CH_2F_2$ optically pumped molecular laser [$\lambda = 214.5$ μm gR(34) $CO_2$ line]. (After Ref. 169.)

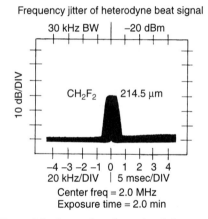

**Figure 9.67**   Frequency jitter of the heterodyne beat signal from two $CH_3F_2$ FIR lasers. (After Ref. 169.)

mesh mirror couples FIR radiation out of a Teflon window whose thickness has been optimized for the HCOOH line at 432.6 μm. The dielectric waveguide consists of a 30-mm-diameter quartz tube with a length of 2.2 m. The $CO_2$ input mirror has a central coupling hole of 1 mm diameter with a radius of curvature of 6 m. Without stabilization, the authors report an FIR power stability of within 10% over an hour. To stabilize the FIR laser actively, a portion of the output is coupled out through a beam splitter, chopped at 20 Hz and then focused onto a sensitive pyroelectric detector. The signal is then introduced into the microprocessor stabilization system shown in Fig. 9.69, which is comprised of a preprogram and a stabilization loop. The former performs such functions as automatically adjusting the $CO_2$ laser resonator length to provide the required FIR laser power. Figure 9.70 displays the results of FIR output

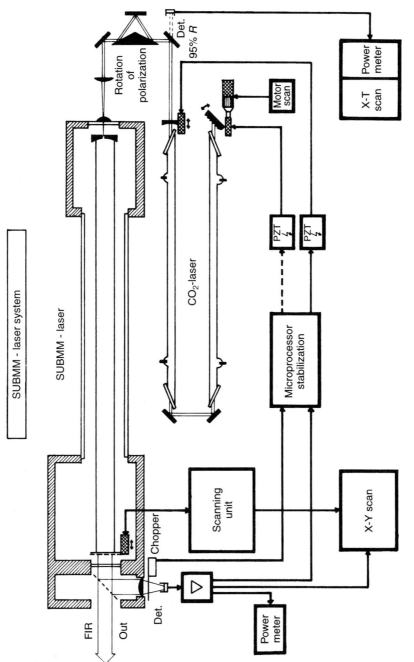

**Figure 9.68** Diagram of FIR laser system. (From Ref. 170.)

581

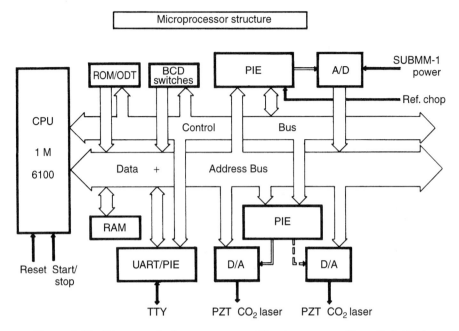

**Figure 9.69** Diagram of microprocessor stabilization system. (From Ref. 170.)

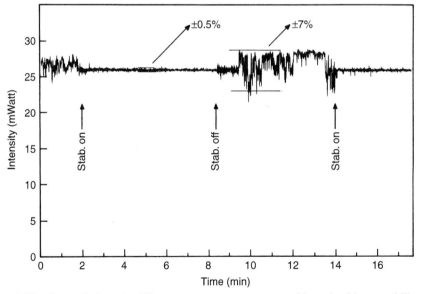

**Figure 9.70** Scan of the submillimeter-wave output power with and without stabilization. (From Ref. 170.)

power stability with and without stabilization. They report that the long-term stability ($\sim$1 h) improves from $\pm$10% without stabilization to $\approx$1.2%. Over shorter time scales of several minutes, the stability is $\approx \pm 0.5\%$. After a warm-up period of 4 h, the frequency drift is $\approx$200 kHz over 1 h.

Another approach to FIR laser stability has been reported by Koepf et al. [171]. First, the stability is improved by reducing the feedback of the IR pump power back into the $CO_2$ laser. To reduce the feedback, an off-axis pumping scheme is employed as illustrated in Fig. 9.71. The F200 input beam makes four round trips with the beam locations on the FIR laser mirrors numbered in sequence with the beam diverging and converging twice in traversing the cavity. The spent beam exits at an angle of 26 arc minutes with respect to the input beam and can, therefore, be readily deflected. In addition to the isolation, the authors note that this configuration eliminates standing waves for the pump radiation, leading to reduced sensitivity to acoustical and thermal effects. In their stabilization scheme, the $CO_2$ laser is locked to the resonance frequency of a temperature controlled mechanically and thermally stable etalon as shown in Fig. 9.72. Here, a 41-Hz dither voltage is applied to the piezoelectric translator in the etalon, with the subsequent power-modulated transmitted beam phase detected with the error signal used to vary the length of the $CO_2$ laser cavity. Using this technique, the FIR output power is reported to be constant to within 4% over periods of several hours. In the frequency drift area, they report mean drifts of 740 Hz and 14 kHz for 1-s and 1-min time intervals, respectively.

An interesting technique that has been used for active stabilization as well as the search for new FIR laser lines makes use of the opto-acoustic effect [172–174]. In this arrangement, a so-called spectrophone is placed in the path of a small chopped portion of the $CO_2$ pump beam [175, 176]. A signal occurs due to the optoacoustic effect when the $CO_2$ laser radiation is absorbed. By utilizing this signal, one may maintain the $CO_2$ laser frequency at the point at which maximum FIR output occurs. Figure 9.73 illustrates the experimental arrangement utilized by Busse et al. to demonstrate the technique [173]. An example of the output signal is shown in Fig. 9.74, where both the microphone signal and $CO_2$ laser output power are displayed as a function of the $CO_2$ laser grating angle, where the FIR gas was $CH_3OH$. The maxima are clearly seen, which correspond to the $CO_2$ pump lines. As noted above, this technique can be employed to search for new FIR lasing transitions. Here the microphone output signal

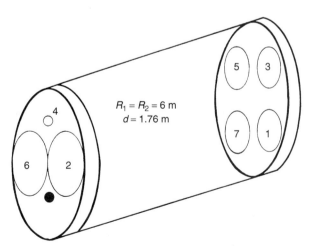

**Figure 9.71**  Pump beam propagation in fourfold degenerate resonator. The beam locations on the mirrors are numbered in sequence. (From Ref. 171.)

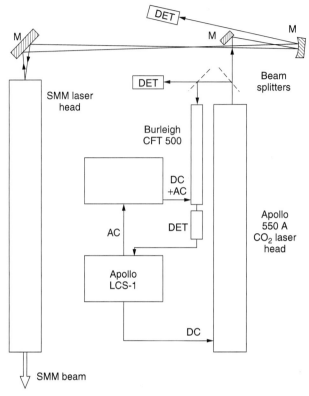

**Figure 9.72**   Active stabilization of pump laser and controlled pump beam propagation by off-axis beam injection. (From Ref. 171.)

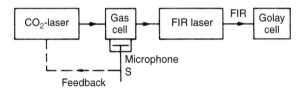

**Figure 9.73**   Experimental arrangement. (From Ref. 173.)

is monitored as the $CO_2$ laser grating is slowly rotated and the $CO_2$ laser frequency is tuned within the emission line.

Mansfield et al. have described an interesting passive isolation scheme which makes use of the polarization of the pump radiation [177, 178]. Figure 9.75 illustrates their approach. Here they employ the combination of a polarizer and a quarter-wave plate as an isolator that is inserted between the $CO_2$ pump and the FIR laser. The principle behind this isolation approach is that the incident $CO_2$ pump is circularly polarized upon entering the FIR cavity, while the feedback beam is converted to linear polarization orthogonal to the input pump beam. The polarizer then eliminates the feedback beam, which results in isolation of the two cavities and therefore increased laser stability. Figure 9.76 displays traces of $CO_2$ and FIR output power with and without isolation,

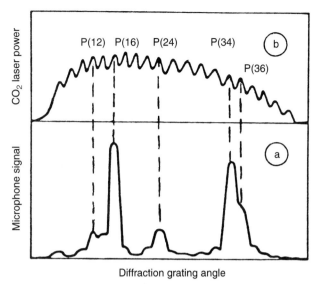

**Figure 9.74** Spectrophone signal (**a**) for $CH_3OH$ (150 mtorr) and $CO_2$ laser power (**b**) in the P branch of the 9.6-$\mu$m band. (To obtain a mean value of the spectrophone signal over the free spectral range of 60 MHz, the resonator length was modulated while the diffraction grating was rotated.) (From Ref. 173.)

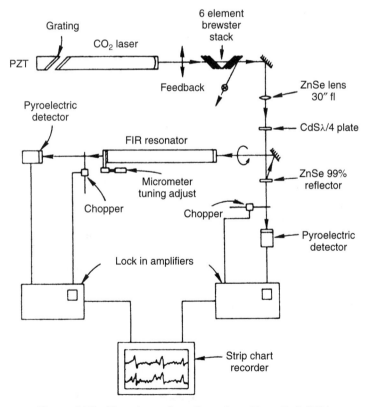

**Figure 9.75** Experimental configuration. (From Ref. 178.)

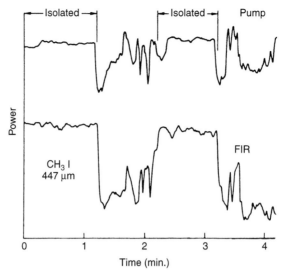

**Figure 9.76**   The pump and FIR power versus time for the 447-$\mu$m line of $CH_3I$. The lock-in integration time was 1 s. (From Ref. 178.)

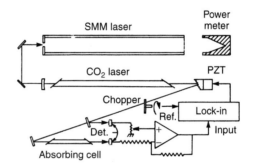

**Figure 9.77**   Schematic view of the $CO_2$ pump stabilization setup. (From Ref. 179.)

showing the power of this approach. However, the change in polarization of the FIR output is not acceptable for some applications.

Lachambre and Gagne also described a stabilization technique that does not require dithering the frequency of the $CO_2$ laser [179]. Their approach utilizes the biased transmission curve of an absorption cell as a discriminator. Figure 9.77 illustrates the experimental arrangement used in their stabilization studies, which were applied to the 214-$\mu$m transition in $CH_2F_2$, which is pumped by the 9R34 $CO_2$ line. Here, the zeroth-order beam from the grating is used as a $CO_2$ power monitor as well as a probe beam that passes through an absorbing cell. Pyroelectric detectors are used to detect the chopped beams with their opposed polarity output fed into an operational amplifier followed by a lock-in amplifier. Figure 9.78 displays their FIR power stability results together with the discriminator output.

An extremely attractive approach to eliminate feedback problems involves the use of a ring resonator [180]. Figure 9.79 schematically illustrates the configuration that was

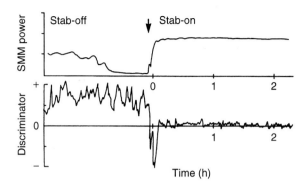

**Figure 9.78**  Effect of the feedback loop on the SMM laser output stability. (From Ref. 179.)

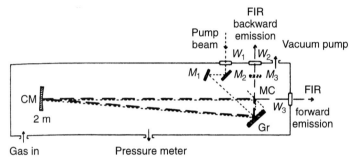

**Figure 9.79**  Experimental setup of FIR ring laser: $W_1$ ZnSe window; $W_2$ and $W_3$ PE windows: $M_1$–$M_3$, mirrors; Gr, grating; CM, curved mirror; MC, mesh coupler. (From Ref. 180.)

employed to demonstrate this concept. Here, the $CO_2$ laser pump beam is coupled into the FIR resonator through the grating operating in first order while serving as a specular reflector to the FIR. The improved amplitude stability of the ring configuration (relative to a linear resonator) is shown in Fig. 9.80. Another interesting feature investigated by Heppner and Weiss concerns the observation of the ring laser tuning characteristics and the switching between forward and backward emission. These results are readily understood by referring to Fig. 9.81, where the "hole" burned into the homogeneous line width at the pump frequency, $v_p$, is clearly indicated. Also shown are the forward and backward emission curves centered, respectively, at $v_f = v_1 - (v_0 - v_p)v_1/v_0$ and $v_b = v_1 + (v_0 - v_p)v_1/v_0$. Away from the overlap near FIR line center, one of the gain curves dominates. The latter has a significant effect on laser performance since the lack of a standing wave (as in a linear resonator) means that spatial hole burning does not occur, and therefore one mode exhausts all the available population inversion. The result of the above is that only one traveling mode can exist at a time and that they switch at line center with the forward mode existing below center.

Another interesting scheme to reduce feedback effects involves the use of rooftop resonators [181], as shown in Fig. 9.82. The isolation arises when the $CO_2$ pump polarization is oriented at $45°$ with respect to the crease in the FIR output rooftop. The reverse propagating feedback beam (which is orthogonal to the pump beam) is prevented from returning to the $CO_2$ laser by a Brewster stack or a polarizer. However,

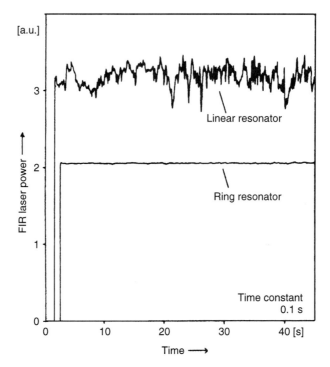

**Figure 9.80** Output power noise of optically pumped FIR lasers. (From Ref. 180.)

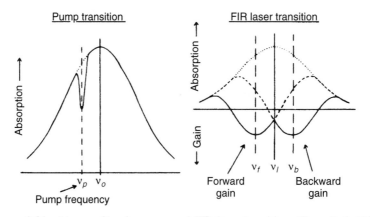

**Figure 9.81** Line profiles for pump and FIR laser transition. (From Ref. 180.)

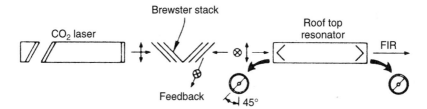

**Figure 9.82** Rooftop resonator. (From Ref. 181.)

the FIR $EH_{11}$ output beam is observed to be circularly polarized, which is not acceptable for some applications, as noted above.

A variety of research areas require high CW FIR output power levels. A specific example involves the use of FIR lasers for fusion plasma diagnostics, such as interferometry, polarimetry, and collective Thomson scattering [28, 59, 60, 182]. This has therefore motivated work on the development of increased performance FIR lasers through a variety of optimization techniques including temperature control and buffer gases [26, 183–185]. The function of the latter is to provide a deexcitation mechanism through collisions with the lasing molecules. Temperature control of the FIR lasing medium, on the other hand, improves performance by optimizing the population distribution for efficient pumping. The importance of a proper temperature FIR lasing medium is readily understood. In thermal equilibrium, the population of a certain energy level $J$ compared with the ground-level population is given by

$$N_J/N_0 = \exp[(E_J - E_0)/KT]$$

where $N_{J,0}$ is the population, $E_{J,0}$ the energy associated with the levels, $K$ is Boltzmann's constant, and $T$ the temperature. A laser is not in thermal equilibrium, but the rate of repopulation of the desired energy level is temperature dependent. Because of this, there is an optimum operation temperature which maximizes the populations and output powers.

This optimum temperature varies with lasing media and energy levels involved in the lasing process. Higher-$J$-level transitions ($J$ = rotational quantum number of the ground state) operate best at higher temperature, while low-$J$-level schemes are optimized at lower temperatures. Calculations of the output variation for varying $J$ levels and temperatures have been performed by Lawandy [186]. In this experiment, output enhancement of as much as a factor of 2 by temperature variation was observed. In the following we describe the UCLA efforts to develop increased power FIR lasers [26, 185].

The UCLA FIR laser developments first made use of the increased $CO_2$ pump power available from the laser described in Section 9.1.3. This provides $\approx 100$ to 170 W on the more commonly employed pump lines. A schematic of the FIR laser arrangement is shown in Fig. 9.83a while Fig. 9.83b contains a photograph of the FIR laser together with the $CO_2$ pump laser. It consists of two vacuum boxes to house the optics connected by circular dielectric waveguide. The waveguide is filled with the lasing gas at low pressures ($< 500$ mT) and propagates the electromagnetic radiation between the two mirrors. Pyrex hollow dielectric waveguide is employed. This waveguide diameter is chosen to minimize propagation losses ($\approx \lambda_0^2/a^3$) and maximize gain ($\approx \lambda_0/a$). The specific attenuation expression is [187]

$$\alpha = 8 \left(\frac{u_{nm}}{2\pi}\right)^2 \frac{\lambda^2}{D^3} \frac{V^2 + 1}{\sqrt{V^2 - 1}}$$

where $u_{nm}$ is the $m$th zero of the Bessell function $J_n$ and $V$ is the index of refraction of the waveguide dielectric. For wavelengths $>1000$ μm, a 51-mm-inner-diameter tube was employed. Below this wavelength, a 38-mm-ID waveguide was required to suppress transverse modes. A precision-bore 38-mm guide was also tested but had no effect on the output power. Several FIR cavity lengths, from 1.2 to 3 m, were

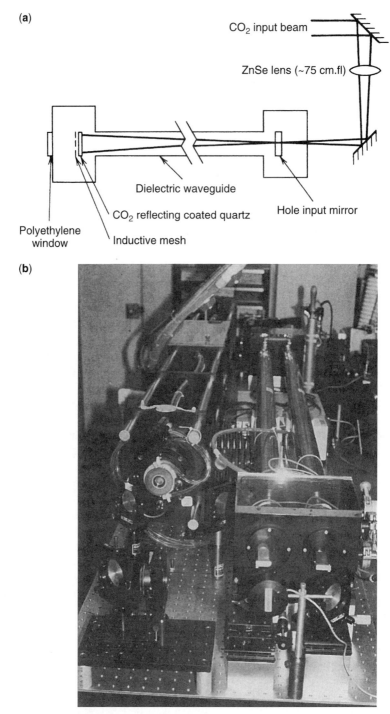

(a)

CO$_2$ input beam

ZnSe lens (~75 cm.fl)

Dielectric waveguide

CO$_2$ reflecting coated quartz

Hole input mirror

Polyethylene window

Inductive mesh

(b)

**Figure 9.83** (a) Schematic of the optically pumped far-infrared (FIR) laser; (b) photograph of FIR laser together with CO$_2$ pump laser.

tested. A 2.4-m cavity length was chosen as a compromise between output power and compactness.

$CO_2$ pump radiation is coupled into the cavity by focusing through a small hole in the copper mirror. A ZnSe lens of 75 cm focal length was found to give the best FIR cavity volume filling factor. That is, the beam expansion after the focus was large enough to pump the gas evenly in the waveguide without expanding into the waveguide walls before one round trip. To investigate $CO_2$ feedback effects on laser stability, two types of input mirrors were tried. The first has the hole drilled on axis, for which the feedback is direct. The second is an off-axis hole mirror. This reduces the amount of reflected $CO_2$ radiation returning directly to the $CO_2$ laser.

Two types of FIR output couplers were also employed. For short wavelengths ($<300$ $\mu$m), the silicon giant hole coupler shown in Fig. 9.61**d** provided the optimum spatial modes and output powers. When this coupler is employed, a quartz window is placed at the output of the FIR laser to reject any $CO_2$ radiation that passes through the 99% reflection-coated silicon. At long wavelengths ($>300$ $\mu$m), a $CO_2$ reflection-coated quartz etalon coupler backed by inductive mesh was used. By changing the mesh, the reflectivity could be optimized.

The input vacuum seal is made by an antireflection-coated ZnSe window. The output window is a 1.6-mm-thick high-density polyethylene sheet. This thin window helps to reduce absorption losses, which can be quite high at short wavelengths. Both the copper mirror and waveguide are watercooled. This was done for two reasons; to provide thermal stability of the FIR cavity, which is heated by the $CO_2$ radiation, and to optimize the temperature of the FIR lasing medium for maximum output power. The water cooler/heater provided temperature variation capability from 2 to 65°C.

Before discussing the power optimization results, it should be stressed that making absolute power measurements and comparisons in the FIR is difficult. Therefore, it should be noted that the output power results quoted in the UCLA studies were the values measured *directly* on a Scientech model 362 thermopile detector, with *no correction factors* applied. These calorimeters were designed for operation in the infrared, and absorption of radiation in the far infrared is not complete. Calibration of this meter has been performed previously [188] and correction factors as large as a factor of 2 are required for $\lambda \sim 500$ $\mu$m. Often, the published results have already accounted for this correction factor. Therefore, when comparing results one must be careful to determine if such corrections have been applied.

Before discussing the details of the optimization studies, it is helpful first to simply list the results. Table 9.14 lists some of the gases investigated, together with the maximum FIR power obtained and the pump power used to achieve these values. As expected, higher output powers are observed at shorter wavelengths. The region from 500 to 1000 $\mu$m is not well covered by known high-power FIR molecules, and outputs in this region are somewhat lower than expected. At several wavelengths, the powers shown are significantly greater than any values reported previously. For example, at 496 $\mu$m the previous maximum was $<20$ mW, whereas the UCLA experiments produced 77 mW output. Powers reported here are also considerably higher for the 393-$\mu$m and 1222-$\mu$m lines.

It is of interest to consider the conversion efficiency of the optically pumped laser. Table 9.15 shows the values of experimentally observed efficiency together with the maximum theoretical efficiency. It is seen that three values of "observed" efficiency have been provided. In all of this work, the pump powers reported are measured directly

**TABLE 9.14   Far-Infrared Output Powers Achieved for the Strong ($\geqslant$ 10 mW) Laser Lines Studied**

| FIR Gas | Emission Wavelength ($\mu$m) | Maximum Output (mW) | Pump Power (W) |
|---|---|---|---|
| $CH_2F_2$ | 185 | 500 (400 $EH_{11}$) | 140 |
| | 215 | 150 | 130 |
| HCOOH | 393 | 150 | 150 |
| | 432 | 65 | 150 |
| | 513 | 32 | 150 |
| | 743 | 12 | 90 |
| $CH_3I$ | 447 | 51 | 130 |
| $CH_3F$ | 496 | 77 | 135 |
| $^{13}CH_3F$ | 1222 | 25 | 140 |
| $CH_3OH$ | 118 | 350 | 140 |

*Source*: Ref. 26.

**TABLE 9.15   Far-Infrared Laser Efficiency for Selected Wavelengths**

| $\lambda_{FIR}$($\mu$m) | $\eta_{theo}$(%) | $\eta_{act}$(%) | $\eta_{corr}$(%) | $\eta_{sci}$(%) | $\eta_{sci}/\eta_{theo}$(%) |
|---|---|---|---|---|---|
| 185 | 2.49 | 0.33 | 0.44 | 0.67 | 26.9 |
| 393 | 1.18 | 0.11 | 0.15 | 0.30 | 25.4 |
| 447 | 1.18 | 0.04 | 0.05 | 0.10 | 8.5 |
| 496 | 0.96 | 0.08 | 0.16 | | 16.7 |
| 1222 | 0.40 | 0.02 | 0.03 | 0.04 | 10.0 |

*Source*: Ref. 26.

at the output of the $CO_2$ laser. This power, $P_{IR}$, is then used to calculate what is called the actual efficiency in Table 9.15 by the relation

$$\eta = P_{FIR}/P_{IR}$$

where $P_{IR}$ is the $CO_2$ laser output power.

However, not all of the $CO_2$ output is delivered into the FIR cavity. Losses in the mirrors, the lens, window, and diffraction at the input hole decrease the infrared power available for pumping the FIR gas. The fraction of $CO_2$ output that was deposited into the FIR laser was determined to be 75.3%. Thus the "corrected efficiency" is greater than the actual measured efficiency by a factor of 1.33.

The "scientific efficiency" is the corrected efficiency with further compensation made for the previously mentioned FIR calorimeter correction factor. This scientific efficiency is the number that should be compared to the maximum theoretical efficiency. This comparison is given in the last column of Table 9.15. The actual efficiency gives a figure of merit for the experimentalist concerned with usable output power. As can be seen, even the scientific efficiency is much less than the theoretical limit, although it is comparable to efficiencies reported elsewhere. The highest scientific efficiency ever reported was 32% of the theoretical maximum at 185 $\mu$m in $CH_2F_2$ [189]. This discrepancy between theoretical and observed efficiencies has not been well explained but is usually attributed to poor absorption of the pump radiation.

Let us now discuss in detail the results of optimization studies for several gases. These include HCOOH (formic acid) at 393, 432, and 743 μm, $CH_3F$ (methyl fluoride) at 496 μm, and $^{13}CH_3F$ at 1222 μm. These gases provided high powers at a variety of wavelengths, as well as an interesting study of the laser physics involved. They also provided coverage of the full wavelength range of interest for plasma diagnostic, radio astronomy, and nondestructive testing applications.

Formic acid provided lasing at three wavelengths, 393, 432, and 743 μm. All were operated with a combination quartz/mesh output coupler giving $EH_{11}$ modes. The 432- and 743-μm lines operated best with an off-axis input mirror and high-reflectivity output couplers. The 432-μm line gave 65 mW output with 90% reflectivity, while the 743-μm line required R 95% to achieve 12 mW. Because it gave the largest output power, up to 150 mW, the 393-μm line was studied in more detail. The $EH_{11}$ mode was the dominant cavity mode while the $LP_{11}$ mode operated at a slightly lower output power of 120 mW.

Relaxation from the lower lasing level to the ground state is dominated by collisions in the HCOOH molecules. For this reason, the formic acid laser (as well as the difluoromethane laser) can be operated over a wider pressure range than diffusion dependent molecules (e.g., $CH_3$, $CH_3I$), as discussed in the preceding. The variation of output with gas pressure displayed in Fig. 9.84 shows operation at pressures from 50 to >400 mT with an optimum pressure of about 270 mT. This optimum pressure increases slightly at higher pump powers, showing the need for more available lasing molecules. It should be noted that this pressure is measured with an uncalibrated thermocouple gauge and is meant to show trends in the power dependence and not absolute values.

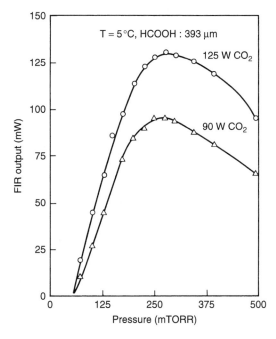

**Figure 9.84**   Far-infrared output power as a function of pressure for the 393-μm HCOOH laser.

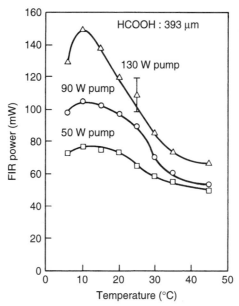

**Figure 9.85** Far-infrared output power as a function of temperature for the 393-$\mu$m HCOOH laser.

The energy levels involved in lasing correspond to $J = 14$, so one would expect lower temperature operation to be more favorable. As Fig. 9.85 shows, the optimum temperature was $10°C$, with enhancement of as much as a factor of 2 over room-temperature ($\approx 25°C$) operation observed. This enhancement was seen to vary with pump power as shown. At higher pump powers, more molecules are needed for the lasing process, and the faster refilling rates are more beneficial. Figure 9.86 shows the variation of FIR power with pump power for three different temperatures. In all cases, as in $CH_2F_2$, the output varies linearly with the $CO_2$ input.

One unfortunate aspect of the 393-$\mu$m laser is that it lies very close to a water-vapor resonance and is attenuated heavily in the atmosphere. Measurement of this attenuation showed a loss of 42% of the power over a 2-m distance in air, while the same measurement for the 432-$\mu$m laser resulted in only an 18% loss. This corre-sponds to an absorption coefficient of 275 dB/km for 393-$\mu$m radiation and 90 dB/km at 432 $\mu$m. This agrees reasonably with published theoretical values of 300 dB/km and 75 dB/km, respectively [190]. This high attenuation makes the 393-$\mu$m laser unattrac-tive for experiments unless it can be propagated through dry nitrogen-filled waveguide.

The temperature variation of the $^{13}CH_3F$ laser was not studied as closely as in other gases. The laser at 1222 $\mu$m used a 51-mm-diameter waveguide and a combination quartz/mesh coupler to achieve 23 mW of output. Also, when helium was added to the lasing medium, the output was boosted to 25 mW. This behavior will be discussed shortly. Again, the FIR output varied linearly with $CO_2$ power. This contrasts with operation in a similar molecule, $CH_3F$, which is discussed later. As a result of this observation, it is concluded that bottlenecking is not a major problem in $^{13}CH_3F$, while it is a problem in the $CH_3F$ laser [191].

The last gas to be studied was $CH_3F$ lasing at 496 $\mu$m. As mentioned earlier, this was the first optically pumped molecule discovered, and a great deal of spectroscopic

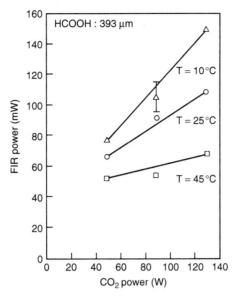

**Figure 9.86**  Far infrared output power as a function of $CO_2$ output power for the 393-$\mu$m HCOOH laser, at three different temperatures.

data is available for it. Lasing in the methyl fluoride laser employs the lowest $J$-value energy level ($J = 12$) of any of the gases studied. As seen in Fig. 9.87, the maximum output power is obtained at lower temperature than other molecules already examined. The output is still increasing slightly as the minimum achievable temperature of 2°C is reached.

The most interesting behavior, observed only in $CH_3F$, is shown in Fig. 9.88, which displays the FIR output power as a function of pump power. Below 100 W of $CO_2$ power, the FIR output increases approximately linearly with increasing $CO_2$ power. However, as the pump power is increased further, the rate of increase in FIR power slows. For $CO_2$ pump power levels above 115 W, no rise in FIR output is seen (i.e., saturation occurs).

This behavior is attributed to bottlenecking in the lasing medium. Bottlenecking occurs when the relaxation rate from the lower lasing level to the ground level is too slow. This causes a buildup of molecules in the lower lasing level, with a resulting decrease in population inversion and thus output power.

In the $CH_3F$ laser, this deexcitation is provided mainly by diffusion to the cavity walls, where energy is deposited by collisions. Because of the $p^{-1}$ ($p = nkT =$ pressure, $n$ = number density) dependence of the diffusion rate, operation of the $CH_3F$ laser is limited to low (<150 mT) pressure operation. At these low pressures, not enough molecules are available to absorb the high pump powers. This is why the roll-off in FIR output with increasing $CO_2$ pump power is observed.

In several gas lasers, including the $CO_2$ laser, buffer gases are used to relieve the bottlenecking. The buffer gases provide a deexcitation mechanism through collisions with the lasing molecules. The buffer gas can then carry the energy to the cavity walls. Helium was the first buffer gas tried in the $CH_3F$ laser. The low mass gives it a high thermal velocity, which provides many collisions with the lasing molecules.

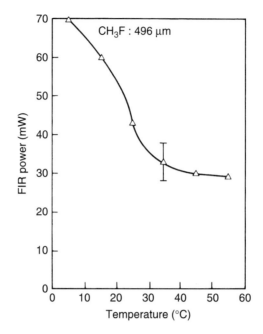

**Figure 9.87**   Far-infrared output power as a function of temperature for the 496-$\mu$m $CH_3F$ laser.

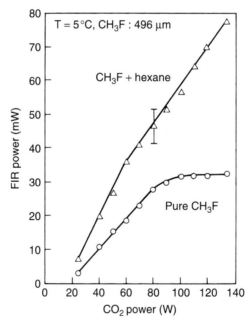

**Figure 9.88**   Far-infrared output power as a function of $CO_2$ output power for the 496-$\mu$m $CH_3F$ laser.

In this experiment, however, the addition of helium did not provide any observable improvement in laser operation.

Pentane and $n$-hexane have also been used as buffer gases in $CH_3F$ lasers [183]. Their operation is attributed to the high vibrational heat capacity of these complex molecules. The large number of vibrational energy levels gives a high probability of energy exchange during a collision with a lasing molecule. A combination of pentane and $n$-hexane was used to relieve the observed bottleneck, with the results shown in Fig. 9.88. At low pump powers, a factor of 1.5 increase in output was seen, which agrees with previous experiments. For higher pump power levels, even greater enhancement is obtained, with a factor of 2.3 improvement achieved at a pump power of 135 W. The maximum 469-$\mu$m power observed was 77 mW, which is more than four times greater than any previously reported values [184]. The $CH_3F$ + hexane combination laser shows a linear increase of FIR power with $CO_2$ input, similar to the other molecules studied. The optimal $CH_3F$: hexane partial pressures are shown in Fig. 9.89. As can be seen, the addition of hexane allows the laser to be operated over a much wider pressure range.

Another buffer gas which has been seen to relieve the bottlenecking is $SF_6$ [184]. The mechanism for operation of the $SF_6$ is a resonant energy exchange with the $CH_3F$ molecules. This gas was also tested in this laser and an improvement in laser performance was seen. Although not studied in detail, $SF_6$ appeared to be at least 90% as effective as the hexane buffer gas.

As mentioned previously, significant information is available concerning the spectroscopic constants for the $CH_3F$ molecule, so that it was possible to carry out a rate equation modeling of this laser [185]. In the following, we describe in detail the modeling results together with detailed comparisons with experiments. Here, we should note that this is a considerably simplified model compared to that described in Section 9.2.1.

To model the molecular laser accurately while taking into account temperature and pressure effects, it is necessary to use a four- or five-level energy scheme. The arrangement for a four-level model is shown in Fig. 9.90. This is similar to a simple three-level scheme, but now account is taken for collisional transfer of energy between the two upper lasing levels and other rotational levels in the upper vibrational manifold. This

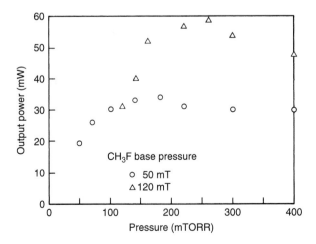

**Figure 9.89**  Far-infrared power for various $CH_3F$/hexane pressure ratios for the 496-$\mu$m laser.

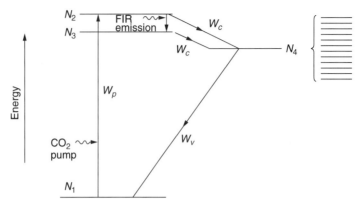

**Figure 9.90** Four-level-energy diagram used for the rate equation modeling of the 496-$\mu$m $CH_3F$ laser. Also shown are the rates associated with the transitions indicated by arrows. (From Ref. 185.)

collision rate, along with the infrared absorption, define the pressure operating regime of the FIR laser. At low pressures, absorption of the $CO_2$ radiation is too low, while at high pressures, collisions destroy the population inversion. For this four-level model, the rate equations are written [192]

$$dN_1/dt = W_p(N_2 - N_1 g_2/g_1) + W_v N_4$$

$$ddN_2/dt = -W_p(N_2 - N_1 g_2/g_1) - B_{23}n(N_2 - N_3 g_2/g_3) + (f_2 N_4 - N_2)W_c$$

$$dN_3/dt = B_{23}n(N_2 - N_3 g_2/g_3) + (f_3 N_4 - N_3)W_c$$

$$dN_4/dt = -W_p(N_2 - N_1 g_2/g_1) - N_4 W_v$$

and

$$dn/dt = b_{23}n(N_2 - N_3 g_2/g_3) - 2\pi n\delta$$

where $n$ is the photon density; $N_i$ the population of the $i$th state; $g_i = 2J + 1$ the degeneracy of the $i$th level; $f_i$ the Boltzmann factor for the $i$th level; $\delta = 1/\tau_c$ the inverse of the FIR cavity lifetime; $B_{23}$ the Einstein coefficient for stimulated emission on the 2–3 transition, which is given by $c^3 g(\nu)(4\pi^2 \nu_p 2t_{sp})^{-1}$; $W_v$ the vibrational deexcitation rate, $W_c$ the rotational energy equilibration rate $= \pi \Delta \nu$; $\Delta \nu$ the Lorentzian line width of the rotational transition; and $W_p$ the pumping rate $= \lambda^2 g(\nu) I (8\pi h n t_{sp})^{-1}$. In the above, $g(\nu)$ is the line width function [193], $I$ the $CO_2$ beam intensity, $h$ is Planck's constant, and $t_{sp}$ is the spontaneous lifetime for the 2–1 transition.

For the case of interest, it is the steady-state solution that is of importance. Then, ignoring time derivatives and making the simplifying assumptions that $g_1 = g_2 = g_3$ (for $CH_3F$ $J = 12$, so $g_1 = g_2 = 25$, $g_3 = 23$), $f_1 = f_2 = f_3$, $f N_4 \ll N_2$, $N_3$, and $W_v < W_c$, these equations reduce to the steady-state expression given by

$$n = \frac{1}{2} \frac{W_c}{B_{23}} \left[ \frac{N B_{23} W_v W_p}{2\pi \delta (W_v + W_p)} - 1 \right]$$

The FIR power is related to the photon density by $P_{FIR} = (h\nu c V t/L)n$, where $L$ is the FIR cavity length, $V$ the active cavity volume, and $t$ the output coupler transmission.

The FIR power for an inhomogeneously broadened oscillator may be written in the form [194]

$$P_{\text{FIR}} = \frac{1}{2} t P_s \left( \frac{2L\alpha}{t+a} - 1 \right)$$

where $\alpha$ is the small-signal gain, $P_s$ the saturation power, and $a$ the loss per round trip. Equating this with the previous solution yields

$$\alpha = \frac{B_{23} N W_v W_p}{c W_c (W_v + W_p)}$$

and

$$P_s = h\nu V c W_c / 2 L B_{23}$$

Although these simple equations do not completely describe the behavior of the FIR laser, a good deal of the physical phenomena can be explained using them. By looking at two limiting cases for the small-signal gain, the mechanism causing bottlenecking is explained. First, when $W_v \gg W_p$ the small-signal gain is given by

$$\alpha \approx B_{23} N W_p / c W_c$$

which is proportional to the pump power as expected. Next, for $W_v \ll W_p$,

$$\alpha \approx B_{23} N W_v / c W_c$$

and now the gain has no dependence on the pump power. This is the bottlenecking which is seen experimentally at high pump powers. A leveling off (saturation) of the FIR power increase was observed experimentally as the pump power was increased, implying that this experiment operates in a region where $W_v \approx W_p$.

The observations concerning the FIR output power dependence on pressures is also clarified by referring to the equation for the FIR power. Writing this as

$$P_{\text{FIR}} = t \frac{h\nu V_c}{4\,L} \left[ \frac{2L W_p W_v N}{c(W_v + W_p)} - \frac{W_c}{B_{23}} \right]$$

we see that the first term varies nearly linearly with pressure because the vibrational rate (which is pressure dependent) appears in both the numerator and denominator. However, the collisional rate, which appears in a negative term, is also proportional to the pressure. This will force the FIR power to decrease and eventually reach threshold at high pressures.

Although not as simple to see physically, information regarding the temperature dependence of the laser output can also be obtained from the steady-state equation. The experiment with methyl fluoride is performed with a constant number density of molecules in the cavity. For this case, the parameters which are temperature dependent are given by [186]

$$W_v = W_{v0} \theta \gamma^{1/2}$$

$$W_p = \frac{W_{p0}}{\gamma^{1/2}} \exp[z(1 - 1/\gamma)]$$

and

$$N = \rho_0 \left( 1 - \frac{E_{02}}{kT_0\gamma} \right) f(\gamma)$$

where

$$T = \gamma T_0 \quad T_0 = 300 \text{ K}$$

$\rho_0$ is the total number of molecules, and

$$z = \frac{4\ln(2)(\nu_p - \nu_{p0})^2}{(\Delta\nu_D)^2}$$

In the above, $\nu_p$ is the pump frequency (tuned to absorption peak), $\nu_{p0}$ the line center frequency of the pump, $\nu_D$ the Doppler line width, and $f(\gamma)\alpha C_1\gamma^{-3/2}\exp[-(C_2/\gamma)]$, where $C_1$ and $C_2$ are constants determined from the molecular parameters.

Using the above, it is predicted that for molecules which lase on low-value $J$ levels, operation at low temperatures is favored, while for high-value $J$-level lasers, operation above room temperature is optimum. For the methyl fluoride laser, the molecular constants were taken primarily from Walzer and Tacke [195]. These include, for the pump transition, $t_{sp} = 2.5$ s, $\Delta\nu_{D0} = 67$ MHz, and $\nu_p - \nu_{P0} = 0$, while for the FIR transition, $t_{sp} = 238$ s and $\Delta\nu_{N0} = 3.75 \times 10^{-2}$ MHz/$\mu$bar. For this experiment, the following values were used $t = 0.1$ (measured), $a = 0.3$ (assumed), $L = 2.4$ m. Also, pump powers that are corrected for mirrors window losses are used in the calculations, but the numbers used in the figures represent the output measured at the $CO_2$ laser.

Some discrepancy was discovered when searching for values of the vibrational relaxation rate $W_{v0}$. Lawandy and Koepf report a value of [184]

$$W_{v0} = 124.0/(P + P_B) + 600.0P + C_3 P_B$$

where $P$ is the pressure (in mtorr) of the laser gas and $P_B$ is the buffer gas pressure. In this equation, the first term is a diffusion term (for a 38-mm-diameter tube), while the second and third terms give the relaxation rates due to collisions involving energy transfer from vibrational levels to translational or kinetic energy ($V - T$ transfer). However, DeTemple and Danielewicz provide the relation [121]

$$W_{v0} = 124.0/(P + P_B) + 1.2 \times 10^5 P + C_3 P_B$$

Here, the much larger coefficient of $CH_3F$ pressure takes into account vibrational energy transfer between two lasing molecules. However, because this involves only a redistribution of the molecules in the excited vibrational states, it does not contribute to the relaxation of molecules to the ground state. Because of this, the first expression for vibrational relaxation was employed. Furthermore, the second expression was tested in the computer model, and the pressure dependence of the FIR output showed no agreement with experimental observations.

To approximate the experimental results, two of the parameters in the model had to be varied from their theoretical values. To gain agreement with the lasing threshold observed at low pump power for the pure $CH_3F$ case, the initial number of molecules in the ground state had to be increased by a factor of 71 over the theoretical value. This is equivalent to saying that a large fraction (two-thirds) of the total number of

$CH_3F$ molecules are in the desired rotational energy level for pumping. This correction may be attributed to the fact that the four-level model employed does not account for refilling of the ground state by thermal equilibration of the rotational levels in the lower vibrational manifold ($W_c$). This process is much faster than the refilling by deexcitation from the upper vibrational level ($W_v$). To check this hypothesis, a modeling using five energy levels, which includes all of the rotational levels in the lower vibrational level, would have to be performed. To achieve agreement with the shape observed on the FIR power versus pump power curve for pure $CH_3F$, the vibrational relaxation rate had to be multiplied by a factor of 33. That is,

$$W_{v0} = 33.0[124.0/(P + P_B) + 600.0P + C_3 P_B]$$

If this factor of 33 is not employed, the roll-off of FIR output power is predicted for much lower pump powers than was observed experimentally. No physical explanation is given for these correction factors, but rather they were used so that the model could reasonably describe the experiment.

No value for the coefficient $C_3$ for hexane buffer gas was found in the literature. By comparing with experimental results, and using the empirical values given in the preceding paragraph, a value of $C_3 = 1.8 \times 10^4$ s$^{-1}$/mT was determined. This is comparable to the value of $C_3 = 3.6 \times 10^4$ s$^{-1}$/mT given by Lawandy and Koepf for $SF_6$ buffer gas [184].

Using these correction factors, comparisons with three experimental observations were performed. First, the variation of FIR power with pressure is shown in Fig. 9.91. Figure 9.92 shows the FIR variation with pump power. The equations predict the roll-off that is seen at high pump power, as well as the relief of the bottlenecking by the addition of hexane. Finally, the temperature dependence of the methyl fluoride laser output is shown in Fig. 9.93. There is some deviation from the experimental results at low temperatures, but overall the agreement is within 10%.

For many applications, such as far-infrared interferometry and scattering in fusion plasmas, one requires an intermediate frequency or beat frequency of the order of

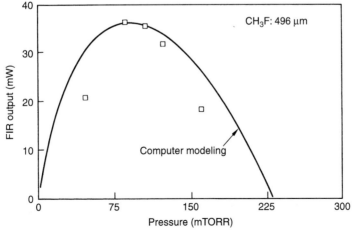

**Figure 9.91** Comparison of rate equation modeling and experimental results for the variation of 496-μm output power as a function of $CH_3F$ pressure.

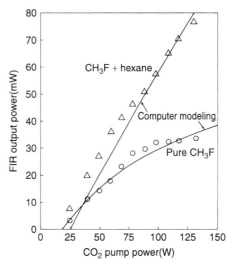

**Figure 9.92** Comparison of rate equation modeling and experimental results for the variation of 496-μm output power as a function of pump power.

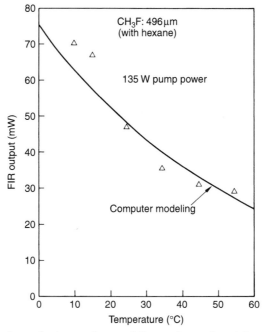

**Figure 9.93** Comparison of rate equation modeling and experimental results for the variation of 496-μm output power as a function of temperature.

1 MHz [28, 59, 60, 182]. One technique that has been employed is to reflect a portion of an FIR beam from a rotating grating and utilize the resultant doppler shift [196]. This, or a similar technique, is essential for electrical discharge lasers. However, in the case of optically pumped lasers, an alternative method exists for obtaining the appropriately frequency-shifted beam. Here, one employs two FIR lasers, whose cavity

lengths are adjusted to provide a frequency offset within the frequency range of the FIR lasing transition [197–200]. A typical arrangement is illustrated in Fig. 9.94, where the output of a single $CO_2$ laser pumps the two FIR lasers employed in the Nagoya, Japan JIPP T-II tokamak interferometer [200]. Such systems can usually provide frequency offsets of the order of several megahertz. Although only a single $CO_2$ pump laser was utilized in the system shown in Fig. 9.94, one may also employ two separate pump lasers to increase the FIR output power level [199]. However, one must then ensure that the two $CO_2$ pump lasers exhibit the proper stability, which is an extremely challenging problem.

To assess the causes of lack of stability in the IF signal, a number of experimental arrangements have been investigated at UCLA [27, 201]. The first approach is to configure the twin laser so that there are still separate cavities but with the optics placed within the same vacuum enclosure. Figure 9.95 illustrates the experimental arrangement. For the moment, the active $CO_2$ laser stabilization portion of the arrangement should be ignored. Such a system has been utilized for heterodyne collective scattering and multichannel interferometry in tokamak fusion plasmas. Output powers of 65/45 mW are available at 430 μm and 11/9 mW at 1.22 mm. The source short- and long-term stability limits are illustrated in Fig. 9.96 for 430 μm operation. The short-term stability is seen to be ±7.5 kHz, whereas there is a slow drift in the center frequency over a period of 30 min (±20 kHz). The latter is not particularly a problem since long term drifts can be readily stabilized or compensating electronics employed. However, the instantaneous jitter sets limits in terms of phase noise in interferometry/polarimetry and frequency resolution in heterodyne scattering.

To reduce the above-mentioned short-term drift, a novel dual-cavity FIR laser was conceived as illustrated in Fig. 9.97. The configuration departs from the normal twin-frequency FIR laser design in that the new design utilizes a single optical cavity with two dielectric waveguides inside a single vacuum vessel. The intermediate frequency is produced by evaporating a small ($\sim$1-μm) step over half the input coupling mirror or by slight horizontal misalignment of the system. Here, it should be noted that a 1-μm step corresponds to an IF frequency of $\approx$1.6 MHz at 100-μm wavelength or $\approx$160 kHz at $\lambda_0 = 1$ mm. The output mode quality obtained from the system is illustrated in Fig. 9.98, where a horizontal scan at 430 μm is displayed. Excellent $EH_{11}$ modes are achieved suitable for the efficient beam propagation required in most applications.

As noted above, the purpose of the novel design was to improve the laser stability. The motivation was the hope that the configuration would reduce differential vibrational and thermal fluctuations between the "two" cavities and thereby reduce the short-term jitter observed in normal twin-frequency sources. However, very similar short-term IF jitter was observed. Investigation of the primary source of IF jitter in this system was traced to feedback of $CO_2$ pump radiation from the FIR laser cavity. The fluctuation in $CO_2$ pump power was found to correlate exactly with the observed variation in IF frequency (monitored using the FM demodulation out of a phase-locked loop). Figures 9.99 and 9.100 show, respectively, the time histories of the $CO_2$ power monitor and the FM demodulation output for the dual-cavity and twin FIR laser systems. Here, the amplitude of an externally applied 100-Hz modulation of the $CO_2$ laser was adjusted so that it was just visible on the photoresistor monitor. Figure 9.101 displays the measured short-term IF jitter frequency ($\Delta F$) as a functions of the frequency of the mixed IF for both of the FIR laser systems. Here, it should be noted that the IF can

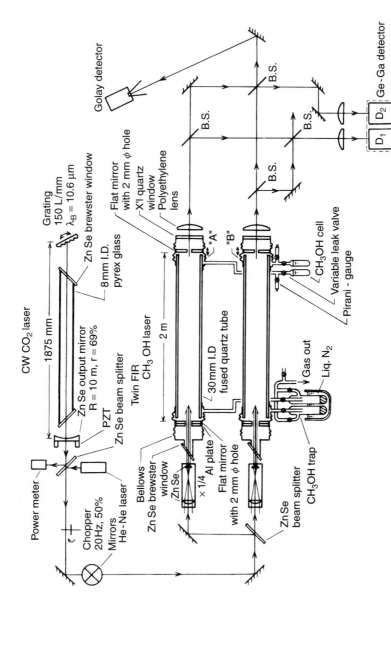

**Figure 9.94**  Schematic of a $CO_2$ optically pumped twin $CH_3OH$ FIR laser operating at 199 μm with greater than 1-MHz modulation frequency. (From Ref. 200.)

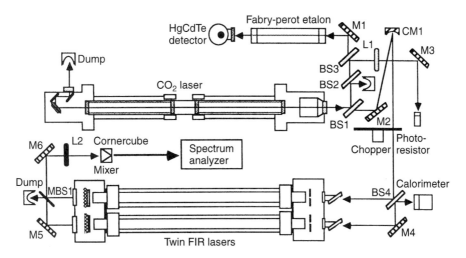

**Figure 9.95**  Twin FIR laser experimental arrangement. (From Ref. 201.)

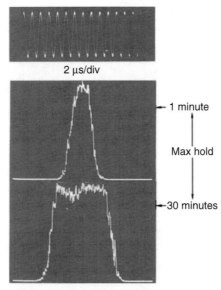

2 μs/div

1 minute

Max hold

30 minutes

700 kHz center frequency, 10 kHz/div,
3 kHz resolution, linear scale

**Figure 9.96**  Intermediate frequency obtained from a twin-frequency FIR laser. The IF drift is indicated on a peak hold spectrum analyzer for periods of 1 and 30 min. 700 kHz center frequency, 10 kHz/div, 3 kHz resolution, linear scale. (From Ref. 27.)

also be changed drastically by feedback of FIR radiation, which can pull the FIR laser frequency. This, therefore, necessitates isolation of the FIR as well as the $CO_2$ laser.

In the preceding, we have discussed relatively high power CW FIR laser systems for use in applications such as plasma diagnostics [28, 59, 60, 182] and nondestructive testing [202, 203]. However, there has also been considerable interest in

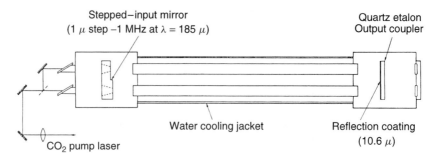

Stepped–input mirror
(1 $\mu$ step –1 MHz at $\lambda$ = 185 $\mu$)

Quartz etalon
Output coupler

Water cooling jacket

$CO_2$ pump laser

Reflection coating
(10.6 $\mu$)

**Figure 9.97** Schematic of novel single-cavity twin-frequency FIR laser. (From Ref. 201.)

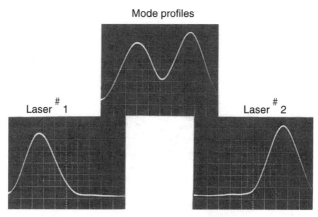

Mode profiles

Laser #1

Laser #2

**Figure 9.98** Horizontal model quality measured 1 m from output. Note that the $EH_{11}$ mode is obtained from both lasers.

developing lightweight systems for both earth-based and airborne radio astronomy applications [204–206]. The system of Roser et al. [205] has been employed for radio astronomical observations from groundbased observatories. The $CO_2$ laser cavity is 2 m long but is folded so that the overall length is $\approx$1 m. With a $CO_2$ pump power level of $\approx$10 to 20 W, the group obtains $\approx$10 to 20 mW of output at 372 $\mu$m. The overall weight of the system is reported to be $\approx$260 lb [206].

*Pulsed Optically Pumped FIR Lasers.* In the preceding, we concentrated on a discussion of CW FIR lasers which provide output powers in the range 1 to 500 mW. However, there are some applications that require pulsed output powers in the range 100 kW to 1 MW. Chief among these is the use of such lasers to determine the ion temperature in magnetic fusion plasmas via collective Thomson scattering [28, 60, 73]. In addition to high power, such applications typically require extremely narrow line widths, therefore dictating single-mode laser operation. In the following, we briefly review pulsed FIR laser developments. More details may be found in a number of reviews [106, 207]. Here, it should be noted that pulsed lasers operate with pressures in the region 1 to 10 torr, compared to the region 30 to 300 mtorr employed with the CW cases. To satisfy the needs of the Thomson scattering diagnostic application, considerable attention has been devoted to the 385-$\mu$m line of $D_2O$ and the 496-$\mu$m line

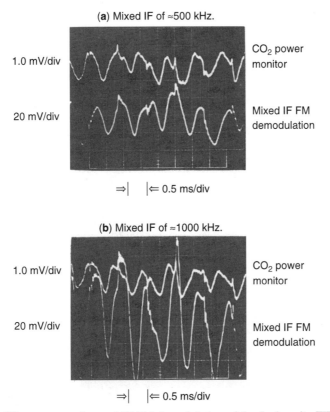

(a) Mixed IF of ≈500 kHz.

1.0 mV/div

$CO_2$ power
monitor

20 mV/div

Mixed IF FM
demodulation

⇒|   |⇐ 0.5 ms/div

(b) Mixed IF of ≈1000 kHz.

1.0 mV/div

$CO_2$ power
monitor

20 mV/div

Mixed IF FM
demodulation

⇒|   |⇐ 0.5 ms/div

**Figure 9.99**   $CO_2$ power monitor and IF FM demodulation of the dual-cavity FIR laser system IF jitter with ≈100 Hz current amplitude modulation on the $CO_2$ laser: **(a)** at a mixed IF of ≈500 kHz; **(b)** at an IF of ≈1000 kHz. (From Ref. 201.)

of $CH_3F$. Both readily produce megawatts of power in the superradiant mode, albeit with too large a bandwidth [208, 209]. To improve the FIR mode quality, a number of oscillator–amplifier configurations have been studied, as described in the following.

Figure 9.102 illustrates the configuration employed by Semet and Luhmann to obtain high-power 496-μm emission [162]. The basic principle of operation for the system depends on the observation that the $CH_3F$ molecule, when excited by a linearly polarized pump, has a tendency to lase with its electric field vector polarized normal to the pump polarization. The output of a grating-tuned TEA $CO_2$ laser oscillator–amplifier configuration is divided by a beam splitter, with part of the output used to pump the FIR oscillator and the remainder used to pump the FIR amplifier. Using an ≈8-J $CO_2$ pump laser, an output power of 156 ± 31 kW was obtained from the amplifier for a duration of ≈20 ns.

An early FIR laser approach that motivated much of the subsequent developments was reported by the Culham Laboratory Group [210]. Figure 9.103 schematically illustrates the details of the oscillator–amplifier configuration. The oscillator is forced to operate on a single longitudinal mode by means of a Fox–Smith interferometer as shown in more detail in Fig. 9.104. The 2.43-m-long oscillator cavity is comprised of a gold-coated rear mirror ($R = 2.46$ m) and the flat copper mesh mirror. Transverse

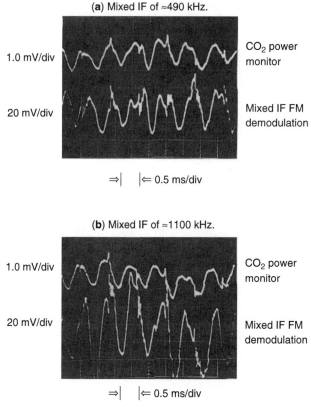

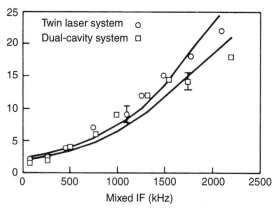

**Figure 9.100** $CO_2$ power monitor and IF FM demodulation of the twin FIR laser system IF jitter with $\approx 100$ Hz current amplitude modulation on the $CO_2$ laser: **(a)** at a mixed IF of $\approx 490$ kHz; **(b)** at an IF of $\approx 1100$ kHz. (From Ref. 201.)

**Figure 9.101** Dual-cavity and twin FIR laser system's short-term IF jitter as a function of mixed IF frequency for the case when $\approx 100$ Hz amplitude modulation is applied to the current of the $CO_2$ pump laser power supply. (From Ref. 201.)

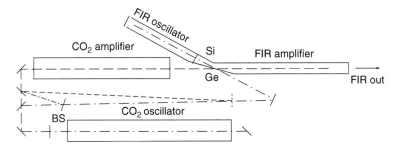

**Figure 9.102** Schematic of $CH_3F$ oscillator–amplifier. (From Ref. 162.)

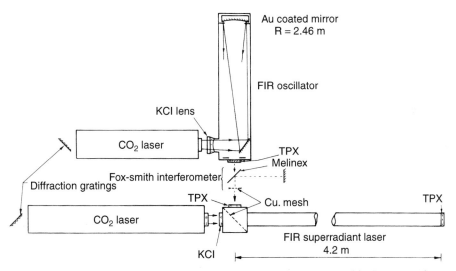

**Figure 9.103** Schematic of the Culham Laboratory injection laser assembly. Important features include FIR oscillator, superradiance tube with its independent $CO_2$ laser pump and Cu mesh mixer, by means of which single-mode FIR radiation from the oscillator is directed into the superradiance tube. (After Ref. 210.)

modes are eliminated with the aperture shown in Fig. 9.104. The pulse shape from the ≈1-kW oscillator is displayed in Fig. 9.105 as obtained using a fast point contact diode detector. The transform of the line width was found to be ≈27 MHz as verified by the Fabry–Perot trace shown in Fig. 9.106.

Without the oscillator, the FIR laser amplifier produces superradiant emission which is quite broad in frequency as evidenced by the ragged, spiky video pulse shown in Fig. 9.107. In contrast, the emission in the presence of the FIR oscillator injection signal is considerably smoother consisting of a low-frequency pedestal upon which small random spikes remain (see Fig. 9.108). Detailed Fabry–Perot scans revealed that the majority of the emission was contained within a bandwidth of ≈55 MHz, although there remained a broad superradiant background out to several hundred megahertz.

As mentioned in the discussion of CW lasers, the standing-wave electric field patterns that exist in linear cavities give rise to spatially inhomogeneous population inversion. This, in turn, provides the environment for simultaneous oscillation even

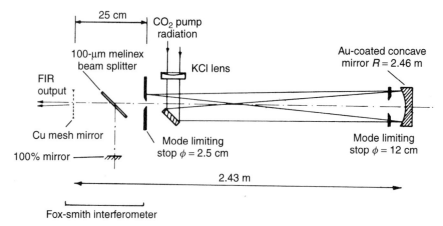

**Figure 9.104**   Schematic of Culham FIR oscillator employing a Fox–Smith interferometer for mode control. (After Ref. 210.)

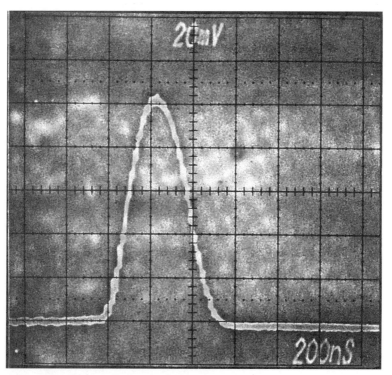

**Figure 9.105**   Video pulse shape of Culham FIR oscillator with Fox–Smith interferometer. (After Ref. 210.)

for the case of a homogeneously broadened transition. However, as noted previously, unidirectional traveling-wave ring lasers eliminate these spatial inhomogenity effects, thereby leading to improved mode quality. In addition, the direct feedback of the $CO_2$ pump is eliminated leading to further improvements.

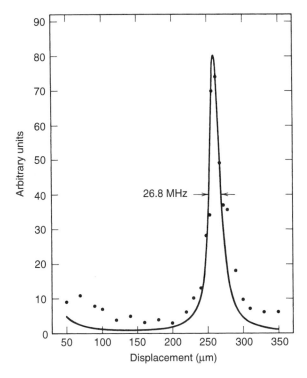

**Figure 9.106** Fabry–Perot scan of output of Culham FIR oscillator. (After Ref. 210.)

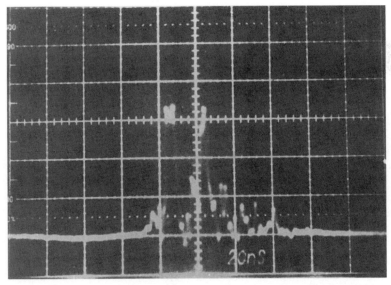

**Figure 9.107** Video pulse shape of output from Culham superradiant FIR Laser. (After Ref. 210.)

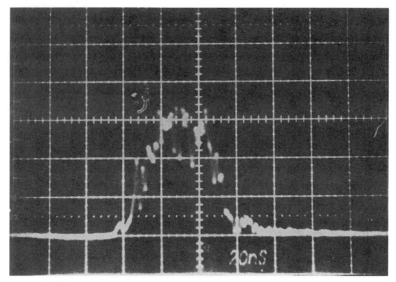

**Figure 9.108**   Video pulse slope of output from Culham FIR laser with narrow-band FIR injection. (After Ref. 210.)

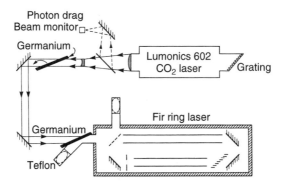

**Figure 9.109**   Experimental arrangement for pulsed FIR ring laser studies. (From Ref. 211.)

Figure 9.109 schematically illustrates the configuration employed by the UCLA group to investigate the properties of the pulsed ring laser and to compare them with those of a folded linear laser [211]. The pump source consisted of a grating-tuned Lumonics model 602 TEA $CO_2$ laser which produced up to $\approx 40$ J on the R22 line ($\lambda = 9.26$ $\mu$m) which is required for 385/359-$\mu$m emission from $D_2O$. The $CO_2$ pump beam impinged on a germanium beam splitter, which upon rotation allowed easy and continuous variation of the incident power. The transmitted $CO_2$ radiation entered the FIR laser through a germanium vacuum window tilted at Brewster's angle. The FIR ring cavity consisted of three flat copper mirrors together with an electroformed mesh (60 lines/cm) output coupler. The total cavity length was $\approx 2$ m, although variation up to $\approx 5$ m was possible. Dielectric waveguide (50 mm diameter, 60 cm long) was used in the long arms of the ring to aid FIR mode control and also confinement of the $CO_2$ pump radiation. To increase diffraction losses for the higher-order transverse modes, a

30-mm aperture was placed on axis in the FIR cavity. Conversion to a folded linear system was easily achieved by a 45° rotation of the output coupler and final copper mirror. In addition, superfluorescent operation could easily be studied by complete removal of the mesh output coupler.

Figure 9.110 displays the video pulse shapes obtained from both the ring and linear cavities utilizing a fast Schottky diode detector and a 400-MHz bandwidth oscilloscope. Both produce output power levels of $\approx 100$ kW. The ring laser is found to possess a smooth output pulse which gives rise to a narrow line output which is essentially transform limited. In contrast, the linear laser clearly exhibits multimode output.

Considerable attention was devoted to the study of the $D_2O$ lasing mechanism by a number of groups. An extremely important result of these investigations was the demonstration by Petuchowski et al. that the 66-$\mu$m line emission observed from $D_2O$ was due to stimulated Raman emission [212]. This work was motivated by the observation that many of the $CO_2$ pump lines were significantly displaced in frequency from their corresponding $D_2O$ absorption lines. First, the observation can possibly be explained by a process of wing absorption followed by FIR emission near line center, which corresponds to off-resonance optical pumping [213]. The second possibility was that stimulated Raman emission was the dominant process. Here, it should be noted that a signature of the stimulated Raman effect is that the FIR emission is off-resonance by an amount equal to the pump detuning. Petuchowski et al. conclusively identified the stimulated Raman nature of the 66-$\mu$m line of $D_2O$ as the major process [212]. Subsequent work by Wiggins et al. provided similar confirmation for the 385-$\mu$m line of $D_2O$ [214].

An important feature of the stimulated Raman process is that the bandwidth of the pump source can then control the bandwidth of the FIR emission, thereby making it possible to realize the goal of narrow-line FIR output. In addition, each absorbed photon can produce an emitted FIR photon, whereas in the standard lasing process only 50% of the excited molecules can emit coherently if the population inversion is to be maintained.

The above-mentioned results had important implications for high-power pulsed $D_2O$ laser developments. Prior to that date, all of the $D_2O$ laser studies yielded broad-bandwidth superradiant emission [210, 215, 216], as was displayed in Fig. 9.107.

Comparison of ring and floded linear oscillators

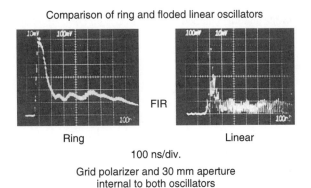

Ring          FIR          Linear

100 ns/div.

Grid polarizer and 30 mm aperture
internal to both oscillators

**Figure 9.110**  Comparison of temporal output from ring and linear cavities. The modulation frequency on the linear emission corresponds to longitudinal mode beating for the linear cavity (grid polarizer and 30-mm aperture internal to both oscillators.) (From Ref. 211.)

However, this is to be expected since the $CO_2$ pump lasers tended to be multimode in nature so that the $D_2O$ emission could be expected to be Raman in origin. An understanding of this fact led to a concentration of the pulsed FIR laser developments on the $D_2O$ molecule [217–224].

The problem of line selection and competition in off-resonantly pumped FIR lasers has been studied in detail by [225]. For off-resonant pumping of moderate intensity, the situation is as shown in Fig. 9.111**a**. Neglecting the interaction between modes within the two resonance lines gives rise to the simplified two-mode model shown in Fig. 9.111**b**. To describe the general case, Dupertuis et al. employ the six-level model illustrated in Fig. 9.112. A single-mode pump field is assumed while the signal fields

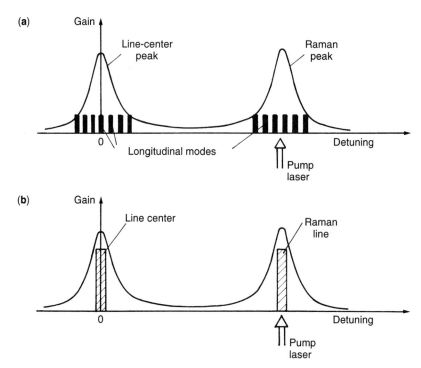

**Figure 9.111** (**a**) Mode and line competition under the gain curve of an off-resonantly pumped FIR laser; (**b**) simplified model. (From Ref. 225.)

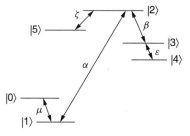

**Figure 9.112** General six level model employed by Dupertuis et al. in their studies [225]. The transition $|1\rangle \rightarrow |2\rangle$ is pumped.

are treated as multimode and a standard density matrix approach is employed. They denote the Rabi frequencies coupling the levels $|i\rangle$ and $|j\rangle$ by

$$\alpha = \frac{\mu_{12}\epsilon_{12}(z,t)}{2\bar{h}} \exp[-i(K_{12}z - \Omega_{12}t)]$$

$$\mu = \frac{\mu_{01}\epsilon_{01}(z,t)}{2\bar{h}} \exp[(-i(K_{01}z - \Omega_{01}t)]$$

$$\beta = \frac{\mu_{23}\epsilon_{23}(z,t)}{2\bar{h}} \exp[-i(K_{23}z - \Omega_{23}t)]$$

$$\epsilon = \frac{\mu_{23}\epsilon_{34}(z,t)}{2\,\text{k bar}} \exp[-iK_{34}z - \Omega_{34}t]$$

and

$$\xi = \frac{\mu_{35}\epsilon_{35}(z,t)}{2\bar{h}} \exp[-i(K_{35}z - \Omega_{35}t)]$$

When only one of the FIR transitions exhibits appreciable gain, one may approximately employ a simplified three-level model [225]. To proceed, the authors assume that the mode $n = -1$ corresponds to the Raman resonance and the mode $n = 0$ to line center. Then, for the case of a weak probe beam and a detuned pump, they find that the probe gain near line center is given approximately by

$$\rho_{23}(1) = -i\frac{|\alpha|^2}{\Delta^2}\frac{\beta_1}{\gamma_{23}}n_1^0 \left(1 - \frac{2\gamma_{21}}{\Gamma_2} + 2\frac{|\beta_0|^2}{\Gamma_3\gamma_{31}} + \frac{|\beta_0|^2}{\gamma_{31}\gamma_{21}}\right)$$

Using the same approximation, they find that near the intense Raman oscillation the gain can be obtained from

$$\rho_{23}(0) = i\frac{|\alpha|^2}{\Delta^2}\frac{\beta_0}{\gamma_{31}}n_1^0$$

When the strong line corresponds to resonance, the probe gain near the Raman peak can be expressed as

$$\rho_{23}(0) = i\frac{|\alpha|^2}{\Delta^2}\frac{\beta_0}{\gamma_{31}}n_1^0 \left(\frac{1}{1 + |\beta_1|^2/\gamma_{31}\gamma_{21}}\right)$$

The equations above clearly illustrate the fact that line center gain can be suppressed by sufficiently intense Raman oscillation. From the expression for $\rho_{23}(1)$, the threshold for suppression of line center gain is given by

$$\beta_{0,\text{th}}^2 = \gamma_{31}\gamma_{21}\frac{(2\gamma_{21}/\Gamma_2) - 1}{(2\gamma_{31}/\Gamma_3) + 1}$$

An examination of the equation above shows that $\beta_{0,\text{th}}^2$ tends to increase with the transverse relaxation rates $\gamma_{31}$ and $\gamma_{21}$. Therefore, Dupertuis et al. point out that the suppression becomes more difficult with increased pump incoherence since this leads to an effective increase in $\gamma_{21}$ and $\gamma_{31}$ [225]. The reader is referred to the work of Dupertuis et al. for discussions of more complicated situations [225].

Let us now proceed to a discussion of various high-power pulsed $D_2O$ systems. The first to be considered is the laser developed for use on the MIT Alcator tokamak [217]. In line with the discussions above, a 50-J single-mode $CO_2$ laser system that could be tuned over a range of $\approx 2$ GHz was employed as a pump source. A grating was employed for $CO_2$ mode selection while a Fox–Smith interferometer provided longitudinal mode control. The longitudinal mode selection was facilitated by choosing a $CO_2$ oscillator cavity length corresponding to a longitudinal mode spacing of $\approx 112$ MHz. The etalon was thermally stabilized and the $CO_2$ oscillator structure fitted with Invar rods for mechanical and thermal stability. The $CO_2$ oscillator was operated at a maximum of 100 mJ per pulse to prevent damage to the etalon. The tunable oscillator was followed by a series of amplifiers comprised of a double-pass Lumonics 103, a triple-pass 601, and finally, two single-pass 601 modules. The $CO_2$ output pulse duration could be varied from 100 ns to $\sim 1$ μs by suitable timing of the oscillator and amplifier modules. The output energy of 50 J was approximately constant for all pulse lengths. To prevent self-oscillation, a formic acid saturable absorber cell (2 to 3 torr) was included between the double-pass stages of the 103 amplifier. In addition, a grating was used which isolated the oscillator output coupler from the 103 except for the 9R22 line. Using the foregoing precautions, the active length of the amplifier chain was able to be maintained at 6.7 m without self-oscillation.

Figure 9.113 schematically depicts the high-power $D_2O$ oscillator developed by Woskoboinikow et al. [217]. An FIR grating (3 lines/mm) is used to ensure that lasing is restricted to the 385-μm transition as well as to strongly polarize the FIR output ($\approx 95\%$). A Fox–Smith interferometer comprised of a quartz plate, a flat mirror, and the FIR grating provides longitudinal mode control. Here, it should be noted that due to the low finesse of the interferometer, the bandwidth of the FIR emission is not appreciably narrowed. Referring back to Fig. 9.113, the $CO_2$ pump beam is coupled into the 4-m-long cavity by an 18-cm-diameter, 0.5-cm-thick crystal-quartz plate which has a reststrahlen resonance at 9.26 μm. This plate provides high reflectivity at the pump wavelength and excellent transmission in the FIR.

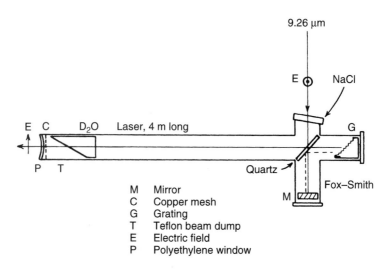

**Figure 9.113**   Details of the 385-μm $D_2O$ oscillator developed at MIT. (After Ref. 217.)

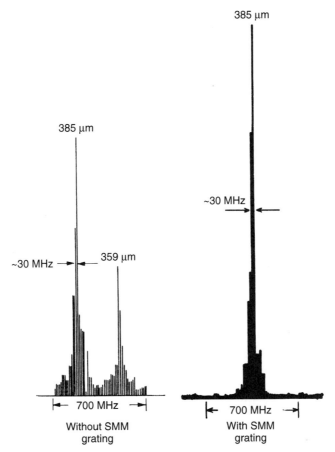

**Figure 9.114** Two Fabry–Perot scans of the $D_2O$ oscillator output. Left: Oscillator output when the main cavity is formed by flat mirrors. Right: Oscillator output when one cavity mirror is grating tuned to the 385-$\mu$m transition. The Fabry–Perot free spectral range is 700 MHz. (After Ref. 217.)

Figure 9.114 contains Fabry–Perot scans of the spectral content of the $D_2O$ oscillator (obtained over many pulses) with and without the FIR grating. The bandwidth of the output is found to be $\approx 30$ MHz (FWHM) and is attributed, as discussed previously, to the single-mode $CO_2$ pump. Output energies of up to $\approx 200$ mJ have been obtained with this configuration [217].

Significantly higher FIR output energies have been reported by Semet et al. [220]. Using a much higher energy $CO_2$ pump laser, they have obtained up to 5 J at 385 $\mu$m with pulse durations in excess of 3 $\mu$s. Figure 9.115 schematically depicts the combined $CO_2$/FIR laser configuration. Here, two Lumonics model 620 TEA $CO_2$ laser modules are employed as a grating-tuned unstable oscillator, while an additional two units comprise the amplifier. In both cases, the gain aperture is $20 \times 25$ cm. The resultant output beam was found to possess a uniform cross section over a $22 \times 20$ cm aperture, excluding the shadows produced by the grating and support structure. To ensure single-mode oscillation, a low-power CW laser is employed to injection lock the high power system (see Fig. 9.115). The low-power unit employed a flat 135 lines/mm

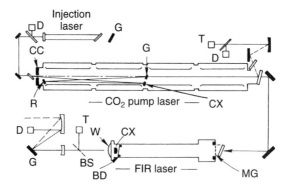

**Figure 9.115** FIR laser system of Semet et al. [220]. BS, beam splitter; BD, beam dump; CC, concave mirror; CX, convex mirror; D, fast detector; G, grating; MG, metal grid; R, relay mirror; T, thermopile; W, window/lens.

grating and an aperture for transverse mode control to ensure its mode purity. Unfortunately, the resultant output of only ≈18 W was barely sufficient to guarantee locking the high-power $CO_2$ laser throughout the entire duration of the pulse.

Referring back to Fig. 9.115, the FIR laser is seen to be comprised of a single unstable oscillator cavity. The $CO_2$ pump beam is coupled into the FIR cavity through a tilted NaCl window that is followed by a $25.5 \times 36.5$ cm free standing wire gird comprised of 25-μm-diameter tungsten wires with a spacing of 100 μm. This grid functions as a plane reflector (≈95%) for the FIR while transmitting (≈75%) the $CO_2$ pump beam. As mentioned previously, the gain of $D_2O$ is larger for the polarization component perpendicular to the $CO_2$ pump field, so that the grid wires are also oriented perpendicular to the pump polarization. The other portion of the unstable FIR oscillator cavity is a 12-cm-diameter convex copper mirror ($R = 15$ m). The divergent output FIR beam is focused by a biconvex polyethylene lens (see Fig. 9.115) at a distance of 1.2 m as well as serving as the output window. The 0.7-mm-thick teflon beam dump BD (see Fig. 9.120) is placed in front or the convex mirror and oriented at a slight angle and serves to reduce $CO_2$ pump feedback.

Figure 9.116 displays time histories of both the $CO_2$ and FIR pulses for pump energies of 60 J and 250 J using short-duration input pulses of ≈100 to 200 ns. The structure in the FIR pulse is quite evident. Figure 9.117 displays the measured FIR output energy as a function of the $D_2O$ fill pressure for several $CO_2$ pump energies for pulses with durations in excess of 3 μs. The authors report a maximum conversion efficiency of 42% of the theoretical limit for the case of a 360-J pump and 3 torr fill pressure and 51.5% at 4 torr and 530 J. A number of FIR and $CO_2$ pulse shapes are displayed in Fig. 9.118 for a variety of time base settings. Fast Fourier transform analysis of the FIR pulse reveals the existence of longitudinal modes at 39 MHz and 78 Hz. Despite these and a smaller occasional signal at 117 MHz, the authors note that the spectral quality was sufficiently good for the proposed scattering application.

As mentioned previously, the Lausanne group has devoted considerable attention to the development of $D_2O$ lasers [221–223]. This has culminated in their recent successful measurement of ion temperature in the TCA tokamak [224]. Figure 9.119 contains a schematic of the $D_2O$ FIR laser system together with the tokamak scattering arrangement. The $CO_2$ pump laser is comprised of a grating-tuned hybrid-TEA oscillator tuned

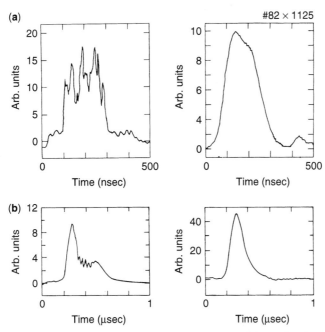

**Figure 9.116**  Left: Pulse-time histories of the FIR emission. Right: Pulse-time histories of the $CO_2$ emission. **(a)** Pump energy 250 J; **(b)** pump energy 60 J. (After Ref. 220.)

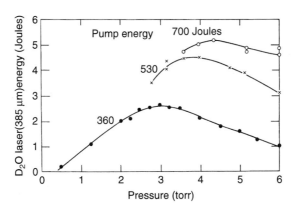

**Figure 9.117**  FIR laser energy versus pressure for "long" pump pulses ($> 3 \mu s$). (After Ref. 220.)

to the 9R22 line which serves as the input to a triple-pass electron beam preionized amplifier. The resultant single-mode output is approximately 600 J with a duration of 1.4 μs. After a propagation distance of 70 m through a beam duct, the pump radiation enters the $D_2O$ laser. The FIR unstable resonator is 4 m long and is configured in an L shape with a wire grid located at the apex to provide efficient coupling of the $CO_2$ pump beam. An output of ≈0.5 J with a pulse duration of 1.4 μs is obtained at a $D_2O$ fill pressure of 6.5 mbar. Figure 9.120 displays typical time histories of the $CO_2$ and FIR laser pulses obtained with fast detectors. As with the system of Semet et al. [220]

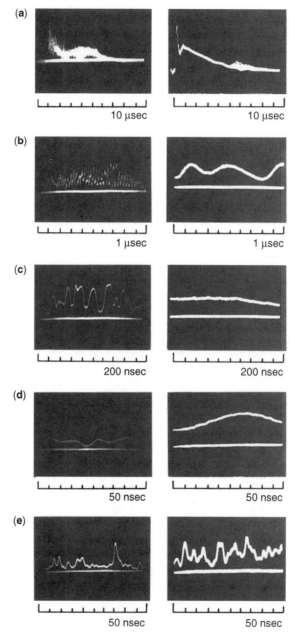

**Figure 9.118** Pulse shapes for "long" pump pulses. Left: FIR pulse shapes. Right: $CO_2$ pulse shapes. **(a)**–**(d)** Normal injection-locked pump pulses; **(e)** superradiant FIR pulse. (After Ref. 220.)

discussed in the preceding, two to three longitudinal modes exist in the FIR laser as evidenced by the observed mode beating.

At the high pump intensities associated with the pulsed FIR lasers, a variety of non-linear effects can occur [214, 226, 227]. Here, we review the work of Peebles et al.,

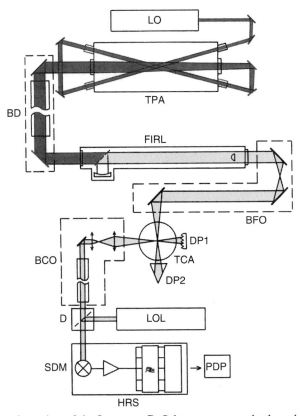

**Figure 9.119** Configuration of the Lausanne $D_2O$ laser system and tokamak scattering experimental system. (From Ref. 224.)

which identified two emission lines (388 μm and 412 μm) in $C^{13}H_3F$ which were due to collisionally coupled two photon absorption [226]. Previous work by Hacker et al. had shown that the "R" branch absorption of $CO_2$ laser radiation excites the methyl fluoride molecules from the ground vibrational state to the first excited state $v_3 = 1$, $J = 5$, resulting in lasing at 1222 μm ($J = 5 \rightarrow 4$ transition) [228]. In addition, three refilling transitions were observed: $v_3 = 0$, $J = 5 \rightarrow 4$(1207 μm), $v_3 = 0$, $J = 6 \rightarrow 5$ (1006 μm), and $v_3 = 0$, $J = 7 \rightarrow 6$ (862 μm). The peak output observed in these measurements was ~5 kW using a $CO_2$ pump power of ~4 MW, which corresponds to a $CO_2$ pump intensity of ≈0.35 MW/cm$^2$ within the isotopic methyl fluoride cell.

Figure 9.121 illustrates the experimental arrangement employed by Peebles et al., which employed up to 80 MW of $CO_2$ pump power in an ≈300-ns-duration pulse [226]. After beam reduction, the ≈50 mm $CO_2$ output beam resulted in pump intensities up to ≈4 MW/cm$^2$. Total FIR output powers of up to ≈75 kW were obtained from the 3.5-m-long FIR laser, which employed the Fabry–Perot as an output coupler. The relative energies of the various lines are shown in Fig. 9.122 as a function of $CO_2$ pump power for two different values of fill pressure. As mentioned above, the 388-μm and 412-μm transitions were identified as due to an R-branch, hot-band absorption of the $CO_2$ pump (see Fig. 9.123). After the absorption, the FIR transitions are $v_3 = 2$, $J = 16 \rightarrow 15$ (388 μm), and the cascade $J = 15 \rightarrow 14$(412) μm.

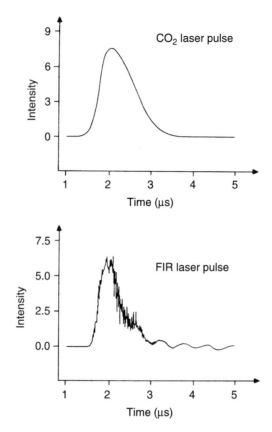

**Figure 9.120**   Typical $CO_2$ and FIR laser pulses obtained with fast detectors. (From Ref. 224.)

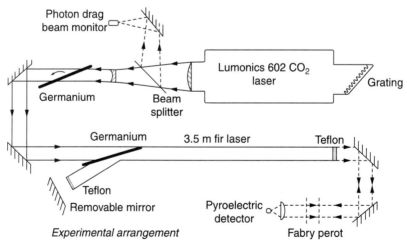

**Figure 9.121**   Experimental arrangement for high-intensity FIR laser pumping studies. (From Ref. 226.)

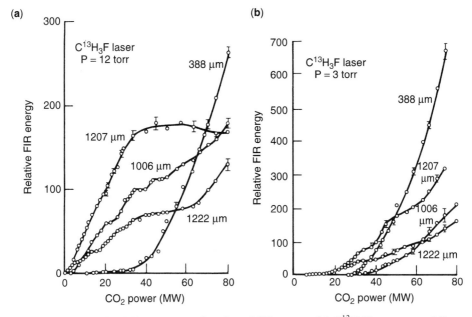

**Figure 9.122**   Relative FIR energy as a function of $CO_2$ power: **(a)** $C^{13}H_3F$ pressure $= 1.2$ torr; **(b)** $C^{13}H_3F$ pressure $= 3$ torr. (From Ref. 226.)

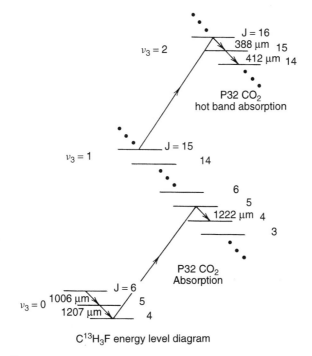

**Figure 9.123**   $C^{13}H_3F$ energy-level diagram indicating $CO_2$ absorption and FIR emission transitions. (From Ref. 226.)

The anharmonicity of the $\nu_3$ mode causes the $\nu_3 = 1 \rightarrow \nu_3 = 2$ hot band to be shifted to lower frequencies with respect to the $\nu_3 = 0 \rightarrow \nu_3 = 1$ fundamental band by about 15.9 cm$^{-1}$. This results in the R(15) $\nu_3 = 1 \rightarrow \nu_3 = 2$ transition having a frequency that coincides closely with the R(5) $\nu_3 = 0 \rightarrow \nu_3 = 1$ transition. Using similar changes in the constants for the $\nu_3 = 2$ vibrational levels of C$^{13}$H$_3$F as for C$^{12}$H$_3$F [139], the calculated frequency for the R(15) $\nu_3 = 1, \rightarrow \nu_3 = 2$ line was in good agreement with the frequency of the CO$_2$ pump laser. However, since the $J = 15$ and $J = 5$ energy levels in the $\nu_3 = 1$ excited vibrational state are separated by 170 cm$^{-1}$, it was necessary to postulate a two-step pumping mechanism involving collisional coupling of the $J = 5$ and $J = 15$ levels to explain the laser behavior. After the initial photon absorption to the $\nu_3 = 1$, $J = 5$ state, collision processes quickly populate the $\nu_3 = 1$, $J = 15$ level and thereby allow significant hot-band absorption to the $\nu_3 = 2$, $J = 16$ state.

## 9.3   EXCIMER LASERS

### 9.3.1   Introduction

Excimers are usually defined as molecules that are bound only in the excited state. They disassemble upon falling into the ground state. To operate a laser, one needs more molecules in the excited state (upper laser level) than in the ground state (lower laser level). Excimers are therefore natural laser candidates. As long as the translational temperature is lower than the electronic temperature (which is the case in glow discharges and electron-beam excited gases), any excitation of molecules results in laser gain.

Many excimer molecules have been observed and used for as lasers media. In the following are some common excimers and the lasers wavelengths they have generated [229, 230].

| | | |
|---|---|---|
| Rare gas dimmers: | Ar$_2$ | 126 nm |
| | Kr$_2$ | 146 nm |
| | Xe$_2$ | 172 nm |
| Rare gas oxides: | KrO | 558 nm |
| | XeO | 537.6 and 544.2 nm |
| Rare gas Halides: | ArF | 193 nm |
| | KrF | 248 nm |
| | XeF | 351 and 353 nm |
| | KrCl | 222 nm |
| | XeCl | 308 nm |
| Metal halides: | Hg Cl | 558 nm |
| | HgBr | 503 and 498 nm |
| | HgI | 444 nm |

We are going to concentrate here on rare gas halides. This type of excimer laser is the most powerful and most commonly used. A typical potential energy curve is shown in Fig. 9.124. In KrF, the low-lying excited states are $B$ and $C$. The ground

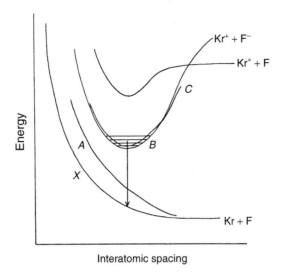

**Figure 9.124**  Energy-level curves for the KrF molecule.

state $X$ is completely unbound, and so is the state $A$. In some excimers, laser gain has been achieved on the $C \rightarrow A$ transition [231]. However, it is usually very low. It is the $B \rightarrow X$ transition that is used for most common excimer lasers.

Rare gas halogen excited molecules can be created by several possible reactions. A typical mechanism for excitation is as follows:

$$Kr + e^- \Rightarrow Kr^* + e^-$$

and

$$Kr^* + F_2 \Rightarrow KrF^* + F$$

In the above interaction, an energetic electron excites a krypton atom. The excited atom has an electron in an outer shell which makes it behave chemically just like a metal. Therefore, it readily reacts with a fluorine molecule generating excited KrF. KrF* can also be created by reactions involving ionized atoms and molecules. The excited dimer (such as KrF*) can lose its energy through radiation (either stimulated or spontaneous emission) or it can be collisionally deexcited. The reaction kinetics can be very complicated due to hundreds of possible alternate paths and detailed discussion can be found in literature [229, 232, 233].

In the following sections, we discuss the technology and applications of the most important types of excimer lasers. We begin with a discussion of the technology and applications of discharge pumped excimer lasers, which are the most commonly used. This is followed by a discussion of electron-beam pumped excimers which can produce very high energy.

### 9.3.2  Discharge Lasers

*Advantages of Discharge Pumped Lasers.*  There are many applications for UV lasers that require modest output energy per pulse ($\leqslant$ joules), medium to high repetition rates ($>10$ Hz), in a compact tabletop system that is inexpensive and reliable. For such laboratory and industrial applications, discharge lasers are the most attractive and frequently used excimer lasers. The electron-beam pumped systems, discussed in the next section, are too costly and complicated for these applications. In addition, it is difficult to produce low-energy electron-beams to couple efficiently to small aperture lasers. On the other hand, as we discuss below, there are limits on the energy and the laser aperture that can be pumped by transverse discharge lasers.

Once the gas is excited and the laser pulse generated, this volume of gas cannot be used for the next laser pulse. The gas must be cooled and under heavy excitation the halogen donor must be replenished. This means that for high repetition rates, it is desirable to flow the gas so that each pulse excites fresh gas. For large apertures, this requires greater flow velocity and more sophisticated fluid dynamical design. Discharge lasers use light buffer gases such as He and Ne, which facilitate this. Electron-beam lasers need heavier buffer gases such as argon. This means that more energy needs to be invested in fluid flow. High-voltage pulse power technology in the tens of kilovolts range, which is used for discharges, is quite mature, due to the enormous effort invested in radar systems. However, compact and efficient pulsers at hundreds of kilovolts, which are necessary for electron-beam excitation of gas, are usually close to state-of-the-art technology and are expensive, have more reliability problems, and cannot easily be operated at high repetition rates.

Discharge excitation works well for modest apertures. A typical commercial excimer laser excites a volume with a cross section of $1 \times 2$ cm and a length of 50 to 100 cm. Thus, the transverse flow needs to clear only about 1 cm. In the transverse discharge case, voltages of 30 to 60 kV are typically used and thyratron circuits can be utilized. For these reasons, discharge pumped lasers, not electron-beam pumped lasers, are used in most laboratories and in industry.

***Glow Discharge Excitation at High Pressure.*** The kinetics of formation of excimer molecules and the relaxation processes in the vibrational manifold in the electronically excited molecules require gas pressures of a few atmospheres for efficient laser operation. This is not a good regime for operating volumetric glow discharges. At such high pressures, discharges are very unstable and have the tendency to collapse rapidly into arcs.

As described briefly in Section 9.1, there exists an established technology for generating large-volume glow discharges for $CO_2$ lasers which also operate at high pressures. Discharges can be kept stable for microseconds with proper grading of electric fields, an external source of ionization, and proper pulse-forming electrical circuits. The most successful $CO_2$ laser technology for very large lasers is the electron-beam sustained discharge. The transverse electric field between the two main electrodes is kept below the self-breakdown value. An electron beam is injected into the laser volume and provides the necessary ionization level while the electrodes independently control the electron temperature. This approach was considered in the early days of excimer lasers. However, the region of stable operation was found to be limited to very low powers for which laser operation was not useful [234].

Another technique for discharge excitation of high-pressure gases involves preionization with either UV, x-rays, or electron-beam, followed by avalanche breakdown and self-sustained discharge between carefully shaped transverse electrodes. Such discharges can remain stable for microseconds in properly designed $CO_2$ systems. This approach was also tried with excimer laser gas mixes with limited success. Typically, discharges remain stable for approximately 20 to 100 ns, depending on the gases used. XeCl laser gas mix usually has longer stable operation than KrF, which in turn is more stable than ArF or XeF. The short duration of the glow discharges means that very fast pulsers have to be used to deliver the energy to the laser on such a time scale.

In $CO_2$ lasers, the collapse of the glow discharge results usually in one bright arc short-circuiting the discharge. In excimer laser discharges, because of the much shorter time scales, the discharge collapses into many small filaments which still have the appearance of a glow discharge but do not generate any laser gain. The duration of the stable discharge is a function of many parameters, such as uniformity of preionization, electric field uniformity, power deposition in the gas, the voltage pulse rise time, and the gas composition.

***Construction of Discharge Lasers.*** Typical cross sections of transverse discharge lasers are shown in Figs. 9.125 and 9.126. A laser body that acts as the insulator between two main electrodes is machined from a plastic material. Early lasers were constructed from plexiglass or similar plastics and they had a very short lifetime. Due to chemical attack of halogens, in particular fluorine, and strong UV fluxes from the discharge, cracks tended to form which propagate rapidly through the body. Only plastics that are resistant to halogens, such as Teflon, or materials such as fiberglass epoxy, in which cracks cannot propagate, can be safely used with fluorine.

With most insulating materials, a surface layer inert to halogens is eventually formed and the chemical reaction stops. Many metals (aluminum, stainless, etc.) also form a pacivation layer in contact with fluorine. In the design of a laser for applications that require little maintenance and long-gas-fill lifetimes, the choice of materials used for laser construction is critical, and in recent years major progress has been made in this area.

Uniform preionization of the discharge volume prior to application of high voltage is critical for reliable operation of this type of laser. One simple approach shown in Fig. 9.125 uses an array of "spark plugs" along the grounded electrode to generate the ultra-violet light, which then ionizes unknown impurities in the breakdown

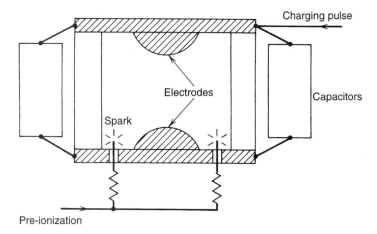

**Figure 9.125** Typical discharge laser with an array of spark plugs for preionization.

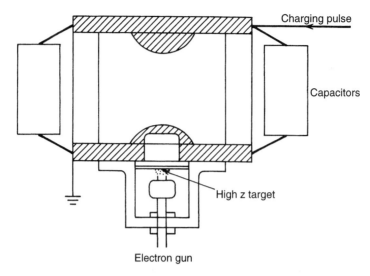

**Figure 9.126**   Discharge laser with x-ray preionization.

region. Another common approach, which provides a more uniform preionization but is mechanically more difficult to implement, involves a screen electrode with a preionization source below it. X-rays can also be used for preionization. The x-ray source is more complex than the simple sparks. However, mechanical design of the laser is simplified significantly since solid metal electrodes can be used as shown in Fig. 9.126.

***Electrical Circuits for Discharge Excitation.***   A commonly used circuit for driving excimer discharge lasers is shown in Fig. 9.127. Between laser pulses, the capacitor $C_1$ is charged to high voltage, on the order of 30 to 60 kV. When the switch $S_1$ closes, the charge starts transferring from $C_1$ to $C_2$. Switch $S_1$ has to be able to handle high currents and have a fast rise time. For low repetition rates, gas-filled triggered spark gaps are often used in "homemade" devices. Spark gaps can be inexpensive, but they require regular maintenance.

Commercial devices usually use hydrogen thyratrons. Thyratrons perform quite well in this circuit; however, their lifetime is significantly reduced from that in usual radar

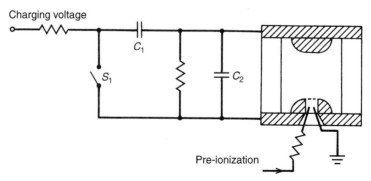

**Figure 9.127**   Electrical circuit commonly used for driving excimer discharge laser. UV preionization from behind a screen electrode is shown here.

circuits operating on a microsecond time scale. The reason for this is the finite time for drop in resistance of the thyratron after triggering. As the thyratron turns on, its resistance drops on the time scale typically of tens of nanoseconds. This turn-on time is comparable to the charging time of $C_2$; therefore, a sizable fraction of available electrical energy can be dissipated in the thyratron. Aside from inefficiency, this power dissipation in the thyratron causes excessive wear and shortens the lifetime of the tube significantly. Adding inductance in series with the thyratron reduces the current rise time and increases thyratron lifetime. Unfortunately, it also causes degradation in laser performance. One solution involves the use of a saturable inductor, which initially has high inductance to protect the thyratron and then switches to a saturated value of low inductance to allow fast charging of $C_2$. Figure 9.128 shows a typical simple electrical circuit utilizing a saturable inductor in such manner. More elaborate circuits utilizing saturable inductors for "magnetic compression" can be used to further reduce the current rise-time requirements on the thyratron [235]. A two-stage magnetic pulse compression circuit is shown on Fig. 9.129. With four stages of magnetic compression, it is possible to build a laser pulser with only solid-state switches such as silicon-controlled rectifiers (SCRs) [236]. This approach would significantly improve the long-term reliability of discharge lasers.

The capacitance of $C_2$ is usually chosen to be smaller then $C_1$, so that it can ring up higher then the dc charge voltage. When the voltage on capacitor $C_2$ is sufficiently high, the gas between the main electrodes breaks down and $C_2$ discharges rapidly. To

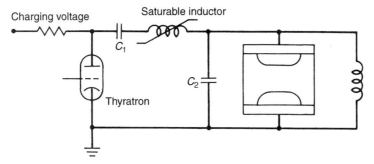

**Figure 9.128** Electrical circuit incorporating saturable inductor for prolonging thyratron lifetime.

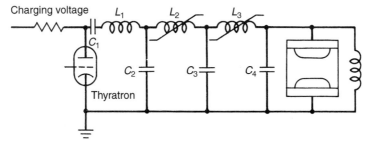

**Figure 9.129** Two-stage magnetic pulse compression circuit.

deposit the energy quickly, the transverse electrode geometry with an array of small "door-knob" capacitors in parallel acting as $C_2$ is used. A typical value of $C_2$ is 10 n$f$.

The time it takes for the voltage on $C_2$ to reach breakdown is very important in determining the behavior of the discharge after breakdown. If this charging time is significantly longer than 100 ns, the discharge will be of very poor quality. There will be low laser output and strong shot-to-shot fluctuations.

Another technique for establishing uniform discharges involves delaying the preionization until after the voltage has been applied across the electrodes. Since unpreionized gas can hold off much higher voltages, the ionization even at very low levels will initiate the discharge. This technique requires very uniform and synchronous ionization of the entire discharge volume [237], and with very uniform hot-cathode x-ray generators [238]. The important advantage of this technique is the reduction in current and rise-time requirement on $S_1$, which translates into much greater switch lifetime. However, the quality of the ionization sources has to be much higher. For that reason, discharges switched in this fashion do not work with common UV sources.

***Long Pulse Lasers.*** The duration of laser gain is very important for laser oscillators, which need well-developed cavity modes. This is the case for oscillators with very narrow spectral line width, high degree of spatial coherence, or with mode locking. For all of these applications, the maximum number of round trips in the laser cavity is desired. In the most common type of discharge laser, which is readily commercially available, the gain duration is 20 ns and one obtains only two or three round trips.

Stable discharges with duration greater than 100 ns can be generated in XeCl lasers by driving the discharge at lower power, which actually allows better impedance matching and by initiating the discharge with a very fast separate high voltage pulser [239]. For small devices, longer-duration pulses were also obtained using segmented inductively isolated electrodes [240] and in electrodeless capillary discharge lasers [241].

***Large Aperture Discharge Lasers.*** In most common excimer lasers, the output beam has a rectangular cross section, with the width of the gain region usually less than half of the electrode separation. When the electrode separation is increased, the ratio between the two dimensions gets larger. Width of the discharge depends on several parameters. Electric field uniformity is very important and is determined by the shape of the electrodes. The gain profile is also sensitive to the uniformity of preionization. Another important parameter is the rise time of the voltage pulse across the electrodes. With a rise time of 100 ns, a preionized discharge will be quite reproducible and yield reasonable laser efficiency (close to 1% in KrF with reasonable gas mix). However, the aspect ratio is such that the width is three or four times narrower than the interelectrode separation. Of course, changing the electrode profiles will have some effect on the aspect ratio, but it is very difficult to obtain wider discharges with this circuit. If the charging time is shortened to below 50 ns, more uniform and wider discharges are obtained, although laser efficiency is not significantly changed. To obtain voltage rise times faster than 50 ns, $S_1$ has to be a multichannel rail gap, or an array of spark gaps or thyratrons, and $C_1$ must be distributed capacitance. Lasers with cross-sectional areas on the order of 50 cm$^2$ have been generated [242].

Using intense UV preionization from creeping discharge along a dielectric, a $13 \times 20$ cm aperture discharge was obtained in XeCl [243]. An output energy of 20 J was reported from this laser. The largest output energy from a discharge excited laser was obtained by a group at the Naval Research Laboratory. Sixty-six joules in a 180-ns pulse was generated in an x-ray preionized XeCl laser [244].

*Optical Configurations.* Discharge excimer lasers are usually very high gain devices. A cavity with one uncoated flat and a high reflectivity mirror has close to optimal output coupling. Energy outputs on the order of a fraction of a joule are obtainable from simple table top oscillators in KrF, ArF, XeF, and XeCl. The output beams are quite divergent because they are basically amplified spontaneous emission which has propagated only two or three passes through the laser. The output is typically quite broadband (on the order of 50 cm$^{-1}$); thus the spatially incoherent beam has excellent uniformity, and this can be of importance in material processing.

To obtain beams with greater spatial coherence, one can decrease the apertures in the cavity until the Fresnel number approaches unity and obtain close to diffraction-limited beams. The power output will also be reduced by two or three orders of magnitude. Beam quality can be significantly enhanced by using an unstable resonator with a high magnification (on the order of 10) with little loss of power. However, the time to reach the steady-state mode is longer than the gain duration and the output will consist of beams with changing divergence, approaching the diffraction limit only at the very end of the laser pulse.

The solution to this, if coherent output is needed, is the oscillator amplifier system. In this system, one laser is used to generate (inefficiently) a high-quality spatial, and narrow-band spectral, output. The second laser is used as a regenerative amplifier which is "seeded" by the first. If the injected "seed" is much greater than the spontaneous emission source in the cavity, the second laser will have the output at the same frequency, and with better beam quality than the first. Figure 9.130 shows a typical optical arrangement for such an oscillator–amplifier configuration.

The main difficulty with operating the oscillator–amplifier system is the need for very strict synchronization between the two lasers. This can be accomplished by using a single switch for circuits driving both lasers. Several commercial systems of this type are presently available.

Special care must be taken in choosing optical materials for components that come in contact with excimer gas mixtures, such as laser windows or internal mirrors. The

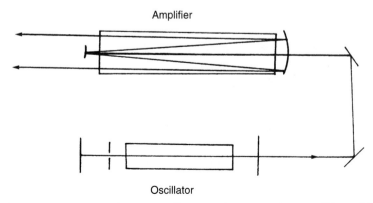

Amplifier

Oscillator

**Figure 9.130** Optical system for an oscillator and regenerative amplifier with unstable resonator.

biggest problem is HF, which is generated whenever $F_2$ comes in contact with moisture or organic materials. If fused silica optics are used, great care must be taken to continuously purify the gas mixture flowing by the windows. Some optical coatings show more resistance to HF. Windows made from $MgF_2$, $BaF_2$, and $CaF_2$ are completely inert from chemical attack. $CaF_2$ is often preferable due to the lack of birefringence and availability of large crystals. However, since it is a soft material, $CaF_2$ optical surfaces are of lower quality than their fused silica counterparts.

***Commercial Discharge Excimer Lasers.*** At present there are several companies that manufacture discharge lasers operating in the self-sustained mode with typical output pulse duration of 20 ns. Some of them are listed below:

| | |
|---|---|
| Lambda Physik | 289 Great Road |
| | Acton, MA 01720 |
| Lumonics | 105 Schneider Road |
| | Kanata, Ontario, K2K 1Y3 |
| | Canada |
| Questek | 44 Manning Road |
| | Billerica, MA 01821 |
| Cymer | 7887 Dunebrook Road |
| | San Diego, CA 92126 |
| Stabelase | 1042 Stage Coach Road |
| | Santa Fe, NM 87501 |
| XMR Inc. | 5403 Betsy Ross Drive |
| | Santa Clara, CA 95054 |

Typical laser outputs are in the range of 0.1 to 1 J per pulse. Highest peak powers are usually obtainable from KrF lasers at 248 nm. Highest electrical efficiency is observed in XeCl lasers at 308 nm. This laser has by far the longest gas-fill lifetimes and because of its wavelength is most commonly used for pumping pulsed dye lasers. The XeF laser at 351 and 353 nm typically has output power lower by a factor of 2 than KrF or XeCl. The most difficult to operate is the ArF laser at 193 nm. When optimized, the output is comparable to XeF; however, it deteriorates rapidly with time due to contamination of gas mix and optics. Pulse repetition rates up to 100 Hz are available on many systems.

For applications requiring high repetition rate and long pulses (100 ns) and low powers, there are other discharge laser designs. Segmented inductively ballasted transverse discharge devices are offered by the Stablex Corporation (New Mexico) with pulse repetition rates up to 1 kHz and pulse energies of tens and millijoules. For higher repetition rates (and lower powers), microwave excitation is used. Potamac Photonics (Maryland) offers lasers with pulse repetition rates up to 10 kHz.

This is a rapidly changing and competitive field, and for details of specific device performance, the interested reader should consult the information readily available from the manufacturers.

***Applications of Discharge Lasers.*** Discharge excimer lasers are rapidly beginning to find applications in diverse areas of technology and medicine. One of the most common uses of the XeCl laser today is as a pump for tunable dye lasers from the near UV through the visible region. Direct use of excimer lasers, however, is also rapidly growing. The main attribute of these lasers which is making them so important is the choice of wavelength in the ultraviolet region. To date, the main obstacle to wider use of excimer lasers in industry is the costly maintenance and the need to handle toxic gases ($F_2$ or HCl). Fortunately, there has been rapid progress in the laser technology, and longer-lifetime, more reliable devices are becoming available, which means that many applications can soon be transferred from the experimental laboratories into the real world.

For material processing, the important parameter is the absorption depth. Ultraviolet light can be absorbed in very shallow depths and combined with nanosecond pulse duration, this can result in very localized heating and melting of a material.

Focusing on the order of a hundred millijoules, even of quite incoherent excimer laser on a surface of any material will result in the generation of very high temperatures with the material melting, vaporizing, and even forming highly ionized plasma. This effect is used to vaporize and sputter highly refractive materials. Thin films of high-temperature superconductors were grown by this process [245].

The broad bandwidth and highly multimode output of excimer lasers result in highly uniform illumination in the near field. For this reason it was discovered that for the process of annealing the surface of semiconductor wafers, excimer lasers worked much better than other types of lasers, such as solid-state lasers [246].

To fabricate submicrometer microelectronic devices with photolithographic techniques, short wavelengths are required. However, for wavelengths below 350 nm, optical components such as lenses can be manufactured presently only from the materials $CaF_2$, $MgF_2$, and synthetic fused silica. Achromatic diffraction limited imaging systems with low f numbers are difficult to construct and are not commercially available. This means that laser line width must be reduced by approximately two orders to magnitude from the freerunning laser case, down to 0.003 nm for the KrF laser, to overcome chromatic aberration in the lenses. Structures down to 0.5 μm in dimensions were fabricated using such a laser [247–250].

A medical application of excimer lasers that is under intense investigation is angioplasty. The output of a XeCl laser coupled through UV fiber bundle is used to clear out the human artery. To propagate maximum energy per pulse through optical fibers, longer-duration pulses are required. Special lasers are being developed for this application [251].

Another medical application involves the ArF laser at 193 nm. When longer-wavelength laser radiation is use for surgery, the irradiated tissue is damaged, due primarily to heating and microexplosions. Because of its short wavelength, the radiation from the ArF laser damages tissue by breaking the molecular bond directly. Perhaps for this reason, cuts made by ArF laser are much cleaner. Recent work shows that it is possible to operate on the surface of the lens of an eye and to preserve its optical quality. This means that the laser can be used for radial keratotomy or perhaps even for direct reshaping of the curvature of the lens [252, 253].

Because of the wide bandwidth of the excimer lasers, they are excellent candidates for generation of ultrashort extremely high power laser radiation. Subpicosecond pulses can be amplified to terawatt power levels in discharge laser amplifiers [254, 255]. These type of systems are expected to play a major role in the development of the next generation of laboratory soft x-ray lasers.

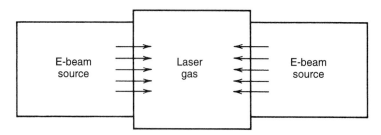

**Figure 9.131**   Typical experimental arrangement for an electron-beam pumped amplifier.

### 9.3.3   Electron-Beam Pumped Excimer Lasers

Electron-beam pumped excimer lasers are among the highest energy pulsed lasers in the ultraviolet wavelength region. Up to 10 kJ has been obtained with KrF and 5 kJ has been obtained with XeCl using electron-beam pumped amplifiers [256, 257]. Energies from single amplifiers up to the 100 to 200 kJ level appear to be feasible [258]. Relatively high efficiencies have also been obtained using these devices. Intrinsic efficiencies (ratio of laser output energy to the electron energy deposited in the gas) above 10% have been obtained routinely using electron-beamed pumped KrF. The combination of high efficiency, high energy, and short wavelength make electron-beam pumped excimer lasers an attractive candidate for the laser-fusion application. In this section, we discuss the physics and technological issues that determine the energy and power capabilities of electron-beam pumped laser systems.

A typical experimental arrangement for an electron-beam pumped laser is shown in Fig. 9.131. An electron beam is accelerated to energies of 200 to 800 keV and is transmitted through a thin foil into the excimer gas mixture. Collisions of the energetic electrons with the gas provide the excitation mechanism for pumping the excimer transition. Use of accelerator potentials near 1 MeV allows electron beams to pump large aperture gas cells ($>1$ m). Present electron-beam technology allows the production of energetic electron beams with pulse durations up to a few microseconds. The combination of the longer pulses and larger apertures that can be obtained with electron-beam pumped lasers in comparison with discharge lasers allow much higher laser energies per pulse. Present systems range from small-aperture (10 cm, 250 keV) commercial systems producing 10 J to large-aperture (100 cm, 800 keV) systems producing multi-kilojoules.

The current state of the art with such amplifiers is summarized below, including the technological and physics limitations. Issues to be considered include (1) the requirements placed on the electron-beam generator by the excimer laser physics, (2) the advantages and disadvantages of using external magnetic fields to guide the electron-beam, and, (3) the means for obtaining short-pulse high power pulses with excimer amplifiers. The discussion will center on KrF system parameters.

***Effect of Laser-Physics Issues.***   Typically, the gas cell for a KrF laser will contain a 0.2 to 1% partial pressure of $F_2$, with the balance of the gas consisting of Kr and a buffer gas, usually argon, with total gas pressures of 1 to 2 atm (760 to 1500 torr). The buffer gas serves to aid in stopping the electron beam and transferring energy to the lasing transition. The kinetics of the interactions in the electron-beam pumped gas have been studied extensively in numerical simultations which give fairly good agreement with experiment [233, 259–261]. Saturation intensities for the laser ($I_s$) of 1 to 2 MW/cm$^2$ are obtained with typical KrF gas mixtures and pressures. The electron beam–gas interaction produces many species, some of which are absorbing at the lasing wavelength. Typically, small-signal gain to nonsaturable-absorption ratios $g_0/\alpha \approx 10$ are obtained for optimized mixtures. The low $I_s$ and the modest $g_0/\alpha$ limit the laser intensity that can be extracted from KrF. Starting with the equation for gain in a homogeneously broadened laser [4],

$$\frac{dI}{dx} = \left( \frac{g_0}{1 + I/I_s} - \alpha \right) I$$

and setting $dI/dx = 0$, one obtains a maximum intensity of

$$I_{\max} = \left( \frac{g_0}{\alpha} - 1 \right) I_s$$

The absolute maximum intensity one can achieve with $g_0/\alpha = 10$ is 9 $I_s \approx 9$ to 18 MW/cm$^2$ for KrF. The optimal local flux $I_{opt}$ for efficient extraction is lower than $I_{\max}$ and is given by [262, 263]

$$I_{opt} = \left[ \left( \frac{g_0}{\alpha} \right)^{1/2} - 1 \right] I_s$$

with a local maximum efficiency given by

$$\eta_{\max} = \left[ 1 - \left( \frac{\alpha}{g_0} \right)^{1/2} \right]^2$$

For typical KrF parameters, $I_{opt} \approx 2$ to 4 MW/cm$^2$ with a local maximum efficiency $\eta_{\max} \approx 50\%$. In actual amplifiers one cannot maintain this optimal extraction intensity, and extraction efficiencies of $\sim 40\%$ are more typically obtained [264, 265]. Figure 9.132 shows the calculated output power for a double-pass laser amplifier as a function of the gain-length product ($g_0 L$) for the case of $g_0/\alpha = 10$ and 20. For large $g_0 L$, the output becomes saturated due to absorptive losses. One obtains reasonable extraction efficiencies in this double-pass configuration for output intensities up to about $0.5 I_s g_0/\alpha$. With the typically realized $g_0/\alpha$ and $I_s \leqslant 2$ MW/cm$^2$, the output is limited to about 10 MW/cm$^2$ with KrF. The potential benefits of $g_0/\alpha$ ratios larger than 10 are obvious from the equations above and Fig. 9.132; methods to obtain larger $g_0/\alpha$ are the subject of current research efforts.

Due to the limitation on intensities that can be efficiently extracted, high energies can be obtained only with excimer lasers by the combination of large apertures and long pulse lengths (rather than long amplifiers and large $g_0 L$). If one wants $>1$ J/cm$^2$ output, the pulse length must be longer than 100 ns duration. High-energy excimer lasers thus require electron beams that can deposit energy over large apertures with $>100$ ns duration pulses.

There are applications, such as laser fusion, which require shorter laser pulse lengths ($\leqslant 1$ ns) than is optimal for amplifier pumping. The deexcitation times for excimer transitions are typically short ($\leqslant 5$ ns). Therefore, excimers cannot store the energy

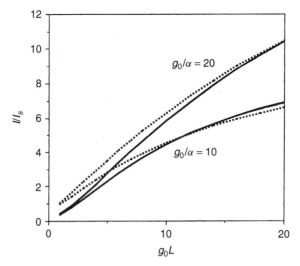

**Figure 9.132**   Calculated output power from a double-pass amplifier as a function of $g_0 L$ for input intensities of $0.1 I_s$ (solid line) and $0.5 I_s$ (dotted line).

over long pumping times, and laser pulse compression schemes must be employed to reduce the duration of the laser pulse for these applications. Several of these schemes will be discussed later.

***Electron-Beam Amplifier Technology.***  In this section, we discuss the electron-beam technology that has been developed to pump excimer lasers. Efficient, uniform deposition of energy in the gas cell calls for two or more sided illumination of the cell with beam energies in the range 200 to 800 keV. Due to various losses, the electron beam energy deposited in the gas must be 8 to 12 times greater than the laser energy. Such high-energy, nearly relativistic electron beams are produced by accelerators that use high-voltage, switched energy-storage devices to drive an electron emitting cathode. For laser energies in the range of a few tens of joules, Marx banks can be attached directly to the cathode to drive the accelerator. It is difficult to obtain low-enough inductance with Marx banks to drive high-power, short-pulse systems. To obtain higher energies and powers, one can use a charged, water-filled transmission line or Blumelein that is impedance matched to the electron beam at the cathode. A typical water-line system is shown in Fig. 9.133. Water is often chosen for the insulating medium in the pulse forming line (PFL) because its high dielectric constant allows high-energy storage density. A major complication in using water is that it cannot withstand high-voltage stresses for more than a few micro-seconds without breakdown. Therefore, the waterline systems require an intermediate speed system, such as a Marx bank, to pulse charge the lines. Some smaller power amplifiers have therefore employed other dielectrics, such as oil, so that the PFL can be dc charged [266].

Large, low-repetition-rate electron-beam pumped lasers typically employ a cold cathode which forms an electron emitting plasma upon the application of high voltage. A cold cathode configuration is shown in Fig. 9.134. The plasma is initiated by field emission from sharp points on the cathode surfaces. Some devices use cold cathodes formed from metal surfaced where sharp points have been placed on the surface; for

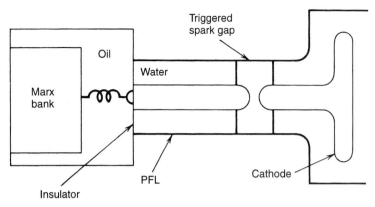

**Figure 9.133** Waterline-driven electron diode typical of those used for electron-beam pumped excimer lasers.

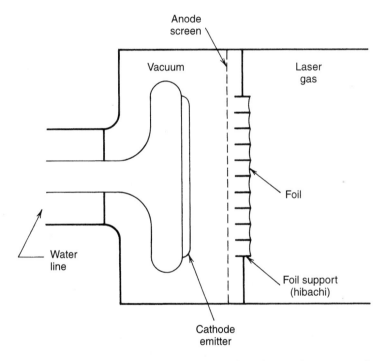

**Figure 9.134** Details of the cold cathode structure of an electron-beam pumped amplifier.

example, a crude but successful technique involved punching many small holes through sheet metal [267]. Various fabrics, such as carbon felt and velvet, have also been used to cover metal cathode surfaces. Broken fibers in the fabric apparently produce the field enhancement for breakdown. Once the plasma has formed, the cathode current is limited by space charge to that given by the Child–Langmuir law for planar geometry [268].

$$J = 2.34 \times 10^{-6} V^{3/2}/s^2 \text{ A/cm}^2$$

Here, $V$ is the anode–cathode voltage and $s$ initially is the anode–cathode separation. One can adjust the emitter impedance by varying the voltage, the anode–cathode spacing, and the total emitting area of the cathode.

There are two phenomena that limit the power and energy in the electron beams produced by cold cathodes. The first is pinching, which limits the current. The second is anode–cathode closure, which limits the duration of the electron beam. We begin with a discussion of pinching. The magnetic self-field produced by the electron beam produces an $eV \times B$ force which deflects the periphery of the beam toward the center and, if large enough, will cause the beam to pinch and distort its spatial profile. Pinching would be prevented in a freely propagating electron beam by the lateral, repulsive electrostatic field. The lateral electrostatic fields are shorted in the electron-beam diode by the proximity of the conducting anode and cathode structures. Pinching becomes an important effect when the electron gyroradius ($R_c$) in the self-magnetic field of the beam approaches the anode–cathode spacing ($s$). For the case of a rectangular beam where the length $L$ is much larger than the width, the peak self-field is given by $B = \mu_0 I / 2L$, and this limit $R_c = s$ is given by

$$I_c = 5.3 \times 10^3 \frac{L[T(T + 2E_0)]^{1/2}}{s} \text{ (Amps)}$$

where $T$ is the electron kinetic energy in MeV, $E_0$ the electron rest mass energy (0.51 MeV), and $I_c$ the critical beam current. One should, in fact, operate with currents several times smaller than $I_c$. Once the electron beam passes through the foil into the gas, pinching is no longer a serious problem because the return current through the ionized gas tends to cancel that due to the electron beam.

Several strategies have been followed to deal with pinching. One is to restrict the current density and dimensions of the cathode so as to remain below the threshold for pinching. Use of a high beam voltage rather than high current to obtain high electron-beam power helps to reduce pinching. The voltage is limited, however, by the requirement for good coupling to the gas cell. Very high laser energies can be obtained in principle by using many cathodes to pump the same gas cell, where the current in each cathode is kept below the pinch limit.

Another method to suppress pinching is to apply an external magnetic field along the axis of the electron beam accelerator. If the amplitude of the external $B$-field is several times larger than that of the self-field, the electrons are guided by the external field and pinching is prevented. An external field has the additional benefit of guiding the beam through the gas and reducing scattering losses to the walls. Guiding by an external magnetic field, however, has the undesirable effect of mapping small-scale nonuniformities on the cathode emitting surface into the gain medium. More attention has to be placed on obtaining uniform cathode emitters with the use of external magnetic fields.

The highest energy reported to date for excimers was 10 kJ for a KrF system using an external magnetic field [256]. Figure 9.135 contains photographs of this 10-kJ, 1-m aperture amplifier, which is located at Los Alamos National Laboratory. Several other systems of electron-beam pumped excimer amplifiers that utilize external magnetic fields have been built with energies in the range 100 to 5000 J [255, 257, 269]. The highest energies obtained with an electron-beam system with multiple cathodes and no external field is in the range 100 to 1000 J [267, 270, 271].

(a)

(b)

**Figure 9.135**  LAM (large-aperture module) electron-beam pumped 10-kJ amplifier of the Aurora laser. (Photographs courtesy of Los Alamos National Laboratory.)

The second important performance-limiting phenomenon occurring in the anode–cathode region is closure of the anode–cathode spacing. This is due to expansion of the electron emitting plasma from the cathode surface [272]. Typical speeds without the magnetic fields are 2 cm/μs. The closure velocity is observed to increase with the external magnetic field amplitude [273]. This enhanced closure velocity may be caused by the magnetic field helping to transport hot spots in the plasma across the anode–cathode spacing. Closure causes the effective anode–cathode spacing to decrease during the electron-beam pulse, and thereby causes the diode impedance to change in time. This complicates matching the impedance of the pulse power to the diode. The pulse lengths are limited by closure to periods short compared to the time for the plasma to expand to the anode. This effect limits the current densities that can be obtained by placing a lower limit on the anode–cathode spacing $s$.

Some fraction of the electron beam transmits through the anode and foil to enter the gas cell. The deposition profiles of the electron energy into the gas vary with gas mixtures, the presence of a magnetic field, the aperture, and the beam voltage. Experiments indicate reasonable agreement between Monte Carlo calculations and observed deposition profiles [274, 275]. The gain, gain-to-loss ratio, saturation fluence, and efficiency vary with the pressure and ratios of the gas mixture. The efficiency is determined by the net effect of the energy channeled into the laser excited state, reduced by de-excitation losses due to collisions. losses due to spontaneous emission, and absorptive loses by the reaction products of the electron beam with the gas mixture. Computer simulations have predicted intrinsic efficiencies as high as 15% for KrF lasers. There is some experimental evidence that such efficiencies can be obtained, although 7 to 10% is more typical [276–278]. Intrinsic efficiency is defined as the ratio of laser output energy to the electron energy deposited in the gas. Comparable efficiencies have been reported with ArF amplifiers [279]. The intrinsic efficiencies reported for XeF and XeCl are about half that of KrF [280–282]. The relatively high efficiency of the KrF laser has fueled interest in its application as a driver for inertial fusion targets. To date, however, the wallplug efficiency that has been obtained with large electron-beam pumped excimer lasers has been more in the range of a few percent. This lower efficiency is due to additional losses in the pulse power lines and transmission losses of the electron beam through the anode and pressure foil and its support structure. At this time, it is not clear exactly how much of this loss can be eliminated by better engineering practices, but efficiencies above 5% appear attainable.

A serious practical problem preventing the routine use of large electron-beam pumped KrF lasers is the problem of the flourine and HF, formed by hydrogen impurities reacting with the flourine, which can damage the surfaces of windows and mirrors. This problem may be solved by the use of flourine resistant coatings [283] and eliminating water and other sources of hydrogen in the gas mixture to eliminate HF. This problem is especially acute for large aperture lasers where flourine-resistant window materials ($CaF_2$, $MgF_2$) do not exist in sufficiently large diameters. The halides ($Cl_2$) used in other excimer lasers are less reactive than $F_2$, and there is much less of a problem with the survival of optics exposed to the gas mixture.

***Short, High-Power Laser Pulses.*** High-energy electron-beam pumped amplifiers require pumping times of many (>100) nanoseconds. The energy storage time for the lasing transitions is only a few nanoseconds. The amplifiers, therefore, naturally

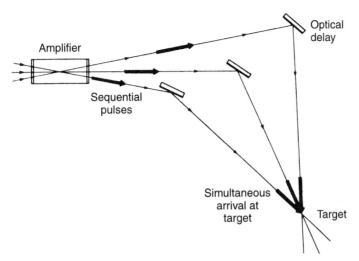

**Figure 9.136**  Angular multiplexing scheme for obtaining short laser pulses. (From Advanced Laser Concepts Technical Summary, Lawrence Livermore National Laboratory, 1980.)

operate with long laser pulses. Several schemes have been developed and tested for obtaining shorter pulses and are discussed below.

The angular multiplexing scheme for pulse compression [284, 285] is shown in Fig. 9.136. Numerous short-duration laser beams are overlapped onto the amplifier. The beams are timed to arrive one after the other, so that the amplifier is presented, effectively, with a single, long-duration extracting laser pulse and the energy can be efficiently extracted. After amplification, the beams are given appropriate delays so that they all arrive at the target at the same time. The simplest configuration is to use individual pulse durations equal to the delay between pulses so that energy is continually being extracted from the amplifier. This has the effect of minimizing losses due to amplified spontaneous emission since the amplifier is continually loaded. However, since the laser amplifier can act as a storage medium over a few nanoseconds, a train of very short laser pulses can also be employed where the period between pulses is comparable to or shorter than the energy storage time. The broad bandwidths available with many of the rare gas excimers allow amplification of subpicosecond pulses. The main problem with using angular multiplexing is the complexity and cost involved with dealing with the numerous beams. Angular multiplexing is limited to applications, such as laser fusion, where numerous beams can be overlapped to obtain the required high energy and power.

Another approach to pulse shortening involves the use of stimulated backward scattering [286]. Stimulated Raman or Brillouin scattering can be used to transfer the energy from a long pump pulse to a shorter counterpropagating Stokes pulse. An important advantage of this technique is that the energy from several poor-quality pump beams, when properly scrambled in a light guide, can be efficiently transferred to one spatially coherent compressed beam [287]. One disadvantage of this technique is the need for relatively narrow bandwidths for the stimulated backward scattering. For large apertures, narrow bandwidth creates problems with parasitic side scattering in the laser windows [288]. Also, as it is discussed below, for laser fusion application, there is an advantage in using broad-bandwidth radiation for uniform target irradiation.

***Application of Electron-Beam Pumped Lasers.*** The main application for large high-energy excimer lasers has been in research. Excimers are the only lasers that have a demonstrated capacity for the direct production of short-wavelength laser light at high energy. The broad bandwidth has fueled interest in their application for producing very short duration ($\leqslant 1$ ps), very high power ($\gg 10^{12}$ W) laser pulses for research applications [255, 289]. Perhaps the most exciting potential application for high-energy excimer lasers, in particular KrF lasers, is laser fusion.

Laser fusion involves the implosion of pellets containing thermonuclear fuel to obtain energy when the fuel is compressed and heated to reaction temperatures [29–31]. For the case of directly illuminated pellets, the implosion is driven by the ablation pressure produced by illuminating the pellet surface with laser beams. High gain can be achieved only with very uniform implosions. This requires highly uniform illumination of the pellet surface. Efficient coupling of the laser energy to the pellet requires that the laser have as short a wavelength as possible. The combination of short wavelength, broad bandwidth, and uniform illumination are advantageous in suppressing instabilities that can occur in the interaction of the high-power laser with the blow-off plasma from the target [290]. It turns out that the KrF laser has many advantages for this application.

The nonuniformity in the pellet illumination must be no larger than a few percent to obtain high-energy gain. Unfortunately, high-power lasers have too many transmissive and reflective components to produce perfect, diffraction-limited beams. Several smoothing techniques have been implemented on high-power Nd glass lasers to solve this problem. These beam-smoothing techniques include the random phase screen (RPS) approach [291], induced spatial incoherence (ISI) [292], lens arrays [293], and a combination of RPS and angular spectral dispersion [294]. All of the techniques above can improve the beam uniformity with the low f-number focusing optics used on most presently operating laser-fusion research facilities. However, high-gain laser fusion will eventually require that the final focusing optics be well removed from the fusion reaction and subtend only a small fraction of the solid angle surrounding the pellet. This requirement implies that the final focusing optics must have large f-numbers, $\geqslant$f20. Of currently available beam-smoothing techniques, only ISI has a demonstrated capability to produce uniform, controlled profiles with large f-number focusing optics. Below, we briefly describe the ISI technology which has been developed for glass lasers, and show the additional options and advantages of applying this type of a beam smoothing to KrF lasers.

Figure 9.137**a** illustrates the induced spatial incoherence technique for obtaining beam smoothing. A broadband laser beam with a short coherence time ($t_c = \Delta\nu$) is broken up into numerous beamlets by a transmissive echelon. The echelon provides differential delays between the beamlets that is longer than $t_c$, thereby rendering the beamlets statistically independent. When these beamlets are focused and overlapped onto a target using a lens, there is an instantaneous interference pattern. This interference pattern fades to a smooth envelope upon time averaging over period long compared to $t_c$. The focal envelope is determined by the diffraction pattern of the individual beamlets provided that the scalelength of aberrations in the incident laser beam are large compared to the lateral dimensions of each beamlet. Figure 9.137**b** shows an implementation of ISI over both lateral dimensions of a laser beam. Figure 9.137**c** shows the improvement in beam uniformity obtained using ISI with a large Nd-glass laser. ISI requires broad laser bandwidth ($t_c \leqslant 1$ ps) in order to have sufficient temporal smoothing during the few nanoseconds it takes for a pellet implosion [295]. This bandwidth is hard to obtain at short wavelengths with frequency-multiplied solid-state lasers due to color dispersion in currently available harmonic crystals [296, 297]. Such bandwidths at short laser wavelength can be obtained easily with KrF lasers.

While conventional ISI using echelons is applicable to the KrF laser, angularly multiplexed KrF lasers have properties that allow a simpler, more flexible beam-smoothing scheme. Figure 9.138 illustrates an echelon-free ISI scheme [298]. The aim of this technique is to obtain a beam whose focal properties are determined by spatial incoherence as in ISI, but without the echelons. In the configuration shown in Fig. 9.138, the beam from a broadband, spatially multimode oscillator uniformly illuminates a filter. The intensity profile of the filter is then imaged onto the target through the amplifier system. The optical information required to produce the focal pattern is carried through the system in small coherence zones. If the laser does not significantly

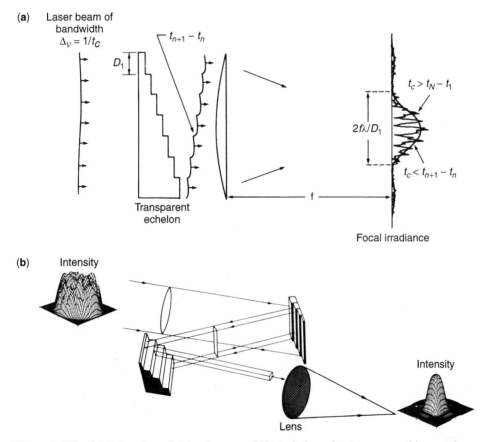

**Figure 9.137**  (a) Induced spatial incoherence (ISI) technique for beam smoothing produces a time-varying interference pattern that approaches a smooth profile when the time averaging interval is long compared to the laser coherence time ($t_C$). (b) A pair of reflecting ISI echelons are shown here producing a two-dimensional array of beamlets. (c) The focal patterns obtained with and without ISI using a frequency-doubled Nd–glass (green) laser. The ISI beam is compared to the $(\sin^2 x)/x^2$ profile predicted theoretically for square beamlets.

(c)

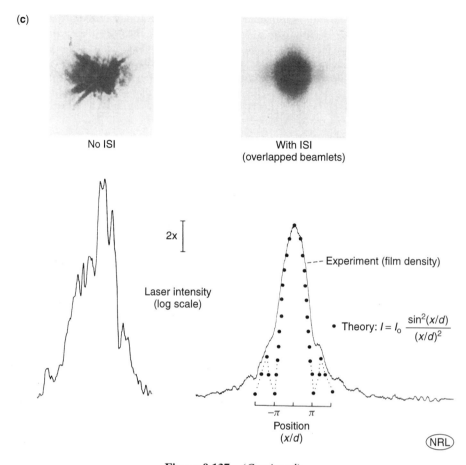

**Figure 9.137** *(Continued)*

distort these coherence zones, the image on the target will faithfully reproduce the pattern determined by the filter. Since the focal distribution is produced by a filter at a low-power stage, one could envision varying the filter with time, allowing the laser to zoom down with the imploding pellet, thereby increasing the uniformity and the efficiency.

This echelon-free beam-smoothing technique requires that the laser have broad bandwidth and that the nonlinear phase shifts are small. Nonlinear phase shifts in the instantaneous intensity hot spots of the incoherent beam profile as it propagates through the laser system will distort (broaden) the focal profile. The requirement for small nonlinear phase shift limits echelon-free ISI to systems where the index of the gain media is small, and the peak power loading on transmissive optics is small. Of currently available high-energy lasers, only angularly multiplexed excimer lasers, in particular KrF, meet these constraints. The low saturation fluence and gaseous amplifying media are advantageous in obtaining low nonlinear phase shifts. The echelon-free ISI scheme has been tested on a small system utilizing a discharge oscillator and amplifier (see Fig. 9.139). Beam uniformities approaching the 1% level have been obtained with this setup. A multi-kilojoule angularly multiplexed KrF laser system is being

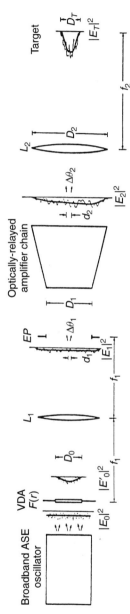

**Figure 9.138** With echelon-free ISI, the smooth pattern produced by the variable-density filter (VDA) is imaged through the laser system and onto the target. (From Ref. 298.)

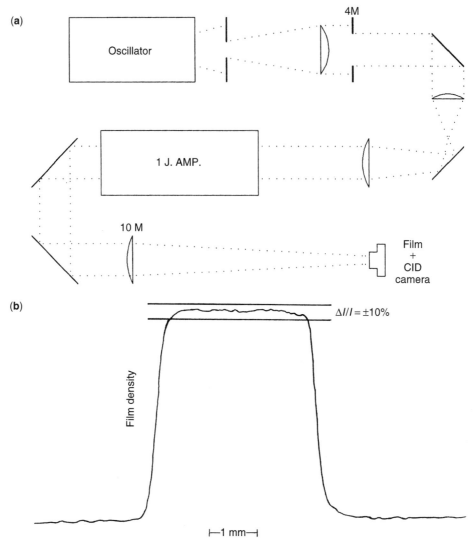

**Figure 9.139**    (a) Experimental setup for testing echelon-free ISI using a KrF discharge pumped oscillator and amplifier. The focal diameter at the camera can be changed by changing the diameter of the aperture after the oscillator. The beam was reflected from several uncoated quartz flats (not shown) after the amplifier to reduce the energy incident on the focal diagnostics. (b) Beam profile at the focus of the 10 m lens obtained from densitometry of photographic film. The reference lines show the excursions in intensity that would occur with $\Delta I/I_0 = \pm 10\%$.

constructed at the Naval Research Laboratory which will implement the echelon-free ISI scheme on a much larger scale for laser-fusion experiments.

KrF is perhaps the most promising laser for the commercial application of laser fusion. In addition to the advantages in beam smoothing and short wavelength, it has potentially high (5 to 10%) wallplug efficiency. Because of the enormous benefits that would occur with success, there is considerable interest in this application for excimer lasers.

# REFERENCES

1. A. Yariv, *Quantum Electronics*, Wiley, New York, 1975.

2. J. T. Verdeyen, *Laser Electronics*, Prentice-Hall, Englewood Cliffs, NJ, 1981.

3. B. A. Lengyel, *Lasers*, Wiley-Interscience, New York, 1971.

4. A. E. Siegman, *Lasers*, University Science Books, Mill Valley, CA, 1986.

5. J. P. Gordon, H. J. Zeiger, and C. H. Townes, *Phys. Rev. Lett.*, **95**, p. 259, 1954.

6. A. L. Schawlow, and C. H., Townes, *Phys. Rev.*, **112**, p. 1940, 1958.

7. T. H. Maiman, *Nature (London)*, **187**, p. 493, 1960.

8. A. Javan, W. R. Bennett, Jr., and D. R. Herriott, *Phys. Rev. Lett.*, **6**, p. 106, 1961.

9. C. K. N. Patel, *Phys. Rev.*, **136A**, p. 1187, 1964.

10. D. Marcuse, *Principles of Quantum Electronics*, Academic Press, New York, 1980.

11. L. I. Schiff, *Quantum Mechanics*, McGraw-Hill, New York, 1955.

12. A. Einstein, *Phys. Z.*, **18**, p. 121, 1917.

13. W. V. Smith, and P. P. Sarokin, *The Laser*, McGraw-Hill, New York, 1966.

14. W. E., Lamb, Jr., *Phys. Rev.*, **134**, p. 1429, 1964.

15. C. S. Willett, *Introduction to Gas Lasers: Population Inversion Mechanisms*, Pergamon Press, Elmsford, NY 1974.

16. J. F. Ready, *Industrial Applications of Lasers*, Academic Press, New York, 1978.

17. G. K. Starkweather, in J. W. Goodman and M. Ross, Eds., *Laser Applications*, Vol. 4, Academic Press, New York, 1980.

18. L. Goldman, Ed., *The Biomedical Laser: Technology and Clinical Applications*, Springer-Verlag, New York, 1981.

19. W. O. N. Guimaraes, C. T., Lin, and A. Mooradian, Eds., *Laser and Applications*, Springer-Verlag, Berlin, 1981.

20. R. Iscoff, *Lasers Optronics*, **7**, p. 46, 1988.

21. J. A. S. Carruth, and A. L. McKenzie, *Medical Lasers, Science, and Clinical Practice*, Adam Hilger, Bristol, England, 1986.

22. W. T. Walter, N. Solimene, M. Piltch, and G. Gould, *IEEE J. Quantum Electron.*, **QE-2**, p. 474, 1966.

23. M. J. Kusher, and B. E. Warner, "Large-Bore Copper-Vapor Lasers: Kinetics and Scaling Issues," *J. Appl. Phys.*, **54**, pp. 2970–2982, 1983.

24. J. D. Anderson, Jr., *Gasdynamic Lasers: An Introduction*, Academic Press, New York, 1976.

25. C. K. N. Patel, *Sci. Am.*, **219**, p. 22, 1968.

26. T. Lehecka, "Development of High Power Optically Pumped Far Infrared Lasers for Fusion Plasma Diagnostics," Master's thesis, University of California–Los Angeles, 1986.

27. T. Lehecka, R. Savage, R. Dworak, W. A. Peebles, N. C. Luhmann, Jr., and A. Semet, "High-Power, Twin-Frequency FIR Lasers for Plasma Diagnostic Applications," *Rev. Sci. Instrum.*, **57**, pp.1986–1988, 1986.

28. N. C. Luhmann, Jr., and W. A. Peebles, in M. Bass and M. L. Stich, Eds., *Laser Handbook Vol. 5*, North-Holland, "Confined Thermonuclear Plasmas, Laser Diagnostics of Magnetically," Amsterdam, 1985, Chap. 5.

29. J. Nuckolls, J. Emmett, and L. Wood, "Laser-Induced Thermonuclear Fusion," *Phy. Today*, pp. 46–53, August 1973.

30. S. E. Bodner, "Critical Elements of High Gain Laser Fusion," *J. Fusion Energy*, **1**, pp. 221–240, 1981.

31. K. A. Brueckner, and S. Jorna, "Laser-Driven Fusion," *Rev. Mod. Phys.*, **46**, pp. 325–367, 1974.

32. C. Joshi, W. B. Mori, T. Katsouleas, J. M. Dawson, J. M. Kindel, and D. W. Forslund, "Ultrahigh Gradient Particle Acceleration by Intense Laser-Driven Plasma Density Waves," *Nature (London)*, **311**, pp. 525–529, 1984.

33. J. F. Figueira, "High-Energy Short-Pulse $CO_2$ Lasers," in P. K. Ches, Ed., *Handbook of Molecular Lasers*, Marcel Dekker, New York, 1987, Chap. 4.

34. A. J. Beaulieu, "Transversely Excited Atmospheric Pressure $CO_2$ Lasers," *Appl. Phys. Lett.*, **16**, p. 504, 1970.

35. R. Dumanchin, and J. Rocca-Serra, "Augmentation de l'énergie et de la puissance fournie par unité de volume dans un laser a $CO_2$ en régime pulse," *C. R. Acad. Sci.*, **269**, p. 916, 1969.

36. H. M. Lamberton, and P. R. Pearson, "Improved Excitation Techniques for Atmospheric Pressure $CO_2$ Lasers," *Electron Lett.*, **7**, p. 141, 1971.

37. J. Jansen, *Review and Status of Antares*, Los Alamos, Natl. Lab. Rep. LA-UR 79–2178, presented at 2nd Int. Pulsed Power Conf., Lubbock, TX, 1979.

38. C. A. Fenstermacher, M. J. Nutter, P. Rink, and K. Boyer, "Electron Beam Initiation of Large Volume Electric Discharges in $CO_2$ Laser Media," *Bull Am. Phys. Soc.*, **16**, p. 42, 1971.

39. J. D. Daugherty, E. Pugh, and D. H. Douglas-Hamilton, "A Stable, Scalable, High Pressure Gas Discharge as Applied to the $CO_2$ Laser," *Bull. Am. Phys. Soc.*, **17**, p. 399, 1972.

40. O. R. Wood, "High-Pressure Molecular Lasers," *Proc. IEEE*, Vol. 62, p. 355, 1974.

41. S. Singer, C. J. Elliott, J. Figueira, I. Liberman, J. V. Parker, and G. T. Schappert, "High Power, Short Pulse $CO_2$ Lasers Systems for Inertial Confinement Fusion," in C. Pellegrini, Ed., *Developments in High-Power Lasers and Their Applications*, Proceedings of the International School of Physics, "Enrico Fermi," Villa Monastero, North-Holland, Amsterdam, 1981.

42. T. F. Stratton, "$CO_2$ Short Pulse Laser Technology," in E. R. Pike, Ed., *High-Power Gas Lasers, 1975*, The Institute of Physics, Bristol, England (1976).

43. J. Ladish, "Helios, a 20 TW $CO_2$ Laser Fusion Facility," *Laser 79' Opto-Electron. Conf.*, Munich, West Germany, 1979.

44. H. Jansen, "A Review of the Antares Laser Fusion Facility," (C. Yamanaka, Ed.), *Proc. IAEA Tech. Comm. Meet. Adv. Inertia Confinement Fusion Res.*, Kobe, Japan, 1984, p. 284.

45. S. Singer, J. S. Parker, and M. J. Nutter, "Cold Cathode Electron Guns in the LASL High-Power Short-Pulse $CO_2$ Laser Program," *Proc. Int. Top. Conf. Electron-Beam Res. Dev.*, Albuquerque, NM, 1975, pp. 274–292.

46. G. V. Loda, and D. A. Meskar, "Repetitively Pulsed Electron-Beam Generators," *Proc. Int. Top. Conf. Electron-Beam Res. Dev.*, Albuquerque, NM, 1975, pp. 252–272.

47. C. R. Mansfield, and W. H., Reichelt, *High-Power $CO_2$ Systems*, D. C. Cartwright, Ed., Los Alamos Natl. Lab. Rep. LA-10380, summary of research for the Inertial Confinement Fusion Program at the Los Alamos National Laboratory, 1985, p. 48.

48. D. Belforte, and M. Levitt, *The Industrial Laser: Annual Handbook*, PennWell, Tulsa, OK, 1986.

49. M., Rosenbluh, R. J. Temkin, and K. J., Button, "Submillimeter Laser Wavelength Tables," *Appl. Opt.* **15**, pp. 2635–2644, 1976.

50. Y. Tsunawaki, M. Yamanaka, and S. Kon, *Rev. Laser Eng.* **10** pp. 78–125, 1982.

51. M. Yamanaka, "Optically Pumped Gas Lasers: A Wavelength Table of Laser Lines," *Rev. Laser Eng. (Japan)*, **3**, pp. 253–294, 1976.

52. J. J. Gallagher, M. D. Blue, B. Bean, and S. Perkowitz, "Tabulations of Optically Pumped Far Infrared Laser Lines and Applications to Atmospheric Transmission," *Infrared Phys.*, **17**, pp. 43–55, 1977.

53. M. S. Tobin, *Opt. Lett.* **7**, p. 322, 1982.

54. H. R. Fetterman, P. E. Tannewald, B. J. Clifton, C. D. Panker, W. D. Fitzgerald, and N. R. Erickson, *Appl. Phys. Lett.*, **33**, p. 151, 1978.

55. V. J. Corcoran, *Soc. Photo. Opt. Laser. Eng.*, **105**, p. 94, 1977.

56. J. Waldman, H. R. Fetterman, W. D. Goodhue, T. G. Bryant, and D. H. Tomme, *Proc. Soc. Photo-Opt. Instrum. Eng.*, **197**, p. 170, 1979.

57. T. S. Hartwick, D. T. Hodges, D. H. Burke, and F. B. Foote, *Appl. Opt.*, **15**, p. 1919, 1976.

58. A. J. Canton, P. K. Chew, M. C. Foster, and L. A. Neuman, *IEEE J. Quantum Electron*, **QE-17**, p. 152, 1981.

59. D. L. Brower, H. K. Park, W. A. Peebles, and N. C. Luhmann, Jr., K. J. Button, Ed., "Multi-channel Far-Infrared Collective Scattering System for Plasma Wave Studies," in *Topics in Millimeter Wave Technology*, Vol. 2, Academic Press, New York, 1988, Chap. 3.

60. N. C. Luhmann, Jr., "Instrumentation and Techniques for Plasma Diagnostics," in K. J. Button, Ed., *Infrared and Millimeter Waves*, Vol. 2, Academic Press, New York, 1979, Chap. 1.

61. R. A. McFarland, W. L. Faust, C. K. N. Patel, and C. G. B. Garrett, *Proc. IEEE*, Vol. 2, p. 318, 1964.

62. A. Crocker, H. A. Gebbie, M. F. Kimmitt, and L. E. S. Mathias, "Stimulated Emission in the Far-Infrared," *Nature (London)*, **201**, pp. 250–251, (1964).

63. H. A. Gebbie, N. W. B. Stone, and F. D. Findlay, *Nature (London)*, **202**, pp. 169–170, 1964.

64. W. J. Witteman, and R. Bleekrode, *Phys. Lett.*, **13**, pp. 126–127, 1964.

65. H. A. Gebbie, N. W. B. Stone, and F. D. Findlay, *Nature (London)*, **202**, p. 685, 1964.

66. K. K. Kneubuhl, and C. K. Sturzenegger, "Electrically Excited Submillimeter Wave Lasers," in K. J. Button, Ed., *Infrared and Millimeter Waves*, Vol. 3, Academic Press, New York, 1980, Chap. 5.

67. H. Steffen, and F. K. Kneubuhl, *IEEE J. Quantum Electron.*, **QE-4**, pp. 922–1008, 1968.

68. R. J. Pressley, *Handbook of Lasers*, Chemical Rubber Company, Cleveland, OH, 1971.

69. J. C. Hassler, G. Hubner, and P. D. Coleman, *J. Appl. Phys.*, **44**, pp. 795–801, 1973.

70. Y. Horiuchi, and A. Murai, *IEEE J. Quantum Electron.*, **QE-12**, pp. 547–549, 1976.

71. T. Y. Chang, and T. J. Bridges, *Opt. Commun.*, **1**, pp. 423–426, 1970.

72. D. J. E. Knight, U.K. Rep. Qu45, 1st rev., *Ordered List of Far Infrared Laser Lines*, National Physics Lab, Teddington, Middlesex, England.

73. N. C. Luhmann, Jr., and W. A. Peebles, *Rev. Sci. Instrum.*, **55**, p. 279, 1984.

74. R. A. McFarlane, W. L. Faust, C. K. N. Patel, and C. G. B. Garrett, *Quantum Electron*, **3**, pp. 573–576, 1964.

75. H. Steffen, J. Steffen, J. F. Moser, and F. K. Kneubuhl, *Phys. Lett.*, **21**, pp. 425–426, 1966.

76. P. Belland, D. Veron, and L. B. Whitbourn, *Appl. Opt.*, **15**, p. 3047, 1976.

77. D. Veron, P. Belland, and M. J. Beccaria, "Continuous 250 mW Gas Discharge DCN Laser at 195 μm," *Infrared Phys.*, **18**, pp. 465–468, 1978.

78. J. L. Bruneau, P. Belland, and D. Veron, "A CW DCN Waveguide Laser of High Volumetric Efficiency," *Opt. Commun.*, **24**, p. 259, 1978.

79. P. Belland, and D. Veron, "Amplifying Medium Characteristics in Optimized 190 μm/ 195 μm DCN Waveguide Lasers," *IEEE J. Quantum Electron.*, **QE-16**, p. 885, 1980.

80. P. Belland, *Appl. Phys.*, **B 27**, p. 123, 1982.

81. H. J. Schotzau, and S. Veprek, *Appl. Phys.*, **1**, p. 271, 1975.

82. H. J. Schotzau, and F. Kneubuhl, *Appl. Phys.*, **6**, p. 25, 1975.

83. H. J. Schotzau, and F. Kneubuhl, *IEEE J. Quantum, Electron.*, **QE-11**, p. 817, 1975.

84. K. Mizuno, R. Kawahara, O. Shimoe, and S. Ono, *J. Appl. Phys.*, **45**, p. 5464, 1974.

85. S. Iwama, N. Satomi, M. Yamanaka, S. Goto, T. Ishimura, and H. Ito, *Int. J. Infrared Millim. Waves*, **2**, p. 1199, 1981.

86. L. O. Hocker, and A., Javan, *Phys. Lett.*, **25A**, pp. 489–490, 1967.

87. D. R. Lide, and A. G. Maki, *Appl. Phys. Lett.*, **11**, pp. 62–64, 1967.

88. J. P. Pichamauthu, J. C. Hassler, and P. D. Coleman, *Appl. Phys. Lett.*, **19**, pp. 510–512, 1971.

89. P. D. Coleman, *IEEE J. Quantum Electron.*, **QE-9**, pp. 130–138, 1973.

90. J. D. Dunning, "Quadrupole Investigation of Far-Infrared Laser Plasma," unpublished, *Master's thesis*, University of California–Los Angeles, 1975.

91. W. S. Benedict, Pollack, and Tomlinson, *IEEE J. Quantum Electron.*, **QE-5**, p. 108, 1969.

92. P. Belland, and J. P. Crenn, *Appl. Opt.*, **18**, p. 1513, 1979.

93. E. A. J. Marcatili, and R. A. Schmeltzer, "Hollow Metallic and Dielectric Waveguides for long Distance Optical Transmission and Lasers," *Bell Syst. Tech. J.*, **43**, p. 1783, 1964.

94. J. J. Degnan, *Appl. Phys.*, **11**, p. 1, 1976.

95. K. K. Kneubuhl, and E. Affolter, "Infrared and Millimeter-Wave Waveguides," in K. J. Button, Ed., *Infrared and Millimeter Waves*, Vol. 1, Academic Press, New York, 1979, pp. 235–278.

96. M. Kawamura, I. Okabayaski, and T. Fukuyama, "A Capacitively Coupled rf-Excited cw-HCN Laser," *IEEE J. Quantum Electron.*, **QE-21**, p. 1833, 1985.

97. R. Turner, and T. O. Poehler, *J. Appl. Phys.*, **42**, pp. 3819–3826, 1971.

98. L. E. Sharp, and A. T. Wetherell, "High Power Pulsed HCN Laser," *Appl Opt.*, **11**, p. 1737, 1972.

99. D. L. Jassby, M. E. Marhic, and P. R. Regan, "High Power Pulsed HCN Laser with Auxiliary DC Discharge," *Appl. Opt.*, **12**, pp. 1403–1404, 1973.

100. M. F. Lam, D. L. Jassby, and L. W. Casperson, "Transverse Excitation Pulsed HCN Laser," *IEEE J. Quantum Electron.*, **QE-8**, pp. 851–852, 1973.

101. B. Adam, H. J. Schotzau, and F. K. Kneubuhl, "Standard Transverse Excitation of the HCN-Laser 337μ Emission," *Phys. Lett.*, **45A**, pp. 365–366, 1973.

102. B. Adam, and F. Kneubuhl, "Transversely Excited 337 μm HCN Waveguide Laser," *Appl. Phys.*, **8**, pp. 281–291, 1975.

103. Ch. Sturzenegger, B. Adam, and F. K. Kneubuhl, *IEEE J. Quantum Electron.*, **QE-13**, pp. 473–475, 1977.

104. Ch. Sturzenegger, H. Vetsch, and F. K. Kneubuhl, *Infrared Phys.*, **19**, pp. 277–296, 1979.

105. T. A. DeTemple, and E. J. Danielewicz, "Continuous -Wave Optically Pumped Lasers," Ed., in K. J. Button, *Infrared and Millimeter Waves*, Vol. 7, Academic Press, New York, pp. 1–41, 1983.

106. T. A. DeTemple, "Pulsed Optically Pumped Far-Infrared Lasers," in K. J. Button, Ed., *Infrared and Millimeter Waves*, Vol. 1, Academic Press, New York, pp. 129–179, 1979.

107. D. T. Hodges, "A Review of Advances in Optically Pumped Far-Infrared Lasers," *Infrared Phys.*, **18**, pp. 375–384, 1978.

108. M. S. Tobin, "Review of Optically Pumped NMMW Lasers," *Proc. IEEE*, Vol. 73, pp. 61–85, 1985.

109. R. W. Waniek, "Far-Infrared Lasers — Two Decades of Progress," *Laser Focus*, **19**, pp. 79–85.

110. T. Y. Chang, "Optical Pumping in Gases," in Y. R. Shen, Ed., *Topics in Applied Physics*, Vol. 16, New York, Springer-Verlag, pp. 215–272, 1977.

111. P. D. Coleman, "Present and Future Problems Concerning Lasers in the Far-Infrared Spectral Region," *J. Opt. Soc. Am.*, **67**, pp. 894–901, 1977.

112. G. Herzberg, *Molecular Spectra and Molecular Structure*, Vol. 2, *Infrared and Raman Spectra of Polyatomic Molecules*, Van Nostrand Reinhold, New York, 1945.

113. H. R. Fetterman, H. R. Schossberg, and C. D. Parker, "CW Submillimeter Laser Generation in Optically Pumped Stark Tuned $NH_3$," *Appl. Phys. Lett.*, **12**, pp. 684–685, 1973.

114. M. Redon, C. Gastaud, and M. Fourrier, "New CW Far-Infrared Lasing in $^{14}NH_3$ Using Stark Tuning," *IEEE J. Quantum Electron.*, **QE-15**, pp. 412–414, 1979.

115. M. S. Tobin, and R. E. Jensen, "Far IR Laser with Metal-Dielectric Waveguide to Observe the Stark Effect," *Appl. Opt.*, **15**, pp. 2023–2024, 1976.

116. K. P. Koo, and P. C. Clasby, "Stark Effects in FIR Lasers," *Conf. Dig. 2nd Int. Conf. SMMW Their Appl.*, San Juan, PR, IEEE Cat. 76 CH1152-8MTT, pp. 171–172, 1976.

117. S. R. Stein, A. S. Risley, H. Van de Stadt, and F. Strumia, "High Speed Frequency Modulation of Far Infrared Lasers Using Stark Effect," *Appl. Opt.*, **16**, pp. 1893–1896, 1977.

118. F. Strumia, and M. Inguscio, "Stark Spectroscopy and Frequency Tuning in Optically Pumped Far Infrared Lasers," in K. J. Button, Ed., *Infrared and Millimeter Waves*, Vol. 5, New York, Academic Press, 1982, pp. 129–213.

119. P. P. Feofilov, *The Physical Basis of Polarized Emission*, Consultant Bureau Enterprises, New York, 1961.

120. D. T. Hodges, and J. R. Tucker, *Appl. Phys. Lett.*, **27**, pp. 667–669, 1975.

121. T. A. DeTemple, and E. J. Danielewicz, *IEEE J. Quantum Electron*, **QE-12**, pp. 40–47, 1976.

122. J. R. Tucker, IEEE Trans. Microwave Technol., **MTT-22**, p. 1117, 1974.

123. C. O. Weiss, "Pump Saturation in Molecular Far-Infrared Lasers," *IEEE J. Quantum Electron.*, **QE-12**, pp. 580–584, 1976.

124. L. A. Gribov, and V. N. Smirnov, *Sov Phys-Usp.*, **4**, pp. 919–946, 1962.

125. L. A. Pugh, and K. N. Rao, K. N. Rao, Ed., *Molecular Spectroscopy: Modern Research*, Vol. 2, Academic Press, New York, 1976.

126. J. D. Lambert, and R. Salter, *Proc. R. Soc. London*, Vol. A253, pp. 277–288, 1959.

127. E. Weitz, and G. W. Flynn, *J. Chem. Phys.*, **58**, pp. 2781–2793, 1973.

128. F. E. Hovis, and C. B. Moore, *J. Chem. Phys.*, **69**, pp. 4847–4950, 1978.

129. R. L. Sheffield, K. Boyer, and A. Javen, *Opt. Lett.*, **5**, pp. 10–11, 1980.

130. S. S. Miljanic, and C. B. Moore, *J. Chem. Phys.*, **73**, pp. 226–229, 1980.

131. E. J. Danielewicz, and C. O. Weiss, *Opt. Commun.*, **27**, pp. 98–100, 1978.

132. T. L. Worchesky, K. J. Ritter, J. P. Sattler, and W. A. Riessler, *Opt. Lett.*, **2**, 1978.

133. A. R. Calloway, and E. J. Danielewicz, *IEEE J. Quantum Electron.*, **QE-17**, pp. 579–581, 1981.

134. G. D. Willenberg, H. Hubner, and J. Heppner, *Opt. Commun.*, **33**, pp. 193–196, 1980.

135. E. M. Frank, C. O. Weiss, K. Siemsen, M. Grinda, and G. D. Willenberg, *Opt. Lett.*, **7**, pp. 96–98, 1982.

136. R. A. Wood, A. Vass, C. R. Pidgeon, M. J. Coller, and B. Norris, *Opt. Commun.*, **33**, pp. 89–90, 1981.

137. H. Jones, and P. B. Davies, *IEEE J. Quantum Electron.*, **QE-17**, pp. 13–14, 1981.

138. D. T. Hodges, F. G. Foote, and R. D. Reel, *IEEE J. Quantum Electron.*, **QE-13**, pp. 491–494, 1977.

139. S. M. Freund, G. Duxbury, M. Romheld, J. T. Tiedje, and T. Oka, "Laser Stark Spectroscopy in the 20 $\mu$m Region the $\nu_3$ Bands of $CH_3F$," *J. Mol. Spectros.*, **52**, pp. 38–57, 1974.

140. M. S. Tobin, J. P. Sattler, and G. C. Wood, *Opt. Lett.*, **4**, pp. 384–386, 1979.

141. M. S. Tobin, and R. D. Felock, *IEEE J. Quantum Electron.*, **QE-17**, pp. 825–826, 1981.

142. M. R. Schubert, M. Durschlag, and T. A. DeTemple, *IEEE J. Quantum Electron.*, **QE-13**, pp. 455–459, 1977.

143. G. Graner, *Opt. Commun.*, **14**, pp. 67–69, 1975.

144. M. S. Tobin and R. D. Felock, *Opt. Lett.*, **5**, pp. 430–432, 1980.

145. L. D. Fesenko, and S. F. Dyubko, *Sov. J. Quantum Electron.*, **6**, pp. 839–943, 1976 [transl. of *Kvantovaya Elektron. (Moscow)*].

146. F. Julien, and J. M. Lourtioz, *Opt. Commun.*, **38**, pp. 294–298, 1981.

147. T. Galantowicz, E. J. Danielewicz, F. B. Foote, and D. T. Hodges, *Proc. Int. Conf. Lasers* McLean, VA, 1979.

148. A. Scalabrin, and K. M. Evenson, *Opt. Lett.*, **4**, pp. 277–280, 1979.

149. R. R. Peterson, A. Scalabrin, and K. M. Evenson, *Int. J. Infrared Millim. Waves*, **1**, pp. 111–116, 1980.

150. T. L. Worchesky, M. S. Tobin, K. J. Ritter, T. W. Daley, and W. J. Lafferty, *Int. J. Infrared Millim. Waves*, **1**, pp. 127–138, 1980.

151. D. T. Hodges, F. B. Foote, and R. D. Reel, *Appl. Phys. Lett.*, **29**, pp. 662–664, 1976.

152. P. K. Cheo, private communication, United Technology Research Center, 1982.

153. J. O. Henningson, *J. Mol. Spectros.*, **83**, pp. 70–93, 1980.

154. S. F. Dyubko, V. A. Svich, and L. D. Fesenko, *Tzv. Vyssh. Uchcbn. Zaued., Radiofiz.*, **18**, pp. 1434–1437, 1975.

155. S. Kon, E. Hagiwara, T. Yano, and H. Hirose, *Jpn. J. Appl. Phys.*, **14**, pp. 731–732, 1975.

156. E. C. C. Vasconcellos, A. Scalabrin, F. R. Petersen, and K. M. Evenson, *Int. J. Infrared Millim. Waves*, **2**, pp. 533–540, 1981.

157. J. Heppner, and D. N. Ghosh Roy, *Infrared Millim. Waves*, **2**, pp. 479–492, 1981.

158. A. Bennett, and H. Herman, *IEEE J. Quantum Electron.*, **QE-18**, pp. 323–325, 1982.

159. E. C. C. Vasconcellos, F. R. Peterson, and K. M. Evenson, *Int. J. Infrared Millim. Waves*, **2**, pp. 705–711, 1981.

160. G. Duxbury, and H. Herman, *J. Phys.*, **11**, p. 419, 1978.

161. F. Julien, and J.-M. Lourtioz, *Int. J. Infrared Millim. Waves*, **1**, p. 175, 1980.

162. A. Semet, and N. C. Luhmann, Jr., "High-Power Narrow-Line Pulsed 496 $\mu$m Laser," *Appl. Phys. Lett.*, **28**, pp. 659–661, 1976.

163. E. J. Danielewicz, and P. D. Coleman, "Hybrid Metal Mesh-Dielectric Mirrors for Optically Pumped Far Infrared Lasers," *Appl. Opt.*, **15**, pp. 761–767, 1976.

164. M. Durschlag, and T. A. DeTemple, *Appl. Opt.*, **20**, p. 1245, 1981.

165. R. Ulrich, *Infrared Phys.*, **7**, p. 37, 1967.

166. J. J. Degnan, "Waveguide Laser Mode Patterns in the Near and Far Field," *Appl. Opt.*, **12**, p. 1026, 1973.

167. M. Yamanaka, "Optically Pumped Waveguide Lasers," *J. Opt. Soc. Am.*, **67**, pp. 952–958, 1977.

168. E. J. Danielewicz, E. L. Fletcher, A. R. Calloway, and D. T. Hodges, *Proc. 3rd APS Topical Conf. High Temp. Plasma Diagn.* University of California–Los Angeles, 1980, Pap. F9.

169. E. J. Danielewicz, E. L. Fletcher, A. R. Calloway, and D. T. Hodges, *Proc. Jpn–USA Workshop Far-Infrared Diagn.* Massachusetts Institute of Technology, Cambridge, MA, 1980.

170. R. Wattenbach, H. P. Roser, and G. V. Schultz, *Int. Infrared Millim, Waves*, **3**, p. 753, 1982.

171. G. A. Koepf, H. R. Fetterman, and N. McAvoy, "A Stable Submillimeter Laser Local Oscillator for Heterodyne Radiometry and Spectroscopy," *Int. Infrared Millim. Waves*, **1**, pp. 597–607, 1980.

172. W. R. Hyarshberger, and M. B. Roin, *Am. Chem. Res.*, **6**, 329, 1973.

173. G. Busse, E. Basel, and A. Pfaller, "Application of the Opto-acoustic Effect on the Operation of Optically Pumped Far-Infrared Gas Lasers," *Appl. Phys.*, **12**, pp. 387–389, 1977.

174. G. Busse, and R. Tharmaier, "Use of the Optoacoustic Effect to Discover CW Far-Infrared Laser Lines," *Appl. Phys. Lett.*, **31**, p. 194, 1977.

175. L. B. Kraeuzer, *J. Appl. Phys.*, **42**, p. 2934, 1971.

176. S. P. Belov, A. V. Burnein, L. J. Gershfein, V. V. Korolikhin, and A. F. Kruprov, *Opt. Spectros.*, **35**, p. 172, 1973.

177. D. K. Mansfield, A. Semet, and L. C. Johnson, "A Lossless, Passive Isolator for Optically Pumped Far-Infrared Lasers," *Appl. Phys. Lett.*, **37**, pp. 688–690, 1980.

178. D. K. Mansfield, G. J. Tesauro, L. C. Johnson, and A. Semet, "Effects of Passive Isolation on Several Optically Pumped Far-Infrared Laser Lines," *Opt. Lett.*, **6**, pp. 230–232, 1981.

179. J. L. Lachambre, and M. Gagne, "Dither-Free Stabilization Scheme for Submillimeter Wave Pumping Sources," *Rev. Sci. Instrum.*, **55**, pp. 1955–1956, 1984.

180. J. Heppner, and C. O. Weiss, "Far-Infrared Ring Laser," *Appl. Phys. Lett.*, **33**, pp. 590–592, 1978.

181. D. K. Mansfield, K. Jones, L. C. Johnson, and A. Semet, *Proc. 6th Int. Conf. Infrared Millim. Waves*, Miami Beach, FL, IEEE Cat 81CH1645-MTT, 1981.

182. N. C. Luhmann, Jr., and W. A. Peebles, "Instrumentation of Magnetically Confined Fusion Plasma Diagnostics," *Rev. Sci. Instrum.*, **55**, p. 279, 1985.

183. T. Y. Chang, and T. Lin, "Effects of Buffer Gases on an Optically Pumped $CH_3F$ FIR Laser," *J. Opt. Soc. Am.*, **66**, pp. 362–369, 1976.

184. N. M. Lawandy, and G. A. Koepf, "Energy Transfer Mechanisms in the $CH_3F : SF_6$ Optically Pumped Laser," *Opt. Lett.*, **5**, pp. 336–338, 1980.

185. T. Lechecka, W. A. Peebles, R. L. Savage, and N. C. Luhmann, "A High-Power $CH_3$ Laser for Plasma Diagnostics," *IEEE J. Quantum Electron.*, **QE-24**, ppp. 5–7, 1988.

186. N. M. Lawandy, *Infrared Phys.*, **19**, pp. 127, 1979.

187. J. P. Crenn, *IEEE Trans. Microwave Theory Tech.* **MTT-27**, p. 573, 1979.

188. F. B. Foote, D. T. Hodges, and H. B. Dyson, *Int. J. Infrared Millim. Waves*, **2**, p. 773, 1981.

189. E. J. Danielewicz, and C. O. Weiss, *IEEE J. Quantum Electron.*, **QE-14**, p. 707, 1978.

190. D. E. Burch, and S. A. Clugh, *Near Millim. Wave Technol. Base Study*, **1**, p. 33, 1979.

191. D. Dangoisse, P. Glorieux, and J. Wascat, *Int. J. Infrared Millim. Waves*, **6**, p. 214, 1981.

192. J. O. Henningsen, and H. G. Jensen, *IEEE J. Quantum Electron.*, **QE-11**, p. 248, 1975.

193. A. Yariv, *Quantum Electronics*, 2nd ed. Wiley, New York, 1980.

194. W. W. Rigrod, *J. Appl. Phys.*, **34**, p. 2602, 1963.

195. K. Walzer, and M. Tacke, *IEEE J Quantum Electron.*, **QE-16**, p. 255, 1980.

196. D. Veron, "Submillimeter Interferometry of High-Density Plasmas" in K. J. Button, Ed., *Infrared and Millimeter Waves*, Vol. 2, Academic Press, New York, 1979, pp. 67–135.

197. S. M. Wolfe, K. J. Button, J. Waldman, and D. R. Cohn *Appl. Opt.* **17**, p. 2645, 1976.

198. D. K. Mansfield, L. C. Johson, and A. Medelsohn, *Int. J. Infrared Millim. Waves*, **1**, p. 631, 1980.

199. C. H. Ma, D. P. Hutchinson, P. A. Stoats, and K. L. Vandersluis, *Int. J. Infrared Millim. Waves*, **3**, p. 263, 1982.

200. M. Yamanaka, Y. Takedo, S. Tanigawa, N. Nishizawa, N. Noda, J. Fujita, M. Takai, M. Shimobayaski, Y. Hayoshi, T. Koizumi, K. Nagasaka, S. Okajima, Y. Tsunawaki, and A. Nagashima, *Int. J. Infrared Millim. Waves*, **1**, p. 57, 1980.

201. R. Dworak, "Dual-Frequency Far-Infared for Fusion Plasma Diagnostics," *M.S. thesis*, University of California–Los Angeles (UCLA Rep. PPG-1047), 1987.

202. P. K. Cheo, "Far-Infrared Laser Scanner for High Voltage Cable Inspection," *Infrared Millim. Waves*, **12**, p. 279, 1984.

203. P. K. Cheo, "Far-Infrared Lasers for Power Cable Manufacturing," *IEEE Circuits Devices Mag.*, **2**, p. 49, 1986.

204. D. Betz, and J. Zmuidzinas, "A 150 $\mu$m to 500 $\mu$m Heterodyne Spectrometer for Airborne Astronomy," *Proc. Airborne Astron. Symp.*, NASA CP-2353, 1984.

205. H. P. Roser, R. Wattenbach, E. J. Durwen, and G. N. Schultz, "A High Resolution Heterodyne Spectrometer from 100 $\mu$m to 100 $\mu$m and the Detection of CO ($J = 7 \to 6$), CO ($J = 6 \to 5$) and $^{13}$CO ($J = 3 \to 2$)." *Astron. Astrophys.*, **165**, p. 287, 1986.

206. G. Chin, "Optically Pumped Submillimeter Gas Lasers the Prospects for Constructing Space-Qualifiable LO Systems," *Int. J. Infrared Millim. Waves*, **8**, pp. 1219–1234, 1987.

207. P. D. Morgan, M. R. Green, M. R. Siegrist, and R. L. Watterson, *Commun. Plasma Phys. Controlled Fusion*, **5**, p. 141, 1979.

208. F. Brown, S. R. Hannau, A. Polevsky, and K. J. Button, *Opt. Commun.*, **9**, p. 28, 1973.

209. D. E. Evans, L. E. Sharp, B. W. James, and W. A. Peebles, *Appl. Phys. Lett.*, **26**, p. 630, 1975.

210. D. E. Evans, L. E. Sharp, W. A. Peebles, and G. Taylor, *IEEE J. Quantum Electron.*, **QE-13**, p. 54, 1977.

211. W. A. Peebles, D. Umstadter, D. L. Brower, and N. C. Luhmann, Jr., "A Unidirectional, Pulsed Far-Infrared Ring Laser," *Appl. Phys. Lett.*, **38**, pp. 851–853, 1981.

212. S. J. Petuchowski, A. T. Rosenberger, and T. A. DeTemple, *IEEE J. Quantum Electron.*, **QE-13**, p. 476, 1977.

213. H. R. Fetterman, H. R. Schlossberg, and S. Waldman, *Opt. Commun.*, **6**, p. 156, 1978.

214. J. D. Wiggins, Z. Drozdowicz, and R. J. Temkin, "Two-photon Transitions in Optically-Pumped Submillimeter Lasers," *IEEE J. Quantum Electron.*, **QE-14, 1**, pp. 23–30, 1978.

215. D. E. Evans, L. E. Sharp, W. A. Peebles, and G. Taylor, *Opt. Commun.* **18**, p. 479, 1976.

216. G. Dodel, and G. Magyar, *Appl. Phys. Lett.*, **32**, p. 44, 1978.

217. P. Woskoboinikow, H. C. Praddaude, W. J. Mulligan, D. R. Cohn, and B. Lax, "High-Power Tunable 385 $\mu$m D$_2$) Vapor Laser Optically Pumped with a Single-Mode Tunable CO$_2$ Laser," *J. Appl. Phys.*, **50**(2), pp. 1125–1127, 1979.

218. P. Woskoboinikow, W. J. Mulligan, and R. Erickson, "385 $\mu$m D$_2$O Laser Linewidth Measurements to $-60$ dB," *IEEE J. Quantum Electron.*, **QE-19**(1), pp. 4–7, 1983.

219. M. R. Green, P. P. Morgan, and M. R. Siegrist, *J. Phys.*, **E11**, p. 389, 1978.

220. A. Semet, L. C. Johnson, and D. K. Mansfield, "A High Energy D$_2$O Submillimeter Laser for Plasma Diagnostics," *Int. J. Infrared Millim. Waves*, **4**(2), pp. 231–246, 1983.

221. R. Behn, I. Kjelberg, P. D. Morgan, T. Okada, and M. R. Siegrist, "A High Power $D_2O$ Laser Optimized for Microsecond Pulse Duration," *J. Appl. Phys.* **54**(6), pp. 2995–3002, 1983.

222. R. Behn, M. A. Dupertuis, I. Kjelberg, P. A. Krug, S. A. Salito, and M. R. Siegrist, "Buffer Gases to Increase the Efficiency of an Optically Pumped Far Infrared $D_2O$ Laser," *IEEE J. Quantum Electron.*, **QE-21**, pp. 1278–1285, 1985.

223. R. Behn, M. A. Deupertuis, P. A. Krug, I. Kjelberg, S. A. Salito, and M. R. Siegrist, *Time-Resolved Linewidth and Lineshape Measurements of a Pulsed Optically Pumped Far-Infrared $D_2O$ Laser*, Ecole Polytechnique Fédérale de Lausanne Rep. LRP 318/87, 1987.

224. R. Behn, D. Dicken, J. Hackman, S. A. Salito, and M. R. Siegrist, *Ion Temperature Measurements of a Tokamak Plasma by Collective Thomson Scattering of $D_2O$ Laser Radiation*, Ecole Polytechnique Fédérale de Lausanne Rep. LRP 357/88, 1988.

225. M. A. Dupertuis, M. R. Siegrist, and R. R. E. Salomaa, *Line Selection in Off-Resonantly Pumped Multi-level Systems*, Ecole Polytechnique Fédérale de Lausanne Rep. LRP 311/87, 1987.

226. W. A. Peebles, D. L. Brower, N. C. Luhmann, Jr., and E. J. Danielewicz, "Pulsed FIR Emission for Isotopic Methyl Fluoride," *IEEE J. Quantum Electron.*, **QE-16**, pp. 505–508, 1980.

227. T. A. DeTemple, and S. J. Petuchowski, "Three-Photon Contributions in Optically Pumped Lasers," *Proc. 4th Int. Conf. Infrared Millim. Waves Their Appl.*, IEEE Cat. 79, CH1384-7-MTT, 1979.

228. M. P. Hacker, Z. Drozdowicz, D. R. Cohn, K. Isobe, and R. J. Temkin, "A High Power, 1.22 mm $^{13}CH_3F$ Laser," *Phys. Lett.*, **57A**, pp. 328–330, June 1976.

229. C. K. Rhodes, Ed., *Excimer Lasers*, Vol. 1, Springer-Verlag, New York, 1984.

230. R. S. Davis, and C. K. Rhodes, Ed., "Electronic Transition Lasers," in M. Weber, *CRC Handbook of Laser Science and Technology*, Vol. 2, *Gas Lasers*, CRC Press, Boca Raton, FL, 1982.

231. W. E. Ernst and F. K. Tittel, *Appl. Phys. Lett.*, **35**, pp. 36–37, 1979.

232. M. Rokni, and J. H. Jacob, "Rare-Gas Halide Lasers," in *Applied Atomic Collision Physics*, Vol. 3, *Gas Lasers*, 1982, Chap. 10.

233. F. Kannari, M. Obara, and T. Fujioka, "Electron-Beam-Excited KrF Lasers Including the Vibrational Relaxation in KrF*(B) and Collisional Mixing of KrF*(B, C)," *J. Appl.*, **57**, pp. 4309–4322, 1985.

234. J. Jacobs and J. Mangano, "Modeling of KrF Laser Discharge," *Appl. Phys. Lett.*, **28**(12), pp. 724–726, 1976.

235. D. L. Birx et al., *Basic Principles Governing the Design of Magnetic Switches*, UCID-18831, Lawrence Livermore National Laboratory, Livermore, CA, 1980.

236. B. Mass, R. Butcher, Hansen, T. Fahlen, and B. Jones, "All Solid State Power Conditioning for 150 W XeCl Laser," *Proc. Lasers Electro-opt. Conf.*, 1987, pap. FC2.

237. L. D. Pleasance, J. R. Murray, J. Goldhar, and L. P. Bradley, "Electron-Beam Switched Discharge for Rapidly Pulsed Lasers," U.S. Patent 4, 308, 507, 1979.

238. E. Muller-Horsche, D. Basting, U. Brinkman, P. Klopotek, P. Oesterlin, and W. Muckenheim, "Pre-ionization Switching for Excimer Lasers," *Proc. Lasers Electro-opt. Conf.*, 1987, pap. FC3.

239. D. E. Rothe, C. Wallace, and T. Petach, "Efficiency Optimization for Discharge Excited High Energy Excimer Laser," in C. K. Rhodes, H. Egger, and H. Pummer, Eds., *Excimer Lasers—1983*, American Institute of Physics, New York, 1983.

240. R. S. Sze, "Inductively Stabilized Rare Gas Halide Minilaser for Long Pulse Operation," *J. Appl. Phys.*, **54**, p. 1224, 1983.

241. C. P. C. Christiansen, C. Gordon III, C. Moutlas, and B. J. Feldman, *Opt. Lett.*, **12**, p. 169, 1987.

242. S. Watanabe, A. Endoh, and M. Watanabe, "Wide Aperture Self-Sustained Discharge KrF and XeCl laser," in C. K. Rhodes, H. Egger, and H. Pummer, Eds., *Excimer Lasers — 1983*, American Institute of Physics, New York, 1983.

243. V. Yu. Baranov, V. M. Borisov, D. N. Molchanov, V. P. Novikov, and O. B. Khristo-forov, *Sov. J. Quantum Electron.*, **17**, p. 978, 1987.

244. L. F. Champagne, A. J. Dudas, and N. W. Harris, "Current Rise Time Limitations of the Large Volume X-Ray Preionized Discharge-Pumped XeCl Laser," *J. Appl. Phys.*, **62**, p. 1576, 1987.

245. D. Dijkkamp, T. Venkatesan, X. D. Wu, S. A. Shahenn, N. Jisravi, Y. H. Min-Lee, W. L. McLean, and M. Croft, *Appl. Phys. Lett.*, **51**, p. 861, 1987.

246. R. T. Young, and R. F. Wood, *Annu. Rev. Mater. Sci.*, **12**, p. 323, 1982.

247. Y. Kawamura, K. Toyoda, and S. Namba, *Appl. Phys. Lett.*, **40**, p. 374, 1982.

248. K. Jain, and R. T. Kerth, "Excimer Laser Projection Lithography," *Appl. Opt.*, **23**, p. 648, 1984.

249. S. Rice, and K. Jain, "Direct High Resolution Excimer Laser Photoetching," *Appl. Phys.*, **A33**, p. 195, 1984.

250. D. J. Erlich, J. Y. Tsao, and C. O. Bozler, "Submicron Paterning by Projected Excimer Laser Beam Induced Chemistry," *J. Vac. Sci. Technol.*, **B3**, p. 1, 1985.

251. J. B. Laundenslager, *Laser Focus*, **24**, p. 57, 1988.

252. S. L. Trokel, R. Srinivasan, and B. Raven, *Am. J. Ophthalmol.*, **96**, p. 710, 1983.

253. V. V. Lantukh, M. M. Pyatin, V. M. Subbotin, V. Ishchenko, S. A. Kochubei, A. M. Razer, and V. M. Chebotayev, *Opt. Spektrosk.*, **63**, p. 1132, 1987.

254. S. Szatmari, F. P. Shafer, E. Muller-Horsche, and W. Muckenheim, "Hybrid Dye Excimer Laser System for Generation of 80 fs, 900 GW Pulses at 248 nm," *Opt. Commun.*, **63**, p. 305, 1987.

255. A. Endoh, M. Watanabe, and S. Watanabe, "Picosecond Amplification in a Wide Aperture KrF Laser," *Opt. Lett.*, **12**, p. 906, 1987.

256. L. A. Rosocha, J. A. Hanlon, J. McLeod, M. Kang, B. L. Kortegaard, M. D. Burrows, and P. S. Bowling, "Aurora Multikilojoule KrF Laser System Prototype for Inertial Confinement Fusion Studies," *Fusion Technol.* **11**, p. 497, 1987.

257. J. R. Oldenettel, and K. Y. Tang, "Multi-kilojoule Narrowband XeCl Laser," *Proc. Soc. Photo-Opt. Instrum. Eng.*, Vol. 710, p. 117, 1986.

258. J. A. Sullivan, "Design of a 100 kJ KrF Power Amplifier Module," *Fusion Technol.*, **11**, p. 684, 1987.

259. T. H. Johnson, L. J. Palumbo, and A. M. Hunter II, "Kinetics Simulation of High-Power Gas Lasers," *IEEE J. Quantum Electron.*, **QE-15**, p. 281, 1979.

260. M. Rokni, J. A. Mangano, J. H. Jacob, and J. C. Hsia, *IEEE J. Quantum Electron.*, **QE-14**, p. 464, 1979.

261. C. B. Edwards, and F. O'Neil, "Computer Modelling of E-Beam-Pumped KrF Lasers," *Laser Part. Beams*, **1**, p. 81, 1983.

262. J. H. Jacob, M. Rokni, R. E. Klinkowstein, and S. Singer, "Expanding Beam Concept for Building Very Large Excimer Amplifiers," *Appl. Phys. Lett.*, **48**, p. 318, 1986.

263. W. W. Rigrod, "Homogeneously Broadened CW Lasers with Distributed Loss," *IEEE J. Quantum Electron.*, **QE-14**, p. 377, 1979.

264. A. H. Hunter, and R. O. Hunter, "Bidirectional Amplification with Nonsaturable Absorption and Amplified Spontaneous Emission," *IEEE J. Quantum Electron.*, **QE-17**, p. 1879, 1981.

265. A. M. Hunter, R. O. Hunter, and T. H. Johnson, "Scaling of KrF Lasers for Inertial Confinement Fusion," *IEEE J. Quantum Electron.*, **QE-22**, pp. 386–404, 1986.

266. R. B. Miller, *Intense Charged Particle Beams*, Plenum Press, New York, 1982.

267. L. G. Schlitt, J. C. Swingle, W. R. Rapoport, J. Goldharand, and J. J. Ewing, *J. Appl. Phys.*, **91**, p. 7, 1981.

268. D. R. Corson, and P. Lorrain, *Introduction to Electromagnetic Fields and Waves*, W. H. Freeman, San Francisco, 1962.

269. Ken-Ichi Ueda, and H. Takuma, *Conf. Lasers Electro-opt., Tech. Dig.*, 1988. Pap. TUB3.

270. C. B. Edwards, O'Neil, and M. J. Shaw, *Appl. Phys. Lett.*, **36**, p. 617, 1980.

271. Y. Owadano, I. Okuda, M. Tanimoto, Y. Matsumoto, T. Kasai, and M. Yano, "Development of a 1-kJ KrF Laser System for Laser Fusion Research," *J. Fusion Technol.*, **11**, p. 486, 1987.

272. R. E. Shefer, L. Friedland, and Klinkowstein, *Phys. Fluids*, **31**, p. 930, 1988.

273. L. A. Rosocha, and B. B. Riepe, "Electron-Beam Sources for Pumping Large Aperture KrF Lasers," *Fusion Technol.*, **11**, p. 576, 1987.

274. A. Mandl, and E. Salesky, "Electron Beam Deposition Studies of the Rare Gases," *J. Appl. Phys.*, **60**, p. 1565, 1986.

275. D. J. Eckstrom, and H. C. Walker, "Determination of Spatial Energy Deposition in E-Beam-Pumped Laser Cells by Pressure Measurements," *J. Appl. Phys.*, **51**, p. 2458, 1980.

276. K. Ueda, H. Nishioka, and H. Takuma, *Proc. Soc. Photo-Opt. Instrum. Eng.* Vol. 710, p. 7, 1986.

277. A. E. Mandl, D. E. Klimek, and E. T. Salesky, "KrF Laser Studies at High Krypton Density," *Fusion Technol.*, **11**, p. 542, 1987.

278. A. Suda, M. Obara, and A. Noguchi, "Atmospheric Operation of a KrF Laser Oscillator and Amplifier with a Krypton-Rich Mixture and a $Kr/F_2$ Mixture," *Fusion Technol.*, **11**, p. 548, 1987.

279. Y. W. Lee, H. Kumagai, S. Ashidate, and M. Obara, *Appl. Phys. Lett.*, **52**, p. 1294, 1988.

280. L. Litzenberger, and A. Mandl, "Increased XeF Laser Efficiency High Pump Rate and Elevated Temperature," *Appl. Phys. Lett.*, **52**, p. 1557, 1988.

281. A. Mandl, and L. Litzenberger, "XeF Laser at High Electron Beam Pump Rate," *Appl. Phys. Lett.*, **51**, p. 955, 1987.

282. N. Nishida, F. K. Tittel, H. Kumagai, Y. Lee, and M. Obara, "Intrinsic Efficiency Comparison in Various Low-Pressure XeF Laser Mixtures Pumped at High Excitation Rates and with Short-Pulse Electron Beam Pumping," *Appl. Phys. Lett.*, **52**, p. 1847, 1988.

283. J. A. Hanlon, and J. McLeod, "The Aurora Laser Optical System," *J. Fusion Technol.*, **11**, p. 1294, 1987.

284. J. J. Ewing, R. A. Haas, J. C. Swingle, E. V. George, and W. F. Krupke, "Optical Compression Systems for Laser Fusion," *IEEE J. Quantum Electron.*, **QE-15**, p. 368, 1979.

285. J. McLeod, "Output Optics for Aurora: Beam Separation, Pulse Stacking, and Target Focusing," *Fusion Technol.*, **11**, p. 654, 1987.

286. J. R. Murray, J. Goldhar, D. Eimerl, and A. Szoke, "Raman Pulse Compression of Excimer Lasers for Application to Laser Fusion," *IEEE J. Quantum Electron.*, **QE-15**, p. 342, 1979.

287. J. Goldhar, M. W. Taylor, and J. R. Murray, "An Efficient Double Pass Raman Amplifier with Pump Intensity Averaging in a Light Guide," *IEEE J. Quantum Electron.*, **QE-20**, p. 772, 1984.

288. J. M. Eggelston, and M. J. Kushner, "Stimulated Brillouin Scattering Parasitics in Large Optical Windows," *Opt. Lett.*, **12**, p. 410, 1987.

289. J. P. Roberts, A. J. Taylor, P. H. Y. Lee, and R. B. Gibson, "High-Irradiance 248 nm Laser System," *Opt. Lett.*, **13**, p. 734, 1988.

290. S. P. Obenschain, J. Grun, M. J. Herbst, K. J. Kearney, C. K. Manka, E. A. McLean, A. N. Mostovych, J. A. Stamper, R. R. Whitlock, S. E. Bodner, J. H. Gardner, and R. H. Lehmberg, *Phys. Rev. Lett.*, **56**, p. 2807, 1986.

291. Y. Kato, K. Mima, M. Miyanka, S. Aringa, Y. Kitagawa, M. Nakatsuka, and C. Yamanaka, "Random Phasing of High-Power Lasers for Uniform Target Acceleration and Plasma Instability Suppression," *Phys. Rev. Lett.*, **53**, p. 1057, 1984.

292. R. H. Lehmberg, and S. P. Obenschain, "Use of Induced Spatial Incoherence for Uniform Illumination of Laser Fusion Targets," *Opt. Commun.*, **46**, p. 27, 1984.

293. Ximing Deng, Xiangchun Liang, Zezun Chen, Wenyan Yu, and Renyong Ma, *Appl. Opt.*, **3**, p. 377, 1986.

294. T. J. Kessler, S. Letzring, S. Skupsky, M. Sheldon, S. Morse, and P. Jaanimagi, "Broadband Phase Conversion of the Frequency Tripled Omega Laser," *Conf. Lasers Electro-opt., Tech. Dig.*, 1989, pap. WD5.

295. R. H. Lehmberg, and J. Goldhar, "Use of Incoherence to Produce Smooth and Controllable Irradiation Profiles with KrF Fusion Lasers," *Fusion Technol.*, **11**, p. 532, 1987.

296. M. S. Pronko, S. P. Obenschain, R. H. Lehmberg, and A. N. Mostovych, "Experimental Studies on the Production of Broadband, High-Peak-Power Laser Radiation," *Conf. Lasers Electron-opt., Conf. Dig.*, 1986, pap. THK33.

297. M. D. Sheldon, T. J. Kessler, R. S. Craxton, S. Skupsky, W. Seka, and J. M. Soures, "Efficient Third Harmonic Generation with a Broadband Laser," *Conf. Lasers Electro-opt., Tech. Dig.*, 1989, pap. WD4.

298. R. H. Lehmberg, A. J. Schmidt, and S. E. Bodner, "Theory of Induced Spatial Incoherence," *J. Appl. Phys.*, **62**, p. 2680, 1987.

# 10

# OPTICAL FIBER TRANSMISSION TECHNOLOGY

CHINLON LIN
*Bell Communications Research (Bellcore)*
*Red Bank, New Jersey*

## 10.1 INTRODUCTION

The idea of using low-loss optical glass fiber waveguides for long-distance optical transmission was first proposed by C. K. Kao in 1966. The realization of low-loss glass optical fibers was first achieved by Corning in 1970; in the very same year, semiconductor diode lasers operating continuously at room temperature were obtained by Bell Labs. The combination of a low-loss compact optical transmission medium and a miniature, directly current-modulated diode laser as the optical source for an optical signal transmitter paved the way to a revolution in communications technology. For the past 20 years or so, there were dramatic progresses made in the field of optical fiber transmission technology in laboratories around the world. Furthermore, the technology has moved from laboratory experiments to actual system applications in the outside world. Optical fiber transmission systems are now widely used in carrying real-life communications traffic throughout the optical-fiber-based long-haul digital telecommunications network.

Clearly optical fiber transmission technology has emerged as the telecommunication system technology of the 1980s and will remain the key communications technology for a long time to come. The key advantages of the optical fiber transmission technology are now well known:

1. small size and light weight;
2. low transmission loss;
3. very high bandwidth possible;
4. immunity to electromagnetic interference;

*Handbook of Optical Components and Engineering*, Edited by Kai Chang
ISBN 0-471-39055-0 © 2003 John Wiley & Sons, Inc.

**659**

5. optical cable flexibility and ruggedness;

6. optical fiber's electrical isolation;

7. signal security within the fiber waveguide;

8. potentially low cost per bandwidth–distance product;

9. optoelectronic devices capable of very high speed operation;

10. technology base shared with other emerging optoelectronic information systems such as laser printers, optical disks, and optical fiber sensors.

Because of these unique advantages, optical fiber transmission systems are widely used not only in the long-haul telephone transmission network but also in specialized data transmission applications within a ship, an aircraft, an industrial plant, an electric power company's monitoring network, along the railroads, and so on. In the near future, the use of fibers for intrabuilding- or intercampus-type local area data communications as well as for broadband video services will certainly see a significant growth. An all-optical-fiber interconnected Broadband Integrated Services Digital Network (BISDN) providing two-way transmission of telephone, data, and high-quality audio and video signals will likely be implemented, eventually making the information age a reality.

This chapter presents a basic review of the essential aspects of optical fiber transmission technology, including optical fibers, optoelectronic devices, and optical components used in optical fiber communication systems. Basic and essential practical information is presented and summarized without the theoretical derivations; relevant emerging technology trends are also briefly discussed. For detailed in-depth theoretical study as well as comprehensive discussions, readers should consult existing books [1–4] and advanced literature [5].

## 10.2   OPTICAL FIBERS FOR TRANSMISSION

Before the first proposal on using optical glass fiber waveguides for long-distance telecommunications, optical fibers have high losses and have been used for short-distance light guiding. The key advance thus was the realization of very low transmission loss in glass fiber waveguides, which enabled the use of optical fibers for long-distance communication applications. Both the optical waveguiding and the material transmittance (loss due to attenuation, scattering, etc.) of a glass optical fiber are therefore equally important. The next few sections discuss the basic waveguiding properties including the transmission loss and dispersion characteristics.

### 10.2.1   Optical Fiber Waveguides

The basic structure of an optical fiber waveguide is a cylindrical glass fiber with two cylindrical layers: the interior region, the core, of refractive index $n_1$, is surrounded by an outer region, the cladding, of refractive index $n_2$, where $n_2 < n_1$. Typically $n_1 - n_2 \ll 1$; the index difference is usually achieved by a slightly different doping concentration in the host material (e.g., $SiO_2$ host with small concentration of $GeO_2$ as dopant). Figure 10.1 shows two such typical step index fiber structures and the corresponding refractive index variations (profiles) along the radial direction. For a typical single-mode fiber the core diameter is about 9 μm, the cladding diameter is

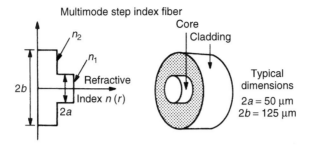

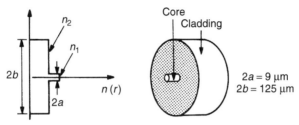

**Figure 10.1**   Refractive index profiles of the core and the cladding of a multimode step index fiber and a single-mode step index fiber. The core diameter is $2a$ and the cladding diameter is $2b$.

125 μm, and the index difference, $\Delta n = n_1 - n_2$, is about 0.0035. For multimode fibers, the core diameter could be 50, 65, 85 μm, or larger depending on the fiber design. Figure 10.1 shows a multimode fiber with 50 μm core diameter and 125 μm cladding diameter, with the index difference being $\Delta n = 0.016$, which is higher than that of the single-mode fiber case.

The rigorous theoretical approaches [2, 3, 6] involve solving Maxwell's equations for the electromagnetic wave propagation in these cylindrically symmetric dielectric waveguide structures. Simplifications in the analysis can be made by noting that the index differences are usually very small ($\Delta n \ll 1$). Nevertheless, an electromagnetic wave propagation analysis is still rather complicated. A physical intuitive understanding of most of the basic propagation properties in optical fiber waveguides could be obtained with a simple ray picture based on geometrical optics of total internal reflection between two dielectric media of different refractive indices. Figure 10.2 illustrates such a case. From Fig. 10.2**a**, Snell's law for refraction leads to

$$n_1 \sin\theta_1 = n_2 \sin\theta_2 \tag{10.1}$$

The expression of the critical angle $\theta_c$ for total internal reflection (the angle of incidence above which the light ray is totally internally reflected rather than refracted into the second medium) is

$$n_1 \sin\theta_c = n_2 \sin\left(\tfrac{1}{2}\pi\right) = n_2 \qquad (n_1 > n_2) \tag{10.2}$$

$$\theta_c = \sin^{-1}\frac{n_2}{n_1}$$

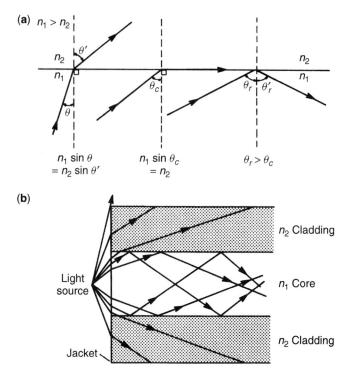

**Figure 10.2** (a) Refraction of light at the interface between media of refractive index $n_1$ and $n_2$, for $n_1 > n_2$. Total internal reflection occurs when incident angle $\theta$ is $\geq \theta_c$, the critical angle. (b) A simple geometrical optics picture of guiding in an optical fiber by total internal reflections at the core–cladding boundaries.

Figure 10.2**b** shows that for incident light rays with different incident angles at the core–cladding boundary, some of them, with an incident angle less than the critical angle for total internal reflection, will be reflected back and thus guided in the core, while some of them will be refracted into the cladding and not be guided in the core. Usually there is a light-absorbing jacket region outside the cladding, so the light scattered out of the cladding will be absorbed. This simple ray picture of total internal reflection can thus explain the basic light-guiding principle.

However, to go one step further to explain the existence of fiber waveguide "modes" based on simple ray optics is not simple. A better description is obtained directly from electromagnetic wave propagation analysis. Such analysis results [2, 3, 6] show that light rays in the optical fiber waveguide can only propagate with a discrete set (and not a continuum) of incident angles, corresponding to the discrete set of waveguide modes, each mode having a specific spatial intensity distribution in the transverse plane [2, 3, 6]. The number of waveguide modes an optical fiber can support depends on the fiber's waveguide parameters such as core diameter, core–cladding refractive index difference, and the light signal wavelength. Optical fibers that can support many waveguide modes are called *multimode fibers*; optical fibers that are designed to support only one mode (the fundamental waveguide mode) are called *single-mode fibers*.

Before we go to the next section, let us first define a few useful parameters. From the optical fiber waveguide dimension $a$ (core diameter), core index $n_1$, and cladding index $n_2$, several commonly used parameters in describing optical fiber waveguides can be defined:

1. The *relative* core–cladding index difference $\Delta$ is given by

$$\Delta = \frac{n_1 - n_2}{n_1} = \frac{\Delta n}{n_1} \tag{10.3}$$

2. The numerical aperture NA, a measure of the light-capturing capability of the optical waveguide, is given by

$$\text{NA} = n_1 \sin\left(\tfrac{1}{2}\pi - \theta_c\right) = n_1 \cos\theta_c = (n_1^2 - n_2^2)^{1/2} \simeq n_1(2\Delta)^{1/2} \tag{10.4}$$

3. The normalized frequency $V$ (also called the $V$-number) is given by

$$V = \frac{2\pi a}{\lambda}(n_1^2 - n_2^2)^{1/2} \simeq \frac{2\pi a}{\lambda} n_1(2\Delta)^{1/2} \tag{10.5}$$

where $\lambda$ is the wavelength of the light. These parameters, defined based on the fiber waveguide structural parameters, are very useful in various discussions on fiber waveguide propagation properties.

### 10.2.2 Multimode and Single-Mode Optical Fibers

A typical multimode fiber has a large core diameter (50–200 μm) and a large index difference between the core and the cladding ($\Delta n = n_1 - n_2 = 0.01$–$0.03$). In this case there is a large number of discrete waveguide modes that can propagate in the fiber.

It can be shown [2, 3, 6] that for a large-core, multimode step index fiber with a large number of modes $M$, $M$ can be approximately expressed by

$$M \cong \frac{V^2}{2} = \frac{1}{2}\left(\frac{2\pi a}{\lambda}\right)^2 (n_1^2 - n_2^2) = 2\pi \frac{A_c}{\lambda^2}(\text{NA})^2 \tag{10.6}$$

where $A_c = \pi a^2$ is the core area. Thus, in multimode step index fibers, the large index differences and large core diameters lead to a large NA and a large $V$-number. Therefore, the number of modes that can propagate in the fiber is large. For example,

$$n_1 = 1.46 \quad n_2 = 1.45 \quad 2a = 50 \ \mu\text{m} \quad \text{NA} = 0.17$$

$$V = 20 \quad \text{at} \quad \lambda = 1.27 \ \mu\text{m} \qquad M \simeq \frac{V^2}{2} = 200$$

The main difference in the waveguide characteristics between the single-mode and the multimode optical fiber waveguides is that the former can only support one waveguide mode while the latter can support propagation of many waveguide modes. This is due to the fact that multimode fibers have larger core diameters and larger refractive index differences between the core and the cladding, while single-mode fibers have much smaller core diameters and index differences that only one waveguide mode can propagate. This in turn leads to a very significant difference in transmission bandwidth and dispersion properties.

It can be shown [2, 3, 6] that in a *step index fiber* the incident light will propagate in only one waveguide mode (the fundamental mode) when the core diameter and the refractive index difference between the core and the cladding are small enough such that the normalized frequency $V$ (the $V$-number) is

$$V = \frac{2\pi a}{\lambda}(n_1^2 - n_2^2)^{1/2} < 2.405 \tag{10.7}$$

This small $V$-number ($V < 2.405$) is to be compared with a typical $V$-number of 20 or more for a large-core, large-index-difference multimode fiber.

The single-mode condition described in the preceding can also be expressed in terms of the wavelength of the propagating light, $\lambda$, with respect to the fiber's *cutoff wavelength* $\lambda_c$,

$$\lambda_c = \frac{2\pi a(n_1^2 - n_2^2)^{1/2}}{2.405} = \frac{V}{V_c}\lambda \tag{10.8}$$

In this case of $\lambda > \lambda_c$, only one mode can be guided in the fiber waveguide (actually two if counting the two possible polarizations). The optical fiber is thus a single-mode fiber waveguide *for light with optical wavelength longer than the cutoff wavelength of the single-mode fiber waveguide*. For example, if a single-mode fiber has a core diameter $2a = 10$ μm and index difference $n_1 - n_2 = 0.003$, then the condition $V < 2.405$ leads to a cutoff wavelength

$$\lambda_c = \frac{\pi(10)(2 \times 1.46 \times 0.003)^{1/2}}{2.405} = 1.22 \; \mu\text{m}$$

With this optical fiber waveguide, an InGaAsP laser or light-emitting diode (LED) at 1.3 μm will propagate in the fiber in the fundamental mode only because $\lambda > \lambda_c$, so the fiber is a single-mode fiber at 1.3 μm wavelength.

It is important to note that the same 10-μm-core-diameter fiber in the preceding example will support higher-order modes (thus becoming a multimode fiber that supports a few higher-order modes) at the He–Ne laser wavelength of 0.63 μm or the GaAlAs laser wavelength of 0.8 μm, because in either case the corresponding $V$-number exceeds 2.405. In terms of the fiber's single-mode cutoff wavelength, both optical sources have wavelengths *shorter than* the fiber's cutoff wavelength ($\lambda < \lambda_c$) and thus will propagate in the fiber as a mixture of higher-order modes. For an optical fiber to be single mode at these shorter wavelengths, a smaller core diameter (5–6.5 μm) is needed to satisfy the same condition of $V < 2.405$. The definition of the single-mode fiber is thus closely tied to the cutoff wavelength and is not independent of the optical source wavelength.

It should be pointed out that in actual single-mode fibers, the practical cutoff wavelength has to be experimentally measured in the fiber cable, because the actual cutoff wavelength depends on the cabling, the fiber/cable length, and the bend radius and could be shorter than the theoretical cutoff wavelength calculated based on (1.7) or (1.8), which is true only for step index single-mode fibers. Table 10.1 illustrates the fiber core diameters and index differences for a typical single-mode (SS) fiber (uncabled) with its cutoff wavelength near 1220 nm, designed for 1300-nm transmission systems. For comparison, parameters for a multimode (MM) fiber are also shown.

**TABLE 10.1  Examples of SM and MM Fiber Parameters**

|  | SM Fiber | MM Fiber |
|---|---|---|
| $2a$ | 10 μm | 50 μm |
| $n$ | 0.003 | 0.01 (NA = 0.17) |
| $V$-number at 1300 nm | <2.4 | 20 |
| Number of modes | 1 | 200 |

### 10.2.3  Graded Refractive Index Profiles and Modal Dispersions

The refractive index difference between the core and the cladding region is responsible, as we already discussed, for the total internal reflection and hence waveguiding in the fiber core region. The index step shown in Fig. 10.1 is only one of the possible index profiles for the core and the cladding regions; in reality, a truly step index profile with sharp boundaries is not always obtained.

There are a variety of refractive index profiles possible for both multimode and single-mode optical fibers; the refractive index profiles together with the doping material properties determine the majority of the optical fiber waveguiding characteristics. Different refractive index profiles have been proposed and realized to achieve various special optical waveguide characteristics. The most well known example is the nearly parabolic graded index profile in multimode fibers for minimizing the intermodal delay distortions (modal dispersion). The other important example is the triangular refractive index profile in single-mode fibers for dispersion shifting (shifting the minimum chromatic dispersion wavelength to the lowest loss region of 1550 nm), which will be discussed in a later section.

Figure 10.3 shows the graded index multimode fiber profile as compared with the step index multimode fiber profile. In a typical step index multimode fiber, the different rays with different angle of incidence will have different travel path lengths in the optical fiber waveguide; this will result in a time-of-arrival difference, also called modal delay spread between the different modes, which causes intermodal delay distortions (modal dispersion). The maximum modal delay spread $\Delta T_{max}$ is that between the ray traveling along the fiber axis (the fastest ray) and the ray traveling at the maximum angle bouncing off the core–cladding interface (the slowest ray), as shown in Fig. 10.4 [1–4]:

$$\Delta T_{max} = t_{2,max} - t_1 = \left.\frac{n_c L}{c \cos\theta}\right|_{max} - \frac{n_c L}{c}$$

$$= \frac{n_c L}{c}\left(\frac{n_c - n}{n}\right) = \frac{L}{c}\frac{\Delta n\, n_c}{n} = \frac{L}{2cn}(\text{NA})^2 \qquad (10.9)$$

Note here that $n_c = n_1$ (core) and $n = n_2$ (cladding) and $c$ is the speed of light. The maximum delay spread due to modal dispersion is therefore proportional to the index difference $\Delta n$ and to the square of the NA. Thus, in general, the larger the index difference (or equivalently the larger the numerical aperture), the larger the maximum modal delay spread. For example, if $n_1 - n_2 = 0.01$, NA = 0.17, the maximum modal delay spread amounts to 50 ns/km. This means for 10-Mb/s optical signals (minimum

(a)

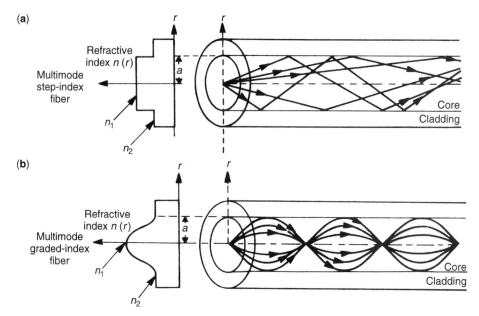

(b)

**Figure 10.3** The refractive index profile and ray path in a multimode step index fiber and those in a multimode graded index fiber. Note the multipath delay difference in the step index fiber is large (large modal dispersion), while that in an ideal graded index fiber is minimized.

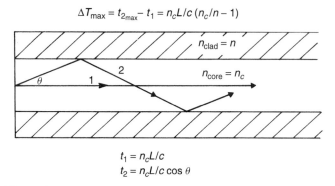

$$\Delta T_{max} = t_{2max} - t_1 = n_c L/c \, (n_c/n - 1)$$

$$t_1 = n_c L/c$$
$$t_2 = n_c L/c \cos \theta$$

**Figure 10.4** The optical path difference between the fastest ray (on axis) and the slowest guided ray leading to modal delay spread in the step index multimode fiber.

pulse spacing 100 ns) the delay spread in a 2-km length of this fiber would be 100 ns, causing the intersymbol interference of the adjacent signal pulses. Such a larger modal delay spread, due to modal dispersion of the fiber, would limit the use of this fiber to <10 Mb/s and 2-km long-distance applications. A higher bit rate or longer distance transmission would be seriously degraded in such a multimode fiber.

To reduce the modal delay spread and improve the performance, instead of using a step-index-profile multimode fiber, graded-index-profile multimode fibers can be fabricated to reduce the modal delay spread (modal dispersion), as shown in Fig. 10.3. With the graded index profile shown, the outer rays, traveling farther from the core center axis and longer overall path, see a lower refractive index (to speed up the propagation time), and the rays closer to the core axis, traveling a shorter distance, see a larger refractive index and travel at a slower speed. In this way the graded index profile is designed to equalize the travel times of all the different rays (modes) so that the overall delay spread is minimized.

The graded index profile in a multimode graded index fiber is often expressed by

$$n(r) = \begin{cases} n_1 \left[ 1 - 2\Delta \left( \dfrac{r}{a} \right)^\alpha \right]^{1/2}, & r < a \ \text{(core)} \\ n_1 (1 - 2\Delta)^{1/2} = n_2, & r \geqslant a \ \text{(cladding)} \end{cases} \tag{10.10}$$

where $\alpha$ is the graded index profile parameter; $\alpha = 2$ corresponds to a parabolic profile; $\alpha = 1$ corresponds to a triangular profile, while the step index profile is the special case of $\alpha = \infty$. Extensive theoretical studies [2, 3, 6] have shown that when $\alpha$ is very close to 2, the fiber has the optimum refractive index profile for minimizing the modal delay spread and thus achieving the smallest modal dispersion multimode fiber propagation. This ideal profile is called a nearly parabolic profile.

Without getting into detailed discussion, the following important facts concerning multimode graded index fibers and modal dispersion characteristics can be summarized [1–4,6]:

1. Actual graded index profile multimode fibers have their profiles made by small step index increments attempting to approximate the ideal near-parabolic index profile; the resultant bandwidth is usually much less than that expected of the ideal profile. Consequently, while ideal profile theoretically can yield 100-GHz-km bandwidths (bandwidth–distance product), the best experimental results are in the 4–6-GHz-km range, while the majority of typical commercial multimode graded index fibers have much smaller bandwidths, in the 400-MHz-km to 1.5-GHz-km range, depending on the NA, the dopants, and the exact index profile.

2. The refractive index profile in a graded index multimode fiber is usually designed to approach the optimum profile *at a given operating wavelength*. Due to "profile dispersion" (wavelength dependence of the index difference), a graded index multimode fiber optimized to have a high bandwidth at one wavelength (e.g., 850 nm) may have a low bandwidth at a different wavelength (e.g., 1300 nm), because the optimum $\alpha$ is wavelength dependent, that is, $\alpha = \alpha(\lambda)$. Figure 10.5 shows examples of such "bandwidth spectra" of two graded index multimode fibers [7]. This spectral dependence of the bandwidth, which is sharply peaked near the optimized wavelength only, makes multimode fibers unattractive for optical fiber systems requiring upgrading by WDM (wavelength division multiplexing).

3. The phenomenon of mode mixing in a multimode fiber leads to a situation where the bandwidth of a concatenated multimode fiber is a nonlinear function of fiber link length, causing some complication in the design of long-link-length multimode fiber transmission systems.

In spite of these serious bandwidth limitations, several early long-haul multimode optical fiber transmission systems were implemented (usually at a lower bit rate such as 45–90 Mb/s) in the 1970s. However, after the 1980s, single-mode fiber transmission systems have been exclusively used for long-distance and high-bit-rate optical

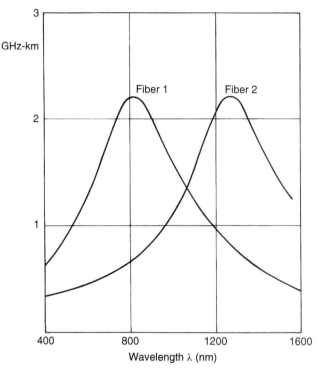

Bandwidth-length product

**Figure 10.5**   Bandwidth spectra of two graded index multimode fibers, fiber 1 was optimized for 800 nm and fiber 2 for 1300 nm.

communication links because multimode fiber-based systems not only have the above-mentioned bandwidth limitations but also are found to have the "modal noise" problem [2, 3] caused by time-varying, fluctuating intensity patterns due to the coherent interference of the fiber modes associated with external disturbances and/or displaced fiber joints. The presence of modal noise is a serious transmission limitation for long-distance transmission systems using multimode fibers; the use of low-coherence optical sources such as LEDs or wide-spectral-width multi-longitudinal-mode laser diodes could reduce the modal noise, but at the expense of increased chromatic dispersion penalty.

Today multimode fibers are principally being used in short-distance (<1 km typically) transmission, although there were longer-distance multimode fiber telecommunication systems installed before 1980. With a cladding diameter of 125 μm, multimode fibers typically have a core diameter of 50, 62.5, or 85 μm, with an NA of 0.2, 0.275, and 0.26, respectively; the choice depends on specific application requirements. In most cases, especially for short-distance, local area network (LAN) communication between computers, the primary concern is usually the power budget, coupling and branching losses, and not the bandwidth or dispersion limitations. For example, with commercial large-NA graded index multimode fibers, the modal dispersion limitation is not serious for a few 100-Mb/s, ≤1-km transmission applications, as the fiber bandwidth is typically in the 200–600-MHz-km range. The high LED light-coupling efficiency, the low fiber jointing (connectors and splices) losses, and low branching losses (through fiber couplers, splitters, etc.) of large-core, large-NA graded index multimode fibers are unique advantages that make it possible to use low-cost optical sources and components for large volumes of LAN applications.

### 10.2.4  Doping and Optical Transmission Loss Spectra

The refractive index profile, whether a step index or a graded index difference between the core and cladding region, is achieved in practice by doping the host material (in our case silica, $SiO_2$) with a dopant that either raises or reduces the effective refractive index of the doped region. For example, $GeO_2$ doping will raise the refractive index above that of the pure silica, while $B_2O_3$ doping reduces the index. Depending on the dopant, the doped region (core or cladding) will have a loss characteristic modified from that of the pure silica. Figure 10.6 illustrates the effect of various dopants on the fiber transmission loss; note some dopants cause the infrared (IR) absorption edge to rise at a shorter wavelength. Similarly, water, not an intentional dopant but practically unavoidable, contributes to the $OH^-$ loss peaks near 1.4, 1.25, and 0.95 μm [1–4]. Likewise, other impurities in the fiber perform fabrication process may cause localized absorption peaks.

Various fiber preform fabrication and fiber drawing processes have been developed for making low-loss glass fibers with the desired refractive index profiles, geometrical parameters, and mechanical properties but with least impurities or OH absorption. For a discussion on the various fiber fabrication processes such as modified chemical vapor deposition (MCVD), outside vapor phase oxidation (OVPO), vapor axial deposition (VAD), and plasma-activated chemical vapor deposition (PCVD), see Refs. 3 and 4.

With the advances in the past 18 years, fiber fabrication and manufacturing technology have now reached a mature stage; commercially available optical silica glass fibers now have in general very low losses. Figure 10.7 shows the loss spectra of a low-OH, low-loss silica glass optical fiber. Typical low-loss fibers can easily have a loss of

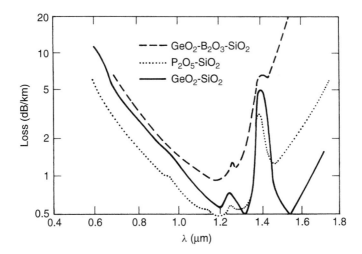

**Figure 10.6**  Effect of various dopants on the silica glass fiber loss spectra.

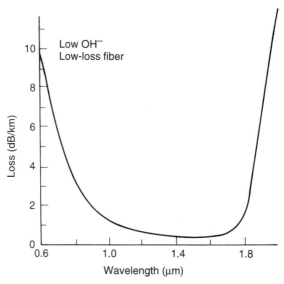

**Figure 10.7**   Loss spectrum of a very low OH content silica glass fiber, showing a single low-loss transmission window.

about 2.4 dB/km at 0.8 μm, 0.4 dB/km at 1.3 μm, and 0.25 dB/km near 1.56 μm. Since fiber loss, or attenuation coefficient $\alpha_l$ in decibels per kilometer, is defined by

$$P(z) = P_0 10^{-\alpha_l z/10} \qquad P_0 = P(z = 0)$$

$$\alpha_l = \frac{1}{L} \left[ 10 \log_{10} \frac{P_0}{P(L)} \right] \qquad (10.11)$$

where $L$ is the fiber length. To appreciate the high transparency of optical glass fibers, note that a loss of 0.25 dB/km means 94.4% of optical transmission after 1 km of glass fiber, an extremely low loss indeed.

While the rising edge in the IR region beyond about 1.7 μm is due to the IR absorption of the host and the dopant glass, the loss curve in the visible–near IR region actually is very close to the intrinsic loss limit due to the Rayleigh scattering associated with microscopic compositional and density fluctuations of the glass [2–6]. Figure 10.8 illustrates the contributions of various loss mechanisms including ultraviolet (UV) and IR absorption (intrinsic, not due to impurity), waveguide imperfection, and Rayleigh scattering [3, 4] and the experimental loss spectra that include the contributions from impurity and OH absorptions. Because Rayleigh scattering is inversely proportional to $\lambda^4$, it is smaller in the IR region. The loss minimum is thus somewhere near the 1.6-μm region bounded by the decreasing Rayleigh scattering loss curve and the rising IR absorption edge. A minimum loss of 0.16 dB/km near 1.58 μm is possible.

In typical low-loss silica glass optical fibers, the transmission window spans from 0.6 to 1.7 μm, with a water peak around 1.4 μm. Although with a low-water-peak fiber the *entire transmission window is useful* for telecommunication applications, due to the availability of laser diode and LED sources, most of the present telecommunication optical fiber systems operate in the transmission loss windows near 0.8, 1.3, and 1.55 μm. In addition to telecommunication applications, optical fibers are also useful

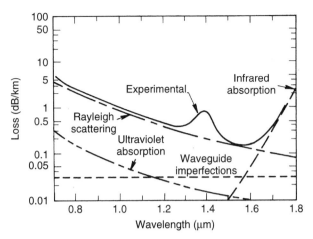

**Figure 10.8**  Figure showing the relative contributions of the various attenuation mechanisms (shown in dashed and dotted lines) to the loss spectra (solid curve) of a low-loss silica glass fiber.

for a variety of very short distance transmission applications such as light delivery and data transmission in an automobile and an airplane, high-power IR and UV laser power delivery, medical applications in a hospital, or robotic material processing applications in a factory.

In a discussion of fiber loss, it should also be mentioned that in addition to low-loss silica-glass-based fibers, there are medium-loss (e.g., 20–40 dB/km) multicomponent glass fibers and higher-loss plastic-based optical fiber waveguides [2, 3], usually for large-core multimode-fiber short-distance transmission applications. In fact, some plastic-based fibers have acceptably low (a few 100 dB/km) loss windows in the visible, allowing the use of inexpensive visible LEDs for some of the very short distance (e.g., automobile) applications previously mentioned.

### 10.2.5  Single-Mode Fiber Dispersion and Dispersion Control

In single-mode optical fibers, there is no group delay spread due to the different modes (modal dispersion) as only one mode is supported. The dispersion limitation in single-mode fibers is due to the glass material's wavelength dispersion (material dispersion) and is associated with the finite (nonzero) spectral width of the optical signals. This is often also referred to as chromatic dispersion, except in fiber waveguides, for which chromatic dispersion includes, in addition to material dispersion, the (usually small) waveguide dispersion due to the wavelength dependence of the propagation constant of the single mode. Actually chromatic dispersion exists in multimode fibers too, but it is usually small compared with modal dispersion, except when very wide spectral width LED sources are used in the fiber transmission. In single-mode fibers the chromatic dispersion becomes the dominant bandwidth limitation; the study, the measurement, and the improvement of chromatic dispersion characteristics have been important aspects of high-bandwidth optical fiber technology research and development [2–8].

Material dispersion in glass materials is due to the variation of the material's refractive index with wavelength. As a result, for a light signal with a finite spectral width

$\Delta\lambda_s$ different spectral components of the signal travel at different speeds in the glass, causing a group delay difference due to chromatic dispersion (such as in a glass prism). The group velocity is determined by the group index $n_g$, that is, $v = c/n_g$, where the group index is related to the refractive index $n$ by

$$n_g = n - \lambda\frac{dn}{d\lambda} \tag{10.12}$$

The most commonly used material dispersion parameter $M$ is defined as

$$M(\lambda) = \frac{d\tau}{d\lambda} = \frac{-1}{c}\frac{dn_g}{d\lambda} = \frac{\lambda}{c}\frac{d^2n}{d\lambda^2} \tag{10.13}$$

For silica glass, material dispersion curves cross zero in the wavelength region near 1300 nm, the exact zero crossing depending on the dopants and the doping concentration [9]. This is the region of zero or minimum material dispersion.

After an optical signal pulse has propagated in a single-mode fiber, there is a group delay spread $\Delta\tau$ due to the different travel speeds of the various spectral components in the optical signal source, where $\Delta\tau$ is given by

$$\Delta\tau = M(\lambda)\Delta\lambda_s L \tag{10.14}$$

where $M$ is a more general chromatic dispersion parameter (usually in ps/nm-km), including both the material dispersion and the waveguide dispersion, discussed later; $\Delta\lambda_s$ is the rms spectral width of the optical signal (in nanometers), and $L$ is the single-mode fiber length (in kilometers). Therefore, it is desirable to have a small-spectral-width optical source such as a narrow-spectral-width laser diode operating near the minimum chromatic dispersion wavelength region ($M$ small) to have the minimum dispersion limitation and higher bandwidth transmission. For example, with a conventional single-mode fiber designed for 1300-nm operation, the chromatic dispersion is typically less than 2 ps/nm-km in the spectral range of 1280–1330 nm. On the other hand, the dispersion is as large as 110 ps/nm-km near 800 nm and 20 ps/nm-km near 1550 nm, limiting the transmission bandwidth distance to a value much less than that achievable near the minimum dispersion region.

Since silica glass fibers have lowest transmission loss in the 1550-nm spectral region rather than in the 1300-nm region, it is desirable to have a fiber with minimum dispersion also in the 1550-nm region. This requires the ideas and techniques for dispersion shifting in single-mode fiber design by controlling the waveguide dispersion. In optical fibers, the wavelength dependence of the propagation constant within an individual mode gives rise to a small spectral-dependent delay difference; this is called waveguide dispersion. Since a single-mode glass fiber is a glass fiber waveguide, the material dispersion effect and the waveguide dispersion effect both need to be considered. The sum is called total chromatic dispersion, or simply, the chromatic dispersion of the single-mode fiber, with the same parameter symbol $M(\lambda)$ used to designate the chromatic dispersion. Waveguide dispersion is usually small compared with material dispersion except in the region of zero material dispersion. In the region near zero material dispersion, the wavelength dependence is such that waveguide dispersion can be used to modify the resultant total dispersion characteristics, such as shifting the zero crossing point to a new wavelength.

The dispersion-shifted single-mode fiber was first realized by

1. using $GeO_2$ doping to increase the material dispersion minimum wavelength and
2. tailoring the waveguide dispersion in a step index single-mode fiber.

The latter was achieved by operating the single-mode fiber at a small $V$-number, achieved by designing a smaller core diameter with a larger index difference for the single-mode fiber. The combination of a small $V$-number and a larger index difference led to the desired waveguide dispersion characteristics for dispersion shifting [10]. In this first experimental demonstration, the dispersion minimum was successfully shifted to the desired 1550-nm region. However, the increased germanium doping caused a higher Rayleigh scattering and hence a higher transmission loss.

Subsequent research and development efforts based on the same idea of waveguide dispersion tailoring have come up with the triangular-shaped index profile single-mode fibers [11–13] that achieved dispersion shifting to the 1550-nm region with a smaller doping requirement (thus keeping the loss lower than if the step index profile is used). Presently commercially available dispersion-shifted single-mode fibers are based on the triangular index profile design and its variations.

To distinguish between the conventional single-mode fibers (minimum dispersion near 1300 nm) and the dispersion-shifted single-mode fibers (minimum dispersion near 1550 nm), we shall use C-SMF to designate the former and DS-SMF for the latter in the rest of this chapter.

It should be pointed out that whether it is C-SMF or DS-SMF, operating at the "zero"-dispersion wavelength does not mean there is no dispersion effect. Theoretical analyses [14–16] have shown that at the minimum dispersion wavelength, only the first-order dispersion vanishes, but the higher-order dispersions may still cause pulse distortions that may lead to intersymbol interference. This represents the ultimate dispersion limitations of single-mode fibers (if there is no polarization dispersion [16]). For the majority of present applications, however, this is of little concern, because these limitations are significant only at 10–100 Gb/s bit rates and over substantial transmission distances [16].

Figure 10.9 shows the chromatic dispersion characteristics of the two types of single-mode fibers: C-SMF and DS-SMF. Note that in addition to the different zero crossing wavelengths, the slopes of the chromatic dispersion curves near the zero crossing point $[S(\lambda_0)]$ are also different; while the dispersion slopes for both types of fibers are less than 0.1 ps/km-nm$^2$, the DS-SMF has a more gradual slope, which is advantageous for multiwavelength WDM applications. Near the minimum dispersion wavelength region the single-mode fiber bandwidth-distance product is $\geqslant 1000$ GHz-km [16]. This is a practically unlimited bandwidth, considering that the wide low loss window allows multiwavelength WDM operation. This is in sharp contrast to the $<10$ GHz-km bandwidth of multimode fibers.

In terms of practical applications, it should be mentioned that the C-SMF fibers have been widely used with 1300-nm laser diode optical transmitters in the existing digital telephone trunk and interoffice transmission network, typically carrying 140 Mb/s, 565 Mb/s, and 1.12 Gb/s traffic over 10–40-km distances. In contrast, DS-SMF fibers, which have become commercially available only in the past few years and typically require 1550-nm laser sources, have only begun to be used in a few long-distance transmission systems. In a later section the different system implementation strategies, the device and fiber requirements, and potential applications will be discussed.

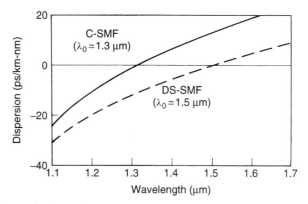

**Figure 10.9**   Chromatic dispersion as a function of wavelength for the C-SMF (conventional single-mode fiber, $\lambda_0 = 1.3$ μm) and the DS-SMF (dispersion-shifted single-mode fiber, $\lambda_0 = 1.5$ μm).

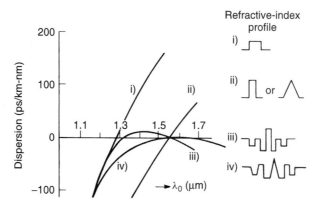

**Figure 10.10**   Dispersion versus wavelength of four types of single-mode fibers with different refractive index profiles: (i) conventional step index; (ii) dispersion-shifted, step index (small core, high index step) or triangular profile; (iii, iv) dispersion-flattened or broadband low-dispersion fibers, with multiple-cladding layers.

In addition to the conventional step index single-mode fiber and the triangular profile dispersion-shifted single-mode fiber, there are also single-mode fibers with various different refractive index profiles in both the core and the cladding regions that help achieve certain desired propagation characteristics. There are a large number of single-mode fiber designs using various index profiles and dopants. A few examples are shown in Fig. 10.10. These include W-profile fibers and doubly clad (DC) and quadruply clad (QC) fibers designed to tailor the waveguide dispersion characteristics such that the total chromatic dispersion either has two zero crossings (two dispersion minima) or has low dispersions (e.g., less than ±2 ps/nm-km) over a broad spectral range (e.g., from 1200 to 1600 nm). These are called *broadband low dispersion* or *dispersion-flattened* single-mode fibers [17]. They are still in the research and exploratory development stage; if these broadband low-dispersion fibers can be made inexpensively with low loss and low dispersion over the 1200–1600-nm spectral range, they would be ideal for broadband WDM applications. Economics and future system needs will determine the practical applications of such advanced single-mode fibers.

### 10.2.6  Spot Size, Mode Field Diameter, and Gaussian Beam in Single-Mode Fibers

One important parameter of single-mode optical fiber waveguides is the mode field diameter, also known as spot size, of the optical field propagating in the single-mode optical fiber waveguide as the fundamental mode (also called $LP_{01}$ mode, see Refs. 2–6). It has been shown [18] that the optical field's intensity distribution in the single-mode optical fiber waveguide can be closely approximated, in many cases except for very small $V$-numbers or unusual index profiles, by a Gaussian intensity distribution:

$$P(r) = P(0)e^{-2(r/\omega_0)^2} \tag{10.15}$$

where the intensity of light is at its maximum at the core center ($r = 0$) and drops to 13.5% of its maximum (i.e., the $1/e^2$ point) at $r = \omega_0$. Here, $\omega_0$ is the mode field radius and $2\omega_0$ is the mode field diameter, or the spot size of the Gaussian intensity distribution. In general, the mode field diameter is not the same as the single-mode fiber core diameter $2a$. Typically for step index single-mode fibers the mode field diameter $2\omega_0$ is between 1 and 1.5 times the fiber core diameter $2a$ [18], depending on the ratio of operating (source) wavelength $\lambda$ to the cutoff wavelength $\lambda_c$.

It is important to note that the optical field in a single-mode fiber waveguide extends well into the cladding region and is not confined to the core region. This of course points out the limitation of the simple ray picture of waveguiding by total internal reflection at the core–cladding boundary as described in Section 10.2.2. Maxwell equations for the electromagnetic waves propagating in a cylindrically symmetrical single-mode optical fiber waveguide have to be solved to obtain the mode field distributions in the core and the cladding regions [18]. Such solutions show that the optical energy of the propagating mode in the single-mode optical fiber that is contained in the core region is only about 80% when the $V$-number [from Eq. (10.7) the $V$-number is a normalized wavelength] is at 2.405, that is, right at the single-mode fiber cutoff condition ($\lambda = \lambda_c$). Figure 10.11 shows the optical power contained in the core of a single-mode optical fiber for different $\lambda/\lambda_c$ or $V$-numbers. The power in the core decreases with a decreasing $V$-number or an increasing ratio of $\lambda/\lambda_c$. When a high percentage of optical energy is propagating in the core region of the waveguide, it is a stronger guiding case. On the other hand, if a high percentage of the energy of the propagating optical wave extends well into the cladding, it is a weaker guiding case. This happens when the optical source wavelength is at such a long wavelength that the corresponding $V$-number is too small. For example, when $V = 1.2$ from Eq. (10.7) [corresponding to $\lambda = 2\lambda_c$ from Eq. (10.8)], it can be shown that the mode field diameter $2\omega_0$ is about 3 times the fiber core diameter $2a$ [18], which means there is weaker guiding with a substantial amount of energy ($>60\%$) in the cladding. In such a case the fiber is more sensitive to bending- and microbending-induced losses [18]. Typically, it is desirable to operate with a $V$-number in the range of 1.8–2.4 for a given single-mode fiber. Equivalently, this corresponds to having the optical source wavelength between $\lambda_c$ and $1.33\lambda_c$ (cutoff wavelength) to reduce sensitivity to bending and microbending

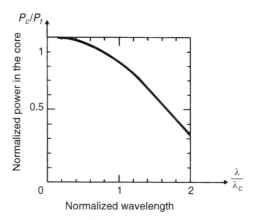

**Figure 10.11** The fundamental mode ($LP_{01}$ mode) power contained in the core, $P_c$, normalized to the total guided power, $P_t$, as a function of normalized wavelength $\lambda/\lambda_c$ ($\lambda_c$ is the cutoff wavelength of the single-mode fiber): $\lambda/\lambda_c = 1$ corresponds to $V = 2.4$; $\lambda/\lambda_c < 1$ corresponds to $V > 2.4$, when the higher-order mode can be guided.

while maintaining a reasonable mode field diameter or spot for ease of fiber coupling and jointing.

### 10.2.7 Optical Fiber Jointing: Splices and Connectors

In optical fiber fabrication, glass preforms are first made by proper deposition of core and cladding materials (host and dopants), with the appropriate refractive index profiles and core–cladding diameter ratios. The preforms are then heated and drawn into thin optical fibers and coated with appropriate coatings in a tall drawing tower. Typical practical fiber length is in the range of 2–10 km, although much longer fibers can be drawn from large preforms. In practical applications, it is often necessary to have the ability to splice and connect the fibers so that different fiber lengths can be jointed together permanently or disconnected and reconnected as needed.

There are several types of splices and splicing techniques [2–5] for permanent or semipermanent fiber jointing as needed in initial system installation or subsequent fiber repair. There are fusion splices and mechanically bonded splices [5]. Fusion splicing machines with fully automated active optical alignment and splicing and with TV camera monitoring of the splicing are available commercially. Fusion splicing techniques use arc fusion or flame fusion of the glass materials to achieve low-loss and high-strength permanent joints. Mechanical splices are also nearly permanent joints except they use bonding materials such as epoxy and index gels and precision mechanical alignment fixtures such as silicon V-grooves for alignment. Both single-fiber and multifiber (such as in a fiber ribbon or array) fusion splicing and mechanical splices are available. Note good cleaved fiber ends are needed for low-loss splicing.

Connectors are intended for connecting and disconnecting the fiber links at the connecting points for system installation, rerouting, and service and maintenance purposes, which may or may not be frequent, although repeated connections are expected. There are many types of optical connectors, but basically they are either fiber butt-joint connectors or lensed expanded-beam connectors [3, 4]. In butt-joint connectors, the fibers are aligned with each other and the fiber ends are butt jointed; the ends may or may not be in physical contact. Biconic connectors and ferrule connectors (FCs) belong to this category and require precision mechanical alignment of the butted fibers. In lensed expanded-beam connectors, collimating lenses are attached to the fiber ends, and the connectors are actually a high-precision imaging system plus stable mechanical fixtures for the optics. The optical fiber alignment tolerances are increased by the magnification of the optical system. Due to the nature of the collimated beam, lensed expanded-beam connectors have several unique features and are useful for device coupling applications such as in-line filters and isolators. However, these connectors are more complicated in optical design.

While insertion loss may be the most important parameter for optical jointing such as splices and connectors, optical reflection at the joint interfaces is becoming more and more of great concern. It is now recognized that even a low level of reflection can induce many undesirable effects in a high-speed optical fiber transmission systems (to be discussed in a later section); for future coherent transmission systems the reflection effect is even more severe. To avoid the reflection-induced noises and optical fiber transmission system penalties, low-loss fusion splices, angled mechanical splices, and low-reflection connectors should be used whenever possible. For optical connectors, several commercially available single-mode optical fiber connectors are either polished to have a convex fiber end surface to guarantee physical contact (such as in FC–PC connectors, where PC stands for physical contact) or polished to have an angle ($5°–12°$) so that reflection is out of the fiber core direction. In lensed expanded-beam connectors, good AR (antireflection) coatings on the lenses are needed to achieve low reflection. These special connectors have low reflections, with a return loss of 30 dB (less than 0.1% reflection) or higher (a higher return loss corresponding to a lower reflection level). This level of return loss (30–40 dB) is found to be desirable for advanced optical fiber transmission systems.

With good fusion splices between identical single-mode fibers (fibers from the same manufacturer and with the same glass composition and fiber parameters), the splice loss is typically less than 0.1 dB. Good mechanical splices have losses in the range of 0.1–0.2 dB. For single-mode fibers of different design or manufacture, the key parameter is the mode field diameter, not the core diameter. Additional splice loss will be incurred if the mode field diameters are different; for example, if one fiber has a mode field diameter of 10 μm while the other has a mode field diameter of 8.7 μm, the additional loss due to mode field mismatch is 0.1 dB. Therefore, it is important to have mode field diameter match for very low loss jointing.

With splices, "splice organizers" are typically used, in which the extra lengths of fibers are coiled and "organized" and properly protected. It is important to use a large enough coil radius (or diameter) for conventional single-mode fibers (with cut-off wavelength and mode field diameter designed primarily for operation at 1300 nm wavelength) to minimize possible bending losses in the 1550-nm-wavelength region if upgrading to include 1550-nm operation is contemplated.

In general, in addition to mode field diameter mismatch between fibers, other common causes of fiber joint losses are due to mechanical misalignments: fiber angular tilt angle $\theta$, axial or lateral offset $d$, and longitudinal end separation $s$ [3, 4, 18]. Figure 10.12 illustrates the different types of fiber end misalignments and effect of lateral and longitudinal misalignments on the fiber jointing loss. For single-mode fibers, the jointing loss is not a very sensitive function of end separation $s$ but increases quadratically with the angular tilt angle $\theta$ and lateral offset $d$ [18]. In practice, the

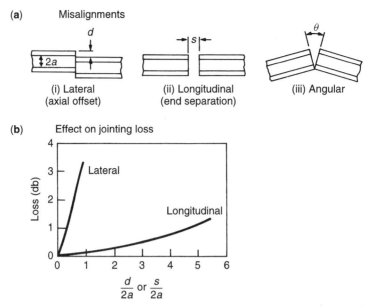

**Figure 10.12**   **(a)** Mechanical misalignments at a fiber-to-fiber joint (connector or splice) due to lateral, longitudinal, or axial offset. **(b)** Curves showing the sensitive effect of axial offset or end separation on the fiber joint loss.

state of the art in fiber jointing technology today is such that single-mode fiber fusion splice loss of $\leqslant 0.1$ dB, mechanical splice loss of $0.1-0.2$ dB, and single-mode fiber connector loss of $0.1-0.3$ dB can all be readily obtained with properly prepared splices and properly designed and assembled connectors.

### 10.2.8   Other Passive Optical Components

In addition to optical fiber splicers and connectors, there are other passive optical components, such as optical attenuators, optical fiber couplers for power mixing and splitting, optical filtering components for spectral separation and combination, and optical fiber taps for power distribution or line monitoring. These can be used in a typical optical fiber transmission system to enhance the system flexibility. For example, optical attenuators are often used to reduce the received optical power to within the dynamic range of the receiver (to avoid saturation and nonlinear distortion), because the optical transmitter–receiver pair could have been designed for a fiber link with a loss higher than the specific one in use. Optical fiber couplers can also be used as optical power splitters (e.g., 50–50% coupler or 10–90% power splitting for distribution bus). Also, as will be discussed in Section 10.3.5, WDM components can be used to multiplex two or more optical source wavelengths for transmission over a single fiber to achieve higher total transmission capacity over an existing fiber. Figure 10.13 illustrates a WDM component for multiplexing 1.3- and 1.55-$\mu$m laser diodes; the component is based on a simple fused biconical tapered fiber coupler (two fibers fused together sideways along the core region). The optical interference filters are added to improve the isolation between the two optical channels and therefore reduce the crosstalk in a WDM transmission system.

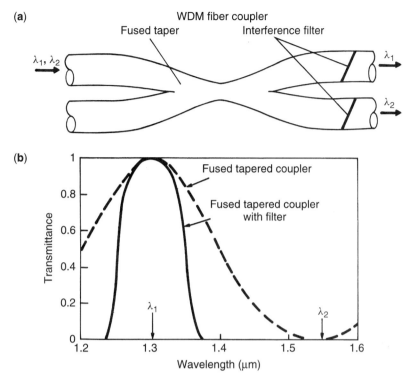

**Figure 10.13**   Example of a simple WDM device based on fused biconical tapered single-mode fiber coupler plus bandpass interference filters for multiplexing and demultiplexing of 1.3 and 1.55-$\mu$m spectral region laser sources onto the same single-mode fiber.

For all passive optical components used in optical fiber transmission systems, in addition to optical characteristics such as insertion loss, spectral dependence, reflection, and durability, there are other concerns, such as long-term stability in uncontrolled environments, cost of technology and installation, and ease of maintenance in mass system deployments; this is especially true for future subscriber loop applications.

## 10.3   OPTICAL FIBER TRANSMISSION TECHNOLOGY

### 10.3.1   Importance of Optical Transmission Technology in Telecommunications

Transmission systems based on optical fiber communication technology have revolutionized the field of communications where there is a change toward a totally digital network for a variety of new functions and features based on the concept of BISDN. The BISDN promises to provide services including high-speed data, low-speed telemetry, high-quality audio, high-definition TV distribution, and two-way video, in addition to the POTS (plain old telephone service) we have today on paired copper wires.

In the following sections we will discuss the basics of optical transmission technology including the choices of fibers, passive optical components, and active devices as well as some of the essential features and limitations of these transmission technologies.

## 10.3.2 Optical Fiber Transmission Systems

Figure 10.14 shows the schematic block diagram of an optical fiber transmission system. There are four important basic building blocks in a typical optical fiber transmission system:

1. transmission medium — the optical fiber/cable;

2. active optical device modules — the laser or LED optical transmitters and photodiode-based optical receivers;

3. passive optical components — connectors, splices, couplers, attenuators, WDM components, and so on; and

4. electronics — electronic multiplexers/demultiplexers, supervisory and maintenance circuits, and so on.

The electronic part of the system is not unique to fiber transmission technology. In contrast, the optoelectronic parts of the system are unique to optical fiber transmission technology. Since we have already discussed fiber characteristics and some of the passive components in the previous sections, the following sections will discuss active devices and the fiber/device choices for different system applications. Note that in an optical fiber transmission system, in general, the characteristics of the active optoelectronic devices and the transmission fibers together determine the most essential features of the particular system. However, the system flexibility is made possible only with the use of various passive interconnection and branching components such as splices, connectors, optical attenuators, fiber couplers, and WDM components.

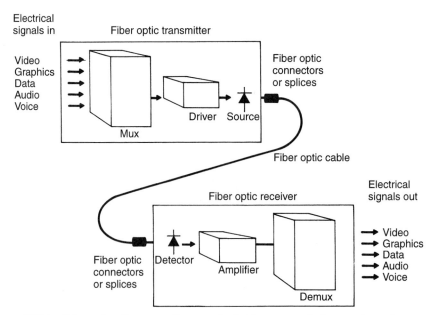

**Figure 10.14** Schematic diagram of an optical fiber transmission system showing the fiber-optical transmitters, receivers, and transmission fiber with fiber jointing components.

### 10.3.3  Multimode Optical Fiber Transmission Systems

Most of the earlier optical fiber transmission systems installed were based on multimode optical fiber technology using, first, short-wavelength (0.8-$\mu$m) AlGaAs semiconductor laser diodes and then, long-wavelength (1.3-$\mu$m) InGaAsP semiconductor laser diodes [19]. However, compared with single-mode optical fiber systems, multimode optical fiber systems have two distinct disadvantages, as already mentioned in Section 1.2.3:

1. limited bandwidth because modal dispersion is large and
2. modal noise, which causes significant system penalty.

Modal noise, due to time-varying modal-distribution-related interference effects, can cause serious system penalty. It is difficult to predict because it depends on the location and characteristics of the interconnections in the multimode fiber link and optical source coherence, requiring extra care in system design, deployment, and maintenance.

Mainly due to these factors, multimode optical fiber systems are typically limited to shorter-distance, lower-bit-rate optical transmission applications. In addition to more complex profile fabrication control and system design, multimode fibers have higher losses than single-mode fibers because of the increased Rayleigh scattering due to higher dopant concentration in the core. Thus, for long-haul transmissions, multimode fibers have distinct disadvantages both in loss and in bandwidth, although the coupling, splicing, and connector losses are lower. Early fiber transmission system installations (pre-1980) used multimode fibers because of the availability of multimode fiber transmission technology. Shortly after 1980, though, there was a clear trend toward using single-mode fiber systems in all long-haul applications as the technology moved toward higher bit rates and longer repeater spacing systems [20].

Nevertheless, there is a strong interest in multimode fiber-based, low-bit-rate, short-distance intrabuilding optical systems because modal dispersion and modal noise do not pose significant problems in these applications, in which the coupling and branching efficiency of large-core multimode fibers are desired. Together with inexpensive LEDs or low-cost lasers, multimode fiber-based systems are finding significant demands for high-speed data transmission in LAN applications.

### 10.3.4  Single-Mode Fiber Transmission Technology

Single-mode fiber transmission technology is clearly the choice for high-bit-rate, long-distance communication systems. Since low-loss optical fiber transmission windows are in the 1.3-$\mu$m region and the broad low-loss window is in the 1.5–1.6-$\mu$m spectral range, these two wavelength regions are used for long-distance single-mode fiber transmission. The minimum loss achievable is about 0.37 dB/km near 1.3 $\mu$m and 0.16 dB/km near 1.58 $\mu$m for high-quality silica glass single-mode optical fibers.

Assuming, for example, a loss budget of 30 dB (considering optical transmitter power, receiver sensitivity, and system margin) for the fiber link alone, the allowable transmission distance is about 80 km near 1.3 $\mu$m and 160 km near 1.56 $\mu$m. This illustrates the clear advantage of using the lowest-loss transmission window with 1.56-$\mu$m-wavelength laser sources in single-mode silica glass optical fibers.

***Loss-limited versus Dispersion-limited Transmission.*** As can be seen from Figure 10.10, the chromatic dispersion of a regular, conventional single-mode fiber

(C-SMF) in the lowest-loss 1.56-μm region is typically 17 ps/nm-km, or more than 10 times that in the 1.3-μm low-dispersion region. The intersymbol interference due to the fiber's chromatic dispersion can limit the high-bit-rate optical signal transmission to a distance much shorter than the loss-limited distance. Since the single-mode fiber's zero-dispersion wavelength is typically around 1.3 μm, optical sources operating in the 1.3-μm loss window region experience much smaller pulse broadening or other dispersion-induced penalty as compared with the 1.56-μm region operation. As a consequence, the dispersion-limited distance in 1.56-μm spectral region transmission is much shorter than that for the 1.3-μm spectral region.

To take advantage of the lowest fiber loss near 1.56 μm without introducing a large dispersion penalty, one can either reduce chromatic dispersion $M(\lambda)$ or reduce the optical spectral width $\Delta\lambda_s$ in Eq. (10.14) by

1. using a dispersion-shifted single-mode fiber (DS-SMF) with its zero-dispersion wavelength shifted to the 1.56-μm spectral region or

2. using a narrow-spectral-width (small $\Delta\lambda_s$) single-longitudinal-mode (SLM) distributed feedback (DFB) laser diode [21] as the optical source.

Either of these two approaches will reduce the chromatic dispersion-related transmission penalties. Figure 10.15 illustrates the spectral characteristics of a SLM DFB semiconductor laser diode as compared with the regular multi-longitudinal-mode (MLM) Fabry–Perot (FP) cavity laser diodes. The broad spectral width (large $\Delta\lambda_s$) of the MLM laser diodes contributes to a large dispersion penalty due to pulse broadening in the transmission fiber, as described by Eq. (10.14) and due to the MLM laser's mode partition noise (MPN), to be discussed later.

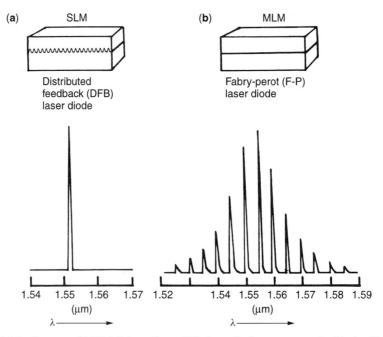

**Figure 10.15**   Spectra of MLM Fabry–Perot (FP) laser diode as compared with the SLM DFB structure laser diode (long-distance transmission).

*Optical Source–Fiber Combinations for Single-Mode Fiber Systems.* Table 10.2 shows some examples of possible source–fiber combinations using different sources (MLM FP laser diodes vs. SLM DFB laser diodes, 1.3 μm versus 1.56 μm wavelength) and fibers (regular 1.3-μm zero-dispersion fiber, C-SMF, versus 1.56-μm dispersion-shifted fiber, DS-SMF) for long-distance single-mode optical fiber transmission systems.

The majority of today's single-mode fiber transmission systems use C-SMF fiber (zero-dispersion, 1.31 μm) and FP-type MLM laser diodes operating near 1.3 μm. However, 1.3 μm SLM DFB lasers are beginning to be used in gigabit-per-second systems. These DFB laser-based single-mode fiber systems represent the first use of SLM (sometimes called single-frequency) laser diodes in commercial optical fiber communication applications. There are also WDM single-mode fiber systems using both the 1.3-μm FP MLM laser diodes and 1.56-μm SLM DFB laser diodes in conventional single-mode fibers (C-SMF) with the zero-dispersion wavelength at 1.3 μm.

The main use of 1.56-μm single-mode fiber systems with MLM laser diode sources are typically for long-repeater-spacing but lower-bit-rate (less than gigabit-per-second) applications. The long-repeater spacing is of particular interest for repeaterless, island-hopping, undersea optical fiber systems [22]. For the next generation of undersea optical fiber systems, 1.56-μm single-mode fiber transmission systems based on DS-SMF or SLM DFB lasers are being developed as such systems promise greatly reduced number of undersea repeaters required [22].

Obviously if one designs a system using *both* a 1.56-μm SLM DFB laser diode source *and* DS-SMF fibers with the minimum dispersion wavelength near 1.56 μm, the loss and dispersion limitations would be simultaneously minimized. The transmission distance/capacity will thus be maximized if we assume that the same optical source power and the receiver sensitivity are available at 1.56 μm as at 1.3 μm. Such DFB

**TABLE 10.2  Examples of Laser Source–Fiber Combinations for Long-Haul System Applications**

| Source Characteristics | | Typical Fiber Characteristics | | |
|---|---|---|---|---|
| Laser Type | Wavelength $\lambda_s$ (nm) | Fiber Type | Zero Dispersion | Loss at $\lambda_s$ (dB/km) |
| 1. MLM F-P LD | 1300 | C-SMF | 1310-nm region | 0.40 |
| 2. SLM DFB LD | 1300 | C-SMF | 1310-nm region | 0.40 |
| 3. SLM DFB-LD | 1560 | C-SMF | 1310-nm region | 0.21 |
| 4. MLM F-P LD | 1560 | C-SMF | 1310-nm region | 0.21 |
| 5. MLM F-P LD | 1560 | DS-SMF | 1550-nm region | 0.24 |
| 6. SLM DFB-LD | 1560 | DS-SMF | 1550-nm region | 0.24 |

*Note*: (i) Combinations 1 and 5 are multi-longitudinal-mode (MLM) lasers operating at the fibers' zero-dispersion wavelength; 2 and 6 are single-longitudinal-mode (SLM) lasers operating at the fibers' zero-dispersion wavelength; combinations 3 and 4 are SLM and MLM lasers, respectively, operating at a high-dispersion region.

(ii) Most of the existing single-mode fiber systems use combinations 1 and 2, at various bit rates, and may use 3 for upgrading these existing systems by WDM. Combinations 4 and 5 are used in much longer repeater spacing or repeaterless island-hopping applications, with 4 limited to low bit rate (e.g., less than 100 Mb/s). Combination 6 is being considered for the next generation of undersea single-mode fiber transmission systems.

laser/DS-SMF systems operating at 1.56 μm are just beginning to be considered for some special transmission applications [23].

Recently, terrestrial single-mode optical fiber transmission systems that operate above 1 Gb/s bit rate have been developed and have become available for commercial applications. Most of these commercial gigabit-per-second, single-mode transmission systems use SLM DFB lasers operating at 1300 nm, although 1300-nm MLM LD (laser diode) is also used in some cases. The use of MLM LD instead of SLM DFB LD for gigabit-per-second system transmission would require a very tight selection of multimode laser spectral width and center wavelength to match very closely the fiber's dispersion minimum, thus minimizing dispersion broadening and MPN. This could mean a less flexible system design or less margin for device degradation.

Beyond the now commercially available 1.12- and 1.7-Gb/s optical single-mode fiber transmission systems, recent advances of very high speed semiconductor laser diodes with a small-signal 3 dB bandwidth up to the 10–15-GHz range [24, 25] has made possible transmission experiments at high gigabit-per-second rates. Laboratory research demonstration of high-speed modulation and transmission experiments at 2.4, 8, 10, and 16 Gb/s [26–28] have been reported; very wide bandwidth (>20-GHz) *pin* photodetectors have also been developed [29]. Therefore, it seems that in the 1990s, multi-gigabit-per-second, high-speed optical transmission technology may become available for a variety of applications, including long-haul telecommunications and short-distance ultrahigh-speed data transmission between supercomputers.

### 10.3.5  Wavelength Division Multiplexing in Single-Mode Optical Fiber Systems

While the development of high-capacity single-mode optical fiber transmission systems continues to move in the direction of higher-speed, high-gigabit-per-second bit rates, to upgrade an existing transmission link capacity, one can either go to a higher transmission bit rate or use additional optical channels sharing the same optical fiber for increased bandwidth transmission. The optical WDM approach [30] uses several simultaneous optical channels transmitting through the same optical fiber. With WDM schemes, for an established route where fiber cable installation has been completed long ago, one can upgrade the system capacity by adding more optical channels without installing new fiber cable. It appears that upgrading by going to higher speeds or higher bit rates (i.e., time division multiplexing) is likely to be considered first, because it is a more straight-forward approach if the higher-speed system is already available. However, when the economic considerations are such that adding another optical channel at the same bit rate is better cost justified, or if the higher-bit-rate system equipments are simply not yet available (e.g., limited by the high-speed electronics required), then WDM may be deployed. This will depend on the relative maturity and cost of WDM technology versus the high-speed optoelectronics technology.

Figures 10.16**a** and 10.16**b** show two typical WDM system configurations. Figure 10.16**a** shows a typical applications, that is, two-channel transmission in one direction, in a single fiber. Figure 10.16**b** shows the case for bidirectional transmission in the same fiber with lasers at different wavelengths. Although only two optical channels (laser wavelengths) are shown in the figures, the number of channels can be as high as 10–20 depending on the channel spacing, the laser spectral properties, and the WDM device characteristics. High-density WDM will require SLM DFB laser diodes.

The technology of WDM devices has been the subject of research and development for years [30, 31]. Several multimode fiber transmission systems have used WDM

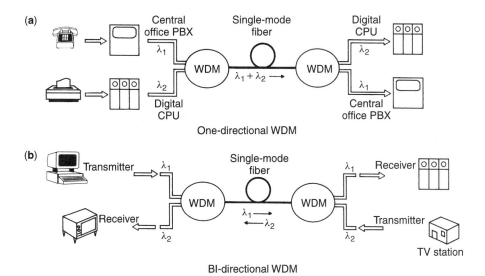

**Figure 10.16** The WDM for increasing the transmission capacity by using more than one optical wavelength channel: **(a)** one-directional WDM; **(b)** bidirectional WDM. Although only 2 optical channels are shown, in principle, 10–50 optical channels per fiber are possible.

technology to increase the total capacity, usually at low transmission bit rates. Single-mode WDM optical fiber transmission systems operating at two to four wavelengths in the 1.3 and 1.55-$\mu$m region have also been developed in the laboratory [31] for multichannel transmission at higher bit rates. There was no significant use of WDM technology until recently when multiplexing 1.3 and 1.55 $\mu$m in an existing 1.3-$\mu$m single-mode transmission link was deployed for upgrading.

Wavelength division multiplexing can be accomplished by using various dispersive optical components such as dielectric interference filters, fused biconical tapered fiber couplers with wavelength-dependent transmission (Fig. 10.13), gratings, and micro-optical components such as graded refractive index (GRIN) rod lenses with wavelength-dependent coatings [30, 31]. Important practical requirements of WDM components include low loss, low crosstalk between optical channels (different wavelength optical sources), low reflection, high stability under operating environmental conditions, small size, and low cost. Specific applications will determine the WDM technology of choice. For example, multiplexing 2 channels such as 1.3 and 1.56 $\mu$m could be accomplished readily with simple fused, biconical, tapered fiber couplers and optical dielectric interference filters. To multiplex 10–20 channels within either the 1.3- or the 1.56-$\mu$m spectral region, on the other hand, may require the use of a more bulky grating multiplexer/demultiplexer with GRIN lenses or equivalent collimating optics for input/output coupling of multi-wavelength channels unless a star coupler is used to mix the optical signal for distribution and tunable narrow-band optical filters are used before the optical detector in the receiver.

It is important to note that WDM technology offers not only the overall bandwidth/bit rate upgrading for long-haul optical fiber transmission systems but also the needed system/network architectural flexibility for future subscriber loop broadband distribution applications. A WDM system, with 4–16 optical channels, for example, allows different broadband services (digital or analog, high-bit-rate or medium-bandwidth, NTSC TV or high-definition TV) to be provided over different optical channels over the same fiber to the customer, depending on the service needs. The high-density multichannel WDM technology requires closely spaced optical channels (2–10 nm); this in turn requires the use of multiwavelength optical sources (such as DFB lasers at different wavelengths) and optical channel selection filters (such as fixed-wavelength optical filters, tunable optical etalons, or FP interferometers). There are currently significant research and development efforts in these areas, which will lead to the realization of multichannel high-density WDM technologies and their applications in future optical systems and networks.

## 10.4   TRANSMISSION LIMITATIONS

As discussed in the preceding, the most obvious transmission limitations are due to fiber transmission loss and fiber dispersion. However, in addition to transmission loss and dispersion of the fibers, in a single-mode fiber transmission system there are other important system degradation factors associated with the interaction of the opto-electronic devices and components with the transmission media that can cause system impairments. In this section we discuss these limitations and a few of the important factors — *mode partition noise* in MLM lasers, *chirping* in SLM lasers, *reflection-induced noise*, and so on — and the corresponding system penalties.

Figure 10.17 shows the general features of single-mode fiber transmission distance versus bit rate bounded by these limiting factors. The individual factors are discussed in the following sections.

### 10.4.1   Transmission Loss and Dispersion

Loss-limited system operation means that the power budgeting consideration limits the transmission distance. For example, if

$$P(\text{transmitter output power}) = P_t \ (\text{dBm})$$

$$P(\text{receiver sensitivity}) = P_r \ (\text{dBm})$$

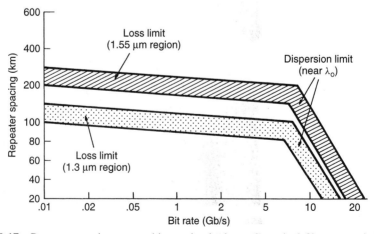

**Figure 10.17**   Repeater spacing versus bit rate in single-mode optical fiber transmission, with boundaries (limits) due to fiber loss and dispersion near $\lambda_0$.

and if there is no dispersion-related penalty or limit, then the maximum transmission distance is

$$L_m = \frac{(P_t - P_r)(\text{in dB})}{\text{average fiber link loss (in dB/km)}} \qquad (10.16)$$

The fiber link loss includes fiber loss, splice loss, and the loss due to connectors, WDM components, fiber couplers, and any other interconnecting components. For example, assuming we have a single-mode fiber system operating at 1300 nm with $P_t = 3$ dBm, $P_r = -34$ dBm, and an average fiber link loss of 0.41 dB/km (C-SMF at 1300 nm), the maximum loss-limited transmission fiber length is $L_m = 37/0.41$ km $= 90$ km. Here, the system margin is not considered.

Dispersion-limited transmission occurs when the power budget allows for transmission over a length $L_m$ but the achievable transmission distance is shorter than the $L_m$ from loss calculation alone because the dispersion-induced signal pulse broadening $\Delta\tau$ is causing significant intersymbol interference between transmitted neighboring pulses. This is usually the case with higher-bit-rate systems with a large overall dispersion. Significant intersymbol interference happens when the pulse broadening $\Delta\tau$ is larger than half of the bit period $T$:

$$\Delta\tau \geqslant \tfrac{1}{2}T \qquad (10.17)$$

where the pulse broadening $\Delta\tau$ is given by Eq. (10.14).

For example, suppose the loss-limited transmission distance is 90 km, as in the preceding example. Now suppose the particular system is to operate at 2 Gb/s (500 ps pulse period) using an MLM laser diode with an rms spectral width of 4 nm at a wavelength of 1280 nm (assuming the laser wavelength does not match the fiber's minimum dispersion wavelength exactly). Since the fiber dispersion $M$ is about 2 ps/nm-km, the dispersion limits the transmission distance to

$$L \cong \frac{500 \text{ ps}/2}{2 \text{ ps/nm} \times 4 \text{ nm}} \text{ km} = \frac{250}{8} \text{ km} = 31 \text{ km}$$

which of course is almost one-third of the 90-km length from the loss consideration alone. As a rule of thumb, loss-limited transmission is usually obtained at relatively low bit rates ($<150$ Mb/s) where the dispersion effects are negligible.

### 10.4.2 MLM Laser Mode Partition Noise

In FP cavity MLM laser diodes, there is a certain degree of power distribution fluctuation among the various longitudinal modes. From theoretical analysis, in MLM lasers the statistical nature of the buildup of the various longitudinal modes leads to a fluctuating power distribution among the modes from one optical pulse to the next in a modulating pulse train. This pulse-to-pulse variation of the optical power distribution means more power is in longitudinal mode $A$ for one pulse while for the next pulse there is more power in longitudinal mode $B$. When coupled with the chromatic dispersion of a long single-mode fiber, the pulse-to-pulse fluctuating optical power distribution will cause pulse-to-pulse signal waveform fluctuations because of the wavelength-dependent delays for different longitudinal modes. This in turn leads to a bit error rate "floor" [32] that is independent of optical power received. This is called mode partition noise, or MPN [32, 33].

If there were no mode partitioning, that is, if the longitudinal-mode power distribution profile were constant from pulse to pulse, then the system penalty would be due to the pulse broadening alone, the broadening being associated with the fixed (not pulse-to-pulse varying) spectral width and the fiber dispersion, as was calculated in the example in the last section. Serious MPN can dominate over the chromatic dispersion broadening in limiting the system transmission [32, 33]. Indeed, this has been the case in several system experiments using MLM laser diodes, where a floor in the bit error rate versus received optical power curve is observed. Figure 10.18 shows a schematic, qualitative diagram of such a bit error rate versus received optical power curve. The pulse broadening due to dispersion causes a dispersion penalty in terms of the power increase needed to get the same bit error rate, while the bit error rate floor is a floor in the bit error rate due to the MPN and is independent of the received optical power.

Because of the MPN limitations, for gigabit-per-second systems, MLM laser diodes should probably only be used near the minimum-dispersion region of the single-mode fiber, except for short-distance transmission. The SLM laser diodes are undoubtedly the primary optical sources of choice for gigabit-per-second long-haul single-mode fiber systems.

### 10.4.3 Chirping in SLM Lasers and System Implications

It has been recognized very early that the broadening of the individual longitudinal modes of a laser diode under fast modulation is due to the transient phenomenon of

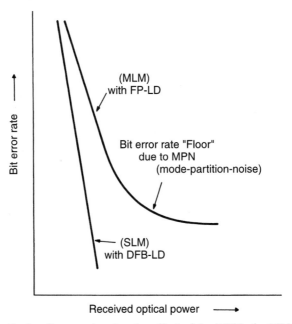

**Figure 10.18** Qualitative diagram showing the effect of the MPN of a MLM laser diode in a single-mode optical fiber transmission system. The bit error rate floor, which cannot be reduced by simply increasing optical signal power, limits the transmission bit error rate. No such floor due to MPN is observed in a good SLM DFB laser diode.

chirping [34, 35]. Chirping refers to a time-dependent frequency (or wavelength) shift during the transient turn on (and off) in a laser diode under fast excitation. The wavelength shift is caused by the time-varying refractive index excursion associated with the carrier density change, which is a natural consequence of nonconstant excitation, such as in short-pulse pumping or high-speed digital modulation. Chirping results in a dynamic line width broadening of the individual longitudinal modes, which is broadened from the continuous-wave (cw) line width of less than 0.001 nm to about 0.2 nm, depending on the biasing and modulation conditions. Since in direct-detection optical transmission systems the laser diode transmitters are directly current modulated, chirping and the resultant line broadening are commonly observed in both MLM and SLM laser diodes. However, in MLM laser diodes, the overall spectral envelope broadening is more important than the individual longitudinal mode broadening, so the latter effect (due to chirping) is usually neglected.

In good SLM lasers, since there is only one dominant longitudinal mode, chirp broadening becomes the most important factor other than the side-mode suppression ratio. Figure 10.19 shows an example of chirp-broadened DFB laser longitudinal mode under high-speed modulation; for comparison, the cw (no modulation) spectra is also shown. This chirped, dynamically broadened line width could impose a significant transmission limitation in multi-gigabit-per-second single-mode fiber transmission systems using SLM DFB laser diodes operating at a dispersive wavelength. If there were no side modes, the limitation is only due to the chirped width of the only longitudinal mode. For transmission near the zero-dispersion wavelength, there is negligible dispersion penalty. However, if the dispersion is, say, 18 ps/nm-km (e.g., with a 1560-nm DFB laser in a C-SMF), then 0.2 nm of chirped laser line width gives rise to 3.6 ps/km of pulse broadening, which for 100 km of low-loss C-SMF transmission amounts to a total pulse broadening of about 360 ps, limiting the transmission bit rate to about 2 Gb/s. So, indeed, this is a significant limitation. Chirp limitation has been observed in several gigabit-per-second system transmission

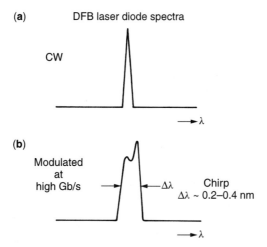

**Figure 10.19** Typical SLM DFB laser diode spectra (**a**) in cw operation and (**b**) under high-speed Gb/s modulation showing the dynamic line broadening due to chirping when modulated.

experiments where a high dc bias was sometimes needed to minimize the chirp-related penalty.

It is possible to reduce the chirp and the chirp-induced penalty in a dispersive transmission by using, for example, an external optical modulator to modulate the laser rather than using dc modulation. Perhaps the simplest solution is to operate the laser wavelength near the zero-dispersion region so that the chirped line width contributes very little to the pulse broadening. However, in a multiwavelength WDM system, there will be some SLM lasers operating in the dispersive regions of the spectrum, and it would be difficult to reduce the chirp penalty for all the laser wavelengths. A broadband low-dispersion single-mode fiber is useful in such applications.

In the meantime, chirping will continue to be an important limitation in SLM-DFB-LD-based gigabit-per-second C-SMF single-mode fiber transmission systems. For this reason, adjusting the laser transmitter bias and modulation conditions to optimize the overall system performance (which could correspond to reduced chirp, increased side-mode suppression, high speed, low pulse patterning, but low extinction ratio) is usually required as part of the system design and testing task.

### 10.4.4  Reflection-induced Noise and Penalty

Depending on the phase and magnitude of optical reflection back into a laser cavity, the laser output characteristics can be disturbed and modified. Reflection-induced noise in semiconductor laser diodes has received much attention, but by nature reflection effects on the laser oscillation behavior are complicated and difficult to quantify.

In direct-detection optical fiber transmission systems, reflection from, for example, the laser transmitter fiber pigtail end and the near end connectors has been observed to introduce noises and system penalties [36]. One can measure directly the induced system penalty for a given reflection feedback noise. Such measurement is most important in single-mode fiber systems using SLM laser diodes. The reflection effect on MLM laser diode characteristics may be just as significant, but the system consequence is less significant in MLM-LD-based systems unless the reflection introduces unpredictable mode jumps.

The experimental results of a 1-Gb/s NRZ (nonreturn to zero) transmission experiment with a 1550-nm DFB laser diode in a conventional single-mode fiber (C-SMF) with zero dispersion at 1310 nm showed that the measured power penalty increased with the increase in the reflection level [36]. Under the particular experimental conditions, a reflection level of $-19$ dB was found to cause a 1 dB penalty; a reflection level of $-10$ dB caused more than a 4-dB penalty. When the optical feedback was large, an increase in noise and jitter and degradation of response speed were observed in the received signal. The effect of the reflection was also dependent on the modulation condition; a larger modulation signal improved the tolerance to reflections.

In general, a reflection level of $-30$ dB (0.1% reflection) is probably satisfactory for most applications; in contrast, a reflection level of $-12$ dB (8% reflection typical of an air gap in noncontacting connectors or other components) or higher will probably always cause a significant system penalty. Figure 10.20 illustrates qualitatively the effect of different levels of reflections on the system bit error rate.

To make sure there is little reflection-induced noise, some recent gigabit-per-second systems use an optical isolator (based on Faraday rotation of the optical polarization in a magnetic field) within the SLM DFB laser package and use low-reflection physical

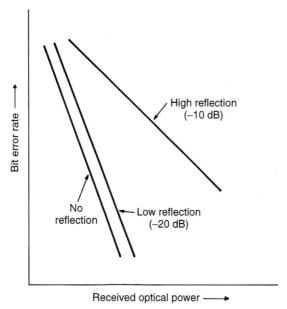

**Figure 10.20**   Bit error rate versus received optical power diagram showing qualitatively the effect of reflection level. High-level reflection induces power penalty.

contact connectors to minimize and prevent reflection penalties. Clearly as the high-speed systems get into high-gigabit-per-second ranges with the increased use of SLM DFB laser diodes, the issues of reflection tolerance in SLM lasers, optical isolator isolation requirement, low-reflection connectors, splices and other branching components, and the effect of multiple reflections will need to be addressed in system engineering.

## 10.5   FUTURE DIRECTIONS IN OPTICAL TRANSMISSION TECHNOLOGY

Since single-mode fibers have practically unlimited bandwidth [16] and very low losses, optical transmission system technology is really not limited by this excellent transmission medium. Rather, the single-mode fiber technology is relatively mature and waiting to be fully utilized. In contrast, the optoelectronics technology and the electronic integrated circuit (IC) technology, impressive as they are today, still cannot fully utilize the remarkable transmission medium — single-mode fiber — to its full potential. Future directions in optical transmission technology are discussed briefly.

### 10.5.1   Very High Speed and High-Density WDM Systems

The rapid progress in both the high-speed optoelectronic and IC technology have been very impressive. The speed limits of the laser diodes and photodetectors seem to be beyond at least 20 Gb/s; the current limitations seem to be mainly in the electronic IC technology. The VHSIC (very high speed IC) technology based on the III–V compound semiconductors and novel structures is under intense research and development and seems to hold great promise for the future. The wide low-loss window of single-mode fibers promises tens or hundreds of simultaneous optical channels, each capable of operating at 10–20 Gb/s bit rates, transmitting through the same fiber by large-channel-number, high-density WDM [37], constituting an ultrahigh-transmission capacity system.

### 10.5.2  Optoelectronic Integrated Circuit Technology

By combining various optical devices and electronic circuits on a single chip to perform complex functions on a monolithic chip or a hybrid chip, optoelectronic IC (OEIC) technology [38] promises to be the equivalent of microelectronic technology in the optoelectronic area in terms of potential applications and impacts on the optoelectronic information transmission and information storage/processing/handling applications. Because of the great importance of OEIC technology, there are significant research and development efforts in various research laboratories worldwide; it is almost certain that in time OEIC technology will be at the center stage of the optoelectronic technology.

### 10.5.3  Coherent Optical Transmission Technology

Based on coherent optical modulation and demodulation (optical heterodyne/homodyne) techniques, one can significantly increase the optical receiver sensitivity over the present direct-detection optical transmission technology as well as achieve very high density multichannel (50–1000 or more channels in principle) optical frequency division multiplexing (OFDM) transmission [39]. The practical deployment of such coherent optical transmission technology depends on future development of stabilized tunable narrowline-width single-frequency lasers for optical sources and local oscillators, polarization control techniques, coherent receiver designs, and so on. Coherent optical fiber transmission systems will most likely be used first in long-haul island-hopping applications; for multichannel OFDM applications the successful deployment may have to depend on the development of OEIC coherent transmitters and receivers.

## 10.6  FUTURE APPLICATIONS: FIBER TO THE HOME AND INTELLIGENT BUILDINGS

So far we have described the important essential aspects of the single-mode fiber transmission technology for actual system applications. In the past 10 years or so, the emphasis was mainly on deploying the single-mode optical fiber transmission systems for the digital long-haul transmission network and on multimode optical fiber transmission technology for business or campus environment local area computer data network.

In the next decade and beyond, the new wave of optical fiber transmission applications will be in the subscriber loop part of the network. This includes both the "loop feeder" (typically 10 km or shorter) and the "distribution" portions of the subscriber network. The distribution portion of the subscriber loop is typically between 1 and 3 km long, with the majority of distribution distance being less than 2 km. Thus, the fiber loss and dispersion considerations take on a different significance; in fact, other

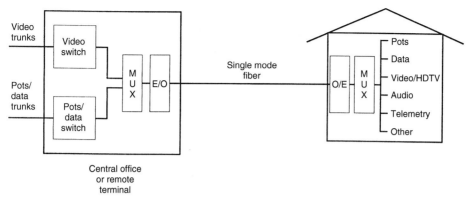

**Figure 10.21** Single-mode optical fiber transmission to the subscriber's home providing the narrow-band and broadband information services.

than the consideration that single-mode fibers should be used for once-for-all installation and nearly unlimited future upgrading, the fiber loss and dispersion are not as important as the network architectural considerations, coupling and distribution losses, maintenance and service concerns, and so on.

There are a number of studies and field trials being conducted worldwide [40, 41] on fiber-to-the-home and fiber-to-the-curb systems and networks to determine the service, architecture, economics, and maintenance/installation issues as well as evolution strategies for deploying the optical fiber transmission technology in subscriber loops to homes and intelligent buildings.

The narrow-band and broadband information could be transmitted to the subscriber's home using various possible architectures and strategies. Figure 10.21 [40] shows a simplified schematic of a single-mode fiber-based transmission approach in which video information (e.g., 600 Mb/s), POTS, and data at a lower bit rate are electronically multiplexed for downstream transmission over a single fiber to the home. Certainly many other versions are possible; for example, two fibers per subscriber instead of one fiber per subscriber, WDM instead of electronic multiplexing, or broadcast video in addition to switched video distribution.

A number of technological, architectural, economic, and legal issues need to be addressed and resolved before single-mode fiber transmission to the subscriber's home can be widely deployed. Nevertheless, with progress in these fiber-to-the-home field trials and future progress in associated optoelectronic and electronic technologies, a broadband integrated services digital network (BISDN) can be envisioned (noting that analog video distribution can be mixed with digital transmission). Such a BISDN, providing a variety of services including POTS, telemetry and remote metering and energy control, high-speed data, high-quality audio and TV distribution, a video library and video-on-demand including HDTV technology, and so on, will probably be realized eventually. This BISDN-based broadband optical network will then be at the center stage of the new information society, which may have significant societal implications. All this is made possible by the dramatic advances in the past 15 years of optoelectronics and optical fiber transmission technology.

## REFERENCES

1. S. D. Personick, *Fiber Optics: Technology and Applications*, Plenum, New York, 1985.
2. J. Gowar, *Optical Communication Systems*, Prentice-Hall, Englewood Cliffs, NJ, 1984.
3. G. Keiser, *Optical Fiber Communications*, McGraw-Hill, New York, 1983.
4. J. Senior, *Optical Fiber Communications: Principles and Practice*, Prentice-Hall, Englewood Cliffs, NJ, 1985.
5. See, e.g., papers published in *IEEE/OSA Journal of Lightwave Technology*, and *Technical Digests of Optical Fiber Communications (OFC) Conferences*.
6. M. J. Adams, *An Introduction to Optical Waveguides*, Wiley, New York, 1981.
7. L. G. Cohen, P. K. Kaiser, and Chinlon Lin, *Proc. IEEE*, **68**, p. 1203, 1980.
8. See, e.g., *Technical Digest, OFC'85*, pp. 92–96, San Diego, CA, 1985.
9. J. W. Fleming, *Electron. Lett.*, **14**, p. 326, 1978.
10. L. G. Cohen, Chinlon Lin, and W. G. French, *Electron. Lett.*, **15**, p. 334, 1979.
11. M. A. Saifi et al., *Opt. Lett.*, **7**, p. 43, 1982.
12. B. J. Ainslie et al., *Tech. Digest, 9th ECOC* (European Conference on Optical Communications), Geneva, p. 53, 1983.
13. C. M. Lemrow and V. A. Bhagavatula, *Laser Focus*, p. 82, March 1985.
14. D. N. Payne and W. A. Gambling, *Electron. Lett.*, **11**, p. 176, 1975; F. P. Kapron, *Electron. Lett.*, **13**, p. 96, 1977.
15. K. Jurgensen, *Appl. Opt.*, **18**, p. 1259, 1979.
16. D. Marcuse and Chinlon Lin, *IEEE J. Quantum Electron.*, **QE-17**, p. 869, 1981.
17. L. G. Cohen, W. L. Mammel, and S. J. Jang, *Electron. Lett.*, **18**, p. 1023, 1982.
18. L. B. Jeunhomme, *Single-Mode Fiber Optics: Principles and Applications*, Marcel Dekker, New York, 1983.
19. T. Li, *IEEE J. Select. Areas Commun.*, **SAC-1**, p. 356, 1983.
20. D. C. Gloge and K. Ogawa, *Tech. Digest, OFC'85*, p. 84, 1985.
21. K. Kobayashi and I. Mito, *IEEE J. Lightwave Tech.*, **LT-3**, p. 1202, 1985.
22. R. E. Wagner, *IEEE J. Lightwave Tech.*, **LT-2**, p. 1007, 1984.
23. L. C. Blank, L. Bickers, and S. D. Walker, *IEEE J. Lightwave Tech.*, **LT-3**, p. 1017, 1985.
24. C. B. Su et al., *Appl. Phys. Lett.*, **46**, p. 344, 1985.
25. J. E. Bowers et al., *Appl. Phys. Lett.*, **47**, p. 78, 1985.
26. A. H. Gnauck et al., *Tech. Digest, OFC'85*, Paper PD-2, 1985.
27. A. H. Gnauck et al., *Tech. Digest, OFC'86*, Paper PDP-9, 1986.
28. Chinlon Lin and J. E. Bowers, *Electron. Lett.*, **21**, p. 906, 1985; A. Gnauck and J. E. Bowers, *Electron. Lett.*, **23**, p. 801, 1987.
29. S. Y. Wang and D. M. Bloom, *Electron. Lett.*, **19**, 1983; J. E. Bowers and C. A. Burrus, *IEEE J. Lightwave Tech.*, **LT-5**, p. 1339, 1987.
30. G. Winzer, *IEEE J. Lightwave Tech.*, **LT-2**, p. 369, 1984.
31. H. Ishio, J. Minowa, and K. Nosu, *IEEE J. Lightwave Tech.*, **LT-2**, p. 448, 1984.
32. Y. Okano, K. Nakagawa, and T. Ito, *IEEE Trans. Commun.*, **COM-28**, p. 238, 1980.
33. K. Ogawa, *IEEE J. Quantum Electron.*, **QE-18**, p. 849, 1982.
34. Chinlon Lin, T. P. Lee, and C. A. Burrus, *Appl. Phys. Lett.*, **42**, p. 141, 1983.
35. S. Yamamoto et al., *IEEE J. Lightwave Tech.*, **LT-5**, p. 1518, 1987.

36. M. Shikada et al., *Tech. Digest, OFC'87*, Paper TuB4, p. 46, 1987.
37. S. Sasaki, H. Nakano, and M. Maeda, *Proc. ECOC'86*, 1986.
38. K. Nosu, H. Toba, and K. Iwashita, *IEEE J. Lightwave Tech.*, **LT-5**, p. 1301, 1987.
39. S. R. Forrest, *IEEE J. Lightwave Tech.*, **LT-3**, p. 1248, 1985.
40. P. Kaiser, *Tech. Digest, ECOC'85*, **II**, p. 125, 1985.
41. B. Catania, *IEEE J. Lightwave Tech.*, **LT-4**, p. 699, 1986.

# 11

## OPTICAL CHANNEL WAVEGUIDES AND WAVEGUIDE COUPLERS

TALAL K. FINDAKLY
*Research Division*
*Hoechst Celanese Corporation*
*Summit, New Jersey*

This chapter deals with the basic structures of integrated optical circuitry, namely, optical channel waveguides and waveguide couplers. The treatment is intended to provide aid and understanding of these components to the designer of such circuits. The properties, behavior, and varieties of planar and channel waveguide structures are presented first. This is followed by a treatment of directional couplers from the standpoint of design and operation for practical applications.

### 11.1 OPTICAL PLANAR WAVEGUIDES

Integrated optical waveguides are simply structures that confine and guide optical waves due to an induced refractive index increase in the guiding region with respect to the surrounding regions. Such waveguides are typically formed at or near the surface of the substrate material by a variety of fabrication techniques. Channel waveguides confine the light in three dimensions, two transverse and one longitudinal, in contrast with the more general form of planar waveguides in which the light is confined in two directions, one transverse and one longitudinal. Since channel waveguides are derived from planar waveguides by providing the extra transverse direction confinement, it is instructive to look first at planar waveguides in order to better understand channel waveguides.

A planar waveguide is typically a thin, flat layer whose refractive index is higher than the two regions that come into immediate contact with it. These regions typically comprise the substrate material and a cover layer, which is often air, but could be any layer of lower refractive index. A typical illustration of such a structure is shown in

*Handbook of Optical Components and Engineering*, Edited by Kai Chang
ISBN 0-471-39055-0 © 2003 John Wiley & Sons, Inc.

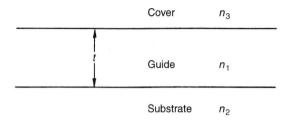

**Figure 11.1**  Planar optical waveguide.

Fig. 11.1. The index profile of planar guiding layers can be either uniform or graded, depending on the method used in forming such layers.

For the designer and maker of integrated optical waveguides, certain information is needed in order to utilize these waveguides for the application intended. Usually, this information includes the number of modes desired for a particular state of polarization, the amount of index change required for a given layer thickness and vice versa, and the effect of the cover layer on such parameters. This information is accurately obtained from solutions to the well-known Maxwell equation leading to a general dispersion equation.

The derivation of the characteristic equation for uniform planar waveguides is briefly given next for TE modes. For a waveguide structure comprising a film of thickness $t$ and refractive index $n_1$, a substrate of index $n_2$, and a cover region of index $n_3$, with $n_1 > n_2, n_3$, as shown in Fig. 11.1, the electric field components are oscillatory in the film and decaying outside of it, leading to solutions of the form

$$E_y = \begin{cases} Ae^{\gamma_3 x} & x \leqslant 0 \qquad\qquad (11.1) \\ Be^{ik_x x} + Ce^{-ik_x x} & 0 \leqslant x \leqslant t \qquad (11.2) \\ De^{-\gamma_2(x-t)} & x \geqslant t \qquad\qquad (11.3) \end{cases}$$

where $A$, $B$, $C$, $D$, $\gamma_2$, $\gamma_3$, and $k_x$ are constants to be determined. The boundary conditions on the electric field component $E_y$ require continuity at $x = 0$ and $x = t$, yielding

$$A = B + C \qquad\qquad (11.4)$$

$$D = Be^{ik_x t} + Ce^{-ik_x t} \qquad\qquad (11.5)$$

Furthermore, the magnetic field component $H_z$ obtained from the wave equation

$$\frac{\partial}{\partial x} E_y = i\omega\mu H_z \qquad\qquad (11.6)$$

must also be continuous at $x = 0$ and $x = t$, leading to

$$\gamma_2 A = ik_x(B - C) \qquad\qquad (11.7)$$

$$-\gamma_3 D = ik_x(Be^{ik_x t} - Ce^{-ik_x t}) \qquad\qquad (11.8)$$

Equations (11.4), (11.5), (11.7), and (11.8) can be used to solve for the characteristic equation by setting the determinate to zero:

$$
\begin{vmatrix}
-\gamma_2 & ik_x & -ik_x & 0 \\
-1 & 1 & 1 & 0 \\
0 & ik_x e^{ik_x t} & ikx e^{-ik_x t} & \gamma_3 \\
0 & e^{ik_x t} & e^{-ik_x t} & -1
\end{vmatrix} = 0
\tag{11.9}
$$

yielding the following characteristic equation:

$$
k_x t = \tan^{-1} \frac{\gamma_2}{k_x} + \tan^{-1} \frac{\gamma_3}{k_x} + m\pi
\tag{11.10}
$$

The eigenvalue constants $k_x$, $\gamma_1$, and $\gamma_2$ are obtained from the wave equation and are related to the propagation wave vector $\beta$ as follows:

$$
k_x = \sqrt{k_0^2 n_1^2 - \beta^2}
\tag{11.11}
$$

$$
\gamma_2 = \sqrt{\beta^2 - k_0^2 n_2^2}
\tag{11.12}
$$

$$
\gamma_3 = \sqrt{\beta^2 - k_0^2 n_3^2}
\tag{11.13}
$$

where $k_0 = 2\pi/\lambda$ and $\beta = (2\pi/\lambda)n_m$, with $n_m$ being the refractive index of the guided mode. In order to simplify the presentation of the dispersion properties of these waveguides, it is convenient to use two normalized parameters that relate to the physical parameters of the structure, namely, a normalized thickness and a normalized propagation wave vector or refractive index. Therefore, we define the normalized thickness as

$$
V = k_0 t \sqrt{n_1^2 - n_2^2}
\tag{11.14}
$$

and the normalized index of the guided mode as

$$
b = \frac{n_m^2 - n_2^2}{n_1^2 - n_2^2}
\tag{11.15}
$$

where $k_0$ is the wave number in free space, $t$ is the guiding layer thickness, $n_1$ is the guiding layer index, and $n_2$ is the substrate index. Since these structures might be asymmetric, that is, the indices of the substrate and cover are different, an asymmetry factor is introduced for the TE and TM modes as

$$
a_{\mathrm{TE}} = \frac{n_2^2 - n_3^2}{n_1^2 - n_2^2}
\tag{11.16}
$$

$$
a_{\mathrm{TM}} = \frac{(n_1^4/n_3^4)(n_2^2 - n_3^2)}{n_1^2 - n_2^2}
\tag{11.17}
$$

The eigenconstants $k_x$, $\gamma_2$, and $\gamma_3$ can now be redefined in terms of $V$, $b$, and $a$ as follows:

$$k_x = \frac{V}{t}\sqrt{1-b} \tag{11.18}$$

$$\gamma_2 = \frac{V}{t}\sqrt{b} \tag{11.19}$$

$$\gamma_3 = \frac{V}{t}\sqrt{b+a} \tag{11.20}$$

Following these normalized definitions, the dispersion equation (11.10) becomes

$$V\sqrt{1-b} = \tan^{-1}\sqrt{\frac{b}{1-b}} + \tan^{-1}\sqrt{\frac{b+a}{1-b}} + m\pi \tag{11.21}$$

From this equation, the cutoff frequency of the $m$th mode as well as the number of modes can be determined as follows:

$$V_m = V_0 + m\pi \tag{11.22}$$

where $V_0$ is the cutoff condition for the fundamental mode given by

$$V_0 = \tan^{-1}\sqrt{a} \tag{11.23}$$

For symmetric waveguides, $a = 0$, and therefore $V_0 = 0$. The solution to Eq. (11.21) is depicted graphically in Fig. 11.2**a**. From this figure, the desired point of operation (number of modes allowed) as well as the degree of optical confinement (especially in the case of single-mode operation) are chosen first. Knowledge of the $V$-number allows for the choice of either the thickness of the layer or the index difference ($\Delta n = n_1 - n_2$), from which the second parameter is easily calculable. The $b$ value also determines the effective index of the guided modes represented for small index difference approximately by

$$n_m = n_2 + b(n_1 - n_2) \tag{11.24}$$

Knowledge of the $V$ and $b$ parameters also allows for the calculation of the eigennumbers ($k_x$, $\gamma_2$, and $\gamma_3$) inside and outside the waveguide as given by Eqs. (11.12)–(11.20).

For guiding layers with nonuniform refractive index distribution, the dispersion relation as well as the field distributions can be obtained from solution to the wave equation with the substitution of the particular index distribution $n(x)$:

$$\frac{d^2 E_y}{dx^2} = [\beta^2 - n^2(x)k_0^2]E_y \tag{11.25}$$

for TE modes and

$$\frac{d^2 H_y}{dx^2} = [\beta^2 - n^2(x)k_0^2]H_y \tag{11.26}$$

for TM modes.

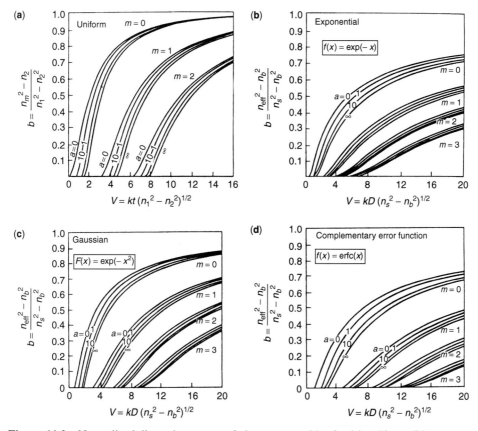

**Figure 11.2** Normalized dispersion curves of planar waveguides for (**a**) uniform, (**b**) exponential, (**c**) Gaussian, and (**d**) complementary error function index profiles.

Such solutions are usually difficult to obtain analytically for complicated index distributions but can be obtained by numerical methods. In practice, fabrication methods employing diffusion techniques lead to a slowly varying index distribution. The relationship between impurity concentration and refractive index distribution justifies the use of Gaussian and error function distributions for diffused waveguides as predicted from diffusion theory. This has been found to be close to actual distributions obtained in the fabrication of Ti:LiNbO$_3$ and glass waveguides, for example. Accordingly, it is useful to generate modal dispersion curves for such distributions. Figures 11.2**b**, **c**, and **d** correspond to normalized dispersion curves obtained numerically for graded index waveguides with exponential, Gaussian, and complementary error function distributions, respectively, represented by the following index distributions:

$$\text{Exponential:} \qquad n(x) = n_s + \Delta n \exp\left(-\frac{x}{D}\right) \qquad (11.27a)$$

$$\text{Gaussian:} \qquad n(x) = n_s + \Delta n \exp\left[-\left(\frac{x}{D}\right)^2\right] \qquad (11.27b)$$

$$\text{Error function:} \qquad n(x) = n_s + \Delta n \,\text{erfc}\,\frac{x}{D} \qquad (11.27c)$$

It can be noticed from Figs. 11.2**b**, **c**, and **d** that the asymmetry in the index profile (maximum at the superstrate interface and decaying to a steady state within the substrate) leads to a cutoff condition for the fundamental mode even when the symmetry parameter $a = 0$, in contrast with the uniform index distribution (Fig. 11.2**a**) where the fundamental mode has no cutoff at $a = 0$.

## 11.2    OPTICAL CHANNEL WAVEGUIDES

Two-dimensional optical confinement can be achieved in a variety of ways. In all cases, the refractive index within the channel waveguide must be larger than that in all the regions around its perimeter. The five basic structures of integrated optical channel waveguides are shown in Fig. 11.3. The strip-loaded waveguide shown in Fig. 11.3**a** consists of a planar film deposited on a substrate of lower index. The channel confinement is provided by depositing a narrow superstrate strip film whose index is higher than air but lower than that of the film. Due to this, the region in the film below the superstrate strip has a higher effective index than the side regions covered with air, and therefore light is confined underneath the strip. The ridge waveguide shown in Fig. 11.3**b** is simply a narrow film deposited on a substrate of lower refractive index, with air covering the top. Embedded channel waveguides are formed by diffusing impurities into a substrate such that the index in the diffused region is higher than the substrate, thus forming a channel guide bound by the substrate on three sides and

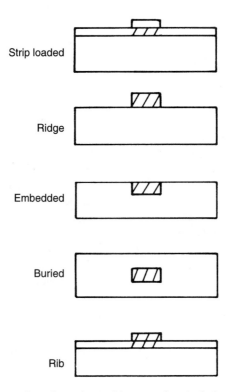

**Figure 11.3**    General configurations of integrated optical channel waveguides.

by air on the fourth, as shown in Fig. 11.3**c**. Buried waveguides are formed when the channel area of higher index is driven into the substrate and is therefore surrounded symmetrically by regions of the same refractive index, as shown in Fig. 11.3**d**. Finally, a rib waveguide may be formed by depositing a planar film layer of higher index than the substrate and then removing part of the film on both sides of a narrow channel, as shown in Fig. 11.3**e**, thus forming a waveguide underneath the rib area. Variations on these structures may be generated, for example, by using a cover layer instead of air.

The formation of such waveguide circuitry requires the use of photolithographic techniques for the definition of the channel. Fabrication techniques include deposition (thermal evaporation, electron beam evaporation, sputtering, spin coating, CVD deposition, etc.), epitaxial growth (sputtering, melting, liquid phase epitaxy, vapor phase epitaxy, metal organic chemical vapor deposition, molecular beam epitaxy, etc.), and modification (out-diffusion, in-diffusion, ion exchange, proton exchange, ion implantation, etc.).

The exact analysis of channel waveguides is more difficult than that of planar waveguides. This is primarily due to the increase in complexity in the boundary conditions where field matching around the waveguide perimeter has to be satisfied. Rigorous and approximation methods have been employed to determine the dispersion properties of rectangular channel waveguides. The more rigorous analysis employed circular harmonic field expansions [1], wave vector optimization for field match [2], and variational methods [3], all of which require computerized numerical computation since closed-form solutions are not easily obtained. The most accurate results have been reported using circular harmonic computer analysis [1]. This method is based on the expansion of the electromagnetic fields in terms of a series of circular harmonics in the form of Bessel functions multiplied by trigonometric functions where the electric and magnetic fields are matched at the boundaries, yielding a set of equations solvable by a computer. Approximation methods have also been introduced to reduce the computational complexities involved in the rigorous solutions. An approximate analytical method for well-guided structures where most of the energy is confined in the channel waveguide assumes oscillatory field solutions inside the channel and decaying fields outside of it both in the horizontal and vertical regions surrounding the channel with the decaying constants being independent of each other [4, 5, 6]. Accordingly, two characteristic equations are derived for each axis and are linked to each other by their relation to the propagation constant in the channel waveguide. The accuracy of these solutions is very good for waveguides well above cutoff. A second approximate but simple method known as the effective index method utilizes the division of the channel waveguide structure into three regions looking like a slab and sandwiched between two side regions. An effective index for each region is found using the simple planar waveguide analysis. From that, the step index planar waveguide treatment is applied to determine the wave vector of the new slab waveguide [6–8]. This method also yields accurate results for waveguides well above cutoff. The accuracy improves with larger width-to-depth aspect ratio. The simplicity and accuracy of this method makes it attractive for a quick prediction of the approximate dispersion properties of a variety of channel waveguide structures. Due to the usefulness of this method, an example is cited to illustrate the approach. Consider the ridge waveguide shown in Fig. 11.4, where the channel waveguide is composed of a raised ridge of width $W$, thickness $t$, and uniform index $n_1$ on a substrate of index $n_2$. The thickness of the

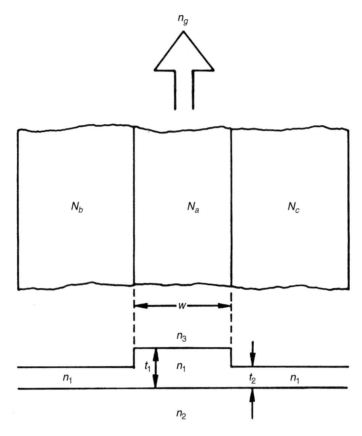

**Figure 11.4**  Equivalent index representation of channel waveguides.

film on both sides of the channel is $t_2$, and the structure is bound on top by a region of index $n_3$. The first step is to determine an effective index for the three regions $a$, $b$, and $c$ corresponding to the channel and its sideway surroundings, respectively, as though they were all planar regions. With the help of Fig. 11.2**a**, the $V$-numbers for the respective regions are first found:

$$V_a = k_0 t_i \sqrt{n_1^2 - n_2^2} \tag{11.28a}$$

$$V_b = V_c = k_0 t_2 \sqrt{n_2^2 - n_2^2} \tag{11.28b}$$

To find the effective indices in each region, the $b$-numbers are found from Fig. 11.2**a** for the appropriate asymmetry factor $a$, as defined in Eqs. (11.16) and (11.17). The corresponding effective indices are found from knowledge of the $b$-numbers, which for small $n_1 - n_2$ are approximately given by

$$N_a^2 = n_2^2 + b_a(n_1^2 - n_2^2) \tag{11.29}$$

$$N_{b,c}^2 = n_2^2 + b_{b,c}(n_1^2 - n_2^2) \tag{11.30}$$

The same exercise is applied to the new slab waveguide. The $V$-number of the channel waveguide is now given by

$$V_g = k_0 W \sqrt{N_a^2 - N_b^2} = k_0 W \sqrt{(n_1^2 - n_2^2)(b_a - b_b)} \qquad (11.31)$$

Due to the symmetry of this structure, $a = 0$, and hence the $b$-number of the waveguide $(b_g)$ is found from the $a = 0$ curve, yielding the effective index of the channel waveguide:

$$n_g = \sqrt{n_2^2 + b_g(b_a - b_b)(n_1^2 - n_2^2)} \qquad (11.32)$$

For a small index difference between the guide and the substrate $(\Delta n)$, Eq. (11.23) is approximated to

$$n_g \cong n_2[1 + b_g(b_a - b_b)\Delta n] \qquad (11.33)$$

While exact solutions to the dispersion relations of channel waveguides involve complicated boundary conditions that require lengthy numerical computations, it is possible to obtain approximate solutions in approximated closed forms by simplification of the boundary conditions. By assuming the waveguide to be of a rectangular geometry, as shown in Fig. 11.5, oscillatory field solutions are assumed in the waveguide region of index $n_1$ in both the $x$ and $y$ directions, and decaying solutions are assumed in the four regions abounding the waveguide on its four sides denoted by indices $n_2, n_3, n_4,$ and $n_5$. For well-guided waveguides, most of the power is confined to region 1, while a small portion propagates in regions 2–5, and even smaller power resides in the excluded areas bound by the corner intersections, thus yielding small error in boundary condition matching. By separation of variables, two characteristic equations are obtained in the $x$ and $y$ directions in much the same way as obtained in the slab waveguides as described earlier, yielding, for TE-like modes, denoted as $E_{MN}^y$, two equations similar to Eq. (11.10):

$$k_x W = \tan^{-1} \frac{\gamma_4}{k_x} + \tan^{-1} \frac{\gamma_5}{k_x} + M\pi \qquad (11.34)$$

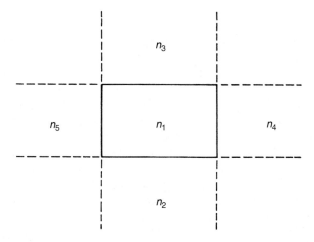

**Figure 11.5**  Cross section and boundaries of channel waveguides.

$$k_x t = \tan^{-1} \frac{\gamma_2}{k_x} + \tan^{-1} \frac{\gamma_3}{k_x} + N\pi \qquad (11.35)$$

with

$$\gamma_{4,5} = \left[ \left( \frac{V_{xi}}{W} \right)^2 - k_x^2 \right]^{1/2} \qquad (11.36)$$

$$\gamma_{2,3} = \left[ \left( \frac{V_{yi}}{t} \right)^2 - k_y^2 \right]^{1/2} \qquad (11.37)$$

$$V_x = \frac{2\pi}{\lambda} W \sqrt{n_1^2 - n_i^2} \qquad (11.38)$$

$$V_{yi} = \frac{2\pi}{\lambda} t \sqrt{n_1^2 - n_i^2} \qquad (11.39)$$

and

$$k_z^2 = k_1^2 - k_x^2 - k_y^2 \qquad (11.40)$$

For well-guided modes, $k_x$ and $k_y$ can be approximated by

$$k_x \cong \frac{m\pi}{W} \left( 1 + \frac{1}{V_{x4}} + \frac{1}{V_{x5}} \right)^{-1} \qquad (11.41)$$

$$k_y \cong \frac{n\pi}{t} \left[ 1 + \left( \frac{n_2}{n_1} \right)^2 \frac{1}{V_{yz}} + \left( \frac{n_3}{n_1} \right)^2 \frac{1}{V_{y3}} \right]^{-1} \qquad (11.42)$$

From the preceding, an approximate closed-form solution of the propagation number or the effective index of the guided mode is obtained as follows:

$$n_{\text{eff}} \cong \left\{ n_1^2 - \frac{M\lambda}{2W} \left( 1 + \frac{1}{V_{x4}} + \frac{1}{V_{x5}} \right)^{-2} - \frac{N\lambda}{2W} \left[ 1 + \frac{n_2}{n_1} \frac{1}{V_{y2}} + \left( \frac{n_3}{n_1} \right)^2 \frac{1}{V_{y3}} \right]^{-2} \right\}^{1/2}$$

A more exact solution is obtained by solving Eqs. (11.34) and (11.35) without using the approximate values for $k_x$ and $k_y$ given in Eqs. (11.41) and (11.42). Such solutions yield accurate results except near cutoff conditions, which are not of great interest in most cases.

Normalized dispersion curves for the various forms of channel waveguides shown in Fig. 11.3 are presented in Figs. 11.6–11.11. These charts were obtained by numerical solutions based on the various methods described earlier. Figure 11.6 shows the normalized waveguide index of the lowest-order mode for a strip-loaded waveguide (Fig. 11.3**a**) for a special case where $W = 2t$ as obtained by the effective index method. Figure 11.7 shows the normalized waveguide index of the ridge channel waveguide (Fig. 11.3**b**) for which $W = 2t$ and $W = 4t$ as obtained also by the effective index method. Figures 11.8 and 11.9 show the normalized waveguide index for embedded channel waveguides (Fig. 11.3**c**) for which $W = t, 2t, 4t$ as obtained by solutions to the transcendental Eqs. (11.34) and (11.35). Figure 11.10 shows the normalized waveguide

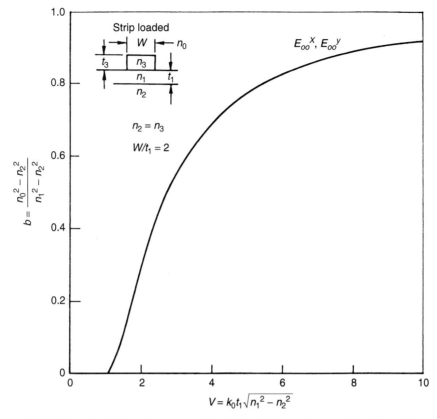

**Figure 11.6**   Normalized dispersion curves of strip-loaded channel waveguides. (From Ungar [2], © 1977, by permission of the Oxford University Press.)

index for buried channel waveguides (Fig. 11.3**d**) for which $W = 2t$ as obtained from exact solutions by the circular harmonic field expansion method. Figure 11.11 shows the normalized waveguide index of rib channel waveguides (Fig. 11.3**e**) for which the rib thickness is 70% of the film thickness and $W = 4t$ as obtained by the effective index method.

## 11.3   OPTICAL DIRECTIONAL COUPLERS

### 11.3.1   General Solution

Optical directional couplers consist of a set of closely spaced optical waveguides whose fields interact with each other by proximity coupling. Coupling is cumulative over a distance (known as the interaction length) along the propagation direction where the waveguides are in sufficient proximity to cause coupling. The coupling strength depends on parameters that can be categorized under three general conditions, namely, synchronism or wave number phase matching, optical confinement, and proximity. Synchronism of the wave vectors of the propagating optical waves in the coupled waveguide set is very important. The coupling strength is highest when the

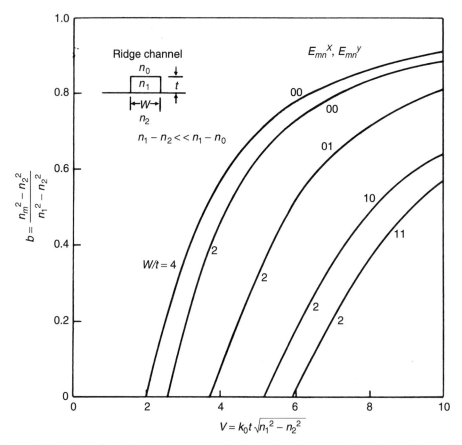

**Figure 11.7**  Normalized dispersion curves of ridge channel waveguides. (From Ungar [2], © 1977, by permission of the Oxford University Press.)

wave vectors of all the interacting waveguides are the same. Conversely, the coupling could be substantially reduced if sufficient imbalance between the wave vectors exists, regardless of proximity and optical confinement. Optical confinement is also important as it determines the extent of field overlap between the interacting waveguides. The optical confinement at a particular wavelength is controlled by the waveguide size (width and height) and index difference with its surrounding ($\Delta n$), which is the make-up of the numerical aperture of the waveguide. Smaller optical confinement (small $\Delta n$ and/or size) implies that a larger portion of the optical fields reside outside the wave-guide, decaying in the transverse plane, yielding greater overlap between the fields of the adjacent waveguides, which increases the coupling strength. Stronger optical confinement (large $\Delta n$ and/or size) yields shorter tails of the decaying fields outside the waveguide and therefore lower field overlap and coupling strength. Finally, the coupling strength is critically dependent on the spacing between the waveguide set. Smaller spacing yields higher coupling strength, and vice versa. In order to translate the preceding discussion into meaningful design parameters, a generalized formulation of the coupling problem is introduced next and then is narrowed down to practical cases of interest.

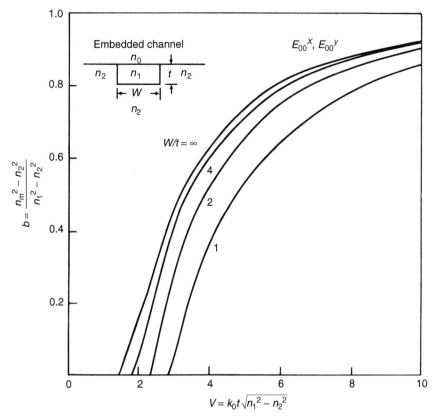

**Figure 11.8** Normalized dispersion curves of embedded channel waveguides. (From Ungar [2], © 1977, by permission of the Oxford University Press.)

Consider a set of $N$ coupled waveguides, as shown in Fig. 11.1. The optical fields in the various waveguides are represented by the following equations:

$$\frac{d}{dz}\begin{bmatrix} A_1(z) \\ A_2(z) \\ \vdots \\ A_n(z) \end{bmatrix} = -i \begin{bmatrix} \beta_1 & K_{12} & \ldots & K_{1n} \\ K_{21} & \beta_2 & & \\ \vdots & & K_{ij} & \vdots \\ K_{n1} & \ldots & \ldots & \beta_n \end{bmatrix} \begin{bmatrix} A_1(z) \\ A_2(z) \\ \vdots \\ A_n \end{bmatrix} \tag{11.43}$$

where

$$A_i(z) = a_i(z)e^{-i\beta \cdot z} \tag{11.44}$$

$\beta_i$ is the wave vector of the $i$th waveguide, and $K_{ij}$ is the coupling coefficient between the $i$th and $j$th waveguides. In most cases of practical interest, the waveguides are made identical, thus having the same $\beta$'s, so that the coupling strength is maximized. Furthermore, the interaction between the nonadjacent waveguides is much smaller than the interaction between the adjacent waveguides due to the substantial increase in spacing. Therefore, it can be assumed that

$$K_{ij} \cong 0 \quad \text{for} \quad |i - j| > 1 \tag{11.45}$$

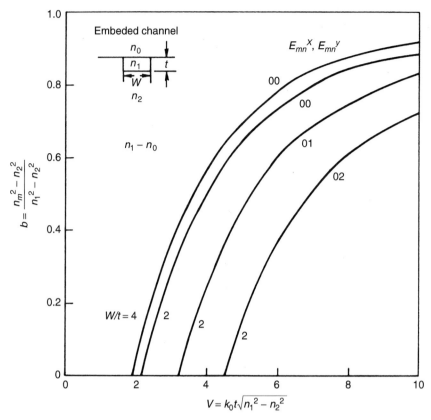

**Figure 11.9**   Normalized dispersion curves of embedded channel waveguides. (From Ungar [2], © 1977, by permission of the Oxford University Press.)

Finally, and assuming lossless conditions, the conservation of power, in normalized form, is defined as

$$\frac{d}{dz} \sum |Ai(z)|^2 = 1.0 \tag{11.46}$$

requiring

$$K_{ij} = K_{ji}^* = K \tag{11.47}$$

under these conditions, and for an equal and uniformly spaced set of waveguides, Eq. (11.1) can be rewritten as

$$\frac{d}{dz} \begin{bmatrix} A_1(z) \\ A_2(z) \\ \vdots \\ A_n(z) \end{bmatrix} = -i \begin{bmatrix} \beta & k & 0 & 0 \\ k & \beta & k & \vdots \\ 0 & k & \vdots & \vdots \\ \ddots & & & \beta \end{bmatrix} \begin{bmatrix} A_1(z) \\ A_2(z) \\ \vdots \\ A_n(z) \end{bmatrix} \tag{11.48}$$

Equation (11.48) can now be solved under the desired initial conditions to determine the power flow in the coupled waveguide set.

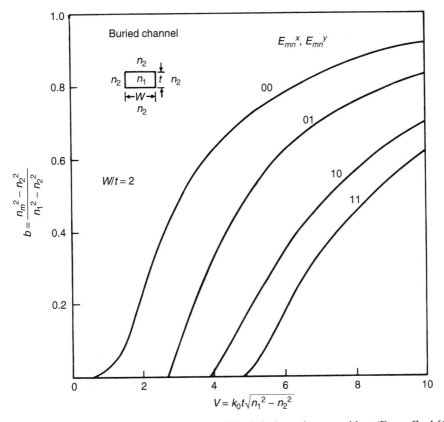

**Figure 11.10**  Normalized dispersion curves of buried channel waveguides. (From Goel [1]. Reprinted with permission. © 1969 AT&T.)

### 11.3.2  Waveguide Pair Directional Coupler

The two-channel directional coupler is an important element in integrated optical circuitry and has been extensively used to build passive couplers, $2 \times 2$ switches, and intensity modulators. For a two-waveguide system in which the waveguides are uniformly spaced throughout the interaction region, as shown in Fig. 11.12, Eq. (11.48), subject to conservation of power, yields the following solution:

$$\begin{bmatrix} a_1(z) \\ a_2(z) \end{bmatrix} = \begin{bmatrix} A & -iB \\ -iB^* & A^* \end{bmatrix} \begin{bmatrix} a_1(0) \\ a_2(0) \end{bmatrix} \tag{11.49}$$

where

$$A = \cos\theta + i\frac{\Delta\beta L}{\pi}\frac{\pi}{2}\frac{\sin\theta}{\theta} \tag{11.50}$$

$$B = \frac{\pi}{2}\frac{L}{l_c}\frac{\sin\theta}{\theta} \tag{11.51}$$

$$\theta = \frac{\pi}{2}\sqrt{\left(\frac{L}{l_c}\right)^2 + \left(\frac{\Delta\beta L}{\pi}\right)^2} \tag{11.52}$$

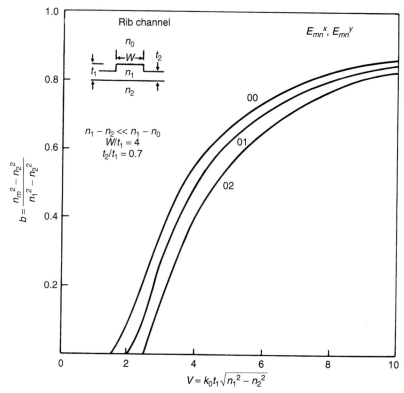

**Figure 11.11**  Normalized dispersion curves of rib channel waveguides. (From Ungar [2], © 1977, by permission of the Oxford University Press.)

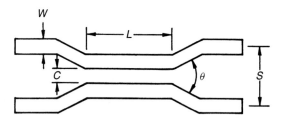

**Figure 11.12**  Integrated optical $2 \times 2$ parallel directional coupler.

$$l_c = \frac{\pi}{2K} \tag{11.53}$$

$$\Delta\beta = \beta_1 - \beta_2 \tag{11.54}$$

where $L$ is the interaction length and $l_c$ is the coupling length. For identical waveguides, $\Delta\beta = 0$, and therefore,

$$A = \cos(Kz) \tag{11.55}$$

$$B = \sin(Kz) \tag{11.56}$$

The amplitudes in the two waveguides in Eq. (2.49) become

$$a_1(z) = a_1(0)\cos(Kz) - ia_2(0)\sin(Kz) \tag{11.57}$$

$$a_2(z) = a_1(0)\sin(Kz) - ia_2(0)\cos(Kz) \tag{11.58}$$

The power flow in the coupled waveguides is therefore

$$|a_1(z)|^2 = |a_1(0)|^2\cos^2 Kz + |a_2(0)|^2\sin^2 Kz - \text{Im}[a_1(0)_{a_2}^*(0)]\sin^2 Kz \tag{11.59}$$

$$|a_2(z)|^2 = |a_1(0)|^2\sin^2 Kz + |a_2(0)|^2\cos^2 Kz - \text{Im}[a_1(0)_{a_2}^*(0)]\sin^2 Kz \tag{11.60}$$

Depending on the initial conditions [$a_1(0)$ and $a_2(0)$], the power in the two waveguides oscillates between maximum and minimum at intervals equal to the coupling length $l_c$, except when $a_1(0) = a_2(0)$, where no coupling or power transfer occurs. Under the condition when all the power is initially excited into one guide, say, for example, $|a_1(0)|^2 = 1.0$ and $|a_2(0)|^2 = 0.0$, the power flow becomes

$$|a_1(z)|^2 = \cos^2 Kz \tag{11.61}$$

$$|a_2(z)|^2 = \sin^2 Kz \tag{11.62}$$

The preceding treatment applies to parallel waveguide directional couplers in which the coupling is uniform throughout. When the coupling within the interaction region is not uniform, such as in nonparallel waveguides, the coupling coefficient is no longer constant with $z$. In this case, under lossless conditions Eq. (11.48) yields the following modified solutions for a two-identical-waveguide coupler:

$$a_1(z) = a_1(0)\cos\int K(z)\,dz - ia_2(0)\sin\int K(z)\,dz \tag{11.63}$$

$$a_2(z) = a_1(0)\sin\int K(z)\,dz - ia_2(0)\cos\int K(z)\,dz \tag{11.64}$$

The oscillatory behavior of the power exchange is now modified depending on the $K(z)$. In certain cases, the oscillatory behavior may have variable periodicity; in others it may not have any oscillatory behavior. The treatment of such cases is presented later in this section for special cases of practical interest.

The coupling coefficient $K$ is a measure of the field overlap between the coupled waveguides. In the case of two adjacent identical waveguides, the coupling coefficient is defined as

$$K = -\frac{iw}{4}\varepsilon_0 \iint_{\infty}^{\infty} a_1(x_1 y)a_2^*(x, y)\Delta\varepsilon_y(x, y)\,dx\,dy \tag{11.65}$$

where the relative dielectric constant difference is $\Delta\varepsilon_y = n_g^2 - n_s^2$ within the waveguide, and $\Delta\varepsilon_y = 0$ outside the waveguide, with $n_g$ and $n_s$ the refractive indices of the guide and its surroundings, respectively. The integration of Eq. (11.65) over the entire cross section of the two-waveguide coupler yields the following expression for the coupling coefficient:

$$K = \frac{2k_x^2 \gamma \exp(-\gamma x)}{\beta(W + 2/\gamma)(k_x^2 + \gamma^2)} \qquad (11.66)$$

where $\beta$ is the wave vector, $k_x$ and $\gamma$ are the eigenconstants inside and outside the waveguides, respectively, $W$ is the guide width, and $c$ is the inner-edge channel separation. By using the normalized slab waveguide parameter notation, the coupling length can be represented in a simple form by utilizing the notation used in Eqs. (11.18) and (11.19):

$$l_c = \frac{\pi^2 n W^2 (V\sqrt{b} + 2)}{2\lambda V^2 b(1 - b)} \exp\left(\frac{V}{W}\sqrt{bc}\right) \qquad (11.67)$$

where

$$V = k_0 W \sqrt{2n\,\Delta n} \qquad (11.68)$$

In most cases of practical interest, a well-confined single-mode optical waveguide is desired. Good single-mode optical confinement is usually achieved at $V \simeq 3$, such that the guide is well above cutoff yet adequately below the cutoff of the next higher-order mode so that any electro-optically induced increase in the index does not push the guide into a two-mode regime. An approximate design chart for Ti:LiNbO$_3$ parallel directional couplers is presented in Fig. 11.13 at various wavelengths for guide widths and $V$-numbers of practical interest. The effect of optical confinement is evident by the longer coupling lengths for higher $V$-numbers and by the longer wavelengths. The spacing is also critical due to the exponential dependence of the coupling factor on spacing. It should be noted that as the spacing is decreased to zero, the formulation used in deriving the coupling efficiency by the perturbation method (which is based on weak coupling) loses accuracy as the situation transforms from weak to strong coupling.

When the waveguide pair are not parallel, the coupling coefficient is no longer constant with distance, and the solution to the coupled-mode equations yield the solutions given in Eqs. (11.63) and (11.64). The argument representing the coupling coefficient–distance product is now an integral of the coupling coefficient over distance. For a linearly changing waveguide spacing under the initial condition where one input guide is excited, the optical powers in the two guides may be obtained by integration over a distance where the spacing is linearly changing with distance to obtain

$$P_1(\infty) = \cos^2 \frac{K_0}{\gamma \tan \alpha} \qquad (11.69)$$

$$P_2(\infty) = \sin^2 \frac{K_0}{\gamma \tan \alpha} \qquad (11.70)$$

where $\alpha$ is the angle between the two guides, $K_0$ is the coupling coefficient at the smallest separation between the two guides as defined in Eq. (11.66), and $\gamma$ is the decaying eigennumber outside the guides.

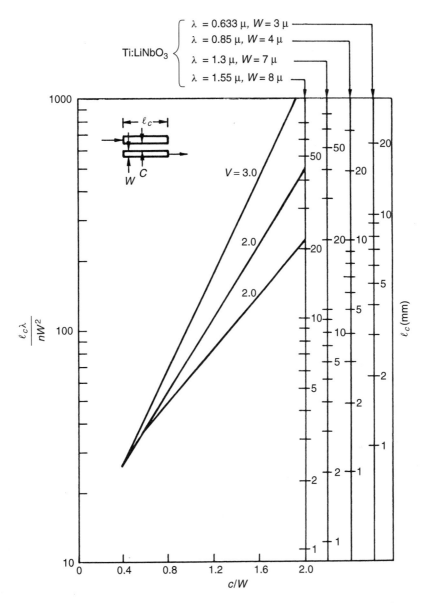

**Figure 11.13**   Calculated parameters for Ti:LiNbO₃ 2 × 2 directional couplers.

The behavior of such directional coupler configuration is shown in Fig. 11.14 for a single-mode Ti:LiNbO₃ coupler at various wavelengths of practical interest. Here, again, the waveguide was assumed to have a single-mode optical confinement at $V = 3$. In this case, the coupling efficiency represents the effect of coupling over infinite distance. Therefore, for very small angles, the power coupling is very sensitive to minor changes in angle as the power is exchanged many times between the two guides in much the same way as in parallel waveguides. As the angle increases, the region where a strong interaction exists becomes shorter and the number of times the power

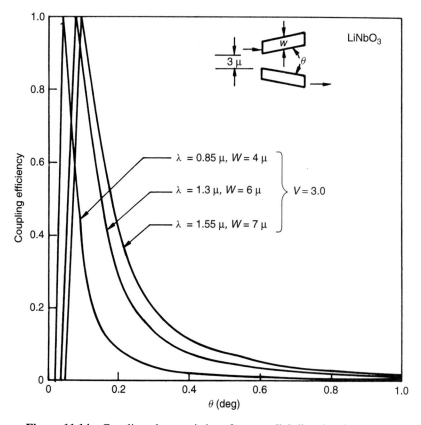

**Figure 11.14**   Coupling characteristics of nonparallel directional couplers.

is exchanged decreases, up to a point where coupling is weak and only a small fraction of the power is allowed to couple to the second waveguide. The behavior is especially important in the design of parallel couplers where the input and output guides have to be separated away from proximity to a wider spacing to allow for fiber coupling, for example. In this case, small angles may cause significant coupling prior to the principal interaction region, thus modifying the assumed initial excitation conditions.

## REFERENCES

1. J. E. Goel, "A Circular-Harmonic Computer Analysis of Rectangular Dielectric Waveguides," *Bell Syst. Tech. J.*, **48**, p. 2071, 1969.

2. H. G. Unger, *Planar Optical Waveguides and Fibers*, Chapter 3, Oxford University Press, 1977.

3. C. B. Shaw, B. T. French, and C. Wagner, "Further Research on Optical Transmission Lines," Science Report No. 2, contract AF449(638)-1504 AD 625 501, Autonetics Report No. C7-929/501, pp. 13–44, 1976.

4. E. A. J. Marcatili, "Dielectric Rectangular Waveguide and Directional Coupler for Integrated Optics," *Bell Syst. Tech. J.*, **48**, p. 2071, 1969.

5. E. A. J. Marcatili, *Bell Syst. Tech. J.*, **53**, p. 645, 1974.

6. R. M. Knox and P. P. Toulios, in J. Fox, Ed., *Proceedings of MRI Symposium of Submillimeter Waves*, Polytechnic, Brooklyn, NY, 1970.

7. V. Rameswamy, *Bell Syst. Tech. J.*, **53**, p. 697, 1974.

8. G. B. Hocker and W. K. Burns, "Mode Dispersion in Diffused Channel Waveguides by the Effective Index Method," *Appl. Opt.*, **113**, 1977.

# 12

# PLANAR OPTICAL WAVEGUIDES AND APPLICATIONS

SHI-KAY YAO

*Optech Laboratory*
*Rowland Heights, California*

Since the invention of lasers, immense progresses have been made in optoelectronics. Optical modulation, beam scanning, frequency mixing, and parametric oscillations have achieved tremendous advances and are employed in various systems applications such as optical sensors, communications, data storage, signal processing, and many other instruments. With the exception of fiber-optics, almost all the applications are implemented in bulk device forms and with a light beam of nearly Gaussian intensity distribution. However, since the introduction of the concept of integrated optics in the late sixties [1–11], the electro-optical community has been in vigorous pursuit of more efficient electro-optical modulators, switches, scanners, spatial light modulators, and nonlinear optical devices in a planar thin-film form. Furthermore, an important question has been raised on the possibility of constructing an optical system or subsystem on the surface of a single substrate using photolithography and planar fabrication processes similar to the integrated electronics circuits. Very compact, efficient, and environmentally stable optical devices could be produced economically using the integrated optical techniques.

Contrary to the one-dimensional nature of electronic systems, classical optical systems tend to be multidimensional. The advances of electro-optics and integrated optics follow both of these two directions. In the areas of optical communications, and optical sensors, the approaches have been analogous to electronic circuits with channel optical waveguides serving as wires. On the other hand, in the area of optical signal processing, multidimensional systems taking advantage of the parallel processing property of Fourier optics are employed most of the time. When reduced to the integrated optics form, a planar optical waveguide is used as a two-dimensional medium in which optical wavefronts are manipulated. Just like in classical optics, the most

*Handbook of Optical Components and Engineering*, Edited by Kai Chang
ISBN 0-471-39055-0   © 2003 John Wiley & Sons, Inc.

important components in the two-dimensional integrated optics are the planar optical waveguides (the optical medium), the waveguide optical lenses (the optical lens), and the spatial light modulators (the optical transparency).

In this chapter, a number of popular planar optical waveguide systems and waveguide optical lenses will be described. General theory, preparation techniques, physical properties, and measurement techniques will be discussed.

## 12.1  PLANAR OPTICAL WAVEGUIDES

### 12.1.1  Introduction

In the late 1960s researchers turned their attention to the intense laser light that can be trapped in an optical thin film and the optical prism film couplers as well as the optical grating film couplers that can bring the laser light into and out of such a thin film. In the meantime many dielectrical optical thin-film systems have been discovered to exhibit good optical waveguiding properties. These include films prepared on glass substrates, semiconductor wafers, and electro-optical crystal surfaces. Optical waveguide effective refractive indices of these waveguiding films are perturbed by additional processes such as thin-film overcoats and thickness modifications over selected areas.

In general, planar optical waveguiding films can be classified as either step index optical waveguides or graded index optical waveguides. Films deposited by the majority of conventional thin-film coating processes such as physical vapor deposition, chemical vapor deposition, vacuum sputtering, and epitaxial crystal growth processes tend to belong to the step index waveguide category. The optical waveguiding layer has a substantially uniform optical index of refraction that is higher than the optical index of refraction of the substrate material. Optical waveguiding is facilitated by the total internal reflection at the thin-film interfaces where a step in the optical index of refraction exists. On the other hand, graded index optical waveguides are made by diffusion type of processes such as in-diffusion, out-diffusion, or ion exchange processes. Again, the waveguiding layer must have a higher index of refraction than its surrounding layers if low waveguide losses are desired. At the graded index interface, the Poynting vector of the optical field is gradually turned back toward the direction where the optical index of refraction is higher. Graded index waveguides are extremely important because they can be fabricated onto dielectric crystal surfaces with a single-crystal waveguide body and because they are relatively tolerant to microscopic local defects.

### 12.1.2  General Theory of Dielectric Planar Waveguides

*Ray Optics of Planar Waveguide.*  The simplest dielectric waveguide is the planar guide shown in Fig. 12.1, where a film of refractive index $n_2$ is sandwiched between a substrate and a cover material with lower refractive indices $n_3$ and $n_1$, respectively. Often the cover material is air, $n_1 = 1$. Due to total internal reflection at the film–substrate and film–cover interfaces, light can be confined in the film layer as guided optical waves. The conditions for waveguiding can be illustrated by considering the optical waves as rays propagating inside a thick slab of glass [9–11], as shown in Fig. 12.2. For example, at the air–film interface, the rays can be turned back completely when the incidence angle is greater than the so-called critical angle $\theta_c$, which is given by

$$\sin \theta_c = \frac{n_1}{n_2} \tag{12.1}$$

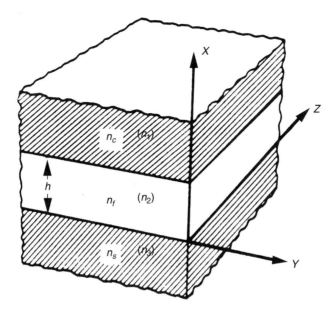

**Figure 12.1**  Cross section of a planar waveguide.

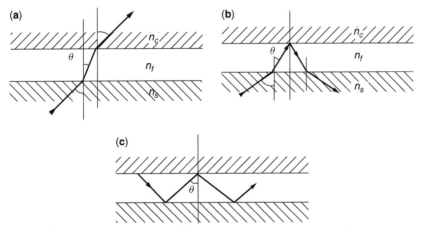

**Figure 12.2**  Zig-zag ray pictures for light waves in a planar waveguiding structure where $n_2 > n_3 > n_1$: (**a**) radiation mode; (**b**) substrate radiation mode; (**c**) guided modes.

In Eq. (12.1), $n_2$ is the refractive index of the medium in the input ray side of the interface. Note that the critical angle exists only when $n_2 > n_1$. Figure 12.2**b** illustrates such a condition in which radiation is allowed in the substrate side only. Further increases in the ray incidence angle beyond the critical angle at the film–substrate interface can cause the total internal reflection to occur at both the interfaces, as shown in Fig. 12.2**c**. Such rays can only propagate in the zig-zag manner within the film region as guided light.

In Fig. 12.2, the rays are shown to actually penetrate slightly into the lower index side of the total internal reflection interfaces while turning back toward the higher index side of the interface. This is due to the presence of evanescent waves on the lower index side of a total internal reflection interface for satisfying the boundary conditions. With the evanescent waves, the energy flow of light actually passes the interface during the total internal reflection and causes a lateral shift in the reflected ray relative to the incident ray. This is the Goos–Hanchen shift, which turned out to be an important element in the understanding of the flow of energy in dielectric waveguides in terms of the ray optics.

Referring to Fig. 12.2c, the zig-zag rays may be considered as two superimposed plane-wave components with wave normals that follow the zig-zag directions and are totally reflected at the film boundaries. These waves are coherent and monochromatic with wavelength $\lambda$. For a guided mode of the planar waveguide, the zig-zag model predicts a propagation constant

$$\beta = kn_2 \sin \theta \tag{12.2}$$

which is the projection of the wave vector of the plane waves, $n_2 k$, in the direction of the waveguiding film. However, only a discrete set of angles that allow the reflected plane waves to interfere constructively will lead to acceptable guided modes. For a film of thickness $h$, the optical phase shift for each transverse passage through the film is simply $n_2 k \cos \theta$. Letting $2\phi_1$ and $2\phi_2$ denote the phase shift on total internal reflections from the air–film interface and the film–substrate interface, respectively, the discrete angles for guided modes is given by the "transverse resonance condition" when the round-trip phase shift of the zig-zag rays equals multiples of $2\pi$:

$$2n_2 kh \cos \theta - 2\phi_1 - 2\phi_2 = 2m\pi \tag{12.3}$$

where $m$ is an integer that identifies the mode number. Equations (12.2) and (12.3) are essentially the dispersion equation for the planar waveguide when the Goos–Hanchen shift angles as a function of the incidence angle $\theta$ are known.

From the preceding equations it is observed that

$$n_3 k < \beta < n_2 k \tag{12.4}$$

or in terms of an *effective waveguide index*, $N = n_2 \sin \theta$,

$$n_3 < N < n_2 \tag{12.5}$$

The dispersion diagram of an asymmetric planar waveguide is illustrated by Fig. 12.3. At the cutoff frequency, the effective waveguide indices assume the value of the lower bound, $n_3$, and as the waveguide thickness $h$ increases (or as the wavelength decreases), the effective waveguide index approaches the upper bound of $n_2$ and more and more waveguide modes are allowed.

To obtain a more precise dispersion diagram for asymmetric planar waveguides in general, the waveguide parameters are rearranged to a number of normalized parameters [12]. The first is a normalized thin-film thickness parameter $V$ such that the effect of optical wavelength is included:

$$V = kh\sqrt{n_2^2 - n_3^2} \tag{12.6}$$

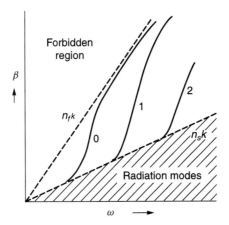

**Figure 12.3**   Typical $\omega-\beta$ diagram of a dielectric waveguide.

and then a normalized guide mode index $b$ is related to the effective mode index $N$ by

$$b = \frac{N^2 - n_3^2}{n_2^2 - n_3^2} \tag{12.7}$$

The index $b$ is zero at cutoff and approaches unity far away from it. Typically, the index difference $(n_2 - n_3)$ is small. Thus,

$$N \approx n_3 + b(n_2 - n_3) \tag{12.8}$$

Finally, we introduce the asymmetry parameter for the waveguide structure as

$$a = \frac{n_3^2 - n_1^2}{n_2^2 - n_3^2} \tag{12.9}$$

for the TE modes, which ranges in value from zero for perfect symmetry $(n_3 = n_1)$ to infinity for strong asymmetry $(n_3 \neq n_1$ and $n_3 \approx n_2)$. Typical values of the asymmetry parameter range from 4 for spattered glass waveguide film over glass substrate to 881 for out-diffused waveguide film on lithium niobate crystal substrate. The asymmetry parameter for the popular titanium diffused waveguide on lithium niobate crystal substrate is 44. For TM modes the asymmetry parameter has larger values of from 27 for the glass waveguide to 21,206 for the out-diffused lithium niobate waveguide. In the form of normalized parameters, the generalized dispersion relation becomes, following Kogelnik and Ramaswamy,

$$V\sqrt{1-b} = m\pi + \tan^{-1}\sqrt{\frac{b}{1-b}} + \tan^{-1}\sqrt{\frac{b+a}{1-b}} \tag{12.10}$$

Figure 12.4 is the normalized dispersion diagram where the normalized effective mode index $b$ is plotted as a function of the normalized frequency $V$ for four values of

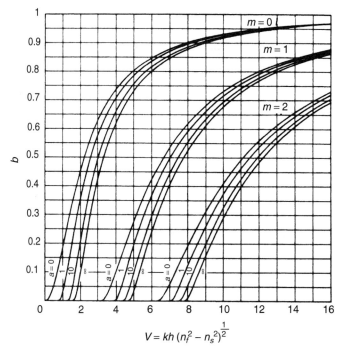

**Figure 12.4** Normalized $\omega - \beta$ diagram of a planar slab waveguide showing the guide index $b$ as a function of the normalized thickness $V$ for various degrees of asymmetry. (After Ref. 12.)

asymmetry parameters $a$ and for the first three modes, $m = 0, 1, 2$. In Eq. (12.10), the cutoff frequency for the $m$th mode, $V_m$, can be obtained by letting $b = 0$:

$$V_m = \tan^{-1}\sqrt{a} + m\pi \tag{12.11}$$

For a symmetric waveguide, $a = 0$, the fundamental mode cutoff frequency $V_0 = 0$. There is always a fundamental mode even for thickness approaching zero. On the other hand, for asymmetric waveguides, a minimum film thickness is needed in order to support any guided mode. From Eq. (12.11), the number of guided modes allowed in a waveguide is

$$m = \frac{2h}{\lambda}\sqrt{n_2^2 - n_3^2} \tag{12.12}$$

The cutoff conditions and dispersion diagrams for the TM mode are very similar to the TE modes, particularly when $n_2 - n_3$ is small. The main difference is in the definition of the asymmetry parameter $a$, which for the TM mode is

$$a_{\text{TM}} = \frac{n_2^4}{n_1^4}\left(\frac{n_3^2 - n_1^2}{n_2^2 - n_3^2}\right) \tag{12.13}$$

***Waveguide Modes for Step Index Guides.*** The waveguide modes are solutions of Maxwell's wave equation [9, 13–15],

$$\nabla^2 E(r) + k^2 n^2(r)E(r) = 0 \tag{12.14}$$

subject to the continuity of the tangential components of $E$ and $H$ at the dielectric interfaces. Consider the step index planar waveguide illustrated in Fig. 12.1. The field components $E_y$ of a TE mode propagating along the $y$ direction must obey the equation

$$\partial^2 E_y / \partial x^2 = (\beta^2 - n^2 k^2) E_y \qquad (12.15)$$

where the $E_y$ for guided modes are, with $p^2 = \beta^2 - n_1^2 k^2$ and $q^2 = \beta^2 - n_3^2$,

$$E_y = \begin{cases} E_1 \exp[-p(x - h)] & \text{for } h < x & (12.16) \\ E_2 \cos(k_2 x - \phi_2) & \text{for } 0 < x < h & (12.17) \\ E_3 \exp(qx) & \text{for } x < 0 & (12.18) \end{cases}$$

From the boundary conditions, the phase shifts and dispersion relation $V$ can be obtained:

$$\phi = \tan^{-1} \frac{p}{k_2} \qquad (12.19)$$

$$= \tan^{-1} \frac{q}{k_2} \qquad (12.20)$$

$$K_2 h - \tan^{-1} \frac{p}{k_2} - \tan \frac{q}{k_2} = m\pi \qquad (12.21)$$

in agreement with the results from the zig-zag model. The relation between the field magnitude is

$$E_2^2 (n_2^2 - N^2) = E_3^2 (n_2^2 - n_3^2) = E_1^2 (n_2^2 - n_1^2) \qquad (12.22)$$

where $N = \beta / K$ is the effective mode index. Finally, the power carried by a mode per unit guide width is

$$P = -2 \int_{-\infty}^{\infty} E_y H_x dx = N \sqrt{\frac{\varepsilon_0}{m_0}} E_2^2 h_{\text{eff}} \qquad (12.23)$$

where

$$h_{\text{eff}} = h + \frac{1}{q} + \frac{1}{p} \qquad (12.24)$$

is the effective thickness of the waveguide.

For the case of TM modes, the boundary conditions require the continuity of $H_y$ and $E_z$ at the interfaces. Assuming the field solutions for the guided modes as

$$H_y = \begin{cases} H_1 \exp[-p(x - h)] & \text{for } h < x & (12.25) \\ H_2 \cos(k_2 x - \phi_2) & \text{for } 0 < x < h & (12.26) \\ H_3 \exp(qx) & \text{for } x < 0 & (12.27) \end{cases}$$

the phase shifts and the dispersion relation are

$$\phi_1 = \tan^{-1} \left(\frac{n_2}{n_1}\right)^2 \frac{p}{k_2} \qquad (12.28)$$

$$\phi_2 = \tan^{-1} \left(\frac{n_2}{n_3}\right)^2 \frac{q}{k_2} \qquad (12.29)$$

and

$$k_2 h - \tan^{-1}\left(\frac{n_2}{n_3}\right)^2 \frac{q}{k_2} - \tan^{-1}\left(\frac{n_2}{n_1}\right)^2 \frac{p}{k} = m\pi \tag{12.30}$$

The magnitude of these field components are related by

$$
\begin{aligned}
\frac{H_2^2(n_2^2 - N^2)}{n_2^2} &= H_3^2(n_2^2 - n_3^2)\frac{(N/n_2)^2 + (N/n_3)^2 - 1}{n_3^2} \\
&= H_1^2(n_2^2 - n_1^2)\frac{(N/n_2)^2 + (N/n_1)^2 - 1}{n_1^2}
\end{aligned}
\tag{12.31}
$$

The power per unit guide width carried by the TM mode is

$$P = 2\int_{-\infty}^{\infty} E_x H_y \, dx = \frac{N\sqrt{\frac{\mu_0}{\varepsilon_0}} H_2^2 h_{\text{eff}}}{n_2^2} \tag{12.32}$$

where the effective guide thickness for the TM mode is

$$h_{\text{eff}} = h + \frac{1}{q\left[\left(\frac{N}{n_2}\right)^2 + \left(\frac{N}{n_3}\right)^2 + 1\right]} + \frac{1}{p\left[\left(\frac{N}{n_2}\right)^2 + \left(\frac{N}{n_1}\right)^2 + 1\right]} \tag{12.33}$$

The dispersion diagrams are given in Fig. 12.4.

***Waveguide Modes for Graded Index Guides.*** Many highly useful optical waveguides are made by diffusion processes that lead to waveguides with refractive indices varying gradually over the cross section. Solution of such graded index waveguides have been done by solving the wave equations for exact solutions for certain index profiles. However, it can also be solved either by the WKB approximation method or by taking the odd symmetric solutions of previously solved Schrödinger equations with symmetric potential profiles. For arbitrary profiles, a numerical method has been developed to obtain the waveguide mode properties.

For TE modes, the wave equations for the $E_y$ component is

$$\frac{d^2 E_y}{dx^2} = (\beta^2 - n^2 k^2) E_y \tag{12.34}$$

which has the same form as the Schrödinger equation of quantum mechanics. The waveguide mode index $N^2 = \beta^2/k^2$ is equivalent to the energy level for each solution to the potential energy well of $n^2(x)$. However, most practical planar waveguides are highly asymmetric with a small gradient into the substrate and a large index step at the air–film interface. The small gradient means close approximation between TE and TM modes. The large step index at the air–film interface allows the assumption that the field in the air region ($x < 0$) is negligibly small. This in turn allows the use of known solutions for Schrödinger equations with odd symmetry that have zero field amplitude at $x = 0$ corresponding to the near-zero amplitude of the guided

**TABLE 12.1**

| | Parabolic Profile | Sech$^2$ Profile | Exponential Profile |
|---|---|---|---|
| Index formula | $n(x) \approx$ $n_2(1 - \frac{1}{2}x^2/x_0^2)$ | $n(x) \approx$ $n_3 + \Delta n\ \mathrm{sech}^2(2x/h)$ | $n(x) \approx n_3 +$ $\Delta n \exp(-2|x|/h)$ |
| Solution | | | |
|   For $x \geqslant 0$ | $E_y \simeq H_{2m+1}(\sqrt{2}x/w)$ $\exp(-x^2/w^2)$ | $E_y = U_{2m+1}(2x/h)$ $\mathrm{sech}^2(2x/h)$ | $E_y = J_p(V\exp[-x/h])$ |
|   For $x < 0$ | $E_y = 0$ | $E_y = 0$ | $E_y = 0$ |
| Polynomial | Hermite polynomials: $H_1 = 2x$ $H_3 = 8x^3 - 12x$ | Hypergeometric functions: $u_1 = \sinh(2x/h)$ $u_3 = \sinh(2x/h)[1 - \frac{2}{3}(s-2)\sinh^2(2x/h)]$ | Bessel function of the first kind with noninteger order $p$ |
| Normalized thickness | — | $V = kh\sqrt{2n_3\Delta n}$ | $V = kh\sqrt{2n_3\Delta n}$ |
| Effective index | $N_m^2 = n_2^2 - (2m + \frac{3}{2})(n_2\lambda/\pi x_0)$ | $N_m^2 = n_3^2 + (s - 2m - 1)^2(\lambda/\pi h)^2$ | $N_m^2 = n_3^2 + (P_m^2/4)(\lambda/\pi h)^2$ |
| Comment | $w = \sqrt{\lambda x_0/\pi n_2}$ is the beam radius | $S = \frac{1}{2}(\sqrt{1+V^2}-1)$ specifies the maximum number of modes $m \leqslant (s-1)/2$ | The order $p$ is solved by finding the Bessel functions such that $J_p(V) = 0$ |

mode at the air–film interface. A number of examples are given in Ref. 15 and are listed here in Table 12.1, which lists the index formulas, approximate wave solutions, and effective mode indices [16–18]. The potential well profiles employed in the Schrödinger equation for these three index profiles are illustrated in Fig. 12.5. Of greater interest is the exponential profile [16,17] of Table 12.1, which is similar to many practical waveguide systems. The mode index is related to the parameter $P$ by $(P_m/V)^2 = (N_m^2 - n_3^2)/(2n_3\Delta n) \approx (N_m - n_3)/\Delta n$. The dispersion relation for the first four modes of such a waveguide is given in Fig. 12.6.

Although these approximate solutions work very well with most asymmetric graded index waveguides, it is sometimes of interest to know the field values at the air–film interface. One then needs to match the boundary conditions by assuming an evanescent field in the air region [15] with the exponential decay constant given by, as usual,

$$l^2 = \beta^2 - n_1^2 k^2 \tag{12.35}$$

The $y$ component of this field is related to the approximate mode solutions by

$$E_y(0) = \frac{1}{l}\left(\frac{dE_y}{dx}\right) x = 0 \tag{12.36}$$

Another example is for the silver ion exchange waveguide on glass that has a waveguide index given in Fig. 12.7 and is

$$n^2 = \begin{cases} n_3^2 - 2n_3\Delta n\left[\dfrac{x}{d} + b\left(\dfrac{x}{d}\right)^2\right] & \text{for } x > 0 \\[2mm] n_1^2 & \text{for } x < 0 \end{cases} \tag{12.37}$$

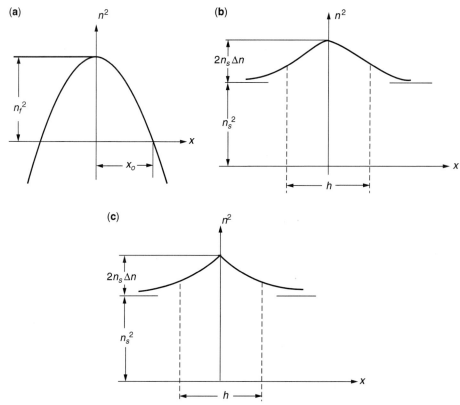

**Figure 12.5**   The potential well profile used to solve the wave equation for (**a**) parabolic index profile, (**b**) $\text{sech}^2$ index profile, and (**c**) exponential index profile.

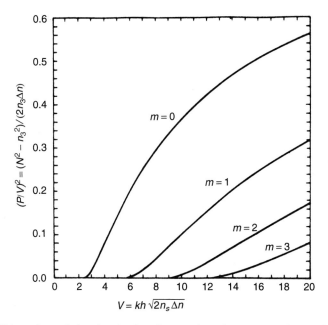

**Figure 12.6**   Dispersion relation for the first four modes of an exponential profile waveguide.

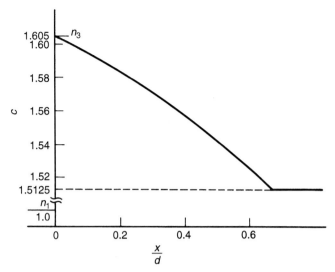

**Figure 12.7** Refractive index profile of a planar waveguide formed by silver ion exchange in glass.

A rigorous solution can be obtained by means of a change of variable. Equations (12.34) and (12.37) can be rewritten in the form [19]

$$\frac{d^2 E_y}{d\xi^2} + \left(\nu + \frac{1}{2} - \frac{\xi}{4}\right) E_y = 0 \tag{12.38}$$

where

$$\xi = \left(\frac{8k^2 n_3 \Delta n \ b}{d^2}\right)^{1/4} (x + a) \tag{12.39a}$$

$$\nu = \left(\frac{8k^2 n_3 \Delta n \ b}{d^2}\right)^{-1/2} \left(k^2 n_3^2 + \frac{k^2 n_3 \Delta n}{b} - \beta^2\right)^{-1/2} \tag{12.39b}$$

and

$$a = \frac{d}{2b} \tag{12.39c}$$

The solutions of Eq. (12.38) are the *parabolic cylinder functions* $D_\nu(\xi)$ and $D_\nu(-\xi)$. For the field to vanish as $\xi$ becomes large, only the $D_\nu(\xi)$'s are retained. Thus

$$E_y(x) = A D_\nu(\xi) \text{ for } x > 0 \tag{12.40a}$$

$$= B \exp(\sqrt{\beta^2 - n_1^2 k^2} x) \text{ for } x < 0 \tag{12.40b}$$

where $A$ and $B$ are arbitrary constants. If the discontinuity in the index at $x = 0$ is large (i.e., highly asymmetric), then the boundary conditions yield

$$D_\nu(\xi_0) = 0 \tag{12.41}$$

where

$$\xi_0 = \xi(x = 0) = \left(kd\sqrt{\frac{n_3 \Delta n}{2b^3}}\right)^{1/2} \tag{12.42}$$

The $\xi_0$ value determines the number of modes allowed in this waveguide. The discrete number of $\nu$ values satisfy Eq. (12.41) leads to the discrete number of solutions. For $n_3 = 1.605$, $n_1 = 1.0$, $\Delta n = 0.0925$, and for a typical value of $kd = 100$, the three lowest-order modes are given in Fig. 12.8.

***The WKB Method.*** The WKB method, which has been thoroughly treated in quantum mechanics, can be applied to obtain approximate solutions of the wave equation (12.34) with slowly varying index profile $n(x)$ [20, 21]. A trial value for the propagation constant $\beta$ (or the mode index $N$) of the waveguide is selected. This leads to the determination of "turning points" $x_{t1}$ and $x_{t2}$, as indicated in Fig. 12.9 and mathematically by

$$n(x_{t1}) = N = \beta/k \tag{12.43}$$

The phase shift at the turning point is $-\pi/2$ each. Thus, if this value $\beta$ satisfies the phase condition

$$\int_{x_{t1}}^{x_{t2}} \sqrt{n^2 k^2 - \beta^2} \, dx = (m + \tfrac{1}{2})\pi \tag{12.44}$$

where $m$ is the mode number, it is the propagation constant for the $m$th waveguide mode. For the mode, the WKB method predicts oscillatory field distribution between $x_{t1}$ and $x_{t2}$ and an exponentially decaying field outside this range. For the cases of highly asymmetric graded index waveguides, $x_{t1}$ is the total internal reflection at the

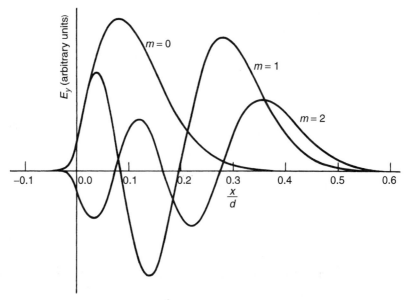

**Figure 12.8** Modal patterns for the three lowest-order TE modes for a typical value of $k_0 d = 100$. (After Ref. 19.)

$$n(x_{t1}) = N = \beta/k$$

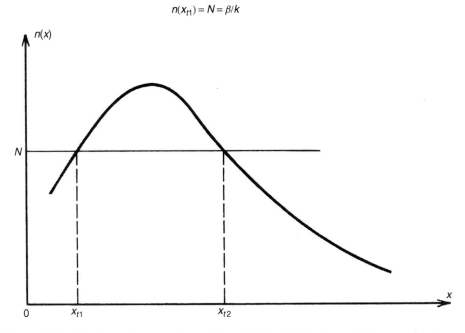

**Figure 12.9**  Turning points are where $n(x) = N$. The field is trapped between $x_{t1}$ and $x_{t2}$.

air–film interface. The phase shift at $x_{t1} = 0$ is given by Eq. (12.19) and is very close to $-\pi$. Thus

$$2 \int_0^{x_{t2}} \sqrt{n^2 k^2 - \beta^2} \, dx = (2 \, \mathrm{m} + \tfrac{3}{2})\pi \tag{12.45}$$

Equation (12.45) can be used to solve for the dispersion relation for each given index profile numerically. Figure 12.10 gives the dispersion curves for two highly useful index profiles [22]. The complimentary error function profile generally describes the diffusion process with infinite surface source. When surface source is depleted with a substantial "drive-in" diffusion process, the index profile will approach a Gaussian function. Figure 12.11 gives the allowed mode number as a function of surface index change $\Delta n$ and diffusion depth $d$ for the two cases [22]. In these figures $B = (N_m^2 - n_3^2)/[\Delta n(2n_3 + \Delta n)] \approx (N_m^2 - n_3^2)/2n_3\Delta n$, $V \cong kd\sqrt{2n_3\Delta n}$, and $x_0 = d/\lambda$.

### 12.1.3  Waveguide Materials and Fabrication Techniques

As described in the previous sections, a planar optical waveguide is a dielectric thin film with its optical index of refraction higher than its substrate as well as cover regions. The index of refraction of the thin-film optical waveguide can be either homogeneous or graded with certain profiles. The index-of-refraction profile is primarily determined by the waveguide fabrication process. Since the name *integrated optics* was coined in 1968, many waveguide fabrication techniques have been proposed and used to form various optical waveguides on a variety of substrates. For instance, organic and inorganic films have been deposited either by physical deposition processes (thermal evaporation, electron beam evaporation, RF sputtering, spin or

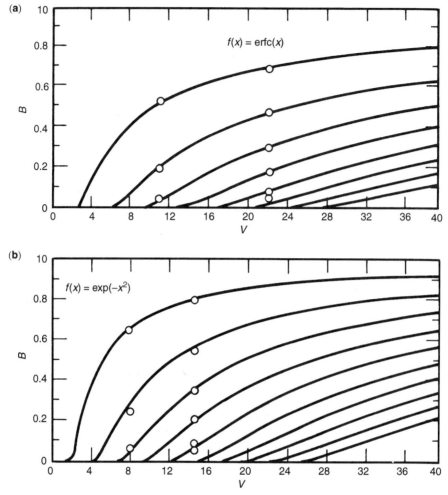

**Figure 12.10**   Normalized dispersion relation for waveguides with (**a**) a complimentary error function index profile and (**b**) a Gaussian index profile; $B \approx (N_m^2 - n_3^2)/2n_3 \Delta n$ and $V \approx kd\sqrt{2n_3\Delta n}$.

dip coating) or by chemical deposition processes. Single-crystal films have been grown on single-crystal substrates by various epitaxial processes. Surface layers with modified optical properties have been made on crystalline and noncrystalline substrates by means of ion exchange, diffusion, or ion implantation processes. The deposition process makes waveguides with physical properties unrelated to the substrate material. Such waveguide materials tend to be amorphous and are useful for passive device applications. On the other hand, the epitaxial processes and the surface modification processes tend to extend certain physical properties from the substrate into the waveguiding layer, making it useful for active device applications. To date, low-loss optical waveguides with propagation loss less than 1 dB/cm have been demonstrated with most processes on selected substrates.

In the meantime, the great variety of potential applications for integrated optical devices have spurred intensive investigation of a large number of waveguide as

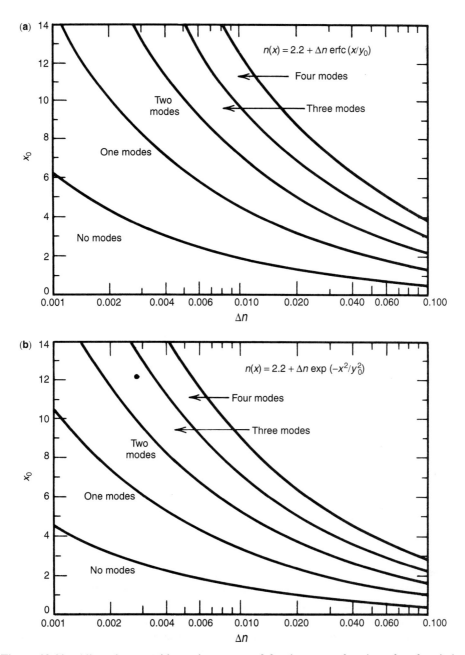

**Figure 12.11**  Allowed waveguide modes on $n_3 = 2.2$ substrate as function of surface index change $\Delta n$ and $v$ normalized diffusion depth $d/\lambda$ for (**a**) an erf $c$ index profile and (**b**) a Gaussian index profile.

well as substrate materials over the past two decades. For instance, semiconductive substrates were studied in the hope of facilitating easy interface with microelectronic or optoelectronic components. Glass substrates were investigated with intended applications as advanced passive fiber-optical components. Electro-optical crystal substrates were explored for the development of optical switches, modulators, and optical signal-processing devices. Nonlinear optical materials were studied in the attempts to make high-efficiency optical harmonic generators. With feasibilities demonstrated, the practicality of each of these potential applications depends on the availability of matured and substrate-compatible process technologies that will provide the required physical properties to the waveguide structure.

In this section, waveguides reported to date are summarized in terms of fabrication technologies with emphasis on the ones that yield low-loss waveguides.

***Deposited Thin-Film Waveguides.*** This section deals with physically or chemically deposited optical waveguides that exhibit either an amorphous structure or a polycrystal structure. The amorphous optical waveguides are passive devices that merely present themselves as light-guiding media. In general, such waveguides can provide very low optical propagation losses when properly prepared. They can serve the purposes of optical interconnecting, branching, combining, wavelength filtering, spatial filtering, and interfacing with optical sources or detectors. On the other hand, polycrystal optical waveguides tend to show considerably higher optical propagation losses. Much of the loss is due to the rough surface texture as the result of anisotropic growth characteristics of each crystallite, and such loss can be improved by surface polishing. Under certain deposition conditions, the crystallites of a polycrystal film can be made to orient all in the same direction. Such oriented film will show some electro-optical or piezoelectric properties and can be utilized to make active devices on a great variety of substrates. However, much work is needed to further reduce the waveguide loss in such oriented polycrystal films in order to fully realize its potential. As a result, to date, the majority of the demonstrated deposited thin-film optical waveguides are passive amorphous waveguides.

A rather straightforward means of waveguide fabrication is by simply depositing a layer of dielectric thin film of high index of refraction over an optically smooth and transparent substrate. This deposition can be done by spin or dip coating, evaporation, RF sputtering, and chemical vapor deposition. The spin or dip coating techniques are popular for the preparation of the delicate organic films while the evaporation, chemical vapor deposition, and the sputtering methods are commonly used for the preparation of inorganic films.

***Organic Thin-Film Waveguides.*** Thin-film optical waveguides have been deposited on glass substrates from liquid solutions. After the solvents are evaporated, a solid film is left on the substrate surface. The liquid film can be coated by spinning the substrates that have been covered by liquid or by dipping into a solution bath. Film thickness is determined by the solid content, viscosity, and spinning or dip withdrawal speed. Various materials were used, including polyurethane, polystyrene, epoxy, photoresist, and organometallic solutions. Usually, coating is done at room temperature followed by curing at elevated temperature. Often the baking is done in several steps with gradual increase in temperature in order to obtain a smooth film surface. For the polymer films VTMS and HMDS, the coated monomer films have to be polymerized by exposure to an RF plasma environment. Table 12.2 lists the waveguide parameters and fabrication procedures of several organic waveguides reported in the literature. Waveguide losses less than 1 dB/cm at the He–Ne laser wavelength of 0.6328 μm can be easily obtained with all the materials except photoresists containing photosensitizer dyes. An interesting

**TABLE 12.2  Organic Thin-Film Waveguides**

| Material | Index | Loss (dB/cm) | Fabrication Method | Comments | References |
|---|---|---|---|---|---|
| Polyurethane | | | | | |
|   9653-1 | 1.555 | 0.8 | Solvent dry and bake | Solvent is toluene/MEK | 23 |
|   LX500 | 1.573 | 4 | Solvent dry and bake | Solvent is toluene/MEK | 23 |
| Epoxy (Araldite) | 1.581 | 0.3 | Solvent and bake | Ethanol | 23 |
| Photoresist | 1.615 | 7 | Solvent and bake, UV | Thinner | 23 |
| KPR | — | 2.2 | Solvent and bake, UV | Thinner, dye removed | 23 |
| AZ 1350 | | 1.6 | Solvent and bake | Dye removed | 24 |
| HMDS, $(CH_3)_3SiOSi(CH_3)_3$ | 1.488 | Low loss | RF plasma, polymerize | $n$ and thickness, reduce by heat | 25–27 |
| VTMS, $CH_2{=}CHSi(CH_3)_3$ | 1.532–1.4797 | 0.04 | RF plasma, polymerize | $n$ and thickness, reduce by heat, mix HMDS and VTMS to adjust $n$ | 25–27 |
| PMMA or Glycidyle MA | 1.515 | 0.1–0.2 | — | UV increase $n$ 1% and thickness | 28–30 |
| PMMA | 1.49–1.56 | Very low | Dip and bake | | |
| Polystyrene | 1.586 TE, 1.589 TM | Low loss | Dip and dry | Stress birefingence | 28 |
| Polyimide | 1.7 | Very low | Spin and bake | 150°C, then 325–400°C | 31 |
| Nitrocellulose | — | Low | Langmuir coating | Isoamyl acetate solution, spread on $H_2O$ surface | 32 |
| MBBA | $1.52 n_0$, $1.72 n_e$ | 500 | Sandwich | Liquid crystal, loss = 30–40 db/cm with field on | 33, 34 |
| Nitrobenzene | 1.55 | — | Sandwich | Electro-optical liquid | 35–37 |

*Note:* Data is for 0.6328 μm optical wavelength.

feature of these films is the possibility of adjusting the film index as well as film thickness by means of mixing or even subsequent heat treatment or UV exposure.

Most organic films are soft and susceptible to chemical attack. The coating technique also lacks the control in film thickness and uniformity needed for waveguide optical devices. However, due to the ease in fabrication, these optical waveguides may be useful for the fabrication of passive components in optical circuits such as wafer scale optical interconnections. In addition, the relatively low film index also make these films useful as overlay coating to reduce the waveguide surface scattering.

Also listed in Table 12.2 are two electro-optically active liquid waveguides, namely the MBBA nematic liquid crystal and nitrobenzene. Unfortunately, the usefulness of these liquid waveguides are hampered by the rather high propagation losses.

*Inorganic Thin-Film Waveguides.* Like solution-coated films, practically all the deposited inorganic films exhibit a step index profile. These waveguides tend to have higher loss than the organic waveguides but are much stronger and durable. The deposition process can be controlled to provide tight thickness as well as uniformity tolerances. Although most deposited films are passive, some are electro-optical. Others have demonstrated frequency doubling and even laser gain.

The deposition of dielectric thin films by evaporation have been known in the optics industry for a long time. However, most evaporated films are not of sufficient quality for optical waveguide applications due to surface roughness scattering and absorption. In Table 12.3, the ZnS film and the Planer CAS 10 are prepared by electron beam evaporation. The ZnS film exhibits high scattering loss unless it is hand polished.

The RF sputtering of dielectric films has been a popular technique for the deposition of low-loss waveguiding films. Note that most of the reported low-loss deposited waveguide films were prepared by the RF sputtering technique. The sputtering process is well understood, and precise control of process parameters is obtainable with commercial sputtering systems. By altering the ground condition, the sputtering can be performed in several modes including RF sputtering, RF sputter-etch, and biased RF sputtering. By admitting reactive gases such as oxygen or nitrogen to the sputtering chamber, reactive sputtering can modify the chemical composition of the deposited films.

Corning 7059 glass (a pyrex glass with composition of $SiO_2$ 50.2%, BaO 25.1%, $B_2O_3$ 13%, $Al_2O_3$ 10.7%, and $As_2O_3$ 0.4%) was the first reported low-loss optical waveguide prepared by the RF sputtering technique. Films must be sputtered with oxygen between 20 and 100% in order to avoid excessive optical absorption in the waveguide due to oxygen deficiency. The waveguide loss at the He–Ne laser wavelength can be lower than 1 dB/cm. However, it was observed that the sputtered 7059 films have a different barium oxide content than the bulk material and that the refractive index of the sputtered film varies as a function of the deposition rate. At 0.6328 µm wavelength, the film refractive index could vary from 1.53 to 1.585 as the deposition rate increases. Another low-loss glass film is the barium silicate glass sputtered from a target formed by a hot-pressed mixture of barium carbonate and silica. By varying the barium oxide content from 0 to 40 wt %, the film index can be varied from 1.48 to 1.62.

Tantalum pentoxide is another popular waveguide material with film index as high as 2.08 at 0.6328 µm. When nitrogen is blended into the sputtering chamber, the resultant film index can vary from 1.85 to 2.13 depending on the $N_2/O_2$ ratio. Optical

**TABLE 12.3 Inorganic Thin-Film Waveguides**

| Material | Index | Loss (dB/cm) | Fabrication Method | Comments | References |
|---|---|---|---|---|---|
| ZnO/glass | 1.973 | 60, 20 | Sputtering | Oriented crystal, after polishing | 38 |
| ZnO | — | 7 | Sputtering | Oriented | 39 |
| ZnO | 1.98 | 0.01 | Sputtering | On oxidized Si, laser annealed | 40 |
| ZnS | 2.342 | 5 | E-Beam | — | 38 |
| Planar CAS 10 | 1.469 | 1.2 | E-Beam | Glass film | 41 |
| 7059 glass | 1.53–1.585 | <1 | Sputtering in oxygen | $n$ depends on deposit rate | 42–45 |
| Ba–Si glass | 1.48–1.62 | <1 | Sputter from silica and $BaCO_3$ | | 46 |
| Nd–Glass | — | 0.5 | Sputter from AO1838 target, anneal 500°C | Gain is 1/cm | 47, 48 |
| Aluminum oxide | 1.66 | 40 | Sputtering | — | 49 |
| $Ta_2O_5$ | 2.08–2.214 | 0.9 | Sputter Ta in $O_2$, then heat 500°C | Amorphous | 50, 51 |
| $Ta_2O_5$ | 1.85–2.13 | Low | Sputter in $O_2$–$N_2$ mix | | 52 |
| $Ta_2O_5$/$SiO_2$ | 1.46–2.08 | 0.8 | Ta from 0–100% adjusted $n$ | Anneal, 450°C for 12 h. | 51 |
| $Nb_2O_5$ | 2.297 | <10 | Sputtering | Low temperature better | 53 |
| SiO/N | 1.45–1.98 | <4 | RF CVD 850°C, 0.2–0.5% NO, 0.02–0.07% Silane, $N_2$ at 1 atm | Low loss for $n = 1.48$–$1.54$ | 54 |

*(Continued)*

737

**TABLE 12.3** (*Continued*)

| Material | Index | Loss (dB/cm) | Fabrication Method | Comments | References |
|---|---|---|---|---|---|
| $Si_3N_4$ | 1.98 | < 0.1 | Low-pressure CVD | On oxidized Si | 55 |
| $LiNbO_3$/Sapphire | 2.247 TM, 2.280 TE | High | RF sputter in $Ar/O_2 = \frac{7}{3}$ | Frequency doubling | 56 |
| $As_2S_3$/Glass | 2.36 | 2.1 at 1.06 μm | RF sputter | 0.4 dB/cm for $TE_0$ on thick film | 57 |
| $Ge_{29.9}Sb_{15.6}Se_{54.5}$ | 2.75 | 10 | RF sputter | Anneal, 265°C, loss is 3.6 dB/cm | 57 |
| $Ge_{34.4}As_{11.1}Se_{54.5}$ | 2.586 | 20 | RF sputter | Anneal improves loss | 57 |
| GaAs | 3.27 | 1–7.4 at 10.6 μm | Platelet single crystal | Electro-optics modulator CdTe or $As_2S_3$ cladding layer | 58 |
| $SiO_2$–$TiO_2$ | — | 0.15 | Flame hydrolysis deposition, then 1250°C heat | Core–cladding index ratio 1.011 | 59, 60 |
| $SiO_2$–$GeO_2$ | — | 1.5 | Same as above | Index ratio 1.009 | 59 |
| Pb–silica 1:2.5 | 1.664 | 0.5 | Solution coat and bake at 60°C | Adjustable by ratio | 23 |

*Note*: Data for 0.6328 μm optical wavelength.

propagation loss of less than 1 dB/cm has been reported. Another way of changing the film index of refraction is by using a target containing $SiO_2$ and $Ta_2O_5$.

Niobium pentoxide must be sputtered at a low substrate temperature in order to keep the waveguide loss low. Further reduction of waveguide loss can be done by laser annealing. With 100% oxygen sputtering, a refractive index of 2.297 was obtained.

Other sputtered films of interest includes ZnO, Nd-doped glass, aluminum oxide, lithium niobate, and several chalcogenide glasses. The ZnO film can be deposited with (002) oriented crystallites and therefore can provide piezoelectric as well as electro-optical properties to an optical waveguide on a glass type of substrate. However, the oriented film also shows high surface scattering loss due to surface roughness unless hand polished. The Nd-doped glass exhibits low loss (0.5 dB/cm) and an optical gain of about 1 $cm^{-1}$. The aluminum oxide film and the lithium niobate film provides possibility for frequency doubling. But the high waveguide loss prevents it from being practical. The chalcogenide films are useful for far-IR waveguiding devices.

A rather interesting deposited optical waveguide is the silicon oxynitride film that results in the chemical vapor deposition process. Silicon oxynitride is a glassy, amorphous, stable silicon–oxygen–nitrogen polymer of variable composition. This film was deposited at 850°C in a RF-heated silica tube reactor. The gas was at 1 atm pressure, typically comprising 0.2–0.5% nitric oxide, 0.02–0.07% silane, and the remainder nitrogen. By controlling the NO-to-Silane concentration ratio, the composition ratio of silicon dioxide ($n = 1.455$) to silicon nitride ($n = 1.98$) may be adjusted. Deposition rate was about 700 Å/min. For the index range between 1.48 to 1.54, this film shows low loss. By using a low-vapor-pressure chemical vapor deposition technique on a thermally oxidized silicon wafer, a waveguide loss of <0.1 dB/cm was observed for the $TE_0$ mode.

Another chemical vapor deposition process with great potential for producing low-loss optical waveguides is the flame hydrolysis deposition technique similar to the OVPO and VAD processes for optical fiber fabrication. To fabricate the $SiO_2$–$TiO_2$ planar waveguide, a mixture of $SiCl_4$–$TiCl_4$ was fed into an oxyhydrogen torch. Fine glass particles synthesized by flame hydrolysis were deposited on the glass substrate on a turntable. Small amounts of $BCl_3$ and $PCl_3$ were added to the gas to lower the melting point of the glass particles. A cladding layer was deposited over the core layer to reduce surface scattering. After the deposition, the samples were heated up to 1250°C for consolidation. The measured waveguide loss was 0.15 dB/cm at 0.6328 μm optical wavelength.

Also included in Table 12.3 are a lapped GaAs platelet as the far-IR waveguide electro-optical modulator and a solution-coated low-loss glass film. The GaAs platelet is coated with CdTe (or $As_2S_3$) cladding layers and supports many waveguide modes. Propagation losses as low as 1–5 dB/cm have been measured on some of the modes.

*Single-Crystal Epitaxial Waveguides.* The previously described waveguide fabrication techniques are useful for the preparation of passive waveguides. However, a major reason for developing integrated optics technology has been the desire to have affordable miniaturized and highly efficient electro-optical devices or subsystems. For active optical power distribution functions as well as optical modulation and switching, the optical waveguide must be constructed with single-crystal materials having large electro-optical coefficients. In order to fabricate a single-crystal waveguiding layer with a higher index of refraction than the substrate material, the epitaxial crystal growth technique appears promising. During the past decade, epitaxial growth technology for the growth of semiconductive thin films has matured rapidly. It has been the major contributor for the recent successes in the laser-diode-related optoelectronics industry. However, the required optical quality in a waveguiding film for integrated optics far exceeds what is needed for laser diodes and photodetectors. This is mainly due to the longer propagation distances as well as angles and bends in a typical integrated optical device. It will continue to be a big challenge to material scientists to refine the epitaxial growth technique such that low-loss optical waveguide devices can be constructed.

In this section, optical waveguides made by epitaxial techniques will be described. Emphasis will be placed on the growth of ferroelectric electro-optical crystals. Although the epitaxial growth technique for compound semiconductor crystal materials is much more mature and is currently advancing toward integrated optoelectronics devices, it will not be the subject of this chapter, which mainly deals with planar optical waveguides and components. Readers interested in the epitaxial growth of compound semiconductive optical waveguides are referred to the chapter on laser diodes and optoelectronic devices.

Table 12.4 lists the reported dielectric waveguiding films prepared by epitaxial techniques. Most of the efforts are for the growth of ferroelectric crystals due to its large electro-optical coefficient. Also included are efforts in ZnO, an electro-optical material, and in a garnet film, a magneto-optical material.

The ZnO has been sputtered as oriented polycrystal films showing electro-optical and electromechanical properties. Early experiments using this film as optical waveguides yielded a waveguide loss greater than 20 dB/cm. Even after surface polishing, the grain boundaries due to the columnar structure as well as voids among crystallites still contributes to large amounts of scattering. Magnetron sputtering produces oriented ZnO films with improved density as well as smoothness. As-grown films on glass substrates with optical waveguide losses of 6–10 dB/cm have been reported. Recently, reduction of optical losses has been observed on sputtered ZnO films after $CO_2$ laser annealing. The laser annealing process is believed to induce coalescence of neighboring crystallites, thus creating a nearly single-crystal film. The waveguide loss for the fundamental mode of a three-mode waveguide was reduced to 0.01 dB/cm. Effects on the electro-optical, electromechanical, and nonlinear optical properties of laser-annealed ZnO films are yet to be determined. Epitaxial growth of ZnO has also been reported on sapphire substrates using RF sputtering and using chemical vapor deposition. The ZnO (0001) and (1-210) epitaxial films were sputter grown on (0001) and (01-12) sapphire substrates, respectively. Demonstrated optical propagation losses were 2 and 1.9 dB/cm, respectively, using a prism-coupled He–Ne laser beam. An electromechanical coupling constant of approximately $k = 0.2$ has been reported. Chemical vapor deposition (CVD) films have shown varied results. One film must be polished in order to conduct measurements yielding a 0.3-dB/cm attenuation value for the $TE_0$ mode. Another as-grown ZnO film gives optical propagation losses for an eight-mode waveguide from 0.7 dB/cm for the $TE_0$ mode to 18.3 dB/cm for the $TE_7$ mode in more or less a linear fashion.

A number of gallium and iron–garnet films developed originally for magnetic bubble memory devices have been tried for optical waveguiding applications. These as-grown garnet films are smooth, uniform, and pin-hole free. But the absorption losses depend on the impurity content in the melt. Two garnet films, Bi0.63 Tm2.3 Fe3.8 Ga1.2 Pb.02 O12 and Bi.95 Yb2.1 Fe3.8 Ga1.1 Pb.03 O12, were grown over the GGG substrate for

**TABLE 12.4  Epitaxial Waveguide**

| Layer | Substrate | Index | Loss (dB/cm) | Fabrication Method | References |
|---|---|---|---|---|---|
| $6Bi_2O_3:TiO_2$ | $B_{12}GeO_{20}$ | — | — | EGM 825–930°C | 61 |
| $LiNbO_3$ | $LiTaO_3$ | $n_0 = 2.288$, $n_e = 2.207$ | Low | EGM 1300° cool at 20°C/h, then polish | 62–64 |
| $LiNbO_3$ | $Z$-$LiTaO_3$ | $n_0 = 2.288$, $n_e = 2.191$ | 5 TM, 11 TE | LPE in $LiO_2$–$V_2O_5$–$Nb_2O_5$ flux at 1100°C with 5:4:1 ratio, film grown at 850°C | 65, 66 |
| Ga, Fe, Garnet | Garnet | — | 1–5 | Grown by LPE | 67 |
| $Eu_3Ge_5O_{15}$ | $Gd_3Sc_2Al_3O_{12}$ | — | — | — | 67 |
| Garnet | GGG | $n = 2.247$ or $n = 2.259$ | 25 | LPE growth for Faraday rotation device | 68 |
| ZnO | Sapphire | $n = 1.98$ | 2 | Sputter grown on sapphire, loss depends on orientation | 39, 69 |
| ZnO | Sapphire | $n = 1.98$ | 0.3 | CVD grown, then hand polish | 70, 71 |

*Note*: Data for 0.6328 μm optical wavelength. Abbreviations: EGM, epitaxial growth by melting; LPE, liquid phase epitaxy.

the fabrication of a Faraday rotation device. The films had a (111) orientation and were grown by horizontal dipping. Strong TE–TM mode conversion was observed as the optical beam propagates in this waveguide. However, the measured optical loss was between 25 and 30 dB/cm. A magneto-optical switch was fabricated on $Eu_3Ge_5O_{12}$ waveguide epitaxially grown over a $Gd_3Sc_2Al_3O_{12}$ substrate. The waveguide optical loss was 1–5 dB/cm. Although garnet films make excellent magneto-optical modulators, the optical losses limits its usefulness to optical wavelengths longer than 1 μm.

Epitaxial growth by melting (EGM) has been used to make $6Bi_2O_3:TiO_2$ waveguides on $Bi_{12}GeO_{20}$ crystal substrate and lithium niobate waveguide on lithium tantalate crystal substrate. The growth of lithium niobate on lithium tantalate is possible because the melting temperature of lithium tantalate is higher by about 300°C than that of lithium niobate. The epitaxial film is grown from bulk melt or melting a powder or lacquer suspension on the substrate. The film grown by EGM always has a transition region that minimizes the lattice mismatch problem and produces a gradient index at the interface. For the growth of lithium niobate on lithium tantalate, lithium niobate ceramics crushed into powder were laid on the polished c plane of the substrate. Then the substrate was heated to about 1300°C to melt the powder and cooled slowly at a rate of about $20°C\ h^{-1}$. The as-grown film had a rough surface requiring hand polish before the optical waveguiding experiments. The waveguide loss for the polished sample was reportedly too small to measure without elaborate set-up. Subsequently, an electro-optical modulator was fabricated. Film uniformity was improved by first suspending the lithium niobate powder in a lacquer and then painting onto the lithium tantalate surface before the melting process.

Another approach of epitaxial growing lithium niobate on lithium tantalate is by the liquid phase epitaxy technique using a $LiO_2–V_2O_5$ flux. The molten mix contains 50 mol % $LiO_2$, 40 mol % $V_2O_5$, and 10 mol % $Nb_2O_5$. The composition is equivalent to 20 mol % lithium niobate in the pseudobinary system. The molten mixture was heated to about 1100°C and cooled slowly to the growth temperature of about 850°C. The lithium tantalate substrate was dipped into the molten mixture for film growth. The growth rate was estimated about 0.1 μm/ min. A 3-μm colorless film was grown supporting seven TE and TM modes. Waveguiding losses of 5 and 11 dB/cm were measured for the $TM_0$ and the $TE_0$ modes, respectively. Other flux systems were tried, resulting in either Li-rich or Nb-rich films.

*Graded Index Optical Waveguides.*   Until now, all the waveguides described belong to the step index family. In this section, optical waveguides with graded index profiles will be described. Graded index waveguides are made by ion implantation, ion exchange, or diffusion processes. The ion implantation process modifies the refractive index of the substrate material by damaging the lattice structure within a certain depth range. The ion exchange process replaces cations in the substrate material with cations from an external source forming a graded impurity concentration near the substrate surface. Ion exchange can be accelerated by the application of an external electrical field. The diffusion process introduces impurities from the exposed surface into the substrate by thermal energy. The impurity concentration profiles for the unassisted ion exchange process and the diffusion process tend to follow a complimentary error function shape unless surface impurity source is depleted. Prolonged heat treatment after the depletion of surface source yields a Gaussian impurity concentration profile. The normalized waveguide mode behavior for an optical waveguide with a refractive index profile of the erfc function and of the Gaussian function has been discussed in an earlier part of this chapter. The impurity profile for an ion-implanted optical waveguide can be tailored by careful programming of the implantation energy as well as dose. The field-assisted ion exchange also has an impurity profile depending on the applied field.

In general, graded index optical waveguides exhibit very low optical scattering as well as absorption losses. The fabrication process simply modifies the substrate material by the introduction of impurities. These processes work on a great variety of substrate materials, glass, or crystalline. In most cases, the bulk physical properties of the substrate material are preserved after the index modification. This is advantageous from device design and fabrication points of view. It has been found that the most popular waveguides used in integrated optics are fabricated using either the ion exchange process or the diffusion process.

*Ion Implantation Waveguide.* Many waveguides have been fabricated by ion implantation. The choice of ion and implant energy mainly affects the penetration depth. Lattice structure is damaged at the penetration depth, resulting in a decrease in the refractive index. The less damaged surface layer is able to support a few guided modes due to the relatively higher refractive index on the substrate surface compared to the buried damaged layer. One of the problems associated with ion-implanted waveguides is the defects that result in scattering losses for the guided modes. Thermal annealing is useful in reducing the damage effect with improved optical propagation loss. However, thermal annealing also reduces the refractive index change produced by the implantation process. Table 12.5 summarizes the results of reported implanted optical waveguides.

Lithium ions were implanted into fused quartz through a PMMA electron-resist mask to form a waveguide. Waveguide propagation loss after annealing was about 3 dB/cm. For lithium niobate substrate, a variety of ions were implanted. The saturation value of the index change due to implantation was approximately $-7$ to $-10\%$ regardless of the choice of ion. Implant energy was adjusted to obtain desired penetration depth. As implanted, the optical waveguide showed very high propagation loss. Annealing at about 200°C for 30 min reduces the optical propagation loss substantially. The index change is also reduced substantially after annealing. The exact behavior of

**TABLE 12.5  Ion Implantation Waveguide**

| Ion Type | Substrate | $\Delta n$ | Loss (dB/cm) | Fabrication Comments | References |
|---|---|---|---|---|---|
| Li | Fused quartz | 0.0375 | 1.8–3 | 200 keV $10^{15}$Li + 7, anneal, 300°C, 1 h | 72–74 |
|  |  | 0.013 | 0.2 |  | — |
| Ne | LiNbO$_3$ | $-0.225$ | — | 60 keV | 75, 76 |
| Ar | LiNbO$_3$ | $-0.225$ | — | 60 keV | 75 |
| He | LiNbO$_3$ | $-0.156--0.25$ | Low | 0.86–1.86 MeV to 2–4 μm below surface, anneal, 200°C, 30 min, $r_{33}$ reduced to 60% | 76, 77 |
| H/He or B | ZnTe | — | 1–4 | Lattice damage | 78 |

*Note*: Data for 0.6328 μm wavelength.

annealing depends somewhat on the choice of ion. The most discouraging finding was the reduction in the electro-optical coefficient, $r_{33}$, to only 60% of its original value in an ion-implanted optical waveguide on lithium niobate substrate.

*Ion Exchange Waveguide.* Ion exchange techniques have been used for more than a century to produce tinted glass and to improve surface mechanical properties of glasses [79]. Recently, ion exchange processes have been employed for the fabrication of low-loss optical waveguides on various glass substrates and on ferroelectric crystal substrates. Due to the ease of fabrication, the low waveguiding loss, and the compatibility with optical fibers, ion exchange optical waveguides are becoming one of the dominant waveguides of choice in integrated optics [80–82].

When a glass containing monovalent cations is placed into a molten salt containing another monovalent cation, ion exchange takes place. A generalized ion exchange reaction can be written

$$\underline{A}\ (\text{glass}) + B\ (\text{salt}) = \underline{B}\ (\text{glass}) + A\ (\text{salt}) \tag{12.46}$$

where $\underline{A}$ and $\underline{B}$ are the counterions in the exchanger phase and A and B are the counterions in the liquid phase. Molten salts are required because of the temperature range needed before the cations in the glass become mobile with reference to the negatively charged oxygens of the rigid, immobile silicate network. The rate at which the exchange occurs is controlled by the diffusion of the ions in the glass. In binary ion exchange, the diffusing species A and B are charged and tend, in general, to diffuse at different rates. Thus, there is a tendency of electrical charge buildup. The gradient in the electrical potential acts to slow down the faster ion and speed up the slower ion, resulting in equal and opposite fluxes for the two ions. Thus, electrical neutrality is preserved. The diffusion coefficient depends on temperature and glass composition. Below the glass transition temperature, the temperature dependence of the diffusion coefficients can be fit to the Arrhenius equation

$$D_i = D_0 \exp\left(-\frac{Q_i}{RT}\right) \tag{12.47}$$

where $Q_i$ is the activation energy (in joules per mole), $R$ is the gas constant 8.314 J/deg mol, and $T$ is temperature in kelvins. For the ion exchange process without an external electrical field, the diffusion process has a solution for the concentration of the B cation in the glass,

$$N_b(x, t) = N_0 \operatorname{erfc} \frac{x}{d_0} \tag{12.48}$$

where $d_0 = 2\sqrt{Dt}$ is the effective depth of diffusion. For example, silver ion in soda–lime glass has $D_0 = 2.26 \times 10^{-6}$ and $Q_i = 8.5 \times 10^4$ J/mol. Tables 12.6 and 12.7 list the ion radius, polarizability, activation energy, and diffusion constant, for a number of commonly used cations. Figure 12.12 is a plot of the effective diffusion constant as a function of the diffusion temperature. The normalized dispersion curves are given by Fig. 12.13. The electric field of the guided modes are given by Fig. 12.8.

When an external electric field is applied, the B cation concentration becomes

$$N_b = 0.5 N_0 (\operatorname{erfc}(x' - r) + \exp(4rx')\operatorname{erfc}(x' - r)) \tag{12.49}$$

**TABLE 12.6  Ion Exchange Waveguide Parameters**

| Ion | Electron Polarizability, $\lambda = D$ (Å$^3$) | Ionic Radius (Å) | Substrate | Salt | Operating Point (°C) | Decomposition Point (°C) | Index Increase | Loss (dB/cm) | References |
|---|---|---|---|---|---|---|---|---|---|
| $Na^+$ | 0.41 | 0.95 | Glass | $NaNO_3$ | 307 | 380 | — | — | 83, 84, 85 |
| $Li^+$ | 0.03 | 0.65 | Soda lime | $LiNO_3$ or | 264 | 600 | 0.01 | > 1 | 85, 86 |
|  |  |  |  | $(LiSO_4)/(K_2SO_4) = 4:1$ | 520–620 | — | 0.015 | — | — |
| $Tl^+$ | 5.2 | 1.49 | Borosilicate | $TlNO_3 + KNO_3 + NaNO_3$ | 206 | 530 | 0.001–0.1 | < 0.1 | 84 |
| $Cs^+$ | 3.34 | 1.65 | Soda lime | $CsNO_3$ | 520 | — | 0.03 | > 1 | 87, 88 |
|  |  |  | BGG21 glass | $CsNO_3 + CsCl$ | 435 | — | 0.043 | — | — |
| $Ag^+$ | 2.4 | 1.26 | Alumino-silicate | $AgNO_3$ | 225–270 | 444 | 0.09–0.13 | 0.1–0.5 | 85, 89–92 |
| $Rb^+$ | 1.98 | 1.49 | Soda lime | $RbNO_3$ | 520 | — | 0.015 | High | 87 |
| $K^+$ | 1.33 | 1.33 | Soda lime | $KNO_3$ | 365 | 400 | 0.009 | 0.2 | 85, 93, 94 |
| $Ag^+$ | 2.4 | 1.65 | Glass | Silver film | — | — | 0.001 | — | 82 |
|  | 2.4 | 1.65 | Glass | Silver film field assisted | — | — | 0.025 | — | 82 |
|  | 2.4 | 1.65 | X-cut $LiNbO_3$ | $AgNO_3$ | 250 | 444 | 0.12 | 6 | 95 |
| $Tl^+$ | 5.2 | 1.49 | $LiNbO_3$ | — | — | — | — | — | 96 |
|  |  |  | $LiTaO_3$ | — | — | — | — | — | 96 |
| $H^+$ | — | — | $LiNbO_3$ | Benzoic Acid | 110–249 | — | 0.2 | ~2 | 97 |
|  |  |  | $Y\text{-}LiNbO_3$ | Diluted benzoic acid | — | — | 0.1 | ~5 | — |

**TABLE 12.7    Diffusion Coefficients of Various Cations in Glass Used for Waveguides**

| Cation | Glass | Temperature (°C) | $D$ $(10^{-14} \text{m}^2/\text{s})$ | $Q$ $(10^4 \text{J/mol})$ | References |
|---|---|---|---|---|---|
| $Tl^+$ | Borosilicate | 530 | 20 | — | 84 |
| $Li^+$ | Soda lime | 575 | 64 | 14.2 | 86 |
| $Ag^+$ | Soda lime | 374 | 0.7 | 9.1 | 90 |
| | Borosilicate | 615 | 0.26 | 9.1 | 98 |
| | Soda lime | 215 | 0.010 | — | 92 |
| | BK-7 | 320 | 0.025 | 9.8 | 91 |
| | Soda lime | 330 | 0.01–0.03 | 8.9 | 99 |
| $K^+$ | Soda lime | 385 | 0.11 | 12.5 | 93 |
| | BK-7 | 385 | 0.14 | — | 94 |
| | Pyrex | 385 | 0.06 | — | 94 |
| $Cs^+$ | BGG21 | 407 | 0.32 | 20 | 88 |
| $Na^+$ | Soda lime | 371 | 12 | 16 | 83 |

*Source*: After Ref. 80.

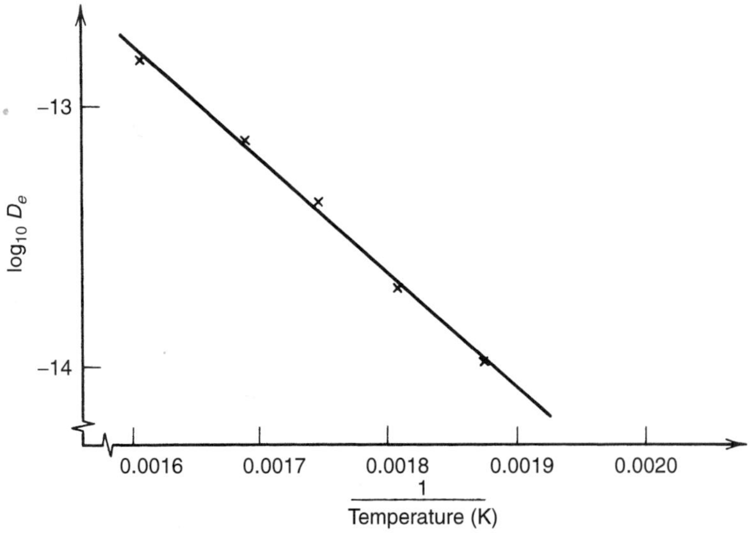

**Figure 12.12**    Dependence of the "effective" diffusion constant on the diffusion temperature.

where $x' = x/d_0$ is the normalized effective depth of diffusion without an external field and $r = \mu Et/d_0$ in which $\mu$ is the electrochemical mobility in $\text{m}^2/\text{V} \cdot \text{s}$ square meters per volt second.

Table 12.6 lists the published waveguide parameters for a number of planar waveguides. The first planar optical waveguide fabricated by this method was by ion exchange from a mixture of thallium, sodium, and potassium salts into a borosilicate glass plate. An external electric field was used to enhance the ion migration rate. The ion exchange occurs between the thallium ions and the sodium as well as potassium ions in the glass. Subsequently, the mixture of salts was replaced by a sodium and potassium

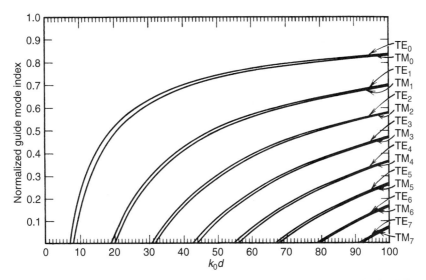

**Figure 12.13**   Theoretical mode dispersion curves for TE and TM modes obtained from the transcendental equation for the second-order polynomial profile.

salts mixture to form a buried waveguide by reversing the ion exchange process. The resultant multimode optical waveguide yielded a total waveguiding loss of less than 0.1 dB/cm.

Ion exchange optical waveguides were also formed on $X$-cut lithium niobate substrate using silver cations. Index change only occurs for the extraordinary rays with a waveguide loss of about 6 dB/cm. No waveguide was observed on $Y$-cut substrates even after prolonged treatment. More recently, proton ion exchange has been employed to form waveguides with index change greater than 0.12 on lithium niobate using benzoic acid ($C_7H_6O_2$) at low temperature (110–249°C). Low-loss optical waveguides (0.5 dB/cm) were observed on $X$-cut and $Z$-cut substrates. However, treatment on $Y$-cut samples resulted in surface damages (pits), and no waveguides were observed. Only when the benzoic acid is substantially diluted with lithium nitride, were guided modes observed on $Y$-cut surfaces.

*Diffused Optical Waveguide.* The ferroelectric crystals lithium niobate and lithium tantalate are among the best materials for electro-optical devices due to large electro-optical, electromechanical, acousto-optical, and nonlinear optical parameters. However, due to the relatively large refractive index of these materials, earlier attempts in making integrated optical devices on these substrates using deposited thin film had only limited successes. The first truly successful optical waveguide on lithium niobate and on lithium tantalate substrates were fabricated by lithium out-diffusion in a vacuum or in an oxygen environment.

Lithium niobate and lithium tantalate crystals can be grown in a slightly nonstoichiometric form with the mole content of the lithium oxide ranging from 0.48 to 0.50. It is known experimentally that for a slightly lithium-deficient crystal, the ordinary refractive index remains unchanged while the extraordinary refractive index increases approximately linearly as the lithium content decreases. For lithium niobate, the extraordinary index increases 1.63% for each percent of lithium oxide loss. The proportional constant is 0.85 for lithium tantalate crystal. Lithium out-diffusion was carried out at high temperature (850–1200°C) in vacuum or in oxygen. If the out-diffusion is done in vacuum, the crystal surface will be darkened due to the excessive loss of oxygen. Reheating the crystal at high temperature in air or in oxygen removes the discoloration. The waveguiding layer had a peak index change on the order of 0.001 and a thickness of several hundred micrometers. Such guides usually support a large number of modes and are not practical for most applications. In fact, the out-diffusion waveguides were quickly replaced by metal in-diffusion waveguides.

Interestingly, as the metal in-diffusion waveguides gained favor, the existence of uncontrolled out-diffusion of lithium oxide during the in-diffusion process became a problem. To suppress the out-diffusion, samples were packed in $Li_2CO_3$ or $LiNbO_3$ powder and annealed after the diffusion. In another approach, a crucible of lithium oxide was placed upstream in the gas flow during the in-diffusion process. More recently, it was found that lithium out-diffusion can be easily suppressed by wetting the incoming gas flow during diffusion.

To date, in-diffusion processes are the most commonly used in the waveguide fabrication of lithium niobate and lithium tantalate crystal substrates. The diffusion source metal, such as Ti, Nb, Mn, Co, Fe, Cu, Zn, and Mg, is deposited on the substrate by evaporation. Table 12.8 lists the reported optical waveguides fabricated by means of diffusion. The diffusion process varies slightly among research laboratories. But the diffusion temperature generally ranges between 850 and 1200°C. By far the most popular metal in-diffused waveguide is the Ti-diffused lithium niobate. Single-mode optical waveguides are easily made by diffusing 200–400 Å of titanium in a wet oxygen environment. Optical waveguiding losses under 1 dB/cm are routinely obtained. The Ti in-diffused lithium niobate waveguide has a larger index change in the extraordinary index than the ordinary index. The copper in-diffused optical waveguide provides an alternative if equal changes in these indices are desired. Another alternative is the zinc in-diffused waveguide. Because of the much larger diffusion coefficient, zinc diffusion may be carried out at lower temperatures (about 800°C) where lithium out-diffusion need not be concerned.

Here one example is given for the fabrication of a single-mode titanium-diffused waveguide on lithium niobate for operation at 0.85 μm wavelength. A titanium thickness of 475 Å is diffused for 5 h at 1000°C in wet oxygen. The wet oxygen can be prepared by feeding through a water bubbler at 95°C. The waveguide loss is sensitive to water bubbler temperature, titanium thickness, and the speed of temperature ramp. A fast temperature ramp is preferred to prevent phase changes in the lithium niobate reported in the 600–900°C range. The resultant waveguide has propagation loss of 0.5–1.5 dB/cm. When the titanium is too thick, there tends to be surface scattering. Slight hand polishing can significantly improve the waveguide loss due to surface scattering.

The lithium tantalate optical waveguide is less popular than the lithium niobate waveguides mainly due to its lower curie temperature (610°C). Unlike lithium niobate, the lithium tantalate crystals must be repoled after diffusion. However, the optical damage threshold of lithium tantalate is two orders of magnitude higher than that of lithium niobate, making it attractive for applications in the visible spectral range.

X-ray photoelectron spectroscopy has been performed to determine the state of in-diffused titanium ions in lithium niobate crystal. It was found that the titanium ions

**TABLE 12.8  Diffused Waveguide Parameters**

| Metal and Substrate | Diffusion | | Temperature (°C) | Time (h) | Diffusion Coefficient | Metal (Å) | Waveguide Depth (µm) | $\Delta n_0$ | $\Delta n_e$ | Number of Modes | Loss (dB/cm) | References |
|---|---|---|---|---|---|---|---|---|---|---|---|---|
| | $E_A$ (eV) | $D_0$ (cm²/s) | | | | | | | | | | |
| Li Out, LiBO$_3$ | — | — | 1100 | 23 | $4.2 \times 10^{-9}_{\perp}$ | — | ~100 | — | 0.003 | Many | <1 | 100, 101 |
| | | | | | $1.5 \times 10^{-9}_{\parallel}$ | — | ~100 | — | 0.003 | Many | <1 | 100, 101 |
| Li Out, LiTaO$_3$ | — | — | 1150 | 3 | $4 \times 10^{-9}_{\perp}$ | — | ~50 | — | 0.002 | Many | <1 | 100, 101 |
| | | | 1400 | 20 min | $25 \times 10^{-9}_{\perp}$ | — | ~30 | — | 0.01 | Many | <1 | 100, 101 |
| V, LiNbO$_3$ | — | — | 950 | 6 | — | 250–500 | 6.2 | 0.0005 | 0.003 | 1 TM, 1 TE | <1 | 102 |
| Ni, LiNbO$_3$ | — | — | 800 | 6 | — | 270 | 2.6–2.9 | 0.007 | 0.005 | 2 TE, 2 TM | <1 | 102 |
| | | | 800 | 6 | — | 500 | 2.8–3.1 | 0.01 | 0.006 | 2–3 TE, 2–3 TM | <1 | 114 |
| Ti, Y-cut LiNbO$_3$ | 2.06 | $1.12 \times 10^{10}$ µm²/h | 1000 | — | $2.2 \times 10^{-10}$ | 500 | 2.6 | — | — | — | — | 114 |
| | 2.2 | $4 \times 10^{-4}$ | 1000 | 10 | $4.6 \times 10^{-13}$ | 25–200 | 3 | — | 0.02 | 1 at 0.87 µm | <1 | 102–110 |
| | | | 970 | 7 | — | 100–500 | 2.5 | — | 0.05 | 4 TE, 1 TM | <1 | 102–110 |
| | | | 970 | 7 | — | 500 | $1.1_{\perp}$ | 0.01 | 0.04 | 1 TE, 5 TM | <1 | 102–110 |
| | | | 960 | 6 | — | | $1.6_{\parallel}$ | 0.006 | 0.025 | — | <1 | 115 |
| | | | | | — | | | | | | — | 115 |
| | | | 1000 | | $9.4 \times 10^{-13}_{\parallel}$ | | | | | | — | 115 |
| | | | | | $1.4 \times 10^{-12}_{\perp}$ | | | | | | — | 115 |
| Cu, LiNbO$_3$ | 1.8 | $4.79 \times 10^{10}$ µm²/h | — | — | — | — | — | — | — | — | — | 111, 114 |
| CuO$_2$, LiTaO$_3$ | — | — | 600 | — | — | — | 6 | 0.0075 | 0.0075 | — | <1 | 112 |
| | | | | | | | | | | | polish | 112 |
| Zn, LiNbO$_3$ | 1.6 | $8.15 \times 10^{7}$ µm²/h | — | — | — | Vapor | — | — | — | — | — | 114 |
| Zn, LiTaO$_3$ (Y-cut) | 1.69 | $5.29 \times 10^{-5}$ | 800 | 6 | $1.4 \times 10^{-12}$ | | 4.2 | 0.0027 | 0.0033 | 1 TE, 1 TM | <1 | 113 |
| | 2.21 | $3.85 \times 10^{-2}_{\parallel}$ | 800 | 6 | | | | | | | | 113 |
| MgO, LiNbO$_3$ | 1.4 | $1.7 \times 10^{-6}$ | 1000 | — | $4.5 \times 10^{-12}$ | 200 | 8.1 | <0 | <0 | — | — | 115 |
| Co, LiNbO$_3$ | 1.34 | $1.4 \times 10^{7}$ µm²/h | 1000 | — | $1.94 \times 10^{-10}$ | — | — | — | — | — | — | 114 |
| Nb$_2$O$_5$, LiTaO$_3$ | — | — | — | — | — | — | — | 0.0006 | 0.0018 | — | <1 | 116, 117 |

*Note:* Data are for 0.62368 µm wavelength.

are fully ionized with no electrons in partially filled $d$ orbitals to absorb light. Optical losses in the titanium in-diffused waveguide are mainly attributed to scattering losses.

***Metal-Clad Optical Waveguide.*** The interest in metal-clad optical waveguides [118–125] stems from the desire to fabricate optical waveguides on ferroelectric electro-optical crystals. In the early 1970s, high-quality optical waveguides were mostly deposited organic or glass films that generally exhibit low refractive indices. In order to bring these low-loss films onto the relatively high refractive index surfaces of lithium niobates and lithium tantalates, a metal buffer layer was recommended. These waveguides have the problem of interacting with the metalized surfaces resulting in substantial waveguide losses. The waveguide losses in such waveguides had been studied using the zig-zag ray model and was found to be proportional to the square of the mode order and inversely proportional to the cube of the waveguide thickness [122]. It was also found that the loss introduced by the metalized surface has different impact on the TE modes and the TM modes. Generally, the TM modes experience greater propagation loss with the presence of metalized surfaces due to the field buildup near conducting walls and due to the existence of an additional surface plasma mode. As a result, these metal-isolated optical waveguides tend to be thick, supporting a large number of modes. The waveguiding loss could be maintained low. For instance, using silver or gold as the cladding metal under a KPR photoresist waveguiding layer [123], the waveguide optical losses were reported between 0.6 and 15 dB/cm. An acousto-optical modulator was demonstrated on such a waveguide using surface acoustic waves generated by the lithium niobate substrate [126].

The advent of out-diffused and metal-diffused optical waveguides on these ferro-electric crystals basically eliminated this approach. However, the problem of guided mode interaction with the metal-cladding layer remains due to the need to have metal electrodes in active waveguide electro-optical devices. For an asymmetric dielectric waveguide with a metal–film outer coating, there is strong differential absorption between the $TE_0$ mode and the TM modes as well as the higher-order TE modes. It appears as if only the lower-order TE modes may be employed for the fabrication of waveguide electro-optical devices. This creates a difficult situation when working with a $C$-cut substrate that provides isotropy for guided modes propagating in all directions on the substrate surface. In order to use the large $r_{33}$ electro-optical coefficient of lithium niobate, a TM mode will be launched that is not compatible with the metal electrodes.

To circumvent this difficulty, a dielectric buffer layer of $SiO_2$ film is deposited between the substrate surface and the metal electrode. The field distributions for the TM modes of a nearly three-mode diffused waveguide under an aluminum electrode are given in Fig. 12.14 with the buffer layer thickness as a parameter. As shown in Fig. 12.14, when there is no buffer layer (top row), the lowest-order TM mode is the highly lossy surface plasma mode, which has maximum field amplitude at the metal–substrate interface. The field distribution of the next TM mode (see Fig. 12.14**b**, top row) actually resembles the $TM_0$ mode in the waveguide region without the top electrode except that a negative dip occurs at the metal–substrate interface. It is expected that at the edge of the electrode area, an incident $TM_0$ mode is decomposed into mainly the combination of a strong $TM_1$ mode (top trace of Fig. 12.14**b**) and a weak plasma mode (top trace of Fig. 12.14**a**). As the buffer layer becomes thicker than a critical thickness, as shown by Fig. 12.14**a** from top row to the bottom row, the

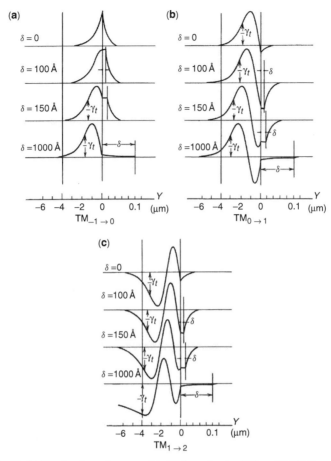

**Figure 12.14** Field distribution of waveguide TM modes in diffused LiNbO$_3$ surface with the buffer SiO$_2$ film thickness of a parameter. With SiO$_2$ of more than 1000 Å thick, the field for all the three lowest-order modes become small at the metal–dielectric interface. (After Ref. 118.)

plasma mode gradually shifts its intensity into the dielectric substrate and becomes less lossy. When the buffer layer is substantially thicker than the critical thickness (bottom trace of Fig. 12.14a), the fundamental TM mode becomes practically the same as the TM$_0$ mode of the waveguide without the top electrode. There is little intensity in the metal–buffer layer interface. At the electrode edge, all the TM modes propagate through with minimum perturbation. The TM mode attenuation under the metalized area is significantly improved when the buffer layer thickness is greater than the critical thickness. Figures 12.14b and 12.14c show the transition of higher-order modes as the buffer layer thickness increases.

Figure 12.15 gives the critical thickness of the buffer layer as a function of the refractive index of the buffer layer, assuming a nearly three-mode Ti-diffused optical waveguide on lithium niobate substrate with a diffusion depth of 4 μm, a peak index change of 0.02, and an aluminum electrode. Figure 12.16 gives the attenuation constant of the three TM modes as a function of the buffer layer thickness. As expected, a lower refractive index buffer layer is much more effective than one with a higher refractive

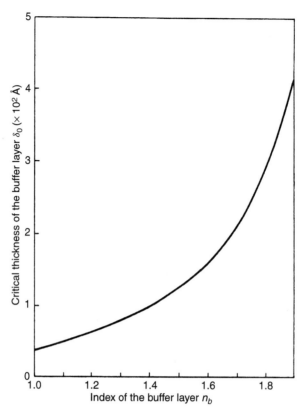

**Figure 12.15** Relation between the critical thickness $\delta_0$ and the index $n_b$ of the buffer layer, calculated from Eq. (3.22), where aluminum is the metal, i.e., $\varepsilon_m/\varepsilon_0 = (1.2 - j7.0)^2$, $\varepsilon_{y0}/\varepsilon_0 = 2.200^2$, and $\Delta\varepsilon_y/\varepsilon_{y0} = 0.018$ for $\lambda = 0.6328$ μm. (After Ref. 118.)

index. Figure 12.17 illustrates the effective indices of the three guided modes in this example. The intersections between the dashed lines and the vertical axis indicate the guided mode indices for the three modes when nothing is coating the crystal surface. The intersections between the solid curves and the vertical axis gives the mode indices when the aluminum top electrode is applied. The rest of the solid curves illustrates how the buildup in buffer layer thickness affects the mode indices. This practice is widely employed to date in fabricating guided-wave electro-optical devices.

***Free-Carrier Optical Waveguide.*** The presence of carrier in a semiconductor lowers the refractive index from that of the pure material. This is primarily due to the negative contribution of the free-carrier plasma to the dielectric constants. Using the free-carrier effective mass in place of the free electron mass, the change in the refractive index due to free carriers is

$$\Delta n_s = -\frac{N\lambda^2 e^2}{\varepsilon_0 n_s 8\pi^2 m^* c^2} \qquad (12.50)$$

where $n$ is the refractive index of the semiconductor, $\lambda$ is the wavelength of light, $e$ is the free carrier charge, $\varepsilon_0$ is the permittivity of the vacuum, and $c$ is the velocity of

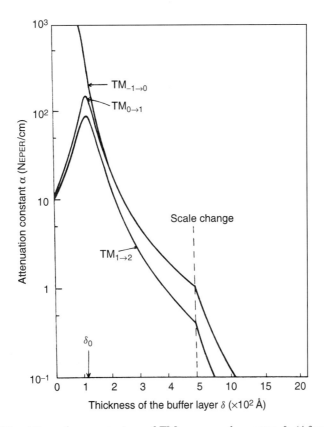

**Figure 12.16**   Attenuation constants $\alpha$ of $TM_{q \to q+1}$ modes versus $\delta$. (After Ref. 118.)

light. For example, in $n$-type GaAs with a carrier concentration of $5 \times 10^{18}$ cm$^{-3}$, the index contribution due to the free carrier is $-0.01$ at an optical wavelength of 1 μm. If a high-resistivity GaAs region of more than 1 μm thick is surrounded by such doped GaAs regions, an optical waveguiding layer can be formed [127]. Light will be confined to the high-resistivity region, which is less lossy than the low-resistivity region due to lack of free-carrier absorption.

Optical waveguides based on this principle can be fabricated by an epitaxially grown high-resistivity layer over a low-resistivity substrate, by a diffused *pin* junction layer, or by ion implantation. Although these waveguides are relatively easy to fabricate, their disadvantages are the inability to independently vary the refractive index and the carrier concentration simultaneously, the inability to make waveguides thinner than 1 μm, and the free-carrier optical losses due to the cladding region. These waveguides are replaced by epitaxially grown ternary and quarternary semiconductor systems.

## 12.2   PASSIVE WAVEGUIDE OPTICAL COMPONENTS

The development of integrated optics requires the integration of a number of waveguide optical components into an optical device substrate. This can be done in two ways. The first approach is to construct a two-dimensional optical lens system on a planar

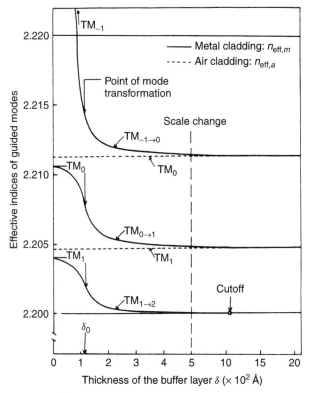

**Figure 12.17**   Effective indices of TM modes versus $\delta$. The solid and dotted lines correspond to the effective indices $n_{\text{eff},m}$ of $TM_{q \to q+1}$ modes for the metal cladding and those $n_{\text{eff},a}$ of $TM_q$ modes for the air cladding, respectively. (After Ref. 118.)

waveguide surface that will manipulate optical wavefronts. The other approach is to construct an electro-optical system using optical channel waveguides to mimic an electronics system. In this section, the passive optical components for the first approach will be discussed.

## 12.2.1   Mode-launching Couplers

The first waveguide optical component to be discussed is the mode-launching couplers that served to interface guided modes with free propagating radiation modes. Mode launching can be achieved by use of a prism coupler, a grating coupler, a tapered thickness coupler, and an end-fire coupler. These couplers will be discussed next.

Prism couplers [5, 128–135] and grating couplers [134–141] are mode converters capable of transferring energy between guided modes and radiation modes. In addition to being convenient devices for launching laser beams into and out of the optical waveguiding films, these devices also provide the means for investigating various properties of guided modes. Of particular importance is the prism coupler, an extremely versatile instrument for bringing light into and out of a waveguiding film. It can be applied almost anywhere on the waveguide substrate surface, requiring no special preparations. It also maps the effective refractive indices of the guided modes into propagation angles of the radiation modes. The only disadvantage is the requirement of relatively bulky mechanical clamping structures. Therefore, the prism couplers are considered a laboratory tool instead of part of an integrated optical device.

On the other hand, the grating couplers must be fabricated to designated areas of the planar waveguides with sophisticated submicrometer lithographic techniques. Once fabricated, the grating couplers cannot be removed or adjusted. The main advantage of a grating coupler is in its planar geometry, which makes it part of the guided wave device instead of just a tool. The problems of a grating coupler are the very high skill level required in fabrication and the difficulty in obtaining input coupling efficiency greater than 50%. Therefore, although grating couplers have been investigated extensively, they are seldom used in integrated optical device research activities other than for grating coupler research. Even their future prospect in packaged integrated optical devices has been shadowed by the alternative of an end-fire coupler.

*The Prism Coupler.*   Figure 12.18 illustrates the physical mechanism of a prism coupler. When a laser beam is incident onto the base of a prism at an angle larger than the total internal reflection angle of the prism–air interface, as shown in Fig. 12.18a, the light will be reflected, yielding a standing wave above the prism–air interface and an evanescent wave below the prism–air interface. The evanescent field decays at the rate of $\exp(-\sqrt{n_4^2 \sin\theta^2 - n_1^2}kx)$ and has a phase constant $n_4 k \sin\theta$ along the $z$ direction. In the meantime, the waveguiding mode of Fig. 12.18a also carries an evanescent wave above the air–guide interface with a phase constant $\beta$ along the $z$ direction. This evanescent wave decays at the rate of $\exp(-\sqrt{n_2^2 - (\beta/k)^2 - 1}kx)$. Figure 12.18b illustrates the field distribution in the air gap between the prism base and the waveguide surface in a magnified scale. In these descriptions, $n_4$ is the refractive index of the prism, $n_2$ is the refractive index of the waveguide material, $n_1$ is the refractive index of the prism–waveguide gap, and $\beta$ is the phase constant of the guided mode of interest. Now, if the gap is reduced such that the two evanescent waves may reach the other interface with appreciable field magnitude, a prism coupling condition is established. The evanescent field component at the other interface simply will leak into the other side of the system, causing the two systems to be coupled.

When the prism coupler is used as an output coupler for an optical waveguide, the guided mode will enter the waveguide area under the prism coupler with its evanescent field reaching the prism–air interface. This in turn induces a field inside the prism with a phase constant in the $z$ direction equal to the phase constant of the guided mode. If the refractive index of the prism is higher than the effective mode index of the guided wave, then this induced field inside the prism is expected to propagate away from the prism–air interface at an angle to the normal,

$$\theta = \sin^{-1}\frac{\beta}{n_4 k} \tag{12.51}$$

Thus, the guided mode is coupled out by this prism coupler with the output angle nearly proportional to the effective mode index. When the optical waveguide supports a number of modes, the output will exhibit a fan of beams from which the waveguide mode indices can be obtained. In case of a poor-quality waveguide that scatters, each of the scattered guided modes will come out of the prism as a line in the three-dimensional space. This is the so-called $m$-lines. Figure 12.18c illustrates the output coupling. Since

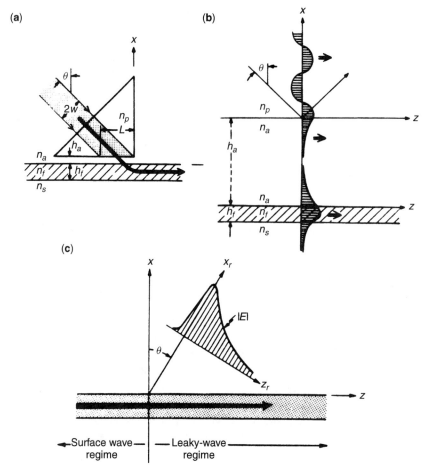

**Figure 12.18**  Prism coupler: (**a**) launching guided wave modes; (**b**) enlarged view of the prism–air gap–waveguide interface and field distribution; (**c**) amplitude variation in the output-coupled field. (After Ref. 135.)

the guided wave is leaking at a constant rate as it enters the prism-covered area of constant air gap, the output beam is expected to have an exponential intensity profile due to depletion of the source. The width of the output beam profile may be adjusted by the height of the air gap and hence the coupling strength or leakage rate.

For the case of input coupling, the laser beam arrives from the inside of the prism toward the air gap, as shown in Fig. 12.18a. The evanescent wave now extends from the prism base to the air–guide interface and induces an optical field in the waveguide having a phase constant equal to $n_4 k \sin\theta$. This phase constant may not correspond to an allowed propagation constant of the particular waveguide of interest. However, this phase constant can be easily adjusted by varying the incidence angle of the laser beam. As long as the prism refractive index $n_4$ is higher than the refractive index of the waveguide, there is always an angle such that the phase constant of the induced field due to the incident laser beam is equal to the phase constant of the guided mode. This angle is again given by Eq. (12.51). The two waves are said to be phase matched when Eq. (12.51) is satisfied and the induced field becomes a guided mode. However, at any arbitrary point of the illuminated prism base of Fig. 12.18**a**, there are four wave components:

1. an incoming evanescent wave from the incident laser beam,

2. an incoming guided wave from the left side due to the induced field,

3. an outgoing guided wave toward the right side that is a combination of the induced field on this site and part of 2, and

4. an outgoing evanescent wave from the reflection of the incident laser beam and the leakage of 2.

Under the phase-matching condition, wave components in 3 will be in phase with each other and the magnitude of the guided wave will increase rapidly as the wave travels toward the $z$ direction. In the meantime, the two components in 4 will interfere destructively to conserve the total power. An ideal condition is for the incident laser beam to have an exponentially increasing magnitude toward the $z$ direction so as to keep up with the increasingly stronger guided wave. Complete power transfer is possible. Interestingly, this ideal condition corresponds to the reciprocal path of the output prism coupler illustrated in Fig. 12.18**c**.

In reality, the input laser beam has a Gaussian distribution instead of the ideal case of exponential distribution. In such a case, the field strength in the guided wave will eventually get to a point such that its leakage equals the power it receives from the incident beam. Optimal power launching occurs when both the illuminated area and the prism base end at that point. This is illustrated in Fig. 12.19. The optimal launch efficiency for a Gaussian laser beam and a prism coupler with constant air gap is 80.1%. In practice, one often adjusts the pressure on the prism to tune the leakage rate for best launch performance. Figure 12.20 gives the calculated maximum possible launch efficiency as a function of the prism coupler leakage rate for a Gaussian beam of width $2W_0 \cos \theta$ with the center of the laser beam offset from the prism corner by an optimal amount $Z_c$. Near-optimal launch performance is achievable over a reasonable range of leakage rate (or laser beam size). Figure 12.21 gives the optimal launch efficiency when the incidence angle of the laser beam is deviated from the phase-matching condition. As expected, when the laser incidence angle error equals the laser beam divergence angle, the launch efficiency will be reduced by a factor of 2.

*The Grating Coupler.* The most important part of a grating coupler design is the periodicity. The grating periodicity provides a wave vector in the $z$ direction for momentum conservation along the $z$ projection between the guided mode and one or more of the radiation modes. This is illustrated in Fig. 12.22. Since the refractive index of the substrate is higher than that of the air, coupling to radiation modes in the air always has a counterpart radiation mode in the substrate. For efficient mode launching, the coupler must not couple the waveguide mode to more than one radiation mode. This requires coupling between a guided mode and a substrate mode. Further complication arises due to the possibility of momentum conservation through higher-order harmonics of the grating wave vector. Therefore, only the reverse grating coupler between the guided modes and the substrate mode can have high efficiency. The problem is in the fabrication of gratings having a wave vector greater than the wave vector of the guided wave itself. The grating periodicity must be less than 0.3 $\mu$m for such grating coupler

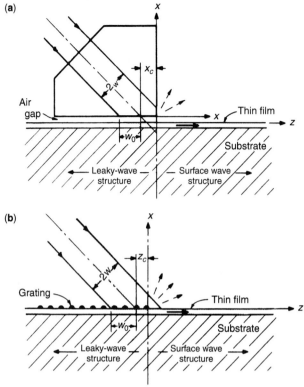

**Figure 12.19**   Alignment of incident beam in prism and grating input couplers for increasing coupling efficiency. Maximum efficiency achievable for Gaussian beams is 80.1%, which is obtained if $\alpha w_0 = 0.68$ and $z_c/w_0 = 0.733$. (After Ref. 134.)

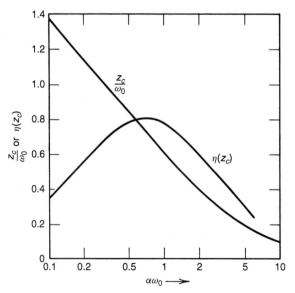

**Figure 12.20**   Optimal beam position offset, $z_c/w_0$, and maximum coupling efficiency at optimal offset, $\eta(z_c)$, as a function of normalized beam diameter, $\alpha w_0$, for a Gaussian beam incident on a prism coupler of leakage rate $\alpha$. (After Ref. 135.)

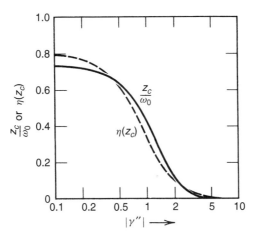

**Figure 12.21**  Optimal beam offset, $z_c/w_0$, and maximum coupling efficiency at optimal offset, $\eta(z_c)$, as a function of normalized beam incidence angle error $\gamma'' = \pi w_0 \Delta/\lambda$. The angular error $\Delta$ is normalized to the laser beam divergence angle $\lambda/\pi w_0$. (After Ref. 135.)

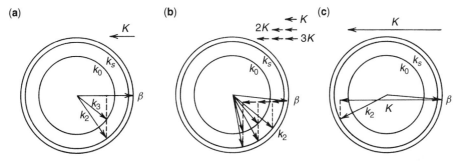

**Figure 12.22**  Grating coupler momentum vector diagrams showing (**a**) waveguide mode vector $\beta$ is scattered by grating vector $K$ producing radiation modes $K_2$ to the substrate and $K_3$ to the air; (**b**) a short grating vector $K$, intended to couple only to one radiation mode $K_2$ to the substrate, may have many radiation modes due to harmonic grating vectors $K$, $2K$, and $3K$; (**c**) a very large grating vector (periodicity much shorter than $\lambda/n$) can scatter the waveguide mode to a single backward substrate mode $K_2$.

to work with the He–Ne laser light on lithium niobate waveguides. In addition, the substrate mode will stay in the substrate unless a prism is mounted to the substrate to bring the light out, as shown in Fig. 12.23. The other important parameter of a grating coupler is the corrugation as well as shape of the phase elements. Due to the very small thickness of the optical waveguide films, there is not much room for the shaping of the phase elements. Most grating couplers reported to date did not attempt to employ shaped geometries.

### 12.2.2  Waveguide Refractive Elements

Optical components are usually made of reflective elements, refractive elements, or diffraction elements. In integrated optics, single-surface reflective elements tend to

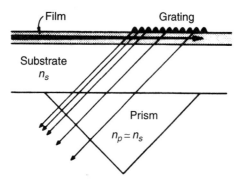

**Figure 12.23**    A prism mounted on the back of substrate serves to bring out the single back scattered output beam from a grating coupler of Fig. 3.22**c**.

have rather limited usefulness. Although waveguides have been etched and waveguide substrates have been cleaved to form clean reflective surfaces, quality as well as reproducibility issues have jeopardized its future. Refractive elements and diffractive elements are far more compatible with integrated optics technologies in general and will be discussed in detail.

Waveguide refractive elements differ from conventional optical refractive elements in many ways. First, the index boundary for waveguide refractive elements tends to be one order of magnitude smaller than for bulk optical refractive elements. Thus, to achieve the same refractive effect, the radius of curvature of refractive boundaries must be considerably smaller than for conventional optics. Second, the material choices as well as considerations are very different. Special considerations must be given to mode conversions at the refractive boundaries and efficiency. Simplicity is a must, and the well-established multielement compound lens concept is not accepted in guided-wave optics. As a result, the waveguide optical refractive lenses used are varieties of fish-eye gradient lenses [141, 142].

The most important part of refractive lens development is the development of waveguide index modification techniques. Two distinctively different techniques have been successfully developed for this purpose: the multilayer waveguide index modification technique and the geodesic curvature waveguide index modification technique. In the following sections, lenses made with these two approaches will be described in detail. Application of these waveguide index modifications to other waveguide refractive elements such as prisms is considered straightforward and will not be discussed here.

*Waveguide Mode Index Modification Techniques.* There are four known methods by which waveguide index modification can be accomplished. Effective index perturbation by waveguide thickness adjustment such as surface corrugation can be obtained by etching or by deposition. Volume index perturbation can be obtained by optical damage in lithium niobate or by ion exchange techniques. The effectiveness of any of these methods in producing a large index perturbation will be affected seriously by the properties of the waveguide employed. Among these techniques, the deposited thin-film overlay is by far the most versatile and thus most widely utilized method. It will be covered with more detail.

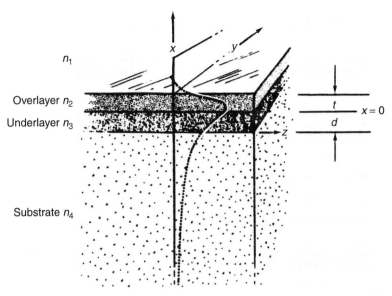

**Figure 12.24**  Cross section of asymmetrical multilayer planar dielectric waveguide showing zero-order mode bound in both layers with evanescent field.

*Multilayer Step Index Waveguide.* Consider the four-layer structure illustrated in Fig. 12.24. It consists of a substrate having a refractive index $n_4$ and superstrate having a refractive index $n_1$, each with an infinite extent with one or two thin films between that serve as the optical waveguides or as a waveguide with an isolation layer. The planar structure considered also has infinite extent in $y$ with optical propagation in the $z$ direction. For wave binding, it is necessary that $n_2$ and/or $n_3$ be greater than $n_1$ and $n_4$. We will not consider the modes bound to the substrate when it is of limited extent. The modes bound to this type of structure may be either transverse electric (TE$_m$) or transverse magnetic (TM$_m$) of order $m$. A transverse field component bounded to both layers (overlayer and underlayer) is illustrated in Fig. 12.24, showing the zero order with the evanescent field extending into the superstrate and substrate. It is possible that the refractive index and the film thickness of these layers be such that wave binding occurs only in that region that has the highest index with an evanescent component in the other layers.

The transverse electric modes (TE) are the eigensolutions of the field equation where the time variation $e^{i\omega t}$ has been suppressed,

$$\left[ \frac{\partial^2}{\partial x^2} + \frac{\partial^2}{\partial z^2} + \omega^2 \varepsilon(x) \mu_0 \right] E_y(x, z) = 0 \qquad (12.52)$$

subject to the following boundary conditions: $E_y \to 0$ as $x \to \pm\infty$, and the tangential components of $E$ and $H$ are continuous across the boundaries. The $z$ dependence will be the same in all four regions for a given mode and can be written as

$$e^{-i\beta_m z}$$

where $m$ is the mode order. Additionally $\varepsilon(x)$ can be expressed as follows

$$\varepsilon(x) = \begin{cases} n_1^2\varepsilon_0 & x \geqslant t \\ n_2^2\varepsilon_0 & t > x > 0 \\ n_3^2\varepsilon_0 & 0 > x > -d \\ n_4^2\varepsilon_0 & -d \geqslant x \end{cases} \tag{12.53}$$

Applying the boundary conditions on $E$ to the solutions of (12.52), the fields in the four regions can be written as

$$E_{m1} = A_m e^{-p_m(x-t)} \sin(h_m t + \phi_m) \sin \gamma_m e^{-i\beta_m z} \tag{12.54a}$$

$$E_{m2} = A_m \sin(h_m x + \phi_m) \sin \gamma_m e^{-i\beta_m z} \tag{12.54b}$$

$$E_{m3} = A_m \sin(l_m x + \gamma_m) \sin \phi_m e^{-i\beta_m z} \tag{12.54c}$$

$$E_{m4} = A_m e^{q_m(x+d)} \sin(-l_m d + \gamma_m) \sin \phi_m e^{-i\beta_m z} \tag{12.54d}$$

where $A_m$ is a normalization constant and $p_m$, $h_m$, $l_m$, and $q_m$ are the transverse components of the propagation vector in regions 1, 2, 3, and 4, respectively. The parameters $p_m$, $h_m$, $l_m$, $q_m$, and $\beta_m$ are related by the following dispersion equations:

$$p_m = (\beta_m^2 - k^2 n_1^2)^{1/2} \tag{12.55a}$$

$$h_m = (k^2 n_2^2 - \beta_m^2)^{1/2} \tag{12.55b}$$

$$l_m = (k^2 n_3^2 - \beta_m^2)^{1/2} \tag{12.55c}$$

$$q_m = (\beta_m^2 - k^2 n_4^2)^{1/2} \tag{12.55d}$$

Application of the boundary conditions on $H$ yields

$$\tan(h_m t + \phi_m) = -\frac{h_m}{p_m} \tag{12.56}$$

$$\tan \phi_m = \frac{h_m}{l_m} \tan \gamma_m \tag{12.57}$$

$$\tan(-l_m d + \gamma_m) = \frac{l_m}{q_m} \tag{12.58}$$

When combined, these equations lead to the following transcendental equation:

$$\frac{1}{l_m} \tan \left( l_m d + \tan^{-1} \frac{l_m}{q_m} \right) + \frac{1}{h_m} \tan \left( h_m t + \tan^{-1} \frac{h_m}{p_m} \right) = 0 \tag{12.59}$$

The roots of this equation are the allowed values of the propagation constants.

The transverse magnetic modes (TM) are the eigensolutions of the field equation

$$\left[ \frac{\partial^2}{\partial x^2} + \frac{\partial^2}{\partial z^2} + \omega^2 \varepsilon(x) \mu_0 \right] H_y(x, z) = 0 \tag{12.60}$$

The development of the TM modes is identical to that for the TE modes, and again the allowed values of the propagation constants are roots of a transcendental equation, which is

$$\frac{n_3^2}{l_m} \tan\left[l_m d + \tan^{-1}\left(\frac{n_4^2}{n_3^2}\frac{l_m}{q_m}\right)\right] + \frac{n_2^2}{h_m}\tan\left[h_m t + \tan^{-1}\left(\frac{n_1^2}{n_2^2}\frac{h_m}{p_m}\right)\right] = 0 \quad (12.61)$$

Tantalum pentoxide and Corning 7059 glass have been employed to form waveguides and lenses on thermally grown $SiO_2$ substrates because they exhibit low scattering loss and may be deposited reproducibly. The single-layer thin-film waveguide dispersion of $Ta_2O_5$ structure on Corning 7440 glass is shown in Fig. 12.25**a** for both TE and TM modes as a function of the normalized film thickness. The thin-film waveguide dispersion of a Corning 7059 glass structure on an $SiO_2$ substrate is shown in Fig. 12.25**b** for both TE and TM modes as a function of the normalized thickness. The effective refractive index approaches the bulk value for thick films and approaches the substrate for thin films. Each of these thin-film waveguide modes exhibits a cutoff where $n_e(m) = 1.47$.

The dispersion for a two-layer structure for only the transverse electric modes is shown in Figs. 12.26**a** and **b**, where $Ta_2O_5$ is employed as the *overlayer* of variable thickness $T$ and where the *underlayer* normalized thickness is constant ($kt = 2.37\pi$), which is equivalent to a 7059 layer having a thickness of 0.75 μm and an optical wavelength of 0.63 μm. Each of the dispersion curves exhibits an inflection representing the transition region where wave binding occurs to only the dense overlayer and where wave binding occurs in both layers. Figure 12.26**b** is an expansion in the region where wave binding occurs in both layers showing the effect of substrate refractive index. The index modification is sufficiently large for practical waveguide refraction element applications.

The dispersion for a two-layer structure for only the transverse electric modes where 7059 is employed as the overlayer of fixed thickness and where the underlayer is $Ta_2O_5$ of variable thickness $T$ is shown in Figs. 12.27**a** and **b**. The fixed-thickness waveguide ($kt = 2.37\pi$) is equivalent to a 7059 layer having a thickness of 0.75 μm for an optical wavelength of 0.63 μm. Although the dispersion data in Figs. 12.26 and 12.27 are similar, the waveguide index modification means by $Ta_2O_5$ underlayer (Fig. 12.27) causes large optical scattering at the refractive boundaries due to the large discontinuity in the waveguiding 7059 glass layer. Therefore, in practice, successful thin-film Luneburg lenses are made by $Ta_2O_5$ overlayer approaches only.

*Overlay Film on Graded Index Waveguide.* Due to the large variations possible in graded index optical waveguides, it is rather difficult to derive the dispersion characteristics for every possible graded index system here. Basically, the mathematical method is similar to the one described in the previous section. One example is given here for the case of TE modes in a graded index optical waveguide with exponential profile.

From Table 12.1, the guided mode is expected to have a field distribution described by the Bessel function of the first kind with noninteger order and with an exponential argument. The higher index overlay film will have an oscillatory field inside the overlay. By matching the fields at the boundaries, the following transcendental equation is

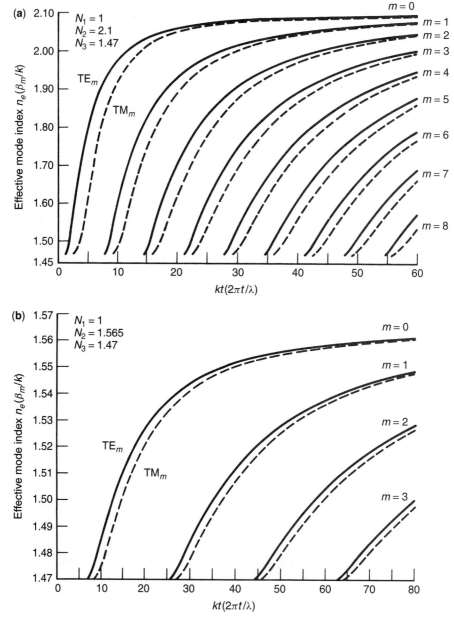

**Figure 12.25** Effective refractive index versus normalized film thickness for (**a**) $Ta_2O_5$ film on $SiO_2$ substrate and (**b**) 7059 glass film on $SiO_2$ substrate: (———) Te modes; (– – –) TM modes.

obtained for the $TE_0$ mode:

$$\frac{\sqrt{n_c^2 - n_b^2}}{\sqrt{n_s^2 - n_b^2}} - \frac{\sqrt{n_c^2 - n_e^2}}{\sqrt{n_s^2 - n_b^2}} \frac{q_2 \sin(q_2 s) + p_1 \cos(q_2 s)}{p_1 \sin(q_2 s) - q_2 \cos(q_2 s)} = \frac{J_{v+1}[(4\pi d/\lambda)/\sqrt{\Delta\varepsilon}]}{J_v[(4\pi d/\lambda)\sqrt{\Delta\varepsilon}]} \quad (12.62)$$

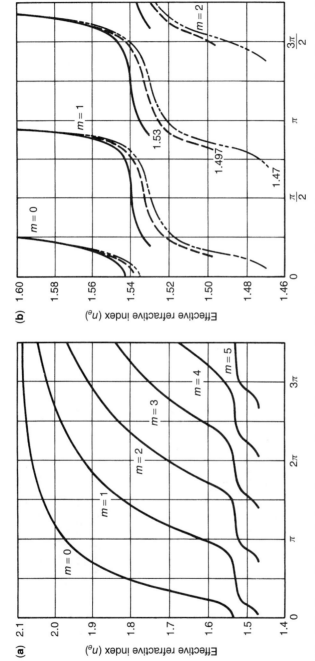

**Figure 12.26** Effective refractive index of composite structure as a function of normalized thickness of the dense overlay film (lens) where the overlay film normalized thickness $kt = 2.37\pi$ for TE modes only (**a**); expanded scale showing effect of three substrates for TE modes only (**b**).

| | |
|---|---|
| Superstrate | $n_1 = 1.0$ |
| Overlayer | $n_2 = 2.1$  *kT* Lens |
| Underlayer | $n_3 = 1.565$  $kt = 2.37\pi$  ($t = 0.75$ μm at $\lambda = 0.63$ μm)  waveguide |
| Substrate | $n_4 = 1.47, 1.497, 1.53$ |

765

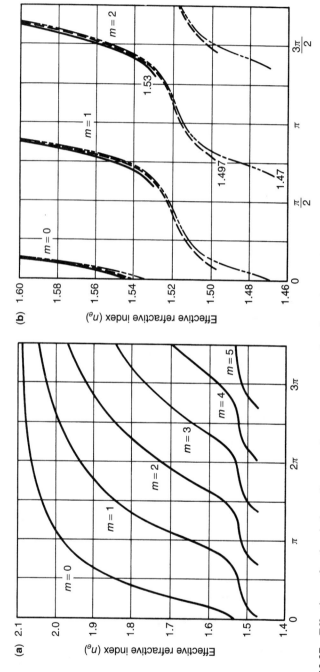

**Figure 12.27** Effective refractive index of composite structure as a function of normalized thickness of the dense underlay film (lens) where the underlay film normalized thickness $kt = 2.37\pi$ for TE modes only (**a**); expanded scale showing effect of three substrates for TE modes only (**b**).

| | | |
|---|---|---|
| Superstrate | $n_1 = 1.0$ | |
| Overlayer | $n_2 = 1.565$ | $kt = 2.37\pi$ ($t = 0.75$ μm at $\lambda = 0.63$ μm) waveguide |
| Underlayer | $n_3 = 2.1$ | $kT$ Lens |
| Substrate | $n_4 = 1.47$ | |

766

where $n_c, n_s$, and $n_b$ are refractive indices for the overlay, the surface of the graded waveguide, and the bulk of the substrate, respectively; $s$ is the overlay thickness; and $d$ is the effective waveguide depth. The modified effective mode index for a Ti in-diffused waveguide on lithium niobate substrate at a wavelength of 0.6328 μm is given in Fig. 12.28 as a function of overlay thickness with the overlay refractive index as a parameter. Note that an overlay thickness of more than 0.1 μm is needed to cause a mode index change of only 0.03. A further increase in overlay thickness results in a multimode waveguide as well as larger mode index modification. This is generally true for index perturbation on a graded index waveguide that has a rather large effective waveguide thickness and a small index difference between the guided mode and the substrate.

In practice, mode index modification on lithium niobate poses special problems due to the difficulty in depositing a film having a higher refractive index than the surface index of metal-diffused lithium niobate [143–145]. Two kinds of films were commonly used for this purpose, namely the $TiO_2$ film of the rutile phase and the $Nb_2O_5$ film. The niobium pentoxide film can be sputtered on lithium niobate surface with low propagation loss (3 dB/cm). However, this film has limited use because its index of refraction is only slightly higher than lithium niobate. The rutile film has a considerably higher refractive index and may be deposited by RF sputtering deposition, reactive sputtering deposition from titanium target, sputtering of titanium followed by oxidation, and E-beam evaporation of titanium followed by oxidation. So far, the first three methods have yielded $TiO_2$ film with either too much attenuation or too low a refractive index. This is reportedly due to the incorporation of contamination in the sputtering process.

Extremely hard and durable transparent films of $TiO_2$ with a refractive index close to that of the pure rutile crystal have been fabricated by E-beam evaporation of a few hundred angstroms thick of Ti under very high vacuum conditions and followed by oxidation in dry oxygen at 450°C for 2–4 hours [143,144]. Experimentally, it was found

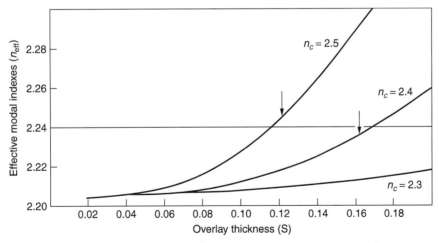

**Figure 12.28**  Effective modal indices of Ti:LiNbO₃ waveguides with a higher index overlay film. The index of the overlay varies from 2.3 to 2.5. Other waveguide parameters are $n_s = 2.24$, $n_b = 2.20$, $d = 0.5$ μm, and wavelength = 0.6328 μm. The arrows indicate when the second-order mode occurs.

that the refractive index of this film depends on the Ti deposition rate as well as on the vacuum background. For a background pressure of $6 \times 10^{-6}$ Torr, the oxidized film index ranges from 2.54 at a slow deposition rate of 0.8 Å/s to 2.72 at a fast deposition rate of 35 Å/s. When the background vacuum pressure is reduced to $6 \times 10^{-7}$ Torr, the oxidized film index becomes a constant value of 2.68 at all deposition rates. This overlay film process has been found compatible with the titanium-diffused waveguide on lithium niobate. However, the waveguide propagation loss in the region covered by 700 Å of this rutile film showed a substantial increase to approximately 20 dB/cm. This relatively high loss and the requirement of heating at 450°C limit the application of the oxidized film to certain diffraction elements on high-temperature metal-diffused waveguides only. Furthermore, the possibility of high waveguide loss due to incomplete oxidation also limits the maximum film thickness to under 0.1 μm, which corresponds to a waveguide propagation loss of about 40 dB/cm.

*Other Index Perturbation Methods.* Two other methods may be used for the creation of index perturbation [145] of optical waveguides in general: the thickness modulation method and the ion exchange or double ion exchange method. These methods can be used to produce high-quality index-modified optical waveguides. The treated waveguide area tends to have a uniform change of waveguide index that limits its application from gradient index type of waveguide optical components such as the Luneburg lenses.

The thickness modulation method is very simple. For practically any waveguide dispersion curve, the guided mode index is a function of waveguide thickness. Therefore, local adjustment of waveguide thickness will produce a corresponding change in the effective waveguide index. This method is extremely easy to implement. A two-step deposition procedure or an etch-back procedure may be employed. Often, the quality of the waveguide is preserved as long as one stays in the single-mode region at all times. However, the magnitude of index modulation is often limited due to the desire to stay in the single-mode region. For step index waveguides, the difference between the effective waveguide index and the substrate refractive index can be large. As a result, significant index perturbation may be obtained. On the other hand, very deep etched steps are required to perturb a graded index waveguide. Due to the small difference between the effective waveguide index and the substrate index, the amount of index modification tends to be very small.

Ion exchange [97] over an existing optical waveguide [146] causes further index perturbation in the volume of the waveguide. The relatively low-temperature process of ion exchange makes it compatible with high-temperature metal-diffused waveguides and many deposited waveguides. The limitation is the relatively small index perturbation that may be achieved by using this method. The one exception is proton ion exchange, which can lead to large index changes in lithium niobate. In this case, index changes greater than 0.1 have been produced.

However, due to the large thickness of graded index waveguides and the volume treatment of the ion exchange process, the proton ion exchange treated graded-index optical waveguide often exhibits multimode behavior unless the amount of dose is significantly reduced. For the purpose of index perturbation, the ion exchange process can be localized by an aluminum mask 0.2 μm thick. Figure 12.29**a** gives experimental results of the refraction index profile of a single-mode titanium in-diffused waveguide on Z-cut lithium niobate substrate after 30 min of benzoic acid (BA) treatment with

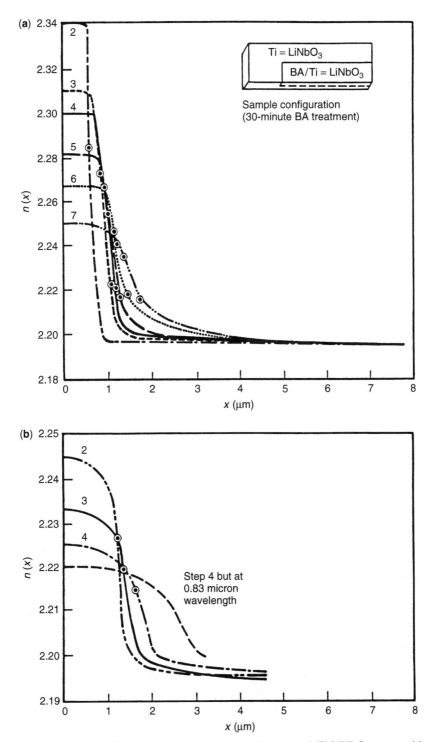

**Figure 12.29** Index profiles and guided modes in a BA-treated Ti:LiNbO$_3$ waveguide with subsequent thermoannealing (**a**); with 10-min BA treatment at 0.63 μm (**b**).

subsequent thermal annealing time as a parameter. The annealing temperature is 240°C. The treated waveguide appears to have two more modes, and the coupling of the titanium waveguide mode and the newer proton ion exchange waveguide modes is very poor. The titanium mode appears unperturbed. Curves 3, 4, 5, 6, and 7 correspond to annealing times of 1, 2, 4, 8, and 16 h, respectively. The annealing process drives the proton ions deeper into the waveguide with reduced surface index value. In the meantime, the titanium mode power becomes more and more coupled to the proton ion exchange waveguide modes that have a higher effective refractive index. After 16 h of annealing, 58% of the titanium waveguide mode power moves into the first proton ion exchange waveguide mode, and the waveguide mode index perturbation is only about 0.03 (from 2.205 for the titanium mode to 2.233 for the ion exchange waveguide mode). Another set of data is given in Fig. 12.29**b**, where the benzoic acid treatment time is reduced to 10 min. At the He–Ne laser wavelength, there are nearly two titanium modes and one ion exchange mode. Curve 2 is the as-perturbed waveguide index profile. Curves 3 and 4 are index profiles after 2 and 8 h of annealing at 240°C. After 8 h of annealing, the conversion efficiency of the titanium waveguide mode to the ion exchange waveguide mode is 74%. Of great interest are the results obtained from the same waveguide at the laser diode wavelength of 0.85 μm. At the longer wavelength, the titanium mode supports only one mode, and over 90% of the titanium mode power is converted to the ion exchange mode when the sample has been annealed for 8 h. However, the index change was only about 0.01.

### Thin-Film Waveguide Luneburg Lens

*Waveguide Luneburg Lens Theory.* Optical lenses with simple form suffer from spherical aberrations particularly when the lenses are thick and the numerical aperture is large. Unfortunately, the desire to develop miniaturized integrated optical devices requires the development of large-numerical-aperture waveguide lenses using thick lenses made of the small index perturbation available on optical waveguides. In order to overcome this serious problem, graded index optical lenses were investigated. It was found that the Luneburg lens can provide large-numerical-aperture imaging without spherical aberration while only a small index perturbation is needed.

The "classical" Luneburg lens [147] is an inhomogeneous positive refractor with radial symmetry that images a plane wavefront to a hemispherical spot located on the opposite boundary of the refractor. Ray traces through the classical Luneburg lens are illustrated in Fig. 12.30. Interest in the Luneburg lens arises because of its $4\pi$ field of view and because it is free of all aberrations except field curvature. The cross section of a dielectric waveguide Luneburg lens is illustrated in Fig. 12.31, where a circular overlay film is employed with a prescribed radial thickness function. For integrated optics, its simplicity of fabrication in a thin-film structure is an important attribute. The circular symmetry is of particular interest where the substrate area is limited because folding of the optical axis may use the same lens for several purposes. The waveguide Luneburg lens has a further advantage that all refractive index variations are continuous, smooth, slow functions with respect to the optical wavelength. Mode conversion in the lens is suppressed by the graded thickness profile and by symmetry. For a classical Luneburg lens the refractive index $n(r, m)$ is

$$n(r, m) = n_e(m)\sqrt{2 - \left(\frac{r}{r_0}\right)^2} \qquad r \leqslant r_0 \qquad (12.63)$$

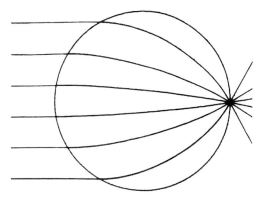

**Figure 12.30**   Ray trajectory through Luneburg lens where $f$-number $= 0.5$, which is equivalent to Morgan's $s = 1$.

**Figure 12.31**   Cross section of a waveguide Luneburg lens using a dense $Ta_2O_5$ overlay lens on a Corning 7059 waveguide on a thermally oxidized silicon substrate.

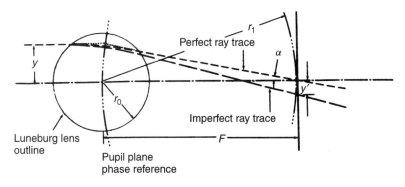

**Figure 12.32**   Generalized Luneburg lens and ray trace definitions.

where $n_e$ is the refractive index of the medium surrounding the lens, $r_0$ is the lens radius, and $r$ is the radial coordinate.

The classical Luneburg lens has been generalized by Morgan [148]. The generalized Luneburg lenses may have incidence rays originated from a point at a finite distance from the lens, and the focused image point is not restricted to the edge of the lens. Figure 12.32 illustrates a generalized Luneburg lens with a focal plane curvature of $r_1$. A perfect ray is shown to arrive at the focal point of the focal plane. An imperfect ray is shown to intercept the focal plane with a ray intercept error of $y'$.

The radial refractive index profile for some generalized Luneburg lenses can be obtained from a paraxial approximation when the $f$-number is over 1.5. The normalized refractive index profile for a focal length of $n$ is given by

$$\frac{n(r, m)}{n_e(m)} = \exp \frac{\sqrt{n_e(m)^2 r_0^2 - r^2 n^2(r, m)}}{\pi n_e(m) r_1}.$$ (12.64)

Since Eq. (12.64) is dependent on the waveguide mode index, a multimode waveguide will have a number of focal planes corresponding to the number of waveguide modes.

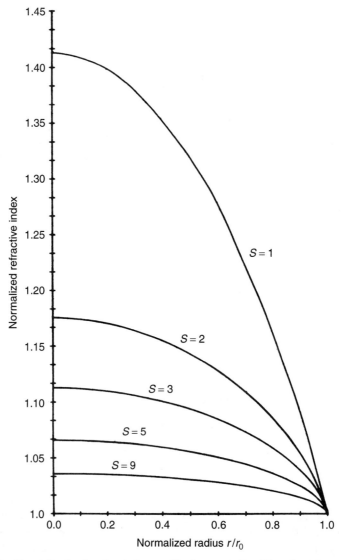

**Figure 12.33** Normalized refractive index profile for generalized Luneburg lenses for various $f$-numbers $= s/2$. (After Ref. 150.)

Only one of these modes will experience perfect focusing for a given index profile. From Eq. (12.64), the normalized refractive index profiles for a number of generalized Luneburg lenses with different $f$-numbers can be calculated and are given in Fig. 12.33.

For the general Luneburg lenses without the restriction that the $f$-number be greater than 1.5, the radially symmetric index profile is the solution of the following integral equations:

$$n = \exp[\omega(\rho, s)] \tag{12.65}$$

where $\rho = rn$ and

$$\omega(\rho, s) = \frac{1}{\pi} \int_{\rho}^{1} \frac{\arcsin(x/s)}{(x^2 - \rho^2)^{1/2}} dx \tag{12.66}$$

Southwell [149] has been able to describe the solution of these integral equations by a power series expansion of only five terms, which is given by Eqs. (12.67). The coefficients for this power series can be determined by least mean squares fit to numerically generated index values from the preceding integral equations. The errors due to the power series approximation is reportedly as small as $10^{-6}$. Table 12.9 lists the coefficients for several generalized Luneburg lenses with different normalized focus values. Table 12.10 gives numerical values for the index profile of four generalized Luneburg lenses:

$$\omega(\rho, s) = p_1(1 - \rho)^{1/2} + p_2(1 - \rho)^{3/2}$$
$$+ p_3(1 - \rho)^{5/2} + p_4(1 - \rho)^{7/2} + p_5(1 - \rho)^{9/2} \tag{12.67a}$$

with

$$p_1 = \frac{\sqrt{2}}{\pi} \arcsin \frac{1}{s} \tag{12.67b}$$

and

$$p_5 = \omega(0, s) - p_1 - p_2 - p_3 - p_4 \tag{12.67c}$$

Once the lens index profile is known, the waveguide mode index of interest must be tailored precisely using index modification techniques. When a high index overlay film is employed for this purpose, the overlay film thickness must be calculated using

**TABLE 12.9  Coefficients for Four Examples of Generalized Luneburg Lenses**

|       | $s = 2$ | $s = 3$ | $s = 5$ | $s = 9$ |
|-------|---------|---------|---------|---------|
| $p_1$ | 0.235687835 | 0.152976535 | 0.0906399959 | 0.0501194645 |
| $p_2$ | $-7.47500358 \times 10^{-2}$ | $-4.22494061 \times 10^{-2}$ | $-2.34300068 \times 10^{-2}$ | $-1.26356394 \times 10^{-2}$ |
| $p_3$ | $6.72894476 \times 10^{-3}$ | $-1.93175172 \times 10^{-3}$ | $-2.51100017 \times 10^{-3}$ | $-1.63934418 \times 10^{-3}$ |
| $p_4$ | $-5.14447054 \times 10^{-3}$ | $-8.25244897 \times 10^{-4}$ | $-1.49192458 \times 10^{-4}$ | $-4.07947091 \times 10^{-5}$ |
| $p_5$ | $-9.89299661 \times 10^{-4}$ | $-1.19124255 \times 10^{-3}$ | $-7.44793717 \times 10^{-4}$ | $-4.11582122 \times 10^{-4}$ |

**TABLE 12.10    Normalized Refractive Index Profiles for Generalized Luneburg Lenses**

| $r$ | $s = 2$ | $s = 3$ | $s = 5$ | $s = 9$ |
|---|---|---|---|---|
| 0 | 1.175311212 | 1.112688200 | 1.065884534 | 1.036025859 |
| 0.05 | 1.174999407 | 1.112507930 | 1.065788087 | 1.035976411 |
| 0.10 | 1.174071705 | 1.111969546 | 1.065500332 | 1.035828873 |
| 0.15 | 1.172521576 | 1.111069129 | 1.065018948 | 1.035581965 |
| 0.20 | 1.170340993 | 1.109801290 | 1.064340603 | 1.035233817 |
| 0.25 | 1.167519780 | 1.108158811 | 1.063460759 | 1.034781859 |
| 0.30 | 1.164044879 | 1.106132212 | 1.062373426 | 1.034222692 |
| 0.35 | 1.159899514 | 1.103709214 | 1.061070836 | 1.033551908 |
| 0.40 | 1.155062156 | 1.100874045 | 1.059543028 | 1.032763858 |
| 0.45 | 1.149505212 | 1.097606509 | 1.057777265 | 1.031851324 |
| 0.50 | 1.143193243 | 1.093880684 | 1.055757219 | 1.030805062 |
| 0.55 | 1.136080437 | 1.089663045 | 1.053461790 | 1.029613132 |
| 0.60 | 1.128106836 | 1.084909633 | 1.050863316 | 1.028259883 |
| 0.65 | 1.119192434 | 1.079561599 | 1.047924746 | 1.026724355 |
| 0.70 | 1.109227400 | 1.073537799 | 1.044594938 | 1.024977603 |
| 0.75 | 1.098054844 | 1.066721643 | 1.040800259 | 1.022977928 |
| 0.80 | 1.085437716 | 1.058935566 | 1.036428193 | 1.020661529 |
| 0.85 | 1.070987513 | 1.049884988 | 1.031290974 | 1.017921692 |
| 0.90 | 1.053981685 | 1.039009952 | 1.025027636 | 1.014552365 |
| 0.91 | 1.050159236 | 1.036524794 | 1.023580201 | 1.013768729 |
| 0.92 | 1.046155078 | 1.033902395 | 1.022045349 | 1.012935510 |
| 0.93 | 1.041942017 | 1.031119755 | 1.020407545 | 1.012043672 |
| 0.94 | 1.037484456 | 1.028146025 | 1.018645667 | 1.011080865 |
| 0.95 | 1.032734249 | 1.024938133 | 1.016729693 | 1.010029408 |
| 0.96 | 1.027623448 | 1.021432662 | 1.014614324 | 1.008862349 |
| 0.97 | 1.022050589 | 1.017529080 | 1.012225136 | 1.007534739 |
| 0.98 | 1.015851602 | 1.013049431 | 1.009422596 | 1.005960167 |
| 0.99 | 1.008726930 | 1.007614146 | 1.005874179 | 1.003921693 |
| 1.00 | 1 | 1 | 1 | 1 |

Eqs. (12.59) and (12.61). For the case of TE modes, Eq. (12.59) is rewritten as

$$ht = \begin{cases} \tan^{-1} \dfrac{h[1 - (l/q)\tan(ld)] + (ph/l)[l/q + \tan(ld)]}{(h^2/l)[l/q + \tan(ld)] - p[1 - (l/q)\tan(ld)]} & n < n_3 \\[2em] \tan^{-1} \dfrac{h[(q + l)e^{2ld} + q - l] + (ph/l)[(q + l)e^{2ld} - q + l]}{(h^2/l)[(q + l)e^{2ld} - q + l] - p[(q + l)e^{2ld} + q - l]} & n \geqslant n_3 \end{cases}$$

$$(12.68)$$

for the TE modes. If the parameters $p$, $h$, $l$, and $q$ are replaced by $p/n_1^2, h/n_2^2, l/n_3^2$, and $q/n_4^2$, Eq. (12.68) can be employed for TM modes.

Figure 12.34 gives the calculated $Ta_2O_5$ overlay film thickness for generalized Luneburg lenses on a 7059 glass waveguide on oxidized silicon substrate. The 7059 glass waveguide has a refractive index of 1.565 and uniform thickness of 1.0665 μm, and the silicon dioxide layer has a refractive index of 1.47. The optical wavelength is assumed to be 0.9 μm. Numerical data are given in Table 12.11. A ray tracing program has been developed for gradient index optical lenses and has been used to evaluate the Luneburg lens with the index profile given in Tables 12.10 and 12.11.

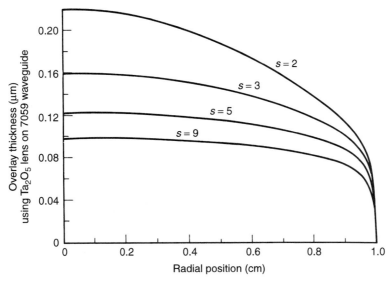

**Figure 12.34**  Waveguide overlay thickness profile for generalized Luneburg lenses using $Ta_2O_5$ on Corning 7059 ($d = 1.0665 \ \mu m$) on $SiO_2$, which is tabulated in Table 3.11. (After Ref. 142.)

Figure 12.35 gives the ray tracing results for a Luneburg lens with $s = 2$, showing good potential performances. For a given thin-film Luneburg lens thickness profile, the focal length may be varied slightly when a thickness error occurs. Figure 12.36 gives the focus sensitivity of a $Ta_2O_5$ thin-film Luneburg lens over a 7059 glass waveguide using computer simulation programs.

*Fabrication of Thin-Film Luneburg Lens.*  One method of forming thin-film waveguide lenses is by shadowed deposition through an appropriate mask. The RF sputter process has been employed throughout for this purpose. The cross section of the RF sputter deposition sample is depicted in Fig. 12.37**a**. Argon ions are accelerated toward the Ta or Nb cathode target where bombardment releases molecules that are oxidized during their flight until collision. In the sputtering environment employed, the mean-free path length is greater than the thickness of the mask, and thus, effects of collision and diffusion are avoided. A Lambertian distribution has been assumed and verified. Thus, the molecular beam arrives at the mask entrance with a predictable angular momentum distribution $A(\theta, v)$. Where the mask produces a shadow, integration over the particle velocity range provides an angular distribution function $A_d(\theta)$. The distribution function $A_d(\theta)$ was experimentally measured with a knife-edge mask confirming the Lambertian source. Based upon this knowledge, the effect of a shadow mask upon sputtering has been predicted with the aid of a computer [150, 151].

The shadow mask computation is relatively simple when the mask has circular symmetry. The coordinates employed to compute the distribution are illustrated in Fig. 12.37**b**. For an arbitrary point on the substrate, $P_0$, the thickness distribution becomes

$$T(x_0, y_0) = \int\int_s \frac{A(\theta)\cos\theta}{R^2} dx \, dy \qquad (12.69)$$

**TABLE 12.11    Overlay Thickness (μm) for Optical Waveguide Luneburg Lenses**

| r (cm) | $s = 2$ | $s = 3$ | $s = 5$ | $s = 9$ |
|--------|---------|---------|---------|---------|
| 0      | 0.2191  | 0.1591  | 0.1223  | 0.0979  |
| 0.05   | 0.2187  | 0.1590  | 0.1222  | 0.0979  |
| 0.10   | 0.2177  | 0.1585  | 0.1220  | 0.0978  |
| 0.15   | 0.2160  | 0.1578  | 0.1216  | 0.0975  |
| 0.20   | 0.2135  | 0.1567  | 0.1211  | 0.0972  |
| 0.25   | 0.2105  | 0.1554  | 0.1204  | 0.0968  |
| 0.30   | 0.2068  | 0.1537  | 0.1196  | 0.0963  |
| 0.35   | 0.2024  | 0.1518  | 0.1186  | 0.0957  |
| 0.40   | 0.1975  | 0.1495  | 0.1174  | 0.0949  |
| 0.45   | 0.1920  | 0.1469  | 0.1160  | 0.0940  |
| 0.50   | 0.1860  | 0.1440  | 0.1141  | 0.0930  |
| 0.55   | 0.1794  | 0.1406  | 0.1126  | 0.0918  |
| 0.60   | 0.1722  | 0.1370  | 0.1105  | 0.0905  |
| 0.65   | 0.1646  | 0.1328  | 0.1081  | 0.0888  |
| 0.70   | 0.1563  | 0.1282  | 0.1054  | 0.0869  |
| 0.75   | 0.1472  | 0.1229  | 0.1022  | 0.0846  |
| 0.80   | 0.1374  | 0.1169  | 0.0983  | 0.0817  |
| 0.85   | 0.1262  | 0.1097  | 0.0935  | 0.0708  |
| 0.90   | 0.1130  | 0.1006  | 0.0870  | 0.0727  |
| 0.91   | 0.1099  | 0.0984  | 0.0853  | 0.0713  |
| 0.92   | 0.1067  | 0.0960  | 0.0835  | 0.0697  |
| 0.93   | 0.1031  | 0.0933  | 0.0814  | 0.0679  |
| 0.94   | 0.0993  | 0.0903  | 0.0790  | 0.0657  |
| 0.95   | 0.0949  | 0.0869  | 0.0762  | 0.0632  |
| 0.96   | 0.0898  | 0.0827  | 0.0728  | 0.0601  |
| 0.97   | 0.0835  | 0.0774  | 0.0682  | 0.0559  |
| 0.98   | 0.0749  | 0.0699  | 0.0616  | 0.0499  |
| 0.99   | 0.0597  | 0.0562  | 0.0496  | 0.0395  |
| 1.00   | 0       | 0       | 0       | 0       |

*Note*: The waveguide parameters are $n_1 = 1, n_2 = 2.1, n_3 = 1.565, n_4 = 1.47, d = 1.0665$ μm, and $\lambda = 0.9$ μm.

where

$$\theta = \cos^{-1} \frac{D}{\sqrt{(x_1 - x_0)^2 + (y_1 - y_0)^2}} \qquad R = \sqrt{(x_1 - x_0)^2 + (y_1 - y_0)^2 + z_1^2}$$

$dx\,dy$ is an infinitesimally small area of the mask entrance window, and $S$ is the portion of the entrance window that is visible from point $P$. The distance $D$ in Fig. 12.37**b** extends from the plane containing the top of the mask to the substrate surface.

The determination of the effective window function, $S$, for a mask contour of a general cylinder is also illustrated by Fig. 12.37. The edge contour is conveniently described by $r = f(z)$. The line of sight between points $P_0$ and $P_1$ is given by the solution of

$$z_1 y + (y_0 - y_1)z - z_1 y_0 = 0$$
$$z_1 x + (x_0 - x_1)z - z_1 x_0 = 0 \qquad (12.70)$$

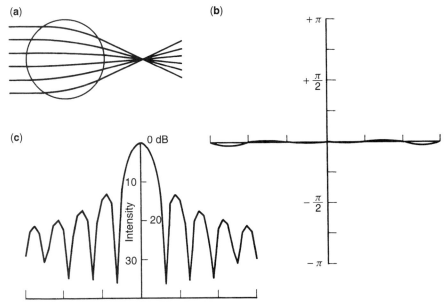

**Figure 12.35**  (a) Ray traces of an $s = 2$ generalized Luneburg lens using the index profile derived from Eq. (12.67), (b) the wavefront phase error in the pupil plane, and (c) the corresponding intensity diffraction pattern using a logarithmic scale in the image plane. (After Ref. 142.)

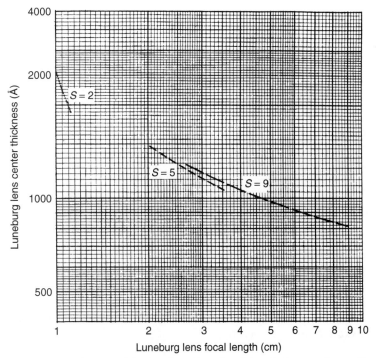

**Figure 12.36**  Calculated waveguide Luneburg lens focal length for $s = 2$, $s = 5$, and $s = 9$ perfect profiles where a fractional thickness error has been introduced. (After Ref. 142.)

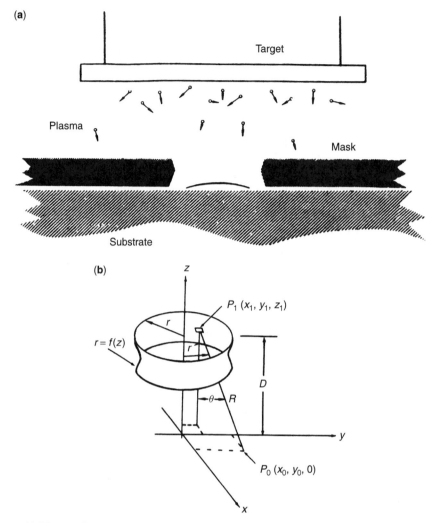

**Figure 12.37** (a) Cross section of RF sputter deposition chamber used to profile Luneburg lenses and taper waveguide films. (After Ref. 152.) (b) Coordinate system for Luneburg lens mask edge synthesis. (After Ref. 152.)

which defines two planes each parallel to the $x$- or $y$-axis while containing the line of sight $P_0 P_1$. At any point on this line, the distance to the axis of rotation of the mask is therefore

$$r' = \left[ \left( \frac{(x_0 - x_1)z}{z_1} - x_0 \right)^2 + \left( \frac{(y_0 - y_1)z}{z_1} - y_0 \right)^2 \right]^{1/2} \tag{12.71}$$

If the line of sight is not interrupted by the edge of the shadow mask, the following function must not have a real solution:

$$(r')^2 - r^2 = \left( \frac{(x_0 - x_1)z}{z_1} - x_0 \right)^2 + \left( \frac{(y_0 - y_1)z}{z_1} - y_0 \right)^2 - f^2(z) = 0$$

For $f(z)$ described by quadratic functions, an analytic result may be easily obtained. For $f(z)$ of higher-order functions, numerical evaluation must be employed. When the preceding function becomes greater than or equal to zero, it means that interference with the line of sight is occurring. Thus, the deposition distribution function becomes

$$T(x_0, y_0) = \int \int_w \frac{A(\theta) \cos \theta \, G}{R^2} dx \, dy \qquad (12.72)$$

where $W$ is the entrance window and $G$ is a logic function that is 1 (or zero) when the line of sight is clear (or interrupted).

With a computer model a conical mask can be computed for a given deposition thickness profile. A cross section of the conical mask and its descriptive parameters are depicted in Fig. 12.38**a**. The effect of changing the conical angle is depicted in Fig. 12.38**b**. The effect of the mask spacing above the substrate is depicted in Fig. 12.38**c**, and the effect of mask aperture diameter is depicted in Fig. 12.38**d**. The accuracy of these computations has been confirmed by a stylus probe, including measurements of the slope and steps for a number of shadow masks.

From these variational effects, depicted in Fig. 12.39, a first-iteration approximation to a Luneburg lens thickness profile was developed using two conical sections, as illustrated by the cross section in Fig. 12.39. The resulting calculated thickness profile is indicated together with an experimentally measured segment near the edge of the deposition. Using the preceding computer simulation programs, the thickness profile is converted to a refractive index profile with ray traces through the lens calculated to find the ray intercept errors at the focus. Figure 12.40 depicts the calculated ray intercepts for the thickness profile of Fig. 12.39 derived from the conical mask. The insert illustrates the ray traces. The experimental points concerning the calculations have been taken from the measured ray traces photographed in Fig. 12.41 using a waveguide Luneburg lens formed with mask SK-1.

A second mask (SK-2) to more closely approximate the ideal lens thickness profile has been developed that employs 12 conical segments to shape the mask edge. The resulting computed thickness profile is depicted in Fig. 12.42 and compared with the profile generated from mask SK-1 and the Luneburg lens ideal thickness profile. A much closer approximation has been achieved. The computed enlarged cross section of the improved mask SK-2 is depicted in Fig. 12.43. Its thickness is 6 mm and its diameter is 8 mm. This computer-generated profile was transferred directly for fabrication of the mask. Figures 12.44**a** and **b** show the measured focal distribution for thin-film Luneburg lenses made by the 2-section mask SK-1 and the 12-section mask SK-2, respectively. The trace of Fig. 12.44**b** indicates diffraction-limited performance.

***Geodesic Lenses.*** Geodesic [153] waveguide lenses have been known and utilized for many years in microwave applications. In waveguide optics form, a shallow spherical depression is drilled into the substrate before the fabrication of a uniform waveguiding layer. As a guided optical ray enters the depression region, its actual path will follow a geodesic over the curved depression as ascertained by Fermat's principle. Thus, for a spherical depression, this is along an arc of a great circle. When this optical path is projected to the plane of the waveguide, the incident rays exhibit a change in direction at the entrance and exit of the waveguide depression region. The behavior is somewhat analogous to a ray plot of a meridional fan through a conventional lens wherein an abrupt change in the direction of the rays occurs at the lens boundaries. However,

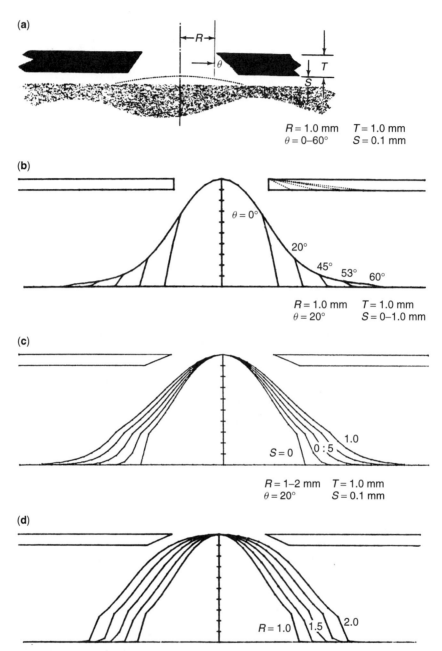

**Figure 12.38** Cross section of mask (**a**) and calculated thin-film deposition profiles for variable cone angle (**b**), variable mask spacing above substrate (**c**), and variable mask aperture diameter (**d**).

although a spherical optical depression is easy to prepare, spherical geodesic lenses behave rather poorly in terms of diffraction-limited performances. One exception is the case of a hemisphere in which all rays from one edge come to perfect focus on the opposite edge. This is the geodesic analog of the Maxwell "fisheye" lens. Unfortunately,

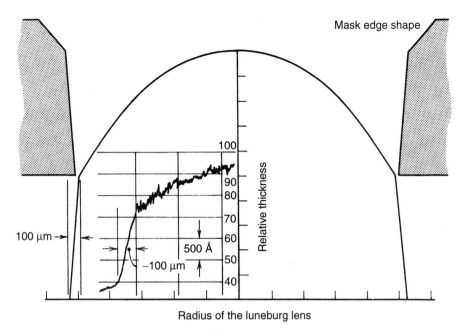

Radius of the luneburg lens

**Figure 12.39**  Computed Luneburg lens thickness profile for SK-1 mask edge shape indicated with a confirming measured stylus thickness profile. (After Ref. 151.)

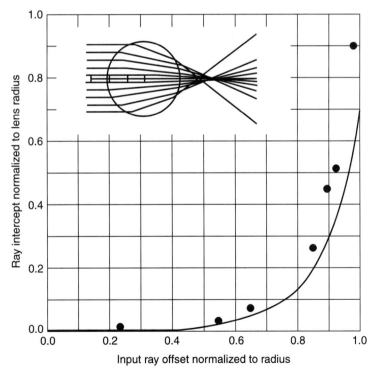

**Figure 12.40**  Calculated and measured ray intercept for waveguide Luneburg lens using mask SK-1. (After Ref. 151.)

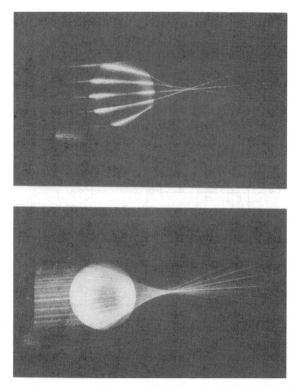

**Figure 12.41**   Measured ray traces for waveguide Luneburg lens formed with mask SK-1. (After Ref. 151.)

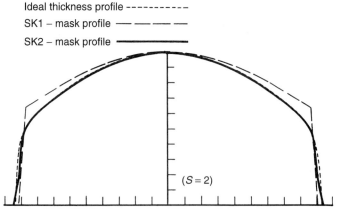

Ideal thickness profile ----------
SK1 – mask profile — — — — —
SK2 – mask profile ▬▬▬▬▬▬

$(S = 2)$

**Figure 12.42**   Sputtered deposition profiles for waveguide Luneburg lens using computer-generated masks. (After Ref. 152.)

the near 90° bend for waveguides at the hemispherical edges is not practical for optical waveguides as it resembles a perfect exit termination.

Various approaches have been investigated for the correction of spherical aberration in a spherical geodesic lens [154–161]. A conical transition region [159] and

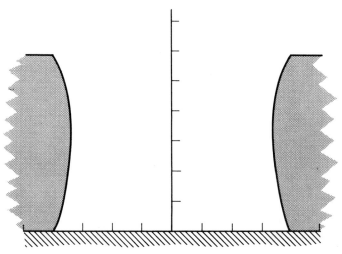

**Figure 12.43**  Computed cross section of SK-2 mask employed for optical waveguide Luneburg lenses which yield diffraction-limited results (mask thickness 6 mm, diameter 8 mm). (After Ref. 152.)

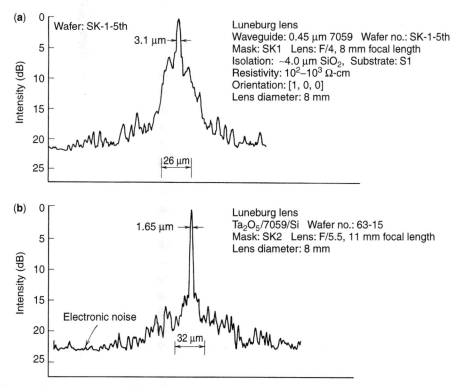

**Figure 12.44**  Measured diffraction pattern: (**a**) Lens using mask SK-1 and Tropel F/3 reimaging lens: lens, F/4, focal length 8 mm; wafer, SK-1-5th. (**b**) Waveguide Luneburg lens fabricated using mask SK-2 using a cleaved exit face at the focus and Zeiss X40, F/.67 microscope objective showing diffraction-limited performance: lens, F/5.5, focal length 11 mm (expanded-scale insert).

a toroidal transition region [157] have been recommended near the boundary of the spherical geodesic depression. When properly designed, the added geodesic transition region serves to not only improve the mode conversion loss at the lens edges but also compensate for part of the spherical aberration in a spherical geodesic lens. Modification of the waveguide parameters on the two sides of the geodesic boundary can also improve lens performance. Another approach calls for the deposition of a phase front correction overlay thin film at one edge of the geodesic lens. This thin film is to have uniform thickness but a carefully designed area shape. A more elegant approach is to generate an aspherical geodesic surface in analogy of the thin-film Luneburg lens. This aspherical geodesic lens is called a generalized Rinehart–Luneburg lens [153]. The issue is in the fabrication of such aspherical geodesic surfaces.

*Spherical Geodesic Waveguide Lenses.* Consider a spherical depression with radius of curvature $R$; the radius of the depression region at the lens contour intersection with the substrate surface, $R_c$; and the polar angle of the lens, $\theta$, where $R_c = R \sin\theta$. The paraxial focus of the spherical geodesic lens can be determined by trigonometry. The normalized focal length, which is defined as the distance between the focus and the center of the depression measured on the substrate plane normalized to the depression radius $R_c$, is simply

$$\frac{f}{R_c} = \frac{1}{2(1 - \cos\theta)} \tag{12.73}$$

The normalized paraxial focal length for a spherical geodesic lens as a function of the spherical geodesic half angle is given in Fig. 12.45. Deeper geodesic depression is required for shorter focal lengths. The ray tracing of a spherical geodesic lens is

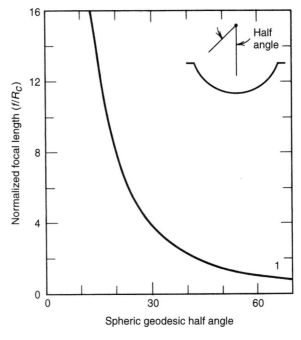

**Figure 12.45**   Spheric geodesic lens normalized focal length as function of half-angular sector.

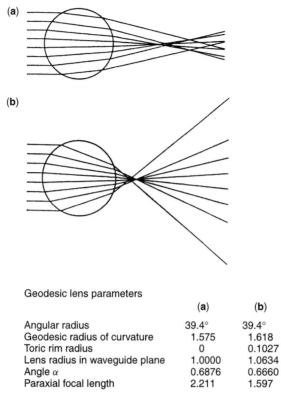

Geodesic lens parameters

|                                | (a)      | (b)      |
| ------------------------------ | -------- | -------- |
| Angular radius                 | 39.4°    | 39.4°    |
| Geodesic radius of curvature   | 1.575    | 1.618    |
| Toric rim radius               | 0        | 0.1027   |
| Lens radius in waveguide plane | 1.0000   | 1.0634   |
| Angle $\alpha$                 | 0.6876   | 0.6660   |
| Paraxial focal length          | 2.211    | 1.597    |

**Figure 12.46**  Computed ray traces through spheric geodesic lens having (**a**) sharp edge and (**b**) toroidal lip for parameters listed having half-angular width 39.4°.

given in Fig. 12.46**a** in which the poor quality of focus is obvious. The rather large longitudinal aberration is displayed in Fig. 12.47 as the curve with $A = 0$.

Transition to the edge of the spherical geodesic lens has been studied for its effect on lens aberrations using ray tracing techniques. Figure 12.48 defines the radius of curvature for a rounded toroidal edge. The ray traces through a spherical geodesic lens with a normalized (relative to $R_c$) toroidal curvature of 0.1 are given in Fig. 12.46**b**, and the longitudinal aberration is plotted in Fig. 12.47 showing remarkable improvement.

The spherical geodesic lens can also be modified by simply making the optical waveguides on the two sides of the geodesic edge different [155]. Figure 12.49**a** shows the improvement in longitudinal aberration when the effective waveguide index in the geodesic depression region is made slightly higher than the waveguide index in the planar region. Experimental data are given in Fig. 12.49**b**. Figure 12.50 gives the optimal index ratio for waveguides inside and outside of the geodesic boundary. The lower trace in Fig. 12.50 compares the normalized focal length for an optimized spherical geodesic lens and a regular spherical geodesic lens. As expected, the higher index waveguide in the geodesic region increases the focusing power slightly.

*The Generalized Rinehart–Luneburg Geodesic Lens.* The modification on the waveguide effective index due to surface curvature can be utilized to generate the effective index profile of a generalized Luneburg lens [153, 159, 162]. This can be accomplished

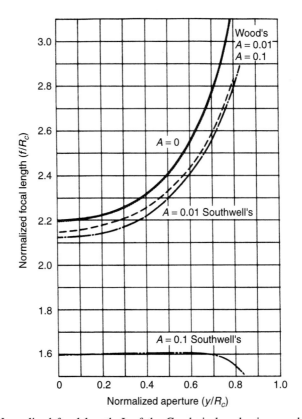

**Figure 12.47**   Normalized focal length $L$ of the Geodesic lens having a spherical depression half angle $\theta_0 = 39.4$ with different amounts of edge rounding. The quantities $L$, $Y$, and $A$ are in units of waveguide lens radius prior to edge rounding. Also shown are the results of Wood's approximation [24]. Note the improved focusing properties of the lens with edge rounding $A = 0.1$.

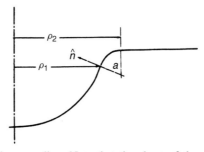

**Figure 12.48**   Toroidal edge rounding. Note that the slope of the geodesic curve must match the slope of the toroidal region at the point of intersection.

by letting the optical path length of a ray path over the assumed aspherical geodesic surface be equal to the optical path length of a corresponding ray through the gradient index thin-film Luneburg lens. Since the thin-film Luneburg lens has no surface depression while the geodesic lens has a constant refractive index, the Fermat principle

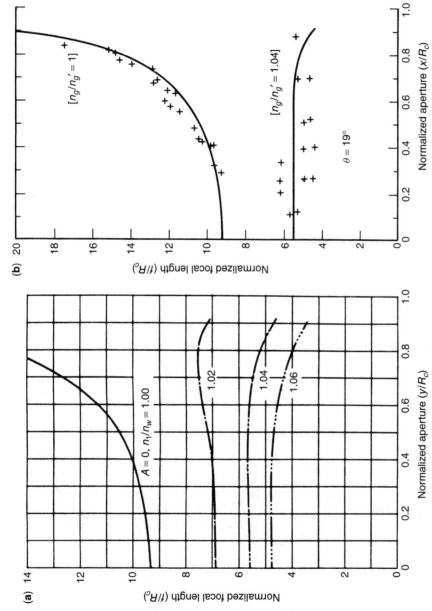

**Figure 12.49** (a) Normalized focal length $f/R_c$ versus normalized distance $y/R_c$ of the incident ray. (After Ref. 155.) (b) Normalized focal length $f/R_c$ versus normalized distance $x/R_c$ of the incident beam from the axis for a spherical depression before and after compensation of the aberrations. Points are experimental; curves are theoretical result ($\theta = 19°$, $R_c = 2.5$ mm, $\lambda = 0.633$ μm). (After Ref. 155.)

787

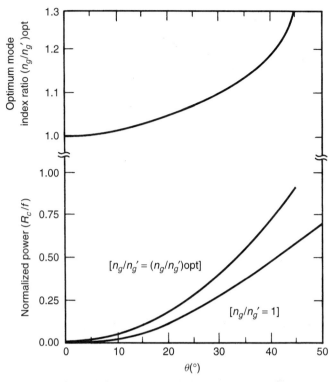

**Figure 12.50**   Ratio of the mode indices $n_g/n'_g$ inside and outside of the depression for best compensation of the aberration and normalized power $R_c/f$ for an uncompensated ($n_g/n'_g = 1$) depression lens and a depression lens with optimum compensation plotted versus the angle $\theta$. (After Ref. 155.)

leads to

$$\rho = rn(r) \qquad (12.74a)$$

and

$$\frac{ds}{dr} = n(r) \qquad (12.74b)$$

where $\rho$ is the radius for an arbitrary point of the geodesic depression projected onto the waveguide plan, $n(r)$ is the refractive index of the Luneburg lens normalized to the waveguide index, and $ds$ is the derivative along the profile of the geodesic curve. The definition of these variables for the geodesic lens are given in Fig. 12.51. From Eq. 12.74a), Eq. (12.74b), and Fig. 12.51**a**, the geodesic generating function $Z(\rho)$ can be obtained as

$$Z = -Z_0 + \int_0^\rho \left[ \frac{\rho}{n} \frac{dn}{d\rho} \left( \frac{\rho}{n} \frac{dn}{d\rho} - 2 \right) \right]^{1/2} d\rho \qquad (12.75)$$

Equation (12.75) may be evaluated numerically. Table 12.12 gives four sets of the aspherical geodesic lens depression depth values for four normalized focal lengths. From these values, the cross-sectional profiles of these four ideal aspherical geodesic lenses (also called generalized Rinehart–Luneburg lenses) are shown in Fig. 12.52.

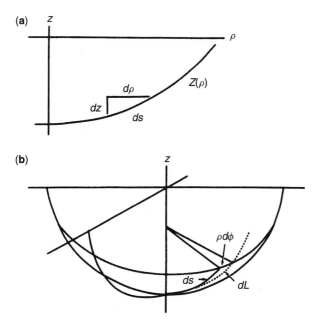

**Figure 12.51** (a) Profile of waveguide depression showing the generating curve $Z(\rho)$ for the surface of revolution. The geodesic lens is represented by rotating $Z(\rho)$ about the $z$-axis. (b) Waveguide depression showing orthogonal surface infinitesimals $ds$ and $\rho\, d\phi$. Distances between neighboring points on the surface are given by $dL = [ds^2 + \rho^2 d\phi^2]^{1/2}$. The angle $\theta$ is between $dL$ and $ds$, so that $\sin\theta = \rho\, d\phi/dL$.

*Fabrication of Geodesic Lenses.* A spherical geodesic lens can be fabricated using universal curve generators similar to the making of spherical lenses. Correction of the spherical geodesic lenses may be done by overlay thin-film deposition on selected areas. A toroidal lip or a conical lip can also be fabricated using conventional polishing techniques. Care must be taken in the depth control [163] because of the sensitivity of focal length to the depression depth, $(df/f)/(dz/z)$, which has been found to be approximately equal to 1.9.

Aspherical geodesic lenses are more difficult to fabricate. The two leading techniques are the ultrasonic impact grinder [160] approach and the diamond machining approach. These techniques provide the proper surface contour, and the substrate is subsequently polished and covered by a waveguide. It is time consuming but relatively straightforward.

Figure 12.53 shows experimental data on the normalized focal length for a Rinehart–Luneburg geodesic lens fabricated by an ultrasonic impact grinder on lithium niobate substrate. The focal spot intensity distribution is given in Fig. 12.54 indicating a diffraction-limited spot size.

### 12.2.3   Waveguide Diffraction Elements

In the section on multilayer step index waveguides, the principles of waveguide index modification using overlay films were described. Modification of the waveguide effective refractive index tends to be small, causing difficulties in fabricating waveguide

**TABLE 12.12 Aspheric Geodesic Lens Depression Depth Values**

| $\rho$ | $L = 1$ | $L = 2$ | $L = 3$ | $L = 5$ |
|---|---|---|---|---|
| 0.0 | −0.63261854 | −0.41516450 | −0.32577595 | −0.24632306 |
| 0.05 | −0.63173418 | −0.41362671 | −0.32497713 | −0.24581597 |
| 0.1 | −0.62907524 | −0.41049277 | −0.32302122 | −0.24442130 |
| 0.15 | −0.62462402 | −0.40601189 | −0.31993946 | −0.24213778 |
| 0.2 | −0.61835026 | −0.40021762 | −0.31572078 | −0.23895513 |
| 0.25 | −0.61021006 | −0.39309853 | −0.31034511 | −0.23485890 |
| 0.3 | −0.60014406 | −0.38462124 | −0.30378519 | −0.22983014 |
| 0.35 | −0.58807494 | −0.37473657 | −0.29600621 | −0.22384467 |
| 0.4 | −0.57390386 | −0.36338075 | −0.28696460 | −0.21687209 |
| 0.45 | −0.55750528 | −0.35047422 | −0.27660619 | −0.20887433 |
| 0.5 | −0.53871973 | −0.33591890 | −0.26486356 | −0.19980375 |
| 0.55 | −0.51734302 | −0.31959338 | −0.25165224 | −0.18960030 |
| 0.6 | −0.49311002 | −0.30134550 | −0.23686495 | −0.17818739 |
| 0.65 | −0.46566912 | −0.28098068 | −0.22036284 | −0.16546551 |
| 0.7 | −0.43453994 | −0.25824265 | −0.20196111 | −0.15130186 |
| 0.75 | −0.39903824 | −0.23277983 | −0.18140406 | −0.13551237 |
| 0.8 | −0.35813042 | −0.20408137 | −0.15831775 | −0.11782775 |
| 0.85 | −0.31011230 | −0.17133889 | −0.13210832 | −0.09782079 |
| 0.9 | −0.25174672 | −0.13308526 | −0.10169755 | −0.07471869 |
| 0.91 | −0.23830090 | −0.12454115 | −0.09494112 | −0.0696055 |
| 0.92 | −0.2240644 | −0.11561444 | −0.08789797 | −0.06428411 |
| 0.93 | −0.20889917 | −0.10624774 | −0.08052637 | −0.05872504 |
| 0.94 | −0.19261701 | −0.09636395 | −0.07277039 | −0.05288894 |
| 0.95 | −0.17494911 | −0.08585453 | −0.06455161 | −0.04672091 |
| 0.96 | −0.15548637 | −0.07455703 | −0.05575321 | −0.04013959 |
| 0.97 | −0.13354614 | −0.06220543 | −0.04618507 | −0.03301353 |
| 0.98 | −0.10780997 | −0.04829728 | −0.03549105 | −0.0250989 |
| 0.99 | −0.07490968 | −0.03158515 | −0.02279784 | −0.01580685 |
| 0.999 | −0.02283366 | −0.00837304 | −0.00572737 | −0.00370575 |
| 0.9999 | −0.00712019 | −0.00245407 | −0.00162569 | −0.00100000 |
| 1 | 0 | 0 | 0 | 0 |

refractive elements. This is particularly true for graded index waveguides. However, large variations in refractive indices are not required in most diffractive optical elements. In general, the maximum optical path length difference in diffractive elements is on the order of micrometers instead of on the order of millimeters for refractive optical elements. On the other hand, the proper operation of diffraction elements requires the use of a highly monochromatic light source, which is generally true in waveguide integrated optics. Therefore, there seems to be a natural compatibility between guided-wave optics and diffraction optical components.

Various types of waveguide optical diffraction elements [164–184] have been reported. Some have been experimentally demonstrated. Some have only been analyzed and published in the open literature. These include lenses, mirror/beam splitters, channel-to-planar waveguide expanders, spatial light modulators, and input/output optical couplers. In this section, attention will be given to lenses and to a lesser extent to mirrors/beam splitters. The waveguide input/output couplers has been discussed before.

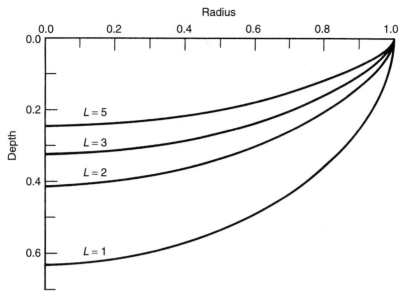

**Figure 12.52** Cross section of four designs of generalized Rinehart–Luneburg geodesic lenses. Focal length $L$ is normalized to geodesic edge radius $R_c$.

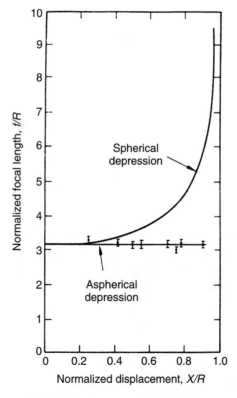

**Figure 12.53** Measured focal distance versus lateral displacement. (After Ref. 160.)

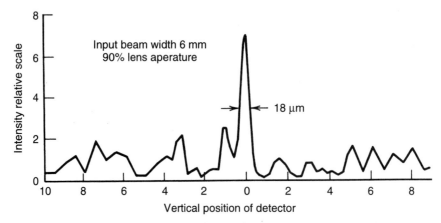

**Figure 12.54**   Intensity distribution at focal plane for a He–He input beam with 6 mm width. (After Ref. 160.)

Channel waveguides and modulators will be discussed in Chapters 11 and 16. The discussion on waveguide diffraction elements will begin with basic design and fabrication considerations for simple waveguide optical gratings.

***Basics of Waveguide Optical Gratings.***   Waveguide optical diffraction elements are volume gratings of different types. In most cases, these two-dimensional volume gratings have very similar behavior as conventional volume gratings. However, due to the uniqueness of planar optical waveguides, there are several special features. First, gratings in optical waveguides can be made in such a way that the grating phase vector is colinear to the optical phase vector, an arrangement more like an optical interference filter than ordinary gratings. Second, these volume gratings can easily be accessed and prepared from the waveguide surface. This feature allows the use of sophisticated planar fabrication technologies such as photo- or E-beam lithography in making grating devices. Third, guided modes can only be manipulated gracefully or it will leak into the substrate. This last feature is very important since it severely limits the design and application of diffraction elements on many integrated optical waveguides [181].

Consider the phase-matching diagram shown in Fig. 12.55. Consider also a grating structure that has a constant periodicity $\Lambda$. The phase-matching condition for efficient volume interaction between the incident guided-wave beam and the diffracted guided-wave beam is satisfied for a given transverse section when the periodicity $\Lambda$ of the grating grooves satisfies the condition $K = 2n_{\text{eff}}k_0 \sin(\theta/2)$, where $K = 2\pi/\Lambda$ and $\theta$ is the total angle of deflection. However, if the projection of the $K$ vector on the direction of the incident beam, $K \sin(\theta/2)$, is large enough such that $K \sin(\theta/2) \geqslant (n_{\text{eff}} - n_s)k_0$, then the incident guided-wave beam will also be coupled to substrate modes that have a propagation wave number equal to or smaller than $n_s k_0$. This may happen despite the fact that the phase-matching condition to the substrate mode may be violated in the direction perpendicular to the direction of the incident beam because each transverse section is very narrow, creating only a very broadband phase-matching condition in the perpendicular direction. Thus, the condition for avoiding the excitation of substrate

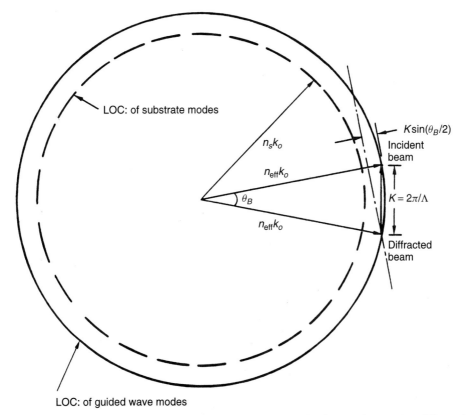

**Figure 12.55**  Phase-matching diagram for the incident guided-wave beam, the diffracted guided-wave beam, and the substrate modes.

mode is the maximum grating diffraction angle

$$\theta_M \leqslant \sqrt{\frac{2(n_{\mathrm{eff}} - n_s)}{n_{\mathrm{eff}}}} \tag{12.76}$$

However, it is believed that this effect will be dependent on optical arrangements and optical beam size. For collimating lenses, this condition implies a limitation on how small an $f$-number may be used without significant reduction of the diffraction efficiency due to coupling into the substrate modes. For step index waveguides, the maximum diffraction angle can be tens of degrees. However, for graded index waveguides such as metal-diffused lithium niobate, the maximum angle can be less than $10°$.

***Waveguide Diffraction Lenses.***  Waveguide diffraction lenses can be made as either a Fresnel lens [165,166] or a chirp grating lens [164]. The Fresnel lens is simply a waveguide optics analogy of the Fresnel zone lens. The basic principle of a zone lens will not be repeated here. Figures 12.56**a** and **b** illustrate a digital zone lens and an analog zone lens, respectively, on planar waveguide. The location of each zone

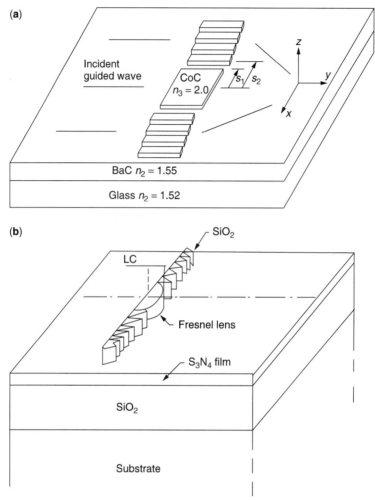

**Figure 12.56** (**a**) Diagram of a surface-deposited Fresnel zone lens in a thin-film waveguide. (After Ref. 166.) (**b**) Schematic diagram of waveguide analog Fresnel lens. (After Ref. 185.)

boundary is given by, for the $m$th zone,

$$S_m = \sqrt{\frac{m\lambda f}{n_{\text{eff}}}} \tag{12.77}$$

where $\lambda/n_{\text{eff}}$ is the optical wavelength in the waveguide and $f$ is the focal length of the zone lens. The digital zone plate has a number of phase shifters, and the output optical phase front usually decomposes into a number of cylindrical components each focusing to a focus. There are convergent as well as divergent wavefronts. If this zone lens is employed to focus a laser beam, the efficiency will be poor. However, most of the unwanted focal spots may be eliminated when the analog zone lens in Fig. 12.56b is employed and when the maximum phase shift is $2\pi$. In both cases, the zone plate has its smallest features at the two ends of the optical aperture. The smallest feature line width can be derived from the equation

$$d = S_m - S_{m-1}$$
$$\cong \frac{F\lambda}{n_{\text{eff}}}, \tag{12.78}$$

which equals the diffraction-limited focal spot of this zone lens. That is, the ideal focal spot size equals the smallest feature size of a waveguide zone lens.

So far, all Fresnel lenses have been made on waveguides that have low index substrates. Their performances are listed by Table 12.13.

Chang and Ashley [186] obtained Fresnel zone lenses on BaO glass waveguides ($n_f = 1.55, n_5 = 1.512$). The phase shift zone pads are obtained by depositing high-index CeO ($n^1 = 2$) on the waveguide. The typical width of the lenses are from 25 to 50 μm, and 23% diffraction efficiency, 3 μm spot size, $F = 5$, 15° angular field of view, and 18 dB of signal-to-noise ratio have been obtained experimentally.

The side lobe was limited to $-12$ dB of the main lobe because of the truncation of the incident beam by the finite aperture size of the lens.

Valette et al. [185] made graded index Fresnel lenses on $Si_3N_4/SiO_2$ waveguides where the phase shift is obtained by deposition of an $SiO_2$ film with a prescribed pattern, as shown in Fig. 12.56**b**. The width of the lens is 20 μm. The focused spot size is 3.6 μm, and the efficiency is 60–70%. The signal-to-noise ratio for a Gaussian incident beam was $-27$ dB at 150 μm off-axis ($\cong 1°$).

Handa et al. [172] have fabricated graded index Fresnel lenses (F/5, 1 mm aperture) in amorphous $As_2S_3$ waveguides on $SiO_2$ substrates. The phase shift pattern was written directly into the $As_2S_3$ waveguide by direct electron beam exposure of the $As_2S_3$. The maximum index change in $As_2S_3$ that can be obtained by this method is $\Delta n \sim 0.06$. By keeping the electron beam current and the electron beam scanning speed fixed, the number of the scanning repetition on a line was varied to give the appropriate does distribution for obtaining the desired index variation. Nearly diffraction-limited focusing characteristics (10 μm spot size) and an efficiency of 48% have been experimentally obtained.

Although these results are impressive, Fresnel zone lenses have a number of problems.

1. Unless a graded analog zone lens is employed, the large number of focal orders means low efficiency as well as high background noise.

2. With graded analog zone lenses, the high efficiency demands a maximum phase shift equal to $2\pi$; otherwise efficiency as well as noise background will deteriorate.

3. They demand a large index change in the phase elements so that the phase elements can be made sufficiently short to maintain a wide acceptance angle.

Another waveguide diffraction lens is the chirp grating lens [164] of Fig. 12.57**b**. The device is a thick grating operating in the Bragg diffraction regime. The grating has a variable diffraction angle over its aperture due to variable grating periodicity. A key feature is the off-axis focusing behavior that separates the focused beam from other grating diffraction orders. As a result, it provides low background noise and insensitivity to fabrication variations and can be quite efficient. The disadvantages of

**TABLE 12.13  Experimental Results on Grating Lenses for Guided-Wave Optics**

| Lens Type | Waveguide/ Substrate Material | Lens Fabrication Technique | Grating Period Λ (μm) | Groove Depth or ΔN | Focal Length f (mm) | Grating Interaction Length (μm) | Focal Spot Size (μm) | f-Number | Acceptance Angle (degrees) | Signal-to-Noise Ratio (dB) | Diffraction Efficiency (%) | Reference |
|---|---|---|---|---|---|---|---|---|---|---|---|---|
| Analog Fresnel lens | $Si_3N_4$ film on $Si - SiO_2$ | $SiO_2$ over layer pattern etched | — | 4000 Å | 8.5 | 20 | 3.6 | — | 10 | 27 | 60–70 | 185 |
| Graded index Fresnel lens | $As_2S_3$ film on $Si - SiO_2$ | Direct electron beam writing | — | 0.05 | 5.0 | 20 | 10.0 | 5 | — | 20 | 48 | 172 |
| Digital Fresnel lens | $Si_3N_4$ film on $Si - SiO_2$ | $SiO_2$ over layer pattern etched | — | 4000 Å | 10.2 | 10 | 4–5 | 8.5 | — | 8 | 19 | 165 |
| Digital Fresnel lens | BaO layer on glass | CeO over layer pattern etched | — | 300 Å | 4.0 | 25–100 | 3.4 | 2.5–5 | 15 | 18 | 23 | 186 |
| Chirp grating lens | Corning 7059 glass on $Si - SiO_2$ | Photolithography and plasma etch | 3.4–6.8 | 400 Å | 32.0 | 150 | 6.1 | 16 | — | −15 | 90 | 164 |
| Chirp grating lens | Corning 7059 glass on pyrex | E-Beam lithography and chemical etch | 3–100 | 3000 Å | 20.0 | 3 mm | — | — | — | — | — | 167 |

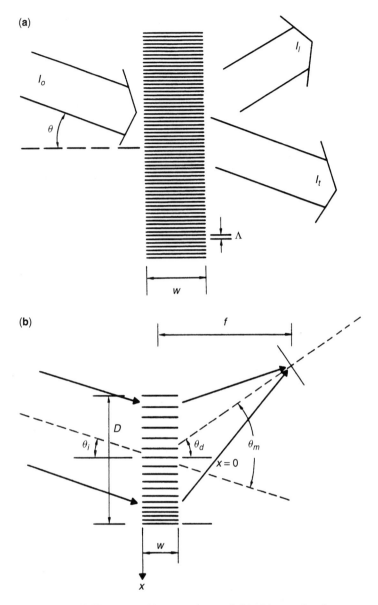

**Figure 12.57** (a) Uniform grating and (b) chirp grating lens.

chirp grating lenses when compared with zone lenses are the need of a smaller line feature width for a given focal spot size, smaller numerical aperture, and narrower field of view. A brief discussion of the theory for chirp grating lenses follows.

To understand the operation of a grating lens, one can start with a simple uniform grating. The angle of deviation between an arbitrary grating diffraction order $m$ and the incident optical beam can be obtained from

$$\theta_d = \sin^{-1}\left(\frac{m\lambda}{\eta\Lambda} - \sin\theta_i\right) \qquad (12.79)$$

$$\theta_m = \theta_i + \theta_d \approx \frac{m\lambda}{\eta\Lambda} \qquad (12.80)$$

where $\lambda$ is the optical wavelength, $n$ is the index of refraction, $\Lambda$ is grating periodicity, and angles $\theta_i$ and $\theta_d$ are given in Fig. 12.57**b**. The error in this approximation is less than 1% when the angle is less than 15°. If the grating spatial frequency $p(x) = 1/\Lambda(x)$ has a nearly linear chirp, the planar optical wave can be diffracted into a converging cone similar to the focusing properties of a thin lens. In practice, the grating spatial frequencies $p(x)$ may be adjusted to compensate the phase front error. To focus a collimated light beam as shown in Fig. 12.57**b**, the following must be satisfied ($\theta_i$ = const):

$$\theta_d(x) = \tan^{-1}\frac{x}{f} \qquad (12.81a)$$

$$\theta_m(x) = \theta_i + \tan^{-1}\frac{x}{f} \qquad (12.81b)$$

The range of $x$ is related to the lens aperture and the $f$-number. Equation (12.79) can be rewritten for grating periodicity as

$$\Lambda(x) = \frac{m\lambda}{2n\sin[\frac{1}{2}\theta_m(x)]\cos[\frac{1}{2}(\theta_i - \theta_d(x))]} \qquad (12.82)$$

which in small-angle approximation is linear to $x$ as shown in the following:

$$p(x) = \frac{n}{m\lambda}\left(\theta_i + \frac{x}{f}\right) \qquad (12.83)$$

Figure 12.58 gives the grating periodicity computed from Eq. (12.82). The similarity between Eq. (12.82) and an optical interferogram allows the use of a holographic

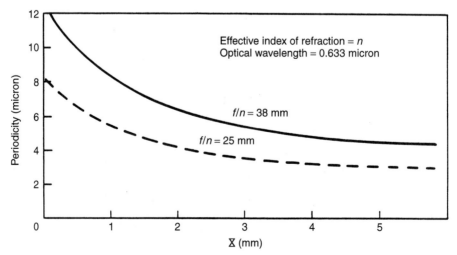

**Figure 12.58** Distribution of grating periodicity for two examples of chirp grating lenses. Focal lengths $f/n$ are 28 and 18.6 mm at 0.85 µm optical wavelength. (After Ref. 187.)

recording process in making this grating lens pattern. However, in such a case, the grating periodicity will be correct only if the recording wavelength in the recording medium is the same as the reconstruction wavelength in the optical waveguide. The recording must be performed in an index matching fluid of proper refractive index.

The lens properties of a chirp grating can be determined by the grating diffraction angles at the extremities of the optical aperture, and the results are

$$f\text{-number} = \frac{n\Lambda_{\max}\Lambda_{\min}}{\lambda(\Lambda_{\max} - \Lambda_{\min})} \tag{12.84}$$

$$\text{Focal spot } \delta \cong \frac{F\lambda}{n} = \frac{\Lambda_{\max}\Lambda_{\min}}{(\Lambda_{\max} - \Lambda_{\min})} \tag{12.85}$$

$$\text{Focal length } f = \frac{W}{2} + \frac{D}{\tan(\theta_d)_{\max} - \tan(\theta_d)_{\min}} \tag{12.86}$$

where $D$ is the optical aperture of the grating lens. The focal length can also be described by the chirp rate as

$$f \approx \frac{n/\lambda}{dp/dx} \tag{12.87}$$

In many practical chirp grating lens designs, the special case of $\Lambda_{\max} = 2\Lambda_{\min}$ corresponding to the octave spatial bandwidth is employed. Thus, Eqs. (12.84) and (12.85) can be simplified to

$$f\text{-number} = \frac{n\Lambda_{\max}}{\lambda} \tag{12.88}$$

$$\text{Focal spot} = \Lambda_{\max} \tag{12.89}$$

Since the optical properties of the chirp grating lens depends only on line geometry, it is very insensitive to fabrication process variations and it is highly predictable.

From small-angle approximations, the focal loci of the chirp grating lens can be determined as the loci of the constant-output beam convergence angle shown in Fig. 12.59. Note that the focal loci is slightly tilted due to the off-axis nature of a chirp grating lens. However, this tilt is small since for all practical applications, all angles in a grating waveguide lens must be small to avoid substrate mode conversion. For a grating lens of $10°$ bend on titanium-diffused lithium niobate waveguides, the smallest focal spot size is approximately 2 μm at the He–Ne laser wavelength.

A very important feature of the chirp grating lens is the separation of the higher-order diffractions from the optical axis of interest. It can be shown using simple geometry that the clear angular field (clear from spurious responses) for the case of $\Lambda_{\max} = 2\Lambda_{\min}$ is related to other lens parameters as

$$\text{Clear angular field} = \frac{\lambda}{n\Lambda_{\max}} = \frac{1}{f\text{-number}} \tag{12.90}$$

$$\text{Clear spatial field} = D \tag{12.91}$$

$$\text{Resolution elements} = \frac{D}{\Lambda_{\max}} \quad \text{(pixels)} \tag{12.92}$$

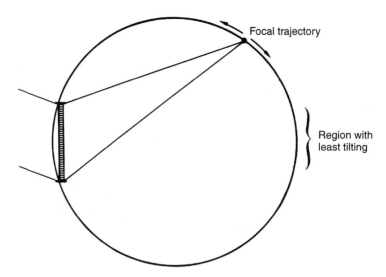

**Figure 12.59**   Focal plane (trajectory) of a diffraction lens (Fresnel as well as chirp grating).

The spatial resolution of such a lens is simply the size of the lens aperture divide by the maximum line width.

Practical waveguide grating lenses have long interaction phase element lengths due to the small index perturbation achievable on optical waveguides. These devices tend to operate in the Bragg diffraction regime like thick gratings. As is known in conventional optics, thick gratings can have very high diffraction efficiency, approaching 100% over a restricted field of view. Therefore, the application of chirp grating lenses must be accompanied with careful design to ensure satisfaction of the Bragg incidence angle condition at the center of the chirp grating. Thus,

$$\theta_i = \sin^{-1} \frac{\lambda}{2n\Lambda_0} \tag{12.93}$$

where

$$\Lambda_0 = \frac{2\Lambda_{\max}\Lambda_{\min}}{\Lambda_{\max} + \Lambda_{\min}} \tag{12.94}$$

Due to the variable grating periodicity in a chirp grating lens, residual angle mismatch exists over the optical aperture:

$$\Delta(x) = \theta_B(x) - \theta_i$$

$$\cong \frac{\lambda}{2n}\left[\frac{1}{\Lambda(x)} - \frac{1}{\Lambda_0}\right] \tag{12.95}$$

Note that the maximum angle mismatch equals $1/4F$. The efficiency of a thick grating with finite angular mismatch is given by

$$\frac{I_d}{I_0} = \frac{\sin^2 \pi \sqrt{(n'W/2\lambda)^2 + (W\Delta/\Lambda)^2}}{1 + (2\lambda\Delta/n'\Lambda)^2} \tag{12.96}$$

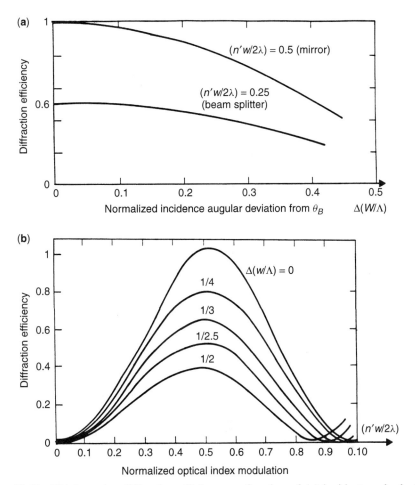

**Figure 12.60**   Thick grating diffraction efficiency as function of (**a**) incident angle deviation and (**b**) phase perturbation.

where $W$ is the grating thickness (interaction length), $n'$ is the magnitude of index perturbation, and $\Delta$ is the angular mismatch, which is a function of $x$ in a chirp grating lens. Figure 12.60 gives the diffraction efficiency of a thick grating as a function of normalized angular mismatch and as a function of normalized optical index modulation over the length of a phase element. It is observed that maximum efficiency occurs when the accumulated index perturbation (i.e., the phase perturbation) $n'W$ equals one optical wavelength and that angular mismatch as high as $0.4\Lambda/W$ may be tolerated. These observations will be extremely valuable for the design of a chirp grating lens with parallel straight-line phase elements.

However, since the chirp grating lens can be produced by advanced lithographic means, the phase elements do not have to be parallel straight lines. When the phase elements are rotated to completely satisfy the Bragg incidence condition over the chirp grating lens aperture, the only angular mismatch will be due to the finite angular spread of the incoming spatially modulated optical wavefront. Figure 12.61 illustrates the phase element design using this concept. From previous discussions on chirp grating

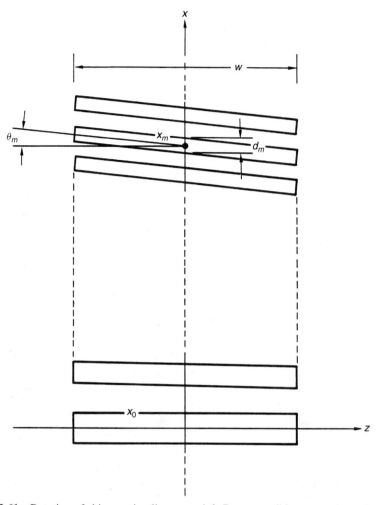

**Figure 12.61**  Rotation of chirp grating lines to satisfy Bragg conditions over the entire aperture of a chirp grating lens.

lens properties, it can be shown that a chirp grating lens with rotated phase elements only has to accommodate half as much angular mismatch as a chirp grating lens with parallel phase elements. Consequently, for a constant normalized mismatch, the grating thickness can be made longer by a factor of 2 using rotated phase elements.

Finally, the field of view of waveguide diffraction elements is limited by the small refractive index difference between the guided mode and the substrate. Figure 12.62 is the momentum vector diagram illustrating the diffraction of an optical wave vector by a grating wave vector $K$. The triangle formed by the dashed lines indicates guided-mode-to-substrate-mode conversion by the same device when the optical incoming wave vector is tilted off-axis, as is required to provide angular field of view. This circumstance may be analyzed by simple geometry, which yields

$$2nk_0 K \ \sin\left(\theta_B + \frac{\Delta'}{2}\right) = (n^2 - n_s^2)k_0^2 + K^2 \tag{12.97}$$

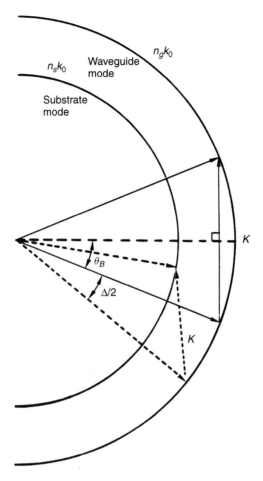

**Figure 12.62** Momentum vector diagram.

which can be reduced to

$$\theta_B \Delta' < \frac{1}{2[1 - (n_s/n)^2]} \qquad (12.98)$$

The constant Bragg angle and mismatch angle product can be very small for a graded index waveguide. For example, this product is about $0.0027$ $rad^2$ for a titanium-diffused single-mode waveguide on lithium niobate. This corresponds to a field of view of about $3.5°$. Although small, this field of view can provide over 400 pixels over an optical aperture of 2 mm. Much better values are achievable on shallower waveguides.

Table 12.13 includes the performance of several chirp grating lenses on glass substrates. The relatively large index step between the glass waveguide and substrate is the major factor for the high performance of these waveguide diffraction lenses. Recent experiments on titanium-diffused waveguides on lithium niobate verified the expected difficulties due to the small index difference between the graded index waveguide and the substrate. However, a diffraction-limited focal spot was routinely obtained. With due carefulness, diffraction efficiencies of 30 and 40% were obtained on titanium-diffused lithium niobate waveguides using benzoic acid treatment and a

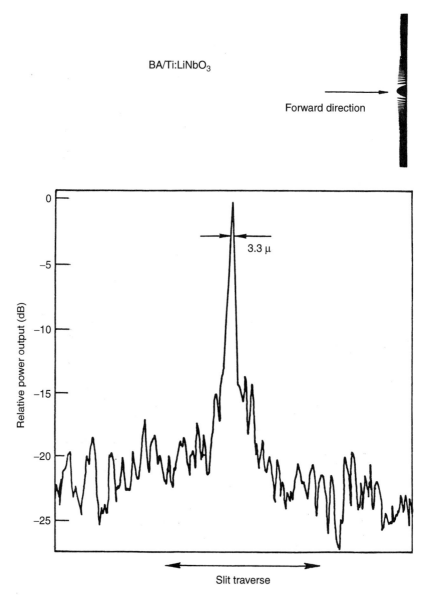

**Figure 12.63**    Focal spot pattern of the shaped-zone Fresnel lens at 0.83 μm wavelength. (After Ref. 145.)

rutile overlay film, respectively, as index perturbation methods. The focal intensity profile of a shaped-zone (analog) Fresnel lens on lithium niobate is given in Fig. 12.63. Figure 12.64 is the focal distribution of a chirp grating lens on lithium niobate with 75% diffraction efficiency.

*Waveguide Diffraction Mirrors and Filters.* For waveguide diffraction lenses, the optical wavefront and the grating phase front are nearly perpendicular to each other. In this section, the waveguide optical components with the optical wavefront parallel

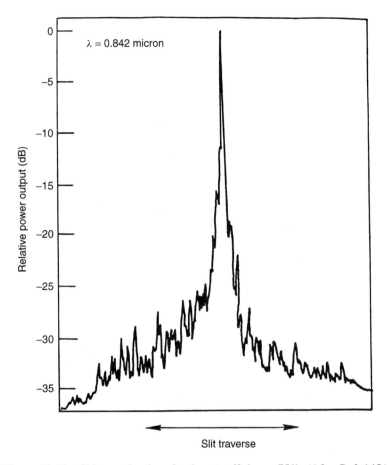

**Figure 12.64**  Chirp grating lens focal spot: efficiency 75%. (After Ref. 145.)

to the grating wavefront are discussed. This parallel arrangement produces waveguide mirrors with spectral response strongly influenced by grating periodicity similar to multilayer interference coatings in conventional optics.

Figure 12.65 illustrates the general condition between the grating and the optical guided mode. The reflectivity from each boundary of the grating lines is very weak. Thus, the response of the grating is simply the algebraic sum of the reflections,

$$r_T = \sum_{n=1}^{N} [-r + re^{j\phi}] B_n e^{j2n\phi} \tag{12.99}$$

where $r_T$ is total reflection, $N$ is number of lines, $r$ is reflectivity at the edge of each line, $\phi = k(\Lambda/2)$ is the propagation delay through each line, and a uniform grating periodicity of $\Lambda$ is assumed with each grating line weighted by $B_n$. The response of the reflector is simply $|r_T|^2$. For instance, if $B_n$ is constant,

$$r_T = \frac{r B_n \sin(N\phi)}{\sin[(\phi/2) + 90°]} \tag{12.100}$$

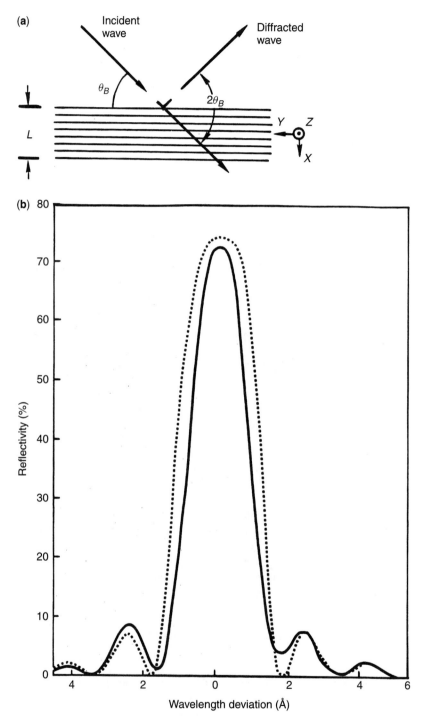

**Figure 12.65**   (a) Schematic diagram of grating waveguide beam splitter. (After Ref. 172.) (b) Reflectivity versus wavelength deviation from the Bragg condition for thin-film grating filter. The solid line is measured and the dotted line is calculated. (After Ref. 180.)

which has a peak value of $N_r B_n$ and a wavelength bandwidth of $\Delta\lambda/\lambda \approx (2N)^{-1}$ centered at $\lambda = \Lambda$ for a $90°$ reflector of projected periodicity $\Lambda$. For control of the passband shape, one simply introduces various weighting coefficients $B_n$ that are generally the Fourier transform of the desired filter passband shape (transversal filter concept). A transmission line model is readily available to take into account multiple reflections within the grating lines. The filter response will be distorted in the case of strong reflection efficiency.

A waveguide grating mirror can be employed as a mirror, a beam splitter, a beam expander, and wavelength filter banks. Table 12.14 lists a number of published grating mirrorlike devices. Because the reflectivity is the algebraic sum of the reflection from a large number of grating lines, small index perturbation can still provide very high reflection efficiency. However, the spectral bandwidth will be narrower and the optical alignment will be more critical for grating reflectors made of lower index perturbation values. Chirp grating has been made to reduce the wavelength selectivity as well as to increase the acceptance angle of a grating reflector with the penalty of reduced efficiency. Figure 12.66 illustrates the use of a long grating as a beam expander. Figure 12.67 illustrates the use of a grating bank to achieve a controlled beam splitter array. Figures 12.68a and b illustrate the use of a variable-periodicity grating reflector as a waveguide optical filter bank, which can be useful for wavelength division multiplexed optical communications. Grating reflectors with curved grating lines can serve as focusing reflectors on optical waveguides. Figure 12.69 illustrates one of the most celebrated use of grating reflectors in waveguide optics, the distributed Bragg reflector (DBR) diode laser. The DBR laser and a similar distributed feedback (DFB) diode laser use the wavelength selectivity of grating reflectors as laser wavelength control devices, resulting in more stable laser performances.

In general, grating reflectors require much smaller line widths than grating lenses. The fabrication of a uniform large-area submicrometer line pattern has been difficult. Imperfections cause scattering and nonuniform responses. Large-angle reflection from a grating reflector is highly polarization dependent, and there is strong TE–TM mode coupling upon such reflections.

### 12.2.4 Waveguide Optical Harmonic Generator

Second-order nonlinear optical harmonic generation has been observed in waveguides on lithium niobate substrates, in GaAs waveguides, and in polycrystalline ZnS thin-film waveguides deposited on BK-7 glass substrate [190–198]. Second-order nonlinear effects in randomly oriented crystalline films exist because there will be a nonvanishing component after the volume average. Recently emphasis is shifted to waveguides made of nonlinear organic materials either in the form of crystalline films or electric-field-oriented polymer films [193, 195, 196].

The theory of nonlinear integrated optics has been treated by Kuhn [199], Anderson et al. [200], and Conwell [201]. More recently, attention has been given to third-order nonlinear integrated optics [202] for potential applications in all-optical-signal processing. Optical waveguides are attractive for efficient nonlinear optical interactions due to (1) the high power density possible with moderate total power due to beam confinement, (2) the diffractionless propagation for long interaction length, and (3) the possibility of phase matching without birefringence using waveguide mode dispersion relations. However, most recent devices tend to use channel waveguides, which provide better field confinement than planar waveguides. Therefore, no detailed treatment will be given here.

**TABLE 12.14  Experimental Results of Wide-Angle In-Plane Bragg Deflection Integrated Optics Devices**

| Waveguide Device | Waveguide/ Substrate Material | Grating Fabrication Technique | Grating Type | Grating Spacing Λ (μm) | Number of Grooves, N | Depth or ΔN | Bragg Angle (degrees) | Wavelength λ (Å) | Diffraction Efficiency (%) | Reference |
|---|---|---|---|---|---|---|---|---|---|---|
| Beam expander | $Cs^+ \leftrightarrow Na^+$ ex. in soda lime glass | Holographic and ion beam mill | Uniform | 0.6 | — | 1000–3000 Å | 70 | 6328 | 16 (total) | 171 |
| Beam deflector splitter | $Ag^+ \leftrightarrow Na^+$ ex. in soda lime glass | Photographic mask inhibits ion exchange | Uniform | 3.0 | — | $2.6 \times 10^{-4}$ | 4 | 6328 | 50–70 | 188 |
| Beam deflector and WDM | $As_2S_3 - SiO_2 -$ $In_2O_3$ – glass | Electron beam lithography | Uniform / Chirped | 0.5 / ~0.5 | 400 / 1000 | $5 \times 10^{-3}$ / $3 \times 10^{-3}$ | 28 / — | 1.153 μm / 1.153 μm | ~100 / ~50 | 172 |
| Beam deflector splitter | PMMA/fused quartz | Electron beam lithography | Uniform | 2.0 / 0.8 | 100 / 375 | $8.4 \times 10^{-4}$ / $8.4 \times 10^{-4}$ | 6.2 / 15.7 | 6328 / 6328 | 5 / 15 | 174 |
| WDM | Corning 7059 glass on glass substrate | Holographic and ion beam mill. | Chirped | 0.298– 0.336 | — | 400–500 Å | 50 | 6030–6300 | ~50 high IL | 177 |
| WDM | Corning 7059 glass on $SiO_2$ | Holographic and AZ1350 grating | Chirped | 0.42– 0.46 | — | 700 Å | ~60 | 6471 | 60–90 | 178 |
| Beam deflector and WDM | Three-layer glass | Holographic ion beam etch | Uniform | 0.237 | — | 120 Å | 39.2 | 5700 | — | 170 |
| Beam deflector | $Ag^+ \leftrightarrow Na^+$ ex. in glass | Holographic and etch | Uniform | 0.29 | — | — | 45–75 | 6328 | — | 175 |
| Multiple beam splitter | $As_2S_3 - SiO_2 -$ $In_2O_3$ – glass | Electron beam lithography | Chirped / Uniform | 0.5 / 0.5 | 500 / 40–200 | $7 \times 10^{-3}$ / $7 \times 10^{-3}$ | 29 / 29 | 1.153 μm / 1.153 μm | 90–100 / 30–100 | 173 |
| Beam deflector | Corning 7059 glass on $SiO_2$ | Holographic and ion beam mill | Chirped | ~0.4 | — | — | — | 8450 | — | 176 |

Abbreviations: WDM, wave division multiplexer; IL, insertion loss.

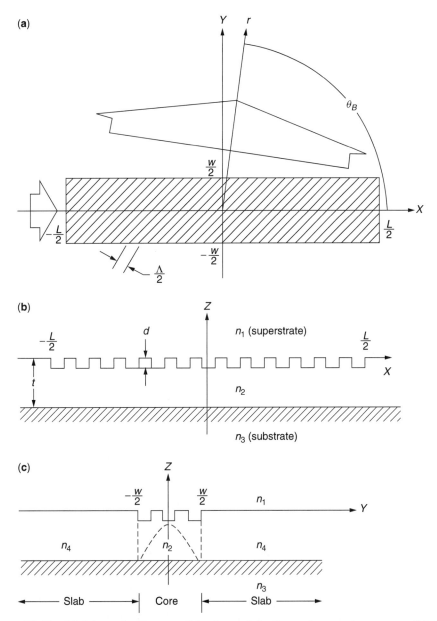

**Figure 12.66**   (**a**) Schematic diagram of the beam deflection and expansion process. (**b**) Side view of the channel waveguide and grating. (**c**) Cross-sectional view of the channel waveguide and grating. (After Ref. 169.)

## 12.3   WAVEGUIDE MEASUREMENT TECHNIQUES AND APPLICATIONS

### 12.3.1   Waveguide Parameter Measurement

When conducting waveguide parameter measurements, very often it is necessary to couple the light into and out of the optical waveguide for observation. The most commonly

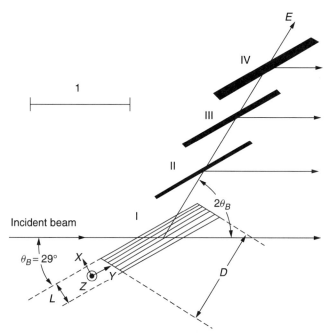

**Figure 12.67**   Schematic diagram of a multiple-grating waveguide beam splitter. (After Ref. 173.)

employed technique is the prism coupler (discussed in an earlier section). A prism with index of refraction greater than the waveguide must be used. Typically, rutile prisms are used for experiments with lithium niobate substrates and dense flint glass prisms are used with glass waveguides. The prism is pressed to the substrate surface to reduce the air gap between the prism base and the substrate surface. Sometimes, a drop of water may be used to make this step easier if contamination of the waveguide surface by water is not a concern. A laser beam is directed toward the $90°$ edge of the prism, and the incidence angle as well as the prism–substrate air gap are adjusted until a streak of light appears either on the waveguide surface of at the exit end of the waveguide/substrate. For substrate modes, the intensity is not sensitive to small changes in incidence angle. For waveguide modes, the intensity flashes as the laser beam incidence angle is adjusted slightly. If a pair of prism couplers are used, the output beam from the second prism can be observed for evidence of waveguiding. Note that an output beam from the second prism does not mean the successful launching of a guided-wave mode. Substrate modes can also be coupled out by the second prism in discrete angles when the substrate mode bounces to the base of the output prism coupler. In most cases, the guided mode exhibits more in-plane scattering than the substrate modes, and the output beam from a guided mode may spread out as a number of nearly evenly spaced $m$-lines. The number of lines and the evenness in line spacing depends on the number of modes supported and the specific index profile of the waveguide being tested.

Sometimes, tests are conducted by end-fire coupling through cleaved waveguide edges or polished waveguide edges [203]. In this case, light is focused to the sharp edges by a high-power microscope objective lens. Success of the launching depends on the quality of the waveguide edges.

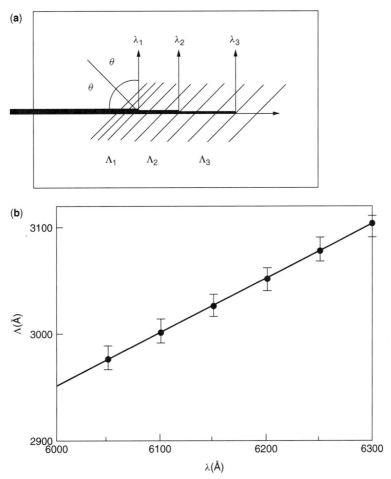

**Figure 12.68** (a) Chirped grating waveguide optical demultiplexer. (After Ref. 178.) (b) Wavelength versus grating periodicity $\Lambda$ in the reflecting region, where $\Lambda$ is a function of distance along the grating. (After Ref. 177.)

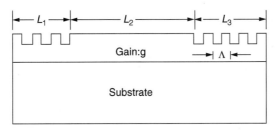

**Figure 12.69** Schematic diagram of a DBR laser. The feedback is supplied by Bragg reflectors outside the gain region of the laser. (After Ref. 189.)

In the following sections, the commonly used method in characterizing waveguide impurity profiles, waveguide index profiles, optical propagation losses, and waveguide device performances are described.

***Impurity Profile Measurement.*** The measurement of an impurity profile under the surface of substrate materials is a useful step in characterizing a newly formed waveguide. The most popular method is the use of an X-ray microanalyzer with an electron microprobe, as illustrated by Fig. 12.70. The substrate with optical waveguide is cleaved or edge polished and is viewed by a scanning electron microscope. The presence of heavy atoms is detected by surface emission of characteristic X-rays. Measuring the intensity of emitted X-rays as the electron beam is focused to different locations yields a plot of impurity concentrations. The limitation of this method is the inability to measure lighter atoms and the lack of accuracy when concentration of the interest species is

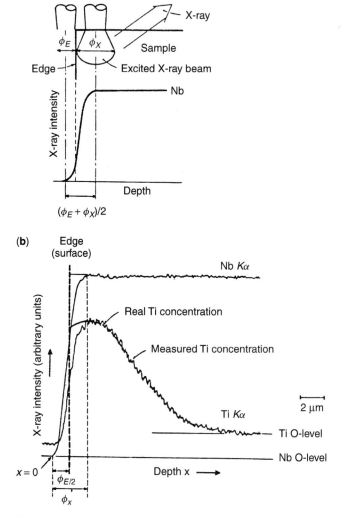

**Figure 12.70**  Illustrated edge correction: (**a**) transient region of measured Nb X-ray intensity for a sample with a steplike Nb concentration: (**b**) relation between real and measured Ti concentration in the LiNbO$_3$ sample diffused at 1000°C for 100 h with 1400-Å-thick and 30-μm-wide strip Ti film. (After Ref. 204.)

low. Resolution is limited by the beam size, beam energy, and penetration. The relative statistical counting uncertainty was believed to be better than 1% with absolute accuracy only $\pm 20\%$. An improved method has been reported by fixing the electron beam while moving the sample at a constant speed [204]. Spatial resolution better than 0.2 μm has been achieved. However, the measured data need to be corrected for the edge effect, as illustrated in Fig. 12.70, due to the finite size of the probe beam. Figure 12.71**a** gives the measured titanium concentration profile in lithium niobate substrate. Figures 12.71**b** and **c** give the relation between titanium concentration and index changes and the relation between maximum index changes and diffusion temperature.

Secondary ion mass spectroscopy (SIMS) and Rutherford alpha particle backscattering are two other methods [205] that have been employed for the measurement of impurity profiles. The SIMS method uses an imaging mass spectrometer. A large area of the sample surface is sputtered away, and the resulting atomic and molecular species are analyzed. It exhibits poor absolute concentration accuracy ($\pm 100\%$) due to uncertainty in component sputtering rates, but it has excellent depth resolution (about 0.01 μm). The Rutherford backscattering method analyzes the alpha particle scattered from nuclei of the sample to determine the energy loss with respect to the incident beam, giving simultaneous information on concentration and depth of the nuclei present. Depth resolution is about 0.01 μm, and absolute concentration accuracy can be as good as 10%. These methods are sometimes used together with the electron microprobe to achieve accuracy in all respects.

***Refractive Index Profile Measurement.*** The refractive index profile of an optical waveguide can be measured by direct viewing of the sample cross section in an optical interferometer or by waveguide mode index measurements and curve fitted by a WKB numerical program. For a step index waveguide, the thickness is measured either by a scanning electron microscope or a high-power microscope at its cross section or by optical thin-film measurement instruments such as an ellipsometer or a spectrometer or simply an etch-and-probe mechanical method. The ellipsometer and the optical spectrometer data also give refractive index values of the thin film as well as the substrate. These measured values are cross checked by correlating effective guided-wave mode indices with predicted values from theory. On the other hand, measurement of a graded index profile is much more involved.

Direct observation of the optical index profile in an optical interferometer setup is conceptually straightforward. The only critical requirement is that sample edges must be polished to perfect sharpness. This method has been very useful, particularly for diffused graded index waveguides of large thickness. However, cutting the sample into thin slices and polishing is a time-consuming process. It is mostly used to provide reference points for other measurements.

By far the most popular method is the use of a prism coupler and the WKB numerical process to determine the refractive index profile of a graded index film. Using a prism coupler, the mode indices of a waveguiding film can be easily determined from the coupling angles of each waveguide modes using the formula

$$N = n_p \sin \left( A + \sin^{-1} \frac{\sin i}{n_p} \right) \tag{12.101}$$

However, there is no data on effective mode depth. The WKB method is used in a numerical program iteratively until an assumed index profile satisfies the measured

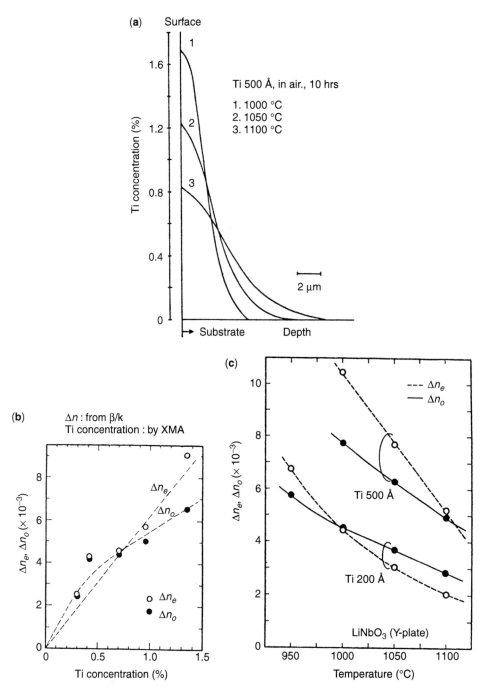

**Figure 12.71**   (**a**) Diffusion profiles for Ti in LiNbO₃ optical waveguides for different diffusion temperatures. (**b**) Refractive index changes in LiNbO₃ optical waveguides as a function of Ti concentration. Refractive index changes are determined independently by using a prism-coupling method. Ti concentrations are measured by an XMA. (**c**) Maximum refractive index changes of LiNbO₃ optical waveguides obtained by the proposed method. Thicknesses of the Ti films are given. (After Ref. 204.)

mode indices. For a single-mode waveguide, the single-mode index value is insufficient to generate both the film index profile and the thickness. One needs to assume a general index profile shape and an effective depth or a surface index value. For a graded index waveguide of known shape containing more than two modes, the index profile can be determined by the least mean-square fit as illustrated in Fig. 12.72. If the index profile is unknown, often a deeply diffused waveguide supporting many modes is fabricated to provide a large number of mode indices for the computation of the index profile. One may assume a surface index slightly higher than the highest-mode index value. Using Eq. (12.45) and assuming a piecewise linear profile, the index profile can be computed. Then, the surface index may be adjusted and more iteration taken to refine the results.

*Optical Loss Measurement.* Optical loss measurement requires the launching of a waveguide mode by whatever means. A highly lossy film can have its loss factor estimated by simply observing the length of the surface streak. Assuming the average dynamic response of the human eye is in the range of 37 dB, the waveguide loss can be estimated as 37 dB divided by the observed streak length. Less lossy films must be measured more carefully.

One of the most commonly used methods is taking output intensity readings from a prism output coupler that is allowed to slide on the surface of the waveguide. This is because a prism output waveguide coupler can be made to couple out nearly 100% of the waveguide mode even when the air gap between the prism base and the substrate surface varies. Plotting the output power relative to the prism location on a log scale

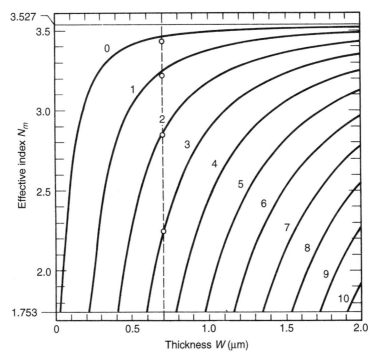

**Figure 12.72**  Waveguide thickness value can be obtained by fitting the measured mode index values to the dispersion diagram.

should form a data distribution that can be approximated by a straight line in the least root-mean-square sense. The slope of this line is the waveguide propagation loss. Figure 12.73 is an example of such waveguide propagation loss measurement for a titanium-diffused waveguide on lithium niobate substrate after several steps of surface polishing. The data tend to be well behaved, and accuracy on the order of a fraction of a decibel per centimeter can be obtained.

Automated waveguide loss measurement can be made by scanning a fiber-optical pick-up probe over the waveguide following the streak of scattered light by the guided mode. Its accuracy is particularly good when a high-quality waveguide is fabricated on an optically absorptive substrate such as oxidized silicon (see Fig. 12.74) propagation loss of 0.2 dB/cm can be measured. It is particularly interesting when waveguide thin-film components are involved that may prevent the sliding of a prism coupler. Figure 12.75 is the surface scattering trace of a glass waveguide with a thin-film Luneburg lens on top of the waveguide. Note that the laser beam is coming from the right on the trace. The scattering of the waveguide mode at lens edges is obvious. The slopes of straight lines I and II are optical losses in the lens region and in the glass waveguide region, respectively. Gap III denotes the insertion loss of the Luneburg lens including mode conversion losses at the two edges and the propagation loss within the lens. However, the problem of the surface probe is the ambiguity to distinguish the

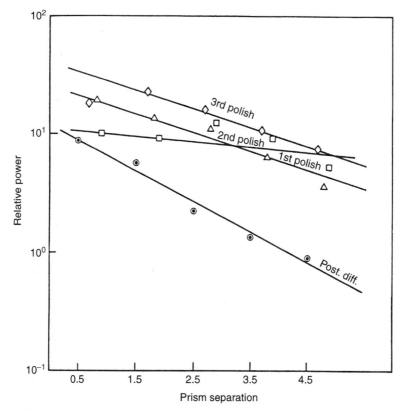

**Figure 12.73**   Waveguide propagation loss data for a Ti:LiNbO$_3$ waveguide with 500 Å starting titanium and 5 h wet oxygen diffusion at 1000°C after several polish steps. (After Ref. 145.)

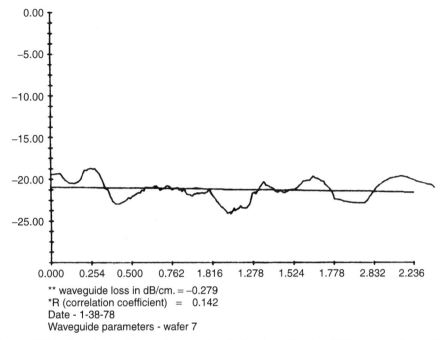

**Figure 12.74**   Longitudinal scan of scattered radiation from Corning 7059 waveguide on silicon water. (After Ref. 142.)

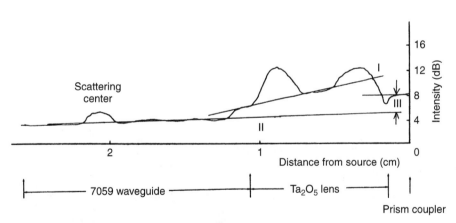

**Figure 12.75**   Waveguide scattering trace across $Ta_2O_5$ thin-film Luneburg lens and subsequent Corning 7059 waveguide. (After Ref. 142.)

cases of normal scattering with high guided-wave intensity and large surface scattering with moderate guided-wave intensity. An example is given in Fig. 12.76 in which the slope may be interpreted in many ways.

For measurement of extremely low-loss waveguides, a pyroelectric probe that will slide on the surface of the waveguide has been developed (see Fig. 12.77). Measurement accuracy of 0.01–0.05 dB/cm has been reported.

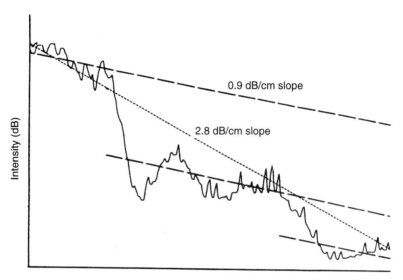

**Figure 12.76**   Measured optical waveguide scattering for Corning 7059 waveguide on a thermally oxidized silicon substrate. Two sharp steps are due to scattering loss and absorption in silicon substrate. Waveguide effective loss is 2.8 dB/cm. Reduced slope lines represent 0.9 dB/cm loss. (After Ref. 142.)

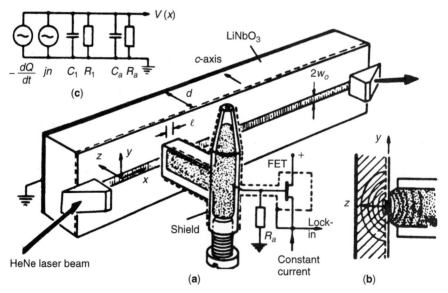

**Figure 12.77**   (a) Schematic arrangement of the Hg electrode; (b) cross section of contact region; (c) equivalent circuit. (After Ref. 206.)

***Waveguide Optical Component Measurement.***   To measure waveguide optical component performances such as focal spots or in-plane scattering, an output coupler must be used to bring the guided mode out for evaluation (see Fig. 12.78). However, care must be taken to consider the effect of reimaging optical quality as the point spread

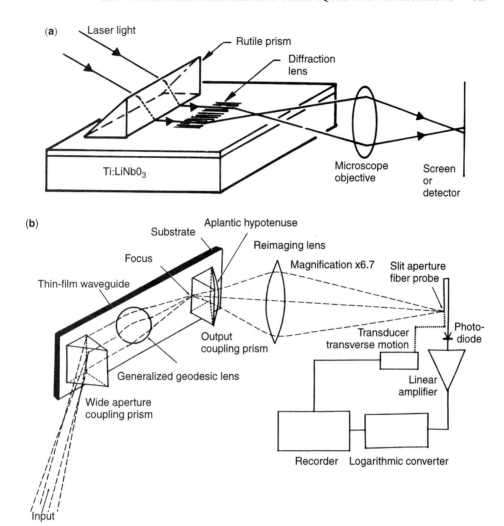

**Figure 12.78** (**a**) Experimental setup for lens evaluation using a polished sharp edge for output coupling. (**b**) Apparatus for evaluation of waveguide geodesic lens diffraction patterns using a prism output coupler.

function of the measurement system will convolve with the beam being evaluated. This is particularly critical for the measurement of diffraction-limited focal spots of a waveguide lens. Figure 12.79 shows the results of two thin-film Luneburg lenses measured by two reimaging optics of different quality. The much better performance SK-2 lens is observed only with the better reimaging optics.

## 12.3.2 Planar Optical Waveguide Applications

This chapter is summarized by several examples of planar waveguide applications. The planar waveguide is suitable for the construction of advanced electro-optical signal-processing subsystems as long as there is no need of a two-dimensional spatial light modulator. Many high-speed signal processors fall into this category. Examples are

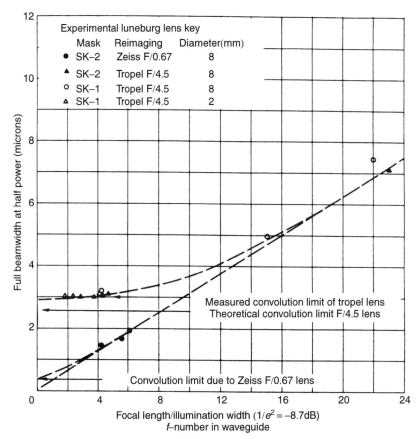

**Figure 12.79**   Measured diffraction pattern beamwidth (3 dB) as a function of the f-number in the waveguide, limited by convolution with the reimaging lens showing experimental results. (After Ref. 152.)

acousto-optical RF spectrum analyzers, and acousto-optical correlators [126, 207–213]. The major components are planar optical waveguides, waveguide optical lenses, and surface acousto-optical modulators [208]. The coupling of light from the laser diode to the photodetectors can be done in several ways including end-fire coupling. An intensive effort has been dedicated by many research organizations over the world on this subject during the past decade. The device is made on lithium niobate substrate that has good surface acoustic wave generation as well as surface acousto-optical interaction capabilities. Most approaches use a geodesic lens fabricated by either impact grinding [212] or diamond machining [209]. Thin-film Luneburg lenses made of niobium pentoxide and chirp grating lenses have also been used in certain cases. Figure 12.80 shows one of the designs. It promises much improved manufacturability and compact device size.

Another planar waveguide device that has been investigated extensively is the wavelength division multiplexer made of waveguide optical filter banks [170–178, 210]. In this case the major candidate substrate material is an oxidized silicon wafer on which photodetector arrays can be fabricated. The optical waveguide is formed by a deposited thin film, and grating reflectors are employed to separate the various optical wavelength components.

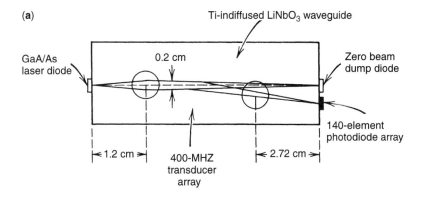

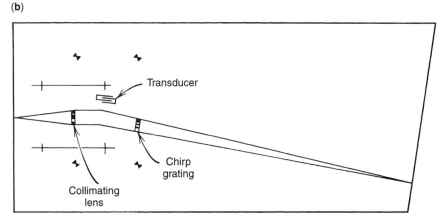

**Figure 12.80** Integrated optical RF spectrum analyzer on LiNbO$_3$ substrate: **(a)** design using geodesic lenses; **(b)** design using diffraction grating lenses. (After Ref. 145.)

Optical waveguide principles have been applied for the characterization of thin-film materials with great success [214–217]. The procedure is simply the measurement of the optical mode indices of the dielectric film of interest with a subsequent fitting to modal dispersion curves. Normally, a pair of prism couplers are used, and the coupling angles for the guided modes are observed. A modified approach uses a single prism coupler for both input and output coupling and a convergent laser beam, as shown in Fig. 12.81. This method eliminates the rotation and observation process, but it does not rely on the actual propagation of the guided modes for appreciable distances. Therefore, it can operate irrespective of the polarization, absorption, or scattering of the guided light. The mode indices appear as dark lines in the reflected beams, indicating the trapping of laser power by the existence of guided modes. The mode indices are fit to normalized dispersion curves by minimizing the least mean-square error for the determination of film thickness and index. Accuracy better than 0.3% is possible.

### 12.3.3 Arrayed Waveguide Grating and Applications

During the 1990s, major progress has been made on planar optical waveguide devices for fiber-optics communications applications. Among the important advances are the

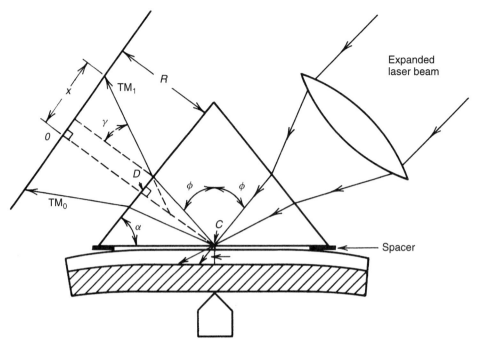

**Figure 12.81** Schematic of prism film arrangement showing expanded collimated laser beam focused onto the prism base and the illuminated area on the screen at distance $R$ from the prism face. A ray incident on prism base at $\phi$ is coupled into guide at angle $\theta$; the "missing ray" makes angle $\gamma$ with normal to prism exit face and appears as a dark line at distance $x$ from the position, $O$, of an undeviated ray. (After Ref. 217.)

arrayed waveguide grating (AWG) technology and applications, [218–220]. Waveguide arrays of up to several hundred channels have been made successfully with excellent quality. The waveguide for these devices are either based on silica or polymers. In the following, a silica-based AWG device is described as an example.

The most prominent feature of the silica waveguides is their simple and well-defined waveguide structures [221, 222]. It is fabricated on silicon or silica substrate by a combination of flame hydrolysis deposition (FHD) and reactive ion etching (RIE). After deposition of under-cladding and waveguiding layers, the substrate is heated to high temperature for consolidation. Then the circuit pattern is fabricated by photolithographic process and reactive ion etching process. Then, the waveguide is covered with over-cladding layer and consolidated again. The properties of the silica waveguide can be adjusted by the index difference between the waveguide core layer and the cladding layers. Lower loss is associated with waveguide made of smaller index difference, [223]. The following table lists the propagation loss and bending radius for several index difference values of the silica waveguide, [220].

The low index difference waveguides are better matched to the standard single-mode optical fibers and thus has low coupling loss. But high index difference waveguides allow tight bending radius and is suited for making highly integrated and large optical circuits such as N × N star couplers, wavelength multiplexers, and optical add and drop multiplexers.

**TABLE 12.15   Propagation Loss and Max Bending Radius for Several Index Difference Values of Silica Waveguide [220]**

|  | Propagation Loss | Bending Radius |
|---|---|---|
| $\Delta n = 2\%$ | 0.1 dB cm$^{-1}$ | R = 2 mm |
| $\Delta n = 0.75\%$ | 0.035 dB cm$^{-1}$ | R = 5 mm |
| $\Delta n = 45\%$ | 0.017 dB cm$^{-1}$ | R = 15 mm |

***Principle of Operation for AWG Devices.*** Arrayed waveguide grating is highly useful in fiber optical communications as wavelength division multiplexing (WDM) devices, N × N star couplers, and optical add/drop multiplexers (OADM), [220, 221]. Figure 12.82 illustrates the configuration of an AWG with the first and the second slab sections magnified, [220]. It consists input and output waveguides, the first and the second focusing slab sections, and a phase array of multiple channel waveguides with constant optical path length difference, $\Delta L$, between adjacent waveguides. As shown, the focusing slab section has the input waveguides and the output waveguides in a curved arrangement. The curvature provides the focusing function similar to curved mirror reflectors or lenses in geometrical optics. The two slab sections may not have to be identical, and they can be each designed for optimal performances.

Due to the constant optical path length difference between the waveguide arrays, the combined output from one set of waveguide channels will interfere at the focal plan similar to a grating diffraction order, or more precisely, like a high spectral resolution etalon. In order to have high spectral resolution, the optical path length differences between adjacent waveguide channels are made to a large number of optical

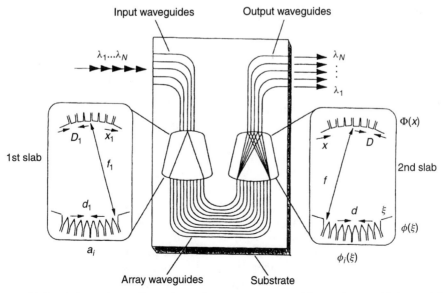

**Figure 12.82**   Schematic configuration of an array waveguide grating multiplexer (after Okamoto).

**TABLE 12.16  Performances of Reported Silica AWG Devices [223, 224]**

| Number of Channels | 16 | 32 | 64 | 128 |
|---|---|---|---|---|
| Diffraction order N | 47 | 59 | 59 | 59 |
| Path length difference | 50.3 micron | 63 micron | 63 micron | 63 micron |
| Channel spacing | 2 nm | 0.8 nm<br>= 100 GHz | 0.4 nm<br>= 50 GHz | 0.2 nm<br>= 25 GHz |
| Slab arc length f | 5.68 mm | 11.35 mm | 24.2 mm | 36.3 mm |
| Number of arrayed<br>  waveguides | 60 | 100 | 160 | 338 |
| On chip insertion loss | 2.3 dB | 2.1 dB | 3.1 dB | 3.5 dB |
| 3-dB bandwidth | 0.74 nm | 40 GHz | 19 GHz | 11 GHz |
| Channel cross-talk | < −29 dB | < −28 dB | < −27 dB | < −16 dB |

wavelengths, $\Delta L = N\lambda$, where $\lambda$ is the optical wavelength of interest and N is the AWG order. For an AWG with M channel of waveguides, the spectral resolving power of an etalon is simply MN, which can be a very large value. For instance, if M = 128 and N = 59, the resolving power is 7552, and the spectral resolution of the AWG shall be $\lambda/MN = \lambda/7552$. At the center of the fiber-optics wavelength range, $\lambda = 1550$ nm, and the AWG resolution is 0.2 nm corresponding to 25 GHz.

Table 12.16 gives some of the reported performances of silica-based AWG devices at 1550-nm center wavelength, [223, 224].

Note that the AWG is often constructed with more waveguide channels than the intended number of wavelength channels for the sake of facilitating tuning of center passband of the AWG channels. When the input and output slabs as well as the channel separations in and out of the slabs are made different, selection of the input–output waveguide pairs can allow vernier adjustment of wavelengths, which is highly useful for compensating of possible center wavelength drift due to fabrication errors. Figure 12.83 gives a typical result of the demultiplexing properties of a 32-channel AWG. Note that the passband of each channel has a sharp tip appearance.

*Flat Spectral Response AWG and Athermal Designs.* A general model of grating type of spectral multiplexers can be established using the correlation operation. Due to spatial dispersion of optical wavelength, the focal image of each of the input channels has a wavelength-dependent shift factor, i.e., smearing according to optical wavelength like a rainbow. Although an output waveguide channel is located at the image plan, where the smeared image of the input waveguide exit aperture is located, only the correct wavelength part of the input image will fall on the output waveguide aperture and be selected. The transfer function of this process is a correlation operation between the input channel image and the output channel image. It is known that the correlation between two circular disks (images of the waveguides) is a sharp-tipped pyramid-like function. Thus, the sharp tip appearance of Fig. 12.83 is easily understood.

However, in a telecommunications system, a signal is required to travel through many multiplexers and demultiplexers. A sharp-tipped channel filter passband shape is highly undesirable because the important signal is contained in the sidebands, which will be lost when traveling through several of such filters. It is necessary to create a flat-top filter response curve if AWG is to be successful as a fiber optical communications device.

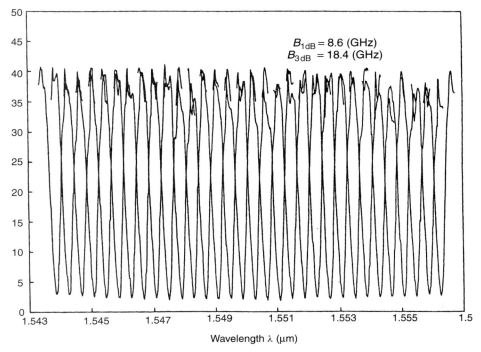

**Figure 12.83**   Demultiplexing properties of a 32-channel AWG, showing only part of the central channels (after Okamoto).

The passband shape of a grating type of device can easily be flattened by means of setting unequal input of output channel apertures. The correlation between a large aperture and a small aperture leads to a flat-topped result with width of the flat top dominated by the size of the larger aperture, and with channel edge slopes defined by the size of the smaller aperture. This can be done by proper adjustment of the channel waveguide input and output channel width inside an AWG, sometimes with the help of a tapered horn structure. However, it is expected that an insertion loss penalty shall come along with this approach because not all of the light can be coupled through the pair of uneven apertures. In the case when the input channels are bigger than the output channels, sure we can understand that only part of the incident light falls on the output channel for transmission, and thus a higher insertion loss shall be incurred according to the ratio of the two channel apertures. In the case of the output channel aperture being the bigger one, all the input channel light falls into the output aperture and be intercepted, but a higher insertion loss will still be there. This is because, in the case of single-mode optical waveguides, the numerical aperture to physical aperture product has to be constant, and approximately equal to the number of waveguide modes supported. So, the larger width aperture of the output channel comes with it a smaller numerical aperture, i.e., angle of acceptance, and much of the light cannot go through without leakages.

Figure 12.84 gives the results for an AWG with a flat top response curve for its channels [225]. The average on-chip insertion loss is about $-7$ dB. Other than the penalty in on-chip insertion loss, the flat top AWG works well and is a highly desirable solution for the DWDM applications. There are other approaches that can be used for

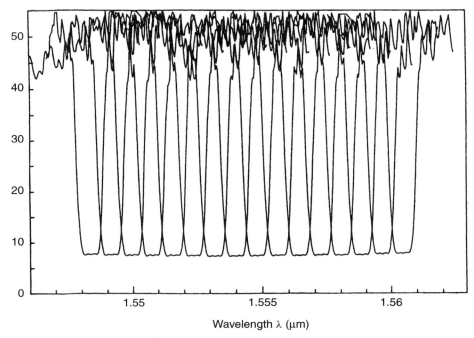

**Figure 12.84**   The response of a 16-channel flat top AWG showing flat top responses, with −7 dB on-chip insertion loss, and −35-dB adjacent channel cross-talk (after Okamoto).

generating a flat top response curve AWG [225–227]. One approach uses s SINC function amplitude distribution at the input waveguide arrays that produces a square-shaped image at the output plan. However, experimentally, the on-chip insertion loss is also about −7 dB, but it is more elaborate in fabrication.

An AWG can be sensitive to temperature due to temperature coefficient of the refractive index of the silica glass. A typical sensitivity number is 0.012 nm per degree Celsius. Because devices in telecommunications applications must endure a wide temperature range of operation, the thermal sensitivity needs be compensated. In one approach, a heater or a Peltier cooler can be constructed into the AWG packaging for temperature control. In another approach, a triangular groove filled with silicone adhesive is inserted into the AWG path, and the negative thermal coefficient of the silicone is employed to compensate for the thermal drift in the AWG. Experimentally, the thermal sensitivity has been reduced about a factor of 20, and the AWG can operate within 0–85°C temperature range with only 0.05 nm of wavelength change. The added insertion loss for the silicone section is about −2 dB.

Phase error in the AWG channels can cause the channel cross-talk to increase due to higher diffraction sidelobes in the focused waveguide images. Phase error in the waveguide channels can be compensated via the photoelastic effect by trimming an amorphous silicon coating on the AWG. Channel cross-talk in a high-resolution AWG has been improved significantly with this technique.

***Dynamic Optical Add/Drop Multiplexer.*** An optical add/drop multiplexer (OADM) is a device that provides simultaneous access to all wavelength channels in a DWDM

Drop port main output

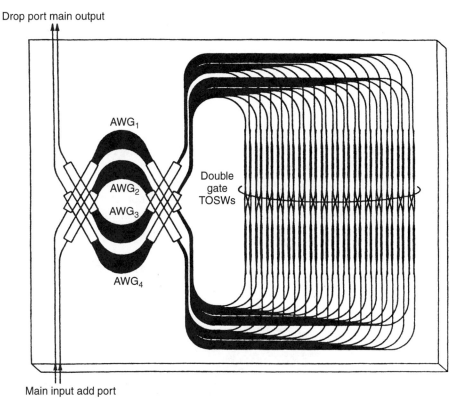

Main input add port

**Figure 12.85**  An OADM can be constructed using four AWG units and a thermal optical waveguide switch array on a single silica waveguide substrate (after Okamoto).

communication system. OADM devices can be constructed using a few AWG units as basic building blocks. The programmable switching function can be provided by integrating thermal optical waveguide switches (TOSW) onto the AWG substrate, or by means of MEMS switch array located at the wavelength domain output plan.

Figure 12.85 illustrates an OADM based on four units of AWG and a thermal optical waveguide switch array on a single silica waveguide substrate [220, 228]. In Fig. 12.85, the PADM has two input ports corresponding to the main communications input port and port for the channels to be added. The input main signal goes through $AWG_1$ and the add signal goes through $AWG_2$, and together they arrive at the thermal optic switch array (TOSWs). Each of the demultiplexed channels goes to one switch element of the thermal optical waveguide switch array. If a channel is to be dropped and added with another channel, the thermal switch corresponding to that channel will be activated and the signal between the main channel and the add channel will be mutated. The add signal is now part of the main channel that goes through $AWG_3$ and reaches the main output port. The original signal in the designated channel coming through the main port is dropped by the thermal optical waveguide switch and goes out to the drop port through $AWG_4$. Thus, by activating a combination of the thermal optical waveguide switch array elements, it is possible to drop and add channels out of and into the main signal path programmably for local use.

### 12.3.4   Multimode Fiber Optics Junctions

Another application of the planar optical waveguide technology is for the construction of multimode optical devices for short haul fiber optics applications. Multimode fiber optics technology used to be active during the 1970s and the early 1980s. As the single-mode fiber-optics technology matured, multimode technology becomes out of favor due to limited bandwidth and transmission capabilities. However, the recent surge of short-distance network applications, for which fiber attenuation and dispersion are not significant issues, has generated a lot of interest in multimode fiber optics devices due to the potential of low-cost implementation. As a result, there is great potential for the development of multimode planar waveguide devices.

Figure 12.86 illustrates multimode Y-junction design parameters [229], Note that a larger distance "d" in the Y-junction corresponds to larger output waveguide width, W. Figure 12.87 gives the amount of optical loss (when total transmitted power becomes

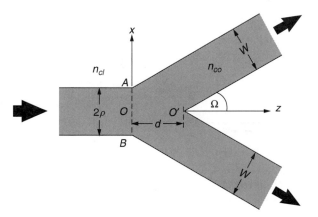

**Figure 12.86**   The Y-junction for multimode fiber-optics applications (after Beltrami).

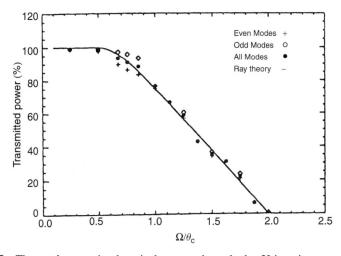

**Figure 12.87**   The total transmitted optical power through the Y-junction, as a function of normalized angle of separation (after Beltrami).

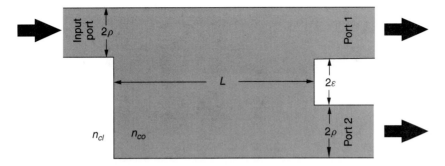

**Figure 12.88**   Schematic of the MMI coupler (after Beltrami).

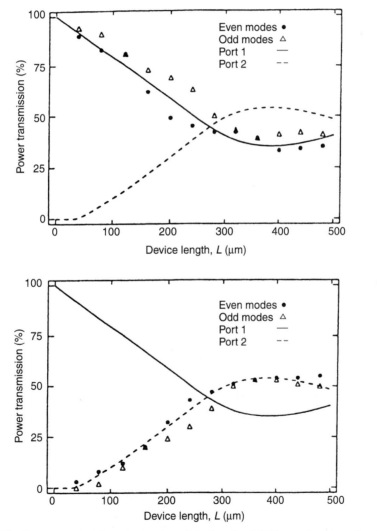

**Figure 12.89**   Power output from ports 1 and 2 for the MMI coupler as a function of mode-mixing length, L.

less than 100%) as a function of the Y-junction angle normalized to the waveguide-critical angle [229]. For small-angle separation, the Y-junction can be lossless and can be a useful device for multimode fiber optics applications.

Figure 12.88 illustrates a multimode interference coupler (MMI) in which a mode-mixing section of length L is provided before one multimode waveguide splitting into two multimode waveguides. The performance of the MMI coupler depends on the mode-mixing length, L. The calculated results for both output ports of the MMI coupler is shown by Fig. 12.89 [229]. With sufficient mode-mixing length, L, the output optical power in both port 1 and port 2 can reach a 50:50 split.

In conclusion, recent advances in fiber-optics applications provide accelerated opportunities for planar optical waveguide technology. For the long-haul fiber-optics applications, planar waveguide technology has been employed for the development of the AWG device technology, which is key to the next-generation DWDM system applications. The AWG technology has become the building block for additional critical-enabling devices such as the OADM for next-generation fiber-optics systems. On the other hand, multimode fiber-optics technology finds new applications in short-distance local networks. Again, multimode planar optical waveguide technology provides several potential low-cost, high-performance devices. Novel phenomena and applications are discovered each year. Future developments will certainly result in more implementation of guided-wave technology in many real-world applications, including commercial applications in many people's home offices.

## REFERENCES

1. D. B. Anderson and R. R. August, *Proc. IEEE*, **54**, p. 657, 1966.
2. A. Yariv and R. C. C. Leite, *Appl. Phys. Lett.*, **2**, p. 55, 1963.
3. W. L. Bond, B. G. Cohen, R. C. C. Leite, and A. Yariv, *Appl. Phys. Lett.*, **2**, p. 57, 1963.
4. D. F. Nelson and F. K. Reinhardt, *Appl. Phys. Lett.*, **5**, p. 148, 1964.
5. H. Osterberg and L. W. Smith, *J. Opt. Soc. Amer.*, **54**, p. 1073, 1964.
6. R. Shubert and J. H. Harris, *IEEE Trans. Microwave Theory Tech.*, **16**, p. 1048, 1968.
7. D. B. Anderson "An Integrated Circuit Approach to Optical Waveguide," *IEEE Microelectron. Symp. Dig.*, St. Louis, MO, 1968.
8. S. E. Miller, *Bell Syst. Tech. J.*, **48**, p. 2059, 1970.
9. P. K. Tien, *Appl. Opt.*, **10**, p. 2395, 1971.
10. S. J. Maurer and L. B. Felson, *Pro. IEEE*, **55**, p. 1718, 1967.
11. H. K. V. Lotsh, *Optik*, **27**, p. 239, 1968.
12. H. Kogelnik and V. Ramaswamy, *Appl. Opt.*, **13**, p. 1857, 1974.
13. D. Marcuse, *Theory of Dielectric Optical Waveguides*, Academic, New York, 1974.
14. J. McKenna, *Bell Syst. Tech. J.*, **46**, p. 1491, 1967.
15. H. Kogelnik, "Theory of Dielectric Waveguides," in T. Tamir, Ed., *Integrated Optics*, Vol. 7, Topics in Applied Physics, Springer-Verlag, New York, Heidelberg, and Berlin, 1975, Chap. 2.
16. J. R. Carruthers, I. P. Kaminov, and L. W. Stulz, *Appl. Opt.*, **13**, p. 2333, 1974.
17. E. M. Conwell, *Appl. Phys. Lett.*, **23**, p. 328, 1973.
18. R. D. Standley and V. Ramaswamy, *Appl. Phys. Lett.*, **25**, p. 711, 1974.
19. A. K. Ghatak, E. Khular, and K. Thyagarajan, *IEEE J. Quantum Electron.*, **QE-14**(6), p. 389, 1978.

20. J. P. Gordon, *Bell Syst. Tech. J.*, **45**, p. 34, 1966.

21. L. B. Felson and N. Marcuvitz, *Radiation and Scattering of Waves*, Prentice-Hall, Englewood Cliffs, NJ, 1973.

22. G. B. Hocker and W. K. Burns, *IEEE J. Quantum Electron.*, **QE-11**(6), p. 270, 1975.

23. R. Ulrich and H. P. Weber, *Appl. Opt.*, **11**(2), p. 428, 1972.

24. D. B. Ostrowsky and A. Jacques, *Appl. Phys. Lett.*, **18**, p. 556, 1971.

25. P. K. Tien, G. Smolinsky, and R. J. Martin, *Appl. Opt.*, **11**, p. 637, 1972.

26. M. J. Vasile and G. Smolinsky, *J. Electrochem. Soc.*, **119**, p. 451, 1972.

27. P. K. Tien, R. J. Martin, and G. Smolinsky, *Appl. Opt.*, **12**, p. 1909, 1973.

28. T. P. Sosnowski and H. P. Weber, *Appl. Phys. Lett.*, **21**, p. 310, 1972.

29. J. P. Kaminow, H. P. Weber, and E. A. Chandross, *Appl. Phys. Lett.*, **18**, p. 497, 1971.

30. E. A. Chandross, C. A. Pryde, W. J. Tomlinson, and H. P. Weber, *Appl. Phys. Lett.*, **24**(2), p. 72, 1974.

31. R. Selvaraj, H. T. Lin, and J. F. McDonald, *IEEE J. Lightwave Technol.*, **6**, p. 1034, 1988.

32. J. Nishigawa and A. Otsuka, *Appl. Phys. Lett.*, **21**, p. 48, 1972.

33. C. Hu and J. R. Whinnery, *J. Opt. Soc. Amer.*, **64**, p. 1424, 1974.

34. J. P. Sheridan, J. M. Schmur, and T. G. Giallorenzi, *Appl. Phys. Lett.*, **22**, p. 56, 1973.

35. Y. A. Bykovskii, A. V. Makovkin, V. L. Smirnov, and V. N. Sorokovikov, *Sov. J. Quantum Electron.*, **5**, p. 1361, 1976.

36. D. P. Gra Russo and J. H. Harris, *Appl. Opt.*, **10**, p. 2786, 1971.

37. L. N. Polky and J. H. Harris, *Appl. Phys. Lett.*, **21**, p. 307, 1972.

38. P. K. Tien et al., *Appl. Phys. Lett.*, **14**, p. 291, 1969.

39. E. L. Paradis and A. J. Shuskus, *Thin Solid Films*, **38**, p. 131, 1976.

40. S. Dutta, H. E. Jackson, J. T. Boyd, F. S. Hickernell, and R. L. Davis, *Appl. Phys. Lett.*, **39**, p. 206, 1981.

41. R. Th. Kersten and W. Rauscher, *Opt. Commun.*, **13**, p. 189, 1975.

42. J. E. Goel and R. D. Standley, *Bell Syst. Tech. J.*, **48**, p. 3445, 1969.

43. P. D. Davidse and L. I. Maissel, *J. Appl. Phys.*, **37**, p. 574, 1966.

44. T. Nishimura, Y. Murayana, K. Dota, and H. Matsumaru, *Dig. 3rd Symp. Deposition of Thin Films by Sputtering*, Rochester, 1969, p. 96.

45. C. W. Pitt, *Optiss Lett.*, **9**, p. 401, 1973.

46. J. E. Goell, *Appl. Opt.*, **12**, p. 737, 1973.

47. H. Yajima, S. Kawase, and Y. Sekimoto, *Appl. Phys. Lett.*, **21**, p. 407, 1972.

48. B. Chen and C. L. Tang, *Appl. Phys. Lett.*, **28**, p. 435, 1976.

49. B. Chen, C. L. Tang, and J. M. Telle, *Appl. Phys. Lett.*, **25**, p. 495, 1974.

50. D. H. Hemster, J. D. Cuthbert, R. J. Martin, and P. K. Tien, *Appl. Opt.*, **10**, p. 1037, 1971.

51. H. Terui and M. Kobayashi, *Appl. Phys. Lett.*, **32**, p. 666, 1978.

52. S. J. Ingrey, W. D. Westwood, Y. C. Cheng, and J. Wei, *Appl. Opt.*, **14**, p. 2194, 1975.

53. S. J. Ingrey and W. D. Westwood, *Appl. Opt.*, **15**, p. 607, 1976.

54. M. J. Rand and R. D. Standley, *Appl. Opt.*, **11**, p. 2482, 1972.

55. W. Stutius and W. Streifer, *Appl. Opt.*, **16**, p. 3218, 1977.

56. G. H. Hewig and K. Jain, *J. Appl. Phys.*, **54**, p. 57, 1983.

57. A. K. Watts, M. deWit, and W. C. Holton, *Appl. Opt.*, **13**, p. 2329, 1974.

58. J. F. Lotspeich, *Appl. Opt.*, **13**, p. 2529, 1974.

59. M. Kawachi, M. Yasu, and T. Edahiro, *Electron. Lett.*, **19**, p. 583, 1983.

60. M. Kawachi, M. Yasu, and M. Kobayashi, *Japan J. Appl. Phys.*, **22**, p. 1932, 1983.

61. A. A. Ballman, H. Brown, P. K. Tien, and R. J. Martin, *J. Crystal Growth*, **20**, p. 251, 1973.

62. S. Miyazawa, *Appl. Phys. Lett.*, **23**, p. 198, 1973.

63. S. Fukunishi, N. Uchinda, S. Miyazawa, and J. Noda, *Appl. Phys. Lett.*, **24**, p. 424, 1974.

64. P. K. Tien, S. Riva-Sanseverino, R. J. Martin, A. A. Ballman, and H. Bro, *Appl. Phys. Lett.*, **24**, p. 503, 1974.

65. S. Miyazawa, S. Fushimi, and S. Kondo, *Appl. Phys. Lett.*, **26**, p. 8, 1975.

66. P. K. Tien and A. A. Ballman, *J. Vac. Sci. Tech.*, **12**, p. 892, 1975.

67. P. K. Tien, R. J. Martin, S. L. Blank, S. H. Wemple, and L. J. Varnerin, *Appl. Phys. Lett.*, **21**, p. 207, 1972.

68. P. G. Van Engen, *J. Appl. Phys. Lett.*, **49**, p. 4660, 1978.

69. N. Chubachi, *Proc. IEEE*, **64**, p. 772, 1976.

70. D. J. Chanin, J. M. Hammer, and M. T. Duffy, *Appl. Opt.*, **14**, p. 923, 1975.

71. T. Shiosaki, S. Ohnishi, Y. Hirokawa, and A. Kawabata, *Appl. Phys. Lett.*, **35**, p. 406, 1978.

72. J. E. Goell, R. D. Standley, W. M. Gibson, and J. W. Rodgers, *Appl. Phys. Lett.*, **21**, p. 72, 1972.

73. A. P. Webb and P. D. Townsend, *J. Phys. D, Appl. Phys.*, **9**, p. 1343, 1976.

74. R. D. Standley, W. M. Gibson, and J. W. Rodgers, *Appl. Opt.*, **11**, p. 1313, 1972.

75. D. T. Y. Wei, W. W. Lee, and L. R. Bloom, *Appl. Phys. Lett.*, **25**, p. 329, 1974.

76. G. L. Destefanis, J. P. Gailliard, E. L. Ligeon, and S. Valette, *J. Appl. Phys.* **50**, p. 7898, 1979.

77. G. L. Destefanis, P. D. Townsend, and J. P. Gailliard, *Appl. Phys. Lett.*, **32**, p. 29, 1978.

78. S. Valette, G. Labrunie, J-C Deutsch, and J. Lizet, *Appl. Opt.*, **16**, p. 1289, 1977.

79. R. F. Bartholomew and H. M. Garfinkel, *Glass: Science and Tech*, Vol. 5, Academic, New York, 1980, Chap. 6.

80. R. V. Ramaswamy and R. Srivastava, *J. Lightwave Tech.*, **6**, p. 984, 1988.

81. C. A. Millar and R. H. Hutchins, *J. Phys. D, Appl. Phys.*, **11**, p. 1567, 1978.

82. T. Findakly, *Opt. Eng.*, **24**, p. 244, 1985.

83. R. H. Doremus, *J. Phys. Chem.*, **68**, p. 2212, 1964.

84. T. Izawa and H. Nakagome, *Appl. Phys. Lett.*, **21**, p. 584, 1972.

85. T. G. Giallorenzi, E. J. West, R. Kirk, R. Ginther, and R. A. Andrews, *Appl. Opt.*, **12**, p. 1240, 1973.

86. G. H. Chartier, P. Jaussaud, A. D. de Oliveira, and O. Parriaux, *Electron. Lett.*, **13**, p. 763, 1977.

87. V. Neuman, O. Parriaux, and L. M. Walpita, *Electron. Lett.*, **15**, p. 704, 1979.

88. L. Ross, H. J. Lilienhof, H. Holscher, H. F. Schlaak, and A. Brandenburg, Topical Meeting Integrated and Guidewave Optics, Atlanta, GA, 1986, pp. 25–26.

89. G. Stewart, C. A. Millar, P. J. R. Laybourn, C. D. W. Wilkinson, and R. M. DeLaRue, *IEEE J. Quantum Electron.*, **QE-13**, p. 192, 1977.

90. G. Stewart and P. J. R. Laybourn, *IEEE J. Quantum Electron.*, **QE-14**, p. 930, 1978.

91. R. G. Eguchi, E. A. Maunders, and I. K. Naik, *Proc. SPIE*, **408**, p. 21, 1983.

92. R. G. Walker, C. D. W. Wilkinson, and J. A. H. Wilkinson, *Appl. Opt.*, **22**, p. 1923, 1983.

93. G. L. Yip and J. Albert, *Opt. Lett.*, **10**, p. 151, 1985.

94. J. E. Gortych and D. G. Hall, *IEEE J. Quantum Electron.*, **QE-22**, p. 892, 1986.

95. M. Shah, *Appl. Phys. Lett.*, **26**, p. 652, 1975.

96. J. Jackel, *Appl. Phys. Lett.*, **37**, p. 739, 1980.

97. J. L. Jackel, C. E. Rice, and J. J. Vaselka, Jr., IEEE/OSA Topical Meeting in Integrated and Guidedwave Optics, Pacific Grove, CA, 1982.

98. H. J. Lilienhof, E. Vogas, D. Ritter, and B. Pantschew, *IEEE J. Quantum Electron.*, **QE-18**, p. 1877, 1982.

99. R. V. Ramaswamy and S. I. Najafi, *IEEE J. Quantum Electron.*, **QE-22**, p. 883, 1986.

100. I. P. Kaminov and J. R. Carruthers, *Appl. Phys. Lett.*, **22**, p. 326, 1973.

101. J. R. Carruthers, I. P. Kaminov, and L. W. Stulz, *Appl. Opt.*, **13**, p. 233, 1974.

102. R. V. Schmidt and I. P. Kaminov, *Appl. Phys. Lett.*, **25**, p. 458, 1974.

103. G. J. Griffith and R. J. Esdail, *IEEE J. Quantum Electron.*, **QE-20**, p. 149, 1984.

104. O. Eknoyan, A. S. Greenblatt, W. K. Burns, and C. H. Bulmer, *Appl. Opt.*, **25**, p. 737, 1986.

105. R. J. Esdaile, *Appl. Phys. Lett.*, **33**, p. 733, 1978.

106. T. R. Ranganath and S. Wang, *Appl. Phys. Lett.*, **30**, p. 376, 1977.

107. J. L. Jackel, *J. Opt. Commun.*, **3**, p. 82, 1982.

108. T. P. Pearsall, S. Chiang, and R. V. Schmidt, *J. Appl. Phys.*, **47**, p. 4794, 1976.

109. A. M. Glass, J. P. Kaminow, A. A. Ballman, and D. H. Olson, *Appl. Opt.*, **19**, p. 276, 1980.

110. M. Minakata, S. Saita, M. Shibata, and S. Miyazawa, *J. Appl. Phys.*, **49**, p. 4677, 1978.

111. J. Noda, T. Saku, and N. Uchida, *Appl. Phys. Lett.*, **25**, p. 308, 1974.

112. Y. Okamura, S. Yamamoto, and T. Makimoto, *Appl. Phys. Lett.*, **32**, p. 161, 1978.

113. O. Eknoyan, D. W. Yoon, and H. F. Taylor, *Appl. Phys. Lett.*, **51**, p. 384, 1987.

114. G. D. Boyd, R. V. Schmidt, and F. D. Storz, *J. Appl. Phys.*, **48**, p. 2880, 1977.

115. J. Noda, M. Fukuma, and S. Saito, *J. Appl. Phys.*, **49**, p. 3150, 1978.

116. M. Minakata, J. Noda, and N. Uchida, *Appl. Phys. Lett.*, **26**, p. 395, 1975.

117. V. Ramaswamy and R. D. Standley, *Appl. Phys. Lett.*, **26**, p. 10, 1975.

118. M. Masuda and J. Koyama, *Appl. Opt.*, **16**, p. 2994, 1977.

119. T. Findakly and C. L. Chen, *Appl. Opt.*, **17**, p. 469, 1978.

120. Y. Suematsu, M. Hakuta, K. Furuya, K. Chiba, and R. Hasumi, *Appl. Phys. Lett.*, **21**, p. 291, 1972.

121. A. Reisinger, *Appl. Opt.*, **12**, p. 1015, 1973.

122. A. Reisinger, *Appl. Phys. Lett.*, **23**, p. 237, 1973.

123. I. P. Kaminov, W. L. Mammel, and H. P. Weber, *Appl. Opt.*, **13**, p. 396, 1974.

124. E. M. Garmire and H. Stoll, *IEEE J. Quantum Electron.*, **QE-8**, p. 763, 1972.

125. H. F. Taylor, W. E. Martin, D. B. Hall, and V. N. Smiley, *Appl. Phys. Lett.*, **21**, p. 95, 197.

126. M. C. Hamilton, D. A. Wille, and W. J. Miceli, *IEEE 1976 Ultrason. Symp. Proc.*, Annapolis, MD, 1976.

127. D. Hall, A. Yariv, and E. Garmire, *Opt. Commun.*, **1**, p. 403, 1970.

128. R. Shubert and J. H. Harris, *IEEE Trans. Microwave Theory Tech.*, **MTT-16**, p. 1048, 1968.

129. D. Marcuse and E. A. J. Marcatili, *Bell Syst. Tech. J.*, **50**, p. 43, 1971.

130. J. H. Harris and R. Shubert, URSI (International Radio Scientific Union) Spring Meeting, Washington DC, 1969, p. 71.

131. P. K. Tien, R. Ulrich, and R. J. Martin, *Appl. Phys. Lett.*, **14**, p. 291, 1969.

132. F. Zernike and J. E. Midwinter, *IEEE J. Quantum Electron.*, **QE-6**, p. 577, 1970.

133. P. K. Tien and R. Ulrich, *J. Opt. Soc. Amer.*, **60**, p. 1325, 1970.

134. T. Tamir and H. L. Bertoni, *J. Opt. Soc. Amer.*, **61**, p. 1397, 1971.

135. T. Tamir, "Beam and Waveguide Couplers," in T. Tamir, Ed., *Integrated Optics*, Vol. 7, Topics in Applied Physics, Springer-Verlag, New York, Heidelberg, and Berlin, 1975, Chap. 3.

136. M. L. Dakss, L. Kuhn, P. F. Heidrich, and B. A. Scott, *Appl. Phys. Lett.*, **16**, p. 523, 1970.

137. H. Kogelnik and T. P. Sosnowski, *Bell Syst. Tech. J.*, **49**, p. 1602, 1970.

138. P. K. Tien, *Appl. Opt.*, **10**, p. 2395, 1971.

139. S. T. Peng and T. Tamir, *Opt. Commun.*, **11**, p. 405, 1974.

140. S. T. Peng, T. Tamir, and H. L. Bertoni, *IEEE Trans. Microwave Theory Tech.*, **MTT-23**, p. 123, 1975.

141. D. B. Anderson, R. L. Davis, J. T. Boyd, and R. R. August, *IEEE J. Quantum Electron.*, **QE-13**, p. 275, 1977.

142. D. B. Anderson, R. R. August, and S. K. Yao, "Waveguide Optics for Coherent Optical Processing," Final Technical Report, Rockwell International, AFAL-TR-78-83, June 1978.

143. J. M. Delavaux, S. Forouhar, W. S. C. Chang, and R. X. Lu, "Experimental Fabrication and Evaluation of Diffraction Lenses in Planar Optical Waveguides," IEEE/OSA Conference on Lasers and Electro-Optics Phoenix, AZ, April 1982.

144. S. Forouhan, C. Warren, R. X. Lu, and W. S. C. Chang, "Tech. for the Fabri. of High Index Overlay Films on LiNbO$_3$," SPIE Technical Symposium, Arlington, VA, April 1983.

145. S. K. Yao and L. Jostad, "Optical Waveguide Diffraction Elements Final Technical Report," TRW, AFWAL-TR-85-1061, May 1985.

146. C. Warren, S. Forouhan, W. S. C. Chang, and S. K. Yao, *Appl. Phys. Lett.*, **43**, p. 424, 1983.

147. R. K. Luneburg, *The Mathematical Theory of Optics*, University of California Press, Berkeley, 1954.

148. S. P. Morgan, *J. Appl. Phys.*, **29**, p. 1358, 1958.

149. W. H. Southwell, *J. Opt. Soc. Amer.*, **67**, p. 1010, 1977.

150. S. K. Yao, *J. Appl. Phys.*, **50**, p. 3390, 1979.

151. S. K. Yao and D. B. Anderson, *Appl. Phys. Lett.*, **33**, p. 307, 1978.

152. S. K. Yao, D. B. Anderson, R. R. August, B. R. Youmans, and C. M. Oania, *Appl. Opt.*, **18**, p. 4067, 1979.

153. R. F. Rinehart, *J. Appl. Phys.*, **19**, p. 860, 1948.

154. G. C. Righini, V. Russo, S. Sottini, and G. T. diFrancia, *Appl. Opt.*, **12**, p. 1477, 1973.

155. E. Spiller and J. S. Harper, *Appl. Opt.*, **13**, p. 2105, 1974.

156. C. M. Verber, D. W. Vahey, and V. E. Wood, *Appl. Phys. Lett.*, **28**, p. 514, 1976.

157. V. E. Wood, *Appl. Opt.*, **15**, p. 2817, 1976.

158. K. S. Kunz, *J. Appl. Phys.*, **25**, p. 642, 1954.

159. G. C. Righini, V. Russo, and S. Sottini, *IEEE J. Quantum Electron.*, **QE-15**, p. 1, 1979.

160. B. Chen, E. Marom, and R. J. Morrison, *Appl. Phys. Lett.*, **33**, p. 511, 1978.

161. G. E. Betts and G. E. Merx, *Appl. Opt.*, **17**, p. 3969, 1978.

162. W. H. Southwell, *J. Opt. Soc. Amer.*, **67**, p. 1293, 1977.

163. S. Sottini, V. Russo, and G. C. Righini, *IEEE Trans. Circ. Syst.*, **CAS-26**, p. 1036, 1979.

164. S. K. Yao and D. E. Thompson, *Appl. Phys. Lett.*, **33**, p. 635, 1978.

165. P. Mottier and S. Valette, *Appl. Opt.*, **20**, p. 1630, 1981.

166. P. R. Ashley and W. S. C. Chang, *Appl. Phys. Lett.*, **33**, p. 491, 1978.

167. G. Hatakoshi and S. Tamaka, *Opt. Lett.*, **2**, p. 142, 1978.

168. P. K. Tien, *Opt. Lett.*, **1**, p. 64, 1977.

169. H. M. Stoll, *Appl. Opt.*, **17**, p. 2562, 1978.

170. K. Wagatsuma, H. Sakaki, and S. Saito, *IEEE J. Quantum Electron.*, **QE-15**, p. 632, 1979.

171. V. Neuman, C. W. Pitt, and L. M. Walpita, *Electron. Lett.*, **17**, p. 165, 1981.

172. Y. Handa, T. Suhara, H. Nishihara, and J. Koyama, *Appl. Opt.*, **9**, p. 2842, 1980.

173. Y. Handa, T. Suhara, H. Nishihara, and J. Koyama, *Opt. Lett.*, **5**, p. 309, 1980.

174. H. Kotani, M. Kawabe, and S. Namba, *Jap. J. Appl. Phys.*, **18**, p. 279, 1979.

175. N. Gremillet, G. Thomin, and J. Marcou, *Opt. Commun.*, **32**, p. 69, 1980.

176. D. A. Bryan and J. K. Powers, *Opt. Lett.*, **5**, p. 407, 1980.

177. A. C. Livanos, A. Katzir, A. Yariv, and C. S. Hong, *Appl. Phys. Lett.*, **30**, p. 519, 1977.

178. T. Fukuzawa and M. Nakamura, *Opt. Lett.*, **4**, p. 343, 1979.

179. A. Yariv and M. Nakamura, *IEEE J. Quantum Electron.*, **QE-13**, p. 233, 1977.

180. D. C. Flanders, H. Kogelnik, R. V. Schmidt, and C. V. Shank, *Appl. Phys. Lett.*, **24**, p. 194, 1974.

181. S. K. Yao, *SPIE Proc.*, **269**, 1981.

182. M. Stockman and W. Beinvogl, *Wave Electron.*, **4**, p. 221, 1983.

183. S. Forouhan, R. X. Lu, W. S. C. Chang, R. L. Davis, and S. K. Yao, *Appl. Opt.*, **22**, p. 3128, 1983.

184. S. Wang, *IEEE J. Quantum Electron.*, **QE-13**, p. 176, 1977.

185. S. Valette, A. Morgue, and P. Mottier, *Electron. Lett.*, **18**, p. 13, 1982.

186. W. S. C. Chang and P. R. Ashley, *IEEE J. Quantum Electron.*, **QE-16**, p. 744, 1980.

187. S. K. Yao and E. H. Young, *SPIE Proc.*, **90**, p. 23, 1976.

188. E. Y. B. Pun and A. Yi-Yan, *Appl. Phys. Lett.*, **38**, p. 673, 1981.

189. H. W. Yen, W. Ng, I. Samid, and A. Yariv, *Opt. Commun.*, **17**, p. 213, 1976.

190. D. B. Anderson and J. T. Boyd, *Appl. Phys. Lett.*, **19**, p. 266, 1971.

191. W. Sohler and H. Suche, *SPIE Proc.*, **408**, p. 163, 1983.

192. R. Regener and W. Sohler, *J. Opt. Soc. Amer. B*, **5**, p. 267, 1988.

193. G. H. Hewig and K. Jain, *Opt. Commun.*, **47**, p. 347, 1983.

194. H. Ito, N. Uesugi, and H. Inaba, *Appl. Phys. Lett.*, **25**, p. 385, 1974.

195. K. Sasaki, T. Kinoshita, and N. Karasawa, *Appl. Phys. Lett.*, **45**, p. 333, 1984.

196. K. D. Singer, J. E. Sohn, and S. J. Lalama, *Appl. Phys. Lett.*, **49**, p. 248, 1986.

197. P. K. Tien, R. Ulrich, and R. J. Martin, *Appl. Phys. Lett.*, **17**, p. 447, 1970.

198. S. Zemon, R. R. Alfano, S. L. Shapiro, and E. Conwell, *Appl. Phys. Lett.*, **21**, p. 327, 1972.

199. L. Kuhn, *IEEE J. Quantum Electron.*, **QE-5**, p. 383, 1969.

200. D. B. Anderson, J. T. Boyd, and J. D. McMullen, *Proc. Symp. Submilliter Waves, MRI Series*, **20**, p. 19, 1971.

201. E. M. Conwell, *IEEE J. Quantum Electron.*, **QE-9**, p. 867, 1973.

202. G. I. Stegeman, E. M. Wright, N. Finlayson, R. Zanoni, and C. T. Seaton, *IEEE J. Lightwave Tech.*, **6**, p. 953, 1988.

203. J. T. Boyd and D. B. Anderson, *Opt. Commun.*, **13**, p. 353, 1975.

204. M. Minakata, S. Saito, M. Shibata, and S. Miyazawa, *J. Appl. Phys.*, **49**, p. 4677, 1978.

205. M. Johnson and C. W. Pitt, *Opt. Commun.*, **23**, p. 121, 1977.

206. K. H. Haegele and R. Ulrich, *Opt. Lett.*, **4**, p. 60, 1979.

207. D. B. Anderson, R. L. Davis, J. T. Boyd, and R. R. August, *IEEE J. Quantum Electron.*, **QE-13**, p. 268, 1977.

208. C. S. Tsai, Le T. Nguyen, S. K. Yao, and M. A. Alhaider, *Appl. Phys. Lett.*, **26**, p. 140, 1975.

    C. J. Li, C. S. Tsai, and C. C. Lee, *IEEE J. Quantum Electron.*, **QE-22**, p. 868, 1986.

    D. A. Wille and M. C. Hamilton, *Appl. Phys. Lett.*, **24**, p. 159, 1974.

209. D. Mergerian, E. C. Malarkey, R. P. Pautienus, J. C. Bradley, G. E. Marx, L. D. Hutcheson, and A. L. Kellner, *Appl. Opt.*, **19**, p. 3033, 1980.

210. J. D. Spear-Zino, R. R. Rice, J. K. Powers, D. A. Bryan, D. G. Hall, E. A. Dalke, and W. R. Reed, *Proc. SPIE*, **239**, p. 293, 1980.

211. C. M. Verber, R. P. Kenan, and J. R. Busch, *J. Lightwave Tech.*, **LT-1**, p. 256, 1983.

212. T. R. Joseph and B. U. Chen, Technical Digest of the Conference on Integrated and Guided wave Optics, Paper ME-2, Incline Village, NV, 1980.

213. C. S. Tsai, *IEEE Trans. Circuits Systems*, **CAS-26**, p. 1072, 1979.

214. M. J. Sun, *Appl. Phys. Lett.*, **33**, p. 291, 1978.

215. M. Olivier and J-C Peuzin, *Appl. Phys. Lett.*, **32**, p. 385, 1978.

216. J. D. Swalen, R. Santo, M. Tacke, and J. Fischer, *IBM J. Res. Dev.*, p. 168, March 1977.

217. J. S. Wei and W. D. Westwood, *Appl. Phys. Lett.*, **32**, p. 819, 1978.

218. H. Takahashi, Y. Hibino, and I. Nishi, *Opt. Lett.*, **17**, p. 499, 1992.

219. M. Zirnigible, C. Dragone, and C.H. Joyner, *IEEE Phot. Tech. Lett.*, p. 1250, 1992.

220. K. Okamoto, *Opt. QE*, **31**, p. 107, 1999.

221. M. Kawachi, *Opt. QE*, **22**, p. 391, 1990.

222. A. Himeno, et al., in *Photonic Networks*, Springer, Berlin, 1997, p. 172.

223. S. Suzuki, et al., *Electron. Lett.*, **28**, p. 1863, 1992.

224. K. Okamoto, et al., *Electron. Lett.*, **32**, p. 1474, 1996.

225. K. Okamoto, et al., *Opt. Lett.*, **20**, p. 43, 1995.

226. M.R. Amersfoort, et al., *Electron. Lett.*, **32**, p. 449, 1996.

227. D. Trouchet, et al., *Proc. OFC '97*, Dallas, TX, 1997, Paper ThM7.

228. K. Okamoto, et al., *Opt. Lett.*, **32**, p. 1471, 1996.

229. D.R. Beltrami, J.D. Love, and F. Ladouceur, *Opt. QE*, **31**, p. 307, 1999.

# 13

# OPTICAL ATTENUATORS, ISOLATORS, CIRCULATORS, AND POLARIZERS

FENG QING ZHOU AND J. J. PAN
*Lightwaves 2020, Inc.*
*Milpitas, California*

## 13.1  INTRODUCTION

This chapter discusses several key optical devices widely used in fiber optic communications, optical testing, and instrumentation, including optical polarizers, optical attenuators, optical isolators, and circulators. Over the past decade, the demand for bandwidth has grown dramatically thanks to the growth of the Internet and voice data traffic. In response, optical communication system manufacturers are turning increasingly to these next-generation devices to meet the demand.

Although most of the devices cover the spectrum from ultraviolet to infrared, we will mainly discuss those that operate in the fiber optic communication wavelength windows, namely, the 900 nm to 1630 nm wavelength range.

Over many years' research and development, so many variations of these devices have been invented, designed, and produced based on different working principles, that it is impossible to cover every aspect of the subject. Therefore, only the most representative and relevant materials and designs will be discussed.

In Section 13.2 of this chapter, optical polarizers will be discussed in detail. Optical polarizers are not new optical devices, but they have existed for centuries. Only with the advancement of the fiber optic communication technologies, however, have they found practical application on a large scale. At the same time, a lot of new approaches have emerged in recent years. In this section, we will first introduce optical polarizers in the strict sense of the word. They are polarizers that generate only one polarization beam component; the other component is absorbed, scattered, or deflected away and dissipated. This kind of polarizer is not used as widely as the ones subsequently discussed.

In reality, most polarizers that are designed for use in fiber optic networks should be called polarization beam splitters. This is because, after passing through a polarizer,

*Handbook of Optical Components and Engineering*,  Edited by Kai Chang
ISBN 0-471-39055-0  © 2003 John Wiley & Sons, Inc.

a single light beam is split into two polarized beam components that can be used. Conversely, these two polarized beam components can be recombined into a single beam by the polarizer. In this sense, a polarizer can also be referred to as a polarization beam combiner. As fiber optic communication devices require polarization independent operation, this kind of polarizer is the most widely used in optical device design. Lastly, we will briefly discuss other types of polarizers, such as fiber polarizers.

Polarization rotators and phase shifters are the topics of Section 13.3. These devices are the basic building blocks of most passive optical devices. Here, we will mainly discuss Faraday rotators and waveplates.

In Section 13.4, various optical attenuators are discussed. Beginning with the properties of Gaussian beams, and then with various coupling loss mechanisms of Gaussian beams, the basic principles of attenuators are described. Following the discussion of Gaussian beams, fixed attenuators, variable optical attenuators (VOA), and EDFA (Erbium Doped Fiber Amplifier) gain equalizers are addressed sequentially.

Section 13.5 introduces the topic of optical isolators, with examples representing different design approaches.

Starting with a single polarization circulator, different circulator designs are discussed in Section 13.6, including those constructed with cube PBSs, walk-off birefringent crystals, planar waveguides, and those that work in reflection mode. Finally, new trends regarding optical isolator and circulator design are reviewed.

At the end of the chapter, references are given. No effort has been made to give a complete list of literature regarding the optical devices discussed in this chapter as the vast amount of existing literature prohibits us from doing so. Throughout the chapter, proposals, designs, and inventions other than the author's own are introduced and discussed. We have done our best to acknowledge the original authors and to present their material clearly and accurately.

## 13.2   OPTICAL POLARIZERS

Optical polarization results from the fact that the optical waves are transverse waves. Standard treatment of optical waves and polarization can be found in any optic classic, such as the book by Max Born and Emil Wolf [1]. For an introductory description, the book by Bahaa E. A. Saleh and Malvin Carl Teich [2] is a good candidate. For an in-depth treatment of optical polarization, please refer to Jean M. Bennett's chapter entitled "Polarization" in volume I of *Handbook of Optics* [3].

An ideal linear polarizer is an optical device that passes the component of the electric field that is vibrating in the direction of its transmission axis and blocks the orthogonal component. An actual linear polarizer, when placed in an unpolarized incident beam, generates a beam of light whose electric field vibrates primarily in one plane, with only a small component vibrating in the plane perpendicular to it.

For a thorough introduction to optical polarizers, please refer to Chapter 3, volume II of *Handbook of Optics*, entitled "Polarizers", by Jean M. Bennett [4].

Polarizers, according to their working principles, can be categorized according to their selective absorption, selective reflection, selective refraction, selective scattering, and selective diffraction. The different categories of polarizers, then, include wire grid and grating polarizers, dichroic polarizers, birefringent crystal polarizers and splitters, and other in-fiber polarizers and waveguide polarizers. Here, however, we will only discuss those polarizers that are frequently found in optical devices used in fiber optic communication systems.

The least expensive polarizers belong to the dichroic category. Polarizers working in the visible region, such as Polaroid™ and Polacoat™, were first made commercially available several decades ago. They are basically polymer based and, as a result, usually cannot meet the stringent requirements of telecommunication equipment.

The most well-known dichroic polarizer used in telecommunication devices is Corning's Polarcor (*http://www.corning.com/photonicmaterials/products/*). Polarcor is a borosilicate glass that contains titanium, aluminum, and silver. The polarizing mechanism consists of elongated silver crystals in the outer 20 to 50 microns of the glass surface and works from 633 nm to 2100 nm. When working in the 1275 nm to 1345 nm or 1510 nm to 1590 nm telecommunication wavelength windows, the largest available dimension for the polarizer is a $15 \times 15$ mm$^2$ glass sheet with a thickness varying from 30 micron to 0.5 mm. The transmittance of the polarizer is 98%, and its residual reflectivity is 0.25%. The typical contrast ratio is as high as 10,000:1 (40 dB). It also has a wide acceptance angle of 30° and can withstand temperatures as high as 400°C for a short period of time.

Wire grid polarizers, on the other hand, were first designed for use with far infrared wavelengths and radio waves in the 1960s. When the wavelength of the light is much longer than the grid pitch, the light beam component with the electric field vibrating perpendicular to the grid wires passes through, while the component with the electric field vibrating parallel to the grid wires is reflected back. With advancements in microphotolithography, the 0.36-micron pitch wire grid manufacturing equipment became commonly available and enabled the wire grid polarizer to go into the near infrared wavelength range, and even into the visible wavelength region in recent years.

Wire grid polarizers have not been widely used in telecommunication devices until now due to the limited aperture size of the telecommunication device, which are mainly constructed by micro-optics. The finite aperture effects of wire grid polarizers are investigated theoretically in a recent publication, which shows that the finite aperture degrades the extinction ratio [5].

Laminated polarizers (also called LAMIPOL) [6], which have a similar structure to wire grid polarizers, have high extinction ratios in a wide wavelength range. The LAMIPOL described in the above-cited publication has an operational wavelength range from 0.8 to 1.55 μm. Components with extinction ratios larger than 50 dB and insertion losses less than 0.4 dB have been developed. Due to their small size, they are often used in telecommunication devices to block the unwanted polarization component.

A typical LAMIPOL structure is depicted in Fig. 13.1. The polarizer consists of alternative laminated layers of transparent and absorptive films. The thickness of the transparent layer is around 1 μm, while the thickness of the absorptive layer is only a few nanometers. SiO$_2$ is the most frequently used transparent material, while metal aluminum is used for longer wavelength polarizers and the semiconductor germanium is often used for shorter wavelength polarizers as absorptive material. The thickness of the LAMIPOL is usually around 30 μm. To increase the working wavelength range and improve the mechanical properties, metal and semiconductor composite materials are used as an absorptive material, such as the composite material described in the above-cited literature, which consists of stainless steel and germanium.

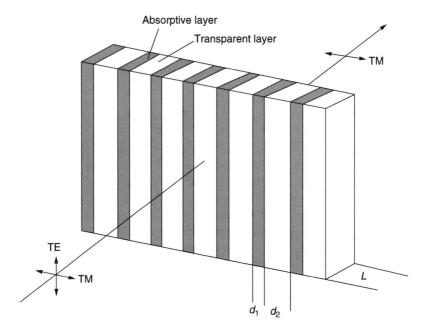

**Figure 13.1** The Schematic structure of the LAMIPOL polarizer.

In the LAMIPOL polarizer, the light beam component with the electric field vibrating perpendicular to the laminated film planes passes through with little loss, while the light beam with the electric field vibrating parallel to the laminated thin film planes is absorbed.

This structure also functions as a beam splitting polarizer when all of the laminated thin film layers are transparent and the light beams enter the structure obliquely [7]. In Fig. 13.2, the beam-splitting polarizer consists of alternating layers of laminated polysilicon and silica thin film. The difference between it and the above mentioned LAMIPOL polarizer is that both polysilicon and silica are transparent layers and they have very large refractive index contrast values. When used in a non-normal incidence (reference to the thin film plane) situation, a significant form birefringence occurs. The optical axis of the laminated thin film stack is perpendicular to the surfaces of the laminated thin films.

The structure is schematically illustrated in Fig. 13.2, in which two types of thin dielectric films, having high ($n_1$) and low ($n_2$) refractive indexes, are alternatively laminated with periodicity. The periodicity is significantly small compared to the wavelength $\lambda$, resulting in a structure that acts as an artificial anisotropic-dielectric material. The optic axis of the form birefringent material is perpendicular to the laminated layers and slanted from the light propagation direction by an angle $\theta$. The input light beam splits into an ordinary wave $E_O$, which is polarized horizontally, and an extraordinary wave $E_e$, which is polarized vertically (in the plane of the page) having a splitting angle $\phi$. The combination of two materials with a small value for the ratio of $n_2/n_1$ and an equal thickness ($d_1 = d_2$) is necessary to attain a large splitting angle.

The laminated polarization splitter mimics the effects of natural birefringent crystals, also called uniaxial crystals, which have double refraction characteristics. This is a

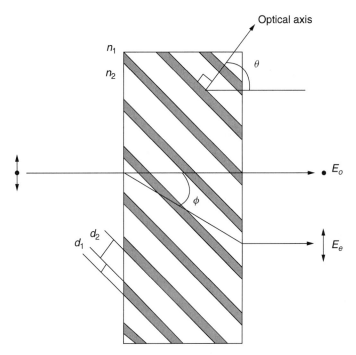

**Figure 13.2** Laminated thin film stack polarization splitter (LPS).

phenomenon where, when an unpolarized light beam passes through these crystals, the incident light beam is refracted into two beams with linear polarizations perpendicular to each other unless the light beam is parallel to the crystal's optic axis. The older generations of polarizers are almost all made from these birefringent crystals. The most frequently used uniaxial crystals are quartz, calcite, rutile, lithium niobate, and yttrium vanadate. Their main optical characteristics are listed in Table 13.1 below.

Quartz has very small birefringence; therefore, it is seldom used as a beam splitter. However, its properties are suitable for making waveplates, which will be discussed further in Section 13.3. Calcite, a natural mineral, is the most frequently used material for making polarizing prisms in the ultraviolet, visible, and near-infrared wavelength range. These prisms include Glan and Nicol type polarizing prisms and Wollaston, Rochon, and Sénarmont type polarizing beam splitter prisms. Of these, Wollaston

**TABLE 13.1   Optical Properties of Several Birefringent Crystals**

| Material | $n_o$ | $n_e$ | $\Delta n$ | Separation Ability (Degree) | Length for 1 mm Separation (mm) |
|---|---|---|---|---|---|
| Quartz | 1.52761 | 1.53596 | 0.00835 | 0.3123 | 183.5 |
| Calcite | 1.6343 | 1.4777 | −0.1566 | 5.733 | 9.96 |
| Rutile | 2.453 | 2.709 | 0.256 | 5.651 | 10.11 |
| Lithium Niobate | 2.2225 | 2.1519 | −0.0706 | 1.848 | 30.99 |
| YVO$_4$ | 1.9447 | 2.1486 | 0.2039 | 5.675 | 10.06 |

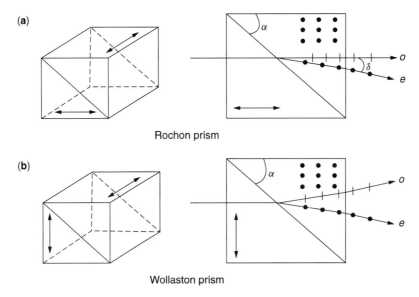

**(a)**

Rochon prism

**(b)**

Wollaston prism

**Figure 13.3**    Three-dimensional and side views of Rochon and Wollaston prisms.

and Rochon prisms are the most frequently used prisms in telecommunication device designs. Figure 13.3 below shows the three-dimensional views and side views of Wollaston and Rochon prisms.

In the case of a Rochon prism, the light beam encounters the entrance face of the prism and travels parallel to the optic axis. As a result, as it passes through the first half of the prism, the ordinary ray and extraordinary ray are unrefracted. Both rays have the same optical refractive index of $n_o$.

The second half of the prism has its optic axis perpendicular to the first half. When the two rays encounter this half, the ordinary ray still travels along without deviation because its refractive index is still $n_o$ while the extraordinary ray has a refractive index of $n_e$. The extraordinary ray will deviate according to Snell's law. If the angle cut of the prism is $\alpha$, the deviation angle is $\delta$, and then, following Snell's law, the deviation angle of the extraordinary ray can be expressed as below with a good approximation:

$$\tan(\alpha) = \frac{n_e - n_o}{\sin \delta} + \frac{\sin \delta}{2n_e} \qquad (13.1)$$

The Wollaston prism, in contrast to the Rochon prism, will deviate both of the transmitted beams. When a light beam encounters the entrance surface of the prism at a normal angle, both the ordinary ray and extraordinary ray will be undeviated. The ordinary ray, however, whose polarization direction is perpendicular to the optic axis, will have a refractive index of $n_o$, and the extraordinary ray, whose polarization direction is parallel to the optic axis, will have a refractive index of $n_e$. When these two rays encounter the second half of the prism, the ordinary ray will become an extraordinary ray and have a refractive index of $n_e$ because the optic axis of the second half of the prism is perpendicular to the first half. At the same time, the extraordinary ray will become an ordinary ray and have a refractive index of $n_o$. As a result, both transmitted beams will deviate an angle of $\delta$ as expressed in formula (13.1).

The two beams emerging from either the Rochon prism or the Wollaston prism have an angle between them. In some applications, parallel beams are desirable. We can put two Rochon prisms or Wollaston prisms opposite each other as shown in Fig. 13.4 below. When the two light beams pass through two prisms, they are parallel to each other and have a separation distance as expressed in formula (13.2) with a good approximation:

$$d = 2D * \tan\{\arcsin[\Delta n * \tan(\alpha)]\} \tag{13.2}$$

where $d$ is the separation of the two transmitted beams and $D$ is the distance between the two prisms. $\Delta n = n_e - n_o$, and $\alpha$ is the prism's angle cut and is the same as defined in formula (13.1).

The beam separator or displacer described above has the advantage of using less material than other designs. It is still more complicated, however, than the polarization beam separator/combiner and displacer depicted in Fig. 13.5.

In the Fig. 13.5 crystal, when the incident light is normal to the left end surface, the component of light with polarization perpendicular to the principal plane (plane

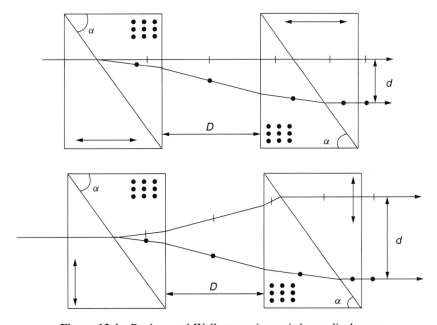

**Figure 13.4**  Rochon and Wollaston prism pair beam displacers.

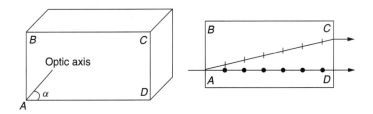

**Figure 13.5**  Birefringent crystal polarization beam splitter/combiner.

ABCD) passes through the crystal without deviation and the component of light with polarization in the principal plane (the angle between the light propagation direction and the optic axis is $\alpha$) is deflected by the crystal. The angle between the two rays can be calculated:

$$\theta = \arctan \left( \frac{(n_e^2 - n_o^2) \sin(2\alpha)}{2(n_e^2 \cos^2 \alpha + n_o^2 \sin^2 \alpha)} \right) \tag{13.3}$$

Then the length of a crystal that is needed to separate two polarization component beams of a light wave for a distance of $d$ is:

$$L = d / \tan \theta \tag{13.4}$$

Another frequently used birefringent crystal beam splitter is the birefringent wedge shown in Fig. 13.6 below. The principal plane of the wedge lies parallel to its rear surface. The optic axis has an angle of 22.5° or 45° with respect to the vertical axis. A light beam encountering the entrance face of the wedge is refracted into two beams with their polarization states perpendicular to each other.

Another kind of widely used polarization beam splitter is the interference polarizer. The ones with a multilayer dielectric thin film stack are the most common. In a typical configuration, the thin film stack is sandwiched between two right angle glass prisms and the incident angle of the thin film stack is 45°. As the finished polarization beam splitter forms a cube shape, it is often referred to as a cube PBS. The conventional cube PBS has a multilayer dielectric thin film stack that is similar in structure to a high reflectivity dielectric multilayer mirror. When a light beam encounters the entrance face of the thin film stack at a 45° incident angle, the reflection peaks for the $S$ polarized beam component and the $P$ polarized beam component shift with respect to each other. As a consequence, at a specific wavelength range, the thin film stack acts as a high reflectivity mirror for $S$ polarized light and acts as an antireflection coating for the $P$ polarized beam component. The $P$ component passes through the PBS undeflected, while the $S$ polarized beam component is reflected back and deflected at a right angle to the incident beam. The reflected beam component and the transmitted beam component form a right angle, as shown in Fig. 13.7.

It is well known that it is hard to fabricate an antireflection thin film coating with very low reflectivity and wide operational wavelength range, while it is relatively easy to

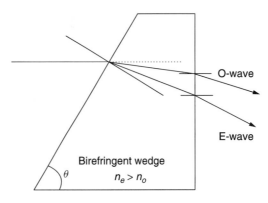

**Figure 13.6**  Birefringent crystal wedge polarization beam splitter/combiner.

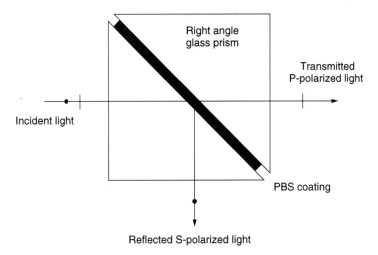

**Figure 13.7**   Thin film coating polarization beam splitter (PBS cube).

fabricate a high reflectivity mirror with a wide bandwidth. Therefore, the conventional cube PBS has a high transmission extinction ratio easily exceeding 40 dB for the $P$ polarization component. At the same time, 30 dB is a very good extinction ratio value for the reflected $S$ polarization component. The conventional thin film interference coating PBS cubes are thoroughly reviewed by Li Li and J. A. Dobrowolski [8] in their paper in which they proposed a new innovative interference coating PBS with greatly improved performance for both the extinction ratios and the acceptance incident angle as well as an extremely wide operational wavelength range. This PBS works at angles greater than the critical angle.

The thin film PBS coating consists of low and high refractive index layers that are sandwiched between two high refractive index substrates, as shown in Fig. 13.8. The thin film stack has a symmetrical structure of ABA around its central layer. The layers are very thin and work beyond the critical incident angle so that the thin film stack is equivalent to an antireflection matching layer for the $S$ polarized light beam component and behaves like a reflective metallic layer for the $P$ polarized light beam component. As a result, it can be fabricated with very high extinction ratios (greater than 50 dB) for both the $S$ and $P$ polarization components. It also has a number of

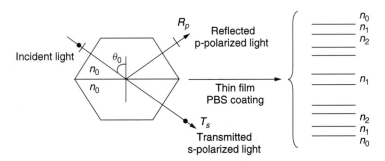

**Figure 13.8**   Thin film PBS operating at angles greater than the critical angle.

performance advantages, including a large incident angular field and extremely wide operational wavelength range. The drawbacks of this design are the strict requirements on the high refractive index of the substrate and the difficulty in making the PBS cube shaped, which is not very convenient for device design.

Other polarizers operating at telecommunication wavelengths and based on various principles have also been proposed. One design by R. M. A. Azzam et al., utilizing silicon prisms and silicon-related material coatings covers both the 1.3-$\mu$m and 1.55-$\mu$m telecommunication windows [9]. A Brewster angle polarizer design proposed by Daniel J. Dummer et al. can cover the wavelength range from 0.4 $\mu$m to beyond 500 $\mu$m with an extinction ratio of 50 dB [10]. There are also polarizers using side polished fibers or waveguides and different cladding overlay layers, such as birefringent polymers, liquid crystals, and various metal layers [11–14]. There are also polarizer designs based on subwavelength diffraction gratings [15]. The various polarizer designs proposed and realized cannot be exhausted in this short introductory text. In addition, new designs are continually emerging.

## 13.3 POLARIZATION ROTATORS AND PHASE RETARDERS

There are two kinds of polarization rotators, namely, reciprocal and nonreciprocal. Faraday rotators, which are nonreciprocal polarization rotators, are the most important devices discussed in this chapter because all optical isolators and circulators described here, as well as some variable optical attenuators (VOA), are based on them.

Faraday rotators are made from ferromagnetic materials. These materials act as polarization rotators when placed in a static magnetic field. The rotation angle of the Faraday rotator is proportional to the thickness of the material. The rotatory power $\rho$ can be expressed as $\rho = VB$, and it has the unit of rotation angle per unit length, where $B$ is the magnetic flux density in the direction of wave propagation and $V$ is the Verdet constant. For $V > 0$, the rotation is in the direction of a right-handed screw pointing in the direction of the magnetic field. The Verdet constant has the unit of rotation angle per unit length and per unit magnetic field. It seems at first glance that the Faraday rotation can be arbitrarily large as long as the magnetic field is strong enough. In fact, this is not the case. The Faraday rotators will be saturated when the magnetic field is large enough. That is to say, the rotation angle will reach a maximum value and remain unchanged thereafter, regardless of how large the magnetic field subsequently gets. In the telecommunication wavelength range, the most commonly used Faraday rotator materials are (1) Yttrium Iron Garnet (YIG, $Y_3Fe_5O_{12}$) single crystal, which is grown using the floating zone method; (2) Bi-substituted iron garnet thick films (($TbBi)_3(FeAl)_5O_{12}$), which are grown using the liquid phase epitaxy (LPE) method. Crystal materials and a $PbO$-$B_2O_3$ flux are heated and made molten in a platinum crucible. Single crystal wafers, such as NGG, CMZ-GGG, YSGG, or a mix of YSGG+GSGG, are then soaked in the molten surface while rotated, which causes a Bi-substituted iron garnet thick film to grow on the wafers. By adding some more rare earth elements to substitute part of the Bi content and some Ga, Al, In, or Sc to substitute part of the Fe content, a self-latching garnet thick film can be fabricated that possesses a saturated Faraday rotation without applying an external magnetic field. These latching garnet thick films have a multicomponent structure, with a form such as $RE1_aRE2_bBi_{3-a-b}Fe_{5-c-d}M1_cM2_dO_{12}$, where RE1 and RE2 are both selected from

**TABLE 13.2   Key Performance Parameters of Three Faraday Rotators**

| Property | Wavelength (nm) | YIG Single Crystal | Bi-substituted Iron Garnet Thick Film (Non-latching) | Bi-substituted Iron Garnet Thick Film (Latching) |
|---|---|---|---|---|
| Material | — | $Y_3Fe_5O_{12}$ | $(TbBi)_3(FeAl)_5O_{12}$ | $RE1_aRE2_bBi_{3-a-b}$ $Fe_{5-c-d}M1_cM2_dO_{12}$ |
| Saturation magnetization (mT) | — | 178 | 60 | N/A |
| Insertion loss (dB) | 1310 | 0.1 | 0.1 | 0.1 |
| Faraday rotation coefficient (deg./cm) | 1310 | 224 | −1570 | −1452 |
|  | 1550 | 175 | −1060 | −978 |
| Faraday rotation temp. Coefficient (deg./°C) | 1310 | 0.034 | 0.054 | −0.076 |
|  | 1550 | 0.042 | 0.062 | −0.078 |
| Faraday rotation wavelength coefficient (deg./nm) | 1310 | 0.056 | 0.089 | −0.086 |
|  | 1550 | 0.040 | 0.064 | −0.061 |
| Extinction ratio (dB) | 1310 | >38 | >41 | >40 |

the lanthanide group. This group includes La, Pr, Nd, Sm, Eu, Gd, Tb, Dy, Ho, Yb, and Lu. M1 and M2 represent Ga, Al, In, or Sc, depending on the element selected. The values of $a$, $b$, $c$, $d$ are adjusted to match the lattice constant between the growing film and the substrate.

Listed in Table 13.2 are the main parameters of the three Faraday rotators described above.

YIG single crystals are better than Bi-substituted iron garnet thick films in terms of their temperature and wavelength coefficients for Faraday rotation (only around 60% of those of Bi-YIG). If these characteristics are of primary consideration in a design, YIG single crystals should be selected as Faraday rotators. On the other hand, the advantages of Bi-substituted iron garnet thick films include (1) lower cost due to a shorter growth period, (2) thinner rotators, which are one-fifth as thick or even less due to increased rotation capacity (around 2.5 mm versus 0.5 mm for 45° rotation at 1550 nm), and (3) low saturation magnetization (1780 Gauss versus 600 Gauss), which relaxes the requirement on the external permanent magnet. These features lead to a price reduction and miniaturization of optical devices.

Latching garnet crystals eliminate the need for an external magnet completely. They are most suitable for compact device designs where space is limited. The drawbacks of the latching garnet thick film Faraday rotator are the relatively large temperature coefficient and considerably higher operating costs.

Besides the nonreciprocal polarization rotators, there are certain materials possessing a property known as optical activity. These materials are normally circularly polarized. Their molecules have inherently helical characteristics; therefore, the light beams with left-handed and right-handed circular polarizations travel at different phase velocities. Materials such as quartz, selenium, and tellurium oxide have optical activity.

It can be shown that an optically active medium with right-handed and left-handed circular polarization phase velocities of $c_0/n_+$ and $c_0/n_-$ acts as a polarization rotator having an angle of rotation $\pi(n_- - n_+)\,d/\lambda_0$, where $d$ is the thickness of the optically active medium. The rotatory power of the optically active medium with a unit of rotation angle per unit length can be expressed as:

$$\rho = \frac{\pi(n_- - n_+)}{\lambda_0} \tag{13.5}$$

The rotation direction is the same as that of the circularly polarized component with greater phase velocity (the component with the smaller refractive index). As most of the available optically active mediums have very small circular birefringence, a 45° rotator will need considerably thick material. As it is quite bulky, this kind of rotator is rarely used in telecommunication devices.

Most of the reciprocal polarization rotator tasks in telecommunication devices are performed by linear phase retarders, especially quarter-wave and half-wave phase retarders, or so-called wave plates.

Wave plates are made from anisotropic materials, which are mostly birefringent crystals. When light beams travel along one of the crystal's principal axes, usually the $z$-axis, the normal modes are linearly polarized waves pointing along the other two principal axes, that is, the $x$- and $y$-axes. If the principal refractive indices of these two axes ($x$- and $y$-axes) are $n_1$ and $n_2$ and $n_1 < n_2$, then the $x$-axis is the fast axis. The phase retardation has the expression of:

$$\Gamma = (n_2 - n_1)k_0\,d = 2\pi(n_2 - n_1)\,d/\lambda_0 \tag{13.6}$$

where $k_0$ is the wave vector of the light beam, $d$ is the thickness of the retarder, and $\lambda_0$ is the wavelength of the light beam.

Although wave plates can be made from any uniaxial crystal, the ones used in telecommunication devices are almost always made from quartz due to its superior chemical and mechanical stability and optical properties. From the above formula, it is clear that waveplates are wavelength-dependent components. For most applications, wavelength dependence is not desirable. Therefore, zero-order wave plates are frequently used. A zero-order half wave plate made from a single crystalline quartz plate for the wavelength of 1.55 μm has a thickness of approximately 91 μm, which is not easy to handle. Another approach for fabricating a zero-order wave plate is to bond two wave plates with the desired small thickness difference together with their fast axes oriented perpendicular to each other.

Generally speaking, wave plates are phase retarders, not polarization rotators. However, a half wave plate can act as a polarization rotator. The working principle can be clearly seen in Fig. 13.9 below.

A linearly polarized light beam with its polarization direction forming an angle $\phi$ with respect to the slow axis of the half wave plate can be decomposed into two components along the wave plate's fast and slow axes. After passing through the half

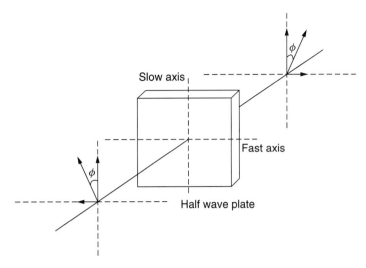

**Figure 13.9** Half wave plate acting as polarization rotator.

wave plate, the component along the fast axis has a $\pi$ phase shift relative to the slow axis. The resulting light beam composed of the two linear polarized components has its polarization direction forming an angle of $\phi$ with respect to the slow axis of the half wave plate, but at the opposite side to the input light beam. The linearly polarized light beam passes through the half wave plate having been rotated an angle of $2\phi$ in total. It is also clear that when the input light beam's polarization aligns with either one of the optical axes of the wave plate, it will not be rotated by the wave plate at all.

Polarization rotators and phase retarders can also be fabricated using liquid crystals. Liquid crystals are organic materials that have elongated molecules (rod shaped) that exhibit orientational order but no positional order. The three common phases of liquid crystals are nematic, smectic, and cholesteric. However, we will discuss mainly the nematic phase liquid crystals. A liquid crystal cell usually has a structure as shown in Fig. 13.10.

The liquid crystal cell is constructed from two identical glass plates. A one-side AR-coated glass plate is first coated with an ITO layer as the transparent electrode, and then spin coated with another polyimide layer. The polyimide layer is rubbed in one direction. Two prepared glass plates are put together with glass rod spacers in between to keep the desired gap. The rubbing directions of the two glass plates are aligned in parallel. After being filling with liquid crystals, the cell is sealed with epoxy.

When there is no voltage being applied to the LC cell, the liquid crystal molecules are aligned along the rubbing direction of the polyimide. The liquid crystal acts like a uniaxial crystal with its optic axis parallel to the molecule orientation. For waves traveling in the $z$ direction (perpendicular to the glass plate surface), the normal modes are linearly polarized in the $x$ and $y$ directions (parallel and perpendicular to the molecular direction). The refractive indices are the extraordinary and ordinary indices $n_e$ and $n_o$. A cell with spacer thickness $d$ will provide a phase retardation of $\Gamma = 2\pi(n_e - n_o)\,d/\lambda_0$.

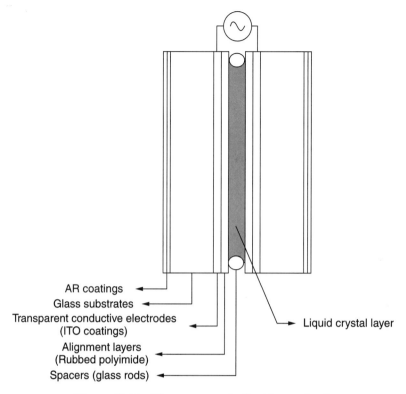

AR coatings
Glass substrates
Transparent conductive electrodes
(ITO coatings)
Alignment layers
(Rubbed polyimide)
Spacers (glass rods)

Liquid crystal layer

**Figure 13.10**    The structure of a liquid crystal cell.

If a voltage is applied to the cell (having a square waveform with frequencies ranging from a few tens of hertz to a few kilohertz), the liquid crystal molecules tend to tilt in the direction of the electric field. If the voltage is high enough, the molecules are almost all aligned with the electric field and the phase retardation decreases to zero. In between, the molecules tilt toward the electric field with an angle $\theta$ with respect to the glass plate surface. For detailed descriptions, please refer to the book by Bahaa E. A. Saleh and Malvin Carl Teich [2].

Polarization rotators can also be fabricated using twisted nematic liquid crystal cells. These cells have the same structure as parallel-aligned nematic liquid crystal cells. The only difference is that the rubbing directions of the polyimide layers on the two glass plates are aligned perpendicular to each other instead of in parallel. When a linearly polarized light beam passes through the twisted nematic liquid crystal cell, it is rotated 90° in the direction of the molecule twist.

## 13.4    OPTICAL ATTENUATORS

An optical attenuator is a device used to reduce the power level of an optical signal in a fiber optic communication system. The basic types of optical attenuator are fixed, step-wise variable and continuously variable. Of these, some operate with multimode fibers and some with single-mode optical fibers. In this section, only optical attenuators working with single-mode optical fibers are discussed.

A light wave in a single-mode fiber has an intensity distribution of a Gaussian function of the radial distance, and it is therefore called a Gaussian beam. Please see "Beam Optics" in [2] for a detailed treatment of Gaussian beams.

The complex electric field amplitude of a Gaussian beam can be expressed as:

$$U(\mathbf{r}) = A_0 \frac{w_0}{w(z)} \exp\left[-\frac{\rho^2}{w^2(z)}\right] \exp\left[-jkz - jk\frac{\rho^2}{2R(z)} + j\zeta(z)\right] \qquad (13.7)$$

where:

$$w(z) = w_0 \left[1 + \left(\frac{z}{z_0}\right)^2\right]^{1/2} \qquad (13.8)$$

$$R(z) = z\left[1 + \left(\frac{z_0}{z}\right)^2\right]$$

$$\zeta(z) = \tan^{-1}\left(\frac{z}{z_0}\right) \qquad (13.9)$$

$$w_0 = \left(\frac{\lambda z_0}{\pi}\right)^{1/2}$$

where $A_0$ is a constant, $w_0$ is the beam waist radius, and $w(z)$ is the beam radius at position $z$. $\rho = \sqrt{x^2 + y^2}$ is the radial distance. $R(z)$ is the wavefront radius of curvature at position $z$. $z_0 = \frac{1}{2}kw_0^2 = \frac{\pi w_0^2}{\lambda}$ is the Rayleigh range and $k = 2\pi/\lambda$.

A lot of attenuators, especially fixed ones, are actually based on the coupling loss between two Gaussian beams. The loss mechanisms of two coupled Gaussian beams include longitudinal separation, lateral offset, mode field diameter mismatch, and oblique coupling, which are shown in Fig. 13.11.

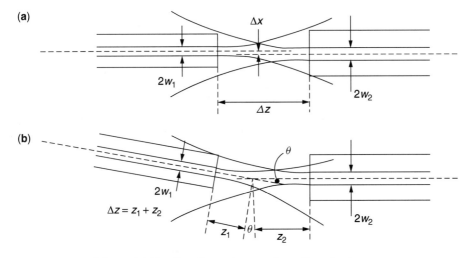

**Figure 13.11**   Coupling schemes of two Gaussian beams.

The coupling loss between the two fibers above is $IL = -10\log_{10}(\eta)$, where $\eta$ is the coupling efficiency. In (**a**), it can be expressed as [16]:

$$\eta = \frac{1}{\dfrac{(w_1^2 + w_2^2)^2}{4w_1^2 w_2^2} + \left(\dfrac{\lambda \Delta z}{2\pi w_1 w_2}\right)^2} \exp\left[-\frac{1}{1 + \left(\dfrac{\lambda \Delta z}{\pi(w_1^2 + w_2^2)}\right)^2} \frac{\Delta x^2}{\left(\dfrac{w_1^2 + w_2^2}{2}\right)}\right]$$

(13.10)

In the case of (**b**), it can be expressed as:

$$\eta = \frac{1}{\dfrac{(w_1^2 + w_2^2)^2}{4w_1^2 w_2^2} + \left(\dfrac{\lambda \Delta z}{2\pi w_1 w_2}\right)^2} \exp\left[-\left[\frac{\dfrac{2z_1^2 w_2^2 + 2z_2^2 w_1^2}{(w_1^2 + w_2^2)^2} + \dfrac{2\pi^2 w_1^2 w_2^2}{\lambda^2(w_1^2 + w_2^2)}}{1 + \left(\dfrac{\lambda \Delta z}{\pi(w_1^2 + w_2^2)}\right)^2}\right]\theta^2\right]$$

(13.11)

Using the above-outlined loss mechanisms, simple fixed optical attenuators are easily made.

In dense wavelength division multiplexing modules made from dielectric thin film interference filters, fixed attenuators are frequently fabricated by splicing machines using the lateral offset of two spliced fiber tips in order to balance the loss level among different channels. A programmable fusion-splicing machine can accurately achieve the desired loss (or attenuation) by setting the alignment and arc parameters.

Compact fused fiber-fixed attenuators are made using biconic fused taper technology. The attenuation ranges from 1 dB to 20 dB.

To combat the attenuation in transmission optical fibers, scientists and engineers have struggled for decades to reduce the loss of the transmission fiber to one tenth of a decibel per kilometer. At the same time, to make fixed attenuators, they also doped absorptive transition metals into the fiber core to form wide flat-band attenuation fibers [17]. The most widely used transition metal element is cobalt. In the above-cited reference, attenuation in the wavelengths from 1530 to 1610 nm was held to within 9.9 to 10.2 dB with only a 3% variation.

With a similar working principle, neutral density filters are optical glass plates coated with absorptive metallic thin film layers made from chrome, nickel, cobalt, or an alloy of them. One obstacle to using neutral density filter attenuators is the attenuation ripple caused by the Fabry–Perot resonance effect that occurs between two weakly reflective surfaces of the glass plate. The reflections arise from the residual reflectivity of the antireflection-coated glass plate surface and the reflection of the metallic layer. To reduce the attenuation ripple, special antireflection coating layers are designed for the reflective metallic layers [18]. Fixed attenuators in the form of fiber optic connectors and adaptors are made from neutral density filters.

Other fixed attenuators include those using fiber bending and prealigned connector adaptor attenuators. Fiber's light guiding capability relies on total internal reflection within the fiber core. When the fiber is bent sharply, the total internal reflection condition may not be satisfied. Light can enter into the cladding layer and be lost.

Beyond fiber coupling and absorption mechanisms, other techniques are used to make VOAs, including light wave scattering; destructive interference; beam blocking; evanescent modes in optical waveguides; Bragg grating diffraction; polarizer-polarization rotator-analyzer structure; and electric absorption. With optical add/drop multiplexers becoming more common in optical networks, channel spacing of DWDM optical networks becoming denser, and dynamic configuration of mesh optical networks becoming a reality, VOAs are being used with increasing frequency in fiber optic communication systems.

Figure 13.12 shows two typical applications of VOAs. The diagram in (**a**) shows a so-called VMUX module. Optical signals from different wavelength transmitters are combined by the multiplexer into a single fiber. Due to the need for accurate wavelength control, the powers of the different wavelength signals are not balanced. Therefore, there is a need to adjust them so that they have a balanced power level. This is done by combining a VOA and a power-monitoring device, as shown in the figure. The second application shown in the figure is an intelligent optical add/drop multiplexer. The power monitor at the drop side detects the power level of the dropped signal so that the added signal can use it as a power level reference. The combination of a VOA and another power monitor is used to adjust the power level of the added signal to the same level as the system signals (following the dropped signal).

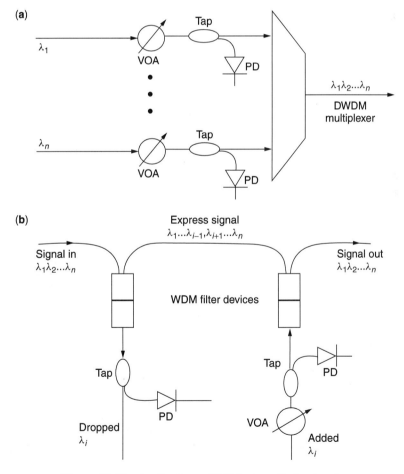

**Figure 13.12**  Applications of VOAs in optical networks.

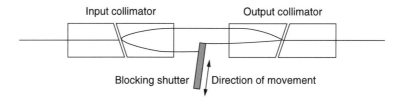

**Figure 13.13** Working principle of an opto-mechanical VOA.

Depending on its actuation mechanisms, VOAs can be categorized as opto-mechanical, thermo-optical, magneto-optical, acousto-optical, electro-optical, or MEMS based. We will introduce these VOAs individually in the following paragraphs.

The opto-mechanical VOAs are actuated mostly by stepping motors. Figure 13.13 below shows a beam blocking type mechanical VOA. Two optical collimators are prealigned and fixed together with a gap between them. A light-blocking shutter plate (the plate's normal is oriented with a small angle with respect to the light propagation axis to avoid back-reflection of light into the input fiber) with a sharp edge actuated by a stepping motor is intruding into the collimated light beam. Part of the light beam is blocked by the intruding shutter (hence, the attenuation of the light beam), which is variable by controlling the stepping motor. The movement mechanism can be a screw and a bolt structure with the shutter plate mounted on the bolt.

The variable beam blocking can also be achieved using a wheel with its axis off-set from the wheel center for a predetermined distance according to the collimated light beam diameter. When a stepping motor rotates the wheel, the collimated light beam is completely blocked by the wheel when it reaches a certain angular position, while in the opposite angular position, the wheel is totally out of the collimated light beam path.

Another mechanical VOA uses longitudinal separation of two aligned fibers [19]. Optical attenuation control is achieved through the axial separation between two tapered single-mode fibers with beveled endfaces aligned within a ceramic sleeve. The attenuation follows the formula listed in (13.10).

Mechanically actuated VOAs using thickness varying metallic thin film neutral density filters are also used widely in the optical industry. The structure of these VOAs is very similar to the beam blocking VOA depicted in the Fig. 13.13. The difference is that the whole light beam always passes through the neutral density filter plate. With the movement of the neutral density filter plate and the varying metallic film thickness across the plate, a light beam experiences different attenuations at different neutral density filter positions.

The advantages of mechanically actuated VOAs are their extremely high attenuation range (larger than 60 dB is available), attenuation latching function, and low insertion loss. The drawbacks are their slow response time and the backlash of the actuation mechanism.

VOAs using the thermo-optical effect are usually in the form of some kind of optical waveguide. The most well known is the Mach–Zehnder interferometer waveguide device as shown in Fig. 13.14. Figure 13.14**a** is a top view of device. This interferometer comprises two 3-dB couplers connected together with two arms. When the two arms are balanced in the light path, light entering the input port comes out at the output port, as indicated in the figure. If a $\pi$ phase retardation is generated by

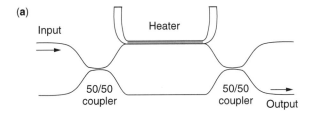

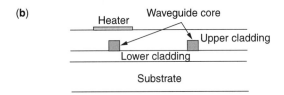

**Figure 13.14** A Mach–Zehnder interferometer thermo-optical VOA.

the heater on the upper arm, then no light comes out at the output port. In other words, the light is completely attenuated. At the states between these two extreme conditions, partial light will come out of the output port. The attenuation in the ideal situation can be easily derived. Suppose the electric field amplitude at the input port is $E = E_0 \cos(\omega t)$. If we neglect the phases that do not affect the results, then the electric field amplitudes of the upper arm and lower arm just after the first 3-dB coupler will be $E_u = \frac{\sqrt{2}}{2} E_0 \cos(\omega t)$ and $E_l = \frac{\sqrt{2}}{2} E_0 \cos(\omega t + \pi)$, respectively. By using $\Phi$ to denote the phase retardation generated by the heater at the upper arm and neglecting identical common phase terms, the electric field amplitudes just before the second 3-dB coupler are $E_u = \frac{\sqrt{2}}{2} E_0 \cos(\omega t + \Phi)$ and $E_l = \frac{\sqrt{2}}{2} E_0 \cos(\omega t + \pi)$ for the upper and lower arms, respectively. When the two light components pass through the second 3-dB coupler, at the output port, the two light components from the upper arm and the lower arm have expressions of $E_{uu} = \frac{1}{2} E_0 \cos(\omega t + \Phi + \pi)$ and $E_{ll} = \frac{1}{2} E_0 \cos(\omega t + \pi)$, respectively. The total electric field amplitude at the output port is $E_{tot} = E_{uu} + E_{ll}$ and has the expression of: $E_{tot} = \frac{1}{2} E_0 (\cos(\omega t + \Phi + \pi) + \cos(\omega t + \pi))$. The light intensities at the input and output port can be expressed using the electric field amplitudes: $I = A E_0^2 < \cos^2(\omega t) >$ and $I_{tot} = \frac{1}{4} A E_0^2 < (\cos(\omega t + \pi + \Phi) + \cos(\omega t + \pi))^2 >$, where $A$ is a constant and $<>$ represents average over time. The attenuation, therefore, can be derived as: Attenuation $= I_{tot}/I = \frac{1}{2}(1 + \cos(\Phi))$ (13.12)

The schematic in Fig. 13.14**b** is a cross-section view of the VOA device shown in (**a**). The substrate material is usually silicon or fused silica. The lower and upper cladding layers are usually pure silica fabricated by PECVD or flame hydrolysis, while the core layer is fabricated by the same process but with $GeO_2$ or $TiO_2$ dopants. If the substrate is pure silica glass, the lower cladding layer is not necessary. The heater is made of metallic thin film and the waveguide pattern is usually fabricated by photolithography and reactive ion etching processes. The advantages of the MZI VOA stem from its manufacturing process, because integrated optical circuit fabrication has the potential of mass production and low cost. Unfortunately, the performance of this type of VOA often suffers from large insertion and polarization-dependent loss. The attenuation range is not very large, and the power consumption is high due to the small thermo-optical coefficient of the fused silica material. To reduce the power consumption level, some manufacturers use polymer materials with large thermo-optical coefficients for the above-described structure.

Another frequently used waveguide approach to fabricating thermo-optical VOAs changes the light wave guiding properties of the waveguide itself. As shown in Fig. 13.15, part of the cladding of a fiber section is removed. To replace the cladding, a polymer material coating is added. On top of the polymer layer, a metallic resistance thin film coating is added and is connected to an electric power supply. The refractive index of the polymer matches that of the original cladding material when the power supply is off. Therefore, the light passes through undisturbed. When the power supply is turned on, the resistance thin film will heat the polymer layer, and due to the thermo-optical effect, the refractive index of the polymer layer will increase. The light guiding capability of the fiber at the polymer section will degrade, and the light will couple into the polymer layer and be dissipated away.

The attenuation is proportional to the temperature of the polymer material, and it depends on the electric current passing through the resistance thin film. This principle also applies to the structure of a planar light wave circuit. When a channel waveguide has a polymer layer as its upper cladding layer, its wave guiding capability will degrade when the upper cladding is heated. A VOA made according to this design will have a very small insertion loss as the light beam never comes out of the fiber for manipulation. The concern is the long-term reliability of the polymer as it is frequently heated to high temperatures and the polymer's temperature resistance is still to be determined.

Magneto-optical VOAs are made using Faraday rotators. The basic structure is a polarizer-rotator-analyzer structure. Shown below in Fig. 13.16 is a polarization independent magneto-optical VOA using two birefringent crystal beam splitters and an electric magnet.

The incoming light is collimated by the input collimator and encounters the first PBS. The optic axis of the birefringent crystal PBS lies in the plane of the page and forms an angle of 45° with respect to the horizontal line. The light beam is then divided into two polarization component beams. The ordinary beam, with its polarization perpendicular to the plane of the page, travels straightforward. The extraordinary beam, with its polarization lying in the plane of the page, is deflected and displaced upward. The extraordinary beam then encounters the half wave plate, which is oriented with a 45° angle with respect to the polarization of the extraordinary beam. The polarization of the extraordinary beam is rotated 90° by the half wave plate so that it is parallel to

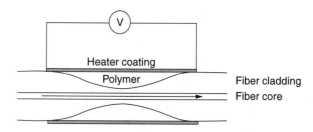

**Figure 13.15**   Thermo-optic VOA having a polymer cladding.

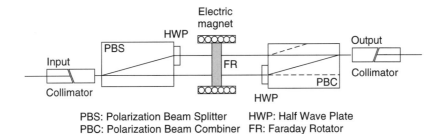

PBS: Polarization Beam Splitter    HWP: Half Wave Plate
PBC: Polarization Beam Combiner    FR: Faraday Rotator

**Figure 13.16**   The structure of a magneto-optic VOA.

the polarization of the ordinary beam. Both beams with parallel polarizations enter into the Faraday rotator. When the power supply of the electric magnet is off, the two beams pass through the Faraday rotator unchanged. The ordinary beam encounters the half wave plate in front of the PBC. This half wave plate is oriented so that it rotates the polarization of the ordinary beam 90° to be oriented with the plane of the page. The PBC (polarization beam combiner) has the same structure as the PBS, but it is inversely positioned. It finally combines the two polarization beams into one, which is then focused by the output collimator into the output fiber without attenuation.

When the power supply of the electric magnet is on, the Faraday rotator rotates the polarization of the incoming beam by an angle of $\theta = VBd$ if the Faraday rotator is not saturated, where $V$ is the Verdet constant, $d$ is the thickness of the Faraday rotator, and $B$ is the magnetic flux density experienced by the Faraday rotator. If the value of $\theta$ is 90°, then the ordinary and extraordinary beam's polarizations are so oriented when they reach the PBC that they will travel along the dotted lines and be totally attenuated. When $\theta$ is less than 90° but larger than zero at the PBC, some of the light beams travel along the dotted lines and the remainder travels in the direction of the solid line and is eventually collected by the output fiber. Thus, the light will be attenuated according to the supplied power strength of the electric magnet.

Magneto-optical VOAs can be designed so that they only require a maximum of 45° polarization rotation of the Faraday rotator or even 22.5° polarization rotation. In fact, almost all optical isolators that are described in Section 13.5 of this chapter can be used as magneto-optical VOAs when the saturated Faraday rotator is replaced by an unsaturated one and the rotation is controlled by an electric magnet.

Depicted in Fig. 13.17 is the working principle of an acousto-optical VOA [20].

Radio frequency power is fed into an acoustic actuator. The actuator generates acoustic waves in a Bragg-mode Acousto-optic Bragg cell. The acoustic waves produce a moving phase grating inside the cell through the acousto-optical effect. When an incident beam travels through the acousto-optic Bragg cell, light is diffracted away by the phase grating.

There are several kinds of VOAs on the market that are based on electro-optic effects, such as VOAs using liquid crystals [21], polymer networked (dispersed) liquid crystals [22], holographic polymer dispersed (nano-sized) liquid crystals, and PLZT ceramics.

As described previously, a liquid crystal cell can act as a variable phase retarder or wave plate. This characteristic makes it very suitable for use in VOA designs. Replacing the variable Faraday rotator in Fig. 13.16 with a liquid crystal cell that has

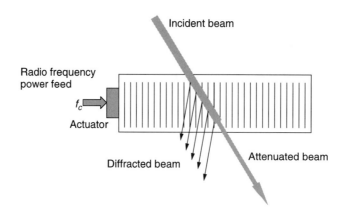

**Figure 13.17**  Working principle of an acousto-optical VOA.

half wave retardation capability, the VOA still works. However, here we will introduce another compact VOA design using a liquid crystal cell that only requires a quarter wave retardation capability, as shown in Fig. 13.18.

In the figure, the birefringent crystal is a parallel plate with a thickness around 300 to 500 microns. The principal plane of the birefringent crystal forms a 45° angle with the alignment (rubbing) direction of the LC cell. When the light beam comes out of the input fiber, the birefringent crystal separates the light beam into two offset, partially overlapping beams. When the LC cell's voltage is off, the LC cell provides quarter-wave retardation for the light beam. Upon the round trip of the light beam through the LC cell, the total half-wave retardation, together with the 45° orientation, will exchange the polarization direction of the separated light beams. This lets the two beams have the correct polarization directions to combine together again in the birefringent crystal and be accepted by the output fiber. When the voltage on the LC cell is large enough, the LC cell provides no phase retardation; therefore, the round tripped light beams experience no polarization changes. This leaves the light beams with incorrect polarization directions upon reaching the birefringent crystal. As a result, the two beams are separated further apart, preventing them from being accepted by the output fiber as the light is totally attenuated. When the LC cell's phase retardation is between 0 and 45°, the light is partially attenuated according to the voltage applied to the LC cell. The polarization state transformations in the above-described LC VOA can be understood fully in Fig. 13.19.

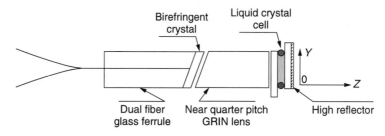

**Figure 13.18**  The structure of a liquid crystal VOA.

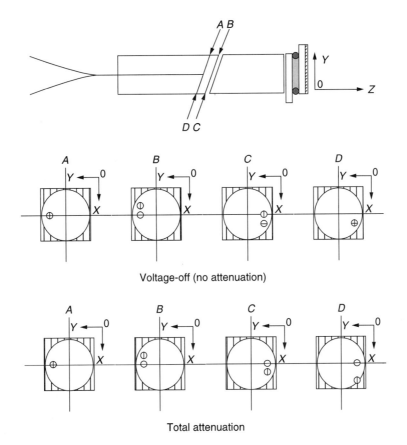

**Figure 13.19**   LC VOA polarization state transformation diagrams.

   Theoretically, the above-introduced VOA has a very large attenuation range. However, due to residual phase retardation of the LC cell when a high voltage is applied, the actual dynamic range of the VOA is limited. Figure 13.20 shows a measured curve of optical attenuation versus the voltage applied to the LC cell.

   PLZT ceramics, which are formed by hot pressing techniques, are the few inorganic materials that present a large electro-optic effect. The quadruple electro-optic coefficient of the PLZT is around $1.2 \times 10^{-16}$ $(m/V)^2$, which makes it possible to generate half wave phase retardation over an optical clear aperture of a few hundred microns (the aperture needed for a GRIN lens collimated light beam) using a 1- or 2-mm-thick PLZT plate with a few hundred volts applied to the aperture. This variable half wave plate can substitute any Faraday rotator or liquid crystal variable wave plate in the above-mentioned VOA designs to form PLZT-based VOA designs. Based on scattering mechanisms, VOAs built from Polymer Dispersed (or networked) Liquid Crystals (PDLC) have simpler structures than LC-based VOA designs. Figure 13.21 below shows a VOA using a PDLC cell. In (**a**), two collimators are aligned with each other and a PDLC cell is sandwiched in between. When a high enough voltage is applied to the PDLC cell [see (**b**)], the molecules inside the liquid crystal droplets are aligned in the direction of the electric field. The liquid crystal's ordinary refractive

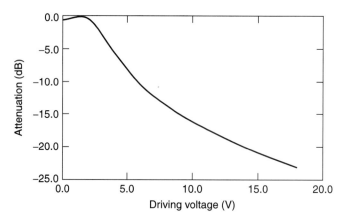

**Figure 13.20**   Attenuation versus driving voltage of the LC VOA.

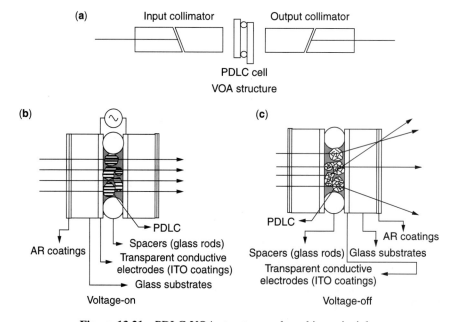

**Figure 13.21**   PDLC VOA structure and working principle.

index is chosen to match the refractive index of the polymer used so that the incident light will not see any refractive index difference across the PDLC cell aperture but, instead, will pass through the cell without attenuation. When the voltage is off [see (**c**)], the molecules inside the liquid crystal droplets are randomly oriented. The extraordinary refractive index presented by liquid crystal droplets is different from the refractive index of the polymer. When the dimensions of the LC droplets are of the same order as the wavelength of the incident light, the incident light experiences a lot of scattering centers inside the PDLC cell and the light beam is greatly attenuated. Between these two extreme conditions, the light beam attenuates according to the voltage applied to the PDLC cell.

Although the performance may not be the best, VOAs built from PDLC have a large potential market due to their simple structure and intrinsically low manufacturing cost.

MEMS (Micro-Electrical-Mechanical-Systems) technology is increasingly found in fiber optic components and subsystems, including optical switches, wavelength routers, dynamic gain equalizers, dynamic optical add/drop multiplexers, and, of course, VOAs [23–26]. Optical MEMS technology is essentially a miniature actuator technology, so that most of the opto-mechanical VOA approaches using stepping motors can be readily redesigned with MEMS actuators. For example, the beam blocking design has been realized using MEMS technology already [24, 25]. Here we need to describe two compact MEMS VOAs in some detail.

The first design uses a MEMS actuated micro-mirror aligned with a dual-fiber collimator, as shown in Fig. 13.22. The optical path of this VOA is very simple. The light from the input fiber is collimated by a lens (usually a GRIN lens), and the collimated light beam impinges on a highly reflective micro-mirror actuated by an MEMS actuator (often, the mirror is a part of the actuator). The light beam is reflected back by the micro-mirror and focused into the output fiber when the actuator is in its normal position. The MEMS actuator of this kind usually operates on electrostatic force. When voltage is applied to the actuator, the micro-mirror tilts and the input fiber and the output fiber are misaligned (in other words, light power is attenuated). As the actuator relies on electrostatic force, it is moisture sensitive and needs a relatively high controlling voltage. As a result, the device is usually assembled in a hermetically sealed package and application-specific integrated circuit (ASIC) chips are used for voltage conversion. The micro-mirror actuator is designed with very high resonance frequencies to protect it from outside vibration interference.

Using the basic structure of the above-described VOA and replacing the micro-mirror actuator by a so-called D-MEMS (diffractive micro-electrical-mechanical systems) actuator, the operational speed of the VOA can be greatly improved. The D-MEMS actuator, which consists of an array of reflective ribbons suspended above a silicon substrate, functions as a mirror with all the reflective ribbons positioned at the same height when the voltage is off. When the voltage is on, half of the ribbons will be pulled downward, so that a diffraction grating is formed. The reflected light beam will be attenuated due to diffraction. When the distance between the two sets of the ribbons reaches a quarter of the incident light wavelength, the light wave reflected from two ribbon surfaces is completely out of phase and destructive interference occurs. At this point, all of the light is totally diffracted and none is reflected.

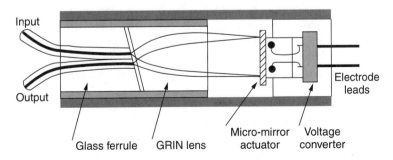

**Figure 13.22** Tilting mirror MEMS VOA structure.

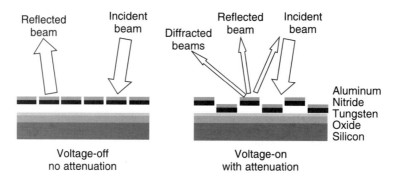

**Figure 13.23**   Working principle of the D-MEMS VOA.

Figure 13.23 shows the structure and working principle of the D-MEMS actuator [27]. A layer of silica is formed on top of the silicon substrate by thermal oxidation. On top of the silica layer is a metallic layer (tungsten or aluminum) acting as one set of electrodes of the actuator. An array of reflective ribbons made of nitride over coated with aluminum is suspended over the substrate. The aluminum serves as reflectors as well as electrodes. One-half of the ribbons are fixed (every other ribbon from the array alternatively). Another half of the ribbons can be pulled downward by electrostatic force when voltage is applied to the electrodes.

The VOAs introduced previously are designed to have an attenuation response over wavelength as flat as possible. However, a new kind of attenuator has emerged in recent years that has wavelength-dependent attenuations. They are called EDFA (erbium doped fiber amplifier) gain flattening filters when the attenuations are fixed, or dynamic gain equalizers when the attenuation and its wavelength dependence can be changed instantly. The fixed EDFA gain flattening filters were invented to meet the needs of current optical networks, which employ a lot of erbium doped fiber amplifiers in their optical transport lines. The EDFA is the enabler of current fiber optic telecommunication networks. It amplifies the attenuated optical signals along a fiber optic transmission line, replacing the expensive electronic regeneration stations.

Unfortunately, the amplification of the EDFA is not flat over the amplification wavelength range. An EDFA operating in C-Band (from 1525 nm to 1565 nm) typically has a wavelength-dependent amplification gain curve as shown in Fig. 13.24**a**. To make things worse, along a fiber optic transmission line, tens of EDFAs must be employed. The cascading of these EDFAs magnifies the gain differentiation. This greatly degrades the performance of fiber optic transmission networks. To flatten the gain, an attenuator with a wavelength-dependent attenuation of the shape in Fig. 13.24**b** (the inverse shape of the gain curve) is desired. This can be done by various technologies, such as dielectric thin film interference filters [28], short period fiber Bragg gratings [29], and long period fiber gratings [30]. Figure 13.24**c** shows a gain curve after flattening.

In next-generation optic networks, reconfigurable optical add/drop nodes and dynamic network configurations will be common. Amplification gain curves change over time, and the need for dynamic gain equalizers arise. Many approaches have been adopted to meet this need, such as cascaded Mach–Zehnder interferometer planar waveguides [31], arrayed waveguide grating pairs with adjustable phase shifters [32],

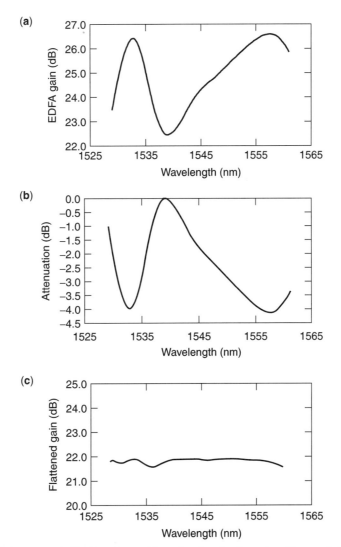

**Figure 13.24** A typical EDFA gain profile, related gain flattening filter and flattened gain curves.

variable reflectivity MEMS mirror arrays coupled with bulk diffraction gratings [33], multistage Fourier filters [34], and acousto-optic tunable filters [35].

## 13.5 OPTICAL ISOLATORS

An optical isolator is a two-port device and is one of the most commonly used nonreciprocal optical devices. An optical isolator passes light waves propagating in one direction with little attenuation and blocks light waves propagating in the opposite direction. This unique characteristic makes it especially suitable for preventing unfavorable back-reflections from entering the active devices in fiber optic communication systems. Active devices, such as laser sources and optical amplifiers, are extremely sensitive to back-reflections from optical network systems. The light beams from semiconductor edge emitting lasers are linearly polarized. Therefore, miniaturized free-space polarization-dependent optical isolators are found inside fiber optic communication laser sources, transmitters, and other semiconductor laser sources. Along with the development and expansion of erbium doped fiber amplifiers, in line fiber optical isolators have experienced unprecedented advancement in performance and deployment volumes in recent years. The optical isolator is now a matured industry in its own right.

Figure 13.25 depicts a typical one-stage erbium doped fiber amplifier. The input tap first couples a small amount of the incoming optical signal (usually around 5%) into a photodiode detector. The remaining signal passes through the input optical isolator. When the signal reaches the WDM device, it is combined with the pump power and enters the erbium doped fiber coils, where the signal is amplified. The amplified signal then passes through an output isolator again. The output tap functions the same way as the input tap, but the tap ratio is smaller (typically 2%).

We can see that there are two optical isolators in this simple EDFA. The input isolator prevents the backward transmission of the ASE noise generated by the EDFA, while the output isolator blocks the back-reflection from the system.

After over 20 years of development, hundreds of designs and inventions for optical isolators have emerged, so it is impossible to describe every design. Our approach is to introduce representative designs only. To reflect the development and application of the optical isolators, we will introduce the polarization-dependent free-space isolator first, and then various in-line optical isolator designs.

Figure 13.26 shows a typical free-space, miniature, polarization-dependent optical isolator and its working principle. It consists of three elements with dimensions of around 1.4 mm × 1.4 mm × 0.5 mm for each element. The polarizer only passes light with vertical polarization. The Faraday rotator garnet rotates the polarization of the light beam 45° counterclockwise, and the analyzer has its passing axis oriented 45° to the vertical, as shown. All three elements can be packaged into a cylindrical housing 2.5 mm in length and 3.0 mm in diameter for use in coaxial DFB lasers and semiconductor amplifiers, or a low height profile package with dimensions of 2.0 mm × 4.0 mm × 1.8 mm for use in 14-pin butterfly configurations. With a one-stage approach (shown in the figure), the insertion loss is around 0.3 dB and the isolation is greater than 25 dB. For a two-stage device, both the insertion loss and

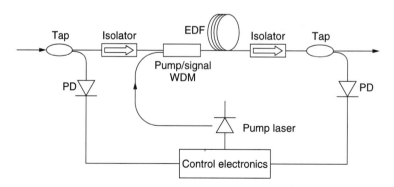

**Figure 13.25** A typical single-stage EDFA.

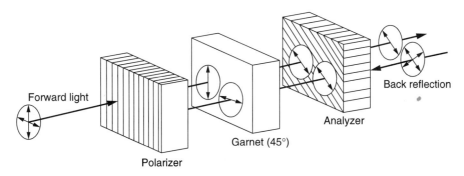

**Figure 13.26**   Structure of a polarization-dependent optical isolator.

isolation increase twofold. We say that the isolator is polarization dependent because, in the forward direction, only the vertical polarization component of the incoming light beam can pass through. The device blocks all back-reflections regardless of their polarization states.

Figure 13.27 shows the most commonly produced in-line optical isolator [36]. The isolator consists of an input optical collimator and an output optical collimator, which comprises a GRIN lens and a fiber holding glass ferrule; a birefringent crystal wedge at the front end and another at the back end; and a Bi-YIG Faraday rotator sandwiched between two wedges with a permanent Sm-Co rare earth element alloy magnet around them. The birefringent crystal wedges can be made from any uniaxial crystals such as calcite, rutile, yttrium vanadate, lithium titanate, and lithium niobate, although isolators

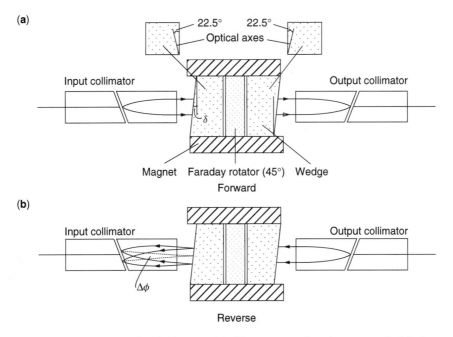

**Figure 13.27**   Structure of a single-stage birefringent crystal wedge type optical isolator.

using wedges made from lithium niobate present the best PMD (polarization mode dispersion) performance. The wedge angle $\delta$ of the lithium niobate wedges is around 12°. The optical axes of the wedges lie in the planes (principal planes of the wedges) perpendicular to the light propagation axis of the device. The optic axes of the wedges form 22.5° angles with respect to the vertical direction, as shown in the figure. The Faraday rotator rotates the polarization 45° clockwise.

Now refer to Fig. 13.28 for the working principle and light beam paths of the isolator shown in Fig. 13.27. (For clarification, the wedge angles are greatly exaggerated; otherwise, the O- and E-beam will crowd together and cannot be distinguished clearly.) When incoming light collimated by the input collimator enters the front wedge, it is divided into two beams, namely, an ordinary beam (O-Wave) and an extraordinary beam (E-Wave), with the polarization of the O-Wave perpendicular to the wedge's optic axis and the polarization of the E-Wave parallel to the wedge's optic axis. When $n_e > n_o$, the two beams are refracted by the wedge, as shown in the figure. The two beams propagate further, with their polarizations rotated 45° clockwise; they encounter the second birefringent crystal wedge at the back end. After the rotation, the O-Wave's polarization is still perpendicular to the second wedge's optic axis so that it remains an O-Wave. Similarly, the E-Wave still remains an E-Wave. Hence, after passing through the second wedge, the two beams become parallel to each other with a small offset (the offset shown in the figure is greatly exaggerated due to the wedge angle exaggeration), and both are focused by the output collimator into the output fiber.

In the reverse direction, due to the nonreciprocal rotation of the Faraday rotator, the O-Wave and E-Wave of the second wedge become the E-Wave and O-Wave of the first wedge, respectively, so that when the two beams come out of the first wedge, they are not parallel to each other. The two beams are refracted downward and upward with an angle of $\Delta\phi$ ($\Delta\phi \approx \Delta n \delta$, where $\delta$ is the wedge angle and $\Delta n = n_e - n_o$) between them. As the optical collimator is very sensitive to the angle deviation, the two beams cannot be focused into the input fiber and are dissipated away.

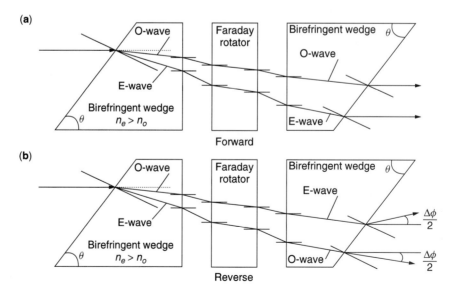

**Figure 13.28** Working principle and light beam paths of the isolator in Fig. 13.27.

As mentioned earlier, the above-mentioned design may have a PDM problem when the incorrect materials are used for wedges. For a design with the wedges made of rutile with a thickness of 0.5 mm, the PMD is around 0.9 picoseconds. This is unacceptably large given that there are many isolators used in an optical transmission line. For comparison, a design using lithium niobate wedges would see its PMD reduced to around 0.2 picoseconds. This is still a large value for some applications. Therefore, PMD compensation designs are desired. The high PMD in the above design is caused by two factors. The main factor is the refractive index difference of the O-Wave and E-Wave. This is why the lithium niobate design has a smaller PMD value, because it has smaller birefringence. Another factor is the path difference between the O-Wave and E-Wave. Compared to the first factor, the effect of this one is considerably smaller. Shown in Figs. 13.29 and 13.30 are several PMD compensated isolator designs [37]. As the Faraday rotator does not introduce PMD, only wedges are shown in the figures.

All designs insert a compensation birefringent crystal plate after the second wedge. Designs in Fig. 13.29 use different materials for compensation. The compensation plate has the same optic axis orientation of that of the second wedge, but the birefringence has the opposite sign. When a material of appropriate thickness is chosen, the design will be PMD free. On the other hand, the designs in Fig. 13.30 use the same birefringent material. However, its optic axis is perpendicular to the optic axis of the second wedge. When the thickness of the compensation plate is approximately the total thickness of the two wedges, the PMD is compensated for fully.

For these single-stage isolator designs, the insertion loss is typically around 0.4 dB with peak isolation around 40 dB. For higher isolation requirements, a two-stage design

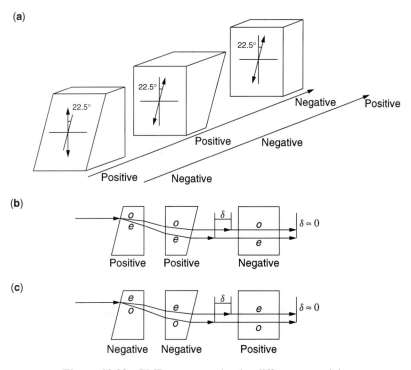

**Figure 13.29** PMD compensation by different materials.

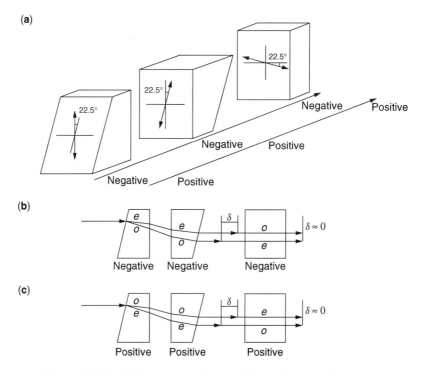

**Figure 13.30**    PMD compensation by different crystal orientations.

is needed, which boosts the isolation to over 60 dB and, at the same time, PMD can also be compensated by design [38]. The only performance sacrifice is the insertion loss, which increases to around 0.8 dB.

Figure 13.31 shows a typical two-stage wedge type isolator design. The first stage is the same as the single-stage design. For the second stage, the optic axes of the two wedges are perpendicular to the optic axes of their counterparts in the first stage. The Faraday rotator, instead of rotating the polarization clockwise, rotates the polarization by 45° counterclockwise. Depicted in (**b**) is the schematic of the light path in the forward direction. The O-Wave in the first stage changes into the E-Wave in the second stage as does the E-Wave. This property makes the design PMD free.

Optical isolators can also be built using a beam displacement design [39]. One benefit of this design is that all parts are in regular cubic or rectangular shape. This makes it easy for handling and assembly.

Figure 13.32 shows a one-stage beam displacement isolator design [40] with three walk-off birefringent crystals. If the thickness of the second and third crystals is annotated as $d$ (they have the same thickness), then the thickness of the first crystal is $\sqrt{2}d$. The principal plane of the first crystal is parallel to the crystal's side surface, and the optic axis is 45° with respect to the vertical edge of the crystal, as shown in the figure. The principal planes of the second and third crystals are parallel to the diagonal planes (the planes indicated by dotted lines) of the crystals and the optic axes of the crystals are 45° with respect to the diagonal lines of the crystals' front and back surfaces. When light travels in the forward direction, the first crystal divides the light beam into two polarization perpendicular beams, with the vertically polarized beam shifted upward.

(a)

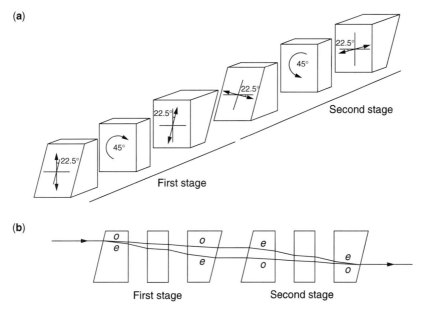

(b)

Figure 13.31   The structure of a two-stage optical isolator with PMD compensation.

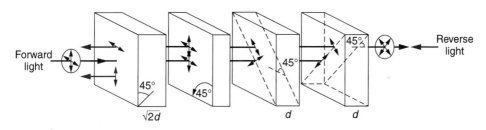

Figure 13.32   Working principle and structure of a walk-off type isolator.

Then, the Faraday rotator rotates the polarizations of the two beams counterclockwise by 45°, making them parallel and perpendicular to the principal planes of the second and third crystals. The second crystal shifts the upper beam downward along its principal plane and the third crystal shifts the lower beam upward along its principal plane. After passing through the two crystals, the light beams combine again, forming an output beam.

For light traveling in the reverse direction, the beam separations and the polarization states are the same as those of the forward traveling light before it reaches the Faraday rotator. The nonreciprocal Faraday rotator rotates the polarization states of the reverse traveling light beam components in such a way that the upper beam has horizontal polarization and the lower beam has vertical polarization, which are just the opposite of the forward traveling light beam polarizations. Therefore, the two beams will be displaced further apart and dissipated away without entering the input fiber.

Other designs utilizing walk-off crystals abound. We will not discuss them further, as their working principles are basically the same as those described here. As increased communication capacity increases the part density in equipment cabinets, compact device designs have become increasingly desirable. Two reflective isolator designs [41, 42] in particular are very compact in form and footprint.

Figure 13.33 shows the top views of a one-stage reflective isolator together with the polarization state transformation diagrams of forward and backward traveling light beams. From left to right, the isolator consists of a dual-fiber ferrule; a thin plate of birefringent crystal proximate to the end faces of the two fibers; a thin glass plate covering the input fiber; a half wave plate covering the output fiber adjacent to the birefringent crystal plate; a near-quarter-pitch GRIN lens; a 22.5° Faraday rotator; and a total reflective mirror.

The principal plane of the crystal is parallel to the horizontal plane, with its optic axis forming a 45° angle with the front surface of the crystal. The half wave plate has its slow axis oriented with a 22.5° angle from the vertical. In other words, when the slow axis of the half wave plate rotates clockwise by 22.5°, it is parallel to the vertical axis. The Faraday rotator, when saturated by the magnet, rotates the polarization 22.5° clockwise. The magnet can be removed by using a latched Faraday rotator here. Figure 13.33a shows the polarization state transformation diagrams.

The incoming light is first divided by the birefringent crystal into two beams with their polarizations perpendicular to each other. The beam with vertical polarization passes through the crystal without deflection. The beam with horizontal polarization shifts a little bit to the right. The two beams making a round trip through the Faraday rotator rotate their polarization states 45° clockwise, while the combination effect of the lens and the mirror exchanges the relative positions of the two polarized beams. Therefore, upon reaching the half wave plate, the two beams have the polarization states shown in polarization state transformation diagram D. The half wave plate then rotates the two polarization states to the horizontal and vertical again, which are then combined by the birefringent crystal into one beam before entering the output fiber.

In the backward direction, the light is first divided into two polarization perpendicular states and the horizontal component is shifted to the right. The half wave plate then rotates the two polarization states into 45° polarization states perpendicular to each other, as shown in polarization state transformation diagram D. The Faraday rotator rotates a total of 45° clockwise for each polarization state, while the lens changes the beam's relative positions horizontally. Finally, the polarization states end up as depicted in polarization state transformation diagram B when the beams reach the birefringent crystal again, because the glass plate does nothing to the polarization states of the traveling beams. The beam component with vertical polarization passes through the crystal without deflection and the horizontal polarized beam component is shifted leftward so that both beams miss the input fiber, as indicated by the dotted circle.

The whole device can be packaged in a cylindrical housing 20 mm in length and 5.5 mm in diameter. However, the isolation of the reflective single-stage isolator can usually only reach 30 dB. To enhance the isolation performance, two-stage reflective isolators are used. Figure 13.34 shows a representative reflective two-stage optical isolator design.

Figure 13.34a shows the top view of the isolator, and Fig. 13.34b is the side view. Annotated in Fig. 13.34c and d are the polarization state transformation diagrams in forward and backward directions, respectively.

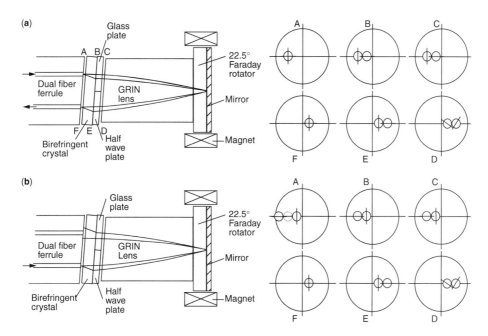

**Figure 13.33** Structure and polarization state transformation diagrams of a reflective isolator.

In Fig. 13.34**a**, from left to right, the isolator has IN and OUT fibers; the first birefringent crystal with a thickness of d; two 45° Faraday rotators with the upper one rotating clockwise and the lower one rotating counterclockwise; the second and third birefringent crystals with a thickness of $\sqrt{2}d$ each; an optical lens; and a total reflective mirror.

The optic axis of the first birefringent crystal lies in the vertical plane [the plane of Fig. 13.34**b**] with a 45° angle relative to the vertical edge of the crystal. The principal planes of the second and third birefringent crystals are parallel to their diagonal planes with the second crystal's principal plane oriented from upper right to lower left and the third crystal's principal plane oriented from upper left to lower right.

The working principle can be understood more clearly by looking at the diagrams in Fig. 13.34**c** and **d**. In Fig. 13.34**c**, light from the IN fiber is divided into two polarization perpendicular beams and the component with vertical polarization is shifted upward while the component with horizontal polarization remains undeflected when passing through the first crystal. The first Faraday rotator then rotates the two beams polarization states 45° clockwise. The second crystal displaces the upper beam downward diagonally to the left of the lower beam with the lower beam remaining unshifted. The lens and the mirror system, in turn, image the two beams to the other side of the device (right side in the diagrams) with their relative positions exchanged. The third crystal shifts the left beam downward diagonally to beneath the other beam component, which is not deflected when passing through the crystal. The second Faraday rotator rotates (45° counterclockwise) the polarization states of the two beams into vertical and horizontal polarization states, respectively. Finally, the first crystal combines the two polarized beams into one beam before it enters the OUT fiber.

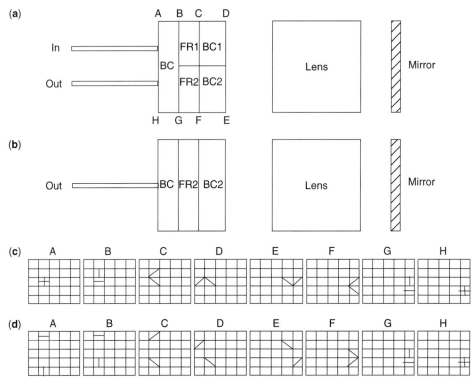

**Figure 13.34**    Structure and polarization state transformation diagrams of a reflective two-stage optical isolator.

In the reverse direction, shown in Fig. 13.34**d**, light from the OUT fiber is divided into two polarization perpendicular beams and the component with vertical polarization is shifted upward when passing through the first crystal. The second Faraday rotator rotates the two polarization states 45° counterclockwise, while the third crystal then shifts the upper beam upward diagonally to the left while the lower beam remains unchanged. The lens and the mirror system images the two beams to the other side of the device (left side of the diagram) with their relative positions exchanged. The second crystal shifts the upper beam upward diagonally to over the lower beam, while the lower beam remains unchanged. The first Faraday rotator rotates the polarization states of the two beams 45° clockwise. When the two beams reach the first crystal, they get the wrong polarization states and are separated further apart and dissipated away. Thus, the light is unable to enter into the IN fiber.

Comparing this isolator's polarization state diagrams to the ones of the one-stage reflective isolator, we can clearly see that the separation of the two polarized beams traveling backward at the IN fiber tip is twice as large as that of the single stage isolator. This suggests that the current design is a two-stage design with high isolation performance.

All isolators that have been introduced until now are based on polarization component separation and combination operations. However, isolators can be built using other principles as well. Fig. 13.35 below shows an isolator design that is based on the light wave interference effect [43]. The isolator introduced is a planar waveguide

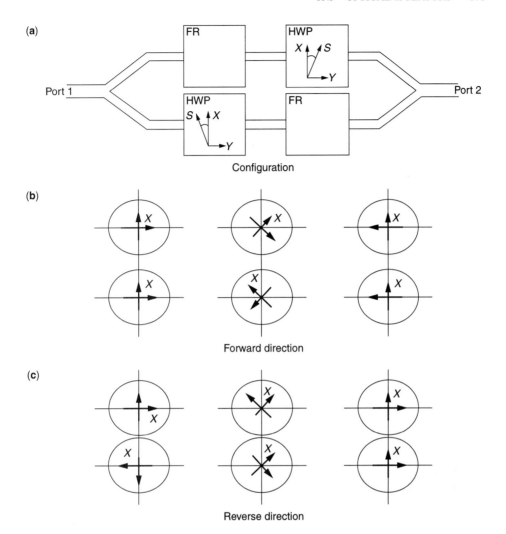

**Figure 13.35**   Working principle of waveguide interference type optical isolator.

device. It consists of two Y-branches connected together to form an interferometer with some polarization rotators inserted into the two arms.

In Fig. 13.35, FR1 and FR2 are Faraday rotators with a 45° clockwise rotation. HWP1 and HWP2 are half wave plates with their slow axes oriented as shown in the figure. The angles between the slow axes and the $x$-axis are 22.5°.

If the interferometer is balanced on other phases, then only the polarization rotators need to be taken into consideration. When light travels in the forward direction (from port 1 to port 2), the polarization components at the upper arm and the polarization components at the lower arm are in phase when they reach port 2 [this can be clearly seen in Fig. 13.35**b**], and thus, the combined beam travels together constructively. On the other hand, when light travels in the reverse direction (from port 2 to port 1), the polarization components at the upper arm and the polarization components at the lower arm are completely out of phase, as can be seen in Fig. 13.35**c**. This will

**(a)**

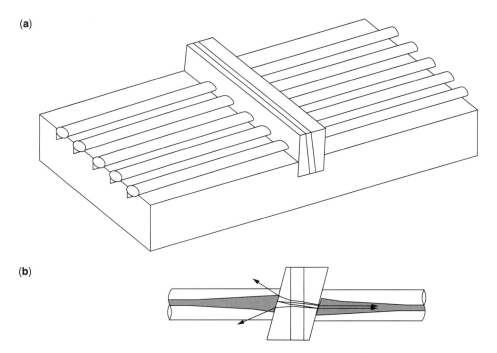

**(b)**

**Figure 13.36**    Schematics of an isolator array device.

cause destructive interference at port 1 so that no light waves can pass through in the backward direction.

To further conserve board space in communication equipment, an array of isolators can be packaged together to form an isolator array device [44, 45]. The design shown in Fig. 13.36 uses the wedge isolator design and thermally expanded core fibers to form a compact arrayed isolator device.

As shown in the Fig. 13.36**a**, V-grooves are made on the substrate first. Then, a slot is cut to accommodate the isolator chip. The isolator chip has the same structure as the single-stage wedge isolator design described previously, but it is thinner and longer. When the chip is inserted into the slot, TEC (thermally expanded core) fibers are placed into the V-grooves to form an isolator array. The working principle of the isolator is shown in Fig. 13.36**b**.

Although the isolators detailed above have attractive performance and compact size, they cannot be easily integrated with planar light wave circuits, i.e., on laser chips. Scientists and engineers are still working on true integrated designs, which only involve the planar waveguides without any discrete parts. A lot of progress has been made so far. However, the devices' performance still cannot compare with that of devices built from discrete components. We will not delve this subject further. Readers interested in learning more about integrated isolators can refer to the references [46, 47].

## 13.6    OPTICAL CIRCULATORS

An optical circulator is an $N$-port ($N > 2$) device in which light enters port $i$ and is transmitted into port $i + 1$. A circulator, therefore, is a nonreciprocal optical device where any two consecutive ports form an optical isolator because light can travel from port $i$ to port $i + 1$, but not in the reverse direction. Optical circulators are frequently used in fiber optic communication systems, testing systems, and instrumentation. Figure 13.37 illustrates two typical applications of optical circulators.

Figure 13.37**a** is an optical add/drop multiplexer used in DWDM optical networks. The multichannel signals coming from port 1 of the left circulator exit port 2. When the signals encounter the FBG (fiber Bragg grating), part of the signals are reflected back and the remaining signals pass through. The reflected signals then enter port 2 again and exit port 3 and become dropped signals. At the same time, the local information is added at port 1 of the right circulator. The added signals exit port 2 and travel leftward. When they encounter the FBG, they are reflected back because they have the same wavelengths as the dropped signals. The back-reflected add signals, together with the remaining signals transmitted from the left circulator, enter port 2 of the right circulator. Finally, they exit port 3 of the right circulator and continue their transmission in the optical network system.

A bidirectional optical transmission system composed of two optical circulators is shown in Fig. 13.37**b**.

Here we must introduce the concept of single polarization circulators. Although they are seldom used in real applications, the single polarization circulator serves as the basis for explaining and analyzing more complex polarization-independent circulator designs. The introduction of several representative circulators that were invented at different time periods then follows just after it.

Figure 13.38**a** shows two single-polarization circulators.

The first single-polarization circulator uses two thin film coated PBS cubes and four Faraday rotators. Faraday rotators are indispensable components in all circulator

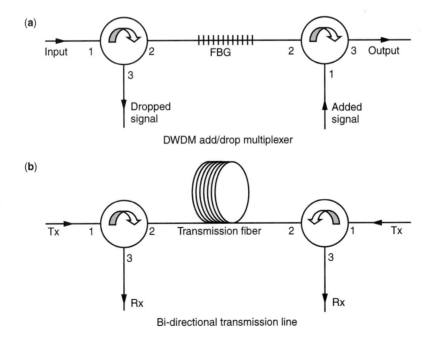

**Figure 13.37** Applications of optical circulators.

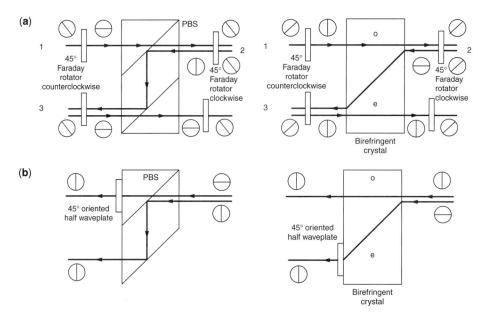

**Figure 13.38**   Working principles of single-polarization optical circulators.

designs discussed here as they provide the nonreciprocal function of the optical circulators. A light wave linearly polarized at 45° enters into port 1 of the circulator, and its polarization state is rotated 45° counterclockwise by the Faraday rotator so that it is transformed into a horizontally polarized light wave. It will then pass through the PBS undisturbed. Rotated by another Faraday rotator 45° clockwise, it passes out of port 2 with a 45° linear polarization again. When the light wave enters port 2 of the circulator, its 45° linear polarization state is rotated 45° clockwise by the Faraday rotator and it becomes a vertically polarized light wave.

When the so polarized light wave encounters the PBS coating, it is reflected and travels downward, is reflected by the second PBS coating again, and travels toward port 3. Its polarization state is rotated another 45° counterclockwise so that when it reaches port 3, the light wave has a 45° linear polarization state once again.

If the light wave with 45° linear polarization enters port 3, then the whole process repeats itself as long as more PBSs are added. The second single-polarization circulator works the same way as the first one except that a birefringent crystal beam splitter replaces the PBS cubes. However, it is much easier to extend to multiport devices as long as the crystal is large enough because there is no need to add anything.

Considering that it is rare to have light waves consisting of only a pure single linearly polarized component, it would seem that the single-polarization circulators are useless after all. In fact, just the opposite is the case. The single-polarization circulator is actually the heart and soul of every polarization-independent circulator. We will fully understand this after we learn the functioning parts of the circulators shown in Fig. 13.38**b**.

The two parts in Fig. 13.38**b** look very similar to the single-polarization circulators in (**a**); however, they function very differently. They are used to transform light waves with arbitrary polarization states into light waves with a single-polarization state. The first part consists of a PBS cube and a half wave plate. Moving from right to the left, the part first decomposes a light wave with an arbitrary polarization state into two predetermined basic polarization states and splits them apart spatially, and then transforms them into the same polarization state. The part on the right works the same way, only replacing the PBS cube with a birefringent crystal beam splitter.

Notice that the two parts do not include nonreciprocal components. Therefore, they are reciprocal and can be used to combine two light wave components with the same polarization state into one light wave. Now that we have these parts, which split a light wave into single-polarization components and combine single-polarization light waves into one, and the single-polarization circulators described previously, making a polarization-independent optical circulator is not difficult.

In the above description, we treated the reciprocal and nonreciprocal polarization rotators separately. In practice, however, they are frequently mixed together. Figure 13.39 below shows common combinations of them.

On the left part of the polarization rotator sets, a half wave plate combines with a 45° clockwise rotation Faraday rotator. The slow axis of the half wave plate forms a 22.5° angle with respect to the vertical axis when the slow axis rotates counterclockwise toward the vertical axis. In the forward direction, this combination will rotate the polarization states 90°, while in the backward direction, it will leave the polarization states unchanged.

Similarly, on the right part of the polarization rotator sets, a half wave plate combines with a 45° clockwise rotation Faraday rotator. The slow axis of the half wave plate forms a 22.5° angle with respect to the vertical axis when the slow axis rotates clockwise toward the vertical axis. In the forward direction, this combination will leave the polarization states unchanged, while in the backward direction, it will rotate the polarization states 90°.

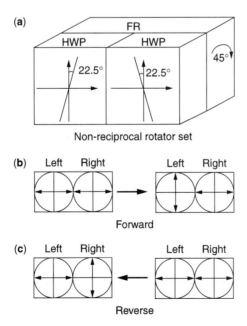

**Figure 13.39**   Combinations of nonreciprocal polarization rotator sets.

Now we are ready to introduce various circulator designs. One of the earliest designs [48] uses only PBS cubes, as shown in Fig. 13.40.

A light beam from the port-1 collimator enters the PBS1 cube and is split into two polarization components. The component with horizontal polarization ($P$ wave) passes through, and the component with vertical polarization ($S$ wave) is reflected upward. The $P$ wave passes through the Faraday rotator (FR) and half wave plate (HWP), where its polarization state is changed into an $S$ wave. The reflected $S$ wave is reflected again by the prism and passes through the FR and HWP, where its polarization state is changed into a $P$ wave. As a result, the two beam components have the right polarization states and are combined by the PBS2 into one beam, which is focused into port-2.

On the other hand, a light beam from the port-2 collimator enters the PBS2 cube and is split into two polarization components. The $P$ wave passes through, and the $S$ wave is reflected downward by the PBS2. The $P$ wave and the $S$ wave then pass the HWP and FR sequentially without changing their polarization states. As a result, they have the right polarization states and are combined by the PBS1 into one beam, which is then focused into port-3. With the same reasoning, a light beam from port-3 will pass into port-4.

This circulator has a very simple structure. However, the alignment is critical because the recombination of the two separated beams relies on accurate alignment. The isolation is also limited by the PBS's performance, which usually only gives larger than 30-dB values.

Circulator designs based on birefringent crystal beam splitters and displacers usually have better isolation performance. The design shown in Fig. 13.41 is from Koga [49]. **a** and **b** are top views. The beam splitters at the two ends are essentially the same. Their principal planes lie parallel to the horizontal plane; therefore, they split and combine light beams horizontally. The birefringent crystal at the center is the beam displacer. Its principal plane is parallel to the vertical plane. Therefore, it displaces beams with polarizations lying vertically in the principal plane. The HWP1, HWP2, and FR are the nonreciprocal rotator set we described previously.

A light beam from port-1 is divided by the splitter into horizontally and vertically polarized beam components. The vertically polarized beam passes straight through,

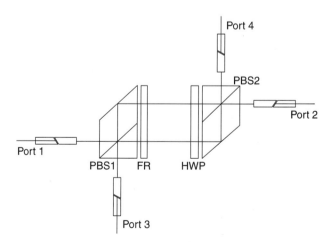

**Figure 13.40**   The structure of a four-port optical circulator.

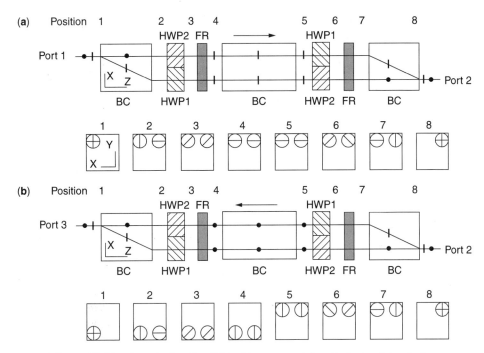

**Figure 13.41** A typical walk-off birefringent crystal optical circulator design.

while the horizontally polarized beam is deflected. After passing through the nonreciprocal rotator set, both beams have horizontal polarization states. Both beams then pass through the beam displacer without deflection. After further passing through another set of nonreciprocal rotators, the polarization states of the two beams are restored to polarization states perpendicular to each other. They have the correct polarization states and are combined by the birefringent crystal into port-2. All of these polarization state transformations can be seen more clearly in Fig. 13.41a.

In Fig. 13.41b, a light beam from port-2 is divided by the splitter into horizontally and vertically polarized beam components. The vertically polarized beam passes straight through, while the horizontally polarized beam is deflected. After passing through the nonreciprocal rotator set, both beams have vertical polarization states. They then pass through the beam displacer and are deflected downward.

After further passing through another set of nonreciprocal rotators, the polarization states of the two beams are once again perpendicular to each other. They have the right polarization states and are combined into port-3, which is beneath port-1. The polarization state transformation diagrams in **b** indicate this more clearly. Optical circulators based on this design are achieved over 60-dB isolation.

Optical circulators can also be built using a hybrid approach, which incorporates both PBS cubes and birefringent crystals. The design shown in the Fig. 13.42 was invented by J. J. Pan et al. [50]. Although it does not improve the isolation much, it decreases the alignment difficulty.

Without using half wave plates, the design uses clockwise and counterclockwise 45° Faraday rotator pairs. The beam splitting and combining crystals have their principal planes parallel to the crystal's diagonal planes (hence, the E-beam displaces diagonally).

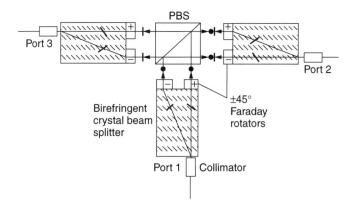

**Figure 13.42**   A circulator consisted of PBS and birefringent crystal beam splitters.

As a result, two Faraday rotators that consist of a Faraday rotator pair are not at the same height.

A light beam from port-1 is first split into two beams with their polarization states perpendicular to each other. The Faraday rotator pair rotates both beams' polarization states until they are vertical. Both beams are reflected by the PBS cube and travel toward port-2. Similarly, a light beam from port-2 is split into two beams with their polarization states perpendicular to each other. Then, the Faraday rotator pair rotates both beams' polarization states to the horizontal (the Faraday rotator pair for port-2 is just the opposite of that of port-1). Both beams will pass through the PBS cube and travel toward port-3.

To save equipment cabinet space and ease fiber routing, reflective type optical circulators are recommended. Figure 13.43 illustrates a design published by Yihao Cheng [51].

The design consists of: a birefringent crystal beam splitter/combiner (BC1); a half wave plate HWP with its slow axis parallel to the horizontal plane; a glass plate; a 45° clockwise Faraday rotator (FR1); a birefringent crystal beam displacer (BC2); a 45° clockwise Faraday rotator (FR2); and a total reflective mirror. The principal plane of BC1 is parallel to the diagonal plane of the crystal (as indicated by dotted lines) so that it deflects the E-beam diagonally upward. The beam displacer BC2 has its principal plane parallel to the horizontal plane. It will deflect the E-beam horizontally from left to right.

Referring to the polarization state diagrams in **b**, when light from port-1 enters the circulator, BC1 will split it into two beams with their polarizations perpendicular to each other. The component with polarization parallel to the diagonal plane is shifted upward diagonally. The upper beam's polarization state is then rotated by the half wave plate so that it is perpendicular to the diagonal plane when it passes through the HWP. The glass plate does not affect the polarization of the lower beam. Faraday rotator FR1 rotates both beams' polarization states clockwise by 45° and makes them vertically polarized and parallel to each other so that both beams are o-beam for BC2 and will pass through BC2 straightforward. Both beams then pass Faraday rotator FR2 and are reflected back by the mirror to pass through the Faraday rotator FR2 again. The round trip changes the beams' polarization states from vertical into horizontal, which is the

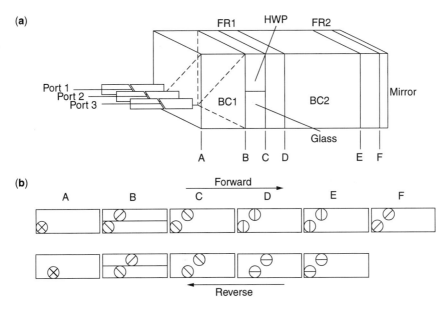

**Figure 13.43** A birefringent crystal, reflective type optical circulator.

E-beam of BC2. Therefore, BC2 will deflect both beams to the right (when looking toward the mirror).

When the two beams pass Faraday rotator FR1 again, both their polarization states are changed into directions perpendicular to the principal plane of BC1. Next, the upper beam's polarization is changed by the HWP so that it is parallel to the principal plane of BC1 when it passes the HWP and the lower beam's polarization state is unchanged by the glass plate. When both beams reach birefringent crystal BC1, they have the right polarizations and will be combined by BC1 into one beam, which is focused into port-2. A light beam transmits from port-2 to port-3 according to exactly the same process as it transmits from port-1 to port-2. This process continues as long as the crystals are large enough.

Although reflective type circulators greatly reduce the footprint of circulator devices, they still do not quite meet the requirements of miniaturization. As most of the circulators used in real applications are three-port devices, a three-port, cylindrically packaged optical circulator is the most desirable. People have tried and succeeded in making such miniature devices. The design introduced in Fig. 13.44 is based on the patent published by Wei-Zhong Li et al. [52].

From left to right, the design consists of: a dual-fiber collimator; a birefringent crystal beam splitter/combiner (BC1); a nonreciprocal rotator set as described previously; a Wollaston prism; a beam displacer birefringent crystal (BC2); another nonreciprocal rotator set; another birefringent crystal beam splitter/combiner (BC3); and a single fiber collimator. The schematics in Fig. 13.44**a** and **b** are top views. BC1 has its principal plane lying horizontally, as does BC3. Actually, BC1 and BC3 are the same part with different orientations. BC2 has its principal plane also parallel to the horizontal plane. The wedge angle of the Wollaston prism is chosen so that the deflection angle for any polarization beam component is equal to the angle of a collimated beam from a dual-fiber collimator with respect to the collimator's optic axis.

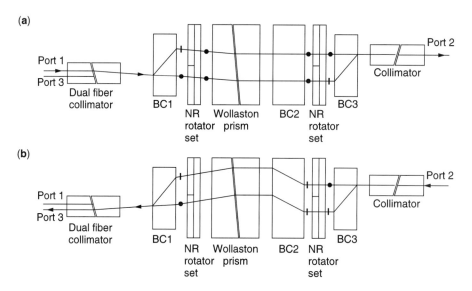

**Figure 13.44**    A miniature 3-port cylindrically packaged optical circulator.

Referring to Fig. 13.44**a**, a light beam from port-1 collimated by the dual-fiber collimator enters BC1 with an angle $\theta$. BC1 splits the beam into two beams horizontally. The beam that is passing through has vertical polarization, and the deflected beam has horizontal polarization. The two beams travel toward the nonreciprocal rotator set with the same angle $\theta$. After passing through the rotator set, both beams have vertical polarizations and, with an angle $\theta$, travel toward the Wollaston prism. As we stated previously, the Wollaston prism is designed so that it converts the obliquely incident single-polarization beam into a beam that exits the prism at a normal angle. Therefore, after passing through the Wollaston prism, the two beams with vertical polarization enter BC2 at a normal incident angle. The two beams will pass BC2 without any change. The nonreciprocal rotator set transforms the polarization states of the two beams into polarization states perpendicular to each other. Finally, the two beams with the right polarization are combined by BC3 into one beam, which is focused into port-2.

Light transmission from port-2 to port-3 can be understood by referring to Fig. 13.44**b**. The circulators made from this design demonstrate a port-to-port insertion loss of 0.7 dB and an isolation of 45 dB with a cylindrical package with dimensions of $\phi 5.5 \times 60$ mm.

Similar to the optical isolator situation, optical circulators can also be designed using waveguides and interference effects [53]. Fig. 13.45 shows the design proposed by J. J. Pan et al. [54]. The proposed optical circulator is actually a Mach–Zehnder interferometer with some polarization rotation elements inserted into its two arms. In this design, inserted into the upper arm is a 90° clockwise Faraday rotator, a slow axis 45° oriented half wave plate and a half wave plate with its slow axis lie horizontally. To balance the phase delay of the two arms, a uniform phase delay plate is inserted into the lower arm.

As discussed in the attenuator section, for a Mach–Zehnder interferometer, if the optical paths of the two arms are balanced, in other words, the two arms introduce the

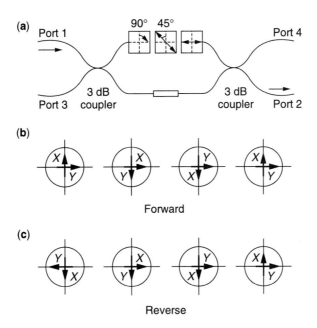

**Figure 13.45**  Mach–Zehnder interferometer type optical circulator.

same phase change, the light from port-1 will travel to port-2 and light from port-2 will also travel to port-1 in reverse. The light from port-3 travels to port-4, and light from port-4 travels to port-3.

The introduction of a Faraday rotator into the upper arm changes the light transmission characteristics in the reverse direction but maintains the same transmission characteristics in the forward direction. This can be understood by looking at the polarization state diagrams in Fig. 13.45**b** and **c**. We denote the transmission direction in the two arms as the z-axis, and then the polarization of the light wave in the two arms can be represented as components in the $x$- and $y$-axes. For the lower arm, because all the materials are homogeneous, a light wave's polarization will not be changed when it travels from the left side to the right side or in the reverse direction. For the upper arm, when a light wave travels from the left side to the right side (forward direction), the polarization components have no direction or phase changes. When a light wave travels in the reverse direction, although the polarization components have no direction change, they get a $\pi$ phase shift. Therefore, after the insertion of the polarization rotators, when a light wave from port-1 travels in the forward direction, it still reaches port-2, as there is no change made by the rotators. When a light wave from port-2 travels in the reverse direction, it will not reach port-1 as before because the $\pi$ phase shift of the upper arm makes the two wave components from the upper arm and the lower arm destructively interfere at port-1. Instead, the light wave goes into port-3 (because the two wave components constructively interfere there). In the same way, a light wave from port-3 will reach port-4 and a light wave from port-4 will reach port-1. All of this makes the structure a true four-port circulator.

The optical circulator is now a very mature commercial product, but scientists and engineers are still working to improve it further. The ultimate goal is to integrate optical circulators into planar light wave circuits [55].

## REFERENCES

1. M. Born and E. Wolf, *Principles of Optics*, 6th ed. Pergamon Press, New York, 1980.

2. Bahaa E. A. Saleh and Malvin Carl Teich, *Fundamentals of Photonics*, 1st ed. Wiley, New York, 1991.

3. J. M. Bennett, "Polarization," in *Handbook of Optics*, Vol. I, M. Bass, ed., McGraw-Hill, New York, 1995, pp. 5.1–5.30.

4. J. M. Bennett, "Polarizers," in *Handbook of Optics,* Vol. II, M. Bass, ed., McGraw-Hill, New York, 1995, pp. 3.1–3.70.

5. M. A. Jensen and G. P. Nordin, "Finite-aperture wire grid polarizers," *J. Opt. Soc. Am. A*, **17**, No. 12, p. 2191, 2000.

6. K. Shiraishi et al., "Laminated Polarizers Exhibiting High Performance Over a Wide Range of Wavelength," *J. Lightwave Technol.*, **15**, No. 6, p. 1042, 1997.

7. Kouichi Muro, et al., "Poly-Si/SiO$_2$ Laminated Walk-Off Polarizer Having a Beam-Splitting Angle of More Than 20°," *J. Lightwave Technol.*, **16**, No. 1, p. 127, 1998.

8. L. Li and J. A. Dobrowolski, "High-performance thin-film polarizing beam splitter operating at angles greater than the critical angle," *Applied Optics*, **39**, No. 16, p. 2754, 2000.

9. R. M. A. Azzam and M. M. K. Howlader, "Silicon-Based Polarization Optics for the 1.30 and 1.55 mm Communication Wavelengths," *J. Lightwave Technol.*, **14**, No. 5, p. 873, 1996.

10. D. J. Dummer, et al., "High-quality Brewster's angle polarizer for broadband infrared application," *Applied Optics*, **37**, No. 7, p. 1194, 1998.

11. S. G. Lee, et al., "Fabrication of a side-polished fiber polarizer with a birefringent polymer overlay," *Optics Letters*, **22**, No. 9, p. 606, 1997.

12. S. P. Ma and S. M. Tseng, "High-performance Side-Polished Fibers and Applications as Liquid Crystal Clad Fiber Polarizers," *J. Lightwave Technol.*, **15**, No. 8, p. 1554, 1997.

13. S. M. Tseng, et al., "Analysis and Experiment of Thin Metal-Clad Fiber Polarizer with Index Overlay," *IEEE Photonics Technology Letters*, **9**. 5, p. 628, 1997.

14. Y. Miyama and H. Nagata, "Optical and Physical Characterization of SiO$_2$-x-Al Thin-Film Polarizer on X-Cut LiNbO$_3$ Substrate," *J. Lightwave Technol.*, **19**, No. 7, p. 1051, 2001.

15. R. C. Tyan, et al., "Design, fabrication, and characterization of form-birefringent multi-player polarizing beam splitter," *J. Opt. Soc. Am. A*, **14**, No. 7, p. 1627, 1997.

16. Y. Long, Internal presentation of E-TEK Dynamics, Inc., 1998.

17. Y. Morishita, et al., "Co$^{2+}$-doped flatband optical fiber attenuator," *Optics Letters*, **26**, No. 11, p. 783, 2001.

18. S. Ajith Kumar, et al., "Low-reflection-loss attenuator optical coatings: theory and experiment," *Applied Optics*, **35**, No. 16, p. 3047, 1996.

19. A. Benner, et al., "Low-Reflectivity In-Line Variable Attenuator Utilizing Optical Fiber Tapers," *J. Lightwave Technol.*, **8**, No. 1, p. 7, 1990.

20. N. A. Riza and Z. Yaqoob, "Submicrosecond Speed Variable Optical Attenuator Using Acoustooptics," *IEEE Photonics Technology Letters*, **13**, No. 7, p. 693, 2001.

21. J. J. Pan, "Fiberoptic liquid crystal on-off switch and variable attenuator," US Patent No. 6181846, 2001.

22. K. Hirabayashi, et al., "Compact optical-fiber variable attenuator arrays with polymer-network liquid crystals," *Applied Optics*, **40**, No. 21, p. 3509, 2001.

23. J. E. Ford, et al., "Micromechanical Fiber-Optic Attenuator with 3 μs Response," *J. Lightwave Technol.*, **16**, No. 9, p. 1663, 1998.

24. B. Barber, et al., "A Fiber Connectorized MEMS Variable Optical Attenuator," *J. Lightwave Technol.*, **10**, No. 9, p. 1262, 1998.

25. C. Marxer, et al., "A Variable Optical Attenuator Based on Silicon Micromechanics," *IEEE Photonics Technology Letters*, **11**, No. 2, p. 233, 1999.

26. N. A. Riza and S. Sumriddetchkajorn, "Digitally controlled fault-tolerant multiwavelength programmable fiber-optic attenuator using a two-dimensional digital micromirror device," *Optics Letters*, **24**, No. 5, p. 282, 1999.

27. D. Bloom, "The Grating Light Valve: Revolutionizing Display Technology," Projection Displays III Symposium, SPIE Proceedings Volume 3013, February 1997.

28. M. Tilsch, et al., "Design and Demonstration of a Thin Film Based Gain Equalization Filter For C-Band EDFAs," NFOEC 1999, Session A7, Chicago, Illinois.

29. C. De Barros, et al., "Tapered slanted Bragg grating for low insertion loss gain equalizing filter," OSA Technical Digest: Conference on Bragg Gratings, Photosensitivity and Poling in Glass Waveguides, Stresa, Italy, July 4–6 (paper JW1), 2001.

30. M. Ibsen, et al., "Custom design of long chirped Bragg gratings: application to gain-flattening filter with incorporated dispersion compensation," *IEEE Photonics Technology Letters*, **12**, No. 5, pp. 498–500, 2000.

31. B. J. Offrein, et al., "Adaptive Gain Equalizer in High-Index-Contrast SiON Technology," *IEEE Photonics Technology Letters*, **12**, No. 5, p. 504, 2000.

32. C. R. Doerr, et al., "Arrayed Waveguide Dynamic Gain Equalization Filter with Reduced Insertion Loss and Increased Dynamic Range," *IEEE Photonics Technology Letters*, **13**, No. 4, p. 329, 2001.

33. J. E. Ford and J. A. Walker, "Dynamic Spectral Power Equalization Using Micro-Opto-Mechanics," *IEEE Photonics Technology Letters*, **10**, No. 10, p. 1440, 1998.

34. T. Huang, et al., "Liquid Crystal Optical Harmonic Gain Equalizer," NFOEC 2001, p. 443, Baltimore, MD, July 8–12, 2001.

35. H. S. Kim, et al., "Actively Gain-Flattened Erbium-Doped Fiber Amplifier Over 35nm by Using All-Fiber Acoustooptic Tunable Filters," *IEEE Photonics Technology Letters*, **10**, No. 6, p. 790, 1998.

36. M. Shirasaki and K. Asama, "Compact Optical Isolator for fibers using birefringent wedges," *Applied Optics*, **21**, No. 23, pp. 4296–4299, 1982.

37. J. J. Pan and M. Shih, "Optical Isolator with Low Polarization Mode Dispersion," US Patent No. 5557692, 1996.

38. J. J. Pan, et. al., "Dual Stage Optical Device with Low Polarization Mode Dispersion," US Patent No. 5566259, 1996.

39. T. Matsumoto, "Polarization-independent isolators for fiber optics," *Trans. IECE Japan*, **62**, pp. 516–517, 1979.

40. K. W. Chang and W. V. Sorin, "Polarization Independent Isolator Using Spatial Walkoff Polarizers," *IEEE Photonics Technology Letters*, **1**, No. 3, p. 68, 1989.

41. M. Tojo, et al., "Optical Passive Components," US Patent No. 5499132, 1996.

42. Y. Cheng and G. S. Duck, "Multi-Stage Optical Isolator," US Patent No. 5768005, 1998.

43. T. Shintaku, et al., "Polarization Independent Optical Nonreciprocal Circuit Based on Even Mode to Odd Mode Conversion," US Patent No. 5905823, 1999.

44. J. J. Pan, "Optical Isolator Array Device," US Patent No. 5706371, 1998.

45. T. Sato, et al., "Lens-free in-line optical isolators," *Optics Letters*, **24**, No. 19, p. 1337, 1999.

46. H. Yokoi, et al., "Demonstration of an optical isolator with a semiconductor guiding layer that was obtained by use of a nonreciprocal phase shift," *Applied Optics*, **39**, No. 33, p. 6158, 2000.

47. J. Fujita, et al., "Polarization-Independent Waveguide Optical Isolator Based on Non-reciprocal Phase Shift," *IEEE Photonics Technology Letters*, **12**, No. 11, p. 1510, 2000.

48. H. Kuwahara, "Optical Circulator," US Patent No. 4650289, 1987.

49. M. Koga and T. Matsumoto, "High-Isolation Polarization-Insensitive Optical Circulator for Advanced Optical Communication Systems," *J. Lightwave Technol.*, **10**, No. 9, p. 1210, 1992.

50. J. J. Pan and Y. Huang, "Compact Fiberoptic Circulator With Low Polarization Mode Dispersion," US Patent No. 5689593, 1997.

51. Y. Cheng and G. S. Duck, "Reflective Optical Non-reciprocal Devices," US Patent No. 5471340, 1995.

52. W. Z. Li, et al., "Optical Circulator," US Patent No. 5909310, 1999.

53. N. Sugimoto, et al., "Waveguide Polarization-Independent Optical Circulator," *IEEE Photonics Technology Letters*, **11**, No. 3, p. 355, 1999.

54. J. J. Pan, et al., "Low Cost Fiber Optic Circulator," US Patent No. 6289156, 2001.

55. K. Matsubara and H. Yajima, "Analysis of Y-Branching Optical Circulator Using Magnetooptic Medium as a Substrate," *J. Lightwave Technol.*, **9**, No. 9, p. 1061, 1991.

# 14

## OPTICAL FILTERS FOR TELECOMMUNICATION APPLICATIONS

C. K. MADSEN

*Bell Laboratories*
*Lucent Technologies*
*Murray Hill, New Jersey*

### 14.1 INTRODUCTION

Optical filters are a key enabling technology for wavelength division multiplexed (WDM) communication systems. Applications span from wavelength management, vis-a-vis multiplexing and channel grooming functions, to signal conditioning in the case of gain equalization and dispersion compensation. These applications are highlighted in a prototypical optical line system as shown for the west-to-east direction in Fig. 14.1. System "hero experiments" with capacities exceeding 10 Tb/s have been reported that contain hundreds of channels operating at bitrates of 10 Gb/s and above. Multiplexers capable of efficiently combining a large number of channels onto a single fiber are required at the transmit side. At the receive side, a demultiplexer separates the channels into separate outputs so that each receiver detects a single channel. Intermediate nodes may drop and add traffic in a channel-specific manner, and optical filters that provide this functionality are called add-drops or wavelength selective switches. If many intermediate nodes are employed, several filters will be cascaded and the effective passband for the cascade will be narrower than for a single filter; consequently, passband width and flatness are important filter characteristics as well as transition band rolloff and stopband rejection.

Optical amplifiers boost the signal to overcome attenuation in the fiber and components; however, their gain response is typically frequency dependent. To maintain a consistent signal-to-noise ratio among the channels, dynamic gain equalization filters

*Handbook of Optical Components and Engineering,* Edited by Kai Chang
ISBN 0-471-39055-0 © 2003 John Wiley & Sons, Inc.

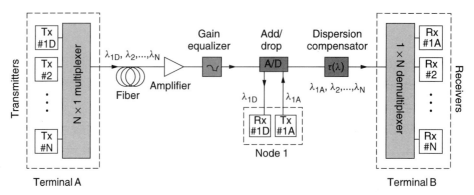

**Figure 14.1**  An optical line system showing a single direction of transmission (west-to-east) and highlighting various optical filter applications.

are employed that use a feedback signal from an optical monitor to minimize the power variation between channels.

As the bitrate per channel increases in optical systems, the allowable tolerance for dispersive effects, such as chromatic and polarization mode dispersion (PMD), decreases substantially. Although dispersion is desired in the optical transmission fiber to minimize nonlinear distortions caused by four-wave mixing, the cumulative path dispersion must be compensated close to zero before the receiver to minimize inter-symbol interference. The tolerance on the allowable cumulative dispersion decreases quadratically as the bitrate increases. Tunable dispersion compensating filters are needed to accommodate statistical and temperature variations in the fiber dispersion. Across many channels, the cumulative dispersion may be quite different if the dispersion slope of the transmission fiber is not compensated. Filters that accomplish this task are dispersion slope compensators. To implement dynamic channel routing in future systems, tunable dispersion compensators that can compensate any channel will be required, because the optimal compensating dispersion will change for different routes and wavelengths.

In addition to evolving towards higher bitrates, system designers also wish to use the available bandwidth most efficiently. Systems with large spectral efficiencies require that the channels be tightly spaced and thereby minimize the total optical amplifier bandwidth, yet demultiplexed without interfering with each other. Signals with bandwidths of 10%, 40% and 80% relative to a constant channel spacing are shown in Fig. 14.2. For a system with a channel spacing of 100 GHz, these spectral efficiencies correspond to bitrates of 10, 40 and 80 Gb/s. Note that a multiplexing filter response must have flatter passbands and sharper rolloffs as the spectral efficiency increases. Besides the magnitude characteristics of the filter response illustrated in Fig. 14.2, the phase response is also important. A simple delay line is characterized by a linear phase response. Deviations from phase linearity introduce dispersion, so it is critical to have a linear-phase response across the passband. Various filter types exhibit different dispersion characteristics, and the dispersion may limit the useable passband instead of the magnitude response as illustrated in Fig. 14.3 [1].

Optical filters for communication applications are based on interference. In this chapter, the underlying types of interference are first described and compared with their digital filter counterparts. The connection between optical waveguide and digital filters was developed by Moslehi et al. [2]. The $Z$ transform and Fourier transform provide

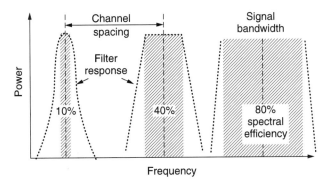

**Figure 14.2** The impact of increasing system spectral efficiency on the desired filter response.

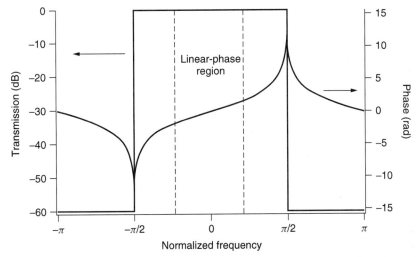

**Figure 14.3** Comparison of a filter magnitude and phase response indicating the region of linear-phase response within the passband [1] © 1998 IEEE.

the mathematical basis for calculating the filter's magnitude and phase response. The Z transform provides a simple means for classifying filters by their phase response and for understanding the connection between a filter's dispersion and magnitude characteristics.

Many diverse technology platforms are used to implement optical filters. Filters based on planar waveguides, diffraction gratings, thin-film filters, and grating-assisted coupled-mode filters, such as Bragg gratings, are discussed along with some of their preeminent applications. Practical considerations such as temperature and polarization dependence, scalability, and integrability are addressed. Tunable filters are critical for next-generation systems to enable dynamic bandwidth management in the optical domain as well as to adaptively compensate for analog impairments such as changes in signal power, chromatic dispersion, and polarization mode dispersion. Dynamic filters for these applications are addressed along with tuning mechanisms such as thermo-optic phase shifters and micro-electro-mechanical (MEM) devices.

## 14.2   OPTICAL FILTER THEORY

In an optical interference filter, the filter response arises from the interference of two or more lightwaves that are delayed relative to each other. The incoming signal is split into multiple paths by a division of the wavefront or amplitude. Diffraction gratings are an example of wavefront division, while directional couplers and partial reflectors are examples of amplitude division. For interference, the optical waves must have the same polarization, the same frequency and be temporally coherent over the longest delay length. When the signals are recombined, their relative phases determine whether they interfere constructively or destructively. Optical interference filters are analogous to digital filters in that they both consist of splitters, weighting coefficients, delay lines and combiners. Optical delay lines implement discrete time delays, and the splitters and combiners are treated as lumped elements. Interference may occur in two ways, feedforward and feedback. Optical and digital filters are easily categorized based on the type of interference. After a brief discussion of optical delays and directional couplers, single-stage optical filters are discussed, followed by a general mathematical description of the magnitude and phase response for multistage filters.

The optical frequency response of a lossless delay line is given by $\exp(-j\beta L)$ where $\beta$ is the propagation constant and $L$ is the delay length. When the delays can be expressed as an integer multiple of a unit delay, denoted by $T$, the filter has a periodic frequency response and is easily described by a $Z$ transform and a discrete impulse response. The period of the frequency response is called the free spectral range (FSR). At the optical radian frequency $\Omega$, the response is expressed by $\exp(-j\Omega T)$, which is periodic in frequency with period FSR $= 1/T$. As the optical frequencies of interest are in the 193-THz range for wavelengths around 1550 nm, it is convenient to describe the frequency response relative to a center optical frequency, $f_0$. For filter design, it is also convenient to normalize the frequency by dividing it by the FSR. The normalized radian frequency is denoted by $\omega = 2\pi(f - f_0)/\text{FSR}$. To normalize in the time domain, the delay is expressed as a multiple of the unit delay. By substituting $z^{-1}$ for $\exp(-j\beta L)$, the $Z$ transform is obtained for a lossless delay with a unit length of $L$. A delay that is $M$ times the unit delay is represented by $z^{-M}$. Loss is neglected to simplify the notation; however, it can easily be included by replacing $z^{-1}$ with $\gamma z^{-1}$, where $20\log_{10}(\gamma)$ is the loss in dB for propagating one unit delay length.

Although there are many other ways to implement optical splitters and combiners, directional couplers are easily fabricated and have a simple mathematical description for their splitting ratios that we will use to calculate the transfer function of multistage optical filters. Directional couplers are created by bringing two waveguides in close proximity as illustrated in Fig. 14.4b. Coupling occurs as a result of the overlap between the evanescent field of one waveguide and the core of the other waveguide. The transfer function relating the input and output electric fields in each arm is described by the $2 \times 2$ matrix in Eq. 14.1 [3].

$$\begin{bmatrix} Y_1 \\ Y_2 \end{bmatrix} = \begin{bmatrix} \cos(\theta) & -j\sin(\theta) \\ -j\sin(\theta) & \cos(\theta) \end{bmatrix} \begin{bmatrix} X_1 \\ X_2 \end{bmatrix} \tag{14.1}$$

where $\theta = \kappa_c L_c$ is the coupling strength $\kappa_c$ times the coupler length $L_c$. Note that the coupler is assumed to be lossless and the matrix is unitary. For simplicity, we assume that the coupling is wavelength independent; however, any wavelength dependence

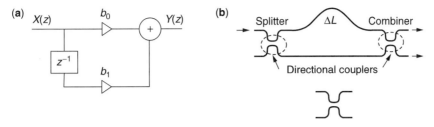

**Figure 14.4**  Single-stage (a) digital and (b) optical FIR filters.

must be included in practice. The abbreviated notation $c = \cos(\theta)$ and $s = \sin(\theta)$ is used below. Partially reflecting mirrors are another example of a splitter described by a $2 \times 2$ matrix. Splitters with a larger number of outputs are formed by multi-mode interference couplers [4–6], star couplers [7–9] and diffraction gratings.

A single-stage optical filter using feedforward interference and its digital filter counterpart are illustrated in Fig. 14.4. The incoming pulse is divided among multiple paths and propagates across each path only once before interfering at the combiner. The $Z$ transform, also called the transfer function, for the digital filter in Fig. 14.4a is $H(z) = b_0 + b_1 z^{-1}$, where the filter coefficients are $b_0$ and $b_1$. The frequency response is obtained by evaluating $H(z)$ at $z = \exp(j\omega)$. The filter has a finite impulse response given by $h(n) = b_0 \delta(n) + b_1 \delta(n-1)$ in this example. Filters with only feedforward interference have a finite impulse response and are called FIR filters.

The optical filter in Fig. 14.4b is commonly known as a Mach–Zehnder interferometer (MZI) [10], where the thick lines indicate an optical waveguide and the regions where the waveguides are brought close together are directional couplers. The MZI is a widely used filter that is easily fabricated in planar waveguides. The Michelson interferometer [11] is also a single-stage FIR filter and can be thought of as an MZI operating in a reflective mode with a mirror placed at the midpoint of the interfering arms. The MZI has two inputs and two outputs, and the transfer function is described by a $2 \times 2$ matrix that is obtained by multiplying the coupler transfer matrices and delay matrix in reverse order of the propagation as follows:

$$\begin{bmatrix} H_{11}(z) & H_{12}(z) \\ H_{21}(z) & H_{22}(z) \end{bmatrix} = \begin{bmatrix} c_1 & -js_1 \\ -js_1 & c_1 \end{bmatrix} \begin{bmatrix} z^{-1} & 0 \\ 0 & 1 \end{bmatrix} \begin{bmatrix} c_0 & -js_0 \\ -js_0 & c_0 \end{bmatrix} \quad (14.2)$$

The common phase due to propagation through the filter is neglected in the above description because it contributes only a constant delay, so $z = \exp(j\beta\Delta L)$ where $\Delta L$ is the path length difference between adjacent interfering paths. Note that an additional common phase term of $\sqrt{z^{-1}}$ could be extracted from the delay matrix to make it unitary. The transfer function for optical FIR filters depends on $\Delta L$ and the coupling coefficients. For example, the cross-port transfer function is $H_{21}(z) = -j(c_0 s_1 z^{-1} + s_0 c_1)$, which is a first-order polynomial in $z^{-1}$. The frequency response of a filter can also be obtained by summing over the contributions from each path through the filter. Constructive interference occurs for positive weighting coefficients when the optical path length difference is equal to an integral number of wavelengths, $n\Delta L = m\lambda$, where $n$ is the refractive index of the medium, or effective index for a waveguide, and $m$ is the diffraction order. The difference between frequencies that constructively interfere for adjacent diffraction orders, such as for order $m$ and $m + 1$, defines the

FSR [3]. For FIR optical filters, the FSR $= c/n_g \Delta L$, where $n_g = n - \lambda \, dn/d\lambda$ is the group index. By introducing a $\pi$-phase shift in the lower interferometer arm, the delay matrix becomes $\begin{bmatrix} z^{-1} & 0 \\ 0 & -1 \end{bmatrix}$ and the transfer function changes to $H_{21}(z) = -j(c_0 s_1 z^{-1} - s_0 c_1)$. A phase shift $\Delta\phi = 2\pi \Delta(nL)/\lambda$ may be realized by changing the index $\Delta n$, for example by the electro-optic or thermo-optic effect, by changing the path length $\Delta L$, or by changing a combination of index and path length $\Delta(nL)$. For a symmetric MZI, the path lengths are nominally identical; i.e., $\Delta L = 0$. An asymmetric MZI has unequal arm lengths, so $\Delta L \neq 0$. When the couplers are 50% splitters so that $s_0^2 = s_1^2 = 0.5$, and a phase shift $\Delta\phi$ is introduced between the arms of a symmetric MZI, $|H_{21}|^2 = \cos^2(\Delta\phi/2)$ and a tunable coupler or switch is realized. Phase shifts are commonly used to implement complex filter coefficients. The delay response for a unit delay path with a phase shift $\phi$ is given by $e^{-j\phi} z^{-1}$.

If a delay is arranged in a feedback topology, an infinite impulse response (IIR) results. A portion of the incoming pulse will traverse the delay infinitely many times in this case. Single-stage digital and optical IIR filter structures are shown in Fig. 14.5. There are two single-stage optical IIR filter implementations, a ring resonator with two directional couplers and a Fabry–Perot filter. The filter output in Fig. 14.5a is described by $y(n) = ay(n-1) + x(n)$, where $a$ is a real number satisfying $0 \leqslant |a| < 1$ and $n \geqslant 0$. Its $Z$ transform is $Y(z) = az^{-1}Y(z) + X(z)$, which yields the transfer function in Eq. 14.3.

$$H(z) = \frac{Y(z)}{X(z)} = \sum_{n=0}^{\infty} a^n z^{-n} = \frac{1}{1 - az^{-1}} \qquad (14.3)$$

The region of convergence is $|z| > a$. For a stable filter, the unit circle defined by $|z| = 1$ must be included in the region of convergence [12]. A ring resonator and a Fabry–Perot etalon [13] filter each have a response that can be expressed in the form of Eq. 14.3. For a passive optical filter, meaning without gain, the magnitude of $a$ will never be larger than one, so the filter will always be stable. In the limit of $a = 1$, the trivial case results where the feedback path is bypassed, so either the partial reflectors are 100% in the etalon case or the cross-port coupling ratio, $\sin\theta$, is zero for the ring resonator.

In general, the $Z$ transform is defined as follows:

$$H(z) = \sum_{n=-\infty}^{\infty} h(n)z^{-n} = \frac{N(z)}{D(z)} \qquad (14.4)$$

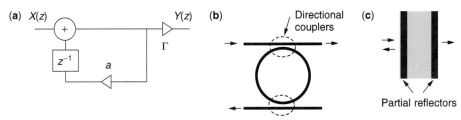

**Figure 14.5**  Single-stage **(a)** digital and optical IIR filters realized using **(b)** a ring resonator with two directional couplers and **(c)** a Fabry-Perot etalon.

where $h(n)$ is the impulse response of a filter, and $z$ is a complex number that may have any magnitude. The Z transform is an analytic extension of the Fourier transform for discrete signals, similar to the relationship between the Laplace transform and the Fourier transform for continuous signals. A delay of one unit results in multiplication by $z^{-1}$. Causality describes the intuitive concept that a filter cannot produce an output before an input signal is applied. Mathematically, a causal filter has $h(n) = 0$ for $n < 0$, so we can modify the lower index on the sum to be $n = 0$ in Eq. 14.4. The transfer function may be expressed as the ratio of two polynomials. The roots of the numerator polynomial $N(z)$ are called "zeros" because they produce nulls in the magnitude response. Multi-stage FIR filters have higher order $N(z)$ polynomials and multiple zeros. The roots of the denominator polynomial $D(z)$ are called "poles," which create peaks in the magnitude response.

A ring resonator with two couplers, as shown in Fig. 14.5, is the simplest optical waveguide filter with a single-pole response. The FSR $= c/n_g L$ where $L$ is the total feedback path length, in contrast to the feedforward filters where only the path length difference between the interferometer arms is critical. The individual contributions from the feedback path form an infinite sum. The cross-port transmission is an all-pole response given by [3]

$$H_{12}(z) = \frac{Y_2(z)}{X_1(z)} = \frac{-s_1 s_2 \sqrt{z^{-1}}}{1 - c_1 c_2 z^{-1}} \tag{14.5}$$

where the pole magnitude is $c_1 c_2$. The bar-port response has both a pole and zero as shown in Eq. 14.6.

$$H_{11}(z) = \frac{Y_1(z)}{X_1(z)} = \left[ \frac{c_1 - c_2 z^{-1}}{1 - c_1 c_2 z^{-1}} \right] \tag{14.6}$$

The pole is identical for both the bar- and cross-ports. For coupled cavities [14] or rings [15, 16], a $2 \times 2$ transfer matrix description is easily applied to describe the response.

A filter's magnitude response is equal to the modulus of its transfer function, $|H(z)|$, evaluated at $z = e^{j\omega}$. When the poles and zeros of $H(z)$ are plotted on the z-plane, only the distance of each pole and zero from the unit circle, i.e., $|e^{j\omega} - z_m|$ or $|e^{j\omega} - p_n|$, affects the magnitude response. Consequently, a zero that is located at the mirror image position about the unit circle, i.e., $1/z_m^*$, cannot be differentiated from $z_m$ based on the magnitude response. Because there are two zero locations that yield the same magnitude response, a naming convention is used to distinguish them. Zeros with a magnitude greater than one are called maximum-phase, and those with magnitudes less than one are called minimum-phase. This nomenclature is illustrated in the pole/zero plot on the z-plane shown in Fig. 14.6**a**. The square magnitude response is easily calculated from the poles and zeros using Eq. 14.7 [17].

$$|H(\omega)|^2 = \frac{|\Gamma|^2 \prod_{i=1}^{M} \{1 - 2r_{zi} \cos(\omega - \phi_{zi}) + r_{zi}^2\}}{\prod_{i=1}^{N} \{1 - 2r_{pi} \cos(\omega - \phi_{pi}) + r_{pi}^2\}} \tag{14.7}$$

where the poles are represented by $p_i = r_{pi} \exp(j\phi_{pi})$ and the zeros by $z_i = r_{zi} \exp(j\phi_{zi})$. A gain factor $\Gamma$ is also included for generality. In practice, $|H(\omega)|$ is

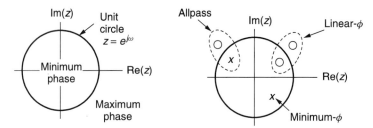

**Figure 14.6** (a) A $z$-plane plot showing the unit circle and regions associated with minimum- and maximum-phase filters. (b) The relative locations for poles and zeros corresponding to allpass, linear-phase and minimum-phase filters.

less than one for optical filters without gain. The filter's group delay is defined as the negative derivative of the phase response with respect to the angular frequency as follows [12]:

$$\tau_n(\omega) = -\frac{d}{d\omega} \tan^{-1} \left[ \frac{\text{Im}\{H(z)\}}{\text{Re}\{H(z)\}} \right]_{z=e^{j\omega}} \tag{14.8}$$

where $\tau_n$ is normalized to the unit delay $T$. The real and imaginary parts of $H(z)$ are denoted by Re{} and Im{}, respectively. The absolute group delay is given by $\tau_g = T\tau_n$. The delay is defined as the slope of the phase response at the frequency where it is being evaluated. For example, a dispersionless, lossless delay line of unit length has a frequency response $H(\omega) = e^{-j\omega}$, a normalized group delay of 1, and an absolute group delay equal to $T$. The group delay for $N$ poles and $M$ zeros can be calculated using Eq. 14.9 [17].

$$\tau_n(\omega) = \sum_{i=1}^{N} \frac{r_{pi}\{\cos(\omega - \phi_{pi}) - r_{pi}\}}{\{1 - 2r_{pi}\cos(\omega - \phi_{pi}) + r_{pi}^2\}} + \sum_{i=1}^{M} \frac{r_{zi}\{r_{zi} - \cos(\omega - \phi_{zi})\}}{\{1 - 2r_{zi}\cos(\omega - \phi_{zi}) + r_{zi}^2\}} \tag{14.9}$$

A plot of the magnitude, phase and group delay response for several pole and zero magnitudes is provided in Fig. 14.7 assuming $\phi_{zi} = \phi_{pi} = 0$. For the zeros, root pairs located reciprocally about the unit circle are calculated for magnitudes $\rho = 0.9$ and 1.1 and $\rho = 0.5$ and 2.0. The filter dispersion, in normalized units, is defined by $D_n \equiv 2\pi \frac{d\tau_n}{d\omega}$. If the phase is linear in $\omega$, then the delay is constant and the dispersion is zero. The filter dispersion is related to the normalized dispersion by $D = -c\left(\frac{T}{\lambda}\right)^2 D_n$. Large filter dispersions are achieved by increasing $T$; however, the FSR decreases accordingly.

A filter's phase response behaves quite differently than an optical fiber's phase response $\phi(\Omega) = -\beta(\Omega)L$, which varies much more slowly with frequency. The fiber's propagation constant $\beta(\Omega)$ is typically expanded in a Taylor series [18]:

$$\beta(\Omega) = \beta(\Omega_c + \Delta\Omega) = \beta(\Omega_c) + \beta'\Delta\Omega + \frac{1}{2!}\beta''\Delta\Omega^2 + \frac{1}{3!}\beta'''\Delta\Omega^3 + \cdots \tag{14.10}$$

where $\Omega_c$ is the center optical frequency, $\Delta\Omega = \Omega - \Omega_c$, and the primes indicate derivatives of $\beta$ with respect to $\Omega$. For optical fibers, dispersion is defined as the

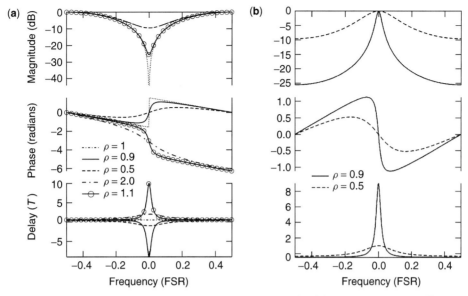

**Figure 14.7**  Magnitude, phase and group delay response for **(a)** a single-zero and **(b)** a single-pole filter with various root magnitudes.

derivative of the group delay with respect to wavelength and normalized with respect to length, $D = \dfrac{1}{L}\dfrac{d\tau_g}{d\lambda} = -\dfrac{1}{L}\dfrac{2\pi c}{\lambda^2}\beta''$, so the units are ps/nm/km. Chromatic dispersion is associated with $\beta''$ in the Taylor series expansion, cubic dispersion with $\beta'''$, and so forth for the nomenclature of higher order dispersion. The units for cumulative path dispersion and filter dispersion are ps/nm. Standard single-mode fiber (SSMF), i.e., nondispersion shifted, has a dispersion of $+17$ ps/nm/km and a dispersion slope of $+0.08$ ps/nm²/km. For a filter with $T = 10$ ps (FSR = 100 GHz) and $\lambda = 1550$ nm, $D = -12.5 D_n$ (ps/nm). For $D_n = -1$, $D$ is equal to 0.7 km of SSMF; however, for a filter with a unit delay of $T = 100$ ps, the corresponding unit dispersion is equal to 70 km of SSMF!

A unique property of FIR filters is that they can be designed to have a linear-phase response. If the zeros lie on the unit circle or occur in pairs located at reciprocal points about the unit circle as indicated by the circled pair in Fig. 14.6**b** labeled linear-$\phi$, then the group delay is independent of frequency. The cancellation of the frequency-dependent group delay by the proper choice of two zeros can be verified by the group delay responses shown for various zero magnitudes in Fig. 14.7. Filters with linear-phase and complex coefficients have an impulse response with Hermitian symmetry [3]. For real coefficients, the impulse response has even or odd symmetry, $h(n) = \pm h(N - n)$.

If all the roots lie within the unit circle, the filter is minimum-phase. A physically realizable all-pole filter is minimum-phase. The magnitude and phase response of a minimum-phase, causal filter form a Hilbert transform pair [19]. Knowledge of one response completely specifies the other. Causality and the minimum-phase condition for continuous-time filters leads to the Kramers–Kronig equations, which relate the real and imaginary parts of the refractive index of a medium [20]. For minimum-phase systems, the phase response is proportional to the change in the magnitude response [21]. This behavior does not describe a mixed-phase system. In particular, linear-phase filters can have very sharp transitions in the magnitude response, although many stages are required, without introducing dispersion [1]. The cross-port response for the ring resonator described by Eq. 14.5 is minimum-phase, but the bar-port response is not, in general. This same situation is found for Fabry–Perot etalons, in which case the transmission response is minimum-phase with a single-pole while the reflection response contains both a pole and zero. Similarly, the transmission response of a Bragg grating is minimum-phase while the reflection response is not. A more detailed discussion of filter classes and their magnitude and phase responses is given in [3].

An important mixed-phase system is an allpass filter, where the amplitude response is constant and the phase response is nonlinear. The allpass filter response consists of pole-zero pairs located at reciprocal points about the unit circle, as indicated in Fig. 14.6b, so that the resulting magnitude response is independent of frequency. The frequency response of an allpass filter is given by $e^{j\Phi(\omega)}$. It has a unity magnitude response, and its phase response can be tailored to approximate any desired response using multiple stages. A single-stage allpass filter has a $Z$ transform expressed as follows, neglecting a common phase factor [3]:

$$H(z) = \frac{-\rho e^{+j\phi} + z^{-1}}{1 - \rho e^{-j\phi} z^{-1}} \tag{14.11}$$

where $\phi$ is a phase shift introduced in the feedback path to change the cavity resonant frequency. There are two optical allpass filter realizations, a ring resonator with a single input and output port and an etalon-based filter known as a Gires–Tournois interferometer (GTI) [22], as shown in Fig. 14.8a and 14.8b, respectively. The back facet of the GTI is a mirror and the front facet is partially reflecting with amplitude reflectance $\rho$. For the ring resonator with a directional coupler having a coupling strength $\theta$, the pole is given by $\rho = \cos\theta$. The phase and group delay responses over two FSRs are shown in Fig. 14.8c and 14.8d, respectively. For $\rho = 0$, the filter is simply a delay line and has linear phase. As $\rho$ approaches unity, the $2\pi$ phase change per period for a single-stage allpass filter is concentrated near the resonance frequency, creating a large but very narrowband delay. Multi-stage allpass filters can be realized by cascading single stages or coupling stages in a lattice architecture. For a higher order allpass filter, the numerator coefficients can be determined directly from the denominator by reversing the order of the coefficients and taking their complex conjugate. When the feedback path is lossy, the magnitude response is dominated by the transfer function's zero, which is given by the ratio $\gamma/\rho$. The magnitude response is calculated for a roundtrip loss of 0.5 dB ($\gamma = 0.94$) and various values of $\rho$ in Fig. 14.8e. For $\rho = 0.9$, the zero is close to the unit circle and a deep null results.

Allpass filters also serve as building blocks for higher order IIR bandpass filters. Jinguji [23] demonstrated that the lattice architecture in Fig. 14.9a is a general multistage IIR filter. The pole and zero locations can be chosen arbitrarily by setting two coupling ratios and two phases for each stage. Although this architecture is complex because of the number of parameters and coupling between them, a related architecture that has been demonstrated in planar waveguides is a single-stage MZI with a single ring in one arm having a feedback path length of two unit delays and a unit delay in the other

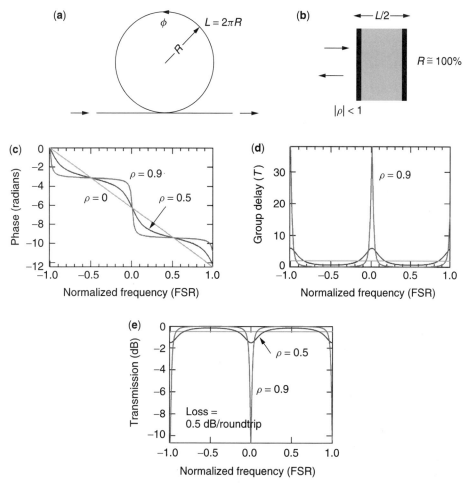

**Figure 14.8** Single-stage optical allpass filters with a feedback path length $L$ implemented with (**a**) a ring-resonator and (**b**) a GTI. The (**c**) phase, (**d**) delay, and (**e**) magnitude response calculated for $\rho = \{0, 0.5, 0.9\}$ over two FSRs.

arm [24, 25] as shown in Fig. 14.9**b**. This architecture exactly realizes a third-order Butterworth filter with the proper choice of the filter parameters. It is worthwhile to adopt a generalized view of this architecture as a sum and difference of two allpass filters [26]. From digital filter theory, it is well known that optimum bandpass filters, such as elliptic, Butterworth, and Chebyshev filters, can be realized as the sum or difference of two allpass functions [27]. A simple optical implementation proposed by Madsen [26] is shown in Fig. 14.9**c** using a MZI with multi-stage allpass filters in each arm. Such allpass filter decomposition architectures may be realized using ring resonator- or etalon-based multi-stage allpass filters.

Although digital filter theory provides a unifying description for optical interference filters, several factors distinguish the practical implementation of optical filters from their digital filter counterparts. First, optical filters have loss. If the filters are too lossy, gain must be introduced, which is expensive relative to the cost of passive optical filters and adds noise to the system. Polarization dependence must be considered as

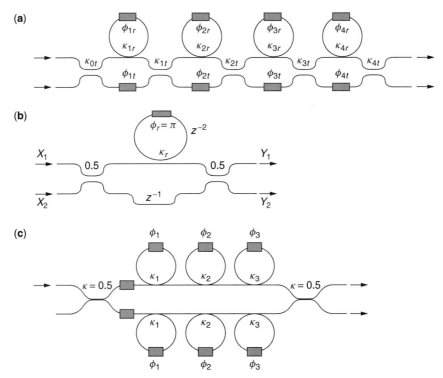

**Figure 14.9** (a) General architecture for realizing an arbitrary pole-zero response [23] © 1996 IEEE, (b) a third-order Butterworth filter [24] © 1988 IEEE, and (c) an architecture employing allpass filter decomposition for realizing optimal bandpass filters [26] © 1998 IEEE.

well as temperature and wavelength dependence of the filter coefficients. The analog realization of optical filter coefficients is limited to a precision on the order of 0.01 compared to the superior precision obtained with floating point arithmetic and digital computers. In practice, the sensitivity of the filter response to variations in the coefficients must be satisfactory from a yield and performance perspective. On the positive side, complex coefficients are easily implemented by varying the optical path length over one wavelength. However, phase errors are introduced if the path lengths are not precise to within a fraction of a wavelength.

## 14.3 FINITE IMPULSE RESPONSE OPTICAL FILTERS

Multi-stage filters are required for filter synthesis to closely approximate a desired response. In a similar fashion to the transversal, cascade, and lattice digital filter architectures, several optical filter architectures have been investigated. Transversal optical planar waveguide filters [28–30] are analogous to the direct form architectures for digital filters. A transversal filter has been used to demonstrate a frequency selector that can add or drop one or more channels [30]; however, it does not scale well to a large number of channels. Coupled Mach–Zehnder interferometers, Solc birefringent filters [31], and codirectional coupled-mode filters are optical FIR lattice filters. For FIR filters, there are several phased-array architectures including diffraction gratings, waveguide grating routers (WGR), and the virtual imaged phased array (VIPA) that inherently scale easily to large numbers of channels. The lattice and phased array architectures have received the most attention and are discussed in more detail below.

### 14.3.1 Lattice Filters

A multi-stage FIR filter can be realized by coupling MZIs in a lattice architecture as shown in Fig. 14.10 [2]. The major advantage of the lattice architecture is that a very low loss passband can be achieved [32]. If each stage has a unit delay or an integer multiple thereof, the transmission is defined by a Fourier series [33]. Applications include bandpass filtering [34], gain equalization [33], and dispersion compensation [35]. Filter synthesis is accomplished for an N-stage filter by finding the coupling ratios $\{\theta_0, \ldots, \theta_N\}$ and phases $\{\phi_1, \ldots, \phi_N\}$ that give a best fit to the desired function using a nonlinear optimization algorithm. The output is the sum of each optical path [33] as illustrated in Fig. 14.10 for the cross-port transmission with all $\phi_n = 0$ for simplicity. Alternatively, $2 \times 2$ transfer matrices and recursion relations to translate between the filter coefficients and coupling ratios, derived by Jinguji and Kawachi [32], may be used. In practice, tunable couplers are implemented instead of directional couplers so that the tap coefficients can be measured and accurately set to their desired values after fabrication. The outputs are power complementary for a lossless filter, so $|t_x(\omega)|^2 + |t_-(\omega)|^2 = 1$. For an $N$th-order lattice filter with identical stages, it can be shown that $|t_x|^2 = \sin^2(N\theta)$ at the maximum transmission frequency and full coupling to the cross-port is achieved if $N\theta = (2m + 1)\dfrac{\pi}{2}$ [3].

Cascaded MZI lattice filters are easily realized in planar waveguides. The path lengths are defined by photolithography and the device is interferometrically stable. The cross section of a planar waveguide is illustrated in Fig. 14.11. A variety of dielectric and semiconductor materials have been used to demonstrate planar waveguide filters; however, silica waveguides provide the most mature platform in terms of manufacturability, low loss, large wafer sizes, and thermo-optic tuning. For silica

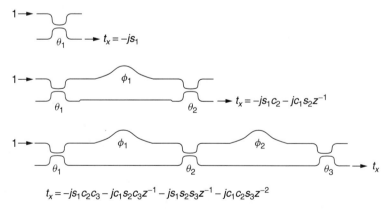

$$t_x = -js_1c_2c_3 - jc_1s_2c_3z^{-1} - js_1s_2s_3z^{-1} - jc_1c_2s_3z^{-2}$$

**Figure 14.10** The transfer function determined from the sum of all optical paths for FIR lattice filters of order 0, 1 and 2 is illustrated for the cross-port and $\phi_n = 0$. The response of a zeroth-order filter is frequency-independent except for any wavelength dependence of the coupler.

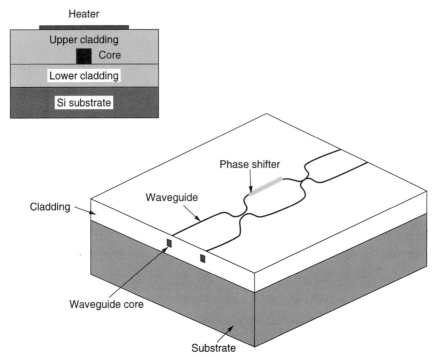

**Figure 14.11**   Top-view of a planar waveguide symmetric Mach–Zehnder interferometer with a thermo-optic phase shifter, and a cross-section through a single waveguide region.

waveguides, a thick lower cladding of silica is deposited on a silicon substrate, followed by the deposition of the core layer which is typically silica doped with either phosphorous or germanium. The core is patterned by exposure of a photoresist through a chrome mask containing the waveguide pattern and etched using reactive ion etching to produce vertical sidewalls. After removing the remaining photoresist, an upper cladding of silica is deposited over the core. Thermo-optic phase shifters are created by depositing a layer of chromium on top of the upper cladding and patterning it to create local heaters [36]. Gold electrode contacts are then fabricated in a similar fashion. The thermo-optic response of silica is dominated by the temperature dependence of the refractive index, which is $dn/dT = 1.1 \times 10^{-5}/^{\circ}\text{C}$. Because of the difference in thermal expansion coefficients between the silicon substrate and silica waveguide and elevated processing temperatures, a stress-induced birefringence on the order of $\Delta n = 10^{-4}$ is typically present. Several techniques to mitigate the birefringence have been explored including inserting waveplates to flip the polarization state within the device [37], changing the glass doping to reduce the stress [38], introducing stress relieving grooves [39], and varying the waveguide width [40].

A six-stage gain-equalizing lattice filter was fabricated using phosphorous-doped silica planar waveguides [41]. The shape of the square magnitude response, shown in Fig. 14.12, was optimized over a 30-nm bandwidth for a 2000-km system containing 40-km spans. Gain equalizers [42] as well as optical add-drop filters [43] have also been realized in silicon oxynitride. In the latter case, one channel out of eight was dropped using an 11-stage filter with 3.2-nm channel spacing.

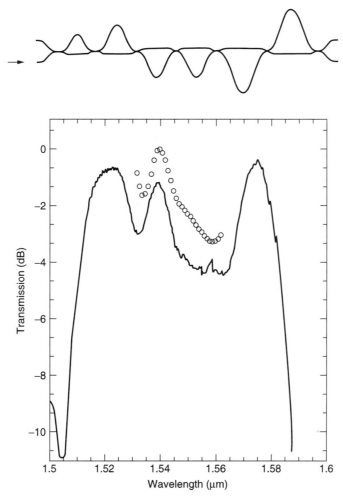

**Figure 14.12**   The waveguide layout for a six-stage, gain equalizing FIR lattice filter and the measured filter response (line) compared to the desired response (circles) [41].

Half-band filters are a special type of lattice filter [44] that find application as interleavers in WDM systems [45]. Interleavers combine two incoming signals with a channel spacing of $2\Delta f$ into a single output signal with a channel spacing of $\Delta f$. De-interleavers operate in the reverse mode. Although the gain equalizer and add-drop filters discussed previously have relatively large FSRs covering tens of channels, an interleaver has an FSR on the order of the channel spacing. As filter dispersion scales inversely to the square of the FSR, more attention must be paid to the filter's dispersion response for smaller FSRs. For lattice filters, it is straightforward to design one of the power complementary responses to be linear-phase; however, requiring both responses to be linear-phase is very constraining. A more practical solution is to cascade two identical filters and interconnect the two different bar-ports (and cross-ports) [45]. The dispersion accumulated in the first filter stage is then cancelled in the second stage because the zeros of the interconnected ports for the first and second stages are mirror images about the unit circle.

For dispersion compensation, a linear group delay response can be approximated using the nonlinear phase response of the zeros. Experimental results were first reported by Takiguchi et al. [46] using 12 stages to compensate the dispersion of 20 km of standard singlemode fiber over a 22-GHz range centered at 1550 nm. A tunable dispersion compensator was demonstrated [47] using 8 stages with tunability from −681 to +786 ps/nm and a 16.3 GHz passband with <1-dB variation over the passband. A dispersion slope equalizer for 40-Gb/s signals has also been reported that is capable of compensating a ±350-ps/nm range over a band of channels [48]. Besides chromatic dispersion compensators, lattice filters combined with tunable couplers, polarization beam splitters, and half wave plates for polarization rotation provide the compensating filter building blocks for integrated polarization mode dispersion compensators [49].

Birefringent plates can also be used to create lattice filters. They are described by $2 \times 2$ unitary Jones matrices that are similar to the $2 \times 2$ transmission matrices for lattice filters used in Eq. (14.2). By cascading birefringent plates between an input and output polarizer, a Solc filter is realized. The angle of each birefringent plate is varied to produce different filter coefficients. The differential delay is proportional to the difference in the refractive index of the ordinary and extraordinary axis. Birefringent filters have been proposed [35] and demonstrated for dispersion compensation applications [50]. Six TiO$_2$ crystals with $n_e = 2.709$ and $n_o = 2.451$ were used to produce a dispersion of 150 ps/nm over a 48.5-GHz bandwidth. Birefringent plates mounted in rotating fixtures have been used to emulate the polarization mode dispersion in optical fibers [51]. Liquid crystals have also been employed as filter elements. A birefringent filter consisting of liquid crystal attenuators and wavelength shifting sections has been used to demonstrate a tunable gain equalizer [52].

Co-directional, coupled-mode filters are a continuous-coupling analog to the FIR filters described so far. Coupling between modes with different propagation constants is assisted by a periodic index perturbation. The difference in propagation constants is matched by the grating spatial frequency, $\Delta \beta = \dfrac{2\pi \Delta n_e}{\lambda_c} = \dfrac{2\pi}{\Lambda}$, where $\Delta n_e$ is the difference in effective indices between the modes at the center wavelength $\lambda_c = \Delta n_e \Lambda$ and $\Lambda$ is the grating period. The coupling strength $\kappa$ depends on the magnitude of the index perturbation and the overlap of the perturbed region with the interacting modes. Analytic solutions exist for uniform coupling using coupled-mode theory [53]. The wavelength dependence is defined by the offset from the center wavelength as $\delta \equiv \pi \Delta n_e \left( \dfrac{1}{\lambda} - \dfrac{1}{\lambda_c} \right)$. The coupling from one mode to the other is $|t_x(\delta)|^2 = \dfrac{\kappa^2}{\gamma^2} \sin^2(\gamma L)$ where $\gamma = \sqrt{\kappa^2 + \delta^2}$. The peak cross-transmission is $|t_x|^2 = \sin^2(\kappa L)$, which is equivalent to the result for the $N$-stage uniform lattice filter with $N\theta$ replaced by $\kappa L$. The bandwidth, defined as the full width between the first zeros, for weak coupling is given by $2/N$, where $N$ is the number of periods in the grating.

Acousto-optic filters use surface acoustic waves (SAWs) in $x$-cut $y$-propagating lithium niobate to create a periodic index grating that couples the two polarizations of the fundamental mode. For a birefringence of $\Delta n = 0.08$ at 1550 nm, the period of the index change is $\Lambda = 20$ μm and the acoustic frequency is 175 MHz [54]. The passband width is limited by the number of periods, so there is a tradeoff between making the device longer to achieve a narrower passband and reducing the device length to obtain faster switching times. For $L = 20$ mm and $\lambda = 1550$ nm, the passband width is $\Delta\lambda = 1.5$ nm and the switching time is $\tau = 6$ ms.

Long period fiber gratings couple light from the fundamental mode to one of the cladding modes, thereby producing a wavelength-dependent loss for the fundamental mode [55]. Typical periods range from 200 to 600 μm. Gratings with different periods are concatenated so that coupling to multiple cladding modes may be used to approximate a desired gain equalizing filter response [56]. A gain response with <1-dB ripple was achieved for an erbium doped fiber amplifier with 22-dB gain over a 40-nm bandwidth using long period grating filters [57]. Six long period grating (LPG) filters were used to equalize the gain in a two-band architecture with a total bandwidth of over 80 nm [58]. In addition, dynamic gain equalizers have been realized by creating an acousto-optically induced long period grating in the fiber with multiple frequencies [59].

### 14.3.2  Phased Arrays

Phased array optical filters implement many stages in parallel and are inherently suited for filtering numerous channels simultaneously. Three implementations are discussed: diffraction gratings, the virtually imaged phased array, and waveguide grating routers.

### 14.3.3  Diffraction Gratings

A transmission grating, shown in Fig. 14.13**a**, is a multi-slit interferometer. To simplify the discussion, let the slits be separated by a distance $d$, the surrounding index be $n_s$, and the source be located at infinity so that the incoming beam is collimated. In the far field of the grating, we sum the diffracted fields from each slit. An interference pattern

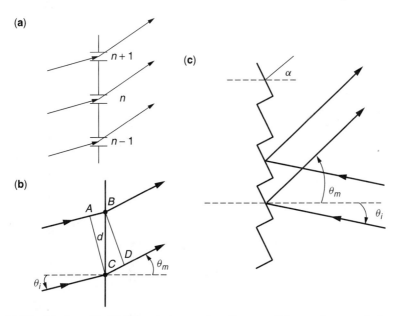

**Figure 14.13**  (**a**) A multi-slit transmission grating diagram, (**b**) an enlargement showing the path length differences between adjacent paths, and (**c**) a reflection grating schematic.

results because the rays from each slit experience different optical path lengths. The angles of incidence and transmission are denoted by $\theta_i$ and $\theta_m$, respectively, measured counter-clockwise from the normal to the grating. The path length difference between adjacent slits is $\Delta L = CD - AB = d[\sin\theta_m - \sin\theta_i]$, as illustrated in Fig. 14.13**b**. At wavelengths where $n_s \Delta L$ is a multiple of $\lambda$, constructive interference occurs. This condition is described by the grating equation

$$m\lambda = n_s d(\sin\theta_m - \sin\theta_i) = n_s \Delta L \qquad (14.12)$$

where $m$ is the diffraction order. Typical diffraction gratings are operated with $m = \pm 1$; however, echelle gratings have courser groove spacings and operate in much larger orders. For WDM filter applications, reflection gratings are typically used. The shape of the grating groove and the blaze angle, $\alpha$, are chosen to produce peak energy in the desired diffraction order. The grating can be folded about its centerline and analyzed as a transmission grating with the angles defined as before. The grating acts like a prism because different wavelengths have different output angles. Unlike a prism, however, there may be multiple diffraction orders for a given wavelength. For uniform illumination, the field just after the grating is

$$E_g(x) = \sum_{n=0}^{N-1} u(x - nd)e^{-jn\beta(n_s d \sin\theta_i)} \qquad (14.13)$$

where $u(x)$ is the aperture function for one slit or reflecting facet. Then, the far field is given by

$$E_f(\theta_m) = U^{env}(\theta_i, \theta_m) \sum_{n=0}^{N-1} e^{-jn\beta n_s d[\sin\theta_i - \sin\theta_m]} = U^{env}(\theta_i, \theta_m)U^{array}(\theta_m, \theta_i) \quad (14.14)$$

There are two contributions to the far field, a slowly varying envelope equal to the diffraction pattern of a single slit or groove and an array factor that depends on the input and output angles as well as the slit spacing and number of slits. The array factor, $U^{array}(\theta_m) = \dfrac{\sin[\pi N\theta_m/\Delta\theta_p]}{\sin[\pi\theta_m/\Delta\theta_p]}$, is periodic with a period $\Delta\theta_p = \dfrac{\lambda}{n_s d}$, assuming a small angle approximation for $\sin(\theta_m) \approx \theta_m$. This angular periodicity results from discretely sampling the incoming field. The width of the main lobe decreases as the number of grooves illuminated increases. The envelope factor determines the grating efficiency and may be quite complicated in practice, depending on the incident and diffracted angles, polarization of the incoming signal, the ratio $\lambda/d$, and the groove shape and material [60].

Diffraction gratings are a fundamental component of optical spectrum analyzers with 0.1-nm resolutions in the near-infrared wavelength region (1000–1700 nm). They are typically combined with other optical components to collimate the incident wave and focus the diffracted waves. To avoid the need for collimating optics, both the dispersion and focusing functions can be performed with a concave grating. An important configuration is the Rowland circle whereby a concave grating forms an arc with a radius of curvature $R$ [61]. The source and detector are placed along a circle of radius $R/2$, called the Rowland circle, which intersects the grating arc at its midpoint [62]. Etched concave gratings in planar waveguide slabs have been demonstrated

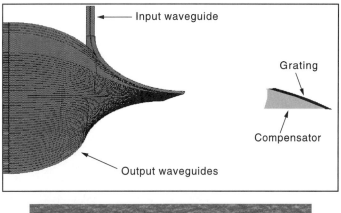

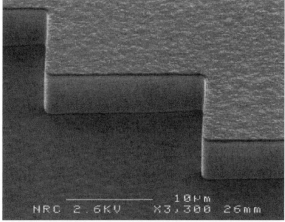

**Figure 14.14**   Top-view of an echelle grating fabricated using planar waveguides and a micrograph of the grating region [65] © 2002 IEEE.

both with the Rowland configuration [63] and other arrangements to obtain a straight focal line [64], which is convenient for attaching a linear fiber or detector array. An echelle grating demultiplexer designed for 40 channels on 100-GHz spacing is shown in Fig. 14.14 [65]. The device can be quite compact and easily scaled to 256 output ports. Note that the slab confines the incident and diffracted waves in the vertical direction. The cross-talk loss may be limited by imperfections in the reflective facets for an etched grating device [66]. Improvements in echelle fabrication have led to vertical sidewalls deviating less than a degree from normal and reduced surface roughness to 50 nm, as illustrated in the micrograph in Fig. 14.14 [65]. Cross-talk of $-35$ dB has been achieved and the polarization wavelength shift reduced to 0.01 nm using a thinned upper cladding region similar to [67]. The Littrow configuration, where $\theta_i \approx \theta_m$, has been used to demonstrate a free-space optics $91 \times 91$ router with 0.33-nm channel spacing [68]. For application in systems with high spectral efficiency, passband broadening techniques employing a double-diffraction-grating architecture and Fabry–Perot etalon have been proposed to increase the spectral efficiency to over 90% [69].

The diffraction grating provides a demultiplexing and multiplexing building block between which an array of tunable elements can be inserted to create a tunable filter

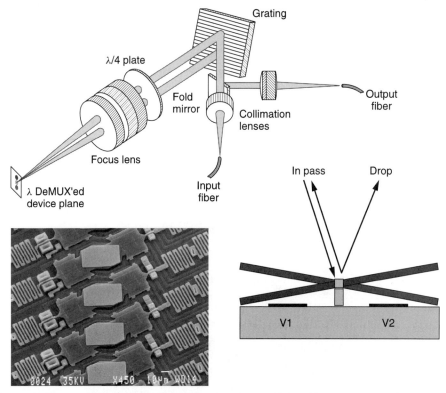

**Figure 14.15**  A 1 × 2 dynamic wavelength add-drop filter with an array of tilting MEMs mirrors located at the demultiplexed plane, a micrograph of the mirror array, and the two states of the tilting mirror to either pass or drop a given channel [70] © 1999 IEEE.

that operates on all channels in parallel. Examples of tuning element arrays are micro-electro-mechanically (MEMs) actuated mirrors [70] and liquid crystals [71]. A 1 × 2 wavelength add-drop filter is shown in Fig. 14.15. Each wavelength is either dropped to an output fiber or returned to the input fiber where it is separated from the incoming signal by a circulator. The mirror tilt is voltage-controlled as shown schematically in Fig. 14.15. Building on the concepts illustrated in the 1 × 2 add-drop filter, an even more powerful architecture is a 1 × K wavelength-selective switch (WSS) capable of switching any wavelength on the incoming port to any of K outputs. A 1 × 4 WSS has been demonstrated for 50-GHz channel spacing that can redirect any of 128 incoming wavelengths on a fiber to any of 4 output fibers [72], as illustrated in Fig. 14.16. The device consists of a section performing spatial-to-angular separation and a frequency-dependent section for spatially separating the channels and controlling their return path with an array of MEMs micromirrors. The tilt-angle for each mirror is voltage-controlled. A micrograph of a single mirror is shown in Fig. 14.16. The large fill factor of the mirror array produces flat-top passbands as shown in the spectral response. An insertion loss of less than 5 dB was achieved. Dynamic gain equalizers have also been demonstrated [73] using interferometric, membrane-MEMs elements [74]. A cavity is formed between the membrane and substrate with a nominal gap of 3λ/4 as shown on the upper right-hand side of Fig. 14.17. By applying a voltage and reducing the gap by

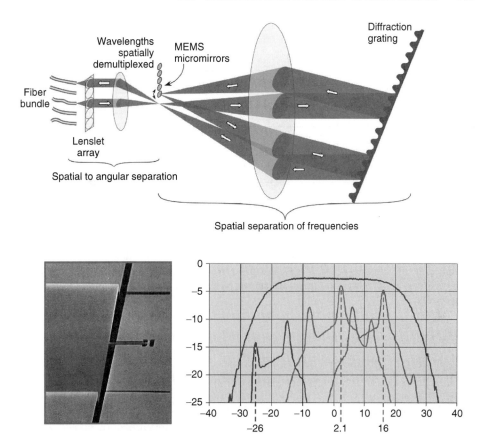

**Figure 14.16** A 1 × 4 wavelength-selective switch schematic, micrograph of a single mirror with a large fill factor, and the spectral response for one channel. Overlaid with the filter response is the spectrum of a 10 Gb/s return-to-zero signal at three different offsets relative to the filter's center frequency [72] © 2002 IEEE.

a quarter-wavelength via the induced electrostatic force, the reflection is varied from maximum to minimum. After passing through a diffraction grating to spatially disperse the incoming light, an array of electrodes controls the induced loss across the spectrum. The resulting filter has a 20-dB dynamic range and a 10-μs response time [74]. The spectral responses before and after equalization of an erbium doped fiber broadband source are shown in Fig. 14.17 along with an enlargement of the remaining ripple on a 0.5-dB scale.

In-fiber diffraction grating devices have also been demonstrated. A chirped grating can angularly disperse light and focus it [75]. A single-mode waveguide with a tilted grating will radiate light over a broad spectrum out of the fundamental mode [76]. By introducing a negative chirp, the radiated light may be focused onto a detector array with the longer wavelengths diffracting at a steeper angle than the shorter wavelengths. A spectrum analyzer using a tilted and chirped UV-induced grating has been demonstrated in fiber [75] and planar waveguides [77]. A resolution of 0.12 nm and a bandwidth of 14 nm have been achieved [75]. The device can be used as an optical channel monitor to provide feedback to dynamic gain equalizing filters.

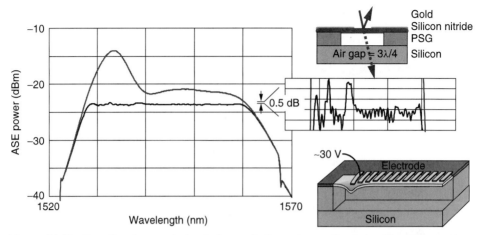

**Figure 14.17**  Broadband transmission before and after gain equalization with a filter employing an array of interferometric MEMs mirrors for dynamic control [73] © 1998 IEEE. The device cross-section and ribbon array are shown on the top and bottom of the right-hand side, respectively.

### 14.3.4   Virtually Imaged Phased Array

The virtually imaged phased array (VIPA) was invented by Shirasaki [78]. As shown in Fig. 14.18, the diffractive element is an etalon coated on the backside with a reflective mirror and on the front facet with a partially reflective coating. The front facet reflectivity is typically higher than 95%. The incoming signal is collimated, focused along a line, and enters the etalon at an angle through an anti-reflection coated window.

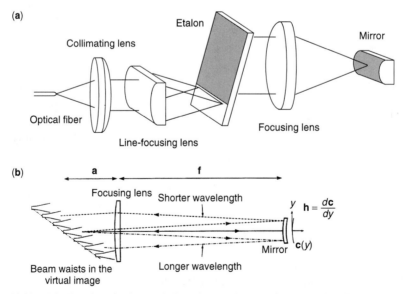

**Figure 14.18**   (a) The virtually imaged phased array in a configuration for dispersion compensation, and (b) a detailed diagram showing the virtual images, the mirror curvature, and paths for extreme wavelengths within one FSR [80] © 2001 IEEE.

Part of the light reflects from the front surface and travels within the etalon creating a series of outputs that form a discrete phased array. The distance traveled between adjacent virtual images is inversely proportional to the FSR. The resulting angular dispersion is converted to a lateral offset at the focusing lens's focal plane. By placing a fiber array at the output, a demultiplexer is formed. If a mirror is used instead, the light reflects back towards the etalon. A curved mirror can be used to make the longer wavelengths travel a shorter distance through the etalon on the return path than shorter wavelengths producing negative dispersion [79].

The dispersion depends on the mirror shape. For constant dispersion, a mirror is required with a curvature of $c(y) = \dfrac{K}{8f^4}y^4 + \dfrac{K\Theta}{2f^3}y^3 + \dfrac{K\Theta^2 - (f-a)}{2f^3}y^2$, where $f$ is the focal length of the focusing lens, $a$ is the distance along the focusing lens axis to the VIPA, and $\Theta$ is the tilt of the etalon as illustrated in Fig. 14.18**b** [80]. The dispersion is given by $D = \dfrac{-2n^4K}{c\lambda}$ where $n$ is the etalon refractive index and $c$ is the velocity of light in vacuum. By designing the mirror curvature to change continuously along the x-axis (out of the paper in Fig. 14.18**b**), tunable dispersion is achieved by moving the mirror in the x-direction. A tradeoff exists between dispersion and bandwidth [81]. A nonuniform design of the partially reflecting mirror can be used to optimize the response for bandwidth and insertion loss. Dispersion slope compensation may be accomplished by introducing a diffraction grating between the etalon and focusing lens to diffract different wavelengths to different x-positions along the mirror [80]. A VIPA tunable dispersion compensator has been used to compensate sixteen, 10-Gb/s channels over 480 km of fiber [82]. In addition, a tuning range of $\pm 800$ ps/nm was demonstrated for 40-Gb/s channels with a non-return-to-zero (NRZ) modulation format [83].

### 14.3.5   Waveguide Grating Routers

An array of waveguides performs the function of a diffraction grating and can easily be integrated using planar waveguides. A waveguide grating was first proposed by Smit [84], and a device with nanometer resolution in the long wavelength window was demonstrated first by Takahashi [85] using free-space optics to couple to the waveguide array. Integration of the waveguide grating with slab couplers to form an $N \times N$ device was proposed by Dragone [86]. In this configuration, a multiplexer/demultiplexer that is easily scaled to a large number of ports is achieved with narrow passbands, low loss, and good cross-talk suppression. The WGR is also referred to as an arrayed waveguide grating (AWG) and a phased array (PHASAR).

The WGR consists of an input waveguide (transmit) array, two slab couplers interconnected by an array of waveguides (the grating array), and an output waveguide (receiver) array as shown in Fig. 14.19. The length between adjacent waveguides in the grating array varies by a constant, $\Delta L$. The diffracted light from the input fiber is Fourier transformed by the first slab coupler, which acts as a lens. The waveguide grating samples the far field like a multi-slit aperture and introduces a differential path length $\Delta L$ between adjacent grating arms upon transmission through the grating. At the input to the second slab coupler, the differential delay produces a wavelength-dependent phase front. This angular dispersion is converted to spatial dispersion by the second slab coupler so that different wavelengths focus to different output waveguides.

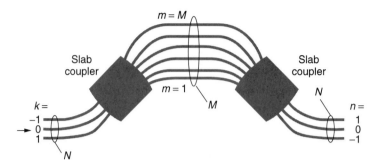

**Figure 14.19**  Schematic of a waveguide grating router with $M$ grating waveguides and $N$ input and output waveguides.

The condition for constructive interference is

$$n_e \Delta L + n_s d[\sin\theta_i + \sin\theta_o] = m\lambda \qquad (14.15)$$

where $m$ is the grating order, $\theta_i \approx kd_i/R$ and $\theta_o \approx nd_o/R$ are the angles for the $k$th input and $n$th output waveguides, respectively, and $R$ is the slab focal length. The pitch of the input and output waveguide arrays is designated by $d_i$ and $d_o$. For a given output port, the response is periodic with FSR $\approx \dfrac{c}{n_g \Delta L}$ for $d(\sin\theta_i + \sin\theta_o) \ll \Delta L$. The spatial dispersion at the receiver array is constant in wavelength, so the demultiplexed outputs are equally spaced in wavelength, not in frequency, as in WDM systems. Three geometrical parameters set the output dispersion $D = -\dfrac{\lambda^2}{c} \dfrac{Rm}{n_s d}$: the focal length $R$, the grating pitch $d$, and the grating order $m$, which is proportional to $\Delta L$.

Typical WGR design requirements are the channel spacing $\Delta f_{ch}$, passband width $\Delta f_L$, cross-talk, loss and loss uniformity, and the number of output channels $N$. To provide some insight into the design process, we look at a simplified case described in [87] where the input, grating and output waveguides are uncoupled and the waveguide modes are defined by a Gaussian distribution $u(x) = \dfrac{1}{w_0 \sqrt{\pi}} \exp\{-(x/w_0)^2\}$ where $w_0$ is the mode field radius. The passband width, defined as the full width at $L$ dB down from the maximum, is $\dfrac{\Delta f_L}{\Delta f_{ch}} \approx 0.96 \dfrac{w_0}{d_o} \sqrt{L}$. The output array pitch divided by the dispersion sets the channel spacing, $\Delta f_{ch} = \dfrac{d_o}{|D|} = \dfrac{d_o d f_c}{R \Delta L} \dfrac{n_s}{n_g}$. The ratio $\dfrac{d_o}{w_0}$ is chosen to satisfy the cross-talk requirement and the passband width. As an example of some design values, reported parameters for a 100-GHz, 16-channel multiplexer are $R = 9.381$ mm, FSR $= 12.8$ nm (1600 GHz), $d = d_o = 25$ μm, $m = 118$, and $w_0 = 4.5$ μm (power half width at $1/e^2$) [87]. By increasing the number of grating waveguides, a narrow array factor can be achieved. Then, power is more localized at the output array and less couples into adjacent outputs. Longer path length differences cause the device to be more susceptible to fabrication variations.

The passband shape of the WGR described so far is Gaussian and can be thought of as the discrete Fourier transform of the grating excitation coefficients convolved with the output waveguide mode [3]. The frequency response can be modified by

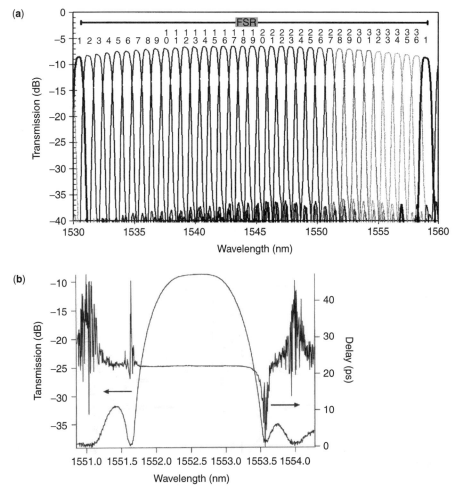

**Figure 14.20** (a) Measured spectral responses for the outputs of a passband flattened $1 \times 36$ WGR [92] © 1998 IEEE, and (b) the measured group delay and magnitude response for a single output of a passband flattened WGR [1] © 1998 IEEE.

changing the input or output waveguide modes or by introducing phase and amplitude filtering in the grating arms [88]. Several techniques have been reported to broaden the passband by widening the input or output waveguide modes using Y-branches [89], MMI couplers [90], and a parabolic waveguide horn [91]. The spectral response for a passband flattened $1 \times 36$ WGR with all the output responses overlaid is shown in Fig. 14.20a [92], and a group delay measurement for a single output waveguide of a flattened passband WGR is shown in Fig. 14.20b [93]. There is no dispersion across the passband. Only in the transition and stop bands, where the zeros are close to the unit circle, does loss begin to introduce a departure from linear phase. For additional information on WGR design, see [94–96].

Many filters using WGRs as building blocks have been realized. A 16-channel, 100 GHz-channel-spacing wavelength add-drop filter is shown in Fig. 14.21a that avoids waveguide crossings and uses a single AWG, an array of $1 \times 2$ thermo-optic

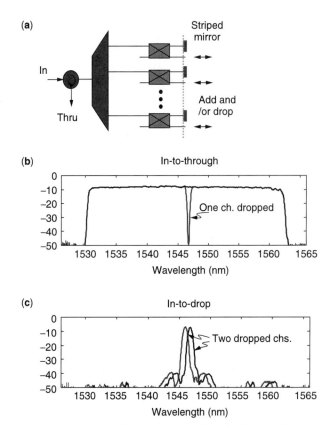

**Figure 14.21** (a) A planar waveguide dynamic add-drop filter architecture using a WGR, an array of 1 × 2 thermo-optic switches, a striped mirror, and an external circulator. (b) The through-channel response for two settings and (c) the drop response for two adjacent channels [97] © 1999 IEEE.

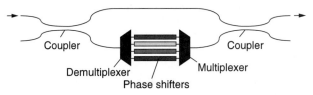

**Figure 14.22** A gain equalization filter using a WGR within a Mach–Zehnder interferometer [98] © 1998 IEEE.

switches, a striped mirror deposited on one waveguide facet, and an external circulator [97]. The through-channel responses for all channels passing through the device and for dropping one channel are shown in Fig. 14.21**b**. The drop response for two channels is shown in Fig. 14.21**c**. A novel gain equalization filter that uses a pair of WGRs and an array of phase shifters within a Mach–Zehnder is shown in Fig. 14.22 [98, 99]. The filter has a 4.5-dB insertion loss and a 14-dB dynamic range. A polarization beam splitter and circulator have been employed for polarization diversity and segmentation of the core in the slab coupler-to-waveguide array transition region have been implemented to reduce the excess loss [100]. A tunable dispersion compensator has been realized by creating a thermally controlled lens between two back-to-back WGRs with FSRs = 200 GHz [101]. By tuning the lens, the excitation of the second grating region is varied so that the longer wavelengths within one FSR can be directed to the longer grating arms and the shorter wavelengths to the shorter arms for positive dispersion or vice versa for negative dispersion. A tuning range of ±80 ps/nm with a passband width sufficient for 40-Gb/s return-to-zero (RZ) modulation format signals and a heater power range of 0 to 6.8W was demonstrated with less than 4.5-dB insertion loss.

## 14.4    INFINITE IMPULSE RESPONSE OPTICAL FILTERS

Optical filters with feedback include etalons, ring resonators, and Bragg gratings. They have at least one pole in their transfer function. Although a sharper transfer function can be realized with one or a few stages, the drawback is that they inherently introduce dispersion.

### 14.4.1    Ring Resonators

Planar waveguides provide a platform for integrating multistage filters using ring resonators; however, ring resonators are more challenging to build than feedforward filters because the FSR depends on the feedback path length and not on the relative path length between interferometer arms as in FIR filters. The feedback path length is inversely proportional to the ring radius. For a given choice of FSR, the core-to-cladding index contrast must be sufficiently high to support low loss fundamental mode propagation and minimize bend loss due to radiation out of the fundamental mode. Typical core-to-cladding index contrasts, $\Delta n = n_{\text{core}} - n_{\text{cladding}}$, for glass waveguides are less than 1% ($\Delta n / n_{\text{core}}$) and limit the FSR to 10 GHz or less. Larger doping concentrations for germanium-doped silica cores and higher index core materials such as silicon oxynitride have been investigated to obtain relative index contrasts up to several percent and FSRs approaching 100 GHz. A compound glass of $Ta_2O_5$-$SiO_2$ with a polymer upper cladding that can be tuned by UV exposure to change the ring's resonant wavelength has been used for bend radii down to 19 μm [102]. Rings have also been fabricated in semiconductors such as AlGaAs on GaAs with FSR = 21.6 nm [103] and Si on $SiO_2$ with bend radii of 10 to 25 microns and FSRs up to 24 nm [104, 105].

A single ring gives a very sharp, but narrow passband filter response. To obtain a flat passband and low loss, the coupled-ring architecture shown in Fig. 14.23**a** has been proposed and demonstrated [15, 16]. Coupled cavities require that the resonant wavelengths (represented by $\phi_n$) for each cavity and the coupling between cavities ($\kappa_n = \sin^2 \theta_n$) is precisely controlled. Vertical coupling provides increased robustness to fabrication variations and shorter couplers without requiring that the small gaps in the coupling region be defined by photolithography [106]. Thermo-optic tuning was used in [107] to align the resonant wavelengths and achieve the third-order, all-pole response shown in Fig. 14.23**b** for orthogonal polarizations. For add-drop functionality and one- or two-stage ring resonator filters, the cross-bar configuration shown in Fig. 14.23**c** has been reported [108]. A two-stage filter response that can be implemented in the cross-bar configuration is shown in Fig. 14.23**d** for one polarization over

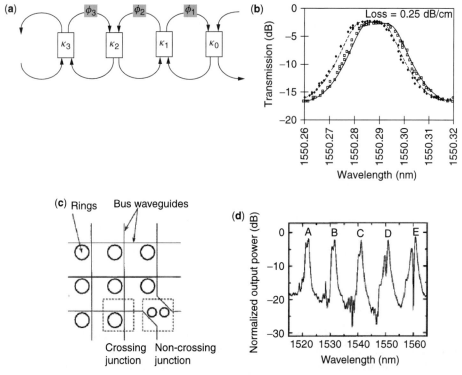

**Figure 14.23**    (**a**) A multi-stage, coupled ring resonator architecture and (**b**) a measured all-pole transmission response for TE and TM polarizations of a coupled three-cavity ring filter [107]. (**c**) A cross-bar wavelength-dependent cross-connect architecture [108] © 2000 IEEE, and (**d**) a two-stage measured response over several periods [109] © 1999 IEEE.

several periods [109]. The dip in the passband is due to misalignment of the two ring resonances, and it is frequency dependent in this example.

A basic allpass filter using a ring resonator is shown in Fig. 14.8**a**. Two parameters control its group delay response: the phase, $\phi$, and the cross-coupling coefficient, $\rho$. The allpass filter architecture shown in Fig. 14.24 replaces the single coupler with a symmetric or asymmetric MZI [110], which is advantageous because complete tunability is realized with two phase shifters and the tolerances on the directional couplers composing the MZI are substantially relaxed compared to the tolerance on $\rho$. A symmetric MZI provides the same dispersion from period-to-period. For a dispersion slope compensating application, an asymmetric MZI may be chosen to produce a slowly varying dispersion from channel to channel. The first application of ring resonators to compensate the dispersion of optical fibers was reported by [111] using a single fiber ring, which is limited by the tradeoff between bandwidth and dispersion. By using a multistage filter where the parameters are chosen optimally for each stage, a constant dispersion (or any desired response) can be approximated over a large portion of the FSR as shown in the four-stage calculated delay response of Fig. 14.25**a** [112]. This yields a large bandwidth utilization (BWU) and decouples the magnitude and phase response in contrast to FIR filters, which trade off bandwidth defined in terms of spectral loss for dispersion. The BWU is defined as the ratio of the passband over which

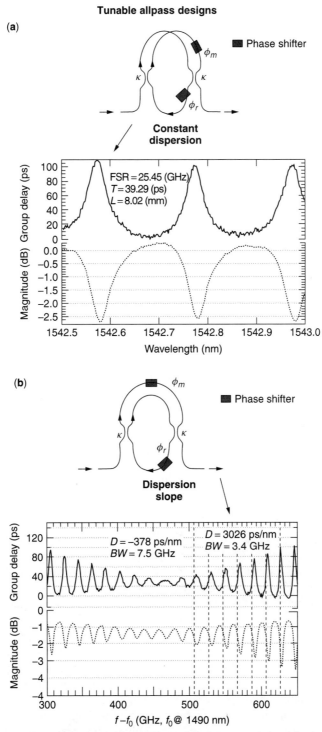

**Figure 14.24**  Tunable single-stage allpass filter architectures consisting of a ring resonator with an embedded **(a)** symmetric MZI and **(b)** asymmetric MZI [110] © 1999 IEEE. The normalized filter loss and delay are shown over several periods for each case.

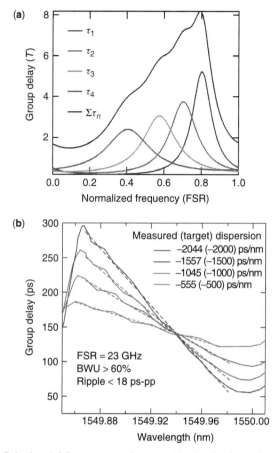

**Figure 14.25**   (**a**) Calculated delay response in normalized units for a dispersion compensating allpass filter showing the individual delay contributions from each ring and (**b**) experimental results using a cascade of four ring-resonator allpass filters.

the dispersion is constant relative to the filter's FSR. Group delay ripple refers to the small deviations of the delay response from a fitted line across the passband. The system impact of group delay ripple is discussed below in the overview of chirped fiber Bragg gratings, where most of the theoretical and simulation work on this subject has been directed. For multichannel dispersion compensation, the filter's FSR is chosen to be an integer multiple of the channel spacing, so that each channel sees the same dispersion over its passband. In the design of multistage allpass filters, tradeoffs are made between the BWU, dispersion and group delay ripple for a given choice of FSR. Because of the large number of channels in dense WDM systems, periodic dispersion compensators, such as allpass filters and the FIR filters discussed previously, are advantageous over Bragg grating approaches that address only a single channel and require a unique device code per channel.

Tunable dispersion compensating filters have been implemented using Ge-doped silica-on-silicon planar waveguides defined by photolithography and reactive ion etching. A core-to-cladding normalized index contrast range of 1.2% to 4% has been investigated, enabling bend radii as small as 350 µm and FSRs up to 80 GHz [113, 114].

The wavelength-dependent filter loss depends on the roundtrip loss [3]. To set the device to a desired dispersion and simultaneously overcome fabrication variations, the spectral magnitude and group delay response are measured at various heater settings. An algorithm then determines the required heater settings for a desired dispersion given requirements on the bandwidth utilization and maximum allowable group delay ripple. By conjugating the filter phases, i.e., $\phi_r \to -\phi_r$, negative dispersion is achieved. Dispersion tuning is demonstrated experimentally in Fig. 14.25**b** for a cascade of four rings with an FSR = 23 GHz. A passband of 16 GHz (60% BWU) is maintained as well as a low group delay ripple over a dispersion range of $\pm 2000$ ps/nm. Bandwidth utilizations up to 80% and passbands up to 60 GHz have been demonstrated, allowing the compensation of spectrally efficient 40-Gb/s NRZ and carrier-suppressed return-to-zero (CSRZ) signals [114]. Roundtrip losses down to 0.4 dB have been measured using the Fig. 14.24**b** architecture. Mode matching of the waveguide to SSMF reduced the coupling loss to 0.8 dB/facet for $\Delta = 2\%$ and 1.1 dB/facet for $\Delta = 4\%$. Tunable dispersion compensators have also been fabricated in silicon oxynitride with a 3% index contrast [115]. Beyond dispersion compensation, allpass filters enable variable-delay lines and first-order polarization mode dispersion compensators [116] to be realized with a solid-state device [117]. Architectures based on allpass filter decomposition allow the realization of optimal bandpass filters [26], multiport interleavers [118], and notch filters [119, 120].

## 14.5 THIN FILM FILTERS

Thin film filters are fabricated by depositing alternating layers of high and low index materials with thicknesses equal to a quarter-wave at the design wavelength. The Fresnel reflections at each interface constructively interfere in the counterpropagating direction at the design wavelength, creating a highly reflective response or photonic bandgap. The analysis of a thin-film stack starts with matching the electric fields, $\widetilde{E}_m^+ + \widetilde{E}_m^- = E_{m+1}^+ + E_{m+1}^-$, and magnetic fields as satisfied by $n_{m+1}(E_{m+1}^+ - E_{m+1}^-) = n_m(\widetilde{E}_m^+ - \widetilde{E}_m^-)$, on both sides of each interface as shown in Fig. 14.26. The resulting transmission and reflection amplitude coefficients are $\tau_m = \tau_m^+ \equiv \dfrac{\widetilde{E}_m^+}{E_{m+1}^+} = \dfrac{2n_{m+1}}{n_m + n_{m+1}}$

and $\rho_m = \rho_m^+ \equiv \dfrac{E_{m+1}^-}{E_{m+1}^+} = \dfrac{n_{m+1} - n_m}{n_m + n_{m+1}}$, where the $+$ superscript denotes that the incident wave is from the left. The relationships for a wave incident from the right are $\tau_m^- = \dfrac{2n_m}{n_m + n_{m+1}}$ and $\rho_m^- = -\rho_m^+$. The input and output fields for a stack with general thicknesses and indices may then be determined by multiplying $2 \times 2$ transfer matrices [3]. The peak reflection for a quarter-wave stack is given by Eq. (14.16) [20]

$$|R|^2 = \tanh^2\left[N \ln \frac{n_H}{n_L} + \frac{1}{2} \ln \frac{n_0}{n_{inc}}\right] \qquad (14.16)$$

where $n_H$ = high refractive index material and $n_L$ = low refractive index. The refractive indices of the incident and output media are denoted by $n_{inc}$ and $n_0$, respectively. The reflectance depends on the ratio of the high and low indices, and it approaches

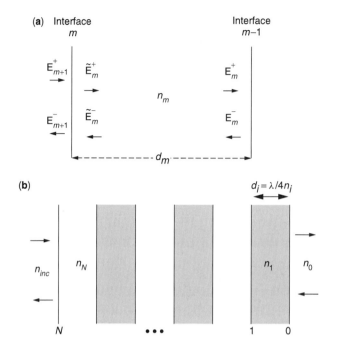

**Figure 14.26**   Thin film schematic (**a**) for a single layer and (**b**) for a quarter-wave stack.

unity as the number of periods $N$ increases. The reflection bandwidth is given by Eq. 14.17 [121]

$$\frac{\Delta\lambda}{\lambda} = \frac{4}{\pi} \sin^{-1} \left( \frac{n_H - n_L}{n_H + n_L} \right) \tag{14.17}$$

The reflection bandwidth depends on the index difference. For alternating layers of silica ($n = 1.44$) and tantalum pentoxide ($n = 2$), the bandgap is several hundred nanometers wide for a design wavelength in the 1550-nm region.

A resonant cavity is created by inserting a multiple of half-wavelength layers between two reflecting stacks. The resulting Fabry–Perot filter has a transmission and reflection transfer function given by [3]

$$T_1^+(z) = T_1^-(z) = \frac{t_0 t_1 \sqrt{z^{-1}}}{1 + \rho_0 \rho_1 z^{-1}} \tag{14.18}$$

$$R_1^+(z) = \frac{\rho_1 + \rho_0 z^{-1}}{1 + \rho_0 \rho_1 z^{-1}} \tag{14.19}$$

and where $t_m = \sqrt{1 - \rho_m^2}$, $z^{-1} = e^{-j\pi\lambda_C/\lambda}$. The reflection for the opposite direction has the same form but with the subscript order exchanged. If each reflector consists of $M$ periods, then the overall structure is conveniently written as $\left[ \frac{L}{2} H \frac{L}{2} \right]^M L \left[ \frac{L}{2} H \frac{L}{2} \right]^M$ where $L$ and $H$ represent a quarter-wave thickness of low and high index material, respectively. By cascading several of these structures, multicavity filters are realized. A calculation of filter responses for 1, 2, and 3 cavities with 10 periods for each reflector

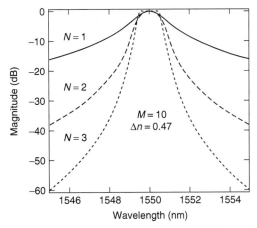

**Figure 14.27** Calculated reflection responses for one, two, and three-cavity filters. Each reflector consists of 10 alternating periods of high and low index materials with an index difference of $\Delta n = 0.47$.

is shown in Fig. 14.27 assuming $n_L = 1.45$ and $n_H = 1.92$. As the number of cavities increases, a flatter passband and sharper rolloff are achieved.

An advantage of thin film filters is that the choice of substrate thermal expansion can be engineered to reduce the temperature dependence of the filter [122]. A temperature dependence of 0.001 nm/°C has been realized [123]. For normal incidence, the polarization dependence is also negligible. Critical fabrication parameters are the loss, thickness control, and number of layers that can be deposited to synthesize a desired filter response. A low-loss 25-GHz-channel-spacing filter has been realized with six cavities and a figure of merit (FOM), defined as the 0.5-dB passband width relative to the 25-dB width, over 50% [124]. The reflection and transmission responses for two filters in cascade are shown in Fig. 14.28. Physical vapor deposition was used to produce 40 alternating layers of tantalum pentoxide and silicon dioxide for each cavity. The stopband in the transmission response has a large rejection; however, there are ripples in the reflection response that limit the stopband rejection as shown in Fig. 14.28. Thin film filters with low passband dispersion for 40-Gb/s applications have been demonstrated by increasing the number of stages to flatten the passband and by moving the transition region, where the dispersion is dominant, away from the passband [125]. It is also possible to cascade a reflective allpass filter with a bandpass filter to reduce the dispersion of their combined response. Wide-band filters with sharp rolloffs, as shown in Fig. 14.29 with a FOM = 83%, have been demonstrated that are capable of demultiplexing bands of WDM channels without sacrificing an intermediate channel to accommodate the filter rolloff. Coupled-cavity filters for interleavers have also been demonstrated [126]; however, they require temperature control because the $dn/dT$ of the cavity cannot be neglected.

A Gires–Tournois interferometer [22] can be thought of as a Fabry–Perot filter with one highly reflecting facet that acts as a mirror so that the device operates in reflection. Coupled-cavities where the last reflector acts as a mirror are multistage allpass filters [112, 3]. The first application of a GTI for dispersion compensation in telecommunications was reported by Gnauck et al. [127]. To improve the bandwidth and dispersion range, multistage etalons are needed [112]. Several multistage architectures

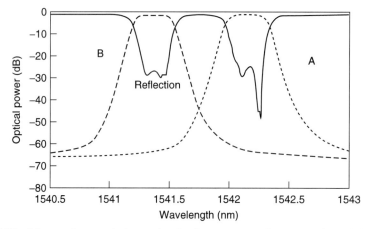

**Figure 14.28** Measured transmission and reflection responses for a two-channel drop module based on thin film filters and designed to separate channels spaced by 25 GHz. Courtesy of Cierra Photonics.

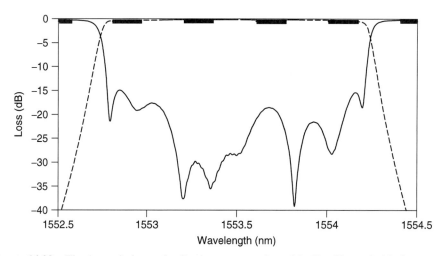

**Figure 14.29** The transmission and reflection response for a thin film filter suitable for separating four 50 GHz-spaced channels as a band. The rolloff is sufficiently steep to avoid interference with adjacent channels in neighboring bands. Courtesy of Cierra Photonics.

have been demonstrated. The first tunable, multistage etalon filter was realized using a combination of MEMs-actuation and thermo-optic control for dispersion compensation [128]. Two cavities with FSR = 100 GHz were cascaded to provide $\pm 100$ ps/nm of dispersion across a 50-GHz passband. A thermo-optically tunable dispersion compensator with FSR = 50 GHz and passbands of 25 GHz has been demonstrated with 4.4-dB loss and a tuning range of $\pm 800$ ps/nm [129]. A tunable cubic dispersion compensator has also been realized using many reflections between parallel cavities and double-passing with a Faraday-rotating mirror [130].

Etalons also provide building blocks for architectures based on allpass filter decomposition [26]. A Michelson interferometer with a GTI in one arm was proposed by

Dingel [131, 132]. The GTI has an air-core cavity and movable reflectors to tune the resonant wavelength of the allpass filter cavity and of the overall interferometer response. An air-core cavity avoids the temperature control issues associated with silica and other materials with non-negligible $dn/dT$.

## 14.6   BRAGG GRATINGS

Since we previously found that codirectional coupled-mode filters are FIR filters, we now investigate Bragg gratings, which are counter-directional coupled-mode filters with infinite impulse responses. For a periodic index modulation with a small index contrast, $n_L \approx n_H$, the reflection and transmission are accurately modeled using coupled-mode theory [53]. When the wavelength is such that the partial reflectances from the index perturbations of a Bragg grating add constructively, a narrowband reflection results. The center wavelength of the reflection is called the Bragg wavelength $\lambda_B$, which is related to the grating period by $\Lambda = \dfrac{\lambda_B}{2n_e}$. The index variation may be induced by ultraviolet (UV) exposure of a photosensitive material through a phase mask or etched. Etched gratings are used extensively in semiconductors for fabricating distributed feedback (DFB) and distributed Bragg reflector (DBR) lasers. The UV-induced Bragg gratings written in optical fibers and planar waveguides are important for realizing low loss, low polarization-dependent, narrowband filters. The photosensitive index changes are less than $\Delta n = 0.01$; however, the filters are typically several millimeters to several centimeters long, consisting of thousands of periods. The frequency dependence is expressed in terms of the detuning parameter $\delta \equiv \beta - \dfrac{\pi}{\Lambda} = 2\pi n_e \left( \dfrac{1}{\lambda} - \dfrac{1}{\lambda_B} \right)$. The grating strength is determined by the coupling coefficient $\kappa = \kappa^* \equiv \dfrac{\pi}{\lambda} \delta \overline{n}_e$ where the average effective index change $\delta \overline{n}_e$ is proportional to the index change times the overlap of the field with the index perturbation. For a uniform grating, let the index perturbation along the length of the grating be described by $\delta n_e(z) = \delta \overline{n}_e (1 + \cos 2\pi z/\Lambda)$, where a unity modulation index has been assumed. Then, the coupled-mode equations have an analytic solution with a reflectance for a grating of length $L$ given by Eq. (14.20) [53].

$$R = |\rho|^2 = \frac{\kappa^2 \sinh^2(\gamma L)}{\kappa^2 \cosh(\gamma L) - \hat{\sigma}^2} \qquad (14.20)$$

where $\sigma = \dfrac{2\pi}{\lambda} \delta \overline{n}_e$, $\hat{\sigma} = \sigma + \delta$, and $\gamma = \sqrt{\kappa^2 - \hat{\sigma}^2}$. The peak reflection is given by $R_{\max} = \tanh^2(\kappa L)$. To quantify the reflection bandwidth, it is defined as the bandwidth between the first zeros of the reflection spectrum [53].

$$\frac{\Delta \lambda}{\lambda_B} = \frac{\delta \overline{n}_e}{n_e} \sqrt{1 + \left( \frac{\lambda_B}{L \delta \overline{n}_e} \right)^2} \qquad (14.21)$$

Bragg gratings are classified in terms of their reflectance as weak or strong gratings. Strong gratings have a reflectance converging to 100% at the Bragg wavelength. The weak grating regime is described by [53] $\delta \overline{n}_e \ll \dfrac{\lambda_B}{L}$. Its reflection bandwidth is given by $\dfrac{\Delta\lambda}{\lambda} = \dfrac{2}{N}$, the same as for weak co-directionally coupled gratings, where $\Delta\lambda$ is the bandwidth and $N$ is the number of periods. The bandwidth is inversely proportional to the grating length for weak gratings and does not depend on the index change. For weak gratings, the reflection spectrum $\rho(\delta)$ is proportional to the Fourier transform of the coupling strength $\kappa(z)$ [133]. A consequence of this approximation is that the reflection response in the weak coupling regime behaves like an FIR filter instead of an IIR response. This behavior has been used to demonstrate a square passband filter response with an approximately constant group delay, i.e., a filter which closely approximates the dispersionless properties of symmetric FIR filters [134]. The drawback is that the peak reflection is $\sim 50\%$; consequently, the remainder of the power is transmitted at the Bragg wavelength. In the strong grating regime $\delta\overline{n}_e \gg \dfrac{\lambda_B}{L}$, and the reflection bandwidth is proportional to the index variation, $\dfrac{\Delta\lambda}{\lambda} = \dfrac{\delta\overline{n}_e}{n_e}$. Increasing the length of a strong grating makes the band edge sharper but does not change the bandwidth.

A key to realizing narrowband reflection filters for WDM applications is control over the coupling strength profile. Tapering, or apodization, of the coupling strength decreases the sidelobes that are present for uniform coupling as shown in Fig. 14.30c. An apodization profile with constant average index reduces the sidelobes symmetrically about the center wavelength as shown in Fig. 14.30d; however, a variation in the average index across the grating profile, produced by exposing with a fixed Gaussian UV beam, for example, will induce an asymmetric response with sidelobes

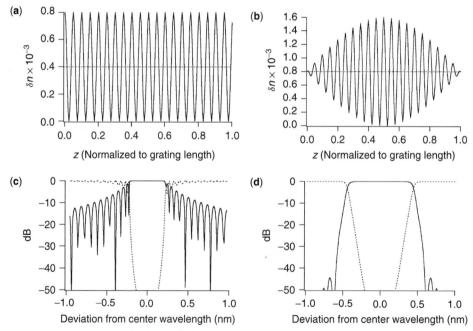

**Figure 14.30**  Index perturbation along the length of a 10 mm-long grating for **(a)** a uniform grating with a maximum index change of $0.8 \times 10^{-3}$, and **(b)** an apodized grating with a maximum index change of $1.6 \times 10^{-3}$. The corresponding reflection (solid line) and transmission (dashed lines) spectra are shown in **(c)** and **(d)**.

on the short wavelength side of the peak. Flat passband filters with sharp transition bands and large stopband rejections have been demonstrated in fibers [135]. To analyze a nonuniform grating, one can solve the coupled-mode equations or segment the grating into approximately uniform sections and multiply the solutions of each section [136, 137]. A discretized propagation analysis, which produces a mathematical description equivalent to the $Z$ transform approach, has been reported for both grating-assisted forward-[138] and reverse-direction [139] coupling.

A sampled or superstructure grating is produced by modulating the coupling strength [140–142]. A sampled grating refers to one whose coupling strength is modulated in a binary fashion to produce multiple passbands; whereas, a superstructure profile may include tapering of the coupling strength within each period of the sample (sub-grating) and tapering of the overall profile (super-grating). The spectral shape for an individual channel is determined by the super-grating, while the overall envelope response is set by the sub-grating, as expected from the Fourier transform relationship between the coupling strength and spectrum in the weak coupling limit. The channel spacing is given by $\Delta\lambda_{ch} = \dfrac{\lambda_B^2}{2\bar{n}\Lambda_{sub}}$, where $\Lambda_{sub}$ is the period of the sub-gratings and $\bar{n}$ is the average index.

As the reflected signal is traveling in the same waveguide as the input, a method to separate the input from the dropped channel is required, and similarly for adding a channel. One solution is to use a circulator for the drop and add channels as shown in Fig. 14.31**a**. An alternative, which is particularly suitable for implementation in planar waveguides, is to write a pair of identical gratings in an MZI [143, 144]. The reflected signal at the Bragg wavelength is then output at the cross-port as shown in Fig. 14.31**b**. The coupling ratios are nominally 50% to achieve ideal performance. Identical gratings must be written across the MZI arms, which has been demonstrated in a single UV exposure [145].

Chirped fiber Bragg gratings where the grating period varies linearly along the length of the grating, as illustrated in Fig. 14.32**a**, are useful for dispersion compensation [146]. The advantages of UV-induced chirped gratings are their low loss, low PMD, and maturity of fabrication. As with other filter-based dispersion compensation

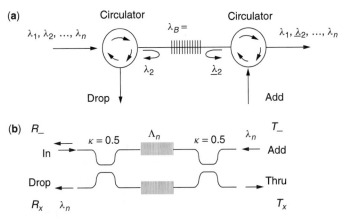

**Figure 14.31**  Bragg grating add/drop filter architectures using (**a**) circulators and (**b**) a Mach–Zehnder interferometer.

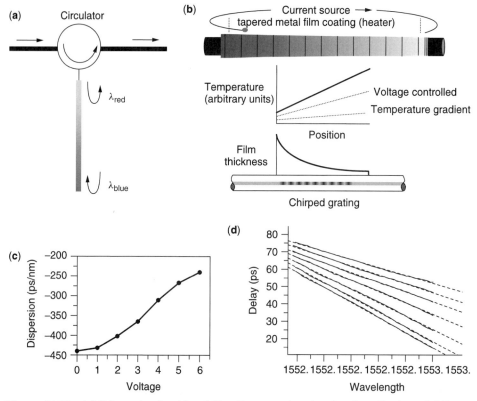

**Figure 14.32** (a) Schematic of a chirped fiber Bragg grating showing the reflection of different wavelengths at different points along the grating, (b) electrically tunable dispersion compensator based on a variable-thickness resistor, (c) dispersion as a function of applied voltage and (d) the corresponding delay response [150] © 2000 IEEE.

solutions, they have negligible nonlinearities compared to dispersion compensating fibers, so the input power can be larger. A very small chirp rate is needed to produce large dispersions, and the required grating length scales with the bandwidth to be compensated. A 17.5-cm-long grating with a 0.06-nm/cm chirp was fabricated to compensate 100 km of conventional fiber (1700 ps/nm) over a 1-nm bandwidth [147]. Typically, one grating covers one channel; however, sampled gratings [148] as well as long gratings have been reported for compensating multiple channels. A 1.3-m fiber grating consisting of thirteen 10-cm gratings was fabricated to compensate for 100 km of conventional fiber over a 10-nm bandwidth [149]. To compensate larger numbers of channels, dispersion compensating filters with periodic responses discussed previously are advantageous.

An electrically tunable dispersion compensator for up to 160-Gb/s signals has been realized using a variable-thickness resistor deposited on the fiber cladding to create a tunable thermal gradient as shown in Fig. 14.32b [150]. As the applied voltage increases, the temperature gradient increases, as does the chirp rate due to the finite $dn/dT$ of silica. The dispersion for a range of applied voltages is plotted in Fig. 14.32c, and the corresponding delay spectrum is given in Fig. 14.32d. As the center wavelength of the grating changes with a change in the average temperature, a second resistor with a silica spacer separating it from the primary resistor is employed to independently vary the dispersion and the center Bragg wavelength.

Slight imperfections in chirped grating fabrication lead to deviations of the delay response from an ideal line for constant dispersion, and the deviation is called the group delay ripple. Depending on the magnitude and period of the group delay ripple, a system penalty may be associated with the resulting pulse distortion and inter-symbol interference. For a given ripple magnitude, the worst-case penalty occurs when the ripple period is roughly equal to the bitrate [151]. For very fast ripples, i.e., having a small period, the pulse averages over the delay variations and only a small penalty is incurred. For long ripple periods relative to the bitrate, the penalty scales in a similar fashion to the cumulative chromatic and cubic dispersion of optical fibers.

## 14.7    CONCLUSION

Optical filters provide enabling functions for WDM systems from bandwidth management to compensation of analog impairments to optical monitoring. Tunable filters, in particular, allow systems to route wavelengths in mesh networks and provide tunable gain equalization and chromatic dispersion compensation. Many advances have been made in just the last few years in new filter architectures, fabrication technologies, and applications. Although many technology platforms are available for filter implementation, continued progress is needed to provide high levels of performance, integration, and functionality at increasingly lower costs. The future looks bright for the field of optical filters to continue playing a major role in the evolution of optical networks.

## REFERENCES

1. G. Lenz, B. Eggleton, C. Giles, C. Henry, R. Slusher, and C. Madsen, "Dispersive properties of optical filters for WDM systems," *J. Quantum. Electron.*, **34**, pp. 1390–1402, 1998.

2. B. Moslehi, J. Goodman, M. Tur, and H. Shaw, "Fiber-Optic Lattice Signal Processing," *Proc. of the IEEE*, **72**(7), pp. 909–930, 1984.

3. C. Madsen and J. Zhao, *Optical Filter Design and Analysis: A Signal Processing Approach*. New York, NY: John Wiley, 1999.

4. L. Soldano, F. Veerman, M. Smit, B. Verbeek, and E. Pennings, "Multimode interference couplers," Proc. Integrated Photonics Research Topical Meeting, Monterey, CA, April, 1991, Poster TuD1.

5. E. Pennings, R. Deri, A. Scherer, R. Bhat, T. Hayes, N. Andreadakis, M. Smit, and R. Hawkins, "Ultra-compact, low-loss directional coupler structures on InP for monolithic integration," Proc. Integrated Photonics Research Topical Meeting, Monterey, CA, April, 1991, Post-deadline PD2.

6. L. Soldano and E. Pennings, "Optical Multi-Mode Interference Devices Based on Self-Imaging: Principles and Applications," *J. Lightw. Technol.*, **13**(4), pp. 615–627, 1995.

7. C. Dragone, "Efficient $N \times N$ star coupler based on Fourier optics," *Electron. Lett.*, **24**(15), pp. 942–944, 1988.

8. C. Dragone, C. Henry, I. Kaminow, and R. Kistler, "Efficient multichannel integrated optics star coupler on silicon," *IEEE Photon. Technol. Lett.*, **1**(8), pp. 241–243, 1989.

9. K. Okamoto, H. Okazaki, Y. Ohmori, and K. Kato, "Fabrication of large scale integrated-optic $N \times N$ star couplers," *IEEE Photon. Technol. Lett.*, **4**(9), pp. 1032–1035, 1992.

10. L. Zehnder, *Zeitschr.f. Instrkde*, **11**, pp. 275, 1891, and L. Mach, *Zeitschr.f. Instrkde*, **12**, pp. 89, 1892.

11. J. Michelson, *Amer. J. Sci.*, **3**(22), pp. 120, 1881.

12. A. Oppenheim and R. Schafer, *Digital Signal Processing*. Englewood, N.J.: Prentice-Hall, Inc., 1975.

13. C. Fabry and A. Perot, *Ann. Chim. Phys.*, **7**(16), pp. 115, 1899.

14. E. Dowling and D. MacFarlane, "Lightwave Lattice Filters for Optically Multiplexed Communication Systems," *J. Lightw. Technol.*, **12**(3), pp. 471–486, 1994.

15. R. Orta, P. Savi, R. Tascone, and D. Trinchero, "Synthesis of Multiple-Ring-Resonator Waveguides," *IEEE Photonics Technol. Lett.*, **7**(12), pp. 1447–1449, 1995.

16. C. Madsen and J. Zhao, "A General Planar Waveguide Autoregressive Optical Filter," *J. Lightw. Technol.*, **14**(3), pp. 437–447, 1996.

17. A. Deczky, "Synthesis of Recursive Digital Filters Using the Minimum $p$-Error Criterion," *IEEE Trans. on Audio and Electroacoustics*, **20**(4), pp. 257–263, 1972.

18. G. Agrawal, *Fiber-optic Communication Systems*. New York, NY: John Wiley & Sons, 1997.

19. A. Papoulis, *The Fourier Integral and Its Applications*. New York: McGraw-Hill, 1962.

20. P. Yeh, *Optical Waves in Layered Media*, New York: Wiley, 1988.

21. H. Bode, *Network Analysis and Feedback Amplifier Design*, New York: Van Nostrand, 1945.

22. F. Gires and P. Tournois, "Interferometre utilisable pour la compression d'impulsions lumineuses modulees en frequence," *C. R. Acad. Sci*, **258**(5), pp. 6112–6115, 1964.

23. K. Jinguji, "Synthesis of Coherent Two-Port Optical Delay-Line Circuit with Ring Waveguides," *J. Lightw. Technol.*, **14**(8), pp. 1882–1898, 1996.

24. K. Oda, N. Takato, H. Toba, and K. Nosu, "A Wide-Band Guided-Wave Periodic Multi/Demultiplexer with a Ring Resonator for Optical FDM Transmission Systems," *J. of Lightw. Technol.*, **6**(6), pp. 1016–1022, 1988.

25. S. Suzuki, M. Yanagisawa, Y. Hibino, and K. Oda, "High-Density Integrated Planar Lightwave Circuits Using $SiO_2$-$GeO_2$ Waveguides with a High Refractive Index Difference," *J. of Lightw. Technol.*, **12**(5), pp. 790–796, 1994.

26. C. Madsen, "Efficient Architectures for Exactly Realizing Optical Filters with Optimum Bandpass Designs," *IEEE Photonics Technol. Lett.*, **10**(8), pp. 1136–1138, 1998.

27. S. Mitra and J. Kaiser, *Handbook for Digital Signal Processing*. New York: John Wiley & Sons, 1993.

28. K. Sasayama, M. Okuno, and K. Habara, "Coherent optical transversal filter using silica-based single-mode waveguides," *Electron. Lett.*, **25**(22), pp. 1508–1509, 1989.

29. K. Sasayama, M. Okuno, and K. Habara, "Coherent Optical Transversal Filter Using Silica-Based Waveguides for High-Speed Signal Processing," *J. Lightw. Technol.*, **9**(10), pp. 1225–1230, 1991.

30. K. Sasayama, M. Okuno, and K. Habara, "Photonic FDM Multichannel Selector Using Coherent Optical Transversal Filter," *J. Lightw. Technol.*, **12**(4), pp. 664–669, 1994.

31. I. Solc, "Birefringent Chain Filters," *J. Opt. Soc. Am.*, **55**, pp. 621, 1965.

32. K. Jinguji and M. Kawachi, "Synthesis of Coherent Two-Port Lattice-Form Optical Delay-Line Circuit," *J. Lightw. Technol.*, **13**, pp. 72–82, 1995.

33. Y. Li and C. Henry, "Silica-based Optical Integrated Circuits," *IEE Proc. Optoelectron.*, **143**(5), pp. 263–280, 1996.

34. M. Kuznetsov, "Cascaded Coupler Mach–Zehnder Channel Dropping Filters for Wavelength-Division-Multiplexed Optical Systems," *J. Lightw. Technol.*, **12**(2), pp. 226–230, 1994.

35. T. Ozeki, "Optical Equalizers," *Opt. Lett.*, **17**(5), pp. 375–377, 1992.

36. M. Okuno, N. Takato, T. Kitoh, and A. Sugita, "Silica-Based Thermo-Optic Switches," *NTT Review*, **7**(5), pp. 57–63, 1995.

37. Y. Inoue, Y. Ohmori, M. Kawachi, S. Ando, T. Sawada, and H. Takahashi, "Polarization Mode Converter with Polyimide Half Waveplate in Silica-Based Planar Lightwave Circuits," *IEEE Photonics Technol. Lett.*, **6**(5), pp. 626–628, 1994.

38. S. Suzuki, S. Sumida, Y. Inoue, M. Ishii, and Y. Ohmori, "Polarization-Insensitive Array-Waveguide Gratings Using Dopant-Rich Silica-Based Glass with thermal Expansion Adjusted to Si Substrate," *Electron. Lett.*, **33**, pp. 1173–1174, 1997.

39. E. Wildermuth, C. Nadler, M. Loaker, W. Hunziker, and H. Melchior, "Penalty-Free Polarization Compensation of SiO2/Si Arrayed Waveguide Grating Multiplexers Using Stress Release Grooves," *Electron. Lett.*, **34**, pp. 1661–1663, 1998.

40. Y. Inoue, M. Itoh, Y. Hashizume, Y. Hibino, A. Sugita, and A. Himeno, "Novel Birefringence Compensating AWG Design." WB4, Optical Fiber Conference, 2001.

41. Y. Li, C. Henry, E. Laskowski, C. Mak, and H. Yaffe, "Waveguide EDFA Gain Equalisation Filter," *Electron. Lett.*, **31**(23), pp. 2005–2006, 1995.

42. B. J. Offrein, F. Horst, G. L. Bona, R. Germann, H. W.M. Salemink, and R. Beyeler, "Adaptive Gain Equalizer in High-Index-Contrast SiON Technology," *IEEE Photon. Technol. Lett.*, **12**, pp. 504–506, 2000.

43. B. Offrein, R. Germann, G. Bona, F. Horst, and H. Salemink, "Tunable Optical Add/Drop Components in Silicon-Oxynitride Waveguide Structures." Proc. 24th European Conf. Opt. Commun. (ECOC). Madrid, Spain, September 20–24, 1998, pp. 325–326.

44. K. Jinguji and M. Oguma, "Optical Half-Band Filters," *J. Lightw. Technol.*, **18**(2), pp. 252–259, 2000.

45. T. Chiba, H. Arai, K. Ohira, H. Nonen, H. Okano, and H. Uetsuka, "Novel Architecture of Wavelength Interleaving Filter with Fourier Transform-Based MZIs." Optical Fiber Communication Conference. Anaheim, California, 2001, p. WB5.

46. K. Takiguchi, K. Okamoto, S. Suzuki, and Y. Ohmori, "Planar Lightwave Circuit Optical Dispersion Equalizer," *IEEE Photonics Technol. Lett.*, **6**(1), pp. 86–88, 1994.

47. K. Takiguchi, K. Jinguji, K. Okamoto, and Y. Ohmori, "Dispersion Compensation Using a Variable Group-Delay Dispersion Equalizer," *Electron. Lett.*, **31**(25), pp. 2192–2194, 1995.

48. K. Takiguchi, K. Okamoto, T. Goh, T. Saida, and M. Itoh, "Integrated-Optic Dispersion Slope Equalizer for $N \times 40$ Gb/s WDM Trasmission," in *7.1.3.* European Conference on Optical Communication. Munich, Germany, 2000.

49. T. Saida, K. Takiguchi, S. Kuwahara, Y. Kisaka, Y. Miyamoto, Y. Hashizume, T. Shibata, and K. Okamoto, "Planar Lightwave Circuit Polarization Mode Dispersion Compensator." European Conference on Optical Communications, 2001, p. 4.

50. M. Sharma, H. Ibe, and T. Ozeki, "Optical Circuits for Equalizing Group Delay Dispersion of Optical Fibers," *J. Lightwave Technol.*, **12**(10), pp. 1759–1765, 1994.

51. J. Damask, "A Programmable Polarization-Mode Dispersion Emulator for Systematic Testing of 10 Gb/s PMD Compensators." Optical Fiber Communications, 2000.

52. J. Chiao, "Liquid-Crystal Optical Harmonic Equalizers," in *WD2.1.* IEEE Leos Summer Topicals, 2001.

53. T. Erdogan, "Fiber Grating Spectra," *J. Lightwave Technol.*, **15**(8), pp. 1277–1294, 1997.

54. D. Smith, R. Chakravarthy, Z. Bao, J. J. Baran, J. L., A. d'Alessandro, D. Fritz, S. Huang, X. Zou, S. Hwang, A. Willner, and K. Li, "Evolution of the Acousto-optic wavelength routing switch," *J. Lightw. Technol.*, **14**(6), pp. 1005–1019, 1996.

55. A. Vengsarkar, P. Lemaire, J. Judkins, V. Bhatia, T. Erdogan, and J. Sipe, "Long Period Fiber Gratings as Band Rejection Filters," *J. Lightwave Technol.*, **14**(1), pp. 58–65, 1996.

56. A. Vengsarkar, J. Pedrazzani, J. Judkins, P. Lemaire, N. Bergano, and C. Davidson, "Long Period Fiber Grating Based Gain Equalizers," *Opt. Lett.*, **21**(5), pp. 336–338, 1996.

57. P. Wysocki, J. Judkins, R. Espindola, M. Andrejco, A. Vengsarkar, and K. Walker, "Erbium-doped fiber amplifier flattened beyond 40 nm using long-period grating," Optical Fiber Communication Conference. Dallas, TX: OSA, February 16–21, 1997, PD2.

58. A. Srivastava, Y. Sun, J. Sulhoff, C. Wolf, M. Zirngibl, R. Monnard, A. Chraplyvy, A. Abramov, R. Espindola, T. Strasser, J. Pedrazzani, A. Vengsarkar, J. Zyskind, J. Zhou, D. Ferrand, P. Wysocki, J. Judkins, and Y. Li, "1 Tb/s transmission of 100 WDM 10 Gb/s channels over 400 km of TrueWave Fiber," in *OSA Technical Digest Series, Vol. 2*, Optical Fiber Communications Conference. San Jose, CA, 1998, p. PD10.

59. H. Kim, S. Yun, H. Kim, N. Park, and B. Kim, "Actively Gain-Flattened Erbium-Doped Fiber Amplifier Over 35 Nm by Using All-Fiber Acoustooptic Tunable Filters," *IEEE Photon. Technol. Lett.*, **10**(6), pp. 790–792, 1998.

60. E. Loewen, M. Neviere, and D. Maystre, "Grating Efficiency Theory as It Applies to Blazed and Holographic Gratings," *Appl. Opt.*, **16**, pp. 2711–2721, 1977.

61. H. Rowland, "On concave gratings for optical purposes," *Amer. J. Sci.*, **3**(26), pp. 87–98, 1883. Also, H. Rowland, *Phil. Mag.*, **13**, p. 467, 1882.

62. M. Klein and T. Furtak, *Optics*. New York: John Wiley & Sons, 1986.

63. J. Soole, A. Scherer, H. Leblanc, N. Andreadakis, R. Bhat, and M. Koza, "Monolithic InP-based grating spectrometer for wavelength-division multiplexed systems at 1.5 μm," *Electron. Lett.*, **27**(2), pp. 132–134, 1991.

64. P. Clemens, G. Heise, R. Marz, H. Michel, A. Reichelt, and H. Schneider, "Flat-field spectrograph in SiO2/Si," *IEEE Photon. Technol. Lett.*, **4**, pp. 886–887, 1992.

65. S. Janz, M. Pearson, B. Lamontagne, L. Erickson, A. Delage, P. Cheben, D.-X. Xu, Gao, A. Balakrishnan, J. Miller, and S. Charbonneau, "Planar Waveguide Echelle Gratings: An Embeddable Diffractive Element for Photonic Integrated Circuits." Optical Fiber Communications Conference. Anaheim, CA, March 19–22, 2002.

66. E. Koteles, J. He, B. Lamontagne, A. Delage, L. Erickson, G. Champion, and M. Davies, "Recent Advances in InP-Based Waveguide Grating Demultiplexers," in *1998 OSA Technical Digest Series Vol. 2*. Optical Fiber Communication Conference. San Jose, CA, February 22–27, 1998, pp. 82–83.

67. J.-J. He, E. Koteles, B. Lamontagne, L. Erickson, A. Delage, and M. Davies, "Integrated Polarization Compensator for WDM Waveguide Demultiplexers," *IEEE Photon. Technol. Lett.*, **11**, pp. 224–226, 1999.

68. E. Churin and P. Bayvel, "Design of Free-Space WDM Router Based on Holographic Concave Grating," *IEEE Photon. Technol. Lett.*, **11**(2), 1999.

69. E. Churin and P. Bayvel, "Passband Flattening and Broadening Techniques for High Spectral Efficiency Wavelength Demultiplexers," *Electronics Lett.*, **35**(1), pp. 27–28, 1999.

70. J. Ford, V. Aksyuk, D. Bishop, and J. Walker, "Wavelength Add-Drop Switching Using Tilting Micromirrors," *J. Lightw. Technol.*, **17**(5), pp. 904–911, 1999.

71. A. Ranalli, B. Scott, and J. Kondis, "Liquid Crystal-Based Wavelength Selectable Cross-Connect." ECOC, 1999.

72. D. Marom, D. Neilson, D. Greywall, N. Basavanhally, P. Kolodner, Y. Low, F. Pardo, C. Bolle, S. Chandrasekhar, L. Buhl, C. Giles, S. Oh, C. Pai, K. Werder, S. Soh,

G. Bogart, E. Ferry, F. Klemens, K. Teffeau, J. Miner, S. Rogers, J. Bower, R. Keller, and W. Mansfield, "Wavelength-Selective $1 \times 4$ Switch for 128 WDM Channels at 50 GHz Spacing," *PD FD7*. Optical Fiber Conference. Anaheim, CA, 2002.

73. J. Ford and J. Walker, "Dynamic Spectral Power Equalization Using Micro-Opto-Mechanics," *IEEE Photon. Technol. Lett.*, **10**(10), pp. 1440–1442, 1998.

74. K. Goossen, J. Walker, and S. Arney, "Silicon Modulator Based on Mechanically-Active Anti-Reflection Layer with 1 Mbit/Sec Capability for Fiber-in-the-Loop Applications," *IEEE Photon. Technol. Lett.*, **6**(9), pp. 1119–1121, 1994.

75. J. Wagener, T. Strasser, J. Pedrazzani, J. DeMarco, and D. DiGiovanni, "Fiber grating optical spectrum analyzer tap," European Conf. on Optical Communications (ECOC), 1997, pp. 65–68, PD V.5.

76. T. Erdogan and J. Sipe, "Tilted Fiber Phase Gratings," *J. Opt. Soc. Am. A*, **13**(2), pp. 296, 1996.

77. C. Madsen, J. Wagener, T. Strasser, M. Milbrodt, E. Laskowski, and J. DeMarco, "Planar Waveguide Grating Optical Spectrum Analyzer," in *Optical Society of America Technical Digest Vol. 4*, Integrated Photonics Research Conf. Victoria, Canada, March 29-April 3, 1998, pp. 99–101.

78. M. Shirasaki, "Large Angular Dispersion by a Virtually Imaged Phased Array and Its Application to a Wavelength Demultiplexer," *Opt. Lett.*, **21**(5), 1996.

79. M. Shirasaki, "Chromatic-Dispersion Compensator Using Virtually Imaged Phased Array," *IEEE Photon. Technol. Lett.*, **9**(12), pp. 1598–1600, 1997.

80. M. Shirasaki and S. Cao, "Compensation of Chromatic Dispersion and Dispersion Slope Using a Virtually Imaged Phased Array." Optical Communication Conference. Anaheim, California, March 19–22, 2001. (TuS1).

81. C. Lin, "Chromatic Dispersion Compensation Using a Virtually Imaged Phased Array (VIPA)," Ph. D. Diss. Massachusetts Institute of Technology, 2000.

82. L. Garrett, A. Gnauck, M. Eiselt, R. Tkach, C. Yang, C. Mao, and S. Cao, "Demonstration of Virtually-Imaged Phased-Array Device for Tunable Dispersion Compensation in $16 \times 10$ Gb/s WDM Transmission Over 480 Km Standard Fiber." Optical Fiber Communications Conference, PD7, 2000.

83. M. Shirasaki, Y. Kawahata, S. Cao, H. Ooi, N. Mitamura, H. Isono, G. Ishikawa, G. Barbarossa, C. Yang, and C. Lin, "Variable Dispersion Compensator Using the Virtually Imaged Phased Array (VIPA) for 40-Gbit/s WDM Transmission Systems." ECOC, PD2.3, 2000.

84. M. Smit, "New Focusing and Dispersive Planar Component Based on an Optical Phased Array," *Electron. Lett.*, **24**(7), pp. 385–386, 1988.

85. H. Takahashi, S. Suzuki, K. Kato, and I. Nishi, "Arrayed-Waveguide Grating for Wavelength Division Multi/Demultiplexer With Nanometre Resolution," *Electron. Lett.*, **26**(2), pp. 87–88, 1990.

86. C. Dragone, "An $N \times N$ Optical Multiplexer Using A Planar Arrangement of Two Star Couplers," *IEEE Photonics Technol. Lett.*, **3**(9), pp. 812–815, 1991.

87. M. Smit and C. Van Dam, "PHASAR-Based WDM-Devices: Principles, Design and Applications," *IEEE J. Selected Topics Quant. Electron.*, **2**(2), pp. 236–250, 1996.

88. K. Okamoto and H. Yamada, "Arrayed-Waveguide Grating Multiplexer with Flat Spectral Response," *Optics Lett.*, **20**(1), pp. 43–45, 1995.

89. C. Dragone, "Frequency routing device having a wide and substantially flat passband," *U.S. Patent*(5,412,744), Issued May 1995.

90. M. Amersfoot, J. Soole, H. Leblanc, N. Andreadakis, A. Rajhel, and C. Caneau, "Passband broadening of integrated arrayed waveguide filters using multimode interference couplers," *Electron. Lett.*, **32**(5), pp. 449–451, 1996.

91. K. Okamoto and A. Sugita, "Flat Spectral Response Arrayed-Waveguide Grating Multi-plexer with Parabolic Waveguide Horns," *Electron. Lett.*, **32**(18), pp. 1661–1663, 1996.

92. Y. Li and L. Cohen, "Planar Waveguide DWDMs for Telecommunications: Design and Tradeoffs." *13th Annual National Fiber Optics Engineers Conference (NFOEC)*. San Diego, CA, September 22–24, 1997.

93. B. Eggleton, G. Lenz, N. Litchinitser, D. Patterson, and R. Slusher, "Implications of fiber grating dispersion for WDM communication systems," *IEEE Photon. Technol. Lett.*, **9**(10), pp. 1403–1405, 1997.

94. K. Okamoto, *Fundamentals of Optical Waveguides*. New York: Academic Press, 2000.

95. P. Munoz, D. Pastor, and J. Capmany, "Modeling and Design of Arrayed Waveguide Gratings," *J. Lightw. Technol.*, **20**(4), pp. 661–674, 2002.

96. C. Doerr, "Planar Lightwave Devices for WDM," in *Optical Fiber Telecommunications IVA*, I. Kaminow and T. Li, Eds. New York: Academic Press, 2002, pp. 405–476.

97. C. Doerr, L. Stulz, J. Gates, M. Cappuzzo, E. Laskowski, L. Gomez, A. Paunescu, A. White, and C. Narayanan, "Arrayed Waveguide Lens Wavelength Add-Drop in Silica," *IEEE Photon. Technol. Lett.*, **11**(5), pp. 557–559, 1999.

98. C. Doerr, C. Joyner, and L. Stulz, "Integrated WDM Dynamic Power Equalizer with Potentially Low Insertion Loss," *IEEE Photon. Technol. Lett.*, **10**, pp. 1443–1445, 1998.

99. C. Doerr, M. Cappuzzo, E. Laskowski, A. Paunescu, L. Gomez, W. Stulz, and J. Gates, "Dynamic Wavelength Equalizer in Silica Using the Single Filtered Arm Interferometer," *IEEE Photon. Technol. Lett.*, **11**(5), pp. 581–583, 1999.

100. C. Doerr, K. Chang, L. Stulz, R. Pafchek, Q. Guo, L. Buhl, L. Gomez, M. Cappuzzo, and G. Bogert, "Arrayed Waveguide Dynamic Gain Equalization Filter with Reduced Insertion Loss and Increased Dynamic Range," *IEEE Photon. Technol. Lett.*, **13**(4), pp. 329–331, 2001.

101. C. Doerr, L. Stulz, S. Chandrasekhar, L. Buhl, and R. Pafchek, "Multichannel Integrated Tunable Dispersion Compensator Employing a Thermooptic Lens." Optical Fiber Communications Conference. Anaheim, CA, March 19–22, 2002.

102. S. Chu, W. Pan, S. Sato, T. Kaneko, B. Little, and Y. Kokubun, "Wavelength Trimming of a Microring Resonator Filter by Means of a UV Sensitive Polymer Overlay," *IEEE Photon. Technol. Lett.*, **11**(6), pp. 688–690, 1999.

103. D. Rafizadeh, J. Zhang, S. Hagness, A. Taflove, K. Stair, and S. Ho, "Nanofabricated Waveguide-Coupled 1.5-Um Microcavity Ring and Disk Resonators with High Q and 21.6-Nm Free Spectral Range." CLEO Conf. Baltimore, MD, May 18–23, 1997, pp. CPD23-2.

104. J. Foresi, B. Little, G. Steinmeyer, E. Thoen, S. Chu, H. Haus, E. Ippen, L. Kimerling, and W. Greene, "Si/SiO2 Micro-Ring Resonator Optical Add/Drop Filters." CLEO Conf. Baltimore, MD, May 18–23, 1997, pp. CPD22-2.

105. B. Little, J. Foresi, G. Steinmeyer, E. Thoen, S. Chu, H. Haus, E. Ippen, L. Kimerling, and W. Greene, "Ultra-Compact Si-SiO2 Microring Resonator Optical Channel Dropping Filters," *IEEE Photonics Technol. Lett.*, **10**(4), pp. 549–551, 1998.

106. S. Chu, W. Pan, T. Kaneko, and Y. Kokubun, "Fabrication of Vertically Coupled Glass Microring Resonator Channel Dropping Filters." Optical Fiber Communication Conference and the International Conference on Integrated Optics and Optical Fiber Communication. San Diego, CA, Feb. 23–26, 1999, p. ThH4.

107. C. Madsen and J. Zhao, "Post-Fabrication Optimization of an Autoregressive Planar Waveguide Lattice Filter," *J. of Applied Optics*, **36**(3), pp. 642–647, 1997.

108. B. Little, S. Chu, and Y. Kokubun, "Microring Resonator Arrays for VLSI Photonics," *IEEE Photon. Technol. Lett.*, **12**(3), pp. 323–325, 2000.

109. S. Chu, B. Little, W. Pan, T. Kaneko, and Y. Kokubun, "Second-Order Filter Response from Parallel Coupled Glass Microring Resonators," *IEEE Photon. Technol. Lett.*, **11**(11), pp. 1426–1428, 1999.

110. C. Madsen, G. Lenz, A. Bruce, M. Cappuzzo, L. Gomez, and R. Scotti, "Integrated Tunable Allpass Filters for Adaptive Dispersion and Dispersion Slope Compensation," *IEEE Photon. Technol. Lett.*, **11**(12), pp. 1623–1625, 1999.

111. S. Dilwali and G. Pandian, "Pulse Response of a Fiber Dispersion Equalizing Scheme Based on an Optical Resonator," *IEEE Photonics Technol. Lett.*, **4**(8), pp. 942–944, 1992.

112. C. Madsen and G. Lenz, "Optical All-Pass Filters for Phase Response Design with Applications for Dispersion Compensation," *IEEE Photonics Technol. Lett.*, **10**(7), pp. 994–996, 1998.

113. C. Madsen, C. K. Madsen, S. Chandrasekhar, E. J. Laskowski, K. Bogart, M. A. Cappuzzo, A. Paunescu, L. W. Stulz, and L. T. Gomez, "Compact Integrated Tunable Chromatic Dispersion Compensator with a 4000 ps/nm Tuning Range." OFC 2001, PD9.

114. C. K. Madsen, S. Chandrasekhar, E. J. Laskowski, M. A. Cappuzzo, J. Bailey, E. Chen, L. T. Gomez, A. Griffin, R. Long, M. Rasras, A. Wong-Foy, L. W. Stulz, J. Weld, and Y. Low, "An Integrated Tunable Chromatic Dispersion Compensator for 40 Gb/s NRZ and CSRZ," OFC 2002, FD9.

115. F. Horst, C. Berendsen, R. Beyeler, G.-L. Bona, R. Germann, H. Salemink, and D. Wiesmann, "Tunable Ring Resonator Dispersion Compensators Realized in High-Refractive-Index Contrast SiON Technology," in *Postdeadline*. European Conference on Optical Communications, 2000.

116. F. Heismann, D. A. Fishman, and D. L. Wilson, "Automatic Compensation of First-Order Polarization Mode Dispersion in a 10 Gb/s Transmission System." ECOC 1998, pp. 529–530.

117. C. K. Madsen, E. J. Laskowski, L. Stulz, A. Griffin, M. A. Cappuzzo, L. Gomez, R. Long, J. Bailey, J. Weld, and P. Oswald, "An Integrated Polarization Mode Dispersion Emulator or Compensator with Tunable Chromatic Dispersion." Integrated Photonics Research Conference, 2002.

118. C. Madsen, "A Multiport Band Selector with Inherently Low Loss, Flat Passbands and Low Crosstalk," *IEEE Photon. Technol. Lett.*, **10**(12), pp. 1766–1768, 1998.

119. C. Madsen, "General IIR Optical Filter Design for WDM Applications Using Allpass Filters," *J. of Lightw. Technol.*, **18**(6), pp. 860–868, 2000.

120. P. Absil, J. Hryniewicz, B. Little, R. Wilson, L. Joneckis, and P.-T. Ho, "Compact Microring Notch Filters," *IEEE Photon. Technol. Lett.*, **12**(4), pp. 398–400, 2000.

121. H. Macleod, *Thin-film Optical Filters*. New York: McGraw-Hill, 1989.

122. H. Takashashi, "Temperature stability of thin-film narrow-bandpass filters produced by ion-assisted deposition," *Appl. Opt.*, **34**(4), pp. 667–675, 1995.

123. M. Scobey and D. Spock, "Passive DWDM components using MicroPlasma optical interference filters," in *Technical Digest*, Optical Fiber Communications 1996, pp. 242–243.

124. M. Scobey, R. Fortenberry, L. Stokes, and W. Kastanis, "Thin Film Interference Filters for 25 GHz Channel Spacing." Optical Fiber Communications Conference, 2002. (ThC5).

125. R. Fortenberry, M. Wescott, L. Ghislain, M. Scobey, and, "Low Chromatic Dispersion Thin Film DWDM Filters for 40 Gb/s Transmission Systems." Optical Fiber Communications Conference, 2002. (WS2).

126. L. Ghislain, R. Sommer, R. Ryall, R. Fortenberry, D. Derickson, P. Egerton, M. Kozlowski, D. Poirier, S. DeMange, L. Stokes, and M. Scobey, "Miniature Solid Etalon Interleaver." 17th Annual Fiber Optic Engineers Conference, 2001.

127. A. Gnauck, L. Cimini, J. Stone, and L. Stulz, "Optical Equalization of Fiber Chromatic Dispersion in a 5-Gb/s Transmission System," *IEEE Photon. Technol. Lett.*, **2**(8), pp. 585–587, 1990.

128. C. Madsen, J. Walker, J. Ford, K. Goossen, and G. Lenz, "A Tunable Dispersion Compensating MARS All-Pass Filter." European Conference on Optical Communications. Nice, France, 1999.

129. D. Moss, S. McLaughlin, G. Randall, M. Lamont, M. Ardekani, P. Colbourne, S. Kiran, and C. Hulse, "Multichannel Tunable Dispersion Compensation Using All-Pass Multicavity Etalons." Optical Fiber Communications Conference. Anaheim, CA, March 19–22, 2002.

130. M. Jablonski, Y. Takushima, K. Kikuchi, and Y. Tanaka, "Adjustable Coupled Two Cavity Allpass Filter for Dispersion Slope Compensation of Optical Fibres," *Electron. Lett.*, **36**(6), pp. 511–512, 2000.

131. B. Dingel and M. Izutsu, "Multifunction Optical Filter with a Michelson-Gires-Tournois Interferometer for Wavelength-Division-Multiplexed Network System Application," *Opt. Lett.*, **23**(14), pp. 1099–1101, 1998.

132. B. Dingel and T. Aruga, "Properties of a Novel Noncascaded Type, Easy-to-Design, Ripple-Free Optical Bandpass Filter," *J. Lightw. Technol.*, **17**(8), pp. 1461–1469, 1999.

133. H. Kogelnik, "Filter Response of Nonuniform Almost-Periodic Structures," *Bell System Techn. J.*, **55**(1), pp. 109–126, 1976.

134. M. Ibsen, M. Durkin, M. Cole, and R. Laming, "Optimized Square Passband Fiber Bragg Grating Filter With In-band Flat Group Delay Response," *Electron. Lett.*, **34**(8), pp. 800–801, 1998.

135. T. Strasser, P. Chandonnet, J. DeMarco, C. Soccolich, J. Pedrazzani, D. DiGiovanni, M. Andrejco, and D. Shenk, "UV-induced Fiber Grating OADM devices for efficient bandwidth utilization," *Optical Fiber Conference*, San Jose, CA, Feb. 25-Mar. 1(PD8, pp. 1–)4, 1996.

136. G. Bjork and O. Nilsson, "A New Exact and Efficient Numerical Matrix Theory of Complicated Laser Structures: Properties of Asymmetric Phase-Shifted DFB Lasers," *J. Lightw. Technol.*, **5**(1), pp. 140–146, 1987.

137. M. Yamada and K. Sakuda, "Analysis of Almost-Periodic Distributed Feedback Slab Waveguides Via a Fundamental Matrix Approach," *Appl. Opt.*, **26**(16), pp. 3474–3478, 1987.

138. R. Feced and M. Zervas, "Efficient Inverse Scattering Algorithm for the Design of Grating-Assisted Codirectional Mode Couplers," *J. Opt. Soc. Am. A*, **17**(9), pp. 1573–1582, 2000.

139. R. Feced, M. Zervas, and M. Muriel, "An Efficient Inverse Scattering Algorithm for the Design of Nonuniform Fiber Bragg Gratings," *IEEE J. of Quantum Electronics*, **35**(8), pp. 1105–1115, 1999.

140. P. Russell, "Optical superlattices for modulation and deflection of light," *J. Appl. Phys.*, **59**, pp. 3344–3355, 1986.

141. P. Russell, "Bragg resonance of light in optical superlattices," *Phys. Rev. Lett.*, **56**, pp. 596–599, 1986.

142. B. Eggleton, P. Krug, L. Poladian, and F. Ouellette, "Long periodic superstructure Bragg gratings in optical fibers," *Electron. Lett.*, **30**, pp. 1620–1622, 1994.

143. D. Johnson, K. Hill, F. Bilodeau, and S. Faucher, "New design concept for a narrowband wavelength-selective optical tap and combiner," *Electron. Lett.*, **23**(13), pp. 668–669, 1987.

144. R. Kashyap, G. Maxwell, and B. Ainslie, "Laser-Trimmed Four-Port Bandpass Filter Fabricated in Single-Mode Photosensitive Ge-Doped Planar Waveguide," *IEEE Photonics Technol. Lett.*, **5**(2), pp. 191–194, 1993.

145. T. Erdogan,   T. Strasser,   M. Milbrodt,   E. Laskowski,   C. Henry,   and   G. Kohnke, "Integrated-Optical Mach–Zehnder Add-Drop Filter Fabricated by a Single UV-Induced Grating Exposure," Optical Fiber Conference. San Jose, CA, Feb., 1996.

146. F. Ouellette, "Dispersion cancellation using linearly chirped Bragg grating filters in optical waveguides," *Opt. Lett.*, **12**, pp. 847–849, 1987.

147. K. Ennser, M. Zervas, and R. Laming, "Optimization of Apodized Linearly Chirped Fiber Gratings for Optical Communications," *IEEE J. Quantum Electron.*, **34**(5), pp. 770–778, 1998.

148. F. Ouellette, P. Krug, T. Stephens, G. Dhosi, and B. Eggleton, "Broadband and WDM dispersion compensation using chirped sampled fibre Bragg gratings," *Electron. Lett.*, **3**(11), pp. 899–901, 1995.

149. R. Kashyap,   A. Ellis,   D. Malyon,   H. Froehlich,   A. Swanton,   and   D. Armes, "Eight wavelength × 10 Gb/s simultaneous dispersion compensation over 100 km single-mode fiber using a single 10 nanometer bandwidth, 1.3 meter long, super-step-chirped fiber bragg grating with a continuous delay of 13.5 nanoseconds," Proc. European Conf. on Optical Communications (ECOC), p. ThB.3.2, 1996.

150. B. Eggleton, A. Ahuja, P. Westbrook, J. Rogers, P. Kuo, T. Nielson, and B. Mikkelsen, "Integrated Tunable Fiber Gratings for Dispersion Management in High-Bit Rate Systems," *J. Lightw. Technol.*, **18**(10), pp. 1418–1432, 2000.

151. K. Ennser, M. Ibsen, M. Durkin, M. Zervas, and R. Laming, "Influence of Nonideal Chirped Fiber Grating Characteristics on Dispersion Cancellation," *IEEE Photon. Technol. Lett.*, **10**(10), pp. 1476–1478, 1998.

# 15

# WAVELENGTH DIVISION MULTIPLEXERS AND DEMULTIPLEXERS

José Capmany and Salvador Sales

*Optical Communications Group*
*IMCO2 Research Institute*
*Universidad Politécnica de Valencia*
*Spain*

## 15.1 INTRODUCTION

Wavelength division multiplexing (WDM) has consolidated over the last years as the preferred transmission technology for the transport network [1–3]. Commercial systems have evolved from the relative modest figures of 1996 (eight wavelengths and 2.5 Gb/s per wavelength) [4] to the current typical values at the time of writing, where a number of products with a channel number in excess of 100 (128 and 256) and bit rates of 10 Gb/s are available in the market [5]. Furthermore, systems providing bit rates per wavelength of 40 Gb/s have been successfully demonstrated at field trial research experiments by different laboratories [6–10], and their commercial deployments are on the line of sight [11].

WDM presents a unique set of advantages that have contributed to its rapid widespread and generalized incorporation to support the need for broadband communications that have been triggered by the increase of data traffic, mainly driven in the last years by Internet applications. The list is extensive, and the interested reader is referred to the literature [12–16] for a comprehensive and detailed description of each one. Among them, the possibility of transmitting signals with independent formats and protocols on each wavelength with negligible or bounded interchannel interaction and, most notably, the possibility for easy capacity upgrade by suitable addition of more wavelengths are to be outlined.

*Handbook of Optical Components and Engineering*, Edited by Kai Chang
ISBN 0-471-39055-0 © 2003 John Wiley & Sons, Inc.

**935**

The consolidation of WDM systems in the present and its subsequent future evolution to optical-based WDM networks has so far been, and will be in the near future, possible as the result of the development and combination of different enabling technologies such as tunable semiconductor lasers, high-performance external modulators, special fibers, optical amplifiers, and signal processing photonics devices. Under the last group, two key components deserve special attention because they are critical for their successful operation; these are the WDM multiplexer/demultiplexer and the optical add drop multiplexer (OADM in short) devices, both of which are the subject of the present chapter.

The outline of this chapter is as follows. Section 15.2 deals with the basic operation principles, characteristics, and applications of both WDM MUX/DEMUX devices as well as of OADMs. We will follow a technology-independent approach, (i.e., a black-box description) first to explain how these devices operate, then to point out the salient features and parameters that must be employed for their complete description, and third to provide a thorough description of the applications of these devices in the context of photonic WDM communication systems and networks.

The different technological options for the implementation of these devices are covered in more detail in Sections 15.3 to 15.6. Each of these sections include specific subsections addressing the description of the technology, the main theoretical and design principles used in the design of the devices, a brief discussion on the performance parameters and the range of values attainable with that particular technology, and finally, a description of the most important device configurations.

In particular, Section 15.3 describes the implementation of WDM MUX/DEMUX devices and OADMs using thin-film technology, Section 15.4 describes the free-space technologies, and finally, Sections 15.5 and 15.6 focus on guided wave implementations, both in fiber (Section 15.5) as well as in integrated optics (Section 15.6) formats.


## 15.2 GENERAL ISSUES ON WDM MUX/DEMUX AND OADM DEVICES

### 15.2.1 Operation Principles

A WDM multiplexer/demultiplexer (WDM MUX/DEMUX in short) is a multiport photonic device that acts as a wavelength multiplexer (combiner) in one direction of propagation and as a wavelength demultiplexer (separator) in the opposite direction [17–22]. Figure 15.1 shows the basic functionality of the device under the two operation regimes and the black-box layout that we will employ in this chapter.

The upper part shows the operation as a multiplexer. Here signals at different wavelengths arrive at the device through different input fiber ports. The device combines all of the input channels and delivers the multiplexed signal to an output signal fiber port. The demultiplexing operation, which is shown in the lower part of Fig. 15.1 is just the opposite operation. Here the compound multiwavelength signal consisting of N different wavelength channels arriving from a single input fiber port is separated into N physically fiber ports (one for each wavelength channel).

Ideally, the combination/separation must not produce any interaction or crosstalk between the information content of the different wavelength channels, and it should be performed, incurring in as low insertion losses per channel as possible. Other properties such as polarization insensitivity, channel granularity, fan-in, fan- out, low cost, and so on, have also to be considered. In practice, however, ideal multiplexing/

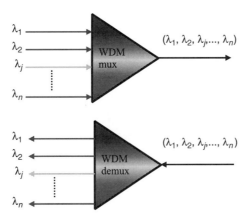

**Figure 15.1** Black-box description of a MUX/DEMUX device: MUX operation (upper); DEMUX operation (lower).

demultiplexing is not possible. Furthermore, there are tradeoffs between the relevant device performance parameters that have to be considered with care, depending on the application [23–27]. In any case, the different technologies available for the implementation of WDM MUX/DEMUX devices offer a variety of ranges for the values of these important parameters, and it is finally the designer who has to choose depending on the specific system's requirements.

An OADM is a device with similar operational characteristics but somehow different applications [28–30]. OADMs are employed in optical communications systems and networks at selected points, where one wants to extract or drop a subset of the wavelengths that propagate through an optical fiber link in order to terminate their traffic at line terminal equipment (LTE) attached to them, leaving the rest of the wavelengths unaltered. At the same time, because the traffic of the extracted wavelengths is terminated, these wavelengths can be reused in order to inject new traffic generated at LTE attached to the OADM into the fiber link; this is the so-called add function. The OADM must therefore have a general input/output fiber port that conveys the in-going/outgoing traffic to/from the device, and a set of the so-called local ports. Local ports come in fiber pairs, and each one is reserved to one of the wavelengths to be extracted/inserted by the OADM. Figure 15.2 shows the operation principle of the OADM. In this case, only two wavelengths are dropped/added.

Although WDM MUX/DEMUX are usually fixed devices, that is, the number of multiplexing/demultiplexing ports N is fixed, ranging from $N = 2$ to $N = 128$ or even higher, OADM can either be fixed or reconfigurable. In this later case, the number of wavelength channels that can be dropped/added can be changed (usually up to a maximum of eight channels) in response to an output control signal.

### 15.2.2 Functional Parameters

The quality of the operation of a WDM MUX/DEMUX or an OADM device can be described in terms of a variety of parameters that can usually be measured in a laboratory environment [31–35]. In order to understand the meaning of some of these parameters, it is instructive to bear in mind that the transfer function of any input–output port configuration either in a multiplexing or a demultiplexing operation

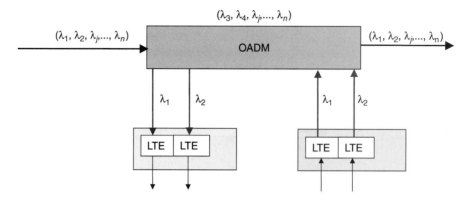

**Figure 15.2** Operation principle of an OADM device with two added/dropped wavelengths.

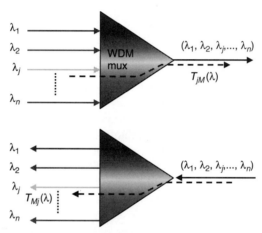

**Figure 15.3** Transfer functions between the input port $j$ and the output multiplexed signal $T_{jM}(\lambda)$ for the MUX operation (upper) and transfer function $T_{Mj}(\lambda)$ between the input multiplexed signal and the output demultiplexed signal at port $j$ for the DEMUX operation (lower).

of a WDM MUX/DEMUX can be represented by an optical bandpass filter placed around at the center wavelength $\lambda_j$ of the channel in question, and the same applies to the drop/add functionality of the OADM. This is shown schematically in Fig. 15.3, where we plot the transfer functions between the input port "j" and the output multiplexed signal $T_{jM}(\lambda)$ for the MUX operation and the transfer function $T_{Mj}(\lambda)$ between the input multiplexed signal and the output demultiplexed signal at port "j" for the DEMUX operation. For a passive device $T_{jM}(\lambda) = T_{Mj}(\lambda)$, and this function corresponds to a bandpass filtering characteristic ideally centered at $\lambda_j$. Therefore, many functional parameters that describe the operation of these devices are identical to those commonly defined to specify the performance of optical filters. The only difference is that whereas for the later case a single input and output port configuration is usual, for WDM MUX/DEMUX and OADM devices, multiple input–output configurations are possible, and thus, a complete characterization requires the measurement of these

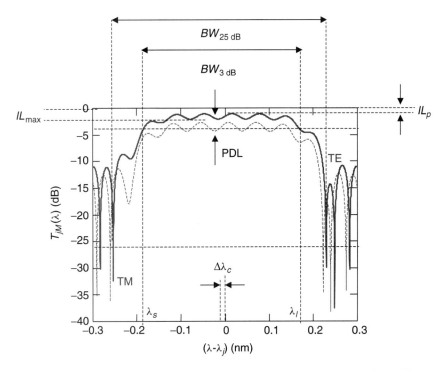

**Figure 15.4**   Modulus of the transfer function of a typical optical bandpass filter.

parameters for all possible input–output port configurations, although only worst-case values are specified in data sheets.

Figures 15.4 and 15.5 show the modulus and the group delay of the transfer function corresponding to a typical optical bandpass filter characteristic representing a given input–output port configuration of a WDM MUX/DEMUX device. Both the response for the transversal electric and transversal magnetic modes is shown in Fig. 15.4 in order to illustrate the polarization sensitivity of the device. The transfer characteristic for the TE mode is used to define some the main parameters of the device. With reference to Fig. 15.4, the following functional parameters can be defined:

1. *Peak Insertion loss* $IL_p$: The lowest insertion loss within the passband of the filter. It should be as small as possible.

2. *Maximum Insertion loss within the Passband* $IL_{max}$: The maximum insertion loss within the passband region. It should be as small as possible.

3. *Ripple* $\Delta IL$: The difference between the peak insertion loss and the maximum insertion loss within the passband region, $|IL_p - IL_{max}|$. Flat type MUX/DEMUX devices should exhibit as small a ripple as possible.

4. *Filter Center Wavelength* $\lambda_c$: Given by $(\lambda_s + \lambda_l)/2$, where $\lambda_s$ is the wavelength for which the value of the transfer function is $IL_p$-3dB on the shorter part of the bandpass, and $\lambda_l$ is the wavelength for which the value of the transfer function is $IL_p$-3dB on the longer part of the bandpass. Ideally, $\lambda_c$ should be equal to the wavelength $\lambda_j$ for which the input–output port configuration is considered.

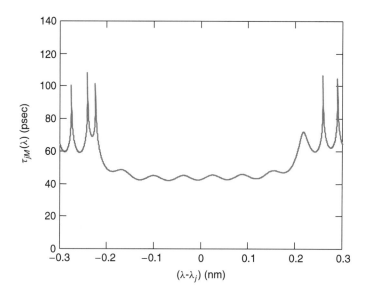

**Figure 15.5**   Group delay of the transfer function of a typical optical bandpass filter.

5. *0.5-dB Filter Center Wavelength* $\lambda'_c$: Same as above, but the wavelength points are those corresponding to $IL_p$-0.5dB on the longer and shorter part of the bandpass.

6. *25-dB Filter Center Wavelength* $\lambda''_c$: Same as above, but the wavelength points are those corresponding to $IL_p$-25 dB on the longer and shorter part of the bandpass.

7. *Center Wavelength Tolerance* $\Delta\lambda_c$: Given by $|\lambda_c - \lambda_j|$.

8. *Filter Asymmetry* $\delta$: Given by $|\lambda'_c - \lambda''_c|$. The smaller the value of $\delta$, the more symmetrical the filter transfer function with respect to $\lambda_c$.

9. *3-dB Bandwidth* $BW_{3dB}$: The filter bandwidth measured at $(IL_p$-3) dB. The result of this measure is usually considered as the passband region of the filter.

10. *25-dB Bandwidth* $BW_{25dB}$: The filter bandwidth measured at $(IL_p$-25) dB.

11. *x-dB Bandwidth* $BW_{xdB}$: The filter bandwidth measured at $(IL_p$-x) dB.

12. *Passband Skirt Parameter* $S_p$: A measure of the transfer function rolloff outside the passband region of the filter. On the shorter wavelength side with respect to $\lambda_c$ is given by $S_{ps} = |T_{Mj}(\lambda_s) - T_{Mj}(\lambda_{s25dB})|$. On the longer wavelength side $S_{pl} = |T_{Mj}(\lambda_l) - T_{Mj}(\lambda_{l25dB})|$, where $\lambda_{s25dB}$ represents the wavelength at which the transfer function is 25 dB below its value at $\lambda_c$ on the shorter wavelength side, and $\lambda_{l25dB}$ represents the same on the longer wavelength side. $S_{p(s/l)}$ is measured in decibels/nanometers. The higher the value of $S_{p(s/l)}$, the sharper the transition between the filter bandpass and the rejected band.

13. *Polarization-Dependent Loss PDL*: Given by the maximum difference between the transfer functions for the orthogonal polarizations corresponding to the TE and TM modes within the filter passband region. The smaller the value of PDL, the more polarization insensitive is the filter.

14. *Coefficient of Thermal Change in Center Wavelength CTW*: The maximum wavelength change within the operating temperature range, divided by the operating temperature range. It is usually expressed in pm/°C. for temperature-insensitive MUX/DEMUX and OADM devices, CTW should be as small as possible.

Relevant parameters can also be defined addressing the group delay characteristic, namely:

15. *Minimum GroupDelay $\tau_{gmin}$*: The lowest value of the group delay within the passband of the filter.

16. *Maximum GroupDelay $\tau_{gmax}$*: The highest value of the group delay within the passband of the filter.

17. *In Band Dispersion D*: Given by $d\tau_g/d\lambda$ within the passband of the filter. Ideally, it should be as close to zero as possible.

All of the parameters described above are important, but only part of the set required to describe the performance of the WDM MUX/DEMUX and OADM devices. Other more global parameters, describing the overall device operation, are also required to fully characterize them. Four of them are especially important and are described using Figs. 15.6 and 15.7 as a reference, where we plot, respectively, the superimposed spectral responses of all the resonances of a typical $1 \times 16$ DEMUX device and the central resonance of the device. For instance, from Fig. 15.6:

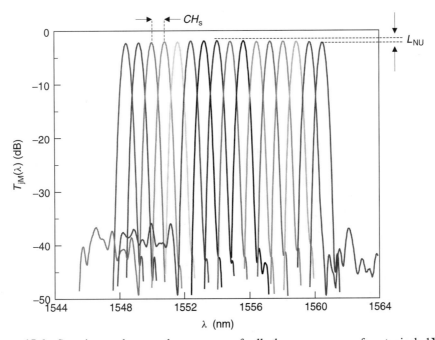

**Figure 15.6** Superimposed spectral responses of all the resonances of a typical 1X16 DEMUX device.

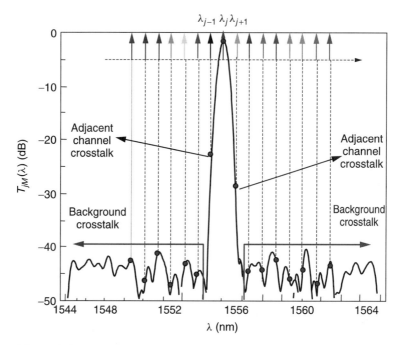

**Figure 15.7**    Central resonance of the spectral characteristic of Fig. 15.6.

1. *Loss Nonuniformity $L_{NU}$*: The difference between the highest and the lowest insertion losses of two bandpass characteristics of the device.

2. *Channel spacing CHs*: The wavelength difference between the center wavelengths of two adjacent passbands of the MUX/DEMUX device. Usually, this is specified by the application. For instance, in WDM systems, the ITU recommendation G.692 specifies possible channel spacings of 200 Ghz (1.6 nm), 100 GHz (0.8 nm), or 50 GHz (0.4 nm) for systems operating in the C band. On the other hand, Coarse WDM (CWDM) systems require channel spacings over 400 GHz (3.2 nm).

Figure 15.7 illustrates the concept of channel crosstalk that gives rise to the two important remaining parameters to be defined for a WDM MUX/DEMUX or an OADM device. The figure represents the filtering action of a particular resonance centered at $\lambda_j$. Under ideal conditions, the rest of the channels (centered at $\lambda_k$   $k \neq j$) should be completely rejected by the filtering action of $T_{Mj}(\lambda)$. In practice, however, energy from the rest or undesired channels leaks into the receiver because $T_{Mj}(\lambda)$ has a finite value at $\lambda = \lambda_k$, $k \neq j$. This energy leak manifests as crosstalk over the desired signal centered at $\lambda_j$. The crosstalk contribution can be divided into two groups: First, the so-called *adjacent channel crosstalk* (ACC) and is due to the energy leaked by $T_{Mj}(\lambda)$ from channels centered at $\lambda_{j-1}$ and $\lambda_{j+1}$. This crosstalk depends on the value of the skirt parameter, which was defined previously and is usually the strongest contribution to the overall crosstalk. The second source is the *background crosstalk* (BC) and is due to the energy leaked by $T_{Mj}(\lambda)$ from the rest of the channels. This crosstalk depends mainly on the background level of $T_{Mj}(\lambda)$ far away from $\lambda_j$.

3. *The Adjacent Channel Crosstalk $ACC_j$*: For the resonance placed at $\lambda_j$ is given in natural units by $ACC_j = (T_{Mj}(\lambda_{j-1}) + T_{Mj}(\lambda_{j+1}))/T_{Mj}(\lambda_j)$.

4. *The Background Crosstalk $BC_j$*: For the resonance placed at $\lambda_j$ is given in natural units by $BC_j = \sum_{k \neq 0,1} T_{Mj}(\lambda_{kj})/T_{Mj}(\lambda_j)$.

### 15.2.3  Applications of WDM MUX/DEMUX and OADMs

WDM MUX/DEMUX and OADM devices have multiple applications that cover different areas, including the design and implementation of more complex subsystems [36], and the use in point-to-point WDM transmission systems design [37], and in the field of optical networks either long-haul [38], metro [39], or access [40]. In this section, we briefly outline the general details of those that are considered more significant. Figure 15.8 shows a possible configuration of a segment within an optical network, which we will employ as a general framework to point out and describe these applications. There are three main subsystems within a WDM optical communications system or network in which WDM MUX/DEMUX devices are critical. These are as follows:

1. Optical Amplifiers (OAs)
2. Optical Line Terminals (OLTs)
3. Optical Cross Connects (OXCs)

Optical Amplifiers (OAs) are widely used in currently deployed optical communication systems to compensate for a variety of losses, including fiber attenuation [41–44], insertion losses at intermediate photonic components inside network nodes [45–47], distribution losses in broadcast-type systems [48], and so on. The most widely used

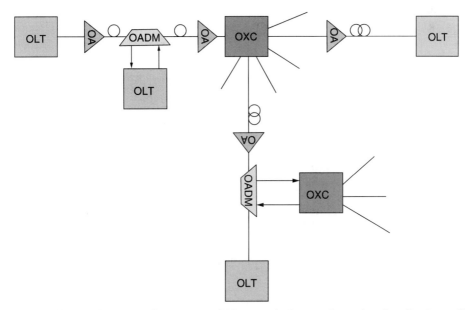

**Figure 15.8**  Configuration of a segment within an optical network, used to describe the applications of MUX/DEMUX and OADM devices.

OA is the standard erbium doped fiber amplifier (EDFA) [48], which is designed to operate in the central wavelength band (C-band), that is, between 1530 and 1565 nm. In standard EDFAs, the active medium is the core of an optical fiber doped with erbium ions. In order to provide gain, ions must be pumped from the ground to an excited state. This is done by means of an external optical power source (usually a laser diode emitting at either at 980 nm or at 1480 nm). At the same time, the signal to be amplified must propagate through the active medium. Thus, it is necessary to employ at least a MUX/DEMUX, in order to couple both radiations into the active medium. There are several options, however, because there is a variety of possibilities, to implement the pumping scheme [49–50], including the codirectional approach, the contradirectional approach, and the bidirectional approach. The details and advantages of each configuration are beyond the scope of this chapter, but the interested reader may refer to [50] for further details. In any case, they are shown schematically in Fig. 15.9.

Note that, depending on the configuration, up to two MUX/DEMUX devices are required. For this particular application, the two channels to be multiplexed are far apart; thus, a coarse MUX/DEMUX device must be employed. For instance, in the case of a pump signal of 980 nm and a C-band EDFA, the typical spectral separation between the pump and any signal to be amplified is over 500 nm! Furthermore, note that all the signals in the C-band must be treated as a "single channel or wavelength" by the MUX device (i.e., exactly the same operation has to be performed by the device on all the wavelengths of the C-band); thus, in this case, the MUX/DEMUX must operate more like a band multiplexer, rather than like a wavelength multiplexer. This concept,

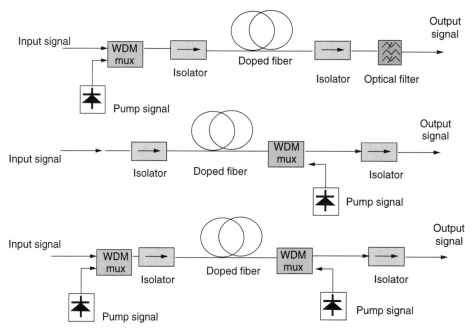

**Figure 15.9** Different alternatives for the implementation of the pumping schemes in EFDAs, including the co-directional (upper), the contra-directional (intermediate), and the bi-directional (lower) approaches.

which is known as waveband division multiplexing [51], will be explained in more detail later in the chapter, because it has important implications in the design of WDM all optical networks. C-band EDFAs provide extremely high, but spectrally nonuniform gains. This means, in practice, that a conventional EDFA is not adequate for its use in WDM systems, unless its gain is equalized [52–56]. Again, WDM MUX/DEMUX devices can find an application here, as shown in Fig. 15.10, where different channels are demultiplexed after amplification and subject to different variable optical attenuators (VOAs) to compensate for the EDFA nonuniform gain before they are multiplexed again.

With the ever increasing demand for more capacity driven by data and Internet applications, there has been a need to expand the usable spectrum in optical fibers beyond the C-band. In the first instance, the wavelengths within the long wavelength, or L-band, which spans from 1565 to 1625 nm, have been the first to be employed [52]–[55]. The use of the L-band allows for practically doubling the capacity of WDM systems if suitable optical amplification can be achieved. Fortunately, EDFA technology can still be employed here to implement the so-called L-band EDFAs. In the long wavelength region, the EDF has a much smaller gain coefficient (around 0.2 dB/m) for an optimum inversion level, between 20% and 30% (that yielding the maximum amplification bandwidth) greater than the gain at the highest peak of the C-band. However, its gain shape is much more uniform, so much less gain equalization is required in an L-band EDFA. Due to the small-gain coefficient, EDF lengths for L-band amplifiers exceed around 5 times that required for C-Band EDFAs. For this reason, one possible way to implement a wideband amplifier (covering C- and L-Bands) is to use a split-band architecture, as shown in Fig. 15.11, where the signal is directed to a different arm (upper or lower), depending on whether it is in the C- or L-band. Each arm then, respectively, implements the C- or L-band EDFA where MUX/DEMUX devices are employed for pump/signal multiplexing into the active medium. Note that the input separation (DEMUX) stage, which is implemented through a combination of a circulator and a fiber Bragg grating filter, could also be implemented by a MUX/DEMUX device performing waveband demultiplexing between the L- and the C-band. The same comment also applies for the output combination (MUX) stage.

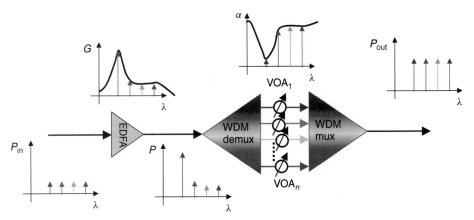

**Figure 15.10**  Configuration of an EDFA gain equalizer based on the use of two MUX/DEMUX devices.

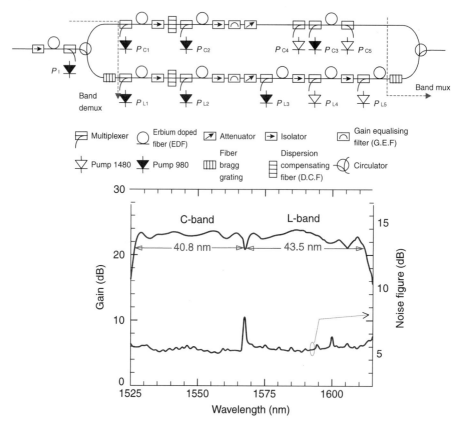

**Figure 15.11**   Split-band architecture for the implementation of C+L Banda EDFAs (upper); typical spectral gain characteristic (lower).

A second alternative to provide wideband amplification that has gained considerable attention in the last years is the use of Raman amplifiers, which can provide gain at any signal wavelength [56]–[63]. This is an extremely attractive feature, because it can continually increase the number of signal wavelengths by extending the bandwidth of amplified transmission. The Raman effect in silica fibers has been widely studied in recent years. Amplification results from stimulated Raman scattering that transfers energy from a pump light to the signal by means of exciting the vibrational modes in the material. The gain coefficient, which is polarization sensitive, can be significant at moderately high pump powers. The Raman gain peak is offset in wavelength from the pump wavelength by a Stokes shift. This means that, as mentioned previously, Raman amplifiers can be implemented at any wavelength just by selecting a suitable pump. In silica fibers, the Stokes shift is around 100 nm (13 THz) at 1550 nm, and the gain spectrum is uniform with a typical bandwidth of 40 nm (5 THz) in the C-band. Raman amplifiers provide lower gains than do those attainable with EDFAs, but they can implement distributed amplification, yielding much lower noise. Broadband flat Raman amplification can be achieved by using multiple pump and equalized wavelengths [64], so it is convenient for the pump source to contain multiple wavelengths. MUX/DEMUX devices are applied here for combining and injecting the multiple optical pump sources into the silica fiber. For instance, Fig. 15.12 shows the largest pump

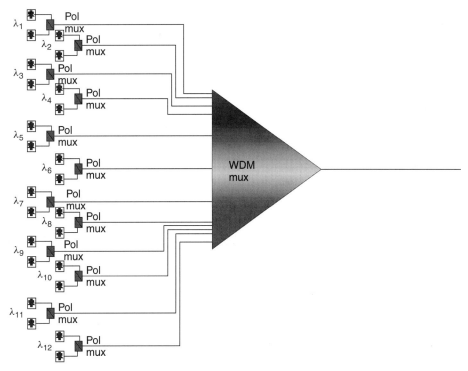

**Figure 15.12**  Pump scheme composed of 24 pump sources at 12 wavelengths (2 polarization multiplexed sources per wavelength) used to achieve very flat, 100 nm-wide Raman gain spectrum.

source so far constructed, composed of 24 pump sources at 12 wavelengths (2 polarization multiplexed sources per wavelength), attaining an output power of 2 W and a very flat, 100-nm–wide Raman gain spectrum.

Perhaps the most straightforward and extended application of WDM MUX/DEMUX devices is in the implementation of optical line terminals (OLT). OLTs are placed at either end of a WDM point-to-point link to multiplex and demultiplex wavelength channels. Figure 15.13 depicts a typical configuration of an OLT showing, where the WDM MUX/DEMUX device is placed. The OLT transforms input signals coming from various electrical equipment (SDH digital cross connects, ATM switches, and IP routers) into wavelength channels with formats compatible to those to be transmitted by the fiber link. These are usually set by ITU recommendation G.692. Wavelengths emitted by the electrical equipment do not match, most of the time, those required by the ITU recommendation, and therefore, intermediate equipment called a transponder is required to change the wavelength emitted by the electrical equipment into another that complies with ITU regulations. In the reverse direction, the opposite operation is performed, and the wavelength carrying a signal destined to an electrical equipment is changed to another that can be input to the latter.

Transponders usually add other functions, such as forward error correcting codes (FECs) and overhead and adaptation bits, for management purposes. The interested reader is referred to Ref. 65 for further information. Some modern equipment is prepared to be directly attached to the WDM MUX/DEMUX device without the need

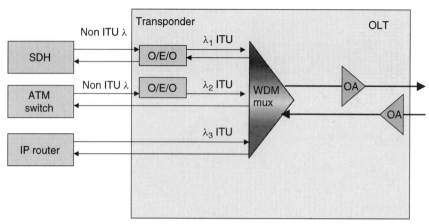

**Figure 15.13**   Application of WDM MUX/DEMUX device for signal combination/separation in Optical Line Terminals (OLTs).

of transponders or just needing one, as shown in the figure. In any case, a WDM MUX/DEMUX is indispensable. The device design requirements for this application are given mainly by the channel separation, which can range typically from 200 GHz for low-count WDM systems, 100 Ghz and 50 Ghz for ITU-regulated DWDM systems, and 20 nm for CWDM applications, and the bandwidth of the resonances, which can range typically from four times the bit rate to be transmitted by each wavelength to half of the channel separation. The reader should also note that optical amplifiers may also be required in OLTs, and this, of course, requires further MUX/DEMUX devices, as discussed earlier.

Optical cross connects (OXCs) are fundamental subsystems for the implementation of advanced all-optical networks, in which switching and routing functions are performed directly in the optical domain. Optical cross connects for WDM networks usually combine two different domains, wavelength and space, to perform switching at a wavelength channel level. The details of possible OXC configurations and their features and advantages can be found elsewhere in the literature [65]–[70], but most of them are composed of three stages, as shown in Fig. 15.14.

The first is the input stage in which input fiber ports, each one carrying WDM signals, are fed to WDM DEMUX devices to separate each wavelength channel. Each demultiplexed optical channel is then fed to the second stage, which is usually a space core switch that can be electrical or fully optical, and routes the channel to a given output fiber port. Output fiber ports can be either local (are connected to electrical terminal equipment attached to the OXC) or nonlocal (they are connected to another OXC). The third stage is composed by MUX devices that multiplex the different wavelength channels that have been switched to the same output fiber port.

The prior description shows that WDM MUX/DEMUX are required at the input and output stages of the OXC. Figure 15.15 shows an example of a possible, all-optical implementation for an OXC using WDM MUX/DEMUX devices.

In certain applications, the routing, switching, and extraction functions are required to be carried at a lower granularity than that provided by a wavelength. In these cases, wavelengths are grouped into bands, and these operations are carried at a band level, rather than at a wavelength level. Coarse WDM MUX/DEMUX devices also can be

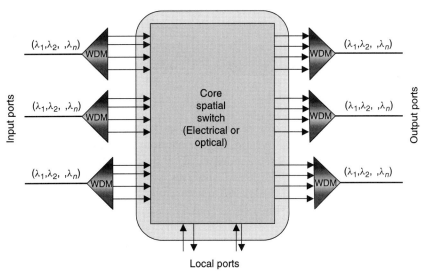

**Figure 15.14**  Three-stage description of an Optical Cross Connect (OXC) showing where MUX/DEMUX devices are needed.

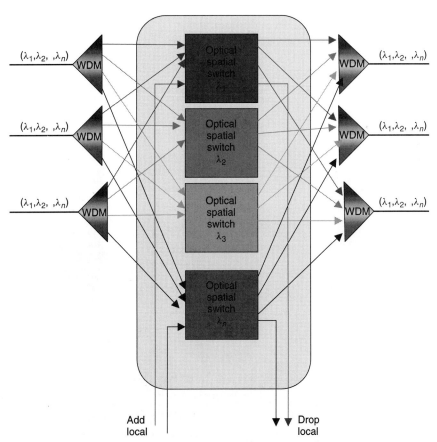

**Figure 15.15**  Example of an implementation of an OXC using WDM MUX/DEMUX devices and optical spatial switches.

applied here as well, because their resonances can be designed to be broad enough to select a certain group of wavelengths (or band). Each wavelength in the band is treated in the same way by the device. For instance, Fig. 15.16 shows an example of a 32-wavelength frequency plan divided into four bands, each one composed of 8 wavelengths. At a certain point, wavelengths in bands C and D need to be fully demultiplexed to have access to each individual wavelength, whereas now, there is no need to access the wavelengths in bands A and B.

A solution is to deploy the subsystem, shown in the figure, where an initial coarse WDM DEMUX device is employed to separate the four bands, and then, in a second stage, finer WDM DEMUX devices are applied to the output ports corresponding to the frequency bands (C and D) whose individual wavelengths need to be accessed. The rest of the bands (A and B) are left untouched and directly remultiplexed by a coarse MUX device. For a typical ITU system with a 100-Ghz separation between wavelengths, the coarse MUX/DEMUX device would require at least 800-GHz channel separation.

Of course, this simple example for a static subsystem can be extended to dynamic OXCs where it is advantageous to carry out some routing and switching operations at a band level in order to save expensive photonic components. For instance, Fig. 15.17 shows a node design for a metropolitan network [66] capable of both waveband and wavelength switching.

In some applications, like, network protection [67], for example, it is interesting to form a signal multiplex by interleaving wavelengths. Here, bands are not formed by contiguous wavelengths, but by wavelengths separated by a fixed amount of times the minimum channel separation in the full multiplex. This concept is shown in Fig. 15.18, where a two stage MUX system is used to interleave two bands composed

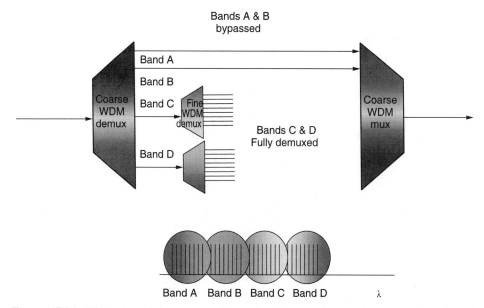

**Figure 15.16** Illustration of the concept of wave-banding and the use of coarse and fine WDM MUX/DEMUX devices to achieve band and wavelength level separation/combination, respectively.

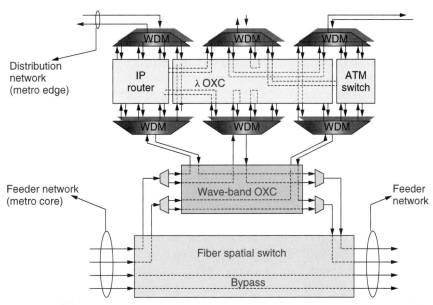

**Figure 15.17**  Node design for a metropolitan network capable of performing waveband and wavelength switching. (Source: [66].)

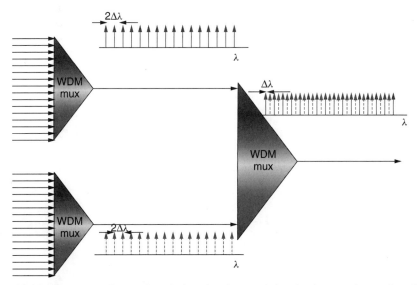

**Figure 15.18**  Concept of wavelength interleaving and its implementation using WDM MUX/DEMUX devices in various stages.

by wavelengths separated by $2\Delta\lambda$ to form a multiplex signal, composed by wavelengths separated by $\Delta\lambda$. This process, of course, can be extended to more stages, because each of the original bands could be formed by a prior interleaving of bands, composed by wavelengths separated by $4\Delta\lambda$, and so forth. The signal demultiplexing is achieved by reversing the direction of propagation.

OADM devices find important applications in the field of optical networks and, more specifically, in the establishment and management of permanent and reconfigurable wavelength-routed connections between nodes [68–75]. Furthermore, OADMs help to save expensive equipment by avoiding the need to convert to electronic-format wavelength channels that can be bypassed at intermediate nodes of the network. OADMs are applied mainly in linear WDM links and rings, just like their SDH counterparts. To illustrate their potential for cost and equipment savings, Fig. 15.19 depicts a simple linear WDM link with three nodes where two wavelengths are required for the traffic between node A and B, six wavelengths are required for the traffic between A and C, and two wavelengths are required to transport the traffic between B and C.

If the link is implemented by means of two point-to-point links (one from A to B, and the second from B to C), then up to 32 electronic-line terminal-equipment cards are required (16 for transmitters and 16 for receivers), as shown in the upper part of Fig. 15.19, because all wavelength channels must be detected and converted to electronic format at node B. If, however, we allow for the use of a fixed OADM with the capability of extracting and inserting two wavelength channels, then the wavelengths used to establish the connections between A and B can be reutilized to establish the link between B and C. Furthermore, the six remaining wavelengths that carry the traffic between A and C do not need to be converted to electronic format, and therefore, we can save 6 receivers and 6 transmitters, for a total of 12 electronic line-terminal equipment cards. A similar procedure can be followed to show equipment savings in ring configurations.

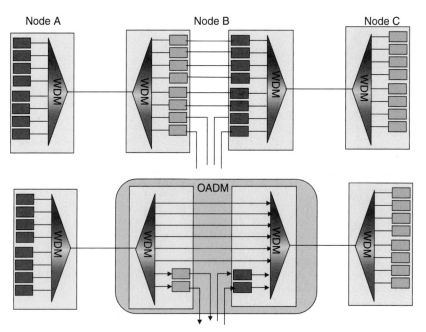

**Figure 15.19** Simple linear WDM link with three nodes. Two wavelengths are required for the traffic between node A and B, 6 wavelengths for the traffic between A and C, and 2 for the traffic between B and C.

OADMS can be implemented in a variety of architectural configurations, either fixed or reconfigurable, and like WDM MUX/DEMUX devices, they can be employed to extract and insert either wavelength channels or wavebands. For instance, the internal configuration for the OADM, shown in Fig. 15.18, corresponds to a parallel configuration. In this particular arrangement, WDM MUX/DEMUX devices are employed as building blocks for OADMS. Figure 15.20 shows typical configurations for parallel fixed OADMs, both wavelength and waveband selective.

The main advantages of these configurations are two fold. First of all, the device loss is independent of the number of wavelengths or wavebands to be added/dropped. Secondly, they lead to very flexible configurations, allowing for the easy implementation of reconfigurable OADMS. For instance, the addition of $2 \times 2$ switches, or even OXCs, in between the DEMUX and MUX devices, enables, as it can be appreciated from Fig. 15.21, the implementation of fully reconfigurable devices.

A second possibility for the implementation of OADM is to use a series arrangement where single channel (SC) OADM devices are placed in tandem. Here again, both fixed and tunable versions can be implemented, either for wavelength channels or for waveband. Series arrangements are particularly interesting for nodes in which a low number of wavelengths are to be inserted/extracted, because losses scale with this number. Furthermore, they are very cost effective, because modules are added as required (pay as you grow). However, they have a number of disadvantages. Most notably, the performance of each SC OADM affects all of those coming after it, so if a particular SC OADM fails, the whole device will come out of service. Figure 15.22 shows the arrangements for fixed and tunable OADMs.

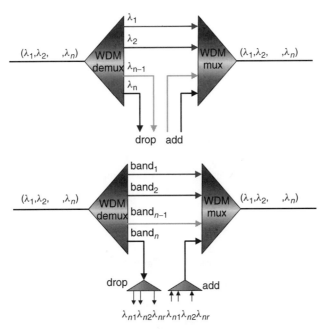

**Figure 15.20**  Typical configurations for parallel fixed OADM devices: wavelength extraction/addition (upper); waveband extraction/addition (lower).

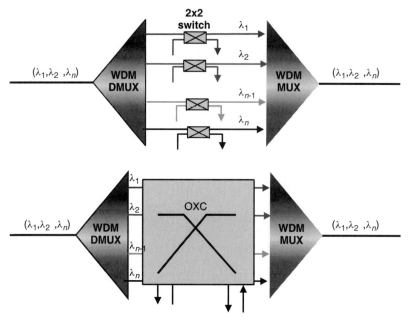

**Figure 15.21** Typical configurations for parallel tunable OADM devices: wavelength extraction/addition (upper); waveband extraction/addition (lower).

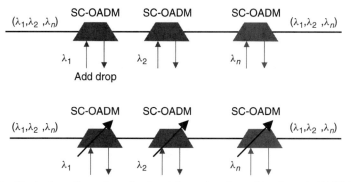

**Figure 15.22** Typical configurations for fixed (upper) and tunable (lower) OADM devices in serial arrangement.

## 15.3    THIN-FILM MUX/DEMUX DEVICES

### 15.3.1    Introduction to Thin-Film Filter Technology

Thin-film filters (TFFs) are composed of one or several coupled Fabry–Perot cavities in which the mirrors that close the etalons are made by stacking multiple dielectric thin-film layers, as shown in Fig. 15.23.

Multiple signals originating from the reflections and the transmissions through the dielectric stacks can combine (interfere) constructively or destructively for given sets of wavelength bands. These devices operate thus as optical bandpass filters allowing the passage through the device of a restricted wavelength band and reflecting the rest of

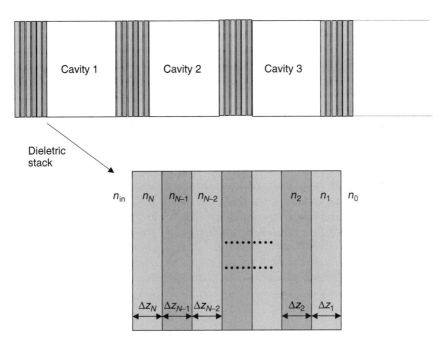

**Figure 15.23** Configuration of a thin film filter.

the wavelength spectrum. Each layer of the stack has a length comparable to (usually a quarter of) that of the center of the wavelength region that the filter is designed to transmit.

The materials most commonly employed in the implementation of the dielectric stacks and their typical refractive indexes are given in Table 15.1.

Physical vapor deposition (PVD) techniques are those most commonly employed for the fabrication of thin-film dielectric stacks. These can be described as vacuum processes in which a solid film condenses from the vapor phase. The easiest method is the one termed as thermal evaporation, and the interested reader can find a thorough description in Ref. 77. Here, the material to be deposited known as the evaporant is heated to a temperature at which it vaporizes. The vapor then condenses as a solid film on the substrates, which are maintained at temperatures below the melting point of the evaporant. Molecules travel virtually in straight lines between the source and the substrate. This method, although simple, has however several disadvantages that stem

**TABLE 15.1**

| Material | Refractive Index $1 \leqslant \lambda \leqslant 2$ nm |
|---|---|
| $SiO_2$ | 1.44 |
| $SiO$ | 1.85 |
| $TiO_2$ | 2.25 |
| $Ta_2O_5$ | 2.09 |
| $Si$ | 3.49 |

from the production of defects of solidity, porosity, scatter, absorption inhomogeneity, and lack of accurate thickness control, which result in several shortcomings regarding poor temperature and environmental stability, high insertion losses, and poor bandpass spectral shape. For WDM and high-quality DWDM applications a different set of methods termed as energetic processes are more widely employed because they lead to the production of high-quality stacks. Energetic processes work by transferring mechanical momentum to the growing film by deliberate bombardment or by an increase in the momentum of the arriving film material that drives the outermost material deeper into the film, thus increasing its solidity [77]. A particularly high-quality process is the MicroPlasma [81] technique, an energetic PVD process that uses refractory metal-oxide coating materials such as silicon dioxide and tantalum pentoxide. The energetic nature of the deposition produces hard, pore-free, amorphous thin-film microstructures with very low absorption and scatter levels. Furthermore, typical temperature coefficients for narrowband filters are on the order of 0.001 nm/°C for temperatures up to 100°C.

### 15.3.2   Theory and Design of Thin-Film Filters

When designing a thin-film filter, one is interested in obtaining the electric field transfer functions for the transmitted and reflected signals. From those, it is straightforward to obtain the intensity transfer functions, that is, the filter reflectance and transmittance. Multiple techniques based on the propagation of plane waves have been developed that allow for the calculation of those quantities, and the interested reader is referred to the existing literature [76]–[80]. A particularly simple and powerful method is that based on the ADBC matrix, which is further developed here.

Figure 15.24 shows an arbitrary dielectric stack and the interfaces and labeling for the electric fields at a given layer (layer $k$) of the stack. At each interface, the electric (magnetic) field is composed by two phasors $E_{k+1}^{+}$ ($H_{k+1}^{+}$) and $E_{k+1}^{-}$ ($H_{k+1}^{-}$) that propagate in the positive and negative z direction, respectively. We define the total transverse electric and magnetic fields at the interface as

$$V_{k+1} = E_{k+1}^{+} + E_{k+1}^{-}$$
$$I_{k+1} = H_{k+1}^{+} + H_{k+1}^{-} \tag{15.1}$$

The application of the continuity and boundary conditions for the electric and magnetic fields yields the following matrix relationship for the dielectric layer:

$$\begin{pmatrix} V_{k+1} \\ I_{k+1} \end{pmatrix} = \begin{pmatrix} \cos\varphi_k & j\dfrac{Z_o}{n_k}\sin\varphi_k \\ j\dfrac{n_k}{Z_o}\sin\varphi_k & \cos\varphi_k \end{pmatrix} \cdot \begin{pmatrix} V_k \\ I_k \end{pmatrix} = M_k \begin{pmatrix} V_k \\ I_k \end{pmatrix} \tag{15.2}$$

where $Z_o$ is the impedance of vacuum and $\varphi_k = (2\pi/\lambda)n_k\Delta z_k$. By an iterative application of the above relationship, we can relate the total transverse fields at the left and right parts of the dielectric stack as

$$\begin{pmatrix} V_{N+1} \\ I_{N+1} \end{pmatrix} = \begin{pmatrix} A & B \\ C & D \end{pmatrix} \cdot \begin{pmatrix} V_0 \\ I_0 \end{pmatrix} = \prod_{j=1}^{N} M_j \begin{pmatrix} V_0 \\ I_0 \end{pmatrix} \tag{15.3}$$

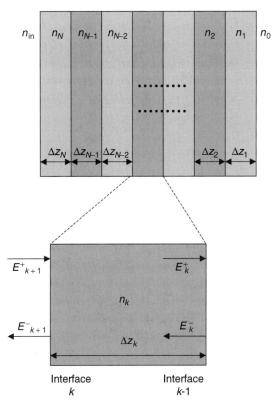

**Figure 15.24**  Arbitrary dielectric stack and the interfaces and labeling for the electric fields at a given layer (layer $k$) of the stack.

Once the coefficients of the overall $ABCD$ matrix are known, it is straightforward to obtain the field reflection and transmission transfer functions:

$$r = \left. \frac{E_{N+1}^-}{E_{N+1}^+} \right|_{E_o^-=0} = \frac{A + \dfrac{Bn_o}{Z_o} - \dfrac{CZ_o}{n_{in}} - \dfrac{Dn_o}{n_{in}}}{A + \dfrac{Bn_o}{Z_o} + \dfrac{CZ_o}{n_{in}} + \dfrac{Dn_o}{n_{in}}} \tag{15.4}$$

$$t = \left. \frac{E_o^+}{E_{N+1}^+} \right|_{E_o^-=0} = \frac{2}{A + \dfrac{Bn_o}{Z_o} + \dfrac{CZ_o}{n_{in}} + \dfrac{Dn_o}{n_{in}}}$$

In general, the dielectric layers are implemented, such that for the design wavelength $\lambda_o$, each layer thickness is equivalent to a quarter wavelength; hence, $\varphi = (\pi/2)(\lambda/\lambda_o)$, and therefore, at that wavelength, the $ABCD$ matrix has the following simple form:

$$\begin{pmatrix} A & B \\ C & D \end{pmatrix} = \begin{pmatrix} 0 & j\dfrac{Z_o}{n_k} \\ j\dfrac{n_k}{Z_o} & 0 \end{pmatrix} \tag{15.5}$$

The stack is then constructed by alternating quarter-wave sections of high ($H$) and low ($L$) refractive index. For example, $(HL)^k$ designates a $2k$ layer stack implemented by alternating $k$ high refractive index quarter-wave layers with $k$ low refractive index quarter-wave layers. Cavities or etalons are implemented using longer sections, usually designed to be half wavelength thick layers at $\lambda_o$. The cavity is, therefore, named by $HH$ or $LL$ depending on whether its refractive index is high or low. The $ABCD$ matrix for such a cavity at $\lambda_o$ is given by

$$\begin{pmatrix} A & B \\ C & D \end{pmatrix} = \begin{pmatrix} -1 & 0 \\ 0 & -1 \end{pmatrix} \tag{15.6}$$

In general, the input and output dielectrics placed before and after the thin-film filter are identical and are composed of a glass with different refractive index to that of the high ($H$) and low ($L$) values of the materials employed in the filter. Thus, it is usual to employ the letter $G$ to designate such dielectrics. For example, $GHLHLLHLHG$ represents a 1-cavity thin-film filter with an internal cavity made of low refractive index material surrounded by mirrors composed of a three-layer stack $HLH$. Figure 15.25 shows its transfer function on transmission as a function of the inverse wavelength normalized by $\lambda_o$ for a typical implementation, where the $H$ sections are made of TiO$_2$ and the $L$ sections are made of SiO$_2$ (see Table 15.1 for the values of the refractive indexes). The Glass substrate $G$ is assumed to have a refractive index of 1.52.

The reader should note that the transfer function is periodic in frequency or in ($\lambda_o/\lambda$). In fact, Fig. 15.25 shows the transfer function for two periods. A narrow

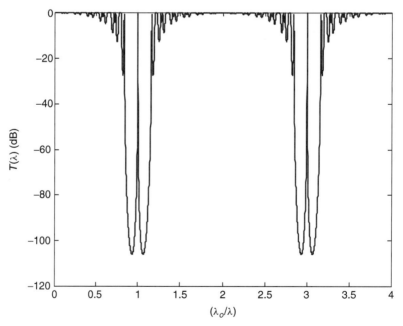

**Figure 15.25** Transfer function on transmission as a function of the inverse wavelength normalized by $\lambda_o$ for a typical HLH filter implementation, where the $H$ sections are made of TiO$_2$ and the $L$ sections are made of SiO$_2$. The Glass substrate $G$ is assumed to have a refractive index of 1.52.

bandpass characteristic is obtained at $(\lambda_o/\lambda = 2\kappa + 1$, with $\kappa = 0, 1, 2, \ldots)$, but it is difficult to appreciate due to the very high-frequency range covered by the plot. A much clearer appreciation of the bandpass nature of the filter is observed if we plot the same transfer function for a narrow spectral region around $\lambda_o$, as it is done in Fig. 15.26.

Using more than one cavity in the thin-film filter structure yields flatter passbands with sharper rolloff characteristics in the stop band. For instance, Fig. 15.26 shows the obtained transfer functions in reflection for two- and three-cavity filters. The same materials as in the case of the single cavity are used, and filter layouts are given, respectively, by $G(HL)^6HLL(HL)^{12}HLL(HL)^6HG$ and $G(HL)^5HLL(HL)^{11}HLL(HL)^{11}HLL(HL)^5HG$.

Thin-film filters can also be designed for non-normal incidence. In this case, the ABDC matrix method can still be used with minor modifications. For instance, the refractive indexes in Eq. (15.2) must be substituted by $n_k \cos \alpha_k$ for TE waves and by $n_k/\cos \alpha_k$ for TM waves, where $\alpha_k$ is the angle of incidence with respect to the normal in the kth layers. Also, the phase shift in the cavity has to be replaced by $\varphi_k = (2\pi/\lambda)n_k \Delta z_k \cos \alpha_k$.

### 15.3.3  Performance Characteristics

We now turn to discuss some of the performance characteristics of individual TFF using some of the functional parameters defined in Section 15.2.2, which will be used as reference values for comparison with other technology options to be discussed in the next sections of the chapter. TFF usually exhibit a 1-dB BW over the nanometer, although in some cases, this figure can be in the subnanometer range with values around 0.4 nm. This implies that a major application field of TFF will be in the field of CWDM because channel spacings that can be resolved will be over 0.8 nm. Nevertheless, as pointed above, recent advances in energetic PVD fabrication techniques are expected

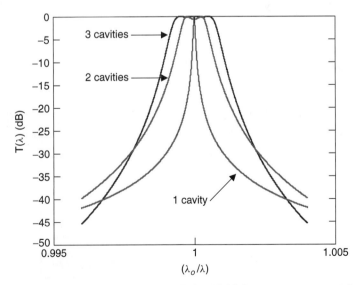

**Figure 15.26**  Same transfer function as that of Fig. 15.25 for a narrower spectral region. Also shown are the spectral responses for two and three coupled cavity filters.

to extend the range of applications to WDM and DWDM. When used under normal incidence, TFF are polarization insensitive, but a typical PDL value around 0.2 dB is obtained in these devices. Also, insertion losses are high, with a typical value around 7 dB due to coupling losses from the input single-mode fiber to the TFF input and from the TFF output to the output fiber. TFF present adjacent channel crosstalk levels (ACC) in the range of −25 dB and an excellent CTW around 0.5 pm/°C for the best cases.

### 15.3.4 Thin-Film MUX/DEMUX Architectures

WDM MUX/DEMUX and OADM devices can be constructed using individual TFF as a basic element in a variety of options, all of which are based on cascading collimated light in reflection from a series of TFF tuned to different center wavelengths. One of such techniques is on the mounting of TFFs between GRIN lens collimators to implement a four-port OADM and subsequently to cascade this module N times to form a N-Port WDM MUX/DEMUX, as shown in the left and right parts of Fig. 15.27, respectively.

A main drawback of this technique is that the loss at the last demultiplexed channel is the sum of the losses of the previous extracted channels, which depends not only on the insertion losses of previous TFFs but also on the collimator losses. These losses might exceed that of the system's requirements for even low-port count devices.

Another option that is shown in Fig. 15.28 is to use a small block to mount the TFF opposing and parallel to each other. GRIN lenses collimate the in–out light before directing the signal at a slight angle to the first bandpass TFF. Wavelengths of light that are in-band of the narrow TFF are transmitted and the rest reflected. In this way of operation, light cascades down the glass block, and with each bounce, a single channel

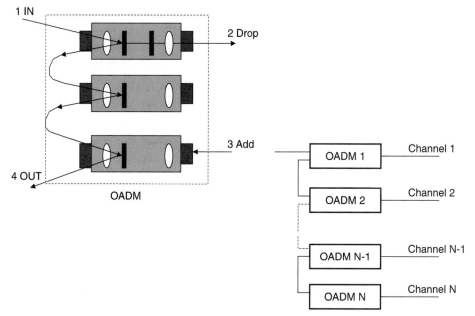

**Figure 15.27** Four-port OADM implemented by the mounting of TFFs between GRIN lens collimators (left). Multiple-port OADM implemented by cascading the former device (right).

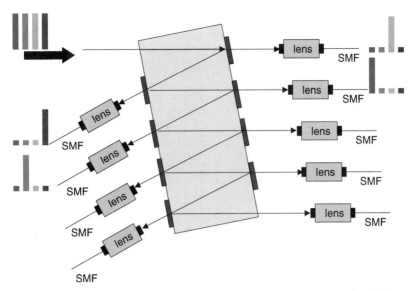

**Figure 15.28**   Multiple-port OADM using a small quartz block to mount the TFFs opposing and parallel to each other.

is removed. If the divergence of the collimators is not excessive, this technique results in lower losses at the last extracted channel.

## 15.4   FREE-SPACE MUX/DEMUX DEVICES

### 15.4.1   Introduction

WDM multiplexers and demultiplexers can be implemented as well by means of exploiting free-space propagation and the use of bulk diffraction gratings. In classic optics, the term grating describes a device whose operation involves the interference among multiple optical signals originating from the same source but showing different phase delays. In free-space propagation, this is achieved by periodically modulating the amplitude or the phase of an incident collimated wave to the device. Gratings have been traditionally employed to separate light into its constituent wavelengths.

Gratings can operate on transmission or in reflection, as shown in the upper part of Fig. 15.29. A transmission grating can be seen as a multislit aperture where very narrow slits are spaced equally apart. The spacing $d$ between adjacent slits is called the grating *pitch*. The light originating from a distant source and impinging on one side of the grating is transmitted by the slits. If the slits are very narrow, then due to diffraction, they behave as secondary light sources, so the light that transmits through them is spread out in all directions. If we consider an observation or *image plane* after the grating, then all of the different contributions from the secondary sources will interfere. If we consider any point in this plane, wavelengths for which the interfering waves are in phase will yield a constructive interference producing an enhancement of the light signal, whereas for wavelengths for which the interfering waves are out of phase, multiple interference will produce signal cancellation. The key point of diffraction gratings is that the interfering waves for different wavelengths add constructively in

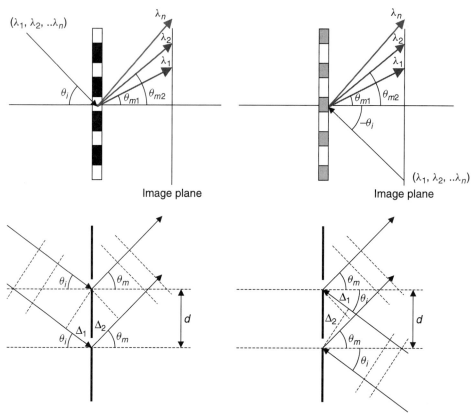

**Figure 15.29** Layout of transmission and reflection bulk gratings (upper). Detail of the interference of transmitted/reflected signals by two consecutive surfaces separated by the grating pitch in transmission/reflection gratings (lower).

different points of the image plane, and therefore, these are effectively separated by the device. Furthermore, in the context of WDM devices, signals from individual wavelength channels can be recovered if optical detectors and optical fibers are placed at those points of the image plane to collect the spatially separated signals.

In the case of reflection gratings, the operation is similar, but in this case, the transmitting slits are replaced by highly reflecting surfaces and the rest of the device surface is nonreflecting. In any case, as it will be shown later, the operation of both configurations is identical.

Diffraction gratings can also be classified into *amplitude and phase gratings* according to the physical nature of the diffracting elements. Amplitude gratings are more known in the literature and are produced by mechanically ruling a thin metallic layer deposited on a glass substrate. Phase gratings do not rely on the mechanical manipulation of a glass-based material, but rather they consist of a periodic variation of the refractive index of the grating material. This variation is usually achieved by writing a volume hologram in a photosensitive medium. The volume hologram is permanently written by exposing this photosensitive glass to the interference pattern coming from two mutually coherent laser beams, by which a periodic modulation of the refractive index of this medium is formed.

### 15.4.2   Theory and Design of Free-Space MUX/DEMUX Devices

The general basic principles governing the behavior of a diffraction grating-based MUX/DEMUX device can be understood by means of a simple relationship known as the grating equation and several related parameters. A deeper understanding of the device operation requires, however, a more thorough analysis that can be performed using the techniques of classic optics [83] or more modern approaches based on Fourier optics [84]. We will first provide the simple description of the basic operation principles of the diffraction grating, and later, a more profound description will be provided using the Fourier optics approach.

The lower part of Fig. 8.29 shows the interference originating from the combination of waves produced by two adjacent slits either in a transmission (left part) or a reflection (right part) grating. Let us consider that the refractive index of the medium surrounding the slits is given by $n$ and that, as mentioned before, $d$ represents the separation between adjacent slits. Interference pattern results because the rays transmitted (or reflected) by each slit suffer different optical path lengths. Referring to the layouts of Fig. 8.29, this path length difference $\Delta L$ is given by

$$\Delta L = \Delta_1 \pm \Delta_2 = d(\sin \theta_m \pm \sin \theta_i) \tag{15.7}$$

where the $+$ sign applies for the transmission grating and the $-$ sign applies for the reflection grating, $\theta_i$ and $\theta_m$ represent the angles of incidence and transmission respectively, measured counterclockwise from the normal to the grating.

At wavelengths where the difference in the phase (or optical path length) experienced by waves transmitted (or reflected) by adjacent slits is a multiple of $\lambda$, then constructive interference occurs. This is expressed by the *grating equation*:

$$m\lambda = nd(\sin \theta_m \pm \sin \theta_i) \tag{15.8}$$

where $m$ is an integer that represents the *diffraction order*. Equation (15.8) states that for a fixed value of $m$ (except for $m = 0$), different wavelengths have different output angles from the diffraction grating, just like what happens when one uses a prism. The angular dispersion that gives the angular separation per unit range of wavelength is given by

$$D_\alpha = \frac{d\theta_m}{d\lambda} = \frac{m}{nd\cos\theta_m} \tag{15.9}$$

It can be observed that the angular dispersion is proportional to the diffraction order $m$, so the higher the value of $m$, the better angular separation. For a positive $m$, the diffraction angle increases as the wavelength increases. Note as well that $m = 0$ results in a null angular dispersion, and therefore, the diffraction grating does not separate the different wavelengths.

The diffraction grating, however, shows distinctive features to that of a prism. For example, for a fixed value of $\lambda$, different output angles $\theta_m$ are possible (one for each possible value of the diffraction order $m$). Thus, the energy that is carried by a given wavelength is spread over the different diffraction orders corresponding to that wavelength. Most of that energy is coupled to the zero-order diffraction ($m = 0$) where, unfortunately, no wavelength angular dispersion is obtained. This problem can

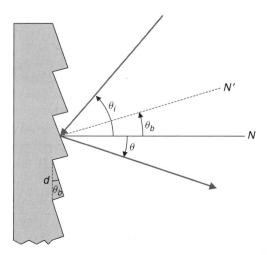

**Figure 15.30**   Example of a blazed reflection grating design.

be solved by using blazed grating designs, such as that shown in Fig. 15.30 for the case of a reflection grating.

Here, the condition for zero path difference is achieved for rays suffering specular reflection referred to the normal N' to the angled groove and not to N. Therefore, maximum power coupling can be achieved for a given wavelength to a diffraction order different from $m = 0$. This requires the proper design of the blaze angle. To do so, we must require that the diffracted beam satisfy both the condition of specular reflection from the groove face and the condition for a principal maximum in the $m$th order ($\theta = \theta_m$). The first condition requires that $2\theta_b = \theta_i - \theta_m$, whereas the second requires the verification of the grating angle. Combining both, we have

$$m\lambda = nd(\sin(2\theta_b - \theta_i) \pm \sin\theta_i) \tag{15.10}$$

From which the desired value of $\theta_b$ can be obtained.

Considerable insight in the operation and design of diffraction gratings for WDM applications can be gained by using a Fourier optics approach for its modeling. Let us assume a diffraction grating composed of N identical slits, each one characterized by a field transmission (reflection) function $u(x)$, illuminated by a plane wave incident at an angle $\theta_i$, as shown in Fig. 15.31.

The field at the plane just after the grating, assuming equal power division of the input wave into the $N$ slits, is given by

$$E(x) = -\frac{1}{N}\sum_{r=0}^{N-1} u(x - rd)e^{-j\frac{2\pi nx\sin\theta_i}{\lambda}} \tag{15.11}$$

The far field under Fraunhofer diffraction regime is given by its Fourier transform, where the spatial frequency variable is $k = n\sin\theta/\lambda$:

$$E_f(k) = -U(k \pm k_i)\frac{1}{N}\sum_{r=0}^{N-1} e^{j2\pi r(k\pm k_i)d} = -U(k \pm k_i)AF(k \pm k_i) \tag{15.12}$$

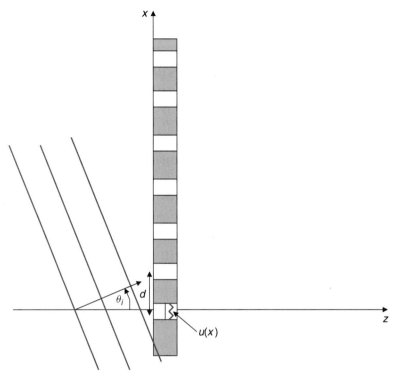

**Figure 15.31**   Grating layout for the description of the device operation by means of Fourier optics.

where $k_i = n \sin \theta_i / \lambda$ and, again, the $+$ and $-$ signs identify the transmission and reflection diffraction gratings, respectively. Note that there are two contributions to the far field. The first one is the far field $U(k)$ of the slit field transmission (reflection) function, which is a slowly varying function (equivalent to the diffraction pattern, under the Fraunhofer approximation, of a single slit). For example, if $u(x) = \prod (x/a)$ with $a < d$, then [83]

$$|U(k)|^2 = \left(\frac{na}{\lambda L}\right)^2 \operatorname{sinc}^2(ka) \tag{15.13}$$

The half-width of this function is $\Delta k_{1/2} = 1/a$.

The second contribution is a fast varying function also known as the *array factor* and depends on the number of slits $N$, the slit separation $d$, and the input and output angles. The array factor for normal incidence is given in terms of the angular direction by

$$AF(\theta) = \frac{1}{N} \sum_{r=0}^{N-1} e^{j \frac{2\pi nrd(\sin \theta)}{\lambda}} \tag{15.14}$$

If $\theta$ is centered around $\theta_m$, then $AF(\theta)$ is periodic with an angular period given by

$$\Delta \theta_p = \frac{\lambda}{nd \cos \theta_m} \tag{15.15}$$

This periodicity is the result of sampling the incident plane wave by the multiple slits of the diffraction grating; furthermore, it can be expressed in terms of wavelengths instead of angles yielding the *spatial Free Spectral Range* of the grating $FSR_s$. Using Eq. (15.9) into Eq. (15.14) [85], [86]:

$$FSR_s = \frac{d\lambda}{d\theta}\Delta\theta_p = \frac{\Delta\theta_p}{D_\alpha} = \frac{\lambda}{m} \tag{15.16}$$

The $FSR_s$ yields the nonoverlapping wavelength range in a particular diffraction order. This overlapping occurs because in the grating equation, the product $d\sin\theta$ may be equal to several possible combinations of $m\lambda$; thus, different wavelengths can be diffracted with the same angle by different diffraction orders. It is interesting to note that $FSR_s$ is inversely proportional to $m$. Thus, increasing the number of the diffraction order used to implement the wavelength separation will decrease the value of the wavelength region that can be actually demultiplexed by the device. In summary, the choice of $m$ is dictated by a tradeoff between the angular dispersion and the spatial Free Spectral Range.

It is customary to express the array factor in a more compact form:

$$AF(k) = \frac{e^{j\pi(N-1)kd}}{N}\left[\frac{\sin(\pi Nkd)}{\sin(\pi kd)}\right] \tag{15.17}$$

This equation shows again the periodic nature of the array factor where resonances or passbands with maximum value of unity placed at $k_v = 2v/(Nd)$ ($v$ integer) and half-width (defined from the maximum to the first zero) given by $\Delta k_{1/2}^{array} = 1/Nd$ are obtained. Note now that $\Delta k_{1/2}/\Delta k_{1/2}^{array} = Nd/a > N$, thus confirming that $U(k)$ is a slowly varying function as compared to $AF(k)$. The resolution of the array factor can be increased (narrower passbands) either by increasing the number of slits $N$ or by increasing the distance $d$ between adjacent slits.

If the values of $\theta_i$ and $\theta$ are fixed, then the array factor can be written as well as a function of the frequency. In this case, we get from this function the information regarding the periodicity in the frequency response at a fixed point in the image plane.

$$AF(f) = \frac{1}{N}\sum_{r=0}^{N-1} e^{j\frac{2\pi rf}{FSR_f}} = \frac{e^{j\frac{(N-1)\pi f}{FSR_f}}}{N}\left[\frac{\sin\left(\frac{\pi Nf}{FSR_f}\right)}{\sin\left(\frac{\pi f}{FSR_f}\right)}\right] \tag{15.18}$$

where

$$FSR_f = \frac{c}{n_g\Delta L} \tag{15.19}$$

represents the period of the frequency response and $n_g$ is the group index. Note that the frequency response has the same functional form as the spatial response. Here the bandpass half-width is given by $\Delta f_{1/2} = FRS_f Nd$. Note however, that the periodic frequency response is due to the constant path length difference $\Delta L$, whereas the periodic angular response is a result of the discrete sampling made by the multiple slits.

The frequency resolution also known as the *resolving power* is given by $\Delta\lambda_{1/2}/\lambda = mN$ and is calculated using the Rayleigh's criterion, that states that the peak transmission of one wavelength must be located at the null of the second wavelength for those two to be resolved.

### 15.4.3  Free-Space MUX/DEMUX Architectures

The implementation of a free-space WDM MUX/DEMUX device usually requires the use of additional devices, most usually lenses, as well as the diffraction grating element. Lenses are used at the input stage of the device to collimate the incident wave to the device and to focus the diffracted waves. Figure 15.32 shows a simplified layout of a free-space MUX/DEMUX device and the typical Gaussian response of a device with 40 ports. In what follows, the device operation is described assuming its operation as a demultiplexer. The operation as multiplexer follows just by considering the propagation in the opposite direction.

The device consists of an input fiber port, a collimating lens, a diffraction grating element, a focusing lens, and a receiving fiber array (output ports). The input port receives the composite multiwavelength signal from the WDM system or network, and the output ports receive the demultiplexed channels signals. At the input, the divergent beam from the input fiber is collimated by the lens placed before the diffraction grating. The parallel beam after the collimating lens is subsequently incident upon the diffraction grating by which different wavelength components are diffracted to different angles, as explained before. Dispersed beams are then focused by a focusing lens placed after the diffraction grating onto different points (a different point for each

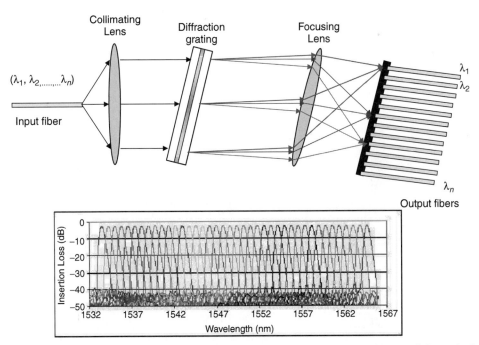

**Figure 15.32**  Simplified layout of a free space MUX/DEMUX device (upper) and the typical Gaussian response of a device with 40 ports (lower).

angle of dispersion) at the focusing plane where a the fiber array is placed. Each fiber thus collects a single wavelength channel. This output fiber assembly can be made by stacking one or more rows of closely spaced, end-flushed, and antireflection-coated optical fibers well aligned in silicon V-grooves.

As the diffraction order $m$ that is used in these devices is relatively low, high values of the spatial Free Spectral Range can be obtained. However, a low angular dispersion is also potentially achieved. An increase in the value of this second parameter can be obtained by means of a double-pass configuration, such as that shown in Fig. 15.33. When the device acts as a multiplexer, an incoming multiwavelength optical signal is transmitted by the input fiber, the end of which is located at the focal plane of a collimating lens. The beam containing the multiplexed wavelengths is collimated and then impinges on the diffraction grating, which spreads and angularly separates the beam into a number of collimated beams. The spatially separated beams are then redirected to the same diffraction grating by means of a 100% reflecting mirror that provides further spatial separation of the individual wavelengths. Finally, the spatially dispersed and collimated beams are focused by the same input lens to its focal plane and received directly by a series of optical fibers.

The configurations described above result in bulky devices with nevertheless quite good performance parameters. For instance, MUX/DEMUX devices with high port count (up to 40 channels) and standard ITU 100-GHz channel separation can be fabricated both with Gaussian and flat-type resonances within the C- or the L-band, providing less than 3.5-dB insertion losses and a PDL of less than 0.3 dB. The typical 3-dB bandwidth values are around 0.4 and 0.5 nm, and the loss nonuniformity is below 0.5 dB. Very low in-band chromatic (less than 5 psec/nm) and polarization mode dispersion (<0.3 psec) are typical. ACC levels are below −30 dB, and background crosstalk values can be made lower than −35 dB. Thermal wavelength stability is very high, with typical values of the CTW parameter around 0.4 pm/°C.

Despite these figures, bulky configurations result in a complicated manufacturing process because of very careful alignment of the input/output fibers and the internal

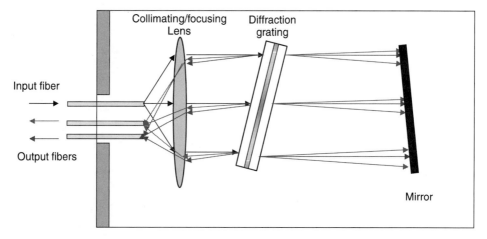

**Figure 15.33**   Double pass configuration for a free space MUX/DEMUX device.

precision optics. Furthermore, performance deviations can be produced due to lens aberrations. To avoid the use of discrete collimating and focusing optics, both the dispersion and focusing functions can be integrated on a single device if concave gratings are employed. This approach is advantageous, especially in integrated optics implementations, although its implementation in fiber format is also possible [87]. Special mountings, including the position of the input signal source, the grating, and the receiving fibers, can be employed to minimize aberrations [88, 89]. An important configuration is the Rowland mounting, which is shown in Fig. 15.34.

In the Rowland mounting [88], a concave grating forms an arc with radius of curvature R. The input and output fibers/waveguides are placed along a circle of radius R/2 called the Rowland circle, which intersects the grating arc at its midpoint. It can be shown [88] that if the input fiber/waveguide is placed on the Rowland circle, the diffracted waves will focus at points in the Rowland circle free from aberrations up to the second order. The Rowland circle provides one stigmatic point (free from all aberrations) that is placed on the axis joining the intersection point between the grating curve and the Rowland circle and the center of the Rowland circle. It has been shown that other mountings with two and even three stigmatic points can substantially reduce the aberrations on the focusing points. The design of such mountings requires, however, a numerical approach.

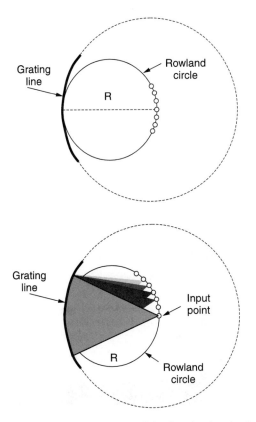

**Figure 15.34**   Configuration of the Rowland mounting.

## 15.5   OPTICAL FIBER-BASED MUX/DEMUX DEVICES

### 15.5.1   Introduction to the Principles and Technology of Fiber Bragg Gratings

A fiber Bragg grating (FBG) is a periodic or almost a periodic perturbation of the refractive index along the fiber length that is formed by exposure of the core to an intense optical interference pattern [91]. The formation of permanent gratings in an optical fiber was first demonstrated by Hill et al. in 1978 [92]. The variation of the refractive index along the length of an FBG can be described as

$$n(z) = \Delta n_{ac}(z) \cos\left(\frac{2\pi}{\Lambda(z)} z + \theta(z)\right) + n_{dc}(z) \tag{15.20}$$

$\Lambda(z)$ is the grating period, $\theta(z)$ is the grating phase, $\Delta n_{ac}(z)$ is the amplitude of the perturbation, and $n_{dc}(z)$ is the average refraction index. As in the thin-film devices, refractive index variations provide small Fresnel reflections (see Fig. 15.35), which create distributed feedback. Light, guiding along the core of an optical fiber, will be scattered by each grating plane. If the Bragg condition is not satisfied, the reflected light from each of the subsequent planes becomes progressively out of phase and will eventually cancel out. If the Bragg condition is satisfied, the contributions of reflected light from each grating plane add constructively in the backward direction to form a backreflected peak. Several theories have been developed to explain the Bragg condition. An intuitive one supposes the propagation of a forward-propagating core-mode and a backward-propagating core-mode. The propagation constant is $\beta = k.n_{eff.} = 2\pi n_{eff}/\lambda$. Thus, the change of phase in FBG within a constant period is $\Phi = \beta\Lambda$. If the reflected contributions have to add in phase $\Phi = m2\pi$, then the strongest interaction or mode-coupling occurs at the Bragg wavelength given by

$$\lambda_B = m2n_{eff}\Lambda \cong m2n_{dc}\Lambda \tag{15.21}$$

The FBG acts as a band-rejection filter passing all wavelengths that do not satisfy the Bragg condition. The reflection spectrum is a band-pass filter, simply the complementary of the transmission spectrum (see Fig. 15.35).

The key characteristics of the FBGs stem from the fact they are made from optical fibers, i.e., low insertion loss, polarization insensitive, easily controllable characteristics, stability, longevity, ease to connect, and ease to manufacture. Typical lengths are from 1 cm to 10 cm, and some types reach 1 m.

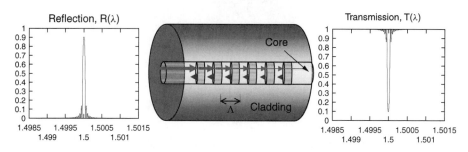

**Figure 15.35**   Fiber Bragg grating layout and behavior in transmission and reflection.

There are several different approaches [93]–[96] to write gratings in fibers. The most common approaches are the two-beam interferometric method [93] and the phase mask method [94]. By exposing the fiber to a UV interference pattern from the side, the pattern is "printed" into the fiber (see Fig. 15.36). Only the core is usually doped (for example, with germanium), and consequently, the grating is only formed in the core and not in the cladding.

In the two-beam interferometric method, the laser beam is split into two components that are subsequently recombined at the fiber to produce an interference pattern. The two-beam interferometric method has the advantage that the period of the interference pattern may be tuned to produce gratings that operate over a wide range of wavelengths, but it requires a high stability and alignment, whereas the phase-mask method is relatively insensitive to vibration and alignment but lacks flexibility to select the grating wavelength.

The phase mask technique is illustrated in Fig. 15.36. The phase mask is made from a flat slab of silica glass that is transparent to ultraviolet light. On one of the surfaces, a one-dimensional periodic structure is etched using photolithographic techniques. The shape of the periodic pattern approximates a square wave.

Ultraviolet light, which is incident normal to the phase mask, passes through and is diffracted by the periodic corrugations of the phase mask. The phase mask is designed to suppress the diffraction into the zero order and to maximize the diffracted light contained in the $\pm 1$ diffracted orders. The two $\pm 1$ diffracted order beams interfere to produce a periodic pattern that photoimprints a corresponding grating in the optical fiber. If the period of the phase mask grating is $\Lambda_{\text{mask}}$, the period of the photoimprinted index grating is $\Lambda_{\text{mask}}/2$. In comparison with the holographic technique, the phase mask technique offers easier alignment of the fiber, reduced stability requirements, and lower coherence requirements of the ultraviolet laser beam. A drawback of the mask technique is that a separate phase mask is required for each different Bragg wavelength.

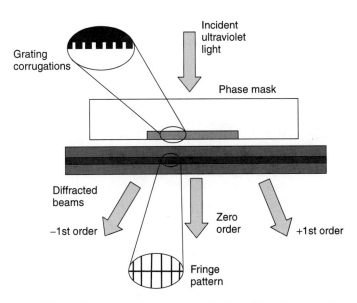

**Figure 15.36**  Phase mask technique to fabricate fiber Bragg gratings.

Improvements in both techniques have attracted considerable interest [97–102]. In order to write nonuniform gratings with advanced characteristics, one can find information in the above cited references or in Refs. 103–106.

The magnitude of the refractive index change ($\Delta n_{ac}$) obtained depends on several different factors, such as the irradiation conditions (wavelength, intensity, and the total dosage of irradiating light), the composition of the material forming the fiber core, and any processing of the fiber prior to irradiation. Usually, $\Delta n_{ac}$ is positive in germanium-doped monomode fibers with a magnitude ranging between $10^{-5}$ and $10^{-3}$. Techniques such as "hydrogen loading" [107] or "flame brushing" [108] are available that can be used to process the fiber prior to irradiation in order to enhance the refractive index change. By the use of hydrogen, loading as high as $10^{-2}$ has been obtained [91].

### 15.5.2   Theory and Design of Fiber Bragg Grating Filters

Theories of FBG are well known. There are already some books in the field on fiber Bragg gratings [109–111], and in addition, there are several review articles; see, e.g., Refs. 91 and 112 and the references therein. Three methods come to mind: the coupled-mode theory, the Bloch-wave analysis [90], and the transfer matrix method.

The **coupled-mode theory** [109–115] is straightforward and intuitive. It accurately models most of the optical properties of the FBGs. The FBG is treated as a perturbation on the fiber along the $z$ axis, and weak guiding is assumed. Thus, the total electric field is defined as a superposition of the forward- and backward-propagating waves

$$E_z(x, y, z) = b_1(z)\Psi(x, y) + b_{-1}(z)\Psi(x, y) \tag{15.22}$$

$b_{\pm 1}$ are dependent on frequency because they include the harmonic propagation factor $\exp(\pm j\beta z)$, being $\beta = \beta(\omega) = n_{eff}\omega/c$, the scalar propagation constant. $\Psi(x, y)$ satisfies the scalar wave of the unperturbed fiber, and the total electric field $E_z$ must satisfy the scalar wave equation for the perturbed fiber. The $z$-dependence of the index perturbation of a FBG is approximately quasi-sinusoidal [Eq. (15.20)]. Thus, solving the wave equation under the condition of Eq. (15.20), we arrive at the couple-mode equations

$$\frac{du}{dz} = +i\delta u + q(z)v$$
$$\frac{dv}{dz} = -i\delta v + q^*(z)u \tag{15.23}$$

where $u$ and $v$ have been defined as

$$b_1(z) = u(z)\exp\left(+i\frac{\pi}{\Lambda}z\right)\exp\left(+i\int_0^z \sigma(z')dz'\right)$$
$$b_{-1}(z) = v(z)\exp\left(-i\frac{\pi}{\Lambda}z\right)\exp\left(-i\int_0^z \sigma(z')dz'\right) \tag{15.24}$$

$\sigma(z)$ is a real and slowly varying function that accounts for the dc index variation, $\Delta\varepsilon_{dc}(z)$ [111, 115]. Also, in Eq. (15.23), we have defined the wavenumber detuning

$\delta = \beta - \pi/\Lambda$ and the coupling coefficient $q$ of the grating:

$$q(z) = ik(z) \exp\left(-2i \int_0^z \sigma(z')dz'\right) \tag{15.25}$$

Any solution $\{u, v\}$ to the coupled-mode equations must satisfy Eq. (15.23). The reflection coefficient of a FBG located in $0 <= z <= L$ can be computed as $r(z_0) = v(z_0)/u(z_0)$. In two cases, it is possible to find simple closed-form solutions to the coupled-mode equations: the *weak gratings* and the *uniform gratings*, where $q(z) =$ constant [116]. The weak grating is studied under the first-order Born approximation [113], which means that the forward-propagating wave $u$ is unaffected by the grating; that is, $u = \exp(i\delta z)$. Then, we obtain

$$r(\delta) = -\frac{1}{2} \int_0^\infty q^*\left(\frac{z}{2}\right) \exp(i\delta z)dz \tag{15.26}$$

This weak grating relation is valid when the top reflectivity of the grating is less than approximately 10% to 40%. Equation (15.26) shows a Fourier transform pair between $r(\delta)$ and $q * (z)$, which is quite intuitive to be able to understand the behavior of the FBGs.

A uniform grating has a constant coupling coefficient over the grating length. In this situation, the coupled-mode equations can be solved analytically. The resulting reflection coefficient is

$$r(\delta) = \frac{-q * \sinh(\gamma L)}{\gamma \cosh(\gamma L) - i\delta \sinh(\gamma L)} \tag{15.27}$$

where $\gamma^2 = |q|^2 - \delta^2$, and the transmission coefficient becomes

$$t(\delta) = \frac{\gamma}{\gamma \cosh(\gamma L) - i\delta \sinh(\gamma L)} \tag{15.28}$$

Recently, a new family of solutions have been published for a set of FBGs [145].

In most cases, Eqs. (15.23) have to be solved using numerical solution techniques. There is a variety of methods to compute the reflection and transmission spectra for nonuniform gratings [125]; the Runge–Kutta method is the most commonly used.

The **transfer matrix method** is also extensively used to analyze the FBGs. It divides the grating into smaller sections [112, 117, 118]. The sections are treated as uniform gratings, yielding the overall spectra by transfer matrix multiplication. Let the section length be $\Delta = L/N$, $N$ the number of sections and $L$ the length of the FBG. By applying the appropriate boundary conditions and solving the coupled-mode equations, the following transfer matrix links the fields at the positions $z$ and $z + \Delta$:

$$\begin{bmatrix} u(z + \Delta) \\ v(z + \Delta) \end{bmatrix} = \begin{bmatrix} \cosh(\gamma\Delta) + i\dfrac{\delta}{\gamma}\sinh(\gamma\Delta) & \dfrac{q}{\gamma}\sinh(\gamma\Delta) \\ \dfrac{q*}{\gamma}\sinh(\gamma\Delta) & \cosh(\gamma\Delta) - i\dfrac{\delta}{\gamma}\sinh(\gamma\Delta) \end{bmatrix} \begin{bmatrix} u(z) \\ v(z) \end{bmatrix} \tag{15.29}$$

Hence, the fields at the two ends of the FBG are linked by

$$\begin{bmatrix} u(L) \\ v(L) \end{bmatrix} = T \begin{bmatrix} u(0) \\ v(0) \end{bmatrix} \quad T = \begin{bmatrix} T_{11} & T_{12} \\ T_{21} & T_{22} \end{bmatrix} \tag{15.30}$$

where $T = T_N.T_{N-1}.\ldots.T_1$ is the overall transfer matrix and $T_j$ is the transfer matrix of each section. The reflection coefficient and the transmission coefficient are calculated as

$$r(\delta) = -\frac{T_{21}}{T_{22}}$$

$$t(\delta) = \frac{1}{T_{22}} \tag{15.31}$$

Instead of using Eq. (15.29), we can discretize the whole grating in a stack of complex discrete reflectors. In this case, the algorithm is in some sense similar to Rouard's method of thin-film optics, which has been applied to corrugated waveguide filters previously [119]. Section 15.3.2 shows the ABCD matrix, which can be used in the cited method. Other quite common matrices are the transfer matrices. In this transfer matrix method, each slab (a fraction of the period of the FBG) is then replaced by two transfer matrices $T^\Delta.T^\rho$ [126, 127], where

$$T^\Delta = \begin{bmatrix} e^{j\beta z} & 0 \\ 0 & e^{-j\beta z} \end{bmatrix}$$

$$T^\rho = \frac{1}{1+\rho} \begin{bmatrix} 1 & \rho \\ \rho & 1 \end{bmatrix} \tag{15.32}$$

$T^\Delta$ is the pure propagation matrix, and $T^\rho$ is the discrete reflector matrix due to the index changes. Using the equations defined by Eq. (15.32), computing the analytical solution is a lengthy process. In Refs. 118 and 127, the matrices $T^\Delta.T^\rho$ are redefined to speed the analysis and later the synthesis approach.

The **synthesis** of FBGs is an important problem well solved recently [117–123]. Synthesis is useful both as a design tool and for characterization of already fabricated gratings with complex profiles. Layer peeling [124], or sometimes called the differential inverse scattering method, is the most common algorithm used. The idea of layer peeling is to calculate the initial part of the FBG from the reflection spectrum. Then, compute the response of the FBG without the initial part of the FBG (layer peeling), taking into account the properties of the transfer matrix and again calculate the initial part of the new FBG, following this procedure until the other end of the FBG. Two types of layer-peeling methods are developed [117]: discrete layer peeling (DLP) and continuous layer peeling (CLP). The DLP algorithm can be described as [117]:

1. Start with a physically realizable reflection coefficient. It must be a casual filter.

2. Calculate the first coefficient of the impulse response of the filter. This coefficient is the reflection coefficient from the first slab of the FBG. It is straightforward to calculate the refraction index of the first slab from the reflection coefficient.

3. Propagate the fields using the transfer matrices [Eq. (15.32)] or better using the matrices of Refs. 117 and 118.

4. Repeat steps 2 and 3 until the entire grating structure is determined.

The CLP [117, 119] method is similar to the DLP. The only difference is in the second step. In the DLP, a discrete Fourier transform is used to calculate the reflection coefficient. In the CLP $q(z)$, the coupling coefficient is calculated evaluating an integral. The CLP method is slower than the DLP method.

### 15.5.3  Performance Characteristics

As seen above, the common reflection response of a FBG is a narrow bandpass filter, $\lambda_B$ [Eq. (15.21)] being the center wavelength. Any change in fiber properties, such as strain, temperature, or polarization, which varies the modal index or grating pitch, will change the Bragg wavelength. The tuning of the FBG with strain and temperature can be modeled

$$\Delta\lambda = \lambda_B \{(1 - \rho_e)\varepsilon + (\alpha_\Lambda + \alpha_n)\Delta T\} \qquad (15.33)$$

where $\rho_e$ is the strain-optic coefficient, $\varepsilon$ the applied strain, $\alpha_\Lambda$ the thermal expansion coefficient, $\alpha_n$ the thermal-optic coefficient, and $\Delta T$ the temperature difference. Typical values are wavelength shift due to strain ranges from 1 and 1.5 pm/μstrain and wavelength shift due to temperature is between 10 and 15 pm/°C.

From Eq. (15.27), it is possible to calculate in a uniform FBG the reflectivity at the Bragg wavelength

$$R(0) = r(0)r^*(0) = \tanh^2(\kappa L) \qquad (15.34)$$

where $\kappa(z) = (\pi/\lambda)\overline{\Delta n}_{ac}(z)\eta$ and $\eta$ is a modal overlap factor. From the former equation, it can be shown that the longer the FBG and/or the higher the value of $\overline{\Delta n}_{ac}(z)$, the higher the value of $R(0)$. This applies also for nonuniform FBG.

A general expression for the approximate full-width-half maximum bandwidth of a grating is given from [91]

$$\Delta\lambda = \frac{s\lambda^2}{\pi n_g}\sqrt{\kappa^2 + \frac{\pi^2}{L^2}} \qquad (15.35)$$

$s$ is a constant that is almost 1 for strong gratings and around 0.5 for weak gratings. So the FWHM is proportional to $\kappa$ and inversely proportional to $L$.

Figure 15.35 shows that the main peak in the reflection spectrum of a finite length Bragg grating with uniform modulation of the index of refraction is accompanied by a series of sidelobes at adjacent wavelengths due to multiple reflections inside the FBG. It is important in some applications to lower and, if possible, eliminate the reflectivity of these sidelobes, or to apodize the reflection spectrum of the grating. In dense wavelength division multiplexing (DWDM), it is important to have very high rejection of the nonresonant light in order to eliminate crosstalk between information channels. Apodization of the refraction index profile is then absolutely necessary. Apodization is accomplished by varying the amplitude of the coupling coefficient along the length of the grating [$\Delta n_{ac}(z)$ in Eq. (15.20) is not constant] [128]. Apodized fiber gratings can have very sharp spectral responses, with channel spacing down to 25–100 GHz [129].

For applications such as add-drop filters, or in demultiplexers, the grating response to less than −30 dB from the maximum reflection is considerable. Figure 15.37 shows the reflection spectrum of two FBGs fabricated with a uniform phase mask and an apodized phase mask (Gaussian profile). The apodized FBG has sidelobes of 40 dB lower than peak reflectivity. This represents a reduction of more than 30 dB in the

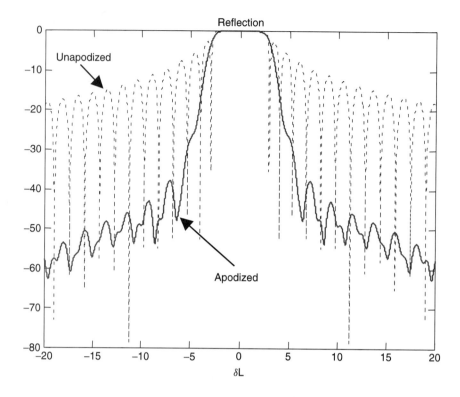

**Figure 15.37**  Comparison of the reflectivity spectrums of unapodized and apodized Fiber Bragg Gratings.

sidelobe levels compared to uniform gratings with the same bandwidth and reflectivity. Another benefit of apodization is the improvement of the dispersion compensation characteristic of chirped Bragg grating for which the group delay becomes linearized and the modulation associated with the presence of sidelobes is eliminated. Chirped Bragg grating has a monotonically varying period. Chirped gratings can be realized by axially varying the period of the grating and the index refraction of the core [130–134]. In this case, the Bragg wavelength is

$$\lambda_B(z) = 2n_{eff}(z)\Lambda(z) \tag{15.36}$$

There are certain characteristic properties offered by monotonically varying the period of gratings that are considered advantages for specific applications in telecommunications and sensor technology, such as dispersion compensation and the stable synthesis of multiple wavelength sources.

### 15.5.4  MUX/DEMUX Structures and Applications

As mentioned in Section 15.5.1, FBGs have outstanding performances; one of them is that the FBG provide excellent filter spectral shapes, allowing almost square-like spectra-filter shapes to be created with a high figure of merit. However, most of the time, FBGs are used with an expensive optical circulator (or a 3-dB coupler with an isolator) to convert the band rejection to band-pass for multiplexing or demultiplexing. To solve this drawback, FBG can be used in an interferometric Mach–Zehnder setup, which requires accurate phase control but suffers from environmental sensitivity.

Figure 15.38 [110] shows a demultiplexer. An eight-wavelength optical input signal passes through an isolator to a 3-dB coupler and is split into two equal output arms. Each of the output arms has four FBGs. The path for each wavelength can be found easily in the figure. This structure can be easily extended to DWDM multiplexers/demultiplexers from 8 channels to 16, 32, and 64, by proportionally adding more FBGs.

The Mach–Zehnder interferometer add/drop filter is shown in Fig. 15.39 [109, 135, 136]. A stream of several wavelengths $\lambda_1, \lambda_2, \ldots, \lambda_N$ is launched into the input port. Assuming the grating resonant wavelength is $\lambda_k$, light at $\lambda_k$ emerges from the drop port and the remaining wavelengths emerge at the output port. If a wavelength $\lambda_k$ is

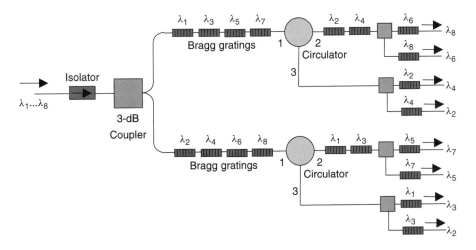

**Figure 15.38** Layout of an eight-channel demultiplexer using fiber Bragg gratings.

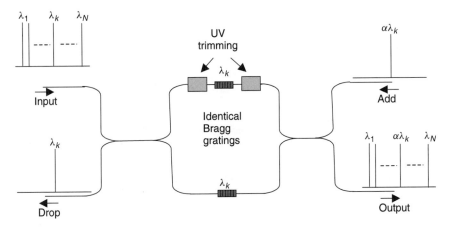

**Figure 15.39** A Mach-Zehnder interferometer add/drop filter using fiber Bragg gratings.

launched in the add port, it is multiplexed with other wavelengths at the output port. Once the device has been fabricated, phase adjustment between the guides to balance the interferometer is a major drawback; this is accomplished by exposing one arm of the interferometer to UV light to photoinduce an average index change. To minimize the balance requirements of a Mach–Zehnder interferometer, a twin-core fiber-based system has been proposed (i.e., two guides are embedded in the same cladding [137]; this scheme is similar to the one shown in Fig. 15.39). Schemes using polarization beam splitters instead of the 3-dB couplers have also been proposed [140].

In Ref. 141, a set of Mach–Zehnder interferometer add/drops are combined to perform a bidirectional OADM. Another add/drop filter, the idea of which seems equal to the last ones is presented in Refs. 138 and 139. The device consists of a mismatched coupler with a Bragg grating written in one core over the coupling region (see Fig. 15.40). The coupler would not normally transfer power from one core to another due to the strong mismatch of the two cores. However, with the existence of the Bragg grating, power transfer of the guided fundamental mode occurs from port 1 to port 4. This allows easy implementation of the device without the need for fine-tuning and is much more stable. The drawback is the small modal overlap, so strong FBGs and a long interaction length are necessary.

Optical add/drop multiplexers can be configured in a different way (see Fig. 15.41) [109]. Channels injected at the input port are reflected by the gratings in between the two circulators and routed to the drop port. All other wavelengths continue to the

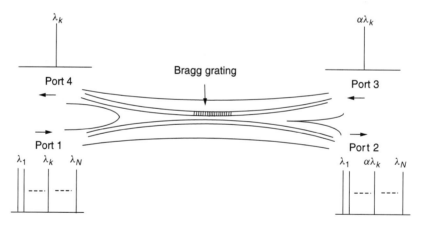

**Figure 15.40**   A fiber Bragg grating written into a coupler to be used as an add/drop filter.

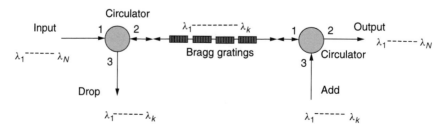

**Figure 15.41**   Layout of a fiber Bragg grating-based $k$-channel add/drop filter.

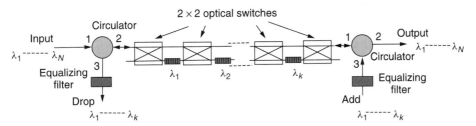

**Figure 15.42**   Layout of a fiber Bragg grating-based $k$-channel add/drop filter with an equalizer to compensate the losses.

output port. If the signals are injected at the add port, the same gratings perform the add function. The reflected channels to the output port, along with the rest of the channels from the input port, are multiplexed.

Dynamic wavelength selective add/drop MUX/DEMUX is often desirable, and several techniques are available to integrate this feature. The scheme shown in Fig. 15.41 is not an interferometric device, so tuning the gratings by stretching/compressing or heating is easily achieved [142, 143]. The drawback is the slow response of this scheme. To overcome this limitation, a reconfigurable OADM using $2 \times 2$ fast optical switches has been proposed (see Fig. 15.40). A set of optical cross-connect switches are connected in series between the two circulators. Fiber gratings at the optical channel wavelengths are connected between one of the output/input ports of the cross-connect switches. This way, the incoming signals may be switched to the grating or bypass it. When switched to the grating, the channel at the grating wavelength $\lambda_1$ is dropped and routed to the drop port. All other channels proceed to the next switch, where the choice is repeated for the other channels. The equalizing filters shown in Fig. 15.42 are intended to compensate for the loss of the switches and the gratings so that all channels suffer the same insertion loss [109, 144].

In Ref. 146, a modification of the reconfigurable OADM is shown. It includes automatic dispersion compensation routing the channels via a "through path."

## 15.6   INTEGRATED OPTICS-BASED MUX/DEMUX DEVICES

### 15.6.1   Introduction to Integrated Optics Technology

Like in the case of fiber technology described in the preceding section, integrated optics brings the possibility of implementing WDM MUX/DEMUX and OADM devices in guided wave format. In this case, the devices consist of planar arrangements of waveguides and active components assembled together on one substrate and can be manufactured in tens, hundreds, and thousands at a time with almost identical performance [147]. Repeatability, reliability, and low cost of fabrication are the main advantages that integrated optics brings over the rest of technologies discussed so far in this chapter. Furthermore, in the context of WDM MUX/DEMUX devices, integrated optics allows for the implementation of several options, including, among others, high port count devices based on the array waveguide grating and low port count devices based on lattice filters [148].

Several material systems can be employed for the implementation of planar light-wave WDM MUX/DEMUX circuits, the most successful currently being silica on silicon, indium phosphide, and polymer.

Silica-on-Silicon circuits are currently the most employed approach for the implementation of passive planar lightwave devices needed for the implementation of WDM MUX/DEMUX components [149]. These circuits are based on optical waveguides that are fabricated in a ridge configuration with square cross sections of typically $6 \times 6$ μm. These waveguides are fabricated on silicon substrates by a combination of flame hydrolysis deposition (FHD) and reactive ion etching (RIE) [150]. The waveguide fabrication process is shown schematically in Fig. 15.43. The first step is to use FHD to deposit two successive glass particle layers that serve as the undercladding and the core. After deposition, the substrate with these two porous glass layers is heated to about $1300°$C for consolidation. The waveguide core ridges are then formed by photolithography and RIE. Finally, FHD is used again to cover the core ridges with an overcladding.

Silica waveguide circuits are ideal for matching to fiber optic systems because the refractive index matches that of the fiber and the propagation loss is potentially very low (<0.02 dB/cm). The typical relative refractive index difference $\Delta$ between the core and the cladding of the waveguide is around 0.7%. This high $\Delta$ value has several advantages, including low loss and its low coupling loss with optical fiber. Researchers have demonstrated losses as low as 0.02 dB/cm in a 10-m-long waveguide and minimum bending radius $r_{min}$ of 5 mm [151].

In an effort to integrate circuits with higher density, researchers have recently improved techniques for fabricating $GeO_2$-doped silica waveguides on Si with a higher $\Delta$ and for their connection to optical fibers [150–152]. High-density PLCs with an $r_{min}$ of 2 mm have been realized using a $GeO_2$-doped silica waveguide with a $\Delta$ of 1.5% [152] known as super-high (SH)-$\Delta$ waveguide. $GeO_2$-doped silica waveguides on Si have been fabricated with a $\Delta$ of ~1.5%, a core size of ~$4.5 \times 4.5$ μm$^2$, and a propagation loss of 0.05 dB/cm from a 40-cm-long waveguide.

Semiconductor integrated optical devices have been traditionally employed for active devices based on semiconductor lasers [153]. Nevertheless, they also have attractive features for the implementation of passive devices as required for WDM MUX/DEMUX applications. One of the most noteworthy properties is the ability to control the refractive index in a wide range by changing material compositions. This leads to the possibility of implementing compact optical circuits because the bending radius can be reduced by making the refractive index difference between the core and cladding materials in large waveguides.

The application of semiconductor materials to passive devices has been limited because of their relatively large propagation and coupling losses and their polarization dependence. However, recent progress in fabrication technologies and device design are enabling the extension of these technologies to passive devices. Fabrication technologies, such as crystal growth and dry etching, enable fabrication of waveguides with uniform and smooth interfaces, which significantly contributes to reducing propagation losses in waveguides while providing a precise control of waveguide thickness that is needed in order to fabricate polarization-insensitive waveguides.

Indium phosphide (InP) is a direct gap semiconductor providing the possibility of giving stimulated emission and optical detection in the wavebands of interest for optical fiber telecommunications. It is routinely employed for the implementation of high-performance tunable DFB and DBR semiconductor lasers and high-performance

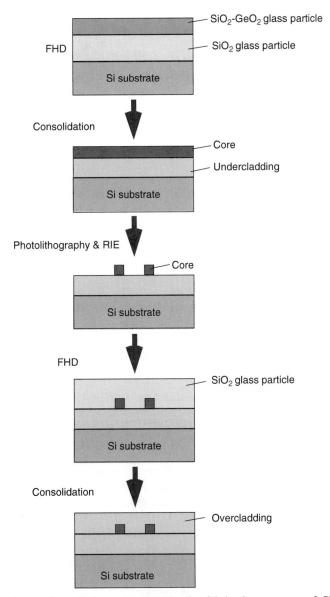

**Figure 15.43**  Illustration of the main steps in the fabrication process of Silica-on-Silicon waveguides.

pin photodiodes, and for high gain semiconductor optical amplifiers. InP is easy to cleave and provides the possibility of linear electrooptic effect.

There are two main alternatives for implementing the waveguides in InP-integrated optic circuits, which are shown in Fig. 15.44 [147]. The first option, shown in the left part of Fig. 15.44, is the buried rib-loaded slab, which generally has a larger bent radius, but lower propagation losses (0.2 dB/cm) and can be wet-etched. The second alternative is the air-clad rib InP waveguide shown in the right part of Fig. 15.44, which needs to be dry-etched and typically present in higher losses (2 dB/cm).

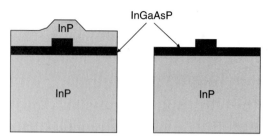

**Figure 15.44**   The two main alternatives for implementing waveguides in InP integrated optic circuits: (left) buried rib-loaded slab; (right) air-clad rib InP waveguide.

Polymer waveguides are of a plastic nature [154] yielding refractive indexes similar to silica. Waveguides are usually of buried type with rectangular cross sections, but the losses are higher due to the multiple absorption peaks that molecular structures in polymers present. Typical values around 0.2 dB/cm in the C-band have been reported. Waveguides can be implemented by directly exposing the polymer with ultraviolet light. The layers are typically spun on in liquid form and then baked. The main advantage of this material is the ease of processing. Another interesting property is that Bragg gratings can be written by exposure to ultraviolet light in both the core and the cladding. Thus, integrated OADMs based on these filters can be implemented using polymer technology. However, they must generally be protected from the environment.

### 15.6.2   Theory and Design of Integrated Optics WDM MUX/DEMUX Devices

There are a number of integrated optic devices that can be employed for WDM applications. These include wavelength $N \times N$ MUX/DEMUX, OADMs, cross-connect switches, multiwavelength sources, programmable lattice filters, and so on. Of these, the arrayed waveguide grating (AWG) multiplexer is currently the most successful for WDM MUX/DEMUX and OADM applications and shall be treated with exclusivity in this section. The interested reader is referred to other sources [78, 155] for descriptions of other components and devices.

The AWG, also known as the waveguide grating router (WGR) or PHASAR, is a device that uses an array of integrated optic waveguides to perform the same function as a free-space diffraction grating. The schematic configuration of the AWG is shown in Fig. 15.45.

The AWG consists of an input waveguide array, two free propagation regions (FPRs), also known as focusing slab waveguides or couplers interconnected by a central array of waveguides, where the length between adjacent waveguides in the array varies by a constant $\Delta L$ (that is, the array operates as the diffraction grating), and finally, an output waveguide array. The device can be operated as a $1 \times N$ device if it has a single input and multiple outputs, a $K \times 1$ device, or, most generally a $K \times N$ device having multiple inputs and outputs.

The evolution toward the final conception of the AWG device is a fascinating and complicated series of proposals and is given in detail in Ref. 147. The central part of the device, that is, the waveguide grating, was first proposed by Smit [156] in the early 1988. Dragone [157] proposed a few months later the planar configuration for the free-space regions acting as star couplers, and subsequently, Smit and Vellenkoop

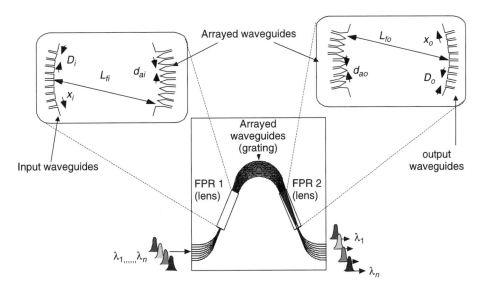

**Figure 15.45**  Schematic configuration of an AWG device.

[158] proposed and demonstrated the converging of the waveguide arrays to points on both sides, terminating the array on an arc before reaching the points with output waveguides on the other side, which was similar to Dragone's coupler. In 1990, Takahashi [159] demonstrated experimentally an array of waveguides similar to Smit's proposal incorporating external bulk lenses acting as free-space optics star couplers to couple the output radiation to fibers. The final version of the AWG as it is known today was due to Dragone [160] in 1991.

The operation of the AWG can be explained in simple terms referring to Fig. 15.45 as follows. We consider the first FPR (input lens) where the input waveguide separation is given by $D_i$, the arrayed waveguide separation is $d_{ai}$, and the focal length of the lens is $L_{fi}$. We will consider that the waveguide parameters in the two FPs (lenses) can be different (although in practice most of the times are equal). Therefore, in the second slab region, the output waveguide separation is $D_o$, the arrayed waveguide separation is $d_{ao}$, and the focal length is given by $L_{fo}$. The input light at position $x_i$ (measured positive in counterclockwise direction) is radiated to the first slab and excites the arrayed waveguides. Within each waveguide in the array, the field amplitude profile has usually a Gaussian distribution. After traveling through the array, light beams interfere constructively at one focal point $x_o$ (measured positive in counterclockwise direction) at the output surface of the second FPR.

The grating equation for this device can be derived following the same procedure employed in Section 15.4.2. For that purpose, let us consider the phase retardation for the two light beams passing through the "$j$" and "$j + 1$" arrayed waveguides, as shown in Fig. 15.46.

If $n_c$ represents the refractive index in the arrayed waveguides and $n_s$ the refractive index at the free propagation regions, then the grating equation is given by the condition that the difference between the total phase retardation experienced by the two light beams originated at $x_i$ and passing through waveguide "$j$" and waveguide "$j + 1$" must be an integer of $2\pi$ in order to achieve constructive interference at $x_o$.

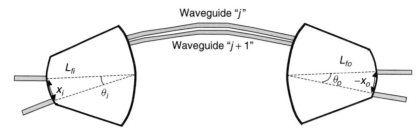

**Figure 15.46**   AWG layout for the calculation of the device grating equation using two adjacent light propagation paths.

In other words,

$$m\lambda = n_c \Delta L + n_s d_{ai} \sin\theta_i - n_s d_{ao} \sin\theta_o \qquad (15.37)$$

Because $\sin\theta_i \approx x_i/L_{fi}$, $\sin\theta_o \approx x_o/L_{fo}$, the former equation can be expressed as

$$m\lambda = n_c \Delta L + \frac{n_s d_{ai} x_i}{L_{fi}} - \frac{n_s d_{ao} x_o}{L_{fo}} \qquad (15.38)$$

Let us assume that the device is designed in such a way that for a given wavelength $\lambda_o$, the constructive interference for a signal injected at $x_i = 0$ is achieved at $x_o = 0$; in other words, $n_c \Delta L = m\lambda_o$. Then, according to Eq. (15.38), if this wavelength is input at position $x_i$, the output position $x_o$ will be given by the following expression:

$$\frac{n_s d_{ai} x_i}{L_{fi}} = \frac{n_s d_{ao} x_o}{L_{fo}} \qquad (15.39)$$

Usually the waveguide parameters for the first and the second free propagation regions are identical as pointed out previously. In this case, Eq. (15.39) transforms into $x_o = x_i$, and in consequence, the input and output distances are the same. This principle can be generalized to show the wavelength routing operation of the device. Consider now another wavelength $\lambda_b$ for which the constructive interference for a signal injected at $x_i = 0$ is achieved at $x_o = r D_o$, with $r$ being an integer. In this case,

$$m\lambda_b = n_c \Delta L - \frac{n_s d_{ao} r D_o}{L_{fo}} \qquad (15.40)$$

If the input signal is now injected at $x_i = l D_i$ with $l$ integer, substitution in Eq. (15.38) with $\lambda = \lambda_b$ and assuming equal value for the parameters in the two free propagation regions and the same waveguide separation at the input and output planes ($D_i = D_o = D$) yields

$$x_o = (l + r)D \qquad (15.41)$$

Equation (15.41) illustrates an important property of AWGs, if the input point for a signal of wavelength $\lambda_b$ is shifted by an integer multiple of the waveguide separation $D$, then output focal point is shifted by the same discrete amount. Figure 15.47 shows a specific example for a $16 \times 16$ AWG. Here, the chosen wavelength is such that if it is

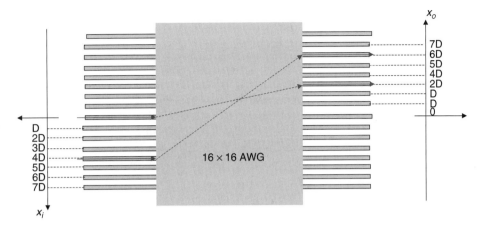

**Figure 15.47**  Wavelength routing functionality of a 16x16 AWG device.

injected in the central port $x_i = 0$, then $x_o = 2D$ (i.e., $r = 2$). If the same wavelength is now injected through the port located at $x_i = 4D$ ($l = 4$), then according to Eq. (15.41), it will be focused at port $x_o = 6D$.

Another parameter of importance, as in the case of free space gratings, is the dispersion of the focal position $x_o$ with respect to the wavelength $\lambda$ for a fixed input position $x_i$. This can be computed by direct differentiation of Eq. (15.38), yielding

$$D_{x_o} = \frac{dx_o}{d\lambda} = \frac{N_c L_f \Delta L}{n_s d\lambda_o} = m\left(\frac{N_c L_f}{n_s n_c d}\right) \tag{15.42}$$

In the above equation, $N_c$ represents the group index corresponding to $n_c$; that is, $N_c = n_c - \lambda dn_c/d\lambda$. Note that the dispersion increases linearly with the value of $m$, as it happened with free-space diffraction gratings. A correct device design for its application as a WDM MUX/DEMUX requires that for a wavelength separation $\Delta\lambda$ corresponding to the fixed ITU channel separation, the spatial dispersion in the focusing plane corresponds to the separation $D$ between adjacent output waveguides. This implies the fulfillment of the following condition:

$$\Delta x_0 = D = D_{x_o}\Delta\lambda = \frac{N_c L_f \Delta L \Delta\lambda}{n_s d\lambda_o} \Rightarrow \Delta\lambda = \frac{n_s d D\lambda_o}{N_c L_f \Delta L} \tag{15.43}$$

As mentioned before, the AWG is conceptually very similar to the free-space diffraction grating discussed in Section 15.4.2. In fact, its equivalent circuit in bulk optics, includes a free-space grating as a central element. Because the input signal expanded by the input FPR is spatially sampled by the grating array, the diffraction pattern of a given wavelength follows a periodic angular pattern in the output plane of the second FPR, each period corresponding to a given order $m$ of diffraction. Therefore, there is a limited spatial region called the spatial free spectral range $FSR_x$ within the output plane where wavelengths can be univocally resolved that is given by the spatial separation between the $m$th and the $m + 1$th focused beams for the same wavelength. From Eq. (15.38), it follows immediately that

$$FSR_x = \frac{\lambda_o L_f}{n_s d} = \frac{n_c \Delta L L_f}{n_s d m} \tag{15.44}$$

It is interesting to note that as in the case of the free-space diffraction grating, the value of $FSR_x$ is inversely proportional to $m$. Thus, increasing the number of the diffraction order used to implement the wavelength separation will decrease the value of the wavelength region that can be actually demultiplexed by the device.

Finally, as it also happens with free-space gratings, the frequency response for a given input–output configuration is periodic. The value of the period or frequency free spectral range, $FSR_f$, is given by

$$FSR_f = \frac{c}{n_c \Delta L} \tag{15.45}$$

The model described so far is based on the propagation of plane waves and provides a useful insight to many of the important properties of AWG. However, this model is only approximate and does not take into account the guided wave nature of the propagation inside the waveguides of the array, the power coupling from the field at the FPR and the waveguides in the array, and finally the coupling between the far field at the second FPR and the output waveguides. Such a description requires an electromagnetic treatment. There have been several authors dealing with such models, and the reader can find detailed treatments in Refs. 161–163. What follows is a resumed version of the technique presented in Ref. 163.

We consider an AWG layout as shown in Fig. 15.48. Because the number of involved parameters is higher than in the plane wave model, we proceed to rename some of them. The inset on the upper left corner of Fig. 15.48 shows the waveguide layout with its corresponding parameters, the waveguide width $W_x$, the gap between waveguides $G_x$, and the waveguide spacing $d_x$. The inset on the upper right side of the figure shows the FPR's layout. It consists of two sets of waveguides positioned over two identical circumferences of radius $L_f$, called the focal length. The centers of these circumferences are separated a distance equal to the focal length.

Consider the field at the output of the central input waveguide (CIW) described by both its slowly varying amplitude and its spatial distribution profile

$$f_{o,t}(x_o) = u(t)e^{j2\pi v_o t} f_o(x_o) \tag{15.46}$$

where $x_o$ is the coordinate over the input plane, as shown in Fig. 15.48, $v_o$ is the optical carrier frequency, and $f(x_o)$ is the spatial field profile. In the temporal frequency domain, Eq. (15.46) can be expressed as

$$f_{o,v}(x_o) = U(v - v_o) f_o(x_o) \tag{15.47}$$

Let us assume that this field is radiated from the CIW to the first FPR. The light spatial distribution in the focal plane can be obtained by the spatial Fourier transform of the input distribution, using the paraxial approximation [84]

$$f_o(x_1) = \left. \frac{1}{\sqrt{\alpha_v}} \Im\{f_o(x_o)\} \right|_{u = x_1/\alpha_v} \tag{15.48}$$

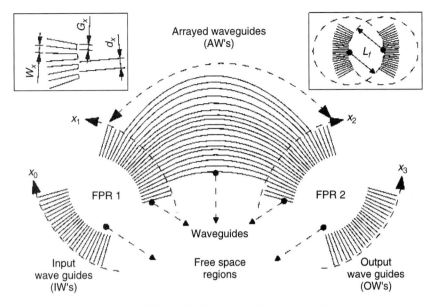

**Figure 15.48**   AWG device layout for electromagnetic analysis.

This approximation holds in the region of Fraunhofer diffraction, which, for the AWG, corresponds to slab lengths, fulfilling

$$L_f \gg \frac{\pi W_x^2}{4\lambda} \tag{15.49}$$

For typical waveguide widths and wavelength values of $W_x = 4.5$ μm and $\lambda = 1550$ nm, respectively, the approximation is valid for $L_f \gg 0.01$ mm. In Eq. (15.48), $u$ is the spatial frequency domain variable of the Fourier transform and $\alpha_v$ is the equivalent to the wavelength focal length product in Fourier optics propagation [84].

$$\alpha_v = \frac{cL_f}{n_s v} = \frac{\lambda L_f}{n_s} \tag{15.50}$$

If $v - v_o \ll v_o$, the approximation $v \approx v_o$ holds, and then $\alpha_v \approx \alpha_{v_o} \approx \alpha$.

The fundamental mode profile in the CIW can be expressed as a power normalized Gaussian function:

$$b_i(x_o) = \sqrt[4]{\frac{2}{\pi w_i^2}} e^{-(x_o/w_i)^2} \tag{15.51}$$

where $w_i$ is the mode field radius related to the waveguide width $W_i$ by

$$w_i = W_i(0.321 + 2.1V^{-3/2} + 4V^{-6}) \tag{15.52}$$

with $V$ being the waveguide normalized frequency [164]. Taking into account Eqs. (15.48) and (15.51), the spatial light distribution impinging on the arrayed waveguides (AWs) will be

$$B_i(x_1) = \sqrt[4]{\frac{2\pi w_i^2}{\alpha^2}} e^{-(\pi w_i x_1/\alpha)^2} \tag{15.53}$$

In order to calculate how much light from Eq. (15.53) couples into one of the AWs, the overlap integral between the illuminating field and the waveguide modes must be solved [164]. If the fundamental mode in the waveguides can be approximated by the following Gaussian function:

$$b_g(x_1) = \sqrt[4]{\frac{2}{\pi w_g^2}} e^{-(x_1/w_g)^2} \tag{15.54}$$

where $w_g$ is modal field radius, and the amount of energy from the illumination that excites the fundamental mode in the waveguide centered at $x_1 = 0$ is [165]

$$a = \int_{-\infty}^{\infty} B_i(x_1) b_g(x_1) dx_1 \tag{15.55}$$

In practice, the following approximation related to the exponents of Eqs. (15.53) and (15.54) holds:

$$\frac{\pi w_i}{\alpha} \ll \frac{1}{w_g} \tag{15.56}$$

which means that the illumination can be considered constant over the width of the modal field of a single AW. Then, the result of Eq. (15.55) can be expressed as follows:

$$a \approx \sqrt[4]{2\pi w_g^2} B_i(0) \tag{15.57}$$

Therefore, the total field distribution for an arbitrary number of illuminated waveguides, spaced by $d_w$ (see the inset in Fig. 15.47), is given by

$$f_1(x_1) = \sqrt[4]{2\pi w_g^2} \sum_r B_i(r d_w) b_g(x_1 - r d_w) \tag{15.58}$$

For an array of $N$ waveguides, Eq. (15.58) can be rewritten as follows:

$$f_1(x_1) = \left[ \prod \left( \frac{x_1}{N d_w} \right) B_i(x_1) \delta_w(x_1) \right] * \sqrt[4]{2\pi w_g^2} b_g(x_1) \tag{15.59}$$

where $\Pi$ is the pi function, given by

$$\prod \left( \frac{x}{a} \right) = \begin{cases} 1 & |x| \leqslant a/2 \\ 0 & \text{otherwise} \end{cases} \tag{15.60}$$

and $\delta_w(x_1)$ is a summation of delta functions

$$\delta_W(x_1) = \sum_{r=-\infty}^{\infty} \delta(x_1 - r d_w) \tag{15.61}$$

Our next step is to compute the field at the end of the waveguide array. The length of the waveguide number $r$ in the array is given by

$$l_r = l_o + \Delta L \left( r + \frac{N}{2} \right) \qquad (15.62)$$

where $l_o$ represents the length of the shortest waveguide, corresponding to $r = -N/2$. As mentioned before, if we set the value of $\Delta \lambda$ to an integer multiple $m$ (the diffraction order) of the design wavelength in the waveguides:

$$\Delta L = \frac{m\lambda_o}{n_c} = \frac{mc}{n_c v_o} \qquad (15.63)$$

We ensure that the lightwave from the CIW $(p = 0)$ focuses on the central output waveguide (COW), where $q = 0$ at the design wavelength $\lambda_o$. The phase change in a waveguide corresponding to $\Delta l$ is

$$\beta \Delta L = \frac{2\pi m v}{v_o} \qquad (15.64)$$

where $\beta$ is the propagation constant and $v$ is the frequency. The corresponding phase shift introduced by waveguide $r$ is thus

$$\Delta \phi_r = \beta l_r = 2\pi \frac{n_c}{c} v l_r \qquad (15.65)$$

This term can be introduced in Eq. (15.59) to yield the field at the output of the waveguide array:

$$f_2(x_2, v) = \left[ \Pi \left( \frac{x_2}{N d_w} \right) B_i(x_2) \delta_w(x_2) \phi(x_2, v) \right] * \sqrt[4]{2\pi w_g^2} b_g(x_2) \qquad (15.66)$$

where the phase factor $\phi(x_2, v)$ is given by

$$\phi(x_2, v) = \psi(v) e^{-j2\pi m(v/v_o)(x_2/d_w)}$$
$$\psi(v) = e^{-j2\pi v[n_c l_o/c + mN/(2v_o)]} \qquad (15.67)$$

To obtain the spatial field distribution at the output plane of the second FPR, coordinate $x_3$, the Fourier transform of Eq. (15.67) is calculated, yielding

$$f_3(x_3, v) = B_g(x_3) \left[ \sqrt[4]{\frac{2\pi w_g^2}{\alpha^2}} \psi(v) \sum_{r=-\infty}^{\infty} f_M \left( x_3 - r\frac{\alpha}{d_w} + \frac{v}{\gamma} \right) \right] \qquad (15.68)$$
$$\gamma = \frac{d_w v_o}{\alpha m}$$

This expression has noticeable similarities with Eq. (15.12) derived for a bulk optics diffraction grating. The field at the output of the second FPR is composed by the product of a slowly varying function proportional to the Fourier transform of the "slit function" of each waveguide in the array, which is represented in this case by the mode profile function $b_g(x)$ and a fast varying array factor, which is given by the term included in brackets in Eq. (15.68). The array factor depends on the $f_M$ function, which is the Fourier transform of a truncated Gaussian function and follows the expression:

$$f_M(x_3) = \sqrt[4]{\frac{\alpha^2}{8\pi w_i^2}} e^{-(x_3/w_i)^2} \left[ erf\left( \frac{\pi w_i N d_w}{2\alpha} + j\frac{x_3}{\alpha} \right) + erf\left( \frac{\pi w_i N d_w}{2\alpha} - j\frac{x_3}{\alpha} \right) \right]$$

(15.69)

As can be observed, the field at the output of the second FPR is composed of a summation of terms. Each one is univocally characterized by an integer value $r$ (the diffraction order). For a fixed value of the diffraction order, the AWG directs each temporal frequency of the input waveform to a different spatial position in the output plane according to the specific value $r$ and $\gamma$, which is actually the frequency spatial dispersion parameter [compare with the definition given in Eq. (15.42)].

Each output waveguide is placed at a selected point of the output plane and selects a temporal frequency band pass, the shape of which is given by the function $f_M$. The term $B_g(x_3)$ is responsible for the loss of nonuniformity, and the term $\psi(v)$ incorporates the minimum delay information to the model.

From Eq. (15.68), it is clear that the same temporal frequency of the input waveform will be directed to different points of the output plane by different diffraction orders. These points follow a periodic spatial pattern; that is, they are separated by a fixed distance that has been previously defined as the spatial free spectral range (SFSR), which is calculated as the distance in the $x_3$ plane corresponding to the focusing points of a fixed input frequency $v$ by two consecutive diffraction orders (i.e., $r$ and $r+1$).

$$FSR_x = \frac{\alpha}{d_w}$$

(15.70)

It is also evident, from Eq. (15.68), that two different frequencies can be focused by two different diffraction orders to the same output point $x_3$. If the diffraction orders are consecutive, then this frequency separation is known as the frequency free spectral range $FSR_f$. From Eq. (15.68), its value is given by

$$FSR_f = \frac{v}{m}$$

(15.71)

The AWG will focus input frequencies separated by this period to the same output port. Using the above definitions, Eq. (15.68) can be rewritten as

$$f_3(x_3, v) = B_g(x_3) \left[ \sqrt[4]{\frac{2\pi w_g^2}{\alpha^2}} \psi(v) \sum_{r=-\infty}^{\infty} f_M\left( x_3 - FSR_x\left( r - \frac{v}{FSR_f} \right) \right) \right]$$

(15.72)

This equation illustrates how, for a given diffraction order $r$, different frequencies focus to different points on the $x_3$ plane. For instance, the reader can check that as requested if $v = v_o$, the order $r = m$ is focused to the COW.

The final step in the model involves the calculation of how much energy from Eq. (15.72) is coupled to excite the fundamental modes in each output waveguide

(OW). This requires the calculation of an overlap integral that finally gives the frequency electric field transfer function from input waveguide $p = 0$ to output waveguide $q$.

$$t_{oq} = \int_{-\infty}^{\infty} f_3(x_3) b_o(x_3 - q d_o) dx_3 \tag{15.73}$$

where $d_o$ represents the spacing between output waveguides and $b_o(x_3)$ represents the spatial profile of the fundamental mode in the output waveguides. The former analysis can be extended to the more general case when the input signal is injected though the input waveguide $p$ (i.e., a waveguide centered at a distance $p d_i$ from the central waveguide, where $d_i$ represents the distance between adjacent input waveguides). In this case,

$$f_{3,p}(x_3, v) = B_g(x_3) \left[ \sqrt[4]{\frac{2\pi w_g^2}{\alpha^2}} \psi(v) \sum_{r=-\infty}^{\infty} f_M \left( x_3 + p d_i - FSR_x \left( r - \frac{v}{FSR_f} \right) \right) \right] \tag{15.74}$$

and

$$t_{pq} = \int_{-\infty}^{\infty} f_{3,p}(x_3) b_o(x_3 - q d_o) dx_3 \tag{15.75}$$

### 15.6.3  Performance Characteristics

Either model of the two presented in the prior section can be employed to describe the operation of the AWG as a MUX/DEMUX device and to outline the most important design parameters involved in their implementation and the main limiting factors. Because the electromagnetic model is more powerful and has been described in detail, we will employ it to describe the main features of AWGs in connection to WDM MUX/DEMUX applications.

The AWG operation as a WDM DEMUX requires in principle a device with only one input waveguide and $N$ output waveguides, where $N$ represents the number of wavelengths to de demultiplexed. If $\Delta v_c$ represents the frequency channel spacing of the WDM grid, then the spacing between the output waveguides must be given by

$$d_o = \frac{\Delta v_c}{\gamma} \tag{15.76}$$

If the above equation is satisfied, then the design frequency $v_o$ is focused to the COW and the center frequencies of the channels focused to the different output waveguides are given by

$$v_{0,q}^{(r)} = v_o - q \Delta v_c + (r - m) FSR_f \tag{15.77}$$

where $r$ represents the diffraction order. Note that for $r = m$, Eq. (15.77) yields $v_{o,q} = v_o - q \Delta v_c$; that is, each wavelength is directed toward a different output waveguide as required. In order for the device to be able to demultiplex all the WDM grid employing a single diffraction order, the spatial free spectral range must satisfy

$$FSR_x \geqslant N d_o \tag{15.78}$$

The above represent the two main requirements to start the design of an AWG-based MUX/DEMUX. Another important requirement is that the band-pass characteristic of every demultiplexed channel must be wide enough to include all the relevant spectral content associated with that particular wavelength. A relatively straightforward relationship can be derived from Eq. (15.73) for the calculation of the 3-dB bandwidth of the AWG resonances:

$$\Delta \nu_{3\ dB} \approx 10.44 \nu_o \gamma \tag{15.79}$$

Indeed, however, as pointed out in Ref. 163, the proper design of the AWG devices involves the calculation of many parameters and therefore requires a highly systematized procedure. The interested reader is referred to Ref. 163 for a full flow-chart description of this procedure.

In practice, there are several factors that degrade the ideal performance of AWG MUX/DEMUX devices. The most important ones will be discussed now together with proposed solutions to overcome them.

Figure 15.49 shows the measured response of a typical AWG MUX/DEMUX device for 16 ITU grid-spaced WDM channels. The first important source of degradation that can be observed are the insertion losses.

Insertion losses in an AWG are due to (1) coupling from fiber to IW and from OW to fibers, (2) propagation in the arrayed waveguides, (3) coupling from the first FPR to the AWs, and (4) spatial diffraction to other orders on the output focal plane.

The first two are covered in detail in Refs. 150 and 166. For instance, coupling losses result from the mode profile mismatch between the fundamental mode of the optical fiber and the fundamental mode of the input/output waveguides due to their different transversal geometry (circular for the fiber and squared for the waveguides). These losses are especially important for high and super-high $\Delta$ waveguides where the mode field diameter (MFD) is reduced. For instance, typical losses of 2 dB/point of connection are observed in 1.5%-$\Delta$ waveguides with a 4.5 × 4.5-$\mu$m core. This leads to a demand for a method that can reduce the coupling loss between SMF and SH- waveguides with small MFDs. A method has been proposed in Ref. 167 to adjust the SMF mode field by providing the waveguides with a field converter prepared by

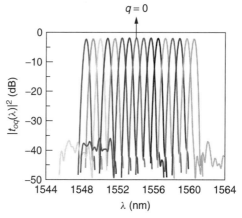

**Figure 15.49** Measured response of a typical AWG MUX/DEMUX device for 16 ITU grid spaced WDM channels.

the TEC technique. Further work has resulted in the development of three methods for reducing the fiber-connection loss in SH-$\Delta$ PLCs, as reported in Ref. 150. Figure 15.50 illustrates these.

In the first method, as shown in Fig. 15.50(**a**), a coupling structure is employed with a field converter on the fiber side. This method connects an SH-waveguide to a high numerical aperture (NA) fiber whose mode field is matched with that of the waveguide. The high-NA fiber is joined to the SMF by fusing and TEC-treating the joints [168]. This high-NA fiber fusion TEC technique is expected to be inexpensive, because no additional PLC treatment is required. Using this method, coupling losses as low as 0.2 dB/point have been reported [150]. In the second method shown in Fig. 15.50(**b**), an intermediate PLC was inserted between an SH-PLC and an SMF to provide a cascaded SSC structure to reduce the coupling loss. The intermediate PLC has an MFD between those of the SH-PLC and the SMF. This method is effective for fabricating large-scale SH-PLC-type devices because no additional process is needed for PLC-type devices. Fiber to waveguide coupling losses as low as 0.1 dB have been reported [150].

In the third method shown in Fig. 15.50(**c**), vertically and laterally tapered waveguides were constructed as the input and output ports in SH-$\Delta$ PLCs, in which both the core height and core width were enlarged. The process requires an optimization procedure reported in Ref 150, which results in typical connection losses of 0.2 dB/point.

Diffraction losses can be estimated as the ratio between the diffraction order focused to $x_3 = 0$ and the sum of energy of all of them [169]. Losses due to coupling from the first FPR to the AWs are identical to the diffraction loss by reciprocity [161]. Then, the total loss due to both effects (3) and (4) can be calculated as

$$l_d = \frac{1}{\left[\displaystyle\sum_{r=-\infty}^{\infty} |B_g(r-m)FSR_x|^2\right]} \qquad (15.81)$$

An approximate formula, taking only into account the energy lost due to transfer to the two adjacent diffraction orders to the order number $m$, can be used to estimate the

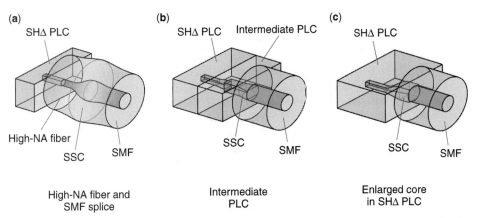

**Figure 15.50**  Three proposed methods for reducing the fiber-connection loss in SH-$\Delta$ PLCs. (Source: [150].)

required ratio $d_w/w_g$:

$$\frac{d_w}{w_g} \approx \frac{\pi\sqrt{2}}{\ln\left[\dfrac{2}{10^{-L_d/20} - 1}\right]} \tag{15.82}$$

where $L_d = 10\log(l_d)$.

A second source of degradation that can be observed in the transfer functions depicted in Fig. 15.49 is loss of nonuniformity, which is due to the weighting of the array factor in Eq. (15.72) by $B_g(x_3)$. This implies that the maximum value of the frequency response will be different depending on the OW. The worst case corresponds to the outermost OW (OOW), located a distance $FSR_x/2$ away from the COW. The loss of nonuniformity $L_u$ can be defined as

$$L_u(dB) = 20\log\left(\frac{B_g(0)}{B_g(FSR_x/2)}\right) \approx 1.08\left(\frac{\pi w_g}{d_w}\right)^2 = 1.08\sigma_g^2 \tag{15.83}$$

The parameter $\sigma_g$ controls the shape of the weighting function $B_g(x_3)$. If the modal field radius $w_g$ in the waveguides decreases, the weighting becomes more uniform, so losses along all the output ports become more uniform, as does the loss difference between the COW and the OOW.

Another important issue regarding the performance of AWG-based WDM MUX/ DEMUX devices is their polarization sensitivity. The propagation characteristics in the arrayed waveguides and, most precisely, the propagation constant for the fundamental modes is different whether they are TE or TM. This results in different spatial focusing points for the same wavelength at the output of the second FPR. To overcome this limitation, a half-wave plate can be inserted in the middle of the arrayed waveguide region, as shown in Fig. 15.51 [170].

The crosstalk is also another important issue. In general, the more uniform the illumination of the AWs is, the higher the side lobes of its Fourier transform will be, as is well known from filter design theory [171]. Equation (15.68) shows how this Fourier transform is replicated over the output plane depending on the light frequency.

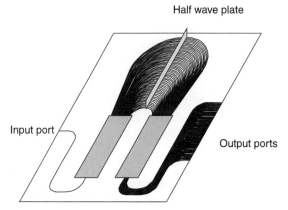

**Figure 15.51**   Half-wave plate insertion solution for combating the polarization dependence of AWG devices. (Source: [170].)

To estimate the frequency response level at the optical frequencies corresponding to the adjacent channels, $\Delta v_c$, Eq. (15.73) with $q = 0$ can be used:

$$t_{00}(v_o + \Delta v_c) = t_{01}(v_o) = \int_{-\infty}^{\infty} f_3(x_3, v_o)b_o(x_3 + d_o)dx_3 \qquad (15.84)$$

Using the normalization procedure described in Ref. 163, it can be shown that the former crosstalk function can be written as a function of two normalized variables $t_{01}(v_o) = t_{01}(\sigma, \sigma_o)$, where

$$\sigma = \frac{\alpha}{\pi N d_w w_o}$$

$$\sigma_o = \frac{d_o}{w_o} \qquad (15.85)$$

$\sigma$ which clearly governs the illumination and uniformity is the inverse of the product between the width of the field distribution over $x_1$ and the total length of the AW interface $N d_w$), whereas $\sigma_o$ controls the energy coupling at the output waveguides. The crosstalk in general increases with $\sigma$ and decreases with $\sigma_o$. It should be pointed here that crosstalk calculations are overestimated when the Gaussian approximation is employed to represent the field modes in the waveguides. Thus, a correct computation requires the knowledge of the exact modal profiles or a better approximation [163]. Typical adjacent crosstalk levels around $-25$ dB have been reported in the literature with background crosstalk levels exceeding $-40$ dB.

In practice, the most important source contributing to the crosstalk is, however, the fabrication errors that produce phase and amplitude deviations in the arrayed waveguides. The interested reader is referred to the literature [172–174] for a detailed treatment. In this case, the compensation of phase errors in all the arrayed waveguides is possible by means of a post-compensation technique using a UV-induced refractive index change [175].

Although most of the reported AWGs show a typical Gaussian-type band-pass characteristic, for WDM applications, it is more desirable to use filters with wider and flat top resonance characteristics. Several techniques have been reported in the literature, including the use of MMI couplers [176], waveguide horns [177], and by suitably apodizing the sampling of the input field to the arrayed waveguides [178]. These designs, however, usually incur in excess insertion losses as compared to Gaussian-type AWGs.

### 15.6.4  MUX/DEMUX Structures and Applications

In principle, the AWG device is in itself a WDM MUX/DEMUX device, so no additional components need to be added to it to perform this operation. Extensive work has been carried out during the last few years to demonstrate the operation of these devices under a very wide range of operating conditions, including increasing higher port count and decreasing channel separation. AWG devices capable of demultiplexing up to 64 channels with standard ITU grid separation of 100 GHz are now currently available. Silica-on-Silicon $64 \times 64$ AWG devices were demonstrated as early as 1995 [179] and on InP in 1997, in both cases, with a channel separation of 50 Ghz. In the Silica-on-Silicon device, a crosstalk value of $-27$ dB was achieved and the on-chip losses from the central channel to channel 32 ranged from 3.1 to 3.4 dB. In the case of the InP device, considerable higher insertion losses were obtained ranging from −14.4 dB in the central port to −16.4 dB in the peripheral ports. The crosstalk level was, however, <−20 dB. A closer channel spacing and a higher port count was reported for Silica-on-Silicon AWGs in 1996, where a 128 × 128 device with channel separation of 25 GHz was presented with crosstalk values of less than −16 dB and on-chip losses between 3.5 and 5.9 dB (loss of nonuniformity of 2.4 dB). A further reduction in the channel spacing to 10 GHz was reported by Yamada and coworkers in a 16 × 16 AWG device [182] where the crosstalk level was reduced using the technique reported in Ref. 175. In this later case, a higher number of ports was reported in Ref. 183, where researchers successfully fabricated a polarization-insensitive 10-GHz-spaced 128-channel arrayed-waveguide grating whose crosstalk ranged from −39 to −36 dB. Again, the optical phases of all arrayed waveguides were adjusted simultaneously by means of a photoinduced refractive-index change under ArF excimer laser irradiation through metal masks [175]. Further efforts have materialized in the report of a 256-port AWG device [184], a 400-channel DEMUX with more than 1000 arrayed waveguides and channel separation of 25 GHz [185], and the implementation of a 16 × 16 AWG with a channel separation of 2 GHz [186]. Despite these figures, higher port count devices have been recently reported by means of a so-called multistage configuration. For instance, in Ref 187, a 320-channel DEMUX device was reported, and in Refs. 188 and 189, a 10-Ghz spacing 1010-channel DEMUX and a 25-Ghz spacing 1080-channel DEMUX have been presented, respectively. In this last case, the MUX/DEMUX device is capable of spawning the C-, L- and S-bands.

The principle of the multistage or tandem configuration to achieve extremely high channel count is shown in Fig. 15.52.

In the tandem configuration, the primary AWG has a wider channel spacing and the secondary AWG has a narrower channel spacing. For example, incoming signals with 1000 different wavelengths are first divided into ten groups by the primary AWG.

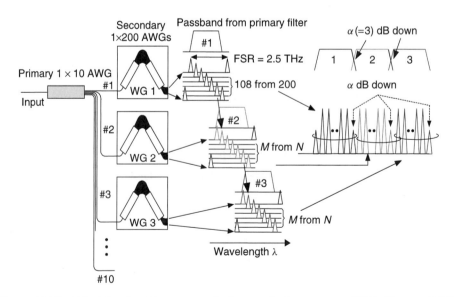

**Figure 15.52** Principle of multistage cascade for very high port-count AWG. (Source: [189]).

The secondary AWGs then further divide each group into 100 signals to achieve a total of 1000 separate signals. As the transmittance of the secondary AWG changes periodically with its $FSR_f$, the $FSR_f$ of the primary AWG should be wider than the whole signal range. If the passband shape of the primary AWG is very steep, the insertion loss around the crossing point with the adjacent channel may be higher and fluctuate. To avoid this problem, the primary AWG needs to have a wider passband in which the transmittance overlaps the adjacent channel around the 3-dB down point. In this case, as the spectral shape of the resultant two-stage MUX/DEMUX is almost the same as that of the secondary AWG, it is necessary to reduce the crosstalk of the secondary AWG, as indicated in Ref. 189.

Apart from the obvious and direct application of the AWG device as MUX/DEMUX, the device can be employed as a building block to implement fixed and/or reconfigurable OADMs and OXCs. For example, Fig. 15.53 shows a possible configuration for a reconfigurable OADM using two AWGs and 2 × 2 spatial switches.

More robust, compact, and economical OADMs can be achieved if the hole component is integrated in a single chip. For instance, in Ref. 190, an integrated OADM has been realized with a silica-based AWG and thermooptic switches. The first AWG demultiplexed the trunk input into 16 channels with a 100-GHz spacing. On each λ channel, a demultiplexer output and an add port were connected to the input branches of a 2 × 2 thermooptic switch. The output branches were connected to a drop port and a port leading to the second AWG. All of the channels that were not dropped were remultiplexed to the trunk line. All of the dropped channels were multiplexed by a third AWG. For this device, a on-off crosstalk of −24 dB and a 7 dB insertion loss were reported. An integrated OADM has also been reported on InP-based AWGs and MZI electrooptic switches [191]. In this case, however, the OADM used a single AWG in loopback configuration, as shown in Fig. 15.54.

For further applications of AWGs to routing and switching, the reader is referred to the literature [192–197].

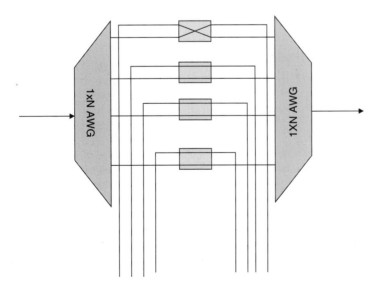

**Figure 15.53**  Layout of a reconfigurable OADM based on two AWGs and spatial switches.

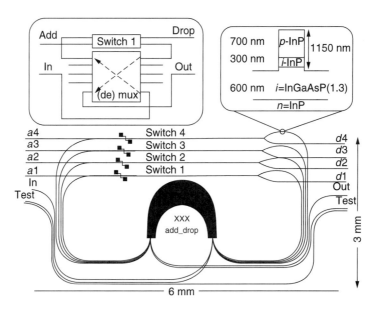

**Figure 15.54** InP-based OADM using a foldback configuration. (After [191].)

## 15.7 EPILOGUE

In this chapter, we have provided a self-contained updated review of the principles, applications, and existing technologies to implement WDM MUX/DEMUX devices. Both consolidated concepts as well as more recent advances have been presented, and interested readers will no doubt have the chance to enlarge their knowledge on the subject by means of the extensive list of references that is provided at the end of the chapter. At the time of writing this chapter, new references of interest have been published, which are also relevant to the subject [198, 199].

Although we have tried to be comprehensive in our treatment, it is also true that we might have skipped some approaches that can be of interest, but that we felt were not essential. Also, the reference list is not complete, and we might also have inadvertedly not included some relevant work by other researchers. We apologize to them in advance.

## REFERENCES

1. I. P. Kaminow and T. L. Koch, Editors, *Optical Fiber Telecommunications III A*, Academic Press, San Diego, CA, 1997.

2. I. P. Kaminow and T. Li, Editors, *Optical Fiber Telecommunications IV B*, Academic Press, San Diego, CA, 2002.

3. T. Stern and K. Bala, *Multiwavelength Optical Networks: A Layered Approach*, Addison Wesley, New York, NY, 1999.

4. R. Ramaswami and K. Syvarajan, *Optical Networks: A Practical Perspective*, 2nd ed., Morgan Kaufman, San Francisco, CA, 2002.

5. www.cisco.com and www.ciena.com.

6. C. M. Weinert, "40 Gb/s and 4 × 40 Gb/s TDM/WDM standard fiber transmission," *IEEE J. Lightwave Technol.*, **17**, pp. 2276–2284, 1999.

7. Y. Inada, et al., "2400-km transmission of 100-GHz-spaced 40-Gb/s WDM signals using a 'double-hybrid' fiber configuration," *Proc ECOC '01*, **3**, pp. 246–247 2001.

8. C. D. Chen, et al., "1-Tb/s (40-Gb/s × 25 ch) WDM transmission experiment over 342 km of TrueWave(R) (non-zero dispersion) fiber," *OFC'99 Tech. Dig.*, pp. PD7/1–PD7/3, 1999.

9. Y. Inada, et al., "32/spl times/40-Gb/s dense WDM transmission over 3000 km using 'double-hybrid' fiber configuration," *IEEE Photon. Technol. Lett.*, **14**, pp. 1366–1368, 2002.

10. J. Yu, et al., "8 × 40 Gb/s 55-km WDM transmission over conventional fiber using a new RZ optical source," *IEEE Photon. Technol. Lett.*, **12**, pp. 912–914, 2000.

11. L. E. Nelson, "Challenges of 40 Gb/s WDM transmission," *OFC 2001 Tech. Dig.*, 2001, Paper ThF1.

12. P. Bonenfant and A. Rodriguez-Moral, "Optical data networking," *IEEE Commun. Mag.*, **38**, pp. 63–70, 2000.

13. J.M.H. Elmirghani and H. T. Mouftah, "Technologies and architectures for scalable dense WDM networks," *IEEE Commun. Mag.*, **38**, pp. 58–66, 2000.

14. B. Rajagopalan, et al., "IP over optical networks: Architectural aspects," *IEEE Commun. Mag.*, **38**, pp. 94–103, 2000.

15. E. L. Goldstein, L. Y. Lin, and R. W. Tkach, "Multiwavelength opaque optical crossconnect networks," *IEICE Trans. Commun.*, **E82-B**, pp. 1095–1104, 1999.

16. R. H. Cardwell, O. J. Wasem, and H. Kobrinski, "WDM architectures and economics in metropolitan areas," *Opt. Networks*, **1**, pp. 41–50, 2000.

17. J. M. Senior and S. D. Cusworth, "Devices for wavelength multiplexing and demultiplexing," *Proc. IEE Pt. J. Optoelectron.*, **136**, pp. 183–202, 1989.

18. E. Pennings, G.-D. Khoe, M. K. Smit, and T Staring, "Integrated-optic versus microoptic devices for fiber-optic telecommunication systems: a comparison," *IEEE J. Sel. Topics Quantum Electron.*, **2**, pp. 151–164, 1996.

19. J. M. Senior, M. R. Handley, and M. S. Leeson, "Developments in wavelength division multiple access networking," *IEEE Commun. Mag.*, **36**, pp. 28–36, 1998.

20. J. P. Laude, *DWDM Fundamentals, Components, and Applications*, Artech House, Boston, MA, 2002.

21. S. Kartalopoulos, *DWDM: Networks, Devices, and Technology*, Wiley, New York, 2002.

22. O. Krauss, *DWDM and Optical Networks: An Introduction to Terabit Technology*, Wiley, New York, 2002.

23. A. Gumaste and T. Anthony, *DWDM Network Designs and Engineering Solutions*, Wiley, New York, 2002.

24. K. Sivalingham and S. Subramanian, Editors, *Optical WDM Networks — Principles and Practice*, Kluwer Academic Press, Amsterdam, The Netherlands, 2000.

25. D. Faulkner and A. Harmer, *WDM and Photonic Networks*, IOP Press, UK, 2000.

26. M.M.K. Liu, *Principles and Applications of Optical Communications*, McGraw-Hill, New York, 1996.

27. S. Kartalopoulos, *Introduction to DWDM Technology: Data in a Rainbow*, Wiley, New York, 2000.

28. G. R. Hill, et al., "A transport network layer based on optical network elements," *IEEE J. Lightwave Technol.*, **11**, pp. 667–679, 1993.

29. S. Okamoto, A. Watanabe, and K. I. Sato, "optical path cross-connect node architectures for photonic transport networks," *IEEE J. Lightwave Technol.*, **14**, pp. 1410–1422, 1996.

30. A.A.M. Saleh and J. Simmons, "Architectural principles for optical regional and metropolitan access networks," *IEEE J. Lightwave Technol.*, **14**, pp. 1349–1355, 1996.

31. D. Derickson, *Fiber Optic Test and Measurement*, Prentice-Hall, New York, 1999.

32. N. Kashima, *Passive Optical Components for Optical Fiber Transmission*, Artech House, Boston, MA, 1995.

33. A. B. Buckman, *Guided Wave Photonics*, Saunders College Publishing, New York, 1992.

34. *WDM Passive Components Test Guide*, HP Lightwave Division, Document 5965-3124E, Santa Rosa, CA, 1996.

35. S. Kartalopoulos, *Fault Detectability in DWDM: Towards Higher Signal Quality and System Reliability*, Wiley, New York, 2001.

36. C. R. Giles and M. Spector, "The wavelength add/drop multiplexer for lightwave communications," *Bell Labs. Tech. J.*, **4**, pp. 207–229, 1999.

37. P. E. Green, *Fiber Optic Networks*, Prentice-Hall, Englewood Cliffs, NJ, 1993.

38. A. M. Hill and A.J.N. Houghton, "Optical networking in the European ACTS programme," *OFC'96 Tech. Dig.*, pp. 238–239, 1996.

39. R. E. Wagner, et al., "MONET: multiwavelength optical networking," *IEEE J. Lightwave Technol.*, **14**, pp. 1349–1355, 1996.

40. N. J. Frigo, et al., "A wavelength-division passive optical network with cost-shared components," *IEEE Photon. Tech. Lett.*, **6**, pp. 1365–1367, 1994.

41. E. Desurvire, J. R. Simpson, and P. C. Becker, "High gain erbium doped travelling wave amplifier," *Opt. Lett.*, **12**, pp. 888–890, 1987.

42. R. I. Laming, et al., "High-sensitivity two-stage erbium doped fiber preamplifier at 10 Gb/s," *IEEE Photon. Tech. Lett.*, pp. 1348–1350, 1992.

43. G. R. Giles and E. Desurvire, "Propagation of signal and noise in concatenated erbium doped fiber optical amplifiers," *IEEE J. Lightwave Technol.*, **9**, p. 147, 1991.

44. K. Inoue, H. Toba, and K. Nosu, "Multichannel amplification utilizing an Er-doped fiber amplifier," *IEEE J. Lightwave Technol.*, **9**, p. 368, 1991.

45. K. Lui and R. Ramaswami, "Analysis of optical bus networks using doped fiber amplifiers," *Proc. IEEE LEOS Topical Meeting in Multiple Access Networks*, 1989.

46. A. F. Elrefaie and S. Zaidi, "Fiber amplifier in close ring WDM networks," *Electron. Lett.*, **28**, pp. 2340–2342, 1992.

47. M. I. Ishid and M. Kavehrad, "Star couplers with gain using fiber amplifiers," *IEEE Photon. Tech. Lett.*, **4**, pp. 58–60, 1992.

48. E. Desurvire, *Erbium Doped Fiber Amplifiers*, Wiley, New York, 1994.

49. P. C. Becker, J. R. Simpson, and N. A. Olsson, *Erbium-Doped Fiber Amplifiers*, Academic Press, San Diego, CA, 1999.

50. A. Bjarklev, *Optical Fiber Amplifiers: Design and System Applications*, Artech House, Boston, MA, 1993.

51. K. Bala, P. Petropoulos, and T. Stern, "Waveband And Channel Routing In A Linear Lightwave Network," *LEOS Summer Topical Meeting on Broadband Analog and Digital Optoelectronics, Optical Multiple Access Networks, Integrated Optoelectronics, Smart Pixels*, pp. A25–A26, 1992.

52. F. A. Flood and O. K. Tonguz, "Individual versus end-to-end gain equalization in erbium-doped fiber amplifier chains: A comparison," *IEEE Global Telecommunications Conference, GLOBECOM '94. "Communications: The Global Bridge"*, 1994, Vol. 2, pp. 1190–1194.

53. F. A. Flood and O. K. Tonguz, "Gain equalization of EDFA cascades," *IEEE J. Lightwave Technol.*, **15**, pp. 1832–1841, 1997.

54. M. Rochette, et al., "Gain equalization of EDFA's with Bragg gratings," *IEEE Photon. Tech. Lett.*, **1**, pp. 536–538, 1999.

55. J. Li, F. Khalehi, and M. Kaverhad, "Gain equalization by mitigating self-filtering effect in a chain of cascaded EDFA's for WDM transmissions," *IEEE J. Lightwave Technol.*, **13**, pp. 2191–2196, 1995.

56. R. H. Stolen and E. P. Ippen, "Raman gain in glass optical waveguides," *Appl. Phys. Lett.*, **22**, pp. 276–278, 1973.

57. Y. Aoki, S. Kishida, K. Washio, and K. Minemura, "Bit error rate evaluation of optical signals amplified via stimulated Raman process in an optical fiber," *Electron. Lett.*, **21**(5), pp. 191–193, 1985.

58. P. B. Hansen, L. Eskildsen, S. G. Grubb, A. J. Stentz, T. A. Strasser, J. Judkins, J. J. DeMarco, R. Pedrazzani, and D. J. DiGiovanni, "Capacity upgrades of transmission systems by Raman amplification," *IEEE Photon. Technol. Lett.*, **9**(2), pp. 262–264, 1997.

59. M. Nissov, C. R. Davidson, K. Rottwitt, R. Menges, P. C. Corbett, D. Innis, and N. S. Bergano, "100 Gb/s (10_10 Gb/s) WDM transmission over 7200 km using distributed Raman amplification," *Proc. Eur. Conf. Optical Commun.*, **5**, pp. 9–12, 1997.

60. T. Tanaka, N. Shimojoh, T. Naito, H. Nakamoto, I. Yokota, T. Ueki, A. Sugiyama, and M. Suyama, "2.1-Tbit/s WDM transmission over 7221 km with 80-km repeater spacing," *Proc. Eur. Conf. Optical Commun.*, 2000, Postdeadline Paper 1.8.

61. J. P. Blondel, F. Boubal, E. Brandon, L. Buet, L. Labrunie, P. Le Roux, and D. Toullier, "Network application and system demonstration of WDM systems with very large spans (Error-free 32 _ 10 Gbit/s 750 km transmission over 3 amplified spans of 250 km)," *Proc. Optical Fiber Commun. Conf.*, 2000, Paper PD31.

62. T. Terahara, T. Hoshida, J. Kumasako, and H. Onaka, "128 × 10.66 Gbit/s transmission over 840-km standard SMF with 140-km optical repeater spacing (30.4-dB loss) employing dual-band distributed Raman amplification," *Proc. Optical Fiber Commun. Conf.*, 2000, Paper PD28.

63. S. Bigo, A. Bertaina, Y. Frignac, S. Borne, L. Lorcy, D. Hamoir, D. Bayart, J. P. Hamaide, W. Idler, E. Lach, B. Franz, G. Veith, P. Sillard, L. Fleury, P. Guenot, and P. Nouchi, "5.12 Tbit/s (128 × 40 Gbit/s WDM) transmission over 3 × 100 km of TeraLight™TM fiber," *Proc. Eur. Conf. Optical Commun.*, 2000, Postdeadline Paper 1.2.

64. S. Yamiki and Y. Emori, "Ultrabroad-Band Raman Amplifiers Pumped and Gain-Equalized by Wavelength-Division-Multiplexed High-Power Laser Diodes," *IEEE J. Sel. T. Quantum Electron.*, **7**, pp. 3–17, 2001.

65. M. Zirngibl, "Applications of Optical switch fabrics," in *Optical Fiber Telecommunications IV*, I. P. Kaminow and T. Li (eds), Academic Press, San Diego (2002).

66. G. Wilfong, et al., "WDM Cross-Connect architectures with reduced complexity," *IEEE J. Lightwave Technol.*, **17** (1999)

67. R. E Wagner, et al., "MONET: Multiwavelength optical networking," *IEEE J. Lightwave Technol.*, **14**, pp. 1349–1355 (1996).

68. E. Almstrom S. N. Larsson, and H. Carlden, "Cascadability of optical add7drop multiplexers," proc. ECOC '98, Madrid, Spain, pp. 589–590, sept. 20–24 (1998).

69. C. R. Doerr, et al., "40-wavelength add-Drop-filter," *IEEE Photon. Tech. Lett.*, **11**, pp. 1437–1439 (1999).

70. S. Y. Kim, et al., "Channel-switching active add/drop multiplexer with tunable gratings," *Electron. Lett.*, **34**, pp. 104–105 (1998).

71. C. R Giles and V. Mizrahi, "Low-loss add/drop multiplexer for WDM lightwave networks," *Tech. Dig, 10 IOOC*, **3**, Paper ThC2-1 (1995).

72. S. Yungfeng, et al., "A novel single-fiber bidirectional optical add/drop multiplexer for distribution networks," Proc. Optical Fiber Communication Conference OFC 2001, Vol: 3, pp. WY5-1–WY5-3, vol. 3 (2001).

73. P. Chuan, et al., "Client-configurable eight-channel optical add/drop multiplexer using micromachining technology," *IEEE Photon. Techn. Lett.*, **12**, pp. 1665–1667 (2000).

74. T. An. Vu, et al., "Reconfigurable multichannel optical add-drop multiplexers incorporating eight-port optical circulators and fiber Bragg gratings," *IEEE Photon. Tech. Lett.*, **13**, pp. 1100–1102 (2001).

75. H. Venghaus, et al., "Optical add/drop multiplexers for WDM communication systems," Proc. Conference on Optical Fiber Communication. OFC 97, 16–21, Feb. 1997, pp. 280–281 (1997).

76. P. Yeh, *Optical Waves in Layered Media*, Wiley, New York, 1988.

77. H. Macleod, *Thin Film Optical Filters*, McGraw-Hill, New York, 1989.

78. C. Masden and J. Zhao, *Optical Filter Design and Analysis: A Signal Processing Approach*, Wiley, New York, 1999.

79. Z. Knittl, *Optics of Thin Films*, Wiley, New York, 1976.

80. S. Ramo, J. R. Winnery, and T. Van Duzer, *Fields and Waves in Communication Electronics*, Wiley, New York, 1993.

81. M. Scobey and D. Spock, "passive DWDM components using microplasma optical interference filters," *Tech. Dig. Optical Fiber Commun. Conf.*, pp. 242–243, 1996.

82. B.E.A. Saleh and M. C. Teich, *Fundamentals of Photonics*, Wiley, New York, 1991.

83. F. Pedrotti and L. Pedrotti, *Introduction to Optics*, Prentice-Hall, New York, 1987.

84. J. W. Goodman, *Introduction to Fourier Optics*, 2nd ed., McGraw-Hill, New York, 1996.

85. E. G. Loewen and E. Popov, *Diffraction Gratings and Applications*, Marcel Dekker, New York, 1997.

86. M. C. Hutley, *Diffraction Gratings*, Academic Press, New York, 1982.

87. F. N. Timofeev, et al., "High performance, free space ruled concave grating demultiplexer," *Electron. Lett.*, **31**, pp. 1466–1467, 1995.

88. R. Marz, *Integrated Optics: Design and Modelling*, Academic Press, San Diego, CA, 1994.

89. R. Marz and C. Cremer, "On the theory of planar spectrographs," *IEEE J. Lightwave Technol.*, **10**, pp. 2017–2022, 1992.

90. E. Peral and J. Capmany, "Generalized Bloch wave analysis for fiber and waveguide Gratings," *IEEE J. Lightwave Technol.*, **15**, pp. 1295–1302, 1997.

91. K. O. Hill and G. Meltz, "Fiber Bragg Grating Technology Fundamentals and Overview," *IEEE/OSA J. Lightwave Technol.*, **15**(8), pp. 1263–1276, 1997.

92. K. O. Hill, Y. Fujii, D. C. Johnson, and B. S. Kawasaki, "Photosensitivity in optical fiber waveguides: Application to reflection filter fabrication," *J. Appl. Phys. Lett.*, **32**, pp. 647–649, 1978.

93. G. Meltz, W. W. Morey, and W. H. Glen, "Formation of Bragg gratings in optical fibers by a transverse holographic method," *Opt. Lett.*, **14**, pp. 823–825, 1989.

94. K. O. Hill, B. Malo, F. Biodeau, D. C. Johnson, and J. Albert, "Bragg gratings fabricated in monomode photosensitive optical fiber by UV exposure through a phase mask," *Appl. Phys. Lett.*, **62**, pp. 1035–1037, 1993.

95. S. J. Mihailov and M. C. Gower, "Recording of efficient high-order Bragg reflectors in optical fibers by mask image projection and single pulse exposure with an excimer laser," *Electron. Lett.*, **30**, pp. 707–709, 1994.

96. B. Malo, K. O. Hill, F. Bilodeau, D. C. Johnson, and J. Albert, "Point-by-point fabrication of micro Bragg gratings in photosensitive fiber using single excimer pulse refractive index modification techniques," *Electron.Lett.*, **29**, pp. 1668–1669, 1993.

97. J. D. Prohaska, E. Snitzer, S. Rishton, and V. Boegli, "Magnification of mask fabricated fiber Bragg gratings," *Electron. Lett.*, **29**, pp. 1614–1615, 1993.

98. A. Othonos and X. Lee, "Novel and improved methods of writing Bragg gratings with phase masks," *IEEE Photon. Technol. Lett.*, **7**, pp. 1183–1185, 1995.

99. Y. Painchaud, A. Chandonnet, and J. Lauzon, "Chirped fiber gratings produced by tilting the fiber," *Electron. Lett.*, **31**, pp. 171–172, 1995.

100. M. C. Farries, K. Sugden, D.C.J. Reid, I. Bennion, A. Molony, and M. J. Goodwin, "Very broad reflection bandwidth (44 nm) chirped fiber gratings and narrow bandpass filters produced by use of an amplitude mask," *Electron. Lett.*, **30**, pp. 891–892, 1994.

101. R. Kashyap, "Assessment of tuning the wavelength of chirped and unchirped fiber Bragg grating with single phase-masks," *Electron. Lett.*, **34**, pp. 2025–2027, 1998.

102. Y. Wang, J. Grant, A. Sharma, and G. Myers, "Modified Talbot interferometer for fabrication of fiber-optic grating filter over a wide range of Bragg wavelength and bandwidth using a single phase mask," *IEEE/OSA J. Lightwave Technol.*, **19**(10), pp. 1569–1573, 2001.

103. M. J. Cole, et al., *Electron. Lett.*, **31**, p. 1488, 1995.

104. W. H. Loh, et al., *Optics Lett.*, **20**, p. 2051, 1995.

105. A. Asseh, H. Storoy, B. E. Sahlgren, S. Sandgren, and R. Stubbe, "A writing technique for long fiber Bragg gratings with complex reflectivity profiles," *J. Lightwave Technol.*, **15**, pp. 1419–1423, 1997.

106. J. Brennan, et al., "Dispersion and dispersion slope correction with a fiber Bragg grating over the full C band," *OFC'2001, PD*, Anaheim, CA, 2001.

107. P. J. Lemaire, R. M. Atkins, V. Mizrahi, and W. A. Reed, "High pressure H2 loading as a technique for achieving ultrahigh UV photosensitivity and thermal sensitivity in GeO2 doped optical fibers," *Electron. Lett.*, **29**, pp. 1191–1193, 1993.

108. F. Bilodeau, B. Malo, J. Albert, D. C. Johnson, K. O. Hill, Y. Hibino, M. Abe, and M. Kawachi, "Photosensitization of optical fiber and silica-on-silicon/silica waveguides," *Opt. Lett.*, **18**, pp. 953–955, 1993.

109. R. Kashyap, *Fiber Bragg Gratings*, Academic Press, San Diego, Ca, 1999.

110. A. Othonos and K. Kalli, *Fiber Bragg Gratings: Fundamentals and Applications in Telecommunications and Sensing*, Artech House, Boston, MA, 1999.

111. A. W. Snyder and D. J. Love, *Optical Waveguide Theory*, Chapman & Hall, 1983.

112. T. Erdogan, "Fiber Grating Spectra," *J. Lightwave Technol.*, **15**, pp. 1277–1294, 1997.

113. H. Kogelnik, "Filter response of nonuniform almost-periodic structures," *Bell Sys. Tech. J.*, **55**, pp. 109–126, 1976.

114. D. Marcuse, *Theory of Dielectric Optical Waveguides*, Academic Press, New York, 1991.

115. L. Poladian, "Resonance Mode Expansions and Exact Solutions for Nonuniform Gratings," *Phys. Rev. E.*, **54**, pp. 2963–2975, 1996.

116. H. Kogelnik, *Theory of Optical Waveguides, Guided-Wave Optoelectronics*, ed. T. Tamir, Springer-Verlag, New York, 1990.

117. J. Skaar, L. Wang, and T. Erdogan, "On the Synthesis of Fiber Bragg Gratings by Layer Peeling," *IEEE J. Quantum Electron.*, **37**(2), pp. 165–173, 2001.

118. R. Feced, M. N. Zervas, and M. A. Muriel, "An efficient inverse scattering algorithm for the design of nonuniform fiber Bragg gratings," *J. Quantum Electron.*, **35**, pp. 1105–1115, 1999.

119. L. Poladian, "Simple grating synthesis algorithm," *Opt. Lett.*, **25**, pp. 787–789, 2000.

120. E. Brinkmeyer, "Simple algorithm for reconstructing fiber gratings from reflectometric data," *Opt. Lett.*, **20**, pp. 810–812, 1995.

121. E. Peral, J. Capmany, and J. Marti, "Iterative solution to the Gel'Fand-Levitan-Marchenko coupled equations and application to the synthesis of fiber gratings," *IEEE J. Quantum Electron.*, **32**, pp. 2078–2084, 1996.

122. M. A. Muriel, J. Azana, and A. Carballar, "Fiber grating synthesis by use of time-frequency representations," *Opt. Lett.*, **23**, pp. 1526–1528, 1998.

123. J. Skaar and K. M. Risvik, "A genetic algorithm for the inverse problem in synthesis of fiber gratings," *J. Lightwave Technol.*, **16**, pp. 1928–1932, 1998.

124. A. M. Bruckstein and T. Kailath, "Inverse scattering for discrete transmission-line models," *SIAM Rev.*, **29**, pp. 359–389, 1987.

125. W. H. Press, S. A. Teukolsky, W. T. Vetering, and B. P. Flannery, *Numerical Recipes in C*, Cambridge University Press, New York, 1992.

126. D. M. Pozar. *Microwave Engineering*, Addison-Wesley, New York, 1990.

127. K. A. Winick, "Effective-index method and couple-mode theory for almost-periodic waveguide gratings: A comparison," *App. Optics*, **31**(6), pp. 757–764, 1992.

128. J. Albert, et al., "Apodisation of the spectral response of fibre Bragg gratings using a phase mask with variable diffraction efficiency," *Electron. Lett.*, **31**, pp. 222–223, 1995.

129. M. Ibsen, "Advanced Bragg Grating Design and Technology," *ECOC'02*, **Short course** (SC-11).

130. F. Ouellete, "Dispersion cancellation using linearly chirped Bragg grating filters in optical waveguides," *Opt. Lett.*, **12**, pp. 847–849, 1987.

131. G. P. Brady, et al., "Extended range, coherent tuned, dual wavelength interferometry using a superfluorescent fibre source and chirped fibre Bragg gratings," *Opt. Commun.*, **134**, pp. 341–346, 1987.

132. K. C. Byron, et al., "Fabrication of chirped Bragg gratings in photosensitive fibre," *Electron. Lett.*, **29**, pp. 1659–1660, 1993.

133. K. Sugden, et al., "Chirped gratings produced in photosensitive optical fibers by fibre deformation during exposure," *Electron. Lett.*, **30**, pp. 440–442, 1994.

134. R. Kashyap, et al., "Novel method of producing all fibre photoinduced chirped gratings," *Electron. Lett.*, **30**, pp. 996–997, 1994.

135. D. C. Johnson, K. O. Hill, F. Bilodeau, and S. Faucher, "New design configuration for a narrow-band wavelength selective optical tap combiner," *Electron. Lett.*, **23**, p. 668, 1987.

136. C. M. Ragdale, T. J. Reid, D.C.J. Reid, A. C. Carter, and P. J. Williams, "Integrated laser and add-drop optical multiplexer for narrowband wavelength division multiplexing," *Electron. Lett.*, **28**, pp. 712–714, 1992.

137. S. Bethuys, et al., "Optical add/drop multiplexer base on UV-written Bragg gratings in twincore fiber Mach-Zehnder interferometer," *Electron. Lett.*, **34**, pp. 1250–1251, 1998.

138. J. L. Archambault, et al., "Grating-frustrated coupler: A novel channel-dropping filter in single mode optical fiber," *Opt. Lett.*, **19**, pp. 180–182, 1994.

139. L. Dong, et al., "Novel add/drop filters for wavelength-division multiplexing optical fiber systems using a Bragg grating assisted mismatched coupler," *IEEE Photon. Technol. Lett.*, **8**, pp. 1656–1658, 1996.

140. K. Y. Se, et al., "Highly stable optical add/drop multiplexer using polarization beam splitters and Fiber Bragg grating," *IEEE Photon. Technol. Lett.*, **9**, pp. 1119–1121, 1997.

141. T. Mizuochi and T. Kitayama, "Interferometric crosstalk-free optical add/drop multiplexer using cascaded Mach-Zehnder fiber gratings," *Technical Proc. OFC'97*, pp. 176–177, 1997.

142. L. Quetel, L. Rivollan, E. Delavaque, E. Gay, and I. Le Gac, "Programmable fibre grating based wavelength demultiplexer," Tech. Dig. OFC'96, Paper WF6, 1996.

143. H. Okayama, Y. Ozeki, and T. Kunii, "Dynamic wavelength selective add/drop node comprising tunable gratings," *Electron. Lett.*, **33**(10), pp. 881–882, 1997.

144. H. Okayama, Y. Ozeki, T. Kamijoh, C. Q. Xu, and I. Asabayashi, "Dynamic wavelength selective add/drop node comprising fibre gratings and switches," *Electron. Lett.*, **34**, pp. 104–105, 1998.

145. D. Shapiro, "Family of exact solutions for reflection spectrum of Bragg grating," *Optics Commun.*, **215**, pp. 295–301, 2003.

146. A. D. Ellis, R. Kashyap, I. Crisp, and D. J. Malylon, "Dispersion compensating, reconfigurable optical add-drop multiplexer using chirped fibre Bragg gratings," *Electron. Lett.*, **34**(17), 1997.

147. C. Doerr, "Planar lightwave Circuits," in I.P. Kaminow and T. Li, *Optical Fiber Telecommunications IV, Part A Devices*, Academic Press, San Diego, CA, 2003.

148. K. Okamoto, "Recent progress of integrated optics planar lightwave circuits," *Opt. Quantum Electron.*, **31**, pp. 107–129, 1999.

149. M. Kawachi, "Silica waveguides on silicon and their application to integrated components," *Opt. Quantum Electron.*, **22**, pp. 391–416, 1990.

150. Y. Hibino, "Recent advances in high density and large scale AWG Milti/Demultiplexers with high index-contrast Silica based PLCs," *IEEE J. Sel. T. Quantum Electron.*, **8**, pp. 115–117, 2002.

151. Y. Hida, Y. Hibino, H. Okazaki, and Y. Ohmori, "10 m long silica-based waveguide with a loss of 1.7 dB/m," *Proc. IPR*, 1995.

152. H. Hibino, et al., "Fabrication of silica on Si waveguide with higher index difference and its application to 256-channel arrayed waveguide multi/demultiplexer," *Tech. Dig. OFC 2000*, **P WH2**, 2000.

153. Y. Yoshikuni, "Semiconductor arrayed waveguide gratings for photonic integrated devices," *IEEE J. Sel. T. Quantum Electron.*, **8**, pp. 1102–1113, 2002.

154. L. Eldada and L. W. Shacklette, "Advances in polymer integrated optics," *IEEE J. Sel. T. Quantum Electron.*, **6**, pp. 54–68, 2000.

155. K. Okamoto, *Theory of Optical Waveguides*. Corona, Tokyo, Japan, 1992, pp. 191–198.

156. M. Smit, "New focusing and dispersive planar component based on an optical phased array," *Electron. Lett.*, **24**, pp. 385–386, 1988.

157. C. Dragone, "Efficient NxN star coupler based on Fourier optics," *Electron. Lett.*, **18**, pp. 942–944, 1988.

158. A. R. Vellenkoop and M. K. Smit, "Four channel integrated optic wavelength demultiplexer with weak polarization dependence," *IEEE J. Lightwave Technol.*, **9**, pp. 310–314, 1991.

159. H. Takahashi, S. Suzuki, K. Kato, and I. Nishi, "Arrayed waveguide grating for wavelength division multi/demultiplexer with nanometer resolution," *Electron. Lett.*, **26**, pp. 87–88, 1990.

160. C. Dragone, C. Edwards, and R. Kistler, "Integrated optics $N \times N$ multiplexer on silicon," *IEEE Photon. Technol. Lett.*, **3**, pp. 896–898, 1991.

161. M. K. Smit and C. van Dam, "PHASAR-based WDM-devices: Principles, design and applications," *IEEE J. Select. Topics Quantum Electron.*, **2**, pp. 236–250, 1996.

162. M. C. Parker and S. D. Walker, "Design of arrayed-waveguide gratings using hybrid Fourier-Fresnel transform techniques," *IEEE J. Select. Topics Quantum Electron.*, **5**, pp. 1379–1384, 1999.

163. P. Muñoz, D. Pastor, and J. Capmany, "Modeling and design of Arrayed Waveguide Gratings," *IEEE J. Lightwave Technol.*, **20**, pp. 661–674, 2002.

164. G. P. Agrawal, *Fiber-Optic Communication Systems*, 2nd ed., Wiley, New York, 1997, p. 103.

165. A. W. Snyder and J. D. Love, *Optical Waveguide Theory*, Chapman & Hall, London, UK, 1983, pp. 420–426.

166. J. C. Chen and C. Dragone, "A proposed design for ultralow-loss waveguide grating routers," *IEEE Photon. Technol. Lett.*, **10**, pp. 379–381, 1998.

167. S. Suzuki, M. Yanagisawa, Y. Hibino, and K. Oda, "High-density integrated planar lightwave circuits using SiO–GeO waveguides with a high refractive index difference," *J. Lightwave Technol.*, **12**, pp. 790–796, 1995.

168. M. Ishii, Y. Hibino, Y. Hida, A. Kaneko, M. Itoh, T. Goh, A. Sugita, T. Saida, A. Himeno, and Y. Ohmori, "Low-loss and compact silica-based 16 channel arrayed-waveguide grating multiplexer module with higher index difference," *Proc. ECOC2000*, **3**, pp. 27–28, 2000.

169. H. Takahashi, S. Suzuki, and I. Nishi, "Wavelength multiplexer based on SiO–Ta O arrayed-waveguide grating," *J. Lightwave Technol.*, **12**, pp. 989–995, 1994.

170. H. Yamada, K. Takada, Y. Inoue, Y. Ohmori, and S. Mitachi, "Statically-compensated 10 GHz-spaced arrayed waveguide grating," *Electron. Lett.*, **32**, pp. 1580–1582, 1996.

171. A. V. Oppenheim, R. W. Schafer, and J. R. Buck, *Discrete-Time Signal Processing*, 2nd ed. Englewood Cliffs, NJ: Prentice-Hall, 1999, Signal Processing Series, pp. 313, 445–317, 457.

172. Y. Chu, et al., "The impact of phase errors on arrayed waveguide gratings," *IEEE J. Sel. T. Quantum Electron.*, **8**, pp. 1122–1129, 2002.

173. P. Muñoz, D. Pastor, J. Capmany, and S. Sales, "Analytical and numerical analysis of phase and amplitude errors in the performance of arrayed waveguide gratings," *IEEE J. Sel. T. Quantum Electron.*, **8**, pp. 1130–1141, 2002.

174. K. Maru, et al., "Statistical analysis of correlated phase error in transmission characteristics of arrayed waveguide gratings," *IEEE J. Sel. T. Quantum Electron.*, **8**, pp. 1142–1148, 2002.

175. K. Takada, et al., "Beam-adjustment free crosstalk reduction in 10 GHz-spaced arrayed waveguide grating via photosensitivity under UV irradiation through metal mask," *Electron. Lett.*, **36**, pp. 60–61, 2000.

176. M. Amersfoot, et al., "Passband broadening of integrated arrayed waveguide filters using multimode interference couplers," *Electron. Lett.*, **32**, pp. 449–451, 1996.

177. K. Okamoto and A. Sugita, "Flat spectral response arrayed waveguide grating multiplexer with parabolica waveguide horns," *Electron. Lett.*, **32**, pp. 1661–1663, 1996.

178. K. Okamoto and H. Yamada, "Arrayed waveguide grating multiplexer with flat spectral response," *Opt. Lett.*, **20**, pp. 43–45, 1995.

179. K. Okamoto, M. Moriwaki, and S. Suzuki, "Fabrication of 64 × 64 arrayed waveguide grating multiplexer on silicon," *Electron. Lett.*, **31**, pp. 184–186, 1995.

180. M. Kohtoku, et al., "InP-based-64-channel arrayed waveguide grating with 50 GHz channel spacing and up to −20 dB Crosstalk," *Electron. Lett.*, **33**, pp. 1786–1787, 1997.

181. K. Okamoto, K. Syuto, H. Takahashi, and Y. Ohmori, "Fabrication of 128-channel arrayed waveguide grating multiplexer with 25 GHz channel spacing," *Electron. Lett.*, **32**, pp. 1474–1476, 1996.

182. H. Yamada, et al., "Statically-phase compensated 10 Ghz spaced arrayed waveguide grating," *Electron. Lett.*, **32**, pp. 1580–1582, 1996.

183. K. Takada, M. Abe, and K. Okamoto, "Low-cross-talk polarization-insensitive 10-GHz-spaced 128-channel arrayed-waveguide grating multiplexer demultiplexer achieved with photosensitive phase adjustment," *Opt. Lett.*, **26**, pp. 64–65, 2001.

184. M. Ishi, et al., "Low-loss fibre-pigtailed 256 channel arrayed-waveguide grating multiplexer using cascaded laterally-tapered waveguides," *Electron. Lett.*, **37**, pp. 1401–1402, 2001.

185. Y. Hida, et al., "400-channel 25-GHz spacing arrayed waveguide grating covering the full range of C and L bands," *Proc. OFC*, **P WB2**, 2001.

186. K. Takada, "Fabrication of 2 GHz-spaced 16-channel arrayed-waveguide grating demultiplexer for optical frequency monitoring applications," *Electron. Lett.*, **36**, pp. 1643–1644, 2000.

187. K. Takada, H. Yamada, and K. Okamoto, "320 channel multiplexer consisting of a 100 GHz-spaced parent AWG and 10 Ghz-spaced subsidiary AWG," *Electrón. Lett.*, **35**, pp. 824–826, 1999.

188. K. Takada, et al., "10 GHz-spaced 1010-channel AWG filter achieved by tandem connection of primary and secondary AWGs," *Proc. ECOC*, Munich, Germany, Postdeadline Paper PD3-8, 2000.

189. K. Takada, M. Abe, T. Shibata, and K. Okamoto, "A 25 GHz spaced 1080-channel tandem multi/demultiplexer covering the S-, C- and L-bands using arrayed waveguide grating with Gaussian passbands as primary filter," *IEEE Photon. Tech. Lett.*, **14**, pp. 648–650, 2000.

190. K. Okamoto, M. Okuno, A. Himeno, and Y. Ohmori, "16 channel optical add/drop multiplexer consisting of arrayed waveguide gratings and double gate switches," *Electron. Lett.*, **32**, pp. 1471–1472, 1996.

191. C.G.M. Vreeburg, et al., "First InP based reconfigurable integrated add/drop multiplexer," *IEEE Photon. Tech. Lett.*, **9**, pp. 188–190, 1997.

192. K. A. McGreer, "Arrayed waveguide Gratings for wavelength routing," *IEEE Communications Magazine*, pp. 62–68, 1998.

193. O. Ishida, H. Takahashi, S. Suzuki, and Y. Inoue, "Multichannel frequency-selective switch employing an arrayed-waveguide grating multiplexer with fold-back optical paths," *IEEE Photon. Technol. Lett.*, **6**, pp. 1219–1221, 1994.

194. S. Suzuki, A. Himeno, Y. Tachikama, and Y. Yamada, "Multichannel optical wavelength selective switch with arrayed-waveguide grating multiplexer," *Electron. Lett.*, **30**, pp. 1091–1092, 1994.

195. Y. Zhao, X. J. Zhao, J. H. Chen, and F. S. Choa, "A novel bidirectional add/drop module using waveguide grating routers and wavelength channel matched fiber gratings," *IEEE Photon. Technol. Lett.*, **11**, pp. 1180–1182, 1999.

196. C. H. Kim, H. Yoon, S. B. Lee, C. H. Lee, and Y. C. Chung, "Optical gain-controlled bidirectional add–drop amplifier using fiber Bragg gratings," *IEEE Photon. Technol. Lett.*, **12**, pp. 894–896, 2000.

197. S. Kim, "Bidirectional Optical Cross Connects for Multiwavelength Ring Networks Using Single Arrayed Waveguide Grating Router," *IEEE J. Lightwave Technol.*, **20**, pp. 188–195, 2002.

198. J. Capmany, C. R. Doerr, K. Okamoto, and M. K. Smit, (eds), "Special issue on arrayed grating routers/WDM MUX/DEMUX and related Applications," *IEEE J. Sel. T. Quantum Electron.*, **8**, pp. 1087–1211, 2002.

199. A. Gopinath, R. M. De la Rue, K. Okamoto, R. A A A. Soref, (eds.), "Special issue on integrated optics and optoelectronics," *IEEE J. Sel. T. Quantum Electron.*, **8**, pp. 1215–1450, 2002.

# 16

# WIDE-BANDWIDTH OPTICAL INTENSITY MODULATORS

G. L. LI

*Semiconductor Photonics Department*
*Agere Systems, Inc.*
*Breinigsville, Pennsylvania*

P. K. L. YU

*Department of Electrical and Computer Engineering*
*University of California, San Diego*
*La Jolla, California*

## 16.1  INTRODUCTION

Lightwave has various characteristics that can be modulated to carry information, including the intensity, phase, frequency, and polarization. Among these, the intensity modulation is the most popular for optical fiber communication systems, primarily due to the simplicity of envelope photo detection. However, the performance requirements for optical modulations are somewhat different between analog systems and digital systems. Analog systems commonly use small-signal modulation, and they are deployed mainly for short distance links, which makes it optimal to use at the 1.32-μm laser wavelength, where fiber has zero dispersion. Therefore, the primary requirements for optical modulation in analog systems are large incremental slope efficiency, wide bandwidth, high linearity and low noise, and so on. Frequency chirp is not a concern due to zero fiber dispersion at 1.32-μm wavelength. In contrast, digital systems usually use on/off modulation format, and they are often deployed in longer distance links, which require laser wavelength around 1.55 μm to keep the fiber loss low. Therefore, the primary requirements for optical modulation in digital systems are large on/off extinction ratio, high data rate (i.e., wide bandwidth), low chirp, large side-mode suppression ratio (SMSR) and signal to noise ratio (SNR), and so on. Nevertheless, both analog and digital systems share some common grounds such as high optical power

*Handbook of Optical Components and Engineering,*  Edited by Kai Chang
ISBN 0-471-39055-0   © 2003 John Wiley & Sons, Inc.

handling ability, small optical loss, polarization insensitivity, and stable performance over ambient temperature variation and time.

The simplest intensity modulation can be implemented by directly modulating the laser sources. Due to the requirements of bandwidth and efficiency, only semiconductor lasers are of practical interest for direct modulation. Modulation bandwidth as high as 25 GHz has been reported for a semiconductor laser operating at 1.55-μm wavelength [1]. However, when the modulation frequency increases toward the relaxation resonance frequency of a semiconductor laser, both the relative intensity noise (RIN) and distortions increase rapidly [2]. This severely limits the feasibility of direct modulation for high-frequency (>20 GHz) links. The large frequency chirp also precludes the direct modulation for long-distance links, especially for digital applications. External modulation can minimize the above effects with the potential disadvantages of adding system complexity and cost. External-modulation link operated at 1.32-μm wavelength can also take advantages of the availability of solid-state (Nd:YAG) lasers with high-power (>300 mW), low relative intensity noise (<−170 dB/Hz), and narrow linewidth (<1 kHz).

Several types of external optical intensity modulators have been developed over several decades for optical fiber communication applications. These include Lithium Niobate modulators, semiconductor electroabsorption modulators (EAMs), semiconductor Mach–Zehnder modulators (MZMs), and polymer modulators. This chapter is intended to provide a fundamental understanding of the operation principles and key performances of these devices. Compared to a review article published earlier [3], this chapter provides more details on the physical models and analysis approaches, especially for analyzing modulation efficiency and bandwidth. We note that all of these modulators share many common aspects, in terms of the performance requirements and operation principles, as well as the approaches employed for their analysis. This chapter also highlights some recent research progress in optical modulators for both analog and digital applications. State-of-the-art device performances and some specific research areas are outlined. At the end of the chapter, some pros and cons of these modulator devices are compared.

## 16.2   PHYSICAL EFFECTS AND MODULATOR MATERIALS

Most of the modern wide-bandwidth modulators are based on two types of physical effects: one is the linear electrooptic (EO) effect,* and the other is the electroabsorption (EA) effects. Both effects depend on the applied electric field, which makes the modulators voltage-controlled devices. Carrier-induced index change in semiconductor materials can also be utilized to make MZMs (controlled by injection current). This kind of index modulation has been proven to be very efficient [4], making it very useful for multisection tunable lasers [5] and other phase tuning devices. However, it is hard to implement fast modulation with this mechanism due to the slow carrier transport process. In this section, we only discuss the linear electro-optic and the electroabsorption effects.

---

*Interested readers are referred to the first edition of this handbook, Chapter 4, "Optical Modulation: Electro-Optical Devices" by S. Thaniyavarn.

### 16.2.1 Electrooptic Effect

The EO effect denotes the change of optical refractive index in nonlinear optical (NLO) crystals due to the presence of electric field. The index change leads to change in optical phase, which can be converted into intensity modulation in a Mach–Zehnder interferometer. A convenient way to visualize the electro-optic effect is to use the *index ellipsoid* of the crystal [6]. The refractive indices for an optical wave propagating in a crystal can be found from the intersection ellipse between the index ellipsoid and the plane that is normal to the propagation direction and through the center of the ellipsoid, as illustrated in Fig. 16.1. The lengths of the two axes of the intersection ellipse are the refractive indices ($n_1$ or $n_2$) for optical waves linearly polarized along these two axes. Optical waves with other polarizations encounter birefringent effect (unless $n_1 = n_2$).

In the principal coordinate system and in the absence of the electric field, the index ellipsoid can be represented by

$$\frac{x^2}{n_x^2} + \frac{y^2}{n_y^2} + \frac{z^2}{n_z^2} = 1 \tag{16.1}$$

where the directions of $x$, $y$, and $z$ are the principal dielectric axes with the corresponding refractive indices $n_x$, $n_y$, and $n_z$. After the electric field is applied, both the size and the orientation of the index ellipsoid are changed. The equation of the index ellipsoid is generally modified to

$$x^2 \left(\frac{1}{n^2}\right)_1 + y^2 \left(\frac{1}{n^2}\right)_2 + z^2 \left(\frac{1}{n^2}\right)_3 + 2yz \left(\frac{1}{n^2}\right)_4 + 2zx \left(\frac{1}{n^2}\right)_5 + 2xy \left(\frac{1}{n^2}\right)_6 = 1 \tag{16.2}$$

Comparing Eq. (16.2) and Eq. (16.1), it is apparent that, without applied electric field,

$$\left(\frac{1}{n^2}\right)_1 = \frac{1}{n_x^2}, \quad \left(\frac{1}{n^2}\right)_2 = \frac{1}{n_y^2}, \quad \left(\frac{1}{n^2}\right)_3 = \frac{1}{n_z^2}, \quad \left(\frac{1}{n^2}\right)_4 = \left(\frac{1}{n^2}\right)_5 = \left(\frac{1}{n^2}\right)_6 = 0 \tag{16.3}$$

After the electric field is applied, the above six terms may be changed. When only the linear EO effect (also known as Pockels effect) is considered, the change induced by

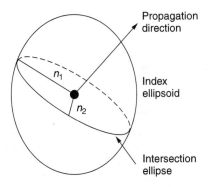

**Figure 16.1** Illustration of index ellipsoid.

the electric field $\mathbf{F} = (F_x, F_y, F_z)$ can be expressed as

$$
\begin{bmatrix}
\Delta\left(\dfrac{1}{n^2}\right)_1 \\[2mm]
\Delta\left(\dfrac{1}{n^2}\right)_2 \\[2mm]
\Delta\left(\dfrac{1}{n^2}\right)_3 \\[2mm]
\Delta\left(\dfrac{1}{n^2}\right)_4 \\[2mm]
\Delta\left(\dfrac{1}{n^2}\right)_5 \\[2mm]
\Delta\left(\dfrac{1}{n^2}\right)_6
\end{bmatrix}
=
\begin{bmatrix}
r_{11} & r_{12} & r_{13} \\
r_{21} & r_{22} & r_{23} \\
r_{31} & r_{32} & r_{33} \\
r_{41} & r_{42} & r_{43} \\
r_{51} & r_{52} & r_{53} \\
r_{61} & r_{62} & r_{63}
\end{bmatrix}
\begin{bmatrix}
F_x \\
F_y \\
F_z
\end{bmatrix}
\tag{16.4}
$$

where the $6 \times 3$ matrix $[r_{ij}]$ is called the *electro-optic tensor*, and its elements are called *electro-optic coefficients*. For various crystals, due to the crystal symmetry, some of the EO coefficients may be zero, and some of them may be equal in value (or opposite in sign) [6]. With the EO tensor determined, the modified index ellipsoid due to arbitrary electric field can be found through Eqs. (16.2)–(16.4), and the refractive indices for arbitrary optical wave can be determined as illustrated in Fig. 16.1.

***Lithium Niobate.*** LiNbO$_3$ is the most widely used material for the manufacture of EO devices, including phase modulators, polarization modulators, Mach–Zehnder intensity modulators, and directional-coupler intensity modulators. It is also a ferroelectric crystal that is commonly used in other types of devices, such as surface-acoustic-wave (SAW) filters. LiNbO$_3$ optical waveguides have the general advantages of high optical coupling efficiency with single-mode optical fiber, low optical loss, and large EO coefficient.

The most popular method of fabricating optical waveguides in LiNbO$_3$ is by titanium (Ti) diffusion [7]. Diffusing Ti into LiNbO$_3$ can cause an increase in the refractive index by $\sim 0.01$, which is ideal for the single-mode optical waveguiding. The dimensions of the optical waveguide can be controlled by properly choosing the initial Ti stripe width, film thickness, and diffusion conditions. Typically, Ti stripes about 3–8 μm wide and $\sim 0.1$ μm thick are patterned on LiNbO$_3$ surface using photolithography and liftoff techniques. The patterned Ti is then driven into LiNbO$_3$ under high temperature around 1000°C for 4–10 hours in a wet argon or oxygen ambient. The resulted single-mode waveguide has a very low optical propagation loss, typically less than 0.2 dB/cm, and its mode size can be matched very well to that of a single-mode optical fiber. There are some other techniques for creating optical index difference in LiNbO$_3$ crystal, including ion exchange, proton exchange, and nickel diffusion. After the optical waveguide is formed, its end surfaces are polished to high optical quality. For $z$-cut LiNbO$_3$, a thin SiO$_2$ buffer layer is often deposited on top of the optical waveguide prior to the deposition of metal electrodes. This is to provide isolation between the metal and the optical mode [see Fig. 16.2b], thus reducing the optical propagation loss.

Without the applied electric field, LiNbO$_3$ is a birefringent crystal. Its extraordinary optical axis is usually chosen as the $z$-axis of the principal coordinate system of the

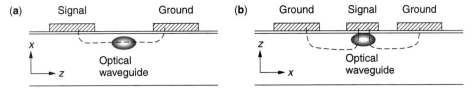

**Figure 16.2**  Electrode configurations for EO modulation with **(a)** $x$-cut and **(b)** $z$-cut LiNbO$_3$ crystal.

index ellipsoid. Therefore, in Eq. (16.1), we have

$$n_x = n_o, \quad n_y = n_o, \quad n_z = n_e \qquad (16.5)$$

where the ordinary index $n_o = 2.29$, and the extraordinary index $n_e = 2.2$.

LiNbO$_3$ belongs to the 3m crystallographic group. Due to the symmetry properties of this group, the corresponding EO tensor takes the form of [6]

$$\begin{bmatrix} 0 & -r_{22} & r_{13} \\ 0 & r_{22} & r_{13} \\ 0 & 0 & r_{33} \\ 0 & r_{51} & 0 \\ r_{51} & 0 & 0 \\ -r_{22} & 0 & 0 \end{bmatrix} \qquad (16.6)$$

Note that in this tensor, $r_{61} = -r_{22}$, $r_{42} = r_{51}$, $r_{12} = -r_{22}$ and $r_{23} = r_{13}$. For LiNbO$_3$ crystal, the EO coefficients are:

$$r_{13} = 8.6 \text{ pm/V}, \quad r_{33} = 30.8 \text{ pm/V}$$
$$r_{22} = 3.4 \text{ pm/V}, \quad r_{51} = 28.0 \text{ pm/V} \qquad (16.7)$$

Combining Eqs. (16.2)–(16.6), the index ellipsoid for LiNbO$_3$ in the presence of the electric field is represented by

$$x^2 \left( \frac{1}{n_o^2} - r_{22}F_y + r_{13}F_z \right) + y^2 \left( \frac{1}{n_o^2} + r_{22}F_y + r_{13}F_z \right) + z^2 \left( \frac{1}{n_e^2} + r_{33}F_z \right)$$
$$+ 2yzr_{51}F_y + 2zxr_{51}F_x - 2xyr_{22}F_x = 1 \qquad (16.8)$$

As $r_{33}$ is the largest EO coefficient, it is desirable to exploit it for EO modulation. This means the electric field is applied along the $z$-direction and the incident optical wave is polarized along the $z$-axis. With the electric field along the $z$-direction, $F_x = F_y = 0$, the index ellipsoid becomes

$$x^2 \left( \frac{1}{n_o^2} + r_{13}F_z \right) + y^2 \left( \frac{1}{n_o^2} + r_{13}F_z \right) + z^2 \left( \frac{1}{n_e^2} + r_{33}F_z \right) = 1 \qquad (16.9)$$

For $x$-cut LiNbO$_3$ crystal, an electric field along the $z$-direction means a horizontal electric field. The electrode configuration is illustrated in Fig. 16.2**a**. In this case, the optical waveguide has to be along the $y$-axis (because the $x$-axis is vertical to the sample surface, and the $z$-axis is the direction of the electric field). Therefore, the intersection ellipse in Fig. 16.1 is obtained by setting $y = 0$ in Eq. (16.9):

$$x^2 \left( \frac{1}{n_o^2} + r_{13} F_z \right) + z^2 \left( \frac{1}{n_e^2} + r_{33} F_z \right) = 1 \qquad (16.10)$$

Therefore, for optical polarization along $x$-axis and $z$-axis, the optical refractive indices become

$$n'_x = \left( \frac{1}{n_o^2} + r_{13} F_z \right)^{-\frac{1}{2}} \approx n_o - \frac{1}{2} n_o^3 r_{13} F_z \qquad (16.11)$$

and

$$n'_z = \left( \frac{1}{n_e^2} + r_{33} F_z \right)^{-\frac{1}{2}} \approx n_e - \frac{1}{2} n_e^3 r_{33} F_z \qquad (16.12)$$

respectively. Equations (16.11) and (16.12) show that for the $x$-cut LiNbO$_3$ modulator, the TE optical mode (polarized along the $z$-axis) is much more efficiently modulated than the TM optical mode (polarized along the $x$-axis), because $r_{33} \gg r_{13}$. For optical polarization other than TE or TM polarization, birefringent effect is expected.

For $z$-cut LiNbO$_3$ crystal, an electric field along $z$-direction means a vertical electric field, as illustrated in Fig. 16.2**b**. In this case, the waveguide can be along either the $x$-axis or the $y$-axis. For example, for optical waveguide fabricated along the $y$-axis, the intersection ellipse is again represented by Eq. (16.10), and the optical refraction indices are the same as that in Eq. (16.11) and (16.12). However, for $z$-cut crystal, the TM optical mode is now polarized along the $z$-axis, and the TE optical mode is polarized along the $x$-axis. Consequently, for the $z$-cut LiNbO$_3$ modulator, the TM mode is much more efficiently modulated than the TE mode.

Some approaches have been demonstrated [8–10] to implement polarization-independent modulators using LiNbO$_3$ crystal. However, their efficiencies are generally compromised (because a smaller EO coefficient is resulted). Special crystal cuts and electrode geometries may also be required, which are not suitable for high-speed modulation.

***III–V Semiconductors.*** III–V compound semiconductors are also suitable candidate materials for EO modulators. Although their EO coefficients are $\sim$20 times smaller than that of LiNbO$_3$, efficient modulation can still be obtained with these materials. This is because that semiconductor crystal growth and fabrication techniques provide great flexibility for waveguide geometry control, so that the optical guided mode can be confined to a very small region (2–3-$\mu$m spot size), and thus a very large electric field can be achieved even with a small voltage applied across the small dielectric gap. Additionally, III–V semiconductors have large optical refractive indices—for example, at an optical wavelength of 1.3–1.6 $\mu$m, the optical index is $\sim$3.2 for InP and $\sim$3.4 for GaAs, compared to $\sim$2.2 for LiNbO$_3$. From Eq. (16.11) and (16.12), this indicates a 3–4 times improvement for the index change. All of these factors make the efficiency of III–V EO modulators comparable to that of LiNbO$_3$ modulators. In addition, III–V EO modulators can potentially be integrated with a wide range of components, such as lasers, semiconductor optical amplifiers, photodetectors, passive optical circuits, and even electronic drivers. This makes III–V EO modulators very attractive.

III–V semiconductor layers can be grown using either molecular beam epitaxy (MBE) or metal organic chemical vapor deposition (MOCVD) techniques. The most popular III–V semiconductor for fabricating EO modulators is the bulk GaAs/AlGaAs waveguide grown on GaAs substrate. This is primarily due to the fact that larger GaAs substrates (4″ diameter) become available first and many devices can be made from each wafer. Similar EO effect also occurs in InP waveguide, and because InP substrates 3″ in diameter are readily available now, EO modulators that are typically 2–3 cm long can be made on InP as well. In the GaAs/AlGaAs waveguide, the GaAs layer has higher optical index than $Al_x Ga_{1-x} As$ layers, and the latter is lattice-matched to GaAs for all $x$ values. By sandwiching the GaAs layer between two $Al_x Ga_{1-x} As$ layers, optical confinement in the vertical growth direction can be achieved. The lateral optical confinement is usually obtained by a material etch to form a mesa structure. Shallow-etched rib waveguide is mostly preferred over other structures (such as buried heterostructure) for this type of devices to achieve single-mode waveguiding (see Section 16.5.5 for further explanation). The resulted optical waveguides may have propagation loss ranging from tenths of decibels/centimeter to a few decibels/centimeter, depending on the waveguide structure and the fabrication process.

III–V semiconductors are Zinc Blende ($\overline{4}3m$) crystals. By themselves, they do not show a birefringent effect. To consider the EO effect, the principal $x$, $y$, and $z$ axes are usually chosen along [100], [010], and [001] crystal orientations, respectively. Due to the symmetry of Zinc Blende crystals, their EO tensor takes the form of [6]

$$
\begin{bmatrix}
0 & 0 & 0 \\
0 & 0 & 0 \\
0 & 0 & 0 \\
r_{41} & 0 & 0 \\
0 & r_{41} & 0 \\
0 & 0 & r_{41}
\end{bmatrix}
\tag{16.13}
$$

Note that in this tensor, we have $r_{41} = r_{52} = r_{63}$. For GaAs layer, the EO coefficient $r_{41}$ is around 1.4 pm/V. Combining Eq. (16.13) with Eq. (16.1)–(16.4), the index ellipsoid for GaAs in the presence of the electric field is:

$$
\frac{x^2 + y^2 + z^2}{n_o^2} + 2r_{41}(yzF_x + zxF_y + xyF_z) = 1
\tag{16.14}
$$

where $n_o \sim 3.4$ for GaAs. As the crystal growth usually proceeds perpendicular to the (001) surface and along the $z$-axis direction, it is most convenient to apply an electric field along the $z$-axis. In this case $F_x = F_y = 0$, and the index ellipsoid becomes

$$
\frac{x^2 + y^2 + z^2}{n_o^2} + 2r_{41}xyF_z = 1
\tag{16.15}
$$

Apparently there is no index modulation if the optical waveguide is along either $x$-axis ([100] direction) or $y$-axis ([010] direction). Due to the off-diagonal component in Eq. (16.15), the principal axes in $xy$ plane are rotated 45° in the presence of the electric field $F_z$, and they are along [110] ($x'$-axis) and [$\bar{1}$10] ($y'$-axis) directions. These crystal orientations and principal axes are illustrated in Fig. 16.3. The coordinate transformation is:

$$x = (x' - y')/\sqrt{2} \qquad y = (x' + y')/\sqrt{2} \tag{16.16}$$

Therefore, the index ellipsoid in the new coordinate system is represented by

$$x'^2 \left( \frac{1}{n_o^2} + r_{41} F_z \right) + y'^2 \left( \frac{1}{n_o^2} - r_{41} F_z \right) + \frac{z^2}{n_o^2} = 1 \tag{16.17}$$

The above equation shows that for waveguide along [110] ($x'$-axis) or [$\bar{1}$10] ($y'$-axis) direction, the index modulations are:

$$n'_{y'} = \left( \frac{1}{n_o^2} - r_{41} F_z \right)^{-\frac{1}{2}} \approx n_o + \frac{1}{2} n_o^3 r_{41} F_z \tag{16.18}$$

or

$$n'_{x'} = \left( \frac{1}{n_o^2} + r_{41} F_z \right)^{-\frac{1}{2}} \approx n_o - \frac{1}{2} n_o^3 r_{41} F_z \tag{16.19}$$

for the TE mode, but the TM mode (polarized along $z$-axis) is not modulated. Therefore, the above GaAs/GaAlAs EO modulator is highly polarization dependent. Also note that the sign of index change depends on waveguide orientation.

With waveguide along [$\bar{1}$10] direction ($y'$-axis), as shown in Fig. 16.3, when a horizontal electric field is applied along $x'$ axis, the principal axes of the intersection ellipse is in the $x'z$ plane, with 45° angle from $x'$ and $z$ axes [11]. The refractive indices along these principal axes are similar to that in Eq. (16.18) and (16.19). This causes birefringent effect to both TE and TM incident waves and, thus, change their polarizations during modulation. This waveguide configuration can be utilized to make an efficient polarization modulator [11]. MZM with both arms designed as the above waveguide and operated in the push-pull mode can provide polarization-independent intensity modulation [12]. However, with the electric field applied in the horizontal direction, the overlap between the electric field and the optical field is relatively small. This makes the polarization-independent MZM less efficient than that using a vertical electric field.

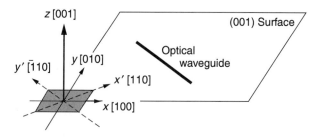

**Figure 16.3**   Crystal orientations and principal axes of index ellipsoid for GaAs.

***Polymers.*** Compared to LiNbO$_3$ and III–V semiconductors, organic polymers are relatively immature EO materials [13], but they also possess a great potential for more advanced modulators. The advantages of EO polymers mostly come from the applicability of spin-coating technique. This not only makes it possible to integrate polymer EO devices with various electronic and optoelectronic components [14], but also creates the opportunity to fabricate multiple devices stacked in the vertical direction. Metal electrodes can be buried between different polymer layers, which makes the electrode design very flexible. To lower the fabrication cost, polymer EO devices can be fabricated directly on top of optical submounts. The optical refractive index of polymers is close to that of single-mode optical fiber. This provides a good match between the polymer waveguide mode and the fiber mode. Besides EO modulators, polymers are also candidate materials for fabricating LEDs, fiber lasers and amplifiers, wavelength converters, variable optical attenuators, tunable filters and optical switches, and so on.

EO polymers are synthetic organic materials mixed with two components: one is the nonlinear optical (NLO) chromophore molecules as active element, and the other is some standard polymers (such as PMMA) as host. NLO chromophores are synthesized organic molecules that possess large electric dipole moment and strong optical nonlinearity. They can be dispersed into the polymer matrix in different ways: either dissolved as guest, or chemically connected to the polymer molecules to form side-chain polymer, cross-linked polymer, or main-chain polymer. When the chromophoric polymer is heated up around 100–200°C, the chromophore molecules become mobile, and their electric dipoles can be aligned to the same direction by applying a strong electric field (100–200 V/μm). After the polymer is cooled down under a high electric field, the alignment of the electric dipoles is "frozen," and macroscopic optical nonlinearity is achieved. This process is called "poling." Many poling techniques have been developed, such as electrode poling (electrodes are deposited for applying the poling electric field), corona (corona discharge is used to create the poling electric field), photo-induced poling, and so on. It is possible to pole adjacent areas in opposite directions, which can facilitate the high-speed push-pull Mach–Zehnder modulator design, as discussed later in Section 16.5.4. To form an optical waveguide, optical confinement in the vertical direction is usually achieved by sandwiching the core EO layer between two cladding layers of different polymer material with lower refractive index ($\Delta n \sim 0.1$). Optical confinement in lateral direction can be achieved by either dry etching or photo-bleaching (exposing the areas outside the waveguide under high-intensity light at certain wavelength to reduce the refractive index). The resulted waveguides typically have propagation loss less than 3 dB/cm, generally due to absorption and scattering.

After poling, EO polymers become birefringent crystal, and they belong to the $m\infty$ symmetry class. The extraordinary axis of their index ellipsoids is along the poling direction, which is chosen as the $z$-axis in the principal coordinate system. The extraordinary and ordinary indices are:

$$n_e = n + 2\Delta, \qquad n_o = n - \Delta \qquad (16.20)$$

where $n \sim 1.6$ is the isotropic refractive index before poling and $\Delta$ is a small number ($\sim 0.1$), depending on poling electric field and material. The corresponding EO tensors take the form of [15]

$$\begin{bmatrix} 0 & 0 & r_{13} \\ 0 & 0 & r_{13} \\ 0 & 0 & r_{33} \\ 0 & r_{13} & 0 \\ r_{13} & 0 & 0 \\ 0 & 0 & 0 \end{bmatrix} \tag{16.21}$$

Note that in the tensor, we have $r_{13} = r_{23} = r_{42} = r_{51}$. For various EO polymers, $r_{33}$ is about 3 times larger than $r_{13}$. The actual value of $r_{33}$ depends on the chromophore type, the mixing density in the polymer, and the poling electric field. Usually, $r_{33}$ increases linearly with the poling field, but so does the optical loss and $\Delta$ in Eq. (16.20). The reported $r_{33}$ values at 1.3–1.6-μm optical wavelength range from a few pm/V to around 100 pm/V, with 15–20 pm/V as the typical value. Compared to LiNbO$_3$ using Eq. (16.12), at the same modulation electric field, $r_{33}$ needs to be around 80 pm/V for the EO polymer to have the same index change as the LiNbO$_3$.

Combining Eqs. (16.1)–(16.4) and (16.21), the index ellipsoid for EO polymers in the presence of electric field is represented by

$$x^2 \left( \frac{1}{n_o^2} + r_{13}F_z \right) + y^2 \left( \frac{1}{n_o^2} + r_{13}F_z \right) + z^2 \left( \frac{1}{n_e^2} + r_{33}F_z \right) + 2yzr_{13}F_y + 2zxr_{13}F_x = 1$$

$$\tag{16.22}$$

As $r_{33}$ is about 3 times larger than $r_{13}$, it is desirable to apply the modulation electric field along the $z$-axis. In this case, $F_x = F_y = 0$, and the resulted index ellipsoid takes the same form as Eq. (16.9). The subsequent analysis for the index change is therefore exactly the same as that for $z$-cut LiNbO$_3$ crystal. The index changes for TE and TM optical modes are represented in Eq. (16.11) and (16.12), respectively. Similar to the $z$-cut LiNbO$_3$ crystal, the TM mode has a much more efficient modulation than does the TE mode in polymer EO waveguides.

Polymer EO modulators are generally polarization dependent, although polarization-independent design has been attempted with compromised efficiency [16]. Major problems with polymer EO devices are the stability issues. Upon thermal aging, the alignment of chromophore molecules can be relaxed, and the polymer materials can be oxidized and become yellowish in color. The above problems can also arise from photo absorption, most severely with blue light that can be generated as harmonics when illuminated with 1.3–1.6-μm light. Humidity in the environment tends to make it even worse, because water incursion enhances the photo absorption in polymer material. Another disadvantage with polymer EO devices is the difficulty in facet cleaving, and facet polishing for polymer waveguides is not as convenient as for LiNbO$_3$ waveguides. These problems, together with the efforts to improve the EO properties of polymers, make the polymer EO device research a very active and intense research area.

### 16.2.2 Electroabsorption Effects

Electroabsorption (EA) effects denote the change of optical absorption coefficient in materials due to the presence of electric field. EA effects in a single optical waveguide directly result into optical intensity modulation. The primary materials for fabricating EAMs working at 1.3–1.60 μm optical wavelengths are currently III–V semiconductors, specifically, ternary and quaternary alloys (including InGaAs, InAlAs, InAsP,

InGaP, InGaAsP, InGaAlAs, etc.) grown on InP substrate. These are also the best materials so far for fabricating many other active optoelectronic components operating in the same wavelength range, including lasers, semiconductors optical amplifiers (SOA), and photodetectors. Hence, EAMs are considered to be the most suitable modulator candidate for integration with these components. The most notable integrated chips include electroabsorption modulated lasers (EML) [17] and tandem EA modulators (often with SOA) for RZ data transmission and wavelength-division demultiplexing [18]. Another attraction of EAMs is their compact size (typically 80–300 μm long) due to high efficiency. This leads to small footprint for a single device and high yield per wafer to lower the fabrication cost. EAMs can also be used as photonic microwave mixer [19] and dual-functional modulator/detector [20].

An EA waveguide usually consists of $p$-$i$-$n$ semiconductor layers, among which the intrinsic layer has higher optical refractive index than do the $p$-type and $n$-type doped layers. This provides vertical optical confinement. The lateral optical confinement is usually achieved by deep mesa etch. The etched deep-ridge waveguide can then be planarized using polyimide or regrown semi-insulating InP. In a typical EA waveguide design, both the optical mode and the applied electric field are tightly confined in a small area around the intrinsic layer, which enables highly efficient modulation. The tight and strong optical confinement sometimes results in multimode waveguide. Note that an EA waveguide can be multimode for analog links, but single-mode waveguide is preferred for digital links to achieve a high on/off extinction ratio. This is further explained in Section 16.3.2. The optical propagation loss in EA waveguides is much larger than that in EO waveguides, typically ranges from 15 dB/mm to 20 dB/mm. This high optical propagation loss is mainly attributed to three sources: the first is the residual absorption loss in the active layer; the second is the interband absorption loss induced by free carriers in the highly doped layers, primarily in $p$-type layers; the third is the scattering loss caused by the roughness of the deep-ridge waveguide sidewalls and defects in the grown materials. Fortunately, EA waveguides are typically very short, so that the overall optical loss is tolerable.

The tight optical confinement results into small, elliptical mode profile in EA waveguides—typically, 2–4-μm lateral and 1–2-μm vertical effective mode sizes, which can have large optical coupling loss to a single-mode optical fiber whose mode size is ∼9 μm in diameter. To overcome this problem, either a micro ball lens is inserted between the EA waveguide and the fiber tip or an optical lens can be directly fabricated on the fiber tip to produce a mode size of 3–4 μm at the focal plane that is ∼20 μm away from the fiber tip. Integrating a spot-size converter (SSC) at both ends of the EA waveguide can further reduce the coupling loss. An SSC is a passive optical waveguide with adiabatic vertical and/or lateral tapers in geometry and/or optical refractive index. It converts the small elliptical EA waveguide mode at one end to a near-circular expanded mode (mode size 3–4 μm) at the other end, or vice versa. Integrating with SSC also makes the EAM longer, which not only makes the device easier to cleave and handle, but also reduces the amount of unguided light directly coupled from the input fiber to the output fiber, resulting in improved on/off extinction ratio for digital links.

Among the $p$-$i$-$n$ layers of an EA waveguide, the intrinsic layer also possesses the smallest energy bandgap. To perform the modulation, photon energy of the incident light has to be slightly less than the bandgap of the intrinsic active layer by an amount called "detuning energy." In the absence of the external electric field, the active layer has little light absorption; when external electric field is applied, the absorption coefficient increases rapidly. This is the so-called "electroabsorption effect." There are two types of electroabsorption effect: one is the Franz–Keldysh effect (FKE) in the bulk active layer, and the other is the quantum-confined Stark effect (QCSE) in multiple-quantum-wells (MQW). Strong index change can be induced by QCSE due to Kramers–Kronig relation, which can be used to make electrorefractive MZMs with proper detuning energy. All of these effects are discussed with more details in the following.

*Franz–Keldysh Effect.* Figure 16.4 shows the band diagram of a bulk semiconductor layer without and with the electric field. Without the electric field, absorption of one photon (with energy $\hbar\omega$) is not enough for an electron in the valence band to make a transition to the conduction band (the detuning energy is usually ~60 meV). However, when the electric field is present, as shown in Fig. 16.4**b**, the energy bands are tilted, and the photon energy $\hbar\omega$ meets the energy difference between the conduction band at location B and the valence band at location A. The electron in valence band at location A can now absorb one photon and make a transition to conduction band at location B. This is sometimes referred to as "photo-assisted interband tunneling." The transition generates a pair of an electron (in conduction band) and a hole (in valence band). As the electric field increases, locations A and B get closer, and the transition rate becomes larger, so that the absorption coefficient is increased.

Using the results of K. Tharmalingam [21], the field-induced optical absorption coefficient for bulk material due to the allowed electron transitions can be expressed as

$$\alpha_{\text{FKE}}(\hbar\omega, F) = \frac{e^2 |P_{cv}|^2}{2\pi \varepsilon_0 \hbar\omega n c m_0^2} \left(\frac{2\mu}{\hbar}\right)^{\frac{3}{2}} \sqrt{\theta_F}(-\beta|Ai(\beta)|^2 + |Ai'(\beta)|^2) \qquad (16.23)$$

where $\hbar\omega$ is the photon energy, $F$ is the electric field, $e$ is the electron charge, $\varepsilon_0$ is the free-space permitivity, $\hbar$ is Plank's constant, $n$ is the optical refractive index, $c$ is the speed of light in free-space, $m_0$ is the free electron mass, $Ai$ denotes the Airy function, and $Ai'$ is the first derivative; $\mu$, $\theta_F$, and $\beta$ are defined by

$$\frac{1}{\mu} = \frac{1}{m_e} + \frac{1}{m_h}, \qquad \theta_F = \left(\frac{e^2 F^2}{2\mu\hbar}\right)^{\frac{1}{3}}, \qquad \beta = \frac{E_g - \hbar\omega}{\hbar\theta_F} \qquad (16.24)$$

where $m_e$ and $m_h$ are the effective masses of electrons and holes, respectively; $E_g$ is the material bandgap energy. The optical matrix element $P_{cv}$ is defined by

$$|P_{cv}|^2 = |\langle u_C|\hat{e} \cdot \vec{p}|u_V\rangle|^2 = M_0 M^2 \qquad (16.25)$$

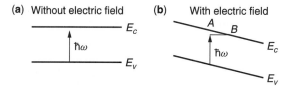

**Figure 16.4**   Band diagram for bulk semiconductor material to show Franz–Keldysh effect.

where $\hat{e}$ is the unit polarization vector of the light, $\vec{p}$ is the momentum operator, $u_C$ and $u_V$ are the Bloch functions in the conduction band and valence bands, the polarization factor $M_0$ takes the value of 1/3 for bulk material [22, 23], and the average matrix element $M^2$ can be calculated using some approximate formula [22 or 26, 31]:

$$M^2 = \left(\frac{m_0}{m_e^*} - 1\right)\frac{(E_g + \Delta)}{2(E_g + \frac{2}{3}\Delta)}m_0 E_g \quad \text{or} \quad M^2 = \frac{m_0^2 E_g(E_g + \Delta)}{12m_e(E_g + \frac{2}{3}\Delta)} \qquad (16.26)$$

where $E_g$ is the bulk material band gap and $\Delta$ is the spin-orbit splitting energy of the valence band. Alternatively, $M^2$ can be evaluated in terms of an associated energy $E_P = 2M^2/m_0$ [22, 23], and $E_P$ for alloy $In_{1-x}Ga_xAs_yP_{1-y}$ can be found using Vegard's law:

$$E_P^{\text{alloy}} = (y - x)E_P^{\text{InAs}} + (1 - y)E_P^{\text{InP}} + xE_P^{\text{GaAs}} \qquad (16.27)$$

where $E_P^{\text{InAs}} = 22.2$ eV, $E_P^{\text{InP}} = 19.7$ eV, and $E_P^{\text{GaAs}} = 28.8$ eV.

Note that the total absorption coefficient is the sum of contributions from the light-hole to conduction band and the heavy-hole to conduction band transitions, i.e.,

$$\alpha = \alpha_{lh} + \alpha_{hh} \qquad (16.28)$$

where $\alpha_{lh}$ and $\alpha_{hh}$ should be calculated as described above using light-hole mass and heavy-hole mass, respectively, for $m_h$ in Eq. (16.24).

The above analysis reveals that the FKE-induced absorption due to the allowed electron transitions is essentially polarization independent. Although the absorption due to forbidden transitions (which become allowed when electric field is present) is polarization dependent [21], it is typically a much smaller effect. However, in an actual FKE-based EA modulator, polarization dependence can arise from different optical confinement for TE and TM optical modes, different facet coupling efficiency, and especially different optical propagation loss. These problems can be minimized by careful optical waveguide design.

***Quantum Confined Stark Effect.*** In a semiconductor quantum well (QW), both electrons and holes are tightly confined in a narrow well, so that the electron and hole energies are quantized to form discrete energy levels called "subbands," and the density of states $\rho(E)$ in a 2D QW becomes a step function, which differs from the square root function of $\rho(E)$ in 3D bulk material. Moreover, the confined electrons and holes are electrically bound as excitons, with a much smaller Bohr radius and a much larger binding energy than that of excitons in bulk material. These properties of QW cause very different optical absorption behavior than that in bulk material. In the idealized case, and if we only consider the electronic transitions from heavy-hole subbands to the conduction band, the absorption spectrum for a QW would look like Fig. 16.5a. The step-like feature is due to the intersubband transitions and determined by the step-like $\rho(E)$ function in QW. The positions of step edges are located at

$$E_{cv} = E_g + E_e + E_h \qquad (16.29)$$

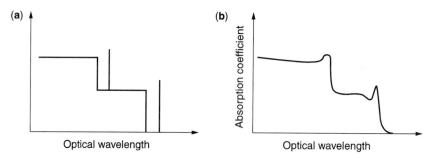

**Figure 16.5**   Illustration of (**a**) idealized and (**b**) broadened optical absorption spectra for a QW. Only heavy-hole exciton resonances and the corresponding non-exciton interband transitions are considered.

where $E_{cv}$ is called intersubband energy gap, $E_g$ is the bandgap of bulk material of the well layer, $E_e$ is the electron energy level in the conduction band, and $E_h$ is the hole energy level in the valence band. The discrete lines are due to the exciton resonances located at

$$E_{ex} = E_{cv} - E_b \qquad (16.30)$$

where $E_{ex}$ is called exciton transition energy, and $E_b$ is the exciton binding energy. Transitions from light-hole subbands result in a similar absorption spectrum, with the step edges and the exciton lines shifted to shorter wavelengths due to the larger $E_{cv}$ and smaller $E_b$. In reality, the steps and the exciton lines are broadened due to various broadening mechanisms, which makes the optical absorption spectrum look like Fig. 16.5**b**. However, even in the broadened absorption spectrum, the absorption edge is still very sharp, so that the detuning energy (defined as $E_{ex} - \hbar\omega$, where $E_{ex}$ is the smallest exciton transition energy) for EA modulators using MQW active layer can be very small (typically, 20–30 meV).

When an electric field is applied across a QW, the band diagram is tilted; thus, the electron and hole confinements are changed, and the electron energy levels $E_e$ and hole energy levels $E_h$ are reduced, as illustrated in Fig. 16.6**b**. As a result, the exciton absorption peaks are shifted to longer wavelengths (red-shift). The exciton linewidth is also further broadened due to the presence of the electric field. These two factors cause a drastic increase in the optical absorption at the longer wavelength side of the band edge. This effect is known as the QCSE. Due to the strong exciton

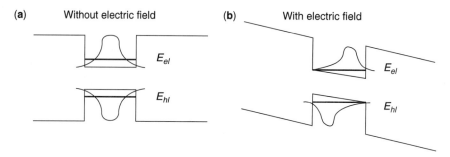

**Figure 16.6**   Band diagram for a single quantum well to show quantum-confined Stark effect.

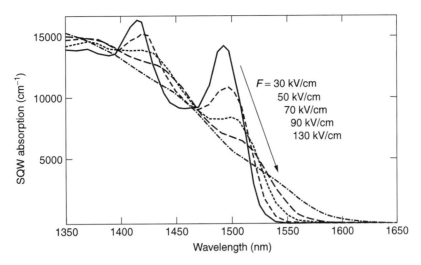

**Figure 16.7**  Simulated absorption spectra for a single QW under different electric field.

resonance, QCSE-induced absorption change is much larger than that of FKE. Exciton binding energy $E_b$ is also reduced as a result of larger electron-hole separation, which causes an opposite effect to QCSE red-shift. However, the decrease of $E_b$ is much smaller than the decrease of $E_{cv}$, so that many researchers assume field-independent $E_b$ in their QCSE simulations. Another important phenomenon is that the decrease of electron-hole overlap in the presence of electric field reduces the exciton oscillation strength, which lowers the exciton peak. Thus, there exists some optimum electric field for maximizing the absorption change. A simulated absorption spectra for a strained single QW in the presence of different electric field is shown in Fig. 16.7. In this example, the best operating wavelength is around 1.55 μm, although it can also work well at other wavelengths in a 20–30-nm range.

The simulation of QCSE consists of two major steps: the first step is to calculate the electron and hole energy levels and their wave functions for the QW in the presence of electric field, and the second step is to calculate the optical absorption coefficient as a function of photon energy and electric field using the results from the first step. To solve the electron and hole states, envelope-wavefunction approximation is usually assumed. Different analysis and numeric methods have been developed, such as transmission matrix method [24, 25], Green's function method [26, 27], and variational method [28]. The goals of the calculation include finding the energy levels of $E_e$ and $E_h$, as well as the overlap integral:

$$M_{cv} = \left| \int_{-\infty}^{\infty} f_e(z) f_h^*(z) \, dz \right| \tag{16.31}$$

where $f_e(z)$ and $f_h(z)$ are the envelope wavefunctions for electron and holes. In addition, exciton binding energy $E_b$ and exciton 2D radius $\lambda_{ex}$ have to be solved [29, 30].

The broadening of the absorption spectrum is a very important concern in the QCSE modeling. The homogeneous broadening, which is related to the lifetime of the excited states, includes phonon scattering, thermionic emission, tunneling through barrier, and electron-hole recombination. The inhomogeneous broadening is mainly caused by well width fluctuation and variation of electric field from well to well. In the earlier research works [24, 31], phonon scattering, which increases with temperature but is independent of electric field, was considered as the most significant among homogeneous broadenings, and the field-induced broadening was attributed only to the inhomogeneous broadenings. However, more recent research works [26, 27, 32] suggested that tunneling through barrier may be the major cause for the field-induced broadening. In the presence of electric field, as shown in Fig. 16.6b, the barrier shape becomes triangular at one side of the well, and the particle wavefunctions, especially of electrons and light-holes, can easily tunnel through this triangular barrier and leak into the semi-infinite space. As a result, the bound electron and hole states become quasi-bound states, and the discrete energy levels become broadened (with finite linewidth). This mechanism contributes to the broadenings of absorption spectrum induced by both exciton resonance and intersubband transition. The field-induced broadening can be calculated analytically using Green's function method or the transmission matrix method.

To calculate the optical absorption coefficient in a QW, the contribution due to the exciton resonance between discrete energy levels can be expressed as [31]

$$\alpha_{ex}(\hbar\omega, F) = \frac{4e^2\hbar|p_{cv}|^2}{\varepsilon_0 cnm_0^2 E_x \lambda_{ex}^2 L_z} M_{cv}^2 B(\hbar\omega, E_{ex}) \tag{16.32}$$

where $E_x$ represents photon energy $\hbar\omega$ [22, 33], or exciton transition energy $E_{ex}$ [31], or bulk material bandgap $E_g$ of the well layer [34]. $L_z$ is the well width. $P_{cv}$ takes the same form as in Eq. (16.25), but with a different polarization factor $M_0$: for heavy-hole exciton, $M_0 = 1$ for TE optical mode and 0 for TM optical mode; for light-hole exciton, $M_0 = 1/3$ for TE mode and $4/3$ for TM mode [22]. As heavy-hole exciton resonance is located at a longer wavelength than light-hole exciton resonance for unstrained QW, the above values of $M_0$ determine that the QCSE effect for TE mode is normally much larger than that for TM mode. $B(\hbar\omega, E_{ex})$ can be a convolution of multiple broadening functions due to different mechanisms. Each broadening function can be either a normalized Gaussian function [31]

$$B(\hbar\omega, E_{ex}) = \frac{1}{\sqrt{2\pi}\xi} \exp[-(\hbar\omega - E_{ex})^2/2\xi^2] \tag{16.33}$$

with a full width at half maximum (FWHM) of $2.35\xi$, or a Lorentzian function [24]

$$B(\hbar\omega, E_{ex}) = \frac{\Gamma}{\pi[(\hbar\omega - E_{ex})^2 + \Gamma^2]} \tag{16.34}$$

with an FWHM of $2\Gamma$. To account for the broadening due to particles tunneling through barriers, Eq. (16.32) should be multiplied by the solved density-of-state functions for electron and holes, and then be integrated over the broadened electron and hole subbands.

The absorption spectrum continuum of a QW due to intersubband transitions between non-exciton states can be expressed as [31]:

$$\alpha_{con}(\hbar\omega, F) = \frac{\mu e^2|p_{cv}|^2 M_{cv}^2}{\varepsilon_0 cnm_0^2\hbar^2\omega L_z} \cdot \frac{2}{1 + \exp(-2\pi\sqrt{(\hbar\omega - E_{cv})/R_y})} \tag{16.35}$$

where $R_y = e^4\mu/(2\varepsilon_0^2\hbar^2)$ is the Rydberg constant. Similarly, to account for the broadening due to particles tunneling through barriers, Eq. (16.35) should be integrated over the broadened electron and hole subbands. The total absorption coefficient in a QW is the sum of $\alpha_{ex} + \alpha_{con}$ due to transitions from all heavy-hole subbands and light-hole subbands.

To achieve a large field-induced absorption change in a QW, it is desirable to have large QCSE red-shift, high exciton peak, and extensive field-induced broadening. A deep and wide well is preferred for large QCSE red-shift, whereas a shallow and narrow well makes carriers easier to tunnel through barriers (or escape out of the well), which results in more broadening. Short carrier escape time is also critical to high-power and high-speed operation. Moreover, a deep and narrow well is ideal for high exciton peak. Consequently, a QW is usually designed to have high electron-barrier, but relatively lower hole-barrier to help heavy holes escape from the well. Typical well width is 80–100 Å.

MQW using InGaAsP well and InGaAlAs barrier can have high electron-barrier and low hole-barrier [35]. High electron-barrier can also be achieved using InGaAsP well and InGaAsP barrier by introducing compressive strain into the well layer and tensile strain into the barrier layer. In strained layers, the bulk-material band positions are shifted due to both composition variations and strain effects [36], as illustrated in Fig. 16.8**a**. However, compressive strain in the well causes split-off between the heavy-hole and light-hole valence bands, and the heavy-hole band is above the light-hole band. This makes the heavy-hole exciton peaks at even longer wavelength than the light-hole exciton peak, which further enhances the polarization dependence. This design also increases the confinement barrier for heavy holes, which may not be suitable for high-optical-power operation.

The above statement infers that by using tensile well and compressive barrier, polarization-insensitive QCSE can be achieved [37], as shown in Fig. 16.8**b**. In this case, the light-hole valence band is above the heavy-hole band in the bulk well material. Using QW confinement, the light-hole subband and heavy-hole subband can be made at a similar level, so that the polarization dependence is compensated. The same

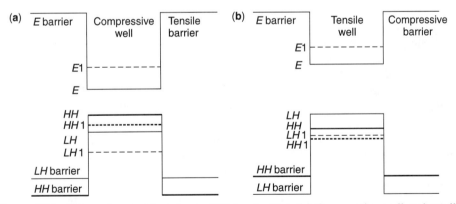

**Figure 16.8**  Effect of material strain on QW band offset. (**a**) Compressive well and tensile barrier, (**b**) tensile well and compressive barrier. *HH* and *LH* are the valence band positions for heavy holes and light holes, *E* is the conduction band position. *HH1*, *LH1*, and *E1* are the first heavy-hole, light-hole, and electron levels in the QW. The band positions are not strictly scaled.

design also lowers the confinement barrier for heavy holes, which is desirable for high-optical-power operation. A more complicated MQW design [38] used tensile (−0.25%) InGaAsP well and tensile (−1.1%) InGaAsP barrier, with a thin compressive (+1.2%) InAsP layer inserted after every 4 QWs to balance the strain. Polarization insensitive modulation has been achieved with >40-mW saturation power [38]. MQW using different well widths for heavy-holes and light-holes has also been demonstrated for polarization-insensitive modulation [39].

***Kramers–Kronig Relations.*** For a passive linear system, there exist general relations between the real part and the imaginary part of its response function. The complex optical index $n + jk$ is a response function, and its imaginary part $k$ is related to the optical absorption coefficient by

$$\alpha = \frac{2\omega}{c}k = \frac{4\pi}{\lambda}k \qquad (16.36)$$

Thus, the optical refractive index $n$ and the optical absorption coefficient $\alpha$ are related to each other, by a set of dispersion relations called Kramers–Kronig relations. The change of $\alpha$ due to electroabsorption effects results in a change of $n$:

$$\Delta n(\hbar\omega, F) = \frac{\hbar c}{\pi} P \int_0^\infty \frac{\Delta\alpha(E, F)}{E^2 - (\hbar\omega)^2} \, dE \qquad (16.37)$$

where $\Delta n$ ($\Delta\alpha$) is the index change (the absorption change) due to the presence of electric field $F$ compared to that at zero electric field. The symbol $P$ denotes the Cauchy principal value of the integral defined as

$$P \int_0^\infty \equiv \lim_{\delta \to 0} \left[ \int_0^{\hbar\omega - \delta} + \int_{\hbar\omega + \delta}^\infty \right] \qquad (16.38)$$

From the Kramers–Kronig relation in Eq. (16.37), the intensity modulation of EA modulators is accompanied by an index (thus, phase) modulation, which in turn results in a broadening of the optical wavelength (or frequency); this is so-called wavelength (or frequency) chirp. Due to the fiber dispersion, the frequency chirping can cause signal distortion. This issue is further discussed in Section 16.3.4.

On the other hand, the phase modulation caused by the EA effects, especially QCSE, can also be utilized to make Mach–Zehnder modulators. For QCSE-based MZM, the detuning energy is typically around 60–80 meV. For example, if the QCSE shown in Fig. 16.7 represents that of the QW in MZM waveguides, the optical wavelength for this MZM will be in the 1600–1650-nm wavelength range. In this range, the optical absorption is very small, while the electric-field-induced index change can be sufficiently large for efficient phase modulation. The ideal QCSE for MZM operation would be large QCSE red-shift, sharp, and high exciton peak, with minimum field-induced broadening. Deeper well is desirable for both electrons and holes, and the carrier escape time is not a concern here due to the small optical absorption. For properly designed MQW structure, the field-induced index change at a small voltage can be larger than $10^{-3}$, which is much more efficient than the index change induced by EO effect.

## 16.3  MODULATION EFFICIENCY AND MODULATOR WAVEGUIDE DESIGN

Modulation efficiency is one of the most important concerns for all types of modulators. In analog applications, it is evaluated in terms of small-signal incremental change of modulated optical power output versus the change of modulation voltage. In digital applications, it is represented by the on/off extinction ratio and the required voltage swing. Modulation efficiency is determined by the modulator transfer function, which, in turn, is determined by the modulator waveguide design.

### 16.3.1  Modulator Transfer Function

For a voltage-controlled optical intensity modulator, its transfer function is the output optical power (or intensity) as a function of the applied voltage. The transfer function not only determines the modulation efficiency, but also determines the linearity of the modulator. The spurious-free dynamic range (SFDR) of the modulator can be calculated from the derivatives of the transfer function [40]. Using the transfer function and the optical phase information, the frequency chirp can also be calculated. The transfer functions for MZMs and EAMs are quite different, and they are analyzed in details below.

The Mach–Zehnder interferometer is the most popular device to implement optical intensity modulation using EO effect. Figure 16.9 is a schematic drawing of such an interferometer. At its optical input port, there is an optical splitter that divides the input optical power into two equal portions. The divided power propagates in two separate waveguides that are often called "two arms." In an MZM, at least one of these two arms is designed as EO waveguide, along which the optical phase can be modulated by an applied voltage. If the optical waves are in-phase after propagating through the two arms, they combine as a single mode in the output optical combiner, which results in a maximum intensity output; whereas if the optical waves are out-of-phase after propagating through the two arms, they combine as a higher order spatial mode near the optical combiner; therefore, most of the optical power becomes unguided wave beyond the combiner and the output intensity is a minimum. If the output Y-branch in Fig. 16.9 is replaced by a $2 \times 2$ coupler, the combiner can be represented as a multi-mode waveguide, and the optical power from each arm is coupled into multi-modes, respectively. All of these multi-modes co-propagate in the combiner, and the power in the output waveguides depend on the net field profile at the end of the combiner.

The optical field amplitude at the output of the MZM can be generally represented by

$$A_{\text{out}} = \frac{\sqrt{2}}{2}(A_1 e^{j\Phi_1} + A_2 e^{j\Phi_2}) \tag{16.39}$$

**Figure 16.9**  Schematic drawing of a Mach–Zehnder interferometer.

where $\Phi_1$ and $\Phi_2$ represents optical phase delays in the two arms, $A_1$ and $A_2$ represent their optical amplitudes. $A_1^2 + A_2^2$ is the total optical input power. The output optical power is

$$P_{\text{out}} = |A_{\text{out}}|^2 = \tfrac{1}{2}(A_1^2 + A_2^2 + 2A_1A_2\cos(\Phi_1 - \Phi_2)) \qquad (16.40)$$

Dividing $P_{\text{out}}$ by the input optical power $P_{\text{in}}(= A_1^2 + A_2^2)$ of the MZM and after some parameter transformation, the optical intensity transmission for the MZM can be written in the form of

$$T_{MZ} = \tfrac{1}{2}[1 + b\cos(\Phi_1 - \Phi_2)] \qquad (16.41)$$

where $b = 2A_1A_2/(A_1^2 + A_2^2)$ is an optical imbalance factor between the two arms, and $b = 1$ for ideally balanced design. Optical losses at the device facets and during the propagation are ignored in the above derivation. The phase difference $\Phi_1 - \Phi_2$ consists of two parts: one is the phase difference $\Phi_0$ at zero applied voltage, and the other is the phase difference $\Delta\Phi$ due to the applied voltage. When only one arm is modulated, the phase difference becomes

$$\Delta\Phi = \Delta n\gamma\frac{2\pi}{\lambda}L \qquad (16.42)$$

where $\Delta n$ is the optical index change in the waveguide active layer; $\gamma$ is the optical confinement factor, defined as the portion of optical mode that is confined in the active layer; $\lambda$ is the optical wavelength; and $L$ is the modulation length. If both of the arms are modulated in a push-pull mode, which means that the phase changes in the two arms are opposite, the overall phase change $\Delta\Phi$ is doubled. If the modulation is based on the EO effect, then

$$\Delta n = \frac{1}{2}n_0^3 r_{ij}\frac{V}{d} \qquad (16.43)$$

where $n_0$ is the optical index of the active layer at zero applied voltage; $r_{ij}$ is the relevant EO coefficient determined by the material, the optical polarization, and the electrode design; $V$ is the applied voltage; and $d$ is the spatial gap between the electrodes across which the voltage $V$ is applied. Combining Eqs. (16.42) and (16.43), we have

$$\Delta\Phi = \frac{\pi}{\lambda}n_0^3 r_{ij}\frac{\gamma L}{d}V = \pi\frac{V}{V_\pi} \qquad (16.44)$$

and

$$V_\pi = \frac{\lambda}{n_0^3 r_{ij}}\cdot\frac{d}{\gamma L} \qquad (16.45)$$

$V_\pi$ is a very important parameter for MZM. It is the voltage value at which the voltage-induced phase difference reaches $\pi$ (or $180°$). Equation (16.41) can now be updated as

$$T_{MZ}(V) = \frac{1}{2}\left[1 + b\cos\left(\pi\frac{V}{V_\pi} + \Phi_0\right)\right] \qquad (16.46)$$

The above equation represents the transfer function for MZM using the EO effect. It is a raised cosine curve, as plotted in Fig. 16.10a, with approximately constant peak values and periodically spaced peaks and valleys. For MZM using the QCSE effect, the relationship between $\Delta n$ and $V$ is nonlinear; thus, its transfer function differs from

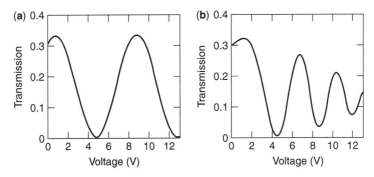

**Figure 16.10** Typical transfer curves for MZMs using **(a)** EO effect and **(b)** QCSE.

Eq. (16.46). The resulted transfer curve is like Fig. 16.10**b**. The peak value and the extinction ratio decrease at higher voltage due to increased optical absorption loss and imbalance; the distance between adjacent peaks also decreases because the index (thus, the phase) changes faster at higher voltage.

The transfer function for an EAM is pretty straightforward:

$$T_{EA}(V) = t_0 e^{-\gamma \alpha(V) L} \qquad (16.47)$$

where $t_0$ is the optical insertion loss of the EAM at zero applied voltage; $\gamma$ is the optical confinement factor; $\alpha(V)$ is the change of optical absorption coefficient due to the applied voltage $V$; and $L$ is the modulation length. Typical transfer curves for EAMs are plotted in Fig. 16.11, in which the transmission $T_{EA}(V)$ has been normalized at zero applied voltage. Typically, EAMs using MQW active layer have sharper transfer curves than do EAMs using bulk active layer. As stated before, for an MQW EAM, there typically exists some voltage point at which the optical absorption coefficient is maximized. On the other hand, for a bulk EAM, optical absorption coefficient typically increases monotonically with applied voltage.

It should be noted that EA effects is accompanied by photocurrent generation. The modulator photocurrent will also change with the applied voltage, as plotted in Fig. 16.11. As discussed in Section 16.2.2, the photocurrent generation may affect the modulator performance at high optical power, including the optical absorption coefficient and the modulation bandwidth. These effects are further discussed in later sections.

### 16.3.2 On/Off Extinction Ratio and Required Voltage Swing

High on/off extinction ratio is desirable for optical modulators used in digital communication systems. For MZMs, the on-state is achieved when the phase difference $\Phi_0 + \Delta\Phi = 0$, and the off-state is achieved when $\Phi_0 + \Delta\Phi = \pi$ (or 180°). The extinction ratio, ER, between the optical intensity outputs at the on-state and at the off-state is determined by the optical balance between the two arms. It can be calculated from Eq. (16.40):

$$ER = \frac{(P_{out})_{max}}{(P_{out})_{min}} = \frac{(A_1 + A_2)^2}{(A_1 - A_2)^2} \qquad (16.48)$$

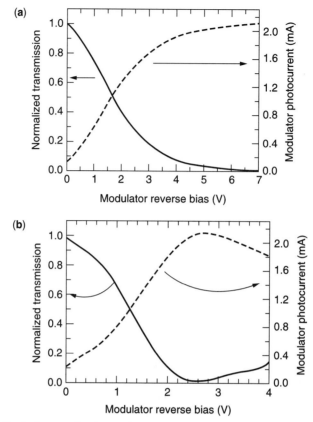

**Figure 16.11**   Typical normalized transfer curves and photocurrent curves for EAMs using (a) bulk active layer and (b) MQW active layer. Both plots have the same left axis and right axis.

The ratio of $(A_1/A_2)^2$ represents the optical power difference between the two arms, which can be caused by different optical loss in the two arms, as well as the imbalances in the optical splitter and combiner. Figure 16.12 plots the extinction ratio versus the power difference $(A_1/A_2)^2$. From this figure, it is seen that in order to achieve more than a 20-dB extinction ratio, the power difference has to be less than 1.7 dB; and a 3-dB optical power difference between the two arms will lead to ~15-dB maximum extinction ratio.

The switching between on- and off-states needs a voltage swing $V_\pi$. A small $V_\pi$ is desirable, especially when the modulator is used at 40-Gb/s data rate. At this high data rate, the electronic driver with a voltage swing larger than 3.5 V can be very expensive. A small $V_\pi$ also implies small power consumption, and this allows smaller heat sink for the electronic circuit, which can make the component footprint smaller. According to Eq. (16.45), $V_\pi$ is proportional to $d/L$, which, in turn, is inversely proportional to the waveguide capacitance. Therefore, a small $V_\pi$ usually results in a large waveguide capacitance and, hence, narrow modulation bandwidth. With a fixed $d/L$ ratio, the electrode gap $d$ should be as small as possible to shorten the waveguide length $L$. However, the design of electrode gap $d$ may affect the confinement factor $\gamma$. According to Eq. (16.45), $\gamma$ should be as large as possible for small $V_\pi$.

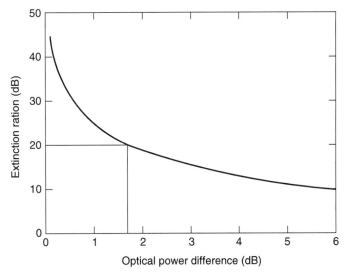

**Figure 16.12** Extinction ratio for an MZM versus the optical power difference between its two arms.

To achieve a high extinction ratio with a small voltage swing using EAM, an MQW active layer is preferred over bulk active layer due to the strong exciton effect in MQW. According to Eq. (16.47), the transfer function for an EAM decays exponentially with a product of three parameters $\gamma$, $\alpha$, and $L$. All three parameters are related to the design of intrinsic active layer thickness $d$ (equivalent to the electrode gap for MZM). As the absorption change $\alpha$ increases rapidly with the electric field, the intrinsic active layer is usually designed pretty thin for EAMs to achieve a large electric field. When $d$ is thin, the optical confinement factor $\gamma$ is roughly proportional to $d$. On the other hand, to meet a certain bandwidth requirement, $d/L$ ratio is fixed; thus, $L$ can also be considered as proportional to $d$. However, before $\alpha$ reaches its maximum value, it increases with electric field in a way faster than square law, and the electric field is inversely proportional to $d$. In this sense, $d$ should be designed as small as possible to achieve large $\gamma\alpha L$ product, which can lead to a sharp transfer curve and a large extinction ratio with a small voltage swing.

It should be watched out that $\alpha$ may reach a maximum value at some voltage point, as plotted in Fig. 16.11**b**. The maximum achievable extinction ratio by applying voltage is thus limited by the maximum $\alpha$ value. If $d$ is too thin, both $\gamma$ and $L$ will be very small; thus, the maximum achievable extinction ratio may be too small. This factor puts a lower limit to the thickness $d$ for QCSE-based EAMs used in digital links, although it may not apply for analog applications. Moreover, this limit does not exist for FKE-based EAMs.

A practical limiting factor is that it is hard to achieve very thin intrinsic layer sandwiched by doped layers using crystal growth techniques. This is due to the difficulty in the control of dopant diffusion, especially the diffusion of $p$-type dopant — Zn. During the growth of $p$-type layers at high temperature ($>500°C$), Zn dopants tend to diffuse into the intrinsic layer. To handle this effect, it is popular to grow an intrinsic InP layer on top of the intrinsic MQW layer to prevent Zn dopants from diffusing into the MQW layer. In addition, a thin lowly doped ($\sim 5 \times 10^{17}$ cm$^{-3}$) $p$-type layer is usually inserted between the highly doped ($>2 \times 10^{18}$ cm$^{-3}$) $p$-type layers and the intrinsic layers. With these designs, the effective intrinsic layer thickness is typically 0.2–0.3 μm, and the MQW active layer thickness is 0.1–0.2 μm. Maximum extinction ratio can be 15–20 dB per 100-μm waveguide length. The control of dopant diffusion in the semiconductor crystal growth is an important research topic.

With the intrinsic layer thickness designed above, the modulation length can be extended as long as the bandwidth requirement is met, so that the modulator transfer curve can be as sharp as possible. The optical waveguide width should be optimized to achieve low waveguide capacitance (which prefers narrow waveguide) and large $\gamma$ (which prefers wide waveguide). The optimized waveguide width is usually considered to be ~2 μm. However, when spot-size converter is not used at both ends of the waveguide for the sake of simplicity, 3-μm waveguide width is a popular choice to facilitate the coupling with lensed fibers. Moreover, single-mode waveguide is preferred for large extinction ratio, because the higher order modes take longer modulation length to decay due to their smaller optical confinement factor $\gamma$ in the active layer. The optimizations of the optical waveguide design usually require complicate optical simulations, and various kinds of commercial software are available to assist the design.

### 16.3.3  Small-Signal Modulation Efficiency

Analog optical systems usually adopt small-signal modulation. RF gain of analog fiber-optic links is one of the most important concerns. It is closely related to the spurious free dynamic range (SFDR) and the noise figure (defined as the ratio of the signal-to-noise ratio at the input over the signal-to-noise ratio at the output) of the link. The link RF gain is determined by the small-signal modulation efficiency of the optical modulator, as analyzed below.

Link RF gain is defined as the ratio of the RF power delivered to the RF load of the photodetector to the available RF power from the RF source that feeds the optical modulator. The available RF power from the source can be represented by

$$P_{\text{RF-in}} = \frac{v_{\text{in}}^2}{R_{\text{in}}} \tag{16.49}$$

where $R_{\text{in}}$ (usually 50 Ω) is the impedance of the transmission line feeding the modulator and $v_{\text{in}}$ is the voltage (rms value) for the microwave going toward the modulator. Due to impedance mismatch and the internal electrical property of the modulator, the actual modulation voltage $v_m$ can be different from $v_{\text{in}}$. The frequency-dependent ratio of $M = (v_m/v_{\text{in}})^2$ is defined as the modulator frequency response, which is the subject for subsequent sections.

The RF power delivered to the load is

$$P_{\text{RF-out}} = i_d^2 R_{\text{out}} = (\eta_d P_{od})^2 R_{\text{out}} \tag{16.50}$$

where $i_d$ is the RF current delivered from the photodetector, $R_{\text{out}}$ is the load impedance (usually 50 Ω), $\eta_d$ is the detector RF optical-to-electrical (O/E) responsivity, and $P_{od}$ is the modulated optical power received by the detector. Therefore, the link RF gain is

$$G = \frac{P_{\text{RF-out}}}{P_{\text{RF-in}}} = M \eta_d^2 R_{\text{in}} R_{\text{out}} \left( \frac{P_{od}}{v_m} \right)^2 \tag{16.51}$$

where $(P_{od}/v_m)^2$ can be obtained from the modulator transfer curve. During modulation, the total modulator voltage consists of the DC bias voltage $V_b$ and the RF modulation voltage $v_m$. Under small signal ($v_m$) approximation, the modulator transfer function can be written as

$$T(V) = T(V_b) + T'(V_b)v_m \qquad (16.52)$$

where the prime superscript represents the first derivative with respect to voltage. Assuming the optical power incident on the modulator input facet is $P_{in}$, the optical power coupled into the output fiber is then $P_{out} = P_{in}T(V)$. If the fiber loss is ignored, $P_{out}$ is also the optical power reaching the photodetector, among which the modulated optical power due to the RF voltage $v_m$ is

$$P_{od} = P_{in}T'(V_b)v_m \qquad (16.53)$$

Substituting Eq. (16.53) into Eq. (16.51), the link RF gain becomes

$$G = M\eta_d^2 R_{in} R_{out}(P_{in}T')^2 \qquad (16.54)$$

The detector responsivity $\eta_d$ has the unit of A/W. The right side of Eq. (16.54) excluding $\eta_d^2$ has the unit of $(W/A)^2$, which is often called *E/O conversion efficiency*. In the above equation, only the two parameters in the bracket are related to the modulator design at low frequency (note that the maximum input optical power $P_{in}$ to the modulator depends on the modulator type and the waveguide design). The term $r_m^2 = (P_{in}T')^2$ is defined as the *small-signal modulation efficiency* of the modulator, with the unit of $(W/V)^2$. Moreover, for the convenience of analysis in analog links, the modulator transfer function is often expressed as the product of the modulator optical insertion loss $t_0$ and the normalized transfer function $T_N(V)$; thus, we have $T'(V) = t_0 T_N'(V)$, and the small-signal modulation efficiency can be written as

$$r_m^2 = (P_{in} t_0 T_N')^2 \qquad (16.55)$$

The derivative $T_N'$ is called *modulator slope efficiency*, and it is bias voltage dependent. To achieve maximum modulation efficiency, the modulator bias voltage should be chosen to maximize $|T_N'|$. This bias voltage is referred as the *best bias point* $V_{bb}$. It has been pointed out that the multioctave SFDR is also optimized at the same bias point $V_{bb}$ [41]. For the transfer function of MZM in Eq. (16.46), because the extinction ratio is not a concern in analog links, and the optical imbalance between the two arms is typically small, it is reasonable to take $b \approx 1$. Therefore, the normalized transfer function for MZM is

$$T_{N-MZ}(V) = \frac{1}{2}\left[1 + \cos\left(\pi\frac{V}{V_\pi} + \Phi_0\right)\right] \qquad (16.56a)$$

and the maximum value of its slope efficiency is $\pi/(2V_\pi)$ when biased at the quadrature point of the raised cosine curve. For EAMs, the normalized transfer function is

$$T_{N-EA}(V) = e^{-\gamma\alpha(V)L} \qquad (16.56b)$$

and the best bias point $V_{bb}$ for achieving the maximum slope efficiency $T'_{N-EA}$ typically makes $T_{N-EA}(V_{bb}) = 0.5$ to $0.7$. The equivalent $V_\pi$ for an EAM is derived from its $T'_{N-EA}(V_{bb})$:

$$V_\pi^{EA} = \frac{\pi}{2|T'_{N-EA}(V_{bb})|} \qquad (16.57)$$

Apparently the maximum slope efficiency should be as high as possible for use in analog optical systems. For MZMs, this is equivalent to achieving a small $V_\pi$, which has already been discussed in the last section; for EAMs, design considerations for achieving large slope efficiency has also been discussed. Single-mode optical waveguide is required for MZMs, because higher order modes have optical refractive indices different from that of fundamental mode, which may cause complicate effects to the overall phase modulation. This should be a concern for the design of III–V MZMs. However, EAMs do not require single-mode optical waveguide for small-signal modulation, because the multimode dispersion does not cause any detrimental effect, and the higher order modes usually hold only small portion of the optical power, which does not significantly impact the slope efficiency.

***Bias Stability.*** Transfer function for any type of optical modulator can change slowly, either due to charging effect, or due to ambient changes such as variations of temperature, optical wavelength, optical power, and polarization. This effect not only causes slow drifting of the optimum bias point for application in analog links, but also can make the voltage point uncertain for achieving high extinction ratio in digital links. A small drift of the optimum bias point can cause significant drop of SFDR in analog links. To maintain the optimum voltage, it is common to employ some voltage control circuitry. The conventional approach is to use a weak low-frequency pilot-tone combined with the modulation signal. A Y-branch coupler has to be inserted after the modulator to tap off a small portion ($\sim 2\%$) of the output and direct it to a photodetector for checking the pilot-tone signal, so that the optimum voltage can be maintained for both analog and digital applications [42].

It is possible to maintain a constant best bias point for thousands of hours for LiNbO$_3$ modulators. However, once there is a sudden voltage change to the device, very slow bias drift (with time constants from hours to days) can occur. This is mainly attributed to the parasitic charge existing at the interface between the SiO$_2$ buffer layer and the LiNbO$_3$ surface. These charges cannot move promptly in response to the sudden change of the voltage, and they will affect the EO modulation [43]. Similarly, in polymer modulators, parasitic charge may also exist at the interface between the polymeric cladding layer and the core layer, because they typically have different conductivity [44]. This kind of bias-drifting problem may have more severe effect for on/off modulation in digital links. This effect can be minimized by matching $\sigma/\varepsilon$ for all layers, where $\sigma$ is the conductivity and $\varepsilon$ is the dielectric constant. LiNbO$_3$ surface treatment can also help to reduce the effect.

So far there is no report on the bias stability issue for III–V MZMs. For EAMs, the transfer function is very sensitive to the ambient variations. To minimize this effect, thermal electric cooler can be used, optical power should be stabilized, and optical wavelength needs to be locked. Moreover, polarization control of the input light is generally required for all types of modulators, unless the modulator is designed to be polarization insensitive. To track the bias point variation for EAMs used in analog links, there are simpler approaches than the conventional one, either based on the correlation between the transfer function and the modulator DC photocurrent [45], or based on the correlation between the transfer function and the modulator second harmonic signal [46].

*Optical Power Handling Capability.* Equations (16.54) and (16.55) indicate that optical input power $P_{in}$ should be as high as possible to achieve a large RF gain for analog optical links. The optical power handling capability is different for different types of modulators. For MZM using $LiNbO_3$ and III–V semiconductors, the maximum input optical power is limited by the catastrophic facet damage. 400-mW optical input power has been demonstrated for a Ti-diffused $LiNbO_3$ modulator after annealing it in $O_2$ atmosphere with high temperature [47]. For polymer modulators, the maximum optical power is limited by the previously mentioned photo absorption problem, which relaxes the alignment of chromophore molecules. The optical power handling capability is usually higher at longer wavelength, and at the same wavelength of 1.32 μm, it varies from 10 mW [48] to 250 mW [49] with different material and fabrication techniques. The optical power handling ability of EAMs is limited by several factors, including electric field screening due to photo-generated carriers, bandgap shrinkage due to heating, and impedance change due to photocurrent resistance. The last factor is discussed in details later. These limiting factors are more severe for MQW EAMs due to stronger absorption and smaller bias voltage $V_{bb}$. Typically, the maximum optical power for MQW EAMs is 10–20 mW [50], while the input optical power can be 40–50 mW for bulk EAMs [51].

*Optical Propagation Loss.* According to Eq. (16.55), the modulation efficiency is proportional to the square of the optical insertion loss. Modulator optical insertion loss consists of the loss at the waveguide facets and the propagation loss in the waveguide:

$$t_0 = C^2 e^{-\alpha_0 L} \tag{16.58}$$

where $C$ is the optical loss factor at each facet, including the mode coupling loss and the reflection loss; $\alpha_o$ is the optical attenuation coefficient in the waveguide (at zero bias for EAMs), including all loss mechanisms such as the scattering loss at the waveguide sidewalls and at the layer interfaces, the free-carrier absorption loss in the doped semiconductor layers, and the photo absorption loss in the EAM active layer. It is desirable to have minimum optical insertion loss for modulators used in both analog and digital optical systems.

The design for minimum facet loss and waveguide propagation loss have been discussed in previous sections. Here we want to show how the waveguide propagation loss can limit the maximum waveguide length. For systems which only require narrow modulation bandwidth (for example, <5 GHz), it seems that the waveguide should be designed very long because the modulator slope efficiency $T_N'$ increases with length $L$. However, the optical propagation loss in the waveguide also increases with the length $L$. For MZMs, the modulation efficiency $r_m^2$ is proportional to $(t_0/V_\pi)^2$, and $V_\pi$ is proportional to $1/L$, therefore $r_m^2$ is proportional to $(Le^{-\alpha_0 L})^2$. Maximizing the value of $(Le^{-\alpha_0 L})^2$ determines the optimum length $L$. For EAMs, by combining the previous equations, we can obtain

$$r_m^2 = (P_{in} C^2 \underline{e^{-\alpha_o L}} \gamma \underline{L\alpha'(V_{bb})} e^{-\gamma\alpha(V_{bb})L})^2 \tag{16.59}$$

The four underlined terms in the above equation have dependence on modulator length $L$. The last exponential term is actually $T_N(V_{bb})$. A longer $L$ leads to a smaller best bias point $V_{bb}$, which in turn results into smaller $\alpha(V_{bb})$. The opposite trends for $L$ and $\alpha(V_{bb})$ make the value of $T_N(V_{bb}) = e^{-\gamma\alpha(V_{bb})L}$ insensitive to the change of $L$. It is observed in practice that $T_N(V_{bb})$ usually falls in the range of 0.5–0.7 with typical waveguide design and detuning energy. Longer $L$ also causes smaller $\alpha'(V_{bb})$ due to smaller $V_{bb}$. For the sake of simplicity, one can also analyze $(Le^{-\alpha_0 L})^2$ first, and put aside $T_N(V_{bb})$ and $\alpha'(V_{bb})$ in Eq. (16.59).

Figure 16.13 plots $(Le^{-\alpha_0 L})^2$ versus $L$ curves for various types of modulators. According to Fig. 16.13**a**, if optical propagation loss in the waveguide is 3 dB/cm (typical for polymer modulators), the waveguide length should not exceed 1.3 cm; likewise, if the propagation loss is 1 dB/cm (typical for III–V EO waveguides), waveguide length more than 2 cm will not improve the modulation efficiency much. For LiNbO$_3$ waveguides, typical optical propagation loss is ~0.2 dB/cm; in this case, the modulation efficiency increases with the waveguide length all the way beyond 6 cm.

The optical propagation loss is much larger for EAMs, typically, 15–20 dB/mm. According to Fig. 16.13**b**, the optimum waveguide length is only 0.2–0.3 mm. Owing

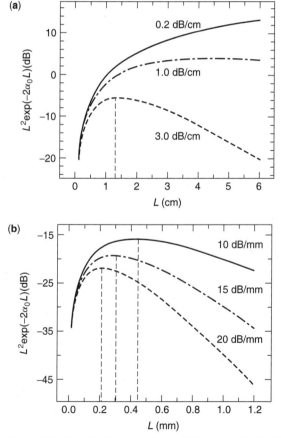

**Figure 16.13** Relative modulation efficiency versus modulator length at different optical propagation losses for (**a**) MZMs and (**b**) EAMs.

to the fact that $\alpha'(V_{bb})$ is smaller for longer $L$, the optimum length will be even shorter. In [52], the optimum modulator length was measured to be 0.17 mm. As a longer modulator length will not improve either the RF gain or the bandwidth, modulator length exceeding the optimum lengths predicted in Fig. 4.13 should be avoided.

### 16.3.4   Chirp Analysis

In general, the intensity modulation is accompanied by the optical phase modulation, which results in optical frequency (wavelength) chirp. This effect is an important concern for digital fiber links, which typically use a 1.55-μm wavelength to minimize optical loss in the long-distance fiber. The frequency chirp may result in increased pulse broadening after propagating through a fiber with nonzero dispersion, and lead to increased bit-error ratio (BER).

It can be shown that each frequency component of the optical field propagates in a single-mode fiber as a guided wave, and it can be represented by [53]

$$\vec{E}(x, y, z, t) = \hat{e} F(x, y) A(z, t) \exp[j(\beta_0 z - \omega_0 t)] \tag{16.60}$$

where $\hat{e}$ is the unit polarization vector; $F(x, y)$ is the field distribution of the optical mode; $A(z, t)$ is the amplitude of the pulse envelope, it varies much slower than the last exponential term in the above equation; $\omega_0$ is the center frequency of the pulse spectrum; and $\beta_0$ is the propagation constant at the center frequency. If the pulse envelope takes a Gaussian shape before it is coupled to the optical fiber ($z = 0$), it can be expressed as

$$A(0, t) = A_o \exp\left[-\frac{1 + iC}{2}\left(\frac{t}{T_o}\right)^2\right] \tag{16.61}$$

where $A_0$ is the peak amplitude; $T_0$ is the half-width at $1/e$-intensity point; and $C$ governs the linear frequency chirp imposed on the pulse, which can be resulted from the modulation process. The frequency chirp can be derived from the phase modulation:

$$\delta\omega = -\frac{\partial\Phi}{\partial t} = \frac{C}{T_0^2}t \tag{16.62}$$

where $\Phi$ is the phase of the complex amplitude $A(0, t)$. If $C$ is positive, then $\delta\omega < 0$ at the rising edge of the pulse (where $t < 0$), and $\delta\omega > 0$ at the falling edge of the pulse ($t > 0$). A negative $C$ will lead to opposite frequency chirp. It has been shown that the Gaussian pulse in Eq. (16.61) will remain in Gaussian shape as it propagates in a dispersive single-mode fiber, and its pulse width $T_1$ changes with propagation distance $z$ as [53]

$$\frac{T_1}{T_0} = \sqrt{\left(1 + \frac{C\beta_2 z}{T_0^2}\right)^2 + \left(\frac{\beta_2 z}{T_0^2}\right)^2} \tag{16.63}$$

where $\beta_2$ is the group-velocity dispersion (GVD) of the fiber. Note that $\beta_2$ is different from the so-called "dispersion parameter" $D$. They are defined and related by

$$\beta_2 = \frac{d^2\beta}{d\omega^2} = \frac{d}{d\omega}\left(\frac{1}{v_g}\right) \qquad D = \frac{d}{d\lambda}\left(\frac{1}{v_g}\right) = -\frac{2\pi c}{\lambda^2}\beta_2 \tag{16.64}$$

where $v_g$ is the optical group velocity and $c$ is the light velocity in free-space. For standard single-mode optical fiber, $\beta_2$ and $D$ are close to zero around 1.32-μm wavelength; at 1.55-μm wavelength, $D$ is 15 to 18 ps/(km-nm) and $\beta_2$ is $-19$ to $-23$ ps$^2$/km.

The change of pulse width results in the change of the peak power of the pulse [53]. Specifically, if the pulse width is increased by a ratio of $T_1/T_0$, the peak amplitude will be decreased by a ratio of $(T_0/T_1)^{1/2}$, and the peak power will be decreased by a ratio of $T_0/T_1$. The dispersion penalty is defined as the required increase in optical power to compensate this peak-power reduction:

$$P = 10\log(T_1/T_o) \tag{16.65}$$

Figure 16.14 plots the dispersion penalty versus the normalized propagation distance $z/L_D$ in the fiber with $\beta_2 < 0$, where $L_D = T_0^2/|\beta_2|$ is called "dispersion length." Fig. 16.14 indicates that there can be an initial negative dispersion penalty (corresponding to an initial compress of the pulse width) for chirped pulse with positive $C$. However, after some critical distance, the dispersion penalty of the chirped pulse quickly increases beyond the penalty of the unchirped pulse. This critical distance is

$$z_0 = \frac{2}{C}L_D = -\frac{2T_0^2}{C\beta_2} \tag{16.66}$$

Larger $C$ value makes $z_0$ shorter. Narrower pulse width $T_0$, corresponding to a higher bit rate $B$, also reduces $z_0$. A simple relationship between $T_0$ and $B$ is $\sqrt{8}BT_0 \leqslant 1$, which ensures at least 95% of the pulse energy remains within the bit slot. It can be calculated that for 10-Gb/s bit rate transmitting in the standard single-mode fiber, $L_D \sim 60$ km, and chirped pulse with $C = 0.7 - 0.8$ would lead to only $\sim$1-dB dispersion penalty for a 80-km fiber link, which is 1 dB less than using unchirped pulse. However, for 40-Gb/s bit rate, $L_D$ is less than 4 km, a 80-km distance makes $z/L_D > 20$. In this case, unchirped pulse is better, and excessive dispersion penalty is inevitable on

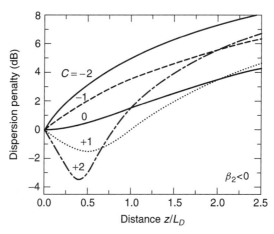

**Figure 16.14**   Dispersion penalty as a function of normalized propagation distance for Gaussian pulses with different chirp parameter $C$.

standard single-mode fiber. Because of this, most long-reach 40-Gb/s systems employ dispersion-compensated transmission links.

However, in actual data transmission systems, the pulses generally have steeper rising and falling edges than the Gaussian shape, and the chirp may not be linear. Therefore, the above analysis is only qualitatively correct. A more general expression for the pulse envelope at the output of an optical intensity modulator (before entering into the fiber) is

$$A(0, t) = \sqrt{I(t)}\, \exp[j\Phi(t)] = A_0 \exp\left(-\frac{2\pi}{\lambda} k(t) L\right) \exp\left(j\frac{2\pi}{\lambda} n(t) L\right) \quad (16.61a)$$

where $I(t)$ is the modulated optical intensity; $\Phi(t)$ is the accompanied phase change; $n(t) + jk(t)$ is the complex optical refractive index in the modulator; and $L$ is the modulation length. The Henry's alpha parameter for evaluating chirp characteristics of modulators is defined as

$$\alpha_H = \frac{\partial n}{\partial k} = -2I \frac{\partial \Phi}{\partial I} \quad (16.67)$$

The above definition of $\alpha_H$ for modulators is similar to the linewidth enhancement factor for lasers. Comparing Eqs. (16.61a) and (16.61), it is seen that if $n(t)$ linearly changes with $k(t)$, then $\alpha_H$ corresponds to $-C$. The actual relationship between $n(t)$ and $k(t)$ is generally not linear. The frequency change is now

$$\delta\omega = -\frac{d\Phi}{dt} = \frac{\alpha_H}{2I}\frac{dI}{dt} \quad (16.62a)$$

At the falling edge of the pulse, $dI/dt$ is negative; thus, negative $\alpha_H$ produces positive frequency chirp at the falling edge and negative frequency chirp at the rising edge, corresponding to positive $C$. Therefore, zero or negative $\alpha_H$ over the entire on/off voltage swing is desirable for optical intensity modulators.

In the analysis of optical modulators, it is often more convenient to replace the time variable $t$ in Eq. (16.61a) with the voltage variable $V$, and the output of a modulator is often expressed in the form of

$$A_{\text{out}} = A_0(V) \exp[j\Phi(V)] \quad (16.68)$$

where $A_0(V) = \sqrt{I}$. In this case, the Henry's alpha parameter is

$$\alpha_H = -A_0 \frac{\partial\Phi/\partial V}{\partial A_0/\partial V} \quad (16.69)$$

*Mach–Zehnder Modulators.* Combining Eqs. (16.39) and (16.42), the amplitude output of an MZM can be generally expressed as

$$A_{\text{out}} = \frac{\sqrt{2}}{2} A_1 \exp\left[j\Delta n_1(V)\gamma\frac{2\pi}{\lambda}L + j\Phi_0\right] + \frac{\sqrt{2}}{2} A_2 \exp\left[j\Delta n_2(V)\gamma\frac{2\pi}{\lambda}L\right]$$

$$(16.70)$$

For balanced arms, $A_1 = A_2$. In the case of single-arm modulation, $\Delta n_2 = 0$; therefore,

$$
\begin{aligned}
A_{\text{out}} &= \sqrt{2} A_1 \cos \left[ \Delta n_1(V) \gamma \frac{\pi}{\lambda} L + \frac{\Phi_0}{2} \right] \exp \left[ j \Delta n_1(V) \gamma \frac{\pi}{\lambda} L + j \frac{\Phi_0}{2} \right] \\
&= \sqrt{2} A_1 \cos \left( \frac{\pi}{2} \frac{V}{V_\pi} + \frac{\Phi_0}{2} \right) \exp \left[ j \left( \frac{\pi}{2} \frac{V}{V_\pi} + \frac{\Phi_0}{2} \right) \right]
\end{aligned}
\tag{16.71}
$$

From Eqs. (16.69) and (16.71), we can obtain

$$
\alpha_H = c \tan \left( \frac{\pi}{2} \frac{V}{V_\pi} + \frac{\Phi_0}{2} \right)
\tag{16.72}
$$

If $\Phi_0 = 0$, then $\alpha_H = +\infty$ at on-state (where $V = 0$), which is an unacceptable scenario for single-arm modulated MZM. One way to avoid this is to use the $\pi$-phase-shifted design to make $\Phi_0 = \pi$; thus, $\alpha_H$ in Eq. (16.72) starts from 0 at off-state (where $V = 0$), and gradually decreases to negative value when the output is changed to on-state [54, 65].

For MZMs using push-pull modulation, $\Delta n_2 = -\Delta n_1$. If $A_1 = A_2$ and $\Phi_0 = 0$, Eq. (16.70) becomes

$$
A_{\text{out}} = \sqrt{2} A_1 \cos \left[ \Delta n_1(V) \gamma \frac{2\pi}{\lambda} L \right] = \sqrt{2} A_1 \cos \left( \frac{\pi}{2} \frac{V}{V_\pi} \right)
\tag{16.73}
$$

Clearly, zero-chirp *at all voltages* is achieved for the above $A_{\text{out}}$, which is a big advantage of push-pull MZMs. However, in actual MZMs, $A_1$ may not be exactly equal to $A_2$, and $\Delta n_2$ may not be exactly equal to $-\Delta n_1$. These factors make the actual $\alpha_H$ for push-pull MZMs slightly deviate from zero. Differential push-pull modulation allows different voltage swings on the two arms (but with opposite phases). By controlling the voltage swings $V_1$ and $V_2$ at the two arms, the chirp can be tuned to be positive, zero, or negative.

***Electroabsorption Modulators.*** To calculate $\alpha_H$ for EAMs, first we have to calculate the optical absorption coefficient $\alpha$ for a wide spectrum using the simulation approaches described in Section 16.2.2, and then calculate the change of optical index using the Kramers–Kronig relation in Eq. (16.37). The above calculations have to be repeated for different bias voltages to obtain $\Delta \alpha(V)$ and $\Delta n(V)$ curves at the desired optical wavelengths, and Henry's alpha parameter $\alpha_H$ can then be calculated by using Eqs. (16.36) and (16.67).

$\alpha_H$ for a typical EAM can take large positive value ($>2$) at low bias, especially when the detuning energy is large. General trends are [55]: $\alpha_H$ decreases as the EAM voltage increases, or as the optical wavelength decreases. To lower $\alpha_H$ values within the on/off voltage swing, one can either choose a larger voltage point as the on-state, or operate the EAM at shorter wavelength, both of them result in larger on-state optical loss. Therefore, there is a tradeoff between the optical loss (perhaps also the extinction ratio) and the $\alpha_H$. Typically, after compromising 3-dB optical power, small positive ($<0.7$) to negative $\alpha_H$ can be achieved within the on/off voltage swing.

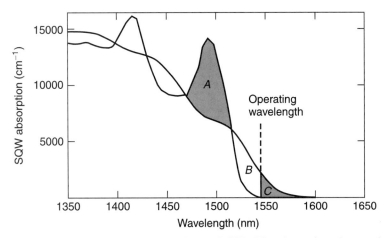

**Figure 16.15**   Illustration of index change due to QCSE. The absorption changes in area $B$ contribute to positive $\Delta n$, while the absorption changes in areas $A$ and $C$ contribute to negative $\Delta n$.

Figure 16.15 can help to understand the behavior of $\alpha_H$ due to QCSE. From Eq. (16.37), it is seen that the absorption changes in area $B$ contribute to positive $\Delta n$, while absorption changes in areas $A$ and $C$ contribute to negative $\Delta n$. When the operating wavelength is shorter, area $B$ is smaller and area $C$ is larger; thus, $\alpha_H$ changes toward negative direction. At low bias, index change from area $B$ may be more significant than that from areas $A$ and $C$, because it is larger than area $C$ and is closer to the operating wavelength than area $A$. This results in positive $\alpha_H$ at low bias. As the bias voltage increases, areas $A$ and $C$ enlarge much faster than area $B$, which also changes $\alpha_H$ toward the negative direction.

Bulk EAMs typically use larger detuning energy, which leads to more positive $\alpha_H$ than that of MQW EAMs under the same electric field. However, after compromising the same amount of on-state optical power, $\alpha_H$ for both types of EAMs may become comparable [56]. This is because the bias voltage can be larger for bulk EAMs for the same reduction of on-state optical power.

## 16.4   LUMPED-ELEMENT MODULATORS

Starting from this section, we focus on wide-bandwidth design and analysis for optical modulators, including lumped-element modulators and traveling-wave modulators. Both EAMs and QCSE-based MZMs can use lumped-element electrode due to their short waveguide lengths, and these two types of devices are fabricated using III–V materials, particularly, on InP substrate. Their frequency responses and modulation bandwidths can be analyzed using the same approach.

The design of lumped-element electrode is simple. Usually, a $p$-type ohmic contact is fabricated directly on top of the optical waveguide. The metal thickness should be designed as the metal skin-depth of the highest modulation frequency. For example, for a 20-GHz bandwidth device, 0.5-$\mu$m-thick gold is sufficient, and thicker gold does not help to reduce resistance due to the microwave skin effect. A metal square, serving as wire-bonding pad or probing pad, is usually fabricated adjacent to the waveguide and is electrically connected to the middle point of the waveguide. This bonding pad contributes parasitic capacitance to the device and causes a reduction of the modulation bandwidth; thus, a smaller pad is desirable to make the parasitic capacitance smaller. However, a minimum area of $30 \times 30 \ \mu m^2$ of the pad is required for easy probing and wire-bonding. To further reduce the parasitic capacitance, it is popular to use thick polyimide or BCB underneath the pad metal. With all of these methods, the parasitic capacitance can be reduced to $\sim 30$ fF. After the waveguide is etched, large-area $n$-type ohmic contact can be fabricated on the front-side, or the backside, of the chip.

### 16.4.1 Frequency Response of Lumped-Element Modulators

The simplest way for analyzing the frequency response of optical modulators is to use the equivalent circuit model. A concise RF equivalent circuit model can facilitate the design of modulator driver and the related RF link performance. As the modulator is considered as lumped-element, the variation of modulation voltage and phase over the waveguide length is ignored. Therefore, the electrical property of the device can be modeled by using a group of lumped circuit elements such as resisters, inductors, and capacitors. However, a valid circuit model should only consist of circuit parameters that carry physical meaning and can be accurately measured.

The key issue for EAM circuit model is to find a way to incorporate its photocurrent effect. Unlike the photocurrent in a photodetector, photocurrent in an EAM is a *current path*, rather than a current source. The current value in this photocurrent path varies with the junction voltage, so that the RF impedance of this photocurrent path can be modeled by a resistance $R_o = (dI_o/dV_J)^{-1}$, where $I_o$ is the EAM DC photocurrent and $V_J$ is the junction DC voltage. The resulted equivalent RF circuit model is depicted in Fig. 16.16, where $V_S$ is the RF source voltage, $Z_S$ is its internal impedance, $R_{SH}$ is an added shunt resistance for broader bandwidth, and $L_{w1}$ and $L_{w2}$ are inductance elements due to wire-bonding for packaged devices. The circuit model for a bared-chip EAM contains only 4 elements: the parasitic capacitance $C_P$, the junction capacitance $C_J$, the device series resistance $R_S$, and the photocurrent resistance

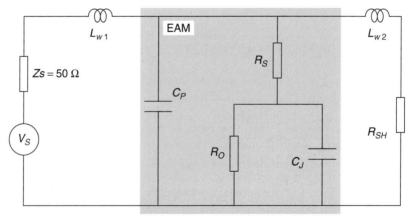

**Figure 16.16** RF equivalent circuit model for lumped-element EAMs. The same model can also work for QCSE-based lumped-element MZMs by replacing $R_o$ with an open-circuit (or setting $R_o = \infty$).

$R_o$ ($R_o$ can also include the junction leakage resistance that is usually $>1$ M$\Omega$). For devices with $C_J \gg 30$ fF, $C_P$ can be ignored. The three major physical parameters in this model, $C_J$, $R_S$, and $R_o$, can be separately calculated or measured using conventional methods: $C_J$ can be either measured from a C–V measurement or calculated using plate-capacitor approximation; $R_S$ can be determined either by measuring the slope of forward I–V curve at large current or by estimating the ohmic resistance at the electrodes plus the bulk resistance in the doped semiconductor layers; $R_o$ can be measured approximately from the slope of reverse I–V curve with optical power input to the EAM. $R_o$ is proportional to the optical power before DC saturation occurs. With detailed analysis of the electroabsorption characteristics, one can establish the theoretical relationship between the EAM photocurrent and the reverse bias voltage [57], from which $R_o$ can also be derived. However, one can also determine all of these parameters in a consistent manner from the $S_{11}$ data using the equivalent circuit model [58].

The above equivalent circuit model also works for QCSE-based lumped-element MZMs. In this case, as the photo-generated current is very small, $R_o$ can be replaced by an electric open-circuit (i.e., set $R_o = \infty$). The circuit model should be considered as the model for one arm of the MZM, and the frequency response for the two arms should be calculated separately if both arms are modulated. However, if the waveguide and electrode designs are symmetrical for the two arms, and if they are modulated in push-pull mode, then the frequency response for one arm is the same as that for the whole device.

The calculation of modulator $S_{11}$ using the above circuit model is straightforward in principle. To obtain a simple mathematical formula, we ignore $R_{SH}$, $L_{w1}$, $L_{w2}$, and $C_P$ in the circuit. This is the case when the device has no shunt resistance and wire-bonding, and $C_J$ is much larger than $C_P$. The microwave impedance for the modulator is then:

$$Z_m = R_S + \frac{R_o}{1 + j\omega C_J R_o} \tag{16.74}$$

where $\omega = 2\pi f$, and $f$ is the microwave frequency. The microwave reflection coefficient $S_{11}$ can be calculated as:

$$S_{11} = \frac{Z_m - 50}{Z_m + 50} \tag{16.75}$$

In the absence of any input optical power, $R_o$ is an open circuit, so that $Z_m$ is also an open circuit at low frequency. Therefore, $S_{11}$ always starts from 1 during the frequency sweep in the $S_{11}$ measurement for the bared chip.

As discussed in Section 16.3.3, the modulator frequency response is $M = (v_m/v_{\text{in}})^2$, where $v_m$ is the RF modulation voltage and $v_{\text{in}}$ is the forward-wave voltage in the 50-$\Omega$ transmission line connecting the RF source and the modulator. From the circuit model, we know that $v_m$ is the RF voltage across $C_J$, and $v_{\text{in}} = V_S/2$. Therefore, the absolute value of the modulator frequency response is (when no shunt resistor is used):

$$M(f) = \left| \frac{2}{1 + j\omega R_{\text{eff}} C_J} * \frac{R_o}{R_o + Z_S + R_S} \right|^2 \tag{16.76}$$

where $R_{\text{eff}} = R_o || (Z_S + R_S)$ represents the resistance of $R_o$ in parallel with $Z_S + R_S$. For MZM, or for EAM with very low input optical power, $R_o$ is very large, and

$R_{eff}$ is simplified to $Z_S + R_S$. In this case, $M = 4$ at $\omega = 0$ Hz. This is known as the voltage-doubling effect at the modulator due to the complete microwave reflection at low frequency. Equation (16.76) shows that the link RF gain monotonically decreases as the modulation frequency increases. From Eq. (16.76), it is easy to derive the 3-dB cutoff frequency, at which $M$ reduces to half compared to its DC value. The frequency range from 0 Hz to this cutoff frequency is defined as the 3-dB modulation bandwidth:

$$f_{3\,dB} = \frac{1}{2\pi R_{eff} C_J} \tag{16.77}$$

The above bandwidth is sometimes referred to as "electrical bandwidth" or "3-dBe bandwidth." This is to differentiate from "3-dB optical bandwidth," as is used by some authors. The 3-dB optical bandwidth, defined as the frequency range in which the modulated output *optical power* reduced to half value, is equivalent to 6-dB electrical bandwidth. It is significantly larger than the 3-dBe bandwidth.

Equation (16.77) clearly states that the modulation bandwidth is limited by the RC time constant. Shunt resistance is often used for a larger bandwidth and for smaller $S_{11}$ value. When a shunt resistance is used, the modulation frequency response becomes (assume no bonding-wire, and $R_o$ is replaced by an electric open-circuit):

$$M(f) = \left| \frac{2r}{1 + j\omega(R_S + rZ_S)C_J} \right|^2 \tag{16.78}$$

where $r = R_{SH}/(R_{SH} + Z_S)$. Therefore, the modulation bandwidth becomes:

$$f_{3\,dB} = \frac{1}{2\pi(R_S + rZ_S)C_J} \tag{16.79}$$

Typically, $R_{SH} = Z_S = 50\ \Omega$, so that $r = 0.5$. In this case, the bandwidth can be increased by nearly a factor of 2, provided that $R_S$ is very small. As a price of a broader bandwidth, the modulation frequency response $M(0)$ is reduced by 6 dB at low frequency when the 50-$\Omega$ shunt resistance is used, as can be seen from Eq. (16.78).

## 16.4.2  Lumped-Element EAM

For EAMs, high optical power makes the photocurrent resistance $R_o$ small. As a consequence, the 3-dB bandwidth will be increased, and the low-frequency gain will be compressed. As an example [58], Table 16.1 lists the measured values of circuit parameters for a MQW EAM at different optical powers. The device had a 180-$\mu$m-long waveguide, with a sharp transfer curve. $R_o$ was reduced from 1400 $\Omega$ at 0-dBm optical power to 180 $\Omega$ at 10-dBm optical power. Correspondingly, $R_{eff} = R_o||(Z_S + R_S)$ is reduced from 52 $\Omega$ to 42 $\Omega$. From Eq. (16.77), the 3-dB bandwidth is increased from 4 GHz to 4.6 GHz. This increase is mainly due to the decrease of $R_o$. On the other hand, from Eq. (16.76), we can derive that $M(0)$ is reduced from 1.9 to 1.5, which results in a low-frequency gain compression of $(1.9/1.5)^2 = 2.1$ dB. This gain-compression is purely due to the decrease of $R_o$ at high optical power, which indicates a new saturation mechanism other than the carrier-screening effect. The gain compression due to the carrier-screening effect was estimated to be 1 dB at 10-dBm optical power.

**TABLE 16.1  Extracted Circuit Parameters for an MQW EAM at Different Optical Powers**

| Optical Power | $R_S$ ($\Omega$) | $C_J$ (pF) | $R_o$ ($\Omega$) |
|---|---|---|---|
| 0  dBm | 3.6 | 0.77 | 1400 |
| 10  dBm | 3.8 | 0.81 | 180 |

Therefore, when the optical power was changed from 0 dBm to 10 dBm, the RF gain was increased by only 16.9 dB, instead of 20 dB, as expected from Eq. (16.54).

The above experimental results and circuit-model analysis show that the EAM bandwidth can be broadened at high optical power. However, high optical power also generates large amount of carriers, and they accumulate in the junction area. As a result, the effective depletion thickness is reduced, and the junction capacitance is increased. From Table 16.1, the increase of $C_J$ was small at 10-dBm optical power. If the optical power is further increased, $C_J$ can be increased drastically, and the bandwidth will start to decrease.

The above device uses very thin intrinsic layer ($\sim$0.1 $\mu$m), which is good for large slope efficiency in analog links. But the small overlap between the optical mode and the modulation layer limits the maximum extinction ratio ($<$10 dB per 100-$\mu$m electrode length), and thus, this may not be suitable for digital applications. The thin intrinsic layer also caused large waveguide capacitance, resulting in narrow modulation bandwidth. As stated in Section 16.3.2, to achieve wide bandwidth and high extinction ratio for digital applications, the EAM can be designed to have 0.2–0.3-$\mu$m intrinsic layer thickness and 2–3-$\mu$m waveguide width. Combined with other experimental results [59], we can derive $R_S \sim 1.0$ $\Omega$-mm and $C_J \sim 1.0$ pF/mm for the EAM waveguide with 0.3-$\mu$m intrinsic layer thickness and 2-$\mu$m waveguide width. The frequency responses for EAMs using the above structure are calculated and plotted in Fig. 16.17. In the calculations, we assumed $C_P = 30$ fF, $R_{SH} = 50$ $\Omega$, $R_o = 1$ M$\Omega$,

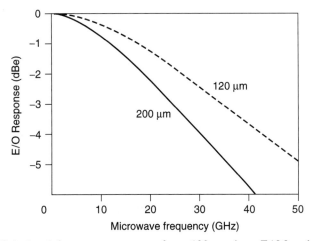

**Figure 16.17**  Calculated frequency responses for a 120-$\mu$m-long EAM and a 200-$\mu$m-long EAM, both devices have a 2-$\mu$m waveguide width and a 0.3-$\mu$m intrinsic layer thickness.

and $L_{w1} = L_{w2} = 0$. The 3-dB bandwidth is ~24 GHz for the 200-μm-long EAM and ~35 GHz for the 120-μm-long EAM. Therefore, the 200-μm-long EAM is ideal for 10-Gb/s application, which can provide a 30–40-dB extinction ratio with ~3-V voltage swing; while the 120-μm-long EAM is suitable for 40-Gb/s application, which provides ~20-dB extinction ratio with a ~3-V voltage swing. To achieve better chirp for the 10-Gb/s EAM, or for the 40-Gb/s EAM used in short-reach (<4 km) data communications, a few decibels of on-state optical power and extinction ratio may need to be compromised, as discussed at the end of Section 16.3.4. In this case, 0-dBm on-state optical power can still be achieved, and +3-dBm on-state optical power is also possible by proper MQW design for high saturation power and employing suitable fabrication techniques for low optical loss. Note that at high optical power, modulation voltage across $C_J$ may be lowered due to the small $R_o$ as well as the carrier screening effect. In this case, the dynamic extinction ratio measured from the eye-diagram may be significantly lowered with the same driver output. For analog applications, the above design may achieve equivalent $V_\pi$ of ~2 V for the 200-μm-long EAM, and ~3 V for the 120-μm-long EAM. The equivalent $V_\pi$ may also be increased at high optical power due to the optical saturation effects.

There are many published experimental demonstrations of wide-bandwidth lumped-element EAMs. In [60], 50-GHz bandwidth was achieved for an EAM with 63-μm active length, 0.23-μm intrinsic layer thickness, and 2.5 μm waveguide width. The whole device structure is illustrated in Fig. 16.18. With passive SSCs integrated at both ends, the total device length was 1.0–1.5 mm. The 12 active QWs had 8-nm $In_{0.48}Ga_{0.52}As$ well with −0.35% tensile strain, and 5-nm $In_{0.60}Al_{0.40}As$ barrier with 0.5% compressive strain, and the MQW photoluminescence was peaked at 1.48-μm wavelength. The device circuit parameters were measured as $C_J + C_P = 0.12$ pF, $R_S = 10\ \Omega$, and $L_{w1} = L_{w2} = 0.25$ nH. At 1.53-μm optical wavelength, this 50-GHz EAM achieved 15-dB DC extinction ratio when the modulator voltage was changed from 0 V to 2.8 V, with an 8.1-dB optical insertion loss at 0 V. Effective chirp parameter $\alpha_H$ was 0.7. The reported device has been well tested for 40-Gb/s modulation.

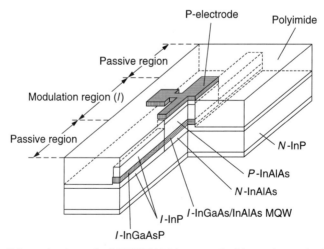

**Figure 16.18**   Schematic view of a MQW EAM integrated with passive optical spot-size converters. [60, Fig. 6.] © 1996 IEEE.

Another earlier work [61] also used 12-InGaAs/InAlAs QWs. The waveguide was designed to be 4 μm wide and 100 μm long, with 0.3-μm thick intrinsic layer. 40-GHz bandwidth was achieved, and the driving voltage was 6 V for 10-dB extinction ratio. InGaAsP MQW was designed in [62], and 40-GHz bandwidth was achieved. In [63], 11 periods of $InAs_{0.41}P_{0.59}/Ga_{0.13}In_{0.87}P$ MQW were designed for analog modulation at 1.32-μm wavelength. For a 100-μm-long device, measurement results indicated 38-GHz bandwidth, 3.2-V equivalent $V_\pi$, and 9-dB optical insertion loss without antireflection coating. With 16-mW input optical power, the link RF gain was measured at −32 dB, the link noise figure was 33 dB, and the spurious-free dynamic range was 105 dB-$Hz^{2/3}$.

The above references all used short waveguide length to achieve wide bandwidth. One can also use thicker intrinsic layer, or use smaller shunt resistance, to increase the bandwidth. However, all of these approaches result in penalty in modulation efficiency. A better way for improving the bandwidth is to use the inductance or transmission line effect, which may not compromise the RF efficiency. When calculating the frequency response in Fig. 16.17, the inductance values of both $L_{w1}$ and $L_{w2}$ were assumed zero. However, properly designed $L_{w1}$ and $L_{w2}$ can substantially increase the modulation bandwidth. With 50-Ω shunt resistance for the previous 120-μm-long EAM as analyzed in Fig. 16.17, the effect of inductance is shown in Fig. 16.19. With $L_{w1} = L_{w2} = 0$, the 3-dB bandwidth is 35 GHz; with $L_{w1} = 0.1$ nH and $L_{w2} = 0.15$ nH, the bandwidth is increased to 53 GHz. This large improvement of bandwidth does not require any compromise of the modulation efficiency. However, further increasing $L_{w1}$ and $L_{w2}$ will not further improve the bandwidth; instead, it causes a hump in the frequency response, which is not desirable for broadband requirement. Although the inductance value of a bonding wire is hard to be controlled accurately, a moderate bandwidth improvement can still be expected. The resulted bandwidth improvement allows sufficient room to tolerate bandwidth reductions due to packaging [60] and due to high optical power effect.

A short-length transmission line can have similar effect to a small value of inductance. The input impedance for a lossless transmission line terminated by 50 Ω is:

$$Z_{in} = Z_o \frac{50 + jZ_o \tan \omega\tau}{Z_o + j50 \tan \omega\tau} \tag{16.80}$$

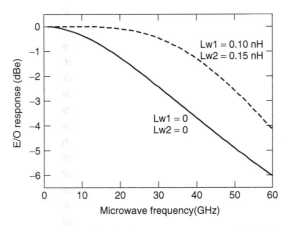

**Figure 16.19**  Inductance effect on the frequency response of the 120-μm-long EAM.

where $Z_o$ is the transmission line impedance and $\tau$ is the transmission line length in electrical delay. For very short transmission line (i.e., $\omega\tau \ll 1$) and with $Z_o > 50\ \Omega$, Eq. (16.80) can be approximated as:

$$Z_{\text{in}} = 50 + j\omega\tau Z_o \qquad (16.81)$$

The impedance in Eq. (16.81) is equivalent to an inductance (with value $\tau Z_o$) plus a 50-$\Omega$ resistance. This effect can be utilized to replace $L_{w2}$ with a short-length transmission line. With equivalent inductance value of 0.15 nH, the transmission line (with 100 $\Omega$ impedance) should be 1.5 ps long, which corresponds to $\sim$200-$\mu$m length of CPW line on top of a semi-insulating InP substrate. This reveals a practical approach for integrating the transmission line with the L-EAM device, and the 50-$\Omega$ thin-film shunt resistor can be soldered at the end of the transmission line.

### 16.4.3  Lumped-Element QCSE-Based MZM

QCSE-based MZMs also use MQW as active modulation layer. As their detuning energy is typically very large, optical absorption in the MQW layers is typically much smaller than that in the MQW layer of the EAMs. The field-induced index change increases with the applied electric field, but not as steep as the increase of the optical absorption coefficient. This is consistent with the fact that the chirp parameter $\alpha_H$ of an EAM decreases with increased bias voltage. Due to this reason, also because of no carrier-escape issue, the active MQW layer of an MZM is usually thicker than that of an EAM. The most popular design is to use 20–30 QWs, with a total intrinsic layer thickness of $\sim$0.4 $\mu$m. The waveguide width is often designed as 2 $\mu$m. To achieve $V_\pi < 5$ V for single-arm modulation, 600-$\mu$m modulation length is required. The electrode design for each arm is similar to that for a lumped-element EAM. A 20-GHz microwave signal propagating in this kind of waveguide will have a wavelength of $\sim$3 mm, which is $\sim$10 times of the half waveguide length. Therefore, the device can be treated as lumped element for microwave frequency up to 20 GHz.

The same equivalent circuit model in Fig. 16.16 also works for lumped-element QCSE-based MZMs. Based on the circuit parameter values in last section, we can derive $C_J \sim 0.75$ pF/mm and $R_S \sim 1.0\ \Omega$-mm for the waveguide with 2 $\mu$m width and 0.4-$\mu$m intrinsic layer thickness. Assuming $C_P = 30$ fF, $R_{SH} = 50\ \Omega$, and $R_o = 1$ M$\Omega$, the frequency response for a QCSE-based MZM with 600-$\mu$m-long modulation length is plotted in Fig. 16.20. Again, the inductance effect can be used to improve the modulation bandwidth. With $L_{w1} = 0.1$ nH and $L_{w2} = 0.5$ nH, the modulation bandwidth is increased from $\sim$13 GHz to $\sim$19 GHz. In general, this type of MZM is hard to achieve more than 20-GHz bandwidth while keeping $V_\pi < 5$ V for single-arm modulation. In digital applications, lumped-element QCSE-based MZM is suitable for 10-Gb/s modulation, but it is hard to meet the requirements of 40 Gb/s with a small $V_\pi$.

Many research works on QCSE-based MZMs have been published. Table 16.2 lists the summary of some published results. Among these works, [64] achieved the widest bandwidth (15 GHz), the highest extinction ratio (18.6 dB at DC), and the smallest optical insertion loss (13 dB). $V_\pi$ for single-arm modulation was 4.5 V. The device structure of [64] is illustrated in Fig. 16.21. Using the same device, 10-Gb/s modulation had been demonstrated in the push-pull mode. With only 2-V voltage-swing per arm (i.e., from $-2$ V to 0 V for one arm, and from $-2$ V to $-4$ V for the other arm), very clear eye-diagram was achieved with a dynamic extinction ratio of 11.5 dB.

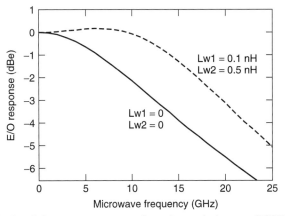

**Figure 16.20**  Calculated frequency response for a lumped-element QCSE-based MZM with 600-μm modulation length, 2-μm waveguide width, and 0.4-μm intrinsic layer thickness. A 50-Ω shunt resistor is used, and 30-fF pad capacitance is included. The effect of inductance is also shown.

**TABLE 16.2  Published Research Works on QSCE-based MZM (Except [66], which was a Phase Modulator). All Parameters are from Measurements. The Cited $V_\pi$ is for Single-arm Modulation**

|  | QW Structure | $\lambda_0$ (μm) | Detuning (meV) | Length (mm) | Thickness (μm) | Width (μm) | $V_\pi$ V | ER (dB) | Optical Loss (dB) | $f_{3\,dB}$ (GHz) |
|---|---|---|---|---|---|---|---|---|---|---|
| [64] | 20 periods InGaAsP/InP 100 Å + 100 Å | 1.56 | 60 | 0.6 | 0.4 | 2 | 4.5 | 18.6 | 13 | 15 |
| [65] | as above | 1.56 | 48.6 | 0.6 | 0.4 | 2 | 4.3 | 13 | N/A | N/A |
| [66] | 30 periods InGaAs/InAlAs 75 Å + 50 Å | 1.554 1.60 | ~40 ~60 | 0.3 | 0.4 | 4 | 2.5 5.5 | PM | N/A ~15 | 10 |
| [67] | 20 periods InGaAs/InAlAs 60 Å + 60 Å | 1.55 1.59 | 43.5 63.7 | 1.45 | 0.25 | 2 | 1.2 2.0 | >15 | 25 20 | N/A |

It should be noted that the results in [66] for a phase modulator are also remarkable. With only 300-μm-long modulation length, the device achieved $V_\pi \sim 5.5$ V at 60-meV detuning energy. This is mainly attributed to the advantage of InGaAs/InAlAs MQW as well as the thinner QW barrier. Using the same active layer design with ~200-μm modulation length and 2-μm waveguide width, the MZM could achieve $V_\pi \sim 5$ V in push-pull modulation, and its modulation bandwidth can be >30 GHz with the aid of inductance effect. This could meet the requirement of 40-Gb/s modulation. The large optical loss (15 dB) in [66] was attributed to 3-dB facet reflection loss, 10-dB mode coupling loss, and 2-dB propagation loss. The mode coupling loss can be greatly

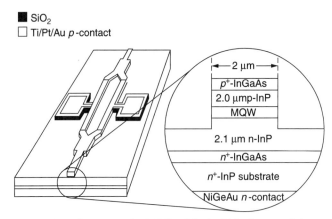

**Figure 16.21**   Schematic diagram of MQW Mach–Zehnder modulator. [64, Fig. 1.] © 1993 IEEE.

reduced by integrating passive SSCs at both ends (this is also true for the device in [64]). The optical propagation loss in the active modulation length is ~5 dB/mm for this type of device, among which ~3 dB/m is due to the free-carrier absorption in the $p$-type layers. Waveguide scattering loss can be minimized by improved fabrication technique, and residual optical absorption in the active layer should be very small at 60-meV detuning energy.

The design of optical splitter and combiner is very critical to the performance of MZM. It affects both the optical insertion loss and the extinction ratio. Multimode interference (MMI) couplers are regarded as the best structure for these purposes [68]. Properly designed $1 \times 2$ and $2 \times 2$ MMIs can work as excellent optical splitter and combiner, with very small optical loss and high extinction ratio. They also have much tolerance to the fabrication error and wavelength deviation [68]. For the device shown in Fig. 16.21, the Y-junction splitter and combiner have a triple-mode section (a 40-μm-long and 4-μm-wide ridge waveguide), followed by two 100-μm-long S-bends separated by 20 μm. To improve the extinction ratio, a thin 0.2-μm $n$-InGaAs absorbing layer was grown between the $n$-InP cladding layer and the substrate to absorb stray light and attenuate high-order modes.

The major limiting factor to the extinction ratio of QCSE-based MZM is the field-induced absorption change. During modulation, no matter it is single-arm or push-pull, the instantaneous voltages on the two arms are different, resulting into different optical loss in the two arms. This can cause a few decibels of optical power difference in the two arms, and limit the maximum extinction ratio, as shown in Fig. 16.12. The optical power difference at the two arms also deteriorates the device's chirp parameter. To improve the chirp parameter for single-arm modulation, $\pi$-phase-shifted waveguide design is often used [54, 65]. Detailed analysis of the device's chirp parameter and extinction ratio as functions of DC bias voltages at the two arms can be found in [69].

## 16.5   TRAVELING-WAVE MODULATORS

MZMs based on linear EO effect typically have a modulation length as long as several centimeters due to the small EO coefficients for various EO materials. For MZMs with wide bandwidth, the modulation length is generally much longer than the microwave wavelength at the 3-dB cutoff frequency. In this case, the microwave voltage varies along the waveguide during propagation, and the modulator electrode can no longer be treated as a lumped-element. In general, traveling-wave design is needed for modulators with long modulation length.

In a traveling-wave modulator, the electrode is designed as a transmission line that typically runs parallel to the optical waveguide. The transmission line is connected to the microwave source at one end, and it is terminated by a matched impedance at the other end, so that microwave propagates mainly toward the forward direction, and the backward-going microwave is largely suppressed. In addition, the forward-going microwave should keep in pace with the modulated optical wave packet, so that the modulation is always being enhanced during the propagation. This requires matching the microwave phase velocity with the optical group velocity [70]. The microwave loss in the transmission line should also be minimized, because it generally increases with the modulation frequency; thereby it is detrimental to the modulation bandwidth. For a traveling-wave modulator, the modulation bandwidth is limited not by the RC-time constant, but by the above-mentioned impedance mismatch, velocity mismatch, and microwave loss.

There are two major types of traveling-wave modulators. One type has microwave transmission line electrode fabricated directly on top of the optical waveguide, and the optical wave is modulated continuously during its propagation. The other type has microwave transmission line electrode separated from the optical waveguide, and the optical wave is periodically modulated during its propagation by the segmented modulation sections. The continuous traveling-wave design has been applied to the traveling-wave EAMs, the traveling-wave $LiNbO_3$ modulators, and the traveling wave polymer modulators. The segmented traveling-wave design has been applied to the traveling-wave III–V bulk MZMs, and it can also be applied to the traveling-wave QCSE-based MZMs and the traveling-wave EAMs.

### 16.5.1  Frequency Response of Continuous Traveling-Wave Modulators

Figure 16.22 is a schematic diagram showing the operation of a continuous traveling-wave modulator. The optical waveguide and the microwave transmission line are represented by the same line from $x = 0$ to $x = L$. Microwave signal and optical carrier are input into the device at the same end of the waveguide, and they copropagate toward the same direction. In this diagram, $\Gamma_S$ and $\Gamma_L$ are the *internal* microwave reflection coefficients in the modulator at the source port ($x = 0$) and the terminator port ($x = L$), respectively; $T$ is the amplitude transmission coefficient at the source port. These coefficients are very important parameters because they represent the quality of impedance matches. Their values depend solely on the modulator transmission

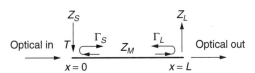

**Figure 16.22**  A schematic diagram to show the operation of a continuous traveling wave modulator.

line impedance $Z_M$ after the source impedance $Z_S$ and the terminator impedance $Z_L$ are determined. They can be calculated by:

$$\Gamma_S = \frac{Z_S - Z_M}{Z_S + Z_M} \quad \Gamma_L = \frac{Z_L - Z_M}{Z_L + Z_M} \quad T = \frac{2Z_M}{Z_S + Z_M} = 1 - \Gamma_S \qquad (16.82)$$

In general, the impedance matches cannot be perfect; thus, $\Gamma_S$ and $\Gamma_L$ are nonzero, and multiple microwave reflections at the two ports are resulted. The microwave voltage on the waveguide can be expressed as [71]

$$v_{ac}(x, t) = \frac{v_{in} T e^{j\omega t}}{1 - \Gamma_L \Gamma_S e^{-2\gamma_\mu L}} \, (e^{-\gamma_\mu x} + \Gamma_L e^{\gamma_\mu x - 2\gamma_\mu L}) \qquad (16.83)$$

where $v_{in}$ is the forward microwave voltage in the source transmission line, $\omega$ is the microwave frequency, and $\gamma_\mu = \alpha_\mu + j\beta_\mu$, where $\alpha_\mu$ is the microwave amplitude-attenuation coefficient and $\beta_\mu = \omega/v_\mu$ is the wave number associated with the microwave phase velocity $v_\mu$. The denominator in Eq. (16.83) is the result of multiple reflections, the first term in the bracket represents the forward-going wave, and the second term represents the backward-going wave.

As an optical wave packet begins to propagate from the microwave source port at time $t_0$ with group velocity $v_o$, it arrives at position $x$ inside the modulator at time $t = t_0 + x/v_o$. This implies that the modulation voltage for this optical wave packet at arbitrary position $x$ inside the waveguide is $v_m(x) = v_{ac}(x, t = t_0 + x/v_o)$, i.e.,

$$v_m(x) = \frac{v_{in} T e^{j\omega t_0}}{1 - \Gamma_L \Gamma_S e^{-2\gamma_\mu L}} \, (e^{(j\beta_0 - \gamma_\mu)x} + \Gamma_L e^{(j\beta_0 + \gamma_\mu)x - 2\gamma_\mu L}) \qquad (16.84)$$

where $\beta_o = \omega/v_o$. As discussed in Section 16.3.3, the modulator frequency response is $M = (v_m/v_{in})^2$. In the case of the traveling-wave modulator, it has to be integrated over the entire modulation length from $x = 0$ to $x = L$. The resulted absolute value of the modulator frequency response is

$$M(f) = \left| \frac{T}{1 - \Gamma_L \Gamma_S e^{-2\gamma_\mu L}} \left\{ \frac{e^{(j\beta_o - \gamma_\mu)L} - 1}{(j\beta_o - \gamma_\mu)L} + \Gamma_L e^{-2\gamma_\mu L} \frac{e^{(j\beta_o + \gamma_\mu)L} - 1}{(j\beta_o + \gamma_\mu)L} \right\} \right|^2 \qquad (16.85)$$

The above equation for frequency response includes the effects of impedance mismatch, velocity mismatch, and microwave loss. For an ideal traveling-wave design with perfect impedance match, perfect velocity match, and zero microwave loss, we have $\Gamma_L = \Gamma_S = 0$, $T = 1$, $\beta_\mu = \beta_o$, and $\alpha_\mu = 0$. In this case, $M(f) \equiv 1$, and the modulation bandwidth is infinitely wide. To evaluate the $M(f)$ for a practical traveling-wave modulator, we have to first quantify the impedance $Z_M$ and the propagation constant $\gamma_\mu = \alpha_\mu + j\beta_\mu$.

### 16.5.2   Traveling-Wave EAM

In Section 16.4.2, we have shown that 40–50-GHz bandwidth can be achieved with a short-length ($\sim$100 µm) lumped-element EAM. However, its modulation efficiency may not be so satisfactory for some applications. Wide-bandwidth analog links may require an equivalent $V_\pi$ smaller than 4 V; 40-Gb/s digital systems may desire a DC

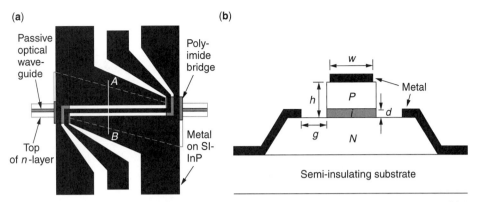

**Figure 16.23**   Schematic drawings for TW-EAM design. **(a)** is the device top view, and **(b)** is the cross-section view along the line $A - B$ in **(a)**.

extinction ratio larger than 20 dB with a small voltage swing. When the system requirements are pushed further, EAMs with wider bandwidth and better modulation efficiency are desirable. In this case, the traveling-wave EAMs (TWEAM) will become increasingly more important.

Figure 16.23 shows a schematic design for a continuous TWEAM. A set of tapered 50-Ω transmission lines are designed at the microwave source port and terminator port to facilitate the microwave probing and wire-bonding. The microwave guide in the modulation section is a slow-wave transmission line, in which the magnetic field is confined by the metal electrodes, whereas the electric field is confined mainly in the semiconductor intrinsic layer. Unfortunately, in this kind of slow-wave transmission line, the microwave impedance is frequency dependent, and it is a complex impedance (due to lossy transmission line) with real part of only 20–30 Ω [72], which is much less than the commonly used 50 Ω. These facts make the impedance-match difficult. The TWEAM electrode also has very slow microwave velocity with phase velocity index ~7, which is a big mismatch with the optical group index ~3.4 [73]. The frequency-dependent microwave loss in the TWEAM transmission line is also exceptionally large [72]. All of these undesirable factors compound the difficulty of TWEAM design. Fortunately, it does not require very long modulation length for a TWEAM to be very efficient. A 200-μm-long TWEAM will improve the modulation efficiency a great deal compared to the 100-μm-long lumped-element EAM. The short TWEAM length makes the velocity mismatch unimportant, and large microwave loss is also tolerable. The only challenge left is to match the impedances. Equation (16.85) shows that if the TWEAM terminator port is well matched so that $\Gamma_L = 0$, the impedance mismatch at the source port will not affect the modulation bandwidth. This infers that low-impedance (20–30 Ω) termination can result in ultrawide bandwidth for a short-length TWEAM.

To evaluate the frequency response for a TWEAM, we start from analyzing the impedance $Z_M$ and the propagation constant $\gamma_\mu$ using an equivalent RF circuit model. If a finite length of TWEAM transmission line is uniformly divided into an infinitive number of small segments, then each small segment can be represented by a lumped-element model [71], as shown in Fig. 16.24. This small-segment lumped-element model relates the microwave properties of the TWEAM transmission line directly to the

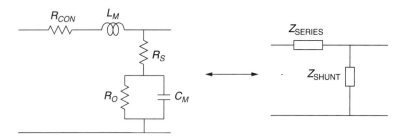

**Figure 16.24** Small-segment lumped-element model for a TWEAM transmission line.

structure design, the material properties, and the fabrication process, which thus can provide more insight about the device. In the circuit model shown in Fig. 16.24, $R_{\text{CON}}$ is the conduction resistance, $L_M$ is the inductance, $R_S$ is the device series resistance, $C_M$ is the junction capacitance, and $R_o = (dI_o/dV_J)^{-1}$ is the differential resistance due to the dependence of the photocurrent $I_o$ on the modulator junction voltage $V_J$. Parasitic capacitance can also be included into this model, although it is typically very small. To facilitate the analysis, we define a series impedance $Z_{\text{SERIES}}$ and a shunt impedance $Z_{\text{SHUNT}}$ for the transmission line model in Fig. 16.24:

$$Z_{\text{SERIES}} = R_{\text{CON}} + j\omega L_M \quad Z_{\text{SHUNT}} = R_S + \frac{R_o}{1 + j\omega R_o C_M} \tag{16.86}$$

From the transmission line theory [71], $Z_M$ and $\gamma_\mu$ can be obtained from:

$$Z_M = \sqrt{Z_{\text{SERIES}} Z_{\text{SHUNT}}} \quad \gamma_\mu = \alpha_\mu + j\beta_\mu = \sqrt{Z_{\text{SERIES}}/Z_{\text{SHUNT}}} \tag{16.87}$$

Note that $Z_{\text{SERIES}}$ has a unit of $\Omega/\text{mm}$, and $Z_{\text{SHUNT}}$ has a unit of $\Omega\text{-mm}$. As the modulation is contributed only by the voltage across the junction capacitance, the voltage drop across $R_S$ should be factored out. Therefore, the frequency response for TWEAM has to be modified as

$$M(f) = \left| \frac{T}{1 - \Gamma_L \Gamma_S e^{-2\gamma_\mu L}} \left\{ \frac{e^{(j\beta_o - \gamma_\mu)L} - 1}{(j\beta_o - \gamma_\mu)L} + \Gamma_L e^{-2\gamma_\mu L} \frac{e^{(j\beta_o + \gamma_\mu)L} - 1}{(j\beta_o + \gamma_\mu)L} \right\} \times \frac{Z_{\text{JUNCT}}}{Z_{\text{SHUNT}}} \right|^2 \tag{16.88}$$

with $Z_{\text{SHUNT}}$ given by Eq. (16.86), and the junction impedance $Z_{\text{JUNCT}}$ equals to $R_o/(1 + j\omega R_o C_M)$. When a short-length TWEAM is open-terminated, it becomes a lumped element EAM, and the frequency response calculated using Eq. (16.88) must agree to the response calculated using Eq. (16.76). This has been verified in [72].

It should be emphasized that only when a TWEAM is both short-length and *open-terminated* is it equivalent to a lumped-element EAM, and the frequency response in Eq. (16.88) can be approximated to the form of Eq. (16.76) only when the conditions $\Gamma_L = 1$, $|\gamma_\mu L| \ll 1$, and $\beta_o L \ll 1$ are all applied. In other words, if a short-length TWEAM is terminated with *matched-impedance* so that $\Gamma_L \approx 0$, the resulted frequency response will be completely different from the response of lumped-element EAM, and its bandwidth will no longer be subject to the RC-time limit.

To show the advantage of a low-impedance terminated TWEAM, we can compare the frequency response of a 200-μm-long TWEAM terminated by $Z_L = 25\ \Omega$ with the frequency response of a 200-μm-long lumped-element EAM shunted by $R_{SH} = 25\ \Omega$. If their optical waveguides are designed the same, their low-frequency responses and efficiencies will be the same; the only difference is their bandwidths. Assuming their waveguides use 0.3-μm intrinsic layer thickness and 2-μm waveguide width, we can derive $R_S \sim 1.0\ \Omega$-mm and $C_J \sim 1.0$ pF/mm, same as the values used in Section 16.4.2. If the metal width is $\sim 3$ μm, according to [59], we will have $L_M = 0.40$ nH/mm and $R_{\text{CON}} = 7.3\ \Omega$-mm$^{-1}$ GHz$^{-1/2}$. Also, assume $R_o \sim 10^6\ \Omega$-mm at low optical power. Using these circuit parameters, the TWEAM frequency response can be calculated by combining Eqs. (16.86), (16.87), (16.82), and (16.88). Frequency response for the lumped-element EAM can be calculated using Eq. (16.78). The results are shown in Fig. 16.25. The 3-dB bandwidth for the TWEAM is 50 GHz, whereas the bandwidth for the lumped-element EAM is only 35 GHz. Their low-frequency efficiencies are the same, except for the 15-GHz bandwidth difference.

The penalty due to the 25-$\Omega$ low-impedance termination is a 3.5-dB reduction of response at low frequency, as shown in Fig. 16.25. This penalty can be reduced by using the transmission-line effect at the terminator port. When a 2-ps-long 50-$\Omega$ transmission line is inserted between the modulation electrode and the terminator, a 30-$\Omega$ terminator impedance can be used and still achieve 50-GHz bandwidth, as shown in Fig. 16.26. In this case, the efficiency penalty is reduced to 2.5 dB. As a comparison, without the 2-ps-long terminator transmission line, the bandwidth is 12 GHz less, as shown in Fig. 16.26. By increasing the terminator impedance from 25 $\Omega$ to 30 $\Omega$, the modulator driver current is also reduced. In addition, [72] showed that increasing the modulator waveguide inductance $L_M$ can improve the TWEAM performance, leading to wider bandwidth or less efficiency penalty. High optical power effects in TWEAM is similar to that in lumped-element EAM [72].

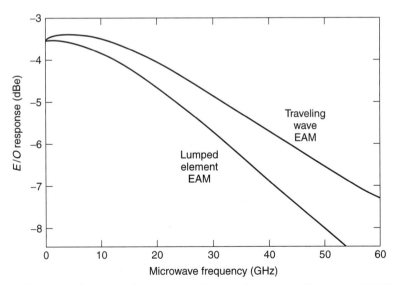

**Figure 16.25** Comparison of frequency responses for a traveling-wave EAM and a lumped-element EAM. Both of them use the same optical waveguide and a 200-μm-long modulation length. The TWEAM is terminated by 25-$\Omega$ impedance, and the lumped-element EAM is shunted by 25-$\Omega$ resistor.

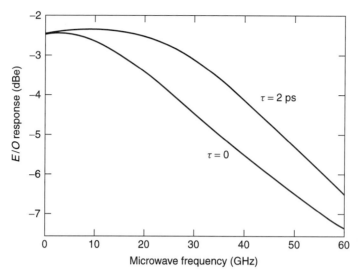

**Figure 16.26**  Frequency responses for a 200-μm-long TWEAM with and without the effects of the 50-Ω terminator transmission line. $\tau$ is the transmission line length. The terminator impedance in both cases are 30 Ω.

Experimental demonstrations of TWEAM have been reported by many researchers [59, 74–76]. Bandwidth beyond 40 GHz was achieved in [59] for a 150-μm-long FKE-based TWEAM using a 26-Ω terminator; for a 200-μm-long TWEAM, the bandwidth was 35 GHz, with equivalent $V_\pi$ of 2.4 V and optical insertion loss of 11 dB, and the optical saturation power was measured at 45 mW. The MQW absorption layer was used for a 330-μm-long TWEAM in [74], in which 40-dB extinction ratio was achieved with voltage swing from 0 V to 2 V. The optical loss was ~18 dB at 0 V, the optical saturation power was measured >14 dBm, and the frequency response was measured up to 20 GHz using a 35-Ω terminator, with 1.5-dB drop at 20 GHz.

### 16.5.3  Traveling-Wave LiNbO₃ Modulator

In order to make an efficient ($V_\pi < 5$ V) LiNbO₃ modulator, at least a 2-cm modulation length is needed. Therefore, traveling-wave electrode must be designed for wide bandwidth. Two types of transmission lines have been used for the traveling-wave LiNbO₃ modulators. One is the asymmetric coplanar strip (A-CPS) or asymmetric strip line (ASL), and the other is the coplanar waveguide (CPW). These electrode configurations on $x$-cut and $z$-cut LiNbO₃ crystals are shown in Fig. 16.27 [77]. It has been shown by different authors [77–79] that CPW design is preferred for modulators fabricated on $z$-cut crystal. If the CPW or the ASL transmission line is designed on the flat top surface of a LiNbO₃ substrate, with very thin metal, it will cause about half of the microwave to propagate in air, and half in the LiNbO₃ substrate. Therefore, the effective microwave velocity index is [71]

$$n_\mu \approx \sqrt{\frac{1 + \varepsilon_{eff}}{2}} \tag{16.89}$$

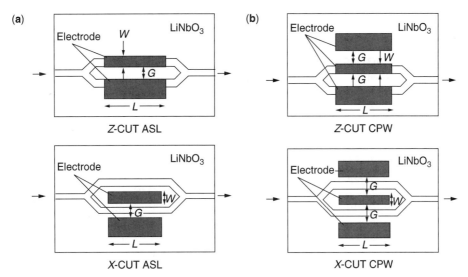

**Figure 16.27**   (a) Asymmetric strip line (ASL) configurations on $x$-cut and $z$-cut LiNbO$_3$ crystals. (b) Co-planar waveguide (CPW) configurations on $x$-cut and $z$-cut LiNbO$_3$ crystals. [77, Fig. 1.] © 1998 IEEE.

where $\varepsilon_{eff}$ is the effective relative dielectric constant in LiNbO$_3$. Note that the relative dielectric constant in LiNbO$_3$ is different at different orientations, i.e., $\varepsilon_z = 28$ and $\varepsilon_x = \varepsilon_y = 43$. For a transmission line running along the $y$-direction, $\varepsilon_{eff} \approx \sqrt{\varepsilon_x \varepsilon_z} \approx 35$ [80]; thus, $n_\mu \sim 4.2$ can be derived from Eq. (16.89). However, the optical group velocity index in a Ti-diffused LiNbO$_3$ waveguide is 2.15, which is much faster than the microwave phase velocity. Therefore, the microwave phase velocity needs to be increased so that it can match with the optical group velocity. In addition, to achieve high efficiency, the electrode gap needs to be small, which often causes the CPW transmission line impedance to be smaller than 50 Ω.

If we use a similar circuit model as in Fig. 16.24 for the transmission line in LiNbO$_3$ modulator and set $R_S = 0$ and $R_o = \infty$, so that only three circuit parameters ($R_{con}$, $L_M$ and $C_M$) are left, then we can derive from Eqs. (16.86) and (16.87) that

$$Z_M = \sqrt{L_M/C_M}, \quad n_\mu = c\sqrt{L_M C_M}, \quad \alpha_\mu = \frac{R_{CON}}{2Z_M} \qquad (16.90)$$

In the derivation, we assumed $R_{con} \ll \omega L_M$. From the above equations, we can see that to reduce the microwave velocity index $n_\mu$, we have to decrease either the capacitance per unit-length $C_M$ or the inductance per unit-length $L_M$. Three methods can be used to reduce $n_\mu$.

The first method is to use a SiO$_2$ buffer layer between the metal electrodes and the LiNbO$_3$ substrate. The SiO$_2$ layer has a relative dielectric constant of 3–4, which is considerably lower than that of LiNbO$_3$. This will reduce $C_M$ without affecting $L_M$ and $R_{CON}$; thus, it will simultaneously increase the transmission line impedance $Z_M$, reduce the microwave phase velocity index $n_\mu$, and reduce the microwave loss $\alpha_\mu$. Another function of the SiO$_2$ buffer layer is to isolate the optical mode from the metal electrodes, because the SiO$_2$ layer has an optical refractive index ($\sim$1.5) smaller than that of LiNbO$_3$. This is very critical in order to keep the optical propagation loss low. However, adding the SiO$_2$ buffer layer causes part of voltage drop across it; thus, it reduces the electric field at the optical waveguide and reduces modulation efficiency. Therefore, this SiO$_2$ buffer layer cannot be too thick; typically, it is less than 2 μm.

The second method is to use thick metal for the electrodes. This causes a large portion of the microwave to propagate in the air; thus, $n_\mu$ can also be reduced. Thick metal actually increases $C_M$; therefore, the reduction of $n_\mu$ is attributed to the decrease of $L_M$. The increased $C_M$ and decreased $L_M$ result in decreased $Z_M$, which may make the impedance mismatch more severe if CPW electrode is used. Therefore, the electrode cannot be too thick, typically, ~20 μm. Such a thick metal is usually fabricated using electrical plating method. It has been shown [81] that to achieve impedance match and velocity match simultaneously by using thick metal in CPW transmission line, the center electrode has to be pretty narrow (~4 μm), and the resulted large microwave loss makes it hard to achieve a wide modulation bandwidth.

The third method is to use ridge structure. By etching waveguide ridge, as illustrated in Fig. 16.28, part of the LiNbO$_3$ material in the electrode gaps is replaced by the SiO$_2$ or air, thus $C_M$ is also reduced, without affecting $L_M$ and $R_{CON}$. Therefore, ridge structure can also simultaneously decrease $n_\mu$, increase $Z_M$, and reduce $\alpha_\mu$. Unlike the effect of the SiO$_2$ buffer layer, the ridge structure may not reduce the electric field in the optical waveguide; instead, it confines more electric field under the ridge due to its larger dielectric constant. This increases the vertical electric field and thus decreases $V_\pi$ for LiNbO$_3$ modulator on $z$-cut substrate. The LiNbO$_3$ waveguide ridge can be

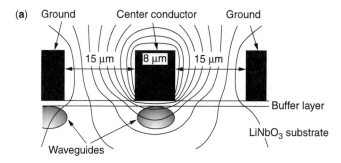

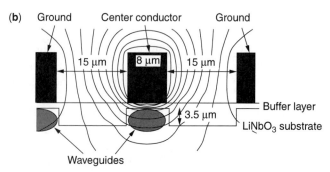

**Figure 16.28**  Cross-sectional view of $z$-cut LiNbO$_3$ modulator using CPW configuration, with the potential distribution around the center electrode of **(a)** planar structure and **(b)** ridge structure. [82, Fig. 3.] © 1993 IEEE.

formed using electron-cyclotron-resonance (ECR) etching with argon and $C_2F_2$ gases. Ti film can be used as etching mask.

Equation (16.85) is still valid for calculating the frequency response of LiNbO$_3$ modulators. However, the transmission line inductance $L_M$ and capacitance $C_M$ cannot be estimated in a simple manner as we did for TWEAM. A quasi-static finite-element method had been used to numerically analyze the performance of LiNbO$_3$ modulators [79]. In the analysis, the electric potential distribution was calculated first from Laplace's equation using finite-element method, and then the capacitances $C_0$ and $C_M$ are obtained by using Gauss's theorem [79], where $C_0$ is the capacitance per unit length for the same CPW transmission line but with all dielectric materials replaced by free-space. In this new CPW line, the microwave velocity is the same as the light velocity $c$ in free-space, and its inductance is the same as $L_M$ of the original CPW line; therefore, $L_M = (c^2 C_0)^{-1}$. Combining with Eq. (16.90), we have

$$Z_M = \frac{1}{c\sqrt{C_0 C_M}}, \quad n_\mu = \sqrt{C_M/C_0}, \quad \alpha_\mu = \frac{R_{CON}}{2Z_M} \qquad (16.91)$$

In [79], $\alpha_\mu$ was calculated using the incremental inductance formula. The calculated $Z_M$, $n_\mu$ and $\alpha_\mu$ can then be substituted into Eqs. (16.82) and (16.85) to obtain the modulator frequency response. The modulator bandwidth is a function of ridge height $t_r$, electrode metal thickness $t_m$, SiO$_2$ buffer thickness $t_b$, electrode gap $G$, and the center electrode width $W$. $V_\pi$ can also be calculated from the overlap integral between the optical-wave and the microwave fields. Figure 16.29a shows the calculated $n_\mu$, $Z_M$, and $\alpha_0$ ($\alpha_\mu$ at $f = 1$ GHz, and $\alpha_\mu \propto f^{1/2}$) as functions of ridge height $t_r$ with predetermined $W = 8$ μm, $G = 15$ μm, $t_b = 1.2$ μm, and $t_m = 10$ μm. It is seen that at $t_r = 3$–4 μm, both impedance match and velocity match can be achieved if CPW electrode is used. Figure 16.29b shows the calculated product of $V_\pi$ and modulation length $L$ as a function of ridge height. The product $V_\pi L$ reaches a minimum at $t_r = 3$–4 μm, and $V_\pi L$ for modulators using CPW electrode is always smaller than that of modulators using A-CPS electrode. Figure 16.29c shows that the modulator's microwave characteristics strongly depend on the SiO$_2$ buffer layer thickness $t_b$ and the electrode metal thickness $t_m$. $V_\pi L$ product linearly increases with $t_b$ but is independent of $t_m$.

Using the ridge structure with SiO$_2$ buffer layer and thick electrode metal, high-performance LiNbO$_3$ modulators have been demonstrated. In [82], 75-GHz optical bandwidth (~35-GHz electrical bandwidth), and $V_\pi \sim 5$ V at 1.5-μm optical wavelength was achieved for a LiNbO$_3$ modulator with $L = 2$ cm, $t_r = 3.3$ μm, $t_m = 10$ μm, $t_b = 1.2$ μm, $G = 15$ μm, and $W = 8$ μm. In [83], $G$ was increased to 25 μm, $t_m$ was increased to 29 μm, $t_b$ was decreased to 0.6 μm, and $t_r = 3.6$ μm, $W = 8$ μm. With modulation length $L = 3$ cm, 30-GHz electrical bandwidth and $V_\pi = 3.5$ V at 1.55-μm wavelength were achieved, DC extinction ratio was 20 dB; when L was decreased to 2 cm, the electrical bandwidth and $V_\pi$ became 70 GHz and 5.1 V. LiNbO$_3$ modulator with 40-GHz bandwidth and $V_\pi \sim 5$ V was also demonstrated in [84].

The bandwidths for these demonstrated modulators were believed to be limited mainly by the microwave loss. From Eq. (16.85), we know that when the impedance is matched, i.e., $\Gamma_L = \Gamma_S = 0$ and $T = 1$, the modulator frequency response can be simplified to

$$M(f) = \left| \frac{e^{(j\beta_o - \gamma_\mu)L} - 1}{(j\beta_o - \gamma_\mu)L} \right|^2 \qquad (16.92)$$

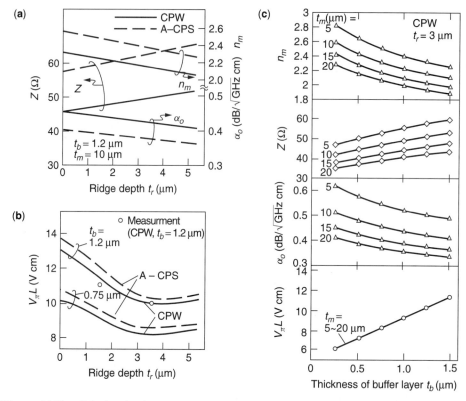

**Figure 16.29**   Calculated microwave characteristics and product of $V_\pi$ and length $L$ of the LiNbO$_3$ modulator as functions of ridge height $t_r$ and SiO$_2$ buffer layer thickness $t_b$. [79, Figs. 2 and 3.] © 1995 IEEE.

If the velocity is also matched, i.e., $\gamma_\mu = \alpha_\mu + j\beta_\mu = \alpha_\mu + j\beta_o$, Eq. (16.92) becomes

$$M(f) = \left| \frac{e^{-\alpha_\mu L} - 1}{\alpha_\mu L} \right|^2, \quad \text{where } \alpha_\mu = \alpha_0 \sqrt{f} \qquad (16.93)$$

Assuming $\alpha_\mu = 0$ when $f = 0$, the above equation determines that the 3-dB bandwidth is at the frequency where $\alpha_\mu L = 0.73$ and the microwave power loss equals to 6.34 dB. Typical microwave loss coefficient in the LiNbO$_3$ electrode ranges from 0.3 to 0.5 dB/cm/GHz$^{1/2}$, corresponding to 4.9–8.2-dB total loss at 30 GHz for a 3-cm-long modulation length. To keep microwave loss low, the SiO$_2$ buffer layer should be thin and of high quality [85], LiNbO$_3$ substrate should be as thin as possible (typically less than 0.5 mm) [86].

The above modulators are all on $z$-cut substrate. From Fig. 16.28**b**, we see that the microwave field in the optical waveguide under the center conductor is larger than that in the other waveguide. This asymmetry limits the extinction ratio, and it makes the chirp parameter deviate from zero. One way to improve this is to use differential push-pull. In this case, two CPW transmission lines have to be designed, and each optical waveguide ridge is under the center conductor of one CPW line. One drawback is that it requires differential driver with dual outputs, and the matching of the output phases is not an easy task. The advantage of $x$-cut LiNbO$_3$ modulators is that they can have ideally symmetrical two arms using CPW transmission line, as seen in Fig. 16.27**b**. This can result in large extinction ratio and zero chirp with single drive. However, in this case, the ridge structure does not provide the same advantage as it does for modulators on $z$-cut substrate. This compromises the modulator $V_\pi$ and bandwidth.

### 16.5.4  Traveling-Wave Polymer Modulator

Polymer modulators usually choose microstrip transmission line as their traveling-wave electrode. This choice is primarily due to two reasons: first of all, a vertical electric field is required to maximize the modulation efficiency, as analyzed in Section 16.2.1; secondly, the dispersion of the dielectric constant of polymer materials is usually very small, i.e., $\varepsilon^{1/2}$ at microwave frequency is very close to the optical group velocity index in the polymer optical waveguide (the difference is <0.1). Microstrip electrode provides the most efficient vertical electric field, and its microwave velocity index is very close to the effective $\varepsilon^{1/2}$ of the filled-in polymer layers, so that close velocity-match can be easily achieved. During fabrication, the large area ground metal is deposited before the spin-coating of the polymeric cladding and core layers. The total polymer stack thickness should be slightly larger than the vertical size of the optical mode to minimize the optical loss due to electrode absorption. After poling, narrow-width metal strip is fabricated on top of the polymer layers to form signal electrode.

Using this electrode configuration, high-performance polymer modulators have been demonstrated. In [87], the polymer modulator was fabricated on a silicon substrate with resistivity of 2000 $\Omega$-cm, the total polymer stack thickness was 6.5 $\mu$m, lateral optical confinement was achieved using photo-bleaching method. The schematic top view of the device is shown in Fig. 16.30. Only one arm of the MZM was modulated. 50-$\Omega$ tapered CPW lines on the silicon substrate were fabricated at both ends of the microstrip line to connect the microwave source and the 50-$\Omega$ termination. With a 12-mm modulation length, the bandwidth was measured as 40 GHz; $V_\pi$ was measured to be 10 V at 1.3-$\mu$m optical wavelength; and extinction ratio was better than 20 dB. For this device, microstrip transmission line impedance was close to 50 $\Omega$, velocity index mismatch was estimated at 0.03, and microwave loss coefficient was 0.75 dB/cm/GHz$^{1/2}$, corresponding to 5.7 dB total loss at 40 GHz for 12-mm modulation length. Therefore, the bandwidth of this device was mostly limited by the microwave loss. The poling

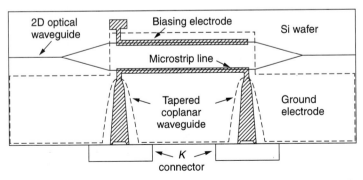

**Figure 16.30**  Schematic top view of a traveling-wave polymer modulator. [87, Fig. 1.] © 1992 American Institute of Physics.

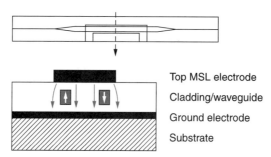

**Figure 16.31**  Top and cross-sectional view of the device showing the push-pull modulation in a traveling-wave polymer MZM using a single microstrip-line electrode. The white and gray arrows represent poling and modulation field directions, respectively. [89, Fig. 1.] © 1999 IEEE.

field of this device was 120 V/$\mu$m. If the EO polymer had been poled at 160 V/$\mu$m, $V_\pi$ would be reduced to 6 V for the same device [87]. Further reduction of $V_\pi$ can be achieved by using push-pull modulation. This can be either differential push-pull [88], or by poling the two optical arms in opposite directions, as reported in [89].

Figure 16.31 shows the schematic view of the device reported in [89]. The device was fabricated on a Cd/Au-coated fused-silica substrate. After spin coating and curing of 4–5-$\mu$m thick lower cladding layer (polyurethane), a 2.5-$\mu$m-thick core layer (LD-3) was spin-coated and vacuum-dried overnight. Two chromium poling electrodes were then fabricated on top of the core layer with a 30-$\mu$m gap. DC voltages of 800 V with opposite signs were applied between the poling electrodes and the ground metal to pole the polymer in opposite directions. To prevent air breakdown between the poling electrodes, a fused-silica cover piece was pressed on top of the poling electrodes to modify the field distribution in the electrode gap, and the poling was performed in an inert gas ($SF_6$) to increase the breakdown voltage. More detailed fabrication process can be found in [89].

To make the single microstrip transmission line 50 $\Omega$, it requires the ratio $w/d$ to be 3–4, where $w$ is the top strip-line-width and $d$ is the total polymer stack thickness. As the poling electrode gap was 30 $\mu$m (smaller gap would lead to easier air breakdown), the separation between the two MZM arms had to be no less than 30 $\mu$m. This makes $w = 30$–40 $\mu$m, and it requires $d \sim 10$ $\mu$m. This $d$ is thicker than the 6.5-$\mu$m polymer stack thickness in [87], which makes $V_\pi$ larger. In addition, the core layer LD-3 polymer has a small EO coefficient (7 pm/V, compared to 16 pm/V in [87]) as the price of better thermal and photo-chemical stabilities, and smaller optical loss. Consequently, for a device of 2-cm-long modulation length, $V_\pi$ was measured at 20 V with push-pull operation at 1.3-$\mu$m optical wavelength. The velocity index mismatch was found negligible ($\sim 0.01$), microwave loss coefficient was 0.55 dB/cm/GHz$^{1/2}$, corresponding to 6-dB total loss at 30 GHz for 2-cm-long length. Optical loss for the 3-cm-long MZM was $\sim 12$ dB, among which $\sim 6$ dB was attributed to the waveguide loss, and $\sim 6$ dB was due to coupling loss with optical fibers at the two end facets.

Recently, the properties of EO polymers have been greatly improved in terms of EO coefficient, optical loss, thermal and photo stabilities. Polymers incorporating CPW-75 type chromophores have shown EO coefficient $> 100$ pm/V. Ridge optical waveguides fabricated from the polycarbonate CPW (PC/CPW) material system exhibited optical loss around 1.7 dB/cm at 1.55-$\mu$m wavelength. A prototype device using PC/CPW core layer and opposite-poling technique has been recently demonstrated in [90]. The device used 1-cm-long modulation length, and lateral optical confinement was formed using reactive-ion etching, and core layer thickness was 3 μm, with 0.4-μm ridge height and 6-μm ridge width. The device was planarized by the upper cladding layer. To avoid a thick polymer stack as required in [89], this device used two 100-Ω microstrip lines for the two arms. A broadband microwave coupler is fabricated at both ends of the modulation length to combine the two 100-Ω lines into one 50-Ω line. In this way, the two MZM arms can have a large separation with thin polymer stack (for example, ~6 μm); the poling field can be high without air breakdown. Under push-pull operation, $V_\pi$ for the demonstrated device was 5.6 V at 1.55-μm wavelength, 3-dB bandwidth was 34 GHz, and the optical insertion loss was 8 dB using lensed fibers.

### 16.5.5  Frequency Response of Segmented Traveling-Wave Modulators

EO coefficients for III–V semiconductors are very small (1.4 pm/V for GaAs). For a III–V MZM based on linear EO effect, more than 1-cm modulation length is required to achieve a $V_\pi < 5$ V under push-pull modulation, and the traveling-wave design is a must to obtain wide bandwidth. However, continuous modulation cannot be implemented in this case due to either velocity-mismatch problem or microwave loss problem. For example, if CPW or CPS transmission line is used for the traveling-wave electrode, the microwave velocity index can be calculated from Eq. (16.89) to be ~2.65 (note that $\varepsilon_{eff}$ is ~13 for both GaAs and InP substrates), which is significantly smaller than the optical group velocity index (3.4–3.6). On the other hand, if microstrip transmission line is used for the traveling-wave electrode, to achieve 50-Ω impedance with a thin dielectric stack (~3 μm), a narrow top-conductor width will be needed, which results in excessive microwave loss and limited bandwidth.

The solution is to design a microwave transmission line separate from the optical waveguide. The transmission line runs parallel with the optical waveguide, and it can be either a CPW line or a coplanar strip (CPS) line, with microwave phase velocity faster than the optical group velocity. Modulation length (and its capacitance) on the optical waveguide is segmented and periodically connected to the microwave transmission line as capacitive-loading, as illustrated in Fig. 16.32. The capacitive-loading lowers the microwave velocity (and the transmission line impedance). For this sake, the above electrode is sometimes called "slow-wave transmission line." The design goal is to match the lowered microwave velocity with the optical group velocity, and match the lowered impedance with 50 Ω. The microwave impedance and the microwave velocity index can be calculated by:

$$\text{before loading: } Z_\mu = \sqrt{L_\mu/C_\mu}, \qquad n_\mu = c\sqrt{L_\mu C_\mu} \qquad (16.94)$$

$$\text{after loading: } \quad Z_0 = \sqrt{L_\mu/(C_\mu + C_L)}, \quad n_0 = c\sqrt{L_\mu(C_\mu + C_L)} \quad (16.95)$$

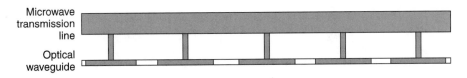

**Figure 16.32**  Schematic diagram to show the segmented traveling-wave electrode design.

where $L_\mu$ and $C_\mu$ are the inductance and capacitance per unit-length for the unloaded transmission line and $C_L$ is the loaded capacitance (due to the modulation sections on the optical waveguide) per unit-length of the transmission line. The requirements of matching $Z_0$ with 50 $\Omega$, and matching $n_0$ with the optical group velocity index $n_o$, determine how much capacitance $C_L$ can be loaded (derived from the above equations):

$$C_L = \frac{n_o^2 - n_\mu^2}{c Z_0 n_o} \tag{16.96}$$

Substituting $n_o = 3.6$, $n_\mu = 2.65$, $Z_0 = 50$ $\Omega$, and $c = 3 \times 10^{10}$ cm/s into the above equation, one can calculate $C_L \sim 1.1$ pF/cm. This loading capability determines how long an electrode is needed to load enough modulation waveguide for a small $V_\pi$. For example, if 1-cm-long modulation length is required and it causes 2.2-pF capacitance, then a 2-cm-long electrode will be needed. In this case, when the 1-cm modulation length is segmented and distributed over a 2-cm-long optical waveguide, half of the waveguide will be active and the other half will be passive, resulting in a fill-factor of 50%. The impedance of the unloaded transmission line should be designed as

$$Z_\mu = Z_0 \times \frac{n_o}{n_\mu} = 50 \times \frac{3.6}{2.65} = 68 \ \Omega \tag{16.97}$$

The frequency response of the segmented traveling-wave MZM can be calculated by using microwave network analysis on the equivalent circuit model, as shown in Fig. 16.33. Each of the periodically loaded active segments (totally $N$ segments) can be represented by an $L_a R_a C_a$ circuit shunted onto the microwave transmission line. The inductance $L_a$ is induced by the connection bridge; the resistance $R_a$ includes the metal bridge resistance and the series resistance of the modulation waveguide; the capacitance $C_a$ is due to the intrinsic active region in the modulation waveguide, and $C_L = C_a/L_0$, where $L_0$ is the length of the microwave transmission line between any two adjacent bridges. The microwave source impedance $Z_S$ and the terminating impedance $Z_L$ are usually 50 $\Omega$.

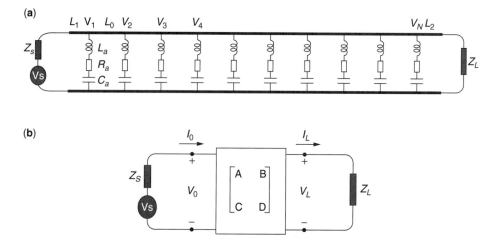

**Figure 16.33** Microwave equivalent circuit model for segmented traveling-wave modulators.

To obtain the modulation frequency response, we can first calculate the microwave voltages ($V_1$, $V_2$, ... $V_N$) across the loaded segments, and then derive the voltages across all the capacitors $C_a$. $V_1$, $V_2$, ... $V_N$ can be easily calculated using the ABCD transmission matrix method. Note that the transmission matrix for a transmission line (with impedance $Z_\mu$, propagation constant $\gamma_\mu = \alpha_\mu + j\beta_\mu$ and length $L_0$) is:

$$T_0 = \begin{bmatrix} \cos h(\gamma_\mu L_0) & Z_\mu \sin h(\gamma_\mu L_0) \\ Z_\mu^{-1} \sin h(\gamma_\mu L_0) & \cos h(\gamma_\mu L_0) \end{bmatrix} \quad (16.98)$$

And the transmission matrix for each of the shunted $L_a R_a C_a$ circuit is:

$$T_a = \begin{bmatrix} 1 & 0 \\ \{R_a + j(\omega L_a - 1/\omega C_a)\}^{-1} & 1 \end{bmatrix} \quad (16.99)$$

The transmission matrix for the whole traveling-wave electrode can be obtained by multiplying the transmission matrices of all segments, i.e., $T_1(T_a T_0)^{N-1} T_a T_2$, where $T_1$ and $T_2$ are the transmission matrices for the transmission line segments at both ends, as shown in Fig. 16.33**a**. Assuming the final result of matrix multiplication is $\begin{bmatrix} A & B \\ C & D \end{bmatrix}$, the equivalent circuit in Fig. 16.33**a** then becomes the equivalent circuit in Fig. 16.33**b**. In this case, the current $I_L$ and the voltage $V_L$ across the terminator can be calculated using the following equations:

$$V_S = V_0 + I_0 Z_S, \quad \begin{bmatrix} V_0 \\ I_0 \end{bmatrix} = \begin{bmatrix} A & B \\ C & D \end{bmatrix} \times \begin{bmatrix} V_L \\ I_L \end{bmatrix}, \quad V_L = I_L Z_L \quad (16.100)$$

and the solution is

$$I_L = \frac{V_S}{AZ_L + B + CZ_S Z_L + DZ_S}, \quad V_L = \frac{V_S Z_L}{AZ_L + B + CZ_S Z_L + DZ_S} \quad (16.101)$$

After $I_L$ and $V_L$ are obtained, $V_N$, $V_{N-1}$, ... $V_1$ can be calculated iteratively using the transmission matrix method. After all $V_1$, $V_2$, ... $V_N$ are obtained, the normalized modulation frequency response of the traveling-wave MZM can be calculated as:

$$M(f) = \left| \frac{2}{N V_S} \sum_{n=1}^{N} V_n e^{j\omega\{L_1+(n-1)L_0\}/v_o} \frac{1}{1 - \omega^2 L_a C_a + j\omega R_a C_a} \right|^2 \quad (16.102)$$

where $v_o$ is the optical group velocity in the optical waveguide. $V_S$ in the above equation will eventually be cancelled; thus, it can be assumed as 1 in the calculation. The dispersion of the microwave transmission line (such as CPS line) can be included because this calculation allows frequency-dependent $Z_\mu$ and $\gamma_\mu$.

The segmented traveling-wave design has been applied to III–V MZM based on linear EO effect, as will be described in the next section. However, it is also possible to design segmented traveling-wave EAM, or segmented traveling-wave MZM based on QCSE. In this case, excellent impedance match and velocity match can be achieved, and microwave loss can be small due to the large area electrode. Ultrawide bandwidth can be achieved together with high efficiency using a traveling-wave electrode 2–5 mm long. However, optical loss may be a very challenging problem because both kinds of optical waveguides are very lossy.

### 16.5.6   Traveling-Wave III–V MZM

Continuous traveling-wave GaAs EO waveguide modulators have been analyzed and demonstrated by several research groups in early times, and the representative performance was reported for a polarization modulator in [11]. Using CPS electrode and 8-mm-long modulation length, 16-GHz bandwidth was achieved with $V_\pi = 11$ V. The device performance was mainly limited by the big velocity mismatch, as analyzed in the last section. Later development for high-speed GaAs EO modulators focused on the segmented traveling-wave design [91–95], and there are mainly two design approaches for GaAs MZM: one uses CPS electrode and implements series push-pull through a thin doped layer [91–93]; the other uses CPW electrode and implements parallel push-pull through undoped layers [94, 95].

Figure 16.34 shows the schematic diagram for the first approach [91]. The transmission line used was an asymmetric CPS line. Active segments on the two optical arms were periodically connected to the two conductors of the CPS line as capacitive loading, as shown in Fig. 16.34a. The cross-section view around the optical waveguides is shown in Fig. 16.34b. The waveguide consisted of a 1-μm-thick GaAs core layer and two 0.5-μm-thick $Al_{0.1}Ga_{0.9}As$ cladding layers. The upper cladding layer was undoped, and the lower cladding layer was partially $n$-type doped at its bottom [92]. This layer structure makes the intrinsic layer gap 1.5–2 μm. Lateral optical confinement was formed by mesa etching. Due to the thick waveguide core, to make the optical waveguide single-mode, shallow-etched rib-loaded waveguides had to be used (in which only upper cladding layer was etched). Waveguide width was 3 μm. Al film was evaporated on top of the waveguides to form Schottky contact with the top AlGaAs layer. During the operation, a large DC voltage was applied to make the electric field in the two optical arms always toward the upper direction, as shown in Fig. 16.34b, while the microwave field was applied across the top metal of the two waveguides. The thin doped layer in the lower cladding helps to enhance vertical microwave field in the waveguides, and this makes the microwave field in the two optical arms always toward opposite directions. This enables push-pull modulation, and it is specifically called "series push-pull" because the two Schottky-$i$-$n$ diodes are series-connected by the doped layer. To implement series push-pull operation, some special on-chip bias

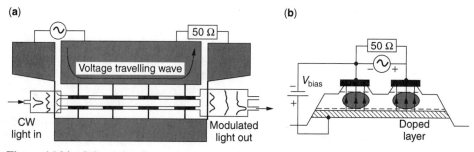

**Figure 16.34**   Schematic diagram of segmented traveling-wave GaAs MZM using series push-pull. (**a**) Device top view and (**b**) cross-section view around the optical waveguides. [91, Fig. 1.] © 1995 IEEE.

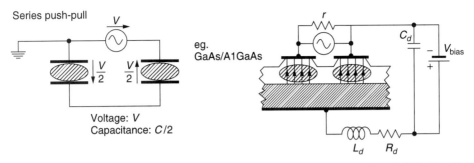

**Figure 16.35**  Illustration of bias and decoupling circuitry for series push-pull. [92, Fig. 3b.] © 1991 IEEE.

and decoupling circuitry is required to isolate the DC and microwave sources [92], as illustrated in Fig. 16.35. This kind of configuration allows small separation between the two optical arms, so that MMIs can be used as optical splitters and combiners, as shown in Fig. 16.34**a**. The advantages of MMIs over conventional X-branch and Y-branch couplers have been mentioned in Section 16.4.3.

Compared to single-arm modulation, series push-pull divides the voltage in halves on each optical arm, which cancels the advantage of two-arm modulation. However, the capacitance per unit length in the active segments is also halved due to the series connection of two capacitors, which, in turn, allows the loaded active length to be doubled for the same velocity match and impedance match. Therefore, series push-pull does provide a factor of 2 improvement in voltage compared to single-arm modulation. The latest performance for GaAs MZM using this kind of configuration was reported in [93]. For the 40-Gb/s device, the total length of traveling-wave electrode was 16 mm, and 33-GHz bandwidth had been achieved with $V_\pi = 4.7$ V at 1.55-μm wavelength. The frequency response of this device can be simulated using the approach in last section. In the simulation, to obtain a good fit with the measurement in [93], we assume 50 active segments with a fill factor of 70%, so that each active segment is 224 μm long. For a pair of series-connected active segments (and two metal bridges), we assume the capacitance $C_a = 35.24$ fF, resistance $r_a = 4\ \Omega$, and inductance $L_a = 0.1$ nH. Also assume source impedance $Z_S = 50\ \Omega$, terminator impedance $Z_L = 40\ \Omega$, unloaded transmission line impedance $Z_\mu \sim 68\ \Omega$, optical group velocity index $n_o \sim 3.6$, microwave phase velocity index $n_\mu = 2.65$, and $\alpha_\mu = 0.52$ dB/cm/GHz$^{1/2}$, The simulated frequency response is plotted in Fig. 16.36, which is very close to the measured result in [94, Fig. 3**a**]. The sharp roll-off after 30 GHz is mostly attributed to the electrical-filter effect due to the periodic loading of active segments [91, 92]. The low-pass filter blocks the transmission of high-frequency microwave signal. To reduce this effect, the active length can be divided into more segments to shorten the loading period, so that the cutoff frequency of the filter is higher, and the sharp roll-off of EO response will be pushed to higher frequency. During the simulation, the inductance $L_a$ is also found to contribute to the roll-off of EO response. Smaller $L_a$ is generally preferred to have wider bandwidth. The modeling in last section can also be used to calculate the microwave properties of the traveling-wave electrode, including its impedance, microwave velocity, and loss. The calculation indicates that after periodic loading, both the microwave impedance and the microwave velocity decrease with frequency, microwave loss is also higher than that of unloaded line. These effects are more pronounced with more

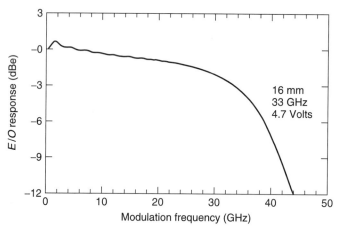

**Figure 16.36** Simulated frequency response for the GaAs traveling-wave MZM reported in [93].

capacitive loading and/or longer loading period. To maximize the modulator performance, the electrode and the periodic loading should be designed in such way that the best velocity match and impedance match are achieved at high frequency, i.e., around the cutoff frequency of 3-dB bandwidth [91].

Another design approach for GaAs MZM uses CPW electrode and implements parallel push-pull through undoped layers. Figure 16.37 illustrates the device structure using this design [94]. Material layers and optical waveguide structure were similar to the device in [91], but there was no intentionally doped layer in this structure. The transmission line used was a CPW line instead of A-CPS line. Active segments on the two optical arms were parallel-connected (instead of series-connected) to the microwave transmission line. With microwave frequency larger than 1 MHz, these semiconductor layers behave like low-loss dielectric material. Therefore, similar to the LiNbO$_3$ modulator shown in Fig. 16.28**b**, electric field in the waveguide ribs will be mostly vertical, and always with opposite signs in the two arms, as illustrated in Fig. 16.37**b**. This enables push-pull operation. Compared to single-arm modulation, the described parallel push-pull structure does not change the voltage across the optical arms. However, the loading capacitance is doubled due to the parallel connection of two capacitors, which, in turn, requires halving the loaded modulation length to keep the same velocity match and impedance match. This factor cancels the advantage of two-arm modulation. Therefore, to the first order of approximation, parallel push-pull does not provide advantage over single-arm modulation in terms of efficiency and bandwidth [92]. The advantage of parallel push-pull is zero-chirp (perhaps also larger extinction ratio), and it does not require any on-chip bias decoupling circuitry. The design in Fig. 16.37 also results in a large separation between the two optical arms (>50 μm, compared to 10–20 μm in the series push-pull design), which may cause more optical loss to the optical splitter and combiner.

For the fabricated device with dimensions shown in Fig. 16.37, modulation bandwidth was measured more than 40 GHz (the EO response dropped 1.5–2 dB at 40 GHz), $V_\pi$ was measured as 16.8 V at 1.55-μm optical wavelength (14 V at 1.3 μm). The large $V_\pi$ was attributed to the poor overlap between the electric field and the optical field. To bring the $V_\pi$ down, the electrode length can be increased to load

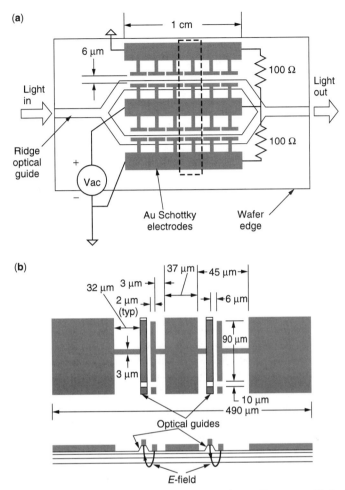

**Figure 16.37**   (**a**) Schematic top view of the segmented traveling-wave GaAs MZM implementing parallel push-pull. (**b**) Enlarged view of the modulator section delineated by the dashed line in  (**a**) together with a schematic cross-sectional view. [94, Figs. 1 and 2.] © 1996 IEEE.

more active segments. For example, if the electrode length is increased from 1.0 cm to 1.5 cm, $V_\pi$ can be decreased from 16.8 V to 11.2 V, while the bandwidth can be possibly maintained at >30 GHz. Further improvement may be made by reducing the gap between the T-rails near the optical waveguide [95]. This gap is currently 6 μm as shown in Fig. 16.37. However, reducing the gap results in increased capacitance of the active segments, which in turn affects the velocity match. In addition, very small gap may cause more tangential component of the electric field in the optical waveguides, thereby not much increase of the vertical field can be expected.

## 16.6   A COMPARISON OF THE AVAILABLE MODULATORS

In the previous sections, we have explored in some details the design, analysis, and state-of-the-art performance for various types of optical intensity modulators. These modulators include lumped-element EAM, lumped-element QCSE-based (or MQW) MZM, traveling-wave EAM, traveling-wave LiNbO$_3$ MZM, traveling-wave polymer MZM, and traveling-wave GaAs MZM. Possibility and issues of segmented traveling-wave QCSE-based MZM and EAM have also been mentioned. Physical effects behind these devices have been described. Our discussions mainly focused on the modulation efficiency and bandwidth, with additional coverage of polarization sensitivity, frequency chirp, optical loss, saturation power, extinction ratio, bias stability, and optical bandwidth. Linearity and noise performance were skipped, material and fabrication issues were only slightly touched, and test and measurement techniques were not included into the scope. These subjects are no less important than those covered in this text.

Most of the above-mentioned modulators are commercially available, some of them have been integrated with other components on the same chip or in the same package. All of these products together provide multiple choices for various applications. To make a comparison, we can use LiNbO$_3$ modulator as the primary benchmark, because it has dominated the market for quite some time; we also use EAM as a secondary benchmark, because it is distinctly different from all other candidates. The pros and cons of these devices are summarized below:

1. *LiNbO$_3$ MZM.* The most remarkable advantages are the lowest optical loss and the highest optical power handling ability. Among all types of MZMs, LiNbO$_3$ modulator is still considered as the device with the best performance. Compared to EAM, LiNbO$_3$ modulator has the additional advantages of broad optical bandwidth, zero or tunable chirp, and temperature insensitivity. The biggest disadvantage for LiNbO$_3$ MZM is the large size, probably the largest among all the modulator candidates. It also has the bias-drifting issue, which requires bias control circuit. Compared to EAM, LiNbO$_3$ modulators are more polarization sensitive, difficult to be integrated with other components, and have a higher fabrication cost in terms of large volume production.

2. *EAM.* The most remarkable advantages are the smallest chip size ($\sim$200 times smaller than LiNbO$_3$ MZM with the same bandwidth and efficiency) and compatibility for integration with other components. It is also the most promising candidate for polarization-insensitive operation with high efficiency. The small chip size not only facilitates many applications, but also greatly lowers the fabrication cost because thousands of chips can be produced on a single wafer. For 40-Gb/s modulation at 1.55-$\mu$m wavelength, EAM is currently the candidate with the smallest driving voltage. Compared to all types of MZMs, the disadvantages of EAM include low saturation power, large chirp, narrow optical bandwidth, and the need of temperature control. The last two problems are due to that its transfer function varies with optical wavelength and temperature, and they become less problematic when the EAM is integrated with a semiconductor laser source, because the laser wavelength is usually locked, and the laser chip itself also needs temperature control.

3. *GaAs MZM.* The performance of GaAs MZM is now getting closer to that of LiNbO$_3$'s. Its modulation efficiency is slightly less — for example, for the 40-Gb/s devices, $V_\pi$ can be $\sim$5 V for GaAs MZM, compared to $\sim$4 V for LiNbO$_3$ MZM; its optical loss is also several decibels higher. However, GaAs MZM has a smaller chip size — for example, for the 40-Gb/s devices, chip size may be

~3 cm long for GaAs MZM, compared to ~5 cm long for LiNbO$_3$ MZM. In addition, no bias-drifting problem has been reported for GaAs MZM. Further advantage for GaAs MZM could be lower fabrication cost. It is possible to fabricate 50–100 chips on a single 4-inch GaAs wafer. Other than these, GaAs MZM possesses the same performance as the LiNbO$_3$ counterpart.

4. *MQW MZM.*  To date, this type of device is limited to 10-Gb/s application in digital systems; in analog links, its bandwidth is hard to exceed 20 GHz with sufficient modulation efficiency. Its chip size is larger than that of EAM, but after packaging, their footprints may be the same (~2 cm long). Compared to 10-Gb/s EAM device, MQW MZM has the advantages of high optical power, small or tunable chirp, and less temperature sensitivity. The disadvantages include higher optical loss (due to optical couplers and longer length) and less extinction ratio (due to optical imbalance of the two arms). In addition, integration with QW laser and other components is more difficult for MQW MZM than for EAM due to the difficulties in butt-joint fabrication. Compared to other types of MZMs, the most notable advantage for MQW MZM is the much smaller chip size. In addition, it is possible to achieve polarization insensitivity with sufficient efficiency, and possible to be integrated with laser and amplifier components on the same chip. The disadvantages may include lower extinction ratio, nonlinear chirp control, higher optical loss, and narrower optical bandwidth.

5. *Polymer MZM.*  This type of device is currently not yet popular in the market. Its chip size, optical loss, and efficiency are now very close to that of GaAs MZM. Polymer MZM is also considered as a candidate to be integrated with a wide range of other components. However, its optical power handling ability (~10 mW) is currently much less than that of other types of MZM, and its thermal and photochemical stabilities need to be improved. These problems could be solved by further research work, and its modulation efficiency can also be greatly improved by the continuous development of new EO polymer materials. In view of these points, polymer MZM may have the greatest potential in the future, just like the development of plastic materials in the history.

Comparisons of the above-mentioned devices are justifiable only in view of some specific applications. For example, in applications where the component footprint is not a concern and high optical power is demanded, LiNbO$_3$ MZM will be the No. 1 choice, and GaAs MZM will be its strong competitor. These applications include various analog and digital distribution links. While in applications that require small size, wide bandwidth, and high efficiency, EAM becomes the best candidate, and MQW MZM can be its competitor for bandwidth below 20 GHz due to its high power handling ability and better chirp. These applications include board-level integration in transponder and transceiver modules. If the modulator is to be integrated with devices on Si substrate (such as micro-electro-mechanical switches, and various passive waveguides), polymer MZM may be the best choice.

Integration is an efficient way to reduce the component cost and the footprint. One of the most notable integrations for optical modulator is to integrate it with a semiconductor laser source, primarily the distributed feedback (DFB) laser. 10-Gb/s electroabsorption modulated laser (EML) chips [17] have been commercialized for several years, and the 40-Gb/s EML chips have also been demonstrated by several vendors since the year of 2001 [96–98]. Copackages of DFB laser with GaAs MZM and MQW MZM have both been commercialized [99]. This kind of copackaging can reduce footprint and facilitate operation, but it may cause higher fabrication cost. On-chip integration of 10-Gb/s MQW MZM with DFB laser has been reported [100, 101], but not commercialized yet, probably due to the difficulties in butt-joint fabrication and the control of optical isolation. After these problems are solved, this integrated chip is expected to become a strong competitor of the 10-Gb/s EML, because it will have higher optical output power and better chirp control. Other integrated chips include tandem EA modulators [18] and polymer MZM integrated with other electronic components [14].

### Acknowledgments

Paul Yu would like to acknowledge the sponsorship of DARPA (RFLICS Program) and Air Force Research Laboratories at Rome for the optical modulator research at UCSD.

### REFERENCES

1. P. A. Morton, R. A. Logan, T. Tanbun-Ek, P. F. Sciortino, Jr., A. M. Sergent, R. K. Montgomery, and B. T. Lee, "25 GHz Bandwidth 1.55 μm GaInAsP p-doped Strained Multiquantum Well Lasers," *Electron. Lett.*, **28**, pp. 2156–2157, 1992.

2. N. K. Dutta, N. A. Olsson, L. A. Koszi, P. Besomi, R. B. Wilson, and R. J. Nelson, "Frequency Chirp under Current Modulation in InGaAsP Injection Lasers," *J. Appl. Phys.*, **56**, pp. 2167–2169, 1984.

3. N. Dagli, "Wide-bandwidth Lasers and Modulators for RF Photonics," *IEEE Trans. Microwave Theory Techniques*, **47**, pp. 1151–1171, 1999.

4. J. G. Mendoza-alvarez, L. A. Coldren, A. Alping, R. H. Yan, T. Hausken, K. Lee, and K. Pedrotti, "Analysis of Depletion Edge Translation Lightwave Modulators," *J. Lightwave Technol.*, **6**, pp. 793–808, 1988.

5. L. A. Coldren and S. W. Corzine, "Continuously-tunable Single-frequency Semiconductor Lasers," *IEEE J. Quantum Electron.*, **QE-23**, pp. 903–908, 1987.

6. A. Yariv and P. Yeh, *Optical Waves in Crystals*, Wiley, New York, 1984; A. Yariv, *Optical Electronics*, 3rd ed., Holt, Rinehart & Winston, New York, 1985.

7. R. V. Schmidt and I. P. Kaminow, "Metal Diffused Optical Waveguides in LiNbO₃," *Appl. Phys. Lett.*, **25**, pp. 458–460, 1974; M. Fukuma and J. Noda, "Optical Properties of Ti-diffused LiNbO₃ Strip Waveguides," *J. Appl. Phys.*, **19**, pp. 4677–4682, 1978.

8. H. F. Taylor, "Polarization Independent Guided Wave Optical Modulators and Switches," *J. Lightwave Technol.*, **LT-3**, pp. 1277–1280, 1985.

9. P. Granestrand, L. Thylen, and B. Stoltz, "Polarization Independent Switch and Polarization Splitter Employing $\Delta\beta$ and $\Delta\kappa$ Modulation," *Electron. Lett.*, **24**, pp. 1142–1143, 1988.

10. L. McCaughan, "Low Loss Polarization Independent Electrooptic Switches," *J. Lightwave Technol.*, **LT-2**, pp. 51–55, 1984.

11. S. Y. Wang and S. H. Lin, "High Speed III–V Electrooptic Waveguide Modulators at $\lambda = 1.3$ μm," *J. Lightwave Technol.*, **6**, pp. 758–771, 1988.

12. R. Spickermann, M. G. Peters, and N. Dagli, "A Polarization Independent GaAs-AlGaAs Electrooptic Modulator," *IEEE J. Quantum Electron.*, **32**, pp. 764–769, 1996.

13. L. A. Hornak, editor, Polymers for Lightwave and Integrated Optics, Marcel Dekker, New York, 1992; D. S. Chemla and J. Zyss, editors, *Nonlinear Optical Properties of Organic Molecules and Crystals*, Academic, New York, 1987.

14. S. Kalluri, M. Ziari, A. Chen, V. Chuyanov, W. H. Steier, D. Chen, B. Jalali, H. Fetterman, and L. R. Dalton, "Monolithic Integration of Waveguide Polymer Electrooptic Modulators on VLSI Circuitry," *IEEE Photon. Technol. Lett.*, **8**, pp. 644–646, 1996.

15. E. V. Tomme, P. P. V. Daele, R. G. Baets, and P. E. Lagasse, "Integrated Optic Devices based on Nonlinear Optical Polymers," *IEEE J. Quantum Electron.*, **27**, pp. 778–787, 1991.

16. A. Donval, E. Toussaere, R. Hierle, and J. Zyss, "Polarization Insensitive Electro-optic Polymer Modulator," *J. Appl. Phys.*, **87**, pp. 3258–3262, 2000.

17. K. Wakita, I. Kotaka, H. Asai, M. Okamoto, Y. Kondo, and M. Naganuma, "High-speed and Low-drive Voltage Monolithic Multiple Quantum-well Modulator/DFB Laser Light Source," *IEEE Photon. Technol. Lett.*, **4**, pp. 16–18, 1992.

18. N. Souli, A. Ramdane, F. Devaux, A. Ougazzaden, and S. Slempkes, "Tandem of Electroabsorption Modulators Integrated with Distributed Feedback Laser and Optical Amplifier for 20-Gbit/s Pulse Generation and Coding at 1.5 μm," *OFC'97 Technical Digest (Optical Fiber Communication)*, **6**, paper WG6, pp. 141–142, 1997.

19. D. S. Shin, G. L. Li, C. K. Sun, S. A. Pappert, K. K. Loi, W. S. C. Chang, and P. K. L. Yu, "Optoelectronic RF Signal Mixing using an Electroabsorption Waveguide as an Integrated Photodetector/Mixer," *IEEE Photon. Technol. Lett.*, **12**, pp. 193–195, 2000.

20. L. D. Westbrook and D. G. Moodie, "Simultaneous Bi-directional Analogue Fiber-optic Transmission using an Electroabsorption Modulator," *Electron. Lett.*, **32**, pp. 1806–1807, 1996. See also [51].

21. K. Tharmalingam, "Optical Absorption in the Presence of a Uniform Field," *Phys. Rev.*, **130**, pp. 2204–2206, 1963.

22. S. W. Corzine, R.-H. Yan, and L. A. Coldren, "Optical gain in III–V bulk and quantum well semiconductors," in *Quantum Well Lasers*, chapter 1, edited by P. S. Zory, Academic Press, San Diego, 1993.

23. P. Lawaetz, "Valence-band Parameters in Cubic Semiconductors," *Phys. Rev. B.*, **4**, pp. 3460–3467, 1971.

24. P. J. Stevens, M. Whitehead, G. Parry, and K. Woodbridge, "Computer Modeling of the Electric Field Dependent Absorption Spectrum of Multiple Quantum Well Material," *IEEE J. Quantum Electron.*, **24**, pp. 2007–2016, 1988.

25. A. K. Ghatak, K. Thyagarajan, and M. R. Shenoy, "A Novel Numerical Technique for Solving the One-dimensional Schrodinger Equation using Matrix Approach — Application to Quantum Well Structures," *IEEE J. Quantum Electron.*, **24**, pp. 1524–1531, 1988.

26. S.-L. Chuang, S. Schmitt-Rink, D. A. B. Miller, and D. S. Chemla, "Exciton Green's-function Approach to Optical Absorption in a Quantum Well with an Applied Electric Field," *Phys. Rev. B*, **43**, pp. 1500–1509, 1991.

27. T. Ikeda and H. Ishikawa, "Analysis of the Attenuation Ratio of MQW Optical Intensity Modulator for 1.55 μm Wavelength taking Account of Electron Wave Function Leakage," *IEEE J. Quantum Electron.*, **32**, pp. 284–292, 1996.

28. G. Bastard, E. E. Mendez, L. L. Chang, and L. Esaki, "Variational Calculations on a Quantum Well in an Electric Field," *Phys. Rev. B*, **28**, pp. 3241–3245, 1983.

29. D. S. Chuu and Y. T. Shih, "Exciton Binding Energy in a GaAs/Al$_x$Ga$_{1-x}$As Quantum Well with Uniform Electric Field," *Phys. Rev. B*, **44**, pp. 8054–8060, 1991.

30. D. A. B. Miller, D. S. Chemla, T. C. Damen, A. C. Gossard, W. Wiegmann, T. H. Wood, and C. A. Burrus, "Electric Field Dependence of Optical Absorption near the Band Gap of Quantum-well Structures," *Phys. Rev. B*, **32**, pp. 1043–1060, 1985.

31. M. Sugawara, T. Fujii, S. Yamazaki, and K. Nakajima, "Theoretical and Experimental Study of the Optical-absorption Spectrum of Exciton Resonance in $In_{0.53}Ga_{0.47}As/InP$ Quantum Wells," *Phys. Rev. B*, **42**, pp. 9587–9597, 1990.

32. T. Yamanaka, K. Wakita, and K. Yokoyama, "Field-induced Broadening of Optical Absorption in InP-based Quantum Wells with Strong and Weak Quantum Confinement," *Appl. Phys. Lett.*, **65**, pp. 1540–1542, 1994.

33. S.-C. Hong, G. P. Kothiyal, N. Debbar, P. Bhattacharya, and J. Singh, "Theoretical and Experimental Studies of Optical Absorption in Strained Quantum-well Structures for Optical Modulators," *Phys. Rev. B*, **37**, pp. 878–885, 1988.

34. W. T. Masselink, P. J. Pearah, J. Klem, C. K. Peng, H. Morkoc, G. D. Sanders, and Y.-C. Chang, "Absorption Coefficients and Exciton Oscillator Strengths in AlGaAs-GaAs Superlattices," *Phys. Rev. B*, **32**, pp. 8027–8034, 1985.

35. F. Devaux, S. Chelles, A. Ougazzaden, A. Mircea, and J. C. Harmand, "Electroabsorption Modulators for High-bit-rate Optical Communications: A Comparison of Strained InGaAs/InAlAs and InGaAsP/InGaAsP MQW," *Semicond. Sci. Technol.*, **10**, pp. 887–901, 1995.

36. T. Ishikawa and J. E. Bowers, "Band Lineup and In-plane Effective Mass of InGaAsP or InGaAlAs on InP Strained-layer Quantum Well," *IEEE J. Quantum Electron.*, **30**, pp. 562–570, 1994.

37. K. Wakita, I. Kotaka, K. Yoshino, S. Kondo, and Y. Noguchi, "Polarization Independent Electroabsorption Modulators using Strain Compensated InGaAs/InAlAs MQW Structures," *IEEE Photon. Technol. Lett.*, **7**, pp. 1418–1420, 1995.

38. A. Ougazzaden and F. Devaux, "Strained InGaAsP/InGaAsP/InAsP Multi-quantum Well Structure for Polarization Insensitive Electroabsorption Modulator with High Power Saturation," *Appl. Phys. Lett.*, **69**, pp. 4131–4132, 1996.

39. T. Yamaguchi, T. Morimoto, K. Akeura, K. Tada, and Y. Nakano, "Polarization Independent Waveguide Modulator using a Novel Quantum Well with Mass Dependent Width," *IEEE Photon. Technol. Lett.*, **6**, pp. 1442–1444, 1994.

40. G. E. Betts, L. M. Walpita, W. S. C. Chang, and R. F. Mathis, "On the Linear Dynamic Range of Integrated Electrooptical Modulators," *IEEE J. Quantum Electron.*, **QE-22**, pp. 1009–1011, 1986.

41. C. K. Sun, S. A. Pappert, R. B. Welstand, J. T. Zhu, P. K. L. Yu, Y. Z. Liu, and J. M. Chen, "High Spurious Free Dynamic Range Fiber Link using a Semiconductor Electroabsorption Modulator," *Electron. Lett.*, **31**, pp. 902–903, 1995.

42. G. E. Betts, C. H. Cox III, and K. G. Ray, "20 GHz Optical Analog Link using an External Modulator," *IEEE Photon. Technol. Lett.*, **2**, pp. 923–925, 1990.

43. S. K. Korotky and J. J. Veselka, "An RC Network Analysis of Long term Ti:LiNbO$_3$ Bias Stability," *J. Lightwave Technol.*, **14**, pp. 2687–2697, 1996.

44. Y. Shi, W. Wang, W. Lin, D. J. Olson, and J. H. Bechtel, "Long Term Stable Direct Current Bias Operation in Electrooptic Polymer Modulators with an Electrically Compatible Multilayer Structure," *Appl. Phys. Lett.*, **70**, pp. 2236–2238, 1997.

45. G. L. Li, R. B. Welstand, W. X. Chen, J. T. Zhu, S. A. Pappert, C. K. Sun, Y. Z. Liu, and P. K. L. Yu, "Novel Bias Control of Electroabsorption Waveguide Modulator," *IEEE Photon. Technol. Lett.*, **10**, pp. 672–674, 1998.

46. G. L. Li, Y. Z. Liu, R. B. Welstand, C. K. Sun, W. X. Chen, J. T. Zhu, S. A. Pappert, and P. K. L. Yu, "Harmonic Signals from Electroabsorption Modulators for Bias Control," *IEEE Photon. Technol. Lett.*, **11**, pp. 659–661, 1999.

47. G. E. Betts, F. J. O'Donnell, and K. G. Ray, "Effect of Annealing Photorefractive Damage in Titanium Indiffused LiNbO$_3$ Modulators," *IEEE Photon. Technol. Lett.*, **6**, pp. 211–213, 1994.

48. Y. Shi, D. J. Olson, and J. H. Bechtel, "Photoinduced Molecular Alignment Relaxation in Poled Electrooptic Polymer Thin Films," *Appl. Phys. Lett.*, **68**, pp. 1040–1042, 1996.

49. Y. Shi, W. Wang, W. Lin, D. J. Olson, and J. H. Bechtel, "Double End Cross Linked Electrooptic Polymer Modulators with High Optical Power Handling Capability," *Appl. Phys. Lett.*, **70**, pp. 1342–1344, 1997.

50. K. K. Loi, J. H. Hodiak, X. B. Mei, C. W. Tu, W. S. C. Chang, D. T. Nichols, L. J. Lembo, and J. C. Brock, "Low-loss 1.3-μm MQW Electroabsorption Modulators for High-linearity Analog Optical Links," *IEEE Photon. Technol. Lett.*, **10**, pp. 1572–1574, 1998.

51. R. B. Welstand, S. A. Pappert, C. K. Sun, J. T. Zhu, Y. Z. Liu, and P. K. L. Yu, "Dual-function Electroabsorption Waveguide Modulator/Detector for Optoelectronic Transceiver Applications," *IEEE Photon. Technol. Lett.*, **8**, pp. 1540–1542, 1996.

52. K. K. Loi, X. B. Mei, J. H. Hodiak, A. N. Cheng, L. Shen, H. H. Wieder, C. W. Tu, and W. S. C. Chang, "Experimental Study of Efficiency-Bandwidth tradeoff of Electroabsorption Waveguide Modulators for Microwave Photonic Links," *Proc. IEEE LEOS*, **1**, pp. 142–143, 1997.

53. G. P. Agrawal, *Fiber-Optic Communications*, John Wiley & Sons, New York, Chapter 2, 1992.

54. J. Yu, C. Rolland, D. Yevick, A. Somani, and S. Bradshaw, "A novel method for improving the performance of InP/InGaAsP multiple-quantum-well Mach-Zehnder modulators by phase shift engineering," *Proc. IPR'96 (Intl. Conf. on InP and Related Materials)*, pp. 376–379, paper 1TuG4-1, 1996.

55. F. Dorgeuille and F. Devaux, "On the Transmission Performances and the Chirp Parameter of a Multiple-quantum-well Electroabsorption Modulator," *IEEE J. Quantum Electron.*, **30**, pp. 2565–2572, 1994.

56. H. Soda, K. Sato, K. Nakai, H. Ishikawa, and H. Imai, "Chirp Behaviour of High-speed GaInAsP/InP Optical Intensity Modulator," *Electron. Lett.*, **24**, pp. 1194–1195, 1988.

57. G. L. Li, R. B. Welstand, J. T. Zhu, and P. K. L. Yu, "Self-bias Control of Electroabsorption Waveguide Modulator," *IEEE MTT-S Digest (Intl. Microwave Symposium)*, **2**, pp. 1007–1010, 1998.

58. G. L. Li, P. K. L. Yu, W. S. C. Chang, K. K. Loi, C. K. Sun, and S. A. Pappert, "Concise RF Equivalent Circuit Model for Electroabsorption Modulators," *Electron. Lett.*, **36**, pp. 818–820, 2000.

59. G. L. Li, S. A. Pappert, P. Mages, C. K. Sun, W. S. C. Chang, and P. K. L. Yu, "High-saturation High-speed Traveling-wave InGaAsP-InP Electroabsorption Modulator," *IEEE Photon. Technol, Lett.*, **13**, pp. 1076–1078, 2001.

60. T. Ido, S. Tanaka, M. Suzuki, M. Koizumi, H. Sano, and H. Inoue, "Ultra-high-speed Multiple-quantum-well Electro-absorption Modulators with Integrated Waveguides," *IEEE J. Lightwave Technol.*, **14**, pp. 2026–2034, 1996.

61. O. Mitomi, I. Kotaka, K. Wakita, S. Nojima, K. Kawano, Y. Kawamura, and H. Asai, "40-GHz Bandwidth InGaAs/InAlAs Multiple Quantum Well Optical Intensity Modulator," *Appl. Opt.*, **31**, pp. 2030–2035, 1992.

62. K. Satzke, D. Baums, U. Cebulla, H. Haisch, D. Kaiser, E. Lach, E. Kuhn, J. Weber, R. Weinmann, P. Wiedemann, and E. Zielinski, "Ultrahigh-bandwidth (42 GHz) Polarization-independent Ridge Waveguide Electroabsorption Modulator Based on Tensile Strained InGaAsP MQW," *Electron. Lett.*, **31**, pp. 2030–2032, 1995.

63. K. K. Loi, X. B. Mei, J. H. Hodiak, C. W. Tu, and W. S. C. Chang, "38 GHz Bandwidth 1.3 mm MQW Electroabsorption Modulators for RF Photonic Links," *Electron. Lett.*, **34**, pp. 1018–1019, 1998.

64. C. Rolland, R. S. Moore, F. Shepherd, and G. Hillier, "10 Gbit/s, 1.56 μm Multiple Quantum Well InP/InGaAsP Mach-Zehnder Optical Modulator," *Electron. Lett.*, **29**, pp. 471–472, 1993.

65. D. Penninck and Ph. Delansay, "Comparison of the Propagation Performance over Standard Dispersive Fiber between InP-based π-phase-shifted and Symmetrical Mach-Zehnder Modulators," *IEEE Photon. Technol. Lett.*, **9**, pp. 1250–1252, 1997.

66. K. Wakita, O. Mitomi, I. Kotaka, S. Nojima, and Y. Kawamura, "High-speed Electrooptic Phase Modulators using InGaAs/InAlAs Multiple Quantum Well Waveguides," *IEEE Photon. Technol. Lett.*, **1**, pp. 441–442, 1989.

67. S. Nishimura, H. Inoue, H. Sano, and K. Ishida, "Electrooptic Effects in an InGaAs/InAlAs Multiquantum Well Structure," *IEEE Photon. Technol. Lett.*, **4**, pp. 1123–1126, 1992.

68. L. B. Soldano and E. C. M. Pennings, "Optical Multi-mode Interference Devices based on Self-imaging: Principles and Applications," *J. Lightwave Technol.*, **13**, pp. 615–627, 1995.

69. J. C. Cartledge, "Optimizing the Bias and Modulation Voltages of MQW Mach-Zehnder Modulators for 10 Gb/s Transmission on Nondispersion Shifted Fiber," *J. Lightwave Technol.*, **17**, pp. 1142–1151, 1999.

70. R. Spickermann, S. R. Sakamoto, and N. Dagli, "GaAs-AlGaAs Traveling Wave Electrooptic Modulators," in *Optoelectronic Integrated Circuits, Proc. SPIE*, **3006**, pp. 272–279, 1997.

71. For example, D. M. Pozar, *Microwave Engineering*, Addison-Wesley, Reading, MA, 1990.

72. G. L. Li, C. K. Sun, S. A. Pappert, W. X. Chen, and P. K. L. Yu, "Ultra High-speed Traveling Wave Electroabsorption Modulator: Design and Analysis," *IEEE Trans. Microwave Theory Techniques*, **47**, pp. 1177–1783, 1999.

73. S. Morasca, F. Pozzi, and C. De Bernardi, "Measurement of Group Effective Index in Integrated Semiconductor Optical Waveguides," *IEEE Photon. Technol. Lett.*, **5**, pp. 40–42, 1993.

74. Y.-J. Chiu, H.-F. Chou, V. Kaman, P. Abraham, and J. E. Bowers, "High Extinction Ratio and Saturation Power Traveling-wave Electroabsorption Modulator," *IEEE Photon. Technol. Lett.*, **14**, pp. 792–794, 2002.

75. S. Irmscher, R. Lewen, and U. Eriksson, "Microwave properties of ultrahigh-speed traveling-wave electro-absorption modulators for 1.55 μm," *IPRM'2001 (Intl. Conf. on InP and Related Materials)*, Nara, Japan, paper IME2-1, May 2001.

76. H. H. Liao, X. B. Mei, K. K. Loi, C. W. Tu, P. M. Asbeck, and W. S. C. Chang, "Microwave Structures for Traveling-wave MQW Electro-absorption Modulators for Wide Band 1.3 mm Photonic Links," in Optoelectronic Integrated Circuits, *Proc. SPIE*, **3006**, pp. 291–300, 1997.

77. K. W. Hui, K. S. Chiang, B. Wu, and Z. H. Zhang, "Electrode Optimization for High-speed Traveling-wave Integrated Optic Modulators," *J. Lightwave Technol.*, **16**, pp. 232–238, 1998.

78. G. K. Gopalakrishnan, W. K. Burns, R. W. McElhanon, C. H. Bulmer, and A. S. Greenblatt, "Performance and Modeling of Broadband LiNbO₃ Traveling Wave Optical Intensity Modulators," *J. Lightwave Technol.*, **12**, pp. 1807–1819, 1994.

79. O. Mitomi, K. Noguchi, and H. Miyazawa, "Design of Ultra Broad Band LiNbO₃ Optical Modulators with Ridge Structure," *IEEE Trans. Microwave Theory Tech.*, **43**, pp. 2203–2207, 1995.

80. B. T. Szentkuti, "Simple Analysis of Anisotropic Microstrip Lines by a Transform Method," *Electron. Lett.*, **12**, pp. 672–673, 1976.

81. X. Zhang and T. Miyoshi, "Optimum Design of Coplanar Waveguide for LiNbO₃ Optical Modulator," *IEEE Trans. Microwave Theory Tech.*, **43**, pp. 523–528, 1995.

82. K. Noguchi, O. Mitomi, H. Miyazawa, and S. Seki, "A Broadband Ti:LiNbO$_3$ Optical Modulator with a Ridge Structure," *J. Lightwave Technol.*, **13**, pp. 1164–1168, 1995.

83. K. Noguchi, O. Mitomi, and H. Miyazawa, "Millimeter-wave Ti:LiNbO$_3$ Optical Modulators," *J. Lightwave Technol.*, **16**, pp. 615–619, 1998.

84. W. K. Burns, M. M. Howerton, R. P. Moeller, R. W. McElhanon, and A. S. Greenblatt, "Low drive voltage, 40 GHz LiNbO$_3$ modulators," *OFC'99 Technical Digest, (Optical Fiber Communication)*, ThT1-1, pp. 284–286, 1999.

85. K. Noguchi, H. Miyazawa, and O. Mitomi, "Frequency Dependent Propagation Characteristics of Coplanar Waveguide Electrode on 100 GHz Ti:LiNbO$_3$ Optical Modulator," *Electron. Lett.*, **34**, pp. 661–663, 1998.

86. G. K. Gopalakrishnan, W. K. Burns, and C. H. Bulmer, "Electrical Loss Mechanism in Traveling Wave Switch/Modulators," *Electron. Lett.*, **28**, pp. 207–209, 1992.

87. C. C. Teng, "Traveling-wave Polymeric Optical Intensity Modulator with more than 40 GHz of 3 dB Electrical Bandwidth," *Appl. Phys. Lett.*, **60**, pp. 1538–1540, 1992.

88. G. D. Girton, S. L. Kwiatkowski, G. F. Lipscomb, and R. S. Lytel, "20 GHz Electro-optic Polymer Mach-Zehnder Modulator," *Appl. Phys. Lett.*, **58**, pp. 1730–1732, 1991.

89. W. Wang, Y. Shi, D. J. Olson, W. Lin, and J. H. Bechtel, "Push-pull Poled Polymer Mach-Zehnder Modulators With a Single Microstrip Line Electrode," *IEEE Photon. Technol. Lett.*, **11**, pp. 51–53, 1999.

90. H. Erlig and H. Fetterman, "Polymer Modulators for the 40-Gbit/sec Market," *Lightwave*, pp. 175–176, August 2001.

91. R. G. Walker, "Electro-optic modulation at mm-wave frequencies in GaAs/AlGaAs guided wave devices," *IEEE LEOS Annual Meeting (LEOS'95)*, San Francisco, CA, pp. 118–119, 1995.

92. R. G. Walker, "High-speed III–V Semiconductor Intensity Modulators," *IEEE J. Quantum Electron.*, **27**, pp. 654–667, 1991.

93. R. G. Walker, R. A. Griffin, R. D. Harris, R. I. Johnstone, N. M. B. Perney, and N. D. Whitbread, "Integrated high-functionality GaAs modulators for 10 & 40 Gb/s transmission," *IPRM'2001 (Intl. Conf. on InP and Related Materials)*, Nara, Japan, paper IME3-1, May 2001.

94. R. Spickermann, S. R. Sakamoto, M. G. Peters, and N. Dagli, "GaAs/AlGaAs Traveling Wave Electro-optic Modulator with an Electric Bandwidth >40 GHz," *Electron. Lett.*, **32**, pp. 1095–1096, 1996.

95. S. R. Sakamoto, R. Spickermann, and N. Dagli, "Narrow Gap Coplanar Slow Wave Electrode for Travelling Wave Electro-optic Modulators," *Electron. Lett.*, **31**, pp. 1183–1185, 1995.

96. H. Takeuchi, "Ultra-fast electroabsorption modulator integrated DFB lasers," *IPRM'2001 (Intl. Conf. on InP and Related Materials)*, paper WA3-1, 2001.

97. H. Kawanishi, Y. Yamauchi, N. Mineo, Y. Shibuya, H. Murai, K. Yamada, and H. Wada, "Over-40-GHz modulation bandwidth of EAM-integrated DFB laser modules," *OFC'01 Technical Digest, (Optical Fiber Communication)*, paper MJ3, 2001.

98. H. Feng, T. Makino, S. Ogita, H. Maruyama, and M. Kondo, "40 Gb/s Electro-absorption-modulator-integrated DFB laser with optimized design," *OFC'02 Technical Digest, (Optical Fiber Communication)*, paper WV4, pp. 340–341, 2002.

99. For copackaged laser and GaAs MZM, please refer to the product catalog of Bookham-Marconi (Northamptonshire, U.K.); for copackaged laser and 10-Gb/s MQW MZM, please refer to the product catalog of Nortel Networks (Canada).

100. D. M. Adams, C. Rolland, N. Puetz, R. S. Moore, F. R. Shephers, H. B. Kim, and S. Bradshaw, "Mach-Zehnder Modulator Integrated with a Gain-coupled DFB Laser for 10 Gb/s 100 km NDSF Transmission at 1.55 μm," *Electron. Lett.*, **32**, pp. 485–486, 1996.

101. S. Lovisa, N. Bouche, H. Helmers, Y. Heymes, F. Brillouet, Y. Gottesman, and K. Rao, "Integrated Laser Mach-Zehnder Modulator on Indium Phosphide Free of Modulated-feedback," *IEEE Photon. Technol. Lett.*, **13**, pp. 1295–1297, 2001.

# 17

# OPTICAL MODULATION: ACOUSTO-OPTICAL DEVICES

CHEN S. TSAI

*Department of Electrical and Computer Engineering*
  *and Institute for Surface and Interface Science*
*University of California*
*Irvine, California*

## 17.1  INTRODUCTION

Acousto-optics broadly refers to the interactions between optical (light) waves and acoustic (sound) waves. In engineering, however, it is now common to refer to Acousto-optics more narrowly as influences of the latter upon the former, influences that have been successfully utilized to construct various types of devices of both scientific and technological importance. Significant influences are possible under certain situations as the refractive index gratings created by acoustic waves will cause diffraction or refraction of an incident light wave. Acousto-optics is further branched into two subareas: bulk-wave and guided-wave acousto-optics. In the former [1–11] both light and sound propagate as unguided (unconfined) columns of waves inside a medium. A great many studies in the area of bulk-wave acousto-optics since the 1960s have resulted in various types of bulk acousto-optical devices and subsystems. In the latter [12–21] both light and sound waves are confined to a small depth in a suitable solid substrate. This subarea has been a subject of intensive study since the early 1970s as an outgrowth of guided-wave optics science and technology [22–32] and surface acoustic wave device technology [33–37], which had undergone intensive research and development a few years earlier. These latest studies on guided-wave acousto-optics have generated many fruitful results [20, 21]. For example, the resulting wide-band planar guided-wave acousto-optical (AO) Bragg modulators and deflectors are now widely used in the development and realization of micro-optical modules for real-time processing of radar signals, for example, integrated optic RF spectrum analyzers. Serious attempts to realize a variety of other integrated AO device modules with applications to optical

*Handbook of Optical Components and Engineering*,   Edited by Kai Chang
ISBN 0-471-39055-0   © 2003 John Wiley & Sons, Inc.

communications, computing, and signal processing are also being made [38–46]. This chapter presents a brief review of bulk-wave acousto-optics and a detailed treatment of guided-wave acousto-optics with emphasis on the principles of AO Bragg diffraction, the resulting wide-band modulators and deflectors, and applications.

In Section 17.2 the geometry, working principles, and device characteristics of a basic bulk-wave AO Bragg modulator together with its applications are reviewed briefly. It suffices to present a brief review on this subarea as a number of excellent review papers and chapters have been written [1–11]. The basic configuration and mechanisms for planar guided-wave AO Bragg diffraction and the resulting diffraction efficiency and frequency response are then analyzed in detail using the coupled-mode technique (Section 17.3). For convenience, the analysis is carried out for the simple but basic case involving a single-surface acoustic wave (SAW) in a LiNbO₃ waveguide. The salient differences between guided-wave AO Bragg diffraction and bulk-wave AO Bragg diffraction are discussed. A comparison of the three major types of AO materials that have been studied is also given in Section 17.3. The key parameters of the resulting AO Bragg modulators and deflectors or Bragg cells and their inherent limitations are then identified and discussed (Section 17.4). Extension of the coupled-mode technique to analysis of AO Bragg diffraction from multiple SAWs, namely, multiple tilted SAWs and phased SAWs, is briefly discussed in Section 17.5. Subsequently, a number of SAW transducer configurations for realization of wide-band Bragg cells are described and compared. Also presented in Section 17.5 are the design, fabrication, testing, and measured performances of wide-band AO Bragg cells in $Y$-cut LiNbO₃ and Z-cut GaAs substrates. Some of the potential applications of wide-band guided-wave AO Bragg cells and modules in optical communications, computing, and RF signal processing are described in Section 17.6.

## 17.2   BULK-WAVE ACOUSTO-OPTICAL BRAGG DIFFRACTION AND APPLICATIONS

As an introduction to this chapter, a brief review of the geometry, working principles, and device characteristics of a basic bulk-wave AO Bragg modulator, as shown in Fig. 17.1, together with its applications now follows. A detailed treatment on application will be given in Section 17.5 since practically all applications of bulk-wave AO devices can be carried over to their guided-wave counterparts. A basic bulk-wave AO Bragg modulator or cell consists of an acoustic cell of sufficiently large aperture and a laser source, both operating at suitable wavelengths. The acoustic cell is in turn made of a column of isotropic or anisotropic material that possesses desirable acoustic, optical, and AO properties. The acoustic waves are generated by applying an RF signal (at a frequency ranging from tens to thousands of megahertz) upon a planar piezoelectric transducer that is bonded on one finely polished end face of the column [37]. A moving optical index grating is induced by the acoustic waves via an elasto-optical effect. We shall now consider isotropic and anisotropic AO Bragg diffraction separately.

### 17.2.1   Isotropic Acousto-optical Bragg Diffraction

For simplicity we shall first consider the case of an isotropic material such as fused quartz in which the refractive index of a light wave is independent of its direction of

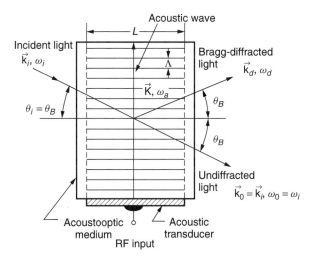

**Figure 17.1**   Basic bulk-wave acousto-optical Bragg modulator.

propagation or polarization. When a light wave is incident upon an acoustic cell at the Bragg angle $\theta_B$ as defined in Eq. (17.1), a portion of the light is diffracted mainly into one side at an angle that is twice the Bragg angle from the undiffracted (incident) light. The intensity of the diffracted light increases linearly with the power of the acoustic or RF driving signal before saturation is reached:

$$\sin \theta_B = \frac{\lambda_0}{2n\Lambda} = \frac{\lambda_0}{2nV_a} f_a \qquad (17.1)$$

In Eq. (17.1), $n$ and $\lambda_0$ designate, respectively, the refractive index of the medium and the wavelength of the light in free space; $\Lambda$, $f_a$, and $V_a$ are, respectively, the wavelength, frequency, and velocity of the acoustic wave.

It has been established that the conservation of both frequency (or energy) and the wave vector (or momentum) as expressed in Eqs. (17.2) and (17.3) is satisfied in AO Bragg diffraction:

$$\omega_d = \omega_i \pm \Omega$$

That is,

$$f_d = f_i \pm f_a \qquad (17.2)$$

and

$$\mathbf{k}_d = \mathbf{k}_i \pm \mathbf{K} \qquad (17.3)$$

where $f_i$, $f_d$, and $f_a$ designate, respectively, the frequencies of the incident light, the diffracted light, and the acoustic wave and $\mathbf{k}_i$, $\mathbf{k}_d$, and $\mathbf{K}$ are the corresponding wave vectors. Note that $\omega_i$, $\omega_d$, and $\Omega$ are the corresponding radian frequencies. For convenience, Eq. (17.3) is often represented as the wave vector diagram shown in Fig. 17.2. It is to be noted that the Bragg angle $\theta_B$ given in Eq. (17.1) is readily derived using Eqs. (17.2) and (17.3) together with the fact that polarization of diffracted light

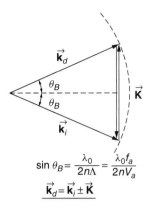

$$\sin \theta_B = \frac{\lambda_0}{2n\Lambda} = \frac{\lambda_0 f_a}{2nV_a}$$

$$\vec{k}_d = \vec{k}_i \pm \vec{K}$$

**Figure 17.2**   Wave vector diagram for isotropic accousto-optical Bragg diffraction.

is identical to that of incidental light and that in practice the acoustic frequency $f_a$ is much lower than the optical frequency $f_i$.

The three desirable characteristics of the basic bulk-wave AO Bragg cell just described, on which numerous applications are based, are now discussed. First, as indicated in Eq. (17.2), the frequency of the diffracted light is either upshifted or downshifted from that of the incident light by the acoustic frequency. The upshift and downshift correspond, respectively, to the cases in which the wave vector of the incident light wave is at an angle greater than and smaller than 90° from the wave vector of the acoustic wave. These frequency shifts can be explained satisfactorily in terms of Doppler effects as well as a heuristic picture [21]. This particular characteristic of the AO Bragg cell has been utilized to perform frequency modulation and frequency shifting of lasers.

Second, as indicated in Eqs. (17.1) and Fig. 17.2, the angle of deflection of the diffracted light varies linearly with the acoustic frequency. Thus, by tuning the frequency of the RF driving signal, the AO Bragg cell can be used to perform light beam deflection/scanning and switching. Finally, as mentioned previously, the intensity of the diffracted light is proportional to the power of the acoustic wave or the RF driving signal, provided the AO Bragg cell is not driven into saturation. This particular characteristic makes the AO Bragg cell one of the most practical intensity modulators for lasers. It should be noted that a variety of real-time signal-processing applications such as spectral analysis, correlation, and convolution of wide-band RF signals have also been realized using simultaneously the second and third characteristics.

### 17.2.2   Anisotropic Acousto-optical Bragg Diffraction

In the preceding discussion we have, for simplicity, considered bulk-wave AO Bragg diffraction in an optically isotropic material. However, since many superior AO materials are optically anisotropic, it is important to summarize the basic characteristics of anisotropic AO Bragg diffraction [6]. Again for simplicity, we shall consider a specific configuration (shown in Fig. 17.3) in which both the incident and the diffracted light as well as the acoustic wave propagate in a plane orthogonal to the $C$-axis of a uniaxial crystal such as LiNbO₃. It has been established that conservation of both frequency and the wave vector as expressed in Eqs. (17.2) and (17.3) still holds. However, through

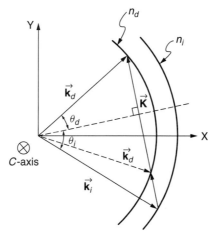

**Figure 17.3**   Wave vector diagram for anisotropic accousto-optical Bragg diffraction in a plane orthogonal to the $C$-axis of a uniaxial crystal.

optical anisotropy both the refractive index and the polarization of the diffracted light differ from that of the incident or undiffracted light. It can be shown that the wave vector diagram depicted in Fig. 17.3 leads to the following expressions for the incident angle $\theta_i$ and the diffraction angle $\theta_d$, which are both measured with respect to the acoustic wavefront:

$$\sin \theta_i = \frac{\lambda_0}{2 n_i V_a} \left[ f_a + \frac{V_a^2}{f_a \lambda_0^2} (n_i^2 - n_d^2) \right] \tag{17.4}$$

$$\sin \theta_d = \frac{\lambda_0}{2 n_i V_a} \left[ f_a - \frac{V_a^2}{f_a \lambda_0^2} (n_i^2 - n_d^2) \right] \tag{17.5}$$

where $n_i$ and $n_d$ designate the refractive indices of the incident and the diffracted light, respectively.

It is readily seen that the general expressions given in Eqs. (17.4) and (17.5) reduce to the simple Bragg condition given by Eq. (17.1) for an isotropic medium in which $n_d = n_i$. Figure 17.4 shows the plots for $\theta_i$ and $\theta_d$ as a function of the acoustic frequency. As a comparison, $\theta_i$ and $\theta_d (= \theta_i)$ for isotropic Bragg diffraction are also plotted as a function of the acoustic frequency in Fig. 17.4.

From Fig. 17.4, we identify two special cases of practical importance (shown in Figs. 17.5**a** and **b**. In the first case, which is commonly called optimized anisotropic interaction, wave vector phase matching may be satisfied for a large range of acoustic frequency. As a result, large-bandwidth light beam deflectors and modulators may be constructed utilizing this particular mode of interaction [10]. For the second case, the collinear geometry (or configuration) involved facilitates construction of electronically tunable AO filters for optical spectra [10b].

The simultaneous availability of a variety of laser sources, piezoelectric transducer technology for efficient generation of very high frequency acoustic waves, and superior solid materials has enabled realization of various types of bulk AO devices including modulators, scanners, deflectors, Q-switches, mode lockers, tunable filters, spectrum

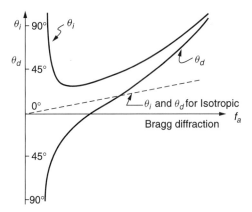

**Figure 17.4**  Variations of incident and diffraction angles versus acoustic frequency in anisotropic acousto-optical Bragg diffraction.

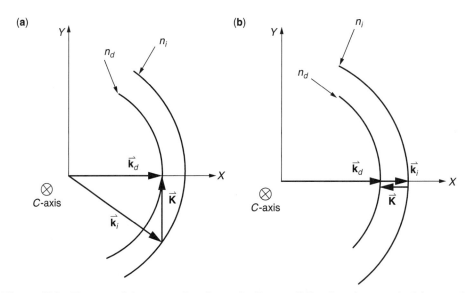

**Figure 17.5**  Two special cases of anisotropic Bragg diffraction in a uniaxial crystal: **(a)** optimized interaction; **(b)** collinear interaction.

analyzers, and correlators [1–11]. Such bulk AO devices have now been deployed in a variety of commercial and military applications.

## 17.3  GUIDED-WAVE ACOUSTO-OPTICAL BRAGG DIFFRACTION IN PLANAR WAVEGUIDES

### 17.3.1  Basic Interaction Configuration and Mechanisms

A basic coplanar guided-wave AO Bragg interaction configuration is shown in Fig. 17.6. Basic elements corresponding to those in Fig. 17.1 for bulk-wave AO Bragg interaction

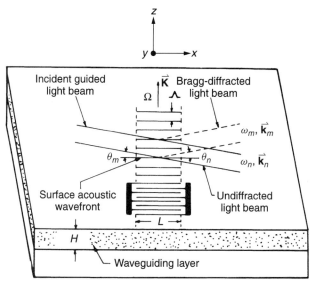

**Figure 17.6**  Guided-wave acousto-optical Bragg diffraction from a single-surface acoustic wave.

are easily identified. The optical waveguide material should possess desirable acoustic, optical, and AO properties. The SAW is commonly excited by an interdigital electrode transducer (IDT) [32–37] deposited upon the optical waveguide. If the substrate material is sufficiently piezoelectric, such as $Y$-cut LiNbO$_3$, the IDT may be deposited directly on it. Otherwise, e.g., GaAs or a piezoelectric film such as ZnO [35, 36, 47] must be deposited either beneath or above the interdigital electrode array. The optical waveguide can be either a graded index layer created beneath the substrate or a step index layer deposited on top of the substrate. Examples of graded index guides are LiNbO$_3$ waveguides formed by out- and in-diffusion techniques [50, 51] while examples of step index guides are glass [12] or As$_2$S$_3$ film [13] waveguides formed by sputtering or deposition. In either case, the waveguide layer is assumed to have an effective thickness or penetration depth $H$. Optical waveguide modes in such structures have been studied in detail [22, 25, 27, 28, 52–56].

Propagation of the SAW creates both a moving optical grating in the optical waveguide and moving corrugations at both the air–waveguide and waveguide–substrate interfaces. The moving grating and corrugations in turn cause diffraction of an incident guided light wave [12–21]. Thus, except for the corrugations, the underlying mechanisms in coplanar guided-wave AO interactions are analogous to those in bulk-wave AO interactions. While the contribution due to the corrugations can be significant for certain ranges of waveguide thickness and acoustic wavelength, in most practical cases, optical grating is the dominant mechanism in diffraction.

As for bulk-wave AO interaction, diffraction may be either Raman–Nath type or Bragg type. Raman–Nath diffraction consists of a number of side orders when the AO parameter $Q \equiv 2\pi \lambda_0 L / n \Lambda^2$ is less than or equal to 0.3. The symbols $\lambda_0$ and $\Lambda$ designate, respectively, the wavelengths of the guided optical wave (in free space) and the SAW, $n$ is the effective refractive index of the guiding medium, and $L$ designates the aperture of the SAW. When the light wave is incident at the Bragg angle defined in

Section 17.2 and $Q$ is larger than $4\pi$, diffraction is of the Bragg type and consists of one side order. We shall limit our discussion to Bragg diffraction because this type of diffraction is capable of higher acoustic center frequency, wider modulation bandwidth, and larger dynamic range and thus greater versatility in application.

It is important, however, to note that while the basic interaction mechanisms for planar guided-wave acousto-optics is analogous to that of bulk-wave acousto-optics, the number of parameters involved in the guided-wave case is greater and the interrelation between them much more complex [20]. For example, the diffraction efficiency is a sensitive function of the spatial distributions of both optical and acoustic waves, which in turn depend, respectively, on the guided optical modes and the acoustic frequency involved. In addition, in the case of a piezoelectric and electro-optical (EO) substrate such as LiNbO$_3$ and ZnO the piezoelectric field accompanying the SAW can be so large that the induced index changes due to the EO effect become very significant [20]. In fact, the contribution from the EO effect that accompanies a $Z$-propagation SAW in a $Y$-cut LiNbO$_3$ substrate is larger than that from the AO effect.

Similar to bulk-wave AO Bragg diffraction, the relevant energy (frequency) and momentum (wave vector) conservation relations between the incident light wave, the diffracted light wave, and the SAW are expressed by Eqs. (17.6a) and (17.6b), where $m$ and $n$ designate the indices of the waveguide modes; $\mathbf{k}_n$, $\mathbf{k}_m$, and $\mathbf{K}$ are, respectively, the wave vectors of the diffracted light, undiffracted light, and the SAW; and $\omega_n$, $\omega_m$, and $\Omega$ are the corresponding radian frequencies:

$$\omega_n = \omega_m \pm \mathbf{\Omega} \tag{17.6a}$$

$$\mathbf{k}_n = \mathbf{k}_m \pm \mathbf{K} \tag{17.6b}$$

Figure 17.7**a** shows the wave vector diagram for the general case in which the diffracted light propagates in a waveguide mode ($n$th mode) that is different from the incident light ($m$th mode). The diffracted light may have a polarization parallel or orthogonal to that of the incident light. Again, as in bulk-wave AO Bragg diffraction, these two classes of interaction are called isotropic and anisotropic AO Bragg diffraction. As a special isotropic case, the diffracted light propagates in the same mode and thus has the same polarization as the incident light, as depicted in Fig. 17.7**b** ($n = m$). Many guided-wave AO Bragg diffraction experiments using LiNbO$_3$ and GaAs belong to this particular case [21]. Figure 17.7**c** depicts a particularly interesting case of anisotropic Bragg diffraction in which the wave vector of the diffracted light is perpendicular or nearly perpendicular to that of the acoustic wave, commonly called optimized anisotropic diffraction [20, 21]. Note that this is a counterpart to Fig. 17.5**a**. Specific examples for both cases in a $Y$-cut LiNbO$_3$ substrate as depicted in Figs. 17.7**a** and **c** have been discussed [20, 21].

The angles of incidence and diffraction measured with respect to the acoustic wavefront, $\theta_m$ and $\theta_n$, are

$$\sin\theta_m = \frac{\lambda_0}{2n_m\Lambda}\left[1 + \frac{\Lambda^2}{\lambda_0^2}(n_m^2 - n_n^2)\right] \tag{17.7a}$$

$$\sin\theta_n = \frac{\lambda_0}{2n_n\Lambda}\left[1 - \frac{\Lambda^2}{\lambda_0^2}(n_m^2 - n_n^2)\right] \tag{17.7b}$$

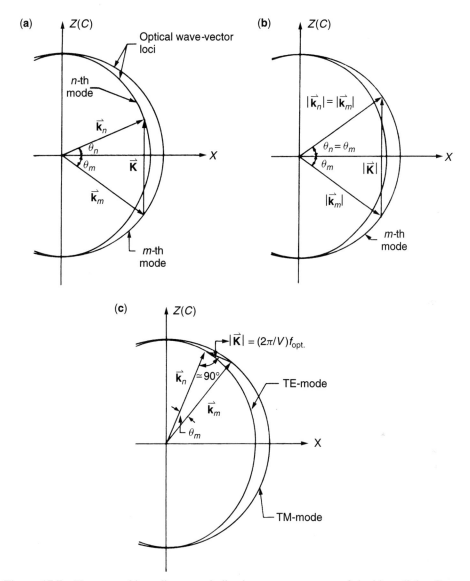

**Figure 17.7**  Phase-matching diagrams indicating wave vectors of incident light, Bragg-diffracted light, and surface acoustic wave in a $Y$-cut LiNbO$_3$ waveguide: (**a**) general case; (**b**) special isotropic case; (**c**) optimized anisotropic case.

where $n_m$ and $n_n$ are the effective refractive indices of the undiffracted and diffracted light waves and, as defined earlier, $\lambda_0$ and $\Lambda$ are the optical wavelength in free space and the wavelength of the SAW. Clearly, when the undiffracted and diffracted light propagate in the same waveguide mode (Fig. 17.7**b**), we have $n_n = n_m$. For this particular case Eqs. (17.7a) and (17.7b) both reduce to the well-known Bragg condition in isotropic diffraction [12], namely, $\sin\theta_n = \sin\theta_m = (\lambda_0/2n_m\Lambda) = (\lambda_0\omega/4\pi n_m V_R)$, where $\theta_n = \theta_m$ is the so-called Bragg angle and $V_R$ designates the propagation velocity of the SAW. Thus, the diffraction angle is identical to the incidence angle, and both angles increase linearly with the acoustic frequency because in practice the latter is much lower than the optical frequency.

### 17.3.2 Coupled-Mode Analysis on Acousto-optical Bragg Diffraction from a Single Surface Acoustic Wave

The most common approach for treatment of AO interactions is the so-called coupled-mode technique [3, 15–18, 20, 21]. This technique has been employed for the analysis of guided-wave AO Bragg diffraction from a single SAW in a LiNbO$_3$ planar waveguide [16]. Such an analysis serves to reveal the physical parameters involved and the key device parameters as well as the performance limitations of the resulting devices. Both the analytical procedures and the methodology for numerical computation developed for this simple case can be conveniently extended to the case involving multiple SAWs [17c] as well as other material substrates. The general case involving $N$ SAWs and the special cases involving multiple tilted SAWs, phased SAWs, and their combinations [17c] are briefly mentioned in Section 17.5.

Refer to the coordinate system of Fig. 17.6 and assume that the medium is lossless both optically and acoustically. First, appropriate forms for electric fields $\hat{E}_m(x, y, z, t)$ and $\hat{E}_n(x, y, z, t)$ of undiffracted and diffracted light waves and the strain field $\hat{S}(x, y, z, t)$ of the SAW and its accompanying piezoelectric field $\hat{E}_p(x, y, z, t)$ are taken. Note that the subscripts $m$ and $n$ designate, respectively, the waveguide modes of the undiffracted and diffracted light waves. Next, all the field quantities are substituted into the wave equation for the amplitude of the normalized electric fields of the optical waves $E_m(x)$ and $E_n(x)$ that depend on the $x$ coordinate only. A set of coupled-wave equations for $E_m(x)$ and $E_n(x)$ is then obtained. Finally, these coupled-wave equations are solved subject to the boundary conditions $E_m(0) \equiv 1$ and $E_n(0) \equiv 0$ at the input $x = 0$, resulting in electric fields of undiffracted and diffracted light waves at the output of the interaction region $x = L$ [16b].

### 17.3.3 Diffraction Efficiency and Frequency Response

From the electric fields of optical waves referred to in Section 17.3.2 the diffraction efficiency $\zeta(f_a)$, defined as the ratio of diffracted light power at the output ($x = L$) and incident light power at the input ($x = 0$) of the interaction region, is found [16b]:

$$\zeta(f_a) = g^2(f_a) \left( \frac{\sin[g^2(f_a) + (K\,\Delta\theta L/2)^2]^{1/2}}{[g^2(f_a) + (K\,\Delta\theta L/2)^2]^{1/2}} \right)^2 \qquad (17.8a)$$

where

$$g^2(f_a) \equiv \frac{\pi^2}{4\lambda_0^2} \frac{n_m^3 n_n^3 |\Gamma_{mn}(f_a)|^2 L^2}{\cos\theta_m \cos\theta_n} \qquad (17.8b)$$

$$|\Gamma_{mn}(f_a)|^2 \equiv \frac{\left| \left[ \int_0^\infty U_m(y) U_n(y)\{P : SU_a(y) + r : E_p U_p(y)\}\,dy \right] \right|^2}{\left| \left[ \int_{-\infty}^\infty U_m^2(y)\,dy \right] \right| \left| \left[ \int_{-\infty}^\infty U_n^2(y)\,dy \right] \right|} \qquad (17.8c)$$

and

$f_a = \Omega/2\pi$ = frequency of the SAW in cycles per second
$K = 2\pi/\Lambda$ = wave number of the SAW
$L$ = aperture of the SAW
$c$ = velocity of light in frequency space
$\Delta\theta$ = deviation of the incidence angle of the light from the Bragg angle

Also, $U_m(y)$, $U_n(y)$, $U_a(y)$, and $U_p(y)$ designate, respectively, the normalized field distributions (along the waveguide thickness) of the light waves, the acoustic wave, and the piezoelectric field. Finally, the following physical constants with suppressed vector and tensor subscripts are designated: $P$ is the relevant photoelastic constant or constants and $r$ is the relevant EO coefficient or coefficients.

The general expression given by Eq. (17.8a) can be used to calculate the diffraction efficiency and its frequency response as a function of both the polarization and the angle of incidence of the light [20]. Equation (17.8a) reduces to the following simple form when the Bragg condition is satisfied ($\Delta\theta \equiv 0$):

$$\zeta(f_a) = \sin^2 g(f_a) \qquad (17.9)$$

It is seen that in contrast to bulk-wave AO Bragg interaction, the diffraction efficiency is a sensitive function of the spatial distributions of the diffracted and the undiffracted light waves and the SAW as determined by the coupling function $|\Gamma_{mn}(f_a)|^2$. Since confinement of the SAW is proportional to the acoustic wavelength, $|\Gamma_{mn}(f_a)|^2$ is also dependent on the thickness of the optical waveguide with respect to the acoustic wavelength. Consequently, $|\Gamma_{mn}(f_a)|^2$ varies strongly with acoustic frequency. An efficient diffraction can occur only in the frequency range for which the confinement of the SAW matches that of the diffracted and undiffracted light waves. The dependence of the diffraction efficiency on acoustic, optical, and AO parameters is further complicated by the accompanying EO effect. Thus, in general, the coupling function $|\Gamma_{mn}(f_a)|^2$ is more complicated than the so-called overlap integral [10, 13, 15–18, 20, 21]. Only for the special case in which the accompanying EO contribution is either negligible (e.g., in a glass waveguide) or proportional to the elasto-optical contribution will Eq. (17.8c) reduce to one that is proportional to the overlap integral.

For the same reason, it is impossible to define an AO figure of merit as simply as in a bulk-wave AO interaction [7, 8]. However, if we again consider the special case just mentioned and employ a simplified model [19], we may define the total power flow of the SAW, $P_a$, as

$$P_a \equiv \tfrac{1}{2}\rho V_R^3 L S^2 \int_{-\infty}^{\infty} U_a^2(y)\, dy \qquad (17.10)$$

where $\rho$ and $V_R$ designate the density and the acoustic propagation velocity of the interaction medium, $S$ designates the acoustic strain, and $\int_{-\infty}^{\infty} U_a^2(y)\, dy$ carries a dimension in length. Equations (17.8b), (17.8c), and (17.10) are now combined to give the following expression for $g^2(f_a)$:

$$g^2(f_a) = \frac{\pi^2}{2\lambda_0^2}\frac{n_m^3 n_n^3 P^2}{\rho V_R^3} C_{mn}^2(f_a)\frac{L}{\cos\theta_m \cos\theta_n} P_a \qquad (17.11a)$$

where the frequency-dependent coupling coefficient $C_{mn}^2(f_a)$ is defined as follows:

$$C_{mn}^2(f_a) \equiv \frac{\left[\int_0^\infty U_m(y)U_n(y)U_a(y)\,dy\right]^2}{\left[\int_{-\infty}^\infty U_m^2(y)\,dy\right]\left[\int_{-\infty}^\infty U_n^2(y)\,dy\right]\left[\int_{-\infty}^\infty U_a^2(y)\,dy\right]} \tag{17.11b}$$

Note that $C_{mn}^2(f_a)$ takes a form similar to the overlap integral, with its value depending upon the optical and the acoustic modes of propagation and equal to unity for bulk-wave AO interactions. Also, the factor $n_m^3 n_n^3 P^2/\rho V_R^3$ is similar to the bulk-wave AO figure of merit $M_2$ [7, 8], henceforth designated by $M_{2mn}$.

From Eqs. (17.9), (17.10), and (17.11a) it is seen that for the case $g^2(f_a) \ll 1$ and $\Delta\theta \equiv 0$, the diffraction efficiency is approximately proportional to the total acoustic power $P_a$. The effect of this quasi-linear dependence on the dynamic range of the resulting AO Bragg modulators and deflectors is discussed in Section 17.4. It is also seen that for a fixed total acoustic power and a relatively low diffraction efficiency (e.g., 60% or less) the diffraction efficiency is linearly proportional to the acoustic aperture.

We shall now examine the quantitative dependence of the diffraction efficiency on the acoustic frequency. From Eqs. (17.8a) and (17.9) it is first noted that in contrast to the bulk-wave AO interaction, even at $\Delta\theta \equiv 0$, the diffraction efficiency is a sensitive function of the acoustic center frequency $f_0$ because of the frequency dependence of the SAW confinement. Thus, the bandwidth of a guided-wave AO Bragg modulator is mainly determined by the frequency dependence of three factors: the transducer conversion efficiency, the Bragg condition (phase matching), and SAW confinement. The individual and combined effects of these three factors on the frequency response can be determined using Eqs. (17.8a) to (17.8c) and a digital computer. For example, to study the effect of SAW confinement alone, we assume that the transducer bandwidth is so large that its frequency dependence can be ignored and the acoustic aperture is so small that the Bragg condition can be satisfied at all frequencies of interest. Similarly, by setting the right side of Eq. (17.8a) equal to one-half of its value at the acoustic center frequency $f_0$ and $\Delta\theta \equiv 0$, the $-3$-dB modulator bandwidth $\Delta f_{-3\,\mathrm{dB,Bragg}}$ as determined by the last two factors can be calculated [16b]. The results for the two types of calculations just described as applied to the $Y$-cut LiNbO$_3$ waveguides will be summarized in Section 17.3.4.

For the purpose of illustration we now calculate the diffraction efficiency at the acoustic center frequency $f_0$ for the special case in which the EO contribution is either negligible or proportional to the elasto-optical contribution. This calculation is applicable to the case involving nonpiezoelectric materials such as glass, oxidized silicon, and As$_2$S$_3$. We further assume that the power of the SAW is uniformly distributed in a depth of one acoustic wavelength [19]. Thus as a special form of Eq. (17.10), we have

$$P_a = \tfrac{1}{2}\rho V_R^3 S^2 L \Lambda_0 = \tfrac{1}{2}\rho V_R^4 S^2 \frac{L}{f_0} \tag{17.12a}$$

or

$$S = \left(\frac{2P_a}{\rho V_R^4}\right)^{1/2}\left(\frac{f_0}{L}\right)^{1/2} \tag{17.12b}$$

where $\Lambda_0$ designates the acoustic wavelength at the center frequency. Substituting Eq. (17.12b) into Eq. (17.8c) and combining Eqs. (17.8b), (17.8c), and (17.9) and restricting to the case of moderate diffraction, we obtain the following expression for the diffraction efficiency:

$$\zeta(f_0) \approx g^2(f_0) = \frac{\pi^2}{2\lambda_0^2} M_{2mn} \frac{f_0 L}{V_R \cos\theta_m \cos\theta_n} P_a \qquad (17.13)$$

Equation (17.13) shows that for this special case the diffraction efficiency is proportional to the product of the center frequency and the aperture of the SAW as well as the total acoustic power.

We now turn to the AO Bragg bandwidth. In contrast to the bulk-wave interaction [4], the AO Bragg bandwidth in general cannot be expressed explicitly in terms of the center frequency and the aperture of the acoustic wave. Only for the special case $g^2(f_a) \ll (K\Delta\theta L/2)^2$ would Eq. (17.8a) lead to an expression similar to that of the bulk-wave case. For this special case the absolute AO Bragg bandwidth for isotropic interaction with the $TE_0$ mode is given as follows:

$$\Delta f_{-3\,\text{dB,Bragg}} \approx \frac{1.8 \; n_0 V_R^2 \cos\theta_0}{\lambda_0 f_0 L} \quad \text{(isotropic)} \qquad (17.14a)$$

or

$$\frac{\Delta f_{-3\,\text{dB,Bragg}}}{f_0} \approx \frac{1.8 \; n_m V_R \cos\theta_0}{\lambda_0 f_0} \frac{\Lambda_0}{L} \quad \text{(isotropic)} \qquad (17.14b)$$

where $n_0$ and $\theta_0$ designate the effective refractive index and the Bragg angle for the $TE_0$ mode, and $\Lambda_0$ again designates the wavelength of the SAW at the center frequency. Earlier measurements in $Y$-cut $LiNbO_3$ waveguides have verified Eqs. (17.14a) and (17.14b) [16b]. The absolute AO Bragg bandwidth is inversely proportional to the product of the center frequency and the aperture of the SAW. Equations (17.13) and (17.14) indicate clearly that the diffraction efficiency and the AO Bragg bandwidth impose conflicting requirements on the acoustic frequency and the acoustic aperture. In fact, the diffraction efficiency–Bragg bandwidth product is a constant that is independent of both the center frequency and the aperture of the SAW.

In summary, because of the complicated spatial distributions of the guided-optical waves (GOW) and the SAW as well as the frequency dependence of the latter, numerical calculations using digital computers are required to determine the efficiency and exact frequency response of a guided-wave AO Bragg modulator. The procedure is as follows:

1. Obtain appropriate analytical expressions for the field distributions of the optical waves and the SAW based on their directions and modes of propagation.

2. Include the frequency dependence of the amplitudes, phases, and penetration depth of the SAW.

3. Identify the relevant photoelastic and EO constants.

4. Calculate the diffraction efficiency versus the acoustic frequency using a digital computer, with the acoustic drive power as a parameter, and inserting the information derived in steps 1–3 together with the remaining optical and acoustic parameters into Eq. (17.8).

Note that the frequency response of the SAW transducer can be incorporated in step 2.

### 17.3.4   Calculated Performances for a $Y$-Cut LiNbO$_3$ Substrate

Detailed numerical calculations and experiments have been carried out for both isotropic (Figs. 17.7**a** and **b**) and anisotropic (Fig. 17.7**c**) Bragg diffraction in $Y$-cut LiNbO$_3$ waveguides. While the latter is discussed in Ref. 18d, some results of the former are presented in this section. The isotropic Bragg diffraction to be given here involves a He−Ne laser light (0.6328 μm) propagating in the TE$_0$ mode in $Y$-cut LiNbO$_3$ in-diffused optical waveguides and a SAW propagating in the $Z(c)$ direction [15]. As indicated earlier, the contribution to Bragg diffraction by the EO effect is very important in a $Y$-cut LiNbO$_3$ substrate. Spatial distributions of optical fields in such waveguides [52−56] and those of strain fields and accompanying piezoelectric fields are taken from the literature [57−60]. Relevant photoelastic constants and EO coefficients are taken from Ref. 61. Specifically, for the case with a $Z$-propagation SAW the appropriate photoelastic constants are $P_{31} = P_{32}$, $P_{33}$. The only relevant EO coefficient is $r_{33}$.

First, the frequency response as determined by confinements of optical waves and the SAW is calculated. This is equivalent to calculating the coupling coefficient $C_{mn}^2(f_a)$ as a function of the acoustic frequency under the assumptions that the transducer bandwidth is so large that it does not introduce any band-limiting effect and that the Bragg condition ($\Delta\theta \equiv 0$) is satisfied at all frequencies of interest. While the first assumption leads to a constant total acoustic power, the second assumption is equivalent to utilization of a sufficiently small acoustic aperture. Subsequently, from Eqs. (17.8a)−(17.9) the relative Bragg diffraction efficiency as a function of acoustic frequency is calculated with the penetration depth of an optical waveguide mode as a parameter. Figure 17.8 shows the calculated results for the case involving a TE$_0$ mode for both diffracted and undiffracted light waves. Note that in these calculated plots the total acoustic power is assumed to be a constant, a consequence of the assumption that the transducer bandwidth is sufficiently large so that it does not introduce any band-limiting effect. These plots clearly show that the smaller the optical penetration depth, the higher will be the optimum acoustic frequency and the diffraction efficiency. It is seen that at the optical wavelength of 0.628 μm the desirable range of optical penetration depth is from 1.0 to 2.0 μm and the optimum range of acoustic frequency should be centered around 700 MHz.

In the preceding discussion, the transducer bandwidth is assumed to be sufficiently large and the acoustic aperture sufficiently small so that no bandwidth limiting could occur. In practice, the finite transducer bandwidth as well as the nonvanishing acoustic aperture must be included in the calculation of the ultimate frequency response of the AO Bragg modulator. For example, a set of frequency responses has been obtained for an optical waveguide of 2.0 μm penetration depth with the fractional transducer bandwidth and the acoustic aperture as parameters [20]. The total acoustic drive power and the peak diffraction efficiency corresponding to each combination of transducer bandwidth and acoustic aperture have also been tabulated [20]. Finally, the corresponding RF drive power is determined by the conversion efficiency of the transducer.

From the aforementioned set of frequency responses a number of conclusions can be drawn. First, for a given acoustic drive power the diffraction efficiency is proportional to the acoustic aperture, and similarly, for a given acoustic aperture the diffraction

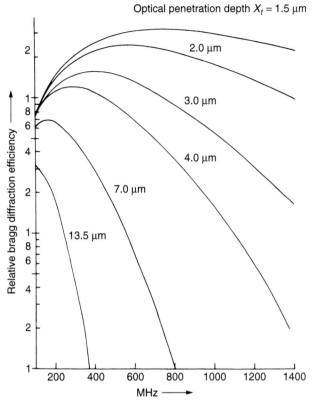

**Figure 17.8**   Frequency response based on coupling coefficient alone with penetration depth of $TE_0$ mode in a $Y$-cut waveguide as a parameter (for $Z$-propagation SAW).

efficiency is proportional to the acoustic drive power. As indicated earlier, this linear dependence is valid for a diffraction efficiency of ph to 60%. Second, the AO Bragg bandwidth decreases as the acoustic aperture increases and may become even smaller than the transducer bandwidth at a sufficiently large acoustic aperture. Also, for a given acoustic aperture the absolute bandwidth of a modulator that utilizes a single SAW of low center frequency is limited mainly by the transducer bandwidth of a periodic inter-digital transducer (IDT). On the other hand, for the same acoustic aperture the absolute bandwidth of a modulator that employs a single SAW of high center frequency is lim-ited mainly by the AO Bragg bandwidth. Clearly, such a set of frequency responses provides the basic data and guidelines required for the design of AO Bragg modulators and deflectors at 0.6328 μm optical wavelength using $Y$-cut $LiNbO_3$ waveguides. A similar set of frequency responses may be generated for AO Bragg modulators and deflectors at the diode laser wavelength.

In summary, the bandwidth of a guided-wave AO Bragg modulator or deflector that employs isotropic Bragg diffraction and a single periodic IDT of relatively large acoustic aperture is rather limited. For example, from the frequency responses referred to in the preceding, a maximum −3-dB modulator bandwidth of only 34 MHz is possible for a device using a $Y$-cut $LiNbO_3$ in-diffused waveguide of 2 μm penetration depth in which the aperture and the center frequency of the $Z$-propagation SAW are, respectively, 3 mm and 700 MHz. This small modulator bandwidth (34 MHz) is caused by the small AO Bragg bandwidth associated with the 3 mm acoustic aperture, irrespective of the fractional transducer bandwidth.

### 17.3.5   Acousto-optical Substrate Materials

Among the many materials that have been explored experimentally for guided-wave AO interactions, $Y$-cut LiNbO$_3$, nonpiezoelectrics such as oxidized Si, and GaAs have demonstrated the highest potential. Relevant physical parameters of these three AO materials together with a few other potential AO materials are listed in Table 17.1. The excitation and propagation of GOWs and SAWs and the AO interactions involved have been studied most thoroughly for $Y$-cut LiNbO$_3$ [20] and to a lesser extent for nonpiezoelectric [47–49] and GaAs [42–64] substrate materials. However, some significant experimental results have only been obtained recently for GaAs [64]. A comparison of these three substrate types now follows.

*LiNbO$_3$ Substrate.* As shown in Section 17.3.4 and also to be discussed in Section 17.5, efficient and wide-band AO Bragg diffraction can be readily realized in a $Y$-cut LiNbO$_3$ substrate, as shown in Fig. 17.6. Aside from a relatively high AO figure of merit, LiNbO$_3$ also possesses desirable acoustic and optical properties. As a result of large piezoelectricity, a SAW can be generated efficiently by directly depositing the IDT on the substrate. The typical propagation loss of the SAW is 1–2 dB/cm at 1.0 GHz, which is by far the lowest among all AO materials that have been studied. Optical waveguides can be routinely fabricated using the well-established titanium in-diffusion technique [51]. The measured optical propagation loss is typically 1.0 dB/cm, again the lowest among all existing AO materials. Furthermore, high-quality LiNbO$_3$ crystals of very large size are commercially available. Consequently, LiNbO$_3$ is at present the best substrate material for the realization of wide-band and efficient planar guided-wave AO Bragg cells at gigahertz center frequencies. Unfortunately, despite the many desirable properties referred to, only a partial or hybrid integration of passive and active components has been realized in LiNbO$_3$ thus far. The ultimate advantages of integrated optics cannot be fully realized until a viable technology has been developed that incorporates other suitable materials and thus lasers and detectors into the LiNbO$_3$ substrate to facilitate total or monolithic integration.

*SiO$_2$, As$_2$S$_3$, or SiO$_2$–Si Substrates.* The second substrate type is comprised of the nonpiezoelectric materials such as fused quartz (SiO$_2$) [12], arsenic trisulfide (As$_2$S$_3$) [13], and oxidized silicon (SiO$_2$–Si) [47–49]. Interest in these substrate materials is based on the fact that the first is a common optical waveguide material, the second is an amorphous material with a very high AO figure of merit, and the third may be used to capitalize on the existing silicon technology to further electrical and optical integrations. While it is common to deposit a piezoelectric zinc oxide (ZnO) film on such substrate materials for the purpose of SAW generation, as depicted in Fig. 17.9, the ZnO–SiO$_2$ composite waveguide has also been utilized to facilitate guided-wave AO interaction because the ZnO film itself also possesses favorable optical waveguiding and AO properties [47–49, 66–68].

Guided-wave AO Bragg diffraction in such substrate materials has also been ana-lyzed in detail using the coupled-mode technique [49c]. The expression for the Bragg

**TABLE 17.1  Relevant Physical Parameters for AO Materials**

| Material | Range of Optical Transmission (μm) | $n_o$ | $n_e$ | $n = (n_o + n_e)/2$ | $\rho$ (g/cm³) | Acoustic Wave Polarization and Direction | $V_s$ (10⁵ cm/s) | Optical Wave Polarization and Direction | $M_2$ (10⁻¹⁸ s³/g) | Photoelastic Coefficients |
|---|---|---|---|---|---|---|---|---|---|---|
| Fused quartz | 0.2–4.5 | — | — | 1.46 | 2.2 | Long. | 6.95 | ⊥ | 1.51 | $P_{11} = 0.121$, $P_{12} = 0.270$, $P_{44} = -0.075$ |
| LiNbO₃ | 0.4–4.5 | 2.29 | 2.20 | 2.25 | 4.7 | Trans. | 3.76 | ‖ or ⊥ | 0.467 | $P_{11} = 0.036$, $P_{12} = 0.072$, $P_{31} = 0.178$, $P_{13} = 0.092$, $P_{33} = 0.088$, $P_{41} = 0.155$ |
|  |  |  |  |  |  | Long. in [11–20] | 6.57 | — | 6.99 |  |
| GaAs | 1.0–11.0 | — | — | 3.37 | 5.34 | Long. in [110] | 5.15 | ‖ | 104 | $P_{11} = -0.165$, $P_{12} = -0.140$, $P_{44} = 0.072$ |
| Gap | 0.6–10.0 | — | — | 3.31 | 4.13 | Trans. in [100] | 3.32 | ‖ or ⊥ in [010] | 46.3 | $P_{11} = -0.151$, $P_{12} = -0.082$, $P_{44} = -0.074$ |
|  |  |  |  |  |  | Long. in [110] | 6.32 | ‖ | 44.6 |  |
| As₂S₃ | 0.6–11.0 | — | — | 2.61 | 3.20 | Trans. in [100] | 4.13 | ‖ or ⊥ in [101] | 24.1 | $P_{11} = 0.277$, $P_{12} = 0.272$ |
|  |  |  |  |  |  | Long. | 2.6 | ⊥ | 433 |  |
| Si | 1.5–12.0 | — | — | 3.435 | 2.33 | — | 8.95 | — | 6.2 | — |
| TeO₂ | 0.35–5.0 | 2.26 | 2.41 | 2.34 | 5.99 | Long. [001] | 4.20 | ⊥ | 34.5 | $P_{11} = 0.0074$, $P_{12} = 0.187$, $P_{13} = 0.340$, $P_{31} = 0.0905$, $P_{33} = 0.240$, $P_{44} = -0.17$, $P_{66} = -0.0463$ |
|  |  |  |  |  |  | Trans. [110] | 0.62 | ‖ | 25.6 |  |
| Tl₃AsS₄ | 0.6–12.0 | — | — | — | — | — | — | — | 800 | — |

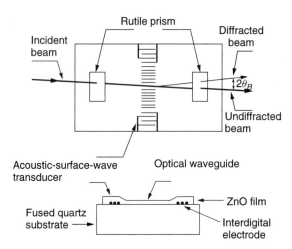

**Figure 17.9**  Guided-wave acousto-optical Bragg diffraction in a $ZnO-SiO_2$ composite waveguide.

diffraction efficiency takes a similar form as that for the $LiNbO_3$ substrate. However, it should be noted that for the case involving the $ZnO-SiO_2$ composite waveguide the distribution of the SAW fields exhibits a marked variation with the thickness of the ZnO film, while the distribution of the GOW fields is practically unaffected. It was shown that the Bragg diffraction efficiency depends strongly on the thickness of the ZnO film and that for each optical waveguide mode there exists a film thickness for which the diffraction efficiency diminishes. Subsequently, a strain-controlling film of fused quartz was introduced on top of the ZnO film to enhance the diffraction efficiency. In a separate study the ZnO film was used solely to generate the SAW for AO Bragg diffraction experiments in the $SiO_2-Si$ substrate at a SAW frequency of up to 700 MHz [68].

At the present time, measured propagation losses of the SAW in the $ZnO-SiO_2-Si$ composite substrate [47–49, 66] are much higher than those in the $LiNbO_3$ substrate and considerably higher than those in the GaAs substrate. This contrast is accentuated as the acoustic frequency goes beyond 200 MHz. Thus, it may be concluded that until a composite substrate with a greatly reduced SAW propagation loss has been developed, guided-wave AO Bragg devices using the foregoing nonpiezoelectric substrate materials have to be limited to a considerably lower RF center frequency and a smaller bandwidth than those using the $LiNbO_3$ substrate.

*GaAs Substrate.* GaAs-based substrate can potentially provide the capability for the total or monolithic integration referred to previously because both the laser sources [69, 70] and the photodetector arrays as well as the associated electronic devices may be integrated in the same GaAs substrate [71–75]. Clearly, one of the key components in such future GaAs-based monolithic integrated AO modules or circuits is an efficient and wide-band AO Bragg modulator or deflector, as shown in Figs. 17.10**a** and 17.10**b**.

The AO Bragg cells that have been realized most recently utilize the interaction configuration shown in Fig. 17.10**a** in which the SAW propagates in the ⟨110⟩ direction [64a]. A previous theoretical study [63] has predicted an AO Bragg bandwidth as

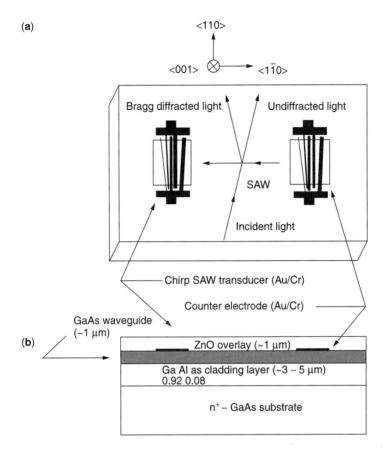

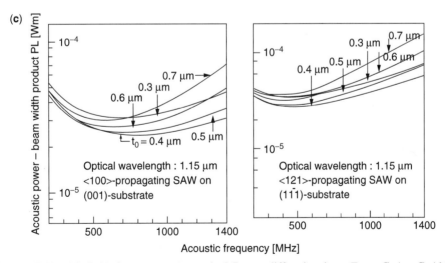

**Figure 17.10**   (**a**) Guided-wave acousto-optical Bragg diffraction in a Z-cut GaAs–GaAlAs planar waveguide (top view). (**b**) Waveguide cross section and ZnO transducer geometry (side view). (**c**) SAW power density required versus acoustic frequency for $TE_0$–$TE_0$ 100% Bragg diffraction in a single-mode waveguide.

large as 1.6 and 1.4 GHz for the $\langle 100 \rangle$ and $\langle 110 \rangle$ propagation SAW, respectively, in a Z-cut GaAs–GaAlAs waveguide that supports a $TE_0$ mode at the optical wavelength of 1.15 μm. Figure 17.10c shows a set of calculated AO frequency responses with the waveguide thickness as a parameter.

In fabrication, a high-quality GaAs–GaAlAs planar optical waveguide of large size (typically $2.2 \times 2.17$ cm$^2$) was first grown using an in-house liquid phase epitaxy (LPE) system in a 0.35-mm-thick silicon-doped GaAs substrate. The ZnO thin film and the interdigital finger electrode array were deposited subsequently. The cross section of the optical waveguide together with the ZnO thin-film overlay and the transducer geometry as well as their typical dimensions are shown in Fig. 17.10**b**.

The measured performance figures of the Bragg cells at the optical wavelengths of 1.15 and 1.30 μm and at various RF frequency bands are summarized in Table 17.2. Consider the Bragg cell that utilized a tilted-finger chirp transducer centered at 485 MHz. A −3-dB acoustic bandwidth of 250 MHz and a −3-dB AO bandwidth of 245 MHz were obtained. The diffraction efficiency was measured to be 5.0% at 1.0 W RF drive power. Next, consider the Bragg cell that utilizes the parallel-finger synchronous transducer centered at 1100 MHz [65]. A high diffraction efficiency was also measured in this case. This represents the highest acoustic frequency that has been employed for AO Bragg diffraction in a GaAs planar waveguide. In order to alleviate the resolution limitations with our existing fabrication facility, a parallel double-electrode synchronous transducer [76] was employed at its third harmonic for this purpose. A diffraction efficiency of 19.2% at 1.0 W RF drive power and an AO bandwidth of 78 MHz were measured. Based on the aforementioned results, it is reasonable to project realization of GaAs planar waveguide AO Bragg cells that operate at a center frequency of 1.0 GHz and a bandwidth of 500 MHz.

Also important for monolithic (total) integration in GaAs are the waveguide microlens and the linear lens array. For this purpose, fabrication of negative-index-change planar waveguide microlenses in both LiNbO$_3$ [77, 78] and GaAs [79] using

**TABLE 17.2   Measured Performances with GaAs Waveguide AO Bragg Cells**

| Acoustic center frequency, MHz | 200 | 360 | 400 | 485 | 600 | 800 | 950 | 1100 |
|---|---|---|---|---|---|---|---|---|
| Saw transducer types: parallel-finger synchronous (I), tilted-finger chirp (II) | I | II | I | II | I | I | I | I |
| Diffraction efficiency (mW acoustic power/mm acoustic aperture), % | 1.0 | 3.2 | 1.9 | — | 2.3 | — | — | 15 |
| −3-dB AO device bandwidth, MHz | 10 | 201 | 37 | 240 | 45 | 35 | 40 | 74 |
| Maximum measured AO diffraction efficiency, % | 65 | 15 | 28 | — | 9 | 9.6 | 18 | 30 |
| Measured or projected AO diffraction efficiency at 1.0 W RF power, % | 40 | 6.5 | 23 | 5.0 | 7.3 | 5.0 | 6.6 | 9 |
| RF drive power per megahertz bandwidth at 1.0% diffraction efficiency, mW/MHz | 2.5 | 0.8 | 1.2 | — | 3.0 | 5.7 | 3.7 | 1.5 |
| SAW propagation loss, dB/μs | 0.32 | 1.4 | 1.5 | 2.0 | 2.8 | — | — | 7.72 |

ion milling was successfully carried out recently. Thus, similar to titanium in-diffusion proton exchange (TIPE) waveguide lenses in LiNbO$_3$, such ion-milled waveguide lenses should facilitate realization of a variety of monolithic IO device modules and circuits in GaAs and related substrates with applications in communications, signal processing, and computing. Naturally, the resulting monolithic IO signal processor modules will include spectrum analyzers and correlators.

## 17.4  KEY PERFORMANCE PARAMETERS OF PLANAR GUIDED-WAVE ACOUSTO-OPTICAL BRAGG MODULATOR AND DEFLECTOR

The planar guided-wave AO Bragg diffraction treated in Section 17.3 can be utilized to modulate and/or deflect a light beam. The resulting light beam modulators and deflectors, commonly called AO Bragg cells, can operate at gigahertz center frequencies and over a wide RF band by using a variety of SAW transducer configurations to be presented in Section 17.5 and thus constitute useful devices for integrated and fiber-optical communication, computing, and signal-processing systems. The key device parameters that determine the ultimate performance characteristics of guided-wave AO modulators and deflectors are bandwidth, time–bandwidth product, acoustic and RF drive power, non-linearity, and dynamic range. A brief discussion of each of these performance parameters now follows.

### 17.4.1  Bandwidth

As shown in Section 17.3, the diffraction efficiency–bandwidth product of a planar guided-wave AO modulator that employs isotropic Bragg diffraction and a single periodic ID SAW transducer [34] is rather limited. However, if the absolute modulator bandwidth is the sole concern, a large bandwidth can be realized by using either a single periodic ID SAW transducer with small acoustic aperture and small number of finger pairs or a single aperiodic ID SAW transducer with small acoustic aperture and large number of finger pairs (chirp transducers) [79, 80]. It should be emphasized, however, that in either case the large bandwidth is obtained at a drastically reduced diffraction efficiency due to the very small acoustic aperture. Consequently, a higher diffraction efficiency will necessarily require large RF or acoustic drive power. Unfortunately, even if the supply of larger RF drive power is not a problem, the risk of transducer failure will increase with the RF drive power. Furthermore, large acoustic drive power will in turn result in an excessive acoustic power density, especially at high acoustic frequency, and thus increase the deleterious effects due to acoustic nonlinearity. A small transducer aperture will also result in a large acoustic radiation impedance [34] and thus in greater complexity in the electrical matching circuit required for optimum electrical-to-acoustic transduction. Therefore, it may be concluded that for applications that require both wide bandwidth and high diffraction efficiency, more sophisticated device configurations must be employed. The unified analysis referred to in Section 17.5 on AO Bragg diffraction from multiple SAWs suggests that this requirement can be met by employing multiple transducers of proper design and placement. Section 17.5.1 lists the five interaction and transducer combinations that have been explored for this purpose. It is now possible to realize high-performance planar guided-wave AO Bragg cells with gigahertz center frequency and gigahertz bandwidth using these wide-band configurations. The design, fabrication, testing, and measured performance figures for a variety of wide-band devices are presented in Sections 17.5.1 and 17.5.2.

### 17.4.2  Time–Bandwidth Product

The time–bandwidth product of an AO modulator, *TB*, is defined as the product of the acoustic transit time across the incident light beam aperture and the modulator bandwidth [1–21]. It is readily shown that this time–bandwidth product is identical to the number of resolvable spot diameters of an AO deflector, $N_R$, which is defined as the total angular scan of the diffracted light divided by the angular spread of the incident light. Thus the following well-known identities hold:

$$TB = N_R = \frac{D}{V_R}\Delta f = \tau\Delta f \tag{17.15a}$$

$$\delta f_R = \frac{V_R}{D} \tag{17.15b}$$

$$\tau = \frac{D}{V_R} \tag{17.15c}$$

where *D* designates the aperture of the incident light beam, $V_R$ the velocity of the SAW, $\Delta f$ the device bandwidth, $\tau$ the transit time of the SAW across the incident light beam aperture, and $\delta f_R$ the incremental frequency change required for deflection of one Rayleigh spot diameter. The acoustic transit time $\tau$ may be considered as the minimum AO switching time if the switching time of the RF driver is sufficiently smaller than the acoustic transit time. The desirable value for $N_R$ depends upon the individual application. For example, in signal processing it is desirable to have this value as large as possible because this value is identical to the processing gain. Thus, for this particular area of application it is also desirable to have a collimated incident light beam of large aperture. The light beam aperture is limited by the quality of the optical waveguide, the excitation mechanism, and the acoustic attenuation in the waveguide. As indicated in Section 17.3.5, acoustic attenuation is only about 1 dB/cm at 1 GHz in a *Y*-cut LiNbO₃ optical waveguide. Using a rutile prism coupler, a guided-light-beam aperture as large as 1.5 cm and good uniformity in this type of waveguide was demonstrated at the author's laboratory. This light beam aperture resulted in an acoustic transit time of about 4.4 μs for a *Z*-propagation SAW ($V_R = 3.488 \times 10^5$ cm/s). It is shown in Section 17.5.2 that a deflector bandwidth of up to 1 GHz can be realized readily using multiple SAW transducers, so that a time–bandwidth product as high as 4400 is achievable. At present the acoustic attenuations measured in all other waveguide materials at 1 GHz are significantly higher and thus considerably limit the maximum time–bandwidth product attainable.

In light modulation and switching applications, on the other hand, it is desirable to have the time–bandwidth product as close to unity as possible so that the highest modulation or switching speed can be achieved. For this purpose, the incident light is focused to a small beam diameter at the interaction region so that the corresponding acoustic transit time is a minimum [81, 82]. Since it is possible to focus both the incident light to a spot size of a few micrometers using "titanium in-diffusion proton-exchange (TIPE)" waveguide lens in the *Y*-cut LiNbO₃ substrate [39, 40] and the

$Z$-propagating SAW in the same substrate using a curved transducer, [83] a switching speed as high as 1 ns can be achieved.

### 17.4.3   Acoustic and RF Drive Power

Using Eqs. (17.10) and (17.12) and following the common practice of specifying the drive power requirement for 50% diffraction, we simply set $\zeta(f_a)$ of Eq. (17.9) equal to 0.50 to arrive at the following expression for the required acoustic drive power at the center frequency $f_0$:

$$P_a(50\% \text{ diffraction}) = \frac{\lambda_0^2 \cos\theta_m \cos\theta_n}{8} \frac{1}{M_{2mn,\text{eff}}} \frac{1}{L} \tag{17.16a}$$

where

$$M_{2mn,\text{eff}} \equiv C_{mn}^2(f_0) M_{2mn} \tag{17.16b}$$

Note that $C_{mn}^2(f_0)$ and $M_{2mn}$ have been defined previously. To determine explicitly the total RF drive power $P_e$, we must first calculate the electrical-to-acoustic conversion efficiency $T_c$ of the transducer used in the modulator or deflector. The frequency response and conversion efficiency of a periodic SAW IDT on LiNbO$_3$ substrates have been studied in detail in terms of one-dimensional equivalent circuits [34]. A detailed treatment is provided in Ref. 34. Accordingly, the total RF drive power at 50% diffraction $P_e(f_0)$ is given by $P_a$ (50% diffraction) divided by $T_c$. Clearly, both the total acoustic and RF drive powers required are inversely proportional to the acoustic aperture as in the bulk-wave modulator. However, in contrast to the bulk-wave case, because of the coupling factor $C_{mn}^2(f_a)$, the frequency dependence of both drive powers must be considered, even if the conversion efficiency of the transducer remains constant over the frequency band of interest.

Finally, the peak electrical-to-acoustic conversion efficiency of the transducer and thus the total RF drive power can be calculated in terms of the synchronous radiation resistance as well as the impedances of the matching network [84] and the RF generator. As shown in Ref. 34, the synchronous radiation resistance is inversely proportional to the product of the synchronous frequency and the acoustic aperture or inversely proportional to the acoustic aperture measured in terms of the synchronous acoustic wavelength. As will be discussed in Section 17.5, this specific dependence imposes a practical limit on the bandwidth of high-efficiency modulators and deflectors that utilize a single acoustic aperture to considerably smaller than 1 GHz. We shall show that using some of the wide-band transducer configurations to be discussed in Section 17.5.1, the AO Bragg cells requiring only milliwatts of electric drive power per megahertz of bandwidth at 50% diffraction efficiency with a bandwidth approaching 1 GHz can be realized.

### 17.4.4   Nonlinearity and Dynamic Range

The ultimate dynamic range of an AO modulator is determined by a number of factors including the nonlinearity inherent in the acoustic wave propagation and AO interaction process, nonlinearity in the RF driver circuit, the optical background noises resulting from scattering in the waveguide and in all passive and active optical components, and the dynamic range of the photodetector. In this chapter the discussion is limited to the acoustic and AO interaction nonlinearities as they are independent of the applications involved.

We again focus attention on the acoustic center frequency $f_0$. It is seen from Eqs. (17.8)–(17.13) that as the diffraction efficiency or the modulation index is increased, the accompanying distortion or nonlinearity is also increased [20]. However, for a very small modulation index it is linearly proportional to the power of the electric drive signal. This linear relationship is the basis of many modulation and signal-processing applications, such as the spectral analysis of RF signals to be discussed in Section 17.6. However, as the RF drive power is sufficiently increased, distortion and nonlinearity become severe and should be included.

In applications such as spectral analysis that involve multiple RF signals the inter- and cross-modulations due to multiple AO diffraction also constitute the sources of nonlinearity. Some measurements on such inter- and cross-modulations in guided-wave diffraction have been made [85]. Note that the accompanying nonlinearity is frequency dependent and thus may be important in applications involving very wide-band RF signals. The AO nonlinearity described sets an upper bound for the strength of the RF signal to be processed and thus limits the dynamic range for large signals. For small signals the dynamic range is limited by the scattered light and the noise current of the photodetector.

Finally, in some cases the effect of acoustic nonlinearity [86] on dynamic range must also be taken into account. Acoustic nonlinearity is particularly important at the higher acoustic frequencies because the absolute penetration depth of the SAW decreases as the frequency increases. As a result, the acoustic power density becomes so large that a significant portion of the fundamental frequency acoustic power is converted to harmonics. In addition, this depletion of fundamental frequency acoustic power increases with the propagation distance. The end result is that an incident light of large aperture will incur significant nonuniformity in the intensity of the diffracted light as well as spurious diffracted light. This in turn degrades the linearity of modulation as well as the spatial resolution in deflection and spectral analysis applications. In fact, this deleterious effect has been observed in an experiment at gigahertz frequencies in a $Y$-cut $LiNbO_3$ waveguide [20].

## 17.5   CONSTRUCTION OF WIDE-BAND GUIDED-WAVE ACOUSTO-OPTICAL BRAGG MODULATORS AND DEFLECTORS

As previously indicated, the planar guided-wave AO Bragg modulator and deflector of Fig. 17.6 constitutes a highly useful component in multichannel integrated and fiber-optical systems. Consequently, it is desirable to realize such modulators and deflectors with as large a bandwidth and as high a diffraction efficiency as possible.

The coupled-mode analysis presented in Sections 17.3.2–17.3.4 has shown that the diffraction efficiency–bandwidth product of a planar AO device that utilizes a single SAW is a constant and rather limited. It is intuitively clear, however, that a larger composite bandwidth and thus a larger diffraction efficiency–bandwidth product can be accomplished by employing multiple SAWs that are properly tailored and configured [5–18, 20]. As treated in detail in Ref. 17c, multiple SAWs of staggered center frequency and tilted propagation direction [15a, 16b] (Fig. 17.11) as well as

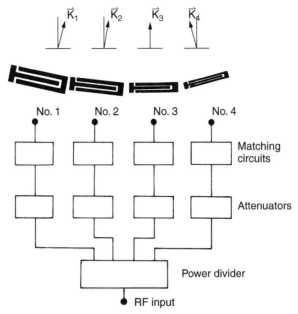

**Figure 17.11**  Multiple tilted SAW transducers of staggered center frequency and RF driver circuits.

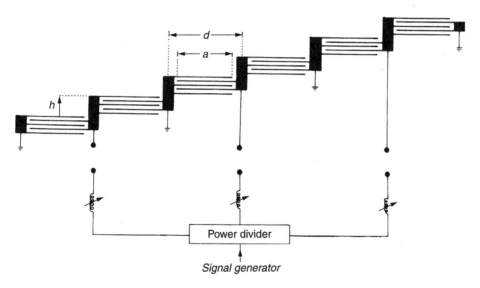

**Figure 17.12**  Phased-array transducer with matching and driver circuits.

phased multiple SAWs of identical center frequency and propagation direction [87] (Fig. 17.12) can be used to achieve this objective. A unified treatment has been developed to analyze the AO Bragg diffraction from $N$ SAWs [17c] (Fig. 17.13). This general approach can be employed to analyze the special cases involving the forementioned multiple tilted and phased SAWs or a combination of both.

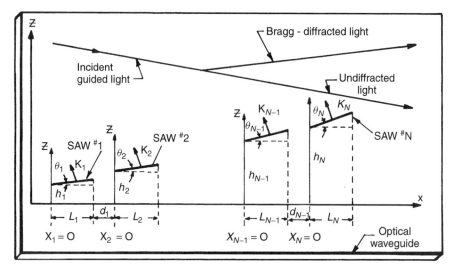

**Figure 17.13**   General interaction configuration for Bragg diffraction from multiple-surface acoustic waves.

In summary, the analysis and methodology for the numerical calculation presented in Sections 17.3.2–17.3.4 can be extended to the case involving $N$ SAWs, and it can be concluded that by using multiple SAWs with proper arrangements in center frequency, propagation direction, and placement, guided-wave AO Bragg devices with a large composite bandwidth and thus a large diffraction efficiency–bandwidth product can be realized. Realization of such wide-band and efficient AO Bragg devices using multiple SAWs is presented in this section.

Five wide-band device configurations have been explored for this purpose. These configurations utilize, respectively, the following interaction and SAW transducer combinations: (1) isotropic diffraction with multiple tilted transducers of staggered center (synchronous) frequency [15–18], (ii) isotropic diffraction with phased-array transducer [87], (iii) isotropic diffraction with a transducer array that combines that of (i) and (ii) [20], (iv) isotropic diffraction with a single tilted-finger chirp transducer [88] or an array of such transducers, and (v) optimized anisotropic diffraction [89] with multiple transducers of staggered center frequency or a parallel-finger chirp transducer. In earlier work these transducer types and arrangements were fabricated in $Y$-cut LiNbO$_3$ waveguides using the well-established photolithographic techniques [90]. In more recent work an electron beam lithographic technique [91, 92] was used to fabricate transducers of gigahertz center frequency in LiNbO$_3$. In even more recent work double-electrode transducers of gigahertz center frequency in GaAs were fabricated using the photolithographic technique [65]. The principle of operation and key design parameters and procedure for each wide-band device configuration have been treated in detail in Ref. 20. Because of space limitations only the first and the fourth wide-band device configurations are treated here. Fabrication, testing, and measured results for specific devices in $Y$-cut LiNbO$_3$ and an update of the performance figures obtained most recently with AO Bragg cells in $Z$-cut GaAs are then described. Some comments on the relative merits of a single transducer versus multiple transducers for wide-band AO Bragg diffraction are also made.

### 17.5.1 Design Guideline and Procedure of Wide-Band Device Configurations

*Isotropic Diffraction with Multiple Tilted Transducers of Staggered Center Frequency.* The transducer arrangement for this wide-band configuration is illustrated in Fig. 17.11. Individual periodic ID transducers are staggered in the center (synchronous) frequency and tilted in the acoustic propagation direction [15–18]. The tilt angle between each pair of adjacent transducers is set equal to the difference of the two Bragg angles at the two corresponding center frequencies. As indicated in Fig. 17.11, each element transducer is incorporated with a matching network, and the transducers are electrically driven in parallel through a power divider. Individual attenuators may also be incorporated between the outputs of the power divider and the inputs of the matching networks [20] to tailor the peak diffraction efficiency at each center frequency. It is clear that the multiple tilted SAWs generated by such a composite transducer satisfy the Bragg condition in each frequency band and thus enable a broad composite frequency response to be realized.

As indicated earlier, a rigorous analysis for this wide-band modulator–deflector configuration using the unified approach [17c] has been presented. Numerical computation of the frequency response based on this unified approach has also been carried out. The key design parameters and procedure that have been identified are now discussed. The key design parameters for the multiple tilted transducers are the center frequency, bandwidth, and aperture of individual element transducers, the tilt angle between adjacent element transducers, and the relative positions of adjacent element transducers. For each pair of adjacent element transducers the two center frequencies are chosen such that the individual frequency responses intersect at −6 dB down from the peak diffraction efficiency. The aperture of each element transducer must be sufficiently large to ensure efficient diffraction and in the meantime presents a suitable acoustic radiation impedance. Furthermore, the number of finger electrode pairs in each element transducer must be sufficiently small to ensure a fractional transducer bandwidth that is consistent with the required fractional AO Bragg bandwidth.

We now turn our attention to the relative positions of the element transducers. As a result of differences in the phase of SAWs generated by adjacent transducers and in the acoustic propagation path as measured from the front edge of the transducers to the interaction region, individual diffracted lights from adjacent SAWs may differ in the phase for the crossover frequencies. However, this phase difference may be compensated by properly configuring the element transducers through proper choice of both the horizontal separation $D_s$ and the vertical step height $h'$ between adjacent element transducers [20]. For the example involving transducers 1 and 2 shown in Fig. 17.14, horizontal separation $D_s$ and vertical step height $h'$ are

$$D_s = M \frac{2\Lambda_i^2}{\lambda} \tag{17.17a}$$

$$h' = \left( \frac{\lambda}{\Lambda_i} - \frac{\Lambda}{2\Lambda_{01}} \right) D_s \tag{17.17b}$$

where      $\lambda$ = wavelength of the light wave in the waveguide

$\Lambda_{01}, \Lambda_{02}$ = wavelength of the SAW at the center frequency $f_{01}$ for transducer 1 and the center frequency $f_{02}$ for transducer 2, respectively

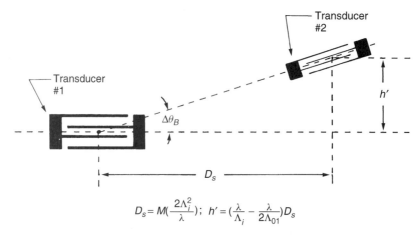

$$D_s = M(\frac{2\Lambda_i^2}{\lambda}); \quad h' = (\frac{\lambda}{\Lambda_i} - \frac{\lambda}{2\Lambda_{01}})D_s$$

**Figure 17.14**  Geometrical configuration of element transducers in a frequency-staggered multiple tilted transducer.

$\Lambda_i$ = wavelength of the SAW at the crossover frequency $(f_{01} + f_{02})/2$
$M$ = integer
$\Delta\theta_B$ = difference of the two Bragg angles at $f_{01}$ and $f_{02}$
  = $\sin^{-1}(\lambda/2\Lambda_{02}) - \sin^{-1}(\lambda/2\Lambda_{01})$

Note that a proper choice of $M$ is dictated by the apertures of the adjacent element transducers and that Eqs. (17.17a) and (17.17b) can be successively applied to determine the relative positions of all element transducers in the array.

Based on the aforementioned design guidelines, useful design procedures may be established. For simplicity, it is assumed that the fractional transducer bandwidths and the corresponding fractional AO Bragg bandwidths are identical for all element transducers but with the latter being slightly smaller than the former. In a more refined design the fractional transducer bandwidth and/or the fractional AO Bragg bandwidth may be varied among the element transducers. Thus, given the upper and lower −3-dB points of the modulator's or the deflector's frequency response, the number of tilted transducers and their center frequencies can be determined once the fractional AO Bragg bandwidth is chosen. Subsequently, the aperture of the element transducers can be determined by substituting the center frequency and the fractional AO Bragg bandwidth into Eq. (17.14). By choosing a fractional transducer bandwidth slightly larger than the fractional AO Bragg bandwidth, the number of finger electrode pairs can be determined. Positioning of the element transducers are then determined using Eqs. (17.17a) and (17.17b). Using the center frequency and the aperture of the individual transducers and the number of finger pairs just determined, electrical parameters such as radiation impedance, static capacitance, and conductive resistance for all individual transducers can be calculated. The matching network [84] and the attenuator for each element transducer can be designed accordingly. Finally, based on specifications for the upper and lower −3-dB frequencies the penetration depth (or the thickness) of the optical waveguide can also be determined using the frequency response plots given in Section 17.3.4. It should be noted that in some cases it may be more expedient to follow the design procedures just presented in reverse order. As discussed in the following section, a deflector of 680 MHz bandwidth was realized in a *Y*-cut titanium-diffused LiNbO$_3$ waveguide using the design procedures and the improved transducer geometry just described.

***Isotropic Diffraction with a Single Tilted-Finger Chirp Transducer or an Array of Such Transducer.*** We now return to Fig. 17.11 and consider the situation involving a large number of element transducers of closely spaced synchronous frequencies and each with a single pair of finger electrodes. It is easily shown that this situation results in a composite transducer of varying finger periodicity and tilted angle, as depicted in Fig. 17.15. This composite transducer is called a *tilted-finger chirp transducer* [88, 93, 94], consistent with the common usage of the term *chirp transducer* for a transducer of parallel finger but linearly varying finger periodicity [79, 80]. Like the conventional chirp transducer the acoustic bandwidth of this composite transducer can be very large. Furthermore, with proper design the wavefront of the SAW generated by this composite transducer can be made to track the Bragg condition for the entire frequency band and thus results in a large AO Bragg bandwidth. Thus, to the extent that the bandwidth imposed by the coupling coefficient $C^2_{mn}(f_a)$ is not the limiting factor, this composite transducer should also be capable of providing a large device bandwidth. Verification of the concept described in the preceding was first carried out using a simple tilted-finger chirp transducer fabricated in a titanium-diffused $Y$-cut LiNbO$_3$ waveguide [88]. Also, transducers of more sophisticated finger electrode arrangements was subsequently realized [20].

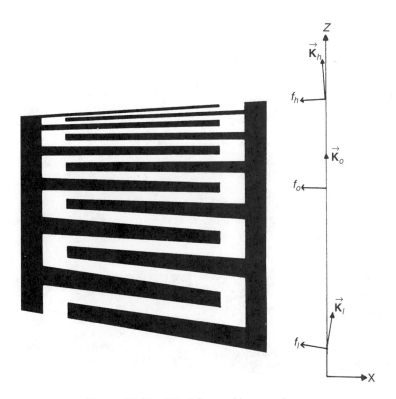

**Figure 17.15**   Tilted-finger chirp transducer.

### 17.5.2    Design, Fabrication, Testing, and Measured Device Performances

*LiNbO₃ AO Bragg Modulators and Deflectors.* Using the design procedures and guidelines presented in Section 17.5.1, a variety of wide-band AO Bragg modulators and deflectors have been designed and fabricated. For $Y$-cut $LiNbO_3$ substrates, single-mode planar optical waveguides were grown using either out-diffusion [50] or in-diffusion [51] and the interdigital SAW transducers were fabricated using the well-established lift-off method [90].

Testing and performance measurement of the devices were carried out using mostly an He–Ne laser light at 0.6328 μm. Excitation of the optical guided waves at $TE_0$ and $TM_0$ modes was accomplished using a rutile prism, and a second rutile prism was used to couple out both the diffracted and the undiffracted light beams for detailed measurement (see Fig. 17.16).

Deflected and undeflected light spots of high beam quality were obtained. For example, Fig. 17.17 shows the undeflected (when no RF power was applied to the device) and deflected light spots, both at the far field, for a device fabricated with the frequency of the driving signal varied from 155 to 410 MHz [16b]. The aperture

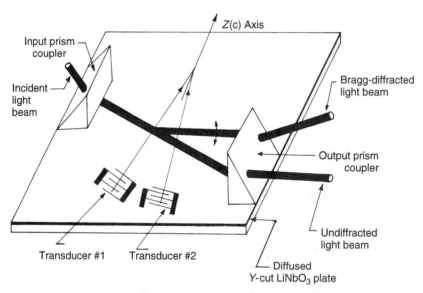

**Figure 17.16**    Basic experimental setup for measurement of planar guided-wave acousto-optical Bragg diffraction using a pair of prism couplers.

**Figure 17.17**    Undeflected light spot (left) and deflected light spots (right) obtained using tilted SAW array. SAW frequency is varied from 155 to 410 MHz at 15 MHz per step.

**Figure 17.18**  Photograph of a guided-wave acousto-optical Bragg deflector using multiple tilted SAW transducers.

of the incident light beam employed for this particular experiment was approximately 0.1 cm. The quality of the undeflected light beam (RF power off) was preserved in the deflected light beam, and deflected light beams of satisfactory quality were achieved. Measured values for the number of resolvable spot diameters $N_R$ and the incremental frequency change $\delta f_R$ required for deflection of one Rayleigh spot diameter were found to be in good agreement with those predicted using Eq. (17.15). For example, 400 resolvable frequency channels and a frequency resolution of approximately 0.8 MHz were measured with a deflector of 358 MHz bandwidth using a uniform light beam aperture of 4.5 mm [17c]. A photograph of one of the complete devices is shown in Fig. 17.18. Note that the prism coupler–LiNbO$_3$ plate combination was sitting in the middle of the brass platform. The input and output RF connectors and the matching circuits for the transducers were located at the right and the left of the combination. The complete device was attached to a precision holder (micromanipulator) to facilitate optical alignment and adjustment. The wide-band device configurations described in Section 17.5.1, each demonstrating certain distinct features, were tested, and the device performance was measured. Again, due to space limitations, detailed design specifications and measured performances are given here only for the first and the fourth wide-band device configurations.

*Isotropic Device with Multiple Tilted Transducers of Staggered Center Frequency.*  Using the improved transducer geometry and related design procedures presented in Section 17.5.1, a deflector of 680 MHz composite bandwidth was realized in a $Y$-cut Ti-diffused LiNbO$_3$ waveguide without an electronic phase shifter [88]. The deflector utilized four tilted transducers with center frequencies of 380, 520, 703, and 950 MHz. The corresponding acoustic apertures were 0.64, 0.47, 0.34, and 0.24 mm, respectively, and the tilt angles between adjacent transducers were 5.9, 8.2, and 11.3 mrad, corresponding to the difference in the Bragg angles at the center frequency of the adjacent transducers. In order to obtain as wide an acoustic bandwidth as possible, the number of finger electrode pairs for each transducer was chosen to be as small as two and a half. The measured acoustic bandwidths of approximately 28–35% of the center frequencies were obtained by inserting a single inductance of proper value to each transducer.

The individual transducers were excited in parallel using power dividers. A deflected light beam of high quality was observed. The measured conversion efficiency of the four element transducers were, respectively, $-7.5, -7.0, -10, -15$ dB. The measured diffraction efficiency was 8% at a total RF drive power of 1 W for the entire $-3$-dB bandwidth of 680 MHz. Since the measured diffraction efficiency of the Bragg cell with only the first three transducers activated was nearly five times higher, a considerably better diffraction efficiency could be expected if the conversion efficiency of the fourth transducer (the one with the highest center frequency) were improved to that of the first three. Also, once a better transducer conversion efficiency is achieved, a Bragg cell with gigahertz bandwidths can be realized by increasing the center frequencies of the element transducers. Based on this projection, a performance figure of approximately 1 mW electric drive power per megahertz bandwidth with 50% diffraction efficiency and 1 GHz bandwidth should be realizable. In conclusion, the experimental study has verified that efficient Bragg modulators and deflectors of large bandwidth at GHz center frequency can be realized using multiple tilted SAW transducers of staggered center frequency.

***Isotropic Device with a Tilted-Finger Chirp Transducer.*** The first tilted-finger chirp transducer was designed and fabricated in a $Y$-cut LiNbO$_3$ waveguide and Bragg diffraction experiments carried out using a 0.6328-μm He–Ne laser light propagating at the TE$_0$ mode [88]. The synchronous frequency of the finger electrodes was designed to vary linearly from 320 MHz ($f_l$) at one end to 630 MHz ($f_h$) at the other. The corresponding finger electrode width at the center of the finger aperture varies from 2.7 to 1.4 μm. The transducer contains a total of 51 finger electrodes each with an aperture of 0.55 mm. The transducer was driven directly with an RF signal generator of 50 Ω source impedance. The measured $-3$-dB transducer bandwidth was 255 MHz. Subsequently, a tilted-finger transducer of improved design using a "dog-leg" configuration was facilitated to realize a deflector bandwidth of 470 MHz [93]. The measured diffraction efficiency was 16% at 200 mW RF drive power. This improved deflector was subjected to 1 W of cw RF drive power without failure. Also, a wide-band transducer of slightly different electrode arrangement with similar performance characteristics was subsequently designed and fabricated [94, 95].

***GaAs AO Bragg Modulators and Deflectors.*** Using the same design procedures and guidelines, a variety of AO Bragg modulators and deflectors have also been designed and fabricated in GaAs waveguides. Fabrication steps involved are described in Section 17.3.5. Excitation of the optical guided waves at TE$_0$ and TM$_0$ modes was facilitated using edge-coupling via cleaved faces of the GaAs samples. Measured performances have been detailed in Section 17.3.5 and summarized in Table 17.2. It suffices to conclude that it is now possible to realize high-performance guided-wave AO Bragg cells in GaAs waveguides at GHz center frequency and octave bandwidth.

### 17.5.3   Relative Merits of a Single Transducer versus Multiple Transducers

As demonstrated in Section 17.5.2, both single transducers and multiple transducers (or array transducers) have been utilized successfully to construct wide-band planar

AO Bragg cells. It is thus appropriate to comment on the relative merits of a single transducer versus multiple transducers. A Bragg cell that employs a single tilted-finger chirp transducer will require no matching circuits and takes less space than the deflector that employs multiple transducers. However, for the Bragg cell that employs multiple transducers, the availability of multiple and independent RF inputs provides the flexibility for compensations and adjustments after the device has been fabricated. Such compensations and adjustments may be required because of the fabrication errors as well as variations in the physical properties of the AO material and in the quality of the waveguide and the transducers. By means of simple electrical attenuators and filter circuits, such compensations and adjustments can be easily made. Thus, it is reasonable to conclude that each wide-band device configuration has its relative merits, and selection of a particular one among the five would largely depend upon the flexibility and performance required in any specific application.

## 17.6 APPLICATIONS OF GUIDED-WAVE ACOUSTO-OPTICAL BRAGG CELLS AND MODULES

As shown in Sections 17.5.1 and 17.5.2, it is now possible to realize high-performance guided-wave AO Bragg modulators and deflectors of gigahertz center frequency and large bandwidth in LiNbO$_3$ and GaAs waveguides. Together with the fabrication of miniature laser sources, waveguide lenses, and photodetector arrays, integration of all passive and active components on a single substrate or a small number of substrates has become a reality [21]. The resulting integrated AO device modules or subsystems possess a number of attractive features such as low electrical drive power, small size, light weight, less susceptibility to environmental effects, and potentially low cost. A number of such integrated AO device modules that are being developed will find a number of unique applications [21] in wide-band multichannel optical communications, signal processing, and computing. In this section some of these device modules and some potential applications are described.

### 17.6.1 Optical Communications

High-speed light beam modulation, deflection and switching, optical frequency shifting, wavelength filtering, and wavelength multiplexing/demultiplexing are among the important functions in future wide-band multichannel integrated and fiber-optical communication systems [20, 21]. With regard to modulation, as indicated in Section 17.4, the existing guided-wave AO Bragg cells may provide a maximum modulation bandwidth of around 1 GHz. High-quality waveguide lenses, such as TIPE lenses [39] developed most recently, can readily produce the small beam waist required. In the meantime, a variety of EO modulators of multigigahertz bandwidth requiring very low drive power have been developed [96–108]. Consequently, existing AO Bragg modulators are in general not as competitive as EO modulators for applications that require nanosecond or sub-nanosecond modulation speed.

However, the situation is quite different with regard to multiport light beam deflection and switching. The AO deflectors discussed herein are capable of deflecting and switching a guided light beam into a large number of ports at moderate speed (microseconds and submicroseconds) and low electrical drive power per port. In contrast, EO deflectors and switches, although capable of a faster switching speed than AO deflectors and switches, can only provide either two ports or a relatively small number of ports per device [20, 21]. Consequently, AO deflectors are unique and superior in this particular area of applications. Like bulk-wave AO deflectors, guided-wave AO deflectors can function under either the digital (random-access) or the analog (sequential) mode of operation [20, 21]. A brief discussion of projected performance figures and potential applications now follows.

***Digital Deflection and Switching.*** In the digital mode of operation, the frequency of the RF signal applied to the deflector is varied in discrete steps to switch the light beam. The number of resolvable beam spots ($N_R$), the frequency step required for deflecting one resolvable spot position ($\delta f_R$) based on Rayleigh criteria, and the minimum switching time ($\tau$) are determined by Eqs. (17.15).

Equation (17.15c) indicates that the instantaneous scan rate $R$, that is, the number of resolvable beam positions or ports scanned per second, is given by $V_R/D$. To appreciate the order of magnitude involved, we consider two specific examples using the LiNbO$_3$ deflector described in Section 17.5.2. We assume that a deflector bandwidth of 500 MHz is used in both examples. In the first example, a light beam aperture of 3.45 mm is used, and we have $N_R = 500$, $\delta f_R = 1$ MHz, and $\tau = 10^{-6}$ s. The corresponding instantaneous scan rate is $10^6$ spots/s. In the second example the same deflector bandwidth is used, but a much smaller light beam aperture of 0.134 mm is assumed. We then have $N_R = 20$, $\delta f_R = 25$ MHz, and $\tau = 4 \times 10^{-9}$ s. The corresponding instantaneous scan rate is $25 \times 10^6$ spots/s. Thus, for a given deflector bandwidth the number of resolvable ports and the scan rate impose a conflicting requirement on the light beam aperture in a digital AO deflector. Nevertheless, the digital deflectors possess the unique capability for medium-speed random-access deflection and switching. One specific application involves a fiber-optical system with a large number of fan-out ports, as depicted in Fig. 17.19. A single-mode optical fiber that carries optical signals is coupled to an optical waveguide. The light beam is expanded and collimated by the TIPE lens and deflected by the AO Bragg cell. By varying the frequency of the RF driving signal the deflected light beams are then routed and

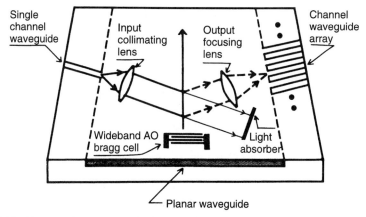

**Figure 17.19**   Integrated acousto-optical space division demultiplexer using a channel-planar composite waveguide in LiNbO$_3$.

coupled via the TIPE microlens array to any element of the optical fiber array that serves as the output terminal.

***Analog Deflection and Switching.*** A guided light beam may also be scanned and switched using an analog [frequency modulated (FM)] mode of operation in which the frequency of the electrical drive signal varies linearly with time. Using this mode of operation, it is possible to achieve a scan rate much greater than the scan rate that can be achieved using a digital mode of operation [20, 21]. This higher scan rate results from the fact that the optical grating created by the linear FM acoustic wave acts as a moving Fresnel zone lens to an incident collimated light beam. Thus, the diffracted light is focused to a small spot size and simultaneously scanned at the acoustic wave velocity to result in a very high scan rate. The relevant parameters include focal length, focused spot size, number of scan spots, and scan rate.

Some potential applications for the analog deflector described in the preceding paragraph include (a) very high data rate multiport switching; (b) very high data rate optical writing and reading in applications such as facsimile; (c) optical pulse compression of radar chirp signal [20]; (d) high-speed parallel-to-serial (spatial-to-temporal) readout for the RF spectrum of integrated optical spectrum analyzers [109]; and (e) time demultiplexing of wide-band multichannel optical pulse trains [20, 110–113]. It is to be noted that the first two applications are self-explanatory.

***Frequency Shifting.*** We now turn to optical frequency shifting. Light sources with electronically tunable frequency offset are needed as local oscillators in heterodyne integrated optical communication and fiber-optical sensor systems [114, 115]. A number of EO schemes for producing such frequency-shifted light sources in integrated optical format have been reported [21].

An alternate integrated optical scheme would be to utilize AO Bragg diffraction from a traveling SAW as the frequency of the diffracted light is shifted from that of the incident light by the acoustic frequency [20]. Earlier, three distinct device configurations, namely, those involving the planar waveguide [20, 21, 116], the channel waveguide, and the spherical waveguide [117, 118], were utilized to implement this AO frequency-shifting scheme. Most recently, a pair of tilted and counterpropagating SAWs in $Y$-cut LiNbO$_3$ planar waveguides were used to facilitate AO Bragg diffractions in cascade, and thus electronically tunable frequency shifting at passband and baseband [119, 120]. The frequency-shifted light propagates in a fixed direction, but spatially resolved from the incident light, irrespective of the magnitude of frequency tuning [119, 120]. These studies have shown that single-sideband frequency shifting using AO Bragg diffraction from the traveling SAW is a viable approach, and the resulting integrated optical modules can be readily and rigidly coupled with single-mode optical fibers to provide simultaneously the unshifted light and the frequency-shifted light sources that are spatially separated.

***Optical Wavelength Filtering and Multiplexing/Demultiplexing.*** With regard to optical wavelength filtering, both passive [121–128] and active [129–132] guided-wave devices have been explored. Active AO filters that utilize guided-wave anisotropic and noncollinear Bragg diffraction have also been reported [131]. Readers are referred to the literature for detail. Also, guided-wave AO wavelength division multiplexing (WDM) optical switches of high capacity can be constructed [133]. Finally, it is to be noted that guided-wave AO diffraction in LiNbO$_3$ was recently utilized to construct an integrated optical bistable device [134].

### 17.6.2 Optical Signal Processing

The block diagram of a typical optical signal processor [114] is shown in Fig. 17.20. It consists of an electrical-to-optical transducer (input transducer) that converts the time window of a wide-band electrical signal to a spatial optical display, an optical system that acts on this display to generate the Fourier transform of the signal or the correlation/convolution of the signal with a reference signal, and an optical-to-electrical transducer (output transducer) that reads out this optical Fourier transform or correlation/convolution as an electrical analog signal or digital bit stream. Optical signal processors can process wide-band signals at very high speeds and offer a significant advantage in hardware performance (equivalent bits per second per dollar) over digital electronic processors [135].

One of the most important classes of optical signal processors is based on the use of coherent AO interactions. A number of bulk-type AO signal processors have been studied and demonstrated [136–138]: spatial-integrating correlators, matched filters for chirp radar, convolvers, time-integrating correlators, snap-shot PROMs, and spectrum analyzers. The AO Bragg cell serves as an input transducer and together with the lens pair provides the optical system for all these optical processors. Finally, a photodetector or photodetector array is used as the output transducer. Most bulk-type AO processors for one-dimensional signal processing can be implemented in a guided-wave format [20] using the wide-band AO Bragg cells described in Section 17.5. As indicated earlier, these guided-wave versions possess a number of attractive features. However, it should be noted that an extension to two-dimensional signal processing is more difficult to implement in the guided-wave format than in the bulk-wave format. Again, because of space limitation, only the spectrum analyzers are discussed further.

As mentioned in Section 17.1, development and implementation of RF spectrum analyzers using monolithic or hybrid integrated optical techniques has already become worldwide [139–152]. A detailed treatment of the LiNbO$_3$-based integrated optical RF spectrum analyzer (IOSA) is given in Refs. 139–153. In this section only the principle of operation and some of the major parameters are discussed.

A schematic of a hybrid integrated optical RF spectrum analyzer is depicted in Fig. 17.21. When a spectrum of RF signals is applied to the transducer, each spectral

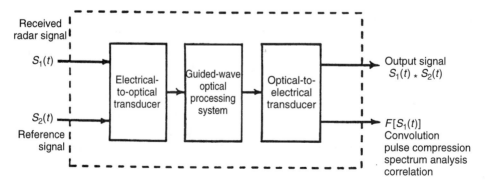

**Figure 17.20**   Block diagram of a basic guided-wave optical signal processor.

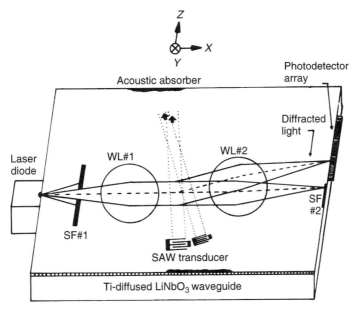

**Figure 17.21**   Acousto-optical spectrum analyzer using hybrid optical waveguide structure.

component generates a SAW that deflects the incident light beam in a corresponding direction. As shown in Section 17.3, the deflection angle and the intensity of the Bragg diffracted light are, respectively, proportional to the frequency and the power of the RF or the acoustic signal. Thus, by measuring the linear positions and intensities of the deflected light spots at the focal plane of the transform lens, the power spectral density of the RF signal of interest may be determined. The number of resolvable frequency channels $N_R$ and the corresponding frequency resolution $\delta f_R$, based on Rayleigh criteria, are also given by Eqs. (17.15a) and (17.15b).

As described in Section 17.5.2, 400 resolvable frequency channels and a frequency resolution of approximately 0.8 MHz were measured using a deflector of 358 MHz bandwidth and a uniform light beam aperture of 4.5 mm [16b]. For the 680-MHz-bandwidth deflector described in Section 17.5.2, the measured frequency resolution was 0.6 MHz for a truncated Gaussian light beam of 6-mm aperture. This resolution was determined by measuring the half-power width of the deflected light spot [16b]. Thus, based on the measured resolution, this deflector, when used as a spectrum analyzer, would provide 1130 resolvable frequency channels.

In addition to the number of resolvable frequency channels and the frequency reso-lution, intermodulation and cross-modulation between different frequency components resulting from multiple AO diffraction are among the other important parameters. The extent of such undesirable modulations was measured using two independent RF signals of varying frequency separation [20, 21, 85]. The measured data show that even for a worst case of as much as 43% diffraction, the intensity of the strongest intermodulation was −38 dB down from those of the two diffracted lights that result from the two funda-mental RF frequencies. For the practical cases the diffraction efficiency involved would be much lower than 43%, and the corresponding inter- and cross-modulation would accordingly be much lower. Thus, the dynamic range of guided-wave AO spectrum analyzers is likely to be limited by background noise (due to light scattering) rather than the inter- and cross-modulations due to multiple AO diffraction.

As previously indicated, development and realization of the integrated optical RF spectrum analyzers have become an international effort. Such integrated optical RF spectrum analyzers, when fully developed, are expected to possess two major advantages: (1) increased performance and reduced cost over both currently employed technology and competing technologies and (2) reduced size and increased compactness.

### 17.6.3    Optical Computing

Realization of optical computing functions in a waveguide substrate has long been considered one of the major objectives of integrated optics. Recently, the prospects for this realization have been significantly advanced through progress in guided-wave microoptic components of active and passive types in both LiNbO$_3$ [39–46] and GaAs [64, 78] substrates. In fact, a variety of integrated optic (IO) device modules have been realized in the LiNbO$_3$ substrate, and their counterparts in GaAs substrate are on the way to be developed. In this section, examples of the LiNbO$_3$ device modules together with applications to systolic array computing and programmable correlation of digital sequences are presented followed by a progress report on GaAs device modules.

*LiNbO$_3$-based Hybrid Multichannel Integrated Optical Device Modules.* Of importance in such device modules are the single-mode microlenses and microlens arrays fabricated in LiNbO$_3$ by the TIPE technique [39, 40]. These microlenses and lens arrays have demonstrated a combination of desirable characteristics. A TIPE microlens array and a large-aperture integrating lens together with a channel waveguide array, a planar waveguide, and an AO Bragg cell were integrated in a LiNbO$_3$ substrate $0.2 \times 1.0 \times 2.0$ cm$^3$ in size [41–45] (see Fig. 17.22).

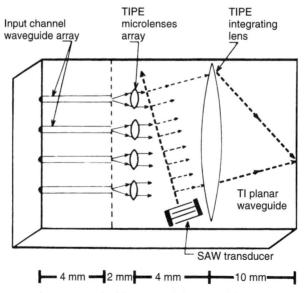

**Figure 17.22**    Integrated acousto-optical Bragg modulator module using a channel-planar composite waveguide in LiNbO$_3$.

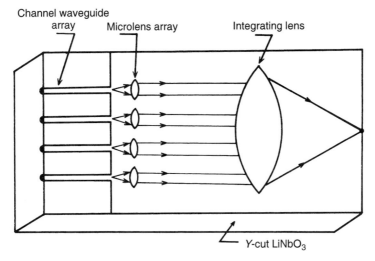

**Figure 17.23**  Linear microlens array–integrating lens combination in LiNbO$_3$ channel-planar composite waveguide.

The versatile structure that facilitates realization of a variety of IO device modules is shown in Fig. 17.23. First, the channel waveguide array was fabricated in a $Y$-cut LiNbO$_3$ substrate using the TIPE process. The channel waveguide array is followed by a TIPE linear microlens array fabricated in the planar waveguide. The microlens array is used to capture, expand, and collimate the multiple light beams from the channel waveguide array before incidence upon the AO, the EO, or the AO–EO Bragg diffraction gratings induced in the planar waveguide. Finally, a large-aperture TIPE lens, which is fabricated using the same masking step as the linear microlens array, serves to collect and focus the multiple Bragg-diffracted light beams upon a single photodetector or a photodetector array.

The integrated AO device module shown in Fig. 17.22 was used successfully to perform optical computing such as optical systolic array processing [41–46, 154] as it can readily perform the two required basic operations, namely, "multiplication" and "addition." For this particular application multiplication is facilitated by AO Bragg diffraction and addition by the integrating lens. Thus by pulsating the digital data sequences separately into the multiple input light beams (through the channel waveguide array), and the SAW high-speed digital filtering as well as matrix–vector and matrix–matrix multiplication can be performed. A simple experiment on matrix–vector multiplication involving a $2 \times 2$ matrix and a two-dimensional vector using two of the channel waveguide arrays and the SAW has been carried out [41–46]. In the following, two of the most advanced versions of the modules that utilize the basic channel-planar composite waveguides together with the microlenses and the active components are described.

Figures 17.24 and 17.25 show the architectures of these two device modules, each fabricated in a $Y$-cut LiNbO$_3$ substrate $0.1 \times 1.0 \times 2.0$ cm$^3$ in size. The width of each channel waveguide and the separation between adjacent waveguides of the 10-element channel waveguide array (only four elements are shown) fabricated are 6 and 200 $\mu$m, respectively. The linear microlens array, which consists of 10 identical lenses each with

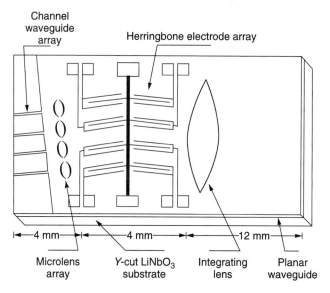

**Figure 17.24**   Multichannel integrated electro-optical device module.

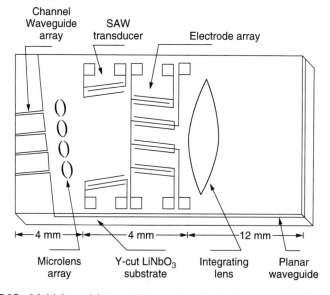

**Figure 17.25**   Multichannel integrated acousto-optical–electro-optical device module.

an aperture of 200 μm, is aligned with the channel waveguide array. Other dimensions along the optical path are shown in Figs. 17.24 and 17.25. A variety of multichannel device modules can be constructed by inserting a single AO Bragg modulator [41], an array of EO Bragg modulators [45], or a combination of the two [46] in the region between the microlens array and the larger-aperture lens. The herringbone Bragg electrode array and the SAW transducer–conventional Bragg electrode array combination used in the two device modules were fabricated using the lift-off technique.

Ideally, the multiple light sources required in the experiments with the two device modules just described should be furnished by a diode laser array. In the absence of such a diode laser array in our laboratory, the multiple light sources were formed by using a large-aperture collimated light beam together with a linear planar microlens array [44]. The separation between adjacent lenses in this planar microlens array was identical to that between adjacent channel waveguides, namely, 200 μm. As noted previously, each of the EO Bragg modulator arrays shown in Fig. 17.24 utilizes a herringbone Bragg electrode that consists of two identical conventional electrode arrays (each with a 13-μm periodicity and 1.9-mm aperture) intersecting at twice the Bragg angle. The two separate electrode arrays facilitate engagement of two independent sets of voltages and thus Bragg diffractions in tandem. Efficient and wide-band double Bragg diffractions, namely, 90% diffraction at a drive voltage of 5 V and a base bandwidth of 840 MHz, was measured at the optical wavelength of 0.6328 μm. A dynamic range of 31 dB was also measured. Therefore, high-performance herringbone Bragg electrode arrays have been fabricated successfully to perform multiplication of two independent sets of voltages. Addition of the resulting multiplication products is facilitated by the large-aperture integrating lens. This multichannel EO device module has been used to perform programmable correlation of binary sequences as well as matrix–vector and matrix–matrix multiplication [41–46, 154]. For example, Figs. 17.26a and b show, respectively, the autocorrelation waveform of the 10-bit binary sequence 1010101010 and the cross-correlation waveform between the sequences 1010101010 and 1000100000 obtained at a data rate of 3 kbits/s. Since the aperture of each EO Bragg modulator can be reduced further from 200 to 100 μm, it should be possible to construct device modules of very large bit capacity (word) (e.g., 100 bits in a LiNbO$_3$ substrate only slightly larger than $0.1 \times 1.0 \times 2.0$ cm$^3$ in size).

Note that the second device module shown in Fig. 17.25 results from the replacement of the first segment of the herringbone Bragg electrode structure by the AO Bragg diffraction grating [20] induced via the SAW, which is in turn generated by a single-SAW transducer. In order to facilitate AO and EO Bragg diffractions in tandem, the interdigital electrodes of the SAW transducer are tilted from the second segment of the herringbone electrode array by an angle that is equal to twice the AO Bragg angle $\theta_B$. A −3-dB AO bandwidth of 45 MHz centered at 260 MHz and a diffraction efficiency of 90% at an RF drive power of 150 mW were measured at the optical wavelength of 0.6328 μm. Thus, double Bragg diffractions of high efficiency and moderate bandwidth

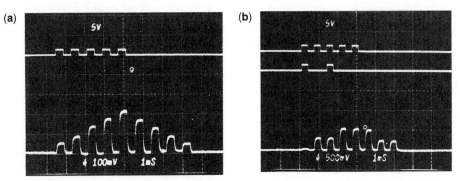

**Figure 17.26** (a) Autocorrelation waveform of 10-bit binary sequence. (b) Cross-correlation waveform of two 10-bit binary sequences.

have also been accomplished. Accordingly, this multichannel AO–EO module is being used to perform systolic array processing and programmable correlation of binary sequences at the data rate of 17.5 Mbits/s via the SAW.

It is important to compare the unique features of the two device modules. The EO module accepts multiple sets of data in parallel format and, therefore, can process the data at a considerably higher rate than is possible with the AO–EO module. However, through the SAW, the AO–EO module accepts data in a pipeline fashion that is desirable in a variety of applications.

***GaAs-based Monolithic Multichannel Integrated Optical Device Modules.*** For the GaAs substrate, as presented in Section 17.3.5, wide-band GaAs waveguide AO Bragg cells that operate in the acoustic frequency range of 300–1200 MHz have been realized. This recent advancement has paved the way for realization of monolithically integrated optical signal processors such as RF spectrum analyzers and correlators in a common GaAs chip. Recently there has also been significant advancement in the realization of multichannel IO modulator modules in GaAs. Specifically, multichannel single-mode EO cutoff modulator arrays [155] and EO Bragg diffraction modulators [156] have been successfully realized in GaAs. Also, waveguide microlenses and linear lens arrays were most recently fabricated in GaAs [78] using the ion milling technique. The wave-guide lenses that have been fabricated and tested include single lenses and analog Fresnel lens arrays, chirp grating, and hybrid analog Fresnel/chirp grating types. Near-diffraction-limited spot sizes and good efficiencies have been obtained. Thus, similar to the TIPE waveguide lenses in LiNbO$_3$, such ion-milled waveguide lenses should facilitate realization of a variety of monolithic IO device modules and circuits in GaAs and related substrates, with applications in communications, signal processing, and computing [157–159].

Most recently, toward eventual realization of monolithic integrated optic RF spectrum analyzers (IOSA), a wide-band AO Bragg cell was integrated with an ion-milled collimation-Fourier transform lens pair in a ZnO-GaAs-Al$_{0.15}$Ga$_{0.85}$ As composite waveguide $7 \times 23$ mm$^2$ in size [157].

The detailed structure of the IOSA is shown in Fig. 17.27. A GaAs guiding layer 1.1 $\mu$m thick and a ZnO layer 0.3 $\mu$m thick were grown to support propagation of the TE$_0$ and the TM$_0$ modes at the optical wavelength of 1.3 $\mu$m and to facilitate efficient and wide-band AO Bragg diffraction centered at 500 MHz. An incident light beam of 10 mW was coupled into the waveguide, and both the diffracted and the undiffracted light were coupled out of the waveguide via cleaved edges of the waveguide. The $f$-number of the collimating lens was 5 with an optical aperture of 1.2 mm, while that of the Fourier lens was 4.4 with an optical aperture of 1.8 mm. A diffraction efficiency of 7% per watt of RF drive power was measured using the TE$_0$ mode of propagation. The dynamic range defined as the diffracted light intensity at 50% diffraction efficiency over background noise level was measured to be higher than 16 dB. A large AO bandwidth of 220 MHz was facilitated by the tilted-finger chirp SAW transducer of 1.1 mm aperture.

The focal spot profiles of the undiffracted and diffracted light after propagating through the waveguide lens pair were measured to be almost identical, namely, a $-3$ dB focal spot width of 4.5 $\mu$m at the optical wavelength of 1.3 $\mu$m. The measured scanning of the focused spot of the diffracted light versus the frequency of the input RF signal shows that a scan of 1.03 $\mu$m per MHz has been demonstrated. Thus, a frequency

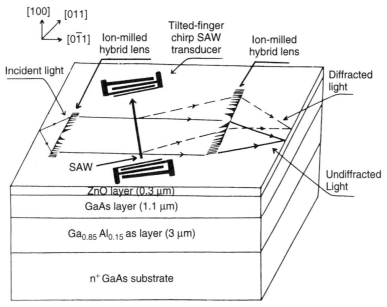

**Figure 17.27**  Acousto-optical RF spectrum analyzer in a ZnO/GaAs/AlGaAs composite waveguide structure.

increment of 4.5 MHz was needed to produce scan of a focused spot size of 4.5 μm. By simultaneously applying two RF signals of various frequency separation to the SAW transducer, the IOSA was found to be capable of resolving a frequency separation 5.5 MHz. Thus, a total of 40 channels for the frequencies centered at 500 MHz at a resolution 5.5 MHz have been accomplished with this first GaAs AO RF spectrum analyzer.

## REFERENCES

1. C. F. Quate, C. D. W. Wilkinson, and D. K. Winslow, "Interaction of Light and Microwave Sound," *Proc. IEEE*, **53**, pp. 1604–1623, 1965.

2. A. J. DeMaria and G. E. Danielson, Jr., "Internal Laser Modulation by Acoustic Lens-Like Effects," *IEEE J. Quantum Electron.*, **QE-2**, pp. 157–164, 1966.

3. A. Korpel, R. Adler, P. Desmares, and W. Watson, "A Television Display Using Acoustic Deflection and Modulation of Coherent Light," *Proc. IEEE*, **54**, pp. 1429–1437, 1966.

4. E. I. Gordon, "A Review of Acoustooptical Deflection and Modulation," *Proc. IEEE*, **54**, pp. 1391–1401, 1966.

5. R. Adler, "Interactions of Light and Sound," *IEEE Spectrum*, **4**, p. 42, 1967.

6. R. W. Dixon, "Acoustic Diffraction of Light in Anisotropic Media," *IEEE J. Quantum Electron.*, **QE-3**, p. 85, 1967.

7. R. W. Dixon, "Photoelastic Properties of Selected Materials and Their Relevance for Applications to Acoustic Light Modulators and Scanners," *J. Appl. Phys.*, **38**, pp. 5149–5153, 1967.

8. D. A. Pinnow, "Guide Lines for the Selection of Acoustooptic Materials," *IEEE J. Quantum Electron.*, **QE-6**, pp. 223–238, 1970.

9. N. Uchida and N. Niizeki, "Acoustooptic Deflection Materials and Techniques," *Proc. IEEE*, **61**, pp. 1073–1092, 1973.

10. I. C. Chang, "Acoustooptic Devices and Applications," *IEEE Trans Sonics Ultrasonics*, **SU-23**, pp. 2–22, 1976.

11. A. Korpel, "Acousto-Optics," in R. Wolfe, Ed., *Applied Solid-State Science*, Vol. 3, Academic, New York, 1972, Chap. 2, pp. 73–179.

12. L. Kuhn, M. D. Dakss, P. F. Heidrich, and B. A. Scott, "Deflection of an Optical Guided Wave by a Surface Acoustic Wave," *Appl. Phys. Lett.*, **17**, p. 265, 1970.

13. Y. Ohmachi, "Acousto-Optical Light Diffraction in Thin Films," *J. Appl. Phys.*, **44**, pp. 3928–3922, 1973.

14. R. V. Schmidt and I. P. Kaminow, "Acoustooptic Bragg Deflection in LiNbO₃ Ti-diffused Waveguides," *IEEE J. Quantum Electron.*, **QE-11**, pp. 57–59, 1975.

15. C. S. Tsai, L. T. Nguyen, S. K. Yao, and M. A. Alhaider, "High-Performance Guided-Light-Beam Device Using Two-Tilted Surface Acoustic Wave," *Appl. Phys. Lett.*, **26**, pp. 140–142, 1975.

16. C. S. Tsai, M. A. Alhaider, Le T. Nguyen, and B. Kim, "Wideband Guided-Wave Acoustooptic Bragg Diffraction and Devices Using Multiple Tilted Surface Acoustic Waves," *Proc. IEEE*, **64**, pp. 318–328, 1976.

17. B. Kim and C. S. Tsai, "High-Performance Guided-Wave Acoustooptic Scanning Devices Using Multiple Surface Acoustic Waves," *Proc. IEEE*, **64**, pp. 788–796, 1976.

18. C. S. Tsai, I. W. Yao, B. Kim, and Le T. Nguyen, "Wideband Guided-Wave Anisotropic Acoustooptic Bragg Diffraction in LiNbO₃ Wave-guides," *International Conference on Integrated Optics and Optical Fiber Communications*, July 18–20, 1977, Tokyo, Japan, Digest of Technical Papers, pp. 57–60.

19. R. V. Schmidt, "Acoustooptic Interactions between Guided Optical Waves and Acoustic Surface Waves," *IEEE Trans. Sonics Ultrasonics*, **SU-23**, pp. 22–23, 1976.

20. C. S. Tsai, "Guided-Wave Acoustooptic Bragg Modulators for Wideband Integrated Optic Communications and Signal Processing," *IEEE Trans. Circuits Syst.*, **CAS-26**, p. 1072, 1979.

21. C. S. Tsai, Ed., *Guided-Wave Acoustooptic Bragg Diffraction, Devices, And Applications*, Springer-Verlag, Berlin, 1990.

22. P. K. Tien, "Light Waves in Thin-Films and Integrated Optics," *Appl. Opt.*, **10**, pp. 2395–2413, 1971.

23. S. E. Miller, "A Survey of Integrated Optics," *IEEE J. Quant. Electron.*, **QE-8**, pp. 199–205, 1972.

24. H. F. Taylor and A. Yariv, "Guided-Wave Optics," *Proc. IEEE*, **62**, pp. 1044–1060, 1974.

25. H. Kogelnik, "An Introduction to Integrated Optics," *IEEE Trans. Microwave Theory Tech.*, **23**, pp. 2–16, 1975.

26. I. P. Kaminow, "Optical Waveguide Modulators," *IEEE Trans. Microwave Theory Tech.*, **23**, pp. 57–70, 1975.

27. P. K. Tien, "Integrated Optics and New Wave Phenomena in Optical Waveguides," *Rev. Mod. Phys.*, **49**, pp. 361–423, 1977.

28. T. Tamir, Ed., *Integrated Optics*, Springer-Verlag, Berlin, 1979.

29. A. Yariv, *Introduction to Optical Electronics*, 2nd ed., Holt, Rinehart and Winston, 1976.

30. R. Hunsperger, *Integrated Optics: Theory and Technology*, Springer-Verlag, Berlin, 1982.

31. H. P. Nolting and R. Ulrich, Ed., *Integrated Optics*, Springer-Verlag, Berlin, 1985.

32. R. M. White, "Surface Elastic Waves," *Proc. IEEE*, **58**, pp. 1238–1276, 1970.

33. A. A. Oliner, Ed., *Surface Acoustic Waves*, Springer-Verlag, Berlin, 1979.

34. R. W. Smith, H. M. Gerard, J. H. Collins, T. M. Reeder, and H. J. Shaw, "Design of Surface Wave Delay Lines with Interdigital Transducers," *IEEE Trans. Microwave Theory Tech.*, **MTT-17**, pp. 856–873, 1969.

35. G. S. Kino and R. S. Wagers, "Theory of Interdigital Couplers on Nonpiezoelectric Substrates," *J. Appl. Phys.*, **44**, pp. 1480–1488, 1973.

36. F. S. Hickernell and J. W. Brewer, "Surface-Elastic-Wave Properties of DC-Triode-sputtered Zinc Oxide Film," *Appl. Phys. Lett.*, **21**, pp. 389–391, 1972.

37. G. S. Kino, *Acoustic Waves: Devices, Imaging, And Analogy Signal Processing*, Prentice-Hall, Englewood-Cliffs, NJ, 1987.

38. C. V. Verber, R. P. Kenan, and J. R. Blusch, "Correlator Based on an Integrated Optical Spatial Light Modulator," *Appl. Opt.*, **20**, p. 1626, 1981.

39. D. Y. Zang and C. S. Tsai, "Single-Mode Waveguide Microlenses and Microlens Arrays Fabrication Using Titanium Indiffused Proton Exchange Technique in LiNbO$_3$", *Appl. Phys. Lett.*, **48**, pp. 703–705, 1985.

40. D. Y. Zang and C. S. Tsai, "Titanium-indiffused Proton-exchanged Waveguide Lenses in LiNbO$_3$ for Optical Information Processing," *Appl. Opt.*, **25**, pp. 2264–2271, 1986.

41. C. S. Tsai, D. Y. Zang, and P. Le, "Guided-Wave Acoustooptic Bragg Diffraction in a LiNbO$_3$ Channel-Planar Waveguide with Application to Optical Computing," *Appl. Phys. Lett.*, **47**, pp. 549–551, 1985.

42. C. S. Tsai, "LiNbO$_3$-based Integrated-Optic Device Modules for Communication, Computing, and Signal Processing," in *1986 Conference on Lasers and Electro-Optics, Technical Digest*, IEEE Cat. No. 86CH2274-9, 44–46, 1986.

43. C. S. Tsai, "Titanium-indiffused Proton-exchanged Microlens-based Integrated Optic Bragg Modulator Modules for Optical Computing," in H. H. Szu, Ed., Special Issue on Optical and Hybrid Computing, *SPIE*, **634**, pp. 409–421, 1987.

44. C. S. Tsai, D. Y. Zang, and P. Le, "Multichannel Integrated Optic Device modules in LiNbO$_3$ for Digital Data Processing," in *1988 Topical Meeting on Integrated and Guided-Wave Optics*, Santa Fe, NM, Technical Digest, pp. TuA5–1–TuA5-4.

45. P. Le, D. Y. Zang, and C. S. Tsai, "Integrated Electrooptic Bragg Modulator Modules for Matrix–Vector and Matrix–Matrix Multiplications", *Appl. Opt.*, Special issue on optical computing, **26**, pp. 1780–1785, 1988.

46. D. Y. Zang, P. Le, G. D. Xu, and C. S. Tsai, "An Integrated Optic Digital Correlator Module Using Both Acoustooptic and Electrooptic Bragg Diffractions" (in press).

47. N. Chubachi, J. Kushibiki, and Y. Kikuchi, "Monolithically Integrated Bragg Deflector for an Optical Guided Wave Made of Zinc-Oxide Film," *Electron. Lett.*, **9**, pp. 193–194, 1973.

48. N. Chubachi, "ZnO Films for Surface Acoustooptic Devices on Nonpiezo-electric Substrates," *Proc. IEEE*, **64**, pp. 772–774, 1976.

49. N. Chubachi and H. Sasaki, "Surface Acousto-Optic Interaction in ZnO Thin Films," *Wave Electron.*, **2**, p. 379, 1976.

50. I. P. Kaminow and J. R. Carruthers, "Optical Waveguiding Layers in LiNbO$_3$ and LiTaO$_3$," *Appl. Phys. Lett.*, **22**, pp. 326–329, 1973.

51. R. V. Schmidt and I. P. Kaminow, "Metal-diffused Optical Wave Guides in LiNbO$_3$," *Appl. Phys. Lett.*, **25**, pp. 459–460, 1974.

52. D. Marcuse, "TE Modes of Graded-Index Slab Waveguides," *IEEE J. Quantum Electron.*, **QE-9**, pp. 1000–1006, 1973.

53. E. M. Conwell, "Modes in Optical Waveguides Formed by Diffusion," *Appl. Phys. Lett.*, **26**, pp. 328–329, 1973.

54. P. K. Tien et al., "Optical Waveguide Modes in Single Crystalline LiNbO$_3$–LiTaO$_3$ Solid-Solution Films," *Appl. Phys. Lett.*, **24**, pp. 503–506, 1974.

55. H. Kogelnik, "Theory of Dielectric Waveguides," in T. Tamir, Ed., *Integrated Optics*, Springer-Verlag, Berlin, 1975, Chap. 2.

56. G. B. Hocker and W. K. Burns, "Modes in Diffused Optical Waveguides of Arbitrary Index Profile," *IEEE J. Quantum Electron.*, **QE-11**, pp. 270–276, June 1975.

57. G. W. Farnell, "Properties of Elastic Surface Waves," in W. P. Mason and R. N. Thurston, Eds., *Physical Acoustics*, Vol. 6, Academic, New York, 1970, pp. 109–166.

58. R. N. Spaight and G. G. Koerber, "Piezoelectric Surface Waves on LiNbO$_3$," *IEEE Trans. Sonics Ultrasonics*, **SU-18**, pp. 237–238, 1971.

59. A. J. Slobodnik, E. D. Conway, and R. T. Delmonico, Eds., *Microwave Acoustics Handbook*, Vol. 1A, *Surface Wave Velocities*, under AFCRL-TR-73-0593, October 1, 1973, Air Force Cambridge Research Laboratories, L.G. Hanscom Field, Bedford, MA.

60. B. A. Auld, *Acoustic Fields and Waves in Solids*, Vol. 2, Wiley, New York, 1973.

61. R. J. Pressley, Ed., *Handbook of Lasers with Selected Data of Optical Technology*, Chemical Rubber Co., Cleveland, OH, 1971.

62. K. W. Loh, W. S. C. Chang, W. R. Smith, and T. Grudkowski, "Bragg Coupling Efficiency for Guided Acoustic Interaction in GaAs," *Appl. Opt.*, **15**, pp. 156–166, 1976.

63. O. Yamazaki, C. S. Tsai, M. Umeda, L. S. Yap, C. J. Lii, K. Wasa, and J. Merz, "Guided-Wave Acoustooptic Interactions in GaAs–ZnO Composite Structure," *1982 IEEE Ultrasonics Symp. Proc.*, IEEE Cat. No. 82CH1823-4, pp. 418–421.

64. C. J. Lii, C. S. Tsai, and C. C. Lee, "Wideband Guided-Wave Acoustooptic Bragg Cells in GaAs–GaAlAs Waveguide," *IEEE J. Quantum Electron.*, **QE-22**, pp. 868–872, 1986.

65. Y. Abdelrazek and C. S. Tsai, "High-Performance Acoustooptic Bragg Cells in ZnO-GaAs Waveguide at GHz Frequencies," *Optoelectronics–Devices and Technologies*, **4**, pp. 33–37, 1989.

66. D. Mergerian, E. Malarkey, B. Newman, J. Lane, R. Weinert, B. R. McAvoy, and C. S. Tsai, "Zinc Oxide Transducer Array for Integrated Optics," *1978 Ultrasonics Symp., Proc.*, IEEE Cat. No. 78-CH1344-ISU, pp. 64–69.

67. S. K. Yao, R. R. August, and D. B. Anderson, "Guided-Wave Acoustooptic Interaction on Nonpiezoelectric Substrates," *J. Appl. Phys.*, **49**, pp. 5728–5730, 1978.

68. N. Mikoshiba, "Guided-Wave Acoustooptic Interactions in ZnO Film on Nonpiezoelectric Substrates," in C. S. Tsai, Ed., *Guided-Wave Acoustooptic Bragg Diffraction, Devices, and Applications*, Springer-Verlag, Berlin, 1988, Chap. 6.

69. A. Yariv and P. Yeh, *Optical Waves in Crystals*, Wiley, New York, 1984, Chap. 11.

70. Y. Suematsu, "Advances in Semiconductor Lasers," *Physics Today*, **32**, pp. 32–39, 1985.

71. J. L. Merz, R. A. Logan, and A. M. Sergent, "GaAs Integrated Optical Circuits by Wet Chemical Etching," *IEEE J. Quantum Electron.*, **QE-5**, pp. 72–82, 1979.

72. Y. R. Yuan, L. Perillo, and J. L. Merz, "Monolithic Integration of Curved Waveguides and Channeled-Substrate DH Lasers by Wet Chemical Etching," *J. Lightwave Tech.*, **LT-1**, pp. 630–636, 1983.

73. N. Bar-Chaim, S. Margalit, A. Yariv, and I. Ury, "GaAs Integrated Optoelectronics," *IEEE Trans. Electron. Devices*, **ED-29**, p. 1372, 1982.

74. A. Yariv, "The Beginning of Integrated Optoelectronic Circuits," *IEEE Trans. Electron. Devices*, **ED-3**, p. 1956, 1984.

75. O. Wada, T. Sakurai, and T. Nakagami, "Recent Progress in Optoelectronic Integrated Circuits," *IEEE J. Quantum Electron.*, **QE-22**, p. 805, 1986.

76. T. W. Bristol, W. R. Jones, P. B. Snow, and W. R. Smith, "Applications of Double Electrodes in Acoustic Surface Wave Device Design," *1972 IEEE Ultrasonics Symp. Proc.*, IEEE Cat. No. 72 CH0708-8SU, p. 343.

77. T. Q. Vu, J. A. Norris, and C. S. Tsai, "Formation of Negative Index Changes Waveguide Lenses in LiNbO$_3$ Using Ion Milling," *Opt. Lett.* **13**, pp. 1141–1143, 1988.

78. T. Q. Vu, J. A. Norris, and C. S. Tsai, "Planar Waveguide Lenses in GaAs Using Ion Milling," *Appl. Phys. Lett.* **54**, pp. 1098–1100, 1989.

79. R. H. Tancrell and M. G. Holland, "Acoustic Surface Wave Filters," *Proc. IEEE*, **59**, pp. 393–409, 1971.

80. W. R. Smith, H. M. Gerard, and W. R. Jones, "Analysis and Design of Dispersive Interdigital Surface-Wave Transducers," *IEEE Trans. Microwave Theory Tech.*, **MTT-20**, pp. 458–471, 1972.

81. H. V. Hance and J. K. Parks, "Wideband Modulation of a Laser Beam Using Bragg-Angle Diffraction by Amplitude Modulated Ultrasonics Waves," *J. Acoust. Soc. Amer.*, **38**, pp. 14–23, 1965.

82. D. Maydan, "Acoustooptical Pulse Modulators," *IEEE J. Quantum Electron.*, **QE-6**, pp. 15–24, 1970.

83. T. Van Duzer, "Lenses and Graded Films for Focusing and Guiding Acoustic Surface Waves," *Proc. IEEE*, **58**, pp. 1230–1237, 1970.

84. G. L. Matthaei, L. Young et al., *Microwave Filters, Impedance Matching Networks and Coupling Structures*, McGraw-Hill, New York, 1964.

85. C. S. Tsai, Le T. Nguyen, B. Kim, and I. W. Yao, "Guided-Wave Acoustooptic Signal Processors for Wideband Radar Systems," Special Issue on Effective Utilization of Optics in Radar Systems, *Proc. SPIE*, **128**, pp. 68–74, 1978.

86. E. G. Lean and C. C. Tseng, "Nonlinear Effects in Surface Acoustic Waves," *J. Appl. Phys.*, **41**, pp. 3912–3917, 1970.

87. Le T. Nguyen and C. S. Tsai, "Efficient Wideband Guided-Wave Acoustooptic Bragg Diffraction Using Phased-Surface Acoustic Wave Array in LiNbO$_3$ Waveguides," *Appl. Opt.*, **16**, pp. 1297–1304, 1977.

88. C. C. Lee, K. Y. Liao, C. L. Chang, and C. S. Tsai, "Wideband Guided-Wave Acoustooptic Bragg Deflector Using a Tilted-Finger Chirp Transducer," *IEEE J. Quantum Electron.*, **QE-15**, pp. 1166–1170, 1979.

89. C. S. Tsai, I. W. Yao, B. Kim, and Le T. Nguyen, "Wideband Guided-Wave Anisotropic Acoustooptic Bragg Diffraction in LiNbO$_3$ Waveguides," *International Conference on Integrated Optics and Optical Fiber Communications*, July 18–20, 1977, Tokyo, Japan, *Digest of Technical Papers*, pp. 57–60.

90. H. I. Smith, F. J. Bachner, and N. Efremow, "A High-Yield Photolithographic Technique for Surface Wave Devices," *J. Electrochem. Soc.*, **118**, pp. 822–825, May 1971.

91. T. Suhara, T. Shiono, H. Nishihara, and J. Koyama, "An Integrated-Optic Fourier Processor Using an Acoustooptic Deflector and Fresnel Lenses in As$_2$S$_3$ Waveguide," *J. Lightwave Technol.*, **LT-1**, pp. 624–630, 1983.

92. C. Stewart, G. Serivener, and W. J. Stewart, "Guided-Wave Acoustooptic Spectrum Analysis at Frequencies Above 1 GHz," *1981 International Conference On Integrated Optics And Optical Fiber Communication, Technical Digest*, April 1981, p. 122.

93. K. Y. Liao, C. L. Chang, C. C. Lee, and C. S. Tsai, "Progress on Wideband Guided-Wave Acoustooptic Bragg Deflector Using a Tilted-Finger Chirp Transducer," *1979 Ultrasonics Symp. Proc.*, IEEE Cat. No. 79CH1482-9SU, pp. 24–27.

94. T. R. Joseph, "Broadband Chirp Transducers for Integrated Optics Spectrum Analyzers," *1979 Ultrasonics Symp. Proc.*, IEEE Cat. No. 79CH1482-9SU, pp. 28–32.

95. D. Gregoris and V. M. Ristic, "Wideband Transducer for Acousto-optic Bragg Deflector," *Can. J. Phys.*, **63**, pp. 195–197, 1985.

96. H. Kogelnik and R. V. Schmidt, "Switched Directional Couplers with Alternating $\Delta\beta$," *IEEE J. Quantum Electron.*, **QE-12**, p. 396, 1976.

97. R. F. Alferness, "Guided-Wave Devices for Optical Communication," *IEEE J. Quantum Electron.*, **QE-17**, p. 946, 1981.

98. C. S. Tsai, B. Kim, and F. R. El-Akkari, "Optical Channel Wave-Guide Switch and Coupler Using Total Internal Reflection," *IEEE J. Quantum Electron.*, **QE-14**, p. 513, 1978.

99. C. L. Chang, F. R. El-Akkari, and C. S. Tsai, "Fabrication and Testing of Optical Channel Waveguide TIR Switching Networks," *Proc. SPIE*, **239**, p. 147, 1981.

100. C. S. Tsai, C. C. Lee, and P. Le, "A 8.5 GHz Bandwidth Single-Mode Crossed Channel Waveguide TIR Modulator and Switch in $LiNbO_3$," *Technical Digest of Postdeadline Papers, 1984 Topical Meeting on Integrated and Guided-Wave Optics*, IEEE Cat. No. 84CH1997-6, pp. PD 5-1–PD 5-4.

101. K. Wasa, H. Adachi, T. Kawaguchi, K. Ohji, and K. Setsune, "Optical TIR Switches Using PLZT Thin Film Waveguides on Sapphire," *International Conference on Integrated Optics And Optical Fiber Communications*, June 27–30, 1983, Tokyo, Japan, *Technical Digest*, pp. 356–357.

102. A. Neyer, "Electro-optic X-switch using single-mode Ti:$LiNbO_3$ channel waveguides," *Electron. Lett.*, **19,** p. 553, 1983.

103. H. Nakajima, I. Sawaki, M. Seino, and K. Asama, "Bipolar Voltage Controlled Optical Switch Using Ti:$LiNbO_3$ Intersecting Waveguides," in *Proc. 4th Int. Conf. IO IOFC*, Tokyo, Japan, 1983, paper 29C4-5, pp. 363–366.

104. M. Papuchon, Y. Comemale, X. Mathieu, D. B. Ostrowsky, L. Reiber, A. M. Roy, B. Sejourne, and M. Werner, "Electrically Switched Optical Directional Coupler: COBRA," *Appl. Phys. Lett.*, **27**, pp. 289–291, 1975.

105. O. Mikami and H. Nakagome, "Waveguided Optical Switch in InGaAs/InP Using Free-Carrier Plasma Dispersion," *Electron. Lett.*, **20**(6), pp. 228–229, 1984.

106. K. Tada and Y. Okada, "Bipolar Transistor Carrier-injected Optical Modulator/Switch: Proposal and Analysis," *IEEE Electron. Device Lett.*, **EDL-7**(11), pp. 605–606, 1986.

107. K. Ishida, H. Nakamura, and H. Matsumura, "In GaAsP/InP Optical Switching Using Carrier Induced Refractive Index Change," *Appl. Phys. Lett.*, **50**, pp. 141–142, 1987.

108. J. P. Lorenzo and R. A. Soref," 1.3 $\mu$m Electro-optic Silicon Switch," *Appl. Phys. Lett.*, **51**, pp. 6–8, 1987.

109. C. C. Lee and C. S. Tsai, "An Acoustooptic Readout Scheme for Integrated Optic RF Spectrum Analyzer," *1978 Ultrasonics Symp. Proc.*, IEEE Cat. No. 78 CH1344-ISU, pp. 79–81.

110. C. S. Tsai, "The Increase of Bragg Diffraction Intensity Due to Acoustic Resonance and Its Application for Demultiplexing and Multiplexing in Laser Communication," *Appl. Opt.*, **10**, pp. 215–218, 1971.

111. S. K. Yao and C. S. Tsai, "Acoustooptic Bragg-Diffraction with Application to Ultrahigh Data Rate Laser Communication Systems, Part I — Theoretical Considerations of the Standing Wave Ultrasonic Bragg Cell," *Appl. Opt.*, **16**, pp. 3032–3043, 1977.

112. C. S. Tsai and S. K. Yao, Part II — Experimental Results of the SUBC and the Acoustooptic Multiplexer/Demultiplexer Terminals," *Appl. Opt.*, **16**, pp. 3044–3060, 1977.

113. C. S. Tsai, S. K. Yao, M. A. Alhaider, and P. Saunier, "High-Speed Guided-Wave Acoustooptic and Electrooptic Switches," *1974 International Electronic Devices Meeting, Technical Digest*, pp. 85–87.

114. J. L. Davis and S. Ezekiel, "Closed-Loop, Low-Noise Fiber-Optic Rotation Sensor," *Opt. Lett.*, **6**, p. 505, 1981.

115. R. F. Cahill and E. Udd, "Phase-Nulling Fiber-Optic Laser Gyro," *Opt. Lett.*, **4**, p. 93, 1979.

116. C. S. Tsai and Z. Y. Cheng, "A Novel Optical Frequency Shifting Scheme using Guided-Wave Acoustooptic Bragg Diffraction in Cascade," *Proc. 1988 IEEE Ultrasonics Symp.*, Oct. 3–5, Chicago, II (in press).

117. C. S. Tsai, C. L. Chang, C. C. Lee, and K. Y. Liao, "Acoustooptic Bragg Deflection in Channel Optical Waveguides," *1980 Topical Meeting on Integrated and Guided-Wave Optics, Technical Digest of Post-Deadline Papers*, IEEE Cat. No. 80CH1489-4QEA, pp. PD7–1–PD7-4, C. S. Tsai, C. T. Lee, and C. C. Lee, "Efficient Acoustooptic Diffraction in Crossed Channel Waveguides and Resultant Integrated Optic Module," *1982 IEEE Ultrasonics Symp. Proc.*, IEEE Cat. No. 82CH1823-4, pp. 422–425.

118. C. S. Tsai and Q. Li, "Wideband Optical Frequency Shifting Using Acoustooptic Bragg Diffraction in a $LiNbO_3$ Spherical Waveguide," *Proc. 5th Int. Conf. Int. Opt. Optic. Fiber Communi.*, Venezia, Italy, October 1–4, 1985, pp. 129–132.

119. C. S. Tsai, and Z. Y. Cheng, "Novel Guided-Wave Acousto-optic Frequency Shifting Scheme Using Bragg Diffractions in Cascade," *Appl. Phys. Lett.*, **54**, pp. 1616–1618, 1989.

120. Z. Y. Cheng and C. S. Tsai, "Baseband Integrated Acoustooptic Frequency Shifter," Presented at *1990 Conference on Lasers and Electrooptics*, Anaheim, CA, May 21–25; *Technical Digest Series*, Vol. 7, pp. 152–153 (Optical Society of America, Washington, D. C.), IEEE Cat. No. 90CH2850–6.

121. D. C. Flanders, H. Kogelnik, R. V. Schmidt, and C. V. Shank, "Grating Filters for Thin Film Optical Waveguides," *Appl. Phys. Lett.*, **24**, pp. 194–196, 1974.

122. A. C. Livanors, A. Katzir, A. Yariv, and C. S. Hong, "Chirped-Grating Demultiplexers in Dielectric Waveguides," *Appl. Phys. Lett.*, **30**, pp. 519–521, 1977.

123. W. J. Tomlinson, "Wavelength Multiplexing in Multimode Optical Fibers," *Appl. Opt.*, **16**, pp. 2180–2194, 1977.

124. H. W. Yen, H. R. Friedrich, R. J. Morrison, and G. L. Tangonan, "Planar Rowland Spectrometer for Fiber-Optic Wavelength Demultiplexing," *Opt. Lett.*, **6**, pp. 639–641, 1981.

125. T. Suhara, Y. Handa, H. Nishihara, and J. Koyama, "Monolithic Integrated Microgratings and Photodiodes for Wavelength Demultiplexing," *Appl. Phys. Lett.*, **40**, pp. 120–122, 1982.

126. E. Voges, "Multimode Planar Devices for Wavelength Division Multiplexing and Demultiplexing," presented at the 1983 International Conference on Integrated Optics and Optical Fiber Communications, Tokyo, Japan, June 27–30, Paper No. 29A1–5.

127. G. Winzer, "Wavelength Multiplexing Components — A Review of Single-Mode Devices and Their Applications," *J. Lightwave Technol.*, **LT-2**, pp. 369–378, 1984.

128. J. Lipson, W. J. Minford, E. J. Murphy, T. C. Rice, R. A. Linke, and G. T. Harvey, "A Six-Channel Wavelength Multiplexer and Demultiplexer for Single Mode Systems," *J. Lightwave Tech.*, **LT-3**, pp. 1159–1163, 1985.

129. L. Kuhn, P. F. Heiderich, and E. G. Lean, "Optical Guided Wave Mode Conversion by an Acoustic Surface Wave," *Appl. Phys. Lett.*, **19**, pp. 428–430, 1971.

130. R. C. Alferness and R. V. Schmidt, "Tunable Optical Waveguide Directional Coupler Filter," *1978 Topical Meeting on Integrated- and Guided-Wave Optics, Technical Digest*, Paper TuA3.

131. R. C. Alferness, "Efficient Electro-optic TE–TM Mode Converter/Wavelength Filter," *Appl. Phys. Lett.*, **36**, pp. 513–515, 1980.

132. Y. Ohmachi and J. Noda, "$LiNbO_3$ Te–TM Mode Converter Colinear Acoustooptic Interaction," *IEEE J. Quantum Electron.*, **QE-13**, pp. 43–46, 1977.

133. B. Kim and C. S. Tsai, "Thin-Film Tunable Optical Filtering Using Anisotropic and Noncollinear Acoustooptic Interaction in $LiNbO_3$ Waveguides", *IEEE J. Quantum Electron.*, **QE-15**, pp. 642–647, 1979.

134. V. P. Hinkov, R. Opitz, and W. Sohler, "Collinear Acoustooptical TM–TE Mode Conversion in Proton Exchanged Ti:LiNbO₃ Waveguide Structures," *J. Lightwave Tech.*, **6**, pp. 903–908, 1988.

135. C. S. Tsai and Z. Y. Cheng, "Wavelength-Division-Multiplexing Optical Switch Using Guided-Wave Acoustooptic Bragg Cells" (unpublished).

136. H. Jerominek, J. Y. D. Pomerleau, P. Tremblay, and C. Delisle, "An Integrated Acoustooptic Bistable Device," *Opt. Commun.*, **51**, pp. 6–10, 1984.

137. K. Preston, Jr., "A Comparison of Analog and Digital Techniques for Pattern Recognition," *Proc. IEEE*, **60**, pp. 1216–1231, 1972.

138. R. W. Damon, W. T. Maloney, and D. H. McMahon, in W. D. Mason and R. N. Thurston, Eds., *Physical Acoustics*, Vol. 7, Academic, New York, 1970, p. 273.

139. R. A. Sprague, "A Review of Acoustooptic Signal Correlators," Special Issue on Acoustooptics, *Opt. Eng.*, **16**, pp. 467–478, 1977.

140. A. Vander Lugt, Ed., Special Section on Acousto-Optic Signal Processing, *Proc. IEEE.*, Vol. 69, pp. 48–92, 1981.

141. M. C. Hamiltion, D. A. Wille, and W. J. Miceli, "An Integrated Optical RF Spectrum Analyzer," *Proc. 1976 Ultrasonics Symp.*, IEEE Cat. No. 74CH11120-5SU, pp. 218–222.

142. D. B. Anderson, J. T. Boyd, M. C. Hamilton, and R. R. August, "An Integrated-Optical Approach to the Fourier Transform" *IEEE J. Quantum Electron.*, **QE-13**, p. 268, 1977.

143. C. S. Tsai, "Guided-Wave Acoustooptic Bragg Modulators for Wideband Integrated Optic Communications and Signal Processing," *IEEE Trans. Circuits Syst.*, **CAS-26**, pp. 1072–1098, 1979.

144. M. K. Barnoski, B. Che, T. R. Joseph, J. Y. M. Lee, and O. G. Ramer, "Integrated-Optic Spectrum Analyzer," *IEEE Trans. Circuits Syst.*, **CAS-26**, pp. 1113–1124, 1979.

145. D. Mergerian, E. C. Malarkey, R. P. Pautienus, J. C. Bradley, G. E. Marx, L. D. Hutcheson, and A. L. Keller, "Operational Integrated Optic RF Spectrum Analyzer," *Appl. Opt.*, **19**, pp. 3033–3034, 1980.

146. E. T. Aksenov, N. A. Esepkina, A. A. Lipovskii, and A. V. Pavenko, "Prototype Integrated Acoustooptic Spectrum Analyzer," *Sov. Tech. Lett. (USA)*, **6**, pp. 519–520, 1980.

147. V. Neuman, C. W. Pitt, and L. M. Walpita, "An Integrated Acoustooptic Spectrum Analyzer Using Grating Components," *Proc. 1st Eur. Conf. Int. Opt.*, London, England, September 14–15, 1981, pp. 89–92.

148. T. Suhara, H. Nishihara, and J. Koyama, "A Folded-Type Integrated Optic Spectrum Analyzer Using Butt-coupled Chirped Grating Lenses," *IEEE J. Quantum Electron.*, **QE-18**, pp. 1057–1059, 1982.

149. V. M. Ristic and S. A. Jones, "Theoretical Considerations Related to Development of an Integrated Acousto-Optic Receiver," *Can. Electr. Eng. J.*, **7**, pp. 7–18, 1982.

150. R. L. Davis and F. S. Hickernell, "Application of Wideband Bragg Cells for Integrated Optic Spectrum Analyzer," *Proc. SPIE, Int. Soc. Opt. Eng.*, **321**, pp. 141–148, 1982.

151. V. M. Ritic, S. A. Jones, and G. R. Dubois, "Evaluation of an Integrated Acousto-Optic Receiver," *Can. Electr. Eng. J.*, **8**, pp. 59–64, 1983.

152. S. Valette, J. Lizet, P. Mottier, J. P. Jadot, S. Renard, A. Fournier, A. M. Grouillet, P. Gidons, and H. Denis, "Integrated Optical Spectrum Analyzer Using Planar Technology on Oxidized Silicon Substrate," *Electron. Lett.*, **19**, pp. 883–885, 1983. Also *IEEE Proc. H.*, **131**, pp. 325–331, 1984.

153. T. Suhara, T. Shiono, H. Nishihara, and J. Koyama, "An Integrated-Optic Fourier Processor Using an Acoustooptic Deflector and Fresnel Lenses in As₂S₃ Waveguide," *J. Lightwave Technol.*, **LT-1**, pp. 624–630, 1983.

154. C. Stewart, G. Serivener, and W. J. Stewart, "Guided-Wave Acoustooptic Spectrum Analysis at Frequencies Above 1 GHz," *100C 1981, Third International Conference on Integrated Optics and Optical Fiber Communication, Technical Digest*, April 1981, p. 122.

155. M. Hamilton and A. Spezio, "Spectrum Analysis Using Integrated Acoustooptics," in C. Tsai, Ed., *Guided-Wave Acoustooptic Interactions, Devices, And Applications*, Springer-Verlag, Berlin, 1990, Chap. 7.

156. H. J. Caulfield, W. J. Rhodes, M. J. Foster, and S. Horvitz, "Optical Implementation of Systolic Array Processing," *Opt. Commun.*, **40**, p. 86, 1981.

157. R. T. Chen and C. S. Tsai, "GaAs–GaAlAs Heterostructure Single-Mode Channel-Waveguide Cutoff Modulator and Modulator Array," *IEEE J. Quantum Electron.*, **QE-23**, pp. 2205–2209, 1987.

158. X. Cheng and C. S. Tsai, "Electrooptic Bragg-Diffraction Modulators in GaAs/AlGaAs Heterostructure Waveguides," *J. Lightwave Tech.*, **6**, pp. 809–817, 1988.

159. Y. Abdelrazek, C. S. Tsai, and T. Q. Vu, "An Integrated Optic RF Spectrum Analyzer in GaAs Waveguide," *1990 Conference on Lasers and Electro-Optics*, May 21–25, Anaheim, CA, *Technical Digest Series*, Vol. 7, pp. 446–447 (Optical Society of America, Washington, D. C.). IEEE Cat. No. 90CH2850–8.

# 18

# OPTICAL MODULATION: MAGNETO-OPTICAL DEVICES

ALAN E. CRAIG*

*U.S. Naval Research Laboratory*
*Washington, D.C.*

## 18.1 INTRODUCTION

Modern uses of magnetic phenomena for optical modulation depend almost exclusively on the Faraday effect (linear with internal magnetic field strength), the phenomenological description of the polarization rotation of linearly polarized light as it propagates through a Faraday-active medium. (The Cotton–Mouton effect also causes optical polarization rotation, with strength depending on the square of the internal magnetic field.) The interaction of spinning, and orbiting, electrons and their response to externally imposed or impinging fields are the bases of all effects that occur in magnetic materials. Only a small subset of magnetic materials exhibits optical properties conductive to propagation of light in the visible or near-infrared spectrum, however. A review of the description of magnetic materials followed by an examination of those that exhibit desirable optical properties will precede discussion of the various magneto-optical interactions applied in devices.

Most devices that incorporate the Faraday effect change their modulation status very slowly (spatial light modulators) or not at all (optical isolators), but some applications invoke magnetic fields that oscillate at microwave frequencies. All three of these types are being developed currently. Intermittently modulated devices envisioned include two-dimensional spatial light modulators for optical linear algebra operations; crossbar switches for communication networks; and perhaps most importantly, erasable disk memories. Planar waveguide optical isolators will undoubtedly

*Present address: Air Force Office of Scientific Research, Bolling Air Force Base, Washington, D.C. 20332-6448

*Handbook of Optical Components and Engineering,* Edited by Kai Chang
ISBN 0-471-39055-0 © 2003 John Wiley & Sons, Inc.

find extensive service in optical fiber telecommunication links where laser diodes need protection from back reflection to ensure stable operation. Magneto-optical modulation at microwave frequencies, where certain magnetic materials support propagation of magnetostatic waves, may supplant acousto-optical devices (which are severely limited by acoustic attenuation at high frequencies) for spectrum analysis and other similar functions.

Some magnetic phenomena are purely quantum mechanical in origin, in particular those linked to the orbital and spin momenta attributed to the electron. However, taking these phenomena and their associated magnetic dipole moments as empirical fact, much of magnetism can be understood in the classical domain; the discussion here will pursue the subject mainly from this point of view. Similarly, optical interactions are in reality quantum mechanical. Considerable understanding can be extracted from classical and macroscopic descriptions of the response of light waves to modification of material properties via models such as the Faraday effect and the optical permeability and permittivity tensors.

## 18.2   UNITS

Although the International System of Units (SI) has attained widespread usage in many branches of electromagnetics, the centimeter-gram-second (cgs) system remains dominant in magnetics literature and will be used here, with exceptions noted.

The definition of units in magnetics is somewhat unintelligible without some rudimentary accompanying physical description. Although the phenomena discussed later derive mostly from quantum-mechanical observations of the behavior of materials, the entities used to describe them originated from observation of a macroscopic phenomenon, namely the attractive force between two parallel wires carrying current in the same direction. This force (in dynes), for two elementary currents (namely, moving charges) is given by the expression

$$\mathbf{F} = \frac{k(q_1 q_2)[\mathbf{v}_1 \times (\mathbf{v}_2 \times \mathbf{n})]}{|\mathbf{r}_1 - \mathbf{r}_2|^2}$$

where $q$ designates the charge of a particle, $\mathbf{v}$ its vector velocity, $\mathbf{r}$ the vector spatial coordinate of the particle, and $\mathbf{n}$ the unit normal vector from particle 1 to particle 2 ($\mathbf{n} = (\mathbf{r}_1 - \mathbf{r}_2)/|\mathbf{r}_1 - \mathbf{r}_2|$). The unit of the elementary charge becomes an issue here: if $k = 1$ is chosen, $q$ is in electromagnetic units (e.m.u.); if $k = 1/c^2$ is chosen ($c = 3 \times 10^{10}$ cm/s), $q$ is in electrostatic units (e.s.u.), the standard in the cgs system; if $k = \frac{1}{100}$ is chosen, $q$ is in coulombs, the mks (or SI) standard (1 e.m.u. = 10 C = $3 \times 10^{10}$ e.s.u.). Using e.s.u., the differential force on a moving charged particle resulting from the presence of a filamentary current element $i d\mathbf{I}$ is

$$d\mathbf{F} = \frac{q_1 \mathbf{v}_1}{c} \frac{i \, d\mathbf{I} \times \mathbf{n}}{c|\mathbf{r}_1 - \mathbf{r}_2|^2}$$

From this expression the magnetic induction $\mathbf{B}$ is defined as

$$\mathbf{B} = \frac{i d\mathbf{I} \times \mathbf{n}}{c|\mathbf{r}_1 - \mathbf{r}_2|^2}$$

where $i$ is in e.s.u. per second and **B** is in gauss. For the special case of the field a distance $r$ from a straight wire carrying current $I$,

$$|\mathbf{B}| = \frac{2I}{c \times 10r} \quad (\text{e.s.u./cm}^2)$$

where $I$ is in e.s.u. per second, or

$$|\mathbf{B}| = \frac{2I}{10r} \quad (\text{A/cm})$$

where $I$ is in amperes; in both cases **B** is in gauss. Figure 18.1 illustrates these fields. Given a current $I$ in a circular wire loop of radius $R$, the field at the center is

$$|\mathbf{B}| = \frac{2\pi I}{cR} \quad (\text{G})$$

where $I$ is in e.s.u. per second, or

$$|\mathbf{B}| = \frac{2\pi I}{10R} \quad (\text{G})$$

where $I$ is in amperes. A solenoid, an infinite stack or spiral of these loops, having radius $R$ and current $I$ per loop, produces a field on its axis $|\mathbf{B}| = 4\pi n(I/c)$ gauss (cgs units), or $|\mathbf{B}| = 4\pi n(I/10)$ gauss ($I$ in amperes), where $n$ is the number of loops or turns per unit length.

Various conventions are adopted by authors for defining the remaining magnetic quantities and their interrelations, in part depending on the system of units adopted. These quantities account for the empirical behavior of materials in magnetic fields and for the response of charged particles moving in materials that have internal magnetization. In the cgs system used here, the magnetic field **H** is substituted for **B** external to magnetic materials. (While **B** is in gauss, the precisely equivalent unit for **H** is

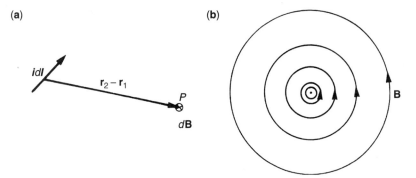

**Figure 18.1**   (a) Current element $i\,d\mathbf{I}$ is situated at the point $\mathbf{r}_1$, and the contribution of this to the induction at the point $P(\mathbf{r}_2)$ is directed normal to the diagram, away from the observer. (b) The induction represented by concentric circles around a current directed toward the observer. (From D. J. Craik, "Magnetostatic Principles," *Structure and Properties of Magnetic Materials*, Pion Limited, London, 1971, p. 1. Copyright 1971, used by permission.)

the oersted.) The preceding equations could justifiably have been written using **H**. In fact, common usage employs $\mathbf{H}_i$, with $\mathbf{B} = \overline{\overline{\mu}}\mathbf{H}_i$, to discuss the magnetic field internal to a magnetic material. Further definition of $\overline{\overline{\mu}}$ is introduced below; the use of $\mathbf{H}_i$ will be seen in the section on spin waves. Inside magnetic materials the magnetization (or magnetic moment per unit volume) **M** contributes to the magnetic induction according to $\mathbf{B} = \mathbf{H} + 4\pi\mathbf{M}$. The factor $4\pi$ arises by definition, but via the concept of equivalent currents.

A current loop placed in a uniform magnetic field **B** experiences a torque $\mathbf{T} = -(I/c)|\mathbf{a}|\hat{\mathbf{a}} \times \mathbf{B}$, in which $I/c$ is the numerical value of the current in the loop ($I$ in e.s.u. per second), $|\mathbf{a}|$ is the scalar area of the loop, and $\hat{\mathbf{a}}$ is the unit vector normal to the plane of the loop. The quantity $(I/c)|\mathbf{a}|\hat{\mathbf{a}}$ is called the magnetic dipole moment of the loop and designated **m**, with units of ergs per gauss (the energy of a magnetic dipole in magnetic induction **B** is $E = -\mathbf{m} \cdot \mathbf{B}$). Magnetic materials also can be described in terms of their magnetic dipole moments because they experience the same torques in the presence of magnetic induction ($\mathbf{T} = -\mathbf{m} \times \mathbf{B}$). Thus any material magnetic dipole moment can be represented by an equivalent current loop. Figure 18.2 shows the field of a magnetic dipole.

The magnetization **M** is just the magnetic dipole moment per unit volume, $\mathbf{m}/V$. A section of magnetic material in which magnetic dipoles, each having scalar area $|\mathbf{a}|$, are aligned in a string back-to-back looks like a solenoid. Then **M** is given by $(\mathbf{m}/|\mathbf{a}|)n$, where $n$ is the number of dipoles per centimeter along the string, or equivalently the number of solenoid turns per centimeter. The current–dipole equivalent is a solenoid with $|\mathbf{M}| = |(\mathbf{m}/|\mathbf{a}|)n| = n(I/c)$, and comparison to the expression for **B** due to a solenoid (see the preceding) necessitates the inclusion of the $4\pi$ factor for the contribution of material magnetization to the magnetic induction.

A few other names are sometimes invoked as magnetic units. The maxwell is a measure of *lines of force*, or magnetic flux; one oersted is equivalent to one maxwell per square centimeter. The weber is also a measure of magnetic flux; one weber equals $10^8$ maxwells. Last, the tesla is equivalent to $10^4$ gauss.

Two parameters, both tensors, describe the response of materials to applied magnetic fields: the susceptibility, $\overline{\overline{\chi}}$, and the permeability, $\overline{\overline{\mu}}$. The susceptibility $\overline{\overline{\chi}}$, with $\mathbf{M} =$

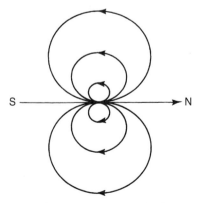

**Figure 18.2**   Field of a magnetic dipole. (From B. D. Cullity, "Definitions and Units," *Introduction to Magnetic Materials*, Addison-Wesley, Reading, MA, 1972, p. 8. Copyright 1972, used by permission.)

$\overline{\overline{\chi}}\mathbf{H}$, characterizes the proclivity of the magnetic dipoles of a material to align with an applied magnetic field. It generally is dependent on both the magnitude and direction of $\mathbf{H}$: $\overline{\overline{\chi}} = \overline{\overline{\chi}}(\mathbf{H})$. For high values of $\mathbf{H}$, $\overline{\overline{\chi}}$ tends asymptotically to zero, and the material is said to saturate. At this point, $\mathbf{M} = \mathbf{M}_s$, the saturation magnetization. ($M_s = |\mathbf{M}_s|$ will denote its magnitude.) The permeability $\overline{\overline{\mu}}$, with $\mathbf{B} = \overline{\overline{\mu}}\mathbf{H}$, characterizes the total flux density produced by a given applied magnetic field. As can be seen by manipulation of $\mathbf{B} = \mathbf{H} + 4\pi\mathbf{M}$, $\overline{\overline{\mu}} = 1 + 4\pi\overline{\overline{\chi}}$; thus it also depends on $\mathbf{H}$: $\overline{\overline{\mu}} = \overline{\overline{\mu}}(\mathbf{H})$. Often $\overline{\overline{\chi}}$ and $\overline{\overline{\mu}}$ are diagonal and isotropic and so can be written as scalars.

## 18.3   MATERIAL PHYSICS OF MAGNETICS

The magnetic moment of a material derives from the consolidated effects of the charge and the angular moments of its electrons. An electron has spin and orbital angular momentum, both of which contribute to its magnetic moment. These properties are fundamentally quantum mechanical; however, a phenomenological description provides a reasonable basis for understanding their magnetic contributions.

The contribution of the orbital angular momentum of an electron to the magnetic moment $\mathbf{m}$ can be thought of classically in terms of a current loop surrounding an area, as in the preceding. For a single electron, the current is $(e/c)v/2\pi r$, where $v$ is the electron velocity around a loop of radius $r$, and the area is $\pi r^2$; so

$$|\mathbf{m}| = \text{current} \times \text{area} = \frac{(e/c)vr}{2}$$

The quantum mechanically allowed values of the angular momentum are $m_e vr = n(h/2)\pi$, in which $m_e$ is the electron mass and $h = 6.626 \times 10^{-27}$ erg-second is Planck's constant. Substituting the allowed values of velocity from this expression, the Bohr magneton is defined as $|\boldsymbol{\mu}_B| = [\mathbf{m}] = (e/c)h/(4\pi m)$ in the orbit for which $n = 1$. (Its value is $|\boldsymbol{\mu}_B| = 0.927 \times 10^{20}$ ergs/Oe. This value turns out to be correct for the orbital magnetic moment even though the model presented here to aid visualization is naive.) There is no reasonable analog to aid envisioning the spin angular momentum contribution of the electron; a spinning top having charge distributed around its equator is cumbersome. However, its value is found experimentally to be exactly the same as that of the orbital magnetic moment.

The magnetic moment contribution of a single electron to an atom is the sum of these orbital and spin moments. For various reasons, either of these moments may be quenched to some degree by the presence of other electrons and by the influence of the atomic environment. The total magnetic moment of an atom results from summing the (modified) contributions of all of its electrons. As already alluded to, the magnetic moments depend on the orbital and spin angular moments, whose allowed values fall under the jurisdiction of quantum mechanics. For free atoms, electrons that inhabit filled orbitals contribute very little since each electron having spin up is matched by one having spin down. The Pauli exclusion principle, which allows only one electron of each spin quantization in any energy level, governs the spin contribution in unfilled levels. In solid materials, electron orbitals are distorted by bonding of neighboring atoms (orbit–lattice coupling) so that the orbital contribution to the magnetic moment often becomes small. That is, electrons that in a free atom might experience distorted orbitals in response to an applied magnetic field will be restrained in a solid by the strong preferential direction of the bonds. Spin–spin and spin–orbit coupling is generally much weaker; consequently spin-originating magnetic dipole moments provide most of an atom's response to a magnetic field in a solid.

A parameter called the $g$-factor has been devised to indicate the degree to which the magnetic moment derives from the electron's spin or orbital angular momentum. The $g$-factor represents the ratio of the number of Bohr magnetons of magnetic moment in an atom to the number of units of angular momentum, measured in $h/2\pi$ ($h$ is Planck's constant, $h = 6.626 \times 10^{-27}$ erg-s). Its value varies from $g = 1$ for orbital dependence alone to $g = 2$ for the solely spin-dependent case. Further details can be obtained in Refs. 1–4.

An external applied magnetic field exerts a torque on the total magnetic dipole. The torque must equal the rate of change of the angular momentum and consequently a change in the magnetic moment of the dipole. (The torque causes no change in momentum in the direction of the field, however.) The equation that describes this process, called the gyromagnetic equation, is $d\mathbf{m}/dt = \gamma(\mathbf{m} \times \mathbf{H})$. Solution of this equation gives sinusoidal precession in the plane perpendicular to the field at the gyromagnetic frequency: $\omega_0 = \gamma|\mathbf{H}|$. (Inside a material, $\mathbf{H}$ would be replaced by $\mathbf{H}_i$.)

With this rudimentary description of electron magnetic dipole moments in mind, the magnetic behavior of solid materials can be addressed.

Only a small subset of magnetic materials is important for optical interactions. A general description of the broad classifications of magnetic materials delineates this subset. The physical mechanisms that produce the various responses will be described; thorough analysis of these mechanisms is deferred to the references [1–3].

Several terms are invoked to describe the macroscopic behavior of materials when subjected to external magnetic fields. These are diamagnetic, paramagnetic, ferromagnetic, antiferromagnetic, and ferrimagnetic. The dipole characteristics that correspond to many of these classifications are illustrated in Fig. 18.3. Although a particular material will respond predominantly according to the features of only one of these classes, more than one mechanism is almost always operating; in addition, the class of response changes with temperature for many materials.

The physical mechanism that produces diamagnetism is the analog at an atomic scale to Lenz's law, which describes the electromotive force (emf) produced in a

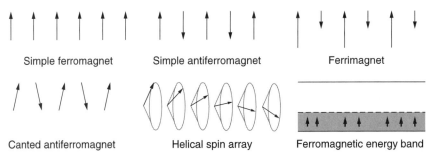

Simple ferromagnet        Simple antiferromagnet        Ferrimagnet

Canted antiferromagnet        Helical spin array        Ferromagnetic energy band

**Figure 18.3** Possible ordered arrangements of electron spins, in which one particular type of cooperative behavior dominates. (From Charles Kittel, "Ferromagnetism and Antiferromagnetism," *Introduction to Solid State Physics*, 4th ed., Wiley, New York, 1971, p. 528. Copyright 1971, used by permission.)

macroscopic wire loop when a change occurs in the magnetic flux encircled (emf $=$ $-d\phi/dt$; $\phi = |\mathbf{H}|A$ volts; $\phi$ in maxwells, $|\mathbf{H}|$ in oersteds, $A$ in square centimeters). The emf establishes a current loop magnetic dipole field that opposes the change in the intercepted flux. An electronic magnetic dipole responds similarly, except that whereas the macroscopic loop current persists only as long as the flux is changing, the electronic dipole state corresponds to the applied magnetic field strength and does not decay, as if its path were without resistance, or superconducting. (This feature is attributable to the quantized nature of the electron orbitals.)

For diamagnetic behavior to dominate in a material a balance must exist between the orientations of dipoles, so that if they are randomly aligned the net magnetic moment is zero; if their alignment is governed (by the orientation of atomic orbitals or by that of interatomic bonds), every spin-up dipole is compensated by one oriented spin down. The Pauli exclusion principle ensures this condition in the inert gas atoms as well as in a great many molecular gases in which atomic orbitals are filled by sharing electrons. Many solids, in which bonding satisfies the full-orbital condition, also exhibit diamagnetic behavior. Regardless of whether diamagnetic behavior is dominant, all materials exhibit diamagnetic response; the other magnetic behaviors are stronger effects, however, and occlude diamagnetism when they are present.

Paramagnetism occurs in materials in which all magnetic moment contributions are not fully compensated, and the atoms, molecules, or crystalline unit cells have net magnetic dipole moments. Many ions are paramagnetic since their valence electrons are unpaired. Metals that have incomplete outer shells also tend to paramagnetism. Most transition elements and rare earths, either in isolation or incorporated in crystal structures, have large intrinsic magnetic dipole moments because of their incomplete inner shells.

Paramagnetism results from the competition between the tendency of the uncompensated magnetic dipoles to coalign with an applied magnetic field and the disorienting effects of thermal phonons or interatomic collisions. The Maxwell–Boltzmann distribution governs the thermal randomization, with the energy term given by the dipole energy in the magnetic field, $E = -\mathbf{m} \cdot \mathbf{H}$. Classically, alignment of a dipole with the field is allowed to occur to any degree whatsoever and to assume any angular orientation; quantum mechanics, permitting only discrete values of angular momentum, constrains the relative orientation angles and prohibits complete alignment. The resulting expression for the dependence of bulk magnetization on the ratio of magnetic field to temperature is called the Langevin function ($\mathbf{M}/M_s = \mathbf{L}[(\mathbf{m} \cdot \mathbf{H})/kT]$, where $\mathbf{L} = \mathbf{L}(\mathrm{x})$ is the Langevin function and $M_s$ is the saturation magnetization) in the case of a classical derivation and the Brillouin function when quantum-mechanical arguments have been used. Curves plotted from the two functions are similar; their slopes, $\frac{1}{3}$ at the origin, decrease gradually with increasing field (or decreasing temperature) to zero, corresponding to an asymptotic approach to magnetic saturation. See Fig. 18.4. The mechanism of paramagnetism, being temperature dependent, obeys the Curie–Weiss law for the susceptibility: $\chi = C/(T - \theta)$, where $C$ is the Curie constant, $T$ is the temperature, and $\chi$ is the scalar magnitude of the susceptibility, assumed isotropic. The parameter $\theta$ has temperature units, and varies from one material to another depending on the interaction between electrons that share orbitals or otherwise pass regularly within close proximity. This form of the law is a low-field, low-temperature linear approximation of the expression derived by either a classical or a quantum-mechanical analysis.

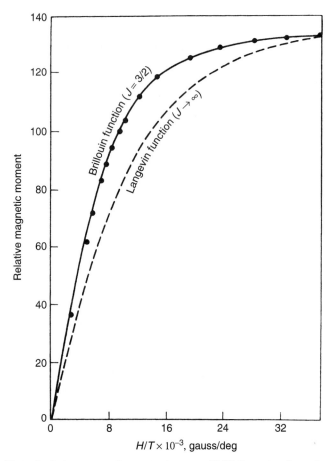

**Figure 18.4**   Plot of relative magnetic moment versus $H/T$ for potassium chromium alum at $T = 1.29$ K. The solid line is a Brillouin curve for $g = 2$ and $J = S = \frac{3}{2}$ (spin only; no orbital contribution). The broken line is a Langevin curve. (From Benjamin Lax and Kenneth J. Button, "Paramagnetism and Ferromagnetism," *Microwave Ferrites and Ferrimagnetics*, McGraw-Hill, New York, 1962, p. 61. Copyright 1962, used by permission.)

Pierre Weiss discovered that the linear susceptibility relationships of paramagnetics in small fields do not intercept the zero magnetization line at the origin in many cases. He postulated that the responsive dipoles do not act totally independently but interact through creation of an interdipole magnetic field he called the molecular field. (In truth, the effect is more nearly atomic in origin.) The Weiss temperature parameter, $\theta$ in the preceding, results from the mathematical introduction and inclusion in the phenomenological theory of this magnetic "molecular field," denoted $\mathbf{H}_m$, which is proportional to the material's magnetization ($\mathbf{H}_m = \lambda \mathbf{M}$); that is, $\mathbf{H}_m$ is proportional to the amount of alignment among the dipoles caused by the external field. The effect is highly cooperative and self-augmenting. When this conceptual molecular field is added to the externally applied field, the actual susceptibility is accurately modeled. The parameter $\theta$ can be positive or negative and is usually small in magnitude for paramagnetics, less than about 10 K.

Weiss's molecular field is explainable in terms of exchange energy. Electrons that share orbitals, more precisely bonding orbitals, are constantly interchanging responsibilities and atom allegiances. An energy is connected to this interchange, called the exchange energy. Generally the exchange energy is negative when the electron spins are antiparallel. Then, as expected from the Pauli exclusion principle applied to the bonding orbitals of molecules, the energy is lower when the electrons involved have opposite spins, and bonding is promoted. Van der Waal forces are the result of exchange, for example. In some materials the exchange energy is negative when the spins of the exchange electrons are parallel, although this situation applies to electrons inhabiting inner, partially filled orbitals. The important class of materials for which this is true is the ferromagnetics, to be discussed.

The expression for the exchange energy is

$$E_{ex} = -2J_e(\mathbf{S}_i \cdot \mathbf{S}_j)$$

where $\mathbf{S}_i$ and $\mathbf{S}_j$ are the spin vectors of two electrons. The exchange integral $J_e$ indicates the overlap of the orbitals of the two electrons (or equivalently, of their charge distributions). The relation between the exchange energy and the molecular field is then $E_{ex} = -\mathbf{m} \cdot \overline{\overline{\mu}} \cdot \mathbf{H}_m = -\mathbf{m} \cdot \overline{\overline{\mu}} \cdot \lambda\mathbf{M}$. Crystals of different elements and compounds have various separations between those constituent atoms responsible for their magnetic character. When two atoms whose electron orbitals exhibit exchange effects are widely separated, the integral is positive but negligibly small. As the atomic separation decreases from specie to specie, the integral grows larger, passes a maximum, and then becomes progressively smaller, turning negative for very small atomic separations. Antiferromagnetic materials are found in this region of negative exchange integral, while ferromagnetics are located in the region of the maximum. Large separation corresponds to paramagnetic response.

Examining once more the dipoles that constitute the paramagnetic material leads to a few other points. First, quantization of the total angular momentum prohibits complete alignment of a magnetic dipole with an applied magnetic field, even at absolute zero temperature. (In fact, because of the angular momentum restriction on dipole orientation, realignment with the field can only occur by means of collisions or phonon interactions, in which energy is exchanged. Otherwise only a precession around the field direction occurs, at the gyromagnetic frequency; it is superimposed on the dipole's orientation, which is determined by its total angular momentum.) A dipole is always free to precess around the field direction, which is driven by the torque of the dipole–field coupling, a useful picture for envisioning the spin waves described in a later section. Second, the contribution of exchange energy—responsible for the molecular field (and the Weiss temperature $\theta$)—to paramagnetic susceptibility, which appears only during the application of an external magnetic field, is not necessarily large compared to the exchange interaction that occurs in an orbital in the absence of the applied field. Exchange is normally acting continuously to affect the strength of interatomic bonding according to the spin orientations of the contributing electrons (or conversely, to determine the spin orientations of electrons in common, bonding orbitals). The exchange that produces the Weiss temperature is generally acting between inner-shell electrons on differing atoms that are barely close enough to exert common influence and do not participate in bonding. It is certainly a secondary effect compared to the paramagnetic realignment of magnetic dipoles in an external magnetic field. Unlike the basic paramagnetic alignment process, however, the dipole may be influenced by exchange without resort to collision or phonon jostling since here the field influence is internal to the atom, changing the energy levels of its orbitals. (Incorporation of the electron magnetic dipole — magnetic field interaction energy into the Hamiltonian perturbs the allowed energy levels of electrons in an atom.) Revisiting the macroscopic picture, paramagnetism is the state assumed by ferromagnetic, antiferromagnetic, and ferrimagnetic materials at elevated temperatures.

Understanding ferromagnetism is a difficult undertaking; experts on magnetism promote various theories that explain, each in part, the mechanisms that cause a material to be ferromagnetic. In fact, a general theory that predicts the ferromagnetic response of all materials may be strictly unattainable. The reality is that ferromagnetism results partly from the interaction of electrons having unpaired spins, and situated on particular atomic sites, via the exchange energy; and partly from the interaction of electrons (sometimes called itinerant) that are not fixed in the material but are to some extent free to move through bands of energy levels, a theory not yet visited here. The aspects of the theories that must be invoked to rationalize the ferromagnetism of any particular material are in some ways unique; the selection depends on the way in which the atomic orbitals of the elements combine into bands of energy states in a solid and to what extent the electrons responsible for the magnetic response occupy the bands.

A qualitative, somewhat cursory description follows. Ferromagnetism is not particularly important for magneto-optics except insofar as ferromagnets normally are used to provide the required biasing fields. For more detailed discussion, read the references [1–3].

Ferromagnetics comprise the three commonly known ferromagnetic transition element metals — iron, cobalt, and nickel — along with a selection of rare earth elements at low temperature (gadolinium below $16°C$) and various compounds and alloys, many of which contain transition elements or rare earths that may not be ferromagnetic in their pure forms.

The classical argument for ferromagnetism provides for a molecular field in the absence of any applied field. This molecular field is very strong, strong enough to cause saturation by coaligning neighboring dipoles. A sample is proscribed from exhibiting bulk magnetization by the subdivision of the coaligned dipoles into small volumes called domains, which orient randomly when no external field is applied. A small applied external field, by aligning the domains, can cause the material to exhibit a very large bulk magnetization. (An applied magnetic field of as low as 50 Oe suffices to achieve technical saturation, all domains aligned, for pure crystalline iron. Fields on the order of a few hundred to at most a few thousand oersteds are required for other ferromagnetics.)

The combination of molecular field theory with the Langevin theory, which describes the thermal statistics of magnetization caused by coaligning dipoles, produces a mathematical argument for the spontaneous magnetization of domains in ferromagnetic materials. Recall that the Langevin function, $\mathbf{M}/M_s = \mathbf{L}[(\mathbf{m} \cdot \mathbf{H})/kT]$, has a slope that equals $\frac{1}{3}$ at the origin and decreases to zero for greater values of the function argument. Meanwhile, the molecular field expression can be inverted to read $\mathbf{M} = \mathbf{H}_m/\lambda$. Whenever $\lambda$ is greater than $3kT/(|\mathbf{m}|M_s)$ the curves representing the two functions intersect. This infers that the molecular field can to a considerable extent align the dipoles of a domain in the absence of an external applied field. For high enough temperature, $\lambda$ is insufficient to ensure intersection of the two curves, and the domains revert to paramagnetic behavior. The temperature at which the transition occurs is called the Curie temperature for a material and corresponds very closely to the Weiss temperature $\theta$ introduced to parameterize the molecular field effect in paramagnetics. For ferromagnetic materials, $\theta$ may be many hundreds of degrees; it is always positive, aiding dipole alignment with the applied field.

The way in which the exchange energy produces the molecular field has been shown in a preceding section. That the exchange energy must be responsible rests on two observations: first, the magnetic dipole moment in ferromagnetics is almost entirely caused by electron spin, not orbital, momentum; second, the internal fields caused by the environment of coaligned dipoles surrounding any particular dipole are simply not strong enough by several orders of magnitude to spontaneously magnetize the domains. As mentioned earlier; for ferromagnetics a negative exchange energy corresponds to parallel dipoles in the domains.

The band theory of solids can help to clarify the details of ferromagnetism. As previously stated, ferromagnetic materials are often formed of elements in which incompletely filled inner shells exist. These are the transition metals and rare earth elements. When these elements form solids, their outer ($s$) shells and inner ($d$ or $f$) shells form bands of allowed states in which the energy levels lose their degeneracy. The energies of the $s$ bands overlap those of the $d$ or $f$ bands. The way that electrons fill the overlapping bands is staggered, and electrons can with relative ease transfer between the states of the two bands. The density of states in the bands is also a factor in determining their electron occupancy; there are five times as many $d$ states as $s$ and seven times as many $f$ states as $s$. In these bands electrons normally would be in the lowest energy configuration if they were paired, one spin up, one spin down, in each level. Exchange forces in the three ferromagnetic metals — iron, cobalt, and nickel — distort the electron occupancy; the $s$ band loses some fraction of its average population and the balance between spins in the $d$ band is destroyed. The exchange energy is least if the $d$ band is half filled with electrons of one spin, the remaining electrons adopting the parity spin and filling states as high as their numbers permit. The end result is ferromagnetism. The theory does not hold for neighboring lighter transition metals in the periodic table (they are not ferromagnetic). Apparently exchange forces can persuade electrons to take on only so much extra potential energy in bands holding progressively fewer of them.

One final clarification: the exchange energy depends strongly on physical separation of the participating electron orbitals. For large separation, the energy is small but is in favor of parallel dipoles. As separation decreases, the energy first increases and then decreases to zero. Even smaller separations see the exchange energy reemerge, but now in favor of antiparallel spins. This arrangement is antiferromagnetism.

Antiferromagnetism depends on forces working in the same way as those that produce ferromagnetism but on a different type of material. Most antiferromagnetics are ionic compounds in which atoms that have intrinsic magnetic moments are not neighbors in the lattice. This separation influences the exchange energy to be negative when nearby spins are antiparallel. For crystalline materials the result is that two interlaced sublattices form, in each of which all dipoles are parallel but which orient in opposite directions as groups.

Above the critical temperature, called the Néel temperature ($T_N$) for antiferromagnetics, each sublattice acts separately as a paramagnetic in the presence of a molecular field, except that the exchange energy influencing an atom in the sublattice originates from the surrounding atoms of the interpenetrating sublattice. (The assumption of neg-ligible interaction with atoms on sites in the same sublattice is not quite accurate. In some crystal structures, some nearest-neighbor atoms are on the same sublattice. How-ever, exchange forces between atoms on differing sublattices dominate those between atoms in the same sublattice for antiferromagnetics.) Since the sublattices are com-posed of identical atoms, the resulting total magnetization is the sum of that of the sublattices individually, that is, twice that for one alone. In this paramagnetic temper-ature region, the material appears to have a negative value of the Weiss temperature parameter. Figure 18.5 is illustrative.

Below the Néel temperature the exchange energy dominates the thermal random-izing effects. Magnetic dipole moments on the two sublattices align antiparallel, and the spontaneous magnetization becomes zero. Small amounts of magnetization may be induced by an external field; the amount depends on the relative orientations of the field and the crystal axis that defines the sublattice orientations. The field causes a small magnetization in each sublattice, which in turn creates a counteracting molec-ular field in the other sublattice; for each sublattice, a complicated reevaluation of the Brillouin function, which governs the balance between the exchange energy and the thermal randomizing statistics, establishes the resultant total magnetization. For a field perpendicular to the crystal axis the dipoles experience a slight rotation; for a parallel field, a small displacement occurs. In the perpendicular case, the susceptibility becomes temperature independent; in the parallel case, it goes to zero with vanishing temperature.

The ferrimagnetic materials remain for consideration. These materials also have characteristics similar to those of ferromagnetics but derive from a different type of

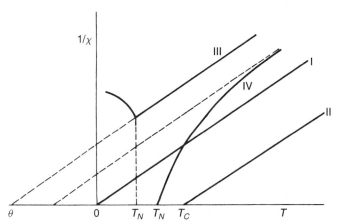

**Figure 18.5**  Reciprocal of susceptibility versus temperature: (I) Simple paramagnetic with no interactions; Curie law. (II) Curie–Weiss paramagnetic with positive interaction term $T_C$, indi-cating ferromagnetic ordering below $T_C$ (the Curie temperature). (III) Antiferromagnetic with negative interaction constant, $\theta$. The Néel temperature below which the interaction is effective in causing antiferromagnetic ordering is $T_N$; $\chi$ falls below $T_N$ because the moments are effec-tively locked into a particular orientation in the lattice and an applied field must overcome this anisotropy as well as the ordering. (IV) Ferrimagnetic. (From D. J. Craik, "Magnetic Dipoles in Applied Fields," *Structure and Properties of Magnetic Materials*, Pion Limited, London, 1971, p. 39. Copyright 1971, used by permission.)

structure. Like the antiferromagnetics, they comprise various complex oxides and are electrically insulating, meaning that the magnetic elements are separated so that bands of electron energies do not form. Local spin interactions dominate, and the molecular field theory provides an accurate description. Unlike the antiferromagnetics, the magnetic ions are of more than one element and are located on more than one type of crystalline site in the material. The overall magnetic character depends on the imbalance of magnetic moments between the elements and on the influence exerted by the site-dependent molecular fields on the intercoupling of the moments.

Several crystal types produce ferrimagnetics. Most are spinel-like structures incorporating iron oxides, known as cubic ferrites. Two types of sites exist for the metallic ions in spinels: one is tetrahedrally coordinated by oxygen atoms ($A$ site); the other is octahedrally coordinated ($B$ site). Materials in which the two different metallic ions are segregated according to site type—divalent on $A$, trivalent on $B$—are called normal spinels. (Ferrites of this structure are generally paramagnetic.) When the divalent ions sit on $B$ sites, forcing the trivalent ions onto either $A$ sites or some mixture of the two sites, the spinel is called inverted; these are often ferrimagnetic.

Hexagonal ferrites are also represented among the ferrimagnetics. This crystal structure differs from the cubic spinel mainly in the stacking of the crystalline planes; face-centered-cubic and hexagonally close-packed structures are otherwise similar. The iron atoms that provide the magnetic moments occupy sites of three different coordinations: tetrahedral, octahedral, and hexahedral.

Another structure that produces ferrimagnetics is garnet. Three different metallic ion sites exist in these materials. They are tetrahedrally, octahedrally, and dodecahedrally coordinated. (See Fig. 18.6. In addition, Ref. 5 exhibits an excellent illustration of the garnet unit cell and the arrangement of eight formula units of atoms that in various orientations comprise it.) Usually, iron ions are found on the first two site types; they interact strongly via exchange to cause antiparallel alignment of the moments on these two sites. The third site type can be occupied by an ion having no magnetic moment. An imbalance of the number of iron ions on its two site types establishes the net moment. The most important of these generally weakly ferrimagnetic materials is yttrium iron garnet (YIG), which propagates several types of magnetic dipole spin waves at microwave frequencies; in particular, magnetostatic waves propagate with relatively low absorption.

Molecular field theory describes ferrimagnetism. Because at least two metallic ions are present, often occupying more than one site type, particular use of the theory is formidable. Marked, yet still useful, simplification results from assuming the presence of only a single ion type. The magnetization of ions on each site type affects the molecular field acting on ions at the same site type, as well as on ions at alternate site types. The algebra is arduous, the result being that above the critical (Curie) temperature, the dependence of the susceptibility on temperature is hyperbolic, appearing paramagnetic at temperatures far above critical but exhibiting increased susceptibility near the Curie temperature (where it is infinite). Below the Curie temperature the ferrimagnetic behavior is the sum of the behavior of the several contributing sublattices. Each of these is determined, as for ferromagnetics and antiferromagnetics, by the simultaneous satisfaction of the thermal distribution of dipole alignment described by the Brillouin function and the spontaneous magnetization determined by the molecular field expression. The sublattice contributions to the magnetization are antiparallel, and various behaviors can be found; in some materials there exists a temperature at

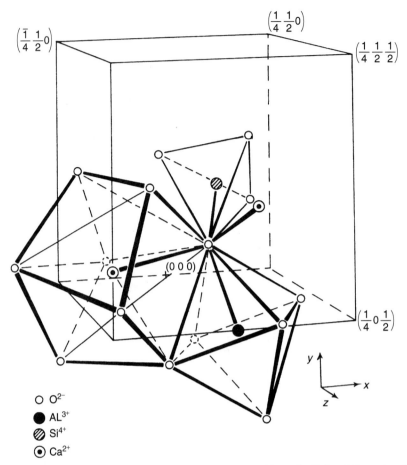

$\left(\dfrac{\overline{1}}{4}\dfrac{1}{2}0\right)$   $\left(\dfrac{1}{4}\dfrac{1}{2}0\right)$   $\left(\dfrac{1}{4}\dfrac{1}{2}\dfrac{1}{2}\right)$   $(0\ 0\ 0)$   $\left(\dfrac{1}{4}0\dfrac{1}{2}\right)$

○ $O^{2-}$
● $AL^{3+}$
◉ $Si^{4+}$
◉ $Ca^{2+}$

**Figure 18.6**   Coordination about an oxygen ion in a garnet. (From S. Geller, "Crystal and Static Magnetic Properties of Garnets," *Physics of Magnetic Garnets*, North-Holland, Amsterdam, 1978, p. 4. Copyright 1978, used by permission.)

which the component contributions exactly cancel, called the compensation point (see Fig. 18.7).

Certain other properties are common to many magnetic materials. Under the influence of an external field, the consequent alignment of magnetic dipoles in a material affects the internal field in a way that depends on the physical shape of the specimen; the diamagnetic response is constrained by the geometry of the specimen's boundary conditions. (An alternative view considers the contribution of fields produced by magnetic poles induced at the specimen boundaries to the total internal field.) This effect reduces the internal field below the applied field value and produces nonuniform (fringing) fields near surfaces and edges. The phenomenon is denoted shape-factor demagnetization. The problem of determining a functional form of the internal field, assuming arbitrary boundary conditions, is realistically intractable, but for ellipsoids of revolution the contribution of surface poles to the internal field is uniform throughout the specimen when the externally applied field is large enough to ensure saturation. The contribution is described by $\mathbf{H}_i = \mathbf{H}_o - (\mathbf{N} \cdot 4\pi\mathbf{M})(\mathbf{M}/M)$. The demagnetizing

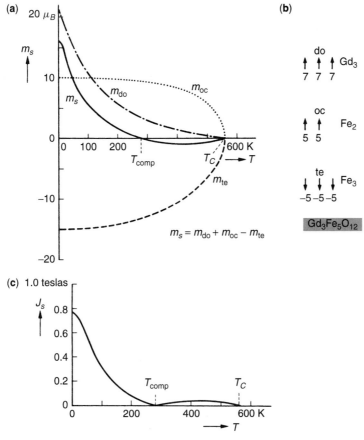

**Figure 18.7**   (**a**) Magnetic saturation moment $m_s$ per unit of $Gd_3Fe_5O_{12}$ (an eighth part of a unit cell) in Bohr magnetons $\mu_B$ as a function of absolute temperature $T$ in a gadolinium–iron garnet film. The saturation moment $m_s$ is the sum of the magnetic moments $m_{oc}$, $m_{te}$, and $m_{do}$ originating from the spins of the $Fe^{3+}$ ions at the octahedral and the tetrahedral lattice sites and of the $GD^{3+}$ ions at the dodecahedral lattice sites. The compensation temperature is $T_{comp}$; when this temperature is exceeded and an external magnetic field is applied, the direction of $m_s$ rotates through $180°$. The Curie point is $T_c$. (**b**) Origin of the total saturation magnetization $m_s$ as the sum of the magnetic moments of the sublattices for a unit $Gd_3Fe_5O_{12}$ at absolute zero. (**c**) Magnetic saturation polarization $J_s$ in teslas (the magnetic saturation moment $j_s = \mu_0 m_s$ weber meters per unit volume) derived from (**a**) for the material as a function of the absolute temperature $T$. The saturation polarization is $J_s = \mu_0 |m_s| \mu_B 8/a^3$ teslas, where $\mu_B = 0.927 \times 10^{-23}$ $Am^2$ is the magnetic moment associated with a Bohr magneton, and $a$ is the lattice constant. (From P. Hansen et al., "Optical Switching with Bismuth-Substituted Iron Garnets," *Philips Tech. Rev.*, **41**, No. 2, 1983/84, p. 37. Published by N. V. Philips. Copyright 1985, used by permission.)

factor **N** has components $N_x$, $N_y$, and $N_z$ that sum to 1 in all cases; for a sphere, $N_x = N_y = N_z = \frac{1}{3}$; for a thin slab with surface perpendicular to the $z$ direction, $N_x = N_y = 0$, $N_z = 1$.

Shape-factor demagnetization contributes to the magnetic anisotropy field when a sample is not saturated. The demagnetizing energy affects the spontaneous domain structure in small (nonsaturating) fields but is not a factor under saturation conditions. In unsaturated planar samples, anisotropy caused by shape-factor demagnetization is in-plane as a consequence of the lower magnetostatic energy for this alignment. This alignment is counter to the contribution of shape-factor demagnetization in saturated samples; in-plane saturation encounters no internal demagnetizing effect, while perpendicular saturation must supersede a demagnetizing internal field of $4\pi M_s$.

The symmetries of the crystallographic structure of a magnetic material generally confer crystalline magnetic anisotropy. It can be uniaxial, as it is in crystals having hexagonal symmetry, but often has much more complicated structure depending on the number of distinguishable directions in the unit cell of the crystal. Anisotropy results from the influence of directional bonding on the orientation of the orbital angular momentum and on the derivative spin dipole orientation caused by spin–orbit coupling. When the internal magnetic field lies parallel to the crystal axis along which the dipoles naturally align (the easy axis), it is enhanced by the anisotropy field. The magnetic anisotropy energy is the difference between the energy required to orient all dipoles along the hard direction and that required to orient them along the easy direction. For a uniaxial crystal the energy follows $E_a = K_1 \sin^2 \theta$ to first order, reflecting the energy acquired by a dipole reorienting in the presence of an aligning torque. In a cubic crystal, with the cosine of the angle between the domain orientation and each crystal axis $i$ given by $\alpha_i$,

$$E_a = K_1(\alpha_1^2\alpha_2^2 + \alpha_2^2\alpha_3^2 + \alpha_3^2\alpha_1^2)$$

In either case, the total anisotropy energy is proportional to the volume of material considered. Susceptibility curves ($B$ vs. $H$) for anisotropic materials vary with specimen alignment in the external field. As discussed further in what follows, the anisotropy energy balances the exchange energy in determining the thickness of the Bloch walls that separate magnetic domains. The anisotropy contribution also changes the (field-dependent) frequency structure of magnetic resonances in a material.

Other sources of magnetic anisotropy can also contribute to material behavior. Strain can be introduced to a sample in several ways. For epitaxial films of magnetic media, crystal mismatch between film and substrate often introduces strain. Alternatively, applying a magnetic field to a sample causes its dimensions to change slightly as a result of the realigning of the dipoles, an effect called magnetostriction. The magnetization of the material is affected by this stress, introducing magnetoelastic energy. For isotropic media the magnetostriction energy contribution to material anisotropy is given by $E_{\mathrm{me}} = \frac{3}{2}\lambda T \sin^2\theta$, in which $\lambda$ is the magnetostriction constant, $T$ the temperature, and $\theta$ the angle between the strain and magnetization directions.

Crystal growers often acknowledge another source of magnetic anisotropy, which they call growth anisotropy. This is apparently caused by imperfect crystallization of a sample during the growth process; disordered or irregular location of large atoms in a crystal, for example, induces local strain fields. Often, postgrowth annealing can relax the irregularities and virtually eliminate this source of anisotropy.

Surface states can inhibit magnetic dipoles from assuming arbitrary orientation. This phenomenon is called spin pinning and is responsible for boundary conditions different from those that would be invoked considering formal truncation of an infinite medium. Spin pinning can contribute strongly to coupling between various modes of the RF magnetic field in finite samples.

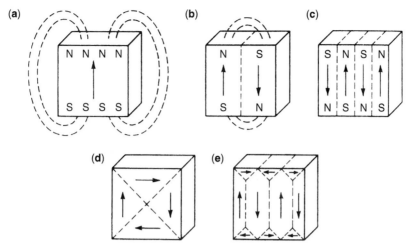

**Figure 18.8**  Several arrangements of domains that may occur in single crystals. The magnetostatic energy associated with the single domain shown in (**a**) is greater than that of (**b**) and (**c**). Domains of closure are shown in (**d**) and (**e**), and the energy of demagnetization is zero. (From B. Lax and K. J. Button, "Paramagnetism and Ferromagnetism," *Microwave Ferrites and Ferrimagnetics*, McGraw-Hill, New York, 1962, p. 76. Copyright 1962, used by permission.)

Magnetic materials that contain unpaired magnetic dipoles, that is, ferromagnetics and ferrimagnetics, generally organize them in coaligned clumps called domains (see Fig. 18.8). Several energy factors contribute to determining the size of the domains, their relative orientation, and the characteristics of the boundaries, called Bloch walls, which separate them. The three most obvious of these are the magnetostatic energy, the exchange energy, and the anisotropy energy. Consider a crystal in which all the magnetic dipoles are aligned in parallel. As previously discussed, the material exhibits an internal magnetic field contribution called the demagnetizing field. The dipoles of the material that align in this field take on an energy, the magnetostatic energy, equal to one-half the dot product of the field and the magnetization:

$$E_m = \tfrac{1}{2}\mathbf{H} \cdot \mathbf{M}$$

The magnitude of $E_m$ when $\mathbf{H}$ is parallel to $\mathbf{M}$ is $E_m = \tfrac{1}{2}NM^2$ ($N$ is the demagnetizing factor; it equals $4\pi$, e.g., when the field is perpendicular to the surface of a thin slab.) The factor of $\tfrac{1}{2}$ discounts the influence of the magnetic dipoles on themselves. This energy can be reduced substantially (by nearly a factor of 2) by dividing the crystal conceptually into halves and reorienting one-half in the antiparallel direction; the total field at any point will be the sum of contributions from the two segments. Calculating the domain magnetostatic energy is difficult, but it is apparently approximately equal to $E_m = LM^2$ per unit area, where $L$ is the sample thickness. When many antiparallel domains are present, the energy per unit area decreases by a factor of about the inverse of the number of domains $n_d$, giving the proportionality dependence $E_m = M^2/n_d$ (see Fig. 6.8). Repeating this conceptual subdivision of the dipoles into counteroriented domains to diminish the internal energy could continue indefinitely except for the influence of the other two energy contributors to domain formation. These (the exchange energy and the anisotropy energy) result from the skew in alignment between neighboring dipoles, and between dipoles and crystallographically preferred directions, and therefore contribute not via the domains themselves but via the interdomain walls. Minimizing the sum of the energy of the domains (according to their size and orientation) and the energy of the domain walls (according to their thickness and separation) governs the magnetic partitioning. The total energy, $E_T = M^2/n_d + n_d E_w$, is minimized for $n_d = M/\sqrt{E_w}$, where $E_w$, the wall energy, is not yet specified.

Consider temporarily a material without anisotropy, so that the effects of exchange energy are isolated. The exchange energy is ferromagnetic and ferrimagnetic materials acts to influence the majority of dipoles to coalign. Reducing exchange energy and reducing magnetostatic energy are thus countervailing criteria, and the size of domains is determined by the configuration of dipoles that minimizes the total energy in a specimen. Recall that the exchange energy for electron-populated orbitals in any two neighboring atoms is given by the product of the exchange integral and the dot product of the spin angular moment of the two electrons. The exchange energy finds the source of its contribution for the domain-organizing problem in the character of the regions that separate antiparallel domains, namely, in the domain walls (Bloch walls). From one domain to its neighbor the dipoles change orientation by 180° (see Fig. 18.9). (Actually, domain wall rotations of 90°, as well as other angles according to the specifics of crystal anisotropy, also occur regularly.) The energy required to make this transition can be decreased if many layers of atoms are invoked, with small rotation between consecutive layers. The exchange energy is then given by

$$E_{\mathrm{ex}} = -\tfrac{1}{2} J_e S^2 \cos^2 \theta$$

for an angle $\theta$ between neighboring spins. For a small angle $\theta$ the difference in $E_{\mathrm{ex}}$ from zero rotation is equal to

$$E_{\mathrm{ex}} = J_e S^2 \theta^2$$

For a 180° wall, $\theta = \pi/n$, where $n$ atomic layers are invoked to constitute the wall. If the atomic separation is $d$, then the total exchange energy per unit area of a wall is $J_e S^2 \pi^2/nd^2$, where $d^2$ represents the atoms per unit area and $n$ is the number of atomic layers through the wall.

The exchange energy could go to zero for very thick walls. This proclivity is tempered by the anisotropy energy, which, per unit area, increases proportionally with the wall thickness. The total wall energy per unit area is

$$E_w = E_{\mathrm{ex}} + E_a = \frac{J_e S^2 \pi^2}{nd^2} + K_1 nd$$

It is minimized when the wall thickness is

$$nd = \sqrt{\frac{J_e S^2 \pi^2}{K_1 d}}$$

The exchange energy equals $kT_c$ at the Curie temperature. (Above the Curie temperature the exchange energy is dominated by thermal fluctuations and the material becomes paramagnetic. Domain structure and domain (Bloch) walls cease to exist.) The wall thickness $nd$ then varies as $\sqrt{T_c/K_1}$.

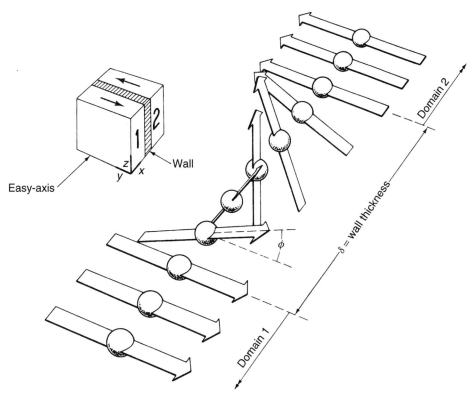

**Figure 18.9**  Structure of a 180° wall. The axis of easy magnetization lies along the *y*-axis. (From B. D. Cullity, "Domains and the Magnetization Process," *Introduction to Magnetic Materials*, Addison-Wesley, Reading, MA, 1972, p. 289. Copyright 1972, used by permission.)

## 18.4  SPIN WAVES

Within a domain or in a material in which the domains have been saturated by a uniform biasing magnetic field (whether an external bias field or an intrinsic field, e.g., an anisotropic field), the magnetic dipoles will precess around the direction of the internal magnetic field. This is because the force of the field on the dipole is in the form of a torque. (Dipoles come near to alignment with the field by means of energy transfer to lattice vibrations. As discussed previously in the section delineating paramagnetic materials, the aligning influence of the torque is balanced by thermal agitation and is ultimately restricted by the quantum-mechanical constraints imposed by the uncertainty principle. The angle between a dipole and the field is never precisely zero.) The coupling between neighboring dipoles — interdipole fields, the source of the magnetostatic energy already discussed, are responsible — tends to cause them to precess with a common phase. This is called the uniform precession mode. The application of an RF electromagnetic field to the material will affect the precession in a way that depends on the directions of propagation and polarization of the electromagnetic disturbance with respect to the magnetization direction in the material. The dipole response to this perturbation is added linearly (for small electromagnetic wave amplitudes) to the intrinsic precessional motion.

The torque-type action of the RF (and bias) magnetic fields on the dipoles is represented in a macroscopic mathematical model by introducing imaginary-valued, off-diagonal elements into the RF permeability (or susceptibility) tensor $\overline{\overline{\mu}}$ (or $\overline{\overline{\chi}}$). Both the diagonal and off-diagonal elements become complex valued in the presence of damping, or loss, in a material.

To examine the reaction of magnetically saturated materials to magnetic field perturbation, assume a linearly polarized electromagnetic wave impinging, of which the magnetic field component is important because of its singular ability to interact with the spin contribution to the dipole moment. A dipole will in general experience a torque coercing it to align with the instantaneous RF magnetic field. Disregarding damping and inertia, the dipole will track the component of the RF magnetic field that is perpendicular to the saturating internal field. (A precessing dipole points nearly in the direction of the saturating field, so it has little moment to respond to a perturbing magnetic field polarized* in the longitudinal direction.) An RF transverse magnetic wave propagating parallel to the constant (or static) internal field drives the angle between the dipole and the bias field to successively greater and smaller values. An RF magnetic wave propagating perpendicular to the bias field can similarly interact with the dipoles if its magnetic polarization is transverse to the bias field. If the wave is polarized so that the magnetic component is parallel to the static internal field, the material response exhibits no magnetic features, and the wave propagates in it as if in a nonmagnetic dielectric. Applying a torque to a vector quantity cannot change its magnitude, only its direction. The locus of the dipole moment therefore oscillates on the surface of a sphere, the same one that delimits the intrinsic precessional motion (see Fig. 18.10).

When the frequency of the magnetic field approximates the natural precession frequency of the dipole (which depends on the bias field strength by way of the gyromagnetic equation), energy can be transferred to the dipole out of the RF field if the required phase relationship exists. The bandwidth of this resonant absorption is governed by a Lorentzian lineshape, which results from the decay of excited spins to lattice phonons, directly or via coupling to the reservoir of other dipoles. Experiments that expose this line width are critical to modeling the attributes of magnetic materials.

The mathematical description of the interaction of magnetic dipoles with RF magnetic fields invokes both Maxwell's equations (viz. $\nabla \times \mathbf{h} = d\mathbf{E}/dt$ and $\nabla \cdot \mathbf{b} = 0$; $\mathbf{b} = \overline{\overline{\mu}} \cdot \mathbf{h}$) and the equation of motion for magnetic dipoles in a magnetic field $[d\mathbf{m}/dt = \gamma (\mathbf{m} \times \mathbf{H}_i)]$.[†] For the appropriate RF magnetic polarization perpendicular to the saturating static magnetic field, the material response depends on the wavelength of the internal RF magnetic field relative to the separation of the contributing magnetic dipoles. (The internal wavelength depends on several parameters: the strength of the static saturating magnetic field, the frequency of the perturbing RF electromagnetic wave, and the material constants, often including the sample's dimensions and orientation in the static field. The wavelength of the vacuum electromagnetic wave is almost

---

* Standard usage connects the term *polarization*, or *polarized*, only with the electric field component of an electromagnetic wave. In this section it will often be used to indicate the direction in which the magnetic field component is pointing, but always with the distinguishing magnetic designation in *magnetic polarization* (or *magnetically polarized*). The text reverts to standard usage in the following section.

† Lowercase representation of the magnetic field vectors **b** and **h** indicate rapidly time-varying components.

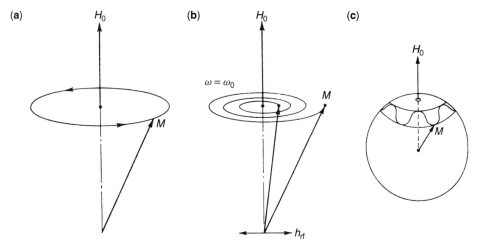

**Figure 18.10**  Precession of the magnetization vector (a) in a constant magnetic field grows in amplitude (b) when the frequency of the alternating magnetic field $h_{rf}$ is near the natural precession frequency. A perspective sketch of the oscillation of the magnetization vector when $\omega \neq \omega_0$ is shown in (c). The magnitude of the magnetization vector is a constant, so that the oscillations shown in this figure actually take place on the surface of a sphere of radius $|M|$. The exaggerated oscillatory component $\omega$ is superimposed on the precessional motion. (From B. Lax and K. J. Button, "Ferromagnetic Resonance," *Microwave Ferrites and Ferrimagnetics*, McGraw-Hill, New York, 1962, p. 148. Copyright 1962, used by permission.)

always much greater than the size of the sample; at 3 GHz the vacuum wavelength is 100 cm.) For a long internal RF magnetic wavelength the material response is minimal, and the propagation of the RF wave occurs as if in a dielectric. Alternatively, for a very short RF magnetic wavelength the driving field varies in strength (via its phase excursion) over the distance separating neighboring dipoles, and the field attempts to modulate neighboring dipoles out of phase. In this case, along with the magneto-static force that acts to coalign the dipoles, the exchange force becomes effective. The restoring force acting to align the dipoles is much stronger in this regime. The resonant frequency shifts upward, but more importantly a new type of wave becomes viable in the material, the exchange-dominated spin wave. An additional term is appended to the right side of the equation of motion, so that now

$$\frac{d\mathbf{m}}{dt} = \gamma(\mathbf{m} \times \mathbf{H}_i) + \gamma H_m d^2 \frac{\mathbf{M} \times \nabla^2 \mathbf{M}}{M},$$

in which $H_m = |\mathbf{H}_m| = \gamma |\mathbf{M}|$ is the magnitude of the exchange field, $\gamma$ is the molecular field coefficient, and $d$ is the separation between neighboring dipoles. The new solution for the resonant frequency is greater by a term $\omega_{ex} d^2 k^2$, in which $\omega_{ex}$ is the frequency corresponding to the exchange energy ($\omega_{ex} = \gamma H_m$), and $k$ is the wave number in the magnetic medium.

If the RF magnetic field is applied to dipoles on one side of a specimen, neighboring dipoles follow the perturbation of their orbits, and the disturbance propagates across the material. (See Fig. 18.11.) In samples of finite dimension, coupling of vacuum RF waves to exchange-dominated spin waves occurs either in the presence of a nonuniform

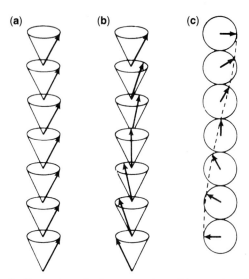

**Figure 18.11**   Schematic illustration of spins precessing with negative phase differences to give spin waves: (**a**) coherent mode; (**b**) one half wavelength; (**c**) spins in plan. (From D. J. Craik, "Effects of Crystal Size and Shape Rotational Processes," *Structure and Properties of Magnetic Materials*, Pion Limited, London, 1971, p. 187. Copyright 1971, used by permission.)

static internal magnetic field, such as can be caused by shape factor demagnetization (recall that only an ellipsoid of rotation has a uniform internal magnetic field), or when surface states cause dipoles near an interface to be held, or "pinned," in a particular orientation so that they are unable to rotate in response to the RF perturbation. In these cases skew between neighboring dipoles is inevitable and exchange-dominated spin waves result.

An intermediate wavelength regime also exists. In this range the exchange field is weak enough compared to the internal field that its influence can be ignored (recall that the exchange energy contribution increases as $k^2$; here $k$ is small). Moreover, in this range the simultaneous solution of Maxwell's equations and the magnetic equation of motion lead to expressions in which the RF electric field components in the material may be very small. Depending on the propagation direction with respect to the static internal magnetic field, the transverse RF magnetic field components may also be negligible, and the wave is essentially longitudinal — magnetic polarized. This behavior will be explained somewhat more thoroughly in what follows.

This type of wave is called, somewhat imprecisely, a magnetostatic wave. The wave field still oscillates with time at RF. But since the magnetic field polarization is quasi-longitudinal, the oscillating component is described approximately by the equation $\nabla \times \mathbf{h} = \mathbf{0}$, which in light of Maxwell's equation, $\nabla \times \mathbf{H} = d\mathbf{E}/dt$, confers the nomenclature magnetostatic wave. Incorporating the magnetostatic approximation makes the simultaneous solution of the gyromagnetic equation of motion and Maxwell's equations more tractable for intermediate values of $k (10 < k < 10^5 \text{ cm}^{-1})$ without severely altering its form compared to the exact solution.

The partitioning of RF magnetic perturbation phenomena outlined in the preceding into uniform precession, magnetostatic, and exchange-dominated spin waves according to wave number magnitude, and in particular the nature of magnetostatic waves, can be difficult to intuit. Propagation of magnetic-field-dominated waves is very sensitive to propagation direction and magnetic polarization. A delineation of the characteristics of RF magnetic plane waves in infinite magnetic media may aid understanding.

The polarization of the magnetic field for which spin waves are stimulated (uniform precession, magnetostatic, or exchange dominated) can be described in the language of optics as the extraordinary wave. The orthogonal magnetic polarization, for which the interaction of the RF magnetic field and the dipole spins are negligible (magnetic polarization parallel to static internal field), is analogous to an ordinary wave. The propagation and dispersion relation for the ordinary wave magnetic polarization are just those of an RF field in a dielectric. (Remember: magnetic fields act on dipoles via torque; no torque exists between coaligned forces and moments.) Ordinary RF magnetic waves will be mentioned only once again, when their straight-line dispersion serves as a reference for the complicated dispersion curves of the extraordinary waves. For extraordinary waves, solution of the preceding equations give rise, in an infinite medium, to two different dispersion relations. The applicable form depends on whether the spin wave propagates parallel or perpendicular to the static internal field. (Intermediate directions exhibit dispersion relations bounded by these orthogonal forms.)

The dispersion relation for the perpendicular propagation direction has two branches separated by a zone that is forbidden in the absence of exchange energy considerations (see Fig. 18.12). A graph of the low-frequency branch is coincident with the dielectric dispersion at the origin but curves gently away to become asymptotic to a line at constant frequency at high wave numbers (the bottom edge of the zone). This asymptote occurs at a frequency that is related to the gyromagnetic frequency, $\omega_0 = \gamma |\mathbf{H}_i|$, where $\mathbf{H}_i$ is the internal field due to an applied external field, and to the equivalent magnetic saturation frequency, $\omega_m = \gamma \times 4\pi |\mathbf{M}_s|$, where $\mathbf{M}_s$ is the material's saturation magnetization, by $\omega = \sqrt{\omega_0(\omega_0 + \omega_m)}$. The graph of the high-frequency branch is coincident with and parallel to the upper edge of the forbidden zone at very low wave numbers and curves gently to become asymptotic to the straight-line dielectric dispersion (the RF magnetic ordinary wave dispersion) at high frequency. The upper edge of the zone occurs at $\omega = \omega_0 + \omega_m$.

The dispersion relation for propagation parallel to the static internal field also has two branches (see Fig. 6.12). Their graphical appearance is very similar to that of the perpendicularly propagating waves, but mathematically their derivation is different, and the physical interpretation of the waves that obey them differs appreciably. One branch has two segments that mimic the appearance of the graph of the perpendicularly propagating wave just described, except that the lower edge of the forbidden zone occurs at a lower frequency, $\omega = \omega_0$. The upper segment appears, at zero wave number, at the same point coincident with the upper edge of the forbidden zone at $\omega = \omega_0 + \omega_m$ as the high-frequency part of the perpendicular dispersion relation. It is asymptotic to the alternate branch of the parallel dispersion relation rather than to the dielectric dispersion. This alternate branch is nearly linear and differs appreciably in appearance from the dielectric dispersion only in that the mean velocity, derivative from its slope, is slightly less.

These two branches of the parallel dispersion relation describe propagation of circularly polarized magnetic waves; the nearly straight branch corresponds to left circularly polarized (LCP) propagation, and the two segments of the other branch correspond to right circularly polarized propagation (RCP). (For RCP propagation, a stationary

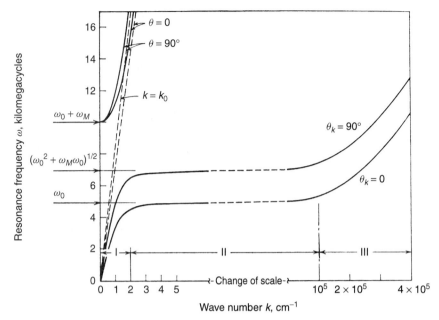

**Figure 18.12** Resonance frequency in the infinite medium as a function of wave number, where $k = 0$ corresponds to the frequency of uniform precession. The regions indicated are (I) electromagnetic propagation, (II) magnetostatic modes, and (III) spin wave modes. The two dashed lines are the ordinary wave dispersion relations. Curves in upper left are upper branches of extraordinary mode. Curves across center are lower branches of extraordinary mode. Here $4\pi M_s = 2000$ Oe. (From B. Lax and K. J. Button, "Plane-Wave Propagation," *Microwave Ferrites and Ferrimagnetics*, McGraw-Hill, New York, 1962, p. 319. Copyright 1962, used by permission.)

observer looking in the direction of propagation would see on a screen in front of him the magnetic field vector rotate in a clockwise sense with increasing time, that is, in the direction that the fingers of a right hand would curl when the right thumb is pointed along the propagation direction.) The LCP straight branch exhibits nearly no interaction with the magnetic dipoles, while the RCP branch is strongly dispersive. This behavior can be anticipated since dipole precession is also clockwise around the static magnetic field. At a certain frequency the RCP wave will be resonant with the dipoles, a situation that can never occur for the LCP wave. These two branches of the parallel propagation dispersion relation have different slopes, demonstrative of different wave velocities for RCP and LCP waves. This phenomenon is commonly known as Faraday rotation. (As will be shown in what follows, its basic mechanism differs from that for optical Faraday rotation, even though the phenomenological effects are similar.)

The mathematics of the derivations that produce these dispersion relations (not reproduced here) show that for perpendicular propagation, at intermediate values of wave number, both transverse components of the RF magnetic field are negligibly small (the component along the static field direction is identically zero), as are all components of the RF electric field. This wave has become essentially longitudinal–magnetic polarized, and the magnetostatic approximation $\nabla \times \mathbf{H} = 0$ is valid. Conversely, the parallel propagating waves have transverse magnetic components that are equal in magnitude and electric field components that are not vanishingly small. These waves do not satisfy the magnetostatic approximation.

Consider the perpendicularly propagating waves. Throughout the magnetostatic region of intermediate wave number magnitudes their dispersion relation is nearly flat, with phase velocity near zero. According to the discussion so far, perpendicular RF magnetic waves exist only along the single dispersion curve below the forbidden zone. Experimentally, perpendicular waves propagate on a manifold of dispersion curves in this region and in the forbidden zone as well. The parallel propagating waves maintain a single dispersion curve, although they too transgress the forbidden zone. The analysis of plane waves in unbounded media requires adjustment to represent physical reality.

The first adjustment reincorporates exchange energy considerations at high wave numbers. Recall that they introduce an additional term in the gyromagnetic equation of motion and an additive term in the resonant frequency expression that depends on $k^2$. This $k^2$ term causes the dispersion curve to turn parabolically across the forbidden zone toward higher frequency at high wave number values ($k > 10^5$ cm$^{-1}$).

The addition of boundary conditions to the problem additionally complicates the mathematics by introducing shape factor demagnetization terms to the internal magnetic field. The shape factor was discussed in Section 18.3 on magnetic materials. For thin planar samples, the demagnetization factor is perpendicular to the surface. When the applied magnetic field (which saturates the material) is also perpendicular to the surface, the effective internal magnetization is reduced by $4\pi \mathbf{M}_s$, and the gyromagnetic resonant frequency is reduced by $\gamma \times 4\pi \mathbf{M}_s$ (i.e., by $\omega_m$). This shifts the dispersion relation and its attendant forbidden zone to lower frequencies. To see how much, replace $\omega_0$ by $\omega_0 - \omega_m$ everywhere it occurs in the preceding discussion. This allows the forbidden zone to be bisected by the frequency of uniform precession at $\omega_0$, and coupling between uniform precession modes and exchange-dominated spin wave modes, for which the dispersion curves bend across the zone, can occur. Of course, if the planar sample is tilted with respect to the applied field, the demagnetization contribution is reduced to $N_i$ times $\omega_m$, where $N_i$ is the demagnetization component in the direction of the field. If the field is in the plane of the specimen, the contribution is zero.

To analyze the propagation of spin wave modes in bounded media, it is convenient to invoke the magnetostatic approximation ($\nabla \times \mathbf{H} = \mathbf{0}$). Then $\mathbf{H} = \nabla \psi$, with $\psi$ the magnetic potential. The gyromagnetic and Maxwell equations become straightforward, if still complicated, differential equations in $\psi$, and propagation in a bounded media becomes a boundary value problem. The mathematical derivations of the various solutions will not be presented here (see Ref. 3 or 4), but a description follows of the types of waves that will propagate for various orientations of a thin, planar sample (or film) in the static field. The eigensolutions are found when the sample is oriented either perpendicular or parallel to the static field (see Fig. 18.13), and only these conditions will be discussed.

Both surface and bulk waves exist. (A preview of the dispersion manifolds for the various modes of each possible wave type appears in Fig. 18.14.) When the static field is in the plane of the sample but perpendicular to the propagation direction, solution of the boundary problem produces surface waves whose magnetic field amplitude decays exponentially with distance away from the surface in both directions. A single dispersion curve applies to these waves for any specified static magnetic field strength.

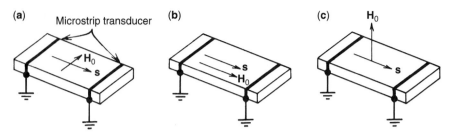

**Figure 18.13**   The required orientation of the static bias magnetic field $\mathbf{H}_0$ for each of the three types of magnetostatic wave. The $s$ vector indicates the MSW propagation direction. Note that surface waves (**a**) are nonreciprocal; they propagate only on the particular surface of the film where $\mathbf{H}_0 \times \hat{\mathbf{n}}$ is parallel to $\mathbf{s}$ ($\hat{\mathbf{n}}$ normal to the surface). Also shown are backward (**b**) and forward (**c**) volume wave modes.

Propagation of the surface waves is nonisotropic; they travel along one surface of the sample in one direction and the opposite surface in the reverse direction, specified according to $\mathbf{H}_i \times \hat{\mathbf{n}}$, where $\hat{\mathbf{n}}$ is the normal vector to the surface under consideration.

The dispersion relation for surface waves is

$$\exp(2\beta_{\text{MSW}}d) = \frac{(2\pi M_s)^2}{(|\mathbf{H}_i| + 2\pi M_s)^2 - \omega/\omega^2}$$

In this equation, $\beta_{\text{MSW}}$ is the component of the MSW wave vector parallel to the film surface, $d$ is the film thickness, $\omega$ is the MSW frequency, and $\gamma$ is the gyromagnetic frequency.

When the static field is in the sample plane but parallel to the propagation, backward-volume waves propagate. These are waveguide-confined, bulklike waves for which the group velocity is negative. A manifold of dispersion curves exists for backward-volume waves that appears approximately as a mirrored replica of the manifold for forward-volume waves, reflected across the center of the frequency bandwidth.

The dispersion relation for backward-volume waves is

$$2\cot(\alpha\beta_{\text{MSW}}d) = \alpha - \frac{1}{\alpha}$$

Here $\alpha = \sqrt{-1/\mu}$, where $\mu$ is the scalar magnitude of the permeability, assumed isotropic; other variables are as previously defined.

Forward-volume waves exist when the static field is perpendicular to the planar sample. Their propagation is isotropic with the direction in the film. The magnetic field amplitude for these waveguide-confined, bulklike, forward-volume waves is a standing-wave pattern across the thin planar sample; the component of the wave vector transverse to the sample is real inside the film. Although the forward-volume waves submit readily to an RF magnetic wave boundary condition analysis, they can also be analyzed by a procedure analogous to that used for optical guided-wave modes. The approach is complicated by the fact that the spin waves in the reflecting-ray or zig-zag ray model propagate neither perpendicular nor parallel to the sample boundary. The applicable dispersion relation depends on the angle between the propagating direction and the surface of the sample, as does the value of the phase change experienced by the

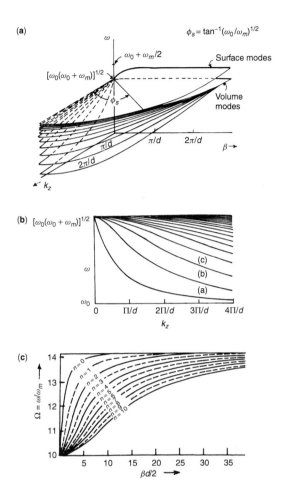

**Figure 18.14** (a) General magnetostatic mode spectrum for a ferrimagnetic slab magnetized parallel to its faces. The surface and bulk modes are simultaneously allowed only when the angle between the directions of propagation and biasing field exceeds $\phi_s = \cot^{-1}(\omega_m/\omega_0)^{1/2} = \tan^{-1}(\omega_0/\omega_m)^{1/2}$. (From M. S. Sodha and N. C. Srivastava, "Magnetostatic Waves in Layered Planar Structures," *Microwave Propagation in Ferrimagnetics*, Plenum, New York, 1981, p. 173. Copyright 1981, used by permission.) (b) Dispersion curves for the case of magnetostatic wave propagation along the biasing field. The curves designated by **a**, **b**, **c**, etc., correspond to $n = 0, 1, 2, \ldots$. The slope of $\omega$-versus-$k_z$ is negative throughout the frequency range for all mode orders. Thus the modes are backward waves, for which the directions of energy flow and phase propagation are mutually opposite. (From M. S. Sodha and N. C. Srivastava, "Magnetostatic Waves in Layered Planar Structures," *Microwave Propagation in Ferrimagnetics*, Plenum, New York, 1981, p. 172. Copyright 1981, used by permission.) (c) Magnetostatic mode spectrum for a normally magnetized slab. The phase velocity is smaller for the higher-order modes. (From M. S. Sodha and N. C. Srivastava, "Magnetostatic Waves in Layered Planar Structures," *Microwave Propagation in Ferrimagnetics*, Plenum, New York, 1981, p. 183. Copyright 1981, used by permission.)

wave on reflection. Either approach results in the existence of a different dispersion curve for each forward-volume wave mode, where a mode is distinguished by the number of standing-wave nulls found between the film surfaces.

The dispersion relation for forward-volume waves is

$$\tan \frac{\beta_{\text{MSW}}}{2\alpha} = \alpha$$

The manifold of dispersion curves resulting from the two bulk-wave solutions (forward- and backward-volume waves) shifts to higher frequency when the static field strength is increased. The frequency bounds for the manifold are

$$\omega_0 < \omega < \sqrt{\omega_0(\omega_0 + \omega_m)}$$

for the upper and

$$\omega_0 = \gamma|\mathbf{H}_i|$$

for the lower. This lower bound is at the gyromagnetic frequency, which is also the lower edge of the forbidden zone for plane RF magnetic waves propagating parallel to the static magnetic field in an unbounded medium. The upper bound is the same as the lower edge of the forbidden zone for propagation perpendicular to the static field. The geometric constraints have induced propagation in frequency ranges that do not support waves in the infinite medium.

The single dispersion curve for the surface wave also shifts to a higher frequency for increased static fields. It is bounded in frequency by

$$\sqrt{\omega_0(\omega_0 + \omega_m)} < \omega < \omega_0 + \tfrac{1}{2}\omega_m$$

This frequency band is adjacent to and above that for the bulk waves, with its upper limit still within the forbidden zones of the analysis of plane waves in an infinite medium.

In recapitulation, three types of spin waves — uniform precession, magnetostatic, and exchange dominated — can be stimulated in saturated magnetic materials by externally impressed, oscillating magnetic (electromagnetic) fields. The uniform precession modes occupy the very low wave number region of frequency–wave number space. Magnetostatic waves are intermediate in wave number. Exchange-dominated spin waves exist only at high wave numbers where the exchange force is appreciable. There exists a forbidden zone for plane waves in an infinite isotropic magnetic medium, but it is breached on inclusion of exchange interactions at high wave numbers and by the effect of shape factor demagnetization on internal fields. Complicated rules emerge from the gyromagnetic and Maxwell equations to govern the propagation of three subtypes of magnetostatic waves in planar film samples.

A few other details merit mentioning. Anisotropy fields in magnetic materials also change the location of the allowed frequency bands for the various spin wave disturbances. Their exact effect depends on the orientation of the anisotropy in the sample and the orientation of the sample in the static field (see Ref. 4).

These spin waves can be stimulated in several ways. Usual approaches comprise two alternatives. A sample can be placed in a microwave waveguide (transmission line) or in a cavity terminating such a line. The degree to which the symmetry of the RF magnetic field in the waveguide matches that required for a particular spin wave mode determines the efficiency of excitation of the mode. Alternatively, for thin-film samples, a simple metallic microstripline antenna carrying RF current and placed near the surface of the sample will stimulate any of the three magnetostatic wave mode types very efficiently.

A technique for determining the attenuation of propagating magnetostatic waves (and exchange-dominated spin waves, perforce, since their resonant frequency regions can overlap) employs the microwave waveguide approach to stimulating the modes in a thin-film sample. Usually the sample is placed near an antinode of the RF magnetic field in a waveguide or cavity, with its surface perpendicular to both the RF magnetic field and an externally imposed static magnetic field (from an electromagnet). A small coil carries a small-amplitude, low-frequency alternating current that modulates the static field slightly. The large field supplied by the electromagnet is swept slowly on a linear ramp, which shifts the magnetostatic and exchange-dominated spin wave spectra commensurately, so that the spin wave modes pass successively through resonance with the RF magnetic field. The tickler magnetic field supplied by the small coil allows lock-in amplifier detection of changes in the RF power in the waveguide after insertion of a small probe. The differential line width of the spin wave mode is detected, from which the propagating absorption can be calculated. (Figure 18.15 illustrates this apparatus schematically.) First the real line width is calculated from the detected derivative line by assuming a Lorentzian lineshape. The absorption is then $\alpha = 76\Delta|\mathbf{H}|$, where $\Delta|\mathbf{H}|$ is the sweep range of the large field over which the resonance occurs, and $\alpha$ is in decibels per microsecond. Exchange-dominated spin wave modes do not always appear if the sample has been grown so that the surface spins remain unpinned and if the sample is situated in the waveguide so that the RF field is uniform over its dimensions.

When the wave numbers of magnetostatic waves overlap those for which uniform precession or, particularly, exchange-dominated waves exist, coupling to them occurs,

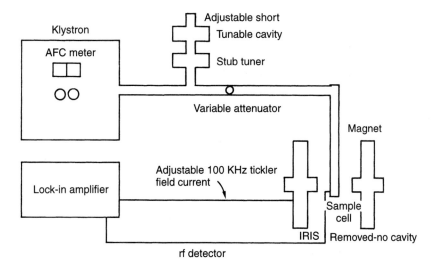

**Figure 18.15**  Ferrimagnet resonance line width measurement.

similar to the coupling between uniform precession and exchange-dominated modes previously mentioned. This is the natural evidence of phase-matched, coupled oscillators. The frequency passband of magnetostatic waves stimulated in a planar sample exhibits notches at the corresponding frequencies.

## 18.5  MATERIAL PHYSICS OF OPTICS

Optical waves are generally recognized as those transverse electromagnetic disturbances governed by Maxwell's equations that have wavelengths in vacuum between approximately 300 and 30,000 nm. Most light-modulating devices are concerned with wavelengths in a considerably narrower range, from 400 to 1600 nm, partly because of the paucity of coherent sources outside this range but more critically because of the reduced efficiency or response speed of detectors and the difficulties encountered in controlling thermal noise for wavelengths beyond these limits on one end or the other. This section presents those characteristics and descriptive models of the propagation of optical waves in materials that pertain particularly to magneto-optical interaction. Many details of optical propagation in materials that are generically useful for description of device operation will be assumed familiar or can be acquired elsewhere. These include, for example, light ray refraction according to Snell's law, as well as the effect on propagation velocity, caused by a changing refractive index; a thorough description, via Bessel function expansion of a sinusoidally varying phasor, of the diffraction of light from thin and thick phase gratings; and a detailed derivation of the equations that delineate optical propagation in planar waveguides. Several books are listed in the bibliography that provide this background material.

Magnetic interactions with light induce changes in the optical polarization. (This is the polarization of the electric field.) The polarization of a transverse electromagnetic wave at some point along its propagation path can be described by any number of pairs of orthogonal states: linear polarizations at any angular orientation; right and left circular polarizations; and elliptical polarizations of opposite handedness, orientation, and ellipticity (elliptical states are complicated, and declaring two elliptical polarizations to be orthogonal requires careful analysis). Generally there is one pair that renders any particular analysis most tractable. Polarization changes can be described (i) as differences in phase change on propagation (i.e., differences in velocity), (ii) by differences in absorption, or (iii) by coupling between the orthogonal states, depending on the mechanism. Some comments on light polarization states follow.

Elliptical polarization is the most general description of polarized light. Linear and circular polarizations are specific cases of elliptical polarization. Elliptical polarization can be constructed from two orthogonal, linearly polarized waves of equal magnitude, one delayed by a fixed phase with respect to the other. This is the easiest construction to visualize. The ellipse drawn on a plane perpendicular to the propagation direction by the tip of the electric field vector will always have its major axis oriented at 45° to the two component linear polarization directions. The ratio of minor to major axis lengths will depend on the phase delay. Zero phase delay makes the ratio zero (ellipticity 0), corresponding to linear polarization; and 90° phase delay produces ellipticity 1, for circular polarization.

Many other pairs of orthogonally polarized waves can replicate any particular ellipse, however, which complicates the mapping of the resultant polarization to a unique description. As an example, consider the following. For any particular ellipticity, the same ellipse can be produced on the perpendicular plane by any pair of linearly polarized waves oriented along the same axes as in the preceding but having different magnitudes in the two component polarizations, provided that the sum of the squares of their magnitudes equals that for the preceding description and their relative phase delay is properly adjusted. Thus a polarization-dependent absorption can appear to produce the same elliptical polarization as a polarization-dependent phase delay. In reality, the electric field vector in the second situation does not rotate at a uniform angular rate, and the appropriate description is not clear. An experimental measurement could easily produce confusing results.

Some authors invoke a geometrical artifice to catalogue and analyze changes in polarization states called the Poincaré sphere. Readers who are avidly interested in polarization states should consult the literature.

The polarization eigenstates of photons emitted by electric dipole transitions are circularly polarized. This physical reality corresponds to the quantum-mechanical rules governing the change of angular momentum that must accompany an electron transition between orbitals in an atom.

Changes in electromagnetic field polarization can be described, rigorously, in terms of either the magnetic permeability tensor or the dielectric permittivity tensor. Mathematically, use of either tensor requires identical manipulation; they occur as multiplicands in the same term of the wave equation. Considering the physical processes that occur in materials, however, it seems reasonable to identify the magnetic permeability with lower-energy magnetic dipole transitions and the dielectric permittivity with higher-energy electric dipole transitions [6]. At optical frequencies there are still residual effects of magnetic exchange interactions that take place at the lower frequencies (in the gigahertz range), discussed in preceding sections, but most features in the optical frequency range are artifacts of resonances that involve electronic orbital transitions. Magnetically induced effects then alter these electronic transition energies. To reflect the nature of the predominant physical processes that produce features in the optical frequency range, the dielectric permittivity tensor will be used in the macroscopic description of magnetooptical effects. In so doing, to avoid confusion, the unitary tensor is assigned to the magnetic permeability tensor. The off-diagonal terms of this tensor govern magnetooptical behavior in materials.

The dielectric permittivity can be separated into a nonmagnetic part and a much smaller part that is dependent on the magnetization vector $\mathbf{M}$ so that

$$\bar{\bar{\varepsilon}} = \bar{\bar{\varepsilon}}_{nm} + \bar{\bar{\varepsilon}}_m$$

Wettling [7] makes a careful argument, introducing Hermitian and anti-Hermitian parts to $\bar{\bar{\varepsilon}}_m$ to discover the symmetries of its real and imaginary parts. Then he invokes the Onsager relation: terms of the permittivity tensor that are symmetrical with respect to the diagonal have dependences on the magnetization $\mathbf{M}$ that differ by a minus sign. Next, the permittivity tensor elements are expanded in a power series in the Cartesian components of $\mathbf{M}$, retaining only the linear and quadratic terms. (The linear terms have anti-Hermitian real and Hermitian imaginary parts; the quadratic terms are real Hermitian and imaginary anti-Hermitian.) Combining the restrictions discovered through the analysis outlined in the preceding, the linear and quadratic magnetic dependence of the nonzero elements of the permittivity tensor can be derived in view of the symmetries of any chosen crystal class. Only results for cubic crystals seem to be extant in the literature; presumably other crystal classes are more difficult to analyze, and the exercise contributes little to understanding.

The resulting tensor for **M** directed along the $z$-axis (the 3 direction) and light also propagating in the $z$ direction is

$$
\bar{\bar{\varepsilon}} = \begin{bmatrix} \varepsilon_{11} + G_{12}M_z^2 & KM_z & 0 \\ -KM_z & \varepsilon_{11} + G_{12}M_z^2 & 0 \\ 0 & 0 & \varepsilon_{11} + G_{11}M_z^2 \end{bmatrix}
$$

where $\varepsilon_{11}$ is the value of the diagonal terms of $\bar{\bar{\varepsilon}}_{nm}$, $KM_z$ is the magnitude of the off-diagonal term that is linearly dependent on **M**, and $G_{12}M_z^2$ and $G_{11}M_z^2$ are the terms quadratic in **M** that can be seen to be diagonal contributors. (Some authors use an alternative convention in which the off-diagonal elements are written with a preceding multiplier $j = \sqrt{-1}$. This difference is resolved by interchanging the assignment of the real and imaginary parts of the tensor elements to phase and absorption effects.)

(Wettling [7] points out some details about optical propagation in other directions that most other sources do not acknowledge: the pertinent $G$ coefficients change with the direction. The parameter $\bar{\bar{G}}$ is really a four-tensor, of which many elements are degenerate, and many zeros. See Ref. 7 for details.)

This tensor is more tractable in a simpler form:

$$
\bar{\bar{\varepsilon}} = \begin{bmatrix} \varepsilon_1 & \varepsilon_2 & 0 \\ -\varepsilon_2 & \varepsilon_1 & 0 \\ 0 & 0 & \varepsilon_3 \end{bmatrix}
$$

The tensor elements generally are complex, reflecting the refractive index and absorption effects of the material. The preceding tensor expression does not include terms for uniaxial or biaxial birefringence or for optical activity, all phenomena that occur in some crystals in the absence of a perturbing imposed field. Birefringence would be evidenced by unequal values of the field-independent terms along the tensor diagonal. Optical activity, also called magnetic optical rotation, contributes to field-independent off-diagonal terms. The symmetry of these field-independent terms is even about the tensor diagonal, in contrast to the behavior of the field-dependent contributions. Here, all materials will be assumed free of these effects unless explicitly stated.

Elements of the tensor that are zero above may be nonzero when static magnetic fields are present along directions that do not correspond to cubic crystal axes or when RF magnetic fields are superimposed. The selection of tensor elements that is effected by superimposed RF magnetic fields depends on the magnetic polarization of those fields. In general, the RF magnetic fields of spin waves have components both transverse and longitudinal to the propagation direction (although none are parallel to the static saturating field). To first order in the RF magnetic fields, the elements of the tensor that are populated above do not change. Each remaining term becomes nonzero: the (1, 3) element (first row, third column) becomes $-Km_y + G_{44}M_z m_x$; the (2, 3) element becomes $+Km_x + G_{44}M_z m_y$; the (3, 1) element becomes $+Km_y + G_{44}M_z m_x$; and the (3, 2) element becomes $-Km_x + G_{44}M_z m_y$ (see Ref. 8).

The microscopic model of the source of the permittivity tensor's off-diagonal elements comprises several different processes. Each of them relies on a difference in the interaction of a particular electronic transition with the RCP and LCP polarization

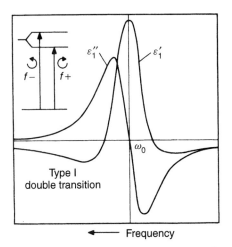 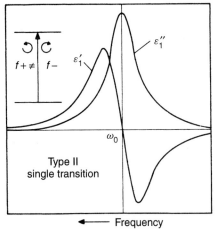

**Figure 18.16**   The calculated spectra of $\varepsilon_1'$ and $\varepsilon_1''$ for a so-called double transition with a "diamagnetic" line shape (type I) and for a so-called single transition with a "paramagnetic" line shape (type II). (From S. Wittekoek et al., "Magneto-Optic Spectra and the Dielectric Tensor Elements of Bismuth-Substituted Iron Garnets at Photon Energies between 2.2–5.2 eV," *Phys. Rev. B*, **12**, No. 7, 1975, p. 2783. Published by the American Physical Society. Copyright 1971, used by permission.)

components of the light. The two transition types, discussed next, are illustrated in Fig. 18.16.

One process, which gives paramagnetic response, derives from an electronic transition in which a single excited state (or energy degenerate pair of excited states) is accessed from a ground state split by the Zeeman effect. This occurs in materials for which the spin–orbit coupling is strong. (The Zeeman effect is phenomenological; application of a magnetic field splits the energy of a transition into two levels, one for RCP and one for LCP light. Classically, this dependence can be calculated by incorporating the Lorentz force into the equation of motion for the electron, with the result that the natural resonance frequency in the expression for the dielectric constant is either augmented or diminished by the Larmor precession frequency, $\omega_L = e\mu_0|\mathbf{H}|/2m_e$, wherein $e$ is the electronic charge, $\mu_0$ is the free-space permeability, and $m_e$ is the electron mass.) The energy difference between states split by the Zeeman effect is actually very small, and its influence on off-diagonal tensor elements, or equivalently on the difference in the refractive index for RCP and LCP light via shifts in their respective resonant frequencies, is dominated by the following quantum-mechanical effect.

Transitions to the excited state can occur for RCP light only when the total magnetic quantum number (spin and orbital quantum numbers summed) is incremented by 1 (the lower energy transition) and for LCP light only when it is decremented by 1 (higher energy transition); a magnetic dipole transition accompanies that of the electric dipole. (These restrictions on allowable transitions are called selection rules.) The average transition probability for each polarization depends on the number of dipoles in each of the two pertinent electronic ground states, that is, the number having the appropriate magnetic dipole state to access the excited state. That number depends on two factors, both of which influence the magnetization of the dipole. The difference in energy between the two Zeeman-split states causes their respective populations to differ, as required by the Boltzmann thermal distribution function at temperatures above absolute zero. The populations also depend on the magnitude of the static internal magnetic field evidenced by the magnetization. These effects on magnetic dipole alignment were discussed earlier and apply to paramagnetic materials — hence the designation of this effect. Since it applies to the balance between two transitions of nearly equivalent energy, the paramagnetic effect exhibits a dispersive lineshape near the transition resonance. Its amplitude is determined by the imbalance and is proportional to the Verdet constant. When the temperature is lowered or the static internal field is raised enough to magnetically saturate the material, the paramagnetic effect also saturates.

The mechanism described in the previous paragraph prevails when the material is characterized by strong spin–orbit coupling. In the case of a saturated material (no further effect of applied field) or for a material in which the spin–orbit coupling is weaker, a paramagnetic response may be caused through a different mechanism. In these materials the orbital angular momentum contributes very weakly to the magnetization, and the magnetic field interacts almost exclusively with the spin. Light interacts only with the electric dipole moment, however. The spin–orbit coupling may still be strong enough to transfer the energy change effected by the magnetic field on the spin levels to the orbital energy levels where it can influence the RCP and LCP electric dipole transitions. In the particular case of ferromagnetics and ferrimagnetics, the exchange energy causes level splitting between the spin states that may exceed the magnetic field effect on the dipole transition energy by a factor of 1000. Spin–orbit coupling plays a vital role in transferring the effect of a magnetic field to the values of the off-axis elements of the permittivity tensor.

A different type of response is elicited when the material has a single ground state and the excited state is split. The magnetic field influences the electric dipole transition via strong spin–orbit coupling, weaker spin–orbit transfer of exchange spin-level splitting, or Zeeman splitting of the upper state of the transition pair. In this case the average probability of a transition is not affected by the field since electrons reside predominantly in the ground state, which is unaffected. Instead, the RCP and LCP resonant frequencies are fractionally offset. The difference in the individually dispersive responses is approximately derivative and has an absorption-type lineshape, which characterizes the off-axis permittivity tensor elements for frequencies in the vicinity of the resonance. This type of response is called diamagnetic; it applies to materials that have no intrinsic magnetic dipole moment (diamagnetic materials) and contributes to the response of saturated magnetic materials.

## 18.6  MAGNETO-OPTICS

Section 18.5 addressed the description of polarized light, outlined the basic physical processes that produce the phenomenological manifestations of magneto-optical interactions, and introduced the tensor formalism that connects the physical processes to empirical observations. This section describes the macroscopic phenomena according to their several nomenclatures and surveys the types of materials in which each effect is important.

Magneto-optical effects in solids can be divided into transmissive and reflective classes. Both depend on the magnitudes of the elements of the dielectric permittivity tensor. Transmissive effects also depend on the thickness of the material transited by the light.

Consider propagation of light through a medium in a direction parallel to the internal magnetization. The effects on its polarization, both in terms of phase retardation and absorption, will be determined by the dielectric permittivity tensor. The simpler of the two forms for the permittivity tensor shown in the preceding will suffice to explain the interaction. Remember that each tensor element is in general a complex value.

Inserting this tensor into the electromagnetic wave equation (obtained from Maxwell's equations) and setting the determinant of coefficients of the Cartesian components of the electric field equal to zero reveals an expression for the refractive index in terms of the tensor elements. It is quartic in $n$, the refractive index. For propagation parallel to the $z$-axis and collinear with the static magnetic field (or magnetization $\mathbf{M}$), two solutions result:

$$n_+^2 = \varepsilon_1 + j\varepsilon_2$$
$$n_-^2 = \varepsilon_1 - j\varepsilon_2$$

Substituting these values of the refractive index into the wave equation produces the respective proper, or eigen-, modes of the electric field:

$$\mathbf{E}_+ = (E_x, E_y, E_z) = E_0(1, j, 0)$$
$$\mathbf{E}_- = (E_x, E_y, E_z) = E_0(1, -j, 0)$$

where $E_0$ is an arbitrary constant. The convention for assigning the nomenclature right and left circularly polarized to a propagating wave in optics is converse to that in radio or microwaves. In optics, RCP is assigned to light whose electric field vector rotates clockwise on a stationary plane placed perpendicular to the propagation direction, viewed into the beam, against the propagation direction. Thus the field labeled with the plus sign is called RCP, and that labeled with the minus sign is the LCP.

The real part of the difference in refractive indices for RCP and LCP propagation through a sample is called magnetic circular birefringence (MCB). The MCB multiplied by the vacuum wave number,

$$k = |\mathbf{k}| = \frac{2\pi}{\lambda_0}$$

(where $\lambda_0$ is the vacuum wavelength), and by the sample thickness $d$ is known as the Faraday rotation $\phi_F$:

$$\phi_F = \frac{2\pi}{\lambda_0}(n_+ - n_-)d$$

where $(n_+ - n_-)(n_+ + n_-) = n_+^2 - n_-^2 = j2\varepsilon_2$, which implies $n_+ - n_- = j\varepsilon_2/n_0$, where $n_0$ is the average refractive index, $n_0 = (n_+ + n_-)/2$. (Here $\varepsilon_2$ is complex. Its imaginary part applies to MCB and the Faraday rotation.) Strictly, the Faraday designation should be reserved for materials in which the MCB is proportional to the applied magnetic field, in which case $\varepsilon_2 = KM_z = K|\chi|\mathbf{H}_z$ (here $|\chi|$ is the magnitude of the diagonal tensor component in an isotropic medium for a static magnetic field, and $K$ is an empirical constant); in this case the equation for the Faraday rotation $\phi_F$ is often written $\phi_F = VH_zd$, where $V$ is the Verdet constant. In practice, the

Faraday designation is also used to describe the rotation experienced by light that transits saturated magnetic materials as well as those that have natural rotatory power (optical activity), in which the phenomenon should be distinguished as magnetic optical rotation (MOR).

Linearly polarized light incident on a material sample that imparts Faraday rotation is decomposed into equal-amplitude components of RCP and LCP light. At any point along its propagation path in the sample the two circular polarizations may be recombined, conceptually, to give a linearly polarized resultant. If the resultant polarization is rotated along a right-handed helical path from its point of entry to the sample, the material is designated dextrorotatory. For left-handed helical propagation the material is designated lavorotatory.

With the sign convention used here, the imaginary part of the off-diagonal tensor element $\varepsilon_2$ gives birefringence (MCB), as just shown. The real part of $\varepsilon_2$ imbues differential absorption to RCP and LCP propagating waves. This effect is called magnetic circular dichroism (MCD) and is responsible for Faraday ellipticity (linearly polarized light is made elliptically polarized, the rotated linear polarization axis becoming the ellipse's major axis).

A similar analysis can be performed for light propagating perpendicular to the magnetic field. There are two eigenmodes. For one the electric field is linearly polarized nearly perpendicular to the magnetic field (its associated electric displacement is exactly perpendicular). If the magnetic field is directed in the $z$ direction and optical propagation is in the $y$ direction, the electric field is given by

$$\mathbf{E}_{\text{perp}} = (E_x, E_y, E_z) = E_0(\varepsilon_1, \varepsilon_2, 0)$$

with the corresponding refractive index given by

$$n_{\text{perp}}^2 = \frac{\varepsilon_1^2 + \varepsilon_2^2}{\varepsilon_1}$$

The second eigenmode is linearly polarized parallel to the magnetic field:

$$\mathbf{E}_{\text{para}} = (E_x, E_y, E_z) = E_0(0, 0, 1)$$

with its corresponding refractive index given by

$$n_{\text{para}}^2 = \varepsilon_3$$

The birefringence implied by the difference in $n_{\text{perp}}$ and $n_{\text{para}}$ applies to linearly polarized eigenmodes and so is called magnetic linear birefringence (MLB):

$$(n_{\text{para}} - n_{\text{perp}})(n_{\text{para}} + n_{\text{perp}}) = n_{\text{para}}^2 - n_{\text{perp}}^2$$

$$= \varepsilon_3 - \frac{\varepsilon_1^2 + \varepsilon_2^2}{\varepsilon_1}$$

which implies

$$n_{\text{para}} - n_{\text{perp}} = \frac{\varepsilon_3 - \varepsilon_1 - \varepsilon_2^2 \varepsilon_1}{2n_0}$$

where $n_0$ is the average refractive index, as before. Since $\varepsilon_3$ and $\varepsilon_1$ vary as the square of the magnetic field, whereas $\varepsilon_2$ varies linearly with $|\mathbf{H}|$, the birefringence has quadratic dependence on the magnetic field.

The MLB depends on the real parts of the permittivity tensor elements (recall that the MCB depends on the imaginary parts). The related quadratic effect to MLB, dependent on the imaginary parts of the tensor elements, is a second form of differential absorption called magnetic linear dichroism (MLD).

Magnetic linear dichroism is the basis in solids (crystals) for the Voigt effect, a response that relies on electronic transitions. The Voigt effect will be revisited later in the discussion of magneto-optical interactions in gases. This quadratic interaction is also often referred to, imprecisely, as the Cotton–Mouton effect. Strictly, the Cotton–Mouton effect applies to realignment of magnetic molecules in liquids; it too is quadratic in dependence on applied magnetic field.

In summary, phenomena that are linear in magnetic field strength occur when light propagation is parallel to the modulating magnetic field component, that is, MCB and MCD for the rotatory and absorptive effects, respectively. The optical normal modes for this interaction are right and left circularly polarized. Phenomena that are quadratic in magnetic field occur when light propagation is perpendicular to the magnetic field component responsible for modulation. These are the MLB and MLD. The normal modes for optical propagation in this case are linearly polarized.

The same magnetic-field-induced differences in refractive index that impart Faraday rotation via MCB and MCD are also responsible for a group of reflection phenomena (see Fig. 18.17). These fall under the designation of Kerr effects, of which there are three types. An analysis by Freiser [9] is instructive. Two significant concepts are proffered. The first concerns the eigenmodes for reflection. As are those for transmission, these eigenmodes are polarized neither circularly nor linearly completely, except at normal incidence angle and when propagating in Cartesian directions with respect to the magnetization. The second concerns the amplitude reflectivity from an interface. The same Fresnel reflection laws apply as for a common dielectric interface, except that the expressions for the refractive indices are modified according to the magneto-optical permittivity.

For external reflection (incident on the magneto-optical material surface from the vacuum or air side), the continuity of tangential electric field across an interface ensures that the angle of reflection equals the angle of incidence regardless of the relative directions of light propagation, sample orientation, and magnetic field. (The component of the propagation $\mathbf{k}$ vector tangential to the interface is always conserved.) Internal reflection presents a more difficult case. The complicated boundary conditions that accompany reflection from a magnetic material interface cause the light's polarization state in the sample to change. The refractive index, which governs the propagation

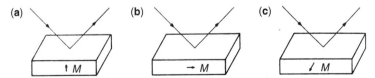

**Figure 18.17** Kerr effect configurations: (**a**) polar; (**b**) longitudinal; (**c**) equatorial. (From M. J. Freiser, "A Survey of Magnetooptic Effects," *IEEE Trans. Magn.*, **MAG-4**, No. 2, 1968, p. 156. Published by the Institute of Electronics Engineers. Copyright 1968, used by permission.)

direction and velocity of the reflected light according to its polarization, does not in general match that which governs the incident wave. After reflection the eigenmodes are generally of mixed polarization and propagate independently away from the interface, just as they would at non-Cartesian directions in transmission.

The analysis of internal reflection is purportedly horrendous (Dillon [10] references a thesis by C. C. Robinson at MIT). Internal incidence reflection will not be described here.

The polar Kerr effect describes reflection from a sample magnetized perpendicular to its surface. For normal incidence RCP light and LCP light are transformed into each other on reflection. Transmission in this orientation occurs with circularly polarized eigenmodes; since these are possible modes of the vacuum, normal reflection can be described in terms of the $n_+$ and $n_-$ previously shown. The reflection coefficient is given in terms of the circular eigenmodes by

$$r_+ = -\frac{n_+ - 1}{n_+ + 1} \quad \text{and} \quad r_- = -\frac{n_- - 1}{n_- + 1}$$

These are complex values; the difference in their phases gives the polar Kerr rotation. Linearly polarized incident light can be analyzed as the sum of equal-amplitude RCP and LCP waves. Then the Kerr rotation is one-half the differential phase retardation accumulated on reflection. Because of the simplicity of these expressions, the polar Kerr effect is most readily related to the values of the permittivity tensor elements.

The longitudinal Kerr effect describes reflection when the magnetization is both in the plane of the surface and in the plane of incidence of the light. At normal incidence the eigenmodes are linearly polarized. At nonnormal incidence elliptical polarization is introduced to linearly polarized incident light on reflection. This effect is important for detection in magnetic disk memories since domains in these materials align in the plane of the surface.

The transverse or equatorial Kerr effect is the third reflection regime. The magnetization is in the plane of the surface but perpendicular to the plane of incidence of the light. For normal incidence the transverse and longitudinal effects are degenerate. For nonnormal incidence, the transverse Kerr effect has minimal effect on light polarized perpendicular to the plane of polarization but produces a small change in reflected amplitude that depends on the direction of the magnetization (to the right or left with respect to the light's incident direction). Magnetization states for strongly magnetized materials can be analyzed without using a polarization analyzer when the equatorial Kerr effect is employed.

Freiser [9] lists the expressions for reflectivity in each of these three cases for nonnormal incidence. Deriving them is a tedious task that originates in assuring the continuity of tangential electric field components and normal magnetic field components across the interface.

Other familiar magneto-optical effects were originally observed in radiating or absorbing gases in applied magnetic fields. The names given them have in some cases also become associated with solids; in either case, the magneto-optical response is determined by magnetic field influences on electronic dipole transitions. The most important of these is the Zeeman effect. The observed traits of the effect change with viewing direction. Viewed parallel to the magnetic field, a natural resonant emission line splits into a doublet, one line up-shifted and LCP, the other down-shifted and RCP. A classical physics description such as that provided by Jenkins and White [11] suffices to explain this splitting. Viewed perpendicular to the magnetic field, a triplet of lines appears. The central line is unshifted in frequency and is linearly polarized along the direction of the magnetic field. The up-shifted and down-shifted lines are linearly polarized perpendicular to the magnetic field. These two lines imply a birefringence for linear polarization; the birefringence is proportional to the square of the magnetic field strength and is called the Voigt effect.

A description of this behavior is straightforward. As already mentioned, in a classical view the Lorentz force of a magnetic field on a moving charge is responsible. In an assemblage of oscillating electric dipoles, in the absence of a magnetic field, linear vibration will occur in evenly distributed random directions. The vibration of any particular dipole may be analyzed into a component parallel to the direction of a magnetic field (to be imposed presently) and two mutually orthogonal components both perpendicular to the field direction. The projection of the vibration along the three directions will vary from dipole to dipole. A sum over the statistical distribution is required to discover the strength of radiation of the assemblage of dipoles, with a particular polarization, in a particular direction. However, the frequency of radiation of any dipole depends only on the polarization and propagation direction of the radiation with respect to the magnetic field, not on its magnitude. Consider the component of a dipole oscillation parallel to a magnetic field (now turned on). The Lorentz force has no effect on this oscillation. Viewed against the direction of the magnetic field, this oscillation has no transverse component; no radiation is observed. Viewed perpendicular to the direction of the magnetic field, linearly polarized light will be seen, and it will have no frequency shift from the natural line center (no magnetic field). (This is the center line of the transverse triplet.) Now consider the component of a dipole oscillation that originally lies along one of the axes perpendicular to the magnetic field. The Lorentz force causes the path of the vibration to curve in the plane perpendicular to the field, counterclockwise, looking against the direction of the field for a negatively charged oscillating particle (an electron). The resulting path is rosette-like, and the resultant emitted ratiation will mimic this behavior. It is plane polarized but continually rotating. This is the behavior of MCB, in which the frequency of the LCP light exceeds that of the RCP light by twice the Larmor frequency (the frequency shift imposed by the Lorentz force; recall $\omega_L = e\mu_0|\mathbf{H}|/2m_e$). Viewed against the direction of the magnetic field, both LCP and RCP light will be seen, the LCP upshifted by the Larmor frequency, the RCP down-shifted by the same amount. Viewed perpendicular to the field, the rosette-like oscillation will still occur at the difference frequency of twice the Larmor frequency. However, the oscillation is in the plane of observation, so the two fields will appear linearly polarized perpendicular to the field. These two lines form the satellites of the transverse triplet, evenly spaced in frequency on either side, and are responsible for the Voigt effect.

Similar effects appear in absorption; these are called the inverse Zeeman effect.

## 18.7  REAL MATERIALS

Materials of importance for magneto-optical devices fall into two basic categories, with some variations and exceptions. The first of these contains materials that are reasonably transparent in at least some portion of the optical spectrum and that have large Verdet constants (when not magnetically saturated) or large values of Faraday rotation (when saturated) or, occasionally, large values of Voigt rotation (which is usually called the Cotton–Mouton effect) in the same wavelength region. The second category contains materials that have high optical reflectivity in at least some segment of the optical spectrum, where they also introduce significant Kerr rotation to reflected incident light.

### 18.7.1  Transmissive Magneto-optical Materials

In transmissive magneto-optical materials, both optical absorption and the magneto-optical effects are generally strongest for optical frequencies (wavelengths) in the vicinity of resonances; longer incident wavelengths are generally accompanied by weaker magneto-optical effects and by weaker absorption (greater transparency). One useful figure of merit is the ratio of Faraday rotation to optical absorption, $Q = \phi_F/\alpha$; this ratio is strongly wavelength dependent near absorption bands in crystalline materials. (Unless otherwise indicated, attention will be confined henceforth to Faraday rotation, the most commonly used transmissive effect.)

Magneto-optically important transmitting materials are of two types: glass doped with paramagnetic or diamagnetic atomic species and crystalline (or polycrystalline) materials, usually garnets. (Some crystals of the ferrite family were investigated previously, but the properties of these materials render them noncompetitive with garnets.) A brief look at glass materials will be indulged first; lower figures of merit, for which a low Verdet constant or Faraday rotation, rather than high absorptivity, is responsible limit their applications somewhat.

A representative paramagnetic glass is terbium-doped borosilicate [12]. Its Verdet constant (0.0058 deg/cm-Oe at 531 nm) is about 200 times that of quartz. (Koshizuka and Okuda indicate that either terbium-doped glass or terbium–aluminum garnet may have a Verdet constant as high as 0.015 deg/cm-Oe at 800 nm. See Ref. 13.) The terbium exists in its $Tb^{3+}$ ionization state in the glass. Its paramagnetic response is evinced by a susceptibility curve that follows $\chi = C/T$ for temperatures of 5–300 K. Judging from data included in the Hoya Corporation magneto-optical products catalogue, which lists Faraday effect isolators and rotators that use paramagnetic and diamagnetic glass, the optical absorption of these glass materials is low, and their refractive indices are high. (Absorption is less than 0.25 dB/cm, and the refractive index is near 3 at a wavelength of 633 nm.) Paramagnetic glass has higher Faraday rotation by a factor of almost 3 but is temperature sensitive, whereas diamagnetic glass is not.

A discussion of garnets, which exhibit Faraday rotation, is incomplete without inclusion of the effects that the parameters and constraints of the growing process impress on their properties. First, the magnetic, optical, and magneto-optical properties that may be important to control are listed. The magnetic properties comprise the magnetic uniaxial and crystalline anisotropy constants $K_1$ and $K_\mu$ and the orientation of the anisotropy; the saturation magnetization $\mathbf{M}_s$; the Curie temperature, the compensation temperature if the material has one, and the rate of variation of the susceptibility in the vicinity of these temperatures; the Bloch wall energy; and the field required to saturate the material in consideration of its saturation magnetization and the magnitude and direction of its anisotropy. Optical properties are optical absorption and its wavelength dependence; optical birefringence; scattering from surface pits; and inclusion defects. Magneto-optical parameters include the Verdet constant and its saturation value, which corresponds to the maximum Faraday rotation, and its wavelength dependence.

Bulk samples of magneto-optical materials may be pulled by the Czochralski method or grown by the float-zone technique. Relatively pure bulk crystals can be expected, with only minute contamination by the crucible (platinum). Devices that can make use of these crystals are relatively large, however, and to accommodate the motivations of laser diode and optical fiber compactness and integrated optical configurations, growing garnets as thin films has been investigated. Viable growth techniques include liquid phase epitaxy (LPE) from fluxes (currently most productive), RF sputtering, and vapor deposition. These are sometimes attended by annealing in a hydrogen, oxygen, or inert atmosphere. The problems encountered in growing garnet films will be discussed briefly here.

To create a garnet that has the desired properties generally entails beginning with a simple, well-known structure and then modifying it by substituting elements that have similar size and chemical proclivities to those of the basic constituents in the crystal. Because of the widely varying sizes of ions that share chemical valences, the degree to which one element substitutes for another may be limited by the integrity of the crystal structure. Examples will be given.

To grow substituted garnets as thin films by LPE, a flux must be found that will hold the appropriate fractions of film constituent atoms in solution and render them to the film in the desired ratio; hopefully the flux material proves reticent to enter the film itself. Among variations in the growth parameters that influence the composition of an LPE film are the fractional quantities of the various elements dissolved in the flux, the temperature of the flux (which determines the degree of supersaturation of the solvents) during film growth, and the nature of agitation, or mixing, of the flux during the growth process.

The films must be grown on a substrate. The substrate must not melt in the flux at the growth temperature. It must have the same crystal structure and lattice constant as the expected film material to prevent stress effects from arising. It should have nearly the same coefficient of expansion so that lattice constants that match at room temperature will also match at growth temperature. For optical waveguide applications, the refractive index of the substrate should be less than that of the film. Generally, the substrate material should be nonmagnetic. It should take a smooth, flat polish while retaining good crystal structure on the surface.

The crystal type that serves as the base structure for almost all significant magnetic and magneto-optical garnets is a rare earth iron garnet (recall Fig. 18.6). (Trivalent iron, a transition metal ion, has a large intrinsic magnetic dipole moment, about 5 Bohr magnetons per ion for temperatures from absolute zero to near 300 K.) Yttrium iron garnet (often denoted YIG), whose chemical formula is $Y_3Fe_5O_{12}$, is representative of this general class of garnets having the composition $R_3A_2B_3O_{12}$. The crystal is cubic. It has three different types of cation sites. The R sites are generally occupied by rare earth ions such as yttrium, lutetium, or gadolinium (large ions). Each of the three trivalent ions located on R sites is surrounded by an irregular dodecahedron (triangular faces) of eight oxygen ions. Each of the two A ions is surrounded by an octahedron (triangular faces) of six oxygen ions, and each of the three B ions by a tetrahedron (also triangular faces) of four oxygens. These sites, like the R sites, are generally and desirably occupied by trivalent ions. (The $Fe^{2+}$ has a somewhat smaller dipole moment than $Fe^{3+}$; more important here, it contributes strongly to optical absorption.) In YIG both the A and B sites are occupied by iron ions. The block of material defined by the 20 ions of the YIG formula has an axis of rotational symmetry through a cube diagonal. Eight of these formula cubes, with their diagonals at various orientations, are stacked together in a larger cube to form the crystallographic unit cell.

In the garnet structure, magnetic ions on the A and B sites are antiferromagnetically coupled. If identical ionic species occupy these sites, as in YIG, their unequal number renders the crystal ferrimagnetic. In YIG the R site is occupied by a nonmagnetic ion, yttrium (no intrinsic magnetic dipole moment), and the entire magnetic behavior is determined by the iron ions. Since the tetrahedrally coordinated B sites outnumber the octahedrally coordinated A sites 3 to 2, the net magnetic moment aligns with the B sites in this crystal. The R site may be occupied by magnetic ions in other similar crystals, however. In this case their dipole moments are influenced to lie canted with respect to the octahedral (A) sites but so that their net contribution aligns with that of the octahedrally coordinated A site ions. The total magnetic dipole moment in the crystal is then determined by summing the contributions of the ions on the three site types. This value generally varies with temperature; the dipole moments of the ions occupying each site type decrease with increasing temperature (all three contributions go to zero at the Curie temperature) but with differing behaviors, depending on the ionic species and the influence its site location in the crystal asserts on the magnetic contribution of its orbital magnetic moment. Many garnets that have magnetic ions on all three sites exhibit a total magnetic moment that passes through zero at a temperature well below the Curie temperature. This is called the compensation point, an important feature for most magneto-optical memory media, as will be discussed later (recall Fig. 18.7).

Yttrium iron garnet is a representative garnet having a simple structure that exhibits good, if not optimum, values of many parameters. Continued discussion of this crystal will illuminate the capabilities, variations, and limitations of the garnet class.

As a thin film YIG is most often grown by LPE methods, although RF-sputtered [14–16], hydrothermal process [17], and chemical vapor deposition (CVD) [18, 19] films have been grown. The most common substrate is gadolinium gallium garnet (GGG), $Gd_3Ga_5O_{12}$, which presents a good lattice match along with good thermal, surface, and optical refractive index characteristics. When grown by LPE, a flux combining lead oxide with boron oxide (PbO and $B_2O_3$) has historically been used, although preventing the incorporation of lead in films grown from this flux is problematic for optical absorption.

Magnetically, YIG has a nominal saturation magnetization value of $4\pi M_s = 1750$ Oe. The crystalline anisotropy field is small, on the order of $2K_1/M_s = 80$ Oe. The crystalline anisotropy constant, $K_1$, in ergs/cm$^3$-G, can be calculated through understanding that $M_s = 1750/4\pi$. (In YIG the saturation field is 1750 Oe; $K_1 = 5570$ ergs/cm$^3$). The anisotropy constant is positive, which imparts in-plane anisotropy to films grown on (111)-oriented substrates. Its Curie temperature is near 600 K, and it has no compensation point in its pure state. Optically, it is very absorptive throughout the visible wavelength region, with peaks broadly localized around 600 and 910 nm. The effects of octahedral and tetrahedral iron crystal field transitions responsible for the absorption dwindle substantially for wavelengths beyond 1100 nm. Faraday rotation is appreciable in various wavelength regions, exceeding 25,000 deg/cm at 430 nm and at 530 nm and nearing 10,000 deg/cm at 680 nm. Unfortunately, in these wavelength regions optical absorption is mostly oppressively high and consequently produces a poor figure of merit. Windows of higher $Q$ occur at 560, 780, and 1100 nm for visible and near-visible wavelengths of interest, but near-IR wavelengths between 1100 and 1500 nm must be considered for best performance.

Another material parameter that is of paramount importance for a particular device application and so should be mentioned here is the ferrimagnetic resonance line width discussed earlier. Pure YIG leads all competitors in exhibiting a narrow line width (indicative of low microwave absorption for spin waves, of which magnetostatic waves are a special case), with values as low as 0.15 Oe occasionally demonstrated at an RF field frequency near 10 GHz and values below 0.8 Oe routinely available. Retaining a low line width value while changing the constituents of the garnet films to enhance other parameters has proved challenging to materials specialists.

A garnet system that vies with YIG in some applications is gadolinium iron garnet, GdIG. (See Ref. 5.) Gadolinium occupies the dodecahedrally coordinated R site in the garnet lattice. Unlike yttrium, gadolinium is paramagnetic, and in the presence of the strongly magnetizing iron garnet crystal structure it displays a large intrinsic magnetic dipole moment, 7 Bohr magnetons per ion at 0 K. This value diminishes rapidly with increasing temperature to about 1.5 Bohr magnetons at room temperature. Combined with the slowly diminishing iron ion moments on the other two lattice sites, a compensation point appears at about room temperature (thus the saturation magnetization is near zero). The Faraday rotation at 546 nm at room temperature for GdIG is 5000 deg/cm.

The problem of substituents will now be superficially addressed. Good reviews of their effects can be found in Refs. 20 and 21. By including various rare earth and transition metal elements in the flux solution, garnets can be grown that incorporate them and whose properties vary from those of YIG or GdIG appreciably. (Substituting the bismuth ion into the garnet structure acts to greatest benefit in many applications.) By intelligent combination of several substituents, many properties of garnet materials can be adjusted simultaneously to fit an application. The saturation magnetization can be lowered. The Faraday rotation can be enhanced. Compensation and Curie temperatures can be adjusted. The anisotropy constant can sometimes be controlled. Of course, efforts to incorporate substituent ions in garnet films require adaptations in flux composition and growth condition parameters and sometimes produce unexpected or undesired changes in the other structurally related parameters as well.

The list of substituents is rather long; it begins with the rare earth elements and seems to effectively end with bismuth, as will be discussed. The radii of the trivalent ions of the rare earths contract fairly smoothly with increasing atomic number, from 1.11 Å for cerium to 0.94 Å for ytterbium (Ref. 22). (Lutetium is also quite small.) Generally one of these is substituted into a garnet structure either to provide a strong magnetic dipole moment (as, e.g., praseodymium or neodymium will contribute) or to compensate by its small size for the large size of another substituent ion (lutetium is commonly employed in this capacity). These ions generally occupy the dodecahedral site in the garnet structure. Cobalt, a transition metal neighboring iron in the periodic table, has been investigated recently for its efficacy in enhancing Faraday rotation in iron garnets. It substitutes for iron on tetrahedrally coordinated sites. Various other elements Ca, Mg, Va, Ga, Ge, Si, and Al have been incorporated into garnets to fine tune the magnetic and magnetooptical properties, provide charge compensation, or stabilize the crystal structure by means of their ionic radii in specific materials development approaches. Bismuth, and to some extent for the same purposes lead, can be substituted into the dodecahedrally coordinated R site, displacing yttrium in YIG or gadolinium in GdIG. Through a complicated interaction with the garnet lattice, these produce the greatest enhancement of the Faraday rotation.

Saturation magnetization at room temperature can be changed by two substitutional approaches. A diamagnetic ion can be included that displaces iron on the tetrahedral site. Gallium is often employed, with aluminum an alternative trivalent choice; germanium or silicon is occasionally used sparingly to give charge compensation. As the number of formula units of this ion is increased from zero to 1, the antiferromagnetically coupled iron ions on the remaining tetrahedrally coordinated B sites and the octahedrally coordinated A sites become more nearly balanced in number. Consequently, the net magnetic moment of the crystal (the difference of the iron contributions when the do-decahedrally coordinated ion is nonmagnetic) approaches zero as the substituent ion concentration nears one formula unit. (Complete compensation is still temperature dependent since the temperature dependence of each iron dipole moment contribution depends on its lattice site.) Conversely, an ion having a permanent intrinsic magnetic dipole moment that will occupy the dodecahedrally coordinated lattice site can be used, provided it aligns parallel to the octahedrally coordinated iron sites. A prominent example of this type of compensation was discussed in the preceding in describing the traits of GdIG. Praseodymium, neodymium, and dysprosium are among other paramagnetic ions used as R site substituents, but not for magnetization compensation. The alignment of their dipoles, unlike that of gadolinium, is parallel to the tetrahedrally coordinated B site iron ions, which outnumber the A site iron ions. These ions therefore increase the room temperature saturation magnetization somewhat.

Considerable effort has been devoted over the years to increasing the magneto-optical effects in garnets: Faraday and Kerr rotation (transmissive and reflective effects, respectively). Substituting paramagnetic rare earth ions in iron garnets is a reasonably useful approach. (The larger of these rare earth ions — those of La, Pr, and Nd, which have lower atomic numbers — cannot form pure iron garnets with the formula $R_3Fe_5O_{12}$ because their large ions distort the would-be crystal structure excessively.) (See Ref. 21.) Praseodymium, Nd, Sm, Dy, and Eu all have been employed with some degree of success. They enter garnets on the dodecahedrally coordinated sites, displacing yttrium in YIG or gadolinium in GdIG. Each of these ions has its own intrinsic magnetic dipole moment. Optical transitions in materials containing these ions therefore exhibit magnetic field dependence, which appears phenomenologically as Faraday, or Kerr, rotation at longer wavelengths. As a starting point for investigating the influence of praseodymium on magneto-optical effects in garnets, the interested reader should consult Ref. 23. Some additionally useful discussion that pertains to the class of rare earth substituted garnets and examines the mechanisms by which their properties are produced is contained in Ref. 21.

Although substituted rare earth ions are effective, the discovery of the effects on magneto-optical parameters of substitution of one of two particular nonmagnetic ions relegated them to a subordinate role. The persuasive ions are those of bismuth and lead. They substitute on the dodecahedrally coordinated garnet sites. The bismuth ion is trivalent, while the lead ion is divalent. (This observation may in part explain why lead incorporation exacerbates optical absorption to the extent that while it enhances Faraday rotation strongly, the quality factor $Q$ for lead-substituted garnets is so poor that it is regarded as an obstinate impurity for most Faraday rotation device applications; Pb is the predominant solvent in the growth flux and thus is difficult to eliminate.)

Substituted in YIG, either of these ions changes the Faraday rotation dramatically. At room temperature and at 633 nm wavelength, Ref. 20 discloses that the Faraday rotation changes at a rate of $-2.06 \times 10^4$ deg/cm per formula unit for bismuth or $-1.84 \times 10^4$ deg/cm per formula unit for lead. It should be noted that the Faraday rotation of unsubstituted YIG is positive, on the order of 200 deg/cm for wavelengths in the near-IR. Substitution of bismuth (or lead) drives the Faraday rotation negative with increasing concentrations; it passes through zero at a small level of substitution and then grows to high negative values as the substitution level nears or exceeds one formula unit. Bismuth makes a similar contribution to Faraday rotation when substituted into GdIG, but lead is not nearly as effective in this host. The details of bismuth-substituted YIG (Bi:YIG) and its variants will be outlined next.

Trivalent bismuth is a relatively large ion, appreciably larger than the yttrium ion it displaces. Bismuth may be substituted to about 1.88 formula units (about 60% of the three R sites) in YIG when grown as a bulk crystal, but to substitute bismuth to greater than about one formula unit in an epitaxial film requires size compensation to prevent stress cracking as the worst case and stress anisotropy and substrate lattice mismatch in any event. To avoid stress anisotropy in an epitaxial film, mismatch of film and substrate lattice constants must not exceed about 1 pm (0.001 nm). Commonly, lutetium is incorporated to relieve the stress; it too occupies the dodecahedrally coordinated R site. In fact, $Bi_x Lu_{3-x} IG$ has been grown, excluding the yttrium component altogether. Alternatively, lattice match with the substrate can be achieved by altering the substrate composition. Gadolinium gallium garnet is the time-tested substrate material. Its lattice constant matches that of YIG almost perfectly at 1.237 nm. Replacing some or all of the gadolinium on the R site in GGG with samarium or neodymium increases its lattice constant: NdGG has a lattice constant of 1.249 nm. Another successful approach is to replace some of the gallium ions with zirconium and magnesium ions. Zirconium is tetravalent while magnesium is divalent, so charge balance is retained. Although this technique produces a larger lattice constant, its value seems to drift as the crystal is grown, perhaps through unequal incorporation of the two ions. Incorporation of an appropriate amount of divalent calcium on the dodecahedrally coordinated site stabilizes the lattice constant; to retain the charge balance, the magnesium content must be reduced commensurately. Varying the amount of zirconium and the relative amounts of calcium and magnesium allows substrates to be grown that have lattice constants throughout the range 1.245–1.25241 nm (Ref. 5).

A brief digression will elucidate the magnetic properties of rare earth gallium garnets vis-à-vis rare earth iron garnets. The rare earth ions contribute paramagnetically to the magnetic properties of materials at room temperature. In the iron garnet crystal, their environment exhibits a strong magnetization because of the ferrimagnetic alignment of the A site and B site iron ions. They align in the magnetization (except for gadolinium, their magnetic moments lie parallel to the magnetization of the tetrahedrally coordinated trio of iron ions) and contribute to the magnetic character of the crystal, as was discussed for GdIG. Now consider the effects of replacing iron ions with trivalent gallium ions. Gallium ions enter the crystal, in small concentrations, on tetrahedrally coordinated sites. (This lowers the Curie temperature by tens of degrees kelvin and raises the compensation temperature dramatically; it disappears in $Gd_3 Fe_{5-y} Ga_y O_{12}$, having surpassed the Curie temperature, when $y$ exceeds about 0.6.) The saturation magnetization of the crystalline environment decreases as the tetrahedrally coordinated iron is replaced. Beyond some gallium concentration, octahedrally coordinated sites are also commandeered, and eventually GGG results from the complete absence of iron ions in the growth melt and the resulting crystal. Paramagnetic ions that are substituted for gadolinium on the R site or replace gadolinium entirely no longer adopt a ferrimagnetic posture or contribute to material magnetization.

Besides the stress-induced anisotropy caused predominantly by film-to-substrate lattice mismatch via magnetostriction, Bi:YIG exhibits magnetic anisotropy induced by other mechanisms. As in YIG itself and most magnetic materials, the bonds that establish the crystal structure influence the orbital angular momentum and through it the magnetization direction according to the degree of spin–orbit coupling in the magnetic ions. The resulting magnetic alignment is called crystalline anisotropy. In garnets, it shows cubic symmetry. Bismuth and yttrium have no intrinsic magnetic moment, so the crystalline anisotropy of Bi:YIG depends only on the spin–orbit coupling of the iron ions, which is weak.

Since YIG is very nearly a perfectly cubic crystal, its uniaxial anisotropy is also very small. The Bi:YIG has a rather large positive uniaxial anisotropy, however, induced in the iron ions by the noncubic (irregular) distribution of the bismuth ions. This is also commonly referred to as growth-induced anisotropy, since the nonuniform distribution of bismuth ions on the R sites, which results from the growth process, is responsible for the distorted bonding orbitals of the iron ions. The possibility of moderating this anisotropy by the inclusion of praseodymium ions, the sole rare earth ion that induces negative anisotropy in garnets, has been explored in Ref. 23. It is apparently very sensitive to constituent concentrations in the growth flux, and to the growth conditions (temperature and sample rotation rate). The effects of substituting particular diamagnetic ions on the tetrahedrally coordinated and octahedrally coordinated sites generally occupied by iron have been studied, showing prospects for controlling anisotropy by this means. (See Ref. 24.) Of course, these diamagnetic substitutions replace iron, the ion that is responsible for the desirably strong magnetic and magneto-optical properties of Bi–YIG.

This uniaxial anisotropy of Bi–YIG may prove troublesome to the reliable manufacture of devices, not so much because of its magnitude as because of the difficulty in accurately controlling, or reproducing, the anisotropy value in the growth process.

Saturation magnetization is not appreciably affected by bismuth substitution in YIG. It increases slightly as bismuth formula fraction increases.

Optically, Bi:YIG has an absorption edge in the near IR, like YIG, but somewhat beyond 1100 nm. Doormann et al. present a conscientious measurement of the refractive index and absorption coefficient of LPE films of YIG substituted with formula fractions of bismuth ranging up to 1.42 [25]. Although their films have lead concentrations up to 0.08 formula fraction, which may contribute significantly to absorption, the indicated trend is that the absorption edge for Bi:YIG films extends slightly farther into the near IR as bismuth concentration is increased. A measurement on one sample in which the bismuth content is quite high, 1.42 formula fraction, shows remanent absorption of about 50 cm$^{-1}$ at a wavelength of 1900 nm. For wavelengths shorter than 1100 nm, bismuth increases the absorption above that of YIG in a complicated way, introducing or enhancing several absorption resonances, most of them intrinsic to iron ion transitions. These transitions are also the root of the Faraday rotation enhancement and appear as the result of a magnetically coupled, second-order effect called super exchange. Measurements of refractive index, optical absorption, and Faraday rotation for various bismuth-substituted garnets are reproduced in Fig. 18.18.

Superexchange is the mechanism by which the magnetically dependent energy levels in one ion influence those of a second ion that is its next-nearest neighbor. The superexchange process, like the simple exchange process, ascribes to an electronic state a magnetic dipole whose strength depends on the spin orientation of the occupying electron. This is equivalent to magnetically induced energy-level splitting of an electronic state. Electronic states involved in superexchange may interact through partial overlap of the bonding orbitals of the affected magnetic ions in the vicinity of the intermediary ion; or they may interact more obliquely by their individual influence on the electron spins in the bonding orbitals of the intermediary ion.

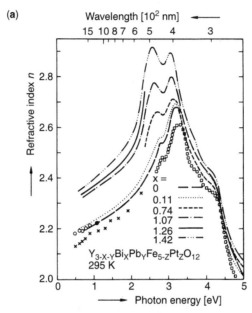

**Figure 18.18** (**a**) Refractive index spectra of Bi:YIG films as measured by thin-film interference for $\lambda \gtrsim 0.52~\mu$m and by ellipsometry for $\lambda \lesssim 0.52~\mu$m. Crosses refer to interference spectrometry on single-crystal wafers, open circles to the refraction of a single-crystal YIG prism, filled circles to $m$-line spectroscopy, and open squares to a Syton polished thick water. For references to original experiments, consult V. Doormann et al., *Appl. Phys.*, **A34**, p. 223, 1984. (From V. Doormann et al., "Measurement of the Refractive Index and Optical Absorption Spectra of Epitaxial Bismuth Substituted Yttrium Iron Garnet Films at uv to Near-ir Wavelengths," *Appl. Phys. A Solids Surf.*, **A34**, No. 4, 1984, p. 227. Published by Springer-Verlag. Copyright 1984, used by permission.) (**b**) Optical absorption spectra of Bi:YIG materials as measured by thin-film interference for $\lambda \gtrsim 0.52~\mu$m and by ellipsometry for $\lambda \lesssim 0.52~\mu$m. The border of applicability of the two spectroscopic techniques is indicated by the horizontal broken lines. Good overlap has been confirmed by ellipsometry on the bulk YIG sample. (From V. Doormann, "Measurement of the Refractive Index and Optical Absorption Spectra of Epitaxial Bismuth Substituted Yttrium Iron Garnet Films at uv to Near-ir Wavelengths," *Appl. Phys. A Solids Surf.*, **A34**, No. 4, 1984, p. 228. Published by Springer-Verlag. Copyright 1984, used by permission.) (**c**) Faraday rotation versus wavelength for (1) $Y_3Fe_5O_{12}$; (2) $Bi_{0.6}Tm_{2.4}Fe_{3.9}Ga_{1.1}O_{12}$ (LPE); (3) $Bi_{0.8}Yb_{2.2}Fe_{3.9}Ga_{1.1}O_{12}$ (LPE); (4) $Bi_{1.0}Tm_{2.0}Fe_{3.9}Ga_{1.1}O_{12}$ (LPE); (5) $Bi_{0.9}Sm_{2.1}Fe_{3.9}Ga_{1.1}O_{12}$; and (6) $Bi_{1.1}Eu_{1.9}Fe_{4.3}Ga_{0.7}O_{12}$. (From G. B. Scott and D. E. Lacklison, "Magnetooptic Properties and Applications of Bismuth Substituted Iron Garnets," *IEEE Trans. Magn.*, **MAG-12**, No. 4, 1976, p. 292. Published by the Institute of Electrical and Electronics Engineers, Copyright 1976, used by permission.)

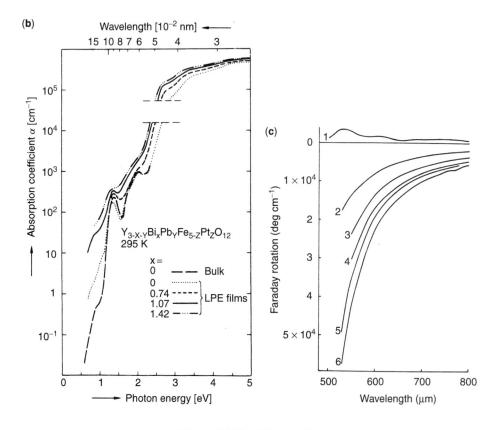

**Figure 18.18** *(Continued)*

In the garnet lattice, superexchange applies to interaction between the energy levels of two iron ions located on differing site types (one on an A site and one on a B site) that are separated by and bonded to an oxygen ion. The presence of bismuth in the lattice structure apparently enhances their magnetic coupling by distorting the angles and ion center-to-center separations along the iron ion–oxygen ion bonds and by introducing a perturbation in the electronic wave function around the oxygen ion with its own bonds to the oxygen, the participating bismuth orbitals being strongly spin–orbit coupled. (That is, bismuth increases the superexchange coupling.) In bismuth-substituted YIG, the superexchange coupling is suspected of promoting two types of transitions. (The paper by Scott and Lacklison contains possibly the most current analysis of the electronic transitions responsible for increased absorption and enhanced Faraday rotation in bismuth-substituted YIG. See Ref. 26.) These are, first, an excitation of a bi-exciton (i.e., two simultaneously created excitons in such close proximity that they interact to lower their total energy, much as molecular hydrogen forms to decrease the total energy of its atomic electrons) in a superexchange coupled electronic transition in next-nearest-neighbor iron ions and, second, a charge transfer transition between states associated with the two differing site iron ions. Each of these mechanisms is responsible for one of the two transitions whose strength increases most dramatically when bismuth is substituted in YIG, with the bi-exciton transition dominating. These transitions occur at wavelengths of 433 nm (the charge transfer transition) and 391 nm (the stronger, bi-exciton transition). Both of these transitions have a diamagnetic lineshape, indicating a split excited state. A curve that is often published (see Fig. 18.19), from work by Wittekoek et al. [27], shows the imaginary part of $\varepsilon_1^2$ between 250 and 550 nm and its dramatic increase near 390 nm for bismuth-substituted films compared to that for pure YIG films.

Controlling the magnitude of the optical absorption has begun to be addressed. As already mentioned, lead incorporated in YIG films contributes strongly to absorption. To eliminate lead from YIG films, fluxes free of PbO and based on $Bi_2O_3$ instead have been used with preliminary indications of success. In addition, divalent and to some extent tetravalent iron ions contribute strongly to optical absorption. Borrowing a technique used to suppress the optical absorption and accompanying optical damage caused by divalent iron ions in lithium niobate, researchers have incorporated divalent magnesium ions into bismuth-substituted LPE films [namely, into LPE $(BiLu)_3Fe_5O_{12}$]. See Ref. 28.

Several paragraphs earlier, the enhancement of Faraday rotation was named as the motivation for considering bismuth substitution in LPE garnet films. Although Bi:YIG demonstrates very high values of Faraday rotation, the values of Faraday rotation for equal substitutional composition in Bi:GdIG are even higher. Consequently, most of the literature that addresses bismuth substitution for enhanced Faraday rotation centers its discussion around GdIG. Bismuth substitution in YIG (or, alternatively, the combined bismuth–lutetium iron garnet system) remains important because these formulations may also prove capable of transmitting microwaves, in the guise of magnetostatic waves, with low absorption, a task for which the GdIG host is apparently less suitable. This aspect will be dealt with later in the discussion of ferrimagnetic resonance line width.

The properties of bismuth-substituted iron garnets are comprehensively presented by Hansen et al. [5], but with emphasis devoted particularly to Bi:GdIG. Two figures from their paper are reproduced here (Fig. 18.20). Figure 18.20**a** shows the variation with wavelength of the Faraday rotation for various substitutional levels of bismuth. Note that the unsubstituted GdIG has positive Faraday rotation at all wavelengths and that bismuth substitution makes it negative, passing through zero for a bismuth content of about 0.1 formula fraction. Unsubstituted YIG also has positive Faraday rotation that becomes negative with bismuth substitution. The peak Faraday rotation is near 530 nm in the Bi:GdIG system; for Bi:YIG it is nearer the superexchange transitions responsible for the large effect, around 400 nm. Although not directly extractable from the figure, the Faraday rotation contribution due to the bismuth incorporation depends linearly on the bismuth fraction, almost uniformly without regard to wavelength (differing slopes apply). Figure 18.20**b** shows the figure of merit $Q$ for Bi:GdIG when the wavelength is 546 nm, as a function of bismuth content. The optimum bismuth content depends on wavelength, but it is broadly peaked, which allows for some adjustment of composition to satisfy other requirements.

Values of Faraday rotation and absorption for wavelengths in the near IR are difficult to ascertain from these curves, and only recently have measurements appeared specifically at wavelengths of lasers that emit in that range: around 1150, 1300, and 1500 nm. At 1150 nm, Ohno and his coworkers [29] discuss a theoretical model for the rate of change of Faraday rotation with bismuth content that is based on observed behavior in the range of 500–800 nm. They estimate $d\phi_F/dX = -1977.2X$ in YIG and

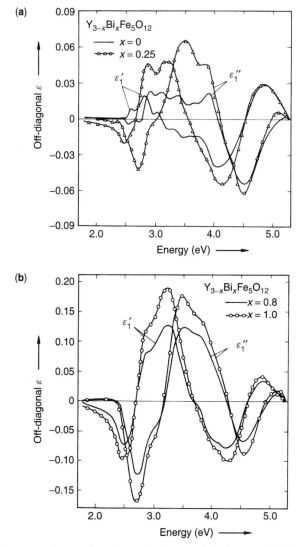

**Figure 18.19**    (**a**) Off-diagonal tensor elements $\varepsilon_1''$ and $\varepsilon_1''$ for $Y_{3-x}Bi_xFe_5O_{12}$ for $x = 0$, 0.25. (From S. Wittekoek et al., "Magneto-Optic Spectra and the Dielectric Tensor Elements of Bismuth-Substituted Iron Garnets at Photon Energies Between 2.2–5.2 eV," *Phys. Rev. B*, **12**, No. 7, 1975, p. 2780. Published by the American Physical Society. Copyright 1975, used by permission.) (**b**) Off-diagonal tensor elements $\varepsilon_1'$ and $\varepsilon_1''$ for $Y_{3-x}Bi_xFe_5O_{12}$ for $x = 0.8$, 1.0. (From S. Wittekoek et al., "Magneto-Optic Spectra and the Dielectric Tensor Elements of Bismuth-Substituted Iron Garnets at Photon Energies Between 2.2–5.2 eV," *Phys. Rev. B*, **12**, No. 7, 1975, p. 2780. Published by the American Physical Society. Copyright 1975, used by permission.)

$d\phi_F/dX = -15,900X$ in GdIG, where $X$ is the formula fraction of bismuth substitution. Since the transitions that produce the large effects are almost exclusively at wavelengths on the short end of this range, the estimate seems reasonable. Experimental corroboration would be useful.

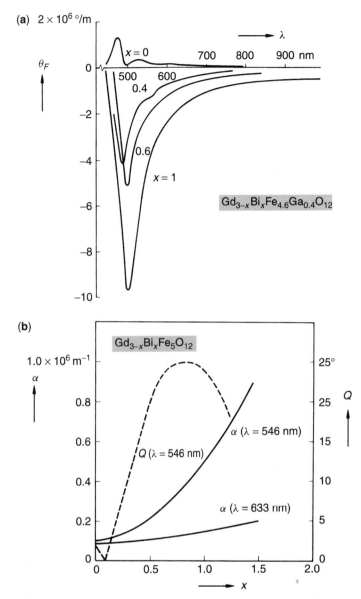

**Figure 18.20** (a) Specific Faraday rotation $\theta_F$ in degrees per meter of $Gd_{3-x}Bi_xFe_{4.6}Ga_{0.4}O_{12}$ as a function of wavelength $\lambda$ in nanometers of incident light, with the bismuth content $x$ as parameter. (From P. Hansen et al., "Optical Switching with Bismuth-Substituted Iron Garnets," *Philips Tech. Rev.*, **41**, No. 2, 1983/84, p. 39. Published by N. V. Philips. Copyright 1985, used by permission.) (b) Results of measurements of the figure of merit $Q = 2|\theta_F|/\alpha$ in degrees (at wavelength of 546 nm) and the absorption coefficient $\alpha$ in reciprocal meters (at wavelengths of 546 and 633 nm) as a function of the bismuth content $x$ in $Gd_{3-x}Bi_xFe_5O_{12}$. The curve for $Q$ is shown dashed. (From P. Hansen et al., "Optical Switching with Bismuth-Substituted Iron Garnets," *Philips Tech. Rev.*, **41**, No. 2, 1983/84, p. 40. Published by N. V. Philips. Copyright 1985, used by permission.)

Hibiya et al. provide some data in Ref. 30. Values of Faraday rotation at 1300 nm as high as $-800$ deg/cm for Bi:YIG with bismuth formula fraction $X = 0.49$ and as high as $-1500$ deg/cm for Bi–GdIG with bismuth formula fraction $X = 0.58$ are shown in one set of figures. Elsewhere in the paper their Figure 7 shows Faraday rotation at 1300 nm of $-2880$ deg/cm for Bi:Gd(FeAlGa)G grown from a different melt composition that produced the bismuth content $X = 1.56$. (This film had Faraday rotation of $-12,800$ deg/cm at 780 nm and $-23,600$ deg/cm at 633 nm.) In addition, Faraday rotation is shown to increase approximately linearly with bismuth content; at 1300 nm the slope $d\phi_F/dX = -2000X$ deg/cm in Bi:GdIG and Bi:Gd(FeAlGa)G. Finally, the authors report the minimum optical absorption value measured on their films at 1300 nm to be 1.6 cm$^{-1}$; this film was of the Bi:Gd(FeAlGa)G variety and had Faraday rotation of $-1880$ deg/cm, so its quality factor $Q = 267$ deg/dB, apparently the highest value for LPE garnets measured to date.

Stress-induced (photoelastic) optical birefringence, caused by lattice constant mismatch of film and substrate, can be untenably large in bismuth-substituted garnets. Presumably, differential thermal expansion rates produce stress on cooling from the oven even with substrates that have adjusted lattice constants. This birefringence has been shown to be linearly proportional both to lattice constant mismatch [31] and to bismuth substitution concentration [32]. Birefringence greater than $\Delta n = 1 \times 10^{-3}$ was reported in each report; in addition, the analysis of Ref. 31 indicates that stress alone cannot account for the measured birefringence in some films.

The ferrimagnetic resonance line width is not often discussed in characterizations of magneto-optical garnet materials; it is a property relevant to the propagation of microwave frequency (1–30 GHz) magnetic spin waves. The interaction of magnetic spin disturbances with light is somewhat removed from the common magneto-optical applications for display, isolation, and low-frequency modulation but holds promise for future high-frequency light modulation applications.

To implement the microwave magnetic modulation, low absorption of spin waves must be assured. As discussed in Section 18.4, microwave magnetic spin waves, in particular the various types of magnetostatic waves (MSWs), require static magnetic bias fields to coalign the internal dipoles. In addition, the dispersion relation for any particular MSW type shifts in frequency as the bias field strength is varied. At any predetermined frequency, the MSW wavelength depends on the strength of the static bias magnetic field. The technique commonly used to ascertain spin wave absorption places a microwave cavity supplied with microwaves from a klystron source (usually at a frequency of about 9.3 GHz) between the poles of a strong electromagnet. (This technique was described in a preceding section, but further details may be of interest. Recall the apparatus of Fig. 18.15.) A carefully prepared rectangular or circular sample of garnet film (or a sphere of bulk material when bulk properties are of interest) is mounted on a nonmagnetic rod in the microwave cavity and oriented so that the static magnetic field direction is appropriate for the type of MSW to be generated (volume waves are generally invoked). The magnitude of the bias field is then slowly swept so that the wavelength of the MSW in the sample varies until it fills the transverse dimension of the sample with a near integral number of half-wavelengths. Near this resonant condition, the energy of the MSW in the garnet sample increases, as does the energy extracted from the microwave cavity via absorption in the sample. This depletion of the cavity RF field can be measured by a microwave probe. The range of MSW wavelengths over which the sample absorbs appreciable energy maps linearly onto the swept bias magnetic field. This range is narrow, as for any Lorentzian resonance, when the absorption is low. Since the range of the bias magnetic field over which the absorption occurs is directly observable, it has become the custom to report the line width in terms of the swept field range $\Delta H$, in oersteds, at 9.3 GHz. Translation of the ferrimagnetic resonance (FMR) line width $\Delta H$ into more traditional terms can be accomplished using $\alpha_{MSW} = 76 \, \Delta H$ decibels per microsecond. The dispersion relation must be consulted for the MSW velocity to convert to decibels per centimeter. Sample thickness and ground plane location play roles in determining the shape of the dispersion curve and thus the dependence of the MSW velocity on frequency; values near 50 mm/µs are typical in the range where MSWs are useful for optical interaction, which gives a rough approximation of absorption as $\alpha_{MSW} = 1.5 \, \Delta H$ decibels per millimeter. It can be appreciated that garnet films that have values of $\Delta H > 2$ Oe are unimportant in this application.

At room temperature, the FMR line width in epitaxial YIG films has been measured as low as about 0.15 Oe, and values around 0.5 Oe are commonplace. This variation depends on a number of factors deriving from the film growth process: the supersaturation temperature of the flux; the base solvent composition and balance of constituents in the flux; the agitation rate of the substrate; the lattice match between film and substrate; and the film surface quality, which depends on the care with which remnant flux is spun or etched off following film growth. No other material can be grown as a thin film with such a low line width, including other garnets as well as ferrites. Yttrium iron garnet films have been grown by methods other than LPE in attempts to find processes more amenable to production, namely sputtering and vapor deposition. Sputtering tends to produce films composed of crystallites in which the line widths are considerably larger, although Faraday rotation is unaffected.

Gadolinium iron garnet has been discussed previously since it assumes the highest values of Faraday rotation, particularly when substituted with bismuth. It has a considerably broader line width than YIG, as do all the garnets that incorporate ions that have intrinsic magnetic dipole moments (Pr, Nd, etc.). The magnetic dipole moment of this type of ion is strongly coupled both to the lattice, via spin–orbit coupling, and to the magnetic dipoles of the neighboring iron ions, by exchange interaction. In addition, the rare earth ions have unusually short relaxation times. In combination, the consequence of these two attributes is that even minute amounts of rare earth ions in garnets increase the spin wave absorption immensely.

Introducing substituents (e.g., Bi) into garnet films, into YIG particularly, can act to broaden the ferrimagnetic resonance line width even when nonmagnetic ions are introduced. Consider an ion that preferentially substitutes for yttrium ions on the R site. Assume, as a best case, that exactly one substituent ion per formula unit is introduced. At least three conditions exist that can produce local distortion in the garnet crystal structure. Substituent ions will not all reside on the sites that they prefer due to the statistics of the crystallization process. Likewise, on a statistical basis, not every unit cell will have its exact apportionment of substituent ions (recall that the unit cell in YIG consists of eight formula units); some will have none, others more than one. Last, the distribution of substituent ions on R sites within unit cells or within formula units need not be exactly lock-step. When substituent ions do not constitute an integral number of formula units, at least some of these events are bound to occur. This atomic-scale local distortion of the lattice broadens the ferrimagnetic resonance line width as it subtly distorts the bond angles and perturbs the bond strengths of the iron ions, consequently changing the exchange and superexchange coupling of their magnetic dipole moments.

In spite of these considerations, substituted garnet films having bismuth concentration exceeding 0.8 formula units have recently been grown with quite narrow FMR line widths [33]. A film having composition $Bi_{0.85}Lu_{2.15}Fe_5O_{12}$ had a line width of 1.5 Oe and that of a film having composition $Bi_{1.4}Y_{1.6}Fe_5O_{12}$ was measured to be 0.73 Oe. The critical procedure for producing films having such low line widths was the use of a lead-free flux consisting of $Bi_2O_3$ alone.

### 18.7.2  Reflective Magneto-optical Materials

Reflective magneto-optical materials are used essentially in a single application, namely, magneto-optical recording in thin film, and their important parameters are defined in terms of the needs of this technology. These parameters are high polar Kerr rotation of the polarization of reflected light; high uniaxial anisotropy perpendicular to the film plane; a moderate value of saturation magnetization far from the compensation temperature (too high and it exceeds the coercivity, too low and the Kerr effect is small, even though the Kerr effect magnitude must depend predominantly on one of the sublattices alone, since magnetization is near zero around $T_{comp}$) and relatively high magnetic remanence and coercivity (also called coercive field, required to demagnetize a domain) near the operating temperature; a magnetic compensation temperature (or a Curie temperature, a rarer occurrence) near the ambient temperature; and the capability of forming small domains to enable high-density recording. (Many of these magnetic properties are indicated in Fig. 18.21.) Thermal properties contribute to the dynamics of domain formation in such a way that low heat capacity and high thermal conductivity are desired to keep data domains small. Substantial optical absorption at the operating wavelength (e.g., 50%) is also important for recording data on magneto-optical media since all writing mechanisms depend on raising the local temperature in the material.

Near-normal light incidence is required for most reflective magneto-optical applications. The polar Kerr effect, rather than either the longitudinal or transverse effect, must be invoked since of the three it alone is nonzero for normal-incidence light. The polarization rotation available from polar Kerr rotation is not large, always less than $1°$ in known materials when used in the absence of enhancing surface coatings (e.g., films in which the incident light is resonant).

Materials that induce high values of polarization rotation in reflected light consist almost exclusively of rare earth–transition metal (RE–TM) alloys. Amorphous films are desired over polycrystalline films since the signal-to-noise ratio of light reflected from oriented domains in this type of film is greater. For example, MnBi was initially considered a promising material since its Kerr rotation is high ($0.7°$) but its polycrystalline films give high background scattering.

The RE–TM alloys are generally Néel-type ferrimagnetic; the RE and TM atoms are coupled antiferromagnetically. The RE magnetization is larger at low temperature while the TM magnetization dominates at higher temperatures. Thus a compensation point exists; its exact value can be adjusted by varying the material composition, particularly in ternary and quaternary alloys.

Near the compensation temperature the magnetization components of the RE and TM sublattices nearly cancel, and the total magnetization is about zero (see Fig. 18.22).

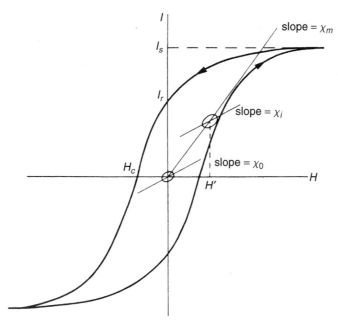

**Figure 18.21** Schematic of $I$ versus $H$ or hysteresis loop for a ferromagnetic (or ferrimagnetic) material, defining the following: $I_s$, saturation magnetization; $I_r$, remanence or remanent magnetization; $H_c$, coercivity or coercive field. The slopes of the lines indicated define: $\chi_m$, maximum susceptibility; $\chi_i$, incremental susceptibility in the bias field $H'$; $\chi_0$, initial susceptibility (for a minor loop of vanishingly small amplitude with no bias field, as found by extrapolation). The differential susceptibility $\chi_d$ is the slope, at any point, of the loop itself. (From D. J. Craik, "Magnetic Dipoles in Applied Fields," *Structure and Properties of Magnetic Materials*, Pion Limited, London, 1971, p. 35. Copyright 1971, used by permission.)

In this condition a saturated domain has very little net magnetic moment, so the torque exerted on the domain by an applied magnetic field is small, and reducing the magnetization of the domain to zero requires a high coercive field. Magnetized domains thus are stable with respect to small variations in local applied magnetic field and temperature when the ambient temperature is in the vicinity of the compensation temperature.

The magnetization of any particular domain can be inverted by changing the local temperature appreciably from the compensation temperature in the presence of a magnetic field. (The field may be externally applied or it may result from a geometric factor, viz., shape factor demagnetization, or another mechanism that induces uniaxial anisotropy, e.g., growth-induced stress.) The coercive field $H_c$ required to reduce domain magnetization to zero decreases with temperature excursion from the compensation temperature [$H_c = H_c(T)$ has a pole at $T = T_{\text{comp}}$]. When the coercive field decreases sufficiently that the local magnetic field exceeds it, the domain orientation may be flipped. Similarly, domain magnetization may be imposed or inverted when the local temperature is raised above the Curie temperature since the domain then loses all intrinsic magnetization. This approach to domain reversal is practical in materials that have been devised to exhibit very low Curie temperatures, within a few hundred degrees of room temperature.

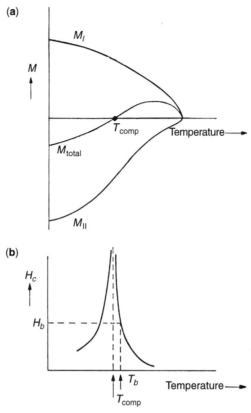

**Figure 18.22** (a) Temperature dependence of the magnetization of a ferrimagnetic garnet and its sublattices and (b) temperature dependence of the coercive force. (From S. Middlehoek, P. K. George, and P. Dekker, "Beam Accessible memories," *Physics of Computer Memory Devices,*" Academic, New York, 1976, p. 354. Copyright 1976, used by permission.)

A high value of uniaxial anisotropy perpendicular to the film surface is important for maintaining high coercivity and rectangular hysteresis loops (magnetization vs. magnetic field) and, when the uniaxial anisotropy field exceeds the saturation magnetization, for inducing domain reorientation when the local temperature is elevated. Perpendicular uniaxial anisotropy may be induced by two types of mechanisms. The first are spin–orbit mechanisms, represented by atomic-level effects like single-ion anisotropy, and anisotropic exchange and magnetostriction. These occur mostly in alloys containing rare earth ions whose outer electrons are not $S$ state (e.g., Tb, which is not a spherically symmetric ion). The second are magnetic dipolar interactions, of which pair ordering (in GdCo) is exemplar and to which growth process stress (differential contraction on cooling or substrate lattice mismatch) is a contributor. These are listed in Ref. 34 and discussed more thoroughly in Ref. 35 and the references therein.

A few other material properties are important in manufacturing magneto-optical films for memory applications. Domain formation and size are determined primarily by the thermal properties of the film material and substrate. Specific heat and thermal conductivity determine the dynamics of raising the local temperature sufficiently to impose domain magnetization and record data, as well as the rate of cooling in the local region, which determines domain size. Equally important is the saturation magnetization, since local demagnetizing fields increase dramatically as the size of the magnetized domain shrinks. The film material must be designed so that this local demagnetizing field does not exceed the coercive field. The saturation demagnetization must not be too large. Finally, for commercial viability the films should be producible using a continuous processing technology and be comprised of affordable elements.

Investigation of several representative reflective magneto-optical materials is reported in the current literature. Most of these are variations of quaternary RE–TM alloys or of their ternary subsets. Achievable values of Kerr rotation near room temperature are in the range $0.20°$–$0.35°$. Uniaxial anisotropy constants can be obtained greater than $5 \times 10^5$ ergs/cm$^3$, equivalent to greater than 7000 Oe anisotropy field in materials that have high values of saturation magnetization ($H_a = 2K_\mu/M_s$). The coercive field can be maintained greater than 300 Oe more than $40°$C from the compensation temperature without unreasonable difficulty.

For extensive details on particular RE–TM materials systems, the following references may be useful. For the GdTbFeCo system, see Refs. 36–43. Several papers discuss the PtMnSb system; see Refs. 44–46. For the (Nd, Pr)–(Fe, Co) system, see Ref. 47. A related system is Tb(Nd, Pr)Co; see Ref. 48. And rejuvenating the MnBi system is Ref. 49.

In addition to these discussions of the properties of specific materials, a good review of the requirements of magneto-optical recording systems is Ref. 50. Also, see Ref. 51.

## 18.8 MODERN MAGNETO-OPTICAL DEVICES

The list of magneto-optical devices currently commercially available or under development in laboratories is not long. By far the greatest resources are apportioned to development of magneto-optical data storage media and system components. Optical addressing of these media allows writing and reading at densities as high as or higher than competing technology (all-magnetic addressing). Prospects for erasure and rewrite capabilities of magneto-optical media are being explored extensively. Second in commitment of resources is the development of magneto-optical polarization-rotation isolators. Their use is envisioned in particular for isolation of laser diodes from reflective feedback of their emitted light, incident on optical fiber end-faces or on lenses positioned to couple light from laser to fiber. Otherwise, commercially proffered devices comprise Faraday rotator modulators and optical isolators, which perform light-ray-controlling functions, and a two-dimensional spatial light modulator for light wavefront addressing and modulating functions. A few other devices are under development in various laboratories. A magneto-optical crossbar switch directs optical signals between multiple communication ports in one system. Several laboratories are investigating the diffraction of wave-guided optical beams from magnetostatic waves in thin garnet films. This new Bragg diffraction technology may provide spectral analysis, convolution functions, and optical modulation at microwave frequencies between 1 and 20 GHz.

Magneto-optical data-recording technology is advancing along two different paths, although both envision erasable media as their competitive advantage over current optical technology and increased data density as the advantage over current nonoptical, magnetic technology. The first of these employs amorphous, or fine-grain polycrystalline, RE–TM films as the magneto-optical medium. The films are generally ternary or quaternary metal alloys; the multiplicity of constituents allows control of various device parameters simultaneously. Most approaches design the films to operate around the ferrimagnetic compensation point, the temperature at which the magnetization components of the antiferromagnetically aligned RE and TM sublattices balance and cause the net internal magnetization to vanish. This temperature is generally near or somewhat above ambient (room) temperature. The coercivity of the films becomes very large in the vicinity of the compensation temperature. (The net magnetic dipole moment disappears; the torque applied to a magnetic domain depends on the dot product of the local magnetic field and the net moment.) The magnetic orientation of a data-storing domain is thus stable with respect to small variations in the environmental magnetic field and temperature. To write data in a magnetic domain, a pulse of light is directed at the domain location, where the energy accrued from absorption of the pulse raises the local temperature. The increased temperature restores the net magnetic dipole moment in the domain and decreases the coercivity. A magnetic field at the domain site that exceeds the coercive field value can then reverse the domain's orientation. Cooling at the domain location freezes in the new data. Figure 18.23 illustrates a domain switching process as well as sketches of several types of domains that can exist in magnetic media under various conditions of material parameters, magnetic field strength, and temperature.

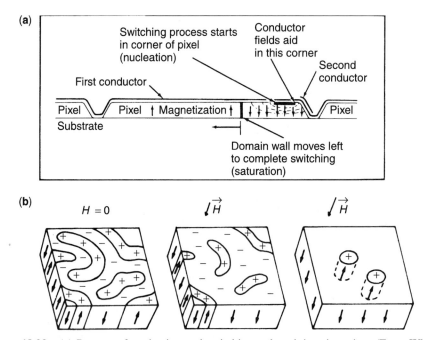

**Figure 18.23**   (a) Process of nucleation and switching a domain's orientation. (From William E. Ross et al., "Two-Dimensional Magneto-Optic Spatial Light Modulator for Signal Processing," *Opt. Eng.*, **22**, No. 4, 1983, p. 486. Published by the Society of Photo-Optical Instrumentation Engineers. Copyright 1983, used by permission.) (b) Equilibrium domain alignment at various applied magnetic field strengths. (From P. Paroli, "Magneto-Optical Devices Based on Garnet Films," *Thin Solid Films*, **114**, No. 1/2, 1984, p. 190. Published by Elsevier Sequoia S.A. Copyright 1984, used by permission.)

The magnetic field at the domain site can be derived from one of a number of sources. In the most general case the magneto-optical film that supports the data storage need not have uniaxial anisotropy. Geometric influences (shape factor anisotropy) cause the magnetic dipoles of this type of film to lie in the film plane. The local magnetic field that drives the domain orientation perpendicular to the film plane is then supplied solely by an external magnet. Usually this is an electromagnet, so that the magnetic field can be modulated on and off to correspond to particular write, read, erase, and overwrite requirements. The direction and magnitude of the magnetic field applied during the optical pulse, when the local temperature is elevated, saturates the dipoles of the region in a perpendicular orientation; the magnetic remanence of the material sustains this orientation when the temperature falls, the coercivity becomes large again, and the external magnetic field is removed. The written domain corresponding to a data bit then appears as a region magnetically oriented perpendicular to the film plane surrounded by material with magnetization in the film plane. In fact, since the ambient temperature is near the compensation temperature, neither the domain itself nor the surrounding in-plane-magnetized dipoles exhibits appreciable magnetization. Employing optical readout requires that the magneto-optical Kerr rotation depend substantially on only one of the ferrimagnetic sublattices in the film. Thus, although the net magnetization of the oriented domain is negligible, the domain may nevertheless be read optically via Kerr rotation. (See Ref. 52, pp. 350–360.)

Alternatively, the film may exhibit strong perpendicular uniaxial anisotropy. The entire film of dipoles then exhibits magnetization perpendicular to the film plane, with various regions aligned antiparallel. If such a film were cooled adiabatically from above the Curie temperature, or annealed, in the absence of an applied magnetic field, many stripe domains would result. The size of the domains in annealed films depends on the balance of energy minimization effects: shape factor anisotropy counteracts the aligning forces of dipole–dipole influences, tending to produce high-density contradirectional domains; uniaxial anisotropy and exchange effects account for the energy and consequent width of domain walls; and the interaction of these energy contributions determine, in the absence of external fields, the density and patterns of domains. These mechanisms have been discussed in Section 6.3. Externally applied perpendicular fields encourage the orientation of domains in one direction over that in the other, causing stripe domains of one magnetization direction to shrink to cylinders for moderate fields and to collapse altogether in higher fields.

Uniaxially anisotropic films for magnetic recording are cooled in the presence of an external magnetic field or saturated by application of a sufficiently high field so that in the unaddressed state all dipoles are aligned in one direction. (In these films the uniaxial anisotropy generally far exceeds the saturation magnetization due to shape factor demagnetization.) With the application of a reversing magnetic field, a small cylindrical domain flips magnetization when heated by an optical pulse. To understand the permanence of these small cylindrical domains, it is important to realize that a true equilibrium condition does not exist over the entire film area. Otherwise, cylindrical domains would not be stable in the absence of an external perpendicular magnetic field; rather, stripes would mediate the domains, as described in the preceding paragraph. Only because the brief optical pulse and the accompanying thermal transient temporarily change the local magnetization and coercivity can these domains be created and persist.

Shieh has devised a procedure to accomplish write, read, and overwrite functions in the absence of any externally applied magnetic field [53]. The shape factor demagnetizing field is invoked to nucleate the cylindrical domain that records a data bit; along with careful control of the optical power incident on the domain site, the various mechanisms of domain dynamics can implement write and erase functions. Once again, domain formation and stability rely on the balance between the domain energy and the domain wall energy. Domain wall energy varies directly as the square root of the product of the exchange constant and the anisotropy constant and inversely as the domain diameter for cylindrical domains. Wall energy thus contributes forces tending to reduce domain size. Both factors in the wall energy product decrease as the temperature is raised; this weakens the retaining forces of the wall. Conversely, demagnetization acts to increase the cylindrical domain size, whereby regions oriented antiparallel may more nearly cancel the total self-induced field. (Recall that the film outside the domain is uniformly magnetized perpendicular to the surface.) This force increases with temperature above the compensation point, where the internal magnetization increases in response to the growing dominance of one of the ferrimagnetic sublattices. So both wall and domain energy contributions act to increase the domain diameter when the local temperature is elevated. When their sum added to that of any applied field exceeds the stabilizing contribution of the coercive field, the domain expands. Expansion is limited first by equilibrium and then by the short duration of the heating optical pulse. When the domain again cools to ambient, the expansive forces diminish, the coercive field reassumes its overriding influence, and the domain stabilizes. Further detail may be discovered in Ref. 53.

Writing a cylindrical domain in the absence of an external magnetic field proceeds in several steps. First, the temperature of the irradiated region of the film increases from below to above the compensation temperature of the material. This causes the local magnetization to change sign with respect to the surrounding saturated medium and the sum of the intrinsic magnetic fields (demagnetizing and anisotropic) to exceed the coercivity. This small (typically $1-2$-$\mu$m-diameter) region establishes its own demagnetizing field, which is now oriented coparallel with the surrounding medium (because of the flip in magnetization of the small region caused by the local temperature's crossing through the compensation temperature). The magnetization in the heated region is greater in the center since more optical power is absorbed there. The demagnetizing field can thus nucleate a domain in the center of the small region, with accompanying cylindrical domain walls. The domain walls expand until they reach the edge of the small, heated region. The region cools, whereupon the temperature falls through the compensation temperature, the local magnetization reverses sign again, and the data-storing domain remains, with magnetization opposite to the surrounding film. Erasure follows an almost identical progression, with the added feature that the second expanding domain wall collides with the first and they coannihilate. The power, or the duration, of the erase pulse must be carefully limited so that a third nucleation is not initiated. For clarifying details, see Ref. 54.

Several other papers extant discuss the desirable material parameters for RE–TM films, fabrication techniques for obtaining them, and the dynamics of domain formation. For example, see Refs. 55 and 56.

The Curie point writing approach to magneto-optical data storage previously mentioned has fallen into disregard. The primary material that can support this approach is MnBi, which presents several problematic characteristics that are difficult to surmount.

First, films of MnBi tend to be polycrystalline rather than amorphous. Polycrystalline films scatter reflected light more strongly, giving high background light levels and poorer signal-to-noise ratio. Second, MnBi has two structural phases. The Curie temperature ($T_C$) for the low-temperature phase is above the phase transition temperature (360°C). The Curie point for the high-temperature phase, however, is below the phase transition temperature (at 190°C). To write a domain by the Curie point technique, the local temperature must be raised above 360°C. The MnBi then becomes paramagnetic, since its temperature is above the Curie temperature of the high-temperature phase. If it is cooled quickly (quenched) to below 190°C, the domain orientation induced in the paramagnetic state will persist. The high-temperature structural phase will be partially frozen in, however, so that a mixed-phase structure results. This structure slowly relaxes to the low-temperature phase. The high-temperature phase demonstrates saturation magnetization 30% lower than the low-temperature phase, so the magneto-optical read signal is initially reduced. Last, MnBi decomposes at above 446°C, only 20% above $T_C$. The MnBi does have a high Kerr rotation, near 0.4°. It can also be used in transmission as a Faraday rotation material, although its high absorptivity requires thin films whose growth must be carefully monitored since pit inclusion is common. (See Ref. 52.) Research on this material continues sporadically. (See Ref. 57.)

The second magneto-optical data-recording technology employs ferrimagnetic garnets, invoking Faraday rotation in transmission. The large possible Faraday rotation in these materials has been discussed in preceding sections. Growing the films by an expeditious technique is problematic; LPE is a slow process for mass production and requires expensive crystalline substrates; sputtering or vapor deposition techniques using glass substrates are preferable provided attractive film properties can be retained. The transmissive geometry is not as complaint as the reflective one to system layout, but it can be accommodated.

Domain writing in ferrimagnetic garnets takes place in the same manner in the vicinity of the compensation temperature as in RE–TM alloys. Iron garnets incorporating bismuth grow with perpendicular uniaxial anisotropy, and the substitution of gallium for some iron allows tailoring of the compensation temperature throughout the range 50–500°C via substitution on the predominating coordination site, as described in Section 18.7.1 The anisotropy is primarily growth induced and is due to the nonuniform distribution of bismuth ions on the dodecahedrally coordinated sites; a smaller contribution to the anisotropy is stress induced, caused by lattice mismatch and differential thermal contraction with respect to the substrate.

Write and read procedures are identical to those employed for reflective media. For analytical discussion of the various transmissive techniques, see Ref. 58. These transmissive media can sometimes be designed to operate in a double-pass configuration incorporating reflection from a back-surface coating and generally with optical access through the substrate. This approach increases the total polarization rotation (Kerr rotation often accompanies the reflection) for a given film thickness and recasts the system layout into a reflective format. For both the transmissive and reflective materials, front-surface quarter-wave film layers have been envisioned to increase the polarization rotation incurred in the magneto-optical interaction. Balasubramanian and Macleod have done computer analysis to demonstrate that Kerr rotation can be enhanced to about 2° using surface films; see Ref. 59.

Representative discussions of garnet films prepared by sputtering techniques are contained in Refs. 60–62. Vapor deposition is a newer approach; see Ref. 63.

   The theoretical aspects of domain formation and stability for magneto-optical data storage are formulated expertly in several papers by Mansipur; see Refs. 64–66. These papers invoke a complicated computer model of domain dynamics over an extended area, incorporating demagnetization, anisotropy, exchange, film thickness, coercivity, and temperature dependencies of all parameters, complete with a graphic visualization.

   Rare earth iron garnets are being investigated for other applications as well. One of these is as Faraday rotators in optical isolators. Both bulk and thin-film waveguide embodiments exist. Bulk isolators are generally constructed of rods of YIG, although paramagnetic Faraday glass is sometimes used at wavelengths shorter than 1.0 μm because of the high absorption of YIG there, saturated along the optical propagation direction in a static magnetic field imposed by a permanent magnet (see Fig. 18.24). The YIG rod in an isolator must be of appropriate thickness to rotate the polarization of incident light by 45°. Since the Faraday rotation of YIG varies with wavelength, these isolators cannot be broadband. Some commercially available isolators can be tuned over a wavelength range of 250 nm by moving the YIG specimen longitudinally into a higher or lower magnetic field. In these devices, the field is not strong enough to saturate the magnetization, and the polarization rotation induced by the crystal is governed by the expression $\theta_F = \int_0^L (VH)\, dx$, in which $\theta_F$ is the total cumulative Faraday rotation, $V$ is the Verdet constant at the wavelength of interest, $H = H(x)$ is the local applied magnetic field strength, and $dx$ and $L$ are the differential and total lengths of the crystal, respectively. At constant temperature, isolation can exceed 20 dB over approximately a 60-nm wavelength excursion, peaking at a little more than 30 dB. The value of the Faraday rotation is temperature sensitive; at the design (center) wavelength, however, isolation can exceed 25 dB over a temperature range of 30°C (from 10 to 40°C).

   A variation in operation of the bulk isolator characteristic produces a magneto-optical modulator (Fig. 18.25). The Faraday rotation of the transmitted optical beam polarization is modulated, via the Verdet dependence on the applied longitudinal magnetic field, by a signal current driving an electromagnet. One version of this modulator operates at rates of up to 10 kHz; it uses paramagnetic or diamagnetic glass, the former for higher rotation (by a factor of almost 3), the latter when stability against temperature drift is required.

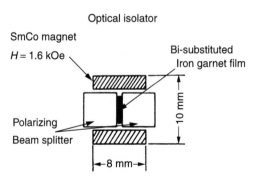

**Figure 18.24** Structure of an optical isolator. (From N. Koshizuka et al., "Application of Bi-Substituted Iron Garnet Films to 0.8 μm Range Optical Isolator," *IEEE Translat. J. Magn. Jpn.*, **TJMJ-I**, No. 1, 1985, p. 103. Published by the Institute of Electrical and Electronics Engineers. Copyright 1985, used by permission.)

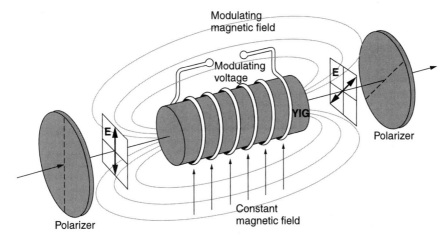

**Figure 18.25**  Faraday effect modulator. (From E. Hecht and A. Zajac, "Polarization," *Optics*, Addison-Wesley, Reading, MA, 1974, p. 263. Copyright 1974, used by permission.)

Optical polarization isolators are less frequently found in a waveguide geometry, particularly with the advent of bismuth-substituted iron garnets (e.g., Bi:YIG and Bi:GdIG). These materials have such large Faraday rotation (near or exceeding 2000 deg/cm at 1300 nm wavelength) and such good figure of merit, particularly at optical wavelengths greater than 1.0 $\mu$m where absorption is low ($\theta_F/\alpha$ greater than 250 deg/dB at 1300 nm), that films only a few hundred micrometers thick can achieve 45° rotation when used at perpendicular incidence. Doping with divalent ions such as calcium and magnesium reduces, optical absorption at 0.8 $\mu$m wavelength by preventing formation of the highly absorptive divalent iron ion; see Ref. 67. Saturation magnetization can be reduced to a few hundred oersteds. Further, techniques have been developed that circumvent the strong temperature dependence of Faraday rotation in bismuth garnets. Among these are the sandwiching of pairs of materials whose change of Faraday rotation with temperature have opposite sign (Ref. 68 suggests several alternatives) and the intermixing of these two kinds of crystal in a kind of solid-solution composite.

Other limitations of optical, waveguide geometry isolators are that they carry only a single spatial mode, restricting polarization isolation to temporally modulated optical beams, and they are somewhat tedious to couple to laser diode sources since use of butt coupling, end firing, or prism or grating surface couplers is required to insert the optical beam into the isolator.

Nevertheless, waveguide isolators continue to be investigated in light of the prospects of integrated optical devices and their anticipated manufacturability. These waveguide isolators must avoid films in which the propagation constants for the equal-mode-number TE and TM optical polarizations are excessively disparate. (Even in the absence of material birefringence, optical waveguide solutions indicate differing velocities along the propagation direction for the TE and TM polarizations. Velocity — or equivalently, wave vector — mismatch impedes the efficiency of polarization conversion because of imperfect phase matching in the coupled-mode interaction.) Intramodal TE–TM phase matching can be approached by making the waveguide weakly guiding, that is, fabricating the waveguide with a small refractive index difference between guide and cladding. Otherwise the wave vector mismatch can often be compensated by material birefringence, which occurs routinely as a growth or structural feature of most unannealed bismuth-substituted garnets.

Several novel concepts that compensate geometrical and material birefringence have been explored for achieving polarization rotation in waveguides (see Refs. 68–70). One conception for a waveguide isolator that combines these effects envisages placing equivalent-rotation sections of Faraday, nonreciprocal rotation material back to back with birefringent, reciprocal-rotation material; the rotation contributions add when the light passes in the direction of the magnetic field applied to saturate the Faraday rotator segment and cancel for reverse propagation. (See Ref. 68.) A second device compensates for the combined birefringence of geometry and materials simultaneously by reversing the in-plane magnetization periodically so that it is alternately parallel, then antiparallel, to the optical propagation direction. The periodicity of the magnetization reversal is established permanently in the magneto-optical waveguide material by local heating (laser light absorption is used) in the presence of a saturating external magnetic field. The field is reversed in direction when magnetizing neighboring regions to produce alternately magnetized strips, in much the same way as domains are written in data-recording applications. The magnetization periodicity equals that of the birefringence phase delay between the TE and TM light; the magnetization direction is switched whenever the phase difference between light in the two polarizations reaches a half wave. Reference 70 describes this device.

Another signal-switching component, a circulator, has been constructed as an optical waveguide device (see Fig. 18.26). This device receives signal data on one (or several) input port(s) and selectively transmits it on a single (or multiple) designated output port(s). Selection of an output port is achieved when signal data arrives there along two different device paths in phase. Combinations of reciprocal and nonreciprocal phase-shifting elements, of which Faraday rotators represent the former, along with mode-selecting waveguide channel junctions can be devised to perform this function. In addition, output port selection can be made flexible by overlaying the Faraday rotating sections of the waveguide structure with metalization to carry current and induce modulating magnetic fields. See Refs. 71 and 72 for details.

Wafers of ferrimagnetic garnet have been incorporated into spatial light modulators capable of binary modulation on a $128 \times 128$ array of pixels. Two different versions have been offered commercially. Both use a bismuth-substituted iron garnet film on which a regular two-dimensional array of mesas has been etched. The iron garnet film exhibits perpendicular magnetic anisotropy, and each mesa can support a single, large, magnetically saturated domain oriented either parallel or antiparallel to the direction of propagation of an incident optical beam whose wavefront illuminates the entire array. According to the magnetization of a mesa pixel, the polarization of transmitted light is rotated clockwise or counterclockwise a few tens of degrees by the Faraday effect. (The exact rotation depends on the optical wavelength since the Faraday rotation is spectrally dependent.) A design compromise is required since thicker films producing higher rotation also inflict higher absorption. The array is sandwiched between polarizers, with the analyzer generally oriented perpendicular to the output polarization produced by one of the pixel magnetization states (see Fig. 18.27); this orientation provides the highest contrast ratio.

The two devices differ in the mechanism employed to switch the pixels; to support each particular switching approach, the characteristics of their garnet films also differ

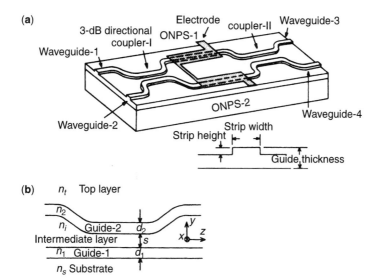

**Figure 18.26**   Schematic drawings of waveguide-type optical circulators. (**a**) Optical waveguide consists of a rib guide, and dc electric current flowing along an electrode yields a magnetic field. (**b**) Light propagates in the $z$-axis direction, and an external magnetic field is applied along the $x$-axis. (From T. Mizumoto et al., "Measurement of Optical Nonreciprocal Phase Shift in a Bi-Substituted $Gd_3Fe_5O_{12}$ Film and Application to Waveguide-Type Optical Circulator," *J. Lightwave Tech.*, **LT-4**, No. 3, 1986, p. 350. Published by IEEE. Copyright 1986, used by permission.)

somewhat. One device switches the mesa domains by the same mechanism discussed for magneto-optical recording in materials for which the compensation temperature is near ambient. The temperature of an addressed pixel is elevated so that the combined magnetic field of the saturation magnetization, the anisotropy, and the external field exceeds the coercivity. Heating results from current conduction through a transparent resistive element that overlies the pixel of interest. The pixel is addressed by a pair of electrodes that, lying in the grooves between rows and columns of mesas, are connected by the resistive element. An externally applied magnetic field is required to reverse the mesa domain orientation for this device, and its direction must be switchable to accommodate the desired orientation of an addressed domain. Operation of this device is described in Ref. 73.

The other device switches the domains using small nucleating magnetic fields produced by circulating currents near the pixel of interest. As in the preceding device, electrodes lie in the channels between rows and columns of mesas. In the activation state the current that passes along one of the wires is insufficient to nucleate a domain alone, but when current passes in the appropriate direction along each electrode of the pair that crosses at a pixel, an inverted domain is nucleated. This nucleation is aided by ion implantation, in the corner of the pixel nearest the crossing electrodes, of impurities that lowers the magnetic anisotropy. Once nucleated, the new domain is driven across the mesa by the nucleating field. Further description of this device may be found in Ref. 74.

The stability of domains in these spatial light modulators is discussed in Ref. 75. As shown therein, the perpendicular anisotropy reduced by the saturation magnetization

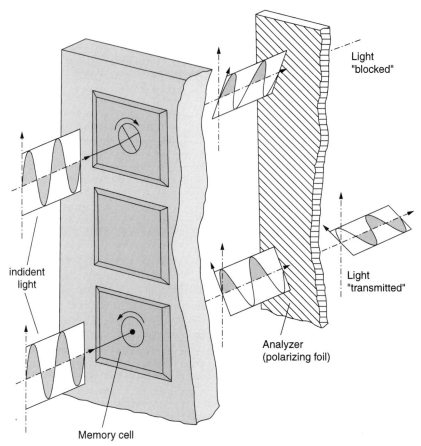

**Figure 18.27**   Viewing of magnetic domains using the Faraday effect and polarization optics. (From B. Hill and K. P. Schmidt, "Fast Switchable Magneto-Optic Memory — Display Components," *Philips J. Res.*, **33**, Nos. 5/6, 1978, p. 214. Published by N. V. Philips. Copyright 1978, used by permission.)

must exceed the field required to saturate the film in order that large domains will be stable. Equivalently, the perpendicular anisotropy must exceed twice the saturation magnetization.

A large-aperture, diffractive, magneto-optical deflector is being developed by a team of scientists at Unisys. The device implementation has seen several revisions, but the physical operation and its foreseen variations are recurrent. A thin film of Bi–YIG is grown with perpendicular anisotropy, so that the dipoles organize naturally into an array of stripe domains, alternatingly oriented up and down with respect to the wafer surface. These stripe domains can be aligned parallel to an externally applied, in-plane, magnetic field. The strength of this field determines the width of the stripes: the stronger the field, the narrower the stripes. Figure 18.28 provides a drawing of the deflector.

The basic device operation relies on the counterrotation of the polarization of light passing through adjacent stripe domains. Light transmitted by a stripe domain acquires a polarization component that lies in the plane orthogonal to that of the incident light

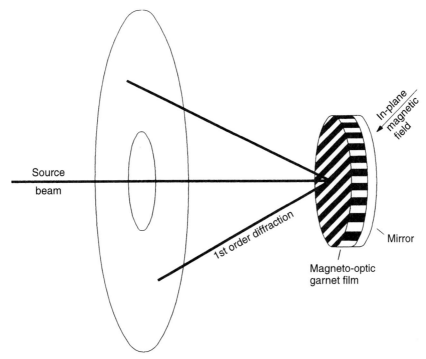

**Figure 18.28** Diffractive magneto-optical deflector. An applied in-plane magnetic field controls the stripe domain grating orientation and periodicity to provide a two-dimensional field of view. (From G. L. Nelson et al., "Stripe Domain Light Deflection for Intersatellite Communications," *Communications Networking in Dense Electromagnetic Environments, Proc. SPIE,* **876**, 1988, p. 122. Published by the Society of Photo-Optical Instrumentation Engineers. Copyright 1988, used by permission.)

polarization. Light transmitted by a neighboring stripe whose polarization is rotated in the opposite sense also acquires a polarization component in that plane but pointing in the opposite direction. The relative phase between light transmitted by neighboring stripes is thus 180° for the polarization component orthogonal to the incident polarization. Ideally the rotation is 90° in each stripe for optimal diffraction efficiency, although considerably smaller rotations can be effective (with lower efficiency). Diffraction as from a phase grating occurs and multiple diffraction orders are produced. Enhanced performance can be obtained by employing nonnormal incidence Bragg diffraction to eliminate all but a single diffracted order. Techniques employing a resonant cavity for either the input light alone or for both the input and diffracted light have been explored that produce increased efficiency; 11% at 800 nm and 14% at 1060 nm have been realized. Rotating the alignment, and consequently the diffraction direction, of the stripe domains and varying the width of the stripes through some small range by adjusting the applied magnetic field strength and orientation provides control for steering of the diffracted beam between multiple ports. For further details see Ref. 76. In this device, as in others described earlier, increasing the ratio of Faraday rotation to absorption is important for high efficiency, and material properties delimit its success. An interconnect switch for an optical communications network is envisioned.

A ferrimagnetic garnet film can also support optical diffraction from a moving magnetic disturbance, such as an MSW. An MSW propagating in a thin film of ferrimagnetic garnet induces a periodic perturbation in the off-diagonal elements of the optical permittivity tensor. The perturbation is thus a traveling-wave, microwave frequency modulation of the Faraday rotation. This type of traveling, periodic, polarization-rotating grating can be produced by each of the three types of magnetostatic waves and can cause Bragg regime diffraction of an intersecting optical wave provided that the process is phase matched. Details of the interaction are somewhat complicated.

The magnetostatic wave modes of a thin film (a waveguide) have been discussed in Section 18.4. Three types of MSW propagate. Volume waves, MSWs whose propagation is isotropic in the plane of the film, are excited when the field that saturates the magnetic dipoles is perpendicular to the film surface. Surface waves are excited when the saturating field is perpendicular to the MSW propagation direction but in the plane of the film; they are localized on the top film surface when propagating in one direction and on the bottom surface when propagating in the opposite direction. Backward-volume waves are excited when the saturating magnetic field is parallel to the propagation direction; the group velocity for these MSWs is antiparallel to the phase velocity. The microwave component of the magnetization for each of these types of waves can be visualized lying perpendicular to the dipole-saturating magnetic field and circulating as the dipole precesses around that direction, approximately circumscribing a circle. Recall that light polarization can be rotated by the Faraday effect only when a component of the magnetic field is parallel to the optical propagation direction. Thus, to first order, forward-volume waves can cause optical polarization rotation at microwave frequencies for light propagating either collinear or perpendicular with respect to the MSW propagation direction, whereas surface and backward waves induce microwave frequency Faraday rotation in light only when its propagation is collinear with the MSW. (Several mechanisms can cause higher-order effects. Magnetostatic wave precession can be saturated or elliptical because of anisotropy fields; high microwave power can introduce the Cotton–Mouton effect or allow multiple optical interactions with various permittivity perturbations to occur.)

Magnetostatic waves propagate as guided waves in thin ferrimagnetic garnet films. Thus for an optical wave to intersect them and diffract from them, the light must also propagate as a mode of the waveguide formed by the garnet film. (An illustration of the layout of magnetic fields, MSW excitation and propagation, and optical insertion, propagation, and diffraction in a YIG film waveguide is provided in Fig. 18.29.) Note that the waveguide criteria for MSWs is not identical to that for waveguiding light. The former requires a low-loss magnetic film medium bounded by nonmagnetic media or, alternatively, media having differing saturation magnetization values from the guiding layer, so that with the predetermined MSW wavevector and external bias field, the excitation lies outside the MSW frequency band in the bounding media. The latter requires that the guiding film have an optical refractive index value greater than that of either bounding medium so that light can be completely trapped inside the film by total internal reflection; i.e., it can be incident on the film surfaces at angles exceeding the critical angle. Fortuitously, a YIG or Bi:YIG film grown on GGG (or one of its variants) satisfies both of these criteria.

In isotropic thin films (no birefringence), light can be guided in one of two polarizations: an electric field parallel to the surface, called transverse electric (TE), or a

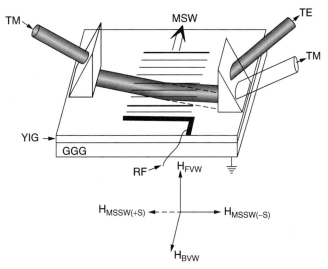

**Figure 18.29**  Bragg diffraction of a guided optical wave by an MSW induces conversion between orthogonally polarized optical modes. Lower section shows the magnetic field bias direction required to produce each of the four types of MSWs. (From A. D. Fisher, "Optical Signal Processing with Magnetostatic Waves," *Circuits, Syst. Signal Process.*, **4**, No. 1–2, 1985, p. 267. Published by Birkhauser Boston. Copyright 1985, used by permission.)

magnetic field parallel to the surface, called transverse magnetic (TM). Guiding is allowed when two conditions are met.

First, light within the film must be incident at each film boundary (or film–cladding interface) at an angle greater than the critical angle, that is, the maximum angle at which light propagating in a higher refractive index medium can be partially transmitted across an interface with a medium of lower refractive index. This causes the light to be totally internally reflected. The value of the critical angle can be derived from Snell's law of refraction:

$$n_1 \sin \theta_1 = n_2 \sin \theta_2$$

where $n_1$ and $n_2$ are the refractive indices of the transmitted light medium and the incident light medium, respectively, with $n_2 > n_1$, and $\theta_1$ and $\theta_2$ are the respective angles of propagation of the light at the interface in the two media, measured with respect to the surface normal. Setting $\theta_1 = 90°$ gives $\theta_2$, the value of the critical angle.

Second, the light must be resonant in the film structure, that is, in propagating from a point in the center of the waveguiding film through a reflection at each of the two interfaces and returning to the center of the film, the total accumulation of phase must be an integral number of wavelengths of the light in the film. Accurately, the phase of interest is the sum of that accrued by the propagation component directed perpendicular to the waveguide surfaces and the phase changes incurred on reflection at each interface. (Constructive interference of overlying optical wavefronts is required.) To better understand this concept, it is instructive to write the optical wave vector as the sum of two orthogonal components: $\mathbf{k} = \boldsymbol{\kappa} + \boldsymbol{\beta}$. In this decomposition $\boldsymbol{\kappa}$ is the component of $\mathbf{k}$ that is perpendicular to the waveguide surface and $\boldsymbol{\beta}$ is the component that is parallel to it: $|\boldsymbol{\kappa}| = |\mathbf{k}| \cos \theta$ and $|\boldsymbol{\beta}| = |\mathbf{k}| \sin \theta$. The characteristic equation that governs the resonant propagation of light in a waveguide film as it reflects alternately from the top and bottom surfaces is

$$2nd \cos\theta + 2\phi_{21} + 2\phi_{23} = m\lambda$$

where $n$ is the refractive index of the film and $d$ is its thickness; $\theta$ is the angle of incidence of the light on the film–boundary interface ($\theta = 0$ is perpendicular to the surface); $\lambda$ is the vacuum optical wavelength; and $m$ is the number of the waveguide mode occupied by the light or, equivalently, the number of nodes of the standing wave formed by the light as it bounces back and forth across the film, propagating down the waveguide. The $\phi$ parameters are somewhat more complicated. When light reflects from an interface at which its incidence angle exceeds the critical angle, it garners a phase increment, as if it had traveled an additional distance. This phase increment depends on the optical polarization, the angle of incidence, and the refractive indices of the interface media; the quantities $\phi_{ij}$ in the preceding expression represent these increments for the two interfaces of the waveguide, the subscript $i$ indicating the waveguide side of the interface and $j$ the cladding side. For a TE mode,

$$\phi_{ij} = \tan^{-1} \frac{\sqrt{n_i^2 \sin^2\theta_i - n_j^2}}{n_i \cos\theta_i}$$

For a TM mode,

$$\phi_{ij} = \tan^{-1} \frac{(n_i^2/n_j^2)\sqrt{n_i^2 \sin^2\theta_i - n_j^2}}{n_i \cos\theta_i}$$

Because these expressions are not identical, the characteristic equation produces differing propagation angles $\theta$ for TE and TM modes having the same mode number $m$. Consequently, their longitudinal propagation velocities (group velocities) in the waveguide differ. Equivalently, the components of their wave vectors parallel to the waveguide surface, conventionally called $\beta$, differ; $\beta_{TM} < \beta_{TE}$ for the same mode number. For waveguide diffraction calculations, these $\beta$s are equivalent to propagation wavevectors for diffraction calculations in bulk devices.

The number of modes supported by a waveguide structure is determined by the values and degree of symmetry of the refractive indices of the guiding film and the cladding media and by the thickness of the guiding film. Waveguide structures that are symmetric in refractive index always support at least a single TE mode, although not always a TM mode. Lower-order modes, those with fewer nodes in the standing-wave pattern across the guiding film, propagate at higher angles, that is, more nearly down the axis of the waveguide. Light in higher-order modes is incident on the boundaries at angles nearer the critical angle. As the mode number is increased, the characteristic equation eventually fails to have a bound solution; this is called the cutoff condition for the waveguide. For YIG films, which form asymmetric refractive index structures on their GGG substrates in air, the lowest-order ($m = 0$) TM mode appears when the thickness exceeds about 0.75 μm (the refractive index values are 2.212 for YIG and 1.94 for GGG at an optical wavelength of 1150 nm). For a review of waveguide optics, see Ref. 77.

Bragg diffraction of waveguide-confined light from an MSW requires that several criteria be met. Because of the magneto-optical nature of the interaction, polarization rotation is induced. However, light propagating in a waveguide is constrained to occupy either a TE or a TM mode; its polarization cannot be arbitrary, as would be the case for bulk implementation. Diffraction therefore requires the waveguide light to transition from a TE to a TM mode polarization, or vice versa. Further, the diffraction mechanism requires that phase-matching criteria be met, or at least approximated, as in any optical interaction, particularly those high-efficiency interactions that depend on long interaction distances. Specifically, the MSW wave vector should be approximately of magnitude and direction such that the vector sum of the TM light wave vector and the MSW wave vector equal the TE wave vector:

$$\boldsymbol{\beta}_{\text{TM}} + \mathbf{K}_{\text{MSW}} = \boldsymbol{\beta}_{\text{TE}}$$

Physically, solutions to this equation are governed by the observation that for the same mode number $|\boldsymbol{\beta}_{\text{TM}}| < |\boldsymbol{\beta}_{\text{TE}}|$. That is, in diffracting, the wave vector of the light changes magnitude. Thus the usual requirement of bulk diffraction (in the Bragg, i.e., phase-matched, regime) that the angle of incidence on the diffraction grating equal the angle of diffraction does not pertain.

From Ref. 78, the intensity diffraction efficiency for low MSW power is given by

$$\frac{I_d}{I_0} = \frac{4\kappa^2}{4\kappa^2 + (\Delta\beta)^2} \sin^2\left[\frac{1}{2}y\sqrt{4\kappa^2 + (\Delta\beta)^2}\right]$$

where $I_0$ and $I_d$ are the initial and diffracted intensities, respectively; $\Delta\beta$ is the phase mismatch; $\kappa$ is the coupling constant; and $y$ is the interaction length. In detail,

$$\Delta\beta = |\boldsymbol{\beta}_{\text{TE}} - (\boldsymbol{\beta}_{\text{TM}} + \mathbf{K}_{\text{MSW}})|$$

is the mismatch magnitude of the sum of wave vectors, which equals zero when the diffraction is perfectly phase matched. Additionally

$$\kappa = \frac{\varepsilon_{31}k_0\sqrt{\varepsilon}}{2\varepsilon_0\varepsilon},$$

where $k_0$ is the free-space optical wave vector; $\varepsilon$ is the permittivity so that $\sqrt{\varepsilon} = n$, the refractive index; $\varepsilon_0$ is the permittivity of free space; and $\varepsilon_{31}$ is the value of the permittivity tensor at the (3, 1) position, which couples the TM and TE modes. These expressions may be recognized as belonging to solutions to a coupled-mode formalism. They show that for perfectly phase-matched interaction, the intensity diffraction efficiency is proportional to $\sin^2(\kappa y)$. For short interaction lengths (or less desirably, for small $\kappa$), $\sin^2(\kappa y)$ can be approximated by $(\kappa y)^2$. So $I_d/I_0$ is proportional to $\kappa^2$, which is in turn proportional to $\varepsilon_{31}$. Rewriting $\varepsilon_{31} = \varepsilon_0 fm$ and substituting for the linear coefficient $f$,

$$f = \frac{2\sqrt{\varepsilon}\phi_F}{k_0 M}$$

where $m$ is the applicable component of the microwave frequency magnetization associated with the MSW, $\phi_F$ is the Faraday rotation coefficient, and $M$ is the strength of the bias magnetic field, a new expression for $\kappa$ can be realized:

$$\kappa = \phi_F \frac{m}{M}$$

The intensity diffraction efficiency is proportional to the square of the Faraday rotation coefficient and to the square of the local microwave frequency magnetization. This realization drives the exploration for materials to support this technology.

First, the material must have a high value of $\phi_F$. Second it must have low MSW attenuation (measured as low ferrimagnetic resonance line width) so that $m$, the RF magnetization, is not negligible in the MSW–optical interaction region. Ancillary concerns are low optical attenuation at the wavelength of interest and repeatable values of saturation magnetization, magnetic anisotropy, and optical birefringence from a particular growth sequence. A technique for measuring MSW attenuation was described in the preceding section treating spin waves. The measurement of $\phi_F$ is straightforward, incorporating an electromagnet to adjust the magnetic field strength, a laser source and detector, and a polarizer–analyzer set of polarizing prisms. Collimated light is transmitted through the polarizer, perpendicularly through the garnet sample, and then through the analyzer and onto the detector. The sample is placed between the poles of the electromagnet so that the optical propagation is parallel to the magnetic field. This apparatus is described in Ref. 79 and illustrated in Fig. 18.30. A technique for measuring optical waveguide attenuation is also described in Ref. 79 (see also Fig. 18.30) in which light scattered from the waveguide surface is detected as a function of propagation distance. Assuming isotropically distributed absorption and scattering centers in the garnet film, an estimate of total waveguide optical attenuation can be derived. Garnet films tend to scatter very little waveguide light since they are nearly perfect crystals, and scattered optical power is often unsatisfactory for accurate measurements. Greater accuracy can be obtained, with care, using the three-prism technique described in Ref. 80 (see Fig. 18.30 again).

The magnitude of the MSW wave vector, $|\mathbf{K}_{MSW}|$, is generally not large. Although some range of selectable values is available at any particular frequency via adjustment of the biasing static magnetic field to shift the MSW dispersion curve, it is common to obtain wave vectors in the vicinity of 500 cm$^{-1}$. (Magnetostatic wave velocity is about $5 \times 10^6$ cm/s; this wave vector value is approximate around 2 GHz frequency.) Since this value is very small compared to the wave vectors for light in either a TE or TM waveguide mode (about $6.7 \times 10^4$ cm$^{-1}$) and is even somewhat smaller than the usual difference between wave vector amplitudes for TE and TM modes having the same mode number, the angle of deflection of the diffracted light is only a few degrees, almost always less than 10°. Also, diffraction in the low MSW power, linear regime is almost without exception between TE and TM modes of the same value of $m$ and the least difference in $\beta$. Low-order modes (between the zeroth and fifth) of waveguides that support 10 or more optical modes are best used. (Such films are about 10 μm thick). Films of thickness on the order of 10 μm constitute a good compromise between the desired attributes of few optical modes, influencing design toward films nearer to 1 μm thick, and that of efficient MSW insertion. An MSW is launched from a microstrip antenna placed near or on the garnet surface; thicker garnet films present a greater capture volume for the microwave field radiating from the antenna.

The efficiency of diffraction in YIG films on GGG substrates has generally been observed to be small when the MSW power has been restrained below the nonlinear,

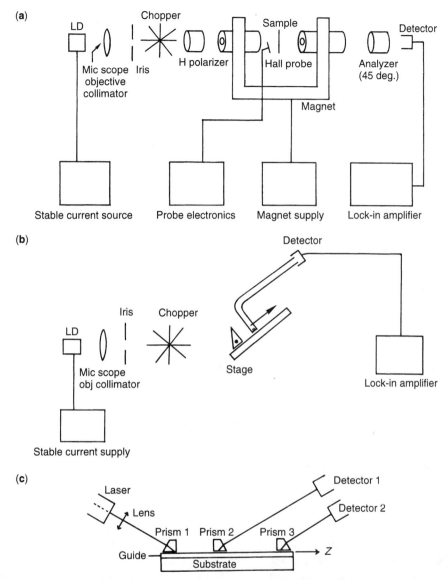

**Figure 18.30** (**a**) Faraday rotation measurement. (**b**) Optical waveguide attenuation measurement. (**c**) Schematic of the three-prism method. [(**c**) From Y. H. Won et al., "Three-Prism Loss Measurements of Optical Waveguides," *Appl. Phys. Lett.*, **37**, No. 3, 1980, p. 269. Published by the American Institute of Physics. Copyright 1980, used by permission.]

saturating region, where only MSW-induced (dynamic) Faraday effects can be invoked to describe the interaction. Diffraction efficiency for transverse interaction of light with low-power, forward-volume MSWs has been in the range of 0.01% (somewhat less than 10% per microwave watt at about 2 mW of inserted MSW power). Diffraction efficiency for collinear interaction of light and surface MSWs having power near the saturation level has reached a few percent (again at about 5% per microwave watt). See Refs. 78, 81, and 82.

The instantaneous bandwidth of the low-power interaction in a uniform bias magnetic field is about 30 MHz. Phase matching is lost over wider bandwidths when the interaction region is long enough to maintain reasonable efficiency; in YIG, the minimum length is about 7 mm. A geometric view of phase matching assists in understanding this limitation. In optically isotropic media, the wave vector loci for TM and TE polarizations are both circles in the plane of the waveguide of radii $|\boldsymbol{\beta}_{TM}|$ and $|\boldsymbol{\beta}_{TE}|$, respectively. Ideally, the tail of the MSW wave vector lies on one circle (e.g., the outer one, the TE circle), touching the head of one optical wave vector, $\boldsymbol{\beta}_{TE}$, while the MSW wave vector head lies on the other circle (the inner one, TM), touching the head of the other optical wave vector, $\boldsymbol{\beta}_{TM}$. The bandwidth of the diffraction interaction reflects the allowable mismatch in magnitude and/or direction of $\mathbf{K}_{MSW}$. As can be seen from the preceding diffraction efficiency equation, for a small coupling constant the efficiency per unit length of interaction depends on $\Delta\beta$ as $\mathrm{sinc}^2(\Delta\beta y/2)$ using the definition sinc $x = (\sin x)/x$. (This expression for diffraction efficiency indicates the roll-off penalty for nonzero mismatch.) The MSW wave vector is generally so short in practice that it connects the two optical wave vectors in a nearly radial direction (hence the small observed diffraction angles). Substantial bandwidth gains could be realized by finding a way for the MSW wave vector to connect the two optical wave vectors while lying perpendicular to the shorter of them ($\boldsymbol{\beta}_{TM}$), that is, tangent to the inner circle; this condition is called tangential phase matching. With this alignment $|\mathbf{K}_{MSW}|$ can vary over a larger range without abandoning the phase-matched region. Tangential phase matching is illustrated in the center diagram of Fig. 18.31. (A means of accomplishing tangential phase matching has been investigated and will be discussed briefly in what follows.) A substantial investigation of the use of gradient bias magnetic fields to invoke different portions of the (nonlinear) MSW dispersion curve across the optical–MSW interaction region has shown that the interaction bandwidth may be increased to about 300 MHz using this technique (again, in YIG). (See Refs. 83 and 84.) Varying relative delays across the band, complicated beam-steering effects on the MSWs, and nonspecific diffraction directionality discourage use of this technique in practical devices, however.

The desirable attributes for garnet materials in which MSW–optical diffraction can occur efficiently were already outlined: high Faraday rotation, low MSW attenuation, and low optical absorption. Recently, scientists have been able to grow bismuth-substituted YIG having these characteristics along with reasonably low optical absorption. Using bismuth-substituted iron garnet films, a research group has recently shown dramatic results in high-efficiency MSW–optical diffraction (Ref. 28). The LPE-grown garnets were grown on GGG and consisted of (BiLu) IG with 0.89 formula units bismuth and 2.11 formula units lutetium. The flux from which the iron garnet films were grown was solely $Bi_2O_3$, relieving the concern of incorporating lead in the films. This proportion of bismuth produced Faraday rotation of $-140.0$ deg/nm at an optical wavelength of 1300 nm. Guided-wave optical attenuation at this wavelength was low in the film for which results are reported (only 0.6 dB/cm), as was MSW attenuation; the FMR line width was 0.7 Oe. Best diffraction results were obtained at a microwave frequency of 4.0 GHz, at which 47% diffraction was achieved with a microwave power of 0.8 W in an interaction region of length 7 mm. Backward-volume magnetostatic waves were employed since forward-volume and surface MSWs were found to couple strongly to the exchange-dominated spin wave manifold in the films used, whereas

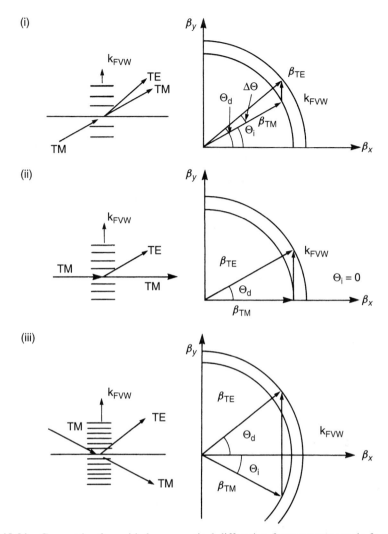

**Figure 18.31** Geometries for guided-wave optical diffraction from magnetostatic forward volume waves. Three different wave vector magnitudes and their corresponding wave vector surface configurations for perfect phase matching are shown. Center sketch shows tangentially phase-matched case.

the backward-wave dispersion relation does not coincide with that for the exchange-dominated spin waves and does not couple.

Development of the highly bismuth-substituted garnet films grown from a lead-free flux to provide low optical attenuation is the primary capability enabling demonstration of this landmark result. The initial report on the material characteristics of the films appears in Ref. 85. The absence of lead apparently promotes low FMR line width as well, possibly because lead may induce 2+ and 4+ ionization states of the iron ions in the lattice. Reference 86 instructs that a few percent MgO was incorporated in films grown later, presumably to maintain the appropriate 3+ ionization of the iron ions. Scientists working on reducing optical damage in lithium niobate, which was believed to be caused by the presence of inappropriately ionized iron, showed several years ago that incorporation of MgO was effective in controlling iron charge states. Perhaps the same type of charge compensation mechanism is responsible for success here.

Higher MSW power (greater than tens of milliwatts) causes multiple interactions to occur, which depend on static Faraday rotation and static and dynamic Cotton–Mouton effects. (In brief, the dynamic Faraday effect is that caused by the microwave frequency magnetic field. The static Faraday effect is caused by the static or biasing magnetic field. Likewise, the dynamic and static Cotton–Mouton effects depend on microwave and static fields.) The static effects occur even in the absence of MSWs. Even though the process is not well phase matched, some conversion of polarization between TM and TE modes of the same mode number occurs in the presence of a static magnetic field. It is not deflected and does not carry the frequency shift (a Doppler shift caused by diffraction from a moving grating) of MSW-diffracted light. The multiple interactions that occur at higher powers seem capable of significantly increasing the instantaneous bandwidth and also possibly of increasing the diffraction efficiency. In addition, interactions that are forbidden or negligible in the linear MSW power regime, such as transverse diffraction from surface MSWs, become possible. This work is described in Refs. 86–89.

At the instigation of this author, Stancil has recently investigated theoretically several concepts that may enhance viability of MSW–optical interactions in the linear power regime [90]. The first relies on two facts. First, an MSW has an evanescent tail that extends several wavelengths into the media bounding its waveguiding host film; this tail can modify the magneto-optical properties of the neighboring medium. Second, the optical phase change accumulated on reflection by an optical beam depends on the refractive index of the reflecting medium as well as that of the medium in which the light propagates. As is evident from the characteristic equation, a change in one of the $\phi_{ij}$ terms may be as effective in perturbing the optical wave as is a change in the propagating phase term caused by a perturbed refractive index. The $\phi_{ij}$ perturbation is particularly sensitive to changes in the interfacial refractive index near the critical angle of incidence (near cutoff). Modulating the refractive index of the reflecting medium can therefore cause diffraction in a similar way as does modulating the refractive index of the propagation medium. The combined impact of these observations is that the optical guided waves and MSWs need not propagate in the same garnet film to effect diffraction but can propagate in neighboring films whose optical and magnetic properties need not be the same. Material requirements for each of the neighboring films can be relaxed considerably. The second concept incorporates periodic variations in the properties of the film that supports the MSW oriented parallel to the MSW propagation direction. The periodic perturbation assists in satisfying the phase-matching criteria for efficient optical diffraction by enhancing the effective wave number magnitude of the MSW; the tangential phase-matching condition may be approached using this technique.

A few other efforts pertain to MSW–optical interactions. Reference 91 discusses the theory of TM to TE mode coupling caused by diffraction from MSW forward-volume waves. In particular, Huahui et al. predict from theory that coupling between differing mode numbers should dominate (over polarization-switching diffraction within the same mode number) at MSW frequencies that invoke large wave vector values, nearer to the high end of the frequency band for a given bias magnetic field value. (These wave vectors are more difficult to excite experimentally.) Further, they write that coupling is restricted to pairs of TM–TE modes that have the same symmetry (odd or even) in the waveguide. This constraint results from the nature of the expression for the coupling constant, which is an overlap integral between the electric field profiles across the waveguide of the polarization-rotated input mode and the diffracted mode. These field profiles are nearly orthogonal for modes of opposite symmetry; the overlap integral and the coupling constant are then negligible.

Probing of MSWs by Brillouin scattering of light incident on garnet films perpendicular to their surfaces has been used as a diagnostic tool for observing MSW propagation and dispersion. See Ref. 92.

Theoretical analysis of the correlation between static and dynamic magneto-optical effects on one hand and elastic and inelastic spin–photon scattering by electric and magnetic dipoles on the other has been presented by Le Gall and associates. For readers engaged in quantum interpretations, these papers provide particular insight to the physics of spin–particle interactions with photons. See Ref. 93 and the references therein.

Finally, Solomko and associates published a succession of papers of both theoretical and more recently experimental focus, detailing their work on MSW–optical interactions. Many of these papers are listed in Refs. 94 and 95.

## REFERENCES

1. B. D. Cullity, *Introduction to Magnetic Materials*, Addison-Wesley, Reading, MA, 1972.
2. C. Kittel, in *Introduction to Solid State Physics*, 4th ed., Wiley, New York, 1971, Chaps. 14–17.
3. B. Lax and K. J. Button, *Microwave Ferrites and Ferrimagnetics*, McGraw-Hill, New York, 1962.
4. M. S. Sodha and N. C. Srivastava, *Microwave Propagation in Ferrimagnetics*, Plenum, New York, 1981.
5. P. Hansen et al., "Optical Switching with Bismuth-Substituted Iron Garnets," *Philips Tech. Rev.*, **41**, p. 33, 1983/84.
6. J. F. Dillon, Jr., "Magneto-Optical Properties of Magnetic Garnets," in A. Paoletti, Ed., *Physics of Magnetic Garnets*, North-Holland, New York, 1978, p. 379.
7. W. Wettling, "Magneto-Optics of Ferrites," *J. Magnet. Magn. Mat.*, **3**, p. 147, 1976.
8. A. D. Fisher, "Optical Signal Processing with Magnetostatic Waves," *Circ., Syst., Sig. Process.*, **4**(1–2), p. 265, 1985.
9. M. J. Freiser, "A Survey of Magnetooptic Effects," *IEEE Trans. Magn.*, **MAG-4**(2), p. 152, 1968.
10. J. F. Dillon, "Magneto-Optical Properties of Magnetic Crystals," in J. Smit, Ed., *Magnetic Properties of Materials*, McGraw-Hill, New York, 1971, Chap. 5.
11. F. A. Jenkins and H. E. White, *Fundamentals of Optics*, 4th ed., McGraw-Hill, New York, 1976.
12. M. A. Butler and E. L. Venturini, "High Frequency Faraday Rotation in FR-5 Glass," *Appl. Opt.*, **26**, p. 1581, 1987.
13. N. Koshizuka and T. Okuda, "The Problems in the Development of Magneto-Optical Devices," *IEEE Trans. J. Magn. Jpn.*, **TJMJ-1**, p. 1044, 1985.
14. J.-P. Krumme, et al., "Bismuth-Substituted Iron Garnet Films Prepared by RF Diode Sputtering," *IEEE Trans. Magn.*, **MAG-20**, p. 983, 1984.

15. J.-P. Krumme et al., "Bismuth Iron Garnet Films Prepared by RF Magnetron Sputtering," *J. Appl. Phys.*, **57**, p. 3885, 1985.

16. T. Okuda et al., "Faraday Rotation in Highly Substituted Yttrium Iron Garnet Films Prepared by Ion Beam Sputtering," *IEEE Trans. Magn.*, **MAG-23**, p. 3491, 1987.

17. Y. Toudic and M. Passaret, "Croissance por Voie Hydrothermale de Films Ferrimagnetiques; Epitaxies sur des Substrats de GdGaG," *J. Cryst. Growth*, **24/25**, p. 621, 1974.

18. G. R. Pulliam, "Chemical Vapor Growth of Single Crystal Magnetic Oxide Films," *J. Appl. Phys.*, **38**, p. 1120, 1967.

19. K. Maeto et al., "Bi-substituted DyIG Thin Films Prepared by Chemical Deposition for Magneto-Optic Recording Medium," *IEEE Trans. J. Magn. Jpn.*, **TJMJ-2**, p. 839, 1987.

20. P. Hansen and J.-P. Krumme, "Magnetic and Magneto-Optical Properties of Garnet Films," *Thin Solid Films*, **114**, p. 69, 1984.

21. S. Geller, "Crystal and Static Properties of Garnets," in A. Paoletti, Ed., *Physics of Magnetic Garnets*, North-Holland, New York, 1978, p. 1.

22. C. Kittel, *Introduction to Solid State Physics*, 4th ed., Wiley, New York, 1971, p. 507.

23. P. Hansen et al., "Magnetic and Magneto-Optic Properties of Praseodymium- and Bismuth-Substituted Yttrium Garnet Films," *J. Appl. Phys.*, **60**, p. 721, 1986.

24. V. J. Fratello et al., "Effect of Diamagnetic Substitution on Growth-Induced Anisotropy in $(YBi)_3Fe_5O_{12}$," *J. Appl. Phys.*, **60**, p. 718, 1986.

25. V. Doormann et al., "Measurement of the Refractive Index and Optical Spectra of Epitaxial Bismuth Substituted Yttrium Iron Garnet Films at UV to Near-IR Wavelengths," *Appl. Phys. A*, **34**, p. 223, 1984.

26. G. B. Scott and D. E. Lacklison, "Magnetooptic Properties and Applications of Bismuth Substituted Iron Garnets," *IEEE Trans. Magn.*, **MAG-12**, p. 292, 1976.

27. S. Wittekoek et al., "Magneto-Optic Spectra and the Dielectric Tensor Elements of Bismuth-Substituted Iron Garnets at Photon Energies between 2.2–5.2 eV," *Phys. Rev. B*, **12**, p. 2777, 1975.

28. H. Tamada et al., "TM–TE Optical-Mode Conversion Induced by a Transversely Propagating Magnetostatic Wave in a $(BiLu)_3Fe_5O_{12}$ Film," *J. Appl. Phys.*, **64**, p. 554, 1988.

29. H. Ohno et al., "Faraday Rotation of LPE Bi-Substituted Magnetic Garnet Films," *IEEE Trans. J. Maan. Jpn.*, **TJMJ-1**, p. 93, 1985.

30. T. Hibiya et al., "Growth and Characterization of Liquid-Phase Epitaxial Bi-Substituted Iron Garnet Films for Magneto-Optic Application," *Japan. J. Appl. Phys.*, **24**, p. 1316, 1985.

31. A. Murata et al., "Optical Birefringence in BiNdLuAlIG Thin Film Waveguides," *IEEE Trans. J. Magn. Jpn.*, **TJMJ-1**, p. 91, 1985.

32. M. Imamura et al., "Bi-Content Dependence of Birefringence in LPE–YIG and GdIG Films," *IEEE Trans. J. Magn. Jpn.*, **TJMJ-1**, p. 95, 1985.

33. H. Tamada et al., "Bi-Substituted LPE Garnet Films with FMR Linewidth as Small as YIG," *IEEE Trans. J. Magn. Jpn.*, **3**, p. 98, 1988.

34. G. Bate, "Materials Challenges in Metallic, Reversible, Optical Recording Media: A Review," *IEEE Trans. Magn.*, **MAG-23**, p. 156, 1987.

35. P. Hansen and M. Hartmann, "Magneto-Optical Properties of Iron Garnets and Amorphous Alloys and Their Use in Device Applications," in J. Rauluszkiewicz et al., Eds., *Proceedings of the Second International Conference on Physics of Magnetic Materials*, World Scientific Publishing, Singapore and Philadelphia, 1985, p. 158.

36. C. D. Wright et al., "Stability Phenomena in Amorphous Rare Earth–Transition Metal Films," *IEEE Trans. Magn.*, **MAG-23**, p. 162, 1987.

37. H. -P. D. Shieh and M. H. Kryder, "Magneto-Optic Recording Materials with Direct Overwrite Capability," *Appl. Phys. Lett.*, **49**, p. 473, 1986.

38. M. H. Kryder et al., "Control of Parameters in Rare Earth–Transition Metal Alloys for Magneto-Optical Recording Media," *IEEE Trans. Magn.*, **MAG-23**, p. 165, 1987.

39. H.-P. D. Shieh and M. H. Kryder, "Operating Margins for Magneto-Optic Recording Materials with Direct Overwrite Capability," *IEEE Trans. Magn.*, **MAG-23**, p. 171, 1987.

40. F. Tanaka et al., "Magneto-Optical Recording Characteristics of TbFeCo Media by Magnetic Field Modulation Method," *Japan. J. Appl. Phys.*, **26**, p. 231, 1987.

41. S. Tsunashima et al., "Magneto-Optic Kerr Effect of Amorphous Gd–Fe Films," *IEEE Trans. Magn.*, **MAG-23**, p. 3205, 1987.

42. F. Kirino et al., "Gd-Base Amorphous Magneto-Optical Films with Large Kerr Rotations," *IEEE Trans. J. Magn. Jpn.*, **TJMJ-2**, p. 1110, 1987.

43. M. Tanaka et al., "Magnetic Properties of Compositionally Modulated Tb/FeCo Films," *IEEE Trans. Magn.*, **MAG-23**, p. 2955, 1987.

44. R. Ohyama et al., "Preparation of PtMnSb Thin Films and Magneto-Optical Properties," *IEEE Trans. J. Magn. Jpn.*, **TJMJ-1**, p. 122, 1985.

45. T. Inukai et al., "Magneto-Optical Properties of Substituted Pt–Mn–Sb Thin Films," *IEEE Trans. J. Magn. Jpn.*, **TJMJ-2**, p. 1102, 1987.

46. A. Ito et al., "Magnetic and Magneto-Optical Properties of PtMnSb Thin Films," *IEEE Trans. J. Magn. Jpn.*, **TJMJ-2**, p. 1100, 1987.

47. T. Suzuki et al., "Magnetic and Magneto-Optical Properties of (Nd,Pr)–(Fe,Co) Alloy Amorphous Films with Huge Perpendicular Magnetic Anisotropy," *IEEE Trans. Magn.*, **MAG-23**, p. 2958, 1987.

48. H.-P. D. Shieh et al., "Magnetic Properties of Amorphous Tb(Nd,Pr)Co Films," *IEEE Trans. Magn.*, **MAG-23**, p. 3208, 1987.

49. M. Masuda et al., "Preparation, Magnetic and Magneto-Optic Properties of Small-Crystalline MnBi Films," *Japan. J. Appl. Phys.*, **26**, p. 707, 1987.

50. M. H. Kryder, "Magneto-Optic Recording Technology," *J. Appl. Phys.*, **57**, p. 3913, 1985.

51. E. Schultheiss et al., "Production Technology for Magnetooptic Data Storage Media," *Solid State Tech.*, p. 107, 1988.

52. S. Middlehoek et al., *Physics of Computer Memory Devices*, Academic, New York, 1976.

53. H.-P. D. Shieh and M. H. Kryder, "Dynamics and Factors Controlling Regularity of Thermomagnetically Written Domains," *J. Appl. Phys.*, **61**, p. 1108, 1987.

54. H.-P. D. Shieh and M. Kryder, "Magneto-Optic Recording Materials with Direct Overwrite Capability," *Appl. Phys. Lett.*, **49**, p. 473, 1986.

55. M. D. Schultz et al., "Performance of Magneto-Optical Recording Media with Direct Overwrite Capability," *J. Appl. Phys.*, **63**, p. 3844, 1988.

56. H.-P. D. Shieh and M. Kryder, "Operating Margins for Magneto-Optic Recording Materials with Direct Overwrite Capability," *IEEE Trans. Magn.*, **MAG-23**, p. 171, 1987.

57. M. Masuda et al., "Preparation, Magnetic and Magneto-Optic Properties of Small-Crystalline MnBi Films," *Japan. J. Appl. Phys.*, **26**, p. 707, 1987.

58. M. Kaneko et al., "Optical Operation of a Magnetic Bubble," *IEEE Trans. Magn.*, **MAG-22**, p. 2, 1986.

59. K. Balasubramanian and A. Macleod, "Performance Calculations for Multi-Layer Thin-Film Structures Containing a Magnetooptical Film," *Abstracts of the 1986 Annual Meeting, Optical Society of America*, p. P28; *J. Opt. Soc. Amer. A*, **3**(13), p. P28, 1986.

60. M. Gomi et al., "RF-Sputtering of Highly Bi-Substituted Garnet Films on Glass Substrates for Magneto-Optic Memory," *IEEE Trans. J. Magn. Jpn.*, **TJMJ-1**, p. 75, 1985.

61. M. Gomi et al., "Bi-Substituted Garnet Films Crystallized During RF Sputtering for M-O Memory," *IEEE Trans. Magn.*, **MAG-23**, p. 2967, 1987.

62. T. Okuda et al., "Faraday Rotation in Highly Bi-Substituted Yttrium Iron Garnet Films Prepared by Ion Beam Sputtering," *IEEE Trans. Magn.*, **MAG-23**, p. 3491, 1987.

63. K. Maeto et al., "Bi-Substituted DyIG Thin Films Prepared by Chemical Deposition for Magneto-Optic Recording Medium," *IEEE Trans. J. Magn. Jpn.*, **TJMJ-2**, p. 839, 1987.

64. M. Mansipur, "Coercivity and Its Role in Thermomagnetic Recording," *J. Appl. Phys.*, **61**, p. 3334, 1987.

65. M. Mansipur, "Magnetization Reversal Dynamics in Magneto-Optic Media" (invited), *J. Appl. Phys.*, **63**, p. 3831, 1988.

66. M. Mansipur, "Magnetization Reversal, Coercivity, and the Process of Thermomagnetic Recording in Thin Films of Amorphous Rare Earth–Transition Metal Alloys," *J. Appl. Phys.*, **61**, p. 1580, 1987.

67. M. Kaneko et al., "A Low Loss 0.8 Micron Band Optical Isolator Using Highly Bi-substituted LPE Garnet Film," *IEEE Trans. Magn.*, **MAG-23**, p. 3482, 1987.

68. K. Matsuda et al., "Bi-Substituted Rare-Earth Iron Garnet Composite Film with Temperature Independent Faraday Rotation for Optical Isolators," *IEEE Trans. Magn.*, **MAG-23**, p. 3479, 1987.

69. H. Hemme et al., "Optical Isolator Based on Mode Conversion in Magnetic Garnet Films," *Appl. Opt.*, **26**, p. 3811, 1987.

70. R. Wolfe et al., "Thin-Film Waveguide Magneto-Optical Isolator," *Appl. Phys. Lett.*, **46**, p. 817, 1985.

71. T. Mizumoto et al., "Measurement of Optical Nonreciprocal Phase Shift in a Bi-Substituted $Gd_3Fe_5O_{12}$ Film and Application to Waveguide-Type Optical Circulator," *J. Lightwave Tech.*, **LT-4**, p. 347, 1986.

72. S. Yamamoto and Y. Okamura, "Magneto-Optical Branching Waveguides and Their Applications to Nonreciprocal Optical Devices," *IEEE Trans. J. Magn. Jpn.*, **TJMJ-1**, p. 1037, 1985.

73. B. Hill and K. P. Schmidt, "Fast Switchable Magneto-Optic Memory — Display Components," *Philips J. Res.*, **33**, p. 211, 1978.

74. W. E. Ross et al., "Fundamental Characteristics of the Litton Iron Garnet Magneto-Optic Spatial Light Modulator," *Advances in Optical Information Processing*, SPIE **388**, p. 55, 1983.

75. G. R. Pulliam et al., "Large Stable Magnetic Domains," *J. Appl. Phys.*, **53**, p. 2754, 1982.

76. G. L. Nelson et al., "Stripe Domain Light Deflection for Intersatellite Communications," *Communication Networking in Dense Electromagnetic Environments*, SPIE **876**, p. 121, 1988.

77. H. Kogelnik, "Theory of Dielectric Waveguides," in T. Tamir, Ed., *Integrated Optics*, Springer-Verlag, New York, 1975, p. 13.

78. A. D. Fisher, "Optical Signal Processing with Magnetostatic Waves," in J. P. Parekh, Ed., *Circuits, Systems, and Signal Processing*, Vol. 4 (Special Issue on Magnetostatic Waves and Applications to Signal Processing), 1985, p. 265.

79. A. E. Craig et al., "Characterization of Ferrimagnetic Garnets for MSW — Optical Diffraction," *Proc. IEEE Ultrasonics Symp.*, p. 174, 1985.

80. Y. H. Won et al., "Three-Prism Loss Measurements of Optical Waveguides," *Appl. Phys. Lett.*, **37**, p. 269, 1980.

81. A. D. Fisher et al., "Optical Guided-Wave Interactions with Magneto-Static Waves at Microwave Frequencies," *Appl. Phys. Lett.*, **41**, p. 779, 1982.

82. A. D. Fisher et al., "Magnetostatic Wave Devices for Integrated-Optical Signal Processing," *Proc. IEEE Ultrason. Symp.*, p. 226, 1983.

83. A. C.-T. Wey et al., "Enhanced-Bandwidth MSFVW — Optical Interaction Employing an Inhomogeneous Bias Field," *Proc. IEEE Ultrason. Symp.*, p. 173, 1986.

84. C.-T. Wey et al., "Inhomogeneous Field MSFVW — Optical Interaction," *Integrated Optical Circuit Engineering*, SPIE **704**, p. 51, 1986.

85. H. Tamada, "Bi-Substituted LPE Garnet Films with FMR Linewidth as Small as YIG," *IEEE Trans. J. Magn. Jpn.*, **3**, p. 98, 1988.

86. C. S. Tsai et al., "Noncollinear Coplanar Magneto-Optic Interaction of Guided Optical Wave and Magnetostatic Surface Waves in Yttrium Iron Garnet — Gadolinium Gallium Garnet Waveguides," *Appl. Phys. Lett.*, **47**, p. 651, 1985.

87. D. Young et al., "Tunable Wideband Guided Wave Magneto-Optic Modulator Using Magnetostatic Surface Wave," *Acousto-Optic, Electro-Optic and Magneto-Optic Devices and Applications*, SPIE **753**, p. 161, 1987.

88. D. Young and T. S. Tsai, "GHz Bandwidth Magneto-Optic Interaction in Yttrium Iron Garnet — Gadolinium Gallium Garnet Waveguide Using Magnetostatic Forward Volume Waves," *Appl. Phys. Lett.*, **53**, p. 1696, 1988.

89. S. H. Talisa, "The Collinear Interaction between Forward Volume Magnetostatic Waves and Guided Light in YIG Films," *IEEE Trans. Magn.*, **MAG-24**, p. 2811, 1988.

90. D. D. Stancil, "Theoretical Investigations for MSW — Optical Interactions," Final Report on U.S. Naval Research Laboratory contract N00173-88-M-X012, September 12, 1988.

91. H. Huahui et al., "Optical Mode Conversion in a Ferrimagnetic Film Containing Magnetostatic Forward Volume Wave," *IEEE Trans. Magn.*, **MAG-23**, p. 3500, 1987.

92. G. Srinivasan et al., "Characterization of Magnetostatic Wave Devices by Brillouin Light Scattering," *IEEE Trans. Magn.*, **MAG-23**, p. 3718, 1987.

93. H. le Gall, "First and Second-Order Inelastic Scatterings of Light by High Amplitudes Spin-Waves in Ferrimagnetic Crystals," in M. Balkanski, Ed., *Proceedings of the Second International Conference on Light Scattering in Solids*, Flammarion Sciences, New York, 1971, p. 170.

94. A. A. Solomko et al., "Interaction of Laser Radiation with Surface Magnetostatic Waves in Iron–Garnet Films," *Opt. Spectrosc. (USSR)*, **59**, p. 381, 1985.

95. A. A. Solomko et al., "Collinear Interaction of Light with Surface Magnetostatic Waves in Ferrite–Garnet Films," *Opt. Spectrosc. (USSR)*, **61**, p. 804, 1986.

## SUGGESTIONS FOR FURTHER READING

### 18.2   Units

D. J. Craik, *Structure and Properties of Magnetic Materials*, Pion Ltd., London, 1971.

B. D. Cullity, *Introduction to Magnetic Materials*, Addison-Wesley, Reading, MA, 1972.

### 18.3   Material Physics of Magnetics

S. Chikazumi, *Physics of Magnetism*, Wiley, New York, 1964.

B. D. Cullity, *Introduction of Magnetic Materials*, Addison-Wesley, Reading, MA, 1972.

C. Kittel, in *Introduction to Solid State Physics*, 4th ed., Wiley, New York, 1971, Chaps. 14–17.

B. Lax and K. J. Button, *Microwave Ferrites and Ferrimagnetics*, McGraw-Hill, New York, 1962.

M. McCaig and A. G. Clegg, *Permanent Magnets in Theory and Practice*, Wiley, New York, 1987.

M. S. Sodha and N. C. Srivastava, *Microwave Propagation in Ferrimagnetics*, Plenum, New York, 1981.

## 18.4 Spin Waves

B. Lax and K. J. Button, *Microwave Ferrites and Ferrimagnetics*, McGraw-Hill, New York, 1962.

J. P. Parekh, Ed., *Circuits, Systems, and Signal Processing* (Special Issue on Magnetostatic Waves and Applications to Signal Processing), **4**(1–2), 1985.

M. S. Sodha and N. C. Srivastava, *Microwave Propagation in Ferrimagnetics*, Plenum, New York, 1981.

D. D. Stancil, *Magnetostatic Waves* (in press).

M. G. Cottam and D. R. Tilley, *Introduction to Surface and Superlattice Excitations*, Cambridge University Press, Cambridge and New York, 1989.

P. E. Tannenwald, "Spin Waves," *Microwave J. for July*, 1959, p. 25.

P. E. Wigen, "Magnetic Excitations," in A. Paoletti, Ed., *Physics of Magnetic Garnets*, North-Holland, New York, 1978, p. 196.

P. E. Wigen, "Microwave Properties of Magnetic Garnet Thin Films," *Thin Solid Films*, **114**(1/2), p. 135, 1984; Special Issue on Magnetic Garnet Films, A. Paoletti, guest editor.

## 18.5 Material Physics of Optics

D. Clarke and J. F. Grainger, *Polarized Light and Optical Measurement*, Pergamon, New York, 1971.

P. Hlawiczka, *Gyrotropic Waveguides*, Academic, New York, 1981.

E. Hecht and A. Zajac, *Optics*, Addison-Wesley, Reading, MA, 1974.

R. G. Hunsperger, Ed., *Integrated Optics: Theory and Technology*, 2nd ed., Springer-Verlag, New York, 1985.

F. A. Jenkins and H. E. White, *Fundamentals of Optics*, 4th ed., McGraw-Hill, New York, 1976.

G. B. Scott and D. E. Lacklison, "Magnetooptic Properties and Applications of Bismuth Substituted Iron Garnets," *IEEE Trans. Magn.*, **MAG-12**(4), p. 292, 1976.

T. Tamir, Ed., *Integrated Optics*, Springer-Verlag, New York, 1975.

W. Wettling, "Magneto-Optics of Ferrites," *J. Magnet. Magn. Mat.*, **3**, p. 147, 1976.

## 18.6 Magneto-optics

J. F. Dillon, Jr., "Magneto-Optical Properties of Magnetic Crystals," in J. Smit, Ed., *Magnetic Properties of Materials*, McGraw-Hill, New York, 1971, Chap. 5, p. 149.

J. F. Dillon, Jr., "Magneto-Optical Properties of Magnetic Garnets," in A. Paoletti, Ed., *Physics of Magnetic Garnets*, North-Holland, New York, 1978, p. 379.

M. J. Freiser, "A Survey of Magnetooptic Effects," *IEEE Trans. Magn.*, **MAG-4**(2), p. 152, 1968.

G. S. Monk, in *Light: Principles and Experiments*, 2nd ed., Dover, New York, 1963, Chap. 16.

G. B. Scott and D. E. Lacklison, "Magnetooptic Properties and Applications of Bismuth Substituted Iron Garnets," *IEEE Trans. Magn.*, **MAG-12**(4), p. 292, 1976.

J. C. Suits, "Faraday and Kerr Effects in Magnetic Compounds," *IEEE Trans. Magn.*, **MAG-8**(1), p. 95, 1972.

## 18.7    Real Materials

J. O. Artman, "Magnetic Anisotropy and Structure of Magnetic Recording Media," *J. Appl. Phys.*, **61**, p. 3137, 1987.

G. Bate, "Materials Challenges in Metallic, Reversible, Optical Recording Media: A Review," *IEEE Trans. Magn.*, **MAG-23**, p. 156, 1987.

A. H. Eschenfelder, *Magnetic Bubble Technology*, Springer-Verlag, New York, 1981.

P. Hansen and J.-P. Krumme, "Magnetic and Magneto-Optical Properties of Garnet Films," *Thin Solid Films*, **114** (Special Issue on Magnetic Garnet Films), p. 69, 1984.

P. Hansen and K. Witter, "Growth-Induced Uniaxial Anisotropy of Bismuth-Substituted Iron-Garnet Films," *J. Appl. Phys.*, **58**, p. 454, 1985.

P. Hansen et al., "Optical Switching with Bismuth-Substituted Iron Garnets," *Philips Tech. Rev.*, **41**, p. 33, 1983/84.

P. Hansen et al., "Magnetic and Magneto-Optical Properties of Bismuth-Substituted Lutetium Iron Garnet Films," *Phys. Rev. B*, **31**, p. 5858, 1985.

M. Kaneko et al., "Optical Operation of a Magnetic Bubble," *IEEE Trans. Magn.*, **MAG-22**, p. 2, 1986.

B. Knorr and W. Tolksdorf, "Lattice Parameters and Misfits of Gallium Garnets and Iron Garnet Epitaxial Layers at Temperatures between 294 and 1300 K," *Mat. Res. Bull.*, **19**, p. 1507, 1984.

G. Nelson and W. A. Harvey, "Optical Absorption Reduction in $Bi_1Lu_2Fe_5O_{12}$ Garnet Magneto-Optical Crystals," *J. Appl. Phys.*, **53**, p. 1687, 1982.

A. Paoletti, Ed., *Physics of Magnetic Garnets*, North-Holland, New York, 1978.

W. Reim et al., "$Tb_xNd_y(FeCo)_{1-x-y}$: Promising Materials for Magneto-Optical Storage?" *J. Appl. Phys.*, **61**, p. 3349, 1987.

B. Strocka et al., "An Empirical Formula for the Calculation of Lattice Constants of Oxide Garnets Based on Substituted Yttrium- and Gadolinium-Iron Garnets," *Philips J. Res.*, **33**, p. 186, 1978.

W. Tolksdorf and C.-P. Klages, "The Growth of Bismuth Iron Garnet Layers by Liquid Phase Epitaxy," *Thin Solid Films*, **114** (Special Issue on Magnetic Garnet Films), p. 33, 1984.

W. H. von Aulock, Ed., *Handbook of Microwave Ferrite Materials*, Academic Press, New York, 1965.

## 18.8    Modern Magneto-optical Devices

G. Bouwhuis et al., *Principles of Optical Disc Systems*, Adam Hilger Ltd., Boston, 1985.

A. D. Fisher, "Optical Signal Processing with Magnetostatic Waves," in J. P. Parekh, Ed., *Circuits, Systems, and Signal Processing* (Special Issue on Magnetostatic Waves and Applications to Signal Processing), **4**, p. 265, 1985.

R. G. Hunsperger, Ed., *Integrated Optics: Theory and Technology*, 2nd ed., Springer-Verlag, New York, 1985.

N. S. Kapany and J. J. Burke, *Optical Waveguides*, Academic, New York, 1972.

N. Koshizuka and T. Okuda, "The Problems in the Development of Magneto-Optical Devices," *IEEE Trans. J. Magn. Jpn.*, **TJMJ-1**, p. 1044, 1986.

M. H. Kryder, "Magneto-Optic Recording Technology," *J. Appl. Phys.*, **57**, p. 3913, 1985.

D. Marcuse, *Theory of Dielectric Optical Waveguides*, Academic, New York, 1974.

S. Middlehoek et al., *Physics of Computer Memory Devices*, Academic, New York, 1976.

P. Paroli, "Magneto-Optical Devices Based on Garnet Films," *Thin Solid Films*, **114** (Special Issue on Magnetic Garnet Films), p. 187, 1984.

W. E. Ross et al., "Two-Dimensional Magneto-Optical Spatial Light Modulator for Signal Processing," *Opt. Eng.*, **22**, p. 485, 1983.

T. Tamir, Ed., *Integrated Optics*, Springer-Verlag, New York, 1975.

K. Tsushima, "Magneto-Optics: Its Past and Present," *IEEE Trans. J. Magn. Jpn.*, **TJMJ-1**, p. 920, 1985.

K. Tsushima and N. Koshizuka, "Research Activities on Magneto-Optical Devices in Japan," *IEEE Trans. Magn.*, **MAG-23**, p. 3473, 1987.

# 19

# OPTICAL DETECTORS

P. K. L. Yu
*University of California, San Diego*
*La Jolla, California*

H. D. Law
*Agura Hill, California*

## 19.1 INTRODUCTION

An optical detector converts photon energy into some other form of energy, mainly electrical energy, and couples it to a receiver circuit. The human eye is an example of an optical detector that excels in its ability to perform parallel signal processing and to adjust itself to the changes in background intensity despite the slow response speed and limited spectral and sensitivity ranges of the photosensitive cells [1]. Presently, optical detectors play an important role in communication systems that employ photons as information carriers. In particular, for digital communication systems, detector devices such as photodiodes, photoconductors, and phototransistors have been developed. For meeting the system requirements, the optical receivers have to be highly sensitive to low-level optical signals, and they need to be high speed and reliable.

For microwave photonics applications, the optical detectors are commonly used in the receiver and sometimes as switching element for microwave signals. Unlike digital communications, the requirements for RF signal receiver stems from those related to analog signal transmission. Therefore, in addition to high sensitivity detection, receivers are required to have large dynamic range and low-noise figure properties [2].

This chapter reviews the basic properties of optical detectors and should be used as introductory reading for engineers using optical receivers. Due to the space limitation, more extensive discussion of the subject is deferred to other texts [3–6]. Section 19.2 describes the terminology for optical detection commonly used in the literature and in commercial specifications. Material issues pertaining to photodetection are also discussed. Section 19.3 provides an update of the characteristics of various types of

*Handbook of Optical Components and Engineering*, Edited by Kai Chang
ISBN 0-471-39055-0 © 2003 John Wiley & Sons, Inc.

photodiodes, including avalanche photodiodes and waveguide photodiode. Section 19.4 describes the characteristics of photoconductors and compares them to photodiodes for microwave switching applications. Section 19.5 describes the nonlinearity in photodiodes, an important consideration for analog link applications.

## 19.2   TERMINOLOGY IN OPTICAL DETECTION AND MATERIAL CONSIDERATIONS

Semiconductor detectors are attractive because they are usually small, require a low bias voltage, and are easily integrated to form an array. Light is absorbed in semiconductor via processes, whereby electrons make a transition from a lower to a higher energy state. These electrons are subsequently detected in an external circuit before they drop back to the lower state. Each semiconductor material has a characteristic absorption spectrum (for example, see Fig. 19.1), and the absorption is much dependent on incident photon energy. The absorptivity, $\alpha_o$, defined as the incremental decrease in optical intensity per unit length per unit incident power, is usually referred to as the absorption coefficient of the material. The photon energy $h\nu$ (where $h$ is the Planck's constant and $\nu$ is the frequency) and the wavelength $\lambda$ are related simply by

$$\lambda = \frac{1.24}{h\nu} \tag{19.1}$$

where $\lambda$ is measured in units of micrometer and $h\nu$ is in electron volt.

$\alpha_o$ describes the power decay of the optical beam as it propagates inside a given material. In general, a large $\alpha_o$ is desirable for optical detection for an efficient optical to electrical conversion. However, too large an $\alpha_o$ can result in a small penetration depth, and many of the photogenerated electrons can be lost via surface (or interface) recombination such that they will not be collected in an external circuit. Optimization of $\alpha_o$ depends on the material and structural aspects of the detector.

An important performance parameter of an optical detector is the external quantum efficiency $\eta$ defined as the ratio of the number of electrons generated to the number of incident photons without photogain. $\eta$ takes into account the surface reflection loss and other losses in the detector. In general, $\eta(\lambda)$ depends on the $\alpha_o(\lambda)$ of the materials and the dimensions of the absorption region [e.g., see Equations (19.8) and (19.22)]. The optical response can be enhanced in the presence of photogain $M$ that results from carrier injection in semiconductor materials, as in photoconductive devices [see Equation (19.19)], or from impact ionization, as in avalanche photodiodes [see Equations (19.12a) and (19.12b)]. The responsivity $R$ of a detector describes the dependence of the output in response to unity input optical power and is expressed as:

$$R = M\eta\frac{q}{h\nu} \quad \text{(A/W)} \tag{19.2}$$

where $q$ is the electronic charge. For many applications, a small optical power reaches the photodetector and the responsivity remains constant as the optical power is raised slightly. For applications where a large power optical beam is incident at the detector, the responsivity can change as a result of the carrier screening of the applied electric field [7]. Such a nonlinear response is not suitable for some analog applications, as discussed in Section 19.5.

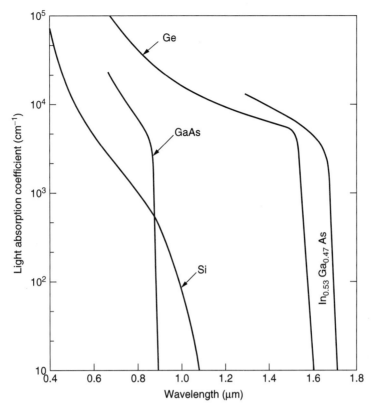

**Figure 19.1**  Absorption coefficient vs. wavelength for silicon, germanium, gallium arsenide, and indium gallium arsenide.

The response speed of a photodetector is another important parameter for many analog and digital applications. It is measured in terms of the characteristics 3-dB frequency, $f_{3\,dB}$, at which the detected electric power is reduced to half of its low frequency value. The $f_{3\,dB}$ is inversely proportional to the response time $\tau$ of the detector. $f_{3\,dB}$, too, can be greatly reduced when detector saturation occurs [see Section 19.5 and Reference 7].

The response time generally depends on the transit time of electrons and holes in the detector, the carrier diffusion process, the carrier multiplication process in the semiconductor, as well as the circuit time ($RC$) constant. In practice, $\tau$ can be determined from the $f_{3\,dB}(= 1/(2\pi\tau))$ measured from a small signal, frequency modulation response of the photodetector. For this measurement, a modulated optical carrier can be generated by direct modulation or external modulation of laser light, or simply by an optical mixing of two laser sources whose emission wavelengths can be separately tuned [8].

Alternatively, $\tau$ can be estimated from the time domain response of the detector when it is excited with ultra-fast optical pulses, such as those generated by mode-locked lasers, or those from a gain switched laser driven by a comb generator [9]. Provided that the rise time and the fall time of the optical pulse are much shorter than those of the detector, $\tau$ can be de-embedded from both the rise time and the fall time of the measured signal current. A high-speed sampling oscilloscope is commonly used to measure the total time response, including the detector response time, the effect of the finite optical pulse width, and the finite response time of the oscilloscope.

For detectors with photogain, an additional figure of merit, the so-called gain-bandwidth product (GB) is used. It is simply the product of the 3-dB bandwidth and the dc gain. However, in many cases, GB depends on device geometry, carrier dynamics, and some material parameters.

The quality of detected signals can be degraded by noise. The noise behavior is crucial in the determination of the overall system dynamic range (which is related to signal-to-noise ratio) for analog applications [2]. Noise can arise at the transmitter, the background environment, the electrical and optical amplifiers, and the receiver. For fiber-optic links, the common noise sources are the laser intensity noise, the Poisson (shot) noise due to the quantized nature of photons, and the finite response time of the detector; dark current noise due to generation and recombination of carriers in the detection region, and thermal (Johnson) noise due to the resistance of the external load of the detector. For infrared focal plane array applications, the background radiation constitutes an additional noise source. Three commonly used measurements of noise performance are the signal-to-noise (S/N) ratio, noise equivalent power (NEP), and specific detectivity ($D^*$) [3]. The S/N ratio (in decibels) is simply the ratio of detected electrical signal power to the noise power at the output of the detector (or receiver). The NEP of a detector is the minimum optical power that must be incident on the detector so that the signal level is raised above the noise floor (or $S/N = 1$) for a given wavelength, detector temperature, and resolution bandwidth. The specific detectivity is the inverse of the NEP normalized to the square root of the detector area $A$ and the bandwidth $B$,[*]

$$D^* = \frac{\sqrt{AB}}{NEP} \tag{19.3}$$

In choosing semiconductor material for effective optical detection, several issues are commonly considered. Materials with good crystal quality and high purity are usually preferred so as to minimize trapping and recombinations of photogenerated carriers, because defects can affect both the near-term and long-term performance of the photodetector. A direct band bandgap material, due to its large absorption coefficient, is preferred over an indirect material. The bandgap energy $E_g$ of the absorption region is usually smaller than the photon energy in order to get a large $\alpha_o$. In cases where heterojunction materials are used, close matching of crystal lattice constants of different materials usually ensures good interface quality. This is a common safeguard against premature failure of detectors, as interfacial states or traps can cause gradual degradation of the dark current characteristics. In addition, two material-related issues are commonly encountered in detector designs: namely, the doping profile in the detection region of the device and the nature of the metal-semiconductor contact. The doping level affects the electric field profile and can be designed to fine tune the carrier transport. A low doping level in the depletion region is often desired for fast carrier sweep out at low bias voltage as well as for stable device junction impedance. The properties of the metal-semiconductor contact can affect carrier injection (and thus

---

[*] Although the noise equivalent power is normally used to characterize the sensitivity of a photodiode, the specific detectivity is commonly used even when the incident flux covers an area larger than the detector.

gain), the response speed, and the optical saturation levels (and thus linearity) of the detector, as the contact itself can be viewed as a potential barrier.

For band-to-band transition of electrons that involves the absorption of a photon — the so-called intrinsic absorption process — the photon energy is larger than that of the bandgap for large $\alpha_o$ [10]. There are many ways that can lead to the absorption of photons with energy less than the bandgap energy. For instance, for far infrared detection, extrinsically doped silicon photoconductor can be used. In impurity band photoconductor, photon absorption occurs as the electron makes a transition from an impurity level within the bandgap to the nearby energy band. Free-carrier absorption can take place when the electron makes a transition within the same energy band while a photon is being absorbed. However, free-carrier absorption becomes significant only in the presence of large electron (or hole) density ($>10^{18}$ cm$^{-3}$). The Franz–Keldysh effect in bulk semiconductor [11, 12] and the quantum confined Stark effect in quantum wells [13] exploit the electric-field dependent shift of the absorption edge for below band-gap absorption. These effects have been investigated for various optical switches [14] as well as for traveling wave waveguide modulators where a small $\alpha_o$ is well suited for a long interaction length [see Chapter 16]. The band-to-band absorption process in semiconductors results in the largest absorptivity simply due to the abundance in initial and final states available for absorptive transition.

In contrast to the direct bandgap semiconductor materials, indirect bandgap materials depend on a phonon or impurity-assisted transition process for absorption near the bandgap energy. This process has a relatively smaller probability to occur. Consequently, for photon near the bandgap energy, a relatively smaller absorption coefficient is observed [15].

### 19.2.1  Digression

The $RC$ limitation is due to the time required to charge and discharge the photodiode capacitance. Capacitance $C$ denotes the sum of the junction capacitance ($C_j$) and other parasitic (stray) capacitance ($C_s$) related to the contact pad and the package; $R$ is the resistance of the load. The junction capacitance $C_j$ of the photodiode is given by

$$C_j = \frac{\varepsilon A}{W} \tag{19.4}$$

where $\varepsilon$ is the dielectric permittivity, A denotes the junction area of the diode, and $W$ denotes the depletion region thickness. In the case of one-sided, abrupt junction, such as $p^+n$ diode, $W$ is related to the doping of the $n$ layer, $N_D$, and the bias voltage $V$ through the Poisson's equation:

$$W \approx \sqrt{\frac{2V\varepsilon}{qN_D}} \tag{19.5}$$

From Equations 19.4 and 19.5, the diode capacitance can be reduced by lowering the doping in the active region. A lower doping in the intrinsic layer may also enhance the optical saturation level of the photodiode [16]. In general, the junction capacitance can be reduced by either reducing the area of the photodiode or by increasing the depletion layer thickness. However, for surface–coupled photodiode, the optical coupling also depends on the area. The intrinsic layer thickness is related to the quantum efficiency and the carrier transit time, so that in general, an optimization of these parameters is needed.

The $RC$ time effect, the transit time, and the carrier diffusion time determines the basic response speed of the detectors. Among these, the diffusion time is usually the slowest. For $n$-type materials, the photogenerated holes take a slow diffusion velocity

$$v_{Diff} = \frac{L_P}{\tau_P} = \frac{D_P}{L_P} \tag{19.6}$$

where $L_p$ is hole diffusion length, $D_p$ is the hole diffusivity, and $\tau_p$ is the (minority carrier) hole lifetime. A similar equation can be written for electrons in $p$-type materials. At low carrier concentration, the diffusion constants are related to the carrier mobility via the Einstein relationship $D = (kT/q)\mu$, where $\mu$ is the corresponding carrier mobility, $k$ is Boltzmann's constant, and $T$ in the temperature measured in Kelvin. For cases where the dimension of the $n$-region, $L$, is smaller than $L_p$, the last part in Equation 19.6 is replaced by $D_p/L$; thus, a higher diffusion velocity is possible with small $L$.

Most direct bandgap III-V materials have short carrier lifetimes and large electron mobility, in addition to a large absorption coefficient ($10^4$–$10^5$ cm$^{-1}$). These properties make them attractive for very high-speed applications. Silicon can be used for 0.8-$\mu$m wavelength where the absorptivity is $\sim 10^3$ cm$^{-1}$ (due to its indirect bandgap). This corresponds to an absorption length of $\sim 10$ $\mu$m; therefore, high-speed and high responsivity detection occurs only for some designs of the detector (such as in Si avalanche photodiode). Figure 19.1 shows the absorptivity versus wavelength for silicon, germanium, GaAs, and InGaAs.

The transit time can be minimized by reducing the thickness of the depletion region. By doing so, the capacitance also increases. Alternatively, one can preferentially use the carrier (typically the electron) with higher drift velocity for signal transport. One example is the uni-traveling carrier photodiode, described at the end of 19.3.1 [17]. In this device, the doping profile and/or the bandgap profile can be tailored to provide a built-in electric field to drift the faster carrier in the right direction.

## 19.3    PHOTODIODES

Conventional photodiodes have an absorption region where a strong electric field is applied to sweep out the photogenerated carriers. Consequently, a current flows in the external electronic circuit. Examples are $pn$ junction photodiodes, $p$-intrinsic-$n$ (PIN) photodiodes, Schottky junction photodiodes, and avalanche photodiodes (APDs).

A $pn$ photodiode usually consists of a $p^+n$ (or $n^+p$) junction in which the optical signal can be incident at the top or the bottom surfaces (see Fig. 19.2). Alternatively, in case of the edge-coupled detectors such as waveguide photodiodes, light is coupled to an optical waveguide parallel to the junction plane [18, 7]. Typically, the diode is reverse biased, and a high electric field is applied across the depletion region. Electron-hole pairs, including those generated due to absorption in the high field region as well as those due to absorption nearby and reaching there via diffusion, are separated by the electric field and drift toward the electrodes. The external circuit detects the signal currents as the photogenerated carriers drift across the depletion region. The induced current stops once the carriers recombine. This phenomenon gives rise to the transit

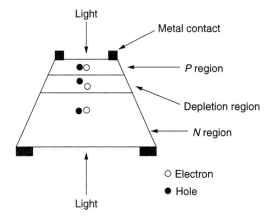

**Figure 19.2** A *pn* junction photodiode with light coupled to top or bottom surface.

time limit of the response speed. For a given velocity distribution $v(x)$ of the carriers and the thickness $(W)$ of the depletion region, the carrier transit time $T_{tr}$ can be obtained from:

$$T_{tr} = \int_0^W \frac{dx}{v(x)} \tag{19.7}$$

From Equation 19.7, it can be seen that the transit time is limited by the slower carriers travel across the high field region.

The quantum efficiency $\eta$ of the photodiode shown in Fig. 19.2 for the case of illumination through the top *p*-region can be approximated by:

$$\eta = (1 - R_o)\left[e^{-\alpha_o W_1}(1 - e^{-\alpha_o W}) + (1 - e^{-\alpha_o W_1})\left(\frac{\alpha_o}{\alpha_o + \dfrac{1}{L_n}}\right)\right.$$

$$\left. + e^{-\alpha_o(W_1+W)}\left(\frac{\alpha_o}{\alpha_o + \dfrac{1}{L_p}}\right)\right] \tag{19.8}$$

where $R_o$ denotes the reflectivity of the air-semiconductor interface, $W_1$ is the thickness of the *p* region, and $L_n$ and $L_p$ are the diffusion lengths of electrons in the *p* region and holes in the *n* regions, respectively. The first term inside the bracket of Equation (19.8) accounts for carriers generated within the depletion region; the second and the third terms account for those in the *p* and *n* regions, respectively. For most III-V photodiodes with a large absorption coefficient, the absorption in the *n* region can be ignored. In the case of a heterojunction photodiode where the *p* region consists of material transparent to incident photons, $\alpha_0$'s in the second term can be effectively set to zero. As can be seen from Equation (19.8), $\eta$ can be enhanced by reducing $R_o$; this is achieved by depositing antireflection coating on the incident facet of the diode. The quantum efficiency in well-optimized heterojunction III-V photodiodes can be as high as 95%.

In junction photodiodes where carriers are swept out by a moderate electric field (typically, $10^3 – 10^4$ V/cm) before avalanche process occurs, there is no photogain in the device. The responsivity can be calculated from Equation (19.2) by setting $M$ to unity. For instance, *pn* junction photodiodes made on silicon with a depletion region about 10 μm thick typically has a responsivity of ~0.4 A/W at 0.8–0.9-μm wavelength [19].

For simple *pn* junction photodiodes, it is difficult to optimize the quantum efficiency and the transit time at the same time due to the fact that $W$ depends on the applied electric field. This is further complicated by the nonlinear relationship between the electron velocity and the electric field.

### 19.3.1  PIN Photodiodes

The PIN photodiode, consisting of a low-doped (either $i$ or $p$) region sandwiched between $p$ and $n$ regions, can be a better alternative to the simple *pn* photodiode. This is because at a relatively low reverse bias, the intrinsic region becomes fully depleted and the total depletion thickness, including those from the $p$ and $n$ regions, remains relatively constant as the voltage is increased. A schematic diagram of a PIN photodiode is shown in Fig. 19.3**a** with light incident from the left; its energy band diagram at reverse bias is depicted in Fig. 19.3**b**. Typically, for a doping concentration of $10^{14}$ cm$^{-3}$ in the intrinsic region, a bias voltage of a few volts is sufficient to deplete several micrometers, and the electron velocity reaches the saturation value. However, as mentioned earlier, transit time is not the only parameter affecting the response speed of the PIN photodiode. Other effects such as carrier diffusion, the $RC$ time constant, and optical saturation can also be important. For $p$ region less than one diffusion

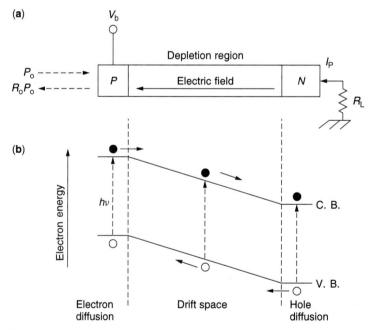

**Figure 19.3**  PIN photodiode schematics. (**a**) Structure: $P_o$, incident power; $V_b$, bias voltage; $R_o$ reflectivity; $I_p$, photocurrent; $R_L$, load resistor. (**b**) Energy band diagram at reverse bias; C.B. stands for conduction band; V.B. stands for valence band.

length, the response time due to diffusion alone is typically 1 ns/μm in $p$-type silicon and about 20–100 ps/μm in $p$-type III-V materials [19]. The corresponding value for $n$-type III-V materials is several nanoseconds per micrometer due to the lower mobility of holes. For absorption in the intrinsic materials, one can minimize the diffusion effect in the $p$ and $n$ regions, by making them very thin; alternatively, one can employ the $p$ or $n$ region transparent to the light, as in the case of heterojunction PIN photodiodes; or a combination of both [20, 21].

For wavelengths in the 1.0–1.55-μm range, which is suitable for optical fiber communication, germanium, and a number of III-V compound semiconductor alloys stand out as candidate materials for PIN photodiodes, primarily because of their large absorption coefficients in this wavelength range. The absorption edge of germanium is near 1.6 μm at room temperature. Its $\alpha_o$ is almost flat and is in the $10^{-4}$-cm$^{-1}$ range over this wavelength region [22]. In comparison, silicon's absorption edge is near 1.1 μm. There are practical problems, however, in using germanium for making PIN photodiodes. The germanium photodiodes typically suffer a large surface leakage current [23], in addition to the dark current arising from generation-recombination across its small bandgap. The large dark current can degrade the signal-to-noise ratio [24]. So far, no satisfactory surface passivation technique is available for germanium PIN photodiodes, and the surface leakage current of unpassivated Ge photodiode tends to be very high and unstable at high ambient temperature [25].

In contrast, III-V compound semiconductor materials are popular detection materials for fiber-optics applications [26, 27]. By properly selecting the material composition of III-V materials, they can be made lattice matched to each other. As noted earlier on heterostructure PIN photodiodes, one can choose material composition of the various layers such that the photon energies lie between that of intrinsic region (smallest) and the $p$ and $n$ regions. This ensures photodiodes with a high quantum efficiency, a high response speed, and a low dark current. The materials of interest are the ternary alloys $In_{0.53}Ga_{0.47}As$ and the quaternary $In_xGa_{1-x}As_yP_{1-y}$ alloys lattice matched to InP. As in the case with germanium, unpassivated InGaAs surfaces also exhibit an unstable dark current. However, the leakage can be reduced by epitaxial regrowth, which buries the absorption region with high bandgap [28] and high resistivity material, for instance, Fe-doped InP [29]. Alternatively it can be reduced by passivating the surface with dielectrics [30]. PIN detectors with excellent performances in the 1–1.6-μm wavelength range have been demonstrated in the InGaAsP/InP material system.

Two generic InGaAs/InP PIN photodiodes, top-illuminated and back-illuminated photodiodes, are commonly used [31, 32]. A simple front-illuminated planar diffused InGaAs/InP homojunction photodiode is shown in Fig. 19.4**a**, where a layer of $In_{0.53}Ga_{0.47}As$ with an absorption edge at 1.65 μm is grown on top of the InP material. The residual background doping level for a metal-organic vapor phase epitaxial (MOVPE) grown InGaAs layer is typically in the low $10^{-14}$ cm$^{-3}$ range. The $p$-type dopant introduced in the $p$-layer during epitaxy can diffuse and raise the background doping concentration of the intrinsic layer. A thin (on the order of 0.5 μm) $p$-type layer can also be formed by shallow zinc diffusion or beryllium ion implementation [31, 33]. Additionally, a mesa (or mesa trench) is subsequently etched to provide electrical isolation and define the contact and the light coupling area. Surface passivation material and anti-reflection coating layer are then deposited. Dark currents below sub-nanoamperes have been obtained for 50-μm-diameter devices. Speeds higher than 20 GHz have been achieved. Devices with long lifetime and good thermal stability

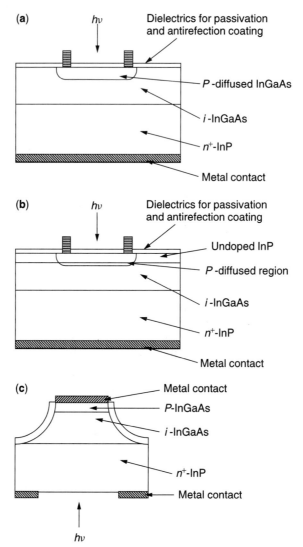

**Figure 19.4**   Schematic diagram of InGaAs/InP PIN photodiode. (**a**) Single heterostructure PIN photodiode with front illumination through the thin *p*-InGaAs layer. (**b**) InP/InGaAs/InP PIN photodiode with a window for front illumination. (**c**) Back-illuminated PIN photodiode with side-wall passivation.

(the dark current versus temperature characteristics has a typical activation energy of 0.42 eV) have been obtained. A double mesa PIN photodiode was popular in the past. The disadvantage of this simple structure is that the second mesa used for contact formation introduced additional junction capacitance and thus dark current to the photodiodes [28].

A variation of the preceding structure is the front-illuminated double-heterojunction photodiode shown in Fig. 19.4**b**. The light coupled from the top is absorbed directly at the InGaAs active layer. This structure has a higher quantum efficiency than that of the homojunction diode.

Backside optical coupling is possible for material systems with substrate transparent to the optical beam. Figure 19.4c illustrates such a structure. In this case, the active-layer thickness is selected such that at the operating bias, the electric field reaches through to the $n^+$ InP material. In principle, this structure has the same basic advantages as the heterojunction front-illuminated photodiode, although the quantum efficiency can be reduced by the free-carrier absorption in the relatively thick substrate. This absorption can be reduced by using a lightly doped or undoped substrate. The main advantage of this configuration is that the front-side contact pad is no longer a concern for light coupling to the photodiode. Therefore, for the same InGaAs optical detection area, the back-illuminated diodes have lower capacitance and lower dark current than those of the front-illuminated diodes.

Planar photodiodes with a $p$-diffused junction have been shown to have low leakage and high reliability. Typically, the heterostructure layers are grown undoped or $n$-doped, and the $p$ dopant, such as zinc, is locally diffused from the topmost layer. A schematic diagram is shown in Fig. 19.4b. A wide mesa outside the junction region is commonly etched to confine the current path. As mentioned, the bandwidth of the conventional surface normal photodiode was largely limited by the tradeoff between the transit time and the $RC$ time. A new approach to enhance the bandwidth of PIN photodiode has been proposed and demonstrated by $T$. Ishibashi [34]. In this approach, the electron-hole pairs are generated in the low-field $p$-region and electrons diffused to the intrinsic region (carrier collection layer in Fig. 19.5) are swept out at high velocity while the holes, as majority carriers, respond quickly to the electron flow to maintain quasi-neutrality. This results in both fast rise and fall time. A faster carrier transport eases the layer thickness design for small capacitance. A further advantage is that, because the hole is absent within the intrinsic region, the carrier screening effect due to hole accumulation at the interface between the $p$ layer and the intrinsic layer can be greatly reduced, and thus, very high optical power operation becomes feasible. Optical saturation power as large as 63 mW has been reported up to 40 GHz, and the optical saturation is attributed to the space charge effect in the collector layer [35]. In the structure shown in Fig. 19.5 where the $p$-InGaAs is the absorption layer and the $i$-InP is the collection (intrinsic) layer, the bandwidth can be limited by the diffusion velocity in the absorption layer. Because diffusion proceeds in both directions

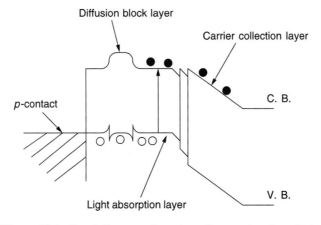

**Figure 19.5**  Band diagram of a unitraveling-carrier photodiode.

(i.e., toward the contact and intrinsic regions), a higher bandgap diffusion barrier near the contact can block the electron diffusion toward the $p$-contact. Using the diffusion blocking layer shown in Fig. 19.5, the average diffusion velocity of electrons in the $p$ layer is enhanced to $\sim 2D_n/L_a$, where $L_a$ is the thickness of the $p$-type absorption layer [see Equation 19.6 and the subsequent comment for fast diffusion velocity]. A graded bandgap or doping profile in the $p$ layer can lead to a small electric field that helps to drift the electron and enhances the bandwidth. Pulsed operation with $f_{3\,dB}$ as high as 310 GHz has been demonstrated in the uni-traveling carrier photodiode [36].

### 19.3.2  Schottky Photodiodes

Schottky photodiodes typically consist of an undoped semiconductor layer (usually $n$ type) on bulk material with a metal layer deposited on top to form a Schottky barrier. An example is shown in Fig. 19.6 [37]. As a strong electric field is developed near the metal–semiconductor interface due to the Schottky barrier, photogenerated holes and electrons are separated and collected at the metal contact and in the bulk semiconductor materials, respectively. By increasing the reverse bias to these diodes, both the peak electric field and the depletion region thickness inside the undoped layer can be increased, and thus, the responsivity can be increased. The complete absorption of optical signals within the high-field region ensures high-speed performance, because absorption outside the high-field region generates slow hole diffusion current. Again, the undoped layer should not be too thick, as the photodiode response speed can then become transit time limited.

In comparison with PIN photodiodes, Schottky photodiodes are simpler in structure. Unfortunately, although good Schottky contacts are readily made on GaAs, for long-wavelength InGaAsP/InP materials, stable Schottky barriers are difficult to make [38]. An alternative way to achieve Schottky photodiode in this material system is to make use of Au-Schottky contact on top of InAlAs lattice-matched to InP. Schottky barrier height of $\sim 0.82$ eV have been obtained, and the resulting photodiode shows $\sim 1$-nA leakage current at 5-V reverse bias [39]. On the other hand, GaAs Schottky photodiodes with a 3-dB bandwidth larger than 100 GHz and operating at less than 4-V reverse bias have been reported [40]. This high speed is attained by restricting the photosensitive area to a very small mesa ($5 \times 5 \ \mu m^2$) to minimize capacitance and by using a semitransparent platinum Schottky contact such that good responsivity can be achieved with top-side illumination.

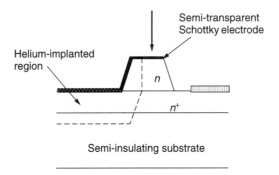

**Figure 19.6**   Schematic diagram of a Schottky photodiode on semi-insulating substrate.

### 19.3.3 Avalanche Photodiodes

Avalanche photodiodes (APDs) are important for imaging applications in which high resolution, high sensitivity, low light level operation are required [41]. APD is similar to PIN photodiode except that it is operated at a high reverse bias such that carrier multiplies through interband impact ionization in the high field region. When free carriers created by photoabsorption are accelerated by the electric field, they gain sufficient energy to excite more electrons from the valence band to the conduction band, thus giving rise to an internal current gain [42–44]. Avalanche photodiodes usually require a higher bias voltage than PIN photodiodes to maintain the high electric field. Also, the current gain is not a linear function of the applied voltage and is sensitive to temperature [45]. Figure 19.7a shows a schematic diagram of a silicon avalanche photodiode that consists of an $n^+$ contact, a $p$-type region, a $p$ drift region, and a $p^+$ contact. The corresponding electric field profile is shown in Fig. 19.7b. In operation, under the high reverse-bias voltage, the depletion region extends from the $n^+$ to the $p^+$ contact. Avalanche multiplication occurs at the $n^+$-$p$ junction. Figure 19.7b shows that when an electron-hole pair is photogenerated, electrons are injected into

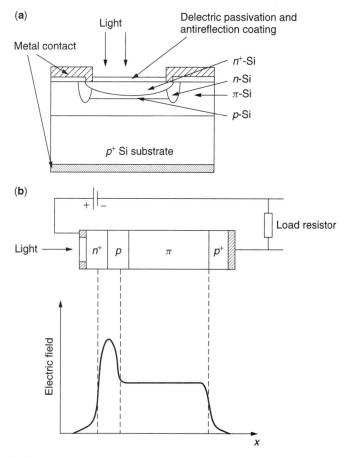

**Figure 19.7** Schematic diagram of a silicon APD (**a**) structure and (**b**) electric field distribution. (From Ref. 44 with permission.)

the multiplication region if light is mostly absorbed to the right of the multiplication region. Similarly, hole injection occurs for light absorbed to the left. Mixed injection of both carriers occurs when light is absorbed within the multiplication region. The doping profile and the thickness of the various layers must be optimized for each material system to achieve the highest performance. It is important that carriers of higher ionization coefficient are injected to the multiplication region in order to have the minimum multiplication noise. For instance, silicon APDs are generally designed so that electrons are the primary carriers that undergo multiplication as electrons in silicon have a larger ionization coefficient than holes. Consequently, both the hole injection or mixed injection are to be avoided [46, 47]. For low noise silicon APDs, a long drift region is incorporated to ensure high quantum efficiency and pure electron injection. The guard ring in the planar structures shown in Fig. 19.7**a** prevents edge breakdown at the perimeter of the multiplying region [19]. Reverse-bias voltage of up to 400 V is common for this kind of diode because of the large voltage drop across the long drift region.

The electric field required for impact ionization depends strongly on the bandgap of the material. The minimum energy needed for impact ionization is known as the ionization threshold energy $E_i$ [48, 49]. $E_i$ of electrons and holes in Si, Ge, GaP, GaAs, and InSb at different crystal direction are listed in [49]. The ionization energy influences strongly the ionization rates (or coefficients) for electrons ($\alpha$) and holes ($\beta$). These quantities are defined as the reciprocal of the average distance traveled by an electron or a hole, measured along the direction of the electric field, to create a secondary electron-hole pair. In other words, $1/\alpha$ and $1/\beta$ are the mean free paths of the secondary ionization of electrons and holes, respectively.

As in the case of the PIN diode, the APD's usefulness for communication critically relies on its noise properties. The noise currents in a photodiode arise from the random motions of charge carriers, thermal noise, and randomness in the arrival times of the carriers at the collecting contacts. These constitute the shot noise. The mean square shot noise $\langle i_s^2 \rangle$ due to the signal current, $I_p$, background radiation current, $I_B$, and dark current, $I_D$, in the presence of an average multiplication, $M$, is given by:

$$\langle i_s^2 \rangle = 2qB(I_P + I_B + I_D)\langle M^2 \rangle \tag{19.9}$$

where $B$ is the bandwidth. As the signal power is proportional to the square of $I_p$ and the square of $\langle M \rangle$, the shot noise limited S/N ratio is determined, as the avalanche gain increases, by the behavior of the statistical variation of $M$. The $\langle M^2 \rangle$ in the above expression is commonly written in terms of $\langle M^2 \rangle$ times an excess noise factor $F$. If there is no variance in the multiplication rate, the excess noise factor is unity. By increasing $\langle M \rangle$, then, the noise will be dominated by the shot noise (rather than the thermal noise). This condition is also known as the quantum noise limited detection. (Note that in this case, the S/N improves as $I_p$ increases, until the noise due to laser intensity fluctuation becomes the dominant noise.)

The multiplication depends on a three-body collision process and is consequently statistical in nature. As a result, it contributes a statistical noise component in addition to the shot noise already present in the diode [47–49]. Randomness in the multiplication process produces the greatest noise when the electron and hole ionization coefficients are equal. This excess noise has been studied for the case of an arbitrary $\alpha/\beta$ ratio for both uniform and arbitrary electric field profiles. For the uniform electric field case, the excess noise factors for electronics, $F_n$, and holes, $F_p$, are given by McIntyre's local field model as [50]:

$$F_n = M_n \left[ 1 - (1-k) \left( \frac{M_n - 1}{M_n} \right)^2 \right] \qquad (19.10a)$$

$$F_p = M_p \left[ 1 - \left( 1 - \frac{1}{k} \right) \left( \frac{M_p - 1}{M_p} \right)^2 \right] \qquad (19.10b)$$

where $k$ is less than 1 and is equal to $\beta/\alpha$ or $\alpha/\beta$ and $M_n$ and $M_p$ are the dc multiplication factors for pure electron and pure hole injection, respectively. For cases where $k$ is not constant, $F_n$ and $F_p$ are given as

$$F_n = k_{eff} M_n + \left( 2 - \frac{1}{M_n} \right) (1 - k_{eff}) \qquad (19.10c)$$

$$F_p = k'_{eff} M_p + \left( 2 - \frac{1}{M_p} \right) (1 - k'_{eff}) \qquad (19.10d)$$

where $k_{eff} \approx k_2$ and $k'_{eff} = k_2/k_1^2$ depending on the spatial averages $k_1$, $k_2$ of $\alpha$ and $\beta$ and

$$k_1 = \frac{\displaystyle\int_0^{W_a} \beta(x) M(x) \, dx}{\displaystyle\int_0^{W_a} \alpha(x) M(x) \, dx} \qquad (19.11a)$$

$$k_2 = \frac{\displaystyle\int_0^{W_a} \beta(x) M^2(x) \, dx}{\displaystyle\int_0^{W_a} \alpha(x) M^2(x) \, dx} \qquad (19.11b)$$

where $W_a$ denotes the thickness of the multiplication region.

Thus, from Equations (19.10a) and (19.10b), to realize low-noise APDs, the ionization rate for one type of carrier must be much greater than that of the other, and the carrier with the larger ionization coefficient must initiate the avalanche process.

The dc avalanche multiplication gain of an APD is expressed as

$$M_n = \frac{1}{1 - \displaystyle\int_o^{W_a} \alpha \exp\left[ - \int_0^{W_a} (\alpha - \beta) \, dx' \right] dx} \qquad (19.12a)$$

$$M_p = \frac{1}{1 - \displaystyle\int_o^{W_a} \beta \exp\left[ \int_0^{W_a} (\alpha - \beta) dx' \right] dx} \qquad (19.12b)$$

where $M_n$ and $M_p$ denote the current gain for electron and hole injection, respectively. For cases where $\beta = 0$, Equation (19.12a) can be reduced to

$$M_n = \frac{1}{1 - \int_o^{W_a} \alpha \exp(-\alpha x)\, dx} \tag{19.13}$$

For a uniform field, $\alpha$ is constant and $M_n$ of Equation (19.13) increases exponentially with the number of ionizing carriers in the high-field region, and there is no sharp breakdown. Figure 19.8a depicts the avalanche buildup process under the $\beta = 0$ condition. A current pulse is induced when the photogenerated electron starts to drift toward the multiplication region. This pulse increases in magnitude during the electron transit time through the high-field region due to the electron impact ionization. Then it is reduced to zero when the last hole is swept outside the high-field region. Neglecting the transit time inside the absorption region, the current pulse width is approximately the average of electron and hole transit times and decreases a little as the gain increases. The corresponding excess noise $F$ [see Equations (19.10a) and (19.10b)] is small because as M increases, statistical variation of the impact ionization

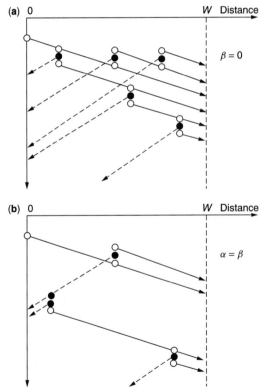

**Figure 19.8**  Physical representation of the avalanche buildup process with **(a)** $\beta = 0$ (electron multiplication only) and **(b)** $\alpha = \beta$ (both carriers multiply). Open circles denote electrons, and filled circles denote holes. (From H. Melchoir, in F. T. Arecchi and E. O. Schulz-Dubois, Eds., *Laser Handbook*, Vol. 1, North-Holland Amsterdam, 1972, with permission.)

process only causes a small fluctuation in the total number of carriers [48]. In fact, when $\beta = 0$ and pure electron injection occurs in the multiplication region, $F_n \approx 2 - 1/M_n$.

When both carriers ionize at the same rate (i.e., $\alpha = \beta$), the APD becomes quite noisy. For the sake of simplicity, in the case of a uniform electric field, the current gain becomes

$$M_n = M_p = \frac{1}{1 - \alpha W_a} \tag{19.14}$$

As $\alpha W_a$ approaches unity, a sharp breakdown ($M$ diverges) occurs and results in unstable device performance. This corresponds to the situation when each electron or hole-injected produces one the average one secondary electron-hole pair during the transit through the high-field region, as depicted in Fig. 19.8**b**. The secondary hole generated transverses in the $-x$ direction and generates further electron-hole pairs, and the electron repeats the process. This process persists for a long time when the gain is large. Theoretically speaking, during breakdown ($M \to \infty$), an optical pulse can produce a dc current. It is clear that for a large gain, there are fewer ionizing carriers in the high-field region than the case when $\alpha \gg \beta$ (or $\beta \gg \alpha$). Thus, any statistical variation in the impact ionization process will produce large fluctuation in the gain and cause considerable excess noise.

The gain can be a function of signal frequency. The overall low-frequency gain, $M_o$, of an APD can be empirically expressed as a function of bias voltage [51]:

$$M_o = \frac{I}{I_p} = \left[ 1 - \left( \frac{V - IR}{V_n} \right)^n \right]^{-1} \tag{19.15}$$

where $I$ is the multiplied diode current, $I_p$ is the primary photocurrent, $V_B$ is the breakdown voltage, $V$ is the applied voltage, and $R$ is the sum of the diode series resistance and the load resistance. For $V_B \gg IR$, a maximum for $M_o$ can be obtained:

$$M_o(\text{max}) = \left( \frac{V_B}{n I_p R} \right)^{1/2} \tag{19.16}$$

The high-frequency gain can be approximated by

$$M(\omega) = \frac{M_o}{[1 + \omega M_o \tau_{eff}]^{1/2}} \tag{19.17}$$

where $\omega$ is the angular frequency and $\tau_{eff}$ is approximately equal to $[\beta\alpha]\tau_{av}\delta$, where $\delta$ is a parameter in the range of 1/3 (at $\beta = \alpha$ to 2 (at $\beta \ll \alpha$) and $\tau_{av}$ is the average transit time. Figure 19.9 depicts the theoretical gain-bandwidth product as a function of $M_o$ for various $\alpha/\beta$ values [42].

The applications of APDs are very much related to the receiver design. Consider the signal-to-noise ratio of an APD receiver:

$$\frac{S}{N} = \frac{\text{signal}_{\text{power}}}{\text{rnoise}_{\text{receiver}} + \text{noise}_{\text{APD}}}$$

$$= \frac{(q\eta P_o M / h\nu)^2}{\dfrac{4kTB}{R_{eq}} + 2qB(I_P + I_B + I_D)M^2 F} \tag{19.18}$$

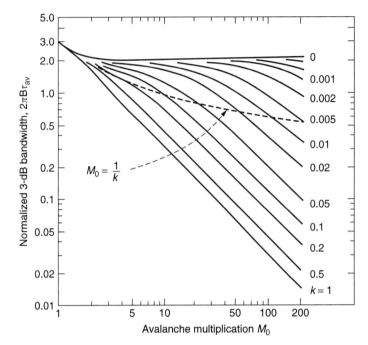

**Figure 19.9** Normalized 3-dB bandwidth as a function of average avalanche multiplication $M_o$ for several values of the ionization rate ratio $k$. For pure injection, $k = \beta/\alpha$; for pure hole injection, $k = \alpha/\beta$; $\tau_{av}$ is the average of the electron and hole transit times in the avalanche region. (From Ref. 42 with permission.)

where $\eta$ is the quantum efficiency at unity gain*, $P_o$ is the optical power incident on the diode, $4kT/R_{eq}$ is the combined Johnson noise of the amplifier, external load, and the junction resistance, and $F$ is the excess noise factor. Using Equations (19.10a) and (19.10b), one can calculate the optimum gain for certain receiver designs from Equation (19.18) and its derivative with respect to $M$. As the noise of an APD goes up faster than the signal term ($M^{2+x}$ versus $M^2$, where $x$ is between 0 and 1), once the APD noise becomes the dominant term in the denominator of Equation (19.18), the signal-to-noise ratio actually degrades with a further increase in gain. This sets the usable gain of the APD.

Silicon APDs are used in the wavelength range of 0.4–1.0 μm because of the low dark current and a large $\alpha/\beta$. The structure shown in Fig. 19.7a was employed in the case of silicon APDs to obtain carrier multiplication with very low excess noise. The high-field $n^+p$ junction served as electron-initiated avalanche region and can be formed by diffusion or ion implantation. At a bias below the breakdown of the $n^+p$ junction, the $\pi$ region will be depleted. Beyond this point, the voltage dropped across the wide $\pi$ region increases with bias faster than that across the $n^+p$, thus increasing the operating bias range for multiplication. The full depletion of the $\pi$ region is essential for high response speed, as photogenerated electrons within the $\pi$ region

---

* For SAM APD, this refers to the $\eta$ at the gain after reach-through occurs (full depletion of the absorption layer).

drift to the $n^+p$ region with high velocity. Holes, on the other hand, drift opposite to the electrons, and thus, a pure electron current is injected into the multiplication region. (Although there exist some hole injections due to the absorption in the $n^+$ region and the avalanche region, they are relatively much smaller than the electron current due to the absorption in the long drift region.) An excess noise factor of 4 at a gain of 100 has been achieved in such an APD [52]. The response speed is limited mostly by the transit time across the $\pi$ region. To increase the response speed without increasing the operating voltage, a built-in field in the $\pi$ region can be achieved by grading the doping profile so as to increase the carrier velocities [53]. Another approach involves the deposition of an antireflective layer at the back contact to increase the quantum efficiency of a relatively narrow $\pi$ region [54]. Response times of 150–200 ps have been achieved.

For long-wavelength application, germanium APDs have been used because silicon cannot respond to wavelengths beyond 1.1 $\mu$m [51]. As can be seen from Fig. 19.10, $\beta/\alpha$ ratio for germanium has a maximum value of 2. A relatively high excess noise factor ($F \approx 0.5$ M), in addition to a relatively high dark-current noise, is observed in germanium APDs due to the small bandgap and large surface leakage. For typical germanium APD receiver, the usable gain is limited to around 10. Beyond this, the signal-to-noise ratio degrades because the signal power increases as $M^2$ while the excess noise increases as $M^3$ [see Equation (19.18)]. By reducing the detection area of the APDs, dark current in the nanoampere range can be achieved.

Attractive alternatives to germanium APDs for long-wavelength applications include III-V APDs because the material bandgap can be tailored to achieve both high absorption and low dark current. Early APDs in the InGaAs and InGaAsP systems have shown low usable gain because the difference in $\alpha$ and $\beta$ in these materials is not as large as that in silicon. For homojunction InGaAs devices, the dark current increases exponentially with reverse bias as the bias approaches avalanche breakdown. This is attributed to the tunneling current, which is more profound in materials with a small bandgap. To reduce the tunneling current, a heterostructure that separates the absorption region from the multiplication region depicted in Fig. 19.11 is commonly adopted [55]. This is known as separate absorption and multiplication (SAM) APD. With this arrangement, the low-bandgap InGaAs material will not experience a high electric field, and thus, the tunneling effect is reduced. The absorption in this structure takes place in the low-bandgap material, which, like the silicon APDs, is low doped, while the multiplication takes place in the wide-bandgap $pn$ junction, which usually consists of InP. Typically, between the multiplication and the absorption region, a charge layer with proper combination of doping concentration and thickness is inserted to regulate the electric field distribution in the APD [56]. Consequently, the noise of the SAM APDs reflects much of the multiplication process in InP [57]. Reasonably small unmultiplied dark currents are observed for APDs with an InGaAsP absorption region (nanoampere range) [58] and InGaAs absorption region (tens of nanoamperes) [59], while the gain is maintained in the range of 10–60. Trapping of holes at the InGaAs/InP heterojunction was observed, which greatly degrades the speed of the APD. To overcome this problem, an additional layer of quaternary InGaAsP with intermediate energy bandgap (or multiple layers of graded bandgap region) is incorporated. A relatively low-noise (as compared with germanium) APD can be achieved with this design. For instance, at a gain of 10, APDs with an excess noise figure of 5, a dark current of 20 nA, and a bandwidth in excess of 8 GHz have been demonstrated.

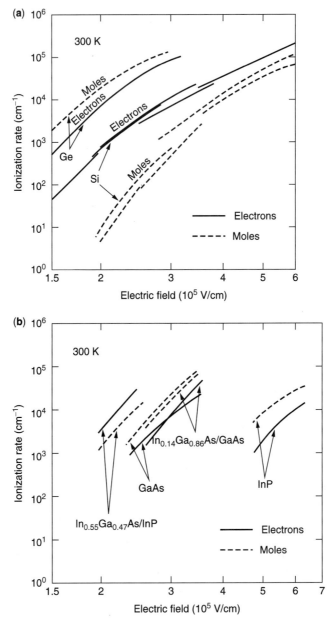

**Figure 19.10** Experimentally obtained carrier ionization rates at 300 K for **(a)** silicon and germanium and **(b)** GaAs, InP and $In_{0.14}Ga_{0.86}As/GaAs$ and $In_{0.53}Ga_{0.47}As/InP/$. (From H. Melchoir, Physics Today, Vol. 30, p. 32, 1977 with permission.)

Alternatives designs to the current III-V SAM APDs to limit the excess noise have been studied. These structures aim at achieving a small effective $k$ values via various means such as utilizing the dead space effect, superlattice structure and heterojunctions [60–62]. A smaller $F$ is demonstrated in structures with thin multiplication layers. The idea is to build up the energy of the carriers (without ionization) in thin

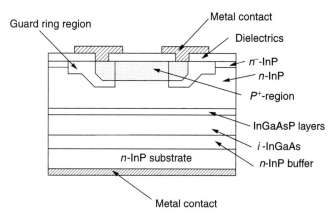

**Figure 19.11**  InGaAs/InP with separated absorption and multiplication regions. The intermediate bandgap InGaAsP layers reduce the hole trapping effect.

layer of high bandgap and high ionization threshold energies before injecting them
into thin layer of small bandgap and low threshold energies (where impact ionization
occurs). This produces carriers with high initial energy and reduced dead space for
multiplication. The localization of the impact ionization reduces the randomness of the
multiplication process and thus reduces $F$.

A simpler approach is to use the wafer fusing technique [63] to form InGaAs/Si
SAM-APD in which the absorption occurs in InGaAs and multiplication occurs in
Si, which has a small $k$ value. A high gain-bandwidth product of 315 GHz had been
reported in the fused APD [64]. Low excess noise performance has been demonstrated
in the wafer-fused InGaAs/Si APD using low stress bonding to reduce the misfit density
at the InGaAs/Si interface [65]. Figure 19.12 shows a schematic layer structure of the
fused InGaAs/Si SAM-APD. The $p$-$n$ junction in silicon is formed by epitaxy or ion
implantation. The sheet charge density of the $p$-Si layer is around $1.8 - 3 \times 10^{12}$ cm$^{-2}$.
During operation, most of the voltage is dropped across the silicon junction where
multiplication takes place, while the electric field in InGaAs is sufficiently high to fully
deplete the layer and to sweep out the photogenerated electron-hole pairs. Electrons
drifting into the multiplication region initiate the impact ionization. Inside the high
field region, electron ionization dominates the impact ionization due to the highly
asymmetric multiplication process in silicon.

Figure 19.13 shows the excess noise factor $F(M)$ extracted from a room temperature
noise measurement of the fused InGaAs/Si SAM-APD (square symbols). The solid lines
are $F(M)$ versus $M$ plots for different $k$'s obtained from McIntyre's model mentioned
above. The $k = 0.01$, and 0.02 and 0.04 curves are typical for conventional silicon
APDs, while the $k = 0.2$ and 0.4 ones are typical for InGaAs/InP SAM-APDs. The
measured excess noise factor of the InGaAs/Si is much smaller than that of conventional
InP-based APDs, and it falls on the curve with effective $k \sim 0.02$, indicative of the
avalanche process inside the silicon $p$-$n$ junction.

Other material systems have also been investigated for avalanche photodiode applications. For instance, GaAlSb/GaSb materials operating at 1.55 um wavelength have
shown very high gain. More interestingly, the bandgap of this alloy is very close
to covalence band spin orbit splitting energy [66]. This greatly enhances the hole

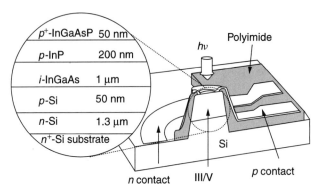

**Figure 19.12**    Schematic diagram of the fused InGaAs/Si APD. The top InGaAsP is transparent to 1.3–1.55-μm light.

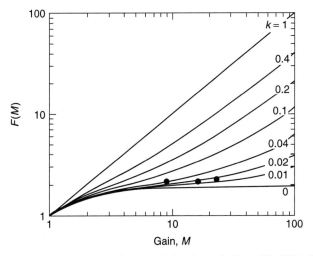

**Figure 19.13**    Measured excess noise factors of the fused InGaAs/Si APD. The squares are the excess noise factor values obtained from measurements, and solid lines are those calculated from McIntyre's model at different $k$'s. (From [65], with permission.)

multiplication process and leads to very small $k = \alpha/\beta (< 0.05)$. However, the dark current (mostly due to surface recombination) of this material system is too high at this moment.

### 19.3.4   Waveguide Photodiodes

Major tradeoffs in the design of analog wide bandwidth photodiode can be related to the intrinsic layer thickness, which links directly to the quantum efficiency, the carrier transit time, and the RC circuit time constant [5]. For conventional surface normal photodiodes, the intrinsic layer thickness needs to be thin for reducing the transit time; however, in doing so, the quantum efficiency is reduced and the capacitance is increased. To reduce the RC time effect, the detection area needs to be reduced

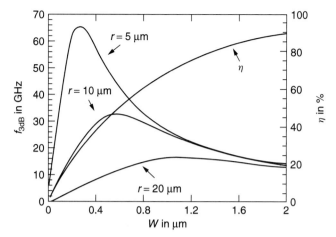

**Figure 19.14** The $f_{3\,dB}$ vs intrinsic layer thickness for PIN photodiode with surface normal illumination; $r$ is the diode radius.

and this can reduce the quantum efficiency and thus the responsivity, as illustrated in Fig. 19.14 [67].

Waveguide photodiode was proposed as a means to ease the tradeoff between the speed and responsivity [68]. In this configuration, the detector is generally edge-illuminated with light guided parallel to the intrinsic layer and incrementally absorbed. The optical waveguiding structure is similar to that of the double heterostructure lasers. An example of the waveguide photodiode for 1.3- and 1.5-$\mu$m detection is shown in Fig. 19.15, where the mesa structure is used to confine the light in the lateral direction. Edge-coupling and optical waveguiding increases the interaction length of the light inside the absorption layer. The responsivity is mainly limited by the coupling efficiency at the fiber-to-waveguide interface, and the scattering loss along the waveguide. In comparison with the surface normal photodiode, the waveguide photodiode typically has a smaller responsivity due to the modal mismatch between the waveguide and the single mode fiber. However, due to the edge-illumination, the responsivity is mainly determined by the waveguide length, and not by absorption layer thickness; therefore, the waveguide detector has a larger bandwidth-efficiency performance than the surface normal photodiode. A quantum efficiency as high as 68% has been achieved for a waveguide photodiode with a $f_{3\,dB}$ greater than 50 GHz [68]. In contrast, a conventional PIN photodiode with a quantum efficiency of 68% will be limited to a $f_{3\,dB}$ of about 30 GHz.

The waveguide photodiode shown in Fig. 19.15 consists of a rib-loaded PIN waveguide integrated with a coplanar waveguide transmission line on a semi-insulating InP substrate. The rib is 5 $\mu$m wide at the top and 20 $\mu$m long, and the intrinsic InGaAs layer is 0.35 $\mu$m thick. For waveguide photodiode, the $f_{3\,dB}$ is limited by the transit time across the intrinsic region and the $RC$ time effect. Using a mushroom-mesa structure to reduce the $RC$ time constant (through reducing the contact resistance while keeping the junction area small), 100-GHz bandwidth waveguide photodiode with a quantum efficiency of 50% has been demonstrated [5].

As noted earlier, the waveguide configuration can pose a limitation to the responsivity due to the fiber coupling. The coupling can be improved through the use of

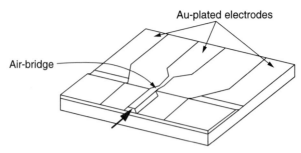

**Figure 19.15**   Schematic diagram of a waveguide photodiode integrated with a coplanar waveguide transmission line. Light is edge-coupled to the waveguide, and the *p*-electrode on top of the waveguide is connected to the central electrode via an air bridge.

lensed fibers, antireflection coating at the waveguide facet, laterally tapered waveguide section, or large optical waveguide [69].

***Traveling Wave Photodiode and Velocity-Matched Photodiode.*** Lumped element designs, including the surface normal and waveguide photodiodes discussed earlier, are limited by the *RC* circuit time. As a means to overcome the *RC* bandwidth limitation, the distributed photodiode was first introduced in 1990 in the form of the "traveling-wave photodetector" for photoabsorption distributed along a microwave transmission line [70]. The optical waveguide structure of the traveling-wave waveguide photodiode is very similar to that of the ordinary wave photodiode described earlier. The device can be viewed as the terminating section of a transmission line with a position-dependent photocurrent source distributed along its length. The transit time limitation to the frequency response still persists, and as a transmission line element, the frequency response is related to the matching between the characteristic impedance of the transmission line and termination impedance. The monolithic design where light is continuously absorbed along the microwave waveguide is commonly referred to as the traveling-wave waveguide photodiode [71]. In an alternative design where light is absorbed in a set of discrete photodiodes distributed along the optical waveguide and the transmission line, the photodiode is referred to as the periodic traveling-wave photodiode [72].

Although the absorption behavior is similar in the two designs of traveling-wave photodiode, the microwave properties can be quite different; in particular, the microwave (phase) wave can be slower than the optical (group) wave in the traveling-wave waveguide photodiode and faster in the periodic traveling-wave photodiode; this mismatch affects the maximum response bandwidth [5]. The bandwidth can also be reduced due to the effect of reflected microwave from the termination of the transmission line. The open termination is bandwidth-limited by the interference between the forward and backward waves. The impedance-matched termination approach leads to a larger bandwidth but the available photocurrent can be reduced a factor of two (and the microwave power by a factor of four). The potential benefit of the traveling-wave photodiode, besides the large bandwidth, is the maximum waveguide length allowable (within the constraint placed by the impedance mismatch and the waveguide scattering loss) for absorption, which eases the design for the modal confinement at the absorber, a feature exploitable for high optical power handling. Bandwidth as high as 172 GHz has been reported for a GaAs/AlGaAs traveling-wave photodiode detecting in the 800–900-nm wavelength range, with an intrinsic GaAs region which is 1 μm wide and 0.17 μm thick [73].

The periodic traveling-wave photodiode can take the form of an array of individual photodiodes serially connected by long, passive waveguides and transmission lines. As the backward microwave seriously disturbs the forward microwave in a long transmission line, a match termination is also required. An advantage of the periodic configuration is the feasibility to perform the velocity matching between the microwave phase velocity and the optical group velocity using electrical delay line — the resulting structure has been referred to as the velocity-matched (VM) waveguide photodiode. The slow-wave line is achieved by periodic capacitive loading using the arrayed photodiodes [72]. The VM photodiode based on the GaAs-MSM photodiode has reached around 50 GHz and that based on InGaAs has reached 78 GHz (with an efficiency of 7.5%) [73].

Avalanche photodiode exploiting the waveguide geometry has recently been studied [74, 75]. The motivation here is again to optimize the 3-dB frequency bandwidth with good quantum efficiency, which is a limitation of the surface normal design. By employing large optical cavity waveguiding layers, a quantum efficiency of 78% and a gain bandwidth product of 180 GHz have been demonstrated [75].

## 19.4  PHOTOCONDUCTIVE DEVICES

Photoconductive devices depend on the change in the electrical conductivity of the material upon irradiation [76, 77]. Although photoconductive effects have been established for a long time, only recently are they employed for high-sensitivity, high-speed detection. In semiconductors, two basic types of photoconductors, extrinsic and intrinsic, are studied. In the intrinsic case, photoconduction is produced by band-to-band absorption. The absorption coefficient can be very large ($\sim 10^4$ cm$^{-1}$) in this case due to the large density of electronic states associated with the conduction and valence bands. In the extrinsic case, photons are absorbed at the impurity level, and consequently, free electrons are created in the $n$-type semiconductor and free holes are created in the $p$-type semiconductor. Extrinsic photoconduction is characterized by lower sensitivity because the absorption is limited by the smaller number of available states. These devices are usually operated at low temperatures (e.g., 77 K) to freeze out impurity carriers (thus reducing the background noise) so that they become available for optical absorption.

Consider the simple photoconductive device shown in Fig. 19.16. A voltage $V$ is applied across the electrodes, and light is illuminated from the top. The signal is detected either as a change in voltage across a resistor in series with the photoconductor or as a change in current through the sample. Frequently, the signal is detected as a change in voltage across a resistor matched to the dark resistance of the detector.

The photoconductive gain $M$ of these devices can be expressed as [77]

$$M = \frac{\tau_n}{t_n} + \frac{\tau_p}{t_p} \tag{19.19}$$

where $\tau_n$ and $\tau_p$ are the lifetimes of the excess electrons and holes, respectively, and $t_n$ and $t_p$ are the corresponding transit times across the device. For cases where electron

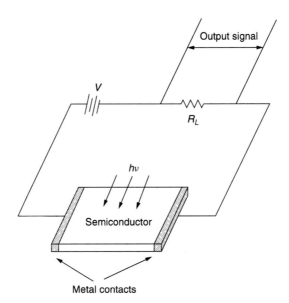

**Figure 19.16**   Schematic diagram of a simple photoconductor; $R_L$ represents the load resistor.

conduction is dominant and the response time of the device is $\tau$, the gain-bandwidth product can be expressed as

$$GB = \frac{\tau_n}{t_n}\frac{1}{\tau} \tag{19.20}$$

For material where carrier trapping is not significant, the response time is determined by carrier lifetimes. However, as the device dimension gets smaller, the response time can be altered by the nature of the metal contact and the method of biasing. When a Schottky barrier is formed at the electrode–semiconductor interface and its depletion region extends across the length of the device to reach the other electrode, the device operates more or less like a photodiode with $\tau_n$ replaced by $t_n$ and its GB value is then equal to $1/\tau$ [78].

When a steady optical beam with power $P_o$ and wavelength $\lambda$ is incident on the photoconductor, the optical generation rate of electron-hole pairs versus the distance $x$ into the sample is given by

$$g(x)\,dx = \frac{(1 - R_o)P_o}{h\upsilon}\alpha_o e^{-\alpha_o x}\,dx \tag{19.21}$$

The quantum efficiency $\eta$ is

$$\eta = (1 - R_o)(1 - e^{-\alpha_o D}) \tag{19.22}$$

where $D$ is the effective sample thickness. For an intrinsic photoconductor at steady state, the average excess electron and hole concentrations are

$$\Delta n = \Delta p = \frac{P_o \eta \tau}{W L D h\upsilon} \tag{19.23}$$

where $W$, $L$ are the respective width and length dimensions of the device.

The change in conductivity, $\Delta\sigma$, due to $P_o$ is given by

$$\Delta\sigma = q(\Delta n\mu_n + \Delta p\mu_p) \qquad (19.24)$$

where $\mu_n$ and $\mu_p$ are the electron and hole mobilities, respectively.

For many signal-processing applications, the ratio of the on-resistance to the off-resistance of the photoconductive device is an important parameter. The off-resistance is primarily limited by the material resistivity in the dark. The on-resistance, as discussed earlier, depends on the carrier mobility and the excess carriers generated.

Due to low surface recombination velocity of InP-related materials, planar photo-conductivity devices on iron-doped semi-insulating InP have been extensively studied [79]. This material has a room temperature resistivity of $10^5$–$10^8$ $\Omega$-cm and an electron mobility of 1500–4500 cm$^2$/V-s depending on the crystal quality and iron concentration. The carrier lifetime ranges from less than 100 ps to 3 ns [79, 80].

A typical photoconductor structure consists of either (1) a single narrow gap made on a metal stripe deposited on semi-insulating InP [81] or (2) interdigitated electrodes deposited on the substrate [80]. Response times as short as 50 ps have been achieved in both devices. As noted earlier, the nature of metal contacts on semiconductors affects much of the device performance. For devices with nonalloyed contacts, a smaller photoconductive gain is observed. However, the response time, especially the fall time, is shorter than that of devices with alloyed contacts. Such alloyed contact devices show lower contact resistance and a higher photoconductive gain. Various contact materials (such as Au/Sn and Ni/Ge/Au) as well as alloying conditions [82] (such as alloying temperature and duration) have been investigated. Compared to photodiodes, the planar photoconductive detector has advantages such as ease of optoelectronic integration, especially with field-effect transistor (FET) circuits. Also, detector arrays are more readily achieved with planar geometry photoconductors.

A photoconductive device can function as a current switch. However, this requires low on-resistance. For alloyed contact devices, the current-voltage characteristics are linear at low voltage, and the current tends to saturate at high voltage, in contrast to Equation (19.26). The current saturation is believed to be caused by electrons transferred from a high-mobility valley to a low-mobility valley, especially at the contact region where the electric field is strong [83]. The photoconductance, on the other hand, can be linearly dependent on the light intensity over several orders of magnitude, as shown in Fig. 19.17.

For 1.0–1.6-$\mu$m wavelength applications, InGaAs lattice matched to InP has been considered for photoconductive detectors [84]. It is an attractive material due to its high-saturation drift velocity. Compared to semi-insulating InP, however, semi-insulating InGaAs materials obtained so far tend to be less resistive ($\sim 10^5$ $\Omega$-cm), the typical dark resistance in the range of k$\Omega$, which compromises some of the device performance.

As InP can be used to detect radiation ranges from the near IR to the UV region with good sensitivity, it becomes very attractive for broadband detection [85, 86]. Together with its small surface recombination velocity, very fast detection (<90-ps rise time) has been achieved at the X-ray wavelengths where $\alpha_o \sim 10$ cm$^{-1}$. Such devices have been studied in conjunction with analog to digital (A/D) conversion where an analog electrical signal is time sampled and coded into a digital format [79]. In this application, photoconductive devices operate as a time switch triggered by laser pulses and serve

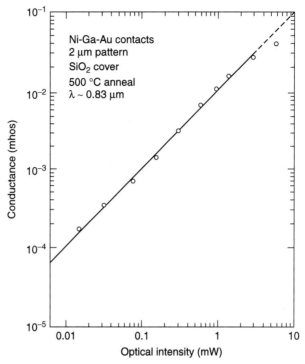

**Figure 19.17**  On-state conductance versus incident optical power ($\lambda \sim 0.85$ mm) for an InP optoelectronic switch with Ni-Ge-Au contacts. This device had a 2-$\mu$m finger pattern and was alloyed at 500°C with SiO$_2$ cover. (From A. G. Foyt and F. J. Leonberger in C. H. Lee, Ed., Picosecond Optoelectronic Devices, Academic Press, Orlando, Fl. 1984, with permission.)

for tracking of the analog signal voltage. When compared to electrical A/D conversion, this can lead potentially to less time jitter as well as higher sampling rates. Up to 6-bit resolution with a sampling rate greater than $1 \times 10^9$ samples per second has been predicted with InP photoconductive switches.

Another important application of the photoconductive device is the generation of high-power, picosecond duration, electrical pulses. In this case, the photoconductor operates as a switching element connected to a low-impedance transmission line and is terminated in a load resistor. With the switch in the off state, one side of the line is charged to a high voltage. Upon laser illumination, the switch is turned on, and a high-power electrical pulse propagates down the transmission line. The on-impedance of the switching element needs to be much smaller than the line impedance so that a small fraction of the voltage (and also power dissipation) dropped across the switching element. The carrier lifetime determines the maximum time window when the switch is in the on state. Silicon photoconductive switches have been used to generate gigawatt pulse power, nanosecond electrical pulses. The pulse duration depends on the length of the charging transmission line [87]. On the other hand, III-V semiconductor materials are inferior to silicon materials because of their relative shorter carrier lifetimes, which limits the duration of the electrical pulse. A photodiode switch has also been investigated as an alternative to the photoconductive switch for this application [88, 89].

## 19.5   NONLINEARITY IN PHOTODETECTORS

For externally modulated analog optical links, both the $RF$ efficiency and the spurious free dynamic range of the link can be improved by using a large optical intensity. In links where a PIN photodiode is used at the receiver, the linearity of the photodiode becomes critical in the consideration of the overall link performance. Consequently, it is of interest to obtain high-speed photodetectors that can be operated without saturation under a large optical intensity. At saturation, the fundamental signal can be accompanied by a high level of nonlinear distortions [16]. Possible causes of the saturation are absorption saturation, and electric field screening, which arise from the space charge effect in the intrinsic layer. The latter can be worsened due to the external circuit effect and the geometric effect.

### 19.5.1   External Circuit Effect

Severe absorption saturation occurs when there is a significant filling of the conduction and valence band states. For InGaAs materials, for instance, this corresponds to an electron concentration in excess of $2 \times 10^{17}$ cm$^{-3}$ and a hole concentration in excess of $10^{18}$ cm$^{-3}$. However, these concentrations are typically orders of magnitude higher than the carrier concentrations at which the space charge screening effect becomes significant [3]. The screening effect is accompanied by a substantial voltage drop across the series resistance outside the junction region of the detector. For example, for load resistance of 50 $\Omega$, a photocurrent of 10 mA can lead to a voltage reduction of 0.5 V across the intrinsic region. The voltage reduction increases as the photodiode current increases until it reaches a significant fraction of the applied bias (5 V or so) across the detector.

Both photodiode and photoconductor operate under a high electric field for sweeping out the photogenerated carriers. However, under a high flux of photogenerated carriers, the applied electric field can be countered by the local dipolar electric field resulting from the separating electrons and holes. The nonlinear response can arise from a redistribution of the electric field within the intrinsic region, which is accompanied by a spatially dependent hole and electron velocity distributions [90]. Consequently, the harmonic distortion levels increase at elevated optical intensities. At even higher intensity, the fundamental RF signal response can be affected. In principle, the screening effect can be neglected when the photodiode is under a large reverse bias. Figure 19.18 demonstrates the effect of bias to the frequency response of a waveguide photodetector at high optical flux. Figure 19.19 shows the fundamental and second harmonic signals at −3 and −5 V as a function of photocurrent. Additional nonlinear effect was observed at high bias that has been attributed to absorption at the highly doped contact region of the photodiode [90].

In the literature, nonlinear distortions of the PIN photodiode have been attributed to the nonlinear transport effects induced by the space charge. Numerical models have been developed to describe the nonlinear transport inside the device [91, 92]. Basically, these solve simultaneously the Poisson equation and the carrier continuity equations under various operating conditions. When the density of photogenerated electron and hole pairs reach a level high enough to partially screen the biasing electric field, a highly nonuniform carrier velocity profile can be resulted, and the photodiode is in severe saturation with high nonlinear distortion levels.

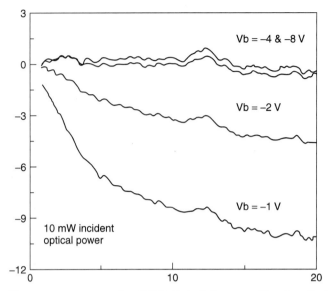

**Figure 19.18**   Frequency response of an InP/InGaAs/InP waveguide photodiode with 10-mW laser power and detector bias voltages of Vb = −8, −4, −2, −1 V. The horizontal axis is the frequency in unit of gigahertz, and the vertical axis is relative response in decibels.

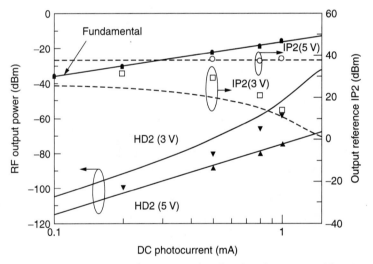

**Figure 19.19**   Fundamental and second harmonic signals of a waveguide photodiode, at 260 MHz and at −3 $(\nu, \tau)$ and −5 V $(\lambda, \sigma)$, respectively, versus photocurrent. The dots and solid curves are the measured and theoretical results, respectively. The calculated (dash line) and extracted output referenced IP2 at −3 V and −5 V are included.

Experimentally, the nonlinear distortion is studied by examining the harmonic levels of the photodiode with the photodiode subject to heterodyned lasers with wavelength slightly offset to generate a RF tone. Alternatively, beat signals are generated at the photodiode using three separate optical beams RF-modulated at frequencies $f_1$, $f_2$,

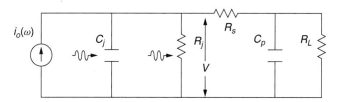

**Figure 19.20**   The small signal equivalent circuit of a photodiode under illumination.

and $f_3$, respectively: the 2nd-order intermodulation distortion signal at $f_1 + f_2$ and the 3rd-order intermodulation distortion at $f_3 - (f_1 + f_2)$ are measured [93].

There is an alternative physical approach to access the photodiode's harmonic and intermodulation distortion levels at different frequencies. This approach relies on the variation of the photodiode microwave impedance at various optical powers for determining the frequency response and nonlinearity behaviors. The microwave impedance can be extracted from measured microwave reflection coefficient ($S_{11}$) of the photodiode. The distortions can therefore be traced back to physical parameters of the photodiode. To illustrate this, an equivalent circuit of the PIN photodiode is shown in Fig. 19.20, where $i_o(\omega)$ represents the photocurrent at angular frequency $\omega$ inside the reverse-biased intrinsic region, $C_j$ and $R_j$ are the respective junction capacitance and the shunt junction resistance of the diode, $C_p$ is the parasitic capacitance, $R_s$ is the series resistance (due to the contacts and the $p$ and $n$ regions of the diode), and $R_L$ is the load resistor. For this circuit, the output current $i(\omega)$ can be expressed as $i(\omega) = i_o(\omega) \cdot H(\omega)$, where $H(\omega)$ is the transfer function of the circuit. The form of $i(\omega)$ suggests that the harmonic distortion can show up in either $i_o(\omega)$ or $H(\omega)$, or both. As mentioned above, the physical origin of the nonlinearity in $i_o(\omega)$ and $H(\omega)$ are attributed to the nonlinear carrier transport and the associated impedance changes induced by the optical input signal. This model can be used to account for the photodiode distortion even at low optical power.

In the analysis, an increase of junction capacitance $C_j$ of a planar PIN photodiode under optical illumination can be paralleled by an increase in the electric polarization due to an increase in photogenerated electron-hole pairs, whose density depends on the net electric field and optical illumination level. In general, both $R_j$ and $C_j$ change with optical illumination. For instance, we consider the effect of $C_j$ and assume it increases linearly with photocurrent:

$$C_j = C_{\text{dark}} + I_{dc} C' \tag{19.25}$$

where $C_{\text{dark}}$ is the dark capacitance of the junction, $I_{dc}$ is the $dc$ photocurrent, and $C'$ is defined as the differential capacitance with respect to the photocurrent. From the equivalent circuit in Fig. 19.20, the equation for the voltage $V$ across the current source satisfies:

$$\frac{d}{dt}\{(C_{jo} + i_o e^{j\omega t} C')V\} + \frac{V}{Z(\omega)} = i_o e^{j\omega t} \tag{19.26}$$

where $C_{jo}$ and $R_{jo}$ are the junction capacitance and resistance at a given $I_{dc}$, $Z(\omega)$ is the equivalent impedance of $C_p$, $R_s$, and $R_L \cdot V$ can be expressed in a harmonic series of $V_1 e^{j\omega t} + V_2 e^{j2\omega t} + V_3 e^{j3\omega t} + \cdots$, where $V_1$, $V_2$, and so on can be determined by comparing the coefficient of the $\omega^n$ term on both sides. In this model, it is observed that a relatively large change in $C_j$ can enhance harmonic levels, while the fundamental power remains largely unchanged. Also, the harmonic distortions have similar $RC$ frequency rolloff as the fundamental signal. At frequency well below the rolloff frequency, harmonic powers are proportional to $\omega^2$ when the variation of $C_j$ dominates; similar observations hold for higher order harmonics and intermodulation distortions [94].

### 19.5.2 Geometrical Effect

Another issue associated with optical saturation can arise from the series resistance associated with geometric design of the photodiode. To illustrate this, consider a surface normal InP/InGaAs/InP PIN diode with a photodetection region of thickness, $d$, and ring electrode, as shown in Fig. 19.21. As mentioned before, at high power, the photocurrent leads to a voltage drop across the photodiode. For the electrode configuration shown in Fig. 19.21, the lateral charging of the distributed capacitors through the resistors can enhance the debiasing effect. A simple distributed circuit model is shown to estimate the effect of high-power optical pulses on the biasing within the photodiode [95]. For the $p^+$ and $n^+$ InP regions, the corresponding ring resistances $R_{ip}$ and $R_{in}$ are obtained as:

$$R_j = \frac{\rho_s}{2\pi} \ln \frac{R + \Delta R}{R} \tag{19.27}$$

where the $\rho_s$ is the resistivity of the layer. Similarly, the capacitance between corresponding rings is obtained from the parallel plate capacitance:

$$C_j = \frac{\varepsilon}{d}\pi \left[(R + \Delta R)^2 - R^2\right] \tag{19.28}$$

For simplicity, the incident light on the photodetector is represented by a Gaussian distribution function. The debiasing effect of the high-power optical pulse is evident in Fig. 19.22 in which 20-ps, 1-W optical pulses cause a bias larger than 10 V at the photodiode, which is opposite to the bias of the photodiode and thus results in the saturation effect as described above. In this model, a high contact resistance can

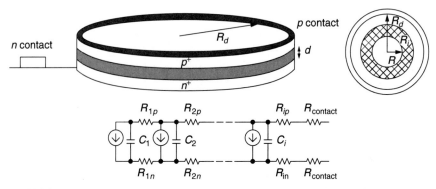

**Figure 19.21**  Distributed detector model of a surface normal photodiode with a ring electrode. The ring capacitance and resistance is represented in the distributed circuit model.

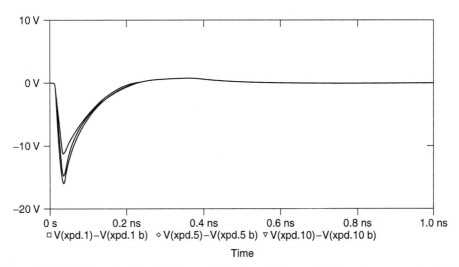

**Figure 19.22** Effect of optical debiasing of the photodiode upon high peak power pulse (1-W, 20-ps duration) excitation. The vertical axis is voltage, and the horizontal is time in nanoseconds. The different curves show the cumulative effect from the innermost ring to the outermost ring (lowest curve), as depicted in Fig. 19.21.

greatly enhance the debiasing effect. To reduce the debiasing of photodiodes, a thick $p^+$ surface layer and distributed contacts should be used to reduce the internal resistance, or alternatively, one can explore the waveguide-coupled configuration where lateral charging effect is minimized because of the continuous metal contact along the absorption waveguide. This geometric effect can be reduced by evenly distributed photocurrent; an example is the waveguide integrated photodiodes where the optical energy is more evenly distributed and absorbed along the waveguide [96]. An example of the waveguide photodiode response to ultra-fast optical pulses is illustrated in Fig. 19.23, where 5-ps light pulses from mode-locked lasers are coupled to a waveguide photodiode incorporated with an index layer to even out the optical absorption along the waveguide.

### 19.5.3 Phototransistor

Previously, most of the work on phototransistors has been based on silicon and germanium [97, 98]. Recent interest in III-V phototransistors is primarily due to the good material quality of heterojunction devices and the higher gain achievable in the heterojunction bipolar transistor. By employing a heterojunction, a wide bandgap emitter-configured phototransistor-heterojunction phototransistor (HPT), is realized, and this greatly improves the emitter efficiency [99]. The heterojunction relieves the restriction on relative dopant concentrations on both sides of the emitter-base junction [100], and a high gain-bandwidth product is possible. HPTs have been considered as useful photodetectors with high intrinsic gain for fiber-optic communication systems. Since the early 1980s, the bandwidth, gain, and noise properties of HPTs were studied with materials based either on GaAs/AlGaAs or on InP/InGaAs. Unlike avalanche photodiodes (APDs), HPTs can provide large photocurrent gain without a high bias voltage and excess avalanche noise in the amplification process.

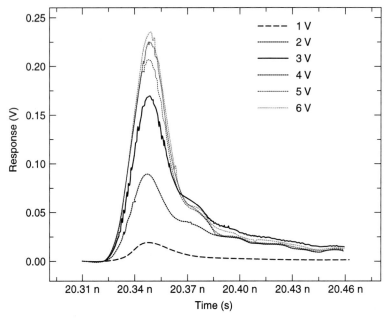

**Figure 19.23** Pulse response of a waveguide photodiode under different bias voltages with the energy per pulse of 2.4 pJ/pulse.

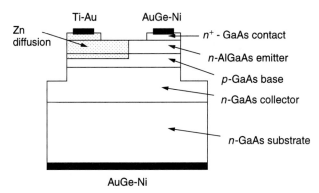

**Figure 19.24** Schematic diagram of an *npn* GaAs/AlGaAs phototransistor with diffused base contact region. (From Scavennec et al., *Electron. Lett.*, Vol. 19, p. 384, 1983.)

For the *npn* phototransistor shown in Fig. 19.24, the base is unbiased, and the voltage is applied between the emitter and collector such that the emitter and collector junctions are forward and reverse biased, respectively. The wide-bandgap emitter is transparent to incident photons, which are mostly absorbed in the base and collector regions. The photogenerated holes accumulated at the base change the potential at the emitter-base junction and cause injection of electrons from the emitter into the base. Current gain is obtained when the electron lifetime in the base is longer than the base transit time. The absence of the base contact is primarily for reducing the capacitance. The frequency response of the phototransistor can be improved by employing designs with a lower emitter capacitance (lightly doped emitter) and a reduced base resistance (heavily doped base). This could have compromised the emitter injection efficiency in the case of homostructure *npn* transistors. However, by employing a wider bandgap material in the emitter, the valence band discontinuity can greatly enhance the emitter injection efficiency [101, 102]. For the GaAs/AlGaAs material system, a phototransistor with a floating base configuration has achieved a current gain of 5000 [101], while the configuration with a diffused base to minimize the emitter-base capacitance has achieved a current gain $>10^4$ [103].

Most of the HPTs are the surface-normal type [104, 105]. As the device dimensions are reduced for high-speed operation, the optical coupling to the device becomes more challenging. Waveguide HPTs are introduced to achieve a good optical coupling with small device dimensions. An edge-coupled InGaAs/InP HPT with a 0.1-$\mu$m-thick base has achieved a DC gain of 270 and a unity gain cutoff frequency $>30$ GHz [106].

As the HPT is structurally compatible with the heterojunction bipolar transistor (HBT), monolithically integrated optical receivers are also considered. A monolithic HPT/HBT optical receiver is fabricated and operated as high as 5 Gb/s [107]. Utilizing a novel base circuit composed of an inductor and a series resistor, an HPT is shown to operate over a 3-dB bandwidth of 0.43–12.1 GHz with $>11$-dB gain over a photodiode with identical quantum efficiency [108]. An integrated HPT/HBT optical receiver from the same design has demonstrated a gain of $>20$ dB [108].

A particular microwave photonics application of the HPT is as an optoelectronic mixer [109–111]. The HPT has a nonlinear transistor gain, which is a function of the incident optical power or the device current. Utilizing this property, the HPT can either mix an optical signal with another optical signal or mix an optical signal with an electrical signal to generate up- or down-converted RF/microwave frequency. For example, in [111], a 10.4-dB intrinsic conversion gain was achieved using an electrical local oscillator signal mixed with an RF-modulated optical signal. Optoelectronic mixing also occurs in the dual-function HPT that incorporates a PIN structure for electroabsorption [112, 113], and the resulting waveguide npin structure can function as an electroabsorption modulator in the modulation mode, and as an HPT in the photodetection mode.

# REFERENCES

1. R. Feymann, *Lecture on Physics*, Vol. 1, Addison Wesley, Reading, HA, 1963, Ch. 35 and 36.

2. R. Simons, *Optical Control of Microwave Devices*, Artech House, 1990.

3. R. J. Keyes, Ed., *Optical and Infrared Detectors, Topics in Applied Physics*, Vol. 19, Springer-Verlag, Berlin, 1980.

4. H. Kressel, *Semiconductor Devices for Optical Communications*, 2nd ed., Springer Verlag, Heidelberg, 1990.

5. K. Kato, *IEEE Transactions on Microwave Theory and Techniques*, **47**, p. 1265, 1999.

6. S. B. Alexander, *Optical Communication Receiver Design, SPIE Tutorial Texts in Optical Engineering*, Vol. TT22, Bellingham, Washington, 1997.

7. A. R. Williams, A. L. Kellner, X. S. Jiang, and P. K. L. Yu, *IEE Electronics Letters*, **28**, p. 2258, 1992.

8. S. Kawanishi, A. Takada, and M. Saruwatari, *J. Lightwave Technol.*, **7**, p. 92, 1989.

9. K. Li, E. Rezek, and H. D. Law, *IEE Electron. Lett.*, **20**, p. 196, 1984.

10. J. I. Pankove, *Optical Processes in Semiconductors*, Prentice Hall, Englewood Cliffs, New Jersey, 1971.

11. W. Franz, Z. Naturforsch, **A13**, p. 484, 1958.

12. L. V. Keldysh, *Zh. Eksp. Teor. Fiz.*, **34**, p. 1138, 1958. *English trans.: Sov. Phys.*, Vol. **JETP** 7, p. 788, 1958.

13. D. S. Chemla, *Helv. Phys. Acta*, **56**, p. 607, 1983.

14. E. W. Jacobs, D. W. Fogliatti, H. Nguyen, D. J. Albares, C. T. Chang, and C. K. Sun, *IEEE Trans. on Microwave Theory and Techniques*, **50**, pp. 413–419, 2002.

15. K. Seeger, *Semiconductor Physics*, 3rd ed., Springer Verlag, 1999, Ch. 11.

16. H. Jiang and P. K. L. Yu, *IEEE Photon. Technol. Lett.*, **10**, p. 1608, 1998.

17. N. Shimuzu, Y. Miyamoto, A. Hirano, K. Sato, and T. Ishibashi, *IEE Electron. Lett.*, **36**, p. 750, 2000.

18. D. Giorgi, P. K. L. Yu, J. R. Long, T. Navapanich, and O. S. F. Zucker, *J. Appl. Phys.*, **63**, p. 930, 1988.

19. S. M. Sze, *Physics of Semiconductor Devices*, Wiley, New York, 1981.

20. H. D. Law, K. Nakano, and L. R. Tomasetta, *J. Quant. Electron.*, **QE-15**, p. 549, 1979.

21. H. Melchoir, Demodulation and Photodetection Techniques, in F. T. Arecchi and E. O. Schulz-Dubois, Eds., *Laser Handbook*, Vol. 1, North-Holland, Amsterdam, 1972, pp. 725–835.

22. H. Ando, H. Kanbe, T. Kimura, T. Yamaoka, and T. Kaneda, *IEEE J. Quant. Electron.*, **QE-14**, p. 804, 1978.

23. R. P. Riesz, *Rev. Sci. Instrum.*, **33**, p. 994, 1962.

24. H. Kanbe, G. Grosskopt, O. Mikami, and S. Machinda, *IEEE J. Quant. Electron.*, **QE-17**, p. 1534, 1981.

25. L, Duraffourg, J. M. Merolla, J. P. Goedgebuer, N. Butterlin, and W. T. Rhodes, *IEEE Journal of Quantum Electronics*, **QE-37**, p. 75, 2001.

26. M. A. Washington, R. E. Nahory, and E. D. Beebe, *Appl. Phys. Lett.*, **33**, p. 854, 1978.

27. A. R. Clawson, W. Y. Lum, G. E. McWilliams, and H. H. Wieder, *Appl. Phys. Lett.*, **32**, p. 549, 1978.

28. C. Fan, P. K. L. Yu, and P. C. Chen, *Electron. Lett.*, **23**, p. 571, 1987.

29. T. Aoyagi, E. Ishimura, S. Funaba, D. Suzuki, T. Kimura, T. Sogo, and M. Aiga, Proceedings of Optical Fiber Communication Conference, p. 37, 1997.

30. J. S. Pereira, P. J. Shieh, A. L. Gobbi, E. Sato, P. Malberti, T. Santos, F. Borin, and N. Patel, Proceedings of IEEE International Reliability Physics Symposium, p. 372, 1993.

31. T. P. Lee, C. A. Burrus, A. Y. Cho, and K. Y. Cheng, *Appl. Phys. Lett.*, **37**, p. 730, 1980.

32. K. Li, E. Rezek, and H. D. Law, *Electron. Lett.*, **20**, p. 106, 1984.

33. H. D. Law, L. R. Tomasetta, and K. Nakano, *Electron. Lett.*, **20**, p. 196, 1984.

34. T. Ishibashi, N. Shimizu, S. Kodama, H. Ito, T. Nagatsuma, and T. Furuta, *Ultrafast Electron. Optoelectron. '97 Conf.*, Incline Village, NV, 1997.

35. N. Shimuzu, Y. Miyamoto, A. Hirano, K. Sato, and T. Ishibashi, *Electron. Lett.*, **36**, p. 750, 2000.

36. T. Furuta, S. Kodama, and T. Ishibashi, *Electron. Lett.*, **36**, p. 1809, 2000.

37. Z. Ray-Noy et al., *Electron. Lett.*, **19**, p. 753, 1983.

38. H. Schumacher, H. P. Leblanc, J. Soole, and R. Blat, *IEEE Electron. Dev. Lett.*, **EDL-9**, p. 607, 1988.

39. K. C. Hwang, S. S. Li, Y. C. Kao, Proceeding of the Second International Conference on Indium Phosphide and Related Materials, p. 372, 1990.

40. S. Y. Wang, and D. M. Bloom, *Electron. Lett.*, **19**, p. 544, 1983.

41. H. Komobuchi, T. A. Ando, *IEEE Trans. on Electron. Dev.*, **37**, p. 1861, 1990.

42. R. B. Emmons, *J. Appl. Phys.*, **38**, p. 3705, 1967.

43. H. W. Ruegg, *IEEE Trans. Electron. Dev.*, **ED-14**, p. 238, 1967.

44. H. Melchoir, A. R. Hartman, D. P. Schinke, and T. E. Seidel, *Bell Syst. Tech. J.*, **57**, p. 1791, 1978.

45. J. Conradi, *Solid-State Electron.*, **17**, p. 99, 1974.

46. K. Nishida, *Electron. Lett.*, **13**, p. 419, 1977.

47. R. J. McIntyre, *IEEE Trans. Electron. Dev.*, **ED-13**, p. 164, 1966.

48. G. E. Stillman, and C. M. Wolfe, in P. K. Williardson and A. C. Beers, Eds., *Semiconductor and Semimetals*, Vol. 12, Academic, New York, 1977, p. 291.

49. C. L. Anderson and C. R. Crowell, *Phys. Rev. B.*, **5**, p. 2267, 1972.

50. R. J. McIntyre, *IEEE Trans. Electron. Dev.*, **ED-19**, p. 703, 1972.

51. H. Melchoir and W. T. Lynch, *IEEE Trans. Electron. Dev.*, **ED-13**, p. 829, 1966.

52. J. Conradi and P. P. Webb, *Proc. Ist Eur. Conf. Opt. Fibre Commun.*, **1**, p. 128, 1975.

53. H. Kanbe, T. Kimura, Y. Mizushima, and K. Kajiyama, *IEEE Trans. Electron. Dev.*, **ED-23**, p. 1337, 1976.

54. J. Muller and A. Ataman, Tech. Dig. Int. Electron Dev. Meet., Washington, DC, IEEE, New York, 1976, p. 416.

55. N. Susa, H. Nakagame, H. Ando, and H. Kanbe, *IEEE J. Quant. Electron.*, **QE-17**, p. 243, 1981.

56. L. E. Tarof, J. Yu, R. Bruce, D. G. Knight, T. Baird, and B. Oosterbrink, *IEEE Photn Tech. Lett.*, **5**, p. 672, 1993.

57. L. W. Cook, G. E. Bulman, and G. E. Stillman, *Appl. Phys. Lett.*, **40**, p. 589, 1982.

58. R. Yeats and K. Von Dessonneck, *IEEE Electron. Dev. Lett.*, **EDL-2**, p. 268, 1981.

59. O. K. Kim, S. R. Forrest, W. A. Bonner, and R. G. Smith, *Appl. Phys. Lett.*, **39**, p. 402, 1981.

60. J. C. Campbell, H. Nie, C. Lenox, G. Kinsey, P. Yuan, A. L. Holmes, and B. G. Streetman, *Technical Digest of Optical Fiber Commun. Conf.* 2000, **37**, p. 114, 2000.

61. T. Kagawa, Y. Kawamura, H. Asai, and M. Naganuma, *Appl. Phys. Lett.*, **57**, p. 1895, 1990.

62. S. Wang, R. Sidhu, X. G. Zheng, X. Li, X. Sun, A. L. Holmes, Jr., and J. C. Campbell, *IEEE Photon. Technol. Lett.*, **13**, p. 1346, 2001.

63. Z. L. Liau and D. E. Mull, *Appl. Phys. Lett.*, **56**, p. 737, 1990.

64. A. R. Hawkins, W. Wu, P. Abraham, K. Streubel, and J. E. Bowers, *Appl. Phys. Lett.*, **70**, p. 303, 1996.

65. Y. Kang, P. Mages, A. R. Clawson, P. K. L. Yu, M. Bitter, Z. Pan, A. Pauchaud, S. Hummel, and Y. H. Lo, *IEEE Photon. Technol. Lett.*, **14**, p. 1593, 2002.

66. O. Hilderbrand, W. Kuebart, K. W. Benz, and M. H. Pilkuhn, *IEEE J. Quant. Electron.*, **17**, p. 284, 1981.

67. J. E. Bowers, and C. A. Burrus, Jr. *J. Lightwave Technol.*, **5**, p. 1339, 1987.

68. K. Kato, S. Hata, K. Kawano, J. Yoshida, and A. Kozen, *IEEE J. Quant. Electron.*, **28**, p. 2728, 1992.

69. L. Giraudet, F. Banfi, S. Demiguel, and G. Herve-Gruyer, *IEEE Photon. Technol. Lett.*, **11**, p. 111, 1999.

70. H. F. Taylor, O. Eknoyan, C. S. Park, K. N. Choi, and K. Chang, *Proc. SPIE on Optoelectronic Signal Processing Phased Array Antenna II*, **1217**, p. 59, 1990.

71. K. S. Giboney, M. Rodwell, and J. Bowers, *IEEE Photon. Technol. Lett.*, **4**. p. 1363, 1992.

72. L. Y. Lin, M. C. Wu, T. Itoh, T. A. Vang, R. E. Muller, D. L. Sivco, and A. Y. Cho, *IEEE Trans. Microwave Theory Tech.*, **45**, p. 1320, 1997.

73. K. S. Giboney, R. Nagarajan, T. Reynolds, S. Allen, R. Mirin, M. Rodwell, and J. E. Bowers, *IEEE. Photon. Technol. Lett.*, **7**, p. 412, 1995.

74. C. Cohen-Jonathan, L. Giraudet, A. Bonzo, and J. P. Praseuth, *Electron. Lett.*, **33**, p. 1492, 1997.

75. T. Nakata, T. Takeuchi, I. Watanabe, K. Makita, and T. Torikai, *Electron. Lett.*, **36**, p. 2034, 2000.

76. R. Bube, *Photoconductivity of Solids*, Wiley, New York, 1960.

77. A. Rose, *Concepts in Photoconductivity and Allied Problems*, Wiley-Interscience, New York, 1963.

78. R. B. Hammond, N. G. Paulter, R. S. Wagner, T. E. Springer, and M. MacRoberts, *IEEE Trans. Electron. Dev.*, **ED-30**, p. 412, 1983.

79. A. G. Foyt and F. J. Leonberger, InP Optoelectronic Switches, In C. H. Lee, Ed., *Picosecond Optoelectronic Devices*, Academic Press, New York, 1984, Ch. 9, p. 271.

80. F. J. Leonburger and P. F. Moulton, *Appl. Phys. Lett.*, **35**, p. 712, 1979.

81. D. H. Auston, *Appl. Phys. Lett.*, **26**, p. 101, 1975.

82. A. G. Foyt, F. J. Leonberger, and R. C. Williamson, *Proc. SPIE*, **269**, p. 109, 1981.

83. J. B. Gunn, *IBM J. Res. Dev.*, **8**, p. 141, 1963.

84. H. J. Klein, R. Kaumanns, and H. Beneking, *Electron. Lett.*, **17**, p. 422, 1981.

85. T. F. Deutsch, F. J. Leonburger, A. G. Foyt, and D. Mills, *Appl. Phys. Lett.*, **41**, p. 403, 1982.

86. R. H. Day, P. Lee, E. B. Solomon, and D. J. Nagel, *Los Alamos Sci. Lab, [REP] LA*, **LA-7941-MS**, 1981.

87. G. Mourou, W. Know, and S. Williamson, in C. H. Lee, Ed., *Picosecond Optoelectronic Devices*, Academic, New York, 1984, Ch. 7, p. 219.

88. D. Giorgi, P. K. L. Yu, J. R. Long, V. D. Lew, T. Navapanich, and O. S. F. Zucker, *J. Appl. Phys.*, **63**, p. 930, 1988.

89. C. K. Sun, P. K. L. Yu, C. T. Chang, and D. J. Albares, *IEEE Trans. Electron Devices*, **39**, p. 2240, 1992.

90. K. J. Williams, *Appl. Phys. Lett.*, **65**, p. 1219, 1994.

91. R. R. Hayes and D. L. Persechini, *IEEE Photonics Technol. Lett.*, **5**, p. 70, 1993.

92. J. Harari, G. Jin, J. P. Vilcot, and D. Decoster, *IEEE Trans. Microwave Theory Tech.*, **45**, p. 4332, 1997.

93. T. Ozeki and E. H. Hara, *IEE Electron. Lett.*, **12**, p. 80, 1976.

94. H. Jiang, D. S. Shin, G. L. Li, J. T. Zhu, T. A. Vang, D. C. Scott, and P. K. L. Yu, *IEEE Photonics. Technol. Lett.*, **12**, p. 540, 2000.

95. D. Ralston, A. Metzger, Yimin Kang, P. Asbeck, and P. Yu, *Proc. SPIE*, **4112**, p. 132, 2000.

96. H. Jiang and P. K. L. Yu, IEEE International Microwave Symposium, *IMS-2000 Digest*, **2**, p. 679, 2000.

97. M. A. Schuster and G. Strull, *IEEE Trans. Electron. Dev.*, **ED-13**, p. 903, 1966.

98. J. N. Shive, *Phys. Rev.*, **76**, p. 575, 1949.

99. Z. L. Alferov, F. Akhmedov, V. Korol'kov, and V. Niketin, *Sov. Phys. Semiconductor*, **7**, p. 780, 1973.

100. W. Shockley, Circuit Element Utilizing Semiconductor Material, U.S. Patent Office, Patent No. 2, 569,347, 1951.

101. M. N. Svilans, N. Crote, and H. Beneking, *IEEE Electron. Dev. Lett.*, **EDL-1**, p. 274, 1980.

102. S. Margalit and A. Yariv, in W. T. Tsang, Ed., *Semiconductor and Semimetal*, Vol. 22, Part E, Academic, New York, 1985.

103. S. C. Lee and G. L. Pearson, *J. Appl. Phys.*, **52**, p. 275, 1981.

104. R. A. Milano, P. D. Dapkus, and G. E. Stillman, *IEEE Trans. Electron. Dev.*, **ED-29**, p. 266, 1982.

105. J. C. Campbell and K. Ogawa, Heterojunction Phototransistors for Long-wavelength Optical Receivers, *J. Appl. Phys.*, **53**, p. 1203, 1982.

106. D. Wake, D. J. Newson, M. J. Harlow, and I. D. Henning, *Electron. Lett.*, **29**, p. 2217, 1993.

107. S. Chandrasekhar, L. M. Lunardi, A. H. Gnauck, R. A. Hamm, and G. J. Qua, *IEEE Photon. Technol. Lett.*, **5**, p. 1316, 1993.

108. H. Kamitsuna, *J. Lightwave Technol.*, **13**, p. 230, 1995.

109. J. Van de Casteele, J. P. Vilcot, J. P. Gouy, F. Mollot, and D. Decoster, *Electron. Lett.*, **32**, p. 1030, 1996.

110. C. P. Liu, A. J. Seeds, and D. Wake, *IEEE Microwave Guided Wave Lett.*, **7**, p. 72, 1997.

111. Y. Betser, D. Ritter, C. P. Liu, A. J. Seeds, and A. Madjar, *J. Lightwave Technol.*, **16**, p. 605, 1998.

112. D. S. Shin, C. K. Sun, W. X. Chen, S. A. Pappert, S. A. Pappert, J. T. Zhu, R. Nguyen, Y. Z. Liu, and P. K. L. Yu, *Proc. SPIE*, **3463**, p. 66, 1998.

113. D. S. Shin, H. Jiang, C. K. Sun, W. X. Chen, J. T. Zhu, S. A. Pappert, and P. K. L. Yu, *Proc. SPIE*, **3631**, p. 108, 1999.

# 20

# ACOUSTO-OPTIC MODULATORS AND SWITCHES

SHI-KAY YAO

*Optech Laboratory*
*Rowland Heights, California*

## 20.1 INTRODUCTION

Acousto-optic modulators certainly have been around for long time, and they are widely available from commercial sources. However, despite numerous publications on this subject since the 1960s, most articles either deal with the fundamental theorem of acousto-optic interactions or deal with some rather advanced device design or special features. Systematic technical discussions on many important and basic properties of rudimentary acousto-optic modulators were not available for people who want to go beyond the entry level and gain solid understanding of the various properties of an acousto-optic modulator as a useful device. Because acousto-optic has been a highly specialized field of engineering, it has been difficult for an optical system engineer who happens to deal with acousto-optic modulator to find answers for the many questions that he encounters daily. Questions that the author is trying to provide an answer to include why and how a laser beam gets slightly distorted after the acousto-optic modulator, why the peak diffraction efficiency does not reach the desired level, why the rise-time and modulation bandwidth sometimes seems less than expected, why the acousto-optic modulator is linear and what to do if not, what effect the acoustic attenuation has on an acousto-optic modulator, and how and why there is thermal drift in acousto-optic devices etc.

For that reason, this chapter is not intended to be a showcase of fascinating acousto-optic device varieties. It is written for engineers who really want to do a good job on a basic acousto-optics design or system applications, and therefore, the book emphasizes basic properties and issues that a good acousto-optic modulator engineer must know. Focus is placed on details of acousto-optic rise-time, laser beam distortion, maximum

*Handbook of Optical Components and Engineering*, Edited by Kai Chang
ISBN 0-471-39055-0 © 2003 John Wiley & Sons, Inc.

achievable diffraction efficiency, the modulation transfer functions, harmonic contents and waveform distortions in the modulated laser, effects of acoustic attenuations, effects of transducer heating and acoustic diffraction inside the modulator, and some basic reasons for the acoustic transducer construction.

On designing of acousto-optic modulators, instead of giving the reader a set of general equations from which they have to struggle for some satisfactory solutions for an application, a unique design procedure based on years of experience by the author in actual practice is provided. This design approach starts from detail device performance requirements and works its way up step by step toward a set of device configuration numbers for a nearly optimal design.

By focusing on basic and fundamental general properties of acousto-optic modulators, the author has to leave out the many fascinating and novel acousto-optic modulator designs and applications for future work. Also, for the same reason, the discussions are focused on behavior and properties of isotropic acousto-optic modulator only, leaving the highly specialized anisotropic acousto-optic modulators and guided wave acousto-optic modulators for those who are willing to consult with additional publications and literatures [1, 3, 4, 6, 7, 8, and 20, 21, 26, 27]. This is because, the author believes, only after having a solid understanding of basic acousto-optic modulator properties of a relatively simple isotropic acousto-optic modulator, can one then move on to the more sophisticated birefringence acousto-optic modulator designs, which requires special knowledge on crystal cuts and orientations and anisotropic optics in birefringent crystals.

### 20.1.1 Background of Acousto-Optic Modulation

When direct current modulation of laser fails to meet the performances required for certain applications, external laser modulation becomes necessary. Most high-speed, high data rate applications of gas lasers and solid state lasers cannot be done without the use of external laser modulators. Specific example, include modulation of continuous wave gas lasers for various forms of laser printing (for example, direct to plate for printing industry, photo-mask generation for micro-electronics industry, and laser direct imaging for electronics industry), laser color separation, laser communications, and laser show applications. Separate laser modulation devices are also widely used inside laser cavities for Q-switching of solid-state lasers, which is the building block for range finders, laser machining, and so on and for active mode locking of gas lasers or dye lasers for generation of sharp pulses and picosecond pulse train.

During the past 40 years since the invention of lasers, acousto-optical modulation of laser beam has enjoyed a special status as the enabling device in many electro-optical systems [1–5]. In general, acousto-optic modulators are slower and more difficult to construct when compared with electro-optic modulators. Its popularity is mainly due to high reliability as well as high optical beam quality that can be maintained through the laser modulation process. High reliability and good optical beam quality are necessary for most laser systems to reach state-of-the-art performances. This is particularly true for high-intensity laser systems and for shorter wavelength laser systems that often have its performances limited by optical damage of components.

In general, the operation of an electro-optical device depends on interferometric balance of optical phase fronts between two paths that are affected by the applied electric field. It is therefore sensitive to other physical parameters that may also affect

**TABLE 20.1   Comparison between E-O and A-O Modulators**

|  | Drive Signal | Typical Rise Time | Extinction Ratio | Comments |
|---|---|---|---|---|
| E-O Modulators | High Voltage | Nanosecond | $\sim$100 to 1 | Bias Level, Optical Damage |
| A-O Modulators | Modulated RF | Microsecond | $>$1000 to 1 | Diffraction Limited |

the optical interferometer such as temperature and stress. In contrary, the operation of an acousto-optic modulator depends on the acoustic power level delivered to the device and does not require critically maintained bias level. As a result, acousto-optic modulators are relatively insensitive to environmental variations.

Unlike the electro-optic parameters that exist in selected materials, among which only a few exhibiting high electro-optics coefficients, acousto-optic modulation relies on photo-elastic properties that exist in almost all materials. Thus, we can design an effective acousto-optic modulator out of a large list of candidate materials and select the one with best optical properties for use. That allows the designer of an acousto-optic modulator to select the most suitable material, trading off drive power requirement with optical quality under specified application.

Table 20.1 summarizes the basic characteristics of acousto-optic modulators and electro-optical modulators.

In this chapter, we will show readers how to design a practical acousto-optical modulator for a given bandwidth and efficiency considerations while keeping the laser beam distortion within bound. The basic considerations in designing acoustic transducers, bonding layers, and top electrodes are also provided. Experimental data are provided from previous published results, and commercial devices are employed where appropriate. A UV acousto-optic modulator design parameter is given as an example.

## 20.2  BASIC PROPERTIES OF ACOUSTO-OPTIC MODULATORS

### 20.2.1  Basic Structure of Acousto-Optic Modulators

Figure 20.1 illustrates the basic construction of an acousto-optic modulator. An acoustic transducer is attached to one side of the acousto-optic interaction medium for launching of modulated acoustic waves into the interaction medium. An electrical impedance-matching circuit must be employed to allow efficient transfer of electrical power to the acoustic transducer that often exhibits a very low input impedance level. Laser beam is allowed to travel through the acousto-optic medium at approximately 90 degrees to the acoustic wave propagation direction. The acoustic wavefront behaves like a traveling optical phase grating to the laser beam and diffracts part of the laser beam off-axis to the diffraction orders, thus facilitating laser modulations [2–5].

The basic principle of acousto-optic interaction has been treated in other chapters of this book and will not be repeated here. However, the design and operation of a well-designed acousto-optic modulator contains far more details than does the acousto-optic diffraction process. To be complete, it must include the associated laser beam forming

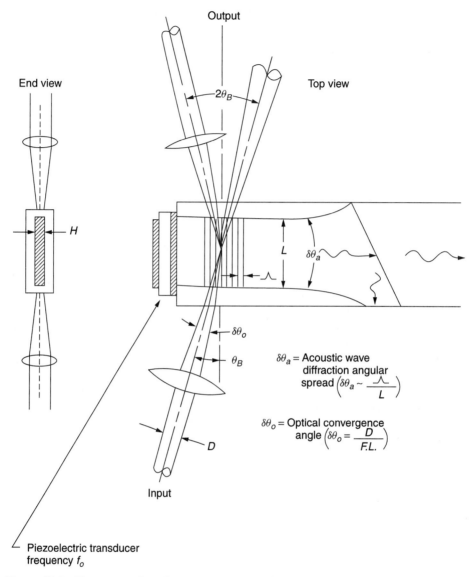

**Figure 20.1**   The construction of an acousto-optic modulator, including the focusing optics that significantly affects modulator performances.

optics and the acoustic launching transducer. The laser beam forming optics is critical to achieve expected modulator performances such as rise time, peak efficiency, and beam distortion [2, 4, 5, 9–15]. The acoustic transducer is an integral part of an acousto-optic modulator and is the most fragile yet most critical element in acousto-optic modulator fabrication. The acoustic transducer must provide the necessary electric-to-acoustic conversion efficiency with proper bandwidth, including proper amplitude as well as phase responses, and it must be well constructed for reliable performances over the life of the device [16]. Often, the quality (or weakness) of an acousto-optic modulator resides in the quality (or weakness) of the acoustic transducer.

Another very important issue for acousto-optic modulators is the acoustic column width, $L$. There is a characteristic thickness value of the acoustic grating, $L_o$, that divides the acousto-optic interaction processes into the so-called Bragg regime and the so-called Raman Nath regime [17].

$$L_o = \Lambda^2 n / \lambda \qquad (20.1)$$

where $\Lambda$ is the acoustic wavelength, $\lambda$ is the optical wavelength, and n is the optical index of refraction in the acousto-optic interaction media. For $L << L_o$, the acoustic grating is considered thin and the process belongs to the Raman Nath regime. The process is insensitive to incidence angle and has a large number of diffraction orders. Acousto-optic modulators are seldom designed into this regime due to poor device throughput efficiency as valuable laser power losses into the unwanted diffraction orders.

For $L >> L_o$, the acoustic column is thick and the process belongs to the Bragg regime. Strong diffraction (energy exchange between grating orders) occurs between only two of the orders when the incident laser beam meets the Bragg incidence angle condition

$$\theta_B = \arcsin(\lambda/(2n\Lambda)) \qquad (20.2)$$

Most acousto-optic modulators are designed to have large acoustic column width, L, and therefore operate in the Bragg regime for the sake of high throughput efficiency.

### 20.2.2 Acousto-Optic Device Rise-Time Considerations

The first requirement of an acousto-optic modulator is the desired frequency response, i.e., the laser modulation rise time. For many electro-optic system designers who simply purchases an acousto-optic modulator for their applications, once an acousto-optic modulator material of choice is selected, often the only consideration needed to complete a subsystem design is to decide the necessary optics before and after acousto-optic modulators so as to provide the desired modulation rise time.

For a sufficiently thin acoustic wave column, the rise time of an acousto-optic modulator is related to the transit time that the acoustic waveform takes to traverse the optical beam diameter [2, 4, 5].

$$Tt = D/v \qquad (20.3)$$

where Tt is the transit time for the acoustic wave to travel across the laser beam diameter, D is the laser beam diameter, and $v$ is the acoustic velocity. For a Gaussian laser beam, D is the effective laser beam diameter at $1/e^2$ of the peak intensity level at the middle of the acousto-optic interaction region. Note that D is related to the beam waist of a focused laser beam, $D = 2\omega_0$, only when the laser beam is properly focused to the center of the acousto-optic modulator.

When the acoustic waveform approximates an ideal step function, the modulation time-response of a diffracted Gaussian laser beam is smoothed by the acoustic transit time, and it assumes the shape of a complimentary error function. The rise time of an acousto-optic modulator with a Gaussian laser beam profile is related to transit time by

$$Tr = Tt/1.5 \qquad (20.4)$$

The modulation frequency bandwidth of an acousto-optic modulator, fm, is related to the rise time, and it is approximately, $f_m = 1/(2\ Tr)$. Note that the modulation bandwidth of an acousto-optic modulator, fm, is related to the frequency bandwidth of the acousto-optic process, $\Delta f$, by a simple relationship, $\Delta f = 2\ fm$, because of a double side-band requirement for intensity modulation or amplitude modulations. Thus,

$$fm = \beta(0.75/Tt)$$

$$= \beta(0.75v/D) \tag{20.5}$$

where $\beta = 1$ for devices in the Raman Nath regime and $\beta <= 1$ for the Bragg regime.

Figure 20.2 illustrates the broadening of laser beam width after the Bragg diffraction that distorts the circular laser beam into elliptical shape and degrades modulation rise time. Simple geometric consideration gives the increase in laser beam width being approximately $(2\ L\theta_B)$, which is the projection of acoustic interaction length, L, at an angle $(\pi/2 - 2\theta_B)$ to the output beam direction. The modulated output beam profile is the convolution of the incident laser beam profile and the projection of acoustic transducer length $(2\ L\ \theta_B)$. As a result, the rise time increases as the laser beam width increases. The modulation bandwidth becomes smaller, corresponding to $\beta$ value smaller than 1.

In addition, the modulated laser waveform can be affected by RF waveform and by distortion in the optics system before the acousto-optic modulator. Saturation of the acousto-optic process can also result in significant distortion of the modulated laser waveform. For instance, offsetting the output optical aperture from the optical axis may cause a modulated square waveform to become the rabbit-ear shape due to imbalance of spectral components in the modulated laser beam. Also, it has been observed that, when an acousto-optic modulator is driven well into its saturation level, significant overshoot at one of the pulse edges may occur depending on the orientation of the incident laser beam due to change in laser incidence condition.

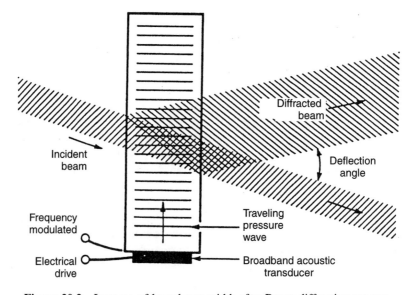

**Figure 20.2**   Increase of laser beam width after Bragg diffraction process.

### 20.2.3 Coupled Mode Solutions and Transfer Functions of Acousto-Optic Process

A rigorous approach to the acousto-optic modulator design must consider the focusing laser beam and the finite width of the acoustic wave column. Using Fourier optics techniques, a focusing laser beam can be decomposed into its spatial frequency components determined by its angular spectrum [18]. Each of the spatial frequency components is a plan wave, for which, the acousto-optic response can be obtained through the acousto-optic coupled mode equations given below [17]. Integrating over the spatial frequency domain after the acousto-optic interaction process yields the resultant output laser beam.

$$dE_0(x)/dx = -(\phi/L)E_1(x) \tag{20.6a}$$

$$dE_1(x)/dx + j(2\Gamma/L)E_1(x) = (\phi/L)E_0(x) \tag{20.6b}$$

$$2\phi = 2\pi n' L/\lambda \tag{20.7}$$

$$n' = -(1/2)n^3 pS \tag{20.8}$$

$$\Gamma = Q(2\alpha - 1)/4$$
$$= \pi(\theta o - \theta_B)L/\Lambda \tag{20.9}$$

$$Q = 2\pi \lambda L/(n\Lambda^2)$$
$$= 2\pi(L/L_o) \tag{20.10}$$

$$\alpha = \sin(\theta o)/2\sin(\theta_B) \tag{20.11}$$

where $E_0$ and $E_1$ are the optical field amplitude in the incident laser beam and the diffracted (output) laser beam, respectively; $x$ is the direction of the optical path parallel to the acoustic wavefront; $2\phi$ is the phase shift due to the acoustic field induced index perturbation, which is the strength of the acousto-optic process; $n'$ is the induced optical index change through photo-elastic coefficient $p$ and acoustic field $S$; $L$ is the acousto-optic interaction length; and $\Gamma$ is the mismatch factor for the Bragg condition. The parameter $Q$ defines the Bragg regime if $Q \gg 1$ and the Raman Nath regime if $Q \ll 1$.

In general, the acousto-optic coupled mode equations include many more equations for other diffraction orders. Klein and Cook have calculated the peak optical intensity values in various diffraction orders numerically, and they found that when $Q$ is substantially larger than 1, the energy transfer into all other modes is small and can be ignored, and only Eqs. (6a) and (6b) need to be concerned [17]. In such case, the solution of the coupled mode equations are given as follows:

$$E_0(x) = E_0 e^{-j\Gamma x/L}[\cos(\sigma x/L) + j(\Gamma/\sigma)\sin(\sigma x/L)] \tag{20.12}$$

$$E_1(x) = E_0 e^{-j\Gamma x/L}(\phi/\sigma)\sin(\sigma x/L) \tag{20.13}$$

Where

$$\sigma = (\phi^2 + \Gamma^2)^{1/2} \tag{20.14}$$

The output laser field in the undiffracted order $E_0$, and in the diffracted order $E_1$, can be obtained from Eqs. (20.12) and (20.13) by letting $x = L$. Note that the exponential

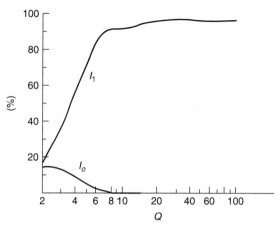

**Figure 20.3**  Zeroth- and first-order light intensities versus $Q$ with optical plan wave incident at Bragg condition (after Klein and Cook).

factor in the laser fields denotes a tilt in the propagation direction in the output laser beams that is the result of the incident laser beam having an error from the Bragg incidence angle when angular mismatch occurs.

Substituting $x = L$ for optical fields exiting the acousto-optic modulator, we obtain the acousto-optical transfer function and the plan wave acousto-optic diffraction efficiency as follows, assuming $Q >> 1$:

$$E_1/E_0 = e^{-j\Gamma}(\phi/\sigma)\sin(\sigma)$$
$$= e^{-j\Gamma}[\phi/(\phi^2 + \Gamma^2)^{1/2}]\sin(\phi^2 + \Gamma^2)^{1/2} \qquad (20.15)$$
$$\eta = |E_1|^2/|E_0|^2$$
$$= (\phi)^2[\sin^2(\phi^2 + \Gamma^2)^{1/2}/(\phi^2 + \Gamma^2)] \qquad (20.16)$$

Figure 20.3 gives the calculated optical power in the zeroth order, $Io = |E_0|^2$, and the optical power in the first order, $I_1 = |E_1|^2$, by Klein and Cook when all diffraction orders are considered. Note, at Bragg angle condition, the optical intensity in the first order exceeds 90% when $Q > 2\pi$, that is, when $L = L_o$.

### 20.2.4   Spatial Frequency Response of Acousto-Optic Modulators:

The mismatch term, $\Gamma$, in the above equations can be rewritten in terms of acoustic diffraction spread angle, $\delta\theta a = \Lambda/L$,

$$\Gamma = \pi(\theta - \theta_B)/\delta\theta a \qquad (20.17)$$

The acousto-optic transfer function, i.e., spatial frequency response if the dependence on angles is rewritten in terms of dependence on spatial frequencies, for optical plane wave near the Bragg angle can be obtained by rewriting Eq. (20.15) as

$$E_1/E_0 = e^{-j\Gamma}\{\phi/[\phi^2 + (\pi(\theta - \theta_B)/\delta\theta a)^2]^{1/2}\}\sin[\phi^2 + (\pi(\theta - \theta_B)/\delta\theta a)^2]^{1/2}$$
$$= e^{-j\Gamma}(\phi)\text{sinc}[\phi^2 + (\pi(\theta - \theta_B)/\delta\theta a)^2]^{1/2} \qquad (20.18)$$

where the spatial frequencies are proportional to the mismatch angle $(\theta - \theta_B)$. In Eq. (20.18), the function sinc$(x)$ is defined by sinc$(x) = [(\sin x)/x]$. In most optical systems, the attention is about optical intensity used for photosensitive media exposure, or photoelectric currents generation. The plan wave acousto-optic diffraction efficiency of an acousto-optic modulator is

$$\eta = |E_1|^2/|E_0|^2$$
$$= (\phi)^2 \{\sin^2(\phi^2 + (\pi(\theta - \theta_B)/\delta\theta a)^2)^{1/2}/[\phi^2 + (\pi(\theta - \theta_B)/\delta\theta a)^2]\} \quad (20.19)$$

For the case of Bragg incidence condition, the mismatch term, $\Gamma$, becomes zero and we have

$$\eta = [\sin(\phi)]^2 \qquad (20.20)$$

Figure 20.4**a** gives the spatial frequency response of the acousto-optic process at several acoustic power levels. Figure 20.4**b** is the plan wave acousto-optic diffraction efficiency as a function of the incidence angle deviation from Bragg condition, i.e., input plan spatial frequency, with acoustic power as a parameter.

These figures indicate that the acousto-optic process serves as an angular (or spatial frequency) bandpass filter for optical plan waves oriented near the Bragg angle. The angular (or spatial frequency) bandwidth is inversely proportional to the acoustic interaction length, L, and in proportion to the acoustic wavelength, $\Lambda$. When the optical incidence angle deviates from the Bragg angle more than the acoustic angular spread, $\delta\theta a$, the Bragg diffraction efficiency decreases to zero. The acousto-optic transfer function, i.e., the spatial frequency response curve, decreases to half its peak value when the incidence angle deviation from Bragg angle is about $+-(0.55\,\delta\theta a)$. In the mean time, the acousto-optic diffraction efficiency decreases to its half peak value when $(\theta - \theta_B)$ is about $+-(0.4\,\delta\theta a)$. As a useful rule of thumb, the acceptance angle for the Bragg process is approximately one acoustic beam spread, $\delta\theta a$ [2, 19].

***20.2.4.1  Optical Beam Distortion.***  As shown in Fig. 20.2, when a laser beam intersects the acoustic wavefront at an angle, and in this case the Bragg angle $\theta_B$, the exiting laser beam after the Bragg diffraction is elongated along the acoustic propagation direction. Assume the acousto-optic modulator is placed at the laser beam waist, and let the laser beam profile at beam waist be F(y), where y is the acoustic wave propagation direction; the laser wavefront can be described by the Fourier integral

$$F(y) = \int_{-\infty}^{\infty} Af(\theta)e^{jky(\sin\theta)}d\theta \qquad (20.21)$$

where $Af(\theta)$ is the angular spectrum of F(y). Employing the acousto-optic transfer function given by Eq. (20.18), the optical wavefront after emerging from the acousto-optic modulator becomes

$$F1(y) = \phi \int_{-\infty}^{\infty} Af(\theta) \quad \text{sinc}[\phi^2 + (\pi(\theta - \theta_B)/\delta\theta a)^2]^{1/2} \quad e^{jky(\theta - \theta_B)}d\theta \qquad (20.22)$$

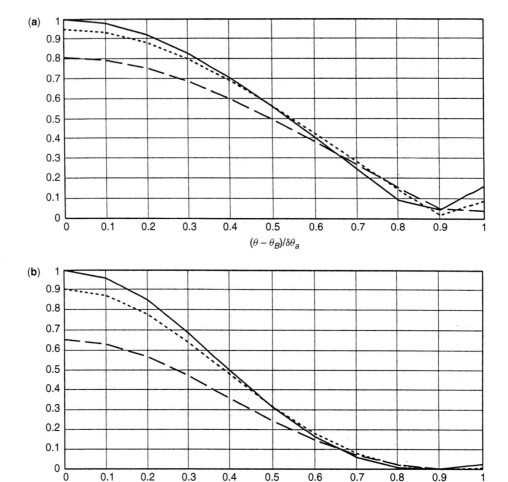

**Figure 20.4** **(a)** Plan wave acousto-optic transfer function, i.e., spatial frequency response curve as function of optical incidence angle deviation from Bragg condition, with acoustic drive power as a parameter. **(b)** Plan wave acousto-optic diffraction efficiency as function of optical incidence angle deviation from Bragg condition, i.e., spatial frequency, with acoustic drive power as a parameter.

When the output optical axis is tilted to the Bragg exit angle, Eq. (20.22) can be rewritten as

$$F1(y) = \delta\theta o\phi \int_{-\infty}^{\infty} Af(\theta/\delta\theta o)\text{sinc}[\phi^2 + (\pi(\theta/\delta\theta o)/R)^2]^{1/2}e^{jky(\theta/\delta o)}d(\theta/\delta\theta o)$$

(20.23)

and

$$R = (\delta\theta a/\delta\theta o)$$
$$= n\pi vD/(4\lambda foL)$$

(20.24)

where $\delta\theta o = (\pi/4)\lambda/(nD)$ is the optical beam spread angle of the Gaussian laser beam, and $R = (\delta\theta a/\delta\theta o)$ is the acoustic-to-optic beam balance ratio (beam balance ratio) defined by the ratio of acoustic beam spread to optical beam spread [9]. In this description, both the Gaussian laser beam angular spectrum, $Af(\theta)$, and the acoustic beam spread in the acousto-optic transfer function are normalized to the optical beam spread, making this equation dimensionless, other than a very important new parameter, R, the acoustic-to-optical beam spread ratio. For a given acousto-optic device used in an optical subsystem, the acoustic-to-optic beam spread ratio R is the most important parameter that dictates many device performances.

Setting $\phi = \pi/2$ corresponding to maximum diffraction efficiency, the integral yields a Gaussian beam profile weighted by a bandpass filter of the sinc function. The result can be considered convolution of the incident Gaussian laser beam with a rectangular-shaped function that can be considered the image of the acoustic transducer with reduced length $(2L \sin\theta_B)$. The output beam will be distorted into elliptical shape as it becomes elongated along the acoustic propagation direction, $y$. Normalizing the elongated dimension to the original laser beam width, we obtain the eccentricity of the output laser spot after the acousto-optic modulator versus the beam spread ratio, $R$; see Fig. 20.5 [9].

At the far field of the acousto-optic exit aperture, the laser output beam is the Fourier transform of the laser profile of Eq. (20.23), and it is simply the integrand of Eq. (20.23). The intensity profile of the far-field distribution is easy to calculate without the need to carry out the integration process, and it is called the "beam" because it is ready to propagate through the optical system [9].

Note that in Fig. 20.5, "spot" means the output laser profile at the exit window of the acousto-optic modulator, whereas "beam" means the far field of the laser output from the acousto-optic modulator. The reason for this definition is, in most optical systems incorporating acousto-optic modulators, it is customary to image the "spot" (i.e., output laser profile at modulator exit window) onto a target or a work surface because the spot is insensitive to thermal gradients inside the acousto-optic modulator as well as fluctuations in acoustic frequency. On the other hand, the "beam" will drift in case of significant thermal gradients or acoustic frequency fluctuations.

*Rise Time and Modulation Bandwidth.* There is another meaning for the curves of Fig. 20.5. As the laser spot size increases after the acousto-optic process, the actual transit time as well as modulation rise time increases in proportion. Thus, the upper curve of Fig. 20.5, "spot," also gives the increase in modulation rise time as a function of the beam balance ratio, $R$. Likewise, the lower curve of Fig. 20.5, "beam," also indicates the reduction of modulation bandwidth, fm, as function of R. Thus, the value $\beta$, which denotes the deterioration of the device modulation bandwidth, Eq. (20.5), is also given by the lower curve of Fig. 20.5.

According to Fig. 20.5, acousto-optic modulators with $R > 2$ will have low laser beam distortion and well-performed modulation bandwidth near its ideal value. Note that, in many prior literatures, the parameter "a" defined by the ratio of optical beam spread to acoustic beam spread is used instead. Although "a" is simply the inverse of R, we prefer to use "R" because from Fig. 20.5, the more useful range with $R > 1$ is shown in better detail.

As an example, the published experimental results of Maydan [12] gave measured rise time at $R \gg 1$ as Tr $= 1.3\omega_0/v = 0.65\ D/v$, where $\omega_0$ is radius of the laser beam

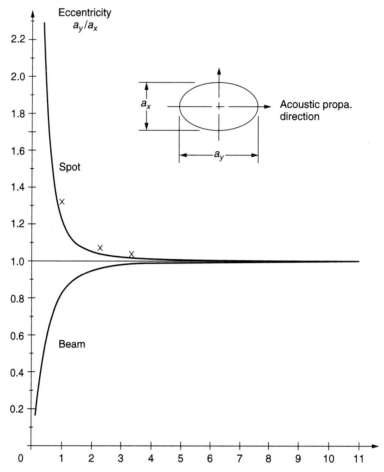

**Figure 20.5**  The distortion of the output laser beam is given by the eccentricity of the output "spot" right after the modulator exit window as a function of the acoustic-to-optic beam spread ratio, R. Fourier transform of the elliptical "spot" gives the "beam" at the far field and is elliptical in the orthogonal direction.

waist. At "$a = 1.5$," which corresponds to $R = 0.66$ in our figures, Maydan measured $Tr = 1.7\omega_0/v = 0.85\ D/v$. Note that, $0.85/0.65 = 1.308$, which is in good agreement with Fig. 20.5.

*Peak Diffraction Efficiency of Practical Acousto-Optic Modulators.* Integrating $|F1(y)|^2$ given by Eq. (20.23) over the y space yields the acousto-optic modulator diffraction efficiency with a realistic input laser beam profile. However, using Praseval's law, the same result can be obtained by integrating the product of $|Af(\theta/\delta\theta o)|^2$ and the plan wave acousto-optic diffraction efficiency, $\eta$ given by Eq. (20.19), over the angular space. The output laser power as a function of the beam balance ratio, $R$, is given by Fig. 20.6. For acousto-optic modulators operating under conditions of small $R$, some of the angular spectrum components of the incident laser beam experience a large mismatch angle resulting in reduced diffraction efficiency. The optical intensity in the reduced diffraction efficiency stays in the zeroth diffraction

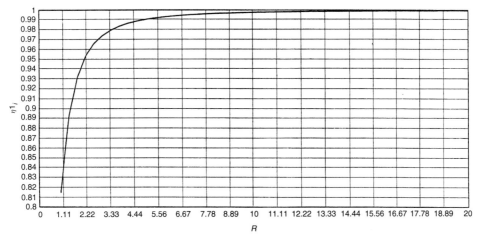

**Figure 20.6**  The peak diffraction efficiency of an acousto-optic modulator as a function of beam balance ratio, $R$, assuming $Q \gg 1$, allowing 100% peak efficiency in the ideal case.

order. Acousto-optic modulator diffraction efficiency into the first order higher than 94% can be obtained for $R > 2$, assuming $Q \gg 1$.

Figure 20.7**a** and **b** are calculated results for the diffracted beam, first order, and the undiffracted beam, zeroth order, observed at the far field of the acousto-optic exit aperture at maximum diffraction efficiency, showing the diffracted order narrows as the beam balance ratio R decreases, and a dark line appears at the far field of the undiffracted order. Experimentally, this is easily observed as one attempts to drive the acousto-optic modulator to its maximum diffraction efficiency. In a well-aligned acousto-optic modulator, a dark line would occur at the center of the zeroth-order beam when the center part of the laser wavefront meets the Bragg condition and is depleted. This is true even with modest values of $Q$ when the depleted laser intensity goes into other diffraction orders. On the other hand, the edges of the laser wavefront mismatch the Bragg condition slightly and cannot be diffracted out of the zeroth order completely, no matter how the acoustic drive power is adjusted. The straightness or curvature of this dark line indicates the quality of the optical wavefront and the acoustic wavefront in the acousto-optic modulator.

Remember the effect of finite value of the $Q$ parameter on peak acousto-optic diffraction efficiency given by Fig. 20.3. A small $Q$ value reduces peak diffraction efficiency because optical power is going into other diffraction orders, notably, the $-1$st order and the $+2$nd order. Although a small $R$ value reduces peak diffraction efficiency due to optical power left in the zeroth order, these two effects can be considered independent to each other. Thus, the peak diffraction efficiency achievable from a practical acousto-optic modulator can be obtained by multiplication of the data from Figs. 20.3 and 20.6. The result is given by Fig. 20.8, which is highly useful in predicting the device efficiency of an acousto-optical modulator design.

### 20.2.5  Temporal Transfer Function and Frequency Response (MTF)

In most applications, the acousto-optic modulator is driven by a modulated RF signal that is time dependent so that the output laser beam can be modulated as a function of

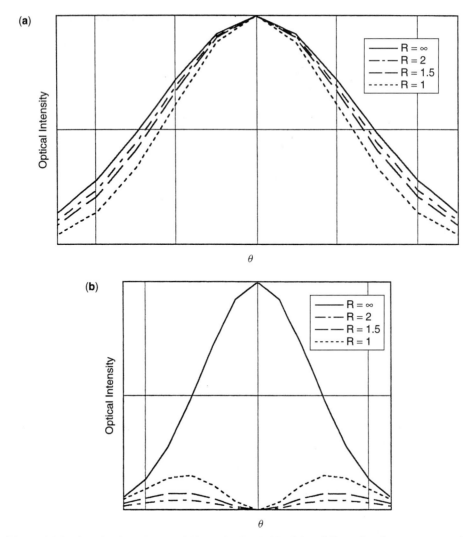

**Figure 20.7**    (a) The laser beam width at the far field of the diffracted order narrows as the beam balance ratio R decreases. (b) A dark line appears at the center of the far field of the undiffracted order when the acousto-optic modulator is operated near its peak efficiency level. The width of the dark line also decreases as the width of the diffracted order decreases when R decreases. Note, $\theta$ is proportional to displacement at far field.

time. In such cases, the acoustic-induced optical index, $n'$, and thus the phase shift term, $\phi$, are time dependent. By substitute $\phi$ with a time-variant expression, the temporal frequency response of an acousto-optic modulator can be obtained from Eqs. (20.15) and (20.16) for amplitude modulation and intensity modulation, respectively.

For the case of amplitude modulation, Fig. 20.9**a** shows (plan wave) acousto-optic diffraction efficiency as a function of applied acoustic field (or amplitude), with Bragg angle mismatch as parameter. It is termed amplitude modulation because the output optical intensity is proportional to the acoustic amplitude. When necessary, a photodetector or other square law devices will convert optical intensity into electrical current.

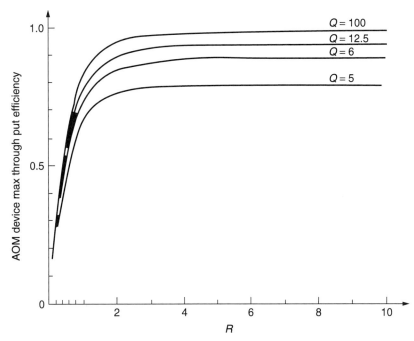

**Figure 20.8**  Maximum achievable diffraction efficiency of an acousto-optic modulator with finite $Q$ and $R$ values, as a function of $R$, the acoustic-to-optic beam balance ratio, with $Q$ as a parameter.

Linear response between output optical intensity and input acoustic amplitude occurs in the middle of the diffraction range, near 50% bias point for the case of no mismatch. Existence of angle mismatch by either focusing laser beam or detuning the incidence angle causes the diffraction efficiency curve to degrade with lower efficiency across the entire range. Due to the sinusoidal nature of these curves, harmonics will be generated when a sinusoidal signal is applied. Details of the nonlinearity and harmonics content will be discussed in the next section.

For intensity modulation, Fig. 20.9**b** shows the acousto-optic diffraction efficiency as a function of applied acoustic power with Bragg angle mismatch as parameter. The small-signal ranges of the diffraction efficiency curves are linearly proportional to the applied acoustic power. At higher drive power levels, the device saturates toward peak diffraction efficiency close to 100% if Bragg condition is satisfied. Once more, existence of mismatch term either due to focusing or due to detuning of the incidence angle causes the peak efficiency to decrease and will never reach 100% regardless of acoustic power applied. Again, due to the nonlinear slope of these curves, harmonics will be generated when a large signal is applied to acousto-optic intensity modulators.

Most acousto-optical modulators are intensity modulators because they do not require a bias level to exhibit a substantially linear response, and because most applications use optical intensity for exposure of photosensitive materials in simple on-off modes. The frequency response of on-off operation is best described by the device rise time. Nonlinearity due to saturation of diffraction efficiency is not a problem for digital on-off type of applications.

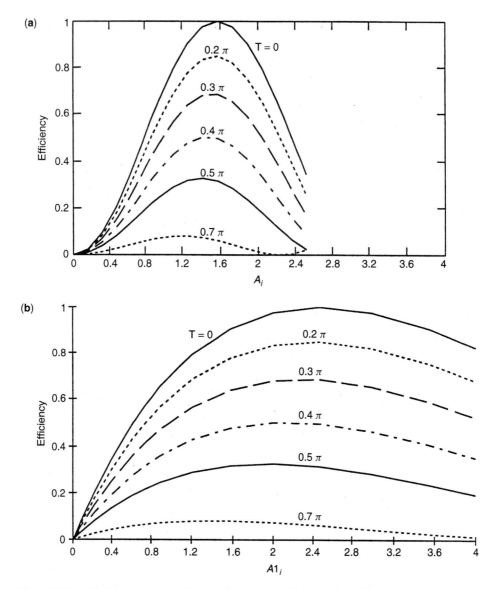

**Figure 20.9** (**a**) For acousto-optic amplitude modulation, the diffraction efficiency of acousto-optic Bragg diffraction is a function of acoustic field applied, with angular mismatch as a parameter. (**b**) For the case of an acousto-optic intensity modulator, the diffraction efficiency of acousto-optic Bragg diffraction is a function of acoustic power applied, with angular mismatch as a parameter.

However, when grayscale exposure is required with precise dosage control, the electric drive voltage (or current) needs to be converted by a square root formula before sending to the acousto-optic modulators. Electronically, this has been done either through a predistortion circuit or through a digital lookup table. Performances of such devices are complicated by the sophisticated electronics circuits and shall not be discussed here.

*Frequency Response (MTF) of Acousto-Optic Amplitude Modulation.* For linear operation of an acousto-optic amplitude modulator, frequency response (the Modulation Transfer Function) and linearity are key performance parameters. Although the rise time of an acousto-optic modulator has been discussed, the relationship between rise time and frequency response is interesting and shall be explored here [9].

To begin with, we shall introduce a time-dependent signal to the acousto-optic process. Noting that the transfer function of an acousto-optic amplitude modulator given by Fig. 20.7 requires a bias level for linearity, we let

$$\phi(t, y) = \pi \gamma [1 + \alpha \cos(\omega_m t - k_m y)] \tag{20.25}$$

where $\phi(t, y)$ is the phase term proportional to the traveling acoustic modulation envelope, $\gamma$ is a constant for setting up a bias level, $\alpha$ is the acoustic modulation index, $\omega_m = (2\pi \text{ fm})$ is the modulation angular frequency, and $k_m = (2\pi \text{ fm/v})$ is the wave vector for the envelope of the modulated acoustic wave that also travels at the speed of sound, v.

The laser output from a acousto-optic modulator is given by Eq. (20.23) for steady state when the acoustic field is constant. But for modulated acoustic field, the acousto-optic process with multiple frequencies belongs to the Hill's problem and is very difficult to solve analytically.

However, one may take advantage of the fact that most modulation envelopes are a slow varying function comparable to or slower than modulator rise time. By breaking up the optical aperture, D, into several smaller subapertures, the acoustic field can be considered uniform within each subaperture. The acousto-optic response of each of the subapertures can be computed using Eq. (20.23). Each subaperture element is elongated upon exiting the acousto-optic modulator. Thus, adjacent subaperture responses overlap. Use of the superposition theorem allows us to sum up these subaperture responses, resulting in the description for the output laser beam with time-dependent acoustic field, which can be written as follows:

$$F1(y, t) = \delta\theta o \int_{-\infty}^{\infty} Af(\theta/\delta\theta o)\phi(y, t)\text{sinc}[\phi(y, t)^2$$
$$+ (\pi(\theta/\delta\theta o)/R)^2]^{1/2}e^{jky}(\theta/\delta\theta o)d(\theta/\delta\theta o) \tag{20.26}$$

Next, we note from Fig. 20.4**a** that the transfer function can be considered nearly constant for a range of small angular mismatch within 20% of the acoustic beam spread. In reality, except some of the highest speed modulators, most acousto-optic modulators operate with relatively large beam balance ratio and fit in this category. Thus, for the cases of $R > 2.5$, we may assume the acousto-optic transfer function being independent of mismatch angle, and it is simply $\sin\phi(y, t)$. Moving the simplified acousto-optic transfer function factor $\sin\phi(y, t)$ out of the integral on $\theta$, and introducing the inverse Fourier transform of $Af(\theta/\delta\theta o)$ into Eq. (20.26), we obtain the following simple approximation:

$$F1(y, t) = F(y) \sin[\pi \gamma (1 + \alpha \cos(\omega_m t - k_m y))] \tag{20.27}$$

Equation (20.27) can be evaluated for the case of a Gaussian input laser beam with beam waist diameter of $D$,

$$F(y) = \exp(-4y^2/D^2) \tag{20.28}$$

The optical intensity of the diffracted light subject to time-dependent acoustic field is, therefore,

$$I = \int |\exp(-4y^2/D^2)\sin[\pi\gamma(1 + \alpha\cos(\omega_m t - k_m y))]|^2 dy \qquad (20.29)$$

Equation (20.29) can be expanded into a series of Bessel functions multiplied by sinusoidal harmonics. The Bessel functions are not time and space dependent and thus can be moved out of the integral, and the rest become Fourier integrals.

For example, let $\gamma = 0.25$, corresponding to a modulator biasing level at 50% diffraction efficiency point. We have

$$I/Io = 1/2 + J_1(\alpha\pi/2)\exp(-\omega_m^2 Tt^2/32)\cos(\omega_m t)$$

$$+ \sum_{n=1}^{\infty} J_{2n+1}(\alpha\pi/2)(-1)^n \exp(-(2n+1)^2 \omega_m^2 Tt^2/32)$$

$$\times \cos((2n+1)\omega_m t) \qquad (20.30)$$

It is noted that the modulated light intensity for a properly biased acousto-optic amplitude modulator contains only odd harmonics. The depth of modulation for the fundamental frequency is given by $J_1(\alpha\pi/2)$. The other higher order Bessel function terms represent harmonics distortions.

From Eq. (20.30), the frequency response for the acousto-optic amplitude modulator biased at 50% point resembles the Fourier transform of the input laser beam profile, and is $\exp(-\omega_m^2 Tt^2/32)$, as shown by Fig. 20.10 [9]. Experimental data from two

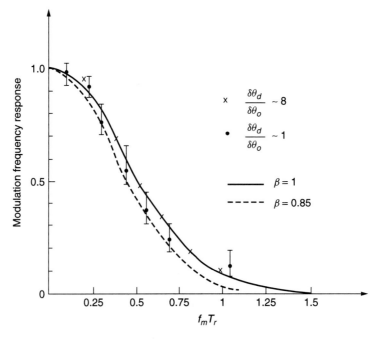

**Figure 20.10** The frequency response (MTF) for an acousto-optic amplitude modulator.

acousto-optic temporal modulators are plotted against the theoretical curve. Note that, for the case of $R = 8$, the data are in excellent agreement with the theoretical model with $\beta \sim 1$. For the $R = 1$ modulator, the rise-time degradation factor is estimated to be $\beta \sim 0.85$. The experimental data also fit well to the degraded frequency response curve, except the last data point when the modulation frequency is way out of the 3-dB bandwidth [9].

The third harmonic intensity content relative to the fundamental frequency intensity is therefore

$$I(3\omega_{\mathrm{m}})/I(\omega_{\mathrm{m}}) = |\,[J_3(\alpha\pi/2)/J_1(\alpha\pi/2)]\exp(-\omega_{\mathrm{m}}^2\,\mathrm{T}t^2/4)\,|^2 \qquad (20.31)$$

which increases with the modulation index, $\alpha$, and decreases with the modulation frequency [9]. Figure 20.11 shows the behavior of fundamental and third harmonic contents as a function of the modulation index, with normalized modulation frequency as a parameter. Note when normalized modulation frequency increases, the fundamental frequency response decreases. But the third harmonic frequency response decreases even faster due to the exponential factor in Eq. (20.31), which behaves as a low-pass filter.

Similarly, one can obtain the frequency response for different bias levels, and even harmonics will appear. For the case of intensity modulators, it can be shown that only even harmonics exist.

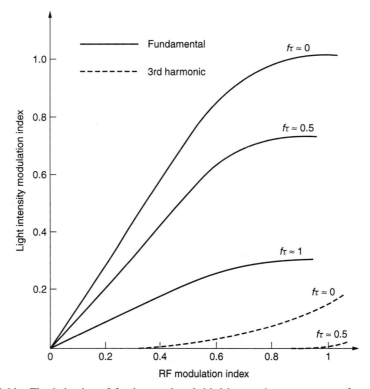

**Figure 20.11**  The behavior of fundamental and third harmonic contents as a function of the modulation index, with normalized modulation frequency as a parameter.

In theory, the frequency response of acousto-optic modulators can be tailored by shaping the input laser beam profile. However, it is difficult to implement such a shaped laser beam due to the need to accommodate diffraction side-lobes at the acousto-optic modulator entrance pupil.

### 20.2.6 Multifrequency Acousto-Optic Beam Deflector and Intermodulations

Another important category of acousto-optic modulators includes acousto-optic spatial light modulators and acousto-optic beam deflectors that are highly useful in optical signal processing, RF spectrum analysis, and laser scanning types of instruments [3, 4]. Often, the useful light output is in the far field of the acousto-optic exit aperture. The input laser beam width, $D$, is large compared with the signal modulation period. Such devices are noted for a very large beam balance ratio, $R \gg 1$.

The resolution of an acousto-optic beam deflector, or the capability of an acousto-optical signal processor, including an acousto-optic spatial light modulator and a RF spectrum analyzer, is determined by the time-bandwidth product, which is [2]

$$TB = (D/v)(\Delta f) \qquad (20.32)$$

Note that $(v/\Delta f)$ is the period of the modulation signal, $\Delta f$, and has the meaning of bit size (or pixel) when the modulated signal is a bit stream. Thus, $TB$ denotes the number of bits, or pixel, covered by the laser beam for signal processing [2]. Rearranging Eq. (20.32), we have

$$TB = (\lambda/\Lambda_1 - \lambda/\Lambda_2)/(\lambda/D) \qquad (20.33)$$

which is the Bragg diffraction angle range for the acoustic bandwidth $(f_1 - f_2) = \Delta f$ divided by the angular spread of the laser beam, and it is the number of resolvable spots covered by the acousto-optic beam deflector.

Here, let us turn our attention to the case when two frequencies are applied to the acousto-optic beam deflector simultaneously. A beat will be formed on the acoustic field. When the two frequency components are equal magnitude, the acoustic wave will show a carrier at the center frequency and a modulation envelope at the difference frequency. Thus,

$$\phi(t, y) = \pi\alpha \cos(\omega_m t - k_m y) \qquad (20.34)$$

As $R \gg 1$, we can assume the acousto-optic transfer function is almost independent of $\theta$. Taking the Fourier transform of $F1(y, t)$ given by Eq. (20.26) and expanding the acousto-optic transfer function containing Eq. (20.34) into the products of Bessel functions and sinusoidal functions, we have

$$F1(\theta) = FT\left\{F(y)\left[2\sum_{n=0}^{\infty}(-1)^n J_{2n+1}(\pi\alpha)\cos[(2n+1)(\omega_m t - k_m y)]\right]\right\}$$

$$= \sum_{n=0}^{\infty}(-1)^n J_{2n+1}(\pi\alpha) FT\{F(y)[\exp(j(2n+1)(\omega_m t - k_m y) + \text{c.c.}]\}$$

$$(20.35)$$

where $FT\{g(y)\}$ denotes the Fourier transform operation on $g(y)$. Because an exponential term in the Fourier transform integral means shifting in the frequency domain, we obtain

$$F1(\theta) = \sum_{n=0}^{\infty} (-1)^n J_{2n+1}(\pi\alpha)[F(\theta - (2n+1)k_m/k_o)\exp(j(2n+1)\omega_m t)$$

$$+ F(\theta + (2n+1)k_m/k_o)\exp(-j(2n+1)\omega_m t)] \qquad (20.36)$$

where $k_o = 2\pi/\lambda$ is the wave vector for the laser light. When the laser beam is collimated with large aperture, $D$, then the Fourier transform of the laser beam, $F(\theta)$, resembles a point function. The first two terms on the right-hand side of Eq. (20.36) with $n = 0$ corresponds to the original two-tone signal terms. All other terms with nonzero value of $n$ are spurious intermodulation terms appearing at displaced locations. The exponential factor is the frequency shift in the laser output spots that is characteristic of acousto-optic processes.

Let the laser output go through a focusing lens of focal length, fl, a typical component of Eq. (20.36), such as $F(\theta - (2n+1)k_m/k_o)$ will be offset from $F(\theta)$ by a distance $-(2n+1)k_m/k_o$ (fl) on the back focal plan. Thus, Eq. (20.36) produces a number of laser spots on the back focal plan at the following locations:

$$x_f(n) = +- (2n+1)k_m fl/k_o$$

$$= +- [(2n+1)fm](\lambda fl/v) \qquad (20.37)$$

which are proportional to the frequencies $(2n+1)$ fm.

Noting that the applied two-tone frequencies, $f1$ and $f2$, are related to the center frequency fo and the beat frequency fm by

$$f1 = fo - fm$$

$$f2 = fo + fm \qquad (20.38)$$

and that the optical axis is defined by the output laser beam for the center frequency fo, the locations for the first two terms of Eq. (20.36) corresponding to $n = 0$ are for the two tones, $f1$ and $f2$. The locations for other terms of Eq. (20.36) are given by

$$fo + (2n+1)fm = (n+1)f2 - n\ f1$$

$$fo - (2n+1)fm = (n+1)f1 - n\ f2 \qquad (20.39)$$

The $n = 1$ terms are the third-order intermodulation products with intensity of the third-order intermodulation products given by $|J_3(\pi\alpha)|^2$. These results are identical to published results by Hecht using far more complicated mathematical manipulations [20]. From $|F1(\theta)|^2$, the following are obtained for two-tone operation of acousto-optic beam deflectors or acousto-optic RF spectrum analyzers:

| | |
|---|---|
| Diffraction efficiency | $|J_1(\pi\alpha)|^2$ |
| Maximum Diff Efficiency | 0.339 at $(\pi\alpha) = 1.841$ |
| Depletion | $1 - |J_0(\pi\alpha)|^2$ |
| Compression | $1 - |J_1(\pi\alpha)/\sin(\pi\alpha/2)|^2$ |
| Third-Order Intermodulation | $|J_3(\pi\alpha)|^2$ |

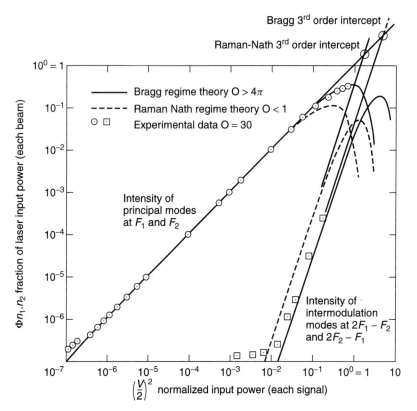

Bragg 3$^{rd}$ order intercept

Raman-Nath 3$^{rd}$ order intercept

———— Bragg regime theory O > 4$\pi$

- - - - Raman Nath regime theory O < 1

○ □    Experimental data O = 30

Intensity of principal modes at $F_1$ and $F_2$

Intensity of intermodulation modes at 2$F_1$ − $F_2$ and 2$F_2$ − $F_1$

$\Phi n_1 \cdot n_2$ fraction of laser input power (each beam)

$\left(\dfrac{V}{2}\right)^2$ normalized input power (each signal)

Two-tone third-order intermodulation.

**Figure 20.12**   Two-tone third-order intermodulation products of acousto-optic modulators (following Hecht [20]).

Figure 20.12 is the published result on two-tone intermodulation product for acousto-optic modulators, including experimental data. The spurious signal level is less than −35 dB when Bragg diffraction efficiency is less than 10%.

Another example is for two unequal tones, for instance, a large tone and a second tone substantially weaker than the first one. In such a case, we arrive at

$$\phi(t, y) = \pi \alpha [1 + B \cos(\omega_m t - k_m y)] \qquad (20.40)$$

which is the same form as Eq. (20.25). The bias point is determined by the strength of the larger tone and is not set to any particular point. The Bessel series expansion will have both even and odd orders, and the transfer function can be expanded into

$$\sin(\phi) = \sin(\alpha \pi) \left[ J_0(\alpha \pi B) + 2 \sum_{n=0}^{\infty} J_{2n}(\alpha \pi B)(-1)^n \cos(2n\omega_m t - 2n\, k_m y) \right]$$

$$+ \cos(\alpha \pi) \left[ 2 \sum_{n=0}^{\infty} J_{2n+1}(\alpha \pi B)(-1)^n \cos((2n + 1)(\omega_m t - k_m y)) \right] \qquad (20.41)$$

Intermodulation terms occur for both even and odd orders with the larger tone being $\sin^2(\alpha\pi)J_0(\alpha\pi B)^2$, and the weak tone being $4\cos^2(\alpha\pi)J_1(\alpha\pi B)^2$, and the m-th intermodulation product being $4\sin^2(\alpha\pi)J_m(\alpha\pi B)^2$ for m even, and $4\cos^2(\alpha\pi)J_m(\alpha\pi B)^2$ for m odd.

## 20.3   DESIGN PROCEDURES OF ACOUSTO-OPTIC MODULATORS

### 20.3.1   Conventional Design Procedures

Although there are a few alternative design approaches [9, 14, 15], the general rule of thumb for the design of an acousto-optic Bragg cell is to begin with a decision of center frequency based on desired rise time or bandwidth considerations. Then, the candidate materials shall be selected. Next, the length of the acoustic transducer is determined by the following [2]:

$$L <= 1.8 \, nv^2/(\lambda \, f_o \Delta f)$$
$$= 1.8 \, L_o(f_o/\Delta f) \qquad (20.42)$$

where $L_o$ is the characteristic length defined by Eq. (20.1), $f_o$ is the center frequency of the acousto-optic modulator, $\Delta f$ is the bandwidth of the modulator defined by the acousto-optic process, and $(\Delta f/f_o)$ is the fractional bandwidth of the modulator. The transducer height, $H$, is then estimated by requiring the acoustic near field to cover the entire optical aperture, $D$. Due to heating generated by power dissipation at the acoustic transducer, it is customary to keep the optical window at least 0.5 mm away from the transducer surface in order to maintain optical beam quality. Thus, the acoustic transducer height must be

$$H >= [(500 + D)/\Lambda_o]^2 \qquad \text{micron} \qquad (20.43)$$

where $\Lambda_o$ is the acoustic wavelength at center frequency fo. Finally, the required acoustic drive power shall be estimated to see if it is feasible.

$$P(@100\%) = 135(H/L/M_2')(\lambda/0.6328)^2 \qquad \text{watts} \qquad (20.44)$$

in which $M2'$ is the relative acousto-optic figure of merit for the selected acousto-optic interaction media compared with fused quartz. Note, the acousto-optic figure of merit is defined by [22]

$$M_2 = n^6 p^2/(\rho v^3) \qquad (20.45)$$

where $\rho$ is density of the acousto-optic material and $(\rho v)$ is the characteristic impedance of the acoustic material. Referring to Eq. (20.8), it is obvious that $M_2$ is related to the acoustic power density required to induce a refractive index change of n'.

Once the acoustic power is acceptable, then the design is deemed completed. The dimensions, L, and H, provide the acoustic transducer top electrode area. Given that, one may move on to estimate the transducer input impedance level and the electrical tuning circuit for impedance matching from a 50-ohm drive amplifier. With the acousto-optic device parameters all decided, then the focal distance of the optical lens shall be determined to yield proper laser beam diameter before entering the acousto-optic modulator. Quality of the device is then reviewed by calculating parameters Q and R to see if satisfactory Bragg interaction can be expected subject to considerations of Section 20.2. Such design procedure is not only time consuming, but also it cannot assure the anticipated performances.

However, it is also possible to design an acousto-optic modulator starting from detail performance parameters and system requirements. The design procedure to be described in the next section will start by specifying the Q and R parameters and work its way to the transducer parameters fo, $L$ and $H$ [9]. In this alternative design approach, which has been used by the author since the late 1970s, design iterations can be shortened and anticipated device performances can be assured.

### 20.3.2   Design Procedures from Specifications and Requirements of Applications

A good modulator design procedure should not be limited to only modulation bandwidth or rise-time considerations. Keeping optical distortion and aberration within tolerance range is also very important in high-performance electro-optical system design. Laser power throughput efficiency is another critical parameter that must be considered due to limited availability of laser power in many system applications.

As discussed in Section 20.2, an acousto-optic modulator always introduces a small amount of laser beam distortion into the system. The amount of distortion for a given modulator rise time can be traded off with other acousto-optic device parameters such as center frequency, peak achievable diffraction efficiency, and drive power requirement. Also, an acousto-optic modulator cannot provide 100% throughput of laser power. However, with a good design, efficiency up to 80% to 95% may be assured in most cases. A logical design procedure is given as follows, for temporal modulators only. Modifications in design considerations for other types of acousto-optic modulators, such as spatial modulators for signal processing or RF spectrum analysis, and acousto-optic beam deflectors will be discussed later.

1. Determine modulation bandwidth, rise time, optical throughput requirements, and optical beam distortion tolerances, together with optical wavelength as well as laser power, for a given application.

2. Determine the acousto-optic modulation rise time and transit time required to provide the necessary bandwidth for the system.

3. Select a list of material candidates for the modulator.

4. Find out the minimum value of R allowed to maintain optical beam quality using Fig. 20.5.

5. Obtain the modulation bandwidth reduction factor, $\beta$, with given $R$ value from Fig. 20.5. Adjust the rise time if necessary.

6. Adjust transit time value, and then calculate the optical aperture $D$.

7. Figure out the minimum value of parameter $Q$ in order to achieve required optical throughput together with the selected $R$ value, using Fig. 20.7.

8. Compute the device center frequency, fo, from $Q$, $R$, and fm.

9. Select the acousto-optic modulator material of highest figure of merit [use $M_2$ times ($nv^{1.5}$) because of dependency on $L$ and $H$] from the group of candidate

materials of suitable optical transparency and adequate acoustic propagation loss factor at fo.

10. Compute the acousto-optic interaction length, $L$, transducer height, and $H$, from $Q$, $R$, fm, and fo.

11. Calculate the acoustic power required to achieve desired diffraction efficiency.

***Design Formulas using Q and R.*** The acousto-optic modulator parameters can be obtained from Eqs. (20.5), (20.10), and (20.24). First, the optical aperture $D$ is obtained from Eq. (20.5). From Eqs. (20.10) and (20.24), we obtain the center frequency fo in terms of $Q$ and $R$. $L$ can be obtained from fo and Eq. (20.24). Noting the approximation, (Tr fm) $\sim$ 0.5, we have

$$D = 0.75\beta v/\text{fm} \tag{20.46}$$

$$\text{fo} = 2\, vQR/(\pi^2 D)$$

$$= 0.27QR\, \text{fm}/\beta \tag{20.47}$$

$$L = n\pi^3 D^2/(8\lambda QR^2)$$

$$= 2.18\beta^2 nv^2/(\lambda QR^2\, \text{fm}^2) \tag{20.48}$$

$$H \geq = [(500 + D)\, v/\text{fo}]^{0.5} \tag{20.49}$$

From Eqs. (20.46) to (20.49), the dimension of the acousto-optic modulator material and the dimension of the acoustic transducer can be calculated, once the choice material is determined.

***Device Design Example.*** As an example, we shall design a acousto-optic UV intensity modulator for a high-performance laser printing system. First, we decide to use fused quartz because of its high optical quality that can tolerate intense UV laser irradiation. Thus, the following material constants are obtained:

| | |
|---|---|
| Optical wavelength | $\lambda = 0.35$ micron |
| Material | Fused Quartz |
| Optical refractive index | $n = 1.477$ |
| Acoustic velocity | $v = 5.95$ mm/micro-sec |

The video bandwidth required is obtained for the specific system of interest, and it is 20 MHz. However, the system can not tolerate a 3-dB degradation in system bandwidth. Instead, a 1.5-dB bandwidth degradation is acceptable. It is estimated that a 3-dB modulation bandwidth of 28 MHz will be required. We would like to give it a bit more bandwidth to be safe, so we let fm be 32.5 MHz. Thus,

Modulation 3-dB bandwidth      fm = 32.5 MHz

Next, we understand the system application, that the writing speed is limited by available UV laser power in many cases. Although continuous UV laser is very expansive, it is also difficult to obtain with reliable high power output. Therefore, high diffraction efficiency in the acousto-optic modulator is highly desirable. But, we also understand that, for a very aggressive diffraction efficiency, the $R$ and $Q$ values will be high, and the device will be small with very high acoustic power density to achieve high diffraction efficiency. Although fused quartz does not have an exceptionally high acousto-optic figure of merit, $M_2' = 1$, the acoustic drive power required may be large. High acoustic power density is not desirable for reasons of thermal stability and ultimate device failure. Thus, we set a minimum goal of achieving 88% diffraction efficiency, and look into Figs. 20.3, 20.6, and 20.7 for recommended $Q$ and $R$ values. In the meantime, we appreciate the importance of minimizing laser spot distortion to print quality, and we decided the laser beam distortion must be kept under 5% in eccentricity. Thus, from Fig. 20.5, we find $R > 2$. And, from Fig. 20.7, we find that $Q > 12.5$ is a necessity to achieve >88% diffraction efficiency. We decide to let $Q$ be a bit smaller and $R$ be a bit larger because of the importance to have a good circular laser beam. Thus,

| | |
|---|---|
| Beam balance ratio | $R = 2.2$ |
| Acousto-optic Q factor | $Q = 12$ |
| Rise-time degradation factor | $\beta = 0.952$ |

where the rise-time degradation factor $\beta$ is the same as the beam eccentricity tolerance value.

Using this set of modulator performance parameters, we calculated the acousto-optic modulator design parameters and dimension values using the equations of Section 3.2.1, and the results are

$$D = 0.13 \text{ mm} \quad \text{or} \quad 130 \text{ micron at } e^{-2} \text{ width}$$

$$\text{fo} = 243.3 \text{ MHz}$$

$$L = 4811 \text{ micron}$$

$$H >= 124 \text{ micron}$$

To determine the transducer height, H, one has to consider the need to accommodate the laser beam diameter, which is 130 micron. Thus, we shall make $H = 250$ micron in order to align the laser beam through the device with sufficient ease. With these numbers, the estimated acoustic drive power needed to achieve maximum diffraction efficiency level will be, using Eq. (20.44),

$$P(\text{max eff.}) = 2.13 \text{ watt}$$

In reality, a 10% larger electrical drive power is needed due to losses in the transmission lines and the impedance-matching circuit elements. The device had been built and performed rather close to the anticipated values.

***Material Selection Considerations.*** Material selection is a critical issue in acousto-optic device design. Although a long list of potential useful materials have been listed on some open literature [22–24], most are not being used because of either availability, not suited in certain optical transmission range, unable to handle large optical power density, or simply difficult to fabricate the device on. Thus, for an experienced acousto-optic device designer who is not in research mode, here there is not really too many material choices. A short list is provided here for a few interesting wavelength ranges. This list is by no means an extensive list for all possible applications, but it is the mostly commonly used materials. Also, the use of crystalline materials may provide many new designs when a special cut and orientation is made, and only the most common cut and orientation is provided here. Care should be taken that most materials require optical input polarization in specific orientation, as shown in Table 20.2. Only a few materials such as $PbMoO_4$, and the shear waves in Ge and in GaP work with random optical polarization.

***Effect of Acoustic Attenuation.*** It is important in material selection to avoid use of material that will exhibit high acoustic attenuation in the frequency band of interest. Diffraction efficiency will decrease, drive power will increase, and in the extreme cases, the diffraction efficiency may exhibit an exponential decay across the optical aperture. In the material list, the acoustic attenuation at 1 GHz is also provided as reference. The acoustic attenuation in most materials increases at the square of acoustic frequency. For most materials in the list, acoustic attenuation will be significant for devices with large optical aperture, i.e., for $D > 10$ mm, when frequency is higher than about 300 MHz. In the case of the highly useful slow shear waves in $TeO_2$, the acoustic attenuation becomes significant at 50 MHz or higher. Thus, often we must live with some acoustic attenuation across the optical aperture when designing high-speed acousto-optic modulators, particularly for the modulators with large optical aperture.

Analytical results have been carried out for optical devices with exponential decay of intensity across the optical aperture for the case of uniform illumination and for

**TABLE 20.2    Properties of Popular A-O Material**

|  | no | ne | $\rho$ g/cm$^3$ | $M_2$ | v m/ms | $\Gamma$ dB/cm/Ghz |
|---|---|---|---|---|---|---|
| *For UV wavelengths* | | | | | | |
| Fused Quartz (longitudinal wave, $\perp$) | 1.457 | — | 2.20 | 1.56 | 5.96 | 12 |
| Fused Quartz (shear wave, // or $\perp$) | 1.457 | — | 2.20 | 0.467 | 3.76 | — |
| Crystalline Quartz (100 longitudinal, $\perp$) | 1.544 | 1.553 | 2.65 | 2.38 | 5.72 | 3 |
| *For visible and near to mid IR wavelengths* | | | | | | |
| $TeO_2$ (001 longitudinal wave, $\perp$) | 2.26 | 2.41 | 6.0 | 34.5 | 4.2 | 15 |
| $TeO_2$ (110 slow shear wave) | 2.26 | 2.41 | 6.0 | 793 | 0.616 | 290 |
| $PbMoO_4$ (001 longitudinal wave) | 2.386 | 2.262 | 6.95 | 36.1 | 3.63 | 15 |
| Dense Flint Glass (SF4) ($\perp$) | 1.616 | — | 3.59 | 4.5 | 3.63 | >200 |
| $LiNbO_3$ (100 longitudinal wave) | 2.29 | 2.20 | 4.64 | 7.0 | 6.57 | 0.15 |
| $LiNbO_3$ (100 35° shear wave, //) | 2.29 | 2.20 | 4.64 | 6.5 | 3.6 | ~1 |
| *For IR wavelengths* | | | | | | |
| Ge (111 longitudinal wave, //) | 4.0 | — | 5.33 | 840 | 5.50 | 30 |
| Ge (100 shear wave) | 4.0 | — | 5.33 | 290 | 3.51 | 9 |
| $As_{12}Se_{55}Ge_{33}$ ($\perp$) | 2.7 | — | 4.4 | 248 | 2.52 | 29 |
| GaP (110 longitudinal wave, //) | 3.31 | — | 4.13 | 45 | 6.32 | 8 |
| GaP (110 shear wave, $\perp$) | 3.31 | — | 4.13 | 24 | 4.13 | 2 |

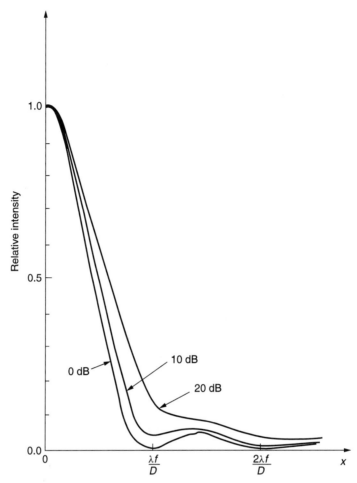

**Figure 20.13** Far-field intensity pattern for a uniformly illuminated acousto-optic modulator aperture, subject to acoustic attenuation across the aperture.

the case of Gaussian illumination [21]. For uniform illumination, the far-field intensity pattern is

$$I/Io = \exp(-\alpha D)|\,\text{sinc}[(k_o\,D\,x/(2\,\text{fl})) - (j\alpha D/2)]|^2 \qquad (20.50)$$

where $\alpha D$ is the acoustic attenuation across the optical aperture and fl is the focal length of the transform lens. As shown by Fig. 20.13, the acoustic attenuation across optical aperture causes larger spot width, higher side-lobes, and is rather undesirable when $\alpha D$ equals or exceeds 10 dB. In addition, the intensity level is reduced because of the exponential factor in Eq. (20.50).

For the more interesting case of calculating the far-field pattern of a Gaussian laser input beam subject to acoustic attenuation, the acoustic attenuation factor, $\exp(-\alpha D)$, can be manipulated with the Gaussian formula and the result in the far-field plan becomes

$$I(x)/Io = (D^2/4)\exp[-\alpha D + (\alpha D)^2/8]\exp[-k_o^2 D^2 x^2/(8\,\text{fl}^2)] \qquad (20.51)$$

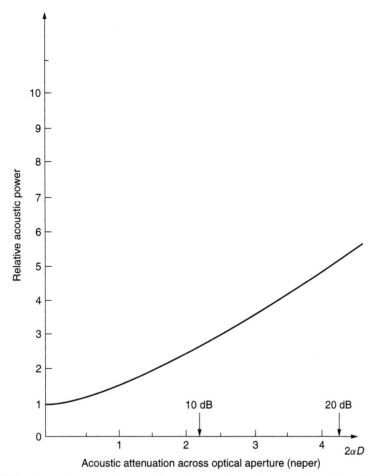

**Figure 20.14**   Acoustic power increases in order to restore the output laser intensity level for Gaussian laser beam after an acousto-optic modulator, subject to acoustic attenuation across the modulator aperture.

where the first exponential factor means reduced optical intensity in the far-field pattern, and the second exponential factor is identical to the laser far-field pattern without acoustic attenuation. Therefore, for the case of a Gaussian laser beam, the acoustic attenuation across the optical aperture only causes optical intensity loss in the far-field pattern. Figure 20.14 gives the increase of acoustic power in order to compensate for the optical intensity loss, assuming the device operates in linear intensity mode. For operations in maximum diffraction efficiency mode, the lost optical intensity cannot be recovered and peak throughput will suffer.

Note the ability for a Gaussian laser beam to hold its shape subject to acoustic attenuation does not mean we can ignore the acoustic attenuation of an acousto-optic modulator in a laser system. The attenuation will cause the effective center location of the laser beam to shift toward the acoustic source, loss of peak diffraction efficiency, and cause thermal heating in the modulator body, which may result in additional laser beam distortion. In particular, the thermal effects must be investigated carefully for each application or adverse effects may arise.

***Q and R Parameters for Other Acousto-Optic Devices.*** There are several important acousto-optic modulators that do not belong to the category of temporal modulator. Examples are spatial light modulators, RF spectrum analyzers, and acousto-optic beam deflectors. All of these devices have large optical aperture, may operate in low diffraction efficiency range, and may demand a very large acoustic bandwidth of operation. The design considerations for these devices shall be discussed in this section. (Note, we do not consider an acousto-optic tunable filter in this chapter, although it can also be used as a laser temporal modulator, albeit being slow, because it operates rather differently and thus demands much more room for discussion.)

The design modifications required for these other types of acousto-optic modulators are mainly in definition of the $Q$ and $R$ parameters. Once the effective $Q$ and $R$ parameters for each application are defined, the design procedure for the acousto-optic temporal modulator can still be employed.

*The Q Parameter for Spatial Modulators and Beam Deflectors.* The $Q$ parameter is important for achieving high optical throughput with Bragg diffraction. However, the operation of RF spectrum analyzers and most spatial light modulators involves many acoustic frequency components and design priority is more about linearity than about throughput efficiency. The multifrequency coupled mode equations for acousto-optic process belongs to the Hill's problem, which is difficult to solve analytically. The linearity issue involving two frequencies has been discussed in Sections 2.5.1 and 2.6. From Figs. 20.11 and 20.12, it is noticed that in order to keep spurious signal under $-20$ dB, the combined signal strength in terms of diffraction efficiency must be under somewhere about 40%.

Figure 20.12 shows the intermodulation product for Raman Nath regime at $Q \sim 1$ is generally higher than the intermodulation product for Bragg regime at $Q \sim 12.5$ by about 10 dB. A large $Q$ value is required for the sake of keeping spurious signal low. Thus, although for different reasons, the desirable $Q$ value remains near $4\pi$ or 12.5 for low diffraction efficiency applications such as spatial light modulators, RF spectrum analyzers, as well as for high diffraction efficiency applications, such as acousto-optic temporal modulators and acousto-optic beam deflectors.

*The R Parameter for Spatial Modulators and Beam Deflectors.* For the cases of spatial light modulators, RF spectrum analyzers, or acousto-optic beam deflectors, the input laser beam is collimated and aligned to the Bragg condition at the center of the frequency bandwidth. Thus, the optical beam spread in the incident light is always much smaller than the acoustic beam spread. Conventional sense of $R$ will be much greater than 1 in all of these cases. However, the collimated input laser beam with large beam diameter, $D$, may be considered as a superposition of many smaller laser beams in close proximity, in a way similar to the subaperture discussion in Section 2.5.1. Here we shall define a pixel dimension being $(v/\Delta f)$, where $\Delta f$ is the bandwidth of the acoustic signal of interest in the acousto-optic modulator. The number of pixel elements within the laser beam diameter $D$ is therefore $(D\Delta f/v) = (D/v)\Delta f$, which is the product of transit time and the bandwidth, the time-bandwidth product.

$$\text{Pixel} = v/\Delta f \tag{20.52}$$

$$\text{TB} = D\Delta f/v \tag{20.53}$$

Note that the definitions of pixel as well as TB have important meaning in signal processing. It is indicative of the information content and the signal processing capability of an optical system. Also, we note that the modulation frequency of an acousto-optic temporal modulator, fm, is related to the acousto-optic bandwidth by, $\Delta f = 2$ fm, due to double side band of modulated electrical signal.

Consider the narrowed laser beam out of each pixel-sized subaperture; the acousto-optic device acts like an acousto-optic temporal modulator, as discussed in previous sections. Each of the pixels comes out of the acousto-optic diffraction process slightly elongated due to the finite value of its beam balance ratio, Rs. Superposition of the elongated pixels together results in an image of the acoustic field of the full optical aperture, D, but it is slightly blurred. As the coherent image transfer function is Fourier transform of the point spread function, the elongated pixels result in degradation of the coherent image transfer function. The image of the acoustic field will be blurred unless the beam balance ratio of the subaperture is kept large, Rs >> 1.

Thus, the performance of an acousto-optic spatial light modulator is defined by the beam balance ratio of the "Pixel" instead of the entire laser beam diameter, D. Note that it is possible to assume each pixel being a Gaussian distribution with $1/e^2$ width defined by Eq. (20.52). Thus,

$$Rs = (\delta\theta a/\delta\theta op)$$

$$= n\pi v^2/(4\lambda \text{ fo}\Delta fL) \tag{20.54}$$

$$Rs = n\pi vD/(4\lambda \text{ fo } L)/TB$$

$$= R/TB \tag{20.55}$$

where $\delta\theta op$ is the optical beam spread of the pixel element and R is the beam balance ratio based on the actual laser beam diameter, D. The design of an acousto-optic spatial light modulator must now determine the appropriate Rs value, Rs = R/TB, for achieving acceptable spatial resolution degradation, or in other words coherent image transfer function degradation. From Fig. 20.5, the degradation of the coherent image transfer function is same as the rise-time degradation factor, $\beta$.

Therefore, if we drive a given acousto-optic spatial light modulator with larger signal bandwidth $(c\Delta f)$, where the constant c is greater than 1, this will result in a larger number of pixels, cTB, but will be subject to more severe pixel elongation as well as resolution degradation factor, $\beta'$, due to smaller beam balance ratio, Rs/c. The resultant spatial resolution for the acousto-optic spatial light modulator should remain unchanged. This is expected for the same acousto-optic spatial light modulator. Also note that, in another interpretation, the blurring of each pixel is due to convolution with the scaled acoustic transducer image. This is because the acousto-optic transfer function is the Fourier transform of the rectangular-shaped acoustic transducer pattern at a reduced scale, $\sin\theta_B$. So, the resolution limit is dictated by the convolution with scaled transducer image regardless of the selection of pixel size.

An acousto-optic RF spectrum analyzer is similar to an acousto-optic spatial light modulator with the observation plan moved to the far field. Note that in a well-designed coherent imaging system, equivalent resolution numbers in the pixel domain and in the spatial frequency domain are identical due to space-bandwidth product conservation. Therefore, the design considerations of beam balance ratio for an acousto-optic spatial light modulator are also applicable to an acousto-optic RF spectrum analyzer.

For the case of an acousto-optic beam deflector, there is usually only one acoustic frequency in the acousto-optic device at one time. (Note, except the case of an acousto-optic device driven by a chirp frequency for which the following discussion is still valid, subject to decomposing the optical aperture into TB number of subapertures.) However, the acoustic frequency changes from time to time, over the bandwidth $\Delta f$. The laser output shall scan over an angular space of $\Delta\theta s = \lambda\Delta f/(nv)$, which in terms of the incident laser beam spread is

$$\Delta\theta s = \lambda\Delta f/(nv)$$
$$= (\pi/4)\text{TB}(\delta\theta o) \tag{20.56}$$

which covers $(\pi/4)$TB number of laser spots in the scan. We shall designate the number of deflector resolution as Ns,

$$\text{Ns} = \Delta\theta s/\delta\theta o$$
$$= (\pi/4)\text{TB} \tag{20.57}$$

The optical incidence angle aligned at the center frequency shall be required to operate over the acoustic bandwidth $\Delta f$. For frequencies at the edges of the bandwidth, the mismatch angle for Bragg condition is, $\Delta\theta\text{in} = \theta\text{in} - \theta_B$,

$$\Delta\theta\text{in} = +- \lambda\Delta f/(4nv) \tag{20.58}$$

This is equivalent to an effective input optical beam divergence angle of $2\Delta\theta\text{in}$, and thus, the effective acoustic to optical beam spread ratio for the acousto-optic beam deflector is

$$\text{Rd} = \delta\theta a/(2\Delta\theta\text{in})$$
$$= 2.546 \text{ R/TB}$$
$$= 2 \text{ R/Ns} \tag{20.59}$$

Comparing Eq. (20.55) to Eq. (20.59), if the same acousto-optic modulator is employed for either the spatial light modulator application or the beam deflector application, we will find the effective beam balance ratio different for the two cases. For the beam deflector application, the effective value of beam balance ratio per system resolution is two times larger than for the same device serving as spatial light modulator. An explanation is required because in the real life of science nothing comes free.

The factor of 2 advantage in Eq. (20.59) is because the scan pixels come one at a time while the acoustic field remains uniform during a scan. Thus, we do not have to be concerned with the elongation of the laser beam at acousto-optic output plan. However, in a sequential scan, the acoustic field applied must be a linear chirp going monotonically up or down the frequency slope. During a scan, the acoustic frequency at each point of the optical aperture is different from any other point. Examining the diffracted output laser light can easily show a focusing or defocusing effect that will blur the focal spots in the scan field. Fortunately, this blurring effect can be compensated by means of focus adjustment that is chirp rate dependent. Whenever the chirp rate changes, the focusing optics needs to be adjusted.

## 20.4 FABRICATION OF ACOUSTO-OPTIC MODULATORS

Fabrication of acousto-optic devices can be rather involved and is usually left to the specialty shops. Here we will discuss some of the basics in acoustic transducer design and fabrication. The acoustic transducer is made of a thin piezoelectric plate attached to the surface of an acousto-optic modulator medium. There are several intermediate layers of metal films between the piezoelectric plate and the acousto-optic medium to facilitate strong bonding of the two materials. The intermediate layers have well-controlled thickness for the purpose of provide acoustic impedance matching between the transducer material and the acousto-optic medium. Then the top of the transducer plate is electroded to activate the acoustic-generating area according to the acousto-optic device design. The thickness of the top electrode material must be carefully done because there is a mass loading effect that will cause the acoustic transducer frequency to move to lower frequencies. Finally, wire is attached to the top electrode from the impedance-matching circuit that allows electricity to flow into the acoustic transducer, which may have very low input impedance level with a large series capacitance. Then, the device needs be packaged properly before put to use.

With all these at work, there are typically five metal layers in the acoustic transducer bonding area, one to two layers for the top electrode, one piezoelectric layer, and one matching network. Making acoustic transducers for low-frequency devices is relatively easy because the large acoustic wavelength is rather tolerating. However, at higher frequencies, the intermediate layers act like acoustic transmission line sections that can have major effects on device performances such as electro-acoustic conversion bandwidth, insertion loss, phase integrity, and the shape of the passband. Numerical modeling is necessary for designing the transducers, and careful fabrication keeping the film thickness under control is important practice.

In this section, we are going to avoid the more academic transducer design considerations, and go right into what is currently being used so the reader can have an understanding about what is in a typical device. Also, we are not going to go into details for some more exotic transducer designs such as multielement array, acoustic quarter-wave intermediate layer impedance matching, and so on. Only the basic acousto-optic fabrication know-how and simplistic modeling results will be described.

### 20.4.1 The Piezoelectric Layer

The piezoelectric layer is the most important part of the acoustic transducer. Nowadays, the piezoelectric layer almost invariably uses a 36-degree rotated Y cut $LiNbO_3$ crystal for longitudinal acoustic wave transducer and uses a 41-degree X plate $LiNbO_3$ crystal for shear wave acoustic transducer. The acoustic parameters for these two materials is listed in Table 20.3.

These materials have a good coupling coefficient for making wideband acoustic transducers. The main task of transducer fabrication is to make a strong bonding to acousto-optic medium substrates, and to design for good acoustic impedance matching between the transducer and the substrate for efficient flow of acoustic power through the interface. The acoustic impedance of the transducer material is given above. For example, the acoustic impedances for most common acousto-optic materials are $TeO_2$ longitudinal wave, $Z = 25.2$, $TeO_2$ slow shear wave, $Z = 3.69$, $PbMoO_4$ longitudinal wave, $Z = 25.2$, and fused quartz, $Z = 13.1$. Comparing with the acoustic impedance

**TABLE 20.3   Properties of A-O Transducer Material**

|  | Mode | Coupling k | v | $Z = \rho v$ | MHz-μm | $\varepsilon$ |
|---|---|---|---|---|---|---|
| 36-degree rotated Y-cut LiNbO$_3$ | L | 0.49 | 7.3 | 33.9 | 3650 | 38.6 |
| 41-degree X plate LiNbO$_3$ | S | 0.68 | 4.8 | 22.3 | 2400 | 44.3 |

of the LiNbO$_3$ transducers, the impedance ratio is anything from 1.34:1 for longitudinal wave TeO$_2$ or PbMoO$_4$, to 2.6:1 for longitudinal wave in fused quartz, and to 6:1 for slow shear wave in TeO$_2$. For reference, the reflectivity for waves on a 1.5:1 interface is about 6%. The reflectivity increases to about 20% on a 2.6:1 interface, and to 51% on a 6:1 interface.

The design goal of transducer intermediate layers is to keep reflectivity as low as possible so that the electrical drive power can be converted to acoustic power efficiently, which means typically <1-dB insertion loss. And that requires good understanding of the acoustic transducer with numerical modeling.

### 20.4.2   Bonding Layers

The purpose of the bonding layer is to facilitate strong bonding between the LiNbO$_3$ plate and the acousto-optic medium, and to enhance acoustic power coupling over a wide bandwidth with minimum rolloff so as to leave all the bandwidth available to the acousto-optic process.

Since the 1970s, the old method of epoxy bond for making of acoustic transducers had been pretty much stopped because the glue thickness is too difficult to control reliably. Instead, an Indium gold eutectic bond has been widely used for acoustic frequencies up to about 1 GHz. Beyond 1 GHz, the indium is considered having too much acoustic loss, and gold bond or aluminum bond has been developed. Here we are only going to discuss the indium gold bond, which is used in most acousto-optic devices to date.

To prepare for the indium-gold bond process, one of the transducer surfaces as well as one of the acousto-optic medium surfaces are both coated with one layer of Cr and one layer of gold in a vacuum system. The Cr layer is generally very thin, about 50 to 300 angstrom thick, and it is only for the purpose of improving the gold thin film adhesion to desired surfaces. Sometimes the Cr layer is replaced with Ti, Ta, or another metal layer with good adhesion properties. The gold layer is typically about 2000 angstrom thick, although the specific thickness must go through numerical evaluation before use, particularly for frequencies higher than about 400 MHz. Once the Cr Au coating is ready, the two pieces are placed in another vacuum system dedicated for indium bonding work to be coated with typically 0.2 to 1 micron thickness of indium, and the two pieces are mated together face to face subject to high pressure at room temperature. Of course, the two surfaces that meet together must be well polished to very good flatness and low scratch and dig count, or otherwise the two pieces simply will refuse to stick together. Also, it is important to avoid dust particles and splattering during any of the coating process as a particle in the middle will definitely ruin the work.

Initially, the LiNbO$_3$ piezoelectric layer may be very thick, typically about 1 mm. The transducer thickness is grinded down to about 30 micron for a 100-MHz device using optical grinding and polishing techniques. The thickness uniformity needs be maintained or the device will not have uniform electric-to-acoustic conversion performances. It is customary to prepare the transducers on a substrate larger than the device so that it can be cut up, making it into several acousto-optic modulator units. RF frequency probe is used to test the thickness of the transducer by placing a metal piece on the LiNbO$_3$ surface and to monitor the RF reflection after the generated acoustic wave reflects back from the opposite surface of the substrate block. The reflection will be strongest where the piezoelectric layer is resonating, indicating the working frequency of the plate. And, it is important to make the piezoelectric plate resonate at a higher frequency than desired in anticipation of mass loading by the top electrode process later. The higher the acoustic frequency, the more severe the mass loading will be, for a given thickness of top electrode of certain material. The frequency downshift in anticipation of mass loading can be numerically calculated or can simply work by experiences. Then, the whole unit is placed inside a vacuum chamber for coating of the top electrode, which is often Cr, Au, or Cr, Al. Thus, the completed bond contains the sequence of layers, Au, Cr, LiNbO$_3$, Cr, Au, In, Au, Cr, and then the acousto-optic medium.

Sittig developed the numerical modeling for acoustic transducers, including intermediate layers in the 1960s [16]. The model used the Mason's equivalent circuit for the transducer and each of the metal layers. Each of the layers is represented in the form of transmission matrices. Multiplication of the transmission matrices gives the results for the entire assembly from which the input impedance, output impedance, conversion loss, and phase and amplitude responses can be obtained. Using this numerical model, the transducer behavior of some simple cases is given in the figures, 15 to 18 for a transducer centered at about 700 to 800 MHz. In these figures, in order to simplify the modeling, the Cr layer is not included. All of these examples are for longitudinal wave transducers. The 800-MHz frequency is higher than most typical acousto-optic devices, but it is selected because at higher frequency, all bonding layers become more sensitive to thickness variations.

Figure 20.15 illustrates thc effect of load impedance on transducer bandwidth. Note the load impedance for fused quartz is about 13, for TeO$_2$ is about 25, and for LiNbO$_3$ is about 35. The larger the impedance mismatch (curve with load impedance being 10), the smaller the fractional bandwidth becomes. In all cases, the insertion loss is high. Except for the case of LiNbO$_3$ load material, which is identical to the transducer material for a perfect match, useful bandwidth is limited, no more than 300 MHz, because we would accept only 1-dB bandwidth rolloff because there will be more band rolloff penalties due to the acousto-optic process.

Figure 20.16 illustrates the effect of an Au-In-Au bond, with TeO$_2$ as the acousto-optic material, and with the indium thickness varying between 0.1 micron and 0.3 micron. Note that gold has acoustic velocity of 3.4 m/ms, and acoustic impedance of 65.5, whereas indium has acoustic velocity of 2.3 m/ms, and acoustic impedance of 17. Thin layer of indium is used because the frequency is high. It is observed that the indium thickness variation as little as 0.05 micron has dramatic effect on the transducer bandwidth and shape, because of the slow acoustic velocity in indium and the low acoustic impedance. No top electrode is added to the transducer yet.

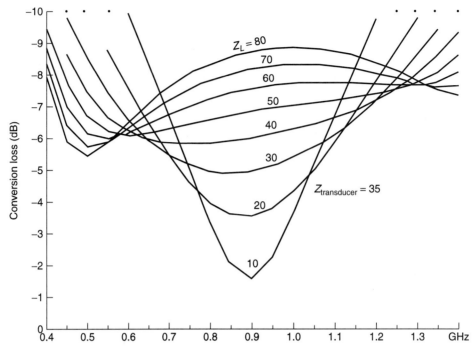

**Figure 20.15**   The transducer frequency response curves for a LiNbO$_3$ longitudinal transducer near 800 MHz, with the load material acoustic impedance as parameter.

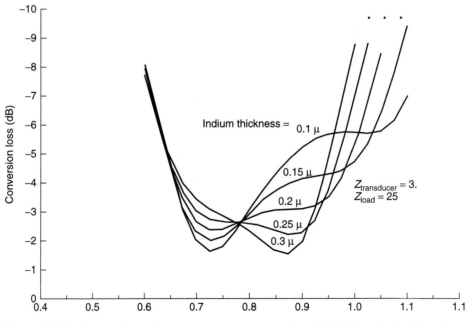

**Figure 20.16**   The transducer frequency response curves for a LiNbO$_3$ longitudinal transducer near 800 MHz, with a gold indium bond, and TeO$_2$ acousto-optic load, and with the indium bond thickness as parameter.

Figures 20.17 and 20.18 illustrate the transducer performances when changing to aluminum and copper bonds. Note that aluminum has an acoustic velocity of 6.4 m/ms, and acoustic impedance of 17.3, and copper has acoustic velocity of 5 m/ms, and acoustic impedance of 40.6. A top electrode of aluminum at 2000 angstrom thickness is added. The aluminum bond and copper bond are not as sensitive to slight thickness variations. Proper thickness control is important to obtain well-behaved bandwidth that can make subsequent impedance-matching circuit easier to do.

### 20.4.3  Top Electrode Thickness, Mass Loading, and Impedance Matching

Figure 20.19 illustrates the effect of top electrode mass loading on a acoustic transducer. With gold top electrode, a thickness change of 0.2 micron can cause a downshift on center frequency by as much as 150 MHz, or 18% for an 800-MHz piezoelectric layer. Although it seems desirable to keep the top electrode thin to minimize mass loading, in reality, a thin top electrode is difficult to attach a wire to with good strength, and the sheet resistance may be too large for low loss electric-to-acoustic power conversion, particularly when the input impedance is low. Therefore, for high-frequency devices, it is desirable to use light density material such as aluminum for a top electrode.

Figure 20.20 illustrates a typical high-frequency acousto-optic transducer bandwidth, when impedance matching is properly done. The example is a transducer made of the previous models starting from near 900-MHz piezoelectric layer resonant thickness and mass loaded down to 750 MHz upon completion. The bonding layers are copper aluminum bond for better bond impedance matching, and the top electrode layer is 0.2 micron aluminum for less mass loading. The impedance tuning is a series coil, parallel

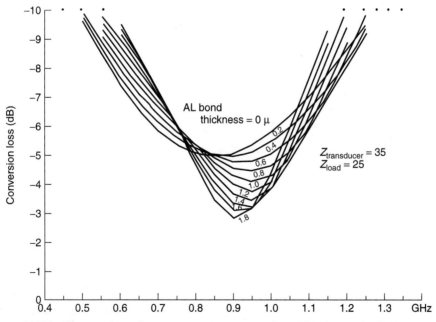

**Figure 20.17**  The transducer frequency response curves for a LiNbO$_3$ longitudinal transducer near 800 MHz, with an aluminum bond, and TeO$_2$ acousto-optic load, and with the aluminum bond thickness as parameter.

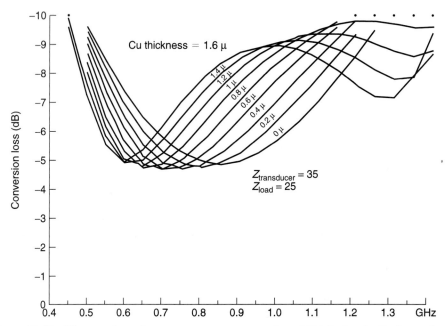

**Figure 20.18**   The transducer frequency response curves for a LiNbO$_3$ longitudinal transducer near 800 MHz, with a copper bond, and TeO$_2$ acousto-optic load, and with the copper bond thickness as parameter.

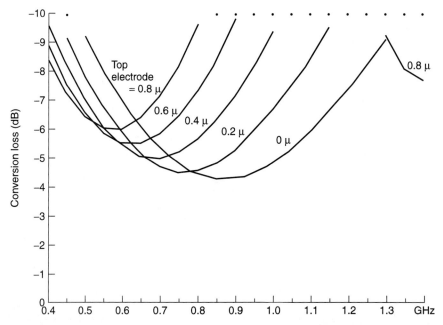

**Figure 20.19**   The transducer frequency response curves for a LiNbO$_3$ longitudinal transducer near 800 MHz, with no bonding layer, and with TeO$_2$ acousto-optic load, and with the gold top electrode layer thickness as parameter showing mass loading causing the center frequency to shift lower.

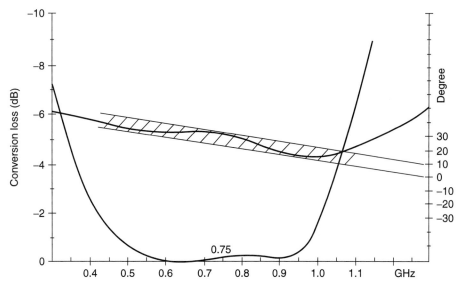

**Figure 20.20** Illustrates a typical high-frequency acousto-optic transducer at 750 MHz, when impedance is matched using series parallel and series tuning circuit.

capacitor, and series coil circuit. Once properly matched, the bandwidth becomes very broad at the passband, with almost no appreciable conversion loss. However, in reality, there are parasitic losses at the tuning circuits, top electrode, and bonding layers, and a <1-dB insertion loss may be expected. Also shown is the calculated phase response of the matched transducer with a ripple of about 10 degrees across the bandwidth. This example provides about 550 MHz of low-loss conversion bandwidth for the acousto-optic modulator, which realistically can only use no more than octave bandwidth, i.e., 500 MHz. So, it shows that well designed and well-fabricated acoustic transducers can place little burden on the acousto-optic modulator performances and is most critical for the making of advanced acousto-optic devices.

### 20.4.4 Transducer Heating

An important performance-limiting issue of acousto-optic modulators is the heating generated at the transducer due to dissipation of electrical and acoustic power [9, 25]. For some of the advanced high-frequency devices, the power density can be very large at the transducer surface, and even a small dissipation factor can cause adverse effects. In this section, the heat equation and typical heat distribution will be discussed to provide a first-order understanding to this problem.

The general two-dimensional heat equation for steady state is the elliptical partial differential equation

$$\frac{d^2 T}{d x^2} + \frac{d^2 T}{d y^2} + \frac{Q}{K} = 0 \tag{20.60}$$

where $T$ is temperature, $Q$ is heat generation, and $K$ is thermal conductivity. Numerical solution is made on the device configuration of Fig. 20.21 where the acousto-optic material is TeO$_2$ with the boundary condition shown. The crystal is sandwiched between

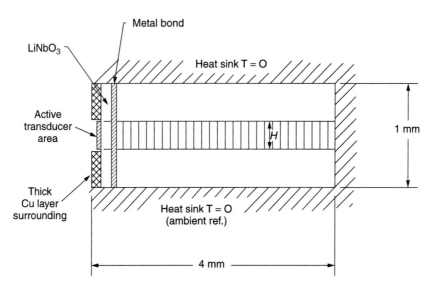

**Figure 20.21**   The acousto-optic modulator heat sink model for numerical analysis.

copper heat sinks. To remove the heating at the transducer surface, the transducer top surface is plated with copper thick film, which surrounded the top electrode pattern. The numerical model includes two heat sources, a RF source of about 1 w with 10% dissipated at the transducer surface, and an acoustic heating due to acoustic attenuation in TeO$_2$ at 0.5 dB/cm. The results are shown in Fig. 20.22, assuming the temperature at the heat sink is zero (reference temperature). Along the acoustic propagation direction, the heat from transducer heating dominates and the temperature profile fades into the crystals rapidly. When the surface copper plating is made thicker, the temperature profile drops significantly, indicating it is rather effective in removing the heat generated from the transducer. In the cross-section view, the heating decreases rapidly to both sides of the acoustic column as expected. This numerical result is in qualitative agreement with our observation when the laser beam is placed near the transducer. In general, the laser beam experiences a prism-like effect along the direction of acoustic beam, and a lens-like focusing effect in the direction perpendicular to the acoustic beam.

### 20.4.5   Top Electrode Geometry

While propagation inside an acousto-optic modulator, the acoustic wave undergoes Fresnel diffractions that cause huge intensity variations inside the acoustic field [26–28] Fresnel diffraction is well known in physical optical and can be found in most optics textbooks. During the transition from near field to far field, the wave intensity can even have deep holes in the middle with strong ripple. The Fresnel diffraction fringes are parallel to the transducer edges, which is troublesome because the laser beam also travels along the same direction. For a tightly focused laser beam in a temporal modulator, the small laser beam can be unstable when unexpected drifts in optics or acoustic field occurs. The problem becomes more severe for acousto-optic spatial light modulators and beam deflectors that incorporate a larger laser beam. The optical

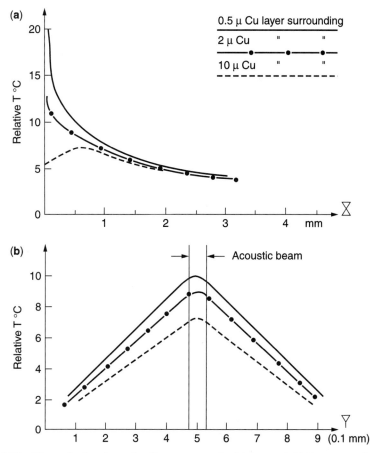

**Figure 20.22** The calculated results for an example high-speed TeO$_2$ temporal modulator showing the temperature profile along the acoustic propagation direction and in the cross section. The copper surface film thickness is used as a parameter, showing it is effective removing the transducer heating.

aperture is filled with an image of the Fresnel diffraction fringes that seriously degrades system performances.

Diamond-shaped or cosine-shaped top electrode pattern has been employed [27] such that a laser beam propagating along the length of a transducer no longer runs parallel to the transducer edges. Experimentally, this technique has been found worthwhile and has become a common practice lately. However, due to the large aspect ratio of acousto-optic modulator transducers, the diamond shape becomes a needle shape in many cases. Optical alignment can be difficult, and effective interaction length can be sensitive to alignment error. And as the laser beam only makes small angle to transducer edges, the diffraction fringes are not completely removed. Figure 20.23 shows a few new types of transducer patterns developed by the author. The transducer edges are either zig-zag shaped or saw tooth shaped. There are several advantages of these types of transducer shape. For one, the optical path will be at a larger angle to the transducer edges so the Fresnel fringes can be averaged out more completely. Next, it no longer produces a needle tip and has a larger acceptance window for the laser beam to align

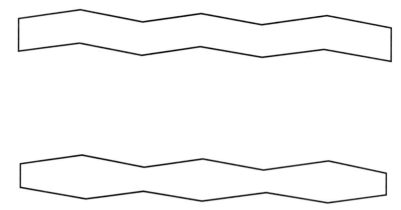

**Figure 20.23**   Top electrode shapes for reduction of Fresnel diffraction fringes from the laser beam.

to. Small lateral drift of the laser beam will be more tolerable than for the needle tip transducer. Computer simulation model results are given by Fig. 20.24, showing it can be effective in removing the diffraction fringes. Many high-speed acousto-optic modulators with this type of top electrode have been employed in high-speed printing equipment since 1996.

## 20.5   SUMMARY

Basic properties of acousto-optic modulators have been discussed. In order to focus on the basic properties and behavior of acousto-optic modulators that an optical engineer may confront in their daily work, the discussions are limited to isotropic acousto-optic modulators only. Technical areas covered include the rise time, laser beam distortion after the modulation, maximum diffraction efficiency of acousto-optic modulators, the modulation transfer functions of various mode of operation, harmonic contents and waveform distortions in the modulated laser, effects of acoustic attenuations, effects of transducer heating and acoustic diffraction, and some basic reasons for the acoustic transducer construction. There are sufficient details in these subjects for an engineer to consult with when in their daily practice involving acousto-optic modulators somehow things do not come out as expected.

On designing acousto-optic modulators, a unique design procedure based on specifying the Q and R parameters for acousto-optic modulators is provided. This design approach starts from detail device performance requirements and works its way up step by step until a set of device configuration and design numbers for a nearly optimal design is reached. Only the most frequently used acousto-optic materials and transducer materials are listed. A zig-zag type of top electrode pattern designed to reduce the adverse effects of acoustic diffraction on laser beam quality is also introduced.

There are many advanced areas the author wishes to cover but regrettably does not. That includes the optical polarization effects through acousto-optic modulators, Q-switches, and other intracavity acousto-optic devices; high-performance and ultra wideband acoustic transducer designs, special power saving techniques, acousto-optic modulators for fiber optical applications, anisotropic acousto-optic beam deflectors such

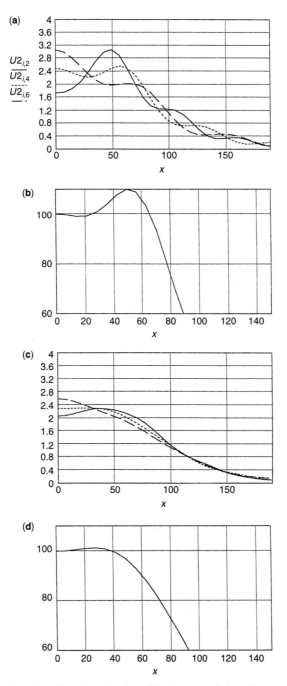

**Figure 20.24**  Calculated results showing the effectiveness of the zig-zag transducer pattern: **(a)** The acoustic field distribution at three distances away from the transducer. **(b)** The effect of lateral displacement for a focused Gaussian laser beam with beam diameter about 25% of the acoustic transducer height, H. The focusing laser beam averaged out the acoustic ripples, but the throughput efficiency is sensitive to lateral displacement within the transducer aperture. **(c)** and **(d)** Corresponding results for the same laser beam when the acoustic transducer employs a zig-zag edge pattern. Sensitivity in laser beam lateral displacement is significantly reduced.

as the important slow shear wave $TeO_2$ devices, wideband birefringent spatial light modulators on $LiNbO_3$, and acousto-optic tunable filters; and phased array wideband beam deflectors and spatial light modulators, acousto-optic signal processing techniques and applications. The discussions presented in this chapter can provide the starting ground for further investigations of these highly specialized or more exotic acousto-optic devices.

# REFERENCES

1. R. W. Dixon, "Acoustic Diffraction of Light in Anisotropic Media," *IEEE J. QE*, **QE-3**, p. 85, 1967.
2. E. I. Gordon, "A Review of Acoustooptical Deflection and Modulation Devices," *Proc. IEEE*, **54**, p. 1391, 1966.
3. S. K. Yao et al., "Current Results and Future Trends in Acousto-Optic Bulk Wave Technologies," *Opt. Eng.*, Sept. 1981.
4. I. C. Chang, "Acoustooptic Devices and Applications," *IEEE Trans. Sonics Ultrasonics*, **SU-23**, p. 2, 1976.
5. E. H. Young and S. K. Yao, "Design Considerations for Acousto-Optic Devices," *Proc. IEEE*, **69**, p. 54, 1981.
6. S. K. Yao, "Wideband Bragg Cell Efficiency Enhancement Techniques," *Proc. SPIE*, **545**, p. 72, 1985.
7. E. Blomme, O. Leroy Jr., and A. Sliwinski, "Acoustically Induced Light Polarization Effects in Isotropic Media," *Proc. SPIE*, **2643**, p. 2, 1995.
8. C. W. Tarn, "Polarization Changing and Beam Profile Deformation Effects of Acousto-Optics," *Proc. SPIE*, **3581**, p. 70, 1998.
9. S. K. Yao, "Analytical Study of Wideband Optical Modulators, Final Technical Report," Harris Corp. Tech. Report TMO-76-002, September 1976.
10. J. Randolph and J. Morrison, "Modulation Transfer Characteristics of an Acoustooptic Deflector," *Appl. Opt.*, **10**, p. 1383, 1971.
11. R. Johnson, "Temporal Response of the Acoustooptic Modulator: Geometrical Optics Model in the Low Scattering Efficiency Limit," *Appl. Opt.*, **16**, p. 507, 1977.
12. D. Maydan, "Acoustooptical Pulse Modulators," *IEEE J. QE*, **QE-6**, p. 15, 1970.
13. D. H. McMahon, "Relative Efficiency of Optical Bragg Diffraction as a Function of Interaction Geometry," *IEEE Trans. Sonics Ultrasonics*, **SU-16**, p. 41, 1969.
14. V. V. Nikulin, V. A. Skormin, and T. E. Busch, "Genetic Algorithm Optimization for Bragg Cell Design," *Opti. Eng.*, **41**, p. 1767, August 2002.
15. Y. H. Li, Y. Cai, G. L. Zheng, X. B. Lin, and W. L. She, "Matrix Series Method of the Raman Nath Equation of Ultrasonic Light Diffraction and its Application to the Optimum Design of Acousto-Optic Modulator," *Chinese J. Lasers*, **A29**, p. 595, July 2002.
16. A. H. Meitzler and E. K. Sittig, "Characterization of Piezoelectric Transducers in Ultrasonic Device Operation Above 0.1 GHz," *J. Appl. Phys.*, **40**, p. 4341, 1969.
17. W. R. Klein and B. D. Cook, "Unified Approach to Ultrasonic Light Diffraction," *IEEE Trans. Sonics and Ultrasonics*, **SU-14**, p. 123, 1967.
18. J. W. Goodman, *Introduction to Fourier Optics*, McGraw Hill, New York, 1968.
19. M. G. Cohen and E. I. Gordon, "Acoustic Beam Probing Using Optical Techniques," *BSTJ*, **44**, p. 693, 1965.
20. D. Hecht, "Multifrequency Acoustooptic Diffraction," *IEEE Trans. Sonics Ultrasonics*, **SU-24**(1), p. 7, 1977.

21. S. K. Yao and C. S. Tsai, "Acousto-Optic Bragg Diffraction with Application to Ultrahigh Data Rate Laser Communication Systems, Part I, Theoretical Considerations of the Standing Wave Ultrasonic Bragg Cell," *Appl. Opt.*, **16**, p. 3032, 1977.

22. R. W. Dixon, "Photoelastic Properties of Selected Materials and Their Relevance for Applications to Acoustic Light Modulators and Scanners," *J. Appl. Phys.*, **38**(13), p. 5149, 1967.

23. N. Uchida and N. Niizeki, "Acoustooptic Deflection Materials and Techniques," *Proc. IEEE*, **61**, p. 1073, 1973.

24. D. A. Pinnow, "Guide Lines for the Selection of Acousto-optic Materials," *IEEE J. QE*, **QE-6**, p. 223, 1970.

25. H. Eschler, "Performance Limits of Acoustooptic Light Deflectors due to Thermal Effects," *Appl. Phys.*, **9**, p. 289, 1976.

26. I. C. Chang and D. L. Hecht, "Characteristics of Acousto-Optic Devices for Signal Processors," *Opt. Eng.*, **21**, p. 76, 1982.

27. D. Pape, P. Wasilousky, and M. Krainak, "A High Performance Apodized Phased Array Bragg Cell," *Proc. SPIE*, **789**, p. 116, 1987.

28. B. D. Cook, "Near Field of Ultrasonic Transducers (and How it Pertains to Acousto-Optics)," *Proc. SPIE*, **2643**, p. 12, 1995.

# 21

# OPTICAL AMPLIFIERS

CHIN B. SU
*Department of Electrical Engineering*
*Texas A&M University*
*College Station, Texas*

This chapter deals with three types of commercially available and commonly used optical amplifiers. They are the erbium-doped fiber amplifier (EDFA), the Raman fiber amplifier, and the semiconductor optical amplifier (SOA). Basic principles governing these amplifiers are first discussed, followed by descriptions of their characteristics and implementations. Selected specification sheets from vendors are presented with discussions and explanations of terms relevant to applications.

## 21.1 ERBIUM-DOPED FIBER AMPLIFIER

### 21.1.1 Spectroscopy of the $Er^{3+}$ Ion

The erbium-doped fiber amplifier (EDFA) is an optical fiber doped with erbium impurities. The erbium atom is responsible for the gain characteristics of the fiber. The spectroscopy of the $Er^{3+}$ ion plays an important role in EDFA technology. It determines the signal and the pump laser wavelength, in turn determining the supportive semiconductor lasers and fiber components technology.

The overall energy level scheme of the $Er^{3+}$ ion from the ground state to an energy corresponding to a wavelength of 0.4 $\mu$m is shown in Fig. 21.1. The measured absorption spectrum is given in Fig. 21.2. For amplification in the 1.5 $\mu$m region, the relevant energy levels are the $^4I_{15/2}$, $^4I_{13/2}$, and $^4I_{11/2}$ levels. These energy levels are shown in Fig. 21.3 as multiple sublevels due to Stark-splitting of degenerate states. The $^4I_{15/2} \rightarrow {}^4I_{11/2}$ transition corresponds to the absorption of the 980-nm pump light. The $^4I_{11/2} \rightarrow {}^4I_{13/2}$ transition is nonradiative and has a very fast lifetime of about 10 $\mu$s. Thus, excited atoms reside mostly in the lower $^4I_{13/2}$ states despite being pumped to the

*Handbook of Optical Components and Engineering,* Edited by Kai Chang
ISBN 0-471-39055-0  © 2003 John Wiley & Sons, Inc.

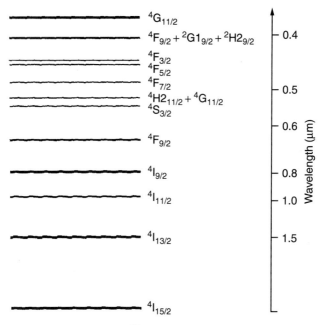

**Figure 21.1** Energy level diagram of $Er^{3+}$ [1]. The wavelength scale represents the transition between the given energy level and the ground state.

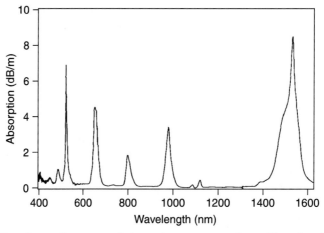

**Figure 21.2** Experimentally measured absorption spectrum of an erbium-doped germano-alumino-silica fiber [1].

$^{4}I_{11/2}$ manifold. The $^{4}I_{13/2} \rightarrow {}^{4}I_{15/2}$ transition is mainly radiative, emitting photons in the 1.5-μm region with a very slow spontaneous emission lifetime of about 10 ms. The slow lifetime is responsible for the beneficial low-channel cross-talk characteristics of EDFA. Absorption of the 1480-nm pump light occurs between the $^{4}I_{15/2}$ levels and the higher levels of the $^{4}I_{13/2}$ manifold. The thermalization time within the same manifold

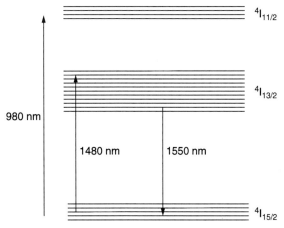

**Figure 21.3**  Energy levels showing sublevels due to Stark-splitting of degenerate states in a given manifold.

is in the picosecond regime so that the erbium-doped fiber amplifier is predominantly a homogeneously broaden gain medium.

### 21.1.2  The McCumber Relationship

The McCumber relationship [1, 2] relates the wavelength-dependent net absorption cross section $\sigma_a(\lambda)$ to the net-stimulated emission cross section $\sigma_e(\lambda)$. The formula for the McCumber relationship is given by

$$\frac{\sigma_e(\lambda)}{\sigma_a(\lambda)} = \exp\left\{ \left( \varepsilon - \frac{hc}{\lambda} \right) /kT \right\} \tag{21.1}$$

where $h$ is the Planck constant and $c$ is the speed of light. $\varepsilon$ is an adjustable parameter determined experimentally. The McCumber relationship is obtained by assuming, under thermal equilibrium condition, a Boltzman statistics of excited erbium population density distribution within the Stark sublevels of each manifold. Equation (21.1) is essentially the Einstein relation modified to account for the Boltzman statistics of atomic population and the summation of multiple transitions between sublevels in different manifolds for a given wavelength.

The spontaneous emission lifetime $\tau$ for the 1.5-$\mu$m emission is related to the absorption cross section $\sigma_a(\lambda)$ by the integral [1, 2]

$$\frac{1}{\tau} = \int \frac{\sigma_a(\lambda)}{\lambda^4} \cdot \exp\left\{ \left( \varepsilon - \frac{hc}{\lambda} \right) /kT \right\} \cdot d\lambda \tag{21.2}$$

Measurements of $\tau$ (10 ms) and $\sigma_a(\lambda)$ allow the determination of $\varepsilon$. Once $\varepsilon$ and $\sigma_a(\lambda)$ are determined, the stimulated emission cross section $\sigma_e(\lambda)$ can be calculated from the McCumber relationship.

The spectra of $\sigma_a(\lambda)$ and $\sigma_e(\lambda)$ are given in Fig. 21.4.

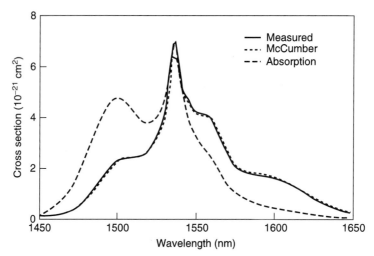

**Figure 21.4**   The spectrum of the absorption cross section $\sigma_a(\lambda)$ is shown by the dashed line. Using the McCumber relationship, the predicted stimulated emission cross section $\sigma_e(\lambda)$ fits the measured $\sigma_e(\lambda)$ (solid line) [1, 3].

### 21.1.3   Rate Equations Model for Erbium-Doped Fiber

Rate equations are used for modeling the performance of erbium-doped fiber. Software packages such as Lucent's OASIX Optical Amplifier Simulation System, Professional Version, are used for comprehensive modeling of EDFA's performance. A student version accompanying Ref. 1 is also available. Formulas used in the software package are described and summarized below.

Material parameters $\sigma_a(\lambda)$, $\sigma_e(\lambda)$, $\tau$; erbium distribution and concentration N; fiber length; and fiber waveguide parameters, such as the mode-field diameter, are, in principle, contained in rate equations for modeling the performance of the erbium-doped fiber amplifier. Complete rate equation formulas can be found in Ref. 1, the result of which is summarized below. Parameters used are defined below.

$\sigma_a(\lambda)$: Net absorption cross section.

$\sigma_e(\lambda)$: Net stimulated emission cross section.

$\tau$: spontaneous lifetime ($\approx 10$ ms).

$\Gamma(\lambda)$: $\Gamma(\lambda)$ gives the lateral overlapped factor (in percentage) between the optical field and the erbium distribution in the fiber.

N: Erbium dopant concentration.

$N_2(z)$: Excited erbium concentration in the $^4I_{13/2}$ manifold. It is position-dependent because of pump and signal depletion effects.

$N_1(z)$: Erbium concentration in the ground state ($^4I_{15/2}$ state).

A: Effective area describing the lateral extent of the erbium distribution.

$\kappa \equiv \tau/(A \cdot N)$.

$g_a(\lambda) \equiv \Gamma(\lambda)\sigma_a(\lambda)N$.

$g_e(\lambda) \equiv \Gamma(\lambda)\sigma_e(\lambda)N$.

$P_p(\lambda_p)$: pump power.

$P_s(\lambda_s)$: signal power.

$P_A(\lambda_j)$: Amplified spontaneous power at wavelength $\lambda_j$.

$P_A^+$: Copropagating (with respect to the signal) amplified spontaneous power per wavelength interval.

$P_A^-$: Counterpropagating amplified spontaneous power per wavelength interval.

$$N_2(z) = \frac{\kappa\left(\frac{\lambda_s}{hc}\right)g_a(\lambda_s)P_s + \kappa\left(\frac{\lambda_p}{hc}\right)g_a(\lambda_p)P_p + \kappa\sum_j\left(\frac{\lambda_j}{hc}\right)g_a(\lambda_j)P_A(\lambda_j)}{\kappa\left(\frac{\lambda_s}{hc}\right)[g_a(\lambda_s)+g_e(\lambda_s)]P_s + \kappa\left(\frac{\lambda_p}{hc}\right)[g_a(\lambda_p)+g_e(\lambda_p)]P_p + \kappa\sum\left(\frac{\lambda_j}{hc}\right)[g_a(\lambda_j)+g_e(\lambda_j)]P_A(\lambda_j)+1} \cdot N \tag{21.3}$$

$$\frac{dP_p}{dz} = \pm\frac{1}{N}\cdot(N_2\cdot g_e(\lambda_p) - N_1\cdot g_a(\lambda_p))\cdot P_p \tag{21.4}$$

A "+" sign is used for copropagating configurations and a negative sign for counterpropagating configurations.

$$\frac{dP_s}{dz} = \frac{1}{N}\cdot(N_2\cdot g_e(\lambda_s) - N_1\cdot g_a(\lambda_s))\cdot P_s \tag{21.5}$$

$$\frac{dP_A^+(\lambda_j)}{dz} = \frac{1}{N}\cdot(N_2\cdot g_e(\lambda_j) - N_1\cdot g_a(\lambda_j))\cdot P_A^+ + \frac{N_2}{N}\cdot g_e(\lambda_j)\cdot P_A^0 \tag{21.6}$$

$$\frac{dP_A^-(\lambda_j)}{dz} = -\frac{1}{N}\cdot(N_2\cdot g_e(\lambda_j) - N_1\cdot g_a(\lambda_j))\cdot P_A^- - \frac{N_2}{N}\cdot g_e(\lambda_j)\cdot P_A^0 \tag{21.7}$$

where, $N = N_1 + N_2$ and $P_A = P_A^+ + P_A^-$.

$P_A^0$ is the spontaneous noise power per wavelength interval and is given by

$$P_A^0 = \frac{2hc^2}{\lambda^3}\cdot\Delta\lambda \tag{21.8}$$

### 21.1.4 Commercial Erbium-Doped Fiber Specifications and Data Sheets

Examples of commercial erbium-doped fiber data sheets are given below. Key design parameters are the erbium concentration, index raising germanium concentration, aluminum codoping (to lessen the high concentration quenching and index depression problems), core radius, refractive index difference between the core, and the cladding. Important parameters from the user standpoint are the mode field diameter (MFD) of the 1.55-μm signal and the maximum absorption and gain, which depends on erbium concentration and other factors.

| Parameter | Value | Tolerance * |
|---|---|---|
| Cutoff wavelength | 860 nm | +/− 5% |
| Loss at 1558.5 nm | 1.76 dB/m | +/− 5% |
| Core radius | 1.77 μm | +/− 5% |
| Refractive index diff (Δn) | 0.0118 | +/− 5% |
| Numerical aperture | 0.186 | +/− 5% |
| Peak absorption (near 1.53 μm) | 5.05 dB/m | +/− 5% |
| Peak absorption (near 980 nm) | 4.20 dB/m | +/− 5% |
| MFD (Petermann 2) 1.55 μm | 7.15 μm | +/− 5% |
| Background loss 1.2 μm | 2.64 dB/km | +/− 5% |
| Est. background 1.55 μm | 1.13 dB/km | +/− 10% |
| Fiber outside diameter | 125 μm | +/− 2 μm |
| Proof test | 200 kpsi | |
| Core eccentricity | <0.3 μm | |
| PMD | <.002 ps/m | |
| Estimated erbium conc. | 7.3e24 m$^{-3}$ | |

\* Bounds based on statistical calculations

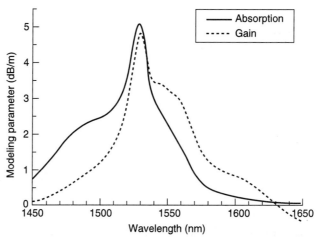

**Figure 21.5** Lucent's EDF-HP980 fiber for high-power operation. This fiber is characterized by a mode field diameter of 7.15 μm.

Data sheets for three types of fiber are given here. They are Lucent's EDF-HP980 for high-power applications, EDF-MP980 for medium-power applications, and EDF-HE980 for high-efficiency low-threshold pump power applications.

From Figs. 21.5–21.7 it is observed that a higher erbium concentration yields a larger gain as expected. However, excessively high erbium concentration does cause a number of undesirable effects, such as a decrease in the lifetime of the $^4I_{13/2} \rightarrow {}^4I_{15/2}$ ($2 \rightarrow 1$) transition due to nonradiative processes [4] and an increase in the up-conversion process due to stimulated photons transferring energy to higher states. Although these undesirable effects are alleviated by codoping with aluminum, generally, the erbium concentration are held to less than 1000 ppm. However, a very high erbium concentration of 8900 ppm

| Parameter | Value | Tolerance * |
|---|---|---|
| Cutoff Wavelength | 895 nm | +/- 5% |
| Loss at 1558.5 nm | 1.95 dB/m | +/- 5% |
| Core radius | 1.47 μm | +/- 5% |
| Refractive index diff (Δn) | 0.0184 | +/- 5% |
| Numerical aperture | 0.233 | +/- 5% |
| Peak absorption (near 1.53 μm) | 5.53 dB/m | +/- 5% |
| Peak absorption (near 980 nm) | 4.05 dB/m | +/- 5% |
| MFD (petermann 2) 1.55 μm | 5.75 μm | +/- 5% |
| Background loss 1.2 μm | 4.08 dB/km | +/- 5% |
| Est. background 1.55 μm | 1.58 dB/km | +/- 5% |
| Fiber outside diameter | 125 μm | +/- 2 μm |
| Proof test | 200 kpsi | |
| Core eccentricity | <0.3 μm | |
| PMD | < .002 ps/m | |
| Estimated erbium conc. | $8.1\ e^{24}\ m^{-3}$ | |

* Bounds based on statistical calculations

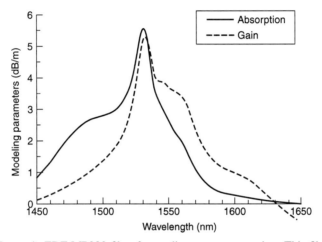

**Figure 21.6**   Lucent's EDF-MP980 fiber for medium-power operation. This fiber is character-ized by a mode field diameter of 5.75 μm.

had been reported [5] with gain of 30 dB/m. Fibers with very high gain find applications mostly in fiber sensors, where, for example, one may wish to build a single-mode fiber laser, requiring a short fiber length.

The mode field diameter of a standard single-mode telecommunication fiber (SMF-28 fiber) is 10.5 μm at 1.55-μm wavelength, larger than the mode-field diameter of erbium-doped fiber. The discrepancy of mode field diameter between the SMF-28 fiber and the erbium-doped fiber introduces splice loss. A useful formula for estimating the splice loss when two fibers with different mode-field diameter is spliced together [6] is

$$\text{Fiber mismatched loss} = 10 \cdot \log \left( \frac{2d_1 d_2}{d_1^2 + d_2^2} \right)^2 \text{ dB} \qquad (21.9)$$

However, it should be noted that by controlling the fusion splice time, the splice loss can be made smaller than the value predicted by the formula given above because

| Parameter | Value | Tolerance * |
|---|---|---|
| Cutoff wavelength | 850 nm | +/− 5% |
| Loss at 1558.5 nm | 1.11 dB/m | +/− 5% |
| Core radius | 1.08 μm | +/− 5% |
| Refractive index diff (Δn) | 0.0308 | +/− 5% |
| Numerical aperture | 0.301 | +/− 5% |
| Peak absorption (near 1.53 μm) | 3.20 dB/m | +/− 5% |
| Peak absorption (near 980 nm) | 2.64 dB/m | +/− 5% |
| MFD (petermann 2) 1.55 μm | 4.84 μm | +/− 5% |
| Background loss 1.2 μm | 12.75 dB/km | +/− 3% |
| Est. background 1.55 μm | 5.73 dB/km | +/− 5% |
| Scattering capture fraction | 0.0024 | − |
| Fiber outside diameter | 125 μm | +/− 2 μm |
| Proof test | >200 kpsi | |
| Core eccentricity | <0.3 μm | |
| PMD | <0.002 ps/m | |
| Estimated erbium conc. | $4.8e^{24}$ m$^{-3}$ | |

• Bounds based on statistical calculations

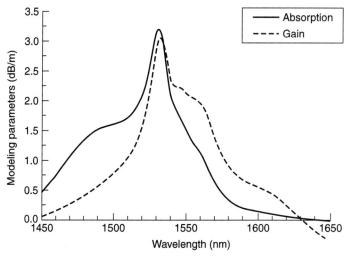

**Figure 21.7**  Lucent's EDF-HE980 fiber for low-threshold pump-power operation. This fiber is characterized by a small mode field diameter of 4.84 μm.

index-altering dopants diffuse faster in smaller core fiber than in higher core fiber, equalizing the mode field diameter [7]. The mode field diameter of a small core fiber can also be increased by tapering the fiber [8].

Figure 21.8 provides an example of a performance evaluation using the Lucent's OASIX Software Package mentioned above.

### 21.1.5   Erbium-Doped Fiber Implementation

Signal and pump power are generally combined by two types of fiber components, namely, the wavelength division multiplexed (WDM) fused fiber coupler and the

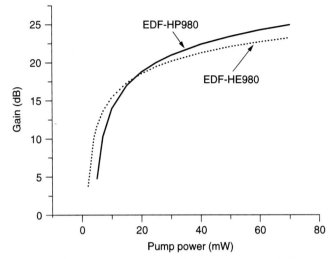

**Figure 21.8**   Small signal gain versus copropagating pump power at 980-nm wavelength. The signal wavelength is 1532 nm, and the fiber length is 10 m. The solid line is for fiber type EDF-HP980, and the dashed line is for fiber type EDF-HE980, indicating a higher efficiency at lower pump power due to core radius and mode-field diameter differences.

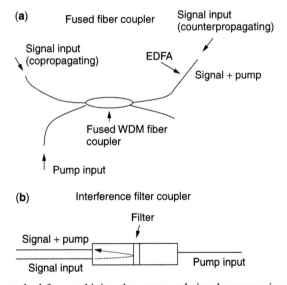

**Figure 21.9**   The method for combining the pump and signal power using **(a)** a Fused WDM coupler and **(b)** an interference filter-type coupler.

interference filter-type WDM coupler. Configurations for both types are given in Fig. 21.9. Pump lasers are usually 980- or 1480-nm semiconductor lasers. If a 980-nm laser is used, one of the fiber is usually of a type (Flexcor-1060, for example) that is single mode at 980 nm. This is because the single mode fiber (SMF-28) is not single mode at 980 nm. In Fig. 21.9, a $2 \times 2$ fused WDM fiber coupler combines the pump laser power and the signal power in the erbium-doped fiber with high efficiency. If the signal input is into the port on the same side as the pump port, the signal is said to be copropagating with the pump light. If the signal input is on the other side (EDFA side) as indicated in Fig. 21.9, the signal is counterpropagating.

WDM coupler data sheet (1480-nm pump, for example) may contain specifications given in Fig. 21.10.

Isolation describes the efficiency combining the pump and signal power. For instance, the isolation for C-P1 is specified as 30-dB which means that at a given wavelength, the C port would have 30-dB more power than the unused port on the same side as the C port. The passband describes the wavelength bandwidth for efficient merging of the 1480/1550-nm pump and signal power in the same fiber. A typical number for the passband is about 40 nm. The insertion loss referring to C-P1 describes the reduced power from P1 port to C port due to scattering and other effects. The directivity refers to the returned power in P2 if the only input power is in P1, whereas the backreflection (return loss) refers to returned power in the same fiber arm as the input arm. PDL is the polarization-dependent loss; an important quantity because the polarization state is random in telecommunication fibers. In general, fused

| Specifications[1] | |
|---|---|
| Passbands | See ordering information |
| Insertion loss, C- P1 | 0.5 dB typ., 0.7 dB max. |
| Insertion loss, C- P2 | 0.4 dB typ., 0.5 dB max. |
| Isolation C- P1 | 30 dB min. |
| Isolation C- P2 | 15 dB typ., 10 dB min. |
| Directivity | 55 dB min. |
| Back-reflection | -50 dB max. |
| PDL | 0.05 dB typ., 0.1 dB max. |
| Thermal stability | 0.005 dB/°C max. |
| Fiber type | 9/125 corning SMF-28 |
| Operating temperature | 0°C to +65°C |
| Storage temperature | -40°C to +85°C |
| Humidity | 85°C/85% RH/14 days |

[1]All specifications referenced without connectors.

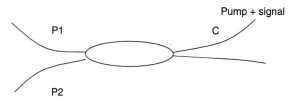

**Figure 21.10**   A specification sheet of a 1480/1550 fused WDM coupler. Ports configuration is included to facilitate the explanation of various terms.

fiber couplers have advantages over interference filter couplers in terms of insertion loss, maximum power capability, reliability, and cost, whereas the interference filter coupler has wider passband and lower polarization-dependent loss.

Pump lasers are necessary for using the erbium-doped fiber. Two types of semiconductor lasers emitting at wavelengths of 980 nm and 1480 nm are used for pumping erbium-doped fiber. They are compact and efficient, and they have high reliability. Typically, they delivered fiber-coupled powers of the order of $70 \sim 100$ mW. Some commercial pump modules operate without a thermal electric cooler, reducing power consumption. The output fiber often consists of a section of fiber Bragg grating for maintaining wavelength stability over a wide temperature range. Figure 21.2 shows that the absorption coefficient at 980 nm is larger than at 1480 nm. However, the bandwidth of the 980-nm band is smaller than the 1480-nm band. Thus, the wavelength tolerance for the 1480-nm pump is better than for the 980-nm pump.

Figure 21.11 shows a specification sheet of the 980-nm pump laser.

| No. | Parameter | Symbol | Test condition | Ratings Min. | Ratings Max. | Unit |
|-----|-----------|--------|----------------|--------------|--------------|------|
| | Part Number | | | | | |
| 1 | SDLO-2200-080 | $P_{kink-free}$ | 25°C, $I_{kink}$ | 80 | - | mW |
| | **Absolute maximum ratings** | | | | | |
| 1 | Operating current | $I_{op}$ | CW | - | 300 | mA |
| 2 | Laser diode reverse voltage | $V_{revLD}$ | - | - | 2.0 | V |
| 3 | Monitor photodiode reverse voltage | $V_{revMPD}$ | - | - | 25 | V |
| 4 | Operating case temperature range | $T_{case}$ | - | 0 | +50 | °C |
| 5 | Storage temperature | $T_{stg}$ | - | −40 | +80 | °C |
| | **Operating characteristics ($T_{case}$ = 0 to 50°C)** | | | | | |
| 1 | Operating current | $I_{op}$ | $P_f$ | - | 225 | mA |
| 2 | Threshold | $I_{th}$ | - | - | 25 | mA |
| 3 | Linearity | $dL/dI$ | - | - | No negative slope | |
| 4 | Laser diode forward voltage | $V_{fwdLD}$ | $I_{op}$ | - | 2.5 | V |
| 5 | Optical power stability | $\Delta P_f/\Delta t$ | $I_{op}$, t = 60 seconds | - | 0.5 | % of $P_f$ |
| 6 | Spectral cutoff (low) | $\lambda_{5\%}$ | $I_{op}$ | 974 | - | nm |
| 7 | Spectral cutoff (high) | $\lambda_{95\%}$ | $I_{op}$ | - | 985 | nm |
| 8 | Spectral shift with temperature | $\Delta\lambda/\Delta T$ | $I_{op}$ | - | 0.02 | nm/°C |
| 9 | Spectrum stability | $\Delta\lambda/\Delta t$ | $I_{op}$, t = 60 seconds $T_{case}$ = constant | - | 0.1 | nm |
| 10 | Monitor photodiode current | $I_{mpd}$ | $I_{op}$ | 100 | - | μA |
| 11 | Thermistor resistance | $R_{them}$ | $T_{laser}$ = 25°C | 9.5 | 10.5 | KΩ |

| No. | Parameter | Specification | Unit |
|-----|-----------|---------------|------|
| | Fiber pigtal specifications | | |
| 1 | Type | SM | - |
| 2 | Mode-field diameter | 6.5 ± 1 | μm |
| 3 | Cladding diameter | 125 ± 2 | μm |
| 4 | Jacket diameter | 250 | μm |

Notes

1.   All measurements taken with back-reflection less than -50dB.
2.   Other optical power ratings are available.
3.   Specifications are flexible. Contact SDL optics for further information.

**Figure 21.11**   Specification sheet of a 980-nm pump laser.

## 21.2 RAMAN FIBER AMPLIFIER

### 21.2.1 Principle of Raman Fiber Amplifier

The Stokes process in Raman scattering involves the absorption of a laser photon at frequency $\nu_0$ creating an optical phonon and a photon at $\nu_0 - \Delta\nu$ [9]. The anti-Stokes process annihilates an optical phonon and simultaneously creates a photon at $\nu_0 + \Delta\nu$. Under thermal equilibrium condition, the Stokes process is larger than the anti-Stoke process by a factor of $\exp(\Delta\nu/kT)$.

If the laser power is high and signal-pump interaction length is long, as in the case using optical fiber, the stimulated Raman effect causes the fiber to exhibit gain at the Stoke frequency [10]. For silica fiber, the optical phonon energy gives a Stokes frequency shift $\Delta\nu$ of about 15 THz, corresponding to a positive wavelength shift of about 120 nm from the pump laser wavelength. The bandwidth of the Raman fiber amplifier system is about 20 THz, providing a bandwidth wider than the erbium-doped fiber amplifier. Some unique features of Raman fiber amplifiers are (1) the use of the transmission fiber itself as the gain medium and (2) the wavelength at which the fiber exhibits gain depends on the wavelength of the pump laser, which means that the wavelength can be tuned by choosing the wavelength of the pump laser. However, high-power pump laser ($\geqslant$500 mW) and long fiber length (tens of kilometers) are needed because the gain coefficient of silica fiber is small.

Figure 21.12 shows a typical Raman gain coefficient as a function of wavelength offset.

### 21.2.2 Equations Governing Raman Amplifiers

Basic equations describing Raman fiber amplifiers neglecting pump depletion effects are given below. The equation describing the signal buildup along the fiber is given by

$$\frac{dIs(z)}{dz} = \gamma \cdot I_p \cdot I_s - \alpha \cdot I_s \tag{21.10}$$

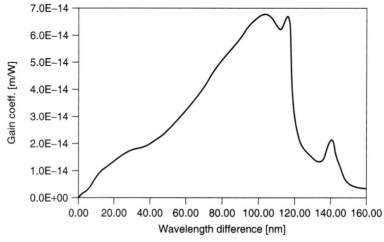

**Figure 21.12** Gain coefficient of silica fiber versus wavelength offset from the pump wavelength.

where $I_s$, $I_p$ are signal and pump intensities, $\gamma$ is the gain coefficient, and $\alpha$ is the fiber loss parameter. Suppose $P_s$ and $P_p$ are signal and pump powers and $I(r)$ describes the mode-field distribution, one writes $I_s = P_s I(r)$ and $I_p = P_p I(r)$. From Eq. (21.10), one obtains

$$\frac{dP_s}{dz} = \gamma \cdot P_s \cdot P_p \cdot \left(\frac{1}{A_{\text{eff}}}\right) - \alpha P_s \tag{21.11}$$

where the effective $A_{\text{eff}}$ is defined by

$$A_{\text{eff}} = \frac{\int r I(r) dr}{\int r I^2(r) dr} \tag{21.12}$$

Suppose the pump intensity decay as $P_p(z) = P_p(0) \exp(-\alpha z)$. By integrating Eq. (21.11) from $z = 0$ to $z = L$, one will obtain the following result:

$$P_s(L) = P_s(0) \cdot \exp\left\{\gamma \left(\frac{L_{\text{eff}}}{A_{\text{eff}}}\right) \cdot P_p(0) - \alpha \cdot L\right\} \tag{21.13}$$

where the effective length $L_{\text{eff}}$ is defined by

$$L_{\text{eff}} = \frac{1 - \exp(-\alpha \cdot L)}{\alpha} \tag{21.14}$$

To achieve a gain of $G$ (in decibels), the required pump power calculated from Eq. (21.13) is

$$P_p(0) = \frac{(0.23\, G + \alpha \cdot L)}{\gamma} \cdot \left(\frac{A_{\text{eff}}}{L_{\text{eff}}}\right) \tag{21.15}$$

Equation (21.15) implies that the gain depends not only on the pump power and fiber length, but also on the fiber type through the effective area $A_{\text{eff}}$. For example, the SMF-28 fiber and dispersion-shifted fiber have effective areas of about 80 $\mu m^2$ and 50 $\mu m^2$, respectively [11]. Using Eq. (21.15), the peak Raman gain as a function of fiber length is calculated and plotted in Fig. 21.13.

The amplified spontaneous emission noise current $I_{\text{spon}}$ within an optical filter bandwidth $B_0$ accounting for two orthogonal polarization modes is given by [12]

$$I_{\text{spon}} = 2 \cdot \eta_{sp} \cdot (G - 1) q B_0 \tag{21.16}$$

with

$$\eta_{\text{sp}} = \frac{1}{1 - \exp(-h\Delta\nu/kT)} \tag{21.17}$$

where $\Delta\nu$ is the Stokes shift. As $h\Delta\nu \gg kT$, $\eta_{\text{sp}} \approx 1$ and the noise is close to the quantum limit. Thus, Raman fiber amplifiers offer lower noise performance than do semiconductor optical amplifiers or erbium-doped fiber amplifiers.

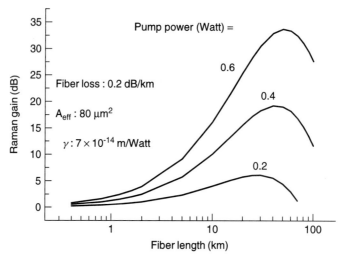

**Figure 21.13**  Peak Raman gain as a function of fiber length for various pump power $P_p(0)$. The fiber loss parameter is 0.2 dB/km near the 1500-nm band ($\alpha = 0.047$/km). The effective area corresponds to a SMF-28 fiber.

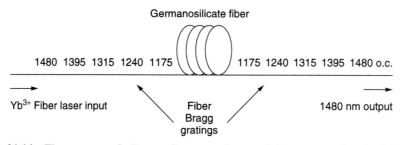

**Figure 21.14**  The structure of a Raman fiber pump laser emitting at a wavelength of 1480 nm. The nested cavities defined by fiber Bragg grating are shown.

### 21.2.3  High-Power Pump Sources and Specification Sheets

High-power pump sources are key components in Raman fiber amplifiers. One type of pump source is the Raman resonator fiber laser. The structure of the Raman resonator fiber laser is shown in Fig. 21.14 [13].

The fiber laser is pumped by a 1100-nm $Yb^{3+}$ pump laser. Multiple pairs of Bragg gratings with each pair forming a laser resonator at the intended wavelength are shown in Fig. 21.14. The innermost resonator at 1175 nm is the first Stokes emission excited by the 1100-nm pump laser. The 1175-nm emission pump is the second resonator at a resonant wavelength of 1240 nm and so on until the 1480-nm wavelength is reached by a cascade of Stokes conversion processes. The 1480-nm emission is then used as a pump source for Raman fiber amplifier in the 1500 communications band. The 1100-nm $Yb^{3+}$ pump laser [14] is a double-clad fiber laser pumped by a high-power GaAs semiconductor diode laser array.

As the range of power levels of semiconductor lasers emitting in the 1480-nm band are in the hundreds of milliwatts, watts of power in the fiber can be attained by multiplexing a few of these laser diodes. Overall, 14-xx-type pump combiners with the following specifications are commercially available for multiplexing. The combiner merges two or more 14-xx-type pump lasers with a combined optical power of up to 2 W. An example of a specification sheet of a combiner is given below. The meaning of the various terms can be found in the section describing the WDM coupler for erbium-doped fiber amplifiers.

Specifications[1]

| | |
|---|---|
| Insertion loss C-P1 | 0.7 dB typ., 0.9 dB max. |
| Insertion loss C-P2 | 0.6 dB typ., 0.7 dB max. |
| Isolation C-P1 | 30 dB[2] |
| Isolation C-P2 | 10 dB min., 12 dB typ. |
| Directivity | 55 dB min. |
| Back-reflection | −50 dB max. |
| PDL | 0.05 dB typ., 0.1 dB max. |
| Optical power | 2.0 W max. |
| Thermal stability | 0.005 dB/°C max. |
| Fiber type | 9/125 corning SMF. 28 |
| Operating temperature | −5°C to +70°C |
| Storage temperature | −40°C to +85°C |

Features:

• High power capabilities up to 2.0 Watts

• Demonstrated long term high power reliability

• Tested to telcordia GR-1221

Applications:

Raman 14-xx pump combiners are used to couple two or more different 14-nm pump lasers in Raman or erbium doped fiber amplifiers with a combined optical power of up to 2.0 W.

1. All specifications referenced without connectors.
2. Isolation depends on bandwidth and channel separation.

**Figure 21.15**  Specification sheet of 14-xx pump combiner. The combiner can handle power up to 2 W.

Specifications

| Parameter | | X-RPU-C (C-band) | X-RPU-L (L-band) |
|---|---|---|---|
| Wavelength range | | 1528 to 1562 nm | 1570 to 1612 nm |
| Pump power output | Minimum | 550 mW | 550 mW |
| Pump degree of polarization | Typical | 5.0% | 5.0% |
| | maximum | 7.5% | 7.5% |
| Insertion loss over C-band | Typical | 0.8 dB | - |
| | maximum | 1.1 dB | - |
| Insertion loss over L-band | Typical | - | 0.8 dB |
| | maximum | - | 1.1 dB |
| Change in insertion loss over T | Maximum | 0.2 dB | 0.2 dB |
| Change in insertion loss over λ | Maximum | 0.2 dB | 0.2 dB |
| Polarization dependent loss | Maximum | 0.1 dB | 0.1 dB |
| Polarization mode dispersion | Maximum | 0.1 ps | 0.1 ps |
| Number of pump lasers | | 4 | 4 |
| Environmental | | | |
| Power dissipation. BOL | Typical | 40 W | 40 W |
| | Maximum | 70 W | 70 W |
| Operating case temperature | | −5 to 70°C | −5 to 70°C |
| Storage temperature | | −40 to 85°C | −40 to 85°C |
| Typical performance | | | |
| Gain in SMF-28 | | 10 to 12 dB | 11 to 13 dB |
| Gain in LEAF | | 15 to 17.5 dB | 14 to 16.5 dB |
| Gain in TW-RS | | 17 to 19 dB | 19 to 22 dB |
| Effective noise figure in SMF-28 | | −0.1 to −0.9 dB | −0.5 to −1.7 dB |
| Effective noise figure in LEAF | | −1 to −2.1 dB | −1.1 to −2.7 dB |
| Effective noise figure in TW-RS | | −1.2 to −2.2 | −1.6 to −3.3 dB |

**Figure 21.16**  Specification sheet for a Raman amplifier pump module.

Figure 21.16 shows the specification sheet of commercial digitally controlled Raman amplifier pump modules for two wavelength bands. The output power is 550 mW obtained by combining powers from four semiconductor lasers. Note the different gain performance for different fiber types.

## 21.3   SEMICONDUCTOR OPTICAL AMPLIFIER

### 21.3.1   Principle and Structure of Semiconductor Optical Amplifiers

The semiconductor optical amplifier (SOA) is a key device for high-speed all-optical routing and switching in wavelength-division-multiplexed (WDM) fiber-optic networks. It is also used as optical preamplifier, as in line amplifiers, and as components for performing all-optical wavelength conversion of WDM channels [15]. The semiconductor optical amplifier is also an important device for building wavelength-tunable external cavity semiconductor ring lasers that operate in a single-mode with narrow linewidth [16].

The operating wavelength of an SOA is determined by the material used in fabricating the device. Overall, 1.55- and 1.3-μm SOA are based on InP technology (the substrate material is InP), and 0.85 μm and 0.98 μm are based on GaAs technology.

The wavelengths of choice for a semiconductor optical amplifier are 1.55 and 1.3 μm. Their availability are determined by commercial factors, the most prominent being applications in telecommunications. Thus, a 1.55-μm wavelength device is more readily available than a 1.3-μm device. Although 0.85-μm wavelength semiconductor optical amplifiers based on GaAs technology are technically feasible, they are essentially not available commercially. An important characteristic is the material gain spectrum under different levels of electron-holes injection into the confining active region (usually quantum-wells). Figure 21.17 shows the gain spectrum of a III-V compound material at various injection levels [17].

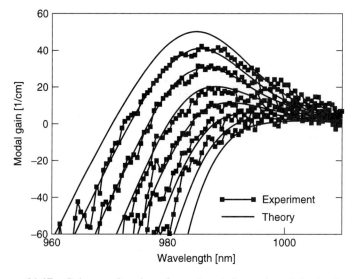

**Figure 21.17**   Gain as a function of wavelength for various injection levels.

The semiconductor optical amplifier is essentially a Fabry–Perot diode laser with very low facet reflectivity. The lower the facet reflectivity, the higher the gain required for lasing because of the higher mirror loss. Thus, for lower facet reflectivity, higher carrier density and, therefore, higher gain can be achieved before the onset of lasing action, which pins the gain of the SOA. By anti-reflection coating and by angle cleaving the diode's facet, low facet reflectivity can be achieved. For finite reflectivity, the device gain is an oscillatory function of wavelength. If the wavelength is in resonant with the Fabry–Perot cavity, the device gain peaks. The ratio of the peak and minimum gain at resonant and at anti-resonant condition describes the gain ripple. With a coating reflectivity of about $10^{-2}$ on both facets, the device gain ripple is about 10 to 20 dB. This type of SOA is referred to as the resonant type SOA. A traveling-wave SOA is realized if the facet reflectivity is $\leqslant 10^{-4}$. The gain ripple is $\leqslant$ a few decibels for a traveling-wave SOA. Very low gain ripple of less than 0.5 dB can be reproducibly achieved often by fabricating a tilted facet of about 7 degrees off the waveguide axis, and by anti-reflective coating the facet. The basic structure of an SOA chip with tilted facet is given in Fig. 21.18.

Figure 21.19 shows a more detail structure of an Alcatel SOA chip.

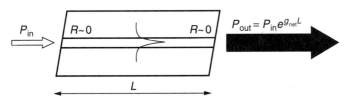

**Figure 21.18**   The basic structure of a traveling-wave SOA chip depicting the input power $P_{in}$ and the amplified output power $P_{out}$. $R$ is the facet reflectivity, $L$ is chip length, and $g_{net}$ is the net gain to be described below (from Jean-Jacques Bernard and Monique Renaud web article on SOA).

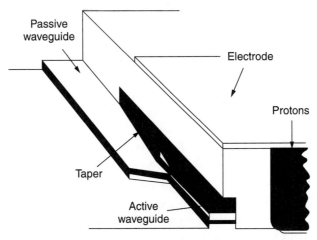

**Figure 21.19**   Alcatel's SOA chip by Jean-Jacques Bernard and Monique Renaud. The following description of the chip structure is based on their web article.

The device consists mainly of three sections: A central active section of around 600 μm in length. There are two passive sections at the input and output sides of the chip. The central active layer is based on a Separate Confinement Heterostructure (SCH) and consists of a 0.2-μm-thick tensile bulk active layer embedded between two quaternary layers 0.1 μm thick. It is tapered over a length of 150 μm, which allows for optical coupling to an underlying passive waveguide. The two passive waveguides are around 100 μm in length. This type of structure leads to a high confinement factor in the gain section together with a large spot size at the facets for achieving a high chip-to-fiber coupling efficiency.

Due to carrier depletion by signal photons through stimulated emission, the carrier density and gain in the active region of an SOA is reduced for finite $P_{in}$. The position dependent amplified power $P(z)$ along the SOA cavity creates a z-dependent gain that involves a complex function of the input power $P_{in}$ and the amplifier's characteristics. As shown in Fig. 21.2, the amplified output power is given by

$$P_{out} = P_{in} \cdot \exp(g_{net}L) \qquad (21.18)$$

The net gain, $g_{net}$, is not only a function of carrier density, but also a function of $P_{in}$ as mentioned above. The net gain $g_{net}$ is modeled by the following equation:

$$g_{net} = g_0 \cdot \exp\left\{-(g_{net} - 1)\frac{P_{in}}{P_{sat}}\right\} \qquad (21.19)$$

where $P_{sat}$ is the power saturation parameter and $g_0$ is the small signal unsaturated gain (note that $P_{in} = 0$, $g_{net} = g_0$).

### 21.3.2  Characteristics and Specifications of Semiconductor Optical Amplifiers

Important design characteristics of a SOA device are as follows:

1. A low residual reflectivity to ensure a low gain ripple.

2. Good material quality with a low optical loss to achieve a high net gain of around 30 dB.

3. A high material gain (utilizing quantum well structure) allowing low current operation (20- to 30-dB fiber-to-fiber gain for a 100-mA drive current).

4. Equalizing the TE and TM mode gain to within 0.5 dB. This is important because the polarization state in a fiber link is random in time (unless polarization maintaining fiber is used as in some sensor applications). If the TE- and TM-mode gain of the SOA are unequal, the amplified optical signal will undergo temporal fluctuations because the input polarization is changing in time.

5. A high output saturation power (defined as the output power for which the gain is reduced by 3 dB from the small signal gain).

6. A high chip-to-fiber coupling efficiency (coupling loss less than 3 dB per facet). This can be achieved by using the mode-expanding tapered waveguides structure mentioned above.

A specification sheet of a commercial fiber-coupled SOA module is given below.

All quantities except the noise figure are discussed previously. The noise figure (NF) is defined as

$$NF = \frac{(SNR)_{in}}{(SNR)_{out}} \qquad (21.20)$$

where $(SNR)_{in}$ is the signal-to-noise ratio at the input fiber (just prior to coupling into the SOA facet) and $(SNR)_{out}$ is the signal-to-noise ratio referring to signal coupled into the output fiber. Thus, NF represents the extra noise due to the amplifier from the standpoint of fiber-to-fiber coupling. The amplified spontaneous emission noise current $I_{spon}$ within an optical filter bandwidth $B_0$ is given for each polarization mode by [18]

$$I_{spon} = \eta_{sp} \cdot (G - 1)qB_0 \qquad (21.21)$$

where $\eta_{sp}$ is the inversion parameter, $G$ is the amplifier gain, and $q$ is the electron charge. For complete inversion, $\eta_{sp} = 1$, giving the lowest possible noise. Generally, for the semiconductor optical amplifier, $\eta_{sp} \approx 2$. The dominant noise factor is the signal-spontaneous beat noise. The mean square current fluctuation $\langle \delta I_{s-sp}^2 \rangle$ due to the signal-spontaneous beat noise is given by

$$\langle \delta I_{s-sp}^2 \rangle = GI_s I_{spon} \left( \frac{B_e}{B_0} \right) \qquad (21.22)$$

where $I_s$ is the signal current, related to the signal power by the factor $q/h\nu$. $B_e$ is the electrical bandwidth of the receiver.

Using Eqs. (21.20)–(21.22), one can show that for high gain, the NF is approximately given by

$$NF = \frac{\eta_{sp}}{\alpha}$$

where $\alpha$ describes the percentage of optical power from the input fiber coupled into the SOA cavity. For $\eta_{sp} = 2$ and $\alpha = 0.5$, typical for semiconductor optical amplifier, one obtains a noise figure of 6 dB, which corresponds quite well to the value given in Fig. 21.20.

| | Standard SOA module | Gain clamped SOA module |
|---|---|---|
| Wavelength of maximum gain | 1540 nm | 1540 nm |
| Fiber-to-fiber gain @ Pin = −25 dBm | 25 dB | 17 dB |
| Gain ripple | 0.5 dB | 0.5 dB |
| Polarization sensitivity | 0.5 dB | 0.5 dB |
| Noise figure nsp/C1 | 7 dB | 7 dB |
| Saturation output power @ 3dB gain compression | 3 dBm | 10 dBm |
| Maximum output power | 10 dBm | 11 dBm |
| 3dB optical bandwidth | 40 nm | 40 nm |
| Lasing wavelength | NA | 1510 nm |
| Driving current | 150 mA | 150 mA |

**Figure 21.20**   A specification sheet for a typical SOA.

## REFERENCES

1. P. C., Becker, N. A., Olsson, and J. R., Simpson, *Erbium-Doped Fiber Amplifiers, Fundamentals and Technology*, Academic Press, New York, 1999.
2. D. E., McCumber, *Phys. Rev.*, **136**: A954 (1964).
3. W. J., Miniscalco and R. S., Quimby, *Opt. Lett.*, **16**: 258 (1991).
4. Y., Mita, T., Yoshida, T., Yagami, and S., Shionoya, *J. Appl. Phys.*, **71**: 938 (1992).
5. Y., Kimura and M., Nakazawa, *Electron. Lett.*, **28**: 1420 (1992).
6. C. M., Miller, S. C., Mettler, and I. A., White, *Optical Fiber Splices and Connectors: Theory and Methods*, Marcel Dekker, New York, 1986.
7. J. T., Krause, W. A., Reed and K. L., Walker, *J. Lightwave. Tech.*, **LT-4**: 837 (1986).
8. W., Zheng, O., Hulten, and R., Rylander, *J. Lightwave Tech.*, **12**: 430 (1994).
9. R. W., Hellworth, *Phys. Rev.*, **130**: 1850 (1963).
10. R. H., Stolen and E. P., Ippen, *Appl. Phys. Lett.*, **22**: 276 (1973).
11. P. C., Becker, N. A., Olsson, and J. R., Simpson, *Erbium-Doped Fiber Amplifiers, Fundamentals and Technology*, Academic Press, New York, 1999, p. 48.
12. P. C., Becker, N. A., Olsson, and J. R., Simpson, *Erbium-Doped Fiber Amplifiers, Fundamentals and Technology*, Academic Press, New York, 1999, p. 205.
13. P. C., Becker, N. A., Olsson, and J. R., Simpson, *Erbium-Doped Fiber Amplifiers, Fundamentals and Technology*, Academic Press, New York, 1999, p. 205.
14. H., Po, J. D., Cao, B. M., Laliberte, R. F., Robinson, B. H., Rockney, R. R., Tricco, and Y. H., Zhang, *Electron Lett.*, **29**: 1500 (1993).
15. B., Ma, M., Saitoh, Y., Nakano, CLEO 2000 Tech. Dig. paper CThG5, p. 422.
16. En T., Peng and C. B., Su, *Optics Lett.*, **17**: 55–58 (1992).
17. J., Hader, A. R., Zakharian, J. V., Moloney, T. R., Nelson, W. J., Siskaninetz, J. E., Ehret, K., Hantke, M., Hofmann, and S. W., Koch, *IEEE Photon. Tech. Lett.*, **14**: 7131 (2002).
18. P. C., Becker, N. A., Olsson, and J. R., Simpson, *Erbium-Doped Fiber Amplifiers, Fundamentals and Technology*, Academic Press, New York, 1999, p. 205.

# INDEX

Abbe nu number, 138

Abbe prism, 132–133

Abbe sine condition, 151

*ABCD* matrix, 25–27. *See also ABCD*
transformation
for common optical elements and media
(table), 28
for quadratic index media, 27–29

*ABCD* transformation, 1, 25–27. *See also*
*ABCD* matrix
for fiber Bragg grating filters, 974
of a Gaussian beam, 33–35
in Gaussian mode matching, 36
for a lens waveguide, 37, 38
thin-film filters and, 957–958

Aberration control, 435–436
with Luneburg lenses, 770–771

Aberrations
chromatic, 137–138
higher-order, 145–146
"longitudinal," 136
"monochromatic," 138–146
of optical lenses, 115
of optical systems, 134–151
spherical, 138, 139–141
of thin lenses, 148–150
third-order, 146–150

Absorbers, for passively mode-locked dye
lasers, 492–496

Absorber switches, saturable, 412–413

Absorption
in avalanche photodiodes, 1233, 1235
in bismuth-substituted YIG, 1179

in optical detectors, 1219
in photoconductive devices, 1239
in waveguide photodiodes, 1238

Absorption coefficient, of optical detectors,
1217

Absorption spectra
of active compounds, 446–456
of erbium-doped
germano-alumino-silica fiber, 1302
of hemoglobin and melanin, 499, 500
of organic dye solutions, 447, 448–455
of quantum wells, 1021–1022, 1023,
1024–1025

Absorptivity, of optical detectors, 1216,
1217

Achromatic doublets, 160, 161

Achromatic optical processing, 217
partially coherent, 218

Achromats, 161

Acoustic attenuation, effect on acousto-optic
modulators, 1281–1283. *See also*
Attenuation entries

Acoustic cells, 1080. *See also* Acousto-optic
Bragg cell; Guided-wave
acousto-optic Bragg cells/modules

Acoustic center frequency, 1102

Acoustic column, 1259

Acoustic drive power, of an acousto-optic
modulator, 1101

Acoustic-induced optical index, 1268

Acoustic power, in acousto-optic design,
1279

*Handbook of Optical Components and Engineering*, Edited by Kai Chang
ISBN 0-471-39055-0   © 2003 John Wiley & Sons, Inc.

**1321**

Acoustic waves, Bragg diffraction from, 1104

Acousto-optic beam deflectors, 1111–1112, 1112–1113
multifrequency, 1274–1277

Acousto-optic Bragg cell, 1081–1082
GaAs in, 1096–1098
key performance parameters of, 1099–1102

Acousto-optic Bragg diffraction
bulk-wave, 1080–1084
coupled-mode analysis of, 1088, 1094–1096
efficiency of and frequency response in, 1088–1092
in planar waveguides, 1084–1099

Acousto-optic Bragg modulators, efficiency and frequency response of, 1088–1092

Acousto-optic (AO) devices, 1079–1129.
*See also* Acousto-optic (AO) modulators
applications of, 1111–1121
bulk-wave acousto-optic Bragg diffraction, 1080–1084
construction of, 1102–1111
$Q$ and $R$ parameters for, 1278, 1279, 1284–1286

Acousto-optic filters, 901–902

Acousto-optic materials, 1278, 1280–1281, 1287–1288
table of popular, 1281

Acousto-optic modulation, background of, 1256–1257

Acousto-optic (AO) modulators, 820, 1255–1299. *See also* Acousto-optic (AO) devices; Planar guided-wave acousto-optic Bragg modulator/deflector; Wide-band guided-wave acousto-optic Bragg modulators/deflectors
applications for, 1256
design of, 1255–1256, 1277–1286
diffraction efficiency–bandwidth product of, 1099–1100
electro-optic modulators versus, 1257
fabrication of, 1287–1296
history of, 1255
materials for, 1280–1281
modulation bandwidth for, 1265–1266
peak diffraction efficiency of, 1266–1267
properties of, 1257–1277

rise-time considerations for, 1259–1260, 1265–1266
spatial frequency response of, 1262–1267
structure of, 1257–1259
temporal transfer function and frequency response in, 1267–1274
time–bandwidth product of, 1100–1101

Acousto-optic RF spectrum analyzers, 1275

Acousto-optics. *See also* Opto-acoustic effect; Surface acoustic waves (SAWs)
coupled mode solutions and transfer functions of, 1261–1262
history of, 1079–1080

Acousto-optic spectrum analyzer, 1114–1115

Acousto-optic substrate materials, 1094–1099
table of, 1095

Acousto-optic switches, 412, 417–418, 1112–1113. *See also* Opto-acoustic effect

Acousto-optic UV intensity modulator, design of, 1279–1280

Acousto-optic VOAs, 854, 857, 858

Acridine dyes, 451

Active compounds, absorption and emission spectra of, 446–456

Active ions, laser action in solid hosts, 371–374

Active layers, in electroabsorption modulators, 1031–1032

Active medium exchange, in liquid lasers, 459–460

Active medium excitation, in liquid lasers, 460–471

Active mode locking. *See* Mode locking

Active optical device modules, in optical fiber transmission systems, 680

Actual image, with optical lenses, 115

Adaptive processing, 228–229

Add-drop filters, 906
with fiber Bragg gratings, 978–979
in waveguide grating routers, 911–912

Add-drop multiplexers, with fiber Bragg gratings, 978–979

Adjacent channel crosstalk (ACC), in multiplexers, 942, 943

Afocal systems, 127

Aging, of lasers, 288

Air-clad rib waveguide, MUX/DEMUX devices and, 981–982

Airy disk, 157, 159. *See also* Spot size
Alacator tokamak, 616
Alcatel SOA chip, 1317
Allpass filters, 896–897, 898
    ring resonators in, 913–914, 915, 916
Aluminum (Al), in avalanche photodiodes,
    1235–1236
Aluminum gallium arsenide (AlGaAs)
    in electro-optic modulators, 1015–1016
    in phototransistors, 1247–1249
    production of, 246
Aluminum gallium arsenide (AlGaAs) diode
    lasers, 241, 242, 244–245
Aluminum gallium nitride (AlGaN) diode
    lasers, 245–246
Ammonia ($NH_3$) maser, 512
Amplification, single- and multiple-pass,
    402, 403
Amplified spontaneous emission (ASE),
    408–409, 410, 478
Amplifier noise, in optical detectors, 1218
Amplifiers. *See also* Optical amplifiers
    electron-beam, 636–640
    in discharge excimer lasers, 631
    fusion, 404–405, 406
    with synchronously pumped dye
        laser, 491
Amplifying media, for lasers, 512, 514
Amplitude gratings, 962
Amplitude modulation (AM)
    in acousto-optic modulators,
        1268–1269, 1270, 1271–1274
    tunable lasers and, 299
Analog deflection/switching, acousto-optic
    devices in, 1113
Analog systems, optical modulation in,
    1009–1010
Analog-to-digital (A/D) conversion, in
    photoconductive devices, 1241–1242
Angioplasty, discharge excimer lasers
    for, 633
Angled-facet tapered amplifier, 289
Angle of incidence, 7, 8, 118
Angular multiplexing, 641, 643
Angular spectral dispersion
    in diffraction gratings, 963
    in laser fusion, 642
Anisotropic acousto-optic Bragg diffraction,
    1082–1084
    in $Y$-cut $LiNbO_3$ substrate, 1092–1094
Anisotropic crystals, waves in, 12–15
Anisotropic diffraction

with multiple transducers of staggered
    center frequency, 1104
with parallel-finger chirp transducer,
    1104
Anisotropy
    of garnet, 1172
    of magnetic fields, 1146, 1147, 1148,
        1158
    in magneto-optic deflectors, 1196
    in reflective magneto-optic materials,
        1186
    in spatial light modulators, 1195–1196
    uniaxial, 1186, 1189
Anisotropy energy. *See* Magnetic anisotropy
    energy
ANN translation, 235
Anode–cathode closure, in electron-beam
    amplifiers, 638
Anorganic laser media, 456
Antares laser, 533, 534–538
Anthracene dyes, 448, 452
Anti-Hermitian part, of permittivity
    tensor, 1161
Antiferromagnetic materials, 1136, 1140,
    1141–1142, 1143
Antiguide, 257, 265, 267
Antimony (Sb), in avalanche photodiodes,
    1235–1236
Antireflection (AR) coatings, on optical
    fiber, 677
Apertures. *See also* Circular aperture; Clear
    aperture; Gaussian aperture; Large
    aperture entries; Numerical aperture
    (NA); Rectangular aperture;
    Semiaperture; Synthesis aperture
    radar (SAR) processing
    in acousto-optic design, 1278, 1279
    numerical, 126
    types of, 48–50
Aperture stop, 117, 123–124, 128
Apodization profile, 922
    for fiber Bragg gratings, 975–976
Apodized fiber Bragg gratings, 975–976
Application-specific integrated circuit
    (ASIC) chips, in MEMS based
    VOAs, 861
Approximation methods, for optical channel
    waveguides, 703
Arclamps, operation of, 380–381
Area arrays, InSb, 90–91
Argon (Ar), in excimer lasers, 626
Argon lasers, 520–524
    mode-locked, 488, 489–490

Argon fluoride (ArF) laser, 627, 632, 640
  applications for, 633
Argus laser, 411, 421
Array architecture
  lasers with, 264–267
  Schottky-barrier, 93–96
Arrayed waveguide grating (AWG)
  as MUX/DEMUX device, 991–992
  operation of, 983–991
Arrayed waveguide grating technology,
    821–827, 909. *See also* Waveguide
    grating routers (WGRs)
  future of, 998
  in MUX/DEMUX integrated optics
    design, 982–991
  principles of, 823–824
Array factor, in diffraction grating analysis,
    965, 966
Arrays, 382. *See also* Diode laser arrays;
    Focal plane arrays; Herringbone
    Bragg electrode array; Infrared focal
    planar array entries; Lens arrays; Line
    arrays; Phased array entries;
    Transducer array; Virtually imaged
    phased array (VIPA)
  index-guided, 264
  of optical isolators, 874
  phase-locked, 264–265
  resonant-optical-waveguide, 267
  reticulated linear, 105
  self-scanning, 103
  in spatial light modulators, 1194–1195
  traveling wave photodiodes as, 1239
  two-dimensional stacked, 290, 291
  VCSEL, 346–347
Array technology
  focal plane architecture in, 69–70
  infrared detectors with, 54–55
Array waveguide grating multiplexer, 823
Arsenic trisulfide ($As_2S_3$), as acousto-optic
    substrate material, 1094–1096
Aspheric coefficients, 117
Aspherics, forms of, 146, 147–148
Aspheric surfaces
  refraction at, 154–155
  transfer between, 153–154
Astigmatic blur length, 142
Astigmatism, 138, 142–144, 436
  in continuously pumped dye lasers,
    483–484
Asymmetric planar waveguide, 722–724,
    727

Asymmetric strip line (ASL) configurations,
    of traveling-wave modulators, 1056,
    1057
Asymmetrical multilayer planar dielectric
    waveguide, 761
Athermal AWG designs, 826
Atmosphere, infrared radiation and, 54,
    55–58
Atmospheric attenuation, 594
Atoms
  antiferromagnetism and, 1141–1142
  ferrimagnetism and, 1143
  in solid-state lasers, 371
  magnetic phenomena and, 1135–1136,
    1138–1139
Attenuation. *See also* Polarization state
    transformation diagrams
  in acousto-optic design, 1281–1283
  atmospheric, 594
  in electro-optic VOAs, 860
  in magneto-optic thin-film waveguides,
    1202, 1203
  of magnetostatic waves, 1159
Attenuation coefficient, 578
Attenuators, 850–863
  variable optical, 846, 852–863
Aurora laser, 639
Automated tunable module (ATM),
    481–482
Automated waveguide loss measurement,
    816–817
Avalanche photodiodes (APDs), 1220,
    1227–1236
  applications for, 1231–1232
  phototransistors versus, 1247
  as waveguide photodiodes, 1239
"Axial color," 137
Axial distances, 118
Axial inhomogeneous media, ray tracing in,
    20–23
Axial ray, 117
Azacoumarin dyes, 453
Azine dyes, 451


Background crosstalk (BC), in multiplexers,
    942, 943
Background noise, in optical detectors, 1218
Background-limited infrared photodetector
    (BLIP), 82
  figures of merit for, 61–63
Backscattering method, measuring impurity
    profiles with, 813

Backside coupling, in PIN photodiodes, 1225

Backward scattering, 641

Backward-volume magnetic waves, 1156
    in magneto-optic thin-film waveguides, 1204–1205

Band diagrams, quantum confined Stark effect in, 1022–1023

Bandgap energy, in optical detectors, 1218–1219, 1220

Bandlimited partially coherent processing, 217–218

Band theory, ferromagnetism and, 1136, 1141

Bandwidth. *See also* Gain bandwidth (GB) product; Modulation bandwidth
    in acousto-optic Bragg diffraction, 1091
    in acousto-optic design, 1278, 1279
    in avalanche photodiodes, 1228, 1232
    in magneto-optic thin-film waveguides, 1204
    in multiplexers, 940
    in optical fiber transmission systems, 681
    optical modulation and, 1010
    in optical systems, 888
    for planar guided-wave acousto-optic Bragg modulator/deflector, 1099–1100, 1100–1101
    of planar photodiodes, 1225
    in traveling-wave modulators, 1059, 1061
    for $Y$-cut $LiNbO_3$ substrate, 1092

Bandwidth utilization (BWU), of ring resonators, 914–916

"Barrel" distortion, 145

Beam attenuators, diagnostics with, 432

Beam combiners, optical filters and, 890–891

Beam deflectors
    acousto-optic, 1274–1277
    $Q$ parameter for, 1284
    $R$ parameter for, 1284–1286

Beam displacement, 130. *See also* Beam displacers
    in optical isolators, 868–869

Beam displacers, 842–843. *See also* Beam displacement
    in optical circulators, 878–879

Beam distortion
    in acousto-optic devices, 1263–1265, 1266
    beam geometry and, 427, 428

Beam monitors, 435

Beam profile, of diode lasers, 260

Beam propagation, in a lens waveguide, 37–38

Beams. *See* Electron-beam entries; Gaussian beams; Laser beams; Molecular beam epitaxy (MBE)

Beam separator, 842–843

Beam-smoothing techniques, in laser fusion, 642–646

Beam splitters. *See also* Beam splitting entries
    diagnostics with, 432
    in optical circulators, 878–879, 879–880
    optical filters and, 890–891
    waveguide diffraction mirrors as, 805, 806, 807, 808, 810

Beam splitting, 130, 132

Beam splitting polarizers, 840, 841–844. *See also* Polarization beam splitters (PBSs)

Bending factors, for thin lenses, 149

Bennett, Jean M., 838

Beryllium aluminate ($BeAl_2O_4$), as solid-state laser host material, 375

Beryllium fluoride ($BeF_2$) glasses, 411

Bessel functions, 1272

Bessel series expansion, 1276–1277

Best bias point voltage, 1033, 1034

BH-LOC devices, 275–277

Bias-drifting, in electro-optic modulators, 1034

Bias stability, 1034–1035. *See also* Best bias point voltage; Debiasing

Biaxial birefringence, 1162

Biaxial crystals, 15

Biaxial medium, 13–14

Biconic connectors, of optical fiber, 677, 679

Bidirectional optical transmission system, optical circulators in, 875–877

Bidirectional WDM, 685

Bi-exciton transition, in bismuth-substituted YIG, 1178

Binary modulation, by spatial light modulators, 1194

Binocular objectives, 160

Binoculars, Porro system for, 133–134

Biperiodic lens sequence, Gaussian beam propagation through, 38

Birefringence, 14, 467, 468, 1162, 1198.
*See also* Birefringent entries;
Extraordinary wave (E-wave);
Magnetic circular birefringence
(MCB); Ordinary wave (O-wave)
in bismuth-substituted YIG, 1182
effects of material properties and
configuration on, 426–427
electro-optic effect and, 1011
of lithium niobate crystals, 1012, 1013,
1014
in magneto-optics, 1165, 1166–1167
in polarizers, 840–841, 841–844
of polymer crystals, 1017–1018
in waveguide isolators, 1194
of Zinc Blende crystals, 1015
Birefringent crystals, 485, 841–844
in electro-optic VOAs, 858
in in-line optical isolators, 865–866
in optical circulators, 878–882
in optical isolators, 867–870, 871–872
Birefringent plates, in lattice filters, 901
Birefringent wedge, 844
in optical isolators, 865–867, 868
Bismuth (Bi), as garnet substituent, 1173,
1174–1175, 1176, 1177–1182,
1183–1184, 1204
Bismuth-substituted iron garnet thick films,
in polarization rotators, 846–847
Bismuth-substituted yttrium iron garnet
(Bi:YIG), 1175, 1176, 1177–1182,
1183–1184. *See also* Yttrium iron
garnet (YIG; $Y_3Fe_5O_{12}$)
magnetostatic waves in, 1198, 1204,
1205
in optical isolators, 1192, 1193, 1194
in spatial light modulators, 1194
Bit-error ratio (BER), 1037
Bitrates, in optical systems, 888
Bloch functions, 1021
Bloch walls, 1146, 1147, 1148, 1149
Bloch-wave analysis, of fiber Bragg grating
filters, 972
Blur function, 140, 141
Bohr magneton, 1135, 1136, 1145
Boltzmann constant, 370
Boltzmann relation, 513
Bonding layers, in acousto-optic
modulators, 1288–1291
Born, Max, 838
Born approximation, 973
Borosilicate glass, ion exchange reactions
and, 746–747

Bottom-emitting VCSEL structure, 326–327
Boundaries, reflection and transmission
at, 7–8
Boundary conditions
for graded index waveguides, 727
magnetism and, 1146
for Maxwell equations, 3
of optical channel waveguides, 705–706
for ray trajectories, 20
for spin waves, 1155
for step index waveguides, 762
for thin-film filters, 956
Bragg conditions, 970
chirp grating lenses and, 801–802, 803
waveguide diffraction filters and, 805,
806, 807, 808
Bragg diffraction, 1269, 1270
anisotropic acousto-optic, 1082–1084
bulk-wave acousto-optic, 1080–1084,
1085–1086
guided-wave acousto-optic, 1084–1099,
1108
in garnet films, 1198, 1199, 1200–1201,
1202–1203
isotropic acousto-optic, 1080–1082,
1091
from multiple-surface acoustic waves,
1104
in planar waveguides, 1084–1099
Bragg diffraction technology, 1187
Bragg electrode arrays, 1118
Bragg gratings, 469, 471. *See also* Fiber
Bragg gratings (FBGs)
analysis of, 921–922
optical filters and, 889, 921–925
Bragg mode acousto-optic Bragg cell, 857
Bragg regime, acousto-optic interaction in,
1259, 1260
Breakdown damage, oscillator–amplifier
laser systems and, 422–423
Brewster angle, 8–9, 418, 468
Brewster windows, 579
in $CO_2$ lasers, 529, 532
Brightness theorem, 150, 151
Brillouin function, 1137, 1138
ferrimagnetism and, 1143
Brillouin scattering, 641
in magneto-optic thin-film waveguides,
1207
Broad-area lasers, 261–267
Broad-area twin-ridge structure (BTRS)
laser, 262
Broadband illumination, 218

Broadband information, in fiber transmission networks, 693

Broadband Integrated Services Digital Network (BISDN), 660, 679, 693

Broadband low dispersion single-mode fibers, 674–675

Buffer gases
  in excimer lasers, 626
  in FIR lasers, 595–597

Buffer layer, in metal-clad optical waveguides, 750–752, 753

Bulk EAMs, 1041

Bulk gratings, 962

Bulk semiconductors, Franz–Keldysh effect in, 1020–1021

Bulk spin waves, 1155–1156

Bulk-wave acousto-optic Bragg diffraction, 1080–1084
  anisotropic, 1082–1084
  isotropic, 1080–1082
  surface acoustic waves and, 1085–1086

Bulk-wave acousto-optic Bragg modulator, 1080
  features of, 1082
  operation and structure of, 1080–1082

Bulk-wave acousto-optics, 1079, 1080

Buried rib-loaded slab, in fabricating integrated optics-based MUX/DEMUX devices, 981–982

Buried waveguides, 702, 703, 707, 711

Burst-mode solid-state lasers, 405–408

Butterworth filters, 897, 898

Butt-joint connectors, of optical fiber, 676–677

BVSIS laser, 262

Calcium (Ca), as garnet substituent, 1175

Canted antiferromagnet, 1136

Capacitance, 320, 321–322, 323
  in optical detectors, 1219–1220
  of photodiodes, 1245

Capacitatively coupled RF-excited systems, 555–556, 557

Capacitors, in discharge excitation circuits, 628–630

Capmany, José, 935

Carbon dioxide ($CO_2$) lasers, 525–538, 579, 595, 596, 621–624
  glow discharge from, 626–627
  invention of, 512
  at UCLA, 589–593

Carbon dioxide ($CO_2$) molecule
  rotational energy levels of, 527–528
  vibrational modes of, 527

Cardinal points, 117, 122–123

Carrier capture time, in high-speed laser design, 313, 318–319

Carrier escape time, in high-speed laser design, 312–314

Carrier noise, in optical detectors, 1218

Carrier photodiodes, 1220

Carrier transport time
  in high-speed laser design, 312, 313, 317–318
  VCSELs and, 343, 344–345

Cartesian coordinate system
  Fabry–Perot resonator analysis in, 180–183
  ray equation in, 18–19

Cascaded Mach–Zehnder interferometer, 899–900

Catastrophic facet damage (COD), 271–274

Cathodes, in electron-beam amplifiers, 636–638

Cations, in ion exchange reactions, 744–746

Causal filters, 893

Cavities, 512. *See also* Dual-cavity FIR lasers; Pumping cavity; Vertical-cavity surface-emitting lasers (VCSELs)
  in continuously pumped dye lasers, 483–484
  with FIR ring lasers, 612–613
  for passively mode-locked dye lasers, 492
  $p$-type doping in, 316
  in stimulating spin waves, 1159
  for synchronously pumped dye lasers, 490
  thin-film filters and, 918–919, 920, 958–959

Cavity dumping, 490–491

C-band EDFA, 944–946

CCD multiplexer, 70

CCD technology. *See also* SPRITE ("signal processing in the element") detectors
  infrared detectors with, 54–55, 69–70, 97, 98, 99
  with pyroelectric detectors, 104–107
  with silicon detectors, 108

$C$-coupled LERF systems, 555–556, 557

$C$-coupled TERF systems, 555, 557

Central input waveguide (CIW), in arrayed waveguide grating, 986, 987, 989

Central output waveguide (COW), in arrayed waveguide grating, 989, 990, 991, 994

Cerium (Ce), as garnet substituent, 1173

$CH_3F$. *See* Methyl fluoride entries

$CH_3I$ FIR lasers, 567

$CH_3OH$ FIR lasers, 567

Chan, Andrew K., 1

Channel spacing
   in multiplexers, 942
   in waveguide grating routers, 910

Channel substrate planar (CSP) laser, 254–258, 262

Channel waveguides, 697, 702–707

Charge, magnetic force and, 1132

Charged-coupled devices (CCDs). *See* CCD entries

Chebyshev filters, 897

Chemical vapor deposition (CVD)
   of garnet, 1172
   waveguide fabrication via, 734

Chemical vapor deposition films, 740

Chief ray, 117, 124, 125, 128

Child–Langmuir law, 637–638

Chirp analysis, 1037–1041

Chirped Bragg gratings, 923–925, 976

Chirp grating lenses, 793, 795, 796–804, 805, 811, 820, 907

Chirping, 495
   optical modulation and, 1009
   in SLM lasers, 686, 688–690

Chirp transducer, single tilted-finger, 1104, 1107, 1110

Chlorophylls, 451

Cholesteric liquid crystals, 849

Chromatic aberrations, 137–138

Chromatic dispersion, in optical fibers, 672, 673, 674, 682

Chromium (Cr)
   in acousto-optic modulators, 1288
   in lattice filters, 900

Circuit model
   of EAMs, 1042–1044, 1045
   of lumped-element QCSE-based MZMs, 1043, 1048
   of segmented traveling-wave modulators, 1064

Circuit time constant, of photodetectors, 1217

Circular aperture
   Fraunhofer diffraction through, 48–49
   power transmission through, 32–33

Circular harmonic analysis, of optical channel waveguides, 703

Circular polarization, 6–7, 1160. *See also* Left circularly polarized (LCP) propagation; Right circularly polarized (RCP) propagation

Circulators, 1194, 1195. *See also* Optical circulators
   optical, 874–883

Cladding, of an optical fiber, 660–661, 662–663

Clear aperture, 117

Cleavage, in NAM laser fabrication, 279

Close-coupled pump cavity, 384

Coarse WDM MUX/DEMUX devices, in optical networks, 948–950

Coaxial flashlamps, for dye lasers, 471–472

Cobalt (Co)
   ferromagnetism of, 1140
   as garnet substituent, 1173
   optical attenuation and, 852

Codirectional coupled-mode filters, 898, 901, 921. *See also* Finite impulse response (FIR) filters

Coefficient of thermal change in center wavelength (CTW), in multiplexers, 941

Coercive field. *See* Coercivity

Coercivity, 1185, 1186
   of reflective magneto-optic materials, 1184, 1185, 1188

Coherent light sources, 199

Coherent optical processing, 210–212

Coherent optical systems, 205–210
   defined, 206–207

Coherent optical transmission technology, 692

Coherent sources, optical diagnostics with, 431–435

Coherent spectrum analyzer, 211–212

Cold cathodes, in electron-beam amplifiers, 636–638

Cold mirror, 130

Collimating lens, in free-space MUX/DEMUX architecture, 967, 968

Collimators, 160–161

Collinear interaction, in anisotropic acousto-optic Bragg diffraction, 1083, 1084

Collision rate, in far infrared lasers, 539

Coma, 138, 141–142, 151, 436. *See also* Elliptical coma

Communications, optical filters in, 888–889. *See also* Optical communication; Telecommunication applications

Compensation temperature, of magneto-optic materials, 1184–1185, 1188, 1190

Complementary error function (erfc) profiles
of graded index waveguides, 731, 732, 733, 743
of planar waveguides, 701–702

Complex amplitude, 208

Complex amplitude transmittance, 212–213

Complex Gaussian parameter, 33–34

Complex light field, 199–204

Complex signal detection, 223–225

Complex spatial filters, synthesis of, 212–215

Compound microscope, 129

Compounds, absorption and emission spectra of, 446–456

Compressive strain
in high-speed laser design, 316
in QW lasers, 253

Concave lens, waveform matrix for, 175

Concave mirrors
Fabry–Perot open resonators with, 176–188
waveform matrix for, 175

Condenser lens, 129, 130

Confinement layers, VCSEL, 328–329

Confinement techniques. *See* Current confinement techniques

Confocal condition, 178

Conic constant, of a surface, 117

Conjugacy, 116

Conjugate factors, for thin lenses, 149

Connectors, optical fiber, 676–678

Conservation of energy, law of, 8

Constant-angle deviation, 130

Constricted DH-structure large optical cavity (CDH-LOC) lasers, 257, 258, 262, 264

Constructive interference, in magneto-optic thin-film waveguides, 1199–1200

Continuous layer peeling (CLP), in fabricating fiber Bragg gratings, 974–975

Continuously pumped dye lasers, 482–486

Continuously variable optical attenuators, 850

Continuous oscillators, 388–390

Continuous traveling-wave modulators, 1053
frequency response of, 1051–1052

Continuous wavelength tuning, 295, 298, 300–303

Continuous-wave (CW) sources, for solid-state lasers, 375, 381. *See also* CW entries

Contrast reversal, 158

Conventional design procedures, for acousto-optic modulators, 1277–1278

Conventional single-mode fibers (C-SMF), 673, 674
in optical fiber transmission systems, 681–682, 683, 689–690

Convex lens, waveform matrix for, 175

Convex mirrors
Fabry–Perot open resonators with, 176
waveform matrix for, 175

Convolution integral, 201, 205, 207, 208

Cooled cylindrical rods, 425

Coordinate systems. *See also* Cartesian coordinate system; Cylindrical coordinate system
for aberration, 135
for diffraction calculations, 156
Kirchhoff diffraction integral and, 199–204
for optical lenses, 117–118
for ray equation, 18–20
for wave equation, 5

Coordination, of atoms in ferrimagnetic materials, 1143, 1144, 1145

Coplanar waveguide (CPW) configurations, of traveling-wave modulators, 1056, 1057, 1058, 1059

Copper (Cu) heat webs, 521, 523

Copper-vapor lasers (CVLs), 524–525, 526
pumping dye lasers with, 476

Core, of an optical fiber, 660–661, 662–663

Cornea, laser light and, 500

Corner cube, 132

Cornu spiral, 45, 46

Corona poling, of polymers, 1017

Cotton–Mouton effect, 1131, 1167, 1170
in magneto-optic thin-film waveguides, 1205

Coumarin dyes, 448, 452

Counter-directional coupled-mode filters, 921. *See also* Bragg gratings

Coupled Mach–Zehnder interferometer, 898, 899

Coupled mode, in acousto-optics, 1261–1262

Coupled-mode analysis, of acousto-optic Bragg diffraction, 1088, 1094–1096

Coupled-mode theory, of fiber Bragg grating filters, 972–973

Coupled waveguides, mathematical analyses of, 709–710, 711–716

Couplers. *See also* Coupling entries; Giant hole coupler; Zinc selenide (ZnSe) coupler
  directional, 890–891
  end-fire, 754, 755, 810
  grating, 754–755, 757–759, 760
  mode-launching, 754–759
  multimode interference, 829, 830, 1067
  optical directional, 707–716
  optical fiber, 678
  prism, 754–755, 755–757, 758, 810, 813, 815–816, 821, 822, 1108, 1109
  tapered thickness, 754
  waveguide pair directional, 711–716

Coupling, backside, 1225

Coupling coefficient, 713–714

Coupling loss, in optical attenuators, 851–852

Craig, Alan E., 1131

CRANK TJS structure, 275–277

*CRC Handbook of Laser Science and Technology*, Vol. 4 (Weber), 431

Critical angle, 8–9
  in dielectric planar waveguides, 721
  optical fibers and, 661
  in prisms, 131
  Snell's law and, 16

Crosstalk
  in AWGs, 994–995
  in multiplexers, 942–943
  in waveguide grating routers, 910

Crystals. *See also* Electro-optic crystal substrates; Ferroelectric electro-optic crystals; Liquid crystal entries; Liquid-crystal light valve (LCLV); Single-crystal epitaxial waveguides; Single-crystal films; Single-crystal infrared detectors; Walk-off crystal optical isolators
  anisotropic, 12–15
  anisotropic acousto-optic Bragg diffraction in, 1083–1084

antiferromagnetism of, 1141–1142

birefringent, 485, 841–844, 858, 865–866

in Faraday rotators, 846, 847

ferrimagnetism of, 1143

growing magneto-optic, 1171

growth anisotropy in, 1146

index ellipsoid of, 1011–1012, 1013

photonic, 471

as solid-state laser host materials, 374–375

wave plates from, 848

Crystal systems, 14–15

Cube corner, 132

Cube PBS, 844, 845
  in optical circulators, 875–877, 878, 879, 880

Cubic ferrites, 1143

Culham Laboratory Group, pulsed optically pumped FIR laser from, 607–612

Curie temperature, 1141, 1142
  ferrimagnetism and, 1143
  magneto-optic data-recording films and, 1189, 1190–1191
  of MnBi films, 1191
  of reflective magneto-optic materials, 1184, 1185

Curie–Weiss law, 1137, 1142

Current. *See* Dark current entries; Electric current; Generation–recombination (G–R) current; Photocurrent generation

Current confinement techniques
  diode laser stripe geometry and, 254–258
  laser problems with, 288
  VCSEL, 331–334

Current path, in EAM circuits, 1042

Current switches, photoconductive devices as, 1241, 1242

Curvature, 117, 118

Curved dielectric interface, ray matrix for, 28

Curved mirror, ray matrix for, 28

Cutoff wavelengths, 664, 675–676
  for HgCdTe detectors, 71, 79

CW diode lasers, pumping by, 486. *See also* Continuous-wave (CW) sources

CW gas lasers, 520, 525–526, 541, 549

CW laser pumping, 383, 486

CW optically pumped lasers, 569–606
  spectral line characteristics of, 570–573

CW oscillators, mode locking in, 393–394

CW passively mode locked dye lasers, 491–493

Cyanine dyes, 448, 449

Cylindrical coordinate system
Fabry–Perot resonator analysis in, 181, 183–186
ray equation in, 19
wave equation in, 5

Cylindrically inhomogeneous media, ray tracing in, 23–24

Cylindrical rods, cooled, 425

Czochralski method, 1171


$D_2O$ FIR lasers. *See* Heavy water ($D_2O$)

Damage limiting, oscillator–amplifier laser systems and, 422–423

Dark current, 88
of HgCdTe detectors, 74
levels in LWIR-PV detectors, 66–67
in PIN photodiodes, 1223–1224

Dark current noise, in optical detectors, 1218

Dark line defects, in lasers, 288

Data extrapolation, from laser literature, 431

Data sheets, for erbium-doped fiber, 1305–1308

Data storage media, magneto-optic, 1187–1192

DCN lasers, 551–557. *See also* Deuterium cyanide (DCN)
energy levels of 544, 545, 549

Debiasing, of photodiodes, 1246–1247

Deep diffused stripe (DDS) lasers, 277

Deflecting mirror surface-emitting lasers, 290–291

Defocus, 137
astigmatism and, 142
diffraction and, 159–160
spherical aberration and, 140
third-order aberrations and, 147

De-interleavers, 901

Delay
in allpass filters, 916
Bragg gratings and, 925
in multiplexers, 939–941
in optical filters, 892, 894, 895, 902
in waveguide grating routers, 911

Delay lines, lossless, 890

Delta function, spatial, 201–202

Demagnetization
of magneto-optic data-recording films, 1190

of reflective magneto-optic materials, 1187
shape-factor, 1144–1146, 1155

Demagnetizing field, 1147

Demultiplexers (DEMUX). *See also* MUX/DEMUX entries; Optical wavelength filtering/multiplexing/demultiplexing; Wavelength division multiplexers/demultiplexers (WDM MUX/DEMUX)
AWG devices as, 824, 825, 827
diffraction gratings as, 905–906
eight-channel fiber Bragg grating, 977
optical filters for, 887

Dense-wavelength division multiplexing (DWDM), 292
dynamic optical add/drop multiplexers and, 826–827
fiber Bragg gratings and, 975
optical attenuators and, 852, 853
in optical circulators, 875

Deposited thin-film overlay, waveguide mode index modification via, 760

Deposited thin-film waveguides, 734. *See also* Thin-film deposition

Deposition processes
for planar optical waveguides, 731–734
waveguide mode index modification via, 760

Depth of focus, 159

Dermatologic surgery, dye lasers for, 500

Design procedures, for acousto-optic modulators, 1277–1286

Detectivity, 55–58, 60–61, 70, 76, 81, 84, 85
of extrinsic silicon detectors, 109
of photo sensors, 112
of SPRITEs, 99–101

Detectors. *See* Optical detectors

Deuterium cyanide (DCN), far infrared lasers using, 541, 543, 544, 545. *See also* DCN lasers

Device modules
GaAs-based monolithic multichannel integrated, 1120–1121
$LiNbO_3$-based hybrid multichannel integrated, 1116–1120

Device parasitics, in high-speed laser design, 313, 314, 320–323

Device physics
of excimer lasers, 634–636
of infrared detectors, 58–70

Diagnostics, optical, 431–435

Diamagnetic materials, 1136–1137, 1164
Dichroic beam splitter, 132
Dielectric. *See* Layered dielectric
Dielectric DBRs, 325
Dielectric films, in polarizers, 840
Dielectric interfaces, ray matrix for, 28
Dielectric permittivity, 1161–1162
   in magneto-optics, 1165
   in optical detectors, 1219
Dielectric planar waveguide theory,
   720–731, 732, 733
Dielectric stack, thin-film filters and,
   956–958
Dielectric waveguides, two-dimensional,
   164–167
Differential gain, in high-speed laser design,
   312, 313, 314–315
Diffraction, 40–50. *See also* Acousto-optic
   Bragg diffraction; Bragg diffraction;
   Bragg regime; Fresnel–Kirchhoff
   diffraction integral; Kirchhoff
   diffraction integral; Raman–Nath
   regime; Reflection high-energy
   electron diffraction (RHEED);
   Waveguide diffraction elements
   Fraunhofer, 42, 43–44, 964–965, 987
   Fresnel, 42, 44–45, 793–795, 796, 800
   in garnet films, 1198, 1199, 1200–1201
   in guided-wave acousto-optic Bragg
      modulators, 1086, 1087
   linear system representation of, 47–48
   Luneburg lens and, 783
   in optical lenses, 155–160
   in periodic structures, 464–466
   in planar guided-wave acousto-optic
      Bragg modulator/deflector,
      1099–1100
   in $Y$-cut $LiNbO_3$ substrate, 1092
Diffraction efficiency
   in acousto-optic modulators,
      1266–1267, 1268, 1269, 1270
   in magneto-optic thin-film waveguides,
      1202–1203, 1204
Diffraction efficiency–bandwidth product,
   of an acousto-optic modulator,
   1099–1100
Diffraction filters, 804–807, 808, 809,
   810, 811
Diffraction grating lenses, 821
Diffraction gratings, 45–47, 903–908. *See
   also* Chirp grating lenses; Fiber
   Bragg gratings (FBGs); In-fiber
   diffraction grating devices;

Waveguide diffraction entries;
   Waveguide optical gratings
amplitude, 962
in $CO_2$ lasers, 529–531, 532
in DFB and DBR lasers, 269–271
in free-space MUX/DEMUX
   architecture, 967, 968, 969
as MUX/DEMUX devices, 961–962
optical filters and, 890
phase, 962
prisms versus, 963–964
reflection, 961, 962
scattering by, 189–190, 191
theory of, 963–967
transmission, 961–962
for tuning dye lasers, 464–466
uniform, 797–799
Diffraction loss, in Fabry–Perot resonators,
   179–186
Diffraction loss and propagation, 186
   in AWGs, 993–994
Diffraction mirrors, 804–807, 808, 809,
   810, 811
Diffraction order, 963
Diffraction point spread functions, 157
Diffractive magneto-optic deflectors,
   1196–1197
Diffractive micro-electrical-mechanical
   systems (D-MEMS), 861. *See also*
   D-MEMS based VOAs
Diffused optical waveguides, 747–750
   table of parameters for, 749
Diffuse reflection hazards, 429–430
Diffusion
   in fabricating planar optical
      waveguides, 732
   in planar photodiodes, 1225–1226
Diffusion coefficients, in ion exchange
   reactions, 744, 746
Digital deflection/switching
   acousto-optic devices in, 1112–1113
   multiport, 1111–1112
Digital filter theory, 897–898
Digital systems, optical modulation in,
   1009–1010
Diode laser arrays, in optical computing,
   1119
Diode laser pumping, 381–382
Diode lasers. *See also* Channel substrate
   planar (CSP) laser; CW diode lasers;
   In-plane diode lasers; Semiconductor
   lasers
applications for, 240–241

characteristics of, 258–260
efficiency of, 260
failure modes of, 288
GaAlAs/GaAs semiconductor, 278–282
GaInAsP/InP, 282–288
multispatial mode, 261–267
reliability of, 288–289
semiconductor materials for, 244–246
single lateral mode high-power, 261
spectral emission range of materials
    in, 242
thermal considerations for, 278–288
Diodes. *See* Photodiode entries
Diopters, 120
Dip coating, waveguide fabrication via, 734
Dipolar interactions, in reflective
    magneto-optic materials, 1186
Dipole moments, 1132. *See also* Magnetic
    moment
defined, 1134
in antiferromagnetic materials, 1142
Dipoles, 1134, 1135, 1137, 1138,
    1139–1140
ferromagnetism and, 1140, 1141
in garnet films, 1198
material magnetic properties and, 1144
precession of, 1149, 1150, 1151, 1152
RF fields and, 1149–1155
superexchange and, 1177
transitions of, 1163–1164
unpaired, 1147–1148
Dirac delta function, 210
Direct band bandgap materials, in optical
    detectors, 1218–1219, 1220
Directional couplers, 890–891
mathematical analysis of, 711–716
Direct modulation, 1010
Dirichlet condition, 41
Discharge excimer lasers. *See also*
    Discharge lasers
applications for, 632–633
commercial, 632
optical configurations for, 631–632
Discharge excitation, electrical circuits for,
    628–630
Discharge lasers, 625–646. *See also*
    Discharge excimer lasers; Discharge
    pumped lasers
construction of, 627–628
large aperture, 630–631
Discharge pumped lasers, advantages of,
    625–626. *See also* Discharge lasers

Discontinuous wavelength tuning, 295,
    298–300
Discrete layer peeling (DLP), in fabricating
    fiber Bragg gratings, 974–975
Discriminator, in stabilizing FIR lasers, 586
Disk amplifiers, 405, 406, 407, 436
limiting processes in, 408–409
medium average power, 408
Dispersing prisms, 131
Dispersion, 130. *See also* Angular spectral
    dispersion; Chromatic dispersion;
    Fiber dispersion; Group-velocity
    dispersion (GVD); Material
    dispersion; Polarization mode
    dispersion (PMD); Profile dispersion
in fiber transmission networks, 692–693
in multilayer step index waveguides,
    763
in optical fiber transmission technology,
    686–687
in virtually imaged phased arrays,
    908–909
in waveguide grating routers, 909–910
Dispersion control, with single-mode fibers,
    671–675
Dispersion curves. *See* Normalized
    dispersion curves
Dispersion diagram
for asymmetric planar waveguide,
    722–724, 728, 732, 733
waveguide thickness values from, 815
Dispersion equation, for planar optical
    waveguides, 700
Dispersion-flattened single-mode fibers,
    674–675
spot size, mode field diameter, and
    Gaussian beam in, 675–676
Dispersion-limited transmission, 681–682,
    687
Dispersion penalty, 1038–1039
in continuous traveling-wave
    electroabsorption modulators,
    1055
Dispersion relation, for magnetic waves,
    1153–1154, 1154–1155, 1155–1156,
    1158
Dispersion-shifted single-mode fibers
    (DS-SMF), 673, 674
in optical fiber transmission systems,
    682, 683–684
Distortion, 139, 144–145
in photodetectors, 1244–1245
Distortion effects, thermal, 424–426

Distributed Bragg reflector (DBR), 164, 193–194, 324–337, 921
phase shift in, 194–195
stopbands in, 190–192
structure of, 269–271
tunable lasers and, 294–295, 297, 298–300, 301–302, 307–308
waveguide grating mirrors and, 807, 811
Distributed feedback, 469–470
Distributed feedback dye lasers (DFDLs), 480–481
Distributed feedback (DFB) resonator, 164, 188–193, 921, 1071–1072
in optical fiber transmission systems, 682, 683–684, 686, 689–690
spectrum of, 195
structure of, 269–270
tunable lasers and, 294–295, 297, 298, 300, 301–303, 304–306, 307
waveguide grating mirrors and, 807
D-MEMS based VOAs, 861–862
Dodecahedral coordination, of atoms in ferrimagnetic materials, 1143, 1144, 1145
Domains
in magnetic materials, 1147–1148, 1149
in magneto-optic data-recording films, 1188–1192
in reflective magneto-optic materials, 1185, 1186–1187
in spatial light modulators, 1195–1196
"Door knob" capacitors, 630
Doped glass, 1170
Döpel, E., 443
Doping, 669–670. See also Doped glass
in PIN photodiodes, 1222
Doping level, in optical detectors, 1218
Double-hetero-structure (DH) configuration, 239, 242, 243, 254, 280
QW lasers and, 247
Double ion exchange method, index perturbation via, 768
Double-layer planar heterostructure (DLPH) HgCdTe detector, 72
Double pass configuration, for free-space MUX/DEMUX device, 968
Double refraction, 14
Dove prism, 132, 133
Down chirp, 495
Dry-etching, in fabricating integrated optics-based MUX/DEMUX devices, 981
Dual-beam system, 533–534, 535

Dual-cavity FIR lasers, 603–606, 607, 608
Dye laser modules, 481–482
Dye lasers, 446–455. See also Lasing dyes
applications for, 444–445, 496–501
continuously pumped, 482–486
dye materials used in, 444–445
flashlamp-pumped, 471–474
frequency selection and tuning of, 463–471
history of, 444
laser-pulse pumped, 475–481
mode locking of, 486–496
passively mode-locked, 491–496
pump sources for, 461–463
synchronously pumped, 488–491
ultra-short light pulses from, 486–488
Dye molecules, in lasers, 443
Dynamic filters, 889
Dynamic gain equalizer filters, 887–888, 906–907, 908. See also EDFA gain equalizer
Dynamic optical add/drop multiplexer, 826–827
Dynamic range, of an acousto-optic modulator, 1101–1102
Dynamic wavelength selective add/drop MUX/DEMUX, 979
Dysprosium (Dy), as garnet substituent, 1174

Eaton lens, ray tracing for, 25
Echelle gratings, 904, 905
Echelon-free beam-smoothing, in laser fusion, 643–646
Echelon ISI, in laser fusion, 642–643
ED-2 silicate glass, 410, 412. See also Lithium-magnesium aluminosilicate glass
pumping of, 376–380
EDFA gain equalizer, 945
EDF-HE980 fiber, commercial specifications for, 1308
EDF-HP980 fiber, commercial specifications for, 1306
EDF-MP980 fiber, commercial specifications for, 1307
Edge-pumped slab amplifiers, 408
Edge rounding, in geodesic lenses, 785, 786, 789, 790, 791
Edgerton, Harold E., 376
Effective index method, for optical channel waveguides, 703–705

Effective index of refraction, 255
  graded index waveguides and, 763, 764,
    765, 766
Effective index perturbation, 760
Effective mode index, 723
Effective waveguide index, in dielectric
    planar waveguides, 722
Effective window function, in Luneburg
    lens fabrication, 776–779, 780, 781,
    782, 783
Efficiency. *See also* Peak diffraction
    efficiency; Quantum efficiency
  of acousto-optic Bragg modulators,
    1088–1092
  of acousto-optic devices, 1259
  of a diode laser, 260
  of electric-to-acoustic conversion, 1101
  of optical modulation, 1010
  of photodetectors, 1243
  of planar guided-wave acousto-optic
    Bragg modulator/deflector,
    1099–1100
  of small-signal modulation, 1032–1037
  of $Y$-cut $LiNbO_3$ substrate diffraction,
    1092
Eigenmode, 163
Eigenstates, polarization, 1161
Eight-channel fiber Bragg grating
    demultiplexer, 977
Eikonal, in geometrical optics, 15
Eikonal equation, 17–18. *See also* Ray
    tracing
Einstein, Albert, 369
Einstein relations, 370–371, 374
EK9740 two-stage saturable absorber, 413
Electrical circuits, for discharge excitation,
    628–630
Electrical discharge FIR lasers, 544–559,
    560, 561
  table of, 542–543
Electrical energy, from optical detectors,
    1215
Electrical pumping. *See also* Discharge
    pumped lasers
  of dye lasers, 445
  of FIR lasers, 539
Electric current
  magnetic force and, 1132–1133
  tunable lasers and, 292
Electric fields
  in avalanche photodiodes, 1227–1228,
    1230–1231
  Franz–Keldysh effect and, 1020–1021

ion exchange reactions and, 744–746
in photodetectors, 1243
quantum wells and, 1022–1023
thin-film filters and, 956–958
Electric-to-acoustic conversion efficiency,
    1101
Electroabsorption (EA) effects, 1010. *See
    also* Franz–Keldysh effect; Quantum
    confined Stark effect (QCSE)
  in modulator materials, 1018–1026
Electroabsorption modulated lasers (EMLs),
    1019
  chips with, 1071
Electroabsorption modulators (EAMs),
    1010, 1019
  advantages and disadvantages of,
    1070
  chirp analysis of, 1040–1041
  circuit model of, 1042–1044, 1045
  lumped-element, 1041, 1044–1048
  optical loss in, 1035, 1036–1037
  optical power handling capability of,
    1035
  transfer function for, 1029, 1030
  traveling-wave, 1052–1056
  voltage swing in, 1030–1032
Electroabsorption waveguides, 1019–1020
Electrodeless discharges, 555, 557
Electrode poling, of polymers, 1017
Electromagnetic field
  energy in, 4
  of light, 1160, 1161
Electromagnetic spectrum, infrared in, 54
Electromagnetic wave propagation, 2–15
Electron-beam amplifiers, 636–640
  large-aperture module, 639
Electron-beam lasers, 626
Electron-beam pumped excimer lasers,
    634–646
  applications for, 642–646
  physics of, 634–636
Electron-beam sustained discharge, 626
Electronic components, of optical fiber
    transmission systems, 680
Electrons
  in avalanche photodiodes, 1227–1228,
    1229, 1230, 1232–1233
  excited states of, 1163–1164
  in ferromagnetic materials, 1140, 1141
  magnetic phenomena and, 1132,
    1135–1136, 1138–1139
  in optical detectors, 1216, 1219
  in photoconductive devices, 1239–1240

Electrons (*continued*)
    in photodetectors, 1243
    superexchange and, 1177
Electro-optic coefficients, 1012
    for $Y$-cut $LiNbO_3$ substrate, 1092
Electro-optic crystal substrates, for planar
    optical waveguides, 733–734
Electro-optic (EO) devices, 199. *See also*
    Multichannel integrated
    acousto-optic–electro-optic device
    module; Pockels cell electro-optic
    switch
    materials in, 414, 723
    in optical fiber transmission systems,
    680
Electro-optic effects, 1011–1018
    linear, 1010, 1011, 1050–1051
    in Mach–Zehnder interferometer,
    1027
    in $Y$-cut $LiNbO_3$ substrate, 1092
Electro-optic (EO) modulators, 412,
    413–416. *See also* Pockels cell
    electro-optic switch
    acousto-optic modulators versus,
    1257
    with III–V compounds, 1014–1016
    with lithium niobate crystals,
    1013–1014
    polymers in, 1017–1018
Electro-optic polymers, 1017–1018
Electro-optic tensor, 1012
    for lithium niobate, 1013
    for polymer crystals, 1017–1018
    for Zinc Blende, 1015
Electro-optic VOAs, 854, 857–861
Electrostatic discharge (ESD) damage
    to lasers, 288
Elliott, C. T., 96
Elliptical coma, 145
Elliptical polarization, 6–7, 1160–1161
Elliptical pump cavity, 384–385
Embedded channel waveguides, 702–703,
    706, 709, 710
Emission spectra
    of active compounds, 446–456
    of argon, 521
End-fire couplers, 754, 755
    for measuring waveguide parameters,
    810
Energy. *See also* Exchange energy
    bulk-wave acousto-optic Bragg
    modulators and, 1081
    from excimer lasers, 638

    from FIR lasers, 621–624
    from HCN lasers, 561
    in magnetic fields, 1146, 1147
    optical detectors and, 1215
    from synchronously pumped dye lasers,
    490–491
    transmission of, 8–9
Energy levels, 1303
    of $Er^{3+}$ ion, 1302
Entrance pupil, 117, 124–125, 127
    distortion of, 155
Entropy, 235
E/O conversion efficiency, 1033
Epitaxial crystal growth technique,
    739–742. *See also* Liquid phase
    epitaxy (LPE); Molecular beam
    epitaxy (MBE)
    in NAM laser fabrication, 279
Epitaxial growth by melting (EGM),
    fabricating optical waveguides via,
    742
Epitaxial waveguides
    single-crystal, 739–742
    table of, 741
Epoxy bond, for acousto-optic modulators,
    1288
Equatorial Kerr effect, 1167, 1168
Equivalent circuit model, for segmented
    traveling-wave modulators, 1064
Equivalent index representation, of optical
    channel waveguides, 704
Erbium (Er)-doped fiber
    commercial specifications and data
    sheets for, 1305–1308
    implementation of, 1308–1311
    rate equations model for, 1304–1305
Erbium-doped fiber amplifier (EDFA),
    1301–1311
    in MEMS based VOAs, 862, 863
    operation of, 864
    in optical networks, 944–946
Erbium ion ($Er^{3+}$), spectroscopy of,
    1301–1303
Erbium lasers, 371–374, 382
Etalons, 466–467, 468, 477, 892, 896,
    920–921
    for gas lasers, 574, 575, 616
    thin-film filters and, 958
    in virtually imaged phased arrays,
    908–909
Etched concave gratings, 904–905
Etched-mesa type VCSELs, 331–332, 334

Etching. *See also* Reactive ion etching (RIE)
  in fabricating integrated optics-based MUX/DEMUX devices, 981
  waveguide mode index modification via, 760
Europium (Eu) ions, in liquid lasers, 474
Evanescent field
  in multilayer step index waveguides, 761
  in prism couplers, 755, 756, 757
Evanescent in-phase mode, 265–267
Evanescent out-of-phase mode, 265–267
Evanescent waves, in dielectric planar waveguides, 722
Evaporation, waveguide fabrication via, 734, 736
Even parity, 131
Excess noise factor, in avalanche photodiodes, 1235, 1236
Exchange-dominated spin waves, 1152, 1153, 1158, 1159–1160
Exchange energy, 1139, 1141, 1147, 1148, 1153
Exchange field, 1151, 1152
Excimer lasers, 538, 624–646
  electron-beam pumped, 634–646
  physics of, 634–636
  pumping dye lasers with, 476–477
Excimers, defined, 624
Excitation
  by arclamps, 380–381
  discharge, 628–630
  by filament lamps, 380–381
  by flashlamp, 376–380
  glow discharge, 626–627
  of solid-state lasers, 375–386
Excited states, 1163–1164
Excitons, quantum confined Stark effect and, 1021–1023. *See also* Bi-exciton transition
Exciton transition energy, 1022
Exit pupil, 117, 124, 126–127, 128, 129
Expanded-beam connectors, of optical fiber, 676–677
Exponential error function index profiles, of planar waveguides, 701–702
Exponential profile, for graded index waveguides, 727, 728
External-cavity tunable (ECT) lasers, 307, 308
External circuit effect, 1243–1246

External quantum efficiency, of optical detectors, 1216
External reflection, in magneto-optics, 1167
Extraordinary wave (E-wave)
  birefringence and, 14, 842–844
  in optical isolators, 866, 868
Extrinsic photon detectors, 58
Extrinsic silicon detectors, 107–109
  with minority carrier CCD readout, 108–109
  performance characteristics of, 109
Eyepiece magnification, 126, 129
Eyepieces, 161
  simple, 127
Eye relief, 129
Eye surgery, discharge excimer lasers for, 633

Fabrication processes
  for geodesic lenses, 789, 790, 791
  for Luneburg lenses, 770–771, 775–779, 780, 781, 782, 783
  for planar optical waveguides, 731–734
Fabry–Perot (FP) diode laser, semiconductor optical amplifier as, 1317
Fabry–Perot etalons. *See* Etalons
Fabry–Perot filters, 892, 918, 919
Fabry–Perot lasers, 260, 267, 269
  in optical fiber transmission systems, 682, 683, 687
Fabry–Perot resonators, 163–164
  mode and diffraction loss in, 179–186
  open, 176–188
  optical attenuation and, 852
  waveguide-type, 164–171
Fabry–Perot scans, of FIR lasers, 609, 611, 617
Face-cooled disk amplifier, 408
Face-pumped disk geometry, 426
Facet damage, to lasers, 288
Facet loss, 1035
Facets
  catastrophic damage to laser, 271–274
  reducing intensity at, 274–275
Fans, 136–137
Faraday-active media, 1131
Faraday effect, 1131, 1132
Faraday rotation, 1154, 1169–1170. *See also* Faraday rotators
  in bismuth-substituted YIG, 1179–1182
  in gadolinium-substituted garnet, 1183
  in garnet, 1174

Faraday rotation, 1154, 1169–1170. *See also* Faraday rotators (*continued*)
  in magneto-optic data-recording films, 1191
  in magneto-optic deflectors, 1197
  in magneto-optics, 1165–1166, 1167
  in magneto-optic thin-film waveguides, 1203, 1204, 1206
  in optical isolators, 1192, 1193, 1194
  in spatial light modulators, 1194, 1196
Faraday rotators, 416–417, 485, 846–847. *See also* Faraday rotation
  in acousto-optic VOAs, 857
  in magneto-optic VOAs, 856–857
  in optical circulators, 875–876, 877, 879–880, 880–881, 882, 883
  in optical isolators, 865, 866, 867, 869, 870, 871, 872, 873
  polarization and, 419
  risk reduction via, 424
Far-field beam monitors, 435
Far-infrared (FIR) lasers, 539–541, 541–624
  applications for, 540–541
  design of, 569–606
  electrical discharge, 544–559, 560, 561
  optically pumped, 559–624
  power output from, 591–606
  pulsed optically pumped, 606–624
  at UCLA, 589–593
Fast Fourier transform (FFT), 48
Feedback
  distributed, 469–470
  randomly distributed, 470–471
  in stabilizing FIR lasers, 583, 584, 603, 605
Female connectors (FCs), of optical fiber, 677
Femtosecond light pulses, 487–488, 492, 494, 495, 496
Fermat's principle, 15–16, 116
Fermi–Dirac statistics, 247
Ferrimagnetic materials, 1136, 1140, 1142–1144, 1147–1149, 1157, 1159, 1164
  hysteresis loop for, 1185, 1186
  reflective magneto-optic materials as, 1184
Ferrimagnetic resonance (FMR) line width, 1173
  of bismuth-substituted YIG, 1182, 1183
Ferrites, 1143
Ferroelectric electro-optic crystals, 740, 744

in metal-clad optical waveguides, 750–752
Ferromagnetic materials, 1136, 1139, 1140–1141, 1147–1149, 1164
  hysteresis loop for, 1185, 1186
  in polarization rotators, 846–847
Fiber amplifiers
  erbium-doped, 1301–1311
  Raman, 1301, 1312–1316
Fiber Bragg grating filters, theory and design of, 972–975
Fiber Bragg gratings (FBGs)
  in optical circulators, 875
  performance characteristics of, 975–976
  principles and technology of, 970–972
  synthesis of, 974–975
Fiber dispersion, single-mode, 671–675
Fiber optics, VCSELs for, 342. *See also* Optical fiber entries
Fiber optics junctions, multimode, 828–830
Fiber to the home, 692–693
Fiber transmission networks, 692–693
Field curvature, 139, 142–144
Field-effect transistor (FET) circuits, 1241
Field lens, 128–129
Field stop, 117, 125–126, 128
Figueroa, Luis, 239
Figures of merit (FOMs), 61–69, 232
  in acousto-optic Bragg diffraction, 1089
  in acousto-optic design, 1278–1279
  of HgCdTe detectors, 74
  for Schottky-barrier detectors, 93–96
Filament lamps, operation of, 380–381
Films, for planar optical waveguides, 720. *See also* Thin-film entries
Filter asymmetry, in multiplexers, 940
Filter center wavelength, in multiplexers, 939–940
Filters. *See also* Gain flattening filters; Optical filters
  add-drop, 906, 911–912
  allpass, 896–897, 898, 913–914, 915, 916
  fiber Bragg grating, 972–975
  lattice, 899–903
  tunable, 905–906
  waveguide diffraction, 804–807, 808, 809, 810, 811
  with waveguide grating routers, 911–913
Filter technology, thin-film, 954–956
Findakly, Talal K., 697

Fine WDM MUX/DEMUX devices, in optical networks, 950

Finite impulse response (FIR) filters, 891, 895, 898–913

First-order properties
of optical lenses, 115, 118–133
of prisms, 130

Fixed optical attenuators, 850, 851, 852

Flame brushing, in fabricating fiber Bragg gratings, 972

Flame hydrolysis deposition (FHD)
for AWG devices, 822
in fabricating integrated optics-based MUX/DEMUX devices, 980
for thin-film waveguides, 739

Flashlamp excitation, 376–380

Flashlamp-pumped dye lasers, 471–474

Flashlamp-pumped rare-earth liquid lasers, 474–475

Flashlamps
failure of, 379–380
for passively mode-locked dye lasers, 491–492

Flat spectral response AWG designs, 824–826

Flat-top filter response curve, 824–826

Flip-chip solder bond interconnection, 106

Fluorene dyes, 452

Fluorescence, 370, 373, 564

Fluorescence bands, in dyes, 463

Fluorescence quenching, 472

Fluorescence spectra, of organic dye solutions, 447

Fluorescence spectroscopy, liquid lasers in, 497–498

Fluorine ($F_2$), in excimer lasers, 627, 631–632, 633, 634, 640

Fluoroberyllate glasses, 411

Fluorophosphate glasses, 411–412

$f$-number, 126, 798, 799, 820
for Luneburg lenses, 773

Focal length, 117, 118, 121
of Fresnel zone plate, 50
of geodesic lenses, 784–785, 786, 787, 789, 791
of Luneburg lenses, 777

Focal plane architecture, 69–70

Focal plane arrays, HgCdTe, 73–82. *See also* Infrared focal planar array entries

Focal points, 117

Focal systems, 126

Focus. *See also* Defocus; Medial focus; Sagittal focus; Self-focusing; Tangential focus
in acousto-optics, 1261
of chirp grating lenses, 797–800, 803, 804
depth of, 159
of shaped-zone Fresnel lens, 804

Focusing lens, in free-space MUX/DEMUX architecture, 967, 968

Foldback configuration, for AWGs, 997, 998

Forbidden zone, for spin waves, 1153, 1155, 1158

Force, magnetic, 1132

Formic acid (HCOOH) FIR lasers, 567, 593–594, 595

Förster mechanism, 459

Forward error correcting codes (FECs), 947

Forward-volume magnetic waves, 1156–1158
in magneto-optic thin-film waveguides, 1204–1205

Four-energy-level system, 517–518, 597–598

Fourier multiplication theorem, 208, 212

Fourier spectra, in white light optical processing, 218–220

Fourier transform, 158, 160, 206, 207–208, 210–212, 213, 214, 221, 224, 989–990, 994–995, 1274–1275
in diffraction grating analysis, 964–965
for fiber Bragg grating filters, 973
of Hermite–Gaussian functions, 39
in linear systems, 47–48
optical filters and, 888–889
in optical systems, 201–204

Four-level lasers, 371–374
dye lasers as, 457–458

Four-port optical circulator, 878, 883

Four-stage amplifier, with synchronously pumped dye laser, 491

Fox–Smith interferometer, 607, 610, 616

Frantz–Nodvik formalism, 400, 401

Franz–Keldysh effect, 1020–1021
in optical detectors, 1219

Fraunhofer approximation, 156, 159–160

Fraunhofer diffraction, 43–44, 964–965, 987
examples of, 48–50

Free-carrier optical waveguide, 752–753

Free-carrier plasma effect, tunable lasers and, 292–293

Free propagation regions (FPRs), in arrayed waveguide grating, 982, 983, 985, 986, 989–990, 994

Free space, waveform matrix for, 175

Free-space MUX/DEMUX devices, 961–969
theory and design of, 963–967

Frequency. *See also* Acoustic center frequency; High-frequency rate equations; Low-frequency noise; Resonant frequencies; Spatial frequency response
in acousto-optic design, 1278, 1279
in avalanche photodiodes, 1231
bulk-wave acousto-optic Bragg modulators and, 1081–1082, 1083
in guided-wave acousto-optic Bragg modulators, 1086–1088

Frequency chirp, 1037–1041

Frequency free spectral range ($FSR_f$)
in arrayed waveguide grating, 986, 990, 997

Frequency jitter, in FIR lasers, 580

Frequency mode selection, in laser oscillators, 395–396

Frequency modulation (FM), tunable lasers and, 299

Frequency response
in acousto-optic Bragg modulators, 1088–1092
in acousto-optic modulators, 1101, 1267, 1271–1274, 1291–1293
in continuous traveling-wave modulators, 1051–1052, 1053–1054, 1055–1056
in lumped-element modulators, 1042–1044
in photodetectors, 1243, 1244
in segmented traveling-wave modulators, 1063–1066
in $Y$-cut LiNbO$_3$ substrate, 1092–1093

Frequency selection
in dye lasers, 463–471
in liquid lasers, 460–471

Frequency shifting, optical, 1113

Frequency-staggered multiple tilted transducer, 1106. *See also* Multiple tilted SAW transducers; Staggered center frequency

Fresnel diffraction, 44–45, 793–795, 796, 800

Fresnel integral, 160, 204
functions with, 45

Fresnel–Kirchhoff diffraction integral, 41–42, 45. *See also* Kirchhoff diffraction integral
simplified, 172

Fresnel–Kirchhoff theory, 199–204

Fresnel number, 45, 173

Fresnel zone lenses, 793–795, 796, 800, 804

Fresnel zone plate, 50. *See also* Zone plate

Fresnel zones, 45

Fringing fields, 1144

Front focal length, 117, 118, 121

Front focal point, 117

Front nodal point, 117, 122

Front principal plane, 117

Full-width-half-width maximum (FWHM) bandwidth, for fiber Bragg gratings, 975

Fused quartz, 1280
as acousto-optic substrate material, 1094, 1095
for acousto-optic modulators, 1289
in bulk-wave acousto-optic Bragg modulators, 1080–1081

Fusion amplifiers, 404–405, 406

Fusion splices, of optical fiber, 677

Gadolinium (Gd)
ferromagnetism of, 1140
as garnet substituent, 1173, 1174, 1175, 1183

Gadolinium–iron garnet film, 1145

Gain
effect of laser glass properties on, 409–412
in gas lasers, 564–566, 568
in high-speed laser design, 312, 313, 314–315
link RF, 1032–1033, 1044
VCSEL, 329–331

Gain bandwidth (GB) product, of optical detectors, 1218

Gain depletion, in passively mode-locked dye lasers, 493–494

Gain flattening filters, in MEMS based VOAs, 862, 863

Galilean telescope, 127, 128

Gallium (Ga)
in avalanche photodiodes, 1235–1236
as garnet substituent, 1174, 1175–1176
in optical waveguides, 740–742

Gallium aluminum arsenide (GaAlAs) lasers, 277, 278–282. *See also* Z-cut GaAs–GaAlAs planar waveguide
power output limit for, 271–272
pumping by, 381
Gallium arsenide (GaAs)
as acousto-optic substrate material, 1095, 1096–1099, 1110
in electro-optic modulators, 1015–1016
in guided-wave acousto-optic Bragg modulators, 1086
in phototransistors, 1247–1249
production of, 246
Gallium arsenide (GaAs) AO Bragg modulators/deflectors, performance of, 1110
Gallium arsenide (GaAs)-based monolithic multichannel integrated optical device modules, 1120–1121
Gallium arsenide (GaAs) lasers, 241, 242, 244–246, 277, 278–282
power output limit for, 271–272
Gallium arsenide (GaAs) MZMs, advantages and disadvantages of, 1070–1071
Gallium indium arsenide phosphide (GaInAsP) diode lasers, 282–288. *See also* Indium gallium arsenide phosphide (InGaAsP) diode lasers
Gallium phosphide (GaP) diode lasers, 245
Garnet
creation of, 1171
ferrimagnetism of, 1143, 1144, 1145
in magneto-optic polarization-rotation isolators, 1192–1194
magneto-optic properties of, 1170
rare-earth-iron, 1171, 1172, 1173
saturation magnetization of, 1174, 1176
structure of, 1172, 1173
substituents in, 1173–1176
superexchange in, 1178
temperature versus magnetization of, 1186
Garnet films
diffraction in, 1198, 1199, 1200–1201
for magneto-optic data-recording, 1191
in optical waveguides, 740–742
in polarization rotators, 846–847
Garnet wafers, in spatial light modulators, 1194
Gas lasers, 511–647
advantages of, 518–519
common, 519–541

excimer lasers, 624–646
far-infrared lasers, 539–541, 541–624
liquid lasers versus, 443–444
properties of, 518–541
properties of gas molecules in, 569, 570–573
solid-state lasers versus, 387
Gauss (unit), 1133–1134
Gaussian aperture, Fraunhofer diffraction through, 49–50
Gaussian beams, 171–176, 1115. *See also* Gaussian modes
in acousto-optic devices, 1259–1260, 1271–1272, 1282–1283
complex representation of, 33
cross-sections of, 32
defined, 30–31
in optical attenuators, 851
prism couplers and, 757, 758
propagation of, 29–39
in single-mode fibers, 675–676
transformation of, 33–35
Gaussian error function index profiles, of planar waveguides, 701–702
Gaussian function, 987, 988, 1024. *See also* Hermite–Gaussian functions
in chirp analysis, 1037, 1039
Gaussian lens formula, 27
Gaussian modes, 1–2, 163, 254. *See also* Gaussian beams
in lens waveguide, 35
matching of, 36–37
in quadratic index media, 35–36
Gaussian noise, 214
in optical information processing, 234
Gaussian profile, for graded index waveguides, 731, 732, 733, 742–743
GdIG. *See* Gadolinium (Gd)
Gemini laser, 533–534, 535
Generalized Rinehart–Luneburg geodesic lens, 785–788, 789, 791
General magnetostatic mode spectrum, 1157
Generation–recombination (G–R) current, surface, 66
Generation–recombination (G–R) noise, in photoconductive detectors, 64
Geodesic lenses, 779–789, 820, 821
fabrication of, 789, 790, 791, 820
Geometrical optics, 15–29, 115–116
Geometry. *See also* Top electrode geometry
beam distortion and birefringence and, 426–427, 428
of laser amplifiers, 402–408

Geometry. *See also* Top electrode geometry
(*continued*)
material magnetic properties and, 1144
of photodiodes, 1246–1247
thermal distortion and, 424–426
Germanium (Ge), in avalanche photodiodes,
1233, 1234
Germanium beam splitter, 612
Germanium extrinsic detectors, 59, 60
Germanium oxide (GeO$_2$)-doped silica
waveguides, in integrated
optics-based MUX/DEMUX
devices, 980
*g*-factor, 1136
GGG (Gd$_3$Ga$_5$O$_{12}$) lasers, 375
Giant hole coupler, for gas lasers, 574–575
Gires–Tournois interferometer (GTI), 896,
897, 919–920, 920–921
Glan prisms, 841
Glan–Thompson polarizer, 418
Glass. *See also* ED-2 silicate glass; Laser
glass properties; Magnifying glasses;
Nd:glass entries; Silica glass fibers
ion exchange reactions and, 744–746
magnetically doped, 1170
in multilayer step index waveguides,
763
in optical fibers, 660
as solid-state laser host material, 375
in thermo-optical VOAs, 855–856
thin-film filters and, 958
Glass substrates
for planar optical waveguides, 733
for thin-film waveguides, 734, 736
Glow discharge excitation, 626–627
Golay cell, 104
Gold (Au)
for acousto-optic modulators, 1288,
1289
in lattice filters, 900
Goldhar, J., 511
Goos–Hanchen shift, 722
Graded-index and separate carrier and
optical confinement hetero-structure
(GRIN-SCH), 243
QW lasers and, 249
Graded-index fiber, ray tracing in, 23–24,
27–29
Graded index optical waveguides, 720,
742–743
modes for, 726–730
optical diffraction elements and, 790
overlay film on, 763–768

Graded refractive index (GRIN) lenses, 685
in in-line optical isolators, 865
in MEMS based VOAs, 861
in thin-film filters, 960–961
Graded refractive index profiles, 665–669
for Luneburg lenses, 773–775
Grating coupler, 754–755, 757–759, 760
Grating equation, 963
for arrayed waveguide grating, 983–991
Grating periodicity, for chirp gratings,
798–799, 800
Grating pitch, 961
Gratings, bulk, 962. *See also* Bragg
gratings; Diffraction gratings; Echelle
gratings; Fiber Bragg gratings
(FBGs); Long period fiber gratings;
Waveguide grating routers (WGRs)
Grating surface-emitting lasers, 290, 291
Grazing incidence grating, 465–466
Green's function, 40, 43
for calculating quantum confined Stark
effect, 1023, 1024
Group delay. *See* Delay
Group-velocity dispersion (GVD), 1037
Growth anisotropy, 1146
GSGG (Gd$_3$Sc$_2$Ga$_3$O$_{12}$) lasers, 375. *See
also* Nd:Cr:GSGG lasers
Guided-optical waves (GOW), 1091, 1094,
1096
Guided-wave acousto-optic Bragg
cells/modules, applications of,
1111–1121
Guided-wave acousto-optic Bragg deflector,
1109
Guided-wave acousto-optic Bragg
diffraction, in planar waveguides,
1084–1099, 1108
Guided-wave acousto-optics, 1079, 1080
Guided-wave mode, of prism couplers,
755–756, 757
Guided-wave optical signal processor, 1114
Gyromagnetic equation of motion, 1150,
1152, 1155, 1158
Gyromagnetic frequency, 1136, 1139

Half-band filters, 901
Half wave plates, 848–849
in AWGs, 994
in optical isolators, 872–874
*Handbook of Optics*, 838
Hankle function, 44
Hänsch-type cavity, 466, 467

Harmonic levels
  in acousto-optic modulators, 1272–1273
  in photodetectors, 1243, 1244–1245
Hazards, reflection, 429–430
HCN lasers, 553, 555, 556, 557–559, 560,
    561. *See also* Hydrogen cyanide
    (HCN)
  energy levels of, 544
HCOOH. *See* Formic acid (HCOOH) FIR
    lasers
Heavy-hole exciton resonances, 1022
Heavy water ($D_2O$), far infrared lasers
    using, 541, 542, 613–614, 616–617,
    618–619, 621
Helios laser, 533–534, 535
Helium (He)
  as buffer gas, 595–597
  in excimer lasers, 626
  far infrared lasers using, 542
Helium–neon (He–Ne) lasers, 519–520
Helmholtz equation, 4
Hemoglobin, absorption spectrum of, 499,
    500
Henry's alpha parameter, 1039
Hermite equation, 31
Hermite–Gaussian functions, 29–30, 31,
    33, 35, 183, 185
  properties of, 39
Hermite–Gaussian modes, 184
Hermite polynomials, 31, 183
  recursion formula for, 39
Hermitian part, of permittivity tensor,
    1161
Herringbone Bragg electrode array, 1118
Hexagonal ferrites, ferrimagnetism of,
    1143
*n*-Hexane, as buffer gas, 597, 601
HgCdTe, in SPRITEs, 96
HgCdTe detectors, 70–86
  focal plane arrays for, 73–82
  performance characteristics of, 70–73
HgCdTe-infrared focal planar arrays
    (IRFPAs), requirements for, 82–86
HgCdTe photoconductive detectors, 61, 62
HgCdTe photodiodes, 72, 76, 77, 78
HgCdTe photovoltaic detectors, 54, 61, 62,
    63, 64–65
  compositional profile of, 73
High-density WDM systems, 691–692
High-efficiency Nd:YAG lasers,
    single-lamp, 384
Higher-order aberrations, 145–146
High-frequency rate equations, 309

Highly coherent sources, optical diagnostics
    with, 431–435
High numerical aperture (NA) fiber, in
    AWGs, 993
High-power lasers
  carbon dioxide, 528, 531, 532–538
  future development of, 350–351
  passively mode locked, 495–496
  short pulses from, 640–641
High-power mode stabilized lasers,
    274–275
High-power pump sources, 1314–1316
High-quality mode locking, 490
High-speed lasers, future development of,
    351
High-speed semiconductor lasers, 308–323
  design parameters for, 312–323
  rate equations for, 309
  small-signal modulation of, 309–312
High-speed vertical-cavity surface-emitting
    lasers, 343–346
Holmium (Ho) lasers, 382
Holography, fabricating fiber Bragg gratings
    via, 971
Homogeneous media, light rays in, 19
Hopfield model, of neural nets, 233
Host materials, solid-state laser, 374–375
Huygens–Fresnel principle, 155
Huygens principle, 42–43, 199–200
Hybrid array detectors, 54
  HgCdTe, 73–82
Hybrid DBRs, 325
Hybrid integrated optical RF spectrum
    analyzer, 1114–1115
Hybrid mode locking, 496
Hybrid optical architecture, for matrix
    multiplication, 227–228
Hybrid optical circulators, 879–880
Hybrid-TEA oscillator, 618–619
Hydrogen cyanide (HCN), far infrared
    lasers using, 541, 543, 544. *See also*
    HCN lasers
Hydrogen fluoride (HF), in excimer lasers,
    640
Hydrogen loading, in fabricating fiber
    Bragg gratings, 972
Hydrogen sulfide ($H_2S$), far infrared lasers
    using, 542
Hydrogen thyratrons, 628–629
Hydrothermal garnet growth, 1172
Hyperbolic secant index media, ray tracing
    in, 23
Hysteresis loop, 1185, 1186

ICBM (intercontinental ballistic missile) detection, 67

Ideal image, with optical lenses, 115

Ideal linear polarizer, 838

ID SAW transducer, 1099. *See also* Interdigital electrode transducer (IDT)

Iga, Kenichi, 163

III–V compound semiconductors, 1014–1016
  in avalanche photodiodes, 1233–1234
  bias stability of, 1034
  electroabsorption effects in, 1018–1019
  electro-optic coefficients for, 1063
  in optical detectors, 1220
  in phototransistors, 1247–1249
  in PIN photodiodes, 1223

III–V Mach–Zehnder modulators
  comparison of different, 1070–1071
  traveling-wave, 1066–1069

Illumination systems, for slide projector, 129–130

Image-forming systems, 126

Image inversion, 130

Image plane, of gratings, 961

Image rotation, 130

Images
  location of, 118–119, 126
  with optical lenses, 115
  parity of, 130–131
  types of, 116

Image sampling, 222

Image space, 116

Immersed thermistor (IT), 104

Impedance
  transmission line, 1047–1048
  in traveling-wave modulators, 1052, 1054–1055

Impedance matching, in acousto-optic modulators, 1291–1293

Impurity profile measurement, 812–813, 814

Incidence angles, 119

Incoherent optical systems, 205–210
  defined, 206–207

Incoherent pump model, for gas lasers, 563–569

Index ellipsoid
  for crystals, 1011–1012
  for GaAs, 1016
  for lithium niobate, 1013
  for polymer crystals, 1018

Index factors, for thin lenses, 149

Index functions, ray tracing for, 22–23

Index-guided arrays, 264. *See also* Antiguide; Negative-index guide; Positive-index guide

Index of refraction. *See* Refractive index

Index perturbation methods, 760, 768–770

Indicatrix, 13–14

Indirect bandgap materials, in optical detectors, 1219

Indium aluminum gallium arsenide (InAlGaAs) diode lasers, 241, 242, 1025

Indium antimonide (InSb), 86

Indium antimonide detectors. *See* InSb entries

Indium antimonide infrared detectors. *See* InSb infrared detectors

Indium arsenide (InAs) diode lasers, 245

Indium gallium aluminum phosphide (InGaAlP) diode lasers, 241, 242, 245–246

Indium gallium arsenide (InGaAs)
  in avalanche photodiodes, 1233, 1235
  in photoconductive devices, 1241
  in phototransistors, 1247, 1249
  in PIN photodiodes, 1223, 1224, 1225

Indium gallium arsenide (InGaAs) diode lasers, 245

Indium gallium arsenide phosphide (InGaAsP), in avalanche photodiodes, 1233, 1235

Indium gallium arsenide phosphide (InGaAsP) diode lasers, 245, 1025, 1026. *See also* Gallium indium arsenide phosphide (GaInAsP) diode lasers

Indium gallium nitride (InGaN) diode lasers, 245–246

Indium-gold bonding, for acousto-optic modulators, 1288, 1289

Indium phosphide (InP)
  in avalanche photodiodes, 1233
  in electro-optic modulators, 1015
  in integrated optics-based MUX/DEMUX devices, 980–982, 997, 998
  in photoconductive devices, 1241–1242
  in phototransistors, 1247, 1249
  in PIN photodiodes, 1223, 1224, 1225

Indium phosphide (InP) diode lasers, 241, 245, 282–288

Induced spatial incoherence (ISI), in laser fusion, 642–646

Inductance, 320, 323
IN fiber, in optical isolators, 871–872
In-fiber diffraction grating devices, 907
Infinite impulse response (IIR) filters, 892,
896, 913–917, 921. *See also* Bragg
gratings
Information, light-carried, 1009
Information processing
large-capacity, 229–232
optical approaches to, 234–235
Infrared (IR) radiation
acousto-optic materials for, 1281
defined, 54
in optical fibers, 670–671
photoconductive-device detection of,
1241–1242
Infrared detectors
history of, 53, 54–58
materials and properties of (table), 59
Infrared dye lasers, pumping of, 486
Infrared focal planar array (IRFPA)
detectors, applications of, 84–85
Infrared focal planar arrays (IRFPAs),
HgCdTe, 82–86
*Infrared Handbook, The*, 53
Infrared techniques, 53–114
applications of, 54
device physics and, 58–70
for extrinsic silicon detectors,
107–109
with HgCdTe, 70–86
history of, 53
with InSb, 86–91
for pyroelectric detectors, 103–107
with Schottky-barrier detectors,
91–96
for SPRITE detectors, 96–103
systems analysis and, 109–113
Inhomogeneous broadening, 446–448
Inhomogeneous media, ray tracing in,
20–25
In-line optical isolators, 865–866
Inorganic thin-film waveguides, 736–739
table of, 737–738
In-plane diode lasers, 241–244
InSb area arrays, 90–91, 92
InSb detectors, 86–91
InSb infrared detectors, 54
InSb line arrays, 89, 91
Insertion loss
in AWGs, 992
modulator optical, 1035
in optical fiber, 677

in optical isolators, 867–868
Integrated acousto-optic Bragg modulator
module, 1116
Integrated acousto-optic space division
demultiplexer, 1112
Integrated circuit (IC) technology, 691
optoelectronic, 692
Integrated optic RF spectrum analyzers
(IOSA), 1120–1121
Integrated optics, 731, 979–998
devices using, 808
nonlinear, 807
technology of, 979–982
Integrated optics-based MUX/DEMUX
devices, 979–998
performance characteristics of, 991–995
theory and design of, 982–991
Intelligent buildings, 692–693
Intensity, linear, 207
Intensity diffraction efficiency, in
magneto-optic thin-film waveguides,
1201–1202
Intensity modulation, 1010, by
acousto-optic modulators, 1269
Interaction length
in acousto-optic design, 1279
in couplers, 707
Interconnections, in neural networks, 233
Interconnection weight matrix (IWM), 233
Interdigital electrode transducer (IDT),
1085, 1093. *See also* ID SAW
transducer
Interdipole fields, 1149
Interface quality, VCSEL, 329
Interference
in magneto-optic thin-film waveguides,
1199–1200
multiple-beam, 466–467
optical filters and, 888–889, 890
optical isolators based on, 872–874
Interferometers. *See also* Fox–Smith
interferometer; Gires–Tournois
interferometer (GTI); Mach–Zehnder
interferometer (MZI); Michelson
interferometer; Multi-slit
interferometer; Twyman–Green
interferogram
in optical isolators, 872–874
ring resonators and, 913
Interleavers, 901
Interleaving wavelengths, in optical
networks, 950–951
Intermodulation products, 1275–1276

Internal magnetic fields
  domains and, 1147–1148
  functional forms of, 1144–1146
  spin waves and, 1149–1160
Internal reflection, in magneto-optics,
    1167–1168
International System of Units (SI), 1132
Intrinsic absorption, in optical detectors,
    1219
Intrinsic photon detectors, 58
Invar support frame, 578–579, 616
Inverted channel substrate planar (ICSP)
    laser, 262–263
Ion exchange, in fabricating planar optical
    waveguides, 732
Ion exchange reactions, 744
Ion exchange techniques, waveguide mode
    index modification via, 760, 768–770
Ion exchange waveguides, 744–747
  table of parameters for, 745, 746–747
Ionic compounds, antiferromagnetism of,
    1141
Ion implantation
  in fabricating planar optical
      waveguides, 732
  in VCSELs, 332–333
Ion implantation waveguides, 743–744
Ionization, in avalanche photodiodes, 1231,
    1234
Ion lasers, 520–524
Ion-milled waveguide lenses, 1099
Ions. *See also* Excimers
  active, 371–374
  ferrimagnetism and, 1143
  paramagnetism of, 1137
Iron (Fe). *See also* Garnet
  ferromagnetism of, 1140
  in garnet, 1171, 1172, 1173, 1191
Iron–garnet films, in optical waveguides,
    740–742
Irradiation, photoconduction and, 1239
Isolators. *See* Optical isolators
Isotope separation, dye lasers in,
    496–499
Isotropic acousto-optic Bragg diffraction,
    1080–1082, 1091
  in $Y$-cut $LiNbO_3$ substrate,
      1092–1094
Isotropic acousto-optic devices,
    performance of, 1109–1110
Isotropic crystals, 15
Isotropic diffraction

  with multiple tilted transducers of
      staggered center frequency, 1104,
      1105–1107, 1109–1110
  with a single tilted-finger chirp
      transducer, 1104, 1107, 1110
  with transducer arrays, 1104, 1107
Isotropic media, 13–14
  thermal distortion effects in, 424–426
Isotropic thin films, light in, 1198–1207
ITU recommendation G.692, 947

Jablonski scheme, 457
Jha, Asu R., 53
Johnson noise
  in avalanche photodiodes, 1232
  in optical detectors, 1218
  in photoconductive detectors, 64
Jones matrices, 902
JTP tracking systems, 229, 230–231
Junction capacitance, in optical detectors,
    1219
Junction-down devices, 291
Junction photodiodes, 1220–1222, 1226
Junctions, multimode fiber-optic, 828–830.
    *See also* Transverse junction stripe
    (TJS) lasers; Y-junctions
Junction temperature, in high-speed laser
    design, 313, 314
Junction-up devices, 291

Kao, C. K., 659
Keplerian telescope, 128, 133
Kerr effect, 1167, 1168
  in garnet, 1174, 1191
  in magneto-optic data-recording films,
      1189
  in MnBi films, 1191
  in reflective magneto-optic materials,
      1184
Kerr lens mode locking, 487
Kirchhoff diffraction integral, 40–41. *See
    also* Fresnel–Kirchhoff diffraction
    integral
  coordinate system of, 199–204
Kramers–Kronig equations, 895, 1020,
    1026, 1040
Krypton (Kr)
  in excimer lasers, 634
  in ion lasers, 520
Krypton fluoride (KrF), molecular energy
    levels of, 625

Krypton fluoride laser, 624–625, 627, 632,
640, 642, 643
in laser fusion, 642–646
physics of, 634–636
Kupferman, Peter N., 53

Lagrange invariant, 126, 127, 150, 151
Laguerre–Gaussian functions, 185
Laguerre–Gaussian modes, 184
Laguerre polynomials, 185
$\lambda_B/4$ phase shift, 194–195
Lambertian distribution, in Luneburg lens
fabrication, 775
Laminated polarizers, 839–841
LAMIPOL polarizer, 839–840
Langevin function, 1137, 1138, 1140
Lanthanide elements, in polarization
rotators, 846–847. *See also* Rare
earth entries
Large-aperture discharge lasers, 630–631
Large-aperture module (LAM)
electron-beam amplifier, 639
Large-capacity information processing,
229–232
Large-volume Nd:glass rod amplifiers,
operating consequences associated
with, 385–386
Larmor precession, 1163, 1169
Laser amplifiers, 163, 399–412
design of, 401–402
geometries of, 402–408
limiting processes for, 408–409
Laser applications, acousto-optic modulators
in, 1256
Laser beams, in acousto-optic devices,
1257–1286
*Laser Crystals* (Kaminskii), 431
Laser diodes (LDs)
in optical fiber transmission systems,
682
semiconductor tunable, 292–308
speed limits on, 691
Laser diode–submount interface problems,
288
Laser fusion, 379
electron-beam pumped excimer lasers
for, 642–646
Laser glass properties, effect on gain and
propagation effects, 409–412
Laser intensity noise, in optical detectors,
1218

Laser literature, data extrapolation from,
431
Laser media, rare-earth, 455–456
Laser media configurations, thermal
distortion effects in, 424–426
Laser oscillators, 163, 386–399. *See also*
Oscillators
Laser paint, 470
Laser-pulse pumped dye lasers, 475–481
Laser pulses. *See also* Femtosecond light
pulses; Pulsed optically pumped FIR
lasers; Short laser pulses;
Subpicosecond light pulses;
Ultra-short light pulses
from FIR lasers, 617, 618, 619, 620,
622
short, high-power, 640–641
Laser pumping
diode, 381–382
of gas lasers, 520–521
pulsed or CW, 383
Laser rod geometries, thermal distortion
and, 424–426. *See also* Cylindrical
rods; Nd:glass rod amplifiers; Rod
amplifiers
Lasers. *See also* Channel substrate planar
(CSP) laser; Diode lasers; Dye lasers;
Fabry–Perot entries; Gas lasers;
Liquid lasers; Semiconductor lasers;
Solid-state lasers
acousto-optic modulators and, 1255,
1256
broad area, 261–267
in bulk-wave acousto-optic Bragg
modulators, 1080
catastrophic facet damage to, 271–274
$CO_2$, 512, 525–538, 579
Cu vapor, 524–525, 526
CW optically pumped, 569–606
discharge, 625–646
electrical discharge, 544–559, 560, 561
excimer, 476–477, 538
far-infrared, 539–541, 541–624
future development of, 350–351
high power mode stabilized, 274–275
history of, 511, 512, 720
in-plane diode, 241–244
ion, 520–524
large aperture discharge, 630–631
long pulse, 630
noble-gas halide, 475–476, 477, 480,
624, 627
optical modulation of, 1010

Lasers. *See also* Channel substrate planar (CSP) laser; Diode lasers; Dye lasers; Fabry–Perot entries; Gas lasers; Liquid lasers; Semiconductor lasers; Solid-state lasers (*continued*)
  optical resonators and, 163–164
  quantum-well, 247–252
  with reduced facet intensity, 274–275
  safety of, 429–431
  single longitudinal mode, 267–270
  single-mode and multimode, 260–271
  strained quantum-well, 252–254
  theory of, 512–518
Laser systems, oscillator–amplifier, 419–424
Lasing dyes, classes of, 449–453
Latching bi-substituted iron garnet thick films, 847
"Lateral color," 137
Lateral mode, 261, 267
Lateral refractive index, 254–255
Lateral structure, in laser structure, 241, 244, 260, 261
Lattice filters, 899–903
Launch conditions, of a ray, 21
Law, H. D., 1215
Law of conservation of energy, 8
Layered dielectric, reflection and transmission from, 10–12
Layer large-optical-cavity structure, 243
Layers, in electroabsorption modulators, 1031–1032, 1045–1046, 1047
L-band EDFAs, 945–946
*L*-coupled systems, 557
Lead (Pb), as garnet substituent, 1174–1175. *See also* Pb entries
Lead oxide vidicon cameras, diagnostics with, 433–434
Lead salts, in infrared detectors, 54
Lead tin telluride infrared detectors. *See* PbSnTe infrared detectors
Lead zirconate (PZ) pyroelectric array, 104, 106
Leaky in-phase mode, 265–267
Leaky out-of-phase mode, 265–267
Leaky-wave coupling, 267
Left circularly polarized (LCP) propagation, 1153–1154. *See also* Left-handed polarization
  of light, 1160, 1162–1163, 1164, 1165, 1166, 1168, 1169
Left-handed optical systems, 130–131

Left-handed polarization, 7. *See also* Left circularly polarized (LCP) propagation
Lehecka, T., 511
Lens arrays, in laser fusion, 642. *See also* Arrays
Lenses. *See also* Optical lenses; Thin lens
  chirp grating, 793, 795, 796–804, 805, 811, 820
  diffraction grating, 821
  in free-space MUX/DEMUX architecture, 967, 968
  Fresnel zone, 793–795, 796, 800, 804
  generalized Rinehart–Luneburg geodesic, 785–788, 789, 791
  geodesic, 779–789, 790, 791, 820, 821
  Luneburg, 768, 770–779, 780, 781, 782, 783, 819–820
  in microscopes, 129
  ray tracing in spherically symmetric, 25
  spherical geodesic waveguide, 784–785
  in telescopes, 127, 128
  waveform matrices for, 175
  waveguide diffraction, 793–804
  in Z-cut GaAs–GaAlAs planar waveguide, 1098–1099
Lens waveguide
  beam propagation in, 37–38
  Gaussian mode in, 29, 35
Lenz's law, 1136–1137
Li, G. L., 1009
Liang, Bing W., 239
Light, 1. *See also* Acousto-optic entries; Electro-optic entries; Magneto-optic entries; Photon entries
  in acousto-optics, 1079–1080, 1261–1262
  in bulk-wave acousto-optic Bragg modulators, 1080–1082, 1083–1084
  Faraday effect and, 1131
  in guided-wave acousto-optic Bragg diffraction, 1085–1086
  in guided-wave acousto-optic Bragg modulators, 1086–1087
  as information carrier, 1009
  in isotropic thin films, 1198–1207
  magnetic interactions with, 1160–1164, 1164–1169
  magnetic polarization of, 1160–1161, 1162–1163, 1164
  in optical communications, 1111, 1131
  path of, 115–116

propagating perpendicular to a magnetic field, 1166–1167
in spatial light modulators, 1194–1195
speed of, 1303
Light-hole exciton resonances, 1022
Light rays, properties of, 19–20
Limiting processes, for laser amplifiers, 408–409
Lin, Chinlon, 659
Linear electro-optic effect, 1010, 1011
in MZMs, 1050–1051
Linear flashlamps, for dye lasers, 471–472
Linear graded index SCH (L-GRINSCH), 318–319
Linear microlens array–integrating lens combination, 1117–1118
Linear polarization, 6–7, 1160, 1166
Line arrays
InSb, 89, 91
for pyroelectric detectors, 103
Linear systems, 199, 201–204
diffraction as, 47–48
Line broadening, 446–448
Lines of force, in magnetic fields, 1134
Line terminal equipment (LTE), OADMs and, 937
Line-width enhancement factor, 253
Link RF gain, 1032–1033, 1044
Liquid crystal cells
in electro-optic VOAs, 857–859
polarization rotators from, 849–850
Liquid-crystal light valve (LCLV), 223–224, 225
Liquid crystals (LCs). See also Polymer dispersed liquid crystals (PDLCs)
in lattice filters, 901
polarization rotators from, 849
Liquid lasers, 443–510
active medium excitation, frequency selection, and tuning in, 460–471
applications for, 444–446, 496–501
comparison with other lasers, 446–460
flashlamp-pumped rare-earth, 474–475
history of, 443–444
operational principles, design, and parameters of, 471–496
pulse-pumped, 471–482
radiative and nonradiative processes of, 456–459
Liquid phase epitaxy (LPE), 246, 264, 382
growing garnet via, 1171, 1172
Lithium (Li) ions, in ion implantation waveguides, 743

Lithium-magnesium aluminosilicate glass, 410–411. See also ED-2 silicate glass
Lithium niobate (LiNbO$_3$), 813, 814. See also Ti:LiNbO$_3$ waveguides; Y-cut LiNbO$_3$ substrate
as acousto-optic substrate material, 1094, 1095, 1096, 1289, 1290, 1291, 1292, 1298
in diffused optical waveguides, 747–750
in electro-optic devices, 414, 723, 1012–1014
in graded index waveguides, 767, 768
in guided-wave acousto-optic Bragg modulators, 1086
III–V compounds versus, 1014–1015
in metal-clad optical waveguides, 750–751
in optical isolators, 866
in optical modulators, 1010
in optical waveguides, 742–744, 747
Lithium niobate (LiNbO$_3$) acousto-optic Bragg modulators/deflectors, performance of, 1108–1109
Lithium niobate (LiNbO$_3$)-based hybrid multichannel integrated optical device modules, 1116–1120
Lithium niobate (LiNbO$_3$) modulator advantages and disadvantages of, 1070
bias stability of, 1034
traveling-wave, 1056–1061
Lithium tantalate (LiTaO$_3$)
in diffused optical waveguides, 747–748, 749
in optical waveguides, 742
Lithium yttrium fluoride (YLF), as solid-state laser host material, 375. See also Nd:YLF lasers
Littrow configuration, 905
Littrow mount, 464, 465, 477
L-I-V curve, of VCSELs, 340–341
"Longitudinal" aberration, 136
Longitudinal astigmatism, 142
Longitudinal chromatic aberration, 137
Longitudinal direction, in laser structure, 241–244, 260
Longitudinal electrodeless RF (LERF) systems, 555–556, 557
Longitudinally integrated tunable lasers, 297, 298–304
Longitudinal Kerr effect, 1167, 1168
Longitudinal mode lasers, 267–270

Longitudinal phase factor, 31

Longitudinal pumping, of dye lasers, 462–463

Longitudinal separation, in opto-mechanical VOAs, 854

Long period fiber gratings, 902

Long-pulsed oscillators, 388–390

Long-pulse lasers, 630

Long-wavelength infrared (LWIR), 54. *See also* LWIR entries

Loop feeder, in fiber transmission networks, 692

Lorentz force, 1163, 1169

Lorentzian resistance, in bismuth-substituted YIG, 1182–1183

Loss. *See also* Insertion loss; Optical propagation loss
  in multiplexers, 939, 940, 942
  in tunable dispersion compensating filters,
  VCSEL, 329–330
  in waveguide grating routers, 910, 911

Lossless delay line, optical frequency response for, 890

Loss-limited transmission, 681–682

Loupes, 161

Low-frequency noise, in LWIR-PV detectors, 66

Low-index difference waveguides, 822

Low-loss silica glass fibers, 669–671

Low-power carbon dioxide lasers, 528–529, 538

Low-scatter reflection, 130

Luhmann, N. C., Jr., 511

Lumonics lasers, 616, 617

Lumped-element electroabsorption modulators, 1041, 1044–1048

Lumped-element electrodes, design of, 1041–1042

Lumped-element modulators, 1041–1050
  frequency response of, 1042–1044

Lumped-element QCSE-based MZMs, 1041, 1048–1050
  circuit model of, 1043, 1048

Luneburg collimating lens. *See* Thin-film waveguide Luneburg lens

Lutetium (Lu), as garnet substituent, 1173, 1175, 1204

LWIR detectors, operating requirements for, 68

LWIR-FPA detectors, capabilities of, 85–86

LWIR hybrid focal planar arrays, 82, 83, 84

LWIR-PV detectors, 65
  low-frequency noise and dark current levels in, 66–67

Lyot filters, 468–469

Mach–Zehnder interferometer (MZI), 1027
  Bragg gratings and, 923
  cascaded, 899–900
  coupled, 898, 899
  with fiber Bragg gratings, 977–978
  in next-generation optical networks, 862
  in optical circulators, 882–883
  in optical filters, 891–892, 896–897
  in ring resonators, 914–916
  in thermo-optical VOAs, 854–856
  in waveguide grating routers, 912

Mach–Zehnder modulator (MZM), 1010, 1016, 1017, 1020, 1026
  bias stability of, 1034–1035
  chirp analysis of, 1039–1040
  comparison of different, 1070–1071
  linear electro-optic effect in, 1050–1051
  lumped-element QCSE-based, 1041, 1048–1050
  on/off extinction ratio in, 1029–1030, 1031
  optical power handling capability of, 1035
  theory of, 1027–1029
  traveling-wave, 1066–1069

Macular degeneration, laser therapy for, 501

Madsen, C. K., 887

Magnesium (Mg), as garnet substituent, 1175

Magnetic anisotropy energy, 1146, 1147, 1148

Magnetic circular birefringence (MCB), 1165, 1166–1167

Magnetic circular dichroism (MCD), 1166

Magnetic dipoles. *See* Dipole moments; Dipoles

Magnetic fields
  defined, 1133–1134
  of dipoles, 1134
  in electron-beam amplifiers, 638
  internal, 1144–1146, 1147, 1149
  in magneto-optic devices, 1131
  paramagnetism and, 1137
  of solenoids, 1133, 1134
  spin waves and, 1149–1160

Magnetic flux, 1134, 1137

Magnetic induction, 1132–1133, 1134

Magnetic linear birefringence (MLB), 1166–1167

Magnetic linear dichroism (MLD), 1167

Magnetic moment, 1135, 1138. *See also* Dipole moments
of an electron, 1135–1136

Magnetic optical rotation (MOR), 1166

Magnetic phenomena, quantum mechanics of, 1132, 1135–1136, 1137, 1163–1164

Magnetic polarization, 1150–1151, 1152, 1153–1154

Magnetic remanence, of reflective magneto-optic materials, 1184

Magnetics
material physics of, 1135–1149
units in, 1132–1135

Magnetic saturation, 1145, 1150. *See also* Saturation magnetization
of garnet, 1174, 1176

Magnetism
domains and, 1147–1148
temperature and, 1138–1139

Magnetization
defined, 1134
of garnet films, 1198, 1200–1207
of magneto-optic data-recording films, 1188–1192
in magneto-optic isolators, 1192–1194
of reflective magneto-optic materials, 1184–1185, 1186–1187
saturation, 1135
in spatial light modulators, 1194–1196

Magnetoelastic energy, 1146

Magneto-optic crossbar switch, 1187

Magneto-optic data-recording technology, 1187–1192

Magneto-optic data storage media, 1187–1192

Magneto-optic deflectors, 1196–1197

Magneto-optic devices, 1131–1214. *See also* Magnetics
important materials for, 1169–1187
material physics of optics and, 1160–1164
modern, 1187–1207
spin waves and, 1149–1160

Magneto-optic interaction, 1160–1164

Magneto-optic materials
reflective, 1184–1187
transmissive, 1170–1184

Magneto-optic modulators, 1192, 1193

Magneto-optic polarization-rotation isolators, 1187, 1192–1194

Magneto-optics, 1164–1169

Magneto-optic spatial light modulator (MOSLM), 223, 224–225. *See also* Spatial light modulator (SLM)
for matrix multiplication, 227–228

Magneto-optic switches, 412, 416–417

Magneto-optic VOAs, 854, 856–857

Magnetostatic energy, 1147, 1149

Magnetostatic waves (MSWs), 1152–1153, 1156, 1157, 1158, 1159–1160. *See also* Spin waves
attenuation of, 1159
in bismuth-substituted YIG, 1182–1183
in garnet films, 1198, 1199–1207

Magnetostriction, 1146
in reflective magneto-optic materials, 1186

Magnetron sputtering, 740

Magnets, in Faraday rotators, 846–847

Magnification, 26, 118, 121, 151
in afocal systems, 127
defined, 123
distortion and, 144–145
in elementary optical systems, 126
of Keplerian telescope, 128
of microscope, 129

Magnifying glasses, 161

Maiman, T. H., 369

Manganese-bismuth (MnBi) films, for magneto-optic data-recording, 1190–1191

Manley–Rowe condition, 539–540

Maréchal's criterion, 159

Marginal ray, 117, 124, 125, 126, 127, 128

Marshak, I. S., 376

Marx banks, 533, 636

Masers, 512

Masking, in Luneburg lens fabrication, 775–779, 780, 781, 782, 783

Mass loading, in acousto-optic modulators, 1291–1293

Master oscillator power amplifiers (MOPAs), 289, 399–400, 478

Matched-impedance TWEAM, 1054

Material dispersion, in optical fibers, 671–673, 674

Material growth technologies, for diode lasers, 246–247

Material physics
of magnetics, 1135–1149
of optics, 1160–1164

Material properties, effects on thermal
distortion and birefringence,
426–427, 428
Materials. *See* Acousto-optic materials;
Crystals; Glass; Host materials;
Pyroelectric materials; Semiconductor
materials; Solid-state laser materials
Matrices, interconnection weight, 233
Matrix multiplication, optical, 226–228
Maximum grating diffraction angle, for
waveguide optical gratings, 793
Maximum insertion loss, in multiplexers,
939
Maxwell (unit), 1134
Maxwell–Boltzmann distribution, 1137
Maxwell equations, 2–3, 11, 1150, 1152,
1155, 1158, 1160, 1165
boundary conditions for, 3
for optical fibers, 661
step index waveguides and, 724–726
Maxwell's fish-eye lens, ray tracing for,
25, 780
McCumber relationship, 1303–1304
McIntyre's local field model, 1229, 1236
McMahon, John M., 369
Mean-square error (MSE), in optical
information processing, 234–235
Mechanical splices, of optical fiber, 676
Media. *See also* Transmission media
axial inhomogeneous, 20–23
cylindrically inhomogeneous, 23–24
homogeneous, 19
hyperbolic secant index, 23
inhomogeneous, 20–25
planar inhomogeneous, 21
quadratic index, 22, 23–24, 27–29,
35–36
ray matrices for (table), 28
spherically symmetric, 19–20
spherically symmetric inhomogeneous,
24–25
Medial focus, 142, 143
Medical applications, dye lasers in,
499–501
Medium average power solid-state lasers,
405–408
Medium-power carbon dioxide lasers, 528,
529–532, 538
Melanin, absorption spectrum of, 499, 500
MEMS based VOAs, 854, 861–863
Mercury (Hg) electrode, 818
Mercury cadmium telluride infrared
detectors. *See* HgCdTe entries

Meridional fan, 136–137
Meridional ray, 22, 23, 151
Merocyanine dyes, 449
Mesa pixels, in spatial light modulators,
1194–1195
Mesa trench, in PIN photodiodes, 1223,
1224, 1225
Mesa type VCSELs, 331–332, 334
Metal–ceramic plasma tube, 521, 522
Metal-clad optical waveguides, 750–752
Metal halides, in excimer lasers, 624
Metallurgical objective, 161
Metal mesh configuration, for gas lasers,
574, 575–577
Metal-organic chemical vapor deposition
(MOCVD), 239, 240, 244, 246, 247,
261, 264, 382
growing III–V semiconductor layers
via, 1015
Metal-organic compounds, in lasers, 443
Metal-organic laser media, 456
Metal-organic vapor phase epitaxy
(MOVPE), 1223
Metals
ferromagnetism of, 1140
paramagnetism of, 1137
Metal vapor lasers, 524–525
Methyl fluoride ($CH_3F$)
energy levels of molecules of, 562,
563–569
far infrared lasers using, 541, 562–569
Methyl fluoride ($CH_3F$) FIR lasers, 541,
562–569, 579, 594–601, 621, 623
from Culham Laboratory Group,
607–612
Michelson interferometer, 920–921
in optical filters, 891
Micro-electrical-mechanical systems
(MEMS), 861. *See also* MEMS based
VOAs
optical filters and, 889, 906, 908
Micro-mirror actuator, in MEMS based
VOAs, 861
MicroPlasma technique, 956
Microprocessor stabilization system, for
FIR lasers, 580–582
Microresonators, dye laser, 469
Microscope objectives, 160–161
Microscopes, 129
Microspheres, 469
Microwaves
phototransistors and, 1249
waveguide photodiodes and, 1238–1239

Microwave signals, optical detectors and, 1215

Microwave transmission line, 1063–1064

Mid-wavelength infrared (MWIR), 54. *See also* MWIR entries

Military systems, VLWIR detectors in, 67–68

Minimum blur circle, 140

Minimum facet loss, 1035

Mirrors, 116. *See also* Concave mirrors; Convex mirrors; Reflection entries
in $CO_2$ lasers, 529–530
cold, 130
in free-space MUX/DEMUX architecture, 968
for gas lasers, 569–574, 575–577
in laser oscillators, 386–387, 388–389
in MEMS based VOAs, 861–862
in optical filters, 906, 907, 908
in optical resonators, 163–164
in unstable resonators, 398–399
waveform matrices for, 175
waveguide diffraction, 804–807, 808, 809, 810, 811

Misalignments, of fiber-optical joints, 677–678

Missile detection, 67

MIT Alcator tokamak, 616

$m$-lines, of prism couplers, 755

MLM laser mode partition noise, 686, 687–688

Modal delay, in optical fibers, 665–667

Modal dispersions, 665–669

Modal noise, in optical fiber transmission systems, 681

Mode confinement factor, 167

Mode-controlled laser, 164

Mode field diameter (MFD), 992–993, 1305
in single-mode fibers, 675–676

MODEIG waveguide program, 256, 257

Mode index, of geodesic lenses, 788

Mode index modification techniques, for waveguides, 760

Mode-launching couplers, 754–759

Mode-locked oscillators, 392–395

Mode locking. *See also* Kerr lens mode locking
of dye lasers, 486–496
hybrid, 496
with synchronously pumped dye lasers, 488–489

Mode matching, 36–37

Mode mixing, in multimode optical fibers, 667

Mode partition noise (MPN), in optical fiber transmission systems, 682, 686, 687–688. *See also* Partition noise

Modes
in Fabry–Perot resonators, 179–186
of optical fibers, 662, 663–665
zero-order, 761

Modified chemical vapor deposition (MCVD), 669

Modular optical design, 160–161

Modular pulsed laser system, 481–482

Modulation. *See also* Optical modulation; Spatial light modulators; Wide-bandwidth optical intensity modulators
efficiency of small-signal, 1032–1037
optical, 1009–1010
small-signal, 309–312

Modulation bandwidth, for acousto-optic modulators, 1265–1266

Modulation transfer function (MTF), 158–159, 208, 225, 1027–1029, 1030
in acousto-optic modulators, 1267, 1271–1274
of SPRITEs, 102–103

Modulator frequency response, 1043

Modulator materials, 1010–1026
electroabsorption effects in, 1018–1026

Modulator optical insertion loss, 1035

Modulators. *See also* Acousto-optic modulators; Magneto-optic modulators; Spatial light modulators
comparison of, 1069–1072
electro-optic, 413–416
electro-optic versus acousto-optic, 1257
lumped-element, 1041–1050
on/off extinction ratio in, 1029–1032
optical propagation loss in, 1035–1037
voltage swing in, 1030–1032

Modulator slope efficiency, 1033–1034

Modulator waveguide design, 1027–1041

Molecular beam epitaxy (MBE), 239, 240, 244, 246–247, 261
growing III–V semiconductor layers via, 1015

Molecular energy levels
of $CO_2$ molecule, 527–528
of KrF, 625
of $N_2$ molecule, 527–528

Molecular field, 1151
antiferromagnetism and, 1141–1142

Molecular field, (*continued*)
  ferrimagnetism and, 1143–1144
  ferromagnetism and, 1140
  magnetic, 1138–1139
Momentum vector
  in bulk-wave acousto-optic Bragg
      modulators, 1081, 1082
  in guided-wave acousto-optic Bragg
      modulators, 1086, 1087
"Monochromatic" aberrations, 138–146
Monochromatic light field, 199–204, 205
Monochromatic point sources, 200–202
Monochromatic properties, of prisms, 130
Monochromator systems, 131
Monolithically integrated flared-amplifier
      master oscillator power amplifiers
      (MFA-MOPAs), 289
Monolithic array detectors, 54
Monolithic surface-emitting-through-
      substrate arrays, 291
Monte Carlo ray tracing, 409, 410
Multichannel integrated
      acousto-optic–electro-optic
      device module, 1118
Multichannel spectrum analyzer,
      211–212
Multifrequency acousto-optic beam
      deflector, 1274–1277
Multilayered medium. *See* Layered
      dielectric
Multilayer step index waveguides,
      761–763, 764, 765, 766
Multi-longitudinal-mode (MLM) laser
      diodes, in optical fiber transmission
      systems, 682, 683–684. *See also*
      MLM laser mode partition noise
Multimode fiber optics junctions, 828–830
Multimode interference (MMI) couplers,
      829, 830, 1067
Multimode lasers, 260–271
Multimode (MM) optical fibers, 662,
      663–665
  modal dispersions in, 665–669
  in optical attenuators, 850
Multimode optical fiber transmission
      systems, 681
Multiple-beam interference, in dye lasers,
      466–467
Multiple-helical-lamp pump geometries, 386
Multiple-linear-lamp pump geometries, 386
Multiple-pass amplification, 402, 403
Multiple-port OADMs, 960–961

Multiple quantum well lasers (MQWs), 241,
      243, 315–316, 1020
  with lumped-element QCSE-based
      MZMs, 1048–1050
  modulators and, 1031–1032,
      1045–1047
  tunneling time in, 319–320
Multiple quantum well MZMs, advantages
      and disadvantages of, 1071–1072
Multiple-surface acoustic waves, Bragg
      diffraction from, 1104
Multiple tilted SAW transducers,
      1102–1103, 1109–1110. *See also*
      Transducer array
  isotropic diffraction with, 1104,
      1105–1107
Multiple transducers, versus single
      transducer, 1110–1111. *See also*
      Transducer array
Multiple transducers of staggered center
      frequency, optimized anisotropic
      diffraction with, 1104
Multiplexers (MUX). *See also* Array
      waveguide grating multiplexer;
      CCD multiplexer; Demultiplexers
      (DEMUX); Dense-wavelength
      division multiplexing (DWDM);
      MUX/DEMUX entries; Optical
      add/drop multiplexers (OADMs);
      Optical frequency division
      multiplexing (OFDM); Optical
      wavelength filtering/multiplexing/
      demultiplexing; Waveguide optical
      demultiplexers; Wavelength division
      multiplexer; Wavelength division
      multiplexers/demultiplexers (WDM
      MUX/DEMUX); Wavelength
      division multiplexing (WDM);
      Wavelength-multiplexed
      reflection-type matched filter
      correlator
  angular, 641, 643
  AWG devices as, 824–826
  diffraction gratings as, 905–906
  optical filters for, 887
Multiplication, in avalanche photodiodes,
      1227–1228, 1228–1233, 1235
Multi-slit interferometer, transmission
      grating as, 903–904
Multi-stage allpass filters, 896, 897
Multi-stage cascade, for AWGs, 996–997
Multi-stage FIR filters, 893
Multi-stage optical filters, 898

Multispatial mode diode lasers, 261–267
Mutual intensity function, 206
MUX/DEMUX applications, 976–979,
    995–998
    future, 998
MUX/DEMUX architectures
    free-space, 967–969
    thin-film, 960–961
MUX/DEMUX devices
    free-space, 961–969
    integrated optics-based, 979–998
    optical fiber-based, 970–979
    thin-film, 954–961
MUX/DEMUX structures, 976–979,
    995–998
    future, 998
MWIR-FPA detectors, capabilities of, 81, 85
MWIR hybrid focal planar arrays, 83,
    84, 85
MWIR photovoltaic detectors, 65

Naphthalene dyes, 452
Narrow-band information, in fiber
    transmission networks, 693
Nd:Cr:GSGG lasers, 388
Nd:glass lasers, in laser fusion, 642–643
Nd:glass oscillators, mode locking in,
    394–395
Nd:glass rod amplifiers, 409
    gain and propagation effects in,
        409–412
    large-volume, 385–386
Nd:YAG lasers, 388. *See also* Nd:YAG
    oscillators; YAG lasers
    diagnostics with, 433
    medical applications of, 499–500
    mode-locked, 488, 489–490
    popularity of, 391–392
    pumping dye lasers with, 476–477, 482
    single-lamp high-efficiency, 384
    single-lamp single-mode, 384–385
Nd:YAG oscillators. *See also* Nd:YAG
    lasers
    frequency mode selection in, 395–396
    mode locking in, 395
Nd:YAG range finders, 424
Nd:YLF lasers, diagnostics with, 433–435
Near-field beam monitors, 435
Near infrared (NIR) lasers, medical
    applications of, 499–500
Néel temperature, 1141–1142
Negative DI lens, waveform matrix for, 175

Negative-index guide, 257, 265
Nematic liquid crystals, polarization rotators
    from, 849–850
Neodymium (Nd), as garnet substituent,
    1173, 1174, 1175
Neodymium ions, in liquid lasers, 474, 475
Neodymium lasers, 371, 382, 443, 475. *See*
    *also* Nd: entries
    saturable absorber switches for, 413
Neon (Ne)
    in Cu vapor lasers, 524
    in excimer lasers, 626
    far infrared lasers using, 542
    in lasers, 519–520
NETD per micrometer, of HgCdTe
    detectors, 80
Networks, fiber transmission, 692–693. *See*
    *also* Optical networks
Neumann condition, 41
Neural networks (*NN*), 233
    optical, 232–234
Nickel (Ni), ferromagnetism of, 1140
Nicol prisms, 841
Niobium pentoxide ($Nb_2O_5$)
    in graded index waveguides, 767
    in thin-film waveguides, 739
Nitrogen lasers, pumping dye lasers with,
    476, 482
Nitrogen molecule ($N_2$), rotational energy
    levels of, 527–528
N × N star couplers, 822, 823
Noble-gas halide lasers, 475–476, 477, 480.
    *See also* Rare gas halide lasers;
    Xenon chloride (XeCl) lasers
    excimer, 538
Noble-gas lasers. *See* Argon (Ar) lasers;
    Helium–neon (He–Ne) lasers
Nodal points, 117, 122
Noise. *See also* Gaussian noise;
    Low-frequency noise; MLM laser
    mode partition noise; Modal noise;
    Partition noise; Relative intensity
    noise (RIN); Signal-to-noise ratio
    (SNR)
    in avalanche photodiodes, 1228–1233,
        1235, 1236
    in CCD images sensor, 99
    in optical detectors, 1218
    in photoconductive detectors, 64
    reflection-induced, 686, 690–691
Noise equivalent power (NEP), in optical
    detectors, 1218

Noise equivalent temperature (NET), 111–112. *See also* NETD per micrometer
Nonabsorbing mirror (NAM) technology, 240, 275–278, 279
Non-exciton interband transitions, 1022
Non-latching bi-substituted iron garnet thick films, 847
Nonlinear integrated optics, 807
Nonlinearity
  acousto-optic modulators and, 1101–1102, 1269
  in photodetectors, 1243–1249
Nonlinear optical crystals (NOCs), 1011
Nonlinear optic (NLO) chromophore molecules, in electro-optic polymers, 1017
Nonparallel directional coupler, 714–716
Nonpiezoelectric substrates, 1094
Nonradiative processes, of liquid lasers, 456–459
Nonreciprocal polarization rotator sets, 877
Nonreciprocal rotation, 1194
Nonuniformity loss, in AWGs, 994
Normalization, of Hermite–Gaussian functions, 39
Normalized detectivity, 61
Normalized dispersion curves
  of optical channel waveguides, 706–707, 708, 709, 710, 711, 712
  of planar waveguides, 701–702, 723–724
Normalized effective mode index, 723–724
Normalized frequency ($V$), of an optical fiber, 663
Normalized guide mode index, 723
Normalized propagation wave vector, 699
Normalized refractive index, 699
Normalized thickness, 699
Normalized thin-film thickness parameter, 722
Nova laser, 404, 411, 412, 421, 423
Novette laser, 421
*npn* phototransistors, 1248–1249
Nucleation, of magnetic domains, 1188, 1190
Numerical aperture (NA)
  in image space, 126
  of an optical fiber, 663
Numerical simulation, 437

OASIX Optical Amplifier Simulation System, 1304, 1308

Obenschain, S. P., 511
Object and pupil coordinates, 135
Objective lens, 129, 160–161
Object space, 116
Oblique spherical aberration, 145
Octahedral coordination, of atoms in ferrimagnetic materials, 1143, 1144, 1145
Odd parity, 131
Oersted (unit), 1133–1134
Off-axis beam injection, in stabilizing FIR lasers, 583, 584
Oligophenylene dyes, 448, 452
One-directional WDM, 685
$1/f$ noise, in photoconductive detectors, 64
One-stage beam displacement optical isolator, 868–869
One-stage reflective optical isolator, 870–871
On/off extinction ratio (ER), 1029–1032
Onsager relation, 1161
Open resonator, 164
Open-terminated TWEAM, 1054
Ophthalmology, dye lasers in, 501
Optical aberration control, 435–436
Optical absorption, in bismuth-substituted YIG, 1179
Optical absorption coefficient, in a quantum well, 1024
Optical add/drop multiplexers (OADMs), 822, 823, 853. *See also* Add-drop entries
  applications of, 943–954
  configuring, 953–954
  dynamic, 826–827
  functional parameters of, 937–943
  history of, 936
  integrated optics in, 979
  multiple-port, 960–961
  operation of, 937
  in optical circulators, 875
  reconfigurable, 979, 997
  single channel, 953–954
Optical amplifiers (OAs), 887–888, 1301–1320. *See also* Erbium doped fiber amplifier (EDFA)
  erbium-doped fiber, 1301–1311
  MUX/DEMUX devices and, 943–947
  Raman fiber, 1312–1316
  semiconductor, 1316–1319
Optical attenuators, 678, 850–863
  theory of, 851–852
  variable, 846, 852–863

Optical beam distortion, in acousto-optic
    devices, 1263–1265
Optical beams, power decay of, 1216
Optical channel waveguides, 697, 702–707
Optical circulators, 874–883. *See also*
    Circulators
    birefringent crystals in, 878–882
    hybrid, 879–880
    reflective, 881–882
Optical communication, acousto-optic
    devices in, 1111–1114, 1131–1132
Optical components, passive, 678–679,
    753–809, 810, 811
Optical computing, acousto-optic devices in,
    1116–1121
Optical confinement techniques
    in high-speed laser design, 313, 314,
        320
    VCSEL, 331–334
Optical correlator, real-time
    microcomputer-based, 223–224
Optical cross connects (OXCs),
    MUX/DEMUX devices and, 943,
    948–950, 997
Optical damage methods, waveguide mode
    index modification via, 760
Optical design, modular, 160–161
Optical detectors, 1215–1253
    applications for, 1215–1216
    nonlinearity in, 1243–1249
    photoconductive devices as, 1239–1242
    photodiodes and, 1219–1220,
        1220–1239
    terminology related to, 1216–1220
Optical device modules
    GaAs-based monolithic multichannel
        integrated, 1120–1121
    LiNbO$_3$-based hybrid multichannel
        integrated, 1116–1120
Optical device substrates, for passive
    waveguides, 753
Optical diagnostics, 431–435
Optical directional couplers, 707–716
Optical elements
    polarizing, 418–419
    ray matrices for (table), 28
Optical fiber-based MUX/DEMUX devices,
    970–979
Optical fiber couplers, 678
Optical fiber jointing, 676–678
Optical fibers
    manufacture of, 669–670

    multimode and single-mode, 662,
        663–665
    in optical attenuators, 850–852
    structure of, 660–661
    telecommunications via, 660
Optical fiber taps, 678
Optical fiber telecommunication links, 1132
Optical fiber transmission technology,
    659–695
    advantages of, 659–660
    future of, 691–693
    history of, 659–660
    in telecommunications, 679
    transmission limitations in, 686–691
Optical fiber waveguides, 660–663
Optical filtering components, 678
Optical filters, 887–933
    applications for, 887–889
    Bragg gratings and, 889, 921–925
    finite impulse response, 898–913
    future of, 925
    implementation of, 889
    infinite impulse response, 892, 896,
        913–917
    interference and, 888–889, 890
    phased array, 903
    single-stage, 891–892
    theory of, 890–898
    thin-film, 917–921
Optical frequency division multiplexing
    (OFDM), 692
Optical frequency shifting, 1113
Optical intensity modulators. *See*
    Wide-bandwidth optical intensity
    modulators
Optical interference filters, 678
Optical isolators, 863–874, 1131–1132. *See
    also* Magneto-optic
    polarization-rotation isolators
    arrays of, 874
    birefringent crystals in, 867–870,
        871–872
    in-line, 865–866
    interferometers in, 872–874
    operation of, 863–864
    polarization dependent, 864–865
    rare earth iron garnets in, 1192–1194
    reflective, 870–872
Optical lenses, 115–162. *See also* Optics
    coordinate systems for, 117–118
    definitions related to, 116–117
    diffraction effects in, 155–160
    first-order properties of, 115, 118–133

Optical lenses, 115–162. *See also* Optics
   (*continued*)
  fundamental concepts related to,
   115–116
  in modular optical design, 160–161
  notational conventions related to,
   117–118
  real ray-trace techniques and, 151–155
Optical line system, 888
  multiplexers in, 943–954
Optical line terminals (OLTs),
   MUX/DEMUX devices and, 943,
   947–948
Optical loss measurement, 815–818
Optically pumped FIR lasers, 539–540,
  559–624
Optical matrix multiplication, 226–228
Optical MEMS technology, 861
Optical modulation, 1009–1010. *See also*
  Acousto-optic (AO) devices;
  Magneto-optic devices;
  Wide-bandwidth optical intensity
  modulators
Optical modulators, 412–419
  materials in, 1010–1026
Optical networks. *See also* Broadband
  Integrated Services Digital Network
  (BISDN); Fiber transmission
  networks
  multiplexers in, 943–954
  next-generation, 862
Optical neural networks, 232–234
Optical path difference (OPD), 116
Optical path length (OPL), 116
Optical planar waveguides. *See* Planar
  optical waveguides
Optical polarizers, 837–838, 838–846
  categories of, 838
  history of, 838
Optical power, 120, 1044, 1045
Optical power handling capability, of
  electro-optic modulators, 1035
Optical processing
  achromatic, 217
  achromatic partially coherent, 218
  adaptive, 228–229
  bandlimited partially coherent, 217–218
  coherent, 210–212
  partially coherent, 215–218
  real-time, 222–228
  spatially partially coherent, 216
  by synthesis aperture radar, 236
  white light, 218–220

Optical propagation loss, 1036–1037
  in electro-optic modulators, 1035–1037
Optical pumping. *See also* Diode laser
  pumping; Laser pumping
  by arclamps and filament lamps,
   380–381
  of dye lasers, 445, 460, 461–463
  of far infrared lasers, 539–540
  of fusion amplifiers, 404–405
  laser excitation via, 371, 375, 376–380
  of long-pulsed lasers, 388–389
Optical reflection, in optical fiber joints, 677
Optical resonators, 163–197
  defined, 163
  distributed Bragg reflector, 191,
   193–195
  distributed feedback resonator, 188–193
  Gaussian beams in, 171–176
  open Fabry–Perot, 176–188
  propagation matrices for, 173–176
  stability of, 177–179
  vertical-cavity surface-emitting laser,
   195–196
  waveguides for, 163–164, 164–168
  waveguide-type Fabry–Perot, 164–171
Optical RF spectrum analyzer, 821
Optical signal processing
  acousto-optic devices in, 1114–1116
  guided-wave, 1114
Optical simulation, advantages of, 437
Optical sine theorem, 150–151
Optical source-fiber combinations, for
  single-mode fiber systems, 683–684
Optical spatial switches, 948, 949
Optical spectrum analyzers, diffraction
  gratings in, 904–905
Optical systems
  aberrations of, 134–151
  coherent and incoherent, 205–210
  elementary, 126–130
  first-order properties of, 120–123
  Fourier transforms of, 201–204
Optical transfer function (OTF), 158, 208
Optical transmission loss spectra, 669–671
Optical transmission technology,
  coherent, 692
Optical tube length, 129
Optical wave propagation, 1–51
  diffraction in, 40–50
  electromagnetic theory of, 2–15
  of Gaussian beam, 29–39
  geometry of, 15–29

Optical wavelength filtering/multiplexing/
demultiplexing, acousto-optic devices
in, 1113–1114
Optic axis, 13–14, 116
Optics. *See also* Magneto-optics
first-order, 118–133
geometrical, 15–29
integrated, 731, 807, 808, 979–998
material physics of, 1160–1164
Optimized anisotropic diffraction, in
guided-wave acousto-optic Bragg
modulators, 1086, 1087
Optimized interaction, in anisotropic
acousto-optic Bragg diffraction, 1083,
1084
Opto-acoustic effect, in stabilizing FIR
lasers, 583–584, 585. *See also*
Acousto-optic switches
Optoelectronic integrated circuit (OEIC)
technology, 692
Optoelectronics, history of, 719–720
Opto-mechanical VOAs, 854
Orbital angular momentum, of an electron,
1135–1136
Orbitals, of atoms, 1135–1136, 1139–1140,
1141
Orbit–lattice coupling, of atoms,
1135–1136
Ordinary wave (O-wave)
birefringence and, 14, 842–844
in optical isolators, 866, 868
Organic dyes, in dye lasers, 444–445,
446–455
Organic luminescence diodes (OLED),
445
Organic thin-film waveguides, 734–736
table of, 735
Orthogonality, of Hermite–Gaussian
functions, 39
Orthogonally polarized waves, 1160–1161
Orthogonal magnetic polarization, 1153
Oscillation, parasitic, 408–409
Oscillator–amplifier laser systems, system
strategies for, 419–424
Oscillators
continuous and long-pulsed, 388–390
in discharge excimer lasers, 631
frequency mode selection in, 395–396
mode-locked, 392–395
*Q*-switch and short-pulse, 390–392
simple, 386–387
spatial mode selection in, 366–399
Outer product, 226–228

OUT fiber, in optical isolators, 871–872
Outside vapor phase oxidation (OVPO), 669
Overlayer, in multilayer step index
waveguides, 761, 763, 764, 765, 766
Overlay film, on graded index waveguides,
763–768
Oxadiazole dyes, 448
Oxazine dyes, 448, 451
Oxazole dyes, 448, 453
Oxide-confined VCSELs, 331, 333–334,
335–337
Oxidized silicon ($SiO_2$–Si), as acousto-optic
substrate material, 1094–1096

Palladium silicide detectors. *See* PdSi
detectors
Pan, J. J., 837
Parabolic cylinder functions, 729–730
Parabolic graded index SCH (P-GRINSCH),
318–319
Parabolic profile, for graded index
waveguides, 727, 728
Parallel directional coupler, 711–714
Parallel double-electrode synchronous
transducer, 1098
Parallel-finger chirp transducer, optimized
anisotropic diffraction with, 1104
Parallel spin wave propagation, dispersion
relation for, 1153–1154, 1154–1155
Parallel-to-serial readout architecture, 70
Paramagnetic glass, 1170
Paramagnetic materials, 1136, 1137–1140,
1142, 1164
antiferromagnetism and, 1141–1142
Parameters, acousto-optic design from,
1278–1286
Parametric equations, for systems analysis,
110–113
Parasitic oscillation, 408–409
Parasitic parameters, 320
Paraxial approximation, 27, 116, 200
Paraxial invariant, 126
Paraxial marginal ray, 124
Paraxial rays, 116–117
Paraxial ray-trace equations, 119–120
Paraxial region, 116
Paraxial wave equation, 30–32
Parity, 130–131
Partial dispersion ratio, 138
Partially coherent optical processing,
215–218
Partition noise, MLM laser mode, 686,
687–688

Passband width
  in multiplexers, 940
  in waveguide grating routers, 910–911
Passive optical components, 678–679
  in optical fiber transmission systems,
    680
Passive optical spot-size converters, 1046
Passive polarization switches, 412, 413
Passive waveguides, optical components of,
    753–809, 810, 811
Passively mode-locked dye lasers, 491–496
Pauli exclusion principle, 1135, 1137
PBS cube. *See* Cube PBS
PbSe, PbS, PbSnTe, and PbTe infrared
    detectors, 54
PC-HgCdTe detectors. *See* HgCdTe
    photoconductive detectors
PdSi detectors, 91–93, 93–96
Peak diffraction efficiency, of acousto-optic
    modulators, 1266–1267, 1268
Peak insertion loss, in multiplexers, 939
Peebles, W. A., 511
Penalties, in optical fiber transmission
    systems, 690–691
Pentane, as buffer gas, 597
Penta prism, 131–132
Periodic slit grating, 47
Periodic structures, diffraction in, 464–466
Permeability tensor, 1134–1135, 1150, 1161
Permittivity
  dielectric, 1161–1162, 1165
  in optical detectors, 1219
Permittivity tensor, 12–13, 1161–1163
  in magneto-optics, 1165
Perpendicular spin wave propagation,
    dispersion relation for, 1153, 1154,
    1155–1156
Perturbation methods. *See* Index
    perturbation methods
Petzval field curvature, 142
Petzval surface, 142
Phase, in magneto-optic thin-film
    waveguides, 1199–1200, 1201, 1204,
    1206
Phased array optical filters, 903
Phased arrays (PHASARs), 909. *See also*
    Waveguide grating routers (WGRs)
  in MUX/DEMUX integrated optics
    design, 982
  virtually imaged, 899, 903, 908–909
Phased-array transducer, 1103
  isotropic diffraction with, 1104
Phase error, in AWG devices, 826

Phase front, for Gaussian beams, 172–173
Phase gratings, 47, 962
Phase-locked arrays, 264–265
Phase mask technique, fabricating fiber
    Bragg gratings via, 971, 975–976
Phase matching
  in waveguide isolators, 1193–1194
  in waveguide optical gratings, 792–793
Phase response, of optical filters, 894–895
Phase retarders, 846–850
Phase shift
  in distributed Bragg reflectors, 194–195
  in Fabry–Perot resonators, 187
Phase transfer function (PTF), 158
Phosphate glasses, 411–412
Phosphorus oxychloride ($POCl_3$), in liquid
    lasers, 475
Photo-induced poling, of polymers, 1017
Photochemical reactions, dye lasers in,
    496–499
Photoconduction, 1239
Photoconductive (PC) detection, 58–61, 62
  figures of merit for, 61–64
Photoconductive devices, 1239–1242
Photoconductive gain, 1239–1240
Photocurrent generation, 1029, 1030
Photodetectors
  nonlinearity in, 1243–1249
  in optical detectors, 1216–1217
Photodiode capacitance, in optical detectors,
    1219–1220
Photodiodes
  avalanche, 1220, 1227–1236
  geometrical effects in, 1246–1247
  HgCdTe, 72, 76, 77, 78
  in optical detectors, 1219–1220,
    1220–1239
  PIN, 1220, 1222–1226, 1227, 1237,
    1243
  planar, 1225–1226
  Schottky, 1220, 1226
  speed limits on, 691
  traveling wave, 1238–1239
  velocity-matched, 1238–1239
  waveguide, 1236–1239
Photodynamic therapy, dye lasers for, 501
Photoelastic birefringence, in
    bismuth-substituted YIG, 1182
Photogain, of optical detectors, 1218
Photolithography, 1104
  discharge excimer lasers for, 633
  fabricating fiber Bragg gratings
    via, 971

of waveguide circuitry, 703
of waveguide optical gratings, 792
Photon density
in high-speed laser design, 314, 323
in laser amplifiers, 400
in $Q$-switch and short-pulse oscillators,
391, 393
Photon detectors, 58
Photon lifetime, in high-speed laser design,
313, 314, 321
Photonic crystals, 471
Photons. *See also* Light; VUV photon
emission
in optical detectors, 1215, 1216, 1219
polarization eigenstates of, 1161
in semiconductor optical amplifiers,
1318
solid-state lasers and, 370
in Stokes process, 1312
Phototransistor-heterojunction
phototransistors (HPTs), 1247–1249
Phototransistors, 1247–1249
Photovoltaic (PV) detection, 58–61, 62
figures of merit for, 61–69
Photovoltaic detectors
HgCdTe, 54, 61, 64–65
LWIR, 65
MWIR, 65
Phthalimide dyes, 448
Phthalocyanine dyes, 448, 449
Physical vapor deposition (PVD)
techniques, thin-film filter fabrication
via, 955–956
Physics. *See* Device physics; Material
physics
Picosecond pulse generation, by
photoconductive devices, 1242
Piezoelectric film, in guided-wave
acousto-optic Bragg diffraction, 1085,
1086
Piezoelectric layer, in acousto-optic
modulators, 1287–1288
Piezoelectric substrates, 1094
Piezoelectric transducer (PZT), 531
Pinching, in electron-beam amplifiers, 638
"Pincushion" distortion, 145
PIN photodiodes, 1220, 1222–1226
avalanche photodiodes versus, 1227
as photodetectors, 1243
Schottky photodiodes versus, 1226
waveguide photodiodes versus, 1237
$p$-intrinsic-$n$ (PIN) photodiodes, 1220,
1222–1226, 1227, 1237, 1243

Piston error, 137, 139
Pitch, of gratings, 961
Pixels, 97, 98
in Schottky-barrier arrays, 93–96
in spatial light modulators, 1194–1195
Plain old telephone service (POTS),
679, 693
Planar guided-wave acousto-optic Bragg
diffraction, 1108
Planar guided-wave acousto-optic Bragg
modulator/deflector, performance
parameters of, 1099–1102
Planar hybrid focal plane array, 69
Planar image, with optical lenses, 115
Planar inhomogeneous media, ray tracing
in, 21
Planar optical waveguides, 697–702,
719–836. *See also* Dielectric planar
waveguide theory
applications for, 819–821, 822
components of passive, 753–809,
810, 811
fabrication of, 731–753
future of, 830
history of, 719, 720
measurement techniques in,
809–830
ray optics in, 720–724
ring resonators and, 913
theory of, 720–731, 732, 733
Planar photodiodes, 1225–1226
Planar waveguide optical isolators,
1131–1132
Planar waveguides, guided-wave
acousto-optic Bragg diffraction in,
1084–1099, 1108
Planck's constant, 1135, 1136, 1303
Plane of incidence, 7, 8
Plane wave, 5, 6
boundary reflection and transmission
of, 7–8
diffraction of, 43, 46–47, 48–50
in layered dielectric, 10–12
Planoconvex lens, 160
Plasma-activated chemical vapor deposition
(PCVD), 669
Platinum silicide detectors. *See* PtSi
detectors
Plössl eyepieces, 161
PLZT ceramics, in electro-optic VOAs,
859
$pn$ junction photodiodes, 1220–1222

Pockels cell electro-optic switch, 390, 395, 413–415
   polarization and, 419
   risk reduction via, 424
Pockels cell magneto-optic switch, 416
Pockels effect, 1011
Poincaré sphere, 1161
Point source, 116
Point spread function (PSF), 216
Poisson noise, in optical detectors, 1218
Polacoat™, 839
Polarcor, 839
Polarization, 418, 564. *See also* Left circularly polarized (LCP) propagation; Magneto-optic polarization-rotation isolators; Polarizers; Right circularly polarized (RCP) propagation
   bulk-wave acousto-optic Bragg modulators and, 1081–1082
   circular, 6–7, 1160
   elliptical, 6–7, 1160–1161
   Faraday effect and, 1131
   in guided-wave acousto-optic Bragg modulators, 1086, 1087
   by half wave plates, 848–849
   left-handed and right-handed, 7
   of light in magnetic fields, 1160–1161, 1162–1163, 1164
   linear, 6–7, 1160, 1166
   of lithium niobate crystals, 1013–1014
   magnetic, 1150–1151, 1152, 1153–1154
   in magneto-optic deflectors, 1196–1197
   in magneto-optics, 1165
   in optical circulators, 875–877
   orthogonal, 1160–1161
   in reflective magneto-optic materials, 1184
   saturation, 1145
   in spatial light modulators, 1194, 1196
   of spin waves, 1153–1154, 1154–1155
   in stabilizing FIR lasers, 584–586, 587–589
   in waveguide grating routers, 912–913
   in waveguide isolators, 1194
   of waves, 5–7
Polarization angle, 9
Polarization beam combiners (PBCs), in magneto-optical VOAs, 857
Polarization beam splitters (PBSs), 837–838, 844–846. *See also* Beam splitting polarizers; Cube PBS

in magneto-optical VOAs, 856–857
Polarization chart, 6–7
Polarization-dependent loss (PDL), in multiplexers, 940
Polarization-dependent optical isolators, 864–865
Polarization eigenstates, 1161
Polarization mode dispersion (PMD), in optical isolators, 866, 867–868
Polarization rotators, 846–850
Polarization rotator sets, 877
Polarization sensitivity, of AWGs, 994
Polarization state, in dye lasers, 467–469
Polarization state transformation diagrams, for VOAs, 859
Polarizers
   aberration control with, 436
   categories of, 838
   laminated, 839–841
   new and future designs of, 846
   optical, 837–838, 838–846
   wire grid, 839
Polarizing optical elements, 418–419
Polar Kerr effect, 1167, 1168
   in reflective magneto-optic materials, 1184
Polaroid™, 839
Polaroid-type polarizers, 419
Poling, of polymers, 1017
Polycrystalline films, for magneto-optic data-recording, 1190–1191
Polymer dispersed liquid crystals (PDLCs), in electro-optic VOAs, 859–861
Polymer modulators, 1010
   traveling-wave, 1061–1063
Polymer MZMs, advantages and disadvantages of, 1071
Polymers, as electro-optic materials, 1017–1018
Population inversion, 514–515
   in argon laser, 520–521
   in FIR lasers, 609–610
Porro system, 133–134
Positive DI lens, waveform matrix for, 175
Positive-index guide, 257, 264–265
Positive lens, Fourier transform and, 201–204
Potassium dideuterium phosphate (KD*P), in electro-optic devices, 414
Potassium dihydrogen phosphate (KDP), in electro-optic devices, 414
Potassium ions, in ion exchange reactions, 746–747

Potential well profiles, for graded index waveguides, 727–731, 732, 733
Power decay, of optical beams, 1216
Power effects, average solid-state laser, 424–428
Power transmission, through a circular aperture, 32–33
Poynting theorem, 4
Praseodymium (Pr), as garnet substituent, 1173, 1174
Precession
    Larmor, 1163, 1169
    of magnetic dipoles, 1149, 1150, 1151, 1152, 1153
Preionization discharge, 626–627, 627–628, 630
Pressure, glow discharge excitation and, 626–627
Principal planes, 117, 121
Principal points, 121–122, 123
Principal ray, 117
Prism beam expander, 465, 468
Prism couplers, 754–755, 755–757, 758, 1108, 1109
    for measuring refractive index profiles, 813
    for measuring waveguide parameters, 810
    in optical loss measurement, 815–816
    in thin-film material characterization, 821, 822
Prismatic effects, 435–436
Prisms, 130–133, 760, 1203
    beam-splitter, 841–844
    in continuously pumped dye lasers, 484–485
    diffraction gratings versus, 963–964
    for dye laser refractive index dispersion, 463–464
    Glan, 841
    Nicol, 841
    Rochon, 841, 842–843
    Sénarmont, 841
    Wollaston, 841–842, 843, 881–882
Profile dispersion, in multimode optical fibers, 667
Projection lens, 129, 130
Projection systems, 129–130
Propagation, effect of laser glass properties on, 409–412
Propagation constant, 171
Propagation matrices, 11–12, 173–176

Proton-implanted VCSELs, 331, 332–333, 334, 335–337, 346, 347
PtSi detectors, 91–93, 93–96
p-type doping, 316
Pulsed laser pumping, 383
Pulsed laser system, modular, 481–482
Pulsed optically pumped FIR lasers, 606–624
Pulsed pump sources
    for liquid lasers, 471, 475–481, 481–482
    for solid-state lasers, 375
Pulsed storage lasers, 376–379
Pulse generation, by photoconductive devices, 1242
Pulse-pumped liquid lasers, 471–482
Pulse shapes
    in laser amplifiers, 400–401
    in Q-switch and short-pulse oscillators, 390–391, 393, 394
Pulse width, in chirp analysis, 1037, 1038–1039
Pump depletion, Raman fiber amplifiers and, 1312
Pumped disk amplifier, 408
Pumping. See also Diode laser pumping; Electrical pumping; Laser pumping; Optical pumping
    of EDFAs, 944
    of Raman amplifiers, 946–947
Pumping cavity, 384–386
Pump sources
    for dye lasers, 461–463
    high-power, 1314–1316
Pupils, 117, 123–126. See also Entrance pupil; Exit pupil; Two-pupil incoherent processing systems
Push-pull modulation, in Mach–Zehnder modulators, 1040, 1066, 1067, 1068–1069
PV-HgCdTe detectors. See HgCdTe photovoltaic detectors
Pyroelectric (PE) detectors, 103–107
Pyroelectric materials, performance of, 104–107
Pyroelectric probe, in optical loss measurement, 817
Pyrylium salts, 448, 453

QCSE red-shift, 1025
Q parameter, in acousto-optic design, 1278, 1279, 1284, 1296

*Q*-switch oscillators, 390–392
    electro-optic modulators for, 413–414
    frequency mode selection in, 395–396, 397
    glasses for, 410–411
    saturable absorber switches as, 412–413
Quadratic index media
    *ABCD* matrix for, 27–29
    Gaussian modes in, 35–36
    Gaussian beams in, 29
    ray tracing in, 22, 23–24
Quantum cascade lasers, 347–348
Quantum confined Stark effect (QCSE), 1021–1026. *See also* Stark-splitting
    in chirp analysis, 1041
    in optical detectors, 1219
    tunable lasers and, 293, 294
Quantum dot (QD) lasers, 348–350
    future development of, 351
Quantum efficiency
    of HgCdTe detectors, 74, 75, 79
    in high-speed laser design, 313, 314
    of optical detectors, 1216
    of photoconductive devices, 1240
    of *pn* junction photodiodes, 1221
    of waveguide photodiodes, 1237
Quantum electronics, 199
Quantum mechanics, of magnetic phenomena, 1132, 1135–1136, 1137, 1163–1164
Quantum noise limited detection, by avalanche photodiodes, 1228
Quantum well (QW) lasers, 239, 240, 243, 247–252, 261, 264. *See also* Multiple quantum well lasers (MQWs); Multiple quantum well MZMs
    future development of, 351
    high-speed, 308–323
    strained, 252–254
    tunable, 294
    VCSEL, 324, 328–331
Quantum wells, quantum confined Stark effect in, 1021–1026
Quantum wire lasers, 348–350
Quarter-wave criterion, 159
Quartz. *See also* Silica substrate
    in bulk-wave acousto-optic Bragg modulators, 1080–1081
    magneto-optic properties of, 1170
    in polarizers, 841–842
    wave plates from, 848
Quasi-continuous wavelength tuning, 298, 303–304

Quasi-two-energy-level system, 517–518
Quenched transient dye lasers (QTDLs), 479–480, 481
Quinolone dyes, 448, 453

Radial keratotomy, discharge excimer lasers for, 633
Radiative processes, of liquid lasers, 456–459
Radius of curvature, 117, 118
    in geometrical optics, 16–17
Raman amplifiers, in optical networks, 946–947
Raman emission, from FIR lasers, 613, 614, 615
Raman fiber amplifiers, 1301, 1312–1316
    equations governing, 1312–1314
    high-power pump sources for, 1314–1316
    specification sheets for, 1314–1316
Raman gain, 1312, 1313, 1314
Raman–Nath diffraction, 1085
Raman–Nath regime, acousto-optic interaction in, 1259, 1260
Raman scattering, 641
Randomly distributed feedback, 470–471
Random phase screen (RPS) approach, to laser fusion, 642
Range equations, 111
Rare earth elements. *See also* Lanthanide elements
    ferromagnetism of, 1140
    as garnet substituents, 1173–1176
    in reflective magneto-optic materials, 1184
Rare earth ions
    in liquid lasers, 455–456
    in solid-state lasers, 371–374
Rare earth iron garnets, 1171, 1172, 1173
    in optical isolators, 1192–1194
Rare earth liquid lasers, flashlamp-pumped, 474–475
Rare gas dimers, in excimer lasers, 624
Rare gas halide lasers, 640. *See also* Noble-gas halide lasers
    excimer, 624–625, 627
Rare gas halides, in excimer lasers, 632
Rare gas oxides, in excimer lasers, 624
Rate equations, 513–518
    high-frequency, 309
Rate equations model, for erbium-doped fiber, 1304–1305

Rate parameters, of lasing dye molecules, 457
Ray angles, 118
    in optical systems, 118–119
Ray equation, 17–18. *See also* Ray tracing
    in three coordinate systems, 18–19
Ray fans, 136
Rayleigh scattering, in optical fibers, 670
Rayleigh's criterion, 159
Ray matrices, 175–176. *See also ABCD* matrix
Ray path, in an optical fiber, 666
Rays, 116
    in birefringence, 842–844
    definitions concerning, 116–117
    in geometrical optics, 15
    in planar optical waveguides, 720–724
Ray tracing, 1
    *ABCD* transformation for, 25–27
    in geodesic lenses, 779–785, 785–788
    in inhomogeneous media, 20–25
    in Luneburg lenses, 25, 771, 775, 777, 780, 781, 782
    Monte Carlo, 409, 410
*RC* limitations
    in optical detectors, 1219–1220
    of photodiodes, 1245–1246
    in waveguide photodiodes, 1236–1237, 1238
Reactive ion etching (RIE). *See also* RIE-etched ridge waveguide laser
    for AWG devices, 822
    in fabricating integrated optics-based MUX/DEMUX devices, 980
Readout structures, for extrinsic silicon detectors, 108–109
Real rays, 117
Real ray-trace techniques, 151–155
Real-time optical processing, 222–228
Rear focal length, 117, 118, 121
Rear focal point, 117, 120–121
Rear nodal point, 117, 122
Rear principal plane, 117
Receiver noise, in optical detectors, 1218
Receivers
    in avalanche photodiodes, 1231–1232
    optical detectors as, 1215
Reciprocal polarization rotator sets, 877
Reciprocal rotation, 1194
Reconfigurable OADMs, 979, 997
Rectangular aperture, Fraunhofer diffraction through, 48, 49
Rectangular pulse function, 46

Recursion formula, for Hermite polynomial, 39
Red VCSELs, 342
Reflection. *See also* Mirrors; Optical reflection
    at a boundary, 7–8
    from layered dielectric, 10–12
    in magneto-optics, 1167–1168
    Snell's law of, 16
    from thin film, 9–10
Reflection bandwidth, of thin-film filters, 918
Reflection coefficient, 9–10, 12
Reflection gratings, 961, 962
Reflection hazards, diffuse and specular, 429–430
Reflection high-energy electron diffraction (RHEED), 246
Reflection-induced noise, 686, 690–691
Reflective magneto-optic materials, 1184–1187
Reflective optical circulators, 881–882
Reflective optical isolators, 870–872
Reflectivity, DBR, 325
Refraction
    in acousto-optic devices, 1259
    at an aspheric surface, 154–155
    at a spherical surface, 152–153
    Snell's law and, 16
Refraction index dispersion, in dye lasers, 463–464. *See also* Refractive index
Refractive elements, in waveguides, 759–789
Refractive index, 7, 8–9, 16, 116, 467–468, 720–724. *See also* Graded refractive index profiles; Step index fiber
    *ABCD* matrix and, 26, 27
    birefringence and, 842–843
    of bismuth-substituted YIG, 1177–1178
    bulk-wave acousto-optic Bragg modulators and, 1080–1081
    of DFB and DBR gratings, 269–271
    effective, 255
    electro-optic effect and, 1011
    free-carrier optical waveguides and, 752–753
    of fused quartz, 1080–1081
    geodesic lenses and, 788
    graded index waveguides and, 763, 764, 765, 766
    grating couplers and, 757
    in magneto-optics, 1165, 1166, 1167–1168

Refractive index, 7, 8–9, 16, 116, 467–468, 720–724. *See also* Graded refractive index profiles; Step index fiber (*continued*)
    lateral, 254–255
    of liquid crystals, 849
    Luneburg lens and, 770–772, 773–775
    of magneto-optic thin-film waveguides, 1199, 1200, 1206
    metal-clad optical waveguides and, 750, 751, 752
    mode-launching couplers and, 754–755
    optical channel waveguides and, 702–707
    optical diffraction elements and, 789–790
    of optical fibers, 660–661, 661–662
    planar optical waveguides and, 697–698, 699
    of polymer crystals, 1017–1018
    prism couplers and, 756
    thin-film filters and, 917
    tunable lasers and, 293, 295
Refractive index change, in fabricating fiber Bragg gratings, 972
Refractive index profile measurement, 813–815
Regenerative amplifiers, in discharge excimer lasers, 631
Relative core, of an optical fiber, 663
Relative intensity noise (RIN), 1010
Relaxation resonance frequency, VCSELs and, 343–344
Relay lenses, 161
Reliability, VCSEL, 335–337
Remanence, 1185
Resistance, 320, 322–323
    in photoconductive devices, 1241
Resistance and capacitance (*RC*) constant, VCSELs and, 343, 344. *See also RC* limitations
Resistance–area product, 87, 88
Resolving power, of diffraction gratings, 966–967
Resonances, in multiplexers, 941–942. *See also* Excitons; Relaxation resonance frequency; Transverse resonance condition
Resonant frequencies
    in distributed feedback resonators, 188–189
    of Fabry–Perot resonators, 186–188
    of optical resonators, 168–171

Resonant mode, 163
Resonant-optical-waveguide (ROW) arrays, 267
Resonant transients, 478–480
    quenched, 479–480
Resonators. *See* Optical resonators
    in discharge excimer lasers, 631
    for FIR lasers, 569–578, 583
    unstable, 398–399
Response speed, of photodetectors, 1217
Response time, of photodetectors, 1217–1218
Responsivity, of waveguide photodiodes, 1237–1238
Reticulated linear array, 105
Retina, laser light and, 500
Retroreflection, 130, 132
Reversion prism, 132
RF (radio frequency) drive power. *See* RF power
RF driver circuits, acousto-optic modulators and, 1101–1102
RF electromagnetic fields
    magnetic dipoles and, 1149–1155
    magnetic waves and, 1158–1159, 1159–1160
RF equivalent circuit model
    of EAMs, 1042–1044, 1045
    of lumped-element QCSE-based MZMs, 1043, 1048
RF optical-to-electrical (O/E) responsivity, 1032–1033
RF power, 1032–1033
    of an acousto-optic modulator, 1099–1100, 1101
RF signals, acousto-optic modulators and, 1102
RF spectrum analyzers, acousto-optic, 1114–1116, 1120–1121
RF sputtering
    Luneburg lens fabrication via, 775–779, 780, 781, 782, 783
    waveguide fabrication via, 734, 736–739
Rhodamine 6G dye laser, 443, 454
Rhodamine 640, 470
Rib waveguides, 702, 703, 707, 712
Ridge structure, in traveling-wave modulators, 1058–1059, 1060
Ridge waveguides, 702, 703–704, 706, 708

RIE-etched ridge waveguide laser, 262–263

Right-angle prism, 132

Right circularly polarized (RCP) propagation, 1153–1154. *See also* Right-handed polarization

of light, 1160, 1162–1163, 1164, 1165, 1166, 1168, 1169

Right-handed optical systems, 130–131

Right-handed polarization, 7. *See also* Right circularly polarized (RCP) propagation

Ring lasers, 492

    FIR, 612–613

    pumping by, 485

Ring node, 145

Ring resonators, 892, 893, 896, 897, 913–917

    bandwidth utilization of, 914–916

    design of, 913–914

    in stabilizing FIR lasers, 586–587

Ring-type reflector, 164

Ripple, in multiplexers, 939

Rise time

    in acousto-optic design, 1278

    for acousto-optic modulators, 1259–1260, 1265–1266

Risk reduction, in oscillator–amplifier laser systems, 423–424. *See also* Hazards; Safety

Rms wavefront variance, 140–141

Rochon prisms, 841, 842–843

Rod amplifiers, 404, 405, 406

    medium average power, 405–408

Rogers, John R., 115

Rooftop resonators, in stabilizing FIR lasers, 587–589

Ross, D., 369

Rotational energy levels

    of carbon dioxide and nitrogen, 527–528

    of $CH_3F$ molecules, 562, 563

    far infrared lasers and, 539

    of water molecules, 545, 546, 550

Rotators, polarization, 846–850. *See also* Faraday rotators

Routers, waveguide grating, 909–913

Rowland circle, 904–905, 969

Rowland mounting, in free-space MUX/DEMUX architecture, 969

$R$ parameter, in acousto-optic design, 1278, 1279, 1284–1286, 1296

Runge–Kutta method, 973

Rutherford backscattering method, measuring impurity profiles with, 813

Rutile film, in graded index waveguides, 767, 768

Safety, of solid-state lasers, 429–430

Sag function, 147

Sagitta ("sag"), of a surface, 117

Sagittal coma, 141, 142

Sagittal focus, 142–144

Saleh, Bahaa E. A., 838, 850

Sales, Salvador, 935

Samarium (Sm), as garnet substituent, 1175

Sampled grating (SG), 923, 924

Sampled grating DBR (AM) tunable lasers, 299

Sapphire ($Al_2O_3$), as solid-state laser host material, 374

Saturable absorber switches, 412–413

Saturation magnetization, 1135, 1137, 1185, 1187. *See also* Magnetic saturation

Saturation polarization, 1145

SAW transducer–conventional Bragg electrode array, 1118

Scalar diffraction theory, 40–41

Scalar wave equation, 5, 17, 30, 40

Scanning beam displacement prism, 131

Scattered guided modes, of prism couplers, 755

Scattered waves

    from distributed Bragg reflectors, 191

    from distributed feedback resonators, 189–190, 191

Schottky-barrier detectors, 91–96

    characteristics of, 91–93

Schottky junction photodiodes, 1220, 1226

Schrödinger equation, graded index waveguides and, 726–727

$Sech^2$ profile, for graded index waveguides, 727, 728

Secondary chromatic coefficients, 147

Secondary ion mass spectroscopy (SIMS), measuring impurity profiles with, 813

Second harmonic generation (SHG), in dye lasers, 486

Segmented traveling-wave modulators, frequency response of, 1063–1066

Seidel aberration coefficients, 436

Selection rules, 1163

Selective zinc diffusion, in NAM laser fabrication, 279

Selenium oxychloride (SeOCl$_2$), in lasers, 443, 475

Self-focusing, in oscillator-amplifier laser systems, 419–422

Self-organized quantum dots, 349

Self-scanning arrays, 103

Semet–Luhmann configuration, 607, 609, 617–618

Semiaperture, 117

Semiconductive substrates, for planar optical waveguides, 733

Semiconductor DBRs, 325

Semiconductor integrated optical devices, in integrated optics-based MUX/DEMUX devices, 980

Semiconductor lasers, 239–368. *See also* III–V compound semiconductors
  applications for, 240–241
  devices using, 241–244
  for diode lasers, 244–246
  future development of, 350–351
  high-power operation of, 271–289
  high-speed, 308–323
  history of, 239–240
  operation of, 241–271
  in optical detectors, 1218–1219, 1220
  quantum cascade lasers, 347–348
  quantum wire and quantum dot lasers, 348–350
  spectra of, 171
  surface-emitting, 290–292
  vertical-cavity surface-emitting, 323–347

Semiconductor optical amplifiers (SOAs), 1301, 1316–1319
  characteristics and specifications of, 1318–1319
  structure of, 1316–1318

Semiconductor optical detectors, 1216–1219

Semiconductors
  bias stability of, 1034
  Franz–Keldysh effect in, 1020–1021
  III–V compound, 1014–1016, 1220, 1223

Semiconductor tunable laser diodes, 292–308

Sénarmont prisms, 841

Sensor trade-off equations, 113

Separate confinement hetero-structure (SCH), 241, 243

carrier transport time in layers of, 317–318

layer designs with, 320

QW lasers and, 247

Series impedance, 1054

Shadow mask, in Luneburg lens fabrication, 775–779, 780, 781, 782, 783

Shape-factor demagnetization, 1144–1146, 1155

SH-$\Delta$ PLCs, in AWGs, 993

Shiva amplifiers, 405, 406, 407, 411

Short-circuit mode, 58–61

Short laser pulses. *See also* Ultra-short light pulses
  generating, 478–479, 1242
  high-power, 640–641

Short-pulse oscillators, 390–392

Short-wavelength infrared (SWIR), 54. *See also* SWIR entries

Shot noise
  in avalanche photodiodes, 1228
  in optical detectors, 1218

Shunt impedance, 1054

Side-mode suppression ratio (SMSR), 1009

Signal detection, complex, 223–225

Signal processing, acousto-optic devices in, 1114–1116

Signal-switching devices, 1194

Signal-to-noise ratio (SNR), 110–111, 218, 1009
  of avalanche photodiodes, 1231–1232

Silica (SiO$_2$)
  as acousto-optic substrate material, 1094–1096
  in lattice filters, 900

Silica-based AWG devices, 822–823, 824

Silica buffer layer, in traveling-wave modulators, 1057–1058

Silica glass
  in D-MEMS based VOAs, 862
  in thermo-optical VOAs, 855–856

Silica glass fibers, 669–671
  dispersion in, 672

Silica-on-silicon circuits, in integrated optics-based MUX/DEMUX devices, 980, 981, 995–996. *See also* Oxidized silicon (SiO$_2$–Si)

Silica substrate, in metal-clad optical waveguides, 750–752, 816, 817, 818. *See also* Quartz

Silica waveguide circuits, in integrated optics-based MUX/DEMUX devices, 980

Silicon (Si). *See also* Extrinsic silicon detectors; Silica entries
    in avalanche photodiodes, 1232–1233, 1234, 1235
    in thermo-optical VOAs, 855–856
Silicon-controlled rectifiers (SCRs), 629
Silicon oxynitride, in thin-film waveguides, 739
Silicon vidicon cameras, diagnostics with, 432–433
Silver ion exchange waveguide, 727, 729
Simple linear WDM link, 952
Simulated annealing algorithms (SAAs), 230–231
Simulation, optical versus numerical, 437
Single-beam system, 533, 534
Single channel (SC) OADMs, 953–954
Single-crystal epitaxial waveguides, 739–742
Single-crystal films, 732
Single-crystal infrared detectors, 54
Single-lamp high-efficiency Nd:YAG lasers, 384
Single-lamp single-mode Nd:YAG lasers, 384–385
Single lateral mode high-power diode lasers, 261
Single longitudinal mode (SLM) lasers, 267–270
    chirping in, 686, 688–690
Single-longitudinal mode laser diodes, in optical fiber transmission systems, 682, 683–684
Single-mode dye lasers, pumping by, 485–486
Single-mode fibers (SMFs). *See also* Conventional single-mode fibers (C-SMF); Dispersion-shifted single-mode fibers (DS-SMF)
    dispersion in, 671–675
    types of, 673–675
Single-mode fiber systems. *See also* Single-mode fiber transmission technology
    optical source-fiber combinations for, 683–684
    wavelength division multiplexing in, 684–686, 992–993
Single-mode fiber transmission technology, 681–684. *See also* Single-mode fiber (SMF) systems
Single-mode lasers, 260–271

Single-mode Nd:YAG lasers, with single lamp, 384–385
Single-mode (SM) optical fibers, 662, 663–665
    in optical attenuators, 850–852
Single-pass amplification, 402, 403
Single-polarization optical circulators, 875–877
Single quantum well lasers (SQWs), 241
    differential gain in, 314–315
Single-slit diffraction, 45, 46–47
Single-stage birefringent crystal wedge type optical isolators, 865–866
Single-stage EDFA, 864
Single-stage optical allpass filters, 897
Single-stage optical filters, 891–892
Single-surface acoustic waves. *See* Surface acoustic waves (SAWs)
Single tilted-finger chirp transducer, isotropic diffraction with, 1104, 1107, 1110
Single transducer, versus multiple transducers, 1110–1111
Sinusoidal amplitude grating, 47
$SiO_2$–Si. *See* Oxidized silicon ($SiO_2$–Si)
Six-energy-level system, 614–615
Six-stage gain-equalizing filter, 900–901
Skew fan, 136–137
Skew ray, 24, 151
Slide projector, 129–130
Slits
    diffraction through single, 45, 46–47
    in gratings, 961
    in Young's experiment, 220–221
Small-signal modulation, 309–312
    efficiency of, 1032–1037
Smectic liquid crystals, 849
Snell's law, 8, 10, 11, 116, 119, 842
    optical fibers and, 661
    statement of, 16
Sodium ions, in ion exchange reactions, 746–747
Solc birefringent filters, 898, 902
Solenoids, 1133
Solid hosts, active-ion laser action in, 371–374
Solid lasers, liquid lasers versus, 443–444
*Solid State Laser Amplifiers* (Brown), 431
*Solid State Laser Engineering* (Koechner), 431
Solid-state laser materials, limits on, 427–428, 429

Solid-state lasers, 369–442
  average power effects, 424–428
  excitation of, 375–386
  history of, 369
  host materials in, 374–375
  laser amplifiers, 399–412
  laser oscillators, 386–399
  medium average power and burst-mode, 405–408
  optical modulators, 412–419
  oscillator-amplifier laser systems, 419–424
  safety of, 429–431
  simple, 386–387
Solute concentration, dye lasers and, 454–455
Solvents, for dye lasers, 454–455, 456
Sommerfeld radiation condition, 41
Sound waves, 1079. *See also* Acoustic waves; Acousto-optic entries; Bragg mode acousto-optic Bragg cell; Opto-acoustic effect; Surface acoustic waves (SAWs)
Source encoding, 220–221
Space systems, VLWIR detectors in, 67–68
Spark plugs, in excimer lasers, 627–628
Sparrow criterion, 159
Spatial coherence function, 206
Spatial delta function, 201–202
Spatial encoding, 220–222
Spatial Free Spectral Range ($FSR_s$, $FSR_x$, SFSR)
  for arrayed waveguide grating, 985–986, 990
  in diffraction grating analysis, 966
Spatial frequency response, of acousto-optic modulators, 1262–1267
Spatial impulse response, 200, 210
Spatial light modulators (SLMs), 199, 1131, 1194–1196. *See also* Magneto-optic spatial light modulator (MOSLM)
  acousto-optic, 1274
  in adaptive optical processing, 228–229
  $Q$ parameter for, 1284
  $R$ parameter for, 1284–1286
Spatially incoherent sources, 207
Spatially inhomogeneous population inversion, in FIR lasers, 609–610
Spatially partially coherent processing, 216
Spatial mode selection, in laser oscillators, 396–399
Specification sheets, Raman fiber amplifier, 1314–1316

Spectra. *See also* Absorption spectra; Angular spectral dispersion; Electromagnetic spectrum; Emission spectra; Fluorescence spectra; Fourier spectra; General magnetostatic mode spectrum; Optical transmission loss spectra
  of active compounds, 446–456
  of DFB lasers, 195
  of distributed feedback resonator, 195
  of semiconductor lasers, 171
Spectral broadening, from quantum confined Stark effect, 1023–1025
Spectral output, of diode lasers, 260
Spectrometer systems, 131
Spectrophone, 583–584, 585
Spectroscopy
  dye lasers in, 496–499
  of $Er^{3+}$ ion, 1301–1303
Spectrum analyzers
  acousto-optic, 1114–1116, 1120–1121
  coherent, 211–212
  multichannel, 211–212
Specular reflection hazards, 429–430
Speed limits, on laser diodes and photodiodes, 691
Spherical aberration, 138, 139–141
  geodesic lens and, 782–784, 785
  Luneburg lens and, 770
Spherical coordinate system
  ray equation in, 19–20
  wave equation in, 5
Spherical geodesic waveguide lenses, 784–785
Spherically symmetric inhomogeneous media, ray tracing in, 24–25
Spherically symmetric media, light rays in, 19–20
Spherical surfaces
  refraction at, 152–153
  transfer between, 152
Spherochromatism, 145
Spin angular momentum, of an electron, 1135–1136, 1139–1140
Spin coating, waveguide fabrication via, 734
Spinels, ferrimagnetism of, 1143
Spin–orbit coupling, in atoms, 1136, 1164
Spin pinning, 1146, 1152
Spin–spin coupling, in atoms, 1136
Spin waves, 1149–1160. *See also* Magnetostatic waves (MSWs)
  bulk, 1155–1156
  polarization of, 1153–1154, 1154–1155

stimulating, 1158–1159
surface, 1155–1156
Splice organizers, for optical fiber, 677
Splices, optical fiber, 676–678
Spontaneous emission, 370, 371
amplified, 408–409
Spontaneous emission lifetime, 1303
Spot diagrams, 139–141
for astigmatism, 142–144
Spot size, 176–177, 178, 179, 180, 182, 183
with acousto-optic modulators, 1265, 1266
for Gaussian beams, 172–173
with integrated optical device modules, 1120–1121
in single-mode fibers, 675–676
Spot-size converter (SSC), 1019
SPRITE ("signal processing in the element") detectors, 96–103
performance of, 101–103
Spurious free dynamic range (SFDR), 1032
Sputtering. *See also* RF sputtering
magnetron, 740
waveguide fabrication via, 734, 736–739
Square-law media. *See* Quadratic index media
Stability
of gas lasers, 578–589, 603–605
of optical resonators, 177–179
Stability diagram, 179
Staggered center frequency, multiple transducers with, 1104, 1105–1107, 1109–1110
Standard single-mode fiber (SSMF), 895
Stark-splitting, energy levels of, 1303. *See also* Quantum confined Stark effect (QCSE)
Step index fiber, 664, 665
Step index optical waveguides, 720
modes for, 724–726
multilayer, 761–763, 764, 765, 766
Stepping motors, in opto-mechanical VOAs, 854
Step-wise variable optical attenuators, 850
Stilbene dyes, 448, 452
Stimulated backward scattering, 641
Stimulated emission, 370, 371, 512, 513–514
Stokes process, 1312, 1313, 1314
Stokes pulse, 641
Stokes shift, 946
Stop shift, 148, 150

Stopbands, in distributed feedback resonators, 190–193
Stops, 123–126. *See also* Aperture stop; Field stop
Straight section, ray matrix for, 28
Straight-through slab amplifier, 408
Strain
compressive, 316
in high-speed laser design, 313, 314
magnetism and, 1146
Strained quantum-well lasers, 252–254
Stransky–Krastanov process, 349
Strehl ratio, 159
Stress-induced birefringence, in bismuth-substituted YIG, 1182
Stripe geometry, 239, 242, 254–258. *See also* CRANK TJS structure; Deep diffused stripe (DDS) lasers; Transverse junction stripe (TJS) lasers
in magneto-optic deflectors, 1196–1197
Strip-loaded channel waveguides, 702, 706, 707
Strobe flashlamps, solid-state laser excitation via, 376
Strong gratings, 921, 922, 975
Su, Chin B., 1301
Subpicosecond light pulses, 486–488, 489, 633
Substituents, in garnet, 1173–1176
Substrates. *See* Acousto-optic substrate materials; Channel substrate planar (CSP) laser; Electro-optic crystal substrates; Glass substrates; Nonpiezoelectric substrates; Optical device substrates; Piezoelectric substrates; Semiconductive substrates; Silica substrate; Titanium-diffused substrate; Twin-channel substrate mesa guide (TCSM) laser; $X$-cut substrates; $Y$-cut $LiNbO_3$ substrate; $Y$-cut substrates
Sulfur dioxide ($SO_2$), far infrared lasers using, 542
Sulfur hexafluoride ($SF_6$), as buffer gas, 597
Superexchange, 1177
in garnet, 1178
Super Oslo Plot, 141
Superposition theorem, 1271
Superradiant emission, from FIR lasers, 609, 611, 620
Superstructure grating (SSG), 923

Superstructure grating DBR (FM) tunable lasers, 299

Surface acoustic waves (SAWs), 1080. *See also* Multiple tilted SAW transducers
  in acousto-optic Bragg diffraction, 1089, 1090, 1091, 1092
  in acousto-optic filters, 901–902
  in coupled-mode Bragg diffraction analysis, 1088
  in guided-wave acousto-optic Bragg diffraction, 1085–1086, 1086–1088
  in $Y$-cut $LiNbO_3$ substrate, 1092–1093
  in $Z$-cut GaAs–GaAlAs planar waveguide, 1097

Surface computations, for third-order aberrations, 146–147

Surface corrugation, waveguide mode index modification via, 760

Surface-emitting lasers, vertical-cavity, 323–347

Surface-emitting semiconductor lasers, 290–292

Surface spin waves, 1155–1156

Surfaces, curvature of, 117, 118

Surface states, magnetism and, 1146

Susceptibility, magnetic, 1137, 1138, 1146

Susceptibility tensor, 1134–1135, 1150

SWIR hybrid focal planar arrays, 82, 83, 84

Switches. *See* Acousto-optic switches; Current switches; Digital deflection/switching; Magneto-optic switches; Passive polarization switches; Pockels cell electro-optic switch; $Q$-switch oscillators; Saturable absorber switches

Synchronization, in couplers, 707–708

Synchronously pumped dye lasers, 488–491

Synthesis aperture radar (SAR) processing, 236

Synthetic discriminant function filter (SDF), 229–230

System penalties, in optical fiber transmission technology, 690–691

Systems analysis, 109–113

Tandem configuration, 996–997

Tangent plane, to a surface, 117

Tangential coma, 142

Tangential focus, 142–144

Tantalum pentoxide ($Ta_2O_5$)

  with Luneburg lenses, 771, 774, 775, 816, 817
  in multilayer step index waveguides, 763, 764, 765, 766
  in thin-film waveguides, 736–739

Tapered thickness coupler, 754

TCA tokamak, 618–620, 621

TEC technique, with AWGs, 993

Teich, Melvin Carl, 838, 850

Telecentric system, 117

Telecommunication applications. *See also* Optical communication; Optical fiber telecommunication links; Optical filters
  optical transmission technology in, 679

Telescope cavity configuration, 473, 474

Telescope-grating configuration, 465

Telescope objectives, 160

Telescopes
  Galilean, 127, 128
  Keplerian, 128, 133

Television, with fiber transmission networks, 693

$TEM_{nm}$ modes, 396–397, 398

Temperature. *See also* Compensation temperature; Curie temperature; Noise equivalent temperature (NET); Thermal entries; Thermo-optical VOAs; Transducer heating
  antiferromagnetism and, 1141–1142
  argon laser and, 521, 523–524
  AWG devices and, 826
  $CO_2$ laser and, 533
  continuously pumped dye lasers and, 483
  diode lasers and, 278–288
  dye lasers and, 472
  with extrinsic silicon detectors, 107
  ferrimagnetism and, 1143–1144
  ferromagnetism and, 1140–1141
  FIR lasers and, 589, 594, 596, 599–600
  HgCdTe detectors and, 71, 77, 85
  in high-speed laser design, 313, 314
  lattice filters and, 900
  magnetism and, 1138–1139
  photoconduction and, 1239
  pyroelectric detectors and, 104
  reflective magneto-optic materials and, 1184, 1185, 1186
  solid-state lasers and, 370
  in thermal annealing, 770
  thin-film filters and, 919
  tunable lasers and, 292, 294–295

VCSELs and, 334–335
Zeeman states and, 1164
Temporal transfer function, in acousto-optic modulators, 1267–1274
Tensile strain, in QW lasers, 253
Terbium–aluminum (TbAl) glass, magneto-optic properties of, 1170
Terbium-doped borosilicate glass, magneto-optic properties of, 1170
Terbium ions, in liquid lasers, 474
Tesla (unit), 1134
Tetrahedral coordination, of atoms in ferrimagnetic materials, 1143, 1144, 1145
Thallium ions, in ion exchange reactions, 746–747
Thermal annealing, index perturbation via, 768–770
Thermal conductivity, VCSELs and, 334–335
Thermal damper sleeve, 385
Thermal distortion
    effects of, 424–426
    effects of material properties and configuration on, 426–427, 428
Thermal noise
    in avalanche photodiodes, 1228
    in optical detectors, 1218
Thermal optical waveguide switches (TOSWs), 827
Thermionic emission time, in high-speed laser design, 319
Thermo-optical VOAs, 854–856
Thermopile, 104
Thick metal electrodes, in traveling-wave modulators, 1058
Thickness modulation method, index perturbation via, 768
Thick slab lasers, 426
Thin-film deposition, for planar optical waveguides, 720. *See also* Deposited thin-film waveguides
Thin-film filters (TFFs), 917–921, 954–956
    advantages of, 919
    analysis of, 917–919
    fabrication of, 919, 955–956
    performance characteristics of, 959–960
    technology of, 954–956
    theory and design of, 956–959
Thin-film materials, characterization of, 821, 822
Thin-film MUX/DEMUX devices, 954–961
Thin-film PBSs, 844, 845–846

Thin films
    growing garnet as, 1171, 1172, 1184
    in magneto-optic data-recording technology, 1187–1192
    in magneto-optic deflectors, 1196–1197
    isotropic, 1198–1207
    reflection from and transmission through, 9–10
    reflective magneto-optic materials as, 1184
    waveguide mode index modification via, 760
Thin-film thickness parameter, 722
Thin-film waveguide Luneburg lens, 768, 770–779, 780, 781, 782, 783, 820. *See also* Generalized Rinehart–Luneburg geodesic lens
    fabrication of, 775–779, 780, 781, 782, 783
    generalized, 773–775
    overlay thicknesses for (table), 776
    ray tracing for, 25, 771, 775, 777, 780, 781, 782
    theory of, 770–775
Thin-film waveguides, 734–739
    Fresnel lenses and, 793–795, 796
    table of inorganic, 737–738
    table of organic, 735
Thin lens, 123
    aberrations of, 148–150
    ray matrix for, 28
    transformation of Gaussian beam by, 34–35
Third-order aberrations, 146–150
Thomson scattering diagnostics, 606–607
Three-dimensional waveguide, modes of, 167–168
Three-energy-level system, 515–517, 518
Three-level lasers, 371–374
Three-port cylindrically packaged optical circulator, 881–882
Thulium lasers, 382
Thyratrons, 628–629
Ti:LiNbO$_3$ coupler, 715–716
Ti:LiNbO$_3$ waveguides, 767, 769, 816
Tilt coefficients, 147
Tilted-finger chirp transducer, 1098, 1104, 1107, 1110
Tilting mirror MEMS VOA structure, 861
Time–bandwidth product (*TB*), 1274
    of an acousto-optic modulator, 1100–1101

Titanium in-diffusion proton exchange (TIPE) waveguide lenses, 1099, 1100, 1116–1117

Titanium oxide (TiO$_2$), in graded index waveguides, 767, 768

Titanium-diffused substrate
in diffused optical waveguides, 748–750
in graded index waveguides, 767, 768–770
for lithium niobate electro-optical devices, 1012
measuring impurity profiles and, 813, 814
in optical loss measurement, 816

Tokamaks
Alcator, 616
TCA, 618–620, 621

Top electrode geometry, in acousto-optic modulators, 1294–1296

Top electrode thickness, in acousto-optic modulators, 1291–1293

Top-emitting VCSEL structure, 326

Toroidal edge rounding, in geodesic lenses, 785, 786, 789, 790, 791

Total chromatic dispersion, in optical fibers, 672

Total internal reflection (TIR), 116
in dielectric planar waveguides, 721–722
in prisms, 131

TQW laser, 262–263

Transducer array, isotropic diffraction with, 1104, 1107. *See also* Multiple tilted SAW transducers; Multiple transducers

Transducer bandwidth, for $Y$-cut LiNbO$_3$ substrate, 1092

Transducer heating, in acousto-optic modulators, 1293–1294

Transducers. *See also* Multiple tilted SAW transducers
in acousto-optic devices, 1258, 1280
efficiency of electric-to-acoustic, 1101
in lithium niobate AO Bragg modulators/deflectors, 1108–1109
single versus multiple, 1110–1111
zigzag, 1296–1297

Transfer functions. *See also* Modulation transfer function (MTF); Temporal transfer function
in acousto-optics, 1261–1262, 1264
bias stability and, 1034–1035
for electroabsorption modulators, 1029, 1030
for electro-optic modulators, 1027–1029, 1030
for multiplexers, 938–941
for thin-film filters, 957, 958–959

Transfer matrix method, for fiber Bragg grating filters, 972, 973–974

Transformation matrix. *See ABCD* matrix

Transients, resonant, 478–480

Transit time limit, of *pn* junction photodiodes, 1220–1221

Transition elements
ferromagnetism of, 1140
in reflective magneto-optic materials, 1184

Transitions, of dipoles, 1163–1164

Transmission
at a boundary, 7–8
through layered dielectric, 10–12
loss-limited versus dispersion-limited, 681–682
through thin film, 9–10

Transmission amplitude gratings, 46–47

Transmission coefficient, 9–10, 12

Transmission gratings, 903–904, 961–962

Transmission limitations, in optical fiber transmission technology, 686–691

Transmission line impedance, 1047–1048

Transmission lines, in stimulating spin waves, 1158–1159

Transmission loss, in optical fiber transmission technology, 686–687

Transmission matrix, 1065
for calculating quantum confined Stark effect, 1023, 1024

Transmission media, in optical fiber transmission systems, 680

Transmission systems
multimode optical fiber, 681
optical fiber, 680
single-mode fiber, 681–684

Transmissive magneto-optic materials, 1170–1184

Transmitter noise, in optical detectors, 1218

Transponders, in optical networks, 947–948

Transversal optical planar waveguide filters, 898

Transversal pumping, of dye lasers, 462, 477–478

Transverse aberration coordinates, 135

Transverse chromatic aberration, 137

Transverse direction, in laser structure, 241, 260–261
Transverse electric (TE) mode, 165, 260–261, 747
  for graded index guides, 726, 730
  in Luneburg lenses, 774
  in magneto-optic thin-film waveguides, 1200, 1201, 1202, 1204, 1206, 1207
  in metal-clad optical waveguides, 750–752
  in multilayer step index waveguides, 761–762, 763, 764, 765, 766
  planar optical waveguides and, 698, 699–700, 723, 724, 725
  in polymer crystals, 1018
  in semiconductors, 1016
Transverse electrodeless (TERF) systems, 555, 557
Transverse junction stripe (TJS) lasers, 277. *See also* CRANK TJS structure
Transverse Kerr effect, 1168
Transversely integrated tunable lasers, 297, 298, 304–306
Transverse magnetic (TM) mode, 165, 260–261, 747
  for graded index guides, 726
  in Luneburg lenses, 774
  in magneto-optic thin-film waveguides, 1198–1199, 1200
  in metal-clad optical waveguides, 750–752, 753
  in multilayer step index waveguides, 761, 762–763, 764
  planar optical waveguides and, 699–700, 723, 724, 725–726
  in polymer crystals, 1018
  in semiconductors, 1016
Transverse mode spacing, 171
Transverse ray deviation, 134, 135
Transverse resonance condition, in dielectric planar waveguides, 722
Traveling-wave electroabsorption modulators (TWEAMs), 1052–1056
  continuous, 1053
Traveling-wave III–V MZMs, 1066–1069
Traveling-wave LiNbO$_3$ modulator, 1056–1061
Traveling-wave modulators, 1050–1069
  continuous, 1051–1052
  segmented, 1063–1066
Traveling-wave photodiodes, 1238–1239

Traveling-wave polymer modulator, 1061–1063
Traveling-wave-type resonator, 164
Triarylmethene dyes, 450
Triglycine sulfate (TGS), for pyroelectric detectors, 103
Trimethylgallium (TMGa), 246
Triplet absorption, in continuously pumped dye lasers, 483
Tsai, Chen S., 1079
Tunable dispersion compensating filters, 916–917, 924–925
Tunable filters, diffraction gratings and, 905–906
Tunable lasers, 292–308. *See also* Tuning
  dye lasers as, 444
  external-cavity, 307, 308
  longitudinally integrated, 297, 298–304
  in photochemistry, 498–499
  structures of, 295–298
  transversely integrated, 297, 298, 304–306
  VCSEL, 307–308
Tunable single-stage allpass filter, ring resonator in, 915
*Tunable Solid State Lasers* (Budgor & Pinto), 431
Tunable twin-guide (TTG) lasers, 297, 304–306, 307
Tungsten disks, 521, 523
Tungsten filament lamps, solid-state laser excitation via, 380–381
Tuning. *See also* Tunable entries
  of dye lasers, 463–471
  of liquid lasers, 460–471
  of passively mode-locked dye lasers, 492–493
  of waveguide grating routers, 913
  of wavelength, 292–295
Tuning wedge, 466, 467, 468
Tunnel diagram, 130
Tunneling time, in high-speed laser design, 319–320
Turning points
  in a ray, 21–22
  with WKB method, 730, 731
Twin-channel substrate mesa guide (TCSM) laser, 262–263
Twin-frequency FIR lasers, 603–606, 607, 608
Twin-ridge structure (TRS) laser, 262
Twisted nematic liquid crystals, polarization rotators from, 850

Two-beam interferometric method, fabricating fiber Bragg gratings via, 971

Two-channel directional couplers, mathematical analysis of, 711–716

Two-dimensional dielectric waveguide, modes of, 164–167

Two-dimensional optical confinement, 702

Two-dimensional stacked array lasers, 290, 291, 292

Two-energy-level system, 513–514, 515, 518

Two-pupil incoherent processing systems, 216

Two-stage magnetic pulse compression circuit, 629

Two-stage reflective optical isolator, 871–872

Two-stage wedge type optical isolator, 868, 869

Twyman–Green interferogram, 137

UCLA
    far infrared laser at, 589–593
    pulsed optically pumped FIR laser at, 612

Ultra-short light pulses, 486–488, 492, 494, 495. See also Femtosecond light pulses; Subpicosecond light pulses

Ultra-small quantum dot devices, 349–350

Ultraviolet (UV) light. See also Acousto-optic UV intensity modulator
    acousto-optic materials for, 1281
    from dye lasers, 444
    from excimer lasers, 538, 625–626
    in fabricating fiber Bragg gratings, 971
    in optical fibers, 670–671
    preionization with, 627–628, 630–631

Unapodized fiber Bragg gratings, 975–976

Underlayer, in multilayer step index waveguides, 761, 763, 764, 765, 766

Uniaxial anisotropy
    in magneto-optic data-recording films, 1189
    in reflective magneto-optic materials, 1186

Uniaxial birefringence, 1162

Uniaxial crystals, 15

Uniaxial medium, 13–14

Uniform error function index profiles, of planar waveguides, 701–702

Uniform gratings, 797–799, 973

Uniform planar waveguides, characteristic equation for, 698–702

Uniform precession mode, of magnetic dipoles, 1149, 1152, 1153, 1158, 1159–1160

Unitraveling-carrier photodiode, 1220, 1225

Units, magnetic, 1132–1135

Unit vectors, in geometrical optics, 16–17

Unpaired magnetic dipoles, 1147–1148

Unstable resonators, mirrors in, 398–399

Vacuum wave number, 1165

Van Cittert–Zernike theorem, 206, 221

Vapor axial deposition (VAD), 669

Vapor deposition techniques. See Chemical vapor deposition; Evaporation; Metal-organic chemical vapor deposition (MOCVD); Modified chemical vapor deposition (MCVD); Physical vapor deposition (PVD) techniques; Plasma-activated chemical vapor deposition (PCVD); RF sputtering; Vapor axial deposition (VAD)

Variable beam blocking, in opto-mechanical VOAs, 854

Variable optical attenuators (VOAs), 846, 852–863
    next-generation, 862–863
    polarization state transformation diagrams for, 859

Variational method, for calculating quantum confined Stark effect, 1023

V-channel inner stripe (VSIS) laser, 262

VCSEL arrays, 346–347

Velocity-matched photodiodes, 1238–1239

Verdet constant, 846, 1169, 1170
    in magneto-optics, 1165
    in optical isolators, 1192

Verneuil process, 387

Vertex, of a surface, 117

Vertical-cavity surface-emitting lasers (VCSELs), 240, 290, 291, 323–347. See also VCSEL arrays
    current and optical confinements in, 331–334
    at different wavelengths, 337–343

850-nm, 324–337, 338
future development of, 351
high-speed, 343–346
history of, 241
optimizing performance of, 339–340, 343–346
quantum wells in, 328–331
reliability of, 335–337
resonators for, 195–196
thermal properties of, 334–335
tunable, 307–308
visible, 341–342
Very high speed integrated circuit (VHSIC) technology, 691
Very high speed WDM systems, 691–692
Very long wavelength infrared (VLWIR), 54. *See also* VLWIR entries
Vibrational modes
of carbon dioxide molecule, 527
of $CH_3F$ molecules, 562, 563
far infrared lasers and, 539
of water molecules, 545–549, 550
Vibronic lasers, 372, 374
Vidicon cameras, diagnostics with, 432–434
Vidicons, for pyroelectric detectors, 103, 104, 105
Vignetting, 117, 125, 128, 155
Virtual image, 116
Virtually imaged phased array (VIPA), 899, 903, 908–909
Visible VCSELs, 341–342
Visible wavelengths, acousto-optic materials for, 1281
VLWIR detectors
performance capabilities of large, 68–69
in space and military systems, 67–68
VLWIR hybrid focal planar arrays, 84
VMUX module, 853
*V*-number
of an optical fiber, 663, 675
planar optical waveguides and, 700
Voigt effect, 1167, 1169, 1170
Voltage, applied to liquid crystal cells, 849–850
Voltage swing, in modulators, 1030–1032
Volume index perturbation, 760

Volume waves, in garnet films, 1198
VUV photon emission, 520–521

Wafer fusing technique, 1235
Walk-off birefringent crystal optical circulator, 879
Walk-off crystal optical isolators, 869–870
Wall energy, in magneto-optic data-recording films, 1190
Water. *See also* Heavy water ($D_2O$)
in electron-beam amplifiers, 636
far infrared lasers using, 541, 542, 543, 544–549, 550
molecular properties of, 545–549, 550
rotational energy levels of molecules of, 545, 546, 550
vibrational modes of molecules of, 545–549, 550
Water cooling, of $CO_2$ lasers, 533
Waterline systems, in electron-beam amplifiers, 636, 637
Wave aberration function, 156
Wave-banding, in optical networks, 948–950
Wave equation, 4–5
Waveform coefficients, transformations of, 173–175
Waveform matrices, 175
Wavefront, 116
Wavefront aberration function, 134–137, 137–139, 139–141
for astigmatism, 142–144
Waveguide $CO_2$ lasers, 528–529. *See also* Carbon dioxide ($CO_2$) lasers
Waveguide diffraction elements, 789–807, 808, 809, 810, 811
Waveguide diffraction filters, 804–807, 808, 809, 810, 811
Waveguide diffraction lenses, 793–804
Waveguide diffraction mirrors, 804–807, 808, 809, 810, 811
Waveguide grating routers (WGRs), 899, 903, 909–913
design of, 909–911
filters using, 911–913
in MUX/DEMUX integrated optics design, 982
Waveguide gratings, scattering by, 189–190, 191
Waveguide isolators, 1193–1194

Waveguide mode index modification techniques, 760

Waveguide modes
for graded index guides, 726–730
for step index guides, 724–726

Waveguide optical component measurement, 818–819, 820

Waveguide optical demultiplexers, with chirped grating, 811

Waveguide optical gratings, 792–793

Waveguide optical harmonic generator, 807–809

Waveguide pair directional coupler, 711–716

Waveguide photodiodes, 1236–1239

Waveguide propagation loss, 1035

Waveguides, 697–716. *See also*
Electroabsorption (EA) waveguides;
Graded index optical waveguides;
Lens waveguide; Planar optical
waveguides; Step index optical
waveguides
applications for, 819–830
deposited thin-film, 734
dielectric planar, 720–731, 732, 733
diffused optical, 747–750
for dye lasers, 469
free-carrier optical, 752–753
for gas lasers, 574, 578, 589
graded index optical, 742–743
for HCN lasers, 560
inorganic thin-film, 736–739
ion exchange, 744–747
ion implantation, 743–744
materials and fabrication techniques for, 731–753
measurement techniques for, 809–819, 820
measuring parameters for, 809–819, 820
metal-clad optical, 750–752
multilayer step index, 761–763, 764, 765, 766
optical channel, 697, 702–707
optical fiber, 660–663
for optical resonators, 163–164, 164–168
organic thin-film, 734–736
planar optical, 697–702
refractive elements in, 759–789
single-crystal epitaxial, 739–742
in stimulating spin waves, 1158–1159
table of epitaxial, 741

thin films as, 1198–1207
three-dimensional, 167–168

Waveguide thickness, 815

Waveguide-type Fabry–Perot resonators, 164

Wave-guiding structures, in dye lasers, 469

Wavelength. *See also* Cutoff wavelength
for far infrared lasers, 540, 541
for VCSELs, 324, 337–343

Wavelength detuning, 316

Wavelength division multiplexer, 820

Wavelength division multiplexers/demultiplexers (WDM MUX/DEMUX), 935–1007. *See also* MUX/DEMUX entries
applications of, 943–954
functional parameters of, 937–943
history of, 935–936
operation of, 936–937

Wavelength division multiplexing (WDM).
*See also* Dense-wavelength division multiplexing (DWDM); Wavelength division multiplexers/demultiplexers (WDM MUX/DEMUX); WDM systems
AWG devices for, 822, 823
Bragg gratings and, 922
devices for, 678, 679
erbium-doped fiber for, 1308–1311
optical filters for, 887
for semiconductor optical amplifiers, 1316
in single-mode optical fiber systems, 684–686

Wavelength interleaving, in optical networks, 950–951

Wavelength-multiplexed reflection-type matched filter correlator, 232

Wavelength-selecting switch (WSS), 906, 907

Wavelength spacing, 171

Wavelength-tunable laser diodes, mechanism comparison for, 296

Wavelength tuning
continuous, 295, 298, 300–303
discontinuous, 295, 298–300
physical mechanisms for, 292–295
quasi-continuous, 298, 303–304

Wavelets, 155

Wave plates, 848–849

Wave propagation, 1, 5–7

Waves, in anisotropic crystals, 12–15

WDM links, 952
WDM systems, very high speed and high-density, 691–692
Weak gratings, 921, 922, 973, 975
Weber (unit), 1134
Wedge. *See* Birefringent wedge; Tuning wedge
Weidner, F., 443
Weiss, Pierre, 1138–1139
Weiss temperature, 1139
Wet-etching, in MUX/DEMUX devices, 981
White light optical processing, 218–220
Wide-band acousto-optic device configurations, 1105–1107
Wide-band guided-wave acousto-optic Bragg modulators/deflectors construction of, 1102–1111 performance of, 1108–1110
Wide-bandwidth optical intensity modulators, 1009–1078 efficiency and waveguide design in, 1027–1041 lumped-element, 1041–1050 physical effects and modulator materials for, 1010–1026 traveling-wave, 1050–1069
Wiener–Hopf solution, to optimize optical information processing, 234–235
Wilhelmi, B., 443
Wire grid polarizers, 839
WKB approximation method, 730–731, 732, 733 for graded index guides, 726 measuring refractive index profiles with, 813–815
Wolf, Emil, 838
Wollaston prisms, 841–842, 843 in optical circulators, 881–882

Xanthene dyes, 448, 450
X-cut substrates, ion exchange waveguides with, 747
Xenon (Xe), in ion lasers, 520
Xenon chloride (XeCl) lasers, 475–476, 477, 480, 624, 627, 632, 640. *See also* Noble-gas halide lasers; Rare gas halide lasers applications for, 633
Xenon flashlamps, solid-state laser excitation via, 376, 380

Xenon fluoride (XeF) laser, 632, 640
*X*-fan, 136–137, 142
X-ray microanalyzer (XMA), measuring impurity profiles with, 812–813, 814
X-ray photoelectron spectroscopy, 748–750
X-ray preionization, 628

YAG lasers, continuous-wave, 381. *See also* Nd:YAG lasers; Yttrium aluminum garnet (YAG)
Yang, Jane J., 239
Yao, Shi-Kay, 719, 1255
*Y*-cut LiNbO$_3$ substrate, 1100–1101, 1104, 1106–1107, 1108, 1109, 1110 calculated performances for, 1092–1094
*Y*-cut substrates, ion exchange waveguides with, 747
*Y*-fan, 136–137, 139, 140, 142
Y-junctions, 828–830
Young's experiment, 220–221
Ytterbium (Yb), as garnet substituent, 1173
Yttrium (Y), as garnet substituent, 1173, 1175
Yttrium aluminate (YALO), as solid-state laser host material, 375
Yttrium aluminum garnet (YAG), as solid-state laser host material, 375. *See also* Nd:YAG entries; YAG entries
Yttrium iron garnet (YIG; Y$_3$Fe$_5$O$_{12}$). *See also* Bismuth-substituted yttrium iron garnet (Bi:YIG) magnetic properties of, 1172–1173 magnetostatic waves in, 1198, 1200 in optical isolators, 1192 structure of, 1171–1172
Yttrium iron garnet (YIG) single crystal ferrimagnetism of, 1143 in polarization rotators, 846–847
Yu, Francis T. S., 199
Yu, P. K. L., 1009, 1215

ZAP code, 409, 410
Z-cut GaAs–GaAlAs planar waveguide, 1097, 1104
Zeeman effect, 1163, 1164, 1168
Zero material dispersion, in optical fibers, 672
Zero-order diffraction, 963–964

Zero-order mode, 761
Zero-order wave plates, 848
Zhou, Feng Qing, 837
Zigzag, rays in, 721–722
Zigzag slab laser, 408, 426, 427
Zigzag transducer pattern, 1296–1297
Zinc Blende crystals, in electro-optic
　modulators, 1015
Zinc diffusion, in NAM laser fabrication,
　279
Zinc oxide (ZnO)
　epitaxial growth of, 740

in guided-wave acousto-optic Bragg
　modulators, 1086
Zinc oxide films, 1094, 1096, 1098
Zinc oxide transducer, 1097
Zinc selenide (ZnSe) coupler, 531, 532
Zirconium (Zr), as garnet substituent, 1175
Zonal spherical aberration, 146
Zone lenses, 793–795, 796, 804
Zone plate, 794. *See also* Fresnel
　zone plate
Z transform, optical filters and, 888–889,
　892–893